GARDNER'S COMMERCIALLY IMPORTANT CHEMICALS

GARDNER'S COMMERCIALLY IMPORTANT CHEMICALS
SYNONYMS, TRADE NAMES, AND PROPERTIES

Edited by

G. W. A. Milne

WILEY-INTERSCIENCE

A JOHN WILEY & SONS, INC., PUBLICATION

Library of Congress Cataloging-in-Publication Data:

Gardner's Commercially Important Chemicals: Synonyms, Tradenames, and Properties. / edited by George W. A. Milne. —
    p. cm.
   Includes index.
   ISBN 13: 978-0-471-73518-2
   ISBN 10: 0-471-73518-3
    1. Chemicals—Dictionaries.   2. Chemicals—Trademarks.   I. Milne, George W. A., 1937–

TP9.G286 2004
660'.03—dc21

98-51143
CIP

CONTENTS

CONTENTS

PREFACE

Through eleven editions, *Gardner's Chemical Synonyms and Trade Names* has become the best-known and most widely used source of information on chemicals in commerce. This companion Book reflects the continuing research underlying this reference work and presents a major expansion of the information provided for individual chemical compounds.

In the preparation of this Book, an effort has been made to include useful information on chemicals that play a significant role in commerce. The US Environmental Protection Agency (EPA), in cooperation with industry, maintains a list of "High Production Volume" (HPV) chemicals whose production volume in the US exceeds one million pounds per year. With the exception of fuels, some 1836 HPV chemicals have been included in this Book. Some other chemicals, notably drugs, are also very important in commerce although they are produced in much smaller quantities. Most pesticides also fail to reach this production level and the 502 most commonly prescribed drugs and 576 most heavily used pesticides have been included in this compilation, along with the more commonly used veterinary medicines and pharmaceutical excipients. In addition, some 2635 chemicals which are not in any of these categories but which are in the 11th. Edition of Gardners have also been included.

This compilation contains 4,174 entries drawn from these various sources. For each chemical, the appropriate identifying information (CAS Registry Number, structure, molecular formula and chemical name is provided and in each case, an exhaustive list of known synonyms for the materials is given. These synonyms include other identifiers, chemical names, tradenames and trivial names, in English and other languages. There are on average, some 18 synonyms for each chemical described in this Edition. Chemical properties of the compounds are given, as is information concerning known uses of the chemical. Biological data, in particular acute toxicity in various species is provided as available and finally, companies that manufacture or supply the material are identified.

The main criterion for inclusion of a material in this handbook is its importance as a significant commercially available chemical. Thus all bulk inorganic chemicals are included, all major pesticides (herbicides, insecticides, antifungal agents, and so on) and many dyestuffs, surfactants, metals and inorganic compounds are described in this book. Only the most commonly used drugs are included and for a more exhaustive coverage of drugs, the reader is referred to a separate publication, *Drugs: Synonyms and Properties*, published by Ashgate. Similarly, a complete coverage of pesticides may be found in the *Handbook of Pesticides*, also from Ashgate.

Almost all the records in this compilation carry the appropriate Chemical Abstracts Service (CAS) Registry Number and the associated EINECS (European Inventory of Existing Commercial Chemical Substances) number. Wherever possible, a chemical is thus tagged with the major American and European identification numbers. In addition, all chemicals in this edition which also appear in the Thirteenth Edition of the Merck Index have the Merck Index Number provided. Identifiers from other databases, such as Beilstein, and files maintained by governmental agencies such as UN, EPA, USDA and NCI are also provided and can be used as links to these databases. A feature of this database is the inclusion of physical properties and use data for pure chemicals. Properties that have been provided as available include the melting point, boiling point, density or specific gravity, optical rotation, ultraviolet absorption, solubility and acute toxicity. The major uses of most of the chemicals are indicated and, where appropriate, regulatory information is also provided.

Details of the structure of a record are provided in the section *How to Use this Book* on page xi.

Proprietary Considerations

Every attempt has been made to ensure the accuracy of the information provided in this Book. However, the publishers cannot be held responsible for the accuracy of the information, and users are expected to bear in mind the following information:

The reporting of a name in this Book cannot imply definitive legality in establishing proprietary usage. Questions concerning legal ownership of a particular name can be resolved by due legal process.

A manufacturer in some countries may manufacture its product under names different from those cited in this Book. Similarly, manufacture or marketing of a product may be licensed to a separate company in another country either under the same or a different name.

We trust that readers will find that this compilation contains a wealth of information which is difficult to obtain from any other source. It is the intention of the publishers to produce regularly updated editions and subsets of this compilation at suitable intervals in both printed and digital form. Companies wishing to submit new or updated material for inclusion in future editions should contact George W A Milne (address below).

Acknowledgements

The Editor would like to acknowledge the skilled programming performed by Dr Ju-yun Li which allowed for accurate formatting and typesetting of this handbook. Finally, I should like to acknowledge the expert assistance and continual support provided by my wife, Kay, without whom none of my deadlines would have been met.

George W A Milne
John Wiley & Sons, Inc.
111 River Street
Hoboken, NJ 07030-5774
USA
E-mail: bill@phm.com

HOW TO USE THIS BOOK

The book is divided into three Sections; a main Section, an Index Section
containing three Indexes and a Directory of Manufacturers and Suppliers.
A brief description of each Section is given below

A. SECTION 1 contains a monograph for every chemical in the database.
Every entry in this Section has a unique Entry Number which is used to
find and refer to the entry it identifies. The Indexes in Section 2 of the
Book make use of this Entry Number to facilitate cross-referencing
back to Section 1.

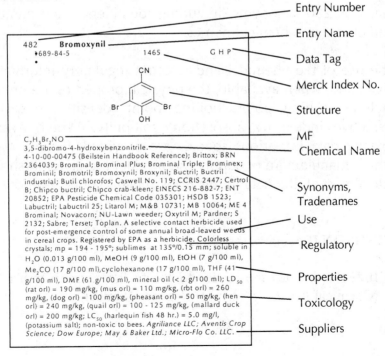

A typical record is shown above. The first line contains the Entry Number for the record (482) and the name of the material (Bromoxynil). The second line gives the Chemical Abstracts Service (CAS) Registry Number (1689-84-5) for the compound, the corresponding Merck Index Entry (1465) and a data tag (G H P) indicating the file(s) which were the source of the compound. The CAS RN, Merck Index Number and Data Tag always appear in the same position (left, centre or right) enabling the reader to determine which source they belong to. The Data Tag may have any of 6 values, defined below:

D – Drug, in US Pharmacopeia
E – Pharmaceutical Excipient
G – In Gardner's (11th. Edition)
H – High Volume Production (HVP) Chemical in EPA HPV List
P – Pesticide
V – Veterinary Medicine

Whenever CAS Registry Numbers are used in the text, they are always enclosed in brackets, for example [122-99-6]. The structure and molecular formula of the compound are provided and the next line carries the chemical name of the compound. This is followed by as many as 200 synonyms, including identifiers, trade names, synonyms and other non-chemical names.

The use of the chemical and relevant regulatory information are then given and, when available, the physical properties are presented. These include melting point, boiling point, density or specific gravity, refractive index, uv absorption, solubility. Where available, acute toxicity data in various species are provided. Finally, the companies which manufacture or supply the product are given. It should be noted that almost all the data in this Book relate to pure compounds.

B. SECTION II contains three Indexes:

i) **Index 1** enables the reader to locate the Entry Number for any CAS (Chemical Abstracts Service) Registry Number.

ii) **Index 2** enables the reader to locate the Entry Number for any EINECS (European Inventory of Existing Commercial Chemical Substances) Number.

iii) **Index 3** contains all the names, synonyms and tradenames and ther identifiers for the compounds in the database. The chemical names of organic compounds are not useful as locators and have therefore been omitted from this Index.

This is probably the most convenient place to start if only a name is known. This thesaurus will refer the reader to the Entry Number in Part I, which relates to the main entry for that chemical.

C. **SECTION III** is a directory of chemical manufacturers and suppliers whose products are described in the Book. The entries are in alphabetical order by company name. Wherever possible, the postal address, telephone number, fax number and website address are provided.

GLOSSARY OF UNITS

Name	Description
Mass	Unless otherwise specified, mass is expressed in a multiple of grams (g), such as micrograms (μg; $= 10^{-6}$ g), milligrams (mg; $= 10^{-3}$ g), grams (g; $= 10^{0}$ g), kilograms (kg; $= 10^{+3}$ g), etc.
Volume	Volume is expressed in litres (l) or millilitres (ml) unless otherwise specified.
Temperature	When no units are cited, the temperature given is in degrees Celsius (°C).
Melting point	Melting points are cited in degrees Celsius (°C) unless otherwise specified.
Boiling point	When measured at atmospheric pressure, boiling points are cited with no pressure, e.g. bp $= 167°$. At other pressures, the pressure is also cited, i.e. $bp_{0.01} = 167°$.
Density	The measurement temperature is given as a superscript; thus a density of 1.123 measured at 25°

will appear as $d^{25} = 1.123$. If the measurement was explicitly referenced to the density of water at 4°, the citation will carry both a superscript and a subscript, as in $d^{25}_4 = 1.123$. Specific gravities are denoted by the abbreviation 'sg'.

Optical rotation
Optical rotations (α) are cited with the measurement temperature superscripted, and the measurement wavelength (often the sodium D line) subscripted, as in $[\alpha]^{25}_D = 105°$. When mutarotation can occur, the starting and an equilibrium values are given, as in $[\alpha]^{25}_D = 105° \pm 91°$.

UV absorption
The ultraviolet absorption maxima given by the material are cited in nanometers (nm = millimicrons, mμ) and the absorptivity E or A, (which are unitless) is also given.

Acute toxicity
Wherever possible the units of toxicity are LD_{50}, i.e. the dose which is lethal to 50% of the test animals. In most cases, acute toxicity is measured with the rat, orally administered, and the result is reported as LD_{50} (rat orl) = 50 mg/kg. Other species (for example, mus = mouse; rbt = rabbit; pgn = pigeon; hmn = human; chd = child; wmn = woman; gpg = guinea pig) are occasionally cited as are other administration routes (sc = subcutaneous; ihl = inhalation; ip = intraperitoneal; iv = intravenous). Chronic toxicity data are not given.

ABBREVIATIONS

Terms	Definition
ABS	acrylonitrile-butadiene-styrene
ACE	acetylcholinesterase
ACN	acrylonitrile
alc.	alcohol
AMP	2-amino-2-methyl-1-propanol
aq.	aqueous
ASA	acrylic-styrene-acrylonitrile
BHA	butylated hydroxyanisole
BHT	butylated hydroxytoluene
BMC	bulk molding compound
bp	boiling point
BP	British Pharmacopeia
BR	butadiene rubbers, polybutadienes
BRN	Beilstein Registry Number
B/S	butadiene/styrene
CAB	cellulose acetate butyrate
CAS	Chemical Abstracts Service
CCRIS	Chemical Carcinogenesis Research Information System (NLM)
CDA	completely denatured alcohol
CFC	Chlorofluorocarbon

C.I.	Color Index
CL	Cyanamid Laboratories
CMC	carboxymethylcellulose, critical micellar concentration
CNS	central nervous system
CPE	chlorinated polyethylene
CPVC	chlorinated polyvinyl chloride
CR	chloroprene rubber, polychloroprene
cs or cSt	centistoke(s)
CTFA	Cosmetic, Toiletries and Fragrance Association
DAP	diallyl phthalate, diammonium phosphate
DB	dichlorophenoxybutyric acid
DEA	diethanolamine, diethanolamide
DEDM	diethylol diethyl
DIBA	diisobutyl adipate
DIDA	diisodecyl adipate
DMC	4,4'-dichloro(methylbenzhydrol)
DMDM	dimethylol dimethyl
DMF	dimethlformamide
DMSO	dimethyl sulfoxide
DNPT	dinitrosopentamethylenetetramine
DOP	dioctyl phthalate
DOT	Department of Transportation
DP acid	diphenolic acid
DPG	diphenylguanidine
DTPA	diethylenetriamine pentaacetic acid
ECTFE	ethylene/chlorotrifluoroethylene copolymer
EDTA	ethylenediamine tetraacetic acid
EINECS	European Inventory of Existing Commercial Chemical Substances
EMC	electromagnetic conductive
EMI	electromagnetic interference
EO	ethylene oxide
EP	extreme pressure
EPA	Environmental Protection Agency
EPDM	ethylene-propylene-diene rubbers
EPM	ethylene-propylene rubbers
EPR	ethylene-propylene rubber
ESCR	environmental stress crack resistance
ETFE	ethylene tetrafluoroethylene

ETU	ethylene thiourea
EVA	ethylene vinyl acetate
F	Fahrenheit
FA	fatty acid
FDA	Food & Drug Administration
FEMA	Federal Emergency Management Agency
FEP	fluorinated ethylene propylene
FFA	free fatty acid
FG	food grade
fp	freezing point
FRP	fiberglass-reinforced plastic(s)
GFRP	glass fiber-reinforced plastic(s)
gran.	granular, granules
GRAS	generally recognized as safe
GRP	glass-reinforced plastics, polyester
HDI	hexamethylene diisocyanate
HDL	high density lipids
HDPE	high density polyethylene
HIPS	high impact polystyrene
HLB	hydrophilic-lipophilic balance
HPLC	high performance liquid chromatography
HSDB	Hazardous Substances Database
HPV	High Production Volume (Chemical)
IC	integrated circuit
ICI	Imperial Chemical Industries
ihl	inhalation
IIR	isobutylene-isoprene rubber
IPA	isopropyl alcohol
IPM	isopropyl myristate
IPP	isopropyl palmitate
IR	(synthetic) isoprene rubber, infrared
IU	international units
iv	intravenous
J	joule(s)
KTPP	potassium tripolyphosphate
LDL	low density lipids
LDPE	low density polyethylene
LED	light-emitting diode
LLDPE	linear low density polyethylene
Ltd	Limited
MA	methacrylic acid

MBCA	4,4'-methylene bis(orthochloroaniline)
MBT	mercaptobenzothiazole
MBTS	2-mercaptobenzothiazole disulfide
MCPA	(4-chloro-2-methylphenoxy) acetic acid
MDI	methylene diphenylene diisocyanate
MDM	monomethylol dimethyl
MDPE	medium density polyethylene
MEA	monoethanolamine, monoethanolamide
MEK	methyl ethyl ketone
MIBK	methyl isobutyl ketone
min.	minute(s), mineral, minimum
MIPA	monoisopropylamine, monoisopropylamide
MKP	monopotassium phosphate
MMW-HDPE	medium molecular weight high density polyethylene
MOCA	methylene bis(orthochloroaniline)
mp	melting point
MPK	methyl propyl ketone
MVTR	moisture vapor transmission rate
mw	molecular weight
N	normal
NBR	nitrile-butadiene rubber
NC	nitrocellulose
NCI	National Cancer Institute (NIH)
NCR	nitrile-chloroprene rubber
NEMA	National Electrical Manufacturers Association
N/F	non-flammable
NF	National Formulary
NIH	National Institutes of Health
NLM	National Library of Medicine (NIH)
NR	natural (isoprene) rubber
NSC	National Service Center (NCI)
NSF	National Science Foundation
NTA	nitrilotriacetic acid
OEM	original equipment manufacturer
OPP	oriented polypropylene
OTC	over the counter (i.e. non-prescription) drug
o/w	oil-in-water
Pa	Pascal
PAN	polyacrylonitrile

PBT	polybutylene terephthalate
pbw	parts by weight
PC	polycarbonate
PCA	2-pyrrolidine-5-carboxylic acid
PCP	pentachlorophenol
PCTFE	polychlorotrifluoroethylene
PE	polyethylene
PEEK	polyetheretherketone
PEG	polyethylene glycol
PEI	polyetherimide
PEK.	polyetherketone
PES	polyether sulfone
PET	polyethylene terephthalate
PFA	perfluoroalkoxy
PG	polypropylene glycol
pH	hydrogen ion concentration as negative logarithm
phr	parts per hundred of rubber or resin
PIB	polyisobutylene
pK	dissociation constant as negative logarithm
PMA	phosphomolybdic acid
PMMA	polymethyl methacrylate
PO	propylene oxide
POE	polyoxyethylene, polyoxyethylated
POM	polyoxymethylene
POP	polyoxypropylene, polyoxypropylated
PP	polypropylene
PPE	polyphenylene ether
PPG	polypropylene glycol
ppm	parts per million
PPO	polyphenylene oxide
PPS	polyphenylene sulfide
PS	polystyrene
PTFE	polytetrafluoroethylene
PTMEG	polytetramethylene ether glycol
PU, PUR	polyurethane
PVA, PVAL	polyvinyl alcohol
PVAc	polyvinyl acetate
PVB	polyvinyl butyral
PVC	polyvinyl chloride
PVDC	polyvinylidene chloride
PVDF	polyvinylidene fluoride

PVE	polyvinyl ethyl ether
PVF	polyvinyl fluoride
PVM	polyvinyl methyl ether
PVP	polyvinyl pyrrolidone
RCRA	Resource Conservation and Recovery Act (EPA)
RFI	radio frequency interference
RIM	reaction injection molded (molding)
RTM	resin transfer molding
RTV	room temperature vulcanizing
RV	recreational vehicle
SAN	styrene-acrylonitrile
S/B	styrene/butadiene
SBR	styrene/butadiene rubber
SBS	styrene-butadiene-styrene
SDA	specially denatured alcohol
SE	self emulsifying
SMA	styrene maleic anhydride
SMC	sheet molding compound
SPF	sun protection factor
SR	styrene rubber
SRF	semi-reinforced furnace
TBHQ	*tert*-butylhydroquinone
TDI	toluene diisocyanate
TEA	triethanolamine, triethanolamide
TFE	tetrafluoroethylene
THF	tetrahydrofuran
TIPA	triisopropanolamine
TMC	thick molding compound
TMPTA	trimethylolpropane triacrylate
TPGDA	tripropylene glycol diacrylate
TPO	thermoplastic polyolefin
UF	urea-formaldehyde
UHF	ultra-high frequency
UHMW	ultra high molecular weight
UHMWPE	ultra high molecular weight polyethylene
UL	Underwriter's Laboratory
UN	United Nations
UPVC	unplasticized polyvinyl chloride
USDA	United States Department of Agriculture

USP	United States Pharmacopeia
uv	ultra-violet
VA	vinyl acetate
VAE	vinyl acetate ethylene
VC	vinyl chloride
VdC, VDC	vinylidenechloride
VHF	very high frequency
VOC	volatile organic compounds
v/v	volume by volume
v/w	volume by weight
w/o	water in oil
w/v	weight by volume
w/w	weight by weight
XLPE	cross-linked polyethylene

PART I

DICTIONARY

Chemical Names
and Synonyms

1   **Abalyn**
127-25-3   6036   G

$C_{21}H_{32}O_2$
Methyl abietate.
4-09-00-02176 (Beilstein Handbook Reference); Abietic acid, methyl ester; AI3-01745; BRN 2702228; EINECS 204-832-7; Methyl 7,13-abietadien-18-oate; Methyl abietate; NSC 2141; Podocarpa-7,13-dien-15-oic acid, 13-isopropyl-, methyl ester. Resin with compatibility, surface wetting properties, viscosity. and tack. Used in lacquers, inks, paper coatings, varnishes, adhesives, sealing compounds, plastics, wood preservatives and perfumes. Pale yellow leaflets; bp = 360-365°, bp_{16} = 225-226°; d^{20} = 1.040; λ_m = 220, 233, 241 nm (ε = 6880, 7100, 7260, MeOH); insoluble in H_2O, soluble in EtOH, AcOH. *Hercules Inc.* 204-832-7

2   **Abietic acid**
514-10-3   5   G

$C_{20}H_{30}O_2$
[1R-(1α,4aβ,4bα,10aα)]-1,2,3,4,4a,4b,5,6,10,10a-Deca-hydro-1,4a-dimethyl-7-(1-methylethyl)-1-phenanthrene-carboxylic acid.
7,13-Abietadien-18-oic acid; Abietate; Abietic acid; I-Abietic acid; AI3-17273; CCRIS 3183; EINECS 208-178-3; 13-Isopropylpodocarpa-7,13-dien-15-oic acid; Kys-elina abietova; NSC 25149; Podocarpa-7,13-dien-15-oic acid, 13-isopropyl-; Sylvic acid. A major active ingredient of rosin; used as varnish driers, in soaps, and in the fermentation industry. Used in lacquers, varnishes, soaps and plastics. Plates; mp = 173.5°; bp_9 = 250°; $[\alpha]_D^{20}$ = -116° (c = 1 EtOH), λ_m = 235, 241.5, 250 nm (ε 19500, 22000, 14300); insoluble in H_2O, very soluble in EtOH, Et_2O, Me_2CO, C_6H_6; LD_{50} (mus iv) = 180 mg/kg. *Hercules Inc.* 208-178-3

3   **Ablumide SDE**
93-82-3   G

$C_{22}H_{45}NO_3$
Stearamide-DEA (1:1).
N,N-Bis(2-hydroxyethyl)octadecanamide; N,N-Bis(2-hydroxyethyl)stearamide; Diethanolamine stearic acid amide; EINECS 202-280-1; Octadecanamide, N,N-bis(2-hydroxyethyl)-;

Stearamide DEA; Stearic acid diethanolamide; Stearoyl diethanolamide; N,N-Bis(hydroxyethyl)octadecanamide; N,N-Bis(2-hydroxy-ethyl)octadecanamide; N,N-Bis(2-hydroxyethyl)stear-amide; N,N-Bis(β-hydroxyethyl)stearamide; Clindrol 868; Clindrol 200-S; Cyclomide SD; Diethanolamine stearic acid amide; Diethanolstearamide; Onyxol 42; Schercomid ST; Stearamide DEA; Stearic acid diethanolamide; Stearic diethanolamide; Stearoyl Diethanolamide; N-Stearoyl diethanolamine; Unamide S. Thickener, emulsifier for mineral and vegetable oils, microcrystalline wax. *Heterene; Taiwan Surfactants.* 202-280-1

4   **Ablunol GML**
142-18-7   E G

$C_{15}H_{30}O_4$
Glyceryl laurate.
Ablunol GML; Aldo® MLD; (±)-2,3-Dihydroxypropyl dodecanoate; 2,3-Dihydroxypropyl laurate; Dodecanoic acid α-monoglyceride; Dodecanoic acid, 2,3-dihydroxypropyl ester; Dodecanoic acid, 2,3-dihydroxypropyl ester; 3-Dodecanoyloxy-1,2-propanediol; EINECS 205-526-6; Glyceryl laurate; 1-Glyceryl laurate; Glycerol 1-laurate; Glycerol α-monolaurate; Glycerin 1-monolaurate; Glyceryl monododecanoate; (±)-Glyceryl 1-monododecanoate; Glyceryl monolaurate; Grindtek ML 90; Lauric acid α-monoglyceride; Lauric acid 1-monoglyceride; Lauricidin; Laurin, 1-mono-; 1-Monododecanoylglycerol; α-Monolaurin; 1-Monolaurin; 1-Monolauroyl-rac-glycerol; 1-Monomyristin; Laurin, 1-mono-2,3-dihydroxypropyl dodecanoate; Glycerin 1-monolaurate; Glycerol 1-dodecanoate; Glycerol 1-laurate; Glycerol 1-monolaurate; Glycerol α-monolaurate; 1-Glyceryl laurate; Glyceryl monododecanoate; Glyceryl monolaurate; Lauric acid 1-monoglyceride; Lauric acid α-monoglyceride; 1-Monolaurin. A dispersant and emulsifier for cosmetic, pharmaceutical and industrial use. Used as a dispersant for food products, oils, waxes, solvents; antifoaming agent; drycleaning soap base. Used in mold release agents. Component of water-oil and oil-water creams, lubricant, antistat, antifogging agent in plastics; antimicrobial effects reported. *Grindsted UK; Henkel/Emery; Lonzagroup; Protameen; Taiwan Surf.*

5   **Ablusol DBD**
26545-53-9   E G

$C_{18}H_{30}O_3S.C_4H_{11}NO_2$
DEA dodecylbenzene sulfonate.
AI3-28007; BAS 089-00E; Benzenesulfonic acid, dodecyl-, compd. with 2,2'-iminobis(ethanol); Caswell No. 413D; DEA-Dodecylbenzenesulfonate; Diethanol-amine do-decylbenzene sulfonate; Diethanolamine do-decyl-benzenesulfonate; Dihydro-3-(tetrapropenyl)-2,5-furan-dione; Dihydro-3-(tetrapropenyl)furan-2,5-di-one; Do-decylbenzenesulfonic acid, diethanolamine salt; Dodecylbenzenesulphonic acid, compound with 2,2'-iminodiethanol (1:1); DSA; DSA (crosslinking agent); EINECS 247-781-6; EINECS 247-784-2; EPA Pesticide Chemical Code 079015; 2,5-Furandione, dihydro-3-(tetrapropenyl)-; 2,5-Furandione, dihydro-3,3,4,4-tetra-1-propenyl-; RD 174; Succinic anhydride, (tetrapropenyl)-;

Tetrapropenylsuccinic anhydride; Bis(2-hydroxyethyl)-ammonium dodecylbenzenesulf-onate; BAS 089-00E; Diethanolamine dodecylbenz-enesulfonate; Diethanol-amine-dodecylbenzenesulf-onic acid adduct; Diethanol-amine-dodecylbenzene-sulfonic acid salt; Diethanol-ammonium dodecylbenz-enesulfonic acid; Dodecyl-benzenesulfonic acid dieth-anolamine salt; Dodecylbenzenesulfonic acid, diethan-olamine salt. Mixture of o, m and p-isomers. Surfactant, emulsifier, wetting agent for bubble baths, shampoos and detergents. *Taiwan Surfactants.*

6   Acephate
30560-19-1   32   H P

C4H10NO3PS

O,S-Dimethyl acetylphosphoramidothioate.

12420; 75 SP; Acephat; Acephate; Acetamidophos; Acetylphosphoramidothioic acid O,S-dimethyl ester; AI3-27822; Asataf; BRN 1936365; Caswell No. 002A; Chevron Orthene; Chevron RE 12,420; Dimethyl acetylphosphoramidothioate; ENT 27822; EPA Pesticide Chemical Code 103301; HSDB 6549; Kitron; N-(Methoxy(methylthio)phosphinoyl)acetamide; Orthene; Orthene-755; Ortho; Ortho 124120; Ortran; Ortril; Phosphoramidothioic acid, N-acetyl-, O,S-dimethyl ester; RE 12420. Registered by EPA as an insecticide. White solid; mp = 64 - 68°; 82 - 93°; soluble in H2O (79 g/100 ml), Me2CO (15.1 g/100 ml), EtOH >10 g/100 ml), EtOAc (3.5 g/100 ml), C6H6 (1.6 g/100 ml), C6H14 (0.01 g/100 ml), poorly soluble in aromatic solvents; LD50 (rat orl) = 700 mg/kg, (mrat orl) = 945 mg/kg, (frat orl) = 866 mg/l, (mus orl) = 361 mg/kg, (dog orl) > 681 mg/kg, (rbt der) > 2000 mg/kg, (mallard duck orl) = 350 mg/kg, (ckn orl) = 852 mg/kg; LC50 (bluegill sunfish 96 hr.) = 2050 mg/l, (rainbow trout 96 hr.) > 1000 mg/l, (channel catfish 96 hr.) = 2230 mg/l, (goldfish 96 hr.) = 9550 mg/l; toxic to bees. *Creative Sales Inc.; Drexel Chemical Co.; Ecolab Inc.; Micro-Flo Co. LLC; Scotts Co.; UAP - Platte Chemical; Valent USA Corp; Value Gardens Supply LLC; Whitmire Micro-Gen Research Laboratories Inc.*

7   Acesulfame
33665-90-6   38   G

C4H5NO4S

6-Methyl-1,2,3-oxathiazin-4(3H0-one 2,2-dioxide.

Acesulfame; Acesulfamo; Acesulfamum; Acetosulfam; 3,4-Dihydro-6-methyl-1,2,3-oxathiazin-4-one-2,2-di-oxide; EINECS 251-622-6; HSDB 3914; 6-Methyl-1,2,3-oxathiazin-4(3H)-on 2,2-dioxid; 6-Methyl-1,2,3-oxathia-zin-4(3H)-one 2,2-dioxide; 6-Methyl-3,4-dihydro-1,2,3-oxathiazin-4-one 2,2-dioxide; 1,2,3-Oxathiazin-4(3H)-one, 6-methyl-, 2,2-dioxide; Sunett. Sweetening agent. Needles; mp = 123-123.5°. AG; Hoechst.

8   Acesulfame potassium
55589-62-3   38   G

C4H5NO4S

6-Methyl-1,2,3-oxathiazin-4(3H)-one 2,2-dioxide.

Acesulfam-K; Acesulfame K; Acesulfame potassium; Acesulfame potassium salt; CCRIS 1032; EINECS 259-715-3; H733293; Hoe 095; 6-Methyl-1,2,3-oxathiazin-4(3H)-one 2,2-dioxide, potassium salt; 1,2,3-Oxathiazin-4(3H)-one, 6-methyl-, 2,2-dioxide, potassium salt; Potassium 6-methyl-1,2,3-oxathiazin-4(3H)-one 2,2-dioxide; Potassium acesulfame. Sweetener used in foods and cosmetics. White solid; dec 225°; d = 1.81; λ_m = 225 nm (ε 10762); soluble in H2O (360 g/l), organic solvents; LD50 (rat orl) = 7431 mg/kg. *Hoechst AG.*

9   Acetal
105-57-7   39   D G

C6H14O2

1,1-Diethoxyethane.

.4-01-00-03103 (Beilstein Handbook Reference); Acetaal; Acetal; Acetal diethylique; Acetaldehyde diethyl acetal; Acetaldehyde ethyl acetal; Acetale; Acetol; Aceton® NS; Acetron®GP; AI3-24135; AT-20GF; BRN 1098310; Diaethylacetal; 1,1-Diaethoxy-aethan; 1,1-Diäthoxy-äthan; 1,1-Diethoxy-ethaan; 1,1-Diethoxyethane; Diethyl acetal; Diethylacetal; 1,1-Dietossietano; EINECS 203-310-6; Ethane, 1,1-diethoxy-; Ethylidene diethyl ether; FEMA No. 2002; HSDB 1635; NSC 7624; UN1088; USAF DO-45 ; Cadco® Acetal; Delrin® 100, 500; Delrin®100ST, 500T; Delrin® 107, 507; Delrin® 150 SA, 550SA; Delrin® 570; Delrin®900; Delrin® AF Blend; Electrafil® J-80/CF/10/TF/10; Thermocomp® KB-1008. Solvent in synthetic perfumes such as jasmine; used for organic syntheses. Acetal resin, 20% glass fiber-reinforced; offers lubricity and chemical and hot water resistance for automotive, hardware, plumbing applications. Acetal compound with solid lubricants; for bearing and wear applications, e.g. bearings, bushings, valve seats, seals, wear surfaces, rollers, gears, cams, liners, tooling fixtures, forming dies. Has been used medically as a sedative/hypnotic. Oil; mp = -100°; bp = 102.2°; d^{20} = 0.8254; soluble in H2O (5 g/100 ml), CHCl3, very soluble in Me2CO, freely soluble in EtOH, Et2O; LD50 (rat orl) = 4.57 gm/kg. *Polymer Ag Inc.*

10   Acetaldehyde
75-07-0   40   G H

C2H4O

Ethyl aldehyde.

Acetaldehyd; Acetaldehyde; Acetic aldehyde; Acetic ethanol; Acetylaldehyde; AI3-31167; Aldehyde acetique; Aldeide acetica; CCRIS 1396; EINECS 200-836-8; Ethanal; FEMA No. 2003; HSDB 230; NCI-C56326; NSC 7594; Octowy aldehyd; RCRA waste number U001; UN1089. Used in manufacture of acetic acid, acetic anhydride, n-butanol, peracetic acid, pentaerythritol, pyridines, 1,3-butylene glycol, trimethylolpropane; synthetic flavors. Volatile liquid; mp = -123°; bp = 20.1°; d^{20} = 0.942; miscible with H2O, EtOH; LD50 (rat orl) = 1930 mg/kg.

11   Acetaldehyde ammonia
75-39-8   41   G

C2H7NO
1-Aminoethanol.
Acetaldehyde ammonia; Al3-52423; Aldamine; EINECS 200-868-2; Ethanol, 1-amino-; Hexahydro-2,4,6-trimethyl-s-triazine trihydrate; UN1841; Velsan. Exists as a trimer which is used in the manufacture of plastics and as a pickling inhibitor for steel, and is a rubber vulcanizing accelerator. A proprietary rubber vulcanization accelerator. Crystals; mp = 97°; bp= 110°; soluble in H2O, less soluble in Et2O.

12   Acetaldoxime
107-29-9   43

C2H5NO
Acetaldehyde oxime.
4-01-00-03121 (Beilstein Handbook Reference); Acetaldehyde, oxime; Acetaldoxime; Al3-52258; Aldoxime; BRN 1209252; CCRIS 1379; EINECS 203-479-6; Ethanal oxime; Ethylidenehydroxylamine; HSDB 2662; Hydroiminoethane; NSC 4974; UN2332; USAF AM-5. Needles; mp = 45°; bp = 115°; d^{20} = 0.9656; λ_m = 190 nm (ε = 5012); freely soluble in EtOH, Et2O, soluble in H2O, CHCl3.

13   Acetamide MEA
142-26-7   G

C4H9NO2
N-Acetyl ethanolamine.
2-Acetamidoethanol; 2-Acetylaminoethanol; N-Acetyl-2-aminoethanol; N-Acetyl ethanolamine; 4-04-00-01535 (Beilstein Handbook Reference); Acetamide MEA; Acetamide, N-(2-hydroxyethyl)-; Acetic acid amide, N-(2-hydroxyethyl)-; Acetylcolamine; Acetyl-ethanolamine; Al3-02836; Amidex AME; BRN 1811708; Carsamide® AMEA; EINECS 205-530-8; N-Ethanolacetamide; HSDB 2713; Hydroxyethyl acetamide; β-Hydroxyethylacetamide; N-(2-Hydroxy-ethyl)acetamide; N-β-Hydroxyethylacetamide; N-2-Hydroxyethylacet-amide; Mackamide AME-75; Mack-amide AME-100; NSC 5999; Foamid AME-70; Foamid AME-75; Foamid AME-100; Hetamide MA; Incro-mectant AMEA-100, AMEA-70; Lipamide MEAA; Mackamide ™ AME-75, AME-100; Schercomid AME; Upamide ACMEA; Witcamide® CMEA. Used as antistat, humectant, conditioner for skin and hair products. Solvent, humectant, skin and hair cond-itioner, intermediate, coupling agent, pigment dispers-ant, clarifying agent for shampoos, moisturizer. Humectant; surfactant, thickener, foam booster/stabilizer for personal care and industrial applications. Solid; mp = 65°; bp8 = 166-167°; d^{25} = 1.1079; slightly soluble in C6H6, ligroin, soluble in Me2CO, freely soluble in H2O. *Chemron; McIntyre*.

14   Acetamide, N-(4-methoxy-3-nitro-phenyl)-
50651-39-3   H

C9H10N2O4
3'-Nitro-p-acetanisidide.
Acetamide, N-(4-methoxy-3-nitrophenyl)-; 4-Acetyl-amino-2-nitroanisole; EINECS 256-686-9; 4'-Methoxy-3'-nitroacetanilide; N-(4-Methoxy-3-nitrophenyl)acet-amide; 3'-Nitro-p-acetanisidide; NSC 14685.

15   Acetaminophen
103-90-2   48   D G

C8H9NO2
4'-Hydroxyacetanilide.
222 AF; A-Per; A.F. Anacin; Abenol; Abensanil; Abrol; Abrolet; Acamol; Acenol; Acertol; Acetagesic; Acetalgin; Acetaminofen; Acetaminophen; Acetamol; Acetanilide, 4'-hydroxy-; Acetofen; Actifed Plus; Actron; Afebrin; Afebryl; Aferadol; Algesidal; Algina; Algomol; Algotropyl; Alpiny; Alvedon; Amadil; Anadin dla dzieci; Anaflon; Analter; Andox; Anelix; Anexsia; Anhiba; Anti-Algos; Antidol; Apacet; Apamide; APAP; Apitrelal; Arfen; Asetam; Asomal; Aspac; Aspirin-Free Anacin; Asplin; Atralidon; Babikan; Bacetamol; Banesin; Bayer Select Allergy-Sinus; Bayer Select Head Cold; Bayer Select Headache Pain; Bayer Select Menstrual Multi-Symptom; Bayer Select Sinus Pain Relief; Ben-u-ron; Benmyo; Biocetamol; Cadafen; Calapol; Calmanticold; Calpol; Capital with Codeine; Captin; Causalon; CCRIS 3; Cefalex; Children's Acetaminophen Elixir Drops; Children's Acetamino-phen Elixir Solution; Children's Acetaminophen Oral Solution; Children's Tylenol Chewable; Claradol Codeine; Clixodyne; Cod-Acamol Forte; Codabrol; Codalgin; Codapane; Codicet; Codisal; Codisal Forte; Codoliprane; Codral Pain Relief; Cofamol; Contac Cough & Sore Throat Formula; Contra-Schmerz P; Coricidin; Coricidin D; Coricidin Sinus; Cosutone; Crocin; Croix Blanche; Cuponol; Curadon; Curpol; Custodial; Dafalgan Codeine; Darocet; Darvocet-N; Datril; Daygrip; DEA No. 9804; Demilets; Deminofen; Democyl; Demogripal; Desfebre; Dhamol; DHCplus; Dial-a-gesic; Dirox; Disprol; Dol-Stop; Dolcor; Dolefin; Dolegrippin; Dolgesic; Doliprane; Dolko; Dolofugin; Doloreduct; Dolorfug; Dolorol Forte; Dolorstop; Dolotec; Dolprone; Dorocoff; Dresan; DRG-0007; Dristan Cold No Drowsiness; Dristancito; Drixoral Cold & Flu; Drixoral Sinus; Duaneo; Duorol; Duracetamol; Durapan; Dymadon; Dymadon Co; Dymadon Forte; Ecosetol; EINECS 203-157-5; Empracet; Endecon; Enelfa; Eneril; Eu-Med; Excipain; Exdol; Fanalgic; Farmadol; Febranine; Febrectal; Febrectol; Febrex; Febricet; Febridol; Febrilix; Febrin; Febro-gesic; Febrolin; Fendon; Fensum; Fepanil; Finimal; Finiweh; Fluparmol; Fortalidon P; Freka-cetamol; Gattaphen T; Gelocatil; Geluprane; Geralgine-P; Gripin Bebe; Grippostad; Gynospasmine; Hedex; Helon N; Homoolan; HSDB 3001; Hy-Phen; Hycomine Compound; Ildamol; Inalgex; Infadrops; Influbene N; Intensin; Junior Disprol; Kataprin; Kinder Finimal; Kratofin simplex; Labamol; Lekadol; Lemgrip; Lemsip; Lestemp; Liquagesic; Liquigesic Co; Liquiprin; Lonarid; Lonarid Mono; Lupocet; Lyteca; Magnidol; Malex N; Malgis; Malidens; Maxadol; Medinol Paediatric; Medocodene; Melabon Infantil; Mexalen; Midol Maximum Strength; Midol PM Night Time Formula; Midol Regular Strength; Midol Teen Formula; Migraleve Yellow; Minafen; Minoset; Miralgin; Mixture Name; Momentum; Mono Praecimed; Multin; N-(4-Hydroxyphenyl)acetamide; N-Acetyl-p-aminophenol; Naldegesic; Naprinol; NCI-C55801; NCX 701; Nebs; Neo-Fepramol; NeoCitran; Neodol; Neodolito; Neuridon; New Cortal for Children; NilnOcen; Nina; No-Febril; Nodolex; Noral; NSC 109028; Oltyl; Oralgan; Ornex Severe Cold Formula; Ortensan;

Oxycocet; p-Acetamidophenol; p-Acetaminophenol; p-Acetylaminophenol; p-Hydroxyacetanilide; Paceco; Pacemo; Pacemol; Pacet; Pacimol; Paedialgon; Paedol; Paldesic; Pamol; Panacete; Panadeine; Panadeine Co; Panadiene; Panado-Co; Panado-Co Caplets; Panadol; Panamax; Panasorbe; Panets; Panodil; Pantalgin; Para-Suppo; Para-Tabs; Paracemol; Paracenol; Paracet; Paracetamol; Paracetamol; Paracetamol; Paracetamol AL; Paracetamol Antipanin P; Paracetamol Basics; Paracetamol BC; Paracetamol Dr. Schmidgall; Paracetamol Fecofar; Paracetamol Genericon; Paracetamol Hanseler; Paracetamol Harkley; Paracetamol Heumann; Paracetamol Hexal; Paracetamol Italfarmaco; Paracetamol Nycomed; Paracetamol PB; Paracetamol Raffo; Paracetamol Ratiopharm; Paracetamol Rosch; Paracetamol Saar; Paracetamol SmithKline Beecham; Paracetamol Stada; Paracetamol von ct; Paracetamol Winthrop; Paracetamolo; Paracetamolo [Italian]; Paracetamolum [INN-Latin]; Paracetol; Paracin; Paracod; Paracodol; Parador; Paradrops; Parakapton; Parake; Paralen; Paralief; Paralink; Paralyoc; Paramol; Paramolan; Paranox; Parasedol; Parasin; Parcetol; Parmol; Parogal; Paroma; PCM Paracetamol Lichtenstein; Pe-Tam; Pediapirin; Pediatrix; Percocet; Percocet-5; Percocet-Demi; Percogesic with Codeine; Perdolan Mono; Phenaphen; Phenaphen W/Codeine; Phenipirin; Phogoglandin; Pinex; Piramin; Pirinasol; Plicet; Polmofen; Predimol; Predualito; Prodol; Prontina; Propacet; Puernol; Pulmofen; Pyregesic-C; Pyrigesic; Pyrinazine; Pyromed; Quiet World; Reliv; Remedol; Rhinex D-Lay Tablets; Robitussin Night Relief; Rockamol Plus; Rounox; RubieMol; Rubophen; Rupemol; Salzone; Sanicet; Sanicopyrine; Scanol; Scentalgyl; Scherzatabletten Rezeptur 534; Schmerzex; Sedalito; Semolacin; Servigesic; Seskamol; Setakop; Setamol; Setol; Sifenol; Sinaspril; Sine-Aid, Maximum Strength; Sine-Off Sinus Medicine Caplets; Sinedol; Sinmol; Sinubid; Spalt fur die nacht; Spalt N; St Joseph Aspirin-Free for Children; St Joseph Aspirin-Free; St. Joseph Cold Tablets for Children; Stanback; Stopain; Sudafed Cold and Cough; Sudafed Severe Cold Formula; Sudafed Sinus; Sunetheton; Supac; Supadol mono; Supofen; Suppap; Supramol-M; Tabalgin; Tachiprina; Tamol; Tapar; Tazamol; Tempra; Termacet; Termalgin; Termalgine; Termofren; TheraFlu; Tiffy; Titralgan; Toximer P; Tralgon; Treupel mon; Treupel N; Treuphadol; Triaminic Sore Throat Formula; Tricoton; Tussapap; Tylenol; Tylenol; Tylenol Allergy Sinus; Tylex; Tylex CD; Tylol; Tylox; Tymol; Upsanol; Utragin; Valadol; Valgesic; Vanquish; Veralgina; Vermidon; Verpol; Viclor Richet; Vicodin; Vips; Viruflu; Vivimed; Volpan; Zatinol; Zolben.; component of: Actifed Plus, Allerest Sinus Pain Formula, Anexsia, Aspirin-Free Anacin, Children's Tylenol Cold Tablets, Contac Cough & Sore Throat Formula, Contac Jr Non-drowsy Formula, Contac Nighttime Cold Medicine, Contac Severe Cold Formula, Coricidin, Darvoset-N, Dristan Cold (Multisymptom), Dristan Cold (No Drowsiness), Empracet, Endecon, Gemnisyn, Headache Strength Allerest, Hycomine Compound, Hy-Phen, Intensin, Liquiprin, Maximum Strength Sine-Aid, Midol, Midol Maximum Strength, Midol PMS, Naldegesic, Naldetuss, Ornex, Percocet, Percogesic with Codeine, Propacet, Quiet World, Rhinex D-Lay Tablets, Sinarest, Sine-Off Maxiumum Strength Allergy/Sinus, Sine-Off Maximum Strength No Drowsiness Formula Caplets, Sinubid, St Joseph's Cold Tablets for Children, Sudafed Sinus, Supac, Teen Midol, TheraFlu, Tylenol Allergy Sinus, Tylenol Cold and Flu Multi-Symptom, Tylenol Cold Medication Caplets, Liquid and Tablets, Tylenol Cold Night TIme Liquid, Tylenol Cold No Drowsiness, Tylenol PM Tablets and Caplets, Tylenol with Codeine, Tylox, Vanquish, Vicodin, Wygesic, Zydone. Analgesic; antipyretic; anti-inflammatory. White solid; mp = 169-170.5°; d_4^{21} = 1.293; λ_m = 250 nm (ε 13800 EtOH); slightly soluble in cold H_2O, Et_2O; more soluble in hot H_2O; soluble in alcohol, dimethylformamide, ethylene dichloride, Me_2CO, EtOAc; nearly insoluble in petroleum ether, pentane, C_6H_6; LD_{50} (mus orl) = 338 mg/kg, (mus ip) = 500 mg/kg. *Bristol Myers Squibb Pharm. Ltd.; Bristol-Myers Squibb Co.; Forest Pharm. Inc.; McNeil Pharm.; Mead Johnson Nutritionals; Novo Nordisk Pharm. Inc.; Parke-Davis; Robins, A. H. Co.; Sterling Health U.S.A.; Warner-Lambert.*

16   Acetanilide-p-sulfonyl chloride
121-60-8   105   H

$C_8H_8ClNO_3S$
p-Acetamidobenzenesulfonyl chloride.
Acetanilide-p-sulfonyl chloride; Acetylsulfanilyl chloride; N-Acetylsulfanilyl chloride; N-Acetylsulphanilyl chloride; ASC; Benzenesulfonyl chloride, 4-(acetylamino)-; Dagenan chloride; EINECS 204-485-1; HSDB 2712; NSC 127860; Sulfanilyl chloride, N-acetyl-. Needles or prisms; mp = 149°; λ_m = 259 nm (ε = 17700, MeOH); soluble in C_6H_6, $CHCl_3$, very soluble in EtOH, Et_2O.

17   Acetic acid
64-19-7   56   G P

$C_2H_4O_2$
Ethanoic acid.
4-02-00-00094 (Beilstein Handbook Reference); Acetasol; Acetic acid; Acetic acid (natural); Acetic acid, glacial; Aci-Jel; Acide acetique; Acido acetico; AI3-02394; Azijnzuur; BRN 0506007; Caswell No. 003; CCRIS 5952; EINECS 200-580-7; EPA Pesticide Chemical Code 044001; Essigsäure; Essigsaeure; Ethanoic acid; Ethylic acid; FEMA Number 2006; Glacial acetic acid; HSDB 40; Kyselina octova; Methanecarboxylic acid; NSC 132953; Octowy kwas; Otic Domeboro; Otic Tridesilon; Pyroligneous acid; Shotgun; TCLP extraction fluid 2; UN2789; UN2790; Vinegar; Vinegar acid; Vosol. Used in manufacture of acetic anhydride, cellulose acetate, vinyl acetate monomer; acetic esters; production of plastics, pharmaceuticals, dyes, insecticides, photographic chemicals, food additives; solvent reagent. Registered by EPA as an antimicrobial and fungicide. FDA GRAS, BP, Europe listed, UK approved, USP/NF, BP compliance. Used as an acidifying agent, buffering agent, solvent, vehicle, flavoring agent, used in ophthalmics and otics. Used in manufacture of acetic anhydride, cellulose acetate, vinyl acetate monomer; acetic esters; production of plastics, pharmaceuticals, dyes, insecticides, photographic chemicals, food additives; solvent reagent. Used with thymol as a fumigant to reduce postharvest brown rot and blue mold rot in cherries. Clear liquid; mp = 16.6°; bp_{765} = 117.9°; d_4^{20} = 1.0492; miscible with H_2O, EtOH, glycerol, Et_2O, insoluble in CS_2. LD_{50} (rat orl) = 3310 mg/kg, TLV = 10 ppm in air. *Albright & Wilson Americas Inc.; Albright & Wilson UK Ltd.; BASF Corp.; BP Chem.; Daicel (U.S.A.) Inc.; Diamalt; Eastman Chem. Co.; Ellis & Everard; Frutarom; General Chem; Harris Keith; Hoechst AG; Hoechst Celanese; Integra; Integrated Ingredients; Janssen Chimica; Mallinckrodt Inc.; Penta Mfg.; Pfalz & Bauer; PMC; Quantum/USI; Ruger; Sibner-Hegner; Sigma-Aldrich Fine Chem.; Spectrum Chem. Manufacturing; Van Waters & Rogers.*

18   Acetic anhydride
108-24-7   57   G H

$C_4H_6O_3$
Acetic acid, anhydride.
4-02-00-00386 (Beilstein Handbook Reference); Acetanhydride; Acetic acid, anhydride; Acetic anhydride; Acetic oxide; Acetyl

acetate; Acetyl anhydride; Acetyl ether; Acetyl oxide; Anhydrid kyseliny octove; Anhydride acetique; Anidride acetica; Azijnzuuranhydride; BRN 0385737; Caswell No. 003A; CCRIS 688; EINECS 203-564-8; EPA Pesticide Chemical Code 044007; Essigsaeureanhydrid; Essigsäureanhydrid; Ethanoic anhydride; HSDB 233; Octowy bezwodnik; UN1715. Used in manufacture of cellulose acetate fibers and plastics and vinyl acetate; dehydrating and acetylating agent for pharmaceuticals, dyes, perfumes, explosives; aspirin; esterifying agent for food starch. Liquid; mp = -73°; bp = 139.5°; d^{20} = 1.082; slightly soluble in CCl_4, soluble in EtOH, $CHCl_3$, C_6H_6; very soluble in H_2O (reacts), freely soluble in Et_2O; LD_{50} (rat orl) = 1780 mg/kg. *Ashland; BP Chem.; Chisso Corp.; Eastman Chem. Co.; Hoechst Celanese; Lancaster Synthesis Co.; Union Carbide Corp.*

19   Acetoacetanilide
102-01-2   59   H

$C_{10}H_{11}NO_2$
2-Acetyl-acetanilide.
 2-Acetyl-acetanilide; Acetoacetamidobenzene; Aceto-acetanilid; Acetoacetanilide; Acetylacetanilide; AI3-00854; Anilid kyseliny acetoctove; Butanamide, 3-oxo-N-phenyl-; Butanoic acid, 3-oxo-, amide, N-phenyl-; CCRIS 4534; EINECS 202-996-4; HSDB 2669; β-Ketobutyranilide; NSC 2656; USAF EK-1239. Prisms or needles; mp = 86°; λ_m = 244 nm (MeOH); slightly soluble in H_2O, soluble in EtOH, Et_2O, C_6H_6, $CHCl_3$, ligroin.

20   Acetoacet-o-anisidide
92-15-9   H

$C_{11}H_{13}NO_3$
Butanamide, N-(2-methoxyphenyl)-3-oxo-.
 4-13-00-00865 (Beilstein Handbook Reference); Acetoacetyl-o-aniside; Acetoacetyl-o-anisidide; AI3-04822; BRN 1459707; Butanamide, N-(2-methoxyphenyl)-3-oxo-; EINECS 202-131-0; o-Methoxyacetoacetanilide; NSC 7563.

21   2-Acetoacetylaminotoluene
93-68-5   H

$C_{11}H_{13}NO_2$
 Butanamide, N-(2-methylphenyl)-3-oxo-.
 4-12-00-01777 (Beilstein Handbook Reference); Aceto-acet-o-toluidide; Acetoaceto-ortho-toluidide; Acetoacetyl-2-methylanilide; 2-Acetoacetylaminotol-uene AI3-08708; BRN 2099098; Butanamide, N-(2-methylphenyl)-3-oxo-; CCRIS 7750; EINECS 202-267-0; 2'-Methylaceto-acetanilide; N-(2-Methylphenyl)-3-oxobutanamide; NSC 7655.

22   Acetochlor
34256-82-1   62   H P

$C_{14}H_{20}ClNO_2$
 2-Chloro-N-(ethoxymethyl)-6'-ethyl-o-acetotoluidide.
 Acenit; Acetal; Acetochlor; Acetochlore; Azetochlor; BRN 2859702; Caswell No. 003B; CCRIS 7709; 2-Chloro-N-(ethoxymethyl)-N-(2-ethyl-6-methylphenyl)-acetamide; 2-Chloro-N-(ethoxymethyl)-6'-ethyl-o-ace-totoluidide; 2-Chloro-2'-methyl-6'-ethyl-N-ethoxy-methyl-acetanilide; CP 55097; Doubleplay; EPA Pesticide Chemical Code 121601; Erunit; Harness; HSDB 6550; MG 02; MON 097; MON-097; Nevirex; Top Hand; Topnotch. Registered by EPA as a herbicide. Straw-colored oil; mp = 0°; $bp_{0.4}$ = 134°; soluble in H_2O (0.0223 g/100 ml at 25°), soluble in Et_2O, Me_2CO, C_6H_6, $CHCl_3$, EtOH, EtOAc, C_7H_8; LD_{50} (rat orl) = 1160 mg/kg, 2953 mg/kg, (rbt der) = 3667 mg/kg, (bobwhite quail orl) = 1260 mg/kg, (bee) = 1.715 mg/bee; LC_{50} (rat ihl 4 hr.) = 3 mg/l air, (rainbow trout 96 hr.) = 0.5 mg/l, (bluegill sunfish 96 hr.) = 1.3 mg/l; toxic to bees. *Dow AgroSciences; Monsanto Co.; Pace International LLC; Syngenta Crop Protection.*

23   Acetone
67-64-1   67   G P

C_3H_6O
 Dimethyl ketone.
 Aceton; Acetone; Acetone pyroacetic ether; AI3-01238; Caswell No. 004; CCRIS 5953; Chevron acetone; Dimethyl formaldehyde; Dimethyl ketone; Dimethylketal; EINECS 200-662-2; EPA Pesticide Chemical Code 004101; FEMA No. 3326; HSDB 41; Ketone, dimethyl; Ketone propane; β-Ketopropane; Methyl ketone; NSC 135802; Propanone; 2-Propanone; Pyroacetic acid; RCRA waste number U002; UN1090. Disinfectant, possibly antibacterial. Registered by EPA (cancelled). FDA listed (30 ppm tolerance in spice oleoresins), FEMA GRAS, Japan approved with restrictions, USP/NF, BP compliance. Used as a solvent and flavoring agent. Used as a solvent for paints, varnishes and lacquers; for cleaning and drying precision equipment; delustrant for cellulose acetate fibers. Clear mobile liquid; mp = -94.8°; bp = 56.0°; d^{20} = 0.7899, d^{25} = 0.788; miscible with H_2O, EtOH, DMF, Et_2O; CCl_4, most oils; LD_{50} (rat orl) = 8300 mg/kg (10.5 ml/kg). *Allied Signal; Ashland; BASF Corp.; BP Chem.; Dow Chem. U.S.A.; Eastman Chem. Co.; Mitsui Petroleum; Montedipe SpA; Ruger; Shell; Sigma-Aldrich Fine Chem.; Spectrum Chem. Manufacturing; Texaco; Union Carbide Corp.*

24   Acetone cyanohydrin
75-86-5   68   H

C_4H_7NO
 2-Hydroxy-2-methyl-propanenitrile.

4-03-00-00785 (Beilstein Handbook Reference); Acetoncianhidrinei; Acetoncianidrina; Acetoncyaan-hydrine; Acetoncyanhydrin; Acetone cyanhydrin; Acet-one cyanohydrin; Acetonecyanhydrine; Acetonkyan-hydrin; AI3-04257; BRN 0605391; CCRIS 4657; Cyanhydrine d'acetone; EINECS 200-909-4; HSDB 971; Lactonitrile, 2-methyl-; NSC 131093; Propanenitrile, 2-hydroxy-2-methyl-; RCRA waste number P069; UN1541; USAF RH-8. Used in chemical synthesis Liquid; mp = -19°; bp$_{23}$ = 82°; d^{19} = 0.932; soluble in H$_2$O and polar organic solvents; LD$_{50}$ (rat orl) = 170 mg/kg. *Lancaster Synthesis Co.; Mallinckrodt Inc.; Sigma-Aldrich Fine Chem.*

25   Acetone dicarboxylic acid
542-05-2   69   G

C$_5$H$_6$O$_5$
1,3-Acetonedicarboxylic Acid.
Acetone-1,3-dicarboxylic acid; Acetonedicarboxylic acid; ADA; EINECS 208-797-9; β-ketoglutaric Acid; 3-Ketoglutaric acid; 3-Oxoglutaric acid; 3-Oxopentanedioic acid; Pentanedioic acid, 3-oxo-. Used in organic synthesis, manufacturing. Needles; mp = 138°; insoluble in C$_6$H$_6$, CHCl$_3$, ligroin, slightly soluble in Et$_2$O, soluble in H$_2$O, EtOH, EtOAc, very soluble in H$_2$O, EtOH, less soluble in organic solvents. *Dajac Labs.; Janssen Chimica; Lonzagroup; Penta Mfg.*

26   Acetone dimethyl ketal
77-76-9   H

C$_5$H$_{12}$O$_2$
2,2-Dimethoxypropane.
Acetone, dimethyl acetal; Acetone dimethyl ketal; AI3-26275; EINECS 201-056-0; NSC 62085; Propane, 2,2-dimethoxy-. Liquid; mp = -47°; bp = 83°; d^{25} = 0.847.

27   Acetonitrile
75-05-8   71   G H

—C≡N

C$_2$H$_3$N
Ethanenitrile.
Acetonitril; Acetonitrile; AI3-00327; CCRIS 1628; Cyanomethane; Cyanure de methyl; EINECS 200-835-2; Ethyl nitrile; HSDB 42; Methane, cyano-; Methane-carbonitrile; Methyl cyanide; Methylkyanid; NCI-C60822; NSC 7593; RCRA waste number U003; UN1648; USAF EK-488. Solvent for hydrocarbon extraction processes, especially for butadiene; intermediate; catalyst; for separation of fatty acids from vegetable oils; manufacture of synthetic pharmaceuticals. Liquid; mp = -43.8°; bp = 81.6°; d$_{20}$ = 0.7857; d$_{30}$ = 0.7712; miscible with H$_2$O, MeOH, organic solvents, immiscible with saturated hydrocarbons; LD$_{50}$ (rat orl) = 3800 mg/kg. *BP Chem.; DuPont UK; E. I. DuPont de Nemours Inc.; Greeff R.W. & Co.; Penta Mfg.*

28   Acetophenone
98-86-2   74   D G H

C$_8$H$_8$O
Ethanone, 1-phenyl-.
Acetofenon; Acetophenon; Acetophenone; Acetyl-benzene; Acetylbenzol; AI3-00575; Benzene, acetyl-; Benzoyl methide; CCRIS 1341; EINECS 202-708-7; Ethanone, 1-phenyl-; FEMA Number 2009; HSDB 969; Hypnone; Ketone, methyl phenyl; Methyl phenyl ketone; NSC 7635; Phenyl methyl ketone; RCRA waste number U004; USAF EK-496. Used in perfumery and as a solvent and an intermediate for pharmaceuticals, resins, etc.; in flavoring; as a polymerization catalyst, in organic synthesis. Has been used medically as a sedative-hypnotic mp = 20.5°; bp = 202°; d$_{15}$ = 1.033; λ$_m$ = 242, 279, 318 nm (ε = 12600, 1050, 60, EtOH); very slightly soluble in H$_2$O, freely soluble in EtOH, CHCl$_3$, Et$_2$O, fatty oils, glycerol; LD$_{50}$ (rat orl) = 0.90 g/kg. *BP Chem.; EniChem Am.; Janssen Chimica; Mitsui Petroleum; Mitsui Toatsu; Montedipe SpA; Penta Mfg.*

29   N-Acetoxyethyl-N-cyanoethylaniline
22031-33-0   H

C$_{13}$H$_{16}$N$_2$O$_2$
Propanenitrile, 3-((2-(acetyloxy)ethyl)phenylamino)-.
3-((2-(Acetyloxy)ethyl)phenylamino)propanenitrile; N-Acetoxyethyl-N-cyanoethylaniline; Aniline, N-acetoxy-ethyl-N-cyanoethyl-; 2-(N-(2-Cyanoethyl)anilino)ethyl acetate; EINECS 244-740-4; Propanenitrile, 3-((2-(acetyloxy)ethyl)phenylamino)-; Propionitrile, 3-(N-(2-hydroxyethyl)anilino)-, acetate (ester).

30   Acetoxytrimethylsilane
2754-27-0   G

C$_5$H$_{12}$O$_2$Si
Trimethylsilylacetate.
(Acetato)trimethylsilane; CT3254; EINECS 220-404-2; NSC 96780; Silanol, trimethyl-, acetate; Trimethyl-acetoxysilane; Trimethylsilyl acetate. Coupling agent, chemical intermediate, blocking agent, release agent, lubricant, primer, reducing agent. *Degussa-Hüls Corp.*

31   Acetrizoate sodium
129-63-5   81   G

C9H5I3NNaO3
3-(Acetylamino)-2,4,6-triiodobenzoic acid sodium salt.

3-Acetamido-2,4,6-triiodobenzoic acid sodium salt; Acetrizoate de sodium; Acetrizoate sodium; Acetrizoato sodico; Acetrizoic acid sodium salt; Acidum acetrizoicum; Benzoic acid, 3-acetamido-2,4,6-triiodo-, monosodium salt; Bronchoselectan; Cystokon; Diaginol; EINECS 204-956-1; Fortombrine-N; Iodopact; Iodopaque; Jodopax; MP 1023; Natrii acetrizoas; Pyelokon-R; Reopak; Salpix; Sodium 3-acetamido-2,4,6-triiodo-benzoate; Sodium 3-acetyl-amino-2,4,6-triiodobenzoate; Sodium acetriz-oate; Sodium urokon; Thixokon; Tri-abrodil; Triiodyl; Triio-trast; 2,4,6-Trijod-3-acetamino-benzosaeure natrium; Triopac; Triopac 200; Triopas; Triumbren; Triurol; Triuropan; Urokon Sodium; Vesamin. A radiopaque medium, used as a diagnostic aid. Marketed as a sterile 30% aqueous solution.

Free acid: mp = 278-283° (dec); soluble in H2O (94 g/100 ml), EtOH; LD50 (rbt iv) = 5200 mg/kg.

32   **Acetulan**
8028-98-6   5374   G

C18H36O2
Cetyl acetate.
Acetol; Acetylated lanolin alcohol; Acetulan®. Acetylated lanolin alcohol; binder for pressed powders; emollient, plasticizer, cosolvent, NV and sebum solvent for personal care products; lubricant for clay, talc, and starch; stabilizer for lanolin; solubilizer in Aerosols®; penetrant and spreading agent. Pale yellow oil; d = 0.867; miscible with mineral oil, castor oil, vegetable oil, i-PrOH, EtOH, isopropyl palmitate, butyl stearate. *Amerchol.*

33   **Acetylacetone**
123-54-6   83   G H

C5H8O2
Pentane-2,4-dione.
4-01-00-03662 (Beilstein Handbook Reference); ACAC; Acetoacetone; Acetyl 2-propanone; Acetyl acetone; AI3-02266; BRN 0741937; CCRIS 3466; Diacetylmethane; EINECS 204-634-0; HSDB 2064; NSC 5575; Pentan-2,4-dione; Pentane-2,4-dione; 2,4-Pentanedione; Pentane-dione; UN2310. Solvent for cellulose acetate; intermediate; chelating agent for metals, paint drier, lubricant additives, pesticides. Liquid; mp = -23°; bp = 138°; d^{25} = 0.9721; λ_m = 272 nm (ε = 6800, MeOH); very soluble inH2O, freely soluble in EtOH, Et2O, Me2CO, CHCl3; LC50 (rat inh) = 4000.ppm/4 hrs. *Penta Mfg.; Sigma-Aldrich Fine Chem.; Union Carbide Corp.; Wacker Chemie GmbH.*

34   **Acetyl chloride**
75-36-5   87   G H

C2H3ClO
Acetic acid, chloride.
4-02-00-00395 (Beilstein Handbook Reference); Acetic acid, chloride; Acetic chloride; BRN 0605303; CCRIS 4568; EINECS 200-865-6; Ethanoyl chloride; HSDB 662; RCRA waste number U006; UN1717. Acetylating agent for organic preparations; used in manufacture of dyestuffs and pharmaceuticals. Liquid; mp = -112.8°; bp = 50.7°; d^{20} = 1.1051; miscible with C6H6, CHCl3, Et2O, AcOH, petroleum ether; reacts with H2O. *Lancaster Synthesis Co.; Sigma-*

Aldrich Fine Chem.

35   **Acetylene**
74-86-2   92   G H

H————H

C2H2Cl2
Ethyne.
Acetylen; Acetylene; EINECS 200-816-9; Ethine; Ethyne; HSDB 166; Narcylen; UN1001. An asphyxiant gas; intermediate for manufacture of vinyl chloride, vinylidene chloride, vinyl acetate, acrylates, acrylonitrile, acetaldehyde, perchloroethylene trichloroethylene, 1,4-butanediol, carbon black, welding and cutting metals. Colorless gas; mp = -81°; d = 1.165 g/l (0°, 760 mm), d of gas = 0.90 (air = 1); soluble in H2O (1:1 v/v), soluble in AcOH, EtOH, Et2O, C6H6, Me2CO; LC (rat inh) = 9000 ppm. *Air Products & Chemicals Inc.; BASF Corp.; BOC Gases; Union Carbide Corp.*

36   **Acetylene dichloride**
540-59-0   94   G

C2H2Cl2
1,2-dichloroethylene.
2-01-00-00158 (Beilstein Handbook Reference); Acetylene dichloride; BRN 1719345; CCRIS 6227; 1,2-DCE; 1,2-Dichlor-aethen; 1,2-Dichlor-äthen; 1,2-Dichloroethene; 1,2-Dichloroethylene; sym-Dichloro-ethylene; Dichloro-1,2-ethylene; Dioform; EINECS 208-750-2; Ethene, 1,2-dichloro-; Ethylene, 1,2-dichloro-; HSDB 149; NCI-C56031; dichloroacetylene. Used as a general solvent for organic materials, dye extraction, perfume, lacquers, thermoplastics, organic synthesis. Liquid; mp = -57°; bp = 48-60°; d = 1.265; insoluble in H2O, soluble in organic solvents; LD50 (rat orl) = 770 mg/kg; LD50 (mus ip) = 2150 mg/kg; gradually decomposes from air, light and moisture forming HCl. *Sigma-Aldrich Fine Chem.*

37   **Acetyl methyl carbinol**
513-86-0   65   G

C4H8O2
3-Hydroxy-2-butanone.
2-01-00-00870 (Beilstein Handbook Reference); Acethoin; Acetoin; Acetyl methyl carbinol; AI3-03314; BRN 0385636; Butan-2-ol-3-one; 2,3-Butanolone; 2-Butanone, 3-hydroxy-; CCRIS 2918; Dimethylketol; 2-Hydroxy-3-butanone; 3-Hydroxy-2-butanone; 1-Hydroxyethyl methyl ketone; EINECS 208-174-1; FEMA No. 2008; γ-Hydroxy-β-oxobutane; HSDB 974; Methanol, acetylmethyl-; NSC 7609; UN2621. Aroma carrier; preparation of flavors and essences. Liquid; mp = 15°; bp = 148°; d_{17} = 0.9972; freely soluble in H2O, EtOH, sparingly soluble in Et2O, petroleum ether. *BASF Corp.; Penta Mfg.*

38   **2-Acetyl pyridine**
1122-62-9   G

9

C7H7NO

Methyl-2-pyridyl ketone.

AI3-52210; 2-Acetopyridine; 2-Acetylpyridine; CCRIS 7784; EINECS 214-355-6; Ethanone, 1-(2-pyridinyl)-; FEMA No. 3251; Ketone, methyl 2-pyridyl; Methyl 2-pyridyl ketone; NSC 15043; 1-(2-Pyridinyl)ethanone; 2-Pyridyl methyl ketone. Chemical intermediate. Crystals; mp = 8-10°; bp = 192°; d = 1.082; λ_m = 226, 258, 266, 276 nm (log ε = 4.43, 2.82, 2.85, 2.60); insoluble in H_2O, slightly soluble in CCl_4, soluble in EtOH, Et2O, AcOH; LD50 (rat orl) = 1495 mg/kg, (rat ip) = 1150 mg/kg. *Lancaster Synthesis Co.; Penta Mfg.; Raschig GmbH; Reilly Ind.; Spectrum Chem. Manufacturing.*

39   Acetyltributylcitrate

77-90-7   G H

C20H34O8

Tributyl 2-(acetyloxy)-1,2,3-propanetricarboxylate.

Acetyl tributyl citrate; Acetylcitric acid, tributyl ester; AI3-01999; Blotrol; BRN 2303316; Caswell No. 005AB; CCRIS 3409; Citric acid, tributyl ester, acetate; Citroflex A; Citroflex A 4; EINECS 201-067-0; FEMA No. 3080; HSDB 656; NSC 3894; Tributyl 2-(acetyloxy)-1,2,3-propanetricarboxylate; Tributyl 2-acetoxy-1,2,3-propanetricarboxylate; Tributyl acetyl citrate; Tributyl citrate acetate; Tributyl O-acetylcitrate. Plasticizer for PVC, PVDC, especially food films, medical articles. Plasticizer for indirect and direct food contact applications; milling lubricant for aluminum foil or sheet steel for use in cans for beverage and food products; in PVC toys, cellulose nitrate films, Aerosol® hair sprays, and dairy product cartons. Colorless oil; mp = -80°; bp = 172-174°; d = 1.048; insoluble in H_2O, soluble in organic solvents. *Yamanouchi Europe B.V.; Yamanouchi Pharma.*

40   Acetyl triethyl citrate

77-89-4   G

C14H22O8

1,2,3-Propanetricarboxylic acid, 2-(acetyloxy)-, triethyl ester.

3-03-00-01105 (Beilstein Handbook Reference); 2-(Acetyloxy)-1,2,3-propanetricarboxylic acid, triethyl ester; Acetyl triethyl citrate; AI3-03574; ATEC; BRN 1804947; Citric acid, acetyl triethyl ester; Citric acid, triethyl ester, acetate; Citroflex A 2; EINECS 201-066-5; NSC 3887; Tricarballylic acid, β-acetoxytributyl ester; Triethyl 2-acetoxy-1,2,3-propanetricarboxylate; Trieth-yl acetyl-citrate; Triethyl citrate, acetate; Triethyl O-acetylcitrate; Triethylester kyseliny acetylcitronove. Plasticizer for cellulosics; used in food packaging materials. Liquid; bp40 = 214°, bp100 = 228-229°; d = 1.136; $n^{20}D$ = 1.4390. *Yamanouchi Europe B.V.; Yamanouchi Pharma.*

41   Acid fuchsine

3244-88-0   109   G

C20H17N3Na2O9S3

2-Amino-5-[(4-amino-3-sulfophenyl)(4-imino-3-sulfo-2,5-cyclohexadien-1-ylidene)methyl]-3-methylbenz-enesulf-onic   acid disodium salt.

Acid fuchsin; Acid fuchsine; Acid fuchsine FB; Acid fuchsine N; Acid fuchsine O; Acid fuchsine S; Acid leather magenta A; Acid magenta; Acid magenta O; Acid rosein; Acid rubin; Acidal fuchsine; Acidal magenta; C.I. Acid Violet 19, disodium salt; 3-(1-(4-Amino-3-methyl-5-sulphonatophenyl)-1-(4-amino-3-sulphonatophenyl)-methylene)cyclohexa-1,4-diene-sulphonic acid; Benzene-sulfonic acid,   2-amino-5-((4-amino-3-sulfophenyl)(4-imino-3-sulfo-2,5-cyclohexa-dien-1-ylidene)methyl]-3-methyl-, disodium salt C.I. Acid Violet 19; EINECS 221-816-5; Fuchsin acid; Fuchsin(E) acid; Fuchsine acid; p-Fuchsine acid; Kiton magenta A; NSC 13979; NSC 56444; Rubine S; Andrade indicator. Used as a pH indicator and a biological stain. Crystals; slightly soluble in H_2O (14.3 g/100 ml), EtOH; λ_m = 540-545 nm (10 mg/l 0.1N HCl).

42   Acid Yellow 23

1934-21-0   9160   G H P

C16H9N4Na3O9S2

Trisodium   5-hydroxy-1-(4-sulphophenyl)-4-(4-sulpho-phenylazo)pyrazole-3-carboxylate.

1310 Yellow; 1409 Yellow; A.F. Yellow No. 4; Acid Leather Yellow T; Acid Yellow 23; Acid Yellow 23; Acid Yellow T; Acilan Yellow GG; Airedale Yellow T; Aizen tartrazine; Amacid Yellow T; Atul Tartrazine; Bucacid tartrazine; C.I. 640; C.I. 19140; C.I. Acid Yellow 23; C.I. Acid Yellow 23, trisodium salt; C.I. Food Yellow 4; Calcocid Yellow; Calcocid Yellow MCG; Calcocid Yellow XX; Canacert tartrazine; CCRIS 2656; Certicol Tartrazol Yellow S; CI 19140; Cilefa Yellow T; Curon Fast Yellow 5G; D and C Yellow No. 5; Dolkwal tartrazine; Dye FD and C Yellow No. 5; Dye Yellow Lake; E 102; Edicol Supra Tartrazine N; Egg Yellow A; EINECS 217-699-5; Erio Tartrazine; Eurocert tartrazine; FD & C Yellow No. 5; FD & C Yellow No. 5 tartrazine; Fenazo Yellow T; Food Yellow 4; Food Yellow 5; HD tartrazine; HD Tartrazine Supra; Hexacert Yellow No. 5; Hexacol tartrazine; Hidazid tartrazine; Hispacid Fast Yellow T; Hydrazine

Yellow; Hydroxine Yellow L; Kako tartrazine; Kayaku Food Colour Yellow No. 4; Kayaku tartrazine; Kca Foodcol Tartrazine PF; Kca Tartrazine PF; Kiton Yellow T; L-Gelb 2; L Yellow Z 1020; Lake Yellow; Lemon Yellow A; Lemon Yellow A Geigy; M-8847; Maple Tartrazol Yellow; Mitsui tartrazine; Naphtocard Yellow O; Neklacid Yellow T; NSC 4760; Oxanal Yellow T; San-ei Tartrazine; Schultz No. 737; Sugai tartrazine; Tartar Yellow; Tartar Yellow FS; Tartar Yellow N; Tartar Yellow PF; Tartar Yellow S; Tartran Yellow; Tartraphenine; Tartrazine; Tartrazine A Export; Tartrazine B; Tartrazine B.P.C.; Tartrazine C; Tartrazine C Extra; Tartrazine Extra Pure A; Tartrazine FD&C Yellow 5; Tartrazine FQ; Tartrazine G; Tartrazine Lake; Tartrazine Lake Yellow N; Tartrazine M; Tartrazine MCGL; Tartrazine N; Tartrazine NS; Tartrazine O; Tartrazine T; Tartrazine XX; Tartrazine XXX; Tartrazine Yellow; Tartrazine Yellow 5; Tartrazol BPC; Tartrazol Yellow; Tartrine Yellow O; Unitertracid Yellow TE; Usacert Yellow No. 5; Vondacid tartrazine; Wool Yellow; Xylene Fast Yellow GT; Y-4; Yellow Lake 69; Yellow No. 5 FDC; Zlut kysela 23; Zlut pigment 100; Zlut potravinarska 4. A dyestuff used to dye silk and wool. Bright orange powder; mp > 300°; freely soluble in H2O. *Aquarium Pharmaceuticals Inc.; Aquashade; Becker Underwood, Inc.*

43   Acifluorfen
50594-66-6   111   G P

C14H7ClF3NO5
5-[2-Chloro-4-(trifluoromethyl)phenoxy]-2-nitrobenzoic acid.
Acifluorfen; Acifluorfene; Benzoic acid, 5-(2-chloro-4-(trifluoromethyl)phenoxy)-2-nitro-; BRN 2953865; CCRIS 3491; EINECS 256-634-5; Tackle. Herbicide. Brown solid; mp = 142 - 160°; soluble in H2O (0.00025 g/100 ml), MeOH (4 g/100 ml), C6H14 (0.3 g/100 ml), C7H8 (33.5 g/100 ml); LD50 (rat, mus orl) > 5000 mg/kg, (rat der) > 5000 mg/kg, (Japanese quail, canary orl) > 15000 mg/kg, (bee orl) > 0.1 mg/bee; LC50 (rainbow trout 96 hr.) = 1.0 - 2.3 mg/l, (carp 96 hr.) = 3.3 mg/l; non-toxic to bees. *BASF Corp.; Rhône-Poulenc.*

44   Aconitic acid
499-12-7   119   G

C6H6O6
Citridic acid.
Achilleaic acid; Achilleic acid; Aconitic acid; AI3-14615; 3-Carboxy-2-pentenedioic acid; 3-Carboxy-glutaconic acid; CCRIS 2507; Citridic acid; Citridinic acid; EINECS 207-877-0; Equisetic acid; FEMA No. 2010; Glutaconic acid, 3-carboxy-; HSDB 635; NSC 7616; Propene-1,2,3-tricarboxylic acid; 1-Propene-1,2,3-tricarboxylic acid; 1,2,3-Propenetricarboxylic acid; 2-Pentenedioic acid, 3-carboxy-; Pyrocitric acid. A plasticizer for buna rubber. Leaflets or plates; dec 198°, 204°, 209°, soluble in H2O (18.2 g/100 ml, 13°, 50 g/100 ml, 25°), in 88% EtOH (50 g/100 ml), slightly soluble in Et2O.

45   Acridinic acid
643-38-9   H

C11H7NO4
2,3-Quinolinedicarboxylic acid.
Acridinic acid; NSC 26342; 2,3-Quinolinedicarboxylic acid.

46   Acrisorcin
7527-91-5   128   D G

C25H28N2O2
4-Hexyl-1,3-benzenediol compd with 9-acridin-amine(1:1).
Acrisorcin; Acrisorcina; Acrisorcinum; Akrinol; 9-Aminoacridine compd. with 4-hexylresorcinol (1:1); 1,3-Benzenediol, 4-hexyl-, compd. with 9-acridinamine (1:1); Bis(9-aminoacridinium) 4-hexyl-1,3-resorcindiolat; Bis(9-aminoacridinium)salz des hexylresorcine; EINECS 231-389-7; 4-Hexylresorcinol compound with 9-aminoacridine (1:1); Sch 7056. Used medicinally as an antifungal agent. Yellow crystals.

47   Acrolein
107-02-8   130   H P

C3H4O
Prop-2-enal.
Acquinite; Acraldehyde; Acrolein; trans-Acrolein; Acrolein, inhibited; Acroleina; Acroleine; Acrylaldehyd; Acrylaldehyde; AI3-24160; Akrolein; Akroleina; Aldehyde acrylique; Aldeide acrilica; Allyl aldehyde; Aqualin; Aqualine; Biocide; Caswell No. 009; CCRIS 3278; Crolean; EPA Pesticide Chemical Code 000701; Ethylene aldehyde; HSDB 177; Magnacide; Magnacide H; Magnacide H and B; NSC 8819; Papite; Propenal; 2-Propenal; Prop-2-enal; Prop-2-en-1-al; Propenaldehyde; 2-Propen-1-one; Propylene aldehyde; RCRA waste number P003; Slimicide; UN1092. Aquatic herbicide and slimicide. Registered by EPA as an antimicrobial, herbicide and fungicide. Pungent liquid; mp = -88°; bp = 52.5°; bp200 = 17.5°, bp100 = 2.5°; soluble in H2O (40 g/100 ml), EtOH, Et2O; d^0 = 0.8621, d^{20} = 0.8389; sg$^{4-20}$ = 0.841; LD50 (rat orl) = 29 mg/kg, (mmus orl) = 13.9 mg/kg, (fmus orl) = 17.7 mg/kg, (rbt orl) = 7.1 mg/kg, (rbt der) = 231 mg/kg, (bobwhite quail orl) = 19 mg/kg, (mallard duck orl) = 9.1 mg/kg; LC50 (rainbow trout 24 hr.) = 0.15 mg/l, (bluegill sunfish 24 hr.) = 0.079 mg/lg, (shiner 24 hr.) = 0.04 mg/l, (mosquito fish 24 hr.) = 0.39 mg/l, (shrimp 48 hr.) = 0.10 mg/l, (oyster 48 hr.) = 0.46 mg/l. *Baker Petrolite Corp.; Generic.*

48   Acrylamide
79-06-1   131   G H

C3H5NO
2-Propenamide.

11

4-02-00-01471 (Beilstein Handbook Reference); AAM; Acrylagel; Acrylamide; Acrylic acid amide; Acrylic amide; AI3-04119; Akrylamid; Amid kyseliny akrylove; Amresco Acryl-40; BRN 0605349; CCRIS 7; EINECS 201-173-7; Ethylenecarboxamide; HSDB 191; NSC 7785; Optimum; Propenamide; Propeneamide; Propenoic acid amide; RCRA waste number U007; UN2074; Vinyl amide. Monomer and chemical intermediate. Used to produce polyacrylamide resins. Crystalline solid; mp = 84.5°; bp = 192.6°, bp_{72} = 103°; Soluble in H_2O 215.5 g/100 ml), MeOH (155 g/100 ml), EtOH (86.2 g/100 ml), Me_2CO (63.1g/100 ml), EtOAc (12.6 g/100 ml), $CHCl_3$ (2.66 g/100 ml), C_6H_6 (0.35 g/100 ml); LD_{50} (mus ip) = 170 mg/kg; highly toxic and irritant, causes CNS paralysis. *Am. Cyanamid; Cyanamid of Great Britain Ltd.*

49   2-Acrylamido-2-methylpropanesulfonic acid
15214-89-8   H

$C_7H_{13}NO_4S$
1-Propanesulfonic acid, 2-acrylamido-2-methyl-.
2-Acrylamido-2-methylpropanesulfonic acid; 2-Acryl-amido-2-methylpropanesulphonic acid; EINECS 239-268-0; 1-Propanesulfonic acid, 2-methyl-2-((1-oxo-2-propenyl)amino)-; 1-Propanesulfonic acid, 2-acrylamido-2-methyl-.

50   Acrylic acid
79-10-7   132   G H

$C_3H_4O_2$
2-Propenoic acid.
4-02-00-01455 (Beilstein Handbook Reference); Acide acrylique; Acido acrilio; Acroleic acid; Acrylic acid; AI3-15717; BRN 0635743; Caswell No. 009A; CCRIS 737; EINECS 201-177-9; Ethylenecarboxylic acid; Glacial acrylic acid; HSDB 1421; Kyselina akrylova; NSC 4765; Propene acid; Propenoic acid; RCRA waste number U008; UN2218; Vinylformic acid. Corrosive liquid used in the manufacture of plastics. Monomer for polyacrylic and polymethacrylic acids, other acrylic acids, acrylic polymers. Used in manufacture of acrylic fiber; In plastics, surface coatings and adhesives industry. mp = 12.3°; bp = 141°; d^{20} = 1.0511; soluble in H_2O, organic solvents; LD_{50} (rat orl) = 2.59 g/kg. *BASF Corp.; Degussa-Hüls Corp.; Hoechst Celanese; Penta Mfg.; Rohm & Haas Co.; Union Carbide Corp.*

51   Acrylic acid, 1,1,1-(trihydroxy--methyl)propane triester
15625-89-5   G H

$C_{15}H_{20}O_6$
2-Ethyl-2-(hydroxymethyl)-1,3-propanediol triacrylate.
Acrylic acid, 1,1,1-(trihydroxymethyl)propane triester; Acrylic acid, triester with 2-ethyl-2-(hydroxymethyl)-1,3-propanediol; Acrylic acid,

triester with 2-ethyl-2-(hydroxymethyl)-1,3-propanediol; Ageflex TMPTA; CCRIS 92; EINECS 239-701-3; 2-Ethyl-2-(((1-oxoallyl)oxy)-methyl)-1,3-propanediyl diacrylate; 2-Ethyl-2-(hydroxy-methyl)-1,3-propanediol triacrylate; M 309; MFM; NK Ester A TMPT; Ogumont T 200; 1,3-Propanediol, 2-ethyl-2-(hydroxymethyl)-, triacrylate; 2-Propenoic acid, 2-ethyl-2-(((1-oxo-2-propenyl)oxy)-methyl)-1,3-propanediyl ester; Saret 351; Sartomer SR-351; Setalux UV 2241; SR 351; TMPTA; Tri-methylolpropane triacrylate; Viscoat 295. With 100-150 ppm hydroquinone inhibitor; high boiling monomeric ester polymerized by common free radical initiators; curing agent. Crosslinker; UV-cured adhesives, wood fillers, inks, coatings, dry film photo polymer resists, flexographic, offset and screen printing inks, vinyl acrylic latex paint, exterior coatings, highly crosslinked polybutadiene rubber. Liquid; bp > 200°; d^{20}_{20} = 1.108; LD_{50} (rat orl) = 5190 mg/kg. *Rit-Chem; Sartomer.*

52   Acrylonitrile
107-13-1   133   G H

C_3H_3N
2-Propenenitrile.
4-02-00-01473 (Beilstein Handbook Reference); Acritet; Acrylnitril; Acrylon; Acrylonitril; AI3-00054; Akrylnitril; Akrylonitryl; BRN 0605310; Carbacryl; Caswell No. 010; CCRIS 8; Cianuro di vinile; Cyanoethylene; Cyanure de vinyle; EINECS 203-466-5; ENT 54; EPA Pesticide Chemical Code 000601; Fumigrain; HSDB 176; Miller's fumigrain; NCI-C50215; Nitrile acrilico; Nitrile acrylique; NSC 6362; Propenenitrile; 2-Propenenitrile; RCRA waste number U009; TL 314; UN 1093; VCN; Ventox; Vinyl cyanide; Vinylkyanid. Monomer for acrylic and modacrylid fibers; in production of ABS and acrylonitrile styrene copolymers. Liquid; mp = -83.5°; bp = 77.3°; d^{20} = 0.8060; λ_m = 206 nm (ε = 6166, MeOH); very soluble in Me_2CO, C_6H_6, Et_2O, EtOH; LD_{50} (rat orl) = 78 mg/kg. *Am. Cyanamid; Asahi Chem. Industry; BP Chem.; DSM Spec. Prods.; DuPont; Mitsui Toatsu; Sigma-Aldrich Fine Chem.*

53   β-(Acryloyloxy)propionic acid
24615-84-7   H

$C_6H_8O_4$
2-Carboxyethyl acrylate.
β-(Acryloyloxy)propionic acid; 3-Acryloyloxypropionic acid; 2-Carboxyethyl acrylate; β-Carboxyethyl acrylate; 2-Carboxyethyl 2-propenoate; EINECS 246-359-9; Hydracrylic acid, acrylate; 2-Propenoic acid, 2-carboxyethyl ester; Sipomer B-CEA.

54   Acticarbone
64365-11-3   G

C
Carbon, activated.
Carbon black; Carbon decolorizing. Powdered and granulated activated carbon; used for purification, decolorization, deodorization, separation and recovery in liquid or gas phase, in the chemical, petrochemical, pharmaceutical and food industries (glucose factories, sugar refiners, oil refining, wine treatment); for the treatment of drinking and industrial water, etc. Catalyst supports. *Elf Atochem N. Am.*

55   Activated charcoal
16291-96-6   H

C

Charcoal.
Charcoal; Charcoal, activated; Charcoal briquettes, shell, screenings, wood, etc.; Charcoal, except activated; EINECS 240-383-3; HSDB 2017; NA1361; Swine fly ash; Whetlerite.

56   Acyclovir
59277-89-3   147   D

$C_8H_{11}N_5O_3$
 2-Amino-1,9-dihydro-9-[(2-hydroxyethoxy)methyl]-6H-purin-6-one.
 Aciclovir; Aciclovirum; Aciclovirum; Acycloguanosine; Acyclovir; 2-Amino-1,9-dihydro-9-((2-hydroxyethoxy)-methyl)-6H-purin-6-one; BW-248U; CCRIS 1953; DRG-0008; EINECS 261-685-1; HSDB 6511; 9-((2-Hydroxy-ethoxy)methyl)guanine; 6H-Purin-6-one, 2-amino-1,9-dihydro-9-((2-hydroxyethoxy)methyl)-; Vipral; Virorax; W-248-U; Wellcome-248U; Zovirax.; Laurocapram; Poviral; Virorax; Zovirax; Vipral; Aciclovir; Acyclo-V; Zyclir. Antiviral agent used in treatment of herpes virus. A nucleoside analog that is preferentially taken up by infected cells and then inhibits viral DNA synthesis by interfering with transcription. mp = 256.5-257°; LD50 (mus orl) >10,000 mg/kg. *Glaxo Wellcome Inc.*

57   Acyclovir Sodium
69657-51-8   D

$C_8H_{10}N_5NaO_3$
 2-Amino-1,9-dihydro-9-[(2-hydroxyethoxy)methyl]-6H-purin-6-one monosodium salt.
 Aciclovir natrium; Acyclovir sodium; Acycloguanosine sodium (Obs.); Acyclovir sodium salt; 2-Amino-1,9-dihydro-9-((2-hydroxyethoxy)methyl)-6H-purin-6-one monosodium salt; BW 248U sodium; 9-((2-Hydroxyethoxy)methyl)guanine monosodium salt; Sodium acyclovir; Zovirax; Zovirax Sterile Powder. Antiviral agent. *Glaxo Wellcome Inc.*

58   Adamsite
578-94-9   7292   G

$C_{12}H_9AsClN$
 10-Chloro-5,10-dihydrophenarsazine.
 4-27-00-09796 (Beilstein Handbook Reference); 5-Aza-10-arsenaanthracene chloride; Adamsit; Adamsite; BRN 0178698; Caswell No. 648; 10-Chloro-5,10-dihydro-arsacridine; 10-Chloro-5,10-dihydrophenars-azine; Chlorodiphenylaminearsine; Diphenylamine chloroarsine; Diphenylaminechloroarsine; DM; EINECS 209-433-1; EPA Pesticide Chemical Code 063901;

Fenarsazinchlorid; NSC 86138; Phenarsazine, 10-chloro-5,10-dihydro-; Phenarsazine, 10-chloro-; Phenarsazine chloride; UN1698. A poison gas, war gas; also used in the formulation of wood-treating solutions against marine borers and similar pests. Crystals; mp = 195°; bp = 410°; insoluble in H_2O, slightly soluble in organic solvents; highly toxic.

59   Adenine
73-24-5   152   G

$C_5H_5N_5$
 1H-purin-6-amine.
 ADE; Adenin; Adenine; Adeninimine; AI3-50679; 6-Aminopurine; 6-Amino-1H-purine; 6-Amino-3H-purine; 6-Amino-7H-purine; 6-Amino-9H-purine; CCRIS 2556; 1,6-Dihydro-6-iminopurine; 3,6-Dihydro-6-iminopurine; EINECS 200-796-1; Leuco-4; NSC 14666; Purine, 6-amino-; 1H-Purin-6-amine; 1H-Purine, 6-amino; 9H-Purine, 1,6-dihydro-6-imino-; USAF CB-18; Vitamin B4. Used in medical and biochemical research. Crystals; mp = 360-365° (dec); λ_m = 207, 260.5 nm (ε 23200, 13400 pH 7.0); soluble in H_2O (0.05 g/100 ml), less soluble in organic solvents; LD50 (rat orl) = 745 mg/kg. *Lonzagroup; Penta Mfg.; Schweizerhall Inc.; Sigma-Aldrich Fine Chem.; U.S. BioChem.*

60   Adenosine triphosphate
56-65-5   156   G

$C_{10}H_{16}N_5O_{13}P_3$
 9-β-D-Arabinofuranosyladenine 5'-triphosphate.
 Adenosine, 5'-(tetrahydrogen triphosphate); Adenosine 5'-triphosphate; Adenosine 5'-triphosphoric acid; Adenosine triphosphate; Adenosintriphosphorsaeure; Adenosintri-phosphorsäure; Adenylpyrophosphoric acid; Adephos; Adetol; Ado-5'-P-P-P; Adynol; Ara-ATP; Atipi; ATP; 5'-Atp; ATP (nucleotide); Atriphos; EINECS 200-283-2; Glucobasin; Myotriphos; Striadine; Striadyne; Triadenyl; Triadenyl; Triphosaden; Triphosphaden; Triphosphoric acid adenosine ester. Organic compound used in biochemical research. Used to inhibit enzymatic browning of raw edible plant materials, such as sliced potatoes and apples. White powder; $[\alpha]_D^{22}$ = -26.7° (c = 3.095); λ_m = 259 nm (a_M = 15400); soluble in H_2O; pH of 1% aq. solution = 2. *Asahi Chem. Industry; Asahi Denka Kogyo; Greeff R.W. & Co.; Penta Mfg.*

61   5'-Adenylic acid
61-19-8   159   G

13

$C_{10}H_{14}N_5O_7P$

Adenosine-5'-monophosphate.

4-26-00-03615 (Beilstein Handbook Reference); 5'-AMP; A-5MP; A5MP; Adeno; Adenosine 5'-(dihydrogen phosphate); Adenosine-5'-phosphoric acid; Adenosine-5-monophosphoric acid; Adenosine 5'-monophosphate; Adenosine 5'-phosphate; Adenosine, mono(dihydrogen phosphate) (ester); Adenosine monophosphate; Adenosine phosphate; Adenosini phosphas; Adenovite; Adenylic acid; AMP; AMP (nucleotide); BRN 0054612; Cardiomone; EINECS 200-500-0; Ergadenylic acid; Fosfato de adenosina; HSDB 3281; Lycedan; Monophosph-adenine; Muscle adenylic acid; Muskeladenosin-phosphorsaeure; Muskeladenosin-phosphor-säure; Muskeladenylsaeure; Muskeladenylsäure; My-B-Den; Myoston; NSC-20264; Phosaden; Phosphaden; Phos-phate d'adenosine; Phosphentaside; Vitamin B_8. Nutrient. Used in biochemical research. White solid; mp = 200° (dec); $[\alpha]_D^{20}$= -47.5° (c = 2 2% NaOH); soluble in H_2O. *U.S. BioChem.*

62   **Adipic acid**
124-04-9   163   G

$C_6H_{10}O_4$

Dicarboxylic acid C_6.

4-02-00-01956 (Beilstein Handbook Reference); Acide adipique; Acifloctin; Acinetten; Adilactetten; Adipate; Adipic acid; Adipinic acid; Adipinsäure; AI3-03700; BRN 1209788; 1,4-Butanedicarboxylic acid; CCRIS 812; EINECS 204-673-3; FEMA Number 2011; Hexanedioic acid; 1,6-Hexanedioic acid; HSDB 188; Kyselina adipova; NSC 7622; Adi-pure®. Used in manufacture of polyurethane foams and nylon; preparation of esters for use as plasticizers and lubricants; food additive (acidulant) for baking powders; flavoring agent; leavening agent, neutralizer, acidulant, pH control agent for baked goods, beverages. For chemical industry, chemical intermediate; used in adhesives, coatings, nylon 66, PU foams/elastomers/fibers, lubricants, textile treatments, cosmetic emollients; pH buffer; food acidulant. Prisms; mp = 153.2°; bp = 337.5°; d_4^{25} 1.360; insoluble in AcOH, ligroin; slightly soluble in H_2O, soluble in Et_2O, very soluble in EtOH; LD_{50} (mus orl) = 1900 mg/kg. *Allied Signal; Asahi Chem. Industry; Monsanto Co.; Penta Mfg.; Rhône-Poulenc; UCB SA.*

63   **Adiponitrile**
111-69-3   H

$C_6H_8N_2$

Hexanedioic acid, dinitrile.

4-02-00-01975 (Beilstein Handbook Reference); Adipic acid dinitrile; Adipinsaeuredinitril; Adipodinitrile; Adiponitrile; Adipyldinitrile; AI3-11080; BRN 1740005; CCRIS 4570; 1,4-Dicyanobutane; EINECS

203-896-3; Hexanedinitrile; Hexanedioic acid, dinitrile; HSDB 627; Nitrile adipico; NSC 7617; Tetramethylene cyanide; Tetramethylene dicyanide; UN2205. Liquid; mp = 1°; bp = 295°; d^{20} = 0.9676; slightly soluble in H_2O, Et_2O, soluble in EtOH, $CHCl_3$.

64   **Agaricic acid**
666-99-9   186   G

$C_{22}H_{40}O_7$

2-Hydroxynonadecane-1,2,3-tricarboxylic acid.

4-03-00-01284 (Beilstein Handbook Reference); Agaric acid; Agaricic acid; Agaricin; Agaricinic acid; Agaricinsaeure; Agaricinsäure; BRN 1729981; α-Cetylcitric acid; EINECS 211-566-5; Laricic acid; n-Hexadecylcitric acid; (-)-2-Hydroxy-1,2,3-nonadecan-tricarbonsaeure; 2-Hydroxy-1,2,3-nonadecanetricarb-oxylic acid; 2-Hydroxynonadecane-1,2,3-tricarboxylic acid; 1,2,3-Nonadecanetricarboxylic acid, 2-hydroxy-; NSC 60429; NSC 65690. A resin acid, obtained by extraction with alcohol of the fruit bodies of *Polyporus officinalis and Agaricus albus.* Used as a febrifuge and antiperspirant Crystalline cream-colored powder; mp = 142° (dec); $[\alpha]_D^{19}$= -8.8° (NaOH); insoluble in C_6H_6, $CHCl_3$, slightly soluble in EtOH, Et_2O, soluble in H_2O; odorless, tasteless.

65   **Agerite® White**
93-46-9   G

$C_{26}H_{20}N_2$

Sym-di-β-Naphthyl-p-phenylenediamine.

4-13-00-00119 (Beilstein Handbook Reference); Aceto DIPP; Agerite White; AgeRite W; AI3-14324; Antigene F; Antioxidant 123; Antioxidant DNP; ASM-DNT; 1,4-Benzenediamine, N,N'-di-2-naphthalenyl-; N,N'-Bis-(2-naftyl)-p-fenylendiamin; 1,4-Bis(2-naphthylamino)-benzene; N,N'-Bis(β-naphthyl)-p-phenylenediamine; N,N'-Bis(2-naphthyl)-p-phenylenediamine; BRN 2224419; CCRIS 6026; Di-β-naphthyl-p-phenyldi-amine; Di-β-naphthyl-p-phenylenediamine; s-Di(β-naphthyl)-p-phenylenediamine; N,N'-Di-β-naphthyl-p-phenylene-diamine; N,N'-Di-2-naphthalenyl-1,4-benz-enediamine; N,N'-Di-2-naphthyl-p-phenylenediamine; Diafen NN; DNPD; Dnpda; Dwu-β-naftylo-p-fenylodwuamina; EINECS 202-249-2; Nonox CL; NSC 3410; p-Phenylenediamine, N,N'-(di-2-naphthyl)-; N,N'-p-Phenylenebis(2-naphthylamine); Santowhite CL; Tisperse MB-2X. Antioxidant. An antidegradant for latex, nitrile rubber, styrene-butadiene and nitrile-butadiene rubber. Solid; mp = 224-230°; d = 1.22-1.28; insoluble in H_2O, EtOH, soluble in organic solvents; LD_{50} (rat orl) = 4500 mg/kg. *Vanderbilt R.T. Co. Inc.*

66   **Alachlor**
15972-60-8   201   G P

C14H20ClNO2

N-(Methoxymethyl)-2,6-diethylchloroacetanilide.

Al3-51506; Ala-Scept; Alachlor; Alachlor technical; Alachlor technical (90% or more); Alachlore; Alamex; Alanex; Alanox; Alatox 480; Alazine; Alochlor; BRN 2944476; Bullet; Cannon; Caswell No. 011; CCRIS 3155; CDMA; Chimiclor; Chloressigsäure-N-(methoxymethyl)-2,6-diäthylanilid; α-Chloro-2',6'-diethyl-N-methoxy-methylacetanilide; 2-chloro-2',6'-diethyl-N-(methoxy-methyl)acetanilide; 2-chloro-N-(2,6-diethylphenyl)-N-(methoxymethyl)acetamide; CP 50144; Crop; EPA Pesticide Chemical Code 090501; Freedom; HSDB 1014; Lariat; Lasagrin; Lasso; Lasso EC; Lasso II; Lasso Micro Tech; Lazo; Metachlor; Methachlor; Methoxymethyl-2',6'-diethylanilide chloroacetate; Micro-tech; Micro-Tech Lasso; Nudor; Partner; Pillarzo; Star. Registered by EPA as a herbicide. Crystals; mp = 40 - 41°; bp$_{0.02}$ = 100°, bp$_{0.3}$ = 135°; d$^{25}_{15.6}$ = 1.133; soluble in H$_2$O (0.0242 g/100 ml), Et$_2$O, Me$_2$CO, C$_6$H$_6$, EtOH, EtOAc; LD$_{50}$ (rat orl) = 930 - 1200 mg/kg, (rbt der) = 13300 mg/kg; LC$_{50}$ (mrat ihl 6 hr.) > 23.4 mg/l air, (mallard duck, bobwhite quail 8 day dietary) > 5000 mg/kg diet, (rainbow trout 96 hr.) = 1.8 mg/l, (bluegill sunfish 96 hr.) = 2.8 mg/l; non-toxic to bees. *Cedar Chemical Corp.; Micro Flo Co. LLC; Monsanto Co.*

67 **Alanine**

56-41-7   203   G

C3H7NO2

2-Aminopropionic acid.

Alanina; Alanine; Alanine, L-; (L)-Alanine; L-(+)-Alanine; (S)-Alanine; Alaninum; L-2-Aminopropanoic acid; (S)-α-Aminopropionsäure; (S)-α-Aminopropion-saeure; (S)-2-Aminopropanoic acid; (S)-2-Amino-propionsäure; (S)-2-Aminopropansaeure; α-Amino-propionic acid; 2-Aminopropanoic acid, L-; 2-Aminopropionic acid; EINECS 200-273-8; HSDB 1801; NSC 206315; Propanoic acid, 2-amino-, (S)-. The natural L-isomer is used in microbiological research, biochemical research and as a dietary supplement. mp = 314-317° (dec); [α]$_D$ = 14° (c = 6 1N HCl); d = 1.401; soluble in H$_2$O (15.8 g/100 ml), cold 80% ethanol; insoluble in ether. *Penta Mfg.; Sigma-Aldrich Fine Chem.; U.S. BioChem.*

68 **Albuterol**

18559-94-9   215   D

C13H21NO3

2-(tert-Butylamino)-1-(4-hydroxy-3-hydroxymethyl-phenyl)ethanol.

Aerolin; AH 3365; Albuterol; dl-Albuterol; Almotex; Alti-Salbutamol; Anebron; Arubendol-Salbutamol; Asmadil; Asmanil; Asmasal; Asmatol; Asmaven; Asmidon; Asmol; Asmol Uni-Dose; Asthalin; BRN 2213614; Broncho-Spray; Broncovaleas; Bronter; Bugonol; Bumol; Butamol; Buto-Asma; Butohaler; Butotal; Butovent; DL-N-tert-Butyl-2-(4-hydroxy-3-hydroxymethylphenyl)-2-hydroxy-ethylamine; Buvent-ol; Cobutolin; Dilatamol; EINECS 242-424-0; Farcolin; Gerivent; Grafalin; Libretin; Medolin; Mozal; Novosalmol; Parasma; Pneumolat; Proventil HFA; Proventil Inhaler; Respax; Respolin; Sabutal; Salamol; Salbetol; Salbron; Salbu-BASF; Salbu-

Fatol; Salbuhexal; Salbulin; Salbupur; Salbusian; Salbutalan; Salbutamol; dl-Salbutamol; Salbutamolum; Salbutan; Salbutol; Salbuven; Salbuvent; Sallbupp; Salmaplon; Salomol; Salvent; Saventol; Servitamol; Spreor; Sultanol; Sultanol N; Suprasma; Suxar; Theosal; Tobybron; Vencronyl; Ventamol; Ventilan; Ventiloboi; Ventolin; Ventolin Inhaler; Ventolin Rotacaps; Ventoline; Volmax; m-Xylene-α,α'-diol, α'-((tert-butylamino)methyl)-4-hydroxy-; Zaperin. Bronchodilator; tocolytic. Crystals; mp = 151°, 157-158°; soluble in most organic solvents. *Allen & Hanbury; Apothecon; Glaxo Wellcome Inc.; Key Pharm.*

69 **Albuterol sulfate**

51022-70-9   215   D

C26H44N2O10S

2-(tert-Butylamino)-1-(4-hydroxy-3-hydroxymethyl-phenyl)ethanol sulfate (2:1).

AccuNeb; Aerotec; Albuterol sulfate; Aloprol; Amocasin; Arm-A-Med; Bis((tert-butyl)(β,3,4-trihydroxyphenethyl)-ammonium) sulphate; Broncho Inhalat; Bronchospray; Broncodil; Combivent; Dipulmin; EINECS 256-916-8; Emican; Epaq; Fartolin; Huma-Salmol; Inspiryl; Loftan; NSC 289928; Proventil Repetabs, Solution, Syrup, and Tablets; Salbutamol hemisulfate; Salbutamol sulfate; (±)-Salbutamol sulfate; dl-Salbutamol sulfate; Sch 13949W sulfate; Venetlin; Ventolin HFA. Bronchodilator; tocolytic. White crystals; soluble in H$_2$O, slightly soluble in EtOH. *Allen & Hanbury; Apothecon; Glaxo Labs.; Key Pharm.; Lemmon Co.; Schering AG.*

70 **Alclometasone dipropionate**

66734-13-2   219   D G

C28H37ClO7

7α-Chloro-11β,17,21-trihydroxy-16α-methylpregna-1,4-diene-3,20-dione 17,21-dipropionate.

Aclovate; Alclometasone 17, 21-dipropionate; Alclo-metasone dipropionate; BRN 2317658; CCRIS 1927; Delonal; EINECS 266-464-3; Sch 22219; Vaderm. Anti-inflammatory (topical). Crystals; mp = 212-216°; [α]$^{26}_D$ = +42.6° (c = 3 DMF); λ$_m$ = 242 nm (ε 15600 MeOH). *Glaxo Labs.; Schering-Plough HealthCare Products.*

71 **Alcohols, C9-C11**

66455-17-2   H

C9-C11 alcohol.

Alcohols, C9-C11; Alcohols, C9-11; C9-C11 alcohol; Dobanol 91; Dobanol 911; EINECS 266-367-6; Linevol 911; Neodol 91; Neodol 91-8T.

72 Alcohols, C$_{12}$-C$_{15}$
63393-82-8   H

C$_{12}$-C$_{15}$ alcohols.
Alcohols, C$_{12}$-C$_{15}$; C$_{12}$-C$_{15}$ alcohol; EINECS 264-118-6.

73 Aldicarb
116-06-3   221   H P

C$_7$H$_{14}$N$_2$O$_2$S
2-Methyl-2-(methylthio)propanal O-[(methylamino)carb-onyl]oxime.
Al3-27093; Aldicarb; Aldicarbe; Carbamyl; Caswell No. 011A; CCRIS 17; EINECS 204-123-2; ENT 27,093; EPA Pesticide Chemical Code 098301; HSDB 1510; Methyl-2-(methylthio)propionaldehyde O-(methylcarb-amoyl)oxime; Methylcarbamic acid, O-((2-methyl-2-(methylthio)propyl-idene)amino) deriv.; NCI-C08640; NSC 379586; OMS-771; RCRA waste number P070; Sulfone aldoxycarb; Temik; Temik 10 G; Temik G; Temik G10; UC-21149; Union carbide 21149. Registered by EPA as an insecticide and acaricide. Crystals; mp = 99-100°; soluble in H$_2$O (0.6 g/100 ml), Me$_2$CO (35 g/100 ml), C$_6$H$_6$ (15 g/100 ml), xylene (5 g/100 ml), CH$_2$Cl$_2$ (30 g/100 ml); LD$_{50}$ (rat orl) = 0.93 mg/kg, (frat orl) = 1 mg/kg, (rbt der) = 5.0 mg/kg, (gpg der) = 2400 mg/kg; LC$_{50}$ (rat ihl 5 min.) < 0.2 mg/l air; (bobwhite quail 7 day dietary) = 2400 mg/kg diet, (rainbow trout 96 hr.) = 0.88 mg/l, (bluegill sunfish 96 hr.) = 1.5 mg/l; toxic to bees. *Agriliance LLC; Aventis Crop Science; Rhône-Poulenc.*

74 Aldicarb oxime
1646-75-9   H

C$_5$H$_{11}$NOS
2-(Methylthio)-2-methylpropionaldehyde oxime.
Aldicarb oxime; CCRIS 660; EINECS 216-709-5; HSDB 5848; 2-(Methylthio)isobutyraldehyde oxime; 2-(Methylthio)-2-methylpropionaldehyde oxime; 2-Methyl-2-(methylthio)propanal oxime; 2-Methyl-2-(methylthio)-propionaldoxime; Propanal, 2-methyl-2-(methylthio)-, oxime; 2-Methyl-2-(methylthio)-propionaldehyde oxime; Propionaldehyde, 2-methyl-2-(methylthio)-, oxime; Temik oxime.

75 Aldrin
309-00-2   225   G P

C$_{12}$H$_8$Cl$_6$
1,2,3,4,10,10-Hexachloro-1α,4α,4aβ,5α,8α,8aβ-hexa-hydro-1,4:5,8-dimethanonaphthalene.
Al3-15949; Aldocit; Aldrex; Aldrex 30; Aldrex 30 E.C.; Aldrex 40; Aldrin; Aldrine; Aldrite; Aldron; Aldrosol; Algran; Altox; Caswell No. 012; CCRIS 18; Compound 118; 1,4:5,8-Dimethanonaphthalene, 1,2,3,4,10,10-hexachloro-1,4,4a,5,8,8a-hexahydro-, endo,exo-; (1R, 4S,4aS,5S,8R,8aR)-1,2,3,4,10,10-Hexachloro-1,4,4a,5, 8,8a-

hexahydro-1,4:5,8-dimethanonaphthalene; (1α, 4α,4aβ,5α,8α,8aβ)-1,2,3,4,10,10-hexachloro-1,4,4a,5, 8,8a-hexahydro-1,4:5,8-dimethanonaphthalene; EINE-CS 206-215-8; ENT 15,949; EPA Pesticide Chemical Code 045101; Hexachlorohexahydro-endo-exo-di-methanonaphthalene; HHDN; HSDB 199; Kortofin; Latka 118; NA2761; NA2762; NCI-C00044; NSC 8937; Octalene; RCRA waste number P004; SD 2794; Seedrin; Soilgrin; Tatuzinho; Tipula. Non-systemic insecticide with contact, stomach and respiratory action. Used for control of soil-dwelling insects, termites and ants and also as a wood preservative. Manufacture and use in the US discontinued. Designated as a Persistent Organic Pollutant (POP) under the Stockholm convention. White solid; mp – 104-105°; bp$_2$ = 145°; insoluble in H$_2$O, moderately soluble in organic solvents; LD$_{50}$ (rat orl) = 38-67 mg/kg.

76 Alendronic acid
66376-36-1   227   D

C$_4$H$_{13}$NO$_7$P$_2$
(4-Amino-1-hydroxybutylidene)bisphosphonic acid.
ABDP; Acide alendronique; Acido alendronico; Acidum alendronicum; Alendronate; Alendronic acid; (4-Amino-1-hydroxybutylidene)diphosphonic acid; 4-Amino-1-hydroxybutylidene-1,1-bis(phosphonic acid); 4-Amino-1-hydroxybutylidene-1,1-bisphosphonate; Arendal; MK 217; Phosphonic acid, (4-amino-1-hydroxybutylidene)bis-. Calcium regulator. White solid; mp = 233-235° (dec). *Inst. Gentili S.p.A.; Merck & Co.Inc.*

77 Alendronic acid trisodium salt
121268-17-5   227   D

C$_4$H$_{12}$NNaO$_7$P$_2$.3H$_2$O
Sodium trihydrogen (4-amino-1hydroxybutylidene)-diphosphonate trihydrate.
Adronat; Alendronate sodium; Alendronate sodium hydrate; Alendronic acid monosodium salt trihydrate; Alendros; Aminohydroxybutylidene biphosphonate monosodium salt trihydrate; Dronal; Elandor; Fosalan; Fosamax; G-704,650; MK-217; Monosodium (4-amino-1-hydroxybutylidene)bisphosphonate trihydrate; Onclast; Phosphonic acid, (4-amino-1-hydroxybutylidene)bis-, monosodium salt, trihydrate; Sodium alendronate hydrate; Sodium trihydrogen (4-amino-1-hydroxy-butylidene)diphosphonate, trihydrate. Calcium regulator. *Inst. Gentili S.p.A.; Merck & Co.Inc.*

78 Alfaprostol
74176-31-1   232   D G

C24H38O5
[1R-[1α(Z),2β(S*),3α,5α]]-7-[2-(5-Cyclohexyl-3-hyd-roxy-1-pentynyl)-3,5-dihydroxycyclopentyl]-5-hepten-oic acid methyl ester. Alfavet; Ro-22-9000; K-11941; Alphacept. Prosta-glandin. Used for estrus control in cows. *Farmitalia Carlo Erba Ltd.*

79   Algin
9005-38-3   240   D E G

(C6H7O6)n
Sodium alginate.
AI3-19772; Algiline; Algin; Algin (Laminaria spp. and other kelps); Algin (polysaccharide); Alginate KMF; Alginic acid, sodium salt; Algipon L-1168; Amnucol; Antimigrant C 45; Cecalgine TBV; Cohasal-IH; Darid QH; Dariloid QH; Duckalgin; FEMA No. 2014; Halltex; HSDB 1909; Kelco Gel LV; Kelcosol; Kelgin; Kelgin F; Kelgin HV; Kelgin LV; Kelgin XL; Kelgum; Kelset; Kelsize; Keltex; Keltone; L'-Algiline; Lamitex; Manucol; Manucol DM; Manucol KMF; Manucol SS/LD2; Manugel F 331; Manutex; Manutex F; Manutex RS 1; Manutex RS-5; Manutex rS1; Manutex SA/KP; Manutex SH/LH; Meypralgin R/LV; Minus; Mosanon; Nouralgine; OG 1; Pectalgine; Proctin; Protacell 8; Protanal; Protatek; Snow algin H; Snow algin L; Snow algin M; Sodium alginate; Sodium polymannuronate; Stipine; Tagat; Tragaya; Colloid 488T; E401; KELCOSOL®; KELGIN® F; KELGIN® HV; KELGIN® LV; KELGIN® MV; KELGIN® XL; KELSET®; KELTONE®; KELTONE® HV; KELTONE® HVCR; KELTONE® LV; KELTONE® LVCR; KELVIS®; Kimitsu Algin I-3; Kimitsu Algin I-7; Kimitsu Algin IS; MANUCOL DM; MANUCOL DMF; MANUCOL LB; MANUCOL LKX; Manugel®; Pronova™ LVG; Pronova™ LVM; Pronova™ MVG; Pronova™ MVM; Pronova™ P LVG; Pronova™ P MVG; Pronova™ UP MVG; Protanal LF 5/60; Protanal LF 10/40; Protanal LF 10/60; Protanal LF 20/200; Protanal LF 120 M; Protanal LF 200; Protanal LF 200 M; Protanal LF 200 RB; Protanal LF 200 S; Protanal LFR 5/60; Protanal SF; Protanal SF 120; Protanal SF 120 RB; Satialgin-H8®; Sobalg FD 100 Range. Stabilizer in manufacture of ice cream; emulsifying agent in foods and paints. Emulsifier; firming agent; formulation aid; processing aid; surfactant. FDA, FEMA GRAS, Japan, UK approved, Europe listed, FDA approved for orals, USP/NF, BP, Ph. Eur. compliance (oral suspensions and tablets). Used as a suspending agent, gellant, thickening and emulsifying agent, excipient, tablet binder, stabilizer, used in orals, film-former for microencapsulation. Cream colored powder; soluble in H2O, insoluble in EtOH, Et2O, CHCl3; dynamic viscosity of 1% aqueous solution = 20-400 mPa. LD50 (rat iv) = 1000 mg/kg, (rat orl) > 5000 mg/kg, (rbt iv) = 100 mg/kg, (cat ip) = 250 mg/kg. *Am. Roland; FMC Corp. Pharm. Div.; Kelco Intl.; Multi-Kem; Sigma-Aldrich Fine Chem.; Spectrum Chem. Manufacturing.*

80   Alginic acid
9005-32-7   239   D E G

(C6H8O6)n;
Polymannuronic acid.
A 2830-9; Acid Algin G 2; Algiline; Alginic acid; CCRIS 6769; EINECS 232-680-1; HSDB 2967; Kelacid; Landalgine; Norgine; Polymannuronic acid; Protanal LF; Satialgine-H 8; Sazzio; Snow acid algin G; Verdyol Super. Polysaccharide composed of β-d-mannuronic acid residues; suspending, thickening, emulsifying, and stabilizing agent. Very slightly soluble in H2O; capable of absorbing 200-300 times *its weight in H2O; soluble in alkaline solutions. Kelco Intl.; Mendell*

Edward; Penta Mfg.; Protan.

81   Alizarin
72-48-0   246   G

C14H8O4
1,2-Dihydroxyanthraquinone.
4-08-00-03256 (Beilstein Handbook Reference); AI3-18244; Alizarin; Alizarin B; Alizarin Red; Alizarina; Alizarine; Alizarine 3B; Alizarine B; Alizarine indicator; Alizarine L paste; Alizarine Lake Red 3P; Alizarine Lake Red IPX; Alizarine Lake Red 2P; Alizarine NAC; Alizarine Paste 20 percent Bluish; Alizarine Red; Alizarine Red B; Alizarine Red B2; Alizarine Red IP; Alizarine Red IPP; Alizarine Red L; Alizarinprimeveroside; Alizerine NAC; Alizerine Red IPP; Anthraquinone, 1,2-dihydroxy-; 9,10-Anthracene-dione, 1,2-dihydroxy-; 1,2-Anthraquinonediol; BRN 1914037; C.I. 58000; C.I. 58000C; C.I. Mordant Red 11; C.I. Mordant Red 11C; C.I. Pigment Red 83; C.I. Pigment Red 83C; CCRIS 3530; Certiqual Alizarine; Certiqual Alizarine D; CI 58000; D and C Orange No. 15; D and C Orange Number 15D; D and C Orange Number 15; Deep Crimson Madder 10821; Deep Crimson Madder 10821E; 1,2-Dihydroxy-9,10-anthraquinone; 1,2-Dihydroxy-9,10-anthracenedione; 1,2-Dihydroxy-anthrachinon; 1,2-Dihydroxyanthraquinone; EINECS 200-782-5; Eljon madder; Eljon Madder M; Mitsui Alizarine B; Mitsui Alizarine BS; Mordant Red 11; NSC 7212; Pigment red 83; Sanyo Carmine L2B; Turkey Red; Turkey Red W. Acid wool dye base. Also used as an acid-base indicator (pH 5.5 yellow; pH 6.8 red) and as a spot-test reagent for Al, In, Hg, Zn and Zr. Orange needles; mp = 290°; bp = 430°; λm = 248, 435 nm (ε 25119 6310 EtOH); slightly soluble in H2O (0.058 g/100 ml); more soluble in organic solvents. *Comlets; Fahlberg-List.*

82   Alkanet
517-88-4   252   D G

C16H16O5
(S)-5,8-dihydroxy-2-(1-hydroxy-4-methyl-3-pentenyl)-1,4-naphthalenedione.
Alkanet extract; Alkanna red; Alkannin; Anchusa acid; Anchusin; C.I. 75530; C.I. Natural Red 20; (S)-5,8-Dihydroxy-2-(1-hydroxy-4-methylpent-3-enyl)-1,4-naph-thoquinone; EINECS 208-245-7; 1,4-Naphth-alenedione, 5,8-dihydroxy-2-(1-hydroxy-4-methyl-3-pentenyl)-, (S)-; NSC 407295. Terms applied to two different plants, *Lawsonia inermis* and *Anchusa tintoria*, whose roots are the source of a red dye, anchusine (alkannin), the name is applied to the dye as well as to the plant. Used as an astringent. Solid; mp = 149°; [α]β0= -165° (C6H6); poorly soluble in H2O, more soluble in organic solvents; LD50 (mmus orl) = 3000 ≅ 1000 mg/kg, (fmus orl) 3100 ≅ 100 mg/kg, (rat orl) > 1000 mg/kg.

83   Alkylbenzyldimethylammonium chloride
8001-54-5   1058   D G P

Benzalkonium chloride.

Alkyl dimethyl ethylbenzyl ammonium chloride; Alkyl dimethylbenzyl ammonium chloride; Alkylbenzyl-dimethylammonium chloride; Alkyldimethyl(phenyl-methyl)quaternary ammonium chlorides; Alkyl-dimethyl-benzylammonium chloride; Ammonium, alkyldimethyl-(phenylmethyl)-, chloride; Ammonium, alkyldimethyl-benzyl-, chloride; Ammonyx; Arquad B 100; Arquad dmmcb-75; Barquat MB-50; Barquat MB-80; Bayclean; Benirol; Benzalconio cloruro; Benzalkonii chloridum; Benzalkonium A; Benzalkonium chloride; Bio-quat 50-24; Bio-quat 50-25; Bio-quat 50-30; Bio-quat 50-40; Bio-quat 50-42; Bio-quat 50-60; Bio-quat 50-65; Bio-quat 80-24; Bio-quat 80-28; Bio-quat 80-40; Bio-quat 80-42; Bionol; BTC 50; BTC 50 USP; BTC 65; BTC 65 USP; BTC 100; BTC 471; BTC 824; BTC 2565; BTC 8248; BTC 8249; BTC E-8358; Capitol; Cequartyl; Chlorure de benzalkonium; Cloruro de benzalconio; Culversan LC 80; Desitin Dabaways; Desitin Skin Care Lotions; Dimanin A; Disinall; Dodigen 226; Drapolene; Drapolex; Drest; E-Pilo Ophthalmic Solution; Enuclen; Enuclene; EPA Pesticide Chemical Code 069105; Epifrin Ophthalmic Solution; Gardiquart SV480; Gardiquat 1450; Genamin KDS; Germ-I-tol; Germicin; Germinol; Germitol; HSDB 234; Hyamine 3500; Intexan LB-50; Kemamine BAC; Leda benzalkonium chloride; Mefarol; Mixture Name; Murine For The Eyes; Muro Tears Ophthalmic Solution; Muro's Opcon A Ophthalmic Solution; Muro's Opcon Ophthalmic Solution; Murocoll-19 Opthalmic Solution; Murocoll-2 Ophthalmic Solution; Neo Germ-I-tol; Onyx BTC (Onyx oil & chem Co); Osvan; Paralkan; Pheneene germicidal solution & tincture; Quaternary ammonium compounds, alkylbenzyldimethyl, chlorides; Quaternium-1; Romergal CB; Sporostacin; Triton K-60; Vikrol RQ; Visalens Soaking/Cleaning Solutions; Visalens Wetting Solution; Visine AC; Zephiral; Zephiran; Zephiran chloride; Benirol; BTC; Capitol; Cequartyl; Drapolene; Drapolex; Enuclen; Germinol; Germitol; Osvan; Paralkan; Roccal; Rodalon; Zephiran Chloride; Zephirol. Alkylammonium chlorides of general formula [C6H5CH2N(CH3)2R]+Cl−; R = C8H17 to C18H37. Germicide, algicide, disinfectant, sanitizer, deodorant; used in pesticides and manufacture of sanitizers; food processing, dairy, restaurant, industrial and household products. Amorphous powder, very soluble in H2O, EtOH, Me2CO, insoluble in C6H6, Et2O; LD50 (rat orl) = 400 mg/kg. *Sherex.*

84 **Allantoin**
97-59-6   254   G

C4H6N4O3
2,5-dioxo-4-imidazolidinyl urea.

5-25-15-00338 (Beilstein Handbook Reference); AI3-15281; Alantan; Allantoin; Allantol; AVC/Dienestrol cream; BRN 0102364; Caswell No. 024; CCRIS 1958; Cordianine; Cutemol emollient; (2,5-Dioxo-4-imidazolidinyl)urea; EINECS 202-592-8; EPA Pesticide Chemical Code 085701; Glyoxyldiureid; Glyoxyl-diureide; Hydantoin, 5-ureido-; NSC 7606; Psoralon; Sebical; Septalan; Uniderm A; Urea, (2,5-dioxo-4-imidazolidinyl)-; 5-Ureido-2,4-imidazolidindion; 5-Ureidohydantoin. A product of animal metabolism, excreted in urine. Used in biochemical research and medicine. Used as a soothing agent and skin protectant; stimulates growth of healthy tissue.
Racemic form: monoclinic plates or prisms; mp=239°; soluble in H2O (0.53 g/100 ml), EtOH (0.2 g/100 ml), almost insoluble in ether; pH of

saturated H2O solution = 5.5. *3V; Alcoa Ind. Chem.; EM Ind. Inc.; Greeff R.W. & Co.; Hommel GmbH; ICI Americas Inc.; Lancaster Synthesis Co.; Penta Mfg.; Sutton Labs.; Tri-K Ind.*

85 **Alloocimene**
673-84-7   H

C10H16
2,6-Dimethyl-2,4,6-octatriene.

AI3-00737; Alloocimene; 2,6-Dimethyl-2,4,6-octatriene; 2,6-Dimethylocta-2,4,6-triene; EINECS 211-614-5; NSC 406263; Ocimene, allo; 2,4,6-Octatriene, 2,6-dimethyl-.

86 **Allopurinol**
315-30-0   276   D

C5H4N4O
1H-Pyrazolo[3,4-d]pyrimidin-4-ol.

Adenock; Ailural; AL-100; Allo-Puren; Allopur; Allopurinol; Allopurinol(I); Allopurinolum; Allozym; Allural; Alopurinol; Aloral; Alositol; Aluline; Anoprolin; Anzief; Apulonga; Apurin; Apurol; Atisuril; B. W. 56-158; Bleminol; Bloxanth; BW 56-158; Caplenal; CCRIS 626; Cellidrin; Cosuric; Dabrosin; Dabroson; DRG-0056; Dura Al; EINECS 206-250-9; Embarin; Epidropal; Epuric; Foligan; Geapur; Gichtex; Gotax; Hamarin; Hexanuret; HPP; 4-HPP; HSDB 3004; Ketanrift; Ketobun-A; Ledopur; Lopurin; Lysuron; Milurit; Miniplanor; Monarch; Nektrohan; NSC 101655; NSC-1390; Progout; Remid; Riball; Suspendol; Takanarumin; Urbol; Uricemil; Uripirim; Uripurinol; Urobenyl; Urolit; Urosin; Urtias; Urtias 100; Xanturat; Zyloprim; Zyloric.; Ketobun-A; Ledopur; Lopurin; Lysuron; Miniplanor; Monarch; Nektrohan; Remid; Riball; Sigaporol; Suspendol; Takanarumin; Urbol; Uricemil; Uripurinol; Urobenyl; Urosin; Urtias; Xanturat; Zyloprim; Zyloric. Xanthine oxidase inhibitor. Used to treat gout. Solid; mp > 350°; λm = 257 nm (ε 7200 0.1N NaOH), 250 nm (ε 7600 0.1N HCl), 252 nm (ε 7600 MeOH); soluble in H2O (0.048 g/100 ml), CHCl3 (0.060 g/100 ml), EtOH (0.030 g/100 ml), DMSO (0.46 g/100 ml), n-octanol (< 0.001 g/100 ml). *Glaxo Wellcome Inc.; Knoll Pharm. Co.*

87 **Alloxan**
50-71-5   279   G

C4H2N2O4
2,4,5,6(1H,3H)-pyrimidinetetrone.

AI3-15282; Alloxan; Alloxan 7169; Alloxane; Barbituric acid, 5-oxo-; EINECS 200-062-0; Mesoxalylcarbamide; Mesoxalylurea; NSC 7169; 2,4,5,6-Pyrimidintetron; 2,4,5,6(1H,3H)-Pyrimidine-tetrone; 2,4,5,6-Tetraoxohexa-hydropyrimidine; Urea, mesoxalyl-. Used in nutrition experiments. Causes diabetes in experimental animals.
Crystals; dec 256°; soluble in H2O. *Wako Pure Chem. Ind.*

88   Allyl alcohol
107-18-6   283   G H

C3H6O
 2-Propen-1-ol.
 AI3-14312; Alcool allilco; Alcool allylique; Allilowy alkohol; Allyl alcohol; Allylalkohol; Allylic alcohol; Caswell No. 026; CCRIS 747; EINECS 203-469-1; EPA Pesticide Chemical Code 068401; HSDB 192; 3-Hydroxypropene; NSC 6526; 2-Propenol; 2-Propen-1-ol; Propen-1-ol-3; Propenyl alcohol; 2-Propenyl alcohol; RCRA waste number P005; Shell unkrautted A; UN1098; Vinyl carbinol; Weed drench. Intermediate for pharmaceuticals and other organic chemicals, herbicide. Liquid; mp = -129°; bp = 97°; d^{20} = 0.8540; λmKs = 273 nm (ε = 4677, H2SO4); freely soluble in H2O, EtOH, Et2O, soluble in CHCl3; LD50 (rat orl) = 64 mg/kg.

89   Allyl chloride
107-05-1   286   G H

C3H5Cl
 3-Chloro-1-propene.
 4-01-00-00738 (Beilstein Handbook Reference); Allile (cloruro di); Allyl chloride; Allylchlorid; Allyle (chlorure d'); Barchlor; BRN 0635704; CCRIS 19; Chlorallylene; Chloro-2-propene; α-Chloropropylene; 3-Chloro-1-propylene; 3-Chloro-1-propene; 3-Chloropropene; 3-Chlorpropen; EINECS 203-457-6; HSDB 178; NCI-C04615; NSC 20939; Propene, 3-chloro-; 1-Propene, 3-chloro-; 2-Propenyl chloride; UN1100. Used in the synthesis of allyl compounds Liquid; mp = -134.5°; bp = 45.1°; d^{20} = 0.9376; λm = 273 nm (ε = 4467); insoluble in H2O, slightly soluble in CCl4, freely soluble in EtOH, Et2O, Me2CO, C6H6, ligroin; LD50 (rat orl) = 460 mg/kg. Lonza Ltd.; Lonzagroup.

90   Allyl glycidyl ether
106-92-3   G H

C6H10O2
 1-(Allyloxy)-2,3-epoxypropane.
 5-17-03-00012 (Beilstein Handbook Reference); AGE; Ageflex® AGE; AI3-37791; Allil-glicidil-etere; Allyl 2,3-epoxypropyl ether; Allyl glycidyl ether; Allylglycidaether; 1-Allilossi-2,3 epossipropano; 1-(Allyloxy)-2,3-epoxy-propane; 1-Allyloxy-2,3-epoxy-propaan; 1-Allyloxy-2,3-epoxypropan; BRN 0105871; CCRIS 2375; EINECS 203-442-4; 1,2-Epoxy-3-allyloxypropane; 2,3-Epoxypropyl-1-allyl ether; Ether, allyl 2,3-epoxypropyl; Glycidyl allyl ether; HSDB 505; M 560; NCI-C56666; NSC 18596; Oxirane, ((2-propenyloxy)methyl)-; Oxyde d'allyle et de glycidyle; Propane, 1-(allyloxy)-2,3-epoxy-; Sipomer® AGE; UN2219. Modifier for elastomer, epoxies, adhesives, fibers; reactive intermediate for coatings, sizing/finishing agent for fiberglass; silane intermediate in electrical coatings. Glutamate decarboxylase inhibitor. Liquid; mp = -100°; bp = 154°; d^{20}_{2} = 0.9698; slightly soluble in H2O, more soluble in organic solvents; LD50 (rat orl) = 1600 mg/kg. Rhône Poulenc Surfactants; Rit-Chem.

91   Allyl methacrylate
96-05-9   H

C7H10O2
 2-Propenoic acid, 2-methyl-, 2-propenyl ester.
 4-02-00-01529 (Beilstein Handbook Reference); Ageflex AMA; AI3-37827; Allyl methacrylate; Allylester kyseliny methakrylove; BRN 1747406; EINECS 202-473-0; HSDB 5297; Methacrylic acid, allyl ester; NSC 18597; Sipomer® AM. Monomer for coatings, elastomers, adhesives, intermediates; contributes hardness and scratch resisance; crosslinker/hardener. Silane monomer intermediate; crosslinker offering two-stage polymerization, abrasion and solvent resistance; polymer modifier for high impact plastics, adhesives, acrylic elastomers, photoresists, optical polymers. Liquid; mp = -65°; bp = 144°; bp^{50} = 67°, bp^{30} = 55°; d^{20}_{2} = 0.9335; soluble in H2O (0.4 g/100 ml), organic solvents; LD50 (rat orl) = 430 mg/kg. Monomer-Polymer & Dajac; Polysciences; Rhône Poulenc Surfactants; Rhône-Poulenc; Richman Chem.; Rit-Chem; Rohm & Haas Co.; Sigma-Aldrich Fine Chem.

92   Allylacetonitrile
592-51-8   H

C5H7N
 1-Cyano-3-butene.
 Allylacetonitrile; Allylmethyl cyanide; 3-Butenyl cyanide; 1-Cyano-3-butene; 4-Cyano-1-butene; EINECS 209-762-0; HSDB 5709; 4-Pentenenitrile; 4-Pentenoic acid, nitrile; 4-Pentenenitrile. Liquid; bp = 140°; d^{25} = 0.712.

93   Allyltrimethoxy silane
2551-83-9   G

C6H14O3Si
 Trimethoxy-2-propenylsilane.
 Allyltrimethoxysilane; CA0567; EINECS 219-855-8; Silane, allyltrimethoxy-. Coupling agent, chemical intermediate, blocking agent, release agent, lubricant, primer, reducing agent. Degussa-Hüls Corp.

94   Allyltrimethylsilane
762-72-1   G

C6H14Si
 Allyltrimethyl silane.
 CA0570; CCRIS 2649; EINECS 212-104-5; Silane, trimethyl-2-propenyl-. Coupling agent, chemical intermediate, blocking agent, release agent, lubricant, primer, reducing agent. Liquid; bp = 84-88°; d = 0.7190. Akzo Chemie.

95   Alprazolam
28981-97-7   310   D

C17H13ClN4
8-Chloro-1-methyl-6-phenyl-4H-s-triazolo[4,3-a][1,4]-benzodiazepine.
Alcelam; Algad; Alpaz; Alplax; Alpram; Alprax; Alprazolam; Alprazolamum; Alpronax; Alprox; Alzam; Alzolam; Alzon; Anpress; Apo-Alpraz; Azor; Bestrol; BRN 1223125; Cassadan; Constan; D 65MT; DEA No. 2882; EINECS 249-349-2; Esparon; Frontal; Gen-Alprazolan; Helex; Intensol; Ksalol; Mialin; Neurol; Novo-Alprazol; Nu-Alpraz; Panix; Pharnax; Prazam; Prazolan; Prinox; Ralozam; Relaxol; Restyl; Solanax; Tafil; Tensivan; Trankimazin; Tranquinal; Tricalma; TUS-1; U 31889; Unilan; Valeans; Xanagis; Xanax; Xanax XR; Xanor; Zacetin; Zanapam; Zaxan; Zenax; Zolan; Zolarem; Zoldac; Zoldax; Zotran. Anxiolytic. Sedative/hypnotic. *Pharmacia & Upjohn.*

96   Alumina
1344-28-1   355   G

Al2O3
Aluminum oxide.
A 1 (Sorbent); A1-0104 T 3/16"; A1-0109 P; A1-1401 P(MS); A1-1404 T 3/16"; A1-3438 T 1/8"; A1-3916 P; A1-3945 E 1/16"; A1-3970 P; A1-3980 T 5/32"; A1-4028 T 3/16"; A1-4126 E 1/16"; A-2; Abramant; Abramax; Abrarex; Abrasit; Activated aluminum oxide; Adamant; Al3-02904; Alcan AA-100; Alcan C-70; Alcan C-71; Alcan C-72; Alcan C-73; Alcan C-75; Alcoa F 1; Almite; Alon C; Aloxite; Alumina; α-Alumina; γ-Alumina; δ-Alumina; Aluminite 37; Aluminium oxide; Aluminum oxide (Al2O3); Aluminum oxide (Brockmann); Aluminum oxide (fibrous forms); Aluminum oxide (ignited); Aluminum sesquioxide; Aluminum trioxide; Alumite (oxide); Alundum; Alundum 600; Bikorit; Brockmann, aluminum oxide; Cab-O-grip; Catapal S; Catapal SB alumina; CCRIS 6605; Compalox; Conopal; Diadur; Dialuminum trioxide; Dispal alumina; Dispal M; Dotment 324; Dotment 358; Dural; EINECS 215-691-6; Eta-alumina; Exolon XW 60; F 360 (Alumina); Faserton; Fasertonerde; Fiber FP; G 0 (Oxide); G 2 (Oxide); GK (Oxide); HSDB 506; Hypalox II; Jubenon R; KA 101; Ketjen B; KHP 2; LA 6; Lucalox; Ludox CL; Martoxin; Microgrit WCA; Neobead C; Poraminar; PS 1 (Alumina); Q-Loid A 30; RC 172DBM; Saffie; T64; T-1061; Versal 150. Used in abrasive industries, super refractories, sandblasting and for safes. Colorant; dispersing agent; used in orals. Polish and abrasive; absorbent, desiccant; filler for paints and varnishes; ceramic materials, electronics and resistors; dental cements; glass, artificial gems; coating for metals. Calcined alumina, a ceramic grade alumina for refractory bricks, whitewares. Used in production of aluminum, abrasives, refractories, ceramics, electrical insulators, catalysts and catalyst supports, paper, spark plugs, crucibles and lab ware, adsorbent for gases/water vapors, chromatographic analysis, heat-resist. fibers. White special fused corundum (crystalline aluminum oxide), used for production of ramming mixes, shape bricks and crucibles for the lining of high temperature furnaces; molding material for precision casting molds. Hard white solid; bp = 2977°; d = 3.54; insoluble in H2O, slightly soluble in mineral acids; toxic by inhalation of dust; eye irritant by mechanical abrasion. *Air Products & Chemicals Inc.; Alcan Chem.; Alcoa Ind. Chem.; Atomergic Chemetals; BA Chem. Ltd.; Degussa AG; Degussa-Hüls Corp.; Ferro/Transelco; LaRoche Ind.; Lonza-Werke GmBH; Nissan Chem. Ind.; Norton Chem. Process Prods.; Rhône-Poulenc; Sigma-Aldrich Fine Chem.; Vista.*

97   Aluminum
7429-90-5   321   G

Al

Al
Aluminum.
A 00; A 95; A 99; A 995; A 999; A99V; AA 1099; Aa1193; AA1199; AD 1; AD1M; ADO; Adom; AE; Alaun; Allbri aluminum paste and powder; Alumina fibre; Aluminium; Aluminium bronze; Aluminium, elementary; Aluminium flake; Aluminum; Aluminum (fume or dust); Aluminum (metal); Aluminum 27; Aluminum A00; Aluminum dehydrated; Aluminum dust; Aluminum, elemental; Aluminum metal; Aluminum metal, pyro powders; Aluminum metal, respirable fraction; Aluminum metal, soluble salts; Aluminum metal, total dust; Aluminum metal, welding fumes; Aluminum, molten [Class 9]; Aluminum powder; Aluminum production; Aluminum, pyro powders; Aluminum soluble salts; Aluminum, welding fumes; AO A1; AR2; AV00; AV000; C-Pigment 1; C.I. 77000; Caswell No. 028A; CI 77000; CI 77000; EINECS 231-072-3; Emanay atomized aluminum powder; EPA Pesticide Chemical Code 000111; HSDB 507; JISC 3108; JISC 3110; L16; Metana; Metana aluminum paste; NA9260; Noral aluminium; Noral aluminum; Noral Extra Fine Lining Grade; Noral ink grade aluminium; Noral Ink Grade Aluminum; Noral non-leafing grade; PAP-1; Pigment metal 1; UN1309; UN1396. Metallic element; building and construction, corrosion-resistant chemical equipment (desalination plants), die-cast auto parts, electrical industry (power transmission lines), photoengraving plates, permanent magnets, cryogenic technology, machinery, tubes for ointments. Malleable metallic solid; mp = 660°; bp = 2327°; d = 2.70. *Alcan Chem.; Alcoa Ind. Chem.; Norsk Hydro AS.*

98   Aluminum butoxide
3085-30-1   H

C12H27AlO3
Aluminum n-butoxide.
Aluminium tributanolate; Aluminum butoxide; Aluminum n-butoxide; 1-Butanol, aluminum salt; EINECS 221-394-2.

99   Aluminum chlorate
15477-33-5   334   D G

AlCl3O9
Chloric acid aluminum salt.
Aluminum chlorate; EINECS 239-499-7; Mallebrin. Occurs as a hexahydrate and a nonahydrate. Used as an antiseptic and astringent. Freely soluble in H2O, soluble in EtOH.

100   Aluminum chloride
7446-70-0   335   D G P

AlCl₃

AlCl3

Aluminum chloride, anhydrous.

Al3-01917; Alluminio(cloruro di); Aluminium, (chlorure d'); Aluminium chloride; Aluminium trichloride; Aluminiumchlorid; Aluminum, (chlorure d'); Aluminum chloride; Aluminum chloride (1:3); Aluminum chloride (AlCl3); Aluminum chloride (anhydrous); Aluminum trichloride; Anhydrol forte; Caswell No. 029; CCRIS 6871; Chlorure d'aluminium; EINECS 231-208-1; EPA Pesticide Chemical Code 013901; HSDB 607; NSC 143015; NSC 143016; Pearsall; Praestol® K2001; TK Flock; Trichloroaluminum; UN1726; UN2581. Ethylbenzene catalyst, dyestuff intermediate, detergent alkylate, ethyl chloride, pharmaceuticals and organics, butyl rubber, petroleum refining, hydrocarbon resins, nucleating agent for titanium dioxide pigments. Coagulant for municipal and industrial water treatment; effective for turbidity reduction, phosphorus removal, water clarification, flotation, oil/water demulsification. Registered by EPA as an antimicrobial and herbicide (cancelled). BP compliance. Used as an antiseptic. The hexahydrate used as a topical astringent. White or colorless crystals; strong irritant; freely soluble in many organic solvents. *Asada Chem. Ind. Ltd.; CK Witco Corp.; Elf Atochem N. Am.; Fluka; Harcros Durham; Sigma-Aldrich Fine Chem.*

101   Aluminum n-decoxide

26303-54-8   H

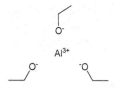

C₃₀H₆₆AlO₃

Aluminium tri(decanolate).

Aluminium tri(decanolate); Aluminum n-decoxide; 1-Decanol, aluminum salt; Decyl alcohol, aluminum salt; EINECS 247-598-1.

102   Aluminum n-docosoxide

67905-30-0   H

C₆₆H₁₃₈AlO₃

1-Docosanol, aluminum salt.

Aluminium tridocosanolate; Aluminum n-docosoxide; 1-Docosanol, aluminum salt; EINECS 267-649-1.

103   Aluminum n-eicosoxide

67905-31-1   H

C₆₀H₁₂₆AlO₃

1-Eicosanol, aluminum salt.

Aluminium tri(icosanolate); Aluminum n-eicosoxide; 1-Eicosanol, aluminum salt; EINECS 267-650-7.

104   Aluminum ethoxide

555-75-9   337   H

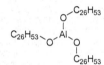

C₆H₁₈AlO₃

Ethanol, aluminum salt.

Aluminium triethanolate; Aluminum ethoxide; Aluminum ethylate; Aluminum triethoxide; EINECS 209-105-8; Ethanol, aluminum salt; Ethyl alcohol, aluminum salt; HSDB 5430; Triethoxyaluminum.

105   Aluminum fluoride

7784-18-1   338   G

AlF3

Aluminum trifluoride.

Aluminium fluoride; Aluminium fluorure; Aluminum fluoride (AlF3); Aluminum trifluoride; CCRIS 2282; EINECS 232-051-1; Fluorid hlinity; HSDB 600. Used mixed with aluminum oxide and silica for use as an electrolyte in reduction of alumina to aluminum metal; as flux in remelting and refining of aluminum and its alloys; opacifier aid in production of ceramic enamels, glass, and glazes. Crystals; sublimes 1272°; soluble in H2O (0.559 g/100 ml). *Alcan Chem.*

106   Aluminum n-hexacosoxide

67905-28-6   H

C₇₈H₁₅₉AlO₃

1-Hexacosanol, aluminum salt.

Aluminium trihexacosanolate; Aluminum n-hexacosoxide; EINECS 267-647-0; 1-Hexacosanol, aluminum salt.

107   Aluminum hexadecan-1-olate

19141-82-3   H

C₄₈H₁₀₂AlO₃

1-Hexadecanol, aluminum salt.

Aluminium hexadecan-1-olate; Aluminum hexadecan-1-olate; Aluminum n-hexadecoxide; EINECS 242-836-0; 1-Hexadecanol, aluminum salt.

108   Aluminum n-hexoxide

23275-26-5   H

C18H39AlO3
1-Hexanol, aluminum salt.
Aluminium tri(hexanolate); Aluminum n-hexoxide; EINECS 245-546-2; 1-Hexanol, aluminum salt; Hexyl alcohol, aluminum salt.

109   Aluminum hydroxide
21645-51-2   342   G

H3AlO3
Aluminum oxide trihydrate.
A 3011; AC 450; AC 714KC; AE 107; AF 260; AKP-DA; Alcoa 331; Alcoa 710; Alcoa A 325; Alcoa AS 301; Alcoa C 30BF; Alcoa C 31; Alcoa C 33; Alcoa C 330; Alcoa C 331; Alcoa C 333; Alcoa C 385; Alcoa H 65; Alhydrogel; Alolt 8; Alolt 80; Alolt 90; Alternagel; ALternaGEL; Alu-Cap; Alugel; Alugelibye; Alumigel; Alumina trihydrate; Aluminic acid (H3AlO3); Aluminium hydroxide; Aluminum hydroxide; Aluminum hydroxide (Al(OH)3); Aluminum hydroxide, dried; Aluminum hydroxide gel; Aluminum oxide trihydrate; Aluminum trihydroxide; Aluminum(III) hydroxide; Alusal; Amberol ST 140F; Amorphous alumina; Amphogel; Amphojel; Antipollon HT; Apyral; Apyral 120; Apyral 120VAW; Apyral 2; Apyral 4; Apyral 8; Apyral 15; Apyral 24; Apyral 25; Apyral 40; Apyral 60; Apyral 90; Apyral B; Arthritis Pain Formula Maximum Strength; Ascriptin; BACO AF 260; Boehmite; British aluminum AF 260; C 31; C-31-F; C-31-F; C 31C; C 31F; C 31F; C 33; C 4D; C.I. 77002; Calcitrel; Calmogastrin; Camalox; Cl 77002; Di-Gel Liquid; Dialume; DRG-0172; EINECS 244-492-7; F-1000 Dried Gel; F-1000®; F-2000® Dried Gel; F-2100® Dried Gel; F-2200® Dried Gel; Gelusil; GHA 331; GHA 332; GHA 431; H 46; Higilite; Higilite H 31S; Higilite H 32; Higilite H 42; HSDB 575; Hychol 705; Hydrafil; Hydral 705; Hydral 710; Hydrated alumina; Hydrated aluminum oxide; Kudrox; Liquigel; Maalox; Maalox HRF; Maalox Plus; Martinal; Martinal A; Martinal A/S; Martinal F-A; Mixture Name; Mylanta; P 30BF; PGA; Reheis F 1000; Simeco Suspension; Tricreamalate; Trihydrated alumina; Trihydroxyaluminum; Trisogel; Wingel. Used in dyes, paints and in textile finishing. As a compressed gel or a powder, is used medicinally as an antacid with good resuspending properties. White refractory powder; insoluble in H2O. *Alcan Chem.; Alcoa Ind. Chem.; Atomergic Chemetals; BA Chem. Ltd.; Nyco Minerals; Reheis; Rhône-Poulenc; Seimi Chem.; Vista; Whittaker Clark & Daniels.*

110   Aluminum hydroxychloride
1327-41-9   343   D G

Al2H5O5Cl.2H2O
Basic aluminum chloride.
ACH 325; ACH 331; ACH 7-321; Aloxicoll; Aluminol ACH; Aluminum chlorhydrate; Aluminum chlorhydroxide; Aluminum chloride, basic; Aluminum chloride hydroxide oxide, basic; Aluminum chloride hydroxide; Aluminum chloride oxide; Aluminum chlorohydrate anhydrous; Aluminum chlorohydrol; Aluminum hydroxide chloride; Aluminum hydroxychloride; Aluminum oxychloride; Aluminum sesquichlorohydrate; Aquarhone 18; Astringen; Astringen 10; Banoltan White; Basic aluminum chloride; Basic aluminum chloride, hydrate; Berukotan AC-P; Cartafix LA; Cawood 5025; Chlorhydrol; Chlorhydrol Micro-Dry SUF; Chlorhydrol Micro-Dry; E 200; E 200 (coagulant); EINECS 215-477-2; Hessidrex WT; HPB 5025; Hydral; Hydrofugal; Kempac 10; Kempac 20; Kemwater PAX 14; Locron; Locron P; Locron S; Nalco 8676; OCAL; Oulupac 180; PAC; PAC (salt); PAC 250A; PAC 250AD; PACK 300M; Paho 2S; PALC; Reach® 101; Reach® 201; Reach® 501; Ritachlor 50%; Wickenol

cps 325. Astringent and antihyperphosphatemic. Used in commercial antiperspirants and deodorants giving increased wetness protection, especially for Aerosols; in water purification; treatment of sewage and plant effluent. A raw material for personal care products. Glassy solid; soluble in H2O (< 55% w/w). *Catomance Ltd.; Giulini; Reheis; Rita.*

111   Aluminum isopropoxide
555-31-7   346   H

C9H24AlO3
Isopropanol aluminum salt.
AI3-14396; Aliso; Aluminium isopropoxide; Aluminium triisopropanolate; Aluminum isoprop-anolate; EINECS 209-090-8; HSDB 5429; Isopropanol aluminum salt; Isopropyl alcohol, aluminum salt; NSC 7604; 2-Propanol, aluminum salt; Triisopropoxy-aluminum; Triisopropyl-oxyaluminum; Tris(isoprop-oxy)aluminum. Used for cosmetics, pharmaceuticals. Solid; mp = 128-133°; bp5.2 =125-130°; d = 1.0350; decomposed by H2O; soluble in organic solvents; LD50 (rat orl) = 11.3 g/kg.

112   Aluminum nitrate
7784-27-2   G

AlN3O9.9H2O
Nitric acid aluminum salt nonahydrate.
Aluminum nitrate nonahydrate; Aluminum trinitrate nonahydrate; Aluminum(III) nitrate, nonahydrate (1:3:9); Nitric acid, aluminum salt, nonahydrate. Occurs mainly as the nonahydrate. Mordant for textiles, leather tanning, manufacture of incandescent filaments, catalyst in petroleum refining, nucleonics, anticorrosion agent, antiperspirant. Crystals; mp = 73°; dec 135°; very soluble in H2O, less soluble in organic solvents; LD50 (rat orl) = 4.28 g/kg. *EM Ind. Inc.; Hoechst Celanese; Sherman Chem. Co.; Sigma-Aldrich Fine Chem.; Spectrum Chem. Manufacturing.*

113   Aluminum nitride
24304-00-5   352   G

Al≡N

AlN
Aluminum nitride.
Aluminum nitride; EINECS 246-140-8. Used as a semiconductor in electronics, nitriding of steel; steel manufacture. Crystals; mp = 2150-2200°; hardness 9 to 10 on Moh's scale. *Atomergic Chemetals; Carborundum Corp.; Mandoval Ltd.; Sigma-Aldrich Fine Chem.*

114   Aluminum n-octacosoxide
67905-27-5   H

C84H171AlO3
1-Octacosanol, aluminum salt.
Aluminium tri(octacosanolate); Aluminum n-octacosoxide; EINECS 267-646-5; 1-Octacosanol, aluminum salt.

115   Aluminum n-octadecoxide
3985-81-7   H

C54H111AlO3
Aluminium octadecan-1-olate.
Aluminium octadecan-1-olate; Aluminum n-octadecoxide; EINECS 223-629-4; 1-Octadecanol, aluminium salt; 1-Octadecanol, aluminum salt.

116   Aluminum oleate
688-37-9   353   G

C54H99AlO6
Aluminum 9-octadecenoate, (Z)-.
AI3-19803; Aluminium trioleate; Aluminum oleate; Aluminum trioleate; EINECS 211-702-3; HSDB 573; 9-Octadecenoic acid, (Z)-, aluminum salt (3:1); 9-Octadecenoic acid (Z)-, aluminum salt; Oleic acid, aluminum salt; Olminat. Olminate is a trade name for a commercial aluminum oleate; contains 1.5% aluminum. Used as lacquer for metals. Insoluble in H_2O, soluble in organic solvents.

117   Aluminum phosphate
7784-30-7   357   7

AlO4P
Aluminum orthophosphate.
Aluminium orthophosphate, natural; Aluminium-phosphat; Aluminophosphoric acid; Aluminum acid phosphate; Aluminum monophosphate; Aluminum phosphate; Aluminum phosphate (1:1); Aluphos; Angelite; Coeruleoactite; EINECS 232-056-9; Evansite; FFB 32; Lucinite; Metavariscite; Monoaluminum phosphate;

Phosphaljel; Phosphalugel; Phosphoric acid, aluminum salt (1:1); Sterretite. Acidic solution. Refractory bonding agent, metal processing. Used as cement admixture with calcium sulfate and sodium silicate; as a flux for cermaics; dental cements and for special glasses. Used as an antacid. White solid; mp > 1460°; d^{23} = 2.56; insoluble in H_2O, AcOH, slightly soluble in concentrated HCl, HNO_3. *Albright & Wilson Americas Inc.; Rasa; Rhône-Poulenc; Superfos Biosector A/S.*

118   Aluminum potassium sulfate
10043-67-1   359   D G

AlKO8S2
Potassium aluminum sulfate.
Alaun; Alum; Alum, N.F.; Alum potassium; Alum, potassium; Alum, potassium anhydrous; Aluminium potassium bis(sulphate); Aluminum potassium alum; Aluminum potassium disulfate; Aluminum potassium sulfate (AlK(SO4)2); Aluminum potassium sulfate (KAl(SO4)2); Aluminum potassium sulfate, alum; Aluminum potassium sulfate, anhydrous; Burnt alum; Burnt potassium alum; CCRIS 6842; Dialuminum dipotassium sulfate; EINECS 233-141-3; Exsiccated alum; HSDB 2685; Potash alum; Potassium alum; Potassium aluminum alum; Potassium aluminum sulfate; Potassium aluminum sulfate (1:1:2); Sulfuric acid, aluminum potassium salt (2:1:1); Tai-Ace K 150; Tai-Ace K 20. Anhydrous or dodecahydrate. Used as an astringent, as a cement hardener, in paper, matches, paints, tanning agents, waterproofing agents and purification of water. Soluble in H_2O (5 g/100 ml at 20°, 100 g/100 ml at 100°), insoluble in EtOH; [dodecahydrate]: mp = 92.5°; d = 1.725; soluble in H_2O (13.9 g/100 ml at 20°, 330 g/100 ml at 100°, insoluble in organic solvents.

119   Aluminum rosinate
61789-65-9   H

Rosin acids, aluminum salts.
Aluminum resinate; Aluminum rosinate; EINECS 263-075-0; Resin acids and rosin acids, aluminum salts; Rosin acids, aluminum salts; UN2715.

120   Aluminum silicate
12141-46-7   361   G

Al2O5Si
Aluminum oxide silicate.
Aciculite; Aluminum oxide silicate (Al2O(SiO4)); Aluminum silicate, natural; Aluminum silicate, synthetic; EINECS 235-253-8; Kaopolite® 1152; Kaopolite® AB; Kaopolite® SF; Kaylene; Kaylene-ol; Tisyn®. A gentle abrasive that speeds cleaning; antiblocking agent for plastic film. Confers mild polishing properties for auto polishes, plastics. Used as a reinforcement for plastic systems, coating, caulks, sealants, and mastics. *Kaopolite.*

121   Aluminum sodium sulfate
10102-71-3   362   D G

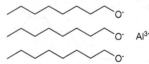

AlNaO8S2
Sodium aluminum sulfate.
  Aluminum sodium sulfate; Aluminum sodium sulfate, NaAl(SO4)2; Aluminum sodium sulfate (AlNa(SO4)2); EINECS 233-277-3; HSDB 571; Kasal; Soda alum; Sodium alum; Sodium aluminum sulfate; Sodium aluminum sulfate (NaAl(SO4)2); Sulfuric acid, aluminum sodium salt (2:1:1). Sodium alum [as dodecahydrate]; Soda alum [as dodecahydrate]. Food additive for baking, cereals, dairy and cheese. Used in textiles as a mordant and for waterproofing, dry colors, ceramics, tanning, paper size precipitant, sugar refining, water purification. An astringent. [dodecahydrate]: mp ≅ 60°; d = 1.61; soluble in H2O (100 g/100 ml), insoluble in EtOH. *Rhône-Poulenc Food Ingredients.*

122   Aluminum stearate
300-92-5   G H

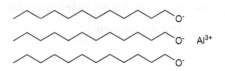

C36H71AlO5
Aluminum hydroxide distearate.
  Aluminum distearate; Aluminum hydroxide distearate; Aluminum, hydroxybis(octadecanoato-O)-; Aluminum, hydroxybis(stearato)-; Aluminum hydroxydistearate; EINECS 206-101-8; HSDB 5487; Hydroxyaluminum distearate; NSC 522176; Special M. Process aid, lubricant for PVC, polyolefins, PS, ABS. *Syn Prods.*

123   Aluminum sulfate
10043-01-3   365   D G P

Al2O12S3
Aluminum(III) sulfate.
  Alum; Aluminium sulfate; Aluminium sulphate; Aluminum alum; Aluminum sesquisulfate; Aluminum sulfate; Aluminum sulfate (Al2(SO4)3); Aluminum sulfate anhydrous; Aluminum sulphate; Aluminum trisulfate; Aluminum(III) sulfate; Cake alum; Caswell No. 031A; Dialuminum sulfate; Dialuminium sulphate; Dialuminum trisulfate; EINECS 233-135-0; EPA Pesticide Chemical Code 013906; Filter Alum; Hi Soft C 2; HSDB 5067; Nalco 7530; NSC 54563; Papermaker's Alum; Pearl Alum; Pickle Alum; Sulfatodialuminum disulfate (Al72(SO4)3); Sulfuric acid, aluminum salt (3:2); Sulfuric acid, aluminum(3+) salt (3:2); Tai-Ace S 150; Tai-Ace S 100. Used in pulp and paper mills, water purification plants, leather, textile, gypsum treatment, in fire retardants; deodorizer, decolorizer, food additive. Used in packing and preserving and to color hydrangeas; also to lower phosphate levels in water. Antiseptic. Registered by EPA as an antimicrobial and herbicide

(cancelled). Solid; mp = 770°; d = 1.61; soluble in H2O (100 g/100 ml), insoluble in EtOH; LD50 (mus orl) = 6207 mg/kg. *Alcan Chem.; Am. Cyanamid; Asada Chem. Ind. Ltd.; Ashland; BA Chem. Ltd.; Ethyl Corp.; General Chem; Herbert; Rasa; Rhône-Poulenc; Sigma-Aldrich Fine Chem.*

124   Aluminum n-tetracosoxide
67905-29-7   H

C72H147AlO3
1-Tetracosanol, aluminum salt.
  Aluminium tetracosanolate; Aluminum n-tetracosoxide; EINECS 267-648-6; 1-Tetracosanol, aluminum salt.

125   Aluminum n-tetradecoxide
67905-32-2   H

C42H90AlO3
1-Tetradecanol, aluminum salt.
  Aluminum n-tetradecoxide; EINECS 267-651-2; 1-Tetradecanol, aluminum salt; Tetradecanolate aluminium.

126   Aluminum tri(octanolate)
14624-13-6   H

C24H51AlO3
1-Octanol, aluminum salt.
  Aluminium tri(octanolate); Aluminum n-octoxide; EINECS 238-666-1; 1-Octanol, aluminum salt; Octyl alcohol, aluminum salt.

127   Aluminum tridodecanolate
14624-15-8   H

C36H75AlO3
1-Dodecanol, aluminum salt.
  Aluminum n-dodecoxide; Aluminium tridodecanolate; 1-Dodecanol, aluminum salt; Dodecyl alcohol, aluminum salt; EINECS 238-667-7.

128   Aluminum trisodium hexafluoride
13775-53-6   G

AlF6Na3

Aluminate(3-), hexafluoro-, trisodium, (OC-6-11)-.

Aluminate(3-), hexafluoro-, trisodium; Aluminum sodium fluoride (Na3AlF6); Aluminum sodium hexafluoride (AlNa3F6); EINECS 237-410-6; Sodium aluminum hexafluoride; Sodium fluoroaluminate (Na3AlF6); Sodium fluoroaluminate(3-); Sodium hexafluoroaluminate; Synkrolith; Synthetic Cryolite; Trisodium aluminum hexafluoride; Trisodium hexafluoroaluminate. Filler for synthetic resin-bonded grinding wheels and brake linings. *Metallgesellschaft GmbH.*

129 Aluminum tristearate
637-12-7 363 G H

$C_{54}H_{105}AlO_6$

Aluminum n-octadecanoate.

AI3-01515; Alugel 34TN; Aluminium stearate; Aluminum octadecanoate; Aluminum(III) stearate; Aluminum tristearate; Aluminium tristearate, pure; EINECS 211-279-5; HSDB 5733; Metasap XX; Monoaluminum stearate; Novogel® ST; Octadecanoic acid, aluminum salt; Rofob 3; SA 1500; Stearic acid, aluminum salt; Synpro® 404; Tribasic aluminum stearate. Solid; mp = 118°; d = 1.01; insoluble in H_2O, soluble in EtOH, petroleum ether; LD_{50} (rat orl) >5 g/kg. *Elf Atochem N. Am.; Ferro/Grant; Magnesia GmbH; Norac; Rhône-Poulenc; Syn Prods.; Witco/Humko.*

130 Ametryn
834-12-8 391 G P

$C_9H_{17}N_5S$

N-Ethyl-N-isopropyl-6-methylthio-1,3,5-triazine-2,4-di-amine.

80W; A 1093; AI3-60365; Amephyt; Ametrex; Ametrine; Ametryn; Ametryne; BRN 0613099; Caswell No. 431; Cemerin; Crisatrine; Doruplant; EPA Pesticide Chemical Code 080801. Selective systemic herbicide used for control of most annual grasses and broad-leaved weeds in pineapples, sugar cane, bananas; citrus fruit, maize, cassava; coffee, tea, cocoa; oil palmsand on non-crop land. An unrestricted, general use pesticide. Registered by EPA as a herbicide. Colorless crystals; mp = 86-88°; d_{20} = 1.19; soluble in H_2O (0.0185 g/100 ml), more soluble in CH_2Cl_2 (60 g/100 ml), Me_2CO (50 g/100 ml), MeOH (45 g/100 ml), C_7H_8 (40 g/100 ml), C_6H_{14} (1.4 g/100 ml); LD_{50} (rat orl) = 1110 mg/kg, (mus orl) = 965 mg/kg, (rbt der) > 8160 mg/kg, (rat der) > 3100 mg/kg; LC_{50} (bobwhite quail 8 day dietary) = 30000 mg/kg diet, (mallard duck 8 day dietary) = 23000 mg/kg diet, (rainbow trout 96 hr.) = 8.8 mg/l, (bluegill sunfish 96 hr.) = 4.1 mg/l, (goldfish 96 hr.) = 14.1 mg/l; low toxicity to bees. *Agan Chemical Manufacturers; Micro-Flo Co. LLC;*

Syngenta Crop Protection.

131 Amicarbalide
3459-96-9 394 G V

$C_{15}H_{16}N_6O$

3,3'-(Carbonyldiimino)bisbenzenecarboximidamide.

Amicarbalida; Amicarbalide; Amicarbalidum; 1,3-Bis(3-amidinophenyl)urea; 3,3'-Diamidinocarbanilide; EINECS 222-402-7; 3,3'-Ureylendibenzamidin. Babesiacide (treatment for piroplasmosis) in cattle. Diampron, the isethionate of amicarbalide, is an antiprotozoan for veterinary use.

132 Amido-G acid
86-65-7 400 G

$C_{10}H_9NO_6S_2$

7-Amino-1,3-naphthalenedisulfonic acid.

4-14-00-02811 (Beilstein Handbook Reference); 2-Amino-6,8-disulfonaphthalene; 2-Aminonaphthalene-6,8-disulfonic acid; 7-Amino-1,3-naphthalenedisulfonic acid; 7-Aminonaphthalene-1,3-disulphonic acid; AI3-28530; Amido-G-acid; Amino-G-Acid; BRN 2669649; EINECS 201-689-2; 1,3-Naphthalenedisulfonic acid, 7-amino-; β-Naphthylamine-6,8-disulfonic acid; 2-Naphthylamine-6,8-disulfonic acid; NSC 4013. Used as an intermediate in manufacture of chemicals, particularly dyestuffs. mp = 274°; soluble in H_2O (92 g/l), EtOH, less soluble in non-polar organic solvents. *Zen Brasal Pvt. Ltd.*

133 Amilperoxy pivalate
29240-17-3 G

$C_{10}H_{20}O_3$

t-Amyl peroxypivalate.

tert-Amyl peroxypivalate; Aztec® t-Amyl peroxypivalate-75 OMS; EINECS 249-530-6; Esperox® 551M; Lupersol 554-M50, 554-M75; tert-Pentyl peroxypivalate; Propaneperoxoic acid, 2,2-dimethyl-, 1,1-dimethylpropyl ester; Trigonox® 125-C75. Initiator used in polymerization of monomers; initiator for bulk, solution, and suspension polymerization. *Catalyst Resources Inc.; Elf Atochem N. Am.*

134 Amines, tris(hydrogenated tallow alkyl)
61790-42-9 H

Tris(hydrogenated tallow alkyl)amines.

Amines, tris(hydrogenated tallow alkyl); EINECS 263-135-6.

135   Aminitrazole
140-40-9   409   D G

C5H5N3O3S
N-(5-Nitro-2-thiazolyl)acetamide.
4-27-00-04676 (Beilstein Handbook Reference); 2-Acetamido-5-nitrothiazole; 4-27-00-04676 (Beilstein Handbook Reference); 5-Nitro-2-acetilaminotiazolo; 5-Nitro-2-acetamidothiazole; Acetamide, N-(5-nitro-2-thiazolyl)-; Acetyl enheptin; Acinitrazol; Acinitrazole; Ametoterina; Aminitrazol; Aminitrozol; Aminitrozole; Aminitrozolum; Amminitrozol; BRN 0167361; CL 5,279; CL 5279; Cyzine premix; EINECS 205-414-7; Enheptin A; Gynofon; Lavoflagin; Nitasol; Nitazol; Nitazole; Nithiamide; N-(5-Nitro-2-thiazolyl)acet-amide; NSC 31539; Pleocide; RP 8243; Thiazole, 2-acetamido-5-nitro-; Torion; Trichlorad; Trichloral; Trichocid; Trichoman; Trichorad; Trichoral; Tricogen; Tricolaval; Tricoral; Tricosil; Tricosteril; Trikolaval; Tritheon. Antiprotozoal (Trichomonas) and an antihistomonad; used with turkeys. Crystals; mp = 264.5°; λm = 236, 340 nm (MeOH); soluble in aqueous alkaline media. *Am. Cyanamid.*

136   3'-Amino-p-acetanisidide
6375-47-9   H

C9H12N2O2
3-Amino-4-methoxyacetanilide.
Acetamide, N-(3-amino-4-methoxyphenyl)-; p-Acet-anisidide, 3'-amino-; 3'-Amino-4'-methoxyacetanilide; 3'-Amino-p-acetanisidide; EINECS 228-938-8.

137   o-Aminoaniline
95-54-5   7368   G H

C6H8N2
1,2-Benzenediamine.
4-13-00-00038 (Beilstein Handbook Reference); AI3-24343; o-Benzenediamine; BRN 0606074; CCRIS 508; CI 76010; CI Oxidation Base 16; o-Diaminobenzene; EINECS 202-430-6; EK 1700; o-Fenylendiamin; HSDB 2893; IK 3; NSC 5354; o-Phenylenediamine; OPDA; Orthamine; Phenylenediamine, ortho-; PODA; SQ 15500; UN1673. Used in dyestuffs manufacture. mp = 102.5°; bp = 257°; λm = 237, 293 nm (ε = 7943, 3981, MeOH); soluble in H2O (1 g/100 ml), more soluble in EtOH, Et2O, C6H6; LD50 (rat orl) = 1070 mg/kg.

138   m-Aminobenzoic acid
99-05-8   420   G

C7H7NO2
3-Aminobenzoic acid.
4-14-00-01092 (Beilstein Handbook Reference); AI3-04707; 3-Aminobenzoic acid; m-Aminobenzoic acid; Aniline-3-carboxylic acid; Benzoic acid, 3-amino-; Benzoic acid, m-amino-; BRN 0471603; 3-Carboxyaniline; m-Carboxyaniline; EINECS 202-724-4; MABA; NSC 15012. Used in chemical synthesis and manufacturing. Crystals; mp = 173°; d^{25} = 1.151; λm = 219, 317 nm (ε = 25800, 1930, MeOH); slightly soluble in H2O (0.59 g/100 ml), EtOH, CHCl3; soluble in Et2O, TFA, very soluble in Me2CO, MeOH; LD50 (mus orl) = 6300 mg/kg. *Lancaster Synthesis Co.; Penta Mfg.*

139   3-Aminobenzotrifluoride
98-16-8   H

C7H6F3N
Benzenamine, 3-(trifluoromethyl)-.
4-12-00-01843 (Beilstein Handbook Reference); m-Abtf; AI3-07422; 1-Amino-3-(trifluoromethyl)benzene; 3-Aminobenzotrifluoride; Benzenamine, 3-(trifluoromethyl)-; BRN 0387672; CCRIS 2812; EINECS 202-643-4; HSDB 4249; NSC 4540; Toluene, 3-amino-α,α,α-trifluoro-; m-Toluidine, α,α,α-trifluoro-; α,α,α-Trifluoro-m-toluidine; 3-(Trifluoromethyl)benzenamine; 3-Trifluoromethyl-aniline; UN2948; USAF MA-4. Liquid; mp = 5.5°; bp = 187°, bp10 = 74-75°; d^{12} = 1.3047; λm = 238, 295 nm (cyclohexane); slightly soluble in H2O, soluble in EtOH, Et2O.

140   2-Aminobutane
13952-84-6   1543   P

C4H11N
2-Butanamine.
2-AB; AI3-35093; 2-Aminobutane; Butafume; 2-Butanamine; Butylamine; (RS)-sec-Butylamine; 2-Butylamine; Caswell No. 125; CCRIS 4757; Deccotane; EINECS 237-732-7; EPA Pesticide Chemical Code 004214; Frucote; HSDB 6312; NSC 8030; Propylamine, 1-methyl; Tutane. Used agriculturally as a fungistat. Liquid; mp = -104°; bp = 63°; d^{20} = 0.724; soluble in H2O, EtOH, LD50 (rat orl) = 380 mg/kg. *DowElanco Ltd.; Pennwalt Corp.*

141   2-Aminobutanol
96-20-8   426   G

C4H11NO
2-Amino-1-butanol.
4-04-00-01705 (Beilstein Handbook Reference); AB; AI3-03357; 2-Amino-1-butanol; 2-Amino-1-hydroxybutane; 2-Amino-n-butyl

alcohol; 2-Aminobutan-1-ol; 2-Aminobutyl alcohol; BRN 1098274; Butanol-2-amine; 1-Butanol, 2-amino-; EINECS 202-488-2; EINECS 235-940-2; 1-(Hydroxymethyl)propylamine; 1-Hydroxy-2-butyl-amine; NSC 1068. Pigment dispersant, neut-ralizing/emulsifying amine; corrosion inhibitor; acid salt catalyst; pH buffer; chemical, pharmaceutical intermediate, solubilizer Liquid; mp = 2°; bp = 176-178°; d = 0.944. *Whittaker Clark & Daniels.*

142 4-Amino-6-tert-butyl-3-mercapto-1,2,4-triazin-5(4H)-one
33509-43-2   H

C7H12N4OS
  as-Triazin-5(4H)-one, 4-amino-6-tert-butyl-3-mercapto-.
  4-Amino-6-tert-butyl-3-mercapto-1,2,4-triazine-5(4H)-one; 4-Amino-6-tert-butyl-3-mercapto-1,2,4-triazin-5(4H)-one; 1,2,4-Triazin-5(2H)-one, 4-amino-6-(1,1-dimethylethyl)-3,4-dihydro-3-thioxo-; as-Triazin-5(4H)-one, 4-amino-6-tert-butyl-3-mercapto-; EINECS 251-548-4.

143 2-Amino-4-chloro-5-methyl-benzenesulfonic acid
88-51-7   H

C7H8ClNO3S
  Benzenesulfonic acid, 2-amino-4-chloro-5-methyl-.
  2B acid; 3-14-00-02216 (Beilstein Handbook Reference); 4-Amino-2-chlorotoluene-5-sulfonic acid; 4-Amino-6-chlorotoluene-3-sulphonic acid; 2-Amino-4-chloro-5-methylbenzenesulfonic acid; 6-Amino-4-chloro-m-toluenesulfonic acid; Benzenesulfonic acid, 2-amino-4-chloro-5-methyl-; Brilliant Toning Red Amine; BRN 2727161; CCRIS 3406; 2-Chloro-4-aminotoluene-5-sulfonic acid; 4-Chloro-6-amino-m-toluenesulfonic acid; 3-Chloro-4-methylaniline-6-sulfonic acid, 2-Chloro-4-toluidine 5-sulfonic acid; FINECS 201-837-6; HSDB 5258; Kyselina 2-chlor-4-toluidin-5-sulfonova; Permanent Red 2B Amine; Red 2B acid; m-Toluenesulfonic acid, 6-amino-4-chloro-.

144 2-Amino-5-chloro-4-methyl-benzenesulf-onic acid
88-53-9   H

C7H8ClNO3S
  Benzenesulfonic acid, 2-amino-5-chloro-4-methyl-.
  AI3-28529; 2-Amino-5-chloro-4-methylbenzenesulf-onic acid; 3-Amino-6-chlorotoluene-4-sulfonic acid; 5-Amino-2-chlorotoluene-4-sulphonic acid; Benzene-sulfonic acid, 2-amino-5-chloro-4-methyl-;

CCRIS 2283; EINECS 201-839-7; HSDB 5259; Lake Red C Amine; Red Lake C amine; p-Toluenesulfonic acid, 2-amino-5-chloro-.

145 Aminodimethylacetonitrile
19355-69-2   H

C4H8N2
  2-Amino-2-methylpropiononitrile.
  Aminodimethylacetonitrile; α-Aminoisobutyronitrile; 2-Amino-2-methylpropanenitrile; 2-Amino-2-methyl-propiononitrile; 2-Aminoisobutyronitrile; 2-Cyano-isopropylamine; EINECS 242-989-3; Propanenitrile, 2-amino-2-methyl-; Propionitrile, 2-amino-2-methyl-; Vazo 64AN.

146 2-Amino-2,3-dimethylbutanenitrile
13893-53-3   H

C6H12N2
  Butanenitrile, 2-amino-2,3-dimethyl-.
  2-Amino-2,3-dimethylbutanenitrile; Butanenitrile, 2-amino-2,3-dimethyl-.

147 4-Aminodiphenylamine
101-54-2   G H

C12H12N2
  1,4-Benzenediamine, N-phenyl-.
  Acna Black DF Base; AI3-15983; N-(4-Aminophenyl)aniline; Azosalt R; N, 4'-Bianiline; Black Base P; CCRIS 513; CI 37240; CI 76085; CI Azoic Diazo Component 22; CI Developer 15; CI Oxidation Base 2; Diphenyl Black; Diphenylamine, 4-amino-; Diphenylamine, p-amino-; EINECS 202-951-9; Fast Blue R Salt; HSDB 2178; Luxan Black R; N-4'-Bianiline; Naphthoelan Navy Blue; NCI-C02233; NSC 3401; Oxy Acid Black Base; Peltol BR; Peltol BR II; p-Phenylenediamine, N-phenyl-; Rodol Gray B base; Semidin; Semidine; p-Semidine; UBOB®; Variamine Blue RT; Variamine Blue Salt RT. Used in chemical manufacturing. Needles; mp = 66°; bp = 354°; λ_m = 286 nm (MeOH); slightly soluble in H2O, CHCl3, soluble in Et2O, ligroin, very soluble in EtOH. *Uniroyal.*

148 Aminoethyl ethanolamine
111-41-1   G H

C4H12N2O
  2-((2-Aminoethyl)amino)ethanol.
  4-04-00-01558 (Beilstein Handbook Reference); AI3-15368; Aminoethyl ethanolamine; N-Aminoethyl-ethanolamine; 2-((2-Aminoethyl)amino)ethanol; BRN 0506012; CCRIS 4825; EINECS 203-867-5; Ethanol, 2-((2-aminoethyl)amino)-; Ethanolethylene diamine; HSDB 2067; Hydroxyethyl ethylenediamine; N-Hydroxyethyl-1,2-ethanediamine; Monoethanolethylenediamine;

NSC 461. Used in textile finishing compounds (antifuming agents, dyestuffs, cationic surfactants), resins, rubber, insecticides, medicinals. Liquid; bp = 239°, bp$_{10}$ = 105°; d^{20} = 1.0286; slightly soluble in C$_6$H$_6$, ligroin, soluble in Me$_2$CO, freely soluble in EtOH, H$_2$O; LD$_{50}$ (rat orl) = 3 gm/kg. *BASF Corp.; Dow Chem. U.S.A.; Lancaster Synthesis Co.; Nippon Nyukazai; Union Carbide Corp.*

149 1-(2-Aminoethyl)imidazolidin-2-one
6281-42-1   G

C$_5$H$_{11}$N$_3$O
1-(2-Aminoethyl)-2-imidazole.
AI3-24564; 1-(β-Aminoethyl)-2-imidazolidone; 1-(2-Aminoethyl)-2-imidazolidinone; 1-(2-Aminoethyl)-imidazolidinone; 1-(2-Aminoethyl)imidazolidin-2-one; DV-2301; EINECS 228-491-9; 2-Imidazolidinone, 1-(2-aminoethyl)-; NSC 5776. Used as a surfactant. *Rhône Poulenc Surfactants.*

150 Aminoethylpiperazine
140-31-8   G H

C$_6$H$_{15}$N$_3$
N-(2-Aminoethyl)piperazine.
5-23-01-00257 (Beilstein Handbook Reference); AI3-52274; Aminoethylpiperazine; N-Aminoethylpiper-azine; N-(β-Aminoethyl)piperazine; N-(2-Aminoethyl)-piperazine; BRN 0104363; CCRIS 6678; EINECS 205-411-0; HSDB 5630; NSC 38968; Piperazine, 1-(2-aminoethyl)-; 1-Piperazineethanamine; UN2815; USAF DO-46. Epoxy curing agent, intermediate for pharmaceuticals, anthelmintics, surface-active agents, synthetic fibers. Solid; mp = -19°; bp = 220°; d^{25} = 0.985; soluble in H$_2$O; LD$_{50}$ (rat orl) = 2140 mg/kg. *Akzo Nobel Chemicals Inc; Dow Chem. U.S.A.; Fabrichem; Texaco; Tosoh; Union Carbide Corp.*

151 2-Amino-2-ethyl-1,3-propanediol
115-70-8   438   G

C$_5$H$_{13}$NO$_2$
1,3-Propanediol, 2-amino-2-ethyl-.
AEPD®; AEPD® 85; AI3-03358; Aminoethyl propanediol; 2-Amino-2-ethylpropanediol; 2-Amino-2-ethyl-1,3-propanediol; EINECS 204-101-2; 2-Ethyl-2-amino-propanediol; NSC 8803; 1,3-Propanediol, 2-amino-2-ethyl-; AEPD®-85; AEPD®. Pigment dispersant, neutralizing amine, corrosion inhibitor, acid-salt catalyst, pH buffer, chemical and pharmaceutical intermediate and solubilizer. Chemical intermediate, formaldehyde scavenger, acid-salt catalyst for permanent-press resins, corrosion inhibitor. Crystals; mp = 37.5°; bp$_{10}$ = 152-153°; d^{20} = 1.0990; freely soluble in H$_2$O; soluble in alcohols; pH 0.1, 1M aqueous solution. *Whittaker Clark & Daniels.*

152 6-Amino-4-hydroxy-2-naphthalenesulfonic acid
90-51-7

C$_{10}$H$_9$NO$_4$S
2-Naphthalenesulfonic acid, 6-amino-4-hydroxy-.
4-14-00-02823 (Beilstein Handbook Reference); AI3-19502; Aminonaphthol sulfonic acid γ; BRN 1821283; C.I. Developer 3; EINECS 202-000-8; γ Acid; NSC 31508; Y acid.

153 Aminomercury chloride
10124-48-8   5902   G

ClH$_2$HgN
Ammoniated mercuric chloride.
Aminomercuric chloride; Aminomercury chloride; Ammoniated mercuric chloride; Ammoniated mercury; EINECS 233-335-8; HSDB 1175; Hydrargyrum ammoniatum; Hydrargyrum praecipitatum album; Hydrargyrum precipitatum album; Lemery's white precipitate; Mercuric amidochloride; Mercuric ammonium chloride; Mercuric chloride, ammoniated; Mercury amide chloride (Hg(NH$_2$)Cl); Mercury amide chloride; Mercury amine chloride; Mercury, ammoniated; Mercury ammonium chloride; Mercury, ammonobasic (HgNH$_2$Cl); Mercury(II) chloride ammonobasic; Quecksilber(II)-amid-chlorid; UN1630; White mercuric precipitate; White mercury precipitated; White precipitate. Mercury ammonium chloride, used for the preparation of cinnabar and in medicine as a topical anti-infective. Powder; d = 5.38; insoluble in H$_2$O, EtOH, soluble in mineral acids.

154 2-Amino-5-methyl-benzenesulfonic acid
88-44-8   H

C$_7$H$_9$NO$_3$S
Benzenesulfonic acid, 2-amino-5-methyl-.
3-14-00-02213 (Beilstein Handbook Reference); AI3-52551; Benzenesulfonic acid, 2-amino-5-methyl-; BRN 2211509; CCRIS 2772; EINECS 201-831-3; Kyselina 4-toluidin-3-sulfonova; NSC 7544; m-Toluenesulfonic acid, 6-amino-; p-Toluidine-m-sulfonic acid; PTMS; PTMSA; Ptmsptmsa; Red 4B acid.

155 2-Amino-2-methyl-1,3-propanediol
115-69-5   447   G

C$_4$H$_{11}$NO$_2$
1,1-Di(hydroxymethyl)ethylamine.
4-04-00-01881 (Beilstein Handbook Reference); AI3-03949; Aminoglycol; Aminomethyl propanediol; 2-Amino-2-methyl-1,3-propandiol; 2-Amino-2-methyl-propanediol; 2-Amino-2-methylpropane-1,3-diol; AMPD; BRN 0635708; 1,1-Di(hydroxymethyl)ethyl-amine; EINECS 204-100-7; Gentimon;

Isobutandiol-2-amine; NSC 6364; Pentaerythritol dichlorohydrin; 1,3-Propanediol, 2-amino-2-methyl-. Pigment dispersant, neutralizing amine, corrosion inhibitor, acid-salt catalyst, pH buffer, chemical and pharmaceutical intermediate; solubilizer or emulsifier system component in personal care products. Crystals; mp = 110°; bp_{10} = 151-152°; soluble in EtOH, very soluble in H_2O, LD_{50} (rat orl) = 17 gm/kg.

156   Aminomethyl propanol
124-68-5   448   G H

C4H11NO
2-Methyl-2-aminopropanol.
4-04-00-01740 (Beilstein Handbook Reference); AI3-03947; Aminomethyl propanol; Aminomethylprop-anol; AMP; AMP 75; AMP 95; AMP Regular; BRN 0505979; Caswell No. 037; Corrguard 75; EINECS 204-709-8; EPA Pesticide Chemical Code 005801; HSDB 5606; Hydroxy-tert-butylamine; Isobutanol-2-amine; KV 5088; NSC 441. Boiler water treatment chemical, corrosion inhibitor, carbon dioxide absorber. Widely used as a buffer and phosphate acceptor in assay of phosphatases. Suitable as buffer for manual and automated determination of alkaline phosphatase using 4-nitrophenyl phosphate as substrate. Solid; mp = 25.5°; bp = 165.5°; d^{20} = 0.934; soluble in CCl_4, freely soluble in H_2O; LD_{50} (rat orl) = 2900 mg/kg. Lancaster Synthesis Co.; Lancaster Synthesis Ltd.

157   Aminonaphthol sulfonic acid J
87-02-5   G H

C10H9NO4S
7-Amino-4-hydroxy-2-naphthalenesulfonic acid.
4-14-00-02823 (Beilstein Handbook Reference); 7-Amino-4-hydroxy-2-naphthalenesulfonic acid; Amino-naphthol sulfonic acid J; BRN 2217192; EINECS 201-718-9; I acid; Isogamma acid; J acid; Kyselina 2-amino-5-naftol-7-sulfonova; Kyselina 6-amino-1-naftol-3-sulfo-nova; Kyselina I; NSC 31510.

158   7-((2-Amino-1-naphthyl)azo)-3-phenyl-2-benzopyrone
67906-30-3   H

C25H17N3O2
2H-1-Benzopyran-2-one, 7-((2-amino-1-naphthyl)azo)-3-phenyl-.
7-((2-Amino-1-naphthyl)azo)-3-phenyl-2-benzopyrone; 2H-1-Benzopyran-2-one, 7-((2-amino-1-naphthyl)azo)-3-phenyl-; 2H-1-Benzopyran-2-one, 7-((2-amino-1-naphth-alenyl)azo)-3-phenyl-; EINECS 267-699-4.

159   2-Amino-5-nitrothiazole
121-66-4   456   G V

C3H3N3O2S
5-Nitro-2-thiazol-amine.
4-27-00-04675 (Beilstein Handbook Reference); AI3-50030; 2-Amino-5-nitrothiazole; Amnizol soluble; BRN 0126797; CCRIS 37; EINECS 204-490-9; Enheptin; Enheptin premix; Enheptin-T; Entramin; HSDB 4022; NCI-C03065; Nitramin IDO; 5-Nitro-2-aminothiazole; 5-Nitro-2-thiazolamine; 5-Nitro-2-thiazolylamine; 5-Nitrothiazol-2-ylamine; NSC 4; 2-Thiazolamine, 5-nitro-; Thiazole, 2-amino-5-nitro-; U 7458; USAF EK-6561. Used as an antihistomonad in turkeys, chickens; for trichomonasis in pigeons Orange-yellow powder; mp = 202° (dec); λ_m = 238, 278 nm (ε = 4250, 15100, MeOH), 224, 263, 446 nm (ε = 6460, 4800, 16000, MeOH-KOH), 236, 378 nm (ε = 4350, 11300, MeOH-HCl); sparingly soluble in H_2O; almost insoluble in $CHCl_3$; soluble in dilute mineral acids. May & Baker Ltd.

160   o-Aminophenol
95-55-6   460   H

C6H7NO
Phenol, 2-amino-.
AI3-09065; o-Aminophenol; BASF ursol 3GA; Benzofur GG; CCRIS 4144; CI 76520; CI Oxidation Base 17; EINECS 202-431-1; Fouramine OP; HSDB 4246; o-Hydroxyaniline; o-Hydroxyphenylamine; Nako Yellow 3GA; Nako Yellow ga; NSC 1534; Paradone Olive Green B; Pelagol 3GA; Pelagol Grey GG; Phenol, 2-amino-; Phenol, o-amino-; Questiomycin B; UN2512; Zoba 3GA. Crystls; mp = 174°, sublimes at 153°; d^{25} = 1.328; λ_m = 233, 285 nm (MeOH); poorly soluble in C_6H_6, TFA, soluble in H_2O, Et_2O, very soluble in EtOH. BASF Corp.

161   p-Aminophenol
123-30-8   461   G H

C6H7NO
4-Aminophenol.
Activol; AI3-14872; p-Aminofenol; p-Aminophenol; Aminophenol, p-; Azol; BASF Ursol P Base; Benzofur P; CCRIS 4146; Certinal; CI 76550; CI Oxidation Base 6; Citol; Durafur Brown RB; EINECS 204-616-2; Fouramine P; Fourrine 84; Fourrine P Base; Furro P Base; HSDB 2640; Nako Brown R; NSC 1545; PAP; Paranol; Pelagol Grey P Base; Pelagol P Base; Phenol, 4-amino-; Phenol, p-amino-; Renal AC; Rodinal; Takatol; Tertral P base; UN2512; Unal; Ursol P; Ursol P base; Zoba Brown P Base. White platelets; mp = 187.5°; $bp_{0.3}$ = 110°; λ_m = 234, 301 nm (ε = 7770, 2290, MeOH); insoluble in C_6H_6, $CHCl_3$, slightly soluble in H_2O, TFA, soluble in alkali, very soluble in EtOH; LD_{50} (rat orl) = 375 mg/kg.

162   2-((p-Aminophenyl)sulphonyl)ethyl   hydrogen sulfate

2494-89-5 H

C8H11NO6S2
Ethanol, 2-((4-aminophenyl)sulfonyl)-, hydrogen sulfate (ester).
2-((p-Aminophenyl)sulphonyl)ethyl hydrogensulphate; EINECS 219-669-7; Ethanol, 2-((4-aminophenyl)sulf-onyl)-, hydrogen sulfate (ester); Ethanol, 2-sulfanilyl-, hydrogen sulfate (ester); 4-((2-Sulfatoethyl)sulfonyl)-aniline.

163   Aminopropyltrimethoxysilane
13822-56-5 G

C6H17NO3Si
3-Aminopropyltrimethoxysilane.
(3-Aminopropyl)trimethoxysilane; 3-Aminopropyltri-methoxysilane; (γ-Aminopropyl)trimethoxysilane; CA 0880; Dynasylan® AMMO; EINECS 237-511-5; KBE 903; NSC 83845; 1-Propanamine, 3-(trimethoxysilyl)-; 3-(Trimethoxysilyl)-1-propanamine; N-(Trimethoxy-silylpropyl)amine; 3-(Trimethoxysilyl)propylamine; Propylamine, 3-(trimethoxysilyl)-; SC 3900; Silane SC 3900; Union Carbide® A-1110. Coupling agent, chemical intermediate, blocking agent, release agent, lubricant, primer, reducing agent. Liquid; bp = 217°; d = 0.942. *Degussa-Hüls Corp.; Fluka; Gelest; PCR; Sigma-Aldrich Fine Chem.; Union Carbide Corp.*

164   11-Aminoundecanoic acid
2432-99-7 H

C11H23NO2
11-Aminoundecylic acid.
4-04-00-02823 (Beilstein Handbook Reference); 11-Aminoundecanoic acid; 11-Aminoundecylic acid; BRN 1767291; CCRIS 39; EINECS 219-397-9; HSDB 4349; NCI-C50613; NSC 240503; Undecanoic acid, 11-amino-.

165   Amitraz
33089-61-1 486 G P V

C19H23N3
1,5-Di(2,4-dimethylphenyl)-3-methyl-1,3,5-triazapenta-1,4-diene.
Acarac; AI3-27967; Amitraz; Amitraz estrella; Ami-traze; Amitrazum; Azadieno; Azaform; Baam; Bipin; Boots BTS 27419; BRN 2946590; BTS 27,419; Caswell No. 374A; CCRIS 1552; 1,5-Di-(2,4-dimethylphenyl)-3-methyl-1,3,5-triazapenta-1,4-diene; Ectodex; Edrizar; EINECS 251-375-4; ENT 27967; EPA Pesticide Chem-ical Code 106201; Formamidine, N-methyl-N'-2,4-xylyl-N-(N-2,4-xylylform-imidoyl)-; Fumilat A; Methan-imidamide, N'-(2,4-di-

methylphenyl)-N-(((2,4-di-methylphenyl)imino)methyl)-N-methyl-; 2-Methyl-1,3-di(2,4-xylylimino)-2-azapropane; 2,4-Xylidine, N,N'-(methyliminodimethylidyne)bis-; Mitaban; Mitaban (Veterinary); Mitac; Mitac 20; N-Methyl-bis(2,4-xylyl-iminomethyl)amin; N-Methyl-N'-2,4-xylyl-N-(N,2,4-xylylformimidoyl)formamidine; N-Methyl-bis(2,4-xylyl-iminomethyl)amine; N'-(2,4-Dimethyl-phenyl)-N-(((2,4-dimethylphenyl)imino)methyl)-N-; N,N-Bis(2,4-xylyliminomethyl)methylamine; N,N'-((Methyl-imino)-dimethylidyne)di-2,4-xylidine; N,N'-Di-2,4-xylid-ine; NSC 324552; Ovasyn; R.D. 27419; Taktic; Triatox; Triatrix; Triazid; U-36059; Upjohn U-36059. An acaricide and insecticide. Used to control mites, bugs, insects and larvae in fruit crops. Used primarily with animals as a non-systemic acaricide and insecticide with contact and respiratory action. Thought to interact with octopamine receptors in the nervous system, causing an increase in nervous activity. Used for control at all stages of mites, pear suckers,scale insects, mealy bugs, whitefly, aphids and lepidoptera. Registered by EPA as an insecticide and acaricide. Colorless, monoclinic needles; mp = 86 - 87°; sg^{25} = 1.128; soluble in H2O (0.0001 g/100 ml), more soluble in Me2CO, C7H8, xylene (all > 30 g/100 ml) and in other organic solvents; LD50 (rat orl) = 800 mg/kg, (mus orl) > 1600 mg/kg, (rbt der) > 200 mg/kg, (rat der) > 1600 mg/kg, (bee orl) = 0.012 mg/bee, (bee contact) = 3600 mg/l; LC50 (rat ihl 6 hr.) = 65 mg/l air, (mallard duck 8 day dietary) = 7000 mg/kg diet, (Japanese quail 8 day dietary) = 1800 mg/kg diet, (bluegill sunfish 96 hr.) = 1.3 mg/l, (harlequin fish 96 hr.) = 3.2 - 4.2 mg/l, (rainbow trout 96 hr.) = 2.7 - 4.0 mg/l. *AgrEvo; Atabay; Aventis Crop Science; Intervet Inc.; Nor-Am; Quimica Estrela; Schering AG; Schering Agrochemicals Ltd.; Virbac AH Inc.*

166   Amitriptyline
50-48-6 487 D

C20H23N
3-(10,11-Dihydro-5H-dibenzo-[a,d]cyclohepten-5-ylidene)-N,N-dimethyl-1-propanamine.
Adepress; Adepril; Amitriptilina; Amitriptylin; Ami-triptyline; Amitriptylinum; BRN 2217885; Damilen; Damitriptyline; EINECS 200-041-6; Elavil; Flavyl; HSDB 3007; Lantron; MK 230; N 750; Proheptadiene; Redomex; Ro 4-1575; Seroten; Triptanol; Triptilin; Triptisol; Tryptanol. A tricyclic antidepressant. *Hoffmann-LaRoche Inc.; Merck & Co.Inc.*

167   Amitriptyline Hydrochloride
549-18-8 487 D

C20H24ClN
3-(10,11-Dihydro-5H-dibenzo-[a,d]cyclohepten-5-ylidene)-N,N-dimethyl-1-propanamine hydrochloride.
Adepril; ADT-Zimaia; Amavil; Ami-Anelun; Amilit; Amineurin;

Amiplin; Amiprin; Amitid; Amitril; Amitrip; Amitriptyline chloride; Amitriptyline hydrochloride; Amyline; Amyzol; Anapsique; Annoyltin; Apo-Ami-triptyline; Belpax; CCRIS 7092; Damilen hydro-chloride; Daprimen; Deprex; Domical; DRG-0169; EINECS 208-964-6; Elatrol; Elatrolet; Elavil hydrochloride; Enafon; Endep; Etrafon; Etrafon; Etravil; Kyliran; Laroxyl; Larozyl; Lentizol; Levate; Limbitrol; Limbitrol DS; Maxivalet; Miketorin; Mitaptyline; NIH 10794; Nornaln; Novoprotect; Novotriptyn; NSC 104210; Oasil-M; Pinsanu; Pinsaun; Proheptadien monohydrochloride; Rantoron; Redomex; Saroten; Saroten Retard; Sarotena; sk-Amitriptyline chloride; Syneudon; Teperin; Trepiline; Triavil; Tridep; Tripta; Triptizol; Trynol; Tryptacap hydrochloride; Tryptine; Tryptizol; Tryptizol retard; Trytomer; Vanatrip; Yamanouchi. Antidepressant. mp = 196-197°; freely soluble in H_2O, $CHCl_3$, alcohol; λ_m = 240 nm; pKa = 9.4; LD_{50} (rat orl) = 380 mg/kg. *Bristol-Myers Squibb Co.; Lemmon Co.; Merck & Co.Inc.; Parke-Davis; Roche Puerto Rico; Schering-Plough HealthCare Products.*

168   Amlodipine
88150-42-9   491   D

$C_{20}H_{25}ClN_2O_5$
 3-Ethyl-5-methyl (±)-2-[[(2-aminoethoxy)methyl]-4-(o-chlorophenyl)-1,4-dihydro-6-methyl-3,5-pyridinedi-carboxylate.
 Amlocard; Amlodipine; Amlodipino; Amlodipinum; Amlodis; Coroval; Lipinox; Norvasc. Antianginal agent. Dihydropyridine calcium channel blocker. Has antihypertensive properties. *Ciba-Geigy Corp.; Pfizer Inc.*

169   Ammohippurate sodium
94-16-6   441   G

$C_9H_9N_2NaO_3$
 p-Aminohippuric acid sodium salt.
 4-Aminohippursaeure, natriumsalz; 4-Aminohippur-säure, natriumsalz; Amino-hippurate-sodium; Ammo-hippurate sodium; p-Aminohippurate sodium; EINECS 202-309-8; Glycine, N-(4-aminobenzoyl)-, mono-sodium salt; Hippuric acid, p-amino-, monosodium salt; Monosodium p-aminohippurate; N-(4-Amino-benzoyl)glycine mono-sodium salt; Natrium 4-aminohippurat; Nephrotest; Paraaminohippurate Injection; Sodium 4-aminohippurate; Sodium p-aminohippurate. Used intravenously to measure effective renal plasma flow and tubular secretory capacity. Crystals; mp = 123-125°. *Merck & Co.Inc.*

170   Ammonia
7664-41-7   492   G P

H_3N
 Ammonia gas.
 AM-Fol; Ammonia; Ammonia, anhydrous; Ammonia gas; Ammonia solution, strong; Ammoniac; Ammoniaca; Ammoniak; Ammoniak Kconzentrierter; Ammoniakgas; Amoniak; Caswell No. 041; CCRIS 2278; EINECS 231-635-3; EPA Pesticide Chemical Code 005302; HSDB 162; Liquid Ammonia; Nitro-sil; R 717; Spirit of hartshorn; UN 1005 (anhydrous gas or >50% solution); UN 2073 (>44% solution); UN 2672 (between 12% and 44% solution). Fertilizers, refrigerant, nitriding of steel, condensation catalyst, neutralizing agent, petroleum industry, latex preservative, explosives. Registered by EPA as an insecticide. Colorless gas; mp = -77°; bp = -33.35°; d = 0.5967; d (gas) = 0.7714 g/l. *Air Products & Chemicals Inc.; Allied Signal; Am. Cyanamid; Asahi Chem. Industry; Chevron; General Chem; LaRoche Ind.; Mitsui Toatsu; Monsanto Co.; Nissan Chem. Ind.; Norsk Hydro AS; OxyChem; PPG Ind.; Unocal.*

171   Ammonium acetate
631-61-8   495   D G

$C_2H_7NO_2$
 Acetic acid ammonium salt.
 Acetic acid, ammonium salt; AI3-26540; Ammonium acetate; EINECS 211-162-9; HSDB 556. Reagent in analytical chemistry, drugs, textile dyeing, preserving meats, foam rubbers, vinyl plastics, and explosives. Used medically as a diuretic. mp = 114°; bp dec; d^{20} = 1.17; freely soluble in H_2O, soluble in EtOH, slightly soluble in Me_2CO; 0.5 M aqueous solution has pH of 7.0. *General Chem; Schaeffer Salt & Chem.; Sigma-Aldrich Fine Chem.; Verdugt BV.*

172   Ammonium alginate
9005-34-9   E G

 $(C_6H_{11}NO_6)_x$
 Ammonium polymannuronate.
 Alginic acid, ammonium salt; Ammonium alginate; Ammonium polymannurate; Analgine; Callatex; Collatex ARM Extra; Digamon; HSDB 1910; Protomon; Superloid; component of: KELTOSE®. Thickening agent and stabilizer in food productsFDA, FEMA GRAS, USDA, Europe listed, UK approved. Used as a thickening agent, gellant and stabilizer in pharmaceuticals. Filamentous, grainy, granular, or powder; colorless or slightly yellow; slowly soluble in H_2O forming viscous solid; insoluble in alcohol; heated to decomposition emits toxic fumes of NO_x. *Kelco Intl.; Spectrum Chem. Manufacturing.*

173   Ammonium alum
7784-25-0   325   D G

$AlH_4NO_8S_2$
 Aluminum Ammonium Sulfate.
 Alum, ammonium; Alum, ammonium anhydrous; Aluminium ammonium bis(sulphate); Aluminum ammonium alum; Aluminum

ammonium disulfate (Al(NH4)(SO4)2); Aluminum ammonium sulfate; Aluminum sulfate compd. with ammonium sulfate (1:1); Ammonium alum; Ammonium aluminum alum; Ammonium aluminum sulfate; Burnt ammonium alum; Caswell No. 041B; Curb; EINECS 232-055-3; EPA Pesticide Chemical Code 098501; Exsiccated ammonium alum; HSDB 611; Monoammonium monoaluminum sulfate; NSC 146176; Sulfuric acid, aluminum ammonium salt (2:1:1). Inorganic salt; mordant in dyeing, water and sewage purification, sizing paper, retanning leather, clarifying agent, food additive, manufacture of lakes and pigments, fur treatment. Used medicinally as an astringent. Crystals; d = 1.65; mp = 94.5°; soluble in H_2O (14.3 g/100 ml at 20°, 200 g/100 ml at 100°, freely soluble in glycerol, insoluble in EtOH.

174 Ammonium bicarbonate
1066-33-7   497   G

CH5NO3
Ammonium hydrogen carbonate.
ABC Trieb; Acid ammonium carbonate; Ammonium acid carbonate; Ammonium bicarbonate; Ammonium hydrogen carbonate; Ammonium hydrogencarbonat; Ammonium hydrogencarbonate; Carbonic acid, monoammonium salt; CCRIS 7327; EINECS 213-911-5; HSDB 491; Monoammonium carbonate. Used for production of ammonium salts, dyes; leavening agent for cookies, crackers; fire-extinguishing compounds; pharmaceuticals, degreasing textiles, blowing agent for foam rubber, boiler scale removal, compost treatment. Baking raising agent. In cooling baths, fire extinguishers, manufacture of porous plastics, ceramics, dyes and fertilizer. White crystals; mp = 60° (dec); d = 1.586; soluble in H_2O at 20° (17.4 g/100 ml); insoluble in EtOH, Me2CO. *BASF AG; General Chem; Nissan Chem. Ind.; Norsk Hydro AS; Rhône-Poulenc.*

175 Ammonium bifluoride
1341-49-7   498   G

F2H5N
Ammonium acid fluoride.
Acid ammonium fluoride; Ammonium acid fluoride; Ammonium bifluoride; Ammonium fluoride; Amm-onium fluoride ((NH4)(HF2)); Ammonium hydro-fluoride; Ammonium hydrogen bifluoride; Ammonium hydrogen fluoride; Ammonium hydrogendifluoride; EINECS 215-676-4; Fluoram; Fluorure acide d'ammonium; HSDB 480; Matt salt; UN1727; UN2817. Used in ceramics, as a chemical reagent, for etching glass, as a sterilizer for breweries and dairies; electroplating, processing beryllium; as a laundry sour. Orthorhombic crystals which readily etch glass; d = 1.5; mp = 124.6°; bp = 235°; soluble in H_2O (630 g/l). *Bayer plc; Hoechst Celanese; Miles Inc.; Solvay Deutschland GmbH*

176 Ammonium bisulfite
10192-30-0   502   G

H5NO3S
Sulfurous acid, monoammonium salt.
Ammonium acid sulfite; Ammonium bisulfite; Ammonium bisulfite (NH4HSO3); Ammonium hydrogen sulfite; Ammonium hydrogensulphite; Ammonium hydrosulfite; Ammonium monosulfite; Ammonium sulfite, hydrogen; EINECS 233-469-7; HSDB 486; Monoammonium sulfite; Sulfurous acid, monoammonium salt.

Preservative. Crystals; mp = 147°; soluble in H_2O (267 g/100 ml 10°; 620 g/100 ml 60°). *Brotherton Ltd.; General Chem; Heico.*

177 Ammonium bromide
12124-97-9   505   D G

$$NH_4^+ \quad Br^-$$

BrH4N
Ammonium bromide.
Ammonii bromidum; Ammonium bromatum; Ammonium bromide; EINECS 235-183-8; FR-1; HSDB 207; Hydrobromic acid monoammoniate; Nervine. Flame retardant for textiles, wood, chipboard, plywood. Used in manufacture photographic films, plates, and papers; in engraving and lithography; as a corrosion inhibitor. Medicinally, a sedative/hypnotic. White hygroscopic crystals; sublimes at high temperature; d^{25} = 2.429; freely soluble in H_2O, EtOH, MeOH, Me2CO; slightly soluble in Et2O; insoluble in EtOAc. *Great Lakes Fine Chem.; Johnson Matthey; Sigma-Aldrich Fine Chem.*

178 Ammonium carbamate
1111-78-0   507   H

CH6N2O2
Ammonium aminoformate.
Ammonium carbamate; Carbamic acid, ammonium salt; Carbamic acid, monoammonium salt; EINECS 214-185-2; HSDB 485. Crystals; bp approx. 60°; freely soluble in H_2O, soluble in EtOH.

179 Ammonium chloride
12125-02-9   510   G

$$NH_4^+ \quad Cl^-$$

ClH4N
Ammonium chloride.
AI3-08937; Amchlor; Ammon Chlor; Ammonchlor; Ammoneric; Ammonii Chloridum; Ammonium Chloratum; Ammonium chloride; Ammonium chloride ((NH4)Cl); Ammonium chloride fume; Ammonium Chloride Injection; Ammonium Chloride Tablets; Ammonium muriate; Ammoniumchlorid; Ammonium-klorid; CCRIS 7262; Chlorammonic; Chloramon; Chlorid ammonia; Chlorid amonny; Cloruro de Amonio; Darammon; EINECS 235-186-4; Elektrolyt; Gen-Diur; HSDB 483; Katapone VV-328; Muriate of Ammonia; PV Tussin Syrup; Sal ammonia; Sal ammoniac; Salammonite; Salmiac. In dry batteries; mordant (dyeing and printing); safety explosives; flux for coating sheet and iron with zinc; manufacture of various ammonia compounds, fertilizer, pickling agent; In washing powders. Used as a flux and cleanser. Medically as a systemic acidifier and in veterinary medicine as an expectorant and diaphoretic. Colorless, odorless crystals; mp = 340°; d^{25} = 1.5274; soluble in H_2O, methanol, ethanol; LD50=im rats 30 mg/kg; incompatible with Ag, Pb salts; LD50 (rat orl) = 1650 mg/kg. *BASF Corp.; Degussa-Hüls Corp.; EM Ind. Inc.; General Chem; Heico; Kemichrom; Sigma-Aldrich Fine Chem.*

180 Ammonium cumenesulfonate
37475-88-0   G H

NH_4^+

C9H15NO3S
 Benzenesulfonic acid, (1-methylethyl)-, ammonium salt.
 ACS 60; Ammonium cumenesulfonate; Ammonium cumenesulphonate; Benzenesulfonic acid, (1-methylethyl)-, ammonium salt; Benzenesulphonic acid, (1-methylethyl)-, ammonium salt; EINECS 253-519-1; Eltesol® AC60; (1-Methylethyl)benzenesulfonic acid, ammonium salt. Surfactant, hydrotrope for agricultural applications. Hydrotrope, solubilizer for personal care applications. *Albright & Wilson UK Ltd.; CK Witco Corp.; Rewo.*

181   Ammonium dichromate
7789-09-5   516   G

$$O=Cr-O-Cr=O \quad 2NH_4^+$$

Cr2H8N2O
 Ammonium dichromate (VI).
 Ammonio (bicromato di); Ammonio (dicromato di); Ammonium (dichromate d'); Ammonium bichromate; Ammonium chromate ((NH4)2Cr2O7); Ammonium dichromate; Ammonium dichromate (VI); Ammonium-bichromaat; Ammoniumdichromaat; Ammonium-dichromat; Bichromate d'ammonium; Chromic acid, diammonium salt; Diammonium dichromate; Dichromic acid, diammonium salt; EINECS 232-143-1; HSDB 481; UN1439. Mordant for dyeing, pigments, manufacture of alizarin, chrome alum, catalysts, oil purification, pickling, leather tanning, synthetic perfumes, photography, lithography, pyrotechnics Bright orange-red crystals; flammable; mp = 170°; d = 2.155, lb/f³; dec 180°; very soluble in H2O. *British Chrome & Chemical; EM Ind. Inc.*

182   Ammonium dithiocarbamate
513-74-6   517   H

$$H_2N \quad SH \quad NH_3$$

CH6N2S2
 Dithiocarbamic acid monoammonium salt.
 Ammonium dithiocarbamate; Ammonium sulfocarb-amate; Carbamic acid, dithio-, monoammonium salt; Dithiocarbamic acid monoammonium salt; EINECS 208-166-8; HSDB 5675; NSC 202959.

183   Ammonium dodecanoate
2437-23-2   H

NH_4^+

C12H27NO2
 n-Dodecanoic acid, ammonium salt.
 AI3-00287; Ammonium laurate; Dodecanoic acid, ammonium salt;

184   Ammonium dodecylbenzene sulfonate
1331-61-9   G

NH_4^+

$C_{12}H_{25}$

C18H33NO3S
 Ammonium dodecylbenzene sulfonate.
 Ablusol DBM; Ammonium dodecylbenzenesulfonate; Ammonium dodecylbenzenesulphonate; Ammonium lauryl benzene sulfonate; Benzenesulfonic acid, dodecyl-, ammonium salt; Dodecylbenzenesulfonic acid, ammonium salt; EINECS 215-559-8; Hetsulf 50A; Nansa® AS 40. Wetting agent, emulsifier, dispersant, for light duty detergent formulations. Used in formulation of domestic and industrial liquid detergents. Surfactant, emulsifier, wetting agent for bubble baths, shampoos and detergents. *Albright & Wilson Americas Inc.; Heterene; Taiwan Surfactants.*

185   Ammonium fluoride
12125-01-8   524   G

$NH_4^+ \quad F^-$

FH4N
 Ammonium Fluoride.
 Ammonium fluoride; Ammonium fluoride ((NH4)F); Ammonium fluorure; CCRIS 2285; EINECS 235-185-9; Fluorek amonowy; Fluorure d'ammonium; Fluoruro amonico; HSDB 6287; Neutral ammonium fluoride; UN2505. Used in the manufacture of fluorides, analytical chemistry, antiseptic in brewing, etching glass, textile mordant, wood preservative, mothproofing agent Crystals; d = 1.009; soluble in H2O (100g/100ml); decomposed by hot H2O into NH3 and ammonium biflouride; incompatible with quinine salts calcium salts; ingestion produces nausea, vomiting, gastroentroenteritlis, convulsions, death. *GE Silicones; GE Specialities; GE; General Chem; Hoechst Celanese; Olin Corp; Sigma-Aldrich Fine Chem.*

186   Ammonium laureth sulfate
67762-19-0   G

 (C2H4O)n.C12H26O4S.H3N (average n = 1-4)
 Ammonium lauryl ether sulfate.
 (C10-C16)-Alkyl alcohol ethoxylate sulfuric acid ammonium salt; (C10-C16) Alkylethoxylate sulfuric acid, ammonium salt; (C10-C16)Alcohol ethoxylate, sulfated, ammonium salt; (C13-C16)Alkyl ethoxylate sulfuric acid, ammonium salt; Avirol® AE 3003; Calfoam NEL-60; Carsonal® SES-A; DeSonol AE; Empicol® EAA, EAB, EAC; Poly(oxy-1,2-ethanediyl), α-sulfo-omega-hydroxy-, C10-16-alkyl ethers, ammonium salts; SDA 15-067-01; Texapon EA-1, NA; Ungerol AM3-75; Witcolate AE; Witcolate LES-60A; Zoharpon LAEA 253; Nonasol N4AS; Nurapon AL1, AL 60; Polystep®B-11; Rhodapex® AB-20, EA, EAY; Standapol® EA-1; Steol® CA-460; Sulfochem EA-1, EA-2, EA-3, EA-60, EA-70; Sulfotex OT. Emulsifier for vinyl acetate copolymers, S/B latexes, vinyl chloride copolymers, acrylate homo-and copolymers. Surfactant; hair and skin detergents; breaks up and holds oils and soil. Liquid; d = 1.02. *Allchem Ind.; Ashland; Clariant Corp.; Great Western; Lonzagroup; Pilot; Rhône Poulenc Surfactants; Stepan; Witco/Oleo-Surf.*

187   Ammonium lauryl sulfate
2235-54-3   G P

C12H29NO4S

Sulfuric acid, monododecyl ester, ammonium salt.

Akyposal als 33; Ammonium dodecyl sulfate; Ammonium dodecyl sulphate; Ammonium lauryl sulfate; Ammonium n-dodecyl sulfate; Caswell No. 044B; Conco sulfate A; Dodecyl ammonium sulfate; Dodecyl sulfate ammonium salt; EINECS 218-793-9; EPA Pesticide Chemical Code 079028; HSDB 2101; Lauryl ammonium sulfate; Lauryl sulfate ammonium salt; Maprofix NH; Montopol LA 20; Neopon LAM; Octosol ALS-28; Presulin; Rhodapon® L-22; Rhodapon® L-22/C; Richonol AM; Sinopon; Sipon LA 30; Siprol 422; Siprol L22; Standapol® A; Sterling AM; Sulfuric acid, lauryl ester, ammonium salt; Sulfuric acid, monododecyl ester, ammonium salt; Texapon A 400; Texapon special; Witcolate NH. Detergent, emulsifier, foaming agent, dispersant, wetting agent; for personal care products, carpet shampoos, firefighting. High foaming detergent, emulsifier for shampoo, bubble bath, pet shampoos, industrial and institutional cleaners, wool scouring, fire fighting foams, assistant for pigment dispersion; emulsion polymerization aid. Registered by EPA as a bactericide, insecticide, fungicide and rodenticide (cancelled). Liquid; anionic detergent. *CK Witco Chem. Corp.; CK Witco Corp.; Henkel/Cospha; Lonzagroup; Rhône Poulenc Surfactants; Rhône-Poulenc; Sandoz; Stepan Canada; Stepan.*

188   Ammonium metavanadate
7803-55-6   572   G

H4NO3V

Ammonium Vanadate (V).

Ammonium metavanadate; Ammonium metavanadate (NH4VO3); Ammonium monovanadate; Ammonium trioxovanadate; Ammonium vanadate; Ammonium vanadate ((NH4)VO3); Ammonium vanadium oxide (NH4VO3); Ammonium vanadium trioxide; CCRIS 4120; EINECS 232-261-3; HSDB 6310; NSC 215196; RCRA waste number P119; UN2859; Vanadic acid (HVO3), ammonium salt; Vanadic acid, ammonium salt. Used in wool dyeing, wood staining, in manufacture of inks, photography and microscopy. White crystalline powder; soluble in H2O (60 g/100 ml); LD50 (rat orl) = 0.16 g/kg. *Kerr-McGee.*

189   Ammonium molybdate
12027-67-7   536   G

H16MoN2O8

Hexaammonium molybdate.

Ammonium heptamolybdate; Ammonium hepta-molybdate ((NH4)6Mo7O24); Ammonium molybdate; Ammonium molybdate(VI); Ammonium paramolybdate; EINECS 234-722-4; Hexaammonium heptamolybdate; Hexaammonium molybdate; Hexammonium hepta-molybdat; Hexammonium tetracosaoxoheptamolybdate; HSDB 1802; Molybdate, hexaammonium; Molybdic acid, hexaammonium salt; PM 20. In soil additives, enamel bonding agents, protective and decorative metal coatings, iron and steel alloys, lubricants, petroleum refining catalysts, pigments, corrosion inhibitors, smoke suppressants, production of molybdenum metal. [tetrahydrate]; colorless or slightly greenish crystals; LD50 (rat orl) = 333 mg/kg. *AAA Molybdenum; Climax Molybdenum Co.; Climax Performance.*

190   Ammonium nitrate
6484-52-2   538   G

H4N2O3

Nitric acid, ammonium salt.

Ammonium nitrate; Ammonium nitrate, urea solution (containing ammonia); Ammonium nitricum; Ammonium saltpeter; Ammonium(I) nitrate (1:1); Ansax; Caswell No. 045; EINECS 229-347-8; EPA Pesticide Chemical Code 076101; German saltpeter; Herco prills; Hero-Prills; HSDB 475; Merco Prills; Nitram; Nitrate d'ammonium; Nitrate of ammonia; Nitrato amonico; Nitric acid, ammonium salt; Norway saltpeter; Old Plantation; UN0222; UN1942; UN2426; Varioform I. Used in fertilizers, explosives, pyrotechnics, herbicides and

insecticides, manufacture of nitrous oxide, absorbent for nitrogen oxides, ingredient of freezing mixtures, oxidizer in solid rocket propellants, nutrient for antibiotics and yeast, catalyst. Transparent, hygroscopic crystals or white granules; mp = 169°; dec at approx. 210° into H2O and N2O; d = 1.7250; soluble in H2O (2 g/ml); pH of 0.1 M solution = 5.43. *Air Products & Chemicals Inc.; Chevron; Faith, Keyes and Clark; LaRoche Ind.; Norsk Hydro AS; Unocal.*

191   Ammonium nonoxynol-4 sulfate
9051-57-4   G

C23H43NO8S

Ammonium salt of sulfated nonylphenoxy POE ethanol.

Alipal CO 436; Alipal EP; Alipal EP 110; Alipal EP 120; Alipal HF-433; Ammonium nonoxynol-4-sulfate; CO 436; Fenopon CO 436; Fenopon EP 110; Fenopon EP 120; Hitenol N 093; HSDB 2687; Newcol 560SF; Nikkol SNP; Nonylphenyl, ethoxylated, monoether with sulfuric acid, ammonium salt; Nonylphenol, ethoxylated, sulfated, ammonium salt; Poly(oxy-1,2-ethanediyl), α-sulfo-ω-(nonylphenoxy)-, ammonium salt; Poly(oxyethylene) nonylphenyl ether ammonium sulfate; Polyethylene glycol nonylphenyl ether ammonium sulfate; Polyethylene glycol nonylphenyl ether ammonium bisulfate; Polyethylene glycol nonylphenyl ether sulfate ammonium salt; Sulfated nonylphenoxypoly(ethylene-oxy)ethanol ammonium salt. High foaming surfactant for emulsion polymerization of acrylic, styrene and vinyl acetate systems, dishwashing detergents, germicides, pesticides, general purpose cleaners and cosmetics. Pale yellow clear liquid; alcoholic odor; soluble in H2O; slightly soluble in organic solvents; d = 8.9 lb/gal; viscosity = 100 cps.; f.p.<0°; flash pt. (PMCC) 83F; pH 6.5-7.5. *Chemron; Cytec; Rhône Poulenc Surfactants; Stepan.*

192   Ammonium oxalate
6009-70-7   G

C2H8N2O4

Ethanedioic acid diammonium salt monohydrate.

Ammonium oxalate; Diammonium oxalate monohydrate; Ethanedioic acid, diammonium salt, monohydrate. Used in analytical chemistry, safety explosives, manufacture of oxalates, rust and scale removal. Orthohombic odorless crystals or granules; mp = 70°; d = 1.50; soluble in H2O (11.8 g/100 ml); poisonous. *Brotherton Ltd.; General Chem; Heico; Rhône-Poulenc.*

193   **Ammonium perchlorate**
7790-98-9   544   G

ClH4NO4
Perchloric acid ammonium salt.
   Ammonium perchlorate; Ammonium perchlorate (NH4ClO4); EINECS 232-235-1; HSDB 474; Perchloric acid, ammonium salt; PKHA; UN0402; UN1442. Manufactured in ordnance and industrial grades. d = 1.95; decomposes when heated. *Kerr-McGee.*

194   **Ammonium perfluorocaprylate**
3825-26-1   G

C8H4F15NO2
Ammonium perfluorooctanoate.
   Ammonium pentadecafluorooctanoate; Ammonium perfluorocaprilate; Ammonium perfluorooctanoate; EINECS 223-320-4; FC-143; Fluorad® FC 143; Fluorad® FC-118; NSC 35120; Octanoic acid, pentadecafluoro-, ammonium salt; Pentadecafluoro-1-octanoic acid, ammonium salt; Perfluoroammonium octanoate; Perfluorooctanoic acid, ammonium salt. Surfactant for emulsion polymerization of fluorinated monomers. *3M Company.*

195   **Ammonium persulfate**
7727-54-0   545   G

H8N2O8S2
Ammonium Peroxydisulfate.
   Ammonium peroxydisulfate; Ammonium persulfate; CCRIS 1430; Diammonium peroxysulfate; Di-ammonium peroxydisulphate; Diammonium persulfate; EINECS 231-786-5; Peroxydisulfuric acid (((HO)S(O)2)2O2), diammonium salt; Peroxydisulfuric acid, diammonium salt; Persulfate d'ammonium; Thioxydant lumire; UN1444. Oxidizer, bleaching agent; photography; etchant for printed circuit boards, copper; electroplating; deodorizing oils; aniline dyes; food preservative; depolarizer in batteries; washing infected yeast; manufacture of other persulfates White crystals; mp = 80° (dec); d = 1.9820; soluble in H2O (80 g/100 ml); LD50 (rat orl) = 689 mg/kg. *Degussa AG; EM Ind. Inc.; FMC Corp.; Sigma-Aldrich Fine Chem.; Solvay Interox Inc.*

196   **Ammonium phosphate dibasic**
7783-28-0   546   G

H9N2O4P
Diammonium hydrogen phosphate.
   AI3-25349; Akoustan A; Ammonium hydrogen phosphate solution; Ammonium monohydrogen orthophosphate; Ammonium orthophosphate dibasic; Ammonium phosphate; Ammonium phosphate ((NH4)2(HPO4)); Ammonium phosphate, dibasic; Ammonium phosphate, secondary; Caswell No. 286C; Coaltrol LPA 445; Diammonium acid phosphate; Diammonium hydrogen orthophosphate; Diammonium hydrogen phosphate; Diammonium monohydrogen phosphate; Diammonium orthophosphate; Diammonium phosphate; Dibasic ammonium phosphate; EINECS 231-987-8; Fyrex; HSDB 301; Hydrogen diammonium phosphate; K2 (phosphate); Pelor; Phos-Chek 202A; Phos-Chek 259; Phosphoric acid, diammonium salt; Secondary ammonium phosphate. Used for fireproofing textiles, as a soldering flux and in dentrifices, corrosion inhibitors and fertilizers. Flame retardant for wood, paper, textiles fertilizer, plant nutrient sol'ns., feed additive; flux for soldering, purifying sugar; in ammoniacal dentifrices; manufacture of yeast, vinegar, bread improvers; foods, pharmaceuticals. Crystals; d = 1.6190; soluble in H2O (59 g/100 ml), insoluble in organic solvents. *Albright & Wilson Americas Inc.; Chisso Corp.; Heico; IMC Fertilizer; LaRoche Ind.; Monsanto Co.; OxyChem; Rhône-Poulenc; Sigma-Aldrich Fine Chem.*

197   **Ammonium phosphate monobasic**
7722-76-1   547   G

H6NO4P
Ammonium dihydrogen phosphate.
   AI3-26062; Ammonium acid phosphate; Ammonium biphosphate; Ammonium diacid phosphate; Ammonium dihydrogen orthophosphate; Ammonium dihydrogen phosphate ((NH4)H2PO4); Ammonium monobasic phosphate (NH4H2PO4); Ammonium monophosphate; Ammonium orthophosphate dihydrogen; Ammonium phosphate (monobasic); Ammonium phosphate, primary; Ammonium primary phosphate; Dihydrogen ammonium phosphate; EINECS 231-764-5; HSDB 1229; Mono-ammonium acid phosphate; Monoammonium di-hydrogen orthophosphate; Monoammonium dihydrogen phosphate; Monoammonium hydrogen phosphate; Monoammonium orthophosphate; Monoammonium phosphate; Monobasic ammonium phosphate; NSC 57633; Phosphoric acid, monoammonium salt; Primary ammonium phosphate; VTI 57. Used in the manufacture of food products, fertilizer, flame retardants, plant nutrient solutions, manufacture of yeast, vinegar, yeast foods and bread improvers, food additive, analytical chemistry White crystalline powder; mp = 190°; d = 1.8030; soluble in H2O (40 g/100 ml), less soluble in organic solvents. *Albright & Wilson Americas Inc.; Chisso Corp.; EniChem Am.; Heico; IMC Fertilizer; Monsanto Co.; OxyChem; Rhône-Poulenc; Showa Denko; Sigma-Aldrich Fine Chem.*

198   **Ammonium picrate**
131-74-8   551   G

C6H6N4O7
2,4,6-Trinitrophenol ammonium salt.
   Ammonium carbazoate; Ammonium picronitrate; EINECS 205-038-3; Explosive D; HSDB 2070; Obeline picrate; Phenol, 2,4,6-trinitro-, ammonium salt; Picratol; Picric acid, ammonium salt; RCRA waste no. P009; 2,4,6-Trinitrophenol ammonium salt; UN0004; UN1310. Used in explsoives, fireworks and rocket propellants. Crystals; d =

1.72; soluble in H_2O (1 g/100 ml), less soluble in organic solvents.

199 Ammonium polyacrylate
9003-03-6

G

$[C_3H_4O_2]_x \cdot xH_3N$
2-Propenoic acid, homopolymer, ammonium salt.
Alcogum 9639; Ammonium polyacrylate; Poly(acrylic acid), ammonium salt; 2-Propenoic acid, homopolymer, ammonium salt. Dispersant for paints and coatings; thickening and stabilizing agent for synthetic latices; used in coatings, adhesives, dipped, cast, and molded goods, cements for rug backimg, spraying, spreading, brushing and extruding compounds.

200 Ammonium stearate
1002-89-7   557

G

$C_{18}H_{39}NO_2$
n-Octadecanoic acid, ammonium salt.
Ammonium stearate; Ammonium stearate, pure; EINECS 213-695-2; Octadecanoic acid, ammonium salt; Stearates; Stearic acid, ammonium salt. Ammonium salt of stearic acid; used in vanishing creams, brushless shaving creams, other cosmetic products, waterproofing of cements, concrete, stucco, paper, textiles. Waxy solid; mp = 38-42°; soluble in H_2O, organic solvents except Me_2CO, CCl_4. *Magnesia GmbH; Original Bradford Soapworks.*

201 Ammonium sulfamate
7773-06-0   558

G P

$H_6N_2O_3S$
Sulfamic acid, ammonium salt.
AI3-17753; Amcide; Amicide; Ammate; Ammate X; Ammate X-NI; Ammonia sulfamate; Ammonium amidosulfate; Ammonium amidosulfonate; Ammonium amidosulphate; Ammonium aminosulfonate; Ammonium sulfamate; Ammonium sulfamidate; Ammonium sulphamate; Ammonium sulphamidate; Ammoniumsalz der amidosulfonsaure; Ammoniumsalz der amidosulfonsäure; AMS; Atlacide; Caswell No. 047; EINECS 231-871-7; EPA Pesticide Chemical Code 005501; Feliderm K; Fyran 200 K; Fyran J 3; HSDB 703; Ikurin; Monoammonium sulfamate; Necco Fire Retardant 2750; Necco Fire Retardant 2578; Necco Fire Retardant 2762; Sepimate; Silvacide; Sulfamate d'ammonium; Sulfamic acid, monoammonium salt; Sulfaminsäure; Sulfaminsaure; monoammonium sulfamate; Amcide; Ikurin. Flameproofing agent for textiles and paper; weed and brush killer; electroplating; generation of nitrous oxide. An inorganic herbicide used to control weeds and grasses in vegetables and ornamentals prior to planting and as a tree-killer. Registered by EPA as a herbicide (cancelled). Large hygroscopic plates; mp = 132-135°; bp = 160° (dec); soluble in H_2O (225 g/100 ml); pH = 5.2; LD50 (rat orl) = 2000 mg/kg. *Battle Hayward & Bower Ltd.; Heico; Nissan Chem. Ind.; Spartan Flame Retardants.*

202 Ammonium sulfate
7783-20-2   559

G

$H_8N_2O_4S$
Sulfuric acid diammonium.
Actamaster; Ammonium hydrogen sulfate; Ammonium sulfate $((NH_4)_2SO_4)$; Ammonium sulfate (2:1); Ammonium sulphate; Caswell No. 048; Diammonium sulfate; Diammonium sulphate; Dolamin; EINECS 231-984-1; EPA Pesticide Chemical Code 005601; HSDB 471; Mascagnite; NSC 77671; Sulfatom ammoniya; Sulfuric acid, diammonium salt. Used in fertilizers, water treatment, fermentation, fireproofing compositions, viscose rayon, tanning, food additive Orthohombic crystals or white granules; d = 1.7690; mp = 280° (dec); soluble in H_2O (77 g/100 ml); insoluble in alcohol, acetone; pH of 0.1 aqueous solution = 5.5; LD50 (rat orl) = 2840 mg/kg. *Accurate Chem. & Sci. Corp.; Allied Signal; BASF Corp.; DSM Fine Chem.; General Chem; Heico; Nissan Chem. Ind.; Schaeffer Salt & Chem.; Showa Denko; Sigma-Aldrich Fine Chem.*

203 Ammonium thiocyanate
1762-95-4   564

G

CH_4N_2S
Ammonium rhodanide.
AI3-08542; Ammonium isothiocyanate; Ammonium rhodanate; Ammonium sulfocyanate; Ammonium sulfocyanide; Ammonium thiocyanate; Ammonium-rhodanid; EINECS 217-175-6; HSDB 701; NSC 31184; Rhodanine, ammonium salt; Thiocyanic acid, ammonium salt; Trans-aid; USAF EK-P-433; Weedazol tl. Analytical chemistry; thiourea; fertilizers; photography; in liquid rocket propellants; fabric dyeing; zinc coating; weed killer, defoliant; adhesives; curing resins; pickling iron and steel; electroplating; polymerization catalyst; metals separation. Crystals; mp = 149°; d = 1.3050; soluble in H_2O (163 g/100 ml); LD50 (rat orl) = 750 mg/kg. *Carbo-Tech GmbH; CK Witco Corp.; Degussa AG.*

204 Ammonium thioglycollate
5421-46-5

H

$C_2H_7NO_2S$
Ammonium mercaptoacetate.
Acetic acid, mercapto-, monoammonium salt; AI3-26246; Ammonium mercaptoacetate; Ammonium thioglycolate; Ammonium thioglycollate; EINECS 226-540-9; Mercaptoacetic acid, monoammonium salt; NSC 6954; Thiofaco A-50; Thioglycolic acid ammonium salt; USAF MO-2.

205 Ammonium thiosulfate
7783-18-8   565

G P

$H_8N_2O_3S_2$
Thiosulfuric acid, diammonium salt.
Ammo hypo; Ammonium hyposulfite; Ammonium thiosulfate; Ammonium thiosulphate; Amthio; Caswell No. 048A; Diammonium thiosulfate; EPA Pesticide Chemical Code 080103; HSDB 2688;

Thio-Sul; Thiosulfuric acid, diammonium salt. Photographic fixing agent; analytical reagent; fungicide; reducing agent; brightener in silver plating baths; cleaning compounds for zinc-base die-cast metals; hair waving preparations; fog screens. Also used as a herbicide. Registered by EPA as a herbicide, fungicide and insecticide. Solid; mp = 150° (dec); soluble in H_2O (64 g/100 ml), insoluble in alcohol, ether; LD_{50} (rat orl) = 2890 mg/kg. *Blythe, Williams Ltd.; DuPont; General Chem; National Chelating Co.*

206   Ammonium tungstate
11120-25-5   567   G

$H_{40}N_{10}O_{41}W_{12}$
Ammonium wolframate.
Ammonium tungstate; Ammonium tungstate(VI); EINECS 234-364-9; Tungstate ($W_{12}(OH)_2O_{41}$-), decaammonium. Preparation of ammonium phosphotungstate and tungsten alloys. Crystals; freely soluble in H_2O. *Climax Molybdenum Co.; Sigma-Aldrich Fine Chem.*

207   Ammonium xylenesulfonate
26447-10-9   G H

$C_8H_{13}NO_3S$
Benzenesulfonic acid, dimethyl-, ammonium salt.
Ammonium xylenesulfonate; Ammonium xylene-sulphonate; Benzenesulfonic acid, dimethyl-, ammonium salt; EINECS 247-710-9; Eltesol® AX 40; Hartotrope AXS; Naxonate® 4AX; Stepanate® AXS; Xylenesulfonic acid, ammonium salt. Hydrotrope, cloud point depressant, stabilizer, solubilizer used in formulating detergents, inks, electroplating baths, dyestuffs, polymers. *Albright & Wilson UK Ltd.; Hart Prod.; Ruetgers-Nease; Stepan Canada; Stepan.*

208   Ammonium zinc edetate-
67859-51-2   H

$C_{10}H_{20}N_4O_8Zn$
Glycine, N,N'-1,2-ethanediylbis(N-carboxymethyl)-, di-ammonium zinc salt.
Ammonium zinc edetate; Diammonium ((N,N'-ethyl-enebis(N-(carboxylatomethyl)glycinato))(4-)-N,N',O,   O',ON,ON')zincate(2-); EINECS 267-400-7; (Ethylene-dinitrilo)tetraacetato zincate(2-), diammonium salt; Glycine, N,N'-1,2-ethanediylbis(N-carboxy-methyl)-, diammonium zinc salt; Zincate(2-), ((N,N'-1,2-ethanediylbis(N-(carboxymethyl)glycinato))(4-)-N,N',O,O',ON,ON')-, diammonium, (OC-6-21)

209   Amoxicillin trihydrate
61336-70-7   582   D

$C_{16}H_{19}N_3O_5S.3H_2O$
[2S-[2α,5α,6β(S*)]]-6-[[Amino(4-hydroxyphenyl)acetyl]-amino]-3,3-dimethyl-7-oxo-4-thia-1-azabicyclo[3.2.0]-heptane-2-carboxylic acid trihydrate.
A-Gram; Alfamox; AM 7; Amodex; Amoksicillin; Amoksicillin forte; Amophar; Amoran; Amoxi-Diolan; Amoxi-wolff; Amoxicillin-ratiopharm; Amoxicillin trihydrate; Amoxicillin; Amoxidal; Amoxil; Amoxillat; Amoxina; Amoxine; Amoxipen; Augmentin; Benzoral; BRL 2333; Ciblor; Clamoxyl; Drg-0075; Dura AX; Flemoxine; Galenamox; Gramidil; Hiconcil; Himinomax; Imacillin; Izoltil; Kentrocyllin; Larotid; Matasedrin; Metifarma; Moxal; Moxaline; Novabritine; Pacetocin; Pamocil; Paradroxil; Polymox; Robamox; Siganopen; Simplamox; Sintopen; Trimox; Uro-clamoxyl; Utimox; Velamox; Wymox; Zamocillin; Zamocilline; Zimox. Antibacterial. Off-white solid; mp - 178-180°; $[α]_D^{20}$= +246° (c = 0.1); $λ_m$ 230, 274 nm (ε 10850, 1400 EtOH); $λ_m$ 229, 272 nm (ε 9500, 1080 0.1N HCl); $λ_m$ = 248, 291 nm (ε 2200, 3000 KOH); soluble in H_2O (0.4 g/100 ml), MeOH (0.75 g/100 ml), EtOH (0.34 g/100 ml); insoluble in C_6H_6, EtOAc, CH_3CN, C_6H_{14}.

210   Ampicillin
69-53-4   591   D G

$C_{16}H_{19}N_3O_4S$
[2S-[2α,5α,6β(S*)]]-6-[(Aminophenylacetyl)amino]-3,3-dimethyl-7-oxo-4-thia-1-azabicyclo[3.2.0]heptane-2-carboxylic acid.
AB-PC; AB-PC Sol; Acillin; Adobacillin; Alpen; Amblosin; Amcill; Amfipen; Amfipen V; Aminobenzylpenicillin; Amipenix S; Ampi-bol; Ampi-Co; Ampi-Tab; Ampichel; Ampicil; Ampicilina; Ampicillin; Ampicillin A; Ampicillin acid; Ampicillin anhydrate; Ampicillina; Ampicilline; Ampicillinum; Ampicin; Ampifarm; Ampikel; Ampimed; Ampipenin; Ampiscel; Ampisyn; Ampivax; Ampivet; Amplacilina; Amplin; Amplipenyl; Amplisom; Amplital; Ampy-Penyl; Austrapen; AY-6108; AY 6108; BA 7305; Bayer 5427; Binotal; Bonapicillin; Britacil; BRL 1341; Campicillin; Cimex; Copharcilin; D-Cillin; Delcillin; Deripen; Divercillin; Doktacillin; Duphacillin; EINECS 200-709-7; Geocillin; Grampenil; Guicitrina; HI 63; HSDB 3009; Lifeampil; Mixture Name; Morepen; Norobrittin; Novo-ampicillin; NSC-528986; Nuvapen; Olin Kid; Omnipen; Orbicilina; P-50; Pen A; Pen Ampil; Penbristol; Penbritin; Penbritin paediatric; Penbritin syrup; Penbrock; Penicline; Penimic; Pensyn; Pentrex; Pentrexl; Pentrexyl; Pfizerpen A; Polyflex (Veterinary); Ponecil; Princillin; Principen; Qidamp; Racenacillin; Ro-Ampen; Rosampline; Roscillin; Semicillin; Semicillin R; Servicillin; SK-Ampicillin; SQ 17382; Sumipanto; Supen; Synpenin; Texcillin; Tokiocillin; Tolomol; Totacillin; Totalciclina; Totapen; Trifacilina; Ultrabion; Ultrabron; Vampen; Viccillin; Viccillin S; WY-5103; Wypicil. Antibacterial. Anhydrous form, mp = 199-202° (dec); $[α]_D^{23}$ = 287.9° (H_2O), soluble in H_2O, DMSO. *Apothecon; Bristol-Myers Squibb Pharm. R&D; Parke-Davis; Wyeth-Ayerst Labs.*

211   Amprolium hydrochloride
121-25-5   595   G V

$C_{14}H_{19}ClN_4$
1-[(4-Amino-2-propyl-5-pyrimidinyl)methyl]-2-picolinium chloride.
1-((4-Amino-2-propyl-5-pyrimidinyl)methyl)-2-picolinium chloride; 1-((4-Amino-2-propyl-5-pyrimidinyl)methyl)-2-methylpyridinium chlorid; Amprol, Veterinary; Amprolio; Amprolium; Amprovine; Corid; EINECS 204-458-4; Pancoxin; 2-Picolinium, 1-((4-amino-2-propyl-5-pyrimidinyl)methyl)-, chloride; Pyridinium, 1-((4-amino-2-propyl-5-pyrimidinyl)methyl)-2-methyl-, chloride . Coccidiostat in veterinary medicine. Also used as an antiparasitic in cattle. Soluble in H_2O, MeOH, EtOH, DMF, insoluble in i-PrOH, BuOH, dioxane, Me_2CO, EtOAc, CH_3CN, isooctane; pH 2.5-3.0.

212   Amsonic acid disodium salt
7336-20-1   H

$C_{14}H_{12}N_2Na_2O_6S_2$
Benzenesulfonic acid, 2,2'-(1,2-ethenediyl)bis(5-amino-, disodium salt.
CCRIS 4430; Diaminostilbene disulphonate disodium salt; 4,4'-Diamino-2,2'-stilbenedisulfonic acid, disodium salt; p,p'-Diaminostilbene-o,o'-disulfonic acid disodium salt; Disodium 4,4'-diaminostilbene-2,2'-disulphonate; 2,2'-Disulfo-4,4'-stilbenediamine disodium salt; EINECS 230-847-3; Flavonic acid disodium salt; 2,2'-Stilbenedisulfonic acid, 4,4'-diamino-, disodium salt.

213   Amyl acetate
628-63-7   G P

$C_7H_{14}O_2$
n-Pentyl acetate.
4-02-00-00152 (Beilstein Handbook Reference); Acetate d'amyle; Acetic acid, amyl ester; Acetic acid, pentyl ester; AI3-02729; Amyl acetate; Amyl acetate, n-; Amyl acetic ester; Amyl acetic ether; Amylazetat; Amylester kyseliny octove; Banana oil; Birnenoel; BRN 1744753; Caswell No. 049A; Chlordantoin; Dymon SWH Wasp & Hornet Spray; EINECS 211-047-3; EPA Pesticide Chemical Code 000169; Holiday Pet Repellent; Holiday Repellent Dust; HSDB 5126; NSC 7923; Octan amylu; Pear oil; Pent-acetate; Pent-acetate 28; Pentyl acetate; Prim-amyl acetate; Primary amyl acetate; UN1104. Solvent for cellulose acetate in lacquers and paints, extraction of penicillin, photographic film, leather and nail polishes, flavoring agent, printing and finishing fabrics, solvent for phosphors in fluorescent lamps. Registered by EPA as an antimicrobial, herbicide and insecticide (cancelled). At one time it was believed to have a preservative effect when applied to leather, but it is much too volatile

to have afforded lasting protection, even if initially effective. Colorless, clear liquid; mp = -70.8°; bp = 149.2°; d^{20} = 0.8756; slightly soluble in H_2O; soluble in CCl_4, freely soluble in EtOH, Et_2O; LD_{50} (rat orl) = 6500 mg/kg. *BP Chem.; Penta Mfg.; Pentagon Chems. Ltd.; Sigma-Aldrich Co.; Sigma-Aldrich Fine Chem.; Union Carbide Corp.*

214   Amyl-m-cresol
53043-14-4   611   G

$C_{12}H_{18}O$
6-n-Amyl-m-cresol.
3-06-00-02005 (Beilstein Handbook Reference); Amilmetacresol; 6-Amyl-m-cresol; 6-n-Amyl-m-cresol; Amylmetacresol; Amylmetacresolum; BRN 2440952; m-Cresol, 6-pentyl-; EINECS 215-094-0; 5-Methyl-2-pentylphenol; 6-n-Pentyl-m-cresol; 6-Pentyl-m-cresol; Phenol, 5-methyl-2-pentyl-. Antiseptic, germicide and mold preventative. Oil; mp = 24°; bp_{15} = 137-139°; insoluble in H_2O, soluble in organic solvents.

215   Amylene
513-35-9   612   G H

C_5H_{10}
1,1,2-Trimethylethylene.
AI3-37711; Amylene; n-Amylene; 2-Butene, 2-methyl-; EINECS 208-156-3; Ethylene, trimethyl-; HSDB 2072; β-Isoamylene; 2-Methyl-2-butene; 2-Methylbut-2-ene; 3-Methyl-2-butene; NSC 74118; Trimethylethylene; 1,1,2-Trimethylethylene; UN2460. Used in organic synthesis, high-octane fuel manufacture. Volatile liquid; mp = -133.7°; bp = 38.58°; d^{20} = 0.6623; λ_m = 205 nm (ε = 851, gas); insoluble in H_2O, soluble in Et_2O, Me_2CO

216   t-Amyl peroxy-2-ethylhexanoate
686-31-7   G

$C_{13}H_{26}O_3$
tert-Pentyl 2-ethylperoxyhexanoate.
t-Amylperoxy 2-ethylhexanoate; EINECS 211-687-3; Hexaneperoxoic acid, 2-ethyl-, 1,1-dimethylpropyl ester; Trigonox® 121. Polymerization initiator. *Akzo Chemie.*

217   4-t-Amylphenol
80-46-6   7220   G P

C11H16O
4-(1,1-Dimethylpropyl)-1-phenol.
AI3-00460; Amilfenol; Amilphenol; Amyl phenol 4T; BRN 1908224; Caswell No. 050; CCRIS 4693; EINECS 201-280-9; EPA Pesticide Chemical Code 064101; HSDB 5236; NSC 403672; Orthophen® 278; Pentaphen; Phenol, 4-(1,1-dimethylpropyl)-; Phenol, p-(tert-pentyl)-; Ptap; tert-Amylphenol; UCAR amyl phenol 4T. Registered by EPA as an antimicrobial and fungicide. Intermediate for chemical specialties; also in manufacture of photographic chemicals, oil demulsifiers, phenolic resins, agricultural surfactants and antiskinning agents, in germicidal formulations. Crystals; mp = 94 - 95°; bp = 262.5°, bp_{740} = 248 - 250°, bp_{15} = 138.5°, bp_3 = 112 - 120°; d_4^{20} = 0.962; insoluble in H_2O, soluble in EtOH, Et_2O, C_6H_6, $CHCl_3$; LD_{50} (rat orl) = 3080 mg/kg. *Athea Laboratories, Inc.; Clariant Corp.; Elf Atochem N. Am.; Huntington Professional Products; Quest Chemical Corp.; Steris Corp.; Walter G. Legge Co Inc.*

218   **Anethole**
    104-46-1   647   G H P

C10H12O
1-Methoxy-4-propenylbenzene.
2-06-00-00523 (Beilstein Handbook Reference); Acintene O; AI3-00380; Anethol; Anethole; Anise camphor; Aniskampfer; Anisole, p-propenyl-; BRN 0774229; Caswell No. 051B; CCRIS 6211; EINECS 203-205-5; EPA Pesticide Chemical Code 015604; FEMA Number 2086; HSDB 1427; Isoestragole; Methoxy-4-propenylbenzene; 1-Methoxy-4-(1-prop-enyl)benzene;   1-Methoxy-4-propenylbenzene; Mona-sirup; Nauli gum; NSC 4018; Oil of aniseed; 1-Prop-ene, 1-(4-methoxyphenyl)-;   4-Propenylanisole;   p-Prop-enylanisole; Propenylanisole. Anise flavoring material used as a rodenticide. Registered by EPA as a rodenticide (cancelled). [Trans isomer]: Crystals; mp = 21.4°; bp = 234°, bp_{12} = 115°, d_4^{20} 0.9883; λ_m = 259 nm (ε = 22,300 EtOH); insoluble in H_2O, very soluble in EtOH, Et_2O, Me_2CO, C_6H_6; LD_{50} (rat ip) = 900 mg/kg. [Cis isomer]: $bp_{2.3}$ = 79-79.5°; d_4^{20} 0.9878; λ_m = 253.5 nm (ε 18,500 EtOH); LD_{50} (rat ip) = 93 mg/kg. *Arizona; Sigma-Aldrich Fine Chem.*

219   **Anethole, trans**
    4180-23-8   647   H

C10H12O
trans-1-(4-Methoxyphenyl)-1-propene.
(E)-Anethol; (E)-Anethole; 4-06-00-03796 (Beilstein Handbook Reference); Anethole; Anethole, trans-; Anisole, p-propenyl-, (E)-; Anisole, p-propenyl-, trans-; Benzene, 1-methoxy-4-(1-propenyl)-, (E)-; BRN 0636190; CCRIS 2481; EINECS 224-052-0; FEMA No. 2086; Methoxy-β-methylstyrene, trans-p-; 1-Methoxy-4-(1-propenyl)benzene, (E)-; 1-p-Methoxyphenylpropene, trans-; NSC 209529; Propenylanisole, p-, (E)-; trans-1-(4-Methoxyphenyl)-1-propene; trans-1-(p-Methoxyphenyl)-1-propene; trans-p-Methoxy-β-methylstyrene. Liquid; mp = 21.35°; bp = 234°, bp_{12} = 115°; d^{20} = 0.9882; λ_m = 259 nm (cyclohexane); slightly soluble in H_2O, soluble

in Me_2CO, $CHCl_3$, very soluble in C_6H_6, freely soluble in EtOH, Et_2O.

220   **Angeliconitrile**
    20068-02-4   H

C5H7N
2-Butenenitrile, 2-methyl-, (Z)-.
Angelic acid nitrile; Angeliconitrile; 2-Butenenitrile, 2-methyl-, (Z)-; 2-Butenenitrile, 2-methyl-, (2Z)-; Crotononitrile, 2-methyl-, (Z)-; EINECS 243-496-6; HSDB 6156; (Z)-2-Methyl-2-butenenitrile; 2-Methyl-cis-2-butenenitrile; cis-2-Methyl-2-butenonitrile.

221   **1,4-Anhydro-D-glucitol 6-dodecanoate**
    5959-89-7   H

C18H34O6
D-Glucitol, 1,4-anhydro-, 6-dodecanoate.
1,4-Anhydro-D-glucitol 6-dodecanoate; D-Glucitol, 1,4-anhydro-, 6-dodecanoate; D-Glucitol, 1,4-anhydro-, 6-dodecanoate; EINECS 227-534-9.

222   **Anilazine**
    101-05-3   659   G P

C9H5Cl3N4
2,4-Dichloro-6-o-chloroanilino-s-triazine.
5-26-08-00019 (Beilstein Handbook Reference); AI3-26058; Anilazin; Anilazine; Aniyaline; B-622; Bortrysan; BRN 0223133; Caswell No. 302; CCRIS 43; (o-Chloroanilino)dichlorotriazine; 2-(2-Chloranilin)-4,6-di-chlor-1,3,5-triazin; 2-Chloro-N-(4,6-dichloro-1,3,5-tri-azin-2-yl)aniline; Dairene®; Dairin®; 2,4-Di-chloro-6-(2-chloroanilino)-1,3,5-triazine; 2,4-Dichloro-6-(o-chloro-anilino)-s-triazine; 4,6-Dichloro-N-(2-chlorophenyl)-1,3,5-triazin-2-amine; Direx; Direz; Dyrene®; Dyrene® 50W; Dyrene® 50W Triazine; Dyrene® Flüssig; EINECS 202-910-5; ENT 26,058; EPA Pesticide Chemical Code 080811; HSDB 1567; Kemate; NCI-C08684; NSC 3851; 1,3,5-Triazin-2-am-ine, 4,6-dichloro-N-(2-chlorophenyl)-; s-Triazine, 2,4-dichloro-6-(o-chloroanilino)-; Triasym; Triasyn; Tri-azin; Triazine Zinochlor; (o-chloro-anilino)dichloro-s-triazine; 2-(2-Chloranilin)-4,6-dichlor-1,3,5-triazin; 2-Chloro-N-(4,6-dichloro-1,3,5-triazin-2-yl)aniline; 2,4-dichloro-6-o-chloroanilino-s-tri-azine; 2,4-Dichloro-6-(2-chloroanilino)-1,3,5-triazine; 4,6-Dichloro-N-(2-chlorophenyl)-1,3,5-triazin-2-amine; Dichloro-6-(o-chloroanilino)-s-triazine; Dichloro-N-(2-chlorophenyl)-1,3,5-triazin-2-amine; Direx; Direz; Dyrene; Dyrene 50W; ENT 26,058; EPA Pesticide Chemical Code 080811; HSDB 1567; Kemate; NCI-C08684; NSC 3851; Triasym; Triasyn; Triazin; Triazine; 1,3,5-Triazin-2-amine, 4,6-dichloro-N-(2-chlorophenyl)-; 1,3,5-Triazine, 2,4-dichloro-6-(o-chloroanilino)-; s-Triazine, 2,4-dichloro-6-(o-chloroanilino)-; Zinochlor. Broad spectrum

fungicide used for tobacco, potatoes, cereals and ornamentals. Non-systemic foliar fungicide with protective action used to control blights of potatoes and tomatoes and leafspot diseases in many crops. Registered by EPA as a fungicide (cancelled). White-tan crystals; mp = 159-160°; insoluble in H_2O (0.0008 g/100 ml), Me_2CO (10 g/100 ml), chlorobenzene (6 g/100 ml); C_7H_8 (5 g/100 ml), xylene (4 g/100 ml); LD_{50} (rat orl) > 5000 mg/kg, (mmus orl) > 5000 mg/kg, (fmus orl) = 3672 mg/kg, (rbt orl) = 460 mg/kg, (cat orl) > 500 mg/kg, (rbt der) > 9400 mg/kg, (rat der) > 5000 mg/kg, (mallard duck orl) > 2000 mg/kg, (Japanese quail orl) =2500 - 3750 mg/kg, (ckn orl) = 3750 - 5000 mg/kg, (canary orl) > 1000 mg/kg; LC_{50} (rat ihl 1 hr.) > 0.228 mg/l air, (bluegill sunfish, goldfish, carp 96 hr.) < 1.0 mg/l, (rainbow trout 48 hr.) = 0.15 mg/l; non-toxic to bees. *Bayer AG; Sigma-Aldrich Co.*

223   Aniline
62-53-3   661   G H

C_6H_7N
Benzenamine.
AI3-03053; Aminobenzene; Aminophen; Anilin; Anilina; Aniline; Aniline oil; Anyvim; Arylamine; Benzenamine; Benzene, amino; Benzidam; Blue Oil; C.I. 76000; C.I. Oxidation Base 1; Caswell No. 051C; CCRIS 44; CI 76000; CI Oxidation Base 1; Cyanol; EINECS 200-539-3; EPA Pesticide Chemical Code 251400; HSDB 43; Huile d'aniline; Krystallin; Kyanol; NCI-C03736; Phenylamine; RCRA waste number U012; UN 1547. Isolated from coal tar. Used as an intermediate in the manufacture of dyestuffs and pharmaceuticals. Oil; mp = -6°; bp = 184°; d = 1.0220; soluble in H_2O (3 g/100 ml), more soluble in organic solvents; LD_{50} (rat orl) = 0.44 g/kg. *Lancaster Synthesis Co.; Mallinckrodt Inc.; Sigma-Aldrich Fine Chem.*

224   1-Anilino-8-naphthalenesulfonic acid
82-76-8   664   G

$C_{16}H_{13}NO_3S$
1-Phenyl-naphthylamine-8-sulfonic acid.
1-Anilino-8-napthalenesulfonate; 1-Anilino-8-naphtha-lene-sulfonate; Anilinonaphthalenesulfonic acid; 8-Anil-ino-1-naphthalenesulfonic acid; 8-Anilinonaphthalene-1-sulphonic acid; ANS; EINECS 201-438-7; 1-Naphthalenesulfonic acid, 8-anilino-; 1-Naphtha-lenesulfonic acid, 8-(phenylamino)-; NSC 1746; Peri acid, phenyl-; 1-(Phenylamino)-8-naphthalenesulfonic acid; Phenylperi acid. Fluorescent probe used in protein conformation studies. Crystals; mp = 215-217°.

225   p-Anisaldehyde
123-11-5   666   G H

$C_8H_8O_2$
4-Methoxybenzaldehyde.
AI3-00223; Anisaldehyde; Anisic aldehyde; Aubepine; Benzaldehyde, 4-methoxy-; Caswell No. 051E; CCRIS 821; Crategine; EINECS 204-602-6; FEMA No. 2670; Formylanisole, p-; HSDB 2641; 4-Methoxybenzaldehyde; NSC 5590; Obepin. Used in perfumery. Liquid; mp = o°; bp = 248°, bp_{12} = 134°; d^{15} = 1.119; λ_m = 273 nm (MeOH); insoluble in H_2O, soluble in C_6H_6, very soluble in Me_2CO, $CHCl_3$, freely soluble in EtOH, Et_2O; LD_{50} (rat orl) = 1510 mg/kg.

226   m-Anisidine
536-90-3   G

C_7H_9NO
3-Methoxy-1-aminobenzene.
4-13-00-00953 (Beilstein Handbook Reference); AI3-52519; 1-Amino-3-methoxybenzene; 3-Aminoanisole; m-Aminoanisole; 3-Anisidine; m-Anisidine; m-Anisylamine; Benzenamine, 3-methoxy-; Benzenamine, 3-methoxy-; BRN 0386119; CCRIS 5886; EINECS 208-651-4; m-Methoxyaniline; 3-Methoxyaniline; 3-Methoxy-benzenamine; NSC 7631. Chemical intermediate. Liquid; mp = -1°; bp = 251°; d^{20} = 1.0960; λ_m = 285 nm (cyclohexane); slightly soluble in H_2O, CCl_4, soluble in EtOH, Et_2O, Me_2CO, C_6H_6. *Penta Mfg.; Rhône-Poulenc; Sigma-Aldrich Fine Chem.*

227   o-Anisidine
90-04-0   G H

C_7H_9NO
Benzenamine, 2-methoxy-.
4-13-00-00806 (Beilstein Handbook Reference); AI3-08584; o-Aminoanisole; o-Anisidine; o-Anisylamine; Benzenamine, 2-methoxy-; BRN 0386210; CCRIS 768; EINECS 201-963-1; HSDB 2073; o-Methoxyaniline; o-Methoxyphenylamine; NSC 3122. Chemical intermediate. Liquid; mp = 6.2°; bp = 224°; d^{20} = 1.0923; λ_m = 237 nm (ε = 8511, cyclohexane); soluble in EtOH, Et_2O, Me_2CO, less soluble in H_2O; LD_{50} (rat orl) = 2 g/kg. *Penta Mfg.; Rhône-Poulenc; Sigma-Aldrich Fine Chem.*

228   p-Anisidine
104-94-9   G

C_7H_9NO
4-Methoxy-1-aminobenzene.
AI3-02392; 4-Aminoanisole; p-Aminoanisole; 1-Amino-4-methoxybenzene; Aniline, 4-methoxy-; Aniline, p-methoxy-; 4-Anisidine; p-Anisidine; Anisole, p-amino-; p-Anisylamine; Benzenamine, 4-methoxy-; CCRIS 917; p-Dianisidine; EINECS 203-254-2; HSDB 1603; 4-Methoxy-1-aminobenzene; 4-Methoxy-aniline; p-Methoxyaniline; 4-Methoxybenzenamine; p-Methoxyphenylamine;

NSC 7921. Intermediate in chemical manufacturing. Plates; mp = 57.2°; bp = 243°; d^{57} = 1.071; λ_m = 235, 300 nm (MeOH); soluble in H_2O, Me_2CO, C_6H_6, very soluble in EtOH, Et_2O; LD_{50} (rat orl) = 1400 mg/kg. *Penta Mfg.; Rhône-Poulenc; Rhone-Poulenc UK; Sigma-Aldrich Fine Chem.*

229   **Anserine**
584-85-0   683   G

$C_{10}H_{16}N_4O_3$
β-Alanyl 3-methyl L-histidine.
  Anserine; EINECS 209-545-0; N-β-Alanyl-3-methyl-L-histidine; L-Anserine; L-Histidine, N-β-alanyl-3-methyl-. A natural peptide from muscle. Used in biochemical research. [L-form]; Needles; mp = 240-242°; $[\alpha]_D^{30}$= +12.3° (c = 5); freely soluble in H_2O, soluble in MeOH, slightly soluble in EtOH.

230   **Anthracene**
120-12-7   687   G

$C_{14}H_{10}$
Paranaphthalene.
  AI3-00155; Anthracen; Anthracene; Anthracin; Bis-alkylamino anthracene; CCRIS 767; Coal tar pitch volatiles: anthracene; EINECS 204-371-1; Green Oil; HSDB 702; NSC 7958; Paranaphthalene; Sterilite Hop Defoliant; Tetra Olive N2G. Anthracene oil; used for chemical stripping in hop vines. Prisms; mp = 215.0°; bp = 339.9°; d^{25} = 1.28; λ_m = 218, 220, 250, 296, 309, 322, 338, 355, 276 nm (ε = 11700, 11800, 20000, 531, 1230, 2750, 5290, 7770, 7590, MeOH); insoluble in H_2O, slightly soluble in EtOH, Et_2O, Me_2CO, C_6H_6, CCl_4, $CHCl_3$, C_7H_8, CS_2.

231   **Anthraflavic acid**
84-60-6   G

$C_{14}H_8O_4$
2,6-Dihydroxyanthraquinone.
  4-08-00-03272 (Beilstein Handbook Reference); 9,10-Anthracenedione, 2,6-dihydroxy-; Anthraflavic acid; Anthraflavin; Anthraquinone, 2,6-dihydroxy-; BRN 2054127; CCRIS 5593; EINECS 201-544-3; NSC-33531. Used as a basis for pigments and dyestuffs. Solid; mp > 320°.

232   **Anthranilamide**
88-68-6   H

$C_7H_8N_2O$
Benzamide, 2-amino-.
  4-14-00-01010 (Beilstein Handbook Reference); AI3-28018; Aminobenzamide; o-Aminobenzamide; Anthra-nilamide; Anthranilimidic acid; Benzamide, 2-amino-; Benzamide, o-amino-; Benzoic acid, 2-amino-, amide; BRN 0508509; EINECS 201-851-2; HSDB 5261; NSC 38768. Has antiviral activity; see USP 5,763,464. Also has anti-coagulant properties (USP 6,498,185). Solid; mp = 110.5°; bp = 300°; λ_m = 213, 248 nm (MeOH); very soluble in EtOAc, soluble in H_2O, EtOH, less soluble in Et_2O, C_6H_6. *BASF Corp.; La-Co Industries; Ludger*

233   **Anthranilic acid**
118-92-3   421   G

$C_7H_7NO_2$
o-Aminobenzoic acid.
  4-27-00-07875 (Beilstein Handbook Reference); AA; AI3-02408; o-Aminobenzoic acid; ortho-Aminobenzoic acid; 2-Aminobenzoic acid; 1-Amino-2-carboxybenzene; Anthranilic acid; o-Anthranilic acid; Benzoic acid, 2-amino-; Benzoic acid, o-amino-; BRN 0471803; 2-Carboxyaniline; o-Carboxyaniline; Caswell No. 033G; CCRIS 49; EINECS 204-287-5; HSDB 1321; Kyselina anthranilova; Kyselina o-aminobenzoova; NCI-C01730; NSC 144; Vitamin L1. Antioxidant for fats, greases, lube oils and polyamides. Sludge preventative in furnace and lube oils. Chelating agent and sequestrant. Corrosion inhibitor. Stabilizer of can lacquers, oils and lubricants. Used as a dye intermediate. mp = 146.5°; d^{20} = 1.412; slightly soluble in C_6H_6, soluble in H_2O (5.7 g/l), EtOH, Et_2O, very soluble in $CHCl_3$, TFA, C_5H_5N; LD_{50} (mus orl) = 1400 mg/kg. *PMC.*

234   **Anthranol**
1143-38-0   689   G

$C_{14}H_{10}O_3$
9(10H)-Anthracenone, 1,8-dihydroxy-.
  4-06-00-07602 (Beilstein Handbook Reference); 9(10H)-Anthracenone, 1,8-dihydroxy-; Anthra-Derm; Anthralin; Anthrone, 1,8-dihydroxy-; Batridol; BRN 2054360; CCRIS 628; Chrysodermol; Cignolin; Cigthranol; 1,8-Dihydroxyanthracen-9(10H)-one; 1,8-Dihydroxy-anthrone; 1,8-Dihydroxy-9-anthrone; Dithranol; Dithranolum; EINECS 214-538-0; NSC 43970; Psoriacid-Stift. Anthranol is a smooth soft ointment containing anthralin (in 0.4, 1.0 and 2.0 strengths) w/w in a base containing cetyl alcohol, liquid paraffin, soft white paraffin and sodium sulfate with salicylic acid; used for the topical treatment of subacute and chronic psoriasis including psoriasis of the scalp. Yellow plates; mp = 179°; λ_m = 256, 288 nm (ε = 11400, 9290, MeOH); insoluble in H_2O, slightly soluble in Et_2O, soluble in EtOH, Me_2CO, C_6H_6, ligroin, dil. NaOH, very

soluble in $CHCl_3$, C_5H_5N. *Stiefel Labs Inc.*

235   Anthraquinone
84-65-1   692   G P

$C_{14}H_8O_2$
9,10-Anthracenedione.
AI3-09073; Anthracene, 9,10-dihydro-9,10-dioxo-; Anthradione; Anthrapel; Anthraquinone; Bis-alkylamino anthraquinone; Caswell No. 052A; CCRIS 649; Corbit; EINECS 201-549-0; EPA Pesticide Chemical Code 122701; Hoelite; HSDB 2074; Morkit; NSC 7957. Bird repellent, used to protect seed stocks. Also used as an intermediate for dyes and organics, discharging auxiliary for textiles and an organic inhibitor. Yellow crystals; mp = 286°; bp = 377°; d^{20} = 1.438; λ_m = 206, 251, 272, 319, 380, 399, 420 nm (ε = 19498, 56234, 16982, 5370, 110, 89, 60 cyclohexane), 205, 252, 273, 325, (ε = 24547, 42658, 14125, 4266 EtOH); slightly soluble in H_2O (0.00006 g/100 ml), soluble in $CHCl_3$ (0.90 g/100 ml), C_6H_6 (0.23 g/100 ml), EtOH (0.35 g/100 ml), C_7H_8 (0.26 g/100 ml), Et_2O (0.08 g/100 ml); LD_{50} (rat, mus orl) > 5000 mg/kg, (rat der) > 5000 mg/kg; LC_{50} (rat ihl 4 hr.) > 1.3 mg/l air. *Bayer Corp.; Buckton Scott Ltd.; Generic; ICI Americas Inc.*

236   Anthrarufin
117-12-4   694   G

$C_{14}H_8O_4$
1,5-Dihydroxyanthraquinone.
4-08-00-03268 (Beilstein Handbook Reference); 9,10-Anthracenedione, 1,5-dihydroxy-; Anthraquinone, 1,5-dihydroxy-; Anthrarufin; BRN 1881718; CCRIS 3150; 1,5-Dihydroxyanthrachinon; 1,5-Dihydroxy-9,10-anthra-quinone; EINECS 204-175-6; NSC 7211. Intermediate in dyestuffs manufacture. Pale yellow platelets; mp = 280° (dec); bp sublimes; λ_m = 250, 420 nm (ε = 39811, 20893, MeOH); insoluble in H_2O, slightly soluble in EtOH, Et_2O, Me_2CO, CS_2; soluble in C_6H_6, conc. H_2SO_4.

237   Antimony
7440-36-0   698   G

Sb

Sb
Antimony.
Antimony; Antimony Black; Antimony, elemental; Antimony, metallic; Antimony powder; Antimony, regulus; Antymon; CI 77050; EINECS 231-146-5; HSDB 508; Regulus of antimony; Stibium; Stibium metallicum; UN2871. Hardening alloy for lead, bearing metal, type metal, solder, collapsible tubes and foil, sheet and pipe, semiconductor technology, pyrotechnics. mp = 630°; bp = 1635°; d =

6.68; LD_{50} (rat orl) = 100 mg/kg. *Amspec Chem.; Atomergic Chemetals; Sigma-Aldrich Fine Chem.*

238   Antimony diamyldithiocarbamate
15890-25-2   H

$C_{33}H_{66}N_3S_6Sb$
Tris(dipentyldithiocarbamato-S,S')antimony.
Antimony diamyldithiocarbamate; Antimony, tris-(dipentylcarbamodithioato-S,S')-, (OC-6-11)-; Carbamic acid, dipentyldithio-, tris(anhydrosulfide) with thioantimonic acid (H_3SbS_3); EINECS 240-028-2; Tris(dipentyldithiocarbamato-S,S')antimony.

239   Antimony oxide
1327-33-9   G

O_3Sb_2
Antimony(III) trioxide.
EINECS 215-474-6; Timonox; Diantimony trioxide; Exitelite; Flowers of antimony; NCI-C55152; Thermoguard B; Thermoguard S. Used as a pigment and flame-proofant and in manufacture of tartar emetic. Also RN 1309-64-4. Solid; mp = 655°; bp = 1425°; d = 5.2000. *Anzon.*

240   Antimony pentachloride
7647-18-9   701   G

Cl_5Sb
Antimony (V) chloride.
Antimoine (pentachlorure d'); Antimonio (pentacloruro di); Antimonpentachlorid; Antimony chloride; Antimony chloride (SbCl5); Antimony pentachloride; Antimony perchloride; Antimony(V) chloride; Antimoonpenta-chloride; CCRIS 4496; EINECS 231-601-8; HSDB 444; Pentachloroantimony; Pentachlorure d'antimoine; Perchlorure d'antimoine; UN1730; UN1731. For analytical testing of alkaloids and cesium; dyeing intermediates; as chlorine carrier in organic chlorinations. Crystals; mp = 2-4°; bp = 140°; Corrosive. *Atomergic Chemetals; Hoechst Celanese; Sigma-Aldrich Fine Chem.*

241   Antimony pentafluoride
7783-70-2   702   G

F_5Sb
Antimony (V) fluoride.
Antimony fluoride; Antimony (V) fluoride; Antimony fluoride (SbF5); Antimony pentafluoride; Antimony(V) fluoride; Antimony(V) pentafluoride; EINECS 232-021-8; HSDB 442; Pentafluoroantimony; UN1732. Catalyst and/or source of fluorine in fluorination reactions.

Moderately viscous liquid; toxic; mp = 8.3°; bp = 141°; $d^{25.8}$ = 3.097. *Allied Signal; Atomergic Chemetals; Elf Atochem N. Am.*

242   Antimony pentasulfide
1315-04-4   703   G

S5Sb2
Antimony(V) sulfide.
Antimonial saffron; Antimonic sulfide; Antimony pentasulfide; Antimony Red; Antimony sulfide; Antimony sulfide (Sb2S5); Antimony sulfide golden; C.I. 77061; Diantimony pentasulphide; EINECS 215-255-5; Golden antimony sulfide; Sulfur gold. Red pigment, used as a pigment; for vulcanizing and coloring rubber and in matches and fireworks. Orange-yellow powder; insoluble in H_2O; soluble in conc. HCl, solns of alkali hydroxides or sulfides; LD50 (rat orl) = 150 mg Sb/100 g. *Atomergic Chemetals.*

243   Antimony pentoxide
1314-60-9   704   G

O5Sb2
Antimony(V) oxide.
A 1530 (metal oxide); A 1550; A 2550; AGO 40; Anchimonzol A 2550; Antimonic oxide; Antimony oxide (Sb2O5); Antimony pentaoxide; Antimony pentoxide; Apox S; CCRIS 4497; Diantimony pentaoxide; Diantimony pentoxide; EINECS 215-237-7; HFR 201; Nyacol 1550; Nyacol A 1590; Nyacol ADP 480; Nyacol ADP 494; Nyacol AGO 40; Sanka Anchimonzol A 2550M; Stibic anhydride; Sun Epoch NA 3080P; Sun Epoch NA 3070P; Sun Epoch NA 100; Suncolloid AME 130; Suncolloid AMT 130; Timonox. Used as a flame retardant in clothing. Yellow powder; d = 3.78; slightly soluble in H_2O; LD50 (rat ip) = 4 g/kg. *Anzon.*

244   Antimony trichloride
10025-91-9   713   G

Cl3Sb
Antimony(III) chloride.
AI3-04463; Antimoine (trichlorure d'); Antimonio (tricloruro di); Antimonous chloride; Antimontrichlorid; Antimony butter; Antimony chloride; Antimony chloride (SbCl3); Antimony trichloride; Antimony(III) trichloride; Antimoontrichlride; Butter of antimony; C.I. 77056; Caustic antimony; CCRIS 4494; Chlorid antimonity; Chlorure antimonieux; CI 77056; EINECS 233-047-2; HSDB 439; Stibine, trichloro-; Trichlorostibine; Trichlorure d'antimoine; UN1733; Alferrlc; Aluminoferric. Used in manufacture of antimony salts, bronzing iron, mordant, manufacture of lakes, chlorinating agent in organic synthesis, pharmaceuticals, fireproofing textiles, analytical reagent. Needles; mp = 73°, bp = 223.5°, bp70 = 143.5°, bp11 = 102°; soluble in H_2O (10 g/100 ml), CHCl3 (22 g/100 ml), EtOH, C6H6, CS2, Et2O, Me2CO, CCl4. *Akzo Chemie; Hoechst Celanese; Nihon Kagaku Sangyo; Sigma-Aldrich Fine Chem.*

245   Antimony trioxide
1309-64-4   717   G

Sb2O3
Antimony(III) oxide.
A 1588LP; Amspec-KR; Amsperse; AN 800; Antimonious oxide; Antimony Bloom 100A; Antimony Bloom 500A; Antimony oxide; Antimony oxide (Sb2O3); Antimony Oxide High Tint, Low Tint, Ultrapure, Very High Tint; Antimony sesquioxide; Antimony trioxide; Antimony White; Antimony(3+) oxide; Antox; AO;A 1582; AP 50; AP 50 (metal oxide); AT 3 (fireproofing agent); AT 3B; Atox B; Atox F; Atox R; Atox S; C.I. 77052; C.I. Pigment White 11; CCRIS 4495; Chemetron fire shield; CI 77052; CI Pigment white 11; Cooksons; Dechlorane A-O; Dechlorane A-O; Diantimony trioxide; EINECS 215-175-0; Exitelite; Fireshield FSPO 405; FireShield H; FireShield LS-FR; Flame Cut 610; Flame Cut 610R; Flameguard VF 59; Flowers of antimony; HM 203P; HSDB 436; LS-FR; LSB 80; MIC 3; Microfine A 05; NCI-C55152; Nyacol A 1510LP; Nyacol A 1530; Octoguard FR 10; Patox C; Patox H; Patox L; Patox M; Patox S; Senarmontite; Stibiox MS; Thermoguard B; Thermoguard L; Thermoguard S; Thermoguard® L; Thermoguard® UF; Timonox; Timonox; Timonox Blue Star; Timonox White Star; Twinkling star; UltraFine® II; Valentinite; Weisspiessglanz; White star; Fireshield® H, HPM, HPM-UF, L; Fyraway; Fyrebloc; KR; Octoguard FR-10; Petcat R-9; Thermoguard® HPM, HPM-UF, L, S, UF; Ultrafine® II. A synergistic agent for fire-retardant plastics. Flame retardant for water-based polymer compounds such as latex adhesives, binders, coatings, foams and for liquid systems, e.g., unsaturated polyesters, epoxies, PU, phenolics, textile treatments; produces minimum loss of physical properties in ABS. Crystals; mp = 655°; bp = 1425°; slightly soluble in H_2O; soluble in KOH, HCl and sulfuric acid, strong alkalis; d = 5.67; LD50 (rat orl) >20 g/kg. *Alfa Aesar; Asarco; Ashland; Fluka; Hoechst AG; Johnson Matthey; Lukens; Miljac; Nihon Kagaku Sangyo; Noah Chem.; Punda Mercantile; Reade Advanced Materials; Royal H. M.; Sigma-Aldrich Fine Chem.*

246   Antimony trisulfide
1345-04-6   G

Sb2S3
Antimony(III) sulfide.
AI3-00983; Antimonous sulfide; Antimony Orange; Antimony sesquisulfide; Antimony sulfide; Antimony sulfide (Sb2S3); Antimony sulphide; Antimony trisulfide; Antimony trisulfide colloid; Antimony vermilion; Antimony(3+) sulfide; Black antimony; C.I. 77060; C.I. Pigment Red 107; CI 77060; CI Pigment Red 107; Crimson antimony; Crimson antimony sulphide; Diantimony trisulfide; EINECS 215-713-4; HSDB 1604; Lymphoscan; Needle antimony; Stibnite. Vermilion or yellow pigment, used to produce antimony salts, in pyrotechnics, matches, percussion caps, camouflage paints and ruby glass. *Atomergic Chemetals; BASF Corp.*

247   Apagallin
2217-44-9   5060   G

C20H8I4Na2O4
Tetraiodophenol-phthalein.
Antinosin; Bilitrast; Cholepulvis; Cholumbrin; EINECS 218-715-3; Foriod; Galisol; Iodeikon; Iodoftaleina sodica; Iodognost;

43

Iodognostum; Iodophene sodium; Iodophtaleine sodique; Iodophthalein sodium; Iodophthaleinum natricum; Iodorayoral; Iodotetragnost; Iodtetragnost; Keraphen; Nosophene sodium; Opacin; Oral-Tetragnost; Phenolphthalein, 3',3",5',5"-tetraiodo-, disodium salt; Photobiline; Piliophen; Radiotetrane; Shadocol; Soluble iodophthalein; Sombrachol; Stipolac; Tetiothalein sodium; Tetraiode; Tetraiodophenolphthalein sodium; Tetraiodophthalein sodium; Tipps; Videopheal. Radiopaque medium, used as a diagnostic aid in cholecystography. Also an antiseptic and acid-base indicator.

248   Arecaidine
499-04-7   782   G

C7H11NO2
1,2,5,6-tetrahydro-1-methyl-3-pyridinecarboxylic acid.
5-22-01-00322 (Beilstein Handbook Reference); Arecaidine; Arecaine; BRN 0112366; Methylguvacine; N-Methylguvacine; Nicotinic acid, 1,2,5,6-tetrahydro-1-methyl-; NSC 76017; 3-Pyridinecarboxylic acid, 1,2,5,6-tetrahydro-1-methyl-; 1,2,5,6-Tetrahydro-1-methylnico-tinic acid; 1,2,5,6-Tetrahydro-1-methyl-3-pyridine-carboxylic acid. Isolated from betel nuts. Plates or tablets, mp = 232° (dec)insoluble in EtOH, Et2O, C6H6, CHCl3, very soluble in H2O.

249   Arecoline
63-75-2   783   G V

C8H13NO2
Methyl 1,2,5,6-tetrahydro-1-methylnicotinate.
5-22-01-00322 (Beilstein Handbook Reference); Arecaidine methyl ester; Arecaline; Arecholine; Arecolin; Arecoline; Arekolin; BRN 0123045; CCRIS 7688; EINECS 200-565-5; Methyl N-methyl-1,2,5,6-tetrahydro-nicotinate; 1-Methyl-$\Delta^{3,4}$-tetrahydro-3-pyridinecarbox-ylate; Methyl N-methyltetrahydronicotinate; Methyl-arecaiden; N-Methyltetrahydronicotinic acid, methyl ester; N-Methyltetrahydropyridine-β-carboxylic acid methyl ester; Nicotinic acid, 1,2,5,6-tetrahydro-1-methyl-, methyl ester; NSC 56321; 3-Pyridinecarboxylic acid, 1,2,5,6-tetrahydro-1-methyl-, methyl ester; 1,2,5,6-Tetrahydro-1-methylnicotinic acid, methyl ester; 1,2,5,6-Tetrahydro-1-methyl-3-pyridinecarboxylic acid methyl ester. Used in veterinary medicine as an anthelmintic effective against Cestodes. Oil; bp = 209°; d^{20} = 1.0495; n_D^{20} = 1.4302; LD50 (mus sc) = 100 mg/kg. Boehringer Ingelheim Vetmedica Inc.

250   Aresin
1746-81-2   G P

C9H11ClN2O2
N'-(4-chlorophenyl)-N-methoxy-N-methylurea.
Aresin; Arezin; Arezine; BRN 2212523; Caswell No. 207D; 3-(4-Chlorophenyl)-1-methoxy-1-methylurea; 3-(4-Chlorophenyl)-1-methoxy-1-methylharnstoff; N-(4-Chloro-phenyl)-N'-methoxy-N-methylurea; EINECS 217-129-5; EPA Pesticide Chemical Code 207500; HOE 02747; HOE 2747; HSDB 5030; Monolinuron; Monorotox; Premalin; Urea, 3-(p-chlorophenyl)-1-methoxy-1-methyl-; Urea, N'-(4-chlorophenyl)-N-methoxy-N-methyl-. Selective systemic herbicide used to control broad-leaved weeds and some annual grasses in vegetable crops such as potatoes and leeks. Crystals; mp = 80-83°; soluble in H2O (735 mg/l), organic solvents; LD50 (rat orl) = 1660 mg/kg. Hoechst UK Ltd.

251   Arova 16
54982-83-1   G

C14H24O4
1,4-Dioxacyclohexadecane-5,16-dione.
4-19-00-01935 (Beilstein Handbook Reference); BRN 0199405; Cyclic ethylene dodecanedioate; 1,4-Dioxacyclohexadecane-5,16-dione; EINECS 259-423-6; Ethylene cyclic dodecanedioate; Ethylenedodecane-dioate; Muskonate. Musk perfume. Oriental; Sandal-wood; Amber; Woody; Musk. Not Found In Nature. Hüls AG.

252   Arsenic
7440-38-2   802   G

As

As
Arsenic.
Arsen; Arsenic; Arsenic-75; Arsenic Black; Arsenic, elemental; Arsenic, inorganic; Arsenicals; CCRIS 55; Colloidal arsenic; EINECS 231-148-6; Fowler's Solution; Gray arsenic; Grey arsenic; HSDB 509; Metallic arsenic; UN1558. Metallic form: alloying additive for metals, especially lead and copper as shot, battery grids, cable sheaths; high-purity semiconductor grade: manufacture of gallium arsenide for dipoles and other electronic devices; doping agent; solders; medicine. mp = 818°; soluble in H2O; d = 4.700. Atomergic Chemetals; Sigma-Aldrich Fine Chem.; Whiting Peter Ltd.

253   Arsenic acid
7778-39-4   803   G P

AsH3O4
Orthoarsenic Acid.
Acide arsenique liquide; Aresenid; Arsenate; Arsenic acid; Arsenic acid (H3AsO4); Caswell No. 056; CCA Type C; Chemonite Part A; Crab grass killer; Desiccant L-10; EINECS 231-901-9; EPA Pesticide Chemical Code 006801; Hi-Yield Dessicant H-10; HSDB 431; Hy-Yield H-10; Orthoarsenic acid; Poly Brand Dessicant; RCRA waste number P010; Scorch; UN1553; UN1554; Zotox; Zotox Crab Grass Killer. Arsenic acid solution; wood preservative and chemical intermediate. Registered by EPA as a herbicide. Solid; mp = 35°; bp = 160°; soluble in H2O (16.7 g/100 ml); LD50 (rbt iv) = 6 mg/kg.

254   Arsenic pentoxide
1303-28-2   808   G P

$$O = As - O - As = O$$

As_4O_{10}
Arsenic oxide.

Anhydride arsenique; Arsenic acid anhydride; Arsenic anhydride; Arsenic oxide; Arsenic oxide (As_2O_5); Arsenic pentaoxide; Arsenic pentoxide; Arsenic(5+) oxide; Arsenic(V) oxide; Caswell No. 057; Diarsenic pentaoxide; Diarsenic pentoxide; EINECS 215-116-9; EPA Pesticide Chemical Code 006802; HSDB 429; RCRA waste number P011; UN1559. Used in insecticides; dyeing and printing; as a weed killer; in colored glass and metal adhesives. Yellow-white solid; mp = 315°; d = 4.3200; soluble in H_2O, alcohol; LD_{50} (rat orl) = 8 mg/kg. *Atomergic Chemetals; Lancaster Synthesis Co.*

255   Arsenic sulfide
1303-34-0   807   G

$$S = As - S - As = S$$

As_2S_5
Arsenic pentasulfide.

Arsenic pentasulfide; Arsenic sulfide; Arsenic sulfide (As_2S_5). Used in paint pigments, light filters and manufacture of other arsenic compounds. Brown solid; Insoluble in H_2O. *Atomergic Chemetals.*

256   Arsenic trichloride
7784-34-1   810   G

$$Cl - As(Cl) - Cl$$

$AsCl_3$
Arsenic(III) chloride.

Arsenic butter; Arsenic chloride; Arsenic chloride ($AsCl_3$); Arsenic trichloride; Arsenic(III) chloride; Arsenic(III) trichloride; Arsenious chloride; Arsenous chloride; Arsenous trichloride; Butter of arsenic; Caustic arsenic chloride; Caustic oil of arsenic; Chlorure arsenieux; Chlorure d'arsenic; EINECS 232-059-5; Fuming liquid arsenic; HSDB 422; Trichloroarsine; Trichlorure d'arsenic; UN1560. Intermediate for organic arsenicals (pharmaceuticals, insecticides), ceramics. Liquid; mp = -9°; bp = 130.21°; d = 2.1497; n_D^{20}=1.6006; reacts with H_2O, soluble in organic solvents. *Atomergic Chemetals; Noah Chem.*

257   Arsenic trifluoride
7784-35-2   811   G

$$F - As(F) - F$$

AsF_3
Arsenic(III) fluoride.

Arsenic fluoride; Arsenic fluoride (AsF_3); Arsenic trifluoride; Arsenous fluoride; Arsenous trifluoride; EINECS 232-060-0; HSDB 421; TL 156; Trifluoroarsine. Used as a fluorinating reagent, catalyst, ion implantation source and dopant. Liquid; mp = -5.95°; bp = 57.8°; d_4^{15} = 2.73; reacts with H_2O, soluble in alcohol, ether, benzene. *Atomergic Chemetals; Elf Atochem N. Am.; Noah Chem.*

258   Arsenic trioxide
1327-53-3   813   G P

As_4O_6
Arsenic(III) oxide.

Acide arsenieux; AI3-01163; Anhydride arsenieux; Arseni trioxydum; Arsenic (III) oxide; Arsenic blanc; Arsenic oxide; Arsenic oxide (3); Arsenic oxide (As_2O_3); Arsenic sesquioxide; Arsenic sesquioxide (As_2O_3); Arsenic trioxide; Arsenic(III) oxide; Arsenicum album; Arsenigen saure; Arsenious acid; Arsenious Acid Anhydride; Arsenious oxide; Arsenious trioxide; Arsenite; Arsenolite; arsenous oxide anhydride; Arsenous acid; Arsenous acid anhydride; Arsenous anhydride; Arsenous oxide; Arsenous oxide anhydride; Arsentrioxide; Arsodent; Caswell No. 059; CCRIS 5455; Claudelite; Claudetite; Crude arsenic; Diarsenic oxide; Diarsenic trioxide; Diarsonic trioxide; EINECS 215-481-4; EPA Pesticide Chemical Code 007001; HSDB 419; Oxyde Arsenieux; Poison flour; RCRA waste number P012; Trisenox; UN 1561; White arsenic. Primary material for all arsenic compounds. Used in chemical manufacturing and also as a parasiticide and rodenticide. Used in pigments, ceramic enamels, aniline colors, decolorizing agent in glass, as an insecticide, rodenticide, herbicide, sheep and cattle dip, hide preservative, wood preservative, in preparation of other arsenic compounds. Crystals; mp = 315°; bp = 465°; soluble in H_2O, dil HCl, alkali hydroxide or carbonate solns; insoluble in EtOH, CHCl3, Et_2O; LD_{50} (rat orl) = 1.46 mg/kg. *Atomergic Chemetals; Noah Chem.; Outokumpu Oy.*

259   Arsine
7784-42-1   818   G

$$H - As(H) - H$$

AsH_3
Arsenic trihydride.

Agent SA; Arsenic hydride; Arsenic hydride (AsH_3); Arsenic trihydride; Arseniuretted hydrogen; Arsenous hydride; Arsenowodor; Arsenwasserstoff; Arsine; EINECS 232-066-3; HSDB 510; Hydrogen arsenide; UN 2188. Used in organic synthesis, as a military poison, doping agent for solid-state electronic components. Gas; mp = -117°; bp = -62.5°; decomposes when heated at 300°; slightly soluble in H_2O. *Air Products & Chemicals Inc.; Atomergic Chemetals.*

260   Asiaticoside
16830-15-2   839   D G

$C_{48}H_{78}O_{19}$

$(2\alpha,3\beta,4\alpha)$-2,3,23-Trihydroxyurs-12-en-28-oic acid O-6-deoxy-α-L-mannopyranosyl-(1→4)-O-β-D-glucopyran-osyl-(1→6)-O-β-D-glucopyranosyl ester.

Asiaticosid; Asiaticoside; Ba 2742; Blastoestimulina; BRN 0078195; Centelase; Dermatologico; EINECS 240-851-7; Emdecassol; Madecassol; NSC 166062. A proprietary preparation extract of *Centella asiatica* in an ointment; used for skin protection. Powder; mp = 230-233°; insoluble in H_2O; soluble in EtOH, C_5H_5N; $[\alpha]_D^{20}$= -14° (EtOH). *Rona Laboratories.*

261   Asparagine
70-47-3   842   G

$C_4H_8N_2O_3$

(S)-2,4-Diamino-4-oxobutanoic acid.

2-Aminosuccinamic acid, L-; -Aminosuccinamic acid; (-)-Asparagine; (S)-Asparagine; Agedoite; Altheine; Asparagine; Asparagine acid; Asparagine, L-; L-Asparagine; L-Asparatamine; L- -Asparagine; Aspar-amide; Aspartamic acid; Aspartic acid -amide; Butanoic acid, 2,4-diamino-4-oxo-, (S)-; Crystal VI; L-2,4-Diamino-4-oxobutanoic acid; (S)-2,4-Diamino-4-oxobutanoic acid; EINECS 200-735-9; NSC 82391. Used in biochemical research for preparation of culture media and in medicine. mp = 234-235°; d_4^{15}= 1.543; $[\alpha]_D^{20}$= -5.42° (c = 1.3); practically insoluble in methanol, ethanol, ether, benzene; soluble in acids and alkalis. *Degussa AG; Penta Mfg.; Sigma-Aldrich Fine Chem.; Spectrum Chem. Manufacturing; Tanabe U.S.A. Inc.; U.S. BioChem*

262   Aspartame
22839-47-0   844   G

$C_{14}H_{18}N_2O_5$

3-Amino-N-(α-methoxycarbonylphenethyl) succinamic acid.

3-Amino-N-(α-methoxycarbonylphenethyl) succinamic acid; 3-Amino-N-(α-carboxyphenethyl)succinamic acid N-methyl ester; APM; Asp-phe-ome; Aspartam; Aspartame; Aspartame, L,L-α-; Aspartamo; Aspartamum; Aspartylphenylalanine methyl ester; L-Aspartyl-L-phenylalanine methyl ester; Canderel; CCRIS 5456; Dipeptide sweetener; EINECS 245-261-3; Equal; HSDB 3915; Methyl aspartylphenylalanate; Methyl L-aspartyl-L-phenylalanine; Methyl L-α-aspartyl-L-phenylalanate; Methyl N-L-α-aspartyl-L-phenylalaninate; 1-Methyl N-L-α-aspartyl-L-phenylalanate; Nutrasweet; L-Phenylalanine, L-α-aspartyl-, 2-methyl ester; L-Phenylalanine, N-L-α-aspartyl-, 1-methyl ester; SC 18862; Succinamic acid, 3-amino-N-(α-carboxyphenethyl)-, N-methyl ester,; Sweet dipeptide; Tri-sweet. A sweetening agent Crystals; mp = 246-247°; $[\alpha]_D^{22}$= -2.3° (1N HCl). *Searle G.D. & Co.*

263   D-Aspartic acid
1783-96-6   845   G

$C_4H_7NO_4$

1-Amino-1,2-carboxyethane.

4-04-00-02998 (Beilstein Handbook Reference); Asp; D-Aspartate; Aspartic acid, D-; Aspartic acid D-form; (-)-Aspartic acid; D-Aspartic acid; (R)-Aspartic acid; BRN 1723529; D; EINECS 217-234-6; NSC 97922. Used in biological and clinical studies, preparation of culture media, organic intermediate, ingredient of aspartame, detergents, fungicides, germicides, metal complexation. L-form is most common, D-form occurs naturally but is less common. Crystals; mp = 270-271°; soluble in H_2O, more soluble in salt solutions, soluble in acid, alkali, insoluble in EtOH; $[\alpha]_D^{20}$= +25° (c = 1.97 6N HCl). *Atomergic Chemetals; Lancaster Synthesis Co.; Penta Mfg.; Tanabe Seiyaku Co. Ltd.; Tanabe U.S.A. Inc.; U.S. BioChem.*

264   Asphalt
8052-42-4   850   H P

Asphalt.

Asphalt; Asphalt (Bitumen)fume as benzene-soluble Aerosol®; Asphalt (cut); Asphalt (cutback); Asphalt (petroleum); Asphalt cements; Asphalt fumes; Asphalt, liquid medium-curing; Asphalt, liquid rapid-curing; Asphalt, liquid slow-curing; Asphaltic bitumen; Asphaltum; Bitumen; Bitumens, asphalt; Bituminous materials, asphalt; Caswell No. 062; Caswell No. 106; EINECS 232-490-9; EPA Pesticide Chemical Code 022002; EPA Pesticide Chemical Code 022001; HSDB 5075; Judean pitch; Mineral pitch; Mineral rubber; NA1999; Petroleum; Petroleum asphalt; Petroleum bitumen; Petroleum pitch; Petroleum refining residues, asphalts; Petroleum roofing tar; Road asphalt; Road tar; Trinidad pitch. Used to make roads and to seal tanks. Registered by EPA as a fungicide (cancelled). Black bituminous pitch; d = 1.00-1.18; insoluble in H_2O, EtOH, acids, alkali, soluble in $CHCl_3$, Et_2O, Me_2CO, CS_2, petroleum ether, C_6H_6.

265   Aspirin
50-78-2   856   D

C9H8O4

Acetylsalicylic acid.

4-10-00-00138 (Beilstein Handbook Reference); A.S.A.; A.S.A. and Codeine Compound; A.S.A. empirin; AC 5230; Acenterine; Acesal; Aceticyl; Acetilsalicilico; Acetilum acidulatum; Acetisal; Acetol; Acetonyl; Acetophen; Acetosal; Acetosalic acid; Acetosalin; Acetylin; Acetylsal; Acetylsalicylate; Acetylsalicylic acid; Acetylsalicylsaeure; Acetylsalicylsäure; Acetylsalycilic acid; Acide acetylsalicylique; Acido acetilsalicilico; Acido O-acetil-benzoico; Acidum acetylsalicylicum; Acimetten; Acisal; Acylpyrin; Adiro; AI3-02956; Anacin; Anacin Maximum Strength; Arthritis Pain Formula Maximum Strength; ASA; Asagran; Asatard; Ascoden-30; Ascriptin; Aspalon; Aspec; Aspergum; Aspirdrops; Aspirin; Aspirina 03; Aspirine; Aspro; Aspro Clear; Asteric; Axotal; Bayer; Bayer Aspirin 8 Hour; Bayer Buffered; Bayer Children's Aspirin; Bayer Enteric 325 mg Regular Strength; Bayer Enteric 500 mg Arthritis Strength; Bayer Enteric 81 mg Adult Low Strength; Bayer Extra Strength Aspirin for Migraine Pain; Bayer Plus; Benaspir; Benzoic acid, 2-(acetyloxy)-; Bi-prin; Bialpirina; BRN 0779271; Bufferin; Caprin; CCRIS 3243; Cemirit; Claradin; Clariprin; Colfarit; Cont-rheuma retard; Coricidin; Crystar; Dasin; DEA No. 9804; Decaten; Delgesic; Dolean pH 8; Duramax; Easprin; ECM; Ecolen; Ecotrin; EINECS 200-064-1; Empirin; Empirin with Codeine; Endydol; Entericin; Enterophen; Enterosarein; Enterosarine; Entrophen; Equagesic; Extren; Globentyl; Globoid; Helicon; HSDB 652; Idragin; Istopirin; Kapsazal; Kyselina 2-acetoxybenzoova; Kyselina acetylsalicylova; Levius; Measurin; Medisyl; Micrainin; Micristin; Neuronika; Norgesic; Novid; NSC 27223; P-A-C Analgesic Tablets; Percodan; Percodan Demi; Persistin; Pharmacin; Phensal; Pirseal; Polopiryna; Pravigard PAC; Premaspin; Rheumin tabletten; Rheumintabletten; Rhodine; Rhonal; Robaxisal; Ronal; S-211; Salacetin; Salcetogen; Saletin; Salicylic acid acetate; Sine-Off Sinus Medicine Tablets-Aspirin Formula; SK-65 Compound; Solfrin; Solprin acid; Solpyron; Soma Compound; SP 189; Spira-Dine; St. Joseph Aspirin for Adults; Supac; Synalgos; Synalgos-DC; Temperal; Triaminicin; Triple-sal; Vanquish; Xaxa; Yasta; ZORprin. Analgesic; antipyretic; anti-inflammatory. Also has platelet aggregation inhibiting, antithrombotic, and antirheumatic properties. mp = 135; d = 1.40; λ_m = 229, 277 nm ($E^{1\%}_{1cm}$ 484, 68 in 0.1N H_2SO_4); pH (25°) = 3.49; somewhat soluble in H_2O (1 g/300 ml at 25°, 1 g/100 ml at 37°); soluble in EtOH (1 g/5 ml), $CHCl_3$ (1 g/17 ml), Et_2O (1 g/10 ml). *Bayer AG; Boots Pharmaceuticals Inc.; Bristol-Myers Squibb Co.; Eli Lilly & Co.; Parke-Davis; Schering-Plough Pharm.; SmithKline Beecham Pharm.; Sterling Health U.S.A.; Sterling Winthrop Inc.; Upjohn Ltd.*

266   **Atenolol**
29122-68-7   863   D V

C14H22N2O3

2-[p-[2-Hydroxy-3-(isopropylamino)propoxy]phenyl]-acetamide.

Aircrit; Alinor; Altol; Anselol; Antipressan; Apo-Atenolol; Atcardil; Atecard; Atehexal; Atenblock; Atendol; Atenet; Ateni; Atenil; Atenol 1A pharma; Atenol acis; Atenol AL; Atenol Atid; Atenol Cophar; Atenol ct; Atenol Fecofar; Atenol Gador; Atenol Genericon; Atenol GNR; Atenol Heumann; Atenol MSD; Atenol NM Pharma; Atenol Nordic; Atenol PB; Atenol Quesada; Atenol Stada; Atenol Tika; Atenol Trom; Atenol von ct; Atenol-Mepha; Atenol-ratiopharm; Atenol-Wolff; Atenolin; Atenolol; Atenololum; Atenomel; Atereal; Aterol; Betablok; Betacard; Betasyn; Betatop Ge; Blocotenol; Blokium; BRN 2739235; Cardaxen; Cardiopress; CCRIS 4196; Corotenol; Cuxanorm; Duraatenolol; Duratenol; EINECS 249-451-7; Evitocor; Farnormin; Felo-Bits; Hipres; HSDB 6526; Hypoten; Ibinolo; ICI 66,082; ICI 66082; Internolol; Jenatenol; Juvental; Loten; Loten; Lotenal; Myocord; Normalol; Normiten; Noten; Oraday;

Ormidol; Panapres; Plenacor; Premorine; Prenolol; Prenormine; Prinorm; Scheinpharm Atenol; Seles beta; Selobloc; Serten; Servitenol; Stermin; Tenidon; Tenobloc; Tenoblock; Tenolol; Tenoprin; Tenoretic; Tenormin; Tenormine; Tensimin; Tredol; Unibloc; Uniloc; Vascoten; Vericordin; Wesipin; Xaten. Antianginal; class II antiarrhythmic agent. Cardioselective β-adrenergic blocking agent. Has antiarrhythmic and antihypertensive properties. Used in preference to propranolol for treatment of bronchospastic disease in small animals. Crystals; mp = 146-148°, 150-152°; λ_m = 225, 275, 283 nm (MeOH); very soluble in MeOH; soluble in AcOH, DMSO; less soluble in Me_2CO, dioxane; insoluble in CH_3CN, EtOAc, $CHCl_3$; LD_{50} (mus orl) = 2000 mg/kg, (mus iv) = 98.7 mg/kg, (rat orl) = 3000 mg/kg, (rat iv) = 59.24 mg/kg. *Apothecon; C.M. Ind.; ICI; Lemmon Co.; Zeneca Pharm.*

267   **Atlas Linuron**
330-55-2   5531   G P

C9H10Cl2N2O2

N'-(3,4-Dichlorophenyl)-N-methoxy-N-methylurea.

Linuron; Methoxydiuron; Arresin; Afalon. A residual urea herbicide for the control of weeds in field crops including potatoes and carrots. Crystals; mp = 93-94°; soluble in H_2O, acetone, alcohol, benzene, toluene, xylene; LD_{50} (rat orl) = 1500 mg/kg. *Agan Chemical Manufacturers; Ashlade Formulations Ltd.; Atlas Interlates Ltd; DuPont UK; Farm Protection Ltd.; Hoechst UK Ltd.; ICI Chem. & Polymers Ltd.; MTM AgroChemicals Ltd.*

268   **Atorvastatin Calcium**
134523-03-8   868   D

0.5Ca^{2+}

C66H68CaF2N4O10

Calcium (βR,δR)-2-(p-fluorophenyl)-β,δ-dihydroxy-5-iso-propyl-3-phenyl-4-(phenylcarbamoyl)pyrrole-1-heptan-oate.

Atorvastatin calcium; CI-981; DRG-0321; Lipitor; PD 134298-38A. Hydroxymethylglutarate co-enzyme A reductase inhibitor. Hypolipidemic agent. Crystals; $[\alpha]_D$ = -7.4° (DMSO c = 1). *Parke-Davis.*

269   **Atrazine**
1912-24-9   874   G H P

C8H14ClN5
6-Chloro-N-ethyl-N'-(1-methylethyl)-1,3,5-triazine-2,4-diamine.
A 361; Aatam; Aatram; Aatram 20G; Aatrex; Aatrex 4L; Aatrex 80W; Aatrex nine-O; Actinite PK; AI3-28244; Akticon; Aktikon; Aktikon PK; Aktinit A; Aktinit PK; Aneldazin; Argezin; Atazinax; Atranex; Atrasine; Atratol; Atratol; Atratol A; Atrazin; Atrazine; Atred; Atrex; Azoprim; BRN 0612020; Candex; Caswell No. 063; CCRIS 1025; Ceasin 50; Cekuzina-T; Chromozin; Crisatrina; Crisazine; Cyazin; EINECS 217-617-8; EPA Pesticide Chemical Code 080803; Farmco atrazine; Fenamin; Fenamine; Fenatrol; G 30027; Gesaprim; Gesaprim 50; Gesaprin; Gesoprim; Griffex; Guardsman; Herbatoxol; HSDB 413; Hungazin; Hungazin PK; Inakor; NSC 163046; Oleogesaprim; Pitezin; Primatol; Primatol A; Primaze; Radazin; Radizin; Shell atrazine herbicide; Strazine; Triazine A 1294; Vectal; Vectal SC; Weedex A; Wonuk; Zeapos; Zeazin; Zeazine; Zeopos; Zeoposs-Triazine. Selective systemic herbicide, absorbed through roots and foliage, inhibits photosynthesis. Used for pre- and post-emergence control of annual grasses and broad-leaved weeds in a variety of crops. Registered by EPA as a herbicide. Colorless crystals; mp = 176°; d = 1.38; soluble in H_2O (0.0028 g/100 ml), DMSO (18.3 g/100 ml), CHCl3 (5.2 g/100 ml), EtOAc (2.8 g/100 ml), MeOH (1.8 g/100 ml), Et2O (1.2 g/100 ml), C5H12 (0.036 g/100 ml); LD50 (rat orl) = 1100, 3080 mg/kg, (mus orl) = 1750 mg/kg, (rbt orl) = 750 mg/kg, (rbt der) = 7500 mg/kg, (rat der) > 3100 mg/kg; LC50 (rat ihl 1 hr.) > 0.71 mg/l air, (bobwhite quail 8 day dietary) = 5760 mg/kg diet, (mallard duck 8 day dietary) = 19650 mg/kg diet, (rainbow trout 96 hr.) = 8.8 mg/l, (bluegill sunfish 96 hr.) = 16 mg/l, (carp 96 hr.) = 76 mg/l, (catfish 96 hr.) = 7.6 mg/l, (perch 96 hr.) = 16 mg/l, (guppy 96 hr.) = 4.3 mg/l; non-toxic to bees. ABM Chemicals Ltd.; Agan Chemical Manufacturers; Agriliance LLC; Albaugh Inc.; Ashlade Formulations Ltd.; Chipman Ltd.; Ciba Geigy Agrochemicals; CIBA plc; Fisons plc, Horticultural Div.; Rhône-Poulenc Environmental Prods. Ltd; Rigby Taylor Ltd.; Syngenta Crop Protection; UAP - Platte Chemical; United Horticultural Supply.

270   **Azadirachtin**
11141-17-6   896   G P

C33H42O13
(2aR-
(2aα,3β,4β(1aR*,2S*,3aS*,6aS*,7S*,7aS*),4aβ,5α,7aS*,8β(E),10β ,10aα,10bβ))-10-(Acetyloxy)octahydro-3,5-dihydr-oxy-4-methyl-8-((2-methyl-1-oxo-2-butenyl)oxy)-4-(3a,6a,7,7a-tetrahydro-6a-hydroxy-7a-methyl-2,7-methanofuro(2,3-b)oxireno(e)oxepin-1a(2H)-yl)-   1H,7H-naphtho(1,8-bc:4,4a-c')difuran-5,10a(8H)-dicarboxylic acid, dimethyl ester.
Azadirachtin; Azadirachtin A; Margosan-O; Azatin. Isolated from the seeds of the neem tree. Used experimentally as an insect control agent. Registered by EPA as an insecticide. Crystalline powder; mp

= 154 - 158°; $[\alpha]_D$ = -53° (c = 0.5 CHCl3); λ_m = 217 nm (ε 9100 MeOH). Agro Logistic Systems Inc.; Biosys; Certis USA LLC; E.I.D. Parry (India) Ltd.; Fortune Biotech Ltd.; PBI/Gordon Corp; PBT International.

271   **Azamethiphos**
35575-96-3   G P

C9H10ClN2O5PS
S-[(6-Chloro-2-oxooxazolo[4,5-b]pyridin-3(2H)-yl)-methyl]   O,O-dimethylphosphorothioate.
AI3-29129; Alfacron; Alfacron 10WP; Alficron; Azamethiphos; BRN 1086470; CGA 18809; 6-Chloro-3-dimethoxyphosphinoylthiomethyl-1,3-oxazolo(4,5-b)-pyridin-2(3H)-one;   S-((6-Chloro-2-oxooxazolo(4,5-b)-pyridin-3(2H)-yl)methyl) O,O-dimethylphosphoro-thioate;   S-((6-Chloro-2-oxooxazolo(4,5-b)-pyridin-3(2H)-yl)methyl) O,O-dimethyl thiophosphate;   S-(6-Chloro-2,3-dihydro-2-oxo-1,3-oxazolo-(4,5-b)pyridin-3-yl)methyl) O,O-di-methyl phosphorothioate; S-6-Chloro-2,3-dihydro-2-oxo-1,3-oxazolo(4,5-b)pyridin-3-ylmethyl O,O-dimethyl phosphorothioate; Ciba-Geigy 18809; Dymos; EINECS 252-626-0; OMS No 1825; Phosphorothioic acid, S-((6-chloro-2-oxooxazolo(4,5-b)-pyridin-3(2H)-yl)methyl) O,O-dimethyl ester; Rubidor;   Snip;   Thiophosphorsaeure-O,O-dimethyl-S-(6-chlor-oxazolo(4,5-b)-pyridin-2(3H)-on-3-yl)methyl-ester; OMS No 1825; S-((6-Chloro-2-oxooxazolo(4,5-b)-pyridin-3(2H)-yl)methyl)   O,O-di-methylphosphoro-thioate; Snip; Thiophosphorsäure-O,O-dimethyl-S-(6-chlor-oxazolo(4,5-b)pyridin-2(3H)-on-3-yl)-methyl-ester.
Insecticide with contact and stomach action; cholinesterase inhibitor, used for control of flies and other insects in animal houses. Used for mosquitoes, tsetse flies and cockroaches. Colorless crystals; mp = 89°; sg^{20} = 1.60; soluble in H_2O (0.11 g/100 ml), CH2Cl2 (8.1 g/100 ml), C6H6 (1.14 g/100 ml), MeOH (0.8 g/100 ml), n-octanol (0.48 g/100 ml); LD50 (rat orl) = 1180 mg/kg, (rat der) > 2150 mg/kg; LC50 (rainbow trout 96 hr.) = 0.2 mg/l, (bluegill sunfish 96 hr.) = 8.0 mg/l, (carp 96 hr.) = 6.0 mg/l; non-toxic to Japanese quail; toxic to bees. Ciba-Geigy Corp.; Novartis.

272   **Azaperone**
1649-18-9   901   D G V

C19H22FN3O
4'-Fluoro-4-[4-(2-pyridinyl)-1-piperazinyl]butyrophenone.
5-23-03-00041 (Beilstein Handbook Reference); Azaperon; Azaperona; Azaperone; Azaperonum; Azeperone; BRN 0565491; CCRIS 1586; EINECS 216-715-8; Eucalmyl; Fluoperidol; NSC 170976; R-1929; Sedaperone vet; Stresnil; Suicalm. Used in veterinary medicine as a sedative and a tranquillizer. Solid; mp = 73-75°. Janssen Pharm. Ltd.

273   **Azelaic acid**
123-99-9   908   D G H

C9H16O4

1,9-Nonanedioic acid.

4-02-00-02055 (Beilstein Handbook Reference); Acide azelaique; Acido azelaico; Acidum azelaicum; AI3-06299; Anchoic acid; Azelaic acid; Azelex; BRN 1101094; EINECS 204-669-1; Emerox 1110; Emerox 1144; Finacea; Heptanedicarboxylic acid; Lepargylic acid; Nonanedioic acid; NSC 19493; ZK 62498. Intermediate in chemical manufacturing of, for example, detergents; also an antiacne agent. Crystals; mp = 106.5°; bp_{100} = 286.5°, bp_{50} = 265°; bp_{15} = 237°; bp_{10} = 225°; slightly soluble in H_2O (2.4g/l), Et_2O, C_6H_6, DMSO, soluble in EtOH; LD_{50} (rat orl) >5 gm/kg. *Henkel/Emery; Schering AG.*

274   Azinphos-ethyl
2642-71-9   G H P

C12H16N3O3PS2

O,O-Diethyl   S-[(4-oxo-1,2,3-benzotriazin-3(4H)-yl)-methyl] phosphorodithioate.

4-26-00-00460 (Beilstein Handbook Reference); AI3-22014; Athylgusathion; Azinfos-ethyl; Azinophos-ethyl; Azinos; Azinphos-äthyl; Azinphos-aethyl; Azinphos ethyl; Azinphos-etile; Azinugec E; BAY 16255; BAY 16259; Bionex; BRN 0297468; Caswell No. 344; Cotnion-ethyl; Crysthion; 3-Diethoxyphos-phinothioylthiomethyl-1,2,3-benzotri-azin-4(3H)-one; 3,4-Dihydro-4-oxo-3-benzotriazinylmethyl O,O-di-ethyl phosphorodithioate; EINECS 220-147-6; ENT 22,014; EPA Pesticide Chemical Code 058002; Ethyl azinphos; Ethyl Gusathion; Ethyl guthion; Ethyl homolog of guthion; Gusathion; Gusathion A; Gusathion A-M; Gusathion H; Gusathion H and K; Gusathion K; Gusathion K forte; Gusation A; Gutex; Guthion (ethyl); HSDB 411; R 1513; Sepizin L; Triazotion. Insecticide with broad spectrum of activity against biting pests such as caterpillars, beetles and their larvae, as well as sucking pests such as aphids, thrips, leafhoppers, etc. on a wide range of crops; side-effect on mites. Colorless needles; mp = 53°; $bp_{0.0013}$ = 111°; sg^{20} = 1.284; soluble in H_2O (0.0004 g/100 ml), freely soluble in polar organic solvents; LD_{50} (rat orl) = 12.5 - 17.5 mg/kg, (mrat der) = 250 mg/kg, (rat ip) > 7.5 mg/kg; LC_{50} (guppy 96 hr.) = 0.01 - 0.10 mg/l, (goldfish 96 hr.) = 0.1 mg/l; toxic to bees. *Bayer Corp.*

275   Azinphos-methyl
86-50-0   915   H P

C10H12N3O3PS2

3-Dimethoxyphosphinothioylthiomethyl-1,2,3-benzo-triazin-4(3H)-one.

4-26-00-00460   (Beilstein Handbook Reference); AI3-23233; Azinfos-methyl; Azinphos-metile; Azin-phosmethyl; BAY 17147; BAY 9027; Bayer 17147; Bayer 9027; Benzotriazinedithiophosphoric acid dimethoxy ester; BRN 0280476; Carfene; Caswell No. 374; CCRIS 63; Cotneon; Cotnion; Cotnion methyl; Crysthion 2L; Crysthyon; S-(3,4-Dihydro-4-oxo- benzo(α)(1,2,3)triazin-3-ylmethyl) O,O-dimethylphos-phorodithioate; S-3,4-Dihydro-4-oxobenzo-(d)(1,2,3)-triazin-3-ylmethyl O,O-dimethylphosphoro-dithioate Dimethyldithiophosphoric acid N-methylbenzazimide ester; O,O-Dimethyl S-(3,4-dihydro-4-keto-1,2,3-benzotriazinyl-3-methyl)-dithiophosphate; O,O-Dimethyl S-(4-oxobenzo-triazino-3-methyl)phosphoro-dithioate; O,O-Dimethyl-S-((4-oxo-3H-1,2,3-benzo-triazin-3-yl)-methyl)-dithiophosphat; O,O-Dimethyl S-(4-oxo-1,2,3-benzotriazino(3)-methyl)thiothionophos-phate; O,O-Dimethyl-S-((4-oxo-3H-1,2,3-benzotriazine-3-yl)-methyl)-dithiofosfaat; O,O-Dimethyl S-4-oxo-1,2,3-benzotriazin-3(4H)-ylmethyl phosphorodithioate; O,O-Dimethyl-S-(4-oxobenzotriazin-3-methyl)-dithio-phosphat; O,O-Dimethyl-S-(benzaziminomethyl) dithiophosphate; O,O-Dimethyl S-(4-oxo-3H-1,2,3-benzotriazine-3-methyl)phosphorodithioate; O,O-Dimethyl S-(4-oxo-1,2,3-benzotriazino(3)-methyl) thio-thionophosphate; O,O-Dimethyl-S-(1,2,3-benzo-triazinyl4-keto)methyl phosphorodithioate; O,O-Dimetil-S-((4-oxo-3H-1,2,3-benzotriazin-3-il)-metil)-ditiofosfato; EINECS 201-676-1; ENT 23,233; EPA Pesticide Chemical Code 058001; EPA Shaughnessy #058001; Gusathion; Gusathion-20; Gusathion 25; Gusathion K; Gusathion M; Gusathion methyl; Guthion; HSDB 1171; 3-(Mercaptomethyl)-1,2,3-benzotriazin-4(3H)-one O,O-dimethylphosphoro-dithioate S-ester; N-Methylbenzazimide, dimethyl-dithiophosphoric acid ester; Methyl guthion; Methylazinphos; Methylgusathion; Metiltriazotion; NCI-C00066; Phosphorodithioic acid, O,O-dimethylS-((4-oxo-1,2,3-benzotriazin-3(4H)-yl)methyl) ester; R 1582. Non-systemic insecticide and acaricide acting on contact or by ingestion. A cholinesterase inhibitor, used for control of chewing and sucking insects and spider mites in fruit and vegetable crops. Registered by EPA as an insecticide. Colorless crystals; mp = 73 - 74°; sg^{20} = 1.44; soluble in H_2O (0.003 g/100 ml at 25°, C_7H_8 (> 20 g/100 ml), CH_2Cl_2 (> 20 g/100 ml); LD_{50} (rat orl) = 4 - 20 mg/kg, (mgpg orl) = 80 mg/kg, (rat der) = 220 mg/kg, (mallard duck orl) = 136 mg/kg; LC_{50} (rat ihl 1 hr.) = 0.385 mg/l air, (bobwhite quail 5 day dietary) = 540 mg/kg diet, (rainbow trout 96 hr.) = 0.02 mg/l, (bluegill sunfish 96 hr.) = 0.0046 mg/l, (guppy 96 hr.) = 1.0 mg/l; toxic to bees. *Bayer Corp.; Gowan Co.; Makhteshim Chemical Works Ltd.; Micro-Flo Co. LLC; UAP - Platte Chemical.*

276   Aziprotryne
4658-28-0   G P

C7H11N7S

4-Azido-N-(1-methylethyl)-6-methylthio-1,3,5-triazin-2-amine.

Azeprotryne; 2-Azido-4-isopropylamino-6-methylthio-s-triazine; 2-Azido-4-isopropylamino-6-methylthio-1,3,5-triazine; 4-Azido-N-(1-methylethyl)-6-(methyl-thio)-1,3,5-triazin-2-amine; 4-Azido-N-isopropyl-6-methylthio-1,3,5-triazin-2-ylamine; Aziprotryn; Bras-oran; Brasoran 50 WP; Brassoron; C 7019; Caswell No. 063B; Ciba C 7019; EINECS 225-101-9; EPA Pesticide Chemical Code 263301; Isopropylamino-4-azido-6-methylthio-1,3,5-triazin; Mesoranil; Mezuron; 1,3,5-Triazin-2-amine, 4-azido-N-(1-methylethyl)-6-(methylthio)-; s-Triazine, 2-azido-4-(isopropylamino)-6-(methylthio)-. A selective herbicide applied pre- or post-emergence and used to control a wide range of annual broad-leaved weeds and some grasses in brassicas. Colorless crystals; mp = 94 - 95°; sg^{20} = 1.40; soluble in H_2O (0.0055 g/100 ml at 25°, Me_2CO (2.1 g/100 ml), CH_2Cl_2 (4.9 g/100 ml), EtOAc (1.1 g/100 ml), i-PrOH (1.1 g1/100 ml), C_6H_6 (0.35 g/100 ml); LD_{50} (rat orl) = 3600 - 5833 mg/kg, (rat der) > 3000 mg/kg; LC_{50} (mallard duck, quail 8 day dietary) > 4000 mg/kg, (bluegill sunfish, largemouth bass 96 hr.) > 1 mg/l; non-toxic to bees. *Ciba Geigy Agrochemicals; Ciba-Geigy Corp.*

277   Azithromycin (anhydrous)
83905-01-5   917   D

$C_{38}H_{72}N_2O_{12}$
[2R-(2R*,3S*,4R*,5R*,8R*,10R*,11R*,12S*,13S*,14R*)]-13-[(2,6-Dideoxy-3-C-methyl-3-O-methyl-α-L-ribohexo-pyranosyl)oxy]-2-ethyl-3,4,10-trihydroxy-3,5,6,8,10,   12-heptamethyl-11-[[3,4,6-trideoxy-3-(dimethyl-amino)-β-D-xylo-hexopyranosyl]oxy]-1-oxa-6-aza-cyclopentadecan-15-one.
Aritromicina; Azenil; Azithromycin; Azithromycine; Azithromycinum; Azitromax; Aziwok; Aztrin; BRN 5387583; CCRIS 1961; DRG-0104; Hemomycin; Misultina; Mixoterin; Setron; Sumamed; Tobil; Tromix; Zeto; Zifin; Zithrax; Zitrim; Zitromax; Zithromax; Zitrotek. Macrolide antibiotic. Crystals; mp = 113-115°; $[\alpha]_D^{20}$ = -37° (c = 1 in CHCl3). *Pfizer Inc.*

278   Azo-bis(cyclohexanecarbonitrile)
2094-98-6   G

$C_{14}H_{20}N_4$
1,1'-Azobis(cyclohexane-1-carbonitrile).
1,1'-Azobis(1-cyclohexanecarbonitrile);   Cyclohexane-carbonitrile, 1,1'-azobis-; EINECS 218-254-8; V-40. Used as a polymerization initiator. Solid; mp = 115-118°. *Wako Pure Chem. Ind.*

279   Azobis(isobutyronitrile)
78-67-1   920   G H

$C_8H_{12}N_4$
2,2'-Dicyano-2,2'-azopropane.
Aceto AZIB; Al3-28716; AIBN; Aivn; Azdh; Azobis(isobutyronitrile); Azobisisobutyrolonitrile; CCRIS 4287; Chkhz 57; EINECS 201-132-3; Genitron; HSDB 5220; NSC 1496; Pianofor an; Poly-Zole AZDN; Porofor 57; Porofor N; Porophor N; Propanenitrile, 2,2'-azobis(2-methyl-; Vazo; VAZO 64. Polymerization initiator for a wide range of monomers. A blowing agent for elastomers and plastic. It is an initiator for free radical reactions. Used as a foaming agent and an inhibitor in plastic materials. A vinyl polymerization catalyst; gives freedom from side reactions and is not readily poisoned. Used as a blowing agent and as an initiator for free radical reactions. Solid; mp = 102-103° (dec); λ_m = 345 nm (EtOH); insoluble in H_2O, soluble in EtOH, MeOH; LD50 (mus orl) = 700 mg/kg. *Schering AG; Wako Pure*

Chem. Ind.

280   2,2'-Azobis(2-methylbutyronitrile)
13472-08-7   H

$C_{10}H_{16}N_4$
Butyronitrile, 2,2'-azobis(2-methyl-.
2,2'-Azobis(2-methylbutyronitrile);   Butanenitrile,   2,2'-azobis(2-methyl-; EINECS 236-740-8.

281   Azocyclotin
41083-11-8
                                                                           G P

$C_{20}H_{35}N_3Sn$
1-(Tricyclohexylstannyl)-1H-1,2,4-triazole.
Azocyclotin; BAY bue 1452; BRN 0621636; EINECS 255-209-1; HSDB 6559; Peropal; Stannane, (1H-1,2,4-triazol-1-yl)tricyclohexyl-; (1H-1,2,4-Triazolyl)tricyclo-hexyl-stannane; 1-(Tricyclohexylstannyl)-1H-1,2,4-tri-azole;   Tri-(cyclohexyl)-1H-1,2,4-triazol-1-yltin;   Tri-(cyklohexyl)-1H-1,2,4-triazol-1-ylstannium. Long-lasting acaricide with contact action, used for control of mobile stages of spider mites on pome and stone fruit, grapes, citrus and vegetables. Colorless crystals; mp = 218°; insoluble in H_2O (< 0.0001 g/100 ml), soluble in CH_2Cl_2 (2 - 5 g/100 ml), i-PrOH (1 - 2 g/100 ml) and other organic solvents; LD50 (mrat orl) = 99 mg/kg, (mus orl) = 410 - 450 mg/kg, (gpg orl) = 261 mg/kg, (mrat der) > 1000 mg/kg, (ckn orl) = 250 - 375 mg/kg; LC50 (carp 96 hr.) = 0.05 - 0.1 mg/l, (goldfish 96 hr.) = 0.01 - 0.1 mg/l, (rainbow trout 96 hr.) = 0.005 - 0.01 mg/l; non-toxic to bees. *Bayer Corp.*

282   Azodicarbonamide
123-77-3   921   G H

$C_2H_4N_4O_2$
1,1'-Azodiformamide.
ABFA;   Al3-52516;   Azobiscarbonamide;   Azodicarb-amide; Azodicarbonamide; Azodicarboxamide; Azo-dicarboxylic acid diamide; Azodiformamide; CCRIS 842; Celogen AZ; Celogen AZ 130; Celogen AZ 199; Celosen AZ; ChKhz 21; ChKhz 21R; ChKhZ 21r; Diazenedicarboxamide; EINECS 204-650-8; Ficel EP-A; Formamide, 1,1'-azobis-; Genitron AC; Genitron AC 2; Genitron AC 4; Genitron EPC; HSDB 1097; Kempore®; Kempore® 125; Kempore® 60/40; Kempore® 60/14FF; Kempore® R 125; Lucel ADA; NCI-C55981; Nitropore; NSC 41038; Pinhole ACR 3; Pinhole AK 2; Poramid K 1; Porofor 505; Porofor adc/R; Porofor Chkhz 21; Porofor Chkhz 21R; Porofor ChKhZ 21r; Porofor DhKhZ 21; Porofor-lk 1074 (Bayer); Santechem 21-21; UN3242; Unifoam AZ; Unifoam

50

AZH 25; Uniform AZ; Yunihomu AZ; 7. Azodicarbonamide based; blowing agent for dynamic foaming processes, e.g., injection molding, extrusion, and calendering. Blowing agent for polyolefins, extruded and injection molded structural foam; eliminates die sink marks in injection molding. Bleaching agent in cereal flour. Solid; mp = 212° (dec);λ_m = 239 nm (ε = 1380, MeOH); soluble in hot H_2O, EtOH. *Elf Atochem N. Am.; Gist-Brocades Intl.; Olin Corp; Schering AG; Uniroyal.*

283   Azo-tert-butane
927-83-3   G

C8H18N2
2,2'-Azobis(2-methylpropane).
2,2'-Azo-2,2'-dimethylpropane; 2,2'-Azoisobutane; Azo-tert-butane; Azoethane, 1,1,1',1'-tetramethyl-; Bis(tert-butyl)diimine; 1,2-Bis(1,1-dimethylethyl)di-azene; Di-tert-butyldiazene; Diazene, bis(1,1-dimethylethyl)-; VR-160. Polymerization initiator. Liquid; bp80 = 47-48°. *Wako Pure Chem. Ind.*

284   Baker's salt
506-87-6   G

2NH3

CH8N2O3
Ammonium carbonate.
AI3-25347; Ammonia sesquicarbonate; Ammonium carbonate; Ammoniumcarbonat; Carbonate d'ammoni-aque; Carbonic acid, diammonium salt; Caswell No. 042; CCRIS 7328; Diammonium carbonate; EINECS 208-058-0; EPA Pesticide Chemical Code 073501; HSDB 6305; Salt of Hartshorn. Also typically contains ammonium carbamate. Crystals; mp = 58° (dec); vaporizes at about 60°; incompatible with acids and acid salts; soluble in H_2O (25 g/100 ml).

285   Barium
7440-39-3   967   G

Ba

Ba
Barium.
Bario; Barium; Baryum; EINECS 231-149-1; HSDB 4481; UN1400. Alkaline-earth element; Alloys in vacuum tubes, deoxidizer for copper, lubricant for anode rotors in x-ray tubes, spark-plug alloys. mp = 725°; bp = 1640°; d = 3.51; reacts with H_2O, producing H_2. *Atomergic Chemetals; Degussa AG; Noah Chem.; Sigma-Aldrich Fine Chem.*

286   Barium acetate
543-80-6   968   G

Ba++

C4H6BaO4
Barium(II) acetate.
Acetic acid, barium salt; Barium acetate; Barium di(acetate); Barium diacetate; Caswell No. 068A; CCRIS 7240; EINECS 208-849-0; NSC 75794; Octan barnaty. Chemical reagent, acetates, textile mordant, catalyst manufacture, paint and varnish driers. White solid; d^{25} = 2.4680; soluble in H_2O; LD50 (rat orl) = 921 mg/kg. *Hoechst Celanese; Mallinckrodt Inc.; Sigma-Aldrich Fine Chem.*

287   Barium bis(dinonylnaphthalenesulphonate)
25619-56-1   H

Ba2+

C56H86BaO6S2
Naphthalenesulfonic acid, dinonyl-, barium salt.
Barium bis(dinonylnaphthalenesulphonate); Barium di-nonylnaphthalenesulfonate; Dinonylnaphthalene sulfonic acid barium salt; EINECS 247-132-7; Naphtha-lenesulfonic acid, dinonyl-, barium salt; NSC 49580

288   Barium bromide
10553-31-8   971   G

BaBr2
Barium dibromide.
Barium bromide; Barium bromide (BaBr2); EINECS 234-140-0. Used in the manufacture of bromides, photographic compounds, phosphors. [dihydrate]; crystals; mp = 850°; very soluble in H_2O, soluble in MeOH, insoluble in Me2CO, EtOAc. *Atomergic Chemetals.*

289   Barium carbonate
513-77-9   972   G

Ba2+

CBaO3
Carbonic acid, barium salt (1:1).
Barium carbonate; Barium carbonate (1:1); Barium monocarbonate; BF 1 (salt); BW-C3; BW-P; C.I. 77099; C.I. Pigment White 10; Carbonic acid, barium salt; Carbonic acid, barium salt (1:1); Caswell No. 069; CI 77099; CI Pigment White 10; Durex White; EINECS 208-167-3; EPA Pesticide Chemical Code 007501; HSDB 950; NSC 83508; Pigment White 10. Used in the treatment of brines in chlorine-alkali cells to remove sulfates, as a rodenticide, in production of barium salts, ceramic flux, optical glass, case-hardening baths, ferrites, in radiation-resistant glass for color television tubes. White powder; mp = 811°; bp = 1450°; d = 4.430;

slightly soluble in H$_2$O (24 mg/l), LD$_{50}$ (rat orl) = 418 mg/kg. *Mallinckrodt Inc.; Solvay Deutschland GmbH.*

290   Barium chloride
10361-37-2   974   G

Cl—Ba—Cl

BaCl$_2$.2H$_2$O
Barium dichloride.
  Ba 0108E; Barium chloride; Barium chloride (BaCl$_2$); Barium dichloride; CCRIS 2286; EINECS 233-788-1; HSDB 2633; NCI-C61074; NSC 146181; SBA 0108E. Production of artificial barium sulfate, other barium salts; reagents; lubrication oil additives; boiler compounds; textile dyeing; pigments; manufacture of white leather. White solid; mp = 963°; d = 3.86; very soluble in H$_2$O, insoluble in organic solvents; LD$_{50}$ (rat orl) = 118 mg/kg. *EM Ind. Inc.; Hoechst Celanese; Mallinckrodt Inc.; Sachtleben Chemie GmbH; Sigma-Aldrich Fine Chem.; Solvay Deutschland GmbH.*

291   Barium chromate
10294-40-3   975   G

O=Cr—O$^-$ Ba^{2+}

BaCrO$_4$
Barium chromate(VI).
  Barium chromate; Barium chromate (1:1); Barium chromate(VI); Barium chromium oxide (BaCrO$_4$); Baryta Yellow; C.I. 77103; C.I. Pigment Yellow 31; CCRIS 7568; Chromic acid (H$_2$CrO$_4$), barium salt (1:1); Chromic acid, barium salt (1:1); Cl 77103; Cl Pigment Yellow 31; EINECS 233-660-5; HSDB 6190; Lemon chrome; Lemon Yellow; Permanent Yellow; Pigment Yellow 31; Ultramarine Yellow. Usd in safety matches, as a corrosion inhibitor in metal-joining compounds, pigment for paints, ceramics, fuses, pyrotechnics, metal primers, ignition control devices. Pigment in anti-corrosion pastes. Yellow monoclinic crystals; d = 4.50; insoluble in H$_2$O. *Atomergic Chemetals; BASF Corp.; Noah Chem.*

292   Barium fluorosilicate
17125-80-3   979   G

F—Si^{2-}—F (with F groups) Ba^{2+}

BaF$_6$Si
Barium hexafluorosilicate.
  Barium fluorosilicate; Barium fluosilicate; Barium hexafluorosilicate; Barium hexafluorosilicate(2-); Barium silicofluoride; Barium silicon fluoride; Barium-silicofluorid; Caswell No. 070; EINECS 241-189-1; EPA Pesticide Chemical Code 075302; Flosol; Hexafluorosilicate(2-) barium (1:1); Silicate(2-), hexafluoro-, barium (1:1); Silicon fluoride barium salt. Colloidal barium silicofluoride, used as a horticultural pesticide. Needles; d$_4^{21}$= 4.29; soluble in H$_2$O (0.0235 g/100 ml).

293   Barium lithol red
1103-38-4   H

C$_{40}$H$_{26}$BaN$_4$Na$_2$O$_8$S$_2$
  Barium bis(2-((2-hydroxynaphthyl)azo)naphthalene-sulfonate).
  1883 red; Barium lithol; Barium lithol red; Calcotone Red B; CCRIS 1355; Cl Pigment Red 49, barium salt (2:1); Cl Pigment Red 49:1; D & C Red No. 12; Dainichi Lithol Red R; EINECS 214-160-6; Eljon Lithol Red MS; HSDB 5523; Irgalite Red BRL; Isol Red 3BK; Isol Red Toner RB; Isol Red Toner GB; Isol Tobias Red 3BK; Isol Tobias Red GB; Isol Tobias Red RB; Light Red RB; Light Red RCN; Lithol Red 18959; Lithol Red 22060; Lithol Red 27965; Lithol Red Barium Toner; Pigment Red 49:1; Poster Red; Red No. 207; Red Toner YTA; Sanyo Fast Red NN; Sanyo Lacquer Red RN; Sanyo Lithol Red R; Symuler Red 2R BA Salt; Vulcanosine Red RBKX.

294   Barium mercuric iodide
10048-99-4   985   G

4I$^-$ Ba^{2+} Hg^{2+}

BaHgI$_4$
Barium tetraiodomercurate.
  Barium mercuric iodide; Barium tetraiodomercurate; EINECS 233-160-7; Rohrbach's solution. A solution of barium and mercuric iodides (100 g barium iodide and 130 g mercuric iodide heated with 20 cc water to 150-200 C.). The solution is allowed to cool, when a double salt is deposited. The liquid is decanted. Used for separating minerals of different densities; also for microchemical detection of alkaloids. Very soluble in H$_2$O.

295   Barium nitrate
10022-31-8   986   G

(nitrate structures) Ba^{2+}

BaN$_2$O$_6$
Barium(II) nitrate.
  Barium dinitrate; Barium nitrate; Barium nitrate (Ba(NO$_3$)$_2$); Barium(II) nitrate (1:2); CCRIS 4140; Dusicnan barnaty; EINECS 233-020-5; HSDB 401; Nitrate de baryum; Nitrato barico; Nitric acid, barium salt; Nitrobarite; UN1446. Used in pyrotechnics, incendiaries, chemicals (barium peroxide), ceramic glazes, rodenticide, electronics. Crystals; mp = 592°; d = 3.23; freely soluble in H$_2$O; LD$_{50}$ (rat orl) = 355 mg/kg. *Berk Pharm. Ltd.; Hoechst Celanese; Noah Chem.; San Yuan Chem. Co. Ltd.; Sigma-Aldrich Fine Chem.*

296   Barium nitrite
13465-94-6   987   G

BaN2O4.H2O
Nitrous acid barium salt monohydrate.
Barium nitrite; EINECS 236-709-9; Nitrous acid, barium salt; Stickstoffoxydbaryt. Used as a chemical intermediate in diazotization reactions. Also used to prevent corrosion of steel and in explosives. [monohydrate]; crystals; d = 3.187; soluble in H2O, insoluble in organic solvents.

297   Barium nonylphenate
28987-17-9   H

C30H46BaO2
Phenol, nonyl-, barium salt.
Barium bis(nonylphenolate); Barium nonylphenate; EINECS 249-359-7; Phenol, nonyl-, barium salt. Mixture of o, m and p-isomers

298   Barium oxide
1304-28-5   989   G

Ba═O

BaO
Barium monoxide.
Barium monoxide; Barium oxide; Barium oxide (BaO); Barium protoxide; Baryta; Calcined baryta; EINECS 215-127-9; Oxyde de baryum; UN1884. Dehydrating agent used as a drying agent for solvents, detergent for lubricating oils. White solid; mp = 1920°; bp = 2000°; d = 5.98; soluble in H2O and dilute acids. *Atomergic Chemetals; Hoechst Celanese; Lancaster Synthesis Co.; Noah Chem.*

299   Barium perchlorate
13465-95-7   990   G

BaCl2O8
Perchloric acid, barium salt.
Barium diperchlorate; Barium perchlorate; Desicchlora; EINECS 236-710-4; Perchloric acid, barium salt; UN1447. A drying agent which can be used in place of calcium chloride, sulfuric acid or potassium hydroxide; absorbs 20% of its weight of water. [trihydrate]; crystals; mp = 505°; d = 3.200; soluble in H2O.

300   Barium peroxide
1304-29-6   992   G

BaO2
Barium binoxide.
Bario (perossido di); Barium binoxide; Barium dioxide; Barium oxide, per-; Barium peroxide; Barium peroxide (Ba(O2)); Barium superoxide; Bariumperoxid; Barium-peroxyde; Dioxyde de baryum; EINECS 215-128-4; HSDB 396; Peroxyde de baryum; UN1449. Used for bleaching, decolorizing glass, thermal welding of aluminum, manufacture of hydrogen peroxide, oxidizing agent, dyeing textiles. Grey powder; insoluble in H2O. *Hoechst Celanese; Lancaster Synthesis Co.*

301   Barium stearate
6865-35-6   G H

C36H70BaO4
n-Octadecanoic acid, barium salt.
Barium distearate; EINECS 229-966-3; Stavinor 40; Stearic acid, barium salt; Synpro® Barium Stearate. Waterproofing agent, lubricant in metalworking, plastics, and rubber; wax compounding; preparation of greases; heat and light stabilizer in plastics. Used as a processing aid, lubricant for PVC, polyolefins, PS, ABS. *Adeka Fine Chem.; CK Witco Corp.; Syn Prods.*

302   Barium sulfate
7727-43-7   997   G

BaO4S
Sulfuric acid barium salt.
Actybaryte; Al3-03611; Albaryt; Artificial barite; Artificial heavy spar; BA147; Bakontal; Bar-Test; Baraflave; Baricon; Baridol; Barıı sulphas; Barite; Baritogen deluxe; Baritop; Baritop 100; Baritop G Powder; Baritop P; Barium 100; Barium Andreu; Barium sulfate; Barium sulfate (1:1); Barium sulfate (BaSO4); Barium sulfuricum; Barium sulphate; Barium sulphate, natural; Barobag; Barocat; Barodense; Baroloid; Barosperse; Barosperse 110; Barosperse II; Barotrast; Baryta White; Barytes; Barytes 22; Barytgen; Baryum (sulfate de); Baryx Colloidal; Baryxine; Basofor; Bayrites; BF 1 (salt); BF 10 (sulfate); Blanc fixe; C.I. 77120; C.I. Pigment White 21; Caswell No. 071B; Citobaryum; Colonatrast; Danobaryt; E-Z Preparations; E-Z-AC; E-Z-Cat Concentrate; E-Z-HD; E-Z-Paque; E-Z-Paste esophageal cream; EINECS 231-784-4; Enamel White; EneCat; EneMark; EneSet; EntroBar; EPA Pesticide Chemical Code 007502; Epi-C; Epi-Stat 57; Epi-Stat 61; Esopho-CAT; Esophotrast; Esophotrast esophageal cream; Eweiss; Finemeal; Fixed white; Gastropaque-S; Gel-Unix; HD 85; HD 200 Plus; HiTone; HSDB 5041; Intropaque; Lactobaryt; Liquibarine; Liquid Barosperse; Liquid E-Z-Paque; Liquid Polibar; Liquid Polibar Plus; Liquid Sol-O-Pake; Liquipake; Macropaque; Microbar; Microfanox; Micropaque; Micropaque RD; Microtrast; Mikabarium B; Mikabarium F; Mixobar;

Mixture III; Neobalgin; Neobar; Novopaque; Oesobar; Oratrast; Permanent White; Pigment White 22; Polibar; Precipitated barium sulphate; PrepCat; Radimix Colon; Radio-Baryx; Radiobaryt; Radiopaque; Raybar; Readi-CAT; Readi-CAT 2; Recto Barium; Redi-Flow; Rugar; Sachtoperse® HU; Sol-O-Pake; Solbar; Sparkle Granules; SS 50; Sulfuric acid, barium salt (1:1); Supramike; Suspobar; Terra ponderosa; Tixobar; TomoCat 1000 Concentrate; TomoCat Concentrate; TonoJug 2000; Tonopaque; Topcontral; Travad; Ultra-R; Umbrasol A; Unibaryt; Unit-Pak; Veri-O-Pake; X-Opac; Xylocaine Viscous. Used as an opaque medium in x-ray diagnosis. Weighting mud in oil drilling, paper coating, paints; filler and delustrant for textiles, rubber, plastics, and lithographic inks; radiation shield. Transparent, conductive pigment. Extender pigment. Filler for paints and coatings. White solid; mp = 1580°; d = 4.500; Insoluble in H_2O. *Cyprus Industrial Min.; Foseco (FS) Ltd; Huber J. M.; Mallinckrodt Inc.; Sachtleben Chemie GmbH.*

303   Barium titanate
12047-27-7   1003   G

BaO3Ti
Barium titanate(IV).
Barium metatitanate; Barium titanate(IV); Barium titanium oxide (BaTiO3); Barium titanium oxide; Barium titanium trioxide; BT 01; BT 02; BT 05; BT 05 (filler); BT 8; BT 10 (titanate); BT 201; BT 204; BT 206; BT 303; BT 100M; BT 100P; BT-HD 9DX; BT-HP 8KB2; EINECS 234-975-0; HBT 3; HPBT 1; Kyorix BT-HD 9DX; Kyorix BT-S; N 220 (titanate); N 5500; Nova Chem AB; Tamtron X 7R262L; Tamtron X 7R302H; TB 1; Ticon 5016; Titanate (TiO3^{2-}) barium (1:1); Titanate barium (1:1); Transelco 219-3; VK 4; VK 4 (oxide); YV 100AN. Ferroelectric ceramics used in storage devices, dielectric amplifiers, digital calculators. Crystals (5 forms); mp = 1625°; d = 6.0800. *Atomergic Chemetals; Ferro/Transelco; Noah Chem.; Sigma-Aldrich Fine Chem.*

304   Bärostab® L 230
557-09-5
                                                              G

Zn^{2+}

C16H30O4Zn
Zinc octoate.
EINECS 209-156-6; NSC 63869; Octanoic acid, zinc salt; Octanoic acid, zinc salt (2:1); Zinc caprylate; Zinc dioctanoate; Zinc dioctylate; Zinc octanoate; Zinc octoate; Zinc octylate; Zinc(2+) octanoate; Zinc(II) octoate. Fast-kicking stabilizer for foam processing in production of sealings for caps; good organoleptic props., excellent color; approved for contact with foodstuffs. Very slightly soluble in H_2O.

305   Behenamide
3061-75-4
                                                              G

C22H45NO
n-Docosanamide.
Behenamide; Behenic acid amide; Docosanamide; EINECS 221-

304-1; Uniwax 1747. Lubricant, mold release agent for EPDM and PVC processing. Solid; mp = 111-112°. *Unichema.*

306   Behenic acid
112-85-6   1024   H

C22H44O2
n-Docosanoic acid.
AI3-52709; Behenic acid; Docosanoic acid; 1-Docosanoic acid; n-Docosanoic acid; Docosoic acid; EINECS 204-010-8; Glycon B-70; HSDB 5578; Hydrofol 2022-55; Hydrofol Acid 560; NSC 32364. Needles; mp = 81°; bp60 = 306°; d^{90} = 0.8223; slightly soluble in H_2O, EtOH, Et2O.

307   Benazepril
86541-75-5   1029   D

C24H28N2O5
(3S)-3-[[(1S)-1-Carboxy-3-phenylpropyl]amino]-2,3,4,5-tetrahydro-2-oxo-1H-1-benzazepine-1-acetic acid 3-ethyl ester.
Benazepril; Benazeprilum; CGS-14824A; Briem; Cibacen; Cibacène; Lotensin. Angiotensin-converting enzyme inhibitor. Used to treat hypertension. Orally active peptidyldipeptide hydrolase inhibitor. Crystals; mp= 148-149°; [α]D = -159° (c = 1.2 EtOH). *Ciba-Geigy Corp.*

308   Benazepril Hydrochloride
86541-74-4   1029   D

H—Cl

C24H29ClN2O5
(3S)-3-[[(1S)-1-Carboxy-3-phenylpropyl]amino]-2,3,4,5-tetrahydro-2-oxo-1H-1-benzazepine-1-acetic acid 3-ethyl ester hydrochloride.
Benazepril hydrochloride; Briem; CGS 14824A HCl; Cibace; Cibacen; Cibacen CHF; Cibacene; HSDB 7081; Labopol; Lotensin; Tensanil; Zinadril.; component of: Lotensin-HCT, Lotrel, Lotrel capsules. Angiotensin-converting enzyme inhibitor. Used to treat hypertension. Orally active peptidyldipeptide hydrolase inhibitor. Crystals; mp = 188-190°; [α]D = -141° (c = 0.9 EtOH). *Ciba-Geigy*

309 Benazolin

3813-05-6   G P

C9H6ClNO3S

4-Chloro-2-oxo-3-(2H)-benzothiazoleacetic acid.

Acetic acid, (4-chloro-2-oxobenzothiazolin-3-yl)-; Asset; BA 7688; Ben-cornox; BEN-30; Benasalox; Benazalox; Benazolin; Benazoline; Benopan; Bensecal; Benzar; 3-Benzothiazolineacetic acid, 4-chloro-2-oxo-; 3(2H)-Benzothiazoleacetic acid, 4-chloro-2-oxo-; BRN 0749788; Chamilox; 4-Chloro-2-oxo-3(2H)-benzothia-zoleacetic acid; 4-Chloro-2-oxobenzothiazolin-3-ylacetic acid; 4-Chloro-2,3-dihydro-2-oxo-1,3-benzothiazol-3-ylacetic acid; Cornox CWK; Cresopur; EINECS 223-297-0; Eunasin; EX 10781; Galipan; Galtak; Grassland weedkiller; Herbazolin; Herbitox; Keropur; Legumex Extra; Ley-Cornox; Leymin; Metizolin; NSC 521058; RD 7693; Springclene; Tillox; Tri-cornox; Tri-cornox special; 3(2H)-Benzothiazoleacetic acid, 4-chloro-2-oxo-; BRN 0749788; Chamilox; 4-Chloro-2-oxobenzothiazolin-3-ylacetic acid; 4-Chloro-2-oxo-3(2H)-benzothiazole-acetic acid; 4-Chloro-2,3-dihydro-2-oxo-1,3-benzo-thiazol-3-ylacetic acid; Cornox CWK; Cresopur; Erbitox; Eunasin; EX 10781; Galipan; Galtak; Grassland weedkiller; Herbazolin; Keropur; Legumex Extra; Ley-Cornox; Leymin; Metizolin; NSC 521058; RD 7693; Springclene; Tillox; Tri-cornox; Tri-cornox special. Selective, systemic growth regulator herbicide Used for control of broad leaved weeds such as black bindweed, chickweed, cleavers and charlock. Colorless crystals; mp = 193°; soluble in H2O (0.06 g/100 ml), Me2CO 13.2 g/100 ml), EtOH (11.1 g/100 ml), CS2 (0.05 g/100 ml); LD50 (rat orl) > 4800 mg/kg, (mus orl) = 3200 mg/kg, (rat der) > 5000 mg/kg, (Japanese quail orl) > 10200 mg/kg; LC50 ((trout 96 hr.) = 27 mg/l, (bluegill sunfish 24 hr.) = 24 mg/l; non-toxic to bees. *Aventis Crop Science; Schering.*

310 Bendiocarb

22781-23-3   1033   G P

C11H13NO4

2,2-Dimethyl-1,3-benzodioxole-N-methylcarbamate.

AI3-27695; Bencarbate; Bendiocarb; Bendiocarbe; Benzodioxole, 2,2-dimethyl-4-(N-methylaminocarb-oxylato)-; 1,3-Benzodioxol-4-ol, 2,2-dimethyl-, methyl-carbamate; 1,3-Benzodioxole, 2,2-dimethyl-4-(N-methylcarbamato)-; 1,3-Benzodioxole, 2,2-dimethyl-4-(N-methylaminocarboxylato)-; BRN 1315404; Carb-amic acid, methyl-, 2,3-(dimethylmethylene-dioxy)-phenyl ester; Carbamic acid, methyl-, 2,3-(isopropyl-idenedioxy)phenyl ester; 2,2-Dimethyl-1,3-benzodiox-ol-4-ol methylcarbamate; 2,2-Dimethylbenzo-1,3-dioxol-4-yl methylcarb-amate; 2,3-Isopropylidene-dioxyphenyl methylcarbamate; Dycarb; Ficam; Ficam 80W; Ficam B; Ficam D; Ficam Plus; Ficam ULV; Ficam W; Ficam Z; Fisons NC 6897; Fuam; Garvox; HSDB 3918; Methylcarbamic acid, 2,3-(dimethylmethylenedioxy)-phenyl ester; Methylcarb-amic acid 2,3-(isopropylidenedioxy)phenyl ester; Mult-amat; Multimet; NC 6897; Niomil; OMS-1394; RCRA waste no. U278; Rotate; Sedox; Seedox;

Seedox 80W; Seedoxin; Tatoo; Tattoo; Turcam. Insecticide with contact action. Registered by EPA as an insecticide. White solid; mp = 129 - 130°; sg20 = 1.25; soluble in H2O (0.004 g/100 ml), Me2CO (20 g/100 ml), CHCl3 (30 g/100 ml), CH2Cl2 (30 g/100 ml), dioxane (20 g/100 ml), C6H6 (4 g/100 ml), EtOH (4 g/100 ml), o-xylene (1 g/100 ml), C6H14 (0.035 g/100 ml), kerosene (< 0.1 g/100 ml); LD50 (rat orl) = 40 - 156 mg/kg, (mus orl) = 45 mg/kg, (gpg orl) = 35 mg/kg, (rbt orl) = 35 - 40 mg/kg, (rat der) = 566 - 800 mg/kg, (bee) = 0.0001 mg/bee; LC50 (rat ihl 4 hr.) = 0.55 mg/l air, (rainbow trout 96 hr.) = 1.55 mg/l; toxic to bees. *AgrEvo; Chemisco; Farnam Companies Inc.; Scotts-Sierra Crop Protection; Value Gardens Supply LLC; Voluntary Purchasing Group Inc.*

311 Benfluralin

1861-40-1   1067   G H P

C13H16F3N3O4

N-butyl-N-ethyl-2,6-dinitro-4-trifluoromethylaniline.

Balan; Balan 1.5 LC; Balan 2.0G; Balan 2.5 G; Balfin; Banafine; Benalan; Benefex; Benefin; Benephin; Benfluralin; Benfluraline; Bethrodine; Binnell; Blulan; Bonalan; BRN 2821329; Carpidor; Caswell No. 130; EINECS 217-465-2; EL-110; Emblem; EPA Pesticide Chemical Code 084301; Flubalex; HSDB 407; L 54521; Pel-Tech; Quilan. A pre-emergence herbicide with a wide range of weed control both of annual grass weeds and broad-leaved weeds. Registered by EPA as a herbicide. Yellow-orange crystals; mp = 65 - 66.5°; bp7 = 148 - 149°; bp0.5 = 121 - 122°; poorly soluble in H2O (< 0.0001 g/100 ml), soluble in Me2CO (65 g/100 ml), dioxane (60 g/100 ml), MEK (58 g/100 ml), CHCl3 (> 50 g/100 ml), DMF (45 g/100 ml), xylene (45 g/100 ml), MeOH (4 g/100 ml), EtOH (2.4 g/100 ml); LD50 (rat orl) > 10000 mg/kg, (mus orl) > 5000 mg/kg, (dog, rbt orl) > 2000 mg/kg, (mallard duck, bobwhite quail, ckn orl) > 2000 mg/kg; LC50 (bluegill sunfish 96 hr.) = 0.064 mg/l. *Agan Chemical Manufacturers; Dow AgroSciences; Earth Care; Riverdale Chemical Co; UAP - Platte Chemical; UAP-West; United Horticultural Supply.*

312 Benodanil

15310-01-7   G P

C13H10INO

2-Iodo-N-phenylbenzamide.

3-12-00-00506 (Beilstein Handbook Reference); Apache; BAS-3170; BAS 3170F; Benefit; Benodanil; Benzamide, 2-iodo-N-phenyl-; Benzanilide, 2-iodo-; BRN 2725018; Calirus; 2-Iodo-N-phenylbenzamide; 2-Iodobenzanilide; 2-Iodobenzoic acid anilide; NSC 100499. Systemic and contact fungicide with protective and curative action. Used for control of rust disease in various crops. Colorless crystals; mp = 137°; soluble in H2O (0.002 g/100 ml), Me2CO (40 g/100 ml), EtOAc (12 g/100 ml), EtOH (9.3 g/100 ml), CHCl3 (7.7 g/100 ml); LD50 (rat, gpg orl) > 6400 mg/kg, (rat der) > 2000 mg/kg; LC50 (trout 96 hr.) = 6.4 mg/l; non-toxic to bees. *BASF*

Corp.

313   Benomyl
17804-35-2   1042   G P

C14H18N4O3

[1-[(Butylamino)carbonyl]-1H-benzimidazol-2-yl]carbamic   acid
methyl ester.

5-25-10-00390 (Beilstein Handbook Reference); Agrocit; Arbortrine;
Arilate; BBC; BC 6597; Benex; Benlate; Benlate 50; Benlate 50 W;
Benlate(R); Benomyl-Imex; Benosan; Benzimidazolecarbamic acid,
1-(butylcarb-amoyl)-, methyl ester; 2-Benzimidazolecarbamic acid,
1-(butylcarbamoyl)-, methyl ester; BNM; BRN 0825455; 1-
(butylcarbamoyl)-1H-benzimidazol-2-yl-,   methyl   ester;   1-
(Butylcarbamoyl)-2-benzimidazolecarbamic acid methyl ester; 1-
(Butylcarbamoyl)-2-benzimidazol-methylcarb-amat;   1-(N-
Butylcarbamoyl)-2-(methoxy-carboxamido)-benzimidazol; Carbamic
acid, methyl-, 1-(butyl-carbamoyl)-2-benzimidazolyl ester; Carbamic
acid, (1-((butylamino)carbonyl)-1H-benzimidazol-2-yl)-, methyl ester;
Caswell No. 075A; CCRIS 773; Chinoin-fundazol; D 1991; Dupont
1991; EPA Pesticide Chemical Code 099101; F1991; Fibenzol;
Fundazol; Fungicide D-1991; Fungochrom; HSDB 1655; Kribenomyl;
MBC;   Methyl   1-((butyl-amino)carbonyl)-1H-benzimidazol-2-
ylcarbamate;   Methyl   1-(butylcarbamoyl)-2-benzimidazolylcarb-
amate; Methyl 1-(butylcarbamoyl)benzimidazol-2-ylcarbamate; NS
02; NS 02 (fungicide); NSC 263489; RCRA waste no. U271; Tersan
1991; Uzgen. Fungicide and acaricide for garden use. Registered
by EPA as a fungicide. White crystals; almost insoluble in H_2O
(0.0002 g/100 ml), soluble in $CHCl_3$ (13.8 g/100 ml), DMF (5 g/100
ml), Me_2CO (1.4 g/100 ml), xylene (0.8 g/100 ml), EtOH (0.3 g/100
ml); LD_{50} (rat orl) > 9590 mg/kg, (rbt der) > 10000 mg/kg, (bee
contact, oral) > 0.01 mg/kg; LC_{50} (rat ihl 4 hr.) > 2 mg/l air, (mallard
duck, bobwhite quail 8 day dietary) > 500 mg/kg diet, (rainbow trout
96 hr.) = 0.17 mg/l, (goldfish 96 hr.) = 4.2 mg/l, (guppy 48 hr.) = 3.4
mg/l, (Daphnia 48 hr.) = 0.64 mg/l; non-toxic to bees. DuPont; Hi-
Yield Chemical Co.; ICI Chem. & Polymers Ltd.

314   Bentazone
25057-89-0   1051   G P

C10H12N2O3S

3-Isopropyl-1H-2,1,3-benzothiadiazin-4(3H)-one-2,2-dioxide.
Adagio; BAS 351-07H; BAS 351H; BAS 351-H; BAS 3510; BAS
3510H; BAS 3512H; BAS 3517H; Basagran; Basagran 480;
Basagran DP; Basagran KV; Basagran M; Basagran-plus;
Bendioxide; Bentazon; Bentazone; 1H-2,1,3-Benzothiadiazin-4(3H)-
one-2,2-dioxide, 3-iso-propyl-; 1H-2,1,3-Benzothiadiazin-4(3H)-one,
3-(1-methyl-ethyl)-, 2,2-dioxide; BRN 0530220; Caswell No. 509C;
CCRIS 6977; EINECS 246-585-8; EPA Pesticide Chemical Code
275200; Graminon-plus; Herbatox; HSDB 3430; 3-Isopropyl-1H-
benzo-2,1,3-thiadiazin-4-one   2,2-dioxide;   3-Isopropyl-1H-2,1,3-
benzothiadiazin-4(3H)-one-2,2-di-oxide;   3-Isopropyl-2,1,3-
benzothiadiazinon-(4)-2,2-; Shell.

315   Bentonite
1302-78-9   1053   G

$Al_2O_3.4SiO_2.H_2O$

Hydrated aluminum silicate.

250M; Accofloc 352; Accugel F; AEG 325; Akajo; Albagel Premium
USP 4444; Altonit SF; Aquagel; Aquagel Gold Seal; Asama;
Askangel; Ben-A-Gel®; Benclay; Bengel; Bengel 11; Bengel 15;
Bengel 23; Bengel A; Bengel FW; Bengel HVP; Bensulfoid;
BentoGrout; Bentolite HS; Bentolite L; Bentone 660; Bentonit T;
Bentonite; Bentonite 2073; Bentonite, calcian-sodian; Bentonite
magma; Bentosolon 82; Black Hills BH 200; Bulgarben BA; CCRIS
3663; Clarit BW 100; Clarit BW 125; Colloidal clay; Cosintam 403;
Culvin; CVIS; Detercal G 1FC; Detercal G 2FC; Detercol P 2;
EINECS 215-108-5; EX 0030 (clay); EX 0276; EX-M 703; Filgel;
Fulbent 570; G 100 (clay); GK 129SA3; GK 129SA5; HI-Jel; HSDB
392; Imvite I.G.B.A; K 129-H; Korthix; Korthix H-NF; Magbond;
Otaylite; Panther creek bentonite; Soap clay; Sodium
montmorillonite; Southern bentonite; Tixoton; Volcaly bentonite BC;
Volclay; Wilkinite; Wilkonite; Wyoming sodium bentonite;
Yellowstone. Native hydrated colloidal aluminum silicate clay; oil-well
drilling fluids, cement slurries for oil-well casings, thickener,
fireproofing, cosmetics, decolorizing agent, filler in ceramics,
emulsifier for oils, suspending agent in pharmaceuticals, base for
plasters. Beneficiated bentonite clay used as thickening, gelling and
emulsifying agent for water systems, a thixotropic agent for water-
based paints, inks, polishes, adhesives, and for household products
a thickener and suspending agent for liquid abrasive cleaning
compounds, water treatment; retention aid for paper, in cosmetics
and pharmaceuticals, steel production and foundries. Powder; forms
gels readily with H_2O. 779; 825; 829; 856; 857; 858; 896.

316   Benzal chloride
98-87-3   1056   H

C7H6Cl2

Benzene, (dichloromethyl)- .
4-05-00-00817 (Beilstein Handbook Reference); AI3-28597; Benzal
chloride; Benzene, dichloromethyl-; Benzyl dichloride; Benzylene
chloride; Benzylidene chloride ; BRN 1099407; CCRIS 959;
Chlorobenzal; Chlorure de benzylidene; Cloruro de bencilideno;
Dichlorophenylmethane; EINECS 202-709-2; HSDB 5322; NSC
7915; RCRA waste number U017; Toluene, α,α-dichloro-; UN1886 .
Liquid; mp = -17°; bp = 205°; d^{25} = 1.26; λ_m = 220, 240, 247, 259,
265, 272, 288 nm (MeOH); very soluble in EtOH, Et_2O.

317   Benzaldehyde
100-52-7   1057   G H

C7H6O

Benzoic aldehyde.
Artificial essential oil of almond; Phenylmethanal; almond artificial
essential oil; artificial almond oil; benzenecarbonal; benzene
carboxaldehyde; oil of bitter almond; Artificial Bitter Almond Oil;
Benzene methylal; Benzoyl hydride. Synthetic oil of bitter almond;
Chemical intermediate for dyes, flavors, perfumes, aromatic
alcohols; solvent for oils, resins, cellulose acetate and nitrate;
manufacture of cinnamic acid, benzoic acid; pharmaceuticals;
photographic chemicals. Oily liquid; mp = -26°; bp = 179°; d^{10} =
1.0415; λ_m = 241, 283, 290 nm (ε = 15849, 1259, 1000, C_6H_{14});

slightly soluble in H2O, soluble in CHCl3, liq. NH3; very soluble in C6H6, ligroin, Me2CO, freely soluble in EtOH, Et2O; LD50 (rat orl) = 1300 mg/kg. 306; 553; 841; 847; 1201; 1281; 3000; 3016.

318 **Benzazimide**
90-16-4   H

C7H5N3O
1,2,3-Benzotriazin-4(1H)-one .
AI3-28017; Benzazimide; Benzazimidone; Benzo-ketotriazine; 1,2,3-Benzotriazin-4-one; 1,2,3-Benzo-triazin-4-ol; 1,2,3-Benzotriazin-4(1H)-one; EINECS 201-971-5; HSDB 5270; NSC 13563; USAF MA-2.

319 **Benzenamine, 2,6-diethyl-N-methylene-**
35203-08-8   H

C11H15N
Benzenamine, 2,6-diethyl-N-methylene-.

320 **Benzenamine, N-(1-ethylpropyl)-4,5-dimethyl-**
56038-89-2   H

C13H21N
Benzenamine, N-(1-ethylpropyl)-4,5-dimethyl-.

321 **Benzene**
71-43-2   1066   G H

C6H6
Benzene.
AI3-00808; (6)Annulene; Benzeen; Benzen; Benzene; Benzin; Benzin (Obs.); Benzine; Benzol; Benzol 90; Benzol diluent; Benzole; Benzolene; Benzolo; Bicarburet of hydrogen; Carbon oil; Caswell No. 077; CCRIS 70; Coal naphtha; Cyclohexatriene; EINECS 200-753-7; EPA Pesticide Chemical Code 008801; Fenzen; HSDB 35; Mineral naphtha; Motor benzol; NCI-C55276; Nitration benzene; NSC 67315; Phene; Phenyl hydride; Polystream; Pyrobenzol; Pyrobenzole; RCRA waste number U019; UN 1114 . Used in manufacture of ethylbenzene, dodecylbenzene, cyclohexane,

phenol, nitrobenzene, chlorobenzene; dyes, medicinals, artificial leather, airplane dopes, lacquers; solvent for waxes, resins, oils. Liquid; mp = 5.5°; bp = 80°; d = 0.8790; λ_m = 243, 248, 254, 260, 268 nm (cyclohexane); highly flammable; LD50 (rat orl) = 3.8 ml/kg. 481; 668; 980; 1159; 1162; 1167; 1281.

322 **Benzene, 1-(bromomethyl)-3-phenoxy-**
51632-16-7   H

C13H11BrO
1-(Bromomethyl)-3-phenoxybenzene.
Benzene, 1-(bromomethyl)-3-phenoxy-; α-Bromo-3-phenoxytoluene; 1-(Bromomethyl)-3-phenoxybenzene; m-(Bromomethyl)phenyl phenyl ether; EINECS 257-327-9; m-Phenoxybenzyl bromide; 3-Phenoxybenzyl bromide.

323 **Benzenepropanoic acid, 3,5-bis(1,1-dimethylethyl)-4-hydroxy-, hydrazide**
32687-77-7   H

C17H28N2O2
Hydrocinnamic acid, 3,5-di-tert-butyl-4-hydroxy-, hydrazide.
Benzenepropanoic acid, 3,5-bis(1,1-dimethylethyl)-4-hydroxy-, hydrazide; 3-(3,5-Di-tert-butyl-4-hydroxy-phenyl)propionohydrazide; 3,5-Di-tert-butyl-4-hydr-oxyhydrocinnamic acid, hydrazide; EINECS 251-155-8; Hydrocinnamic acid, 3,5-di-tert-butyl-4-hydroxy-, hydrazide.

324 **Benzenesulfonamide**
98-10-2   G

C6H7NO2S
Benzenesulfonamide.
4-11-00-00050 (Beilstein Handbook Reference); AI3-04492; Benzenesulfonamide; Benzenesulphonamide; Benzolsulfonamide; Benzosulfonamide; BRN 1100566; EINECS 202-637-1; M and B 7973; NSC 5341. Chemical intermediate. Crystals; mp = 150-152°; soluble in H2O (0.43 g/100 ml), organic solvents; LD50 (rat orl) = 991 mg/kg. *Unitex.*

325 **Benzenesulfonic acid**
98-11-3   1070   H

C6H6O3S

Benzenesulfonic acid.

17-120A; 4-11-00-00027 (Beilstein Handbook Reference); AI3-15297; Benzene sulphonic acid; Benzenemonosulfonic acid; Benzenesulfonic acid; Benzenesulphonic acid; Besylic acid; BRN 0742513; CCRIS 4595; EINECS 202-638-7; HSDB 2642; Kyselina benzensulfonova; Phenylsulfonic acid. Used in manufacture of phenol. Needles; mp = 65°; λ_m = 235, 252, 258, 262, 265, 269 nm (ε = 731, 218, 273, 308, 265, 253, MeOH); insoluble in Et2O, CS2, slightly soluble in C6H6, soluble in AcOH, very soluble in H2O, EtOH.

326   **Benzenesulfonic acid butyl amide**
3622-84-2   G H

C10H15NO2S

N-Butylbenzene sulfonamide.

4-11-00-00051 (Beilstein Handbook Reference); AI3-08011; Benzenesulfonamide, N-butyl-; Benzenesulfonic acid butyl amide; BM 4 (sulfonamide); BRN 2725965; Cetamoll BMB; Dellato® BBS; EINECS 222-823-6; N-Butylbenzenesulfonamide; N-Butylbenzenesulphon-amide; NSC 3536; Plasthall BSA; Plastomoll® BMB; Uniplex 214. Plasticizer for polymide 6, 66, 11, and 12 and copolymides; also for flexibilizing cellulose derivatives, especially flame-retardant cable coatings based on cellulose acetate and cellulose acetobutyrate. BASF Corp.; Bayer AG; Unitex.

327   **Benzenesulfonic acid, 2-methyl-5-nitro-**
121-03-9   H

C7H7NO5S

4-Nitro-2-toluenesulfonic acid.

3-11-00-00173 (Beilstein Handbook Reference); 4-Nitro-2-toluenesulfonic acid; 4-Nitrotoluene-2-sulphonic acid; 5-Nitro-2-methylbenzenesulfonic acid; 5-Nitro-o-toluenesulfonic acid; Benzenesulfonic acid, 2-methyl-5-nitro-; BRN 2216176; EINECS 204-445-3; HSDB 5470; Kyselina 4-nitrotoluen-2-sulfonova; p-Nitrotoluene-o-sulfonic acid; NSC 9580; Pntos.

328   **Benzenesulfonyl chloride**
98-09-9   1072   H

C6H5ClO2S

Benzenesulfonyl chloride.

4-11-00-00049 (Beilstein Handbook Reference); AI3-18043; Benezenesulfochloride; Benzene sulfochloride; Benzene sulfonechloride; Benzene sulfonyl chloride; Benzenesulfon chloride; Benzenesulfonic acid chloride; Benzenesulfonyl chloride; Benzenesulphonyl chloride; Benzenosulfochlorek; Benzenosulfochloride; Benzeno-sulphochloride; Benzolsulfochloride; BRN 0606926; BSC-refine D; EINECS 202-636-6; HSDB 6004; NSC 2864; Phenylsulfonyl chloride; RCRA waste number U020; UN2225.

Solid; mp = 14.5°; bp = 251° (dec); d^{15} = 1.3470; λ_m = 210, 270 nm (ε = 12023, 1259, EtOH); insoluble in H2O, soluble in EtOH, Et2O, C6H6.

329   **1,2,4-Benzenetricarboxylic acid, decyl octyl ester**
67989-23-5   H

C27H42O6

1,2,4-Benzenetricarboxylic acid, decyl octyl ester.

1,2,4-Benzenetricarboxylic acid, decyl octyl ester; EINECS 268-007-3.

330   **Benzethonium chloride**
121-54-0   1074   G P V

C27H42ClNO2

N,N-Dimethyl-N-[2-[2-[4-(1,1,3,3-tetramethylbutyl)-phenoxy]ethoxy]ethyl]benzenemethaminium chloride.

Anti-germ 77; Antiseptol; Antiseptol (quaternary com-pound); Banagerm; Benzethoni chloridum; Benzethonii chloridum; Benzethonium; Benzethonium chloride; Benzetonio cloruro; Benzetonium chloride; BZT; Caswell No. 614B; CCRIS 4748; Chlorure de benzethonium; Cloruro de benzetonio; Diapp; Diisobutyl phenoxy ethoxy ethyl dimethyl benzyl ammonium chloride; Diisobutylphenoxyethoxyethyl-dimethyl benzyl ammon-ium chloride; Diisobutyl-phenoxyethoxyethyl dimethyl benzyl ammonium chloride; p-Diisobutyl phenoxy-ethoxyethyl dimethyl benzylammonium chloride; (2-(2-(4-Diisobutyl-phen-oxy)ethoxy)ethyl)dimethylbenzyl-ammonium chloride; 2-(2-(p-(Diisobutyl)phenoxy)-ethoxy)ethyl dimethyl benzyl ammonium chloride; Disilyn; EINECS 204-479-9; EPA Pesticide Chemical Code 069154; Formula 144; HSDB 567; Hyamine; Hyamine 1622; Inactisol; Kylacol; NCI-C61494; NSC 20200; p-tert-Octyl-phenoxyethoxyethyldimethylbenzyl-ammonium chloride; Phemeride; Phemerol chloride; Phemithyn; Polymine D; QAC; Quatrachlor; Sanizol; Solamin; Solamine. A topical anti-infective, used primarily in cosmetics for its antimicrobial and cationic surfactant properties. Bactericide, deodorant, anti-infective; antiseptic (veterinary); preservative for veterinary and pharmaceutical products. Registered by EPA as an antimicrobial and disinfectant. Colorless crystals; mp = 160 - 165°; soluble in H2O, EtOH, CHCl3, slightly soluble in Et2O; LD50 (mus iv) = 29.5 mg/kg. Chemical Packaging Corp.; Earth Care; Gold Coast Chemical Products; Lonza Ltd.; Lonzagroup; Metrex Research Corp; Microban Systems Inc.; Parke Davis & Co. Ltd.; Parke-Davis; Wepak Corp.

331   **Benzhydrol**
91-01-0   1091   G

C13H12O
Diphenyl methanol.
4-06-00-04648 (Beilstein Handbook Reference); AI3-03066; Benzenemethanol, α-phenyl-; Benzhydrol; Benzhydryl alcohol; Benzohydrol; BRN 1424379; Diphenyl carbinol; Diphenylcarbinol; Diphenylmethanol; Diphenylmethyl alcohol; EINECS 202-033-8; Hydroxy-diphenylmethane; NSC 32150; α-Phenylbenzenemethanol. Used in organic synthesis Solid; mp = 69°; bp = 298°, bp20 = 180°; λ_m = 242, 253, 258, 264 nm (ε = 304, 472, 520, 395, MeOH); slightly soluble in H2O, soluble in organic solvents; LD50 (rat orl) = 5000 mg/kg. *Alfa Aesar; Purecha Group.*

332   Benzimidazole
51-17-2   1082   G

C7H6N2
1,3-benzodiazole.
5-23-06-00196 (Beilstein Handbook Reference); AI3-03737; 3-Azaindole; Azindole; Benzimidazole; o-Benzimidazole; 1H-Benzimidazole; 1,3-Benzodiazole; Benzimidazole; Benziminazole; Benzoglyoxaline; Benzoimidazole; BRN 0109682; BZI; CCRIS 5967; 1,3-Diazaindene; EINECS 200-081-4; HSDB 2797; N,N'-Methenyl-o-phenylenediamine; NSC 759. Used in chemical synthesis, particularly of pesticides. Crystals; mp = 172-174°; soluble in H2O, organic solvents; LD50 (mus orl) = 2910 mg/kg. *Alcoa Ind. Chem.; Naturex; Penta Mfg.; Schweizerhall Pharma.*

333   Benzisothiazolin-3-one
2634-33-5   G P

C7H5NOS
1,2-Benzisothiazol-3(2H)-one.
1,2-Benzisothiazol-3(2H)-one; 1,2-Benzisothiazolin-3-one; 1,2-Benzisothiazoline-3-one; Caswell No. 079A; Caswell No. 513A; CCRIS 6369; EINECS 220-120-9; EPA Pesticide Chemical Code 098901; IPX; Nipacide® BIT; Proxan; Proxel PL. Used as a preservative in over 150 products, including cleaning agents, polishes,and paints. Used as a preservative in cooling fluids, paints, adhesives, paper and in the textile industry. The concentration is less than 0.01% in 46% of products and greater than 0.1% in 24%. The concentration is lowest in shampoos, skin care agents, cleaning agents, printing inks, and polishes. Sensitized patients have reacted to 0.0099% concentration in aqueous and alcohol. Registered by EPA as an antimicrobial and fungicide. Solid; mp = 154-158°. *Avecia Inc.; Clariant Corp.; Creanova Inc.; Nipa; Ondeo Nalco Co.; PMC; Sanitized Inc; Thor; Troy Chemical Corp; Trutec Industries Inc; Zeneca Pharm.*

334   Benzoguanamine
91-76-9   1090   H

C9H9N5
1,3,5-Triazine-2,4-diamine, 6-phenyl-.
4-26-00-01244 (Beilstein Handbook Reference); AI3-22162; Benzoguanamine; Benzoguanimine; BRN 0153223; EINECS 202-095-6; ENT 60118; HSDB 5275; NSC 3267; s-Triazine, 2,4-diamino-6-phenyl-; USAF RH-5. Used in manufacture ofdyestuffs, pesticides, drugs and polymers. Solid; mp = 226.5°; d^{25} = 1.40; λ_m = 243 nm (ε = 27900 MeOH); soluble in EtOH, Et2O, CF3COOH. *Alfa Aesar; Manuel Vilaseca SA; Solutia, Inc.; Wuhan Youji Industries Co. Ltd.*

335   Benzoic acid
65-85-0   1092   G P

C7H6O2
Benzenecarboxylic acid.
Acide benzoique; Acido benzoico; AI3-0310; AI3-03710; Benzenecarboxylic acid; Benzeneformic acid; Benzenemethanoic acid; Benzenemethonic acid; Benzoate; Benzoesäure; Benzoesäure GK; Benzoesäure GV; Benzoesaeure; Benzoesaeure GK; Benzoesaeure GV; Benzoic acid; Carboxybenzene; Caswell No. 081; CCRIS 1893; Diacylic acid; Dracylic acid; E 210; EINECS 200-618-2; EPA Pesticide Chemical Code 009101; FEMA No. 2131; Flowers of Benjamin; Flowers of benzoin; HA 1; HA 1 (acid); HSDB 704; Kyselina benzoova; NSC 149; Oracylic acid; Phenylcarboxylic acid; Phenylformic acid; Retarder BA; Retardex; Salvo; Salvo liquid; Solvo powder; Tenn-Plas; Tennplas; Unisept BZA; Used in the manufacture of sodium and butyl benzoates, plasticizers, benzoyl chlorides, food preservatives, flavors, perfumes, antifungal agent. Fungicide. Registered by EPA as a fungicide and insecticide (cancelled). FDA GRAS, included in the FDA Inactive Ingredients Guide (im and iv injections, irrigation and oral solutions, suspensions, syrups and tablets, rectal, topical and vaginal preparations). Used as an antimicrobial preservative in cosmetics, foods and pharmaceuticals. Used in the manufacture of sodium and butyl benzoates, plasticizers, benzoyl chlorides, food preservatives, flavors, perfumes, antifungal agent. White crystals; mp = 122.4°; bp = 249°; d^{24} = 1.311; soluble in H2O (0.29 g/100 ml), EtOH (37.0 g/100 ml at 15°, 45.4 g/100 ml at 37°), Me2CO (43.5 g/100 ml), C6H6 (10.6 g/100 ml), CHCl3 (22.2 g/100 ml), Et2O (33.3 g/100 ml) and other organic solvents; LD50 (rat orl) = 1700 mg/kg, 2530 mg/kg, (mus orl) = 1940 mg/kg, (mus ip) = 1460 mg/kg, (dog orl) = 2000 mg/kg, (cat orl) = 2000 mg/kg. *Ashland; E. Merck; Elf Atochem N. Am.; Fluka; Greeff R.W. & Co.; Integra; Jan Dekker; Mallinckrodt Inc.; Mitsubishi Corp.; Napp Labs Ltd; Nipa; Penta Mfg.; Sigma-Aldrich Fine Chem.; Spectrum Chem. Manufacturing; Velsicol.*

336   Benzoic   acid,   3-(2-chloro-4-(trifluoromethyl)phenoxy)-
63734-62-3   H

$C_{14}H_8ClF_3O_3$

3-(2-Chloro-4-(trifluoromethyl)phenoxy)benzoic acid.
Benzoic acid, 3-(2-chloro-4-(trifluoromethyl)phenoxy)-; 3-(2-Chloro-4-(trifluoromethyl)phenoxy)benzoic acid; EINECS 264-433-9.

337   Benzoin
119-53-9   1094   D G

$C_{14}H_{12}O_2$

α-Hydroxybenzyl phenyl ketone.
4-08-00-01279 (Beilstein Handbook Reference); Aceto-phenone, 2-hydroxy-2-phenyl-; AI3-00851; Benzoin; (±)-Benzoin; Benzoin extract (resinoid); Benzoylphenyl-carbinol; Bitter-almond-oil camphor; BRN 0391839; CCRIS 75; CCRIS 6696; EINECS 204-331-3; EINECS 209-441-5; Ethanone, 2-hydroxy-1,2-diphenyl-; FEMA No. 2132; Fenyl-α-hydroxybenzylketon; HSDB 384; 2-Hydroxy-2-phenylacetophenone; 2-Hydroxy-1,2-di-phenylethanone; (1)-2-Hydroxy-1,2-diphenylethan-1-one;   α-Hydroxy-α-phenylacetophenone; α-Hydroxybenzyl phenyl ketone; Ketone, α-hydroxybenzyl phenyl; NCI-C50011; NSC 8082; Phenylbenzoyl carbinol; Phenyl-α-hydroxybenzyl ketone; WY-42956. Organic synthesis; intermediate; photopolymerization catalyst. Crystals; mp = 134-136°; bp12 = 194°; slightly soluble in H_2O, more soluble in organic solvents; LD50 (rat orl) = 10 gm/kg. *Janssen Chimica; Madis Dr. Labs; Sigma-Aldrich Fine Chem.; Snia UK; Spectrum Chem. Manufacturing.*

338   Benzonitrile
100-47-0   1098   G

C_7H_5N

Phenyl cyanide.
AI3-24184; Benzene, cyano-; Benzenecarbonitrile; Benzenenitrile; Benzoic acid nitrile; Benzonitrile; CCRIS 3184; Cyanobenzene; EINECS 202-855-7; Fenylkyanid; HSDB 45; NSC 8039; Phenyl cyanide; UN2224. Manufacture of benzoquanamine; intermediate for rubber chemicals; solvent for nitrile rubber, specialty lacquers, resins and polymers, anhydrous metallic salts. Liquid; mp = -12.7°; bp = 191.1°; d15 = 1.0093; λm = 222, 230, 263, 277 nm (ε = 10600, 9100, 804, 932, MeOH); slightly soluble in H_2O (1.0 g/100 ml), soluble in CCl4, very soluble in Me2CO, C6H6, freely soluble in EtOH, Et2O; LD50 (mus orl) = 971 mg/kg. *Penta Mfg.; PMC; Spectrum Chem. Manufacturing.*

339   Benzophenone
119-61-9   1099   G H

$C_{13}H_{10}O$

Diphenyl ketone.
ADK STAB 1413; AI3-00754; Benzene, benzoyl-; Benzoylbenzene; Caswell No. 081G; CCRIS 629; Diphenyl ketone; Diphenylmethanone; EINECS 204-337-6; EPA Pesticide Chemical Code 000315; FEMA No. 2134; HSDB 6809; Ketone, diphenyl; Methanone, diphenyl-; NSC 8077; Phenyl ketone. UV absorber for polyolefins, PVC, etc.; good compatibility with polymers. Prisms; mp = 47.8°; bp = 305.4°; λm = 252, 331 nm (ε = 18600, 167, MeOH); insoluble in H_2O, soluble in C6H6, MeOH, very soluble in EtOH, Et2O, Me2CO, CHCl3, CS2, AcOH. *Asahi Denka Kogyo.*

340   Benzophenone-1
131-56-6   1107   G H

$C_{13}H_{10}O_3$

2,4-Dihydroxybenzophenone.
4-08-00-02442 (Beilstein Handbook Reference); 4-Benzoyl resorcinol; Advastab 48; Benzophenone-1; Benzophenone, 2,4-dihydroxy-; Benzoresorcinol; BRN 1311566; Dastib 263; Eastman Inhibitor DHPB; EINECS 205-029-4; HHB; HSDB 5617; Inhibitor DHBP; Methanone, (2,4-dihydroxyphenyl)phenyl-; NSC 38555; Quinsorb 010; Resbenzophenone; Syntase 100; UF 1; USAF DO-28; USAF ND-54; UV 12; Uvinul® 400; Uvistat 12. UV absorber; used for polyester, acrylics, PS, in outdoor paints and coatings, varnishes, colored liquid toiletries and cleaning agents, filters for photographic color films and prints, and rubber-based adhesives. Crystals; mp = 144-147°; bp1 = 194°; insoluble in H_2O, soluble in organic solvents. *BASF Corp.; EM Ind. Inc.; Ferro/Bedford; Greeff R.W. & Co.; Haarmann & Reimer GmbH; Hoechst Celanese; Quest Intl.; Sartomer.*

341   Benzophenone-2
131-55-5   G

$C_{13}H_{10}O_5$

2,2',4,4'-Tetrahydroxybenzophenone.
4-08-00-03505 (Beilstein Handbook Reference); Benzophenone-2; Benzophenone,   2,2',4,4'-tetrahydroxy-;   Bis(2,4-dihydroxyphenyl)methanone; BRN 1914746; EINECS 205-028-9; Methanone, bis(2,4-dihydroxy-phenyl)-; NSC 38556; 2,2',4,4'-Tetrahydroxybenzo-phenone; 2,4,2',4'-Tetrahydroxybenzophenone; Uvinol D-50; Uvinul D-50. Commercial UV absorber with the broadest UV absorption spectrum; retards fading of pigments and dyestuffs; prolongs the life of polymeric materials; photostabilizes cosmetic formulations; and minimizes discoloration of synthetic rubber of plastic latices. Crystals; mp = 200-203°; LD50 (rat orl) = 1220 mg/kg. *BASF Corp.; EM Ind. Inc.; Ferro/Bedford; Greeff R.W. & Co.; Haarmann & Reimer GmbH; Hoechst Celanese; Quest Intl.;*

342   Benzophenone-6
131-54-4   1100   G

C15H14O5
2,2'-Dihydroxy-4,4'-dimethoxybenzophenone.
4-08-00-03505 (Beilstein Handbook Reference); 4,4'-Dimethoxy-2,2'-dihydroxybenzophenone; Benzophen-one, 2,2'-dihydroxy-4,4'-dimethoxy-; Benzophenone-6; Bis(2-hydroxy-4-methoxyphenyl)methanone; BRN 1887087; Caswell No. 353C; Cyasorb UV 12; 2,2'-Dihydroxy-4,4'-dimethoxybenzophenone; EINECS 205-027-3; Methanone, bis(2-hydroxy-4-methoxyphenyl)-; NSC 40149; Uvinul® D 49. Organic benzophenone derivative; UV light absorber, used especially in paints and plastics. Most economical of the near-uv absorbers; greater heat stability and more solubility (in chlorinated and aromatic solvents); gives broad protection to plastics, coatings, textiles such as PVC, chlorinated polyesters, epoxies, acrylics, urethanes and cellulosics. Crystals; mp = 139.5°; λ_m = 284, 340 nm (log ε 4.12, 4.12). *BASF Corp.; EM Ind. Inc.; Ferro/Bedford; Greeff R.W. & Co.; Haarmann & Reimer GmbH; Hoechst Celanese; Quest Intl.; Sartomer.*

343   Benzophenone-9
3121-60-6   G

C15H13NaO8S
Sodium 2,2'-dihydroxy-4,4'-dimethoxy-5-sulfobenzo-phenone.
Benzenesulfonic acid, 4-hydroxy-5-(2-hydroxy-4-methoxybenzoyl)-2-methoxy-, monosodium salt; Benzophenone-9; EINECS 221-498-8; NSC 76059; Sodium 2,2'-dihydroxy-4,4'-dimethoxy-5-sulfobenzo phenone; Sodium 4-hydroxy-5-(2-hydroxy-p-anisoyl)-2-methoxybenzenesulphonate; Uvinul® DS 49. Organic benzophenone derivative; a sulfonated derivative of Uvinul D-49 [131-54-4]; UV absorber in cosmetic formulations to prevent fading of colors and viscosity changes caused by UV light. *BASF Corp.; EM Ind. Inc.; Ferro/Bedford; Greeff R.W. & Co.; Haarmann & Reimer GmbH; Hoechst Celanese; Quest Intl.; Sartomer.*

344   Benzophenone-12
1843-05-6   6774   D G H

C21H26O3
[2-hydroxy-4-(octyloxy)phenyl]phenylmethanone.
Aduvex 248; Advastab 46; Anti-UV P; Benzon OO; Benzophenone-12; Benzophenone, 2-hydroxy-4-(octyl-oxy)-; 2-Benzoyl-5-octyloxyphenol; Biosorb 130; BRN 1915198; Carstab 700;

Chimassorb 81; Cyasorb® UV 531; EINECS 217-421-2; HSDB 5858; (2-Hydroxy-4-(octyloxy)phenyl)phenylmethanone; 2-Hydroxy-4-(n-octyloxy)benzophenone; 2-Hydroxy-4-(octyloxy)benz-ophenon; 2-Hydroxy-4-(octyloxy)benzophenone; 2-Hydroxy-4-octoxybenzophenone; 2-Hydroxy-4-(n-octoxy)benzophenone; 2-Hydroxy-4-oktyloxybenzo-fenon; Mark 1413; Methanone, (2-hydroxy-4-(octyl-oxy)phenyl)phenyl-; NSC 163400; Octabenzona; Octabenzone; Octabenzonum; 4-(Octyloxy)-2-hydroxybenzophenone; 4-n-Octyloxy-2-hydroxy-benzophenone; 4-Octoxy-2-hydroxybenzophenone; Rhodialux P; Sanduvor 3035; Seikalizer E; Spectra-sorb UV 531; Sumisorb 130; UF 4; UV 1; UV 531; UV Absorber HOB; UVA 1; Uvinul® 408; Uvinul® M 408; Viosorb 130; Zislizer E. Light stabilizer and UV absorber for plastics and coatings, e.g., polyethylene, PP, PVC, and EVA; uses include pipe, storage tanks, and auto, marine, garden products, auto refinish and industrial coatings, adhesives and sealants. UV absorber and stabilizer for polyethylene, PP, plasticized and rigid PVC, and other polymers; offers good compatibility, maximum protection and minimum color and a low order of toxicity. Crystals; mp = 48.5°; λ_m = 243, 290, 326 nm (ε = 10965, 15849, 10715, EtOH). *Am. Cyanamid; BASF Corp.*

345   Benzotetraoxatriplumbacycloundecin-1,9-dione
17976-43-1   H

C8H4O6Pb3
2,4,6,8,3,5,7-Benzotetraoxatriplumbacycloundecin-3,5,7-triylidene,1,9-dihydro-1,9-dioxo-.
$3\lambda2,5\lambda2,7\lambda2$-2,4,6,8,3,5,7-Benzotetraoxatriplumba-cycloundecin-1,9-dione; Cyclo-di-μ-oxo(μ-phthalato)tri-lead; EINECS 241-894-4; Lead, (μ-(1,2-benzene-dicarboxylato(2-)-O1:O2))di-μ-oxotri-, cyclo-; Lead, di-μ-oxo(μ-phthalato)tri-, cyclo-.

346   Benzothiazole, 2-((chloromethyl)thio)-
28908-00-1   H

C8H6ClNS2
2-((Chloromethyl)thio)benzothiazole.
Benzothiazole, 2-((chloromethyl)thio)-; 2-((Chloro-methyl)thio)benzothiazole; EINECS 249-306-8.

347   4-(2-Benzothiazolylthio)morpholine
102-77-2   G H

C11H12N2OS2
Benzothiazole, 2-(4-morpholinylthio)-.
4-27-00-01868 (Beilstein Handbook Reference); Accel NS; Accelerator MF; AI3-27134; Amax®; Amax® No 1; Benzothiazole, 2-(4-morpholinylthio)-; 4-(2-Benzo-thiazolylthio)morpholine; BRN

61

0191684; CCRIS 4911; Cure-rite OBTS; Delac MOR; EINECS 203-052-4; HSDB 2867; MBS; Meramide M; Morpholine, 4-(2-benzothiazolylthio)-; Nobs special; NSC 70078; S; OMTS; Perkacit® MBS; Santocure MOR; Sulfenamide M; Sulfenax mob; Sulfenax MOR; USAF CY-7; Vulcafor BSM; Vulcafor SSM; Vulkacit MOZ. A primary and secondary accelerator for rubber; safe at processing temperature and active over a wide curing range; particularly advantageous in styrene/butadiene rubber tires compounded with fine particle fumace blacks. The most delayed action accelerator offered by Uniroyal; activated by thiurams, dithiocarbamates, BIK, guanidines and aldehyde-amines; nondiscoloring and nonstaining to rubber stocks and materials in contact with them; used in tire treads, carcass, mechanicals and wire jackets. *Akrochem Chem. Co.; Akzo Chemie; Uniroyal; Vanderbilt R.T. Co. Inc.*

348   Benzotriazole
95-14-7   1109   G H

$C_6H_5N_3$
1H-Benzotriazole.
4-26-00-00093 (Beilstein Handbook Reference); AI3-15984; Azimidobenzene; Aziminobenzene; Benzene azimide; Benzisotriazole; Benztriazole; Benztriazole; BRN 0112133; CCRIS 78; Cobratec 99; EINECS 202-394-1; HSDB 4143; NCI-C03521; NSC-3058; Preventol® CI 8; U-6233. Chelating agent and sesquestrant for copper ions; corrosion inhibitor for copper, brass, bronze; used in antifreeze, cleaners, coatings, detergents, functional fluids, metalworking fluids, packaging materials, polishes. Corrosion inhibitor particularly suitable for antifreezes, coolants, cutting fluids and hydraulic fluids. Needles; mp = 100°; bp_{15} = 204°; λ_m = 253, 258, 275 nm (MeOH); soluble in H_2O (2.0 g/100 ml), EtOH, C_6H_6, $CHCl_3$, C_7H_8, DMF; LD_{50} (rat orl) = 600 mg/kg. *Alcoa Ind. Chem.; Bayer AG; Miles Inc.; PMC; Sandoz.*

349   2-(2H-Benzotriazol-2-yl)-4,6-bis(1-methyl-1-phenylethyl)phenol
70321-86-7   H

$C_{30}H_{29}N_3O$
Phenol, 2-(2H-benzotriazol-2-yl)-4,6-bis(1-methyl-1-phenylethyl)-.
2-(2H-Benzotriazol-2-yl)-4,6-bis(1-methyl-1-phenyl-ethyl)-phenol; EINECS 274-570-6; Phenol, 2-(2H-benzotriazol-2-yl)-4,6-bis(1-methyl-1-phenylethyl)-.

350   Benzotrichloride
98-07-7   1110   H

$C_7H_5Cl_3$
Benzene, (trichloromethyl)-.
4-05-00-00820 (Beilstein Handbook Reference); AI3-02583; Benzene, (trichloromethyl)-; Benzenyl trichloride; Benzoic trichloride; Benzotrichloride; Benzotricloruro; Benzyl trichloride; Benzylidyne chloride; BRN 0508152; CCRIS 1292; Chlorure de benzenyle; Chlorure de benzylidyne; EINECS 202-634-5; HSDB 2076; NSC 14663; Phenyl chloroform; Phenyltrichloromethane; RCRA waste number U023; Toluene, α,α,α,-trichloro-; Toluene trichloride; Trichloormethylbenzeen;   Trichlor-methylbenzol; Trichloromethylbenzene;   Trichlorophenyl-methane; Triclorometilbenzene;   (Trichloromethyl)benz-ene;   1-(Trichloromethyl)benzene; α,α,α-Trichlorotoluene; Triclorotoluene; UN2226. Oil; mp = -5°; bp = 221°; d^{20} = 1.3723; λ_m = 225, 261, 267, 274 nm (ε = 6910, 499, 558, 450, MeOH); insoluble in H_2O, soluble in EtOH, Et_2O, C_6H_6.

351   Benzoyl chloride
98-88-4   1113   H

C_7H_5ClO
Benzoyl chloride.
4-09-00-00721 (Beilstein Handbook Reference); Benzaldehyde, α-chloro-; Benzenecarbonyl chloride; Benzoic acid, chloride; Benzoyl chloride; BRN 0471389; CCRIS 802; EINECS 202-710-8; HSDB 383; UN1736. Liquid; mp = -1°; bp = 197.2°; bp_9 = 71°; d^{20} = 1.2120; λ_m = 241, 281 nm (cyclohexane); very soluble in Et_2O, soluble in C_6H_6, CCl_4.

352   Benzoyl chloride, 3,5-dichloro-
2905-62-6   H

$C_7H_3Cl_3O$
3,5-Dichlorobenzoyl chloride.
Benzoyl chloride, 3,5-dichloro-; 3,5-Dichlorobenzoyl chloride; EINECS 220-813-6; RH-24,299.

353   Benzoyl tert-butyl peroxide
614-45-9   G H

$C_{11}H_{14}O_3$
Peroxybenzoic acid, t-butyl ester.
4-09-00-00715 (Beilstein Handbook Reference); AI3-06625; Aztec® t-Butyl Perbenzoate; Benzenecarbo-peroxoic acid, 1,1-dimethylethyl ester; Benzoyl tert-butyl peroxide; BRN 1342734; t-Butyl perbenzoate; t-Butyl peroxy benzoate; terc.Butylester kyseliny peroxy-benzoove; terc.Butylperbenzoan; tert-Butyl perbenzoate; tert-Butyl peroxybenzoate; CCRIS 6217; Chaloxyd tbpb; EINECS 210-

382-2; Esperox 10; HSDB 2891; Novox; NSC 674; Perbenzoate de butyle tertiaire; Perbenzoic acid, tert-butyl ester; Perbutyl Z; Peroxybenzoic acid, t-butyl ester; Polyvel CR-5T; Trigonox® 93; Trigonox® C. Initiator for elevated-temperature polyester cures, for LDPE polymerization, and high-temperature polymerization of acrylic emulsion polymers; used in BMC and SMC molding in temperature range 275-325°F. For polymerization and/or crosslinking of monomers and unsaturated polymers. Additive for grafting and extrusion reactions. Liquid; mp = 8°; bp = 113°, bp2 = 75-76°; d^{25} = 1.021; insoluble in H_2O, soluble in organic solvents; LD50 (rat orl) = 1012 mg/kg. *Akzo Chemie; Catalyst Resources Inc.; Elf Atochem N. Am.; International Hormones; Norac; Polyvel.*

354   3-Benzoyloxy-2,2,4-trimethylpentyl iso-butyrate
22527-63-5   H

C19H28O4
 Isobutyric acid, 3-hydroxy-2,2,4-trimethylpentyl ester benzoate.
 3-Benzoyloxy-2,2,4-trimethylpentyl isobutyrate; EINECS 245-054-8; Isobutyric acid, 3-hydroxy-2,2,4-trimethylpentyl ester benzoate; Propanoic acid, 2-methyl-, 3-(benzoyloxy)-2,2,4-trimethylpentyl ester.

355   Benztropine Mesylate
132-17-2   1123   D

C22H29NO4S
 3α-(Diphenylmethoxy)-1αH,5αH-tropane methane-sulfonate.
 Benzatropine methanesulfonate; Benzotropine mesyl-ate; Benzotropine methanesulfonate; Benztropine mes-ilate; Benztropine mesylate; Benztropine methane-sulfonate; Cobrentin methanesulfonate; Cogentin; Co-gentin mesylate; Cogentin methanesulfonate; DRG-0198; EINECS 205-048-8; MK 02; NSC 169913. Anticholinergic. Antiparkinsonian. mp = 143°; λm = 259 nm (E_M 437); soluble in H_2O. *Apothecon; Merck & Co.Inc.*

356   Benzyl acetate
140-11-4   1125   G H

C9H10O2
 Acetic acid, phenylmethyl ester.
 Acetic acid, benzyl ester; Acetic acid, phenylmethyl ester; AI3-01996; Benzyl acetate; Benzyl ethanoate; Benzylester kyseliny octove; Caswell No. 081EA; CCRIS 1423; EINECS 205-397-6; FEMA No. 2135; HSDB 2851; NCI-C06508; NSC 4550; Phenylmethyl acetate; Plastolin I. A natural constituent of several essential oils and flower absolutes extracted from jasmine, hyacinth,

gardenia, tuberose, ylang-ylang, cananga, and neroli. Commercial benzyl acetate, a liquid prepared synthetically from benzyl chloride, acetic acid, and triethylamine is used primarily as a component of perfumes for soaps and as a flavoring ingredient. Constituent of artificial jasmine and other perfumes; used as a soap perfume; flavors; solvent and high boiler for cellulose acetate and nitrate, natural and synthetic resins; oils; lacquers; polishes; printing inks; varnish removers. Liquid; mp = -51.3°; bp = 213°; d^{20} = 1.0550; λm = 252, 257, 263, 267 nm (ε = 174, 214, 174, 114, MeOH); slightly soluble in H_2O, soluble in Et2O, Me2CO, CHCl3, freely soluble in EtOH. *Bush Boake Allen   (IFF bought 2000); Haarmann & Reimer GmbH; Janssen Chimica; MTM AgroChemicals Ltd.; Penta Mfg.; Pentagon Chems. Ltd.; Quest Intl.*

357   Benzyl acrylate
2495-35-4   G

C10H10O2
 Acrylic acid benzyl ester.
 Acrylic acid, benzyl ester; AI3-03836; Benzyl acrylate; EINECS 219-673-9; Melcril 4085; NSC 20964; 2-Propenoic acid, phenylmethyl ester; Sartomer SR 432; SR 432. Used as a monomer. Liquid; bp = 228°, bp8 = 110-111°; d^{20} = 1.0573; λm = 251, 257, 263, 267 nm (cyclohexane); insoluble in H_2O, soluble in EtOH, Et2O, Me2CO, CCl4. *Danbert Chemical Co.*

358   Benzyl alcohol
100-51-6   1126   G P

C7H8O
 Phenylmethanol.
 4-06-00-02222 (Beilstein Handbook Reference); AI3-01680; Alcohol bencilico; Alcoholum benzylicum; Alcool benzilico; Alcool benzylique; Bentalol; Bentanol; Benzal alcohol; Benzenecarbinol; Benzenemethanol; Benzoyl alcohol; Benzyl alcohol; Benzylicum; BRN 0878307; Caswell No. 081F; CCRIS 2081; EINECS 202-859-9; EPA Pesticide Chemical Code 009502; Euxyl K 100; FEMA No. 2137; HSDB 46; HSDB 46; (Hydroxymethyl)benzene; α-Hydroxytoluene; Hydroxy-toluene; Itch-X; Methanol, phenyl-; NCI-C06111; NSC 8044; Phenylcarbinol; Phenylcarbinolum; Phenyl-methanol; Phenylmethyl alcohol; α-Toluenol; Perfumes; flavors; photographic developer; dyeing nylon, textiles, sheet plastics; solvent for dyestuffs, cellulose esters, casein, waxes; heat-sealing polyethylene films; intermediate for benzyl esters and ethers; bacteriostat; cosmetics; inks. Used in perfumes; flavors; photographic developer; dyeing nylon, textiles, sheet plastics; solvent for dyestuffs, cellulose esters, casein, waxes; heat-sealing polyethylene films; intermediate for benzyl esters and ethers; bacteriostat; cosmetics; inks. Registered by EPA as a fungicide and insecticide (cancelled). Colorless liquid; mp = -15°; bp = 205°; d^{20}_4 = 1.04535; λm = 247, 252, 257, 263, 267 nm (ε = 150, 233, 229, 165, 112, MeOH); freely soluble in CHCl3, EtOH, Et2O, fixed and volatile oils, soluble in 95% EtOH (40 g/100 ml), H_2O (4 g/100 ml at 25°, 7.1 g/100 ml at 90°); LD50 (rat orl) = 1230 mg/kg, (rat iv) = 60 mg/kg, (rat ip) = 40 mg/kg, (rbt orl) = 1040 mg/kg, (mus orl) = 1580 mg/kg, (mus iv) = 320 mg/kg. *Ashland; Bush Boake Allen   (IFF bought 2000); E. Merck; Elf Atochem N. Am.; Givaudan Iberica SA; Greeff R.W. & Co.; Haarmann & Reimer GmbH; Janssen Chimica; Mallinckrodt*

Inc.; Osaka Org. Chem. Ind.; Penta Mfg.; Quest Chemical Corp.; Quest Intl.; Ruger; Sigma-Aldrich Fine Chem.; Spectrum Chem. Manufacturing; Takasago Intl. Corp.; Tosoh.

359   Benzyl benzoate
120-51-4   1129   D G P

C14H12O2

Benzoic acid phenylmethyl ester.
4-09-00-00307 (Beilstein Handbook Reference); Al3-00523; Antiscabiosum; Ascabin; Ascabiol; Benylate; Benzoesaeurebenzylester; Benzoesäurebenzylester; Benzoic acid, benzyl ester; Benzoic acid, phenyl-methyl ester; Benzyl alcohol benzoic ester; Benzyl benzenecarboxylate; Benzyl benzoate; Benzyl phenylformate; Benzylbenzenecarboxylate; Benzyl-ester kyseliny benzoove; Benzylets; Benzylis benzoas; Benzylum benzoicum; BRN 2049280; Caswell No. 082; Colebenz; EINECS 204-402-9; EPA Pesticide Chemical Code 009501; FEMA Number 2138; HSDB 208; Novoscabin; NSC 8081; Peruscabin; Peruscabina; Phenylmethyl benzoate; Scabagen; Scabanca; Scabide; Scabiozon; Scabitox; Scobenol; Spasmodin; Vanzoate. The active constituent of Peru balsam, used in the same manner, and for the same purpose as Peruol. Fixative and solvent for musk in perfumes and flavors; external medicine; plasticizer for nitrocellulose and cellulose acetate; miticide. Also used to treat scabies and pediculosis. Needles or leaflets, mp = 21°; bp = 323.5°; λ_m = 230, 257, 263, 267, 272, 280 nm (ε = 1470, 952, 952, 954, 968, 771, MeOH); insoluble in H2O, soluble in EtOH, Et2O, Me2CO, C6H6, CHCl3, MeOH, petroleum ether; LD50 (rat orl) = 1.7 g/kg. *Haarmann & Reimer GmbH; Janssen Chimica; Lancaster Synthesis Co.; May & Baker Ltd.; Penta Mfg.; Pentagon Chems. Ltd.*

360   Benzylbis(hydrogenated   tallowalkyl)-methyl, quaternary ammonium chlorides
61789-73-9   H

Quaternary ammonium compounds, benzylbis-(hydrogenated tallowalkyl)methyl, chlorides.
EINECS 263-082-9; Quaternary ammonium comp-ounds, benzylbis(hydrogenated tallowalkyl)methyl, chlorides.

361   Benzyl chloride
100-44-7   1131   H

C7H7Cl

Benzene, (chloromethyl)-.
Al3-15518; Benzene, (chloromethyl)-; Benzile (cloruro di); Benzyl chloride; Benzylchlorid; Benzylchloride; Benzyle (chlorure de); CCRIS 79; Chloromethylbenzene; Chlorophenylmethane; Chlorure de benzyle; EINECS 202-853-6; HSDB 368; NCI-C06360; NSC 8043; RCRA waste number P028; Toluene, α-chloro-; Tolyl chloride; UN1738. Liquid; mp = -45°; bp = 179°; d^{20} = 1.1004; λ_m = 217, 254, 259, 265 nm (ε = 6970, 190, 231, 211, MeOH); insoluble in H2O, slightly soluble in CCl4, freely soluble in EtOH, Et2O, CHCl3.

362   Benzyldimethylamine
103-83-3
                                                    G

C9H13N

N,N-dimethylbenzenemethanamine.
Al3-26794; Araldite accelerator 062; BDMA; Benz-enemethanamine, N,N-dimethyl-; Benzyl-N,N-dimethyl-amine; Benzylamine, N,N-dimethyl-; Benzyldimethyl-amine; N-Benzyldimethylamine; N-Benzyl-N,N-dimethyl-amine; CCRIS 6693; Dabco® B-16; Dimethyl-benzylamine; N,N-Dimethylbenzylamine; N,N-Dimethyl-N-benzylamine; N,N-Dimethylbenzenemethanamine; EINECS 203-149-1; NSC 5342; Pentamin BDMA; N-(Phenylmethyl)dimethylamine; Sumine 2015; UN2619. Amine-based catalyst for flexible slabstock PU foam. Used as a polyurethane catalyst and an epoxy curing agent. Liquid; mp = -75°; bp = 181°; d^0 = 0.915; λ_m = 252, 258, 265 nm (cyclohexane); slightly soluble in H2O, freely soluble in EtOH, Et2O; LD50 (rat orl) = 265 mg/kg. *Air Products & Chemicals Inc.; Pentagon Chems. Ltd.* -

363   Benzylidene acetone
122-57-6   1139   G

C10H10O

Benzalacetone.
2-07-00-00287 (Beilstein Handbook Reference); Aceto-cinnamone; Al3-00944; Benzalaceton; Benzalacetone; Benzilidene acetone; Benzilideneacetone; Benzylidene-acetone; BRN 0742046; 3-Buten-2-one, 4-phenyl-; CCRIS 5319; EINECS 204-555-1; FEMA No. 2881; Ketone, methyl styryl; Methyl 2-phenylvinyl ketone; Methyl styryl ketone; Methyl β-styryl ketone; NSC 5605; 4-Phenylbutenone; 4-Phenyl-3-buten-2-one; 2-Phenylvinyl methyl ketone; Styryl methyl ketone. Used in organic synthesis and perfumery; used as a fixative and in flavors. Solid; mp = 41.5°. *Lancaster Synthesis Co.; Penta Mfg.; Raschig GmbH.*

364   Benzyl iodide
620-05-3
                                                    G

C7H7I

Iodomethylbenzene.
Benzene, (iodomethyl)-; Benzyl iodide; EINECS 210-623-1; Fraissite; α-Iodotoluene; UN2653. Used in chemical synthesis. Yellow needles; mp = 24.5°; bp10 = 93°; d^{25} = 1.7335; very soluble in C6H6, Et2O, EtOH.

365   Benzyl   3-isobutyryloxy-1-isopropyl-2,2-dimethylpropyl phthalate
16883-83-3
                                                    H

C27H34O6
Phthalic acid, benzyl 3-hydroxy-1-isopropyl-2,2-dimethylpropyl ester isobutyrate.
Benzyl 3-isobutyryloxy-1-isopropyl-2,2-dimethylpropyl phthalate; EINECS 240-920-1; Phthalic acid, benzyl 3-hydroxy-1-isopropyl-2,2-dimethylpropyl ester isobutyrate.

366 Benzyl mercaptan
100-53-8   9394   H

C7H8S
Benzenemethanethiol.
4-06-00-02632 (Beilstein Handbook Reference); AI3-22955; Benzenemethanethiol; Benzyl hydrosulfide; Benzyl mercaptan; Benzylhydrosulfide; Benzylthiol; BRN 0605864; EINECS 202-862-5; FEMA No. 2147; HSDB 2105; Methanethiol, phenyl-; NSC 41897; Phenylmethanethiol; Phenylmethyl mercaptan; Thiobenzyl alcohol; Toluene, α-mercapto-; Toluene-α-thiol; USAF ex-1509. Oil; mp = -30°; bp = 194.5°; d^{20} = 1.058; λ_m = 263, 269 nm (cyclohexane); insoluble in H_2O, slightly soluble in CCl_4, soluble in CS_2; very soluble in EtOH, Et_2O.

367 Benzylparaben
94-18-8   G

C14H12O3
p-Hydroxybenzoic acid benzyl ester.
AI3-02955; Benzoic acid, 4-hydroxy-, phenylmethyl ester; Benzoic acid, p-hydroxy-, benzyl ester; Benzyl 4-hydroxybenzoate; Benzyl p-hydroxybenzoate; Benzyl parahydroxybenzoate; Benzyl Parasept; Benzyl Tegosept; Benzylparaben; EINECS 202-311-9; Nipabenzyl; NSC 8080; Parosept; Phenylmethyl 4-hydroxybenzoate; Solbrol Z. Among the most widely used preservatives in food, drugs, and cosmetics. Preservative, bactericide, fungicide for pharmaceuticals, cosmetics, foods, medicinal preparations, industrial applications. Crystals; mp = 110-112°. *Nipa.*

368 Benzyl salicylate
118-58-1   1146   H

C14H12O3
Phenylmethyl 2-hydroxybenzoate.
4-10-00-00157 (Beilstein Handbook Reference); AI3-00517; Benzoic acid, 2-hydroxy-, phenylmethyl ester; Benzyl o-hydroxybenzoate; Benzyl salicylate; BRN 2115365; CCRIS 4749; EINECS 204-262-9; FEMA No. 2151; NSC 6647; Phenylmethyl 2-hydroxybenzoate; Salicylic acid, benzyl ester; Salicylsaeurebenzylester; Salicylsäurebenzylester. Solid; bp = 320°; d^{20} = 1.1799; λ_m = 238, 306 nm (ε = 10700, 4550, MeOH); slightly soluble in H_2O, soluble in EtOH, Et_2O, CCl_4.

369 Benzyl trimethyl ammonium chloride
56-93-9   G

C10H16ClN
Trimethylbenzylammonium chloride.
Ammonium, benzyltrimethyl-, chloride; Benzene-methanaminium, N,N,N-trimethyl-, chloride; Benzyl-trimethylammonium chloride; BTM; CCRIS 4587; EINECS 200-300-3; HSDB 4196; Tmbac; N,N,N-Trimethyl-benzenemethanaminium chloride; Trimethylbenzyl-ammonium chloride; Variquat®. Quaternary ammonium salt; dispersant, dye leveler and retarder, emulsifier used in textile industry; solvent for cellulose; gelling inhibitor in polyester resins; intermediate. Used as a dispersant, dye leveler and retarder, as an emulsifier used in textile industry. White solid; mp = 239° (dec). *Janssen Chimica; Pentagon Chems. Ltd.; Sherex; Sybron.*

370 Benzyltriethyl ammonium chloride
56-37-1   G

C13H22ClN
N,N,N-triethylbenzenemethanaminium chloride.
AI3-14906; Ammonium, benzyltriethyl-, chloride; Benzenemethanaminium, N,N,N-triethyl-, chloride; Benzyl triethyl ammonium chloride; Benzyltriethyl-ammonium chloride; BTEAC; EINECS 200-270-1; Sumquat® 2355; TEAC; TEBA; TEBAC; Triethyl-benzylammonium chloride; N,N,N-Triethylbenzene-methanaminium chloride. Quaternary ammonium salt; solvent for cellulose; gelling inhibitor in polyester resins; intermediate in chemical manufacturing. Also used as a fibre dyeing auxilliary and phase transfer catalyst. Solid; mp = 185°; soluble in H_2O (170 g/100 ml). *Hexcel; Janssen Chimica; Zeeland Chem. Inc.*

371 Berylla
1304-56-9   1176   G

Be=O

BeO
Beryllium oxide.

Beryllia; Beryllium monoxide; Beryllium oxide; Beryllium oxide (BeO); Bromellete; CCRIS 83; EINECS 215-133-1; Glucina; HSDB 1607; Natural bromellite; Thermalox; Thermalox 995. Used in manufacture of beryllium oxide ceramics, glass in nuclear reactor fuels and moderators; electrically resistive; catalyst for organic reactions. Electrical conductor but thermal insulator. Light amorphous powder; mp = 2530°; very sparingly soluble in H_2O.

372   Beryllium
7440-41-7   1164   G

Be

Be
Beryllium.
Beryllium; Beryllium-9; Beryllium dust; Beryllium, elemental; Beryllium metal; Beryllium, metal powder; Beryllium metallic; Beryllium powder; CCRIS 81; EINECS 231-150-7; Glucinium; Glucinum; HSDB 512; RCRA waste number P015; UN1566; UN1567. Metallic element; Structural material in space technology; moderator in nuclear reactors; source of neutrons; windows for x-ray tubes; in gyroscopes, computer parts, inertial guidance systems; additive in solid-propellant rocket fuels; beryllium-copper alloys. mp = 1287°; bp = 2500°; d = 1.8477. 2973; *Degussa AG; Noah Chem*

373   Betaine
107-43-7   1182   D G

$C_5H_{11}NO_2$
1-Carboxy-N,N,N-trimethylmethanaminium inner salt.
α-Earleine; Betaine; 4-04-00-02369 (Beilstein Handbook Reference); Abromine; AI3-24187; AI3-52598; BRN 3537113; EINECS 203-490-6; Jortaine; Loramine AMB 13; Lycine; NSC 166511; Oxyneurine; Rubrine C. Used as a soldering flux, in chemical synthesis and as a hepatoprotectant. Solid; Dec 293°; soluble in H_2O (160 g/100 ml), MeOH (55 g/100 ml), EtOH (8.7 g/100 ml); sparingly soluble in Et_2O. *Sterling Winthrop Inc.*

374   Betamethasone Sodium Phosphate
151-73-5   1183   D G V

$C_{22}H_{28}FNa_2O_8P$
(11β,16β)-9-Fluoro-11,17,21-trihydroxy-16-methylpregna-1,4-diene-3,20-dione 21-phosphate di-sodium salt.
Beta-Methasone phosphate; Betasone (Veterinary); Betavet Soluspan (Veterinary); BDP; Bentelan; Bentelan; Betnesol; Betsolan; Celestone Phosphate; Celestone Phosphate Injection; Celestone Soluspan; Disodium Betamethasone 21-phosphate; EINECS 205-797-0; Linolosal; Linosal; NSC 90616.; Bentalan; Betnersol Injectable; Durabetason; Vista-Methasone. A glucocorticoid used especially in veterinary medicine. *Glaxo Labs.; Merck & Co.Inc.; Roussel-UCLAF; Schering AG.*

375   Betazole hydrochloride
138-92-1   1187   G

$C_5H_{11}Cl_2N_3$
1H-Pyrazole-3-ethanamine dihydrochloride.
3-(2-Aminoethyl)pyrazole dihydrochloride; 3-β-Amino-ethylpyrazole dihydrochloride; Ametazole dihydro-chloride; Betazole dihydrochloride; Betazole hydro-chloride; EINECS 205-345-2; Gastramine; Histalog; Histalog dihydrochloride; Histimin; Pyrazole, 3-(2-aminoethyl)-, dihydrochloride; 1H-Pyrazole-3-ethan-amine, dihydrochloride; 3,5-β-Pyrazoleethylamine dihydrochloride. A stimulant of gastric secretion, used as a diagnostic aid. Crystals; mp = 224-226°; soluble in H_2O, insoluble in EtOH.

376   Bifenox
42576-02-3   1214   G P

$C_{14}H_9Cl_2NO_5$
5-(2,4-Dichlorophenoxy)-2-nitrobenzoic acid methyl ester.
Benzoic acid, 5-(2,4-dichlorophenoxy)-2-nitro-, methyl ester; Bifenox; BRN 2170169; Caswell No. 561AA; CCRIS 7158; 5-(2,4-Dichlorophenoxy)-2-nitrobenzoic acid methyl ester; MC-4379; Modown; Modown 4 Flowable; EPA Pesticide Chemical Code 104301; MC-4379; Methyl 5-(2,4-dichlorophenoxy)-2-nitrobenz-oate; Modown. Selective herbicide used for control of annual broad-leaved weeds and some grasses in cereals, maize, sorghum, soybeans, rice and some other crops. Registered by EPA as a herbicide (cancelled). Yellow crystals; mp = 84-86°; poorly soluble in H_2O (0.000035 g/100 ml), more soluble in Me_2CO (31.5 g/100 ml), chlorobenzene (44.3 g/100 ml), xylene (25.8 g/100 ml), EtOH (< 4 g/100 ml); slightly soluble in aliphatic hydrocarbons; LD_{50} (rat orl) > 6400 mg/kg, (mus orl) = 4556 mg/kg, (rbt der) > 20000 mg/kg; LC_{50} (rat ihl) > 200 mg/l air, (duck, pheasant 8 day dietary) > 5000 mg/kg diet, (rainbow trout 96 hr.) = 0.87 mg/l, (bluegill sunfish 96 hr.) = 0.64 mg/l. *Aventis Crop Science.*

377   Bioresmethrin
28434-01-7   1230   G P

C22H26O3

(5-Benzyl-3-furyl)methyl (+)-trans-2,2-dimethyl-3-(2-methylpropenyl)cyclopropanecarboxylate.

AI 3-27662; Benzyl-3-furylmethyl (+)-trans-chrysanth-emate; 5-Benzyl-3-furylmethyl(+)-trans-chrys-anthem-ate; 5-Benzyl-3-furylmethyl-d-trans-crysanthe-mate; d-trans-((5-Benzyl-3-furyl)methyl)chrysanthe-mumate; d-trans-(5-Benzyl-3-furyl)methyl chrysanth-emate; 5-Benzyl-3-furylmethyl (1R)-trans-chrysanthemate (8CI); 5-Benzyl-3-furylmethyl (1R,3R)-2,2-dimethyl-3-(2-methylprop-1-enyl)cyclopropanecarboxylate; (5-Benzyl-3-furyl)methyl (+)-trans-2,2-dimethyl-3-(2-methylpropenyl)cyclopropanecarboxylate; 5-Benzyl-3-furylmethyl (1R)-trans-2,2-dimethyl-3-(2-mthylprop-1-enyl)cyclopropanecarboxylate; Biobenzyfuroline; Bio-resmethrin; Bioresmethrin; Bioresmethrine; Biores-methrinum; Bioresmetrina; CCRIS 5401; d-trans-Chrysron; Combat White Fly Insecticide; Cyclo-propanecarboxylic acid, 2,2-dimethyl-3-(2-methyl-propenyl)-, (5-benzyl-3-furyl)methyl ester, d-trans-; Cyclopropanecarboxylic acid, 2,2-dimethyl-3-(2-methyl-1-propenyl)-,(5-(phenylmethyl)-3-furanyl)-methyl ester, (1R-trans)-; EINECS 249-014-0; FMC 18739; 3-Furanmethanol, 5-benzyl-, 2,2-dimethyl-3-(2-methyl-propenyl)cyclopropanecarboxylate, t-(+)-; Isathrine; Isatrin; NIA-18739; NRDC 107; Penick 1390; Pyrethroid NRDC 107; Pyrethroids; Resbuthrin; Reslin; (+)-trans-Resmethrin; d-trans-Resmethrin; 1R-trans-Resemethrin; RU-11484; SBP-1390; d-trans-((5-Benzyl-3-furyl)methyl)-chrysanthemumate; d-trans-(5-Benzyl-3-furyl)methyl chrysanthemate; d-trans-Chrysrond-trans-Resmethrin; FMC 18739; 3-Furanmethanol, 5-benzyl-,2,2-dimethyl-3-(2-methylpropenyl)cycloprop-anecarboxylate, t-(+)-; Isathrine; Isatrin; NIA-18739; NRDC 107; Penick 1390; Pyrethroid NRDC 107; Resbuthrin; Resmethrin isomer; (+)-trans-Resmethrin; 1R-trans-Resemethrin; RU-11484; SBP-1390. Reg-istered by EPA as an insecticide. Liquid; mp = 32°; bp0.01 = 181°; $[\alpha]_D^{20}$ = -7.8° (c = 5 Me2CO); insoluble in H2O (< 0.00003 g/100 ml), soluble in most organic solvents; LD50 (mrat orl) = 1244 mg/kg, (frat orl) = 1721 mg/kg, (rat orl) = 7070 - 8000 mg/kg, (frat der) > 10000 mg/kg, (ckn orl) > 10000 mg/kg, (bee orl) = 3 ng/bee; LC50 (rat ihl 4 hr.) = 5.2 mg/l air, (harlequin fish 96 hr.) = 0.014 mg/l, (guppy 96 hr.) = 0.5 - 1.0 mg/l; toxic to bees. AgrEvo; Aventis Environmental Science USA LP; FMC Corp.; Sumitomo Corp.; Valent Biosciences Corp.

378 Biphenyl
92-52-4   3346   G P

C12H10
Phenylbenzene.
AI3-00036; Bibenzene; Biphenyl; Biphenyl, 1,1-; Carolid AL; Caswell No. 087; CCRIS 935; CP 390; Diphenyl; EINECS 202-163-5; EPA Pesticide Chemical Code 017002; FEMA No. 3129; HSDB 530; Lemonene; MCS 1572; NSC 14916; Phenador-X;

Phenylbenzene; PHPH; Tetrosin LY; Xenene. Registered by EPA as an antimicrobial and fungicide (cancelled). Fungicide used in packaging citrus fruit and in plant disease control. Used as a heat transfer agent and in organic synthesis. Colorless crystals; mp = 69°; bp = 256.1°; d20 = 1.04; λ_m = 247 nm (ε = 19300, EtOH); insoluble in H2O, readily soluble in EtOH, Et2O, other organic solvents; LD50 (rat orl) = 3280 mg/kg, (rbt orl) = 2400 mg/kg, (cat orl) > 2600 mg/kg; exposure to > 0.005 mg/l air is dangerous to humans. Coalite Chem. Div.; Fluka; Generic; Monsanto Co.; Sigma-Aldrich Fine Chem.; Sybron.

379 Biphenylyl phenyl ether
28984-89-6   H

C18H14O
1,1'-Biphenyl, phenoxy-.
1,1'-Biphenyl, phenoxy-; Biphenylyl phenyl ether; EINECS 249-357-6; Ether, biphenylyl phenyl; Phenoxy-1,1'-biphenyl; Phenyl biphenyl ether. Mixture of o, m and p-isomers.

380 Bipyridyl
553-26-4   3383   H

C10H8N2
4,4'-Bipyridyl.
5-23-08-00028 (Beilstein Handbook Reference); AI3-00138; 4,4-Bipyridyl; BRN 0113176; CCRIS 2363; 4,4-Dipyridyl; EINECS 209-036-3; 'σ,σ'-Bipyridyl; 'σ,σ'-Dipyridyl; NSC 404423; 4-(4-Pyridyl)pyridine. Needles; mp = 111°; bp = 305°; λ_m = 240 nm (ε = 14900, MeOH); slightly soluble in H2O, very soluble in EtOH, Et2O, C6H6, ligroin.

381 Biguanide
56-03-1   1219   G

C2H7N5
Imidodicarbonimidic diamide.
AI3-52571; Amidinoquanidine; Baquacil; Baquagold; Biguanide; Diquanide; EINECS 200-251-8; Imido-dicarbonimidic diamide; Quanyl-quanidine. Polymeric biguanide used as a swimming pool sanitizer. Salts of biguanide are used in assays for copper and nickel. mp = 130°; dec = 142°; soluble in H2O, alcohol, insoluble in organic solvents. ICI Chem. & Polymers Ltd.

382 α-Bisabolol
515-69-5   1241   D G

C15H26O

α,4-Dimethyl-α-(4-methyl-3-pentenyl)-3-cyclohexene-1-methanol.
Bisabolol; α-Bisabolol; Camilol; 3-Cyclohexene-1-methanol, α,4-dimethyl-α-(4-methyl-3-pentenyl)-; 3-Cyclohexene-1-methanol, α,4-dimethyl-α-(4-methyl-3-pentenyl)-, (R*,R*)-; 3-Cyclohexene-1-methanol, α,4-dimethyl-α-(4-methyl-3-pentenyl)-, (αR,1R)-rel-; (R*, R*)-(1)-α,4-Dimethyl-α-(4-methyl-3-pentenyl)cyclohex-3-ene-1-methanol; (R*,R*)-α,4-Dimethyl-α-(4-methyl-3-pent-enyl)cyclohex-3-ene-1-methanol; EINECS 208-205-9; EINECS 246-973-7; 5-Hepten-2-ol, 6-methyl-2-(4-methyl-3-cyclohexen-1-yl)-; Hydagen® B; 6-Methyl-2-(4-methyl-3-cyclohexen-1-yl)-5-hepten-2-ol. Anti-inflammatory agent for emulsions, oils, lotions, and oral hygiene preparations. Cosmetic ingredient. Oil; bp$_{12}$ = 155-157°; d$_4^{23}$ = 0.9223; insoluble in H$_2$O, soluble in organic solvents. *Henkel/Cospha; Henkel/Organic Prods.; Henkel.*

383 Bis(4-aminocyclohexyl)methane
1761-71-3   H

C13H26N2

Di(p-aminocyclohexyl)methane.
1,4-Bis(aminocyclohexyl)methane; Bis(4-aminocyclo-hexyl)methane; BRN 2079179; CCRIS 4777; Cyclo-hexanamine, 4,4'-methylenebis-; 4,4'-Diaminodicyclo-hexylmethane; Di(p-aminocyclohexyl)methane; p,p'-Diaminodicyclohexylmethane; EINECS 217-168-8; HLR 4219; HLR 4448; Methylenebis(4-aminocyclo-hexane); 4,4'-Methylenebis(cyclohexylamine); PACM 20; Wandamin HM. Liquid; mp = 15°; bp = 320°; d^{75} = 0.92.

384 Bis(6-aminohexyl)amine
143-23-7   H

C12H29N3

Dihexylamine, 6,6'-diamino-.
Bis(6-aminohexyl)amine; Bis(hexamethylene)triamine; Dihexylamine, 6,6'-diamino-; Dihexylenetriamine; EINECS 205-593-1; HSDB 5646; N-(Aminohexyl)-1,6-hexanediamine; N-(6-Aminohexyl)-1,6-hexanedi-amine; NSC 92231.

385 Bis(adiponitrile)bis(cyanotriphenyl-borato) nickel
83864-02-2   H

C50H46B2N6Ni

Nickel, bis[(cyano-C)triphenylborato(1-)-N]bis(hexane-dinitrile-N,N')-.
Bis(adiponitrile)bis(cyanotriphenylborato)nickel; Nick-el, bis((cyano-C)triphenylborato(1-)-N)bis(hexanedinit-rile-N,N')-.

386 Bis(4-aminophenyl) ether
101-80-4   H

C12H12N2O

4,4'-Diaminodiphenyl ether.
4-13-00-01038 (Beilstein Handbook Reference); AI3-18375; Aniline, 4,4'-oxydi-; Benzenamine, 4,4'-oxybis-; Bis(4-aminophenyl) ether; Bis(p-aminophenyl) ether; BRN 0475735; CCRIS 491; Dadpe; Di-(4-aminophenyl)ether; Diaminodiphenyl ether; EINECS 202-977-0; Ether, 4,4'-diaminodiphenyl; HSDB 1316; NCI-C50146; NSC 37075; p,p'-Oxydianiline. Crystals; mp = 189°.

387 Bis(2,4-di-tert-butylphenyl)pentaerythritol diphosphite
26741-53-7   H

C33H50O6P2

Phosphorous acid, cyclic neopentanetetrayl bis(2,4-di-tert-butylphenyl) ester.
Bis(2,4-di-tert-butylphenyl)pentaerythritol diphosphite; 3,9-Bis(2,4-di-tert-butylphenoxy)-2,4,8,10-tetraoxa-3,9-diphosphaspiro(5.5)undecane; BRN 4773772; EINECS 247-952-5; MARK PEP 24; Phosphorous acid, cyclic neopentanetetrayl bis(2,4-di-tert-butylphenyl) ester; 2,4,8,10-Tetraoxa-3,9-diphosphaspiro(5.5)un-decane, 3,9-bis(2,4-bis(1,1-dimethylethyl)phenoxy)-; Ultranox 624; Ultranox 626; Weston 626; Weston MDW 626.

388 Bis(2-benzothiazyl) disulfide
120-78-5   3406   G H

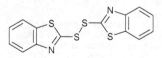

68

$C_{14}H_8N_2S_4$

2-Mercaptobenzothiazole disulfide.

4-27-00-01862 (Beilstein Handbook Reference); Accel TM; AI3-07662; Altax; Benzothiazole, 2,2'-dithiobis-; Benzthiazole disulfide; BRN 0285796; Caswell No. 408A; CCRIS 4637; Dithiobis(benzothiazole); Dwusiarczek dwubenzotiazylu; EINECS 204-424-9; Ekagom GS; EPA Pesticide Chemical Code 009202; HSDB 1137; MBTS; MBTS rubber accelerator; Mercaptobenzthiazyl ether; NSC 2; Perkacit® MBTS; Pneumax DM; Royal MBTS; Thiofide; USAF CY-5; USAF EK-5432; Vulcafor MBTS; Vulkacit® DM; Vulkacit® DM/C; Vulkacit® DM/mgc. Dibenzothiazyl disulfide, mineral oil coated; delayed action, semi-ultra accelerator for NR, IR, BR, SBR, NBR, CR, IIR, and chlorosulfonated polyethylene; used for mechanical goods, tires, conveyor belts, cables, hoses, rubber footwear, expanded rubber goods. A component of mercapto mix.

Crystals; mp = 168°; d = 1.5; insoluble in H_2O (<0.1 mg/ml), slightly soluble in organic solvents; LD_{50} (rat orl) > 12 g/kg. Akzo Chemie; Bayer AG; Polysar.

389 Bis(butoxyethoxyethoxy)methane
143-29-3   H

$C_{17}H_{36}O_6$

Bis(2-(2-butoxyethoxy)ethoxy)methane.

Bis(butoxyethoxyethoxy)methane; Bis(butoxyethoxyethyl)-formal; Bis(butylcarbitol)formal; BRN 2450899; Cryoflex; Dibutoxyethoxyethyl formal; EINECS 205-598-9; 5,8,11,13,16,19-Hexaoxatricosane; HSDB 5647; NSC 7185; TP 90B.

390 Bis(4-t-butylcyclohexyl) peroxydicarbonate
15520-11-3   G

$C_{22}H_{38}O_6$

Peroxydicarbonic acid, bis(4-(1,1-dimethylethyl)cyclo-hexyl) ester.

Bis(4-tert-butylcyclohexyl)peroxydicarbonate; EINECS 239-557-1; Perkadox® 16; Peroxydicarbonic acid, bis(4-(1,1-dimethylethyl)cyclohexyl) ester. Ultrafast initiator for polyester cure above 180°F; pultrusion, matched die molding; short ambient temperature compound shelf life. Akzo Chemie.

391 Bisbutyl peroxy trimethylcyclohexane
6731-36-8   G

$C_{17}H_{34}O_4$

1,1-bis(t-butylperoxy)-3,3,5-trimethylcyclohexane.

Aztec®; Aztec® 1,1-Bis(t-Butylperoxy)-3,3,5-Trimethyl Cyclohexane; 1,1-Bis(t-butylperoxy)-3,3,5-trimethyl-cyclohexane; BRN 5932965; Di-tert-butyl 3,3,5-trimethyl-cyclohexylidene diperoxide; Di-tert-butyl-peroxy-3,3,5-trimethylcyclohexane peroxide; EINECS 229-782-3; Luperco 231G; Luperco 231XL; Luperco 231XLP; Luperox 231; Lupersol 231; Perhexa 3M; Perhexa 3M40; Peroxide, (3,3,5-trimethylcyclohexyl-idene)bis(tert-butyl); Peroxide, (3,3,5-trimethylcyclo-hexylidene)bis((1,1-di-methylethyl); Peroximon® S-164/40P; Trigonox® 29; Trigonox® 29/40; Trigonox® 29/40mb; Trigonox® 29b50; Trigonox® 29b75; Trigonox® 29c75; Varox 231xl. Polymerization initiator and catalyst for polymerization reactions. Akrochem Chem. Co.; Akzo Chemie; Catalyst Resources Inc.

392 Bis(2-chloroethyl) (2-(((2-chloro-ethoxy)(2-chloroethyl)phosphinyl)oxy)ethyl)-phosphonate
58823-09-9   H

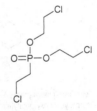

$C_{10}H_{20}Cl_4O_6P_2$

Bis(2-chloroethyl) (2-(((2-chloroethoxy)(2-chloroethyl)-phosphinyl)oxy)ethyl)phosphonate.

Bis(2-chloroethyl) (2-(((2-chloroethoxy)(2-chloroethyl)-phosphinyl)oxy)ethyl)phosphonate; (2-(((2-Chloroeth-oxy)-(2-chloroethyl)phosphinyl)oxy)ethyl)-phosph-onic acid, bis(2-chloroethyl) ester; EINECS 261-459-2; Phos-phonic acid, (2-(((2-chloroethoxy)(2-chloroethyl)-phos-phinyl)-oxy)ethyl)-, bis(2-chloroethyl) ester.

393 Bis(2-chloroethyl) 2-chloroethylphos-phonate
6294-34-4   H

$C_6H_{12}Cl_3O_3P$

Phosphonic acid, (2-chloroethyl)-, bis(2-chloroethyl) ester.

AI3-25413; Antiblage 78; Bis-chloroethyl 2-chloro-ethanephosphonate; Bis(β-chloroethyl) β-chloroethyl-phosphonate; Bis(2-chloroethyl) (2-chloroethyl)phos-phonate; Ethanol, 2-chloro-, (2-chloroethyl)phosphonate (2:1); NSC 9297; Phosphonic acid, (2-chloroethyl)-, bis(2-chloroethyl) ester. Liquid; bp_5 = 170.2°.

394 Bis(2-chloroethyl) ether
111-44-4   3091   H

$C_4H_8Cl_2O$

Di(2-chloroethyl) ether.

4-01-00-01375 (Beilstein Handbook Reference); AI3-04504; BCEE; Bis(β-chloroethyl) ether; Bis(2-chloroethyl) ether; Bis(chloro-2-ethyl)

69

oxide; BRN 0605317; Caswell No. 309; CCRIS 88; Chlorex; Chloroethyl ether; Clorex; DCEE; Di(β-chloro-ethyl)ether; Di(2-chloroethyl) ether; Dichloroether; Dichloroether; Dichloroethyl ether; Dichloroethyl oxide; Dicholoroethyl ether; Dwuchlorodwuetylowy eter; EINECS 203-870-1; ENT 4,504; EPA Pesticide Chemical Code 029501; Ethane, 1,1'-oxybis(2-chloro-; Ether, bis(2-chloroethyl); Ether dichlore; HSDB 502; Khloreks; NSC 406647; Oxyde de chlorethyle; RCRA waste number U025; sym-Dichloroethyl ether; UN1916. Liquid; mp = -51.9°; bp = 178.5°; d^{20} = 1.22; insoluble in H_2O, soluble in EtOH, Et_2O, Me_2CO, freely soluble in C_6H_6, MeOH.

395 Bis(3,5-di-tert-butyl-4-hydroxyhydro-cinnamoyl)hydrazine
32687-78-8   H

$C_{34}H_{52}N_2O_4$
N,N'-Bis(3-(3,5-di-tert-butyl-4-hydroxyphenyl)propionyl)-hydrazine.
2-(3-(3,5-Bis(1,1-dimethylethyl)-4-hydroxyphenyl)-1-oxo-propyl)hydrazide;   Bis(3,5-di-tert-butyl-4-hydroxy-hydrocinnamoyl)hydrazine; 2',3-Bis((3-(3,5-di-tert-but-yl-4-hydroxyphenyl)propionyl))propionohydrazide; 3, 5-Bis-(1,1-dimethylethyl)-4-hydroxybenzenepropionic acid; EINECS 251-156-3; Irganox 1024; Irganox MD 1024; MD 1024; N,N'-Bis(3-(3',5'-di-tert-butyl-4'-hydroxyphenyl)-propionyl)hydrazine.

396 2,2-Bis(4-(3,4-dicarboxyphenoxy)-phenyl)propane dianhydride
38103-06-9   H

$C_{31}H_{20}O_8$
4,4'-((Isopropylidene)bis(p-phenyleneoxy))diphthalic dianhydride.
2,2-Bis(4-(3,4-dicarboxyphenoxy)phenyl)propane di-anhydride; EINECS 253-781-7; 1,3-Isobenzofuran-dione, 5,5'-((1-methylethylidene)bis(4,1-phenylene-oxy))bis-; 4,4'-((Isopropylidene)bis(p-phenyleneoxy))-diphthalic dianhydride.

397 Bis(2-(dimethylamino)ethyl) ether
3033-62-3   H

$C_8H_{20}N_2O$
2,2'-Oxybis(N,N-dimethylethylamine).
2,2'-Oxybis(N,N-dimethylethylamine); 4-04-00-01441 (Beilstein Handbook Reference); A 99 (Amine); Bis(2-(dimethylamino)ethyl) ether; BRN 1739668; Dabco BL 11; Dabco BL 19; Dabco BL 19I; EINECS 221-220-5; Ethylamine, 2,2'-oxybis(N,N-dimethyl-; Kalpur PC; Niax A 1; Niax A 4; Niax A 99; Niax catalyst al; NSC 109887; Texacat ZF 20; Toyocat ET; Toyocat ETS.

398 Bis(1,3-dimethylbutyl) dithiophosphate
6028-47-3   H

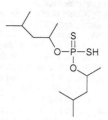

$C_{12}H_{27}O_2PS_2$
2-Pentanol, 4-methyl-, hydrogen phosphorodithioate.
Bis(1,3-dimethylbutyl) dithiophosphate; O,O-Bis(1,3-dimethylbutyl) phosphorodithioate; EINECS 227-900-8; 2-Pentanol, 4-methyl-, hydrogen phosphorodithioate; 2-Pentanol, 4-methyl-, O,O-diester with phosphorodithioic acid.

399 1,7-Bis(1,3-dimethylbutylidene)diethylenetriamine
10595-60-5   H

$C_{16}H_{33}N_3$
N,N'-Bis(1,3-dimethylbutylidene)-2,2'-iminobis(ethyl-amine).
1,7-Bis(1,3-dimethylbutylidene) diethylenetriamine; N,N'-Bis(1,3-dimethylbutylidene)-2,2'-iminobis(ethylamine); Diethylenetriamine, 1,7-bis(1,3-dimethylbutylidene)-; EINECS 234-205-3.

400 Bis(1,3-dimethylbutyl) maleate
105-52-2   H

$C_{16}H_{28}O_4$
Bis(1,3-dimethylbutyl)-2-butenedioate (2Z)-.
Bis(1,3-dimethylbutyl)-2-butenedioate (2Z)-; Bis-(1,3-dimethylbutyl)ester kyseliny maleinove; Bis(1,3-dimethylbutyl) maleate; 2-Butenedioic acid, bis(1,3-dimethylbutyl) ester; Di(1,3-dimethylbutyl)maleate; Di(4-methyl-2-amyl) maleate; Di(4-methyl-2-pentyl) maleate; DMAM; EINECS 203-304-3; Maleic acid, bis(1,3-dimethylbutyl) ester; Maleic acid, di(1,3-dimethylbutyl) ester.

401 Bis(dimethyldithiocarbamato)zinc
137-30-4   10225   G P

$C_6H_{12}N_2S_4Zn$
Bis(dimethylcarbamodithioato-S,S')zinc.
AAProtect; Aaprotent; Aavolex; Aazira; Accelerator L; Accelerator MZ Powder; Aceto ZDED; Aceto zdmd; AI3-00988; Alcobam ZM; Antene;   Bis-dimethyldithio-carbamate   de   zinc; Bis(dimethylcarbamodithioato-S,S')zinc; Bis(dimethyldithiocarbamato)zinc;   (T-4)-Bis-(dimethyldithiocarbamato-S,S')zinc;   Bis(N,N-dimetil-ditiocarbammato) di zinco; Carbamic acid, dimethyl-dithio-, zinc salt; Carbamodithioic acid, dimethyl-, zinc salt; Carbazinc; Caswell No. 931; CCRIS 625; Corona corozate; Corozate; Crittam; Crittan;

70

Cuman; Cuman L; Cymate; Dimethylcarbamodithioic acid, zinc salt; Dimethyldithiocarbamate zinc salt; Drupina 90; EINECS 205-288-3; ENT 988; EPA Pesticide Chemical Code 034805; Eptac 1; Fuclasin; Fuclasin Ultra; Fuklasin; Fungostop; Hermat ZDM; Hexazir; HSDB 1788; Karbam White; Methasan; Methazate; Methyl zimate; Methyl zineb; Methyl Ziram; Mezene; Micosin F30; Milbam; Milban; Molurame; Mycronil; NCI-C50442; Nocceler pz; NSC 737; Orchard brand ziram; Pomarsol Z forte; Pomarzol Z-forte; Prodaram; Ramedit; RCRA waste no. P205; Rhodiacid; Rodisan; Sabceler PZ; Soxinal PZ; Soxinol PZ; Tricarbamix Z; Trikagol; Triscabol; Tsimat; Tsiram; Ultra Zinc DMC; USAF P-2; Vancide MA-96; Vandice® MZ-96; Vulcacure ZM; Vulkacit L; Vulkacite L; Weisstaub; Z 75; Z-C spray; Zarlate; Zerlate; Zimate®, methyl; Zinc, bis(dimethylcarbamodithioato-S,S')-; Zinc bis(dimethyl-dithiocarbamoyl)disulphide; Zinc bis(di-methyldithio-carbamate); Zinc bis(dimethylthiocarbamoyl) disulfide; Zinc dimethyldithiocarbamate; Zink-bis(N,N-dimethyl-dithiocarbamaat); Zinkcarbamate; Ziradin; Ziram; Ziram Granuflo; Ziram technical; Zirame; Zirame; Ziramvis; Zirasan; Zirasan 90; Zirberk; Ziretec; Zirex 90; Zirex Fungicide; Ziride; Zirthane; Zitox; Vancide MZ96; Zincmate; Ziram; Ziram F4; Ziram W76; Ziramvis; Zirasan 90; Zirex 90; Ziride; Zitox; Fuclasin Ultra; Accelerator L; Methazate; Aceto ZDED; Aceto ZDMD; Methyl Ziram; Mexene; Molurame; Orchard Brand Ziram; Pomarsol Z Forte; Corona Corozate; Cymate; EPTAC 1; Rhodiacid; Soxinal PZ; Soxinol PZ; Tsimat; Vulcacure; Vulcacure ZM; Vulkacite L; Z 75; ZC; Z-C Spray; Zirasan; Aavolex; Aazira; Antene; Amyl Zimate; Ciram; Cuman L; Hermat Zdm; Methyl Zineb; Mycronil; Zarlate; Zirthane; Alcobam Zm;milam; Pomarsolz; Vancide. Non-discoloring, nonstaining accelerator for natural rubber, isoprene rubber, BR, styrene/butadiene rubber, isobutylene-isoprene rubber, and ethylene-propylene-diene rubbers and natural rubber latex. Rubber vulcanization accelerator and agricultural fungicide. Antimicrobial, preservative for starch and synthetic latex adhesives and food packaging adhesives. Used as a bird and animal repellent. mp = 250°; d4²⁵ = 1.66; insoluble in H₂O (0.006 g/100 ml), slightly soluble in organic solvents; LD₅₀ (rat orl) = 1.4 g/kg. *Akrochem Chem. Co.; Universal Crop Protection Ltd.; Vanderbilt R.T. Co. Inc.*

402 Bis(2-ethylhexyl) dithiophosphate
5810-88-8   H

C₁₆H₃₅O₂PS₂
Phosphorodithioic acid, O,O-bis(2-ethylhexyl) ester.
Bis(2-ethylhexyl) dithiophosphate; Bis(2-ethylhexyl) hydrogen dithiophosphate; Bis(2-ethylhexyl) phosphoro-dithioate; BRN 1913062; Di-2-ethylhexylphosphoro-dithioic acid; EINECS 227-376-0; HSDB 2612; Phoslex DT 8; Phosphorodithioic acid, O,O-bis(2-ethylhexyl) ester.

403 Bis(2-ethylhexyl) peroxydicarbonate
16111-62-9   H

C₁₈H₃₄O₆
Peroxydicarbonic acid, bis(2-ethylhexyl) ester.
Bis(2-ethylhexyl) perdicarbonate; Bis(2-ethylhexyl)

peroxydicarbonate; BRN 5964960; Di(2-ethylhexyl) peroxydicarbonate; EINECS 240-282-4; Lupersol 223; Peroxydicarbonic acid, bis(2-ethylhexyl) ester; Peroxydicarbonic acid, di(2-ethylhexyl) ester; Peroyl OPP; Peroyl OPP 60E; Trigonox® EHP.

404 Bis(2-ethylhexyl) terephthalate
6422-86-2   G H

C₂₄H₃₈O₄
1,4-Benzenedicarboxylic acid, bis(2-ethylhexyl) ester.
4-09-00-03306 (Beilstein Handbook Reference); AI3-17104; BRN 2302822; CCRIS 7052; Di-(2-ethylhexyl) terephthalate; EINECS 229-176-9; HSDB 6150; Kodaflex® DOTP; Terephthalic acid, bis(2-ethylhexyl) ester. Plasticizer used with PVC resins, in PVC plastisols, rubber; application including wire coatings, automotive and furniture upholstery; compatible with acrylics, CAB, cellulose nitrate, polyvinyl butyral, styrene, oxidizing alkyds, nitrile rubber. Solid; mp = 30-34°; bp = 400°; d = 0.980. *Eastman Chem. Co.*

405 Bis(2-ethylhexyl) tetrabromophthalate
26040-51-7   H

C₂₄H₃₄Br₄O₄
Phthalic acid, tetrabromo-, di(2-ethylhexyl) ester.
1,2-Benzenedicarboxylic acid, 3,4,5,6-tetrabromo-, bis(2-ethylhexyl) ester; Bis(2-ethylhexyl) tetrabromophthalate; EINECS 247-426-5; Phthalic acid, tetrabromo-, bis(2-ethylhexyl) ester.

406 Bis(2-ethylhexyl)phthalate
117-81-7   2884   G H

C₂₄H₃₈O₄
Di-(2-ethylhexyl) phthalate.
4-09-00-03181 (Beilstein Handbook Reference); AI3-04273; Bis(2-ethylhexyl) 1,2-benzenedicarboxylate; Bis-(2-ethylhexyl)ester kyseliny ftalove; Bis(2-ethylhexyl) phthalate; Bisoflex 81; Bisoflex DOP; BRN 1890696; Caswell No. 392K; CCRIS 237; Celluflex DOP; Compound 889; DEHP; Di-(2-ethylhexyl) phthalate; Di-sec-octyl phthalate; Diacizer DOP; Dioctyl phthalate; DOF; EINECS 204-211-0; EPA Pesticide Chemical Code 295200; Ergoplast FDO; Ergoplast FDO-S; Etalon; Ethyl hexyl phthalate; Eviplast 80; Eviplast 81; Fleximel; Flexol DOP; Flexol Plasticizer DOP; Good-rite GP 264;

Hatcol DOP; Hercoflex 260; HSDB 339; Jayflex DOP; Kodaflex® DEHP; Kodaflex® DOP; Mollan O; Monocizer DOP; NCI-C52733; NSC 17069; Nuoplaz DOP; Octoil; Octyl phthalate; Palatinol AH; Phthalic acid, bis(2-ethylhexyl) ester; Pittsburgh PX-138; Plasthall DOP; Platinol AH; Platinol DOP; PX-138; RC Plasticizer DOP; RCRA waste number U028; Reomol D 79P; Reomol DOP; Sansocizer DOP; Sansocizer R 8000; Sconamoll DOP; Sicol 150; Staflex DOP; Truflex DOP; Vestinol AH; Vinicizer 80; Witcizer 312. All-purpose plasticizer used with PVC resins incl. film and sheeting for upholstery, clothing, food pkg., paper coatings, molded vinyl products, electrical wire insulation; compatible with PS, methylmethacrylate, chlorinated rubber, NC, and CAB; low odor, relatively low toxicity and low volatility. Also used in vacuum pumps. Registered by EPA (cancelled). Light-colored liquid; mp = -55°; bp = 384°, bp$_5$ = 231°; d$_{20}^{20}$ = 0.9861; viscosity (centistokes) = 386 at 0°, 22 at 20°, 31 at 37.8°, 5 at 100°; insoluble in H$_2$O, slightly soluble in CCl$_4$; soluble in mineral oil. LD$_{50}$ (rat orl) = 30600 mg/kg; TLV = 5 mg/m^3. *Aristech; BASF Corp.; BP Chem.; Chemisphere; Chisso Corp.; Daihachi Chem. Ind. Co. Ltd; Degussa-Hüls Corp.; Eastman Chem. Co.; Hall C.P.; Mitsubishi Gas; Sigma-Aldrich Co.; UCB SA.*

407   Bis(3-glycidoxypropyl)tetramethyl-disiloxane
126-80-7   G

C$_{16}$H$_{34}$O$_5$Si$_2$
1,3-Bis(3-(glycidyloxy)propyl)-1,1,3,3-tetramethyldi-siloxane.
1,3-Bis(3-(2,3-epoxypropoxy)propyl)-1,1,3,3-tetramethyl-disiloxane; 1,3-Bis(3-(2,3-epoxypropoxy)-propyl)tetra-methyldisiloxane; Disiloxane, 1,1,3,3-tetramethyl-1,3-bis(3-(oxiranylmethoxy)propyl)-; Disil-oxane, 1,3-bis(3-(2,3-epoxypropoxy)propyl)-1,1,3,3-tetramethyl-; EINECS 204-803-9; NSC 93976. Coupling agent, chemical intermediate, blocking agent, release agent, lubricant, primer, reducing agent. *Degussa-Hüls Corp.*

408   Bis(2-hydroxy-3-tert-butyl-5-methyl-phenyl)methane
119-47-1   G H

C$_{22}$H$_{32}$O$_2$
2,2'-Methylenebis(6-tert-butyl-p-cresol).
2246; 4-06-00-06801 (Beilstein Handbook Reference); 6,6'-Di-tert-butyl-2,2'-methylenedi-p-cresol; A 22-46; Advastab 405; Al3-18027; Alterungsschutzmittel BKF; Antage W 400; Anti Ox; Antioxidant 1; Antioxidant 2246; Antioxidant BKF; Antioxidant NG-2246; AO 1 (Antioxidant); AO 2246; AO1; Bis(2-hydroxy-3-tert-butyl-5-methylphenyl)methane; Bisaklofen BP; BKF; BRN 2062676; Calco 2246; CAO 14; CAO 5; Catolin 14; CCRIS 4919; Chemanox 21; Cyanox® 2246; EINECS 204-327-1; HSDB 5585; Lederle 2246; Methane, 2,2'-bis(6-t-butyl-p-cresyl)-; Methane, 2,2'-bis(6-tert-butyl-p-cresyl)-; Methylene bis methyl butyl phenol; 2,2'-Methylenebis(4-methyl-6-tert-butylphen-ol); 2,2'-Methylenebis(6-(1,1-dimethylethyl))-p-cresol); 2,2'-Methylenebis(6-tert-butyl-p-cresol); NG 2246; Nocrac NS 6; NSC 7781; Oxy Chek 114; Plastanox 2246; Plastanox 2246 Antioxidant; S 67; Sumilizer MDP; Synox 5LT; Ultranox® 2246; Vulkanox® BKF. Antioxidant preventing thermal oxidation of ABS, polyethylene, PP, and EVA; oxidation inhibitor for fats, oils, and paraffin wax; polymerization inhibitor in chemical processes; for ABS, hot-melt adhesives, latex carpet backing, and speciality olefin

applications. Stabilizer, antioxidant for PP, polyethylene, Polyoxymethylene copolymer, polyoxymethylene homopolymer, styrenics, and rubber. Nonstaining antioxidant for natural and synthetic rubbers; used in transparent, bathing, surgical techniques and latex goods, fabric proofings; stabilizer for hot-and cold-polymerized emulsion-SBR, BR, and NBR, and solution-BR, IR, and ABS. Solid; mp = 123-127°. *Am. Cyanamid; Bayer AG; Ciba-Geigy Corp.; GE Silicones; Polysar.*

409   Bis(2-hydroxyethyl)lauramide
120-40-1   G P

C$_{16}$H$_{33}$NO$_3$
N,N-Bis(2-hydroxyethyl)lauramide.
4-04-00-01539 (Beilstein Handbook Reference); Ablumide LDE; Al3-09484; Alkamide® 327; Alkamide LE; Amide, dodecyl-,; Aminon L 02; Amisol LDE; Bis(2-hydroxyethyl)lauramide; BRN 1791417; Caswell No. 519A; CCRIS 4662; Chemistat 2500; Chemstat LD 100; Clindrol 100L; Clindrol 101CG; Clindrol 200L; Clindrol 203CG; Clindrol 210CGN; Clindrol superamide 100L; Coco diethanolamide; Coconut oil amide of diethanolamine; Comperlan LD; Condensate PL; Crillon LDE; Diethanol lauric acid amide; Diethanolamine lauric acid amide; Diethanolamine lauroylamide; Diethanollauramide; Dodecanamide, N,N-bis(2-hydroxyethyl)-; Duspar LA 2000; EINECS 204-393-1; EMID 6511; Empilan LDE; EPA Pesticide Chemical Code 079018; Ethylan MLD; Hetamide ML; HSDB 5586; Incromide LR; Lalmin D; Lankrostat JP; Lauramide DEA; Lauramido DEA; Lauric acid diethanolamide; Lauric acid diethanolamine condensate (1:1); Lauric diethanolamide; Lauroyl diethanolamide; Lauryl diethanolamide; LDA; LDE; Mackamide LL; Mackamide LLM; Methyl laurate-diethanolamine condensate; Monamid 150-LW; NCI-C55323; Ninol 4821; Ninol AA-62 Extra; Ninol AA62; Ninol P-621; Onyxol 345; Rewomid DL 203/S; Rewomid DLMS; Richamide 6310; Richamide STD; Rolamid CD; Standamid LD; Standamidd LD; Steinamid DL 203 S; Stepan LDA; Super amide L-9A; Super amide L-9C; Synotol L-60; Unamide J-56; Varamid ML 1; Varamide ML 1; Witcamide® 5195. Mixture of ethanolamides of lauric acid; foam booster/stabilizer, detergency and viscosity builder, emulsifier, wetting agent for personal care products, household and institutional detergents. Viscosity modifier, foam stabilizer, lubricant, conditioner, lubricant, emulsifier, wetting agent, thickener, penetrant, dye dispersant, scouring aid, antistat; cosmetics and toiletries; industrial foamer and stabilizer; metal processing; textile surfactant; Aerosol®® formulations. Widely used in cosmetics, shampoos, soaps, and related consumer products, to which there is extensive human exposure. Registered by EPA as an antimicrobial, fungicide, insecticide and rodenticide (cancelled).

410   2,2-Bis(hydroxymethyl)propionic acid
4767-03-7   3279   H

C$_5$H$_{10}$O$_4$
α,α-Bis(hydroxymethyl)propionic acid
α,α-Bis(hydroxymethyl)propionic acid; 2,2-Bis(-hydroxymethyl)propionic acid; 2,2-Dimethylolprop-ionic acid; Dimethylolpropionic acid; EINECS 225-306-3; Hydracrylic acid, 2-(hydroxymethyl)-2-methyl-; NSC 96616; Propanoic acid, 3-hydroxy-2-(hydroxy-methyl)-2-methyl-; Propionic acid, 2,2-bis(hydroxy-

72

methyl)-. Solid; mp = 190°.

411 Bis(2-hydroxypropyl) ether
110-98-5   G H

$C_6H_{14}O_3$
2,2'-Dihydroxydipropyl ether.
4-01-00-02473 (Beilstein Handbook Reference); 4-Oxaheptane-2,6-diol; AI3-01233; Bis(2-hydroxypropyl) ether; BRN 1698372; 2,2'-Dihydroxydipropyl ether; 1,1'-Dimethyldiethylene glycol; EINECS 203-821-4; HSDB 2854; NIAX catalyst D-19; NSC 8688; 1,1'-Oxybis-2-propanol; 1,1'-Oxydipropan-2-ol; 2-Prop-anol, 1,1'-oxydi-; 2-Propanol, 1,1'-oxybis-. Used as a solvent; in polyester and alkyd resins, reinforced plastics, plastics, plasticizers and solvents. Liquid; bp = 233°; d = 1.023; soluble in H2O, organic solvents; irritant, mildly toxic by ingestion. *Allchem Ind.; Arco; Ashland; Berje; Brown Chem. Co.; Coyne; Great Western; Olin Corp; PMC; Sigma-Aldrich Fine Chem.; Texaco.*

412 Bis(isobutyl)aluminum chloride
1779-25-5   H

$C_8H_{18}AlCl$
Chlorobis(2-methylpropyl)aluminum.
Alluminio diisobutil-monocloruro; Aluminum, chloro-bis-(2-methylpropyl)-; Aluminum, chlorodiisobutyl-; Bis-(isobutyl)aluminum chloride; Chlorobis(2-methyl-propyl)-aluminum; Chlorodiisobutylaluminium; Dibac; Diiso-butylaluminum chloride; Diisobutylaluminum mono-chloride; Diisobutylchloroaluminum; EINECS 217-216-8; HSDB 5856.

413 Bis(1-methylamyl) Sodium Sulfosuccinate
6001-97-4   1254   G

$C_{16}H_{29}NaO_7S$
Sulfobutanedioic acid 1,4-bis(1-methylpentyl) ester sodium salt.
Alphasol MA; Aerosol®® MA; Bis(1-methylamyl) sodium sulfosuccinate; Butanedioic acid, sulfo-, 1,4-bis(1-methylpentyl) ester, sodium salt; Dihexyl sodium sulfosuccinate; EINECS 227-847-0; Lankropol® KMA; Sodium dihexyl sulfosuccinate; Sodium 1,4-diisohexyl 2-sulphosuccinate. Emulsifier, wetting agent, especially in solutions of electrolytes; used as a solubilizer for soaps, emulsion polymerization aid. Solid; soluble in H2O, pine oil, oleic acid, Me2CO, kerosene, CCl4, EtOH, C6H6, hot olive oil, glycerol; insoluble in liquid petrolatum. *Harcros.*

414 Bis(o-methylbenzoyl)peroxide
3034-79-5   G

$C_{16}H_{14}O_4$
Di(2-methylbenzoyl) peroxide.
Bis(o-toluoyl) peroxide; EINECS 221-231-5; Perkadox® 20; Peroxide, bis(2-methylbenzoyl). Initiator. *Akzo Chemie.*

415 Bis(1-methylheptyl) adipate
108-63-4   H

$C_{22}H_{42}O_4$
Bis(1-methylheptyl) hexanedioate.
Adipic acid, bis(1-methylheptyl) ester; Bis(1-methylheptyl) adipate; EINECS 203-601-8; Bis(1-methylheptyl) hexanedioate; Hexanedioic acid, bis(1-methylheptyl) ester; NSC 6197.

416 m-Bis(1-methylvinyl)benzene
3748-13-8   H

$C_{12}H_{14}$
1,3-Bis(1-methylethenyl)benzene.
Benzene, 1,3-bis(1-methylethenyl)-; Benzene, m-diisopropenyl-; 1,3-Bis(1-methylethenyl)benzene; BRN 2073277; 1,3-Diisopropenylbenzene; EINECS 223-146-9; m-Bis(1-methylvinyl)benzene; m-Diisopropenylbenzene; m-DIPEB.

417 Bis(4-nitrophenyl) ether
101-63-3   H

$C_{12}H_8N_2O_5$
4,4'-Dinitrodiphenyl oxide.
4-06-00-01290 (Beilstein Handbook Reference); AI3-08831; Benzene, 1,1'-oxybis(4-nitro-; Bis(4-nitrophenyl) ether; BRN 2058965; 4,4'-Dinitrodiphenyl ether; p,p'-Dinitrodiphenyl ether; Di-4-nitrophenyl ether; EINECS 202-961-3; Ether, bis(4-nitrophenyl); NSC 8740; p-Nitrophenyl ether; Oxybis(4-nitrobenzene).

418 2,5-Bis(tert-nonyldithio)-1,3,4-thia-diazole
89347-09-1   H

C20H38N2S5
1,3,4-Thiadiazole, 2,5-bis(tert-nonyldithio)-.
2,5-Bis(tert-nonyldithio)-1,3,4-thiadiazole; EINECS 289-493-3;
1,3,4-Thiadiazole, 2,5-bis(tert-nonyldithio)

419   Bis(nonylphenyl)amine
36878-20-3   H

C30H47N
Benzenamine, ar-nonyl-N-(nonylphenyl)-.
Benzenamine, ar-nonyl-N-(nonylphenyl)-; Bis(nonyl-phenyl)amine;
EINECS 253-249-4.

420   Bis(4-nonylphenyl) chlorophosphite
63302-49-8   H

C30H746ClO2P
Phosphorochloridous acid, bis(4-nonylphenyl) ester.
Bis(4-nonylphenyl) chlorophosphite; EINECS 264-082-1;
Phosphorochloridous acid, bis(4-nonylphenyl) ester.
421   Bis(p-octylphenyl)amine
421   Bis(p-octylphenyl)amine
101-67-7   G H

C28H43N
4,4'-Dioctyldiphenylamine.
AI3-17279; Benzenamine, 4-octyl-N-(4-octylphenyl)-; Bis(4-
octylphenyl)amine; Bis(p-octylphenyl)amine; CCRIS 6029; Di-n-octyl
diphenylamine; p,p-Dioctyl-diphenylamine; Diphenylamine, 4,4'-
dioctyl-; EINECS 202-965-5; HSDB 5341; Naugalube® 438; NSC
79268. Antioxidant in automatic transmission fluids, turbine oil, and
synthetic lubricants used in jet turbine engines; thermal stabilizer for
automatic transmission fluid at high temperatures. Crystals; mp= 96-
97°; insoluble in H2O (<1 mg/ml); LD50 (rat orl) = 7.58 g/kg. *Uniroyal.*

**422   N,N'-Bis(2,2,6,6-tetramethyl-4-piperidinyl)-1,6-
hexanediamine**
61260-55-7   H

C24H50N4
1,6-Hexanediamine, N,N'-bis(2,2,6,6-tetramethyl-4-pip-eridinyl)-.
A 31; A 31 (stabilizer); N,N'-Bis(2,2,6,6-tetramethyl-piperidin-4-
yl)hexane-1,6-diamine; EINECS 262-679-1; 1,6-Hexanediamine,

N,N'-bis(2,2,6,6-tetramethyl-4-pip-eridinyl)-.

423   Bis(2,2,6,6-tetramethyl-4-piperidinyl) sebacate
52829-07-9   G H

C28H52N2O4
Decanedioic acid, bis(2,2,6,6-tetramethyl-4-piperid-inyl) ester.
ADK Stab LA 77; Bis(2,2,6,6-tetramethyl-4-piperidinyl)
decanedioate; Bis(2,2,6,6-tetramethyl-4-piperidinyl)-sebacate;
Decanedioic acid, bis(2,2,6,6-tetramethyl-4-piperidinyl) ester;
EINECS 258-207-9; HALS 1; HALS 770; Lowilite® 77; LS 770; Mark
LA 77; Sanol; Sanol 770; Sanol LS 700; Sanol LS 770; Sumisorb
577; T 770; TIN 770; Tinuvin 770; Tinuvin 770DF; Tinuvin 770LS;
TK-10665; Uvaseb 770. Light stabilizer for polyolefins, ABS, PUR
acrylics, PS and styrene copolymers, thermoplastic elastomers,
ethylene-propylene copolymers, polyamides. *Ciba-Geigy Corp.;
EniChem SpA; Lowi.*

424   1,2-Bis(2,4,6-tribromophenoxy)ethane
37853-59-1   H

C14H8Br6O2
1,1'-(1,2-Ethanediylbis(oxy))bis(2,4,6-tribromobenzene).
Benzene, 1,1'-(1,2-ethanediylbis(oxy))bis(2,4,6-tri-bromo-; 1,2-
Bis(2,4,6-tribromophenoxy)ethane; 1,2-Bis(tribromo-
phenoxy)ethane; CCRIS 4752; EINECS 253-692-3; Ethane, 1,2-
bis(2,4,6-tribromophenoxy)-; 1,1'-(1,2-Ethane-diylbis(oxy))bis(2,4,6-
tribromo-benzene); 1,1'-(Ethane-1,2-diylbisoxy)bis(2,4,6-tri-
bromobenzene); FF 680; FF 680; FireMaster 680; FireMaster FF
680; HSDB 6099.

425   Bis(tributyltin) Oxide
56-35-9   H P

C24H54OSn2
5,5,7,7-Tetrabutyl-6-oxa-5,7-distannaundecane.
AF-SeafloZ-100; AI3-24979; AW 75-D; Biomet; Biomet 66; Biomet
75; Biomet TBTO; BioMeT SRM; Bis-(tri-n-butylcin)oxid; Bis(tri-n-
butyltin) oxide; Bis(tri-n-butylzinn)-oxyd; Bis(tri-n-tributyltin)oxide;
Bis(tri-butyloxide) of tin; Bis(tributylstannium) oxide; Bis(tri-
butylstannyl)oxide; Bis(tributyltin) oxide; BKA007; BMC; BTO;
Butinox; C-Sn-9; Caswell No. 101; CCRIS 3697; Distannoxane,
hexabutyl-; EINECS 200-268-0; ENT 24,979; EPA Pesticide
Chemical Code 083001; HBD; Hexabutyldistannioxan;
Hexabutyldistannoxane; Hexa-butylditin; HSDB 6505; Interlux

74

Micron; Interswift; Kyslicnik tri-n-butylcinicity; L.S. 3394; Lastanox F; Lastanox Q; Lastanox T; Lastanox T 20; MBC; Mykolastanox F; Navicote 2000; NSC 22332; OTBE; Oxybis(tributyltin); Oxyde de tributyletain; Sigmaplane 7284; Stannane, tri-n-butyl-, oxide; Stannicide A; Super Sea Jacket; TBOT; TBTO; Tin, bis(tributyl)-, oxide; Tin, oxybis(tributyl-; Tributyltin oxide; Vikol AF-25; Vikol LO-25; ZK 21995. Registered by EPA as a fungicide. Primary use is as a fungicide, secondary uses are as algicide, molluscicide and bactericide. Colorless - light yellow liquid; bp$_2$ = 180°; mp = -45°; d$_4^{20}$ = 1.14; soluble in H$_2$O (0.01 g/100 ml at 25°), in sea water (0.006 g/100 ml), soluble in most organic solvents. *Akcros Chemicals America; Akzo Nobel Chemicals Inc; Betzdearborn; Biosentry, Inc.; CK Witco Chem. Corp.; Elf Atochem N. Am.; Eprova AG; Flexabar Corp.; H & S Chemicals Division; IBC Manufacturing Co.; International Paint, Inc.; Janssen Chimica; La Littorale; PPG Ind.; Thomson Research Associates; Vinings Industries Inc.*

426   Bis(tridecyl) adipate
16958-92-2   H

C32H62O4
Hexanedioic acid, ditridecyl ester.
Adipic acid, ditridecyl ester; Bis(tridecyl) adipate; Dilinoleic acid, ditridecyl ester; Dimer acid, ditridecyl ester; Ditridecyl adipate; Ditridecyl dimer dilinoleate; Ditridecyl dimerate; Ditridecyl hexanedioate; EINECS 241-029-0; Hexanedioic acid, ditridecyl ester.

427   Bis(3,5,5-trimethylhexanoyl) peroxide
3851-87-4   G

C18H34O4
Peroxide, bis(3,5,5-trimethyl-1-oxohexyl).
Bis(1-oxo-3,5,5-trimethylhexyl)peroxide;   Bis(3,5,5-tri-methyl-1-oxohexyl) peroxide; EINECS 223-356-0; Peroxide, bis(3,5,5-trimethyl-1-oxohexyl); 3,5,5-Tri-methylhexanoyl peroxide; USP® -355M. Used as an initiator for radical polymerization. *Ethitek Pharmaceuticals Co.*

428   Bismuth
7440-69-9   1256   G

Bi

Bi
Bismuth.
Bismuth; Bismuth-209; Bismuth, elemental; EINECS 231-177-4; HSDB 2078. Used in the synthesis of pharmaceuticals and medicinals, cosmetics, alloys, catalyst in making acrylonitrile, additive, coating selenium. Metal; mp = 271°; bp = 1420°. *Asarco; Atomergic Chemetals; Frys Metals Ltd; Noah Chem*

429   Bismuth dimethyldithiocarbamate
21260-46-8   G

C9H18BiN3S6
Bismuth, tris(dimethylcarbamodithioato-S,S')-, (OC-6-11)-
Bismate®; Bismet; Bismuth dimethyldithiocarbamate; Bismuth N,N-dimethyldithiocarbamate; Bismuth, tris-(dimethylcarbamodithioato); Bismuth, tris(dimethyl-carbamodithioato-S,S')-; Bismuth tris(dimethyldithio-carbamate); Carbamic acid, dimethyldithio-, bismuth salt; Carbamic acid, dimethyldithio-, bismuth(3+) salt; Carbamodithioic acid, dimethyl-, bismuth salt; EINECS 244-299-8; NSC 70079; Tris(dimethyldithiocarbam-ato)-bismuth. Ultra accelerator for NR, IR, BR, and SBR; for high temperature, high speed vulcanization of compounds which must be high-temperature cured; nonstaining. *Akrochem Chem. Co.; Vanderbilt R.T. Co. Inc.*

430   Bismuth nitrate
10361-46-3   G

Bi(NO3)3
Bismuth (III) nitrate, basic.
Bismuthyl nitrate; Bismuthoxy nitrate; Bismuth trinitrate; (Nitrooxy)oxobismuthine; EINECS 233-792-3. Preparation of other bismuth salts, bismuth luster on tin, luminous paints and enamels, precipitation of alkaloids. Crystals; d = 2.83; soluble in H$_2$O containing nitric acid. *Atomergic Chemetals; Mallinckrodt Inc.; Nihon Kagaku Sangyo; Noah Chem.*

431   Bismuth oxide
1304-76-3   1273   D G

Bi2O3
bismuth trioxide.
Bismuth oxide; Bismuth oxide (Bi2O3); Bismuth sesquioxide; Bismuth sesquioxide (Bi2O3); Bismuth trioxide; Bismuth Yellow; Bismuth(3+) oxide; Bismuth(III) oxide; Bismuthous oxide; C.I. 77160; Dibismuth trioxide; EINECS 215-134-7; Flowers of Bismuth. Used in enameling cast iron, ceramic, porcelain colors. In disinfectants, magnets, glass, rubber vulcanization, fireproofing papers and polymers. An astringent, also used in disinfectants, magnets and catalysts. Yellow powder, practically insoluble in H$_2$O. *Atomergic Chemetals; Ferro/Transelco; Mallinckrodt Inc.; Nihon Kagaku Sangyo.*

432   Bismuth oxychloride
7787-59-9   1262   D G

BiClO
Bismuth(III) oxychloride.
Basic bismuth chloride; BIJU; Bismuth chloride oxide; Bismuth oxychloride; Bismuth subchloride; Bismuthine, chlorooxo-; Bismuthyl chloride; Blanc d'Espagne; Blanc de perle; Chlorbismol;

Chlorooxobismuthine; CI 77163; EINECS 232-122-7; Flake white; Mearlite; Pearl white; Pigment white 14; U Pink 113, U Violet 109; Ultramarine pink. Bismuth oxychloride in liquid suspension; in face powders; as pigment; used in the manufacture of artificial pearls, dry-cell cathodes. A basic bismuth nitrate, Bi(OH)2NO3 is also known as pearl white, a pigment. The term is somctimes used in connection with a white lead which has been tinted with Paris blue or indigo. Has been used to treat syphilis. Fine powder or crystals; d = 7.72; melts at low red heat; insoluble in H2O; soluble in HCl, HNO3. Atomergic; British; Chemetals; Drug; Dyk; Houses; Mallinckrodt; Mearl; Van.

433   Bismuth subcarbonate
5892-10-4   1283   D G V

CBi2O5
Bismuth oxycarbonate.
Amforol (Veterinary); Basic bismuth carbonate; Bismuth subcarbonate; Dibismuth carbonate dioxide; 2,4-Dioxa-1,5-dibismapentane, 1,3,5-trioxo-; EINECS 227-567-9. Bismuth compounds, cosmetics, opacifier in x-ray diagnosis, enamel fluxes, ceramic glazes. FDA permanently listed. Used as a skin protectant and as a color additive for externally applied pharmaceuticals. Odorless, tasteless powder; insoluble in H2O, EtOH; d = 6.860. *Atomergic Chemetals; Mallinckrodt Inc.; Spectrum Chem. Manufacturing.*

434   Bismuth subnitrate
1304-85-4   1285   D G

H9Bi5N4O22
Bismuth hydroxide nitrate oxide.
Basic bismuth nitrate; Bismuth hydroxide nitrate oxide (Bi5(OH)9(NO3)4O); Bismuth magistery; Bismuth nitrate, basic; Bismuth oxide (Bi2O3), compd with nitrogen oxide (N2O5) (6:5); Bismuth oxynitrate; Bismuth paint; Bismuth subnitrate; Bismuth subnitricum; Bismuth White; Bismuthyl nitrate; Blanc de fard; C.I. 77169; C.I. Pigment White 17; Cosmetic White; EINECS 215-136-8; Flake White; HSDB 1608; Magistery of bismuth; Mammol; Novismuth; Paint White; Snowcal 5 SW; Snowcal 5SW; Spanish White; Vicalin; Vikaline. Used in cosmetics, ceramic glazes, enamel fluxes and as an antacid. Odorless, tastless, slightly hydroscopic microcrystalline powder; practically insoluble in H2O and alcohol. *Atomergic Chemetals; Celtic Chem. Ltd.; Greeff R.W. & Co.; Mallinckrodt Inc.*

435   Bismuth subsalicylate
14882-18-9   1286   D G

C7H5BiO4
(2-Hydroxybenzoato-O^1)-oxobismuth.
Basic bismuth salicylate; Bismogenol tosse inj; Bismuth, (2-hydroxybenzoato-O1,O2)oxo-; Bismuth oxide salicylate; Bismuth, oxo(salicylato)-; Bismuth oxy-salicylate; Bismuth salicylate, basic; Bismuth subsalicylate; CCRIS 4751; EINECS 238-953-1; HSDB 332; (2-Hydroxybenzoato-O^1)-oxobismuth; 2-Hydroxy-benzoic acid bismuth(3+) salt; 2-Hydroxybenzoic acid bismuth (3+) salt, basic; Salicylic acid basic bismuth salt; Salicylic acid, bismuth basic salt;

Stabisol. Used in surface-coating plastics and copying paper. Imparts pearly surface to cellulose-base, polystyrene and phenol-formaldehyde resins. Used medicinally as an antidiarrheal, antacid and antiulcerative. White solid; almost insoluble in H2O, EtOH. *Atomergic Chemetals; Greeff R.W. & Co.; Mallinckrodt Inc.; Mobay; Spectrum Chem. Manufacturing.*

436   Bisoprolol
66722-44-9   1294   D G

C18H31NO4
(±)-1-[[α(2-Isopropoxyethoxy)-p-tolyl]oxy]-3-(isoprop-yl-amino)-2-propanol.
Bisoprolol; Bisoprololum; CL 297,939; EMD 33 512. Cardioselective β1-adrenergic blocker. Antihyper-tensive. *E. Merck.*

437   Bisoprolol Hemifumarate
104344-23-2   1294   D

C20H33NO6
(±)-1-[[α(2-Isopropoxyethoxy)-p-tolyl]oxy]-3-(isopropyl-amino)-2-propanol hemifumarate.
Biso-Puren; Bisobloc; Bisomerck; Bisoprolol fumarate; Bisoprolol hemifumarate; CL 297,939; Concor; Concor Plus Forte; Detensiel; Emcor; EMD 33512; Emvoncor; Euradal; Eurtadal; Fondril; Godal; Isoten; Maintate; Monocor; Soprol; TA-4708; Zebeta; Ziac.; component of: Ziac. Cardioselective β1-adrenergic blocker. Anti-hypertensive. Crystals; mp = 100°; soluble in EtOH. *E. Merck.*

438   Bisphenol A
80-05-7   1296   G H

C15H16O2
Bis(4-hydroxyphenyl) dimethylmethane.
4-06-00-06717 (Beilstein Handbook Reference); AI3-04009; Biphenol A; Bis(4-hydroxyphenyl) dimethyl-methane; Bis(4-hydroxyphenyl)propane; Bisferol A; Bis-phenol; Bisphenol A; BRN 1107700; CCRIS 95; DIAN; Diano; Dimethyl bis(p-hydroxyphenyl)methane; Di-methylmethylene-p,p'-diphenol; Diphenylolpropane; EINECS 201-245-8; HSDB 513; Ipognox 88;

NCI-C50635; NSC 1767; Parabis A; Phenol, (1-methylethylidene)bis-; Phenol, 4,4'-(1-methylethyl-idene)bis-; Phenol, 4,4'-isopropylidenedi-; Pluracol 245; Propane, 2,2-bis(p-hydroxyphenyl)-; Rikabanol; Ucar bisphenol A; Ucar bisphenol HP. Intermediate in manufacture of epoxy, polycarbonate, phenoxy, polysulfone, polyester resins, flame retardant, rubber chemicals; fungicide. Solid; mp = 147.9°; bp_{10} = 250°; λ_m = 226, 248, 277 nm (ε 15000, 25900, 1310, MeOH); insoluble in H_2O; soluble in organic solvents; LD_{50} (rat orl) = 3250 mg/kg. *Aristech; Mitsui Petroleum; Mitsui Toatsu; Shell.*

439 Bisphenol A diglycidyl ether
1675-54-3   H

C22H14O4
Bis(4-glycidyloxyphenyl)dimethylmethane.
2,2-Bis(p-(2,3-epoxypropoxy)phenyl)propane; 4,4'-Bis-(2,3-epoxypropoxy)diphenyldimethylmethane; 2,2-Bis-(4-(2,3-epoxypropyloxy)phenyl)propane; Bis(4-glyc-idyloxy-phenyl)dimethylmethane; 2,2-Bis(4-glycidyl-oxyphenyl)-propane; 2,2-Bis(p-glycidyloxyphenyl)-propane; Bis(4-hydroxyphenyl)dimethylmethane di-glycidyl ether; 2,2-Bis(4-hydroxyphenyl)propane, di-glycidyl ether; 2,2-Bis(p-hydroxyphenyl)propane, di-glycidyl ether; Bisphenol A diglycidyl ether; Bpdge; CCRIS 1965; D.E.R. 332; Dian-bis-glycidylether; Dian diglycidyl ether; Diglycidyl bisphenol A ether; Diglycidyl bisphenol A; Diglycidyl diphenylolpropane ether; Diglycidyl ether of 4,4'-isopropylidenediphenol; Diglycidyl ether of 2,2-bis(p-hydroxyphenyl)propane; Diglycidyl ether of 2,2-bis(4-hydroxyphenyl)propane; Diglycidyl ether of bisphenol A; 4,4'-Dihydroxy-diphenyldimethylmethane diglycidyl ether; p,p'-Di-hydroxydiphenyldimethylmethane diglycidyl ether; Diomethane diglycidyl ether; EINECS 216-823-5; EP 274; Epi-Rez 508; epi-Rez 510; Epi-Rez 510; Epophen EL 5; Epotuf 37-140; Epoxide A; ERL-2774; GY 6010; HSDB 331; 4,4'-Isopropylidenebis(1-(2,3-epoxy-prop-oxy)benzene); 4,4'-Isopropylidenediphenol diglyc-idyl ether; 2,2'-((1-Methylethylidene)bis(4,1-phenylene-oxymethylene))bisoxirane; NSC 5022; Oligomer 340; Oxirane, 2,2'-((1-methylethylidene)bis(4,1-phenylene-oxy-methylene))bis-; Propane, 2,2-bis (p-(2,3-epoxy-propoxy)phenyl); Propane, 2,2-bis(4-(2,3-epoxy-prop-oxy)phenyl)-.

440 Bisphenol A sodium salt
2444-90-8   H

C15H14Na2O2
4,4'-(1-Methylethylidene)bisphenol disodium salt.
Bisphenol A disodium salt; Bisphenol A sodium salt; Diphenylolpropane disodium salt; Disodium 4,4'-isopropylidenediphenolate; EINECS 219-488-3; HSDB 5660; Phenol, 4,4'-(1-methylethylidene)bis-, disodium salt; Phenol, 4,4'-isopropylidenedi-, disodium salt; Sodium, (isopropylidenebis(p-phenyleneoxy))di-.

441 Bisphenol-A polycarbonate
25037-45-0   H

(C15H16O2.CH2O3)x
Carbonic acid, polyester with 4,4'-isopropylidenediphenol.
Bisphenol-A-polycarbonate; Carbonic acid, polyester with 4,4'-isopropylidenediphenol; Carbonic acid, polymer with 4,4'-(1-methylethylidene)bis(phenol).

442 Bistrichloromethylsulfone
3064-70-8   G P

C2Cl6O2S
Sulfonylbis(trichloromethane).
4-03-00-00285 (Beilstein Handbook Reference); Amerstat® 294; Bis(trichloromethyl)sulfone; Bis(trichloro-methyl) sulphone; BRN 1783352; Caswell No. 104; CCRIS 809; Chlorosulfona; Clorosulfona; EINECS 221-310-4; EPA Pesticide Chemical Code 035601; Hexachlorodimethyl sulfone; HSDB 7126; Methane, sulfonylbis(trichloro-; N 1386; N-1386 biocide; NSC 21713; Slimicide E; Stauffer N-1386®; Sulfone, bis(trichloromethyl); Trichloromethylsulfone. Industrial biocide for control of algae, bacteria and fungi. Used to control bacteria and fungi in wastewater, as a slimicide for paper/paperboard production; a preservative for adhesives, latexes, secondary oil well recovery. Registered by EPA as an algicide, fungicide, herbicide and antimicrobial. *Akzo Chemie; Betzdearborn; Drew Industrial Div.; Hercules Inc.; Ondeo Nalco Co.; Verichem Inc.*

443 Bistrimethylsilyl acetamide
10416-59-8   G

C8H21NOSi2
N,O-Bis(trimethylsilyl)acetamide.
Acetimidic acid, N-(trimethylsilyl)-, trimethylsilyl ester; BRN 1306669; CB2500; Dynasylan® BSA; EINECS 233-892-7; Ethanimidic acid, N-(trimethylsilyl)-, trimethylsilyl ester; Trimethylsilyl N-trimethylsilylacetamidate; N-(Trimethylsilyl)acetimidic acid, trimethylsilyl ester; N,O-Bis(trimethylsilyl)acetamide. Coupling agent, chemical intermediate, blocking agent, release agent, lubricant, primer, reducing agent. Liquid, d = 0.832; bp = 71-73°; n_D^{20}= 1.418; flash pt=11°; moderately toxic by intraperitoneal route. *Degussa-Hüls Corp.*

444 Bitertanol
55179-31-2   1301   G P

C20H23N3O2
β-[(1,1'-Biphenyl)-4-yloxy]-α-(1,1-dimethylethyl)-1H-1,2,4-triazole-1-ethanol.
5-26-01-00134 (Beilstein Handbook Reference); Bay KWG 0599; Baycor; Baycor 25 WP; Baymat-spray; Biloxazol; Bitertanol; BRN 0620948; EINECS 259-513-5; KWG 0599; Sibutol. Broad spectrum fungicide. Mixture of diastereoisomers. Systemic foliar fungicide used especially to control scab on apples and black spot on roses. Crystals; mp = 140° (diastereoisomer A), 146° (diastereoisomer B), 118° (Eutectic mixture); soluble in H_2O (0.0005 g/100 ml), C_6H_{14} (0.1 - 1.0 g/100 ml), CH_2Cl_2 (10 - 20 g/100 ml), i-PrOH (3 - 10 g/100 ml), cyclohexanone (5 - 10 g/100 ml), C_7H_8 (1 - 3 g/100 ml); LD_{50} (rat orl) > 5000 mg/kg, (mus orl) = 4200 - 4500 mg/kg, (dog orl) > 5000 mg/kg, (rat der) > 5000 mg/kg; LC_{50} (carp 48 hr.) = 2.5 mg/l, (rainbow trout 96 hr.) = 2.2 - 2.7 mg/l; non-toxic to bees. *Bayer AG; Bayer Corp., Agriculture.*

445   Biuret
108-19-0   1305   H

C2H5N3O2
Dicarbamylamine.
AI3-14905; Allophanamide; Allophanic acid amide; Allophanimidic acid; Biuret; Carbamoylurea; Carbamyl-urea; Caswell No. 106A; Caswell No. 159A; Di-carbamylamine; EINECS 203-559-0; EPA Pesticide Chemical Code 206200; HSDB 5382; Imidodicarbonic diamide; Isobiuret; NSC 8020; Urea, (aminocarbonyl)-; Ureidoformamide. Crystals; mp = 190° (dec); λ_m = 216 nm (ε = 5623, pH 13); insoluble in Et_2O, slightly soluble in H_2O, very soluble in EtOH.

446   Bone charcoal
8021-99-6   H

Bone charcoal.
Bone black; Bone charcoal; Charcoal, bone; CI 77267; EINECS 232-421-2; Pigment black 9.

447   Borethyl
97-94-9   G

C6H15B
Triethylboron.
4-04-00-04359 (Beilstein Handbook Reference); Borane, triethyl-; Boron ethyl; Boron triethyl; BRN 1731462; EINECS 202-620-9; HSDB 897; Triethylborane; Triethylborine; Triethylboron. Used as an initiator of radical reactions. Liquid; mp = -93°; bp = 95°; d^{23} = 0.700;

soluble in EtOH, Et_2O; LD_{50} (rat orl) = 235 mg/kg.

448   Boric acid
10043-35-3   1326   D G P

H3BO3
Orthoboric acid.
AI3-02406; Ant Flip; Basilit B; Bluboro; Boracic acid; Boric acid; Borofax; Boron hydroxide; Boron trihydroxide; Borsaure; Caswell No. 109; CCRIS 855; Collyrium Eye Wash; Collyrium Fresh-Eye Drops; component of Aci-Jel; Dr.'s 1 Flea Terminator DTPBO; Dr.'s 1 Flea Terminator DF; Dr.'s 1 Flea Terminator DFPBO; Dr.'s 1 Flea Terminator DT; EINECS 233-139-2; EPA Pesticide Chemical Code 011001; Flea Prufe; Homberg's salt; HSDB 1432; NCI-C56417; NSC 81726; Orthoboric acid; Sal Sedativus; Super Flea Eliminator; Three elephant; Trihydroxyborone.; component of: Bluboro, Borofax, Collyrium Eye Wash, Collyrium Fresh-Eye Drops. Used for waterproofing wood; fireproofing fabrics; preservative; manuf.cements, ceramics, glass, hats, soaps, leather, borates, enamels, crockery; cosmetics; printing, painting, and dyeing; photography; hardening steel. Active ingredient in ant bait; for household, hospital, restaurant use for controlling pharoah and other sweet eating ants; workers carry bait back to nest to kill the colony. Fungicide. Registered by EPA as a fungicide, herbicide and insecticide. Used medicinally as an astringent and antiseptic. Transparent crystals, mp = 171°; pH of 0.1M solution = 5.1; soluble in H_2O (5.5 g/100 ml at 20°, 25 g/100 ml at 100°), EtOH (16.6 g/100 ml at 78°), glycerol (25 g/100 ml); LD_{50} (rat orl) = 5140 mg/kg. *Allergan Herbert; Amvac Chemical Corp.; Arch Wood Protection Inc.; Drexel Chemical Co.; Glaxo Wellcome Inc.; J.C. Chemical Co.; Janssen Chimica; OxyChem; Qualis Inc.; U.S. Borax Inc.; Waterbury Companies Inc.; Willert Home Products Inc.; Wyeth-Ayerst Labs.*

449   Boric oxide
1303-86-2   1327   G P

B2O3
Boron trioxide.
AI3-51845; Boric acid (HBO_2), anhydride; Boric anhydride; Boric oxide; Boric oxide (B_2O_3); Boron oxide; Boron sesquioxide; Caswell No. 109B; Diboron trioxide; EINECS 215-125-8; Fused boric acid; HSDB 1609. Used in production of boron, heat-resistant glassware, fire-resistant additive for paints, electronics, liquid encapsulation techniques, herbicide. White crystals; mp = 450°; bp = 1860°; d = 2.46; soluble in H_2O (3 g/100 ml); LD_{50} (mus orl) = 3163 mg/kg. *Atomergic Chemetals; Lancaster Synthesis Co.; Noah Chem.; U.S. Borax Inc.*

450   Borneol
507-70-0   1328   G

C10H18O
endo-1,7,7-Trimethylbicyclo[2.2.1]heptan-2-ol.
AI3-00116; Baros; Baros camphor; Bhimsaim camphor; Bicyclo(2.2.1)heptan-2-ol, 1,7,7-trimethyl-, endo-; Bicyclo(2.2.1)heptan-2-ol, 1,7,7-trimethyl-, (1R,2S,4R)-rel-; 2-

Bornanol, endo-; Borneo camphor; Borneol; DL-Borneol; trans-Borneol; Bornyl alcohol; 2-endo-Bornyl alcohol; Camphane, 2-hydroxy-; 2-Camphanol; Camphol; CCRIS 7300; Dryobalanops camphor; EINECS 208-080-0; FEMA No. 2157; HSDB 946; 2-Hydroxybornane; 2-Hydroxycamphane; 2-Hydroxy-1,7,7-trimethylnor-bornane, endo-; Malayan camphor; Sumatra camphor; 1,7,7-Trimethyl-bicyclo(2.2.1)heptan-2-ol, endo-; UN1312. A terpene from *Dryobalanops camphora*; used in perfumery, in celluloid manufacture, and in medicine as an antiseptic. Crystals; mp = 208°; bp = 212°; insoluble in H_2O, soluble in organic solvents; LD_{50} (rat orl) = 500 mg/kg.

451   Boron
7440-42-8   1333   G

B

B
Boron.
Boron; Boron, metallic; EINECS 231-151-2; HSDB 4482. Nonmetallic element. Used in special-purpose alloys; cementation of iron; neutron absorber in reactor controls; oxygen scavenger for copper and other metals; fibers and filaments in composites; semiconductors; rocket propellant mixtures. mp = 2200°; insoluble in H_2O. *Atomergic Chemetals; Noah Chem.*

452   Boron bromide
10294-33-4   1337   G

BBr3
Boron tribromide.
Borane, tribromo-; Boron bromide; Boron bromide (BBr3); Boron tribromide; EINECS 233-657-9; HSDB 327; Tribromoborane; Tribromoboron; Trona; UN2692. Used as a catalyst in organic synthesis, in the manufacture of diborane and boron. Liquid; mp = -46.0°; bp = 91°; d_0 = 2.650; reacts with H_2O. *Janssen Chimica; Kerr-McGee; Sigma-Aldrich Fine Chem.*; 3973Aldrich; Atomergic; Chemetals; Chimica; Janssen; Kerr-McGee.

453   Boron chloride
10294-34-5   1338   G

BCl3
Boron trichloride.
Borane, trichloro-; Boron chloride (BCl3); Boron trichloride; Chlorure de bore (French); EINECS 233-658-4; HSDB 326; Trichloroborane; Trichloroboron; UN 1741. Catalyst in organic synthesis; source of boron compounds; refining of alloys; soldering flux; electrical resistors; extinguishing magnesium fires in heat-treating furnaces; manufacture of diborane and boron. Gas; mp = -107°; bp = 12.5°; d_0 = 0.7380. *Air Products & Chemicals Inc.; Atomergic Chemetals; Kerr-McGee; Sigma-Aldrich Fine Chem.*

454   Boron nitride
10043-11-5   1336   G

B≡N

BN
Boron mononitride.

BN 40SHP; Borazon; Boron nitride; Boron nitride (BN); BZN 550; Denka boron nitride GP; Denka GP; EINECS 233-136-6; Elbor; Elbor LO 10B1-100; Elbor R; Elbor RM; Elboron; Geksanit R; Hexanit R; Hexanite R; KBN-H10; Kubonit; Kubonit KR; Sho BN; Sho BN HPS; SP 1; SP 1 (Nitride); Super mighty M; UHP-Ex; Wurzin. Used in manufacture of alloys, in semiconductors, nuclear reactors, lubricants. Solid; mp = 3000°. *Atomergic Chemetals; Carborundum Corp.; New Metals & Chems. Ltd; Noah Chem.*

455   Boron trifluoride
7637-07-2   1339   G

BF3
Boron fluoride.
Anca 1040; Borane, trifluoro-; Boron fluoride; Boron fluoride (BF3); Boron trifluoride; EINECS 231-569-5; Fluorure de bore; HSDB 325; Trifluoroborane; Trifluoroboron; UN1008. Lewis acid catalyst in organic synthesis; production of diborane; instruments for measuring neutron intensity; soldering fluxes; gas brazing. Gas; bp = -127.1°; d = 0.8700. *Air Products & Chemicals Inc.; Akzo Chemie; Allied Signal; Atomergic Chemetals; Sigma-Aldrich Fine Chem.*

456   Bretonite
3019-04-3   G

C3H5IO
Iodoacetone.
EINECS 221-161-5; Iodoacetone; 1-Iodoacetone; 1-Iodo-2-propanone; 2-Propanone, 1-iodo-.

457   Brodifacoum
56073-10-0   1363   G P

C31H23BrO3
3-(3-(4'-Bromo(1,1'-biphenyl)-4-yl)-1,2,3,4-tetrahydro-1-naphthalenyl)-4-hydroxycoumarin.
2H-1-Benzopyran-2-one,   3-(3-(4'-bromo(1,1'-biphen-yl)-4-yl)-1,2,3,4-tetrahydro-1-naphthalenyl)-4-hydroxy-; BFC; Brodifacoum; Brodifakum; Bromfenacoum; 3-(3-(4'-Bromobiphenyl-4-yl)-1,2,3,4-tetrahydronaphth-1-yl)-4-hydroxycoumarin;   3-(3-(4'-Bromo-1,1'-biphenyl-4-yl)-1,2,3,4-tetrahydro-1-naphthyl)-4-hydroxycoum-arin; 3-(3-(4'-Bromo-(1,1'-biphenyl)-4-yl)-1,2,3,4-tetra-hydro-1-napthalenyl)-4-hydroxycoumarin;   3-(3-(4'-Bromo(1,1'-biphenyl)-4-yl)-1,2,3,4-tetrahydro-1-naph-thalenyl)-4-hydroxy-2H-1-benzopyran-2-one; Caswell No. 114AAA; Coumarin, 3-(3-(4'-bromo-1,1'-biphenyl-4-yl)-1,2,3,4-tetrahydro-1-naphthyl)-4-hydroxy-; EINECS 259-980-5; EPA Pesticide Chemical Code 112701; HSDB 3916; 4-Hydroxy-3-(3-(4'-bromo-4-bi-phenylyl)-1,2,3,4-tetrahydro-1-naphthyl)coumarin; Klerat; PP 581; Ratak; Super warfarin; Talon; Talon rodenticide; Volid; WBA 8119; 3-(3-(4'-Bromobiphenyl-4-yl)-1,2,3,4-tetrahydronaphth-1-yl)-4-hydroxycoumarin; Caswell No. 114AAA; EPA Pesticide Chemical

79

Code 112701; Havoc; HSDB 3916; Klerat; PP581; Ratak; Ratak Plus; Talon; Talon Rat Bait; Talon rodenticide; Volid; WBA 8119. Rodenticide. Registered by EPA as a rodenticide and insecticide. Off-white powder; mp = 228-230°; insoluble in H_2O (< 0.001 g/100 ml at 20°, pH 7), slightly soluble in EtOH, C_6H_6 (< 0.0006 g/100 ml), soluble in Me_2CO (0.002 g/100 ml), $CHCl_3$ (0.0003 g/100 ml); LD_{50} (mrat orl) = 0.16 mg/kg, 0.27 mg/kg, (frat orl) = 0.18 mg/kg, (mrbt orl) = 0.3 mg/kg, (mmus orl) = 0.4 mg/kg, (fgpg orl) = 2.8 mg/kg, (cat orl) = 25 mg/kg, (dog orl) = 0.25 - 1.0 mg/kg, (rat der 6 hr.) = 200 mg/kg, (ckn orl) = 4.5 mg/kg, (mallard duck orl) = 2 mg/kg; LC_{50} (bluegill sunfish 96 hr.) = 0.165 mg/l, (rainbow trout 96 hr.) = 0.051 mg/l. *Bell Laboratories Inc.; Hacco Inc.; Reckitt Benckiser Inc.; Sorex Ltd.; Syngenta Crop Protection.*

458   Bromacil
314-40-9   1365   G H P

$C_9H_{13}BrN_2O_2$
5-Bromo-6-methyl-3-(1-methylpropyl)-2,4(1H,3H)pyr-imidinedione.
5-24-07-00040 (Beilstein Handbook Reference); Borea; Borocil; BRN 0647896; Bromacil; Bromax; Bromazil; Caswell No. 111; Cynogan; Du Pont herbicide 976; Eerex Granular Weed Killer; Eerex Water Soluble Concentrate Weed Killer; EINECS 206-245-1; EPA Pesticide Chemical Code 012301; Herbicide 976; Hibor; HSDB 1522; Hyvar; Hyvar X; Hyvar X-L; Hyvar-X Weed Killer; Hyvar X-WS; Hyvar-xl Weed Killer; Hyvarex; Istemul; Krovar I; Krovar I Weed Killer; Krovar II; Krovar II Weed Killer; Nalkil; Rout G-8; Uragan; Uragon; Ureabor; Urox B; Urox B water soluble concentrate weed killer; Urox ha; Urox HX; Urox HX granular weed killer; Weed Killer. A versatile herbicide for control of established annual and perennial broadleaf weeds and grasses and brush. Photosynthesis inhibitor, used as a herbicide for total weed and brush control on non-crop land and selective control of annual and perennial weeds and grasses in citrus and pineapple plantations. Registered by EPA as a herbicide. Colorless crystals; mp = 158-159°; d^{25} = 1.55; soluble in H_2O (0.0815 g/100 ml), EtOH (13.4 g/100 ml), Me_2CO (16.7 g/100 ml), CH_3CN (7.1 g/100 ml), xylene (3.2 g/100 ml), 3% aq. NaOH (8.8 g/100 ml); LD_{50} (rat orl)= 5200 mg/kg, (rbt der) > 5000 mg/kg; LC_{50} (rat ihl 4 hr.) > 4.8 mg/l air, (mallard duck, bobwhite quail 8 day dietary) > 10000 mg/kg diet, (rainbow trout 48 hr.) = 75 mg/l, (bluegill sunfish 48 hr.) = 71 mg/l, (carp 48 hr.) = 164 mg/l; non-toxic to bees. *Agan Chemical Manufacturers; DuPont UK; DuPont; Selectokil Ltd (Velpar).*

459   Bromacil, sodium salt
69484-12-4
                                                                 P

$C_9H_{12}BrN_2NaO_2$
5-Bromo-6-methyl-3-(1-methylpropyl)-2,4(1H,3H)-pyrimidinedione, sodium salt.
5-Bromo-6-methyl-3-(1-methylpropyl)-2,4(1H,3H)-pyrimidinedione sodium salt; 5-Bromo-3-sec-butyl-6-methyluracil, sodium salt; Bromacil, dimethylamine salt; 2,4(1H,3H)-Pyrimidinedione, 5-bromo-6-methyl-3-(1-methylpropyl)-,sodium salt; Sodium bromacil.

Registered by EPA as a herbicide (cancelled).

460   Bromadiolone
28772-56-7   1366   G P

$C_{30}H_{23}BrO_4$
3-[3-(4'-Bromo[1,1'-biphenyl]-4-yl)-3-hydroxy-1-phenylpropyl]-4-hydroxy-2H-1-benzopyran-2-one.
2H-1-Benzopyran-2-one, 3-(3-(4'-Bromo(1,1'-biphenyl)-4-yl)-3-hydroxy-1-phenylpropyl)-4-hydroxy-; Boldo; Boot hill; BRN 1335336; Bromadiolone; Bromatrol; Bromone; 3-(3-(4'-Bromo-(1,1'-biphenyl)-4-yl)-3-hydroxy-1-phenyl-propyl)-4-hydroxycoumarin; 3-(3-(4'-Bromo(1,1'-bi-phenyl)-4-yl)3-hydroxy-1-phenylpropyl)-4-hydroxy-2H-1-benzopyran-2-one; 3-(3-(4'-Bromo(1,1'-biphenyl)-4-yl)-3-hydroxy-1-phenylpropyl)-4-hydroxy-2-benzopyrone; 3-(3-(4'-Bromobiphenyl-4-yl)-3-hydroxy-1-phenylpropyl)-4-hydroxycoumarin; 3-(α-(p-(p-Bromophenyl)-β-hydroxy-phenethyl)benzyl)-4-hydroxycoumarin; Bromore; Bropro-difacoum; Canadien 2000; Caswell No. 486AB; Contrac; Contrax; Coumarin, 3-(α-(p-(p-bromophenyl)-β-hydroxy-phenethyl)benzyl)-4-hydroxy-; Coumarin, 3-(3-(4'-bromo-1,1'-biphenyl-4-yl)-3-hydroxy-1-phenylpropyl)-4-hydroxy-; EINECS 249-205-9; EPA Pesticide Chemical Code 112001; Eradic; HSDB 6458; (Hydroxy-4 coumarinyl 3)-3 phenyl-3 (bromo-4 biphenylyl-4)-1 propanol-1; 3-(4-Hydroxycoumarin-3-yl)-3-phenyl-1-(4-bromobiphenyl)propan-1-ol; 3-(4-Hydroxy-2-oxo-chrom-en-3-yl)-3-phenyl-1-(4-bromobiphenyl)propan-1-ol; LM-637; Maki; Rafix; Ratimus; Rotox; Sup'operats; Super-caid; Super-Cald; Super-rozol; Temus; Topidion; Bromo-4-biphenylyl)ethyl)benzyl)-4-hydroxycoumarin; LM-637; Super-caid; 3-(3-(4'-Bromo-(1,1'-biphenyl)-4-yl)-3-hydroxy-1-phenylpropyl)-4-hydroxycoumarin; Broma-diolone; Canadien 2000; Contrac; Maki; Ratimus; Tamogam; Boldo. An anticoagulant rodenticide as a concentrated bait. A ready-to-use bait for the control of rats and house mice. Registered by EPA as a rodenticide. Yellow powder; mp = 200-210°; λ_m = 260 nm ($E^{1\%}_{cm}$ = 538-582 EtOH); soluble in H_2O (0.0019 g/100 ml), soluble in DMF (73 g/100 ml), EtOAc (2.5 g/100 ml), EtOH (0.82 g/100 ml), soluble in Me_2CO, slightly soluble in $CHCl_3$, insoluble in Et_2O, C_6H_{14}; LD_{50} (rat orl) = 1.125 mg/kg, (mus orl) = 1.75 mg/kg, (rbt orl) = 1.0 mg/kg, (dog orl) > 10 mg/kg, (cat orl) > 25 mg/kg, (quail orl) = 1600 mg/kg, (rbt der) = 2.1 mg/kg; LC_{50} (rainbow trout 96 hr.) = 1.4 mg/l; non-toxic to bees. *Bell Laboratories Inc.; Ciba Geigy Agrochemicals; Lipha Pharm. Inc.; Liphatech Inc.; Rentokil Ltd.*

461   Bromcresol purple
115-40-2   1371   G

C21H16Br2O5S
Dibromo-o-cresolsulfonphthalein.
Bromcresol purple; Bromocresol purple; EINECS 204-087-8; NSC 374134; Phenol, 4,4'-(3H-2,1-benzoxathiol-3-ylidene)bis(2-bromo-6-methyl-, S,S-dioxide; Phenol, 4,4'-(1,1-dioxido-3H-2,1-benzoxathiol-3-ylidene)bis(2-bromo-6-methyl. An acid-base indicator. Crystals; mp = 241.5°; irritant; practically insoluble in H2O, soluble in EtOH.

462   Bromeikon
860-07-1   9262   G

2Na+

C20H8Br4Na2O4
Sodium tetrabromophenolphthalein.
Brom-tetragnost; Cholegnostyl; Tetrabrom. Radio-opaque medium, used as a diagnostic aid. Solid; soluble in H2O, insoluble in organic solvents.

463   Brometone
76-08-4   9685   G

C4H7Br3O
2-Tribromomethyl-2-propanol.
Acetonbromoform; Acetone-bromoform; Al3-14655; Brometon; Bromobutanol; EINECS 200-931-4; HSDB 5212; NSC 2865; 2-Propanol, 1,1,1-tribromo-2-methyl-; Tribromo-tert-butyl alcohol; 2-Tribromomethyl-2-propanol; 1,1,1-Tribromo-2-methyl-2-propanol; 1,1,1-Tribromo-2-methylpropan-2-ol. Modifier in the polymerization of vinyl chloride. Crystalline solid; mp = 169°; slightly soluble in H2O, more soluble in organic solvents. Uniroyal.

464   Bromine
7726-95-6   1378   G P

Br—Br

Br2
Bromine.
Brom; Brome; Bromine; Bromo; Broom; Caswell No. 112; Dibromine; EINECS 231-778-1; EPA Pesticide Chemical Code 008701; HSDB 514; UN1744. Used in the manufacture of ethylene dibromide, organic synthesis; in water disinfection, bleaching fibers, fire retardants, medicinals and dyestuffs. Registered by EPA as an antimicrobial, fungicide and insecticide. In the US, use of bromine is estimated as: fire retardants, 27%; agriculture, 15%; petroleum additives, 15%; well drilling fluids, 10%; sanitary preparations, 5%; and other uses, 28%. Other uses included intermediate chemicals used in the manufacture of other products and bromide solutions used alone or in combination with other chemicals. Brown gas; mp = -7.25°; bp = 59.47°; soluble in H2O, EtOH, Et2O; critical temperature = 315; critical pressure = 102 atm; Ingestion of solution may cause severe gastroenteritis and/or death; burns and blisters the skin. *Bromine & Chems. Ltd; Ethyl Corp.; Great Lakes Fine Chem.;*

Janssen Chimica; Spectrum Chem. Manufacturing. [

465   Bromoacetanilide
103-88-8   1383   D G

C8H8BrNO
N-(4-Bromophenyl)acetamide.
4-12-00-01504 (Beilstein Handbook Reference); Al3-01799; Antisepsin; Asepsin; BRN 2208091; Bromoanilide; Bromoantifebrin; EINECS 203-154-9; NSC 105442; USAF DO-40. Used as an analgesic and antipyretic. Solid; mp = 168°; d^{25} = 1.717; λ_m = 257 nm (ε = 19953, MeOH); insoluble in H2O; slightly soluble in Et2O, C6H6, soluble in EtOH, CHCl3. *Monsanto Co.*

466   m-Bromobenzaldehyde
3132-99-8   H

C7H5BrO
3-Bromobenzaldehyde.
Benzaldehyde, 3-bromo-; Benzaldehyde, m-bromo-; 3-Bromobenzaldehyde; m-Bromobenzaldehyde; EINECS 221-526-9; NSC 66828. Liquid; bp = 234°; λ_m = 245, 291 nm (ε = 6500, 790, MeOH); insoluble in H2O, slightly soluble in CCl4, very soluble in EtOH, Et2O.

467   Bromochloro dimethyl hydantoin
126-06-7   G

C5H6BrClN2O2
1-Bromo-3-chloro-5,5-dimethyl hydantoin.
BRN 0780014; 3-Bromo-1-chloro-5,5-dimethylimidazol-idine-2,4-dione; 3-Bromo-1-chloro-5,5-dimethyl-2,4-imidazolidinedione; 3-Bromo-1-chloro-5,5-dimethyl-hydantoin; Dantoin® BCDMH; Dantoin® GSD-550; EINECS 204-766-9; Halobrom; HSDB 5608; Hydantoin, 3-bromo-1-chloro-5,5-dimethyl-; 2,4-Imidazolidinedione, 3-bromo-1-chloro-5,5-dimethyl-; Quesbrom; Imidazol-idinedione, 3-bromo-1-chloro-5,5-dimethyl-. Broad spectrum biocide for control of algae, bacterial and fungal slimes in swimming pools and industrial water systems; nonflammable. An extremely effective microbiocial bactericide, fungicide and algicide used in cooling towers, once-thru and closed loop water system. Low temperature industrial bleach. *Dead Sea Bromine; Lonzagroup; Quest Intl.*

468   Bromochlorodifluoromethane
353-59-3   H

CBrClF₂

CBrClF2

Chlorodifluoromonobromomethane.

4-01-00-00075 (Beilstein Handbook Reference); BCF; BRN 1732514; Bromochlorodifluoromethane; CCRIS 5663; Chlorodifluobromometano; Chlorodifluorobromo-methane; Chlorodifluoromonobromomethane; Daiflon 12B1; Dwufluorochlorobromometan; EINECS 206-537-9; Flugex 12B1; Fluorocarbon 1211; Freon 12B1; Halon 1211; HSDB 6784; Methane, bromochlorodifluoro-; R 12B1; Refrigerant gas R 12B1; UN1974. Gas; mp = -159.5°; bp = -3.7°.

469   Bromochloromethane
74-97-5   H

CH₂BrCl

Methane, bromochloro-.

4-01-00-00074 (Beilstein Handbook Reference); AI3-15514; BRN 1730801; Bromochloromethane; CCRIS 817; Chlorobromomethane; Chloromethyl bromide; EINECS 200-826-3; Fluorocarbon 1011; Halon 1011; HSDB 2520; Methane, bromochloro-; Methylene chlorobromide; Mil-B-4394-B; Mono-chloro-mono-bromo-methane; NSC 7294; UN1887. Has been used in refrigeration and fire retardants. Has been identified as an ozone-threatening chemicals and subjected to restrictions. Liquid; mp = -87.9°; bp = 68.0°; d^{20} = 1.9344; poorly soluble in H₂O, soluble in organic solvents.

470   Bromociclen
1715-40-8   G P V

C₈H₅BrCl₆

5-bromomethyl-1,2,3,4,7,7-hexachloro-2-norbornene.

AI3-23393; Alugan; Bicyclo(2.2.1)hept-2-ene, 5-(bromomethyl)-1,2,3,4,7,7-hexachloro-; Bromociclen; Bromociclene; Bromocicleno; Bromociclenum; Bro-mocyclen; Bromocyclene; Bromodan; Bromomethyl-hexachlorobicycloheptene; 5-(Bromomethyl)-1,2,3,4,7, 7-hexachloro-2-norbornene; 5-(Bromomethyl)-1,2,3, 4,7,7-hexachlorobicyclo(2.2.1)hept-2-ene; Caswell No. 116; EINECS 216-996-7; EPA Pesticide Chemical Code 008705; 2-Norbornene, 5-(bromomethyl)-1,2,3,4,7,7-hexachloro-. A veterinary pesticide. Also used as a brominating agent. Fisons plc, Horticultural Div.

471   Bromocresol green
76-60-8   1370   G

C₂₁H₁₄Br₄O₅S

m-Cresol, 4,4'-(3H-2,1-benzoxathiol-3-ylidene)bis(2,6-di-bromo-, S,S-dioxide.

5-19-03-00460 (Beilstein Handbook Reference); BCG; BRN 0372527; Bromcresol green; EINECS 200-972-8; NSC 7817; Phenol, 4,4'-(1,1-dioxido-3H-2,1-benz-oxathiol-3-ylidene)bis(2,6-dibromo-3-methyl-; Phenol, 4,4'-(3H-2,1-benzoxathiol-3-ylidene)bis(2,6-dibromo-3-methyl-, S,S-dioxide; o-Toluenesulfonic acid, α-hydroxy-, gamma-sultone; Tetrabromo-m-cresolphthalein sulfone. An acid indicator pH 3.8 (yellow)-5.4 (blue-green). Red-yellow crystals; mp = 218.5°; λm = 211, 278, 338, 426 nm (ε 54100, 8500, 7550, 17100 MeOH), 312, 394, 620 nm (ε 13300, 8520, 40100 MeOH/KOH); insoluble in H₂O, soluble in organic solvents. Lancaster Synthesis Co.; Mallinckrodt Inc.; Sigma-Aldrich Fine Chem.

472   1-Bromodecane
112-29-8   G

C₁₀H₂₁Br

n-Decyl bromide.

AI3-28586; 1-Bromodecane; Decane, 1-bromo-; Decyl bromide; 1-Decyl bromide; n-Decyl bromide; EINECS 203-955-3; NSC 8780. Chemical intermediate and solvent. Liquid; mp = -29.2°; bp = 240.6°; d^{20} = 1.0702; insoluble in H₂O, soluble in CCl₄, very soluble in EtOH, CHCl₃. Ethicon Inc.; Great Lakes Fine Chem.; Humphrey.

473   6-Bromo-2,4-dinitroaniline
1817-73-8   H

C₆H₄BrN₃O₄

2-Bromo-4,6-dinitroaniline.

4-12-00-01734 (Beilstein Handbook Reference); Aniline, 2-bromo-4,6-dinitro-; Benzenamide, 2-bromo-4,6-dinitro-; BRN 0916722; 6-Bromo-2,4-dinitroaniline; Bromo DNA; CCRIS 820; EINECS 217-329-2; HSDB 5453; NCI-C60844; NSC 16572. Yellow needles; mp = 153.5°; bp sublimes; λm = 229, 268, 336 nm (ε = 10100, 8700, 12000, MeOH); soluble in AcOH, very soluble in EtOH, Me₂CO.

474   Bromoethane
74-96-4   3807   H

C₂H₅Br

Ethane, bromo-.

AI3-04462; Bromic ether; Bromoethane; Bromure d'ethyle; CCRIS 2504; EINECS 200-825-8; Ethane, bromo-; Ethyl bromide; Etylu

82

bromek; Halon 2001; HSDB 532; Hydrobromic ether; Monobromoethane; NCI-554813; NCI-C55481; NSC 8824; UN1891.

475 p-Bromofluorobenzene
460-00-4   H

C_6H_4BrF
1-Fluoro-4-bromobenzene.
1-Bromo-4-fluorobenzene; 4-Bromofluorobenzene; 1-Fluoro-4-bromobenzene; 4-Fluoro-1-bromobenzene; 4-Fluorobromobenzene; 4-Fluorophenyl bromide; Benzene, 1-bromo-4-fluoro-; EINECS 207-300-2; NSC 10268; p-Bromofluorobenzene; p-Fluorobromobenzene; p-Fluorophenyl bromide. Liquid; mp = -17.4°; bp = 151.5°; d^{15} = 1.593; λ_m = 221, 265, 271, 278 nm (ε = 7079, 794, 1047, 891, cyclohexane); insoluble in H_2O, soluble in EtOH, Et_2O, $CHCl_3$.

476 Bromofluoroform
75-63-8   H

$CBrF_3$
Bromotrifluoromethane.
Bromofluoroform; Bromotrifluoromethane; Carbon monobromide trifluoride; Daiflon 13B1; EINECS 200-887-6; F 13B1; FC 13B1; FE 1301; Flugex 13B1; Fluorocarbon 1301; Freon 13B1; Halocarbon 13B1; Halon 1301; HSDB 141; Khladon 13B1; Methane, bromotrifluoro-; Monobromotrifluoromethane; R 13B1; Refrigerant 13B1; Trifluorobromomethane; Trifluoro-methyl bromide; Trifluoromonobromomethane; UN1009. Used as an Aerosol®® propellant. Presumed to be a threat to the ozone later. Gas; bp = -58°; mp = -168°; d = 1.5; LC_{50} (rat ihl) = 84000 ppm, (mus ihl) = 381 g/m^3, (gpg ihl) = 88000 ppm.

477 Bromoform
75-25-2   1404   G

$CHBr_3$
Tribromomethane.
4-01-00-00082 (Beilstein Handbook Reference); AI3-28587; BRN 1731048; Bromoform; Bromoforme; Bromoformio; CCRIS 98; EINECS 200-854-6; HSDB 2517; Methane, tribromo-; Methenyl tribromide; Methyl tribromide; NCI-C55130; NSC 8019; RCRA waste number U225; Tribrommethaan; Tribrommethan; Tribromomethan; Tribromomethane; UN2515. Used for gem and mineral testing and for generation of dibromocarbene. Used medically as sedative, hypnotic and antitussive. Liquid; mp = 8°; bp = 149.1°; d^{15} = 2.899; poorly soluble in H_2O, soluble in organic solvents; LD 50 (rat orl) = 933 mg/kg. Geoliquids Inc.

478 Bromonitroform
560-95-2   G

$CBrN_3O_6$
Bromotrinitromethane.
3-01-00-00116 (Beilstein Handbook Reference); BRN 1787170; Bromotrinitromethane; Methane, bromotrinitro-

479 Bromopicrin
464-10-8   G

CBr_3NO_2
Tribromonitromethane.
4-01-00-00106 (Beilstein Handbook Reference); BRN 1756139; Bromopicrin; CCRIS 5977; EINECS 207-348-4; Methane, nitrotribromo-; Methane, tribromonitro-; Nitrobromoform; Nitrotribromomethane; Tribromo-nitromethane. A military poison, used in organic synthesis. Liquid; mp = 10°; bp^{18} = 127°; d^{12} = 2.811; λ_m = 240, 260, 300 nm (ε = 1906, 1000, 79, petroleum ether); insoluble in H_2O, soluble in EtOH, Et_2O, very soluble in Me_2CO, C_6H_6, AcOH; LD_{50} (mus ipr) = 15 mg/kg.

480 α-Bromopropionic acid
598-72-1   H

$C_3H_5BrO_2$
2-Bromopropionic acid.
AI3-52314; α-Bromopropionic acid; 2-Bromopropanoic acid; 2-Bromopropionic acid; EINECS 209-947-6; NSC 172; Propanoic acid, 2-bromo-; Propionic acid, 2-bromo

481 Bromotrimethylsilane
2857-97-8   G

C_3H_9BrSi
Trimethylbromosilane.
CT2928; EINECS 220-672-0; NSC 139857; Silane, bromotrimethyl-; Trimethylbromosilane; Trimethylsilicon bromide; Trimethylsilyl bromide. Coupling agent, chemical intermediate, blocking agent, release agent, lubricant, primer, reducing agent. Liquid; mp = -43°; bp = 79°; d = 1.1600. Degussa-Hüls Corp.

482 Bromoxynil
1689-84-5   1465   G H P

C7H3Br2NO
3,5-dibromo-4-hydroxybenzonitrile.
4-10-00-00475 (Beilstein Handbook Reference); Brittox; BRN 2364039; Brominal; Brominal Plus; Brominal Triple; Brominex; Brominil; Bromotril; Bromoxynil; Broxynil; Buctril; Buctril industrial; Butil chlorofos; Caswell No. 119; CCRIS 2447; Certrol B; Chipco buctril; Chipco crab-kleen; EINECS 216-882-7; ENT 20852; EPA Pesticide Chemical Code 035301; HSDB 1523; Labuctril; Labuctril 25; Litarol M; M&B 10731; MB 10064; ME 4 Brominal; Novacorn; NU-Lawn weeder; Oxytril M; Pardner; S 2132; Sabre; Terset; Toplan. A selective contact herbicide used for post-emergence control of some annual broad-leaved weeds in cereal crops. Registered by EPA as a herbicide. Colorless crystals; mp = 194 - 195°; sublimes at 135°/0.15 mm; soluble in H_2O (0.013 g/100 ml), MeOH (9 g/100 ml), EtOH (7 g/100 ml), Me_2CO (17 g/100 ml),cyclohexanone (17 g/100 ml), THF (41 g/100 ml), DMF (61 g/100 ml), mineral oil (< 2 g/100 ml); LD_{50} (rat orl) = 190 mg/kg, (mus orl) = 110 mg/kg, (rbt orl) = 260 mg/kg, (dog orl) = 100 mg/kg, (pheasant orl) = 50 mg/kg, (hen orl) = 240 mg/kg, (quail orl) = 100 - 125 mg/kg, (mallard duck orl) = 200 mg/kg; LC_{50} (harlequin fish 48 hr.) = 5.0 mg/l, (potassium salt); non-toxic to bees. *Agriliance LLC; Aventis Crop Science; Dow Europe; May & Baker Ltd.; Micro-Flo Co. LLC.*

483   Bromoxynil butyrate
3861-41-4   H P

C11H9Br2NO2
3,5-Dibromo-4-hydroxybenzonitrile butyrate.
BRN 2742676; Bromoxynil butyrate; Butanoic acid, 2,6-dibromo-4-cyanophenyl ester; Butyric acid, ester with 3,5-dibromo-4-hydroxybenzonitrile; Caswell No. 119E; 2,6-Dibromo-4-cyanophenyl butyrate; EPA Pesticide Chemical Code 035303. Selective contact herbicide with some systemic activity. Used for post-emergence control of broad-leaved weeds. Registered by EPA as a herbicide (cancelled). *Aventis Crop Science.*

484   Bromoxynil heptanoate
56634-95-8   H P

C14H15Br2NO2
2,6-Dibromo-4-cyanophenyl heptanoate.
Bromoxynil heptanoate; 3,5-Dibromo-4-hydroxybenzo-nitrile heptanoate; Heptanoic acid, 2,6-dibromo-4-cyanophenyl ester. Registered by EPA as a herbicide. *Agsco Inc.; Aventis Crop Science.*

485   Bromoxynil octanoate
1689-99-2   H

C15H17Br2NO2
Octanoic acid 2,6-dibromo-4-cyanophenyl ester.
BRN 2756636; Brominal W; Bromoxynil octanoate; Buctril; Caswell No. 119A; EINECS 216-885-3; EPA Pesticide Chemical Code 035302; M&B 10731; NPH 1320; RP-16272. Registered by EPA as a herbicide. Colorless crystals; mp = 45 - 46°; insoluble in H_2O, soluble in xylene (70 g/100 ml), EtOH (10 g/100 ml), n-PrOH (12 g/100 ml), Me_2CO (10 g/100 ml), cyclohexanone (55 g/100 ml), EtOAc (62 g/100 ml), DMF (70 g/100 ml), $CHCl_3$ (80 g/100 ml), CCl_4 (50 g/100 ml); LD_{50} (rat orl) = 365 mg/kg, (mus orl) = 306 mg/kg, (rbt orl) = 325 mg/kg, (rat der) > 2000 mg/kg, (rbt der) = 1675 mg/kg, (pheasant orl) = 175 mg/kg; LC_{50} (rainbow trout 96 hr.) = 0.15 mg/l; non-toxic to bees. *Agsco Inc.; Albaugh Inc.; Aventis Crop Science; Micro-Flo Co. LLC; Nufarm Americas Inc.*

486   Bronopol
52-51-7   1432   G

C3H6BrNO4
2-Bromo-2-nitropropane-1,3-diol.
4-01-00-02501 (Beilstein Handbook Reference); AI3-61639; Bioban®; BNPD-40; BRN 1705868; β-Bromo-β-nitrotrimethyleneglycol; 2-Bromo-2-nitropropane-1,3-diol; Bronidiol; Bronocot; Bronopol; Bronopolu; Bronopolium; Bronosol; Bronotak; Caswell No. 116A; EINECS 200-143-0; EPA Pesticide Chemical Code 216400; Lexgard bronopol; NSC 141021; Onyxide 500; 1,3-Propanediol, 2-bromo-2-nitro-; UN3241. Broad spectrum antimicrobial agent; preservative for topical cosmetics, pharmaceuticals, toiletries. Used as a preservative for metalworking fluids. mp = 120-122°; soluble in H_2O, EtOH, EtOAc; slightly soluble in chloroform, acetone, ether, benzene; LD_{50} (rat orl) = 350 mg/kg. *Inolex; Knoll AG; Whittaker Clark & Daniels.*

487   Budesonide
51333-22-3   1454   D

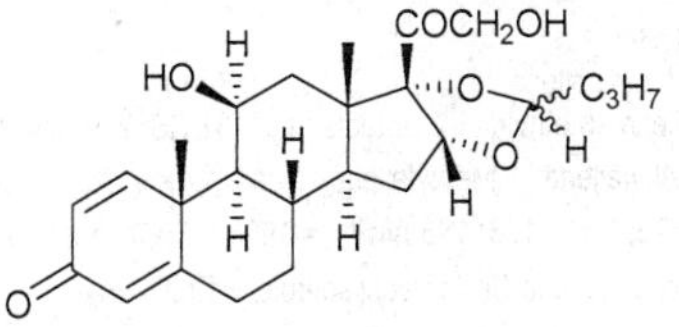

C$_{25}$H$_{34}$O$_6$

(R,S)-11β,16α,17,21-Tetrahydroxypregna-1,4-diene-3,20-dione cyclic 16,17-acetal with butyraldehyde.

Bidien; Budeson; Budesonide; Budesonido; Budes-onidum; CCRIS 5230; Cortivent; EINECS 257-139-7; Entocort; Entocort EC; Micronyl; Preferid; Pulmicort; Respules; Rhinocort; Rhinocort α; S 1320; Spirocort. Anti-inflammatory; antiasthmatic. Non-halogenated gluco-corticoid related to triamcinolone hexacetonide with a high ratio of topical to systemic activity. A mixture of two isomers in which the S-isomer varies between 40-51%. Crystals; mp = 221-232° (dec); [α]$_D^{25}$= 98.9 (c = 0.28 CH$_2$Cl$_2$). *Astra Chem. Ltd.; Bofors.*

488   R-Budesonide
51372-29-3   D

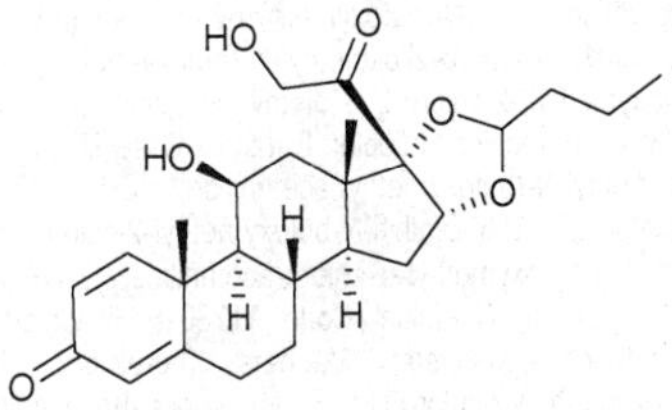

C$_{25}$H$_{34}$O$_6$

11β,16α(R),17,21-Tetrahydroxypregna-1,4-diene-3,20-dione cyclic 16,17-acetal with butyraldehyde.

Budesonide; EINECS 257-161-7. Anti-inflammatory.

489   S-Budesonide
51372-28-2   D

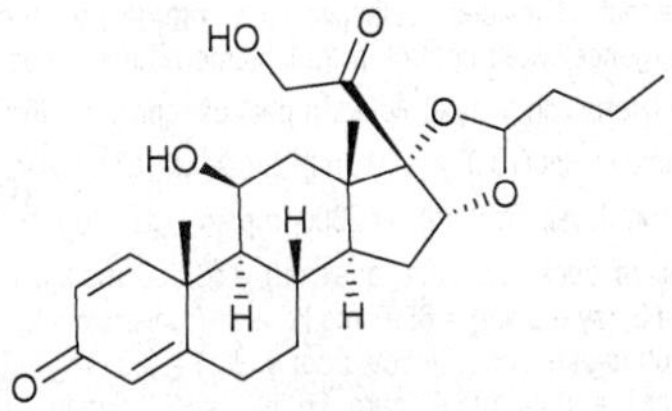

C$_{25}$H$_{34}$O$_6$

11β,16α(S),17,21-Tetrahydroxypregna-1,4-diene-3,20-dione cyclic 16,17-acetal with butyraldehyde.
EINECS 257-160-1. Anti-inflammatory.

490   Bufexamac
2438-72-4   1461   D G

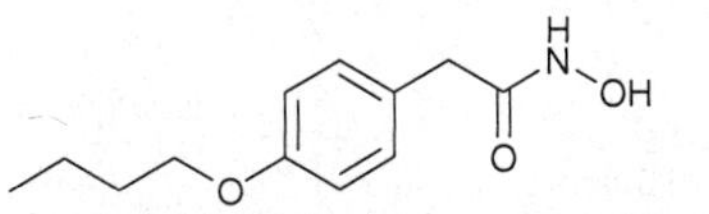

C$_{12}$H$_{17}$NO$_3$

2-(p-Butoxyphenyl)acetohydroxamic acid.
BRN 2646848; Bufessamac; Bufexamac; Bufexamaco;

Bufexamacum; Bufexamic acid; CP-1044; CP-1044-J3; Droxarol; Droxaryl; EINECS 219-451-1; Feximac; Flogicid; Flogocid N plastigel; J3; Malipuran; Mofenar; Norfemac; Parfenac; Parfenal. Anti-inflammatory; analgesic; antipyretic. Used as a skin cream. Crystals; mp = 153-155°; nearly insoluble in H$_2$O; LD$_{50}$ (rat orl) >4 g/kg, (mus orl) > 8 g/mg. *C.M. Ind.; Pfizer Inc.*

491   Bunamidine hydrochloride
1055-55-6   1473   G

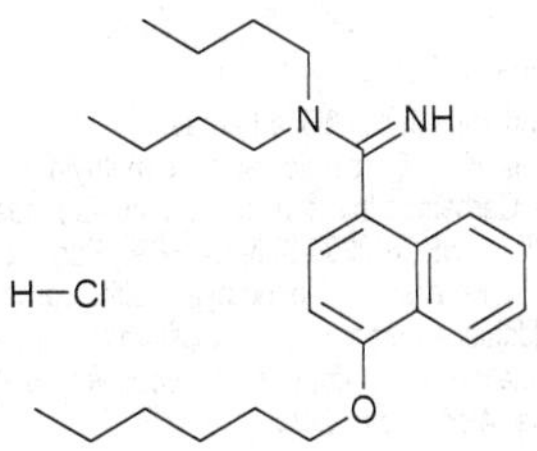

C$_{25}$H$_{39}$ClN$_2$O

N,N-Dibutyl-4-(hexyloxy)-1-naphthalenecarboximidamide hydrochloride.

Buban; Bunamidine hydrochloride; Bunamidine, monohydrochloride; N,N-Dibutyl-4-hexyloxy-1-naph-thamidine hydrochloride; N,N-Dibutyl-4-(hexyloxy)-naphthalene-1-carboximidamide hydrochloride; EINECS 213-890-2; 1-Naphthalenecarboximidamide, N,N-dibutyl-4-(hexyloxy)-, hydrochloride; 1-Naph-thamidine, N,N-dibutyl-4-hexyloxy-, monohydro-chloride; NSC-106571; Scolaban. Anthelmintic effec-tive against Cestodes. Crystals; mp = 214-215°; LD$_{50}$ (rat orl) = 591 mg/kg. *Burroughs Wellcome Inc.; Wellcome Foundation Ltd.*

492   Bupropion
34911-55-2   1487   D

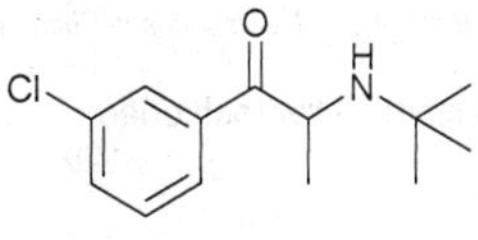

C$_{13}$H$_{18}$ClNO

(±)-2-(tert-Butylamino)-3'-chloropropiophenone.

Amfebutamona; Amfebutamone; Amfebutamonum; BRN 2101062; Bupropion; (±)-Bupropion. Anti-depressant with action similar to that of the tricyclic antidepressants. Pale yellow oil; bp = 52°; soluble in MeOH, EtOH, Me$_2$CO, Et$_2$O, C$_6$H$_6$. *Burroughs Wellcome Inc.*

493   Bupropion hydrochloride
31677-93-7   1487   D

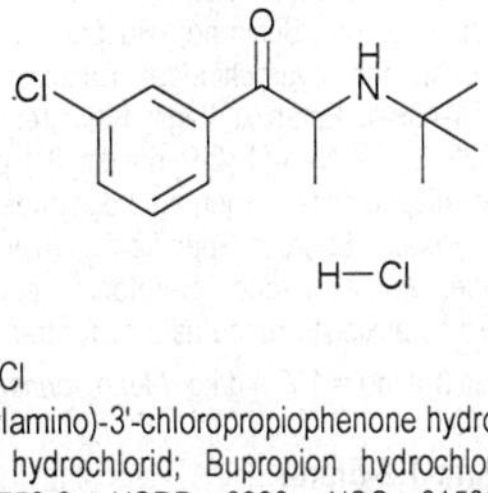

C$_{13}$H$_{18}$ClNO.HCl

(±)-2-(tert-Butylamino)-3'-chloropropiophenone hydrochloride.

Amfebutamon hydrochlorid; Bupropion hydrochloride; BW 323; EINECS 250-759-9; HSDB 6988; NSC 315851; Wellbutrin; Wellbutrin SR; Wellbutrin XL; Zyban; Zyban (pharmaceutical),. Antidepressant. Also used in smoking cessation therapy. Crystals; mp = 233-234°; solubility (in H$_2$O) = 32 mg/ml, (in alcohol) = 193 mg/ml, (in 0.1 N HCl) = 333 mg/ml; LD$_{50}$ (mus ip) = 230 mg/kg, (rat ip) = 210 mg/kg, (mus orl) = 575 mg/kg, (rat orl) = 600 mg/kg. *Burroughs Wellcome Inc.*

494   Busan 85
128-03-0   G

$C_3H_6KNS_2$
Potassium dimethyl dithiocarbamate.
Aquatreat KM; Carbamic acid, dimethyldithio-, potassium salt, hydrate; Carbamodithioic acid, dimethyl-, potassium salt; Caswell No. 691; EINECS 204-875-1; EPA Pesticide Chemical Code 034803; Potassium dimethyl dithiocarbamate; Potassium dimethyldithiocarbamate; Potassium dimethylthio-carbamate. Polymerization short stops in the copolymerization of styrene and butadiene. *Alco Chem. Corp.*

495   Buspirone
36505-84-7   1492   D

$C_{21}H_{31}N_5O_2$
N-[4-[4-(2-Pyrimidinyl)-1-piperazinyl]butyl]-1,1-cyclopentanediacetamide hydrochloride.
5-25-10-00059 (Beilstein Handbook Reference); Ansial; BRN 0964904; Buspirona; Buspirone; Buspironum; EINECS 253-072-2. Serotonin receptor agonist. A non benzodiazepine anxiolytic, used as a minor tranquillizer. *Mead Johnson Pharmaceuticals.*

496   Buspirone hydrochloride
33386-08-2   1492   D

$C_{21}H_{32}ClN_5O_2$
N-[4-[4-(2-Pyrimidinyl)-1-piperazinyl]butyl]-1,1-cyclopentanediacetamide hydrochloride.
Ansial; Ansiced; Ansitec; Anxinil; Anxiolan; Apo-Buspirone; Axoren; Barpil; Bespar; Biron; Busirone; Buspar; Buspimem; Buspimen; Buspinol; Buspirone hydrochloride; Buspisal; Censpar; Effiplen; EINECS 251-489-4; Establix; Itagil; Kallmiren; Lucelan; Mabuson; MJ 9022-1; N-(4-(4-(2-Pyrimidinyl)-1-piperazinyl)butyl)-1,1-cyclopentanediacetamide monohydrochloride; Narol; Nerbet; Normaton; Relac; Sburol; Spamilan; Travin; Tutran; Uspirone hydrochloride. An azapirone. Serotonin receptor agonist. Non-benzodiazepine anxiolytic, used as a minor tranquillizer. mp = 201.5-202.5°; LD50 (rat ipr) = 136 mg/kg. *Mead Johnson Pharmaceuticals.*

497   Buta-1,2-diene
590-19-2   H

C_4H_6
1-Methylallene.
Allene, methyl-; 1,2-Butadiene; EINECS 209-674-2; HSDB 5705; Methylallene; 1-Methylallene. Gas; mp = -136.2°; bp = 10.9°; d^0 = 0.676; λ_m = 178, 186 nm (ε = 19953, 3982, gas); insoluble in H_2O, very soluble in C_6H_6, freely soluble in EtOH, Et2O.

498   Butachlor
23184-66-9   1497   G H P

$C_{17}H_{26}ClNO_2$
N-(Butoxymethyl)-2-chloro-N-(2,6-diethylphenyl)acetamide.
Acetamide, N-(butoxymethyl)-2-chloro-N-(2,6-diethyl-phenyl)-; Acetanilide, 2-chloro-2',6'-diethyl-N-(butoxy-methyl)-; Acetanilide, N-(butoxymethyl)-2-chloro-2',6'-diethyl-; Amichlor; Bilchlor; BRN 2873811; Butachlor; Butaclor; Butanex; Caswell No. 119B; 2-Chloro-2',6'-diethyl-N-(butoxymethyl)acetanilide; CP 53619; Del-chlor; Delchlor 5G; 2',6'-Diethyl-N-butoxymethyl-2-chloroacetanilide; 2',6'-Diethyl-N-butoxymethyl-α-chloro-acetanilide; EINECS 245-477-8; EPA Pesticide Chemical Code 112301; Hiltachlor; HSDB 6865; Mach-Mach; Machete; Machete (herbicide); Machette; N-(Butoxymethyl)-2-chloro-N-(2,6-diethylphenyl)acetamide; N-(Butoxymethyl)-2-chloro-2',6'-diethylacetanilide; N-Butoxymethyl-α-chloro-2',6'-diethylacetanilide; NSC 221683; Pillarsete; Rasayanchlor; SHA 112301; Weedout; N-(Butoxymethyl)-2-chloro-2',6'-diethyl-acetanilide; N-(Butoxymethyl)-2-chloro-N-(2,6-diethyl-phenyl)acetamide; Caswell No. 119B; 2-Chloro-2',6'-diethyl-N-(butoxymethyl)acetanilide; CP 53619; Del-chlor; Delchlor 5G; 2',6'-Diethyl-N-butoxymethyl-2-chloroacetamide; EPA Pesticide Chemical Code 112301; Hiltachlor; HSDB 6865; Mach-Mach; Machete; Machette; NSC 221683; Pillarsete; Rasayanchlor; SHA 112301; Weedout. Provides selective pre-emergence and early post-emergence weed control in transplanted, direct seeded and upland rice. Herbicide. Amber liquid; mp < -5°; bp0.5 = 156°; sg25 = 1.070; soluble in H_2O (0.002 g/100 ml), Et2O, Me2CO, C_6H_6, EtOH, EtOAc, C_6H_{14}; LD50 (rat orl) = 2000 mg/kg, (rbt der) > 13000 mg/kg, (mallard duck, bobwhite quail orl) > 10000 mg/kg; LC50 (bobwhite quail 8 day dietary) = 6597 mg/kg diet, (mallard duck 8 day dietary) > 10000 mg/kg diet, (rainbow trout 96 hr.) = 0.52 mg/l, (bluegill sunfish 96 hr.) = 0.44 mg/l, (carp 96 hr.) = 0.32 mg/l. *Agan Chemical Manufacturers; Comlets; Crystal; Hindustan Insecticides; Krishi Rasayan; Makhteshim-Agan; Monsanto Co.; Shen Hong; Shinung; Sintesul; Sundat.*

499   Butadiene
106-99-0   1498   H

C_4H_6
1,3-Butadiene.
Biethylene; Bivinyl; Butadieen; Buta-1,3-dieen; Butadien; Buta-1,3-dien; Butadiene; Buta-1,3-diene; 1,3-Butadiene; α,σ-Butadiene; Butadiene monomer; CCRIS 99; Divinyl; EINECS 203-450-8; Erythrene; HSDB 181; NCI-C50602; UN 1010\r\n; Vinylethylene. Monomer for BUNA rubbers. Gas; mp = -108.9°; bp = -4.4°; λ_m =

217 nm (ε = 20893, EtOH); insoluble in H_2O, soluble in EtOH, Et_2O, C_6H_6, very soluble in Me_2CO.

500   Butamisole hydrochloride
54400-62-3   1504   G V

$C_{15}H_{19}N_3OS.HCl$
2-methyl-N-[3-(2,3,5,6-tetrahydroimidazo[2,1-b]thiazol-6-yl)phenyl]propanamide hydrochloride.
CL-206214; Styquin. Anthelmintic. *Am. Cyanamid.*

501   Butane
106-97-8   1505   G H

C_4H_{10}
n-Butane.
A 17; A 21; Bu-Gas; Butane; n-Butane; Butanen; Butani; Butyl hydride; CCRIS 2279; Diethyl; EINECS 203-448-7; HC 600; HC 600 (hydrocarbon); HSDB 944; Liquefied petroleum gas; LPG; Methylethylmethane; R 600; UN1011. Used as producer gas and propellant; raw material for motor fuels, in the manufactur of synthetic rubbers. May be narcotic in high concentrations. A simple asphyxiant. Gas; mp = -138.2°; bp = -0.5°; d = 2.046 (air=1); soluble in H_2O, very soluble in EtOH, Et_2O, $CHCl_3$. *Air Prods & Chems; Electrochem. Ltd.; Monsanto Co.; Phillips 66.*

502   Butanediol
110-63-4   G H

$C_4H_{10}O_2$
1,4-Dihydroxybutane.
4-01-00-02515 (Beilstein Handbook Reference); Agrisynth B1D; AI3-07553; 1,4-BD; BRN 1633445; Butanediol; 1,4-Butanediol; Butane-1,4-diol; 1,4-Butylene glycol; CCRIS 5984; Dabco® BDO; 1,4-Dihydroxybutane; Diol 14B; EINECS 203-786-5; HSDB 1112; NSC 406696; Sucol B; Tetramethylene 1,4-diol; Tetramethylene glycol; 1,4-Tetramethylene glycol. Curative, chain extender; provides reactive H-source in prepolymer production; used to provide hard segments in PU. Used as an intermediate and in polyurethane formulation in the hard segment as a curative. Crystals; mp = 20.1°; bp = 235°; d^{20} = 1.0171; λ_m = 291 nm (ε = 200, H_2SO_4); freely soluble in H_2O, soluble in EtOH, slightly soluble in DMSO, Et_2O; LD_{50} (rat orl) = 1525 mg/kg. *Air Products & Chemicals Inc.; Arco; BASF Corp.; Degussa-Hüls Corp.; DuPont; Intl. Specialty Products Inc. (ISP).*

503   Butanethiol
109-79-5   1576   P

$C_4H_{10}S$
n-Butyl mercaptan.
4-01-00-01555 (Beilstein Handbook Reference); AI3-22954; Bear skunk; BRN 1730908; Butanethiol; 1-Butanethiol; n-Butanethiol; Butyl mercaptan; 1-Butyl mercaptan; n-Butyl mercaptan; Butylthiol; Caswell No. 119D; EINECS 203-705-3; EPA Pesticide Chemical Code 125001; FEMA No. 3478; HSDB 290; 1-Mercaptobutane; NCI-C60866; Normal butyl thioalcohol; Thiobutyl alcohol. Registered by EPA as an animal repellent and insecticide. Used as an animal repellent. Mobile, foul-smelling liquid; mp = -115.7°; bp = 98.5°; d^{20} = 0.8416; λ_m = 277 nm (ε = 158, isooctane); slightly soluble in H_2O, $CHCl_3$, very soluble in EtOH, Et_2O. *Sigma-Aldrich Fine Chem.*

504   n-Butanol
71-36-3   1539   G H

$C_4H_{10}O$
1-Butanol.
AI3-00405; Alcool butylique; Butanol; Butanol, 1-; Butanolen; Butanolo; Butyl alcohol; Butyl hydroxide; Butylowy alkohol; Butyric alcohol; CCRIS 4321; CCS 203; EINECS 200-751-6; FEMA Number 2178; Hemostyp; HSDB 48; Methylolpropane; Normal primary butyl alcohol; NSC 62782; Propyl carbinol; Propylmethanol; RCRA waste number U031; Tebol 88; Tebol 89. Solvent, cosolvent, compatibilizer, coupling agent, processing aid for pharmaceuticals, personal care products, aqueous coatings and adhesives, agricultural formulations, polymer processing, cleaners/disinfectants; chlorinated hydrocarbon stabilizer. Clear liquid; mp = -89.8°; bp = 117.7°; d^{20} = 0.8098; soluble in H_2O (9.9 g/100 ml), more soluble in organic solvents; LD_{50} (rat orl) = 4.36 g/kg. *Arco*

505   sec-Butanol
78-92-2   1540   H

$C_4H_{10}O$
Butan-2-ol.
2-01-00-00400 (Beilstein Handbook Reference); AI3-24189; Alcool butylique secondaire; BRN 0773649; Butan-2-ol; Butanol secondaire; Butyl alcohol, sec-; Butylene hydrate; Caswell No. 119C; CCS 301; EINECS 201-158-5; EINECS 240-029-8; EPA Pesticide Chemical Code 001502; Ethyl methyl carbinol; HSDB 674; Methyl ethyl carbinol; NSC 25499; s-Butanol; s-Butyl alcohol; sec-Butanol; sec-Butyl alcohol. Used in manufacture of flavors, perfumes, dyestuffs and wetting agents. A general purpose solvent, used with natural oils. Liquid; mp = -114.7°; bp = 99.5°; d^{20} = 0.8063; soluble in H_2O, organic solvents; LD_{50} (rat orl) = 6480 mg/kg. *Ashland; BASF Corp.; Callery Chem. Co. UK; Mitsui Toatsu.*

506   2-Butanone oxime
96-29-7   G H

C_4H_9NO
2-Butanone, oxime.
4-01-00-03250 (Beilstein Handbook Reference); BRN 1698241; Butanone oxime; CCRIS 1382; EINECS 202-496-6; Ethyl methyl

ketone oxime; Ethyl methyl ketoxime; Ethyl-methylketonoxim; MEK-oxime; Methyl ethyl ketone oxime; Methyl ethyl ketoxime; Methyl ethyl ketoxime; NSC 442; Skino 2; Troykyd anti-skin B; USAF AM-3; USAF EK-906. Antiskinning agent for paint industry. Yellow oil; mp = -29.5°; bp = 152.5°; d^{20} = 0.9232; soluble in H_2O, very soluble in C_6H_6, EtOH, Et_2O; LD_{50} (mus ip) = 200 mg/kg. *Akzo Chemie; Allied Signal; La Littorale.*

507   2-Butanone, O,O',O''-(methylsilylidyne)-trioxime
22984-54-9   H

$C_{13}H_{27}N_3O_3Si$
2-Butanone, O,O',O''-(methylsilylidyne)trioxime.
2-Butanone, O,O',O''-(methylsilylidyne)trioxime; EINECS 245-366-4.

508   1-Butene
106-98-9   1512   H

C_4H_8
But-1-ene.
'α-Butene; 1-Butene; α-Butylene; 1-Butylene; But-1-ene; Butene, 1-; EINECS 203-449-2; Ethylethylene; HSDB 179. Gas; mp = -185.3°; bp = -6.2°; λ_m = 162, 175, 187 nm (ε = 10000, 15849, 12589, gas); insoluble in H_2O, soluble in C_6H_6, very soluble in EtOH, Et_2O.

509   2-Butene
107-01-7   1513   H

C_4H_8
But-2-ene.
3-01-00-00732 (Beilstein Handbook Reference); BRN 1718755; Butene-2; Butene, 2-; 2-Butene; β-Butylene; Butylene-2; Dimethylethylene; EINECS 203-452-9; HSDB 180; Pseudobutylene.

510   cis-But-2-ene
590-18-1   H

C_4H_8
1,2-Dimethylethylene (Z).
2-Butene, (2Z)-; (Z)-But-2-ene; 2-Butene-cis; cis-2-Butene; cis-Butene; β-cis-Butylene; cis-1,2-Dimethylethylene; EINECS 209-673-7; High-boiling butene-2; HSDB 5704. Gas; mp = -138.9; bp = 3.7°; d^{25} = 0.616; λ_m = 160, 175, 196, 200 nm (ε = 12589, 19953, 1000, 501, gas); insoluble in H_2O, soluble in C_6H_6, very soluble in EtOH, Et_2O.

511   Butenediol
110-64-5   H

$C_4H_8O_2$
2-Butene-1,4-diol.
Agrisynth B2D; AI3-07551; Butenediol; 1,4-Butenediol; 2-Buten-1,4-diol; 2-Butene-1,4-diol; 2-Butene, 1,4-dihydroxy-; Caswell No. 120; 1,4-Dihydroxy-2-butene; EINECS 203-787-0; EPA Pesticide Chemical Code 220100; HSDB 5540; NSC 1260; Penitricin C.

512   Butoxycarbonylmethyl butyl phthalate
85-70-1   G

$C_{18}H_{24}O_6$
Butyl phthalyl butyl glycolate.
4-09-00-03256 (Beilstein Handbook Reference); AI3-01793; AI3-01793 (USDA); 1,2-Benzenedicarboxylic acid, 2-butoxy-2-oxoethyl butyl ester; BRN 2007363; Butyl carbobutoxymethyl phthalate; Butyl glycolyl butyl phthalate; Butyl phthalyl butyl glycolate; Butylcarbobutoxymethyl phthalate; Caswell No. 131B; Dibutyl O-(o-carboxybenzoyl) glycolate; EINECS 201-624-8; Glycolic acid, butyl ester, butyl phthalate; Glycolic acid, phthalate, dibutyl ester; HSDB 284; Phthalic acid, butoxycarbonylmethyl butyl ester; Phthalic acid, butyl ester, butyl glycolate; Reomol 4pg; Santicizer B-16. Used as a plasticizer in food-packaging material, PVC tubing and dental cushions. Oil; bp^{10} = 220°; almost insoluble in H_2O. *CIBA plc; Morflex.*

513   Butoxyethanol
111-76-2   1558   G P

$C_6H_{14}O_2$
O-Butyl ethylene glycol.
4-01-00-02380 (Beilstein Handbook Reference); AI3-0993; AI3-09903; BRN 1732511; BUCS; Butoksyetylowy alkohol; 2-Butossi-etanolo; 2-Butoxy-äthanol; 2-Butoxy-aethanol; Butoxyethanol; β-Butoxyethanol; 2-n-Butoxy-1-ethanol; 2-Butoxy-1-ethanol; 2-Butyl cellosolve®; Butyl cellu-sol; Butyl glycol; Butyl oxitol; Butylcelosolv; Butylglycol; Caswell No. 121; CCRIS 5985; Chimec NR; Dowanol EB; EGBE; EGMBE; EINECS 203-905-0; Ektasolve® EB; EPA Pesticide Chemical Code 011501; Eter monobutilico del etilenglicol; Ethanol, 2-butoxy-; Ether monobutylique de l'ethyleneglycol; Ethylene glycol butyl ether; Ethylene glycol mono-n-butyl ether; Ethylene glycol monobutyl ether; Ethylene glycol monobutyl ether [Keep away from food]; Ethylene glycol n-butyl ether; Gafcol EB; Glycol butyl ether; Glycol ether EB; Glycol monobutyl ether; HSDB 538; Jeffersol EB; Monobutyl ether of ethylene glycol; Monobutyl ethylene glycol ether; n-Butoxyethanol; NSC 60759; Poly-Solv EB; UN2369. Solvent for nitrocellulose resins, spray lacquers, co-solvent, gas chromatography. Used in alkyd, phenolic, maleic, and cellulose nitrate resins; excellent retarder for lacquers, improving gloss and flow-out, blush resistance, and reducing the formation of orange peel; useful in formulating hot-spray, brushing, flow-coat, and Aerosol®® lacquers. Registered by EPA as an antimicrobial and fungicide (cancelled). Liquid: mp = -74.8°; bp = 168.4°; bp_{12} = 67°; d_4^{20}= 0.9012, d^{20} = 0.9015, d_{20}^{20} = 0.9020; soluble in H_2O (5 g/100

ml), freely soluble in EtOH, Et2O, slightly soluble in CCl4; LD50 (rat orl) = 1480 mg/kg. *Arco; Eastman Chem. Co.; ICI; Sigma-Aldrich Fine Chem.; Union Carbide Corp.*

514   Butoxyethyl stearate
109-38-6   G

C24H48O3
2-Butoxyethyl stearate.
AI3-04494; 2-Butoxyethyl stearate; Dermol BES; EINECS 203-668-3; KP-23; Octadecanoic acid, 2-butoxyethyl ester. Used as a solvent. *Akzo Chemie.*

515   N-(Butoxymethyl)acrylamide
1852-16-0   H

C8H15NO2
2-Propenamide, N-(butoxymethyl)-.
Acrylamide, N-(butoxymethyl)-; BRN 1906225; N-(Butoxymethyl)-2-propenamide; N-(Butoxymethyl)-acrylamide; N-Butoxymethylakrylamid; EINECS 217-442-7; HSDB 5859; 2-Propenamide, N-(butoxy-methyl)-.

516   Butoxypropanol
5131-66-8   H

C7H16O2
1,2-Propylene glycol 1-monobutyl ether.
4-01-00-02471 (Beilstein Handbook Reference); AI3-18549; BRN 1733910; Butoxypropanol; n-Butoxypropanol; n-Butoxy-2-propanol; 1-Butoxy-2-propanol; 1-Butoxypropan-2-ol; EINECS 225-878-4; 2-Hydroxy-3-butoxypropane; NSC 2211; 2-Propanol, 1-butoxy-; Propasol solvent B; Propylene glycol monobutyl ether; Propylene glycol n-butyl ether; 1,2-Propylene glycol 1-monobutyl ether. Liquid; bp = 171.5°, bp20 = 71°; d^{20} = 0.882; soluble in EtOH, MeOH, Et2O, C6H6, CCl4.

517   Butoxytriethylene glycol
143-22-6   H

C10H22O4
2-(2-(2-Butoxyethoxy)ethoxy)ethanol.
4-01-00-02402 (Beilstein Handbook Reference); AI3-30236; BRN 1750600; Butoxytriethylene glycol; Butoxytriglycol; Dowanol tbat; EINECS 205-592-6; Ethanol, 2-(2-(2-butoxyethoxy)ethoxy)-; HSDB 5645; NSC 164915; Poly-Solv TB; Triethylene glycol butyl ether. Liquid; bp = 278°; d^{20} = 0.9890; very soluble in EtOH, MeOH.

518   Butter yellow
60-11-7   3256   G

C14H15N3
4-Dimethylaminoazobenzene.
AI3-08903; Aniline, N,N-dimethyl-p-phenylazo-; Atul Fast Yellow R; Azobenzene, p-dimethylamino-; Benzenamine, N,N-dimethyl-4-(phenylazo)-; Benzene-azodimethylaniline; Brilliant Fast Oil Yellow; Brilliant Fast Spirit Yellow; Brilliant Oil Yellow; Butter Yellow; Butyro flavine; C.I. 11020; C.I. Solvent Yellow 2; CCRIS 251; Cerasine Yellow GG; Dab; Dab (Carcinogen); Dimethyl aminoazobenzene; 4-(Dimethylamino)azobenzene; p-Dimethylamino-azobenzen; p-Dimethylaminoazobenzene; p-Dimethylamino-azobenzol; Dimethyl Yellow; Dimethyl Yellow Analar; Dimethyl Yellow N,N-dimethylaniline; DMAB; EINECS 200-455-7; Enial Yellow 2G; Fast Oil Yellow B; Fast Yellow; Fat Yellow; Fat Yellow A; Fat Yellow AD OO; Fat Yellow ES; Fat Yellow Extra Conc.; Fat Yellow R; Fat Yellow R (8186); Grasal Brilliant Yellow; HSDB 2692; Iketon Yellow Extra; Jaune de beurre; Methyl Yellow; NSC 6236; Oil Yellow; Oil Yellow 20; Oil Yellow 2625; Oil Yellow 7463; Oil Yellow 2G; Oil Yellow BB; Oil Yellow D; Oil Yellow DN; Oil Yellow FF; Oil Yellow FN; Oil Yellow G; Oil Yellow G-2; Oil Yellow GG; Oil Yellow GR; Oil Yellow II; Oil Yellow N; Oil Yellow Pel; Oil Yellow S; Oleal Yellow 2G; Organol Yellow ADM; Orient Oil Yellow GG; P.D.A.B.; Petrol Yellow Wt; 4-(Phenylazo)-N,N-dimethylaniline; RCRA waste number U093; Resinol Yellow GR; Resoform Yellow GGA; Silotras Yellow T2G; Solvent Yellow 2; Somalia Yellow A; Stear Yellow JB; Sudan GG; Sudan Yellow; Sudan Yellow GG; Sudan Yellow GGA; Toyo Oil Yellow G; USAF EK-338; Waxoline Yellow AD; Waxoline Yellow ADS; Yellow G soluble in grease; Zlut maselna; Zlut rozpoustedlova 2. Azo dyestuffs based on dimethylaminoazobenzene were formerly used for coloring butter and oils. Acid-base indicator; red at pH 2.9, yellow at pH 4.0 Crystals; mp = 111°; bp = 200°; LD50 (rat orl) = 200 mg/kg. *Lancaster Synthesis Co.; Sigma-Aldrich Fine Chem.*

519   Butyl acetate
123-86-4   1534   H

C6H12O2
n-Butyl acetate.
4-02-00-00143 (Beilstein Handbook Reference); Acetate de butyle; Acetic acid, butyl ester; AI3-00406; BRN 1741921; Butile (acetati di); Butyl acetate; Butyl ethanoate; Butylacetat; Butylacetaten; Butyle (acetate de); Butylester kyseliny octove; CCRIS 2287; EINECS 204-658-1; HSDB 152; NSC 9298; Octan n-butylu; UN1123. Mobile liquid; mp = -78°; bp = 126.1°; d^{20} = 0.8825; λ_m = 202, 214 nm (ε = 51, 50, isooctane), 208 nm (ε = 56, MeOH); slightly soluble in H2O, soluble in Me2CO, CHCl3, freely soluble in EtOH, Et2O.

520   tert-Butyl Acetate
540-88-5   1536   G

C6H12O2
Acetic acid 1,1-dimethylethyl ester.
4-02-00-00151 (Beilstein Handbook Reference); Acetic acid, 1,1-dimethylethyl ester; Acetic acid, tert-butyl ester; BRN 1699506; t-Butyl acetate; tert-Butyl acetate; tert-Butyl ethanoate; EINECS 208-760-7; HSDB 835; NSC 59719; Texaco lead appreciator; TLA; UN1123. Used as a gasoline additive. Liquid; bp = 95.1°; d^{20} =

0.8665; λ_m = 213 nm (ε = 55, MeOH); insoluble in H_2O; soluble in EtOH, Et2O, CHCl3, AcOH. *Lancaster Synthesis Co.; Sigma-Aldrich Fine Chem.*

521 Butyl acetyl ricinoleate
140-04-5   G

$C_{24}H_{44}O_4$

Butyl 12-(acetyloxy)-9-octadecenoate.

AI3-00400; Bakers P-6; Baryl; Butyl 12-acetoxyoleate; Butyl O-acetylricinoleate; EINECS 205-393-4; Flexricin P 6; NSC 2319; 9-Octadecenoic acid, 12-(acetyloxy)-, butyl ester; 9-Octadecenoic acid, 12-(acetyloxy)-, butyl ester, (R-(Z))-; 9-Octadecenoic acid, 12-(acetyloxy)-, butyl ester, (R-(Z))- (9CI); 9-Octadecenoic acid, 12-(acetyloxy)-, butyl ester, (9Z,12R)-; Ricinoleic acid, butyl ester, acetate. A plasticizer used with vinyl polymers. *Harcros.*

522 Butyl acrylate
141-32-2   1538   G H

$C_7H_{12}O_2$

n-Butyl 2-propenoate.

4-02-00-01463 (Beilstein Handbook Reference); Acrylic acid butyl ester; Acrylic acid n-butyl ester; AI3-15739; BRN 1749970; Butyl 2-propenoate; Butyl acrylate; Butylester kyseliny akrylove; CCRIS 3401; EINECS 205-480-7; HSDB 305; NSC 5163; UN2348. Intermediate in organic synthesis, polymers and copolymers for solvent coatings, adhesives, paints, binders, emulsifiers. Liquid; mp = -64.6°; bp = 145°; d^{20} = 0.8898; λ_m = 242 nm (isooctane); insoluble in H_2O, slightly soluble in CCl4, soluble in EtOH, Et2O, Me2CO; LD50 (rat orl) = 3.73 g/kg. *BASF Corp.*

523 tert.-Butyl Alcohol
75-65-0   1541   P

$C_4H_{10}O$

1,1-Dimethylethanol.

4-01-00-01609 (Beilstein Handbook Reference); AI3-01288; Alcool butylique tertiaire; Arconol; BRN 0906698; Butanol tertiaire; t-Butanol; tert-Butanol; t-Butyl hydroxide; tert-Butyl alcohol; tertiary-Butanol; Caswell No. 124A; CCRIS 4755; Dimethylethanol; EINECS 200-889-7; HSDB 50; Methanol, trimethyl-; Methyl-2-propanol; NCI-C55367; TBA; Trimethyl carbinol; Trimethyl methanol. Registered by EPA as an antimicrobial, fungicide, herbicide and insecticide (cancelled). Crystals; mp = 25.4°; bp = 82.4°; d^{20} = 0.7887; LD50 (rat orl) = 3500 mg/kg, pulmonary toxicity at 25 ppm. *Lancaster Synthesis Co.; Mallinckrodt Inc.; Sigma-Aldrich Fine Chem.*

524 tert-Octylamine
107-45-9   H

$C_8H_{19}N$

1,1,3,3-Tetramethylbutylamine.

4-04-00-00769 (Beilstein Handbook Reference); AI3-52247; BRN 1732753; Butylamine, 1,1,3,3-tetramethyl-; Butylamine, bis(1,3-dimethyl)-; EINECS 203-491-1; NSC 33852; tert-Octylamine; 2-Pentanamine, 2,4,4-trimethyl-; 1,1,3,3-Tetramethyl-butanamine; 2,4,4-Trimethyl-2-pentylamine.

525 Butylamine
109-73-9   1542   G H

$C_4H_{11}N$

n-Butylamine.

AI3-24197; Amine-C4; 1-Amino-butaan; 1-Aminobutan; 1-Aminobutane; 1-Butanamine; n-Butilamina; n-Butylamin; Butylamine; n-Butylamine; CCRIS 4756; EINECS 203-699-2; FEMA Number 3130; HSDB 515; Mono-n-butylamine; Monobutilamina; Monobutylamine; Norvalamine; NSC 8029; UN1125. Chemical intermediate. Liquid; mp = -49.1°; bp = 77.0°, d^{20} = 0.7414; freely soluble in H_2O, soluble in EtOH, Et2O; LD50 (rat orl) = 366 mg/kg. *Akzo Nobel Chemicals Inc; Akzo Nobel.*

526 t-Butylamine
75-64-9   1544   H

$C_4H_{11}N$

Trimethylaminomethane.

AI3-24036; Butylamine, tert; CCRIS 4758; CP 11093; EINECS 200-888-1; Erbumine; HSDB 5209; NSC 9571; t-Butylamine; tert-Butylamine; Trimethylamino-methane. Used in chemical synthesis and analysis. Liquid; mp = -66.9°; bp = 44°; d^{20} = 0.6958; very soluble in H_2O, organic solvents. *Lancaster Synthesis Co.; Mallinckrodt Inc.; Sigma-Aldrich Fine Chem.*

527 tert-Butylaminoethyl methacrylate
3775-90-4   G

$C_{10}H_{19}NO_2$

2-[(1,1-Dimethylethyl)amino]ethyl 2-methyl-2-prop-enoate.

4-04-00-01509 (Beilstein Handbook Reference); Ageflex FM-4; AI3-61480; BRN 1761825; 2-(tert-Butylamino)ethyl methacrylate; N-tert-Butylaminoethyl methacrylate; 2-((1,1-Dimethylethyl)amino)ethyl 2-methyl-2-propenoate; EINECS 223-228-4; Ethanol, 2-(tert-butylamino)-, methacrylate (ester); HSDB 6111; Methacrylic acid, 2-(tert-butylamino)ethyl ester; Methacrylic acid, 2-(tert-butylamino)ethyl ester; 2-Methyl-2-propenoic acid, 2-((1,1-dimethylethyl)-amino)ethyl ester; 2-Propenoic acid, 2-methyl-, 2-((1,1-dimethylethyl)amino)ethyl ester. Used in automotive dip tanks, coatings, industrial/consumer adhesives and coatings, dye and lube oil additives, intermediate for water treatment chemicals, oil-water separations. Liquid; d_{20}^{20} = 0.914. *Rit-Chem.*

528 Butylated hydroxyanisole
25013-16-5   1546   G

$C_{11}H_{16}O_2$
 BHA.
 (1,1-Dimethylethyl)-4-methoxyphenol; Anisole, butylated hydroxy-; Antioxyne B; Antrancine 12; BHA; BOA (antioxidant); Butyl hydroxyanisole; Butylated hydroxyanisole; Butylhydroxyanisole; 2-terc.Butyl-4-methoxyfenol; 2(3)-t-Butyl-4-hydroxyanisole; 3-t-Butyl-4-hydroxyanisole; t-Butyl hydroxyanisole; tert-Butyl-4-hydroxyanisole; Butylohydroksyanizol; CCRIS 102; EEC No. E320; EINECS 246-563-8; Embanox; FEMA No. 2183; HSDB 3913; Nepantiox 1-F; Nipantiox 1-F; Phenol, (1,1-dimethylethyl)-4-methoxy-; Phenol, tert-butyl-4-methoxy-; Protex; Sustane® 1-F; Tenox BHA. Mixture of isomers of t-butyl-substituted 4-methoxyphenols; antioxidant and preservative for foods mp = 48-55°; bp$_{733}$ = 264-270°; insoluble in H_2O, soluble in organic solvents; LD$_{50}$ (rat orl) = 2200 mg/kg. *Eastman Chem. Co.; Nipa; Penta Mfg.; UOP.*

529 p-t-Butylbenzaldehyde
939-97-9   H

$C_{11}H_{14}O$
 Benzaldehyde, 4-(1,1-dimethylethyl)-.
 AI3-37199; Benzaldehyde, 4-(1,1-dimethylethyl)-; Benzaldehyde, p-tert-butyl-; p-t-Butylbenzaldehyde; 4-tert-Butylbenzaldehyde; EINECS 213-367-9.

530 n-Butyl benzoate
136-60-7   1551   G

$C_{11}H_{14}O_2$
 Benzoic acid butyl ester.
 4-09-00-00290 (Beilstein Handbook Reference); AI3-00521; Anthrapole AZ; Benzoic acid, butyl ester; Benzoic acid n-butyl ester; BRN 1867073; Butyl benzoate; n-Butyl benzoate; Butylester kyseliny benzoove; Dai Cari XBN; EINECS 205-252-7; Hipochem B-3-M; HSDB 2089; Marvanol® Carrier BB; NSC 8474. Emulsified butyl benzoate; self-emulsifying carrier used in dyeing of polyester and triacetate fibers. Thick, oily liquid; mp = -22.4°; bp = 250.3°; d^{20} = 1.0000; λ$_m$ = 228, 272, 279 nm (ε = 12200, 874, 716, MeOH); insoluble in water; soluble in Me$_2$CO, CCl$_4$, freely soluble in EtOH, Et$_2$O; LD$_{50}$ (rat orl) = 5.14 g/kg. *Pentagon Chems. Ltd.; PMC; Raschig GmbH; Uniroyal.*

531 N-t-Butyl-2-benzothiazolesulfenamide
95-31-8   G H

$C_{11}H_{14}N_2S_2$
 2-Benzothiazolesulfenamide, N-(1,1-dimethylethyl)-.
 4-27-00-01866 (Beilstein Handbook Reference); Accel BNS; BBTS; Benzothiazolyl-2-t-butylsulfenamide; BRN 0158370; Delac NS; EINECS 202-409-1; HSDB 5288; Nocceler NS; NSC 84176; Pennac Tbbs; Perkacit® TBBS; Santocure NS; Santocure NS vulcanization accelerator; Vanax NS; Vulkacit NZ. A delayed action accelerator very safe at processing temperatures producing lower scorch and faster cure than Durax but producing high modulus stocks at curing temperatures; activated by thiurams, dithiocarbamates, aldehyde amines, guanidines and BIK; nondiscoloring and nonstaining to rubber stocks and materials in contact with them; used in tire treads, carcass, mechanicals and wire jackets. Solid; mp=105°; d = 1.28; practically insoluble in H$_2$O. *Akrochem Chem. Co.; Akzo Chemie; Monsanto Co.; Uniroyal; Vanderbilt R.T. Co. Inc.*

532 Butylbenzyl phthalate
85-68-7   H

$C_{19}H_{20}O_4$
 Butyl phenylmethyl 1,2-benzenedicarboxylate.
 4-09-00-03218 (Beilstein Handbook Reference); AI3-14777; BBP; Benzyl butyl phthalate; Benzyl-butylester kyseliny ftalove; Benzyl n-butyl phthalate; BRN 2062204; Butyl benzyl phthalate; Butyl phenylmethyl 1,2-benzenedicarboxylate; Butylbenzyl phthalate; Caswell No. 125G; CCRIS 104; EINECS 201-622-7; HSDB 2107; NCI-C54375; NSC 71001; Palatinol BB; Phthalic acid, benzyl butyl ester; Santicizer 160; Sicol; Sicol 160; Unimoll BB. Used as a plasticizer for polyvinyl and cellulosic resins, and as an organic intermediate. Oil; mp = -35°; bp = 370°; d = 1.100; slightly soluble in H$_2$O (0.269 mg/100 ml), more soluble in organic solvents; LD$_{50}$ (rat orl) = 2330 mg/kg. *Ashland; Bayer AG; Monsanto Co.*

533 Butylbiphenyl
41638-55-5   H

$C_{16}H_{18}$
 Butyl-1,1'-biphenyl.
 1,1'-Biphenyl, butyl-; Butyl-1,1'-biphenyl; Butylphenyl; EINECS 255-471-7.

534 Butyl cellosolve acetate
112-07-2   G H

C8H16O3
2-Butoxyethyl acetate.
4-02-00-00215 (Beilstein Handbook Reference); Acetic acid, 2-butoxyethyl ester; BRN 1756960; 2-Butoxyethanol acetate; n-Butoxyethanol acetate; Butoxyethyl acetate; 2-Butoxyethyl acetate; 2-Butoxyethylester kyseliny octove; Butyl cellosolve acetate; Butyl glycol acetate; Butylcelosolvacetat; Butylglycol acetate; Diethylene glycol butyl ether acetate; EGBEA; EINECS 203-933-3; Ektasolve EB acetate; Ethanol, 2-butoxy-, acetate; Ethylene glycol butyl ether acetate; Ethylene glycol monobutyl ether acetate; Glycol monobutyl ether acetate; HSDB 435. Solvent for high-solids coatings Liquid; bp = 192.3°. *Arco; Eastman Chem. Co.; OxyChem; Union Carbide Corp.*

535   Butyl chloride
109-69-3   1559   H

C4H9Cl
n-Butyl chloride.
4-01-00-00246 (Beilstein Handbook Reference); AI3-15309; BRN 1730909; Butane, 1-chloro-; Butyl chloride; n-Butyl chloride; CCRIS 1389; 1-Chlorobutane; Chlorobutane, 1-; Chlorure de butyle; EINECS 203-696-6; HSDB 4167; NBC wormer; NCI-C06155; NSC 8419; n-Propylcarbinyl chloride; Sure Shot; UN 1127. Liquid; mp = -123.1°; bp = 78.6°; d^{20} = 0.8862; insoluble in H2O, slightly soluble in CCl4, freely soluble in EtOH, Et2O.

536   6-t-Butyl-3-chloromethyl-2,4-xylenol
23500-79-0   H

C13H19ClO
3-(Chloromethyl)-6-(1,1-dimethylethyl)-2,4-dimethyl-phenol.
6-t-Butyl-3-chloromethyl-2,4-xylenol; 3-(Chloro-methyl)-6-(1,1-dimethylethyl)-2,4-dimethylphenol; 2,4-Dimethyl-3-(chloromethyl)-6-t-butylphenol; EINECS 245-697-4; Phenol, 6-t-butyl-3-(chloromethyl)-2,4-dimethyl-; Phenol, 3-(chloromethyl)-6-(1,1-dimethyl-ethyl)-2,4-dimethyl-.

537   o-t-Butyl-p-cresol
2409-55-4   G H

C11H16O
4-Methyl-2-(1,1-dimethylethyl)phenol.
4-06-00-03397 (Beilstein Handbook Reference); BRN 1817645; Cao; CCRIS 4045; EINECS 219-314-6; HSDB 5875; Lowinox® MBPC; 4-Methyl-2-butylphenol; 4-Methyl-2-t-butylphenol; NSC 60301; o-t-Butyl-p-cresol; p-Cresol, 2-t-butyl-; Phenol, 2-(1,1-dimethylethyl)-4-methyl-. Intermediate used in manufacture of antioxidants for food, plastic, rubbers, and other general purpose requirements and of UV absorbing chemicals. Crystals; mp = 51-52°;

bp = 237°; d^{25} = 0.9247; λ_m = 275, 277, 284 nm (cyclohexane); slightly soluble in H2O, soluble in Me2CO, C6H6, CHCl3; LD50 (rat orl) = 2500 mg/kg. *Lowi; PMC; PMS Specialities Group Inc.*

538   Butyl cumyl peroxide
3457-61-2   G

C13H20O2
1,1-Dimethylethyl 1-methyl-1-phenylethyl peroxide.
tert-Butyl α,α-dimethylbenzyl peroxide; EINECS 222-389-8; Luperco 801-XL; Peroxide, 1,1-dimethylethyl 1-methyl-1-phenylethyl; Trigonox® T. Used as an initiator for high-temperature cure of polyester resins, curing elastomers, and polymer modification thermoplastic cross-linking. *Elf Atochem N. Am.*

539   4-t-Butylcyclohexanol
98-52-2   H

C10H20O
Cyclohexanol, 4-(1,1-dimethylethyl)-.
1-06-00-00018 (Beilstein Handbook Reference); AI3-02503; BRN 1902277; 4-t-Butylcyclohexanol; p-t-Butylcyclohexanol; Cyclohexanol, 4-t-butyl-; Cyclohexanol, 4-(1,1-dimethylethyl)-; EINECS 202-676-4; NSC 404197; Padaryl; USAF DO-20.

540   p-tert-Butylcyclohexyl acetate
32210-23-4   H

C12H22O2
Cyclohexanol, 4-tert-butyl-, acetate.
Acetic acid, (4-tert-butylcyclohexyl) ester; AI3-36523; 4-t-Butylcyclohexyl acetate; 4-tert-Butyl cyclohexyl acetate; 4-tert-Butylcyclohexanol acetate; 4-tert-Butylhexahydrophenyl acetate; Cyclohexanol, 4-(1,1-dimethylethyl)-, acetate; Cyclohexanol, 4-tert-butyl-, acetate; Cyclohexanol, 4-tert-butyl-, acetate (8CI); EINECS 250-954-9; NSC 163103; p-tert-Butyl cyclohexyl acetate; p-tert-Butylcyclohexyl acetate; Vertenex.

541   Butyl diethylene glycol acetate
124-17-4   G H

C10H20O4
2-(2-Butoxyethoxy)ethanol acetate.
3-02-00-00308 (Beilstein Handbook Reference); Acetic acid 2-(2-butoxyethoxy)ethyl ester; AI3-00170; BRN 1771533; Butoxyethoxyethyl acetate; 2-(2-Butoxyethoxy)ethanol acetate; 2-(2-Butoxyethoxy)ethyl acetate; 2-(2-Butoxyethoxy)ethylester kyseliny

octove; Butyl carbitol acetate; Butyl diethylene glycol acetate; Butyl diglycol acetate; Butylkarbitolacetat; Diethylene glycol butyl ether acetate; Diglycol monobutyl ether acetate; EINECS 204-685-9; Ektasolve DB acetate; Ethanol, 2-(2-butoxyethoxy)-, acetate; Glycol ether DB aceatate; HSDB 334; NSC 5175. Used as a solvent. Liquid; mp = -32°; bp = 245°; d^{20} = 0.985; soluble in H_2O, very soluble in Me_2CO, EtOH, Et_2O; LD$_{50}$ (rat orl) = 6500 mg/kg. *Allchem Ind.; Arco; Ashland; Eastman Chem. Co.; Fluka; Occidental Chem. Corp.; OxyChem; Sigma-Aldrich Fine Chem.; Spectrum Chem. Manufacturing.*

542   Butylene
25167-67-3   H

C4H8
But-1-ene.
Butene; But-1-ene; n-Butene; Butylene; n-Butylene; EINECS 246-689-3; HSDB 5127; UN1012.

543   2-Butylene dichloride
764-41-0   H

C4H6Cl2
1,4-Dichloro-2-butene.
BRN 1361446; 2-Butene, 1,4-dichloro-; 2-Butylene dichloride; 1,4-DCB; 1,4-Dichloro-2-butene; 1,4-Dichlorobut-2-ene; 1,4-Dichlorobutene-2; 1,4-Dichloro-2-butylene; EINECS 212-121-8; HSDB 6008; NSC 9452; RCRA waste number U074.

544   Butylene Glycol
107-88-0   1566   H

C4H10O2
1,3-Dihydroxybutane.
0-01-00-00477 (Beilstein Handbook Reference); AI3-11077; BD; BRN 1731276; 1,3-Butandiol; 1,3-Butanediol; Butane-1,3-diol; 1,3-Butylene glycol; β-Butylene glycol; 1,3-Butylenglykol; 1,3-Dihydroxybutane; Caswell No. 128GG; EINECS 203-529-7; HSDB 153; Methyltrimethylene glycol; NSC 402145. Registered by EPA (cancelled). Viscous liquid; mp < 50°; bp = 207.5°; d^{20} = 1.005 (8.398 lb/gal); viscosity (centistokes) 24.6 at 50°, 96 at 25°; 590 at 0°, 3253 at -17.7°, 6059 at -23°; soluble in H_2O, Me_2CO, MEK, EtOH, dibutyl phthalate, castor oil, insoluble in hydrocarbons, CCl4, ethanolamines, mineral oil, linseed oil; LD$_{50}$ (rat orl) = 22500 mg/kg. *Sigma-Aldrich Co.*

545   sec-Butyl ether
6863-58-7   H

C8H18O
2,2'-Oxybisbutane.
0-01-00-00372 (Beilstein Handbook Reference); Bis(2-butyl)ether; BRN 1733014; Butane, 2,2'-oxybis-; sec-Butyl ether; Di-sec-butyl ether; EINECS 229-961-6. Liquid; bp = 121°; d^{25} = 0.756.

546   Butylethylamine
13360-63-9   H

C6H15N
N-Ethyl-n-butylamine.
4-04-00-00547 (Beilstein Handbook Reference); BRN 1731324; 1-Butanamine, N-ethyl-; Butyl(ethyl)amine; Butylamine, N-ethyl-; CCRIS 4815; EINECS 236-415-0; Ethylbutylamine; N-Butylethylamine; N-Ethyl-N-butylamine.

547   Butyl Ethyl Cellosolve
112-34-5   1556   G H

C8H18O3
Diethylene glycol mono-n-butyl ether.
4-01-00-02394 (Beilstein Handbook Reference); AI3-01954; BRN 1739225; BUCB; Butoxy diethylene glycol; Butoxydiglycol; Butoxyethoxyethanol; 2-(2-Butoxyethoxy)ethanol; Butyl carbitol; Butyl diglycol; Butyl digol; Butyl dioxitol; Butyl Ethyl Cellosolve; Caswell No. 121B; Caswell No. 125H; CCRIS 5321; Diethylene glycol butyl ether; Diethylene glycol monobutyl ether; Diglycol monobutyl ether; Dowanol DB; EINECS 203-961-6; Ektasolve® DB; EPA Pesticide Chemical Code 011502; Ethanol, 2-(2-butoxyethoxy)-; Glycol ether DB; HSDB 333; Jeffersol DB; NSC 407762; Poly-Solv DB. A lacquer solvent boiling at 230°. Liquid; mp = -68°; bp = 231°; d^{20} = 0.9553; freely soluble in H_2O, very soluble in EtOH, Et_2O, Me_2CO, soluble in C_6H_6; LD$_{50}$ (rat orl) = 6.56 g/kg. *Eastman Chem. Co.*

548   Butyl 2-ethylhexyl phthalate
85-69-8   H

C20H30O4
1,2-Benzenedicarboxylic acid, butyl 2-ethylhexyl ester.
1,2-Benzenedicarboxylic acid, butyl 2-ethylhexyl ester; 2-Ethylhexyl butyl phthalate; Butyl 2-ethylhexyl phthalate; EINECS 201-623-2; HSDB 5251; Phthalic acid, butyl 2-ethylhexyl ester.

549   tert-Butyl 2-ethylperoxyhexanoate
3006-82-4   G H

C12H24O3
Hexaneperoxoic acid, 2-ethyl-, 1,1-dimethylethyl ester.
Aztec® t-Butyl Peroctoate; Aztec® t-Butyl Peroctoate-50 OMS; tert-Butyl 2-ethylperoxyhexanoate; EINECS 221-110-7; Hexaneperoxoic acid, 2-ethyl-, 1,1-dimethylethyl ester; Peroxyhexanoic acid, 2-ethyl-, tert-butyl ester; Trigonox® 21. Initiator for acrylates, styrenics, and LDPE polymerization; used where presence of water is

objectionable. *Akzo Chemie; Catalyst Resources Inc.*

550 Butyl glycidyl ether
2426-08-6   G H

C7H14O2
1-Butoxy-2,3-epoxypropane.
5-17-03-00011 (Beilstein Handbook Reference); Ageflex BGE; BGE; BRN 0103668; 1-Butoxy-2,3-epoxypropane; Butyl 2,3-epoxy propyl ether; Butyl glycidyl ether; CCRIS 828; EINECS 219-376-4; ERL 0810; Ether, butyl 2,3-epoxypropyl; Ether, butyl glycidyl; Glycidyl butyl ether; HSDB 299; n-Butyl glycidyl ether; NSC 83413; Oxirane, (butoxymethyl)-; Propane, 1-butoxy-2,3-epoxy-; Sipomer BGE; TK 10408. Reactive diluent in epoxy resins, laminating, flooring, electrical casting and encapsulants. Liquid; bp = 169°, bp_{26} = 75°; d^{20} = 0.918, d_4^{25} = 0.908; soluble in H_2O (1.0-2.0 g/100 ml), more soluble in organic solvents; LD_{50} (rat orl) = 2050 mg/kg. *Rit-Chem.*

551 t-Butyl glycidyl ether
7665-72-7   G

C7H14O2
((1,1-Dimethylethoxy)methyl)oxirane.
5-17-03-00011 (Beilstein Handbook Reference); Ageflex TBGE; t-Bge; BRN 0103483; (tert-Butoxymethyl)oxirane; Butyl glycidyl ether, tert-; tert-Butyl glycidyl ether; CCRIS 2632; 1,1-Dimethylethyl glycidyl ether; EINECS 231-640-0; 2,3-Epoxypropyl-t-butyl ether; Oxirane, ((1,1-dimethylethoxy)methyl)-; Propane, 1-tert-butoxy-2,3-epoxy-. Reactive diluent in epoxy resins, corrosion inhibitor in some solvents, modifier for amines, acids and thiols. Liquid; mp = -70°; bp = 152°, 164-166°; d_{20}^{20} = 0.898; TWA 25ppm; moderately toxic by ingestion, skin contact, intraperitoneal route; mildly toxic by inhalation. *Rit-Chem.*

552 t-Butylhydroperoxide
75-91-2   1569   G H

C4H10O2
1,1-Dimethylethyl hydroperoxide.
4-01-00-01616 (Beilstein Handbook Reference); AI3-50541; BRN 1098280; Cadox TBH; Caswell No. 130BB; CCRIS 5892; DE 488; EINECS 200-915-7; HSDB 837; Hydroperoxide, 1,1-dimethylethyl-; Hydroperoxide, tert-butyl; Hydroperoxyde de butyle tertiaire; NSC 672; Perbutyl H; Slimicide; Slimicide DE-488; t-Butylhydroperoxide; TBHP-70; terc. Butylhydroperoxid; tert-Butyl hydrogen peroxide; tert-Butyl hydroperoxide; Tertiary butyl hydroperoxide; Trigonox A-75; Trigonox A-W70; USP®-800. Polymerization initiator and slimicide. Liquid; mp = 6°; bp_{17} = 36°, bp = 89° (dec); d^{20} = 0.8960; moderately soluble in H_2O, organic solvents; LD_{50} (rat orl) = 406 mg/kg. *Akzo Chemie; Catalysts & Chem. Ind.; CK Witco Chem. Corp.*

553 t-Butylhydroquinone
1948-33-0   G

C10H14O2
2-(1,1-Dimethylethyl)-1,4-benzenediol.
4-06-00-06013 (Beilstein Handbook Reference); AI3-61039; Banox 20BA; 1,4-Benzenediol (1,1-di-methylethyl)-; 1,4-Benzenediol, 2-(1,1-dimethylethyl)-; BRN 0637923; tert-Butyl-1,4-benzenediol; 2-tert-Butyl-1,4-benzenediol; Butylhydroquinone, tert-; t-Butyl hydroquinone; tert-Butylhydroquinone; tertiary-Butylhydroquinone; 2-t-Butylhydroquinone; 2-tert-Butylhydroquinone; 2-tert-Butyl(1,4)hydroquinone; CCRIS 1447; 2-(1,1-Dimethylethyl)-1,4-benzenediol; Eastman® MTBHQ; EINECS 217-752-2; HSDB 838; Hydroquinone, t-butyl-; Hydroquinone, tert-butyl-; Mono-tert-butylhydroquinone; Mono-tertiary butyl-hydroquinone; MTBHQ; NSC 4972; Sustane®; Sustane® TBHQ; TBHQ; Tenox TBHQ. Preservative and antioxidant for foodstuffs and meat products; color stable and useful as substitute for reactive antioxidants that tend to form purple complexes with iron or copper. White to light tan crystal solid; soluble in organic solvents; slightly soluble in water; mp = 126.5-128.5; LD_{50} (rat orl) = 700 mg/kg. *AC Ind. Inc.; Aceto Corp.; Allchem Ind.; Charkit; Eastman Chem. Co.; Lancaster Synthesis Co.; Penta Mfg.; Showa Denko; UOP.*

554 Butyl-3-iodo-2-propynylcarbamate
55406-53-6   H

C8H12INO2
Carbamic acid, butyl-3-iodo-2-propynyl ester.
BRN 2248232; Butyl-3-iodo-2-propynylcarbamate; Carbamic acid, butyl-3-iodo-2-propynyl ester; Caswell No. 501A; EINECS 259-627-5; EPA Pesticide Chemical Code 107801; Iodopropynyl butylcarbamate; 3-Iodo-2-propynyl butyl carbamate; Troysan KK-108A; Troysan polyphase anti-mildew; Woodlife.

555 Butyl isocyanate
111-36-4   1574   H

C5H9NO
Isocyanic acid, butyl ester.
BIC; Butane, 1-isocyanato-; Butyl isocyanate; EINECS 203-862-8; HSDB 5548; 1-Isocyanatobutane; Isocyanic acid, butyl ester; n-Butyl isocyanate; UN2485.

556 Butyl lactate
138-22-7   G

C7H14O3
n-Butyl lactate.
4-03-00-00649 (Beilstein Handbook Reference); AI3-00397; BRN

1721597; Butyl α-hydroxypropionate; Butyl 2-hydroxypropanoate; Butyl lactate; n-Butyl lactate; Butylester kyseliny mlecne; EINECS 205-316-4; FEMA No. 2205; 2-Hydroxypropanoic acid butyl ester; Lactic acid, butyl ester; NSC 6533; Propanoic acid, 2-hydroxy-, butyl ester. Solvent, used with dipped latex products, coatings, specialty finishes, inks, diluents, adhesives, intermediates, photoresist solvents, screen printing of electronic parts, flavors and fragrances, solvents for nitrocellulose, antiskinning agent, in dry cleaning fluids. Liquid; mp = -28°; bp = 185-187°; d = 0.968; n_D^{20}= 1.4220; LD50 (rat orl) >5 gm/kg. *Penta Mfg.; Purac Biochem. BV; Rit-Chem.*

557   Butyllithium
109-72-8   H

C4H9Li
n-Butyl lithium.
Butyllithium; EINECS 203-698-7; Lithium, butyl-.

558   t-Butyl mercaptan
75-66-1   1578   G H

C4H10S
2-Methyl-2-propanethiol.
4-01-00-01634 (Beilstein Handbook Reference); BRN 0505947; EINECS 200-890-2; HSDB 1611; t-Butyl mercaptan; tert-Butanethiol; tert-Butyl mercaptan; tert-Butylthiol; tertiary-Butyl mercaptan. Has skunk-like odor. Used as additive to natural gas for leak detection. Liquid; mp = -0.5°; bp = 64.3°; d^{25} = 0.7943; soluble in organic solvents, insoluble in H2O. *Lancaster Synthesis Co.; Mallinckrodt Inc.; Sigma-Aldrich Fine Chem.*

559   Butyl methacrylate
97-88-1   G H

C8H14O2
2-Propenoic acid, 2-methyl-, butyl ester.
4-02-00-01525 (Beilstein Handbook Reference); AI3-25420; BRN 0773960; Butil metacrilato; Butyl 2-methacrylate; Butyl 2-methyl-2-propenoate; Butyl methacrylate; n-Butyl methacrylate; Butylester kyseliny methakrylove; Butylmethacrylaat; Butylmethacrylate; CCRIS 4760; EINECS 202-615-1; HSDB 289; Methacrylate de butyle; Methacrylic acid, butyl ester; Methacrylsaeurebutylester; Methacrylsäurebutylester; NSC 20956; UN2227. Used as a monomer in resins, solvent coatings, adhesives, oil additives; emulsions for textiles, leather, paper finishing. Liquid; bp = 160°; d^{20} = 0.8936; very soluble in EtOH, Et2O; Irritant; LD50 (rat orl) = 18000 mg/kg. *Mitsubishi Gas; Rohm & Haas Co.*

560   Butyl methoxydibenzoylmethane
70356-09-1   801   G

C20H22O3
4-t-Butyl-4'-methoxy-dibenzoylmethane.
Avobenzona; Avobenzone; Avobenzonum; Butyl methoxydibenzoylmethane; 1-(p-t-Butylphenyl)-3-(p-methoxyphenyl)-1,3-propanedione; 1-(4-(1,1-Di-methylethyl)phenyl)-3-(4-methoxyphenyl)propane-1,3-dione; 1-(4-(1,1-Dimethylethyl)phenyl)-3-(4-methoxy-phenyl)-1,3-propanedione; EINECS 274-581-6; Parsol® A; Parsol® 1789; Photoplex; 1,3-Propanedione, 1-(4-(1,1-dimethylethyl)phenyl)-3-(4-methoxyphenyl)-. UV-A absorber for production photostability; used in sunscreens. An ultraviolet-A absorbing agent, used in cosmetics, lipsticks and lip balms, moisturizers, nail polishes, shampoos and other hair care products and sunscreens. Crystals; mp = 83°. *Bernel; Givaudan.*

561   2-t-Butyl-5-methylphenol
88-60-8   H

C11H16O
Phenol, 2-(1,1-dimethylethyl)-5-methyl-.
4-06-00-03400 (Beilstein Handbook Reference); BRN 1908225; m-Cresol, 6-t-butyl-; EINECS 201-842-3; HSDB 5260; NSC 48467; Phenol, 2-(1,1-dimethylethyl)-5-methyl-; Phenol, 2-t-butyl-5-methyl-. Crystals; mp = 46.5°; bp11 = 127°; d^{80} = 0.922; insoluble in H2O, soluble in EtOH, Et2O, Me2CO.

562   Butyl myristate
110-36-1   G

C18H36O2
n-butyl tetradecanoate.
4-02-00-01132 (Beilstein Handbook Reference); AI3-07958; BRN 1782649; Bumyr; Butyl myristate; Butyl tetradecanoate; Butyl n-tetradecanoate; EINECS 203-759-8; Myristic acid, butyl ester; NSC 4814; Tetradecanoic acid, butyl ester; Wickenol 141. Emollient, solubilizer, and lubricant for use in cosmetic and toilet preparations; plasticizer. White liquid; soluble in H2O, organic solvents; practically odorless; d = 0.860. *Amerchol; CasChem.*

563   2-Butyloctanol
3913-02-8   G

C12H26O
2-Butyl-1-octanol.
4-01-00-01855 (Beilstein Handbook Reference); AI3-19958; BRN

1738522; Butyloctanol; 2-Butyloctanol; 2-Butyl-1-octanol; 2-Butyloctan-1-ol; 2-Butyloctyl alcohol; EINECS 223-470-0; 5-(Hydroxymethyl)-undecane; Isodecyl alcohol; Michel XO-150-12; NSC 2414; 1-Octanol, 2-butyl-. Used in organic synthesis. *Michel M.*

564   Butyl octyl phthalate
84-78-6

G

C$_{20}$H$_{30}$O$_4$
1,2-Benzenedicarboxylic acid, butyl octyl ester.
BOP; BRN 2289337; Butyl octyl phthalate; EINECS 201-562-1; NSC 69894; Octyl butyl phthalate; Phthalic acid, butyl octyl ester; Plasticizer BOP; Plasticizer OBP; PX 914; Staflex BOP; Truflex OBP. Reagent grade plasticizer. Used as a plasticizer. *Aristech; BASF Corp.*

565   Butyl oleate
142-77-8

G H

C$_{22}$H$_{42}$O$_2$
9-Octadecenoic acid, butyl ester (Z)-.
4-02-00-01653 (Beilstein Handbook Reference); Advaplast 42; Al3-00660; BRN 1728057; Butyl 9-octadecenoate; Butyl cis-9-octadecenoate; Butyl oleate; EINECS 205-559-6; Hallco C 503; Hallco C-503 Plasticizer; HSDB 5483; Kessco 554; Kesscoflex BO; NSC 6700; Oleic acid, butyl ester; Plasthall® 503; Plasthall 914; Uniflex BYO; Vinicizer 30; Wilmar Butyl Oleate; Witcizer 100; Witcizer 101. Textile surface finisher, softener, thread lubricant and antistat; plasticizer. Liquid; mp = -26.4°; bp$_{15}$ = 227-228°; d^{15} = 0.8704; very soluble in EtOH. *Ferro/Keil; Hall C.P.; Inolex; Sybron; Witco/Humko.*

566   Butylparaben
94-26-8   1619

G P

C$_{11}$H$_{14}$O$_3$
Butyl p-hydroxybenzoate.
4-10-00-00375 (Beilstein Handbook Reference); 4-Hydroxybenzoic acid butyl ester; Al3-02930; Aseptoform butyl; Benzoic acid, 4-hydroxy-, butyl ester; Benzoic acid, p-hydroxy-, butyl ester; BRN 1103741; Butoben; 4-(Butoxycarbonyl)phenol; Butyl butex; Butyl chemosept; Butyl 4-hydroxybenzoate; Butyl p-hydroxybenzoate; Butyl paraben; Butyl parahydroxybenzoate; Butyl-Parasept; Butyl tegosept; Butylparaben; Caswell No. 130A; CCRIS 2462; EINECS 202-318-7; EPA Pesticide Chemical Code 061205; FEMA Number 2203; HSDB 286; n-Butyl hydroxybenzoate; n-Butyl p-hydroxybenzoate; n-Butyl parahydroxybenzoate; Nipabutyl; NSC 8475; p-Hydroxybenzoic acid butyl ester; Nipabutyl; Preserval B; Solbrol B; SPF; Tegosept B; Tegosept butyl. Antifungal agent used as a preservative in pharmaceuticals and foods. Solid; mp = 68.5°; λ$_m$ = 256 nm (ε = 2470, MeOH); soluble in H$_2$O, more soluble in

EtOH, CCl$_4$; LD$_{50}$ (mus orl) = 13.2 g/kg. *Inolex; Nipa; Penta Mfg.*

567   t-Butyl peracetate
107-71-1

G

C$_6$H$_{12}$O$_3$
t-Butyl peroxyacetate.
4-02-00-00391 (Beilstein Handbook Reference); Aztec® t-butyl peracetate-50 OMS, 60 OMS, 75 OMS; BRN 1701510; t-Butyl peroxyacetate; tert-Butyl peracetate; tert-Butyl peroxyacetate; EINECS 203-514-5; Ethaneperoxoic acid, 1,1-dimethylethyl ester; Lupersol 70; NSC 118417; Peroxyacetic acid, tert-butyl ester; Trigonox F-C50. Used as a catalyst and initiator. *Catalyst Resources Inc.*

568   tert-Butyl peroxyneodecanoate
26748-41-4

H

C$_{14}$H$_{28}$O$_3$
Neodecaneperoxoic acid, 1,1-dimethylethyl ester.
tert-Butyl peroxyneodecanoate; EINECS 247-955-1; Neodecaneperoxoic acid, 1,1-dimethylethyl ester; Peroxyneodecanoic acid, tert-butyl ester.

569   t-Butyl peroxypivalate
927-07-1

G H

C$_9$H$_{18}$O$_3$
tert-Butyl trimethylperoxyacetate.
4-02-00-00912 (Beilstein Handbook Reference); Aztec® t-Butyl Peroxypivalate-75 OMS; BRN 1704745; EINECS 213-147-2; Esperox 31M; Lupersol 11; Peroxypivalic acid, tert-butyl ester; Propane- peroxoic acid, 2,2-dimethyl-, 1,1-dimethylethyl ester; t-Butyl peroxypivalate; tert-Butyl peroxypivalate; tert-Butyl trimethylperoxyacetate; Trigonox 25/75; Trigonox® 25-C75. An initiator for LDPE polymerization, styrenics, and PVC. Supplied as a solution in odorless mineral spirits. Liquid; mp = -17°; insoluble in H$_2$O. *Akzo Chemie; Catalyst Resources Inc.*

570   tert-Butylphenol
98-54-4   1584

G P

C$_{10}$H$_{14}$O
4-t-Butylphenol.
Al3-00126; Butylphen; p-t-Butyl phenol; p-terc.Butylfenol; p-tert-Butylphenol; Caswell No. 130E; EINECS 202-679-0; EPA Pesticide Chemical Code 064113; NSC 3697; Phenol, 4-(1,1-dimethylethyl)-;

Phenol, p-(tert-butyl)-; PTBP; Terbutol; Ucar butylphenol 4-T; Ucar butylphenol 4-T flake. Antioxidant; chemical intermediate for synthetic resins, plasticizers, surface active agents. Intermediate in the manf. of varnish and lacquer, as a soap antioxidant, in motor oil additives. Used as an intermediate in manufacture of varnishes, as a soap antioxidant, in demulsifiers and as a motor oil additive. Registered by EPA as an antimicrobial and fungicide (cancelled). Needles; mp = 98°; bp = 237°; d_4^{14} 0.9081; λ_m = 220, 276 nm (cyclohexane); insoluble in cold H_2O, soluble in EtOH, Et_2O; LD_{50} (rat orl) = 3250 mg/kg. *ICI Chem. & Polymers Ltd.; Janssen Chimica; PMC; Sigma-Aldrich Co.*

571   2-sec-Butylphenol
89-72-5   H

$C_{10}H_{14}O$
Phenol, 2-(1-methylpropyl)-.
3-06-00-01852 (Beilstein Handbook Reference); BRN 1210026; o-sec-Butylphenol; CCRIS 6218; EINECS 201-933-8; HSDB 5266; Phenol, 2-(1-methylpropyl)-; Phenol, o-sec-butyl-. Solid; mp = 16°; bp = 228°; bp_{21} = 116°; d^{25} = 0.9804; λ_m = 216, 274 nm (ε = 6500, 2250, MeOH).

572   2-t-Butylphenol
88-18-6   H

$C_{10}H_{14}O$
Phenol, 2-(1,1-dimethylethyl)-.
4-06-00-03292 (Beilstein Handbook Reference); AI3-26292; Benzene, 1-t-butyl-2-hydroxy-; BRN 1907120; o-t-Butylphenol; o-t-Butylphenol; CCRIS 5825; EINECS 201-807-2; HSDB 5255; Phenol, 2-(1,1-dimethylethyl)-; Phenol, o-t-butyl-. Oil; mp = -6.8°; bp = 223°; d^{20} = 0.9783; λ_m = 216, 273, 279 nm (ε = 5830, 2260, 2000, MeOH), 213, 245, 273, 279 nm (ε = 6090, 726, 2170, 2000, MeOH-KOH); very soluble in Et_2O, less soluble in EtOH, CCl_4.

573   t-Butylphenyl diphenyl phosphate
56803-37-3   G H

$C_{22}H_{23}O_4P$
Phosphoric acid, (1,1-dimethylethyl)phenyl diphenyl ester.
t-Butylphenyl diphenyl phosphate; tert-Butylphenyl diphenyl phosphate; CCRIS 4761; EINECS 260-391-0; HSDB 6102; Phosphate, tert-butylphenyl diphenyl; Phosphoric acid, (1,1-dimethylethyl)phenyl diphenyl ester; Syn-O-Ad® 8478; Syn-O-Ad®

8485. Antiwear and EP agents in non-crankcase lubricants; ferrous metal passivator; fluid base stock where inhibition of flame propagation is desired. Liquid; bp_5 = 245-260°; SG_{25} = 1.15; insoluble in H_2O, soluble in organic solvents. *Akzo Chemie.*

574   Butyl propionate
590-01-2   1587   H

$C_7H_{14}O_2$
Propionic acid n-butyl ester.
4-02-00-00708 (Beilstein Handbook Reference); AI3-24352; BRN 1700932; Butyl propanoate; Butyl propionate; EINECS 209-669-5; FEMA No. 2211; n-Butyl propanoate; n-Butyl propionate; NSC 8449; Propanoic acid, butyl ester; Propionic acid, butyl ester; UN1914. Liquid; mp = -89°; bp = 146.8°; d^{20} = 0.8754; slightly soluble in H_2O, CCl_4, freely soluble in EtOH, Et_2O.

575   Butyl rubber
9010-85-9   G

[-C(CH3)2CH2-]x-[CH2CH=C(CH3)CH2-]y

Poly(isobutylene-co-isoprene).
1,3-Butadiene, 2-methyl-, polymer with 2-methyl-1-propene; Butyl rubber; Isobutylene/isoprene co-polymer; Kalar® 5214; Kalar® 5263; Kalene® 800; 2-Methyl-1,3-butadiene polymer with 2-methyl-1-propene; Poly(isobutylene-co-isoprene). Cross-linked butyl composition; produces nonsagging butyl-based sealants, e.g., automotive windshield tape, hot melt sealant; as base for butyl mastics Viscosity (Mooney, ML 1+8, 100°) = 42-52. *Hardman.*

576   Butyl salicylate
2052-14-4   G

$C_{11}H_{14}O_3$
2-Hydroxy-benzoic acid, butyl ester.
4-10-00-00153 (Beilstein Handbook Reference); AI3-00512; Benzoic acid, 2-hydroxy-, butyl ester; BRN 2208904; Brunol; Butyl (2-hydroxyphenyl)formate; Butyl 2-hydroxybenzoate; Butyl o-hydroxybenzoate; n-Butyl o-hydroxybenzoate; n-Butyl salicylate; EINECS 218-142-9; FEMA No. 3650; NSC 1511; NSC 403676; Salicylic acid, butyl ester. Used as a fragrance. Liquid; mp = -5.9°; bp = 271°; d^{20} = 1.0728; λ_m = 238, 306 nm (ε = 8990, 4210, MeOH); slightly soluble in CCl_4. *Catomance Ltd.*

577   Butyl stearate
123-95-5   1589   G H

$C_{22}H_{44}O_2$
n-Butyl octadecanoate.
4-02-00-01219 (Beilstein Handbook Reference); ADK STAB LS-8; AI3-00398; APEX 4; BRN 1792866; BS; Butyl octadecanoate; n-Butyl octadecanoate; Butyl octadecylate; Butyl stearate; n-Butyl stearate; EINECS 204-666-5; Emerest 2325; Estrex 1B 54, 1B 55;

FEMA Number 2214; Groco 5810; HSDB 942; Kemester® 5510; Kessco® BS; Kessco BSC; Kesscoflex BS; NSC 4820; Octadecanoic acid, butyl ester; Polycizer 332; Radia® 7051; RC plasticizer B-17; Starfol BS-100; Stearic acid, butyl ester; Tegester butyl stearate; Uniflex® BYS-Tech; Unimate® BYS; Wickenol 122; Wilmar butyl stearate; Witcizer 200; Witcizer 201; Witconol 2326. Has low viscosity, good color, low odor for plasticizers, textile fiber lubricants, metalworking oils. A biodegradable replacement for mineral oil; used as lubricant in textile spin finish, coning oils, carding, dye bath. Used as a solvent, spreading and softening agent in plastics, textiles, cosmetics, rubbers. Emollient for cosmetics; lubricant, plasticizer. Chemical intermediate, used in chemical synthesis; lubricant in mineral, cutting, lamination, and textile oils, rust inhibitors; textile and leather application. mp = 27°; bp = 343°; d^{25} = 0.854; insoluble in H_2O, soluble in EtOH, very soluble in Me_2CO. Amerchol; Asahi Denka Kogyo; CK Witco Corp.; Fina Chemicals; Hall C.P.; Henkel/Organic Prods.; Inolex; Mosselman NV; Penta Mfg.; Stepan Canada; Stepan; Union Camp; Witco/Humko.

578   Butyltin tris(isooctyl mercaptoacetate)
25852-70-4   H

$C_{34}H_{66}O_6S_3Sn$
Butyltris(2-ethylhexyloxycarbonylmethylthio)stannane.
Acetic acid, ((butylstannylidyne)trithio)tri-, triisooctyl ester; Acetic acid, 2,2',2''-((butylstannylidyne)-tris(thio))tris-, triisooctyl ester; Butyltin tris(isooctyl mercaptoacetate); Butyltris(2-ethylhexyloxycarbony-methylthio)stannane; CCRIS 4592; EINECS 247-295-4; Stannane, butyltris((carboxymethyl)thio)-, triisooctyl ester; Stannane, butyltris(carboisooctoxymethylthio)-; Stannane, butyltris(isooctyloxycarbonylmethylthio)-; Stannane, n-butyltris(carboisooctoxymethylthio)-; Tin, butyl-, tris(isooctylthioglycollate); Triisooctyl 2,2',2''-((butylstannylidyne)tris(thio))triacetate.

579   Butyl toluene
98-51-1   G

$C_{11}H_{16}$
p-t-Butyl toluene.
4-05-00-01130 (Beilstein Handbook Reference); 1-t-Butyl-3,5-dimethylbenzene; 5-t-Butyl-1,3-dimethyl-benzene; 5-t-Butyl-m-xylene; Al3-02460; Benzene, 1-(1,1-dimethylethyl)-3,5-dimethyl-; Benzene, 1,3-dimethyl-5-(1,1-dimethylethyl)-; Benzene, 5-t-butyl-1,3-dimethyl-; BRN 1853314; 1,3-Dimethyl-5-t-butylbenzene; EINECS 202-647-6; Lowinox® PTBT; NSC 11016; TBT; m-Xylene, 5-t-butyl-. Antioxidant. mp = -52°; bp = 190°; d^{20} = 0.8612; λ_m = 212, 259, 264, 266, 272 nm (ε = 8313, 282, 407, 355, 347 isooctane); insoluble in H_2O, slightly soluble in EtOH, soluble in Me_2CO, C_6H_6, $CHCl_3$, very soluble in Et_2O. Hall C.P.

580   sec-Butylurea
689-11-2   H

$C_5H_{12}N_2O$
Urea, (1-methylpropyl)-.
EINECS 211-709-1; HSDB 5742; sec-Butylurea; Secondary butylurea; N-sec-Butylurea; NSC 27458; Urea, (1-methylpropyl)-; Urea, sec-butyl-; Urea, 1-sec-butyl-.

581   Butyl vinyl ether
111-34-2   G

$C_6H_{12}O$
1-(Ethenyloxy)butane.
4-01-00-02052 (Beilstein Handbook Reference); Agrisynth BVE; Al3-24225; BRN 1560217; Butane, 1-(ethenyloxy)-; Butil vinil eter; Butoxyethene; Butoxyethylene; Butyl vinyl ether; n-Butyl vinyl ether; BVE; EINECS 203-860-7; Ethenyl n-butyl ether; Ether, butyl vinyl; Ether butylvinylique; HSDB 6384; NSC 8264; Polyvinox; Shostakovsky Balsam; UN2352; Vinyl butyl ether; Vinyl n-butyl ether. A proprietary preparation of synthetic vinyl butyl ether. Liquid; mp = -92°; bp = 94°; d^{20} = 0.7888; λ_m = 192 nm (ε = 9550, C_7H_{16}); insoluble in H_2O, slightly soluble in ethylene glycol, soluble in C_6H_6; very soluble in EtOH, Me_2CO, freely soluble in Et_2O.

582   Butyl-m-xylene
98-19-1   G

$C_{12}H_{18}$
t-Butyl-m-xylene.
4-05-00-01130 (Beilstein Handbook Reference); 5-t-butyl-m-xylene; 5-t-Butyl-1,3-dimethylbenzene; 1-t-Butyl-3,5-dimethylbenzene; 5-t-Butyl-m-xylene; Al3-02460; Benzene, 1-(1,1-dimethylethyl)-3,5-dimethyl-; Benzene, 1,3-dimethyl-5-(1,1-dimethylethyl)-; Benz-ene, 5-t-butyl-1,3-dimethyl-; BRN 1853314; 1,3-Dimethyl-5-t-butylbenzene; EINECS 202-647-6; Lowinox® TBMX; NSC 11016; m-Xylene, 5-t-butyl-. Intermediate for perfumes and fragrances. Oil; mp = -18°; bp = 207°; λ_m = 331 nm (ε = 4230, MeOH); insoluble in H_2O, slightly soluble in EtOH, soluble in Et_2O, $CHCl_3$. Hall C.P.

583   Butyl zimate
136-23-2   G H

$C_{18}H_{36}N_2S_4Zn$
Bis(N,N-dibutyldithiocarbamato)zinc.
Accel BZ; Aceto ZDBD; Al3-14880; Bis(dibutyl-dithiocarbamato)zinc; Bis(N,N-dibutyldithiocarb-amato)zinc; Butasan; Butazate; Butazate 50-D; Butyl zimate®; Butyl ziram;

Carbamic acid, dibutyldithio-, zinc complex; Carbamodithioic acid, dibutyl-, zinc salt; Dibutyldithiocarbamic acid zinc salt; EINECS 205-232-8; HSDB 2906; Nocceler BZ; NSC 36548; Octocure ZDB-50; Perkacit® ZDBC; Soxinol BZ; USAF GY-5; Vulcacure; Vulcacure ZB; Vulkacit Idb/C; Vulkacit LDB; Zimate, butyl; Zinc bis(dibutyldithiocarbamate); Zinc dibutyldithio-carbamate. Ultra accelerator for EPDM and natural and synthetic latexes; provides fast, flat cures in SBR, nitrile, and neoprene latexes; functions as nondiscoloring antioxidant in noncuring applications and stabilizer in IIR; antioxidant in thermoplastic rubbers and hot melts. Latex and rubber accelerator. An activator, antidegradant, and accelerator for natural rubber, butadiene, styrene-butadiene, nitrile-butadiene, butyl rubber, and ethylene-propylenediene terpolymers. *Akzo Chemie; Vanderbilt R.T. Co. Inc.*

584 2-Butynediol
110-65-6   G H

C4H6O2
2-Butyne-1,4-diol.
4-01-00-02687 (Beilstein Handbook Reference); Agrisynth B3D; AI3-61467; Bis(hydroxymethyl)-acetylene; BRN 1071237; 2-Butin-1,4-diol; 1,4-Butinodiol; Butynediol; 2-Butynediol; Butynediol-1,4; But-2-yne-1,4-diol; 1,4-Butynediol; 2-Butyne-1,4-diol; 1,4-Dihydroxy-2-butyne; EINECS 203-788-6; HSDB 2004; NSC 834; UN2716. Corrosion inhibitor in acid pickles and cleaners. Platelets; mp = 50°; bp = 238°; d^{20} = 1.4804; very soluble in H2O, EtOH, MeOH, Me2CO; slightly soluble in CHCl3, Et2O, insoluble in C6H6, petroleum ether; LD50 (rat orl) = 105 mg/kg. *BASF Corp.; Intl. Specialty Products Inc. (ISP).*

585 Butynorate
77-58-7   3063   G

C32H64O4Sn
Dibutyltin dillaurate.
AI3-26331; Bis(lauroyloxy)di(n-butyl)stannane; Butyn-orate; Cata-Chek 820; CCRIS 4786; Davainex; DBTL; Di-n-butyltin dilaurate; Dibutyl-tin-dilaurate; Dibutyl-zinn-dilaurat; Dibutylbis(1-oxododecyl)oxy)stannane; Dibutylbis(laurato)tin; Dibutylbis(lauroxy)stannane; Dibutylbis(lauroyloxy)tin; Dibutylstannium dilaurate; Dibutylstannylene dilaurate; Dibutyltin didodecanoate; Dibutyltin dilaurate; Dibutyltin laurate; Dibutyltin n-dodecanoate; DXR 81; EINECS 201-039-8; Fomrez sul-4; HSDB 5214; Kosmos 19; KS 20; Lankromark LT 173; Laudran di-n-butylcinicity; Lauric acid, dibutylstannylene salt; Lauric acid, dibutyltin deriv.; Laustan-B; Mark 1038; Mark BT 11; Mark BT 18; Neostann U 100; NSC 2607; Ongrostab BLTM; SM 2014C; Stabilizer D-22; Stanclere DBTL; Stannane, bis(dodecanoyloxy) di-n-butyl-; Stannane, bis(lauroyl-oxy)dibutyl-; Stannane, dibutylbis((1-oxododecyl)oxy)-; Stannane, dibutylbis(lauroyloxy)-; Stannane, dibutyl-bis(lauroyloxy)-; Stavincor 1200 SN; Stavinor 1200 SN; T 12; T 12 (catalyst); Therm chek 820; Thermolite T 12; Tin, di-n-butyl-, di(dodecanoate); Tin dibutyl dilaurate; Tin, dibutylbis(lauroyloxy)-; Tinostat; TN 12; TN 12 (catalyst); TVS Tin Lau; TVS-TL 700. Metal-based catalyst for PU flexible molded foam, coatings. Standard catalyst for PU elastomer applications; also for slabstock, semirigid,

and rigid applications. PVC stabilizer providing good heat and weathering resistance; improves processability in rigid transparent formulation. Also used as an anthelmintic, against tapeworms in chickens. Solid; mp = 22-24°; d = 1.066; insoluble in H2O, soluble in organic solvents; LD50 (rat orl) = 175 mg/kg. *Air Prods & Chems; Air Products & Chemicals Inc.; Asahi Denka Kogyo; CK Witco Corp.; Elf Atochem N. Am.; Ferro/Bedford; La Littorale; Tosoh.*

586 Butyraldehyde
123-72-8   1591   G H

C4H8O
n-Butyraldehyde.
4-01-00-03229 (Beilstein Handbook Reference); AI3-24198; Aldehyde butyrique; Aldeide butirrica; BRN 0506061; Butal; Butaldehyde; Butalyde; Butanal; Butanaldehyde; Butylaldehyde; Butyral; Butyraldehyd; Butyraldehyde; Butyric aldehyde; CCRIS 3221; EINECS 204-646-6; FEMA Number 2219; HSDB 2798; NCI-C56291; NSC 62779; UN1129. Used in manufacture of plasticizers, rubber accelerators, solvents and high molecular weight polymers. Liquid; mp = -99°; bp = 74.8°; d^{20} = 0.8016; λ_m = 225, 283 nm (ε = 12, 13, H2O); slightly soluble in CHCl3, soluble in H2O, soluble in Me2CO, C6H6, freely soluble in EtOH, Et2O; LD50 (rat orl) = 5.89 g/kg. *Eastman Chem. Co.; Hoechst Celanese; Neste UK; Penta Mfg.; Union Carbide Corp.*

587 4-Butyramidophenol
101-91-7   4839   G

C10H13NO2
N-Butyryl-p-aminophenol.
Butanamide, N-(4-hydroxyphenyl)-; 4-Butyramido-phenol; Butyranilide, 4'-hydroxy-; N-Butyroyl-p-aminophenol; N-Butyryl-p-aminophenol; p-(Butyryl-amino)phenol; p-Butyramidophenol; EINECS 202-988-0; 4'-Hydroxybutyranilide; NSC 166351; Suconox-4. Antioxidant and processing aid for thermoplastics. Needles; mp = 139.5°; very soluble in H2O, EtOH. *Zeeland Chem. Inc.*

588 Butyric acid
107-92-6   1593   H

C4H8O2
n-Butyric acid.
4-02-00-00779 (Beilstein Handbook Reference); AI3-15306; BRN 0906770; Butanic acid; Butanoic acid; Buttersaeure; Buttersäure; Butyrate; Butyric acid; 1-Butyric acid; n-Butyric acid; CCRIS 6552; EINECS 203-532-3; Ethylacetic acid; FEMA Number 2221; HSDB 940; Kyselina maselna; NSC 8415; 1-Propanecarboxylic acid; Propylformic acid; UN2820. Liquid; mp = -5.7°; bp = 163.7°; d^{20} = 0.9577; λ_m = 280 nm (ε = 50, H2O); slightly soluble in CCl4, freely soluble in H2O, EtOH, Et2O.

589 Butyric anhydride
106-31-0   1594   H

$C_8H_{14}O_3$
n-Butyric acid anhydride.
4-02-00-00802 (Beilstein Handbook Reference); Anhydrid kyseliny maselne; BRN 1099474; Butanoic acid, anhydride; Butanoic anhydride; Butyranhydrid; Butyric acid anhydride; Butyric anhydride; Butyryl oxide; Caswell No. 132A; EINECS 203-383-4; EPA Pesticide Chemical Code 077705; HSDB 5369; UN2739. Liquid; mp = -75°; bp = 200°; d^{20} = 0.9668; λ_m = 282 nm (ϵ = 126, C_6H_6); soluble in Et_2O, slightly soluble in CCl_4.

590   Butyroin
496-77-5   1595   G

$C_8H_{16}O_2$
5-Hydroxy-4-octanone.
AI3-05612; Butyroin; EINECS 207-830-4; FEMA No. 2587; 5-Hydroxyoctan-4-one; 5-Hydroxy-4-octanone; NSC 1479; Octan-4-ol-5-one; 5-Octanol-4-one; 4-Octanone, 5-hydroxy-. Chemical intermediate. Liquid; mp = -10°; bp = 185°; d_4^{16} 0.9107.

591   Butyrolactam
616-45-5   8106   G H V

C_4H_7NO
2-Pyrrolidone.
4-Aminobutyric acid lactam; Butanoic acid, 4-amino-, lactam; Butyrolactam; EINECS 210-483-1; σ-Aminobutyric acid lactam; σ-Aminobutyric lactam; σ-Aminobutyrolactam; HSDB 2652; 2-Ketopyrrolidine; LAM; NSC 4593; NSC 8413; 2-Oxopyrrolidine; 2-Pyrol; α-Pyrrolidinone; 2-Pyrrolidinone; Pyrrolidon; Pyrrolidone; 2-Pyrrolidone. Solvent for polymers, pesticides and sugars. Used in printers inks and as a plasticizer and coalescing agent for floor polishes. Plasticizer and coalescing agent; solvent for veterinary medicine. Solid; mp = 25°; bp = 251°, bp12 = 133°; d^{20} = 1.120; λ_m = 190 nm (ϵ = 7079, H_2O); very soluble in H_2O, EtOH, Et_2O, C_6H_6, $CHCl_3$, CS_2. Akzo Chemie; BASF Corp.; Intl. Specialty Products Inc. (ISP); UCB SA.

592   4-Butyrolactone
96-48-0   1596   G H

$C_4H_6O_2$
2(3H)-Furanone, dihydro-.
'σ-6480; Agrisynth BLO; Agsol Ex BLO; AI3-28121; σ-BL; BLO; BLON; Butanoic acid, 4-hydroxy-, σ-lactone; Butyric acid, 4-hydroxy-, σ-lactone; Butyric acid lactone; Butyrolactone; σ-Butyrolactone; Butyryl-actone; C-1070; Caswell No. 132B; CCRIS 2924; Dihydro-2-furanone; Dihydro-2(3H)-furanone; EINECS 202-509-5; EPA Pesticide Chemical Code 122303; FEMA No. 3291; HSDB 4290; σ-Hydroxybutyric acid cyclic ester; σ-Hydroxybutyric acid lactone; σ-

Hydroxybutyrolactone; NCI-C55878; No Go; NSC 4592; Tetrahydro-2-furanone. Solvent for PAN, PS, fluorinated hydrocarbons, cellulose triacetate, shellac; used in paint removers; petroleum processing, hectograph process, specialty inks; intermediate for aliphatic and cyclic compounds; reaction and diluent solvent for pesticides; used in dyeing of acetate, wetting agent for cellulose acetate films, fibers, solvent welding of plastic films in adhesive applications. Liquid; mp = -43.3°; bp = 204°; d^{16} = 1.1284; λ_m = 209 nm (ϵ = 43, MeOH); soluble in H_2O, Me_2CO, C_6H_6, Et_2O, EtOH; LD50 (rat orl) = 19.4 g/kg. BASF Corp.; ISP; Janssen Chimica; Sigma-Aldrich Fine Chem.; Spectrum Chem. Manufacturing; UCB SA.

593   Butyrone
123-19-3   3379   G

$C_7H_{14}O$
Di-n-propyl ketone.
4-01-00-03323 (Beilstein Handbook Reference); AI3-15181; BRN 1699049; Di-n-propyl ketone; Dipropyl ketone; EINECS 204-608-9; FEMA No. 2546; GBL; Heptan-4-one; 4-Heptanone; NSC 8692; Propyl ketone; UN2710. Solvent, chemical intermediate. Liquid; mp = -33°; bp = 144°; d^{20} = 0.8174; λ_m = 283 nm (ϵ = 25, MeOH), 274 nm (ϵ = 60, H_2O), 280 nm (ϵ = 24, EtOH), 283 nm (ϵ = 21, C_6H_{14}); insoluble in H_2O, soluble in CCl_4, freely soluble in EtOH, Et_2O; LD50 (rat orl) = 3.04 g/kg.

594   Butyronitrile
109-74-0   1597   G H

C_4H_7N
n-Butyronitrile.
4-02-00-00806 (Beilstein Handbook Reference); AI3-08778; BRN 1361452; Butane nitrile; Butanenitrile; Butyric acid nitrile; Butyronitrile; Butyrylonitrile; 1-Cyanopropane; EINECS 203-700-6; HSDB 5013; Nitrile c4; NSC 8412; Propyl cyanide; n-Propyl cyanide; Propylkyanid; UN2411. Basic material in industrial, chemical and pharmaceutical intermediates and products, poultry medicines. Liquid; mp = -111.9°; bp = 117.6°; d^{20} = 0.7936; slightly soluble in H_2O, CCl_4, soluble in C_6H_6, freely soluble in EtOH, Et_2O; LD50 (rat orl) = 140 mg/kg. Air Products & Chemicals Inc.; Eastman Chem. Co.; Janssen Chimica; Lonzagroup.

595   C.I. Direct Black 22
6473-13-8   H

$C_{44}H_{32}N_{13}Na_3O_{11}S_3$
Trisodium   6-((2,4-diaminophenyl)azo)-3-((4-((4-((7-((2,4-diaminophenyl)azo)-1-hydroxy-3-sulphonato-2-naphthyl)azo)phenyl)amino)-3-sulphonatophenyl)azo)-4-hydroxynaphthalene-2-sulphonate.
CCRIS 1390; C.I. Direct Black 22; EINECS 229-326-3; 2-Naphthalenesulfonic acid, 6-((2,4-diaminophenyl)-azo)-3-((4-((4-((7-((2,4-diaminophenyl)azo)-1-hydroxy-3-sulfo-2-naphthalenyl)azo)phenyl)amino)-3-sulfo-phenyl)azo)-4-hydroxy-,

100

trisodium salt; Trisodium 6-((2,4-diaminophenyl)azo)-3-((4-((4-((7-((2,4-diamino-phenyl)azo)-1-hydroxy-3-sulphonato-2-naphthyl)azo)-phenyl)amino)-3-sulphonatophenyl)azo)-4-hydroxy-naphthalene-2-sulphonate.

596 C.I. Pigment Green 51
68553-01-5   H

Victoria green garnet.
C.I. Pigment Green 51; EINECS 271-385-2; Victoria green garnet.

597 C.I. Pigment Red 48, calcium salt
7023-61-2   H

$C_{18}H_{11}CaClN_2O_6S$
Calcium 4-((5-chloro-4-methyl-2-sulphonatophenyl)-azo)-3-hydroxy-2-naphthoate.
4-((5-Chloro-2-sulfo-p-tolyl)azo)-3-hydroxy-2-naphth-alenecarboxylicacid, calcium salt (1:1); EINECS 230-303-5.

598 Cacodyl
471-35-2   1603   G

$C_4H_{12}As_2$
Tetramethyldiarsine.
Alkarsin; Cacodyl; Diarsine, tetramethyl-; EINECS 207-440-4; Tetramethyldiarsenic; Tetramethyldiarsine. Chemical intermediate.
Platelets; mp = -6°; bp = 165°; d^{15} = 1.447; slightly soluble in H_2O, very soluble in EtOH, Et_2O.

599 Cadaverine
462-94-2   1609   G

$C_5H_{14}N_2$
Pentamethylenediamine.
4-04-00-01310 (Beilstein Handbook Reference); AI3-26937; Animal coniine; BRN 1697256; Cadaverin; Cadaverine; Cadaverine; Cadaverine; 1,5-Diamino-pentane; EINECS 207-329-0; Pentamethylenediamine; 1,5-Pentamethylenediamine; 1,5-Pentanediamine. A base found in ergot and formed by bacterial decomposition of human and animal tissues; used in the preparation of high polymers and in biomedical research. Oil; mp = 9°; bp = 179°; d^{25} = 0.873; pKa1 = 10.25, pKa2 = 9.13; soluble in H_2O, EtOH, slightly soluble in Et_2O.

600 Cadmium chloride
10108-64-2   1617   G P

$CdCl_2$
Cadmium(II) chloride.
AI3-09096; Caddy; Cadmium chloride; Cadmium chloride ($CdCl_2$); Cadmium dichloride; Caswell No. 135; CCRIS 114; Dichlorocadmium; EINECS 233-296-7; EPA Pesticide Chemical Code 012902; HSDB 278; Kadmiumchlorid; NSC 51148; VI-Cad. Used in photography, dyeing and printing and as a fungicide. Preparation of cadmium sulfide, analytical chemistry, photography, dyeing and calco printing, in electroplating baths and tinning solutions, manufacture of special mirrors, cadmium yellow. Registered by EPA as a fungicide (cancelled). Hygroscopic rhombohedral crystals; mp = 568°; bp = 960°; d = 4.05; freely soluble in H_2O, soluble in Me_2CO, slightly soluble in MeOH, EtOH, insoluble in Et_2O; LD_{50} (rat orl) = 88 mg/kg. *Cleary W. A.; Fluka.*

601 Cadmium 2-ethylhexanoate
2420-98-6   H

$C_{16}H_{30}CdO_4$
Cadmium 2-ethylhexanoate.
Cadmium 2-ethylhexanoate; Cadmium 2-ethylhexoate; Cadmium bis(2-ethylhexanoate); Cadmium di-2-ethylhexylate; Cadmium ethylhexanoate; EINECS 219-346-0; Hexanoic acid, 2-ethyl-, cadmium salt; HSDB 6140.

602 Cadmium fluoride
7790-79-6   1619   G

CdF_2
Hydrofluoric acid cadmium salt.
Cadmium difluoride; Cadmium fluoride; Cadmium fluoride (CdF_2); Cadmium fluorure; EINECS 232-222-0. Used in electronic and optical applications, high-temperature dry-film lubricants, starting material for crystals for lasers, phosphors. mp = 1049°; bp = 1748°; d = 6.33; soluble in H_2O 4.3 g/100ml; soluble in HF, mineral acids, insoluble in EtOH, NH_3. *Atomergic Chemetals; Cerac; Noah Chem.*

603 Cadmium hydroxide
21041-95-2   1620   G

$Cd(OH)_2$
Cadmium hydrate.
AI3-09097; Cadmium dihydroxide; Cadmium hydroxide; Cadmium hydroxide ($Cd(OH)_2$); EINECS 244-168-5. Used in preparation of cadmium salts, in cadmium plating and storage-battery electrodes. Trigonal or hexagonal crystals; d = 4.79; dec <200°; insoluble in H_2O, slightly soluble in alkali, acid. *Noah Chem.*

604   Cadmium iodide
7790-80-9   1621   G

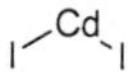

I—Cd—I

CdI$_2$
Hydriodic acid cadmium salt.
AI3-02411; Cadmium diiodide; Cadmium iodide; Cadmium iodide (CdI$_2$); EINECS 232-223-6. Used in photography, process engraving and lithography, analytical chemistry, electroplating, lubricants, phosphors, nematocide. mp = 388°; bp = 787°; d = 5.67; soluble in H$_2$O, Et$_2$O, EtOH, Me$_2$CO. *Atomergic Chemetals; Cerac; Spectrum Chem. Manufacturing.*

605   Cadmium oxide
1306-19-0   1623   G

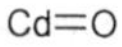

Cd=O

CdO
Cadmium oxide brown.
Aska-Rid; Cadmium fume; Cadmium monoxide; Cadmium oxide; Cadmium oxide (CdO); Caswell No. 136AA; CCRIS 115; EINECS 215-146-2; EPA Pesticide Chemical Code 236200; HSDB 1613; Kadmu tlenek; NCI-C02551. Used in cadmium plating baths, electrodes for storage batteries, cadmium salts, catalyst, ceramic glazes, phosphors, nematocide. Dark brown heavy powder; d = 8.15; insoluble in H$_2$O, soluble in acids, ammonium salts; LD$_{50}$ (rat orl) = 72 mg/kg. *Amax; Asarco; Atomergic Chemetals; Chemisphere; Mallinckrodt Inc.; Nihon Kagaku Sangyo.*

606   Caffeine
58-08-2   1636   D H V

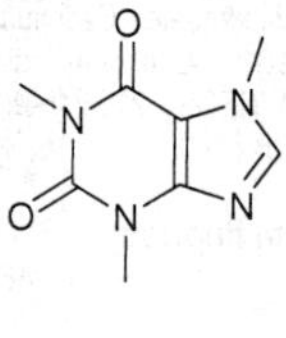

O=, N, N... (structure)

C$_8$H$_{10}$N$_4$O$_2$
3,7-Dihydro-1,3,7-trimethyl-1H-purine-2,6-dione.
5-26-13-00558 (Beilstein Handbook Reference); A.S.A. and Codeine Compound; AI3-20154; Alert-pep; Anacin; Anacin Maximum Strength; Anhydrous caffeine; Bayer Select Headache Pain; BRN 0017705; Cafamil; Cafecon; Cafeina; Cafergot; Caffedrine; Caffein; Caffeina; Caffeine; Caffine; Cafipel; CCRIS 1314; Coffein; Coffeine; Coffeinum; Dasin; Dexitac; DHCplus; EINECS 200-362-1; Eldiatric C; FEMA No. 2224; Guaranine; HSDB 36; Hycomine Compound; Kofein; Koffein; Mateina; Methyltheobromide; Methyltheobromine; Methylxanthine theophylline; Midol Maximum Strength; NCI-C02733; Nix Nap; NO-Doz; Nodaca; Norgesic; NSC 5036; Organex; P-A-C Analgesic Tablets; Phensal; Propoxyphene Compound 65; Quick-Pep; Refresh'N; SK-65 Compound; Stim; Synalgos; Synalgos-DC; Thein; Theine; Theobromine, 1-methyl-; Theophylline, 7-methyl-; Tirend; Vanquish; Vivarin; Wigraine; Xanthine, 1,3,7-trimethyl. Occurs naturally in tea, coffee. A CNS and respiratory stimulant, used in veterinary medicine as a stimulant. Crystals; mp = 238°; d = 1.23; λ$_m$ = 272 nm (ε 8510 MeOH); soluble in H$_2$O, EtOH, Me$_2$CO, CHCl$_3$, AcOH, C$_5$H$_5$N, CS$_2$; LD$_{50}$ (mmus orl) = 127 mg/kg, (fmus orl) = 137 mg/kg, (mrat orl) = 355 mg/kg, (frat orl) = 247 mg/kg, (mrbt orl) = 246 mg/kg, (frbt orl) = 224 mg/kg.

607   Calcium acetate
62-54-4   1645   G H

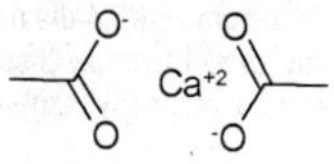

(structure: acetate with Ca^{+2})

C$_4$H$_8$CaO$_4$
Acetic acid, calcium salt.
Acetate of lime; Acetic acid, calcium salt; AI3-02903; Brown acetate; Brown Acetate of Lime; Calcium acetate; Calcium diacetate; CCRIS 4921; EINECS 200-540-9; FEMA No. 2228; Gray acetate; Gray Acetate of Lime; HSDB 928; Lime acetate; Lime pyrolignite; PhosLo; Sorbo-calcion; Teltozan; Vinegar salts. Used in manufacture of acetic acid, acetone; in tanning, as food stabilizer, corrosion inhibitor. Used as a mordant in dyeing and printing of textiles, a stabilizer in resins, an additive to calcium soap lubricants, a food additive and a corrosion inhibitor. Solid; dec 160°; d = 1.50; soluble in H$_2$O, slightly soluble in MeOH, insoluble in EtOH, Me$_2$CO, C$_6$H$_6$; LD$_{50}$ (rat orl) = 4280 mg/kg. *General Chem; Mallinckrodt Inc.; Mechema Chemicals Ltd.; Niacet; Verdugt BV.*

608   Calcium arsenate
7778-44-1   1648   G P

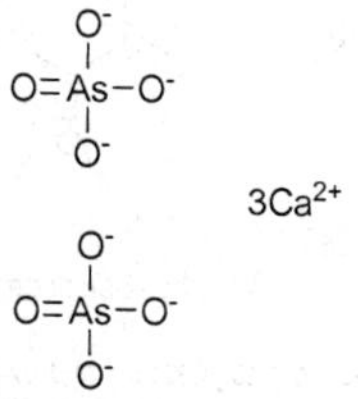

(structure: O=As–O$^-$ groups with 3Ca^{2+})

As$_2$Ca$_3$O$_8$
Calcium(II) arsenate.
AI3-24838; Arseniate de calcium (French); Arsenic acid (H$_3$AsO$_4$), calcium salt (2:3); Arsenic acid, calcium slt (2:3); Calcium arsenate; Calcium-o-arsenate; Calcium orthoarsenate; Chip-Cal Granular; Cucumber dust; EINECS 233-287-8; HSDB 1433; Kalziumarseniat; Pencal; Spra-cal; Tricalcium arsenate; Calcars. Used as insecticide and molluscicide. Registered by EPA as a fungicide, herbicide and insecticide (cancelled). Solid; slightly soluble in H$_2$O and dilute acids; LD$_{50}$ (rat orl) = 298 mg/kg. *Mechema Chemicals Ltd.*

609   Calcium bis(dinonylnaphthalene-sulfonate)
57855-77-3   H

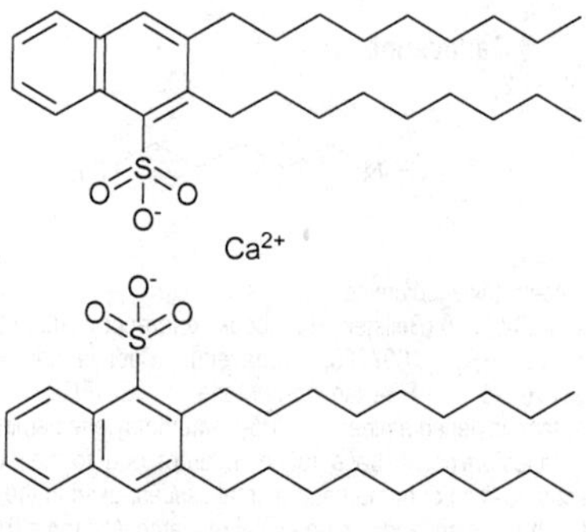

(structure with Ca^{2+})

C$_{56}$H$_{86}$CaO$_6$S$_2$
Dinonylnaphthalenesulfonic acid, calcium salt.
Calcium bis(dinonylnaphthalenesulphonate); Dinonyl-naphthalenesulfonic acid, calcium salt; EINECS 260-991-2; Naphthalenesulfonic acid, dinonyl-, calcium salt. -

610   Calcium borate

12007-56-6   1652   G

CaB_4O_7

Calcium tetraborate.

Boric acid ($H_2B_4O_7$), calcium salt (1:1); Boron calcium oxide (B_4CaO_7); Boron calcium oxide; Calcium borate; EINECS 234-511-7; Meyerhofferite. An artificially prepared mineral. Used as a flux, in antifreeze preparations and in fire retardant paint. Slightly soluble in H_2O. -

611   Calcium carbide

75-20-7   1657   G H

$$^-C\equiv C^- \quad Ca^{2+}$$

C_2Ca

Calcium acetylide.

Acetylenogen; Al3-03101; Calcium acetylide; Calcium acetylide (Ca(C_2)); Calcium carbide; Calcium carbide (CaC_2); Calcium dicarbide; Carbure de calcium; Carburo calcico; EINECS 200-848-3; Ethyne, calcium deriv.; HSDB 1434; UN1402. Used in generation of acetylene gas for welding, chloroethylenes, vinyl acetate monomer, acetylene chemicals, reducing agent. mp = 2300°; d = 2.22; soluble in H_2O, MeOH, EtOH, slightly soluble in Me_2CO, insoluble in $CHCl_3$, Et_2O. *Spectrum Chem. Manufacturing.*

612   Calcium carbonate

471-34-1   1658   D G

Ca^{+2}

$CCaO_3$

Carbonic acid calcium salt (1:1).

Aeromatt; Akadama; Albacar; Albacar 5970; Albafil; Albaglos; Albaglos SF; Allied whiting; Atomit; Atomite; AX 363; BF 200; Brilliant 15; Brilliant 1500; Britomya M; Britomya S; C.I. 77220; C.I. Pigment White 18; Cal-Light SA; Cal-Sup; Calcene CO; Calcene NC; Calcene TM; Calcicoll; Calcidar 40; Calcilit 100; Calcilit 8; Calcitrel; Calcium carbonate; Calcium carbonate, precipitated; Calcium carbonate slurry; Calcium monocarbonate; Calibrite; Calmos; Calmote; Calofil A 4; Calofil B 1; Calofil E 2; Calofor U 50; Calofort® S; Calofort® T; Calofort® U; Calopake® F; Calopake® FS; Calopake® H; Calopake® high opacity; Calopake® PC; Calseeds; Caltec; Caltrate; Calwhite; Camalox; Camel-CAL®; Camel-CARB®; Camel-FIL; Camel-FINE; Camel-tex; Camel-wite; Carbital® 50; Carbital® 90; Carbium; Carbium MM; Carbonic acid calcium salt (1:1); Carbonic acid, calcium salt (1:1); Carborex 2; Carusis P; Caswell No. 139; CCC G-white; CCC No.AA oolitic; CCR; CCRIS 1333; CCW; Chemcarb; Chooz; Cl 77220; Clefnon; CP Filler; Crystic prefil S; Dacote; Di-Gel Tablets; DOMAR; Duramite; Durcal 10; Durcal 2NH; Durcal 40; Durcal C 640305; EGRI M 5; EINECS 207-439-9; EPA Pesticide Chemical Code 073502; Eskalon 100; Eskalon 1500; Eskalon 200; Eskalon 400; Eskalon 800; Filtex White Base; Finncarb 6002; Garolite SA; Gilder's whiting; Hakuenka CC; Hakuenka CCR; Hakuenka DD; Hakuenka O; Hakuenka PX; Hakuenka PZ; Hakuenka R 06; Hakuenka T-DD; Homocal D; HSDB 927; Hydrocarb; Hydrocarb 60; Hydrocarb 65; K 250; Kalvan; Kotamite®; Kredafil 150 Extra; Kredafil RM 5; KS 1300; KS 1500; KS 1800; KS 2100; KS 500; KULU 40; Levigated chalk; Marble white; Marblewhite 325; Marfil; MC-T; Microcarb; Micromic CR 16; Micromya; Microwhite 25; Mild lime; Monocalcium carbonate; MSK-C; MSK-G; MSK-K; MSK-P; MSK-PO; MSK-V; Multifex MM; Multiflex MM; Multiflex SC; Mylanta Gelcaps; Mylanta Soothing Lozenges; N 34; N 43; NCC 45; NCC-P; Neoanticid; Neolite F; Neolite SP; Neolite

TPS; Non-Fer-Al; NS; NS 100; NS 200; NS 2500; NS 400; P-Lite 500; P-Lite 700; Pigment white 18; Precipitated calcium carbonate; Precipitated chalk; R Jutan; Sandscale; Statuary marble; Strucal®; T 130-2500; Titralac; Tylenol Headache Plus.; component of: Bufferin Arthritis Strength, Bufferin Extra Strength, Bufferin Regular, Calcitrel, Camalox, Di-Gel Tablets, Mylanta Gelcaps, Mylanta Tablets, Titralac, Tylenol Headache Plus. Used for sealants, plastics, rubber, as a filler in paper, water-based paint, inks and PVC pipe, putty, caulk, ceramics, adhesives, linoleum; floor tile, textile coatings. Fine, wet-ground filling pigment offering excellent retention in the sheet, excellent optical properties; also for matte or dull; offers easy dispersion in plastic compounds, e.g., polyolefins, rigid and flexible PVC; for wire and cable insulation compounds, improved impact properties in PP. Used in rubber to give wear resistance. Marble, with a crystalline or saccharoid structure. Used medically as a calcium replenisher, antacid and dentrifice. White solid; mp = 825°, 1339° (102.5 atm); $d_{25.2}$ = 2.711; insoluble in H_2O. *3M Pharm.; Bristol-Myers Squibb HIV Products; ECC Intl.; Genstar Stone Prods.; Johnson & Johnson-Merck Consumer Pharm.; McNeil Consumer Products Co.; Rhône-Poulenc Rorer Pharm. Inc.; Rhône-Poulenc; Schering-Plough Pharm.; Sterling Winthrop Inc.*

613   Calcium carbonate

1317-65-3   G

$CCaO_3$

Calcium(II) carbonate.

Agricultural limestone; Agstone; Bell mine pulverized limestone; Calcichew; Calcidia; Calcium carbonate; Chalk; Citrical; Domolite; EINECS 215-279-6; Franklin; Ground limestone; Limestone; Lithograpic stone; Marble; Marble dust; Natural calcium carbonate; Portland stone; Sohnhofen stone; Aragonite; Aeromatt; Albacar; Purecal; Chalk; English White; Paris White; Carbonic acid, calcium salt; precipitated calcium carbonate, commercial form; prepared calcium carbonate, native purified form. Source of lime; neutralizing agent; opacifying agent in paper; fortification of bread; putty; tooth powders. Used in medicine, polishing powders, silicate cement. Building stone, metallurgy (flux), manufacture of lime; source of CO_2; Portland and natural cement; removal of sulfur dioxide from stack gases and sulfur from coal. Crystalline solid; mp = 825° (dec); d = 2.83; insoluble in H_2O, soluble in dilute acids. *BASF Corp.; ECC Intl.; EM Ind. Inc.; Genstar Stone Prods.; Georgia Marble; Huber J. M.; Mallinckrodt Inc.; Nichia Kagaku Kogyo; Pfizer Inc.; Whittaker Clark & Daniels.*

614   Calcium chloride

10043-52-4   1660   D G P

$CaCl_2 \cdot 2H_2O$

Hydrochloric acid calcium salt.

Al3-02239; Anhydrous calcium chloride; Calcium chloride; Calcium chloride anhydrous; Calcium chloride, dihydrate; Calcium chloride solution; Calcium dichloride; Calcosan; Caloride; Calplus; Caltac; CCRIS 1334; Dowflake; EINECS 233-140-8; HSDB 923; Intergravin-orales; Liquidow; Peladow; Snomelt; Superflake anhydrous; Uramine MC. Inorganic salt; deicing and dust control of roads, drilling muds, dustproofing, freezeproofing, and thawing coal, coke, stone, sand, ore, concrete conditioning; drying and desiccating agent; sequestrant in foods. Registered by EPA as an antimicrobial, fungicide, herbicide, insecticide and molluscicide (cancelled). Calcium replenisher. Solid; mp = 772°; bp > 1600°; d_4^{25} 2.152; freely soluble in H_2O (exothermic), EtOH; LD_{50} (mus iv) = 42.2 mg/kg. *Akzo Chemie; Allied Signal; EM Ind. Inc.; Gist-Brocades Intl.; Jarchem.; Kemira Kemi UK; Mallinckrodt Inc.; Nichia Kagaku Kogyo; OxyChem; Schaeffer Salt & Chem.*

615 Calcium citrate

813-94-5   1662   G

$C_{12}H_{10}Ca_3O_{14} \cdot 4H_2O$

2-Hydroxy-1,2,3-propanetricarboxylic acid, calcium salt (2:3).
Acicontral; Calcium citrate; Calcium citrate, tribasic; Calcium 2-hydroxy-1,2,3-propanetricarboxylate (3:2); Citracal; Citric acid, calcium salt (2:3); Citrical; EINECS 212-391-7; HSDB 5756; 2-Hydroxy-1,2,3-propanetricarboxylic acid calcium salt (2:3); Lime cirate; 1,2,3-Propanetricarboxylic acid, 2-hydroxy-, calcium salt (2:3); Tribasic calcium citrate; Tricalcium dicitrate. Dietary supplement, sequestrant, buffer, and firming agent in foods. Crystals; soluble in H_2O (0.095 g/100 ml 20º), insoluble in EtOH. *EM Ind. Inc.; Rit-Chem; Rottapharm SpA.*

616 Calcium cyanide

592-01-8   1665   G

C_2CaN_2

Hydrocyanic acid calcium salt.
3-02-00-00061 (Beilstein Handbook Reference); BRN 3926685; Calcid; Calcium cyanide; Calcium cyanide (Ca(CN)2); Calcyan; Calcyanide; Caswell No. 142; Cyanide of calcium; Cyanide salts; Cyanogas; Cyanolime; Cyanure de calcium; Degesch Calcium Cyanide A-Dust; EINECS 209-740-0; EPA Pesticide Chemical Code 074001; HSDB 242; NSC 74784; RCRA waste number P021; UN1575. Dissolves in H_2O liberating HCN. Used as a source of HCN to exterminate rodents. Fumigant and rodenticide. Also used in stainless steel manufacture and extraction of precious metals.
Soluble in H_2O (liberates HCN), EtOH; LD50 (rat orl) = 39 mg/kg. *Am. Cyanamid; Mechema Chemicals Ltd.*

617 Calcium dioxide

1305-79-9   1694   D G

CaO_2

Calcium peroxide.
Calcium peroxide; Calcium peroxide (Ca(O2)); Calper; Calper G; EINECS 215-139-4; HSDB 965; Oxy-Gro; UN1457. For agricultural and horticultural uses. Use as a rubber stabilizer and as an antiseptic. White solid; slightly soluble in H_2O. *Solvay Interox Inc.*

618 Calcium disodium edetate

62-33-9   3542   G

$C_{10}H_{12}CaN_2Na_2O_8$

Calcium disodium ethylenediamine tetraacetate.
Acetic acid, (ethylenedinitrilo)tetra-, calcium disodium salt; Adsorbonac; Antalin; Antallin; Calciate(2-), ((ethylenedinitrilo)tetraacetato)-, disodium; Calci-edetate de sodium; Calcioedetato sodico; Calcitetracemate disodium; Calcium disodium (ethylenedinitrilo)tetraacetate; Calcium disodium edathamil; Calcium disodium EDTA; Calcium disodium versenate; Calcium edetate de sodium; Calcium-EDTA; Calcium sodium EDTA; Calcium titriplex; CCRIS 3657; Chelaton; Dinatrium calcium edeticum; Disodium ((ethylenedinitrilo)tetra-acetato)calciate; Disodium calcium EDTA; Disodium calcium ethylenediaminetetraacetate; Disodium monocalcium EDTA; Edathamil calcium disodium; Edetamin; Edetamine; Edetate calcium; Edetate calcium disodium; Edetato sodico calcico; Edetic acid calcium disodium salt; Edta-calcium sodium; Edtacal; Edta,calcium disodium; EINECS 200-529-9; Ethylene diamine tetraacetic acid, disodium calcium salt; Glycine, N,N'-1,2-ethanediylbis(N-(carboxymethyl)-, calcium sodium salt (1:1:2); HSDB 4072; Ledclair; Monocalcium disodium EDTA; Mosatil; Natrii calcii edetas; NSC 310151; Rikelate calcium; Sodium (calciedetate de); Sodium calcium edetate; Sormetal; Tetacin; Tetacin-calcium; Tetazine; Versene; Versene CA.; calcium disodium (ethylenedinitrilo)tetraacetate; calcium disodium ethylenediamine tetraacetate; EDTA calcium; edathamil calcium disodium; calcium disodium edetate; edetic acid calcium disodium salt; sodium calciumedetate; Calcitetracemate Disodium; Calcium Disodium Versenate; Ledclair; Mosatil; Antallin; Sormetal. Chelating agent. Used to sequester lead, which displaces the calcium in the compound. Soluble in H_2O, insoluble in organic solvents. *Dow Europe; Dow UK.*

619 Calcium dodecylbenzenesulfonate

26264-06-2   H

$C_{36}H_{58}CaO_6S_2$

Dodecylbenzenesulfonic acid calcium salt.
1371A; Benzenesulfonic acid, dodecyl-, calcium salt; Calcium alkylbenzenesulfonate; Calcium bis(dodecyl-benzenesulfonate); Calcium dodecylbenzenesulfonate; Calcium dodecylbenzenesulphonate; Casul 70HF; Dodecylbenzenesulfonic acid calcium salt; EINECS 247-557-8; HSDB 996; Sinnozon NCX 70; Soprofor S 70; Wettol EM 1.

620 Calcium 2-ethylhexanoate

136-51-6   H

C16H30CaO4

Calcium bis(2-ethylhexanoate).
Calcium 2-ethylhexanoate; Calcium bis(2-ethyl-hexanoate); EINECS 205-249-0; Hexanoic acid, 2-ethyl-, calcium salt.

621   Calcium fluoride
7789-75-5   1669   G

CaF2

Hydrofluoric acid calcium salt.
Acid-spar; Al3-01166; Calcium difluoride; Calcium fluoride; Calcium fluoride (CaF2); EINECS 232-188-7; Fluorite; Fluorspar; HSDB 995; Irtran 3; Liparite; Met-spar; Natural fluorite. Source of fluorine, flux, ceramics, phosphors, paint pigment, catalyst in wood preservative, spectroscopy, electronics, lasers, high-temperature dry-film lubricants. Solid; mp = 1403°; bp = 2500°; d = 3.18; insoluble in H2O; LD50 (gpg orl) > 5 g/kg. *Cerac; GE Silicones; Noah Chem.; Solvay Deutschland GmbH.*

622   Calcium formate
544-17-2   1670   G H

C2H4CaO4

Formic acid, calcium salt.
Calcium diformate; Calcium formate; Calcium formate (Ca(HCO2)2); Calcoform; EINECS 208-863-7; Formic acid, calcium salt; HSDB 5019; Latibon®; Mravencan vapenaty. As feed additive and as silage additive; binder for fine-ore briquets. Crystalline powder; d = 2.02; soluble in H2O; insoluble in EtOH. *Bayer AG.*

623   Calcium gluconate
299-28-5   1671   D G

C12H22CaO14

Calcium D-gluconate.
Calcet; Calcicol; Calciofon; Calcipur; Calcium gluconate; Calcium hexagluconate; Calglucol; Calglucon; CCRIS 1336; Dragocal; Ebucin; EINECS 206-075-8; Glucal; Glucobiogen; Gluconate de calcium; Gluconato di calcio; Gluconic acid, calcium salt; Gluconic acid, calcium salt (2:1), D-; D-Gluconic acid, calcium salt (2:1); HSDB 994; Kalpren; Neo-Calglucon; Novocal. Calcium replenisher. Used in sewage purification and in coffee powders as an anticaking agent. $[\alpha]_D^{20} = 10°$ (c = 1 H2O); soluble in H2O (3.3 g/100 ml at 20°,

20 g/100 ml at 100°), insoluble in organic solvents. *Astra Chem. Ltd.; Mission Pharmacal Co.; Parke-Davis; Sandoz Pharm. Corp.*

624   Calcium hydride
7789-78-8   1674   G

CaH2

Calcium hydride.
Calcium dihydride; Calcium hydride; Calcium hydride (CaH2); EINECS 232-189-2; Hydrolete. Reacts with water to produce hydrogen. Powerful reducing agent, used to prepare rare metals from their oxides, as a drying agent or as a source of hydrogen. Solid; mp = 816°; d = 1.900.

625   Calcium hydroxide
1305-62-0   1675   D G

CaH2O2

Calcium hydrate.
Al3-02602; Bell mine; Biocalc; Calbit; Calcium dihydroxide; Calcium hydrate; Calcium hydroxide; Calcium hydroxide (Ca(OH)2); Calcium hydroxide slurry; Caldic 1000; Calvit; Carboxide; Caswell No. 144; Edelwit; EINECS 215-137-3; EPA Pesticide Chemical Code 075601; HSDB 919; Hydralime; Hydrated lime; Kalkhydrate; Kemikal; Kentoku K 100; Limbux; Lime hydrate; Lime milk; Lime water; Milk of lime; NICC 3000; Rhenofit CF; Sa 074; Slaked lime; Super Microstar; Trulime; Yukijirushisakanyo. Inorganic base; mortar, plaster, cements, calcium salts, disinfectant, food additives, lubricants, pesticides, manufacture of paper pulp, in SBR vulcanization. Used in chemical industries, drinking water treatment, waste water treatment. Hydrated lime, used for horticulture/agriculture, alkalinity, building industry, civil engineering (soil stabilization and soil modification), leather processing, organic and inorganic chemicals, petrochemicals, plasterwork, sewage and water treatment. White powder; d = 2.08-2.34; slightly soluble in H2O; LD50 (rat orl) = 7.34 g/kg. *EM Ind. Inc.; Janssen Chimica; Mallinckrodt Inc.; Pfizer Inc.; U.S. Gypsum.*

626   Calcium 12-hydroxystearate
3159-62-4   H

C36H70CaO6

12-Hydroxy-n-octadecanoic acid, calcium salt.
Calcium(2+) 12-hydroxyoctadecanoate; EINECS 221-605-8; Octadecanoic acid, 12-hydroxy-, calcium salt (2:1).

627   Calcium hypochlorite
7778-54-3   1676   G P

105

$$Cl-O^-$$
$$Cl-O^- \quad Ca^{2+}$$

CaCl2O2

Hypochlorous Acid, Calcium Salt.

B-K powder; Bleaching powder; Calcium chlorohydrochlorite; Calcium chlorohypochloride; Calcium hypochlorite; Calcium oxychloride; Caporit; Caswell No. 145; CCH; Chemichlon G; Chemichlor G; Chloride of lime; Chlorinated lime; Chlorine of lime; Chlorolime chemical; EINECS 231-908-7; EPA Pesticide Chemical Code 014701; Hipoclorito calcico; HSDB 914; HTH; HTH (bleaching agent); HY-Chlor; Hypochlorite de calcium; Hypochlorous acid, calcium salt; Induchlor; Lime chloride; LO-Bax; Losantin; NSC 21546; Pittabs; Pittchlor; Pittcide; Pittclor; Prestochlor; Repak; Sentry; Solvox KS; T-Eusol; UN 1748; Chemichlor G; Chloride of lime; Chlorinated lime; Chlorine of lime; Chlorolime chemical; EPA Pesticide Chemical Code 014701; Hipoclorito calcico; HSDB 914; HTH; HTH (bleaching agent); HY-Chlor; Hypochlorite de calcium; Hypochlorous acid, calcium salt; Lime chloride; LO-Bax; Losantin; NSC 21546; Perchloron; Pittchlor; Pittcide; Pittclor; Sentry; Solvox KS; T-Eusol; UN1748. Algicide, bactericide, deodorant, potable water purification, disinfectant for pools, bleaching agent, oxidizing agent; commercial grade usually 50% or more Ca(OCl)2. Used as an algicide. Registered by EPA as an antimicrobial and fungicide. Commercial product is typically <70% Ca(OCl)2 and usually contains impurities of CaCl2, CaCO3, Ca(OH)2 and Ca(ClO4)2. *Alliance Packaging Inc; Arch Chemicals Inc; Jonas; Nippon Soda Co. Ltd.; Nissho Iwai American Corp.; Olin Corp; PPG Ind.; Qualco Inc.*

628   Calcium iodate

7789-80-2   1678   D G

CaI2O6

Iodic acid calcium salt.

Autarite; Calcium iodate; Calcium iodate (Ca(IO3)2); EINECS 232-191-3; HSDB 986; Iodic acid, calcium salt; Iodic acid (HIO3), calcium salt; Lautarite. Used in deodorants, mouthwashes, food additives and as a dough conditioner. Antiseptic and nutritional source of iodine in foods and feedstuffs. Crystals; mp = 740°; bp = 1100°; stable below 540°; d_4^{14} 4.519; sensitive to reducing agents; slightly soluble in H2O; more soluble in aqueous solutions of iodides and amino acids; soluble in nitric acid; insoluble in alcohol. *Atomergic Chemetals; Blythe, Williams Ltd.; Greeff R.W. & Co.; Mitsui Toatsu; Spectrum Chem. Manufacturing.*

629   Calcium metasilicate

10101-39-0   1707   G P

CaO3Si

Calcium metasilicate.

Baysical K; β-Calcium silicate; C.I. 77230; C.I. Pigment White 28; Calcium metasilicate; Calcium silicate; Calcium silicon oxide; CS (cement component); EINECS 233-250-6; Florite; Kemolit ASB 8 K;

Microcal; Okenite; Pigment White 28; Silasorb; Silicic acid (H2SiO3), calcium salt (1:1). Used as a filler and in building materials. An adsorbent used to control free fatty acids. Registered by EPA as an antimicrobial, fungicide and insecticide (cancelled). FDA GRAS (limited to 2% in table salt, 5% in baking powder), Europe listed, UK approved. USP/NF compliance. FDA approved for orals. Used as a filler, glidant and anti-caking agent in orals and as an antacid. Insoluble in H2O, forms gel with mineral acids; d = 2.10 (15-16 lb/ft3); pH 8.4 - 10.2. Nontoxic orally. *Celite; Crosfield; Degussa Ltd.; Great Western; Huber J. M.; Kraft; Vanderbilt R.T. Co. Inc.*

630   Calcium molybdate

7789-82-4   1686   G

CaMoO4

Molybdic acid calcium salt.

Calcium molybdate; Calcium molybdate(VI); Calcium molybdenum oxide (CaMoO4); EINECS 232-192-9; Molybdate (MoO42-), calcium (1:1); Molybdate, calcium; Molybdic acid (H2MoO4), calcium salt (1:1). Alloying agent in production of iron and steel, crystals in optical and electronic applications, phosphors. Solid; d = 4.350; insoluble in H2O, alcohols, soluble in concentrated mineral acids. *AAA Molybdenum; Atomergic Chemetals; Cerac; Noah Chem.*

631   Calcium nitrate

10124-37-5   1687   G

CaN2O6

Nitric acid calcium salt.

Calcium dinitrate; Calcium nitrate; Calcium saltpeter; Calcium(II) nitrate (1:2); EINECS 233-332-1; HSDB 967; Lime nitrate; Lime saltpeter; Nitric acid, calcium salt; Nitrocalcite; Norge saltpeter; Norway saltpeter; Norwegian saltpeter; Saltpeter; Synfat 1006; UN1454; Wall saltpeter. Usually crystallizes as the tetrahydrate [13477-34-4] or the trihydrate [15842-29-2]. Used to support combustion in matches, explosives, as a fertilizer and corrosion inhibitor. Solid; mp = 560°; very soluble in H2O, MeOH, EtOH, Me2CO, insoluble in other organic solvents.

632   Calcium oleate

142-17-6   1689   G

C36H66CaO4

9-Octadecenoic acid calcium salt.

AI3-19804; Calcium dioleate; Calcium 9-octadecenoate, (Z)-; Calcium oleate; EINECS 205-525-0; 9-Octadecenoic acid (9Z)-, calcium salt; 9-Octadecenoic acid, (Z)-, calcium salt (2:1); Oleic acid, calcium salt. Used as a thickening lubricating grease, waterproofing agent, emulsifier. Solid; mp > 140° (dec); insoluble in H2O and organic solvents.

633 Calcium oxalate
563-72-4   1690   H

C2CaO4
Oxalic acid calcium salt.
Calcium oxalate; EINECS 209-260-1; Ethanedioic acid, calcium salt (1:1); Oxalic acid, calcium salt (1:1). Cubic crystals; d = 2.2; insoluble in H2O, AcOH, soluble in dilute HCl, HNO3.

634 Calcium oxide
1305-78-8   1691   G P

CaO
Lime.
Airlock; Bell CML(E); Bell CML(P); Burnt lime; Calcia; Calcia (CaO); Calcium monoxide; Calcium oxide; Caloxal CPA; Caloxol CP2; Caloxol W3; Calx; Calx usta; CALX; Calxyl; Caswell No. 147A; CCRIS 7496; Chaux vive; CML 21; CML 31; CML 35; Desical P; EINECS 215-138-9; EPA Pesticide Chemical Code 075604; Gebrannter kalk; HSDB 1615; KM Pebble Lime; Lime; Lime, burned; Oxyde de calcium; QC-X; Quick lime; Quicklime; Rhenosorb C; Rhenosorb F; UN1910; Unslaked lime; Vesta PP; Wapniowy tlenek; CaO. Inorganic oxide; refractory; sewage treatment, insecticides, fungicides, manufacture of steel and aluminum; flotation of nonferrous ores; manufacture of glass, paper, Ca salts; in drilling fluids, lubricants; laboratory. A dessicant used with rubber. White solid; mp = 2572°; bp = 2850°; d = 3.32-3.35. *Cerac; Degussa-Hüls Corp.; Kerr-McGee; Mallinckrodt Inc.; Pfizer Inc.; Rhône-Poulenc; U.S. Gypsum.*

635 Calcium palmitate
542-42-7   1692   H

C32H62CaO4
Palmitic acid calcium salt.
Calcium dipalmitate; Calcium hexadecanoate; EINECS 208-811-3; Hexadecanoic acid, calcium salt; Palmitic acid, calcium salt. Used to thicken lubricating oil and for waterproofing fabrics and lubricating greases. Also used as a corrosion inhibitor with halohydrocarbons. White powder; dec. 155°; slightly soluble in CHCl3, C6H6, AcOH, insoluble in H2O, EtOH, Et2O, Me2CO, petroleum ether.

636 Calcium d-pantothenate
137-08-6   7985   D G

C18H32CaN2O10
(R)-N-(2,4-dihydroxy-3,3-dimethyl-1-oxobutyl)-β-alanine calcium salt.
β-Alanine, N-(2,4-dihydroxy-3,3-dimethyl-1-oxobutyl)-, calcium; Bis(pantothenato)calcium; Ca-HOPA; Calcii pantothenas; Calcium, bis(pantothenato)-; Calcium D-pantothenate (1:2); Calcium N-(2,4-dihydroxy-3,3-dimethyl-1-oxobutyl)-β-alanine; Calcium panthothen-ate; Calcium pantothenate, D-form; Calcium panto-thenate (2:1); Calpan; Calpanate; CCRIS 1338; Cris Pan; Dextro calcium pantothenate; N-(2,4-Dihydroxy-3,3-dimethylbutyryl)-β-alanine calcium; EINECS 205-278-9; NSC 36292; Pancal; Panthoject; Pantholin; Pantotenato calcico; Pantothenate calcium; Pantothenate de calcium; Pantothenic acid, calcium salt (2:1), (+)-; Pantothenic acid, calcium salt (2:1), D-; Stuartinic; Vitamin B5, calcium salt. Vitamin (enzyme cofactor). An essential dietary factor. Solid; mp = 195-196° (dec); [α]$_D^{20}$ = 28.2° (c = 5); soluble in H2O (0.36 g/ml), poorly soluble in organic solvents. *BASF Corp.; Eli Lilly & Co.; Johnson & Johnson-Merck Consumer Pharm.; Pharmacia & Upjohn.*

637 Calcium permanganate
10118-76-0   1693   G

CaMn2O8
Permanganic acid calcium salt.
Acerdol; Calcium permanganate; Calcium permanganate (Ca(MnO4)2); EINECS 233-322-7; HSDB 966; Monol; Permanganate de calcium; Permanganato calcico; Permanganic acid (HMnO4), calcium salt; UN1456. Used to treat gastro-enteritis and diarrhea. Antiseptic, deodorizer, disinfectant. Used in the textile industry. Freely soluble in H2O, decomposes in alcohol.

638 Calcium phosphate
7757-93-9   1697   D G

CaHO4P
Calcium hydrogen phosphate.
A-Tab; Anhydrous Emcompress; Biofos; Calbrite; Calcium acid phosphate; Calcium dibasic phosphate; Calcium hydrogen orthophosphate; Calcium hydrogen phosphate; Calcium monohydrogen phosphate; Calcium monohydrogen phosphate anhydrous; Calcium phosphate (1:1); Calcium phosphate (CaHPO4); Calcium phosphate, dibasic; Calcium phosphate dibasic anhydrous; Calcium phosphate, dibasic, dental grade; Calcium secondary phosphate; CCRIS 1337; D.C.P.; DCP; DCP-O; Dicafos AN; Dicalcium orthophosphate; Dicalcium phosphate; EINECS 231-826-1; Fujicalin S; HSDB 992; Ipifosc 20; Monetite; Monocalcium acid phosphate; Monocalcium phosphate; Monohydrogen calcium phosphate; Phosphoric acid, calcium salt (1:1); secondary Calcium phosphate. Foods, pharmaceuticals, dentifrice, medicine, glass, fertilizer, stabilizer for plastics, dough conditioner, yeast food. Used medicinally as a calcium replenisher. Triclinic crystals; d = 2.31; insoluble in H2O, EtOH, soluble in mineral acids. *Albright & Wilson Americas Inc.; EM Ind. Inc.; FMC Corp.; GE Specialities; Janssen Chimica; Mallinckrodt Inc.; OxyChem; Rhône-Poulenc.*

639 Calcium phosphate dihydrate
7789-77-7   G

$$\text{O=P} \begin{array}{c} \text{O}^- \\ | \\ \text{O}^- \\ | \\ \text{OH} \end{array} \quad Ca^{2+} \quad 2H_2O$$

$CaH_5O_6P \cdot 2H_2O$
Dibasic calcium phosphate dihydrate.
Calcium monohydrogen phosphate dihydrate; Calcium phosphate, dibasic, dihydrate; Dicalcium phosphate dihydrate; Phosphoric acid, calcium salt (1:1), dihydrate. Dibasic calcium phosphate dihydrate USP/BP; excipient for production of pharmaceutical tablets by direct compression process. *Penwest Pharmaceuticals Co.*

640   Calcium phosphate (monobasic)
7758-23-8   1698   G

$CaH_4O_8P_2$
Phosphoric acid, calcium salt, monobasic.
Acid calcium phosphate; C 38 (phosphate); Calcium biphosphate; Calcium bis(dihydrogen phosphate); Calcium dihydrogen orthophosphate; Calcium dihydrogen phosphate; Calcium diorthophosphate; Calcium hydrogen phosphate $(Ca(H_2PO_4)_2)$; Calcium monobasic phosphate; Calcium phosphate (1:2); Calcium phosphate, monobasic; Calcium superphosphate; Calcium tetrahydrogen orthophosphate; Calcium tetrahydrogen phosphate; EINECS 231-837-1; HSDB 1441; Monobasic calcium phosphate; Monocalcium orthophosphate; Monocalcium phosphate, monobasic; Phosphoric acid, calcium salt (2:1); Primary calcium phosphate; V 90®; acid calcium phosphate; calcium biphosphate; monocalcium orthophosphate; primary calcium phosphate; calcium superphosphate; MCP. Leavening acid in food products, mineral supplement, fertilizer, stabilizer for plastics, to control pH in malt, glass manufacture, firming agent. White solid; mp = 200° (dec); d_4^{18} =2.220; moderately soluble in H_2O, soluble in acids. *Albright & Wilson Americas Inc.; FMC Corp.; Kemira Kemi UK; Mallinckrodt Inc.; Monsanto Co.; Nichia Kagaku Kogyo; Rhône-Poulenc.*

641   Calcium phosphate (tribasic)
7758-87-4   1699   D G

$Ca_3O_8P_2$
Phosphoric acid, calcium salt, tribasic.
Bonarka; Calcigenol simple; Calcium orthophosphate; Calcium phosphate; Calcium phosphate (3:2); Calcium phosphate $(Ca_3(PO_4)_2)$; Calcium phosphate, tribasic; Calcium tertiary phosphate; Caswell No. 148; CCRIS 3668; EINECS 231-840-8; EPA Pesticide Chemical Code 076401; FEMA No. 3081; HSDB 879; Natural whitlockite; Phosphoric acid, calcium salt (2:3); Phosphoric acid calcium(2+) salt (2:3); Synthos; Tertiary calcium phosphate; Tribasic calcium phosphate; Tricalcium diphosphate; Tricalcium orthophosphate; Tricalcium phosphate; tricalcium orthophosphate; tertiary calcium phosphate; Calcigenol Simple; oxydapatit; voelicherite; whitlockite; bone ash. Used in foods, pharmaceuticals, polystyrene, ceramics, mordant, fertilizers, dentifrices, stabilizer for plastics, in meat tenderizer, anticaking agent, nutrient supplement and calcum replenisher. White powder; mp = 1670°; d = 3.14; insoluble in H_2O, EtOH, soluble in mineral acids. *Albright & Wilson Americas Inc.; FMC Corp.; Mallinckrodt Inc.; Monsanto Co.; Rhône-Poulenc.*

642   Calcium phosphide
1305-99-3   1700   G P

Ca_3P_2
Tricalcium diphosphide.
Calcium phosphide; Calcium phosphide (Ca_3P_2); Calcium photophor; EINECS 215-142-0; HSDB 963; Photophor®; UN1360. Used for signal fires and as a rodenticide. Solid; mp = 1600°; d = 2.51; decomposed by H_2O.

643   Calcium polysulfide
1344-81-6   G P

CaS_x
Calcium polysulfide.
Calcium polysulfide; Calcium polysulphide; Calcium sulfide $(Ca(S_x))$; Caswell No. 150; Eau grison; EINECS 215-709-2; EPA Pesticide Chemical Code 076702; HSDB 5826; Lime sulfur; Lime sulphur; Neviken; Orthorix; Polysulfure de calcium; Sulka; Zolfosol 25. Lime-sulfur fungicide. Acts directly and alo by decomposition to sulfur which is also a fungicide. Used for control of powdery mildews, anthracnose, scab and other diseases in benas, clover and fruits. Control of insects and spider mite eggs in fruit trees. Registered by EPA as a fungicide and insecticide. Deep orange liquid; d15.6 > 1.28; soluble in H_2O; skin and eye irritant. *Bonide Products, Inc.; Earth Care; Generic; PBI/Gordon Corp; Scotts Co.; Value Gardens Supply LLC; Western Farm Service, Inc.*

644   Calcium propionate
4075-81-4   1703   D G P

$C_6H_{10}CaO_4$
Propionic acid, calcium salt.
Bioban-C; Calcium dipropionate; Calcium propanoate; Calcium propionate; Caswell No. 151; EINECS 223-795-8; EPA Pesticide Chemical Code 077701; HSDB 907; Mycoban; Propanoic acid, calcium salt; Propionic acid, calcium salt. Mold-inhibitor; food additive; antifungal agent; improves scorch resistance and processibility of butyl rubber. Registered by EPA as a fungicide (cancelled). Monoclinic crystals; soluble in H_2O, slightly soluble in MeOH, EtOH, insoluble in Me_2CO, C_6H_6. *Gist-Brocades Intl.; Niacet; Riedel de Haen (Chinosolfabrik); Verdugt BV.*

645   Calcium silicate
1344-95-2   1707   G

$CaSiO_3$, Ca_2SiO_4, Ca_3SiO_5
Silicic acid, calcium salt.
Calcium hydrosilicate; Calcium metasilicate; Calcium monosilicate; Calcium polysilicate; Calcium silicate; Calcium silicate, synthetic;

Calcium silicate, synthetic nonfibrous; Calflo E; Calsil; CS lafarge; EINECS 215-710-8; Florite R; HSDB 705; Marimet 45; Micro-cel; Micro-cel A; Micro-cel B; Micro-cel C; Micro-cel E; Micro-cel T; Micro-cel T26; Micro-cel T38; Micro-cel T41; Microcal 160; Microcal ET; Promaxon P60; Silene EF; Silicic acid, calcium salt; Silmos T; Solex; Stabinex NW 7PS; Starlex L; SW 400; Toyofine A; afwillite; akermanite; calcium pectolith; centrallasite; crestmoreite; eaklite; Cecasil; foshagite; foshallasite; gjellebaekeite; grammite; gyrolite; hillebrandite; larnite; okenite; parawollastonite; pseudo-wollastonite; riversideite; table spate; tobermorite; wollastonite; xonaltite; xonotlite. A constituent of lime glass and portland cement; used as a reinforcing filler in elastomers and plastics; absorbent for liquids, gases, vapors; anticaking agent, suspending agent. Functional filler and reinforcing agent, bright color used in adhesives, ceramics, elastomers, insulating materials, cosmetics, plastics, paint, resins, sealants, and wallboards. Cream-colored powder; d^{25} = 2.10. *Celite; Crosfield; Degussa AG; Vanderbilt R.T. Co. Inc.*

646   Calcium silicate
13983-17-0   G

CaSiO3;Ca2SiO4; Ca3SiO5
 Silicic acid calcium salt.
 350 Wollastokup 10014; 1250 Wollastokup 10224; Aedelforsite; Bistal; Bistal W; Bistal W 101; Cab-O-lite; Cab-O-lite 100; Cab-O-lite 130; Cab-O-lite 160; Cab-O-Lite P 4; Calcium silicate; Casiflux; Casiflux VP 413-004; CCRIS 4763; CHC 62N10; CHC 74N; CHD 62N10; Dab-O-lite P 4; EINECS 237-772-5; Fibernite HG; FPW 400; FPW 800; Fuwalip; FW 50; FW 200 (mineral); FW 325; Gillebachite; HN 3 (mineral); HSDB 4294; Hycon A 60; K 160 (mineral); Kemolit A-60; Kemolit ASB; Kemolit ASB-3; Kemolit ASB 8; Kemolit-N; NCI-C55470; Nyad 1250; Nyad 1250T; Nyad Denacup 100G; Nyad G; Nyad G 400; Nyad G Wollastocoat 1001; Nyad G Wollastocoat 2075; NYAD 10; NYAD 325; NYAD G; NYCO 1250; Nycor 200; Nycor 300; Okenite; Rivaite; Schalstein; Tabular spar; Tremin (mineral); Vansil® W 10; Vansil® W 20; Vansil® W 30; Vilnite; Wollastokup; Wollastonite; Wollastonite (Ca(SiO3)). Extender pigment for solvent-thinned and latex paints with high aspect ratio grades, milled grades and fine particle size grades; CaSiO3; used in polymer composites, coatings, adhesives, elastomers and friction products. *Nyco Minerals; Vanderbilt R.T. Co. Inc.*

647   Calcium stearate
 1592-23-0   1708   G H

C36H70CaO4
 n-Octadecanoic acid, calcium salt.
 Afco-Chem CS; AI3-01335; Aquacal; Calcium bis(stearate); Calcium distearate; Calcium octadecanoate; Calcium stearate; Calstar; EINECS 216-472-8; Flexichem; Flexichem CS; G 339 S; G 339S; HSDB 905; Nopcote C 104; Octadecanoic acid, calcium salt; Stavinor 30; Stearates; Stearic acid, calcium salt; Synpro stearate; Witco G 339S. Water repellent; flatting agent in paints, emulsions; release agent for plastic molding powds.; stabilizer for PVC resins; lubricant for paper coatings; in pencils and crayons; food grade as conditioning agent in foods. Lubricant for metal sintering. Lubricant and stabilizer for resins. Pigment dispersant, mold releasing agent, waterproofing agent and lubricant additive. Solid; mp = 179.5°; slightly soluble in H2O, EtOH, Et2O. *Adeka Fine Chem.; CK Witco Corp.; Eka Nobel Ltd.; Elf Atochem N. Am.; Ferro/Grant; Henkel/Organic Prods.; Mallinckrodt Inc.; PPG Ind.; Vanderbilt R.T. Co. Inc.; Witco/Oleo-Surf.*

648   Calcium sulfate
7778-18-9   1711   G P

CaO4S
 Calcium sulfate (anhydrous).
 AI3-02330; Anhydrite; Anhydrous calcium sulfate; Anhydrous gypsum; Anhydrous sulfate of lime; Calcium sulfate; Calcium sulfate, anhydrous; Calcium sulphate, natural; Caswell No. 152; CCRIS 3666; Crysalba; Drierite; EINECS 231-900-3; EPA Pesticide Chemical Code 005602; GIBS; Gypsum; HSDB 902; Karstenite; Muriacite; Natural anhydrite; NSC 529649; Pearl Hardening; Pigment white 25; Plaster of paris; Plaster of Paris, anhydrite; Satin White; Sulfuric acid, calcium salt; Sulfuric acid, calcium salt (1:1); Surfuric acid calcium(2+) salt (1:1); Terra Alba; Thiolite. Used in cement formulations, as paper filler; soluble anhydride: drying agent, desiccant. FDA GRAS, BATF limitation 16.69 lbs/1000 gallons, Japan approved with restrictions, Europe listed, UK approved, FDA approved for orals, USP/NF compliance. Used as an excipient, tablet/capsule diluent, dessicant, used in tablets prepared by direct compression, orals, abrasive and firming agent in tooth powders. (EPA cancelled). White solid; d = 2.96; soluble in H2O (0.2%). *U.S. Gypsum.*

649   Calcium sulfate dihydrate
 10101-41-4   G

CaO4S.2H2O
 Calcium sulfate dihydrate.
 Alabaster; Annaline; C.I. 77231; C.I. pigment white 25; Calcium sulfate, dihydrate; Calcium(II) sulfate, dihydrate (1:1:2); Compactrol; Light spar; Magnesia White; Mineral White; Native calcium sulfate; Precipitated calcium sulfate; Satin spar; Satinite; Sulfuric acid, calcium(2+) salt, dihydrate; Terra alba. Tablet and capsule filler for pharmaceutical tablets manufacturing by direct compression. FDA approved for orals, USP/NF, BP, JP compliance. Used as an excipient, filler, dessicant, tablet/capsule diluent, used in orals. Off-white, odorless, tasteless powder; mp = 1450°; d = 2.308; soluble in H2O, insoluble in most organic solvents. Non-toxic. *EM Ind. Inc.; Fluka; Franklin Ind. Mins.; Lohmann; Penwest Pharmaceuticals Co.; Ruger; Sigma-Aldrich Fine Chem.; Spectrum Chem. Manufacturing.* 9

650   Calcium sulfite
 10257-55-3   1713   G

CaO3S
 Sulfurous acid calcium salt.
 Calcium sulfite; Calcium sulphite; EINECS 233-596-8; Katarsit;

Sulfurous acid, calcium salt (1:1). Used as a dechlorinating agent for water. Used to preserve fruit juices, in paper processing and sugar manufacture. [dihydrate]; slightly soluble in H_2O, EtOH.

651   Calcium thiobis(dodecylphenolate)

26998-97-0   H

$C_{36}H_{56}CaO_2S$

Phenol, thiobis(dodecyl-, calcium salt (1:1).
Calcium salt of thiobis(C12-alkylated phenol); Calcium thiobis(dodecylphenolate); EINECS 248-159-7; Phenol, thiobis(dodecyl-, calcium salt (1:1).

652   Calcium titanate

12049-50-2   G

CaO_3Ti

Calcium titanate(IV).
Calcium titanate; Calcium titanium oxide (CaTiO3); Calcium titanium trioxide; EINECS 234-988-1; RC 17; Titanate (TiO3^{2-}), calcium (1:1); Perovskite; Titanium calcium oxide. Inorganic compound; used in electronic devices. *Atomergic Chemetals; Cerac; Ferro/Transelco; Tam Ceramics.*

653   Calcium verate

5793-94-2   1709   G

Ca^{2+}

$C_{48}H_{86}CaO_{12}$

Calcium 2-stearoyl lactylate.
Calcium 2-(1-carboxyethoxy)-1-methyl-2-oxoethyl-octadecanoate; Calcium bis(2-(1-carboxylatoethoxy)-1-methyl-2-oxoethyl) disterarate; Calcium stearoyl lactylate; Calcium stearoyl-2-lactylate; Calcium α-(α-(stearoyloxy)propionyloxy)propionate; Crolactil CSL; EINECS 227-335-7; Octadecanoic acid, 2-(1-carboxyethoxy)-1-methyl-2-oxoethyl ester, calcium salt; Stearic acid, ester with lactate of lactic acid, ca salt; Stearoyl-2-lactylic acid; Verv®. Water-oil emulsifier, emollient for skin care and treatment products; ingredients in food products. Starch and protein complexing agent, softener for use in yeast-leavened bakery products; conditioning agent in dehydrated potatoes. *Croda Surfactants; Patco.*

654   Camite

12070-12-1   9945   G

WC
Tungsten carbide.
EINECS 235-123-0; NCI C61198; Tungsten carbide; Tungsten carbide (WC); Tungsten, containing > 2% cobalt [Tungsten and cemented tungsten carbide]; Tungsten, containing > 3% nickel [Tungsten and cemented tungsten carbide]; Tungsten monocarbide. Proprietary trade name for tungsten carbide materials.

655   Camphene

79-92-5   1737   H

$C_{10}H_{16}$

2,2-Dimethyl-3-methylene-bicyclo(2.2.1)heptane.
AI3-01775; Bicyclo(2.2.1)heptane, 2,2-dimethyl-3-methylene-; Camphene; CCRIS 3783; 2,2-Dimethyl-3-methylene-norbornane; EINECS 201-234-8; EINECS 209-275-3; FEMA No. 2229; HSDB 900; NSC 4165. In essential oil of nutmeg. Used in manufacture of isobornyl esters and camphor and other chemicals related to the cosmetics and perfumery industries. It is also used as a starting material in the synthesis of terpene phenolic resins, as a plasticizer for resins, paints and as a starting material in the synthesis of acrylates and methacrylates. [dl-form]: Crystals; mp = 51 - 5°; bp = 158.5 - 159.5°; d^{54} = 0.8422; insoluble in H_2O, moderately soluble in organic solvents; [d-form]: Crystals; mp = 52°; $[\alpha]_D^{17}$ = +103.5° (c = 9.67 Et2O); d^{50} = 0.8486; [l-form]: Crystals; mp = 52°; $[\alpha]_D^{21}$ = -119.11 (c = 2.33 C6H6); d^{54} = 0.8422. *Camphor and Allied Products.*

656   Camphor

76-22-2   1739   G

$C_{10}H_{16}O$

1,7,7-Trimethylbicyclo(2.2.1)-2-heptanone.
0-07-00-00135 (Beilstein Handbook Reference); 1,7,7-Trimethylbicyclo(2.2.1)heptan-2-one; 1,7,7-Trimethyl-norcamphor; 4-07-00-00213 (Beilstein Handbook Reference); AI3-18783; Alphanon; Bicyclo(2.2.1)-heptan-2-one, 1,7,7-trimethyl-; Bornane, 2-oxo-; 2-Bornanone; DL-Bornan-2-one; BRN 1907611; BRN 3196099; 2-Camphanone; Campho-Phenique Cold Sore Gel; Campho-Phenique Gel; Campho-Phenique Liquid; Camphor; (±)-Camphor; DL-Camphor; Camphor, synthetic; dl-Camphor; Caswell No. 155; EINECS 200-945-0; EINECS 244-350-4; EPA Pesticide Chemical Code 015602; Gum camphor; Heet; HSDB 37; Huile de camphre; 2-Kamfanon; Kampfer; 2-Keto-1,7,7-trimethylnorcamphane; Matricaria camphor; Norcamphor, 1,7,7-trimethyl-; Root bark oil; Sarna; Spirit of camphor; UN2717. Ordinary camphor, which separates from the essential oil of *Laurus camphora.* Used in organic synthesis, as a moth repellant, odorant and flavoring agent, plasticizer and preservative. Medical use as a carminative, antipruritic and antiseptic. White crystals; mp = 179°; bp = 209°; λ_m = 292 nm (CHCl3); soluble in H_2O (0.12 g/100 ml), EtOH (100 g/100 ml), Et2O (1 g/100 ml); CHCl3 (2 g/100 ml), other organic solvents; LD50 (mus orl) = 1300 mg/kg. *Sigma-Aldrich Fine Chem.*

657   d-Camphorsulfonic acid

3144-16-9   1741   G

$C_{10}H_{16}O_4S$

7,7-Dimethyl-2-oxobicyclo[2.2.1]heptane-1-methane-sulfonic acid.
AI3-23953; Bicyclo(2.2.1)heptane-1-methanesulfonic acid, 7,7-dimethyl-2-oxo-, (1S,4R)-; 10-Bornane-sulfonic acid, 2-oxo-, (1S,4R)-(+)-; Campher-sulfosaeure; Camphersulfosäure; Camphorsulfonic acid; d-Camphorsulfonic acid; d-10-Camphorsulfonic acid; D-Camphor-10-sulfonic acid; Camphostyl; Camsylate; CSA; EINECS 221-554-1; 2-Oxobornane-10-sulphonic

acid; Reychler's acid   Reychler's acid;. Used for resolution of optical isomers. Prisms; mp = 193-195° (dec); λ_m = 285 nm (ε = 35, H_2O); $[\alpha]_D^{20}$= +43.5° (c = 4.3 EtOH); very soluble in H_2O, soluble in EtOH, poorly soluble in other organic solvents.

658   Candicidin
1403-17-4   1748   D G

C47H84N2O18 (Candicidin D)
Vanobid.
Levorin; Candeptin; Candimon. Candicidin; antifungal. The major of four components of Candicin is Candicidin D ([1403-17-4]). White powder; λ_m = 403, 380 (E$_{1cm}^{1\%}$ 1150), 360 nm; insoluble in H_2O, most organic solvents; soluble in DMF, DMSO, AcOH; LD50 (mus ip) = 14 mg/kg. Key Pharm.; Marion Merrell Dow Inc.

659   Canthaxanthin
514-78-3   1758   G

C40H52O2
β,β-Carotene-4,4'-dione.
4-07-00-02680 (Beilstein Handbook Reference); BRN 1898520; C.I. Food Orange 8; Cantaxanthin; Cantaxanthine; Canthaxanthin; Canthaxanthine; Carophyll Red; Carotene-4,4'-dione, β-; all-trans,β-Carotene-4,4'-dione;   β-Carotene-4,4'-dione,   all-trans-;   β,β-Carotene-4,4'-dione; CCRIS 3276; CI 40850; 4,4'-Dioxo-β-carotene; E 161 G; EINECS 208-187-2; Food orange 8; L-Orange 7; NSC 374110; Orobronze; Roxanthin Red 10; Carophyll. A pigment used in animal feedstuffs. Crystals; mp = 217° (dec); λ_m = 470 nm (E$_{1cm}^{1\%}$ = 2250); soluble in CHCl3, oils. Roche Products Ltd.

660   Capramide DEA
136-26-5   G

C14H29NO3
N,N-Bis(2-hydroxyethyl)decanamide.
3-04-00-00706 (Beilstein Handbook Reference); Amidex CP; N,N-Bis(2-hydroxyethyl)decan-1-amide; BRN 1785093; Capramide DEA; Capric acid diethanolamide; Decanamide, N,N-bis(2-hydroxy-ethyl)-; EINECS 205-234-9; Standamid® CD; Upamide CD. Capramide DEA (2:1) - diethanolamine; detergent, foam enhancer for anionic systems; solubilizes fragrances into hydroalcoholic systems;

secondary emulsifier in oil-water systems, perfume stabilizer; personal care products, industrial cleaners. Chemron; Henkel/Cospha; Henkel.

661   Caproic acid
142-62-1   1765   G

C6H12O2
n-Hexanoic acid.
4-02-00-00917 (Beilstein Handbook Reference); AI3-07701; BRN 0773837; Butylacetic acid; Caproic acid; n-Caproic acid; Capronic acid; CCRIS 1347; EINECS 205-550-7; FEMA No. 2559; Hexacid 698; Hexanoic acid; n-Hexanoic acid; 1-Hexanoic acid; Hexoic acid; n-Hexoic acid; n-Hexylic acid; HSDB 6813; Kyselina kapronova; NSC 8266; Pentanecarboxylic acid; 1-Pentanecarboxylic acid; Pentiformic acid; Pentylformic acid; UN2829. Used in analytical chemistry, flavors, manufacture of rubber chemicals, varnish driers, resins, pharmaceuticals. Liquid; mp = -3°; bp = 205.2°; d^{20} = 0.9274; insoluble in H_2O (1.082 g/100g), soluble in EtOH, Et2O, CHCl3; LD50 (rat orl) = 3.0 g/kg. Chisso Corp.; Janssen Chimica; Lancaster Synthesis Co.; Penta Mfg.; Sigma-Aldrich Fine Chem.; Aldrich; Am.; Chimica; Chisso; Janssen; Mfg.; Penta; Schweizerhall.

662   Caprolactam
105-60-2   1767   H

C6H11NO
6-Aminohexanoic acid cyclic lactam.
5-21-06-00444 (Beilstein Handbook Reference); A1030; AI3-14515; Aminocaproic lactam; 6-Aminocaproic acid lactam; 6-Aminohexanoic acid cyclic lactam; BRN 0106934; Caprolactam; 6-Caprolactam; Caprolattame; Capron PK4; CCRIS 119; Cyclohexanone iso-oxime; ε-Kaprolaktam; EINECS 203-313-2; Epsilon kaprolaktam; Extrom 6N; Hexahydro-2H-azepin-2-one; Hexamethylenimine, 2-oxo-; 6-Hexanelactam; Hexanoic acid, 6-amino-, lactam; Hexanolactam; Hexanone isoxime; Hexanonisoxim; HSDB 187; Kapromine; NCI-C50646; NSC 117393; Stilon. Leaflets; mp = 69.3°; bp = 270°; λ_m = 198 nm (ε = 7762, H_2O); very soluble in H_2O, C6H6, EtOH, CHCl3.

663   Caprolactone
502-44-3   G H

C6H10O2
Hexanoic acid, 6-hydroxy-, ε-lactone.
5-17-09-00034 (Beilstein Handbook Reference); BRN 0106919; Caprolactone; EINECS 207-938-1; ε-Kaprolakton; ε-Caprolactone; Hexamethylene oxide, 2-oxo-; 6-Hexanolactone; Hexan-6-olide; 1,6-Hexanolide; HSDB 5670; 6-Hydroxyhexanoic acid lactone; 2-Oxepanone; Placcel M. Intermediate in adhesives, urethane coatings, elastomers; solvent; diluent for epoxy resins; synthetic fibers; organic synthesis. Liquid; mp = -18°; bp = 215°, bp10 = 96-98°; d^{20} = 1.0693; soluble in EtOH, Et2O, Me2CO; LD50 (rat orl) =

4290 mg/kg. *Union Carbide Corp.*

664　Caprylic alcohol

123-96-6　　　　　　6785　　　　　　H

OH

$C_8H_{18}O$

2-Octyl alcohol.

AI3-05598; Capryl alcohol; Caprylic alcohol, secondary; sec-Caprylic alcohol; EINECS 204-667-0; EINECS 223-938-4; FEMA Number 2801; Hexyl methyl carbinol; HSDB 5601; Methyl hexyl carbinol; n-Octan-2-ol; NSC 14759; 2-Octyl alcohol; 2-Octanol; s-Octyl alcohol.

665　Caprylic/capric triglyceride

65381-09-1　　　　　　　　　　　　H

COOH

COOH

OH

HO　　OH

$C_{10}H_{20}O_2.xC_8H_{16}O_2.xC_3H_8O_3$

Decanoic acid, ester with 1,2,3-propanetriol octanoate.

Caprylic acid, capric acid triglyceride; Caprylic/capric triglyceride; Decanoic acid, ester with 1,2,3-propanetriol octanoate; EINECS 265-724-3; Octanoic/decanoic acid triglyceride.

666　Captafol

2425-06-1　　　　　　1777　　　　　　G P

$C_{10}H_9Cl_4NO_2S$

3a,4,7,7a-Tetrahydro-2-[[(1,1,2,2-tetrachloroethyl)thio]-1H-isoindole-1,3(2H)-dione.

5-21-10-00136 (Beilstein Handbook Reference); Alfloc 7020; Alfloc 7046; Arborseal; BRN 1543712; Captafol; Captaspor; CCRIS 848; CS 5623; 4-Cyclohexene-1,2-dicarboximide, N-((1,1,2,2-tetrachloroethyl)thio)-; Di-folatan; Difolatan 4F; Difolatan 4F1; Difolatan 80W; Difolatan BOW; Difosan; EINECS 219-363-3; Folcid; Foltaf; Haipen; Haipen 50; HSDB 340; 1H-Isoindole-1,3(2H)-dione, 3a,4,7,7a-tetrahydro-2-((1,1,2,2-tetra-chloroethyl)thio)-; Kenofol; Merpafol; N-1,1,2,2-Tetra-chloroethylmercapto-4-cyclohexene-1,2-carb-oximide; N-((1,1,2,2-Tetrachloroethyl)sulfenyl)-cis-4-cyclohex-ene-1,2-dicarboximide; N-((1,1,2,2-Tetrachloroethyl)-thio)-4-cyclohexene-1,2-dicarboximide; N-(1,1,2,2-Tetrachloraethylthio)-cyclohex-4-en-1,4-diacarbox-imid; N-(1,1,2,2-Tetrachloroaethylthio)-tetrahydro-phthalamid; N-(Tetrachloroethylthio)tetrahydro-phthal-imide; N-(1,1,2,2-Tetrachloroethylthio)-Δ[4]-tetrahydro-phthalimide; Nalco 7046; Ortho 5865; Ortho Difolatan 4 Flowable; Ortho Difolatan 80W; Proxel EF; Sanspor; Santar SM; Sulfonimide; Sulpheimide; Terrazol; Tetrachloroethylthiotetrahydrophthalimide; 1,2,3,6-Tetrahydro-N-(1,1,2,2-tetrachloroethylthio)-phthalimide; 3a,4,7,7a-Tetrahydro-2-((1,1,2,2-tetra-chloroethyl)thio)-1H-isoindole-1,3(2H)-dione; 3a,4,7, 7a-Tetrahydro-N-(1,1,2,2-tetrachloroethanesulph-enyl)phthalimide; [2939-80-2]. Fungicide, used with potatoes. Registered by EPA as a fungicide (cancelled). Colorless to pale

yellow crystals; mp = 160-162°; slightly soluble in H_2O (0.00014 mg/100 ml), soluble in i-PrOH (1 g/100 ml), C_6H_6 (2.2 g/100 ml), C_7H_8 (1.5 g/100 ml), xylene (8.6 g/100 ml), Me_2CO (3.4 g/100 ml), MEK (3.5 g/100 ml), DMSO (18.7 g/100 ml); LD$_{50}$ (rat orl) = 6200 (maize oil), 4200 mg/kg (aqueous), (rbt der) = 15400 mg/kg; LC$_{50}$ (pheasant 10 day dietary) > 23070 mg/kg diet, (mallard duck 10 day dietary) > 1010700 mg/kg diet, (rainbow trout 96 hr.) = 0.5 mg/l, (goldfish 96 hr.) = 3 mg/l, (bluegill sunfish 96 hr.) = 0.15 mg/l; non-toxic to bees. *Generic; Monsanto (Solaris); Rallis India Ltd.*

667　Captamine Hydrochloride

13242-44-9　　　　　　　　　　　　D G

HS　　　N(CH₃)₂ · HCl

$C_4H_{12}ClNS$

2-(Dimethylamino)ethanethiol hydrochloride.

Captamine hydrochloride; EINECS 236-221-6; MEDA; NSC-45463; Thiofluor™. Nucleophile used in preparation of o-phthaldehyde reagent for fluorescence detection in HPLC. Metal complexing agent and depigmentor. Crystals; mp = 158-160°. *Schering-Plough HealthCare Products.*

668　Captan

133-06-2　　　　　　1778　　　　　　G P

$C_9H_8Cl_3NO_2S$

2-[(Trichloromethyl)thio]-1H-isoindole-1,3(2H)-dione.

5-21-10-00136 (Beilstein Handbook Reference); Aacaptan; Agrosol S; Agrox 2-way and 3-way; AI3-26538; Amercide; Bangton; Bean Seed Protectant; Bonide Captan 50%WP; BRN 0023177; Buvisild K; Captab; Captadin; Captaf; Captaf 50W; Captaf 85W; Captan; Captan 50W; Captan-streptomycin 7.5-0.1 potato seed piece protectant; Captancapteneet 26,538; Captane; Captanex; Captazel; Captec; Capteneet 26,538; Captex; Captol; Caswell No. 159; Catanex; CCRIS 120; Dangard; EINECS 205-087-0; ENT 26,538; EPA Pesticide Chemical Code 081301; Esso fungicide 406; Flit; Flit 406; Fungus ban type II; Glyodex 3722; Granox PFM; Gustafson captan 30-DD; Hexacap; HSDB 951; Isotox Seed Treater D; Isotox Seed Treater F; Kaptan; Kaptazor; LE Captane; Malipur; Merpan; Merpan 90; Merphan; Meteoro; Micro-check 12; NCI-C00077; Neracid; NSC 36726; Orthocide; Orthocide 7.5; Orthocide 50; Orthocide 83; Orthocide 406; Ortocid 50; Ososcide; Phytocape; Pillarcap; Sepicap; Sorene; SR 406; Stauffer Captan; Trichlormethyl-thioamid kyseliny 1,2,3,6-tetrahydroftalove; Trichlor-methylthio-1,2,5,6-tetrahydrophthalamide; Trimegol; Ugecap; Ugecap 83; Vancide 89; Vancide 89RE; Vancide P-75; Vangard K; Vanguard K; Vanicide; Vondcaptan. Fungicide and bactericide, used in seed treatment, in plants, plastics, leather, fabrics, and fruit preservation; gas odorant. Used as a fungicide for natural and synthetic rubber compounds containing susceptible plasticizers, as an industrial preservative for vinyl, polyethylene, paint, lacquer, soap, wallpaper flour paste and as a bacteriostat in soap. Registered by EPA as a fungicide. Colorless crystals; mp = 178°; d = 1.74; slightly soluble in H_2O (0.00033 g/100 ml), xylene (1.7 g/100 ml), CHCl₃ (11.5 g/100 ml), Me_2CO (1.65 g/100 ml), cyclohexanone (4.7 g/100 ml), dioxane (4.8 g/100 ml), C_6H_6 (1.8 g/100 ml), C_7H_8 (0.59 g/100 ml), i-PrOH (0.13 g/100 ml), EtOH (0.23 g/100 ml), Et₂O (0.18 g/100 ml); LD$_{50}$ (rat orl) = 9000 mg/kg, (rbt der) > 4500 mg/kg, (mallard duck, pheasant orl) > 5000 mg/kg, (bobwhite quail orl) = 2000 - 4000 mg

/kg; LC50 (mus ihl 4 hr.) = 4 mg/l air, (bluegill sunfish 96 hr.) = 0.072 mg/l, (harlequin fish 96 hr.) 0.3 mg/l; non-toxic to bees. *Agriliance LLC; Arvesta Corp.; Bonide Products, Inc.; Drexel Chemical Co.; Generic; Gustafson LLC; ICI Agrochemicals; Makhteshim Chemical Works Ltd.; Micro-Flo Co. LLC; Monsanto (Solaris); Murphy Chemical Co Ltd.; Trace Chemicals LLC; UAP - Platte Chemical; Vanderbilt R.T. Co. Inc.*

669 Captopril
62571-86-2   1780   D

C9H15NO3S
1-[(2S)-3-Mercapto-2-methylpropionyl]-L-proline.
Acediur; Aceplus; Acepress; Acepril; Alopresin; Capoten; Capozide; Captolane; Captopril; L-Captopril; Captoprilum; Captopryl; Captoril; Cesplon; Dilabar; EINECS 263-607-1; Garranil; HSDB 6527; Hypertil; Lopirin; Lopril; SA 333; SQ 14,225; Tenosbon; Tensoprel.; component of: Capozide, Acezide, Captea, Ecazide. Angiotensin-converting enzyme inhibitor. Used to treat hypertension. Orally active peptidyldipeptide hydrolase inhibitor. Crystals; mp = 103-104°, 86°, 87-88°, 104-105°; $[\alpha]_D^{22}$ = -131.0° (c = 1.7 EtOH); freely soluble in H2O, EtOH, CHCl3, CH2Cl2; LD50 (mus iv) = 1040 mg/kg, (mus orl) = 6000 mg/kg. *Apothecon; Bristol-Myers Squibb Co.; Squibb E.R. & Sons.*

670 Carbadox
6804-07-5   1787   D G

C11H10N4O4
Methyl 3-(2-quinoxalinylmethylene)carbazate N^1,N^4-dioxide.
Carbadox; Carbadoxum; CCRIS 3002; EINECS 229-879-0; Fortigro; Getroxel; GS-6244; HSDB 7028; Karbadox; Mecadox. Antimicrobial, used in swine. Crystals; mp = 239-240°; λm = 236, 251, 303, 366, 373 nm (ε = 11000, 10900, 36400, 16100, 16200 H2O); almost insoluble in H2O. *Pfizer Intl.*

671 Carbamepazine
298-46-4   1788   D

C15H12N2O
5H-Dibenz[b,f]azepine-5-carboxamine.
5-20-08-00247 (Beilstein Handbook Reference); Amizepin; Biston; BRN 1246090; Calepsin; Carbamazepen; Carbamazepina; Carbamazepine; Carbamazepinum; Carbamezepine; Carbazepine;

Carbelan; EINECS 206-062-7; Epitol; Finlepsin; G-32883; Geigy 32883; HSDB 3019; Karbamazepin; Lexin; Neurotol; NSC 169864; Sirtal; Stazepine; Tegretal; Tegretol; Telesmin; Timonil. Analgesic; anticonvulsant. mp = 190.2°; λm = 214, 238, 288 nm (ε = 27542, 14125, 11220, MeOH); soluble in alcohol, Me2CO, propylene glycol; nearly insoluble in H2O; LD50 (mus orl) = 3750 mg/kg, (rat orl) = 4025 mg/kg. *Ciba-Geigy Corp.; Lemmon Co.*

672 Carbanilide
102-07-8   1792   G

C13H12N2O
Diphenylurea.
4-12-00-00741 (Beilstein Handbook Reference); Acardite; Acardite I; AD 30; AI3-52320; BRN 0782650; Carbanilide; CCRIS 4634; N,N'-Difenylmocovina; Diphenylcarbamide; 1,3-Diphenyl-carbamide; Diphenylurea; 1,3-Diphenylurea; N,N'-Diphenylurea; s-Diphenylurea; sym-Diphenylurea; EINECS 203-003-7; HSDB 2757; Karbanilid; NSC 227401; N-Phenyl-N'-phenylurea; Urea, 1,3-diphenyl-; Urea, N,N'-diphenyl-; USAF EK-534; Zeonet U. Retarder to slow down the cure of Nipol AR and HyTemp 4050 series elastomers. Also used in organic synthesis. Prisms; mp = 236°; bp = 260° (dec); λm = 210, 258 nm (ε = 8913, 24547, EtOH); insoluble in C6H6, slightly soluble in H2O (0.15 g/l), Me2CO, EtOH, CHCl3, AcOH,. *Nippon Zeon.*

673 Carbaryl
63-25-2   1794   G H

C12H11NO2
Methylcarbamic acid, 1-naphthyl ester.
4-06-00-04219 (Beilstein Handbook Reference); AI3-23969; Arilat; Arilate; Arylam; Atoxan; Bercema NMC50; BRN 1875862; Bug; Bug master; Caprolin; Carbamec; Carbamic acid, methyl-, 1-naphthyl ester; Carbamine; Carbaril; Carbarilo; Carbarilum; Carbaryl; Carbatox; Carbatox-60; Carbatox-75; Carbavur; Carbomate; Carpolin; Carylderm; Caswell No. 160; CCRIS 850; Cekubaryl; Clinicide; Compound 7744; Crag sevin; Crunch; Denapon; Derbac; Devicarb; Dicarbam; Dicarbament 23,969; Dyna-carbyl; EINECS 200-555-0; ENT 23,969; EPA Pesticide Chemical Code 056801; Experimental Insecticide 7744; Gamonil; Germain's; Hexavin; HSDB 952; Karbaryl; Karbaspray; Karbatox; Karbatox 75; Karbatox zawiesinowy; Karbosep; Latka 7744; Master; Menaphtam; Methylcarbamate 1-naphthalenol; Methylcarbamate 1-naphthol; Methylcarbamic acid, 1-naphthyl ester; Monsur; Mugan; Murvin; NAC; NAC (insecticide); Naphthyl methylcarbamate; Naphthyl N-methylcarbamate; NMC 50; Noflo 5 vet; NSC 27311; Olititox; Oltitox; OMS-29; Padrin; Panam; Patrin; Pomex; Prosevor 85; Ravyon; Rylam; Savit; Seffein; Septene; Septene; Sevimol; Sevin; Sevin 4; SOK; Suleo; Tercyl; Thinsec; Tornado; Toxan; Tricarnam; UC 7744; Union carbide 7,744; Vetox; Vioxan. Registered by EPA as an insecticide. Used as an insecticide. Used in lice infestation. Crystals; mp = 142°; d_4^{20} = 1.232; soluble in H2O (0.004 g/100 ml), DMF (40 - 45 g/100 ml), Me2CO (20 - 30 g/100 ml), cyclohexanone (20 - 25 g/100 ml), i-PrOH (10 g/100 ml), xylene (10 g/100 ml), isophorone; LD50 (rat orl) = 250 mg/kg, (mrat orl) = 850 mg/kg, (frat orl) = 500

mg/kg, (rbt orl) = 710 mg/kg, (mallard duck orl) > 2179 mg/kg, (pheasant orl) > 2000 mg/kg, (Japanese quail orl) = 2230 mg/kg, (pgn orl) = 1000 - 3000 mg/kg, (rat der) > 4000 mg/kg, (rbt der) > 2000 mg/kg; LC50 (rat ihl) = 206.1 mg/l air, (rainbow trout 96 hr.) = 1.3 mg/l, (bluegill sunfish 96 hr.) = 10 mg/l; toxic to bees. *Agriliance LLC; Amvac Chemical Corp.; Aventis Crop Science; Aventis Environmental Science USA LP; Bonide Products, Inc.; Gowan Co.; ICI Agrochemicals; Napp Labs Ltd; PET Chemicals; Rhône-Poulenc; Southern Agricultural Insecticides, Inc.; Value Gardens Supply LLC.*

674   Carbendazim
10605-21-7   1799   G P

C9H9N3O2
1H-Benzimidazol-2-ylcarbamic acid methyl ester.
2-MBC; A 118 (pesticide); Agrizim; Antibac MF; BA 67054F; BAS-3460; BAS 3460 F; BAS 67054F; Battal; Bavistan; Bavistin; Bavistin 3460; BCM; BCM (fungicide); Bengard; Bercema-Bitosen; Bitosen; BMK; BMK (fungicide); Carben VL; Carbendazim; Carbendazime; Carbendazol; Carbendazole; Carbendazym; Carbenzazim; CCRIS 1553; CTR 6669; Custos; Delsene; Derosal; E-965; EINECS 234-232-0; EK 578; EPA Pesticide Chemical Code 128872; Equitdazin; Falicarben; Funaben; Funaben 3; Funaben 50; Fungisol; G 665; Garbenda; HOE 17411; HSDB 6581; IPO-1250; IPO Y; Jkatein; Karben; Kemdazin; Kolfugo; Kolfugo 25 FW; Kolfugo Extra; MBC (VAN); Mecarzole; Medamine; Myco; NSC 109874; Pillarstin; Preparation G 665; Preventol BCM; RCRA waste no. U372; Spin; Stein; Stempor; Supercarb; Thicoper; Triticol; U-32.104. Systemic fungicide with protective and curative action for use in rape, sunflower, strawberries, vegetables. Controls a wide variety of fungal diseases in cerals, fruit and vegetables. Used against Dutch elm disease. Registered by EPA as a fungicide. Light gray powder; mp = 302-307° (dec); slightly soluble in H2O (0.0028 g/100 ml at pH 4, (0.0008 g/100 ml at pH 7, 0.0007 g/100 ml at pH 8), soluble in DMF (0.5 g/100 ml), Me2CO (0.03 g/100 ml), EtOH (0.03 g/100 ml), CHCl3 (0.01 g/100 ml), EtOAc (0.0135 g/100 ml), CH2Cl2 (0.0068 g/100 ml), C6H6 (0.0036 g/100 ml), cycohexane (< 0.001 g/100 ml), Et2O (< 0.001 g/100 ml), insoluble in C6H14; pKa = 4.48; LD50 (rat orl) > 15000 mg/kg, (dog orl) > 2500 mg/kg, (mrat ip) = 7320 mg/kg, (frat ip) = 15000 mg/kg, (rbt der) > 10000 mg/kg, (rat der) > 2000 mg/kg, (quail orl) > 10000 mg/kg; LC50 (carp 96 hr.) = 261 mg/kg, (rainbow trout 96 hr.) = 2.4 mg/l; non-toxic to bees. *Aventis Crop Science; BASF AG; DuPont; Farmers Crop Chemicals Ltd.; Fisons plc; ICI Agrochemicals; ICI Chem. & Polymers Ltd.; J.J. Mauget Co.; Pan Britannica Industries Ltd.; Rigby Taylor Ltd.; Troy Chemical Corp.*

675   Carbendazim Phosphate
52316-55-9   P

C9H12N3O6P
Methyl (1H-benzimidazol-2-yl)-carbamate, phosphate (1:1).
Carbendazim phosphate; Caswell No. 525BB; Correx; EPA Pesticide Chemical Code 099102; Lignasan BLP; Lignosan BLP;

MBC-P; Methyl 2-benzimidazole-carbamate phosphate. Registered by EPA as a fungicide. *Elm Research Institute.*

676   Carbetamide
16118-49-3   G P

C12H16N2O3
(R)-N-Ethyl-2-[[(phenylamino)carbonyl]oxy]propanamide.
11,561 RP; 2-Phenyl-carbamoyloxy-N-äthyl-propion-amid; BRN 2983326; Carbanilic acid, (1-ethylcarbamoyl)ethyl ester, D-(-)-; Carbetamex; Carbetamid; Carbetamide; Carbethamide; Caswell No. 159B; D-(-)-1-(Ethylcarbamoyl)ethyl phenylcarbamate; D-(-)-N-Ethyl-2-(phenylcarbamoyloxy)propionamide; D-N-Ethylacetamide carbanilate; (R)-1-(Ethylcarb-amoyl)ethyl carbanilate; (R)-1-(Ethylcarbamoyl)ethyl phenylcarbamate; D-N-Ethyllactamide carbanilate; (R)-N-Ethyl-2-(((phenylamino)carbonyl)oxy)propanamide; EPA Pesticide Chemical Code 259200; Lactamide, N-ethyl-, carbanilate (ester), D-; Legurame; Legurame PM; N-Phenyl-1-(ethylcarbamoyl-1)-ethylcarbamate, D isomer; (Phenylcarbamoyloxy)-2-N-ethylpropion-amide; Propanamide, N-ethyl-2-(((phenylamino)-carbonyl)oxy)-, (R)-. Herbicide. Colorless crystals; mp = 119°; soluble in H2O (0.35 g/100 ml), freely soluble in Me2CO, CH2Cl2, DMF, EtOH, MeOH; LD50 (rat orl) = 11000 mg/kg, (mus orl) = 1250 mg/kg, (dog orl) = 1000 mg/kg, (rbt der) > 500 mg/kg; LC50 (rat ihl 4 hr.) > 0.13 mg/l air; non-toxic to bees. *Aventis Crop Science; Rhône-Poulenc.*

677   Carbidopa
28860-95-9   1806   D

C10H14N2O4.H2O
(-)-L-α-Hydrazino-3,4-dihydroxy-α-methylhydrocinn-amic   acid monohydrate.
Carbidopa; Carbidopum; CCRIS 5093; EINECS 249-271-9; HMD; Hydrazino-α-methyldopa; α-Methyldopahydrazine; Lodosin; Lodosyn; MK 486; MK-486; N-Aminomethyldopa.; component of: Sinemet. Decarboxylase inhibitor. Used as an antiparkinsonian agent. mp = 203-205° (dec), 208°; [α]D = -17.3 (MeOH); [DL-form]: mp = 206-208° (dec); λm = 282.5 nm (ε 2940 MeOH). *DuPont-Merck Pharm.; Lemmon Co.*

678   Carbobenzoxy-L-aspartic acid
1152-61-0   H

C12H13NO6

N-Benzyloxycarbonyl-L-aspartic acid.

Aspartic acid, N-(benzyloxy)carbonyl-; Aspartic acid, N-carboxy-, N-benzyl ester, L-; L-Aspartic acid, N-((phenylmethoxy)carbonyl)-; N-Benzyloxycarbonyl-L-aspartic acid; N-Benzyloxycarbonylaspartic acid; Carbobenzoxy-L-aspartic acid; N-Carbobenzoxy-L-aspartic acid; EINECS 214-568-4; NSC 9972; NSC 88479.

679   Carbobenzoxy chloride
501-53-1   1810   G H

C8H7ClO2

Formic acid, chloro-, benzyl ester.

Benzyl carbonochloridate; Benzyl chlorocarbonate; Benzyl chloroformate; Benzylcarbonyl chloride; Benzyloxycarbonyl chloride; Carbobenzoxy chloride; Carbonochloridic acid, phenylmethyl ester; CCRIS 2599; Chloroformic acid, benzyl ester; EINECS 207-826-2; Formic acid, chloro-, benzyl ester; HSDB 364; NSC 83466; UN1739. Used to provide the carbobenzoxy group, a protecting group widely used in peptide synthesis. Oily liquid; bp20 = 103°; d^{25} = 1.1950; soluble in Et2O, Me2CO, C6H6. *Janssen Chimica; Pentagon Chems. Ltd.; PPG Ind.*

680   Carbodiimide
622-16-2   G

C13H10N2

Diphenylcarbodiimide.

Benzenamine, N,N'-methanetetraylbis-; Carbodiimide, diphenyl-; Diphenylcarbodiimide;   EINECS   210-721-4;   N,N'-Diphenylcarbodiimide; NSC 159432; Stabilizer 2013-P®. Activator; improves resistance to hydrolysis by acids, bases, and hot water in vulcanizates based on millable PU. Solid; mp = 169°; bp = 331°. Bp20 = 175°; slightly soluble in H2O. EtOH, Et2O, soluble in C6H6. *TSE Industries.*

681   Carbofuran
1563-66-2   1813   G H P

C12H15NO3

2,2-Dimethyl-2,2-dihydrobenzofuranyl-7 N-methylcarbamate.

5-17-04-00048 (Beilstein Handbook Reference); AI3-27164; BAY 70143; BAY 78537; Brifur; BRN 1428746; C2292-59A; Carbodan; Carbofuran; Carbofurane; Caswell No. 160A; CCRIS 4017; Chinufur; Crisfuran; Curaterr; D 1221; EINECS 216-353-0; ENT 27,164; EPA Pesticide Chemical Code 090601; FMC 10242; Furacarb; Furadan; Furadan 3G; Furadan 4F; Furadan G; HSDB 1530; Karbofuranu; Kenofuran; ME F248; NIA 10242; Niagara 10242; Niagara NIA-10242; NSC 167822; OMS-864; Pillarfuran; Rampart; Yaltox. Granular soil-applied insecticide for use in sugar beet, tobacco, maize, rice, sugar cane and vegetables. Cholinesterase inhibitor. A systemic carbamate insecticide and nematode. Registered by EPA as an miticide and insecticide. Used as a miticide and nematicide. White crystals; mp = 150-153°; sg^{20} = 1.18; soluble in H2O (0.032 g/100 ml); Me2CO (11.8 g/100 ml), CH3CN (10.9 g/100 ml); CH2Cl2 (16 g/100 mol); cyclohexanone (8.46 g/100 ml); C6H6 (3.52 g/100 ml), EtOH (3.16 g/100 ml), DMSO (27.5 g/100 ml), DMF (25.4 g/100 ml); LD50 (rat orl) = 8.2 - 14.1 mg/kg, (mus orl) = 2 mg/kg, (dog orl) = 19 mg/kg, (rat der) > 3000 mg/kg, (ckn orl) = 25 - 39 mg/kg; LC50 (gpg ihl) = 0.043 - 0.053 mg/l air, (pheasant 10 day dietary) = 960 mg/kg diet, (rainbow trout 96 hr.) = 0.28 mg/l, (bluegill sunfish 96 hr.) = 0.24 mg/l; toxic to bees. *Bayer AG; Bayer Corp.; Chemisco; FMC Corp.; Griffin LLC; Makhteshim Chemical Works Ltd.; Sipcam; Universal Crop Protection Ltd.; Whitmire Micro-Gen Research Laboratories Inc.*

682   Carbofuran 7-phenol
1563-38-8   H

C10H12O2

2,3-Dihydro-2,2-dimethyl-7-benzofuranol.

AI3-27488; 7-Benzofuranol, 2,3-dihydro-2,2-dimethyl-; Carbofuran 7-phenol; Carbofuran phenol; EINECS 216-350-4; HSDB 5839; NIA 10272; RCRA waste no. U367.

683   Carbon
7440-44-0   1817   D P

C

C

Carbon.

Acticarbone; Activated carbon; Activated carbon, decolorizing; Activated charcoal; Adsorbit; AG 3 (Adsorbent); AG 5; AG 5 (Adsorbent); AK (Adsorbent); Amoco PX 21; Anthrasorb; Aqua nuchar; AR 3; ART 2; AU 3; BAU; BG 6080; Black 140; Black pearls; Calcotone Black; Canesorb; Carbolac; Carbon; Carbon-12; Carbon, activated; Carbon, Activated or Decolorizing; Carbon, animal or vegetable origin; Carbon, colloidal; Carbon-^{12}C (amorphous) (^{13}C-depleted); Carbopol Extra; Carbopol M; Carbopol Z 4; Carbopol Z Extra; Carboraffin; Carborafine; Carbosieve; Carbosorbit R; Caswell No. 161; Cecarbon; CF 8 (Carbon); Charcoal; Charcoal, activated; CLF II; CMB 50; CMB 200; Coke powder; Colgon BPL; Colgon PCB 12X30; Colgon PCB-D; Columbia LCK; Conductex; CUZ 3; CWN 2; Darco; EINECS 231-153-3; EPA Pesticide Chemical Code 016001; Filtrasorb; Filtrasorb 200; Filtrasorb 400; Graphite, synthetic; Grosafe; HSDB 5037; Hydrodarco; Irgalite 1104; Jado; K 257; MA 100 (Carbon); Norit; Nuchar; OU-B; Pelikan C 11/1431a; SKG; SKT; SKT (adsorbent); SU 2000; Suchar 681; Supersorbon IV; Supersorbon S 1; U 02; UN1362; Watercarb; Witcarb 940; XE 340; XF 4175L. Registered by EPA as an insecticide and plant growth regulator. FDA GRAS; Used in medicine as an antidote, adsorber, for odor control and refining of pharmaceutical chemicals. Carbon black banned by FDA for use in foods, drugs or cosmetics. Insoluble in H2O, organic solvents. *Allied Signal; Calgon Carbon; Ceca SA; Elf Atochem N. Am.; Norit Am.; Sigma-Aldrich Fine Chem.; United Catalysts Inc.; Westvaco.*

684 Carbon black
1333-86-4   H

C
Carbon, amorphous.
Acetylene black; Animal bone charcoal; Aroflow; Arogen; Arotone; Arovel; Arrow; Atlantic; Black Kosmos 33; Black pearls; Cancarb; Carbodis; Carbolac; Carbolac 1; Carbomet; Carbon, amorphous; Carbon black; Carbon black, acetylene; Carbon black, channel; Carbon black, furnace; Carbon black, lamp; Carbon black, thermal; Carbon Black BV and V; CCRIS 7235; Channel Black; Char, from refuse burner; CI 77266; CI Pigment Black 6; CI Pigment Black 7; CK3; Collocarb; Columbia carbon; Conductex 900; Continex; Corax A; Corax P; Croflex; Crolac; Degussa; Delussa Black FW; Durex O; Eagle Germantown; EINECS 215-609-9; ELF 78; Elftex; Essex; Excelsior; Explosion Acetylene Black; Explosion Black; Farbruss; Fecto; Flamruss; Furnace black; Furnal; Furnex; Furnex N 765; Gas Black; Gas-furnace black; Gastex; HSDB 953; Huber; Humenegro; Impingement Black; Impingement carbons; Ketjenblack EC; Kosmink; Kosmobil; Kosmolak; Kosmos; Kosmotherm; Kosmovar; Lamp black; Magecol; Metanex; Micronex; Miike 20; Modulex; Mogul; Mogul L; Molacco; Monarch 1300; Monarch 700; Neo-spectra II; Neo Spectra Beads AG; Neo-Spectra Mark II; Neotex; Niteron 55; Oil-furnace Black; P 33 (carbon black); P68; P1250; Peach black; Pelletex; Permablak 663; Philblack; Philblack N 550; Philblack N 765; Philblack O; Pigment Black 6; Pigment Black 7; Printex; Printex 60; Raven; Raven 30; Raven 420; Raven 500; Raven 8000; Rebonex; Regal; Regal 99; Regal 300; Regal 330; Regal 400R; Regal 600; Regal SRF; Regent; Royal spectra; Sevacarb; Seval; Shawinigan Acetylene Black; Shell carbon; Special Black 1V & V; Special schwarz; Spheron; Spheron 6; Statex; Statex N 550; Sterling MT; Sterling N 765; Sterling NS; Sterling SO 1; Super-carbovar; Super-spectra; Superba; Therma-atomic Black; Thermal Acetylene Black; Thermal black; Thermatomic; Thermax; Thermblack; Tinolite; TM 30; Toka Black 4500; Toka Black 5500; Toka Black 8500.

685 Carbon dioxide
124-38-9   1819   D G P

O=C=O

CO2
Carbonic acid anhydride.
AER Fixus; After-damp; Anhydride carbonique; Carbon dioxide; Carbon Oxide; Carbon Oxide, di-; Carbonic acid anhydride; Carbonic acid gas; Carbonic anhydride; Carbonica; Caswell No. 163; Dioxido de carbono; Dioxyde de carbone; Dry ice; EINECS 204-696-9; EPA Pesticide Chemical Code 016601; HSDB 516; Khladon 744; Kohlendioxyd; Kohlensäure; Kohlensaure; Makr carbon dioxide; R 744; UN1013; UN1845; UN2187. Refrigeration, carbonated beverages, Aerosol®® propellant, chemical intermediate, fire extinguishing, inert atmospheres, municipal water treatment, medicine, oil wells, mining, blowing agent. Registered by EPA as an insecticide and plant growth regulator. Used medically as a respiratory stimulant. Colorless, odorless, non-combustible gas; bp = -56.6°; d^{25} (gas) = 0.742; solid carbon dioxide is stable at temperatures < -78.5°, where it sublimes; soluble in H2O (171 ml gas/100 ml H2O at 0°, 88 ml/100 ml at 20°, 36 ml/100 ml at 60°), less soluble in EtOH, other organic solvents. Regarded as essentially non-toxic, but has long-term exposure limits of 5000 ppm (8 hour TWA) and 15000 ppm short-term (10 minutes) in the UK, and 5000 (long term), 10000 (short term) and 30000 (short term maximum) ppm in the US. *ADM; Air Products & Chemicals Inc.; Biosensory,*

Inc.; BOC Gases; Cytec; D.I.E. Corp; G.H.G. Co. Inc.; Nissan Chem. Ind.; Norsk Hydro AS; Praxair Inc; Showa Denko.

686 Carbon disulfide
75-15-0   1821.   G H

S=C=S

CS2
Carbon bisulfide.
4-03-00-00395 (Beilstein Handbook Reference); AI3-08935; BRN 1098293; Carbon bisulfide; Carbon bisulphide; Carbon disulfide; Carbon disulphide; Carbone (sulfure de); Carbonio (solfuro di); Caswell No. 162; CCRIS 5570; Dithiocarbonic anhydride; EINECS 200-843-6; EPA Pesticide Chemical Code 016401; HSDB 52; Kohlendisulfid (schwefel-kohlenstoff); Koolstofdisulfide (zwavelkoolstof); NCI-C04591; RCRA waste number P022; Schwefel-kohlenstoff; Solfuro di carbonio; Sulfure de carbone; Sulphocarbonic anhydride; Sulphuret of carbon; UN 1131; Weeviltox; Wegla dwusiarczek. Used in manufacture of viscose rayon, cellophane, carbon tetrachloride and flotation agents, used as a solvent. Volatile liquid; mp = -111.5°; bp = 46°; d^{20} = 1.2632; soluble in H2O, more soluble in organic solvents. *Akzo Chemie; PPG Ind.; Rhône Poulenc Surfactants; Rhône-Poulenc Environmental Prods. Ltd.*

687 Carbon monoxide
630-08-0   1823   H

$^-C{\equiv}O^+$

CO
Carbon monoxide.
Carbon monoxide; Carbon oxide (CO); Carbone (oxyde de); Carbonic oxide; Carbonio (ossido di); EINECS 211-128-3; Exhaust gas; Flue gas; HSDB 903; Kohlenmonoxid; Kohlenoxyd; Koolmonoxyde; NA9202; Oxyde de carbone; UN1016; Wegla tlenek. Gas; mp = -205°; bp = -191.5°; d^{-19} = 0.7909, slightly soluble in H2O, soluble in C6H6, AcOH.

688 Carbon tetrachloride
56-23-5   1826   H

Cl
|
Cl—C—Cl
|
Cl

CCl4
Tetrachloromethane.
AI3-04705; Benzinoform; Carbon chloride (CCl4); Carbon tetrachloride; Carcona; Caswell No. 164; CCRIS 123; Chlorid uhlicity; Czterochlorek wegla; EINECS 200-262-8; ENT 27164; ENT 4,705; EPA Pesticide Chemical Code 016501; Fasciolin; Flukoids; Freon 10; Halon 1040; HSDB 53; Methane tetrachloride; Methane, tetrachloro-; Necatorina; Necatorine; NSC 97063; Perchloromethane; R 10; R 10 (Refrigerant); RCRA waste number U211; Tetrachloorkoolstof; Tetrachloormetaan; Tetrachlor-kohlenstoff, tetra; Tetrachlormethan; Tetrachloro-methane; Tetrachlorure de carbone; Tetraclorometano; Tetracloruro di carbonio; Tetrafinol; Tetraform; Tetrasol; UN1846; Univerm; Vermoestricid. Used as a refrigerant, in metal degreasing and dry-cleaning, as an agricultural fumigant, in chlorinating organic compounds, production of semiconductors and solvents. Mobile liquid; mp = -23°; bp = 76.7°; d_4^{25} = 1.589; soluble in H2O (1 ml/2000 ml), soluble in organic solvents; LC50 (mus inh) = 9528 ppm. *Ashland; Janssen Chimica; Kemichrom; Mitsubishi Chem. Corp.;*

689 Carbonochloridothioic acid, S-(phenyl-methyl) ester
37734-45-5   H

C_8H_7ClOS
Carbonochloridothioic acid, S-(phenylmethyl) ester.

690 Carborundum
409-21-2   8566   G

Si—C

SiC
Silicone carbide.
Annanox CK; Betarundum; Betarundum ST-S; Betarundum UF; Betarundum ultrafine; Carbofrax M; Carbogran; Carbogran E; Carbogran UF; Carbolon; Carbomant; Carbon silicide; Carborex; Carborundeum; Carborundum; Carsilon; CCRIS 7813; Crystar; Crystolon 37; Crystolon 39; Densic C 500; DU-A 1; DU-A 2; DU-A 3; DU-A 3C; DU-A 4; EINECS 206-991-8; GC 10000; Green densic; Green densic GC 800; Hitaceram SC 101; HSDB 681; KZ 3M; KZ 5M; KZ 7M; Meccarb; Nicalon; SC 201; SC 9; SC 9 (Carbide); SCW 1; SD-GP 6000; SD-GP 8000; Silicon carbide; Silicon monocarbide; Silundum; Tokawhisker; UA 1; UA 2; UA 3; UA 4; UF 15; YE 5626. Used in abrasive, refractories and electrical industries. UF grade used in sintered ceramics and for abrasion resistant surfaces. Green-black solid; d = 3.23. *Lonzagroup.*

691 Carbostyril
59-31-4   1835   G

C_9H_7NO
2-Hydroxyquinoline.
AI3-00782; o-Aminocinnamic acid lactam; Carbostyril; CCRIS 4327; EINECS 200-420-6; α-Hydroxyquinoline; 2-Hydroxyquinoline; NSC 156783; 2-Quinolinol; 2-Quinolinone; α-Quinolone; 2-Quinolone; 2(1H)-Quinolinone; 2(1H)-Quinolone. Crystals; mp = 199-200°; slightly soluble in H_2O; soluble in EtOH, Et_2O, HCl. *Rutgers Organics GmbH.*

692 Carbosulfan
55285-14-8   1836   G P

$C_{20}H_{32}N_2O_3S$
2,3-Dihydro-2,2-dimethyl-7-benzofuranyl[(dibutylamino)thio]methyl carbamate.
Advantage; AI3-29259; BRN 1397995; Carbamic acid, ((dibutylamino)thio)methyl-, 2,2-dimethyl-2,3-dihydro-7-benzofuranyl ester; Carbosulfan; Caswell No. 463C; Dibutylaminosulfenylcarbofuran; ((Dibutylamino)thio)-methylcarbamic acid, 2,2-dimethyl-2,3-dihydro-7-benzofuranyl ester; 2,3-Dihydro-2,2-dimethyl-7-benzofuranyl (di-n-butylaminosulfenyl)methylcarb-amate; 2,3-Dihydro-2,2-dimethyl-7-benzofuryl((di-butylamino)thio)methylcarbamate; EINECS 259-565-9; EPA Pesticide Chemical Code 090602; FMC 35001; Marshal; Marshall; Posse; RCRA waste no. P189; ((Dibutylamino)thio)methylcarbamic acid 2,3-dihydro-2,2-dimethyl-7-benzofuranyl ester; 2,3-Dihydro-2,2-dimethyl-7-benzofuranyl ((dibutylamino)thio)methyl-carbamate; 2,3-Dihydro-2,2-dimethyl-7-benzofuranyl (di-n-butylaminosulfenyl)methylcarbamate; EPA Pesticide Chemical Code 090602; FMC 35001; Marshal; Marshall; Posse; RCRA waste no. P189. Systemic insecticide with contact and stomach action. Used for control of soil-dwelling and foliar insects in a wide variety of crops. Brown liquid; bp = 124 - 128°; sg^{20} = 1.956; almost insoluble in H_2O (0.000003 g/100 ml), soluble in xylene, C_6H_{14}, $CHCl_3$, CH_2Cl_2, MeOH, Me_2CO and most other solvents; LD_{50} (mrat orl) = 250 mg/kg, (frat orl) = 185 mg/kg, (rbt der) > 2000 mg/kg, (mallard duck orl) = 8.1 mg/kg, (quail orl) = 82 mg/kg, (pheasant orl) = 26 mg/kg; LC_{50} (bluegill sunfish 96 hr.) = 0.015 mg/l, (trout 96 hr.) = 0.042 mg/l; toxic to bees. *FMC Corp.*

693 Carboxin
5234-68-4   1837   G P

$C_{12}H_{13}NO_2S$
1,4-Oxathiin-3-carboxamide, 5,6-dihydro-2-methyl-N-phenyl-.
5-19-07-00251 (Beilstein Handbook Reference); BRN 0983249; Carboxanilido-2,3-dihydro-6-methyl-1,4-oxathiin; 5-Carboxanilido-2,3-dihydro-6-methyl-1,4-oxathiin; Carboxin; CCRIS 5217; Carbathiin; Carboxin; Carboxine; Caswell No. 165ACerevaxCerevax Extra; D-735; DCMO; 5,6-Dihydro-2-methyl-1,4-oxathiin-3-carboxanilide; 5,6-Dihydro-2-methyl-N-phenyl-1,4-oxathiin-3-carboxamide; DMOC; DMOC (VAN); Dual Murganic RPB; Enhance; Enhance Plus; EPA Pesticide Chemical Code 090201; F 735; Flo Pro V Seed Protectant; Germate Plus; HSDB 1532; Kemikar; Kisvax; Krisvax; Mist-O-Matic Murganic; Murganic; NSC 263492; 1,4-Oxathiin, 2,3-dihydro-5-carboxanilido-6-methyl-; 1,4-Oxathiin-3-carbox-anilide, 5,6-dihydro-2-methyl-; 1,4-Oxathiin-3-carboxamide, 5,6-dihydro-2-methyl-N-phenyl-; Oxatin; Pro-Gro; RTV Vitavax; V 4X; Vitaflo 250; Vitaflow; Vitavax; Vitavax 30-C; Vitavax 34; Vitavax 75W; Vitavax 100; Vitavax-200; Vitavax-300; Vitavax 735D; Vitavax-Thiram-Lindane; Vivatex. Systemic fungicide used as a seed treatment for control of smuts, bunts and seedling diseases. Registered by EPA as a fungicide. Colorless crystals, two forms; mp = 93-95°, and 98 -100°; soluble in H_2O (0.0175 g/100 ml, at 25°); soluble in DMSO (165 g/100 ml), Me_2CO (47 g/100 ml), MeOH (16.8 g/100 ml), C_6H_6 (13.2 g/100 ml), EtOH (8.7 g/100 ml); LD_{50} (rat orl) = 3820 mg/kg, (rbt der) > 8000 mg/kg; LC_{50} (rat ihl 1 hr.) > 20 mg/l air, (mallard duck 8 day dietary) > 4640 mg/kg diet, (bobwhite quail 8 day dietary) > 10000 mg/kg diet, (rainbow trout 96 hr.) = 2 mg/l, (bluegill sunfish 96 hr.) = 1.2 mg/l; non-toxic to bees. *Crompton Corp.; Diachem; Gustafson LLC; Jin Hung; Kemira Agro OY; Kemira Kemi UK; Trace Chemicals LLC; Uniroyal.*

694 Carboxymethylcellulose sodium
9004-32-4   1840   G

R_nOCH_2COONa
Carboxymethyl ether cellulose sodium salt.
7H3SF; AC-Di-sol. NF; Aku-W 515; Aquaplast; Avicel RC/CL; B 10; B 10 (Polysaccharide); Blanose BS 190; Blanose BWM; Camellose

117

gum; Carbose 1M; Carboxymethyl cellulose; Carboxymethyl cellulose, sodium salt; Carmellose gum; Carmethose; CCRIS 3653; Cekol; Cellofas; Cellofas B; Cellofas B5; Cellofas B50; Cellofas B6; Cellofas C; Cellogel C; Cellogen 3H; Cellogen PR; Cellogen WS-C; Cellpro; Cellufix FF 100; Cellufresh; Cellugel; Cellulose, carboxymethyl ether; Cellulose carboxymethyl ether, sodium salt; Cellulose glycolic acid, sodium salt; Cellulose gum; Cellulose sodium glycolate; Celluvisc; CM-Cellulose sodium salt; CMC; CMC 2; CMC 3M5T; CMC 41A; CMC 4H1; CMC 4M6; CMC 7H; CMC 7H3SF; CMC 7L1; CMC 7M; CMC 7MT; CMC sodium salt; Collowel; Copagel PB 25; Courlose A 590; Courlose A 610; Courlose A 650; Courlose F 1000G; Courlose F 20; Courlose F 370; Courlose F 4; Courlose F 8; Daicel 1150; Daicel 1180; Edifas B; Ethoxose; Fine Gum HES; Glikocel TA; KMTs 212; KMTs 300; KMTs 500; KMTs 600; Lovosa; Lovosa 20alk.; Lovosa TN; Lucel (polysaccharide); Majol PLX; Modocoll 1200; NaCm-cellulose salt; Nymcel S; Nymcel slc-T; Nymcel ZSB 10; Nymcel ZSB 16; Orabase; Orabase; Polyfibron 120; S 75M; Sanlose SN 20A; Sarcell tel; Sodium carboxmethylcellulose; Sodium cellulose glycolate; Sodium CM-cellulose; Sodium CMC; Sodium CMC; Sodium glycolate cellulose; Sodium salt of carboxymethylcellulose; Tylose 666; Tylose C; Tylose C 1000P; Tylose C 30; Tylose C 300; Tylose C 600; Tylose CB 200; Tylose CB series; Tylose CBR 400; Tylose CBR series; Tylose CBS 30; Tylose CBS 70; Tylose CR; Tylose CR 50; Tylose DKL; Unisol RH; Thylose; Xylomucine; Tylose MGA; Cellolax; Polycell; CMC; cellulose gum (CTFA); sodium carboxymethylcellulose; sodium CMC. Thickening agent. Used in oil well drilling muds, detergents as a soil-suspending agent, wallpaper adhesives, ceramic glazes, emulsion paints, emulsion paints, adhesives, inks, textile sizes; as a protective colloid. White solid; mp >300°. *Berol Nobel AB; Courtaulds Water Soluble Polymers; Delavau J. W. S.; FMC Corp.; Hercules Inc.*

695 Cardolite® NC-507
501-24-6   G

C21H36O
3-(n-Pentadecyl)phenol.
Anacardol, tetrahydro-; Cyclogallipharaol; EINECS 207-921-9; Hydrocardanol; Hydroginkgol; 3-Pentadecylphenol; 3-n-Pentadecylphenol; m-Pentadecylphenol; NSC 9781; Phenol, 3-pentadecyl-; Phenol, m-pentadecyl-; Tetrahydroanacardol. Starting raw material for surfactants, antioxidants, anticorrosives; lubricant additive; cosolvent for insecticides, germicides; resin modifier. Needles, mp = 53.5°; bp$_8$ = 230°, bp$_1$ = 190-195°; very soluble in Me2CO, C6H6, EtOH. *Whitecourt.*

696 Carisoprodol
78-44-4   1854   D

C12H24N2O4
2-Methyl-2-propyl-1,3-propanediol carbamate iso-propylcarbamate.
4-04-00-00520 (Beilstein Handbook Reference); Apesan; Artifar; Arusal; Atonalyt; Brianil; BRN 1791537; Calenfa; Caprodat; Caridolin; Carisol; Carisoma; Carisoprodate; Carisoprodatum; Cariso-prodol; Carisoprodolo; Carisoprodolum; Carlsodol; Carlsoma;

Carlsoprol; Carsodal; Carsodol; CB 8019; CCRIS 4764; Chinchen; Diolene; Domarax; EINECS 201-118-7; Fibrosona; Flexagilt; Flexagit; Flexal; Flexartal; Flexidon; HSDB 3021; Isobamate; Isomeprobamate; Isopropyl meprobamate; Isoprotan; Isoprotane; Isoprothane; Izoprotan; Listaflex; Mediquil; Meprodat; Mioartrina; Miolisodal; Mioratrina; Mioril; Mioriodol; Muslax; NCI-C56235; Neotica; Nospasm; NSC-172124; Rela; Relasom; Sanoma; Sch 7307; Scutamil-C; Skutamil; Soma; Somadril; Somalgit; Somanil; Stialgin. Skeletal muscle relaxant. Crystals; mp = 92-93°; poorly soluble in H2O (0.03 g/100 ml at 25°, 0.140 g/100 ml at 50°), soluble in organic solvents; LD$_{50}$ (mus orl) = 2340 mg/kg, (mus ip) = 980 mg/kg, (rat orl) = 1320 mg/kg, (rat ip) = 450 mg/kg. *Schering-Plough HealthCare Products; Wallace Labs.*

697 Carmine
1390-65-4   G

C23H22O13
Alum carmine.
B Rose liquid; Carmine; Carmine (Coccus cacti L.); CCRIS 1204; Cochineal (Coccus cacti L.); Cochineal extract lake; EINECS 215-724-4; FEMA No. 2242; FEMA No. 2330. Aluminum lake of the coloring agent, cochineal; cochineal is a natural pigment derived from the dried female insect Coccus cacti; dyes, inks, indicator in chemical analysis, coloring food, medicine. *Aceto Corp.; Greeff R.W. & Co.; Penta Mfg.; Warner-Jenkinson.*

698 Carminic acid
1260-17-9   1855   G

C22H20O13
7-α-D-Glucopyrosanosyl-9,10-dihydro-3,5,6,8-tetra-hydroxy-1-methyl-9,10-dioxo-2-anthracenecarboxylic acid.
AI3-18242; 2-Anthracenecarboxylic acid, 7-beta-D-glucopyranosyl-9,10-dihydro- 3,5,6,8-tetrahydroxy-1-methyl-9,10-dioxo; 2-Anthroic acid, 7-D-glucopyrosyl-9,10-dihydro-3,5,6,8-tetrahydroxy-1-methyl-9,10-dioxo-; C.I. 75470; C.I. Natural red 4; Carmine; Carminic acid; CCRIS 1397; CI 75470; CI Natural Red 4; Coccus cacti extract; Cochineal extract; Cochineal tincture; E 120; EINECS 215-023-3; 7-D-Glucopyranosyl-9,10-dihydro-3,5,6,8-tetrahydroxy-1-methyl-9,10-dioxo-2-anthroic acid; 7β-D-Gluco-pyranosyl-9,10-dihydro-3,5,6,8-tetrahydroxy-1-methyl-9,10-dioxo-2-anthracenecarboxylic acid; HSDB 912; Natural red 4; NSC 326224; San-Ei Gen San Red 1; Sanred 1; Sun Red 1; Sun Red No. 1. Aluminum lake of the coloring agent, cochineal; cochineal is a natural pigment derived from the dried female insect Coccus cacti; dyes, inks, indicator in chemical analysis, coloring food, medicine. Red prisms; mp = 136° (dec); [α]$_{D}^{15}$50 = +51.6°; (C = 1, H2O); λ$_m$ = 222, 277, 312, 499 nm (ε =

13900, 24100, 7630, 5610, MeOH), 300, 336, 560 nm (ε = 15400, 13800, 5790, MeOH-KOH); insoluble in C_6H_6, $CHCl_3$, slightly soluble in Et_2O, soluble in H_2O, EtOH. *Aceto Corp.; Greeff R.W. & Co.; Penta Mfg.*; Aceto; Greef; Mfg.; Penta; R.; RITA; W.; Warner-Jenkinson.

699   Carnidazole
42116-76-7   1861   G D V

$C_8H_{12}N_4O_3S$
O-Methyl [2-(2-methyl-5-nitroimidazol-1-yl)ethyl]-thiocarbamate.
5-23-05-00062 (Beilstein Handbook Reference); BRN 0620260; Carnidazol; Carnidazole; Carnidazolum; EINECS 255-663-0; ME 108; NSC 293873; R 25831; Spartrix. Antiprotozoal against Trichomonas. Used in veterinary medicine, especially with pigeons. Crystals; mp = 142.4°. *Abbott Laboratories Inc.; Janssen Chimica.*

700   Caro's acid
7722-86-3   1864   G

H_2O_5S
Peroxymonosulfuric acid.
Caro's acid; EINECS 231-766-6; Peroxymonosulfuric acid; Peroxomonosulphuric acid. Used in dye manufacture, oxidizing agent, bleaching. Viscous liquid.

701   Carotene
7235-40-7   1866   D G

$C_{40}H_{56}$
β-Carotene.
Betacaroteno; Betacarotenum; 'β Carotene; β-Carotene, all-trans-; trans-β-Carotene; β,β-Carotene; all-trans-β-Carotene; C.I. 75130; C.I. Food Orange 5; Carotaben; CCRIS 3245; CI 40800; CI 75130; Cyclohexene, 1,1'-(3,7,12,16-tetramethyl-1,3,5,7,9, 11,13,15,17-octadecanonaene-1,18-diyl)bis(2,6,6-tri-methyl-, (all-E)-; Diet,β-carotene supplementation; E160A; EINECS 230-636-6; Food orange 5; HSDB 3264; Karotin; KPMK; Natural Yellow 26; NSC 62794; Provatene; Provitamin A; Serlabo; Solatene; Solatene (caps); 1,1'-(3,7,12,16-Tetramethyl-1,3,5,7,9,11,13, 15,17-octadecanonaene-1,18-diyl)bis(2,6,6-trimethylcyclohexene), (all E)-; Zlut prirodni 26. Color additive for foods; vitamin A precursor; UV screen. Purple crystals; mp = 183°; λ_m 497, 466 nm; soluble in CS_2, C_6H_6, $CHCl_3$; moderatley soluble in Et_2O, petroleum ether, oils; soluble in C_6H_{14} (0.11 g/100 ml at 0°); poorly soluble in MeOH, EtOH; insoluble in H_2O. *BASF Corp.; Hoffmann-LaRoche Inc.; Lancaster Synthesis Co.; Penta Mfg.; Warner-Jenkinson.*

702   Carvacrol
499-75-2   1887   D G

$C_{10}H_{14}O$
2-Methyl-5-(1-methylethyl)phenol.
AI3-03438; Antioxine; BRN 1860514; Carvacrol; Caswell No. 511; CCRIS 7450; EINECS 207-889-6; EPA Pesticide Chemical Code 022104; FEMA No. 2245; HSDB 906; Isopropyl-o-cresol; Karvakrol; NSC 6188; Oxycymol; o-Thymol. Found in oil of origanum; thyme, marjoram, summer savory. Used in perfumes; fungicides, disinfectant, flavoring, organic synthesis.Anthelmintic. Targets nematodes. Used as a general disinfectant and in organic synthesis. Liquid; mp 0°; bp = 237-238°; bp_{18} = 118-122°; bp_3 = 93°; d_4^{20} = 0.976; d_{25}^{25} = 0.9751; λ_m = 277.5 (log ε 3.262 in EtOH); volatile with steam; nearly insoluble in H_2O; soluble in alcohol, Et_2O; LD_{50} (rbt orl) = 100 mg/kg.

703   Carvediol
72956-09-3   1888   D

$C_{24}H_{26}N_2O_4$
(±)-1-(Carbazol-4-yloxy)-3-[[2-(o-methoxyphenoxy)ethyl]amino]-2-propanol.
BM 14190; Carvedilol; Carvedilolum; Coreg; Dilatrend; Dimitone; DQ 2466; Eucardic; HSDB 7044; Kredex; Querto; SKF 105517. Non-selective β-adrenergic blocker with vasodilating activity. Antianginal; antihypertensive. Crystals; mp = 114-115°. *Boehringer Mannheim GmbH; SmithKline Beecham Pharm.*

704   Carvone
99-49-0   1889   G

$C_{10}H_{14}O$
2-Methyl-5-(1-methylethenyl)-2-cyclohexene-1-one.
4-07-00-00316 (Beilstein Handbook Reference); 6,8(9)-p-Menthadien-2-one; AI3-08877; BRN 1364206; Carvol; Carvone; 1-Carvone; 2-Cyclohexen-1-one, 2-methyl-5-(1-methylethenyl)-; d-p-Mentha-1(6),8-dien-2-one; Δ^1-Methyl-4-isopropenyl-6-cyclo-hexen-2-one; EINECS 202-759-5; FEMA Number 2249; HSDB 707; Karvon; p-Mentha-6,8-dien-2-one; 2-Methyl-5-(1-methylethenyl)-2-cyclohexen-1-one; 2-Methyl-5-isopropenyl-2-cyclohexenone; NCI-C55867; NSC 6275; $\Delta^{6,8}$-(9)-Terpadienone-2; DL - [99-49-0], D-(+) - [2244-16-8], L-(-) - [6485-40-1]. Used as a flavorant in liqueurs,in perfumery and soaps. D-(+): bp = 228-230°; d = 0.962; L-(-): bp = 228-230°; d = 0.959; DL: bp = 230-231°; d = 0.9645. *Lancaster Synthesis Co.; Penta Mfg.*

705 Cashew nut oil
8007-24-7   H

Cashew nut oil.
Cashew nut oil; Cashew nut shell liquid; Cashew nut shell oil (untreated); Cashew, nutshell liq.; CCRIS 7898; EINECS 232-355-4; Oils, cashew nuts

706 Cassella's acid
92-40-0   G

$C_{10}H_8O_4S$
7-Hydroxy-2-naphthalenesulfonic acid.
Cassella's acid; EINECS 202-153-0; F Acid; 2-Hydroxynaphthalene-7-sulfonic acid; 7-Hydroxy-2-naphthalenesulfonic acid; 7-Hydroxynaphthalene-2-sulphonic acid; Mono Acid F; Mono F Acid; Monosulfonic acid F; β-Naphthol-7-sulfonic acid; 2-Naphthalenesulfonic acid, 7-hydroxy-; 2-Naphthol-7-sulfonic acid; 2-Naphthol, 7-sulfo-; NSC 1704. Used as an intermediate for azo dyes.
Crystals; mp 89°; soluble in H_2O, EtOH, insoluble in C_6H_6, Et_2O.

707 Castor Oil
8001-79-4   D G P

Castor oil.
Aceite de ricino; Aromatic castor oil; Castor oil; Castor oil (*Ricinus communis* L.); Castor oil aromatic; Castor oil [Oil, edible]; Caswell No. 165B; CCRIS 4596; Cosmetol; Crystal O; DB Oil; EINECS 232-293-8; EPA Pesticide Chemical Code 031608; FEMA No. 2263; Gold bond; HSDB 1933; Huile de ricin; LAVCO; NCI-C55163; Neoloid; Oil from seeds of *Ricinus communis.*; Oil of Palma christi; Oleum Ricini; Olio di ricino; Palma christi oil; Phorbyol; Ricini oleum; Ricinol; Ricinus communis oil; Ricinus oil; Tangantangan oil; Triglyceride of ricinoleic, oleic, linoleic, palmitic, stearic acids; Vegetable oil; Viscotrol C. Registered by EPA as an animal repellent and rodenticide. Used as an animal repellent. FDA, FEMA GRAS, FDA approved for injectables, orals, topicals, USP/NF, BP, JP, Ph.Eur. compliance. Used as an oleaginous vehicle, emollient, lubricant, plasticizer, laxative, solvent in intramuscular injectables, solid oral dosage forms, topical pharmaceuticals, capsules and emulsions. Soothing to the eyes, cathartic and purgative. Pale yellow, viscous liquid; mp = -12°; bp = 313°; d = 0.961; soluble in EtOH, AcOH, $CHCl_3$, Et_2O, insoluble in H_2O. Moderately toxic by ingestion. *Acme-Hardesty Co.; Amber Syn.; Arista Ind.; Ashland; CasChem; Charkit; Climax Performance; Degen; Fanning; Greeff R.W. & Co.; Harcros; Lipo; Norman-Fox; Ruger; Sigma-Aldrich Fine Chem.; Spectrum Chem. Manufacturing; Süd-Chemie; Van Waters & Rogers; Zeochem.*

708 Catechol
120-80-9   8089   G H

$C_6H_6O_2$
1,2-Dihydroxybenzene.
4-06-00-05557 (Beilstein Handbook Reference); AI3-03995; Benzene, o-dihydroxy-; 1,2-Benzenediol; BRN 0471401; Catechol; CCRIS 741; CI 76500; CI Oxidation Base 26; 1,2-Dihydroxybenzene; ortho-Dihydroxybenzene; Durafur developer C; EINECS 204-427-5; Fouramine PCH; Fourrine 68; HSDB 1436;

Kachin; Katechol; NCI-C55856; NSC 1573; Oxyphenic acid; Pelagol Grey C; Phthalhydroquinone; Pyrocatechin; Pyrocatechine; Pyrocatechinic acid; Pyrocatechol; Pyrocatechuic acid; Pyrokatechin; Pyrokatechol. Used as an antiseptic, in photography, dyestuffs, electroplating; specialty inks, light stabilizers. Crystals; mp = 105°; bp = 245°, bp400 = 221.5°; bp200 = 197.7°; bp100 = 176°; bp60 = 161.7°; bp40 = 150.6°; bp20 = 134°; bp10 = 118.3°; bp5 = 104°; d^{22} = 1.1493; λ_m = 278 nm (MeOH); very soluble in H_2O, EtOH, Et_2O, C_6H_6. *Madis Dr. Labs; Sigma-Aldrich Fine Chem.; Snia UK; Spectrum Chem. Manufacturing.*

709 Caustic baryta
17194-00-2   980   G

BaH_2O_2
Barium hydroxide.
Barium dihydroxide; Barium hydroxide; Barium hydroxide (Ba(OH)$_2$); Barium hydroxide lime; Caustic baryta; EINECS 241-234-5; HSDB 1605. Used in glass manufacture, rubber vulcanization, corrosion inhibitors, drilling fluids and lubricants. [monohdrate]; White powder; d = 3.743; slightly soluble in H_2O, soluble in acids; [octahydrate]; crystals; mp = 78°; freely soluble in H_2O, MeOH, soluble in EtOH, insoluble in Me_2CO.

710 Cefadroxil
66592-87-8   1925   G D

$C_{16}H_{17}N_3O_5S \cdot H_2O$
[6R-[6α,7β(R*)]]-7-[[Amino-(4-hydroxyphenyl)acetyl]-amino]-3-methyl-8-oxo-5-thia-1-azabicyclo[4.2.0]oct-2-ene 2-carboxylic acid monohydrate.
Baxan; Baxan Kapseln; Bidocef; Bidocel; BL-S 578; Cefa-Drops; Cefadroxil; Cefadroxil monohydrate; Cefamox; Ceforal; Cefos Granulat; Cephos; Duracef; Duricef; Kefroxil; Kefroxil Kapseln; MJF 11567-3; Moxacef Kapseln; Omnidrox Kapseln; Oracéfal; Sedral; Ultracef. Antibacterial. A semisynthetic cephalosporin antibiotic intended for oral admin-istration. It is a white to yellowish-white crystalline powder. It is soluble in water and it is acid-stable. Powder, mp = 197° (dec). *Bristol-Myers Squibb Pharm. R&D; Mead Johnson Labs.*

711 Cefprozil [anhydrous]
92665-29-7   1955   D

$C_{18}H_{19}N_3O_5S$
[6R-[6α,7β(R*)]]-7-[[Amino(4-hydroxyphenyl)acetyl]-amino]-8-oxo-3-(1-propenyl)-5-thia-1-azabicyclo-[4.2.0]oct-2-ene-2-carboxylic acid.

Arzimol; BMY 28100; Brisoral; Cefprozil; Cefprozil anhydrous; Cefprozilo; Cefprozilum; Cronocef; Procef; Serozil. Antibacterial. *Bristol Laboratories.*

712 Cefprozil
121123-17-9   1955   D

$C_{18}H_{19}N_3O_5S.H_2O$
[6R-[6α,7β(R*)]]-7-[[Amino(4-hydroxyphenyl)acetyl]-amino]-8-oxo-3-(1-propenyl)-5-thia-1-azabicyclo-[4.2.0]oct-2-ene-2-carboxylic acid monohydrate.
BBS-1067; BMY-28100 [Z form]; BMY-28100-03-800; BMY-28167 [E form]; Cefprozil hydrate; Cefzil; Procef. Antibacterial. [Z Form]: mp = 218-220° (dec); λm 228, 279 nm (ε 12300, 9800 pH 7 phosphate buffer); [E Form]: mp = 230° (dec); λm 228, 292 nm (ε 13000, 16900 pH 7 phosphate buffer). *Bristol Laboratories.*

713 Celecoxib
169590-42-5   1968   D

$C_{17}H_{14}F_3N_3O_2S$
4-[5-(4-Methylphenyl)-3-(trifluoromethyl)-1H-pyrazol-1-yl]benzenesulfonamide.
Celebra; Celebrex; Celecoxib; HSDB 7038; SC 58635; YM 177; YM177. A COX-2 inhibitor. Anti-inflammatory, used as an NSAID, especially in pain control in arthritis. *Searle G.D. & Co.*

714 Celluflow C-25
9004-34-6   1977   E G

$(C_6H_{10}O_5)x$
Cellulose.
α Cel PB 25; α-Cellulose; β-Amylose; Abicel; Arbocel; Arbocel BC 200; Arbocell B 600/30; Avicel; Avicel 101; Avicel 102; Avicel CL 611; Avicel PH; Avicel PH 101; Avicel PH 105; Avicel RC/CL; CCRIS 6600; Cellex MX; Cellulose; Cellulose 248; Cellulose crystalline; Cellulose, microcrystalline; Cellulose regenerated; Cellulose, respirable fraction; Cellulose, total dust; Celufi; Cepo CFM; CEPO; CEPO S 20; CEPO S 40; Chromedia CC 31; Chromedia CF 11; Cupricellulose; EINECS 232-674-9; Elcema F 150; Elcema G 250; Elcema P 050; Elcema P 100; Fresenius D 6; Heweten 10; Hydroxycellulose; Kingcot; LA 01; Microcrystalline cellulose; MN-Cellulose; Onozuka P 500; Pyrocellulose; Rayon; Rayon flock; Rayophane; Rayweb Q; Rexcel; Sigmacell; Solka-fil; Solka-floc; Solka-floc BW 20; Solka-floc BW 100; Solka-floc BW; Solka-floc BW 2030; Solka-floc BW 200; Spartose OM-22; Sulfite cellulose; Tomofan; Tunicin; Whatman CC-31; Wood pulp, bleached. Cosmetic ingredient with excellent oil absorbency, moisture retention, high lubricity for powders, emulsions and anhydrous systems. FDA GRAS, Europe listed, UK approved, USP/NF, BP, Ph.Eur., Japan

compliance. Used as a tablet binder and disintegrant, lubricant, capsule/tablet filler and diluent, filler, filter aid, sorbent, suspending agent, viscosity-increasing agent. Insoluble in H_2O and organic solvents. *Presperse.*

715 Cephalexin [anhydrous]
15686-71-2   1986   D

$C_{16}H_{17}N_3O_4S$
[6R-[6α,7β(R*)]]-7-[(Aminophenylacetyl)amino]-3-methyl-8-oxo-5-thia-1-azabicyclo[4.2.0]oct-2-ene-2-carboxylic acid.
Alcephin; Alexin; Alsporin; Biocef; BRN 0965503; Carnosporin; Cefa-iskia; Cefablan; Cefadal; Cefadin; Cefadina; Cefaleksin; Cefalessina; Cefalexin; Cefalexina; Cefalexine; Cefalexinum; Cefalin; Cefaloto; Cefaseptin; Cefax; Ceforal; Cefovit; Celexin; Cepastar; Cepexin; Cephacillin; Cephalexin; Cephalexine; Cephalexinum; Cephanasten; Cephaxin; Cephin; Cepol; Ceporex; Ceporex Forte; Ceporexin; Ceporexin-E; Ceporexine; Check; Cophalexin; Durantel; Durantel DS; Ed A-Ceph; EINECS 239-773-6; Erocetin; Factagard; Felexin; Fexin; HSDB 3022; Ibilex; Ibrexin; Inphalex; Kefalospes; Keflet; Keflex; Kefolan; Keforal; Kekrinal; Kidolex; L-Keflex; Lafarine; Larixin; Lenocef; Lexibiotico; Lilly 66873; Lonflex; Lopilexin; Madlexin; Mamalexin; Mamlexin; Medoxine; Neokef; Neolexina; Novolexin; Nufex; Oracef; Oriphex; Oroxin; Ortisporina; Ospexin; Palitrex; Pectril; Pyassan; Roceph; Roceph Distab; S 6437; Sanaxin; Sartosona; Sencephalin; Sepexin; Servispor; Sialexin; Sinthecillin; Sporicef; Sporidex; SQ 20248; Syncl; Syncle; Synecl; Tepaxin; Tokiolexin; Uphalexin; Voxxim; Winlex; Zozarine.; Alfaspoven [as sodium salt]. Antibacterial. White solid; λm = 260 nm (ε 7750). *Eli Lilly & Co.*

716 Cephalexin hydrochloride
105879-42-3   1986   D

$C_{16}H_{20}ClN_3O_5S$
[6R-[6α,7β(R*)]]-7-[(Aminophenylacetyl)amino]-3-methyl-8-oxo-5-thia-1-azabicyclo[4.2.0]oct-2-ene-2-carboxylic acid hydrochloride monohydrate.
Cephalexin hydrochloride; Cephalexin monohydro-chloride monohydrate; Keftab; LY061188. Anti-bacterial. Crystals; LD50 (mus orl) = 1600-4500 mg/kg, (mus ip) = 400-1300 mg/kg, (rat orl) > 5000 mg/kg, (rat ip) > 3700 mg/kg. *Eli Lilly & Co.*

717 Cephalexin monohydrate
23325-78-2   1986   D

C16H17N3O4S.H2O

[6R-[6α,7β(R*)]]-7-[(Aminophenylacetyl)amino]-3-methyl-8-oxo-5-thia-1-azabicyclo[4.2.0]oct-2-ene-2-carboxylic acid monohydrate.

Adcadina; Ambal; Beliam; Cefa-Iskia; Cefacet; Cefalekey; Cefalexgobens; Cefalexin generics; Cefalexin Scand Pharm; Cefalexina Northia; Cefalexina Richet; Cefalival; Ceffanex; Cefibacter; Cephalex von ct; Cephalexin monohydrate; Cephalobene; Ceporex; Ceporexin; Cusisporina-Cefalox; Domucef; Doriman; Efemida; Henina Oral; Karilexina; Keflet; Keflex; Keforal; Maksipor; Medolexin; Oracef; Ortisporina; Prindex; Rilexine; Sartosona; Servicef; Servispor; Sintolexyn. Antibacterial. λm = 260 nm (ε = 7750); pKa = 5.2, 7.3; [monohydrate]: LD50 (rat orl) > 5000 mg/kg. *Bristol-Myers Oncology; Eli Lilly & Co.; Lemmon Co.*

718   Cerium
7440-45-1   2003   G

Ce

Ce
Cerium.
Cerium; EINECS 231-154-9; UN3078. A rare-earth element; used in manufacture of cerium salts, cerium-iron alloys, ignition devices, military signaling, illuminant in photography, reducing scavenger, catalyst, alloys for jet engines, solid state devices, rocket propellants, vacuum tubes, Metal; mp = 815°; bp = 3257°; d = 6.77. *Cerac; Ferro/Transelco; Rhône-Poulenc; Sigma-Aldrich Fine Chem.*

719   Cerium sulfate
10294-42-5   G

CeO8S2.4H2O
Ceric sulfate tetrahydrate.
Sulfuric acid, cerium(4+) salt (2:1), tetrahydrate. Dyeing and printing textiles, analytical reagent, waterproofing, mildewproofing. *Atomergic Chemetals; Noah Chem.; Rhône-Poulenc.*

720   Cerous acetate
537-00-8   H

C6H9CeO6
Acetic acid, cerium(3+) salt.
Acetic acid, cerium(3+) salt; Cerium acetate; Cerium triacetate; Cerium(3+) acetate; Cerium(III) acetate; Cerous acetate; EINECS 208-654-0.

721   Cesium
7440-46-2   2018   G

Cs

Cs
Cesium.
Caesium; Cesium; Cesium-133; EINECS 231-155-4; UN1407. An alkali-metal (Group I) element; Cs; used in photoelectric cells, vacuum tubes, hydrogenation catalyst, ion propulsion systems, rocket propellant, heat transfer fluid, thermochemical reactions. Metal; mp = 28.5°, bp = 706°. *Atomergic Chemetals; Cabot Carbon Ltd.; Cerac; Sigma-Aldrich Fine Chem*

722   Cesium bromide
7787-69-1   2019   G

Cs—Br

CsBr
Hydrobromic acid cesium salt.
Caesium bromide; Cesium bromide; Cesium bromide (CsBr); EINECS 232-130-0; NSC 84269; Tricesium tribromide. Crystals for infrared spectroscopy, scintillation counters, fluorescent screens. Crystals; mp = 636°; bp = 1300°; d = 4.44; very soluble in H2O, soluble in EtOH, insoluble in Me2CO; LD50 (rat ip) =1.4 g/kg. *Cabot Carbon Ltd.; Cerac; Noah Chem.*; 3973Atomergic; Cabot; Cerac; Chemetals; Chemical; Noah.

723   Cesium carbonate
534-17-8   2020   G

CCs2O3
Cesium(I) carbonate.
Carbonic acid, dicesium salt; Cesium carbonate (Cs2CO3); Dicesium carbonate; EINECS 249-784-8; NSC 112218. Used in brewing, mineral waters, in specialty glasses, as a polymerization catalyst for ethylene oxide. Soluble in H2O, organic solvents. *Cabot Carbon Ltd.; Cerac; Sigma-Aldrich Fine Chem.*

724   Cesium chloride
7647-17-8   2021   G

Cs—Cl

CsCl
Hydrochloric acid cesium salt.
Caesium chloride; Cesium chloride; Cesium chloride (CsCl); Cesium monochloride; Dicesium dichloride; EINECS 231-600-2; NSC 15198; Tricesium trichloride. Used in brewing; preparation of cesium compounds; mineral waters; evacuation of radio tubes; in ultracentrifuge separations; fluorescent screens; contrast medium. Crystals; mp = 646°; bp = 1303°; d = 3.99; LD50 (rat ip) = 1.5 g/kg. *Accurate Chem. & Sci. Corp.; Atomergic Chemetals; Cabot Carbon Ltd.; Cerac; Janssen Chimica.*

725   Cesium fluoride
13400-13-0   G

Cs—F

CsF
Hydrofluoric acid cesium salt.

122

Caesium fluoride; Cesium fluoride; Cesium fluoride (CsF); Cesium monofluoride; Dicesium difluoride; EINECS 236-487-3; NSC 84270; Tricesium trifluoride. Used in optics, catalysis, specialty glasses. *Atomergic Chemetals; Cerac; Noah Chem.; Spectrum Chem. Manufacturing.*

726   Cesium hydroxide
21351-79-1   2022   G

Cs—OH

CsOH
Cesium hydroxide.
Caesium hydroxide; Caesium hydroxide, solution; Cesium hydrate; Cesium hydroxide; Cesium hydroxide (Cs(OH)); EINECS 244-344-1; NSC 121987; UN2681; UN2682. Electrolyte in alkaline storage batteries (especially at subzero temperatures), polymerization catalyst for siloxanes. Yellow-white crystals; mp = 272°; d = 3.68; LD_{50} (rat ip) = 100 mg/kg. *Atomergic Chemetals; Cabot Carbon Ltd.; Cerac; Noah Chem.; Sigma-Aldrich Fine Chem.*

727   Cesium iodide
7789-17-5   2023   G

Cs—I

CsI
Hydriodic acid cesium salt.
Caesium iodide; Cesium iodide; Cesium iodide (CsI); Cesium monoiodide; Dicesium diiodide; EINECS 232-145-2; NSC 15199; Tricesium triiodide. Crystals for infrared spectroscopy, scintillation counters, fluorescent screens. Crystals; mp = 621°, bp = 1280°; d = 4.5; LD_{50} (rat ip) = 1.4 g/kg. *Atomergic Chemetals; Cabot Carbon Ltd.; Cerac; Noah Chem.*

728   Cesium nitrate
7789-18-6   2024   G

CsNO3
Nitric acid cesium salt.
Caesium nitrate; Cesium nitrate; Cesium(I) nitrate (1:1); EINECS 232-146-8; Nitric acid, cesium salt; NSC 84271; UN1451. Crystals; mp = 414°; d_4^{20} 3.64-3.68; LD_{50} (rat ip) = 1.2 g/kg. *Cabot Carbon Ltd.; Cerac; Noah Chem.*

729   Cesium sulfate
10294-54-9   2025   G

Cs2O4S
Sulfuric acid cesium salt.
Caesium sulphate; Cesium sulfate; Dicesium sulfate; EINECS 233-662-6; Sulfuric acid, dicesium salt. Used in brewing, mineral waters, for density gradient in ultracentrifuge separation. Crystals; mp = 1019°; d = 4.24; soluble in H_2O, insoluble in organic solvents. *Atomergic Chemetals; Cabot Carbon Ltd.; Cerac; Sigma-Aldrich Fine Chem.*

730   Cetalkonium chloride
122-18-9   2026   D G

$C_{25}H_{46}ClN$
Benzylhexadecyldimethylammonium chloride.
Acinol; AI3-03464; Ammonium, benzyldimethyl-hexadecyl-, chloride; Ammonyx G; Ammonyx T; Baktonium; Banicol; Benzaletas; Benzenemethan-aminium, N-hexadecyl-N,N-dimethyl-, chloride; Benzylcetyldimethylammonium chloride; N-Benzyl-N-cetyldimethylammonium chloride; Benzyldimethyl-cetylammonium chloride; Benzyldimethylhexadecyl-ammonium chloride; Benzylhexadecyldimethyl-ammonium chloride; Bicetonium; Bonjela; Cdbac; Cetalconio cloruro; Cetalkonii chloridum; Cetalkonium chloride; Cetol; Cetyl dimethyl benzyl ammonium chloride; Cetyl zephiran; Cetylbenzyldimethyl-ammonium chloride; Cetylon; Chlorure de cetalkonium; Cloruro de cetalconio; Dehyquart CBB; Dehyquart CDB; Dimethylbenzylcetylammonium chloride; Dimethylbenzylhexadecylammonium chloride; Dmcbac; EINECS 204-526-3; Hexadecyl-benzyldimethylammonium chloride; Hexadecyl-dimethylbenzylammonium chloride; n-Hexadecyl-dimethylbenzylammonium chloride; N-Hexadecyl-N,N-dimethylbenzenemethanaminium chloride; NSC 32942; Pharycidin concentrate; Rodalon; Spilan; Sumquat® 6050; Tetraseptan; WIN 357; Winzer solution; Zettyn. Cationic quaternary ammonium surfactant germicide and fungicide. Used as a topical anti-infective. White solid; mp = 59°; soluble in H_2O, EtOH, Me_2CO, EtOAc, propylene glycol, sorbitol solutions, glycerol, Et_2O, CCl_4; pH of aqueous solutions 7.2. *I.G. Farben; ICI; Sterling Winthrop Inc.; Zeeland Chem. Inc.*

731   Cetene
629-73-2   G H

$C_{16}H_{32}$
Hexadec-1-ene.
AI3-06556; Cetene; 1-Cetene; Cetylene; Dialene 16; EINECS 211-105-8; Gulftene 16; α-Hexadecene; 1-Hexadecene; 1-n-Hexadecene; n-Hexadec-1-ene; Hexadecylene-1; α-Hexadecylene; HSDB 5730; NSC 60602. Intermediate for biodegradable surfactants and specialty industrial chemicals. Liquid; mp = 2.1°; bp = 284.9°; d^{20} = 0.7811; insoluble in H_2O, soluble in EtOH, Et_2O, CCl_4, petroleum ether. *Shell.*

732   Ceteth-15
9004-95-9   E G

$C_{56}H_{114}O_{21}$
POE (15) cetyl ether.
Alfonic 16-8; Akyporox RC 200; Atlas G 3802; Atlas G 3816; BC 2; BC 7; BC 10; BC 20; BC 20 TX; BC 30 TX; BC 8SY; Berol 28; Blaunon CH 340; BRIJ 38; BRIJ 52; BRIJ 56; BRIJ 58; BRIJ W1; C 30 (polyoxyalkylene); CA 16; Ceteth; Ceteth-1; Ceteth-2; Ceteth-3; Ceteth-4; Ceteth-5; Ceteth-6; Ceteth-12; Ceteth-14; Ceteth-15; Ceteth-16; Ceteth-20; Ceteth-24; Ceteth-25; Ceteth-30; Ceteth-45; Cetocire; Cetomacrogol; Cetomacrogol 1000; Cetomacrogol 1000 BPC; Cetomacrogolum 1000; Cetyl alcohol, ethoxylated; Cetyl poly(oxyethylene) ether; Cirrasol ALN-WF; Collone AC; Emalex 103; Emalex 105; Emalex 110; Emalex 115; Emalex 120; Emulgen 210; Emulgen 2200; Ethosperse CL20; Ethoxylated cetyl alcohol; Ethoxylated hexadecyl alcohol; Ethoxylated(20) cetyl alcohol; Ethylene glycol monocetyl ether; G 3802; G 3802POE; G 3804POE; G 3816; G 3820; Glycols, polyethylene, monohexadecyl ether (8CI); 1-Hexadecanol, monoether with polyethylene glycol; α-Hexadecyl-ω-hydroxy-poly(oxy-1,2-ethanediyl); α-Hydroxy-ω-cetylpoly(ethylene oxide); Hexadecylpoly-(ethyleneoxy) ethanol;

Lipocol C2; Lubrol W; Nikkol BC; Nikkol BC 15TX; Nikkol BC 20TX; Nikkol BC 30; Nikkol BC 40; Nissan nonion P 208; Nissan nonion P 210; Nissan nonion P 213; Nissan nonion P 220; Nonion P 208; Nonion P 210; NTS 40; OS 30; OS 55; OTsS 14; OTsS 20; Oxyethylenated hexadecyl alcohol; PEG-1 Cetyl ether; PEG-3 Cetyl ether; PEG-4 Cetyl ether; PEG-5 Cetyl ether; PEG-6 Cetyl ether; PEG-12 Cetyl ether; PEG-14 Cetyl ether; PEG-15 Cetyl ether; PEG-16 Cetyl ether; PEG-20 Cetyl ether; PEG-24 Cetyl ether; PEG-25 Cetyl ether; PEG-30 Cetyl ether; PEG-45 Cetyl ether; Poly(oxy-1,2-ethanediyl), α-hexadecyl-ω-hydroxy-; Poly(oxyethylene)cetyl ether; Poly(oxy-ethylene)hexadecyl ether; Poly(oxyethylene)monocetyl ether; Poly(oxyethylene)palmityl ether; Polyethylene glycol (3) cetyl ether; Polyethylene glycol (5) cetyl ether; Polyethylene glycol (14) cetyl ether; Polyethylene glycol (15) cetyl ether; Polyethylene glycol (16) cetyl ether; Polyethylene glycol (24) cetyl ether; Polyethylene glycol (25) cetyl ether; Polyethylene glycol (30) cetyl ether; Polyethylene glycol (45) cetyl ether; Polyethylene glycol 200 cetyl ether; Polyethylene glycol 300 cetyl ether; Polyethylene glycol 600 cetyl ether; Polyethylene glycol 1000 cetyl ether; Polyethylene glycol 1000 monocetyl ether; Polyethylene glycol cetyl ether; Polyethylene glycol monocetyl ether; Polyethylene glycol palmityl ether; Polyethylene glycols monohexadecyl ether; Polyethylene oxide cetyl ether; Polyethylene oxide hexadecyl ether; Polyoxyethylated cetyl alcohol; Polyoxyethylene (3) cetyl ether; Polyoxyethylene (4) cetyl ether; Polyoxyethylene (5) cetyl ether; Polyoxyethylene (6) cetyl ether; Polyoxyethylene (12) cetyl ether; Polyoxyethylene (14) cetyl ether; Polyoxyethylene (15) cetyl ether; Polyoxyethylene (16) cetyl ether; Polyoxyethylene (20) cetyl ether; Polyoxyethylene (24) cetyl ether; Polyoxyethylene (25) cetyl ether; Polyoxyethylene (30) cetyl ether; Polyoxyethylene (45) cetyl ether; Polyoxyethylene monohexadecyl ether; Romopal O; S 30; Ts 14; Ts 20; Ts 30; Ts 35; Ts 35 (polymer); Ts 40; Ts 55; Ts 62; Ts A 16. Solubilizer for cosmetic products. Emulsifier for pharmaceuticals. *Chem-Y GmbH.*

733   Cetirizine Dihydrochloride
83881-52-1   2030   D

C21H25ClN2O3.2HCl
(±)-[2-[4-(p-Chloro-α-phenylbenzyl)-1-piperazinyl]ethoxy]acetic acidetirizine hydrochloride.
Alercet; Alergex; Alerid; Alerlisin; Alertisin; Cetirizine hydrochloride; Cetriler; Cetrine; Cetzine; Cezin; Formistin; P 071; Reactine; Ressital; Riztec; Ryzen; Salvalerg; Setiral; Stopaler; Triz; UCB-P 071; Virdos; Virlix; Xero-sed; Zirtek; Zirtin; Zyrlex; Zyrtec; Zyrtec-D; Zyrzine. Antihistaminic. Crystals; mp = 225°. *Pfizer Inc.*

734   Cetrimonium bromide
57-09-0   2034   G

C19H42BrN
Cetyl trimethyl ammonium bromide.
(1-Hexadecyl)trimethylammonium bromide; 1-Hexadecanaminium, N,N,N-trimethyl-, bromide; Acetoquat CTAB; AI3-12209; Ammonium, hexadecyltrimethyl-, bromide; Bromat; Bromure de cetrimonium; Bromuro de cetrimonio; C.T.A.B.; Caswell No. 167; Cee dee; Centimide; Cetab; CETAB; Cetaflon; Cetarol; Cetavlex; Cetavlon;

Cetavlon bromide; Cetrimide; Cetrimide bp; Cetrimonii bromidum; Cetrimonio bromuro; Cetylamine; Cetyltrimethyl-ammonium bromide; N-Cetyl-trimethylammonium bromide; Cirrasol OD; CTAB; CTABr; Ctmab; Cycloton V; EINECS 200-311-3; EPA Pesticide Chemical Code 069117; Hexadecyl trimethyl ammonium bromide; n-Hexadecyltrimethylammonium bromide; HTAB; Lauroseptol; Lissolamin V; Lissolamine; Lissolamine A; Lissolamine V; Mical; Micol; NSC 32927; Palmityltrimethyl ammonium bromide; Pollacid; Quamonium; Rhodaquat® M242B/99; Softex KW; Sumquat® 6030; Suticide; Trimethylcetylammonium bromide; Trimethylhexa-decylammonium bromide; Varisoft® CTB-40. Germ-icide, sanitizing agent. Quaternary ammonium comp-ound used in conditioners to increase antistatic and comb out effects. Cationic detergent and antiseptic. Also used as a laboratory reagent. mp = 237-243°; soluble in 10 parts H2O, freely soluble in EtOH, sparingly soluble in Me2CO, insoluble in Et2O, C6H6; LD50 (mus, iv) = 32.0, 44.0 mg/kg. *Aceto; Hexcel; ICI Chem. & Polymers Ltd.; Rhône Poulenc Surfactants.*

735   Cetrimonium chloride
112-02-7   G H

C19H42ClN
N,N,N-Trimethyl-1-hexadecanaminium chloride.
Adogen 444; Aliquat 6; Ammonium, hexadecyl-trimethyl-, chloride; Ammonyx Cetac 30; Arquad 16; Arquad 16-25LO; Arquad 16-25W; Arquad 16-26; Arquad 16/28; Arquad 16-29; Arquad 16-29W; Arquad 16-50; Barquat® CT 29; Carsoquat CT 429; Caswell No. 167A; Catinal CTC 70ET; Cation PB 40; CETAC; Cetrimonium chloride; Cetyltrimethyl ammonium chloride; Dehyquart A; Dehyquart A-CA; Dodigen 1383; EINECS 203-928-6; EPA Pesticide Chemical Code 069133; FSM 28; Genamin CTAC; HDTMA-Cl; Hexadecyltrimethylammonium chloride; N-Hexadecyltrimethylammonium chloride; HSDB 5553; HTAC; Intexan CTC 29; Intexsan CTC 29; Intexsan CTC 50; Lebon TM 16; Lebon TM 60; Morpan CHA; Nissan Cation PB 40; Palmityltrimethylammonium chloride; PB 40; Pionin B 611; Quartamin 60W; Quatramine C 16/29; Radiaquat 6444; Rhodaquat® M242C/29; Surfroyal CTAC; Swanol CA 2350; Trimethylcetylammonium chloride; N,N,N-Trimethyl-1-hexadecanaminium chloride; Trimethylhexadecyl-ammonium chloride; Variquat E 228; Varisoft® 250; Varisoft® 300; Varisoft® 355. Softener, antistat, detergent for textile, leather, detergent, and cosmetic industries; clay modifier. Surfactant with conditioning and emolliency effect on hair, used as a base for hair conditioners/creme rinses; antistat, surfactant, emulsifier; hair grooming aids; imparts softness and manageability without greasiness. Coagulating agent in manufacturing of antibiotics. Solid; d = 0.968. *CK Witco Corp.; Fina Chemicals; Lonzagroup; Rhône Poulenc Surfactants.*

736   Cetrimonium tosylate
138-32-9   2034   G

C26H49NO3S
Cetrimonium p-toluene sulfonate.
Cetats®; Cetyl trimethyl ammonium p-toluene sulfonate; EINECS 205-324-8; Hexadecyltrimethyl-ammonium toluene-p-sulphonate; 1-Hexadecan-aminium, N,N,N-trimethyl-, salt with 4-methyl-benzenesulfonic acid (1:1); N,N,N-Trimethyl-1-hexadecanaminium salt with 4-methylbenzenesulfonic acid; Trimethylcetylammonium p-

toluenesulfonate. Surfactant and topical antiseptic. *Zeeland Chem. Inc.*

737 Cetyl acetate
629-70-9   G

$C_{18}H_{36}O_2$
n-Hexadecyl acetate.
4-02-00-00171 (Beilstein Handbook Reference); Acelan A; Acetic acid, hexadecyl ester; 1-Acetoxyhexadecane; AI3-01025; BRN 1782695; Cetyl acetate; EINECS 211-103-7; ENT 1025; 1-Hexadecanol, acetate; n-Hexadecyl ethanoate; NSC 8492; Palmityl acetate. Liquid; mp = -18.5°; bp_{205} = 220-225°; d^{25} = 0.8574; insoluble in H_2O, slightly soluble in EtOH, soluble in CCl_4. *Fabriquimica.*

738 Cetyldimethylethylammonium bromide
124-03-8   2038   G

$C_{20}H_{44}BrN$
N-Ethyl-N,N-dimethyl-1-hexadecanaminium bromide.
AI3-12210; Alkenyl dimethyl ethyl ammonium bromide; Ammonium, ethylhexadecyldimethyl-, bromide; Ammonyx DME; Bretol; Caswell No. 165E; CDA; Cetethyldimonium bromide; Cetyl dimethyl ethyl ammonium bromide; Cetylcide; Cetyldimethylethyl-ammonium bromide; Dimethylethylhexadecyl-ammonium bromide; EINECS 204-672-8; EPA Pesticide Chemical Code 069156; Ethyl cetab; Ethyl hexadecyl dimethyl ammonium bromide; Ethylhexadecyldimethylammonium bromide; 1-Hexadecanaminium, N-ethyl-N,N-dimethyl-, bromide; Hexadecyldimethylammonium bromide; NSC 8494; Quaternium-17; Radiol germicidal solution; Sumquat® 6020. Used as a disinfectant, laboratory reagent and antiseptic in detergents. Solid; mp = 178-186° (dec); soluble in H_2O, EtOH, less soluble in organic solvents; LD_{50} (rat orl) = 500 mg/kg. *Hexcel.* -

739 Cetyl ethylene
3452-07-1   G

$C_{20}H_{40}$
1-Eicosene.
AI3-36496; Cetyl ethylene; α-Eicosene; 1-Eicosene; EINECS 222-374-6; Icos-1-ene; Neodene® 20; NSC 77138. Intermediate for biodegradable surfactants and specialty industrial chemicals. Solid; mp = 29°; bp = 342°, bp_2 = 151°; d^{30} = 0.7882; insoluble in H_2O, soluble in C_6H_6, petroleum ether. *Shell.*

740 Cetyl lactate
35274-05-6   2039   G

$C_{19}H_{38}O_3$
2-Hydroxypropionic acid hexadecyl ester.
AI3-14483; Cetyl lactate; EINECS 252-478-7; n-Hexadecyl-2-hydroxypropanoate; Hexadecyl lactate; n-Hexadecyl lactate; Propanoic acid, 2-hydroxy-, hexadecyl ester; Ceraphyl® 28. A non-ionic emollient binder for pressed powders, lipsticks, hair products Solid; mp = 41°; bp_1 = 170°. *Am. Biorganics; Van Dyk.*

741 Cetyl palmitate
540-10-3   2040   G

$C_{32}H_{64}O_2$
n-Hexadecyl hexadecanoate.
4-02-00-01168 (Beilstein Handbook Reference); BRN 1805188; Cetin; Cetyl palmitate; EINECS 208-736-6; Hexadecanoic acid, hexadecyl ester; Hexadecyl palmitate; Kemester® CP; Kessco® 653; Palmitic acid, hexadecyl ester; Palmityl palmitate; Radia® 7500; Rewowax CG; Standamul 1616; Waxenol® 815; Crodamol CP. Emollient for replacing spermaceti wax. Used as a base for ointments and in manufacture of candles and soaps, as a consistency factor in creams, ointmnets, liquid emulsions and fatty make-ups. Leaflets; mp = 54°; d^{20} = 0.989; insoluble in H_2O, very soluble in EtOH, Et_2O. *Croda Surfactants; Henkel/Cospha; Henkel; Sherex; Smith Werner G.; Stepan; Witco/Humko.*

742 Cetylpyridinium bromide
140-72-7   G P

$C_{21}H_{38}BrN$
1-Hexadecylpyridinium bromide.
Acetoquat CPB; Bromocet; Caswell No. 166; Cetapharm; Cetasol; Cetazol; Cetazolin; Cetyl pyridinium bromide; Cetylpyridinium bromide; 1-Cetylpyridinium bromide; N-Cetylpyridinium bromide; Cetylpyridine bromide; Cetylpyridinium bromide; EINECS 205-428-3; EPA Pesticide Chemical Code 069118; Fixanol C; Hexadecylpyridinium bromide; N-Hexadecylpyridinium bromide; Hexadecylpyridine bromide; HSDB 6859; Morpan CBP; Nitrogenol; NSC 8495; Pyridinium, 1-hexadecyl-, bromide; Seprisan; Sterogenal; Sterogenol; TsPB. Germicide, sanitizing agent. Crystals; mp = 67-69°. *Aceto Corp.*

743 Cetylpyridinium chloride [anhydrous]
123-03-5   2041   G D

$C_{21}H_{38}ClN$
1-Hexadecylpyridinium chloride.
Acetoquat CPC; AI3-15070; Aktivex; Ammonyx CPC; Biosept; Caswell No. 166A; Ceeprin chloride; Ceepryn; Ceepryn chloride; Cepacol; Cepacol chloride; Ceprim; Cetamium; Cetilpiridinio cloruro; Cetyl pyridinium chloride; Cetylpyridini chloridum; Cetylpyridinium chloride; Cetylpyridinium chloride anhydrous; Chlorure de

cetylpyridinium; Cloruro de cetilpiridinio; Dobendan; EINECS 204-593-9; EPA Pesticide Chemical Code 069160; Fixanol C; Hexadecylpyridinium chloride; HSDB 38; Intexsan CPC; Medilave; Merocet; NSC 14864; Pristacin; Pyrisept; Quaternario CPC; Swabettes Hoechst; Tserigel. Antiseptic, disinfectant and germicide. Used as a sanitizing agent. White powder; mp = 80°; very soluble in H_2O, $CHCl_3$; LD_{50} (rat orl) = 200 mg/kg. *Aceto Corp.; Hexcel; Lancaster Synthesis Co.; Marion Merrell Dow Inc.; Spectrum Chem. Manufacturing; Weiders Farmasoytiske A/S; Zeeland Chem. Inc*

744   Cetyl stearate

1190-63-2   G

$C_{34}H_{68}O_2$
n-Hexadecyl stearate.
 Cetyl stearate; EINECS 214-724-1; Hexadecyl stearate; 1-Hexadecyloctadecanoate; n-Hexadecyl stearate; Octadecanoic acid, 1-hexadecyl ester; Octadecanoic acid, hexadecyl ester. Used as an emollient. Leaflets; mp = 57°; very soluble in Me_2CO, Et_2O, $CHCl_3$. *Koster Keunen; Sherex.*

745   Chalkone

94-41-7   2045   G

$C_{15}H_{12}O$
1,3-Diphenyl-2-propen-1-one.
 Acrylophenone, 3-phenyl-; AI3-00946; Benzalaceto-phenone; 2-Benzalacetophenone; 1-Benzoyl-1-phenylethene; 1-Benzoyl-2-phenylethene; Benzylidene acetophenone; 2-Benzylideneacetophenone; α-Benzyl-ideneacetophenone; β-Benzoylstyrene; CCRIS 2213; Chalcone; Chalkone; Cinnamophenone; 1,3-Diphenyl-1-propen-3-one; 1,3-Diphenylpropenone; EINECS 202-330-2; NSC 26612; β-Phenylacrylophenone; 3-Phenylacrylophenone; 1-Phenyl-2-benzoylethylene; Phenyl 2-phenylvinyl ketone; Phenyl styryl ketone; 2-Propen-1-one, 1,3-diphenyl-; Styryl phenyl ketone. Pale yellow leaflets; mp = 59°; bp = 346° (dec), bp_{25} = 208°; d^{62}_2 = 1.0712; insoluble in H_2O, soluble in EtOH, Et_2O. *Showa Denko.*

746   Chavicol

501-92-8   2058   G

$C_9H_{10}O$
4-(2-Propenyl)phenol.
 4-Allylphenol; p-Allylphenol; CCRIS 3208; Chavicol; EINECS 207-929-2; p-Hydroxyallylbenzene; σ-(p-Hydroxyphenyl)-α-propylene; NSC 290195; Phenol, 4-(2-propenyl)-; Phenol, p-allyl-. Occurs in many essential oils such as volatile betel oil. Liquid; mp = 15.8°; bp = 238°, bp_{16} = 123°; insoluble in H_2O, very soluble in EtOH, Et_2O, $CHCl_3$.

747   Chevron 100

68602-80-2   H

 Aromatic petroleum distillate.
 Aromatic oil distillate; Aromatic petroleum distillate; (C_{12}-C_{30}) Aromatic oil; Caswell No. 055; Chevron 100; Distillates (petroleum), C_{12-30} aromatic; Distillates, aromatic petroleum; EINECS 271-620-9; EPA Pesticide Chemical Code 006601; Espesol 3A; Petroleum derived aromatic hydrocarbons. Registered by EPA as an insecticide. *Amvac Chemical Corp.; Quest Chemical Corp.; Valent USA Corp.*

748   Chimassorb® 119FL

106990-43-6   G

$C_{124}H_{234}N_{32}$
1,3,5-Triazine-2,4,6-triamine.
 N,N',N",N'''-Tetrakis(4,6-bis(butyl-(N-methyl-2,2,6,6-tetramethylpiperidin-4-yl)-amino)triazin-2-yl)-4,7-diazadecane-1,10-diamine. Light and thermal stabilizer for PP fiber applications in automotive, marine and residential carpets, agricultural films, fertilizer bags, thick section pigmented applications, rotational molding applications. *Ciba-Geigy Corp.*

749   Chitin

1398-61-4   2072   G P

$(C_8H_{13}NO_5)_n$
 Poly(N-acetyl-1,4-β-D-glucopyranosamine).
 Chitan, N-acetyl-; Chitin; Chitin Tc-L; Chitina; Clandosan; EINECS 215-744-3; EPA Pesticide Chemical Code 128991; Kimitsu Chitin; Regitex FA. A glucosamine polysaccharide. Principal component of exoskeletons. Used in biological research, source of chitosan. Amorphous white solid; insoluble in H_2O, organic solvents. *Ajinomoto Co. Inc.; Ajinomoto USA Inc.; Amerchol; Atomergic Chemetals; Tri-K Ind.*

750   Chloral

75-87-6   9699   H

C_2HCl_3O
 Trichloroacetaldehyde.
 4-01-00-03142 (Beilstein Handbook Reference); Acetaldehyde, trichloro-; Anhydrous chloral; BRN 0506422; CCRIS 852; Chloral; Chloral, anhydrous, inhibited; Cloralio; EINECS 200-911-5; Grasex; HSDB 2557; RCRA waste number U034; Trichloro-acetaldehyde; Trichloroethanal; UN2075. Used in chemical synthesis and as a source of the drug chloral hydrate [302-17-0]. mp = -90°; bp = 65.6°; d^{20} = 1.273; soluble in H_2O and organic solvents. *Bristol-Myers Squibb Pharm. R&D; SmithKline Beecham Pharm.*

751   Chloral Hydrate

302-17-0   2080   D

$C_2H_3Cl_3O_2$
 2,2,2-Trichloro-1,1-ethanediol.
 4-01-00-03143 (Beilstein Handbook Reference); AI3-00082; Aquachloral; Bi 3411; BRN 1698497; Caswell No. 168; CCRIS 4142; Chloradorm; Chloral hydrate; Chloral monohydrate; Chloraldural; Chloraldurat; Chloralex; Chlorali Hydras; Chloralvan; Cohidrate; DEA No. 2465; Dormal; EINECS 206-117-5; EPA Pesticide Chemical Code 268100; Escre; Felsules; HSDB 222; Hydrate de chloral; Hynos; Kessodrate; Kloralhydrat; Knockout drops; Lorinal; Noctec;

Novochlorhydrate; NSC 3210; Nycoton; Nycton; Oradrate; Phaldrone; Rectules; SK-Chloral Hydrate; Somni SED; Somnos; Sontec; Tosyl; Trawotox. Sedative/hypnotic. mp = 57°; d = 1.91; bp = 98°; freely soluble in H_2O (240 g/100 ml at 0°, 830 g/100 ml at 25°, 1430 g at 40°), EtOH (76.9 g/100 ml), $CHCl_3$ (50 g/100 ml), Et_2O (66.6 g/100 ml), glycerol (200 g/100 ml), CS_2 (1.47 g/100 ml); sparingly soluble in CCl_4, C_6H_6, C_7H_8; LD_{50} (rat orl) = 479 mg/kg. *Bristol-Myers Squibb Pharm. R&D; SmithKline Beecham Pharm.*

752   Chloramine B
127-52-6   2084   D G

$C_6H_5ClNNaO_2S$
N-Chlorobenzenesulfonamide sodium salt.
AI3-16452; Annogen; Benzene chloramine; Benzene-sulfo-sodium chloramide; Benzenesulfochloramide, N-chloro-, sodium salt; Benzenesulfochloramide sodium; Benzenesulfonamide, N-chloro-, sodium salt; Caswell No. 169; Chloramin B; Chloramine B; Chlordetal; Chlorogen; N-Chlorobenzenesulfonamide sodium; N-Chloro-N-sodiobenzenesulfonamide; EINECS 204-847-9; EPA Pesticide Chemical Code 076501; HSDB 3422; Khloramin B; Monochloramine B; Neomagnol; NSC 75446; Sodium, (chloro(phenylsulfonyl)amino)-; Sodium, (N-chlorobenzenesulfonamido)-; Sodium benzenesulfochloramine; Sodium benzosulfo-chloramide; Sodium N-chlorobenzenesulfonamide. Used like Chloramine T. Soluble in H_2O (5 g/100 ml), EtOH (4 g/100 ml), insoluble in organic solvents.

753   Chloramine T
127-65-1   2085   D G

$C_7H_7ClNNaO_2S$
N-Chloro-4-methylbenzenesulfonamide sodium salt.
Acti-chlore; AI3-18426C; Aktiven; Aktivin; Anexol; Aseptoclean; Benzenesulfonamide, N-chloro-4-methyl-, sodium salt; Berkendyl; Caswell No. 170; Chloralone; Chloramin Dr. Fahlberg; Chloramin Heyden; Chlor-amine; Chloramine T; Chlorasan; Chloraseptine; Chlor-azan; Chlorazene; Chlorazone; Chlorina Aktivin; N-Chloro-4-methylbenzenesulfonamide sodium salt; N-Chloro-4-methylbenzylsulfonamide sodium salt; Chlor-osol; N-Chlorotoluenesulfonamide sodium salt; N-Chloro-p-toluenesulfonamide sodium; Chlorozone; Chlorseptol; Cloramine T; Clorina; Clorosan; Des-infect; EINECS 204-854-7; EPA Pesticide Chemical Code 076502; Euclorina; Gansil; Gyneclorina; Halamid; Heliogen; HSDB 4303; Ketjensept; Kloramin; Kloramine-T; Mannolite; Mianin; Mianine; Mono-chloramine T; Multichlor; NSC 36959; Pyrgos; Sodium chloramine T; Sodium N-chloro-p-toluenesulfonamide; Sodium p-toluenesulfon-chloramide; Sodium p-toluenesulfonylchloramide; Sodium tosylchloramide; Tampules; Tochlorine; Tolamine; p-Toluenesulfon-amide, N-chloro-, sodium salt; Tosilcloramida sodica; Tosylchloramid-natrium; Tosylchloramide sodique; Tosylchloramide sodium; Tosylchloramidum natricum. An antiseptic used in medicine; solutions of Chloramine T are also used as detergents and bleaching agents. Crystals; mp = 167-170°; Soluble in H_2O, insoluble in organic solvents. *Akzo Chemie.*

754   Chloranil
118-75-2   2088   G

$C_6Cl_4O_2$
2,3,5,6-Tetrachloro-2,5-cyclohexadiene-1,4-dione.
AI3-03797; 1,4-Benzoquinone, 2,3,5,6-tetrachloro-; p-Benzoquinone, 2,3,5,6-tetrachloro-; Caswell No. 171; CCRIS 7155; Chloranil; p-Chloranil; Chloranile; Coversan; 2,5-Cyclohexadiene-1,4-dione, 2,3,5,6-tetrachloro-; Dow Seed Disinfectant No. 5; EINECS 204-274-4; ENT 3,797; EPA Pesticide Chemical Code 079301; G-444E; G-25804; Geigy-444E; HSDB 1533; Khloranil; NSC 8432; Psorisan; Quinone tetrachloride; Reranil; Spergon; Spergon I; Spergon technical; 2,3,5,6-Tetrachlorobenzoquinone; Tetrachloro-1,4-benzoquinone; 2,3,5,6-Tetrachloro-1,4-benzoquinone; Tetrachloro-p-benzoquinone; 2,3,5,6-Tetrachloro-p-benzoquinone; 2,3,5,6-Tetrachloro-2,5-cyclohexadi-ene-1,4-dione; Tetrachloro-p-quinone; 2,3,5,6-Tetrachloroquinone; Vulklor. Used as an agricultural fungicide, dye intermediate, electrodes for pH measurement, reagent. Used in tyre industry: combines with R6 to form an effective bonding system for compounds featuring steel cord reinforcement; used to activate GMF; also functions as a vulcanizing agent without sulfur; used in natural, SBR, nitrile, butyl and chlorobutyl rubbers. mp = 290°; bp sublimes; λ_m = 286, 364 nm (ε = 12600, 248, MeOH); insoluble in H_2O, ligroin, slightly soluble in EtOH, $CHCl_3$, CS_2, soluble in Et_2O; LD_{50} (rat orl) = 4.0 g/kg. *Ashland; Hall C.P.; Unitex.*

755   Chlordane
57-74-9   2096   P

$C_{10}H_6Cl_8$
1,2,4,5,6,7,8,8-Octachloro-2,3,3a,4,7,7a-hexahydro-4,7-Methano-1H-indene.
1068; α,σ-Chlordane; Aspon-chlordane; Belt; BELT; BRN 1915474; Caswell No. 174; CCRIS 127; CD 68; Chloordaan; Chlor Kil; Chlordan(e); Chlordane, α & γ isomers; Chlorindan; Chlortox; Clordan [Italian]; Clordano; Compound K; Corodane; Cortilan-neu; Dichlorochlordene; Dow-klor; Dowchlor; ENT 25,552-X; ENT 9,932; EPA Pesticide Chemical Code 058201; Gold Crest; HCS 3260; HSDB 802; Intox; Intox 8; Kilex; Kypchlor; Latka 1068; M 140; 4,7-Methano-1H-indene, 1,2,4,5,6,7,8,8-octachloro-2,3,3a,4,7,7a-hexahydro-; 4,7-Methanoindan, 1,2,4,5,6,7,8,8-octachloro-3a,4,7,7a-tetrahydro-; NCI-C00099; NSC 8931; Octa-Klor; Octachlor; Octachloro-4,7-methanotetrahydro-indane; Octachlorodihydrodicyclopentadiene; Octa-chlorohexahydromethanoindene; 1,2,4,5,6,7,8,8-Octa-chloor-3a,4,7,7a-tetrahydro-4,7-endo-methano-indaan; 1,2,4,5,6,7,8,8-Octachloro-3a,4,7,7a-tetrahydro-4,7-endo-methano-indan; 1,2,4,5,6,7,8,8-Octachloro-4,7-methano-3a,4,7,7a-tetrahydroindane; 1,2,4,5,6,7,10,10-Octachloro-4,7,8,9-tetrahydro-4,7-methylene-indane; 1,2,4,5,6,7,8,8-Ottochloro-3a,4,7,7a-tetraidro-4,7-endo-metano-indano[Italian]; 1-Exo,2-endo,4,5, 6,7,8,8-octachloro-2,3,3a,4,7,7a-hexahydro-4,7-meth-anoindene; Oktaterr; OMS 1437; Ortho-Klor; RCRA waste number U036; SD 5532; Shell SD-5532; Starchlor; Syndane; Synklor; Synklor; TAT chlor 4; Termi-ded;

127

Topichlor 20; Topiclor; Topiclor 20; Toxichlor; Unexan-koeder; Velsicol 1068; [5103-71-9]; [12789-03-6]. Registered by EPA as an insecticide (cancelled). Designated as a Persistent Organic Pollutant (POP) under the Stockholm convention. Brown solid - amber-colored liquid; mp = 106 - 107° (cis-form), 104 - 105° (trans-form); $bp_{1.0}$ = 175°; d^{25} = 1.59 - 1.63; viscosity = 69 poises at 25°, much lower at temperatures above 50° when it can be sprayed directly; insoluble in H_2O (< 0.00001 g/100 ml), soluble in hydrocarbon solvents; LD_{50} (rat orl) = 457 - 590 mg/kg, (mus orl) = 430 mg/kg, (rbt orl) = 300 mg/kg, (mrat ip) = 343 mg/kg, (rbt der) = 200 - 2000 mg/kg, (rat der) = 217 mg/kg, (bobwhite quail orl) = 83 mg/kg; LC_{50} (bobwhite quail 8 day dietary) = 421 mg/kg diet, (mallard duck 8 day dietary) = 795 mg/kg diet, (rainbow trout 96 hr.) = 0.09 mg/l, (bluegill sunfish 96 hr.) = 0.07 mg/l; toxic to bees. *Novartis*.

756   Chlordene
3734-48-3   H

$C_{10}H_6Cl_6$
4,5,6,7,8,8-Hexachloro-3a,4,7,7a-tetrahydro-4,7-meth-ano-1H-indene.
Addukt hexachlorcyklopentadienu S cyklopenta-dienem; AI3-15150; BRN 5749626; Chlordene; Chlordene 50; EINECS 223-096-8; 4,5,6,7,8,8-Hexa-chlor-$\Delta^{1,5}$-tetrahydro-4,7-methanoinden; 4,5,6,7,8,8-Hexachloro-3a,4,7,7a-tetrahydro-4,7-methanoindene; 4,5,6,7,8,8-Hexachloro-3a,4,7,7a-tetrahydro-4,7-meth-ano-1H-indene; 4,7-Methano-1H-indene, 4,5,6,7,8,8-hexachloro-3a,4,7,7a-tetrahydro-; 4,7-Methanoindene, 4,5,6,7,8,8-hexachloro-3a,4,7,7a-tetrahydro-; 4,7-Methanoindene, 4,5,6,7,8,8-hexachloro-$\Delta^{1,5}$-tetra-hydro-; HSDB 2800.

757   Chlorendic acid
115-28-6   H

$C_9H_4Cl_6O_4$
Hexachloro-5-norbornene-2,3-dicarboxylic acid.
CCRIS 896; Chlorendic acid; EINECS 204-078-9; HET acid; Hexachloro-endo-methylenetetrahydrophthalic acid; Hexachloroendomethylenetetrahydrophthalic acid; HSDB 2915; Kyselina 1,2,3,4,7,7-hexachlorbicyklo(2,2,1)hept-2-en-5,6-dikarboxylova; NCI-C55072; NSC 22231.

758   Chlorendic anhydride
115-27-5   2101   H

$C_9H_2Cl_6O_3$
Hexachloro-5-norbornene-2,3-dicarboxylic anhydride.
4-17-00-06070 (Beilstein Handbook Reference); BRN 0092693; Chloran 542; Chlorendic anhydride; EINECS 204-077-3; HET anhydride; Hexachloro-5-norbornene-2,3-dicarboxylic anhydride; Hexachloroendo-methylene tetrahydrophthalic anhydride; HSDB 2920; NSC 22229.

759   Chlorfenvinphos
470-90-6   2105   G P

$C_{12}H_{14}Cl_3O_4P$
Phosphoric acid 2-chloro-1-(2,4-dichlorophenyl)-ethenyl diethyl ester.
2-Chloro-1-(2,4-dichlorophenyl)-vinyl diethyl phos-phate; β-2-Chloro-1-(2',4'-dichlorophenyl)vinyl diethyl phosphate; Benzyl alcohol, 2,4-dichloro-α-(chloro-methylene)-, diethyl phosphate; Birlan; Birlane; Birlane 10G; C 8949; C-10015; Clofenvinphos; Chloro-fenvinphos; Clofenvineosum; Clofenvinfos; CVP; SD-7859; Compd. 4072; Dermaton; Sapecron; Steladone; Supona. An insecticide and acaricide with contact and stomach action. Used in seed treatment or soil application for control of root flies, rootworms, fruit flies and Colorado beetles. Used as a liquid seed dressing for winter wheat. Oil; mp = -22 - -16°; $bp_{0.05}$ = 170°; soluble in H_2O (0.015 g/100 ml), freely soluble in EtOH, Me_2CO, propylene glycol; LD_{50} (rat orl) = 9.6 mg/kg. *ICI Chem. & Polymers Ltd.*

760   Chlorhexidine acetate
56-95-1   2108   G

$C_{22}H_{30}Cl_2N_{10}.2C_2H_4O_2$
Bis(p-chlorophenyldiguanidohexane) diacetate.
10,040 Diacetate; Arlacide A; Bactigras; Biguanide, 1,1'-hexamethylenebis(5-(p-chlorophenyl)-, diacetate; 1,6-Bis(p-chlorophenylbiguanido)hexane di-acetate; N,N'-Bis(4-chlorophenyl)-3,12-diimino-2,4,11,13-tetraazatetradecanediimidamide, diacetate; Caswell No. 481E; Chlorasept 2000; Chlorhexidine acetate; Chlorhexidine diacetate; 1,6-Di(4'-chlorophenyl-diguanidino)hexane diacetate; EINECS 200-302-4; EPA Pesticide Chemical Code 045502; Hexamethylene-bis(5-(p-chlorophenyl)biguanide) diacetate; 1,1'-Hexa-methylene bis(5-(p-chloro-phenyl)biguanide) diacetate; Hibitane diacetate; Nolvasan; NSC 526936; Tetraaza-tetradec-anediimidamide, N,N''-bis(4-chlorophenyl)-3,12-diimino-, diacetate;

2,4,11,13-Tetraazatetradec-ane-diimidamide, N,N'-bis(4-chlorophenyl)-3,12-di-imino-, diacetate. Antiseptic and disinfectant. Crystals; mp = 154-155°; soluble in H$_2$O (1.9 g/100 ml), alcohols; LD$_{50}$ (mus orl) = 2 g/kg. *Hopkins.*

761   Chlorhexidine gluconate
18472-51-0   2108   D E G

C$_{34}$H$_{54}$Cl$_2$N$_{10}$O$_{14}$
1,1'-Hexamethylenebis[5-(p-chlorophenyl)biguanide]   di-D-gluconate.
Abacil; Arlacide G; Bacticlens; Caswell No. 481G; Chloraprep; Chlorhexamed; Chlorhexidin glukonatu; Chlorhexidine D-digluconate; Chlorhexidine digluconate; Chlorhexidine gluconate; Corsodyl; Disteryl; DRG-0091; Dyna-hex; EINECS 242-354-0; EPA Pesticide Chemical Code 045504; Gingisan; Hibiclens; Hibidil; Hibiscrub; Hibistat; Hibital; Hibitane; Hibitane 5; Orahexal; Peridex; Peridex (antiseptic); Periogard; pHiso-Med; Plac Out; Plurexid; Rotersept; Secalan; Septeal; Sterilon; Unisept.; component of: Hibistat. A bisbiguanide with bacteriostatic activity. Antiseptic; disinfectant. Used in dental hygene and for the prophylaxis of puerperal mastitis. Solid; soluble in H$_2$O at 20°; LD$_{50}$ (mus iv) 22 mg/kg, (mus orl) 1800 mg/kg. *Boots Pharmaceuticals Inc.; Fair Products; ICI; Poythress.*

762   Chloridazon
1698-60-8   G P

C$_{10}$H$_8$ClN$_3$O
5-Amino-4-chloro-2-phenyl-3(2H)-pyridazinone.
5-25-14-00196 (Beilstein Handbook Reference); 5-Amino-4-chloro-2,3-dihydro-3-oxo-2-phenylpyrid-azine; 5-Amino-4-chloro-2-phenylpyridazinone; 5-Amino-4-chloro-2-phenylpyridazin-3(2H)-one; 5-Amino-4-chloro-2-phenyl-3-pyridazinone; 5-Amino-4-chloro-2-phenyl-3(2H)-pyridazinone; BAS 13033; BAS 1191611; Better Flowable; BRN 0397241; Burex; Caswell No. 714C; Chloridazon; Chloridazone; Clorizol; Curbetan; EINECS 216-920-2; EPA Pesticide Chemical Code 069601; 1-Fenyl-4-amino-5-chlor-6-pyridazinon; HS 119-1; HSDB 1759; PCA; Phenazon; Phenazon (herbicide); Phenazone; Phenazone (herbicide); Phenosane; 1-Phenyl-4-amino-5-chloro-pyridaz-6-one; 1-Phenyl-4-amino-5-chlorpyridaz-6-one; 1-Phenyl-4-amino-5-chlorpyridazin-6(-6); 1-Phenyl-4-amino-5-chloropyridazin-6-one; Pyramin (herbicide); Pyramin RB; Pyramin(e); Pyramine; Pyrazon; Pyrazone; 3(2H)-Pyridazinone, 5-amino-4-chloro-2-phenyl-; Suzon; Better; Brek; Curbetan; Gladiator; H 119;

Hyzon; Pyramin; Pyrazol; Silver; Starter; Trojan. Selective systemic herbicide, rapidly absorbed by roots, used for control of annual broad-leaved weeds in sugar beets, fodder beet and beetroot. Used for pre- and post-emergence weed control in onions, leeks, chives, and flower bulbs. Crystals; mp = 205-206° (dec); soluble in H$_2$O (400 mg/l), more soluble in organic solvents; LD$_{50}$ (mrat orl) = 3830 mg/kg, (frat orl) = 2140 mg/kg, (mus orl) = 2500 mg/kg, (rbt orl) = 1250 mg/kg, LC$_{50}$ (rat ihl 4hr.) > 30.8 mg/l, (trout 96hr.) = 27mg/l; not toxic to bees. *Agrichem (International) Ltd.; Agrimont; Atlas Interlates Ltd; BASF AG; BASF Corp.; BASF plc; Chiltern Farm Chemicals Ltd; Farmers Crop Chemicals Ltd.; MTM AgroChemicals Ltd.; Portman Agrochemicals Ltd.; Rhône-Poulenc Environmental Prods. Ltd; Schering Agrochemicals Ltd.; Sipcam; Tripart Farm Chemicals Ltd; Truchem; Wacker Chemie GmbH.*

763   Chlorinated paraffin waxes
63449-39-8   H

Chlorinated paraffins.
A 70 (wax); Adekacizer E 410; Adekacizer E 450; Adekacizer E 470; ADK Cizer 450; ADK Cizer 470; ADK Cizer E 410; Aquamix 108; Arubren; Arubren CP; CCRIS 1420; CCRIS 4770; Cerechlor 54; Cereclor; Cereclor 30; Cereclor 42; Cereclor 48; Cereclor 50LV; Cereclor 51L; Cereclor 52; Cereclor 54; Cereclor 56L; Cereclor 63L; Cereclor 65L; Cereclor 70; Cereclor 70L; Cereclor S 42; Cereclor S52; Cereclor S70; Chlorcosane; Chlorez 700; Chlorez 700hmp; Chlorinated paraffin; Chlorinated paraffin waxes; Chlorinated paraffins; Chlorinated wax; Chlorinated waxes; Chloroflo 35; Chloroflo 40; Chloroflo 42; Chloroparaffine 40G; Chlorowax; Chlorowax 170; Chlorowax 40; Chlorowax 40-40; Chlorowax 45AO; Chlorowax 50; Chlorowax 70; Chlorowax 70-5; Chlorowax 70S; Chlorowax 500C; Chlorowax S 70; Clorafin; Crechlor S 45; Creclor S 45; EINECS 264-150-0; Flexchlor; HSDB 4214; NCI-C53587; Paraffin, chlorinated; Paraffin wax, chlorinated; Paroil chlorez; Unichlor; Unichlor 50 .

764   Chlorine
7782-50-5   2112   G P

Cl—Cl

Cl$_2$
Chlorine.
Bertholite; Bertholite /warfare gas/; Caswell No. 179; CCRIS 2280; Chloor; Chlor; Chlore; Chlorinated water (chlorine); Chlorine; Chlorine gas; Cloro; Dichlorine; EPA Pesticide Chemical Code 020501; HSDB 206; Molecular chlorine; UN1017. Used in the manufacture of CCl$_4$, trichloroethylene, chlorinated hydrocarbons, neoprene, PVC; water purification and disinfection; shrinkproofing wool; in flame retardant compounds; food processing; bleaching. Registered by EPA as an antimicrobial and disinfectant. Green gas; mp = -101°; bp = -34.05°; d = 2.48 (air = 1); soluble in H$_2$O. *Air Products & Chemicals Inc.; Asahi Chem. Industry; Asahi Denka Kogyo; BASF Corp.; Cerexagri Decco Inc.; Dow Chem. U.S.A.; Elf Atochem N. Am.; Georgia Pacific; Hill Brothers Chemical Co; JCI Jones Chemicals, Inc; LaRoche Ind.; Olin Corp; OxyChem; PPG Ind.; Showa Denko; SSC Chemical Corp; Vulcan.*

765   Chlorine dioxide
10049-04-4   2113   G P

ClO$_2$
Chlorine peroxide.
Alcide; Anthium dioxcide; Caswell No. 179A; Chlorine dioxide; Chlorine oxide; Chlorine oxide (ClO2); Chlorine peroxide; Chlorine(IV) oxide; Chloroperoxyl; Chloryl radical; Doxcide 50; EINECS 233-162-8; EPA Pesticide Chemical Code 020503; HSDB

517. Stabilized chlorine dioxide; broad spectrum biocide, preservative; deodorant designed to remove odors caused by residual monomers in resins; works as an oxidizer. Used to bleach wood pulp, fats and oils; biocide; odor control; water purification; oxidizing agent; bactericide, antiseptic. Registered by EPA as an antimicrobial, fungicide, herbicide and insecticide. Yellow-red gas; unstable to light; mp = -59°; bp = 11°; d^0 = 1.642; soluble in H_2O (0.302 g/100 ml at 25°). *Bio-Cide International Inc.; Chemical Cealin Co. Inc.; Conseal International Inc.; Drew Industrial Div.; International Dioxcide Inc.*

766   Chlormequat
999-81-5   2120   H P

Cl⁻   [structure]

[C5H13Cl2N]⁺
  (2-Chloroethyl)trimethylammonium.
  5C Cycocel; 60-CS-16; AC 38555; AI3-26939; Ammonium, (2-chloroethyl)trimethyl-, chloride; Antywylegacz; Arotex Extra; Barleyquat; Barleyquat B; BAS 06200W; Bercema CCC; Bettaquat; Brevis; Caswell No. 191; CCC; CCC (plant growth regulant); CCRIS 135; Cecece; CeCeCe; Chlorcholinchlorid; Chlorcholine chloride; Chlormequat; Chlormequat chloride; Chlorocholine chloride; Chloroethyl-trimethylammonium chloride; Choline dichloride; Cyclocel; Cycocel 460; Cycocel C 5; Cycocel Extra; Cycogan; Cycogan extra; Cyocel; EI 38,555; EINECS 213-666-4; EPA Pesticide Chemical Code 018101; Ethanaminium, 2-chloro-N,N,N-trimethyl-, chloride; Farmacel; Halloween; Helstone; Hico CCC; Hormocel; Hormocel-2CCC; HSDB 1541; Hyquat; Increcel; Lihocin; Mirbel 5; NCI-C02960; NSC 34858; Premix; Retacel; Stabilan; Titan; Trimethyl-β-chlorethyl-ammoniumchlorid; Trimethyl-β-chloroethyl-ammon-ium chloride; TUR; WR 62; Zar. Registered by EPA as an antimicrobial, fungicide and herbicide. Used as a plant growth regulator. Improves resistance against lodging of oats, rye, and wheat White crystals; mp = 239-243° (dec); soluble in H_2O, lower alcohols, in-soluble in Et_2O, hydrocarbons; LD_{50} (rat orl) = 600 mg/kg, (rat ip) = 64 mg/kg, (mus orl) = 54 mg/kg, (mus ip) = 62 mg/kg, (mus iv) = 7 mg/kg. *ABM Chemicals Ltd.; Agan Chemical Manufacturers; BASF AG; BASF Corp.; Farmers Crop Chemicals Ltd.; Sigma-Aldrich Co.*

767   Chloroacetic acid
79-11-8   2129   G H

[structure]

C2H3ClO2
  Monochloroacetic acid.
  4-02-00-00474 (Beilstein Handbook Reference); Acetic acid, chloro-; Acide chloroacetique; Acide monochloracetique; Acidomonocloroacetico; AI3-25035; BRN 0605438; Caswell No. 179B; CCRIS 2117; Chloroacetic acid; Chloroacetic acid, molten; Chloroacetic acid, solid; Chloroacetic acid solution; Chloroethanoic acid; EINECS 201-178-4; EPA Pesticide Chemical Code 279400; HSDB 939; Kyselina chloro-ctova; MCA; Monochloorazijnzuur; Monochloracetic acid; Monochloressigsäure; Monochloressigsaeure; Monochloroacetic acid; Monochloroethanoic acid; NCI-C60231; NSC 142; UN1750; UN1751; UN3250. Herbicide, preservative, bacteriostat; intermediate in production of carboxymethylcellulose, ethyl chloro-acetate, glycine, synthetic caffeine, sarcosine, thio-glycolic acid, EDTA, 2,4-D, 2,4,5-T. Has three crystal forms; mp = 50°, 55 - 56°, 63°; bp = 189°; very soluble in H_2O, soluble in EtOH, C_6H_6, $CHCl_3$. Et_2O; LD_{50} (rat orl) = 650 mg/kg (sodium salt), (mus orl) = 165 mg/kg; LC_{50} (rainbow trout 48

hr.) = 900 mg/l; toxic to poultry; toxic to bees. *Atlas Interlates Ltd; E. I. DuPont de Nemours Inc.; Rhône Poulenc Crop Protection Ltd.; Rhône Poulenc Rorer.*

768   Chloroacetone
78-95-5   2131   H

[structure]

C3H5ClO
  Chloromethyl methyl ketone.
  4-01-00-03215 (Beilstein Handbook Reference); A-Stoff; Acetone, chloro-; Acetonyl chloride; BRN 0605369; CCRIS 1943; Chloracetone; Chloro-2-propanone; Chloroacetone; Chloromethyl methyl ketone; Chloropropanone; EINECS 201-161-1; HSDB 1070; Monochloracetone; Monochloropropanone; NSC 30673; Tonite; UN1695. Has been considered as a tear gas. Used in chemical manufacturing and as a catalyst. Liquid; mp = -44.5°; bp = 119°; d^{20} = 1.15; moderately soluble in H_2O, organic solvents; LD_{50} (mus orl) = 127 mg/kg, (rat orl) = 100 mg/kg, LC_{50} (rat ihl 1 hr.) = 262 ppm. *Lancaster Synthesis Co.; Mallinckrodt Inc.; Sigma-Aldrich Fine Chem.*

769   Chloroacetophenone
532-27-4   2132   G

[structure]

C8H7ClO
  2-Chloro-1-phenylethanone.
  4-07-00-00641 (Beilstein Handbook Reference); Acetophenone, 2-chloro-; AI3-52322; BRN 0507950; Caswell No. 179C; CCRIS 2370; Chemical mace; Chloroacetophenone; α-Chloroacetophenone; 1-Chloroacetophenone; 2-Chloro-1-phenylethanone; 2-Chloroacetophenone; Chloromethyl phenyl ketone; CN; CN (lacrimator); EINECS 208-531-1; EPA Pesticide Chemical Code 018001; Ethanone, 2-chloro-1-phenyl-; HSDB 972; Mace (lacrimator); NCI-C55107; NSC 41666; ω-Chloroacetophenone; Phenacyl chloride; Phenacylchloride; Phenylchloromethylketone; UN1697. Pharmaceutical intermediate. A lachrymator, is the primary component of Mace, a riot-control gas. Plates, mp = 56.5°; bp = 247°; d^{15} = 1.324; insoluble in H_2O; soluble in Me_2CO, petroleum ether, very soluble in EtOH, Et_2O, C_6H_6, CS_2; LD_{50} (rat orl) = 127 mg/kg. *Janssen Chimica; Lancaster Synthesis Co.; Penta Mfg.*

770   2-Chloro-1-(3-acetoxy-4-nitrophenoxy)-4-(trifluoromethyl)benzene
50594-44-0   H P

[structure]

C15H9ClF3NO5
  Phenol, 5-(2-chloro-4-(trifluoromethyl)phenoxy)-2-nitro-, acetate.
  2-Chloro-1-(3-acetoxy-4-nitrophenoxy)-4-(trifluoro-methyl)benzene;

5-(2-Chloro-4-(trifluoromethyl)phen-oxy)-2-nitrophenyl acetate; EINECS 256-632-4; Phenol, 5-(2-chloro-4-(trifluoromethyl)phenoxy)-2-nitro-, acetate.

771 Chloroacetyl chloride
79-04-9   2074   H

C2H2Cl2O
 Chloroacetic acid chloride.
 4-02-00-00488 (Beilstein Handbook Reference); Acetyl chloride, chloro-; BRN 0605439; Chloracetyl chloride; Chlorid kyseliny chloroctove; Chloroacetic acid chloride; Chloroacetic chloride; Chlorure de chloracetyle; EINECS 201-171-6; HSDB 973; Monochloroacetyl chloride; UN1752. Used in organic synthesis. Liquid; mp = -22°; bp = 220° (explosive); very soluble in organic solvents.

772 Chloroalkanes
61788-76-9   H

 Alkanes, chloro [Chlorinated paraffins].
 Alkanes, chloro; Alkanes, chloro [Chlorinated paraffins]; EINECS 263-004-3.

773 Chloroallyl-3,5,7-triaza-1-azonia-adamantane Chloride
4080-31-3   2135   G P

C9H16Cl2N4
 1-(3-Chloro-2-propenyl)-3,5,7-triaza-1-azoniatricyclo-[3.3.1.1(3,7)]decane chloride.
 Caswell No. 181; CCRIS 1398; Cinartc 200; Dowco 184; Dowicide 184; Dowicide Q; Dowicil® 75; Dowicil® 100; EINECS 223-805-0; EPA Pesticide Chemical Code 017901; Hexamethylenetetramine chloroallyl chloride; HSDB 6820; Methenamine 3-chloroallylochloride; NSC 172971; Quaternium 15; XD-1840. Used with sodium bicarbonate as a bactericide and stabilizer; preservative for aqueous end products such as adhesives, latex emulsions, paints, metal-cutting fluids, drilling muds, biodegradable detergents, and paper coatings; antimicrobial activity. Registered by EPA as an antimicrobial and disinfectant (cancelled). Cream-colored powder; freely soluble in H_2O (< 25 g/100 ml), less soluble in organic solvents; LD50 (rat orl) = 500 mg/kg. Dow Chem. U.S.A.

774 m-Chloroaniline
108-42-9   2136   G

C6H6ClN
 m-Aminochlorobenzene.
 AI3-12126; 1-Amino-3-chlorobenzene; m-Amino-chlorobenzene; meta-Aminochlorobenzene; Aniline, m-chloro-; Benzenamine, 3-chloro-; CCRIS 3402; 3-Chloroanilinen; 3-Chloroaniline; m-Chloraniline; 3-Chlorobenzenamine; m-Chloroaminobenzene; 3-Chlorophenylamine; m-Chlorophenylamine; 3-Cloroaniline; EINECS

203-581-0; Fast Orange GC Base; HSDB 2046; NSC 17581; Orange GC Base. Intermediate for azo dyes and pigments, pharmaceuticals, insecticides, agricultural chemicals. Liquid; mp = -10.4°; bp = 230.5°; d$_4^{20}$ 1.2161; λ_m = 242, 294 nm (MeOH); insoluble in H_2O, soluble in $CHCl_3$, freely soluble in EtOH, Et2O, Me2CO, C6H6, CCl4. DuPont; Janssen Chimica; Lancaster Synthesis Co.

775 o-Chloroaniline
95-51-2   2136   G H

C6H6ClN
 Benzenamine, 2-chloro-.
 AI3-16321; o-Aminochlorobenzene; Aniline, o-chloro-; Benzenamine, 2-chloro-; CCRIS 2880; o-Chloraniline; o-Chloroaminobenzene; o-Chloroaniline; EINECS 202-426-4; Fast Yellow GC Base; HSDB 2045; NSC 6183. Dye intermediate, standards for colorimetric apparatus, manufacture of petroleum solvents and fungicides. mp = -14°; bp = 208.8°. bp11 = 95-97°; d = 1.213; insoluble in H_2O, soluble in Eto, Me2CO, freely soluble in EtOH. E. I. DuPont de Nemours Inc.

776 p-Chloroaniline
106-47-8   2136   G H

C6H6ClN
 4-Chloroaniline.
 AI3-14908; p-Aminochlorobenzene; Aniline, 4-chloro-; Aniline, p-chloro-; Benzenamine, 4-chloro-; Benzenamine, p-chloro-; Caswell No. 182; CCRIS 131; p-Chloroaniline; p-Chlorophenylamine; EINECS 203-401-0; EPA Pesticide Chemical Code 017203; HSDB 2047; NCI-C02039; NSC 36941; p-CA; para-Chloroaniline; RCRA waste number P024. Used in manufacture of dyestuffs, pharmaceuticals and agricultural chemicals. Crystals; mp = 72.5°; bp = 232°; d^{19} = 1.429; λ_m = 243, 296 nm (MeOH); soluble in H_2O, EtOH, Et2O, CHCl3; LD50 (rat orl) = 0.31 g/kg. DuPont; Janssen Chimica; Mitsui Toatsu.

777 Chlorobenzaldehyde
35913-09-8   G

C7H5ClO
 Chlorobenzaldehyde.
 Benzaldehyde, chloro-; Chlorobenzaldehyde; NSC 174140. Intermediate for triphenylmethane and related dyes, organic intermediate. Hoechst Celanese; Janssen Chimica; Penta Mfg.; Rit-Chem.

778 Chlorobenzene
108-90-7   2139   G

C6H5Cl

Phenyl chloride.

Abluton T30; AI3-07776; Benzene chloride; Benzene, chloro-; Caswell No. 183A; CCRIS 1357; Chloor-benzeen; Chlorbenzene; Chlorbenzol; Chlorbenzen; Chlorobenzene; Chlorobenzene, mono-; Chloro-benzenu; Chlorobenzol; Clorobenzene; CP 27; EINECS 203-628-5; EPA Pesticide Chemical Code 056504; HSDB 55; I P Carrier T 40; MCB; Monochloorbenzeen; Monochlorbenzene; Monochlorbenzol; Monochloro-benzene; Monoclorobenzene; NCI-C54886; NSC 8433; Phenyl chloride; RCRA waste no. U037; Tetrosin SP; UN1134. Carrier for textile dyeing with low stain to wool. Pesticide intermediate; manufacture of phenol, aniline, DDT; solvent carrier for methylene diisocyanate; solvent for paints; heat transfer medium. Liquid; mp = -45.2°; bp = 131.7°; d$_4^{20}$ = 1.1058; λ$_m$ = 263 nm (ε = 191, H2O); insoluble in H2O, very soluble in C6H6, CCl4, CHCl3, CS2; freely soluble in EtOH, Et2O; LD50 (rat orl) = 1110 mg/kg. *Elf Atochem N. Am.; Janssen Chimica; Monsanto Co.; PPG Ind.; Sigma-Aldrich Fine Chem.; Taiwan Surfactants.*

779   p-Chlorobenzotrichloride
5216-25-1   H

C7H4Cl4

1-Chloro-4-(trichloromethyl)benzene.

4-05-00-00823 (Beilstein Handbook Reference); AI3-02822; Benzene, 1-chloro-4-(trichloromethyl)-; BRN 1866549; 1-Chloro-4-(trichloromethyl)benzene; 4-Chlorobenzotrichloride; EINECS 226-009-1; p-Chlorobenzotrichloride; p-Chlorophenyltrichloro-methane; p-Trichloromethylchlorobenzene; p,α,α,α-Tetrachlorotoluene; α,α,α-Trichloro-4-chlorotoluene; Toluene, α,α,α,p-tetrachloro-; Toluene, p,α,α,α-tetrachloro-. Liquid; bp = 245°; d20 = 1.4463; λsim = 233, 266, 278 nm (ε = 11800, 495, 264, MeOH); very soluble in Me2CO, Et2O.

780   2-Chlorobenzyl chloride
611-19-8   H

C7H6Cl2

1-Chloro-2-(chloromethyl)benzene.

4-05-00-00816 (Beilstein Handbook Reference); AI3-14885; Benzene, 1-chloro-2-(chloromethyl)-; BRN 0471700; 1-Chloro-2-(chloromethyl)benzene; 2-Chlorobenzyl chloride; α,2-Dichlorotoluene; α,o-Dichlorotoluene; EINECS 210-258-8; NSC 8446; o-Chlorobenzyl chloride; Ortho-α-dichlorotoluene; Toluene, α,o-dichloro-; Toluene, o,α-dichloro-. Liquid; mp = -17°; bp = 217°; d0 = 1.2699; λm = 271, 278 nm (ε = 339, 263, MeOH); insoluble in H2O, slightly soluble in EtOH, CCl4, soluble in CS2, very soluble in Et2O, C6H6, AcOH.

781   Chlorobromoform
124-48-1   2154   G

CHBr2Cl

Chlorodibromo methane.

4-01-00-00081 (Beilstein Handbook Reference); BRN 1731046; CCRIS 938; CDBM; Chlorodibromomethane; Dibromochloromethane; EINECS 204-704-0; HSDB 2763; Methane, chlorodibromo-; Methane, dibromochloro-; Monochlorodibromomethane; NCI-C55254. Used in organic synthesis. Liquid; mp = -20°; bp = 120°; d20 = 2.451; insoluble in H2O, soluble in EtOH, Et2O, Me2CO, C6H6, CCl4; LD50 (rat orl) = 370 mg/kg.

782   2-Chloro-N-(chloromethyl)-N-(2,6-diethylphenyl)acetamide
40164-69-0   H

C13H17Cl2NO

Acetamide, 2-chloro-N-(chloromethyl)-N-(2,6-diethyl-phenyl)-.

Acetamide, 2-chloro-N-(chloromethyl)-N-(2,6-diethyl-phenyl)-; 2-Chloro-N-(chloromethyl)-N-(2,6-diethyl-phenyl)acetamide; EINECS 254-817-4.

783   Chlorodiethylaluminum
96-10-6   H

C4H10AlCl

Aluminum, chlorodiethyl-.

Aluminum, chlorodiethyl-; Aluminum, dichloro-tetraethyldi-; Aluminum diethyl monochloride; Chlorodiethylaluminum; Deac; Deak; Diethyl-aluminium chloride; Diethylaluminum monochloride; Diethylchloroaluminum; EINECS 202-477-2; HSDB 5299.

784   Chlorodifluoromethane
75-45-6   H

CHClF2

Difluorochloromethane.

4-01-00-00032 (Beilstein Handbook Reference); Algeon 22; Algofrene 22; Algofrene type 6; Arcton 22; Arcton 4; BRN 1731036; CCRIS 858; CFC 22; Chlorodifluoromethane; Chlorofluorocarbon 22; Daiflon 22; Difluorochloromethane; Difluoromono-chloromethane; Dymel 22; EINECS 200-871-9; Electro-CF 22; Eskimon 22; F 22; F 22 (halocarbon); FC 22; FKW 22; Flugene 22; Fluorocarbon 22; Forane 22; Forane 22 B; Freon 22; Frigen; Frigen 22; Genetron 22;

Haltron 22; HCFC 22; HFA - 22; HSDB 143; Hydrochlorofluorocarbon 22; Isceon 22; Isotron 22; Khaladon 22; Khladon 22; Methane, chlorodifluoro-; Monochlorodifluoromethane; Propellant 22; R-22; R 22 (refrigerant); R22 [Nonflammable gas]; Racon 22; Refrigerant R 22; Ucon 22; UN1018. Used as a refrigerant. Gas; mp = -157.4°; bp = -40.7°; d^{-69} = 1.4909; soluble in H_2O and in organic solvents. *DuPont; Honeywell; PPG Ind.*

785   Chlorodimethyl ether
107-30-2   2165   H

C_2H_5ClO
 Methyl chloromethyl ether.
 4-01-00-03046 (Beilstein Handbook Reference); BRN 0505943; CCRIS 138; Chlordimethylether; Chlorodimethyl ether; Chloromethoxymethane; Chloromethyl methyl ether; α,α-Dichlorodimethyl ether; Dimethylchloroether; EINECS 203-480-1; Ether, chloromethyl methyl; Ether, dimethyl chloro; Ether methylique monochlore; HSDB 908; Methane, chloromethoxy-; Methoxychloromethane; Methoxy-methyl chloride; Methyl chloromethyl ether; Monochlorodimethyl ether; Monochloromethyl methyl ether; NSC 21208; RCRA waste number U046; UN1239. Liquid; mp = -103.5°; bp = 59.5°; d^{10} = 1.063; soluble in EtOH, Et2O, Me2CO, CHCl3 .

786   Chlorodimethyloctadecylsilane
18643-08-8   G

$C_{20}H_{43}ClSi$
 Dimethyloctadecylchloro silane.
 CD5636; Dimethyloctadecylchlorosilane; EINECS 242-472-2; Octadecyldimethylchlorosilane; Silane, chlorodimethyloctadecyl-. Coupling agent, chemical intermediate, blocking agent, release agent, lubricant, primer, reducing agent. Solid; mp = 28-30°; bp = 300°. *Degussa-Hüls Corp.*

787   Chlorodimethylsilane
1066-35-9   G H

C_2H_7ClSi
 Dimethylchlorosilane.
 CD5470; Chlorodimethylsilane; Dimethylchlorosilane; DMCS; EINECS 213-912-0; Silane, chlorodimethyl-. Coupling agent, chemical intermediate, blocking agent, release agent, lubricant, primer, reducing agent. Volatile liquid; mp = -111°; bp = 35°; d = 0.8520. *Hüls Am.*

788   Chloroethane
75-00-3   3817   G H

C_2H_5Cl
 Ethane, chloro-.
 Aethylchlorid; Äthylchlorid; Aethylchloride; Äthyl-chloride; Aethylis chloridum; AI3-24474; Anodynon; CCRIS 3349; Chelen; Chloorethaan; Chlorene; Chlorethyl; Chloroaethan; Chloroäthan;

Chloroethane; Chlorure d'ethyle; Chloryl; Chloryl anesthetic; Cloretilo; Cloroetano; Cloruro di etile; Dublofix; EINECS 200-830-5; Ethane, chloro-; Ether chloratus; Ether chloridum; Ether hydrochloric; Ether muriatic; Ethyl chloride; Etylu chlorek; HSDB 533; Hydrochloric ether; Kelene; Monochlorethane; Monochloroethane; Muriatic ether; Narcotile; NCI-C06224; UN1037. Used as a refrigerant, and solvent and as a chemical feedstock. A hazardous air pollutant identified by the US government as a toxic air contaminant. Gas; mp = -138.7°; bp = 12.3°; d^{25} = 0.8902; soluble in H_2O (0.574 g/100 ml), more soluble in organic solvents.

789   2-Chloroethylmethyldichlorosilane
7787-85-1   G

$C_3H_7Cl_3Si$
 2-Chloroethylmethyldichlorosilane.
 CC3005; 2-Chloroethylmethyldichlorosilane; Di-chloro(2-chloroethyl)methylsilane; EINECS 232-134-2; Silane, dichloro(2-chloroethyl)methyl-. Coupling agent, chemical intermediate, blocking agent, release agent, lubricant, primer, reducing agent. Liquid; bp744 = 157°; n_D^{20} = 1.4580. *Degussa-Hüls Corp.*

790   Chloroform
67-66-3   2160   G H

$CHCl_3$
 Trichloromethane.
 4-01-00-00042 (Beilstein Handbook Reference); AI3-24207; BRN 1731042; Caswell No. 192; CCRIS 137; Chloroform; Chloroforme; Cloroformio; EINECS 200-663-8; EPA Pesticide Chemical Code 020701; Formyl trichloride; Freon 20; HSDB 56; Methane trichloride; Methane, trichloro-; Methenyl chloride; Methenyl trichloride; Methyl trichloride; NCI-C02686; NSC 77361; R 20 (Refrigerant); RCRA waste number U044; Trichloormethaan; Trichlormethan; Trichloromethane; Triclorometano; UN1888. Used in manufacture of fluorocarbon plastics, as a solvent, in analytical chemistry, as a fumigant, in insecticides. Clear mobile liquid; mp = -63.5°; bp = 61.1°; d^{20} = 1.4832; LD50 (rat orl 14 day) = 1330 - 3230 mg/kg (0.9 - 2.18 ml/kg). *Degussa-Hüls Corp.; Elf Atochem N. Am.; Mallinckrodt Inc.; Mitsui Toatsu; OxyChem.*

791   3-Chloro-2-hydroxypropylammonium chloride
3327-22-8   G H

$C_6H_{15}Cl_2NO$
 Trimethyl(2-hydroxy-3-chloropropyl)ammonium chloride.
 Ammonium, (3-chloro-2-hydroxypropyl)trimethyl-, chloride; Catiomaster C; CHPTA 65%; Dextrosil; Dextrosil KA; Dowquat 188; EINECS 222-048-3; NSC 51216; NT 21; QUAB; Quab 188; Quat 188; Reagens-CF2; Trimethyl(2-hydroxy-3-chloropropyl)ammonium chloride; Verolan KAF. Cationizing reagent for natural and synthetic polymers; starch modifier for textiles. Used for quaternization of compounds with hydroxyl, amino, and other functional groups, especially corresponding polymers, production of cationic polyelectrolytes. Crystals; mp = 191-193°; d = 1.154. *Chem-Y GmbH; Degussa AG; Synthetic Chemicals Ltd.*

792   Chloromethane
74-87-3   6069   G H

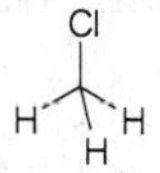

CH₃Cl → CH_3Cl

Methane, chloro-.
Al3-01707; Artic; Caswell No. 557; CCRIS 1124; Chloor-methaan; Chlor-methan; Chloromethane; Chlorure de methyle; Clorometano; Cloruro di metile; EINECS 200-817-4; EPA Pesticide Chemical Code 053202; HSDB 883; Methane, chloro-; Methyl chloride; Methylchlorid; Methylchloride; Metylu chlorek; Monochloromethane; R 40; RCRA waste number U045; UN1063. Catalyst carrier in low-temperature polymerization, tetramethyl lead, silicones, refrigerant, fluid for thermometric/ thermostatic equipment, methylating agent in organic synthesis, extractant and low-temperature solvent, herbicide, topical anesthetic. Colorless gas; mp = -97°; bp = -24°; d = 0.92; insoluble in H2O, soluble in organic solvents; LC₁₀₀ (mus ihl) = 3146 ppm. *Air Products & Chemicals Inc.; Mitsui Toatsu; OxyChem.*

793   Chloromethylbenzene
25168-05-2   G H

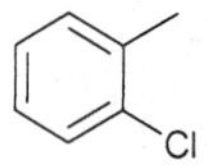

C₇H₇Cl
Toluene, ar-chloro-.
4-05-00-00809 (Beilstein Handbook Reference); BC; Benzene, chloromethyl-; BRN 0471308; Chloro-methylbenzene; Chlorotoluene; EINECS 246-698-2; Halso® 99; Oxsol 10; Toluene, ar-chloro-. Extender, diluent or substitute for other organic solvents; for dye carrier, fuel oil additive, sludge solvent, in paint thinners, metal parts cleaners, adhesives. *OxyChem.*

794   Chloromethyl tert-butyl ketone
13547-70-1   H

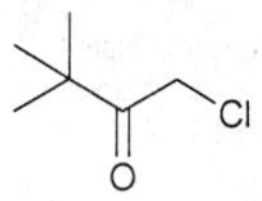

C₆H₁₁ClO
1-Chloro-3,3-dimethyl-2-butanone.
2-Butanone, 1-chloro-3,3-dimethyl-; tert-Butyl chloro-methyl ketone; 1-Chloro-3,3-dimethyl-2-butanone; 1-Chloro-3,3-dimethylbutan-2-one; Chloromethyl tert-butyl ketone; α-Chloropinacolin; α-Chloropinacoline; 1-Chloropinacolone; Chlorpinakolin; EINECS 236-920-6; 1-Monochloropinacoline.

795   Chloromethyldimethylchlorosilane
1719-57-9   G

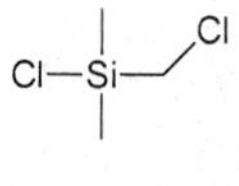

C₃H₈Cl₂Si
Chloromethyldimethylchlorosilane.
CC3270; Chloro(chloromethyl)dimethylsilane; EINECS 217-006-6; Silane, chloro(chloromethyl)dimethyl-. Coupling agent, chemical intermediate, blocking agent, release agent, lubricant, primer, reducing agent. Liquid; bp = 115.5°; bp₇₅₂ = 114°; d²⁰ = 1.0865.

Degussa-Hüls Corp.

796   Chloromethyldiphenylsilane
144-79-6   H

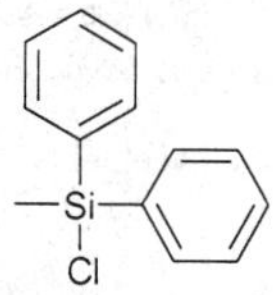

C₁₃H₁₃ClSi
Methyldiphenylchlorosilane.
Chloro(methyl)diphenylsilane; Chloromethyldiphenyl-silane; EINECS 205-639-0; Methyldiphenylchloro-silane; NSC 93961; Silane, chloromethyldiphenyl-. Liquid; bp = 295°; d²⁰ = 1.1277.

797   (Chloromethyl)ethylbenzene
26968-58-1   H

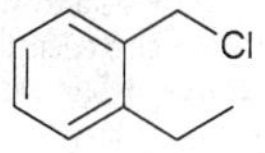

C₉H₁₁Cl
Toluene, α-chloro-ar-ethyl-.
Benzene, (chloromethyl)ethyl-; (Chloromethyl)ethyl-benzene; E!NECS 248-148-7; Toluene, α-chloro-ar-ethyl-. Mixture of o, m and p-isomers.

798   3-Chloromethyl-4-keto-1,2,3-benzo-triazine
24310-41-6   H

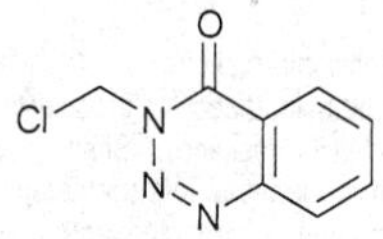

C₈H₆ClN₃O
1,2,3-Benzotriazin-4(3H)-one, 3-(chloromethyl)-.
1,2,3-Benzotriazin-4(3H)-one, 3-(chloromethyl)-; 3-(Chloromethyl)-1,2,3-benzotriazin-4(3H)-one; 3-Chloromethyl-4-keto-1,2,3-benzotriazine; EINECS 246-152-3.

799   Chloromethylphthalimide
17564-64-6   H

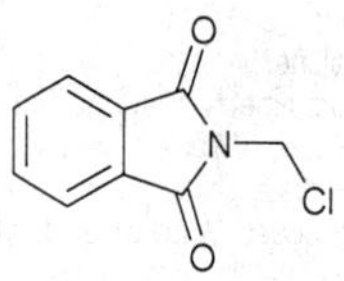

C₉H₆ClNO₂
1H-Isoindole-1,3(2H)-dione, 2-(chloromethyl)-.
CCRIS 7991; Chloromethylphthalimide; N-(Chloro-methyl)phthalimide; N-Chloromethyltrimellitimide; EINECS 241-541-4; 1H-Isoindole-1,3(2H)-dione, 2-(chloromethyl)-; NSC 29558; Phthalimide, N-(chloromethyl)-; Phthalimide, N-chloromethyl-.

800   Chloromethyltrimethyl silane
2344-80-1   G

C4H11ClSi
Chloromethyltrimethyl silane.
CC3285; Chloromethyltrimethylsilane; EINECS 219-058-5; Silane, (chloromethyl)trimethyl-. Coupling agent, chemical intermediate, blocking agent, release agent, lubricant, primer, reducing agent. Reagent for the Peterson olefination and the homologation of ketones and aldehydes via α, β-epoxysilanes. Liquid; bp = 98-99°; d^{25} = 0.8790; n_D^{20} = 1.4175. *Degussa-Hüls Corp.; Janssen Chimica.*

801 Chloroneb
2675-77-6   G P

C8H8Cl2O2
Benzene, 1,4-dichloro-2,5-dimethoxy-.
4-06-00-05772 (Beilstein Handbook Reference); AI3-52191; Benzene, 1,4-dichloro-2,5-dimethoxy-; BRN 1952749; Caswell No. 198; CCRIS 5996; Chloroneb; Chloroneb 65W Fungicide; Chloroneb Systemic Flowable Fungicide; Chloronebe; Demasan; Demosan; Demosan 10D; Demosan 65W; Demosan 65W Termec SP; 1,4-Dichloro-2,5-dimethoxybenzene; EINECS 220-222-3; EPA Pesticide Chemical Code 027301; Flo Pro D Seed Protectant; HSDB 1542; NSC 151546; Nuflo D; Soil fungicide 1823; Terraneb; Terraneb B; Terraneb SP; Terraneb SP Turf Fungicide; Tersan SP. Systemic soil- and seed fungicide, absorbed by the roots. Used for control seedling diseases like snow mold (*Typhula*) and pythium blight. Seed treatment to suppress soreshin and pre- and post-emergence damp-off caused by *rhizoctonia solani, pythium spp and sclerotium rolfsii* on cotton, beans, soybeans and sugar beets. Registered by EPA as a fungicide. Crystals; mp 133-135°; bp = 268°; soluble in H2O (0.0008 g/100 ml), Me2CO (9 g/100 ml), DMF (11.1 g/100 ml), xylene (7.65 g/100 ml), CH2Cl2 (17.7 g/100 ml); LD50 (rat orl) > 11000 mg/kg, (rbt orl) > 5000 mg/kg, (mallard diuck, Japanese quail orl) > 5000 mg/kg; LC50 (bluegill sunfish 46 hr.) > 4200 mg/l. *Agriliance LLC; Gustafson LLC; Kincaid Enterprises; Micro-Flo Co. LLC; PBI/Gordon Corp; Wilbur Ellis Co.*

802 1-Chloro-2-nitro-benzene
88-73-3   2170   H

C6H4ClNO2
Benzene, 1-chloro-2-nitro-.
AI3-15385; Benzene, 1-chloro-2-nitro-; CCRIS 141; Chloronitrobenzene, ortho, liquid; o-Chloronitro-benzene; EINECS 201-854-9; HSDB 1322; o-Nitro-chlorobenzene; NSC 36934; ONCB; UN1578. Used in manufacture of dyestuffs. Needles; mp = 32.5°; bp = 245.5°; d^{242} = 1.368; λ_m = 215, 270 nm (ε = 8140, 10300, MeOH); soluble in EtOH, C6H6, Et2O.

803 4-Chloronitrobenzene
100-00-5   2170   H

C6H4ClNO2
Benzene, 1-chloro-4-nitro-.
AI3-15387; Benzene, 1-chloro-4-nitro-; CCRIS 142; Chloronitrobenzene, para, solid; p-Chloronitro-benzene; EINECS 202-809-6; HSDB 1666; p-Nitrochloorbenzeen; p-Nitrochlorobenzene; p-Nitro-chlorobenzol; p-Nitroclorobenzene; p-Nitrophenyl chloride; NSC 9792; PNCB; UN1578. Prisms; mp = 83.5°; bp = 242°; d^{90} = 1.2979; λ_m = 215, 270 nm (ε = 8140, 10300, MeOH); insoluble in H2O, slightly soluble in EtOH, soluble in Et2O, CHCl3, CS2

804 o-Chloro-p-nitrotoluene
121-86-8   H

C7H6ClNO2
2-Chloro-1-methyl-4-nitrobenzene.
4-05-00-00855 (Beilstein Handbook Reference); Benzene, 1-chloro-2-methyl-5-nitro-; Benzene, 2-chloro-1-methyl-4-nitro-; BRN 1817924; 2-Chloro-1-methyl-4-nitrobenzene; 2-Chloro-4-nitrotoluene; CNT; EINECS 204-501-7; HSDB 5592; NSC 60111; o-Chloro-p-nitrotoluol; Toluene, 2-chloro-4-nitro-. Needles; mp = 66.5°; bp = 260°; λ_m = 254, 304 nm (MeOH); slightly soluble in H2O, CHCl3, soluble in EtOH, Et2O, AcOH.

805 1-Chloro-8-(p-octylphenoxy)-3,6-dioxa-octane
66028-01-1   H

C20H33ClO3
1-(2-(2-(2-Chloroethoxy)ethoxy)ethoxy)-4-octyl-benzene.
Benzene, 1-(2-(2-(2-chloroethoxy)ethoxy)ethoxy)-4-octyl-; 1-(2-(2-(2-Chloroethoxy)ethoxy)ethoxy)-4-octylbenzene; 1-Chloro-8-(p-octylphenoxy)-3,6-dioxa-octane; EINECS 266-079-0.

806 Chloropentafluoroethane
76-15-3   H

C2ClF5
Pentafluoroethyl chloride.
4-01-00-00129 (Beilstein Handbook Reference); BRN 1740329; CFC-115; Chloropentafluoretano; Chloro-pentafluorethane; Chloroperfluoroethane; EINECS 200-938-2; Ethane, chloropentafluoro-; F-115; FC 115; FKW 115; Fluorocarbon-115; Freon 115; Genetron 115; Halocarbon 115; HSDB 147; Monochloropentafluoroethane; Pentafluorochloro-ethane; Pentafluoroethyl chloride; Perfluoroethyl chloride; Propellant 115; R 115; Refrigerant 115; UN1020. Has been used as a refrigerant Gas; mp = -99.4°; mp = -37.9°; d^{-42} = 1.5678; insoluble in H2O, soluble in organic solvents. *Lancaster Synthesis Co.; Mallinckrodt Inc.; Sigma-*

807   Chlorophacinone

3691-35-8   2171   G P

$C_{23}H_{15}ClO_3$

2-((p-Chlorophenyl)phenylacetyl)-1,3-indandione.

Actosin C; Afnor; Baraage; BRN 2063081; CAID; Caswell No. 211C; Chloorfacinon; 2(2-(4-Chloor-fenyl-2-fenyl)-acetyl)-indaan-1,3-dion; Chlorfacinon; Chloro-phacinone; Chlorphacinon; ((4-Chlorphenyl)-1-ph-enyl)-acetyl-1,3-indandion; 1-(4-Chlorphenyl)-1-phen-ylacetylindan-1,3-dion; 2(2-(4-Chlor-phenyl-2-phenyl)-acetyl)indan-1,3-dion; 2(2-(4-Chlorophenyl)-2-phenyl)indan-1,3-dione; 2-((4-Chlorophenyl)-phenyl-acetyl)-1H-indene-1,3(2H)-dione; 2-((p-Chlorophenyl)-phenylacetyl)-1,3-indandione; 2-(2-(4-Chlorophenyl)-2-phenylacetyl)indan-1,3-dione; 2-(-p-Chlorophenyl-acetyl)indane-1,3-dione; 2-(-p-Chlorophenyl- -phenyl-acetyl)indane-1,3-dione; 2(2-(4-Cloro-fenil-2-fenil)-acetil)indan-1,3-dione; Drat; DRAT; EINECS 223-003-0; EPA Pesticide Chemical Code 067707; HSDB 6432; Indandione, 2-((p-chlorophenyl)phenylacetyl)-; 1,3-Indandione, 2-((p-chlorophenyl)phenylacetyl)-; Indene-1,3(2H)-dione, 2-((4-chlorophenyl)phenylacetyl)-; 1H-Indene-1,3(2H)-dione, 2-((4-chlorophenyl)phenyl-acetyl)-; Karate; Lepit; Liphadione; LM 91; LM-91; Microzul; Muriol; Orcomolebait; Partox; 2-(2-Phenyl-2-(4-chlorophenyl)-acetyl)-1,3-indandione; Quick; Rat-indan 3; Ratomet; Raviac; Razol; Redentin; Ridene; Rozol; Sakarat Special; Saviac; Skaterpax; Topitox. A powerful anticoagulant rodenticide; controls black rats, brown rats, house mice, long-tailed field mice, voles and musk rats. Registered by EPA as a rodenticide. Silky yellow needles; mp = 138°, 140°; λm = 325 nm (Me2CO); soluble in H2O (0.01 g/100 ml), MeOH, EtOH, Me2CO, AcOH, EtOAc, C6H6, oil; LD50 (rat orl) = 20 mg/kg; LC50 (rat ihl 1 hr.) > 3 mg/kg; low toxicity to birds, non-toxic to bees. JT Eaton & Co. Inc.; Killgerm Chemicals Ltd.; Lever Industrial Ltd.; Liphatech Inc.; Rhône-Poulenc Environmental Prods. Ltd; Rhône-Poulenc.

808   1-(4-Chlorophenyl)-4,4-dimethyl-3-pentanone

66346-01-8   H

$C_{13}H_{17}ClO$

3-Pentanone, 1-(4-chlorophenyl)-4,4-dimethyl-.

1-(4-Chlorophenyl)-4,4-dimethyl-3-pentanone; 3-Pentanone, 1-(4-chlorophenyl)-4,4-dimethyl-; HWG 1608-Alkylketon.

809   Chlorophyll

1406-65-1   2174   G

$C_{55}H_{72}MgN_4O_5$

Amplex.

C.I. 1956; Chlorofolin; Chlorofyl; Chlorophyll; Chlorophylls; Deodophyll; E 140; EINECS 215-800-7; L-Gruen No. 1; L-Grün No. 1. The green pigments of plants and green algae contain chlorophyl

a and chlorophyl b in a ratio of aprox 3:1; A proprietary preparation of chlorophyll; a deodorant. Used to color soaps, oils, fats, perfumes and sensitizing color film. Source of phytol. Green solid; soluble in oils. Ashe Chemicals.

810   Chloropicrin

76-06-2   2175   G P

CCl_3NO_2

Nitrotrichloromethane.

4-01-00-00106 (Beilstein Handbook Reference); Acquinite; AI3-00027; Aquinite; BRN 1756135; Caswell No. 214; CCRIS 146; Chloorpikrine; Chlor-O-pic; Chloroform, nitro-; Chloropicrin; Chloropicrin mixtures, UN1583; Chloropicrin [UN1580] [Poison]; Chloropicrine; Chlorpikrin; Cloropicrina; Dojyopicrin; Dolochlor; EPA Pesticide Chemical Code 081501; G 25; HSDB 977; KLOP; Larvacide; Larvacide 100; Methane, trichloronitro-; Microlysin; NCI-C00533; Nemax; Nimax; Nitrochloroform; Nitrotrichloro-methane; NSC 8743; OG 25; Pic-Chlor; Pic-Clor; Picfume; Picride; Profume A; PS; S 1; Timberfume; Tri-clor; Tri-Con; Trichloornitromethaan; Trichlornitro-methan; Trichloronitromethane; Tricloro-nitro-metano; UN1580; UN1583. Registered by EPA as an insecticide. Used as an insecticide and nematicide and in organic synthesis, dyestuffs, fumigants, insecticides, rat extermination, tear gas. Oil; mp = -64°; bp757 = 112°; d420 = 1.6558, d425 = 1.6483; soluble in H2O (0.23 g/100 ml at 0°, 0.16 g/100 ml at 25°), freely soluble in C6H6, CS2, Et2O, EtOH. Albemarle Corp.; AmeriBrom Inc.; Dow AgroSciences; Great Lakes Chemical Corp; La-Co Industries; Niklor Chem Co. Inc.; Osmose Inc.; Reddick Fumigants, Inc.; Soil Chemicals Corp.; Trical; Trinity Manufacturing Inc.

811   Chloropivaloyl chloride

4300-97-4   H

$C_5H_8Cl_2O$

3-Chloro-2,2-dimethylpropanoyl chloride.

β-Chloropivaloyl chloride; 3-Chloro-2,2-dimethylpropanoyl chloride; Chloropivaloyl chloride; 3-Chloropivaloyl chloride; EINECS 224-311-8; NA9263; Propanoyl chloride, 3-chloro-2,2-dimethyl-.

812   Chloroprene

126-99-8   G H

C_4H_5Cl

2-Chloro-1,3-butadiene.

1,3-Butadiene, 2-chloro-; Bayprene® 110; CCRIS 873; 2-Chloor-1,3-butadieen; 2-Chlor-1,3-butadien; Chloro-butadiene; 2-Chloro-1,3-butadiene; 2-Chloro-buta-diene; 2-Chlorobuta-1,3-diene; Chloropreen; Chloro-pren; Chloroprene; Cloroprene; 2-Cloro-1,3-butadiene; EINECS 204-818-0; HSDB 1618; Neoprene; NSC 18589; UN1991; β-Chloroprene. Synthetic polymer used for moldings and extrudates, reinforced hoses, roll covers, belting, cable sheathings and insulation, sponge rubber, sheeting, fabric proofings, footwear, food-contact goods, adhesives for footwear, furniture, building industry. Used as a protective coating and in synthetic rubber manufacture by polymerization. The polymerized product bears the name of Neoprene. Liquid; mp = -130°; bp = 59°; d^{20} = 0.956; insoluble in H_2O (0.211 g/100 ml), soluble in EtOH, Me_2CO, very soluble in Et_2O, $CHCl_3$; LD_{50} (rat orl) = 450 mg/kg. *Bayer AG; Miles Inc.*

813 3-Chloropropylmethyldimethoxysilane
18171-19-2   G

$C_6H_{15}ClO_2Si$
(3-Chloropropyl)methyldimethoxysilane.
CC3290; EINECS 242-056-0; Silane, (3-chloropropyl)dimethoxymethyl-. Coupling agent, chemical intermediate, blocking agent, release agent, lubricant, primer, reducing agent. *Degussa-Hüls Corp.*

814 Chloropropyltrichlorosilane
2550-06-3   G H

$C_3H_6Cl_4Si$
Trichloro(3-chloropropyl)silane.
CC3291; (3-Chloropropyl)trichlorosilane; Chloropropyltrichlorosilane; EINECS 219-844-8; NSC 139831; Silane, 3-chloropropyltrichloro-; Silane, trichloro(3-chloropropyl)-; Trichloro(3-chloropropyl)-silane. Coupling agent, chemical intermediate, blocking agent, release agent, lubricant, primer, reducing agent. Liquid; bp = 181-183°; d^{20} = 1.3590. *Degussa-Hüls Corp.*

815 3-Chloropropyltriethoxysilane
5089-70-3   G

$C_9H_{21}ClO_3Si$
(3-Chloropropyl)triethoxysilane.
CC3292; (3-Chloropropyl)triethoxysilane; EINECS 225-805-6; NSC 252156; Silane, (3-chloropropyl)triethoxy-; Triethoxy(γ-chloropropyl)silane. Coupling agent, chemical intermediate, blocking agent, release agent, lubricant, primer, reducing agent. *Degussa-Hüls Corp.; Union Carbide Corp.*

816 2-Chloropyridine
109-09-1   G H

C_5H_4ClN
2-Chloropyridine.
AI3-19231; α-Chloropyridine; 2-Chloropyridine; o-Chloropyridine; CCRIS 1724; EINECS 203-646-3; NSC 4649; Pyridine, 2-chloro-; UN2822. Intermediate used in manufacture of antihistamines, germicides, pesticides and agricultural chemicals. Liquid; mp = -46°; bp = 170°; d^{15} = 1.205; λ_m = 208, 257, 263, 270 nm (MeOH);

slightly soluble in H_2O (27 g/l), soluble in EtOH, Et_2O; LD_{50} (mus orl) = 110 mg/kg. *EXPANSIA; Lancaster Synthesis Co.; Olin Corp; Penta Mfg.*

817 4-Chlororesorcinol
95-88-5   G

$C_6H_5ClO_2$
4-Chloro-1,3-dihydroxybenzene.
4-06-00-05684 (Beilstein Handbook Reference); AI3-03873; 1,3-Benzenediol, 4-chloro-; BRN 2042864; 4-Chloro-1,3-benzenediol; 4-Chloro-1,3-dihydroxy-benzene; 4-Chlororesorcin; 4-Chlororesorcinol; 6-Chlororesorcinol; p-Chlororesorcinol; CI 76510; 2,4-Dihydroxychlorobenzene; EINECS 202-462-0; NSC 1569; Resorcinol, 4-chloro-. Used in chemical manufacturing and synthesis. Solid; mp = 106.5-108°; bp_{18} = 147°. *Rit-Chem.*

818 N-Chlorosuccinimide
128-09-6   2183   G

$C_4H_4ClNO_2$
1-Chloro-2,5-pyrrolidinedione.
5-21-09-00543 (Beilstein Handbook Reference); AI3-52304; BRN 0113915; Caswell No. 807; 1-Chloro-2,5-pyrrolidinedione; Chlorosuccinimide; N-Chloro-succinimide; EINECS 204-878-8; EPA Pesticide Chemical Code 077301; HSDB 5407; NCS; NSC 8748; 2,5-Pyrrolidinedione, 1-chloro-; Succinchlorimide; Succinic acid, imide, N-chloro-; Succinic N-chloroimide; Succinimide, N-chloro-; Succinochlorimide. Chlorinating agent, disinfectant for swimming pools, bactericide. Plates; mp 150°; d^{25} = 1.65; soluble in H_2O (1.4 g/100ml), EtOH, C_6H_6, $CHCl_3$, ligroin, soluble in AcOH, Me_2CO; MLD (rat orl) = 2.7 g/kg. *Janssen Chimica; Lancaster Synthesis Co.; Penta Mfg.*

819 1-Chlorotetradecane
2425-54-9   H

$C_{14}H_{29}Cl$
1-Chloro-n-tetradecane.
1-Chlorotetradecane; EINECS 219-368-0; Tetradecane, 1-chloro-. Liquid; mp = 2.5°; bp = 292°; d^{20} = 0.8665; insoluble in H_2O, slightly soluble in CCl_4, soluble in EtOH, $CHCl_3$, very soluble in Me_2CO, C_6H_6.

820 Chlorothalonil
1897-45-6   2185   G P

C8Cl4N2

2,4,5,6-Tetrachloroisophthalonitrile.

AI3-28721; Black Leaf Lawn & Garden Fungicide; Bombardier; Bonide; Bravo; Bravo 6F; Bravo 500; Bravo-W-75; BRN 1978326; Caffaro; Caswell No. 215B; CCRIS 150; Chloroalonil; Chlorothalonil; Chlrthalonil; Clorthalonil; Clorto; Clortocaf ramato; Clortosip; DAC 2787; Dacobre; Daconil; Daconil 2787; Daconil Flowable; Daconil M; Dacosoil; Dexol Fungicide Containing Daconil; Dragon; Echo 75; EINECS 217-588-1; EPA Pesticide Chemical Code 081901; Evade; Exotherm; Exotherm Termil; Faber; Farber; Ferti-lome; Forturf; Green Charm Multi-Purpose Fungicide; Green Thumb Lawn & Garden Fungicide; HSDB 1546; Jupital; NCI-C00102; Nopccocide; Nopccocide N-96; Nopcocide N40D & N96; Ole; Ortho Multi-Purpose Fungicide Daconil 2787; Pennington's Pride Multi-Purpose; Pillarich; Pro-Care Multi-Purpose Fungicide; Repulse; Rigo's Best Lawn & Garden Fungicide; SA Lawn Ornamental & Vegetable; Security Fungi-Gard; Sweep; Taloberg; TCIN; m-TCPN; Termil; Terraclactyl; Tetrachlorisoftalonitril; Tetra-chloroisophthalonitrile; m-Tetrachlorophthalonitrile; TPN; Tuffcide; Vanox. A fungicide for a wide range of agricultural crops. Antimicrobial for marine antifouling coatings. Registered by EPA as a fungicide. Colorless crystals; mp = 250°; bp = 350°; d = 1.800; insoluble in H_2O (0.00006 g/100 ml), more soluble in xylene (6.9 g/100 ml), cyclohexanone (2.8 g/100 ml), DMF (2.8 g/100 ml), Me_2CO (1.58 g/100 ml), DMSO (2.2 g/100 ml), kerosene (< 1 g/100 ml); LD_{50} (rat orl) = > 10000 mg/kg, (dog orl) > 5000 mg/kg, (rbt der) > 10000 mg/kg, (mallard duck orl) > 4640 mg/kg; LC_{50} (rat ihl 1 hr.) > 4.7 mg/l air, (mallard duck 8 day dietary) > 21500 mg/kg diet, (bobwhite quail 8 day dietary) = 5200 mg/kg diet, (rainbow trout 96 hr.) = 0.25 mg/l, (bluegill sunfish 96 hr.) = 0.39 mg/l, (channel catfish 96 hr.) = 0.43 mg/l; non-toxic to bees. *BASF plc; Brown Butlin Ltd; Caffaro SpA; Chiltern Farm Chemicals Ltd; Fermenta Animal Health Co.; Fermenta; GB Biosciences Corp.; Griffin LLC; Henkel; Hüls Am.; ICI Agrochemicals; Shell UK; Snia UK; Syngenta Crop Protection; Thor; Value Gardens Supply LLC.*

821   **Chlorothiolalcohol**

693-07-2   H

C4H9ClS

2-Chloroethyl ethyl sulfide.

4-01-00-01407 (Beilstein Handbook Reference); BRN 0635672; Chlordiethylsulfid; 1-Chloro-2-(ethyl-thio)ethane; 2-Chloroethyl ethyl sulfide; 2-Chloroethyl ethyl sulphide; Chlorothiolalcohol; EINECS 211-742-1; Ethane, 1-chloro-2-(ethylthio)-; Ethyl β-chloroethyl sulfide; Ethyl 2-chloroethyl sulfide; β-Ethylmerkapto-ethylchlorid; 2-(Ethylthio)chloroethane; 2-Ethylthio-ethyl chloride; h-Mg; h-MG; Half-mustard gas; HSDB 5744; NSC 10977; Sulfide, chloroethyl ethyl; Sulfide, 2-chloroethyl ethyl.

822   **2-Chlorothiophene**

96-43-5   G

C4H3ClS

2-Thienyl chloride.

EINECS 202-505-3; NSC 8747; Thiophene, 2-chloro-. Used in

organic synthesis. Liquid; mp = -71°; bp = 128.3°; d^{20} = 1.2863; λ_m = 235 nm (MeOH); insoluble in H_2O, slightly soluble in $CHCl_3$, very soluble in EtOH, Et_2O. *Janssen Chimica.*

823   **o-Chlorotoluene**

95-49-8   2190   H

C7H7Cl

Benzene, 1-chloro-2-methyl-.

4-05-00-00805 (Beilstein Handbook Reference); AI3-15912; Benzene, 1-chloro-2-methyl-; BRN 1904175; CCRIS 7785; o-Chlorotoluene; EINECS 202-424-3; Halso 99; HSDB 5291; NSC 8766; Toluene, o-chloro-; o-Tolyl chloride; UN2238. Oil; mp = -35.6°; bp = 159.0°; d^{20} = 1.0825; λ_m = 212, 259, 265, 273 (ε = 9390, ?, 288, 250, MeOH); insoluble in H_2O, soluble in EtOH, C_6H_6, freely soluble in Et_2O, Me_2CO, $CHCl_3$, CCl_4.

824   **p-Chlorotoluene**

106-43-4   2190   H

C7H7Cl

4-Chlorotoluene.

AI3-16045; Benzene, 1-chloro-4-methyl-; CCRIS 6000; 1-Chloro-4-methylbenzene; 4-Chloro-1-methyl-benzene; 4-Chlorotoluene; p-Chlorotoluene; EINECS 203-397-0; HSDB 1343; 1-Methyl-4-chlorobenzene; NSC 404114; Toluene, p-chloro-; UN2238. Liquid; mp = 7.5°; bp = 162.4°; d^{20} = 1.0697; λ_m = 220, 263, 269, 277 nm (ε = 11000, 388, 527, 507, MeOH); insoluble in H_2O, soluble in EtOH, Et_2O, CCl_4, $CHCl_3$, AcOH.

825   **3-Chloro-p-toluidine**

95-74-9   H

C7H8ClN

Benzenamine, 3-chloro-4-methyl-.

4-12-00-01985 (Beilstein Handbook Reference); 4-Amino-2-chlorotoluene; Benzenamine, 3-chloro-4-methyl-; BRN 0636511; CCRIS 152; DRC 1339; DRC 1347; EINECS 202-446-3; HSDB 2060; 4-Methyl-3-chloroaniline; NCI-C02040; NSC 96620; p-Toluidine, 3-chloro-. Solid; mp = 26°; bp = 243°; λ_m = 237, 296 nm (cyclohexane); soluble in EtOH, poorly soluble in CCl_4.

826   **Chlorotoluron**

15545-48-9   2191   H

C10H13ClN2O
N'-(3-Chloro-4-methylphenyl)-N,N-dimethylurea.
4-12-00-01986 (Beilstein Handbook Reference); BRN 2647688; C 2242; Caswell No. 216D; CGA 15646; 3-(3-Chlor-4-methylphenyl)-1,1-dimethylharnstoff; 3-(3-Chloro-4-methylphenyl)-1,1-dimethylurea; N-(3-Chloro-4-methylphenyl)-N',N'-dimethylurea; N'-(3-Chloro-4-methylphenyl)-N,N-dimethylurea; Chloro-toluron; 3-(3-Chloro-p-tolyl)-1,1-dimethylurea; Clor-tokem; Deltarol; Dicuran; Dicurane; Dikurin; N,N-Dimethyl-N'-(3-chloro-4-methylphenyl)urea; EINECS 239-592-2; EPA Pesticide Chemical Code 216500; Higaluron; Highuron; HSDB 2760; Ludorum; Talisman; Tolurane; Tolurex; Toro; Urea, 3-(3-chloro-p-tolyl)-1,1-dimethyl-; Urea, N'-(3-chloro-4-methyl-phenyl)-N,N-dimethyl-. Selective pre-and post-emergence herbicide absorbed by foliage and roots. Used with winter cereals for control of annual grasses and broad leaved weeds. Inhibits photosynthesis. Colorless crystals; mp = 147 - 148°; soluble in H2O (0.007 g/100 ml), Me2CO (3.9 g/100 ml), CH2Cl2 (5.7 g/100 ml), C6H6 (2.1 g/100 ml), EtOAc (1.8 g/100 ml), i-PrOH (1.2 g/100 ml); LD50 (rat orl) > 10000 mg/kg, (rat der) > 2000 mg/kg; LC50 (rainbow trout 96 hr.) = 20 - 35 mg/l, (bluegill sunfish 96 hr.) = 40 - 50 mg/l, (crucian carp 96 hr.) > 100 mg/l; non-toxic to birds; non-toxic to bees. *Agan Chemical Manufacturers; Agrolinz; Ciba-Geigy Corp.; Diachem; Farmers Crop Chemicals Ltd.; Hightex; Kemichrom; Makhteshim-Agan; Syngenta Crop Protection.*

827 2-Chloro-5-trichloromethylpyridine
69045-78-9   H

C6H3Cl4N
Pyridine, 2-chloro-5-(trichloromethyl)-.

828 Chlorotriethylsilane
828 Chlorotriethylsilane
994-30-9   G

C6H15ClSi
Triethylchlorosilane.
CT2520; EINECS 213-615-6; Silane, chlorotriethyl-; Triethylchlorosilane. Coupling agent, chemical intermediate, blocking agent, release agent, lubricant, primer, reducing agent. Liquid; bp = 142-144°; d = 0.8980. *Degussa-Hüls Corp.*

829 Chlorotrifluoromethane
75-72-9   G

CClF3
Monochlorotrifluoromethane.
4-01-00-00034 (Beilstein Handbook Reference); Arcton; Arcton 3; BRN 1732392; CFC; CFC-13; Chlorotrifluoromethane; Chlorotrifluoromethane or Refrigerant gas R 13; EINECS 200-894-4; F 13; FC 13; Fluorocarbon 13; Freon; Freon 13; Frigen; Frigen 13; Genetron; Genetron 13; Halocarbon 13; Halocarbon 13/ucon 13; HSDB 140; Methane, chlorotrifluoro-; Methane, monochlorotrifluoro-; R-13; Trifluorochloro-methane; Trifluoromethyl chloride; Trifluoromono-chlorocarbon; UN1022. Refrigerant, dielectric and aerospace chemical, hardening of metals, pharm-aceutical processing. As Aerosol® propellant. Low temperature refrigerant used in low stage of cascade systems to provide evaporator temperatures of - 100°F. Colorless gas; mp = -181°; bp = -81.4°. *Allied Signal; Elf Atochem N. Am.; PCR.*

830 3-(2-Chloro-4-(trifluoromethyl)-phenoxy)phenol acetate
50594-77-9   H

C15H10ClF3O3
Phenol, 3-(2-chloro-4-(trifluoromethyl)phenoxy)-, acetate.
2-Chloro-1-(3-acetoxyphenoxy)-4-(trifluoromethyl)-benzene; 2-Chloro-4-trifluoromethyl-3'-acetoxy-di-phenyl ether; 3-(2-Chloro-4-(trifluoromethyl)-phenoxy)-phenol acetate; EINECS 256-635-0; Phenol, 3-(2-chloro-4-(trifluoromethyl)phenoxy)-, acetate

831 Chlorotrifluoroethylene
79-38-9   H

C2ClF3
1-Chloro-1,2,2-trifluoroethylene.
4-01-00-00704 (Beilstein Handbook Reference); BRN 1740373; CFE; Chlorotrifluoroethylene; Chlortrifluor-aethylen; Chlortrifluoräthylen; CTFE; Daiflon; EINECS 201-201-8; Ethene, chlorotrifluoro-; Ethylene, chlorotrifluoro-; Ethylene, trifluorochloro-; Fluoroplast 3; Genetron 1113; HSDB 2806; Monochlorotri-fluoroethylene; Trifluorchlorethylen; Trifluorochloro-ethylene; Trifluoromonochloroethylene; Trifluorovinyl chloride; Trithene; UN1082. Used as a monomer in production of fluoropolymers, particularly Ethylene-Chlorotrifluoroethylene Copolymer (ECTFE). Gas; mp = -158°; bp = -27.8°; d-60 = 1.54; soluble in C6H6, CHCl3. *Allied Signal; Fluorochem Ltd.*

832 4-Chloro-α,α,α-trifluorotoluene
98-56-6   2145   H

C7H4ClF3
Benzene, 1-chloro-4-(trifluoromethyl)-.
4-05-00-00815 (Beilstein Handbook Reference); BRN 0510203; CCRIS 720; EINECS 202-681-1; HSDB 4251; NSC 10309; p-Chloro-α,α,α-trifluorotoluene; p-Chlorobenzotrifluoride; p-Chlorotrifluoromethyl-benzene; p-Trifluoromethylphenyl chloride; Toluene, p-chloro-α,α,α-trifluoro-; p-(Trifluoromethyl)chloro-benzene. Used in manufacture of pesticides and as a dielectric fluid and solvent. Oil; mp = -33°; bp = 138.5°; d25 = 1.3340; λm = 219, 255, 261, 266, 272 nm (ε = 6680, 184, 207, 160, 117, MeOH); almost insoluble in H2O (0.0029 g/100 ml).

833 Chlorotriphenyltin
639-58-7   H

C8H15ClSn

Triphenyltin chloride.

AI3-25207; Aquatin; Aquatin 20 EC; Brestanol; Caswell No. 896D; CCRIS 6325; Chlorotriphenylstannane; EINECS 211-358-4; EPA Pesticide Chemical Code 496500; Fentin chloride; GC 8993; General chemicals 8993; HOE 2872; HSDB 6404; LS 4442; NSC 1214; NSC 43675; Phenostat-C; Stannane, chlorotriphenyl-; Tinmate; TPTC; Triphenylchlorostannane; Triphenylchlorotin; Triphenyltin chloride. Solid; mp = 103.5°; λ_m = 245, 251, 257, 261, 267 nm (ε = 620, 830, 1040, 887, 636, MeOH); soluble in CHCl₃.

834   Chloroxuron
1982-47-4   G P

C15H15ClN2O2

N-[4-(4-chlorophenoxy)phenyl]-N,N-dimethylurea.

BRN 2814275; C-1933; C 1983; Caswell No. 217B; 3-(4-(4-Chloor-fenoxy)-fenyl-1,1-dimethylureum; 3-(4-(4-Chloor-fenoxy)-fenyl)-1,1-dimethylureum; 1-(4-(4-Chloro-phenoxy)phenyl)-3,3'-methyluree; 3-(4-(4-Chlorophenoxy)phenyl)-1,1-dimethylurea; 3-(4-(4-Chlorphenoxy)-phenyl)-1,1-dimethylharnstoff; 3-(p-(p-Chlorophenoxy)phenyl)-1,1-dimethylurea; 3-(4-(4-Chlorofenossil)fenil)-1,1-dimetil-urea; 3-(4-(4-Cloro-fenossil)-fenossil)-1,1-dimetil-urea; Chloroxifenidim; Chloroxuron; Chlorphencarb; Ciba 1983; EINECS 217-843-7; EPA Pesticide Chemical Code 025501; Gesa-moos; HSDB 980; N'-(4-Chlorophenoxy)phenyl-N,N-dimethylurea; N'-4-(4-Chlorophenoxy)phenyl-N,N-di-methylurea; Norex; Tenoran; Urea, 3-(p-(p-chloro-phenoxy)phenyl)-1,1-dimethyl-; Urea, N'-(4-(4-chloro-phenoxy)phenyl)-N,N-dimethyl-. Selective urea herb-icide, inhibits photosynthesis. Used for pre- and post-emergence control of annual broad-leaved weeds and some grasses in peas, beans, carrots, celery, onions, leeks, garlic, chives, fennel, parsley, dill, tomatoes, cucurbits, soya beans, strawberries and ornamentals. Crystals; mp = 151-152°; soluble in H₂O (3.7 mg/l), more soluble in organic solvents; LD50 (rat orl) = 3700 mg/kg. Ciba Geigy Agrochemicals.

835   Chloroxylenol
88-04-0   2194   D G

C8H9ClO

4-Chloro-3,5-xylenol.

4-06-00-03152 (Beilstein Handbook Reference); AI3-08632; Benzytol; BRN 1862539; Caswell No. 218; Chloroxylenol; Chloroxylenolum; Chlorxylenolum; Clorossilenolo; Cloroxilenol; Desson; Dettol; Dettol, liquid antiseptic; EINECS 201-793-8; EPA

Pesticide Chemical Code 086801; Espadol; Husept Extra; Nipacide MX; NSC 4971; Ottasept; Ottasept Extra; Parametaxylenol; PCMX; RBA 777; Septiderm-Hydrochloride; Willenol V. Antiseptic, antibacterial and germicide. Used as a mildewcide, a topical urinary bactericide and a germicide. A preservative for disinfectant, algicide, slimicide, and water treatment pesticide products, polymer emulsions, adhesives, latex paints, metalworking cutting fluids. Solid; mp = 115°; bp = 246°; λ_m = 279, 287 nm (ε = 2089, 2089, cyclohexane); volatile with steam; slightly soluble in H₂O; more soluble in organic solvents; LD50 (rat orl) = 3830 mg/kg. Nipa.

836   Chlorphenesin
104-29-0   2196   D G

C9H11ClO3

3-(4-Chlorophenoxy)-1,2-propanediol.

4-06-00-00831 (Beilstein Handbook Reference); Adermykon; AI3-24623; BRN 2210845; Chlorophenesin; Chlorphenesin; Chlorphenesine; Chlorphenesinum; Clorfenesina; Demykon; EINECS 203-192-6; Gecophen; Mycil; NSC 6401. A medical antifungal agent. Crystals; mp = 78°; bp19 = 214-215°; λ_m = 228, 281, 288 nm (ε = 12400, 1580, 1280, MeOH); insoluble in H₂O (< 1 g/100 ml), soluble in C₆H₆, very soluble in EtOH, Et₂O. British Drug Houses.

837   Chlorpropham
101-21-3   2206   G P

C10H12ClNO2

Carbamic acid, (3-chlorophenyl)-, 1-methylethyl ester.

4-12-00-01149 (Beilstein Handbook Reference); AI3-18060; Beet-Kleen; BRN 2211397; Bud Nip; Bygran; Carbamic acid, (3-chlorophenyl)-, 1-methylethylester; Carbanilic acid, m-chloro-, isopropyl ester; Caswell No. 510A; N-(3-Chloor-fenyl)-isopropyl carbamaat; Chlor IFK; Chlor IPC; Chlor-IFC; 3-Chlorocarbanilic acid, isopropyl ester; N-(3-Chloro phenyl) carbamate d'isopropyle; N-(3-Chlorophenyl)carbamic acid, isopropylester; N-(3-Chlorophenyl)isopropyl carb-amate; N-(3-Cloro-fenil)-isopropil-carbammato; N-3-Chlorophenylisopropyl-carbamate; N-(3-Chlor-phenyl)-isopropyl-carbamat; Chloro-IFK; Chloro-IPC; (3-Chlorophenyl)carbamic acid, 1-methylethylester; Chloropropham; Chlor-propham; Chlorprophame; Cl-IFK; Cl-IPC; CIPC; EINECS; Elbanil; ENT 18,060; EPA Pesticide Chemical Code 018301; Fasco WY-hoe; Furloe; Furloe 3EC; Furloe 4EC; HSDB 981; Iso-propyl N-chlorophenylcarbamate; Isopropyl 3-chloro-carbanilate; Isopropyl-m-chlorocarbanilate; Isopropyl 3-chlorophenylcarbamate; Isopropyl metachloro-carbanilate; Isopropyl N-(m-chlorophenyl)carbamate; Isopropyl-N-(3-chlorophenyl)carbamate; Isopropyl-N-(3-chlorphenyl)-carbamat; Isopropyl-N-m-chloro-phenylcarbamate; Jack Wilson chloro 51(oil); Keim-stop; Liro CIPC; 1-Methylethyl(3-chlorophenyl)-carbamate; Metoxon; Mirvale; Nexoval; NSC 29466; O-Isopropyl N-(3-chlorophenyl)carbamate; Preventol; Preventol 56; Preweed; Sprout nip; Sprout-nip EC; Spud-Nic; Spud-Nie; Stopgerme-S; Taterpex; Triherbicide CIPC; Unicrop CIPCY. Selective systemic herbicide and growth regulator. Used for pre-emergence control of annual grasses and broad-leaved weeds. Registered by EPA as a herbicide. Colorless crystals; mp = 41°; bp2 = 149°; d^{30} = 1.180; soluble in H₂O (0.0089 g/100 ml), more soluble in organic solvents;

LD50 (rat orl) = 5000 - 7500 mg/kg, (rbt orl) = 5000 mg/kg, (mallard duck orl) > 2000 mg/kg; LC50 (bluegill sunfish 48 hr.) = 12 mg/l, (bass 48 hr.) = 10 mg/l; non-toxic to bees. *Aceto; Atlas Interlates Ltd; Cerexagri, Inc.; Dean AG; Hortichem Ltd.; ICI Chem. & Polymers Ltd.; Mirfield Sales Services Ltd; MTM AgroChemicals Ltd.; Pfizer Intl.; UAP - Platte Chemical; UAP-West.*

838   Chlorpyrifos
2921-88-2   2208   G H P

OH

C9H11Cl3NO3PS

O,O-Diethyl O-(3,5,6-trichloro-2-pyridinyl) phos-phorothioate.
AI3-27311; Bonidel; BRN 1545756; Brodan; Caswell No. 219AA; CCRIS 7144; Chlorpyrifos; Chlorpyrifos-ethyl; Chlorpyriphos; Chlorpyriphos-ethyl; Coroban; Crossfire; Danusban; Detmol U.A.; Dhanusban; Dowco 179; Durmet; Dursban; Dursban 10CR; Dursban 44; Dursban 4E; Dursban F; Dursban R; EINECS 220-864-4; Empire 20; ENT 27311; EPA Pesticide Chemical Code 059101; Equity; Ethyl chlorpyriphos; Geodinfos; Grofo; HSDB 389; Killmaster; Lentrek; Lock-On; Lorsban; Lorsban 50SL; OMS-0971; O,O-Diaethyl-O-3,5,6-trichlor-2-pyridyl-monothiophosphat; O,O-Dithyl-O-3,5,6-trichlor-2-pyridylmonothiophosphat; O,O-Diethyl O-3,5,6-trichloro-2-pyridyl phosphorothioate; O,O-Diethyl O-(3,5,6-trichloro-2-pyridinyl)phosphorothioate; O,O-Diethyl O-(3,5,6-trichloro-2-pyridyl) phsophorothioate; Phosphorothioic acid, O,O-diethyl O-(3,5,6-trichloro-2-pyridinyl) ester; Phosphorothioic acid, O,O-diethyl O-(3,5,6-trichloro-2-pyridyl) ester; Piridane; 2-Pyridinol, 3,5,6-trichloro-, O-ester with O,O-diethyl phosphorothioate; Pyrinex; Radar; Radar (fungicide); Spannit; Stipend; suSCon Blue; suSCon Green; Suscon; Tafaban; Talon; Terial; Terial 40L; Trichlorpyrphos; XRM 429; XRM 5160; Zidil; 3,5,6-trichloro-2-pyridinol O-ester with O,O-diethyl phosphorothioate; AI3-27311; BRN 1545756; Brodan; Caswell No. 219AA; CCRIS 7144; Chloropyrifos; Chloropyriphos; Chlor-pyrifos-ethyl; Chlorpyrifos 4E-AG-SG; Chlorpyrifos ethyl; Chlorpyriphos-ethyl; Chlorpyritos; Coroban; Danusban; Detmol; Detmol U.A.; O,O-Diäthyl-O-3,5,6-trichlor-2-pyridylmonothiophosphat; Diethyl O-(3,5,6-trichloro-2-pyridyl) phosphorothioate; Diethyl O-(3,5,6-trichloro-2-pyridinyl) phosphorothio-ate; Dowco 179; Durmet; Dursban; Dursban 10cr; Dursban 4E; Dursban f; Dursban 44; Dursban 4E; Dursban F; Dursban HF; Dursban/Lorsban; Dursban(R); ENT 27311; EPA Pesticide Chemical Code 059101; Eradex; Ethion, dry; HSDB 389; Killmaster; Lorsban; Lorsban 50SL; Lorsban 4E-SG; OMS-0971; Phosphorothioic acid O,O-diethyl O-(3,5,6-trichloro-2-pyridinyl) ester; Phosphorothioic acid, O,O-diethyl O-(3,5,6-trichloro-2-pyridyl) ester; Piridane; Pyrindol, 3,5,6-trichloro-, O-ester with O,O-diethylphos-phorothioate; Pyrinex; Stipend; Super I.Q.A.P.T; Suscon; Terial; Terial 40l; Trichlorpyrphos; u.a; Zidil. Non-systemic insecticide. A broad-spectrum insect-icide with many crop uses, an organophosphorus insecticide used for control of soil insects. Registered by EPA as a herbicide. Insecticides containing chlorpyrifos are used to control ticks on cattle, mosquitoes, and other insects. Colorless crystals; mp = 41-42°; λm = 208, 230, 290 nm; insoluble in H2O (0.0002 g/100 ml), soluble in C6H6 (695 g/100 ml), Me2CO (512 g/100 ml), CHCl3 (932 g/100 ml), CS2 (743 g/100 ml), Et2O (358 g/100 ml), xylene (344 g/100 ml), CH2Cl2 (532 g/100 ml), isooctane (55 g/100 ml), MeOH (36 g/100 ml); LD50 (rat orl) = 135 - 163 mg/kg, (gpg orl) = 504 mg/kg, (rbt orl) = 1000 - 2000 mg /kg, (rbt der) = 2000 mg/kg, (ckn orl) = 32 mg/kg, (bee contact) = 59 ng/bee, (bee orl) = 250 ng/bee; LC50 (rat ihl 4 hr.) > 0.2 mg/l air, (rainbow

trout 96 hr.) = 0.003 mg/l, (goldfish 24 hr.) = 0.18 mg/l; toxic to crustaceans, toxic to bees. *Diachem; Dow AgroSciences; Dow UK; Farmers Crop Chemicals Ltd.; FCC; Frowein; Makhteshim-Agan; Pan Britannica Industries Ltd.; Planters Products; Rhône Poulenc Crop Protection Ltd.; Siapa; TopPro Specialties; United Horticultural Supply; W Neudorff GMBH KG; Wilbur-Ellis Co.*

839   Chlorpyrifos methyl
5598-13-0   2208   G P

C7H7Cl3NO3PS
2-(3-Chlorophenoxy)propanamide.
AI3-27520; BRN 1541078; Caswell No. 179AA; Chloropyriphos-methyl; Chlorpyriphos-methyl; Chlor-pyrifos O,O-dimethyl analog; Chlorpyriphos-methyl; Dowco 214; Dursban methyl; EINECS 227-011-5; ENT 27520; EPA Pesticide Chemical Code 059102; HSDB 6981; M 3196; Methyl chlorpyrifos; Methyl chlor-pyriphos; Methyl dursban; Methylchlorpyrifos; Noltran; NSC 60380; OMS-1155; O,O-Dimethyl O-(3,5,6-trichloro-2-pyridyl)phosphorothioate; O,O-Dimethyl O-(3,5,6-trichloro-2-pyridinyl) phosphoro-thioate; Phosphorothioic acid, O,O-dimethyl O-(3,5,6-trichloro-2-pyridinyl) ester; Phosphorothioic acid, O,O-dimethyl O-(3,5,6-trichloro-2-pyridyl) ester; Reldan; Reldan 50 EC; Trichlormethylfos; Tumar; Zertell; O,O-Dimethyl-O-(3,5,6-trichloro-2-pyridyl)-phosphorothioate; Dowco 214; Dursban methyl; EPA Pesticide Chemical Code 021203; Methyl chlorpyrifos; Methyl Dursban; Propionamide, 2-(m-chlorophenoxy)-; Reldan; Trichlormethylfos. An organophosphate insecticide used to control pests in stored grain and oilseed rape. Registered by EPA as an insecticide and acaricide. Crystals; mp = 45.5 - 46.5°; slightly soluble in H2O (0.0005 g/100 ml), Me2CO (504 g/100 ml), C66 (458 g/100 ml), Et2O (336 g/100 ml), CHCl3 (5180 g/100 ml); MeOH (31 g/100 ml), C6H14 (15 g/100 ml); LD50 (rat orl) = 1630 - 2140 mg/kg, (mus orl) = 1100 - 2250 mg/kg, gpg orl) = 2250 mg/kg, (rbt orl) = 2000 mg/kg, (rbt der) > 2000 mg/kg, (rat der) > 3700 mg/kg, (ckn orl) > 7950 mg/kg; LC50 (rat ihl 4 hr.) > 0.67 mg/l, (mallard ducks 8 day dietary) = 2500 - 5000 mg/kg, (rainbow trout 96 hr.) = 0.301 mg/l; toxic to crustacea. *Dow AgroSciences; DowElanco Ltd.; Gustafson LLC; Wellcome Foundation Ltd.*

840   Cholebrine
16034-77-8   5032   G

C12H13I3N2O3
Iocetamic acid.
3-(Acetyl-(3-amino-2,4,6-triiodophenyl)amino)-2-methylpropanoic acid; Acide iocetamique; Acido iocetamico; Acidum iocetamicum; Acidum iocetamicum; β-Alanine, N-acetyl-N-(3-amino-2,4,6-triiodophenyl)-2-methyl-; BRN 2171366; Cholebrine; Cholimil; Colebrina; DRC 1201; EINECS 240-173-1; HSDB 3344; Iocetamic acid; MP 620; N-Acetyl-N-(3-amino-2,4,6-triiodophenyl)-2-methyl-β-alanine; N-Acetyl-N-(3-amino-2,4,6-triiodophenyl)-β-aminoiso-butyric acid; N-Acetyl-N-(2,4,6-triiodo-3-amino-phenyl)-β-aminoisobutyric acid; Propanoic acid, 3-(acetyl(3-amino-2,4,6-triiodophenyl)amino)-2-methyl-; Propionic acid, 3-(acetyl-(3-amino-2,4,6-triiodo-phenyl)amino)-2-methyl-. Diagnostic aid in cholecystography. Cream-white powder; mp = 224-225°; insoluble in H2O, slightly

soluble in Et$_2$O, C$_6$H$_6$,EtOH, Me$_2$CO, CHCl$_3$; LD$_{50}$ (rat orl) = 7.1 g/kg, (rat iv) = 700 mg/kg. *Mallinckrodt Inc.*

841 Cholesterol
57-88-5   2221   D G

C$_{27}$H$_{46}$O
 Cholest-5-en-3-β-ol.
 AI3-03112; CCRIS 2834; Δ^5-Cholesten-3-β-ol; Cholesterin; Cholesterine; Cholesterol; Cholesterol base H; Cholesteryl alcohol; Cholestrin; Cholestrol; Cordulan; Dusoline; Dusoran; Dythol; EINECS 200-353-2; HSDB 7106; Hydrocerin; Kathro; Lanol; Lidinit; Lidinite; Nimco cholesterol base H; NSC 8798; Provitamin D; Super hartolan; Tegolan; Wool alcohols B. P. Found in all body tissues; used as an emulsifying agent in cosmetics and pharmaceutical products; source of estradiol and other steroidal drugs. White solid; mp = 148.5°; bp = 360° (dec); d^{18} = 1.052; [α]$^{20}_D$ = -31.5° (c = 2, Et$_2$O); insoluble in H$_2$O, soluble in organic solvents. *Croda; Enzypharm BV; Schweizerhall Inc.; Solvay Duphar Labs Ltd.; U.S. BioChem.*

842 Choline chloride
67-48-1   2226   G H P

C$_5$H$_{14}$ClNO
 Ethanaminium, 2-hydroxy-N,N,N-trimethyl-, chloride .
 AI3-18302; Ammonium, (2-hydroxyethyl)trimethyl-, chloride; Bilineurin chloride; Biocolina; Biocoline; CCRIS 3716; Chloride de choline; Chlorure de choline; Choline chlorhydrate; Choline chloride; Choline hydrochloride; Cholini chloridum; Cholinium chloride; Cloruro de colina; Colina cloruro; EINECS 200-655-4; Ethanaminium, 2-hydroxy-N,N,N-trimethyl-, chloride; Hepacholine; Hormocline; HSDB 984; Lipotril; Luridin chloride; Neocolina; NSC 402838; Paresan; Trimethyl(2-hydroxyethyl)ammonium chloride .
 Animal feed additive. Freely soluble in H$_2$O, EtOH; LD$_{50}$ (rat orl) = 6.64 g/kg. *Am. Biorganics; Mitsubishi Gas; Penta Mfg.; Tanabe U.S.A. Inc.; UCB Res. Inc.*

843 Chromated copper arsenate
37337-13-6   H

 Chromated copper arsenate.
 Arsenic acid (H$_3$AsO$_4$), copper(2+) salt (2:3), chromated; Chromated copper arsenate. Used as an antifungal agent, to protect wood structures. Use to impregnate building materials, including wood.

844 Chrome black
12018-10-9   2662   G

 Cr$_2$Cu$_2$O$_5$
 Cupric chromite(III).
 Chromium copper oxide (Cr$_2$CuO$_4$); Copper chromite; Copper chromium oxide; Copper dichromium tetraoxide; Cupric chromite; EINECS 234-634-6. Used as a hydrogenation catalyst. Black

tetragonal crystals; dec > 900°; insoluble in H$_2$O, acids.

845 Chrome red
18454-12-1   5424   G

CrPb$_2$O$_5$
 Lead chromate(VI) oxide.
 American vermilion; Arancio cromo; Austrian cinnabar; Basic chromium lead oxide (CrPb$_2$O$_5$); CCRIS 6187; Chinese red; Chrome carmine, chinese scarlet; Chrome cinnabar; Chrome garnet; Chrome orange; Chrome ruby; Chromic acid, lead(2+) salt (1:2); Chromium dilead pentaoxide; Chromium lead oxide; Derby red; Dilead chromate oxide; EINECS 242-339-9; HSDB 6185; Lead chromate oxide; Lead chromate oxide (Pb$_2$(CrO$_4$)O); Lead chromate(IV); Lead chromate(VI) oxide; Persian red; Victoria red; Vienna red. Pigments consisting of basic lead chromate, Red powder; insoluble in H$_2$O.

846 Chromic acetate
1066-30-4   2239   G

C$_6$H$_9$CrO$_6$
 Chromium(III) acetate .
 Acetic acid, chromium(3+) salt; CAC 10; Chromic acetate; Chromic(III) acetate; Chromium acetate; Chromium acetate (Cr(AcO)$_3$); Chromium triacetate; Chromium(III) acetate; EINECS 213-909-4; HSDB 985. Used as a textile mordant, tanning, polymerization and oxidation catalyst, emulsion hardener. [Hydrate] gray-green powder; slightly soluble in H$_2$O; MLD (frg iv) = 6185 mg/kg, (mus iv) = 2290 mg/kg, (rbt iv) = 1604 mg/kg. *Alcoa Ind. Chem.; Noah Chem.; Spectrum Chem. Manufacturing.*

847 Chromic acid
7738-94-5   G P

CrH$_2$O$_4$
 Chromium trioxide.
 Acide chromique; AI3-51760; Caswell No. 221; Chromic acid; Chromic acid (H$_2$CrO$_4$); Chromic(VI) acid; Chromium hydroxide oxide; EINECS 231-801-5; EPA Pesticide Chemical Code 021101; HSDB 6769. Used in manufacture of chemicals (chromates, oxidizing agents, catalysts) and medicines, in process engraving, anodizing, ceramic glazes, colored glass, metal cleaning, inks, tanning, paints, textile mordant, etchant for plastics. Registered by EPA as a fungicide. *British Chrome & Chemical; Elf Atochem N. Am.; OxyChem; Rit-Chem; Sigma-Aldrich Fine Chem.; Spectrum Chem. Manufacturing.*

848 Chromic chloride
10025-73-7   2242   G

Cl3Cr
Chromium(III) chloride anhydrous.
C.I. 77295; CCRIS 5557; Chrometrace; Chromic chloride; Chromic chloride anhydrous; Chromic(III) chloride; Chromium chloride (CrCl3); Chromium trichloride; Chromium(3+) chloride; Chromium(III) chloride; Chromium(III) chloride, anhydrous; Cl 77295; EINECS 233-038-3; HSDB 6341; Puratronic chromium chloride; Trichlorochromium. Used to prepare chromium salts, intermediates, textile mordant, chromium plating, preparation of sponge chromium, catalyst for polymerizing olefins, waterproofing. The hexahydrate [10025-73-7] is used as a dietary supplement. Crystals; mp = 1152°; d^{25} = 2.87; MLD (mus, iv) = 801 mg/kg. *Armour Pharm. Co. Ltd.; Atomergic Chemetals; Cerac; Hoechst Celanese.*

849   Chromic fluoride
7788-97-8   2243   G

CrF3
Chromium(III) fluoride.
Chrome fluorure; Chromic fluoride; Chromic trifluoride; Chromium fluoride (CrF3); Chromium trifluoride; Chromium(III) fluoride; Fluorchrome; EINECS 232-137-9; UN1756; UN1757. A mordant. Used in printing, dyeing woolens, coloring marble, treating silk and polishing metals. Solid; mp = 1100°; d = 3.8; insoluble in H2O, EtOH.

850   Chromic oxide
1308-38-9   2247   G

Cr2O3
Chromium (III) oxide.
11661 Green; Amdry 6410; Amperit 704.0; Anadonis Green; Anhydride chromique; C-Grun; C.I. 77288; C.I. Pigment Green 17; Casalis Green; CCRIS 3182; Chrome Green; Chrome Green F 3; Chrome Ocher; Chrome Ochre; Chrome oxide; Chrome oxide (Cr2O3); Chrome Oxide Green GN-M; Chrome Oxide Green BX; Chrome Oxide Green GP; Chromia; Chromic Acid Green; Chromic oxide; Chromium oxide; Chromium oxide (Cr2O3); Chromium oxide (Cr8O12); Chromium Oxide Green; Chromium Oxide Pigment; Chromium Oxide X1134; Chromium sesquioxide; Chromium trioxide; Chromium(3+) oxide; Chromium(3+) trioxide; Chromium(III) oxide; Chromium(III) oxide (2:3); Chromium(III) sesquioxide; Cl 77288; Cl Pigment Green 17; Cosmetic hydrophobic Green 9409; Cosmetic micro blend chrome oxide 9229; Dichromium trioxide; EINECS 215-160-9; Green Chrome Oxide; Green chromic oxide; Green chromium oxide; Green cinnabar; Green Oxide of Chromium; Green rouge; HSDB 1619; Kromex U 1; Leaf green; Levanox Green GA; M100; Oil Green; OKhP1; Oxide of chromium; P 106F10; Pigment green 17; PK 5304; Pure Chromium Oxide Green 59. Used in metallurgy, green paint pigment, ceramics, catalyst in organic synthesis, green granules in asphalt roofing, component of refractory brick, abrasive. Green solid; mp = 2435°; bp = ca. 3000°; d^{25} = 5.22; insoluble in H2O, EtOH, Me2CO, soluble in alkalis, acids. *Atomergic Chemetals; British Chrome & Chemical; Noah Chem.*

851   Chromic potassium sulfate
10141-00-1   2250   G

CrKO8S2
Potassium chromium(III) sulfate.
0% Basicity chrome alum; CCRIS 7532; Chrome alum; Chrome potash alum; Chromic potassium sulfate; Chromic potassium sulphate; Chromium potassium bis(sulphate); Chromium potassium sulfate (1:1:2); Chromium potassium sulfate (CrK(SO4)2); Chromium potassium sulphate; Chromium(III) potassium sulfate; Crystal Chrome Alum; EINECS 233-401-6; Potassium chromic sulfate; Potassium chromic sulphate; Potassium chromium alum; Potassium chromium disulfate (KCr(SO4)2); Potassium disulphatochromate (III); Sulfuric acid, chromium(3+) potassium salt (2:1:1). Mordant for dyeing fabrics uniformly [dodecahydrate]; mp = 89°; d^{25} = 1.83; soluble in 4 parts H2O.

852   Chromium
7440-47-3   2252   G

Cr

Cr
Chromium.
CCRIS 159; Chrom; Chrome; Chromium; Chromium, elemental; Chromium metal; EINECS 231-157-5; HSDB 910. Metallic element; Cr; alloying and plating element for corrosion resistance, stainless steels, protective coatings, nuclear and high-temperature research, constituent of inorganic pigments. Metal; mp = 1903°; bp = 2642°; d^{20} = 7.14. *Atomergic Chemetals; Cerac; Noah Chem.; Sigma-Aldrich Fine Chem.*

853   Chromium phosphate
7789-04-0   2248   G

CrO4P
Phosphoric acid chromium (III) salt.
Arnaudon's Green; Chromic phosphate; Chromium monophosphate; Chromium orthophosphate; Chromium phosphate; EINECS 232-141-0; Phosphoric acid, chromium(3+) salt (1:1); Plessy's green. Chromic phosphate, used as a pigment.

854   Chromium sulfate
15244-38-9   G

Cr2O12S3.H2O
Chromium(III) sulfate hydrate.
CCRIS 6186; Chromic sulfate; Chromium sulfate hydrate; Santotan KR. Peach-colored solid, occurring as several hydrated forms. Used in leather tanning. Solid; d = 3.012; MLD (mus iv) = 247 mg/kg.

855   Chromium trioxide
1333-82-0   2256   G

143

CrO3

Chromium(VI) trioxide.

Anhydride chromique; Anidride cromica; CCRIS 160; Chrome (trioxyde de); Chromia (CrO3); Chromic acid, solid; Chromic anhydride; Chromic oxide; Chromic trioxide; Chromic(VI) acid; Chromium anhydride; Chromium oxide; Chromium oxide (Cr4O12); Chromium oxide (CrO3); Chromium trioxide; Chromium(6+) oxide; Chromium(6+) trioxide; Chromium(VI) oxide; Chromsaeureanhydrid; Chromtrioxid; Chroomtrioxyde; Chroomzuur-anhydride; Cromo(triossido di); EINECS 215-607-8; HSDB 518; Monochromium trioxide; Puratronic chromium trioxide; Red oxide of chromium; Sintered chromium trioxide; UN1463. Dark red crystals, powerful oxidizing agent; used in chromium plating, copper stripping; aluminum anodizing; photography, corrosion inhibition, hardening microscopical preparations, oxidizing agent in organic chemistry. Red crystals; mp = 197°; d = 2.70; very soluble in H_2O.

856   Chromotrope acid
148-25-4   2261   G

$C_{10}H_8O_8S_2$

4,5-Dihydroxynaphthalene-2,7-disulfonic acid.

AI3-18239; Chromotropic acid; 4,5-Dihydroxy-2,7-naphthalenedisulfonic acid; 4,5-Dihydroxynaphth-alene-2,7-disulphonic acid; EINECS 205-712-7; 2,7-Naphthalenedisulfonic acid, 4,5-dihydroxy-. An azo dye for wool. Needles or leaflets, λ_k = 235, 312, 332, 348 nm (ε = 79433, 6310, 7943, 10000, pH 0.75); soluble in H_2O, alkali, insoluble in EtOH, Et_2O.

857   Chromous chloride
10049-05-5   2264   G

Cl_2Cr

Chromium (II) chloride anhydrous.

Chromium chloride; Chromium chloride (CrCl2); Chromium dichloride; Chromium(II) chloride; Chromous chloride; EINECS 233-163-3; HSDB 988. Reducing agent, catalyst, reagent, chromizing agent. Solid; mp = 824°; d_4^{14} 2.751; soluble in H_2O; LD50 (rat orl) = 1.87 g/kg. Cerac; Noah Chem.; 3973Atomergic; Cerac; Chemetals; Chemical; Noah.

858   Chromous sulfate
13825-86-0   2267   G

CrO_4S

Chromium(II) sulfate pentahydrate.

Chromium sulfate (CrSO4); Chromous sulfate; Chromous(II) sulfate; Sulfuric acid, chromium(2+) salt (1:1). Oxygen scavenger, reducing agent, analytical reagent. [pentahydrate]; blue crystals; soluble in H_2O, EtOH, insoluble in less polar organic solvents.

859   Chrysene
218-01-9   2276   G

$C_{18}H_{12}$

1,2-Benzphenanthrene.

AI3-00867; 1,2-Benzphenanthrene; Benz(a)phen-anthrene; Benzo(a)phenanthrene; 1,2-Benzophen-anthrene; CCRIS 161; Chrysene; Coal tar pitch volatiles: chrysene; 1,2,5,6-Dibenzonaphthalene; EINECS 205-923-4; HSDB 2810; NSC 6175; RCRA waste number U050. Found in coal tar. Used in chemical synthesis. mp = 258.2°; bp = 448°; d_4^{20} 1.274; λ_m = 257, 267, 294, 306, 319 nm (ε = 92000, 164000, 11400, 12100, 12200, MeOH); insoluble in H_2O, slightly soluble in EtOH, Et_2CO, Me_2CO, C_6H_6, CS_2, AcOH, soluble in C_7H_8.

860   Chrysoidine
532-82-1   2279   G

$C_{12}H_{13}ClN_4$

4-(Phenylazo)-1,3-benzenediamine hydrochloride.

Astra Chrysoidine R; Atlantic Chrysoidine Y; Basic Orange 2; Basonyl Orange 200; 1,3-Benzenediamine, 4-(phenylazo)-, monohydrochloride; Brasilazina Orange Y; Brilliant Oil Orange Y Base; C.I. 11270; C.I. Basic Orange 2; C.I. Basic Orange 3; C.I. Basic Orange 2, monohydrochloride; C.I. Solvent Orange 3; Calcozine Chrysoidine Y; Calcozine Orange YS; CCRIS 162; Chrysoidin; Chrysoidin FB; Chrysoidin Y; Chrysoidin YN; Chrysoidine; Chrysoidine (II); Chrysoidine A; Chrysoidine B; Chrysoidine C crystals; Chrysoidine Crystals; Chrysoidine G; Chrysoidine GN; Chrysoidine GS; Chrysoidine HR; Chrysoidine J; Chrysoidine M; Chrysoidine Orange; Chrysoidine PRL; Chrysoidine PRR; Chrysoidine SL; Chrysoidine special (biological stain and indicator); Chrysoidine SS; Chrysoidine Y; Chrysoidine Y Crystals; Chrysoidine Y EX; Chrysoidine Y Special; Chrysoidine YGH; Chrysoidine YL; Chrysoidine YN; Chrysoidine(II); Chryzoidyna F.B.; CI 11270; 2,4-Diaminoazobenzene hydrochloride; Diazocard chrysoidine G; EINECS 208-545-8; Elcozine chrysoidine Y; HSDB 5491; Leather Orange HR; 4-(Phenylazo)-1,3-phenylenediamine, monohydrochloride; 4-(Phenylazo)-m-phenylene-diamine, monohydrochloride; 4-Phenylazophenylene-1,3-diamine monohydrochloride; m-Phenylene-diamine, 4-(phenylazo)-, monohydrochloride; m-Phenylenediamine, 4-(phenylazo)-, hydrochloride; Nippon Kagaku Chrysoidine; NSC 152834; Pure chrysoidine YD; Pure Chrysoidine YBH; Pyracryl Orange Y; Sugai Chrysoidine; Tertrophene Brown CG; Verona Chrysoidine GN; Chrysoldine crystal. A dyestuff. It consists of the hydrochloride of phenylazo-m-phenylenediamine, with some of the homologues from o-and p-toluidine. Dyes wool and silk orange. The citrate is used as an antiseptic. Reddish-brown crystals; mp = 118.5°; soluble in H_2O (5.5%, 15°), very soluble in Me_2CO, insoluble in C_6H_6.

861   CI Direct Blue 15
2429-74-5   H

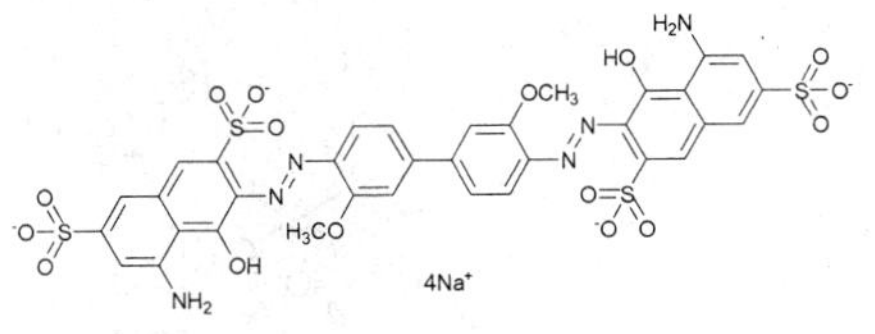

C34H24N6Na4O16S4

2,7-Naphthalenedisulfonic acid, 3,3'-((3,3'-dimethoxy-(1,1'-biphenyl)-4,4'-diyl)bis(azo))bis(5-amino-4-hydroxy-, tetrasodium salt.

Airedale Blue D; Aizen Direct Sky Blue 5B; Aizen Direct Sky Blue 5BH; Amanil Sky Blue; Atlantic Sky Blue A; Atul Direct Sky Blue; Azine Sky Blue 5B; Belamine Sky Blue A; Benzanil Sky Blue; Benzo Sky Blue A-CF; Benzo Sky Blue S; Cartalsol Blue 2GF; Cartasol Blue 2GF; CCRIS 2419; Chloramine Sky Blue 4B; Chloramine Sky Blue A; Chrome Leather Pure Blue; CI 24400; CI Direct Blue 15; CI Direct Blue 15, tetrasodium salt; Cresotine Pure Blue; Diacotton Sky Blue 5B; Diamine Blue; Diamine Blue 6B; Diamine Sky Blue; Diamine Sky Blue CI; Diaphtamine Pure Blue; Diazol Pure Blue 4B; Diphenyl Brilliant Blue; Diphenyl Sky Blue 6B; Direct blue 15; Direct Blue 10G; Direct Blue 15; Direct Blue HH; Direct Pure Blue M; Direct Pure Blue; Direct Pure Blue N; Direct Sky Blue 5B; Direct Sky Blue; Direct Sky Blue A; EINECS 219-385-3; Enianil Pure Blue AN; Fenamin Sky Blue; Hispamin Sky Blue 3B; HSDB 4227; Kayafect Blue Y; Kayaku Direct SKH Blue 5B; Kayaku Direct Sky Blue 5B; Mitsui Direct Sky Blue 5B; Modr Prima 15; Naphtamine Blue 10G; NCI C61290; Niagara Blue 4B; Niagara Sky Blue; Nippon Direct Sky Blue; Nippon Sky Blue; Nitsui Direct Sky Blue 5B; Nitto Direct Sky Blue 5B; NSC 47757; NSC 9616; Oxamine Sky Blue 5B; Paper Blue S; Phenamine Sky Blue A; Pontacyl Sky Blue 4BX; Pontamine Sky Blue 5 BX; Pontamine Sky Blue 5BX; Shikiso Direct Sky Blue 5B; Sky Blue 4B; Sky Blue 5B; Tertrodirect Blue F; Tetrasodium 3,3'-((3,3'-dimethoxy(1,1'-biphenyl)-4,4'-; Vondacel Blue HH.

862   CI Disperse Blue 79:1
3618-72-2   H

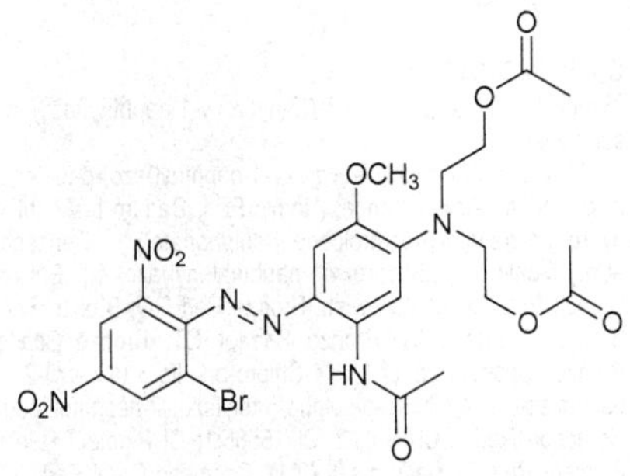

C23H25BrN6O10

3-(2,4-Dinitro-6-bromophenylazo)-4-acetamido-6-(N,N-bis(acetoxyethyl)amino)anisole.

4-(6-Bromo-2,4-dinitrophenylazo)-3-acetylamino-6-methoxy-N-bis(acetoxyethyl)aniline; C.I. Disperse Blue 79:1; 3-(2,4-Dinitro-6-bromophenylazo)-4-acetamido-6-(N,N-bis(acetoxyethyl)amino)anisole; EINECS 222-813-1.

863   CI Fluorescent brightener
16090-02-1   H

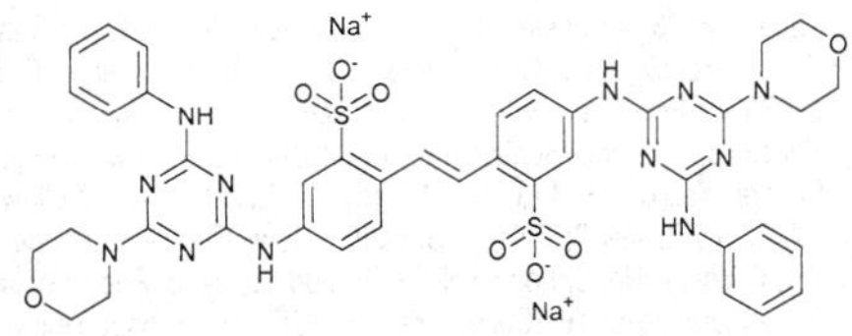

C40H38N12Na2O8S2

Disodium   4,4'-bis((4-anilino-6-morpholino-1,3,5-tri-azin-2-yl)amino)stilbene-2,2'-disulphonate.

4,4'-Bis((4-anilino-6-morpholino-s-triazin-2-yl)-amino)-2,2'-stilbenedisulfonate; 4,4'-Bis((4-anilino-6-morpho-lino-1,3,5-triazine-2-yl)amino)stilbene; Belotex KD; Blankophor BBH; Blankophor MBBH; C.I. Fluorescent Brightener 71; C.I. Fluorescent Brightener 244; C.I. Fluorescent Brightener 250; C.I. Fluorescent Brightener 260; Calcofluor White RC; CCRIS 3187; CI Fluorescent Brightener 260; Disodium 4,4'-bis((4-anilino-6-morpholino-1,3,5-triazin-2-yl)amino)stilbene-2,2'-di-sulfonate; EINECS 240-245-2; FBA-260; Heliofor 3BC; Heliofor BDC; Hiltamine Arctic White DML; HSDB 5061; Leukopur PAM; MBBH 766; Mikephor TB; Phorwite MBBH 766; Stralex MD; Tinopal AMS; Tinopal dms-x; Tinopal DMS-X; Tinopal EMS.

864   CI Food Red 17
25956-17-6   281   H

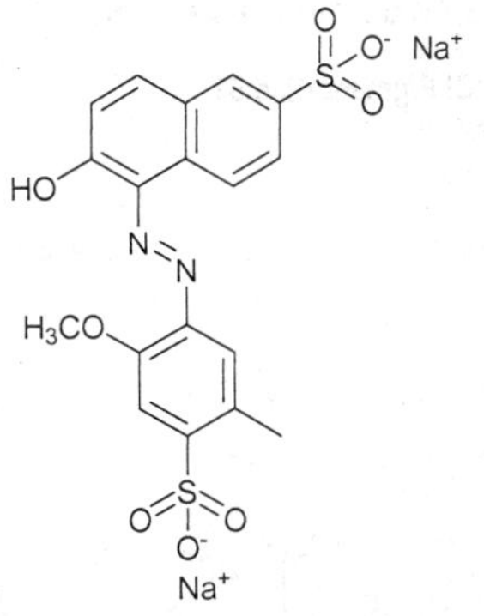

C18H14N2Na2O8S2

Disodium   6-hydroxy-5-((6-methoxy-4-sulfo-m-tolyl)-azo)-2-naphthalenesulfonate.

Allura Red; Allura Red AC Dye; Allura Red AC; CI Food Red 17; CCRIS 3493; CI 16035; Curry red; EINECS 247-368-0; FD&C Red No. 40; Food Red 17; Food Red No. 40; Red No. 40.

865   CI Food Yellow 3
2783-94-0   9091   H

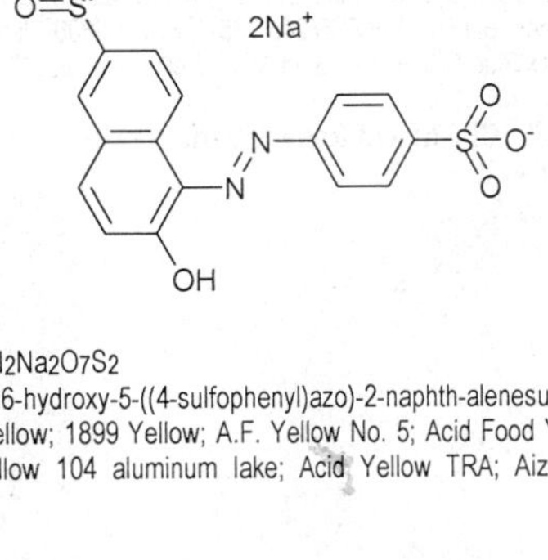

C16H10N2Na2O7S2

Sodium 6-hydroxy-5-((4-sulfophenyl)azo)-2-naphth-alenesulfonate.

1351 Yellow; 1899 Yellow; A.F. Yellow No. 5; Acid Food Yellow 3; Acid Yellow 104 aluminum lake; Acid Yellow TRA; Aizen Food

Yellow No. 5; Alabaster No. 3; Atul Sunset Yellow FCF; Canacert Sunset Yellow FCF; CCRIS 170; Certicol Sunset Yellow CFS; CI 15985; CI Food Yellow 3, disodium salt; Cilefa Orange S; D and C Yellow 6; Dolkwal Sunset Yellow; Dye FDC Yellow No. 6; Dye Sunset Yellow; E 110; E 110 (dye); Edicol Supra Yellow FC; EINECS 220-491-7; Eniacid Sunset Yellow; Eurocert Orange FCF; FD&C Yellow No. 6; Food Yellow 3; Food Yellow 6; Food Yellow No. 5; Gelborange-S; HD Sunset Yellow FCF; HD Sunset Yellow FCF Supra; Hexacol Sunset Yellow FCF; Hexacol Sunset Yellow FCF Supra; HSDB 4136; KCA Foodcol Sunset Yellow FCF; L. Orange Z2010; Maple Sunset Yellow FCF; NCI-C53907; Orange II R; Orange PAL; Orange Yellow S; Orange Yellow S. AF; Orange Yellow S. FQ; Orange Yellow S.fq; Para Orange; Solar radiation color; Standacol Sunset Yellow FCF; Sun Orange A Geigy; SUN Yellow; SUN Yellow A-CE; SUN Yellow A-FDC; SUN Yellow Extra Conc. A Export; SUN Yellow Extra Pure A; SUN Yellow FCF; Sunlight Yellow FCF; Sunset Yellow; Sunset Yellow 6; Sunset Yellow FCF; Sunset Yellow FCF 6; Sunset Yellow FCF Supra; Sunset Yellow FU; Sunset Yellow FU Supra; Sunset Yellow Lake; Twilight Yellow; Usacert FD&C Yellow No. 6; Usacert Yellow No. 6; Yellow No. 6; Yellow Orange S; Yellow Orange S Specially Pure; Yellow Orange Specially Pure 85; Yellow S; Yellow sun; Yellow SY for food; Zlut potravinarska 3.

866   CI Leuco Sulphur Black 1
66241-11-0   H

CI Leuco Sulphur Black 1.
CI 53185; CI Leuco Sulphur Black 1; EINECS 266-273-5.

867   CI Pigment Blue 61
1324-76-1   H

C37H29N3O3S
((4-((4-(phenylamino)phenyl)(4-(phenylimino)-2,5-cyclohexadien-1-ylidene)methyl)phenyl)amino)-benzenesulfonic acid.
((4-((4-(Anilino)phenyl)(4-(phenylimino)-2,5-cyclohexadien-1-ylidene)methyl)phenyl)amino)-benzenesulphonic   acid; Benzenesulfonic acid, ((4-((4-(4-(phenylamino)phenyl)(4-(phenylimino)-2,5-cyclohexa-dien-1-ylidene)methyl)phenyl)amino)-; CI 42765; CI Pigment Blue 61; EINECS 215-385-2; HSDB 5809; Pigment Blue 61; Reflex Blue AGL; Reflex Blue AGM; Reflex Blue R.

868   CI Pigment Red 48, barium salt
7585-41-3   H

C18H11BaClN2O6S
Barium 4-((5-chloro-4-methyl-2-sulphonatophenyl)azo) -3-hydroxy-2-naphthoate.
Bon Red Yellow Shade; Bright Red G Toner; C.I. Pigment Red 48:1; EINECS 231-494-8; Eljon Rubine BS; Isol Bona Red NR barium salt; Isol Bona Red N 5R barium salt; Lithol Scarlet K 3700; Permanent Red BB; Permanent Red BBa; Pigment Red 48:1; Resino Red K; Rubine Toner B; Rubine Toner BA; Rubine Toner BT; Sanyo Fast Red 2BE; Sanyo Fast Red 2B; Segnale Red GS; SeikaFast Red 8040; Symuler Red 3023; Symuler Red NRY; Watchung Red Y.

869   CI Pigment Red 53:1
5160-02-1   H

C34H24BaCl2N4O8S2
Barium   bis(2-chloro-5-((2-hydroxy-1-naphthyl)azo)-toluene-4-sulphonate).
1860 Red; 5-Chloro-2-((2-hydroxy-1-naphthyl)azo)-p-toluenesulfonic acid, barium; Astro Orange; Atomic Red; Barium bis(2-chloro-5-((2-hydroxy-1-naphthyl)-azo)toluene-4-sulphonate); Benzenesulfonic acid, 5-chloro-2-((2-hydroxy-1-naphthalenyl)azo)-4-; Brilliant Red; Bronze Red 16913 Yellowish; Bronze Red RO; Bronze Scarlet CA; Bronze Scarlet CBA; Bronze Scarlet CT; Bronze Scarlet CTA; Bronze Scarlet Toner; 1-(4-Chloro-o-sulfo-5-tolylazo)-2-naphthol, barium salt; 1-(4-Chloro-o-sulpho-5-tolylazo)-2-naphthol, barium salt; Carnation Red; CCRIS 172; CI 15585:1; CI Pigment Red 53:1; CI Pigment Red 53, barium salt (2:1); Cosmetic Coral Red KO Bluish; Cosmetic DVR; Cosmetic Pigment Yellow Red DVR; D&C Red 9; Dainichi Lake Red C; Desert Red; EINECS 225-935-3; Eljon Lake Red C; Hamilton Red; Helio Red Toner LCLL; HSDB 4138; Irgalite Red CBN; Irgalite Red CBR; Irgalite Red CBT; Irgalite Red MBC; Isol Lake Red LCS 12527; Isol Lake LCR 2517; Japan Red 204; Japan Red No. 204; Lake Red 1520; Lake Red C 27217; Lake Red C Toner 8366; Lake Red C; Lake Red C 27200; Lake Red C 27218; Lake Red C 18287; Lake Red C 21245; Lake Red C Toner 8195; Lake Red C Barium Toner; Lake Red C RLC-232 (barium); Lake Red CBA; Lake Red CC; Lake Red CCT; Lake Red CR; Lake Red CRLC-232 (barium); Lake Red GB barium salt; Lake Red RRG; Lake Red Toner C; Lake Red Toner LCLL; Lake Red ZHB; Latexol Scarlet R Solupowder; Latexol Scarlet R; Ld Rubber Red 16913; Lutetia Red CLN; Lutetia Red CLN-ST; Microtex Lake Red CR; Mohican Red A-8008; NCI-C53792; Paridine Red LCL; Pigment Lake Red CD; Pigment Lake Red LC; Pigment Lake Red BFC; Pigment Red (53:1) barium salt; Pigment Red 53:1; Pigment Red CD; Potomac Red; Recolite Red Lake CR; Recolite Red Lake C;

146

Red 16913H; Red 1860; Red For Lake C; Red For Lake C Toner RA-5190; Red For Lake Toner RA-5190; Red Lake C Toner; Red Lake C; Red Lake C Toner RA-5190; Red Lake C Toner 20-5650; Red Lake CM 20-5650; Red Lake CR-1; Red Lake R-91; Red No. 204; Red Scarlet; Red Toner Z; Red ZhB; Rubber Red 16913R; Sanyo Lake Red C; Scarlet Toner Y; Segnale Red LC; Segnale Red LCG; Segnale Red LCL; Sico Lake Red 2L; Sumikaprint Red C-O; Superol Red C RT-265; Symuler Lake Red C; Termosolido Red LCG; Texan Red Toner D; Toner Lake Red C; Transparent Bronze Scarlet; Vulcafix Scarlet R; Vulcafix Scarlet R-D masse; Vulcafor Red 2R; Vulcan Red LC; Vulcol Fast Red L; Wayne Red X-2486.

870   CI Pigment Red 57:1
5281-04-9   H

$C_{18}H_{12}CaN_2O_6S$
3-Hydroxy-4-((4-methyl-2-sulfophenyl)azo)-2-naphth-alenecarboxylic acid, calcium salt.
C.I. Pigment Red 57:1; C.I. Pigment Red 57, calcium salt (1:1); Calcium   3-hydroxy-4-((4-methyl-2-sulph-onatophenyl)azo)-2-naphthoate; CCRIS 4903; CI 15850:1 (Ca salt); D&C Red No. 7; EINECS 226-109-5; 3-Hydroxy-4-((4-methyl-2-sulfophenyl)azo)-2-naphth-alenecarboxylic acid, calcium salt; 3-Hydroxy-4-((2-sulfo-p-tolyl)azo)-2-naphthalenecarboxylic acid, calcium salt (1:1); Lithol rubin B ca; 2-Naphthalenecarboxylic acid, 3-hydroxy-4-((4-methyl-2-sulfophenyl)azo)-, calcium salt.

871   CI Pigment Yellow 12
6358-85-6   H

$C_{32}H_{26}Cl_2N_6O_4$
Acetoacetanilide, 2,2'-((3,3'-dichloro-4,4'-biphenylyl-ene)diazo).
Acetoacetanilide,   2,2'-((3,3'-dichloro-4,4'-biphenyl-ene)diazo)bis-; Amazon Yellow X2485; Benzidene Yellow; Benzidene Yellow ABZ-245; Benzidene Yellow WD-266 (Water Dispersible); Benzidene Yellow YB-1; Benzidine Lacquer Yellow G; Benzidine Yellow; Benzidine Yellow 1178; Benzidine Yellow 45-2685; Benzidine Yellow 45-2650; Benzidine Yellow 45-2680; Benzidine Yellow ABZ-245; Benzidine Yellow E; Benzidine Yellow G; Benzidine Yellow GF; Benzidine Yellow GR; Benzidine Yellow GT; Benzidine Yellow GTR; Benzidine Yellow HG; Benzidine Yellow HG PLV; Benzidine Yellow Toner; Benzidine Yellow Toner YT-378; Benzidine Yellow Toner YA-8081; Benzidine Yellow WD-266 (water dispersible); Benzidine Yellow YB-1; Benzidine Yellow YB 5722; Brilliant Yellow Slurry; C.I. Pigment Yellow 12; Carnelio Yellow GX; CCRIS 203; CI 21090; Dainichi Benzidine Yellow GRT; Dainichi Benzidine Yellow G; Dainichi Benzidine Yellow GT; Dainichi Benzidine Yellow GY;

Dainichi Benzidine Yellow GYT; Daltolite Fast Yellow GT; Diarylanilide Yellow; Diarylanilide Yellow; Diarylide Yellow AAA; Diarylide Yellow YT 553D; diyl)bis(azo)bis(3-oxo-N-phenyl-; EINECS 228-787-8; Eljon Yellow BG; Graphtol Yellow A-HG; Hancock Yellow 10010; Helic Yellow GW; Helio Yellow GWN; Helioyellow GW; HSDB 2926; Irgalite Yellow BO; Irgalite Yellow BST; Irgalite Yellow BTR; Isol Benzidine Yellow G Special; Isol Benzidine Yellow G 2537; Isol Benzidine Yellow Gapropyl; Isol Benzidine Yellow Gbpropyl; Isol Benzidine Yellow G; Kromon Yellow GXT Conc; Kromon Yellow MTB; Light Yellow JB; Light Yellow JBO; Light Yellow JBT; Lodestone Yellow YB-57; Monolite GT; Monolite Yellow 2GRA; Monolite Yellow GRA; Monolite Yellow GT; Monolite Yellow GTA; Monolite Yellow GTN; Monolite Yellow GTNA; Monolite Yellow GTS; NCI-C03269; No. 49 Conc. Benzidine Yellow; NSC 521237; Permanent Yellow DHG; Permanent Yellow GHG; Pigment Yellow 12; Pigment Yellow 12; Pigment Yellow GT; Rangoon Yellow; Recolite Yellow BG; Recolite Yellow BGT; Sanyo Benzidine Yellow-B; Segnale Light Yellow 2GR; Segnale Light Yellow 2GRT; Segnale Light Yellow 2 GR; Segnale Yellow HG; Seikafast Yellow 2300; Siliton Yellow 3GX; Siliton Yellow GTX; Siloton Yellow 3GX; Siloton Yellow GTX; Sumikarrint Yellow 3A-0; Symuler Fast Yellow 219; Symuler Fast Yellow 224; Symuler Fast Yellow 4078; Symuler Fast Yellow GF; Verona Yellow X-1791; Vulcafor Fast Yellow GTA; Vulcafor Fast Yellow GT; Vulcol Fast Yellow GR; Yellow 205; Yellow No. 205 (Japan); Zlut pigment 12.

872   CI Pigment Yellow 14
5468-75-7   H

$C_{34}H_{30}Cl_2N_6O_4$
2,2'-((3,3'-Dichloro(1,1'-biphenyl)-4,4'-diyl)bis(azo))-bis(N-(2-methylphenyl)-3-oxobutyramide).
Aaot Yellow; Atul Vulcan Fast Pigment Oil Yellow T; Benzidene Yellow ABZ-249; Benzidene Yellow G; Benzidine Yellow (Grease proof); Benzidine Yellow AAOT; Benzidine Yellow ABZ 249; Benzidine Yellow G; Benzidine Yellow GGT; Benzidine Yellow L; Benzidine Yellow OT; Benzidine Yellow OT (6Cl); Benzidine Yellow OTYA 8055; C.I. 21095; C.I. Pigment Yellow 14; C.I. Pigment Yellow 14; Calcotone Yellow GP; Diarylide Yellow AAOT; EINECS 226-789-3; Graphtol Yellow GXS; Hostaperm Yellow GT; Hostaperm Yellow GTT; Irgalite Yellow BR; Irgalite Yellow BRE; Isol Benzidine Yellow GO; Lake Yellow GA; Light Yellow JBV; Light Yellow JBVT; Lionol Yellow GGR; No. 55 Conc. Pale Yellow SF; No.55 Conc. Pale Yellow SF; NSC 15087; Permagen Yellow; Permagen Yellow GA; Permanent Yellow G; Permanent Yellow Light; Pigment Fast Yellow 2GP; Pigment Fast Yellow GP; Pigment Yellow 14; Pigment Yellow 2G; Pigment Yellow GGP; Pigment Yellow GPP; Plastol Yellow GG; Plastol Yellow GP; Radiant Yellow; Recolite Fast Yellow B 2T; Resamine Fast Yellow GGP; Resamine Yellow GP; Resorcin Brown R; Rubber Fast Yellow GA; Sacandaga Yellow X 2476; Sanyo Benzidine Yellow; Segnale Yellow 2GR; Seikafast Yellow 2200; Silogomma Fast Yellow 2G; Silotermo Yellow G; Sumatra Yellow X 1940; Sumikaprint Yellow GFN; Symuler Fast Yellow 4090G; Symuler Fast Yellow 5GF; Tertropigment Fast Yellow VG; Tertropigment Yellow BG; Tetropigment Fast Yellow VG; Tetropigment Fast Yellow BG; Versal Fast Yellow PG; Vulcafor Fast Yellow 2G; Vulcan Fast Yellow G; Vynamon Yellow 2G; Vynamon Yellow 2GE.

873   CI Solvent Black 7
8005-02-5   H

CI Solvent Black 7.
CCRIS 4718; CI Solvent Black 7.

874 Cinchophen

132-60-5   2309   D G

$C_{16}H_{11}NO_2$
2-Phenylquinoline-4-carboxylic acid.
5-22-03-00484 (Beilstein Handbook Reference); Acifenokinolin; Aciphenochinoline; Aciphenochin-olinium; Aciphenochinolinum; Agotan; Al3-15400; Alutyl; Alutyo; Aminophan; Artam; Artamin; Artexin; Atigoa; Atocin; Atofan; Atophan; BRN 0192803; Cinchoninic acid, 2-phenyl-; Cinchopen; Cinchophen; Cinchophene; Cinchophenic acid; Cinchophenum; Cincofene; Cincofeno; Cinconal; Cincophen; Cincosal; EINECS 205-067-1; HSDB 2085; Ikterosan; Mylofanol; NSC 2617; Phenophan; Phenoquin; Polyphlogin; Quinofen; Quinophan; Quinophen; Rhematan; Rheumatan; Rheumin; Tervalon; Tophol; Tophosan; Traubofan; Vantyl; Viophan. Quinoline derivative formerly used in treatment of chronic gout. Used experimentally to induce ulcers. Formerly used as an analgesic. Crystals; mp = 213-216°; nearly insoluble in H_2O; more soluble in $CHCl_3$ (0.25 g/100 ml), Et_2O (1 g/100 ml), alcohol (0.83 g/100 ml). *Parke-Davis-*

875 Cinnamal

104-55-2   2319   G H

C_9H_8O
3-Phenyl-2-propenal.
2-07-00-00273 (Beilstein Handbook Reference); Abion CA; Acrolein, 3-phenyl-; Al3-00473; Aldehyd skoricovy; Benzylideneacetaldehyde; BRN 0605737; Cassia aldehyde; Caswell No. 221A; CCRIS 6222; Cinnamal; Cinnamaldehyde; Cinnamic aldehyde; Cinnamyl aldehyde; EINECS 203-213-9; EPA Pesticide Chemical Code 040506; FEMA No. 2286; FEMA Number 2286; Hefty Dog and Cat Repellent; HSDB 209; NCI-C56111; NSC 16935; Phenyl-2-propenal; Phenylacrolein; β-Phenylcrolein; Zimtaldehyde. Insecticide. Registered by EPA as an insecticide and rodenticide. Also used in flavors and perfumery. Oily liquid; mp = -7°; bp = 246°, bp_{400} = 222°, bp_{200} = 199°, bp_{100} = 178°, bp_{60} = 164°, bp_{40} = 152°, bp_{20} = 136°, bp_{10} = 120°, bp_5 = 106°, bp_1 = 76°; d = 1.048 - 1.052; soluble in 60% EtOH (14.3 g/100 ml), Et_2O, $CHCl_3$, fixed oils, slightly soluble in H_2O (0.14 g/100 ml). LD_{50} (rat orl) = 2220 mg/kg. Allergenic. *Florida Treatt; Penta Mfg.; Quest Intl.*

876 Cinnamic acid

621-82-9   2321   G

$C_9H_8O_2$
3-Phenyl-2-propenoic acid.
Al3-00891; Benzenepropenoic acid; Benzylidene-acetic acid; BRN 0507757; Cinnamic acid; Cinnamylic acid; EINECS 210-708-3; FEMA No. 2288; Kyselina skoricove; NSC 9189; Phenylacrylic acid; 3-Phenylacrylic acid; 3-Phenylpropenoic acid; 3-Phenyl-2-propenoic acid; 2-Propenoic acid, 3-phenyl-; Zimtsaeure; Zimtsäure. A chemical intermediate used in perfumes and as an anthelmintic. Monoclinic crystals; mp = 133°; bp = 300°; λ_m = 273 nm (EtOH); soluble in H_2O (0.5 g/l), more soluble in organic solvents. *Aceto Corp.; Degussa-Hüls Corp.; Penta Mfg.; Raschig GmbH.*

877 Cinnamoyl chloride

102-92-1   2323   G

C_9H_7ClO
3-Phenyl-2-propenoyl chloride.
Cinnamic acid chloride; Cinnamic chloride; Cinnamoyl chloride; EINECS 203-065-5; NSC 4683; β-Phenylacryloyl chloride; 3-Phenyl-2-propenoyl chloride; 2-Propenoyl chloride, 3-phenyl-. Reagent for determination of small amounts of water, chemical intermediate. Yellow crystals; mp = 36°; bp = 257°, bp_2 = 101°; d^{45} = 1.1617; very soluble in ligroin. *ICI Spec.; Janssen Chimica; Lancaster Synthesis Co.; Penta Mfg.; Pfalz & Bauer.*

878 Cinnamyl alcohol

104-54-1   2325   G

$C_9H_{10}O$
3-phenyl-2-propen-1-ol.
1-06-00-00281 (Beilstein Handbook Reference); Al3-00949; Alkohol skoricovy; BRN 1903999; CCRIS 3191; Cinnamic alcohol; Cinnamyl alcohol; EINECS 203-212-3; FEMA No. 2294; 3-Fenyl-2-propen-1-ol; HSDB 5011; NSC 8775; Peruviol; 3-Phenylallyl alcohol; σ-Phenylallyl alcohol; Phenyl-2-propen-1-ol; 1-Phenylprop-1-en-3-ol; 3-Phenyl-2-propen-1-ol; 3-Phenyl-2-propenol; 2-Propen-1-ol, 3-phenyl-; 2-Propen-y1-ol, 3-phenyl-; Propenoic acid, 3-phenyl-, (trans)-; Styrone; Styryl alcohol; Styryl carbinol; Zimtalcohol. Used in perfumery and in deodorants. Crystals; mp = 33-35°; bp = 249-250°; d^{35}_{35} = 1.0397; soluble in H_2O, organic solvents.

879 Ciprofloxacin

85721-33-1   2337   D

$C_{17}H_{18}FN_3O_3$
1-Cyclopropyl-6-fluoro-1,4-dihydro-4-oxo-7-(1-piperazinyl)-3-quinolinecarboxylic acid.
Alcon Cilox; Bacquinor; Baflox; BAY q 3939; Bernoflox; Bi-Cipro; BRN 3568352; CCRIS 5241; Cifloxin; Cilab; Ciplus; Ciprecu; Cipro; Cipro IV; Cipro XR; Ciprobay; Ciprobay Uro; Ciprocinol; Ciprodar; Ciproflox; Ciprofloxacin; Ciprofloxacina; Ciproflox-acine; Ciprofloxacino; Ciprofloxacinum; Ciprogis; Ciprolin; Ciprolon; Cipromycin; Ciproquinol; Ciprowin; Ciproxan; Ciproxina; Ciproxine; Ciriax; Citopcin; Cixan; Corsacin; Cycin; Eni; Fimoflox; HSDB 6987;

Ipiflox; Italnik; Loxan; Probiox; Proflaxin; Proflox; Proksi 250; Proksi 500; Quinolid; Quintor; Rancif; Roxytal; Septicide; Sophixin Ofteno; Spitacin; Superocin; Unex; Zumaflox. Quinolone antibiotic. Solid; Dec 255-257°. *Bayer Corp. Pharm. Div.*

880   Ciprofloxacin monohydrochloride monohydrate
86393-32-0   2337   D

C17H19ClFN3O3.H2O
1-Cyclopropyl-6-fluoro-1,4-dihydro-4-oxo-7-(1-piperazinyl)-3-quinolinecarboxylic acid monohydro-chloride monohydrate.
Bay o 9867 monohydrate; Baycip; Belmacina; Catex; Cenin; Ceprimax; Ciflan; Ciflosin; Ciflox; Cilox; Ciloxan; Cipad; Ciprinol; Cipro; Cipro HC Otic Suspension; Ciprobay; Ciprocinal; Ciprofloxacin hydrochloride; Ciprofloxacin hydrochloride monohydrate; Ciprofur; Ciproktan; Cipronex; Cipropol; Ciproxan; Ciproxin; Citeral; Cunesin; Disfabac; DRG-0110; Felixene; Flociprin; Floxacipron; Flunas; Globuce; Inkamil; Keefloxin; Loxacid; Lypro; Megaflox; Microgan; Nixin; Novidat; Novoquin; Ofitin; Oftacilox; Ophaflox; Phaproxin; Piprol; Plenolyt; Proxacin; Quinoflox; Quipro; Renator; Roflazin; Sepcen; Septicide; Siprogut; Sophixin; Strox; Suiflox; Supraflox; Velmonit. Quinolone antibiotic. Crystals; mp = 318-320°. *Alcon Labs; Bayer Corp. Pharm. Div.*

881   Citalopram Hydrobromide
59729-32-7   2342   D

C20H22BrFN2O
1-[3-(Dimethylamino)propyl]-1-(p-fluorophenyl)-5-phthalancarbonitrile hydrobromide.
Celexa; Cipramil; Citalopram hydrobromide; EINECS 261-890-6; HSDB 7042; Lu 10-171-B. A serotonin uptake inhibitor. Antidepressant. Crystals; mp = 182-183°. *Kefalas A/S.*

882   Citral
5392-40-5   2346   G H

C10H16O
3,7-Dimethyl-trans-2,6-octadienal.
3-01-00-03053 (Beilstein Handbook Reference); AI3-01011; BRN 1721871; Caswell No. 221B; CCRIS 1043; Citral; 2,6-Dimethyloctadien-2,6-al-8; 3,7-Dimethyl-2,6-octadienal; 3,7-Dimethyl-trans-2,6-octadienal; EINECS 226-394-6; EPA Pesticide

Chemical Code 040510; FEMA Number 2303; HSDB 993; Lemsyn GB; NCI-C56348; NSC 6170; 2,6-Octadienal, 3,7-dimethyl-. Mixture of two geometrical isomers, geranial and neral. Perfumes, flavoring agents, intermediate for other fragrances, vitamin A synthesis. Fragrance and flavoring; fresh, lemon-like, green, slightly lime-like. Natural citral is a mixture of the cis-isomer (neral) and the trans-isomer (geranial). Oil; bp2.6 = 91-93°; d4^{20} = 0.8869-0.8888; insoluble in H2O, soluble in organic solvents. *BASF Corp.; Lancaster Synthesis Co.; Lucta SA; Penta Mfg.; SCM Glidco Organics.*

883   Citric acid
77-92-9   2350   G H P

C6H8O7
2-Hydroxy-1,2,3-propanetricarboxylic acid.
4-03-00-01272 (Beilstein Handbook Reference); Aciletten; AI3-06286; Anhydrous citric acid; BRN 0782061; Caswell No. 221C; CCRIS 3292; Chemfill; Citralite; Citretten; Citric acid; Citric acid, anhydrous; Citro; Descote® Citric Acid; EINECS 201-069-1; EPA Pesticide Chemical Code 021801; F 0001 (polycarboxylic acid); FEMA Number 2306; HSDB 911; Hydrocerol A; Hydroxytricarballylic acid; 'β-Hydroxytricarballylic acid; K-Lyte; K-Lyte/Cl; K-Lyte DS; Kyselina 2-hydroxy-1,2,3-propantrikarbonova; Kyselina citronova; NSC 30279; Preparation of citrates, flavoring extracts, confections, soft drinks; antioxidant in foods; sequestering agent; detergent builder; metal cleaner. Registered by EPA as an antimicrobial and fungicide. FDA GRAS, approved for injectables, buccals, inhalants, nasals, ophthalmics, topicals, orals, otics. Acidifier, buffering agent, pH adjuster, flavorant; anticoagulant. Solid; mp = 153°; d = 1.665; pK1 = 3.128, pK2 = 4.761, pK3 = 6.396; soluble in H2O (59.2% w/w 20°), moderately soluble in organic solvents. LD50 (rat orl) = 6730 mg/kg, LD50 (rat ip) = 975 mg/kg. *ADM; Albright & Wilson Americas Inc.; Am. Roland; Ashland; Browning; Cargill; Ellis & Everard; Frutarom; Greeff R.W. & Co.; Haarmann & Reimer GmbH; Hoffmann-LaRoche Inc.; Integra; Jungbunzlauer; Norman-Fox; Penta Mfg.; Pfizer Food Science; PMC; Ruger; Schweizerhall Inc.; SCI Natural Ingredients; Sigma-Aldrich Fine Chem.; Spectrum Chem. Manufacturing; U.S. Petrochem. Ind.*

884   Citronellal
106-23-0   2353   G

C10H18O
3,7-dimethyl-6-octenal.
AI3-00203; Citronella; Citronellal; Citronellel; 2,3-Dihydrocitral; 3,7-Dimethyl-6-octenal; 3,7-Dimethyl-6-octen-1-al; D-Rhodinal; EINECS 203-376-6; FEMA No. 2307; HSDB 594; NSC 46106; 6-Octenal, 3,7-dimethyl-; Rhodinal. Used in soap perfumery, manufacture of hydroxycitronellal, an insect repellant. Oil; bp1 = 47°; d = 0.848-0.856; [α]D^{25} = 11.50°. *Lancaster Synthesis Co.; Penta Mfg.; PMC.*

885   Citronellol
106-22-9   2354   G H

C10H20O

3,7-Dimethyl-6-octen-1-ol.
4-01-00-02188 (Beilstein Handbook Reference); AI3-25080; BRN 1721507; CCRIS 7452; Cephrol; Citronellol; 2,6-Dimethyl-2-octen-8-ol; 3,7-Dimethyl-6-octen-1-ol; EINECS 203-375-0; Elenol; FEMA No. 2309; NSC 8779; 6 Octen-1-ol, 3,7-dimethyl-; Rhodinol; Rodinol. Used in perfumery as a source of floral odors. bp = 221-222°; d = 0.858; n_D^{20} = 1.4560; $[\alpha]_D^{20}$ = 5.22°; slightly soluble in H2O, soluble in organic solvents. *Bush Boake Allen; IFF; Penta Mfg.; SCM Glidco Organics.*

886   **Citronellyl acetate**
150-84-5   G

C12H22O2

3,7-Dimethyl-6-octen-1-yl acetate.
1-02-00-00065 (Beilstein Handbook Reference); Acetic acid, 3,7-dimethyl-6-octen-1-yl ester; Acetic acid, citronellyl ester; 1-Acetoxy-3,7-dimethyloct-6-ene; AI3-02039; BRN 1723886; Cephreine; Citronellol acetate; Citronellyl ethanoate; 3,7-Dimethyl-6-octen-1-ol acetate; 3,7-Dimethyl-6-octen-1-yl ethanoate; (1)-3,7-Dimethyloct-6-enyl acetate; EINECS 205-775-0; EINECS 266-837-0; FEMA No. 2311; FEMA No. 2981; Natural rhodinol, acetylated; NSC 4893; 2-Octen-8-ol, 2,6-dimethyl-, acetate; 6-Octen-1-ol, 3,7-dimethyl-, acetate; Rhodinyl acetate. Used in perfumery. Oil; bp = 229°; d = 0.8900; insoluble in H2O, soluble in organic solvents; LD50 (rat orl) = 6800 mg/kg. *Bush Boake Allen.*

887   **Citrus oil, limonene fraction**
65996-98-7   H
Citrus oil, limonene fraction.
Citrus stripper oil; EINECS 266-034-5; Sulfate turpentine, distilled; Terpenes and Terpenoids, limonene fraction.

888   **Clarithromycin**
81103-11-9   2362   D

C38H69NO13

6-O-Methylerythromycin.
A-56268; Abbotic; Abbott-56268; Adel; Astromen; Biaxin; Biaxin HP; Bicrolid; Clacine; Clambiotic; Claribid; Claricide; Clarith; Clarithromycin; Clarithromycine; Clarithromycinum; Claritromicina; Clathromycin; Cyllid; DRG-0099; Helas; Heliclar; Klacid; Klaciped; Klaricid; Klaricid H.P.; Klaricid Pediatric; Klarid; Klarin; Klax; Kofron; Mabicrol; Macladin; Maclar; Mavid; Naxy; TE-031; Veclam; Zeclar. Macrolide antibiotic. White powder; mp = 217-220° (dec), 222-225°; λ_m = 288 nm (CHCl3); $[\alpha]_D^{34}$ = -90.4° (c = 1 CHCl3); LD50 (mmus orl) = 2740 mg/kg, (mmus ip) = 1030, (mmus sc) > 5000 mg/kg, (fmus orl) = 2700 mg/kg, (fmus ip) = 850 mg/kg, (fmus sc) > 5000 mg/kg, (mrat orl) = 3470 mg/kg, (mrat ip) = 669 mg/kg, (mrat sc) > 5000 mg/kg, (frat orl) = 2700 mg/kg, (frat ip) = 753 mg/kg, (frat sc) > 5000 mg/kg. *Abbott Labs Inc.*

889   **Cleve's acid**
119-28-8   2373   G

C10H9NO3S

8-Amino-2-naphthalenesulfonic acid.
1-Aminonaphthalene-7-sulfonic acid; 8-Amino-naphthalene-2-sulfonic acid; 8-Aminonaphthalene-2-sulphonic acid; 1-Amino-7-sulfonaphthalene; Cassella's acid; C.I. 26135; Cleve's acid; 1,7-Cleve's acid; Cleve's theta-acid; Delta acid; EINECS 204-311-4; F acid; J acid; 2-Naphthalenesulfonic acid, 8-amino-; 1-Naphthylamine-7-sulfonic acid; 8-Naphthylamine-2-sulfonic acid; NSC 4983. Used as an intermediate for azo dyes. Crystallizes as the monohydrate, soluble in H2O (0.45 g/100 ml).

890   **Cleve's β acid**
119-79-9   2372   G

C10H9NO3S

α-naphthylamine-6-sulfonic acid.
4-14-00-02804 (Beilstein Handbook Reference); 5-Amino-2-naphthalenesulfonic acid; 5-Amino-naphthalene-2-sulphonic acid; 1-Amino-6-sulfo-naphthalene; 1-Amino-6-naphthalenesulfonic acid; BRN 1819887; 1,6-Cleve's acid; Cleve's β-acid; Cleve's acid-1,6; Cleve's acid, mixed; EINECS 204-351-2; Kyselina 1-naftylamin-6-sulfonova; Kyselina cleve; 2-Naphthalenesulfonic acid, 5-amino-; 1-Naphthylamine-6-sulfonic acid; 5-Naphthylamine-2-sulfonic acid; NSC 31506. Intermediate in synthesis of dyestuffs. Solid; slightly soluble in 1000 parts H2O (0.1 g/100 ml), insoluble in EtOH, Et2O.

891   **Clindamycin**
18323-44-9   2377   D

C18H33ClN2O5S

Methyl   7-chloro-6,7,8-trideoxy-6-(1-methyl-trans-4-propyl-L-2-pyrrolidinecarboxamido)-1-thio-L-threo-α-D-galactooctopyranoside.
Antirobe; 7-CDL; Chlolincocin; Cleocin; Clindamicina; Clindamycin; Clindamycine; Clindamycinum; Clinimycin; Dalacin C; Dalacine; DRG-0011; EINECS 242-209-1; HSDB 3037; Klimicin; Sobelin; U-21251. Antibacterial. Yellow solid; $[\alpha]_D$ = 214° (CHCl3). *Pharmacia & Upjohn.*

892   **Clindamycin hydrochloride mono-hydrate**
58207-19-5   2377   D

150

C18H33ClN2O5S.H2O

Methyl   7-chloro-6,7,8-trideoxy-6-(1-methyl-trans-4-propyl-L-2-pyrrolidinecarboxamido)-1-thio-L-threo-α-D-galactooctopyranoside hydrochloride monohydrate.

7-CDL; Chlolincocin; 7-Chloro-7-deoxylincomycin; 7(S)-Chloro-7-deoxylincomycin; 7-Chlorolincomycin; 7-Deoxy-7(S)-chlorolincomycin; Cleocin; Clinda-micina; Clindamycin; Clindamycine; Clindamycinum; Clinimycin; Dalacin C; DRG-0011; EINECS 242-209-1; HSDB 3037; Sobelin; U-21,251; U-21251;. Antibacterial. White powder; mp = 141-143°; [α]D = 144° (H2O); pKa = 7.6; soluble in H2O, C5H5N, EtOH, DMF; LD50 (mus iv) = 245 mg/kg, (mus ip) = 361 mg/kg, (mus orl) = 2618 mg/kg. *Pharmacia & Upjohn.*

893   Clofazimine
2030-63-9   2396   D G

C27H22Cl2N4

3-(p-Chloroanilino)-10-(p-chlorophenyl)-2,10-dihydro-2-(isopropylimino)phenazine.

4-25-00-03033 (Beilstein Handbook Reference); B-663; BRN 0060420; Chlofazimine; Clofazimina; Clofazimine; Clofaziminum; DRG-0067; EINECS 217-980-2; G-30320; Lampren; Lamprene; NSC-141046. Antibacterial (tuberculostatic). Also leprostatic. Crystals; mp = 210-212°; λm 284, 486 nm (abs. 1.30, 0.64, 0.01M HCl/MeOH); soluble in AcOH, DMF, CHCl3 (6.6 g/100 ml), EtOH (0.14 g/100 ml), Et2O (0.1 g/100 ml); insoluble in H2O; LD50 (mus, rat, gpg orl) > 4000 mg/kg. *Ciba-Geigy Corp.*

894   Clonazepam
1622-61-3   2413   D

C15H10ClN3O3

5-(o-Chlorophenyl)-1,3-dihydro-7-nitro-2H-1,4-benzo-diazepin-2-one.

5-24-00-00351 (Beilstein Handbook Reference); Alti-Clonazepam; Antelepsin;   BRN   0759557;   Chlon-azepam;   Cloazepam; Clonazepam; Clonazepamum; Clonex; Clonopin; DEA No. 2737; EINECS 216-596-2; HSDB 3265; Iktorivil; Kenoket; Klonopin; Landsen; Lonazep; NSC 179913; Paxam; Ravotril; Rivatril; Rivoril; Rivotril; Ro 4-8180; Ro-5-4023; Ro 5-4023/B-7; Solfidin. Antiepileptic agent with anxiolytic and antimanic properties. Used as an an anticonvulsant. Crystals; mp = 236.5-238.5°; λm = 248, 310 nm (ε 14500 11600 MeOH/iPrOH); soluble in Me2CO (3.1 g/100 ml), CHCl3 (1.5 g/100 ml), MeOH (0.86 g/100 ml), Et2O (0.07 g/100 ml), C6H6 (0.05 g/100 ml), H2O (< 0.01 g/100 ml); LD50 (mus orl) > 4000 mg/kg. *Hoffmann-LaRoche Inc.*

895   Clonidine
4205-90-7   2414   D

C9H9Cl2N3

2-[(2,6-Dichlorophenyl)imino]imidazolidine.

734571A; Adesipress; Catapres-TTS; Catarpres; Catarpresan; CCRIS 7787; Chlornidinum; Clonidin; Clonidina; Clonidine; Clonidinum; Duraclon; EINECS 224-119-4; HSDB 3040; M-5041T; ST-155-BS. An α2-adrenergic agonist used as an antihypertensive and antidyskinetic. Crystals; mp = 130°. *Boehringer Ingelheim GmbH.*

896   Clonidine hydrochloride
4205-91-8   2414   D

C9H10Cl3N3

2-[(2,6-Dichlorophenyl)imino]imidazolidine hydro-chloride.

Apo-Clonidine; Atensina; Barclyd; Capresin; Caprysin; Catanidin; Catapres; Catapresan; Chlophazolin; Clofelin; Clonid-Ophal; Clonidil-Riker;   Clonidine   hydrochloride;   Clonidine monohydrochloride; Clonilou; Clonisin; Clonistada; Combipres; DCAI; Dispaclonidin; Dixarit; Duraclon; Edolglau; EINECS 224-121-5; Glausine; Haemiton; Hemiton; Iporel; Isaglaucon; Isoglaucon; Katapresan; Klophelin; Normopresan; Novo-Clonidine; Nu-Clonidine; ST-155.; component of: Combipres. An α2-adrenergic agonist used as an antihypertensive and antidyskinetic. Crystals; mp = 305°; soluble in H2O (7.7 g/100 ml at 20°, 16.6 g/100 ml at 60°), MeOH (17.25 g/100 ml), EtOH (4 g/100 ml), CHCl3 (0.02 g/100 ml); λm = 213, 271, 302 nm (ε 8290, 713, 340 H2O); LD50 (mus orl) = 328 mg/kg, (mus iv) = 18 mg/kg, (rat orl) = 270 mg/kg, (rat iv) = 29 mg/kg. *Boehringer Ingelheim GmbH; Parke-Davis.*

897   Clonitralide
1420-04-8   6543   G P

$C_{13}H_8Cl_2N_2O_4$

2',5-Dichloro-4'-nitrosalicylanilide compound with 2-aminoethanol (1:1).

2-Aminoethanol salt of 2',5-dichloro-4'-nitro-salicylanalide; Bay 73; BAY 6076; Bayer 25648; Bayer 73; Baylucit; Bayluscid; Bayluscide; Benzamide, 5-chloro-N-(2-chloro-4-nitrophenyl)-2-hydroxy-, compd with 2-aminoethanol (1:1); Caswell No. 314; CCRIS 178; 5-Chloro-N-(2-chloro-4-nitrophenyl)-2-hydroxy-benzamide 2-aminoethanol salt; Clonitralid; Clonitralide; 2',5-Dichloro-4'-nitrosalicylanilide; 2',5-Dichloro-4'-nitrosalicylanilide, 2-aminoethanol salt; 2',5-Dichloro-4'-nitrosalicyloylanilide ethanolamine salt; EINECS 215-811-7; EPA Pesticide Chemical Code 077401; Ethanolamine salt of 5,2'-dichloro-4'-nitrosalicylicanilide; HL 2448; HSDB 4045; M 73; Molluscicide Bayer 73; Mollutox; NCI-C00431; Niclosamide; Niclosamide-(2-hydroxyethyl)ammon-ium; Niclosamide ethanolamine salt; Phenasal ethanolamine salt; Phenasal ethanolamine salt; Salicylanilide, 2',5-dichloro-4-nitro-, ethanolamine salt; Salicylanilide, 2',5-dichloro-4'-nitro-, compd with 2-aminoethanol; SR 73; Yomesan. A molluscicide. Registered by EPA as an insecticide. Yellow crystals; mp = 225 - 230°; sparingly soluble in H_2O (0.023 g/100 ml), sparingly soluble in EtOH, $CHCl_3$, Et_2O; LD_{50} (mrat orl) > 3710 mg/kg, (frat orl) = 3710 mg/kg, (rat der) > 1000 mg/kg; LC_{50} (rainbow trout 48 hr.) = 0.05 mg/l, (carp 48 hr.) = 0.235 mg/l, toxic to frogs, salamanders, plankton. *Bayer AG; U.S. Fish and Wildlife Services.*

898   Clopidogrel
113665-84-2   2421   D

$C_{16}H_{16}ClN_2OS$

Methyl (+)-(S)-α-(o-chlorophenyl)-6,7-dihydrothieno-[3,2-c]pyridine-5(4H)-acetate.

SR-25990. Platelet inihibtor; antithrombotic. Crystals; $[α]_D^{20}$= 51.52° (c = 1.61 MeOH). *Sanofi Winthrop.*

899   Clopidogrel hydrogen sulfate
135046-48-9   2421   D

$C_{16}H_{18}ClN_2O_5S_2$

Methyl (+)-(S)-α-(o-chlorophenyl)-6,7-dihydrothieno-[3,2-c]pyridine-5(4H)-acetate sulfate (1:1).

SR-25990C; Plavix. Platelet inihibtor; antithrombotic. Crystals; mp =

184°; $[α]_D^{20}$= 55.10° (c = 1.891 MeOH). *Sanofi Winthrop.*

900   Clopyralid
1702-17-6   2426   G P

$C_6H_3Cl_2NO_2$

3,6-Dichloropyridine-2-carboxylic acid.

5-22-02-00046 (Beilstein Handbook Reference); Acide 3,6-dichloropicolinique; Acide dichloro-3,6 picolinique; Benzalox; BRN 0473755; Campaign; Caswell No. 323H; Cirtoxin; Cliophar; Clopyralid; Clopyralide; Crusader S; Cyronal; 3,6-Dichloro-picolinic acid; 3,6-Dichloro-2-pyridinecarboxylic acid; 3,6-Dichloropyridine-2-carboxylic acid; Dowco 290; EINECS 216-935-4; EPA Pesticide Chemical Code 117401; Format; HSDB 6593; Huiloralid; IWD 3523; Kyselina 3,6-dichlorpikolinova; Loncid; Lontrel 3; Lontrel 100; Lontrel 300; Lontrel L; Lontrel SF 100; Matrigon; Picolinic acid, 3,6-dichloro-; 2-Pyridinecarboxylic acid, 3,6-dichloro-; Reclaim; Shield; Stinger; Transline; Versatill; XRM 3972. Used for post-emergence control of broad-leaf weeds of *Polygonaceae, Compositae, Leguminosae and Umbelliferae.* Provides good control of creeping thistle, sow thistle, coltsfoot, mayweeds and *Polygonum* spp. mp = 151-152°; soluble in H_2O (9 g/l), soluble in organic solvents; LD_{50} (rat orl) >4300 mg/kg. *DowElanco Ltd.*

901   Closantel
57808-65-8   2437   G V

$C_{22}H_{14}Cl_2I_2N_2O_2$

N-[5-Chloro-4-[(4-chlorophenyl)cyanomethyl]-2-methylphenyl]-2-hydroxy-3,5-diiodobenzamide.

Benzamide, N-(5-chloro-4-((4-chlorophenyl)cyano-methyl)-2-methylphenyl)-2-hydroxy-3,5-diiodo-; 5'-Chloro-$α^4$-(p-chlorophenyl)-$α^4$-cyano-3,5-diiodo-2',4'-salicyloxylidide; N-(5-Chloro-4-((4-chlorophenyl)-cyanomethyl)-2-methylphenyl)-2-hydroxy-3,5-diiodobenzamide; Closantel; Closantelum; EINECS 260-967-1; Flukiver; R-31520; Seponver. A proprietary preparation containing closantel; a veterinary anthelmintic (flukicide). Crystals; mp = 217.8°. *Janssen Chimica.*

902   Coal Tar
8007-45-2   P

Coal tar extract.

Alphosyl; Carbo-cort; Caswell No. 223; Coal tar; Coal tar extract; Coal tar ointment; Coal tar pitch volatiles; Coal tar solution USP; Coal tar [Soots, tars, and mineral oils]; Coal Tar Pitch; Coal-tars; Coke oven emissions; Coke-oven tar; Coking tar; Crude coal tar; Egopsoryl TA; EINECS 232-361-7; EPA Pesticide Chemical Code 022003; Estar; Estar (skin treatment); Fongitar; HSDB 5050; Impervotar; Ionil T plus; KC 261; Lavatar; Linotar Gel; Picis carbonis; Pixalbol; Polytar bath; psoriGel; RT 7; RT 7 (coal tar); Tar; Tar, coal; Tar, coal, colorless purified; Tar, coal, purified colorless; Tar, coking; Tarcron 180; Tarcron 180L; Tarcron 230; Zetar. Registered by EPA as a fungicide. *KMG-Bernuth Inc.*

903   Coal Tar/Creosote
8001-58-9   2600   D G P

Creosote oil.
AWPA 1; Brick oil; Caswell No. 225; CCRIS 6005; Coal tar creosote; Coal tar creosote oils; Coal tar oil; Coal tars (during destructive distillation); Creosote; Creosote (coal tar); Creosote (coal) [Soots, tars, and mineral oils]; Creosote P1; Creosotes; Creosotum; Cresylic creosote; Dead oil; EINECS 232-287-5; EPA Pesticide Chemical Code 025004; Heavy oil; HSDB 6299; Liquid pitch oil; Naphthalene oil; Original Carbolineum; Osmoplastic-D; Petroleum creosote; Preserv-O-sote; RCRA waste number U051; Sakresote 100; Smoplastic-F; Tar oil; Wash oil. Registered by EPA as a fungicide. Black oily liquid; d > 1.0; d_4^{38} = 1.06; insoluble in H2O. *Coopers Creek Chemical Corp.; KMG-Bernuth Inc.; Koppers Industries Inc.; Osmose Inc.*

904   Cobalt
7440-48-4   2452   G

Co

Co
Cobalt.
Aquacat; C.I. 77320; CCRIS 1575; Cobalt; Cobalt-59; Cobalt, elemental; Cobalt fume; Cobalt metal, dust and fume; Cobalt metal powder; EINECS 231-158-0; HSDB 519; Kobalt; NCI-C60311; Super cobalt. Oxidizing agent, lamp filaments, in manufacture of cobalt steel; in porcelain, glass, pottery, enamels. *Atomergic Chemetals; Cerac; Noah Chem.; Sigma-Aldrich Fine Chem.*

905   Cobalt acetate
71-48-7   2458   H

C4H8CoO4
Acetic acid, cobalt(2+) salt.
Acetic acid, cobalt(2+) salt; Bis(acetato)cobalt; Cobalt acetate (Co(OAc)2); Cobalt diacetate; Cobalt(2+) acetate; Cobalt(II) acetate; Cobaltous acetate; Cobaltous diacetate; EINECS 200-755-8; HSDB 997; NSC 78410. Bleaching agent and drier in lacquers; esterification and oxidation catalyst, foam stabilizer in beers. Pink crystals; freely soluble in H2O. [Tetrahydrate]; red crystals; d = 1.705; becomes anhydrous at 140°; soluble in H2O, alcohols, pentyl acetate.

906   Cobalt bromide
7789-43-7   2460   G

Br2Co
Cobalt(II)Bromide.
Cobalt bromide; Cobalt bromide (CoBr2); Cobalt dibromide; Cobalt(II) bromide; Cobaltous bromide; EINECS 232-166-7; HSDB 998. Solid; mp = 678°; d_4^{25} = 4.909; soluble in H2O, organic solvents. *Mechema Chemicals Ltd.*

907   Cobalt chloride
7646-79-9   2462   D G

Cl2Co
Cobaltous chloride anhydrous.
CCRIS 4224; Cobalt chloride; Cobalt chloride (CoCl2); Cobalt dichloride; Cobalt muriate; Cobalt(II) chloride; Cobaltous chloride; Cobaltous dichloride; Dichlorocobalt; EINECS 231-589-4; HSDB 1000; Kobalt chlorid; NSC 51149. Absorbent for ammonia, gas masks, electroplating, inks, hygrometers, manufacture of vitamin B12, flux for magnesium refining, solid lubricant, dye mordant, catalyst, barometers, laboratory reagent, fertilizer additive. Usd medicinally as a hematinic agent. Pale blue leaflets; mp = 735°; bp = 1049°; d^{25} = 3.367; soluble in H2O, EtOH, MeOH, Me2CO, Et2O, C5H5N; LD50 (mus orl) = 360 mg/kg, (rat orl) = 171 mg/kg, (mus ip) = 92.6 mg/kg, (rat ip) = 36.9 mg/kg, (mus iv) = 23.3 mg/kg, (rat iv) = 4.3 mg/kg. *Celtic Chem. Ltd.; Mallinckrodt Inc.; Nihon Kagaku Sangyo; Spectrum Chem. Manufacturing.*

908   Cobalt 2-ethylhexanoate
136-52-7   H

C16H32CoO4
Cobalt bis(2-ethylhexanoate).
8SEH-Co; C 101; C 101 (catalyst); CO 12; Cobalt 2-ethylhexoate; Cobalt bis(2-ethylhexanoate); Cobalt octoate; Cobalt(II) 2-ethylhexanoate; Cobaltous 2-ethylhexanoate; Cobaltous octoate; EINECS 205-250-6; Hexanoic acid, 2-ethyl-, cobalt(2+) salt; HSDB 5621; NL 49P; NL 51P; NL 51S; Versneller NL 49.

909   Cobalt naphthenate
61789-51-3   H

Naphthenic acids, cobalt(II) salts.
Caswell No. 226; CCRIS 772; Cobalt naphthenate; Cobalt naphthenates, powder; Cobalt(II) naphthenate; Cobaltous naphthenate; EINECS 263-064-0; EPA Pesticide Chemical Code 025101; Naftolite; Naphtenate de cobalt; Naphthenic acid, cobalt salt; UN2001.

910   Cobalt neodecanoate
27253-31-2   H

C20H38CoO4
Neodecanoic acid, cobalt salt.
Cobalt neodecanoate; EINECS 248-373-0; Neodeca-noic acid, cobalt salt.

911   Cobalt(2+) neodecanoate
52270-44-7   H

153

C20H38CoO4
Neodecanoic acid, cobalt(2+) salt.
Cobalt(2+) neodecanoate; EINECS 257-798-0; Neodecanoic acid, cobalt(2+) salt.

912 Cobalt nitrate
10141-05-6   2469   G

CoN2O6
Cobalt(II) nitrate.
Cobalt bis(nitrate); Cobalt dinitrate; Cobalt nitrate; Cobalt nitrate (Co(NO3)2); Cobalt(2+) nitrate; Cobalt(II) nitrate; Cobaltous nitrate; EINECS 233-402-1; HSDB 238; Nitric acid, cobalt(2+) salt. Oxidizing agent, dangerous fire risk in contact with organic materials. Pale red powder; dec 100°; d = 2.49; LD50 (rbt orl) = 250 mg/kg, (rbt sc) = 75 mg/kg. [hexahydrate]; mp = 55°; d = 1.88; very soluble in H2O, EtOH, organic solvents; LD50 (rat orl) = 691 mg/kg. *Atomergic Chemetals; Celtic Chem. Ltd.; Mallinckrodt Inc.; Nihon Kagaku Sangyo; Noah Chem.*

913 Cobalt oleate
14666-94-5   H

C36H66CoO4
9-Octadecenoic acid (9Z)-, cobalt salt.
Cobalt oleate; EINECS 238-709-4; 9-Octadecenoic acid (9Z)-, cobalt salt; Oleic acid, cobalt salt.

914 Cobalt oxide monohydrate
12016-80-7   2456   G

CoHO2
Cobaltic hydroxide.
Cobalt hydroxide oxide; Cobalt hydroxide oxide (CoO(OH)); Cobalt oxide hydroxide; Cobalt oxyhydroxide; Cobaltic oxide monohydrate; Cobalt(III) oxide monohydrate; EINECS 234-614-7. Oxidation catalyst Dark brown-black powder; insoluble in H2O.

915 Cobalt sulfate
10124-43-3   2473   G

CoO4S
Cobalt(II) sulfate.
Cobalt (2+) sulfate; Cobalt Brown; Cobalt monosulfate; Cobalt sulfate; Cobalt sulfate (1:1); Cobalt sulfate (CoSO24); Cobalt sulphate; Cobalt(2+) sulfate; Cobalt(II) sulfate; Cobalt(II) sulphate; Cobaltous sulfate; Cobaltous sulfate salt (1:1); EINECS 233-334-2; HSDB 240; Sulfuric acid, cobalt(2+) salt (1:1). Ceramics, pigments, glazes, in plating baths for cobalt, additive to soils, catalyst, paint and ink drier, storage batteries. Red crystals; d:s25 = 3.71; slightly soluble in boiling H2O. *Atomergic Chemetals; Elf Atochem N. Am.; Johnson Matthey; Mallinckrodt Inc.; Nihon Kagaku Sangyo.*

916 Cobalt tallate
61789-52-4   H

Tall oil fatty acids, cobalt salts.
Cobalt tallate; EINECS 263-065-6; Fatty acids, tall-oil, cobalt salts; Tall oil fatty acids, cobalt salts. 263-065-6

917 Cobalt trifluoride
10026-18-3   2455   G

F3Co
Cobalt(III) fluoride.
Cobalt fluoride (CoF3); Cobalt trifluoride; Cobaltic fluoride; EINECS 233-062-4. Used as a fluorinating agent in the fluorination of hydrocarbons (Fowler process). Solid; d = 3.88; reacts with air and H2O. *Atomergic Chemetals; Cerac; Elf Atochem N. Am.*

918 Cobaltic oxide
1308-04-9   G

Co2O3
Cobalt (III) oxide.
C.I. 77323; Cobalt black; Cobalt oxide; Cobalt oxide (Co2O3); Cobalt peroxide; Cobalt sesqioxide; Cobalt sesquioxide; Cobalt trioxide; Cobalt(3+) oxide; Cobalt(III) oxide; Cobaltic oxide; Dicobalt oxide; Dicobalt trioxide; EINECS 215-156-7. A pigment used in coloring enamels and glazing pottery. Black solid; insoluble in H2O, soluble in H2SO4. *Atomergic Chemetals; Cerac; Noah Chem.*

919 Cobaltous acetate
6147-53-1   G

C4H6CoO4.4H2O
Cobalt(II) acetate tetrahydrate.
Acetic acid, cobalt(2+) salt, tetrahydrate; Bis-(acetato)tetraquacobalt; Cobalt acetate tetrahydrate; Cobalt diacetate tetrahydrate; Cobalt(II) acetate tetrahydrate; Cobaltous acetate tetrahydrate; Octan kobaltnaty. Used in inks, paints and varnish driers, catalyst, anodizing, mineral supplement in feed additives, foam stabilizer. [Tetrahydrate]; d = 1.705; soluble in H2O and EtOH. *Atomergic Chemetals; Celtic Chem. Ltd.; Mallinckrodt*

920 Cobaltous carbonate
513-79-1   2461   G

CCoO3

Cobalt(II) carbonate.

C.I. 77353; Carbonic acid, cobalt(2+) salt (1:1); Carbonic acid, cobalt(2+) salt; CI 77353; Cobalt carbonate (1:1); Cobalt carbonate (CoCO3); Cobalt monocarbonate; Cobalt spar; Cobalt(2+) carbonate; Cobaltous carbonate; EINECS 208-169-4; HSDB 999; NSC 112219; Sphaerocobaltite. Used in the manufacture of cobaltous oxide, cobalt pigments, cobalt salts; as a chemical intermediate. Powder; d= 4.13; insoluble in H2O, EtOH, EtOAc. Celtic Chem. Ltd.; Nihon Kagaku Sangyo; Noah Chem.

921 Cobaltous oxide
1307-96-6   2471   G

$$Co{=}O$$

CoO

Cobalt(II) oxide.

C.I. 77322; C.I. Pigment Black 13; CCRIS 4229; Cobalt Black; Cobalt monooxide; Cobalt monoxide; Cobalt oxide; Cobalt oxide (CoO); Cobalt(2+) oxide; Cobalt(II) oxide; Cobaltous oxide; EINECS 215-154-6; HSDB 239; Monocobalt oxide; Zaffre. Readily absorbs oxygen at room temperature. Used as pigment in ceramics. Pigment, coloring enamels, glazing pottery. Solid; mp = 1935°; d = 5.7 - 6.7; LD50 (rat orl) = 1.70 g/kg. Atomergic Chemetals; Cerac; Noah Chem.

922 Cocamide DEA
8051-30-7   E H

Coconut oil diethanolamide.

Coconut oil amides; Coconut oil, condensed with diethanolamine; Coconut oil diethanolamine; Coconut oil, diethanolamine condensate; Coconut oil diethanolamide; Coconut oil, reaction products with diethanolamine; Comperlan® COD; Diethanolamine condensate of coconut oil; Diethanolamine, coconut oil condensate; EINECS 232-483-0; Witcamide 128t. FDA (0.2% maximum), FDA approved for topicals. Used as a mild surfactant, thickener, emulsifier and solubilizer in topical dermatological products. Insoluble in H2O. Aquatec Quimica SA.

923 Cocamidopropyl betaine
61789-40-0   G

RCO-NH(CH2)3N⁺(CH3)CH2COO⁻

Cocamidopropyl dimethyl glycine.

CADG; Abluter BE, CPB; Amido betaine C, C-45; Amonyl® 380BA, 440 NI; Ampholak BCA-30; Ampholan® 197; Ampholyt™ JB 130; Rewoteric® AMB-14, AMB-14S, AMB-15, AM B45; AM B 50; Ritataine; Rititaine B; Schercotaine CAB, CAB-G; Swanol AM-3130N; AMPHOSOL® CA, CG; Amphotensid B4, B4 LS; Caltaine C35; Chembetaine® C, CGF, CL; Dehyton® PK, 3016 B, K, KE, KE 3016, PK 45; Deriphat® BAW; DeTAINE CAPB-35, CAPB-35HV; Tego® betaine C, E, F, F50, L-7, L-7F, L-5351, S, ZF; Velvetex® BA-35,; Emcol® 6748, Coco betaine; Emcol® DG, NA30; Emery® 6744; Empigen ® BS/AU, BS/F, BS/FA, BS/H, BS/P; Euroquat C45, CPB, K, LA, LAC; Foamtaine CAB, CAB-G; Genagen CA 818, CAB; Incronam 30; Lebon 200; Lonzaine® C, CO; Mackam™ 35, 35HP, J, L; Manroteric CAB; Miratiane BD-J, BET-C-30, BET-W, CAB, CAB-

A, CAB-O, CB, CBC, CB/M, CBR; Monoteric ADA, CAB, CAB-LC, CAB-XLC, COAB; Naxaine® C, CO, Cocbetaine; Nutrol betaine MD 3863, OL 3798; Proteric CAB, COAB; Ralufon® 414; Rewoteric AMB-13,. High foaming, conditioning detergent for mild shampoos; solubilizer for lauryl sulfates in concentrated shampoos; thickener.

Gardner 4 liquid; soluble in H2O; disperses in glycerol trioleate; density = 8.8 lb/gal; visc 7cSt (100°F); sp gr = 1.03; clear pt = 0°. Chemron; Goldschmidt; Henkel/Emery; Henkel/ Organic Prods.; Huntingdon Labs.; Inolex; Lonza-group; McIntyre; Mona; Scher; Sherex; Taiwan Surfactants; Witco/Oleo-Surf.

924 Cocamine acetate
2016-56-0   G

C14H31NO2

n-Dodecylamine acetate salt.

Acetamin 24; Acetic acid, dodecylamine salt; AI3-14974; Armac 12D; Dodecanamine acetate; 1-Dodecanamine, acetate; 1-Dodecanamine acetate (salt); Dodecylamine, acetate; 1-Dodecylamine acetate; Dodecylammonium acetate; EINECS 217-956-1; Laurylamine acetate; Nopcogen 16L; NSC 97256. Surface coating agent for pigments, anticaking agent for fertilizer; emulsifier, dispersant, and softening agent for textiles; mineral flotation reagent. White solid; soluble in H2O. Kao Corp. SA.

925 Cocamine oxide
61788-90-7   G

R-NH-CHCH2COOHCH3

(R represents the coconut radical)

Coco dimethylamine oxide.

Aminoxid A 4080; Aromax® DMC; Aromax® DMC-W; Barlox® 12; Chemoxide® WC; Empigen® OB/AU; Genaminox KC; Mackamine™ CO; Noxamine CA 30; Rhodamox® C; Schercamox DMC; Sochamine OX 30Empigen® 5083. Foam booster/stabilizer and viscosity modifier for shampoos, foam baths, cleaners; improves conditioning in shampoos; solubilizer for liquid bleach products. Gardners 1 max liquid; sp gr = 0.89; HLB 18.6; flash point = 21°; pH = 6-9; prolonged contact may cause severe burns to eyes and severe skin irritation; flammable. Albright & Wilson UK Ltd.

926 Coceth-27
61791-13-7   G

R(OCH2CH2)nOH

(R represents coconut radical, avg n=27)

Coceth 10.

Dehydol LT 3; Genapol® C-050; Genapol® GC-050. Emulsifier, solubilizer for solvents, oils; base for production of ether sulfates; raw material for low-foaming detergents, dishwashing, cleansing agents, cold cleaners; superfatting agent. Liquid; soluble in oil; HLB 7.3; hyd no. 163-174; pH = 6.0-7.5.

927 Cocoamidopropylbetaine
61789-40-0   H

N-(3-Cocoamidopropyl)-N,N-dimethyl-N-carboxymethyl betaine.

Alkateric Cab-A; Coco Amido βine; Cocoamido-propylbetaine; N-(3-Cocoamidopropyl)-N,N-dimethyl-N-carboxymethyl betaine; N-(3-Cocoamidopropyl)-N,N-dimethyl-N-carboxymethylammonium hydroxide, inner salt; N-(Coco alkyl) amido propyl dimethyl betaine; N-(Cocoamidopropyl)-N,N-dimethyl-N-carb-oxymethyl ammonium, betaine; N-Cocamidopropyl-N,N-dimethylglycine, hydroxide, inner salt; Coconut oil amidopropyl betaine; EINECS 263-058-8; 1-Propanaminium, 3-amino-N-(carboxymethyl)-N,N-di-methyl-, N-coco acyl derivs, inner salts.

928   Cocoamine
61788-46-3   H

Amines, coco alkyl.
Amines, coco alkyl; Cocamine; Cocoamine; Coconut amine; EINECS 262-977-1.

929   Coconitriles
61789-53-5   H

Nitriles, coco.
EINECS 263-066-1; Nitriles, coco.

930   Coconut acid
61788-47-4   H

Coconut fatty acids.
Coco fatty acid; Coconut oil acids; Distilled Whole Coconut Oil 6226 6222' Low I.V. Coconut Oil 6227 6228; Split Coconut Oil 6254 6255; Stripped Coconut Oil 6212 6256; component of: Hystrene® 1835. Used as an emulsifier in topicals. FDA approved for topicals. *Abitec Corp.; Akzo Nobel; Norman-Fox; Procter & Gamble Pharm. Inc.; Witco/Oleo-Surf.*

931   Coconut Oil
8001-31-8   2486   H

Coconut oil.
Acidulated coconut soapstock; Cobee 76; Cocoanut oil; Coconut acidulated soapstock; Coconut Oil® 76; Coconut Oil® 92; Cocos nucifera oil; Copra oil; Pureco® 76; Trifat C-24. FDA GRAS, approved for orals, topicals, BP compliance. Used as an emollient, emulsifier, excipient, ointment base for pharmaceuticals and coating agent. Used in orals, topicals and massage creams. Yellow oil; mp = 21-25°; d^0_4 = 0.903; insoluble in H_2O, 95% EtOH, slightly soluble in EtOH, very soluble in CS_2, $CHCl_3$, Et_2O; non-toxic. *Aarhus; Abitec Corp.; Akzo Chemie; Amber Syn.; Arista Ind.; British Arkady; Calgene; Charkit; Gittens Edwd.; Integra; Penta Mfg.; Ruger; Spectrum Chem. Manufacturing; Spectrum Naturals; Tri-K Ind.*

932   Coconut oil acids, diethanolamine salt
61790-63-4   H

Fatty acids, coco, compds. with diethanolamine.
Coconut oil acids, diethanolamine salt; Coconut oil fatty acids, diethanolamine salt; EINECS 263-153-4; Fatty acids, coco, compds. with diethanolamine

933   Codeine
6059-47-8   2488   D

C18H21NO3.H2O
7,8-Didehydro-4,5α-epoxy-3-methoxy-17-methyl morphinan-6α-ol monohydrate.
Brontex; Codeine; DEA No. 9050; 7,8-Didehydro-4,5α-epoxy-3-methoxy-17-methylmorphinan-6α-ol monohydrate; Morphinan-6-ol,

7,8-didehydro-4,5-epoxy-3-methoxy-17-methyl-,   monohydrate, (5α,6α)-. Analgesic, narcotic; antitussive. Federally controlled substance (opiate). Crystals; mp = 154-156°; d^{20}_4 1.32; $[\alpha]^{55}_D$ = -136° (c= 2, EtOH); pK = 6.05; soluble in H_2O (0.8 g/100 ml); more soluble in organic solvents

934   Codeine [anhydrous]
76-57-3   2488   D

C18H21NO3
7,8-Didehydro-4,5α-epoxy-3-methoxy-17-methyl morphinan-6α-ol.
CCRIS 7555; Codeine; I-Codeine; Codeine anhydrous; Codicept; Coducept; EINECS 200-969-1; HSDB 3043; Methylmorphine; Morphine-3-methyl ether; Morphine monomethyl ether. Analgesic, narcotic; antitussive. Federally controlled substance (opiate).

935   Colfosceril palmitate
63-89-8   2501   D G

C40H80NO8P
1,2-Dipalmitoyl-sn-glycero-3-phosphocholine.
Choline hydroxide, dihydrogen phosphate, inner salt, ester with L-1,2-dipalmitin; Colfosceril palmitate; Colfoscerili palmitas; Colfoseril palmitate; 1,2-Dipalmitoyl-sn-glycero-3-phosphocholine; EINECS 200-567-6; Exosurf; Palmitate de colfosceril; Palmitato de colfoscerilo. Pulmonary surfactant, used to treat infant respiratory distress symptoms. Also used as an aid in the diagnosis of fetal lung immaturity. White solid; mp = 234-235°; $[\alpha]^{23}_D$= 6.6° (c = 4.2 CHCl3-EtOH); soluble in organic solvents. *Burroughs Wellcome Inc.* 200-567-6

936   Collidine
108-75-8   9789   G

C8H11N
2,4,6-Trimethylpyridine.
5-20-06-00093 (Beilstein Handbook Reference); AI3-10050; BRN 0107283; α,γ,α'-Collidine; γ-Collidine; 2,4,6-Collidine; s-Collidine; EINECS 203-613-3; HSDB 57; 2,4,6-Kollidin; NSC 460; Pyridine, 2,4,6-trimethyl-; 2,4,6-Trimethylpyridine. Used as chemical inter-mediate and dehydrohalogenating agent. Liquid; mp = -46°; bp =

170.6°; d^{22} = 0.9166; λ_m = 208, 264 nm (MeOH); soluble in H_2O (35 g/l), EtOH, Et_2O, Me_2CO, CCl_4; LD_{50} (rat orl) = 400 mg/kg.

937   **Copper**
7440-50-8   2545   G P

Cu

Cu
Copper.
1721 Gold; Allbri Natural Copper; ANAC 110; Anode copper; Arwood copper; Blister copper; Bronze powder; C 100 (metal); C.I. 77400; C.I. Pigment Metal 2; Caswell No. 227; Cathode copper; CCRIS 1577; CDA 101; CDA 102; CDA 110; CDA 122; CDX (metal); CE 1110; CE 7 (metal); Cl Pigment metal 2; Copper; Copper-airborne; Copper bronze; Copper dust; Copper, dusts and mists; Copper, elemental; Copper, fume; Copper M 1; Copper metal powder; Copper, metallic powder; Copper-milled; Copper powder; Copper precipitates; Copper slag-airborne; Copper slag-milled; CU M3; Cuprum; Cutox 6010; Cutox 6030; E 115 (metal); EINECS 231-159-6; Electrolytic refinery billet copper; Electrolytic refinery wirebar copper; EPA Pesticide Chemical Code 022501; Gold bronze; HSDB 1622; Kafar copper; M1 (Copper); M2 (Copper); M3 (Copper); M3R; M3S; M4 (Copper); OFHC Cu; Pigment metal 2; Rame; Raney copper. Electric wiring, switches, plumbing, heating, roofing; chemical and pharmaceutical machinery; alloys; coatings; cooking utensils. Fungicide. Registered by EPA as a fungicide, herbicide and insecticide. Metal; mp = 1083°; bp = 2595°; d = 8.94. *Asarco; M&T Harshaw; Noah Chem.; Sigma-Aldrich Fine Chem.*

938   **Copper ammonium carbonate**
33113-08-5   G P

$C_2H_4CuN_2O_6$
Cupric ammonium carbonate.
Copper carbonate, basic; copper ammonium carbonate; Carbonic acid, ammonium copper salt; Copper count N; Croptex Fungex. A protectant fungicide. Registered by EPA as a fungicide and herbicide. *Central Garden & Pet; Chemical Specialties Inc.; Hortichem Ltd.*

939   **Copper carbonate**
12069-69-1   2658   G

$CuCO_3 \cdot Cu(OH)_2$
Copper(II) carbonate, basic.
Basic copper carbonate; Basic copper(II) carbonate; Basic cupric carbonate; (Carbonato(2-))dihydroxy-dicopper; (Carbonato)dihydroxydicopper; Carbonic acid, copper complex; Carbonic acid, copper(2+) salt (1:1), basic; Caswell No. 235; Cheshunt compound; Copper, (carbonato)dihydroxydi-; Copper carbonate, basic; Copper carbonate hydroxide; Copper hydroxide carbonate ($Cu_2(OH)_2CO_3$); Copper(II) carbonate copper(II) hydroxide (1:1); Copper(II) carbonate hydroxide (2:1:2); Copper(II) carbonate hydroxide; Copper rust; Cupric carbonate, basic; Cupric carbonate hydroxide ($CuCO_3.Cu(OH)_2$); Dicopper dihydroxy-carbonate; EINECS 235-113-6; EPA Pesticide Chemical Code 022901; Green spar; Green verditer; Kop karb; Malachite; Mountain green; Verditer blue; Bremen green; Carbonato(2-(O:O'))dihydroxydicopper; Basic copper carbonate ($Cu_2(OH)_2CO_3$). Used in pigments, pyrotechnics, insecticides, copper salts. An

astringent in pomades, antidote for phosphorus poisoning, smut preventive, fungicide for seed treatment, feed additive. Green crystals; mp = 200° (dec); d = 4.000; LD_{50} (rat orl) = 1350 mg/kg. *Am. Chemet; Boliden Intertrade; Nihon Kagaku Sangyo.*

940   **Copper dimethyldithiocarbamate**
137-29-1   G

$C_6H_{12}CuN_2S_4$
Cupric N,N-dimethyldithiocarbamate.
AI3-14693; Bis(dimethylcarbamodithioato-S,S')copper; Bis(dimethyldithiocarbamate)Cu (II) complex; Carb-amic acid, dimethyldithio-, copper(II) salt; Compound-4018; Copper, bis(dimethylcarbamodithioato-S,S')-; Copper dimethyldithiocarbamate; Copper(2+) di-methyldithiocarbamate; Copper(II) dimethyldithio-carbamate; Cumate; Dimethyldithiocarbamatocopper; Dimethyldithiocarbamic acid copper salt; EINECS 205-287-8; Hermat Cu; NSC 32947; Wolfen. Accelerator for butyl rubber, ethylene-propylene-diene rubbers, for high-speed vulcanization of SBR, IIR, EPDM. Solid; mp >300°; d = 1.75; slightly soluble in organic solvents. *Akrochem Chem. Co.; Akzo Chemie; Vanderbilt R.T. Co. Inc.*

941   **Copper gluconate**
527-09-3   2667   G V

$C_{12}H_{22}CuO_{14}$
Copper(II) gluconate.
Bis(D-gluconato-O1,O2)copper; Bis(D-gluconato)-copper; CCRIS 3652; Chelates of copper gluconate; Copper, bis(D-gluconato)-; Copper, bis(D-gluconato-O1,O2)-; Copper D-gluconate (1:2); Copper di-D-gluconate; Copper gluconate; Copper(2+) D-gluconate, (1:2); Copper(II)gluconate; Cupric gluconate; Cupric gluconate monohydrate; EINECS 208-408-2; Gluconal® CU; Gluconic acid, copper(2+) salt; Gluconic acid, copper(2+) salt (2:1), d-; Gluconic acid, copper(2+) salt (2:1), D-; HSDB 261. Pharmaceutical/food grade mineral source for human and veterinary pharmaceutical preparations, dietary supplements, fortified foods and animal feed. Also used as an oral deodorant. Solid; d = 1.710 mg/kg; soluble in H_2O (30 g/100 ml); slightly soluble in EtOH, insoluble in organic solvents. *Akzo Chemie; Glucona; Spectrum Chem. Manufacturing.*

942   **Copper hydroxide**
20427-59-2   2670   G P

CuH_2O_2
Copper(II) hydroxide.
Blue Copper; Blue Shield; Blue Shield DF; Caswell No. 242; Champ; Champ Formula II; Chiltern Kocide 101; Comac Parasol; Copper blue; Copper dihydroxide; Copper hydroxide; Copper

hydroxide (Cu(OH)2); Copper(2+) hydroxide; Copper(II) hydroxide; Criscobre; Cudrox; Cuidrox; Cupravit blau; Cupravit Blue; Cupravit blue; Cupric hydroxide; Cupric Hydroxide Formulation Grade Agricultural Fungicide; Cuzin; EINECS 243-815-9; EPA Pesticide Chemical Code 023401; Funguran OH; HSDB 262; Hydrocop T; Kocide; Kocide 101; Kocide 101PM; Kocide 220; Kocide 404; Kocide 2000; Kocide Copper Hydroxide Antifouling Pigment; Kocide Cupric Hydroxide Formulation Grade; Kocide DF; Kocide LF; Kocide SD; KOP Hydroxide; KOP Hydroxide WP; Kuprablau; Nu-Cop; Parasol; Schweitzer's reagent; Spin Out FP; Technical Hydrox; Wetcol. Copper salts, mordant, paper staining; pesticide, fungicide, catalyst and solvent for cellulose. Registered by EPA as a fungicide and herbicide. Light blue crystals; d = 3.37; slightly soluble in H2O (0.00029 g/100 ml at pH 7)readily soluble in aqueous ammonia, insoluble in organic solvents; LD50 (rat orl) = 1000 mg/kg, (rbt der) > 3160 mg/kg, (bobwhite quail orl) = 3400 mg/kg, (mallard duck orl) > 5000 mg/kg; LC50 (ihl) > 2 mg/l air, (rainbow trout 24 hr.) = 0.08 mg/l, bluegill sunfish 96 hr.) > 180 mg/l; non-toxic to bees. *Agriliance LLC; Agtrol International; Am. Chemet; Chiltern Farm Chemicals Ltd; Cuproquim Corp.; Drexel Chemical Co.; Faesy & Besthoff Inc.; Generic; Gowan Co.; Griffin LLC; Kop-Coat Inc.; McKechnie Chemicals Ltd.; Micro-Flo Co. LLC; Spiess-Urania Chemicals Gmbh.*

943   Copper naphthenate
1338-02-9   P

Naphthenic acid, copper salt.
Caswell No. 245; CCRIS 1401; Chapco Cu-nap; CNC; Copper naphthenate; Copper uversol; Cunapsol; Cuprinol; EINECS 215-657-0; EPA Pesticide Chemical Code 023102; HSDB 245; Naphtenate de cuivre; Naphthenic acids, copper salts; Troysan; Troysan copper 8%; Wiltz-65; Wittox C. Fungicide. Registered by EPA as a fungicide, herbicide and insecticide. Viscous, green oil; insoluble in H2O, moderately soluble in petroleum oils, most organic solvents; LD50 (rat orl) = 110 mg/kg, (mus orl) = 6400-7200 mg/kg, (rbt der) > 2000 mg/kg; LC50 (rat ihl 1 hr.) > 5.5 mg/kg. *Chemical Specialties Inc.; Generic; IBC Manufacturing Co.; Kop-Coat Inc.; Lanco Manufacturing Corp.; Merichem Chemicals & Refinery Services LLC; OMG Americas Inc.; Osmose Inc.; Sherwin-Williams Co.*

944   Copper oleate
10402-16-1   P

C36H66CuO4
Oleic acid, copper(II) salt.
Copper oleate; EINECS 233-866-5; 9-Octadecenoic acid (Z)-, copper salt; (Z)-9-Octadecenoic acid, copper salt; Oleic acid, copper salt. Fungicide. Registered by EPA as a fungicide and molluscicide.

945   Copper oxychloride
1332-40-7   G P

H3ClCu2O3
Copper(II) chloride hydroxide.
Agrizan; Areecop; Basic copper chloride; Blitox; Blitox 50; Blue Copper; Blue Copper 50; Bordeaux A; Bordeaux Z; Caswell No. 249; ChemNut 50; Chemocin; Cobox; Cobox Blue; Cobrex; Colloidox; Copen; Copper chloride, basic; Copper chloride, mixed with copper oxide, hydrate; Copper chloride oxide, hydrate; Copper chloroxide; Copper OC fungicide; Copper oxychloride; Coppercide;

Copperthom; Coppesan; Coppesan Blue; Coprantol; Coprex; Coprosan Blue; Cozib 62; Cupral 45; Cupramer; Cuprargos; Cuprasol; Cupravit; Cupravit-Forte; Cupravit Green; Cupric oxide chloride; Cupricol; Cupritox; Cuprokylt; Cuprokylt L; Cuprosan Blue; Cuprovit; Cuprox; Cuproxol; Demildex; Dicopper chloride trihydroxide; EPA Pesticide Chemical Code 023501; Faligruen; Funguran; Fycol 8; Fycop; Fycop 40A; Fytolan; H 200A; Hokko Cupra Super; Kauritil; KT 35; Kupferoxychlorid; Kupricol; Kuprikol; Maccopper; Microcop; Miedzian; Miedzian 50; Ob 21; Oxicob; Oxivor; Oxychlorue de cuivre; Oxychlorure de cuivre; Oxyclor; Oxycur; Parrycop; Peprosan; Pol-kupritox; Recop; Tamraghol; Tricop 50; Turbair Copper Fungicide; Viricuivre; Vitigran; Vitigran Blue. Copper containing spray used for control of fungal diseases like downy mildews, late blight, early blight, apple scab and various leaf spot diseases on a wide range of crops. Effective against *Phytophthora infestans*, the cause of potato blight. Registered by EPA as a fungicide and herbicide. Blue-green powder; dec. > 220°; almost insoluble in H2O (< 10^{-9} g/100 ml at 20°, pH 7), insoluble in organic solvents; LD50 (rat orl) = 1440 mg/kg; LC50 (carp 48 hr.) = 2.2 mg/l; non-toxic to bees. *Agrichem (International) Ltd.; Agtrol International; All-India Medical; BASF Corp.; Bayer AG; Caffaro SpA; Cequisa; Ciba-Geigy Corp.; Continental Sulfur Co. LLC; Crystal; Cuproquim Corp.; Diachem; Drexel Chemical Co.; Generic; Griffin LLC; Hoechst AG; Industrias Quimicas del Valles SA; Industrie Chimche Caffaro; Phibro-Tech Inc.; Rhône Poulenc Crop Protection Ltd.; Sandoz; Universal Crop Protection Ltd.*

946   Copper phthalocyanine
147-14-8   2546   H

C32H16CuN8
Tetrabenzo-5,10,15,20-diazaporphyrinephthalocyan-ine.
Accosperse Cyan Blue GT; AI3-26192; Aqualine Blue; Arlocyanine Blue PS; Bahama Blue BC; Bahama Blue BNC; Bahama Blue Lake NCNF; Bahama Blue WD; Bermuda Blue; Blue GLA; Blue pigment; Blue Toner GTNF; Calcotone Blue GP; Ceres Blue BHR; Chromatex Blue BN; Chromofine Blue BX; Chromophtal 4G Blue; Chromophthal GF Green (VAN); CI 74160; CI Ingrain Blue 2; CI Pigment Blue 15; CI Pigment Blue 15:1; CI Pigment Blue 15:3; CI Pigment Blue 15:4; Congo Blue B-4; Copper (II) phthalocyanine; Copper Phthaloycanine Blue; Copper tetrabenzoporphyrazine; Copper(2+) phthalocyanine; Copper(II) phthalocyanine; Cromophtal Blue 4G; Cromophtal Blue 4GN; Cromophtal Blue GF; Cyan Blue BNC 55-3745; Cyan Blue BNF 55-3753; Cyan Blue GT 55-3295; Cyan Blue GT 55-3300; Cyan Blue GT; Cyan Blue GTNF; Cyan Blue Toner GT 3000; Cyan Blue XR 55-3758; Cyan Blue XR 55-3760; Cyan Peacock Blue G; Cyanine Blue BB; Cyanine Blue BF; Cyanine Blue Bnrs; Cyanine Blue C; Cyanine Blue GNPS; Cyanine Blue HB; Cyanine Blue LBG; Cyanine Blue Rnf; Dainichi Cyanine Blue B; Dainichi Cyanine Blue BX; Dainichi Cyanine Blue FPG; Dainichi Cyanine Blue PG; Daltolite Fast Blue B; Duratint Blue 1001; EINECS 205-685-1; Euvinyl Blue 702; Fastogen Blue 5007; Fastogen Blue 5110; Fastogen Blue 5120; Fastogen Blue B; Fastogen Blue BS; Fastogen Blue BSF; Fastogen Blue FP; Fastogen Blue GNPT; Fastogen Blue GR; Fastogen Blue GS; Fastogen Blue TGR; Fastolux Blue; Fastolux Peacock Blue; Fenalac Blue B Disp; Franconia Blue A-4431; Graphtol Blue 2GLS; Graphtol Blue BL; Graphtol Blue BLF; Helio Blue B; Helio Fast Blue BRN; Helio Fast

Blue HG; Helio Fast Blue BT; Helio Fast Blue B; Heliogen blue (VAN); Heliogen blue 7080; Heliogen Blue 6902K; Heliogen Blue 6960; Heliogen Blue 7044T; Heliogen Blue 7100; Heliogen Blue A; Heliogen Blue B; Heliogen Blue BA-CF; Heliogen Blue BKA-CF; Heliogen Blue BNC; Heliogen Blue BR; Heliogen Blue BV; Heliogen Blue BWS; Heliogen Blue BWS Extra; Heliogen Blue BWSN; Heliogen Blue IBG; Heliogen Blue K; Heliogen Blue LBG; Heliogen Blue LBGN; Heliogen Blue LBGO; Heliogen Blue LBGT; Heliogen Blue NCB; Heliogen Blue WX; Hostaperm Blue A2R; Hostaperm Blue A3R; Hostaperm Blue AFN; Hostaperm Blue B3G; Hostaperm Blue BG; HSDB 2925; Indolen Blue 3G; Irgalite Blue BLP; Irgalite Blue GLSM; Irgalite Blue LGLD; Irgalite Fast Brilliant Blue BL; Irgaplast Blue RBP; Isol Fast Blue Toner BT; Isol Fast Blue B; Isol Phthalo Blue B; Isol Phthalo Blue FCB; Isol Phthalo Blue BT; Isol Phthalo Blue CBG; Isol Phthalo Blue E 7543; LBX 5; Linnol Blue KLG; Lionol Blue ER; Lionol Blue KLG; Lumatex Blue B; Lutetia Fast Cyanine B; Lutetia Percyanine BRS; Monarch Blue GX-2480; Monarch Blue NCNF X-2658; Monarch Blue Toner NCNF X-2810; Monarch Blue Toner X-2303; Monarch Blue Toner X-2763; Monarch Blue Toner NC X-2371; Pigment Blue 15.

947   Copper sulfate monohydrate
10257-54-2   P

CuO4S.H2O
Copper(II) sulfate, monohydrate.
Caswell No. 257; Copper sulfate monohydrate; EPA Pesticide Chemical Code 024402. Registered by EPA as a fungicide (cancelled). Blue - white crystals; pH 3.7 - 4.5; LD_{50} (rat orl) = 300 mg/kg. Generic.

948   Copper sulfate pentahydrate
7758-99-8   2682   G P

CuSO4.H2O
Copper(II) sulfate, pentahydrate.
Blue copper AS; Blue Copper; Blue Copperas; Blue stone; Blue Vicking; Blue Vitriol; Bluestone; Calcanthite; Caswell No. 256; CCRIS 5556; Copper sulfate (CuSO4) pentahydrate; Copper(2+) sulfate (1:1) pentahydrate; Copper(II) sulfate, pentahydrate; Cupric sulfate (pentahydrate); EPA Pesticide Chemical Code 024401; Hi-Chel; HSDB 2968; Kocide® Copper Sulfate Pentahydrate Crystals; Kupfersulfat-pentahydrat; Kupfervitriol; Roman vitriol; Salzburg vitriol; Sulfuric acid, copper(2+) salt (1:1), pentahydrate; Triangle; Vencedor. Used as a fungicide to control plant diseases, in fertilizers to correct copper deficiencies in soils. Registered by EPA as a fungicide. Blue triclinic crystals; becomes anhydrous at 250°; $d^{15.6}_4$ = 2.286; very soluble in H2O, soluble in MeOH, glycerol, slightly soluble in EtOH; pH of 0.2M aqueous solution = 4.0; LD_{50} (rat orl) = 960 mg/kg. Aquatronics; Biotex Laboratories International Ltd.; Cheltec Inc.; CNM Technologies; Earth Care; Environ Intercontinental Ltd.; Florida Water Works, Inc.; Frank Miller & Sons, Inc.; Generic; Griffin LLC; IBC Manufacturing Co.; Jungle Lab.; Marjon & Associates; Phibro-Tech Inc.; Roebic Labs, Inc.; Voluntary Purchasing Group Inc.

949   Corn Oil
8001-30-7   2559   H

Oil of corn.
Corn germ oil; Lipex 104; Lipomul; Maise oil; Maize oil; Maydol; Mazola oil; Oils, glyceridic, corn; Oils, corn. FDA GRAS, FDA approved for injectables, orals, topicals, USP/NF, BP, JP compliance. Pale yellow oil; mp = -10°; d = 0.914 - 0.921; insoluble in H2O, soluble in Et2O, CHCl3, amyl acetate, C6H6, CS2, slightly soluble in EtOH; viscosity = 37.36 - 38.83 mPa. Non-toxic. Abitec Corp.; ADM; Arista Ind.; British Arkady; Calgene; Cargill; Columbus Foods; Grain Processing; Penta Mfg.; Ruger; Spectrum Chem. Manufacturing; Spectrum Naturals; Tri-K Ind.; Vandemoortele Professional; Wensleydale Foods.

950   Corn Syrup
8029-43-4   E H

Syrups, hydrolyzed starch, obtained by partial hydrolysis of corn starch.
C*Pharm 01165; Corn sugar syrup; Corn syrup; EINECS 232-436-4; Glucose syrup; Mixture of D-glucose, maltose and maltodextrins; Syrups, corn; Syrups, hydrolyzed starch.; refined maize oil; Stripped corn oil; Super Refined® Corn Oil. FDA GRAS, cleared by USDA to flavor sausage, hamburger, meat loaf, luncheon meat, chopped or pressed ham, limit 2% maximum, approved for orals. Used as a sweetener in orals, texturizer and carrying agent, used in aspirin. Non-toxic. ADM; Am. Maize Products; Gist-Brocades Intl.; Jungbunzlauer; MLG Enterprises; Staley, A. E. Mfg.

951   Cottonseed Oil
8001-29-4   2580   E P

Oil of cottonseed.
Aceite de Algodon; Caswell No. 259A; Cotton oil; Cotton seed oil [Oil, edible]; Cottonseed acidulated soapstock; Cottonseed oil; Cottonseed oil, fatty acid; Cottonseed oil, winterized; Deodorized winterized cottonseed oil; EINECS 232-280-7; EPA Pesticide Chemical Code 031602; HSDB 913; Lipomul IV; NCI-C50168; Oils, cottonseed; Prime bleachable summer yellow cottonseed oil; Solvent extracted crude cottonseed oil; Oleaginous vehicle and solvent. Registered by EPA as an antimicrobial, fungicide, herbicide, insecticide and rodenticide (cancelled). FDA GRAS, FDA approved for intramuscular injectable, orals, USP/NF compliance. Used as an oleaginous vehicle, as a solvent for intramuscular preparations and solid oral dosage forms, topical formulations and intravenous nutrients. Pale yellow oil; d = 0.915 - 0.921; insoluble in H2O, slightly soluble in EtOH, soluble in Et2O, C6H6, CHCl3, CS2, petroleum ether; viscosity < 70.4 mPa at 20°. Non-toxic, but used as an experimental tumorigen and teratogen. Abitec Corp.; Amber Syn.; Arista Ind.; British Arkady; Cargill; Columbus Foods; Good Food; National Cottonseed; Penta Mfg.; Ruger; Sigma-Aldrich Fine Chem.; Spectrum Chem. Manufacturing; Tri-K Ind.

952   Coumalic acid
500-05-0   2583   G

C6H4O4
2-Oxo-1,2H-pyran-5-carboxylic acid.
Coumalic acid; Cumalic acid; EINECS 207-899-0; NSC 22978; 2-Oxopyran-5-carboxylic acid; 2-Pentenedioic acid, 4-(hydroxymethylene)-, delta-lactone; α-Pyrone-5-carboxylic acid; 2H-Pyran-5-carboxylic acid, 2-oxo-. Used in chemical synthesis. Crystals; mp= 203-205° (dec); bp120 = 218°; λ_m = 240, 290 nm (ε =

7943, 3981, MeOH); insoluble in C_6H_6, $CHCl_3$, ligroin, slightly soluble in H_2O, Et_2O, Me_2CO, EtOAc, soluble in EtOH, AcOH.

953 Coumaphos
56-72-4 2584 G V

$C_{14}H_{16}ClO_5PS$
3-Chloro-4-methyl-7-coumarinyl diethyl phosphoro-thioate.
4-18-00-00344 (Beilstein Handbook Reference); Agridip; Al3-17957; Asunthol; Asuntol; Azunthol; BAY 21/199; Bayer 21/199; Baymix; Baymix 50; BRN 0327083; Caswell No. 335; CCRIS 180; 3-Chloro-4-methyl-7-hydroxycoumarin diethyl thiophosphoric acid ester; 3-Chloro-7-diethoxyphosphinothioyloxy-4-methylcoumarin; 3-Chloro-7-hydroxy-4-methyl-coumarin O,O-diethyl phosphorothioate; Co-Ral; Coumafos; Coumafosum; Coumaphos; Coumaphos; Cumafos; Cumafosum; Diethyl 3-chloro-4-methylumbelliferyl thionophosphate; Diethyl-O-(3-chloro-4-methyl-7-coumarinyl) phosphorothioate; Diolice; EINECS 200-285-3; ENT 17,957; EPA Pesticide Chemical Code 036501; HSDB 249; Meldane; Meldone; Muscatox; NCI-C08662; Negashunt; Negasunt; NSC 8944; OMS 485; OMS 495; Perizin®; Phosphorothioic acid, O-(3-chloro-4-methyl-2-oxo-2H-1-benzopyran-7-yl) O,O-diethyl ester; Resitox; Suntol; Thiophosphate de O,O-diethyle et de O-(3-chloro-4-methyl-7-coumarinyle) Umbethion. Non-systemic insecticide used for the control of ectoparasites (Varroa jacobsoni) in bees. Also used as a general insecticide and nematicide and, in veterinary medicine as an anthelmintic. Off-white crystals; mp = 95°; d = 1.474; soluble in H_2O (1.5 mg/l), poorly soluble in organic solvents; LD_{50} (mrat orl) = 41 mg/kg, (frat orl) = 16 mg/kg. *Bayer AG.*

954 Coumarin
91-64-5 2588 H

$C_9H_6O_2$
2H-1-Benzopyran-2-one.
5-17-10-00143 (Beilstein Handbook Reference); Al3-00753; Benzo-α-pyrone; BRN 0383644; Caswell No. 259C; CCRIS 181; Cinnamic acid, o-hydroxy-, δ-lactone; cis-o-Coumaric acid anhydride; Coumarin; cis-o-Coumarinic acid lactone; Coumarinic anhydride; Coumarinic lactone; Cumarin; EINECS 202-086-7; EPA Pesticide Chemical Code 127301; HSDB 1623; o-Hydroxycinnamic lactone; o-Hydroxyzimtsaure-lacton; Kumarin; NCI C07103; NSC 8774; Rattex; Tonka bean camphor. Used in pharmaceuticals as a flavorant. Crystals; mp = 71°; bp = 301.7°; d^{20} = 0.935; λ_m = 273, 311 nm (ε = 11200, 5710, MeOH); moderately soluble in H_2O, EtOH, more soluble in Et_2O, $CHCl_3$, C_5H_5N. *Alfa Aesar; Rhodia Inc.; Rhodia Organique Fine Ltd.*

955 Creosol
93-51-6 2599 G

$C_8H_{10}O_2$
1-Hydroxy-2-methoxy-4-methylbenzene.
4-06-00-05878 (Beilstein Handbook Reference); Al3-15891; BRN 1862340; Creosol; p-Creosol; p-Cresol, 2-methoxy-; Cresolum drudum; EINECS 202-252-9; FEMA No. 2671; Homocatechol monomethyl ether; Homoguaiacol; 4-Hydroxy-3-methoxy-1-methyl-benzene; 4-Hydroxy-3-methoxytoluene; Kreosol; 2-Methoxy-p-cresol; 2-Methoxy-4-cresol; 2-Methoxy-4-methylphenol; 3-Methoxy-4-hydroxytoluene; 4-Methyl-2-methoxyphenol; 4-Methyl guaiacol; 4-Methylguaiacol; p-Methylguaiacol; NSC 4969; Phenol, 2-methoxy-4-methyl-; Phenol, 3-methoxy-4-methyl-; Rohkcrsol. Constituent of creosote; used as wood preservative, disinfectant. Prisms; mp = 5.5°; bp = 221°, bp_{15} = 104-105°; d^{20} = 1.098; Soluble in EtOH, Et_2O; LD_{50} (rat orl) = 740 mg/kg. Frinton; *Midori Kagaku; Protocol Analytical Supplies.*

956 Cresidine
120-71-8 H

$C_8H_{11}NO$
1-Amino-2-methoxy-5-methylbenzene.
3-13-00-01577 (Beilstein Handbook Reference); Azoic Red 36; Benzenamine, 2-methoxy-5-methyl-; BRN 0637071; C.I. Azoic Red 83; CCRIS 183; Cresidine; p-Cresidine; EINECS 204-419-1; HSDB 4107; p-Kresidin; Krezidin; Krezidine; 2-Methoxy-5-methylaniline; NCI-C02982; 5-Methyl-o-anisidine; NSC 406904. Solid; mp = 53°; bp = 235°; λ_m = 236, 290 nm (ε = 7120, 3620, MeOH); slightly soluble in H_2O, $CHCl_3$, soluble in EtOH, Et_2O, C_6H_6.

957 Cresol
1319-77-3 2604 D G H P

C_7H_8O
Cresylic acid (mixed isomers).
4-06-00-02035 (Beilstein Handbook Reference); Acede cresylique; Acide cresylique; Al3-02360; Bacillol; BRN 0506719; CCRIS 6006; Coal tar phenols; Cresol; Cresol (isomers and mixture); Cresol (mixed isomers); Cresol, all isomers; Cresol, pure; Cresoli; Cresols; Cresols / cresylic acid (isomers and mixture); Cresols, mixed isomers; Cresolum crudum; Cresylic acid; Cresylic acid (isomers and mixture); Cresylic acid, dephenolized; EINECS 215-293-2; HSDB 250; Hydroxymethylbenzene; Hydroxytoluole; Kresole; Kresolen; Kresolum venale; Krezol; Methyl phenol; Methylphenol, mixed; Mixed cresols; Phenol, methyl-; Phenol, methyl-. mixed; RCRA waste number U052; Tekresol; ar-Toluenol; Tricresol; Tricresolum; Trikresolum; UN2022. Mixture of o, m and p-isomers. Used in manufacture of commercial mixtures of phenolic materials boiling above the cresol range; phosphate esters, phenolic resins, wire enamel solvent, plasticizers, gasoline additives, laminates, coating for magnet wire, disinfectants, metal cleaning, flotation agents,

Liquid; mp = 1-2°; bp = 88-94°, d = 1.04; soluble in H$_2$O (1.932 g/l); more soluble in organic solvents; LD$_{50}$ (rat orl) = 1454 mg/kg. *Allchem Ind.; Crowley Tar Prods.; PMC.*

958   m-Cresol
108-39-4   2604   G P

C$_7$H$_8$O
3-Hydroxytoluene.
AI3-00136; Bacticin; Caswell No. 261A; CCRIS 645; Celcure Dry Mix (chemicals for wood preserving); Cresol, all isomers; Cresol, m-; m-Cresol; m-Cresylic acid; EINECS 203-577-9; EPA Pesticide Chemical Code 022102; FEMA Number 2337; FEMA Number 3530; Franklin Cresolis; Gallex; HSDB 1815; Hydroxy-3-methylbenzene; m-Hydroxytoluene; m-Kresol; Metacresol; Methylphenol, 3-; m-Methylphenol; NSC 8768; m-Oxytoluene; Phenol, 3-methyl-; RCRA waste number U052; Rover's Dog Shampoo; m-Toluol; UN2076. Used in disinfectants, fumigants, in photographic developers, explosives; phenolic resins, ore flotation; textile scouring agent; manufacture of coumarin and salicylaldehyde; herbicides, surfactants. Registered by EPA as a fungicide. Liquid; mp = 11.8°; bp = 202.2°; d^{20}= 1.0341; slightly soluble in H$_2$O (2.5 g/100 ml), freely soluble in EtOH, CHCl$_3$, Et$_2$O, Me$_2$CO, C$_6$H$_6$, CCl$_4$; LD$_{50}$ (rat orl) = 2020 mg/kg. 11763451; *Allchem Ind.; Lancaster Synthesis Co.; Spectrum Chem. Manufacturing.*

959   o-Cresol
95-48-7   2604   G H

C$_7$H$_8$O
Phenol, 2-methyl-.
AI3-00137; CCRIS 646; Cresol, o-; Cresol, o-isomer; Cresol, ortho-; o-Cresylic acid; EINECS 202-423-8; FEMA No. 3480; HSDB 1813; o-Hydroxytoluene; o-Kresol; o-Methylphenol; o-Methylphenylol; o-Oxytoluene; NSC 23076; Phenol, 2-methyl-; RCRA waste number U052; o-Toluol; UN2076. Used as a disinfectant, in phenolic resins, manufacture of tricresyl phosphate, ore flotation, textile scouring, organic intermediate, manufacture of salicylaldehyde, coumarin, and herbicides, surfactant. Crystals; mp = 29.8°; bp = 191°; d^{25} = 1.135; λ$_m$ = 214, 273 nm (ε = 6030, 1820, MeOH); moderately soluble in H$_2$O, freely soluble in EtOH, Et$_2$O, Me$_2$CO, C$_6$H$_6$, CCl$_4$, LD$_{50}$ (rat orl) = 121 mg/kg. *Allchem Ind.; Crowley Tar Prods.; Penta Mfg.; PMC.*

960   p-Cresol
106-44-5   2604   G H

C$_7$H$_8$O
4-Methylphenol.
AI3-00150; CCRIS 647; Cresol, p-; 4-Cresol; p-Cresol; Cresol, p-isomer; Cresol, para-; p-Cresylic acid; EINECS 203-398-6; FEMA Number 2337; HSDB 1814; 1-Hydroxy-4-methylbenzene; 4-Hydroxytoluene; p-Hydroxytoluene; p-Kresol; 1-Methyl-4-hydroxy-

benzene; 4-Methylphenol; NSC 3696; p-Oxytoluene; Phenol, 4-methyl-; p-Toluol; p-Tolyl alcohol; RCRA waste number U052; UN2076. Disinfectant, phenolic resins, ore flotation; textile scouring agent; manufacture of coumarin and salicylaldehyde; herbicides, surfactants, synthetic food flavors. Liquid; mp = 35.5°; bp = 201.9°; d^{40} = 1.0185; λ$_m$ = 279 nm (MeOH); slightly soluble in H$_2$O (2 g/100 ml), freely soluble in EtOH, Et$_2$O, Me$_2$CO, C$_6$H$_6$, CCl$_4$; LD$_{50}$ (rat orl) = 207 mg/kg. *Allchem Ind.; Am. Biorganics; Penta Mfg.; PMC; Spectrum Chem. Manufacturing*

961   Cresol purple
2303-01-7   2606   G

C$_{21}$H$_{18}$O$_5$S
Phenol, 4,4'-(3H-2,1-benzoxathiol-3-ylidene)bis[3-methyl-, S,S-dioxide.
m-Cresol, 4,4'-(3H-2,1-benzoxathiol-3-ylidene)di-, S,S-dioxide; Cresol purple; 3-Cresol purple; m-Cresol purple; m-Cresolsulfonephthalein; EINECS 218-960-6; NSC 9607; Phenol, 4,4'-(3H-2,1-benzoxathiol-3-ylidene)bis(3-methyl-, S,S-dioxide; Phenol, 4,4'-(1,1-dioxido-3H-2,1-benzoxathiol-3-ylidene)bis(3-methyl-. Used as an indicator. Solid; slightly soluble in EtOH; pH 1.2 red, pH 2.8 yellow, pH 7.4 yellow, pH 9.0 purple.

962   Cresol red
1733-12-6   2607   G

C$_{21}$H$_{18}$O$_5$S
4,4'-(3H-2,1-Benzoxathiol-3-ylidene)bis(2-methyl-phenol) S,S-dioxide.
o-Cresol, 4,4'-(3H-2,1-benzoxathiol-3-ylidene)di-, S,S-dioxide; Cresol Red; o-Cresol Red; o-Cresolsulfonephthalein; Cresolsulfophthalein; 3',3''-Dimethylphenolsulfonephthalein; EINECS 217-064-2; NSC 7224; Phenol, 4,4'-(3H-2,1-benzoxathiol-3-ylidene)bis(2-methyl-, S,S-dioxide. A pH-sensitive dye, used as an indicator, range 7.2 (yellow), 8.8 (red), 2-3 (orange). Red-brown crystals; mp = 290° (dec); very soluble in H$_2$O, EtOH; pK 8.3.

963   m-Cresyl Acetate
122-46-3   2611   G D

C$_9$H$_{10}$O$_2$
Acetic acid 3-methylphenyl ester.

161

Acetic acid, 3-methylphenyl ester; Acetic acid m-tolyl ester; m-Acetoxytoluene; AI3-04169; Cresatin; Cresatin Metacresylacetate; Cresatin-Sulzberger; m-Cresol acetate; m-Cresyl acetate; EINECS 204-546-2; Kresatin; 3-Methylphenyl acetate; m-Methylphenyl acetate; Metacresol acetate; NSC 4795; m-Tolyl acetate. Antiseptic (topical); antifungal. Used as an external antiseptic and analgesic. Oily liquid; bp = 212°; bp_{13} = 99°; d_4^{26} 1.048; λ_m = 262.5, 269.5 nm (MeOH); insoluble in H_2O, glycerol; very soluble in alcohol, H_2O, $CHCl_3$, petroleum ether, C_6H_6; soluble in petrolatum (5%), cottonseed oil. *Merck & Co.Inc.*

964   Cresyl phosphate
1330-78-5   9832   G H

C21H21O4P
Phosphoric acid, tris(methylphenyl) ester.
AI3-16771; Caswell No. 884; CCRIS 5947; Celluflex 179C; Cresyl phosphate; Disflamoll TKP; Durad; EINECS 215-548-8; EPA Pesticide Chemical Code 083401; Flexol Plasticizer TCP; Fyrquel 150; HSDB 6774; Imol S-140; IMOL S 140; Kronitex; Lindol; NCI-C61041; Phosflex 179A; Phosphate de tricresyle; Phosphoric acid, tris(methylphenyl) ester; Phosphoric acid, tritolyl ester; Syn-O-Ad® 8484; TCP; Thiorthocresyl phosphate; Tricresilfosfati; Tricresyl phosphate; Tricresylfosfaten; Trikresylfosfat; Trikresyl-phosphate; Tris(methylphenyl) phosphate; Tris(tolyl-oxy)phosphine oxide; Tritolyl phosphate; Tritolylfosfat; UN2574. Mixture of o, m and p-cresyl isomers. Antiwear and EP agents in non-crankcase lubricants; ferrous metal passivator; fluid base stock where inhibition of flame propagation is desired. A plasticizer for cellulose lacquers, and polyvinyl chloride; it has a specific gravity of 1.185-1.189, a boiling range of 430-440° and a flash point of 215°; mixture of o, m and p isomers (1:65:35). Liquid; bp_{10} = 265°; d_{25}^{25} = 1.16; insoluble in H_2O, soluble in organic solvents. *Akzo Chemie.*

965   Crimidine
535-89-7   2615   G

C7H10ClN3
2-chloro-N,N,6-trimethyl-4-pyrimidinamine.
4-25-00-02184 (Beilstein Handbook Reference); 4-Pyrimidinamine, 2-chloro-N,N,6-trimethyl-; AI3-61801; BRN 0127995; Castrix; Castrix Grains; Caswell No. 188; 2-Chloor-4-dimethylamino-6-methyl-pyrimidine; 2-Chlor-4-dimethylamino-6-methylpyrimidin; 2-Chloro-4-dimethylamino-6-methylpyrimidine; 2-Chloro-4-methyl-6-dimethylaminopyrimidine; 2-Chloro-N,N-6-trimethyl-4-pyrimidinamine; 2-Chloro-N,N,6-trimethylpyrimidin-4-ylamine; 2-Cloro-4-di-metilamino-6-metil-pirimidina; Crimidin; Crimidina; Crimidine; EINECS 208-622-6; EPA Pesticide Chemical Code 288200; HSDB 2812; NSC 2017; Pyrimidine, 2-chloro-4-(dimethylamino)-6-methyl-; UN 2588; W 491. Rodenticide; used for control of field and house mice. Mode of action is similar to that of the strychnine alkaloids. Pyridoxine [58-56-0] is an antidote for crimidine poisoning in mice. Brown wax; mp = 87°; bp = 140-147°;

insoluble in H_2O, very soluble in EtOH; LD_{50} (mus orl) = 1.8 mg/kg. *Bayer AG.*

966   Crotonaldehyde
4170-30-3   2624   H

C4H6O
2-Butenaldehyde.
AI3-18303; CCRIS 909; Crotonal; Crotonaldehyde; Crotonic aldehyde; Crotylaldehyde; EINECS 224-030-0; 1-Formylpropene; HSDB 252; Krotonaldehyd; β-Methylacrolein; 3-Methylacrolein, inhibited; Methyl-propenal; NSC 56354; Propylene aldehyde; RCRA waste number U053; UN 1143. Liquid; mp = -76.5°; bp = 104°; d_{20}^{20} = 0.853; soluble in H_2O (18.1 g/100 ml 20°, 19.2 g/100 ml 5°), LD_{50} (rat orl) = 300 mg/kg.

967   Crotonic acid
107-93-7   2625   G

C4H6O2
β-Methylacrylic acid.
4-02-00-01498 (Beilstein Handbook Reference); BRN 1719943; 2-Butenoic acid, (E)-; (E)-2-Butenoic acid; trans-2-Butenoic acid; (E)-Crotonic acid; Crotonic acid, (E)-; trans-Crotonic acid; EINECS 203-533-9; NSC 8751. Synthesis of resins, polymers, plasticizers, drugs. Prisms or needles; mp = 72°; bp = 184.7°; d^{77} = 0.9604; very soluble in H_2O (54.6 g/l), EtOH, soluble in Et_2O, Me_2CO, ligroin; LD_{50} (rat orl) = 1.0 g/kg. *Alcoa Ind. Chem.; Allchem Ind.; Chisso Corp. USA; Eastman Chem. Co.; Sigma-Aldrich Fine Chem.*

968   Cryolite
15096-52-3   2634   G P

AlF6Na3
Aluminum sodium fluoride.
Aluminum sodium fluoride; Caswell No. 264; Cryolite; Cryolite (AlNa3F6); Cryolite (Na3(AlF6)); EINECS 239-148-8; ENT 24,984; EPA Pesticide Chemical Code 075101; Greenland spar; HSDB 1548; ICE Spar; Icetone; Koyoside; Kriolit; Kryalith; Kryocide; Kryolith; Natriumaluminiumfluorid; Natriumhexafluoro-alum-inate; Sodium aluminum fluoride; Kryocide; Kryolith]; Natriumaluminiumfluorid; Natriumhexa-fluoroalum-inate; ·sodium aluminofluoride; Sodium aluminum fluoride; Sodium fluoaluminate; trisodium hexa-fluoroaluminate(3-), (OC-6-11)-. Found in deposits in Greenland and Russia. Used in the aluminum industry as a source of the metal. Also used as an insecticide. Registered by EPA as an insecticide. Snow white monoclinic crystals; mp = 1000°; d = 2.95. *Amvac Chemical Corp.; Cerexagri, Inc.; Elf Atochem N. Am.; Gowan Co.*

969   Cumene hydroperoxide
80-15-9   G H

C9H12O2
1-Methyl-1-phenylethyl hydroperoxide.
4-06-00-03221 (Beilstein Handbook Reference); BRN 1908117; CCRIS 3801; CHP; Cumeenhydroperoxyde; Cumene hydroperoxide; Cumenyl hydroperoxide; Cumolhydroperoxid; Cumolhydroperoxide; Cumyl hydroperoxide; EINECS 201-254-7; HSDB 254; Hydroperoxide, α,α-dimethylbenzyl; Hydroperoxide de cumene; Hydroperoxyde de cumene; Hydroperoxyde de cumyle; Idroperossido di cumene; Idroperossido di cumolo; Isopropylbenzene hydroperoxide; Kumenylhydroperoxid; RCRA waste number U096. Initiator for vinyl monomers and copolymers and the crosslinking of unsaturated polyester resins. Facilitates room temperature curing of vinyl ester and other unsaturated polyester resins. Liquid; mp = -40°; bp = 127° (dec); d = 1.024, insoluble in H2O, soluble in organic solvents; LD50 (rat orl) = 382 mg/kg. *Catalyst Resources Inc.; CK Witco Chem. Corp.*

970   Cuminaldehyde
122-03-2   2648   G

C10H12O
p-Isopropylbenzaldehyde.
4-07-00-00723 (Beilstein Handbook Reference); Al3-01853; Benzaldehyde, 4-(1-methylethyl)-; Benz-aldehyde, p-isopropyl-; BRN 0636547; Cumaldehyde; Cumic aldehyde; p-Cumic aldehyde; Cuminal; Cuminaldehyde; Cuminic aldehyde; Cuminyl aldehyde; EINECS 204-516-9; FEMA No. 2341; 4-Isopropylbenzaldehyde; p-Isopropylbenzaldehyde; 4-Isopropylbenzenecarboxylate; p-Isopropylbenzene-carboxaldehyde; 4-(1-Methylethyl)benzaldehyde; NSC 4886. Intermediate for pharmaceuticals, perfumes. Liquid; bp = 235.5°; d^{20} = 0.9755; λ_m = 255 nm (MeOH); insoluble in H2O, slightly soluble in CCl4, soluble in EtOH, Et2O; LD50 (rat orl) = 1390 mg/kg. *Mitsubishi Gas.*

971   p-Cumyl phenol
599-64-4   G H

C15H16O
2-Phenyl-2-(p-hydroxyphenyl)propane.
4-06-00-04761 (Beilstein Handbook Reference); Al3-08269; BRN 1870517; 4-(α,α-Dimethylbenzyl)phenol; EINECS 209-968-0; 4-Hydroxydiphenyldimethyl-methane; 4-(1-Methyl-1-phenylethyl)phenol; NSC 6237; p-(α-Cumenyl)phenol; p-(α,α-Dimethylbenzyl)-phenol; p-Cumyl phenol; p-Cumylphenol; Phenol, 4-(1-methyl-1-phenylethyl)-; Phenol, p-(α,α-dimethyl-benzyl)-; 2-Phenyl-2-(p-hydroxyphenyl)propane. Intermediate for resins, insecticides, lubricants. Crystals; mp = 70-73°; bp = 335°. *Degussa-Hüls Corp.; ICI Spec.; PMC; Schenectady.*

972   Cupferron
135-20-6   2649   G

C6H9N3O2
N-Hydroxy-N-nitrosobenzenamine ammonium salt.
Al3-63016; Ammonium cupferron; Ammonium N-nitrosophenylhydroxylamine; Ammonium nitroso-β-phenylhydroxylamine; Benzenamine, N-hydroxy-N-nitroso-, ammonium salt; CCRIS 184; Cupferon; Cupferron; Cupferron, ammonium salt; EINECS 205-183-2; HSDB 4109; Hydroxylamine, N-nitroso-N-phenyl-, ammonium salt; Kupferon; Kupferron; N-Hydroxy-N-nitroso-benzenamine, ammonium salt; N-Nitroso-N-phenylhydroxylamine, ammonium salt; N-Nitrosofenylhydroxylamin amonny; N-Nitrosophenyl-hydroxylamin ammonium salz; NCI-C03258; NSC 112124. The ammonium salt is used as a precipitating agent for copper, in the determination of copper; used as an analytical reagent, especially for the separation and precipitation of metals, e.g., copper, iron, vanadium. Crystals; mp = 163.5°; λ_m = 221, 292 nm (MeOH); slightly soluble in DMSO, soluble in H2O, EtOH.

973   Cupric acetate
142-71-2   2651   G

C4H6CuO4
Copper(II) acetate.
Acetate de cuivre; Acetic acid, copper (2+) salt; Acetic acid, copper(II) salt (2:1); Acetic acid, cupric salt; Al3-01379; CCRIS 5286; Copper acetate; Copper acetate (Cu(C2H3O2)2); Copper acetate (Cu(OAc)2); Copper diacetate; Copper(2+) acetate; Crystallized verdigris; Crystals of venus; Cupric acetate; Cupric acetate; Cupric diacetate; EINECS 205-553-3; HSDB 915; Neutral verdigris; NSC 75796; Octan mednaty; Venus copper. Basic copper acetate; it is usually a mixture of mono-, di- and tri-acetates of copper. Green verdigris consists chiefly of the basic acetate, 2(C2H3O2)2Cu2O. Blue verdigris consists mainly of the basic acetate, (C2H3O2)2Cu2O. Used as a pesticide, catalyst, fungicide, pigments, manufacture of Paris green. Solid; mp = 115°; d = 1.882; soluble in H2O, slightly soluble in organic solvents; LD50 (rat orl) = 0.71 g/kg. *Celtic Chem. Ltd.; Mallinckrodt Inc.; Nihon Kagaku Sangyo.*

974   Cupric acetoarsenite
12002-03-8   2653   G

C4H6As6Cu4O16
(Acetato)trimetaarsenitodicopper.
(Acetato-O)(trimetaarsenito)dicopper; (Acetato)trimeta-arsenitodicopper; Acetoarsenite de cuivre; Basle Green; C.I. 77410; C.I. Pigment Green 21; Caswell No. 229A; CCRIS 4771; CI 77410; Copper acetate arsenite; Copper aceto-arsenite; Copper acetoarsenite; Copper, bis(acetato)hexametaarsenitotetra-; Copper(II) acetate meta-arsenate; Cupric acetoarsenite; ENT 884; EPA Pesticide Chemical Code 022601; French Green; Genuine Paris Green; HSDB 1824; Imperial Green; King's Green; Meadow Green; Mineral Green; Mitis Green; Moss Green; Mountain Green; Neuwied Green; New Green; Ortho P-G bait; Paris Green; Parrot Green; Patent Green; Powder Green; Schweinfurt Green;

Schweinfurtergrün; Schweinfurth Green; Sowbug & cutworm bait; Sowbug cutworm control; Swedish Green; UN1585; Vienna Green; Wuerzberg Green; Zwickau Green. Used as a pigment, particularly in ships and submarines, and as a wood preservative and insecticide. Green crystalline powder, insoluble in H_2O, unstable in acid or base and with H_2S; LD_{50} (rat orl) = 100 mg/kg.

975   Cupric arsenate
7778-41-8   G

As2Cu3O8
Copper(II)arsenate.
Arsenic acid (H_3AsO_4), copper(2+) salt (2:3); Cupar; Copper arsenate. Protective fungicide. *Mechema Chemicals Ltd.*

976   Cupric arsenite
10290-12-7   2654   G P

AsCuHO3
Arsonic acid copper(2+) salt (1:1).
acid copper arsenite; Arsenious acid, copper(2+) salt (1:1); Arsonic acid, copper(2+) salt (1:1); Copper(I) arsenite; Cupric arsenite; Cupric arsenite ($CuHAsO_3$); Pickle green; Scheele's green; Scheele's mineral. Used as a pigment, wood preservative, insecticide, fungicide, and rodenticide. Registered by EPA as a rodenticide (cancelled). Insoluble in H_2O, organic solvents.

977   Cupric bromide
7789-45-9   2656   G

Br2Cu
Copper(II) bromide.
Copper bromide ($CuBr_2$); Copper dibromide; Copper(II) bromide; Cupric bromide; EINECS 232-167-2; HSDB 257. Used in photography (intensifier), organic synthesis (brominating agent), battery electrolyte, wood preservative. Crystals; mp = 498°; bp = 900°; d_4^{20} = 4.710. *Atomergic Chemetals; Cerac; Hoechst Celanese; Mallinckrodt Inc.; Sigma-Aldrich Fine Chem.*

978   Cupric chloride
7447-39-4   2660   G

Cl2Cu.2H2O
Copper(II) chloride dihydrate.
AI3-01658; CCRIS 6883; Coclor; Copper bichloride; Copper chloride; Copper chloride ($CuCl_2$); Copper dichloride; Copper(2+) chloride; Copper(II) chloride; Copper(II) chloride (1:2); Coppertrace; Cupric chloride; Cupric dichloride; EINECS 231-210-2; HSDB 259;

NSC 165706. Supplement, trace mineral. Used as a catalyst, deodorant and desulfurizer. Solid; mp = 100°; d = 2.51; freely soluble in H_2O, EtOH, less soluble in Me_2CO, EtOAc. *Armour Pharm. Co. Ltd.; Mechema Chemicals Ltd.*

979   Cupric formate
544-19-4   2666   G

C2H2CuO4
Copper(II) formate.
Copper diformate; Copper formate; Cupric diformate; Cupric formate; EINECS 208-865-8; Formic acid, copper(2+) salt (1:1); Formic acid copper(2+) salt; HSDB 260; NSC 112232; Tubercuprose; Cufor. Antibacterial agent used to treat cellulose.
Light blue, royal blue or turquoise crystals; soluble in H_2O, insoluble in organic solvents. *Mechema Chemicals Ltd.*

980   Cupric nitrate
3251-23-8   2671   G P

CuN2O6
Cupper(II) nitrate.
Caswell No. 246; Claycop; Copper dinitrate; Copper nitrate; Copper(2+) nitrate; Copper(II) nitrate; Cupric dinitrate; Cupric nitrate; EINECS 221-838-5; EPA Pesticide Chemical Code 076102; HSDB 264; Nitric acid, copper(2+) salt. Used in light-sensitive papers, as an analytical reagent, textile dyeing mordant, nitrating agent, insecticide, coloring copper black, electroplating, paints, varnishes, enamels, pharmaceuticals, catalyst. Has been used as a fungicide. Registered by EPA as a fungicide and herbicide (cancelled). Large blue-green orthorhombic crystals; mp = 255 - 256°; sublimes 150 - 225°; soluble in H_2O, EtOAc, dioxane, Et_2O (reacts); [trihydrate]: blue rhombic plates; mp = 114.5°; d = 2.05; freely soluble in H_2O, EtOH, insoluble in EtOAc; pH of 0.2M aqueous solution = 4.0; LD_{50} (rat orl) = 940 mg/kg; [hexahydrate]: blue prismatic crystals; dec. (to trihydrate) 6.4°; freely soluble in H_2O, soluble in EtOH. *Blythe, Williams Ltd.; Mallinckrodt Inc.*

981   Cupric oxalate
814-91-5   2673   G

C2CuO4
Oxalic acid copper(II) salt.
Caswell No. 248A; Copper oxalate; Copper oxalate (CuC_2O_4); Copper(II) oxalate; Crow Chex; Cupric oxalate; EINECS 212-411-4; EPA Pesticide Chemical Code 023305; Ethanedioic acid copper salt; Ethanedioic acid, copper(2+) salt (1:1); HSDB 265; NSC 112246;

NSC 86015; Oxalic acid, copper(2+) salt (1:1). Active ingredient: copper oxalate; seed protectant to prevent sprout pulling by birds in newly planted corn. Solid; mp = 310° (dec); insoluble in H$_2$O, organic solvents. *Borderland Products Inc.*

982   Cupric oxide
1317-38-0   2674   G P

$$Cu=O$$

CuO
 Copper(II) oxide.
 Banacobru ol; Black copper oxide; Boliden-CCA Wood Preservative; Boliden Salt K-33; C.I. 77403; C.I. Pigment Black 15; Caswell No. 265; CCA Type C Wood Preservative; Chrome Brown; Copacaps; Copper Brown; Copper monooxide; Copper oxide; Copper oxide (CuO); Copper(2+) oxide; Copper(II) oxide; Copporal; Cupric oxide; EINECS 215-269-1; EPA Pesticide Chemical Code 042401; Farboil Super Tropical Anti-Fouling 1260; HSDB 266; Natural tenorite; NSC 83537; Osmose K-33 Wood Preservative; Osmose K-33-A Wood Preservative; Osmose K-33-C Wood Preservative; Osmose P-50 Wood Preservative; Paramelaconite; Wolmanac concentrate. Ceramic colorant, reagent in analytical chemistry, fungicide, insecticide, catalyst, purification of hydrogen, batteries and electrodes, electroplating, solvent, desulfurizing oils, rayon, metallurgical and welding fluxes, antifouling paints, phosphors. Brown crystalline powder; d$_4^4$= 6.315; insoluble in H$_2$O, EtOH. *Am. Chemet; Cerac; Chemisphere; Nihon Kagaku Sangyo; Noah Chem.; Sigma-Aldrich Fine Chem.*

983   Cupric sulfate
7758-98-7   2682   D G P

$$O=\overset{\overset{\displaystyle O}{\|}}{\underset{\underset{\displaystyle O^-}{|}}{S}}-O^- \quad Cu^{2+}$$

CuO4S
 Copper(II) sulfate anhydrous.
 All Clear Root Destroyer; Aqua Maid Permanent Algaecide; Aquatronics Snail-A-Cide Dri-Pac Snail Powder; BCS copper fungicide; Blue Copper; Blue stone; Blue vitriol; Bonide Root Destroyer; CCRIS 3665; Copper(II) sulfate; Copper(II) sulfate (1:1); Copper monosulfate; Copper sulfate; Copper sulphate; Copper(2+) sulfate; Cupric sulfate; Cupric sulfate anhydrous; Cupric sulphate; Delcup; EINECS 231-847-6; Granular Crystals Copper Sulfate; HSDB 916; Hylinec; Incracide 10A; Incracide E 51; Kupfersulfat; MAC 570; Monocopper sulfate; NSC 57630; Phelps Triangle Brand Copper Sulfate; Roman vitriol; Sa-50 Brand Copper Sulfate Granular Crystals; Snow Crystal Copper Sulfate; Sulfate de cuivre; Sulfuric acid, copper(2+) salt (1:1); Tobacco States Brand Copper Sulfate; Trinagle. Occurs as the mineral chalcanthite. Used as a soil additive, pesticides, feed additive, germicides, textile mordant, leather, pigments, batteries, electroplated coatings, copper salts, analytical reagent, medicine, wood preservative, lithography, ore flotation, petroleum, rubber, steel. Registered by EPA as a fungicide. Used as a fungicide and in manufacture of agricultural fungicides and many industrial products. Mixed with lime, effective against *Phytophthora infestans*, the cause of potato blight. Used medicinally as an antiseptic; antifungal(topical); antidote to phosphorus. Used in veterinary medicine as a mineral source, anthelmintic, emetic and fungicide. Gray-green-white rhombic crystals; dec. 360°; sg$^{15.6}$ = 2.286; soluble in H$_2$O (14.8 g/100 ml at 0°, 23 g/100 ml at 25°, 33.5 g/100 ml at 50°, 73.6 g/100 ml at 100°), MeOH (15.6 g/100 ml), glycerol, insoluble in EtOH; LD$_{50}$ (rat orl) = 960 mg/kg; toxic to fish. *Allchem Ind.; Farleyway Chem. Ltd.; Generic; Helena Chemical Co.; Industrias Quimicas del Valles SA; McKechnie Chemicals Ltd.; Seaboard Ind.*

984   Cupric sulfate, basic
1332-14-5   2683   G

Cu4H6O10S
 Copper(II) hydroxide sulfate.
 Brochantite; Caldo Bordeles Valles; Cupric sulfate, basic; Cusatrib; EINECS 215-568-7; Langite; Sulfuric acid, copper(2+) salt, basic; Tribasic copper sulfate. Bordeaux mixture plus adjuvants; wettable powder used as protective fungicide for foliage application to ornamental and crop plants. Used as a fungicide for plants, seed treatment. Blue-green crystals; insoluble in H$_2$O. *Industrias Quimicas del Valles SA; Mechema Chemicals Ltd.*

985   Cuprous bromide
7787-70-4   2689   G

$$Cu-Br$$

CuBr
 Copper(I) bromide.
 Copper bromide (CuBr); Copper monobromide; Copper(1+) bromide; Copper(I) bromide; Cuprous bromide; EINECS 232-131-6; HSDB 270. Used as a catalyst in organic reactions. Solid; mp = 504°; bp = 1345°; d$_4^{25}$= 4.72; slightly soluble in H$_2$O. *Atomergic Chemetals; Cerac; Hoechst Celanese; Sigma-Aldrich Fine Chem.*

986   Cuprous chloride
7758-89-6   2690   G P

$$Cu-Cl$$

ClCu
 Copper(I) chloride.
 Chlorid medny; Copper chloride; Copper chloride (CuCl); Copper monochloride; Copper(1+) chloride; Copper(I) chloride; Cuprous chloride; Dicopper dichloride; EINECS 231-842-9. Catalyst, preservative and fungicide, desulfurizing and decolorizing agent in petroleum industry, absorbent for carbon monoxide. Registered by EPA as a fungicide. Used in non-food applications. White crystalline powder; mp = 430°; d$_4^{25}$= 4.14; slightly soluble in H$_2$O. *Atomergic Chemetals; Cerac; Hoechst Celanese; Mallinckrodt Inc.; Mechema Chemicals Ltd.; Nihon Kagaku Sangyo; ZinderSpA.*

987   Cuprous iodide
7681-65-4   2692   G

$$Cu-I$$

CuI
 Copper(I) iodide.
 Copper (I) iodide; Copper iodide (CuI); Copper monoiodide; Copper(1+) iodide; Copper(I) iodide; Cuprous iodide; EINECS 231-674-6; EPA Pesticide Chemical Code 108301; HSDB 271; Hydro-Giene; Natural marshite. Used as a feed additive, in table salt as source of dietary iodine, catalyst, in cloud seeding. Solid; mp = 588-606°; bp= 1290°; d$_4^{25}$= 5.63; insoluble in all solvents. *Atomergic Chemetals; Blythe, Williams Ltd.; Cerac; Greeff R.W. & Co.; Mitsui Toatsu; Nihon Kagaku Sangyo; Sigma-Aldrich Fine Chem.*

988   Cuprous oxide
1317-39-1   2694   G

$$Cu^{\diagup O}\diagdown Cu$$

Cu2O
 Copper(I) oxide.
 Brown copper oxide; C.I. 77402; Caocobre; Caswell No. 266; CCRIS 2289; CI 77402; Cobre; Copox; Copper (I) oxide; Copper

hemioxide; Copper nordox; Copper oxide; Copper oxide (Cu2O); Copper oxide, red; Copper protoxide; Copper Sandoz; Copper sardez; Copper suboxide; Copper(1+) oxide; Copper(I) oxide; CP Cuprous Oxide Pigment Grade; Cupper oxide; Cupramar; Cupridan; Cuprocide; Cuprous oxide; Cuprous Oxide, AA Grade; Cuprous Oxide Type Two; Cuprox; Dicopper monoxide; Dicopper oxide; EINECS 215-270-7; Fungi-rhap Cu-75; Fungimar; HSDB 1549; Kupferoxydul; Kuprite; Nordox; NSC 83538; Oleo nordox; Oleocuivre; Oxyde cuivreux; Perecot; Perenex; Perenox; Purple Copp 92; Purple Copp 97; Purple Copp 97N; Red Copp 92; Red Copp 97; Red Copp 97N; Red copper oxide; Ruby ore; Violet copper; Yellow compound; Yellow cuprocide; Copper oxide. Fungicide. Cubic crystals; mp = 1232°; bp = 1800°; d = 6.000; LD50 (rat orl) = 0.47 g/kg. *Makhteshim Chemical Works Ltd.*

989   Cuprous potassium cyanide
13682-73-0   2695   G

$$N\equiv C-Cu^{-} \; C\equiv N \quad K^{+}$$

C2CuKN2
 Copper(I) potassium cyanide.
 Copper(I) potassium cyanide; Cuprate(1-), dicyano-, potassium; Cuprous potassium cyanide; EINECS 237-192-2; Potassium copper(I) cyanide; Potassium cuprocyanide; Potassium dicyanocuprate; Potassium dicyanocuprate(1-); UN1679. A double cyanide of potassium and copper used in electroplating copper and brass. Prisms; d = 2.38; insoluble in H2O, soluble in DMSO.

990   Cuprous thiocyanate
1111-67-7   2699   G

$$N\equiv\!\!\!\!=\!\!\!-S^{-} \quad Cu^{+}$$

CCuNS
 Copper(I) thiocyanate.
 Caswell No. 266A; Copper thiocyanate; Copper(1+) thiocyanate; Cuprous thiocyanate; Cusyd; EINECS 214-183-1; EPA Pesticide Chemical Code 025602; Thiocyanic acid, copper(1+) salt. Used as marine anti-fouling agent. Solid; d = 2.85; insoluble in H2O, organic solvents. *Mechema Chemicals Ltd.*

991   Cyanazine
21725-46-2   2714   G H P

C9H13ClN6
 Propanenitrile, 2-[[4-chloro-6-(ethylamino)-1,3,5-tri-azin-2-yl]amino]-2-methyl.
 5-26-08-00471 (Beilstein Handbook Reference); Bladex; Bladex 80WP; Blanchol; BRN 0615509; Caswell No. 188C; CCRIS 6823; 2-((4-Chloro-6-(ethylamino)-s-triazin-2-yl)amino)-2-methylpropio-nitrile; 2-(4-Chloro-6-ethylamino-1,3,5-triazin-2-yl-amino)-2-methylpropionitrile;   2-Chloro-4-(1-cyano-1-methylethylamino)-6-ethylamine-1,3,5-triazine; 2-Chloro-4-(1-cyano-1-methylethylamino)-6-ethylamino-s-triazine; Cyanazin; Cyanazine; DW3418; EPA Pesticide Chemical Code 100101; Fortrol; HSDB 6842; Payze; Propanenitrile,2-((4-chloro-6-(ethylamino)-1,3,5-triazin-2-yl)amino)-2-methyl-;   Propionitrile,   2-((4-chloro-6-(ethylamino)-s-triazin-2-yl)amino)-2-methyl-; SD 15418; s-Triazine, 2-chloro-4-ethylamino-6-

(1-cyano-1-methyl)ethylamino-; WL 19805. Selective systemic herbicide, absorbed by roots and foliage. Used for control of annual grass and broad-leaved weeds. Registered by EPA as a herbicide (cancelled). White crystals; mp = 167.5-169°; soluble in H2O (0.0171 g/100 ml), methylcyclohexanone (21 g/100 ml), CHCl3 (21 g/100 ml), Me2CO (19.5 g/100 ml), EtOH (4.5 g/100 ml), C6H6 (1.5 g/100 ml), C6H12 (1.5 g/100 ml), CCl4 (< 1 g/100 ml); LD50 (rat orl) = 182 - 334 mg/kg, (mus orl) = 380 mg/kg, (rbt orl) = 141 mg/kg, (rat der) >1200 mg/kg, (rbt der) > 2000 mg/kg, (mallard duck orl) > 2000 mg/kg, (quail orl) = 400 mg/kg; LC50 (harlequin fish 48 hr.) = 10 mg/l, (sheepshead minnow 48 hr.) = 18 mg/l; non-toxic to bees. *DuPont; Shell UK; Shell.*

992   Cyanoacetic acid
372-09-8   2718   H

C3H3NO2
 Monocyanoacetic acid.
 4-02-00-01888 (Beilstein Handbook Reference); Acetic acid, cyano-; Acide cyanacetique; AI3-15026; BRN 0506325; Cyanessigsäure; Cyanoacetic acid; EINECS 206-743-9; HSDB 272; Kyselina kyanoctova; Malonic acid mononitrile; Malonic mononitrile; Monocyanoacetic acid; NSC 5571; USAF KF-17. Solid; mp = 66°; = 160° (dec), bp0.15 = 108°; λ_m = 208 nm (ε = 40, EtOH); slightly soluble in CHCl3, AcOH, soluble in H2O, EtOH, Et2O, DMSO.

993   β-Cyanoethyltriethoxysilane
919-31-3   H

C9H19NO3Si
 3-(Triethoxysilyl)propionitrile.
 4-04-00-04271 (Beilstein Handbook Reference); AI3-51457; BRN 1776392; EINECS 213-050-5; HSDB 5768; NSC 77092; Propanenitrile, 3-(triethoxysilyl)-; Silane, (2-cyanoethyl)triethoxy-; Triethoxy-2-kyanethylsilan; 3-(Triethoxysilyl)propionitrile.

994   Cyanogen bromide
506-68-3   2723   G

$$N\equiv C-Br$$

CBrN
 Bromine cyanide.
 4-03-00-00092 (Beilstein Handbook Reference); AI3-28715; BRN 1697296; Bromine cyanide; Bromocyan; Bromocyane; Bromocyanide; Bromocyanogen; Bromure de cyanogen; Campilit; Cyanobromide; Cyanogen bromide; Cyanogen monobromide; EINECS 208-051-2; HSDB 708; NSC 89684; RCRA waste number U246; TL 822; UN1889. Organic synthesis, parasiticide, fumigating compositions, rat extermination, cyaniding reagent in gold extraction processes. Needles; mp = 52°; bp = 61-62°; d_4^{20} = 2.015; soluble in H2O, EtOH, Et2O. *Alcoa Ind. Chem.; Eastman Chem. Co.; Janssen Chimica; Sigma-Aldrich Fine Chem.*

995   Cyanoguanidine
461-58-5   3119   G H

C2H4N4
Dicyanodiamide.
  ACR-H 3636; AI3-14632; Araldite HT 986; Araldite XB 2879B; Araldite XB 2979B; Bakelite VE 2560; CCRIS 3478; Cyanoguanidine; N-Cyanoguanidine; Dicyan-diamide; Dicyandiamido; Dicyandiamin; Dicyanodi-amide; EINECS 207-312-8; Epicure DICY 15; Epicure DICY 7; Guanidine, cyano-; HSDB 2126; NSC 2031; Pyroset DO; XB 2879B. Used in fertilizers; nitrocellulose stabilizer; organic synthesis, especially of melamine, barbituric acid, and guanidine salts; explosives; catalyst for epoxy resin; pharmaceuticals; fireproofing compounds; stabilizer in detergents; modifier for starch production. Crystals; mp = 211°; (eutectic with cyanamide at 35.6°); d^{14} = 1.404; λ_m = 217 nm (ε = 1500, MeOH), 217 nm (ε = 14700, MeOH-KOH, 217 nm (ε = 14900, MeOH-HCl); insoluble in C6H6, CS2, CCl4, CHCl3, Et2O, slightly soluble in EtOAc, soluble in H2O, EtOH, Me2CO. *Akzo Chemie; Andrulex Trading Ltd.*

996   p-Cyanophenyl acetate
13031-41-9   H

C9H7NO2
4-(Acetyloxy)benzonitrile.
  4-(Acetyloxy)benzonitrile; Benzonitrile, 4-(acetyloxy)-; Benzonitrile, p-hydroxy-, acetate (ester); 4-Cyanophenyl acetate; p-Cyanophenyl acetate; EINECS 235-893-8.

997   3-Cyanopropyltrichlorosilane
1071-27-8   G

C4H6Cl3NSi
Trichloro-3-cyanopropylsilane.
  4-04-04272 (Beilstein Handbook Reference); BRN 0906924; Butanenitrile, 4-(trichlorosilyl)-; Butyro-nitrile, 4-(trichlorosilyl)-; CC3555; 4-Cyanobutyl-trichlorosilane; (3-Cyanopropyl) trichlorosilane; EINECS 213-990-6; σ-Cyanopropyltrichlorosilane; Silane, (3-cyanopropyl)trichloro-; Silane, trichloro(3-cyanopropyl)-; Trichlor-3-kyanpropylsilan; 4-(Trichlorosilyl)butyronitrile; 4-Trichlorosilylbutane-nitrile; CC3555. Coupling agent, chemical intermediate, blocking agent, release agent, lubricant, primer, reducing agent. Liquid; bp8 = 94°; d = 1.2800; LD50 (rat orl) = 2830 mg/kg. *Degussa-Hüls Corp.*

998   3-Cyanopyridine
100-54-9   H

C6H4N2
3-Pyridinecarbonitrile.
  5-22-02-00115 (Beilstein Handbook Reference); AI3-15766; BRN 0107711; EINECS 202-863-0; HSDB 5335; Nicotinic acid nitrile; Nicotinonitrile; Nitryl kwasu nikotynowego; NSC 17558. Crystals; mp = 51°; bp = 206.9°, bp300 = 170°; d^{25} = 1.1590; λ_m = 216, 225, 258, 264, 270 nm (MeOH); slightly soluble in ligroin, soluble in CHCl3, very soluble in C6H6, EtOH, Et2O.

999   4-Cyanothiazole
1452-15-9   H

C4H2N2S
4-Thiazolecarbonitrile.
  4-Cyanothiazole; 4-Thiazolecarbonitrile; EINECS 215-919-4; HSDB 5829.

1000   Cyanox® 425
88-24-4   G

C25H36O2
2,2'-Methylenebis (4-ethyl-6-*tert*-butylphenol).
  4-06-00-06806 (Beilstein Handbook Reference); Agidol 7; AI3-25275; Antage W 500; Antioxidant 425; AO 425; Bis(2-hydroxy-3-t-butyl-5-ethylphenyl)-methane; Bis(3-t-butyl-5-ethyl-2-hydroxyphenyl)-methane; 6,6'-Di-t-butyl-4,4'-diethyl-2,2'-methylene-diphenol; BRN 2016207; CCRIS 7788; Chemanox 22; Cyanox 425; EINECS 201-814-0; HSDB 5257; 2,2-Methylenebis(4-ethyl-6-t-butylphenol); 2,2'-Methylene-bis(4-ethyl-6-t-butylphenol); Nocrac NS 5; NSC 7782; Phenol, 2,2'-methylenebis(6-(1,1-dimethylethyl)-4-ethyl-; Phenol, 2,2'-methylenebis(6-t-butyl-4-ethyl-; Plastanox 425; Plastanox 425 antioxidant; USAF CY-6; Yoshinox 425. Antioxidant for impact molding resins; stabilizes acrylics and ABS; end-uses include extruded and molded products. *Am. Cyanamid; Ciba-Geigy Corp.*

1001   Cyanuric acid
108-80-5   2728   G P

C3H3N3O3
1,3,5-Triazine-2,4,6(1H,3H,5H)-trione.

AI3-26483; Caswell No. 862; CCRIS 5895; CP 4789; Cyanuric acid; s-Cyanuric acid; EINECS 203-618-0; EPA Pesticide Chemical Code 081402; HSDB 2818; Isocyanurate acid; Isocyanuric acid; Kyselina kyanurova; normal Cyanuric acid; NSC 6284; Pseudocyanuric acid; sym-Triazinetriol; Tricarbimide; Tricyanic acid; Trihydroxy-1,3,5-triazine; Trihydroxycyanidine; Tri-hydroxytriazine; Zeonet A. Intermediate for chlorinated bleaches, selective herbicide, whitening agents. Used to stabilize chlorine solutions in swimming pools; bleaches, sanitizers. Curing agent for Nipol AR and HyTemp 4050 series polyacrylate elastomers. Registered by EPA as an antimicrobial and disinfectant (cancelled). White crystals; mp > 330°; pKa_1 = 6.88, pKa_2 = 11.40, pKa_3 = 13.50; λ_m = 214 nm (log ε 3.38 0.1M HCl), 214 nm (log ε 4.00 0.1N phosphate buffer, pH 7), 213 nm (log ε 4.64 0.1M NaOH); slightly soluble in Me_2CO, Et_2O, C_6H_6, EtOH, C_6H_{14}, DMF (7.2 g/100 ml), DMSO (17.4 g/100 ml), H_2O (0.2 g/100 ml at 25°, 2.6 g/100 ml at 90°, 10.0 g/100 ml at 150°; d^{25} = 1.75 (anhydrous), 1.66 (dihydrate); LD_{50} (rat orl) > 5000 mg/kg. 3V; Allchem Ind.; Fluka; Monsanto Co.; Nippon Zeon.

1002 Cyanuric chloride
108-77-0   H

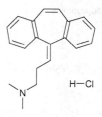

$C_3Cl_3N_3$
1,3,5-Trichlorotriazine.
5-26-01-00311 (Beilstein Handbook Reference); AI3-17788; BRN 0124246; Chlorotriazine; Cyanur chloride; Cyanurchloride; Cyanuric acid chloride; Cyanuric acid trichloride; Cyanuric chloride; Cyanuryl chloride; EINECS 203-614-9; HSDB 2905; Kyanurchlorid; NSC 3512; 1,3,5-Triazine, 2,4,6-trichloro-; s-Triazine, 2,4,6-tri-chloro-; Trichlorocyanidine; Tricholorotriazine; Trichloro-s-triazine; 1,3,5-Trichlorotriazine; 2,4,6-Trichlorotriazine; 2,4,6-Trichloro-1,3,5-triazine; 2,4,6-Trichloro-s-triazine; Tricyanogen chloride; UN2670. Crystals; mp = 154°; bp = 192°; very soluble in EtOH.

1003 Cyclamic acid
100-88-9   2731   G

$C_6H_{13}NO_3S$
Cyclohexylsulfamic acid.
4-12-00-00102 (Beilstein Handbook Reference); BRN 2208885; Cyclamate; Cyclamic acid; Cyclohexane-sulfamic acid; Cyclohexanesulphamic acid; Cyclohexyl-amidosulfuric acid; Cyclohexylamidosulphuric acid; Cyclohexylamine sulfamic acid; Cyclohexylaminesulfonic acid; Cyclohexylaminesulphonic acid; Cyclohexylsulf-amic acid; Cyclohexylsulphamic acid; EINECS 202-898-1; Hexamic acid; HSDB 275; N-Cyclohexylsulfamic acid; N-Cyclohexylsulphamic acid; NSC 220327; Polycat 200; Sucaryl; Sucaryl acid; Sulfamic acid, cyclohexyl-; Sulfuric acid monoamide, N-cyclohexyl-. Used as a non-nutritive sweetener and acidulant. mp = 169.5°; sparingly soluble in H_2O, soluble in alkaline solutions; LD_{50} (rat orl) = 15.25 g/kg.

1004 Cyclobenzaprine
303-53-7   2742   D

$C_{20}H_{21}N$

N,N-Dimethyl-5H-dibenzo[a,d]cycloheptene-$\Delta^{5,\gamma}$-propyl-amine. 9715 R.P.; Ciclobenzaprina; Cyclobenzaprine; Cyclo-benzaprinum; EINECS 206-145-8; MK-130; Proeptatriene; Proheptatrien; Proheptatriene; Ro 4-1577; RP-9715; Yurelax. Anxiolytic; skeletal muscle relaxant. bp1 = 175-180°; λ_m = 224, 289 nm (log ε = 4.57, 4.02). Apothecon; Hoffmann-LaRoche Inc.; Merck & Co.Inc.

1005 Cyclobenzaprine Hydrochloride
6202-23-9   2742   D

$C_{20}H_{22}ClN$

N,N-Dimethyl-5H-dibenzo[a,d]cycloheptene-$\Delta^{5,\gamma}$-propyl-amine hydrochloride.
10,11δ-Amitriptyline hydrochloride; Cloben; Cyben; Cyclobenzaprine hydrochloride; Cycloflex; 3-(5H-Dibenzo(a,d)cyclohepten-5-ylidene)propyl(dimethyl)-ammonium chloride; 5H-Dibenzo(a,d)cycloheptene-Δ^5-γ-propylamine, N,N-dimethyl-, hydrochloride; 5-(3-Dimethylaminopropylidene)-5H-dibenzo-(a,d)cyclo-heptene hydrochloride; N,N-Dimethyl-5H-dibenzo-(a,d)cycloheptene-$\Delta^{5,\gamma}$-propylamine hydrochloride; EINECS 228-264-4; Flexeril; Flexeril hydrochloride; Flexiban; Lisseril; MK-130 HCl; MK 130 hydrochloride; Novo-Cycloprine; NSC 78206; NSC 169900; NSC 173379; Proheptatrien monohydrochloride; Proheptatriene hydrochloride; Proheptatriene monohydrochloride; 1-Propanamine, 3-(5H-dibenzo(a,d)cyclohepten-5-ylid-ene)-N,N-dimethyl-, hydrochloride; Tensodox. Anxiolytic; skeletal muscle relaxant. mp = 216-218°; soluble in H_2O (> 20 g/100ml); soluble in MeOH, EtOH; less soluble in iPrOH, $CHCl_3$, CH_2Cl_2; insoluble in hydrocarbons; λ_m = 226, 295 nm (ε 52300, 12000); LD_{50} (mus iv) = 35 mg/kg, (mus orl) = 250 mg/kg. Apothecon; Hoffmann-LaRoche Inc.; Merck & Co.Inc.

1006 Cyclocarboxypropyloleic acid
53980-88-4   H

$C_{21}H_{36}O_4$
5(or 6)-Carboxy-4-hexyl-2-cyclohexene-1-octanoic acid.
C21-Dicarboxylic acid; 5(or 6)-Carboxy-4-hexyl-2-cyclo-hexene-1-octanoic acid; Cyclocarboxypropyloleic acid; 2-Cyclohexene-1-octanoic acid, 5(or 6)-carboxy-4-hexyl-; EINECS 258-897-1.

1007   Cyclododecalactam
947-04-6   H

C₁₂H₂₃NO
12-Aminododecanoic acid lactam.
5-21-06-00566 (Beilstein Handbook Reference); 12-Aminododecanoic acid lactam; Azacyclotridecan-2-one; BRN 0122031; Cyclododecalactam; ω-Dodecalactam; Dodecane-12-lactam; Dodecyllactam; EINECS 213-424-8; HSDB 5774; Laurin lactam; Laurinolactam; ω-Laurolactam; Lauryl lactam; NSC 77100.

1008   Cyclododecane
294-62-2   8076   H

C₁₂H₂₄
Cyclododecane.
4-05-00-00169 (Beilstein Handbook Reference); BRN 1901008; Cyclododecane; EINECS 206-033-9; HSDB 5557. Needles; mp = 60.4°; bp = 247°; d⁸⁰ = 0.82

1009   Cyclododecanol
1724-39-6   H

C₁₂H₂₄O
Cyclododecanol.
Cyclododecanol; EINECS 217-031-2; HSDB 5850; NSC 524960.

1010   Cyclododecanone
830-13-7   H

C₁₂H₂₂O
Cyclododecanone.
Cyclododecanone; EINECS 212-595-6; HSDB 5762; NSC 77116. Crystals; mp = 59°; bp₁₂ = 126-128°; d⁶⁶ = 0.9059. 6

1011   1,5,9-Cyclododecatriene
4904-61-4   H

C₁₂H₁₈
1,5,9-Cyclododecatriene.
(Z,E,E)-CDT; 1,5,9-Cyclododecatriene; 1,5,9-Cyclo-dodecatriene (Z,E,E); AI3-26695; CDT; EINECS 225-533-8; HSDB 6481; NSC 72433; UN2518. Liquid; mp = -17°; bp = 240°; d¹⁰⁰ = 0.84.

1012   Cyclohex-1,2-ylenediamine
694-83-7   G H

C₆H₁₄N₂
1,2-Diaminocyclohexane.
0-13-00-00001 (Beilstein Handbook Reference); 1,2-Cyclohexanediamine; 1,2-Diaminocyclohexane; BRN 0506142; Cyclohex-1,2-ylenediamine; EINECS 211-776-7; HSDB 5748. High-quality polyamine; epoxy curing agent; chelating agent for oilfield, textile, water treatment, detergent fields; herbicide intermediate; polyamide resins for adhesives, films, plastics, inks and corrosion inhibitors. Used as a chemical intermediate. *DuPont; Milliken Chemical; Pacific Anchor Co.; Sigma-Aldrich Fine Chem.*

1013   Cyclohex-3-enylmethyl   cyclohex-3-ene-carboxylate
2611-00-9   H

C₁₄H₂₀O₂
3-Cyclohexenylmethyl 3-cyclohexenecarboxylate.
Cyclohex-3-enylmethyl cyclohex-3-enecarboxylate; 3-Cyclohexene-1-carboxylic acid, 3-cyclohexen-1-ylmethyl ester; 3-Cyclohexenyl 3-cyclohexene 1-carboxylate; 3-Cyclohexenylmethyl 3-cyclohexenecarboxylate; Diene 221; EINECS 220-031-5; HSDB 5887; NSC 49615.

1014   Cyclohexane
110-82-7   2752   G H

C₆H₁₂
Cyclohexane.
AI3-08222; Benzene, hexahydro-; Benzenehexahydride; Caswell No. 269; CCRIS 3928; Cicloesano; Cyclohexaan; Cyclohexan; Cyclohexane; Cykloheksan; EINECS 203-806-2; EPA Pesticide Chemical Code 025901; Hexahydrobenzene; Hexamethylene; Hexanaphthene; HSDB 60; NSC 406835; RCRA waste number U056; UN1145. Used in manufacture of nylon; solvent for cellulose ethers, fats, oils, waxes; paint and varnish remover; glass substitutes; in analytic chemistry; chemical intermediate; in fungicidal formulations. Mobile liquid; mp = 6.6°; bp = 80.7°; d²⁰ = 0.7785; insoluble in H₂O, freely soluble in EtOH; Et₂O, Me₂CO, C₆H₆, CCl₄,

1015 Cyclohexane-1,4-dimethanol
105-08-8   H

$C_8H_{16}O_2$
1,4-Cyclohexanedimethanol.
AI3-28853; 1,4-Bis(hydroxymethyl)cyclohexane; BRN 1902271; CHDM; 1,4-Chidm; Cyclohex-1,4-ylenedimethanol; Cyclohexane-1,4-dimethanol; 1,4-Cyclohexamethylenebis methylol; 1,4-Cyclohexane-dimethanol; 1,4-Dimethylolcyclohexane; EINECS 203-268-9; HSDB 5364; NSC 44508; Rikabinol DM.

1016 Cyclohexanesulfenyl chloride
17797-03-4   H

$C_6H_{11}ClS$
Cyclohexanesulfenyl chloride.

1017 Cyclohexanethiol
1569-69-3   H

$C_6H_{12}S$
Cyclohexyl mercaptan.
4-06-00-00072 (Beilstein Handbook Reference); BRN 1236342; Cyclohexanethiol; Cyclohexyl mercaptan; Cyclohexyl thiol; Cyclohexylmercaptan; Cyclohexylthiol; Cyklohexanthiol; Cyklohexylmerkaptan; EINECS 216-378-7; HSDB 5840; NSC 59723; UN3054. Liquid; bp = 158.9°; d^{20} = 0.9782; λ_m = 230 nm (ε = 155, cyclohexane); freely soluble in Me_2CO, C_6H_6, EtOH. Et_2O.

1018 Cyclohexanol
108-93-0   2754   G H

$C_6H_{12}O$
Hydroxycyclohexane.
4-06-00-00020 (Beilstein Handbook Reference); Adronal; Adronol; AI3-00040; Anol; BRN 0906744; CCRIS 5896; Cicloesanolo; Cyclohexanol; Cyclohexyl alcohol; Cykloheksanol; EINECS 203-630-6; Hexahydrophenol; Hexalin; HSDB 61; Hydralin; Hydrophenol; Hydroxycyclohexane; Naxol; NSC 403656; Phenol, hexahydro-. Used in soap making and manufacture of phenolic insecticides; source of adipic acid for nylon, textile finishing; solvent for alkyd and phenolic resins and rubber. Solid; mp = 25.4°; bp = 160.8°; d^{20} = 0.9624; slightly soluble in $CHCl_3$, soluble in H_2O, EtOH, Et_2O, Me_2CO, freely soluble in C_6H_6, CS_2; LD_{50} (rat orl) = 2.06 g/kg.

1019 Cyclohexanone
108-94-1   2755   G H

$C_6H_{10}O$
Cyclohexanone.
AI3-00041; Anon; Anone; Caswell No. 270; CCRIS 5897; Cicloesanone; Cyclohexanon; Cyclohexanone; Cyclohexyl ketone; Cykloheksanon; EINECS 203-631-1; EPA Pesticide Chemical Code 025902; Hexanon; HSDB 186; Hytrol O; Ketocyclohexane; Ketohexamethylene; Nadone; NCI-C55005; NSC 5711; Oxocyclohexane; Pimelic ketone; Pimelin ketone; RCRA waste number U057; Sextone; UN1915. Used as paint and varnish remover; solvent for cellulose acetate, nitrocellulose, natural resins, vinyl resins, rubber, waxes, fats; in production of adipic acid for nylon, cyclohexanone resins. Registered by EPA as an insecticide (cancelled). Clear liquid; mp = -31°; bp = 155.4°; bp_{400} = 132.5°, bp_{200} = 110.3°, bp_{100} = 90.4°, bp_{60} = 77.5°, bp_{40} = 67.8°, bp_{20} = 52.5°, bp_{10} = 38.7°, bp_5 = 26.4°, bp_1 = 1.4°; d^{20} = 0.9478; λ_m = 276, 280 nm (ε = 26, 27, MeOH); soluble in H_2O (8.7 g/100 ml), EtOH, Et_2O, Me_2CO, C_6H_6; $CHCl_3$, CCl_4; LD_{50} (rat orl) = 1620 mg/kg. *Allied Signal; BASF Corp.; DSM Spec. Prods.; Fluka; Union Carbide Corp.*

1020 Cyclohexyl acetate
622-45-7   G

$C_8H_{14}O_2$
Acetic acid, cyclohexyl ester.
4-06-00-00036 (Beilstein Handbook Reference); Adronal acetate; AI3-03294; BRN 1906543; Cyclohexane acetate; Cyclohexanol, acetate; Cyclohexanolazetat; Cyclohex-anyl acetate; Cyclohexyl acetate; Cyclohexylester kyseliny octove; EINECS 210-736-6; FEMA No. 2349; H.A. Solvent; Hexalin acetate; HSDB 2820; NSC 8772; UN2243. A proprietary solvent for cellulose acetate and nitrate, rosin, rubber, oils, and metallic resinates. Liquid; bp = 173°, bp_{75} = 96°; d^{20} = 0.968; very soluble in EtOH, Et_2O.

1021 Cyclohexyl chloride
542-18-7   2761   G

$C_6H_{11}Cl$
Chlorocyclohexane.
4-05-00-00048 (Beilstein Handbook Reference); AI3-23841; BRN 1900796; Chlorocyclohexane; Cyclohexane, chloro-; Cyclohexyl chloride; EINECS 208-806-6; HSDB 2801; Monochlorocyclohexane; NSC 8434. Solvent and chemical intermediate. Liquid; mp = -44°; bp = 142°; d^{20} = 1.000; insoluble in H_2O, slightly soluble in CCl_4, very soluble in $CHCl_3$, freely soluble in EtOH, Et_2O, Me_2CO, C_6H_6. *Degussa-Hüls Corp.; Janssen Chimica.*

1022   Cyclohexyl methacrylate
101-43-9   G

C10H16O2
2-Methyl-2-propenoic acid, cyclohexyl ester.
Ageflex CHMA; AI3-33324; Cyclohexyl methacrylate; EINECS 202-943-5; HSDB 5340; Methacrylic acid, cyclohexyl ester; NSC 20968; 2-Propenoic acid, 2-methyl-, cyclohexyl ester. Clear colorless monomeric liquid. Polymer modifier for optical lens coatings, adhesives, floor polishes, vinyl polymerization, anaerobic adhesives. Liquid; bp = 210°; d^{20} = 0.9626; Insoluble in H2O. *Rit-Chem.*

1023   Cyclohexyl pyrrolidone
6837-24-7   G

C10H17NO
N-Cyclohexyl-2-pyrrolidone.
5-21-06-00331 (Beilstein Handbook Reference); BRN 0121832; CHP; 1-Cyclohexyl-2-pyrrolidinone; N-Cyclohexylpyrrolidinone; N-Cyclohexylpyrrolidone; N-Cyclohexyl-2-pyrrolidone; EINECS 229-919-7; 2-Pyrrolidinone, 1-cyclohexyl-. Solvent, used as a reaction intermediate, textile auxiliary, cosmetic ingredient. Liquid; bp7 = 154°; d = 1.0070; LD50 (rat orl) = 370 mg/kg. *ISP.*

1024   Cyclohexylamine
108-91-8   2758   G H

C6H13N
Aminocyclohexane.
4-12-00-00008 (Beilstein Handbook Reference); AI3-15323; Aminocyclohexane; Aminohexahydrobenzene; Aniline, hexahydro-; Benzenamine, hexahydro-; BRN 0471175; CCRIS 3645; CHA; Cyclohexanamine; Cyclohexylamine; EINECS 203-629-0; Hexahydroaniline; Hexahydrobenzenamine; HSDB 918; UN 2357. Boiler water treatment, rubber accelerator, intermediate in organic synthesis. Liquid; mp = -17.7°; bp = 134°; d^{20} = 0.8191; soluble in H2O, CCl4, very soluble in EtOH, freely soluble in Et2O, Me2CO, C6H6; LD50 (rat orl) = 0.61 mg/kg. *Air Products & Chemicals Inc.; BASF Corp.; Elf Atochem N. Am.; PMC.*

1025   Cyclohexylidenebis(t-butyl) peroxide
3006-86-8   G

C14H28O4
Peroxide, cyclohexylidenebis(1,1-dimethylethyl).
1,1-Bis(tert-butyl peroxy)cyclohexane; Chaloxyd P 1250AL; EINECS 221-111-2; Peroxide, cyclo-hexyl-idenebis((1,1-dimethylethyl); USP®-400P; 1,1-Di(tert-butylperoxy)cyclohexane; Lupersol 331; Lupersol 331-80B; Trigonox 22B50. Iinitiator useful in heat-cured polyester resin systems where improved flow and pot life are critical. *Ethitek Pharmaceuticals Co.*

1026   Cyclohexylisocyanate
3173-53-3   H

C7H11NO
Isocyanic acid, cyclohexyl ester.
AI3-28283; Cyclohexane, isocyanato-; Cyclohexyl isocyanate; Cyclohexylisocyanate; EINECS 221-639-3; Isocyanatocyclohexane; Isocyanic acid, cyclohexyl ester; NSC 87419; UN2488. Liquid; bp = 172°; d^{25} = 0.98

1027   Cyclohexylthiophthalimide
17796-82-6   H

C14H15NO2S
2-(Cyclohexylthio)-1H-isoindole-1,3(2H)-dione.
5-21-11-00117 (Beilstein Handbook Reference); BRN 0613992; CP 29242; Cyclohexylthiophthalimide; N-(Cyclohexylthio)phthalimide; N-Cyclohexylsulfenyl-phth-alimide; EINECS 241-774-1; 2-(Cyclohexylthio)-1H-iso-indole-1,3(2H)-dione; Phthalimide, N-(cyclohexylthio)-; Santogard PVI; Santogard PVI-DS.

1028   Cyclonite
121-82-4   2765   H

C3H6N6O6
1,3,5-Triaza-1,3,5-trinitrocyclohexane.
4-26-00-00022 (Beilstein Handbook Reference); BRN 0288466; CX 84A; Cyclonite; Cyclotrimethylene-nitramine; Cyklonit; EINECS 204-500-1; Esaidro-1,3,5-trinitro-1,3,5-triazina; Geksogen; Heksogen; Hexahydro-1,3,5-trinitro-s-triazine; Hexahydro-1,3,5-trinitro-1,3,5-triazine; Hexogeen; Hexogen (Explosive); Hexogen 5W; Hexolite; HSDB 2079; KHP 281; NSC 312447; PBX (af) 108; PBXW 108(E); PE 4; Perhydro-1,3,5-trinitro-1,3,5-triazine; RDX; 1,3,5-Triazine, hexahydro-1,3,5-trinitro-; Trimethyleentrinitramine; Trimethylenetrinitramine; UN0072; UN0391; UN0483. Crystals; mp = 205.5°; d^{20} = 1.82; λ_m = 213 nm (ε = 10965, EtOH); insoluble in H2O, EtOH, C6H6, CCl4, slightly soluble in Et2O, MeOH, C7H8; soluble in Me2CO, AcOH.

1029   Cyclopentadiene
542-92-7   2767   H

C5H6
1,3-Cyclopentadiene.
Cyclopentadiene; 1,3-Cyclopentadiene; EINECS 208-835-4; HSDB 2514; R-Pentine; Pentole; Pyropentylene. Liquid; mp = -85°; bp = 41°; d^{20} = 0.8021; λ_m = 240, 246, 252, 256, 262 nm (ϵ = 3236, 2818, 2138, 1445, 741, C6H14); insoluble in H2O, soluble in Me2CO, freely soluble in EtOH, Et2O, C6H6.

1030   Cyclopentane
287-92-3   2769   H

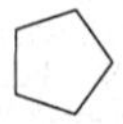

C5H10
Cyclopentane.
Cyclopentane; EINECS 206-016-6; HSDB 62; NSC 60213; Pentamethylene; UN1146. Liquid; mp = -93.8°, bp = 49.3°; d^{20} = 0.7457; insoluble in H2O, freely soluble in EtOH, Et2O, Me2CO, C6H6, CCl4, petroleum ether.

1031   Cyclopropanecarbonyl   chloride,   3-(2,2-dichloroethenyl)-2,2-dimethyl-
52314-67-7   H

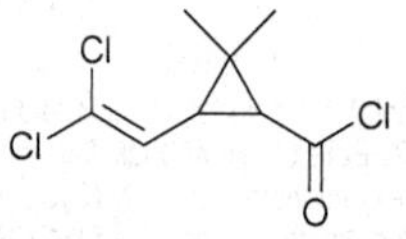

C8H9Cl3O
3-(2,2-Dichlorovinyl)-2,2-dimethylcyclopropanecarbonyl chloride.
Cyclopropanecarbonyl chloride, 3-(2,2-dichloroethenyl)-2,2-dimethyl-; 3-(2,2-Dichloroethenyl)-2,2-dmethyl-cyclopropanecarbonyl chloride; 3-(2,2-Dichlorovinyl)-2,2-dimethylcyclopropanecarbonyl chloride; EINECS 257-840-8.

1032   Cyclopropanecarboxylic   acid,   3-(2,2-dichloroethenyl)-2,2-dimethyl-
55701-05-8   H

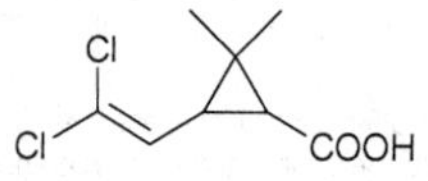

C8H10Cl2O2
3-(2,2-Dichlorovinyl)-2,2-dimethylcyclopropanecarboxylic acid.
Cyclopropanecarboxylic acid, 3-(2,2-dichloroethenyl)-2,2-dimethyl-; 3-(2,2-Dichlorovinyl)-2,2-dimethylcyclo-propanecarboxylic acid; EINECS 259-768-2.

1033   Cycloxydim
101205-02-1
                                                                       G P

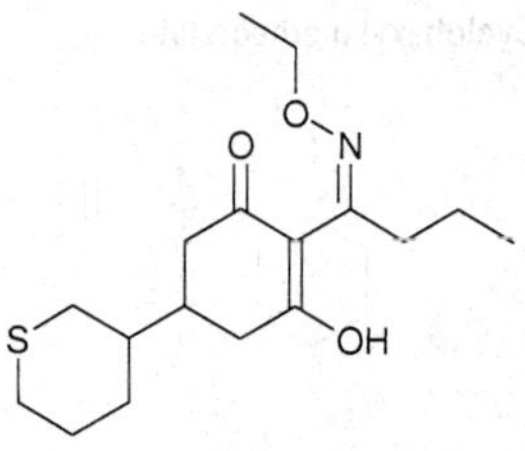

C17H27NO3S
2-[1-(Ethoxyimino)butyl]-3-hydroxy-5-(tetrahydro-2H-thiopyran-3-yl)-2-cyclohexene-1-one.
BAS 517; BAS 517-02H; BAS 51701H; BAS 51704H; BAS 517H; 2-Cyclohexen-1-one, 2-(1-(ethoxyimino)butyl)-3-hydroxy-5-(tetrahydro-2H-thiopyran-3-yl)-; Cycloxydim; (±)-2-(1-(Ethoxyimino)butyl)-3-hydroxy-5-thian-3-cyclo-hex-2-enone; 2-(1-(Ethoxyimino)butyl)-3-hydroxy-5-(tetra-hydro-2H-thiopyran-3-yl)-2-cyclohexen-1-one; Focus Ultra; Laser; Stratos. Selective herbicide, absorbed primarily by leaves. Inhibits mitosis. Used for post-emergence control of annual and perennial grasses in broad-leaved crops, such as beans and potatoes. Colorless crystals; mp = 36°; soluble in H2O (0.0088 g/100 ml), Me2CO, EtOH, CH2Cl2, Et2O, C7H8 (all > 100 g/100 ml); LD50 (rat orl) = 3940 mg/kg, (rat dr) > 2000 mg/kg, (quail orl) > 2000 mg/kg; LC50 (trout 96 hr.) = 220 mg/l, (bluegill sunfish 96 hr.) > 100 mg/l; non-toxic to bees. *BASF Corp.*

1034   Cyfluthrin
68359-37-5   2785   G P

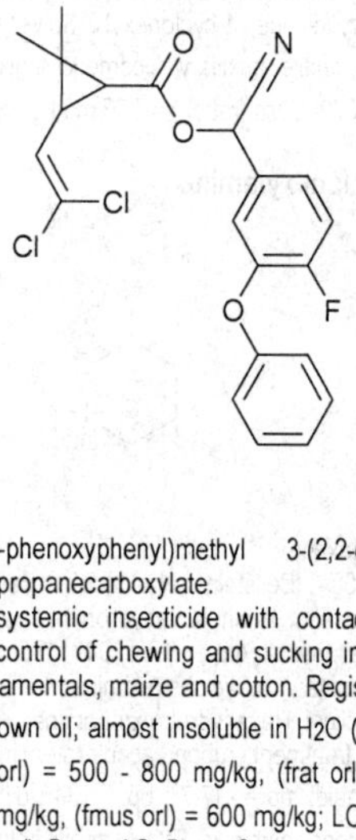

C22H18Cl2FNO3
Cyano(4-fluoro-3-phenoxyphenyl)methyl 3-(2,2-dichloro-ethenyl)-2,2-dimethylcyclopropanecarboxylate.
Baythroid. Non-systemic insecticide with contact and stomach action. Used for control of chewing and sucking insects on oilseed rape, cereals, ornamentals, maize and cotton. Registered by EPA as an insecticide. Brown oil; almost insoluble in H2O (0.0000002 g/100 ml); LD50 (mrat orl) = 500 - 800 mg/kg, (frat orl) = 1200 mg/kg, mmus orl) = 300 mg/kg, (fmus orl) = 600 mg/kg; LC50 (rainbow trout 96 hr.) = 0.0006 mg/l. *Bayer AG; Bayer Corp.*

1035   Cyhalothrin K
91465-08-6   2787   G P

$C_{23}H_{19}ClF_3NO_3$

(1-α(S*),3-α(Z))-($\pm$)-Methyl 3-(2-chloro-3,3,3-trifluoro-1-propenyl)-2,2-dimethyl-, cyano (3-phenoxyphenyl)-cyclopropanecarboxylate.

Charge; Commodore; λ-Cyalothrin; λ-Cyhalothrin; Cyhalothrin K; EPA Pesticide Chemical Code 128897; Excalibur; Hallmark; Icon; Karate; Karate C50; Matador; Ninja 10WP; PP 321; PP321; Saber; Sentinel. Non-systemic insecticide with contact and stomach action. Used for control of a wide variety of insects. Registered by EPA as an insecticide. Colorless solid; mp = 49.2°; poorly soluble in H2O (0.0000005 g/100 ml)readily soluble in Me2CO, EtOH, MeOH, C7H8, EtOAc, C6H14 (all > 50 g/100 ml); LD50 (mrat orl) = 79 mg/kg, (frat orl) = 56 mg/kg (in corn oil), (mrat der) = 632 mg/kg, (frat der) = 696 mg/kg (in propane-1,2-diol), (mallard duck orl) > 3950 mg/kg, (bee orl) = 38 ng/bee, (bee contact) 909 ng/bee; LC50 (quail dietary) > 5000 mg/kg, (bluegill sunfish) = 0.00021 mg/l, (rainbow trout) 0.00024 mg/l. Chemisco; Schering-Plough Veterinary Operations, Inc.; Syngenta Crop Protection; Zeneca Pharm.

1036 Cyhexatin
13121-70-5   2789   G P

$C_{18}H_{34}OSn$

Tricyclohexylhydroxystannane.

tricyclohexylstannoltricyclohexyltin hydroxide; Acarstin; Aracnol F; ENT-27395; Dowco 213; Mitacid; Plictran; TCTH; Tetran; Triran; Caswell No. 884A; CCRIS 6824; Cyhexatin; Dowco-213; ENT 27395; ENT 27,395-X; EPA Pesticide Chemical Code 101601; HSDB 1782; Hydroxytricyclohexylstannane; M 3180; NSC 179742; Ortho Plictran 50 wettable miticide; Plictran; Plictran 5OW miticide; Plyctran; Stannane, tricyclohexylhydroxy-; Tin, tricyclohexylhydroxy-; Tricyclohexylhydroxytin; Tri-cyclohexylhydroxystannane; Tricyclohexylstannanol; Tri-cyclohexylstannium hydroxide; Tricyclohexyltin hydr-oxide; Tricyclohexylzinnhydroxid. Non-systemic acar-icide with contact action. Used for control of mites on fruit crops. Registered by EPA as an acaricide and insecticide (cancelled). Colorless crystals; mp = 195-198°; insoluble in H2O (<0.0001 g/100 ml), soluble in CHCl3 (32 g/100 ml), MeOH (3.0 g/100 ml), CH2Cl2 (4.5 g/100 ml), CCl4 (4.5 g/100 ml), C6H6 (1.4 g/100 ml, C7H8 (0.9 g/100 ml), xylene (0.31 g/100 ml), Me2CO (0.1 g/100 ml); LD50 (rat orl) = 540 mg/kg, (rbt orl) = 500 - 1000 mg/kg, (gpg orl) = 780 mg/kg, (rbt der) > 2000 mg/kg, (ckn orl) = 650 mg/kg, (bee der) = 0.032 mg/bee; LC50 (mallard duck 8 day dietary) = 3189 mg/kg diet, (bobwhite quail 8 day dietary) = 520 mg/kg diet, (large mouth bass 24 hr.) = 0.06 mg/l, (goldfish 24 hr.) = 0.55 mg/l; non-

toxic to bees. *Elf Atochem N. Am.*

1037 Cymoxanil
57966-95-7   2794   G P

$C_7H_{10}N_4O_3$

2-Cyano-N-[(ethylamino)carbonyl]-2-(methoxyimino)-acetamide.

Acetamide, 2-cyano-N-((ethylamino)carbonyl)-2-(meth-oxyimino)-; Curzate; 1-(2-Cyano-2-methoxyiminoacetyl)-3-ethylurea; 2-Cyano-N-((ethylamino)carbonyl)-2-(meth-oxyimino)acetamide; 2-Cyano-N-(ethylcarbamoyl)-2-(methoxyimino)acetamide; Cymoxanil; DPX 3217; DPX 3217M; HSDB 6914; INT-3217-49. Foliar fungicide with protective and curative action. Used for control of Peronosporates, particularly *Peronospora, Phytophthora and Plasmopara* species. Registered by EPA as a fungicide. Used in potatoes. Crystals; mp = 160 - 161°; sg^{25} = 1.31; soluble in H2O (0.1 g/100 ml), DMF (17.4 g/100 ml), Me2CO (8.3 g/100 ml), CHCl3 (15.2 g/100 ml), MeOH (3.3 g/100 ml), C6H6 (0.18 g/100 ml), C6H14 (< 0.07 g/100 ml); LD50 (mrat orl) = 1196 mg/kg, (frat orl) = 1390 mg/kg, (gpg orl) = 1096 mg/kg, (mrbt, dog der) > 3000 mg/kg; LC50 (bobwhite quail 8 day dietary) = 2847 mg/kg diet, (Mallard duck 8 day dietary) > 10000 mg/kg diet; (rainbow trout 96 hr.) = 18.7 mg/l, (bluegill sunfish 96 hr.) = 13.5 mg/l; non-toxic to bees. *DuPont.*

1038 Cypermethrin
52315-07-8   2796   G P

$C_{22}H_{19}Cl_2NO_3$

Cyclopropanecarboxylic acid, 3-(2,2-dichloro ethenyl)-2,2-dimethyl-, cyano (3-phenoxyphenyl) methyl ester.

Agrothrin; AI3-29295; Ambush C; Ambush CY; Ammo; Ammo (pesticide); Antiborer 3767; Ardap; Arrivo; Asymmethrin; Barricade; Barricade (insecticide); Bas-athrin; β-cypermethrin; BRN 2422506; Caswell No. 268AA; CCN 52; CCRIS 2499; Chinmix; Colt; Creokhin; ($\pm$)-α-cyano-3-phenoxybenzyl ($\pm$)-cis,trans-3-(2,2-di-chlorovinyl)-2,2-dimethylcyclopropane carboxylate; (RS)-α-Cyano-3-Phenoxybenzyl (1RS)-cis,trans-3-(2,2-di-chlorovinyl)-2,2-dimethylcyclopropanecarboxylate; α-Cyano(3-phenoxyphenyl)methyl($\pm$)cis,trans-3-(2,2-di-chlorovinyl)-2,2-dimethylcyclopropanecarboxylate; Cymbush; Cymbush 2E; Cymbush 3E; Cympa-Ti; Cymperator; Cynoff; Cypercare; Cyperco; Cypercopal; Cyperil; Cyperkill; Cypermar; Cypermethrin; Cyper-methrin-25EC; cis-Cypermethrin; Cypermethrine; Cypermetryna; Cypersect; Cypor; Cyrux; Demon; Demon TC; 3-(2,2-Dichloroethenyl)-2,2-dimethylcyclopropane-carboxylic acid cyano(3-phenoxyphenyl)methyl ester; Drago; Dysect; Ectomin; Ectopor; EPA Pesticide Chemical Code 109702; Excis; EXP 5598; Fendona; Fenom; Fenom (pesticide); Flectron; FMC 30980; FMC 45497; FMC 45806; Folcord; Fury; Hilcyperin; HSDB 6600; Imperator; JF 5705F; KafilSuper; Kalif Super; Kefil Super; Kordon; Kreokhin; Mustang; Neramethrin; Neramethrin EC 50; NRDC 149; Nurele; Nurelle;

Parasol; Polytrin; PP 383; Ripcord; RU 27998; Rycopel; SF 06646; Sherpa; Siperin; Supercypermethrin; Supercypermethrin forte; Super-methrin; Topclip; Toppel; Ustaad; Vucht 424; WL 8517; WL 43467; Wrdc149; YT 305. Non-systemic insecticide with contact and stomach action. Used to control wide range of insects, e.g. *lepidoptera, coleoptera, diptera and hemiptera*. Registered by EPA as an insecticide. Yellow liquid - colorless crystals; mp = 60-80°; d^{20} = 1.25; insoluble in H_2O (< 0.000001 g/100 ml), soluble in Me_2CO, $CHCl_3$, cyclohexanone, xylene (all > 45 g/100 ml), EtOH (33.7 g/100 ml), C_6H_{14} (10.3 g/100 ml); LD_{50} (rat orl) = 25 - 4150 mg/kg, (mus orl) = 138 mg/kg, (rat der) > 1600 mg/kg, (rbt der) > 2400 mg/kg, (mallard duck orl) > 10000 mg/kg, (ckn orl) > 2000 mg/kg; LC_{50} (brown trout 96 hr.) = 0.0020 - 0.0028 mg/l; toxic to bees. *Am. Cyanamid; Aventis Environmental Science USA LP; Bonide Products, Inc.; Chinoin; Control Solutions, Inc.; CTX-Cenol Inc.; Farnam Companies Inc.; FMC Corp.; Novartis; S.C. Johnson & Son, Co; Sumitomo Corp.; Syngenta Crop Protection; Unicorn Laboratories; Valent Biosciences Corp.; Waterbury Companies Inc.; Zeneca Pharm.*

1039 Cyproconazole
94361-06-5   2800   P

$C_{15}H_{18}ClN_3O$
α-(4-Chlorophenyl)-α-(1-cyclopropylethyl)-1H-1,2,4-triazole-1-ethanol.
Alto; Alto 100SL; Atemi; Atemi C; Cyproconazole; SAN 619F; Sentinel Turf Fungicide; SN 108266. Registered by EPA as a fungicide. Colorless crystals, mp = 103-105°; poorly soluble in H_2O (0.0140 g/100 ml), soluble in Me_2CO (> 23 g/100 ml), EtOH (> 23 g/100 ml), xylene (12 g/100 ml), DMSO (>18 g/100 ml); LD_{50} (mrat orl) = 1020 mg/kg, (frat orl) = 1330 mg/kg, (rat der) 2000 mg/kg, (carp) = 18.9 mg/l, (trout) = 7.2 mg/l, (bluegill sunfish) = 6.0 mg/l, (bobwhite quail orl) = 150 mg/kg. *Janssen Pharmaceutical, Belgium; Osmose Inc.; Syngenta Crop Protection.*

1040 Cysteine
52-90-4   2810   G

$C_3H_7NO_2S$
L-(+)-2-amino-3-mercaptopropionic acid.
L-Alanine, 3-mercapto-; 2-Amino-3-mercaptopropanoic acid, (R)-; 2-Amino-3-mercaptopropionic acid; L-2-Amino-3-mercaptopropanoic acid; (R)-2-Amino-3-mercaptopropanoic acid; α-Amino-β-thiolpropionic acid; α-Amino-β-thiolpropionic acid, L-; α-Amino-β-mercaptopropanoic acid, L-; AI3-26559; CCRIS 912; Cisteina; Cisteinum; Cystein; Cysteine; L-Cysteine; L-(+)-Cysteine; Cysteine, L-; (R)-Cysteine; Cysteinum; EINECS 200-158-2; FEMA No. 3263; Half cystine; HSDB 2109; NSC-8746; Thioserine. A nonessential amino acid; biochemical and nutrition research, reducing agent in

bread doughs. *Diamalt; Nippon Rikagakuyakuhin; Showa Denko.*

1041 Cysteine hydrochloride
52-89-1   2810   G

$C_3H_7NO_2S.HCl$
L-2-Amino-3-mercaptopropanoic acid monohydro-chloride.
AI3-18781; CCRIS 3613; Cystein chloride; Cysteine chlorhydrate; Cysteine hydrochloride; Cysteine hydro-chloride anhydrous; L-Cysteine hydrochloride; L-(+)-Cysteine hydrochloride; (R)-Cysteine hydrochloride; Cysteine monohydrochloride; Cysteine, L-, mono-hydrochloride; Cysteine, L-, hydrochloride; EINECS 200-157-7; EK 2367; 3-Mercaptoalanine hydrochloride; NSC 8746; WR 348. Essential amino acid. Used as a dough conditioner. Crystals; mp = 175-178° (dec); $[α]_D^{25}$= 5° (5N HCl); soluble in H_2O, organic solvents. *Bretagne Chimie Fine SA; Degussa AG; EM Ind. Inc.; Greeff R.W. & Co.; Nippon Rikagakuyakuhin; Penta Mfg.; Tanabe U.S.A. Inc.; U.S. BioChem.*

1042 Cystine
56-89-3   2811   G

$C_6H_{12}N_2O_4S_2$
Di(α-amino-β-thiolpropionic acid.
4-04-00-03155 (Beilstein Handbook Reference); AI3-09064; Alanine, 3,3'-dithiobis-; Alanine, 3,3'-dithiodi-; L-Alanine, 3,3'-dithiobis-; Bis(β-amino-β-carboxyethyl)-disulfide; BRN 1728094; CCRIS 5822; Cysteine disulfide; L-Cysteine disulfide; Cystin; L-Cystin; Cystine; L-Cystine; 1-Cystine; Cystine (L)-; Cystine acid; Dicysteine; α-Diamino-lb-dithiolactic acid; β,β'-Diamino-β,β'-dicarboxydiethyl disulfide; (R-(R*,R*))-3,3'-Dithiobis(2-amino-propanoic acid); β,β'-Dithioalanine, L-; β,β'-Dithio-bisalanine; 3,3'-Dithiobis(2-aminopropanoic acid), (R-(R*,R*))-; 3,3'-Dithiodialanine; EINECS 200-296-3; Gelucystine; Nephrin; NSC 13203; Oxidized L-cysteine; Propanoic acid, 3,3'-dithiobis(2-amino-, (R-(R*,R*))-. Biochemical and nutrition research, nutrient and dietary supplement. Crystals; mp = 260-261° (dec); $[α]_D^{20}$= -223.4° (1.0N HCl); soluble in H_2O (0.011 g/100 ml at 25°, 0.024 g/100 ml at 50°, 0.052 g/100 ml at 75°, 0.114 g/100 ml at 100°), insoluble in EtOH. *Am. Biorganics; Bretagne Chimie Fine SA; Degussa AG; Greeff R.W. & Co.; Sigma-Aldrich Fine Chem.; Tanabe U.S.A. Inc.; U.S. BioChem.*

1043 Cytisine
485-35-8   2818   G

$C_{11}H_{14}N_2O$
1,2,3,4,5,6-Hexahydro-1,5-methano-8H-pyrido[1,2-a]-[1,5]diazocin-8-one.
5-24-02-00535 (Beilstein Handbook Reference); Bapti-toxin; Baptitoxine; BRN 0083882; Citizin; Cystisine; Cytisine; Cytiton;

Cytitone; Cytizin; EINECS 207-616-0; 1,2,3,4,5,6-Hexahydro-1,5-methano-8H-pyrido(1,2-a)-(1,5)diazocin-8-one; HSDB 3560; Laburnin; 1,5-Methano-8H-pyrido(1,2-a)(1,5)diazocin-8-one, 1,2,3,4,5, 6-hexahydro-; 1,5-Methano-8H-pyrido(1,2-a)-(1,5)diazocin-8-one, 1,2,3,4,5,6-hexahydro-, (1R,5S)-; NSC 407282; Sophorin; Sophorine; Tabax; Tabex; Tsitizin; Ulexin; Ulexine. An alkaloid found in laburnum and furze. Crystals; mp= 152-153°; bp2 = 218°; $[\alpha]_D^{17}$= -120°; soluble in H_2O (77 g/100 ml), more soluble in organic solvents; LD_{50} (mus orl) = 101 mg/kg.

1044   2,4-D, Sodium Salt
2702-72-9   H P

$C_8H_5Cl_2NaO_3$
Sodium 2,4-dichlorophenoxyacetate.
Acetic acid, (2,4-dichlorophenoxy)-, sodium salt; Agrion; Caswell No. 315D; CCRIS 1433; 2,4-D sodium salt; 2,4-Dichlorophenoxyacetic acid, sodium salt; Diconirt; Diconirt D; Dikonirt; Dikonirt D; EINECS 220-290-4; EPA Pesticide Chemical Code 030004; Fernoxene; Fernoxone; Hormit; Pielik E; Pielika; Sodium 2,4-D; Sodium 2,4-dichlorophenoxyacetate; Spray-hormite; Spritz-hormit; U-46-D-Fluid. Herbicide. Registered by EPA as a herbicide and fungicide. *Aquacide Co.; Farnam Companies Inc.; Nufarm Americas Inc.; Riverdale Chemical Co.*

1045   DAB
91-95-2   G

$C_{12}H_{14}N_4$
3,3'-Diaminobenzidine.
3,3',4,4'-Biphenyltetramine. A proprietary intermediate for various high-temperature plastics, used to make polypyrones and polyquinoxalines. Crystals; mp = 172-174°; soluble in H_2O (0.55 g/l), more soluble in organic solvents; LD_{50} (mus orl) = 1834 mg/kg, carcinogen. *Upjohn Ltd.*

1046   2,4-DB
94-82-6   2855   G P

$C_{10}H_{10}Cl_2O_3$
4-(2,4-Dichlorophenoxy)butanoic acid.
4-06-00-00927 (Beilstein Handbook Reference); BRN 1976809; Buratal; Butanoic acid, 4-(2,4-dichloro-phenoxy)-; Butirex; Butormone; Butoxon; Butoxone; Butoxone amine; Butoxone ester; Butyrac; Butyrac 118; Butyrac 200; Butyrac ester; Butyric acid, 4-(2,4-dichlorophenoxy)-; Campbell's DB Straight; Caswell No. 316; CCRIS 1021; DB; 2,4-DB; 2,4-D butyric acid; 2,4-(Dichlorophenoxy)butyric acid; 4-(2,4-Dichloro-phen-oxy)butyric acid; 4-(2,4-Dichlorophenoxy)butanoic acid; 2,4-DM; EINECS 202-366-9;

Embutone; Embutox; EPA Pesticide Chemical Code 030801; γ-(2,4-Dichloro-phenoxy)butyric acid; HSDB 6603; Kyselina 4-(2,4-dichlorfenoxy)maselna; Legumex; Legumex D; M&B 2878; NSC 70337; Sys 67 Buratal; Venceweed. Selective systemic hormone type herbicide. Used for post-emergence control of many annual and perennial broad-leaved weeds in lucerne, clovers, cereals, grassland, forest legumes, soybeans and ground nuts. mp = 117-119°; soluble in H_2O (4.6 g/100 ml), readily soluble in organic solvents; LD_{50} (rat orl) = 370-700 mg/kg. *ICI Chem. & Polymers Ltd.; May & Baker Ltd.; MTM AgroChemicals Ltd.*

1047   Dacthal
1861-32-1   2896   G P

$C_{10}H_6Cl_4O_4$
2,3,5,6-Tetrachloro-1,4-benzenedicarboxylic acid dimethyl ester.
3-09-00-04257 (Beilstein Handbook Reference); Al3-52124; 1,4-Benzenedicarboxylic acid, 2,3,5,6-tetra-chloro-, dimethyl ester; BRN 1888840; Caswell No. 382; Chlorothal; Chlorthal-dimethyl; Chlorthal dimethyl ester; Chlorthal-methyl; DAC 4; DAC 893; Dacthal; Dacthalor; Daktal; DCP; DCPA; Dimethyl tetrachloro-1,4-benzenedicarboxylate; Dimethyl 2,3,5,6-tetra-chloro-benzene-1,4-dicarboxylate; Dimethyl 2,3,5,6-tetrachloro-terephthalate; Dimethyl ester of tetrachloro-terephthalic acid; Dimethyl tetrachloroterephthalate; Dimethylester kyseliny tetrachlortereftalove; EINECS 217-464-7; EPA Pesticide Chemical Code 078701; Fatal; HSDB 358; NSC 155745; Rid; TCTP; Terechloroterephthalic acid dimethyl ester; Terephthalic acid, tetrachloro-, dimethyl ester; Terephthalic acid, 2,3,5,6-tetrachloro-,dimethyl ester; Tetrachloro-terephthalic acid dimethyl ester; 2,3,5,6-tetrachloro-1,4-benzenedicarboxylate; 2,3,5,6-Tetra-chlorphthalsäure-dimethylester; 2,3,5,6-Tetrachlor-phthalsaeure-dimethyl-ester; 2,3,5,6-Tetrachlorotere-phthalic acid dimethyl ester; Tetral; Vegetable Turf and Ornamental Weeder. A preemergence sprayable herbicide that can be used in vegetable gardenson ornamentals and in turf areas; controls spurge. Crystals; mp = 155-156°; soluble in H_2O, organic solvents; LD_{50} (rat orl) >3000 mg/kg. *Fermenta; Lawn & Garden Products Inc.*

1048   Dahl's acid II
85-74-5   6427   G

$C_{10}H_9NO_6S_2$
α-Naphthylamine-4,6-disulfonic acid.
4-Aminonaphthalene-1,7-disulphonic acid; EINECS 201-629-5; 1-Naphthylamine-4,6-disulfonic acid; 1,7-Naphthalenedisulfonic acid, 4-amino-. Intermediate in synthesis of dyestuffs. Readily soluble in H_2O, EtOH

1049   Dahl's acid III
85-75-6   6428   G

C10H9NO6S2

α-naphthylamine-4,7-disulfonic acid.
4-Aminonaphthalene-1,6-disulphonic acid; 1-Naphthyl-amine-4,7-disulfonic acid; 1,6-Naphthalenedisulfonic acid, 4-amino-. Intermediate in synthesis of dyestuffs. Soluble in H_2O, insoluble in EtOH.

1050   Dalapon
75-99-0   2830   G P

C3H4Cl2O2
2,2-Dichloropropanoic acid.
4-02-00-00753 (Beilstein Handbook Reference); AI3-28206; Alatex; Basfapon B; Basfapon/Basfapon N; Basinex P; BH Dalapon; BRN 1750149; Caswell No. 273; CCRIS 7752; Crisapon; Dalapon; Dalapon 85; Dalascam; Dawpon-Rae; D-Granulat; α,α-Dichloropropionic acid; 2,2-Dichloropropanoic acid; 2,2-Dichloropropionic acid; 2,2-DPA; Dowpon M; Dowpon NF; DPA; EINECS 200-923-0; EPA Pesticide Chemical Code 028901; HSDB 4010; Kenapon; Kyselina 2,2-dichlorpropionova; Liropon; NSC 56352; Propanoic acid, 2,2-dichloro-; Propionic acid, 2,2-dichloro-; Proprop; Revenge; S 1315; S 95 (herbicide); Sys-Omnidel; Tripon; Unipon. Selective systemic herbicide absorbed by roots and leaves. Used for control of annual and perennial grasses on non-crop land and also orchards and vineyards. Registered by EPA as a herbicide (cancelled). Soluble powder containing dalapon; used for control of grasses in crop and noncrop areas. Selective systemic herbicide absorbed by roots and leaves; used for control of annual and perennial grasses on non-crop land and also orchards and vineyards. Liquid; bp18 = 98-100°; d = 1.4014; LD50 (rat orl) = 7126 mg/kg. Rhône-Poulenc Environmental Prods. Ltd; Synchemicals Ltd; Vitax Ltd.

1051   Dalapon sodium
127-20-8   2830   G P

C3H3Cl2NaO2
2,2-dichloropropanoic acid, sodium salt.
Antigramigna; Basfapon; Caswell No. 273A; Dalapon sodium; Dalapon sodium salt; α-α-Dichloropropionic acid sodium salt; 2,2-Dichloropropionic acid, sodium salt; 2,2-Dichlorpropionsaeure natrium; 2,2-Dichlor-propionsäure natrium; 2,2-DPA Na salt; Dikopan; Dowpon; Dowpon S; EINECS 204-828-5; EPA Pesticide Chemical Code 028902; Gramevin; Hico DCPAS; Natriumsalz der 2,2-dichlorpropionsaure; Omnidel; Omnidel Spezial; Propanoic acid, 2,2-dichloro-, sodium salt; Propinate; Propionic acid, 2,2-dichloro-, sodium salt; Radapon; Sodium α,α-dichloropropionate; Sodium 2,2-dichloropropionate; Sodium Dalapon; SYS 67 Omnidel; Tafapon. Post-emergence systemic herbicide for control of grasses in annual and perennial crops, used on nonagricultural land, in ditches, and

pastures. Used primarily in sugar cane, sugar beets, orchards and also in noncrop applications such as railroads and rubber plantations. Registered by EPA as a herbicide (cancelled). Crystals; dec. 166.5°; soluble in H_2O (90 g/100 ml), EtOH (14.6 g/100 ml), MeOH (14.3 g/100 ml), Me2CO (0.011 g/100 ml), C6H6 (0.002 g/100 ml), Et2O (0.011 g/100 ml); LD50 (rat orl) = 9330 mg/kg, (frat orl) = 7570 mg/kg, (fmus orl) > 4600 mg/kg, (fgpg orl) = 3860 mg/kg, (frbt orl) = 3860 mg/kg, (cattle orl) > 4000 mg/kg, (rbt der) > 2000 mg/kg, (ckn orl) = 5660 mg/kg; LC50 (mallard duck, Japanese quail, pheasant 5 day dietary) > 5000 mg/kg diet, (rainbow trout, goldfish, channel catfish 96 hr.) > 100 mg/l, (carp 96 hr.) > 500 mg/l, (guppy 96 hr.) > 1000 mg/kg; non-toxic to bees. BASF Corp.; Bayer AG; Dow AgroSciences; Dow UK.

1052   Daminozide
1596-84-5   2836   G P

C6H12N2O3
Succinic acid N,N-dimethylhydrazide.
AI3-50727; Alar 85; ALAR; Aminozide; B 995; B-Nine; Bernsteinsäure-2,2-dimethylhydrazid; Bernstein-saeure-2,2-dimethylhydrazid; BRN 1863230; Butanedioic acid mono(2,2-dimethylhydrazide); Caswell No. 808; CCRIS 191; Daminozide; Dimas; N-Dimethylamino-+lb-carbamylpropionic acid; Dimethylaminosuccinamic acid; N-(Dimethylamino)succinamic acid; N-Dimethylamino-succinamidsa+a5ure; N-Dimethylamino-succinamid-saeure; 2,2-Dimethylhydrazid kyseliny jantarove; DMASA; DYaK; EINECS 216-485-9; EPA Pesticide Chemical Code 035101; HSDB 1769; Kylar; NCI-C03827; Nine; SADH; Succinic 1,1-dimethyl hydrazide; Succinic acid 2,2-dimethylhydrazide; Succinic acid N,N-dimethylhydrazide; Succinic N',N'-dimethylhydrazide; Succinic acid, mono(2,2-dimethylhydrazide). Plant growth regulator, absorbed by leaves with translocation throughout the plant. Used on apples to restrict vegetative growth. Registered by EPA as a herbicide and plant growth regulator. Crystals; mp = 154 - 155°; soluble in H_2O (10 g/100 ml), Me2CO (2.0 g/100 ml), MeOH (4 g/100 ml); LD50 (rat orl) = 8400 mg/kg, (rbt der) > 5000 mg/kg; LC50 (rat ihl 1 hr.) > 147 mg/l air, (mallard duck, bobwhite quail 8 dau dietary) > 10000 mg/kg diet, (rainbow trout 96 hr.) = 149 mg/l, (bluegill sunfish 96 hr.) = 423 mg/l; non-toxic to bees. Crompton Corp.; Uniroyal.

1053   Daphnetin
486-35-1   2843   G

C9H6O4
7,8-Dihydroxy-2H-1-benzopyran-2-one.
5-18-03-00211 (Beilstein Handbook Reference); 2H-1-Benzopyran-2-one, 7,8-dihydroxy-; BRN 0009372; Coumarin, 7,8-dihydroxy-; Daphnetin; Daphnetol; 7,8-Dihydroxy-2H-1-benzopyran-2-one; 7,8-Dihydroxy-coumarin; EINECS 207-632-8. Used in manufacture of dyes and pharmaceuticals. Pale yellow needles; mp= 262° (dec); bp sublimes; λm = 258, 335 nm (ε = 6310, 12589, EtOH); slightly soluble in Et2O, C6H6, CS2, CHCl3, soluble in H_2O, EtOH, AcOH.

1054   Dapsone
80-08-0   2847   G

C12H12N2O2S
4,4'-Sulfonylbisbenzeneamine.
1358F; 4-13-00-01306 (Beilstein Handbook Reference); Acedapsone; AI3-08087; 4-Aminophenyl sulfone; p-Aminophenyl sulfone; Aniline, 4,4'-sulfonyldi-; Araldite HT; Araldite HT 976; Avlosulfon; Avlosulfone; Benzenamine, 4,4'-sulfonylbis-; Bis(4-aminophenyl) sulfone; Bis(4-aminophenyl)sulphone; BRN 0788055; CCRIS 192; Croysulfone; Croysulphone; DADPS; Dapson; Dapsona; Dapsone; 4,4'-Dapsone; 4,4-Diaminodifenylsulfon; 4,4'-Diaminodiphenyl sulfone; 4,4'-Diaminodiphenyl sulphone; Dapsone; Dapsonum; DDS; DDS (pharmaceutical); Di(4-aminophenyl)sulfone; Di(4-aminophenyl)sulphone; Di(p-aminophenyl) sulfone; Di(p-aminophenyl)sulphone; p,p-Diaminodiphenyl sulphone; Diamino-4,4'-diphenyl sulfone; Diamino-4,4'-diphenyl sulphone; Diaminodifenilsulfona; Diamino-diphenyl sulfone; Diaphenylsulfon; Diaphenylsulfone; Diaphenylsulphon; Diaphenylsulphone; Diphenasone; N,N'-Diphenyl sulfondiamide; Diphone; Disulone; DRG-0036; DSS; Dubronax; Dumitone; EINECS 201-248-4; Eporal; F 1358; Hardener HT 976; HSDB 5073; HT 976; HY 976; ICI; Metabolite C; NCI-C01718; Novophone; NSC-6091; 1,1'-Sulfonylbis(4-amino-benz-ene); 1,1'-Sulphonylbis(4-aminobenzene); 4,4'-Sulfonyl-bisaniline; 4,4'-Sulfonylbisbenzenamine; 4,4'-Sulfonylbis-benzenamine; 4,4'-Sulfonyldianiline; 4,4'-Sulphonyl-dianiline; p,p-Sulfonylbisbenzamine; p,p-Sulfonylbis-benzenamine; p,p-Sulphonylbisbenzamine; p,p-Sulph-onylbisbenzenamine; p,p'-Sulfonyldianiline; p,p-Sulph-onyldianiline; Sulfadione; Sulfanonamae; Sulfon-mere; Sulfona; Sulfona-Mae; Sulfone, diphenyl, 4,4'-diamino-; Sulfone UCB; Sulfonyldianiline; Sulphadione; Sulphonyl-dianiline; Sumicure S; Tarimyl; Udolac; WR 448. Curing agent for epoxy resins; antibacterial. mp = 175.5°; soluble in EtOH, MeOH, Me2CO, HCl, insoluble in DMSO, H2O; LD50 (rat orl) = 1000 mg/kg. *Crown Metro.*

1055   Dazomet
533-74-4   2854   G H P

C5H10N2S2
Tetrahydro-3,5-dimethyl-2H-1,3,5-thiadiazine-2-thione.
4-27-00-07436 (Beilstein Handbook Reference); Amer-stat® 233; Basamid; Basamid G; Basamid-Granular; Basamid P; Basamid-Puder; Basamid - Purder; BRN 0116039; Carbothialdin; Caswell No. 840; Crag 85W; Crag 974; Crag fungicide 974; Crag nemacide; D 35; Dazomet; Dazomet-Powder BASF; Dimethylformo-carbothialdine; 3,5-Dimethyl-1,2,3,5-tetrahydro-1,3,5-thiadiazinethione-2; 3,5-Dimethyl-1,3,5-(2H)-tetrahydro-thiadiazine-2-thione; 3,5-Dimethyl-1,3,5-thiadiazinane-2-thione; 3,5-Dimethyl-2-thionotetrahydro-1,3,5-thiadiaz-ine; 3,5-Dimethylperhydro-1,3,5-thiadiazin-2-thion; 3,5-Dimethyltetrahydro-1,3,5-2H-thiadiazine-2-thione; 3,5-Dimethyltetrahydro-1,3,5-thiadiazine-2-thione; 3,5-Di-methyltetrahydro-2H-1,3,5-thiadiazine-2-thione; 3,5-Di-metil-peridro-1,3,5-tiadiazin-2-tione; DMTT; EINECS 208-576-7; EPA Pesticide Chemical Code 035602; Fennosan B 100; HSDB 1642; Mico-fume; Mylon; Mylone; Mylone 85; N 521; N 521® Biocide; Nalcon 243; Nefusan; NSC 4737; Prezervit; Salvo; Stauffer N 521; Tetrahydro-2H-3,5-dimethyl-1,3,5-thiadiazine-2-thi-one; Tetrahydro-3,5-dimethyl-2H-1,3,5-thiadiazine-2-thi-one; 2H-1,3,5-Thiadiazine-2-thione, tetrahydro-3,5-dimethyl-; 2-Thio-3,5-dimethyltetrahydro-1,3,5-thiadiaz-ine; Thiadiazin (pesticide); Thiazon; Thiazone; Tiazon;

Troysan 142; UCC 974. Used as a soil fumigant; acts by release of methyl isothiocyanate. Antimicrobial (slimicide) in industrial water systems, e.g. paper mills, preservative in aqueous systems. Needles; mp = 106-107°; soluble in EtOH, decomposed by EtOH, H2O; λ_m = 242, 289 nm (ε 7150, 9900, cyclohexane); soluble in H2O (0.3 g/100 ml), cyclohexane (31.1 g/100 ml), CHCl3 (57.9 g/100 ml), Me2CO (10.5 g/100 ml), C6H6 (4.5 g/100 ml), EtOH (1.2 g/100 ml); Et2O (0.42 g/100 ml); LD50 (rat orl) = 320 - 620 mg/kg, (mus orl) = 180 mg/kg, (mmus orl) = 455 mg/kg, (fmus orl) = 710 mg/kg, (gpg orl) = 160 mg/kg, (rbt orl) = 320 - 620 mg/kg, (rat ip) = 87 mg/kg, (rbt ip) = 127 mg/kg, (dog ip) = 47 - 63 mg/kg, (rat der) > 2000 mg/kg, (rbt der) = 7100 mg/kg; LC50 (rat ihl 4 hr.) = 84 mg/l air; toxic to fish; non-toxic to bees. *BASF Corp.; Bos Chemicals Ltd.; Diachem; DowElanco Ltd.; Drew Industrial Div.; Rhône-Poulenc.*

1056   DDT
50-29-3   2861   G

C14H9Cl5
1,1'-(2,2,2,-Trichloroethylidene)bis[4-chlorobenzene].
4-05-00-01885 (Beilstein Handbook Reference); 4,4'-DDT; 4,4'-Dichlorodiphenyltrichloroethane; Aavero-ex-tra; Agritan; AI3-01506; Anofex; Arkotine; Azotox M 33; Benzene, 1,1'-(2,2,2-trichloroethylidene)bis(4-chloro-; Benzochloryl; α,α-Bis(p-chlorophenyl)-β,β,β-trichloro-ethane; 1-Bis-(p-chlorophenyl)-2,2,2-trichloroethane; 2,2-Bis(p-chlorophenyl)-1,1,1-trichloroethane; Bosan Supra; Bovidermol; BRN 1882657; Caswell No. 308; CCRIS 194; Chlofenotan; Chlorophenothan; Chloro-phenothane; Chlorophenotanum; Chlorophenothanum technicum; Chlorophenotoxum; Chlorphenothan; Chlorphenotoxum; Citox; Clofenotane; Clofenotane technique; Clofenotano; Clofenotanum; D.D.T. technique; DDT; DDT 50 WP; DDT, p,p'-; p,p'-DDT; Deoval; Detox (pesticide); Detoxan; Dibovin; Dichlorodiphenyltrichloroethane; p,p'-Dichlorodi-phenyltrichloroethane; Dicophane; Didigam; Didimac; Dodat; Dykol; EINECS 200-024-3; ENT 1,506; EPA Pesticide Chemical Code 029201; Estonate; Ethane, 1,1,1-trichloro-2,2-bis(4-chlorophenyl)-; Genitox; Gesafid; Gesarol; Guesarol; Gyron; Hildit; HSDB 200; Ivoran; Ixodex; Klorfenoton; Kopsol; Mutoxan; NCI-C00464; Neocid; Neocidol; NSC 8939; OMS 0016; OMS 16; Parachlorocidum; PEB1; Pentachlorin; Pentech; Penticidum; RCRA waste number U061; Rukseam; Santobane; Tafidex; Trichlorobis(4-chlorophenyl)ethane; 1,1,1-Trichlor-2,2-bis(4-chloor fenyl)-ethaan; 1,1,1-Trichlor-2,2-bis(4-chlor-phenyl)-aethan; 1,1,1-Trichloro-2,2-di(4-chlorophenyl)-ethane; 1,1,1-Trichloro-2,2-bis-(4,4'-dichlorodiphenyl)ethane; 1,1,1-Tricloro-2,2-bis(4-cloro-fenyl)-etano; 1,1,1-Tricloro-2,2-bis(4-cloro-fenil)-etano; 1,1'-(2,2,2-Trichloroethylidene)bis(4-chloro-benz-ene); p'-Zeidane; Zerdane. A powerful polychlorinated, nondegradable pesticide. One of the Dirty Dozen pesticides. mp = 108.5-109°; λ_m= 236 nm; insoluble in H2O; soluble in acetone (58 mg/100ml), benzene (78 mg/100ml), carbon tetrachloride (45 mg/100ml); LD50 (rat orl) = 113 mg/kg. 3975; no longer sold in the US.

1057   Deanol
108-01-0   2863   D G H

C4H11NO
2-Dimethylaminoethanol.
4-11-00-00122 (Beilstein Handbook Reference); AI3-09209; Amietol M 21; Bimanol; BRN 1209235; Dabco® DMEA; Deanol; Dimethyl(2-hydroxyethyl)amine; Dimethyl(hydroxyethyl)amine; Di-methylaethanolamin; Dimethylaminoaethanol; Dimethyl-aminoethanol; β-Dimethylaminoethyl alcohol; 2-(Di-methylamino)-1-ethanol; 2-(N,N-Dimethylamino)ethanol; 2-Dimethylaminoethanol; 2-Dwumetyloaminoetanolu; Dimethylmonoethanolamine; DMAE; EINECS 203-542-8; Ethanol, 2-(dimethylamino)-; HSDB 1329; Kalpur P; Liparon; Norcholine; NSC 2652; Propamine A; Tegoamin® DMEA; Texacat® DME; UN2051; Varesal. Solubilizer of synthetic resins for water soluble paints, raw material for ion exchange resins and coagulants. Used in synthesis of dyestuffs, pharmaceuticals and textile auxiliaries. Amine-based catalyst for polyurethane rigid foam, flexible slabstock. Used medically as a CNS stimulant. Liquid; mp = -59°; bp = 134°; d4^20 = 0.8866; freely soluble in H2O, EtOH, Et2O, soluble in CHCl3; LD50 (rat orl) = 2 gm/kg. *Air Products & Chemicals Inc.; BASF Corp.; Elf Atochem N. Am.; Goldschmidt; Nippon Nyukazai; Texaco; Union Carbide Corp.*

1058   **Decabromobiphenyl ether**
1163-19-5   G H

C12Br10O
Bis(pentabromophenyl) ether.
1-06-00-00108 (Beilstein Handbook Reference); Adine 505; AFR 1021; AI3-27894; BDE 209; Berkflam B 10E; Bis(pentabromophenyl)ether; BR 55N; BRN 2188438; Bromkal 82-0DE; Bromkal 83-10DE; Caliban F/R-P 39P; CCRIS 1421; DB 10; DB 101; DB 102; De 83R; DE 83; Decabrom; Decabromobiphenyl ether; Decabromo-diphenyl oxide; Decabromophenyl ether; DP 10F; EB 10; EB 10FP; EB 10W; EB 10WS; EBR 700; EINECS 214-604-9; F/R-P 53; Fire Cut 83D; Flame Cut 110R; Flame Cut Br 100; FR 10; FR 10 (ether); FR 300; FR 300BA; FR-PE; FR-PE(H); FRP 53; HSDB 2911; NCI-C55287; Nonnen DP 10; Nonnen DP 10(F); NSC 82553; Octoguard FR-01; PBED 209; Pentabromophenyl ether; Planelon DB; Planelon DB 100; Planelon DB 101; Plasafety EB 10; Plasafety EBR 700; Saytex® 102; Saytex® 102E; Tardex 100; Thermoguard® 505. Flame retardant used in thermoplastics and fibers, including HIPS, glass-reinforced thermoplastic polyester molding resins, LDPE extrusion coatings, PP (homo and copolymers), ABS, nylon, PBT, PET, PU, SBR latex, textiles, rubber. Used in water-based polymers such as latex adhesives, binders, coatings, and foams. Electrical grade flame retardant for wire and cable insulation applications. Solid; mp = 300-310°; bp = 425°; d = 3.00; insoluble in H2O (< 0.1 g/100 ml); LD50 (rat orl) > 5000 mg/kg; experimental carcinogen with teratogenic and reproductive effects. *Albemarle Corp.; Allchem Ind.; AmeriBrom Inc.; Dead Sea Bromine; Elf Atochem N. Am.; Ethyl Corp.; Fluka; Great Lakes Fine Chem.; Rohm & Haas Co.; Sigma-Aldrich Fine Chem.*

1059   **Decalin®**
91-17-8   2866   G

C10H18
Decahydronaphthalene.
3-05-00-00245 (Beilstein Handbook Reference); AI3-01256; Bicyclo(4.4.0)decane; BRN 0878165; CCRIS 3410; De-kalin; DEC; Decahydronaphthalene; Decalin; Dekalin; Dekalina; EINECS 202-046-9; HSDB 287; Naphthalane; Naphthalene, decahydro-; Naphthan; NSC 406139; Perhydronaphthalene; UN1147. A paint and resin solvent; used as turpentine substitute. Liquid; mp = -43°; bp = 155.5°, 194°; d22 = 0.8965; insoluble in H2O; very soluble in organic solvents; LD50 (rat orl) = 4.2 g/kg. *DuPont.*

1060   **Decamethylcyclopentasiloxane**
541-02-6   2868   H

C10H30O5Si5
Cyclic dimethylsiloxane pentamer.
4-04-00-04128 (Beilstein Handbook Reference); BRN 1800166; CCRIS 1328; CD3770; Cyclic dimethylsiloxane pentamer; Cyclopentasiloxane, decamethyl-; D3770; Decamethylcyclopentasiloxane; Dekamethylcyklopenta-siloxan; Dimethylsiloxane pentamer; Dow Corning 345 fluid; Dow Corning 345; EINECS 208-764-9; HSDB 5683; KF 995; NUC silicone VS 7158; Polydimethylsiloxane; SF 1202; Silicon SF 1202; Union carbide 7158 silicone fluid; VS 7158. Coupling agent, chemical intermediate, blocking agent, release agent, primer, reducing agent. As cleaning, polishing, and damping media; offers low toxicity, inertness. *Degussa-Hüls Corp.*

1061   **Decamethyltetrasiloxane**
141-62-8   2870   G

C10H30O3Si4
Tetrasiloxane, decamethyl-.
CD3780; D3780; Decamethyltetrasiloxane; EINECS 205-491-7. Coupling agent, chemical intermediate, blocking agent, release agent, lubricant, primer, reducing agent. Used in cleaning, polishing and damping media; offers low toxicity, inertness. Liquid; mp = -76°; bp = 194°; d25 = 0.8536; insoluble in H2O, slightly soluble in EtOH, soluble in C6H6, petroleum ether. *Degussa-Hüls Corp.*

1062   **Decane**
124-18-5   H

C10H22
n-Decane.
4-01-00-00464 (Beilstein Handbook Reference); AI3-24107; BRN 1696981; CCRIS 653; Decane; n-Decane; Decyl hydride; EINECS 204-686-4; HSDB 63; NSC 8781; UN2247. Liquid; mp = -29.7°; bp = 174.1°; d20 = 0.7300; insoluble in H2O, slightly soluble in CCl4, soluble in Et2O, freely soluble in EtOH.

1063   **Decanenitrile**
1975-78-6   G H

$C_{10}H_{19}N$
n-Decanonitrile.
AI3-11101; Caprinitrile; 1-Cyanononane; Decanenitrile; Decanonitrile; EINECS 217-830-6; Nitrile 10 D; NSC 6085. Chemical intermediate. Liquid; mp = -17.9°; bp = 243°, bp_{10} = 106°; d^{20} = 0.8199; very soluble in Me_2CO, EtOH, Et_2O, $CHCl_3$. *Berol Nobel AB.*

1064   Decanoic acid
334-48-5   1764   G H

$C_{10}H_{20}O_2$
n-Decanoic acid.
4-02-00-01041 (Beilstein Handbook Reference); AI3-04453; BRN 1754556; C_{10} fatty acid; Capric acid; n-Capric acid; Caprinic acid; Caprynic acid; CCRIS 4610; Decanoic acid; n-Decanoic acid; Decoic acid; Decylic acid; Econosan Acid Sanitizer; EINECS 206-376-4; EPA Pesticide Chemical Code 128955; Fatty acid(C_{10}); FEMA No. 2364; Hexacid 1095; HSDB 2751; Neo-fat 10; NSC 5025. Used to manufacture esters used in perfumes and flavors; a base for wetting agents; an intermediate for chemical synthesis. Needles; mp = 31.9°; bp = 268.7°; d^{40} = 0.8858; insoluble in H_2O, very soluble in Me_2CO, C_6H_6, Et_2O, EtOH; LD_{50} (mus iv) = 129 mg/kg. *Akzo Chemie; Henkel/Emery; Mirachem Srl.; Procter & Gamble Co.; Sigma-Aldrich Fine Chem.; Witco/Humko.*

1065   Decanol
112-30-1   2875   G H

$C_{10}H_{22}O$
n-Decyl alcohol.
4-01-00-01815 (Beilstein Handbook Reference); Agent 504; AI3-02173; Alcohol C_{10}; Alfol 10; Antak; BRN 1735221; C 10 alcohol; C_{10} alcohol; Capric alcohol; Caprinic alcohol; Caswell No. 275A; CCRIS 654; Conol 10N; Decanol; 1-Decanol; Decan-1-ol; n-Decanol; n-Decan-1-ol; Decyl alcohol; Dytol S-91; EINECS 203-956-9; EPA Pesticide Chemical Code 079038; Epal 10; Exxal® 10; Fatty alcohol (C_{10}); FEMA Number 2365; HSDB 1072; 1-Hydroxydecane; Kalcohl 1098; Kalcohl 10H; Lorol 22; Nacol 10-99; Nonylcarbinol; n-Nonylcarbinol; NSC 406313; Primary decyl alcohol; Royaltac; Royaltac-85; Royaltac M-2; Sipol L10; T; T-148. Plant growth regulator. Registered by EPA as a herbicide and insecticide. Also used as a solvent. Viscous liqiud; mp = 6.9°; bp = 231.1°, bp_{15} = 115 - 120°, bp_8 = 109.5°; d^{40} = 0.8297; insoluble in H_2O, slightly soluble in CCl_4, freely soluble in EtOH, Et_2O, Me_2CO, C_6H_6, $CHCl_3$. *Coastal Chemical Corp.; Drexel Chemical Co.; Exxon; Generic; Uniroyal.*

1066   Decanox-F
14156-10-6   G

$C_{10}H_{20}O_3$
Decanoyl peroxide.
Peroxydecanoic acid. Initiator for bulk, solution, and suspension polymerization, curing elastomers, and high-temp. cure of polyester

resins. *Elf Atochem N. Am.*

1067   Decanoyl chloride
112-13-0   G

$C_{10}H_{19}ClO$
Capric acid chloride.
Capric acid chloride; Caprinoyl chloride; Decanoyl chloride. Intermediate, polymerization initiator. Liquid; mp = -34.5°; bp = 95°; d^{25} = 0.919; reacts with H_2O, soluble in Et_2O, CCl_4. *Elf Atochem N. Am.; Janssen Chimica.*

1068   Dec-1-ene
872-05-9   G H

$C_{10}H_{20}$
n-Dec-1-ene.
CCRIS 5718; α-Decene; 1-Decene; 1-n-Decene; Dec-1-ene; Decene, n-; Decylene; Dialene 10; EINECS 212-819-2; Gulftene 10; HSDB 1073; Neodene® 10; NSC 62122. Used as an intermediate in the manufacture of biodegradable surfactants and specialty industrial chemicals. Liquid; mp = -66.3°; bp = 170-171°; d^{20} = 0.7408; insoluble in H_2O, freely soluble in EtOH, Et_2O. *Shell UK; Shell.*

1069   Deceth-6
26183-52-8   G

$C_{22}H_{46}O_7$
3,6,9,12,15,18-Hexaoxaoctacosan-1-ol.
Bio-soft® FF 400; Chemal DA-4; Deceth-4; Decyl alcohol, ethoxylated; α-Decyl-ω-hydroxypoly(oxy-1,2-ethanediyl); Decylpolyethyleneglycol 300; Desonic® DA-4; Desonic® DA-6; Emulphogene DA 630; Ethal DA-040; Genapol® DA-040; Iconol DA-4; Iconol DA-6; Iconol DA-9; Marlipal® 1012/4; Oxetal D 104; PEG-4 Decyl ether; Poly(oxy-1,2-ethanediyl), α-decyl-ω-hydroxy-; Polyethylene glycol 200 decyl ether; Polyoxyethylated (6) isodecyl alcohol; Polyoxyethylene (4) decyl ether; Polyoxyethylene monodecyl ether; Prox-onic DA-1/04; Rhodasurf® DA-4; 3,6,9,12-Tetraoxadocosan-1-ol; Trycol® 5950; Surfonic® DA-6; Synthrapol KB; PEG-6 decyl ether. Wetting agent for built scour systems. Detergent; penetrant; emulsifier for textiles, clay soils, fire fighting; surfactant. Oil; d = 1.0014; viscosity = 109 cps; HLB 12.5; hyd no 132; pour point =12.8°; cloud point = 42°. *Rhône Poulenc Surfactants.*

1070   Dechlorane® Plus
13560-89-9   2871   G H

$C_{18}H_{12}Cl_{12}$
Dodecachlorododecahydrodimethanodibenzocyclooctane.
4-05-00-01783 (Beilstein Handbook Reference); AI3-27889; BRN 2195908; Dechloran A; Dechlorane 605; Dechlorane Plus; Dechlorane Plus 25; Dechlorane Plus 35; Dechlorane Plus 515; Dechlorane Plus 1000; Dechlorane Plus 2520; 1,4:7,10-

Dimethanodibenzo-(a,e)cyclooctane, 1,2,3,4,7,8,9,10,13,13,14,14-dodeca-chloro-; 1,4,4a,5,6,6a,7,10,10a,11,12,12a-dodecahydro-1,4:7,10-Dimethanodibenzo(a,e)cyclooctene,; EINECS 236-948-9. Chlorine-containing cycloaliphatic comp-ound. Used as a flame retardant in polymer systems (thermoplastics, thermosets and elastomers); usually combined with antimony oxide as a synergist. Colorless crystals; mp >325°; soluble in o-dichlorobenzene; LD50 (rat orl) = 25 g/kg. OxyChem.

1071   Decylamine
2016-57-1   G H

C10H23N
n-Decylamine.
4-04-00-00783 (Beilstein Handbook Reference); AI3-52306; Amine 10; Amine 12-98D; Aminodecane; 1-Aminodecane; BRN 1735220; 1-Decanamine; Decylamine; 1-Decylamine; n-Decylamine; EINECS 217-957-7; Kemamine P 190D; Lauramine; Monodecylamine. Emulsifier and chemical intermediate; ethoxylated end products used in detergent, cosmetic, and agricutural applications. Liquid; mp = 16°; bp = 211°; d = 0.794. Kao Corp. SA.

1072   Decyl benzene sodium sulfonate
1322-98-1   G

C16H26NaO3S
Sodium decylbenzenesulfonate.
Benzenesulfonic acid, decyl-, sodium salt; Decylbenzenesulfonic acid, sodium salt; EINECS 215-347-5; Santomerse D; Sodium decylbenzene sulfonate; Sodium decylbenzenesulfonamide; Sodium decyl-benzenesulphonate; Ultrawet DS; Witconate DS; Sodium decylbenzenesulfonate; Ultrawet DS. Mixture of o, m and p-isomers. Detergent, foaming agent, and wetting agent. CK Witco Corp.

1073   Decyl chloride
1002-69-3   H

C10H21Cl
n-Decyl chloride.
1-Chlorodecane; 4-01-00-00469 (Beilstein Handbook Reference); BRN 1735224; Decane, 1-chloro-; Decyl chloride; EINECS 213-691-0; n-Decyl chloride; NSC 6088. Liquid; mp = -31.3°; bp = 223.4°; d20 = 0.8705; insoluble in H2O, soluble in CCl4, very soluble in Et2O, CHCl3. Lancaster Synthesis Co.; Sigma-Aldrich Fine Chem.

1074   Decyl octyl adipate
110-29-2   H

C24H46O4
n-Decyl n-octyl hexanedioate.
Adipic acid, decyl octyl ester; Adipol ODY; Decyl octyl adipate; n-

Decyl n-octyl hexanedioate; EINECS 203-754-0; Hercoflex 290; Hexanedioic acid, decyl octyl ester; HSDB 5397; Monoplex noda; Octyl decyl adipate; Px-202; Staflex NODA; Truflex 146.

1075   Decyl octyl phthalate
119-07-3   G

C26H42O4
Octyldecyl phthalate.
4-09-00-03186 (Beilstein Handbook Reference); 1,2-Benzenedicarboxylic acid, decyl octyl ester; BRN 2009141; Decyl octyl phthalate; n-Decyl n-octyl phthalate; Decyl Octyl 1,2-benzenedicarboxylate; Dinopol 235; EINECS 204-295-9; Good-rite® GP-265; HSDB 1242; Octyl decyl phthalate; n-Octyl-n-decyl phthalate; Phthalic acid, decyl octyl ester; Polycizer 532; Polycizer 562; Staflex 500. Plasticizer used with vinyl polymers. Liquid; mp = -50°; bp4 = 239°; d20 = 0.980; LD50 (rat orl) = 45 g/kg. Goodrich B.F. Co.

1076   Decyl oleate
3687-46-5   G

C28H54O2
9-Octadecenoic acid(Z), decyl ester.
rCeraphyl® 140; Decyl 9-octadecenoate; Decyl oleate; EINECS 222-981-6; 9-Octadecenoic acid, decyl ester; 9-Octadecenoic acid (Z)-, decyl ester; 9-Octadecenoic acid (9Z)-, decyl ester; Oleic acid, decyl este. Emollient; binder for pressed powders; pigment dispersant; co-solvent. Van Dyk.

1077   3-Decyloxysulfolane
18760-44-6   H

C14H28O3S
Thiophene, 3-(decyloxy)tetrahydro-, 1,1-dioxide.
3-Decyloxysulfolane; 3-(Decyloxy)tetrahydrothiophene 1,1-dioxide; EINECS 242-556-9; Tetrahydro-3-(decyl-oxy)thiophene 1,1-dioxide; Thiophene, 3-(decyloxy)-tetrahydro-, 1,1-dioxide; Thiophene, tetrahydro-3-(decyloxy)-, 1,1-dioxide.

1078   Decylphenoxybenzene
69834-17-9   H

C22H30O
Benzene, decylphenoxy-.
Benzene, decylphenoxy-; (Decylphenoxy)benzene;

Decylphenoxybenzene; EINECS 274-140-8.

1079   Decyl(sulfophenoxy)benzenesulfonic acid
70191-75-2   H

$C_{22}H_{30}O_7S_2$
Benzenesulfonic acid, decyl(sulfophenoxy)-.
Benzenesulfonic acid, decyl(sulfophenoxy)-; Benzene-sulfonic acid,
decyl(sulphophenoxy)-; Decyl(sulfophen-oxy)benzenesulfonic acid.

1080   DEDM hydantoin
26850-24-8   G

$C_9H_{16}N_2O_4$
1,3-Bis(2-hydroxyethyl)-5,5-dimethylimidazolidine-2,4-dione.
1,3-Bis(2-hydroxyethyl)-5,5-dimethyl-2,4-imidazolidinedione;
Dantocol® DHE; DEDM Hydantoin; Di-(2-hydroxyethyl)-5,5-dimethyl
hydantoin; 1,3-Di-(hydroxyethyl)-5,5-dimethylhydantoin; Diethylol
dimethyl hydantoin; EINECS 248-052-5; 2,4-Imidazolidinedione, 1,3-
bis(2-hydroxyethyl)-5,5-dimethyl-. Resin crosslinker in coatings and
polymers; intermediate for epoxies, urethane resins, and antistatic
lubricants for the textile and plastics industries. Crystals; mp = 63°;
pH 6.5 (5%). *Lonzagroup.*

1081   Dehydrated castor oil
64147-40-6   H

Castor oil, dehydrated.
Castor oil, dehydrated; Dehydrated castor oil; EINECS 264-705-7.

1082   Dehydroabietylamine
1446-61-3   G P

$C_{20}H_{31}N$
1R-1α,2,3,4,4aβ,9,10,10aα-Octahydro-1,4a-dimethyl-7-(1-
methylethyl)-1-phenanthrenemethanamine.
4-12-00-03005 (Beilstein Handbook Reference); Amine D; BRN
3084620; Caswell No. 276; Dehydro-abietylamine; EPA Pesticide
Chemical Code 004206; HSDB 5665; 13-Isopropylpodocarpa-
8,11,13-trien-15-amine; Podocarpa-8,11,13-trien-15-amine, 13-
isopropyl-. Used as asphalt additive, as cationic collectors for calcite,
sylrite, mica, feldspar, vermicilulite and phosphate rock
concentration operations. Registered by EPA as an antimicrobial
and fungicide (cancelled). Crystals; $[\alpha]_D^{20}$= +56.10 (c=2.4, pyridine).
Fluka; Hercules Inc.

1083   Dehydroacetic acid
520-45-6   2885   G

$C_8H_8O_4$
3-acetyl-6-methyl-2H-pyran-2,4(3H)-dione, ion(1-).
4-17-00-06699 (Beilstein Handbook Reference); Acetic acid,
dehydro-; 2-Acetyl-5-hydroxy-3-oxo-4-hexenoic acid, delta-lactone;
3-Acetyl-4-hydroxy-6-methyl-2H-pyran-2-one; 3-Acetyl-6-
methyldihydropyrandione-2,4; 3-Acetyl-6-methyl-2,4-pyrandione; 3-
Acetyl-6-methyl-2H-pyran-2,4(3H)-dione; 3-Acetyl-6-methyl-2H-
pyran-2,4(3H)-dione. enol form; 3-Acetyl-6-methyl-2H-pyran-2,4(3H)-
dione, ion(1-),; AI3-01464; Biocide 470F; BRN 0006129; Caswell No.
278; Dehydracetic acid; Dehydroacetic acid; DHA; DHAA; DHS;
EINECS 208-293-9; EPA Pesticide Chemical Code 027801; 4-
Hexenoic acid, 2-acetyl-5-hydroxy-3-oxo-, delta-lactone; HSDB 291;
Kyselina dehydroacetova; Methylaceto-pyronone; NSC 8770; 2H-
Pyran-2-one, 3-acetyl-4-hydroxy-6-methyl-; 2H-Pyran-2,4(3H)-dione,
3-acetyl-6-methyl-. A fungicide, bactericide, plasticizer and chemical
intermediate. Used in medicated toothpastes. Crystals; mp = 109°;
bp = 270°; very soluble in H_2O, Et_2O, slightly soluble in EtOH,
$CHCl_3$; LD50 (rat orl) = 570 mg/kg.

1084   Dehydrothio-p-toluidinesulfonic acid.
130-17-6   G

$C_{14}H_{12}N_2O_3S_2$
2-(4-Aminophenyl)-6-methyl-7-benzothiazolesulfonic acid.
2-27-00-00492 (Beilstein Handbook Reference); 2-(4-Aminophenyl)-
6-methylbenzothiazole-7-sulphonic acid; 2-(p-Aminophenyl)-6-
methyl-7-benzothiazolesulfonic acid; 2-(p-Aminophenyl)-6-
methylbenzothiazolyl-7-sulf-onic acid; 7-Benzothiazolesulfonic acid,
2-(p-aminophenyl)-6-methyl-; BRN 0313390; Dehydrothio-p-
toluidinesulfonic acid; D.T.S.; EINECS 204-979-7; NSC 203387; p-
(6-Methyl-7-sulfobenzothiazole)aniline.
Chemical intermediate. Crystals; mp > 300°; soluble in DMSO,
insoluble in H_2O, organic solvents; LD50 (mus iv) = 178 mg/kg.

1085   Dehyquart LDB
139-07-1   G

$C_{21}H_{38}ClN$
Lauralkonium chloride.
Ammonium, benzyldimethyldodecyl-, chloride; Amoryl BR 1244; Bas
2631; Benzenemethanaminium, N-dodecyl-N,N-dimethyl-, chloride;
Benzenemethan-aminium, N,N-dimethyl-N-dodecyl-, chloride;
Benzododecinii chloridum; Benzododecinio cloruro;
Benzododecinium chloride; Benzyldimethyldodecyl-ammonium
chloride; Benzyldodecyldimethylammonium chloride; N-Benzyl-N-
dodecyl-N,N-dimethylammonium chloride; Benzyl-
lauryldimethylammonium chloride; Caswell No. 073A; Caswell No.
416C; Catigene OM; Catinal CB 50; Catiogen PAN; Catiolite BC 50;
Cequartyl A; Chlorure de benzododecinium; Cloruro de

benzododecinio; Dehyquart LDB; Dimethylbenzyl-dodecylammonium chloride; Dimethylbenzyllauryl-ammonium chloride; Dimethyldodecylbenzylammonium chloride; Dodecylbenzyldimethylammonium chloride; Dodecyldimethylbenzylammonium chloride; n-Dodecyl-dimethylbenzylammonium chloride; N-Dodecyl-N,N-dimethylbenzenemethananiminium chloride; N-Dodecyl-N,N-dimethyl-N-benzylammonium chloride; N-Dodecyl-N,N-dimethylbenzenemethananiminium chloride; N-Dodecyldimethylbenzylammonium chloride; DYK 1125; EINECS 205-351-5; EPA Pesticide Chemical Code 069124; Lauralkonium chloride; Lauryl dimethyl benzyl ammonium chloride; Laurylbenzalkonium chloride; Laurylbenzyldimethylammonium chloride; Lauryldi-methylbenzylammonium chloride; Loraquat B 50; N-Lauryldimethylbenzylammonium chloride; Noramium DA 50; NSC 85508; Orthosan HM; QBA 1211; Retarder N; Rewoquat B 50; Rolcril; Swanol CA 100; Swanol CA 101; Tetranil BC 80; Texnol R 5; Triton K60; Vantoc CL; Zephirol. Bactericide and fungicide for disinfectants; emulsifier; external antistat for plastics. *Henkel/Organic Prods.*

1086   Deltamethrin
52918-63-5   2902   G P

C22H19Br2NO3
(1R-(1α(S*),3α))- 3-(2,2-Dibromoethenyl)-2,2-dimethyl-, cyano(3-phenoxyphenyl)methyl cyclopropanecarboxy-late.
AI3-29279; Butoflin; Butoss; Butox; CCRIS 3704; Cislin; Crackdown; (S)-α-cyano-3-phenoxybenzyl (1R)-3-(2,2-dibromovinyl)-2,2-dimethyl-cyclopropane carboxylate; Decamethrin; Decamethrine; Decis; Decis 0.5ULV; Decis 1.5ULV; Decis 2.5ULV; Dekametrin (Hungarian); Delsekte; Deltacide; DeltaGard; Deltagran; Deltamethrin; Deltamethrine; Deltametryna; EINECS 258-256-6; EPA Pesticide Chemical Code 978051; Esbecythrin; FMC 45498; Glossinex 200; HSDB 6604; IPO 8831; K-Obiol; K-Otek; K-Othrin; K-Othrine; New Musigie; NRDC 161; OMS 1988; Phagase 1; RU 22974; RUP 987; Stricker; Suspend; Thripstick®; Zodiac; Zorcis. A fast-acting non-systemic pyrethroid insecticide with contact and stomach action. Used to control many species of insect in many crops. Non-phytotoxic. Registered by EPA as an insecticide. Crystals; mp = 98 - 101°; insoluble in H_2O (< 0.0000002 g/100ml), soluble in dioxane (90 g/100 ml), cyclohexanone (75 g/100 ml), CH_2Cl_2 (70 g/100 ml), Me_2CO (50 g/100 ml), C_6H_6 (45 g/100 ml), DMSO (45 g/100 ml), xylene (25 g/100 ml), EtOH (1.5 g/100 ml), i-PrOH (0.6 g/100 ml); LD50 (mrat orl) = 128 mg/kg, (frat orl) = 139 mg/kg (in vegetable oil), (frat iv) = 4 mg/kg, (rat, rbt der) > 2000 mg/kg, (mallard duck orl) > 4640 mg/kg; LC50 (rat ihl 6 hr.) = 0.6 mg/l air, (mallard duck 8 day dietary) > 4640 mg/kg diet, (quail 8 day dietary) > 10000 mg/kg diet, (fish) 0.001 - 0.01 mg/l; toxic to bees, also repellant. *AgrEvo; Aventis Environmental Science USA LP; Bonide Products, Inc.; Rhône-Poulenc; S.C. Johnson & Son, Co.*

1087   Demeton-S-methyl
919-86-8   6077   G P

C6H15O3PS2
S-[2-(Ethylthio)ethyl] O,O-dimethyl phosphorothioate.
Bay 18436; Bay 25/154; DSM; Duratox; Metasystox 55; Metasystox I; Methyl demeton; Methylmercaptofostiol; Mifatox; Persyst; Power DSM. Systemic insecticide and acaricide with contact and stomach action. Used for control of aphids and other insects in a wide variety of crops. Emulsifiable concentrate containing 580 g de-meton-S-methyl per liter is used as a systemic organophosphorus insecticide and acaricide. Pale yellow oil; bp1= 118°; d^{20} = 1.207; soluble in H_2O (0.33 g/100 ml), more soluble in organic solvents; LD50 (rat orl) = 40 - 106 mg/kg, (rat der) = 30 mg/kg; LC50 (rat ihl 4 hr.) = 0.5 mg/l air, (rainbow trout, carp, Japanese killifish 48 hr.) > 10 mg/l; toxic to bees. *Bayer Corp.*

1088   Dequalinium Chloride
522-51-0   2930   D G

C30H40Cl2N4
1,1'-(1,10-Decanediyl)bis[4-amino-2-methylquinolinium chloride].
BADQ-10; Chlorure de dequalinium; Cloruro de decalinio; Decamine; Decaminum; Decatylen; Dekadin; Dekamiln; Dekamin; Dequadin; Dequadin Chloride; Dequafungan; Dequafungen; Dequalin chloride; Dequalinii chloridum; Dequalinio cloruro; Dequalinium chloride; Dequavagyn; Dequavet; Dynexan-mhp; Efisol; EINECS 208-330-9; Eriosept; Erosept; Evazol; Grocreme; Ivazil; Labosept; NSC 166454; Optipect; Oralgol; Phylletten; Polycidine; Rumilet; Sanoral; Sentril; Sorot. Antibacterial agent. Solid; mp = 326°; soluble in H_2O.

1089   Desloratadine
100643-71-8   2939   D

C19H19ClN2
8-Chloro-6,11-dihydro-22-(4-piperidinylidene)-5H-benzo(5,6)cyclohepta(1,2-b)pyridine.
Clarinex; Descarboethoxyloratadine; Desloratadine; Sch 3411/. FDA-approved 21 December, 2001 for the relief of the nasal and non-nasal symptoms of seasonal allergic rhinitis in patients 12 years of age and older. USP 4,659,716 (21 April, 1987) to Schering Corp. Crystals; mp = 150-151°. *Schering-Plough Pharm.*

1090   Desmeninol
583-91-5   6005   G V

C5H10O3S

2-Hydroxy-4-(methylthio)-butanoic acid.
Alimet; Butanoic acid, 2-hydroxy-4-(methylthio)-; Butyric acid, 2-hydroxy-4-(methylthio)-; EINECS 209-523-0; γ-(Methylthio)-α-hydroxybutyric acid; α-Hydroxy-γ-(methylmercapto)butyric acid; α-Hydroxy-γ-(methylthio)butyric acid; 2-Hydroxy-4-(methylthio)butyric acid; 2-Hydroxy-4-(methylthio)butanoic acid; HSDB 5700; Methionine hydroxy analog; MHA; MHA acid; MHA-FA. Used as a feed additive for livestock, especially poultry. *Monsanto Co.*

1091   **Desmetryn**
1014-69-3   G P

C8H15N5S

N-Methyl-N'-(1-methylethyl)-6-(methylthio)-1,3,5-triazine-2,4-diamine.
BRN 0612017; Caswell No. 509B; Desmetryn; Desmetryne; EPA Pesticide Chemical Code 080810; G 34360; GS 34360; HSDB 1729; 2-(Isopropylamino)-4-(methylamino)-6-(methylthio)-s-triazine; 2-Isopropyl-amino-4-methylamino-6-methylthio-1,3,5-triazine; 2-Isopropilamino-4-metilamino-6-metiltio-1,3,5-triazina; 2-Isopropylamino-4-methylamino-6-methylmercapto-s-triazine; 2-Methylamino-4-methylthio-6-isopropylamino-1,3,5-triazine; 2-(Methylthio)-4-(methylamino)-6-(iso-propylamino)-s-triazine; 2-Methylthio-4-isopropylamino-6-methylamino-s-triazine; 2-Methylmercapto-4-methyl-amino-6-isopropylamino-s-triazine; Methylmercapto-4-isopropylamino-6-methylamino-s-triazine; N-Methyl-N'-(1-methylethyl)-6-(methylthio)-1,3,5-triazine-2,4-diamine; N2-Isopropyl-N4-methyl-6-methylthio-1,3,5-triazine-2,4-diamine; Norametryne; Samuron; Semeron; Semeron-25; Semeron 250; Topusyn; s-Triazine, 2-(isopropylamino)-4-(methylamino)-6-(methylthio)-; 1,3,5-Triazine-2,4-diam-ine, N-methyl-N'-(1-methylethyl)-6-(methylthio)-.
Selective systemic herbicide, inhibits photosynthesis. Used for post-emergence control of broad-leaved weeds and some grasses in brassicas, herbs, onions, leeks and conifer seed beds. Crystals; mp = 84 - 86°; soluble in H2O (0.058 g/100 ml), MeOH (30 g/100 ml), Me2CO (23 g/100 ml), CH2Cl2 (20 g/1200 ml), C7H8 (20 g/100 ml), C6H14 (0.26 g/100 ml); LD50 (rat orl) = 1390 mg/kg, (rat der) > 1000 mg/kg; toxic to fish; non-toxic to bees. *Ciba-Geigy Corp.; Syngenta Crop Protection.*

1092   **Dexibuprofen**
51146-56-6   4906   D

C13H18O2

(+)-(S)-p-Isobutylhydratropic acid.
Dexibuprofen; Dexibuprofene; Dexibuprofeno; Dexi-buprofenum; d-Ibuproten. A cyclooxygenase inhibitor used as an analgesic and anti-

inflammatory. *Merck & Co.Inc.*

1093   **Dexibuprofen lysine**
141505-32-0   4906   D

C19H34N2O5

(+)-(S)-p-Isobutylhydratropic acid L-lysine salt monohydrate.
Ibuprofen S-form L-lysine salt; ML-223; MK-223; Doctrin. S-Form of ibuprofen L-lysine salt. Cyclooxygenase inhibitor. Analgesic; anti-inflammatory. *Merck & Co.Inc.*

1094   **Dexibuprofen lysine [anhydrous]**
141505-32-0   4906   D

C19H3N2O4

(+)-(S)-p-Isobutylhydratropic acid L-lysine salt.
L-669445. S-Form of ibuprofen L-lysine salt. Cyclooxygenase inhibitor.Analgesic; anti-inflammatory. *Merck & Co.Inc.*

1095   **Dexpanthenol**
81-13-0   2964   G

C9H19NO4

d-2,4-Dihydroxy-N-(3-hydroxypropyl)-3,3-dimethylbutanamide.
4-04-00-01652 (Beilstein Handbook Reference); Alcopan-250; Bepanthen; Bepanthene; Bepantol; BRN 1724947; Butanamide, 2,4-dihydroxy-N-(3-hydroxypropyl)-3,3-dimethyl-, (R)-; Butanamide, 2,4-dihydroxy-N-(3-hydroxypropyl)-3,3-dimethyl-, (θ)-; Butyramide, 2,4-di-hydroxy-N-(3-hydroxypropyl)-3,3-dimethyl-, D-(+)-; CCRIS 3947; Cozyme; Dexpantenol; Dexpanthenol; Dexpanthenolum; Dextro pantothenyl alcohol; D-(+)-2,4-Dihydroxy-N-(3-hydroxypropyl)-3,3-dimethylbutyramide; (R)-2,4-Dihydroxy-N-(3-hydroxy-propyl)-3,3-dimethyl-butanamide; 2,4-Dihydroxy-N-(3-hydroxypropyl)-3,3-dimethylbutanamide, (R)-; 2,4-Dihydroxy-N-(3-hydroxy-propyl)-3,3-dimethylbutyramide, D-(+)-; D-P-A Injection; EINECS 201-327-3; HSDB 296; Ilopan; Intrapan; Motilyn; NSC 302962; Panadon; Pantenyl; (+)-Panthenol; d(+)-Panthenol; D-Panthenol; D-Panthenol 50; Panthoderm; Pantol; Panthotenol; Pantothenol, D-; d-Pantothenol; Pantothenyl alcohol; d-Pantothenyl alcohol; D-Pantothenyl alcohol; D(+)-Pantothenyl alcohol; Pantothenylol; Penthenol; Propanolamine, N-pantoyl-; Provitamin B; Provitamin B5; Ritapan D; Synapan; Thenalton; Urupan; Zentinic. Nutrient, humectant for hair and skin care formulations; hair repair agent; soothing to skin. Nutritional factor, source of pantothenic acid. Liquid; bp0.02 = 118-120°; d$_{20}^{20}$ = 1.2; [α]$_D^{20}$= 30° (c = 5); soluble in H2O, EtOH. *Hoffmann-LaRoche Inc.; Roche Vitamins Inc.*

1096   Dextran
9004-54-0   2965   D E G

(C6H10O5)n

Dextran.
CCRIS 2469; Dextran; Dextran 40; Dextran 70; Dextrano; Dextrans; Dextranum; Dextranum 40,70; Dextraven; EINECS 232-677-5; Eudextran; Gentran; Gentran 40; Hemodex; Intradex; LMD; LMWD; Lomodex 40; LVD; Macrodex; Macrose; Onkotin; Plavolex; Polyglucin; Polyglucinum; Promit; Rheomacrodex; Rheopolyglucine; Rheotran. Polysaccharide composed primarily of D-glucose units linked α-D-(1→6); produced by *Leuconostoc mesenteroides, L. dextranicum (Lactobacteriaceae)*. Used as a blood plasma substitute or extender, in confections, lacquers, oil-well drilling muds, filtration gel, food additive. *Accurate Chem. & Sci. Corp.; Am. Biorganics; Baxter Healthcare Systems; CIBA Vision AG; Lancaster Synthesis Co.; McGaw Inc.; Pharmacia & Upjohn AB; Pharmacia; Spectrum Chem. Manufacturing; U.S. BioChem.*

1097   Diacetazotol
83-63-6   G

C18H19N3O4

N,N-Diacetyl-o-tolylazo-o-toluidine.
3-16-00-00387 (Beilstein Handbook Reference); Acetamide, N-acetyl-N-(2-methyl-4-((2-methylphenyl)-azo)phenyl)-; N-Acetyl-N-(2-methyl-4-((2-methylphenyl)-azo)phenyl)acetamide; BRN 0686326; Dermagan; Dermagen; Diacetazotol; Diacetotoluide; o-Diacetotoluidide, 4''-(o-tolylazo)-; Diacetylaminoazotoluene; Diamazo; Dimazon; EINECS 201-490-0; Epidermol; Epithelone; Granulin; NSC 6509; Pellidol; Pellidole; Periphermin; 4-o-Tolylazo-o-diacetotoluide; 4'-(o-Tolylazo)-o-diacetotoluidide; N-(4-o-Tolylazo-o-tolyl)diacetamide. A red dye used in ointment or as a dusting powder. Crystallizes in two modifications: brick-red needles, mp = 65°, or stout, red prisms, mp = 75°; insoluble in H2O; soluble in organic solvents.

1098   Diacetin
25395-31-7   2980   G

C7H12O5

Glycerol diacetate.
Acetin, di-; AI3-00676; Diacetin; Diacetylglycerol; EINECS 246-941-2; Estol 1582; Estol 1583; Glycerin diacetate; Glycerine diacetate; Glycerol diacetate; Glycerol 1,3-di(acetate); 1,2,3-Propanetriol, diacetate. Auxiliary for use in foundries. mp = -30°; bp = 280°; d = 1.18; LD50 (mus orl) = 8500 mg/kg. *Bayer AG.*

1099   Diacetone alcohol
123-42-2   2983   G H

C6H12O2

4-Hydroxy-4-methylpentan-2-one.
4-01-00-04023 (Beilstein Handbook Reference); Acetonyldimethylcarbinol; AI3-00045; BRN 1740440; Caswell No. 280; CCRIS 6177; Diacetonalcohol; Diacetonalcool; Diacetonalkohol; Diacetone; Diacetone alcohol; Diacetone-alcool; Diacetonyl alcohol; Diketone alcohol; Dimethyl acetonyl carbinol; EINECS 204-626-7; EPA Pesticide Chemical Code 033901; HSDB 1152; 4-Hydroxy-4-methyl-pentan-2-on; 4-Hydroxy-4-methyl-pentan-2-one; 4-Idrossi-4-metil-pentan-2-one; NSC 9005; 2-Pentanone, 4-hydroxy-4-methyl-; Pyranton; Pyranton A; Tyranton; UN1148. Solvent for nitrocellulose, cellulose acetate, oils, resins, waxes, fats, dyes, tars, lacquers, dopes, coatings, wood preservatives, rayon, artificial leather, metal cleaning; laboratory reagent; hydraulic fluids; textile stripping agent. Liquid; mp = -44°; bp = 167.9°; d^{20} = 0.9387; λ_m = 238, 281 nm (MeOH); soluble in CHCl3, freely soluble in H2O, EtOH, Et2O; LD50 (rat orl) = 4.0 g/kg. *Allchem Ind.; BP Chem.; Elf Atochem N. Am.; Hoechst Celanese; Shell; Union Carbide Corp.*

1100   3-Diacetoxyethylamino-4-methoxy-acetanilide
23128-51-0   H

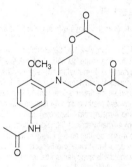

C17H24N2O6

Acetamide, N-(3-(bis(2-(acetyloxy)ethyl)amino)-4-meth-oxyphenyl)-.
Acetamide, N-(3-(bis(2-(acetyloxy)ethyl)amino)-4-meth-oxyphenyl)-; 2,2'-((5-Acetamido-2-methoxyphenyl)-imino)diethyl diacetate; p-Acetanisidide, 3'-(bis(2-hydroxyethyl)amino)-, diacetate (ester) N-(3-(Bis(2-acetoxyethyl)amino)-4-methoxyphenyl)acetamide; EINECS 245-441-1; 3-Diacetoxyethylamino-4-methoxy-acetanilide.

1101   Diacetyl
431-03-8   2985   G

C4H6O2

2,3-Butanedione.
4-01-00-03644 (Beilstein Handbook Reference); AI3-03313; Biacetyl; BRN 0605398; Butadione; 2,3-Butadione; Butanedione; 2,3-Butanedione; CCRIS 827; Diacetyl; 2,3-Diketobutane; 2,3-Dioxobutane; Dimethyl diketone; Dimethyl glyoxal; Dimethylglyoxal; EINECS 207-069-8; FEMA No. 2370; Glyoxal, dimethyl-; HSDB 297; NSC 8750; UN2346. Aroma carrier in food products. Liquid; mp = -1.2 °; bp = 88°; d^{15} = 0.9808; λ_m = 290 nm (MeOH); soluble in C6H6, CCl4, very soluble in H2O, Me2CO, freely soluble in EtOH, Et2O; LD50 (rat orl) = 1580 mg/kg. *Penta Mfg.; Sigma-Aldrich Fine*

1102   N,N-Diallyldichloroacetamide
37764-25-3   H

C8H11Cl2NO

2,2-Dichloro-N,N-di-2-propenylacetamide.

4-04-00-01064 (Beilstein Handbook Reference); Acet-amide, 2,2-dichloro-N,N-di-2-propenyl-; Acetamide, N,N-diallyl-2,2-dichloro-; BRN 1768843; Compound R-25788; Dichlormid; 2,2-Dichloro-N,N-di-2-propenyl-acetamide; EINECS 253-658-8; N,N-Diallyl-2,2-dichloroacetamide; N,N-Diallyldichloroacetamide; R-25788; Stauffer R-25788.

1103   Diallyldimethylammonium chloride
7398-69-8   G H

C8H16ClN

2-Propen-1-aminium, N,N-dimethyl-N-2-propenyl-, chloride.

Ageflex mDMDAC; Ageflex NB-50; Agestat 41; Ammonium, diallyldimethyl-, chloride; Dimethyl-diallylammonium chloride; N,N-Dimethyl-N-2-propenyl-2-propen-1-aminium chloride; EINECS 230-993-8; NSC 59284. Monomer for synthesis of homo and copolymers used as coagulant and flocculants for water treatment, mineral processing, demulsifier for petrol, recovery, electrically conductive paper and coatings, wet and dry strength resins, antistatic additives and coatings; cosmetic additives in hair conditioners, biocides, detergent additives, water-soluble polymers and electrographic paper and film. Paper industry retention aid, pigment dispersion, drainage aid, fiber dewatering, stabilizer for sizes, electroconductive polymer, recycling operations, raw and waste water clarification. Solid; soluble in H2O; d = 1.04; mildly corrosive; nonhazardous. *Rit-Chem.*

1104   Diallyl glycol carbonate
142-22-3   H

C12H18O7

Diethylene glycol, bis(allyl carbonate).

01M; 4-03-00-00012 (Beilstein Handbook Reference); Al3-07498; Allyl diglycol carbonate; BRN 1803874; Diethylene glycol, bis(allyl carbonate); High ADC-CR 39; HSDB 5638; Nouryset 200; NSC 5246; RAV 7N; Transallyl CR 39; TS 16.

1105   Diallyl isophthalate
1087-21-4   G

C14H14O4

Isophthalic acid, di-(2-propenyl) ester.

4-09-00-03295 (Beilstein Handbook Reference); Al3-16904; 1,3-Benzenedicarboxylic acid, di-2-propenyl ester; BRN 2055011; Dapon M; Dappu 100; Di-2-propenyl isophthalate; EINECS 214-122-9; Isophthalic acid, diallyl ester; NSC 6098. Used as a molding material. *FMC Corp.*

1106   Diallyl maleate
999-21-3   G

C10H12O4

2-Butenedioic acid, di-2-propenyl ester, (Z)-.

4-02-00-02214 (Beilstein Handbook Reference); Al3-02531; BRN 1725954; 2-Butenedioic acid (2Z)-, di-2-propenyl ester; Diallyl maleate; Diallylester kyseliny maleinove; EINECS 213-658-0; Maleic acid, diallyl ester; NSC 4799; Sipomer DAM. Used in manufacture of polymers, copolymers and insecticides. Liquid; bp10 = 129°, bp3 = 109°; d^{20} = 1.075; soluble in CHCl3. *Aceto Corp.; Ashland.*

1107   Diallyl phthalate
131-17-9   G H

C14H14O4

Di-2-propenyl 1,2-benzenedicarboxylate.

4-09-00-03188 (Beilstein Handbook Reference); Al3-02574; Allyl phthalate; BRN 1880877; CCRIS 1361; Dapon 35; Dapon R; Decobest DA; Di-2-propenyl 1,2-benzenedicarboxylate; Diallyl phthalate; Diallylester kyseliny ftalove; EINECS 205-016-3; HSDB 4169; NCI-C50657; Nonflammable Decobest DA; NSC 7667; Phthalic acid, diallyl ester; RX® 1-501N. A mineral reinforced; thermoset molding material used in decorative laminates. Liquid; mp = -70°; bp5 = 165-167°; d = 1.1200; n$_D^{20}$ = 1.5194; slightly soluble in H2O, more soluble in organic solvents; LD50 (rat orl) = 656 mg/kg. *Allchem Ind.; Arco; BP Chem.; Hall C.P.; OxyChem; Sumitomo Corp.; Van Waters & Rogers.*

1108   3,3'-Diaminodiphenylsulfone
599-61-1   G

C12H12N2O2S
Bis(m-aminophenyl) sulfone.
4-13-00-01009 (Beilstein Handbook Reference); AI3-52564; Aniline, 3,3'-sulfonyldi-; Benzenamine, 3,3'-sulfonylbis-; BRN 2215938; 3,3'-Diaminodifenylsulfon; 3,3'-Diaminodiphenyl sulfone; 3,3'-Diaminophenyl sulfone; EINECS 209-967-5; NSC 20610; Sulfone, bis(3-aminophenyl); Sulfone, bis(m-aminophenyl); 3,3'-Sulfonylbis(aniline); 3,3'-Sulfonylbisbenzenamine; 3,3'-Sulfonyldianiline; 3,3'-Sulphonyldianiline; 3,3'-sulfonyldianiline; 3-Aminophenyl sulphone. Crystals; mp = 171-172°; irritant. *BASF Corp.; Lancaster Synthesis Co.; Mitsui Toatsu.*

1109   2,6-Diaminopyridine
141-86-6   G

C5H7N3
2,6-Pyridinediamine.
5-22-11-00255 (Beilstein Handbook Reference); AI3-18054; BRN 0108513; CCRIS 6682; DAP; DAP (amine); 2,6-Diaminopyridine; EINECS 205-507-2; NSC 1921; 2,6-Pyridinediamine; Pyridine, 2,6-diamino-; Pyridine-2,6-diyldiamine. Used as a chemical intermediate. Crystals; mp = 121.5°; bp = 285°; bp5 = 148°; λ_m = 244, 308 nm (MeOH); slightly soluble in H2O, Me2CO. *Cilag-Chemie Ltd.; Janssen Chimica; Lancaster Synthesis Co.; Reilly Ind.*

1110   2,4-Diaminotoluene
95-80-7   H

C7H10N2
1,3-Benzenediamine, 4-methyl-.
4-13-00-00235 (Beilstein Handbook Reference); 5-Amino-o-toluidine; AI3-03717; Benzenediamine, ar-methyl-; Benzofur MT; BRN 2205839; Brown for Fur T; CCRIS 202; CI Oxidation Base 35; Developer 14; Developer B; Developer DB; Developer DBJ; Developer MC; Developer MT; Developer MT-CF; Developer MTD; Developer T; EINECS 202-453-1; Eucanine GB; Fouramine J; Fourrine 94; Fourrine M; HSDB 2849; 4-Methyl-1,3-benzenediamine; 4-Methyl-m-phenylene-diamine; Nako TMT; NCI-C02302; Pelagol Grey J; Pelagol J; Pontamine developer TN; RCRA waste number U221; Renal MD; TDA; Tertral G; Toluenediamine; Toluenediamine, o-; Toluene-2,4-diamine; m-Toluene-diamine; m-Toluylendiamin; Tolylene-2,4-diamine; m-Tolylenediamine; 4-m-Tolylenediamine; meta-Tolyl-enediamine; UN1709; Zoba GKE; Zogen Developer H. Needles; mp = 99°; bp = 292°; λ_m = 294 nm (MeOH); very soluble in H2O, EtOH, Et2O, C6H6, soluble in CHCl3.

1111   Diammonium dithiocarbazate
20469-71-0   H

CH4N2S2.H4N2
Dithiocarbonic acid monohydrazide hydrazine salt.
Carbazic acid, dithio-, compd. with hydrazine (1:1); Carbazic acid, dithio-, hydrazine (salt); Carbazic acid, dithio-, hydrazinium salt;

Diammonium dithiocarbazate; Dithiocarbazic acid, compound with hydrazine (1:1); Dithiocarbazic acid hydrazine (salt); Dithiocarbonic acid monohydrazide hydrazine salt; EINECS 243-844-7; Hydrazinecarbodithioic acid, compd. with hydrazine (1:1); Hydrazinium dithiocarbazate; NSC 512576.

1112   Diammonium EDTA
20824-56-0   H

C10H22N4O8
Ethylenediaminetetraacetic acid, diammonium salt.
Acetic acid, (ethylenedinitrilo)tetra-, diammonium salt; Diammonium edetate; Diammonium EDTA; Di-ammonium N,N'-1,2-ethanediylbis(N-(carboxymethyl)-glycine); Diammonium ethylene diamine tetraacetate; Diammonium dihydrogen ethylenediaminetetraacetate; Edetate diammonium; EINECS 244-063-4; Glycine, N,N'-1,2-ethanediylbis(N-(carboxymethyl)-, diammonium salt; Versene Diammonium EDTA. Chelating agent, drug stabilizing agent.

1113   Diamylamine
2050-92-2   H

C10H23N
Di-n-Pentylamine.
4-04-00-00676 (Beilstein Handbook Reference); AI3-15326; Amine, dipentyl; BRN 0906746; CCRIS 6225; Diamyl amine; Diamylamine; Di-n-amylamine; Di-n-Pentylamine; Dipentylamine; EINECS 218-108-3; HSDB 5864; NSC 6329; 1-Pentanamine, N-pentyl-; Pentyl-amine, pentyl-; UN2841. Liquid; bp = 202.5°; d^{20} = 0.7771; slightly soluble in H2O, soluble in Me2CO, very soluble in EtOH, freely soluble in Et2O.

1114   2,5-Di-t-amylhydroquinone
79-74-3   3336   G

C16H26O2
2,5-Di-t-pentylhydroquinone.
3-06-00-04748 (Beilstein Handbook Reference); AI3-61041; 2,5-Di-t-amylbenzene-1,4-diol; 2,5-Di-t-amylhydroquinone; Antage DAH; 1,4-Benzenediol, 2,5-bis(1,1-dimethylpropyl)-; 2,5-Bis(1,1-dimethylpropyl)-1,4-benzenediol; 2,5-Bis(1,1-dimethylpropyl)hydroquinone; BRN 2214556; Dahq; Diamylhydroquinone; EINECS 201-222-2; HSDB 5231; Hydroquinone, 2,5-di-t-amyl-; Hydroquinone, 2,5-di-t-pentyl-; NSC 455; 2,5-Di-t-pentylhydroquinone; 2,5-Di-t-pentylhydroquinone; 2,5-Di-t-pentylbenzene-1,4-diol; Santouar A; Santovar A; USAF B-21. Antioxidant for unvulcanized rubber. Used to protect rubber from

staining. Solid; mp = 179-180°. *Monsanto Co.*

1115 Di-t-amylphenol
120-95-6   H

C16H26O
2,4-Di-t-pentylphenol.
3-06-00-02085 (Beilstein Handbook Reference); 2,4-Di-t-amylphenol; BRN 2370274; Di-t-amylphenol; EINECS 204-439-0; HSDB 5588; NSC 158351; 2,4-Di-t-pentylphenol; Phenol, 2,4-bis(1,1-dimethylpropyl)-; Phenol, 2,4-di-t-pentyl-; Prodox 156.

1116 Dianilinomethane
101-77-9   3000   G H

C13H14N2
4,4'-Diaminodiphenylmethane.
4-13-00-00390 (Beilstein Handbook Reference); AI3-02615; Ancamine TL; Aniline, 4,4'-methylenedi-; Araldite hardener 972; Avaldite HT 972; Benzenamine, 4,4'-methylenebis-; Bis-p-aminofenylmethan; Bis(4-amino-phenyl)methane; Bis(p-aminophenyl)methane; BRN 0474706; CCRIS 1010; Curithane; Dadpm; Di-(4-aminophenyl)methane; Diaminodiphenylmethane; Dianilinemethane; Dianilinomethane; EINECS 202-974-4; Epicure DDM; Epikure DDM; HSDB 2541; HT 972; Jeffamine AP-20; MDA; MDA; Methylenebis(aniline); Methylenedianiline; NCI-C54604; NSC 4709; Sumicure M; Tonox; UN2651. Crude methylene dianiline; used as a curing agent for epoxy resins. Solid; mp = 91-92°; bp = 398-399°; slightly soluble in H2O, very soluble in organic solvents. *Uniroyal.*

1117 Diatrizoate meglumine
131-49-7   5830   G

C18H26I3N3O9
1-Deoxy-1-(methylamino)-D-glucitol   3,5-bis(acetyl-amino)-2,4,6-triiodobenzoate salt.
Amidotrizoate meglumine; Angiografin; Angiovist 282; Benzoic acid, 3,5-diacetamido-2,4,6-triiodo-, compd. with 1-deoxy-1-(methylamino)-D-glucitol; 3,5-Bis(acet-amido)-2,4,6-triiodbenzoesaeure, 1-desoxy-1-methyl-amino-D-glucit-salz; Cardiografin; Cystografin; Cysto-graphin Dilute; 1-Deoxy-1-(methylamino)-D-glucitol 3,5-diacetamido-2,4,6-triiodobenzoate (salt); Diatrizoate meglumine; Diatrizoate methylglucamine; EINECS 205-024-7; Gastrografin; Gastrografin Oral (Veterinary); Glucitol, 1-deoxy-1-(methylamino)-, 3,5-diacetamido-2,4,6-triiodobenzoate (salt), D-; D-Glucitol, 1-deoxy-1-(methylamino)-, 3,5-bis(acetylamino)-2,4,6-triiodo-benzoate (salt); Hypaque 13.4; Hypaque 60; Hypaque Cysto; Hypaque-DIU; Hypaque M 30; Hypaque Meglumine; MD 60; Meglumine amidotrizoate; Meglumine diatrizoate; Methylglucamine diatrizoate; Reno-M; Reno-M-Dip;

RENO M 60; Renocal; Renograffin M-76; Renografin; Renovist; Renurix; Sinografin; Unipaque; Urovist. Diagnostic aid. Crystals; mp = 189-193°; soluble in H2O (89 g/100 ml). *Bristol-Myers Squibb Co.*

1118 Diatrizoate sodium
737-31-5   3016   D G

C11H8I3N2NaO4
3,5-bis(acetylamino)-2,4,6-triiodobenzoic acid sodium salt.
Amidotrizoate de sodium; Amidotrizoato sodico; Benzoic acid, 3,5-diacetamido-2,4,6-triiodo-, monosodium salt; Benzoic acid, 3,5-diacetamido-2,4,6-triiodo-, sodium salt; Benzoic acid, 3,5-bis(acetylamino)-2,4,6-triiodo-,; Conray 35; 3,5-Diacetamido-2,4,6-triiodobenzoic acid, sodium salt; 3,5-Diacetylamino-2,4,6-trijodbenzosäure natrium; 3,5-Diacetylamino-2,4,6-trijodbenzosaeure natrium; Diatrizoate sodium; Diatrizoic acid sodium salt; EINECS 212-004-1; Gastrografin; Histopaque; Hypaque; Hypaque 50; Hypaque Cysto; Hypaque-DIU; Hypaque sodium; Iothalamate; MD 50; Meglumine diatrizoate; Methalamic acid; MI 216; Monosodium 3,5-diacetamido-2,4,6-triiodobenzoate; Natrii amidotrizoas; NSC 61815; Renografin 76 Injectable (Veterinary); Sodium 3,5-diacetamido-2,4,6-triiodobenzoate; Sodium amido-trizoate; Sodium diacetyldiaminetriiodobenzoate; Sodium diatrizoate; Triombrin; Triombrine; Urografic acid, sodium salt; Urovist Sodium 300; Vascoray; WIN 8308-3. Radio-opaque medium used as a diagnostic acid. Crystals; mp = 261-262°; soluble in H2O (60 g/100 ml), less soluble in organic solvents; LD50 (rat iv) = 14.7 g/kg.

1119 Diaveridine
5355-16-8   3017   D G V

C13H16N4O2
5-[(3,4-Dimethoxyphenyl)methyl]-2,4-pyrimidine-diamine.
5-25-13-00391 (Beilstein Handbook Reference); AI3-23935; BRN 0258464; BW 49-210; CCRIS 3784; Diaveridin; Diaveridina; Diaveridine; Diaveridinum; EGIS 5645; EINECS 226-333-3; NSC 408735. Used with sulfaquonoxaline [59-40-5] as Darvisul, a veterinary antiprotozoal agent. White powder; mp = 233°. *Glaxo Wellcome Inc.*

1120 Diazabicycloundecene
6674-22-2   G

C9H16N2
2,3,4,6,7,8,9,10-Octahydropyrimido[1,2-a]azepine.
1,5-Diazabicyclo(5.4.0)undec-5-ene; 1,8-Diazabicyclo-(5.4.0)undec-

7-ene; DBU; EINECS 229-713-7; NSC 111184; 2,3,4,6,7,8,9,10-Octahydropyrimido(1,2-α)aze-pine; Pyrimido(1,2-a)azepine, 2,3,4,6,7,8,9,10-octa-hydro-. Can be used as a biotin substitute as cofactor in a number of enzymatic carboxylation reactions. Liquid; bp$_{0.6}$ = 80-83°; d = 1.018. *BASF; Chem; Fluka; Prods; Schweizerhall.*

1121   Diazepam
439-14-5   3018   D

C16H13ClN2O
7-Chloro-1,3-dihydro-1-methyl-5-phenyl-2H-1,4-benzo-diazepin-2-one.
Alboral; Aliseum; Amiprol; An-Ding; Ansilive; Ansiolin; Ansiolisina; Antenex; Anxicalm; Anxionil; Apaurin; Apo-diazepam; Apozepam; Armonil; Arzepam; Assival; Atensine; Atilen; Azedipamin; Baogin; Bensedin; Benzopin; Best; Betapam; Bialzepam; Britazepam; BRN 0754371; Calmaven; Calmocitene; Calmociteno; Calmod; Calmpose; Caudel; CB 4261; CCRIS 6009; Centrazepam; Cercine; Ceregulart; Chuansuan; Condition; DEA No. 2765; Desconet; Desloneg; Diacepan; Diaceplex; Dialag; Dialar; Diapam; Diapax; Diapine; Diaquel; Diastat; Diatran; Diazem; Diazemuls; Diazepam; Diazepam Dak; Diazepam Desitin; Diazepam Elmu; Diazepam-Eurogenerics; Diazepam Fabra; Diazepam-Lipuro; Diazepam Nordic; Diazepam-ratiopharm; Diazepam Rectubes; Diazepam Stada; Diazepamu; Diazepamum; Diazepan; Diazepan leo; Diazepin; Diazetard; Dienpax; Dipam; Dipaz; Dipezona; Disopam; Dizac; Domalium; Doval; Drenian; Ducene; Duksen; Dupin; Duxen; DZP; EINECS 207-122-5; Elcion CR; Eridan; Euphorin P; Eurosan; Evacalm; Faustal; Faustan; Freudal; Frustan; Gewacalm; Gihitan; Gradual; Gubex; Horizon; HSDB 3057; Iazepam; Jinpanfan; Kabivitrum; Kiatrium; Kratium; Kratium 2; LA III; Lamra; Lembrol; Levium; Liberetas; Lizan; Lovium; Mandro; Mandro-Zep; Mandrozep; Medipam; Mentalium; Metamidol; Methyl diazepinone; Methyldiazepinone; Metil Gobanal; Morosan; Nellium; Nerozen; Nervium; Neurolytril; Nivalen; Nixtensyn; Noan; Notense; Novazam; Novodipam; NSC-77518; NSC-169897; Ortopsique; Paceum; Pacitran; D-Pam; Paralium; Paranten; Parzam; Pax; Paxate; Paxel; Paxum; Placidox 2; Placidox 5; Placidox 10; Plidan; Pomin; Pro-Pam; Propam; Prozepam; Psychopax; Q-Pam; Quetinil; Quiatril; Quievita; Radizepam; Relaminal; Relanium; Relax; Reliver; Renborin; Ro 5-2807; Ruhsitus; Saromet; Sedapam; Sedipam; Seduksen; Seduxen; Serenack; Serenamin; Serenzin; Servizepam; Setonil; Sibazon; Sico Relax; Simasedan; Sipam; Solis; Sonacon; Stesolid; Stesolin; Tensopam; Tranimul; Trankinon; Tranqdyn; Tranquase; Tranquirit; Tranquo-Puren; Tranquo-Tablinen; Trazepam; Umbrium; Unisedil; Usempax AP; Valaxona; Valeo; Valiquid; Valitran; Valium; Valium Injectable; Valrelease; Valuzepam; Vanconin; Vatran; Vazen; Velium; Vival; Vivol; Winii; WY-3467; Zepaxid; Zipan.; Calmpose; Ceregulart; Dialar; Diazemuls; Dipam; Eridan; Eurosan; Evacalm; Faustan; Gewacalm; Horizon; Lamra; Lembrol; Levium; Mandrozep; Neurolytril; Noan; Novazam; Paceum; Pacitran; Paxate; Paxel; Pro-Pam; Q-Pam; Relanium; Sedapam; Seduxen; Servizepam; Setonil; Solis; Stesolid; Tranquase; Tranquo-Puren; Tranquo-Tablinen; Unisedil; Valaxona; Valiquid; Valium; Valium Injectable; Valrelease; Vival; Vivol. Anxiolytic, skeletal muscle relaxant and sedative/hypnotic. Also used as an intravenous anesthetic. Crystals; mp = 132°; λ$_m$ = 230, 315 nm (ε = 30903, 2188, EtOH); soluble in DMF, CHCl3, C6H6; Me2CO, EtOH; slightly soluble in H2O; LD50 (rat orl) = 710 mg/kg. *Berk Pharm. Ltd.;*

Hoffmann-LaRoche Inc.

1122   Diazinon Liquid
333-41-5   3043   G H P

C12H21N2O3PS
O,O-diethyl-O-(6-methyl-2-(1-methylethyl)-4-pyrimid-inyl)phosophorothioate.
5-23-11-00187 (Beilstein Handbook Reference); Ag-500; AI3-19507; Alfa-tox; Antigal; Bassadinon; Basudin; Basudin 10 G; Basudin S; Bazuden; BRN 0273790; Caswell No. 342; CCRIS 204; Ciazinon; Compass; Compass (insecticide); Cooper's Flystrike Powder; Dacutox; Dassitox; Dazzel; Delzinon; Diagran; Dianon; Diaterr-fos; Diazajet; Diazide; Diazinon; Diazinon ag 500; Diazinone; Diazitol; Diazol; Dicid; Dimpilato; Dimpylat; Dimpylate; Dimpylatum; Dipofene; Disonex; Dizictol; Dizinil; Dizinon; Drawizon; Dyzol; EINECS 206-373-8; Ektoband; ENT 19,507; EPA Pesticide Chemical Code 057801; Exodin; Fezudin; Flytrol; G 301; G-24480; Galesan; Garden Tox; Geigy 24480; HSDB 303; Isopropylmethylpyrimidyl diethyl thiophosphate; Kayazinon; Kayazol; KFM Blowfly Dressing; Kleen-Dok; Knox-out; Knox Out 2FM; Knox Out Yellow Jacket Control; Meodinon; NCI-C08673; Neocidol; Neodinon; Nipsan; NSC 8938; Nucidol; Oleodiazinon; OMS 469; Optimizer; PT 265; Sarolex; Spectracide; Srolex; Terminator; Diazide; Diazol; Diethyl Dimpylatum; Dizinon; Drawizon; Dyzol; Exodin; Fezudin; Flytrol; Galesan; Kayazol; Knox Out 2FM; Neocidol; Nipsan; Nucidol; Sarolex. Insecticide. Registered by EPA as an insecticide and fungicide. Oil; bp0.002 = 83 - 84°, bp1 = 125°; dec. > 120°; d4^{20} 1-116 - 1.118; slightly soluble in H2O (0.004 g/100 ml at 20°), EtOH, Et2O, petroleum ether, cyclohexane, C6H6, C7H8; LD50 (rat orl) = 300 - 400 mg/kg, (mrat orl) = 250 mg/kg, (frat orl) = 285 mg/kg, (mus orl) = 80 - 135 mg/kg, (gpg orl) = 250 - 355 mg/kg, (mallard duckling orl) = 3.5 mg/kg, (pheasant orl) = 4.3 mg/kg, (rat der) > 2150 mg/kg, (rbt der) = 540 - 650 mg/kg; LC50 (rat ihl 4 hr.) = 3.5 mg/l air, (bluegill sunfish 96 hr.) = 16 mg/l, (rainbow trout 96 hr.) = 2.6 - 3.2 mg/l; highly toxic to bees. *DowElanco Ltd.; Farnam Companies Inc.; Novartis; Scotts Co.; Syngenta Crop Protection.*

1123   Diazolidinyl urea
78491-02-8   3024   G

C8H14N4O7
Urea, N-(1,3-bis(hydroxymethyl)-2,5-dioxo-4-imidazol-idinyl)-N,N'-bis(hydroxymethyl)-.
1-(1,3-Bis(hydroxymethyl)-2,5-dioxoimidazolidin-4-yl)-1,3-bis(hydroxymethyl)urea; Diazolidinyl urea; EINECS 278-928-2; Germall® 11; Imidazolidinyl urea 11; N-(1,3-Bis(hydroxymethyl)-2,5-dioxo-4-imidazolidinyl)-; N-(1,3-Bis(hydroxymethyl)-2,5-dioxo-4-imidazolidinyl)-N,N'-bis-(hydroxymethyl)urea; N-(Hydroxymethyl)-N-(1,3-di-hydroxymethyl-2,5-dioxo-4-imidazolidinyl)-; N-(Hydro-xymethyl)-N-(1,3-dihydroxymethyl-2,5-dioxo-4-imidazol-idinyl)-N'-

(hydroxymethyl) urea; N,N'-Bis-(hydroxy-methyl) urea; Urea, N-(1,3-bis(hydroxymethyl)-2,5-dioxo-4-imidazolidinyl-N,N'-bis(hydroxymethyl)-. Broad-spect-rum antimicrobial preservative for cosmetics and toil-etries. *Sutton Labs.*

1124   Dibenzosuberone
1210-35-1   G

C$_{15}$H$_{12}$O
10,11-Dihydro-5H-dibenzo[a,d] cyclohepten-5-one.
CCRIS 2780; Dibenzo(a,d)cycloheptadien-5-one; Di-benzo(a,d)cyclohepta(1,4)dien-5-one; 5H-Dibenzo-(a,d)cyclohepten-5-one, 10,11-dihydro-; Dibenzo-(b,f)cycloheptan-1-one; Dibenzocycloheptenone; Di-benzosuberan-5-one; Dibenzosuberone; 2,3:6,7-Di-benzosuberone; Dibenzsuberone; 10,11-Dihydro-5H-dibenzo(a,d)cyclohepten-5-one; 10,11-Dihydrodibenzo-(a,d)cycloheptanone; 10,11-Dihydrodibenzo(a,d)-cyclohepten-5-one; EINECS 214-912-3; NSC 49727. Used in chemical synthesis. Crystals; mp = 30°; bp$_7$ = 203-204°; bp$_{0.3}$ = 148°; d^{20} = 1.1635; LD$_{50}$ (mus orl) = 2.1 g/kg. *Lancaster Synthesis Co.; Lonzagroup; Penta Mfg.; Sandoz.*

1125   Dibenzoyl peroxide
94-36-0   1117   D G H

C$_{14}$H$_{10}$O$_4$
Peroxide, dibenzoyl.
Abcat 40; Abcure S-40-25; Acetoxyl; Acne-Aid Cream; Acnegel; Akneroxid 5; Akneroxide L; Aksil 5; Asidopan; Aztec BPO; B 75W; Benbel C; Benox 50; Benoxyl; Benoxyl (5&10) Lotion; Benzac; Benzac W; Benzagel; Benzagel 10; Benzaknen; Benzamycin; Benzashave; Benzoic acid, peroxide; Benzol peroxide; Benzoperoxide; Benzoyl peroxide; Benzoyl superoxide; Benzoylperoxid; Benzoylperoxyde; BPO; Brevoxyl; BZF-60; Cadat BPO; Cadox 40E; Cadox B; Cadox B 40E; Cadox B 50P; Cadox B 70W; Cadox B-CH 50; Cadox BS; CCRIS 630; Chaloxyd BP 50FT; Clear By Design; Clearasil Antibacterial Acne Lotion; Clearasil BP Acne Treatment Cream; Debroxide; Desanden; Desquam E; Desquam X; Dibenzoyl peroxide; Dibenzoylperoxid; Dibenzoyl-peroxyde; Diphenylglyoxal peroxide; Dry and Clear; Duresthin 5; EINECS 202-327-6; Eloxyl; Epi-Clear; Epi Clear Antiseptic Lotion; Fostex; Fostex BPO; G 20; Garox; HSDB 372; Incidol; Loroxide; Lucidol; Lucidol (peroxide); Lucidol 50P; Lucidol-70; Lucidol B 50; Lucidol G 20; Lucidol KL 50; Luperco AA; Luperco AST; Luperox FL; Luzidol; Mytolac; Nayper B and BO; Nayper bo; Norox bzp-250; Norox bzp-C-35; Novadelox; NSC 671; Oxy-10; Oxy-10 Cover; Oxy-5; Oxylite; PanOxyl; Perossido di benzoile; Peroxide, dibenzoyl; Peroxyde de benzoyle; Persa-Gel; Persadox; pHisoAc BP; Quinolor compound; Resdan Akne; Stri-dex B.P.; Sulfoxyl; Superox 744; Theraderm; Topex; Vanoxide; Xerac; Xerac BP. Catalyst, initiator, oxidizing agent and curing agent. An alternative to MEK peroxide. Benzoyl peroxide pastes or granules are used as initiators for catalysis of unsaturated polyester resins and are well-suited for spray applications. Used medically as a keratolytic in the treatment of *acne vulgaris*. Crystals; mp = 103-106°, explosive at higher temperatures; λ$_m$ = 235, 275 nm (dioxane); sparingly soluble in H$_2$O, EtOH, soluble in C$_6$H$_6$, CHCl$_3$, Et$_2$O, CS$_2$ (2.5 g/100 ml); olive

oil (2 g/100 ml). *Bristol-Myers Squibb Co.; Catalyst Resources Inc.; Dermik Labs. Inc.; Galderma Labs Inc.; Hyland Div. Baxter; Ortho Pharm. Corp.; SmithKline Beecham Pharm.; Sterling Winthrop Inc.; Stiefel Labs Inc.; Taiho; Westwood-Squibb Pharm. Inc.*

1126   Dibenzoyl-p-quinone-dioxime
120-52-5   G

C$_{20}$H$_{14}$N$_2$O$_4$
1,4-Bis(benzoyloxyimino)cyclohexa-2,5-diene.
Benzoquinone dioxime dibenzoate; p-Benzoquinone bis(O-benzoyloxime); p-Benzoquinone dioxime dibenzoate; 2,5-Cyclohexadiene-1,4-dione, bis(O-benzoyloxime); Dibenzo G-M-F; Dibenzoyl-p-benzoquinone dioxime; Dibenzoylquinone dioxime; p-Dibenzoylquinone dioxime; p,p'-Dibenzoylquinone dioxime; Dilbenzo P Q D; EINECS 204-403-4; NSC 113483; Quinone dioxime dibenzoate; p-Quinone dioxime dibenzoate; Rhenocure BQ; Vulnoc DGM. A non-sulfur vulcanizing agent for natural, SBR, butyl and EPDM rubber; used to impart heat resistance to tyre curing bags, gaskets and wire insulation, also used as a coagent with peroxide curatives. *Uniroyal.*

1127   Dibenzylamine
103-49-1   3035   G

C$_{14}$H$_{15}$N
N-(Phenylmethyl)benzenemethanamine.
Accelerator DBA; AI3-15327; Benzenemethanamine, N-(phenylmethyl)-; (N-Benzylaminomethyl)benzene; Bi-benzylamine; DBA; Dibenzylamine; Dibenzylamine; EINECS 203-117-7; N-Benzylbenzylamine; N,N-Dibenylamine; NSC 4811. A proprietary rubber vulcanizing accelerator. Crystals; mp = -26°; bp = 300° (dec), bp$_{250}$ = 270°, bp$_{10}$ = 160-163°; d^{22} = 1.0256; λ$_m$ = 252 nm (MeOH); insoluble in H$_2$O, soluble in CCl$_4$, very soluble in EtOH, Et$_2$O.

1128   Dibenzylsulfoxide
621-08-9   G

C$_{14}$H$_{14}$OS
Bis(phenylmethyl) sulfoxide.
4-06-00-02651 (Beilstein Handbook Reference); AI3-62190; Benzene, 1,1'-(sulfinylbis(methylene))bis-; Benzyl sulfoxide; BRN 2049262; Dibenzyl sulfoxide; Dibenzyl sulphoxide; EINECS 210-668-7; NSC 55; Prevento® Cl 5; Sulfoxide, dibenzyl; Tardiol D. An organic inhibitor for use in cleansing acids and in the surface treatment of metals. Leaflets; mp = 134°; bp = 210° (dec); λ$_m$ = 220, 253, 260, 266, 270 nm (MeOH); insoluble in H$_2$O, soluble in Et$_2$O,

very soluble in EtOH. *Bayer AG.*

1129 1,4-Dibromobutane
110-52-1   G

$C_4H_8Br_2$

Tetramethylene dibromide.
4-01-00-00267 (Beilstein Handbook Reference); AI3-14617; BRN 1071199; Butane, 1,4-dibromo-; DBB; 1,4-Dibrombutan; 1,4-Dibromobutane; α,ω-Dibromobutane; EINECS 203-775-5; NSC 71435; Tetramethylene dibromide; Tetramethylenebromide. Used in chemical synthesis. Liquid; mp = -20°; bp = 197°; d = 1.8080; insoluble in H_2O, slightly soluble in CCl_4, soluble in $CHCl_3$.organic solvents. *Humphrey; Janssen Chimica.*

1130 Dibromo-3-nitrilopropionamide
10222-01-2   G P

$C_3H_2Br_2N_2O$

2,2-Dibromo-2-cyanoacetamide.
3-02-00-01641 (Beilstein Handbook Reference); Acetamide, 2-cyano-2,2-dibromo-; Acetamide, 2,2-dibromo-2-cyano-; Amerstat® 300; Biosperse® 240; BRN 1761192; Caswell No. 287AA; Dbnpa; 2,2-Dibromo-2-carbamoylacetonitrile; Dibromocyanoacetamide; 2,2-Dibromo-2-cyanoacetamide; 2,2-Dibromo-3-nitrilo-propionamide; EINECS 233-539-7; EPA Pesticide Chemical Code 101801; HSDB 6982; NSC 98283; XD-1603; XD-7287I Antimicrobial; XD 7287L; 2,2-Dibromo-2-carbamoylacetonitrile; 2,2-Dibromo-2-cyanoacet-amide; 2,2-Dibromo-3-nitrilopropionamide; NSC 98283; Slimicide 508; XD-1603; XD-7287I Antimicrobial; XD 7287L. Paper mill slimicide, antimicrobial agent for enhanced oil recovery systems; preservative for metal working fluids containing water, antimicrobial for control of bacteria and algae in recirculating cooling water systems, air washer systems, evaporative condensers. Fungicide. Registered by EPA as an antimicrobial, fungicide and herbicide. *AmeriBrom Inc.; Aqua-Serv Engineers Inc.; Baker Petrolite Corp.; Betzdearborn; Buckman Laboratories Inc.; Chemtreat Inc; Clearwater Inc.; Dow Chem. U.S.A.; Drew Industrial Div.; Hercules Inc.; Nalco Diversified Technologies, Inc.; Ondeo Nalco Co.; Verichem Inc.; Walling Water Management.*

1131 Dibromopentaerythritol
3296-90-0   G H

$C_5H_{10}Br_2O_2$

2,2-Dibromomethyl-1,3-propanediol.
4-01-00-02554 (Beilstein Handbook Reference); BRN 1304582; CCRIS 5972; Dibromoneopentyl glycol; Dibromopentaerythritol; DPNG; EINECS 221-967-7; Emery® 9336; FR-522; FR-521; FR 1138; HSDB 4184; NCI-C55516; NSC 9001; Pentaerythritol dibromide; Pentaerythritol dibromohydrin. Flame retardant for unsaturated polyesters, rigid PU and foams; flame retardant intermediate. Crystals; mp=109°; d = 2.23; LD_{50} (rat orl) = 3450 mg/kg. *Albemarle Corp.; AmeriBrom Inc.; Amerihaas; Dead Sea*

Bromine; Sigma-Aldrich Fine Chem.

1132 2,4-Dibromophenol
615-58-7   G

$C_6H_4Br_2O$

Phenol, 2,4-dibromo-.
4-06-00-01061 (Beilstein Handbook Reference); AI3-15480; BRN 1861291; DBP; 2,4-Dibromophenol; EINECS 210-436-5; Emery® 9331; FR-612; NSC 6213. Flame retardant for epoxy resins, phenolic resins, polyester resins; flame retardant intermediate. Needles; mp = 38°; bp = 238.5°; d^{20} = 2.07; λmLks = 223, 289 nm (ε = 9780, 2610, MeOH); slightly soluble in H_2O, CCl_4, very soluble in EtOH, Et_2O, C_6H_6, CS_2; LD_{50} (mus orl) = 282 mg/kg. *AmeriBrom Inc.; Dead Sea Bromine; Esprit Chem. Co.; Fluka; Great Lakes Fine Chem.; Rohm & Haas Co.; Sigma-Aldrich Fine Chem.*

1133 Dibromopropanol
96-13-9   G

$C_3H_6Br_2O$

2,3-Dibromo-1-propanol.
4-01-00-01446 (Beilstein Handbook Reference); AI3-26304; allyl alcohol dibromide; BRN 1719127; Brominex 257; CCRIS 940; 1,2-Dibromopropan-3-ol; 2,3-Dibromo-1-propanol; 2,3-Dibromopropan-1-ol; 2,3-Dibromo-propanol; 2,3-Dibromopropyl alcohol; DBP (flame retardant); Dibromopropanol; EINECS 202-480-9; HSDB 2879; NCI-C55436; NSC 6203; 1-Propanol, 2,3-dibromo-; USAF DO-42. Intermediate in preparation of flame retardants, insecticides, and pharmaceuticals. Liquid; bp_{10} = 94-97°; d = 2.1200. *ICI Americas Inc.; Lancaster Synthesis Co.; Sigma-Aldrich Fine Chem.*

1134 Di-t-butoxydiacetoxysilane
13170-23-5   G

$C_{12}H_{24}O_6Si$

Di-*t*-butoxydiacetoxysilane.
Acetic acid, CD4153; dianhydride with silicic acid (H_4SiO_4) bis(1,1-dimethylethyl) ester; Di-t-butoxy-diacetoxysilane; Diacetoxydi-t-butoxysilane; Dynasylan® BDAC; EINECS 236-112-3; Silane, di(tert-butoxy)-diacetoxy-. Coupling agent, chemical intermediate, blocking agent, release agent, lubricant, primer, reducing agent. *Degussa-Hüls Corp.*

1135 Dibutoxyethoxyethyl adipate
141-17-3   H

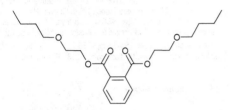

$C_{22}H_{42}O_8$
Bis(2-(2-butoxyethoxy)ethyl) adipate.
3-02-00-01718 (Beilstein Handbook Reference); Bisoflex 111; BRN 1808453; Dibutoxyethoxyethyl adipate; EINECS 205-465-5; HSDB 5480; Plasthall 226S; Plasthall DBEEA; Reomol BCD; RX 11806; Thiokol TP 759; Thiokol TP 95; TP 759; TP-95; Wareflex.

1136   Dibutoxyethyl phthalate
117-83-9   G

$C_{20}H_{30}O_6$
n-Butyl glycol phthalate.
4-09-00-03242 (Beilstein Handbook Reference); AI3-00524; 1,2-Benzenedicarboxylic acid, bis(2-butoxyethyl) ester; Bis(2-butoxyethyl) phthalate; BRN 2006754; β-Butoxyethyl phthalate; Butyl cellosolve phthalate; Butyl glycol phthalate; Di-(2-Butoxyethyl) fster kyseliny ftalove; Di-(2-butoxyethyl)ester kyseliny ftalove; Di(2-butoxyethyl) phthalate; Dibutoxyethyl phthalate; Dibutyl cellosolve phthalate; Dibutylcellosolve ftalat; Dibutylglycol phthalate; EINECS 204-213-1; Ethanol, 2-butoxy-, phthalate (2:1); HSDB 5402; Kesscoflex; Kesscoflex BCP; Kronisol; NSC 4840; Palatinol; Palatinol K; Phthalic acid, bis(2-butoxyethyl) ester; Phthalic acid, dibutoxyethyl ester; Plasthall® 200; Plasthall® DBEP. Plasticizer for PVC, polyvinyl acetate, and other resins. Viscous liquid; bp = 270°; d = 1.06. Ashland; Hall C.P.; Unitex.

1137   Dibutylamine
111-92-2   3058   H

$C_8H_{19}N$
Di-(n-butyl)amine.
4-11-00-00122 (Beilstein Handbook Reference); AI3-15329; AI3-52649; BRN 0506001; 1-Butanamine, N-butyl-; Dibutilamina; Dibutylamine; Di-n-butylamine; Di-(n-butyl)amine; EINECS 203-921-8; HSDB 310; UN2248. Liquid; mp = -62°; bp = 159.6°; d^{20} = 0.7670; λ_m = 30 nm (ε = 5, EtOH); soluble in H_2O, Me_2CO, C_6H_6, very soluble in EtOH, Et_2O.

1138   2,4-Di-t-butyl-6-(5-chlorobenzotriazol-2-yl)phenol
3864-99-1   G

$C_{20}H_{24}ClN_3O$
2-(3',5'-Di-t-butyl-2'-hydroxyphenyl)-5-chlorobenzo-triazole.
2-(3',5'-Di-t-butyl-2'-hydroxyphenyl)-5-chloro-2H-benzotriazole; EINECS 223-383-8; Phenol, 2-(5-chloro-2H-benzotriazol-2-yl)-4,6-bis(1,1-dimethylethyl)-; Uvazol 237. Efficient UV absorber used to protect polyolefins, PVC, unsaturated polyester, acrylics, ABS. Crystals; mp = 150-153°. EniChem SpA.

1139   Di-t-butyl dicarbonate
24424-99-5   G

$C_{10}H_{18}O_5$
BOC anhydride.
Bis(tert-butoxycarbonyl)oxide; CCRIS 2598; Di-t-butyl dicarbonate; Di(tert-butyl) carbonate; Dicarbonic acid, bis(1,1-dimethylethyl) ester; EINECS 246-240-1. Plasticizer. mp = 23°; bp0.5 - 56-57°. Fluka; Lancaster Synthesis Co.; Sigma-Aldrich Fine Chem.

1140   Dibutyldichlorotin
683-18-1   H

$C_8H_{18}Cl_2Sn$
Di-n-butyltin dichloride.
CCRIS 6321; Chlorid di-n-butylcinicity; Di-n-butyl-zinn-dichlorid; Di-n-butyltin dichloride; Dibutyldichloro-stannane; Dibutyldichlorotin; Dibutylstannium dichloride; Dibutyltin chloride; Dibutyltin dichloride; Dichlorodibutylstannane; Dichlorodibutyltin; EINECS 211-670-0; HSDB 6071; NSC 2604; Stannane, dibutyldichloro-; Tin, dibutyl-, dichloride.

1141   Di-t-butyl diperphthalate;
2155-71-7   G

$C_{16}H_{22}O_6$
Di-t-butyl peroxyphthalate.
1,2-Benzenedicarboperoxoic acid, bis(1,1-dimethylethyl) ester; Di-(terc-butilperoxi)ftalato; Di-t-butyl diperoxy-phthalate; Di-t-butyl perphthalate; Diperoxyphtalate de tert-butyle; Diperoxyphthalic acid, di-t-butyl ester; EINECS 218-454-5; NSC 43583; Perbutyl MA; Peroxyphthalic acid, di-t-butyl ester; Trigonox® 111-B40.

191

Polymerization initiator. *Akzo Chemie.*

1142   Dibutyl ether
142-96-1   1568   H

C8H18O
Di-n-butyl ether.
4-01-00-01520 (Beilstein Handbook Reference); AI3-00402; BRN 1732752; Butane, 1,1'-oxybis-; Butyl ether; Butyl oxide; CCRIS 6010; Dibutyl ether; Di-n-butyl ether; EINECS 205-575-3; Ether butylique; HSDB 306; NSC 8459. Liquid; mp = -95.2°; bp = 140.2°; d^{20} = 0.7684; insoluble in H2O, slightly soluble in CCl4, very soluble in Me2CO, freely soluble in EtOH, Et2O.

1143   Dibutyl fumarate
105-75-9   G

C12H20O4
2-Butenedioic acid (E)-, dibutyl ester.
4-02-00-02210 (Beilstein Handbook Reference); AI3-09505; BRN 1726635; 2-Butenedioic acid (2E)-, dibutyl ester; 2-Butenedioic acid, dibutyl ester, (E)-; Butyl fumarate; Dibutyl fumarate; Dibutylester kyseliny fumarove; EINECS 203-327-9; Fumaric acid, di-n-butyl ester; Fumaric acid, dibutyl ester; NSC 140; RC Comonomer DBF; Stafex DBF; Staflex DBF. Used in monomeric plasticizers copolymers and as a chemical intermediate. Liquid; mp = -13.5°; bp = 285°, bp4 = 150°; d^{20} = 0.9775; insoluble in H2O, soluble in CHCl3, Me2CO. *AC Ind. Inc.; Monomer-Polymer & Dajac; Penta Mfg.; Unitex.* 203-327-9

1144   Dibutyl hydrogen phosphite
1809-19-4   G H

C8H19O3P
Phosphonic acid, dibutyl ester.
4-01-00-01525 (Beilstein Handbook Reference); BRN 1099706; Butyl alcohol, hydrogen phosphite; Butyl phosphonate ((BuO)2HPO); Di-n-butyl hydrogen phos-phite; Dibutoxyphosphine oxide; Dibutyl hydrogen phosphite; Dibutyl hydrogen phosphonate; Dibutyl phosphite; Dibutyl phosphonate; Dibutylfosfit; EINECS 217-316-1; HSDB 2597; Mobil DBHP; NSC 2668; Phosphonic acid, dibutyl ester; Phosphorous acid, dibutyl ester; Syn-O-Ad® P-316. Used as amine salts in mineral and synthetic base stocks; as load-carrying additives with secondary activity as low-temperature stabilizers and metal deactivators. Liquid; bp11 = 118-119°; d = 0.9950. *Akzo Chemie.*

1145   Dibutylhydroquinone
88-58-4   G

C14H22O2
2,5-Di-*t*-butylhydroquinone.
4-06-00-06074 (Beilstein Handbook Reference); AI3-16630; Antage DBH; 1,4-Benzenediol, 2,5-bis(1,1-dimethylethyl)-; 2,5-Bis(1,1-dimethylethyl)-1,4-benzene-diol; BRN 2049542; CCRIS 5218; DBH; 1,4-Dihydroxy-2,5-di-t-butylbenzene; 2,5-Di-t-butylhydroquinone; 2,5-Di-t-butylbenzene-1,4-diol; 2,5-Di-t-butyl-1,4-benzene-diol; 2,5-Di-t-butyl-1,4-benzohydroquinone; 2,5-Di-t-butyl-1,4-hydroquinone; 2,5-Di-t-butylhydroquinone; 2,5-Di-t-butylquinol; Di-t-butylhydroquinone; Dibug; DTBHQ; Dybug; Eastman® DTBHQ; EINECS 201-841-8; Hydroquinone, 2,5-di-t-butyl-; Naugard 451; Nocrac NS 7; Nonflex Alba; NSC 11; Santovar O. Antioxidant for rubber, polyesters. Polymerization inhibitor, stabilizer against UV deterioration of rubber. Solid; mp=215-220°; insoluble in H2O, soluble in organic solvents; flash point = 216°. *Eastman Chem. Co.; Fluka.*

1146   Dibutyl maleate
105-76-0   G H

C12H20O4
2-Butenedioic acid (2Z)-, dibutyl ester.
4-02-00-02209 (Beilstein Handbook Reference); AI3-00644; Bisomer DBM; BRN 1726634; Butyl maleate; CCRIS 4136; DBM (VAN); Dibutyl maleate; Dibutylester kyseliny maleinove; Dibutylmaleate; EINECS 203-328-4; Maleic acid, dibutyl ester; NSC 6711; Octomer DBM; PX-504; RC Comonomer DBM; Staflex DBM. A plasticizer and intermediate in the manufacture of copolymers and plasticizers. Oil; bp4 = 129°; d = 0.9930; LD50 (rat orl) = 3730 mg/kg. *Aristech; BP Chem.; Penta Mfg.; Pentagon Chems. Ltd.; Unitex.*

1147   Di-t-butyl-p-methylphenol
128-37-0   1547   G H

C15H24O
2,6-Di-t-butyl-4-methylphenol.
Advastab 401; Agidol; Agidol 1; AI3-19683; Alkofen BP; Antioxidant 4; Antioxidant 30; Antioxidant 29; Antioxidant 264; Antioxidant 4K; Antioxidant DBPC; Antioxidant KB; Antioxidant MPJ; Antioxidant T 501; Anti-Oxydant Bayer; Antox QT; Antrancine 8; AO 4; AO 4K; AO 29; AOX 4; AOX 4K; BAT; BHT; BHT 264; BUKS; Butylated hydroxytoluene; Butylated hydroxytoluol; Butylhydroxytoluene; Butylohydroksytoluenu; CAO 1; CAO 3; Caswell No. 291A; Catalin antioxydant 1; Catalin cao-3; CCRIS 103; Chemanox 11; Dalpac; DBPC; Deenax; Dibunol; Dibutylated hydroxytoluene; EINECS 204-881-4; EPA Pesticide Chemical Code 022105; FEMA No. 2184; HSDB 1147; Impruvol; Ionol; Ionol (antioxidant); Ionol 1; Ionol CP; Ionol CP-antioxidant; Ionole; Kerabit; NCI-C03598; Nocrac 200;

Nonox TBC; NSC 6347; Oxyguard; P21; Parabar 441; Paranox 441; Ralox® BHT food grade; Stavox; Sumilizer BHT; Sustane; Sustane BHT; Swanox BHT; Tenamen 3; Tenamene 3; Tenamine 3; Tenox BHT; Tonarol; Topanol; Topanol O; Topanol OC; Toxolan P; Vanlube PC; Vanlube PCX; Vanox® PCX; Vianol; Vulkanox® KB. Antioxidant and antiskinning agent for foods, animal feed, petrol. products, synthetic rubbers, plastics, soaps; antiskinning agent in paints and inks. A nonstaining antioxidant for plastics in contact with food, rubber, latex, adhesives, hot melts, mineral oil processing (lubricants), petrol., feedstuffs. Oxidation inhibitor in soaps and cosmetics. Nondiscoloring/nonstaining antioxidant for light colored and transparent natural and synthetic rubber goods; used in fabric proofings, toys, bathing, latex, and dipped goods; stabilizer in emulsion and solution-polymerized elastomers (SBR, NBR, BR, IR); costabilizer for NBS. Crystals; mp = 71°; bp = 265°; d^{75} = 0.8937, d_4^{20} 1.048; λ_m = 227, 277, 283 nm (ε = 5623, 2188, 2188, isooctane); insoluble in H_2O, alkali, soluble in EtOH, Me_2CO, C_6H_6, petroleum ether; LD_{50} (mus orl)= 1040 mg/kg. *Bayer AG; Eastman Chem. Co.; ExxonMobil Chem. Co.; Naugatuck; Penta Mfg.; PMC; Raschig GmbH; Uniroyal; Vanderbilt R.T. Co. Inc.*

1148   Di-t-butyl peroxide
110-05-4   3494   G H

$C_8H_{18}O_2$
Bis(1,1-dimethylethyl)peroxide.
Aztec® Di-*t*-Butyl Peroxide; Bis(1,1-dimethylethyl)-peroxide; Bis(t-butyl)peroxide; Bis(tert-butyl) peroxide; t-Butyl peroxide; tert-Butyl peroxide; Cadox; Cadox TBP; CCRIS 4613; Di-t-butyl peroxide; Di-t-butyl peroxide; Di-t-butyl peroxyde; Di-t-Butyl hydroperoxide; Di-t-butylperoxi; DTBP; EINECS 203-733-6; HSDB 1326; Interox DTB; Kayabutyl D; NSC 673; Perbutyl D; Perossido di butile terziario; Peroxide, bis(1,1-dimethylethyl); Peroxyde de butyle tertiaire; Trigonox® B. Polymerization catalyst for resins (e.g., olefins, styrene, styrenated alkyds, silicones); ignition accelerator for diesel fuel; organic synthesis; intermediate. Initiator for LDPE polymerization; lower molecular weight at higher temperatures.; high-temperature peroxide used as finishing initiator for styrenics; crosslinking agent for olefin copolymers; volatile liquid best suited when used in injected extruder application. Liquid; mp = -40°; bp = 111° (dec explosively), bp_{284} = 80°; d^{20} = 0.704; insoluble in H_2O (0.01 g/100 ml), soluble in CCl_4, ligroin, freely soluble in Me_2CO; LD_{50} (rat orl) = 7710 mg/kg. *Akzo Chemie; Catalyst Resources Inc.; CK Witco Chem. Corp.; Elf Atochem N. Am.*

1149   Di-t-butylphenol
128-39-2   G H

$C_{14}H_{22}O$
2,6-Bis(1,1-dimethylethyl)phenol.
AI3-26293; AN 701; CCRIS 5828; Di-t-butylphenol; EINECS 204-884-0; Ethanox 701; Ethyl 701; Ethyl AN 701; Hitec 4701; HSDB 5616; Isonox 103; NSC 49175; Phenol, 2,6-bis(1,1-dimethylethyl)-; Phenol, 2,6-di-t-butyl-. Used as an intermediate; antioxidant, stabilizer for foods and in aviation and other gasolines. Prisms; mp = 39°; bp = 253°, bp_{50} - 161°, bp_{20} = 133°; d^{20} = 1.5001; λ_m = 271 nm (MeOH); insoluble in alkali, slightly soluble in EtOH, soluble in CCl_4;

LD_{50} (mus orl) = 120 mg/kg. *Allchem Ind.; Daicel (U.S.A.) Inc.; Ethicon Inc.; Penta Mfg.*

1150   2,4-Di-t-butylphenol
96-76-4   H

$C_{14}H_{22}O$
Phenol, 2,4-bis(1,1-dimethylethyl)-.
4-06-00-03493 (Beilstein Handbook Reference); Antioxidant No. 33; BRN 1910383; EINECS 202-532-0; NSC 174502; Phenol, 2,4-bis(1,1-dimethylethyl)-; Phenol, 2,4-di-t-butyl-; Prodox 146; Prodox 146A-85X. Solid; mp = 56.5°; bp = 263.5°; λ_m = 218, 225, 277, 283 nm (ε = 6740, 6940, 2360, 2130, MeOH); poorly soluble in CCl_4.

1151   N,N'-Di-s-butyl-p-phenylenediamine
101-96-2   G H

$C_{14}H_{24}N_2$
N,N'-Di-sec-butyl-p-phenylenediamine.
4-13-00-00111 (Beilstein Handbook Reference); Antioxidant 22; BRN 2805827; CCRIS 4603; CP 40182; Du Pont Gasoline Antioxidant No. 22; EINECS 202-992-2; HSDB 5343; Kerobit BPD; Naugalube® 403; NSC 68417; N,N'-Di-sec-butyl-p-phenylenediamine; p-Phenylenediamine, N,N'-di-sec-butyl-; Santoflex 44; Tenamene 2; Topanol® M; UOP 5. Acrylic polymerization inhibitor; gasoline antioxidant/sweetener. Solid; mp = 18°; insoluble in H_2O (<1 mg/ml), soluble in organic solvents; LDLo (rat orl)= 200 mg/kg. *ICI; ICI Americas Inc.; ICI Chem. & Polymers Ltd.; Uniroyal.*

1152   Dibutyl phenyl phosphate
2528-36-1   H

$C_{14}H_{23}O_4P$
Phosphoric acid, dibutyl phenyl ester.
4-06-00-00710 (Beilstein Handbook Reference); BRN 2140519; Butyl phenyl phosphate ((BuO)2(PhO)PO); CCRIS 4604; Di(n-butyl) phenyl phosphate; Dibutyl phenyl phosphate; EINECS 219-772-7; HSDB 2604; Phosphoric acid, dibutyl phenyl ester.

1153   Dibutyl phthalate
84-74-2   1586   G H

C16H22O4

Di-n-butyl 1,2-benzenedicarboxylate.

4-09-00-03175 (Beilstein Handbook Reference); AI-3-00283; Araldite 502; Benzene-o-dicarboxylic acid di-n-butyl ester; Benzenedicarboxylic acid, dibutyl ester; BRN 1914064; Butyl phthalate; Caswell No. 292; CCRIS 2676; Celluflex DPB; DBP; DBP (ester); Di-n-butyl phthalate; Di-n-butylester kyseliny ftalove; Dibutyl 1,2-benzenedicarboxylate; Dibutyl-o-phthalate; Dibutyl Phthalate; EINECS 201-557-4; Elaol; EPA Pesticide Chemical Code 028001; Ergoplast FDB; Ersoplast FDA; Genoplast B; Hatcol DBP; Hexaplas M/B; HSDB 922; Kodaflex DBP; NSC 6370; Palatinol C; Phthalate, di-n-butyl; Phthalate, dibutyl-; Phthalic acid, dibutyl ester; Polycizer DBP; PX 104; RC Plasticizer DBP; RCRA waste number U069; Staflex DBP; Uniflex DBP; Unimoll db; Witcizer 300. Plasticizer in nitrocellulose, lacquers, elastomers, explosives, nail polishes; solvent for perfumes, oils; perfume fixative; textile lubricating agent. Used in coatings industry as primary plasticizer-solvent for nitrocellulose lacquers; for rubbers and CAB, ethyl cellulose, PVAc, and synthetic resins; solvent for oil-soluble dyes, insecticides, peroxides, and organic compounds; antifoamer and fiber lubricant in textile manufacturing. Also used as an insect repellent. Registered by EPA (cancelled). Oily liquid; mp = -35°; bp = 340°; d^{20} = 1.0465; slightly soluble in H2O (0.04 g/100 ml), very soluble in EtOH, Et2O, C6H6, Me2CO; LD50 (rat orl) = 8000 mg/kg. *Aristech; Bayer AG; BP Chem.; Chisso Corp. USA; Daihachi Chem. Ind. Co. Ltd; Hall C.P.; Mitsubishi Gas; Sigma-Aldrich Co.; Sigma-Aldrich Fine Chem.; Unitex.*

1154   Dibutyl sebacate
109-43-3   E G

C18H34O4

Di-n-butyl sebacate.

4-02-00-02081 (Beilstein Handbook Reference); AI3-00393; Bis(n-butyl) sebacate; BRN 1798308; Butyl sebacate; Decanedioic acid, dibutyl ester; Di-n-butyl sebacate; Dibutyl 1,8-octanedicarboxylate; Dibutyl decanedioate; Dibutyl sebacate; Dibutyl sebacinate; Dibutylester kyseliny sebakove; EINECS 203-672-5; FEMA No. 2373; HSDB 309; Kodaflex DBS; Monoplex DBS; NSC 3893; Plasthall® DBS; Polycizer DBS; PX 404; Reomol DBS; Sebacic acid, dibutyl ester; Staflex DBS; Uniflex® DBS. Monomeric plasticizer for plastics and rubber (cellulosics, PVC food wraps, nitrite and neoprene rubbers). Used as an excipient in various pharmaceutical coating formulations. Viscous liquid; mp = -10°; bp = 344.5°, bp3 = 180°; d^{15} = 0.9405; insoluble in H2O, soluble in Et2O, CCl4. *Hall C.P.; Union Camp.*

1155   Di-t-butyl   1,1,4,4-tetramethyltetra-methylene diperoxide
78-63-7   G H

C16H34O4

2,5-Dimethyl-2,5-di(tert-butylperoxy)hexane.

AD; AD 40C; BRN 1863369; CCRIS 4626; CR 05; CT 8 (crosslinking agent); Di-t-butyl 1,1,4,4-tetramethyl-tetramethylene diperoxide; EINECS 201-128-1; HC 4 (peroxide); Hexane, 2,5-dimethyl-2,5-di(t-butylperoxy)-; Interox DHBP; Interox DHBP 45IC/G; Kayahexa AD; Kayahexa AD 40; Kayahexa AD 40C; Luperco 101X45; Luperco 101XL; Luperco 101; Lupersol 101; Lupersol 101XL; NSC 38203; Perhexa 2.5B; Perhexa 2.5B40; Perhexa 3M40; Peroxide, (1,1,4,4-tetramethyl-1,4-butanediyl)bis((1,1-dimethylethyl); Peroxide, (1,1,4,4-tetramethyltetramethylene)bis(tert-butyl; RC 4 (peroxide); TC 8 (catalyst); Trigonox 101; Trigonox 101-101/45; Trigonox XQ 8; Triqanox XQ 8; Varox; Varox 50; Varox DBPH 50; Varox Liquid. Initiator for polyester when molding at elevated temperatures; also used for crosslinking of olefin copolymers, chlorinated polyethylene, EPDM, SBR, and vinyls; initiator for free radical polymerizations of styrene; crosslinking agent for olefin copolymers. Flow modifier and processing aid for PP. Crosslinking agent for elastomers and thermoplastic resins. Liquid; mp = 8°; bp = 50-52°; insoluble in H2O; LD50 (mus ip) = 1700 mg/kg. *Akzo Chemie; Elf Atochem N. Am.*

1156   N,N'-Dibutylthiourea
109-46-6   G

C9H20N2S

1,3-dibutylthiourea.

4-04-00-00585 (Beilstein Handbook Reference); AI3-08621; BRN 0507434; EINECS 203-674-6; 1,3-Dibutylthiourea; 1,3-Di-n-butyl-2-thiourea; 1,3-Dibutyl-2-thiourea; NSC 3735; Pennzone B; Pennzone B 0685; Thiate U; Thiourea, N,N'-dibutyl-; Urea, 1,3-dibutyl-2-thio-; Urea, 1,3-di-N-butyl-2-thio-; USAF EK-2138. Accelerator for mercaptan-modified chloroprene rubber, an activator for ethylenepropylenediene terpolymers and natural rubber, an antidegradant for natural rubber-latex and thermoplastic styrene-butadiene rubber. Crystals; mp = 63-65°. *Elf Atochem N. Am.*

1157   Dibutyltin diacetate
1067-33-0   G

C12H24O4Sn

Diacetoxydibutylstannane.

BA 2726; Bis(acetyloxy)dibutylstannane; Caswell No. 293A; CCRIS 218; Diacetoxybutyltin; Diacetoxydibutyl-stannane; Diacetoxydibutyltin; Dibutyl tin diacetate; Di-n-Butyltin diacetate; Dibutylstannium diacetate; Dibutyltin diacetate; EINECS 213-928-8; Fomrez sul-3; HSDB 4115; Metacure® T-1; NCI-C02028; NSC 8786; Stannane, bis(acetyloxy)dibutyl-; T 1; Tin, dibutyl-, diacetate. Catalyst for use in production of PU coatings, adhesives, and sealants. Liquid; mp < 12°; bp5 = 139°; d = 1.3100; insoluble in H2O; LD50 (rat orl) = 32 mg/kg. *Air Prods & Chems.*

1158   Dibutyltin diisooctylthioglycolate
25168-24-5   H

C28H56O4S2Sn

Tin, dibutylbis((carboxymethyl)thio)-, diisooctyl ester.

Acetic acid, ((dibutylstannylene)dithio)di-, diisooctyl ester; Acetic acid, 2,2'-((dibutylstannylene)bis(thio))bis-, diisooctyl ester; Bis(2-ethylhexyloxycarbonylmethylthio)-dibutylstannane; BS 8T; BTS 70; BTS 8; Dibutyltin diisooctylthioglycolate; Dibutylzinn-S,S'-bis(isooctylthio-glycolat); Diisooctyl 2,2'-((dibutylstannylene)bis(thio))-diacetate; EINECS 246-703-8; Irgastab 17M; Mark 292; Nitto 1360; Ongrostab BS 8T; Stannane, bis(isooctyloxycarbonylmethylthio)dibutyl-; T 101; T 101 (accelerator); Thermolite 31; Tin, dibutyl-, bis(isooctylthioglycollate); Tin, dibutylbis((carboxy-methyl)thio)-, diisooctyl ester; TVS 1360.

1159   Dibutyltin maleate
78-04-6   G

C12H20O4Sn

2,2-Dibutyl-1,3,2-dioxastannepin-4,7-dione.

Advastab DBTM; Advastab T290; Advastab T340; BRN 4138538; BT 31; 2,2-Dibutyl-1,3,2-dioxastannepin-4,7-dione; Dibutyl(maleoyldioxy)tin; Dibutylstannylene maleate; Dibutyltin maleate; 1,3,2-Dioxastannepin-4,7-dione, 2,2-dibutyl-; EINECS 201-077-5; Irgastab DBTM; Irgastab T 4; Irgastab T 150; Irgastab T 290; KK 3206; KS 4B; MA300A; Mark BT 31; Mark BT 58; Markure UL2; Nuodex V 1525; OTS 227; Stancleret 157; Stann BM; Stann RC 40F; Stann RC 71A; Stavinor 1300SN; Stavinor Sn 1300; TK-10422; TN 1000; TN 3J; TVS 86SP5; TVS-MA 300; TVS-N 2000E. Used as a stabilizer for PVC resins and as a condensation catalyst. Solid; mp = 110°; insoluble in H2O; soluble in organic solvents. Elf Atochem N. Am.; Ferro/Bedford; Gelest.

1160   Dibutyltin oxide
818-08-6   H

C8H18ON

Dibutyloxostannane.

BRN 4126243; DBOT; Di-n-butyl-zinn-oxyd; Di-n-butyltin oxide; Dibutyloxide of tin; Dibutyloxostannane; Dibutyloxotin; Dibutylstannane oxide; Dibutylstannium oxide; Dibutyltin oxide; EINECS 212-449-1; Kyslicnik di-n-butylcinicity; NSC 28130; Stannane, dibutyloxo-; Tin, dibutyl-, oxide; Tin, dibutyloxo-.

1161   Dicamba
1918-00-9   3065   G H P

C8H6Cl2O3

3,6-Dichloro-2-methoxybenzoic acid.

AI3-27556; Banex; Banlen; Banvel; Banvel 480; Banvel 70WP; Banvel CST; Banvel herbicide; Banvel II herbicide; Banvel SGF; BRN 2453039; Brush buster; Caswell No. 295; CCRIS 1471; Compound B; Compound B dicamba; Dianat; Dicamba; EINECS 217-635-6; EPA Pesticide Chemical Code 029801; HSDB 311; Kyselina 3,6-dichlor-2-methoxybenzoova; MDBA; Mediben; Vanquish; Velsicol 58-CS-11; Velsicol compound R. Selective systemic herbicide which acts as an auxin-like growth regulator. Used for control of annual and perennial broad-leaved weeds and brush sp-ecies in cereals, maize, sorghum, sugar cane, asparagus, seed pastures, turf, pastures and rangeland. Used to control bracken. Registered by EPA as a herbicide. Colorless crystals; mp = 114-116°; sg^{25} = 1.57; soluble in H2O (0.65 g/100 ml), EtOH (92.2 g/100 ml), cyclohexanone (91.6 g/100 ml), Me2CO (81 g/100 ml), CH2Cl2 (26 g/100 ml), C7H8 (13 g/100 ml), xylene (7.8 g/100 ml); LD50 (rat orl) = 1707 mg/kg, (rbt der) > 2000 mg/kg, (mallard duck orl) = 2000 mg/kg; LC50 (mallard duck, bobwhite quail 8 day dietary) > 10000 mg/kg diet, (rainbow trout, bluegill sunfish 96 hr.) = 135 mg/l; non-toxic to bees. Diachem; Sandoz; Shell UK; Syngenta Crop Protection; Syngenta Professional Prod.

1162   Dicamba, isopropylamine salt
55871-02-8   P

C11H15Cl2NO3

Isopropylamine 3,6-dichloro-o-anisate.

Benzoic acid, 3,6-dichloro-2-methoxy-, compd with 2-propanamine (1:1); Dicamba, isopropylamine sal; Isopropylamine 3,6-dichloro-o-anisate; Propanamine, 3,6-dichloro-2-methoxybenzoate. Registered by EPA as a herbicide. Albaugh Inc.; BASF Corp.; Monsanto Co.

1163   Dicamba, potassium salt
10007-85-9   H P

C8H5Cl2KO3

Potassium 3,6-dichloro-2-methoxybenzoate.

Dicamba K; Dicamba-potassium; Dicamba, potassium salt; EINECS 233-002-7; Potassium 3,6-dichloro-o-anisate. Registered by EPA as a herbicide. Albaugh Inc.; BASF Corp.; Gharda USA Inc.; Micro-Flo Co. LLC.

1164   Dicapryl adipate
105-97-5   G

C26H50O4

Didecyl hexanedioate.

Adipic acid didecyl ester; Dicapryl adipate; Didecyl adipate; Di-n-Decyl adipate; Didecyl hexanedioate; EINECS 203-349-9; Hexanedioic acid, didecyl ester; NSC 4445; Polycizer 632; Uniflex® DCA; Unimate® DCA. Monomeric plasticizer for plastics and rubber. *Union Camp.*

1165　Dicapryl phthalate

131-15-7　　　　　　　　　　　　　　　　　　G

C24H38O4

Di-(2-octyl)phthalate.

4-09-00-03181 (Beilstein Handbook Reference); 1,2-Benzenedicarboxylic acid, bis(1-methylheptyl) ester; Bis(1-methylheptyl) phthalate; Bis(2-octyl)phthalate; Bis-(2-oktyl)ester kyseliny ftalove; BRN 2005093; Capryl o-phthalate; DCP; Dicapryl 1,2-benzenedicarboxylate; Dicapryl phthalate; Dioctanol-2-phthalate; Di-sec-octyl phthalate; EINECS 205-014-2; Genomoll 100; Monoplex DCP; NSC 3926; Phthalic acid, bis(1-methylheptyl) ester; Phthalic acid, bis(2-octyl) ester; Phthalic acid, dicapryl ester; Phthalic acid, di-2-octyl ester; Uniflex® DCP. A plasticizer for vinyl plastics and cellulosic resins. Oil; bp = 227-234°; d = 0.965; insoluble in H2O. *Union Camp.*

1166　Dichlofluanid

1085-98-9　　　　　　　　3070　　　　　　　G P

C9H11Cl2FN2O2S2

1,1-Dichloro-N-[(dimethylamino)sulfonyl]-1-fluoro-N-phenylmethanesulfenamide.

Aniline, N-((dichlorofluoromethyl)thio)-N-((dimethyl-amino)sulfonyl)-; BAY 47531; BRN 2947992; Caswell No. 297A; CCRIS 3647; Dichlofluanid; Dichlofluanide; Dichlorfluanid; 1,1-Dichloro-N-((dimethylamino)sulf-onyl)-1-fluoro-N-phenylmethanesulfenamide; Diparen; Elvaron; EPA Pesticide Chemical Code 128844; Eparen; Euparen; Euparene; HSDB 1565; KU 13-O32-C; KUE 13032c; Methanesulfenamide,1,1-dichloro-N-((dimethyl-amino)sulfonyl)-1-fluoro-N-phenyl-; N-(Dichlor-fluor-methyl-thio)-N',N'-dimethyl-N-phenyl-schwefel-säurediamid; N-(Dichlorofluoromethylthio)-N-(dimethyl-sulfamoyl)-aniline; N-Dichlorfluormethylthio-N',N'-di-methylaminosulfonsäureanilid; N-Dichloroacetyl-N-phenylsemicarbazide; N-Dichlorofluoromethane-sulphenyl-N',N'-dimethyl-N-phenylsulphamide; N-Di-chlorofluoromethylthio-N',N'-dimethyl-N-phenylsulph-amide; N,N-Dimethyl-N'-phenyl-N'-fluorodichloro-methylthiosulfamide; NSC 218451; Oiparen; Pecudin; Sulfamide, N-((dichlorofluoromethyl)thio)-N',N'-di-methyl-N-phenyl-. Fungicide with specific action against *Botrylis*. Colorless powder; mp

= 106°; soluble in H2O (0.00013 g/100 ml), CH2Cl2 (> 20 g/100 ml), C7H8 (10 - 20 g/100 ml), xylene (7 g/100 ml), i-PrOH (1 - 2 g/100 ml), MeOH (1.5 g/100 ml), C6H14 (0.2 - 0.5 g/100 ml; LD50 (rat orl) = 5000 mg/g, (mus orl) = 5464 - 5597 mg/kg, (fgpg orl) = 945 mg/kg, (frbt orl (3500 mg/lg), (rat der) > 5000 mg/kg, (Japanese quail, ckn orl) > 5000 mg/kg; LC50 (rat ihl 4 hr.) > 0.3 mg/l air, (goldfish 48 hr.) = 0.3 mg/l, (carp 48 hr.) = 0.25 mg/l, (Japanese ricefish 48 hr.) = 0.85 mg/l, (rainbow trout 96 hr.) = 0.05 mg/l, (golden orfe 96 hr.) = 0.12 mg/l; non-toxic to bees. *Bayer AG; Bayer Corp., Agriculture.*

1167　Dichlone

117-80-6　　　　　　　　3071　　　　　　　G P

C10H4Cl2O2

2,3-Dichloro-1,4-naphthalenedione.

AI3-03776; Algistat; Caswell No. 298; CCRIS 6667; CNQ; Compound 604; Dichlone; 2,3-Dichlor-1,4-naftochinon; 2,3-Dichlor-1,4-naphthochinon; 2,3-Dichloro-1,4-naphthalenedione; Dichloronaphtho-quinone; 2,3-Dichloronaphthoquinone; 2,3-Dichloro-1,4-naphthoquinone; Diclone; EINECS 204-210-5; ENT 3,776; EPA Pesticide Chemical Code 029601; HSDB 313; Latka 604; 1,4-Naphthalenedione, 2,3-dichloro-; 1,4-Naphthoquinone, 2,3-dichloro-; NSC 537; Phygon; Phygon paste; Phygon seed protectant; Phygon XL; Quintar; Quintar 540F; Sanquinon; Uniroyal; U.S. Rubber 604; USR 604. A fungicide used as a seed dressing, insecticide, organic catalyst. Registered by EPA as a fungicide (cancelled). Yellow needles; mp = 193°; insoluble in H2O, soluble in xylene (4 g/100 ml), o-dichlorobenzene (4 g/100 ml), Me2CO, Et2O, C6H6, dioxane; LD50 (rat orl) = 1300 mg/kg.

1168　Dichloramine T

473-34-7　　　　　　　　3073　　　　　　　D G

C7H7Cl2NO2S

N, N-Dichloro-p-toluenesulfonamide.

AI3-15331; Benzenesulfonamide, N,N-dichloro-4-methyl-; Benzyl p-sulfondichloramide; Dichloramine T; EINECS 207-462-4; N,N-Dichloro-p-toluolsulfonamide; N,N-Dichloro-p-toluenesulfonamide; N,N-Dichloro-toluene-4-sulphonamide; NSC 1130; p-Toluene-sulfonamide, N,N-dichloro-; p-Toluenesulfone dichlor-amide; Peraktivin. Sulfonamide antibiotic. Used as a disinfectant, germicide and antibacterial. Prisms; mp = 83°; λm = 228, 274 nm (MeOH); insoluble in H2O, soluble in EtOH, Et2O, C6H6, CCl4, CHCl3, AcOH, petroleum ether. *Shell.*

1169　Dichloroacetyl chloride

79-36-7　　　　　　　　3078　　　　　　　H

C2HCl3O

2,2-Dichloroacetyl chloride.

4-02-00-00504 (Beilstein Handbook Reference); Acetyl chloride, dichloro-; BRN 1209426; CCRIS 6011; Chlorid kyseliny dichloroctove; Chlorure de dichloracetyle; Dichloracetyl chloride; Dichloroacetyl chloride; Dichloroethanoyl chloride; EINECS 201-199-9; HSDB 5229; UN1765. Liquid; bp = 108°; d^{16} = 1.5315; very soluble in H2O, EtOH, Et2O; LD50 (rat orl) = 2460 mg/kg. *Clariant Corp.; Hoechst AG.*

1170   3,4-Dichloroaniline
95-76-1   3079   H

C6H5Cl2N
  Benzenamine, 3,4-dichloro-.
  4-12-00-01257 (Beilstein Handbook Reference); 4-Amino-1,2-dichlorobenzene; Aniline, 3,4-dichloro-; Benzenamine, 3,4-dichloro-; BRN 0636837; CCRIS 2395; DCA; 3,4-DCA; 4,5-Dichloroaniline; EINECS 202-448-4; HSDB 1319; LY 004892; NSC 247. Needles; mp = 72°; bp = 272°; λ_m = 247 nm (MeOH); soluble in EtOH, slightly soluble in C6H6, Et2O, CHCl3.

1171   2,5-Dichloroanisole
1984-58-3   H

C7H6Cl2O
  1,4-Dichloro-2-methoxybenzene.
  Anisole, 2,5-dichloro-; Benzene, 1,4-dichloro-2-methoxy-; 1,4-Dichloro-2-methoxybenzene; 2,5-Dichloroanisole; EINECS 217-852-6.

1172   m-Dichlorobenzene
541-73-1   3080   H

C6H4Cl2
  1,3-Dichlorobenzene.
  AI3-15517; Benzene, 1,3-dichloro-; Benzene, m-dichloro-; CCRIS 4259; m-DCB; m-Dichlorobenzene; m-Dichlorobenzol; EINECS 208-792-1; HSDB 522; NSC 8754; m-Phenylene dichloride; RCRA waste number U071. Liquid; mp = -24.8°; bp = 173°; d^{20} = 1.2884; λ_m= 216, 250, 256, 263, 270, 278 nm (ε = 12800, 830, 1420, 2330, 3340, 2800, MeOH); insoluble in H2O, soluble in EtOH, Et2O, C6H6, freely soluble in Me2CO, CCl4, ligroin.

1173   o-Dichlorobenzene
95-50-1   P

C6H4Cl2
  1,2-Dichlorobenzene.
  AI3-00053; Benzene, 1,2-dichloro-; Benzene, o-dichloro-; Caswell No. 301; CCRIS 1360; Chloroben; Chloroden; Cloroben; DCB; Dichloricide; Dichlorobenzene; 1,2-Dichlorobenzene; 2-Dichlorobenzene; Dichlorobenzene, 1,2-; o-Dichlorobenzene; Dichlorobenzene, ortho, liquid; Dichlorobenzol; o-Dichlorbenzol; Dilantin DB; Dilatin DB; Dizene; Dowtherm E; EINECS 202-425-9; EPA Pesticide Chemical Code 059401; HSDB 521; NCI-C54944; NSC 60644; ODB; ODCB; Ortho-dichlorobenzene; Orthodichlorobenzol; RCRA waste no. U070; Special termite fluid; Termitkil; UN1591. Registered by EPA as an antimicrobial and disinfectant (cancelled). Used as a solvent, deodorizer, degreasing agent, in metal polishes and as a heat transfer medium. Liquid; mp = -17°; bp = 180.5°; d_4^{20}= 1.3059, d_4^{25}= 1.3003; logP = 3.65; slightly soluble in H2O (0.0145 g/100 ml), soluble in EtOH, Et2O, C6H6. *Fluka; Sigma-Aldrich Co.*

1174   p-Dichlorobenzene
106-46-7   3082   G H P

C6H4Cl2
  1,4-Dichlorobenzene.
  AI3-0050; AI3-00050; Benzene, 1,4-dichloro-; Benzene, p-dichloro-; Caswell No. 632; CCRIS 307; DCB; Di-chloricide; Dichlorobenzene; Dichlorocide; Dicloro-benzene; EINECS 203-400-5; EPA Pesticide Chemical Code 061501; Evola; Globol; HSDB 523; Kaydox; NCI-C54955; NSC 36935; p-Chlorophenyl chloride; p-Dichloorbenzeen; p-Dichlorbenzol; p-Dichlorobenzene; p-Dichlorobenzol; p-Diclorobenzene; Para crystals; Para-zene; PARA; Paracide; Paradi; Paradichlorbenzol; Paradichlorobenzene; Paradichlorobenzol; Paradow; Paramoth; Paranuggets; PDB; PDCB; Persia-perazol; RCRA waste number U071; RCRA waste number U070; RCRA waste number U072; Santochlor. Used as an insecticidal fumigant and as a room deodorant. Intermediate in plastics manufacture. Registered by EPA as a bactericide, fungicide, insecticide and rodenticide. Used in the manufacture of pharmaceuticals and intermediates. Crystals; mp = 52.7°; bp = 174°; λ_m = 224, 264, 272, 280 nm (ε = 12800, 287, 377, 301, MeOH); insoluble in H2O, soluble in Et2O, C6H6, CHCl3, CS2, freely soluble in EtOH; LD50 (rat orl) = 500 mg/kg. *Ashland; Mitsubishi Gas; Monsanto Co.; Murphy Chemical Co Ltd.; PPG Ind.; Ruger; Sigma-Aldrich Fine Chem.; Spectrum Chem. Manufacturing.*

1175   3,3'-Dichlorobenzidine dihydrochloride
612-83-9   3084   H

C12H12Cl4N2
  (1,1'-Biphenyl)-4,4'-diamine, 3,3'-dichloro-, dihydro-chloride.
  AI3-22046; Benzidine, 3,3'-dichloro-, dihydrochloride; (1,1'-Biphenyl)-4,4'-diamine, 3,3'-dichloro-, dihydro-chloride; CCRIS 5899; 3,3'-Dichlorobenzidine dihydrochloride; EINECS 210-323-0; HSDB 5715; NSC 3524.

1176   3,4-Dichlorobenzotrifluoride
328-84-7   H

C7H3Cl2F3
 3,4-Dichloro-1-(trifluoromethyl)benzene.
 Benzene,   1,2-dichloro-4-(trifluoromethyl)-;   3,4-Dichloro-1-
(trifluoromethyl)benzene; 3,4-Dichloro-α,α,α-trifluoro-toluene; 3,4-
Dichlorobenzotrifluoride;   3,4-Dichloro-phenyltrifluoromethane;
EINECS 206-337-1; HSDB 6126; Toluene, 3,4-dichloro-α,α,α-
trifluoro-.

1177   2,4-Dichlorobenzyl alcohol
1777-82-8   3110   G P

C7H6Cl2O
 2,4-Dichlorobenzenemethanol.
 4-06-00-02597 (Beilstein Handbook Reference); Al3-20619;
Benzenemethanol, 2,4-dichloro-; Benzyl alcohol, 2,4-dichloro-; BRN
1448652; 2,4-Dichlorobenzyl alcohol; Dybenal; EINECS 217-210-5;
Myacide® SP; NSC 15635; Rapidosept. Antifungal agent, antiseptic
and preservative. Crystals; mp = 59.5°; bp25 = 150°; soluble in
CHCl3. *Boots Co.; Boots Pharmaceuticals Inc.; Inolex.*

1178   1,2-Dichloro-3-butene
760-23-6   H

C4H6Cl2
 (±)-1-Butene, 3,4-dichloro-.
 4-01-00-00772 (Beilstein Handbook Reference); BRN 1739135; (±)-
1-Butene, 3,4-dichloro-; (±)-1,2-Dichloro-3-butene; 1,2-Dichloro-3-
butene; 3,4-Dichloro-1-butene; EINECS 212-079-0; HSDB 5751.
Liquid; mp = -61°; bp = 116°; d^{20} = 1.1170; insoluble in H2O, soluble
in EtOH, Et2O, CCl4; very soluble in C6H6, CHCl3.

1179   2,3-Dichlorobutadiene
1653-19-6   H

C4H4Cl2
 2,3-Dichloro-1,3-butadiene.
 4-01-00-00986 (Beilstein Handbook Reference); BRN 1698807; 1,3-
Butadiene, 2,3-dichloro-; 2,3-Dichlor-1,3-butadien; 2,3-Dichloro-1,3-
butadiene; 2,3-Dichlorobuta-1,3-diene; 2,3-Dichlorobutadiene; 2,3-
Dichloro-butadiene-1,3; EINECS 216-721-0. Liquid; bp = 98°; d^{20} =

1.1829; very soluble in CHCl3.

1180   Dichlorobutane
616-21-7   H

C4H8Cl2
 1,2-Dichlorobutane.
 Butane, 1,2-dichloro-; 1,2-Dichlorobutane; EINECS 210-469-5;
HSDB 5717; NSC 93880. Liquid; bp = 124.1°; d^{25} = 1.1116;
insoluble in H2O, slightly soluble in CCl4, soluble in Et2O, CHCl3.

1181   1,4-Dichlorobutene
110-57-6   H

C4H6Cl2
 trans-1,4-Dichloro-2-but-2-ene.
 4-01-00-00787 (Beilstein Handbook Reference); Al3-52332; BRN
1719693; 2-Butene, 1,4-dichloro-, (E)-; CCRIS 2458; 1,4-Dichloro-
trans-2-butene;   trans-1,4-Dichloro-2-butene;   trans-1,4-
Dichlorobutene; EINECS 203-779-7; HSDB 1501. Liquid; mp = -1°;
bp = 155.4°; d^{25} = 1.183; very soluble in Me2CO, C6H6, EtOH, Et2O.

1182   1,4-Dichloro-cis-2-butene
1476-11-5   H

C4H6Cl2
 2-Butene, 1,4-dichloro-, (2Z)-.
 2-Butene, 1,4-dichloro-, (2Z)-; 2-Butene, 1,4-dichloro-, (Z)-; 2-
Butene, 1,4-dichloro-, cis-; CCRIS 2651; 1,4-Dichloro-cis-2-butene;
cis-1,4-Dichloro-2-butene; EINECS 216-021-5; HSDB 5832. Liquid;
mp = -48°; bp = 152.5°; d^{25} = 1.188; very soluble in Me2CO, EtOH,
Et2O, C6H6.

1183   1,3-Dichloro-2-butene
926-57-8   H

C4H6Cl2
 1,3-Dichlorobut-2-ene.
 BRN 0906754; 2-Butene, 1,3-dichloro-; 1,3-Dichloro-2-butene; 1,3-
Dichlorobut-2-ene;   1,3-Dichlorobutene-2;   1,3-Dichlorobutylene;
EINECS 213-138-3; HSDB 1500; NSC 8749.

1184   Dichloro(chloromethyl)methylsilane
1558-33-4   G H

C2H5Cl3Si
 Silane, dichloro(chloromethyl)methyl-.
 CC3275;   Chloromethyldichloromethylsilane;   Dichloro-

198

(chloromethyl)methylsilane; EINECS 216-319-5; Silane, dichloro(chloromethyl)methyl-. Coupling agent, chemical intermediate, blocking agent, release agent, lubricant, primer, reducing agent. Liquid; bp = 121.5°; d^{20} = 1.2858. *Degussa-Hüls Corp.*

1185 Dichlorodifluoromethane
75-71-8   3089   H

CCl_2F_2

Dichlorodifluoromethane.

AI3-01708; Algofrene type 2; Arcton 6; Arcton 12; Caswell No. 304; CCRIS 3501; CFC 12; Chlorofluorocarbon 12; Dichlorodifluoromethane; Diclorodifluometano; Difluorodichloromethane; Dwu-chlorodwufluorometan; Dymel 12; EINECS 200-893-9; Electro-CF 12; EPA Pesticide Chemical Code 000014; Eskimon 12; F 12; FC 12; FCC 12; FKW 12; Fluorocarbon-12; Forane 12; Freon 12; Freon F-12; Frigen 12; Genetron 12; Halocarbon 12; Halon; Halon 122; HSDB 139; Isceon 122; Isotron 2; Isotron 12; Kaiser chemicals 12; Ledon 12; Methane, dichlorodifluoro-; Propellant 12; R-12; R 12 (refrigerant); R12 [Nonflammable gas]; RCRA waste number U075; Refrigerant 12; Refrigerant R12; Ucon 12; Ucon 12/halocarbon 12; UN 1028. Registered by EPA as an insecticide (cancelled). Colorless gas; mp = -158°; bp = -29.8°, bp_{2atm} = -12.2°, bp_{5atm} = 16.1°, bp_{10atm} = 42.4°, bp_{20atm} = 74.0°, bp_{30atm} = 95.6°; stable below 550°; $d^{-29.8}_{liq}$ = 1.486; soluble in EtOH, Et2O, H2O.

1186 1,3-Dichloro-5,5-dimethyl hydantoin
118-52-5   3090   G

$C_5H_6Cl_2N_2O_2$

1,3-Dichloro-5,5-dimethyl-2,4-imidazolidinedione.

5-24-05-00373 (Beilstein Handbook Reference); AI3-23669; BRN 0146013; Caswell No. 306; CCRIS 5900; Dactin; Daktin; Dantochlor; Dantoin; Dantoin® DCDMH; DDH; Dichlorantin; 1,3-Dichloro-5,5-dimethyl hydantoin; 1,3-Dichloro-5,5-dimethyl-2,4-imidazolidine-dione; 1,3-Dichloro-5,5'-methylhydantoin; Dwuchlor-antyny; 1,3-Dwuchloro-5,5-dwumetylohydantoina; EINECS 204-258-7; EPA Pesticide Chemical Code 028501; Halane; HSDB 4373; Hydan; Hydan (antiseptic); Hydantoin, dichlorodimethyl-; Hydantoin, 1,3-dichloro-5,5-dimethyl-; 2,4-Imidazolidinedione, 1,3-dichloro-5,5-dimethyl-; NCI-C03054; NSC 33307; Omchlor; Sulfochloranthine . Intermediate for custom chemical synthesis, laundry bleach formulations, and automatic dishwashing compounds; disinfectant, industrial deodorant; stabilizer for vinyl chloride polymers; polymerization catalyst. Crystals; mp = 132°; soluble in H2O, soluble in chlorinated and highly polar solvents; LD50 (rat orl) = 542 mg/kg. *Lonzagroup.*

1187 Dichlorodimethylsilane
75-78-5   H

$C_2H_6Cl_2Si$

Dimethyldichlorosilane.

4-04-00-04111 (Beilstein Handbook Reference); AI3-51462; BRN 0605287; CCRIS 783; Dichloro-(dimethyl)silane; Dichlorodimethylsilane; Dichloro-dimethylsilicon; Dimethyl-dichlorsilan; Dimethyldi-chlorosilane; Dimethylsilane dichloride; EINECS 200-901-0; HSDB 361; Inerton AW-DMCS; Inerton dw-dmc; LS 130; NSC 77070; Repel-Silan; Silane, dichlorodimethyl-; UN1162. Coupling agent, chemical intermediate, blocking agent, release agent, lubricant, primer, reducing agent. Liquid; mp = -16°; bp = 70.3°; d^{25} = 1.064; soluble in H2O and in organic solvents. *Degussa-Hüls Corp.*

1188 Dichlorodimethyltin
753-73-1   H

$C_2H_6Cl_2Sn$

Dimethyltin dichloride.

CCRIS 6320; Cotin 210; Dichlorid dimethylcinicity; Dichlorodimethylstannane; Dichlorodimethyltin; Di-methyldichlorostannane; Dimethyldichlorotin; Dimethyl-tin dichloride; Dimethylzinndichloride; EINECS 212-039-2; NSC 55159; Stannane, dichlorodimethyl-; Tin, dimethyl-, dichloride.

1189 Dichlorodiphenylsilane
80-10-4   G H

$C_{12}H_{10}Cl_2Si$

Diphenyldichlorosilane.

4-16-00-01526 (Beilstein Handbook Reference); AI3-51461; BRN 0609882; CD5950; Dichlor-difenylsilan; Dichloro(diphenyl)silane; Dichlorodiphenylsilane; Di-phenyl dichlorosilane; Diphenyldichlorosilane; Diphenylsilicon dichloride; Diphenylsilyl dichloride; EINECS 201-251-0; HSDB 316; NSC 77110; Silane, dichlorodiphenyl-; TSL 8062; UN1769. Coupling agent, chemical intermediate, blocking agent, release agent, lubricant, primer, reducing agent. Liquid; mp = -22°; bp = 305°; d^{25} = 1.2040; soluble in EtOH, Et2O, Me2CO, C6H6, CCl4. *Degussa-Hüls Corp.*

1190 4,4'-Dichlorodiphenyl sulfone
80-07-9   H

$C_{12}H_8Cl_2O_2S$

4,4'-Dichlorodiphenyl sulfone.

4-06-00-01587 (Beilstein Handbook Reference); 4,4'-Dichlorodiphenyl sulfone; AI3-01386; AI3-02901; Benzene, 1,1'-sulfonylbis(4-chloro-; Bis(4-chlorophenyl) sulfone; Bis(4-chlorophenyl) sulphone; Bis(p-chlorophenyl) sulfone; BRN 2052955; Di-p-chlorophenyl sulfone; EINECS 201-247-9; HSDB 5233; NSC

23899; Sulfone, bis(p-chlorophenyl).

1191 1,1-Dichloroethane
75-34-3   3843   H

C2H4Cl2
Ethylidene chloride.
 Aethylidenchlorid; Äthylidenchlorid; CCRIS 224; Chlorinated hydrochloric ether; Chlorure d'ethylidene; Cloruro di etilidene; DCE; α,α-Dichloroethane; 1,1-Dichloorethaan; 1,1-Dichloraethan; 1,1-Dichlorethane; 1,1-Dichloroethane; 1,1-Dicloroetano; 1,1-Ethylidene dichloride; EINECS 200-863-5; Ethane, 1,1-dichloro-; Ethylidene chloride; Ethylidene dichloride; HSDB 64; NCI-C04535; RCRA waste number U076; UN2362. Used to make flexible plastic food wrap and flame retardant fabrics, in piping, coating for steel pipes, and adhesive applications. Found in many food and other packaging materials. Liquid; mp = -96.9°; bp = 57.4°; d^{20} = 1.1757; soluble in organic solvents. Dow Chem. U.S.A.; PPG Ind.

1192 1,1-Dichloroethene
75-35-4   10057   H

C2H2Cl2
Vinylidene chloride.
 AI3-28804; CCRIS 622; Chlorure de vinylidene; 1,1-Dichloroethene; 1,1-Dichloroethylene; as-Dichloro-ethylene; asym-Dichloroethylene; EINECS 200-864-0; Ethene, 1,1-dichloro-; Ethylene, 1,1-dichloro-; HSDB 1995; NCI-C54262; RCRA waste number U078; UN1303; VDC; Vinylidene chloride; Vinylidene dichloride. Used as a monomer in production, for example, of Saran. Volatile liquid; mp = -122.5°; bp = 31.6°; d^{20} = 1.213; insoluble in H2O, soluble in organic solvents. Dow Chem. U.S.A.; Dow UK.

1193 Dichloroethylaluminum
563-43-9   H

C2H5AlCl2
Dichloromonoethylaluminum.
 Aluminum, dichloroethyl-; Dichloroethylaluminum; Dichloromonoethylaluminum; EINECS 209-248-6; Ethyl aluminum dichloride; Ethylaluminium dichloride; Ethyldichloroaluminum; HSDB 317.

1194 Dichlorofluoroethane
1717-00-6   G

C2H3Cl2F
1,1-Dichloro-1-fluoroethane.
 4-01-00-00134 (Beilstein Handbook Reference); BRN 1731585; CCRIS 7208; CFC-141; CFC 141b; Dichlorofluoroethane; 1,1-Dichloro-1-fluoroethane; Ethane, 1,1-dichloro-1-fluoro-; Freon-141; Freon 141b; Genetron® 141b; HCFC 141b; HSDB 6757; Isotron

141b; R 141b; Refrigerant 141b; Solkane 141b . Blowing agent (replacement for CFCs) in foam applications, rigid board, flexible foam. Liquid; mp = -103.5°; bp = 32°; d^{10} = 1.250. Allied Signal.

1195 Dichloromethoxy ethane
111-91-1   H

C5H10Cl2O2
Bis(2-chloroethoxy)methane.
 4-01-00-03028 (Beilstein Handbook Reference); AI3-01455; Bis(β-chloroethyl) formal; Bis(2-chloro-ethoxy)-methane; BRN 1698909; Dichlorodiethyl formal; Dichlorodiethyl methylal; Dichloromethoxy ethane; EINECS 203-920-2; Ethane, 1,1'-(methylenebis(oxy))bis(2-chloro-; Formaldehyde bis(2-chloroethyl) acetal; HSDB 1333; Methane, bis(2-chloroethoxy)-; NSC 5212; RCRA waste number U024.

1196 Dichloromethyl-3,3,3-trifluoropropylsilane
675-62-7   H

C4H7Cl2F3Si
(3,3,3-Trifluoropropyl)methyldichlorosilane.
 4-04-00-04169 (Beilstein Handbook Reference); BRN 1744964; Dichloromethyl(3,3,3-trifluoropropyl)silane; EINECS 211-623-4; Silane, dichloromethyl(3,3,3-trifluoropropyl)-; (3,3,3-Trifluoropropyl)methyldichloro-silane.

1197 Dichloromethylvinylsilane
124-70-9   G H

C3H6Cl2Si
Dichloro(methyl)(vinyl)silane.
 4-04-00-04184 (Beilstein Handbook Reference); AI3-51464; BRN 1740822; CCRIS 2457; CV-4772; Dichloro(methyl)(vinyl)silane; Dichloromethylvinylsilane; EINECS 204-710-3; Methyldichlorovinylsilane; NSC 96610; Silane, dichloroethenylmethyl-; Silane, dichloro-methylvinyl-; Vinylmethyldichlorosilane. Coupling agent, chemical intermediate, blocking agent, release agent, lubricant, primer, reducing agent. Liquid; bp = 92.5°; d^{20} = 1.0868; freely soluble in H2O. Degussa-Hüls Corp.

1198 1,2-Dichloro-4-nitrobenzene
99-54-7   H

C6H3Cl2NO2
Benzene, 1,2-dichloro-4-nitro-.

4-05-00-00726 (Beilstein Handbook Reference); Al3-03268; Benzene, 1,2-dichloro-4-nitro-; BRN 1818163; CCRIS 3097; DCNB; EINECS 202-764-2; HSDB 4252; NSC 6295. Needles, mp = 43°; bp = 255.5°; d^{75} = 1.4558; λ_m = 225, 287 nm (ε = 6130, 11400, MeOH); insoluble in H_2O, slightly soluble in CCl_4, soluble in EtOH, Et_2O.

1199 Dichlorophen
97-23-4   3096   D F G

$C_{13}H_{10}Cl_2O_2$
2,2'-Methylenebis(4-chlorophenol).
4-06-00-06658 (Beilstein Handbook Reference); Acticide DDM; Al3-02370; Algafen; Algofen; Anthipen; Anthiphen; Antifen; Antiphen; BRN 1884514; Caswell No. 563; CCRIS 6060; Cordocel; DDDM; DDM; Di-phentane-70; Di-phenthane-70; Dicestal; Dichloorfeen; Dichlorofen; Dichlorophen; Dichlorophen B; Dichlorophene; Dichlorophene 10; Dichlorophenum; Dichlorphen; Diclorofeno; Didroxan; Didroxane; Difentan; Diphenthane 70; EINECS 202-567-1; Embephen; EPA Pesticide Chemical Code 055001; Fungicide GM; Fungicide M; G-4; Gefir; Gingivit; Giv Gard G 4-40; Halenol; HSDB 6033; Hyosan; Korium; NSC-38642; Nuophen; Palacel; Panacide; Parabis; Plath-Lyse; Prevental; Preventol G-D; Preventol GD; Preventol GDC; Sandocide; Sindar G 4; Super mosstox; Taeniatol; Teniathane; Teniatol; Teniotol; Trivex; Vermithana; Wespuril. Agricultural fungicide, antimicrobial and germicide. Bactericide/fungicide used in textiles, cellulose solutions, proteins, adhesives and soaps; bactericide/algicide for water treatment. Fungicide for use on textiles, cordage, and hair felt. Registered by EPA as an antimicrobial, fungicide and insecticide (cancelled). Used medically as an anthelmintic targeting cestodes. Colorless crystals; mp = 177 - 178°; soluble in H_2O (0.003 g/100 ml), EtOH (53 g/100 ml), Me_2CO (80 g/100 ml), MeOH, isopropyl ether, petroleum ether; LD_{50} (mrat orl) = 1506 mg/kg, (frat orl) = 1683 mg/kg, (mus orl) = 1000 mg/kg, (gpg orl) = 1250 mg/kg, (dog orl) = 2000 mg/kg; toxic to fish. Aventis Crop Science; Degussa-Hüls Corp.; Hüls Am.; May & Baker Ltd.; Sigma-Aldrich Co.; Thor.

1200 Dichlorophenol
120-83-2   3098   . H

$C_6H_4Cl_2O$
2,4-Dichlorophenol.
4-06-00-00885 (Beilstein Handbook Reference); 4,6-Dichlorophenol; Al3-00078; BRN 0742467; CCRIS 657; 2,4-DCP; 2,4-Dichlorohydroxybenzene; 2,4-Dichloro-phenol; EINECS 204-429-6; HSDB 1139; 1-Hydroxy-2,4-dichlorobenzene; NCI-C55345; NSC 2879; Phenol, 2,4-dichloro-; RCRA waste number U081. Needles; mp = 45°; bp = 210°; λ_m = 211, 229, 287 nm (ε = 6580, 6550, 2450, MeOH); slightly soluble in H_2O, soluble in EtOH, Et_2O, C_6H_6, $CHCl_3$.

1201 2,5-Dichlorophenol
583-78-8   H

$C_6H_4Cl_2O$
Phenol, 2,5-dichloro.
4-06-00-00942 (Beilstein Handbook Reference); BRN 1907692; CCRIS 5903; 2,5-Dichlorophenol; EINECS 209-520-4; HSDB 4287; NSC 6296; Phenol, 2,5-dichloro-. Prisms; mp = 59°; bp = 211°; λ_m = 225, 282, 289 nm (ε = 4250, 1560, 1380, MeOH); slightly soluble in H_2O, soluble in C_6H_6, petroleum ether, very soluble in EtOH, Et_2O.

1202 2,4-Dichlorophenoxyacetic acid
94-75-7   2825   H P

$C_8H_6Cl_2O_3$
(2,4-Dichlorophenoxy)acetic acid.
2,4-D; 2,4-D acid; 2,4-D Amine No. 4; 2,4-D LV6; 2,4-D, salts and esters; 2,4-PA; 4-06-00-00908 (Beilstein Handbook Reference); Acetic acid, (2,4-dichloro-phenoxy)-; Acide 2,4-dichloro phenoxyacetique; Acido(2,4-dicloro-fenossi)-acetico; Acme LV 4; Acme LV 6; Agricorn D; Agrotect; Al3-08538; Amidox; Amine 4 2,4-D Weed Killer; Amoxone; Asgrow; Asgrow Aqua KD; B-Selektonon; Barrage; BH 2,4-D; BRN 1214242; Brush-rhap; Butoxy-D 3: 1 Liquid emulsifiable Brushkiller LV96; Caswell No. 315; CCRIS 949; Chipco turf herbicide D; Chloroxone; Citrus fix; Crop Rider; Croprider; Crotilin; De-Pester Ded-Weed LV-2; Debroussaillant 600; Decamine; Ded-Weed LV-69; Deherban; Dichloro-phenoxyacetic acid; 2,4-Dichlorophenoxyacetic acid; Dicopur; Dicotox; DMA 4; Dormon; Dormone; 2,4-Dwuchlorofenoksyoctowy kwas; EINECS 202-361-1; Emulsamine; Emulsamine BK; Emulsamine E-3; ENT 8,538; Envert DT; EPA Pesticide Chemical Code 030001; Esteron 44 weed killer; Esteron 76 BE; Estone; Fernesta; Fernimine; Ferxone; Foredex 75; Formula 40; Formula 40 4L; Green Cross Weed-No-More 80; Hedonal; Hedonal (the herbicide); Helena 2,4-D; Herbidal; Hivol-44; HSDB 202; Ipaner; Kwas 2,4-dwuchlorofenoksyoctowy; Kwasu 2,4-dwuchlorofenoksyoctowego; Kyselina 2,4-dichlor-fenoxyoctova; Lawn-keep; Low Vol 4 Ester Weed Killer; Macondray; Macrondray; MCP; Miracle; Monosan; Mota Maskros; Moxon; Moxone; Netagrone; Netagrone 600; NSC 2925; Pennamine D; Phenox; Phenoxyacetic acid, 2,4-dichloro-; Pielik; Planotox; Plantgard; R-H Weed Rhap 20; RCRA waste number U240; Red Devil Dry Weed Killer; Rhodia; Scott's 4-XD Weed Control; Silvaprop 1; Standard 2,4-D Amine; Superormone concentre; Tributon; U 46DP; U-5043; Vergemaster; Verton; Verton 2D; Verton 38; Vidon 638; Visko-rhap low volatile 4l; Visko-rhap low drift herbicides; Weed-Ag-Bar; Weed-B-gon; Weed-Rhap; Weed-Rhap A-4; Weed-Rhap B-4; Weed-Rhap B-266; Weed-Rhap I-3.34; Weed-Rhap LV-4-0; Weed Tox; Weedar 64; Weedar 64A; Weedatul; Weedez; Weedez Wonder BAR; Weedone; Weedone 100 Emulsifiable; Weedone-2,4-DP; Weedone LV4; Weedtrol; Wonder BAR. A selective weed killer, used for the control of broad leafed weeds on amenity areas, golf courses, playing fields, etc. A translocatable herbicide for cereals and established grassland. Registered by EPA as a herbicide and fungicide. Crystals; mp = 140.5°; $bp_{0.4}$ = 160°; sg^{30} = 1.565; soluble in H_2O (0.062 g/100 ml), EtOH (125 g/100 ml), Et_2O (24.3 g/100 ml), C_7H_{16} (0.11 g/100 ml), C_7H_8 (0.67 g/100 ml), xylene (0.58 g/100 ml); LD_{50} (rat orl) = 375 mg/kg, (rbt der) > 1600 mg/kg, (wild duck orl) > 1000 mg/kg, (Japanese quail, pigeon orl) = 668 mg/kg,

1203 Dichloropinacolin
22591-21-5   H

$C_6H_{10}Cl_2O$

1,1-Dichloro-3,3-dimethylbutan-2-one.
4-01-00-03313 (Beilstein Handbook Reference); BRN 1752918; 2-Butanone, 1,1-dichloro-3,3-dimethyl-; 1,1-Dichloro-3,3-dimethyl-2-butanone; 1,1-Dichloro-3,3-dimethylbutan-2-one; Dichloromethyl tert-butyl ketone; Dichloropinacolin; Dichloropinakolin; ω,ω-Dichlorpinakolin; EINECS 245-111-7.

1204 1,3-Dichloro-2-propanol
96-23-1   3100   H

$C_3H_6Cl_2O$

2-Propanol, 1,3-dichloro-.
4-01-00-01491 (Beilstein Handbook Reference); AI3-14899; BRN 1732063; CCRIS 953; Dichlorohydrin; s-Dichloroisopropyl alcohol; sym-Dichloroisopropyl alcohol; Sym-dichloroisopropyl alcohol; EINECS 202-491-9; Enodrin; Gdch; Glycerol α,γ-dichlorohydrin; Glycerol 1,3-dichlorohydrin; s-Glycerol dichlorohydrin; Sym-glycerol dichlorohydrin; HSDB 5302; NSC 70982; α-Propenyldichlorohydrin; Propylene dichlorohydrin; U 25,354. Liquid; bp = 176°; d^{17} = 1.3506; Very soluble in H_2O, EtOH, Et_2O, soluble in Me_2CO, $CHCl_3$.

1205 1,3-Dichloropropene
542-75-6   3101   G H

$C_3H_4Cl_2$

1,3-Dichloro-1-propylene.
3-01-00-00704 (Beilstein Handbook Reference); BRN 1719556; Caswell No. 324A; CCRIS 955; 3-Chloroallyl chloride; 3-Chloropropenyl chloride; D-D 92; DD 95; Di-Trapex CP; Dichloro-1,3 propene; Dichloropropene; Dorlone; Dorlone II; EINECS 208-826-5; EPA Pesticide Chemical Code 029001; HSDB 1109; NCI-C03985; Nematox; Nemex; NSC 6202; Propene, 1,3-dichloro-; RCRA waste number U084; Telone; Telone C; Telone C17; Telone II; Telone II-B; Telone II soil fumigant; Vorlex 201; Z-isomer [10061-01-5]; E-isomer [10061-02-6]. Used in soil fumigants added to soil prior to planting to control soil pests such as nematodes which feed on the roots of plants and reduce yields. Registered by EPA as a fungicide and nematicide. Used as a nematicide. Liquid; mp < -50°; bp = 104.3°; d^{25} = 1.220; LD50 (rat orl) = 150 mg/kg, (mrat orl) = 713 mg/kg, (frat orl) = 470 mg/kg, (rat der) = 1200 mg/kg, (rbt der) = 504 mg/kg; LC50 (rat ihl 90 day) = 0.05 mg/l air, (rat ihl 2 year) = 0.099 mg/l air, (mus ihl 2 year) > 0.05 mg/l air, (mallard duck, bobwhite quail 8 day dietary) > 10000 mg/kg diet, (rainbow trout 96 hr.) = 3.9 mg/l, (bluegill sunfish 96 hr.) = 7.1 mg/l, (*Daphnia* 48 hr.) = 6.2 mg/l; non-toxic to bees. *Dow AgroSciences; DowElanco Ltd.; Soil Chemicals*

Corp.; Trical.

1206 Dichlorotetrafluoroethane
76-14-2   2633   G H

$C_2Cl_2F_4$

1,2-Dichloro-1,1,2,2-tetrafluoroethane.
4-01-00-00137 (Beilstein Handbook Reference); Arcton 33; Arcton 114; BRN 1740333; Caswell No. 326A; CFC 114; Criofluorano; Cryofluorane; Cryofluoranum; Dichlorotetrafluoroethane; EINECS 200-937-7; EPA Pesticide Chemical Code 326200; Ethane, 1,2-dichloro-1,1,2,2-tetrafluoro-; Ethane, 1,2-dichlorotetrafluoro-; F 114; F 114 (halocarbon); FC 114; FKW 114; Fluorane 114; Fluorocarbon 114; Freon 114; Frigen 114; Frigiderm; Genetron 114; Genetron 316; Halocarbon 114; HSDB 146; Ledon 114; Propellant 114; R 114; R 114 (halocarbon); Refrigerant 114; sym-Dichlorotetrafluoroethane; Ucon 114. Used with centrifugal compressors for higher capacities or for lower evaporator temperature process applications; also for foam applications. Gas; mp = -94°; bp = 3.8°; d^{25} = 1.455; soluble in organic solvents. *Allied Signal.*

1207 Dichlorotetrafluoroethane
374-07-2   H

$F_3C{-}CFCl_2$

$C_2Cl_2F_4$

1,1-Dichlorotetrafluoroethane.
Dichlorotetraflueroethane; 1,1-Dichloro-1,2,2,2-tetra-fluoroethane; 1,1-Dichlorotetrafluoroethane; EINECS 206-774-8; Ethane, 1,1-dichloro-1,2,2,2-tetrafluoro-; Ethane, 1,1-dichlorotetrafluoro-; Frigen 114A; HSDB 5564; 1,1,1,2-Tetrafluoro-2,2-dichloroethane. Gas; mp = -56.6°; bp = 4°; d^{25} = 1.455; very soluble in C_6H_6, EtOH, Et_2O.

1208 1,3-Dichlorotetraisopropyl-disiloxane
69304-37-6   G

$C_{12}H_{28}Cl_2OSi_2$

1,3-Dichloro-1,1,3,3-tetraisopropyldisiloxane.
CD4368; 1,3-Dichlorotetraisopropyldisiloxane; 1,3-Di-chloro-1,1,3,3-tetraisopropyldisiloxane; 1,3-Dichloro-tetraisopropylsiloxane; Disiloxane, 1,3-dichloro-1,1,3,3-tetrakis(1-methylethyl)-; 1,1,3,3-Tetraisopropyl-1,3-di-chlorodisiloxane; Tipdsicl2; TIPSCl. Coupling agent, chemical intermediate, blocking agent, release agent, lubricant, primer, reducing agent. Liquid; bp15 = 120°; d = 1.0010; n_D^{20} = 1.4543. *Degussa-Hüls Corp.*

1209 Dichloro(trifluoromethyl)methane
306-83-2   G H

$F_3C{-}CHCl_2$

$C_2HCl_2F_3$

2,2-Dichloro-1,1,1-trifluoroethane.
4-01-00-00135 (Beilstein Handbook Reference); BRN 1736763; CCRIS 7216; CFC-123; Chlorofluorocarbon 123; Dichloro(trifluoromethyl)methane; Dichloro-trifluoroethane; EINECS

206-190-3; Ethane, 2,2-dichloro-1,1,1-trifluoro-; FC 123; Freon 123; Genetron® 123; HCFC 123; HFA 123; HSDB 6752; R 123; Refrigerant HCFC-123; Refrigerant R 123; Solkane 123. Centrifugal refrigerant with low operating pressures. *Allied Signal.*

1210 Dichloroxylenol
133-53-9   3102   G

$C_8H_8Cl_2O$

2,4-Dichloro-3,5-dimethylphenol.

AI3-24011; Benzene, 2,4-dichloro-1,3-dimethyl-5-hydroxy-; DCMX; Decasept; 2,4-Dichloro-3,5-dimethylphenol; Dichloroxylenol; Dichloro-m-xylenol; 2,4-Dichloro-3,5-xylenol; 2,4-Dichloro-m,5-xylenol; Dichloroxylenolum; Dichlorxylenolum; Dicloroxilenol; Dicloroxilenolo; 3,5-Dimethyl-2,4-dichlorophenol; Dixol; EINECS 205-109-9; Hewsol; HSDB 2778; Nipacide® DX; NSC 9774; Ottacide; Phenol, 2,4-dichloro-3,5-dimethyl-; Prinsyl; 3,5-Xylenol, 2,4-dichloro-. Bacteriostat and preservative. mp = 83°; soluble in H_2O (200 mg/l), soluble in Et_2O. *Nipa.*

1211 Dichlorvos
62-73-7   3105   G

$C_4H_7Cl_2O_4P$

Dimethyl 2,2-dichlorovinyl phosphate.

4-01-00-02063 (Beilstein Handbook Reference); AI3-20738; Apavap; Astrobot; Atgard; Atgard C; Atgard V; BAY-19149; BAY-b 4986; Bayer 19149; Benfos; Bibesol; Brevinyl; Brevinyl E50; BRN 1709141; Canogard; Caswell No. 328; CCRIS 230; Chlorvinphos; Cypona; DDVP; DDVP (insecticide); Dedevap; Denkavepon; Deriban; Derribante; Des (phosphate); Devikol; Dichlofos; Dichloorvo; Dichlorfos; Dichlorman; Dichlorophos; Dichlorovos; Dichlorphos; Dichlorvos; Dichlorvosum; Diclorvos; Dimethyl 2,2-dichloroethenyl phosphate; Dimethyldichlorovinyl phosphate; Divipan; EINECS 200-547-7; ENT 20738; EPA Pesticide Chemical Code 084001; Equigard; Equigel; Estrosel; Estrosol; Fecama; Fekama; Fly-Die; Fly fighter; Herkal; Herkol; HSDB 319; Insectigas D; Krecalvin; Lindanmafu; MAFU strip; Marvex; Mopari; NCI-C00113; Nefrafos; Nerkol; NO-Pest; NO-Pest strip; Nogos; Nogos 50; Nogos 50 EC; Nogos G; Novotox; NSC-6738; Nuvan; Nuvan 100EC; Nuvan 7; OMS 14; Panaplate; Phosphate de dimethyle et de 2,2-dichlorovinyle; Phosphoric acid, 2,2-dichlorovinyl dimethyl ester; Phosvit; SD-1750; Szklarniak; TAP 9VP; Task; Tenac; Tetravos; Unifos; Unifos (pesticide); Unifos 50 EC; Uniphos; Unitox; Vapona; Vapona insecticide; Vaponite; Vapora II; Vinyl alcohol, 2,2-dichloro-, dimethyl phosphate; Vinylofos; Vinylophos; Winylophos; XLP 30. An insecticide used for spray control of sucking, biting and mining insects in greenhouses. Liquid; $bp_{20} = 140°$; $bp^{18} = 93 - 94°$; $d^{20} = 1.120$; $d^{25} = 1.415$; soluble in H_2O (10-20 mg/ml), very soluble in EtOH, Me_2CO, ligroin; soluble in other organic solvents; LD_{50} (mrat orl) = 80 mg/kg (frat orl) = 56 mg/kg. *Bayer AG.*

1212 Diclofenac
15307-86-5   3108   D

$C_{14}H_{11}Cl_2NO_2$

2-[(2,6-Dichlorophenyl)amino]benzeneacetic acid.

BRN 2146636; Dichlofenac; Diclofenac; Diclofenac acid; Diclofenaco; Diclofenacum; Diclophenac; EINECS 239-348-5; Pennsaid; ProSorb-D. Anti-inflammatory. Crystals; mp = 156-158°. *Ciba-Geigy Corp.*

1213 Diclofenac Sodium
15307-79-6   3108   D

$C_{14}H_{10}Cl_2NNaO_2$

2-[(2,6-Dichlorophenyl)amino]benzeneacetic acid mono-sodium salt.

Allvoran; Assaren; Benfofen; Cataflam; CCRIS 1909; Delphimix; Dichronic; Diclo-Phlogont; Diclo-Puren; Diclobenin; Diclofenac sodium; Diclophenac sodium; Diclord; Dicloreum; Dolobasan; Duravolten; Effekton; EINECS 239-346-4; Evofenac; GP-45840; Kriplex; Neriodin; Novapirina; Orthophen; Ortofen; Primofenac; Prophenatin; Rhumalgan; Sodium diclofenac; Solaraze; Tsudohmin; Valetan; Voldal; Voltaren; Voltaren ophthalmic; Voltarol; Xenid. Anti-inflammatory. Crystals; mp = 283-284°; $\lambda_m = 283$ nm (ϵ 10500 MeOH); soluble in H_2O, MeOH; less soluble in Me_2CO, CH_3CN, cyclohexane; LD_{50} (mus orl) 390 mg/kg, (rat orl) = 150 mg/kg. *Ciba-Geigy Corp.*

1214 Dicloran
99-30-9   G P

$C_6H_4Cl_2N_2O_2$

2,6-Dichloro-4-nitroaniline.

4-12-00-01681 (Beilstein Handbook Reference); AI3-08870; AL-50; Allisan; Aniline, 2,6-dichloro-4-nitro-; Batran; Benzenamine, 2,6-dichloro-4-nitro-; Bortran; Botran; Botran 45W; Botran 75W; BRN 1459581; Caswell No. 311; CCRIS 3111; CDNA; CNA; DCNA; DCNA (fungicide); Dichloran; Dichloran (amine fungicide); Dicloran; Dicloron; Ditranil; EINECS 202-746-4; EPA Pesticide Chemical Code 031301; HSDB 1570; Kiwi Lustr 277; NSC 218; RD-6584; Resisan; Resissan; U-2069. Used in smoke fungicides; for use in enclosed areas on protected crops. Protective fungicide used for control of *Botrytis, Monilinia, Rhizopus, Sclerotinia and Sclerotium* species in fruits and vegetables. Registered by EPA as a fungicide. Yellow crystals; mp = 191°; soluble in H_2O (0.00063 g/100 ml), Me_2CO (3.4 g/100 ml), dioxane (4 g/100 ml), $CHCl_3$ (1.2 g/100 ml), EtOAc (1.9

g/100 ml), C6H6 (0.46 g/100 ml), xylene (0.36 g/100 ml), cyclohexane (0.006 g/100 ml); LD50 (rat orl) = 4040 mg/kg, (mus orl) = 1500 - 2500 mg/kg, (gpg orl) = 1450 mg/kg, (mus der) > 5000 mg/kg, (rbt der) > 2000 mg/kg, (mallard duck orl) = 2000 mg/kg, (bee contact) = 0.18 mg/bee; LC50 (mallard duck 5 day dietary) = 5960 mg/kg diet, (bobwhite quail 5 day dietary) = 1435 mg/kg diet; (rainbow trout 96 hr.) = 1.6 mg/l, (bluegill sunfish 96 hr.) = 37 mg/l, (goldfish 96 hr.) = 32 mg/l; essentially non-toxic to bees. *Boots Co.; Britz Fertilizers Inc.; Gowan Co.; Octavius Hunt Ltd.*

1215   Dicloxacillin sodium
13412-64-1   3111   D G

C19H16Cl2N3NaO5S

[2S-[2α,5α,6β]]-6-   [[[3-(2,6-Dichlorophenyl)-5-methyl-4-isoxazolyl]carbonyl]amino]-3,3-dimethyl-7-oxo-4-thia-1-azabicyclo[3.2.0]heptane-2-carboxylic acid sodium salt.
BLP-1011; Brispen; BRL-1702; Constaphyl; Dichlor-Stapenor; Dichlorstapenor sodium; Diclocil; Dicloxacillin sodium; Dicloxacillin sodium hydrate; Dycill; Dynapen; MDI-PC; Noxaben; P-1011; Pathocil; Pen-Sint; sodium dicloxacillin; Sodium dicloxacillin hydrate; Sodium dicloxacillin monohydrate; Stampen; Staphcillin A banyu; Syntarpen; Veracillin.; Anhydrous form [343-55-5]. Antibacterial. White powder; dec 222-225°; [α]$_D^{20}$= 127.2° (H2O); soluble in H2O, MeOH; less soluble in butanol; slightly soluble in Me2CO and the usual organic solvents; LD50 (mus iv) = 900 mg/kg, (rat ip) = 630 mg/kg, (rat orl) > 5000 mg/kg. *Bristol Laboratories; SmithKline Beecham Pharm.*

1216   Dicocodimonium chloride
61789-77-3   H

Dicoco dimethyl ammonium chloride.
Arquad 2C; C8-C18 dialkyldimethylammonium chloride; Di(coco alkyl) dimethyl ammonium chloride; Dialkyl(C8-C18) dimethylammonium chloride from coconut oil; Dicoco dimethyl ammonium chloride; Dicocoalkyldi-methylquaternary ammonium chloride compound; Dicocodimethylammonium chloride; Dicocodimonium chloride; Dicoconutdimethylammonium chloride; Dimethyldialkyl(C8-18)ammonium chloride; Dimethyl-dicocoammonium chloride; EINECS 263-087-6; Noramium M 2C; Quaternary ammonium compounds, dicoco alkyldimethyl, chlorides; Quaternium 34; Variquat K 300.

1217   Dicofol
115-32-2   3113   G P

C14H9Cl5O
1,1-Bis(chlorophenyl)-2,2,2-trichloroethanol.

4-06-00-04722 (Beilstein Handbook Reference); Acarin; AI3-23648; Benzenemethanol, 4-chloro-α-(4-chloro-phenyl)-α-(trichloromethyl)-; Benzhydrol, 4,4'-dichloro-α-(trichloromethyl)-; 1,1-Bis(4-chlorophenyl)-2,2,2-tri-chloroethanol; 1,1-Bis(chlorophenyl)-2,2,2-trichloro-ethanol; 1,1-Bis(p-chlorophenyl)-2,2,2-trichloroethanol; BRN 1886299; Carbax; Caswell No. 093; CCRIS 231; Cekudifol; 4-Chloro-α-(4-chloro-phenyl)-α-(trichloro-methyl)benzenemethanol; 4-Chloro-α-(4-chlorophenyl)-α-(trichloromethyl)benzyl alcohol; CPCA; Di-(p-chlorophenyl)trichloromethylcarbinol; Dichlorokelthane; 4,4'-Dichloro-α-(trichloromethyl)benzhydrol; Dicofol; Dicomite; DTMC; EINECS 204-082-0; ENT 23,648; Ethanol, 2,2,2-trichloro-1,1-bis(p-chlorophenyl)-; FW 293; Hilfol; Hilfol 18.5 EC; HSDB 631; Kelthane; para,para'-Kelthane; Kelthane A; Kelthane dust base; Kelthanethanol; Milbol; Mitigan; NCI-C00486; 2,2,2-Trichloor-1,1-bis(4-chloor-fenyl)-ethanol; 1,1,1-Trichlor-2,2-bis(4-chlorphenyl)-aethanol; 2,2,2-Trichlor-1,1-bis(4-chlor-phenyl)-aethanol; 2,2,2-Trichloro-1,1-bis(p-chloro-phenyl)ethanol; 2,2,2-Trichloro-1,1-di-(4-chlorophenyl)-ethanol; 2,2,2-Trichloro-1,1-bis(4-cloro-fenil)-etanolo; Tricofol; 4-Chloro-α-(4-chlorophenyl)-α-(trichloro-methyl)benzyl alcohol; CPCA; Di-(p-chlorophenyl)-trichloromethylcarbinol; Dichlorokelthane; 4,4'-Dichloro-α-(trichloromethyl)benzhydrol; Dicofol; Dicomite; DTMC; ENT 23,648; Ethanol, 2,2,2-trichloro-1,1-bis(p-chlorophenyl)-; FW 293; Hilfol; Hilfol 18.5 EC; HSDB 631; Kelthane; Kelthane A; Kelthane dust base; Kelthanethanol; Milbol; Mitigan; NCI-C00486; para,para'-Kelthane; 2,2,2-Trichloor-1,1-bis(4-chloor-fenyl)-ethanol; 1,1,1-Trichlor-2,2-bis(4-chlorphenyl)-äthanol; 2,2,2-Trichlor-1,1-bis(4-chlor-phenyl)-äthanol; 2,2,2-Trichloro-1,1-bis(4-chlorophenyl)-ethanol; 2,2,2-Trichloro-1,1-di-(4-chlorophenyl)ethanol; 2,2,2-Trichloro-1,1-bis(4-cloro-fenil)-etanolo. An acaricide and miticide. Registered by EPA as an acaricide and insecticide. Crystals; mp = 77.5°; bp = 225°, bp360 = 193°; d^{10} = 1.1234; [α]D = 100°; insoluble in H2O (0.00008 g/100 ml), soluble in Me2CO (40 g/100 ml), EtOAc (40 g/100 ml), C7H8 (40 g/100 ml), MeOH (3.6 g/100 ml), C6H14 (3 g/100 ml), i-PrOH (3 g/100 ml), λm = 226, 258, 266, 276 nm (logε = 4.43, 2.82, 2.85, 2.60); LD50 (mrat, orl) = 595 mg/kg, (frat orl) = 578 mg/kg, (gpg orl) = 1810 mg/kg, (rbt orl) = 1870 mg/kg, (rat dr) > 5000 mg/kg, (rbt der) = 2000 - 5000 mg/kg; LC50 (rat ihl 4 hr.) > 5 mg/l air, (bluegill sunfish 96 hr.) = 0.52 mg/l, (channel catfish 96 hr.) = 0.36 mg/l, (rainbow trout 24 hr.) = 0.11 mg/l; non-toxic to bees. *Agan Chemical Manufacturers; Dow AgroSciences; Gowan Co.; Griffin LLC; Makhteshim Chemical Works Ltd.; Makhteshim-Agan; Micro-Flo Co. LLC; Rohm & Haas Co.; Rohm & Haas UK.*

1218   Dicumene
1889-67-4
G

C18H22
2,3-Dimethyl-2,3-diphenylbutane.
AI3-23740; Benzene, 1,1'-(1,1,2,2-tetramethyl-1,2-ethanediyl)bis-; Bibenzyl, α,α,α',α'-tetramethyl-; Butane, 2,3-dimethyl-2,3-diphenyl-; α,α'-Dicumyl; 2,3-Dimethyl-2,3-diphenylbutane; EINECS 217-568-2; NSC 34859; Perkadox® 301,1'-(1,1,2,2-Tetramethylethylene)di-benzene. *Akzo Chemie.*

1219   Dicumyl peroxide
80-43-3
G H

204

C18H22O2

Bis(1-methyl-1-phenylethyl) peroxide.

4-06-00-03225 (Beilstein Handbook Reference); Active dicumyl peroxide; Aztec® DCP-R; Bis(α,α-dimethylbenzyl) peroxide; Bis(1-methyl-1-phenylethyl) peroxide; Bis(2-phenyl-2-propyl) peroxide; BRN 2056090; CCRIS 4616; Cumene peroxide; Cumyl peroxide; Di-cup 40C; Di-cup 40haf; Di-cup 40ke; Di-cup R; Di-cup T; Di-Cup; Di-Cup 40 KE; Di-cupr; Dicumene hydroperoxide; Dicumenyl peroxide; Dicumyl peroxide; Dicup 40; DiCup 40KE; Diisopropylbenzene peroxide; EINECS 201-279-3; HSDB 320; Isopropylbenzene peroxide; Kayacumyl D; Luperco; Luperco 500-40C; Luperco 500-40KE; Luperox; Luperox 500; Luperox 500R; Luperox 500T; Lupersol 500; NSC 56772; Percumyl D; Percumyl D 40; Perkadox 96; Perkadox B; Perkadox BC; Perkadox BC 40; Perkadox BC 9; Perkadox BC 95; Perkadox SB; Peroxide, bis(α,α-dimethylbenzyl); Peroxide, bis(1-methyl-1-phenylethyl); Peroximon® DC-40; Polyvel PCL-20; Samperox DCP; Varox dcp-R; Varox dcp-T. Polymerization initiation catalyst and vulcanizing agent; crosslinking agent for olefinic polymers. Liquid; bp = 130°; insoluble in H_2O (<1 mg/ml), soluble in organic solvents; LD_{50} (rat orl) = 4100 mg/kg. *Akrochem Chem. Co.; Akzo Chemie; British Traders & Shippers; Catalyst Resources Inc.; Elf Atochem N. Am.; Hercules Inc.; Mitsui Petroleum; Polyvel; Vanderbilt R.T. Co. Inc.*

1220   Dicyanoethyl sulfide
111-97-7   G H

C6H8N2S

3,3'-Thiodipropiononitrile.

4-03-00-00749 (Beilstein Handbook Reference); AI3-16840; Bis(2-cyanoethyl) sulfide; BRN 1701139; 2-Cyanoethyl sulfide; Di(2-cyanoethyl)sulfide; Dicyanoethyl sulfide; β,β'-Dicyanodiethyl sulfide; EINECS 203-926-5; Nitril kyseliny β,β'-thiodipropionove; NSC 2040; Propanenitrile, 3,3'-thiobis-; Propionitrile, 3,3'-thiodi-; Sulfide, bis(2-cyanoethyl); 2,2'-Thiodiethylkyanid; 3,3'-Thiodipropiononitrile; β,β'-Thiodipropionitrile; USAF HA-5.

1221   Dicyclohexyl adipate
849-99-0   G

C18H30O4

Hexanedioic acid, dicyclohexyl ester.

4-06-00-00041 (Beilstein Handbook Reference); Adipic acid, dicyclohexyl ester; AI3-01028; BRN 1997888; Dicyclohexyl adipate; EINECS 212-702-6; Ergoplast ADC; NSC 4199; Sipalin AOC. A solvent for cellulose nitrate.

1222   Dicyclohexylamine
101-83-7   3122   H

C12H23N

Cyclohexanamine, N-cyclohexyl-.

4-12-00-00022 (Beilstein Handbook Reference); AI3-15334; BRN 0605923; CCRIS 6228; Cyclohexanamine, N-cyclohexyl-; DCH; DCHA; Dicha; Dicyclohexylamine; Dicyklohexylamin; Dodecahydrodiphenylamine; EINECS 202-980-7; HSDB 4018; NSC 3399; UN2565. Liquid; mp = -0.1°; bp = 256° (dec), bp9 = 114°; d^{20} = 0.9123; slightly soluble in H_2O, CCl_4, soluble in EtOH, Et_2O, C_6H_6.

1223   Dicyclohexylamine nitrite
3129-91-7   G

C12H24N2O2

Cyclohexanamine, N-cyclohexyl-, nitrite.

AI3-19769; CCRIS 4617; Cyclohexanamine, N-cyclohexyl-, nitrite; Dechan; Diana; Dichan; Dichau; Dicyclohexylamine nitrite; Dicyclohexylaminonitrite; Dicyclohexylammonium nitrite; Dicyklohexylamin nitrit; Dicynit; Dodecahydrophenylamine nitrite; Dusitan dicyklohexylaminu; EINECS 221-515-9; Leukorrosin C; N-Cyclohexylcyclohexanamine nitrite; NDA; NSC 49612; UN2687; VP 1260; VPI 260; VPT-260. Vapor-phase corrosion inhibitor for ferrous metals; designed for items enclosed in packaging, e.g., hot water heating systems, nuclear reactor heat-exchange units, gas recovery systems, jet aircraft engine compressors, internal combustion engines, welding electrodes and double-walled pipes. Solid; mp = 182-183°; oxidizer; soluble in H_2O, EtOH, less soluble in organic solvents; LD_{50} (rat orl) = 284 mg/kg. *Olin Corp.*

1224   Dicyclohexyl-2-benzothiazylsulfenamide
4979-32-2   H

C19H26N2S2

N,N-Dicyclohexylbenzothiazole-2-sulphenamide.

Accelerator DZ; 2-Benzothiazolesulfenamide, N,N-dicyclohexyl-; BRN 0621701; Dicyclohexyl-2-benzothiazylsulfenamide; N,N-Dicyclohexyl-2-benzo-thiazolesulfenam; N,N-Dicyclohexylbenzothiazole-2-sulphenamide; N,N-Dicyklohexylbenzthiazolsulfenamid; EINECS 225-625-8; M 181; Meramid DCH; Rhodifax 30; Soxinol DZ; Sulfenamid DC; Vulkacit DZ.

1225   Dicyclohexyl carbodiimide
538-75-0   3123   G

C13H22N2

N,N'-Methanetetraylbiscyclohexanamine.
4-12-00-00072 (Beilstein Handbook Reference); AI3-08191; Bis(cyclohexyl)carbodiimide; BRN 0610662; Carbodicyclohexylimide; Carbodiimide, dicyclohexyl-; Cyclohexanamine, N,N'-methanetetraylbis-; DCC; DCCD; DCCI; Dicyclohexylcarbodiimide; 1,3-Dicyclohexylcarbodiimide; N,N'-Dicyclohexylcarbo-diimide; EINECS 208-704-1; N,N'-Methanetetraylbiscyclohexaamine; NSC 30022. A chemical intermediate and coupling agent used in peptide synthesis. Crystals; mp = 34.5°; bp11 = 154-156°, bp6 = 123°, bp0.5 = 99°; reacts with H2O; contact allergen. *Janssen Chimica; Lancaster Synthesis Co.*

1226 Dicyclohexylmethane 4,4'-diisocyanate
5124-30-1   H

C15H22N2O2

1,1-Methylenebis(4-isocyanatocyclohexane).
BRN 2217800; CP 99173; Cyclohexane, 1,1'-methylenebis(4-isocyanato-; Dicyclohexylmethane 4,4'-diisocyanate; EINECS 225-863-2; Hydrogenated MDI; Hylene W; Isocyanic acid, methylenedi-4,1-cyclo-hexylene ester; Methylene bis-(4-cyclohexylisocyanate); Nacconate H 12.

1227 Dicyclohexyl phthalate
84-61-7   G H

C20H26O4

Dicyclohexyl 1,2-benzenedicarboxylate.
4-09-00-03189 (Beilstein Handbook Reference); AI3-00515; BRN 1889288; CCRIS 6190; DCHP; Dicyclohexyl 1,2-benzenedicarboxylate; Dicyclohexyl phthalate; EINECS 201-545-9; Ergoplast FDC; HF 191; Howflex CP; HSDB 5246; KP 201; NSC 6101; Phthalic acid, dicyclohexyl ester; Unimoll 66. A plasticizer for nitrocellulose, ethylcellulose, chlorinated rubber, PVAc, PVC, and other polymers. Used in formulation of delayed tack heat sealable coatings, it is a heat activated plasticizer for heat seal applications such as food wrappers/labels, pharmaceutical labels and other applications where delayed heat activated adhesive is required; used in manufacture of printing ink formulations for paper, vinyl, textiles, and other substrates. Solid; mp = 66°; d20 = 1.383; insoluble in H2O, soluble in EtOH, Et2O, CHCl3. *Bayer AG; Lancaster Synthesis Co.; Miles Inc.; Morflex; Unitex.*

1228 Dicyclopentadiene
77-73-6   H

C10H12

3a,4,7,7a-Tetrahydro-4,7-Methano-1H-indene.
2-05-00-00391 (Beilstein Handbook Reference); AI3-03386; Bicyclopentadiene; Biscyclopentadiene; BRN 1904092; CCRIS

4790; Cyclopentadiene dimer; Dicyclopentadiene; α-Dicyclopentadiene; Dicyklo-pentadien; Dimer cyklopentadienu; EINECS 201-052-9; HSDB 321; NSC 7352; Tricyclo(5.2.1.0)-3,8-decadieneUN2048. Component of inks, adhesives and polyester resins for molded parts such as tub and shower stalls and boat hulls. Solid; mp = 33.6°; bp = 170.7°; d35 = 0.977; almost insoluble in H2O (0.002 g/100 ml), soluble in organic solvents; EC50 (Selenastrum) = 27 mg/l, 72 hr., (Daphnia magna) = 8 mg/l, 48 hr., (Oryzias latipes) = 4.3 mg/l, 96 hr. *Chevron; Equistar Chemical Products; Lancaster Synthesis Co.; Mallinckrodt Inc.; Phillips 66; Sigma-Aldrich Fine Chem.*

1229 Didecylamine
1120-49-6   G

C20H43N

Di-n-decylamine.
Armeen® 2-10; 1-Decanamine, N-decyl-; Didecylamine; EINECS 214-312-1; Radiamine 6310. Industrial surfactant and intermediate for wide range of chemicals used as fabric softeners, household products and disinfectants. Solid; mp = 42-45°. *Akzo Chemie; Fina Chemicals.*

1230 Didecyldimethylammonium chloride
7173-51-5   3126   G P

C22H48ClN

Didecyl dimethyl ammonium chloride.
Aliquat 203; Ammonium, didecyldimethyl-, chloride; Arquad 10; Arquad 210-50; Astop; Bardac 22; Bardac 2250; Bardac 2270E; Bardac 2280; Bio-dac 50-22; Bio-Dac; Britewood Q; BTC 99; BTC 1010; BTC® 1010-80; BTCO 1010; Calgon H 130; Caswell No. 331A; D 10P; DDAC; DDC 80; 1-Decanaminium, N-decyl-N,N-dimethyl-, chloride; N-Decyl-N,N-dimethyl-1-decan-aminium chloride; Dimethyldidecylammonium chloride; Dodigen 1881; EINECS 230-525-2; EPA Pesticide Chemical Code 069149; H 130 (molluscicide); Maquat 4480E; Nissan Cation 2DB; Odex Q; Quartamin D 10E; Quartamin D 10P; Quaternium 12; Querton 210CL; Querton 210Cl-50; Radiaquat 6410; Radiaquat 6412; Rewoquat B 10; Slaoff 91; Timbercote 2000; Tret-O-Lite XC 507. Fungicide for hard-surface disinfection and sanitization; algicide in swimming pool and industrial water treatment; deodorizer. Bactericide, fungicide for food processing industry, breweries, catering and hospitals; in detergent sanitizers, used in commercial laundries. Disinfectant for cleaners, e.g., for dairies and food industry. Registered by EPA as a fungicide, herbicide and antimicrobial. Used as a slimicide. Solid; soluble in Me2CO, C6H6, insoluble in C6H14. *Berol Nobel AB; Fina Chemicals; Rewo; Stepan Canada; Stepan.*

1231 Di(decyl)methylamine
7396-58-9   G H

C21H45N

1-Decanamine, N-decyl-N-methyl-.
Amine M210D; Armeen® M2-10D; Dama® 1010; Didecylamine, N-

methyl-; EINECS 230-990-1; Methyl decyl-1-amino decane; N-Methyldidecylamine; Radia-mine 6310. Intermediate in the manufacture of quaternary ammonium compounds for biocides, textile chemicals, oil field chemicals, amine oxides, betaines, polyurethane foam catalysis, epoxy curing agent; in fabric softeners, disinfectants and laundry detergents. Solid; d = 0.807. *Ethyl Corp.*

1232   Didecyl phthalate
84-77-5   G H

$C_{28}H_{46}O_4$
Didecyl 1,2-benzenedicarboxylate.
4-09-00-03186 (Beilstein Handbook Reference); AI3-02692; Bis(n-decyl) phthalate; BRN 1893077; Decyl phthalate; Di-n-decyl phthalate; Didecyl 1,2-Benzenedicarboxylate; Didecyl phthalate; EINECS 201-561-6; HSDB 5248; NLA-40; NSC 15319; Phthalic acid, didecyl ester; Vinicizer 105; Vinysize 105. Plasticizer. Clear oil; mp = 4°; d = 0.9700. *National Lead Co.*

1233   Dieldrin
60-57-1   3129   P

$C_{12}H_8Cl_6O$
1,2,3,4,10,10-Hexachloro-1R,4S,4aS,5R,6R,7S,8S,8aR-octahydro-6,7-epoxy-1,4:5,8-dimethanonaphthalene.
AI3-16225; Aldrin epoxide; Alvit; Alvit 55; Caswell No. 333; CCRIS 233; Compound 497; Dieldren; Dieldrex; Dieldrin; Dieldrina; Dieldrine; Dieldrinum; Dieldrite; Dielmoth; Dildrin; Dorytox; EINECS 200-484-5; ENT 16,225; EPA Pesticide Chemical Code 045001; HEOD; Hexachloroepoxyoctahydro-endo, exo-dimethano-naphthalene; HSDB 322; Illoxol; Insecticide No. 497; Insectlack; Kombi-Albertan; Latka 497; Moth Snub D; NA2761; NCI-C00124; NSC 8934; Octalox; Panoram D-31; RCRA waste number P037; Red Shield; SD 3417; Shelltox; Termitox. Has been used as an insecticide. No longer used in Europe and the US but is used widely in Africa and S. America for termite control. Crystals; mp = 176 - 177°; d²⁵ = 1.75; soluble in organic solvents, insoluble in H₂O; LD₅₀ (rat orl) = 46 mg/kg. (terminated); *Shell.*

1234   Dienochlor
2227-17-0   3131   G P

$C_{10}Cl_{10}$
1,1',2,2',3,3',4,4',5,5'-decachlorobi-2,4-cyclopentadien-1-yl.
4-05-00-01542 (Beilstein Handbook Reference); AI3-25718; Bi-2,4-cyclopentadien-1-yl, 1,1',2,2',3,3',4, 4',5,5'-decachloro-; Bi-2,4-cyclopentadien-1-yl, deca-chloro-; Bis(pentachlor-2,4-cyclopentadien-1-yl); Bis(pentachloro-2,4-cyclopentadien-1-yl);

Bis(penta-chlorocyclopentadienyl); BRN 2064747; Caswell No. 274; Decachlor; Decachlorobi-2,4-cyclopentadien-1-yl; Decachlorobis(2,4-cyclopentadiene-1-yl); 1,1',2,2',3,3', 4,4',5,5'-Decachlorobi-2,4-cyclopentadien-1-yl; Dieno-chlor; Dienochlore; EINECS 218-763-5; ENT 25,718; EPA Pesticide Chemical Code 027501; Hooker hrs-16; Hooker HRS 1654; HRS-16; HRS 1654; HRS 16A; HSDB 1557; NSC 26106; NSC 41880; Pentac; Pentac (50WP); Pentac SP; Pentac WP; Perchloro-1,1'-bicyclopenta-2,4-diene. Acaricide used to control mites on ornamental plants. Solid; mp = 121-122°; λm = 330 nm (ε 2950). *Hooker Chem.*

1235   Diethanolamine
111-42-2   3134   G H

$C_4H_{11}NO_2$
2,2'-Dihydroxydiethylamine.
4-04-00-01514 (Beilstein Handbook Reference); AI3-15335; Bis(2-hydroxyethyl)amine; BRN 0605315; CCRIS 5906; Dabco DEOA-LF; DEA; Di(β-hydroxyethyl)amine; Di(2-hydroxyethyl)amine; Diaethanolamin; Diäthanol-amin; Diethanolamin; Diethanolamine; Diethylolamine; 2,2'-Dihydroxydiethylamine; Diolamine; EINECS 203-868-0; Ethanol, 2,2'-iminobis-; Ethanol, 2,2'-iminodi-; HSDB 924; Iminodiethanol; N,N'-Iminodiethanol; NCI-C55174; Niax DEOA-LF; NSC 4959. Used in gas scrubbing; as a rubber chemicals intermediate; in the manufacture of surfactants for textiles, herbicides, petroleum demulsifier; emulsifier and dispersant in agriculture, cosmetics, pharmaceuticals, textile lubricants, humectant, softening agents, in organic synthesis. Solid; mp = 28°; bp = 268.8°; d²⁰ = 1.0966; slightly soluble in Et₂O, C₆H₆, very soluble in H₂O, EtOH; LD₅₀ (rat orl) = 12.76 g/kg. *Degussa-Hüls Corp.*

1236   Diethanolchloroanilide
92-00-2   H

$C_{10}H_{14}ClNO_2$
Ethanol, 2,2'-[(3-chlorophenyl)imino]bis-.
4-12-00-01141 (Beilstein Handbook Reference); AI3-12121; Aniline, m-chloro-N,N-bis(2-hydroxyethyl)-; BRN 2806249; m-Chloro-N,N-bis(2-hydroxyethyl)aniline; Di-ethanolaminochlorobenzene; Diethanolchloroanilide; N,N-Diethanolanilide, 3-chloro-; N,N-Dihydroxyethyl-3-chloroaniline; EINECS 202-115-3; Emery 5715; Emery 5717; Ethanol, 2,2'-((3-chlorophenyl)imino)bis-; NSC 58170.

1237   Diethanolmethylamine
105-59-9   H

$C_5H_{13}NO_2$
Bis(2-hydroxyethyl)methylamin.
4-04-00-01517 (Beilstein Handbook Reference); Bis(2-hydroxyethyl)methylamine; BRN 1734441; CCRIS 4843; Diethanolmethylamine; EINECS 203-312-7; Ethanol, 2,2'-(methylimino)bis-; MDEA; Methyl diethanolamine;

Methyldiethanolamine; Methyliminodiethanol; NSC 11690; USAF DO-52. Liquid; mp = -21°; bp = 247°; d^{25} = 1.038; very soluble in H_2O.

1238 Diethoxy(3-(glycidyloxy)propyl)methyl silane;
2897-60-1   G

C11H24O4Si
(3-Glycidoxypropyl)-methyl-diethoxysilane.
CG6710; EINECS 220-780-8; (3-(2,3-Epoxypropoxy)-propyl)diethoxymethylsilane; (γ-Glycidoxypropyl)methyl-diethoxysilane; GP-137; 3-(Methyldiethoxysilyl)propyl glycidyl ether; NSC 252159; Silane, (3-(2,3-epoxy-propoxy)propyl)diethoxymethyl-; Silane, diethoxymethyl-(3-(oxiranylmethoxy)propyl)-. Coupling agent, chemical intermediate, blocking agent, release agent, lubricant, primer, reducing agent. Used as a coupling agent between inorganic fillers and epoxy, melamine, phenolic and urethane resins, PS, acrylic sealants, butyl rubber. *Degussa-Hüls Corp.; Genesee Polymers.*

1239 Diethoxydiphenylsilane
2553-19-7   G

C16H20O2Si
Diphenyldiethoxy silane.
CD6000; Diphenyldiethoxysilane; EINECS 219-860-5; NSC 77127; Silane, diethoxydiphenyl-. Coupling agent, chemical intermediate, blocking agent, release agent, lubricant, primer, reducing agent. Liquid; bp = 302°; bp15 = 167°; d^{20} = 1.0329; λ_m = 254, 259, 265, 272 nm (ε = 363, 513, 562, 407, EtOH). *Degussa-Hüls Corp.*

1240 3-(Diethoxymethylsilyl)propylamine
3179-76-8   G

C8H21NO2Si
3-Aminopropylmethyldiethoxy silane.
4-04-00-04201 (Beilstein Handbook Reference); (3-Aminopropyl)diethoxymethylsilane; BRN 1744264; CA0742; Dynasylan 1505; Dynasylan 1506; EINECS 221-660-8; σ-Aminopropylmethyldiethoxysilane; KBE 902; 1-Propanamine, 3-(diethoxymethylsilyl)-; Propylamine, 3-(diethoxymethylsilyl)-; Silane, (3-aminopropyl)diethoxy-methyl-. Coupling agent, chemical intermediate, blocking agent, release agent, lubricant, primer, reducing agent. *Degussa-Hüls Corp.*

1241 Diethoxythiophosphoryl chloride
2524-04-1   H

C4H10ClO2PS
Diethyl chlorothiophosphate.
4-01-00-01352 (Beilstein Handbook Reference); BRN 0471434; Diethoxythiophosphoryl chloride; O,O-Diethyl chloridophosphorothioate; O,O-Diethyl chloridothiono-phosphate; Diethyl chlorothiophosphate; Diethyl phosphorochloridothioate; Diethyl phosphorochloro-thioate; Diethyl phosphorothiochloridate; Diethyl-chlorthiofosfat; Diethylthiophosphoryl chloride; EINECS 219-755-4; Ethyl PCT; HSDB 2627; NSC 43776; Phosphorochloridothioic acid, O,O-diethyl ester; UN2751. Liquid; bp3 = 45°; soluble in CCl4.

1242 Diethylamine
109-89-7   3138   G H

C4H11N
N,N-Diethylamine.
AI3-24215; CCRIS 4792; DEA; Diaethylamin; Diäthylamin; Diethamine; Diethylamine; Dietilamina; Dwuetyloamina; EINECS 203-716-3; Ethanamine, N-ethyl-; N-Ethylethanamine; HSDB 524; N,N-Diethylamine; UN1154. Used in rubber chemicals, textile specialties, selective solvent, dyes, flotation agents, resins, pesticides, polymerization inhibitor, pharmaceuticals, petroleum chemicals, electroplating, corrosion inhibitors. Liquid; mp = -49.8°; bp = 55.5°; d^{20} = 0.7056; λ_m = 194, 222 nm (ε = 2951, 295, gas); soluble in Et2O, CCl4, very soluble in H_2O, freely soluble in EtOH; LD50 (rat orl) = 540 mg/kg. *Air Products & Chemicals Inc.; Allchem Ind.; BASF Corp.; Elf Atochem N. Am.; Union Carbide Corp.*

1243 2-Diethylaminoethanol
100-37-8   3139   G H

C6H15NO
Ethanol, 2-(diethylamino)-.
AI3-16309; CCRIS 4793; DEAE; Diaethylaminoaethanol; Diäthylaminoaethanol; Diethyl(2-hydroxyethyl)amine; Diethylamino ethanol; Diethylaminoethanol; N-Diethyl-aminoethanol; Diethylethanolamine; Diethylmono-ethanolamine; EINECS 202-845-2; Ethanol, 2-(diethylamino)-; HSDB 329; NSC 8759; Pennad 150; UN2686. Used in making pharmaceuticals, pesticides, and other chemicals. Hygroscopic liquid; mp = -70°; bp = 163°; d^{20} = 0.8921; freely soluble in H_2O, soluble in EtOH, Et2O, Me2CO, C6H6, petroleum ether, slightly soluble in CCl4; LD50 (rat orl) = 1100 mg/kg.

1244 Diethylaminoethyl acrylate
2426-54-2   G

C9H17O2N
2-Propenoic acid, 2-(diethylamino)ethyl ester.
4-04-00-01477 (Beilstein Handbook Reference); Acrylic acid, 2-(diethylamino)ethyl ester; Acrylic acid, N,N-diethylaminoethyl ester; Ageflex FA-2; AI3-03840; BRN 1099061; Diethylaminoethyl acrylate; 2-(Diethyl-amino)ethyl acrylate; N,N-Diethylaminoethyl acrylate; 2-Diethylaminoethylester kyseliny akrylove; EINECS 219-378-5; HSDB 5460; NSC 3118. Used in industrial and automotive coatings, electronic photo resists, dye additives, lube oil additives; intermediate for water treatment chemicals, silane coupling agents, conductive paper coatings; retention aids for paper manufacturing; flocculant and coagulant. Liquid; mp < -60°; bp$_{10}$ = 81°; d^{20} = 0.937; corrosive; severe eye and skin irritant; harmful if swallowed or inhaled. *Rit-Chem. -*

1245   Diethylaminoethyl methacrylate
105-16-8   H

C10H19NO2
2-(N,N-Diethylamino)ethyl methacrylate.
4-04-00-01477 (Beilstein Handbook Reference); BRN 1761794; Daktose B; Diethylaminoethyl methacrylate; EINECS 203-275-7; Ethanol, 2-(diethylamino)-, methacrylate (ester); HSDB 5365; Methacrylic acid, 2-(diethylamino)ethyl ester; N,N-Diethylaminoethyl methacrylate; NSC 14490.

1246   N,N-Diethyl-m-aminophenol
91-68-9   H

C10H15NO
Phenol, 3-(diethylamino)-.
4-13-00-00969 (Beilstein Handbook Reference); AI3-14895; BRN 0908212; CCRIS 4615; m-(Diethylamino)phenol; N,N-Diethyl-3-hydroxyaniline; N,N-Diethyl-3-aminophenol; EINECS 202-090-9; NSC 93934; Phenol, 3-(diethylamino)-; Phenol, m-(diethylamino)-. Intermediate in manufacture of dyestuffs. Crystals; mp = 78°; bp = 276°; bp^{15} = 17-°; λm = 209, 259, 299 nm (ε = 23800, 13400, 3080, MeOH); soluble in H2O, EtOH, Et2O, CS2.

1247   o-(4-(Diethylamino)salicyloyl)benzoic acid
5809-23-4   H

C18H19NO4
Benzoic acid, 2-(4-(diethylamino)-2-hydroxybenzoyl)-.
Benzoic acid, 2-(4-(diethylamino)-2-hydroxybenzoyl)-; 2-(4-(Diethylamino)-2-hydroxybenzoyl)benzoic acid; o-(4-

(Diethylamino)salicyloyl)benzoic acid; EINECS 227-370-8.

1248   (Diethylamino)trimethylsilane
996-50-9   G

C7H19NSi
N,N-Diethylaminotrimethyl silane.
CD4450; N,N-Diethyl-1,1,1-trimethylsilylamine; EINECS 213-637-6; NSC 377650; Silanamine, N,N-diethyl-1,1,1-trimethyl-. Coupling agent, chemical intermediate, blocking agent, release agent, lubricant, primer, reducing agent. Liquid; bp = 125-126°; d = 0.7670; n$_D^{20}$ = 1.4081. *Degussa-Hüls Corp.*

1249   N,N-Diethylaniline
91-66-7   3140   H

C10H15N
Benzenamine, N,N-diethyl-.
AI3-52227; Aniline, N,N-diethyl-; Benzenamine, N,N-diethyl-; CCRIS 2847; DEA; Diäthylanilin; Diäthylanilin; Diethylaniline; N,N-Diethylaniline; N,N-Diethylbenzen-amine; Diethylphenylamine; EINECS 202-088-8; HSDB 1639; NSC 7205; Phenyldiethylamine; UN2432. Used in chemical manufacturing and as an intermediate in the dyestuffs industry. Yellow oil; mp = -38.8°; bp = 216.3°; d^{20} = 0.9307; λm = 260, 304 nm (ε = 1390, 195, MeOH); poorly soluble in H2O (1.43 g/100 ml), soluble in EtOH, Et2O, Me2CO, CCl4, CHCl3. *Baiye; Buffalo Color Corp.; Callery Chem. Co. UK; Callery Chem. Co.; Kunshan Chemical Material Co. Ltd.*

1250   2,6-Diethylaniline
579-66-8   H

C10H15N
2,6-Diethylbenzenamine.
4-12-00-02841 (Beilstein Handbook Reference); AI3-26297; Aniline, 2,6-diethyl-; Benzamine, 2,6-diethyl-; Benzenamine, 2,6-diethyl-; BRN 1423626; CCRIS 2688; 2,6-Diethylaniline; 2,6-Diethylbenzenamine; EINECS 209-445-7; HSDB 5699. Liquid; mp = 1.5°; bp = 243°; d^{25} = 0.906.

1251   Diethylbenzene
25340-17-4   H

C10H14
Diethylbenzenes (mixed isomers).
4-05-00-01067 (Beilstein Handbook Reference); AI3-15336; Benzene, diethyl-; BRN 1903396; Diethyl-benzene; Diethylbenzenes

(mixed isomer); Diethyl-benzol; EINECS 246-874-9; NSC 405068; UN2049.

1252 p-Diethylbenzene
105-05-5   H

$C_{10}H_{14}$
1,4-Diethylbenzene.
Benzene, 1,4-diethyl-; Benzene, p-diethyl-; EINECS 203-265-2; HSDB 4083; 1,4-Diethylbenzene; p-Diethylbenzene; p-Ethylethylbenzene. Liquid; mp = -42.8°; bp = 183.7°; d^{20} = 0.8620; λ_m = 218, 259, 264, 267, 273 nm (ε = 7980, 418, 503, 481, 513, MeOH); insoluble in H_2O, freely soluble in EtOH, Et_2O, C_6H_6, Me_2CO, CCl_4, ligroin.

1253 Diethyldithiocarbamate sodium
148-18-5   3414   G

$C_5H_{10}NNaS_2$
Sodium diethyldithiocarbamate.
AI3-14688; Carbamic acid, diethyldithio-, sodium salt; Carbamodithioic acid, diethyl-, sodium salt; CCRIS 235; Cupral; DDTC; DEDC; DeDTC; Diethylcarbamodithioic acid, sodium salt; Diethyldithiocarbamate sodium; Diethyldithiocarbamic acid, sodium; N,N-Diethyl-dithiocarbamic acid, sodium salt; Diethyl sodium dithiocarbamate; Dithiocarb; Ditiocarb sodium; Ditiocarbe sodique; Ditiocarbo sodico; Ditiocarbum natricum; DRG-0066; DTC; EINECS 205-710-6; GS 694A; HSDB 4091; Imuthiol; Kupral; Na-ddtc; NCI CO2835; Nocceler SDC; NSC 38583; Octopol SDE-25; Sodium DEDT; Sodium diethylaminocarbodithioate; Sodium diethylcarbamodithioate; Sodium diethyldithio-carbamate; Sodium N,N-diethyldithiocarbamate; Sodium salt of N,N-diethyldithiocarbamic acid; Soxinol ESL; Thiocarb; USAF EK-2596; Diethyldithiocarbamic acid sodium salt; sodium salt of N,N-diethyldithiocarbamic acid; dithiocarbamate sodium; DEDK; sodium dedt; cupral; Sodium N,N-diethyldithiocarbamate trihydrate. Natural rubber latex preservative; precipitant for heavy metals in waste water treatment. Crystals; mp = 94-102°; soluble in H_2O, polar organic solvents; λ_m = 257, 290 nm (ε 1200, 13000 EtOH); LD50 (rat orl) = 2830 mg/kg. *Titarco Chemical.*

1254 Diethyl dithiophosphate
298-06-6   G H

$C_4H_{11}O_2PS_2$
Dithiophosphoric acid, O,O-diethyl ester.
4-01-00-01354 (Beilstein Handbook Reference); BRN 0507407; Di-O-ethyl dithiophosphate; Diethyl dithiophosphate; Diethyl phosphorodithioate; Dithiophosphoric acid, O,O-diethyl ester; EINECS 206-055-9; EP-1; HSDB 5558; Kyselina O,O-diethyldithiofosforecna; NSC 171184; Phosphorodithioic acid, O,O-diethyl ester. Used as an intermediate in manufacture of organophosphorus insecticides. *A/S Cheminova.*

1255 Diethylene glycol
111-46-6   3146   G H

$C_4H_{10}O_3$
Bis(β-hydroxyethyl) ether.
4-01-00-02390 (Beilstein Handbook Reference); AI3-08416; Bis(β-hydroxyethyl) ether; Bis(2-hydroxyethyl) ether; Brecolane ndg; BRN 0969209; Caswell No. 338A; CCRIS 2193; Deactivator E; DEG; D.E.H. 52; D.E.H. 20; Dicol; Diethylene glycol; Diethylenglykol; Digenos; Diglycol; Digol; Dihydroxydiethyl ether; Dissolvant APV; EINECS 203-872-2; EPA Pesticide Chemical Code 338200; Ethanol, 2,2'-oxybis-; Ethylene diglycol; Glycol ether; Glycol ethyl ether; HSDB 69; NSC 36391; TL4N. Used in the production of polyurethane and unsaturated polyester resins, triethylene glycol; textile softener; solvent for nitrocellulose, dyes and oils; dehydration of natural gas, elasticizers, and surfactants; humectant for tobacco, casein, and synthetic sponges. Liquid; mp = -10.4°; bp = 245.8°; d^{15} = 1.1197; soluble in H_2O, EtOH, Et_2O, $CHCl_3$; LD50 (rat orl) = 20.76 g/kg. *BASF Corp.; BP Chem.; DuPont; Eastman Chem. Co.; Hoechst Celanese; Mitsui Petroleum; Olin Corp; OxyChem; Shell; Texaco; Union Carbide Corp.*

1256 Diethylene glycol chloroformate
106-75-2   H

$C_6H_8Cl_2O_5$
Formic acid, chloro-, diester with diethylene glycol.
4-03-00-00029 (Beilstein Handbook Reference); AI3-26267; BRN 1812829; Carbonochloridic acid, oxydi-2,1-ethanediyl ester; Diethylene glycol, bischloroformate; Diethylene glycol chloroformate; Diglycol chlorformate; Diglycol chloroformate; EINECS 203-430-9; Formic acid, chloro-, diester with diethylene glycol (6CI); Formic acid, chloro-, oxydiethylene ester; HSDB 5374; NSC 2346; Oxydiethylene bis(chloroformate); Oxydiethylene chloro-formate; Oxydiethylenebis(chloroformate).

1257 Diethylene glycol dibenzoate
120-55-8   G H

$C_{18}H_{18}O_5$
Benzoyloxyethoxyethyl benzoate.
4-09-00-00356 (Beilstein Handbook Reference); AI3-02293; Benzo Flex 2-45; Benzoyloxyethoxyethyl benzoate; BRN 2509507; Dibenzoyldiethyleneglycol ester; Diethylene glycol dibenzoate; EINECS 204-407-6; HSDB 5587; Polyethylene glycol 100 dibenzoate; Polyoxyethylene (2) dibenzoate. A chemical intermediate. Solid; mp = 33.5°; bp24 = 280°, bp1 = 250°; d^{15} = 1.1690; very soluble in H_2O, EtOH.

1258 Diethylene glycol dimethacrylate
2358-84-1   G

C12H18O5
3-Oxapentane-1,5-diyl dimethacrylate.
Ageflex DEGDMA; CCRIS 7050; DGM 2; Diethylene glycol bis(methacrylate); Diethylene glycol di-methacrylate; EINECS 219-099-9; HSDB 5458; Meth-acrylic acid, oxydiethylene ester; MFM-418; Oxydi-2,1-ethanediyl bismethacrylate; Oxydiethylene methacrylate; 2-Propenoic acid, 2-methyl-, oxydi-2,1-ethanediyl ester; TGM 2. Crosslinker for rubber vulcanization, moisture barrier films and coatings, photopolymer printing plates and letterpress inks, conversion coatings and adhesives. Liquid; bp > 200°, bp8 = 150°, bp2 = 134°; d = 1.082. *Monomer-Polymer & Dajac; Polysciences; Rit-Chem; Rohm Tech.; Sartomer; Sigma-Aldrich Fine Chem.*

1259   Diethylene glycol dinitrate
693-21-0   H

C4H8N2O7
Ethanol, 2,2'-oxybisdinitrate.
4-01-00-02412 (Beilstein Handbook Reference); Bis(hydroxyäthyl)-äther-dinitrat; BRN 1791998; CCRIS 3415; Di(hydroxyethyl) ether dinitrate; Diethylene glycol dinitrate; Diethylenglykoldinitrate; Diglycol (dinitrate de); Diglycoldinitraat; Diglykoldinitrat; Dinitrate de diethylene-glycol; Dinitrodiglicol; EINECS 211-745-8; Ethanol, 2,2'-oxybis-, dinitrate; Ethanol, 2,2'-oxybis-dinitrate; Oxydiethylene dinitrate; UN0075.

1260   Diethylene glycol ethyl ether
111-90-0   1809   G H

C6H14O3
2-(2-Ethoxyethoxy) ethanol.
4-01-00-02393 (Beilstein Handbook Reference); Aethyldiaethylenglycol; Äthyldiäthylenglycol; AI3-01740; BRN 1736441; Carbitol®; Carbitol cellosolve; Carbitol solvent; DEGMEE; Diethoxol; Diethylene glycol ethyl ether; Diglycol monoethyl ether; Dioxitol; Dowanol 17; Dowanol DE; EINECS 203-919-7; Ektasolve® DE; Ethanol, 2-(2-ethoxyethoxy)-; Ethoxy diglycol; 2-(2-Ethoxyethoxy) ethanol; Ethyl carbitol; Ethyl diethylene glycol; Ethyl digol; Ethyl di-Icinol; Ethylene diglycol monoethyl ether; HSDB 51; Karbitol; Losungsmittel APV; NSC 408451; PM 1799; Poly-Solv DE; Solvolsol; Transcutol; Transcutol P. A solvent for cellulose nitrate, shellac, copal, rosin; used in dopes for artificial leather and is added to brushing lacquers. Used in protective coatings, inks, cleaning products, agricultural chemicals; aids wetting, penetration, and soil removal; coupling solvent. Solvent used for active ingredients in pharmaceutical preparations; co-surfactant for microemulsions. Liquid; bp = 196°; d^{20} = 0.9885; freely soluble in H2O, EtOH, Me2CO, C6H6, very soluble in Et2O; LD50 (rat orl) = 11 g/kg. *Allchem Ind.; Ashland; Eastman Chem. Co.; Fluka; Gattefosse SA; Great Western; ICI Chem. & Polymers Ltd.; ICI Surfactants; Occidental Chem. Corp.; Oxiteno; Sigma-Aldrich Fine Chem.; Spectrum Chem. Manufacturing; Union Carbide Corp.*

1261   Diethylene glycol hexyl ether
112-59-4   H

C10H22O3
Ethanol, 2-(2-(2-hexyloxy)ethoxy)-.
4-01-00-02396 (Beilstein Handbook Reference); AI3-00301; BRN 1743959; Diethylene glycol hexyl ether; Diethylene glycol n-hexyl ether; EINECS 203-988-3; Ethanol, 2-((2-hexyloxy)ethoxy)-; Hexol carbitol; Hexylkarbitol; HSDB 5571; NSC 403666.

1262   Diethylene glycol monooleate
106-12-7   G

C22H42O4
2-(2-Hydroxyethoxy)ethyl oleate.
AI3-00971; Diethylene glycol monooleate; EINECS 203-364-0; 9-Octadecenoic acid (9Z)-, 2-(2-hydroxy-ethoxy)ethyl ester; 9-Octadecenoic acid, 2-(2-hydroxy-ethoxy)ethyl ester; PEG-2 Oleate; Polyethylene glycol 100 monooleate; Polyoxyethylene (2) monooleate. Emulsifier, dispersant, antistat for cosmetic, textile, paper processing, cutting oils, polishes, emulsion cleaners, rubber latex, wool lubricants; leather softener. *Henkel/Emery.*

1263   Diethylene glycol stearate
106-11-6   G

C22H44O4
2-(2-Hydroxyethoxy)ethyl stearate.
4-02-00-01222 (Beilstein Handbook Reference); AI3-00977; Aqua Cera; Atlas G 2146; BRN 1800568; Cerasynt; Cerasynt special; Clindrol SDG; Diethylene glycol, monoester with stearic acid; Diethylene glycol monostearate; Diethylene glycol sesquistearate; Diethylene glycol stearate; Diglycol monostearate; Diglycol stearate; EINECS 203-363-5; Emcol CAD; Emcol DS-50 CAD; Emcol ETS; Glyco stearin; Nonex 411; NSC 404230; Octadecanoic acid, 2-(2-hydroxyethoxy)ethyl ester; PEG-2 Stearate; Polyethylene glycol 100 monostearate; Polyoxyethylene (2) monostearate; Promul 5080; Stearic acid, 2-(2-hydroxyethoxy)ethyl ester; USAF KE-8. Emulsifier, plasticizer, lubricant, wetting agent, binding and thickening agent, dispersant, antistat, opacifier, pearlescent, stabilizer used in cosmetics, dry cleaning, leather, textile industries, paper processing, rubber. *Henkel/Emery.*

1264   Diethylenetriamine
111-40-0   G H

C4H13N3
Bis(β-aminoethyl)amine.
4-04-00-01238 (Beilstein Handbook Reference); Aminoethylethandiamine; N-(2-Aminoethyl)-1,2-ethane-diamine; Ancamine DETA; Barsamide 115; Bis(β-aminoethyl)amine; Bis(2-aminoethyl)amine; N,N-Bis(2-aminoethyl)amine; BRN 0605314; CCRIS 4794; ChS-P 1; DEH 20; Diethylamine, 2,2'-diamino-; Diethylene triamine; Diethylenetriamine; EINECS 203-865-4; Epicure T; Epon 3223; Ethylamine, 2,2'-iminobis-; H 9506; HSDB 525; Imino-bis-ethylamine; NSC 446; UN2079. A general purpose aliphatic polyamine curing agent used with epoxy resins, for civil engineering, adhesives, grouts, casting and electric encapsulation. Also used as a solvent for sulfur, acid gases, various resins, dyes; saponification

agent for acidic materials; fuel component. Liquid; mp = -39°; bp = 207°; d^{20} = 0.9569; slightly soluble in Et2O, soluble in ligroin; freely soluble in EtOH, H2O; LD50 (rat orl) = 1080 mg/kg. *Allchem Ind.; Dow Chem. U.S.A.; Janssen Chimica; Tosoh; Union Carbide Corp.*

1265 Diethylenetriamine, pentamethylenepentaphosphonic acid
15827-60-8   H

C9H28N3O15P5
Diethylenetriaminepenta(methylenephosphonic) acid.
Diethylenetriamine, pentamethylenepentaphosphonic acid; Diethylenetriaminepenta(methylenephosphonic) acid; EINECS 239-931-4; Phosphonic acid, (((phosphono-methyl)imino)bis(2,1-ethanediylnitrilobis(methylene)))tetrakis-; (((Phosphonomethyl)imino)bis-(ethane-2,1-diylnitrilobis(methylene)))tetrakisphosphonic acid.

1266 Diethylenetriamine, tall oil fatty acids reaction product
61790-69-0   H

Tall oil fatty acids, reaction product with diethylene-triamine.
Diethylenetriamine, tall oil fatty acids reaction product; EINECS 263-160-2; Fatty acids, tall-oil, reaction products with diethylenetriamine; Tall oil fatty acids, reaction product with diethylenetriamine.

1267 Diethyl ether
60-29-7   3840   D G H V

C4H10O
Ethane, 1,1'-oxybis-.
Aether; Äther; AI3-24233; Anaesthetic ether; Anesthesia ether; Anesthetic ether; Diaethylaether; Diäthyläther; Diethyl ether; Diethyl oxide; Dwuetylowy eter; EINECS 200-467-2; Etere etilico; Ethane, 1,1'-oxybis-; Ether; Ether, ethyl; Ether ethylique; Ethoxyethane; Ethyl ether; Ethyl oxide; HSDB 70; NSC 100036; Oxyde d'ethyle; Pronarcol; RCRA waste number U117; Solvent ether; UN1155. Used in organic synthesis and in smokeless powders; as an industrial solvent; in analytical chemistry; as an extractant. Used as an inhalation anesthetic and, in veterinary medicine as a stimulant and to treat colic. Clear mobile liquid; mp = -116°; bp = 34.5°; d^{20} = 0.7138; slightly soluble in H2O, very soluble in organic solvents. *ExxonMobil Chem. Co.; Hüls Am.; Lancaster Synthesis Co.; Mallinckrodt Inc.; Quantum/USI; Sigma-Aldrich Fine Chem.; Spectrum Chem. Manufacturing.*

1268 Diethyl fumarate
623-91-6   H

C8H12O4
2-Butenedioic acid (E)-, diethyl ester.
4-02-00-02207 (Beilstein Handbook Reference); AI3-05613; Anti-Psoriaticum; BRN 0775347; 2-Butenedioic acid (E)-, diethyl ester; 2-Butenedioic acid (2E)-, diethyl ester; 2-Butenedioic acid, diethyl ester, (E)-; trans-2-Butenedioic acid diethyl ester; Diethyl fumarate; Diethylester kyseliny fumarove; EINECS 210-819-7; Fumaric acid, diethyl ester; HSDB 5722; NSC 20954. Liquid; mp = 0.8°; bp = 214°; d^{20} = 1.0452; λ_m = 213 nm (MeOH); insoluble in H2O, soluble in Me2CO, CHCl3.

1269 Di(2-ethylhexyl)phosphate
298-07-7   H

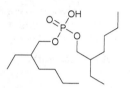

C16H35O4P
Phosphoric acid, bis(2-ethylhexyl) ester.
4-01-00-01786 (Beilstein Handbook Reference); AI3-23072; Bis(2-ethylhexyl) hydrogen phosphate; Bis(2-ethylhexyl) phosphate; BRN 1712988; D 2EHPA; Dehpa extractant; EINECS 206-056-4; Escaid 100; Hdehp; HSDB 341; Hydrogen bis(2-ethylhexyl) phosphate; Kyselina di-(2-ethylhexyl)fosforecna; Phosphoric acid, bis(2-ethyl-hexyl) ester.

1270 Diethyl isophthalate
636-53-3   H

C12H14O4
1,3-Benzenedicarboxylic acid, diethyl ester.
4-09-00-03294 (Beilstein Handbook Reference); 1,3-Benzenedicarboxylic acid, diethyl ester; BRN 2052705; Diethyl isophthalate; EINECS 211-260-1; HSDB 5732; Isophthalic acid, diethyl ester; NSC 249815. Liquid; mp = 11.5°; bp = 302°; d^{17} = 1.1239; λ_m= 225, 274 nm (ε = 8310, 1270, MeOH); insoluble in H2O, soluble in Me2CO, C6H6, CCl4, freely soluble in EtOH, Et2O.

1271 Diethyl ketone
96-22-0   3148   G H

C5H10O
3-Pentanone.
AI3-24337; DEK; Diethyl ketone; Diethylcetone; Dimethylacetone; EINECS 202-490-3; Ethyl ketone; Ethyl propionyl; HSDB 5301; Metacetone; Methacetone; NSC 8653; Pentanone-3; Propione; UN1156. Used in pharmaceutical and chemical manufacturing. Clear mobile liquid; mp = -39°; bp = 101.9°; soluble in H2O (4 g/100 ml), EtOH, Et2O, CCl4; LD50 (rat orl) = 2100 mg/kg. *BASF Corp.; Janssen Chimica; Penta Mfg.; Union Carbide Corp.*

1272   Diethyl oxalate

95-92-1   3152   G

$C_6H_{10}O_4$

Ethyl oxalate.

4-02-00-01848 (Beilstein Handbook Reference); BRN 0606350; Diethyl ethanedioate; Diethylester kyseliny stavelove; EINECS 202-464-1; Ethanedioic acid, diethyl ester; Ethyl oxalate; HSDB 2131; NSC 8851; Oxalic acid, diethyl ester; Oxalic ether; UN2525. Used as a solvent and intermediate in manufacture of chemicals and pharmaceuticals. Liquid; mp = -40.6°; bp = 185.7°; d^{20} = 1.0785; poorly soluble in H_2O, soluble in EtOH, Et_2O, Me_2CO, CCl_4.

1273   Diethyl sulfate

64-67-5   3156   H

$C_4H_{10}O_4S$

Sulfuric acid, diethyl ester.

Al3-15355; CCRIS 242; DES; Diaethylsulfat; Diäthylsulfat; Diethyl sulfate; Diethyl sulphate; Diethyl tetraoxosulfate; Diethylester kyseliny sirove; Diethylsulfate; EINECS 200-589-6; Ethyl sulfate; HSDB 1636; NSC 56380; Sulfuric acid, diethyl ester; UN1594. Used as an ethylating agent and a sulfonation catalyst and as an accelerator in sulfation of ethylene. Liquid; mp = -24°; bp = 208°; d^{25} = 1.172; slightly soluble in H_2O, very soluble in organic solvents. *Sigma-Aldrich Fine Chem.*

1274   Diethyl sulfide

352-93-2   3885   P

$C_4H_{10}S$

Diethyl thioether.

Al3-18785; Bear skunker; Diethyl sulfide; Diethyl sulphide; Diethylsulfid; Diethylthioether; EINECS 206-526-9; Ethyl monosulfide; Ethyl sulfide; Ethyl Thioether; Ethylthioethane; HSDB 5563; NSC 75157; Sulfodor; Thioethyl ether; UN2375. Used as a bear and animal repellent. Registered by EPA as an animal repellent (cancelled). Liquid; mp = -103.9°; bp = 92.1°; d^{20} = 0.8362; λ_m = 194, 215 nm (ϵ = 4786, 1585, C_7H_{16}); slightly soluble in H_2O, CCl_4, soluble in EtOH, Et_2O. *Fluka.*

1275   Diethyl toluamide

134-62-3   2876   G H P

$C_{12}H_{17}NO$

N,N-Diethyl-3-methylbenzamide.

4-09-00-01716 (Beilstein Handbook Reference); Al 3-22542; Amincene C 140; Amincene C-EM; Autan; Benzamide, N,N-diethyl-3-methyl-; Bepper DET; BRN 2046711; Caswell No. 346; CCRIS 6018; Deet; Delphene; DET (insect repellant); DETA; DETA-20; Detamide; Dieltamid; Diethyl-m-toluamide; Diethyl toluamide; Diethyltoluamidum; Dietiltoluamida; EINECS 205-149-7; ENT 20,218; ENT 22,542; EPA Pesticide Chemical Code 080301; Flypel; FlypelHSDB 1582; HSDB 1582; M-Det; Metadelphene; MGK; MGK diethyltoluamide; Muscol; Naugatuck DET; NSC 33840; Off; Repel; Repper DET; Repudin-Special. Insect repellant, resin solvent, film formers. Registered by EPA as an insecticide. Liquid; bp$_{19}$ = 160°; bp$_1$ = 111°; d$_4^{20}$ = 0.996; very soluble in H_2O, EtOH, Et_2O, sparingly soluble in petroleum ether; LD$_{50}$ (rat orl) = 2000 mg/kg. *Chase Products Co.; DuPont; Honeywell & Stein; Morflex; S.C. Johnson & Son, Co; Spectrum Chem. Manufacturing; WPC Brands Inc.*

1276   Diethylhydroxylamine

3710-84-7   G H

$C_4H_{11}NO$

N,N-Diethylhydroxylamine.

4-04-00-03304 (Beilstein Handbook Reference); Al3-28026; BRN 1731349; CCRIS 964; Diethylhydroxyl-amine; N,N-Diethylhydroxyamine; EINECS 223-055-4; Ethanamine, N-ethyl-N-hydroxy-; Hydroxylamine, N,N-diethyl-; N-Hydroxydiethylamine; Pennstop® 1866. Free radical scavenger used by the rubber industry as an emulsion polymerization inhibitor; vapor phase inhibitor for olefin or styrene monomer recovery systems; in-process inhibitor for production of styrene, divinyl benzene, butadiene and isoprene. Liquid; mp = 10°; bp = 133°; d^{20} = 0.8669. *Elf Atochem N. Am.*

1277   Diethylthiourea

105-55-5   G

$C_5H_{12}N_2S$

1,3-Diethylthiourea.

4-04-00-00375 (Beilstein Handbook Reference); Al3-14636; BRN 0773905; CCRIS 243; N,N'-Diethyl-thiocarbamide; 1,3-Diethylthiourea; 1,3-Diethyl-2-thiourea; N,N'-Diethylthiourea; N,N-Diethyl-2-thiourea; EINECS 203-308-5; HSDB 4106; NCI-C03816; NSC 3507; Pennzone E; Thiate H; Thiourea, N,N'-diethyl-; U 15030; Urea, 1,3-diethyl-2-thio-; USAF EK-1803. Accelerator for mercaptan-modified chloroprene rubber. Antidegradant for natural, nitrile-butadiene, styrene-butadiene, and chloroprene rubbers. Crystals; mp = 78°; bp dec.; λ_m = 234, 265 nm (ϵ = 6310, 7244, MeOH); slightly soluble in CCl_4, soluble in H_2O (0.1 - 0.5 g/100 ml), EtOH, very soluble in Et_2O; LD$_{50}$ (rat orl) = 316 mg/kg. *Elf Atochem N. Am.*

1278   Difenacoum

56073-07-5   G P

213

C31H24O3
Biphenyl-4-yl-1,2,3,4-tetrahydro-1-naphthyl)-4-hydroxy-1(2H)-benzopyran-2-one.
2H-1-Benzopyran-2-one, 3-(3-(1,1'-biphenyl)-4-yl-1,2,3, 4-tetrahydro-1-naphthalenyl)-4-hydroxy-; 3-(3-(1,1'-Bi-phenyl)-4-yl-1,2,3,4-tetrahydro-1-naphthalenyl)-4-hydroxy-2H-1-benzopyran-2-one; 3-Biphenyl-4-yl-1,2, 3,4-tetrahydro-1-naphthyl)-4-hydroxycoumarin; Coumarin, 3-(3-(4-biphenylyl)-1,2,3,4-tetrahydro-1-naph-thyl)-4-hydroxy-; Difenacoum; Difenakum; Diphena-coum; 3-(3-p-Diphenyl-1,2,3,4-tetrahydronaphth-1-yl) -4-hydroxycoumarin; EINECS 259-978-4; HSDB 6609; Neosorexa; Neosorexa PP580; Ratak; WBA 8107; Killgerm® Ratak Cut Wheat Rat Bait; Neosorexa; Ratak; Matikus; Talon. Anti-coagulant, used as a rodenticide. Difenacoum; a ready-to-use anticoagulant rodenticide. Registered by EPA as a rodenticide. Colorless crystals,; mp = 215 - 217°; slightly soluble in H2O (< 0.001 g/100 ml at pH 7), slightly soluble in alcohols, soluble in Me2CO (> 5 g/100 ml), CHCl3 (> 5 g/100 ml), EtOAc (0.2 g/100 ml), C6H6 (0.006 g/100 ml); LD50 (rat orl) = 1.8 mg/kg, (mmus orl) = 0.8 mg/kg, (rbt orl) = 2.0 mg/kg, (fgpg orl) = 50 mg/kg, (dog orl) > 50 mg/kg, (cat orl) = 100 mg/kg, (pig orl) > 50 mg/kg, (rat der) > 50 mg/kg, (rbt der) = 1000 mg/kg, (rat orl 5 days) = 0.16 mg/kg/day, (ckn orl) > 50 mg/kg. *Ace Chemicals Ltd.; ICI Agrochemicals; ICI Chem. & Polymers Ltd.; ICI Garden Products; Killgerm Chemicals Ltd.; Lever Industrial Ltd.; Sorex Ltd.*

1279   Diflubenzuron
35367-38-5   3166   G P

C14H9ClF2N2O2
1-(4-Chlorophenyl)-3-(2,6-difluorobenzoyl)urea.
AI3-29054; Astonex; Benzamide, N-(((4-chlorophenyl)-amino)carbonyl)-2,6-difluoro-; BRN 2162461; Caswell No. 346A; 1-(4-Chlorophenyl)-3-(2,6-difluorobenzoyl)-urea; Dimilin; ENT-29054; Micromite; Difluron; DU 112307; Duphacid; Largon; OMS-1804; PDD 6040-I; PH-60-40; TH-6040; 1-(p-Chlorophenyl)-3-(2,6-difluorobenzoyl)urea; Diflubenzuron; Difluron; Dimilin; Dimilin G1; Dimilin G4; Dimilin ODC-45; Dimilin WP-25; DU 112307; Duphacid; ENT 29054; EPA Pesticide Chemical Code 108201; HSDB 6611; Largon; Larvakil; Micromite; N-(((4-Chlorophenyl)amino)carbonyl)-2,6-difluorobenzamide; N-(4-Chlorophenylcarbamoyl)-2,6-difluorobenzamide; OMS 1804; PDD 6040I; PH 60-40; Philips-duphar PH 60-40; TH 6040; Thompson Hayward 6040; Urea, 1-(p-chlorophenyl)-3-(2,6-difluorobenzoyl)-. Registered by EPA as an insecticide and larvicide. Used in bait stations for below-ground termite control. Yellow-brown crystalline solid; mp = 230 - 232°, 239°; poorly soluble in H2O (0.00002 g/100 ml at 20°), soluble in Me2CO (0.65 g/100 ml at 20°), DMF (10.4 g/100 ml), dioxane (2 g/100 ml), moderately soluble in more polar solvents; LD50 (rat orl) = 10000 mg/kg, (mus orl) = 4640 mg/kg (formulation with 50% kaolin), (rbt der) > 2000 mg/kg, (mus ip) > 2150 mg/kg; LC50 (mallard duck, bobwhite quail 8 day dietary) > 4640 mg/kg diet, (rainbow trout 96 hr.) = 140 mg/l, (bluegill sunfish 96 hr.) = 1365 mg/l; non-toxic to bees. *Ensystex Inc.; Uniroyal.*

1280   Diflufenican
83164-33-4   3168   G P

C19H11F5N2O2
N-(2,4-Difluorophenyl)-2-[3-(trifluoromethyl)phenoxy]-3-pyridinecarboxamide.
BRN 4212494; Diflufenican; Diflufenicanil; M&B 38544; N-(2,4-Difluorophenyl)-2-(3-(trifluoromethyl)phenoxy)-3-pyridinecarboxamide. Selective contact and residual herbicide, used for control of broad-leaved weeds and some grasses, particularly *Galium, Veronica and Viola* species in cereal crops. Colorless crystals; mp = 161 - 162°; poorly soluble in H2O (0.000005 g/100 ml), soluble in Me2CO (7.9 g/100 ml), DMF (9.4 g/100 ml), acetophenone (5.2 g/100 ml), cyclohexanone (4.7 g/100 ml), isophorone (3.2 g/100 ml), xylene (1.7 g/100 ml), cyclohexane (< 0.78 g/100 ml), kerosene (< 0.8 g/100 ml); LD50 (rat orl) > 2000 mg/kg, (mus orl) > 1000 mg/kg, (rbt orl) > 5000 mg/kg, (rat der) > 2000 mg/kg, (quail orl) > 2150 mg/kg, (mallard duck orl) > 4000 mg/kg; LC50 (rat ihl 4 hr.) > 2.34 mg/l air, (rainbow trout 96 hr.) = 56 - 100 mg/l, (carp 96 hr.) = 105 mg/l. *Aventis Crop Science; Rhône-Poulenc.*

1281   Difluoroethane
75-37-6   H

C2H4F2
1,1-Difluoroethane.
4-01-00-00120 (Beilstein Handbook Reference); Algofrene type 67; BRN 1696900; Difluoroethane; Dymel 152; Dymel 152A; EINECS 200-866-1; Ethane, 1,1-difluoro-; Ethylene fluoride; Ethylidene difluoride; Ethylidene fluoride; FC 152a; Fluorocarbon 152a; Freon 152a; Genetron 100; Genetron 152a; Halocarbon 152a; HFC 152a; HSDB 5205; Hydrofluorocarbon 152a; Propellant 152a; R 152a; Refrigerant 152a. Used as a refrigerant. Gas; mp = -117°; bp = -24.9°; d^{20} = 0.896. 3902; *Dow Chem. U.S.A.; Fluorochem Ltd.; Molekula Fine Chemicals.*

1282   Difluoroethene
75-38-7   H

C2H2F2
1,1-Difluoroethene.
CCRIS 6744; EINECS 200-867-7; Ethene, 1,1-difluoro-; Ethylene, 1,1-difluoro-; Genetron 1132a; Halocarbon 1132A; HSDB 5206; NCI-C60204; NCI-C60208; UN1959; VDF; Vinylidene difluoride; Vinylidene fluoride. Used as a refrigerant. Gas; mp = -144°; bp = -85.7°; very soluble in organic solvents. *Dow Chem. U.S.A.; DuPont.*

1283   4-(Diglycidylamino)phenyl glycidyl ether
5026-74-4   H

C15H19NO4

p-(2,3-Epoxypropoxy)-N,N-bis(2,3-epoxypropyl)aniline.

Aniline, p-(2,3-epoxypropoxy)-N,N-bis(2,3-epoxypropyl); BRN 1393091; 4-(Diglycidylamino)phenyl glycidyl ether; EINECS 225-716-2; N-(4-(Oxiranylmethoxy)phenyl)-N-(oxiranylmethyl)oxiranemethanamine; Oxiranemeth-anamine, N-(4-(oxiranylmethoxy)phenyl)-N-(oxiranyl-methyl)-; p-(2,3-Epoxypropoxy)-N,N-bis(2,3-epoxy-propyl)aniline; TK 12759.

1284 Diglycolamine
929-06-6   G H

C4H11NO2

2-(2-Aminoethoxy)ethanol.

4-04-00-01412 (Beilstein Handbook Reference); 2-Aminoethoxyethanol; 2-(2-Aminoethoxy)ethanol; 5-Aminoethyl 2-hydroxyethyl ether; 2-Amino-2-hydroxydiethyl ether; 5-Hydroxy-3-oxapentylamine; BRN 0906728; Diethylene glycol amine; Diethylene glycol monoamine; Diglycolamine; Diglycolamine® Agent (DGA®); EINECS 213-195-4; Ethanol, 2-(2-aminoethoxy)-; HSDB 5770; 2-(2-Hydroxyethoxy)ethylamine; NSC 86108; UN3055. Solvent for removal of CO2 or H2S from gases, for recovery of aromatics from refinery streams, for preparation of foam stabilizers, wetting agents, emulsifiers, and condensation polymers. Liquid; mp = -12.5°; bp = 218-224°; d = 1.0500; soluble in H2O, organic solvents; LD50 (rat orl) = 5660 mg/kg. Texaco.

1285 Diglyme
111-96-6   3187   G H

C6H14O3

Diethylene glycol dimethyl ether.

4-01-00-02393 (Beilstein Handbook Reference); AI3-28583; Bis(2-methoxyethyl) ether; BRN 1736101; CCRIS 6212; Dariloid® 100; Di(2-Methoxyethyl) ether; Diethylene glycol dimethyl ether; Diglycol methyl ether; Diglyme; Dimethoxydiglycol; Dimethyl carbitol; Dricoid® 200; EINECS 203-924-4; Ethane, 1,1'-oxybis(2-methoxy-; Ether, bis(2-methoxyethyl); Glyme 2; HSDB 72; Methyldiglyme; NSC 59726; Poly-Solv. A solvent for polystyrene PVC/PVA copolymer and polymethyl methacrylate. Liquid; mp = -68°; bp = 162°; d^{20} = 0.9434; freely soluble in H2O, EtOH, Et2O. ICI Chem. & Polymers Ltd.

1286 Digoxin
20830-75-5   3189   D

C41H64O14

3-[(O-2,6-Dideoxy-β-D-ribohexopyranosyl-(1→4)-O-2,6-dideoxy-β-D-ribohexopyranosyl-(1→4)-2,6-dideoxy-β-D-ribohexopyranosyl)oxy]-14-hydroxycard-20(22)-enolide.

5-18-04-00381 (Beilstein Handbook Reference); Acygoxin; BRN 0077011; Cardigox; Cardiogoxin; Cardioxin; Cardoxin; Chloroformic digitalin; Coragoxine; Cordioxil; Davoxin; Digacin; Digoksyna; Digomal; Digon; Digonix; Digos; Digosin; Digossina; Digoxigenin-tridigitoxosid; Digoxin; Digoxin Nativelle; Digoxin-Sandoz; Digoxin-Zori; Digoxina-Sandoz; Digoxina; Digoxine; Digoxine Nativelle; Digoxinum; Dilanacin; Dimecip; Dixina; Dokim; Dynamos; EINECS 244-068-1; Eudigox; Fargoxin; Grexin; Hemigoxine Nativelle; Homolle's digitalin; HSDB 214; Lanacordin; Lanacrist; Lanatilin; Lanicor; Lanikor; Lanorale; Lanoxicaps; Lanoxin; Lanoxin PG; Lenoxicaps; Lenoxin; Lifusin; Longdigox; Mapluxin; Natigoxin; Neo-Lanicor; NeoDioxanin; Novodigal [inj.]; NSC 95100; Purgoxin; Rougoxin; Saroxin; SK-Digoxin; Stillacor; Vanoxin. Secondary glycoside from Digitalis purpurea L. Scrophulariaceae. Used as a cardiotonic. Dec = 230-265°; $[\alpha]_{Hg}^{25}$ = 13.4 - 13.8° (c = 10 C5H5N); λm = 220 nm (ε 12800 EtOH); soluble in EtOH, C5H5N; insoluble in CHCl3, Me2CO, EtOAc, H2O, Et2O. Glaxo Wellcome Inc.; Wyeth-Ayerst Labs.

1287 Dihexyl adipate
110-33-8   G H

C18H34O4

Di-n-hexyl hexanedioate.

4-02-00-01963 (Beilstein Handbook Reference); Adimoll® DH; Adipic acid, dihexyl ester; AI3-07963; BRN 1798668; Di-n-hexyl adipate; Dihexyl adipate; Dihexyl hexanedioate; EINECS 203-757-7; Hexanedioic acid, dihexyl ester; Di-n-hexyl hexanedioate; Plasticizer dihexyl adipate. Monomeric plasticizer. Bayer AG.

1288 Dihydroabietyl alcohol
26266-77-3   G

215

C20H34O
 Dodecahydro-1,4a-dimethyl-7-(1-methylethyl)-1-phenanthrenemethanol.
 Abietyl alcohol, dihydro-; Abitol® E; Arbitol E; Dihydroabietyl alcohol; Dodecahydro-1,4a-dimethyl-7-(1-methylethyl)-1-phenanthrenemethanol; (1R-(1α,4Aβ, 4bα,10aα))-Dodecahydro-7-isopropyl-1,4a-dimethyl-phenanthren-1-methanol; EINECS 247-574-0; Hydro-abietyl alcohol; 1-Phenanthrenemethanol, 1,2,3,4,4a,4b, 5,6,7,9,10,10a-dodecahydro-1,4a-dimethyl-7-(1-methyl-ethyl)-, (1R,4aR,4bS,10aR)-; 1-Phenanthrenemethanol, dodecahydro-1,4a-dimethyl-7-(1-methylethyl)-; Abietyl alcohol, dihydro-; Dihydroabietyl alcohol; Hydroabietyl alcohol; 1-Phenanthrenemethanol, dodecahydro-1,4a-dimethyl-7-(1-methylethyl)-. Resinous plasticizer and tackifier in plastics, lacquers, inks and adhesives; chemical intermediate. Hercules Inc.

1289   Dihydrogenated tallow dimethyl ammonium methosulfate
    61789-81-9   H

 Quaternary ammonium compounds, bis(hydrogenated tallowalkyl)dimethyl, methyl sulfates.
 Dihydrogenated tallow dimethyl ammonium methosulfate; EINECS 263-091-8; Quaternary ammonium compounds, bis(hydrogenated tallowalkyl)dimethyl, Me sulfates; Quaternium-18 methosulfate.

1290   Dihydrogenated tallow methylamine
    61788-63-4   H

 Amines, bis(hydrogenated tallow alkyl)methyl.
 Amines, bis(hydrogenated tallow alkyl)methyl; Dihydrogenated tallow methylamine; EINECS 262-991-8.

1291   Dihydroisophorone
    873-94-9   G

C9H16O
 3,5,5-Trimethylcyclohexanone.
 AI3-33978; Cyclohexanone, 3,3,5-trimethyl-; Dihydro-isophorone; Dihydroisophorone; EINECS 212-855-9; 3,3,5-Trimethylcyclohexanone; 3,3,5-Trimethylcyclo-hexan-1-one. A high boiling point ketone solvent used for surface coatings. Yellow oil; bp = 189°; d19 = 0.8919.

1292   Dihydrojasmone
    1128-08-1   G

C11H18O
 2-Pentyl-3-methyl-2-cyclopenten-1-one.
 4-07-00-00230 (Beilstein Handbook Reference); AI3-15185; 2-Amyl-3-methyl-2-cyclopenten-1-one; BRN 1906471; 2-Cyclopenten-1-one, 3-methyl-2-pentyl-; Dihydrojasmone; EINECS 214-434-5; FEMA No. 3763; Jasmone, dihydro-; 3-Methyl-2-(n-pentanyl)-2-cyclopenten-1-one; 3-Methyl-2-pentyl-2-cyclopenten-1-one; 3-Methyl-2-pentylcyclopent-2-enone; NSC 71928; 2-Pentyl-3-methyl-2-cyclopenten-1-one. Jasmine, fresh, fruity odor. Used in floral & citrus perfumes. Liquid; bp22 = 143°, bp12 = 116°; d18 = 0.9165. Quest

Intl.; Quest UK.

1293   3,4-Dihydro-2-methoxy-2H-pyran
    4454-05-1   H

C6H10O2
 2,3-Dihydro-2-methoxy(4H)pyran.
 5-17-03-00196 (Beilstein Handbook Reference); BRN 0107274; 2,3-Dihydro-2-methoxy(4H)pyran; 3,4-Di-hydro-2-methoxy-2H-pyran; EINECS 224-698-3; MDP; 2-Methoxy-3,4-dihydro-2H-pyran; NSC 44974; 2H-Pyran, 3,4-dihydro-2-methoxy-.

1294   6,13-Dihydroquinacridone
    5862-38-4   H

C20H14N2O2
 5,6,12,13-Tetrahydroquino(2,3-b)acridine-7,14-dione.
 6,13-Dihydroquinacridone; EINECS 227-508-7; Quino-(2,3-b)acridine-7,14-dione, 5,6,12,13-tetrahydro-; 5,6,12, 13-Tetrahydroquino(2,3-b)acridine-7,14-dione.

1295   Dihydroxyacetone
    96-26-0   3204   G

C3H6O3
 1,3-Dihydroxy-2-propanone.
 4-01-00-04119 (Beilstein Handbook Reference); AI3-24477; BRN 1740268; CCRIS 4899; Chromelin; Dihydroxyacetone; 1,3-Dihydroxyacetone; 1,3-Di-hydroxydimethyl ketone; 1,3-Dihydroxypropanone; Dihydroxyacetone; Dihyxal; EINECS 202-494-5; Ketochromin; NSC-24343; Otan; Oxantin; Oxatone; 2-Propanone, 1,3-dihydroxy-; Prtotsol; Soleal; Triulose; Viticolor. Used as cosmetic stain for suntan lotion and as a reagent in chemical synthesis; as a humectant, plasticizer, in fungicides. mp = 90°; soluble in H2O (100 g/100 ml), EtOH, Et2O, Me2CO. EM Ind. Inc.; Gist-Brocades Intl.; Janssen Chimica; Penta Mfg.; Spectrum Chem. Manufacturing.

1296   Di(4-hydroxy-3,5-di-t-butylphenyl)methane
    118-82-1   H

C29H44O2
 4,4'-Methylenebis(2,6-di-t-butylphenol).
 4-06-00-06811 (Beilstein Handbook Reference); 4,4'-Methylenebis(2,6-bis(1,1-dimethylethyl)phenol); 4,4'-

216

Methylenebis(2,6-di-t-butylphenol); Antioxidant E 702; Bimox M; Binox M; BRN 1916919; CCRIS 5836; Di(4-hydroxy-3,5-di-t-butylphenyl)methane; E 702; EINECS 204-279-1; Ethyl 702; Etil 702; Ionox 220; Ionox 220 antioxidant; L 3MB1; LZ-MB 1; MB 1 (Antioxidant); NSC 30551; Phenol, 4,4'-methylenebis(2,6-di-t-butyl-.

1297   Dihydroxyethyl cocamine oxide
61791-47-7   H

N,N-Bis(2-hydroxyethyl)cocamine oxide.
Amines, coco alkyl dihydroxyethyl, oxides; Bis(2-hydroxyethyl) cocoamine oxide; N,N-Bis(2-hydroxy-ethyl)cocamine oxide; Coco di-(hydroxyethyl) amine oxide; Dihydroxyethyl cocamine oxide; EINECS 263-180-1; Ethanol, 2,2'-iminobis-, N-coco alkyl derivs., N-oxides; 2,2'-Iminobisethanol, N-coco alkyl, N-oxide.

1298   Diiodomethyl-p-tolyl Sulfone
20018-09-1   G P

$C_8H_8I_2O_2S$
4-Methylphenyl diiodomethyl sulfone.
Amical 48; Benzene, 1-((diiodomethyl)sulfonyl)-4-methyl-; BRN 2212639; Caswell No. 353B; Diiodomethyl p-tolyl sulfone; 1-(Diiodomethyl)sulfonyl-4-methyl benzene; p-((Diiodomethyl)sulphonyl)toluene; EPA Pesticide Chem-ical Code 101002. Mildewcide, fungicide for latex paints, emulsions, caulks, adhesives and sealants, and in lumber, construction, home improvement, textile, and automotive industries. Registered by EPA as an antimicrobial, fungicide and herbicide. *Angus; Dow Chem. U.S.A.; Kop-Coat Inc.; Thomson Research Associates.*

1299   3,5-Diiodosalicylic acid
133-91-5   3212   G

$C_7H_4I_2O_3$
2-Hydroxy-3,5-diiodobenzoic acid.
4-27-00-07537 (Beilstein Handbook Reference); AI3-33355; Benzoic acid, 2-hydroxy-3,5-diiodo-; Benzoic acid, 3,5-diiodo-2-hydroxy-; BRN 2615358; 3,5-Diiodo-2-hydroxybenzoic acid; 3,5-Diiodosalicylate; Diosal; EINECS 205-124-0; 2-Hydroxy-3,5-diiodobenzoic acid; 2-Hydroxy-3,5-diiodobenzenecarboxylic acid; 2-Hydroxy-3,5-diiodobenzoate; NSC 6303; Salicylic acid, 3,5-diiodo-. An intermediate in the manufacture of thyroxine. Needles; mp = 235.5°; λ_m = 222, 274 nm (MeOH); insoluble in C_6H_6, $CHCl_3$, slightly soluble in H_2O, very soluble in EtOH, Et_2O.

1300   Diisobutyl adipate
141-04-8   G

$C_{14}H_{26}O_4$
Bis(2-methylpropyl) hexanedioate.

4-02-00-01962 (Beilstein Handbook Reference); Adipic acid bis(2-methylpropyl) ester; Adipic acid, diisobutyl ester; AI3-00995; Bis(2-methylpropyl) hexanedioate; BRN 1791084; DBE-IB; DIBA; Dibutyl adipate; Diisobutyl adipate; Diisobutyl hexanedioate; EINECS 205-450-3; Ftaflex DIBA; Hexanedioic acid, bis(2-methylpropyl) ester; Hexanedioic acid, diisobutyl ester; HSDB 345; Isobutyl adipate; NSC 6343; Plasthall® DIBA. Used as a plasticizer. Liquid; bp = 293°. Bp 15 = 187°; d^{19} = 0.9543; insoluble in H_2O, soluble in most organic solvents; LD_{50} (rat ip) = 5950 mg/kg. *Aceto Corp.; Hall C.P.; Sigma-Aldrich Fine Chem.*

1301   Diisobutylamine
110-96-3   H

$C_8H_{19}N$
N,N-Bis(2-methylpropyl)amine.
4-11-00-00122 (Beilstein Handbook Reference); AI3-15330; Amine, diisobutyl-;   Bis(β-methylpropyl)amine;   N,N-Bis(2-methylpropyl)amine; BRN 1209251; CCRIS 6232; Diisobutylamine; EINECS 203-819-3; HSDB 5543; 2-Methyl-N-(2-methylpropyl)-1-propanamine; 1-Prop-anamine, 2-methyl-N-(2-methylpropyl)-; UN2361. Liquid; mp = -73.5°; bp = 139.6°; d^{20} = 1.4090; soluble in Et_2O, Me_2CO, C_6H_6, slightly soluble in H_2O, CCl_4.

1302   Diisobutylene
107-39-1   H

C_8H_{16}
2,4,4-Trimethyl-1-pentene.
4-01-00-00892 (Beilstein Handbook Reference); AI3-30049; BRN 1098309; Diisobutylene; EINECS 203-486-4; HSDB 1442; 1-Methyl-1-neopentylethylene; NSC 73942; 1-Pentene, 2,4,4-trimethyl-; 2,2,4-Trimethyl-4-pentene; 2,4,4-Trimethyl-1-pentene. Liquid; mp = -93.5°; bp = 101.4°; d^{20} = 0.7150; λ_m = 193 nm (ε = 7079, C_6H_{12}); insoluble in H_2O, soluble in Et_2O, C_6H_6, CCl_3, $CHCl_3$, ligroin.

1303   Diisobutylene
25167-70-8   H

C_8H_{16}
Pentene, 2,4,4-trimethyl-.
Diisobutylene; EINECS 246-690-9; Pentene, 2,4,4-trimethyl-; 2,4,4-Trimethylpentene; UN2050.

1304   Diisobutyl ketone
108-83-8   G H

$C_9H_{18}O$
2,6-Dimethylheptan-4-one.
4-01-00-03360 (Beilstein Handbook Reference); AI3-11270; BRN 1743163; Caswell No. 355B; CCRIS 6233; Di-isobutylcetone; DIBK; Diisobutilchetone;   Diisobutyl-keton;   Diisobutyl ketone; Diisopropylacetone; s-Diiso-propylacetone; sec-Diisopropyl acetone;

sym-Diiso-propylacetone; 2,6-Dimethyl-4-heptanone; 2,6-Dimethyl-heptan-4-on; 2,6-Dimetil-eptan-4-one; EINECS 203-620-1; FEMA No. 3537; Heptanone, 2,6-dimethyl-, 4-; 4-Heptanone, 2,6-dimethyl-; HSDB 527; Isobutyl ketone; Isovalerone; NSC 15136; UN1157; Valerone. Solvent for nitrocellulose, rubber, synthetic resins; lacquers, coatings, organic synthesis, roll-coating inks, stains. Registered by EPA (cancelled). Liquid; mp = -41.5; bp = 169.4°; λ_m = 233 nm (ε = 287, cyclohexane); insoluble in H_2O; soluble in CCl_4, freely soluble in EtOH, Et_2O; LD_{50} (rat orl) = 4300 mg/kg. *Allchem Ind.; Degussa-Hüls Corp.; Eastman Chem. Co.; Sigma-Aldrich Co.; Union Carbide Corp.*

1305 Diisobutyl nonyl phenol
4306-88-1   G

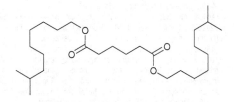

$C_{23}H_{40}O$
2,6-Di-t-butyl-4-nonylphenol.
EINECS 224-320-7; Phenol, 2,6-bis(1,1-dimethylethyl)-4-nonyl-; Uvi-Nox 1494. *Rhône Poulenc Surfactants.*

1306 Diisobutyl phthalate
84-69-5   G H

$C_{16}H_{22}O_4$
Diisobutyl 1,2-benzenedicarboxylate.
4-09-00-03177 (Beilstein Handbook Reference); AI3-04278; BRN 2054802; CCRIS 6193; Di-iso-butyl phthalate; Di(i-butyl)phthalate; DIBP; Diisobutyl phthalate; Diisobutylester kyseliny ftalove; EINECS 201-553-2; Hatcol DIBP; Hexaplas M/1B; HSDB 5247; Isobutyl phthalate; Kodaflex DIBP; NSC 15316; Palatinol IC; Phthalic acid, diisobutyl ester. Plasticizer Oil; mp = -64°; bp = 296.5°, bp4 = 159°; fp = 185°; d^{15} = 1.0490; soluble in CCl_4. *Eastman Chem. Co.*

1307 Diisobutyl sodium sulfosuccinate
127-39-9   3216   G

$C_{12}H_{21}NaO_7S$
Sulfo-butanedioic acid 1,4-bis(2-methylpropyl)ester sod-ium salt.
Aerosol® IB-45; AI3-18859; 1,4-Bis(2-methylpropyl)-sulfobutanyedioate, sodium salt; Butanedioic acid, sulfo-, 1,4-bis(2-methylpropyl) ester, sodium salt; EINECS 204-839-5; Rewopol® SBDB 45; Sodium 1,2-diisobutoxycarbonylethanesulphonate; Sodium 1,4-diisobutyl sulfosuccinate; Succinic acid, sulfo-, 1,4-diisobutyl ester, sodium salt; Sulfosuccinic acid, diisobutyl ester,

sodium salt. Emulsifier, wetting agent. Used in emulsion polymerization of styrene, butadiene and copolymers; dye and pigment dispersant. Used for leaching and electroplating. Biodegradable. *Am. Cyanamid; Rewo.*

1308 2,4'-Diisocyanatodiphenylmethane
5873-54-1   H

$C_{15}H_{10}N_2O_2$
Isocyanic acid, diester with 2,4'-methylenediphenol.
Benzene, 1-isocyanato-2-((4-isocyanatophenyl)methyl)-; 2,4'-Diisocyanatodiphenylmethane; 2,4'-Diphenyl-methanediisocyanate; EINECS 227-534-9; o-(p-Isocyanatobenzyl)phenyl isocyanate; Isocyanic acid, diester with 2,4'-methylenediphenol.

1309 Diisodecyl adipate
27178-16-1   H

$C_{26}H_{50}O_4$
Hexanedioic acid, diisodecyl ester.
Adipic acid, diisodecyl ester; Diisodecyl adipate; EINECS 248-299-9; Hexanedioic acid, diisodecyl ester.

1310 Diisodecyl azelate
28472-97-1   H

$C_{29}H_{56}O_4$
Azelaic acid, diisodecyl ester.
Azelaic acid, diisodecyl ester; Diisodecyl azelate; EINECS 249-044-4; Nonanedioic acid, diisodecyl ester.

1311 Diisodecyl phenyl phosphite
25550-98-5   H

C26H47O3P
Phosphorous acid, diisodecyl phenyl ester.
Diisodecyl phenyl phosphite; Diisodecylphenyl phos-phite; EINECS 247-098-3; Phosphorous acid, diisodecyl phenyl ester.

1312 Diisodecyl phthalate
26761-40-0   H

C28H46O4
1,2-Benzenedicarboxylic acid, diisodecyl ester.
1,2-Benzenedicarboxylic acid, diisodecyl ester; Bis(isodecyl) phthalate; BRN 2171889; CCRIS 6194; Di(i-decyl) phthalate; Didp; DIDP (plasticizer); Diisodecyl phthalate; EINECS 247-977-1; HSDB 930; Palatinol Z; Phthalic acid, bis(8-methylnonyl) ester; Phthalic acid, diisodecyl ester; Plasticized ddp; PX 120; Sicol 184; Vestinol DZ.

1313 Diisodecyl phthalate
68515-49-1   G

C28H46O4
1,2-Benzenedicarboxylic acid, diisodecyl ester.
1,2-Benzenedicarboxylic acid, di-C9-11-branched alkyl esters, C10-rich; Di(C9-C11) branched alkyl phthalate; DIDP; Diplast® R; EINECS 271-091-4; Jayflex® DIDP; Nuoplaz® DIDP; Palatinol® DIDP; Phthalic acid, di-C9-11-branched alkyl esters, C10-rich; Plasthall® DIDP-E; PX-120. Used as a plasticizer. Liquid; bp = 250-257°; d = 0.966; insoluble in H2O, soluble in organic solvents; fp = -50°; viscosity = 108 cps; flash point = 450°F. *Allchem Ind.; Aristech; Ashland; BASF Corp.; Coyne; Exxon; Hall C.P.; Harwick; Hatco; Hoechst AG; OxyChem.*

1314 Diisononyl adipate
33703-08-1   G H

C24H46O4
Adipic acid, diisononyl ester.
Adimoll® DN; Adipic acid, diisononyl ester; CCRIS 2004; Diisononyl adipate; Diisononyl hexanedioate; EINECS 251-646-7; Hexanedioic acid, diisononyl ester; PX-209. A reagent grade, monomeric plasticizer. *Aristech; Bayer AG.*

1315 Diisononylnaphthalene
63512-64-1   H

C28H44
Naphthalene, diisononyl-.

Diisononylnaphthalene; EINECS 264-290-2; Naph-thalene, diisononyl-.

1316 Diisononyl phthalate
28553-12-0   3319   H

C26H42O4
1,2-Benzenedicarboxylic acid, diisononyl ester.
Baylectrol 4200; 1,2-Benzenedicarboxylic acid, diisononyl ester; CCRIS 6195; Diisononyl phthalate; DINP; DINP2; DINP3; EINECS 249-079-5; Enj 2065; ENJ 2065; HSDB 4491; Isononyl alcohol, phthalate (2:1); JAY-DINP; Jayflex DINP; Palatinol DN; Palatinol N; Phthalic acid, diisononyl ester; Phthalisocizer DINP; Sansocizer DINP; Vestinol 9; Vestinol NN; Vinylcizer 90; Witamol 150.

1317 Diisononyl phthalate
68515-48-0   G

C26H42O4
Di(isononyl) phthalate branched.
1,2-Benzenedicarboxylic acid, di-C8-10-branched alkyl esters, C9-rich; CCRIS 7927; Di(C8-C10) branched alkyl phthalate; Di(isononyl) phthalate branched; Diisononyl phthalate, technical grade; EINECS 271-090-9. Plasticizer. *BASF Corp.; Chemisphere; Chisso Corp.; Hall C.P.*

1318 Diisooctyl adipate
1330-86-5   G H

C22H42O4
Adipic acid, diisooctyl ester.
Adipic acid, diisooctyl ester; Adipol 10A; Ageflex FA-10; DB 32 (ester); Diisooctyl adipate; Dimethylheptyl adipate; Dioa; EINECS 215-553-5; Hatcol 2906; Hexanedioic acid, diisooctyl ester; HSDB 5813; HSDB 615; Monoplex® DIOA; Isooctyl adipate; Isooctyl alcohol, adipate (2:1); Plasthall® DIOA; PX 208. Used as a plasticizer. *Hall C.P.*

1319 Diisooctyl phosphite
36116-84-4   G

C16H35PO3
Diisooctyl phosphonate.
Di-isooctyl phosphite; Diisooctyl phosphonate; Doverphos® DIOP;

EINECS 252-873-4; Phosphonic acid, diisooctyl ester. Stabilizer for PVC, PP; lubricant additive. Liquid; bp = 129°; d = 0.925-0.933; flash point = 146°. *Dover.*

1320 O,O'-Diisoctylphosphorodithioic acid
26999-29-1   H

C16H35O2PS2

Phosphorodithioic acid, O,O-diisooctyl ester.
EINECS 248-161-8; O,O-Diisooctyl hydrogen dithiophosphate; O,O'-Diisooctylphosphorodithioic acid; O,O'-Diisooctyl phosphorodithioic acid; Phosphoro-dithioic acid, O,O-diisooctyl ester.

1321 Diisooctyl phthalate
27554-26-3   G H

C24H38O4

Phthalic acid, bis(6-methylheptyl)ester.
AI3-27697-X (USDA); 1,2-Benzenedicarboxylic acid, diisooctyl ester; Corflex 880; Di-iso-octyl phthalate; Diisooctyl 1,2-benzenedicarboxylate; Diisooctyl phthalate; Diop; EINECS 248-523-5; Flexol Plasticizer DIOP; Genomoll 100; Hexaplas M/O; HSDB 588; Isooctyl phthalate; NSC 6381; Phthalic acid, bis(6-methylheptyl)ester; Phthalic acid, diisooctyl ester; Witcizer 313. General purpose plasticizer.

1322 Diisopropanolamine
110-97-4   3218   G H

C6H15NO2

N,N-Bis(2-hydroxypropyl)amine.
3-04-00-00761 (Beilstein Handbook Reference); Bis(2-hydroxypropyl)amine; N,N-Bis(2-hydroxypropyl)amine Bis(2-propanol)amine; BRN 0605363; CCRIS 6234; Di-2-propanolamine; Diisopropanolamine; DIPA; Dipropyl-2,2'-dihydroxy-amine; EINECS 203-820-9; HSDB 338; NSC 4963. Emulsifying agent used in polishes, textile specialties, leather compounds, insecticides, cutting oils, aqueous paints. Crystals; mp = 44.5°; bp = 250°; bp23 = 151°; d^{20} = 0.989; soluble in EtOH, H2O, slightly soluble in Et2O. *BASF Corp.*

1323 Diisopropyl adipate
6938-94-9   G H

C12H22O4

Hexanedioic acid, bis(1-methylethyl) ester.
4-02-00-01961 (Beilstein Handbook Reference); Adipic acid, diisopropyl ester; AI3-06066; Bis(1-methylethyl)hexanedioate; BRN 1785346; Ceraphyl 230; Crodamol DA; β-dia; EINECS 230-072-0; Iso-adipate 2/043700; Isopropyl adipate; NSC 56587; Prodipate; Schercemol DIA; Standamul DIPA; Tegester 504-D; Unimate® DIPA; Wickenol 116. Used in cosmetics. Liquid; mp = -1.1°; bp6.5 = 120°; d^{20} = 0.9569; very soluble in Me2CO, EtOH, Et2O. *Union Camp.*

1324 Diisopropylamine
108-18-9   3219   G H

C6H15N

N-Isopropyl-1-amino-2-methylethane.
AI3-15345; CCRIS 6235; Diisopropylamine; DIPA; EINECS 203-558-5; HSDB 931; N-(1-Methylethyl)-2-propanamine; N-Isopropyl-1-amino-2-methylethane; NSC 6758; UN1158. Intermediate, catalyst. Liquid; mp = -61°; bp = 83.9°; d^{20} = 0.7153; λ_m = 301 nm (ε = 7, gas); very soluble in EtOH, Et2O, Me2CO, C6H6; LD50 (rat orl) = 770 mg/kg. *Air Prods & Chems; BASF Corp.; Elf Atochem N. Am.; Sigma-Aldrich Fine Chem.; Union Carbide Corp.*

1325 Diisopropylbenzene
25321-09-9   H

C12H18

Diisopropylbenzenes.
Benzene, bis(1-methylethyl)-; Benzene, diisopropyl-; Bis(1-methylethyl)benzene; Diisopropylbenzene; Diiso-propylbenzenes; EINECS 246-835-6; HSDB 6500. Mixture of o, m and p-isomers.

1326 m-Diisopropylbenzene
99-62-7   G H

C12H18

Benzene, 1,3-bis(1-methylethyl)-.
4-05-00-01125 (Beilstein Handbook Reference); AI3-08224; Benzene, 1,3-bis(1-methylethyl)-; Benzene, m-diisopropyl-; BRN 1905828; m-Diisopropylbenzene; m-Diisopropylbenzol; EINECS 202-773-1; HSDB 5325. Used as a solvent and chemical intermediate. mp = -63.1°; bp = 203.2°; d = 0.8560; λ_m = 211, 257, 263, 270 nm (ε = 8970, 232, 282, 236, MeOH); insoluble in H2O, freely soluble in EtOH, Et2O, Me2CO, C6H6, CCl4; LD50 (rat orl) = 7400 mg/kg. *Robert Koch Industries.*

1327   p-Diisopropylbenzene
100-18-5   G H

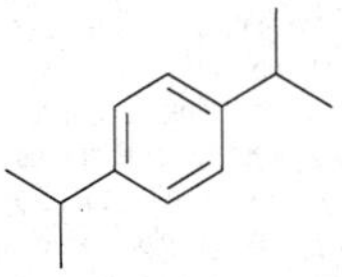

C12H18

Benzene, 1,4-bis(1-methylethyl)-.

4-05-00-01126 (Beilstein Handbook Reference); Benzene, 1,4-bis(1-methylethyl)-; Benzene, p-diisopropyl-; BRN 1854739; p-Diisopropylbenzene; p-Diisopropylbenzol; para-Diisopropylbenzene; EINECS 202-826-9; HSDB 5331; NSC 84198. Used as a solvent and chemical intermediate. Oil; mp = -17°; bp = 210.3°; d^{20} = 0.8568; λ_m = 212, 217, 263, 271 nm (ϵ = 27700, 26500,1460, 1440, MeOH); insoluble in H_2O, freely soluble in EtOH, Et_2O, Me_2CO, C_6H_6, CCl_4; LD_{50} (mus orl) = 3400 mg/kg. *Robert Koch Industries.*

1328   N,N-Diisopropyl-2-benzothiazolesulfen-amide
95-29-4   H

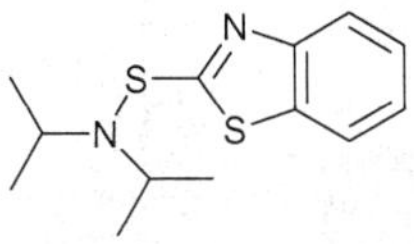

C13H18N2S2

2-Benzothiazolesulfenamide, N,N-bis(1-methylethyl)-.

4-27-00-01866 (Beilstein Handbook Reference); N,N-Bis(isopropyl)-2-benzothiazolesulfenamide; BRN 0177776; DIBS; N,N-Diisopropyl-2-benzothiazolesulfen-amide; Dipac (accelerator); DIPAC; DIPAK; EINECS 202-407-0; HSDB 5287; Meramid P; Santocure IPS.

1329   Diisopropylbiphenyl
36876-13-8   H

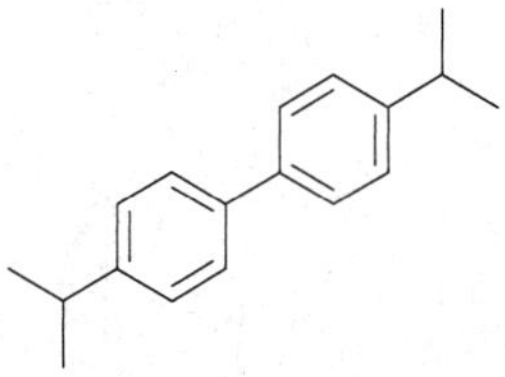

C18H22

1,1'-Biphenyl, ar,ar'-bis(1-methylethyl)-.

1,1'-Biphenyl, ar,ar'-bis(1-methylethyl)-; Dicumyl; Diisopropylbiphenyl; EINECS 253-247-3; HSDB 6166; Isopropyl(isopropylphenyl)benzene. Mixture of o, m and p-isomers.

1330   Diisopropylbiphenyl
69009-90-1   G

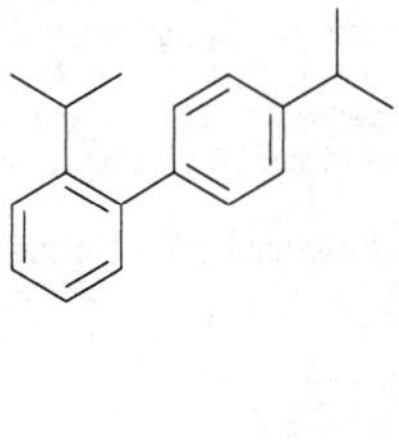

C18H22

Bis(methylethyl)-1,1'-biphenyl.

1,1'-Biphenyl, bis(1-methylethyl)-; Diisopropyl-1,1'-biphenyl; Diisopropylbiphenyl; EINECS 273-683-8; Nusolv ABP-103. Mixture of o, m and p-isomers. Solvent possessing excellent solvency, chemical and thermal stability, nonvolatility; for Aerosols®, adhesives, electronic parts manufacturing and cleaning, industrial cleaning, inks and paper coatings, metal cleaning, paints, plastics, sealants, tile manufacturing. *Ridge Tech.*

1331   Diisopropyl ether
108-20-3   5232   H

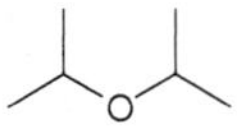

C6H14O

2,2'-Oxybispropane.

AI3-24270; Bis(isopropyl) ether; Diisopropyl ether; Diisopropyl oxide; EINECS 203-560-6; Ether, isopropyl; Ether isopropylique; HSDB 624; Isopropyl ether; Izopropylowy eter; Propane, 2,2'-oxybis-; UN1159. Liquid; mp = -86.8°; bp = 68.5°; d^{20} = 0.7241; slightly soluble in H_2O, soluble in Me_2CO, CCl_4, freely soluble in EtOH, Et_2O.

1332   Diisopropyl peroxydicarbonate
105-64-6   H

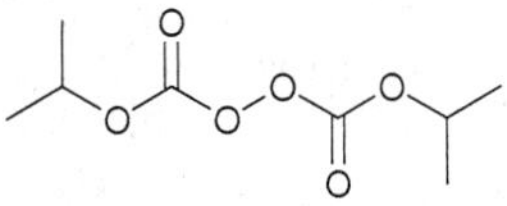

C8H14O6

Peroxydicarbonic acid, bis(1-methylethyl) ester.

4-03-00-00019 (Beilstein Handbook Reference); Bisisopropyl peroxydicarbonate; BRN 1786996; Diisopropyl perdicarbonate; Diisopropyl peroxy-dicarbonate; Diisopropyl peroxydiformate; EINECS 203-317-4; HSDB 349; Isopropyl percarbonate; Isopropyl peroxydicarbonate; Luperox IPP; Peroxydicarbonate d'isopropyle; Peroxydicarbonic acid, bis(1-methylethyl) ester; Peroxydicarbonic acid, diisopropyl ester.

1333   Diisopropyl phosphorodithioate
107-56-2   H

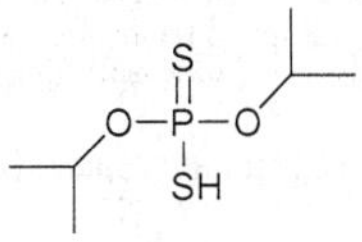

C6H15O2PS2

Phosphorodithioic acid, O,O-diisopropyl ester.

Aerofloat 211; CP 41348; Diisopropyl phosphoro-dithioate; EINECS 203-492-7; HSDB 5379; Isopropyl aerofloat; Isopropyl phosphorodithioate; NSC 15258; O,O-Diisopropyl dithiophosphate; Phosphorodithioic acid, O,O-diisopropyl ester; Phosphorodithioic acid, O,O-bis(1-methylethyl) ester; Phosphorodithioic acid, O,O-diisopropyl ester.

1334   Diisopropyl sebacate
7491-02-3   G

C16H30O4

Decanedioic acid, bis(1-methylethyl) ester.

Bis(1-methylethyl)decanedioate; Diisopropyl sebacate; EINECS 231-306-4; Unimate® DIPS; Wickenol 117. Skin emollient, solubilizer, coupler, spreading agent for creams, lotions, bath oils, Aerosol® toilet preparations. *Union Camp.*

1335   Diketene
674-82-8   H

C4H4O2

4-Methylene-2-oxetanone.

5-17-09-00115 (Beilstein Handbook Reference); Acetyl ketene; BRN 0104541; But-3-en-3-olide; 3-Buteno-β-lactone; 3-Butenoic acid, 3-hydroxy-, β-lactone; Diketene; EINECS 211-617-1; Ethenone, dimer; HSDB 2063; Ketene, dimer; 4-Methylene-2-oxetanone; NSC 93783; 2-Oxetanone, 4-methylene-; UN2521; Vinyl-aceto-β-lactone. Liquid; mp = -6.5°; bp = 126.1°; d^{20} = 1.0877.

1336   Dilauryl phosphite
21302-09-0   G

C24H51O3P

Di(n-dodecyl)hydrogen phosphate.

BRN 2056801; Chelex H 12; Didodecyl phosphite; Di-n-dodecyl phosphite; Didodecyl phosphonate; Dilauryl hydrogen phosphite; Dilauryl phosphite; Doverphos® 271L; Doverphos® 274; Duraphos™ AP-230; EINECS 244-325-8; JP 212; NSC 41924; Phosphonic acid, didodecyl ester; Weston® DLP. Antioxidant and EP additive for lubricants; catalyst in polymerization of unsaturated compounds; stabilizer. Oil; d = 0.898-0.906; flash point = 138°; LD_{50} (rat orl) >10 g/kg. *Albright & Wilson Americas Inc.; CK Witco Chem. Corp.; GE Specialities; Nichia Kagaku Kogyo.*

1337   Dilauryl 3,3'-thiodipropionate
123-28-4   G H

C30H58O4S

Propanoic acid 3,3'-thiobis(didodecyl) ester.

4-03-00-00744 (Beilstein Handbook Reference); Advastab 800; AI3-25277; Antioxidant AS; Antioxidant LTDP; Argus DLTDP; Bis(dodecyloxycarbonylethyl) sulfide; BRN 1808848; Carstab® DLTDP; CCRIS 3936; Cyanox® LTDP; D 1 (antioxidant); Didodecyl 3,3'-thiodipropio-nate; Dilauryl thiodipropionate; Dilauryl 3,3'-thiodi-propionate; Dilaurylester kyseliny β',β'-thiodipropionove; Dilaurylthiodipropionate; DLT; Dltdp; DLTP; Dmptp; EINECS 204-

614-1; HSDB 353; Ipognox 89; Irganox PS 800; Lauryl 3,3'-thiodipropionate; Lusmit; Milban F; Neganox DLTP; NSC 65494; Plastanox LTDP; Plastanox LTDP Antioxidant; Propanoic acid, 3,3'-thiobis-, didodecyl ester; Stabilizer DLT; Thiobis(dodecyl propionate); Thiodipropionic acid, dilauryl ester; Tyox B. Antioxidant used for polyolefins, thermoplastic elastomers synthetic rubber, used in cosmetics and pharmaceuticals and as a secondary antioxidant in ABS, PP, and polyethylene, in food packaging materials, automotive, appliance, battery casing, pipe; stabilization of oils, lubricants, sealants, and adhesives. A heat aging stabilizer used in conjunction with primary antioxidants; for PP, HDPE. Solid; mp = 40°; bp = 240°; d = 0.975; insoluble in H2O; soluble in organic solvents; LD_{50} (rat orl) >10.3 g/kg. *Am. Cyanamid; CK Witco Corp.; Evans Chemetics; Morton Intn'l.*

1338   Dilinoleic acid
6144-28-1   G

C36H64O4

9,12-Octadecadienoic acid, dimer.

Dimer acid; Empol® 1016; Empol® 1020; Empol® 1022, 1004, 1026; Industrene® D; Pripol 1017, 1022, 1025. Dicarboxylic acid formed by the catalytic dimerization of linoleic acid; used as a lubricant, corrosion inhibitor, mildness additive in household detergents, plastics, and protective coatings. *Arizona; Henkel/Emery; Sherex; Union Camp; Witco/Humko.*

1339   Diltiazem
42399-41-7   3226   D

C22H26N2O4S

(+)-5-[2-(Dimethylamino)ethyl]-cis-2,3-dihydro-3-hydroxy-2-(p-methoxyphenyl)-1,5-benzothiazepin-4(5H)-one acetate (ester).

Acalix; BRN 3573079; Dilcontin; Dilta-Hexal; Diltiazem; d-cis-Diltiazem; Diltiazemum; Dilticard; Dilzen; EINECS 255-796-4; Endrydil; HSDB 6528; Incoril AP. Antianginal; antihypertensive; antiarrhythmic (class IV). Calcium channel blocker with coronary vasodilating properties. *Bristol Myers Squibb Pharm. Ltd.; Forest Pharm. Inc.; Hoechst AG (USA); Lemmon Co.; Rhône-Poulenc Rorer Pharm. Inc.; Shionogi & Co. Ltd.; Tanabe Seiyaku Co. Ltd.*

1340   d-cis-Diltiazem hydrochloride
33286-22-5   3226   D

C22H27ClN2O4S

(+)-5-[2-(Dimethylamino)ethyl]-cis-2,3-dihydro-3-hydroxy-2-(p-methoxyphenyl)-1,5-benzothiazepin-4(5H)-one acetate (ester) monohydrochloride.

Adizem; Adizem-CD; Altiazem; Altiazem Retard; Altiazem RR; Anginyl; Angiotrofin; Angiotrofin Retard; Angitil; Angizem; Anzem; Apo-diltiazem; Bi-Tildiem; Britiazim; Bruzem; Calcicard; Calnurs; Cardiazem; Cardil; Cardil Retard; Cardizem; Cardizem CD; Cardizem Retard; Carex; Carzem; Cirilen; Cirilen AP; Citizem; Citizen; Clarute; Coras; Corazet; Cormax; CRD-401; Dazil; Deltazen; Diacor; Diatal; Dil-Sonaramia; Dilacor XR; Dilacor XR Extended Release Capsules; Diladel; Dilatam; Dilatam 120; Dilatame; Dilcard; Dilcor; Dilem; Dilfar; Dilgard; Dilicardin; Dilpral; Dilren; Dilrene; Dilsal; Dilso; Diltahexal; Diltam; Diltan; Diltan SR; Diltelan; Diltiasyn; Diltiazem AWD; Diltiazem Basics; Diltiazem chloridrate; Diltiazem-Cophar; Diltiazem Eu Rho; Diltiazem GNR; Diltiazem-GRY; Diltiazem HCl; Diltiazem Henning; Diltiazem hydrochloride; Diltiazem-Isis; Diltiazem-Mepha; Diltiazem Merck; Diltiazem MSD; Diltiazem Stada; Diltiazem UPSA; Diltiazem Verla; Diltikor; Diltime; Dilzem; Dilzem Retard; Dilzem RR; Dilzene; Dilzereal 90 Retard; Dilzicardin; Dinisor; Dinisor Retard; Doclis; Dodexen; Dodexen A.P.; Dyalac; EINECS 251-443-3; Entrydil; Etizen; Etyzen; Farmabes; Gadoserin; Hart; Helsibon; Herben; Herbesser; Herbesser 60; Herbesser 90 SR; Herbesser 180 SR; Hesor; Incoril; Iski; Iski-90 SR; Kaltiazem; Kardil; Lacerol; Levozem; Longazem; Lytelsen; Masdil; Metazem; Miocardie; Mono-Tildiem; Myonil; Myonil Retard; Novo-Diltazem; Oxycardil; Pazeadin; Pentilzeno; Poltiazem; Presoken; Presokin A. P.; RG-83606; Slozen; Surazem; Syn-Diltiazem; Tazem; Taztia XT; Tiadil; Tiaves; Tiazac; Tilazem; Tilazem 90; Tilazem AS 60; Tilazem AS 90; Tildiem; Tildiem CR; Tildiem LA; Tildiem Retard; Trumsal; Ubicor; Uni Masdil; Viazem SR; Viazem XL; WL Diltiazem; Zildem; Zilden; Ziruvate. Antianginal; antihypertensive; class IV antiarrhythmic. Calcium channel blocker with coronary vasodilating properties. mp = 207.5-212°; $[\alpha]_D^{24}$ = +98.3 ± 1.4° (c = 1.002 in MeOH); soluble in H2O, MeOH, CHCl3; slightly soluble in absolute EtOH; practically insoluble in C6H6; LD50 (mmus iv) = 61 mg/kg, (mmus sc) = 260 mg/kg, (mmus orl) = 740 mg/kg, (fmus iv) = 58 mg/kg, (fmus sc) = 280 mg/kg, (fmus orl) = 640 mg/kg, (mrat iv) = 38 mg/kg, (mrat sc) = 520 mg/kg, (mrat orl) = 560 mg/kg, (frat iv) = 39 mg/kg, (frat sc) = 550 mg/kg, (frat orl) = 610 mg/kg. *Bristol Myers Squibb Pharm. Ltd.; Forest Pharm. Inc.; Hoechst AG (USA); Lemmon Co.; Rhône-Poulenc Rorer Pharm. Inc.; Tanabe Seiyaku Co. Ltd.*

1341   Diltiazem malate
144604-00-2   D

C26H32N2O9S

(+)-5-[2-(Dimethylamino)ethyl]-cis-2,3-dihydro-3-hydroxy-2-(p-methoxyphenyl)-1,5-benzothiazepin-4(5H)-one acetate (ester) (S)-malate (1:1) monohydrochloride.

Diltiazem malate; MK-793. Calcium channel blocker with coronary vasodilating activity. Class IV antiarrhythmic. Antianginal; antihypertensive; class IV antiarrhythmic. *Bristol Myers Squibb Pharm. Ltd.; Forest Pharm. Inc.; Hoechst AG (USA); Lemmon Co.; Rhône-Poulenc Rorer Pharm. Inc.*

1342   Dimefuron
34205-21-5   G P

C15H19ClN4O3

N'-[3-Chloro-4-[5-(1,1-dimethylethyl)-2-oxo-1,3,4-oxa-diazol-3(2H)-yl]phenyl]-N,N-dimethylurea.

BRN 0628305; 3-(4-(5-t-Butyl-2,3-dihydro-2-oxo-1,3,4-oxadiazol-3-yl)-3-chlorophenyl)-1,1-dimethylurea; 3-(4-(5-(tert-Butyl)-2-oxo-1,3,4-oxadiazol-3(2H)-yl)-3-chloro-phenyl)-1,1-dimethylurea; N'-(3-Chloro-4-(5-(1,1-di-methylethyl)-2-oxo-1,3,4-oxadiazol-3(2H)-yl)phenyl)-N,N-dimethylurea; Dimefuron; Legurame TS; Pradone kombi; Pradone plus; Pradone TS; RP 23465; Urea, 3-(4-(2-t-butyl-5-oxo-Δ^2-1,3,4-oxadiazolin-4-yl)-3-chloro-phenyl)-1,1-dimethyl-; Vt 2809. Herbicide. Crystals; mp = 193°; soluble in H2O (0.0016 g/100 ml), very soluble in CHCl3, soluble in Me2CO, CH3CN, acetophenone, EtOH, slightly soluble in C6H6, C7H8, xylene; LD50 (rat orl) > 2000 mg/kg, (mus orl) > 10000 mg/kg, (dog orl) > 2000 mg/kg, (rbt der) > 1000 mg/kg, (rat ip) = 4000 mg/kg; LC50 (rainbow trout, bluegill sunfish 96 hr.) > 1000 mg/l; non-toxic to bees. *Aventis Crop Science; May & Baker Ltd.; Rhône-Poulenc.*

1343   Dimethirimol
5221-53-4   3243   G P

C11H19N3O

5-Butyl-2-(dimethylamino)-6-methyl-4(1H)-pyrimidinone.

BRN 0746334; 5-Butyl-2-dimethylamino-4-hydroxy-6-methylpyrimidine; 5-Butyl-2-(dimethylamino)-6-methyl-4-pyrimidinol; 5-Butyl-2-(dimethylamino)-6-methyl-4(1H)-pyrimidinone; 5-n-Butyl-2-dimethylamino-4-hydroxy-6-methylpyrimidine; Caswell No. 128A; Dimethirimol; 2-(Dimethylamino)-4-hydroxy-5-N-butyl-6-methylpyrimidine; 2-Dimethylamino-4-hydroxy-5-n-butyl-6-methyl-pyrimidine; Dimethyrimol; EINECS 226-021-7; EPA Pesticide Chemical Code 228200; Melkeb; Methyrimol; Milcurb; NSC 263490; PP 675; 4-Pyrimidinol, 5-butyl-2-dimethylamino-6-methyl-; 4(1H)-Pyrimidinone, 5-butyl-2-(dimethylamino)-6-methyl-; 2-Dimethylamino-4-methyl-5-n-butyl-6-hydroxypyrimidine; Dimethyrimol; EPA Pesticide Chemical Code 228200; Melkeb; Methyrimol; Milcurb; NSC 263490; PP 675; 4-Pyrimidinol, 5-butyl-2-dimethylamino-6-methyl-; 4(1H)-Pyrimidinone, 5-butyl-2-(dimethylamino)-6-methyl-. Systemic fungicide with protective and curative action. Used as soil application for control of powderyt mildews in curcubits, tobacco, capsicum, tomatoes and some ornamentals. Colorless needles; mp = 102°; soluble in H_2O (0.12 g/100 ml), $CHCl_3$ (120 g/100 ml), xylene (36 g/100 ml), EtOH (6.5 g/100 ml), Me_2CO (4.5 g/100 ml); LD_{50} (rat orl) = 2350 mg/kg, (mus orl) = 800 - 1600 mg/kg, (gpg orl) = 500 mg/kg, (rat ip) = 200 - 400 mg/kg, (hen orl) = 4000 mg/kg; LC_{50} (brown trout 24 hr.) = 42 mg/l, (brown trout 48 hr.) = 33 mg/l, (brown trout 96 hr.) = 28 mg/l; non-toxic to bees. *ICI Agrochemicals; Syngenta Crop Protection.*

1344 Dimethoate
60-51-5 3246 G H P

$C_5H_{12}NO_3PS_2$
O,O-Dimethyl-S-(N-methylcarbamoylmethyl) phosphorodithioate.
4-04-00-00252 (Beilstein Handbook Reference); 8014 Bis HC; Aadimethoal; AC-12880; AC-18682; Acetic acid, O,O-dimethyldithiophosphoryl-, N-monomethylamide salt; AI3-24650; American Cyanamid 12880; BI-58; BI 58 EC; BRN 1785339; Caswell No. 358; CCRIS 245; Cekuthoate; CL 12880; Cygon; Cygon 2-E; Cygon 4E; Cygon 400; Cygon insecticide; Daphene; De-fend; Defend; Demos-L40; Devigon; Dimate 267; Dimetate; Dimethoaat; Dimethoat; Dimethoate; Dimethoate 30 EC; Dimethogen; Dimethyl phosphorodithioate,; Dimethyl S-((methylcarbamoyl)methyl) phosphorodithioate; Dimeton; Dimevur; EI-12880; EINECS 200-480-3; ENT 24,650; EPA Pesticide Chemical Code 035001; Experimental insecticide 12,880; Ferkethion; FIP; Fortion NM; Fosfamid; Fosfatox R; Fosfotox; Fosfotox R; Fosfotox R 35; Fostion MM; HSDB 1586; L-395; Lurgo; NCI-C00135; OMS 94; PEI 75; Perfecthion; Perfekthion; Phosphamid; Phosphamide; Racusan; RCRA waste number P044; Rebelate; Rogodan; Rogor; Rogor 20L; Rogor 40; Rogor L; Rogor P; Roxion; Roxion UA; Salut; Sevigor; Sinoratox; Sistemin; Solut; Systemin: Systoate; Tara; Tara 909; Trimetion; Turbair. Registered by EPA as an avicide and insecticide. An insecticide and acaricide with contact and plant systemic activity suitable for protection against a broad range of insects and mites. Grey-white crystals; mp = 52-53°; $bp_{0.05}$ = 107°; d = 1.281; sg^{65} = 1.277; soluble in H_2O (2.5 g/100 ml), alcohols, ketones, C_6H_6, C_7H_8, $CHCl_3$, CH_2Cl_2 (all > 25 g/100 ml), slightly soluble in xylene, CCl_4, aliphatic hydrocarbons; LD_{50} (rat orl) = 290 - 325 mg/kg, (mus orl) = 160 mg/kg, (rbt orl) = 400 - 500 mg/kg, (gpg orl) = 600 mg/kg, (mpheasant orl) = 15 mg/kg, (quail orl) = 84 mg/kg, (ckn orl) = 108 mg/kg, (fmallard duck orl) = 40 mg/kg, (bee orl) = 0.00015 mg/bee, (bee contact) = 0.00012 mg/bee), (rat der) > 800 mg/kg, (gpg der) > 1000 mg/kg; LC_{50} (rat ihl 4 hr.) > 0.2 mg/l air, (mosquito fish 96 hr.) = 40 - 60 mg/l, (rainbow trout 96 hr.) = 6.2 mg/l, (bluegill sunfish 96 hr.) = 6 mg/l; toxic to bees. *A/S Cheminova; Agriliance LLC; Am. Cyanamid; Amvac Chemical Corp.; BASF Corp.; Celaflor Gmbh; Dragon Chemical Corp.; Drexel Chemical Co.; Gowan Co.; Helena Chemical Co.; Micro-Flo Co. LLC; Pan Britannica Industries Ltd.; Southern Agricultural Insecticides, Inc.; Uniroyal; Universal Cooperatives Inc.; Value Gardens Supply LLC; Wilbur Ellis Co.*

1345 Dimethoxane
828-00-2 3250 G P

$C_8H_{14}O_4$
Dimethyl-m-dioxan-4-ol acetate.
4-19-00-00641 (Beilstein Handbook Reference); Acetic acid, 2,6-dimethyl-m-dioxan-4-yl ester; Acetic acid, ester with 2,6-dimethyl-m-dioxan-4-ol; Acetomethoxane; 6-Acetoxy-2,4-dimethyl-m-dioxane; 6-Acetoxy-2,4-di-methyl-1,3-dioxane; AI3-51032; BRN 0128710; Caswell No. 363; CCRIS 246; Dimethoxane; 1,3-Dioxan-4-ol, 2,6-dimethyl-, acetate; m-Dioxan-4-ol, 2,6-dimethyl-, acetate; Dioxin (bactericide) (Obs.); 2,4-Dimethyl-6-m-dioxanyl acetate; 2,4-Dimethyl-6-acetoxy-1,3-dioxane; 2,6-Di-methyl-1,3-dioxan-4-ol acetate; 2,6-Dimethyl-1,3-dioxan-4-yl acetate; 2,6-Dimethyl-m-dioxan-4-ol acetate; 2,6-Dimethyl-m-dioxan-4-yl acetate; EPA Pesticide Chemical Code 001001; G1V Gard DXN; GIV Gard DXN-CO; GIV Gard DXN; HSDB 4311; NCI-C56213. Preservative and gasoline additive. Fungicide. Registered by EPA as an antimicrobial and fungicide. Liquid; bp_6 = 74-75°; d_4^{20}= 1.0655; soluble in H_2O, organic solvents; LD_{50} (rat orl) = 1930 mg/kg. *Dow Chem. U.S.A.*

1346 4,5-Dimethoxy-1,3-bis(methoxymethyl)-2-imidazolidinone
4356-60-9 H

$C_9H_{18}N_2O_5$
4,5-Dimethoxy-1,3-bis(methoxymethyl)imidazolidin-2-one.
4,5-Dimethoxy-1,3-bis(methoxymethyl)-2-imidazolidin-one; EINECS 224-431-0; 2-Imidazolidinone, 4,5-dimethoxy-1,3-bis(methoxymethyl)-.

1347 Dimethoxydimethylsilane
1112-39-6 G

$C_4H_{12}O_2Si$
Dimethyldimethoxysilane.
AY 43-004; CD5605; EINECS 214-189-4; KBM 22; NSC 93882; Silane, dimethoxydimethyl-; TSL 8112; TSL 8117. Coupling agent, chemical intermediate, blocking agent, release agent, lubricant, primer, reducing agent. Liquid; bp = 82°; d^{20} = 0.8646; freely soluble in H_2O. *Degussa-Hüls Corp.*

1348 Dimethoxydiphenylsilane
6843-66-9 G

C14H16O2Si
Diphenyldimethoxy silane.
AY 43-047; CD6010; Diphenyldimethoxysilane; EINECS 229-929-1; KBM 202; KBM 202LS5300; NSC 93509; Silane, dimethoxydiphenyl-; TSL 8172; Z 6074. Coupling agent, chemical intermediate, blocking agent, release agent, lubricant, primer, reducing agent. Liquid; bp15 = 161°; d^{20} = 1.0771. *Degussa-Hüls Corp.*

1349   Dimethoxyethyl phthalate
117-82-8   G

C14H18O6
1,2-Benzenedicarboxylic acid, bis(2-methoxyethyl) ester.
4-09-00-03241 (Beilstein Handbook Reference); AI3-01366; 1,2-Benzenedicarboxylic acid, bis(2-meth-oxyethyl) ester; Bis(methoxyethyl) phthalate; Bis(2-methoxyethyl) phthalate; BRN 2056929; Di-(2-Methoxyethyl) ester kyseliny ftalove; Di(2-meth-oxyethyl)phthalate; Dimethoxyethyl phthalate; Dimethyl cellosolve phthalate; Dimethylglycol phthalate; DMEP; EINECS 204-212-6; HSDB 5016; Kesscoflex MCP; Kodaflex® DMEP; Methox; 2-Methoxyethyl phthalate; Methyl glycol phthalate; NSC 2147; Phthalic acid, bis(2-methoxyethyl) ester; Phthalic acid, di(methoxyethyl) ester; Reomol P. A chemical bonding agent for cellulose acetate staple fiber. *CIBA plc.*

1350   Dimethoxymethane
109-87-5   6042   G H

C3H8O2
Methylene dimethyl ether.
AI3-16096; Anesthenyl; Bis(methoxy)methane; Dimeth-oxymethane; Dimethyl formal; 2,4-Dioxapentane; EINECS 203-714-2; Formal; Formaldehyde dimethyl-acetal; Formaldehyde methyl ketal; HSDB 1820; Methane, dimethoxy-; Methoxymethyl methyl ether; Methylal; Methylene dimethyl ether; Metylal; UN1234. Used as a solvent; in organic synthesis; perfumes, adhesives, protective coatings; special fuel. Liquid; mp = -104.8°; bp = 42°; d^{20} = 0.8593; soluble in H2O (30 g/100 ml), very soluble in EtOH, Et2O, Me2CO, C6H6; LD50 (rbt orl) = 5708 mg/kg.

1351   N-(3-(Dimethoxymethylsilyl)propyl)-1,2-ethanediamine
3069-29-2   G

C8H22N2O2Si
N-(2-Aminoethyl)-3-aminopropylmethyldimethoxysilane.
3-(N-(β-Aminoethyl)amino)propylmethyldimethoxysilane; CA0699; N-(3-(Dimethoxymethylsilyl)propyl)ethylene-diamine; Dynasylan 1411; EINECS 221-336-6;. Coupling agent, chemical intermediate, blocking agent, release agent, lubricant, primer, reducing agent. *Degussa-Hüls Corp.*

1352   Dimethoxy polyethylene glycol
24991-55-7   H

(C2H4O)n.C2H6O
Poly(oxy-1,2-ethanediyl), α-methyl-ω-methoxy-.
Carpol CLE 1000; Dimethoxy polyethylene glycol; Genosorb 175; Genosorb 300; Glycols, polyethylene, dimethyl ether; Glyme-23; α,ω-Methoxypoly(ethylene oxide); α-Methyl-ω-methoxypoly(oxy-1,2-ethanediyl); Nissan Unisafe MM 400; Nissan Unisafe MM 1000; PEG-DME 2000; Poly(oxy-1,2-ethanediyl), α-methyl-ω-methoxy-; Polyethylene glycol dimethyl ether; Polyoxyethylene dimethyl ether; Sanfine DM 200; Sanfine DM 400; Sanfine DM 1000; Selexol; U-Nox DM 200; U-Nox DM 1000; Varonic DM 55.

1353   2,2-Dimethoxypropane
4744-10-9   G

C5H12O2
1,1-Dimethoxypropane.
AI3-28233; DMP; EINECS 225-258-3; Propanal dimethyl acetal; Propane, 1,1-dimethoxy-. Chemical intermediate for pharmaceuticals; dehydrating agent. Mobile liquid; bp = 87°. *Schering AG.*

1354   Dimethoxy tetrahydrofuran
696-59-3   G

C6H12O3
Tetrahydro-2,5-dimethoxyfuran.
AI3-61757; Dimethoxytetrahydrofuran; 2,5-Dimethoxy-tetrahydrofuran; EINECS 211-797-1; Furan, tetrahydro-2,5-dimethoxy-; NSC 7911; Protectol® DMT; Tetrahydro-2,5-dimethoxyfuran. Biocide used in disinfectants Liquid; mp = -45°; bp = 145.7°; d^{25} = 1.0200. *BASF Corp.; Bristol-Myers Squibb Co.; Fluka; Sigma-Aldrich Fine Chem.*

1355   Dimethylacetal
534-15-6   3253   H

C4H10O2
1,1-Dimethoxyethane.
4-01-00-03103 (Beilstein Handbook Reference); Acetaldehyde dimethyl acetal; Acetaldehyde methyl acetal; AI3-24137; BRN 1697039; 1,1-Dimethoxyethane; Dimethyl acetal; Dimethyl aldehyde; EINECS 208-589-8; Ethane, 1,1-dimethoxy-; Ethylidene dimethyl ether; FEMA No. 3426; HSDB 5427; Methyl formyl; UN2377. Liquid; mp = -113.2°; bp = 64.5°; d^{20} = 0.8501; λ_m = 290 nm; soluble in

H2O, EtOH, Et2O, CCl4, CHCl3; very soluble in Me2CO.

1356   Dimethylacetamide
127-19-5   3254   H

C4H9NO
N,N-Dimethylacetamide.
Acetamide, N,N-dimethyl-; Acetdimethylamide; Acetic acid, dimethylamide; Acetyldimethylamine; AI3-15276; CBC 510337; CCRIS 4623; Dimethyl acetamide; Dimethylamid kyseliny octove; Dimethylamide acetate; DMA; DMAc; EINECS 204-826-4; HSDB 74; NSC 3138; SK 7176; U-5954. Liquid; mp = -20°; bp = 165°; d^{25} = 0.9366; λ_m = 218 nm (ε = 1000, C7H16); freely soluble in H2O, EtOH, Et2O, Me2CO, C6H6, CHCl3.

1357   N,N-Dimethylacetoacetamide
2044-64-6   H

C6H11NO2
N,N-Dimethyl-3-oxobutyramide.
4-04-00-00260 (Beilstein Handbook Reference); Acetoacetamide, N,N-dimethyl-; Acetylacetamide, N,N-dimethyl-; BRN 1755038; Butanamide, N,N-dimethyl-3-oxo-; Dimethylamid kyseliny acetoctove; EINECS 218-059-8; N,N-Dimethyl-3-oxobutanamide; N,N-Dimethyl-3-oxobutyramide; N,N-Dimethylacetoacetamide; NSC 524755.

1358   Dimethyl adipate
627-93-0   G H

C8H14O4
Hexanedioic acid, dimethyl ester.
4-02-00-01959 (Beilstein Handbook Reference); Adipic acid, dimethyl ester; AI3-00668; BRN 1707443; DBE 6; Dimethyl adipate; Dimethyl hexanedioate; 1,6-Dimethylhexanedioate; Dimethyladipate; EINECS 211-020-6; Hexanedioic acid, dimethyl ester; HSDB 5021; Methyl adipate; NSC 11213. Used as a solvent and plasticizer. Liquid; mp = 10.3°; bp_{13} = 115°; d^{20} = 1.0600; insoluble in H2O, soluble in EtOH, Et2O, CCl4, AcOH. DuPont; Morflex; UCB SA.

1359   Dimethylamine
124-40-3   3255   H

C2H7N
N,N-Dimethylamine.
AI3-15638-X; CCRIS 981; Dimethylamine; EINECS 204-697-4; HSDB 933; Methanamine, N-methyl-; N-Methylmethanamine; N,N-Dimethylamine; NSC 8650; RCRA waste number U092; UN1032;

UN1160. Gas; mp = -92.2°; bp = 6.8°; d^0 = 0.6804; λ_m = 191, 222 nm (ε = 3236, 100, gas); soluble in EtOH, Et2O, very soluble in H2O.

1360   Dimethylamine hydrochloride
506-59-2   3255   G H

C2H8ClN
N-Methylmethanamine hydrochloride.
AI3-52357; Dimethylamine, hydrochloride; Dimethyl-ammonium chloride; EINECS 208-046-5; Hydrochloric acid dimethylamine; Methanamine, N-methyl-, hydro-chloride; N-Methylmethanamine hydrochloride. Used as a vulcanization accelerator and in manfacture of soaps. Crystals; mp = 160°; LD50 (rat orl) = 1070 mg/kg. Janssen Chimica; Lancaster Synthesis Co.; Penta Mfg.

1361   Dimethylaminobenzaldehyde
100-10-7   3257   G

C9H11NO
4-Dimethylaminobenzaldehyde.
AI3-15337; Benzaldehyde, 4-(dimethylamino)-; Benz-aldehyde, p-(dimethylamino)-; p-(Dimethylamino)-benzaldehyde; 4-(Dimethylamino)benzaldehyde; 4-Di-methylaminobenzaldehyde; 4-Dimethylaminobenzene-carbonal; p-(N,N-Dimethylamino)benzaldehyde; N,N-Dimethyl-p-aminobenzaldehyde; Ehrlich's reagent; EINECS 202-819-0; NSC 5517; p-Formyl-N,N-dimethylaniline; p-Formyldimethylaniline; Reagens ehrlichovo. Used in manufacture of dyestuffs, as a reagent and in medicine. Leaflets; mp = 74.5°; bp_{17} = 176-177°; λ_m = 242, 340 nm (MeOH); soluble in H2O (0.03 g/100 ml), CHCl3, more soluble in EtOH, Et2O, Me2CO, C6H6. Aceto Corp.; BASF Corp.; Lancaster Synthesis Co.; Penta Mfg.

1362   4-(Dimethylamino)benzoic acid
619-84-1   3259   G

C9H11NO2
N,N-Dimethyl-p-aminobenzoic acid.
Benzoic acid, 4-(dimethylamino)-; Benzoic acid, p-(dimethylamino)-; EINECS 210-615-8; 4-(Dimethyl-amino)benzoic acid; 4-Dimethylaminobenzoic acid; p-Dimethylamino benzoic acid; p-N,N-(Dimethyl-amino)benzoic acid; N,N-Dimethyl-4-aminobenzoic acid; NSC 16596; Solarchem® O. Esters used as ultraviolet screens. Needles; mp = 242.5°; λ_m = 308 nm (ε = 25119, EtOH); slightly soluble in Et2O, soluble in EtOH. CasChem; ISP Van Dyk Inc.; Lipo.

1363   4-(Dimethylamino)benzoic acid 3-methyl-butyl ester
21245-01-2   3259   G

C14H21NO2

4-(Dimethylamino)benzoic acid isoamyl ester.

Benzoic acid, 4-(dimethylamino)-, 3-methylbutyl ester; EINECS 244-288-8; Escalol® 506; Escalol® 507; Isoamyl-p-N, N-dimethylaminobenzoate; Padimate; Padimate A; Padimate O; Padimate; Padimatum; Solarchem® OSpectraban. Isoamyl-p-N,N-dimethylaminobenzoate in ethanol; a lotion used to protect skin from UV light. A PABA derivative that is used as a sun-screening agent. Of the PABA group it is the most commonly used agent in sunscreens marketed in the US. It is also used in cosmetics for skin, hair, and nails such as moisturizers and lipsticks and lip balms. *CasChem; Lipo; Stiefel Labs Inc.; Van Dyk.*

1364 Dimethylaminoethoxyethanol

1704-62-7   G H

C6H15NO2

2-(2-(Dimethylamino)ethoxy)ethanol.

3-04-00-00648 (Beilstein Handbook Reference); AI3-18588; BRN 1209271; Dimethylaminoethoxyethanol; 2-(2-Dimethylaminoethoxy)ethanol; N,N-Dimethyl-diglycolamine; EINECS 216-940-1; Ethanol, 2-(2-(dimethylamino)ethoxy)-; NSC 3146; Texacat® ZR-70. Catalyst for polyurethane molded high resilience flexible foam, ether slabstock, packaging foam. *Texaco.*

1365 Dimethylaminoethyl acrylate

2439-35-2   G H

C7H13NO2

2-Propenoic acid, 2-(dimethylamino)ethyl ester.

4-04-00-01431 (Beilstein Handbook Reference); Acrylic acid, 2-(dimethylamino)ethyl ester; Adame; Ageflex FA-1; AI3-08751; BRN 1099119; CCRIS 4797; Dimethyl-aminoethyl acrylate; 2-(Dimethylamino)ethyl acrylate; EINECS 219-460-0; NSC 20952; 2-Propenoic acid, 2-(dimethylamino)ethyl ester. Adhesion promoter in UV- and EB-cured coatings for metals, plastic, paper and wood surfaces. Catalyst for epoxy molding and extrusion resins, intermediate for water treatment chemicals, quaternary monomers; silane coupling agents, conductive paper coatings. Liquid; mp < -60°; bp50 = 94°; d20 = 0.940; insoluble in H2O, slightly soluble in organic solvents. *Rit-Chem; TIC Gums.*

1366 2-(Dimethylamino)ethyl acrylate metho-chloride

44992-01-0   G H

C8H16ClNO2

Ethanaminium, N,N,N-trimethyl-2-((1-oxo-2-propenyl)-oxy)-, chloride.

(2-(Acryloyloxy)ethyl)trimethylammonium chloride; ADAMQUAT 80 MC; Ageflex FA-1Q75MC; 2-(Dimethylamino)ethyl acrylate methochloride; EINECS 256-176-6; Ethanaminium, 2-((1-oxo-2-propenyl)oxy)-N,N,N-trimethyl-, chloride; 2-((1-Oxo-2-propenyl)oxy)-N,N,N-trimethylethanaminium chloride. Quaternary antistatic finish for polyester fibers, flocculant and coagulant for industrial process water treatment, flocculant for mineral recovery, ion exchange resins, adhesives, acid dye receptivity, electrostatic coatings on wodd, retention aids for paper. *Rit-Chem.*

1367 Dimethylaminoethyl chloride hydro-chloride

4584-46-7   G

C4H11Cl2N

1-Chloro-2-dimethylaminoethane hydrochloride.

AI3-22778; Bis(methyl)-2-chloroethylamine hydro-chloride; (β-Chloroethyl)dimethylamine hydrochloride; (2-Chloroethyl)dimethylamine, hydrochloride; 2-Chloro-N,N-dimethylethylamine hydrochloride; Chloro(dimethyl-amino)ethane hydrochloride; 1-Chloro-2-(dimethyl-amino)ethane hydrochloride; 2-Chloroethyl dimethyl ammonium chloride; 2-Chloroethyldimethylammonium chloride; Dimethyl-β-chloroethylamine hydrochloride; Dimethyl(2-chloroethyl)amine hydrochloride; 2-Dimethylaminochloroethane hydrochloride; Dimethyl-aminoethyl chloride hydrochloride; 2-(Dimethyl-amino)ethyl chloride hydrochloride; DMC; EINECS 224-970-1; Ethanamine, 2-chloro-N,N-dimethyl-, hydro-chloride; N-(2-Chloroethyl)-N,N-dimethylammonium chloride; N,N-Dimethyl-2-chloroethylamine hydro-chloride; N,N-Dimethyl-N-(2-chloroethyl)amine hydro-chloride; NSC 1917; NSC 111230. Used in the manufacture of antihistamines and other pharmaceuticals; organic intermediate for introduction of β-dimethylaminoethyl radical. Crystals; mp = 199°, 205-208°; slightly soluble in H2O; irritant. *ICI Spec.; Janssen Chimica; Lancaster Synthesis Co.; Lonzagroup; Sigma-Aldrich Fine Chem.*

1368 2-Dimethylamino-2-methyl-1-propanol

7005-47-2   G

C6H15NO

2-Dimethylamino-2-methyl-1-propanol.

4-04-00-01740 (Beilstein Handbook Reference); AI3-28043; BRN 1732837; 2-(Dimethylamino)-2-methyl-propanol; 2-(Dimethylamino)-2-methylpropan-1-ol; 2-(N,N-Dimethylamino)-2-methyl-1-propanol; DMAMP; DMAMP-80; EINECS 230-279-6; NSC 17706; 1-Propanol, 2-(dimethylamino)-2-methyl-; USAF CS-1. Amine solubilizer for resins in aqueous coatings; emulsifier for waxes; vapor-phase corrosion inhibitor; urethane catalyst; titanate solubilizer raw material for synthesis. Liquid; mp = 19°. *Angus.*

1369 p-(Dimethylamino)nitrosobenzene

138-89-6   6672   G

C8H10N2O

p-Nitroso-N,N-dimethylaniline.

4-12-00-01558 (Beilstein Handbook Reference); Accelerine; AI3-15393; Aniline, N,N-dimethyl-p-nitroso-; Benzenamine, N,N-dimethyl-4-nitroso-; BRN 0607293; CCRIS 3057; p-(N,N-Dimethylamino)nitrosobenzene; N,N-Dimethyl-4-nitrosobenzenamine; N,N-Dimethyl-4-nitrosoaniline; N,N-Dimethyl-p-nitrosoaniline; Dimethyl-(p-nitrosophenyl)amine; 4-(Dimethylamino)-nitroso-benzene; EINECS 205-343-1; HSDB 1320; NCI-C01821; NDMA; 4-Nitrosodimethylaniline; 4-Nitroso-N,N-di-methylaniline; p-Nitroso-N,N-dimethylaniline; p-Nitroso-dimethylaniline; p-Nitrosodimethylanilide; Paranitrosodi-methylaniline; NSC 2775; Ultra Brilliant Blue P; UN1369; Vulcaniline. An intermediate for dyes. Also used as an accelerator in vulcanizing. Solid; mp = 92.5°; d^{20} = 1.145; λ_m = 273 nm (MeOH); slightly soluble in H_2O, soluble in EtOH, Et2O, CHCl3, formamide.

1370 Dimethylaminopropionitrile
1738-25-6   H

C5H10N2
3-(N,N-Dimethylamino)propionitrile.
4-04-00-02533 (Beilstein Handbook Reference); AI3-25451; BRN 0773779; Dimethylaminopropionitrile; N,N-(Dimethylamino)-3-propionitrile; Dmapn; EINECS 217-090-4; HSDB 5029; NSC 232; Propanenitrile, 3-(dimethylamino)-; Propionitrile, 3-(dimethylamino)-. Liquid; bp = 173°; d^{20} = 0.8705.

1371 γ-(Dimethylamino)propylamine
109-55-7   H

C5H14N2
3-Dimethylaminopropylamine.
4-04-00-01259 (Beilstein Handbook Reference); AI3-25441; 1-Amino-3-dimethylaminopropane; BRN 0605293; CCRIS 4799; EINECS 203-680-9; HSDB 5391; 1-(Dimethylamino)-3-aminopropane; 3-(Dimethylamino)-propylamine; γ-(Dimethylamino)propylamine; N,N-Di-methyl-1,3-diaminopropane; NSC 1067; 1,3-Propane-diamine, N,N-dimethyl-; Propylamine, 3-(N,N-dimethyl-amino)-.

1372 Dimethylaminotrimethylsilane
2083-91-2   G

C5H15NSi
N,N-Dimethylaminotrimethylsilane.
CD5400; EINECS 218-222-3; Pentamethylsilylamine; Silanamine, pentamethyl-. Coupling agent, chemical intermediate, blocking agent, release agent, lubricant, primer, reducing agent. Liquid; bp = 84°; d^{20} = 0.7400. Degussa-Hüls Corp.

1373 Dimethylaniline
121-69-7   3261   H

C8H11N
N,N-Dimethylaniline.
AI3-17284; Aniline, N,N-dimethyl-; Benzenamine, N,N-dimethyl-; CCRIS 2381; Dimethylaniline; N,N-Dimethylaniline; Dwumetyloanilina; EINECS 204-493-5; HSDB 1179; NCI-C56428; NL 63-10P; NSC 7195; UN2253; Versneller NL 63/10. Pale yellow liquid; mp = 2.4°; bp = 194.1°; d^{20} = 0.9557; λ_m = 251, 298 nm (ε = 11600, 2290, MeOH); slightly soluble in H_2O; soluble in EtOH, Et2O, Me2CO, C6H6, CCl4, very soluble in CHCl3.

1374 Dimethyl anthranilate
85-91-6   G

C9H11NO2
2-Methylamino methyl benzoate.
4-14-00-01016 (Beilstein Handbook Reference); AI3-03340; Anthranilic acid, N-methyl-, methyl ester; Benzoic acid, 2-(methylamino)-, methyl ester; BRN 0607217; CCRIS 2846; Dimethyl anthranilate; EINECS 201-642-6; FEMA No. 2718; HSDB 2784; 2-Methylaminomethyl benzoate; N-Methylanthranilic acid, methyl ester; Methyl 2-(methylamino)benzoate; Methyl methanthranilate; Methyl methylaminobenzoate; Methyl methylanthranilate; Methyl N-methyl anthranilate; Methyl o-(methylamino)benzoate; NSC 9406. Used in perfumes, flavoring, drugs. Crystals; mp = 19°; bp = 255°; d^{15} = 1.120; λ_m = 220, 252 nm (cyclohexane); insoluble in H_2O, soluble in organic solvents. Am. Bio-Synthetics; Bell F & F; Penta Mfg.

1375 Dimethyl azelate
1732-10-1   G

C11H20O4
Nonanedioic acid dimethyl ester.
AI3-06080; Azelaic acid dimethyl ester; Dimethyl azelate; Dimethyl nonanedioate; EINECS 217-060-0; Emery® 2914; Methyl azelate; Nonanedioic acid, dimethyl ester; NSC 59040. Synthetic lubricant. Liquid; mp = -0.8°; bp20 = 156°; d^{20} = 1.0082; insoluble in H_2O, soluble in EtOH, Me2CO, C6H6, CCl4. Henkel/Emery.

1376 Dimethylbenzenesulfonic acid
25321-41-9   H

C8H10O3S
Xylenesulfonic acids.
Benzenesulfonic acid, dimethyl-; Dimethylbenzene-sulfonic acid; EINECS 246-839-8; Xylene sulfonic acid; Xylenesulphonic acid. Mixture of o, m and p-isomers.

1377 Dimethyl benzyl hydrogenated tallow ammonium chloride
61789-72-8   H

Quaternary ammonium compounds, benzyl(hydrogen-ated tallowalkyl)dimethyl, chlorides.

Dimethyl benzyl hydrogenated tallow ammonium chloride; EINECS 263-081-3; Quaternary ammonium compounds, benzyl(hydrogenated tallowalkyl)dimethyl, chlorides.

1378 Dimethyl cellosolve
110-71-4   3251   G H

$C_4H_{10}O_2$
1,2-Dimethoxyethane.
4-01-00-02376 (Beilstein Handbook Reference); AI3-61007; Ansol E-121; Ansul ether 121; BRN 1209237; 1,2-Dimethoxyethane; Dimethyl Cellosolve; DME (glycol ether); 2,5-Dioxahexane; Egdme; EINECS 203-794-9; Ethane, 1,2-dimethoxy-; 1,2-Ethanediol dimethyl ether; Ethylene dimethyl ether; Ethylene glycol dimethyl ether; Glycol dimethyl ether; Glyme; Hisolve MMM; HSDB 73; Monoethylene glycol dimethyl ether; Monoglyme; NSC 60542; UN2252. Used as a general purpose solvent. Liquid; mp = -58°; bp = 85°; d^{20} = 0.8691; soluble in H_2O, EtOH, Et_2O, Me_2CO, C_6H_6, CCl_4, $CHCl_3$. *BASF Corp.*

1379 N,N-Dimethylcyclohexylamine
98-94-2   H

$C_8H_{17}N$
Cyclohexylamine, N,N-dimethyl.
4-12-00-00018 (Beilstein Handbook Reference); BRN 1919922; Cyclohexanamine, N,N-dimethyl-; N-Cyclo-hexyldimethylamine; Cyclohexyldimethylamine; N,N-Dimethylaminocyclohexane; Dimethylcyclo-hexyl-amine; N,N-Dimethyl-N-cyclohexylamine; EINECS 202-715-5; HSDB 5323; NSC 163904; Polycat 8; UN2264. Liquid; bp = 162°.

1380 N,N-Dimethyldecanamide
14433-76-2   H

$C_{12}H_{25}NO$
N,N-Dimethyldecan-1-amide.
AI3-34960; BRN 1906042; Decanamide, N,N-dimethyl-; EINECS 238-405-1; N,N-Dimethylcapramide; N,N-Dimethylcapylamide; N,N-Dimethyldecanamide; N,N-Dimethyldecan-1-amide; N,N-Dimethyldecanoamide; NSC 131411.

1381 2,5-Dimethyl-2,5-di(t-butylperoxy)hexyne-3
1068-27-5   G

$C_{16}H_{30}O_4$
(1,1,4,4-tetramethyl-2-butyne-1,4-diyl)bis((1,1-dimethylethyl) peroxide.
4-01-00-02701 (Beilstein Handbook Reference); Aztec® 2,5-Tri; BRN 1711920; 2,5-Dimethyl-2,5-di(t-butyl-peroxy)hexyne-3; Di-t-butyl 1,1,4,4-tetramethylbut-2-yn-1,4-ylene diperoxide; EINECS 213-944-5; 3-Hexyne, 2,5-dimethyl-2,5-di(t-butylperoxy)-; Polyvel CR-L10. Initiator and additive for roto molding and crosslinking. *Catalyst Resources Inc.; Polyvel.*

1382 Dimethyldiethoxy silane
78-62-6   G

$C_6H_{16}O_2Si$
Diethoxydimethylsilane.
4-04-00-04101 (Beilstein Handbook Reference); BRN 1736110; CCRIS 1321; CD5600; Diethoxy(dimethyl)-silane; Diethoxydimethylsilane; Dimethyl-diethoxysilan; Dimethyldiethoxysilane; Dimethylsilicondiethoxide; EINECS 201-127-6; EXP-49; KBE 22; NSC 77085; Silane, diethoxydimethyl-; UN2380. Coupling agent, chemical intermediate, blocking agent, release agent, lubricant, primer, reducing agent. Intermediate for blocking hydroxyl and amino groups in organic synthesis reactions; also for preparing hydrophobic and release materials and enhancing flow of powders. Liquid; mp = -87°; bp = 114°; d^{25} = 0.8650; LD_{50} (rat orl) = 9280 mg/kg. *Degussa-Hüls Corp.; Genesee Polymers.*

1383 Dimethyldioctadecylammonium chloride
107-64-2   G H

$C_{38}H_{80}ClN$
Di-n-octadecyldimethylammonium chloride.
Adogen® TA 100; Aliquat 207; Ammonium, dimethyldioctadecyl-, chloride; Ammonyx 2200P100; Arosurf TA 100; Arquad 218-100; Arquad 218-100P; Arquad R 40; CA 3475; Cation DS; Cedequat TD 75; Dehyquart DAM; Di-n-octadecyldimethylammonium chloride; Dimethyl dioctadecyl ammonium chloride; Dimethyldistearylammonium chloride; N,N-Dimethyl-N-octadecyl-1-octadecanaminium chloride; N,N-Diocta-decyl-N,N-dimethylammonium chloride; Distearyldi-methylammonium chloride; Distearyldimonium chloride; DODA(Cl); DODAC; EINECS 203-508-2; Genamin DSAC; HSDB 5380; KD 83; Kemamine Q 9702CLP; NSC 61374; 1-Octadecanaminium, N,N-dimethyl-N-octa-decyl-, chloride; Prepagen WK; Q-D 86P; Quartamin D 86; Quartamin DM 86P; Quaternium 5; Sokalan 9200; Sumquat® 6045; Surfroyal DSAC; Talofloc; Varisoft 100; Varisoft TA 100. Textile softener producing very slick cationic hand and maximum softness. Good non-yellowing and high absorbency, can replace wax in sizing formulations. Used as a conditioner, anti-stat and softener. A cationic detergent and antiseptic. Also used as a laboratory reagent. *Hexcel; Hoechst AG; Sherex.*

1384 Dimethyldodecylamine N-oxide
1643-20-5   G H

C14H31NO

N,N-Dimethyl-1-dodecanamine-N-oxide.

4-04-00-00798 (Beilstein Handbook Reference); Ablumox LO; Ammonyx® AO; Ammonyx® LO; Aromox DMCD; Aromox dmmc-W; BRN 1769927; Conco XAL; DDNO; Dimethyllaurylamine oxide; Dimethyldodecylamine N-oxide; Dimethyldodecylamine oxide; EINECS 216-700-6; Empigen® OB; HSDB 5451; Lauramine oxide; Oxyde de dimethyllaurylamine; Refan; Rhodamox® LO. Foamer/foam stabilizer, wetting agent, viscosity builder, grease emulsifier for shampoos, bath products, fine fabric cleaners, hard surface cleaners containing acids or bleach, dishwash, shaving creams, lotions; textile lubricant, emulsifier, dye dispersant. Detergent, antistat, foam booster/stabilizer and viscosity modifier for personal care products, surgical scrubs, fire fighting foam concentrates, foamed rubbers, bleach additive; solubilizer. Foaming agent/stabilizer, thickener, emollient for shampoos, bath products, dishwash, fine fabric detergents, shaving creams, lotions, textile softeners, foam rubber, in electroplating, paper coatings; used in toiletries for mildness. Solid; mp = 130-131°; LD$_{50}$ (rat orl)= 1000 mg/kg. *Albright & Wilson Americas Inc.; Albright & Wilson UK Ltd.; Rhône Poulenc Surfactants; Rhône-Poulenc; Stepan Canada; Stepan; Taiwan Surfactants.*

1385 Dimethyl ether
115-10-6   6096   G H

C2H6O

Dimethyl oxide.

Demeon D; Dimethyl ether; DME; Dymel A; EINECS 204-065-8; Ether, dimethyl; Ether, methyl; HSDB 354; Methane, oxybis-; Methoxymethane; Methyl ether; Methyl oxide; Oxybismethane; UN1033; Wood ether. Used as a solvent, motor fuel and Aerosol® propellant, in refrigerants, as an extraction agent, a catalyst and stabilizer in polymerization. Gas; mp = -141.5°; bp = -24.8°; λ_m = 163, 184 nm (ε = 3981, 2512, gas); slightly soluble in C$_6$H$_6$, soluble in H$_2$O, EtOH, Et$_2$O, Me$_2$CO.

1386 Dimethylethylamine
598-56-1   G H

C4H11N

N,N-Dimethylethylamine.

Al3-52225; Dimethylethylamine; N,N-Dimethylethyl-amine; EINECS 209-940-8; Ethanamine, N,N-dimethyl-; Ethylamine, N,N-dimethyl-; Ethyldimethylamine; N-Ethyldimethylamine; HSDB 5712; Methanamine, N-ethyl-N-methyl-. Used in chemical synthesis. Liquid; mp = -140°; bp = 36.5°; d^{20} = 0.675.

1387 Dimethylethynylmethanol
115-19-5   6061   G H

C5H8O

2-Methyl-3-butyn-2-ol.

4-01-00-02229 (Beilstein Handbook Reference); Al3-23121; BRN 0635746; 1-Butyn-3-ol, 3-methyl-; 3-Butyn-2-ol, 2-methyl-; Carbavane; Dimethylacetylenecarbinol; Dimethylacetylenylcarbinol; Dimethylethynylcarbinol; Dimethylethynylmethanol; α,α-Dimethylpropargyl alcohol; EINECS 204-070-5; Ethynyldimethylcarbinol; 2-Methyl-2-butynol; 2-Methyl-3-butyn-2-ol;

2-Methylbut-3-yn-2-ol; NSC 523. Corrosion inhibitor; reactive intermediate in manufacture of pharmaceuticals, plastics, rubbers, fragrances, agriculture. Used as solvent, acid inhibitor, viscosity reducer, and stabilizer in vinyl plastisols, platinum catalyst blocker for silicones. Liquid; mp = 1.5°; bp = 104°; d^{20} = 0.8618; insoluble in H$_2$O, soluble in organic solvents; LD$_{50}$ (rat orl) = 1950 mg/kg. *Air Prods & Chems; BASF Corp.*

1388 Dimethylformamide
68-12-2   3269   G H

C3H7NO

Formamide, N,N-dimethyl-.

Al3-03311; Caswell No. 366A; CCRIS 1638; Dimethyl formamide; Dimethylamid kyseliny mravenci; Di-methylformamid; Dimethylformamide; N,N-Dimethyl-methanamide; N,N-Dimetilformamida; Dimetilform-amide; Dimetylformamidu; DMF; DMFA; Dwumetylo-formamid; EINECS 200-679-5; EPA Pesticide Chemical Code 366200; Formamide, N,N-dimethyl-; Formic acid, amide, N,N-dimethyl-; N-Formyldimethylamine; HSDB 78; NCI-C60913; NSC 5356; U-4224; UN2265. Used as a solvent in vinyl resins and acetylene, butadiene, acid gases; polyacrylic fibers; catalyst in carboxylation reactions, organic synthesis; carrier for gases; chemical reagent. Liquid; mp = -60.4°; bp = 153°; d^{25} = 0.944; soluble in H$_2$O, organic solvents; LD$_{50}$ (rat orl) = 7600 mg/kg, (mus orl) = 6800 mg/kg, (rat ip) = 4700 mg/kg, (mus ip) = 6200 mg/kg. *Aceto Corp.; Air Products & Chemicals Inc.; Ashland; Baker J. T.; BASF Corp.; Brown Chem. Co.; Browning; Chemcentral; Coyne; E. I. DuPont de Nemours Inc.; ICI Spec.; Mallinckrodt Inc.; Mitsubishi Gas; Sigma-Aldrich Fine Chem.*

1389 Dimethyl glutarate
1119-40-0   H

C7H12O4

Glutaric acid, dimethyl ester.

Al3-06026; DBE 5; Dimethyl glutarate; Dimethyl pentanedioate; EINECS 214-277-2; Glutaric acid, dimethyl ester; HSDB 5789; NSC 58578; Pentanedioic acid, dimethyl ester. Liquid; mp = -42.5°; bp = 214°, bp$_{21}$ = 109°; d^{20} = 1.0876; soluble in CHCl$_3$, very soluble in EtOH, Et$_2$O.

1390 Dimethylglyoxime
95-45-4   3272   G

C4H8N2O2

2,3-Butanedionedioxime.

3-01-00-03105 (Beilstein Handbook Reference); Al3-14925; Biacetyl dioxime; BRN 0506731; Butanedione dioxime; 2,3-Butanedione, dioxime; Chugaev's reagent; Diacetyldioxime; 2,3-Diisonitrosobutane; Dimethyl-glyoxime; EINECS 202-420-1; Glyoxime, dimethyl-; NSC 9. Used in analytical chemistry, especially as a reagent for nickel; also used in biochemical research. Crystals; mp = 238-240°; insoluble in H$_2$O, soluble in organic solvents. *Alcoa Ind. Chem.; Pfalz & Bauer.*

1391   Dimethyl hexahydroterephthalate
94-60-0   H

C10H16O4
1,4-Cyclohexanedicarboxylic acid, dimethyl ester.
AI3-28580; 1,4-Cyclohexanedicarboxylic acid, dimethyl ester;
Dimethyl 1,4-cyclohexanedicarboxylate; Dimethyl cyclohexane-1,4-
dicarboxylate; Dimethyl hexahydro-terephthalate; EINECS 202-347-
5; HSDB 5284.

1392   Dimethylhexanediol
110-03-2   H

C8H18O2
2,5-Dimethyl-2,5-hexanediol.
AI3-20685; Dimethylhexanediol; 2,5-Dimethyl-2,5-hexanediol; 2,5-
Dimethylhexane-2,5-diol; EINECS 203-731-5; 2,5-Hexanediol, 2,5-
dimethyl-; HSDB 5395; NSC 5595; 1,1,4,4-Tetramethyl-1,4-
butanediol. Prisms; mp = 92°; bp = 214°; d^{20} = 0.898; soluble in
H2O, very soluble in EtOH, C6H6, CHCl3.

1393   Dimethylhexynediol
142-30-3   H

C8H14O2
2,5-Dimethyl-3-hexyne-2,5-diol.
Acetylenepinacol; AI3-14500; D 43; D 43 (VAN);
Dimethylhexynediol; 2,5-Dimethyl-3-hexyne-2,5-diol; EINECS 205-
533-4; 3-Hexyne-2,5-diol, 2,5-dimethyl-; HSDB 5639; Kemitracin-50;
NSC 117261; Olfine Y; Tetramethyl-2-butynediol;
Tetramethylbutynediol.
Crystals; mp = 95°; bp = 205°; d^{20} = 0.947; soluble in H2O, CHCl3,
very soluble in C6H6, EtOH, Et2O, Me2CO.

1394   3,5-Dimethyl 1-hexyn-3-ol
107-54-0   G

C8H14O
3,5-Dimethylhex-1-yn-3-ol.
AI3-23126; 3,5-Dimethyl-1-hexyn-3-ol; EINECS 203-500-9; 1-
Hexyn-3-ol, 3,5-dimethyl-; NSC 978; Surfynol®; Surfynol® 61.
Surfactant, wetting agent used for paper coatings, inks, floor
polishes, and glass cleaning formulations; cleaner in silicon wafer
industry. bp = 150°; d= 0.859; n_D^{20} = 1.4340. Air Products &
Chemicals Inc.

1395   Dimethylhydantoin
77-71-4   G H

C5H8N2O2
5,5-Dimethyl- 2,4-Imidazolidinedione.
5-24-05-00348 (Beilstein Handbook Reference); AI3-61127; BRN
0002827; Dantoin 736; Dantoin DMH; Dimethylhydantoin; DM
Hydantoin; DMH; EINECS 201-051-3; HSDB 5216; Hydantoin, 5,5-
dimethyl-; NSC 8652; T10. Intermediate for textiles and other
applications. Prisms; mp = 178°; soluble in H2O and organic
solvents; λ_m = 219 (ϵ 6607 EtOH/KOH). Great Lakes Fine Chem.;
Janssen Chimica; Lonzagroup.

1396   Dimethyl hydrogenated tallowamine
61788-95-2   H

Amines, (hydrogenated tallow alkyl)dimethyl.
Amines, (hydrogenated tallow alkyl)dimethyl; Dimethyl
hydrogenated tallowamine; EINECS 263-022-1; Hydrogenated tallow
dimethyl amine.

1397   Dimethyl hydrogen phosphite
868-85-9   H

C2H7O3P
Dimethoxyphosphine oxide.
Bis(hydroxymethyl)phosphine oxide; CCRIS 1354;
Dimethoxyphosphine oxide; Dimethyl acid phosphite; Dimethyl
hydrogen phosphite; Dimethyl hydrogen phosphonate; Dimethyl
phosphite; Dimethyl phosphonate; Dimethylester kyseliny fosforite;
Dimethylfosfit; Dimethylfosfonat; Dimethylhydrogen-phosphite;
EINECS 212-783-8; HSDB 2593; Hydrogen dimethyl phosphite;
Methyl phosphonate ((MeO)2HPO); NCI-C54773; Phosphonic acid,
dimethyl ester. Liquid; bp = 170.5°; d^{20} = 1.2002; slightly soluble in
CCl4, soluble in EtOH, C5H5N.

1398   Dimethyl isophthalate
1459-93-4   G H

C10H10O4
1,3-Benzenedicarboxylic acid, dimethyl ester.
4-09-00-03293 (Beilstein Handbook Reference); AI3-02247; BRN
1912251; Dimethyl 1,3-benzenedi-carboxylate; Dimethyl
isophthalate; Dimethyl m-phthalate; Dimethylester kyseliny
isoftalove; EINECS 215-951-9; HSDB 6138; Isophthalic acid
dimethyl ester; Methyl 3-(carbomethoxy)benzoate; Methyl
isophthalate; Morflex 1129; NSC 15313; Uniplex 270. Chemical
intermediate in the synthesis of polyesters. Used as a plasticizer,
modifies clarity and melting point of PET resins used in films, blow-
molded bottles, and similar products. Needles; mp = 67.5°; bp =
282°; d^{20} = 1.194; λ_m = 280, 288 nm (MeOH); slightly soluble in
H2O. Morflex; Unitex.

1399 Dimethyl methylphosphonate
756-79-6   3425   G H

C3H9O3P
Methanephosphonic acid dimethyl ester.
4-04-00-03499 (Beilstein Handbook Reference); AI3-08678; BRN 0878263; CCRIS 876; Dimethoxymethyl-phosphine oxide; Dimethyl methanephosphonate; Di-methyl methylphosphonate; DMMP; EINECS 212-052-3; Fyrol® DMMP; HSDB 2590; Methyl phosphonic acid, dimethyl ester; NCI-C54762; NSC 62240; Phosphonic acid, methyl-, dimethyl ester; Pyrol dmmp. Flame retardant for applications where high phosphorus content, good solvency, and low viscosity are desired; lowers viscosity of epoxy resins and unsaturated polyesters filled with hydrated alumina oxide. Liquid; bp = 181°, bp20 = 79.5°; d30 = 1.4099; λ_m = 217 nm (ε = 13, EtOH); soluble in H2O, Et2O, EtOH; LD50 (rat orl) > 5000 mg/kg. *Akzo Chemie.*

1400 Dimethyl myristamine
112-75-4   G H

C16H35N
N,N-Dimethyltetradecanamine.
Adma® 14; Armeen® DM 14D; Dimethyl myristamine; Dimethyl-n-tetradecylamine; Dimethyltetradecylamine; N,N-Dimethyltetradecylamine; EINECS 204-002-4; Genamin 14R302D; HSDB 2785; IPL 30; Myristyl dimethyl amine; NSC 78319; Onamine 14; 1-Tetradecanamine, N,N-dimethyl-; Tetradecylamine, N,N-dimethyl-; Tetradecyldimethylamine. Intermediate in the manufacture of surfactants, antioxidants, oil and grease additives. *Asahi Denka Kogyo.*

1401 1,2-Dimethyl-4-nitrobenzene
99-51-4   H

C8H9NO2
Benzene, 1,2-dimethyl-4-nitro-.
Benzene, 1,2-dimethyl-4-nitro-; CCRIS 3118; EINECS 202-761-6; HSDB 5324; NSC 66555; 4-Nitro-o-xylene; p-Nitro-o-xylene; para-Nitro-ortho-xylene; o-Xylene, 4-nitro-. Crystals; mp = 30.5°; bp = 251°, bp21 = 143°; d15 = 1.112; λ_m = 215 nm (ε = 13183, EtOH), 211 nm (ε = 15136 C6H14); insoluble in H2O, freely soluble in EtOH.

1402 Dimethyloctanamide
1118-92-9   H

C10H21NO
N,N-Dimethyl-n-octanamide.
3-04-00-00128 (Beilstein Handbook Reference); AI3-26660; BRN 1754903; EINECS 214-272-5; N,N-Dimethylcaprylamide; N,N-Dimethyloctanamide; N,N-Dimethyl-n-octanamide; Octanamide, N,N-dimethyl-.

1403 Dimethylol dihydroxyethyleneurea
1854-26-8   H

C5H10N2O5
4,5-Dihydroxy-1,3-bis(hydroxymethyl)imidazolidin-2-one.
5-25-02-00369 (Beilstein Handbook Reference); Arkofix; Arkofix NG; BRN 0881343; Cassurit LR; CCRIS 4804; Depremol G; 4,5-Dihydroxy-1,3-bis(hydroxymethyl)-imidazolidin-2-one; 4,5-Dihydroxy-1,3-bis(hydroxy-methyl)-2-imidazolidinone; Dimethylol dihydroxy-ethyleneurea; Dimethyloldihydroxyethyleneurea; Di-methyloldihydroxyethylene urea; N,N'-Dimethylol-4,5-dihydroxyethyleneurea; Dimethylolglyoxalurea; N,N'-Dimethylolglyoxal monoureine; Dmdheu; EINECS 217-451-6; Firmatex RK; Fixapret CP; Fixapret CP 40; Fixapret CPK; Fixapret CPN; Fixapret CPNS; HSDB 4358; Hylite LF; 2-Imidazolidinone, 1,3-bis(hydroxymethyl)-4,5-dihydroxy-; 2-Imidazolidinone, 4,5-dihydroxy-1,3-bis(hydroxymethyl)-; Knittex LE; NCI-C60322; Neuperm GFN; NS 11; Permafresh 113B; Permafresh 183; Permafresh LF; Permafresh LH; Permafresh LKS; Protocol C; PROX DW; Readpret KPN; Reapret KPN; Sarcoset GM; Sumitex FSK; Sumitex NS; Sumitex NS 1 SPE; Sumitex NS 2; Verapret DH; Verapret DKh; WNM.

1404 Dimethyloxazolidine
51200-87-4   G

C5H11NO
4,4-dimethyl oxazolidine.
Amine CS-1135®; Bioban CS-1135; BRN 0969276; Canguard® 327; Caswell No. 374AB; Cosan 101; Dimethyl-1-oxa-3-aza-cyclopentane; Dimethyloxazolid-ine; Dimethyl oxazolidine; 4,4-Dimethyloxazolidine; 4,4-Dimethyl-1,3-oxazolidine; EINECS 257-048-2; EPA Pesticide Chemical Code 114801; Nuosept 101; Oxaban®-A; Oxadine A; Oxazolidine, 4,4-dimethyl-; Oxazolidine A; Troysan 192. Cosmetic preservative, antimicrobial, emulsifying amine, corrosion inhibitor, alkaline pH stabilizer; for metalworking fluids and aqueous systems. A preservative used in cooling fluids and paints. Bioban CS 1135 is a preservative used in cooling fluids and paints. In-can preservative for water-containing systems, e.g., latex paint, resin emulsions, caulks, adhesives; broad spectrum antimicrobial activity; emulsifying capability when used with fatty acids. *Angus.*

1405 Dimethyl palmitamine
112-69-6   G H

232

C18H39N
Dimethyl-n-hexadecylamine.
4-04-00-00818 (Beilstein Handbook Reference); Adma® 16; AI3-16727; Armeen DM16D; Bairdcat B16; BRN 1755921; Cetyldimethylamine; Crodamine 3.A16D; Dimethyl-n-hexadecylamine; Dimethyl palmitamine; Di-methyl palmitylamine; Dimethylcetylamine; Dimethyl-hexadecylamine; Dimethylpalmitylamine; EINECS 203-997-2; Farmin DM 60; Genamin 16R302D; 1-Hexadecanamine, N,N-dimethyl-; Hexadecylamine, N,N-dimethyl-; Hexadecyldimethylamine; IPL 67; NSC 404177; Palmityl dimethyl amine. Emulsifier for herbicides, ore flotation, pigment dispersion; auxiliary for textiles, leather, rubber, plastics, and metal industries. Chemical intermediate. *Asahi Denka Kogyo; Croda.*

1406   Dimethylphenol
1300-71-6   10137   G H

C8H10O
Xylenol (mixed isomers).
AI3-00056; Dimethylhydroxybenzene; Dimethylphenol; Dimethylphenols, liquid or soild; EINECS 215-089-3; HSDB 363; Hydroxydimethylbenzene; Phenol, dimethyl-; Stericol; Sudol; UN2261; Xilenole; Xilenoli; Xylenol; Xylenol (mixed isomers); Xylenolen; Xylenols. Used in disinfectants, solvents, pharmaceuticals, insecticides, fungicides, plasticizers, rubber chemicals, additives to lubricants and gasoline, manufacture of polyphenylene oxide, wetting agents, and dyestuffs. Solid; slightly soluble in H_2O, freely soluble in organic solvents. *Coalite Chem. Div.; Crowley Tar Prods.*

1407   3,4-Dimethylphenol
95-65-8   H

C8H10O
Phenol, 3,4-dimethyl-.
4-06-00-03099 (Beilstein Handbook Reference); AI3-01552; BRN 1099267; CCRIS 723; 4,5-Dimethylphenol; EINECS 202-439-5; FEMA No. 3596; HSDB 5294; 4-Hydroxy-1,2-dimethylbenzene; NSC 1549; Phenol, 3,4-dimethyl-. Liquid; mp = 60.8°; bp = 227°; d^{20} = 0.9830; λ_m = 218, 270 nm (ϵ = 6230, 1890, MeOH); slightly soluble in H_2O, soluble in EtOH, very soluble in Et_2O, CCl4.

1408   Dimethylphenylmethanol
617-94-7   H

C9H12O
2-Phenylisopropanol.
4-06-00-03219 (Beilstein Handbook Reference); AI3-05532; Benzenemethanol, α,α-dimethyl-; Benzyl alcohol, α,α-dimethyl-; BRN 1905012; α-Cumyl alcohol; α,α-Dimethylbenzenemethanol; α,α-Dimethylbenzyl alcohol; Dimethylphenylcarbinol; Dimethylphenylmethanol; EINECS 210-539-5; HSDB 5718; 1-Hydroxycumene; NSC 1261; NSC 212537; Phenyldimethylcarbinol; 2-Phenyl-2-propanol. Prisms; mp = 36°; bp = 202°; d^{20} = 0.9735; λ_m = 210, 243, 247, 253, 258, 265, 268 nm (MeOH); insoluble in H_2O, soluble in EtOH, Et_2O, C6H6, AcOH.

1409   Dimethyl phosphorochloridothioate
2524-03-0   G H

C2H6ClO2PS
O,O-Dimethyl chlorothiophosphate.
4-01-00-01263 (Beilstein Handbook Reference); BRN 0471300; Chlorodimethoxyphosphine sulfide; Dimeth-oxythiophosphonyl chloride; Dimethyl chlorothio-phosphate; Dimethyl phosphorochlorothioate; Dimethyl thiophosphorochloridate; Dimethyl thiophosphoryl chloride; Dimethylchlorthiofosfat; Dimethylthionochloro-phosphate; EINECS 219-754-9; HSDB 2603; Methyl PCT; MP-2; NSC 132984; Phosphorochloridothioic acid, O,O-dimethyl ester. Used in the production of organophosphorus insecticides. Liquid; bp16 = 66-67°; d = 1.3220. *A/S Cheminova.*

1410   Dimethyl phosphorodithioate
756-80-9   H

C2H7O2PS2
O,O-Dimethyl dithiophosphoric acid.
4-01-00-01264 (Beilstein Handbook Reference); BRN 0635993; Dimethyl phosphodithionate; O,O-Dimethyl dithiophosphate; O,O-Dimethyl dithiophosphoric acid; O,O-Dimethyl hydrogen dithiophosphate; O,O-Dimethyl phosphorodithioate; EINECS 212-053-9; HSDB 5750; Kwas dwumetylo-dwutiofosforowy; Kyselina O,O-dimethyldithiofosforcna; Methyl phosphorodithioate; Phosphorodithioic acid, O,O-dimethyl ester.

1411   Dimethyl phthalate
131-11-3   3281   G H P

C10H10O4
Dimethyl 1,2-benzenedicarboxylate.
AI3-00262; Avolin; Caswell No. 380; CCRIS 2674; Dimethyl 1,2-benzenedicarboxylate; Dimethyl benzene-o-dicarboxylate; Dimethyl o-phthalate; Dimethyl phthalate; Dimethylester kyseliny ftalove; DMF (insect repellent); DMP; EINECS 205-011-6; ENT 262; EPA Pesticide Chemical Code 028002; Fermine; HSDB 1641; Kemester® DMP; Kodaflex® DMP; Mipax; NSC 15398; NTM; Palatinol M; Phtalate de dimethyle; Phthalic acid, dimethyl ester; Phthalsäuredimethylester; RCRA waste number U102; Repeftal; Solvanom; Solvarone; Unimoll® DM. Solvent for resins, plasticizer for cellulose acetate and nitrocellulose lacquers; plastics, rubber; coating agents; safety glass. Has high solvent power for cellulose acetate extrusion compounds; compatible with ethyl cellulose, CAB, PS, PVAc, polyvinyl butyral, and PVC; used in NC-based printing inks. Emollient used in cosmetics, textiles, metalworking lubricants. Fixative for perfumes. Registered by EPA (cancelled). Pale yellow oil; mp = 5.5°; d_4^{20} = 1.1905; d_4^{25} = 1.189; bp760 = 283.7°; bp400 = 257.8°; bp200 = 232.7°; bp100 = 210.0°; bp40 = 182.8°; bp20 = 164°; bp10 = 147.6°; bp5 = 131.8°; bp1 = 100.3°; λ_m = 277 nm (E 1%cm 57.7

EtOH); poorly soluble in H₂O (430 mg/100 ml), soluble in organic solvents; LD$_{50}$ (rat orl) = 8.2 g/kg. *Allchem Ind.; BASF Corp.; Bayer AG; CK Witco Corp.; Daihachi Chem. Ind. Co. Ltd; Degussa-Hüls Corp.; Eastman Chem. Co.; Fluka; Morflex Inc.; UCB Pharma.*

1412   N,N'-Dimethyipiperazine
106-58-1   G

C$_6$H$_{14}$N$_2$
1,4-Dimethylpiperazine.
CCRIS 6690; 1,4-Dimethylpiperazine; EINECS 203-412-0; Lupetazine; N,N'-Dimethylpiperazine; NSC 41177; Piperazine, 1,4-dimethyl-; Texacat® DMP. Catalyst for polyurethane molded high resilience flexible foam, ester slabstock. Liquid; bp = 131°; d^{20} = 0.8600; very soluble in H₂O, EtOH, Et₂O. *Texaco.*

1413   Dimethyl sebacate
106-79-6   H

C$_{12}$H$_{22}$O$_4$
Dimethyl decanedioate.
AI3-00662; Decanedioic acid, dimethyl ester; Dimethyl decanedioate; Dimethyl octane-1,8-dicarboxylate; Dimethyl sebacate; EINECS 203-431-4; Methyl sebacate; NSC 9415; Sebacic acid, dimethyl ester. Solid; mp = 38°; bp$_{20}$ = 175°, bp$_5$ = 144°; d^{28} = 0.9882; insoluble in H₂O, soluble in EtOH, Et₂O, Me₂CO, CCl₄.

1414   Dimethyl soyamine
61788-91-8   H

Amines, dimethylsoya alkyl.
Amines, dimethylsoya alkyl; Dimethyl soyamine; EINECS 263-017-4; Soya dimethyl amine.

1415   Dimethyl succinate
106-65-0   G H

C$_6$H$_{10}$O$_4$
Dimethyl butanedioate.
AI3-02480; Butanedioic acid, dimethyl ester; CCRIS 4803; Dimethyl butanedioate; Dimethyl succinate; EINECS 203-419-9; FEMA No. 2396; HSDB 5370; Methyl butanedioate; Methyl succinate; NSC 52209; Succinic acid, dimethyl ester. Light and heat stabilizer for polyolefins, ABS polymer systems, flexible PVC, food packaging; chemical intermediate. Liquid; mp = 19°; bp = 196.4°; d^{20} = 1.0020; very soluble in Me₂CO, C₆H₆, Et₂O. *Ashland; Chemie Linz N. Am.; DuPont; Fluka; Lancaster Synthesis Co.; Penta Mfg.; Sigma-Aldrich Fine Chem.*

1416   Dimethyl succinylsuccinate
6289-46-9   H

C$_{10}$H$_{12}$O$_6$
Dimethyl 2,5-dioxocyclohexane-1,4-dicarboxylate.
AI3-14663; 1,4-Cyclohexanedicarboxylic acid, 2,5-dioxo-, dimethyl ester; Dimethyl 2,5-dioxocyclohexane-1,4-dicarboxylate; Dimethyl 2,5-dioxo-1,4-cyclohex-anedicarboxylate; Dimethyl succinylsuccinate; EINECS 228-528-9; NSC 5670; NSC 122567; Succinosuccinic acid, dimethyl ester.

1417   Dimethyl sulfate
77-78-1   3282   H

C$_2$H$_6$O$_4$S
Sulfuric acid, dimethyl ester.
4-01-00-01251 (Beilstein Handbook Reference); AI3-52118; BRN 0635994; CCRIS 265; Dimethyl monosulfate; Dimethyl sulfate; Dimethyl sulphate; Dimethylester kyseliny sirove; Dimethylsulfaat; Dimethylsulfat; Dimetilsolfato; DMS (methyl sulfate); Dwumetylowy siarczan; EINECS 201-058-1; HSDB 932; Methyl sulfate; Methyle (sulfate de); NSC 56194; RCRA waste number U103; Sulfate de dimethyle; Sulfate dimethylique; Sulfato de dimetilo; Sulfuric acid, dimethyl ester; UN1595. Liquid; mp = -27°; bp$_{15}$ = 76°, bp = ¨188° (dec); d^{20} = 1.3322; moderately soluble in H₂O, organic solvents.

1418   Dimethyl 5-sulfoisophthalate
138-25-0   H

C$_{10}$H$_{10}$O$_7$S
Isophthalic acid, 5-sulfo-, 1,3-dimethyl ester.
1,3-Benzenedicarboxylic acid, 5-sulfo-, 1,3-dimethyl ester; Dimethyl 5-sulphoisophthalate; EINECS 205-320-6; Isophthalic acid, 5-sulfo-, 1,3-dimethyl ester.

1419   Dimethyl sulfoxide
67-68-5   3285   G H V

C$_2$H$_6$OS
Methane, sulfinylbis-.
A 10846; AI3-26477; Caswell No. 381; CCRIS 943; Deltan; Demasorb; Demavet; Demeso; Demsodrox; Dermasorb; Dimethyl sulfoxide; Dimethyl sulphoxide; Dimethyli sulfoxidum; Dimethylsulfoxide; Dimethyl-sulfoxyde; Dimetil sulfoxido; Dimetilsolfossido; Di-mexide; Dipirartril-tropico; DMS-70; DMS-90; DMSO; Dolicur; Doligur; Domoso; Dromisol; Durasorb; EINECS 200-664-3; EPA Pesticide Chemical Code 000177; Gamasol 90; HSDB 80; Hyadur; Infiltrina; M 176; Methane, sulfinylbis-; Methyl sulfoxide; Methylsulfinyl-methane; NSC-763; Rimso-5; Rimso-50; Somipront; SQ 9453; Sulfinylbis(methane); Syntexan; Topsym. Solvent for polymerization; analytical reagent; industrial cleaners, pesticides,

paint stripping, hydraulic fluids, medicine (anti-inflammatory), veterinary medicine, plant pathology and nutrition, pharmaceuticals, spinning synthetic fibers. Aprotic solvent used as reaction medium in the manufacture of pesticides, pharmaceuticals, dyes, inks; solvent in polymers, electronics, refining; chemical intermediate for pesticides. Clear, viscous liquid; mp = 18.5°; bp = 189°; d^{20} = 1.1014; soluble in H_2O; insoluble in organic solvents; LD_{50} (rat, orl) = 19,700 mg/kg. *Allchem Ind.; Elf Atochem N. Am.; Fluka; Gaylord; Howard Hall; Itochu; Monomer-Polymer & Dajac; Sigma-Aldrich Fine Chem.; Spectrum Chem. Manufacturing; Toray Ind. inc.*

1420 Dimethyl terephthalate
120-61-6   H

$C_{10}H_{10}O_4$
Dimethyl 1,4-benzenedicarboxylate.
AI3-02246; CCRIS 266; Dimethyl 1,4-benzene-dicarboxylate; Dimethyl p-benzenedicarboxylate; Di-methyl 4-phthalate; Dimethyl p-phthalate; Dimethyl terephthalate; Dimethylester kyseliny tereftalove; DMT; EINECS 204-411-8; HSDB 2580; NCI-C50055; NSC 3503. A Plasticizer. Solid; mp = 141°; bp = 288°; d^{141} = 1.075; λ_m = 240, 285 nm (ϵ = 20300, 1680, MeOH); slightly soluble in H_2O, EtOH, MeOH, soluble in Et_2O, $CHCl_3$.

1421 1,2-Dimethyltetrachlorodisilane
4518-98-3   H

$C_2H_6Cl_4Si_2$
1,1,2,2-Tetrachloro-1,2-dimethyldisilane.
1,2-Dimethyltetrachlorodisilane; Disilane, 1,1,2,2-tetra-chloro-1,2-dimethyl-; EINECS 224-844-6; 1,1,2,2-Tetrachloro-1,2-dimethyldisilane. Liquid; bp = 154°.

1422 1,2-Dimethyl-3-tetrapropylenebenzene
66697-27-6   H

$C_{20}H_{34}$
Benzene, 1,2-dimethyltetrapropylene-.
1423 Dimethyl
1423 Dimethyl thiodipropionate
4131-74-2   H

$C_8H_{14}O_4S$
Dimethyl 3,3'-thiobispropionate.
Bis(2-methoxycarbonylethyl)sulfide; β,β'-Dicarbometh-oxydiethyl sulfide; Dimethyl 3,3'-thiobispropionate; Dimethyl thiodipropionate; Dimethyl 3,3'-thiodi-propionate; EINECS 223-948-9; NSC 2244;

NSC 35224; Propanoic acid, 3,3'-thiobis-, dimethyl ester; Propionic acid, 3,3'-thiodi-, dimethyl ester. Liquid; bp_{18} = 162°, bp_{18} = 148°; d^{20} = 1.1559.

1424 Dimethyltin-bis(isooctylthioglycolate)
26636-01-1   H

$C_{22}H_{44}O_4S_2Sn$
Stannane, bis(isooctyloxycarbonylmethylthio)dimethyl-ester.
Acetic acid, 2,2'-((dimethylstannylene)bis(thio))bis-, di-isooctyl ester; Advastab TM 181; Advastab TM 181S; Bis((((isooctyloxy)carbonyl)methyl)thio)dimethyltin; Diiso-octyl ((dimethylstannylene)dithio)diacetate; Diiso-octyl 2,2'-((dimethylstannylene)bis(thio))diacetate; Di-methyltin bis(isooctyl mercaptoacetate); Dimethyltin-bis(isooctyl-thioglycolate); Dimethyltin S,S'-bis(isooctyl mercapto-acetate); Dimethyltinbis(isooctylmercaptoacetate); Di-methylzinn-S,S'-bis(isooctylthioglycolat); EINECS 247-862-6; HSDB 6097; Stannane, bis(isooctyloxycarbonyl-methylthio)dimethyl-ester; Tin, bis((carboxymethyl)thio)-dimethyl-, diisooctyl ester; Tin, dimethyl-, bis(isooctylthioglycollate); TM 181S.

1425 N,N-Dimethylurea
96-31-1   H

$C_3H_8N_2O$
Urea, N,N'-dimethyl-.
4-04-00-00207 (Beilstein Handbook Reference); AI3-24386; BRN 1740672; CCRIS 2509; N,N-Dimethylurea; N,N'-Dimethylharnstoff; N,N'-Dimethylurea; sym-Di-methylurea; Symmetric dimethylurea; DMU; EINECS 202-498-7; HSDB 3423; NSC 14910; Urea, 1,3-dimethyl-; Urea, N,N'-dimethyl-. Crystals; mp = 108°; bp = 269°; d^{25} = 1.142; very soluble in H_2O, EtOH, poorly soluble in $CHCl_3$, insoluble in Et_2O.

1426 Dimethylvinylchlorosilane
1719-58-0   G

C_4H_9ClSi
Vinyldimethylchlorosilane.
AI3-25204; Chlorodimethylvinylsilane; CV-4720; EINECS 217-007-1; Silane, chloroethenyldimethyl-; Vinyldi-methylchlorosilane. Coupling agent, chemical intermediate, blocking agent, release agent, lubricant, primer, reducing agent. Liquid; bp = 83.5°; d^{20} = 0.8744. *Degussa-Hüls Corp.*

1427 Dimetridazole

551-92-8   3292   G V

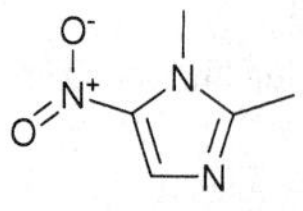

C5H7N3O2

1,2-Dimethyl-5-nitro-1H-imidazole.

5-23-05-00058 (Beilstein Handbook Reference); 8595 R.P.; Al3-27217; BRN 0130665; Caswell No. 371A; CCRIS 997; 1,2-Dimethyl-5-nitroimidazole; 1,2-Dimethyl-5-nitro-1H-imidazole; Dimetridazol; Dimetrid-azole; Dimetridazolo; Dimetridazolum; EINECS 209-001-2; Emtryl; Emtrylvet; Emtrymix; EPA Pesticide Chemical Code 371200; Imidazole, 1,2-dimethyl-5-nitro-; 1H-Imidazole, 1,2-dimethyl-5-nitro-; 5-Nitro-1,2-dimethyl-imidazole; NSC 226253; RP 8595; Unizole; Unizole Soluble. A veterinary antiprotozoan (Histomonas). Needles; mp = 138-139°; sparingly soluble in H_2O, very soluble in EtOH, Et_2O.

1428 2,4-Dinitroaniline

97-02-9   3298   H

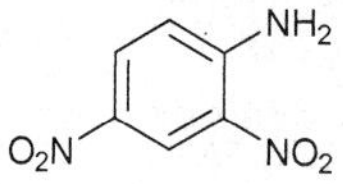

C6H5N3O4

Benzenamine, 2,4-dinitro-.

4-12-00-01689 (Beilstein Handbook Reference); Al3-02920; Aniline, 2,4-dinitro-; Benzenamine, 2,4-dinitro-; BRN 0982999; CCRIS 2872; EINECS 202-553-5; HSDB 1142; NCI-C60753; NSC 8731. Crystals; mp = 180°; d^{14} = 1.615; λ_m = 210, 225, 257, 336 nm (ε = 1120, 1110, 969, 1480, MeOH); insoluble in H_2O, slightly soluble in EtOH, Me_2CO, HCl.

1429 Dinitrobenzene

528-29-0   3301   H

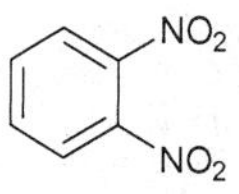

C6H4N2O4

o-Dinitrobenzene.

Al3-15338; Benzene, 1,2-dinitro-; Benzene, o-dinitro-; CCRIS 3091; 1,2-Dinitrobenzene; 1,2-Dinitrobenzol; Dinitrobenzene, o-; o-Dinitrobenzene; EINECS 208-431-8; HSDB 4486; NSC 60682; UN1597. Needles or plates; mp = 118.5°; bp = 318°; bp30 = 194°; insoluble in H_2O, slightly soluble in DMSO, soluble in EtOH, C_6H_6, $CHCl_3$, MeOH, EtOAc, C_7H_8.

1430 Dinitrobenzoic acid

99-34-3   3304   G

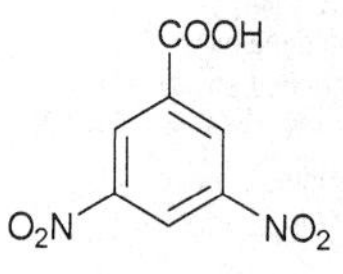

C7H4N2O6

3,5-dinitrobenzoic acid.

Al3-01801; Benzoic acid, 3,5-dinitro-; 3-Carboxy-1,5-dinitrobenzene; CCRIS 3129; Dinitrobenzoic acid; 3,5-Dinitrobenzoic acid; DNBA; EINECS 202-751-1; NSC 8732. Used in chemical analysis, in identification of alcohols and chromatographic determination of essential oil constituents. Crystals; mp = 205°; sublimes; λ_m = 228 nm (ε 19700, MeOH); soluble in H_2O (2 g/100 ml), EtOH and glacial AcOH, less soluble in Et_2O, CS_2, C_6H_6. *Akzo Nobel; Eka Nobel Ltd.; Lancaster Synthesis Co.*

1431 2,4-Dinitrochlorobenzene

97-00-7   2155   H

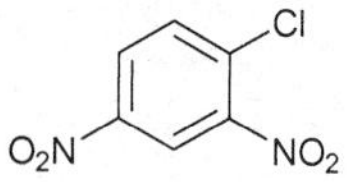

C6H3ClN2O4

Benzene, 1-chloro-2,4-dinitro-.

Al3-01053; Benzene, 1-chloro-2,4-dinitro-; Caswell No. 389C; CCRIS 1799; CDNB; Chlorodinitrobenzene; Dinitrochlorobenzene; Dinitrochlorobenzol; DNCB; DRG-0080; EINECS 202-551-4; EPA Pesticide Chemical Code 055102; HSDB 5306; NSC 6292. Yelow rhomboid crystals; mp = 53°; bp = 315°; d^{75} = 1.4982; λ_m = 206, 238 nm (MeOH); insoluble in H_2O, slightly soluble in EtOH, soluble in Et_2O, C_6H_6, CS_2.

1432 Dinitro-p-cresol

609-93-8   H

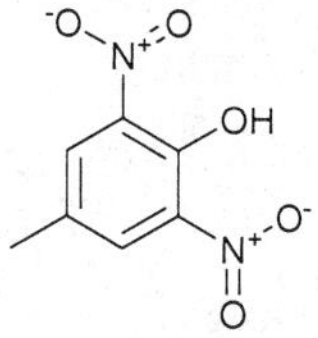

C7H6N2O5

2,6-Dinitro-4-methylphenol.

4-06-00-02152 (Beilstein Handbook Reference); Al3-24606; BRN 1978786; p-Cresol, 2,6-dinitro-; Dinitro-p-cresol; 2,6-Dinitro-4-methylphenol; 2,6-Dinitro-p-cresol; DNPC; EINECS 210-203-8; HSDB 5434; 4-Methyl-2,6-dinitrophenol; NSC 33870; Phenol, 4-methyl-2,6-dinitro-; Toluene, 3,5-dinitro-4-hydroxy-; Victoria Orange; Victoria Yellow.

1433 4,6-Dinitrocresol

534-52-1   3307   G P

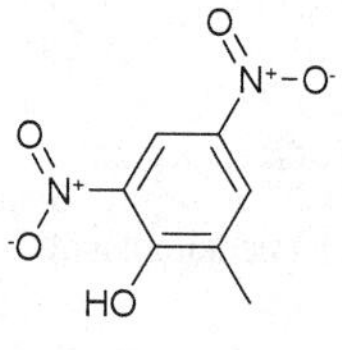

C7H6N2O5

3,5-Dinitro-2-Hydroxytoluene.

Antinonin; Antinonnin; Dekryll; Dekrysil; Detal; Dinitrocresol; Dinitro-o-cresol; 4,6-Dinitrocresol; Dinitrol; Dinitrosol; Ditrosol; DNC; DNOC; 4,6-DNOC; Effusan; Elgetol; Elgetol 30; Elgetox; Extar A; Extar Lin; K III; K IV; Lipan; Nitrador; Prokarbol; Sandoline; Selinon; Sinox; Trifocide; Trifrina. Selective herbicide and insecticide. Yellow prisms or needles; mp = 86.5°; λ_m = 215, 263 nm (MeOH); slightly soluble in H_2O, petroleum ether, soluble in EtOH, Et_2O, Me_2CO,

CHCl3.slightly soluble in H2O, soluble in organic solvents; LD50 (rat orl) = 7 mg/kg. *Bayer AG; Pennwalt Holland; PMC; Sandoz.*

1434   2,4-Dinitrophenol
51-28-5   3309   H P

C6H4N2O5
1-Hydroxy-2,4-Dinitrobenzene.
2,4-DNP; AI3-01535; Aldifen; Camello mosquito coils; Caswell No. 392; CCRIS 3102; Chemox PE; Cobra salts (Impregna salts); Dinitra; Dinitrofenolo; Dinitrophenol, 2,4-; Dinofan; DNP; EINECS 200-087-7; EK 102; EPA Pesticide Chemical Code 037509; Fenoxyl; Fenoxyl Carbon N; HSDB 529; Maroxol-50; Nitro kleenup; Nitrophen; Nitrophene; NSC 1532; Osmoplastic-R; Osmotox-Plus; Phenol, 2,4-dinitro-; RCRA waste number P048; Shirakiku brand mosquito coils; Solfo Black 2B Supra; Solfo Black B; Solfo Black BB; Solfo Black G; Solfo Black SB; Tertrosulphur Black PB; Tertrosulphur PBR; X 32. Fungicide. Registered by EPA as a fungicide and insecticide (cancelled). Used in manufacture of dyestuffs, as a wood preservative and as an acid-base indicator (pH 2,6 colorless, pH 4.4 yellow). Yellow orthorhombic crystals; mp = 112 - 114°; d = 1.683; soluble in H2O (0.137 g/100 ml at 54.5°, 0.301 g/100 ml at 75.8°, 0.587 g/100 ml at 87.4°, 1.22 g/100 ml at 96.2°), EtOAc (15.55 g/100 ml), Me2CO (35.9 g/100 ml), CHCl3 (5.39 g/100 ml), C5H5N (20 g/100 ml), CCl4 (0.42 g/100 ml), C7H8 (6.36 g/100 ml), all at 15°, soluble in EtOH, C6H6; LD50 (rat orl) = 30 mg/kg. *Sigma-Aldrich Co.*

1435   3,5-Dinitro-o-toluamide
148-01-6   3297   G

C8H7N3O5
2-Methyl-3,5-dinitrobenzamide.
3-09-00-02316 (Beilstein Handbook Reference); Benzamide, 2-methyl-3,5-dinitro-; BRN 1990738; Caswell No. 932; Coccidine A; Coccidot; Dinitolmida; Dinitolmide; Dinitolmidum; 3,5-Dinitro-o-toluamide; EINECS 205-706-4; EPA Pesticide Chemical Code 037510; 2-Methyl-3,5-dinitrobenzamide; Zoalene; Zoamix. Coccidiostat; often used on an exchange program with Coyden (clopidol); the main purpose in rotating Zoamix coccidiostat with Coyden coccidiostat is to prevent poultry from developing a resistance to the latter product. Crystals; mp = 181°. *Alpharma; Dow UK; Hoechst Roussel Veterinary; Pharmacia & Upjohn Co.*

1436   Dinitrotoluene
25321-14-6   H

C7H6N2O4
Dinitrotoluene (mixed isomers).
Benzene, methyldinitro-; Binitrotoluene; Dinitrophenyl-methane; Dinitrotoluene; Dinitrotoluene (mixed isomers); Dinitrotoluenes; Dinitrotoluol; DNT; EINECS 246-836-1; HSDB 3907; Methyldinitrobenzene; Toluene, ar,ar-dinitro-; Toluene, dinitro-; UN1600; UN2038. Mixture of o, m and p-isomers.

1437   2,3-Dinitrotoluene
602-01-7   H

C7H6N2O4
1-Methyl-2,3-dinitrobenzene.
4-05-00-00865 (Beilstein Handbook Reference); Benzene, 1-methyl-2,3-dinitro-; BRN 2212428; CCRIS 2837; 2,3-Dinitrotoluene; 2,3-Dinitrotoluol; 2,3-DNT; EINECS 210-013-5; HSDB 5499; 1-Methyl-2,3-dinitrobenzene; Toluene, 2,3-dinitro-.

1438   2,4-Dinitrotoluene
121-14-2   G H

C7H6N2O4
2,4-Dinitrotoluene.
AI3-15342; Benzene, 1-methyl-2,4-dinitro-; CCRIS 268; Dinitrotoluene; 2,4-Dinitrotoluene; 2,4-Dinitrotoluol; DNT; 2,4-DNT; EINECS 204-450-0; HSDB 1144; NCI-C01865; NSC 7194; RCRA waste number U105; Toluene, 2,4-dinitro-. Used in synthesis of toluidines, dyes and explosives. Yellow needles; mp = 71°; bp = 300° (dec); d^{71} = 1.3208; λm = 234 nm (cyclohexane); insoluble in H2O, soluble in EtOH, Et2O, C6H6, CHCl3, C7H8, CS2, EtOAc; very soluble in Me2CO, C5H5N; LD50 (rat orl) = 268 mg/kg.

1439   2,5-Dinitrotoluene
619-15-8   H

C7H6N2O4
2-Methyl-1,4-dinitrobenzene.
4-05-00-00866 (Beilstein Handbook Reference); Benzene, 2-methyl-1,4-dinitro-; BRN 2213718; CCRIS 2838; 2,5-Dinitrotoluene; 2,5-DNT; EINECS 210-581-4; HSDB 5504; 2-Methyl-1,4-dinitrobenzene; NSC 159120; Toluene, 2,5-dinitro-. Needles; mp = 52.5°; d^{111} = 1.282; λm = 267 nm (ε = 11482, 5% EtOH); soluble in EtOH, C6H6, very soluble in CS2.

1440   2,6-Dinitrotoluene
606-20-2   H

$C_7H_6N_2O_4$
2-Methyl-1,3-dinitrobenzene.
4-05-00-00866 (Beilstein Handbook Reference); Benzene, 2-methyl-1,3-dinitro-; BRN 2052046; CCRIS 1006; 2,6-Dinitromethylbenzene; 2,6-Dinitrotoluene; 2,6-DNT; EINECS 210-106-0; HSDB 2931; 2-Methyl-1,3-dinitrobenzene; RCRA waste number U106; Toluene, 2,6-dinitro-. Rhombic needles; mp = 66°; d^{111} = 1.2833; λ_m = 233 nm (ϵ = 9670, MeOH); soluble in EtOH, $CHCl_3$.

1441   3,4-Dinitrotoluene
610-39-9   H

$C_7H_6N_2O_4$
4-Methyl-1,2-dinitrobenzene.
Benzene, 4-methyl-1,2-dinitro-; CCRIS 2839; 3,4-Dinitrotoluene; 3,4-DNT; EINECS 210-222-1; HSDB 5501; 4-Methyl-1,2-dinitrobenzene; NSC 52216; Toluene, 3,4-dinitro-. Yellow needles; mp = 58.3°; d^{111} = 1.2594; λ_m = 216, 262 nm (ϵ = 12400, 5900, MeOH); insoluble in H_2O, slightly soluble in $CHCl_3$, soluble in Et_2O, CS_2.

1442   Dinobuton
973-21-7   3315   G P

$C_{14}H_{18}N_2O_7$
Carbonic acid 1-methylethyl 2-(1-methylpropyl)-4,6-dinitrophenyl ester.
Acrex; Al3-27244; Akrex; BRN 2065340; (2-sek.Butyl-4,6-dinitrofenyl)-isopropylkarbonat; 2-sec-Butyl-4,6-di-nitrophenyl isopropyl carbonate; Carbonic acid, 1-methylethyl 2-(1-methylpropyl)-4,6-dinitrophenyl; Carb-onic acid, 2-sec-butyl-4,6-dinitrophenyl isopropyl ester; Caswell No. 128F; Dessin; Dinitro-sec-butylphenyl isopropyl carbonate; 2,4-Dinitro-6-sec-butylphenyl iso-propyl carbonate; 2,4-Dinitro-6-sek.butyl-isopropyl-phenylcarbonat; Dinobuton; Dinofen; Drawinol; DS 18302; EINECS 213-546-1; ENT 27,244; EPA Pesticide Chemical Code 228700; HSDB 1527; Isophen (pesticide); Isopropyl 2,4-dinitro-6-sec-butylphenyl carbonate; Isopropyl-2-(1-methyl-n-propyl)-4,6-dinitrophenyl carb-onate; Kasebon; MC 1053; 1-Methylethyl 2-(1-methylpropyl)-4,6-dinitrophenyl carbonate; 2-(1-Methyl-2-propyl)-4,6-dinitrophenyl isopropylcarbonate; OMS 1056; Phenol, 2-sec-butyl-4,6-dinitro-, isopropyl-carbonate; Sytasol; Talan; UC 19786; Union carbide 19786. Miticide. Crystals; mp = 56-57°; LD_{50} (rat orl) = 59 mg/kg. *Murphy Chemical Co Ltd.*

1443   Dinocap
39300-45-3   3316   G P

$C_{18}H_{24}N_2O_6$
Crotonic acid, 2(or 4)-(1-methylheptyl)-4,6(or 2,6)-dinitrophenyl ester.
Actual dinocap; Arathane; 2-Butenoic acid, 2(or 4)-isooctyl-4,6(or 2,6)-dinitrophenyl ester; 2-Butenoic acid 2-(1-methylheptyl)-4,6-dinitrophenyl ester; Caprane; Capryldinitrophenyl crotonate; 2-Capryl-4,6-dinitro-phenyl crotonate; Caratan; Carathane; Caswell No. 391D; CR 1639; Crotonate de 2,4-dinitro 6-(1-methyl-heptyl)-phenyle; Crotonic acid 2-(1-methylheptyl)-4,6-dinitro-phenyl ester; Crotonic acid, 2(or 4)-(1-methylheptyl)-4,6(or 2,6)-dinitrophenyl ester; Crotonic acid 2,4-dinitro-6-(2-octyl)phenyl ester; Crotonic acid 2,4-dinitro-6-(1-methylheptyl)phenyl ester; Crotothane; Dinitro methyl-heptyphenyl crotonate; Dinitro(1-methylheptyl)phenyl crotonate; 4,6-Dinitro-2-caprylphenyl crotonate; 4,6-Dinitro-2-(2-capryl)phenyl crotonate; 4,6-Dinitro-2-(1-methylheptyl)phenyl crotonate; 2,4-Dinitro-6-(1-methyl-heptyl)phenyl crotonate; 2,4-Dinitro-6-(1-methylheptyl)-phenylcrotonat; 2,4-Dinitro-6-octyl-phenyl crotonate; 2,4-Dinitro-6-(2-octyl)phenyl crotonate; 2,6-Dinitro-4-octyl-phenyl crotonate; Dinitrocaprylphenyl crotonate; Dinocap; Dinokap; Dinokapu; DNOCP; DPC; DPC (pesticide); EINECS 254-408-0; ENT 24727; EPA Pesticide Chemical Code 036001; HSDB 1597; Isocothane; Karatan; Karathane; Karathane 25; Karathane FN 57; Karathane WD; (6-(1-Methyl-heptyl)-2,3-dinitro-phenyl)-crotonat; (6-(1-Methyl-heptyl)-2,4-dinitro-fenyl)-croton-aat; (6-(1-Metil-epitl)-2,4-dinitro-fenil)-crotonato; 2-(1-Methylheptyl)-4,6-dinitrofenylester kyseliny krotonove; 2-(1-Methylheptyl)-4,6-dinitrophenyl crotonate; Mildex; Mixed dinitrooctylphenylcrotonates and dinitrooctyl-phenols; Phenol, 2-(1-methylheptyl)-4,6-dinitro-, croton-ate (ester); Crotonic acid 2,4-dinitro-6-(1-methyl-heptyl)phenyl ester; Crotonic acid 2,4-dinitro-6-(2-octyl)phenyl ester; Crotothane; Dinitro methylheptyl-phenyl crotonate; Dinitro(1-methylheptyl)phenyl crotonate; Dinitrocaprylphenyl crotonate; 2,4-Dinitro-6-(1-methylheptyl)phenyl crotonate; 2,4-Dinitro-6-(1-methylheptyl)-phenylcrotonat; 2,4-Dinitro-6-octyl-phenyl crotonate; 2,4-Dinitro-6-(2-octyl)phenyl crotonate; 2,6-Dinitro-4-octyl-phenyl crotonate; 4,6-Dinitro-2-caprylphenyl crotonate; 4,6-Dinitro-2-(2-capryl)phenyl crotonate; Dinocap; Dinokap; Dinokapu; DPC; DPC (pesticide); ENT 24727; EPA Pesticide Chemical Code 036001; HSDB 1597; Isocothane; Karatan; Karathane; Karathane 25; Karathane FN 57; Karathane WD; (6-(1-Methyl-heptyl)-2,4-dinitro-fenyl)-crotonaat; (6-(1-Methyl-heptyl)-2,3-dinitro-phenyl)-crotonat; (6-(1-Metil-epitl)-2,4-dinitro-fenil)-crotonato; 2-(1-Methylheptyl)-4,6-dinitro-fenylester kyseliny krotonove; 2-(1-Methylheptyl)-4,6-dinitrophenyl crotonate; Mildex; Mixed dinitrooctylphenylcrotonates and dinitrooctylphenols; Phenol, 2-(1-methylheptyl)-4,6-dinitro-, crotonate (ester). Acaricide. Used to control mildew in fruit, chrysanthemums and roses. Registered by EPA as an fungicide and insecticide. Brown liquid; $bp_{0.05}$ = 138-140°; almost insoluble in H_2O (< 0.00001 g/100 ml), soluble in most organic solvents; LD_{50} (mrat orl) = 980 mg/kg, (frat orl)= 1190 mg/kg, (dog orl) = 100 mg/kg, (rbt der) > 4700 mg/kg, (mrat iv) = 23 mg/kg; LC_{50} (rat ihl 4 hr.) = 0.36 mg/l air; toxic to fish; non-toxic to bees. *Dow AgroSciences; Rohm & Haas UK.*

1444   Dinonyl phenol
1323-65-5   G H

C24H42O

Phenol, dinonyl-.

Dinonyl phenol; Dinonylphenol; EINECS 215-356-4; NSC 2431; Phenol, dinonyl-. Mixture of dialkyl substituted phenols; used as a solvent. Colorless liquid; insoluble in H2O; soluble in organic solvents. *Allchem Ind.; Texaco.*

1445   Dinonyl phthalate
84-76-4   G

C26H42O4

Di-n-nonyl 1,2-benzenedicarboxylate.

4-09-00-03183 (Beilstein Handbook Reference); 1,2-Benzenedicarboxylic acid, dinonyl ester; Bisoflex 91; Bisoflex DNP; Bisolflex 91; BRN 1916263; Di-n-nonyl phthalate; Di-n-nonylphthalate (DnNP); Dinonyl phthalate; EINECS 201-560-0; HSDB 365; Nonyl phthalate; Phthalic acid, dinonyl ester; Unimoll DN. A linear plasticizer with good low temperature, permanence, processibility and low volatility, high efficiency, good low temperature performance and wide processing range. Used as a plasticizer for vinyl polymers. Clear liquid; d^{20} = 0.969. *BDH Laboratory Supplies; ExxonMobil Chem. Co.; Koninklijke Nederlandsche Gist-En Spiritusfabriek; National Lead Co.*

1446   Dinonylnaphthalenesulphonic acid
25322-17-2   H

C28H44O3S

Naphthalenesulfonic acid, dinonyl-.

Dinonylnaphthalenesulphonic acid; EINECS 246-841-9; Naphthalenesulfonic acid, dinonyl-. Mixture of o, m and p-isomers.

1447   Dinoseb
88-85-7   3317   H P

C10H12N2O5

2-(1-Methylpropyl)-4,6-dinitrophenol.

4-06-00-03279 (Beilstein Handbook Reference); Aatox; AI3-01122; Basanite; Blaartox; BNP 20; BNP 30; BRN 3211812; Butaphen; Butaphene; 2-sec-Butyl-4,6-dinitrophenol; Caldon; Caswell No. 392DD; Chemox general; Chemox P.E.; DBNF; Desicoil; Dibutox; Dibutox 20CE; Dinitrall; Dinitro-ortho-sec-butyl phenol; Dinitrobutylphenol; 4,6-Dinitro-2-sec-butylfenol; 2,4-Dinitro-6-sec-butylphenol; 4,6-Dinitro-2-sec-butylphenol; 4,6-Dinitro-o-sec-butylphenol; 2,4-Dinitro-6-(1-methyl-propyl)phenol; 4,6-Dinitro-2-(1-methyl-propyl)phenol; Dinoseb; Dinosebe; DN 289; DNBP; DNOSBP; DNSBP; Dow General Weed Killer; Dow Selective Weed Killer; Dytop; Elgetol 318; ENT 1,122; EPA Pesticide Chemical Code 037505; Gebutox; Hel-Fire; Hivertox; HSDB 1445; Kiloseb; Knox-weed; Ladob; Laseb; Liro DNBP; 6-(1-Metil-propil)-2,4-dinitro-fenolo; 2-(1-Methylpropyl)-4,6-dinitrophenol; 6-(1-Methyl-propyl)-2,4-dinitrofenol; Nitropone C; NSC 202753; Phenol, 2-(1-methylpropyl)-4,6-dinitro-; Phenol, 2-sec-butyl-4,6-dinitro-; Phenotan; Premerge; RCRA waste number P020; Sparic; Spurge; Subitex; Unicrop DNBP; Vertac Dinitro Weed Killer; Vertac General Weed Killer; Vertac Selective Weed Killer; WSX 8365. Herbicide and insecticide. Orange crystals; mp = 38 - 42°; sg[45] = 1.265; soluble in H2O (0.0052 g/100 ml), EtOH (38 g/100 ml), C7H16 (18 g/100 ml), soluble in most organic solvents; LD50 (rat orl) = 58 mg/kg, (gpg orl) = 25 mg/kg, (rbt der) = 80 - 200 mg/kg, (gpg der) = 500 mg/kg, (ckn orl) = 26 mg/kg; LC50 (Japanese quail 5 dat dietary) = 409 mg/kg diet, (ring-necked pheasant 5 day dietary) = 515 mg/kg diet; highly toxic to fish; toxic to bees. *DowElanco Ltd.; Hoechst AG; La Littorale; La Quinoleine S.A.*

1448   Dioctanoyl peroxide
762-16-3   G

C16H30O4

Caprolyl peroxide.

Capryl peroxide; Caprylyl peroxide; Dicaprylyl peroxide; Dioctanoyl peroxide; EINECS 212-094-2; HSDB 2752; Octanoyl peroxide; Perkadox® SE-8; Peroxide, bis(1-oxooctyl). Radical polymerization initiator. *Akzo Chemie.*

1449   Dioctyl adipate
103-23-1   G H

C22H42O4

Di-(2-ethylhexyl) adipate.

4-02-00-01964 (Beilstein Handbook Reference); Adimoll® DO; Adipic acid, bis(2-ethylhexyl) ester; Adipic acid, di(2-ethylhexyl) ester; Adipol 2EH; ADO (lubricating oil); AI3-28579; Arlamol DOA; Beha; BEHA; Bis-(2-ethylhexyl)ester kyseliny adipove; Bis(2-ethylhexyl) adipate; Bisoflex DOA; BRN 1803774; CCRIS 236; Crodamol DOA; DEHA; Di-(2-ethylhexyl) adipate; Dioctyl adipate; DOA; Effomoll DOA; Effomoll DA; Effomoll DOA; EINECS 203-090-1; Ergoplast Addo; Flexol A 26; Flexol plasticizer 10-A; Flexol plasticizer A-26; Good-rite® GP-223; Hatcol 2908; Hexanedioic acid, bis(2-ethylhexyl) ester; Hexanedioic acid, dioctyl ester; HSDB 343; Jayflex® DOA; K 3220; Kemester 5652; Kodaflex® DOA; Lankroflex DOA; Mollan S; Monoplex® DOA; Morflex 310; NCI-C54386; NSC 56775; Octyl adipate; Palatinol® DOA; Plasthall® DOA; Plastomoll® DOA; Polycizer 332; PX-238; Reomol DOA; Rucoflex Plasticizer DOA; Sansocizer DOA; Sicol 250; Staflex DOA; Truflex DOA; Uniflex® DOA; USS 700; Vestinol OA; Vistone A 10; Wickenol® 158; Witamol 320; Witcizer 412. Plasticizer providing flexibility at low temperatures to vinyl products; used in unfilled garden hose, clear sheeting, electrical insulation. Emollient, moisturizer, pigment wetter/dispersant, cosolvent; increases water vapor porosity of fatty components used in cosmetic and topical pharmaceutical preparations. Liquid; mp = -67.8°; bp = 417°, bp5 = 214°; d^{25} = 0.922; n_{20D} = 1.4472; insoluble in H_2O; very soluble in EtOH, Et2O, Me2CO; LD50 (rat orl) = 9110 mg/kg. *Aristech; BASF Corp.; CasChem; Chisso Corp.; Degussa-Hüls Corp.; Eastman Chem. Co.; Goodrich B.F. Co.; Hall C.P.; Harwick; Inolex; Monsanto Co.; Union Camp.*

1450   Dioctyl adipate
123-79-5   H

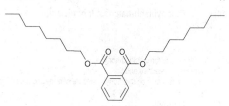

$C_{22}H_{42}O_4$
Bis(2-ethylhexyl) hexanedioate.
Adimoll DO; Adipic acid, dioctyl ester; AI3-17824; Bis(2-ethylhexyl) hexanedioate; Dicaprylyl adipate; Dioctyl adipate; Di-n-octyl adipate; EINECS 204-652-9; Hexanedioic acid, dioctyl ester; HSDB 366; NSC 16201; Octyl adipate.

1451   Dioctyl azelate
103-24-2   G H

$C_{25}H_{48}O_4$
Bis(2-ethylhexyl) azelate.
3-02-00-01787 (Beilstein Handbook Reference); AI3-07965; Azelaic acid, bis(2-ethylhexyl) ester; Bis-(2-ethylhexyl)ester kyseliny azelaove; Bis(2-ethylhexyl) azelate; BRN 1806182; Di-2-ethylhexyl azelate; Dioctyl azelate; Dioctyl nonanedioate; DOZ; EINECS 203-091-7; Emery 2958; Emolien 2986; HSDB 2859; Octyl azelate; Plasthall® DOZ; Plastolein 9058; Plastolein 9058 DOZ; Plastolein 9058DOZ; Sansocizer DOZ; Staflex DOX; Truflex DOX. Plasticizer. Liquid; mp = -78°; bp5 = 237°; d^{25} = 0.915; insoluble in H_2O, slightly soluble in CCl4, soluble in EtOH, Me2CO, C6H6. *Hall C.P.*

1452   Dioctyl maleate
142-16-5   H

$C_{20}H_{36}O_4$
Di-(2-ethylhexyl)maleate.
4-02-00-02211 (Beilstein Handbook Reference); AI3-07870; Bis-(2-ethylhexyl)ester kyseliny maleinove; Bis(2-ethylhexyl) maleate; BRN 1729133; Di-2-ethylhexyl maleate; Dioctyl maleate; EINECS 205-524-5; HSDB 5481; Maleic acid, bis(2-ethylhexyl) ester; RC Comonomer DOM.

1453   Dioctyl maleate
1330-76-3
                                                                       G

$C_{20}H_{36}O_4$
Diisooctyl maleate.
2-Butenedioic acid (2Z)-, diisooctyl ester; 2-Butenedioic acid (Z)-, diisooctyl ester; Diisooctyl maleate; DIOM; EINECS 215-547-2; HSDB 5812; Maleic acid, diisodecyl ester; Octomer DIOM; NSC 6373; RC Comonomer DIOM. Plasticizer, chemical intermediate. *Tiarco.*

1454   Dioctyl maleate
2915-53-9   G H

$C_{20}H_{36}O_4$
2-Butenedioic acid (2Z)-, dioctyl ester.
Bis(1-octyl) maleate; 2-Butenedioic acid (2Z)-, dioctyl ester; 2-Butenedioic acid (Z)-, dioctyl ester; Di-n-octyl maleate; Dioctyl maleate; EINECS 220-835-6; Maleic acid, dioctyl ester; Octomer DOM; PX-538. Reagent grade plasticizer and chemical intermediate, e.g. for the production of sulfosuccinates. *Aristech; Tiarco.*

1455   Di-n-octyl phthalate
117-84-0   H

$C_{24}H_{38}O_4$
Di-n-octyl phthalate.
4-09-00-03180 (Beilstein Handbook Reference); AI3-15071 (USDA); Benzenedicarboxylic acid di-n-octyl ester; 1,2-Benzenedicarboxylic acid, dioctyl ester; BRN 1915994; CCRIS 6196; Celluflex DOP; Di-n-octyl phthalate; Dicapryl phthalate; Dinopol NOP; Dioctyl 1,2-benzenedicarboxylate; Dioctyl o-benzenedicarboxylate; Dioctyl phthalate; Dioktylester kyseliny ftalove; DNOP; EINECS 204-214-7; HSDB 1345; NSC 15318; Octyl phthalate; Phthalic acid, dioctyl ester; Polycizer 162; Px-138; RCRA waste number U107; Vinicizer 85. Liquid; mp = 25°; λ_m = 223, 274 nm (cyclohexane).

1456   Dioctyl phosphite
3658-48-8   G

C16H35O3P

Di-2-ethylhexyl phosphite.

4-01-00-01785 (Beilstein Handbook Reference); Bis(2-ethylhexyl) hydrogen phosphite; Bis(2-ethylhexyl) phosphite; Bis(2-ethylhexyl) phosphonate; Bis(2-ethylhexyl) phosphonic acid; BRN 1711893; Chelex H 8; Di-2-ethylhexyl hydrogen phosphite; Diisooctyl phosphite; EINECS 222-904-6; HSDB 2607; NSC 2664; Phosphonic acid, bis(2-ethylhexyl) ester; Phosphorous acid, bis(2-ethylhexyl) ester; Syn-O-Ad® P-310. Used as amine salts in mineral and synthetic base stocks; as load-carrying additives with secondary activity as low-temperature stabilizers and metal deactivators. *Akzo Chemie.*

1457   Dioctyl sebacate
122-62-3   1251   G H

C26H50O4

Bis(2-ethylhexyl) decanedioate.

4-02-00-02083 (Beilstein Handbook Reference); AI3-09124; Bis-(2-ethylhexyl)ester kyseliny sebakove; Bis(2-ethylhexyl) decanedioate; Bis(2-ethylhexyl) sebacate; Bisoflex; Bisoflex DOS; BRN 1806504; CCRIS 6191; Decanedioic acid, bis(2-ethylhexyl) ester; Di(2-ethylhexyl)sebacate; Dioctyl sebacate; DOS; Edenol 888; Edenor DEHS; EINECS 204-558-8; Ergoplast SDO; HSDB 2898; Monoplex DOS; NSC 68878; Octoil S; Plasthall® DOS; Plexol; Plexol 201J; PX 438; Reolube DOS; Reomol DOS; Reomol DDS; Sansocizer DOS; Staflex DOS; Uniflex® DOS. Monomeric plasticizer used in vinyls, plastics, rubber, and lubricants. Liquid; mp = -48°; bp5 = 256°; d^{25} = 0.912; very soluble in EtOH, Me2CO, C6H6. *Hall C.P.; Union Camp.*

1458   Dioxane
123-91-1   3330   G H

C4H8O2

1,4-Dioxane.

5-19-01-00016 (Beilstein Handbook Reference); AI3-01055; BRN 0102551; CCRIS 269; Di(ethylene oxide); Diethylene dioxide; Diethylene ether; 1,4-Diethylene dioxide; Dioksan; Diossano-1,4; Dioxaan-1,4; 1,4-Dioxacyclohexane; Dioxan; p-Dioxan; Dioxan-1,4; Dioxane; 1,4-Dioxane; p-Dioxane; Dioxanne; Dioxyethylene ether; EINECS 204-661-8; Glycol ethylene ether; Glycolethylenether; HSDB 81; NCI-C03689; NE 220; NSC 8728; RCRA waste number U108; Tetrahydro-1,4-dioxin; Tetrahydro-p-dioxin; Tetrahydro-para-dioxin; UN 1165. Stabilizer for chlorinated hydrocarbons; solvent for adhesives, dyes, cellulose, lacquer, wax, pharmaceuticals, and coatings. Clear, mobile liquid; mp = 11.8°; bp = 101.5°; d^{20} = 1.0337; λ_m = 180 nm (ε = 6310, gas); soluble in CCl4, freely soluble in H2O, EtOH, Et2O, Me2CO, C6H6, AcOH; LD50 (rat orl) = 5.1 gm/kg. *Ashland; BASF Corp.; Ferro/Grant; Mallinckrodt Inc.; Rit-Chem; Union Carbide Corp.*

1459   Dioxybenzone
131-53-3   3334   D G

C14H12O4

·2,2'-Dihydroxy-4-methoxybenzophenone.

4-08-00-03163 (Beilstein Handbook Reference); Advastab 47; AI3-25363; Benzophenone-8; BRN 2055461; CCRIS 6231; Cyasorb® UV 24; Dioxibenzona; Dioxibenzonum; Dioxybenzon; Dioxybenzone; Dioxybenzonum; EINECS 205-026-8; NSC-56769; Solaquin; Spectra-sorb UV 24; UF 2; UV 24.; component of: Solaquin. Light stabilizer and UV absorber for coatings and plastics, e.g., alkyds, phenolics, PU coatings;; stabilizer for polyester film and PVC formulations. Used as an ultraviolet screen in cosmetics and toiletries. Solid; mp = 68°; bp = 170-175°; soluble in EtOH (21.8 g/100 ml), iPrOH (17 g/100 ml), propylene glycol (6.2 g/100 ml), ethylene glycol (3.0 g/100 ml), n-C6H14 (1.5 g/100 ml). *Am. Cyanamid; ICN Pharm. Inc.*

1460   Dipentaerythritol
126-58-9   G H

C10H22O7

2,2',2'-Tetrakis(hydroxymethyl)-3,3'-oxydipropan-1-ol.

Bis(pentaerythritol); Dipentaerythritol; Dipentek; EINECS 204-794-1; Hercules® Tech Di-PE;; HSDB 5610; NSC 65881; 1,3-Propanediol, 2,2'-(oxybis(methylene))bis(2-(hydroxymethyl)-; 2,2,2',2'-Tetrakis(hydroxymethyl)-3,3'-oxydipropan-1-. Used as intermediate in manufacture of alkyds and drying oils, paints and coatings. Crystals; mp = 212-220°; d = 1.33. *Allchem Ind.; Honeywell & Stein.*

1461   Dipentamethylenethiuram tetrasulfide
120-54-7   G

C12H20N2S6

Bis(piperidinothiocarbonyl) tetrasulfide.

4-20-00-01016 (Beilstein Handbook Reference); AI3-28516; Bis(pentamethylenethiuram)-tetrasulfide; Bis-(piperidinothiocarbonyl) tetrasulphide; BRN 0298051; Dipentamethylenethiuram tetrasulfide; Di-N,N'-penta-methylenethiuram tetrasulfide; EINECS 204-406-0; Nocceler TRA; Noksera TRA; NSC 4823; Perkacit® DPTT; Piperidine, 1,1'-(tetrathiodicarbonothioyl)bis-; Sanceler TRA; Soxinol TRA; Sulfads®; Tetrasulfide, bis(pentamethylenethiuram)-; Tetrasulfide, bis(piperidino-thiocarbonyl); Tetron A; Tetrone A; Thiuram MT; Thiuram tetrasulfide, bis(piperidinothiocarbonyl); USAF B-31. Essentially dipentamethylene thiuram hexasulfide; ultra accelerator for NR and synthetic rubbers; vulcanizing agent. *Akzo Chemie; Vanderbilt R.T. Co. Inc.*

1462   Dipentene
138-86-3   5515   G

C10H16
1-Methyl-4-(1-methylethenyl)cyclohexene.
Achilles Dipentene; Acintene DP; Acintene DP dipentene; AI3-00739; Cajeputen; Cajeputene; Caswell No. 526; Cinen; Cinene; Cyclohexene, 1-methyl-4-(1-methylethenyl)-; Dipanol; Dipenten; Dipentene; (±)-Dipentene; Dipentene 200; 4-Isopropenyl-1-methyl-1-cyclohexene; DL-4-Isopropenyl-1-methylcyclohexene; d(R)-4-Isopropenyl-1-methylcyclohexene; (±)-α-Limon-ene; α-Limonene; dl-Limonene; d,l-Limonene; DL-Limonene; EINECS 205-341-0; EINECS 231-732-0; EPA Pesticide Chemical Code 079701; Eulimen; Flavor orange; Goldflush II; HSDB 1809; Inactive limonene; Kautschin; Limonen; Limonene; Limonene, dl-; Mentha-1,8-diene (DL); 1,8-p-Menthadiene; 1,8(9)-p-Mentha-diene; Di-p-mentha-1,8-diene; 1-Methyl-4-isopropenyl-1-cyclohexene; (1)-1-Methyl-4-(1-methylvinyl)cyclohexene; 1-Methyl-4-(1-methylethenyl)cyclohexene; Nesol; NSC 844; NSC 21446; Orange flavor; p-Mentha-1,8-diene, (±)-; p-Mentha-1,8-diene; p-Mentha-1,8-diene, DL-; PC 560; $\Delta^{1,8}$-Terpodiene; UN2052; Unitene.
Solvent for oleoresinous products, rosin, ester gum, etc.; rubber compounding. and reclaiming; dispersant for oils, resins and combinations, pigments, driers, paints, enamels, lacquers: general wetting; printing inks, perfumes, flavors, waxes, polishes. Liquid; mp = -95°; bp = 176°; d^{21} = 0.8402; insoluble in H_2O, soluble in CCl_4, freely soluble in EtOH, Et_2O. Arizona; Hercules Inc.; Langley Smith Ltd.; Penta Mfg.; SCM Glidco Organics; Veitsiluoto Oy.

1463   Dipentite
144-35-4   H

C17H18O6P2
Diphenyl pentaerythritol diphosphite.
Dipentite; Diphenyl pentaerythritol diphosphite; EINECS 205-625-4; Pentaerythritol bis(phenyl phosphite); Pentaerythritol, cyclic bis(phenyl phosphite); Phos-phorous acid, cyclic neopentanetetrayl diphenyl ester.

1464   Diphemanil methylsulfate
62-97-5   3337   D G V

C21H27NO4S
4-(Diphenylmethylene)-1,1-dimethylpiperidinium methyl sulfate.
Ban-Guard (Veterinary); Demotil; Diathal (Veterinary); Diphemanil; Diphemanil methosulfate; Diphemanil methyl sulfate; Diphemanil methylsulfate; Diphemanil methylsulphate; Diphemanil metilsulfate; Diphemanili metilsulfas; Diphemanilum; Diphenmanil methyl sulfate; Diphenmethanil; Diphenmethanil methyl sulfate; 4-(Diphenylmethylene)-1,1-dimethylpiperidinium methyl-sulfate; EINECS 200-552-4; Metilsulfate de diphemanil; Metilsulfato de difemanilo; Nivelon; Nivelona; NSC 41725; Piperidinium, 4-(diphenylmethylene)-1,1-dimethyl-, methyl sulfate; Prantal; Talpran; Vagophemanil methyl sulfate; Variton. Anticholinergic. Used to treat excessive perspiration. Crystals; mp = 194-195°; soluble in H_2O; LD_{50} (rat orl) = 1107 mg/kg, (gpg orl) = 404 mg/kg, (mus orl) = 64 mg/kg. Schering-Plough Pharm.

1465   Diphenyl acetonitrile
86-29-3   G

C14H11N
α-Phenylbenzylcyanide.
4-09-00-02505 (Beilstein Handbook Reference); Acetonitrile, diphenyl-; AI3-17436; Benzeneacetonitrile, α-phenyl-; Benzhydrylcyanide; Benzyhydrylcyanide; BRN 1911160; Caswell No. 396; α-Cyanodiphenylmethane; Difenylacetonitril; Dipan; Diphenatrile; Diphenyl-α-cyanomethane; Diphenylacetonitrile; Diphenylmethyl-cyanide; EINECS 201-662-5; EPA Pesticide Chemical Code 037901; NSC 130268; α-Phenylbenzene-acetonitrile; α-Phenylphenylacetonitrile; USAF KF-13. Used in preparation of diphenylacetic acid and in synthesis of antispasmodics and herbicides. Prisms; mp = 74.3°; bp_{16} = 184°; λ_m = 247, 252, 258, 264 nm (cyclohexane); soluble in EtOH, Et_2O, less soluble in $CHCl_3$, poorly soluble in ligroin; LD_{50} (rat orl) = 3500 mg/kg. Andeno; Greeff R.W. & Co.; Janssen Chimica; Lancaster Synthesis Co.

1466   Diphenylamine
122-39-4   3349   H P

C12H11N
N-Phenylbenzenamine.
AI3-00781; Aniline, N-phenyl-; Anilinobenzene; Benzenamine, N-phenyl-; Benzene, (phenylamino)-; Benzene, anilino-; Big Dipper; C.I. 10355; Caswell No. 398; CCRIS 4699; CI 10355; Deccoscald 282; DFA; Difenylamin; Diphenylamine; DPA; EINECS 204-539-4; EPA Pesticide Chemical Code 038501; HSDB 1108; Naugalube 428L; No-Scald; No-Scald DPA 283; NSC 215210; Phenylaniline; Scaldip; Shield DPA. Used as an insecticide. Registered by EPA as a fungicide and herbicide. Colorless crystals; mp = 52.9°; bp = 302°; d = 1.16; insoluble in H_2O, very soluble in EtOH (45.5 g/100 ml), n-PrOH (22.2 g/100 ml), freely soluble in C_6H_6, Et_2O, AcOH, CS_2. Cerexagri, Inc.; Decco; Generic; Pace International LLC.

1467   Diphenylamine, 4,4'-bis(α,α-dimethyl-benzyl)-
10081-67-1   H

242

C30H31N

Benzenamine, 4-(1-methyl-1-phenylethyl)-N-(4-(1-meth-yl-1-phenylethyl)phenyl)-.

Benzenamine, 4-(1-methyl-1-phenylethyl)-N-(4-(1-meth-yl-1-phenylethyl)phenyl)-; Diphenylamine, 4,4'-bis(α,α-dimethylbenzyl)-; EINECS 233-215-5; 4-(1-Methyl-1-phenylethyl)-N-(4-(1-methyl-1-phenylethyl)phenyl)-aniline.

1468   Diphenylcresyl phosphate
26444-49-5   G

C19H17O4P

Cresyl diphenyl phosphate.

Al3-07853; CCRIS 4773; Cresol diphenyl phosphate; Cresyl diphenyl phosphate; Diphenyl cresol phosphate; Diphenyl cresyl phosphate; Diphenyl tolyl phosphate; Disflamoll® DPK; EINECS 247-693-8; HSDB 6096; Kronitex CDP; Methylphenyl diphenyl phosphate; Methylphenyldiphenyl phosphate; Monocresyl diphenyl phosphate; Phosflex 112; Phosflex® CDP; Phosphoric acid, cresyl diphenyl ester; Phosphoric acid, diphenyl tolyl ester; Phosphoric acid methylphenyl diphenyl ester; Phosphoric acid, methylphenyldiphenyle; Santicizer 140; Tolyl diphenyl phosphate; o-Tolyldiphenylphosphate. Mixture of o, m and p-cresyl isomers. Flame retardant plasticizer for plasticized PVC products; used in air ducts, tarpaulins, driving and conveyor belts, imitation leather, coatings, hoses and extruded goods, cable sheathing and insulation, soles and injection molded items. Liquid; mp = -38°; d = 1.204-1.208; flash point = 233-237° (COC); LD50 (rat orl) = 6400 mg/kg. *Bayer AG; FMC Corp.; Miles Inc.; Polysar; Velsicol.*

1469   Diphenyldichloromethane
2051-90-3   G

C13H10Cl2

p-Dichlorodiphenylmethane.

Benzene, 1,1'-(dichloromethylene)bis-; Benzophenone dichloride; Dichloro(diphenyl)methane; Dichlorodi-phenyl methane; Dichlorodizane; α,α-Dichlorodi-phenylmethane; DPM (halocarbon); EINECS 218-134-5; Methane, dichlorodiphenyl-; NSC 37425. A chemical intermediate. Liquid; bp = 305° (dec), bp21 = 190°; d18 = 1.235; λm = 250 nm (ε = 5120, MeOH); soluble in Et2O, C6H6, CCl4. *Bakelite Corp.*

1470   Diphenyl disulfide
882-33-7   H

C12H10S2

Phenyldithiobenzene.

Al3-02911; Biphenyl disulfide; Diphenyl disulfide; Diphenyl disulphide; Disulfide, diphenyl; EINECS 212-926-4; FEMA No. 3225; NSC 2689; Phenyl disulfide; Phenyldithiobenzene; USAF E-1. Needles; mp = 62°; bp = 310°; d^{20} = 1.353; λm = 239 nm (ε = 1682, 90% EtOH); insoluble in H2O, soluble in EtOH, Et2O, C6H6, CS2.

1471   Diphenyl ether
101-84-8   7372   G H

C12H10O

Biphenyl oxide.

4-06-00-00568 (Beilstein Handbook Reference); Al3-00749; Benzene, 1,1'-oxybis-; Benzene, phenoxy-; Biphenyl oxide; BRN 1364620; CCRIS 5912; Chemcryl JK-EB; Diphenyl ether; Diphenyl oxide; EINECS 202-981-2; Ether, diphenyl-; FEMA No. 3667; Geranium crystals; HSDB 934; NSC 19311; Oxybisbenzene; Oxydiphenyl; Phenoxybenzene; Phenyl ether; Phenyl oxide. Used in perfumery, soaps; as a heat-transfer medium and chemical intermediate for halogenation, acylation, alkylation. Solid; mp = 26.8°; bp = 258°; d^{30} = 1.0661; λm = 207, 217, 248, 280, 286, 295 nm (ε = 39600, 39000, 19500, 17400, 16100, 9170, MeOH); insoluble in H2O, slightly soluble in CHCl3, soluble in EtOH, Et2O, C6H6, AcOH; LD50 (rat orl) = 3370 mg/kg. *Monsanto Co.; Penta Mfg.*

1472   Diphenyl 2-ethylhexyl phosphate
1241-94-7   G H

C20H27O4P

Phosphoric acid, 2-ethylhexyl diphenyl ester.

4-06-00-00718 (Beilstein Handbook Reference); Al3-16360; BRN 2568983; CCRIS 6199; Diphenyl-2-ethylhexyl phosphate; EINECS 214-987-2; 2-Ethylhexyldiphenylphosphate; HSDB 370; Octicizer; Phosphoric acid, 2-ethylhexyl diphenyl ester; Santicizer 141. Flame retardant plasticizer for type PVC applications, dip, rotationally, extruded and injection molded parts, mechanical foam. Liquid; bp5 = 232°; d^{25} = 1.090; LD50 (mus ip) = 930 mg/kg. *Akzo Nobel; Ashland; Bayer AG; Harwick; Miles Inc.; Monsanto Co.; Polysar.*

1473   Diphenylguanidine
102-06-7   3356   G H

C13H13N3
1,3-diphenylguanidine.
Accelerator D; AI3-00225; Akrochem® DPG; CCRIS 1395; Denax; Denax DPG; DFG; 1,3-Difenylguanid; Diphenylguanidine; 1,3-Diphenylguanidine; N,N'-Diphenylguanidine; s-Diphenylguanidine; DPG; DPG accelerator; Dwufenyloguanidyna; Dynamine; EINECS 203-002-1; Guanidine, 1,3-diphenyl-; Guanidine, N,N'-diphenyl-; HSDB 5345; Melaniline; NCI-C60924; Nocceler D; NSC 3272; Perkacit® DPG; USAF B-19; USAF EK-1270; Vanax® DPG; Vulcafor DPG; Vulcaid DPG; Vulkacit D; Vulkacit D/C; Vulkacite D; Vulkazit. Accelerator for NR and SR; secondary accelerator. A medium accelerator for use with thiazoles and sulfenamides in various rubber products. Needles; mp = 150°; bp = 170° (dec); d^{20} = 1.13; λ_m = 242 nm (MeOH); slightly soluble in H_2O, soluble in EtOH, Et2O, CCl4, CHCl3, C7H8.

1474   Diphenyl isodecyl phosphite
26544-23-0   G H

C22H31O3P
Isodecyl diphenyl phosphite.
Chelex MD; Diphenyl isodecyl phosphite; DPDP; EINECS 247-777-4; Isodecyl alcohol, diphenyl phosphite; Isodecyl diphenyl phosphite; Mark 135A; Phoselere T 26; Phosphorous acid, isodecyl diphenyl ester; Weston DPDP. Chelating agent with metal carboxylates as polymer additives, especially for chlorinated polymers such as PVC and chlorinated PE; improves color, heat and light stability. Liquid; bp = 190°; d = 1.022-1.032; flash point = 154°. Akzo Chemie; Dover.

1475   Diphenylmethyl diisocyanate
101-68-8   G H

C15H10N2O2
4,4'-Methylenediphenylene diisocyanate.
4-13-00-00396 (Beilstein Handbook Reference); AI3-15256; Benzene,   1,1'-methylenebis(4-isocyanato-;   Bis(1,4-isocyanatophenyl)methane; BRN 0797662; Caradate 30; CCRIS 2303; Desmodur 44; Di-(4-isocyanatophenyl)methane; Difenil-metan-diisocianato; Difenylmethaan-dissocyanaat; Diphenyl methane   diisocyanate;   Diphenylmethane   diisocyanate; Diphenylmethyl diisocyanate; EINECS 202-966-0; HSDB 2630; Hylene M50; Isocyanic acid, ester with diphenylmethane; Isocyanic acid, methylenedi-p-phenylene ester; Isonaphthol; Isonate; Isonate 125 MF; Isonate 125M; MDI; MDR; Methylbisphenyl isocyanate; Nacconate 300; NCI-C50668; NSC 9596; Rubinate® 44; Rubinate® LF-168; UN2489. A raw material in polyurethane resins. The polyurethanes are of the thermosetting type of plastic and the polymerization is a polyaddition reaction. Monomer used in the preparation of polyurethane resin; bonding rubber to rayon and nylon. Used in manufacturing of high performance polyurethane elastomeric materials including reaction injection molding processed

and cast elastomers, sealants, coatings, and adhesives. Solid; mp = 37-39°; bp5= 194°; d = 1.1900; LD50 (mus orl) = 2200 mg/kg. Allchem Ind.; BASF Corp.; ICI; ICI Chem. & Polymers Ltd.; Miles Inc.

1476   Diphenyloxazole
92-71-7   G

C15H11NO
2,5-Diphenyl-1,3-oxazole.
4-27-00-01435 (Beilstein Handbook Reference); BRN 0157021; 2,5-Diphenyloxazole; DPO (scintillator); EINECS 202-181-3; NSC 24856; Oxazole, 2,5-diphenyl-; PPO (scintillator); Tritosol; USAF EK-6775; DPO. Primary fluor used in scintillation counters or in wavelength shifters. Solid; mp = 74°; bp = 360°; d^{100} = 1.0940; λ_m = 205, 220, 302 nm (MeOH); insoluble in H_2O, soluble in EtOH, Et2O, CHCl3. E. I. DuPont de Nemours Inc.; Penta Mfg.; Spectrum Chem. Manufacturing.

1477   N,N'-Diphenyl-4-phenylenediamine
74-31-7   3361   G

C18H16N2
4,4'-Diphenyl-p-phenylenediamine.
4-13-00-00116 (Beilstein Handbook Reference); AgeRite DPPD; AI3-14323; Altofane DIP; Antage DP; Antigene P; Antioxidant H; 1,4-Benzenediamine,   N,N'-diphenyl-;   1,4-Bis(phenylamino)benzene; BRN 2215944; CCRIS 3500; DFFD; Diafen FF; 1,4-Dianilinobenzene; Diphenyl-p-phenylenediamine; DPPD; EINECS 200-806-4; Ekaland DPPD; Flexamine G; HSDB 2894; N,N-Diphenyl-1,4-benzenediamine; N,N'-Difenyl-p-fenylendiamin; N,N'-Diphenyl-1,4-phenylenediamine; Nocrac DP; Nonflex H; Nonox DPPD;   NSC   5761;   p-Bis(phenylamino)benzene;   p-Phenylaminodiphenylamine;   4-Phenylaminodiphenyl-amine;   p-Phenylenediamine, N,N'-diphenyl-; Permanax 18; Permanax DPPD; Stabilizer DPPD; USAF GY-2. An antioxidant for use in rubber, polyethylene, petroleum and vegetable oils; in natural rubber, it protects against copper and manganese and gives protection against outdoor flexing and static weather cracking; protects against thermal oxidation in polyethylene, inhibits gum formation and degradation at elevated temperatures in petroleum oils. Solid; mp = 144-153°; bp0.5 = 220-225°; d = 1.28; insoluble in H_2O, soluble in Me2CO, C7H8, CHCl3; LD50 (rat orl) = 2370 mg/kg. Akzo Chemie; Uniroyal; Vanderbilt R.T. Co. Inc.

1478   Diphenyl phosphite
4712-55-4   G

C12H11O3P
Phosphonic acid, diphenyl ester.

244

Diphenoxyphosphine oxide; Diphenyl hydrogen phosphite; Diphenyl phosphite; Diphenyl phosphonate; Doverphos® DPP; Doverphos® 213; EINECS 225-202-8; NSC 43786; Phenyl phosphonate; Phenyl phosphonate, (PhO)2HPO; Phosphonic acid, diphenyl ester; Weston® DPP; DPP. Stabilizer used in epoxies, hot-melt adhesives, PU, polyester, SBR, PP; in molding and extrusion of PP, HDPE, LDPE, HIPS, PC, ABS, PVC, polyesters, in calendering of ABS, PVC; in film applications of PP, PE, PVC; fiber applications of PP, polyesters. Used in the synthesis of organophosphorous compounds; color stabilizer for unsaturated polyesters; stabilizer for PVC; antioxidant for PP. Liquid; mp = 12°; bp26 = 218-219°; d = 1.2230; insoluble in H_2O. *Dover; Fluka; GE Specialities; Janssen Chimica; Sigma-Aldrich Fine Chem.; Spectrum Chem. Manufacturing.*

1479   Diphenylsilanediol
947-42-2   G

C12H12O2Si
Dihydroxydiphenylsilane.
4-16-00-01523 (Beilstein Handbook Reference); AI3-51470; BRN 2523445; CD6150; Difenyl-dihydroxysilan; Diphenyldihydroxysilane; Diphenylsilanediol; EINECS 213-427-4; NSC 12561; Silane, dihydroxydiphenyl-; Silanediol, diphenyl-. Coupling agent, chemical intermediate, blocking agent, release agent, lubricant, primer, reducing agent. Liquid; LD50 (mus orl) = 2150 mg/kg. *Degussa-Hüls Corp.*

1480   Diphenylthiourea
102-08-9   3366   G

C13H12N2S
1,3-Diphenylthiourea.
A-1 Thiocarbanilide; Activit; AI3-00852; Akrochem® Thio No. 1; Carbanilide, thio-; CCRIS 5941; DFT; 1,3-Difenylthiomocovina; N,N'-Diphenylthiocarbamide; Di-phenylthiourea; 1,3-Diphenyl-2-thiourea; N,N'-Diphenyl-thiourea; s-Diphenylthiocarbamide; sym-Diphenylthiourea; EINECS 203-004-2; 2-Fenylotiomocznik; HSDB 2758; Nocceler C; NSC 28134; Rhenocure CA; Stabilisator C; Sulfocarbanilide; Thiocarbanilide; Thiokarbanilid; Thiourea, N,N'-diphenyl-; Thiourea, s-diphenyl-; Thiourea, sym-phenyl-; Urea, 1,3-diphenyl-2-thio-; USAF EK-245; Vulkacit CA. Vulcanization accelerator for fast-curing repair stocks; neoprene latex, natural rubber latex and cements; an activator for thiazole accelerators. Accelerator for CR latex, natural rubber latex, and cements, ethylene-propylene-diene rubbers sponge compounds; activates thiazole accelerators; essentially nondiscoloring. Used as a vulcanizing accelelrator and in manufacture of sulfur dyes. Crystals; mp = 154.5°; d^{25} = 1.32; λm:ls = 274 nm (MeOH); slightly soluble in H_2O, very soluble in EtOH, Et2O, CHCl3, oils; MLD (rbt orl) = 1.5 g/kg. *Akrochem Chem. Co.; Monsanto Co.*

1481   Dipicrylamine
131-73-7   3369   G

C12H5N7O12
2,4,6-Trinitro-N(2,4,6-trinitrophenyl)benzenamine.
4-12-00-01737 (Beilstein Handbook Reference); Aurantia; Benzenamine, 2,4,6-trinitro-N-(2,4,6-trinitrophenyl)-; Bis(2,4,6-trinitrophenyl)amine; BRN 0735589; C.I. 10360; CCRIS 5347; Diphenylamine, hexanitro-; Diphenylamine, 2,2',4,4',6,6'-hexanitro-; Dipicrylamine; Dipikrylamin; EINECS 205-037-8; Esanitrodifenilamina; Hexamine; Hexamine (potassium reagent); Hexa-nitrodifenylamine; Hexanitrodiphenylamine; 2,2',4,4',6, 6'-Hexanitrodifenylamin; Hexanitrodiphenylamine; 2,2', 4,4',6,6'-Hexanitrodiphenylamine; 2,4,6,2',4',6'-Hexa-nitrodiphenylamine; Hexil; Hexite; Hexyl; HSDB 2873; NSC 1786; 2,4,6-Trinitro-N-(2,4,6-trinitrophenyl)benzen-amine; UN0079. Used as a booster explosive and in gravimetric analysis for potassium. Pale yellow prisms; mp = 244° (dec); λm = 205, 384 nm (ε = 34200, 18900, MeOH); insoluble in H_2O, EtOH, C6H6, CCl4, C7H8, C5H5N, alkali, slightly soluble in Et2O, Me2CO.

1482   Dipotassium 3,6-dichlorosalicylate
68938-80-7   H

C7H2Cl2K2O3
Benzoic acid, 3,6-dichloro-2-hydroxy-, dipotassium salt.
Benzoic acid, 3,6-dichloro-2-hydroxy-, dipotassium salt; Dipotassium 3,6-dichlorosalicylate; EINECS 273-146-8.

1483   Dipotassium tin bis(sulfate)
27790-37-0   G

K2O12S3Sn
Potassium stannosulfate.
Dipotassium tin bis(sulphate); EINECS 248-659-5; Marignac's salt; Potassium stannosulfate; Sulfuric acid, potassium tin(2+) salt (2:2:1). Solid; decomposed by H_2O.

1484   Dipropylamine
142-84-7   3377   H

C6H15N
Di-n-propylamine.
4-11-00-00122 (Beilstein Handbook Reference); AI3-24037; BRN 0505974; CCRIS 4805; Dipropylamine; Di-n-propylamine; n-Dipropylamine; EINECS 205-565-9; HSDB 2644; RCRA waste number U110; UN2383. Liquid; mp = -63°; bp = 109.3°; d^{20} = 0.7400; soluble in H_2O, EtOH, very soluble in Me2CO, C6H6.

245

1485 Dipropylene glycol
25265-71-8   H

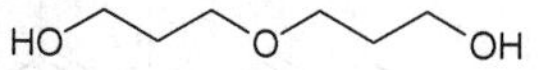

HO—CH₂CH₂CH₂—O—CH₂CH₂CH₂—OH (structure)

$C_6H_{14}O_3$
4-Oxa-2,6-heptandiol.
Caswell No. 399E; CCRIS 475; Dipropylene glycol; Dipropylenglykol; EINECS 246-770-3; EPA Pesticide Chemical Code 068604; HSDB 2658; 4-Oxa-2,6-heptandiol; Oxybispropanol; 1,1'-Oxybis(2-propanol); Oxydipropanol; Propanol, oxybis-.

1486 Dipropylene glycol butoxy ether
29911-28-2   H

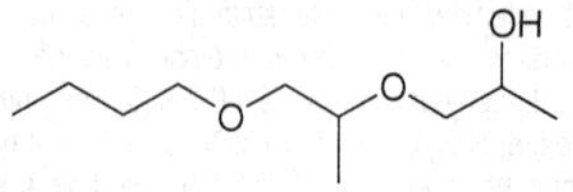

(structure)

$C_{10}H_{22}O_3$
2-Propanol, 1-(2-butoxy-1-methylethoxy)-.
4-01-00-02474 (Beilstein Handbook Reference); BRN 1743918; 1-(2-Butoxy-1-methylethoxy)-2-propanol; Butyl dipropasol solvent; Dipropylene glycol butoxy ether; Dipropylene glycol monobutyl ether; Dipropylene glycol monobutyl ether; Dowanol 54B; EINECS 249-951-5; 2-Propanol, 1-(2-butoxy-1-methylethoxy)-.

1487 Dipropylene glycol, dibenzoate
27138-31-4   H

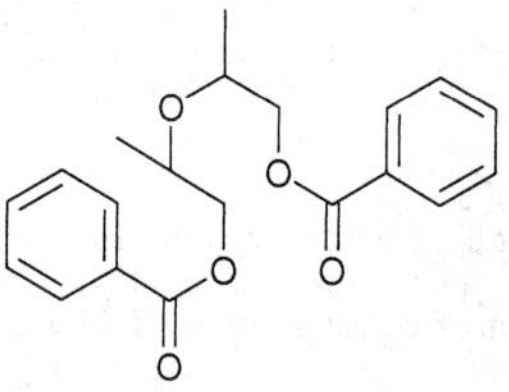

(structure)

$C_{20}H_{22}O_5$
Polypropylene glycol (2) dibenzoate.
Benzoflex 9-88; Dipropylene glycol dibenzoate; EINECS 248-258-5; Finsolv PG 22; Oxybispropanol dibenzoate; Oxydipropyl dibenzoate; Polypropylene glycol (2) dibenzoate; PPG-2 Dibenzoate; Propanol, oxybis-, dibenzoate.

1488 Dipropylene glycol monomethyl ether
34590-94-8   3378   G H

$C_7H_{16}O_3$
(2-Methoxymethylethoxy)propanol.
Arcosolv DPM; Bis(2-(methoxypropyl) ether; 1,4-Dimethyl-3,6-dioxa-1-heptanol; Dipropylene glycol methyl ether; Dipropylene glycol, monomethyl ether; Dowanol-50B; Dowanol DPM; DPGME; EINECS 252-104-2; Glysolv DPM; HSDB 2511; Icinol DPM; Kino-red; (2-Methoxymethylethoxy)propanol; 1-(2-Methoxyiso-propoxy)-2-propanol; Poly-Solv® DPM; PPG-2 methyl ether; Propanol, (2-methoxymethylethoxy)-; Propanol, 1(or 2)-(2-methoxymethylethoxy)-; Ucar solvent 2LM. Solvent for use in protective coatings, inks, cleaning products, agricultural chemicals; aids wetting, penetration, and soil removal; coupling solvent. Used in brake fluids, hard-surface cleaners, leather dyeing, paints, coatings, printing inks, textile vat dyeing and printing, adhesives, antifreeze, floor waxes/polishes, insect repellents; solubilizer for dyes; plasticizer. Liquid; mp = -83°; bp = 190°; d_4^{25} 0.948; n_D^{25} 1.419; soluble in H_2O, organic solvents; LD$_{50}$ (rbt orl) = 5.4 mg/kg. *ICI; Olin Corp.*

1489 Dipyrithione
3696-28-4   D G

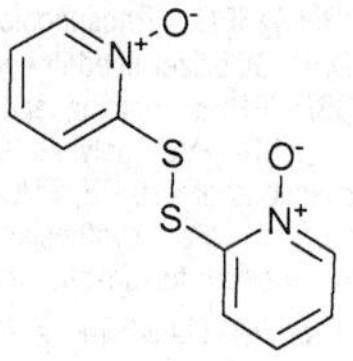

(structure)

$C_{10}H_8N_2O_2S_2$
Bispyrithione.
5-21-02-00055 (Beilstein Handbook Reference); Bispyrithione; BRN 0217725; Dipiritiona; Dipyrithione; Dipyrithionum; EINECS 223-024-5; NSC 241716; NSC 84740; Omadine disulfide; Omadine DS; Omadine® MDS; OMDS; OSY 20. An antidandruff agent for nonalkaline hair care products; antimicrobial agent for Gram-negative and Gram-positive bacteria; also inhibits the growth of fungi. Formulated with MgSO₄. Crystals; mp = 205° (dec). *Olin Corp.*

1490 Diquat
2764-72-9   H P

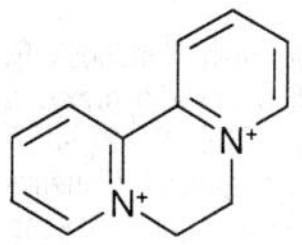

(structure)

$(C_{12}H_{12}N_2)^{2+}$
6,7-Dihydropyrido[1,2-a:2',1'-c]pyrazinediium.
9,10-Dihydro-8a,10a-diazoniaphenanthrene; 9,10-di-hydro-8a,10a-diazoniaphenanthrene ion; 6,7-Dihydro-dipyrido(1,2-a:2',1'-c)pyrazinediylium; 6,7-Dihydrodi-pyrido(1,2-a:2',1'-c)pyrazine-5,8-di-ium; 6,7-Dihydrodi-pyrido-(1,2-a:2',1'-c)pyrazinediium; Dipyrido(1,2-a:2',1'-c)pyrazinediium, 6,7-dihydro-; Diquat; Diquat dication; EINECS 220-433-0; 1,1'-Ethylene-2,2'-bipyridylium ion; 1,1'-Ethylene-2,2'-bipyridyldylium ion. Non-selective contact herbicide. Absorbed by foliage. Used for pre-harvest dessication of many crops, for control of emergent and submerged aquatic weeds, weed control on non-crop land and in sugar cane. Yellow crystals; mp >300° (dec); d^{20} = 1,24; very soluble in H_2O (70 g/100 ml), slightly soluble in EtOH, Et₂O, Me₂O, insoluble in C₆H₆, petroleum ether; LD₅₀ (rat orl) = 231 mg/kg, (mus orl) = 125 mg/kg, (rbt orl) = 187 mg/kg, (dog orl) = 1000-200 mg/kg, (cow orl) = 37 mg/kg, (hen orl) = 200-400 mg/kg, (partridge orl) = 295 mg/kg; LC₅₀ (rainbow trout 96 hr.) = 21 mg/l, (mirror carp 96 hr.) = 67 mg/l; non-toxic to bees. *ICI Agrochemicals.*

1491 Diquat dibromide
85-00-7   3386   G H

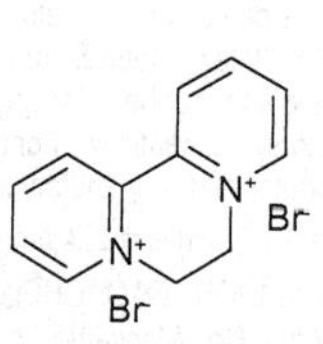

(structure)

$C_{12}H_{12}Br_2N_2$
6,7-Dihydropyrido[1,2-a:2',1'-c]pyrazinediium dibromide.
1,1'-Äthylen-2,2'-bipyridinium-dibromid; Aquacide; Aquicide; Caswell No. 402; Cleansweep; Deiquat; Deiquat dibromide; Detrone; Dextrone; Dihydro-8a,10a-diazoniaphenanthrene dibromide; Katalon; Dihydro-dipyrido(1,2-a:2',1'-c)pyrazinediium dibromide; 6,7-Dihydro-Dipyrido[1,2-a:2',1-c]pyrazinediium dibromide; 6,7-

dihydropyrido(1,2-a:2',1'-c)pyrazinediium dibromide; 9,10-Dihydro-8a,10a-diazoniaphenanthrene dibromide; 9,10-Dihydro-8a,10a-diazoniaphenanthrene; (1,1'-ethyl-ene-2,2'-bipyridylium)dibromide; Dipyrido(1,2-a:2',1'-c)-pyrazinediium, 6,7-dihydro-, dibromide; Diquat; Diquat dibromide; EPA Pesticide Chemical Code 032201; [6385-62-2] - diquat dibromide monohydrate. Registered by EPA as a herbicide and plant growth regulator. Colorless-yellow crystals; mp = 337°; d^{20} = 1.24; λ_m = 308.31 nm (ϵ 18000); very soluble in H_2O (70 g/100 ml), slightly soluble in alcohols, polar solvents; insoluble in non-polar solvents; LD_{50} (rat orl) = 231 mg/kg, (mus orl) = 125 mg/kg, (rbt orl) = 187 mg/kg, (dog orl) = 100 - 200 mg/kg, (cow orl) = 37 mg/kg, (rbt der) > 750 mg/kg, (hen orl) = 200 - 400 mg/kg, (partridge orl) = 295 mg/kg; LC_{50} (rainbow trout 96 hr.) = 21 mg/l, (mirror carp 96 hr.) = 67 mg/l; non-toxic to bees. *ABC Compounding Co.; Frank Miller & Sons, Inc.; Makhteshim Chemical Works Ltd.; Monsanto Co.; Syngenta Crop Protection; Value Gardens Supply LLC.*

1492   1,3-Di-6-quinolylurea
532-05-8   3387   G V

$C_{19}H_{14}N_4O$
Bis(6-quinolyl)urea.
Acaprin®; Atral; Baburan; 1,3-Di-6-quinolylurea; sym-Di-(6-quinolyl)urea; 6,6'-Diquinolinylurea; EINECS 208-525-9; N,N'-Di-6-quinolinylurea; Pirevan; Pyroplasmin; SN 5870; Zothelone. Chemotherapeutic against piroplasmosis (babesiasis); used in veterinary medicine. Crystals; mp = 237° (dec). *Bayer AG.*

1493   Disodium 4,4'-dinitro-2,2'-stilbenedisulfon-ate
3709-43-1   H

$C_{14}H_8N_2Na_2O_{10}S_2$
2,2'-(1,2-Ethenediyl)bis(5-nitrobenzenesulfonic acid), di-sodium salt.
Benzenesulfonic acid, 2,2'-(1,2-ethenediyl)bis(5-nitro-, disodium salt; 4,4'-Dinitrostilbene-2,2'-disulfonic acid, disodium salt; Disodium 4,4'-dinitro-2,2'-stilbene-disulfonic acid; Disodium 4,4'-dinitro-2,2'-stilbene-disulfonate; Disodium 4,4'-dinitrostilbene-2,2'-disulphon-ate; EINECS 223-051-2; NSC 163175; 2,2'-Stilbenedisulfonic acid, 4,4'-dinitro-, disodium salt.

1494   Disodium   decyl(sulphonatophenoxy)-benzenesulfonate
36445-71-3   H

$C_{22}H_{28}Na_2O_7S_2$
Decyl phenoxybenzenedisulfonic acid, disodium salt.
Benzenesulfonic acid, decyl(sulfophenoxy)-, disodium salt; Decyl phenoxybenzenedisulfonic acid, disodium salt; Decyl(sulfophenoxy)benzenesulfonic acid, disodium salt; Disodium decyl(sulphonatophenoxy)benzene-sulphonate; EINECS 253-040-8.

1495   Disodium   dodecyl(sulphonatophenoxy)-benzenesulfonate
28519-02-0   H

$C_{24}H_{32}Na_2O_7S_2$
Benzenesulfonic acid, dodecyl(sulfophenoxy)-, disodium salt.
Benzenesulfonic acid, dodecyl (sulfophenoxy)-, disodium salt; Benzenesulfonic acid, dodecyloxydi-, disodium salt; Disodium dodecyl(sulphonatophenoxy)benzenesulphon-ate; Dodecylphenoxybenzenedisulfonic acid, disodium salt; EINECS 249-063-8.

1496   Disodium iminodiacetate
928-72-3   H

$C_4H_5NNa_2O_4$
Glycine, N-(carboxymethyl)-, disodium salt.
Acetic acid, iminodi-, disodium salt; AI3-52517; Disodium iminodiacetate; EINECS 213-181-8; Glycine, N-(carboxymethyl)-, disodium salt; Iminodiacetic acid, disodium salt; Iminodiacetic acid sodium salt; Iminodioctan disodny; Iminodioctan sodny; NSC 147496.

1497   Disodium inosinate
4691-65-0   G

$C_{10}H_{13}N_4O_8P \cdot 2Na$
Sodium inosinate.
CCRIS 6560; Disodium inosinate; Disodium 5'-inosinate; Disodium inosine 5'-monophosphate; Disodium inosine 5'-phosphate; EINECS 225-146-4; FEMA No. 3669; Gluxor® 1626; IMP disodium salt; IMP sodium salt; 5'-Imp disodium salt; 5'-Inosinic acid, disodium salt; Inosin-5'-monophosphate disodium; Inosine-5'-monophosphate disodium; Inosine 5'-monophosphate disodium salt hydrate; Inosine-5'-monophosphoric acid disodium salt; Luxor® 1639; NSC 20263; Sodium inosinate; Sodium 5'-inosinate IMP sodium; Disodium IMP; Sodium 5-inosinate; Disodium 5'-inosinate; Inosine 5'-disodium phosphate. A 5'-nucleotide derived from seaweed or dried fish; Flavor potentiator in foods. Solid; soluble in H_2O; slightly soluble in alcohol; insoluble in ether; LD_{50} (rat orl) = 15,900 mg/kg. *Lancaster Synthesis Co.; Penta Mfg.*

1498   Disodium laneth-5 sulfosuccinate
68890-92-6   G

$C_{14}H_{26}O_S \cdot 2Na$
Sulfobutanedioic acid, 4-isodecyl ester, sodium salt.
Butanedioic acid, sulfo-, laneth-5 ester, disodium salt; Disodium laneth-5 sulfosuccinate; Incrosul LAFS; Lanolin alcohol, ethoxylated, sulfosuccinate, disodium salt; Poly(oxy-1,2-ethanediyl), α-(3-carboxy-1-oxosulfopropyl)-ω-hydroxy-, ethers with lanolin alcs, disodium salts; Rewolan® 5; Succinic acid, sulfo-, laneth-5 ester, disodium salt; Sulfobutanedioic acid, laneth-5 ester, disodium salt.

Mild, low foaming, conditioning surfactant with good emulsifying properties used in personal care products Liquid; pH = 6.5-7.5. *Croda; Witco/Oleo-Surf.*

1499   Disodium methanearsonate
144-21-8   G H P

Na$^+$
O$^-$
|
—As=O
|
O$^-$
Na$^+$

CH3AsNa2O3
Methylarsonic acid, disodium salt.
Arrhenal; Ansar 184; Ansar 8100; Ansar DSMA Liquid; Arrhenal; Arsinyl; Arsonic acid, methyl-, disodium salt; Arsynal; Cacodyl New; Calar-E-Rad; Caswell No. 405; Chipco Crab Kleen; Crab-3-rad 100; Cralo-E-rad; Dal-E-Rad 100; Di-tac; Diarsen; Dimet; Dinate; Disodium methanarsonate; Disodium methanearsenate; Disodium methanearsonate; Disodium methyl arsenate; Disodium methyl arsonate; Disomear; DMA; DMA 100; Drexel DSMA liquid; DSMA; DSMA liquid; EINECS 205-620-7; EPA Pesticide Chemical Code 013802; HSDB 1701; Jon-trol; Maa sodium salt; Methanearsonic acid, disodium salt; Methar; Methar 30; Metharsan; Metharsinat; Methylarsonat disodny; Methylarsonic acid, disodium salt; Namate; Neo-Asycodile; NSC 5270; Sodar; Sodium methanearsonate; Sodium metharsonate; Sodium methylarsonate; Somar; Stenosine; Tonarsan; Tonarsin; Versar DSMA LQ; Weed broom; Weed-E-rad 360; Weed-E-Rad; Weed-E-Rad DMA Powder; Weed-hoe. Selective contact herbicide. Registered by EPA as a herbicide, rodenticide and insecticide. [hexahydrate]: Colorless crystals; mp = 132 -139°; freely soluble in H2O (27.9 g.100 ml, anhydrous), soluble in MeOH, insoluble in most other organic soplvents; LD50 (rat orl) = 1800 mg/kg, (rbt der) = 10000 mg/kg. *Bonide Products, Inc.; Cleary Chemical Corp.; Cleary W. A.; Drexel Chemical Co.; Earth Care; KMG-Bernuth Inc.; Luxembourg-Pamol, Inc.; PBI/Gordon Corp; UAP - Platte Chemical.*

1500   Disodium octaborate tetrahydrate
12280-03-4   G P

B8Na2O13
Disodium octaborate.
Boric acid (H2B8O13), disodium salt, tetrahydrate; Boric acid, disodium salt, tetrahydrate; Boron sodium oxide, tetrahydrate; Caswell No. 406B; Disodium octaborate tetrahydrate; EPA Pesticide Chemical Code 011103; FLEABOR; Polybor 3; Tim-Bor; Timbor. Fire retardant for treatment of lumber. Registered by EPA as an insecticide, fungicide and herbicide. *Borax Consolidated Ltd; Environmental Laboratories, Inc.; Honolulu Wood Treating Co.; Osmose Inc.; U.S. Borax.*

1501   Disodium phosphate
7558-79-4   8733   D G P

OH
|
O=P—O$^-$ 2Na$^+$
|
O$^-$

HNa2O4P
Sodium phosphate, dibasic, anhydrous.
Acetest; Caswell No. 778; CCRIS 5931; Dibasic sodium phosphate; Disodium acid orthophosphate; Disodium acid phosphate; Disodium hydrogen phosphate; Disodium hydrogenorthophosphate; Disodium hydro-phosphate; Disodium monohydrogen phosphate; Disodium orthophosphate; Disodium phosphate, anhydrous; Disodium

phosphoric acid; DSP; EINECS 231-448-7; EPA Pesticide Chemical Code 076403; Exsiccated sodium phosphate; FEMA Number 2398; HSDB 376; Natriumphosphat; Phosphate of soda; Phosphoric acid, disodium salt; Salt Perlate; secondary Sodium phosphate; Soda phosphate; Sodium acid phosphate, anhydrous; Sodium hydrogen phosphate; Sodium monohydrogen phosphate (2:1:1); Sodium orthophosphate, secondary; Sodium phosphate; Sodium phosphate (NaHPO4); Sodium phosphate, dibasic; Sodium phosphate, exsiccated; Tasteless salts. Inorganic salt of phosphoric acid, disodium salt controls pH in mildly alkaline solutions; used in food products, water treatment, animal feed, textiles, pharmaceuticals, chemicals, fertilizers, detergents. Used as a chelator, emulsifier and buffer. Also as a cathartic or (vet) laxative. Registered by EPA as an antimicrobial, fungicide and herbicide (cancelled). FDA GRAS, Japan approved, FDA approved for buccals, parenterals, ophthalmics, vaginals, USP/NF, BP, Ph.Eur. compliant. Used in buccals, parenterals, ophthalmics, vaginals and enemas. Used in chemicals, fertilizers, pharmaceuticals, textiles, food additives, used in boiler water treatment, fireproofing wood and paper, detergents, ceramic glaze, soldering elements and tanning. Hygroscopic powder; on exposure to air absorbs 2-7 moles H2O; soluble in H2O (12 g/100 ml), insoluble in alcohol; pH of 1% aq soln = 9.1. *Albright & Wilson Americas Inc.; Monsanto Co.; Rhône-Poulenc; U.S. BioChem.; Whiting Peter Ltd.*

1502   Disodium phosphate
10028-24-7   8733   D G

OH
|
O=P—O$^-$ 2Na$^+$ 2H2O
|
O$^-$

HNa2O4P.2H2O
Sodium phosphate, dibasic dihydrate.
DSP-2; Disodium hydrogen phosphate, dihydrate; Metro I.V.; Sodium phosphate; Sorenson's phosphate; Sorensen's sodium phosphate. Controls pH in mildly alkaline solutions. Used medicinally as a laxative-cathartic. [heptahydrate]; crystals or granular powder; d = 1.7; soluble in H2O (25 g/100 ml), insoluble in alcohol; pH = 9.5; LD50 (rat orl) = 12.93 g/kg. *Albright & Wilson Americas Inc.; McGaw Inc.; Monsanto Co.; Rhône-Poulenc; U.S. BioChem.*

1503   Disofenin
65717-97-7   3393   G

(((((2,6-Diisopropylphenyl)carbamoyl)methyl)imino)di-acetic acid.
Disofenin; Disofenine; Disofenino; Disofeninum; Glycine, N-(2-((2,6-bis(1-methylethyl)phenyl)amino)-2-oxoethyl)-N-(carboxymethyl)-; Hepatolite. The 99mTc-complex is used as a radioactive imaging agent. As a diagnostic aid, the radioactive agent is used to measure liver function. White crystals; mp = 196-198°; insoluble in H2O. *DuPont.*

1504   Distearyl pentaerythritol diphosphite
3806-34-6   G H

C18H37—O—P(...)P—O—C18H37

C41H82O6P2

2,4,8,10-Tetraoxa-3,9-diphosphaspiro(5.5)undecane, 3,9-bis(octadecyloxy)-.

2,4,8,10-Tetraoxa-3,9-diphosphaspiro(5.5)undecane, 3,9-bis(octadecyloxy)-; EINECS 223-276-6; O,O'-Dioctadecylpentaerythritol bis(phosphite). Color stabilizer and melt flow aid for polymer processing; antioxidant. Crystals; mp = 40-70°; d = 0.920-0.935. Dover; GE Specialities; Sigma-Aldrich Fine Chem.

1505 Distearyl thiopropionate
693-36-7

G H

C42H82O4S

Dioctadecyl 3,3'-thiobispropanoate.

Advastab 802; Advastab PS 802; Alkanox® 240-3T; Antage STDP-N; Antiok S; Antioxidant STDP; Arbestab DSTDP; BRN 2229929; Carstab® DSTDP; Cyanox-stdp; Cyanox STDP; Dioctadecyl 3,3'-thiobispropanoate; Dioctadecyl thiodipropionate; Distearyl β,β'-thiodi-propionate; Distearyl 3,3'-thiodipropionate; DSTDP; DSTP; EINECS 211-750-5; Evanstab® 18; Hostanox SE 2; Hostanox SE 4; Hostanox VP-SE 2; HSDB 5746; Irganox PS 802; Lusmit SS; Naugard DSTDP; NSC 65493; Plastanox STDP; Plastanox STDP Antioxidant; PS 802; Seenox DS; Sumilizer TPS; Thio 1; Thiodipropionic acid, distearyl ester; Varox DSTDP; Yoshinox DSTDP. An antioxidant, plasticizer, softening agent; antioxidant for cosmetics, pharmaceuticals; stabilizer for plastics and elastomers. Long-term heat aging stabilizer in conjunction with primary antioxidants; for PP, HDPE. Used in food-pkg. materials and edible fats and oils. Solid; mp = 58-62°; bp = 360°; insoluble in H2O, soluble in organic solvents; LD50 (rat orl) > 2500 mg/kg. Am. Cyanamid; CK Witco Chem. Corp.; Cytec; EniChem Am.; Evans Chemetics; Hampshire; Hoechst AG; ICI Americas Inc.; Morton Intn'l.; Sigma-Aldrich Fine Chem.

1506 Disulfiram
97-77-8   3399

D G H

C10H20N2S4

Thioperoxydicarbonic diamide ([(H2N)C(S)]2S2), tetra-ethyl-.

Abstensil; Abstensyl; Abstinil; Abstinyl; Accel TET; Accel TET-R; Al3-27340; Akrochem® TETD; Alcophobin; Alk-aubs; Antabus; Antabuse; Antadix; Antaenyl; Antaethyl; Antäthyl; Antaetil; Antalcol; Antetan; Antetil; Anteyl; Anti-ethyl; Antiaethan; Antiäthan; Anticol; Antietanol; Antietil; Antikol; Antivitium; Antivitium (Spain); Aversan; Averzan; Bis((diethylamino)thioxomethyl) disulfide; Bis((diethylamino)thioxomethyl)disulphide; Bis(diethylthiocarbam-oyl) disulfide; Bis(diethylthiocarbamoyl)disulphide; Bonibal; CCRIS 582; Contralin; Contrapot; Cronetal; Dicupral; Disetil; Disulfan; Disulfide, bis(diethylthio-carbamoyl); Disulfiram; Disulfirame; Disulfiramo; Disulfiramum; DRG-0068; Dupon 4472; Dupont fungicide 4472; EINECS 202-607-8; Ekagom DTET; Ekagom TEDS; Ekagom TETDS; ENT 27,340; Ephorran; Esperal; Etabus; Ethyl thiram; Ethyl thiurad; Ethyl tuads; Ethyl TUEX; Ethyldithiourame; Ethyldithiurame; Exhorran; HOCA; Hocakrotenalnci-C02959; HSDB 3317; Krotenal; NCI-C02959; Nocbin; Nocceler TET; Nocceler TET-G; Noxal; Noxal (VAN); NSC 190940; NSC 25953; Perkacit® TETD; Refusal; Refusal; Ro-Sulfiram; Ro-Sulfram-500 (USA); Sanceler TET; Sanceler TET-G; Soxinol TET; Stopetyl; TATD; Tenurid; Tenutex;

TETD; Tetidis; Tetradine; Tetraethylthioperoxydicarbonic diamide; Tetraethylthiram disulfide; Tetraethylthiram disulphide; Tetraetil; Teturam; Teturamin; Thiosan; Thioscabin; Thiuram disulfide, tetraethyl-; Thiuram E; Thiuranide; Tiuram; TTD; TTS; Tuads, ethyl; USAF B-33. Ulta-accelerator and vulcanizing agent for rubber. Has also been used as a fungicide and as an alcohol deterrent. Crystals; mp = 71.5°; bp17 - 117°; d = 1.30; poorly soluble in H2O (0.02 g/100 ml); soluble in EtOH (3.82 g/100 ml), Et2O (7.14 g/100 ml), Me2CO, C6H6, CHCl3, CS2; LD50 (rat orl) = 8600 mg/kg. Abbott Labs Inc.; Akrochem Chem. Co.; Akzo Chemie; Mitchell Cotts Chem.; Monsanto Co.; Naugatuck; Sigma-Aldrich Fine Chem.; Uniroyal; Vanderbilt R.T. Co. Inc.; Wyeth Labs.

1507 Ditalimfos
5131-24-8

G P

C12H14NO4PS

O,O-Diethyl (1,3-dihydro-1,3-dioxo-2H-isoindol-2-yl)-phosphonothioate.

5-21-11-00141 (Beilstein Handbook Reference); BRN 1542822; Diethyl (1,3-dihydro-1,3-dioxo-2H-isoindol-2-yl)phosphonothioate; Diethyl phthalimidophosphono-thioate; (1,3-Dihydro-1,3-dioxo-2H-isoindol-2-yl)phos-phonothioic acid O,O-diethylester; Ditalimfos; Dital-imphos; Dowco 199; Frutogard; Laptran; Leucon; M-2452; Millie; O 199; Plondrel; O,O-Diäthyl- -phtalimido-thiophosphat; O,O-Diethyl (1,3-dihydro-1,3-dioxo-2H-isoindol-2-yl)phosphonothioate; O,O-Diethyl phthalim-ido-phosphonothioate; O,O-Diethyl phthalimidothio-phosphate; Ortho 199; Phosphonothioic acid, (1,2-dihydro-1,3-dioxo-2H-isoindol-2-yl)-,O,O-diethyl ester; Phosphonothioic acid, phthalimido-, O,O-diethyl ester; Plondrel; RE 199; SF 101. Fungicide used to control powdery mildew and scab. Registered by EPA as a fungicide (cancelled). ICI Chem. & Polymers Ltd.

1508 Ditallow dimethyl ammonium chloride
61789-80-8

H

Dimethyl di(hydrogenated tallow) ammonium chloride.

Adogen 442; Adogen 442-100 P; Adogen 442E83; Adogen 448; Adogen 448E; Aliquat 264; Aliquat 2HT; Ammonyx 2200; Arquad 2HD75; Arquad 2HT; Arquad 2HT75; Arquard HC; Carsosoft V 100; Carsosoft V 90; chlorides; EINECS 263-090-2; Jet Quat 2HT75; Kemamine Q 9702; Kemamine Q 9702C; Kemamine QSML 2; Kemamine qsml2; Kemamium M 2SH; Kemamium M 2SH15; Noramium M 2SH; Noramium M 2SH15; Prapagen WK; Prapagen WKT; Quaternium-18; Querton 442; Radiaquat 6442; Varisoft 3262; Varisoft 442-100P; Varisoft DHT.

1509 Dithianone
3347-22-6   3404

G P

C14H4N2O2S2

5,10-Dihydro-5,10-dioxonaphtho[2-3-b]-1,4-dithiin-2,3-dicarbonitrile. 5-19-08-00189 (Beilstein Handbook Reference); 5,10-Dihydro-5,10-dioxonaphtho(2,3-b)-1,4-dithiin-2,3-di-carbonitrile; 5,10-Dihydro-5,10-dioxonaphtho(2,3-b)-p-dithiin-2,3-dicarbonitrile; Al3-28720; BRN 1325563; Caswell No. 349B; Delan; Delan (fungicide); Delancol; Delan WP; 2,3-Dicarbonitrilo-1,4-diathiaanthrachinon; 2,3-Dicyano-1,4-dithia-anthraquinone; 2,3-Dinitrilo-1,4-dithia-9,10-anthraquinone; 2,3-Dinitrilo-1,4-dithia-anthraquinone; 2,3-Dinitrilo-1,4-dithioanthrachinon; 1,4-Dithiaanthraquinone-2,3-dicarbonitrile; 1,4-Dithiaanthra-quinone-2,3-dinitrile; Dithianon; Dithianone; DTA; EINECS 222-098-6; EPA Pesticide Chemical Code 099201; HSDB 1583; IT 931; Merkdelan; MV 119A; Naphtho(2,3-b)-1,4-dithiin-2,3-dicarbonitrile, 5,10-di-hydro-5,10-dioxo-; NSC 218452; Stauffer MV-119A; Thynon . Fungicide used for control of scab in fruit apples and pears. Dark brown crystals; mp = 225°; soluble in H2O (0.00005 g/100 ml), CHCl3 (1.2 g/100 ml), Me2CO (1 g/100 ml), C6H6 (0.8 g/100 ml); LD50 (rat orl) = 638 mg/kg, (gpg orl) = 115 mg/kg, (rat der) > 2000 mg/kg, (mquail orl) = 280 mg/kg, (fquail orl) = 430 mg/kg, (bee contact) > 0.1 mg/bee; LC50 (rat ihl 4 hr.) ≅3 mg/l air; toxic to fish; non-toxic to bees. *BASF AG; ICI Chem. & Polymers Ltd.; Shell UK.*

1510   1,4-Dithiothreitol
3483-12-3   3412   G

C4H10O2S2

(R*,R*)-1,4-Dimercapto-2,3-butanediol. 3-01-00-02360 (Beilstein Handbook Reference); Al3-62064; BRN 1719757; (±)-2,3-Butanediol, 1,4-dimercapto-, (R*,R*)-; 2,3-Butanediol, 1,4-dimercapto-, (2R,3R)-rel-; 2,3-Butanediol, 1,4-dimercapto-, D-threo-; 2,3-Butanediol, 1,4-dimercapto-, DL-threo-; CCRIS 3617; Cleland reagent; (R*,R*)-(±)-1,4-Dimercapto-2,3-butanediol; (R*,R*)-1,4-Dimercaptobutane-2,3-diol; D-threo-1,4-Dimercapto-2,3-butanediol; Dithiothreitol; 1,4-Dithiothreitol; D-1,4-Dithiothreitol; rac-Dithiothreitol; Dithiotreitol; DL-threo-1,4-Dimercapto-2,3-butanediol; L-DTT; EINECS 222-468-7; EINECS 248-531-9; Sputolysin; Threitol, 1,4-dithio-; Threitol, 1,4-dithio-, DL-; WR 34678. Reducing agent for proteins and enzymes, used in biochemical research. Crystals; mp = 42-43°; soluble in H2O, organic solvents. *Bio-Rad Labs.; Biosynth AG; Lancaster Synthesis Co.; Sigma-Aldrich Fine Chem.; U.S. BioChem.*

1511   Dithioxamide
79-40-3   8360   G

C2H4N2S2

Ethanedithioamide. Dithiooxamide; Dithioxamide; EINECS 201-203-9; Ethanedithioamide; HSDB 5230; Hydrorubeanic acid; NSC 1893; Oxaldiimidic acid, dithio-; Oxalic acid, dithiono-, diamide; Oxamide, dithio-; Rubean; Rubeane; Rubeanic acid; RVK; USAF B-43; USAF EK-4394; USAF MK-6. The amide of dithiooxalic acid, a chelating agent, especially for copper, cobalt and nickel.Also used to stabilize ascorbic acid solutions. Red crystals; mp = 170° (dec); λm = 312 nm (ε = 10300 MeOH); slightly soluble in H2O, EtOH, insoluble in organic solvents.

1512   O,O-Ditolyl phosphorodithioate
27157-94-4   H

C14H15O2PS2

Phosphorodithioic acid, O,O-ditolyl ester. Cresyl aerofloat; EINECS 248-273-7; HSDB 2623; O,O-Bis(methylphenyl) hydrogen dithiophosphate; O,O-Ditolyl phosphorodithioate; Phosphorodithioic acid, O,O-bis(methylphenyl) ester; Phosphorodithioic acid, O,O-ditolyl ester. Mixture of o, m and p-isomers.

1513   Di-o-tolylthiourea
137-97-3   G

C15H16N2S

2,2'-Dimethylthiocarbanilide. Accelerator A22; Al3-03718; N,N'-Bis(2-methylphenyl)-thiourea; 1,3-Bis(o-tolyl)thiourea; Carbanilide, 2,2'-dimethylthio-; 2,2'-Dimethylthiocarbanilide; Di-o-toluylthiourea; Di-o-tolylthiourea; 1,3-Di-o-tolylthiourea; N,N'-Di-o-tolylthiourea; 1,3-Di-o-tolyl-2-thiourea; 1,3-Di-o-tolylthiomocovina; EINECS 205-309-6; NSC 119321; Thiourea, N,N'-bis(2-methylphenyl)-; Urea, 1,3-bis(o-tolyl)-2-thio-; USAF EK-1651. A proprietary rubber vulcanizing accelerator. Solid; mp = 157-159°. *Lancaster Synthesis Ltd.*

1514   1,3-Ditolylguanidine
97-39-2   G H

C15H17N3

Guanidine, N,N'-bis(2-methylphenyl)-. 4-12-00-01764 (Beilstein Handbook Reference); Al3-14630; Akrochem® DOTG; BRN 2653884; CNS 1001; Di-o-tolylguanidine; Diorthotolylguanidine; DOTG; DOTG accelerator; DTG; EINECS 202-577-6; Eveite DOTG; Guanidine, 1,3-di-o-tolyl-; Guanidine, 1,3-di(2-tolyl)-; Guanidine, N,N'-bis(2-methylphenyl)-; HSDB 5307; Nocceler DT; NSC 132023; Perkacit® DOTG; Sanceler DT; Soxinol DT; USAF A-6598; Vanax® DOTG; Vulcafor DOTG; Vulkacit dotg/C; Vulkacit DOTG; Vulkacite DOTG. A rubber vulcanization accelerator; provides activation for MBTS, MBT, and other thiazole accelerators in natural rubber, styrene/butadiene rubber, NBR, and CR. Crystals; mp = 179°; d20 = 1.10; poorly soluble in H2O, EtOH, TFA, soluble in CHCl3, very soluble in Et2O. *Akrochem Chem. Co.; Akzo Chemie; Vanderbilt R.T. Co. Inc.*

1515   Ditridecyl phthalate
119-06-2   G H

C34H58O4

1-Tridecanyl phthalate.

Bis(tridecyl) phthalate; BRN 2023076; CCRIS 6197; Ditridecyl 1,2-benzenedicarboxylate; Ditridecyl phthalate; DTDP; EINECS 204-294-3; HSDB 381; Jayflex® DTDP; Nuoplaz; Phthalic acid, ditridecyl ester; Polycizer 962-bpa; Polycizer 962BPA; PX-126; Staflex DTDP; 1-Tridecanyl phthalate; Truflex DTDP. Reagent grade plasticizer Liquid; bp$_5$ > 285°; d^{20} = 0.951. *Aristech; Exxon.*

1516 Ditridecyl thiodipropionate

10595-72-9   H

C32H62O4S

di(tridecyl) thiodipropionate.

Di(tridecyl) thiodipropionate; Di(tridecyl) 3,3'-thiodipropionate; EINECS 234-206-9; Propanoic acid, 3,3'-thiobis-, ditridecyl ester; Propionic acid, 3,3'-thiodi-, ditridecyl ester; 3,3'-Thiobispropanoic acid, ditridecyl ester. Used as a secondary antioxidant preservative in pharmaceuticals. bp$_{0.25}$ = 265°; insoluble in H$_2$O, slightly soluble in MeOH, soluble in most organic solvents. LD$_{50}$ (rat orl) > 9000 mg/kg, (rbt der) > 4500 mg/kg. *Cytec; Hampshire; Sigma-Aldrich Fine Chem.; Witco/PAG.*

1517 Diundecyl phthalate

3648-20-2   G H

C30H50O4

1,2-Benzenedicarboxylic acid, diundecyl ester.

1,2-Benzenedicarboxylic acid, diundecyl ester; BRN 1894045; CCRIS 6198; Diundecyl phthalate; Di-n-undecyl phthalate; EINECS 222-884-9; HSDB 5132; Jayflex® DUP; Phthalic acid, diundecyl ester; PX-111; Santicizer 711. Plasticizer. Oil; d = 0.954; insoluble in H$_2$O, soluble in organic solvents. *Aristech; ExxonMobil Chem. Co.*

1518 Diuron

330-54-1   3418   G H P

C9H10Cl2N2O

1-(3,4-Dichlorophenyl)-3,3-dimethylurea.

4-12-00-01263 (Beilstein Handbook Reference); AF 101; AI3-61438; Anduron; Ansaron; Bioron; BRN 2215168; Caswell No. 410; CCRIS 1012; Cekiuron; Crisuron; Dailon; DCMU; DCMU 99; Di-on; Diater; Dichlorfenidim; Direx 4L; Dirurol; Ditox-800; Diurex; Diuron; Diuron 4L; Diuron 900; Diuron Nortox; DMU; DP Hardener 95; Drexel; Duran; Durashield; Dynex; EINECS 206-354-4; EPA Pesticide Chemical Code 035505; Farmco; Farmco diuron; Herbatox; Herburon; HSDB 382; HW 920; Karmex; Karmex D; Karmex Diuron Herbicide; Karmex DW; Lucenit; Marmer; NSC 8950; Preventol A 6; Seduron; Sup'r flo; Telvar diuron weed killer; Unidron; Urox D; USAF P-7; USAF XR-42; Vonduron. Pre-emergent herbicide, sugar cane flowering suppressant. Inhibits photosynthesis. Used for total control of weeds and mosses in non-crop areas and selectiver control of germinating grass and broad-leaved weeds in many crops. A flowable herbicide for control of many weeds and grasses in a variety of crops. Effective against a wide range of both broadleaf weeds and annual grasses. Registered by EPA as a herbicide. Crystals; mp = 158 - 159°; soluble in H$_2$O (0.0042 g/100 ml at 25°), Me$_2$CO (4.2 g/100 ml), butyl stearate (0.11 g/100 ml), C$_6$H$_6$ (0.11 g/100 ml), poorly soluble in hydrocarbon solvents; LD$_{50}$ (rat orl) = 3400 mg/kg, 437 mg/kg; LC$_{50}$ (bobwhite quail 8 day dietary) = 1730 mg/kg diet, (Japanese quail, mallard duckling, pheasant chick 8 day dietary) > 5000 mg/kg diet, (rainbow trout 96 hr.) = 5.6 mg/l, (bluegill sunfish 96 hr.) = 5.9 mg/l, (guppy 96 hr.) = 25 mg/l; non-toxic to bees. *Aceto Agriculture Chemicals Corp.; Agan Chemical Manufacturers; BASF Corp.; Bayer AG; Drexel Chemical Co.; DuPont UK; DuPont; Griffin LLC; Helena Chemical Co.; Rhône Poulenc Crop Protection Ltd.; Rhône-Poulenc Environmental Prods. Ltd; Riverdale Chemical Co; Rohm & Haas UK; Syngenta Crop Protection; Thor; Troy Chemical Corp.*

1519 Divalproex Sodium

76584-70-8   9979   D

(C16H31NaO4)n

Sodium hydrogen bis(2-propylvalerate) oligomer.

Abbott 50711; Delepsine; Depakote; Depakote ER; Divalproex sodium; Epilex; Epival; Sodium hydrogen bis(2-propylvalerate); Sodium hydrogen bis(2-propylpentanoate); Sodium hydrogen divalproate; Sprinkle; Valcote; Valparin; Valproate semisodique; Valproato semisodico; Valproatum seminatricum; Valproic acid semisodium salt (2:1). Anticonvulsant; epileptic. *Abbott Labs Inc.*

1520 Divinylbenzene

1321-74-0   H

C10H10

Vinylstyrene.

Benzene, diethenyl-; Benzene, divinyl-; CCRIS 4809; Diethenylbenzene; Divenylbenzene; Divinyl benzene; Divinylbenzene-55; Divinylbenzene, pure; DVB-22; DVB-27; DVB-55; DVB-80; DVB-100; EINECS 215-325-5; HSDB 5449; NSC 4833; Vinylstyrene. Mixture of o, m and p-isomers. Mixture of isomers of divinylbenzene.

1521 Divinyltetramethyldisiloxane
2627-95-4 G

C8H18OSi2
1,1,3,3-Tetramethyl-1,3-divinyldisiloxane.
Bisvinyltetramethyldisiloxane; CD62101,3-Diethenyl-1,1,3,3-tetramethyl disiloxane; Disiloxane, 1,3-diethenyl-1,1,3,3-tetramethyl-; Divinyltetramethyldisiloxane; 1,1'-Divinyltetramethyldisiloxane; 1,3-Divinyltetramethyldisiloxane; EINECS 220-099-6; sym-Tetramethyldivinyl-disiloxane. Coupling agent, chemical intermediate, blocking agent, release agent, lubricant, primer, reducing agent. Liquid; mp = -99.7°; bp = 30°; d^{20} = 0.811. *Degussa-Hüls Corp.*

1522 DMDM hydantoin
6440-58-0 G

C7H12N2O4
1,3-Bis (hydroxymethyl)-5,5-dimethyl-2,4-imidazolidine-dione.
1,3-Bis(hydroxymethyl)-5,5-dimethylhydantoin; 1,3-Bis-(hydroxymethyl)-5,5-dimethylimidazolidine-2,4-dione; 1,3-Bis(hydroxymethyl)-5,5-dimethyl-2,4-imidazolidine-dione; BRN 0882348; Caswell No. 273AB; Dantion DMDMH 55; Dantoguard; Dantoin DMDMH; Dantoin DMDMH 55; Dimethyloldimethyl hydantoin; Dimethyl-ol-5,5-dimethylhydantoin; 1,3-Dimethylol-5,5-dimethyl-hydantoin; DMDMH; DMDMH 55; DMDM hydantoin; EINECS 229-222-8; EPA Pesticide Chemical Code 115501; Glycoserve-DMDMH; Glydant; Glydant Plus; Glydant®; Hydantoin, 1,3-bis(hydroxymethyl)-5,5-di-methyl-; 2,4-Imidazolidinedione, 1,3-bis(hydroxymethyl)-5,5-dimethyl-; Mackgard DM; Mackstat® DM; Nipaguard DMDMH. A preservative and formaldehyde donor; broad spectrum antimicrobial for cosmetics and toiletries; effective against Gram-positive and Gram-negative bacteria, fungi, and yeast. *Lonzagroup; McIntyre; Nipa.*

1523 DMP 30
90-72-2 G H

C15H27N3O
Phenol, 2,4,6-tris[(dimethylamino)methyl]-.
4-13-00-01946 (Beilstein Handbook Reference); Actiron NX 3; AI3-03346; Ancamine K 54; Anchor K 54; Araldite DY 061; Araldite DY 064; Araldite Hardener HY 960; Araldite HY 960; BRN 0795751; Capcure EH 30; Dabco TMR 30; DMF 3; DMP 30; DY 061; EH 30; EINECS 202-013-9; Epilink 230; NSC 3257; Phenol, 2,4,6-tris((dimethylamino)methyl)-; S 41028-4; Sumicure D; UP 606/2. Accelerator, catalyst, and hardener for epoxy resins. *Polysciences; Protex SA.*

1524 Docosanol
661-19-8 3433 G H

C22H46O
n-Docosanol.
4-01-00-01906 (Beilstein Handbook Reference); AI3-36489; Behenic alcohol; Behenyl alcohol; BRN 1770470; 1-Docosanol; n-Docosanol; Docosyl alcohol; EINECS 211-546-6; HSDB 5739; IK 2; Lidavol; NSC 8407; Tadenan. Used in synthetic fibers, lubricants, as an evaporation retardant on water surfaces. Crystals; mp = 72.5°; bp0.22 = 180°; slightly soluble in H2O, Et2O, soluble in CHCl3, very soluble in EtOH, MeOH, petroleum ether. *Lancaster Synthesis Co.; Michel M.; Sherex; Vista.*

1525 Docusate sodium
577-11-7 3435 D G H

C20H38NaO7S
Bis(2-ethylhexyl)sodium sulfosuccinate.
Alphasol OT; Adekacol EC 8600; Aerosol A 501; Aerosol AOT; Aerosol GPG; Aerosol OT; Aerosol OT 70PG; Aerosol OT 75; Aerosol OT-A; Aerosol OT-B; AI3-00239; Alcopol O; Alkasurf SS-O 75; Berol 478; Celanol DOS 65; Celanol DOS 75; Clestol; Colace; Complemix; Constonate; Coprol; Correctol Caplets; Correctol Extra Gentle Tablets; D-S-S Plus; D-S-S; Defilin; DESS; Dialose; Dialose Plus; Dioctlyn; Dioctyl-medo forte; Dioctyl sodium sulfosuccinate; Dioctyl sodium sulphosuccinate; Dioctylal; Diomedicone; Diosuccin; Diotilan; Diovac; Diox; Disonate; Discol DFW; Docusate sodique; Docusate sodium; Docusato sodico; Docusatum natricum; Dorbantyl; Doxan; Doxinate; Doxol; DSS; Dulsivac; Duosol; EINECS 209-406-4; Empimin® OP70; Feen-a-Mint Pills; Ferro-Sequels; Geriplex; HSDB 3065; Humifen WT 27G; Konlax; Kosate; Laxcaps; Laxinate; Laxinate 100; Manoxal OT; Mervamine; Modane Plus; Modane Soft; Molatoc; Molcer; Molofac; Monawet MD 70E; Monawet MO-70; Monawet MO-70 RP; Monawet MO-84 R2W; Monoxol OT; Natrii dioctylsulfosuccinas; Nekal WT-27; Nevax; Nikkol OTP 70; Norval; Obston; Octowet 40; Peri-Colace; Phillips Gelcaps; Rapisol; Regutol; Requtol; Revac; Sanmorin OT 70; SBO; Senokap DSS; Senokot S; Sobital; Softil; Soliwax; Solusol-100%; Solusol-75%; Sulfimel DOS; SV 102; Tex-Wet 1001; Triton GR-5; Triton GR 7; Unilax; Vatsol OT; Velmol; Waxsol; Wetaid SR. Used as a wetting agent, lubricant and detergent for dry cleaning, corrosion resistant lubricants, agricultural emulsions and organic solvent systems. Used when a higher flash is required. A fast wetting agent for denim finishing and continuous carpet dyeing. Used in textile washing and dyeing operations, agriculture, mining, paper and printing. Ingredient in laxative preparations. Waxy solid; mp = 153-157°; soluble in H2O (1.5 g/100 ml 25°, 23 g/100 ml 40°, 30 g/100 ml 50°, 55 g/100 ml 70°), CCl4, petroleum ether, Me2CO, EtOH. *3M Pharm.; Albright & Wilson UK Ltd.; Am. Cyanamid; Ascher B.F. & Co.; Bristol-Myers Squibb Pharm. R&D; Hoechst Roussel Pharm. Inc.; Johnson & Johnson-Merck Consumer Pharm.; Mallinckrodt Inc.; Parke-Davis; Purdue Pharma L.P.; Roberts Pharm. Corp.; Savage Labs; Schering-Plough Pharm.; Sterling Health U.S.A.*

1526   Dodecamethylcyclohexasiloxane
540-97-6   3437   H

C12H36O6Si6
Cyclohexasiloxane, dodecamethyl-.
Cyclohexasiloxane,   dodecamethyl-;   Dodecamethylcyclo-hexasiloxane; EINECS 208-762-8. Liquid; mp = -1.5°; bp = 245°; d^{25} = 0.9672; insoluble in H_2O.

1527   Dodecane
112-40-3   H

C12H26
n-Dodecane.
4-01-00-00498 (Beilstein Handbook Reference); Adakane 12; Ba 51-090453; Bihexyl; BRN 1697175; CCRIS 661; Dihexyl; Dodecane; n-Dodecan; n-Dodecane; EINECS 203-967-9; HSDB 5133; NSC 8714. Liquid; mp = -9.6°; bp = 216.3°; d^{20} = 0.7487; insoluble in H_2O, very soluble in EtOH, Et2O, Me2CO, CCl4, CHCl3.

1528   Dodecanedioic acid
693-23-2   H

C12H22O4
1,12-Dodecanedioic acid.
Decamethylenedicarboxylic acid; 1,10-Decanedi-carboxylic acid; 1,10-Dicarboxydecane; Dodecanedioic acid; 1,12-Dodecanedioic acid; EINECS 211-746-3; HSDB 5745; NSC 400242. Crystals; mp = 128°; bp^{25} = 222°; d^{25} = 1.15; soluble in TFA.

1529   Dodecanenitrile
2437-25-4   G H

C12H23N
n-Dodecanenitrile.
4-02-00-01104 (Beilstein Handbook Reference); AI3-00311; BRN 0970348; 1-Cyanoundecane; Decyl-acetonitrile; Dodecane nitrile; Dodecanenitrile; EINECS 219-440-1; Lauric acid nitrile; Lauric nitrile; Lauronitrile; Nitrile 12; NSC 1804; Undecyl cyanide. Chemical intermediate. Liquid; mp = 4°; bp = 277°, bp_{100} = 198°; d^{20} = 0.8240; insoluble in H_2O, freely soluble in EtOH, Et2O, Me2CO, C6H6, CHCl3. *Berol Nobel AB.*

1530   tert-Dodecanethiol
25103-58-6   H

C12H26S
2,3,3,4,4,5-Hexamethyl-2-hexanethiol.
3-01-00-01794 (Beilstein Handbook Reference); BRN 1738382; CCRIS 6030; t-DDM; t-Dodecanethiol; tert-Dodecyl mercaptan; terc.Dodecylmerkaptan;   tert-Dodecanethiol;   tert-Dodecylthiol; EINECS 246-619-1; 2,3,3,4,4,5-Hexamethyl-2-hexanethiol; Sulfole 120.

1531   1-Dodecene
112-41-4   G H

C12H24
n-Dodec-1-ene.
Adacene 12; α-Dodecene; 1-Dodecene; α-Dodecylene; n-Dodec-1-ene; EINECS 203-968-4; HSDB 1076; Neodene® 6/12; Neodene® 12; Neodene® 1012; NSC 12016. Used as an intermediate for biodegradable surfactants, specialty industrial chemicals, flavors, perfumes, medicines, oils, dyes, resins. mp = -35.2°; bp = 213.8°; d^{20} = 0.7584; insoluble in H_2O, soluble in EtOH, Et2O, Me2CO, CCl4, petroleum ether. *Ethicon Inc.; Monsanto (Solaris); Shell.*

1532   Dodecenylsuccinic anhydride
25377-73-5   H

C16H26O3
2,5-Furandione, 3-(dodecenyl)dihydro-.
CCRIS 698; DDSA; Dodecenylsuccinic acid anhydride; Dodecenylsuccinic anhydride; 2-(Dodecyl)succinic anhydride; EINECS 246-917-1; 2,5-Furandione, 3-(dodecenyl)dihydro-; HSDB 4374; K 12; K 12 (anhydride); Rikacid DDSA; Succinic anhydride, dodecenyl-.

1533   Dodecylamine
124-22-1   G H

C12H27N
n-Dodecanamine.
4-04-00-00794 (Beilstein Handbook Reference); AI3-15083; Alamine 4; Amine 12; Amine BB; 1-Aminododecane; Armeen® 12; Armeen® 12D; BRN 1633576; Dodecanamine; 1-Dodecanamine; Dodecyl-amine; 1-Dodecylamine; EINECS 204-690-6; Farmin 20D; HSDB 2645; Kemamine P690; Lauramine; Laurinamine; Laurylamine; n-Laurylamine; Monododecyl-amine; Nissan amine BB; Radiamine 6164. Chemical intermediate; ethoxylated end products used in detergent, cosmetic, and agricultural applications. Used in mineral flotation, as a corrosion inhibitor, pigment dispersant; cosmetics; lubricant and mold release for hard rubber, textile chemical, chemical synthesis; antistat and antifog additive for plastic foils. Solid; mp = 28.3°, bp = 259°; d^{20} = 0.8015; slightly soluble in H_2O, freely soluble in EtOH, Et2O, C6H6, CHCl3, CCl4. *Akzo Chemie; Fina Chemicals.*

1534   Dodecylbenzene
123-01-3

H

C$_{18}$H$_{30}$

1-Phenyldodecane.

4-05-00-01200 (Beilstein Handbook Reference); AI3-00435; Alkane; Alkylate P 1; Benzene, dodecyl-; BRN 1909107; CCRIS 2291; Detergent alkylate; Detergent Alkylate No. 2; Dodecane, 1-phenyl-; Dodecylbenzene; n-Dodecylbenzene; EINECS 204-591-8; HSDB 937; Laurylbenzene; Marlican; Nalkylene 500; NSC 102805; Phenyldodecan; Ucane alkylate 12. Liquid; mp = 3°; bp = 328°; d^{20} = 0.8551; insoluble in H$_2$O.

1535   Dodecylbenzenesulfonic acid
27176-87-0

G P

C$_{18}$H$_{30}$O$_3$S

Laurylbenzenesulfonic Acid.

Arylsulfonat BASF; Benzenesulfonic acid, dodecyl; Biosoft S-100; Calsoft las 99; Calsoft LAS 99; Caswell No. 413C; Conco AAS-985; Conoco 597; Conoco SA 597; DDBSA; Dobanic acid 83; Dobanic acid JN; n-Dodecylbenzenesulfonate; Dodecylbenzene sulfonic acid; Dodecylbenzenesulphonic acid; E 7256; EINECS 217-555-1; EINECS 248-289-4; Elfan WA sulphonic acid; EPA Pesticide Chemical Code 098002; HSDB 6285; Laurylbenzenesulfonic acid; Marlon AS 3; NA2584; Nacconol 98SA; Nansa 1042P; Nansa SSA; Pentine Acid 5431; Richonic acid B; Sulframin acid 1298; Witco® 1298 sulfonic acid. Mixture of o, m and p-isomers. Detergent intermediate, oil-water emulsifier, solubilizer, wetting agent, and detergent for household products, metal cleaning; emulsion polymerization surfactant for latex stabilization and pigment dispersion. Base for dishwashing and laundry detergents, industrial and institutional cleaners. Bactericide used to disinfect brewery and dairy equipment. Registered by EPA as an antimicrobial, fungicide and insecticide. Liqiud; mp = 10°; bp = 315°; d = 1.2000; LD$_{50}$ (rat orl) = 650 mg/kg. *Anderson Chemical Co; Chem-Serv Inc; CK Witco Corp.; Clough; Devere Chemical Co.; Diverseylever; Drexel Chemical Co.; Ecolab Inc.; Guaranteed Chemical Co; Hydrite Chemical Co; Kay Chemical Co; Shephard Bros Inc; US Chemical Corp; West Agro Inc.; ZEP Manufacturing Co.*

1536   tert-Dodecyl 2-hydroxypropyl sulfide
67124-09-8

H

C$_{15}$H$_{32}$OS

2-Propanol, 1-(tert-dodecylthio)-.

EINECS 266-582-5; tert-Dodecyl 2-hydroxypropyl sulfide; 1-(tert-Dodecylthio)propan-2-ol; 2-Propanol, 1-(tert-dodecylthio)-.

1537   Dodecyl mercaptan
112-55-0

H

C$_{12}$H$_{26}$S

n-Dodecyl mercaptan.

4-01-00-01851 (Beilstein Handbook Reference); AI3-07577; BRN 0969337; CCRIS 743; 1-Dodecanethiol; Dodecyl mercaptan; 1-Dodecyl mercaptan; EINECS 203-984-1; HSDB 1074; Lauryl mercaptan; Lauryl mercaptide; 1-Mercaptododecane; NCI-C60935; NSC 814; Pennfloat M; Pennfloat S. White solid; mp = -6.7°; bp = 277°, bp$_{15}$ = 143°; d^{20} = 0.844; soluble in EtOH, Et$_2$O, CHCl$_3$, insoluble in H$_2$O. LD$_{50}$ (rat orl) = 12800 mg/kg.

1538   Dodecyl methacrylate
142-90-5

G H

C$_{16}$H$_{30}$O$_2$

Dodecyl 2-methyl-2-propenoate.

4-02-00-01529 (Beilstein Handbook Reference); Acrylic acid, 2-methyl-, dodecyl ester; Ageflex FM-12; Ageflex FM 246; AI3-08765; BRN 1708160; Caswell No. 521; Dodecyl 2-methyl-2-propenoate; Dodecyl methacrylate; EINECS 205-570-6; EPA Pesticide Chemical Code 053101; GE 410; GE 410 (methacrylate); HSDB 5417; LAMA; Lauryl methacrylate; Laurylester kyseliny methakrylove; Metazene; Methacrylic acid, dodecyl ester; Methacrylic acid, lauryl ester; NSC 5188; Sipomer LMA; SR 313. Used in lube oil additives, coatings for nonwoven fiber, floor waxes, paints, adhesives, varnishes, sealants, caulks, stabilizer for nonaq. dispersions and inks. *Rit-Chem.*

1539   Dodecyl sodium sulfoacetate
1847-58-1

H

C$_{14}$H$_{27}$NaO$_5$S

Sodium 2-(dodecyloxy)-2-oxoethane-1-sulfonate.

Acetic acid, sulfo-, dodecyl ester, S-sodium salt; Acetic acid, sulfo-, 1-dodecyl ester, sodium salt; Dodecyl sodium sulfoacetate; Dodecyl sulfoacetate S-sodium salt; EINECS 217-431-7; Lathanol; Lathanol-lal 70; Lathanol LAL; Nacconol LAL; Sodium 2-(dodecyloxy)-2-oxoethane-1-sulphonate; Sodium lauryl sulfoacetate; Sulfoacetic acid 1-dodecyl ester, sodium salt; Sulfoacetic acid dodecyl ester S-sodium salt.

1540   Dodecylphenols
27193-86-8

H

C$_{18}$H$_{30}$O

Phenol, dodecyl-, mixed isomers.

Dodecyl phenol; Dodecylphenol; Dodecylphenol (mixed isomers); Dodecylphenols; EINECS 248-312-8; HSDB 386; NSC 6812; Phenol, dodecyl-; Phenol, dodecyl-, mixed isomers; T-Det. Mixture of o, m and p-isomers.

1541   p-Dodecylphenol
104-43-8   H

C18H30O
4-Dodecylphenol.
EINECS 203-202-9; p-Dodecylphenol; Phenol, 4-dodecyl-; Phenol, p-dodecyl-.

1542   Dodecylthioethanol
1462-55-1   G

C14H30OS
2-(Dodecylthio)ethanol.
2-(Dodecylthio)ethanol; DV-1936; EINECS 215-969-7. *Rhône Poulenc Surfactants.*

1543   Dodemorph Acetate
31717-87-0   3441   G P

C20H39NO3
4-Cyclododecyl-2,6-dimethylmorpholine acetate.
Acetate de dodemorphe; BAS 2382 F; BAS 238F; BASF mehltaumittel; Caswell No. 268C; Caswell No. 268E; Cyclododecyl-2,6-dimethylmorpholine acetate; 4-Cyclo-dodecyl-2,6-dimethylmorpholine acetate; Cyclododecyl-2,6-dimethylmorpholinium acetate; 4-Cyclododecyl-2,6-dimethylmorpholinium acetate; Cyclomorph; Dodemorfe; Dodemorph acetate; EINECS 250-778-2; EPA Pesticide Chemical Code 110401; EPA Pesticide Chemical Code 213600; Mehltaumittel; Meltox; Milban; Morpholine, N-cyclododecyl-2,6-dimethyl-, acetate; N-Cyclododecyl-2,6-dimethylmorpholinacetat; N-Cyclododecyl-2,6-dimethylmorpholinium acetate; Cyclomorph; Dodemorfe; Dodemorph acetate; EPA Pesticide Chemical Code 110401; EPA Pesticide Chemical Code 213600; Mehltaumittel; Meltatox; Meltox; Milban; Morpholine, N-cyclododecyl-2,6-dimethyl-, acetate; N-Cyclododecyl-2,6-dimethylmorpholinium acetate; N-Cyclododecyl-2,6-dimethylmorpholinacetat. Systemic fungicide with protective and curative action. Used for control of powdery mildews on roses. Registered by EPA as a fungicide (cancelled). Yellow liquid - colorless solid; mp = 63 - 64°; d = 0.93; sparingly soluble in H2O (< 0.01 g/100 ml), soluble in CHCl3 (> 148 g/100 ml), C6H6 (> 88 g/100 ml), EtOH (13.4 g/100 ml); LD50 (mrat orl) = 3944 mg/kg, (frat orl) = 2645 mg/kg, (mus ip) = 100 mg/kg, (rat der) > 4000 mg/kg; LC50 (rat ihl 4 hr.) = 5 mg/l air, (guppy 96 hr.) = 40 mg/l; non-toxic to bees. *BASF Corp.*

1544   Dodine
2439-10-3   3442   G P

C15H33N3O2
1-Dodecylguanidinium acetate.
Aadodin; AC 5223; American Cyanamid 5223; Carpene; Caswell No. 419; Curitan; Cyprex; Cyprex 65W; Dodecylguanidine acetate; n-Dodecylguanidine acetate; Dodecylguanidine monoacetate; 1-Dodecylguanidinium acetate; Dodin; Dodine; Dodine, mixture with glyodin; Doguadine; EINECS 219-459-5; ENT 16,436; EPA Pesticide Chemical Code 044301; Experimental fungicide 5223; Guanidine, dodecyl-, monoacetate; Guanidine, dodecyl-, acetate; HSDB 1705; Karpen; Kyselina 3-dodecylguanidinooctova; Laurylguanidine acetate; Melprex; Melprex 65; Melprex 65W; Melprex Liquid Dodine; Questuran; Radspor; Syllit; Syllit 65; Tsitrex; Venturol; Vondodine; Dodecylguanidine monoacetate; Dodin; Dodine; Doguadine; ENT 16,436; EPA Pesticide Chemical Code 044301; Experimental fungicide 5223; Guanidine, dodecyl-, acetate; HSDB 1705; Karpen; Kyselina 3-dodecylguanidinooctova [Czech]; Laurylguanidine acetate; Melprex; Melprex 65; Melprex 65W; Melprex Liquid Dodine; Questuran; Radspor; Syllit; Syllit 65; Tsitrex; Venturol; Vondodinen-Dodecylguanidine acetate. A foliar fungicide with some protective and curative action; used for the control of scab in apples and pears. Registered by EPA as an antimicrobial and fungicide. Waxy solid; mp = 132 - 135°; soluble in H2O).063 g/100 ml), EtOH, MeOH, insoluble in most other organic solvents; LD50 (mrat orl) = 1000 mg/kg, (rbt der) > 1500 mg/kg, (rat der) > 6000 mg/kg, (mallard duck orl) = 1142 mg/kg, (bee topical) > 0.011 mg/bee; LC50 (harlequin fish) = 0.53 mg/l; non-toxic to bees. *Am. Cyanamid; BASF Corp.; Stepan; Truchem; UAP - Platte Chemical.*

1545   Donepezil
120014-06-4   3453   D

C24H29NO3
(±)-2-[(1-Benzyl-4-piperidyl)methyl]-5,6-dimethoxy-1-indanone.
Donepezil; 1H-Inden-1-one, 2,3-dihydro-5,6-dimethoxy-2-((1-(phenylmethyl)-4-piperidinyl)methyl)-. Nootropic. A reversible acetylcholine esterase inhibitor used to treat Alzheimer's disease. *Pfizer Inc.*

1546   Donepezil hydrochloride
120011-70-3   3453   D

C24H30ClNO3
(±)-2-[(1-Benzyl-4-piperidyl)methyl]-5,6-dimethoxy-1-indanone hydrochloride.
Aricept; BNAG; Donepezil hydrochloride; E-2020. Nootropic. A reversible acetylcholine esterase inhibitor used to treat Alzheimer's

disease. An adjunct in treatment of dimentia symptoms. A cognition adjuvant. Solid; Freely soluble in CHCl3; soluble in H2O, glacial acetic acid; slightly soluble in EtOH, acetonitrile; practically insoluble in EtOAc, n-hexane. *Pfizer Inc.*

1547 Dowicide 3
85-97-2   7364   G P

C12H9ClO
(1,1'-Biphenyl)-2-ol, 3-chloro-.
AI3-03271; 2-Biphenylol, 3-chloro-; Caswell No. 211; 2-Chloro-6-phenylphenol; 2-Hydroxy-3-chlorobiphenyl; 2-Phenyl-6-chlorophenol; 3-Chloro-(1,1'-biphenyl)-2-ol; 3-Chloro-2-biphenylol; 3-Chlorobiphenyl-2-ol; 6-Chloro-2-phenylphenol; Dowcide 31; Dowcide 32; EPA Pesticide Chemical Code 062210; NSC 2600. Used as an antiseptic and fungicide. Registered by EPA as an antimicrobial and disinfectant (cancelled). Yellow, viscous liquid; mp = 6°; bp = 317 - 318° (dec); d$_4^{25}$ 1.24; insoluble in H2O, soluble in alkali and in most organic solvents. *Dow Chem. U.S.A.*

1548 Dowicide 6
25167-83-3   G P

C6H2Cl4O
2,3,4,5-Tetrachlorophenol.
AI3-01682; Caswell No. 832; EPA Pesticide Chemical Code 063004; NSC 406124; Phenol, tetrachloro-; Tetrachlorophenol; Tetrachlorophenol, isomer unspec-ified; Tetrachlorophenols. A proprietary trade name for tetrachlorophenol; an antiseptic. Registered by EPA as a fungicide (cancelled). *Dow Chem. U.S.A.*

1549 Doxazosin
74191-85-8   3466   D

C23H25N5O5
1-(4-Amino-6,7-dimethoxy-2-quinazolinyl)-4-(1,4-benzodioxan-2-ylcarbonyl)piperazine.
Doxazosin; Doxazosina; Doxazosine; Doxazosinum; UK 33274. Selective α-adrenergic blocker related to Prazosin. Antihypertensive. Also used in the treatment of benign prostatic hyperplasia. [monohydrochloride]: mp = 289-290°. *Pfizer Intl.; Roerig Div. Pfizer Pharm.*

1550 Doxazosin monomethanesulfonate
77883-43-3   3466   D

C24H29N5O8S
1-(4-Amino-6,7-dimethoxy-2-quinazolinyl)-4-(1,4-benzo-dioxan-2-ylcarbonyl)piperazine monomethanesulfonate.
Alfadil; Benur; Cardenalin; Cardoral; Cardoxan; Cardran; Cardular; Cardular PP; Cardular Uro; Cardura; Cardura XL; Carduran; Dedralen; Diblocin; Diblocin PP; Diblocin Uro; Doksura; Doxaben; Doxazomerck; Doxazosin AZU; Doxazosin mesylate; Doxazosin monomethanesulfonate; Doxolbran; HSDB 7082; Kaltensif; Normathen; Normothen; Progandol; Prostadilat; Supressin; Tensiobas; Tonocardin; UK 33,274-27. Selective α-adrenergic blocker related to Prazosin. Antihypertensive. Also used in the treatment of benign prostatic hyperplasia. *Pfizer Intl.; Roerig Div. Pfizer Pharm.*

1551 Doxycycline
564-25-0   3474   D

C22H24N2O8
4-(Dimethylamino)-1,4,4a,5,5a,6,11,12a-octahydro-3,5,10,12,12a-pentahydroxy-6-methyl-1,11-dioxo-2-naphthacenecarboxamide.
Azudoxat; Deoxymykoin; Dossiciclina; Doxiciclina; Doxiciclina; Doxitard; Doxivetin; Doxy-Caps; Doxy-Puren; Doxy-Tabs; Doxycen; Doxycycline; Doxycyclinum; Doxytetracycline; EINECS 209-271-1; GS-3065; HSDB 3071; Investin; Liviatin; Monodox; Nordox; Oxytetracycline, 6-deoxy-; Ronaxan; Spanor; Vibra-Tabs; Vibramycin; Vibramycine; Vibravenos. Tetracycline antibiotic. *Oclassen Pharm. Inc.; Pfizer Intl.*

1552 Doxycycline hyclate
24390-14-5   3474   D G

256

H—Cl
H—Cl

H₂O (shown as H-O-H)

ethanol (CH₃CH₂OH)

$C_{46}H_{56}Cl_2N_4O_{17}.H_2O$

4-(Dimethylamino)-1,4,4a,5,5a,6,11,12a-octahydro-3,5,10,12,12a-pentahydroxy-6-methyl-1,11-dioxo-2-naphthacenecarboxamide monohydrochloride compound with ethyl alcohol (2:1) monohydrate. Atridox; Azudoxat; Bassado; Clinofug; Diocimex; Doryx; Doxatet; Doxichel hyclate; Doxicrisol; Doxycycline hyclate; Doxycycline hydrochloride hemiethanolate hemihydrate; Doxylar; Doxytem; Duradoxal; Granudoxy; Hydramycin; Mespafin; Nordox; Paldomycin; Periostat; Retens; Ronaxan; Sigadoxin; Spanor; Tetradox; Unacil; Vibra-Tabs; Vibramycin Hyclate; Vibraveineuse; Vibravenös; Vivox; Zadorin. Tetracycline antibiotic. Solid; dec 201°; $[\alpha]_D^{25}$ = -110° (c = 1 in 0.01N HCl/MeOH); λm 267, 351 nm (log ε 4.24 4.12 0.01N HCl/MeOH); soluble in H₂O; LD₅₀ (rat ip) = 262 mg/kg. *Apothecon; Bristol-Myers Squibb Pharm. R&D; Elkins-Sinn; Lemmon Co.; Parke-Davis; Pfizer Intl.*

1553 Drazoxolon
5707-69-7   3478   G P

$C_{10}H_8ClN_3O_2$

3-Methyl-4,5-isoxazoledione 4-[(2-chlorophenyl)hydrazone].
Caswell No. 207C; 4-(2-Chlorophenylhydrazone)-3-methyl-5-isoxazolone; 4-((o-Chlorophenyl)hydrazono)-3-methyl-2-isoxazolin-5-one; Drazoxolon; Drazoxolone; EINECS 227-197-8; EPA Pesticide Chemical Code 207400; Ganocide; 4,5-Isoxazoledione, 3-methyl-, 4-((o-chlorophenyl)hydrazone); 3-Methyl-2-isoxazoline-4,5-dione 4-((2-chlorophenyl)hydrazone); 3-Methyl-4-(o-chlorophenylhydrazono)-5-isoxazolone; 3-Methyl-4-((o-chlorophenyl)hydrazone)-4,5-isoxazoledione; 3-Methyl-4,5-isoxazoledione 4-((2-chlorophenyl)hydrazone); Mil-col; PP781; Saisan; Sopracol; Sopracol 781; 3-Methyl-2-isoxazoline-4,5-dione 4-((2-chlorophenyl)hydrazone); 3-Methyl-4-((o-chlorophenyl)hydrazone)-4,5-isoxazole-dione; 3-Methyl-4,5-isoxazoledione 4-((2-chlorophenyl)-hydrazone); Mil-col; PP781; Saisan; Sopracol; Sopracol 781. Fungicide. Yellow crystals; mp = 167°; insoluble in H₂O, soluble in EtOH, CHCl₃, ketones, aromatic hydrocarbons; LD₅₀ (rat orl) = 126 mg/kg, (mus orl) = 129 mg/kg, (rbt orl) = 100 - 200 mg/kg, (gpg orl) = 12.5 - 25 mg/kg, (dog orl) = 17 mg/kg, (cat orl) = 50 - 100 mg/kg, (hen orl) ≅100 mg/kg; LC₅₀ (brown trout 96

hr.) = 0.55 mg/l. *ICI Agrochemicals.*

1554 Drocarbil
900-77-6   783   G V

$C_{16}H_{23}AsN_2O_7$

Methyl 1,2,5,6-tetrahydro-1-methylnicotinate with mono-((3-acetamido-4-hydroxyphenyl)arsonate).
Arecoline-acetarsol; Drocarbil; EINECS 212-983-5; Nemural. Drocarbil is a formulation of cestarsol with acetarsone and is used in veterinary medicine as an anthelmintic (Cestodes) and a cathartic. Pale yellow powder; freely soluble in H₂O. *May & Baker Ltd.*

1555 Drometrizole
2440-22-4   3482   D G H

$C_{13}H_{11}N_3O$

2-(2H-Benzotriazol-2-yl)-p-cresol.
Benazol P; Benazol P; 2-(2H-Benzotriazol-2-yl)-4-methylphenol; 2-(2H-Benzotriazol-2-yl)-p-cresol; 2-Benzotriazolyl-4-methylphenol; BRN 0615546; p-Cresol, 2-(2H-benzotriazol-2-yl)-; Drometrizol; Drometrizole; Drometrizolum; EINECS 219-470-5; 2-(2-Hydroxy-5-methylphenyl)-2H-benzotriazole; 2-(2-Hydroxy-5-methyl-phenyl)benzotriazole; NSC 91885; Phenol, 2-(2H-benzotriazol-2-yl)-4-methyl-; Porex P; Tin P; Tinuvin P; Topanex 100BT; UV Absorber-1; Uvazol P. UV absorber, light stabilizer, antioxidant protecting plastics (PVC, styrenics, acrylics, unsaturated polyesters), lacquers; effective at 290-380 nm. Used medicinally as an ultraviolet screen. Crystals; mp = 131-133°; bp₁₀ = 225°; soluble in EtOAc, Me₂CO, dioctyl phthalate; caprolactam. *Ciba-Geigy Corp.; EniChem Am.; ICI Americas Inc.*

1556 Dulcin
150-69-6   3497   G

$C_9H_{12}N_2O_2$

(4-Ethoxyphenyl)urea.
4-13-00-01154 (Beilstein Handbook Reference); AI3-08931; BRN 2096445; CCRIS 5913; Dulcin; Dulcine; Dulein; EINECS 205-767-7; NCI-C02073; NSC 1839; p-Aethoxyphenylharnstoff; p-Ethoxyfenylmocovina; 4-Ethoxyphenylurea; p-Ethoxyphenylurea; N-(4-Ethoxy-phenyl)urea; p-Phenethylurea; p-Phenetolcarbamid; p-Phenetolcarbamide; p-Phenetylurea; Phenethylcarbamid; Phenetolcarbamide; Sucrol; Suesstoff; Urea, (4-ethoxyphenyl)-; Urea, (p-ethoxyphenyl)-; Valzin. Used as a sweetening substance; 200 times sweeter than cane sugar. Use of this material in foods is prohibited by the USA FDA. Crystals; mp = 173-174°; soluble in H₂O, EtOH; TDLo (wmn orl) = 600 mg/kg, LDLo (chd orl) = 400

mg/kg.

1557 Durene
95-93-2   3501   G

$C_{10}H_{14}$

1,2,4,5-Tetramethylbenzene.
AI3-25182; Benzene, 1,2,4,5-tetramethyl-; Durene; Durol; EINECS 202-465-7; NSC 6770; 1,2,4,5-Tetramethylbenzene; p-Xylene, 2,5-dimethyl-. Used in organic synthesis, plasticizers, polymers, fibers. An agricultural chemical used as a fungicide, bactericide and wood preservative. Crystals; mp = 79.3°; bp = 196.8°; d^{81} = 0.8380; λ_m = 278 nm (cyclohexane); insoluble in H_2O, soluble in EtOH, Et_2O, Me_2CO, C_6H_6, CCl_4, petroleum ether; LD_{50} (rat orl) = 6989 mg/kg.

1558 Dymanthine
124-28-7   3508   D G H

$C_{20}H_{43}N$

N,N-Dimethyl-1-octadecanamine.
3-04-00-00433 (Beilstein Handbook Reference); ADMA 18; Adogen 342D; Adogen® MA-108 SF; AI3-09118; Armeen DM 18D; BRN 1763346; Crodamine 3.A18D; Dimantina; Dimantine; Dimantinum; Dimethyloctadecyl-amine; Dimethyl-N-octadecylamine; Dimethyl stear-amine; Dimethyl stearylamine; DM 18D; Dymanthine; EINECS 204-694-8; Farmin DM 80; HSDB 5472; Kemamine 9902D; Kemamine® T-9902; 1-Octadecanamine, N,N-dimethyl-; Octadecylamine, N,N-dimethyl; Octadecyldimethylamine; Onamine 18; Stearyldimethylamine. Emulsifier for herbicides, ore flotation, pigment dispersion; auxiliary for textiles, leather, rubber, plastics, and metal industries. Chemical intermediate; personal care additive. Also used as an anthelmintic. Intermediate for quats; used in cosmetics and textiles; acid scavenger in petrol, products. Intermediate in the synthesis of surfactants, antioxidants, oil and grease additives. Both Dymanthine and its hydrochloride salt Thelmesan ($C_{20}H_{43}N \cdot HCl$) are used as anthelmintics. Liquid; mp = 23°; d = 0.800. CK Witco Corp.; Croda; Lonzagroup; Sherex.

1559 Edetate Trisodium
150-38-9   3545   D G P

$C_{10}H_{13}N_2Na_3O_8$

Trisodium hydrogen(ethylenedinitrilo)tetraacetate.
AI3-18050; CCRIS 294; Edetate Trisodium; EDTA; EDTA trisodium salt; EINECS 205-758-8; Limclair; NCI-C03974; Nevanaid-B powder; Perma kleer 50, trisodium salt; Sequestrene Na3; Sequestrene NA3; Sequestrene trisodium; Sequestrene trisodium salt; Trilon AO; Trisodium edetate; Trisodium EDTA; Trisodium ethylenediaminetetraacetate; Trisodium ethylenediamine-tetraacetate trihydrate; Trisodium versenate; Versene-9. Chelating

agent. Crystals; mp > 300°; very soluble in H_2O. *Grace W.R. & Co.*

1560 Edetic acid
60-00-4   3546   G P

$C_{10}H_{16}N_2O_8$

Ethylenediaminetetraacetic acid.
4-04-00-02449 (Beilstein Handbook Reference); Acetic acid, (ethylenedinitrilo)tetra-; Acide edetique; Acide ethylenediaminetetracetique; Acido edetico; Acidum edeticum; AI3-17181; BRN 1716295; Caswell No. 438; CCRIS 946; Celon A; Celon ATH; Cheelox; Cheelox BF acid; Chelest 3A; Chemcolox 340; Clewat TAA; Complexon II; Dissolvine E; Edathamil; Edetic acid; EDTA; EDTA acid; EINECS 200-449-4; Endrate; EPA Pesticide Chemical Code 039101; Ethylene diamine tetraacetic acid; Ethylenebisiminodiacetic acid; Gluma cleanser; Glycine, N,N'-1,2-ethanediylbis(N-(carboxy-methyl)-; Hamp-ene acid; Havidote; HSDB 809; ICRF 185; Komplexon II; Kyselina ethylendiamintetraoctova; Metaquest A; Nervanaid B acid; Nullapon B acid; Nullapon BF acid; Permakleer 50 acid; Quastal Special; Questex 4H; SEQ 100; Sequestrene AA; Sequestric acid; Sequestrol; Tetrine acid; Titriplex; Titriplex II; Trilon BS; Trilon BW; Universene acid; Versene; Versene acid; Vinkeil 100; Warkeelate acid; YD 30. Registered by EPA as an antimicrobial, fungicide, herbicide and insecticide (cancelled). FDA approved for otics, rectals, topicals, USP/NF compliance. Used as a chelating, metal complexing agent, excipient and preservative, used in otics, rectals, topicals, ophthalmics, ear/nose/eye drops, local anesthetics, antibiotics, antihistamines, in diagnosis and treatment of heavy metal poisoning and iron overload. Also used as a food additive. A chelating agent; used where sodium ion is undesirable; for soaps, detergents, water treatment, metal finishing and plating, pulp and paper manufacturing, synthesis of polymers, photographic products, textiles, chemical cleaning for scale removal. Irritant, toxic if ingested. Solid; mp > 220° (dec); soluble in H_2O (200 mg/100 ml); LD_{50} (rat ip) = 397 mg/kg, (mus orl) = 30 mg/kg, (mus ip) = 25 mg/kg. *Akzo Nobel; Allchem Ind.; Allied Colloids; Chemplex; Hampshire; Protex SA; Showa Denko; Sigma-Aldrich Fine Chem.; Spectrum Chem. Manufacturing.*

1561 Edetol
102-60-3   3630   G H

$C_{14}H_{32}N_2O_4$

Tetrahydroxypropyl ethylenediamine.
4-04-00-01685 (Beilstein Handbook Reference); Adeka Quadrol; BRN 1781143; Edetol; Edetolum; EINECS 203-041-4; Entprol; HSDB 5349; Neutrol TE; NSC 369219; Quadrol; Quadrol L; Tetrahydroxypropyl ethylene-diamine; Tetrakis(2-hydroxypropyl)ethylenedi-amine; THPE. Chelating agent; intermediate used in resins, emulsifiers, surfactants, pharmaceuticals, herbicides, fungicides, insecticides, adhesives, and

plasticizers. Liquid; bp0.8 = 175-181°, d25 = 1.030; slightly soluble in CHCl3, soluble in H2O, EtOH, Et2O.

1562   Edifenphos
17109-49-8   3547   G P

C14H15O2PS2
O-Ethyl S,S-diphenyl phosphorodithioate.
O-Aethyl-S,S-diphenyl-dithiophosphat; O-Äthyl-S,S-diphenyl-dithiophosphat; Bay-hinosan; BAY 78418; Bayer 78418; BRN 1988797; Caswell No. 434B; Dithio-phosphorsaeure-O-aethyl-S,S-diphenylester; Dithiophos-phorsäure-O-aethyl-S,S-diphenylester; EDDP; EDDP (pesticide); Edifenphos; Ediphenphos; EINECS 241-178-1; EPA Pesticide Chemical Code 434300; O-Ethyl-S,S-diphenyl phosphorodithioate; O-Ethyl S,S-diphenyl dithiophosphate; Hinosan; Phosphorodithioic acid, O-ethyl-S,S-diphenyl ester; SRA 7847. Foliar fungicide with protective and curative action. Used for control of rice blast. Also for blight diseases, stem rot and *Fusarium* leaf spot in rice. Light brown liquid; bp0.01 = 154°; sg20 = 1.23; soluble in H2O (0.0056 g/100 ml), freely soluble in MeOH, Me2CO, C6H6, xylene, CCl4, dioxane; LD50 (mrat orl) = 212 - 340 mg/kg, (frat orl) = 100 - 150 mg/kg, (mus orl) = 218 - 670 mg/kg, (gpg, rbt orl) = 350 - 400 mg/kg, (mrat der) = 700 - 800 mg/kg, (hen orl) = 750 mg/kg; LC0 (rat ihl 4 hr.) = 0.32 - 0.36 mg/l air; (mirror carp 48 hr.) = 1.3 mg/l; non-toxic to bees. *Bayer Corp., Agriculture; Mobay; Nihon Nohyaku Co. Ltd.*

1563   Eicosanol
629-96-9   H

C20H42O
n-1-Eicosanol.
Al3-36485; Arachic alcohol; Arachidic alcohol; Arachidyl alcohol; 1-Eicosanol; Eicosyl alcohol; EINECS 211-119-4; HSDB 5731; Icosan-1-ol; n-1-Eicosanol; n-Eicosanol; NSC 120887. Solid; mp = 66.1°; bp = 309°, bp3 = 222°; D20 = 0.8405; insoluble in H2O, slightly soluble in EtOH, CHCl3, soluble in C6H6, petroeum ether, very soluble in Me2CO.

1564   Eicosyl methacrylate
45294-18-6   H

C24H46O2
2-Propenoic acid, 2-methyl-, eicosyl ester.
Eicosyl methacrylate; EINECS 256-220-4; Icosyl methacrylate; 2-Propenoic acid, 2-methyl-, eicosyl ester.

1565   Eiver-Pick acid
525-37-1   6399   G

C10H8O6S2
1,6-Naphthalene-disulfonic acid.
N-1,6-Dsa; 1,6-Naphthalenedisulfonic acid. Chemical intermediate. Orange prisms, mp = 125° (dec); λm = 237, 273, 342 nm (ε = 77625, 4169, 1698, 5% HCl-EtOH); insoluble in Et2O, soluble in EtOH, very soluble in H2O.

1566   Enalapril
75847-73-3   3599   D

C20H28N2O5
1-[N-[(S)-1-Carboxy-3-phenylpropyl]-L-alanyl]-L-proline   1'-ethyl ester.
Bonuten; Enalapril; Enalapril Richet; Enalaprila; Enalaprilum; Gadopril; HSDB 6529; Kinfil. Angiotensin-converting enzyme inhibitor. Used to treat hypertension. Orally active peptidyldipeptide hydrolase inhibitor. *Merck & Co.Inc.*

1567   Enalapril Maleate
76095-16-4   3599   D

C24H32N2O9
1-[N-[(S)-1-Carboxy-3-phenylpropyl]-L-alanyl]-L-proline   1'-ethyl ester maleate (1:1).
A-Rin; Acetensil; Alphrin; Amprace; Analept; Atens; Baripril; Benalapril; Benalipril; Biocronil; Bitensil; BQL; Cardiovet; Controlvas; Converten; Convertin; Coprilor; Coroldil; Crinoren; Dabonal; Defluin; Denapril; Ecapril; Ednyt; EINECS 278-375-7; Elfonal; Enacard; Enaladil; Enalapril maleate; Enalasyn; Enaloc; Enap; Enapren; Enapres; Enapril; Enaprin; Enarenal; Enaril; Envas; Eupressin; Feliberal; Glioten; Herten; Hipoartel; Hipten; Hytrol; Innovace; Innovade; Inoprilat; Insup; Invoril; Kenopril; Konveril; Lapril; Lipraken; Lotrial; Mapryl; Mepril; Minipril; MK-421; MK-421 maleate; Naprilene; Naritec; Neotensin; Norpril; Nuril; Olivin; Palane; Pres; Presil; Pressotec; Pril; Prilenap; Pulsol; Reca; Regomed; Renavace; Renitec; Renitek; Reniten; Renivace; Repantril; Ristalen; Sintec; Tenace; Tensazol; Unaril; Unipril; Vapresan; Vaseretic; Vasopril;

Vasotec; Xanef.; component of: Vaseretic, Acesistem, Co-Renitec, Innozide, Renacor, Xynertec. Angiotensin-converting enzyme inhibitor. Used to treat hypertension. Orally active peptidyldipeptide hydrolase inhibitor. Crystals; mp = 143-144.5°; $[\alpha]_D^{25}$ = -42.2° (c = 1 MeOH); soluble in H_2O (2.5 g/100 ml), EtOH (8 g/100 ml), MeOH (20 g/100 ml). *Merck & Co.Inc.*

1568   Endobil
31127-82-9   G

C26H26I6N2O10
Iodoxamic acid.
Acide iodoxamique; Acido iodossamico; Acido iodoxamico; Acidum iodoxamicum; B 10610; BC-17; Benzoic acid, 3,3'-((1,16-dioxo-4,7,10,13-tetraoxahexa-decane-1,16-diyl)diimino)bis(2,4,6-triiodo-; BRN 2801368; Cholevue; EINECS 250-478-1; Endomirabil; 3,3'-(Ethylenebis(oxyethyleneoxyethylenecarbonyl-imino))bis(2,4,6-triiodobenzoic acid); Iodoxamic acid; Iodoxamsaeure; Iodoxamsäure; SQ 21982; 4,7,10,13-Tetraoxahexadecane-1,16-dioyl-bis(3-carboxytriiodo-anilide); Videocolangio. Diagnostic aid. *Bracco Diagnostics Inc.*

1569   Endosulfan
115-29-7   3608   G H P

C9H6Cl6O3S
1,4,5,6,7,7-Hexachloro-5-norbornene-2,3-dimethanol cyclic sulfite.
AI3-23979; Benzodioxathiepin-3-oxide; Benzoepin; Beosit; BIO 5462; Caswell No. 420; CCRIS 275; Chlorthiepin; Crisulfan; Cyclodan; Devisulfan; Devisulphan; EINECS 204-079-4; Endocel; Endosalfan and metabolites; Endosulfan; Endosulfan 35 EC; Endosulfan and metabolites; Endosulphan; Endotaf; ENT 23,979; EPA Pesticide Chemical Code 079401; FMC 5462; Goldenleaf tobacco spray; Hexachlorohexa-hydromethano 2,4,3-benzodioxathiepin-3-oxide; Hildan; HOE 2,671; HSDB 390; Insectophene; Kop-thiodan; Malix; NCI-C00566; NIA 5462; Niagara 5,462; OMS 570; PFF Thiodan 4E; Phaser; Rasayansulfan; RCRA waste number P050; SD-4314; Sialan; Thifor; Thimul; Thiodan; Thiodan 35; Thiodan 4E Insecticide Liquid; Thiodan 4EC; Thiodan 4EC Insecticide; Thiodan 50 WP; Thiodan 50 WP Insecticide; Thiodan Dust Insecticide; Thiofor; Thiomul; Thionate; Thionex; Thiosulfan; Thiosulfan tionel; Thiotox; Thiotx (insecticide); Tionex; Tiovel. Non-systemic insecticide and acaricide with contact and stomach action. Used in control of sucking, chewing and boring insects and mites on a wide variety of crops. The commercial product is a mixture of an α-isomer (mp = 108-110°) and a β-isomer. Registered by EPA as an insecticide and acaricide. Crystals; mp = 106° (α-form) or 213° (β form); bp0.7 = 106°; slightly soluble in H_2O

(0.000032 g/100 ml), soluble in EtOAc (20 g/100 ml), CH2Cl2 (20 g/100 ml), C7H8 (20 g/100 ml), EtOH (6.5 g/100 ml), C6H14 (2.4 g/100 ml); LD50 (rat orl) = 70 - 110 mg/kg, (frat orl) = 18 mg/kg, (dog orl) = 77 mg/kg, (rbt der) = 359 mg/kg, (mallard duck orl) = 205 - 245 mg/kg; LC50 (rat ihl 1 hr.) = 21 mg/l air, (golden orfe 96 hr.) = 0.002 mg/l; non-toxic to bees. *AgrEvo; Agriliance LLC; Aventis Crop Science; Cape Fear Chemicals Inc; Dragon Chemical Corp.; FMC Corp.; Gowan Co.; Helena Chemical Co.; Hoechst UK Ltd.; Makhteshim Chemical Works Ltd.; Micro-Flo Co. LLC; Universal Cooperatives Inc.; Value Gardens Supply LLC.*

1570   Enquik
21351-39-3   H

CH6N2O5S
Urea Sulfate.
EINECS 244-343-6; Enquik; EPA Pesticide Chemical Code 128961; Monocarbamide dihydrogen sulfate; Monourea sulfuric acid adduct; N-Tac; Sulfuric acid, monourea adduct; Tac dessicant; Urea sulfate; Uronium hydrogen sulphate. Registered by EPA as a herbicide. *Entek Corp.*

1571   Eosin
15086-94-9   G

C20H6O5Br4Na2
2',4',5',7'-Tetrabromofluorescein.
4-19-00-02917 (Beilstein Handbook Reference); acid eosin; BRN 0063410; Bromeosin; Bromoeosin; C.I. 45380:2; C.I. Solvent Red 43; CCRIS 4904; CI 45380:2; D & C Red no. 21; D and C Red No. 21; D&C Red No. 21; 2-(3,6-Dihydroxy-2,4,5,7-tetrabromoxanthen-9-yl)-benzoic acid; 3,6-Dihydroxy-2,4,5,7-tetrabromospiro-(xanthene-9,3'-phthalide); EINECS 239-138-3; Eosin; Eosin A; Eosin acid; Eosin 3J; Eosin 4J extra; Eosin A extra; Eosin B; Eosin C; Eosin DH; Eosin G; Eosin G Extra; Eosin GGF; Eosin JJS; Eosin KS; Eosin yellowish; Eosin Y spirit soluble; Eosine acid; Fluorescein, 2',4',5',7'-tetrabromo-; Japan Red 223; Japan Red No. 223; NSC 244436; Red No. 223; Solvent Red 43; Spiro(isobenzofuran-1(3H),9'-(9H)-xanthen)-3-one, 2',4',5',7'-tetrabromo-3',6'-di-hydroxy-; Water soluble eosin. The alkali salts of tetrabromo-fluoresceine, dyes wool and silk yellowish red; used as a microscopic stain and a fluorescent tracer dye; red writing ink; cosmetic products and a colorant for motor fuel. Red solid.

1572   Epichlorohydrin
106-89-8   3642   G H

C3H5ClO
1-Chloro-2,3-epoxypropane.
5-17-01-00020 (Beilstein Handbook Reference); AI3-03545; BRN

0079785; Caswell No. 424; CCRIS 277; Chloromethyloxirane; Chloropropylene oxide; EINECS 203-439-8; EPA Pesticide Chemical Code 097201; epi-Chlorohydrin; Epichloorhydrine; Epichlorhydrin; Epi-chlorhydrine; Epichlorohydryna; Epichlorophydrin; Epi-cloridrina; Glycerol epichlorhydrin; Glycerol epichloro-hydrin; Glycidyl chloride; HSDB 39; NCI-C07001; NSC 6747; Oxirane, (chloromethyl)-; Oxirane, 2-(chloro-methyl); Propane, 1-chloro-2,3-epoxy-; RCRA waste number U041; Skekhg; UN2023. Major raw material for epoxy and phenoxy resins, manufacture of glycerol, curing propylene-based rubbers; solvent for cellulose esters and ethers; high wet-strength resins for paper industry. Liquid; mp = -25.6°; bp = 117.9°, $bp_{400} = 98°$, $bp_{200} = 79.3°$, $bp_{100} = 62.0°$, $bp_{40} = 42.0°$, $bp_{10} = 16.6°$, $bp_{1.0} = -16.5°$; d_4^{20} 1.1812; insoluble in H_2O, very soluble in EtOH, Et_2O, $CHCl_3$, CCl_4; LD_{50} (rat orl) = 0.09 g/kg.

1573   Epoxiconazole
133855-98-8   3659   P

C13H17ClFN3O
(±)-1-((3-(2-Chlorophenyl)-2-(4-fluorophenyl)oxiranyl)-methyl)-1H-1,2,4-triazole cis-.
(2RS,   35R)-1-(3-(2-Chlorophenyl)-2,3-epoxy-2-(4-fluoro-phenyl)propyl)-1H-1,2,4-triazole;   1-((3-(2-Chlorophenyl)-2-(4-fluorophenyl)oxiranyl)methyl)-1H-1,2,4-triazole   cis-(±)-; Epoxiconazol;   Epoxiconazole;   1H-1,2,4-Triazole,   1-((3-(2-chlorophenyl)-2-(4-fluorophenyl)oxiranyl)methyl)-, cis-(±)-; 1H-1,2,4-Triazole,   1-(((2R,3S)-3-(2-chloro-phen-yl)-2-(4-fluorophenyl)oxiranyl)methyl)-, rel-. Fungicide. Crystals; mp = 136.2°; insoluble in H_2O (0.000663 g/100 ml), soluble in Me_2CO (18 g/100 ml), CH_2Cl_2 (14 g/100 ml), C_7H_{16} (< 0.1 g/100 ml); LD_{50} (rat orl) > 5000 mg/kg, (rat der) > 2000 mg/kg; LC_{50} (rat ihl 4 hr.) > 5.3 mg/l. BASF Corp.

1574   Epoxidized linseed oil
8016-11-3   H

Epoxidized linseed oil.
EINECS 232-401-3; Epoxidized linseed oil; Linseed oil, epoxidized.

1575   Epoxidized soybean oil
8013-07-8   H

Epoxidized soybean oil.
EINECS 232-391-0; Epoxidized soybean oil; Fatty acid, soybean oil, epoxidized; Flexol EPO; Oils, soybean, epoxidized; Paraplex G-60; Paraplex G-62; PX-800; Soybean oil, epoxidized.

1576   Epoxycyclohexylethyl trimethoxy silane
3388-04-3   G

C11H22O4Si
2-(3,4-Epoxycyclohexyl) ethyltriacetoxysilane.
A 186 (coupling agent); A 186 (heterocycle); AI3-52751; CCRIS

3047;   CE   6250;   E   6250;   EINECS   222-217-1; ((Epoxycyclohexyl)ethyl)trimethoxy   silane;   (3,4-Epoxy-cyclohexyl)ethyltrimethoxysilane;   (2-(3,4-Epoxycyclo-hexyl)ethyl)trimethoxysilane;   β-(3,4-Epoxycyclohexyl)-ethyltrimethoxy silane; KBM 303; NSC 139838; NUCA 186; S 530; Sila-Ace S 530; Silane, (β-(3,4-epoxycyclo-hexyl)ethyl)trimethoxy-; Silane, β-(3,4-epoxycyclohexyl)-ethyltrimethoxy-; Silane A 186; Silane, trimethoxy(2-(7-oxabicyclo(4.1.0)hept-3-yl)ethyl)-; Silane Y-4086; Silic-one A-186; 3-(2-(Trimethoxysilyl)ethyl)-7-oxabicyclo-(4.1.0)heptane;   4-(2-(Trimethoxysilyl)ethyl)-7-oxabicyclo-(4.1.0)heptane; UC-A 186; Union carbide® A-186; Y 4086. Coupling agent, chemical intermediate, blocking agent, release agent, lubricant, primer, reducing agent. Used as a crosslinking agent, adhesion promoter for coatings; provides durability. Liquid; bp = 310°; d = 1.065. Degussa-Hüls Corp.; Union Carbide Corp

1577   3,4-Epoxycyclohexylmethyl-3,4-epoxycyclohexanecarboxylate
2386-87-0   H

C14H20O4
7-Oxabicyclo(4.1.0)hept-3-ylmethyl   7-oxabicyclo(4.1.0)-heptane-3-carboxylate.
BRN 1381750; Chissonox 221 monomer; EINECS 219-207-4; 3,4-Epoxycyclohexylmethyl 3,4-epoxycyclo-hexane carboxylate; ERL-4221;   HSDB   5873;   7-Oxabicyclo(4.1.0)hept-3-ylmethyl   7-oxabicyclo(4.1.0)-heptane-3-carboxylate; UT 632. Chisso Corp. USA.

1578   2,3-Epoxypropyl neodecanoate
26761-45-5   G H

C13H24O3
Neodecanoic acid, 2,3-epoxypropyl ester.
Cardura E 10; CCRIS 2627; EINECS 247-979-2; 2,3-Epoxypropyl neodecanoate; Glycidyl ester of neodecanoic acid; Glycidyl neodecanoate; Glydexx N 10; Neodecanoic acid, 2,3-epoxypropyl ester; Neodecanoic acid, oxiranylmethyl ester; 1-Propanol, 2,3-epoxy-, neodecanoate. Polyfunctional monomer; in coatings to improve adhesion to substrate and solvent resistance. Liquid; bp = 260°; d = 0.97; insoluble in H_2O, soluble in organic solvents; LD_{50} (rat orl) > 9.59 g/kg. ExxonMobil Chem. Co.

1579   EPTC
759-94-4   3669   G P

C9H19NOS
S-Ethyl dipropylcarbamothioate.
4-04-00-00487 (Beilstein Handbook Reference); S-Aethyl-N,N-dipropylthiolcarbamat;   S-Äthyl-N,N-dipropylthiol-carbamat;   Alirox;

BRN 1762751; Carbamic acid, di-propylthio-, S-ethyl ester; Carbamothioic acid, dipropyl-, S-ethyl ester; Caswell No. 435; CCRIS 6035; Dipropylcarbamothioic acid S-ethyl ester; Dipropyl-thiocarbamic acid S-ethyl ester; N,N-Dipropylthio-carbamic acid S-ethyl ester; EINECS 212-073-8; EPA Pesticide Chemical Code 041401; Eptam; Eptam 6E; EPTC; Ethyl dipropylthiocarbamate; S-Ethyl dipropylthio-carbamate; S-Ethyl N,N-dipropylthiocarbamate; S-Ethyl N,N-di-n-propylthiocarbamate; FDA 1541; Genep; Genep EPTC; HSDB 394; Niptan; NSC 40486; R 1608; Stauffer R 1608; Torbin; Witox. Selective systemic herbicide, used to control annual and perennial grasses in many crops. Liquid; bp_{20} = 127°; d^{30} = 0.9546; soluble in H_2O (36.5 g/100 ml 20°, 37.5 g/100 ml 25°), freely soluble in C_6H_6, C_7H_8, C_8H_{10}; LD_{50} (rat orl) = 1630 mg/kg.

1580   Erbium

7440-52-0   3675   G

Er

Er
Erbium.
EINECS 231-160-1; Erbium. Rare-earth element; used in nuclear controls, special alloys, room-temperature laser. Metal; mp = 1529°; bp = 2868°; d = 9.066. *Atomergic Chemetals; Cerac; Noah Chem.; Rhône-Poulenc.*

1581   Erbium oxide

12061-16-4   G

Er_2O_3
Erbia.
Dierbium trioxide; EINECS 235-045-7; Erbia; Erbium oxide; Erbium oxide (Er_2O_3); Erbium sesquioxide; Erbium trioxide; Erbium(3+) oxide; Erbium(III) oxide. Phosphor activator, infrared-absorbing glass. Solid; d = 8.64; changing into crystals on heating at 1300°; soluble in acids, H_2O. *Atomergic Chemetals; Cerac; Noah Chem.; Rhône-Poulenc.*

1582   Ergocalciferol

50-14-6   10078   G

$C_{28}H_{44}O$
(3β,5Z,7E,22E)-9,10-Secoergosta-5,7,10(19)-,22-tetraen-3-ol.
Activated ergosterol; D-Arthin; Buco-D; Calciferol; Calciferolum; Calciferon 2; Condocaps; Condol; Crystallina; Daral; Davitamon D; Davitin; De-rat concentrate; Decaps; Dee-Osterol; Dee-Ron; Dee-Ronal; Dee-Roual; Deltalin; Deratol; Detalup; Diactol; Divit urto; Doral; Drisdol; EINECS 200-014-9; Ergocalciferol; Ergocalciferolo; Ergocalciferolum; Ergorone; Ertron; Fortodyl; Geltabs; Haliver; Hl-Deratol; HSDB 819; Hyperkil; Infron; Metadee; Mina D2; Mulsiferol; Mykostin; Novovitamin-D; NSC 62792; Oleovitamin D2; Ostelin; Radiostol; Radsterin; Rodine C; Rodinec; Shock-ferol; Sorex C.R.; Sterogyl; D-Tracetten; Vigantol; Vio-D; Viosterol; Vitamin D2; (+)-Vitamin D2; Vitavel-D. A fat soluble vitamin, deficiency of which

causes rickets, in children and osteomalacia in adults; $C_{28}H_{44}O$; used as a dietary supplement and as a rodenticide. Crystals; mp= 115-118°; $[\alpha]_D^{25}$= +83° (c= 2 Me2CO); λ_m = 264.5 nm ($E^{1\%}_{1cm}$ 459, hexane); insoluble in H_2O, soluble in organic solvents.

1583   Erucamide

112-84-5   G H

$C_{22}H_{43}NO$
13-Docosenamide, (13Z)-.
Armid® E; 13-Docosenamide; 13-Docosenamide, (13Z)-; 13-Docosenamide, (Z)-; (Z)-Docos-13-enamide; EINECS 204-009-2; Erucamide; Erucic acid amide; Erucyl amide; Erucylamide; HSDB 5577; Petrac® Eramide®; Unislip 1753. Mold release agent for rubber and plastics; auxiliary for processing rubber. Foam stabilizer; solvent for waxes and resins, emulsions; slip/antiblock agent for polyethylene. slip, release, antitack, and/or internal mold release agent; used in PP for extrusion of sheets, in injection molding; antistat; in polyvinyls for films and sheeting; in polyethylene it imparts slip and antiblock characteristics in film application; internal mold release agent in molded products, lamination of PE to cellophane and in PE extrusion coatings; withstands high processing temperatures, food contact applications. Solid; mp = 79°; insoluble in H_2O, soluble in organic solvents. *Chemax; CK Witco Corp.; Croda; Syn Prods.; Syntetics Products Co; Unichema.*

1584   Erucic acid

112-86-7   3707   G H

$C_{22}H_{42}O_2$
13-Docosenoic acid, (Z)-.
AI3-18180; 13-Docosenoic acid, (Z)-; 13-Docosenoic acid, (13Z)-; 13-cis-Docosenoic acid; cis-13-Docosenoic acid; (Z)-Docos-13-enoic acid; EINECS 204-011-3; Erucic acid; HSDB 5015; Hystrene 2290; NSC 6814. Major fatty acid in mustard, rapeseed and wallflower seed. Used as a chemical intermediate. Needles; mp = 33.5°; bp = 381°, bp_{15} = 265°; d^{55} = 0.860; insoluble in H_2O, soluble in EtOH, CCl_4, very soluble in Et_2O, MeOH.

1585   Erythrosin

16423-68-0   3727   G

$C_{20}H_6O_5I_4Na_2$
Disodium 2',4',5',7'-tetraiodofluorescein.
1427 Red; 1671 Red; Acid Red 51; AI3-09094; Aizen erythrosine; Aizen Food Red 3; C.I. 45430; C.I. Acid Red 51; C.I. Food Red 14;

Calcocid erythrosine N; Canacert erythrosine BS; 9-(o-Carboxyphenyl)-6-hydroxy-2,4,5,7-tetraiodo-3H-xanthene-3-one disodium salt monohydrate; Caswell No. 425AB; CCRIS 892; Cerven kysela 51; Cerven potravinarska 14; CI 45430; Cilefa Pink B; D&C Red No. 3; 3',6-Dihydroxy-2',4',5',7'-tetraiodospiro(isobenzofuran-1(3H),9'(9H)xanthen)-one disodium salt; Disodium 9-(O-carboxyphenyl)-6-hydroxy-2,4,5,7-tetra-iodo-3H-xanthen-3-one monohydrate; Disodium 3',6'-dihydroxy-2',4',5',7'-tetraiodospiro(isobenzofuran-1(3H), 9'-(9H)xanthen)-3-one; Disodium 2-(2,4,5,7-tetraiodo-6-oxido-3-oxoxanthen-9-yl)benzoate; Disodium 2',4',5',7'-tetraiodofluorescein; Dolkwal erythrosine; Dye FD and C Red No. 3; E 127; Edicol Supra Erythrosin AS; Edicol Supra Erythrosine A; EINECS 240-474-8; EPA Pesticide Chemical Code 120901; Erythrosin; Erythrosin B; Erythrosin B sodium salt; Erythrosin BS; Erythrosine; Erythrosine 3B; Erythrosine B; Erythrosine B (biological stain); Erythrosine B-FO (biological stain); Erythrosine bluish; Erythrosine bluish (biological stain); Erythrosine Bluish; Erythrosine BS; Erythrosine Extra; Erythrosine Extra Bluish; Erythrosine Extra Conc. A Export; Erythrosine Extra Pure A; Erythrosine I; Erythrosine K-FO (biological stain); Erythrosine Lake; Erythrosine sodium; Erythrosine sodium (close form); Erythrosine TB; Erythrosine TB Extra; FD&C Red No. 3; FDC Red 3; FDC Red 3 dye; Fluorescein, 2',4',5',7'-tetraiodo-, disodium salt; Food Color Red 3; Food Dye Red 3; Food Red 14; Food Red 3; Food Red No. 3; Hexacert Red No. 3; Hexacol erythrosine BS; LB-Rot 1; Maple erythrosine; New Pink Bluish Geigy; Red Dye No. 3; Schultz No. 887; Sodium erythrosin; Spiro(isobenzofuran-1(3H),9'-(9H)xanthen)-3-one, 3',6'-dihydroxy-2',4',5',7'-tetraiodo-, disodium salt; 2,4,5,7-Tetraiodofluorescein disodium salt; 2',4',5',7'-Tetraiodofluorescein, disodium salt; Tetraiodofluorescein sodium salt; Usacert Red No. 3. The salt of tetraiodofluorescein, Dyes silk and wool bluish-red; used for paper staining. Brown powder; λ_m = 524 nm (H2O), 531 nm (95% EtOH); LD50 (mrat ip) = 340 mg/kg, (frat ip) = 370 mg/kg, (mmus ip) = 400 mg/kg, (fmus ip) = 320 mg/kg, (mrat orl) = 7400 g/kg, (frat orl) = 6800 mg/kg, (mmus orl) = 6700 mg/kg, (fmus orl) = 6900 mg/kg.

1586   Esfenvalerate
66230-04-4   4038   P

C25H22ClNO3
[S-(R*,R*)]-4-Chloro-α-(1-methylethyl)benzeneacetic acid cyano(3-phenoxyphenyl)methyl ester.
A Alpha; Asana; Asana XL; Benzeneacetic acid, 4-chloro-α-(1-methylethyl)-, cyano(3-phenoxyphenyl)methyl ester, (S-(R*,R*))-; BRN 4275674; Cyano-3-phenoxybenzyl (S)-2-(4-chlorophenyl)isovalerate; Cyano-3-phenoxybenzyl (S)-2-(4-chlorophenyl)-3-methylbutyrate; (S-(R*,R*))-cyano(3-phenoxyphenyl)methyl 4-chloro-α-(1-methyl-ethyl)benzeneacetate; (S)-α-Cyano-3-phenoxybenzyl(S)-2-(4-chlorophenyl)-3-methylbutyrate; DPX-YB656-84; Du-Pont Asana SP; Esfenvalerate; Fenvalerate (S,S)-isomer; Fenvalerate α; Fenvalerate A α; Halmark; HSDB 6625; OMS 3023; S-1844; S-5620A α; S-fenvalerate; SS-pydrin; Sumi-α; Sumi-alfa; Sumiciclin Aα; Sumicidin A α. Registered by EPA as an insecticide. White crystalline solid; mp = 59 - 60.2°; bp = 151 - 167°; d$_{23}^{23}$ = 1.163; [α]$_D^{25}$= -15.0° (c = 2.0 MeOH); soluble in H2O (0.00003 g/100 ml) CH3CN (> 47 g/100 ml), CHCl3 (> 89 g/100 ml), DMSO (> 66 g/100 ml), DMF (> 56 g/100 ml), EtOAc (> 54 g/100 ml), Me2CO (> 47 g/100 ml), ethyl Cellosolve (30 - 40 g/100 ml), C6H14 (0.7 - 3.3 g/100 ml), kerosene (< 0.8 g/100 ml), MeOH (5.6 - 8.0 g/100 ml), xylene (> 52 g/100 ml); LD50 (rat orl)

= 75 - 458 mg/kg, (rbt der) > 2000 mg/kg; LC50 (fathead minnow 96 hr.) = 690 ng/l. *Bonide Products, Inc.; DuPont; Mclaughlin Gormley King Co; Shell; Speer Products Inc.; Sumitomo Corp.*

1587   Estoral
53370-45-9   5864   G

C10H21BO3
Menthyl borate.
Cyclohexanol, 5-methyl-2-(1-methylethyl)-, monoester with boric acid; (H3BO3), (1R,2S,5R)-rel-; Cyclohexanol, 5-methyl-2-(1-methylethyl)-, monoester with boric acid; (H3BO3), (1α,2β,5α)-; Menthyl borate. Decomposes in solution into its components. Source of menthol. Insoluble in H2O, EtOH, soluble in organic solvents.

1588   Estradiol
50-28-2   3738   D

C18H24O2
(17β)-Estra-1,3,5(10)-triene-3,17-diol.
Aerodiol; Alora; Altrad; Aquadiol; Bardiol; CCRIS 280; Climaderm; Climara; Climara Forte; Compudose; Compudose 200; Compudose 365; Corpagen; Dermestril; Dihydrofollicular hormone; Dihydrofolliculin; Dihydro-menformon; Dihydrotheelin; Dihydroxyestrin; Dihydroxy-oestrin; Dimenformon; Diogyn; Diogynets; Divigel; EINECS 200-023-8; Encore; Estra-1,3,5(10)-triene-3,17-β-diol; Estrace; Estracomb TTS; Estraderm; Estraderm MX; Estraderm TTS; Estraderm TTS 50; Estradiol; α-Estradiol; β-estradiol; d-Estradiol; D-3,17-β-Estradiol; D-Estradiol; Estradiol-17β; Estradiol-3,17β; Estradiolo; Estradiolum; Estraldine; Estrapak 50; Estrasorb; Estreva; Estrifam; Estring; Estring Vaginal Ring; Estroclim; Estroclim 50; Estrodiolum; Estrofem 2; Estrofem Forte; Estrogel; Estrogens, esterified; Estrovite; Evorel; Femestral; Femogen; Fempatch; Femtran; Follicyclin; Ginedisc; Ginosedol; Gynergon; Gynestrel; Gynoestryl; GynPolar; HSDB 3589; Lamdiol; Macrodiol; Macrol; Menest; Menorest; Microdiol; Nordicol; NSC-9895; NSC-20293; Oesclim; Oestergon; Oestra-1,3,5(10)-triene-3,17-β-diol; Oestradiol; d-Oestradiol; D-Oestradiol; D-3,17-β-Oestradiol; Oestradiol-17β; Oestradiol Berco; Oestradiol R; Oestradiolum; Oestrogel; Oestroglandol; Oestrogynal; Ovahormon; Ovasterol; Ovastevol; Ovociclina; Ovocyclin; Ovocycline; Perlatanol; Primofol; Profoliol; Profoliol B; Progynon; Progynon-DH; Sandrena Gel; Sisare Gel; SK-Estrogens; Syndiol; Systen; Theelin, dihydro-; Tradelia; Trial SAT; Trocosone; Vagifem; Vivelle; Zerella; Zumenon.
Estrogen. mp = 173-179°; [α]$_D^{25}$= 76° to 83° (dioxane); λ_m = 225, 280 nm; insoluble in H2O; soluble in EtOH, Me2CO, dioxane. *Apothecon; Berlex Labs Inc.; Ciba-Geigy Corp.; Marion Merrell Dow Inc.; Mead Johnson Labs.; Pfizer Inc.; Pharmacia & Upjohn; Schering-Plough HealthCare Products.*

1589 Estragole

140-67-0   3740   G H

$C_{10}H_{12}O$

4-Allyl-1-methoxybenzene.

4-06-00-03817 (Beilstein Handbook Reference); AI3-16052; 4-Allylanisole; 4-Allyl-1-methoxybenzene; Allyl-phenyl methyl ether, p-; Anisole, p-allyl-; Benzene, 1-methoxy-4-(2-propenyl)-; BRN 1099454; CCRIS 1317; Chavicol methyl ether; Chavicol, O-methyl-; Chavicyl methyl ether; EINECS 205-427-8; EPA Pesticide Chemical Code 062150; Esdragol; Esdragole; Esdragon; Estragole; FEMA Number 2411; HSDB 5412; Isoanethole; Methyl chavicol; NCI-C60946; NSC 404113; Tarragon; Terragon. Used as a perfume and flavorant. bp = 215.5°; d^{25} = 0.965; λ_m = 226, 255, 284 nm (MeOH); very soluble in EtOH, $CHCl_3$; LD_{50} (rat orl) = 1820 mg/kg.

1590 Estrogens, conjugated

2533   D

Premarin; component of: PMB-2000, PMB-4000. Estrogen. *Wyeth-Ayerst Labs.*

1591 Ethane

74-84-0   3758   H

C_2H_6

Ethane.

Bimethyl; Dimethyl; EINECS 200-814-8; Ethane; Ethyl hydride; HSDB 941; Methylmethane; UN1035; UN1961. Used in the manufacturing of chemical intermediates; in ionization chambers used in subatomic species measurement and in the manufacturing of calibration mixtures. Colorless gas; mp = -182.8°; bp = -88.6°; soluble in hydrocarbon solvents. *BOC Gases*; Mesa Specialty Gases & Equipment.

1592 1,2-Ethanediamine, N-(2-aminoethyl)-N'-(2-(2-(8-heptadecenyl)-4,5-dihydro-1H-imidazol-1-yl)ethyl)-, (Z)-

65817-50-7   H

$C_{26}H_{53}N_5$

1,2-Ethanediamine, N-(2-aminoethyl)-N'-(2-(2-(8Z)-8-heptadecenyl-4,5-dihydro-1H-imidazol-1-yl)ethyl)-.

(Z)-N-(2-Aminoethyl)-N'-(2-(2-(8-heptadecenyl)-4,5-dihydro-1H-imidazol-1-yl)ethyl)ethylenediamine; EINECS 265-935-0; 1,2-Ethanediamine, N-(2-aminoethyl)-N'-(2-(2-(8Z)-8-heptadecenyl-4,5-dihydro-1H-imidazol-1-yl)ethyl)-.

1593 Ethanol

64-17-5   3795   G P

C_2H_6O

1-Hydroxyethane.

Äthanol; Äthylalkohol; Absolute ethanol; Aethanol; Aethylalkohol; AI3-01706; Alcohol; Alcohol, anhydrous; Alcohol dehydrated; Alcohol, diluted; Alcohol, ethyl; Alcool ethylique; Alcool etilico; Algrain; Alkohol; Alkoholu etylowego; Anhydrol; Caswell No. 430; CCRIS 945; Cologne Spirit; Denatured alcohol; Denatured alcohol; Distilled spirits; EINECS 200-578-6; EPA Pesticide Chemical Code 001501; Etanolo; Ethanol; Ethanol absolute; Ethanol solution; Ethanol, undenatured; Ethyl alcohol; Ethyl alcohol; Ethyl hydroxide; Ethylol; EtOH; Etylowy alkohol; FEMA Number 2419; Fermentation alcohol; Grain alcohol; HSDB 82; Hydroxyethane; Jaysol; Jaysol S; Methylcarbinol; Molasses alcohol; NCI-C03134; NSC 85228; Potato alcohol; Ru-Tuss Expectorant; Ru-Tuss Hydrocodone Liquid; Ru-Tuss Liquid; SD Alcohol 23-hydrogen; SDM No. 37; Spirits of wine; Synasol; Tecsol; Tecsol C. Registered by EPA as an antimicrobial, fungicide, herbicide and insecticide. In alcoholic beverages. Used as a solvent, in pharmaceuticals, perfumery and organic synthesis. Also as an additive in gasolines; and an antiseptic. Clear liquid; mp = -114°; bp = 78.2°; $d^{15.5}$ = 0.816, d^{20} = 0.7893; miscible with H_2O, MeOH, $CHCl_3$, Me_2CO, Et_2O. TLV = 1000 ppm in air; moderately toxic if ingested. Depressant drug. *ADM Ethanol; BP Chem.; Condea Vista Co.; Coyne; Eastman Chem. Co.; Georgia Pacific; Gist-Brocades Intl.; Grain Processing; Great Western; Quantum/USI; Spectrum Chem. Manufacturing; Union Carbide Corp.*

1594 Ethanol, 2,2'-iminobis-, N-coco alkyl derivs.

61791-31-9   H

N,N-Bis(hydroxyethyl)cocoamine.

N,N-Bis(2-hydroxyethyl)(coconut oil alkyl)amine; N,N-Bis(hydroxyethyl)cocoamine; Coconut fatty acid diethanolamide; Diethanolamine coconut fatty acid condensate; EINECS 263-163-9; Ethanol, 2,2'-iminobis-, N-coco alkyl derivs.

1595 Ethanol, 2,2'-iminobis-, N-tallow alkyl derivs.

61791-44-4   H

Ethanol, 2,2'-iminobis-, N-tallow alkyl derivs.
EINECS 263-177-5; Ethanol, 2,2'-iminobis-, N-tallow alkyl derivs.

1596 Ethanolamine

141-43-5   3762   G P

C_2H_7NO

1-Amino-2-hydroxyethane.

Äthanolamin; Aethanolamin; AI3-24219; Aminoethanol; Caswell No. 426; CCRIS 6260; Colamine; EINECS 205-483-3; EPA Pesticide Chemical Code 011601; Etanolamina; Ethanol, 2-amino-; Ethanolamine; Ethylolamine; Glycinol; HSDB 531; Kolamin; MEA (alcohol); Monoäthanolamin; Monoaethanolamin; Monoethanolamine; Olamine; Thiofaco M-50; UN2491; USAF EK-1597. Scrubbing acid gases, especially in synthesis of ammonia; nonionic detergents for dry cleaning wool treatment, emulsion paints, polishes, agricultural sprays; chemical intermediate; pharmaceuticals; corrosion inhibitor; rubber accelerator. Registered by EPA as an antimicrobial and fungicide (cancelled). Colorless clear liquid; mp = 10.5°; bp = 171°; d^{20} = 1.0180; freely soluble in H_2O, EtOH, MeOH, Me_2CO, $CHCl_3$, glycerin, slightly soluble in C_6H_6, Et_2O, C_7H_{16}. LD_{50} (rat orl) = 1720 mg/kg, (rat sc) = 1500 mg/kg, (rat iv) = 230 mg/kg, (rat ip) = 70 mg/kg, (rat im) = 1750 mg/kg, (rbt orl) = 1000 mg/kg, (mus orl) = 700 mg/kg, (mus ip) = 50 mg/kg, gpg orl) = 620 mg/kg. TLV:TWA 3 ppm. *BP Chem.; OxyChem; Sigma-Aldrich Fine Chem.; Texaco; Union Carbide Corp.*

1597 **Ethanolaminedi(methylenephosphonic acid)**
5995-42-6   H

$C_4H_{13}NO_7P_2$
2-Hydroxyethyliminodimethanephosphonic acid.
EINECS 227-833-4; Ethanolaminedi(methylenephos-phonic acid); (((2-Hydroxyethyl)imino)bis(methylene))-bisphosphonic acid; 2-Hydroxyethylbis(phosphono-methyl)amine; 2-Hydroxyethyliminobis(methylene phos-phonic acid); 2-Hydroxyethyliminodimethanephosphonic acid; Phosphonic acid, (((2-hydroxyethyl)imino)di-methylene)di-; Phosphonic acid, (((2-hydroxyethyl)-imino)bis(methylene))bis-; Wayplex 61A.

1598 **Ethenediyl diphenylene bisbenzoxazole**
1533-45-5   G

$C_{28}H_{18}N_2O_2$
2,2'-(1,2-Ethenediyldi-4,1-phenylene)bisbenzoxazole.
Benzoxazole, 2,2'-(1,2-ethenediyldi-4,1-phenylene)bis-; Eastobrite® OB-1; EINECS 216-245-3; 2,2'-(1,2-Ethene-diyldi-4,1-phenylene)bisbenzoxazole; 2,2'-(Vinylenedi-4-phenylene)bis(benzoxazole). Optical brightener; fluor-escent whitening agent for use in linear polyester, PET, nylon fibers. Solid; d = 1.39. *Eastman Chem. Co.*

1599 **Ethephon**
16672-87-0   3767   G P

$C_2H_6ClO_3P$
2-Chloroethyl phosphonic acid.
3-04-00-01780 (Beilstein Handbook Reference); Acide chloro-2-ethyl-phosphonique; Amchem 68-250; BRN 1751208; Bromeflor; Bromoflor; Camposan; Caswell No. 426A; CEP; CEPA; 2-Cepa; Cepha; Cepha 10LS; Cerone; Chipco Florel PRO; 2-Chloroethyl-phosphonsaeure; 2-Chloräthyl-phosphonsaeure; 2-Chloroethanephosphonic acid; Chlorethephon; Chloroethylphosphonic acid; (2-Chloroethyl)phosphonic acid; 2-Chloroethylphosphonic acid; EINECS 240-718-3; EPA Pesticide Chemical Code 099801; Ethefon; Ethel; Ethephon; Ethepon; Etheverse; Ethrel; Ethrel C; Flordimex; Florel; Florel Plant Growth; Floridex; G 996; Gagro; HSDB 2618; Kamposan; Phosphonic acid, (2-chloroethyl)-; Prep; Roll-Fruct; Terpal; Tomathrel. Plant growth regulator used for winter barley. Manufactured by Union Carbide. Registered by EPA as a herbicide and plant growth regulator. Hygroscopic needles; mp = 74 - 75°; freely soluble in H_2O (100 g/100 ml), EtOH, Me_2CO, glycols, slightly soluble in C_6H_6, C_7H_8, insoluble in petroleum ether; LD$_{50}$ (mus orl) = 2850 mg/kg, 4229 mg/kg, (rbt der) = 5730 mg/kg, (bobwhite quail orl) =1000 mg/kg; LC$_{50}$ (mallard duck 8 day dietary) > 10000 mg/kg diet, (bluegill sunfish 96 hr.) = 300 mg/l, (rainbow trout 96 hr.) = 350 mg/l; non-toxic to bees. *A H Marks & Co LTD; Aventis Crop Science; Embetec Crop Protection Ltd.; ICI Agrochemicals; ICI Chem. & Polymers Ltd.*

1600 **Ethion**
563-12-2   3772   G P V

$C_9H_{22}O_4P_2S_4$
O,O,O',O'-Tetraethyl-S,S'-methylene di(phosphorodithio-ate).
Ethanox; Ethiol; FMC 1240; Hylemox; Rhodiacide; Rhodocide; RP-Thion; Vegfru-Fosmite. Has both acaricidal and insecticidal properties; its acaricidal action is widely used in the abatement of cattle ticks; as a non-systemic insecticide it is used on citrus, deciduous fruits, tea, cotton and ornamental plants. Liquid; mp = -13°; bp0.3 = 165°; d^{20} = 1.22; slightly soluble in H_2O, more soluble in organic solvents; LD$_{50}$ (rat orl) = 208 mg/kg. *A/S Cheminova.*

1601 **Ethofumesate**
26225-79-6   3778   G P

$C_{13}H_{18}O_5S$
3-Ethoxy-2,3-dihydro-3,3-dimethyl-5-benzofuranyl methanesulfonate.
5-Benzofuranol, 2-ethoxy-2,3-dihydro-3,3-dimethyl-, methanesulfonate, (±)-; BRN 5759730; Caswell No. 427BB; CR 14658; 2-Ethoxy-2,3-dihydro-3,3-dimethyl-5-benzofuranyl methanesulfonate; (±)-2-Ethoxy-2,3-dihydro-3,3-dimethyl-5-benzofuranol methanesulfonate; (±)-2-Ethoxy-2,3-dihydro-3,3-dimethylbenzofuran-5-yl methanesulphonate; EPA Pesticide Chemical Code 110601; Ethofumesate; NC 8438; Nortron; Nortron (new); Progress; Tramat. Herbicide used for weed control in field crops. Registered by EPA as a herbicide. White crystalline solid; mp = 70 - 72°; soluble in H_2O (11.0 g/100 ml), more soluble in organic solvents; LD$_{50}$ (rat orl) = 1130 mg/kg. *Agvalue Inc.; Aventis Crop Science; Schering Agrochemicals Ltd.*

1602 **Ethopabate**
59-06-3   3781   G V

$C_{12}H_{15}NO_4$
Methyl 4-acetamido-2-ethoxybenzoate.
4-Acetamido-2-methoxybenzoic acid methyl ester; Amprol Plus (veterinary); Benzoic acid, 4-(acetylamino)-2-ethoxy-, methyl ester; EINECS 200-414-3; Ethopabate; Ethyl pabate; pancoxin. Coccidiostat, used in poultry, often as Amprol Plus, a combination with amprolium. Crystals; mp = 148-149°; λ_m = 298, 267 nm (A$_{1 cm}^{1\%}$ 805, 365); slightly soluble in H_2O, more soluble in organic solvents.

1603 **Ethoxydiglycol acetate**
112-15-2   H

265

C8H16O4
 2-(2-Ethoxyethoxy)ethanol acetate.
 3-02-00-00308 (Beilstein Handbook Reference); Acetic acid 2-(2-ethoxyethoxy)ethyl ester; Al3-01953; BRN 1764643; Carbitol acetate; Diethylene glycol ethyl ether acetate; Diethylene glycol monoethyl ether acetate; Diglycol monoethyl ether acetate; EINECS 203-940-1; Ektasolve de acetate; Ethanol, 2-(2-ethoxyethoxy)-, acetate; Ethoxydiglycol acetate; Glycol ether de acetate; HSDB 5555; Karbitolacetat; NSC 8702. Liquid; mp = -25°; bp = 218.5°; d^{20} = 1.0096; very soluble in H2O, Me2CO, EtOH, Et2O.

1604   Ethoxydimethylsilane
14857-34-2   G

C4H12OSi
 Dimethylethoxysilane.
 4-04-00-03991 (Beilstein Handbook Reference); BRN 1731481; CD5635; Dimethylethoxysilane; EINECS 238-921-7; Ethoxydimethylsilane; Silane, ethoxydimethyl-. Coupling agent, chemical intermediate, blocking agent, release agent, lubricant, primer, reducing agent. Degussa-Hüls Corp.

1605   Ethoxyethanol acetate
111-15-9   3787   G H

C6H12O3
 2-Ethoxyethanol acetate.
 4-02-00-00214 (Beilstein Handbook Reference); Acetate d'ethylglycol; Acetate de cellosolve; Acetate de l'ether monoethylique de l'ethylene-glycol; Acetato di cello-solve; Acetic acid, 2-ethoxyethyl ester; Aethylenglykol-aetheracetat; Äthylenglykolätheracetat; Al3-01955; BRN 1748677; Cellosolve acetate; Celosolvacetat; Diethylene glycol ethyl ether acetate; EGEEA; EINECS 203-839-2; Ektasolve EE acetate solvent; Ethanol, 2-ethoxy-, acetate; Ethoxyethanol, 2-, acetate; Ethoxyethanol acetate; Ethoxy-ethyl acetate; Ethyl cellosolve acetaat; Ethyl cellosolve acetate; Ethylene glycol ethyl ether acetate; Ethylglycol acetate; Ethylglykolacetat; 2-Etossietil-acetato; Glycol ether EE acetate; Glycol monoethyl ether acetate; HSDB 539; NSC 8658; Octan etoksyetylu; Oxitol acetate; Oxytol acetate; Poly-Solv EE acetate; UN 1172. Solvent used in automobile lacquers; retards evaporation and imparts high gloss. Liquid; mp = -61,7°; bp = 156.4°; d^{20} = 0.9740; λ_m = 214 nm (ϵ = 83, EtOH); very soluble in H2O, Me2CO, EtOH, Et2O; LD50 (rat orl) = 2700 mg/kg. Allchem Ind.; Arco; Eastman Chem. Co.; OxyChem; Union Carbide Corp.

1606   Ethoxylated nonylphenol phosphate
51811-79-1   H

(C2H4O)x(C15H24O)y. H3O4P
 Nonylphenol, ethoxylated, phosphate ester.
 Ethoxylated nonylphenol phosphate; Nonylphenol, ethoxylated and phosphated; Nonylphenol, ethoxylated, phosphate ester; Phosphated, ethoxylated nonylphenol; Phosphoric ester of poly(oxyethylene) nonylphenol ether; Poly(oxy-1,2-ethanediyl), α-(nonylphenyl)-ω-hydroxy-, phosphate.

1607   Ethoxyquin
91-53-2   3789   G P

C14H19NO
 6-Ethoxy-1,2-dihydro-2,2,4-trimethylquinoline.
 Al3-17715; Alterungsschutzmittel EC; Amea 100; Antage AW; Antioxidant EC; Antox; Aries Antox; Caswell No. 427D; CCRIS 2513; Dawe's nutrigard; Dihydro-6-ethoxy-2,2,4-trimethylquinoline; EINECS 202-075-7; EMQ; EPA Pesticide Chemical Code 055501; EQ; Ethoxychin; Ethoxyquin; Ethoxyquine; HSDB 400; Niflex; Niflex D; Nix-Scald; Nocrac AW; Nocrack AW; NSC-6795; Permanax 103; Polyflex; Quinol ED; Quinoline, 6-ethoxy-1,2-dihydro-2,2,4-trimethyl-; Santoflex; Santoflex A; Santoflex AW; Santoquin; Santoquine; Stop-Scald; USAF B; USAF B-24. Used as an antioxidant in feed and food; antidegradation agent for rubber. Registered by EPA as a herbicide and plant growth regulator. Yellow liquid; bp2 = 123 - 125°; d^{25} = 1.026; LD50 (rat orl) = 1920 mg/kg, (mus orl) = 1730 mg/kg, (rat orl) = 1925 mg/kg. Cerexagri Decco Inc.; Naugatuck; Wrap Pack Inc.

1608   Ethoxytrimethylsilane
1825-62-3   G

C5H14OSi
 Trimethylethoxysilane.
 4-04-00-03994 (Beilstein Handbook Reference); BRN 1731950; CT2970; EINECS 217-370-6; Ethyl trimethylsilyl ether; EXP-51; NSC 43345; Silane, ethoxytrimethyl-; Silane, trimethylethoxy-; Trimethylethoxysilane; Trimethylsilyl ethyl ether. Coupling agent, blocking agent, release agent, lubricant, primer, reducing agent. Chemical intermediate, useful for blocking hydroxyl or amino groups in order to perform reactions on multifunctional organic compounds or polymers; also for deactivating glass surfaces used in gas chromatographic applications. Liquid; bp = 76°; d^{20} = 0.7573; insoluble in H2O, soluble in EtOH, Et2O, Me2CO. Degussa-Hüls Corp.; Genesee Polymers.

1609   Ethyl Acetate
141-78-6   3792   G P

C4H8O2
 Ethyl acetic ester.
 Äthylacetat; Acetate d'ethyle; Acetato de etilo; Acetic acid, ethyl ester; Acetic ester; Acetic ether; Acetidin; Acetoxyethane; Aethylacetat; Al3-00404; Caswell No. 429; CCRIS 6036; EINECS 205-500-4; EPA Pesticide Chemical Code 044003; Essigester; Ethyl acetate; Ethyl acetic ester; Ethyl ester; Ethyl ethanoate; Ethylacetaat; Ethyle (acetate d'); Ethylester kyseliny octove: Etile (acetato di); EtOAc; FEMA No. 2414; HSDB 83; NSC 70930; Octan etylu; RCRA waste number U112; UN1173; Vinegar naphtha. A general solvent in coatings and plastics, organic synthesis,

smokeless powders, artificial leather, photographic films and plates, pharmaceuticals, synthetic fruit essences; cleaning textiles. Registered by EPA as an antimicrobial, herbicide and insecticide (cancelled). FDA, FEMA GRAS, Japan approved with restrictions, FDA approved for ophthalmics, orals, topicals, USP/NF, BP, Ph.Eur. compliance. Used in ophthalmics, topicals and orals as a pineapple-like flavoring agent and solvent. Colorless liquid with a pleasant, fruity odor detectable at 7 to 50 ppm; mp = -83.6; bp = 77.1°; d^{20} = 0.9003; λ_m = 209 nm (ϵ = 72, MeOH); soluble in H_2O (8 g/100 ml), very soluble in Me_2CO, C_6H_6, freely soluble in EtOH, Et_2O, $CHCl_3$; LD_{50} (rat orl) = 5620 mg/kg, TLV = 400 ppm in air. *Allchem Ind.; Berje; BP Chem.; Brown; Chisso Corp.; Daicel (U.S.A.) Inc.; Eastman Chem. Co.; Hoechst Celanese; Hüls Am.; Lonzagroup; Mallinckrodt Inc.; Monsanto Co.; Penta Mfg.; Sigma-Aldrich Fine Chem.; Tokuyama Petrochem.; Union Carbide Corp.*

1610   Ethyl acetoacetate

141-97-9   3793   H

$C_6H_{10}O_3$
  Ethyl β-ketobutyrate.
  Acetoacetic acid, ethyl ester; Acetoctan ethylnaty; Active acetyl acetate; AI3-00066; Butanoic acid, 3-oxo-, ethyl ester; CCRIS 1343; Diacetic ether; EAA; EINECS 205-516-1; Ethyl β-ketobutyrate; Ethyl 3-oxobutanoate; Ethyl 3-oxobutyrate; Ethyl acetoacetate; Ethylester kyseliny acetoctove; FEMA No. 2415; HSDB 402; NSC 8657. Liquid; mp = -45°; bp = 180.8°; d^{10} = 1.0368; λ_m = 243 nm (cyclohexane); slightly soluble in CCl_4, soluble in $CHCl_3$, C_6H_6, very soluble in H_2O, freely soluble in EtOH, Et_2O.

1611   Ethyl acrylate

140-88-5   3794   H

$C_5H_8O_2$
  Ethyl 2-propenoate.
  4-02-00-01460 (Beilstein Handbook Reference); Acrylate d'ethyle; Acrylic acid, ethyl ester; Acrylsäureäthylester; Äthylacrylat; AI3-15734; Akrylanem etylu; BRN 0773866; Carboset 511; CCRIS 248; EINECS 205-438-8; Ethoxycarbonylethylene; Ethyl 2-propenoate; Ethyl acrylate; Ethyl propenoate; Ethylacrylaat; Ethylakrylat; Ethylester kyseliny akrylove; Etil acrilato; Etilacrilatului; FEMA Number 2418; HSDB 193; NCI-C50384; NSC 8263; RCRA waste number U113; UN1917. Liquid; mp = -71.2°; bp = 99.4°; d^{20} = 0.9234; λ_m = 208 nm (ϵ = 6918, EtOH); slightly soluble in H_2O, DMSO, soluble in $CHCl_3$, freely soluble in EtOH, Et_2O.

1612   Ethyl aluminum sesquichloride

12075-68-2   H

$C_6H_{15}Al_2Cl_3$
  Trichlorotriethyldialuminum.
  Aluminum, dichloroethyl-, mixt with chlorodiethyl-aluminum; Aluminum, trichlorotriethyldi-; EINECS 235-137-7; Ethyl aluminum sesquichloride; HSDB 2013; Sesquiethylaluminum chloride;

Trichlorotriethyldi-aluminium; Triethylaluminum sesquichloride; Triethyl-dialuminium trichloride; Triethyldialuminum trichloride; Triethyltrichlorodialuminum.

1613   Ethylamine

75-04-7   3797   H

C_2H_7N
  Ethane, amino-.
  Aethylamine; Äthylamine; AI3-24228; Aminoethane; CCRIS 6261; EINECS 200-834-7; Ethanamine; Ethylamine; Etilamina; Etyloamina; HSDB 803; Mono-ethylamine; UN1036; UN2270. Used as a stabilizer for rubber latex; an intermediate in dyestuffs manufacture, in manufacture of medicinals and in chemical synthesis. Gas; mp = -80.5°; bp = 16.5°; d^{25} = 0.677; very soluble in H_2O and organic solvents; LD_{50} (rat orl) = 400 mg/kg. *ICI; ICI Spec.; Union Carbide Corp.*

1614   Ethylaniline

103-69-5   3800   H

$C_8H_{11}N$
  N-Ethylaniline.
  4-12-00-00250 (Beilstein Handbook Reference); Aethylanilin; Äthylanilin; AI3-15346; Aniline, N-ethyl-; Anilinoethane; Benzenamine, N-ethyl-; BRN 0507468; CCRIS 4641; EINECS 203-135-5; Ethylaniline; Ethylphenylamine; HSDB 5354; N-Ethyl-N-phenylamine; N-Ethylaminobenzene; N-Ethylaniline; N-Ethylbenzen-amine; N-Ethylbenzenamino; N-Ethylbenzeneamino; NSC 8736; UN2272. Liquid; mp = -63.5°; bp = 203°; d^{20} = 0.9625; λ_m = 246, 295 nm (MeOH); insoluble in H_2O, soluble in CCl_4, very soluble in Me_2CO, C_6H_6, freely soluble in EtOH, Et_2O.

1615   Ethylbenzene

100-41-4   3801   H

C_8H_{10}
  Benzene, ethyl-.
  Aethylbenzol; Äthylbenzol; AI3-09057; Benzene, ethyl-; CCRIS 916; EB; EINECS 202-849-4; Ethyl benzene; Ethylbenzeen; Ethylbenzene; Ethylbenzol; Etilbenzene; Etylobenzen; HSDB 84; NCI-C56393; NSC 406903; Phenylethane; UN1175. Liquid; mp = -94.9°; bp = 136.1°; d^{20} = 0.8670; λ_m = 208, 255, 261, 270 nm (ϵ = 7520, 168, 200, 142, MeOH); insoluble in H_2O, slightly soluble in $CHCl_3$, freely soluble in EtOH, Et_2O.

1616   p-Ethylbenzaldehyde

4748-78-1   G

$C_9H_{10}O$
  4-Ethylbenzaldehyde.

Benzaldehyde, 4-ethyl-; Benzaldehyde, p-ethyl-; Ebal; EINECS 225-268-8; 4-Ethylbenzaldehyde; Ethylbenzalde-hyde, p-; FEMA No. 3756. Additive for resins; intermediate for pharmaceuticals, fragrances. Liquid; bp = 221°; d^{20} = 0.9790. *Mitsubishi Gas*.

1617 Ethylbutylcarbinol
589-82-2   G

$C_7H_{16}O$
sec-Heptyl alcohol.
4-01-00-01741 (Beilstein Handbook Reference); AI3-21994; BRN 1719067; Butyl ethyl carbinol; EINECS 209-661-1; Ethyl butyl carbinol; FEMA No. 3547; 3-Heptanol; 3-Hydroxyheptane; NSC 2586. Chemical intermediate and solvent. Liquid; mp = -70°; bp_{20} = 66°; d = 0.8180; slightly soluble in H_2O, more soluble in organic solvents; LD_{50} (rat orl) = 1870 mg/kg.

1618 Ethyl cadmate
14239-68-0   G

$C_{10}H_{20}CdN_2S_4$
Cadmium, bis(diethylcarbamodithioato-S,S')-, (T-4)-.
AI3-14696; Bis(diethyldithiocarbamato)cadmium; Cad-mate; Cadmium, bis(diethylcarbamodithioato-S,S')-, (β4)-; Cadmium, bis(diethylcarbamodithioato-S,S')-, (T-4)-; Cadmium, bis(diethyldithiocarbamato)-; Cadmium bis(diethyldithiocarbamate); Cadmium diethyldithio-carbamate; Carbamic acid, diethyldithio-, cadmium salt; Cd diethyldithiocarbamate; Diethyldithiocarbamic acid, cadmium salt; EINECS 238-113-4; Hoernesite; HSDB 2941; Natural roesslerite; NSC 154470. Activated cadmium diethyldithiocarbamate; primary accelerator for NR and synthetic rubbers; used with a thiazole; gives heat resistance and low compression set properties. *Vanderbilt R.T. Co. Inc.*

1619 Ethyl chloroacetate
105-39-5   3818   H

$C_4H_7ClO_2$
Chloroacetic acid, ethyl ester.
4-02-00-00481 (Beilstein Handbook Reference); Acetic acid, chloro-, ethyl ester; AI3-19743; BRN 0506455; CCRIS 7747; Chloroacetic acid, ethyl ester; EINECS 203-294-0; Ethyl chloracetate; Ethyl chloroacetate; Ethyl α-chloroacetate; Ethyl chloroethanoate; Ethyl monochloracetate; Ethylester kyseliny chloroctove; HSDB 408; NSC 8833; UN1181. Liquid; mp = -21°; bp = 144.3°; d^{20} = 1.1585; insoluble in H_2O, soluble in $CHCl_3$, C_6H_6, very soluble in EtOH, Et_2O, Me_2CO.

1620 Ethyl chloroformate
541-41-3   3819   H

$C_3H_5ClO_2$
Chloroformic acid ethyl ester.
4-03-00-00023 (Beilstein Handbook Reference); AI3-19852; BRN 0385653; Carbonochloridic acid, ethyl ester; Cathyl chloride; Chlorameisensäureäthylester; Chloro-carbonate d'ethyle; Chlorocarbonic acid ethyl ester; Chloroformic acid ethyl ester; Cloroformiato de etilo; ECF; EINECS 208-778-5; Ethoxycarbonyl chloride; Ethyl carbonochloridate; Ethyl chlorocarbonate; Ethyl chloro-formate; Ethyl chloromethanoate; Ethylchloorformiaat; Ethyle, chloroformiat d'; Ethylester kyseliny chlor-mravenci; Etil clorocarbonato; Etil cloroformiato; Formic acid, chloro-, ethyl ester; HSDB 409; TL 423; UN1182. Liquid; mp = -80.6°; bp = 95°; very soluble in C_6H_6, Et_2O. $CHCl_3$.

1621 Ethyl cyanoacetate
105-56-6   3821   H

$C_5H_7NO_2$
Cyanoacetic acid ethyl ester.
4-02-00-01889 (Beilstein Handbook Reference); Acetic acid, cyano-, ethyl ester; AI3-19027; BRN 0605871; Cyanacetate ethyle; Cyanoacetic acid ethyl ester; Cyanoacetic ester; EINECS 203-309-0; Estere cianoacetico; Ethyl cyanacetate; Ethyl cyanoacetate; Ethyl cyanoethanoate; Ethylester kyseliny kyanoctove; HSDB 2769; Malonic acid ethyl ester nitrile; NSC 8844; UN2666; USAF KF-25. Liquid; mp = -22.5°; bp = 205°; d^{20} = 1.0654; very soluble in EtOH, Et_2O.

1622 Ethyl cyanoacrylate
7085-85-0   H

$C_6H_7NO_2$
2-Propenoic acid, 2-cyano-, ethyl ester.
Acrylic acid, 2-cyano-, ethyl ester; CCRIS 1693; EINECS 230-391-5; Ethyl cyanoacrylate; Ethyl 2-cyanoacrylate; 2-Propenoic acid, 2-cyano-, ethyl ester.

1623 Ethyl cyanomethylcarbamate
60754-24-7   H

$C_5H_8N_2O_2$
Carbamic acid, cyanomethyl-, ethyl ester.
Carbamic acid, cyanomethyl-, ethyl ester; Ethyl cyanomethylcarbamate.

1624 Ethyl (dimethylamidino)methylcarbamate, hydrochloride
65206-90-8   H

$C_7H_{16}ClN_3O_2$

Carbamic acid, (aminoiminomethyl)methyl-, dimethyl deriv., ethyl ester, monohydrochloride.

Carbamic acid, (aminoiminomethyl)methyl-, dimethyl deriv., ethyl ester, monohydrochloride; EINECS 265-620-8; Ethyl (dimethylamidino)methylcarbamate, hydro-chloride; Ethyl amidinomethylcarbamate monohydro-chloride, dimethyl derivative.

1625 Ethyl 4-dimethylaminobenzoate
10287-53-3   G

$C_{11}H_{15}NO_2$

Ethyl-p-dimethyl aminobenzoate.

Benzoic acid, 4-(dimethylamino)-, ethyl ester; EINECS 233-634-3; Ethyl 4-dimethylaminobenzoate; Parbenate; Speedcure EDB. UV photoinitiator. Crystals; mp = 63-65°; d = 1.0610; insoluble in H_2O, soluble in organic solvents. *Aceto Corp.; Lambson Ltd.*

1626 Ethyl (diphenylmethylene)cyanoacetate;
5232-99-5   D G

$C_{18}H_{15}NO_2$

2-Propenoic acid, 2-cyano-3,3-diphenyl-, ethyl ester.

4-09-00-03640 (Beilstein Handbook Reference); Acrylic acid, 2-cyano-3,3-diphenyl-, ethyl ester; Acrylonitrile, β, β-biscyclopropyl-α-carbethoxy-; Acrylonitrile, 3,3-di-cyclopropyl-2-(ethoxycarbonyl)-; BRN 1885803; CE 2; α-Carbethoxy-β,β-biscyclopropyl acrylonitrile; α-Cyano-β-phenylcinnamic acid, ethyl ester; EINECS 226-029-0; Ethyl (diphenylmethylene)cyanoacetate; Ethyl α-cyano-β,β-diphenylacrylate; Ethyl 2-cyano-3,3-diphenyl-2-prop-enoate; Ethyl 2-cyano-3,3-diphenyl-acrylate; Etocrilene; Etocrileno; Etocrilenum; Etocrylene; NSC 52678; 2-Propenoic acid, 2-cyano-3,3-diphenyl-, ethyl ester; USAF A-15972; UV Absorber-2; Uvinul N-35; Uvinul® N 35. Noncolor-contributing UV absorber; does not contain aromatic hydroxyl groups; effective under varying pH conditions; for NC lacquers and PVC; used in alkaline systems such as urea-formaldehyde and epoxyamine formulations, and in cosmetics. Crystals; mp = 110.5°. *BASF Corp.*

1627 Ethylene
74-85-1   3825   H

C_2H_4

Ethene.

Acetene; Athylen; Bicarburretted hydrogen; Caswell No. 436; EINECS 200-815-3; Elayl; EPA Pesticide Chemical Code 041901; Ethene; Ethylene; Etileno; HSDB 168; Liquid ethylene; Olefiant gas; UN1038; UN1962. Registered by EPA as a herbicide. Colorless gas; mp = -169°; bp_{700} = -102°; d^0 = 1.260 g/l gas; gas is soluble in H_2O (0.25 v/v), EtOH (2 v/v), Et_2O (0.2 v/v); soluble in Me_2CO, C_6H_6; LC (mus ihl) = 960,000 ppm. *Air Liquide America Corp.; Air Products & Chemicals Inc.; Livingston Group Inc.; Permviro Systems Inc.; Praxair Inc.*

1628 Ethylene bis(bis(2-chloroethyl)phosphate)
33125-86-9   H

$C_{10}H_{20}Cl_4O_8P_2$

Phosphoric acid, ethylene tetrakis(2-chloroethyl) ester.

EINECS 251-384-3; Ethylene bis(bis(2-chloroethyl)-phosphate); Phosphoric acid, 1,2-ethanediyl tetrakis(2-chloroethyl)ester; Phosphoric acid, ethylene tetrakis(2-chloroethyl) ester.

1629 Ethylenebis(chlorodimethylsilane)
13528-93-3   G

$C_6H_{16}Cl_2Si_2$

1,1,4,4-Tetramethyldichlorodisilethylene.

CT2015; EINECS 236-871-0; Ethylenebis(chlorodimethyl-silane); Silane, 1,2-ethanediylbis(chlorodimethyl-; 1,1,4,4-Tetramethyl-1,4-dichlorodisilethylene. Coupling agent, chemical intermediate, blocking agent, release agent, lubricant, primer, reducing agent. *Degussa-Hüls Corp.*

1630 Ethylene bis(tetrabromophthalimide)
32588-76-4   G

$C_{18}H_4Br_8N_2O_4$

2,2'-(1,2-Ethanediyl)bis(4,5,6,7-tetrabromo-1H-isoindole-1,3(2H)-dione)

1,2-Bis(tetrabromophthalimide)ethane; BT 93; BT-93D; BT 93W; CCRIS 6188; Citex BT 93; EINECS 251-118-6; 2,2'-(1,2-Ethanediyl)bis(4,5,6,7-tetrabromo-1H-isoindole-1,3(2H)-dione); N,N'-Ethylenebis(3,4,5,6-tetrabromo-phthalimide); 1H-Isoindole-1,3(2H)-dione, 2,2'-(1,2-ethanediyl)bis(4,5,6,7-tetrabromo-]; Phthalimide, N,N'-ethylenebis(tetrabromo-; Saytex® BT 93; Saytex® BT 93W. Flame retardant for high-impact PS, polyethylene, PP, thermoplastic polyesters, nylon, EPDM, rubbers, PC, ethylene copolymers, ionomer resins, textile treatment. Solid; mp = 446°; SG = 2.67;

insoluble in H$_2$O (<1 mg/ml). *Ethyl Corp.*

1631 Ethylene brassylate
105-95-3
G

C$_{15}$H$_{26}$O$_4$
Ethylene undecane dicarboxylate.
4-19-00-01936 (Beilstein Handbook Reference); Al3-24589; Astratone; BRN 0212344; Cyclo-1,13-ethylenedioxytridecan-1,13-dione; 1,4-Dioxacyclohepta-decane-5,17-dione; EINECS 203-347-8; Emeressence 1150; Ethyl brassylate; Ethylene brassylate; Ethylene glycol brassylate, cyclic diester; Ethylene glycol, cyclic tridecanedioate; Ethylene undecane dicarboxylate; FEMA No. 3543; Musk T; NSC 46155; Tridecanedioic acid, cyclic ethylene ester; 1,1'-Undecanedicarboxylic acid, ester with ethylene glycol. Musk chemical for fragrance of odor masking applications. *Henkel/Emery.*

1632 Ethylene carbonate
96-49-1
G H

C$_3$H$_4$O$_3$
1,3-Dioxolan-2-one.
Al3-18365; Carbonic acid, cyclic ethylene ester; CCRIS 293; Cyclic ethylene carbonate; Dioxolone-2; EINECS 202-510-0; Ethylene carbonate; Ethylene carbonic acid; Ethylene glycol carbonate; Ethylene glycol, cyclic carbonate; Ethylenester kyseliny uhlicite; Glycol carbonate; HSDB 6803; NSC 11801; Texacar® EC. Solvent for organic and inorganic material; EPA Rule 66 exempt; also used as reactant and plasticizer in fibers and textiles, plastics and resins, aromatic hydrocarbon extraction, electrolytes, hydraulic brake fluids. Plates; mp = 36.4°; bp = 248°; d^{39} = 1.3214; freely soluble in H$_2$O, EtOH, Et$_2$O, C$_6$H$_6$, CHCl$_3$, EtOAc, AcOH. *Texaco.*

1633 Ethylenediamine
107-15-3
3829
G P

C$_2$H$_8$N$_2$
1,2-Diaminoethane.
4-04-00-01166 (Beilstein Handbook Reference); Äthaldiamin; Äthylenediamin; Aethaldiamin; Aethyl-enediamin; Al3-24231; Algicode 106L; Amerstat 274; β-Aminoethylamine; Aminophylline Injection; BRN 0605263; Caswell No. 437; CCRIS 5224; 1,2-Diamino-ethaan; 1,2-Diamino-ethano; 1,2-Diaminoäthan; 1,2-Diaminoaethan; Diaminoethane; 1,2-Diaminoethane; Dimethylenediamine; EDA; EINECS 203-468-6; EN; EPA Pesticide Chemical Code 004205; Ethanediamine; Ethane-1,2-diamine; 1,2-Ethanediamine; 1,2-Ethylene-diamine; Ethyleendiamine; Ethylendiamine; Ethylene-diamine; Ethylenediamine; HSDB 535; NCI-C60402; UN1604. Used as a fungicide, in manufacture of chelating agents, dimethylolethylene-urea resins, chemical intermediate, solvent, emulsifier, textile lubricants, antifreeze inhibitor. Registered by EPA as an antimicrobial, fungicide and

herbicide (cancelled). Colorless clear liquid; mp = 8.5°; bp = 116 - 117°; d$_4^{25}$ = 0.898; freely soluble in H$_2$O, EtOH, less soluble in Et$_2$O, C$_6$H$_6$; LD$_{50}$ (rat orl) = 11600 mg/kg. *Allchem Ind.; BASF Corp.; Sigma-Aldrich Co.; Sigma-Aldrich Fine Chem.; Texaco; Union Carbide Corp.*

1634 Ethylenediamine bisstearamide
110-30-5
G H

C$_{38}$H$_{76}$N$_2$O$_2$
N,N'-Distearoylethylenediamine.
Abluwax EBS; Abril wax 10ds; Acrawax® C; Acrawax CT; Acrowax C; Advawachs 280; Advawax®; Advawax® 275; Advawax® 280; Advawax® 290; Al3-08515; Armowax EBS-P; 1,2-Bis(octadecanamido)ethane; Carlisle 280; Carlisle Wax 280; CCRIS 2293; Chemetron 100; EINECS 203-755-6; Ethylene distearamide; Ethylenebisstearoamide; Ethylenediamine bisstearamide; Glycowax® 765; HSDB 5398; Kemamide® W-39; Kemamide® W 40; Lubrol EA; Microtomic 280; N,N'-Distearoylethylenediamine; N,N'-Ethylene bisstearamide; N,N'-Ethylene distearylamide; Nopcowax 22-DS; NSC 83613; Octadecanamide, N,N'-1,2-ethanediylbis-; Plastflow; Stearic acid, ethylenediamine diamide; Uniwax 1760; Wax C. Internal and surface lubricant in resins and plastics; processing aid; plasticizer for resin; flow improver; pigment dispersant; used in hot-melt adhesives and coatings; powdered grade used as lubricant, processing aid, detackifier, mold release, and antiblocking agent. Synthetic wax used as plastics processing lubricant and release agent; melting point modifier for waxes and resin blends and industrial asphalt and tar; pigment dispersing agent for resin systems; paper-making defoamer; used in adhesive tapes, coatings, food packaging materials. Additive in pulp and paper defoamer formulations; lubricant, plasticizer, antistat, pigment dispersant for resins and plastics. High melting synthetic wax; binder, thickener for latex formulation, coatings, adhesives; used in powder metallurgy as internal lubricant. Solid; mp = 140-145°; insoluble in H$_2$O. *Armour Pharm. Co. Ltd.; CK Witco Chem. Corp.; Henkel Surface Tech; Hess & Clark Inc.; Lonzagroup; Morton Intn'l.; Rhône Poulenc Surfactants; Taiwan Surfactants; Unichema.*

1635 Ethylenediamine Tetraacetic acid, Di-sodium Salt
139-33-3
3543
G P

C$_{10}$H$_{14}$N$_2$Na$_2$O$_8$
Disodium dihydrogen ethylenediaminetetraacetate.
Al3-18049; CBC 50152966; CCRIS 3658; Cheladrate; Chelaplex III; Chelaton 3; Chelaton III; Chelest 200; Chelest B; Clewat N; Complexon III; Dinatrium ethylendiamintetraacetat; Diso-Tate; Disodium edathamil; Disodium edetate; Disodium edta, anhydrous; Disodium EDTA; Disodium salt of EDTA; Disodium sequestrene; Disodium tetracemate; Disodium versenate; Disodium versene; Dotite 2NA; DR-16133; E.D.T.A. disodique; Edathamil disodium;

Edetate disodium; EDTA disodium; EINECS 205-358-3; Endrate disodium; F 1; F 1 (complexon); Glycine, N,N'-1,2-ethanediylbis(N-(carboxymethyl)-, disodium salt; Kiresuto B; Komplexon III; Mavacid ED 4; Metaquest B; NSC 2760; Perma kleer 50 crystals disodium salt; Perma kleer di crystals; Selekton B 2; Sequestrene sodium 2; Sodium ethylenediaminetetraacetate; Sodium versenate; Tetracemate disodium; Titriplex III; Trilon BD; Triplex III; Veresene disodium salt; Versene Na2; Versonol 120; Zonon D. Food preservative, chelating and sequestering agent; anticoagulant; pharmaceutic aid. Chelating agent for use in mildly acidic dry formulations. Registered by EPA as an antimicrobial, fungicide, herbicide and insecticide (cancelled). [dihydrate]: Crystals; mp = 252° (dec); pH = 5.3; soluble in H_2O; LD_{50} (rat orl) = 2000 mg/kg. *3M Company; 3M Pharm.; BASF Corp.; Clough; Grace W.R. & Co.; Greeff R.W. & Co.; Hampshire.*

1636 Ethylenediamine, N-(3-(trimethoxysilyl)-propyl)-
1760-24-3   G H

C8H22N2O3Si
N-(3-Trimethoxysilylpropyl)-ethylenediamine.
A 0700; AAS-M; Aminoethylaminopropyltrimethoxy silane; AP 132; BRN 0636230; CA0700; Dow corning Z-6020 silane; Dynasylan® DAMO; EINECS 217-164-6; en-APTAS; Ethylenediamine, N-(3-(trimethoxysilyl)propyl)-; GF 91; KBM 603; NUCA 1120; Prosil® 3128; SH 6020; Silane, (3-(2-aminoethyl)aminopropyl)trimethoxy-; Sil-icone A-1120; Z 6020; N-(3-Trimethoxysilylpropyl)-ethylenediamine; Union Carbide® A-1120. Coupling agent for epoxies, phenolics, melamines, nylons, PVC, urethanes, acrylics. Chemical intermediate, blocking agent, release agent, lubricant, primer, reducing agent. Crosslinking agent, adhesion promoter for coatings; features active hydrogen reaction. Liquid; d = 1.0100; bp^{15} = 146°; LD_{50} (rat orl) = 7460 mg/kg. *Degussa-Hüls Corp.; Dow Corning; PCR.*

1637 Ethylenediamine, N-ethyl-N-m-tolyl-
19248-13-6   H

C11H18N2
1,2-Ethanediamine, N-ethyl-N-(3-methylphenyl)-.
N-(2-Aminoethyl)-N-ethyl-m-toluidine; EINECS 242-914-4; 1,2-Ethanediamine, N-ethyl-N-(3-methylphenyl)-; N-Ethyl-N-(β-aminoethyl)-m-toluidine; Ethylenediamine, N-ethyl-N-m-tolyl-; N-Ethyl-N-(m-tolyl)ethylenediamine; NSC 151043.

1638 Ethylenediaminetetraacetonitrile
5766-67-6   H

C10H12N6
N,N,N',N'-Tetracyanomethylethylenediamine.
4-04-00-02453 (Beilstein Handbook Reference); Aceto-nitrile, (ethylenedinitrilo)tetra-; Acetonitrile, 2,2',2'',2'''-(1,2-

ethanediyldinitrilo)tetrakis-; AI3-23663; BRN 1711285; EDTN; EINECS 227-290-3; Ethylenediamine-tetraacetonitrile; N,N,N',N'-Tetracyanomethyläthylene-diamin; N,N,N',N'-Tetracyanomethylethylenediamine; NSC 49104.

1639 Ethylenediaminetetracetic acid, ferric ammonium salt
21265-50-9   H

C10H16FeN3O8
Ferrate(1-), ((ethylenedinitrilo)tetraacetato)-, ammonium.
Ammonium ((N,N'-ethylenebis(N-(carboxymethyl)glycin-ato))(4-)-N,N',O,O',ON,ON')ferrate(1-); EINECS 244-302-2; Ethylenediaminetetracetic acid, ferric ammonium salt; Ferrate(-1), ((N,N'-1,2-ethanediylbis(N-(carboxymethyl)-glycinato))(4-)-N,N',O,O',ON,ON')-, ammonium; Ferrate-(1-), ((ethylenedinitrilo)tetraacetato)-, ammonium.

1640 Ethylene dibromide
106-93-4   3830   H P

C2H4Br2
Ethylene dibromide.
4-01-00-00158 (Beilstein Handbook Reference); Äthylenbromid; Aadibroom; Aethylenbromid; AI3-15349; BRN 0605266; Bromofume; Bromuro di etile; Caswell No. 439; CCRIS 295; Celmide; DBE; Dibromoethane; Dibromoethane, 1,2-; α,β-Dibromo-ethane; α,ω-Dibromoethane; 1,2-Dibromäthan; 1,2-Dibromaethan; 1,2-Dibromoetano; 1,2-Dibromoethane; 1,2-Dibroom-ethaan; Dibromoethylene; Dibromure d'ethylene; Dowfume; Dowfume 40; Dowfume EDB; Dowfume W-8; Dowfume W-85; Dowfume W-90; Dowfume W-90D; Dowfume W-100; Dwubromoetan; E-D-Bee; Edabrom; EDB; EDB-85; EINECS 203-444-5; ENT 15,349; EPA Pesticide Chemical Code 042002; Ethane, 1,2-dibromo-; Ethylene bromide; Ethylene Bromide; Ethylene dibromide; 1,2-Ethylene dibromide; Fumo-gas; Garden; Glycol Bromide; Glycol Dibromide; HSDB 536; Iscobrome D; Kopfume; NCI-C00522; Nefis; Nephis; Pestmaster; Pestmaster edb-85; RCRA waste number U067; Sanhyuum; Soilbrom; Soilbrom-40; Soilbrom-85; Soilbrom-90; Soilbrom 90EC; Soilbrom-100; Soilfume; sym-dibromoethane; UN 1605; Unifume; W 85; Wubromoetan. Registered by EPA as an insecticide (cancelled). Liquid; mp = 9°; bp = 131 - 132°; d$_4^{25}$ = 2.172; soluble in EtOH, Et2O, H_2O (0.4 g/100 ml); LD_{50} (mus ip) = 220 mg/kg. *Sigma-Aldrich Co.*

1641 Ethylene dichloride
107-06-2   3831   G H P

C2H4Cl2
1,2-Dichloroethane.
Aethylenchlorid; Äthylenchlorid; AI3-01656; 1,2-Bichloroethane; Bichlorure d'ethylene; Borer sol; Brocide; Caswell No. 440; CCRIS 225; Chlorure d'ethylene; Cloruro di ethene; DCE; Destruxol borersol; Di-chlor-mulsion; Dichlor-Mulsion; Dichloremulsion; α,β-Di-chloroethane; 1,2-Dichloorethaan; 1,2-Dichlor-aethan; 1,2-Dichlor-äthan; 1,2-Dichlorethane; sym-Dichloro-ethane; Dichloro-1,2-ethane; Dichloroethylene; Dichlor-ure d'ethylene; Dutch liquid; Dutch oil;

271

EDC; EDC (halocarbon); EINECS 203-458-1; ENT 1,656; EPA Pesticide Chemical Code 042003; Ethane, 1,2-dichloro-; Ethane dichloride; Ethyleendichloride; Ethylene chloride; Ethylene dichloride; Freon 150; Glycol dichloride; HCC 150; HSDB 65; NCI-C00511; RCRA waste number U077; UN1184. Used in production of vinyl chloride, trichloroethylene, vinylidene chloride, trichloroethane; lead scavenger in gasoline; paint, varnish remover; metal degreasing; soaps; wetting/penetrating agents; organic synthesis; ore flotation; solvent; fumigant. Registered by EPA as an insecticide and fungicide (cancelled). Liquid; mp = -35.5°; bp = 83.5°; d_4^{20} = 1.2351; freely soluble in Et_2O, very soluble in EtOH, soluble in Me_2CO, C_6H_6, CCl_4, $CHCl_3$, slightly soluble in H_2O; LD_{50} (rat orl) = 770 mg/kg. *Albright & Wilson Americas Inc.; Albright & Wilson UK Ltd.; Ashland; BASF Corp.; BP Chem.; Ethicon Inc.; Georgia Gulf; Norsk Hydro AS; OxyChem; PPG Ind.; Sigma-Aldrich Co.*

1642 Ethylene dimethacrylate
97-90-5   H

$C_{10}H_{14}O_4$
2-Propenoic acid, 2-methyl-, 1,2-ethanediyl ester.
4-02-00-01532 (Beilstein Handbook Reference); Ageflex EGDM; Ageflex EGDMA; BRN 1776663; CCRIS 179; Diglycol dimethacrylate; EINECS 202-617-2; Ethanediol dimethacrylate; Ethyldiol metacrylate; Ethyldiol methacrylate; Ethylene dimethacrylate; Ethylene glycol bis(methacrylate); Ethylene glycol dimethacrylate; Ethylene methacrylate; Ethylenedimethyacrylate; Glycol dimethacrylate; HSDB 5313; Methacrylic acid ethylene ester; MFM-416; NSC 24166; Perkalink® 401; Sartomer SR 206; SR 206. Crosslinker and modifier of ABS, acrylic and PVC, ion exchange resins, encapsulation of smokeless powder, glaze coatings, dental polymers, paper processing aids, rubber modifier, adhesives, optical polymers, leather finishing, moisture barrier films; fiberglass-reinforced polyesters, emulsion polymerization. Co-agent to improve efficiency of peroxideinduced cross-linking of rubber; sensitizer for radiationcured compounds. Liquid; mp = -40°; bp = 260°; d^{20} = 1.053; soluble in H_2O, freely soluble in C_6H_6, EtOH, ligroin; LD_{50} (rat orl) = 3300 mg/kg. *Akzo Chemie; Rit-Chem; Sartomer.*

1643 Ethylene fluoride - hexafluoropropene copolymer
9011-17-0   G
(-CH2CF2-)x[-CF2CF(CF3)-]y
Poly(vinylidene fluoride-co-hexafluoropropylene).
Akeogard CO; 1,1-Difluoroethyl - hexafluoropropylene copolymer; 1,1-Difluoroethylene-1,1,2,3,3,3-hexafluoro-1-propene copolymer; 1,1-Difluoroethylene - hexafluoropropene copolymer; 1,1-Difluoroethylene - perfluoropropropene copolymer; Ethene, 1,1-difluoro-, polymer with 1,1,2,3,3,3-hexafluoro-1-propene; F 26; F 26L; Fluoroplast 26; Fluorovinylidene – hexafluoro-propylene copolymer; Ftorlon F 26; Ftoroplast 26; Ftoroplast 26L; Ftoroplast F 26; Hexafluoropropene - vinylidene fluoride copolymer; Hexafluoropropylene - vinylidene difluoride copolymer; Hexafluoropropylene - vinylidene fluoride polymer; Hylar 2800; Hylar FXH 6; KF 2000; KF 2300; KF polymer 2300; KF polymer T 2300; Kymar 1800; Kymar 2750; Kymar 2751; Kymar 2800; Kymar 2801; Kymar 2801F; Kymar 2812; Kymar 2822; Kymar 2850; Kymar 2850-04; Kymar 2950-05; Kymar Flex 2750; Kymar Flex 2751; Kymar Flex 2801-00; Kymar Flex 2801GL; Kymar Flex 2821; Kymar Flex 2850; 1-Propene, 1,1,2,3,3,3-hexafluoro-, polymer with 1,1-difluoroethene. Used for various applications especially in wire and cable jacketing. Tm = 140-145°. *Elf Atochem N. Am.*

1644 Ethylene Glycol
107-21-1   3832   G P

$C_2H_6O_2$
1,2-Dihydroxyethane.
146AR; 2-Hydroxyethanol; Äthylenglykol; Aethylen-glykol; AI3-03050; Caswell No. 441; CCRIS 3744; 1,2-Dihydroxyethane; Dowtherm SR 1; EINECS 203-473-3; EPA Pesticide Chemical Code 042203; 1,2-Ethanediol; Ethylene alcohol; Ethylene dihydrate; Ethylene glycol; Fridex; Glycol; Glycol alcohol; Glycol, ethylene-; HSDB 5012; Lutrol-9; M.E.G.; Macrogol 400 BPC; Monoethylene glycol; NCI-C00920; Norkool; NSC 93876; Ramp; Tescol; Ucar 17; Union Carbide XL 54 Type I De-icing Fluid; Zerex. Used as antifreeze in cooling and heating systems; in hydraulic brake fluids; industrial humectant; solvent in paints, plastics, inks; softening agent for cellophane; stabilizer; in explosives, alkyd resins, elastomers, synthetic fibers and waxes; asphalt. Registered by EPA as an antimicrobial, fungicide and insecticide (cancelled). FDA listed, approved for topicals, BP compliance. Used as a solvent in topicals, eardrops. Viscous liquid; mp = -13°; bp = 197.3°; d^{20} = 1.1088; freely soluble in H_2O, EtOH, Me_2CO, AcOH, soluble in Et_2O, $CHCl_3$, slightly soluble in C_6H_6; LD_{50} (rat orl) = 4700 mg/kg. *Ashland; BASF Corp.; Eastman Chem. Co.; Hoechst Celanese; Mitsui Petroleum; Mitsui Toatsu; Olin Mathieson; Shell; Sigma-Aldrich Fine Chem.; Spectrum Chem. Manufacturing; Texaco; Union Carbide Corp.*

1645 Ethylene glycol diphenyl ether
104-66-5   H

$C_{14}H_{14}O_2$
1,2-Diphenoxyethane.
AI3-00789; Benzene, 1,1'-(1,2-ethanediylbis(oxy))bis-; EINECS 203-224-9; Ethane, 1,2-diphenoxy-; Ethylene glycol diphenyl ether; NSC 6794. Leaflets; mp = 98°; bp_{12} = 182°; λ_m = 220, 270, 277 nm (ε = 18000, 3330, 2770, MeOH); insoluble in H_2O, slightly soluble in EtOH, soluble in Et_2O, $CHCl_3$.

1646 Ethylene glycol distearate
627-83-8   H

$C_{38}H_{74}O_4$
Octadecanoic acid, 1,2-ethanediyl ester.
Alkamuls® EGDS; EINECS 211-014-3; Elfan L 310; Emerest 2355; 1,2-Ethanediyl bis(octadecanoate); Ethylene distearate; Ethylene glycol dioctadecanoate; Ethylene glycol distearate; Ethylene stearate; Glycol distearate; NSC 6820; Octadecanoic acid, 1,2-ethanediyl ester; Rewopal® PG 280; Rita EDGS; Stearic acid, ethylene ester. A thickener, opacifier, pearlizing agent used in shampoos and cosmetic lotions. An intermediate, lubricant, emulsifier, emollient; for emulsion shampoos and foam baths. A viscosity builder for cosmetic systems. Leaflets; mp = 79°; bp^{20} =

241°; d^{78} = 0.8581; insoluble in H_2O, EtOH, very soluble in Et_2O, Me_2CO. *Rewo; Rhône Poulenc Surfactants; Rita.*

1647 Ethylene glycol ethyl ether
110-80-5   3786   G H

$C_4H_{10}O_2$
2-Methoxyethanol.
AI3-01236; Äthylenglykol-monoäthyläther; Bikanol E 1; CCRIS 2294; Cellosolve; Cellosolve solvent; Celosolv; Dowanol 8; Dowanol EE; EE Solvent; EGEE; EINECS 203-804-1; Ektasolve EE; Emkanol; Eter monoetilico del etilenglicol; Ethanol, 2-ethoxy-; Ether monoethylique de l'ethylene-glycol; β-Ethoxyethanol; Ethoxyethanol, 2-; 2-Ethoxyethanol; 2-Ethoxyethyl alcohol; Ethyl cellosolve; Ethyl ethylene glycol; Ethylene glycol ethyl ether; Ethylene glycol monoethyl ether; Ethylethylene glycol; Ethyl Icinol; Etoksyetylowy alkohol; Glycol monoethyl ether; HSDB 54; Hydroxy ether; Jeffersol EE; NCI-C54853; NSC 8837; Oxitol; Plastiazan 60; Poly-Solv® EE; RCRA waste number U359; RCRA waste number U227; Solvid; Solvulose; UN1171. Solvent for nitrocellulose, lacquer, lacquer thinners, dyeing and printing textiles, varnish removers, cleaning solutions. Used in protective coatings, inks, cleaning products, agricultural chemicals; aids wetting, penetration, and soil removal; as a coupling solvent, in brake fluids, hard-surface cleaners, leather dyeing, paints, coatings, printing inks, textile vat dyeing and printing, adhesives, antifreeze, floor waxes/polishes, insect repellents; solubilizer for dyes; plasticizer. Liquid; mp = -70°; bp = 135°; d^{20} = 0.9297; very soluble in Me_2CO, Et_2O, etOH, H_2O; LD_{50} (rat orl) = 2125 mg/kg, 3000 mg/kg. *Arco; Ashland; Eastman Chem. Co.; Olin Corp; OxyChem; Union Carbide Corp.*

1648 Ethylene glycol monopropyl ether
2807-30-9   G H

$C_5H_{12}O_2$
Ethanol, 2-propoxy-.
4-01-00-02379 (Beilstein Handbook Reference); AI3-30229; BRN 1731983; EINECS 220-548-6; Ektasolve® EP; Ethanol, 2-propoxy-; Ethylene glycol mono-n-propyl ether; HSDB 6499; n-Propoxyethanol; Propoxyethanol; Propyl cellosolve; n-Propyl Oxitol glycol. Slow evaporating solvent used in coatings; useful in waterborne coating systems; coupling solvent for resin/water systems; controls viscosity of waterborne resins; effective for NC, acrylic, epoxy, polyamide, and alkyd resins; retarder in coating systems. Liquid; mp = -70°; bp = 149°; d^{20} = 0.9112; soluble in H_2O, very soluble in EtOH, Et_2O; LD_{50} (rat orl) = 3089 mg/kg. *Ashland; Eastman Chem. Co.; Sigma-Aldrich Fine Chem.*

1649 Ethylene glycol monostearate
111-60-4   G H

$C_{20}H_{40}O_3$
2-Hydroxyethyl n-octadecanoate.
4-02-00-01222 (Beilstein Handbook Reference); Ablunol EGMS; Alkamuls® SEG; BRN 1794033; Clindrol SEG; EINECS 203-886-9; Emerest 2350; Empilan 2848; Ethylene glycol monostearate; Ethylene glycol stearate; Glycol monostearate; Glycol stearate; 2-Hydroxyethyl octadecanoate; 2-Hydroxyethyl stearate; Ivorit; Lipo EGMS; Monthybase; Monthyle; Octadecanoic acid, 2-hydroxyethyl

ester; Parastarin; Prodhybas N; Prodhybase ethyl; S 151; Sedetol; Stearic acid, 2-hydroxyethyl ester; Stearic acid, monoester with ethylene glycol; Tego-stearate; USAF KE-11. Opacifier and pearling agent for shampoos, creams, liquid hand soaps, liquid detergents; emulsion stabilizer, viscosity builder. Solid; mp = 60.5°; bp_3 = 189-191°; d^{60} = 0.8780; soluble in Et_2O, slightly soluble in EtOH. *Rhône Poulenc Surfactants; Taiwan Surfactants.*

1650 Ethylene Oxide
75-21-8   3836   G P

C_2H_4O
1,2-Epoxyethane.
Äthylenoxid; Aethylenoxid; AI3-26263; Amprolene; Anprolene; Anproline; Caswell No. 443; CCRIS 297; Dihydro-Oxyfume; Dihydrooxirene; Dimethylene oxide; E.O.; EINECS 200-849-9; ENT-26263; EPA Pesticide Chemical Code 042301; Epoxyethane; Ethene oxide; Ethox; Ethyleenoxide; Ethylene (oxyde d'); Ethylene oxide; Etilene (ossido di); ETO; Etylenu tlenek; FEMA No. 2433; HSDB 170; Merpol; NCI-C50088; Oxacyclopropane; Oxane; Oxidoethane; Oxiraan; Oxiran; Oxirane; Oxirene, dihydro-; Oxyfume; Oxyfume 12; Qazi-ketcham; RCRA waste number U115; T-Gas; UN 1040. Fungicide. Registered by EPA as an antimicrobial, fungicide, insecticide and rodenticide. Used in manufacture of ethylene glycol and higher glycols, surfactants, acrylonitrile, ethanolamines; petroleum demulsifier; fumigant; rocket propellant; industrial sterilant (medical plastic tubing); fungicide. Colorless gas; mp = -111°; bp = 10.7°; d^4_4 = 0.891, d^7_7 = 0.887, d^{10}_{10} = 0.882; soluble in H_2O, EtOH, Et_2O. *3M Company; Andersen Sterilizers Inc.; ARC Specialty Products; Balchem Corp.; Honeywell; Pennsylvania Engineering Co.; Praxair Inc.*

1651 Ethylene thiourea
96-45-7   3837   G

$C_3H_6N_2S$
2-Mercapto-2-imidazoline.
AI3-16292; Akrochem etu-22; Aperochem ETU-22; CCRIS 298; 4,5-Dihydro-2-mercaptoimidazole; 4,5-Dihydroimidazole-2(3H)-thione; EINECS 202-506-9; Ethylenethiourea; Ethylene thiourea; Ethylene thiourea; l'Ethylene thiouree; 1,3-Ethylene-2-thiourea; 1,3-Ethylenethiourea; ETU; HSDB 1643; Imidazole-2(3H)-thione, 4,5-dihydro-; Imidazolidinethione; Imidazoline, 2-mercapto-; Imidazoline-2-thiol; 2-Imidazoline-2-thiol; 2-Imidazolidinethione; Imidazoline-2(3H)-thione; Mercaptoimidazoline; 2-Merkaptoimidazolin; 2-Mercaptoimidazoline; Mercazin I; NA-22; NA-22-D; NCI-C03372; N,N'-Ethylenethiourea; Nocceler 22; Pennac CRA; Perkacit® ETU; RCRA waste number U116; Rhenogran etu; Rhodanin S 62; Robac 22; Rodanin S-62; Sanceller 22; Sodium-22 neoprene accelerator; Soxinol 22; Tetrahydro-2H-imidazole-2-thione; 2-Thiol-dihydro-glyoxaline; Thiourea, N,N'-(1,2-ethanediyl)-; Urea, 1,3-ethylene-2-thio-; USAF EL-62; Vulkacit NPV/C; Vulkacit NPV/C2; Warecure C.An accelerator in synthetic rubber productions and a degradation product of ethylenebisdithiocarbamate fungicides such as mancozeb, maneb, and zineb. Used in electroplating baths; as an intermediate for antioxidants, insecticides, fungicides, synthetic resins, vulcanization accelerators, dyes. Needles; mp = 203°; soluble in H_2O (0.1 g/100 ml), EtOH, insoluble in Et_2O, C_6H_6, $CHCl_3$; LD_{50} (rat orl) = 1832 mg/kg. *Akzo Chemie; Faesy & Besthoff Inc.; Ore & Chem.*

1652   O-Ethyl ethanephosphonochloridothioate
1497-68-3   H

C$_4$H$_{10}$ClOPS
Phosphonochloridothioic acid, ethyl-, O-ethyl ester.
EINECS 216-095-9; Ethanephosphonochloridothioic acid, O-ethyl ester; O-Ethyl ethanephosphonochloridothioate; Ethylthionophosphonic acid O-ethyl ester chloride; O-Ethyl ethylthiophosphonyl chloride; HSDB 5833; Phosphonochloridothioic acid, ethyl-, O-ethyl ester.

1653   Ethyl 3-ethoxypropionate
763-69-9   G H

C$_7$H$_{14}$O$_3$
Ethyl β-ethoxypropionate.
4-03-00-00697 (Beilstein Handbook Reference); AI3-03254; BRN 1751976; EINECS 212-112-9; Ethoxypropionic acid, ethyl ester; 3-Ethoxypropionic acid ethyl ester; Ethyl β-ethoxypropionate; Ethyl 3-ethoxypropanoate; Ethylester kyseliny 3-ethoxypropion-ove; NSC 8870; Propanoic acid, 3-ethoxy-, ethyl ester. Liquid; mp = -75°; bp = 166°, bp$_5$ = 48°; d^{20} = 0.9490; soluble in H$_2$O (1.6 g/100 ml), more soluble in organic solvents; LD$_{50}$ (rat orl) = 5 gm/kg.

1654   Ethyl hexanoic acid
149-57-5   G H

C$_8$H$_{16}$O$_2$
2-Ethylhexanoic acid.
3-10-00-01641 (Beilstein Handbook Reference); AI3-01371; BRN 1750468; Butylethylacetic acid; CCRIS 3348; EINECS 205-743-6; α-Ethylcaproic acid; Ethyl hexanoic acid; Ethyl hexanoic acid, 2-; Ethylhexoic acid; Hexanoic acid, 2-ethyl-; HSDB 5649; Kyselina 2-ethylkapronova; Kyselina heptan-3-karboxylova; NSC 8881. Paint and varnish drying agents (metallic salts); esters as plasticizers. mp = -59°; bp = 228°, bp$_{13}$ = 120°;; d^{25} = 0.9031; soluble in H$_2$O (2 g/l), Et$_2$O, CCl$_4$, slightly soluble in EtOH; LD$_{50}$ (rat orl) = 3 g/kg. *Ashland; BASF Corp.; Eastman Chem. Co.; Neste UK; Sigma-Aldrich Fine Chem.; Union Carbide Corp.*

1655   2-Ethylhexanoyl chloride
760-67-8   H

C$_8$H$_{15}$ClO
2-Ethylhexanoic acid chloride.
EINECS 212-081-1; 2-Ethylcaproyl chloride; 2-Ethylhexanoic acid

chloride; 2-Ethylhexanoyl chloride; Hexanoyl chloride, 2-ethyl-; NSC 87892. Liquid; bp$_{40}$ = 101°, bp$_{11}$ = 67°; d^{25} = 0.939.

1656   2-Ethylhexenal
26266-68-2   H

C$_8$H$_{14}$O
2-Ethyl-3-propylacrolein.
CCRIS 4644; EINECS 247-571-4; 2-Ethyl-3-propylacrolein; 2-Ethylhexenal; Hexenal, 2-ethyl-; PM 2015.

1657   2-Ethylhexyl chloroformate
24468-13-1   H

C$_9$H$_{17}$ClO$_2$
Carbonochloridic acid, 2-ethylhexyl ester.
Carbonochloridic acid, 2-ethylhexyl ester; EINECS 246-278-9; 2-Ethylhexyl chloroformate; UN2748.

1658   2-Ethylhexyl 10-ethyl-4,4-dimethyl-7-oxo-8-oxa-3,5-dithia-4-stannatetradecanoate
57583-35-4   H

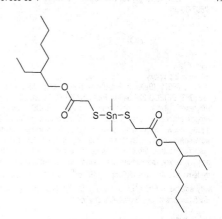

C$_{22}$H$_{44}$O$_4$S$_2$Sn
8-Oxa-3,5-dithia-4-stannatetradecanoic acid, 10-ethyl-4,4-dimethyl-7-oxo-, 2-ethylhexyl ester.
EINECS 260-829-0; 2-Ethylhexyl 10-ethyl-4,4-dimethyl-7-oxo-8-oxa-3,5-dithia-4-stannatetradecanoate; 8-Oxa-3,5-dithia-4-stannatetradecanoic acid, 10-ethyl-4,4-dimethyl-7-oxo-, 2-ethylhexyl ester.

1659   2-Ethylhexyl methacryate
688-84-6   H

C12H22O2

2-Propenoic acid, 2-methyl-, 2-ethylhexyl ester.
4-02-00-01528 (Beilstein Handbook Reference); AI3-03266; BRN 1769420; EINECS 211-708-6; 2-Ethylhexyl methacrylate; HSDB 5440; Methacrylate, 2-ethylisohexy; Methacrylic acid, 2-ethylhexyl ester; NSC 24173; NSC 32647; 2-Propenoic acid, 2-methyl-, 2-ethylhexyl ester. Liquid; bp_{18} = 120°, bp_{14} = 110°; d^{25} = 0.880.

1660 Ethylhexyl nitrate
27247-96-7   G H

C8H17NO3

Nitric acid, 2-ethylhexyl ester.
EINECS 248-363-6; Ethylhexyl nitrate; 2-Ethylhexyl nitrate; Exchem GO-1; Nitric acid, 2-ethylhexyl ester; Nitronal.

1661 Ethylhexyl palmitate
29806-73-3   H

C24H48O2

Hexadecanoic acid, 2-ethylhexyl ester.
AI3-31580; Ceraphyl 368; EINECS 249-862-1; Elfacos EHP; Ethylhexyl palmitate; 2-Ethylhexyl hexadecanoate; 2-Ethylhexyl palmitate; Hexadecanoic acid, 2-ethylhexyl ester; Octyl palmitate; Palmitic acid, 2-ethylhexyl ester; Unimate® EHP; Wickenol® 155. Used in sunscreens, antispersipirants, bath oils, liquid make-up; imparts gloss; binder for pressed powders; solubilizer for benzophenone-3. Emollient, moisturizer, pigment wetter/dispersant; increases water vapor porosity of fatty components used in cosmetic and topical pharmaceutical preparations. Skin emollient for lotions and creams. *CasChem; Inolex; Union Camp; Van Dyk.*

1662 2-Ethylhexyl phosphate
12645-31-7   G H

C8H21O5P

Phosphoric acid, esters with 2-ethylhexanol.
EINECS 235-741-0; 2-Ethylhexanol phosphate; 2-Ethylhexyl phosphate; Phosphated 2-ethyl hexanol; Phosphoric acid, 2-ethylhexyl ester; Phosphoric acid, 2-ethylhexyl esters; Phosphoric acid, esters with 2-ethylhexanol; Rhodafac® PEH. Detergent, dispersant, and wetting agent in textile wet processing. *Rhône Poulenc Surfactants.*

1663 2-Ethyl hexylamine
104-75-6   G H

C8H19N

2-Ethyl-n-hexylamine.
4-04-00-00766 (Beilstein Handbook Reference); AI3-26287; 1-Amino-2-ethylhexan; Armeen® L8D; BRN 1209249; CCRIS 4646; EINECS 203-233-8; 1-Hexanamine, 2-ethyl-; β-Ethylhexylamine; 2-Ethyl hexylamine; 2-Ethylhexanamine; 2-Ethylhexylamine; Hexylamine, 2-ethyl-; UN2276. A chemical intermediate for vapor phase corrosion inhibitors. *Akzo Chemie.*

1664 2-Ethyl-1,3-hexanediol
94-96-2   3780   G H P

C8H18O2

2-Ethyl-1,3-hexanediol.
4-01-00-02597 (Beilstein Handbook Reference); 6-12; 6-12-Insect repellent; AI3-00375; BRN 1735324; Carbide 6-12; Caswell No. 445; CCRIS 4034; Compound 6-12 insect repellent; Diol-Kyowa 8; EH diol; EHD; EINECS 202-377-9; ENT 375; EPA Pesticide Chemical Code 041001; Ethohexadiol; Ethyl hexanediol; Ethyl hexylene glycol; HSDB 1716; Latka 612; NSC 3881; Octylene glycol; Repellent 6-12; Rutgers 6-12. Insect repellent, cosmetics, vehicle and solvent in printing inks, medicine, chelating agent for boric acid. Registered by EPA as an insecticide (cancelled). Colorless, viscous liquid; mp = -40°; bp = 244°, bp_{50} = 163°, bp_{10} = 129°; bp_3 = 102°, $bp_{0.5}$ = 94 - 96°; d^{20}_{20} = 0.9422, d^{22}_4 = 0.9325; d^{22} = 0.9325; absolute viscosity = 271 cps; soluble in H2O (0.6 g/100 ml), EtOH, i-PrOH, propylene glycol, castor oil; LD_{50} (mrat orl) = 9260 mg/kg, (frat orl) = 4625 mg/kg. *Degussa-Hüls Corp.; Fluka; Union Carbide Corp.*

1665 Ethyl p-hydroxybenzoate
120-47-8   3869   D G

C9H10O3

4-hydroxybenzoic acid ethyl ester.
4-10-00-00367 (Beilstein Handbook Reference); AI3-30960; Aseptin A; Aseptoform E; Bonomold OE; BRN 1101972; Carbethoxyphenol; Caswell No. 447; Easeptol; EINECS 204-399-4; EPA Pesticide Chemical Code 061202; Ethyl butex; Ethyl paraben; Ethyl Parasept; Ethylparaben; HSDB 938; Mekkings E; Mycocten; Napagin A; Nipagin A; Nipazin A; NSC 23514; Sobrol A; Solbrol A; Tegosept E. Antifungal, antiseptic preservative used with pharmaceuticals. Crystals; mp = 117°; bp = 297.5°; λ_m = 256 nm (ε = 16400, MeOH); insoluble in CS2, slightly soluble in H2O, CHCl3, TFA, ligroin, very soluble in alcohol, Et2O, Et2O.

1666 Ethylidene acetate
542-10-9   3844   H

C6H10O4
1,1-Diacetoxyethane.
Al3-24218; 1,1-Diacetoxyethane; EINECS 208-800-3; 1,1-Ethanediol, diacetate; Ethylidene acetate; Ethylidene di(acetate); NSC 8852. Liquid; mp = 18.9°; bp = 169°; d^{25} = 1.3985; very soluble in EtOH, Et2O.

1667   2,2'-Ethylidenebis(4,6-bis(1,1-dimethyl-ethyl)-phenol
35958-30-6   G

C30H46O2
2,2'-Ethylidene bis(4,6-di-t-butylphenol).
2,2'-Ethylidenebis(4,6-di-t-butylphenol); EINECS 252-816-3; Phenol, 2,2'-ethylidenebis(4,6-bis(1,1-dimethyl-ethyl)-; Vanox® 1290. Antioxidant; oxidative inhibitor for polymers; process stabilizer for polyolefins; stabilizer for PU and PS. Crystals; mp = 162-164°. Vanderbilt R.T. Co. Inc.

1668   Ethylidenenorbornene
16219-75-3   H

C9H12
Bicyclo(2.2.1)hept-2-ene, 5-ethylidene-.
Bicyclo(2.2.1)hept-2-ene, 5-ethylidene-; BRN 2039935; CCRIS 4816; EINECS 240-347-7; 5-Ethylidene-2-norbornene; 5-Ethylidene-8,9,10-trinorborn-2-ene; 5-Ethylidenebicyclo(2.2.1)hept-2-ene; ENB; Ethylidene norbornene; Ethylidenenorbornene; HSDB 1160; 2-Norbornene, 5-ethylidene-.

1669   Ethyl isothiocyanate
542-85-8   3848   G

C3H5NS
Isothiocyanic acid, ethyl ester.
Al3-18428; CCRIS 7323; EINECS 208-831-2; Ethane, isothiocyanato-; Ethyl isothiocyanate; Ethyl mustard oil; Isothiocyanatoethane; NSC 84212. Used in chemical synthesis and as a military poison gas. Liquid; mp = -5.9°; bp = 131.5°; d^{20} = 0.9990; λ_m = 244 nm (MeOH); insoluble in H2O, freely soluble in EtOH, Et2O.

1670   Ethyl lactate
97-64-3   3850   G

C5H10O3
Ethyl L-(-)-lactate.
4-03-00-00643 (Beilstein Handbook Reference); Actylol; Acytol; Al3-00395; BRN 1209448; EINECS 202-598-0; Ethyl α-hydroxypropionate; Ethyl 2-hydroxypropanoate; Ethyl lactate; Ethylester kyseliny mlecne; Eusolvan; FEMA No. 2440; HSDB 412; 2-Hydroxypropanoic acid ethyl ester; Lactate d'ethyle; Lactic acid, ethyl ester; NSC 8850; Propanoic acid, 2-hydroxy-, ethyl ester; Solactol; UN1192. Used in manufacture of dipped latex products, coatings, specialty finishes, inks, diluents, adhesives, intermediates, photoresist solvents, screen printing of electronic parts, flavors and fragrances, solvent resins. mp = -26°; bp = 154°; d_4^{14} = 1.0420; $[\alpha]_D^{14}$ = -10°slightly soluble in H2O, soluble in organic solvents. Farleyway Chem. Ltd.; Penta Mfg.; Rit-Chem.

1671   Ethyl mercaptan
75-08-1   3761   H

C2H6S
Ethanethiol.
Aethanethiol; Aethylmercaptan; Al3-26618; EINECS 200-837-3; Etantiolo; Ethaanthiol; Ethanethiol; Ethyl hydrosulfide; Ethyl mercaptan; Ethyl sulfhydrate; Ethyl thioalcohol; Ethylmercaptaan; Ethylmerkaptan; Etilmerc-aptano; HSDB 814; LPG ethyl mercaptan 1010; Mercaptoethane; 1-Mercaptoethane; NSC 93877; Thio-ethanol; Thioethyl alcohol; UN2363. Used to impart odor to natural gas, in manufacture of pesticides, plastics and antioxidants. Volatile liquid; mp = -147.8°; bp = 35.1°; d^{25} = 0.8315; slightly soluble in H2O, more soluble in organic solvents. Atofina; Sigma-Aldrich Fine Chem.; U.S. BioChem.

1672   Ethyl methacrylate
97-63-2   G H

C6H10O2
2-Propenoic acid, 2-methyl-, ethyl ester.
4-02-00-01523 (Beilstein Handbook Reference); Al3-25421; BRN 0471201; CCRIS 4817; EINECS 202-597-5; Ethyl α-methylacrylate; Ethyl 2-methacrylate; Ethyl 2-methyl-2-propenoate; Ethyl 2-methylacrylate; Ethyl methacrylate; Ethylester kyseliny methakrylove; HSDB 1332; Methacrylic acid, ethyl ester; NSC 24152; RCRA waste number U118; Rhoplex AC-33; Rhoplex AC-33 (Rohm and Haas); UN2277. Used in manufacture of polymers and chemical intermediates. Liquid; mp = -75°; bp = 117°; d^{20} = 0.9135; soluble in CHCl3, H2O (0.4 g/100 ml), more soluble in EtOH, Et2O; LD50 (rat orl) = 14800 mg/kg. Rohm & Haas Co.; Rohm & Haas UK.

1673   N-Ethyl-2-methylallylamine
18328-90-0   H

276

C6H13N

Allylamine, N-ethyl-2-methyl-.

Allylamine, N-ethyl-2-methyl-; EINECS 242-217-5; N-Ethyl-2-methylallylamine; N-Ethyl-2-methyl-2-propen-1-amine; N-Ethylmethacrylamine; N-Ethylmethallylamine; N-Ethylmethylallylamine; 2-Propen-1-amine, N-ethyl-2-methyl-.

1674 N-Ethyl-3-methyl-2-butanamine
2738-06-9   H

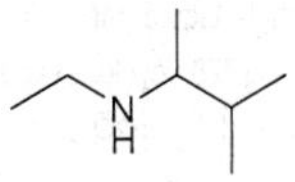

C7H17N

N-Ethyl-1,2-dimethylpropylamine.

2-Butanamine, N-ethyl-3-methyl-; N-Ethyl-1,2-dimethyl-propylamine; N-Ethyl-3-methyl-2-butanamine; Propyl-amine, N-ethyl-1,2-dimethyl-.

1675 N-Ethylheptadecafluoro-N-(2-hydroxy-ethyl)octanesulphonamide
1691-99-2   H

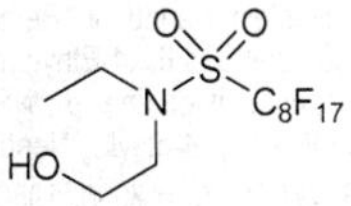

C12H10F17NO3S

N-Ethyl-1,1,2,2,3,3,4,4,5,5,6,6,7,7,8,8,8-heptadeca-fluoro-N-(2-hydroxyethyl)-1-octanesulfonamide.

N-Ethylheptadecafluoro-N-(2-hydroxyethyl)octane-sulphonamide; 1-Octanesulfonamide, N-ethyl-1,1,2,2,3, 3,4,4,5,5,6,6,7,7,8,8,8-heptadecafluoro-N-(2-hydroxy-ethyl)-; EINECS 216-887-4.

1676 Ethylmethyl imidazole
931-36-2   G

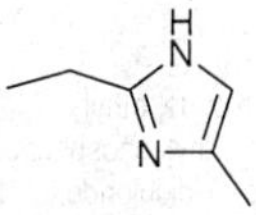

C6H10N2

2-Ethyl-4-methylimidazole.

EINECS 213-234-5; EMI-24; 2-Ethyl-4-methylimidazole; Imidazole, 2-ethyl-4-methyl-; 1H-Imidazole, 2-ethyl-4-methyl-; 4-Methyl-2-ethylimidazole; NSC 82315. A curing agent for epoxy resins used in low proportions thus improving chemical resistance. Crystals; mp = 36-42°; bp = 292-295°; d = 0.9750; soluble in H2O (18.0 g/100 ml), organic solvents. *Lancaster Synthesis Co.*

1677 (Ethyl(3-methylphenyl)amino)acetonitrile
63133-74-4   H

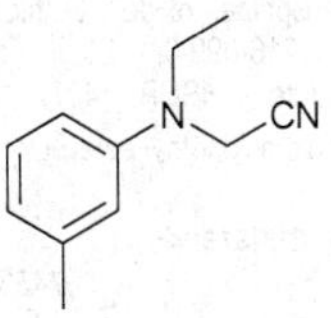

C11H14N2

Acetonitrile, (ethyl(3-methylphenyl)amino)-.

Acetonitrile, (ethyl(3-methylphenyl)amino)-; EINECS 263-891-7;

(Ethyl(3-methylphenyl)amino)acetonitrile.

1678 2-Ethyl-6-methylphenylazomethine
35203-06-6   H

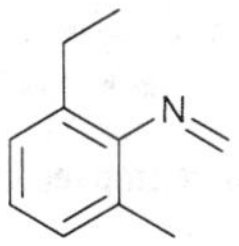

C10H13N

2-Ethyl-6-methyl-N-methylenebenzenamine.

Benzenamine, 2-ethyl-6-methyl-N-methylene-; EMA2O; Ethylmethylazomethine; 2-Ethyl-6-methylphenylazo-methine; 2-Ethyl-6-methyl-N-methylenebenzenamine; N-Methylene-6-ethyl-2-methylaniline.

1679 N-Ethyl morpholine
100-74-3   G

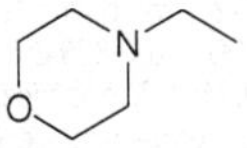

C6H13NO

4-Ethylmorpholine.

4-27-00-00023 (Beilstein Handbook Reference); AI3-24288; BRN 0102969; CCRIS 4818; EINECS 202-885-0; N-Ethylmorfolin; Ethylmorpholine; 4-Ethylmorpholine; N-Ethylmorpholine; HSDB 1644; Morpholine, 4-ethyl-; NEM; NSC 6110; Texacat® NEM; Toyocat® -NEM. Catalyst for PU molded flexible, semirigid, and rigid applications. Amine-based catalyst for flexible slabstock PU foam. Liquid; mp = -63°; bp = 138.5°; d = 0.9050; freely soluble in H2O, EtOH, Et2O, soluble in Me2CO, C6H6; LD50 (rat orl) = 1780 mg/kg. *Texaco; Tosoh.*

1680 Ethyl naphthaleneacetate
2122-70-5   G P

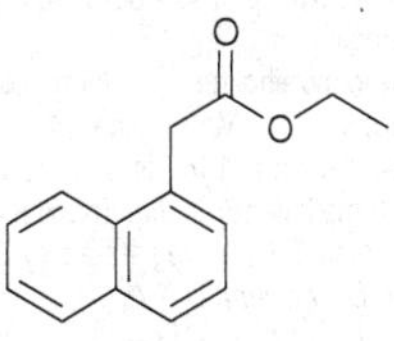

C14H14O2

Ethyl-1-naphthyl acetate.

4-09-00-02425 (Beilstein Handbook Reference); AI3-02254; BRN 1106730; Caswell No. 589AA; EINECS 218-332-1; EPA Pesticide Chemical Code 056008; Ethyl 1-naphthaleneacetate; Ethyl 1-naphthylacetate; 1-Naphthaleneacetic acid, ethyl ester; 2-(1-Naphthyl)acetic acid ethyl ester; NSC 74497; Tre-Hold. Plant growth regulator. Used to inhibit sprouting at pruning points. Oil; bp20 = 222°, bp13 = 118°, bp3 = 158-160°; d25 = 1.106; insoluble in H2O, soluble in EtOH, Et2O; LD50 (rat orl) = 3850 mg/kg. *Marks A. H. & Co. Ltd.*

1681 Ethyl nitrate
625-58-1   G

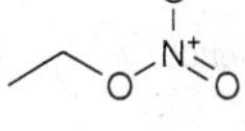

C2H5NO3

Nitric acid, ethyl ester.
4-01-00-01327 (Beilstein Handbook Reference); AI3-24234; BRN 1700275; EINECS 210-903-3; Ethyl nitrate; Ethylester kyseliny dusicne; HSDB 415; Nitric ether; NSC 8826. Liquid; mp = -94.6°; bp = 87.2°; d^{20} = 1.1084; λ_m = 260 nm (ε = 13, petroleum ether); soluble in H_2O, freely soluble in EtOH, Et_2O.

1682   2-Ethyl-2-nitropropanediol
597-09-1   G

C5H11NO4

2-Nitro-2-ethyl-1,3-propanediol.
4-01-00-02550 (Beilstein Handbook Reference); AI3-02257; BRN 1705448; 2-Ethyl-2-nitropropane-1,3-diol; 2-Ethyl-2-nitro-1,3-propanediol; EINECS 209-893-3; NEPD; 2-Nitro-2-ethyl-1,3-propanediol; NSC 2024; 1,3-Propanediol, 2-ethyl-2-nitro-. Chemical and pharmaceutical intermediate, used in tire cord adhesives, as formaldehyde release agents, deodorants, antimicrobials. Needles, mp = 57.5°; bp dec; λ_m = 278 nm (MeOH); very soluble in H_2O, EtOH, Et_2O. *Whittaker Clark & Daniels.*

1683   4,4'-(2-Ethyl-2-nitrotrimethylene)-dimorph-oline
1854-23-5   G P

C13H25N3O4

Ethyl-(2-nitrotrimethylene)dimorpholine.
Bioban P 1487; EINECS 217-450-0; 4,4'-(2-Ethyl-2-nitropropane-1,3-diyl)bismorpholine;   4,4'-(2-Ethyl-2-nitrotrimethylene)dimorpholine;   Morpholine,   4,4'-(2-ethyl-2-nitrotrimethylene)di-;   Morpholine,   4,4'-(2-ethyl-2-nitro-1,3-propanediyl)bis-. Bioban P 1487 is a preservative composed of 4,4-(2-Ethyl-2-nitro-trimethylene) dimorpholine (30%) and 4-(2-Nitrobutyl)morpholine (70%). Registered by EPA as an antimicrobial and disinfectant. *Dow Chem. U.S.A.*

1684   Ethyl oxirane
106-88-7   H

C4H8O

1,2-Monoepoxybutane.
5-17-01-00056 (Beilstein Handbook Reference); BRN 0102411; Butane, 1,2-epoxy-; CCRIS 1015; EINECS 203-438-2; DL-1,2-Epoxybutane; Ethylene oxide, ethyl-; Ethyl ethylene oxide; Ethylethylene oxide; Ethyloxirane; (±)-2-Ethyloxirane; (±)-Ethyloxirane; HSDB 2855; n-Butene-1,2-oxide; NCI-C55527; NSC 24240; Oxirane, ethyl-; UN3022. Liquid; bp = 63.3°; d^{20} = 0.8297; very soluble in EtOH, Me_2CO, freely soluble in Et_2O.

1685   m-Ethylphenol
620-17-7   H

C8H10O

3-Ethylphenol.
AI3-19938; Benzene, 1-ethyl-3-hydroxy-; EINECS 210-627-3; HSDB 5720; 1-Ethyl-3-hydroxybenzene; 3-Ethylphenol; m-Ethylphenol; meta-Ethylphenol; 1-Hydroxy-3-ethylbenzene; NSC 8873; Phenol, 3-ethyl-; Phenol, m-ethyl-. Liquid; mp = -4°; bp = 218.4°; d^{20} = 1.0283; λ_m = 273 nm (ε = 1778, cyclohexane); slightly soluble in H_2O, CHCl3, very soluble in EtOH, Et_2O.

1686   p-Ethylphenol
123-07-9   H

C8H10O

4-Ethylphenol.
4-06-00-03020 (Beilstein Handbook Reference); AI3-26063; BRN 1363317; EINECS 204-598-6; 4-Ethylphenol; FEMA No. 3156; HSDB 5598; Hydroxyphenylethane, p-; NSC 62012; p-Ethylphenol; para-Ethylphenol; Phenol, 4-ethyl-. Needles; mp = 45.0°; bp = 217.9°; λ_m = 223, 278 nm (ε = 7090, 1950, MeOH); slightly soluble in H_2O, CHCl3, slightly soluble in Me_2CO, very soluble in EtOH, Et_2O, C6H6, CS2.

1687   Ethylphosphonothioic dichloride
993-43-1   H

C2H5Cl2PS

Phosphonothioic dichloride, ethyl-.
EINECS 213-609-3; Ethyl phosphonothioic dichloride, anhydrous; Ethylphosphonothioic dichloride; NA2927; Phosphonothioic dichloride, ethyl-.

1688   Ethyl phosphorodichloridate
1498-51-7   H

C2H5Cl2O2P

Dichlorophosphoric acid, ethyl ester.
Dichloroethoxyphosphine oxide; Dichlorophosphoric acid, ethyl ester; EINECS 216-099-0; Ethyl dichlorophosphate; Ethyl phosphorodichloridate; HSDB 417; NA2927; NSC 87531; Phosphorodichloridic acid, ethyl ester. Liquid; bp10 = 60-65°.

1689   Ethyl phthalate
84-66-2   7456   G H

C12H14O4

Diethyl 1,2-benzenedicarboxylate.

4-09-00-03172 (Beilstein Handbook Reference); AI3-00329; Anozol; Benzenedicarboxylic acid, diethyl ester; BRN 1912500; CCRIS 2675; DEP; Diethyl 1,2-benzenedicarboxylate; Diethyl o-phenylenediacetate; Diethyl o-phthalate; Diethyl phthalate; Diethylester kyseliny ftalove; DPX-F5384; EINECS 201-550-6; Estol 1550; Ethyl phthalate; HSDB 926; Kodaflex DEP; NCI-C60048; Neantine; NSC 8905; Palatinol A; Phthalic acid, diethyl ester; Phthalol; Phthalsäurediäthylester; Phthalsaeurediaethylester; Placidol E; RCRA waste number U088; Solvanol; Unimoll DA. Solvent for nitrocellulose and cellulose acetate; plasticizer, wetting agent, insectidal preparations; in perfumery as solvent and fixative; plasticizer in solid rocket propellants. Plasticizer; wetting agent in grinding pigments; pigment-dispersing medium in cellulose acetate solutions and plastics, and solvent for natural resins and polymers and PVC products due to relatively high volatility. Registered by EPA (cancelled). Colorless oil; mp = -40.5°; bp = 295°; d_4^{14}= 1.232; insoluble in H2O, soluble in EtOH, Et2O and most other organic solvents; LD50 (rat ip) = 7.06 ml/kg (8700 mg/kg). *Allan Chem. Corp.; Allchem Ind.; Berje; BP Chem.; Daihachi Chem. Ind. Co. Ltd; Degussa-Hüls Corp.; Eastman Chem. Co.; Morflex; Penta Mfg.; Sigma-Aldrich Co.; Sigma-Aldrich Fine Chem.; Unitex.*

1690   2-Ethyl-3-propylacrolein
645-62-5   H

C8H14O

2-Ethyl-2-hexenal.

Acrolein, 2-ethyl-3-propyl-; AI3-08105; CCRIS 4645; EINECS 211-448-3; 2-Ethylhexenal; 2-Ethyl-2-hexenal; 2-Ethylhex-2-enal; 2-Ethyl-2-hexen-1-al; 2-Ethylhex-2-en-1-al; α-Ethyl-β-propylacrolein; 2-Ethyl-3-propylacryl-aldehyde; 2-Ethyl-3-propylacrolein; 2-Hexenal, 2-ethyl-; HSDB 1120; NSC 4787.

1691   N-Ethylpyrrolidinone
2687-91-4   G

C6H11NO

N-Ethyl-2-pyrrolidone.

5-21-06-00328 (Beilstein Handbook Reference); Agsol Ex 2; BRN 0107971; EINECS 220-250-6; 1-Ethyl-2-pyrrol-idinone; 1-Ethylpyrrolidin-2-one; N-Ethylpyrrolidinone; N-Ethylpyrrolidone; NEP; 2-Pyrrolidinone, 1-ethyl-. Solvent. Liquid; bp20= 97°; d = 0.9920; LD50 (rat orl) = 1350 mg/kg. *ISP.*

1692   5-Ethylquinolinic acid
102268-15-5   H

C9H9NO4

2,3-Pyridinedicarboxylic acid, 5-ethyl-.

C-80185; CL 271,191; 5-Ethylpyridine-2,3-dicarboxylic acid; 5-Ethylquinolinic acid; 2,3-Pyridinedicarboxylic acid, 5-ethyl-.

1693   Ethyl salicylate
118-61-6   3881   G V

C9H10O3

2-Hydroxybenzoic acid ethyl ester.

4-10-00-00149 (Beilstein Handbook Reference); AI3-00513; Benzoic acid, 2-hydroxy-, ethyl ester; BRN 0907659; EINECS 204-265-5; o-(Ethoxycarbonyl)phenol; Ethyl 2-hydroxybenzoate; Ethyl o-hydroxybenzoate; Ethyl salicylate; FEMA No. 2458; Mesotol; NSC 8209; Sal ether; Sal ethyl; Salicylic acid, ethyl ester; Salicylic ether; Salicylic ethyl ester; Salotan. Used in perfumery; has been used in veterinary medicine as a counter-irritant. Crystals; mp = 45°; bp = 231-234°, bp10 = 150-151°; d^{20} = 1.1326; λ_m = 238, 305 nm (ε = 13800, 6590, MeOH); insoluble in H2O, soluble in CCl4, very soluble in Et2O, freely soluble in EtOH.

1694   Ethyl selenac
5456-28-0   G

C20H40N4S8Se

Selenium diethyl dithiocarbamate.

EINECS 226-713-9; Ethyl selenac; Ethyl seleram; Selenium, tetrakis(diethyldithiocarbamato)-; Selenium tetrakis(diethyldithiocarbamate); Selazate; Tetrakis-(diethylcarbamodithioato-S,S')selenium; Tetrakis(diethyl-dithiocarbamato)selenium. Rubber accelerator for NR, SBR and IIR; vulcanizing agent; effective in low sulfur and sulfurless heat-resistant compounds. *Naugatuck; Vanderbilt R.T. Co. Inc.*

1695   Ethyl stearate
111-61-5   G

C20H40O2
Ethyl n-octadecanoate.
4-02-00-01218 (Beilstein Handbook Reference); AI3-01781; BRN 1788183; EINECS 203-887-4; Ethyl octadecanoate; Ethyl stearate; FEMA No. 3490; NSC 8919; Octadecanoic acid, ethyl ester; Radia® 7185; Stearic acid, ethyl ester. Chemical intermediate, chemical synthesis; lubricant in mineral, cutting, lamination, textile oils, and rust inhibitors; textile and leather application; also as emollient, plasticizer, solubilizer of active components in cosmetics and pharmaceuticals. Solid; mp = 33°; bp$_{15}$ = 213-215°, bp$_{10}$ = 199°; insoluble in H$_2$O, soluble in EtOH, Et$_2$O, CHCl$_3$, very soluble in Me$_2$CO. *Fina Chemicals.*

1696   Ethylstyrene
28106-30-1   H

C10H12
Styrene, ar-ethyl-.
Benzene, ethenylethyl-; EINECS 248-846-1; Ethylstyrene; Styrene, ar-ethyl-. Mixture of o, m and p-isomers.

1697   Ethylsuccinonitrile
17611-82-4   H

C6H8N2
Ethylbutanedinitrile.
Butanedinitrile, ethyl-; EINECS 241-587-5; Ethyl-succinonitrile; Succinonitrile, ethyl-.

1698   Ethyl Tellurac®
20941-65-5   G

C20H40N4S8Te
Tellurium diethyldithiocarbamate.
Akrochem® TDEC;Carbamodithioic acid, diethyl-, tetrakis(anhydrosulfide) with thiotelluric acid; CCRIS 304; Diethyldithio carbamic acid tellurium salt; EINECS 244-121-9; Ethyl tellurac; HSDB 4123; NCI-C02857; Nocceler TTTE; Perkacit® TDEC Tellurac; Tellurium bis(diethyldithiocarbamate); Tellurium diethyldithiocarb-amate; Tellurium, tetrakis(diethylcarbamodithioato-

S,S')-, (DD-8-111"1"1'1'1'"1'")-; Tellurium, tetrakis(diethyl-dithiocarbamato)-; Tellurium(IV) diethyl dithiocarbamate; Tetrakis(diethylcarbamodithioato-S,S')tellurium; Tetrakis-(diethyldithiocarbamato-S,S')tellurium; tellurac; diethyl-dithiocarbamic acid tellurium salt; tellurium, tetrakis(diethyldithiocarbamate)-; Tellurium(IV) diethyl-dithiocarbamate. Fast-curing primary or secondary accelerator for use in natural rubber, styrene/butadiene rubber, NBR, ethylene-propylene-diene rubbers, and butyl; ultra accelerator for NR, SBR, NBR, EPDM; used with thiazole modifiers; produces high modulus vulcanization; particularly active in IIR compounds. Solid; mp = 108-118°; SG = 1.44; insoluble in H$_2$O (<0.1 mg/ml), soluble in organic solvents. *Akrochem Chem. Co.; Akzo Chemie; Vanderbilt R.T. Co. Inc.*

1699   Ethyl thiochloroformate
2941-64-2   H

C3H5ClOS
Formic acid, chlorothio-, S-ethyl ester.
Carbonochloridothioic acid, S-ethyl ester; CCRIS 4642; EINECS 220-928-1; Ethyl chlorothioformate; Ethyl thiochloroformate; Ethylthiocarbonyl chloride; Ethylthiol chloroformate; Formic acid, chlorothio-, S-ethyl ester; HSDB 5906; S-(Ethyl)chlorothioformic acid; S-Ethyl carbonochloridothioate; S-Ethyl chlorothiocarbonate; UN2826.

1700   Ethyl toluene
25550-14-5   H

C9H12
Toluene, ar-ethyl-.
Benzene, ethylmethyl-; EINECS 247-093-6; Ethyl toluene; Ethyltoluene; Toluene, ar-ethyl-.

1701   Ethyltoluene-2-sulfonamide
1077-56-1   G

C9H13NO2S
N-Ethyl o-toluene sulfonamide.
Benzenesulfonamide, N-ethyl-2-methyl-; EINECS 214-073-3; N-Ethyl-o-toluenesulfonamide; N-Ethyltoluene-2-sulphonamide; HSDB 5782; o-Toluenesulfonamide, N-ethyl-; Uniplex 108. Plasticizer for nylon, shellac, cellulose acetate, protein materials, PVAc adhesives, and nitrocellulose lacquers. *Unitex.*

1702   Ethyl p-toluenesulfonate
80-40-0   3889   G

C9H12O3S
Ethyl p-methyl benzenesulfonate.
4-11-00-00248 (Beilstein Handbook Reference); AI3-08004; Benzenesulfonic acid, 4-methyl-, ethyl ester; BRN 0611213; CCRIS 1028; EINECS 201-276-7; Ethyl p-methylbenzenesulfonate; Ethyl p-

tosylate; Ethyl p-TS; Ethyl PTS; Ethyl toluene-4-sulphonate; Ethyl tosylate; Ethylester kyseliny p-toluensulfonove; HSDB 5235; NSC 8887; p-Toluenesulfonic acid, ethyl ester; p-Toluolsulfonsaeure aethyl ester; p-Toluolsulfonsäure aethyl ester. Used in ethylation reactions. Solid; mp = 34.5°; bp$_{15}$ = 173°; d^{48} = 1.166; λ_m = 223, 255, 260, 266, 271 nm (MeOH); insoluble in H_2O, soluble in organic solvents. *Acros Organics - USA; Lancaster Synthesis Co.; Mallinckrodt Inc.; Sigma-Aldrich Fine Chem.*

1703   N-Ethyl-m-toluidine
102-27-2   H

$C_9H_{13}N$
 N-Ethyl-3-methyl-benzenamine.
 4-12-00-01816 (Beilstein Handbook Reference); BRN 0742170; EINECS 203-019-4; N-Ethyl-3-methylaniline; HSDB 5347; m-Methyl-N-ethylaniline; m-Toluidine, N-ethyl-; NSC 8624; Toluene, 3-(ethylamino)-. Liquid; bp = 221°; d^{15} = 0.9263; λ_m = 247, 295 nm (cyclohexane); soluble in EtOH, Et$_2$O.

1704   Ethyltriacetoxysilane
17689-77-9   G H

$C_8H_{14}O_6Si$
 Silanetriol, ethyl-, triacetate.
 CE6345; Dynasylan® ETAC; EINECS 241-677-4; Ethyltriacetoxysilane; Silanetriol, ethyl-, triacetate; Triacetoxyethylsilane. Coupling agent, chemical intermediate, blocking agent, release agent, lubricant, primer, reducing agent. *Degussa-Hüls Corp.*

1705   Ethyl tosylamide
80-39-7   G

$C_9H_{13}NO_2S$
 p-Toluenesulfonyl-N-ethylamide.
 Benzenesulfonamide, N-ethyl-4-methyl-; CCRIS 6037; EINECS 201-275-1; Ethyl tosylamide; N-Ethyl-4-methylbenzenesulfonamide; N-Ethyl-p-methylbenzene-sulfonamide; N-Ethyl-4-toluenesulfonamide; N-Ethyl-p-toluenesulfonamide; N-Ethyl-p-tolylsulfonamide; N-Ethyltoluene-4-sulphonamide; NSC 68803; p-Toluenesulfonyl-N-ethylsulfonamide; p-Toluenesulfonamide, N-ethyl-; N-Tosylethylamine; Santicizer 3; Uniplex 108. Mixture of isomers of aromatic amides; plasticizer for shellac, cellulose acetate, and protein materials; used in NC lacquers, cellulose acetate compositions, PVAc emulsion adhesives, synthetic polyamides; food packaging applications. Used as a plasticizer for nylon, shellac, cellulose acetate, protein materials, PVAc adhesives, and nitrocellulose lacquers. Crystals; mp = 63-65°; bp$_{245}$ = 208°. *ICI*

Spec.; Rit-Chem; Unitex.

1706   Ethyl vanillin
121-32-4   3890   G

$C_9H_{10}O_3$
 3-Ethoxy-4-hydroxybenzaldehyde.
 4-08-00-01765 (Beilstein Handbook Reference); 4-Hydroxy-3-ethoxybenzaldehyde; AI3-00786; Benzalde-hyde, 3-ethoxy-4-hydroxy-; Bourbonal; BRN 1073761; CCRIS 1346; EINECS 204-464-7; Ethavan; Ethovan; 2-Ethoxy-4-formylphenol; 3-Ethoxy-4-hydroxybenzalde-hyde; 3-Ethoxyprotocatechualdehyde; Ethyl protal; Ethyl vanillin; Ethylprotal; Ethylprotocatechualdehyde-3-ethyl ether; Ethylprotocatechuic aldehyde; Ethylvanillin; FEMA No. 2464; HSDB 945; NSC 1803; Protocatechuic aldehyde ethyl ether; Quantrovanil; Rhodiarome; Vanbeenol; Vanilal; Vanillal; Vanillin, ethyl-; Vanirom; Vanirome. Vanilla flavor for baking, cereal, diary, cheese, processed foods, beverages, confections. Used by flavor manufacturers to replace part of the vanillin to give bouquet to the finished flavor or fragrance. mp = 77.5°; bp = 285°; λ_m = 231, 277, 308 nm (ε = 15600, 10200, 9310, MeOH); slightly soluble in H_2O, soluble in EtOH, Et$_2$O, C$_6$H$_6$, CHCl$_3$; LD$_{50}$ (rat orl) = 1590 mg/kg. *Boehringer Mannheim GmbH; Bush Boake Allen (IFF bought 2000); Lancaster Synthesis Co.; Maggioni Farmaceutici S.p.A.; Penta Mfg.; Rhône-Poulenc Food Ingredients; Rhone-Poulenc UK.*

1707   Ethyl Ziram
14324-55-1   G

$C_{10}H_{20}N_2S_4Zn$
 Zinc N,N-diethyldithiocarbamate.
 AI3-14877; Ancazate ET; Bis(diethyldithiocarbamato)-zinc; Carbamodithioic acid, diethyl-, zinc salt; CCRIS 4908; Diethyldithiocarbamic acid zinc salt; EINECS 238-270-9; Ethazate; Ethyl cymate; Ethyl zimate; Ethyl Ziram; Ethylzimate; Hermat ZDK; HSDB 2907; Nocceler EZ; NSC 177699; Octocure ZDE-50; Perkacit® ZDEC; Soxinol EZ; Vulcacure ZE; Vulkacit LDA; Vulkacit ZDK; Zimate, ethyl; Zinc bis(diethyldithiocarbamate); Zinc, bis(diethyldithiocarbamato); Zinc, bis(diethylcarbamo-dithioato-S,S')-; Zinc diethyldithiocarbamate; Zinc N,N-diethyldithiocarbamate; Zinc, tetrakis(diethylcarbamo-dithioato)di-. Latex and rubber accelerator. An accelerator and activator for natural rubber, styrene-butadiene, nitrile-butadiene and butyl rubber. *Akzo Chemie; Tiarco.*

1708   Etidronic acid
2809-21-4   3894   D G H

$C_2H_8O_7P_2$
 (1-Hydroxyethylidene)diphosphonic acid.
 0-02-00-00171 (Beilstein Handbook Reference); 1000SL; Acetodiphosphonic acid; Acide etidronique; Acido etidronico; Acidum etidronicum; Briquest® ADPA-60A' BRN 1789291; Dequest 2010; Dequest 2015; Dequest Z 010; Diphosphonate (base); EHDP; EINECS 220-552-8; Etidronic acid; Etidronsaeure; Ferrofos 510;

HEDP; HSDB 5898; NSC 227995; RP 61; Turpinal SL; Unihib® 106. Dispersant, scale inhibitor for cooling tower, boiler treatment, oilfield applications. Used medically as a calcium regulator. Used for water treatment and oil-drilling muds, in powder detergents and photographic applications, sequestering agent for calcium carbonate. Crystals; very soluble in H2O (69 g/100 ml), insoluble in AcOH. *Albright & Wilson Americas Inc.; Lonzagroup; Procter & Gamble Pharm. Inc.*

1709   Etifenin
63245-28-3   G

C16H22N2O5

Glycine, N-(carboxymethyl)-N-(2-((2,6-diethylphenyl)-amino)-2-oxoethyl)-.

(((((2,6-Diethylphenyl)carbamoyl)methyl)imino)diacetic acid; EHIDA Kit; EINECS 264-041-8; Etifenia; Etifenin; Etifenine; Etifenino; Etifeninum; Glycine, N-(carboxymethyl)-N-(2-((2,6-diethylphenyl)amino)-2-oxo-ethyl)-. Diagnostic aid. Solid; mp = 186-189°; soluble in H2O. *Amersham Corp.*

1710   Etridiazole
2593-15-9   G P

C5H5Cl3N2OS

5-Ethoxy-3-(trichloromethyl)-1,2,4-thiadiazole.

Aaterra; Al3-29280; Banrot; BRN 1074817; Caswell No. 428; Dwell; Echlomezol; Echlomezole; EINECS 219-991-8; EPA Pesticide Chemical Code 084701; Etcmtb; Ethazol; Ethazole (fungicide); ETMT; Etridiazol; Etridiazole; HSDB 1709; Koban; MF-344; NSC 524929; Olin mathieson 2,424; OM 2424; OM 2425; Pansoil; Planvate; Terrachlor-super X; Terracoat; Terracoat L21; Terraflo; Terrazole; Truban. Protective fungicide which is incorporated into soil or compost. Registered by EPA as a fungicide. Pale yellow liquid; mp = 19.9°; bp[1] = 95°; sg[25] = 1.503; soluble in H2O (0.005 g/100 ml at 25°), soluble in common organic solvents; LD50 (mrat orl) = 1100 mg/kg, (mus orl) = 2000 mg/kg, (rbt der) = 1366 mg/kg, (mallard duck orl) = 1640 mg/kg; LC50 (rainbow trout 96 hr.) > 4 mg/l, (bluegill sunfish 96 hr.) > 7.5 mg/l; non-toxic to bees. *Crompton Corp.; ICI Agrochemicals; Scotts-Sierra Crop Protection; Uniroyal.*

1711   Etrimfos
38260-54-7   3936   G P

C10H17N2O4PS

O-(6-ethoxy-2-ethyl-4-pyrimidinyl) O,O-dimethyl phos-phorothioate. Satisfar; Ekamet; Ekamet G; Ekamet ULV; Etrimphos.

Organophosphorus insecticide with stomach and contact action. Used to control pests in stored grain. Liquid; mp = -3°; d = 1.195; n$_D^{20}$ = 1.5068; soluble in H2O (40 mg/l), more soluble in organic solvents; LD50 (rat orl) = 1800 mg/kg. *Nickerson Seeds Ltd.*

1712   Euparen® M
731-27-1   G P

C10H13Cl2FN2O2S2

1,1-Dichloro-N-((dimethylamino)sulfonyl)-1-fluoro-N-(4-methylphenyl)methanesulfonamide.

Al3-27470; Bayer 49854; Bayer 5212; Bayer 5712a; BRN 2949607; Dichlofluanid-methyl; Dichloro-N-((dimethyl-amino)sulphonyl)fluoro-N-(p-tolyl)methanesulphenamide; 1,1-Dichloro-N-((dimethylamino)sulfonyl)-1-fluoro-N-(4-methylphenyl)methanesulfenamide; N-Dichlorofluoro-methanesulphenyl-N',N'-dimethyl-N-p-tolylsulphamide; N-Dichlorofluoromethylthio-N',N'-dimethyl-N-p-tolyl-sulphamide; N'-Dichlorofluoromethylthio-N,N-dimethyl-N'-(4-tolyl)sulfamide; N,N-Dimethyl-N-(4-tolyl)-N-(di-chlorofluor-methylthio)-sulfamide; N,N-Dimethyl-N'-(4-tolyl)-N'-(dichlorfluormethylthio)-sulfamid; EINECS 211-986-9; Euparen M; KUE 13183b; Methanesulfenamide, 1,1-dichloro-N-((dimethylamino)sulfonyl)-1-fluoro-N-(4-methylphenyl)-; Sulfamide, N-((dichlorofluoromethyl)-thio)-N',N'-dimethyl-N-(p-tolyl)-; Tolyfluanid; Tolylfluan-ide; Tolylfluanid; Tolylfluanide. Broad spectrum fungicide; effective against *Botrytis*. Pale yellow powder; mp = 95-97°; soluble in H2O (0.4 g/100 ml), MeOH (4.6 g/100 ml), C6H6 (57 g/100 ml); LD50 (rat orl) > 5000 mg/kg, (mus orl) > 1000 mg/kg, (rat sc) > 5000 mg/kg, (fcanary orl) = 1000 mg/kg, LC50 (goldfish 96 hr.) = 1.0 mg/l, (carp, roach 96 hr) = 0.25 - 0.5 mg/l, (golden orfe 96 hr) = 0.07 - 0.25 mg/l; non-toxic to bees. *Bayer AG.*

1713   Extracts, petroleum, gas oil solvent
65652-41-7   H

Extracts, (petroleum), gas oil solvent.
EINECS 265-211-4; Extracts, (petroleum), gas oil solvent; Extracts, petroleum, gas oil solvent.

1714   Famotidine
76824-35-6   3961   D

C8H15N7O2S3

[1-Amino-3-[[[2-[(diaminomethylene)amino]-4-thiazolyl]-methyl]thio]propylidene]sulfamide.

Amfamox; Antodine; Apo-Famotidine; Apogastine; Bestidine; Blocacid; Brolin; Cepal; Confobos; Cronol; Cuantin; Dibrit 40; Digervin; Dinul; Dipsin; Dispromil; Dispronil; Duovel; Durater; Evatin; Fadin; Fadine; Fadyn; Fagastine; Famo; Famocid; Famodar; Famodil; Famodin; Famodine; Famogard; Famonit; Famopsin; Famos; Famosan; Famotal; Famotep; Famotida; Famotidine; Famotidinum; Famotin; Famovane; Famowal; Famox; Famoxal; Famoxal; Famtac; Famulcer; Fanobel; Fanosin; Fanox; Farmotex; Ferotine; Fibonel; Fudone; Ganor; Gaster; Gastridan; Gastridin;

Gastrion; Gastro; Gastrodomina; Gastrofam; Gastropen; Gastrosidin; H2 Bloc; Hacip; HSDB 3572; Huberdina; Ifada; Ingastri; Invigan; L 643341; Lecedil; Logos; Mensoma; Midefam; MK-208; Mosul; Motiax; Muclox; Muclox; Mylanta AR; Neocidine; Nevofam; Notidin; Novo-Famotidine; Nu-Famotidine; Nulceran; Nulcerin; Panalba; Pepcid; Pepcid AC; Pepcid®; Pepcidac; Pepcidin; Pepcidin Rapitab; Pepcidina; Pepcidine; Pepdif; Pepdine; Pepdul; Pepfamin; Peptan; Peptidin; Peptifam; Pepzan; Purifam; Quamatel; Quamtel; Renapepsa; Restadin; Rogasti; Rubacina; Sedanium-R; Sigafam; Supertidine; Tairal; Tamin; Tipodex; Topcid; Ulcatif; Ulceprax; Ulcetrax; Ulcofam; Ulfagel; Ulfam; Ulfamid; Ulfinol; Ulgarine; Vagostal; Weimok; Whitidin; Yamarin; YM 11170. Antiulcerative. Histamine H_2 receptor antagonist. Used for short term treatment of active duodenal ulcers. Crystals; mp = 163-164°; soluble in DMF (80 g/100 ml), AcOH (50 g/100 ml), MeOH (0.3 g/100 ml), H_2O (0.1 g/100 ml); insoluble in EtOH, EtOAc, $CHCl_3$; LD_{50} (mus iv) = 244.4 mg/kg. *Johnson & Johnson-Merck Consumer Pharm.; Merck & Co.Inc.*

1715   Farnesol
4602-84-0   3968   G P

$C_{15}H_{26}O$
3,7,11-Trimethyl-2,6,10-dodecatrien-1-ol.
AI3-44561; Dihydrofarnesol; 2,6,10-Dodecatrien-1-ol, 3,7,11-trimethyl-; EINECS 225-004-1; EPA Pesticide Chemical Code 128911; Farnesol; Farnesyl alcohol; FCI 119a; FEMA No. 2478; HSDB 445; NSC 60597; Stirrup; Stirrup-A/WF; Stirrup-CRW; Stirrup-H; Stirrup-HB; Stirrup-TPW; Trimethyl dodecatrienol; 3,7,11-Trimethyl-2,6,10-dodecatrienol; 3,7,11-Trimethyl-2,6,10-dodecatrien-1-ol; 2,6,10-Trimethyl-2,6,10-dodecatrien-12-ol. A sesquiterp-ene alcohol prepared from nerolidol; a perfume. Reg-istered by EPA as an insecticide. Oil; bp4 = 149°, bp0.35 = 111°; d = 0.8880; λ_m = 192 - 196 nm (ε 28500); insoluble in H_2O, soluble in organic solvents; LD_{50} (rat orl) = 6 g/kg. *Troy Biosciences Inc.*

1716   Fatty acid, C$_{18}$-unsaturated, dimers
61788-89-4   H

Fatty acid, C$_{18}$-unsaturated, dimers.
Fatty acid, C$_{18}$-unsaturated, dimers; Fatty acids, C$_{18}$-unsatd., dimers.

1717   Fatty acids, dehydrated castor-oil
61789-45-5   H

Dehydrated castor-oil fatty acids.
EINECS 263-061-4; Fatty acids, dehydrated castor-oil. 263-061-4

1718   Fenamiphos
22224-92-6   3984   P

$C_{13}H_{22}NO_3PS$
Ethyl 4-(methylthio)-m-tolyl isopropylphosphoramidate.
AI3-27572; BAY 68138; BAY sra 3886; BRN 4752893; Caswell No. 453A; CCRIS 4694; EINECS 244-848-1; ENT 27572; EPA Pesticide Chemical Code 100601; Fenamiphos; HSDB 6452; Methaphenamiphos; Nemacur; Nemacur P; NSC 195106; Phenamiphos; SRA 3886. Cholinesterase inhibitor. Systemic nematicide with contact action.Absorbed by the roots and translocated to the leaves. Used for control of ecto- and endoparasitic, free-living, cyst-forming and root-knotnematodes. Registered by EPA as an insecticide and fungicide. Used as a nematicide. mp = 49.2°; sg^{20} = 1.15; soluble in H_2O (0.07 g/100 ml), readily soluble in CH_2Cl_2, i-PrOH, C_7H_8, sparingly soluble in C_6H_{14}; LD_{50} (mrat orl) = 15.3 mg/kg, (frat orl) = 19.4 mg/kg, (mus orl) = 22.7 mg/kg, (dog,cat orl) = 10 mg/kg, (mrbt der) = 225 mg/kg, (frbt der) = 178 mg/kg, (rat der) = 80 mg/kg, (bobwhite quail orl) = 1.6 mg/kg, (hen orl) = 12 mg/kg; LC_{50} (rat ihl 4 hr.) = 80 mg/l air Aerosol®®, (mallard duck 8 day dietary) = 316 mg/kg diet, (bobwhite quail 5 day dietary) = 38 mg/kg diet, (bluegill sunfish 96 hr.) = 0.0096 mg/l, (rainbow trout 96 hr.) = 0.121 mg/l, (goldfish 96 hr.) = 3.2 mg/l. *Bayer Corp.; Mobay.*

1719   Fenazaquin
120928-09-8   P

$C_{20}H_{22}N_2O$
4-[[4-(1,1-Dimethylethyl)phenyl]ethoxy]quinazoline.
EL-193136; EL-436; Fenazaquin; Quinazoline, 4-((4-(1,1-dimethylethyl)phenyl)ethoxy)-. Acaricide. Registered by EPA as an fungicide and insecticide. Colorless crystals; mp = 70 - 71°; slightly soluble in H_2O (0.000022 g/100 ml), soluble in MeOH (5 g/100 ml), i-PrOH (5 g/100 ml), Me$_2$CO (40 g/100 ml), CH$_3$CN (3.3 g/10 ml), C_6H_{14} (3.3 g/100 ml), $CHCl_3$ (> 50 g/100 ml), C_7H_8 (5 g/100 ml); LD_{50} (mrat orl) = 50 - 500 mg/kg, (mus orl) > 500 mg/kg, (bobwhite quail gvg) > 2000 mg/kg. *Dow AgroSciences; DowElanco Ltd.*

1720   Fenbendazole
43210-67-9   3987   G

$C_{15}H_{13}N_3O_2S$
[5-(phenylthio)-1H-benzimidazol-2-yl]carbamic acid methyl ester.
2-Benzimidazolecarbamic acid, 5-(phenylthio)-, methyl ester; Carbamic acid, (5-(phenylthio)-1H-benzimidazol-2-yl)-, methyl ester; CCRIS 7309; EINECS 256-145-7; Fenbendazol; Fenbendazole; Fenbendazolum; Hoe 881v; 2-(Methoxycarbonylamino)-5-(phenylthio)benzimidazole; Methyl 5-(phenylthio)-2-benzimidazolecarbamate; Meth-yl (5-(phenylthio)-1H-benzimidazol-2-yl)carbamate; Pa-nacur; Phenbendasol. Anthelmintic (nematodes). Crystals; mp = 233° (dec); insoluble in H_2O, organic solvents, freely soluble in DMSO. *Hoechst AG.*

1721   Fenbutatin Oxide
13356-08-6   3991   G P

C60H78OSn2

Hexakis(2-methyl-2-phenylpropyl)distannoxane.

AI3-27738; Bendex; Bis(trineophyltin) oxide; Bis(tris(β,β-dimethylphenethyl)tin)oxide; Bis(tris(2-methyl-2-phenyl-propyl)tin)oxide; BRN 4097400; Caswell No. 481DD; Di(tri-(2,2-dimethyl-2-phenylethyl)tin)oxide; Distannox-ane, hexakis(β,β-dimethylphenethyl)-; Distannoxane, hexakis(2-methyl-2-phenylpropyl)-; EINECS 236-407-7; ENT 27738; EPA Pesticide Chemical Code 104601; Fenbutatin oxide; Fenbutatin-oxyde; Fenylbutatin oxide; Fenylbutylstannium oxide; Hexakis; Hexakis (2-methyl-2-phenylpropyl)-distannoxane; Hexakis(β,β-dimethylphen-ethyl)-distannoxane; Hexakis(2-methyl-2-phenylpropyl)-distannoxane; HSDB 6632; 2-(Methyl-2-phenylpropyl)-distannoxane; Neostanox; Osdaran; SD 14114; Shell SD-14114; Torque; Vendex; AI3-27738; Bendex; Bis(trineo-phyltin) oxide; Bis(tris(2-methyl-2-phenylethyl)tin)oxide; Bis(tris(β,β-dimethylphenethyl)tin)oxide; BRN 4097400; Caswell No. 481DD; Di(tri-(2,2-dimethyl-2-phenylethyl)-tin)oxide; Distannoxane, hexakis(2-methyl-2-phenyl-propyl)-; ENT 27738; EPA Pesticide Chemical Code 104601; Fenbutatin oxide; Fenbutatin-oxyde; Fenyl-butatin oxide; Fenylbutylstannium oxide; Hexakis(2-methyl-2-phenylpropyl)distannoxane; Hexakis(β,β-di-methylphenethyl)-distannoxane; HSDB 6632; 2-(Methyl-2-phenylpropyl)distannoxane; Neostanox [ISO:PROP]; Osdaran; SD 14114; Shell SD-14114; Torque; Vendex. A non-systemic acaricide with contact and stomach action. Used for control of phytophagous mites in fruit crops. Registered by EPA as an acaricide and insecticide. White powder; mp = 138-139°; poorly soluble in H2O (0.0000005 g/100 ml), more soluble in Me2CO (0.6 g/100 ml), C6H6 (14 g/100 ml), CH2Cl2 (38 g/100 ml), slightly soluble in hydrocarbon solvents; LD50 (rat orl) = 2631 mg/kg, (mus orl) = 1450 mg/kg, (dog orl) > 1500 mg/kg, (rbt der) > 2000 mg/kg, (rat der) > 1000 mg/kg, (bee orl) > 0.1 mg/bee; LC50; (bobwhite quail 8 day dietary) = 5056 mg/lg diet, (rainbow trout 48 hr.) = 0.27 mg/l; not toxic to bees. *Am. Cyanamid; DuPont; Griffin LLC; ICI Agrochemicals; Scotts Co.*

1722   Fenitrothion
122-14-5   4003   G P

C9H12NO5PS

Dimethyl-(3-methyl-4-nitrophenyl)thiophosphate.

8057HC; AC-47300; Accothion; Agria 1050; Agriya 1050; Agrothion; AI3-25715; Akotion; American Cyanamid CL-47,300; Arbogal; Bayer 41831; Bayer S 5660; BRN 1887367; C-9146; Caswell No. 373; CCRIS 1559; Cekutrothion; CL 47300; CP 47114; m-Cresol, 4-nitro-, O-ester with O,O-dimethyl phosphorothioate; Cyfen; Cytel; Dimethyl-(3-methyl-4-nitrophenyl)thio-phosphate; Dimethyl 3-methyl-4-nitrophenyl phosphoro-thionate; O,O-Dimethyl-O-(3-methyl-4-nitrofenyl)-mono-thiofosfaat; O,O-Dimethyl O-(3-methyl-4-nitrophenyl) phosphorothionate; O,O-Dimethyl O-(3-methyl-4-nitrophenyl) thiophosphate; O,O-Dimethyl O-(3-methyl-4-nitrophenyl) phosphorothioate; O,O-Dimethyl-O-(3-methyl-4-nitro-phenyl)-monothiophosphat; O,O-Dimetil-O-(3-metil-4-nitro-fenil)-monotiofosfato; O,O-Dimetil-O-(3-metil-4-nitrofenil) fosforotioato; O,O-Dimethyl O-(4-nitro-3-methylphenyl) thiophosphate; O,O-Dimethyl-O-(4-nitro-5-methylphenyl)-thionophosphat; O,O-Dimethyl O-4-nitro-m-tolyl phosphorothioate; Dimethyl 4-nitro-m-tolyl phosphorothionate; EI 47300; EINECS 204-524-2; ENT 25,715; EPA Pesticide Chemical Code 105901; Falithion; Fenition; Fenitox; Fenitrothion; Fenitrotion; Fentrothione; Folithion; Folithion EC 50; HSDB 1590; Insectigas F; Kotion; Macbar; MEP (pesticide); Metathion; Metathion E 50; Metathione; Metathionine; Metathionine E50; Metation; Metation E50; Methylnitrophos; 3-Methyl-4-nitrophenyl dimethyl phosphorothioate; Mglawik F; Monsanto CP 47114; Nitrophos; Novathion; Nuvanol; Oleosumifene; OMS 43; Ovadofos; Owadofos; Owadophos; Pennwalt C-4852; Phenitrothion; Phosphorothioic acid, O,O-dimethyl O-(4-nitro-m-tolyl) ester; Phosphorothioic acid, O,O-dimethyl O-(3-methyl-4-nitrophenyl) ester; S-1102A; S 112A; Sumifene; Sumithion; Sumitomo S-1102A; Super Sumithion; Thiophosphate de O,O-dimethyle et de O-(3-methyl-4-nitrophenyle); Tionofosforan O,O-dwumetylo-O-(3-metylo)-4-nitrofenylowy; Verthion; Dimethyl 4-nitro-m-tolyl phosphorothionate; Dimethyl O-(3-methyl-4-nitrophenyl) phosphorothioate; Dimethyl O-(4-nitro-m-tolyl) phosphorothioate; O,O-Dimethyl O-(3-methyl-4-nitrophenyl) phosphorothionate; O,O-Dimethyl-O-(3-methyl-4-nitrofenyl)-monothiofosfaat; O,O-Dimethyl O-(3-methyl-4-nitrophenyl) thiophosphate; O,O-Dimethyl-O-(3-methyl-4-nitro-phenyl)-monothiophosphat O,O-Dimetil-O-(3-metil-4-nitro-fenil)-monotiofosfato; O,O-Dimetil-O-(3-metil-4-nitrofenil) fosforotioato; O,O-Dimethyl-O-(4-nitro-5-methylphenyl)-thionophosphat; O,O-Dimethyl O-4-nitro-m-tolyl thiophosphate; O,O-Dimethyl O-4-nitro-m-tolyl phosphorothioate; O,O-Dimethyl O-(4-nitro-3-methylphenyl) thiophosphate; Dybar; EI 47300; ENT 25,715; EPA Pesticide Chemical Code 105901; Falithion; Fenition; Fenitox; Fenitrothion; Fenitrotion; Fentrothione; Folithion; Folithion EC 50; HSDB 1590; Insectigas F; Kotion; Macbar; MEP (pesticide); Metathion; Metathion E 50; Metathione; Metathionine; Metathionine E50; Metation; Metation E50; 3-Methyl-4-nitrophenyl dimethyl phosphorothioate; O-(3-methyl-4-nitrophenyl) phosphorothioate; Methyl-nitrophos; Mglawik F; Monsanto CP 47114; Nitrophos; Novathion; Nuvanol; Nuvanol/N-20; Oleosumifene; OMS 43; Ovadofos; Owadofos; Owadophos; Pennwalt C-4852; Phenitrothion; Phosphorothioic acid, O,O-dimethyl O-(3-methyl-4-nitrophenyl) ester; Phospho-rothioic acid, O,O-dimethyl O-(4-nitro-m-tolyl) ester; S 112AS-1102A; Sumifene; Sumithion; Sumitomo S-1102A; Super SumithionThiophosphate de O,O-dimethyle et de O-(3-methyl-4-nitrophenyle); Tionofosforan O,O-dwumetylo-O-(3-metylo)-4-nitrofenylowy; Verthion. All-round low toxic insecticide for forest protection, agriculture and public health; used especially where long term effect is desired. Registered by EPA as an insecticide and acaricide. Yellow oil; bp1 = 164°, bp0.05 = 118°; d4²⁵ = 1.3227; λm = 269.5 nm (ε 6756); almost insoluble in H2O (0.003 g/100 mlat 21°), poorly soluble in aliphatic hydrocarbons, in C6H14 (2.4 g/100 ml at 24°, soluble in other organic solvents; LD50 (rat orl) = 570 - 800 mg/kg, (mus orl) = 870 mg/kg, (gpg orl) = 500 mg/kg, (mrat der) = 890 mg/kg (frat der) = 1200 mg/kg, (mus der) > 3000 mg/kg, (quail orl) = 23.6 mg/kg, (mallard duck orl) = 1190 mg/kg; LC50 (rat ihl 4 hr.) = 1.2 mg/l air, (carp 48 hr.) = 4.1 mg/l, (bluegill sunfish 96 hr.) = 3.8 mg/l, (brook trout 96 hr.) = 1.7 mg/l; toxic to bees. *A/S Cheminova; Am. Cyanamid; Bayer Corp.; Sumitomo Corp.; Walco-Linck Co.*

1723   Fenofibrate
49562-28-9   4005   D

C20H21ClO4

Isopropyl 2-[p-(p-chlorobenzoyl)phenoxy]-2-methyl-propionate.
Ankebin; Elasterin; Fenobrate; Fenofibrate; Fenotard; LF-178;
Lipanthyl; Lipantil; Lipidil; Lipoclar; Lipofene; Liposit; Lipsin; Nolipax;
Procetofen; Procetofene; Procetoken; Protolipan; Secalip.
Hypolipidemic agent. A fibric acid; reduces plasma triglycerides by
lowering plasma levels of very-low-density lipoprotein. Crystals; mp
= 80-81°; insoluble in H_2O; slightly soluble in EtOH, MeOH; more
soluble in non-polar solvents; LD50 (mus orl) = 1600 mg/kg.

1724   Fenoxaprop-P
113158-40-0   P

C16H12ClNO5

(R)-2-[4-[(6-Chloro-2-benzoxazolyl)oxy]phenoxy]-propanoic acid.
(R)-2-(4-((6-Chloro-2-benzoxazolyl)oxy)phenoxy)-propanoic   acid;
Fenoxaprop-P; HOE 046360; Propanoic acid, 2-(4-((6-chloro-2-
benzoxazolyl)oxy)phenoxy)-, (R)-. Herbicide. Hoechst AG.

1725   Fenoxaprop-P-ethyl
71283-80-2   P

C18H16ClNO5

Ethyl (R)-2-(4-((6-chloro-2-benzoxazolyl)oxy)phenoxy)-propanoate.
Fenoxaprop-p-ethyl; Fenoxaprop-p ethyl ester; Propanoic acid, 2-(4-
{(6-chloro-2-benzoxazolyl)oxy}phenoxy)-,   ethyl   ester,   (R)-.
Registered by EPA as a herbicide. Colorless solid; mp = 84 - 85°;
poorly soluble in H_2O (0.00009 g/100 ml), soluble in Me2CO (>39
g/100 ml), EtOH (> 0.8 g/100 ml), cyclohexane (> 0.78 g/100 ml), n-
octanol (> 0.8 g /100 ml), EtOAc (> 18 g/100 ml), C7H8 (> 25.8 g/100
ml); LD50 (mrat orl) = 3040 mg/kg, (frat orl) = 2090 mg/kg, (mus orl)
> 5000 mg/kg, (rat der) > 2000 mg/kg, (bobwhite quail orl) > 2510
mg/kg; LC50 (rat ihl) = 6.04 mg/l air. Aventis Crop Science; Aventis
Environmental Science USA LP; Syngenta Crop Protection.

1726   Fenpropathrin
39515-41-8   4033   G P

C22H23NO3

cyano(3-phenoxyphenyl)methyl   2,2,3,3-tetramethyl-
cyclopropanecarboxylate.
AI3-29234;   Caswell   No.   273H;   CCRIS 7727;   Cyano-3-
phenoxybenzyl-2,2,3,3-tetramethylcyclopropane-carboxylate;   α-
Cyano-3-phenoxybenzyl   2,2,3,3-tetra-
methylcyclopropanecarboxylate;   Cyclopropanecarbox-ylic   acid,
2,2,3,3-tetramethyl-,   cyano(3-phenoxyphenyl)-methyl   ester;
Danimen; Danitol; Danitol Fiori; EINECS 254-485-0; EPA Pesticide
Chemical Code 127901; Fenpropanate; Fenpropathrin; (±)-
Fenpropathrin;   2,2,3,3-Tetramethylcyclopropanecarboxylic   acid
cyano-3-phen-oxyphenyl)methyl ester; Fenpropathrine; Herald;
Kilumal; Meiothrin; Meothrin; Miothrin; OMS-1999; Ortho Danitol;
Rody; S 3206; SD 41706; Smash; Tame; WL 41706; XE-938. A
pyrethroid-based acaricide and insecticide with contact, stomach
and repellent action. Used for control of mites and insects in fruit and
vegetable crops. Liquid; mp = 45-50°; d^{25} = 1.15; almost insoluble in
H_2O (0.33 mg/l), more soluble in organic solvents; LD50 (rat orl) = 71
mg/kg. Shell UK.

1727   Fenpropidin
67306-00-7   4020   P

C19H31N

(±)-1-[3-[4-(1,1-Dimethylethyl)phenyl]-2-methylpropyl]-piperidine.
5-20-02-00071 (Beilstein Handbook Reference); BRN 1245248;
CGA 114900; Fenpropidin; Fenpropidine; Patrol; Piperidine, 1-(3-(4-
(1,1-dimethylethyl)phenyl)-2-methylpropyl)-; Ro 12-3049. Inhibitor of
ergosterol biosynthesis. Used as a fungicide. Pale yellow, viscous
liquid; bp0.004 = 100°; soluble in H_2O (0.035 g/100 ml at 25° and pH
7), Me2CO, CHCl3, EtOH, EtOAc, C7H16, xylene; LD50 (rat orl) =
1800 mg/kg, (mus orl) > 3200 mg/kg, (rat der) > 1800 mg/kg, (rat ip)
= 350 mg/kg, (mallard duck orl) = 1900 mg/kg, (pheasant orl) = 370
mg/kg, (bee orl 48 hr.) > 0.01 mg/bee, (bee contact 48 hr.) = 0.046
mg/bee; LC50 (rat ihl) = 1.22 mg/l air, (rainbow trout 96 hr.) = 2.6
mg/l, (mirror carp 96 hr.) = 3.6 mg/l, (bluegill sunfish 96 hr.) = 1.8
mg/l; (Daphnia 48 hr.) = 0.5 mg/l; toxic to bees. ICI Agrochemicals;
Maag; Syngenta Crop Protection.

1728   Fenpropimorph
67564-91-4   4021   P

C20H33NO

cis-4-[3-[4-(1,1-Dimethylethyl)phenyl]-2-methylpropyl]-2,6-dimethylmorpholine.

ACR-3320; BAS 42100F; BAS 421F; Corbel; EINECS 266-719-9; Fenpropimorph; cis-Fenpropimorph; Fenpropi-morph cis-form; Fenpropimorphe; Forbel; Mistral; Morpholine, 2,6-dimethyl-4-(3-(4-(1,1-dimethylethyl)-phenyl)-2-methylpropyl)-, cis-; Ro 14-3169; Task. Fungicide. Colorless oil; $bp_{0.005}$ = 120°; sg^{20} = 0.931; soluble in H_2O (0.00043 g/100 ml at 25° and pH 7), Me_2CO, $CHCl_3$, EtOAc, cyclohexane, C_7H_8, Et_2O, EtOH (all > 80 g/100 ml); LD_{50} (rat orl) = 3515 mg/kg, (mus orl) = 5980 mg/kg, (rat der) > 4000 mg/kg; LC_{50} (rat ihl 4 hr.) = 2.9 mg/l air, (rainbow trout 96 hr.) = 9.5 mg/l, (bluegill sunfish 96 hr.) = 3.9 mg/l, (carp 96 hr.) = 3.2 mg/l; non-toxic to bees. *BASF Corp.; Maag; Rhône-Poulenc; Syngenta Crop Protection.*

1729   Fenpyroximate
134098-61-6   4024   H

C24H27N3O4

4-[[[(E)-[(1,3-Dimethyl-5-phenoxy-1H-pyrazol-4-yl)meth-ylene]amino]oxy]methyl]benzoic acid 1,1-dimethylethyl ester.

Fenpyroximate; HOE-555-02A; NNI 850; Sequel. Registered by EPA as an insecticide and acaricide. Used on ornamentals. White crystals; mp = 101 - 102°; almost insoluble in H_2O (0.0.0000015 g/100 ml at 25°), soluble in MeOH (1.5 g/100 ml), C_6H_{14} (0.4 g/100 ml), xylene (17.5 g/100 ml); LD_{50} (mrat orl) = 480 mg/kg, (frat orl) = 245 mg/kg, (mfrat der) > 2000 mg/kg; LC_{50} (carp 48 hr.) = 0.0061 mg/l, (*Daphnia pulex* 3 hr.) = 0.085 mg/l. *AgrEvo; Hoechst AG.*

1730   Fensulfothion
115-90-2   4028   G P

C11H17O4PS2

O,O-Diethyl O-[4-(methylsulfinyl)phenyl] phosphoro-thioate.

AI3-24945; BAY 25141; Bayer 25141; BRN 2219515; Caswell No. 343; Chemagro 25141; Daconit; Dasanit; Diethyl p-methylsulfinylphenyl thiophosphate; p,O-Diethyl O-p-(methylsulfinyl)phenyl thiophosphate; EINECS 204-114-3; ENT 24,945; EPA Pesticide Chemical Code 032701; Fensulfothion; HSDB 1580; OMS 37; O,O-Diaethyl-O-4-methylsulfinyl-phenyl-monothiophos-phat; O,O-Diäthyl-O-4-methylsulfinyl-phenyl-monothio-phosphat; O,O-Diethyl O-p-(methylsulfinyl)phenyl thiophosphate; O,O-Diethyl O-4-methylsulphinylphenyl phosphorothioate; O,O-Diethyl O-(p-(methylsulfinyl)-phenyl) phosphorothioate; O,O-Diethyl O-(4-(methylsulfinyl)phenyl) phosphorothioate; Phenol, p-(methylsulfinyl)-, O-ester with O,O-diethyl phosphoro-thioate; Phosphorothioic acid, O,O-diethyl O-(4-(methylsulfinyl)phenyl) ester; S 767; Terracur® P; VUAgT 96;

VUAgT 108. Granular insecticide; used for treatment of biting insects and nematodes. Nematicide and insecticide with primarily contact action. Cholinesterase inhibitor. Used for control of nematodes and soil-dwelling insects. Registered by EPA as an insecticide and fungicide. Used as an insecticide and nematicide (cancelled). Yellow oil; $bp_{0.01}$ = 138-141°; d_{20} = 1.202; soluble in H_2O (0.154 g/100 ml), soluble in most organic solvents; LD_{50} (mrat orl)= 10.5 mg/kg, (frat orl) = 2.2 mg/kg, (gpg orl) = 9 mg/kg, (mrat der) = 30 mg/kg, (frat der) = 3.5 mg/kg, (bobwhite quail orl) = 40 mg/kg; LC_{50} (rat ihl 1 hr.) = 0.113 mg/l air, (bobwhite quail 5 day dietary) = 35 mg/kg diet, (mallard duck 5 day dietary) = 43 mg/kg diet, (bluegill sunfish 96 hr.) = 0.12 mg/l, (rainbow trout 96 hr.) = 8.6 mg/l, (golden orfe 96 hr.) = 6.8 mg/l; toxic to bees. *Bayer AG.*

1731   Fenthion
55-38-9   4030   G V

C10H15O3PS2

O,O-Dimethyl O-(4-methylthio-3-methylphenyl) thio-phosphate.

AI3-25540; B 29493; Bay 29493; Bay-Bassa; BAY 29493; Baycid; Baycid; Bayer 29493; Bayer 9007; Bayer S-1752; Baytex; BRN 1974129; Caswell No. 456F; CCRIS 310; EINECS 200-231-9; ENT 25,540; Entex; EPA Pesticide Chemical Code 053301; Fenthion; Fenthion 4E; Fenthion-methyl; Fenthione; Figuron; HSDB 1403; Lebaycid; Mercaptophos; 4-Methylmercapto-3-methylphenyl di-methyl thiophosphate; MPP; MPP (pesticide); NCI-C08651; OMS 2; Phenthion; Phosphorothioic acid, O,O-dimethyl O-(4-(methylthio)-m-tolyl) ester; Phosphoro-thioic acid, O,O-dimethyl O-(3-methyl-4-(methylthio)-phenyl) ester; Pro-Spot; Queletox; S 1752; Spotton; Sulfidophos; Talodex; Thiophosphate de O,O-dimethyle et de O-(3-methyl-4-methylthiophenyle); Tiguvon. Veterinary preparation; used on domestic animals against warble infestation and lice. Also used in agriculture as an insecticide with stomach and contact action. Oil; mp = 7.5°; $bp_{0.01}$ = 87°; d_{20} = 1.246; soluble in H_2O (2 mg/l), more soluble in organic solvents; LD_{50} (mrat orl) = 215 mg/kg, (frat orl) = 245 mg/kg. *Bayer AG; Bayer Corp., Ag. Div., Animal Health.*

1732   Fentin acetate
900-95-8   H P

C20H18O2Sn

Triphenyltin acetate.

Acetate de triphenyl-etain; Acetato di stagno trifenile; Acetatotriphenylstannane; Acetoxy-triphenyl-stannan; Acetoxy-triphenylstannane; Acetoxytriphenyltin; (Acetyl-oxy)triphenylstannane; AI3-25208; Batasan; Brestan; Brestan 60; BRN 4146107; Caswell No. 896C; EINECS 212-984-0; ENT 25208; EPA Pesticide Chemical Code 496700; Fenolovo acetate; Fentin acetaat; Fentin acetat; Fentin acetate; Fentin azetat; Fentinacetat; Fentine acetate; Fintin acetato; GC 6936; HOE-2824; HSDB 1783; Liromatin; Lirostanol; NSC 76068; OMS 1020; Phenostat A; Phentin acetate; Phentinoacetate; Stannane, (acetyloxy)triphenyl-; Stannane, acetoxytriphenyl-; Suzu; Tin, acetoxytriphenyl-; Tin triphenyl acetate; Tinestan; Tinestan 60 WP; Tinestan WP 20; Tinestan WP 60; TPTA; Tpza; Trifenil stagno acetato; Trifenyltinacetaat; Triphenyl-zinnacetat; Triphenylaceto stannane; Triphenylstannium acetate; Triphenyltin acetate; Triphenyltin(IV) acetate; Tubotin; VP 19-40; Acetate de triphenyl-etain; Acetato di stagno trifenile; Acetatotriphenylstannane; Acetoxy-triphenyl-stannan; Acetoxy-triphenylstannane; Acetoxytriphenyltin; Al3-25208; Batasan; Brestan; Brestan 60; BRN 4146107; Caswell No. 896C; ENT 25208; EPA Pesticide Chemical Code 496700; Fenolovo acetate; Fentin acetaat; Fentin acetat; Fentin acetate; Fentin azetat; Fentinacetat; Fentine acetate; Fintin acetato; GC 6936; HOE-2824; HSDB 1783; Liromatin; Lirostanol; NSC 76068; OMS 1020; Phenostat A; Phentin acetate; Phentinoacetate; Stannane, (acetyloxy)triphenyl-; Stannane, acetoxytriphenyl-. Fungicide, algicide and molluscicide. Colorless crystals; mp = 121.5°; sg^{20} = 1.5; poorly soluble in H_2O (0.0009 g/100 ml at 20° and pH 5), soluble in EtOH (2.2 g/100 ml), EtOAc (8.2 g/100 ml), C_6H_{14} (0.5 g/100 ml), CH_2Cl_2 (46 g/100 ml), C_7H_8 (8.9 g/100 ml); LD$_{50}$ (rat orl) = 140 - 298 mg/kg, (gpg orl) = 20 mg/kg, (rbt orl) = 30 - 50 mg/kg, (rat der) = 450 mg/kg, (mus der) = 350 mg/kg; LC$_{50}$ (mrat ihl 4 hr.) = 0.044 mgl air, (frat ihl 4 hr.) = 0.069 mg/l air, (carp 48 hr.) = 0.32 mg/l; non-toxic to bees. Aventis Crop Science; Hoechst AG.

1733   Fentin hydroxide
76-87-9   9816   P

$C_{18}H_{16}OSn$
Triphenyltin hydroxide.
Al3-28009; Brestan H 47.5 WP fungicide; Brestan R; BRN 4139186; Caswell No. 896E; CCRIS 612; Dowco 186; Du-Ter; Du-Ter Fungicide; Du-Ter Fungicide Wettable Powder; Du-Ter PB-47 Fungicide; Du-Ter W-50; Du-Tur Flowable-30; Duter extra; EINECS 200-990-6; ENT 28009; EPA Pesticide Chemical Code 083601; Erithane; Fenolovo; Fentin; Fentin hydroxide; Fentine; Fintin hydroxid; Fintin hydroxyde; Fintin idrossido; Fintine hydroxide; Fintine hydroxyde; Flo Tin 4l; Flo-Tin 4L; Haitin; Haitin WP 20 (fentin hydroxide 20%); Haitin WP 60 (fentin hydroxide 60%); HSDB 1784; Hydroxyde de triphenyl-etain; Hydroxytriphenylstannane; Hydroxytriphenyltin; Ida; Idrossido di stagno trifenile; Imc Flo-Tin 4L; K 19; NCI-C00260; NSC 113243; OMS 1017; Phenostat-A H; Phenostat-H; Stannane, hydroxytriphenyl-; Stannol, triphenyl-; Sunitron H; Super tin 4L; Super Tin; Super Tin 4L Gardian Flowable Fungicide; Suzu H; Tenhide; Tin, hydroxytriphenyl-; Tpth Technical; TPTH; Tptoh; Trifenyl-tinhydroxyde; Trifenylstanniumhydroxid; Triphenyl tin hydroxide; Triphenyl-zinnhydroxid; Tri-phenylhydroxytin; Triphenylstannanol; Triphenylstann-ium hydroxide; Triphenyltin hydroxide; Triphenyltin oxide; Triphenyltin(IV) hydroxide; Triple Tin; Triple Tin 4l; Tubotin; Vancide KS; Vito Spot Fungicide; Wesley Technical Triphenyltin Hydroxide. Registered by EPA as a fungicide. Suspension concentrate containing 625 g triphenyltin hydroxide per liter; used for prevention of potato blight and disease control in sugar beet. Flowable fungicide for pecans, potatoes, sugarbeets; restricted use. Crystals; mp = 119°, 122-122.5°; bp = 124.3°; d^{20} = 0.9943; poorly soluble in H_2O (0.0001 g/100 ml at 20°, pH 7), slightly soluble in EtOH (1 g/100 ml), CH_2Cl_2 (17.1 g/100 ml), Et$_2$O (2.8 g/100 ml), Me$_2$CO (5 g/100 ml), C_7H_8; LD$_{50}$ (mrat orl) = 171 mg/kg, (frat orl) = 110 mg/kg, (mmus orl) = 225 mg/kg, (fmus orl) = 209 mg/kg, (mgpg orl) = 27.1 mg/kg, (fgpg orl) = 31.1 mg/kg; LC$_{50}$ (bobwhite quail dietary 8 day) = 38.5 mg/kg diet; (guppies 48 hr.) = 0.054 mg/l, (carp 48 hr.) = 0.05 mg/l, (golden orfe (0.11 mg/l), (Japanese killifish 48 hr.) = 0.072 mg/l, (harlequin fish 48 hr.) = 0.042 mg/l; non-toxic to

bees. Agtrol International; Ashlade Formulations Ltd.; Aventis Crop Science; Cerexagri, Inc.; Chiltern Farm Chemicals Ltd; Dow AgroSciences; Griffin LLC; ICI Chem. & Polymers Ltd.

1734   Fenvalerate
51630-58-1   4038   G P

$C_{25}H_{22}ClNO_3$
4-Chloro-α-(1-methylethyl)benzeneacetic   acidcyano(3-phenoxyphenyl)methyl ester.
Agrofen; Al3-29235; Aqmatrine; Belmark; Benzeneacetic acid, 4-chloro-α-(1-methylethyl)-, cyano(3-phenoxy-phenyl)methyl ester; BRN 2025982; Caswell No. 077A; CCRIS 311;S-5602; Pydrin; Pyridin; SD-43775; Sumicidin; Tirade; WL-43775; (+)-α-Cyano-3-phenoxy-benzyl-(+)-α-(4-chlorophenyl)isovalerate; (RS)-α-Cyano-3-phenoxybenzyl(RS)-2-(4-chlorophenyl)-3-methylbutyrate; α-Cyano-3-phenoxybenzyl α-isopropyl-4-chlorophenyl-acetate; Cyano(3-phenoxybenzyl)methyl 2-(4-chloro-phenyl)-3-methylbutyrate; Cyano-(3-phenoxyphenyl)-methyl 4-chloro-α-(1-methylethyl)benzene acetate; α-Cyano-(3-phenoxyphenyl)methyl-4-chloro-α-(1-methyl-ethyl)benzeneacetate; Ectrin; EPA Pesticide Chemical Code 109301; EPA Shaughnessy Code: 109301; Evercide 2362; Evercide fenvalerate; Extrin; Fenaxin; Fenkem; Fenkill; Fenoxin; Fenval; Fenvalerate; Furitrothion; Gold crest tribute; HSDB 6640; Insectral; OMS-2000; Phenoxin; Phenvalerate; Pydrin; Pyridin; S 5602; Sanmarton; SD 43775; Sumibac; Sumicide; Sumicidin; Sumicidin 20E; Sumifleece; Sumifly; Sumipower; Sumitick; Sumitox; Sumkidin; Tirade; Tribute; WL 43775. Synthetic pyethroid insecticide lacking the cyclopropane ring. Registered by EPA as an insecticide. Viscous yellow liquid; d^{23} = 1.17; insoluble in H_2O (< 0.0001 g/100 ml, soluble in Me$_2$CO (> 79 g/100 ml), CHCl$_3$ (> 150 g/100 ml), EtOH (> 79 g/100 ml), MeOH (> 36 g/1200 ml), C_6H_{14} (10.2 g/100 ml); LD$_{50}$ (rat orl) = 451 mg/kg, (rat der) > 5000 mg/kg, (rbt der) = 2500 mg/kg, (domestic fowl orl) > 1600 mg/kg, (mallard duck orl) = 9932 mg/kg, LC$_{50}$ (quail dietary) > 10000 mg/kg, (mallard duck dietary) = 5500 mg/kg; very toxic to fish, LC$_{50}$ (rainbow trout 96 hr.) = 0.0036 mg/l; highly toxic to bees (LD$_{50}$ (bee contact) = 0.00023 mg/bee. Mclaughlin Gormley King Co; PET Chemicals; Sumitomo Corp.; Whitmire Micro-Gen Research Laboratories Inc.

1735   Ferric chloride
7705-08-0   4048   D G

FeCl$_3$
Iron(III) chloride.
Al3-51902; Caswell No. 459; CCRIS 2299; Chlorure ferrique; EINECS 231-729-4; EPA Pesticide Chemical Code 034901; Ferric chloride; Flores martis; HSDB 449; Iron chloride; Iron (III) chloride; Iron chloride (FeCl3); Iron perchloride; Iron sesquichloride; Iron trichloride; Natural molysite; NSC 51150; NSC 135798; Perchlorure de fer; UN1773; UN2582. Used in the treatment of sewage and industrial wastes; etching agent, mordant, disinfectant, pigment, feed additive. Has been used as an astringent. Dark hexagonal leaflets; mp = 300°; bp = 316°; d = 2.90; readiliy absorbs water; readily soluble in H_2O, EtOH, Et$_2$O, Me$_2$CO; LD$_{50}$ (rat orl) = 450 mg/kg. Asahi Denka Kogyo; BASF Corp.; Eka Nobel Ltd.; Mallinckrodt Inc.;

1736 Ferric hydroxide
20344-49-4   4055   G

FeHO$_2$
Ferric hydroxide oxide.
EINECS 243-746-4; Ferric hydroxide; Ferrugo; Hydrated ferric oxide; Iron hydroxide oxide (Fe(OH)O); Iron hydroxide oxide; Iron oxides. Red-brown powder or crystals. Used as a catalyst, pigment and in water purification. Red-brown crystals; d = 3.4-3.9; insoluble in H$_2$O, EtOH, soluble in mineral acids.

1737 Ferric nitrate
10421-48-4   4057   G

FeN$_3$O$_9$
Iron(III) nitrate.
EINECS 233-899-5; Ferric nitrate; HSDB 451; Iron nitrate; Iron (III) nitrate, anhydrous; Iron nitrate (Fe(NO$_3$)$_3$); Iron trinitrate; Nitric acid, iron(3+) salt; UN1466. Used in dyeing, tanning and analytical chemistry. [nonahydrate]; d = 1.68; mp = 47°; dec below 100°; d^{21} = 1.68; freely soluble in water, alcohol, acetone; LD$_{50}$ (rats orl) = 3.25 g/kg. General Chem; Hoechst Celanese; Mallinckrodt Inc.; Sherman Chem. Co.; Sigma-Aldrich Fine Chem.

1738 Ferric oxide
1309-37-1   4058   G

Fe$_3$O$_4$
Iron(III) oxide.
11554 Red; Anchred standard; Anhydrous iron oxide; Anhydrous oxide of iron; Armenian bole; Bauxite residue; Blackox; Black oxide of iron; Blended red oxides of iron; Burnt sienna; Burnt umber; Burntisland Red; C.I. 77491; C.I. Pigment Red 101; C.I. Pigment Red 102; C.I. Pigment Red 101 and 102; Calcotone Red; Caput mortuum; CCRIS 7330; CI 77491; Colcothar; Colloidal ferric oxide; Crocus martis adstringens; Deanox; Deanox DNX Pigments; Diiron trioxide; EINECS 215-168-2; Eisenoxyd; English Red; Ferric oxide; Ferric oxide (colloidal); Ferric oxide [Haematite and ferric oxide]; Ferric sesquioxide; Ferrous-ferric oxide; Ferroxide; Ferrugo; Foundrox; Hematite mineral; HSDB 452; Indian red; Iron oxide; Iron oxide (Fe$_2$O$_3$); Iron oxide dust; Iron oxide fume; Iron oxide pigments; Iron Oxide Red; Iron oxides; Iron Red; Iron sesquioxide; Iron trioxide; Iron(III) oxide; Jeweler's rouge; Kroma red; Levanox Red 130A; Light Red; Manufactured iron oxides; Mars Brown; Mars Red; Natural hematite; Natural iron oxides; Natural Red Oxide; Ocher; Ochre; OSO® 440; OSO® 1905; Pigment black 11; Pigment brown 6; Pigment brown 7; Pigment Red 101; Prussian Brown; Quick rouge; Raddle; Red Iron Oxide; Red ochre; Red oxide; Red oxide D3452; Red oxide D6984; Red oxide of iron; Rouge; Rubigo; Sal mineral; Sienna; Sienna brown; Specular iron; Specularite; Stone Red; Supra; Synthetic iron oxide; Trip; Venetian Red; Vitriol Red; Vogel's Iron Red; Yellow ferric oxide; Yellow oxide of iron. Foundry grade black iron oxide for core and mold use with particular emphasis on elimination of sub-surface porosity and carbon streaking when phenolic urethane sand binder systems are in use. Red-black powder; mp = 1538°; d = 5.24; insoluble in H$_2$O, organic solvents. &; Co; Color; DCS; Inc; Supply.

1739 Ferric sodium edetate
15708-41-5   4062   D G H

C$_{10}$H$_{12}$FeN$_2$NaO$_8$
Sodium [(ethylenedinitrilo)tetraacetato]ferrate(1-).
Calmosine; CCRIS 6795; Edathamil monosodium ferric salt; EINECS 239-802-2; Feredato sodico; Feredetate de sodium; Ferisan; Ferric sodium edetate; Ferric sodium EDTA; Ferrostrane; Ferrostrene; Monosodium ferric EDTA; Natrii feredetas; NSC 5237; Sequestrene Na Fe iron chelate; Sequestrene NaFe; Sodium feredetate; Sodium ferric EDTA; Sodium Iron EDTA; Sodium ironedetate; Sybron; Sytron. Used as a source of iron, a hematinic. Crystals.

1740 Ferric sulfate
10028-22-5   4065   G H P

Fe$_2$O$_{12}$S$_3$
Iron (III) sulfate.
CCRIS 7885; Coquimbite mineral; Diiron tris(sulphate); Diiron trisulfate; EINECS 233-072-9; Elliott's Lawn Sand; Ferric persulfate; Ferric sesquisulfate; Ferric sulfate; Ferric tersulfate; HSDB 6311; Iron persulfate; Iron sesquisulfate; Iron sulfate (2:3); Iron sulfate (Fe$_2$(SO$_4$)$_3$); Iron tersulfate; Iron(3+) sulfate; Iron(III) sulfate; Maxicrop Moss Killer & Conditioner; Sulfuric acid, iron(3+) salt (3:2); Vitax Turf Tonic; Hart Lawn Sand; Hart Moss Killer; Maxicrop Moss Killer & Conditioner; Taylors Lawn Sand; Vitax Micro Gran; Vitax Turf Tonic; Walkover Moss Killer; Elliott's Lawn Sand; Elliott's Moss Killer; Green-up Mossfree; Greenmaster Autumn; Greenmaster Mosskiller. Used for moss control in turf; polymerization catalyst; coagulant in water purification and sewage treatment; in etching aluminium. In pigments, reagent, etching aluminum, disinfectant, textile dyeing and printing, flocculant in water and sewage purification, soil conditioner, polymerization catalyst, metal pickling, chelated iron products, chemical intermediate. Registered by EPA as a herbicide. Gray-white powder or crystals; d^{18} = 3.097; slowly soluble in H$_2$O, sparingly soluble in EtOH, insoluble in Me$_2$CO, EtOAc. Boliden Intertrade; Central Garden & Pet; Faezy & Besthoff; Fisons plc, Horticultural Div.; Hart Prod.; Maxicrop International Ltd; Rhône-Poulenc; Thomas Elliott Ltd; UAP - Platte Chemical; Vitax Ltd.

1741 Ferron
547-91-1   4851   D G

C9H6INO4S
8-Hydroxy-7-iodo-5-quinolinesulfonic Acid.
5-22-07-00582 (Beilstein Handbook Reference); Anayodin; BRN 0223832; EINECS 208-938-4; Ferron; Ferron (analytical reagent); 8-Hydroxy-7-iodo-5-quinolinesulfonic acid; 8-Hydroxy-7-iodoquinoline-5-sulphonic acid; 8-Hydroxy-7-iodoquinolinesulfonic acid; 5-Iodo-8-quinolinol-5-sulfonic acid; 7-Iodo-5-sulfonic acid-8-hydroxyquinoline; 7-Iodo-8-hydroxylquinoline-5-sulfonic acid; 7-Iodooxine-5-sulfonic acid; Loretin; Loretine; Meditrene; NSC 3784; Quiniophen; 5-Quinolinesulfonic acid, 8-hydroxy-7-iodo-; Quinoxyl; Sefona; 5-Sulfo-7-iodo-8-quinolinol; 5-Sulfo-7-iodo-8-hydroxyquinoline; Yatren; Yellon; 8-hydroxy-7-iodoquinoline-5-carboxylic acid. Antiamebic and antiseptic. Used as a germicide. Yellow prisms; mp = 260° (dec); λ_m = 327, 355 nm (ε = 2512, 2512, 1N HCl), 335, 362 nm (ε = 15849, 63096, pH 9); soluble in H_2O (0.2 g/100 ml at 25°, 0.59 g/100 ml at 100°), slightly soluble in EtOH, insoluble in other organic solvents.

1742   Ferrous acetate
3094-87-9   G

C4H6FeO4
Iron (II) acetate.
Acetic acid, iron(2+) salt; Black liquor; Black mordant; EINECS 221-441-7; Ferrous acetate; Iron di(acetate); Iron diacetate; Iron(2+) acetate; Iron liquor; Liqueur de ferraile; Pyrolignite of iron. Prepared by the action of acetic acid upon iron turnings. The solution also contains ferric acetate and is used in calico printing, and in dyeing, for the preparation of blue, violet, black, and brown colors. Crystals; mp= 190°.

1743   Ferrous ammonium sulfate
10045-89-3   523   G P

FeH8N2O8S2
Ammonium iron(II) sulfate.
Ammonium ferrous sulfate; Ammonium iron sulfate; Ammonium iron sulfate (2:2:1); Ammonium iron(II) sulfate; Caswell No. 459B; Diammonium ferrous sulfate; Diammonium iron bis(sulphate); Diammonium iron sulfate; EINECS 233-151-8; EPA Pesticide Chemical Code 050506; Ferroammonsulfat; Ferrous ammonium sulfate; Ferrous ammonium sulphate; Ferrous diammonium disulfate; HSDB 456; Mohr's salt; Mohr'sches salz; Sulfuric acid, ammonium iron(2+) salt (2:2:1); Vitaferro; Diammonium iron bis(sulphate); EPA Pesticide Chemical Code 050506; Ferrous ammonium sulfate; Ferrous ammonium sulphate; Ferrous diammonium disulfate; HSDB 456; Mohr's salt; Sulfuric acid,

ammonium iron(2+) salt (2:2:1). Used in analytical chemistry, photography, as a polymerization catalyst and in metallurgy. Registered by EPA as a herbicide (cancelled). Blue-green crystals; d_4^{20}= 1.86; soluble in H_2O, insoluble in EtOH; LD50 (rat orl) = 3250 mg/kg. Sigma-Aldrich Fine Chem.

1744   Ferrous chloride
7758-94-3   4077   G

Cl2Fe
Iron(II) chloride.
EINECS 231-843-4; Ferro 66; Ferrous chloride; Ferrous dichloride; HSDB 459; Iron chloride; Iron chloride (FeCl2); Iron dichloride; Iron protochloride; Iron(2+) chloride; Iron(II) chloride; NA1759; NA1760; Natural lawrencite. Water treatment chemical. Used in metallurgy; as reducing agent; in pharmaceutical preparations; as mordant in dyeing. White-green crystals; mp = 674°; bp = 1023°; d^{25} = 2.90; freely soluble in H_2O, EtOH, Me_2CO, slightly soluble in C_6H_6, practicallly insoluble in Et_2O; LD50 (rat orl) = 900 mg/kg. Rheox Inc.

1745   Ferrous sodium HEDTA
16485-47-5   H

C10H15FeN2NaO7
Sodium   (N-(2-(bis(carboxymethyl)amino)ethyl)-N-(2-hydroxyethyl)glycinato(3-))ferrate(1-).
EINECS 240-541-1; Ferrate(1-), ((N-(carboxymethyl)-N'-(2-hydroxyethyl)-N,N'-ethylenediglycinato)(3-))-, sodium; Ferrate(1-), (N-(2-(bis(carboxymethyl)amino)ethyl)-N-(2-hydroxyethyl)glycinato(3-))-, sodium; Ferrous sodium HEDTA; Sodium   (N-(2-(bis(carboxymethyl)amino)ethyl)-N-(2-hydroxyethyl)glycinato(3-))ferrate(1-).

1746   Ferrous sulfate
7720-78-7   4091   G P

FeO4S
Iron(II) sulfate.
AI3-51903; CCRIS 6796; Combiron; Copperas; Duretter; Duroferon; EINECS 231-753-5; Exsiccated ferrous sulfate; Exsiccated ferrous sulphate; Feosol; Feospan; Fer-In-Sol; Ferobuff; Ferolix; Ferralyn; Ferro-gradumet; Ferro-Theron; Ferrosulfat; Ferrosulfate; Ferrous sulfate; Fersolate; Ferulen; Ferusal; Green Salts; Green Vitriol; HSDB 465; Iron monosulfate; Iron sulfate (1:1); Iron sulfate (FeSO4); Iron sulphate; Iron vitriol; Iron(2+) sulfate; Iron(2+) sulfate (1:1); Iron(II) sulfate; Irospan; Kesuka; Mol-Iron; NSC 57631; NSC 146177; Odophos; Quickfloc (salt); Sal Chalybis; SFE 171; Slow Fe; Sorbifer durules; Sulferrous; Sulfuric acid, iron(2+) salt (1:1). Iron oxide pigment, other iron salts, ferrites, water and sewage treatment, catalyst especially for synthetic ammonia, fertilizer, feed additive, flour enrichment, reducing agent, herbicide, wood preservative, process engraving. Green crystals. EM Ind. Inc.; Huber J. M.;

Mallinckrodt Inc.

1747   Ferrous sulfate heptahydrate
7782-63-0   4091   D G P

$$O=\overset{\displaystyle O^-}{\underset{\displaystyle O}{\overset{\displaystyle |}{\underset{\displaystyle |}{S}}}}-O^- \quad Fe^{2+} \quad 7H_2O$$

$FeO_4S.7H_2O$

Iron(II)Sulfate Heptahydrate.

Caswell No. 460; CCRIS 7331; EPA Pesticide Chemical Code 050502; Ferrous sulfate, heptahydrate; Fesofor; Fesotyme; Haemofort; Iron protosulfate; Iron sulfate heptahydrate; Iron(2+) sulfate heptahydrate; Iron(II) sulfate heptahydrate; Iron vitriol; Ironate; Irosul; Melanterite mineral; Presfersul; Sal Martis; Salt of steel; Shoemaker's Black; Siderotil mineral; Sulfuric acid, iron(2+) salt (1:1), heptahydrate; Szomolnikite mineral; Tauriscite mineral. As a chemical intermediate, in electroplating, as a pesticide and medicinally as a hematinic. Registered by EPA as a herbicide (cancelled). Blue-green monoclinic crystals; d = 1.897; loses H_2O to form the tetrahydrate at 56.6°, and the monohydrate at 65°; soluble in H_2O, insoluble in EtOH; LD$_{50}$ (mus iv) = 65 mg/kg, (mus orl) = 1520 mg/kg. *Generic; Sigma-Aldrich Co.*

1748   Ferrous sulfide
1317-37-9   4092   G

$$Fe{=}S$$

FeS

Iron(II) sulfide.

Black iron sulfide; CI 77540; EINECS 215-268-6; Ferrous monosulfide; Ferrous sulfide (FeS); HSDB 5803; Iron monosulfide; Iron monosulfide (FeS); Iron protosulfide; Iron sulfide (FeS); Iron sulfuret; Iron sulphide; Magnetkies; Pyrrhotine; Troillite. A variety of iron pyrites. Used as a source of H_2S, in ceramics, as a pigment, in anodes and lubricant coatings. Colorless-grey crystals; mp = 1194°; d = 4.84; insoluble in H_2O, soluble in acids.

1749   Ferrous sulfide
12068-85-8   G

$$^-S{-}S^- \quad Fe^{2+}$$

FeS_2

Iron(II) sulfide.

Coxcomb pyrites; EINECS 235-106-8; Iron disulphide; Iron sulfide (FeS$_2$); Marcasite; Radiated pyrites; White iron pyrites. Iron disulfide, a mineral. The term marcasite is also occasionally applied to bismuth.

1750   Fexofenadine hydrochloride
153439-40-8   4101   D

$C_{32}H_{40}ClNO_4$

(±)-p-[1-Hydroxy-4-[4-(hydroxydiphenylmethyl)piperid-ino]butyl]-α-methylhydratropic acid hydrochloride.

Allegra; Fexofenadine hydrochloride; MDL 16,455A; Telfast. Antihistaminic. *Hoechst AG (USA).*

1751   Fichtelite
2221-95-6   4108   G

$C_{19}H_{34}$

[1S-(1α,4aα,4bβ,7β,8aα,10aβ)-tetradecahydro-1,4a-dimethyl-7-(1-methylethyl)phenanthrene.

(1S-(1α,4aα,4bβ,7β,8aα,10aβ))-Tetradeca-hydro-1,4a-dimethyl-7-(1-methylethyl) phenanthrene; Fichtelite; 18-Norabietane. A hydrocarbon, $C_{19}H_{34}$, found in fossil coniferous resins. Crystals; mp = 45-46°; bp$_{43}$ = 235-236°; d$_4^{22}$ 0.9380; n$_D^{20}$ 1.5052; [α]$_D$ = 19°.

1752   Flamprop-M-isopropyl (D-)
63782-90-1   G P

$C_{19}H_{19}ClFNO_3$

N-Benzoyl-N-(3-chloro-4-fluorophenyl)-D-alanine iso-propyl ester.

Barnon plus; BRN 2899801; Commando; D-Alanine, N-benzoyl-N-(3-chloro-4-fluorophenyl)-, isopropyl ester; Effix; Flame; Flamprop-m-isopropyl; Gunner; Power Flame; Isopropyl N-benzoyl-N-(3-chloro-4-fluorophenyl)-D-alaninate; Mutaven L; Suffix BW; Super Barnon; Super Suffix; Superbarnon; WL 43 423; WL 43 425. Herbicide. Used for control of wild oats in cereal crops. Colorless crystals; mp = 72 - 75°; soluble in H_2O (0.001 g/100 ml at 20°), Me$_2$CO (> 40 g/100 ml), cyclohexanone (68 g/100 ml), EtOH (15 g/100 ml), xylene (≅50 g/100 ml), C$_6$H$_{14}$ (≅0.6 g/100 ml); LD$_{50}$ (rat, mus orl) > 4000 mg/kg, (rat der) > 1600 mg/kg, (rat ip) > 1200 mg/kg, (fowl orl) > 2000 mg/kg; LC$_{50}$ (rainbow trout 96 hr.) = 3.3 mg/l; non-toxic to bees. *BASF Corp.; Shell.*

1753   Flamprop-M-isopropyl (L-)
57973-67-8   G P

C19H19ClFNO3
N-Benzoyl-N-(3-chloro-4-fluorophenyl)-L-alanine isopropyl ester.
EINECS 261-051-4; Isopropyl N-benzoyl-N-(3-chloro-4-fluorophenyl)-L-alaninate. Herbicide. Used for control of wild oats in cereal crops. Crystals; mp - 72 - 75°; soluble in H2O (0.001 g/100 ml), Me2CO (> 40 g/100 ml), cyclohexanone (68 g/100 ml), EtOH (15 g/100 ml), xylene (≅50 g/100 ml), C6H14 (0.6 g/100 ml); LD50 (rat, mus orl) > 4000 mg/kg, (rat der) > 1600 mg/kg, (rat ip) > 1200 mg/kg, (domestic fowl orl) > 2000 mg/kg; LC50 (rainbow trout 96 hr.) = 3.3 mg/l; non-toxic to bees. *BASF Corp.; Shell.*

1754   Flavone
525-82-6   4120   G

C15H10O2
2-phenylbenzopyran-4-one.
5-17-10-00552 (Beilstein Handbook Reference); Asmacoril; 4H-1-Benzopyran-4-one, 2-phenyl-; BRN 0157598; CCRIS 4288; Chromocor; Cromaril; Cromarile; EINECS 208-383-8; Flavone; NSC 19028; 2-Phenyl-4-benzopyron; 2-Phenyl-4H-1-benzopyran-4-one; 2-Phen-yl-γ-benzopyrone; Phenylchromone; 2-Phenylchromone; 2-Phenyl-4-chromone; 2-Phenyl-4H-chromen-4-on. One of a group of flavonoid plant pigments existing as colorless needles, insoluble in water and melting at about 100°. The flavones produce ivory and yellow colors in plants and flowers. Needles; mp = 100°; λ_m = 250, 350 nm (ε = 12589, 15849, MeOH); insoluble in H2O, soluble in EtOH, Et2O, Me2CO,C6H6, CHCl3, ligroin; LD50 (mus orl) = 2500 mg/kg.

1755   Fluazifop-butyl
69806-50-4   4142   G P

C19H20F3NO4
2-[4-[[5-(Trifluoromethyl)-2-pyridinyl]oxy]phenoxy]propanoic   acid butyl ester.
BRN 1510062; (±)-Butyl-2-(4-(((5-trifluoro-methyl)-2-pyridinyl)oxy)phenoxy)propanoate; Butyl (RS)-2-(4-((5-(tri-fluoromethyl)-2-pyridinyl)oxy)phenoxy)propanoate; Butyl 2-(4-(5-trifluoromethyl-2-pyridinyloxy)phenoxy)propan-oate; Butyl 2-(4-(5-trifluoromethyl-2-pyridyloxy)phenoxy)-propionate; Caswell No. 460C; EPA Pesticide Chemical Code 122805; Fluazifop butyl ester; Fluazifop butyl; Fusilade; Hache uno super; Halokon; HSDB 6644; IH 773B; Onecide; Onecide EC; PP 009; Propanoic acid, 2-(4-((5-(trifluoromethyl)-2-pyridinyl)oxy)phenoxy)-, butyl ester; Propionic acid, 2-(p-((5-(trifluoromethyl)-2-pyridyl)oxy)phenoxy)-, butylester; SL-236; TF 1169; 2-(4-((5-(Trifluoromethyl)-2-pyridinyl)oxy)phenoxy)propanoic acid butylester; TS-7236. Used for grass weed control for broad-leaved crops. Registered by EPA as a herbicide (cancelled). Straw-colored liquid; mp = 13°; bp0.5 = 170°; sg^{20} = 1.21; poorly soluble in H2O (0.0002 g/100 ml), soluble in Me2CO, cyclohexanone, C6H14, MeOH, CH2Cl2, xylene; LD50 (rat orl) = 3328 mg/kg, (mmus orl) = 1490 mg/kg, (fmus orl) = 1770 mg/kg, (mgpg orl) = 2659 mg/kg, (rbt orl) = 621 mg/kg, (rat der) >

6050 mg/kg, (rbt der) >2420 mg/kg, (rat ip) = 1761 mg/kg, (mallard duck orl) > 17000 mg/kg; LC50 (mallard duck dietary 5 day) > 25000 mg/kg diet (ring-necked pheasant dietary 5 day) > 18500 mg/kg diet; LC50 (rainbow trout 96 hr.) = 1.37 mg/l, (mirror carp 96 hr.) = 1.31 mg/lg, (bluegill sunfish 96 hr.) = 0.53 mg/lg, low toxicity towards bees. *Dow UK; ICI Chem. & Polymers Ltd.; Ishihara Sangyo; Syngenta Crop Protection.*

1756   Flubendazole
31430-15-6   4144   D G V

C16H12FN3O3
Methyl 5-(p-fluorobenzoyl)-2-benzimidazolecarbamate.
CCRIS 4480; EINECS 250-624-4; Flubendazol; Flubendazole; Flubendazolum; Flubenol; Flumoxal; Flumoxane; Fluvermal; NSC 313680; R 17,889. Anthelmintic; antiprotozoal. Used in veterinary medicine. mp = 260°; LD50 (mus, rat, gpg orl) > 2560 mg/kg. *Abbott Laboratories Inc.; Janssen Chimica.*

1757   Flubenzimine
37893-02-0   4145   G

C17H10F6N4S
N-(3-Phenyl-4,5-bis((trifluoromethyl)imino)-2-thiazol-idinylidene)benzenamine.
BAY slj 0312; Benzenamine, N-(3-phenyl-4,5-bis((tri-fluoromethyl)imino)-2-thiazolidinylidene)-; BRN 24036-90; Cropotex®; (2Z,4E,5Z)-N2,3-Diphenyl-N4,N5-bis(tri-fluoromethyl)-1,3-thiazolidine-2,4,5-triylidene-triamine; EINECS 253-703-1; Flubenzimine; N-(3-Phenyl-4,5-bis((trifluoromethyl)imino)thiazolidin-2-ylidene)-aniline; N-(3-Phenyl-4,5-bis((trifluoromethyl)imino)-2-thiazolid-inylidene)benzenamine; SLJ 0312. Contact acaricide for control of spider mites on pome and stone fruit, plums, damsons, citrus fruit and citrus rust mites. *Bayer AG.*

1758   Fluconazole
86386-73-4   4148   D

C13H12F2N6O

2,4-Difluoro-α,α-bis(1H-1,2,4-triazol-1-ylmethyl)benzyl alcohol. Afungil; Alflucoz; Baten; Biocanol; Biozolene; Canzol; CCRIS 7211; Cryptal; Diflucan; Dimycon; DRG-0005; Elazor; Flucazol; Fluconazol; Fluconazole; Flucon-azolum; Flukezol; Flunizol; Fluzone; Forcan; Fungata; Mutum; Oxifugol; Pritenzol; Syscan; Triflucan; UK 49858; Zemyc; Zonal. Antifungal. Crystals; mp = 138-140°. *Roerig Div. Pfizer Pharm.*

1759   Fludioxonil
131341-86-1   4155   P

C12H6F2N2O2

4-(2,2-Difluoro-1,3-benzodioxol-4-yl)-1H-pyrrole-3-carbonitrile. CGA 100-TLO; CGA-173506; Fludioxonil; Maxim. Fungicide. Colorless crystals; mp = 199°; soluble in H2O (0.000153 g/100 ml at 20°), Me2CO (9.5 g/100 ml), cyclohexanone (16.9 g/100 ml), N-methylpyrrolidone (48 g/100 ml), MeOH (< 1.6 g/100 ml), CH2Cl2 (< 2.7 g/100 ml), C7H8 (< 17.2 g/100 ml), C6H14 (< 13.2 g/100 ml); LD50 (rat orl) > 2000 mg/kg, (rat der) > 2000 mg/kg; LC50 (rat ihl 4 hr.) > 2.6 mg/l air; toxic to fish; non-toxic to bees. *Ciba-Geigy Corp.; Syngenta Crop Protection.*

1760   Fluometuron
2164-17-2   4180   H P

C10H11F3N2O

1,1-Dimethyl-3-(α,α,α-trifluoro-m-tolyl)urea. BRN 2217354; C 2059; Caswell No. 460A; CCRIS 314; Ciba 2059; Cotogard; Cotoran; Cotoran multi 50WP; Cottonex; EINECS 218-500-4; EPA Pesticide Chemical Code 035503; Fluometuron; Herbicide C-2059; Higalcoton; HSDB 1721; Lanex; NCI-C08695; Pakhtaran. A flowable herbicide used to control annual grasses and broadleaf weeds in cotton and sugarcane. Fluometuron is a residual herbicide effective against a wide range of both annual broadleaf weeds and grasses. Registered by EPA as a herbicide. Crystals; mp = 163 - 164.5°; soluble in H2O (0.0105 g/100 ml at 25°), MeOH (11 g/100 ml), Me2CO (10.5 g/100 ml), CH2Cl2 (2.3 g/100 ml), n-octanol (2.2 g/100 ml), C6H14 (0.017 g/100 ml); LD50 (rat orl) = 6416 - 8900 mg/kg, (rat der) > 2000 mg/kg, (rbt der) > 10000 mg/kg; LC50 (Japanese quail 8 day dietary) = 4620 mg/kg diet, (mallard duck 8

day dietary) = 4500 mg/kg diet, (ringneck pheasant 8 day dietary) = 3150 mg/kg diet; (rainbow trout 96 hr.) = 47 mg/l, (bluegill sunfish 96 hr.) = 96 mg/l, (catfish 96 hr.) = 55 mg/l, (crucian carp 96 hr.) = 170 mg/l; non-toxic to bees. *Agan Chemical Manufacturers; Agriliance Crop Nutrient; Agriliance LLC; Cedar Chemical Corp.; Griffin LLC; Helena Chemical Co.; Makhteshim-Agan; Micro-Flo Co. LLC; Syngenta Crop Protection.*

1761   Fluoranthene
206-44-0   G

C16H10

Benzo(jk)fluorene. 1,2-Benzacenaphthene; Benzene, 1,2-(1,8-naphthylene)-; Benzene, 1,2-(1,8-naphthalenediyl)-; CCRIS 1034; EINECS 205-912-4; Fluoranthene; HSDB 5486; Idryl; 1,2-(1,8-Naphthalenediyl)benzene; 1,2-(1,8-Naphthylene)-benzene; NSC 6803; RCRA waste number U120. Used in chemical research. Crystals; mp = 109-111°; bp = 384°; insoluble in H2O, soluble in organic solvents; LD50 (rat orl) = 2 gm/kg.

1762   Fluorescein
2321-07-5   4185   G

C20H12O5

3',6'-Dihydroxyspiro[isobenzofuran-1(3H),9'-[9H]xanth-en]-3-one. 11712 Yellow; 5-19-06-00456 (Beilstein Handbook Reference); Benzoic acid, 2-(6-hydroxy-3-oxo-3H-xanthen-9-yl)-; Benzoic acid, o-(6-hydroxy-3-oxo-3H-xanthen-9-yl)-; BRN 0094324; 9-(o-Carboxyphenyl)-6-hydroxy-3-isoxanthenone; 9-(o-Carboxyphenyl)-6-hydroxy-3H-xanthen-3-one; C.I. 45350:1; C.I. Solvent Yellow 94; CCRIS 7076; D & C Yellow no. 7; 3',6'-Dihydroxyfluoran; 3',6'-Dihydroxyspiro(isobenzofuran-1(3H),9'-(9H)xanthen)-3-one; 3,6-Dihydroxyspiro-(xanthene-9,3'-phthalide); EINECS 219-031-8; Fluoran, 3',6'-dihydroxy-; 3,6-Fluorandiol; Fluorescein; Fluor-escein acid; Fluorescein Red; Fluoresceine; Hidacid fluorescein; HSDB 2128; 2-(6-Hydroxy-3-oxo-(3H)-xanthen-9-yl)benzoic acid; Japan Yellow No. 201; Resorcinolphthalein; Soap Yellow F; Solvent Yellow 94; Spiro(isobenzofuran-1(3H),9'-(9H)xanthen)-3-one, 3',6'-dihydroxy-; Yellow fluorescein; 3H-Xanthen-3-one, 9-(o-carboxyphenyl)-6-hydroxy-; Zlut kysela 73. Used to dye seawater for spotting purposes, as a tracer to locate impurities in wells, in dyeing silk and wool, as a diagnostic aid in ophthalmology, an indicator and a reagent for bromine. Crystals; mp = 314-316°; slightly soluble in H2O, C6H6, CHCl3, Et2O, soluble in hot EtOH, AcOH, alkali hydroxides or carbonates. *EM Ind. Inc.; Greeff R.W. & Co.; Hilton Davis; Kraeber GmbH; U.S. BioChem.*

1763   Fluorescein disodium salt
518-47-8   4185   G

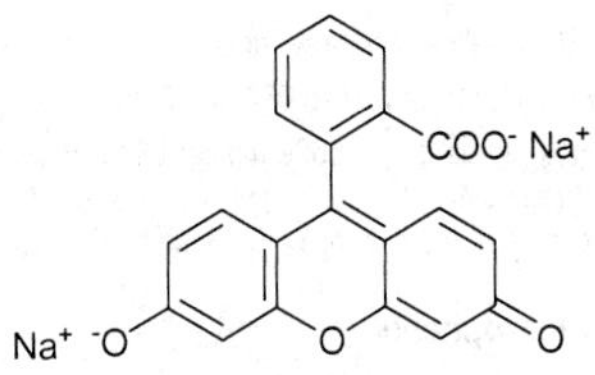

C20H10Na2O5

3',6'-Dihydroxyspiro[isobenzofuran-1(3H),9'-[9H]xanth-en]-3-one disodium.

11824 Yellow; 12417 Yellow; Acid Yellow 73; Aizen uranine; Basacid Yellow 226; C.I. 45350; C.I. Acid Yellow 73; Calcocid uranine B4315; 9-o-Carboxyphenyl-6-hydroxy-3-isoxanthone, disodium salt; CCRIS 6239; Certiqual fluoresceine; CI 45350 (Na salt); D&C Yellow No. 8; 3',6'-Dihydroxyspiro(isobenzofuran-1(3H),9'-(9H)xanthen)-3-one, disodium salt; Disodium 6-hydroxy-3-oxo-9-xanthene-o-benzoate; Disodium 2-(3-oxo-6-oxidoxanthen-9-yl)benzoate; EINECS 208-253-0; Fluor-I-Strip; Fluor-I-strip A.T.; Fluorescein disodium salt; Fluorescein LT; Fluorescein sodium; Fluorescein, soluble; Fluorescite; Flurenate; Fluress; Ful-Glo; Funduscein; Furanium; Hidacid uranine; Japan Yellow 202(1); NCI-C54706; Obiturine; Resorcinol phthalein sodium; Sodium fluorescein; Sodium fluoresceinate; Sodium salt of hydroxy-o-carboxy-phenyl-fluorone; Soluble fluorescein; Soluble Fluoresceine BPS; Spiro(isobenzofuran-1(3H),9'-(9H)xanthen)-3-one, 3',6'-dihydroxy-, disodium salt; Uranin; Uranin A; Uranin S; Uranine; Uranine A; Uranine A Extra; Uranine O; Uranine SS; Uranine WSS; Uranine Yellow. Used to dye silk and wool yellow. Used as a diagnostic aid in corneal examination. mp = 314-316°; λ_m = 460, 493.5 nm; slightly soluble in H_2O, alcohols; LD50 (rat orl) = 6721 mg/kg. *Wyeth Laboratories.*

1764 Fluorescent whitening agent (stilbene-based)

16470-24-9   H

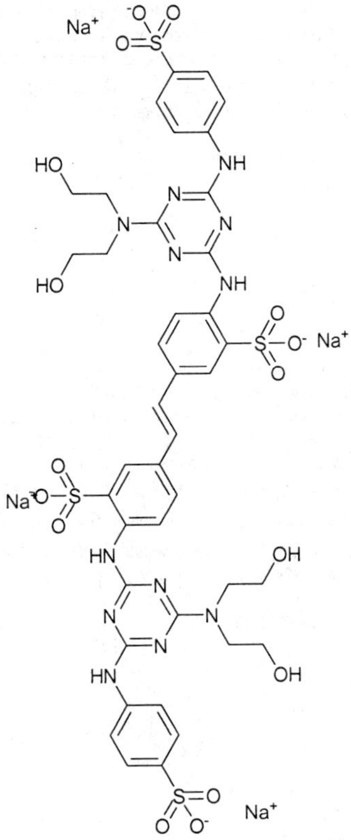

C40H40N12Na4O16S4

2,2'-Stilbenedisulfonic acid, 4,4'-bis((4-(bis(2-hydroxy-ethyl)amino)-6-(p-sulfoanilino)-s-triazin-2-yl)amino)-, tetrasodium salt.

Blancofor BBU; Blankophor BBU; C.I. Fluorescent Brightener 220; EINECS 240-521-2; Fluorescent Brightener 220; Fluorescent whitening agent (stilbene-based); Leucophor·U.

1765 Fluorfolpet

719-96-0   G P

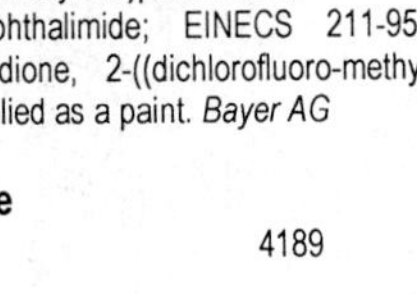

C9H4Cl2FNO2S

N-(Fluordichloromethylthio) phthalimid.

N-(Dichlorofluoromethylthio)phthalimide; Dichloro-fluoromethyl)thio)phthalimide; EINECS 211-952-3; Fluorofolpet; Isoindole-1,3(2H)-dione, 2-((dichlorofluoro-methyl)thio)-; Preventol A3. Fungicide applied as a paint. *Bayer AG*

1766 Fluorine

7782-41-4   4189   G

$$F\text{—}F$$

F2

Fluorine.

Bifluoriden; EINECS 231-954-8; Fluor; Fluorine; Fluoro; Fluorures acide; Fluoruri acidi; HSDB 541; RCRA waste number P056; Saeure fluoride; Säure fluoride; UN1045. Nonmetallic halogen element; Used in the production of metallic and other fluorides, fluorocarbons, fluoridating compounds for drinking water and toothpaste. Gas; mp = -219°; bp = -188.13°; d (liquid) = 1.5127; LD$_{50}$ (rat ihl 1hr) = 185 ppm. *Air Prods & Chems; Allied Signal; Solvay Deutschland GmbH.*

1767 Fluorobenzene
462-06-6   4917   G H

C$_6$H$_5$F
Monofluorobenzene.
AI3-28560; Benzene, fluoro-; EINECS 207-321-7; Fluorobenzene; Monofluorobenzene; NSC 68416; Phenyl fluoride; UN2387. Intermediate in the manufacture of insecticides, larvicides and pharmaceuticals, identification reagent for plastic or resin polymers. Liquid; mp = -42.2°; bp = 84.7°; d^{20} = 1.0225; λ_m = 259 nm (ε = 513, H$_2$O); insoluble in H$_2$O, very soluble in C$_6$H$_6$, Et$_2$O, EtOH, ligroin; LD$_{50}$ (rat orl) = 4399 mg/kg. *Hoechst Celanese; ICI Americas Inc.; Lancaster Synthesis Co.*

1768 Fluoroform
75-46-7   4200   G H

CHF$_3$
Trifluoromethane.
4-01-00-00024 (Beilstein Handbook Reference); Arcton 1; BRN 1731035; Carbon trifluoride; EINECS 200-872-4; FC 23 (fluorocarbon); Fluoroform; Fluoryl; Freon 23; Freon F-23; Freon R 23; Genetron-23; Halocarbon 23; HSDB 5207; Methane, trifluoro-; Methyl trifluoride; Propellant 23; R 23; Refrigerant 23; Trifluoromethane; UN1984; UN3136. Low temperature refrigerant replacing CFC-13 and R-503 in low stage of cascade systems. Used as a refrigerant, intermediate in organic synthesis, direct coolant for infrared detector cells and blowing agent for infrared foams. Gas; mp = -155.1°; bp = -82.1°; d^{25} = 0.673; soluble in organic solvents, less soluble in H$_2$O; non-toxic. *Allied Signal.*

1769 Fluoromide
41205-21-4   G P

C$_{10}$H$_4$Cl$_2$FNO$_2$
3,4-Dichloro-1-(4-fluorophenyl)-1H-pyrrole-2,5-dione.
BRN 1534427; 2,3-Dichloro-N-4-fluorophenylmaleimide; 3,4-Dichloro-1-(4-fluorophenyl)-1H-pyrrole-2,5-dione; Fluoroimide; MK 23; N-p-Fluorophenyl-2,3-dichloro-maleimide; 1H-Pyrrole-2,5-dione, 3,4-dichloro-1-(4-fluorophenyl)-; Sparticide; Spat. Foliar fungicide with protective action. Used for control of scab, *Alternaria* leaf spot and powdery mildew in apples, scab of citrus fruit and coffee berry

disease. Pale yellow crystals; mp = 240 - 242°; soluble in H$_2$O (0.00059 g/100 ml), MeOH (0.067 g/100 ml); LD$_{50}$ (rat, mus orl) > 15000 mg/kg, (mus der) > 5000 mg/kg; LC$_{50}$ (mrat ihl 4 hr.) = 0.57 mg/l air, (frat ihl 4 hr.) = 0.72 mg/l air, (pheasant 5 day dietary) > 27000 mg/kg diet, (carp 48 hr.) = 5 - 6 mg/l. *Mitsubishi Chem. Corp.*

1770 Fluorophenol
371-41-5   G

C$_6$H$_5$FO
p-Fluorophenol.
4-06-00-00773 (Beilstein Handbook Reference); BRN 1362752; CCRIS 665; EINECS 206-736-0; 4-Fluorophenol; NSC 10295; Phenol, 4-fluoro-; Phenol, p-fluoro-. Fungicide, intermediate for pharmaceuticals. mp = 48°; bp = 185.5°; d^{56} = 1.1889; λ_m = 211 nm (ε = 4169, cyclohexane); slightly soluble in H$_2$O, soluble in Me$_2$CO, petroleum ether. *ICI Americas Inc.; Lancaster Synthesis Co.; PCR.*

1771 Fluorotributyltin
1983-10-4   G P

C$_{12}$H$_{27}$FSn
Tributyltin fluoride.
BioMeT TBTF; BRN 4125568; Caswell No. 867C; EINECS 217-847-9; EPA Pesticide Chemical Code 083112; Fluorotributylstannane; NSC 179737; NSC 195319; Stannane, fluorotributyl-; Stannane, tributylfluoro-; Tri-n-butylstannyl fluoride; Tin, tributyl-, fluoride; Tin, tributylfluoro-; Tributylfluorostannane; Tributylstannane fluoride; Tributyltin fluoride. Antifoulant used in marine paints. Registered by EPA as a fungicide (cancelled). Crystals; mp = 270° (dec). *Elf Atochem N. Am.; Fluka.*

1772 Fluoxetine
54910-89-3   4211   D

C$_{17}$H$_{18}$F$_3$NO
(±)-N-Methyl-3-phenyl-3-[(α,α,α-trifluoro-p-tolyl)oxy]-propylamine.
Animex-On; Benzenepropamine, N-methyl-γ-(4-(trifluoromethyl)phenoxy)-, (±)-; Benzenepropanamine, N-methyl-γ-(4-(trifluoromethyl)phenoxy)-, (±)-; Eufor; Flu-oxetina; Fluoxetine; Fluoxetinum; HSDB 6633; (+) or (-)-N-Methyl-3-phenyl-3-((α,α,α-trifluoro-p-tolyl)oxy)propyl-amine; (±)-N-Methyl-3-phenyl-3-((α,α,α-trifluoro-p-tolyl)-oxy)propylamine; (+) or (-)-N-Methyl-γ-(4-(trifluoro-meth-yl)phenoxy)benzenepropanamine; (±)-N-Methyl-γ-(4-(trifluoromethyl)phenoxy)benzenepropanamine; N-Methyl-3-(p-trifluoromethylphenoxy)-3-phenylpropyl-amine; Por-tal; Prozac; Pulvules; dl-3-(p-Trifluoro-methylphenoxy)-N-methyl-3-phenylpropylamine. Antidepressant. Serotonin uptake inhibitor. *C.M.*

1773   Fluoxetine hydrochloride
56296-78-7   D

C17H19ClF3NO

(±)-N-Methyl-3-phenyl-3-[(α,α,α-trifluoro-p-tolyl)oxy]propylamine hydrochloride.

Adofen; Affectine; Alzac 20; Ansilan; CCRIS 6150; Deproxin; EINECS 260-101-2; Erocap; Fluctin; Fluctine; Fludac; Flufran; Flunil; Fluox-Puren; Fluoxac; Fluoxeren; Fluoxetine hydrochloride; Fluoxil; Flutin; Flutine; Fluxen; Fluxil; Fontex; Foxetin; Lilly 110140; Lorien; Lovan; LY 110140; Margrilan; Modipran; Neupax; Nopres; Nuzak; Oxedep; Pragmaten; .Prizma; Proctin; Prodep; Prozac; Prozac 20; Reneuron; Rowexetina; Sanzur; Sinzac; Zactin; Zepax. Serotonin uptake inhibitor. Antidepressant. Crystals; mp = 158-159°; slightly soluble in H2O (1-2 mg/ml), ethyl acetate, toluene, CHCl3, C6H14 (0.5-0.77 mg/ml); soluble in MeOH, ETOH (>100 mg/mg), acetonitrile (33-100 mg/ml), C6H6; λm = 227, 264, 268, 275 nm (E$_1^{1\%}$ cm 372, 29,29,22 MeOH); LD50 (rat orl) = 452 mg/kg. *C.M. Ind.; Eli Lilly & Co.*

1774   Flupirtine
56995-20-1   4216   G D

C15H17FN4O2

[2-Amino-6-[[(4-fluorophenyl)methyl]amino]-3-pyr-idinyl]carbamic acid ethyl ester.

D-9998; EINECS 260-503-8; Flupirtine; Flupirtino; Flupirtinum; Katadolon. Substituted pyridine with central analgesic properties. Analgesic (CNS). Crystals; mp = 115-116°; LD50 (mus orl) = 617 mg/kg, (rat orl) = 1660 mg/kg. *Carter-Wallace; Degussa AG; Degussa-Hüls Corp.*

1775   Flupirtine maleate
75507-68-5   4216   G D

C15H17FN4O2.C4H4O4

[2-Amino-6-[[(4-fluorophenyl)methyl]amino]-3-pyridin-yl]carbamate maleate (1:1).

EINECS 278-225-0; Flupirtine maleate; Katadolon; W 2964M. Analgesic (CNS). Crystals; mp = 175.5-176°. *Carter-Wallace; Degussa AG; Degussa-Hüls Corp.*

1776   Fluroxypyr
69377-81-7   4229   G P

C7H5Cl2FN2O3

[(4-Amino-3,5-dichloro-6-fluoro-2-pyridinyl)oxy]acetic acid.

Acetic acid, ((4-amino-3,5-dichloro-6-fluoro-2-pyridinyl)-oxy)-; Advance; ((4-Amino-3,5-dichloro-6-fluoro-2-pyridinyl)oxy)acetic acid; 4-Amino-3,5-dichloro-6-fluoro-2-pyridyloxyacetic acid; Dowco 433; EF 689; FF4014; Fluroxypyr; HSDB 6655; Starane. Selective systemic herbicide. Used for post-emergence control of broad-leaved weeds such as *Galium aparine and Stella media* and some deep-rooted perennial weeds. Used in wheat, barley and oats. Registered by EPA as a herbicide. Colorless crystals; mp = 232 - 233°; soluble in H2O (0.0091 g/100 ml), Me2CO (6.5 g/100 ml), MeOH (4.4 g/100 ml), EtOAc (1.2 g/100 ml), i-PrOH (1.2 g/100 ml), CH2Cl2 (0.01 g/100 ml), C7H8 (0.1 g/100 ml), xylene (0.03 g/100 ml); LD50 (rat orl) = 2405 mg/kg, (rbt dr) > 5000 mg/kg, (mallard duck, bobwhite quail orl) > 2000 mg/kg, (bee 48 hr) > 0.1 mg/bee; LC50 (rat ihl 4 hr.) > 0.296 mg/l air, (rainbow trout, golden orfe 96 hr.) > 100 mg/l; non-toxic to bees. *Dow AgroSciences; DowElanco Ltd.*

1777   Fluroxypyr-meptyl
81406-37-3   4229   G P

C15H21Cl2FN2O3

1-Methylheptyl ((4-amino-3,5-dichloro-6-fluoropyridin-2-yl)oxy)acetate.

Dowco 433; EINECS 279-752-9; Fluroxypyr-(1-methylheptyl); Fluroxypyr 1-methylheptyl ester; Fluroxypyr meptyl; Starane; Starane 2; Starane 250; XRM 5084. Herbicide. Colorless crystals; mp = 56 - 57°; soluble in H2O (0.00009 g/100 ml), Me2CO (87 g/100 ml), MeOH (47 g/100 ml), EtOAc (79 g/100 ml), CH2Cl2 (90 g/100 ml), C7H8 (74 g/100 ml), xylene (64 g/100 ml), C6H14 (4.5 g/100 ml); LD50 (rat orl) = 5000 mg/kg, (rat, rbt der) > 2000 mg/kg, (mallard duck, bobwhite quail orl) > 2000 mg/kg, (bee) > 0.1 mg/bee; LC50 (rainbow trout, golden orfe 96 hr.) > 0.7 mg/l; non-toxic to bees. *DowElanco Ltd.*

1778   Flusilazole
85509-19-9   4232   G P

$C_{16}H_{15}F_2N_3Si$

1-[[Bis(4-fluorophenyl)methylsilyl]methyl]-1H-1,2,4-tri-azole.
BRN 5824097; Caswell No. 419K; DPX 6573; DPX-H 6573; EPA Pesticide Chemical Code 128835; Flusilazol; Flusilazole; Fluzilazol; Nustar; Olymp; Punch; Sanction. Foliar, systemic fungicide with protective and curative action. Used to control *Ascomycetes, Basdiomycetes and Deuteromycetes* in cereals, apples, vines and sugar beet. Colorless crystals; mp = 53°; soluble in H_2O (0.0054 g/100 ml at 20°, pH 7.2, 0.09 g/100 ml at 20°, pH 1.1), readily soluble (>200 g/100 ml) in common organic solvents; LD_{50} (mrat orl) = 1100 mg/kg, (frat orl) = 674 mg/kg, (rbt der) > 2000 mg/kg, (mallard duck orl) > 1590 mg/kg, (bee) > 0.15 mg/bee; LC_{50} (rainbow trout 96 hr.) = 1.2 mg/l, (bluegill sunfish 96 hr.) = 1.7 mg/l, (*Daphnia* 48 hr.) = 3.4 mg/l; non-toxic to bees. *DuPont.*

1779   Fluticasone
90566-53-3   D

$C_{25}H_{31}F_3O_5S$

S-Fluoromethyl   6α,9-difluoro-11β,17-dihydroxy-16α-methyl-3-oxoandrosta-1,4-diene-17β-carbothioic acid.
Fluticasona; Fluticasone; Fluticasonum. Derivative of flumethasone. Antiallergic; anti-inflammatory. *Glaxo Labs.*

1780   Fluticasone propionate
80474-14-2   4237   D

$C_{22}H_{27}F_3O_4S$

S-Fluoromethyl   6α,9-difluoro-11β,17-dihydroxy-16α-methyl-3-oxoandrosta-1,4-diene-17β-carbothioate 17-propionate.
Asmatil; Atemur; Axotide; Brethal; CCI 18781; Cutivate; Flixonase; Flixonase Nasal Spray; Flixotide; Flixotide Inhaler; Flonase; Flovent; Fluinol; Flunase; Flusonal; Fluspiral; Fluticasone; Fluticasone propionate; Fluticasonpropionat Allen; Flutide; Flutivate; Inalacor; Rinosone; Trialona; Ubizol; Zoflut. Antiallergic; anti-inflammatory. Crystals; mp = 272-273° (dec); $[\alpha]_D$ = +30° (c = 0.35). *Glaxo Labs.*

1781   Flutriafol
76674-21-0   4240   G P

$C_{16}H_{13}F_2N_3O$

(±)-α-(2-Fluorophenyl)-α-(4-fluorophenyl)-1H-1,2,4-tri-azole-1-ethanol.
(RS)-2,4'-Difluoro-α-(1H-1,2,4-triazol-1-ylmethyl)benz-hydryl alcohol; α-(2-Fluorophenyl)-α-(4-fluorophenyl)-1H-1,2,4-triazole-1-ethanol; Flutriafol; Flutriafol; Flutriafol; Impact (pesticide); PP 450; R 152450; 1H-1,2,4-Triazole-1-ethanol, α-(2-fluorophenyl)-α-(4-fluoro-phenyl)-; 1H-1,2,4-Triazole-1-ethanol, α-(2-fluoro-phenyl)-α-(4-fluorophenyl)-; Flutriafol; Impact (pesticide); PP 450; R 152450. Fungicide. Colorless crystals; mp = 130°; soluble in H_2O (0.013 g/100 ml at 20°, pH 7), Me_2CO (19 g/100 ml), CH_2Cl_2 (15 g/100 ml), MeOH (7 g/100 ml), xylene (1.2 g/100 ml), C_6H_{14} (0.03 g/100 ml); LD_{50} (mrat orl) = 1140 mg/kg, (frat orl) = 1480 mg/kg, (rat der) > 1000 mg/kg, (rbt der) > 2000 mg/kg, (fmallard duck orl) > 5000 mg/kg, (bee orl) > 0.005 mg/bee; LC_{50} (rainbow trout 96 hr.) = 61 mg/l, (*Daphnia* 48 hr.) = 78 mg/l; toxic to bees. *ICI Agrochemicals; Syngenta Crop Protection.*

1782   Folic acid
59-30-3   4247   D G

$C_{19}H_{19}N_7O_6$

N-[p-[[(2-Amino-4-hydroxy-6-pteridinyl)methyl]amino]-benzoyl]-L-glutamic acid.
Acfol; Acide folique; Acido folico; Acidum folicum; Acifolic; AI3-26387; Antianemia factor; CCRIS 666; Cytofol; EINECS 200-419-0; Facid; Factor U; Folacid; Folacin; Folaemin; Folan; Folasic; Folate; Folbal; Folcidin; Folcysteine; Foldine; Folettes; Foliamin; Folic acid; Folicet; Folico; Folina; Folipac; Folsan; Folsaure; Folsav; Folvite; HSDB 2002; Incafolic; Kyselina listova; Liver Lactobacillus casei factor; Millafol; Mission Prenatal; Mittafol; Nifolin; Novofolacid; NSC 3073; PGA; Plastulen-N.; Pteglu; Pteroylglutamic acid; Pteroylmono-glutamate; Pteroylmonoglutamic acid; USAF CB-13; Vitamin B9; Vitamin B11; Vitamin Bc; Vitamin Be; Vitamin M.; component of: Mission Prenatal, Plastulen-N. Vitamin, vitamin source. Considered a member of the vitamin B complex; used in medicine, nutrition, and as a food additive. Used as a vitamin (hematopoietic) and nutritional factor. White solid; mp > 250°; $[\alpha]_D^{25}$= 23° (c = 0.5 in 0.1N NaOH); λ_m = 256, 283 368 nm (log ε 4.43, 4.40, 3.96 pH 13); slightly soluble on H_2O (0.00016 g/100 ml at 20°, 1 g/100 ml at 100°), MeOH; less soluble in EtOH, BuOH; insoluble in $CHCl_3$, Me_2CO, Et_2O, C_6H_6. *Bristol-Myers Squibb Pharm. R&D; Lederle Labs.; Mission Pharmacal Co.*

1783   Folpet
133-07-3   4250   G P

C9H4Cl3NO2S
N-(Trichloromethylthio)phthalimide.
5-21-11-00118 (Beilstein Handbook Reference); Acryptan; Al3-26539; BRN 0193373; Caswell No. 464; CCRIS 1036; Cosan I; EINECS 205-088-6; ENT 26539; EPA Pesticide Chemical Code 081601; Faltan; Faltex; Folnit; Folpan; Folpel; Folpet; Ftalan; Fungitrol; Fungitrol 11; Fungitrol II; HSDB 2651; Intercide TMP; Murphy's rose fungicide; Ortho phaltan 50W; Orthophaltan; Phaltan; Phthalimide, N-((trichloromethyl)thio)-; Phthaltan; Spolacid; Thiophal; Trichlormethylthioimid kyseliny ftalove; Trichloromethylthiophthalimide; Troysan anti-mildew O; Vinicoll. Agricultural fungicide. Used for control of downy mildews, powdery mildews, leaf spot diseases, scab, etc. in fruit and vegetable crops.Registered by EPA as a fungicide. Colorless crystals; mp = 177°; insoluble in H_2O (< 0.0001 g/100 ml at 25°), slightly soluble in organic solvents; $CHCl_3$ (8.7 g/100 ml), C_6H_6 (2.2 g/100 ml), i-PrOH (1.25 g/100 ml); LD_{50} (rat orl) > 10000 mg/kg, (rbt der) > 22600 mg/kg, (mallard duck orl) = 2000 mg/kg; toxic to fish; non-toxic to bees. *Bayer Corp., Agriculture; Makhteshim Chemical Works Ltd.; Monsanto (Solaris); Sherwin-Williams Co.*

1784   Foral 85
8050-31-5   H

Gum rosin, glyceryl ester.
Disproportionated rosin, glycerol ester; EINECS 232-482-5; Ester gum; Foral 85; Glycerol ester of rosin; Glycerol ester of disproportionated rosin; Glycerol, rosin polymer; Gum rosin, glyceryl ester; Resin acids and Rosin acids, esters with glycerol; Rosin, glycerin ester; Rosin, glycerine ester; Rosin, glycerol ester.

1785   Formal glycol
646-06-0   H

C3H6O2
1,3-Dioxacyclopentane.
CCRIS 4912; 1,3-Dioxacyclopentane; Dioxalan; 1,3-Dioxolan; Dioxolane; 1,3-Dioxolane; 1,3-Dioxole, dihydro-; EINECS 211-463-5; Ethylene glycol formal; Formal glycol; Glycol formal; Glycol methylene ether; Glycolformal; HSDB 5737; UN1166. Liquid; mp = -95°; bp = 78°; d^{20} = 1.060; freely soluble in H_2O, soluble in EtOH, Et2O, Me2CO.

1786   Formaldehyde
50-00-0   4259   G

CH2O
Methanal.
Al3-26806; Aldehyd mravenci; Aldehyde C1; Aldehyde formique; Aldeide formica; BFV; Caswell No. 465; CCRIS 315; Dormol; EINECS 200-001-8; EPA Pesticide Chemical Code 043001; Fannoform; Formaldehyd; Formaldehyde; Formalin; Formalin 40; Formalin-lösungen; Formalin-loesungen; Formalina; Formaline; Formalith; Formic aldehyde; Formol; FYDE; Hercules® 37M6-8;

HSDB 164; Karsan; Lysoform; Methaldehyde; Methanal; Methyl aldehyde; Methylene oxide; Morbicid; NCI-C02799; NSC 298885; Oplossingen; Oxomethane; Oxymethylene; Paraform; RCRA waste number U122; Superlysoform; UN 1198; UN 2209. Used in urea and melamine resins, polyacetal resins, phenolic resins, fertilizers, preservatives, reducing agent, corrosive inhibitor. Used in manufacture of synthetic resins by reaction with phenols, urea, melamines for molded goods, electrical insulation, binders, plywood adhesives, varnishes, wet-strength resins for paper and textiles. A formalin preparation is used as an antimicrobial and antiseptic. A soil sterilant and fumigant for glass houses. A solution of formaldehyde in alcoholic potash soap solution; a disinfectant much like Lysol. Registered by EPA as an antimicrobial and fungicide. FDA approved for topicals, BP compliance (solutions). Used as an antimicrobial in biologics, topicals, hepatitis B vaccine, sterilizer for kidney dialysis membranes. Colorless gas; mp = -92°; bp = -19.5°; soluble in water, alcohol, ether. LD_{50} (rat orl) = 550 - 800 mg/kg (formalin); TLV = 1 ppm in air, (rbt der) = 270 mg/kg (formalin); LC_{50} (rat ihl 0.5 hr.) = 0.82 mg/l air; toxic to fish. *Aqualon; Baker Petrolite Corp.; Champion Technologies Inc; DuPont; E. I. DuPont de Nemours Inc.; Farleyway Chem. Ltd.; Georgia Pacific Resins; Georgia Pacific; Hess & Clark Inc.; Hoechst Celanese; Mallinckrodt Inc.; Mitsubishi Gas; Monsanto Co.; Ruger; Sigma-Aldrich Fine Chem.; Spectrum Chem. Manufacturing; Vineland Lab.*

1787   Formaldehyde sodium sulfoxylate
149-44-0   8693   G H

CH3NaO3S
Hydroxymethanesulfinic acid sodium salt.
Al3-23202; Aldanil; Bleachit D; Discolite; EINECS 205-739-4; Formapon; Formopan; HSDB 5648; Hyraldite C Ext; Hydrolit; Hydrosulfite AWC; Hydroxymethanesulfinic acid sodium salt; Hydroxymethansulfinsaeure, natrium-salz; Leptacid; Leptacit; NSC 4847; Oxymethan-sulfinsaeuren natrium; Redol C; Rodite; Rongalit; Rongalit C; Rongalite; Rongalite C; Sodium formaldehyde sulfoxylate; Sodium hydroxymethanesulfinate; Sodium hydroxymethanesulphinate; Superlite C. Used as a reducing agent in calico printing, in textile printing, emulsion polymerization; treatment of mercury poisoning. Crystals; mp = 63-64°; soluble in water; insoluble in EtOH, Et2O, C6H6; LD_{50} (mus, sc) = 4.0 g/kg.

1788   Formamide
75-12-7   4261   G H

CH3NO
Formic acid amide.
Al3-15357; Amid kyseliny mravenci; Carbamaldehyde; CCRIS 6240; EINECS 200-842-0; Formamide; Formic acid, amide; Formimidic acid; HSDB 88; Methanamide; Methanoic acid, amide; NSC 748. Solvent, softener, intermediate in organic synthesis. Liquid; mp = 2.55°; bp = 220°; d^{20} = 1.1334; soluble in H_2O, organic solvents; LD_{50} (rat orl) = 5570 mg/kg. *BASF Corp.; Fluka; Penta Mfg.; Sigma-Aldrich Fine Chem.*

1789   Formamidine sulfinic acid
1758-73-2   G

297

CH4N2O2S
Thiourea dioxide.
AIMSA; Aminoiminomethanesulfinic acid; Aminoimino-methanesulphinic acid; EINECS 217-157-8; Formamidine sulfinic acid; Formamidinesulfinic acid; Manofast; Methanesulfinic acid, aminoimino-; NSC 34540; NSC 226979; Thiourea dioxide; Thiourea S,S-dioxide. Reducing agent for dyes. Used in organic synthesis to reduce ketones to secondary alcohols. Solid; mp = 126° (dec); soluble in H2O (3.0 g/100 ml). *Allchem Ind.; Cia-Shen; Degussa AG; Rhone-Poulenc UK.*

1790   Formic acid
64-18-6   4265   G H P

CH2O2
Methanoic acid.
Acide formique; Acido formico; Add-F; AI3-24237; Ameisensäure; Ameisensaeure; Aminic acid; Bilorin; C1 acid; CCRIS 6039; Collo-bueglatt; Collo-didax; EINECS 200-579-1; EPA Pesticide Chemical Code 214900; FEMA No. 2487; Formic acid; Formira; Formisoton; Formylic acid; HSDB 1646; Hydrogen carboxylic acid; Kwas metaniowy; Kyselina mravenci; Methanoic acid; Mierenzuur; Myrmicyl; RCRA waste number U123; Spirit of formic acid; UN1779. Reducing agent, used in dyeing and finishing of textiles, leather treatment, chemical manufacturing, manufacture of fumigants, insecticides, refrigerants, solvents for perfumes, lacquers; electroplating; silvering glass; ore flotation. Registered by EPA as an insecticide. Used as a miticide. Colorless liquid; mp = 8.3°; bp = 101°; d^{20} = 1.220; pKa = 3.75; viscosity = 1.784 cP at 20°; freely soluble in H2O, Et2O, Me2CO, EtOAc, MeOH, EtOH, partially soluble in C6H6, C7H8, xylene; LD50 (mus orl) = 1100 mg/kg, (mus iv)= 145 mg/kg. *BASF AG; BP Chem.; Hoechst Celanese; Lancaster Synthesis Co.; Mallinckrodt Inc.; Norsk Hydro AS; Sigma-Aldrich Co.; Sigma-Aldrich Fine Chem.*

1791   4-Formylcyclohexene
100-50-5   H

C7H10O
3-Cyclohexene-1-carboxaldehyde.
AI3-21661; BRN 0774001; 3-Cyclohexene-1-carbox-aldehyde; Cyclohex-3-ene-1-carbaldehyde; Cyclohexene-4-carboxaldehyde; Δ^1-Tetrahydrobenzaldehyde; EINECS 202-858-3; HSDB 5334; NSC 16241; UN2498. Liquid; mp = 1°; bp = 105°; d^{20} = 0.9692; slightly soluble in CCl4, soluble in MeOH, Me2CO.

1792   Fosinopril
98048-97-6   4279   D

C30H46NO7P
(4S)-4-Cyclohexyl-1-[[(R)-[(S)-1-hydroxy-2-methyl-propoxy](4-phenylbutyl)phosphinyl]acetyl-L-proline propionate (ester).
Fosenopril; Fosinopril; L-Proline, 4-cyclohexyl-1-(((2-methyl-1-(1-oxopropoxy)propoxy)(4-phenylbutyl)phos-phinyl)acetyl)-, (1(S*(R*)),2α,4β)-. An angiotensin-converting enzyme inhibitor used as an antihypertensive. [diacid (SQ-27519)]: mp = 149-153°; [α]D= -24° (c = 1 MeOH). *Squibb E.R. & Sons.*

1793   Fosinopril sodium
88889-14-9   4279   D

C30H45NNaO7P
(4S)-4-Cyclohexyl-1-[[(R)-[(S)-1-hydroxy-2-methyl-propoxy](4-phenylbutyl)phosphinyl]acetyl-L-proline propionate (ester) sodium salt.
Acecor; Acenor-M; Dynacil; Eliten; Fosinil; Fosinopril sodium; Fosinorm; Fosipres; Fositen; Fositens; Foziretic; Fucithalmic; Hiperlex; Monopril; Newace; Sapril; Secorvas; SQ 28555; Staril; Tenso Stop; Tensogard. Angiotensin-converting enzyme inhibitor. Anti-hypertensive. *Mead Johnson Labs.; Mead Johnson Pharmaceuticals.*

1794   Fospirate
5598-52-7   4281   G P

C7H7Cl3NO4P
Dimethyl-(3,5,6-trichloro-2-pyridyl) phosphate.
AI3-27521; BRN 1541077; Caswell No. 465CC; Chlorpyrifos-methyl oxon; Dimethyl 3,5,6-trichloro-2-pyridyl phosphate; Dimethyl-3,5,6-trichlor-2-pyridyl-fosfat; O,O-Dimethyl O-(3,5,6-trichloro-2-pyridyl) phos-phate; Dowco 217; ENT 27521; EPA Pesticide Chemical Code 103501; Fospirat; Fospirate; Fospirate methyl; Fospirato; Fospiratum; NSC 195058; OMS 1168; Phosphoric acid, dimethyl 3,5,6-trichloro-2-pyridyl ester; Phosphoric acid, dimethyl 3,5,6-trichloro-2-pyridinyl ester; Torelle; Dimethyl 3,5,6-trichloro-2-pyridyl phosphate; O,O-Dimethyl O-(3,5,6-trichloro-2-pyridyl) phosphate;

Dowco 217; ENT 27521; EPA Pesticide Chemical Code 103501; Fospirat; Fospirate; Fospirate methyl; Fospirato; Fospiratum; NSC 195058; OMS 1168; Phosphoric acid, dimethyl 3,5,6-trichloro-2-pyridyl ester; Torelle. Anthelmintic. Registered by EPA as an insecticide (cancelled). Crystals; mp = 86.5 - 88°. *Dow AgroSciences; Dow UK.*

1795 Fremy's salt
7789-29-9   7692   G

$$K^+ \quad F^- \quad HF$$

F$_2$HK
Potassium hydrogen fluoride.
Bifluorure de potassium; EINECS 232-156-2; Fremy's salt; Hydrogen potassium fluoride; Potassium acid fluoride; Potassium bifluoride; Potassium fluoride (K(HF$_2$)); Potassium hydrogen difluoride; Potassium hydrogen fluoride; Potassium hydrogendifluoride; Potassium monohydrogen difluoride; UN1811. Used as an electrolyte and in manufacture of frosted glass. Also as a flux for silvr solders and as an organic catalyst. Crystals; mp = 238°; d = 2.37; soluble in H$_2$O (39 g/100 ml).

1796 Fructose
57-48-7   4295   D G

C$_6$H$_{12}$O$_6$
D-(-)-Levulose.
AI3-23514; Arabino-Hexulose; CCRIS 3335; EINECS 200-333-3; Fructose; Fructose, D-; D-Fructose; D-(-)-Fructose; Fructose, pure; Fructose solution; Fruit sugar; Furucton; Hi-Fructo 970; Krystar 300; Levulose; Nevulose; Sugar, fruit. D-fructose, sugar occurring in fruit and honey; Used as a sweetener in food stuffs, medicine, and as a preservative. Prisms; dec 103 - 105°; mutarotates; $[\alpha]_D^{20}$= -132° → -92° (c = 2); soluble in H$_2$O, EtOH (6.7 g/100 ml), MeOH (7.1 g/100 ml), Me$_2$CO, C$_5$H$_5$N, methylamine, ethylamine. *Am. Maize Products; Corn Products; Laevosan GmbH; Pfanstiehl Labs Inc.; Staley, A. E. Mfg.*

1797 Fuberidazole
3878-19-1   G P

C$_{11}$H$_8$N$_2$O
2-(2-Furanyl)-1H-benzimidazole.
B-33172; BAY 33172; Benzimidazole, 2-(2-furyl)-; 1H-Benzimidazole, 2-(2-furanyl)-; Fuberidatol; Fuberidazol; Fuberidazole; Fuberisazol; Fubridazole; 2-(2-Furanyl)-1H-benzimidazole; Furidazol; Furidazole; 2-(2-Furyl)benzi-midazole; NSC 72670; Voronit; Voronite; W VII/117. Systemic fungicide used as a seed treatment for control of *Fusarium* spp. in cereals. Crystals; mp = 284 - 288° (dec); soluble in H$_2$O (0.0078 g/100 ml), i-PrOH (3.9 g/100 ml), C$_7$H$_8$ (0.8 g/100 ml), CH$_2$Cl$_2$ ($\cong$1.3 g/100 ml), petroleum ether ($\cong$0.7 g/100 ml); LD$_{50}$ (rat orl) = 500 mg/kg, (mus orl) = 825 mg/kg, (rat der) > 1000 mg/kg, (rat ip) = 100 mg/kg; LC$_{50}$ (rat ihl 4 hr.) > 0.3 mg/l air; (mosquito fish, *Poecilia latipinna* 96 hr.) > 1 mg/l;

non-toxic to bees. *Bayer AG; Bayer Corp., Agriculture.*

1798 Fuel gases, producer gas
8006-20-0   H

Fuel gases, producer gas.
EINECS 232-344-4; Fuel gases, low and medium B.T.U.; Fuel gases, producer gas; Gas, producer.

1799 Fumaric acid
110-17-8   4308   G H

C$_4$H$_4$O$_4$
2-Butenedioic acid (2E)-.
4-02-00-02202 (Beilstein Handbook Reference); AI3-24236; Allomaleic acid; Allomalenic acid; Boletic acid; BRN 0605763; Butenedioic acid, (E)-; 2-Butenedioic acid, (E)-; trans-Butenedioic acid; Caswell No. 465E; CCRIS 1039; EINECS 203-743-0; EPA Pesticide Chemical Code 051201; 1,2-Ethenedicarboxylic acid, trans-; 1,2-Ethylenedicarboxylic acid, (E); trans-1,2-Ethylenedi-carboxylic acid; FEMA Number 2488; Fumaric acid; HSDB 710; Kyselina fumarova; Lichenic acid; NSC-2752; Tumaric acid; U-1149; USAF EK-P-583. Modifier for polyester, alkyd and phenolic resins; paper sizing resins; plasticizers, rosin esters and adducts, alkyd resin coatings, upgrading natural drying oils, food additive, acidulant, flavoring agent, mordant, organic synthesis, inks. Prisms or leaflets; mp = 287° (dec); bp sublimes 165°; d^{20} = 1.635; λ_m = 207 nm (ε = 14400, MeOH); slightly soluble in H$_2$O, Et$_2$O, Me$_2$CO, CCl$_4$, CHCl$_3$; soluble in conc. H$_2$SO$_4$; EtOH. *Chemie Linz N. Am.; Haarmann & Reimer GmbH; Lonzagroup; Mitsubishi Gas; Monsanto Co.; Schaeffer Salt & Chem.*

1800 Fuming sulfuric acid
8014-95-7   9064   G

H$_2$O$_4$S
Oleum.
Fuming sulfuric acid; HSDB 1236; Nordhausen acid; Oleum; Oleum iodisum; Pyrosulfuric acid; Sulfur trioxide, mixt. with sulfuric acid; Sulfuric acid (fuming); Sulfuric acid fuming; Sulfuric acid mixture with sulfur trioxide; UN1831. Consists of sulfur trioxide dissolved in sulfuric acid. The commonest fuming acid contains 55% sulfuric acid, and 45% sulfur trioxide; used as a sulfating and sulfonating agent, dehydrating agent in nitrations, dyes, and explosives. Oily liquid; d = 1.84; bp = 290°; dec 340°; mp = 10°; miscible with H$_2$O and alcohol; LD$_{50}$ (rat orl) = 2.14 g/kg.

1801 Fungisterol
53260-54-1   4312   G

C28H44O

5α-Ergosta-6,8,22E-trien-3β-ol.

Crystals; mp = 147.5°; [α]$_D^{15}$ = -21.9° (chloroform

1802   Furalaxyl

57646-30-7   G P

C17H19NO4

N-(2,6-Dimethylphenyl)-N-(2-furanylcarbonyl)-DL-alan-ine   methyl ester.

A 5430; CGA 38140; Fonganil; Fongarid; Furalaxyl; Methyl N-(2,6-dimethylphenyl)-N-(2-furanylcarbonyl)-DL-alaninate; Methyl N-(2,6-dimethylphenyl)-N-(2-furoyl)-DL-alaninate; Methyl N-(2-furoyl)-N-(2,6-xylyl)-DL-alaninate. Systemic fungicide with protective and curative action for ornamentals. Used for control of soil diseases caused by *Phytophthora and Pythium* spp. Used to combat potato blight. Colorless crystals; 2 forms, mp = 70°, 84°; sg^{20} = 1.22; soluble in H$_2$O (0.023 g/100 ml), CH$_2$Cl$_2$ (80 g/100 ml), Me$_2$CO (41 g/100 ml), MeOH (40 g/100 ml), C$_6$H$_6$ (42 g/100 ml), C$_6$H$_{14}$ (0.26 g/100 ml); LD$_{50}$ (rat orl) = 940 mg/kg, (rat der) > 3100 mg/kg, (rbt der) = 5508 mg/kg; LC$_{50}$ (rainbow trout 96 hr.) = 32.5 mg/l, (crucian carp 96 hr.) = 38.4 mg/l; non-toxic to birds; non-toxic to bees. *Ciba Geigy Agrochemicals; Syngenta Crop Protection.*

1803   Furfural

98-01-1   4325   G H

C5H4O2

2-Furancarboxaldehyde.

5-17-09-00292 (Beilstein Handbook Reference); AI3-04466; Ant Oil, artificial; Artificial ant oil; Artificial oil of ants; Bran oil; BRN 0105755; Caswell No. 466; CCRIS 1044; EINECS 202-627-7; EPA Pesticide Chemical Code 043301; FEMA No. 2489; Fufural; Fural; Furaldehyde; Furale; Furancarbonal; Furrural; Furfuraldehyde; α-Furfuraldehyde; Furfurale; Furfurol; Furfurole; Furfurylaldehyde; Furol; Furole; α-furole; Furyl-methanal; HSDB 542; NCI-C56177; NSC 8841; Pyromucic aldehyde; Quakeral; RCRA waste number U125; UN1199. Chemical intermediate for manufacture of derivatives (furan, THF); solvent for petroleum lube, nitrocellulose; wetting agent; in manufacture of furfuralphenol plastics; vulcanization accelerator; insecticide, fungicide, germicide; reagent in analytical chemistry. Registered by EPA as an insecticide. Colorless liquid; mp = -36.5°; bp = 161.8°, bp$_{100}$ = 103°, bp$_{20}$ = 67.8°, bp$_{1.0}$ = 18.5°; d$_4^{25}$ = 1.1563; soluble in H$_2$O (9.1 g/100 ml), very

soluble in EtOH, Et$_2$0; LD$_{50}$ (rat orl) = 127 mg/kg. *Allchem Ind.; Great Lakes Fine Chem.; QO; Sigma-Aldrich Fine Chem.*

1804   Furfuramide

494-47-3   4812   G

C15H12N2O3

2-(Bis(furfurylidenamino))methylfuran.

5-17-09-00306 (Beilstein Handbook Reference); AI3-04501; BA 51-90222; BRN 0028784; EINECS 207-790-8; 2-Furanmethanediamine, N,N'-difurfurylidene-; Hydro-furamide; Methanediamine, 1-(2-furanyl)-N,N'-bis(2-furanylmethylene)-; Methanediamine, N,N-difurfuryl-idene-1-(2-furyl)-; NSC 49110; 1,3,5-Tri(2-furyl)-2,4-diazapenta-1,4-diene; Vulcazol.

A vulcanization accelerator. Needles; mp = 117°; bp = 250° (dec); λ$_m$ = 259, 215 nm (log ε 4.18, 4.16); insoluble in H$_2$O, very soluble in EtOIH, Et$_2$O.

1805   Furfuryl alcohol

98-00-0   4326   G

C5H6O2

2-Furanmethanol.

5-17-03-00338 (Beilstein Handbook Reference); AI3-01171; BRN 0106291; CCRIS 2922; EINECS 202-626-1; FEMA No. 2491; 2-Furancarbinol; 2-Furanmethanol; 2-Furanylmethanol; Furfural alcohol; Furfuralcohol; Furfuranol; Furfuryl alcohol; 2-Furfuryl alcohol; Furfurylalkohol; α-Furylcarbinol; 2-Furylcarbinol; 2-Furylmethanol; HSDB 711; 2-Hydroxymethylfuran; 5-Hydroxymethylfuran; Methanol, (2-furyl)-; NCI-C56224; NSC 8843; UN2874. Used as a wetting agent, in furan polymers, as a solvent for dyes and resins, and in flavoring. Pale yellow liquid; mp = -31°; bp = 171°; d$_4^{20}$ = 1.1296; λ$_m$ = 270 nm (MeOH); soluble in H$_2$O (9 g/100 ml), EtOH, Et$_2$O, CHCl$_3$; LD$_{50}$ (rat orl) = 127 mg/kg. *Allchem Ind.; Great Lakes Fine Chem.; QO; Sigma-Aldrich Fine Chem.*

1806   2-Furoic acid

88-14-2   4328   G

C5H4O3

2-Furancarboxylic acid.

5-18-06-00102 (Beilstein Handbook Reference); AI3-16500; BRN 0110149; 2-Carboxyfuran; CCRIS 2157; EINECS 201-803-0; α-Furancarboxylic acid; 2-Furancarboxylic acid; α-Furoic acid; 2-Furoic acid; Kyselina 2-furoova; Kyselina pyroslizova; NSC 8842; Pyromucic acid. Preservative, bactericide, furoates for perfume and flavoring, fumigant, textile processing, chemical intermediate. Needles; mp = 133.5°; sublimes at 130-140°; bp = 231°; λ$_m$ = 247 nm (12300, MeOH); soluble in alcohol, ether, water. *Greeff R.W. &*

1807   Furosemide

54-31-9   4330   D

$C_{12}H_{11}ClN_2O_5S$
4-Chloro-N-furfuryl-5-sulfamoylanthranlic acid.
5-18-09-00555 (Beilstein Handbook Reference); Aisemide; Aldic; Aluzine; Anfuramaide; Apo-Frusemide; Apo-Furosemide; Aquarid; Aquasin; Arasemide; Beronald; Bioretic; BRN 0840915; CCRIS 1951; Cetasix; Depix; Desal; Desdemin; Dirine; Disal; Discoid; Disemide; Diural; Diurapid; Diuretic salt; Diurin; Diurolasa; Diusemide; Diusil; Diuzol; Dranex; Dryptal; Durafurid; Edemid; Edenol; EINECS 200-203-6; Eliur; Endural; Errolon; Eutensin; Farsix; Fluidrol; Fluss; Franyl; Frumex; Frumide; Frumil; Frusedan; Frusema; Frusemid; Frusemide; Frusemin; Frusenex; Frusetic; Frusid; Fulsix; Fuluvamide; Fuluvamine; Furanthril; Furanthryl; Furantril; Furanturil; Furesis; Furetic; Furex; Furfan; Furix; Furmid; Furo-Basan; Furo-Puren; Furobeta; Furocot; Furodiurol; Furodrix; Furomen; Furomex; Furomide M.D.; Furorese; Furosan; Furose; Furosedon; Furosemid; Furosemide; Furosemide mita; Furosemidu; Furosemidum; Furosemix; Furoside; Furosifar; Furosix; Furoter; Furovite; Fursemid; Fursemida; Fursemide; Fursol; Fusid; Golan; Hissuflux; HSDB 3086; Hydrex; Hydro; Hydro-rapid; Hydroled; Impugan; Jenafusid; Katlex; Kofuzon; Kolkin; Kutrix; Lasemid; Lasex; Lasiletten; Lasilix; Lasix; Lasix Retard; Laxur; Lazix; LB 502; less Diur; Liside; Logirene; Lowpston; Lowpstron; Luscek; Macasirool; Marsemide; Mirfat; Mirfat; Mita; Moilarorin; Nadis; NCI-C55936; Nelsix; Neo-renal; Nephron; Nicorol; Novosemide; NSC 269420; Odemase; Odemex; Oedemex; Osyrol; Polysquall A; Prefemin; Profemin; Promedes; Promide; Protargen; Puresis; Radisemide; Radonna; Radouna; Retep; Rosemide; Rosis; Rusyde; Salinex; Salix; Salurex; Salurid; Seguril; Selectofur; Sigasalur; Spirofur; Synephron; Transit; Trofurit; Uremide; Uresix; Urex; Urex-M; Urian; Uridon; Uritol; Urosemide; Vesix; Yidoli; Zafimida. Used as a diuretic and antihypertensive agent. Used with controlled release potassium chloride to control edema. Crystals; mp = 206°; λ_m = 288, 276, 336 nm ($E^{1\%}_{1cm}$ 945, 588, 144 95% EtOH); soluble in Me_2CO, MeOH, DMF; less soluble in EtOH, H_2O, $CHCl_3$, Et_2O; LD_{50} (frat orl) = 2600 mg/kg, (mrat orl) = 2820 mg/kg. *Astra USA Inc.; Elkins-Sinn; Fermenta Animal Health Co.; Hoechst Roussel Pharm. Inc.; Parke-Davis; Rhône-Poulenc Rorer Pharm. Inc.*

1808   Fusel oil

8013-75-0   H

Fusel oil.
Amyl alcohol, commercial; EINECS 232-395-2; FEMA No. 2497; Fusel oil; Fusel oil, refined; Fusel oil, refined (mixed amyl alcohols); Fusel oil, sugar beet; Fuselöl; Huile de fusel; Isoamyl fusel oil; UN1201. 232-395-2

1809   Gabapentin

60142-96-3   4342   D

$C_9H_{17}NO_2$
1-(Aminomethyl)cyclohexaneacetic acid.
Aclonium; BRN 2359739; CCRIS 7210; CI 945; EINECS 262-076-3; Gabapentin; Gabapentine; Gabapentino; Gabapentinum; Gabapetin; Go 3450; GOE 2450; Neurontin. An amino acid related to σ-aminobutyric acid (GABA). An anticonvulsant. Crystals; mp = 162-166°, 165-167°; soluble in H_2O (> 10 g/100 ml at pH 7.4). *Parke-Davis.*

1810   Galactitol

608-66-2   4353   G

$C_6H_{14}O_6$
Dulcitol.
AI3-19423; Dulcite; Dulcitol; D-Dulcitol; Dulcose; EINECS 210-165-2; Euonymit; Galactitol; Melampyrin; Melampyrit; NSC 1944. Occurs in Madagascar manna. Crystals; mp = 189.5°; bp_1 = 275-280°; d^{20} = 1.47; soluble in H_2O (3 g/100 ml), slightly soluble in EtOH, C_5H_5N, insoluble in Et_2O, C_6H_6.

1811   Galaxolide

1222-05-5   H

$C_{18}H_{26}O$
Hexahydro-4,6,6,7,8,8-hexamethylcyclopenta-γ-2-benzo-pyran.
Cyclopenta(g)-2-benzopyran, 1,3,4,6,7,8-hexahydro-4,6,6,7,8,8-hexamethyl-; EINECS 214-946-9; Galaxolide; Hexahydrohexamethyl cyclopentabenzopyran.

1812   Gallic acid

149-91-7   4366   G

$C_7H_6O_5$
3,4,5-Trihydroxybenzoic acid.
3-10-00-02070 (Beilstein Handbook Reference); AI3-16412; Benzoic acid, 3,4,5-trihydroxy-; BRN 2050274; CCRIS 5523; EINECS 205-749-9; Gallic acid; HSDB 2117; Kyselina 3,4,5-trihydroxybenzoova; Kyselina gallova; NSC 20103; 3,4,5-Trihydroxybenzoic acid. Photography, writing ink, dyeing; manufacture of pyrogallol, tannins, paper; tanning agent; pharmaceuticals, engraving, lithography; analytical reagent. Prisms, sublimes at 210° giving stable form with mp = 258-265°; λ_m = 275 nm (EtOH), slightly soluble in H_2O (1.15 g/100 ml); LD_{50} (rbt orl) = 5.0 g/kg. *Fuji Chem. Ind.; Mallinckrodt Inc.; Penta Mfg.; U.S. BioChem.*

1813 Gallicin
99-24-1 4366 G

C8H8O5

Methyl gallate.

4-10-00-01998 (Beilstein Handbook Reference); AI3-00861; Benzoic acid, 3,4,5-trihydroxy-, methyl ester; BRN 2113180; CCRIS 5567; EINECS 202-741-7; Gallic acid methyl ester; Methyl 3,4,5-trihydroxybenzoate; Methyl gallate; Methylgallate; NSC 363001. Used by ophthalmologists as an antiseptic in conjunctivitis. Crystals; mp = 202°; soluble in hot H2O, EtOH, MeOH, Et2O.

1814 Gallium
7440-55-3 4367 G

Ga

Ga

Gallium.

EINECS 231-163-8; Gallium; Gallium, elemental; HSDB 6956; UN2803. Compounds used as semiconductors. Metallic element Metal; mp = 29.78°; bp = 2400°. *Atomergic Chemetals; Cerac; Eagle-Picher; International Gallium; Rhône-Poulenc; Sigma-Aldrich Fine Chem.*

1815 Gasoline
8006-61-9 H

Gasoline.

Antiknock gasoline; Benzin; Casing head gasoline; Cracked gasoline; EINECS 232-349-1; Gasoline; Gasoline (casinghead); Gasoline, natural; High-octane gasoline; HSDB 6477; Light gasoline; Natural gasoline; Natural gasoline (natural gas); Petrol (British); Petrol, natural; Petroleum distillates; Polymer gasoline; Pyrolysis gasoline; Reformed gasoline; Straight-run gasoline; UN 1203; UN1203; Unleaded gasoline (wholly vaporized); White gasoline.

1816 Gemfibrozil
25812-30-0 4399 D

C15H22O3

2,2-Dimethyl-5-(2,5-xylyloxy)valeric acid.

Apo-Gemfibrozil; Ausgem; Bolutol; BRN 1881200; Brozil; CCRIS 318; Cholespid; CI 719; Clearol; Decrelip; EINECS 247-280-2; Elmogan; Fetinor; Fibratol; Fibrocit; Gem-S; Gemd; Gemfibril; Gemfibromax; Gemfibrozil; Gemfibrozilo; Gemfibrozilum; Gemlipid; Genlip; Gevilon; Gevilon Uno; Gozid; Hidil; Hipolixan; Innogen; Ipolipid; Jezil; Lanaterom; Lifibron; Lipazil; Lipigem; Lipira; Lipizyl; Lipur; Litarek; Lopid; Lopizid; Low-Lip; Micolip; Normolip; Pilder; Polyxit; Progemzal; Reducel; Regulip; Renabrazin; Sinelip; Synbrozil; Taborcil; Tentroc; Trialmin; WL-Gemfibrozil. A fibric acid; reduces plasma triglycerides by lowering plasma levels of very-low-density lipoprotein. Hypolipidemic agent. Serum lipid regulator. mp = 61-63°; bp0.02 = 158-159°; LD50 (mus orl) = 3162 mg/kg, (rat orl) = 4786 mg/kg. *Parke-Davis.*

1817 Gentian violet
548-62-9 4406 D G

C25H30ClN3

N-[4-[bis[4-(dimethylamino)phenyl]methylene]-2,5-cyclohexadien-1-ylidene]-N-methylmethanaminium chloride.

12416 Violet; Adergon; Aizen Crystal Violet Extra Pure; Aizen Crystal Violet; Aniline Violet; Aniline Violet Pyoktanine; Atmonil; Avermin; Axuris; Badil; Basic Violet 3; Basic Violet BN; Bismuth Violet; Blaues pyoktanin; Brilliant Violet 5B; C.I. 42555; C.I. Basic violet 3; C.I. Basic Violet 1; Calcozine Violet 6BN; Calcozine Violet C; Caswell No. 264A; CCRIS 2464; Chlorure de methylrosanilinum; CI 42555; CI Basic Violet 3; Cloruro de metilrosanilina; Crystal Violet; Crystal Violet 10B; Crystal Violet 5BO; Crystal Violet 6B; Crystal Violet 6BO; Crystal Violet AO; Crystal Violet AON; Crystal Violet BP; Crystal Violet BPC; Crystal Violet chloride salt; Crystal Violet Chloride; Crystal Violet Extra Pure APNX; Crystal Violet Extra Pure; Crystal Violet Extra Pure APN; Crystal Violet FN; Crystal Violet HL2; Crystal Violet O; Crystal Violet Pure DSC Brilliant; Crystal Violet Pure DSC; Crystal Violet SS; Crystal Violet Technical; Crystal Violet USP; EINECS 208-953-6; EPA Pesticide Chemical Code 039502; Gentersal; Gentian Violet; Gentiaverm; Genticid; Gentioletten; Hecto Violet R; Hectograph Violet SR; Hexamethyl Violet; Hidaco Brilliant Crystal Violet; Hidaco Crystal Violet; HSDB 4366; Kristall-violett; Meroxyl; Meroxyl-wander; Meroxylan; Meroxylan-wander; Methyl Violet 10B; Methyl Violet 10BD; Methyl Violet 10BK; Methyl Violet 10BN; Methyl Violet 10BNS; Methyl Violet 10BO; Methyl Violet 5BNO; Methyl Violet 5BO; Methyl Violet 6B (biological stain); Methylrosanilinchlorid; Methylrosaniline chloride; Methylrosanilini chloridum; Methylrosanilinii chloridum; Methylrosanilinum chloratum; Methylviolett; Metilrosanilinio cloruro; Mitsui Crystal Violet; NCI-C55969; NSC 3090; Oxiuran; Oxycolor; Oxyozyl; Paper Blue R; Plastoresin Violet 5BO; Pyoktanin; Pyoverm; Vermicid; Vianin; Viocid; Violet 5BO; Violet 6BN; Violet CP; Violet cristallise; Violet gencianova; Violet krystalova; Violet xxiii; Violet zasadita 3.; aizen crystal violet; aniline violet pyoktanine; basic violet BN; bismuth violet; brilliant violet 58; calcozine violet C; calcozine violet 6BN; crystal violet O; crystal violet 5BO; crystal violet 6B; crystal violet 6BO; crystal violet 10B; crystal violet AO; crystal violet AON; crystal violet base; crystal violet BPC; crystal violet FN; crystal violet HI2; crystal violet SS; gentersal; hectograph violet SR; hecto violet R; hidaco crystal violet; meroxylanwander; methyl violet 5BNO; methyl violet 5BO; methyl violet 10B; methyl violet 10BD; methyl violet 10BK; methyl violet 10BN; methyl violet 10BO. A dye for wood, paper and silk; used in inks and as a biological stain. Grey powder; mp = 215° (dec); very soluble in H2O, CHCl3, EtOH (10 g/100 ml), glycerin (6.7 g/100 ml); LD50 (mus orl) = 1200 mg/kg, (rat orl) = 1000 mg/kg.

1818 Gentisin
437-50-3 4410 G

C14H10O5

1,3,7-Trihydroxyxanthone-3-methyl ether.

5-18-04-00497 (Beilstein Handbook Reference); BRN 0384788; CCRIS 3151; 1,7-Dihydroxy-3-methoxyxanth-en-9-one; EINECS 207-114-1; Gentianic acid; Gentianin; Gentisin; Gentisine; Xanthen-9-one, 1,7-dihydroxy-3-methoxy-; 9H-Xanthen-9-one, 1,7-dihydroxy-3-methoxy-. The yellow pigment of *Gentiana lutea*. Yellow rhombic crystals; mp = 266-267°; λ_m = 235, 260, 315 nm (ϵ = 39811, 50119, 15849, EtOH); insoluble in H2O, Me2CO, soluble in C5H5N, very soluble in EtOH.

1819 Geraniol
106-24-1   4415   G H

C10H18O

(E)-3,7-Dimethyl-2,6-octadien-1-ol.

4-01-00-02277 (Beilstein Handbook Reference); AI3-00206; BRN 1722456; CCRIS 7243; (E)-3,7-Dimethyl-2,6-octadien-1-ol; trans-3,7-Dimethyl-2,6-octadien-1-ol; 2-trans-3,7-Dimethyl-2,6-octadien-1-ol; 3,7-Dimethyl-trans-2,6-octadien-1-ol; (E)-Geraniol; β-Geraniol; trans-Geraniol; EINECS 203-377-1; EPA Pesticide Chemical Code 597501; FEMA No. 2507; Geraniol; HSDB 484; Lemonol; Meranol; (E)-Nerol; 2,6-Octadien-1-ol, 3,7-dimethyl-, (E)-; NSC 9279. A high-quality rose petal odor used in fragrances in perfumery, as a constituent of synthetic fragrances and with synthetic linalool. Oil; bp = 229-230°; d = 0.8750; λ_m = 190-195 nm (ϵ 18000); insoluble in H2O, soluble in organic solvents. *Bush Boake Allen; Lancaster Synthesis Co.; Penta Mfg.; SCM Glidco Organics.*

1820 Geranyl acetate
16409-44-2   G

C14H20O2

3,7-Dimethylocta-2,6-dienyl acetate.

3,7-Dimethylocta-2,6-dienyl acetate; EINECS 240-458-0; Geraniol acetate. Used In perfumery and flavoring. Oil; bp12 = 115°; d^{15} = 0.9163. *Firmenich; IFF; Penta Mfg.; SCM Glidco Organics.*

1821 Geranyl acetate, trans
105-87-3   G

C12H20O2

(E)-3,7-dimethyl-2,6-octadien-1-yl acetate.

4-02-00-00204 (Beilstein Handbook Reference); Acetic acid, geraniol ester; AI3-00207; Bay pine (oyster) oil; BRN 1722815; CCRIS 877; 2,6-Dimethyl-2,6-octadiene-8-yl acetate; 3,7-Dimethyl-2-trans, 6-octadienyl acetate; 3,7-Dimethyl-2,6-octadien-1-ol acetate; 3,7-Dimethyl-2,6-octadien-1-yl ethanoate, trans-; trans-2,6-Dimethyl-2,6-octadien-8-yl ethanoate; trans-3,7-Dimethyl-2,6-octadien-1-ol, acetate; EINECS 203-341-5; FEMA Number 2509; Geraniol acetate; HSDB 586; Meraneine;NCI-C54728; NSC 2584; 2,6-Octadien-1-ol, 3,7-dimethyl-, acetate,(E)-; 2,6-Octadien-1-ol, 3,7-dimethyl-, acetate, (2E)-; 2,6-Octadien-1-ol, 3,7-dimethyl-, acetate, trans-. Rosy, green, slightly lavendaceous odor. Used in perfume blends & fruit flavors. Oil; bp = 242°; d = 0.911; insoluble in H2O, soluble in organic

solvents; LD50 (rat orl) = 6330 mg/kg. *Bush Boake.*

1822 Germall® 115
39236-46-9   4939   G

C11H16N8O8

Imidazolidinyl urea.

Abiol; Biopure 100; EINECS 254-372-6; Germall 115; Imidazolinidyl urea; Imidurea; Imidurea NF; Methane-bis(N,N'-(5-ureido-2,4-diketotetrahydro-imidazole)-N,N-dimethylol); 1,1'-Methylenebis(3-(3-(hydroxymethyl)-2,5-dioxo-4-imidazolidinyl)urea); N,N'-Methylenebis(N'-(1-(hydroxymethyl)-2,5-dioxo-4-imidazolidinyl)urea); Sept 115; Tristat 1U; Unicide U-13; Urea, N,N''-methylene-bis(N'-(1-(hydroxymethyl)-2,5-dioxo-4-imidazolidinyl)yl)-urea. Antimicrobial preservative used in cosmetics. *3V; Sutton Labs.*

1823 Germanium
7440-56-4   4419   G

Ge

Ge
Germanium.

EINECS 231-164-3; Germanium; Germanium element; Germanium, metal powder; HSDB 2118. Nonmetallic element; used in electronics: manufacture of Ge diodes, transistors, solid state electronic devices, semiconducting applications, brazing alloys, phosphors, gold and beryllium alloys, infrared-transmitting glass; dental alloys. Metal; mp = 937°; bp = 2700°; poor electrical conductor. *Atomergic Chemetals; Cabot Carbon Ltd.; Cerac; Eagle-Picher; New Metals & Chems. Ltd; Noah Chem.*

1824 Germanium hydride
7782-65-2   4418   G

GeH4
Germanium tetrahydride.

EINECS 231-961-6; Germane; Germanium hydride; Germanium tetrahydride; Monogermane; UN2192. A flammable, toxic, colorless gas used for the deposition of epitaxial and amorphous silicon - germanium alloys. Gas; mp = -165°; bp = -88°; d = 2.600; LD50 (mus orl) = 1250 mg/kg.

1825 Gibberellic acid
77-06-5   4430   G P

303

$C_{19}H_{22}O_6$

(3S,3AS,4S,4aS,6S,8aR,8bR,11S)-6,11-Dihydroxy-3-methyl-12-methylene-2-oxo-4a,6-ethano-3,8b-prop-1-enoperhydroindeno(1,2-b)furan-4-carboxylic acid.
5-18-09-00269 (Beilstein Handbook Reference); Acide gibberellique; Activol; Activol GA; AI3-52922; Berelex; Brellin; BRN 0054346; Caswell No. 467; CCRIS 4820; Cekugib; EPA Pesticide Chemical Code 043801; GA; GA3; Gib-Sol; Gib-Tabs; Gibb-3-ene-1,10-dicarboxylic acid, 2,4a,7-trihydroxy-1-methyl-8-methylene-, 1,4a-lactone; Gibberelic acid; gibberellin x; Gibberellin; Gibberellin 1; Gibberellin A3; Gibberellin X; Gibberellins; Gibberellins A4A7; Gibbrel; Gibefol; Giberellin; Gibrel; Gibrescol; Gibreskol; Grocel; HSDB 712; 2β-Hydroxygibberellin 1; NCI-C55823; NSC 14190; Pgr-iv; Pro-Gibb; Pro-Gibb Plus; Pro-Gibb Plus; Regulex; Ryzup; Trihydroxy-1-methyl-8-methylenegibb-3-ene-1,10-dicarboxylic acid, 1,4a-lactone. A plant growth regulator; increases cropping in apples and pears. Registered by EPA as a plant growth regulator. Crystals; mp = 233-235°; $[\alpha]_D^{19}$ = +86° (c = 2.12); pK = 4.0; freely soluble in MeOH, EtOH, Me2CO, slightly soluble in H2O, Et2O, EtOAc. *American Camellia Society; Fine Agrochemicals Ltd.; Griffin LLC; ICI Agrochemicals; La-Co Industries; Micro-Flo Co. LLC; Syngenta Crop Protection; Valent Biosciences Corp.*

1826 Girard's reagent T
123-46-6   G

$C_5H_{14}ClN_3O$

2-hydrazino-N,N,N-trimethyl-2-oxoethanaminium chloride.
AI3-61529; Ammonium, (carbazoylmethyl)trimethyl-, chloride; Ammonium, (carboxymethyl)trimethyl-, chloride, hydrazide; Betaine hydrazide hydrochloride; Carbazoylmethyltrimethylammonium chloride; Carb-azoylmethyltrimetylammonium chloride; (Carboxy-methyl)trimethylammonium chloride hydrazide; EINECS 204-629-3; Ethanaminium, 2-hydrazino-N,N,N-trimethyl-2-oxo-; Girard reagent T; Girard T reagent; Girard's Reagent T; (Hydrazinocarbonylmethyl)trimethyl-ammonium chloride; NSC 9242; Trimethylacethydrazide ammonium chloride; Trimethylaminoacetohydrazide chloride; Trimethylammonium acetyl hydrazide chloride; Trimethylammonium chloride acethydrazide. Reacts with ketones to form water-soluble hydrazones. Used in the isolation of ketosteroids. Girard's reagent P [1126-58-5], used in the same way, is 1-(2-hydrazino-2-oxoethyl)pyridinium chloride. Crystals; mp = 192°; soluble in H2O, polar organic solvents, insoluble in non-polar organic solvents.

1827 Glass H
50813-16-6   8741   G

$Na_6O_{18}P_6$
Sodium hexametaphosphate.
EINECS 256-779-4; Grahamsches salz; Metaphosphoric acid, sodium salt; Natrium polymetaphosphat; Sodium metaphosphate; Sodium metapolyphosphate; Sodium polymetaphosphate; Sodium polyphosphate. Water softener and detergent. Solid; mp = 628°; d = 2.181; soluble in H2O, insoluble in organic solvents; LD50 (rat orl) = 6200 mg/kg. FMC.

1828 Glipizide
29094-61-9   4455   D

$C_{21}H_{27}N_5O_4S$

1-Cyclohexyl-3-[[p-[2-(5-methylpyrazinecarboxamido)-ethyl]phenyl]sulfonyl]urea.
Aldiab; Antidiab; Apamid; BRN 0903495; CP 28,720; Digrin; Dipazide; EINECS 249-427-6; Glibenese; Glibetin; Glican; Glide; Glidiab; Glipid; Glipizida; Glipizidum; Gluco-Rite; Glucolip; Glucotrol; Glucotrol XL; Glucozide; Glupitel; Glupizide; Glyde; Glydiazinamide; K 4024; Melizide; Metaglip; Mindiab; Minidiab; Minodiab; Napizide; Ozidia; Sucrazide; TK 1320. Antidiabetic agent. Crystals; mp = 208-209°; LD50 (rat ip) = 1.2 g/kg, (mus ip) > 3 g/kg. *Apothecon; Pratt Pharm.*

1829 Gluconal® ZN
4468-02-4   4469   G V

$C_{12}H_{22}O_{14}Zn$
Zinc gluconate.
BioZn-AAS; Bis(D-gluconato-O^1,O^2)zinc; Bis(D-gluco-nato-O1,O2) zinc; EINECS 224-736-9; HSDB 1054; Rubozine; Zinc gluconate; Zinc D-gluconate (1:2); Zinc bis(D-gluconato-O1,O2); Zinc, bis(D-gluconato-O^1,O^2)-. Pharmaceutical/food grade mineral source for human and veterinary pharmaceutical preparations, dietary supplements, fortified foods and animal feed. White solid; freely soluble in H2O (100 g/l); insoluble in organic solvents; LD50 (rat orl) > 5,000 mg/kg. *Akzo Chemie.*

1830 D-Gluconic acid
526-95-4   4469   D G P

$C_6H_{12}O_7$
2,3,4,5,6-Pentahydroxyhexanoic acid.
4-03-00-01255 (Beilstein Handbook Reference); BRN 1726055; Dextronic acid; EINECS 208-401-4; Glosanto; Gluconic acid; D-Gluconic acid; Glycogenic acid; Glyconic acid; HSDB 487; Maltonic acid; NSC 77381; Pentahydroxycaproic acid. Used in pharmaceuticals and food products, cleaning and pickling metals, sequestrant, cleansers, catalyst in textile printing. Medically, an antispasmodic agent. Registered by EPA as an antimicrobial and

disinfectant (cancelled). FDA GRAS, Japan approved, FDA approved for orals. Used as in orals a chelating agent, sequestrant and dietary supplement. Crystals; mp = 131°; d_4^{25} 1.24; $[\alpha]_D^{20}$ -6.7° (c = 1); freely soluble in H_2O, slightly soluble in EtOH, insoluble in Et_2O and most other organic solvents. *Akzo Chemie; Am. Biorganics; Faezy & Besthoff; Glucona; Jungbunzlauer; Pfizer Intl.; PMP Fermentation Prods.; Sigma-Aldrich Fine Chem.; Spectrum Chem. Manufacturing.*

1831   δ-D-Gluconolactone

90-80-2   4470   D G H

$C_6H_{10}O_6$

D-Gluconic acid, δ-lactone.

AI3-19578; Deltagluconolactone; EINECS 202-016-5; GDL; Gluconic acid, δ-lactone, D-; Gluconic acid lactone; Gluconic δ-lactone; D-Gluconic acid lactone; D-Gluconic acid-δ-lactone; Gluconolactone; Glucono δ-lactone; D-Glucono-1,5-lactone; δ-D-Gluconolactone; δ-Gluconolactone; D-δ-Gluconolactone; HSDB 488. Can behave as a chelator and is used in metal cleaning materials. mp = 153° (dec); $[\alpha]_D^{20}$ 61.7° (c = 1), mutarotates to +6.2°; soluble in H_2O (59 g/100 ml), less soluble in organic solvents. *Pfizer Intl.*

1832   Glucose

50-99-7   4472   H

$C_6H_{12}O_6$

α-D-glucose.

4-01-00-04302 (Beilstein Handbook Reference); AI3-09328; Anhydrous dextrose; Blood sugar; BRN 1724615; Cartose; CCRIS 950; Cerelose; Cerelose 2001; Corn sugar; 'α-D-Glucopyranose D-Glucose; Dextropur; Dextrose; Dextrose, anhydrous; Dextrose solution; Dextrosol; EINECS 200-075-1; Glucolin; Glucose; D-Glucose; D-Glucose, anhydrous; D(+)-Glucose; Glucosteril; Goldsugar; Grape sugar; HSDB 489; Maxim Energy Gel; NSC 406891; Sirup; Staleydex 111; Staleydex 333; Sugar, grape; Tabfine 097(HS); Traubenzucker; Vadex. Sugar obtained from the hydrolysis of starch; (anhydrous), (hydrous); Confectionery, foods, medicine, brewing, baking, canning. mp = 83°; $[\alpha]_D$ = 102.0°, mutarotates to 47.9° (H_2O); soluble in H_2O (1 g/ml), less soluble in organic solvents; LD_{50} (rbt iv) = 35 g/kg. *Avebe BV; Corn Products; Hightex; Lanaetex; Mallinckrodt Inc.; Mendell Edward; Sigma-Aldrich Fine Chem.; U.S. BioChem.*

1833   Glutaraldehyde

111-30-8   4485   G P

$C_5H_8O_2$

1,5-Pentanedial.

4-01-00-03659 (Beilstein Handbook Reference); Aldehyd glutarowy; Aldesan; Aldesen; Alhydex; BRN 0605390; Caswell No. 468; CCRIS

3800; Cidex; Coldcide-25; Dialdehyde; Dioxopentane; EINECS 203-856-5; EPA Pesticide Chemical Code 043901; GKN-O Microbiocide Concentrate; Glutaral; Glutaraldehyd; Glutaraldehyde; Glutaralum; Glutardialdehyde; Glutaric acid dialdehyde; Glutaric aldehyde; Glutaric dialdehyde; Glutarol; Gluteraldehyde; Hospex; HSDB 949; Microbiocide; NCI-C55425; NSC 13392; Pentanedial; Pentane-1,5-dial; Sonacide; Sporicidin; Ucarcide® 225; Veruca-sep. Preservative, antimicrobial for cosmetic, toiletry, and chemical specialty products. Used as an intermediate, fixative for tissues, for crosslinking protein and polyhydroxy materials, tanning of soft leathers. Registered by EPA as an antimicrobial and fungicide. Oil; mp = -14°; bp = 188° (dec), bp_{50} = 106 - 108°, bp_{10} = 71 - 72°; soluble in C_6H_6, freely soluble in H_2O, EtOH; LD_{50} (rat orl) = 134 mg/kg, LD_{50} of 25% aq. solution (rat orl) = 2.38 ml/kg, (rbt der) = 2.56 ml/kg,. *Baker Petrolite Corp.; BASF Corp.; Buckman Laboratories Inc.; CID Lines NV; Colcide Inc.; Dow Chem. U.S.A.; M & S Research; Medical Chemical Corp; Metrex Research Corp; Multisorb Technologies Inc; Ondeo Nalco Co.*

1834   Glutaric acid

110-94-1   4486   H

$C_5H_8O_4$

1,5-Pentanedioic acid.

4-02-00-01934 (Beilstein Handbook Reference); AI3-24247; BRN 1209725; EINECS 203-817-2; Glutaric acid; HSDB 5542; NSC 9238; Pentandioic acid; Pentanedioic acid; 1,5-Pentanedioic acid; 1,3-Propanedicarboxylic acid; n-Pyrotartaric acid. Needles; mp = 97.8°; bp = 303° (dec); d^{15} = 1.429; very soluble in H_2O, EtOH, Et_2O, soluble in $CHCl_3$, conc. H_2SO_4; slightly soluble in DMSO, ligroin, insoluble in C_6H_6.

1835   Glyburide

10238-21-8   4491   D

$C_{23}H_{28}ClN_3O_5S$

5-Chloro-N-[2-[4-[[[(cyclohexylamino)carbonyl]amino]-sulfonyl]phenyl]ethyl]-2-methoxybenzamide.

Abbenclamide; Adiab; Apo-Glibenclamide; Azuglucon; Bastiverit; Benclamin; Betanase; Betanese 5; BRN 2230085; Calabren; Cytagon; Daonil; Debtan; Dia-basan; Diabeta; Diabiphage; Dibelet; Duraglucon; EINECS 233-570-6; Euclamin; Euglucan; Euglucon; Euglucon 5; Euglykon; GBN 5; Gen-Glybe; Gewaglucon; Gilemal; Gl; Glamide; Gliban; Gliben; Gliben-Puren N; Glibenbeta; Glibenclamid AL; Glibenclamid Basics; Glibenclamid-Cophar; Glibenclamid Fabra; Glibenclamid Genericon; Glibenclamid Heumann; Glibenclamid-Ratiopharm; Glibenclamid Riker M.; Glibenclamida; Glibenclamide; Glibenclamidum; Glibenil; Glibens; Glibesyn; Glibet; Glibetic; Glibil; Gliboral; Glicem; Glidiabet; Glimel; Glimide; Glimidstada; Glisulin; Glitisol; Glubate; Gluben; Gluco-Tabinen; Glucobene; Glucohexal; Glucolon; Glucomid; Glucoremed; Glucoven; Glyben; Glyben-clamide; Glybenzcyclamide; Glyburide; Glycolande; Glycomin; HB 419; Hemi-Daonil; Hexaglucon; Humedia;

Lederglib; Libanil; Lisaglucon; Maninil; Med-Glionil; Melix; Micronase; Miglucan; Nadib; Neogluconin; Norglicem 5; Normoglucon; Novo-Glyburide; Orabetic; Pira; Praeciglucon; Prodiabet; Renabetic; Semi-Daonil; Semi-Euglucon; Semi-Gliben-Puren N; Sugril; Suraben; Tiabet; U 26452; UR 606; Wuglucon; Yuglucon. Antidiabetic agent. Second generation sulfonylurea with hypoglycemic activity. Crystals; mp = 169-170°, 172-174°; poorly soluble in H_2O, soluble in organic solvents; LD_{50} (mus orl) > 20 g/kg, (mus ip) > 12.5 g/kg, (mus sc) > 20 g/kg. (rat orl) > 20 g/kg, rat ip) > 12.5 g/kg, (rat sc) > 20 g/kg. *Abbott Labs Inc.*

1836 Glycarsamide
144-87-6   4489   G

C8H10AsNO5
4-(Glycolloylamino)phenylarsonic acid.
Astryl; Glycarsamide; EINECS 205-643-2. Anthelmintic. Rhône-Poulenc; UK.

1837 Glycerin tricaprylate
7360-38-5   G

C27H50O6
Glyceryl trioctanoate.
Caprylic acid triglyceride; Caprylin; EINECS 230-896-0; Emalex O.T.G; Glycerin tricaprylate; Glycerol trioctanoate; Glyceryl tri(2-ethylhexanoate); Glyceryl trioctanoate; Hexanoic acid, 2-ethyl-, 1,2,3-propanetriyl ester; MCT; Octanoic acid, 1,2,3-propanetriol ester; Panacete 800; Propane-1,2,3-triyl 2-ethylhexanoate; RATO; Tricaprylin; Tricapryloylglycerol; Tricaprylyl glycerin; Trioctanoin; tricaprylic glyceride; trioctanoyl-glycerol. Oil-phase cosmetic ingredient; emollient. mp = 6°; bp = 233°; d = 0.9530; insoluble in H_2O, soluble in organic solvents; LD_{50} (rat orl) = 33300 mg/kg. *Nihon Nohyaku Co. Ltd.*

1838 Glycerol
56-81-5   4497   G H

C3H8O3
1,2,3-Propanetriol.
4-01-00-02751 (Beilstein Handbook Reference); 90 Technical glycerine; Aci-Jel; AI3-00091; Auralgan; BRN 0635685; Caswell No. 469; CCRIS 2295; Citifluor AF 2; Clyzerin, wasserfrei; Collyrium Fresh-Eye Drops; Dagralax; EINECS 200-289-5; EPA Pesticide Chemical Code 063507; FEMA No. 2525; Glicerina; Glicerol; Glycerin; Glycerin (mist); Glycerin, anhydrous; Glycerin, synthetic; Glycerine; Glycerine mist; Glycerinum; Glyceritol; Glycerol;

Glycerolum; Glycyl alcohol; Glyrol; Glysanin; Grocolene; HSDB 492; IFP; Incorporation factor; MOON; NSC 9230; Ophthalgan; Optim; Osmoglyn; Polyhydric alcohols; Propanetriol; Star; Synthetic glycerin; Synthetic glycerine; Trihydroxy-propane; Vitrosupos. Used in manufacture of alkyd resins, dynamite, ester gums, pharmaceuticals, cosmetics, perfumery, lubricants, softener, bacteriostat, penetrant, solvent, humectant, plasticizer, emollient; antifreeze; in production of antibiotics. Registered by EPA (cancelled). Used as a diagnostic aid in ophthalmology and as an emollient and demulcent. Also used as a humectant, solvent, in cosmetics, liquid soaps, confections, inks, and lubricants; used as a chemical intermediate in polyester and PU formulations. Viscous, colorless oil; mp = 17.8°; bp = 290°; d^{15} = 1.2656, d^{20} = 1.2636, d^{s25} = 1.2620; freely soluble in EtOH, MeOH, H_2O, slightly soluble in Me2CO, soluble in Et2O (0.2 g/100 ml), EtOAc (9 g/100 ml), insoluble in C6H6, CHCl3, oils; viscosity of aqueous solutions ranges from 1.143 mPa (5% w/w) to 111 mPa (83% w/w). LD_{50} (rat orl) = 12600 mg/kg, (rat sc) = 100 mg/kg, (rat iv) = 5600 mg/kg, (rat ip) = 8300 mg/kg, (mus orl) = 4100 mg/kg, (mus sc) = 90 mg/kg, (mus iv) = 6200 mg/kg, (mus ip) = 8980 mg/kg, (gpg orl) = 7750 mg/kg. *Alba Intl. Inc.; Asahi Denka Kogyo; Farleyway Chem. Ltd.; Fina Chemicals; Henkel/Emery; Lonza Ltd.; Procter & Gamble Co.; Sigma-Aldrich Co.; Unichema; Witco/Humko.*

1839 Glycerol 1,2-dichlorohydrin
616-23-9   H

C3H6Cl2O
1,2-Dichloro-3-propanol.
4-01-00-01442 (Beilstein Handbook Reference); AI3-61542; BRN 1732060; CCRIS 954; α,β-Dichlorohydrin; 1,2-Dichloro-3-propanol; 1,3-Dichloropropanol-2; 2,3-Dichloro-1-propanol; 2,3-Dichloropropan-1-ol; 2,3-Dichloropropanol; EINECS 210-470-0; Glycerol α,β-dichlorohydrin; Glycerol 1,2-dichlorohydrin; HSDB 2743; 1-Propanol, 2,3-dichloro-; UN2750. Liquid; bp = 184°; d^{20} = 1.3607; slightly soluble in H_2O, ligroin, freely soluble in EtOH, Et2O, Me2CO, C6H6.

1840 Glycerol 1-monooleate
25496-72-4   H

C21H40O4
9-Octadecenoic acid (Z)-, monoester with 1,2,3-propanetriol.
Adchem GMO; AJAX GMO; Aldo 40; Aldo MO-FG; Alkamuls GMO 45LG; Arlacel 129; Atmer 1007; Canamex Glicepol 182; Dimodan GMO 90; Dimodan LSQK; Dur-Em 204; Dur-EM 114; Edenor GMO; EINECS 247-038-6; Emalsy MO; Emalsy OL; Emasol MO 50; Emcol O; Emerest 2400; Emerest 2421; Emery oleic acid ester 2221; Emrite 6009; Emuldan RYLO-MG 90; Excel O 95F; Excel O 95N; Excel O 95R; Glycerin monooleate; Glycerine monooleate; Glycerol monooleate; Glycerol, 1-mono (9-octa-decenoate); Glycerol 1-monooleate; Glycerol oleate; Glyceryl monooleate; Glyceryl oleate; Glycolube 100; GMO 8903; Harowax L 9; HSDB 493; Kemester 2000; Kessco GMO; Loxiol G 10; Mazol GMO; Monoglyceryl oleate; Monomuls 90018; Monolein; Monoolein; Monooleoylglycerol; Nikkol MGO; 9-Octadecenoic acid (Z)-, monoester with 1,2,3-propanetriol (9CI); OL 100; Oleic acid glycerol monoester; Oleic acid, monoester

with glycerol; Oleic acid monoglyceride; Oleic monoglyceride; Olein, mono-; Oleoylglycerol; Oleylmonoglyceride; Olicine; Peceol; Rikemal O 71D; Rikemal ol 100; S 1096; S 1096R; S 1097; Sinnoester ogc; Sunsoft O 30B; Supeol.

1841 Glycerol monostearate
31566-31-1   4502   G H

$C_{21}H_{42}O_4$
n-Octadecanoic acid, monoester with 1,2,3-propanetriol.
Abracol S.L.G.; Admul; Advawax 140; Al3-00966; Aldo-28; Aldo-72; Aldo HMS; Aldo MS; Arlacel 161; Arlacel 169; Armostat 801; Atmos 150; Atmul 124; Atmul 67; Atmul 84; Cefatin; Celinhol - A; Cerasynt 1000-D; Cerasynt S; Cerasynt SD; Cerasynt SE; Cerasynt WM; Citomulgan M; Cyclochem GMS; Dermagine; Distearin; Drewmulse TP; Drewmulse V; Drumulse AA; EINECS 250-705-4; Emcol CA; Emcol MSK; Emerest 2400; Emerest 2401; EMUL P.7; Estol 603; Glycerin monostearate; Glycerol monostearate; Glyceryl monostearate; Glyceryl stearate; Grocor 5500; Grocor 6000; Hodag GMS; Imwitor 191; Imwitor 900K; Kessco 40; Lipo GMS 410; Lipo GMS 450; Lipo GMS 600; Monelgin; Monoglyceryl stearate; Monostearate (glyceride); Monostearin; Monostearin; Myvaplex 600; Octadecanoic acid, monoester with 1,2,3-propanetriol; Ogeen 515; Ogeen GRB; Ogeen M; Ogeen MAV; Orbon; Protachem GMS; Sedetine; Starfol GMS 450; Starfol GMS 600; Starfol GMS 900; Stearates; Stearic acid, monoester with glycerol; Stearic monoglyceride; Stearoylglycerol; Tegin; Tegin 503; Tegin 515; Unimate GMS; USAF KE-7; Witconol 2400, Witconol MS; Witconol MST, Witconol RHT. Oil-water emulsifier, lubricant. Emulsifier for cosmetic, pharmaceutical, Aerosol® formulations; internal lubricant, plasticizer, and emulsifier in industrial applications; flow control agent for polymerization reactions; dispersant. Lubricant, plasticizer, oil-water emulsifier used in industrial applications. Waxy solid; mp = 56-58°; insoluble in H_2O, soluble in organic solvents. *CK Witco Corp.*

1842 Glyceryl behenate
18641-57-1   G

$C_{69}H_{134}O_6$
Docosanoic acid, 1,2,3-propanetriyl ester.
2-02-00-00374 (Beilstein Handbook Reference); Behenic acid, 1,2,3-propanetriol ester; BRN 1811476; Compritol 888; Docosanoic acid, 1,2,3-propanetriyl ester; Docosanoin, tri-; EINECS 242-471-7; Glyceryl tribehenate; 1,2,3-Propanetriol tridocosanoate; Propane-1,2,3-triyl tridocosanoate; Syncrowax HR-C; Tribehenin; Tribehenoyl glycerol; Tridocosanoin. Emulsifier, emollient, opacifier; also suspending agent, thickener, gloss improver used in personal care products. *Croda.*

1843 Glyceryl caprylate
26402-26-6   G

$C_{11}H_{22}O_4$
Octanoic acid monoester with 1,2,3-propanetriol.
Capmul MCM-C 8; Caprylic acid monoglyceride; EINECS 247-668-1; Glycerin monocaprylate; Glycerol monocaprylate; Glyceryl caprylate; Glyceryl monocaprylate; Imwitor 308; Monocaprylin; Monooctanoin; Octanoic acid, monoester with glycerol; Octanoic acid, monoester with 1,2,3-propanetriol; Octanoin, mono-; Poem M 100; Sefsol 318; Sunsoft 700P; Sunsoft 700p2; Witafrol® 7420; DL-glyceryl-1-mono-octanoate; caprylic/capric glycerides. Surfactant for pharmaceutical, cosmetic, nutritional fields; as emulsifier, solubilizer, dispersant, plasticizer, lubricant, consistency regulator, skin/mucous membrane protectant, refatting agent, penetrant, carrier, adsorption promoter, antifoaming agent. Solid; mp = 39.5-40.5°. *Degussa-Hüls Corp.*

1844 Glyceryl monooleate
37220-82-9   G

$C_{21}H_{40}O_4$
9-Octadecenoic acid (Z)-, ester with 1,2,3-propanetriol.
EINECS 253-407-2; Glycerine oleate; PEG-15 Tallow polyamine; 9-Octadecenoic acid (9Z)-, ester with 1,2,3-propanetriol; 9-Octadecenoic acid (Z)-, ester with 1,2,3-propanetriol; Polyethylene glycol (15) tallow polyamine; Polyoxyethylene (15) tallow polyamine. 1- or 2-Monoester of glycerin and oleic acid; Emulsifier, coemulsifier, stabilizer, wetting agent, lubricant, and antistat; used in cosmetic, pharmaceutical, industrial, and food applications. *Calgene; Ferro/Keil; Grindsted UK; Henkel/Emery; ICI Spec.; Inolex; Karlshamns; Lonzagroup; Mona; Patco; Stepan; Witco/Humko.*

1845 Glyceryl monoricinoleate
141-08-2   G

$C_{21}H_{40}O_5$
12-Hydroxy-9-octadecenoic acid, monoester with 1,2,3-propanetriol.
2,3-Dihydroxypropyl 12-hydroxy-9-octadecenoate; EINECS 205-455-0; Glyceryl monoricinoleate; Glyceryl ricinoleate; 12-Hydroxy-9-octadecenoic acid, monoester with 1,2,3-propanetriol; Monoricinolein; 9-Octadecenoic acid, 12-hydroxy-, monoester with 1,2,3-propanetriol; 9-Octadecenoic acid, 12-hydroxy-, 2,3-dihydroxypropyl ester, (9Z,12R)-; Ricinolein, 1-mono-. Used as an emulsifying agent. *CasChem; Lonzagroup.*

1846 Glyceryl stearate
123-94-4   G

$C_{21}H_{42}O_4$
Monostearin.

Aldo 33; Aldo 75; Aldo MSD; Aldo MSLG; Arlacel 165; (1)-2,3-Dihydroxypropyl stearate; EINECS 204-664-4; EINECS 245-121-1; Emerest 2407; FEMA No. 2527; Glycerin 1-monostearate; Glycerin 1-stearate; Glycerol α-monostearate; Glycerol 1-monostearate; Glycerol 1-stearate; Glyceryl monostearate; Glyceryl 1-monostearate; 1-Glyceryl stearate; NSC 3875; Octadecanoic acid, 2,3-dihydroxypropyl ester; α-Monostearin; 1-Monostearin; 1-Monostearoylglycerol; Sandin EU; Stearic acid α-monoglyceride; Stearic acid 1-monoglyceride; Stearin, 1-mono-; 3-Stearoyloxy-1,2-propanediol; Tegin 55G; Octadecanoic acid, monoester with 1,2,3-propanetriol. Nonionic secondary o/w emulsifier for creams and lotions; viscosity booster for emulsions. *Eastman Chem. Co.; Goldschmidt; Grindsted UK; Hart Prod.; Henkel/Emery; ICI Spec.; Inolex; Karlshamns; Lanaetex; Lipo; Lonzagroup; MTM Spec. Chem. Ltd.; Patco; Van Dyk; Witco/Humko.*

1847   Glyceryl stearate citrate
91744-38-6   G

C27H48O10
2-Hydroxy-1,2,3-propanetricarboxylic acid, monoester with 1,2,3-propanetriol monooctadecanoate.
EINECS 294-600-1; Glycerides, C16-18 mono-, di- and tri-, hydrogenated, citrates, potassium salts; Imwitor® 370. Food emulsifier; oil-water emulsifier for very polar oils, fats and liquid wax esters in cosmetics. *Degussa-Hüls Corp.*

1848   Glyceryl thioglycolate
30618-84-9   H

C5H10O4S
Mercaptoacetic acid, monoester with 1,2,3-propanetriol.
Acetic acid, mercapto-, monoester with 1,2,3-propanetriol; Acetic acid, mercapto-, monoester with glycerol; EINECS 250-264-8; Glycerol monomercapto-acetate; Glyceryl monothioglycolate; Glyceryl thio-glycolate; Mercaptoacetic acid, monoester with propane-1,2,3-triol; Mercaptoacetic acid, monoester with 1,2,3-propanetriol. May also contain the isomeric 2'-ester.

1849   Glyceryl triacetyl ricinoleate
101-34-8   G H

C63H110O12
9-Octadecenoic acid, 12-(acetyloxy)-, 1,2,3-propanetriyl ester, (9Z,9'Z,9"Z,12R,12'R,12"R)-.
Baker P-8; EINECS 202-935-1; Glyceryl triacetyl ricinoleate; 9-Octadecenoic acid, 12-(acetyloxy)-, 1,2,3-propanetriyl; 1,2,3-Propanetriyl 12-(acetyloxy)-9-octa-decenoate; Ricinolein, tri-, triacetate. A plasticizer for vinyl polymers and GR/S, neoprene GN, ethyl cellulose, and perbunan. Used for vinyl wire jacketing and semi-rigid vinyls; as an emollient and stabilizer for anhydrated

pigmented systems. Viscous oil; d = 0.967; iodine no 76; saponification no 300; insoluble in H2O. *CasChem; Sigma-Aldrich Fine Chem.*

1850   Glycidoxypropyltrimethyoxysilane, γ
2530-83-8   G H

C9H20O5Si
3-(2,3-Epoxypropoxy)propyltrimethoxysilane.
A 187; AI3-52752; Aktisil EM; BRN 4308125; CCRIS 3044; CG6720; DZ 6040; EINECS 219-784-2; Glycidyl 3-(trimethoxysilyl)propyl ether; Glycidyloxypropyltrimeth-oxysilane; KBM 403; KBM 430; NSC 93590; NUCA 187; Prosil® 5136; Silan A 187; Silane A 187; Silane, trimethoxy(3-(oxiranylmethoxy)propyl)-; Silane-Y-4087; Silane Z 6040; Silicone A-187; Silicone KBM 403; Union carbide A-187; Y 4087; Z 6040. Coupling agent for epoxies, urethanes, acrylics and polysulfides; filler for thermosets, chemical intermediate, blocking agent, release agent, lubricant, primer and reducing agent. Liquid; bp2 = 120°; d = 1.0700; LD50 (rat orl) = 23 g/kg. *Degussa-Hüls Corp.; Hoffmann Mining; PCR.*

1851   Glycidyl acrylate
106-90-1   G

C6H8O3
Acrylic acid 2,3-epoxypropyl ester.
4-17-00-01005 (Beilstein Handbook Reference); Acrylic acid, 2,3-epoxypropyl ester; Acrylic acid glycidyl ester; BRN 0109092; CCRIS 2625; EINECS 203-440-3; 2,3-Epoxypropyl acrylate; Glycidyl acrylate; Glycidyl propenate; Glycidylester kyseliny akrylove; HSDB 5376; M 581; Methacyclic acid, 2,3-epoxypropyl ester; NSC 24151; 1-Propanol, 2,3-epoxy-, acrylate; 2-Propenoic acid, oxiranylmethyl ester. Polyfunctional monomer; in coatings to improve adhesion to substrate and solvent resistance. Oil; bp10 = 53°; d20 =1.1109; very soluble in C6H6. *Estron Chemical, Inc.*

1852   Glycidyl isocyanurate
2451-62-9   H

C12H15N3O6
s-Triazine-2,4,6(1H,3H,5H)-trione, tris(2,3-epoxypropyl)-.
Araldite PT 810; BRN 0765833; CCRIS 6112; EINECS 219-514-3; Glycidyl isocyanurate; NSC 296964; TGT; s-Triazine-

2,4,6(1H,3H,5H)-trione, 1,3,5-tris(2,3-epoxy-propyl)-; Tri(epoxypropyl)isocyanurate; Triglycidyl isocyanurate; N,N',N''-Triglycidyl isocyanurate; Tris(2,3-epoxypropyl)isocyanurate; XB 2615.

1853   Glycidyl methacrylate
106-91-2   G

$C_7H_{10}O_3$
Methacrylic acid 2,3-epoxypropyl ester.
5-17-03-00035 (Beilstein Handbook Reference); Acriester G; Blemmer G; Blemmer GMA; BRN 0002506; CCRIS 2626; CP 105; EINECS 203-441-9; 2,3-Epoxypropyl methacrylate; Glycidol methacrylate; Glycidyl α-methyl acrylate; Glycidyl methacrylate; HSDB 494; Light Ester G; Methacrylic acid, 2,3-epoxypropyl ester; 2-((Methacryloxy)methyl)oxirane; NSC 24156; 1-Propanol, 2,3-epoxy-, methacrylate; 2-Propenoic acid, 2-methyl-, oxiranylmethyl ester; SR 379; SY-Monomer G. Polyfunctional monomer; in hydrogels for contact lenses and membranes, molding and casting compounds, impregnating paper, concrete, wood, coatings, printing inks, adhesives, sealants, elastomers. With 50 ppm inhibitor; polyfunctional monomer polymerized by applications of heat, heat and peroxidic catalysts, and irradiation by UV, β, σ, or x-ray; in hydrogels for contact lenses and membranes, molding and casting compounds; impregnating paper, concrete and wood, coatings and printing inks, adhesives, sealants and elastomers. Liquid; bp = 189°, bp10 = 75°; d^{20} = 1.042; soluble in H_2O, very soluble in C_6H_6, Et_2O, EtOH; LD50 (rat orl) = 597 mg/kg. *Estron Chemical, Inc.; Mitsubishi Gas; Sartomer; Gas; Mitsubishi; Sartomer.*

1854   Glycidyl 2-methylphenyl ether
2210-79-9   G H

$C_{10}H_{12}O_2$
((2-Methylphenoxy)methyl)oxirane.
4-17-00-00993 (Beilstein Handbook Reference); AI3-13110; Araldite DY 023; BRN 0004585; CCRIS 2636; o-Cresol glycidyl ether; Cresyl glycidyl ether, o-; DY 023; EINECS 218-645-3; 2,3-Epoxypropyl o-tolyl ether; Glycidyl 2-methylphenyl ether; Glycidyl o-methylphenyl ether; Glycidyl o-tolyl ether; o-Kresol-glycidaether; NSC 11571; NSC 20291; Oxirane, ((2-methylphenoxy)methyl)-; Propane, 1,2-epoxy-3-(o-tolyloxy)-. Diluent for epoxy resins and cements. Liquid; bp = 259°; d = 1.09 g/ml; insoluble in H_2O, soluble in organic solvents. *Ciba-Geigy Corp.*

1855   Glycidyl trimethyl ammonium chloride
3033-77-0   G

$C_6H_{14}ClNO$
2,3-Epoxypropyltrimethylammonium chloride.
Ammonium, (2,3-epoxypropyl)trimethyl-, chloride; CCRIS 4128; EINECS 221-221-0; β,σ-Epoxypropyltrimethyl-ammonium chloride; (2,3-Epoxypropyl)trimethyl-ammonium chloride; G-MAC; Glycidyl-trimethyl-ammonium chloride; N-Glycidyl-N,N,N-trimethyl-ammonium chloride; Glytac; Glytac A 100; NSC 51213; Ogtac 85 V; Oxiranemethanaminium, N,N,N-trimethyl-, chloride; 1-(Trimethylammonio)-2,3-epoxypropane chloride; Trimethylglycidylammonium chloride; N,N,N-Trimethyloxiranemethanaminium chloride. Intermediate for cationic surfactants; modifier for synthetic polymers such as starch, cellulose, gelatins, polyacrylic acids, epoxy resins; used for production of emulsion layers on photographic plates. *Chem-Y GmbH.*

1856   Glycine
56-40-6   4504   H

$C_2H_5NO_2$
Aminoacetic acid.
Acetic acid, amino-; Acide aminoacetique; Acido aminoacetico; Acidum aminoaceticum; Aciport; AI3-04085; Aminoacetic acid; 2-Aminoacetic acid; Aminoazijnzuur; Aminoethanoic acid; Amitone; CCRIS 5915; Corilin; EINECS 200-272-2; FEMA No. 3287; Glicina; Glicoamin; GLY; Glycin; Glycine; L-Glycine; Glycine, non-medical; Glycinum; Glycocoll; Glycolixir; Glycosthene; Hampshire glycine; HSDB 495; Leimzucker; NSC 25936; Padil; Sucre de gelatine. Used in organic synthesis, nutrient, biochemical research, buffering agent and as a chicken-feed additive. Reduces bitter taste of saccharin, retards rancidity in animal and vegetable fats. dec 233°; d = 1.1607; soluble in H_2O (25 g/100 ml), less soluble in organic solvents. *Allchem Ind.; Degussa Ltd.; Degussa-Hüls Corp.; Grace W.R. & Co.; Sigma-Aldrich Fine Chem.; U.S. BioChem.*

1857   Glycol diacetate
111-55-7   3833   G H

$C_6H_{10}O_4$
1,2-Ethanediol diacetate.
4-02-00-00217 (Beilstein Handbook Reference); AI3-08223; Aptex Donor H-plus; BRN 1762308; 1,2-Diacetoxyethane; EINECS 203-881-1; Ethanediol diacetate; 1,2-Ethanediol, diacetate; Ethylene di(acetate); Ethylene diacetate; Ethylene diethanoate; Ethylene glycol, diacetate; Glycol diacetate; HSDB 430; NSC 8853. Used as an extraction solvent, in foundry resins, as a perfume fixative, solvent for coatings. Liquid; mp = -31°; bp = 190°; d^{20} = 1.1043; very soluble in H_2O, freely soluble in EtOH, Et_2O, Me_2CO, C_6H_6, CCl_4, CS_2, AcOH; LD50 (rat orl) = 6.86 g/kg. *Eastman Chem. Co.; Rit-Chem.*

1858   Glycol dilaurate
624-04-4   4510   G

$C_{26}H_{50}O_4$
Dodecanoic acid 1,2-ethanediyl ester.
AI3-03485; Dodecanoic acid, 1,2-ethanediyl ester; EINECS 210-827-0; Emalex EG-di-L; Ethylene dilaurate; Ethylene glycol didodecanoate; Ethylene glycol dilaurate; Ethylene laurate; Glycol

dilaurate; Kemester® EGDL; Lauric acid, 1,2-ethanediyl ester; Lauric acid, ethylene ester; NSC 406565 . Oil-phase cosmetic ingredient, emulsifier for creams, milky lotions, hair conditioners; cleaner, superfattening agent, thickener, reforming agent. Used in cosmetics and pharmaceuticals. Viscous liquid; mp = 56.6°; bp20 = 188°; insoluble in H_2O, very soluble in EtOH, Et₂O. *Nihon Nohyaku Co. Ltd.; Witco/Humko.*

1859   Glycol dimercaptoacetate
123-81-9   G

$C_6H_{10}O_4S_2$
Ethylene glycol bisthioglycolate.
Acetic acid, mercapto-, 1,2-ethanediyl ester; Acetic acid, mercapto-, ethylene ester; AI3-26087; BRN 1948305; EINECS 204-653-4; Ethylene bis(mercaptoacetate); Ethylene bis(thioglycolate); Ethylene di(S-thioacetate); Ethylene glycol bis(mercaptoacetate); Ethylene glycol bis(thioglycolate); Ethylene glycol bis(thioglycolic ester); Ethylene mercaptoacetate; GDMA; Glycol bis(mercaptoacetate); Glycol dimercaptoacetate; NSC 30032. Crosslinking agent for rubbers, accelerator in curing epoxy resins. Oil; bp1 = 137-139°; d = 1.3170; soluble in H_2O (20 g/l), organic solvents; LD50 = 330 mg/kg. *Evans Chemetics; Janssen Chimica.*

1860   Glycol dinitrate
628-96-6   3834   H

$C_2H_4N_2O_6$
Ethylene glycol dinitrate.
4-01-00-02413 (Beilstein Handbook Reference); BRN 1709055; Dinitrate d'ethylene glycol; Dinitrato de etilenglicol; Dinitroglicol; Dinitroglycol; EGDN; EINECS 211-063-0; Ethanediol dinitrate; 1,2-Ethanediol, dinitrate; Ethylene dinitrate; Ethylene glycol dinitrate; Ethylene nitrate; Ethylenglykoldinitrat; Glycol (dinitrate de); Glycol dinitrate; Glycoldinitraat; Glykoldinitrat; HSDB 537; Nitroglycol; Nitroglykol. Liquid; mp = -22.3°; bp = 198.5°; d^{20} = 1.4918; very soluble in EtOH, Et₂O.

1861   Glycolic acid
79-14-1   4511   P

$C_2H_4O_3$
2-Hydroxyacetic acid.
4-03-00-00571 (Beilstein Handbook Reference); Acetic acid, hydroxy-; AI3-15362; BRN 1209322; Caswell No. 470; EINECS 201-180-5; EPA Pesticide Chemical Code 000101; Glycolic acid; Glycollic acid; HSDB 5227; Hydroxyacetic acid; Hydroxyethanoic acid; Kyselina glykolova; Kyselina hydroxyoctova; NSC 166. Registered by EPA as an antimicrobial and disinfectant. Used in leather dyeing and tanning; textile dyeing; cleaning, polishing and soldering compounds; copper pickling; adhesives; electroplating; petroleum demulsifier; chelating agent for iron; chemical milling; pH control. Hygroscopic crystals; mp = 80°; soluble in H_2O, organic

solvents; pH of aqueous solutions = 2.5 (0.5%), 2.33 (1%), 2.16 (2%), 1.91 (5%), 1.73 (10%); LD50 (rat orl) = 1950 mg/kg. *DuPont; Greeff R.W. & Co.; Hoechst Celanese.*

1862   Glycolonitrile
107-16-4   H

C_2H_3NO
2-Hydroxyacetonitrile.
4-03-00-00598 (Beilstein Handbook Reference); Acetonitrile, hydroxy-; AI3-23958; BRN 0605328; CCRIS 4655; Cyanomethanol; EINECS 203-469-1; Formaldehyde cyanohydrin; Glycolic nitrile; Glycollonitrile; Glycolo-nitrile; Glykolonitril; HSDB 2123; Hydroxyacetonitrile; α-Hydroxymethylcyanide; 2-Hydroxyacetonitrile; Hydroxy-azetonitril; Hydroxymethylkyanid; Hydroxymethylnitrile; Methylene cyanohydrin; NSC 1790; USAF A-8565. Liquid; mp <-72°; bp = 183° (dec), bp24 = 119°; freely soluble in H_2O, EtOH, Et₂O, insoluble in C_6H_6, CHCl₃.

1863   Glycol ricinoleate
106-17-2   G

$C_{20}H_{38}O_4$
Ethylene glycol monoricinoleate.
Cithrol EGMR N/E; EINECS 203-369-8; 1,2-Ethanediol monoricinoleate; Ethylene glycol monoricinoleate; Flexricin® 15; Glycol monoester ricinoleate; Glycol monoricinoleate; Glycol ricinoleate; 2-Hydroxyethyl 12-hydroxy-9-octadecenoate; 2-Hydroxyethyl ricinoleate; NSC 7394; 9-Octadecenoic acid, 12-hydroxy-, 2-hydroxyethyl ester; 9-Octadecenoic acid, 12-hydroxy-, 2-hydroxyethyl ester, (9Z,12R)-; 9-Octadecenoic acid, 12-hydroxy-, 2-hydroxyethyl ester, (R-(Z))- (9CI); Ricinoleic acid, 2-hydroxyethyl ester; S152. Wetting agent, plasticizer, textile, household, and cosmetic applications, rewetting dried skins; chemical intermediate. Oil; d = 0.965; insoluble in water, miscible with most organic solvents. *CasChem.*

1864   Glycols, polyethylene, monooleate
9004-96-0   E G H

Poly(oxyethylene) monooleate.
Ablunol 200MO; Akyporox O 50; Atlas G-2142; Atlas G-2144; Cemulsol 1050; Cemulsol A; Cemulsol C 105; Cemulsol D-8; Chemester 300-OC; Cithrol PO; Crodet O 6; Emanon 4115; Emcol H-2A; Emcol H 31A; Emerest 2646; Emerest 2660; Empilan BP 100; Empilan BQ 100; Emulphor A; Emulphor UN-430; Emulphor VN 430; Ethofat O; Ethofat O 15; Ethylan A3; Ethylan A6; Extrex P 60; hydroxy-; Ionet MO-400; Lannagol LF; Lipal 30W; Lipal 400-ol; Macrogol oleate 600; Nikkol MYO 10; Nikkol MYO 2; Noigen ES 160; Nonex 25; Nonex 30; Nonex 52; Nonex 64; Nonion 06; Nonion O2; Nonion O4; Nonisol 200; Nopalcol 1-0; Nopalcol 4-O; Nopalcol 6-0; OK 7; Oleic acid, ethylene oxide adduct; Oleic acid poly(oxyethylene) ester; Oleox 5; Olepal I; Olepal III; PEG 200MO; PEG 600MO; PEG 1000MO; PEG-3 Oleate; PEG-4 Oleate; PEG-5 Oleate; PEG-6 Oleate; PEG-7 Oleate; PEG-8 Oleate; PEG-9 Oleate; PEG-10 Oleate; PEG-11 Oleate; PEG-12 Oleate; PEG-14 Oleate; PEG-15 Oleate; PEG-20 Oleate; PEG-32 Oleate; PEG-36 Oleate; PEG-75 Oleate; PEG-150 Oleate; Pegosperse 400MO; Poly(ethylene oxide) oleate; Poly(oxyethylene) mono-oleate; Poly(oxyethylene) oleic acid ester; POOA; Prodhyphore B; Rokacet; Rokacet O 7; S 1006; S 1132; Slovasol A; Trydet OS series; Unisol 4-O; Witco 31; X-539-R. Emulsifier, lubricant, dispersing and leveling

agent used in cosmetic, textile, leather, paint and other industrial uses. *Taiwan Surfactants*.

1865   Glyconyl
122-87-2   4863   G

C8H9NO3
N-(4-Hydroxyphenyl)glycine.
AI3-15473; 4-(Carboxymethylamino)phenol; EINECS 204-580-8; Glitsin; Glycin; Glycine, N-(4-hydroxyphenyl)-; Glycine, N-(p-hydroxyphenyl)-; p-Hydroxyanilinoacetic acid; p-Hydroxyphenylaminoacetic acid; Hydroxyphenyl glycine; p-Hydroxyphenyl glycine; N-(4-Hydroxy-phenyl)glycine; N-(p-Hydroxyphenyl)glycine; Iconyl; Monazol; NSC 9267; Photoglycine. A photographic developer, with p-hydroxyphenylglycine as the active constituent. Leaflets or plates; mp = 245-247°; λ_m = 240, 306 nm (ε = 9460, 2340 MeOH), 245, 311nm (ε = 9310, 2980, MeOH-KOH), 223, 276 nm (ε = 7760, 1740, MeOH-HCl); insoluble in Et2O, slightly soluble in H2O, EtOH, Me2CO, C6H6; soluble in EtOAc, CHCl3.

1866   Glycoserve
116-25-6   4856   G

C6H10N2O3
1-(Hydroxymethyl)-5,5-dimethylhydantoin.
Dantoin® MDMH; EINECS 204-132-1; EPA Pesticide Chemical Code 121301; GlycoServe; Hydantoin, 1-(hydroxymethyl)-5,5-dimethyl-; 1-(Hydroxymethyl)-5,5-dimethyl-2,4-imidazolidinedione; 1-(Hydroxymethyl)-5,5-dimethylhydantoin; 2,4-Imidazolidinedione, 1-(hydroxy-methyl)-5,5-dimethyl-; MDM Hydantoin; MDMH; Monomethylol dimethyl hydantoin; 1-Monomethylol-5,5-dimethylhydantoin; NSC 9185. Intermediate for cosmetics and other applications; preservation and gelation agent. Registered by EPA as an antimicrobial and disinfectant. Crystals; mp = 100°; insoluble in Et2O, CCl4, freely soluble in H2O, MeOH, EtOH, Me2CO. *Lonzagroup*.

1867   N-Glycylglycine
556-50-3   4516   G

C4H8N2O3
2-(Aminoacetamido)acetic acid.
AI3-62521; Diglycine; Diglycocoll; EINECS 209-127-8; Gly2; Glycine dipeptide; Glycine, glycyl-; Glycine, N-glycyl-; Glycylglycine; α-Glycylglycine; N-Glycylglycine; NSC 49346. Used in peptide synthesis. Solid; mp = 215° (dec); soluble in H2O, EtOH, less soluble in organic solvents. *Am. Biorganics; Lancaster Synthesis Co.; Penta Mfg.; Spectrum Chem. Manufacturing*.

1868   Glyoxal
107-22-2   4522   G H

C2H2O2
1,2-Ethanedione.
4-01-00-03625 (Beilstein Handbook Reference); Aerotex glyoxal 40; AI3-24108; Biformal; Biformyl; BRN 1732463; CCRIS 952; Diformal; Diformyl; EINECS 203-474-9; Ethanedial; 1,2-Ethanedione; Glyoxal; Glyoxylaldehyde; HSDB 497; Oxal; Oxalaldehyde. Used in textiles, glues, biocides, paper and oil field industries and in organic synthesis. Solid; mp = 15°; bp = 50.4°; λ_m = 268 nm (H2O); very soluble in H2O, soluble in EtOH, Et2O; LD50 (rat orl) = 2020 mg/kg. *Am. Cyanamid*.

1869   Glyoxylic acid
298-12-4   4524   H

C2H2O3
α-Ketoacetic acid.
4-03-00-01489 (Beilstein Handbook Reference); Acetic acid, oxo-; BRN 0741891; CCRIS 1455; EINECS 206-058-5; Formylformic acid; Glyoxalic acid; Glyoxylic acid; HSDB 5559; α-Ketoacetic acid; Kyselina glyoxylova; NSC 27785; Oxalaldehydic acid; Oxoacetic acid; Oxoethanoic acid. Prisms; mp = 98°; λ_m = 565, 576, 580, 584 nm (ε = 1, 8, 10, 8, H2O pH 7.1); slightly soluble in EtOH, Et2O, C6H6, very soluble in H2O.

1870   Glyphosate
1071-83-6   4525   G H P

C3H8NO5P
N-(Phosphonomethyl)glycine.
BRN 2045054; Caswell No. 661A; CCRIS 1587; CP 67573; EPA Pesticide Chemical Code 417300; Glyphosate; Glyphosphate; HSDB 3432; Isopropylamine glyphosate; MON 0573; MON 2139; NSC 151063; Phosphonomethyliminoacetic acid; N-(Phosphono-methyl)glycine; Pondmaster. Broad spectrum trans-locatable herbicide. Registered by EPA as a herbicide and plant growth regulator. White solid; mp = 200° (dec); soluble in H2O (1.2 g/100 ml), insoluble in common organic solvents; LD50 (rat mus orl) = 5600 mg/kg, (rbt der) > 5000 mg/kg, (bobwhite quail orl) = 3850 mg/kg, (bee contact, orl) > 0.1 mg/bee; LC50 (rat ihl 4 hr.) = 12.2 mg/l , (quail, duck 8 day dietary) > 4640 mg/kg diet, (trout 96 hr.) = 86 mg/l, (bluegill sunfish 96 hr.) = 120 mg/l. *Albaugh Inc.; BASF Corp.; Chemisco; Dow AgroSciences; Entek Corp.; Micro-Flo Co. LLC; Riverdale Chemical Co; Syngenta Crop Protection*.

1871   Gold
7440-57-5   4529   G

Au

Au
Gold.

Burnish gold; C.I. 77480; C.I. Pigment Metal 3; Colloidal gold; EINECS 231-165-9; Gold; Gold-197; Gold, colloidal; Gold flake; Gold leaf; Gold powder; HSDB 2125; Magnesium gold purple; Pigment metal 3; Shell Gold. Metallic element; Au; infrared reflectors; electrical contact alloys; brazing alloys; laboratoryware; decorative arts; electronics; dental alloys; jewelry; colloidal dispersions for coloring glass, as nucleating agent, for specialized medical treatments. Metal; mp = 1064.76°; bp = 2700°; d = 19.3. *Cerac; Degussa AG; Noah Chem.; Robert Koch Industries.*

1872   Gold trichloride
13453-07-1   4536   G

AuCl3
Gold(III) chloride.
Auric chloride; Auric trichloride; EINECS 236-623-1; Gold chloride; Gold chloride (AuCl3); Gold trichloride; Gold(III) chloride. Laboratory reagent. [dihydrate]; Orange-red crystals; d = 3.9; bp = 229°; sublimes at 180° (760mm); soluble in water, alcohol, ether; LD50 (mus sc) = 1.5 g/kg. *Degussa AG; Spectrum Chem. Manufacturing.*

1873   Graphite
7782-42-5   4554   G

C
Graphite.
Aerodag G; AG 1500; Aquadag; AS 1; AT 20; ATJ-S; ATJ-S graphite; Black lead; C.I. 77265; C.I. Pigment Black 10; Canlub; Cb 50; CB 50; Ceylon Black Lead; CPB 5000; DC 2; EG 0; EINECS 231-955-3; Electrographite; EXP-F; Fortafil 5Y; GK 2; GK 3; GP 60; GP 60S; GP 63; Grafoil; Grafoil GTA; Graphite; Graphite, natural; Graphite, synthetic; Graphitic acid; Graphnol N 3M; GS 2; GY 70; H 451; Hitco HMG 50; IG 11; Korobon; MG 1; Mineral carbon; MPG 6; Papyex; PG 50; Plumbago; Plumbago (graphite); Pyro-Carb 406; Rocol X 7119; Rollit; S 1; S 1 (Graphite); Schungite; Shungite; Silver graphite; SKLN 1; Stove Black; Swedish Black Lead; Ucar 38; VVP 66-95. Graphite powder; used as mandrel bar lubricant. Soft black crystals; d = 2.09-2.23. *Lonzagroup.*

1874   Grotan
4719-04-4   G P

C9H21N3O3
2,2',2''-(Hexahydro-1,3,5-triazine-1,3,5-triyl)triethanol.
Actane; BRN 0124982; Busan 1060; Caswell No. 481C; CCRIS 6246; EINECS 225-208-0; EPA Pesticide Chemical Code 083301; ETA 75; Grotan; Grotan B; Grotan BK; Grotan HD; Kalpur TE; KM 200 (alcohol); Miliden X-2; Nipacide® BK; NSC 516387; Onyxide 200; Ottaform 204; Rancidity control agent; Roksol T 1-7; Tris(2-hydroxyethyl)hexahydro-s-triazine.
Preservative, bacteriostat, fungistat for soluble cutting fluids, coolants and other products. Registered by EPA as an antimicrobial and disinfectant. *Arch Chemicals Inc; Avecia Inc.; BASF Microcheck Ltd.; Buckman Laboratories Inc.; Clariant Corp.; Dow Chem. U.S.A.; Henkel Surface Tech; Nalco Diversified Technologies, Inc.; Nipa;*

Stepan; Surety Laboratories Inc.; Troy Chemical Corp.

1875   Guanidine nitrate
506-93-4   4578   G H

CH6N4O3
Guanidine mononitrate.
AI3-15039; EINECS 208-060-1; Guanidine mononitrate; Guanidine nitrate; Guanidinium nitrate; HSDB 5671; NSC 7295; UN1467. Used in manufacture of explosives, disinfectants, photographic chemicals. Leaflets; mp = 217°; bp dec; very soluble in H2O (13 g/100 ml), EtOH; LD50 (rat orl) = 730 mg/kg. *Dajac Labs.; Greeff R.W. & Co.; Spectrum Chem. Manufacturing.*

1876   Guanidine thiocyanate
593-84-0   G

C2H6N4S
Guanidine hydrothiocyanate.
AI3-18430; EINECS 209-812-1; Guanidine, monothiocyanate; Guanidine thiocyanate; Guanidinium thiocyanate; Isothiocyanic acid, compd. with guanidine (1:1); NSC 2119; Thiocyanic acid, compd. with guanidine (1:1); USAF EK-705. Potent protein denaturant used in isolation of intact DNA, RNA. Crystals; mp= 117°. *Dajac Labs.; Eastman Chem. Co.; Fluka; U.S. BioChem.*

1877   Guanine
73-40-5   4580   G

C5H5N5O
2-Amino-1,7-dihydro-6H-purin-6-one.
AI3-24393;   2-Amino-1,7-dihydro-6H-purin-6-one;   2-Amino-6-hydroxypurine; 2-Amino-6-purinol; 2-Amino-hypoxanthine; CI 75170; CI Natural white 1; Dew Pearl; EINECS 200-799-8; Guanin; Guanine; Guanine enol; HSDB 2127; 6-Hydroxy-2-aminopurine; Hypoxanthine, 2-amino-; Mearlmaid; Natural Pearl Essence; Natural White 1; Naturon; Pathocidin; Pearl essence; 6H-Purin-6-one, 2-amino-1,7-dihydro-; Stella Polaris. Naturally occuring purine; used in biochemical research, as a pigment and in cosmetics. dec above 360°; λm = 246, 275 nm (ε = 10700, 8100); soluble in ammonia water, aqueous KOH solns, dilute acids; sparingly soluble in alcohol, ether; almost insoluble in water. *Greeff R.W. & Co.; Henley; Janssen Chimica; Mearl; Penta Mfg.*

1878   Guanylurea Sulfate
591-01-5   3120   G

C4H14N8O6S
(Aminoiminomethyl)urea sulfate.
Bis(amidinourea) sulphate; Dicyanodiamidine sulfate; EINECS 209-697-8; Grossman's reagent; Guanylurea sulfate; N-Guanylurea sulfate; Urea, (aminoiminomethyl)-, sulfate (2:1); Urea, (aminoiminomethyl)-, sulfate (2:1). An ammoniacal solution of dicyandiamidine sulfate, a reagent for nickel, cobalt. Crystals; mp = 190°; d = 1.61; soluble in H_2O (5-30%), EtOH, insoluble in organic solvents.

1879   Guazatine
13516-27-3   P

C18H41N7
N,N-(Iminodi-8,1-octanediyl)bisguanidine.
Bis(8-guanidinooctyl)amine; Caswell No. 471D; EINECS 236-855-3; EM-379; EPA Pesticide Chemical Code 498200; Guanidine, 1,1'-(iminobis(octamethylene))di-; Guanidine, N,N'-(iminodi-8,1-octanediyl)bis-; Guazat-ine; Iminobis(octamethylene)diguanidine; Iminoctadine; Kenopel; MC 25; Panoctine; Panolil; Radam; Rappor. Fungicide and bird repellent. Brown solid; mp $\cong$ 60°; sg^{20} = 1.09; soluble in H_2O (> 300 g/100 ml), MeOH (> 300 g/100 ml), EtOH ($\cong$20 g/100 ml), N-methylpyrrolidone (100 g/100 ml), DMSO (50 g/100 ml), DMF (50 g/100 ml), sparingly soluble in hydrocarbon solvents; LD50 (rat orl) = 227 - 300 mg/kg, (rat der) > 1000 mg/kg, (rbt der) = 1176 mg/kg; non-toxic to bees. Aventis Crop Science; Dow AgroSciences; DowElanco Ltd.; KenoGard; Rhône-Poulenc.

1880   Guvacine
498-96-4   4598   G

C6H9NO2
1,2,5,6-tetrahydropyridine-3-carboxylic acid.
5-22-01-00322 (Beilstein Handbook Reference); BRN 111256; Guvacine; 3-Pyridinecarboxylic acid, 1,2,5,6-tetrahydro-; 1,2,5,6-Tetrahydro-3-pyridinecarboxylic acid; 1,2,5,6-Tetrahydronicotinic acid. A pyridine alkaloid derived from betel nuts. Its methyl ester is also called guvacine. Prisms or rods; mp = 295° (dec); very soluble in H_2O, insoluble in organic solvents.

1881   Gypsum
13397-24-5   1711   G

CaO4S.2H2O
Calcium sulfate.
Alabaster; Annaline; GIPS; Gypsite; Gypsum; Gypsum (Ca(SO4).2H2O); Gypsum stone; Landplaster; Lenzit; Light spar; Phosphogypsum; Satin spar; Satinite; Selenite; Terra alba. A mineral used in plasters, in Portland cement, paints, and as a filler for paper and cotton. White powder; d = 2.32; soluble in H_2O, insoluble in organic solvents.

1882   Halethazole
15599-36-7   D G

C19H21ClN2OS
5-Chloro-2-[p-(2-diethylaminoethoxy)phenyl]benzo-thiazole.
5-Chlor-2-[4-(2-diethylaminoethoxy)phenyl)benzothiazol; 5-Chloro-2-(p-(2-diethylaminoethoxy)phenyl)benzothi-azole; Episol; Haletazol; Haletazolum; Halethazole. Antiseptic and antifungal. mp = 93-94°; [citrate]: mp = 167°. Crookes Healthcare Ltd.

1883   Halofuginone hydrobromide
64924-67-0   4627   G V

C16H18Br2ClN3O3
(±)-trans-7-Bromo-6-chloro-3-[3-(3-hydroxy-2-piperidyl)-acetonyl]-4(3H)-quinazolinone monohydrobromide.
(±)-trans-7-Bromo-6-chloro-3-(3-(3-hydroxy-2-piperidyl)-acetonyl)-4(3H)-quinazolinone monohydrobromide; Halofuginone hydrobromide; 4(3H)-Quinazolinone, 7-bromo-6-chloro-3-(3-(3-hydroxy-2-piperidinyl)-2-oxo-propyl)-, hydrobromide, trans-(±)-; 4(3H)-Quinazolinone, 7-bromo-6-chloro-3-(3-(2R,3S)-3-hydroxy-2-piperidinyl)-2-oxopropyl)-, monohydrobromide, rel-; Ru 19110; Stenorol; Tempostatin. Antiprotozoal - coccidiostat Solid; mp = 247° (dec).

1884   Haloxon
321-55-1   4621   G V

C14H14Cl3O6P

Phosphoric acid bis(2-chloroethyl) 3-chloro-4-methyl-2-oxo-2H-1-benzopyran-7-yl ester.

96H60; AI3-50680; O,O-Bis(2-chloroethyl) O-(3-chloro-4-methyl-7-coumarinyl) phosphate; BRN 1271357; 3-Chloro-4-methyl-umbelliferone bis(2-chloroethyl)phos-phate; 3-Chloro-7-hydroxy-4-methylcoumarin bis(2-chlo-roethyl)phosphate; 3-Chloro-7-hydroxy-4-methyl-2H-1-benzopyran-2-one bis(chloroethyl)phosphate; Coumarin, 3-chloro-7-hydroxy-4-methyl-, bis(2-chloroethyl) phos-phate; Di-(2-chloroethyl) 3-chloro-4-methylcoumarin-7-yl phosphate; Di-(2-chloroethyl)-3-chloro-4-methyl-7-coum-arinyl phosphate; O,O-Di(2-chloroethyl)-O-(3-chloro-4-methylcoumarin-7-yl)-phosphate; EINECS 206-289-1; Ethanol, 2-chloro-, phosphate diester, ester with 3-chloro-7-hydroxy-4-methylcoumarin; Ethanol, 2-chloro-, hydro-gen phosphate, ester with 3-chloro-7-hydroxy-4-methylcoumarin; Eustidil; Galloxon; Galoxone; Haloxon; Haloxona; Haloxone; Haloxonum; Helmirane; Helmiron; Helmirone; Loxon; Luxon; Phosphoric acid, bis(2-chloroethyl) 3-chloro-4-methyl-2-oxo-2H-1-benzopyran-7-yl ester; Loxon; 3-chloro-7-hydroxy-4-methylcoumarin bis(2-chloroethyl) phosphate; 3-chloro-7-hydroxy-4-methyl-2H-1-benzopyran-2-one bis(chloroethyl) phos-phate; 3-chloro-4-methylumbelliferone di(2-chloroethyl) phosphate; galoxone; helmirone; 96-H-60; Galloxon; Luxon. A proprietary preparation of haloxon; a veterinary anthelmintic. Used in combination with oxyclozanide. Crystals; mp = 91°; LD50 (rat orl) = 900 mg/kg. *Schering-Plough Veterinary Operations, Inc.*

1885   Heavy aromatic distillate (petroleum)
67891-79-6   H

Distillates, (petroleum), heavy arom.

Distillates, (petroleum), heavy arom.; Distillates, petroleum, heavy arom.; EINECS 267-563-4; Heavy aromatic distillate (petroleum).

1886   Hematin
15489-90-4   4652   G

C34H33FeN4O5

(SP-5-13)-[7,12-diethenyl-3,8,13,17-tetramethyl-21H,23H-porphine-2,8-dipropanoato(4-)-N^{21},N^{22},N^{23},N^{24}]hydroxyferrate(2-) dihydrogen.

Bovine hemin; EINECS 239-518-9; Ferrihemate; Ferriheme; Ferriheme hydroxide; Ferrihemic acid; Ferriporphyrin hydroxide; Ferriprotoporphyrin basic; Ferriprotoporphyrin IX hydroxide; Hematin; Hematin; Hydroxy(dihydrogen protoporphyrin IX-ato^{2-})iron; Hydroxyhemin; Phenodin; Protohematin. Used in biochemical research. Crystals; λ$_m$ = 580 nm; LD50 (rat iv) = 43.2 mg/kg.

1887   Heptachlor
76-44-8   4675   P

C10H5Cl7

1(3a),4,5,6,7,8,8-Heptachloro-3a(1),4,7,7a-tetrahydro-4,7-methanoindene.

Aahepta; Agroceres; AI3-15152; Arbinex 30TN; Basaklor; Caswell No. 474; CCRIS 324; 3-Chlorochlordene; Drinox H-34; E 3314; ENT 15,152; EPA Pesticide Chemical Code 044801; Eptacloro; 1,4,5,6,7,8,8-Eptacloro-3a,4,7,7a-tetraidro-4,7-endo-metano-indene; Gold Crest H-60; GPKh; H-60; H-34; Hepta; Hepta; Heptachloor; Heptachlor; Heptachlorane; Heptachlore; Heptachloro-tetrahydro-4,7-methanoindene; 3,4,5,6,7,8,8a-hepta-chloro-dicyclopentadiene; 3,4,5,6,7,8,8A-heptachloro-α-dicyclopentadiene; 3,4,5,6,7,8,8a-Heptachlorodicyclo-pentadiene; 1,4,5,6,7,10,10-heptachloro-4,7,8,9-tetra-hydro-4,7-endomethyleneindene; 1,4,5,6,7,8,8-hepta-chloro-3a,4,7,7,7a-tetrahydro-4,7-methyleneindene; 1,4,5,6,7,8,8-Heptachlor-3a,4,7,7,7a-tetrahydro-4,7-endo-methano-inden; 1,4,5,6,7,8,8-Heptachloro-3a,4,7, 7a-tetrahydro-4,7-methanol-1H-indene; 1,4,5,6,7,8,8-Heptachloor-3a,4,7,7a-tetrahydro-4,7-endo-methano-indeen; Heptagran; Heptagranox; Heptaklor; Heptamak; Heptamul; Heptasol; Heptox; HSDB 554; Latka 104; 4,7-Methanoindene, 1,4,5,6,7,8,8-hepta-chloro-3a,4,7,7a-tetrahydro-; 4,7-Methano-1H-indene, 1,4,5,6,7,8,8-heptachloro-3a,4,7,7a-tetrahydro-; MS 193; NCI-C00180; NSC 89300; RCRA waste number P059; Rhodiachlor; Soleptax; Technical heptachlor; Termide; Velsicol heptachlor; Velsicol 104. Registered by EPA as an insecticide (cancelled). Designated as a Persistent Organic Pollutant (POP) under the Stockholm convention. Colorless crystals - waxy solid; mp = 95 - 96°; bp = 136 - 145°; sg^{25} = 1.65 - 1.67; insoluble in H2O (0.000056 g/100 ml), soluble in Me2CO (75 g/100 ml), C6H6 (106 g/100 ml), xylene (102 g/100 ml), cyclohexanone (119 g/100 ml), CCl4 (113 g/100 ml), EtOH (4.5 g/100 ml); LD50 (rat orl) = 147 - 220 mg/kg, (gpg orl) = 116 mg/kg, (mus orl) = 68 mg/kg, (rbt der) > 2000 mg/kg, (rat der) = 119 - 250 mg/kg, (mallard duckling orl) > 2000 mg/kg; LC50 (bobwhite quail 8 day dietary) = 450 - 700 mg/kg diet, (Japanese quail 8 day dietary) = 80 - 95 mg/kg diet, (pheasant 8 day dietary) = 250 - 275 mg/kg diet, (rainbow trout 96 hr.) = 0.007 mg/l, (bluegill sunfish 96 hr.) = 0.026 mg/l, (fathead minnow 96 hr.) = 0.078 - 0.130 mg/l. *Novartis.*

1888   Heptacosafluorotributylamine
311-89-7   H

C12F27N

1,1,2,2,3,3,4,4,4-Nonafluoro-N,N-bis(nonafluorobutyl)-1-butanamine.

4-02-00-00819 (Beilstein Handbook Reference); AI3-16951; BRN 1813883; EINECS 206-223-1; FC 43; FC 47; Fluorinert FC 43; Fluorocarbon FC 43; Fluosol 43; Heptacosafluorotributylamine; HSDB 7103; Mediflor FC 43; NSC 3501; Perfluorotributylamine; Tri(nonafluoro-butyl)amine; Tri(perfluorobutyl)amine; Tributylamine, heptacosafluoro-. Oil; bp = 178°; d^{25} = 1.884; soluble in Me2CO.

1889   Heptadecafluoro-N-(2-hydroxyethyl)-N-methyloctanesulphonamide

24448-09-7 H

C11H8F17NO3S
1,1,2,2,3,3,4,4,5,5,6,6,7,7,8,8,8-Heptadecafluoro-N-(2-hydroxyethyl)-N-methyl-1-octanesulfonamide.
EINECS 246-262-1; Heptadecafluoro-N-(2-hydroxyethyl)-N-methyloctanesulphonamide; 1,1,2,2,3,3,4,4,5,5,6,6,7, 7,8,8,8-Heptadecafluoro-N-(2-hydroxyethyl)-N-methyl-1-octanesulfonamide; 1-Octanesulfonamide, 1,1,2,2,3,3,4, 4,5,5,6,6,7,7,8,8,8-heptadecafluoro-N-(2-hydroxyethyl)-N-methyl-.

1890   Heptanal
111-71-7   4677   H

C7H14O
n-Heptyldehyde.
4-01-00-03314 (Beilstein Handbook Reference); AI3-02066; Aldehyde C7; BRN 1560236; CCRIS 6041; EINECS 203-898-4; Enanthal; Enanthaldehyde; Enanthic aldehyde; Enanthole; FEMA Number 2541; Heptaldehyde; Heptanal; Heptylaldehyde; HSDB 6026; n-Heptaldehyde; n-Heptanal; n-Heptylaldehyde; NSC 2190; Oenanthal; Oenanthaldehyde; Oenanthic aldehyde; Oenanthol; UN3056. Liquid; mp = -43.3°; bp = 152.8°; d^{25} = 0.8132; slightly soluble in H_2O, CCl_4; freely soluble in EtOH, Et_2O.

1891   Heptane
142-82-5   4679   G H

C7H16
n-Heptane.
AI3-28784; Dipropylmethane; EINECS 205-563-8; Eptani; Exxsol® Heptane; Gettysolve-C; Heptan; Heptane; n-Heptane; Heptanen; Heptyl hydride; HSDB 90; NSC 62784; Skellysolve C; UN1206. Hydrocarbon derived chiefly from petroleum. Used as a solvent. Liquid; mp = -90.6°; bp = 98.5°; d^{20} = 0.6837; insoluble in H_2O, soluble in CCl_4; very soluble in EtOH; freely soluble in Et_2O, Me_2CO, C_6H_6, $CHCl_3$, petroleum ether. Ashland; Exxon; Humphrey; Phillips 66; Texaco.

1892   Heptanoic acid
111-14-8   4680   G H

C7H14O2
n-Heptanoic acid.
4-02-00-00958 (Beilstein Handbook Reference); AI3-02073; BRN 1744723; CCRIS 6042; EINECS 203-838-7; Enanthic acid; Enanthylic acid; FEMA No. 3348; Heptanoic acid; n-Heptanoic acid; Heptoic acid; Heptylic acid; Hexacid C-7; 1-Hexanecarboxylic acid; HSDB 5546; NSC 2192; Oenanthic acid; Oenanthylic acid. Liquid; mp = -7.5°; bp = 222.2°; d^{20} = 0.9181; slightly soluble in H_2O, CCl_4, soluble in EtOH, Et_2O,Me_2CO; LD_{50} (mus iv) = 1200 mg/kg.

1893   Heptanol
111-70-6   4681   G

C7H16O
n-Heptyl alcohol.
4-01-00-01731 (Beilstein Handbook Reference); AI3-15363; Alcohol C7; l'Alcool N-heptylique primaire; BRN 1731686; C7 alcohol; EINECS 203-897-9; Enanthic alcohol; Enanthyl alcohol; Fatty alcohol (C7); FEMA Number 2548; Gentanol; Heptanol; 1-Heptanol; Heptanol; Heptyl alcohol; Heptyl alcohol, primary; Hexyl carbinol; HSDB 1077; Hydroxy heptane; 1-Hydroxyheptane; n-Heptan-1-ol; n-Heptanol; n-Heptyl alcohol; NSC 3703; Pri-n-heptyl alcohol. Organic intermediate, solvent, cosmetic formulations. Liquid; mp = -34°; bp = 176.4°; d^{20} = 0.8219; slightly soluble in H_2O (1 g/l), CCl_4, freely soluble in EtOH, Et_2O. Elf Atochem N. Am.; Penta Mfg.; Suchema.

1894   2-Heptanone
110-43-0   4683   G H

C7H14O
Methyl n-amyl ketone.
4-01-00-03318 (Beilstein Handbook Reference); AI3-01230; Amyl-methyl-cetone; Amyl methyl ketone; BRN 1699063; Butylacetone; EINECS 203-767-1; FEMA Number 2544; 2-Heptanone; HSDB 1122; Ketone C-7; Ketone, methyl pentyl; MAK; Methyl-amyl-cetone; Methyl amyl ketone; Methyl-n-amyl ketone; Methyl n-pentyl ketone; Methyl pentyl ketone; NSC 7313; Pentyl methyl ketone; UN1110. Industrial solvent for nitrocellulose lacquers, synthetic flavoring and perfumery. Liquid; mp = -35°; bp = 151.0°; d^{20} = 0.8111; λ_m = 274 nm (ε = 22, MeOH); soluble in EtOH, Et_2O, very soluble in H_2O; LD_{50} (rat orl) = 1670 mg/kg. Ashland; Eastman Chem. Co.; Union Carbide Corp.

1895   Heptene
25339-56-4   H

C7H14
Heptene, mixed isomers.
EINECS 246-871-2; Heptene; Heptene (mixed cis and trans); Heptylene.

1896   Heptenophos
23560-59-0   4684   G P

C9H12ClO4P
7-Chlorobicyclo[3.2.0]hepta-2,6-dien-6-yl dimethyl phos-phate.
AI3-27500; BRN 1978448; Caswell No. 721C; 7-Chlorobicyclo(3.2.0)hepta-2,6-dien-6-yl dimethyl phos-phate; O,O-Dimethyl-O-(6-chlorobicyclo(3.2.0)hepta-dien-1,5-yl)phosphate; 5-(O,O-Dimethylphosphoryl)-6-chlorobicyclo(3.2.0)hepta-1,5-dien; EINECS 245-737-0; EPA Pesticide Chemical Code 215600; Heptenophos; HOE 02982; HOE 2982; HOE 2982 OJ; Hostaquick; Hostavik; Phosphonic acid, (7-chlorobicyclo(3.2.0)hepta-3,6-dien-6-

yloxy)-, dimethyl ester; Phosphoric acid, (7-chloro-bicyclo(3.2.0)hepta-2,6-dien-6-yl) dimethyl ester; Ragadan; XOE 2982; Hostaquick; Hostavik; OMS 1845; O,O'-Dimethyl-O-(6-chlorobicyclo(3.2.0)heptadiene-1,5-yl)phosphate; Phosphoric acid, (7-chloro-bicyclo-(3.2.0)hepta-2,6-dien-6-yl) dimethyl ester; Ragadan; XOE 2982. Systemic insecticide with contact, stomach and respiratory action. Used for control of sucking insects, particularly aphids, in a variety of crops. Pale-brown liquid; bp$_{0.1}$ = 64°; sg^{20} = 1.294; soluble in H_2O (0.22 g/100 ml at 23°), C_6H_{14} (13 g/100 ml), Me_2CO, MeOH, xylene (all > 100 g/100 ml); LD$_{50}$ (mrat orl) = 121 mg/kg, (frat orl) = 96 mg/kg, (dog orl) > 500 mg/kg, (frat der) = 2925 mg/kg, (Japanese quail orl) = 17 - 50 mg/kg; LC$_{50}$ (rat ihl 4 hr.) = 0.4 mg/l air; (guppy 96 hr.) = 13.1 g/l, (carp 96 hr.) = 46 mg/l; toxic to bees. *AgrEvo; Hoechst AG; Hoechst UK Ltd.*

1897 Heptylparaben
1085-12-7   G

C$_{14}$H$_{20}$O$_3$
n-Heptyl p-hydroxybenzoate.
4-10-00-00378 (Beilstein Handbook Reference); Benzoic acid, 4-hydroxy-, heptyl ester; Benzoic acid, P-hydroxy-, heptyl ester; BRN 2726540; EINECS 214-115-0; Heptyl 4-hydroxybenzoate; Heptyl p-hydroxybenzoate; n-Heptyl p-hydroxybenzoate; Heptyl paraben; p-Hydroxybenzoic acid heptyl ester; Nipaheptyl; NSC 309818; p-Oxybenzoesaureheptylester; Staypro WS 7. Preservative, bactericide, fungicide for pharmaceuticals, cosmetics, foods, medicinal preparations and industrial applications. *Nipa.*

1898 Herborane
54546-26-8   G

C$_{11}$H$_{22}$O$_2$
2-Butyl-4,4,6-trimethyl-1,3-dioxane.
2-Butyl-4,4,6-trimethyl-1,3-dioxane; 1,3-Dioxane, 2-butyl-4,4,6-trimethyl-; EINECS 259-210-8. A flavorant. *Quest UK.*

1899 Hetacillin
3511-16-8   4691   D G

C$_{19}$H$_{23}$N$_3$O$_4$S
Versapen.
BL-P 804; BRL 804; 6-(2,2-Dimethyl-5-oxo-4-phenyl-1-imidazolidinyl)penicillanic acid; 6-(2,2-Dimethyl-5-oxo-4-phenyl-1-imidazolidinyl)-3,3-dimethyl-7-oxo-4-thia-1-azabicycloheptane-2-

carboxylic acid; EINECS 222-512-5; Etacillina; Hetacilina; Hetacillin; Hetacilline; Hetacill-inum; N,N'-Isopropylidene-A-amino-benzyl penicillin; Penplenum; Phenazacillin; 4-Thia-1-azabicyclo(3.2.0)-heptane-2-carboxylic acid, 6-(2,2-dimethyl-5-oxo-4-phenyl-1-imidazolidinyl)-3,3-dimethyl-7-oxo-, (2S-(2α, 5α,6β(S*)))-; Versapen; Versatrex. Antibacterial. Solid; dec 182.8-183.9°, 189.2-191.0°; [α]$_D^{25}$= +366° (C$_5$H$_5$N). *Bristol Laboratories; Bristol-Myers Squibb Pharm. R&D.*

1900 Hetacillin potassium
5321-32-4   4691   D G V

C$_{19}$H$_{22}$KN$_3$O$_4$S
[2S-[2α,5α,6β(S*)]]-6-(2,2-Dimethyl-5-oxo-4-phenyl-1-imidazolidinyl)-3,3-dimethyl-7-oxo-4-thia-1-azabicyclo-[3.2.0]heptane-2-carboxylic acid potassium salt.
EINECS 226-182-3; H-K Mastitis (Veterinary); Hetacillin potassium; Hetacin-K (Veterinary); Natacillin; Potassium 6-(2,2-dimethyl-5-oxo-4-phenyl-1-imidazolidinyl)-3,3- di-methyl-7-oxo-4-thia-1-azabicyclo(3.2.0)heptane-2-carb-oxylate; Potassium (2S-(2α,5α,6β(S*)))-6-(2,2-di-methyl-5-oxo-4-phenylimidazolidin-1-yl)-3,3-dimethyl-7-oxo-4-thia-1-azabicyclo(3.2.0)heptane-2-carboxylate; Potassium hetacillin; 4-Thia-1-azabicyclo(3.2.0)heptane-2-carboxyl-ic acid, 6-(2,2-dimethyl-5-oxo-4-phenyl-1-imidazolidinyl) -3,3-dimethyl-7-oxo-, monopotassium salt, (2S-(2α,5α, 6β(S*)))-; 4-Thia-1-azabicyclo(3.2.0)heptane-2-carboxylic acid, 6-(2,2-dimethyl-; 5-oxo-4-phenyl-imidazolinyl)-3,3-dimethyl-7-oxo-, monopotassium salt. Antibacterial. Used to treat mastitis in cattle. *Bristol-Myers Squibb Pharm. R&D.*

1901 Hexabromocyclododecane
3194-55-6   G H

C$_{12}$H$_{18}$Br$_6$
1,2,5,6,9,10-Hexabromocyclododecane.
Cyclododecane, 1,2,5,6,9,10-hexabromo-; EINECS 221-695-9; Hexabromocyclododecane; HSDB 6110; Saytex® HBCD-LM. Low melting flame retardant Solid; mp = 173-177°. *Ethyl Corp.*

1902 Hexabromocyclododecane
25637-99-4   G

C12H18Br6

1,2,5,6,9,10-Hexabromocyclododecane.

CCRIS 4821; Cyclododecane, hexabromo-; EINECS 247-148-4; FR-1206; Great Lakes CD-75P™; Great Lakes SP-75™; Hexabromocyclododecane; Saytex® 60006L; Saytex® BCT-610; Saytex® HBCD-LM. Fire retardant for wide range of plastics, textiles, adhesives, and coatings; esp. for styrene-based systems. Solid; mp = 188-191; d = 2.36; soluble in common solvents; LD_{50} (rat orl) >10,000 mg/kg, (rbt dermal) >10,000 mg/kg. *Amerihaas; Dead Sea Bromine; Great Lakes Fine Chem.*

1903 Hexachlornaphthalene

1335-87-1   G

C10H2Cl6

Naphthalene, hexachloro-.

EINECS 215-641-3; Halowax 1014; Hexachlornaftalen; Hexachloronaphthalene; HSDB 2509. Used to insulate electrical equipment, in flameproofing and water proofing, and as a lubricant additive. Crystals; mp = 137°; bp = 343-387°. *Bakelite Corp.*

1904 Hexachlorobenzene

118-74-1   4696   G P

C6Cl6

Pentachlorophenyl chloride.

4-05-00-00670 (Beilstein Handbook Reference); AI 3.01719; Amatin; Anticarie; Benzene, hexachloro-; BRN 1912585; Bunt-cure; Bunt-no-more; Caswell No. 477; CCRIS 325; CEKU C.B.; CO-OP Hexa; EINECS 204-273-9; ENT-1719; EPA Pesticide Chemical Code 061001; Esaclorobenzene; Granox; Granox NM; HCB; Hexa C.B.; Hexa CB; Hexachlorbenzol; Hexachlorobenzene; HSDB 1724; Julin's carbon chloride; NO Bunt; NO Bunt 40; NO Bunt 80; NO Bunt liquid; NSC 9243; Perchlorobenzene; Phenyl perchloryl; RCRA waste number U127; Saatbeizfungizid; Sanocid; Sanocide; Smut-Go; Snieciotox; UN2729; Voronit C. Different from benzenehexachloride (lindane). Used in organic synthesis. Has been used in agriculture as a fungicide. Designated as a Persistent Organic Pollutant (POP) under the Stockholm convention. Colorless crystals; mp = 231.8°; bp = 325°;λ_m = 291, 301 nm (ϵ = 245, 209, EtOH); insoluble in H2O, slightly soluble in EtOH, soluble in Et2O, CHCl3, very soluble in C6H6; LD_{50} (rat orl) = 10 g/kg. *Cequisa; Compania Quimica; Generic; Hightex.*

1905 Hexachlorocyclopentadiene

77-47-4   H

C5Cl6

Hexachloro-1,3-cyclopentadiene.

4-05-00-00381 (Beilstein Handbook Reference); AI3-15558; BRN 0976722; C 56; C-56; Caswell No. 478; CCRIS 5919; Cyclopentadiene, hexachloro-; EINECS 201-029-3; EPA Pesticide Chemical Code 027502; Graphlox; Hexachlorcyklopentadien; Hexachloro-1,3-cyclopenta-diene; Hexachlorocyclopentadiene; HRS

1655; HSDB 4011; NCI-C55607; NSC 9235; Perchloro-1,3-cyclopentadiene; Perchlorocyclopentadiene; RCRA waste number U130; UN2646. Pesticide. No longer used. Yellow-green liquid; mp = -9°; $bp^{0.3}$ = 48°, bp = 239°; d^{25} = 1.7019; λ_m = 323 nm (ϵ 1549). *Sigma-Aldrich Fine Chem.; Spectrum Chem. Manufacturing.*

1906 Hexachlorocyclopentene

72030-26-3   H

C5H2Cl6

Cyclopentene, hexachloro-.

Cyclopentene, hexachloro-; Hexachlorocyclopentene.

1907 Hexachloroethane

67-72-1   4698   H

C2Cl6

Ethane, hexachloro-.

4-01-00-00148 (Beilstein Handbook Reference); AI3-00633; Avlothane; BRN 1740341; Caswell No. 479; CCRIS 330; Distokal; Distopan; Distopin; Egitol; EINECS 200-666-4; EPA Pesticide Chemical Code 045201; Ethane hexachloride; Ethane, hexachloro-; Ethylene hexa-chloride; Falkitol; Fasciolin; Fron 110; Hexachlor-aethan; Hexachlor-äthan; Hexachlorethane; Hexachloro-ethane; Hexachloroethylene; HSDB 2033; Mottenhexe; NCI-C04604; NSC 9224; Perchloroethane; Phenohep; RCRA waste number U131. Used as a solvent, in smoke generators, fire extinguishers, in refining of aluminum and magnesium as a polymer additive and vulcanizing agent. Solid; mp = 187° (triple point); d^{20} = 2.091; slightly soluble in H2O, very soluble in organic solvents; MLD (dog iv) = 325 mg/kg. *Lancaster Synthesis Co.; Penta Mfg.; Sigma-Aldrich Fine Chem.*

1908 Hexachlorophene

70-30-4   4699   G

C13H6Cl6O2

2,2'-Methylenebis(3,4,6-trichlorophenol).

4-06-00-06659 (Beilstein Handbook Reference); Acigena; AI3-02372; Almederm; At-17; AT 7; B & b Flea Kontroller for Dogs Only; B32; Bilevon; Bis-2,3,5-trichloro-6-hydroxyfenylmethan; Bis(2-hydroxy-3,5,6-trichloro-phenyl)methane; Bis(3,5,6-trichloro-2-hydroxyphenyl)-methane; Bivelon; Blockade Anti Bacterial Finish; Brevity Blue Liquid Sanitizing Scouring Cream; Brevity Blue Liquid Bacteriostatic Scouring Cream; BRN 2064407; Caswell No. 566; CCRIS 331; Compound G-11; Cotofilm; Dermadex; Dihydroxy-3,3'5,5',6,6'-hexachlorodiphenyl-methane; 2,2'-Dihydroxy-3,5,6,3',5',6'-hexachlorodi-phenylmethane; Distodin; EINECS 200-733-8; Eleven; En-Viron D Concentrated Phenolic Disinfectant;

Enditch Pet Shampoo; EPA Pesticide Chemical Code 044901; Esaclorofene; Exofene; Fesia-sin; Fomac; Fostril; G-11; G-Eleven; Gamophen; Gamophene; Germa-medica; German-Medica; HCP; Hexabalm; 2,2',3,3',5,5'-Hexa-chloro-6,6'-dihydroxydiphenylmethane; Hexa-chlorofen; Hexachlorophen; Hexachlorophene; Hexachlorophenum; Hexaclorofeno; Hexafen; Hexaphene-LV; Hexide; Hexophene; Hexosan; Hilo Cat Flea Powder; Hilo Flea Powder; HSDB 224; Methane, bis(2,3,5-trichloro-6-hydroxyphenyl)-; Methylene-bis(3,4,6-trichlorophenol); Nabac; Nabac 25 EC; NCI-C02653; Neosept V; NSC 49115; Pedigree Dog Shampoo Bar; Phenol, 2,2'-methylenebis(3,4,6-trichloro)-; Phisodan; pHisoHex; RCRA waste number U132; Ritosept; Septisol; Septofen; Soy-Dome; Staphene O; Steral; Steraskin; Surgi-Cen; Surgi-cin; Surofene; Tersaseptic; Thera-Groom Pet Shampoo for Dogs for Veterinary Use Only; Trichlorophene; Turgex; UN2875. Used in manufacture of germicidal soaps. mp = 166.5°; insoluble in H2O, soluble in organic solvents; LD50 (mrat orl) = 66 mg/kg, (frat orl) = 67 mg/kg. *Sanofi Synthelabo; SmithKline Beecham Pharm.*

1909   Hexachloro-p-xylene
68-36-0   1299   H

C8H4Cl6
Benzene, 1,4-bis(trichloromethyl)-.
4-05-00-00968 (Beilstein Handbook Reference); AI3-02587; Benzene, 1,4-bis(trichloromethyl)-; Bitriben; BRN 2051282; Chloksil; Chloxil; Chloxyl; Cloxil; EINECS 200-686-3; Hetol; Hexachloro-4-xylene; Hexachloro-paraxylol; Hexichol; HSDB 5202; Khloksil; Khloxil; NSC 41883; p-Xylene, α,α'-hexachloro-; p-Xylene, α,α,α,α',α',α'-hexachloro-. Used as an insecticide and an anthelmintic effective against trematodes. White solid; mp = 109°; soluble in CHCl3; λm = 236, 273, 281 nm (MeOH).

1910   Hexacosyl alcohol
506-52-5   H

C26H54O
n-Hexacosanol.
EINECS 208-044-4; 1-Hexacosanol; n-Hexacosanol; Hexacosyl alcohol; NSC 4058. Rhombic plates, mp = 80°, bp20 = 305° (dec); insoluble in H2O, soluble in EtOH, Et2O.

1911   Hexadecane
544-76-3   H

C16H34
n-Hexadecane.
4-01-00-00537 (Beilstein Handbook Reference); AI3-06522; BRN 1736592; CCRIS 5833; Cetane; n-Cetane; EINECS 208-878-9; Hexadecane; n-Hexadecane; HSDB 6854; NSC 7334. Liquid; mp = 18.1°; bp = 286.8°; d^{20} = 0.7733; insoluble in H2O, slightly soluble in EtOH, soluble in CCl4, freely soluble in Et2O.

1912   Hexadecanol
36653-82-4   2037   G H

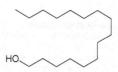

C16H34O
1-Hexadecanol.
4-01-00-01876 (Beilstein Handbook Reference); Adol; Adol 52; Adol 52 NF; Adol 520; Adol 54; Adol® 52 NF; AI3-00755; Alcohol C-16; Aldol 54; Alfol 16; Atalco C; BRN 1748475; C16 alcohol; Cachalot C-50; Cachalot C-51; Cachalot C-52; Cachalot® C-50 NF; Caswell No. 165D; Cetaffine; Cetal; Cetalol CA; Cetanol; 1-Cetanol; Cetostearyl alcohol; Cetyl alcohol; n-Cetyl alcohol; Cetylic alcohol; Cetylol; CO-1670; CO-1695; Crodacol C; Crodacol-cas; Crodacol-cat; Cyclal cetyl alcohol; Dytol F-11; EINECS 253-149-0; Elfacos C; EPA Pesticide Chemical Code 001508; EPAL 16NF; Ethal; Ethol; Fatty alcohol(C16); FEMA Number 2554; Hexadecanol; n-Hexadecanol; n-1-Hexadecanol; Hexadecan-1-ol; 1-Hexadecanol; Hexadecyl alcohol; n-Hexadecyl alcohol; 1-Hexadecyl alcohol; HSDB 2643; Hyfatol; Hyfatol 16; Lanol C; Lorol 24; LorolL 24; Loxanol K; Loxanol K Extra; Loxanwachs SK; NSC 4194; Palmityl alcohol; Product 308; Rita CA; Siponol CC; Siponol wax-A; SSD; SSD RP. Coemulsifier, lubricant, foam control agent, cosolvent; plasticizer, stabilizer, emollient, intermediate for metal lubricants, inks, textiles, emulsions, paper, cosmetics, mineral processing, oil field chemicals and fabric softeners. Used in cosmetics. Raw material for cosmetics & pharmaceuticals, as an auxiliary emulsifier and for viscosity control. Secondary emulsifier, thickener, opacifier, and structural agent in anhydrous stick systems. Cetyl alcohol NF is a primary structural agent in antiperspirant sticks. Solid; mp = 48-50°; bp15 = 189-190°; d = 0.8180; insoluble in H2O, soluble in organic solvents; LD50 (rat orl) = 5 g/kg. *Aarhus Oliefabrik A/S; Amerchol; Chemron; Croda; Ethyl Corp.; Lipo; Lonzagroup; Michel M.; Norman-Fox; Procter & Gamble Co.; Rita; Stepan; Vista.*

1913   Hexadecenylsuccinic anhydride
32072-96-1   H

C20H34O3
2,5-Furandione, 3-(hexadecenyl)dihydro-.
EINECS 250-911-4; 2,5-Furandione, 3-(hexadecenyl)-dihydro-; Hexadecenylsuccinic anhydride; Succinic anhydride, hexadecenyl-.

1914   Hexadecyl chloride
4860-03-1   H

C16H33Cl
1-Chloro-n-hexadecane.
Cetyl chloride; 1-Chlorohexadecane; EINECS 225-461-7; Hexadecane, 1-chloro-; Hexadecyl chloride; n-Hexadecyl chloride; NSC 57106. Liquid; mp = 17.9°; bp = 322°; d^{20} = 0.8652; insoluble in H2O.

1915   Hexadecyl methacrylate
2495-27-4   H

C20H38O2
2-Propenoic acid, 2-methyl-, n-hexadecyl ester.
Cetyl methacrylate; EINECS 219-672-3; Hexadecyl 2-methyl-2-propenoate; Hexadecyl methacrylate; HSDB 5883; Methacrylic acid, hexadecyl ester; 2-Propenoic acid, 2-methyl-, hexadecyl ester.

1916   Hexadecylbetaine
693-33-4   H

C20H41NO2
N-(Carboxymethyl)-N,N-dimethyl-1-hexadecanaminium hydroxide inner salt.
4-04-00-02387 (Beilstein Handbook Reference); Ammonium, (carboxymethyl)hexadecyldimethyl-, hydroxide, inner salt; BRN 3676787; (Carboxymethyl)-hexadecyldimethylammonium hydroxide, inner salt; C16BET; Cetyl betaine; EINECS 211-748-4; Hexadecylbetaine; Lonzaine 16S; N-(Carboxymethyl)-N,N-dimethyl-1-hexadecanaminium hydroxide inner salt; Palmityldimethylbetaine; Product HDN.

1917   Hexadecylene oxide
7320-37-8   H

C16H32O
1,2-Epoxyhexadecane.
5-17-01-00169 (Beilstein Handbook Reference); AI3-32877; BRN 0110428; CCRIS 2618; EINECS 230-786-2; Hexadecane, 1,2-epoxy-; 1,2-Hexadecane oxide; Hexadecene epoxide; 1,2-Hexadecene epoxide; HSDB 4187; NCI-C55538; Oxirane, tetradecyl-; Tetradecyl-oxirane.

1918   Hexa-1,4-diene
592-45-0   H

C6H10
1-Allylpropene.
Allylpropenyl; 1-Allylpropene; EINECS 209-756-8; 1,4-Hexadiene; HSDB 5708. Liquid; bp = 65°; d^{20} = 0.7000; insoluble in H2O, very soluble in Et2O.

1919   Hexafluoropropylene
116-15-4   H

C3F6
1,1,2,3,3,3-Hexafluoro-1-propene.

EINECS 204-127-4; Hexafluoropropene; 1,1,2,3,3,3-Hexafluoro-1-propene; Hexafluoropropylene; HSDB 5582; Perfluoro-1-propene; Perfluoropropene; Perfluoro-propylene; Propene, hexafluoro-; 1-Propene, 1,1,2,3,3,3-hexafluoro-; R1216; UN1858. Gas; mp = -156.5°; bp = -29.6°; d^{-40} = 1.583.

1920   Hexahydroazepine
111-49-9   H

C6H13N
1-Azacycloheptane.
5-20-04-00003 (Beilstein Handbook Reference); AI3-26610; Azacycloheptane; 1-Azacycloheptane; Azepine, hexahydro-; 1H-Azepine, hexahydro-; BRN 0001084; CP 18407; Cycloheptane, 1-aza-; Cyclohexamethylenimine; EINECS 203-875-9; G 0; Hexahydroazepine; Hexahydro-1H-azepine; Hexamethylenimine; HMI; Homopiperidine; HSDB 562; NSC 16236; Perhydroazepine; UN2493. Liquid; bp = 138°; d^{22} = 0.8643; soluble in H2O, very soluble in EtOH, Et2O.

1921   Hexahydrophthalic anhydride
85-42-7   H

C8H10O3
Cyclohexane-1,2-dicarboxylic anhydride.
Araldite HT 907; Cyclohexane-1,2-dicarboxylic anhydride; EINECS 201-604-9; Hexahydrophthalic acid anhydride; Hexahydrophthalic anhydride; HHPA; Lekutherm Hardener H; NSC 8622; NT 907. Widely used as an epoxy hardener. It may also be used in the manufacture of alkyd resins, plasticizers, and polyesters. Crystals; mp = 32°; bp18 = 145°. *Buffalo Color Corp.*

1922   Hexa(methoxymethyl)melamine
3089-11-0   H

C15H30N6O6
Hexakis(methoxymethyl)-s-triazine-2,4,6-triamine.
4-26-00-01274 (Beilstein Handbook Reference); AI3-26859; BRN 0356813; EINECS 221-422-3; Hexa(meth-oxymethyl)melamine; Hexakis(methoxy-methyl)-s-triaz-ine-2,4,6-triamine; Hexamethyl methyl-olmelamine; Hexamethylol-melamin-hexa-methyläther; Hexamethylol-melamine hexamethyl ether; LK 36; Malamine, hexakis(methoxymethyl); Pidifix 330; TLF-3617.

319

1923 Hexamethylcyclotrisilazane
1009-93-4 G

C6H21N3Si3
1,1,3,3,5,5-Hexamethylcyclotrisilazane.
4-04-00-04114 (Beilstein Handbook Reference); BRN 0774739; Cyclotrisilazane, 2,2,4,4,6,6-hexamethyl-; Di-methylsilazane trimer; EINECS 213-773-6; Hexamethyl-cyclotrisilazane; 2,2,4,4,6,6-Hexamethyl-cyclotrisilazane; NSC 139842. Difunctional blocking agent; reagent for cyclosilylation. Liquid; mp= -10°; bp = 188°; d^{20} = 0.9196. *Hüls Am.*

1924 Hexamethylcyclotrisiloxane
541-05-9 H

C6H18O3Si3
Dimethylsiloxane cyclic trimer.
AI3-62005; CCRIS 1326; CH7260; Cyclotrisiloxane, hexamethyl-; DC 246; Dimethylsiloxane cyclic trimer; EINECS 208-765-4; Hexamethylcyclotrisiloxane; LS 8120; SDK 10. Coupling agent, chemical intermediate, blocking agent, release agent, lubricant, primer, and reducing agent. Solid; mp = 64.5°; bp = 134°; d^{20} = 1.1200; insoluble in H2O. *Degussa-Hüls Corp.*

1925 Hexamethyldisilane
1450-14-2 G

C6H18Si2
Permethyldisilane.
CH7280; Disilane, hexamethyl-; EINECS 215-911-0; Hexamethyldisilane; Permethyldisilane. Coupling agent, chemical intermediate, blocking agent, release agent, lubricant, primer, and reducing agent. Starting material for preparation of trimethylsilyl alkali compounds. Liquid; mp = 15°; bp = 112-114°; d = 0.7247; λ_m = 198 nm (ε = 8511, cyclohexane); insoluble in H2O, soluble in Et2O, Me2CO, C6H6, freely soluble in alkali. *Degussa-Hüls Corp.; Greeff R.W. & Co.*

1926 Hexamethyldisiloxane
107-46-0 G H

C6H18OSi2
Bis-trimethylsilyl oxide.
AI3-51466; Belsil DM 0.65; Bis-trimethylsilyl oxide; Bistrimethylsilyl ether; CCRIS 1325; CH7310; Disiloxane, hexamethyl-; EINECS 203-492-7; Fluka AG; H7250; H7310; HMDSO; Hexamethyldisiloxane; HMS; HSDB 5378; KF 96L; NSC 43346; OS 10; Oxybis(trimethyl-silane); Silane, oxybis(trimethyl-; SWS-F 221. Coupling agent,

chemical intermediate, blocking agent, release agent, lubricant, primer, and reducing agent. Used as cleaning, polishing and damping media; offers low toxicity, inertness. Liquid; mp = -66°; bp = 99°; d^{20} = 0.7638; insoluble in H2O. *Hüls Am.*

1927 Hexamethylene bis(3,5-di-t-butyl-4-hydroxyhydrocinnamate)
35074-77-2 H

C40H62O6
1,6-Hexamethylenebis(3,5-t-butyl-4-hydroxyhydro-cinnamate).
Benzenepropanoic acid, 3,5-bis(1,1-dimethylethyl)-4-hydroxy-, 1,6-hexanediyl ester; EINECS 252-346-9; Hexamethylene bis(3-(3,5-di-t-butyl-4-hydroxyphenyl)-propionate); Hexamethylene bis(3,5-di-t-butyl-4-hydroxy-hydrocinnamate); 1,6-Hexamethylenebis(3,5-t-butyl-4-hydroxyhydrocinnamate); 1,6-Hexanediyl 3,5-bis(1,1-dimethylethyl)-4-hydroxybenzenepropanoate; Irganox 249; Irganox 259.

1928 Hexamethylene diisocyanate
822-06-0 H

C8H12N2O2
Hexamethylene-1,6-diisocyanate.
4-04-00-01349 (Beilstein Handbook Reference); AI3-28285; BRN 0956709; 1,6-Diisocyanatohexane; EINECS 212-485-8; HDI; Hexamethylene-1,6-diisocyanate; Hexamethylene diisocyanate, 1,6-; 1,6-Hexamethylene diisocyanate; Hexamethylendiisocyanat; Hexane 1,6-diisocyanate; Hexane, 1,6-diisocyanato-; 1,6-Hexanediol diisocyanate; 1,6-Hexylene diisocyanate; HMDI; HSDB 6134; Isocyanic acid, diester with 1,6-hexanediol; Isocyanic acid, hexamethylene ester; Metyleno-bis-fenyloizocyjanian; NSC 11687; Szesciometylenodwu-izocyjanian; TL 78; UN2281. Liquid; bp_{10} = 122°, bp_1 = 94°; d^{20} = 1.0528.

1929 Hexamethylene glycol
629-11-8 4709 G H

C6H14O2
1,6-Hexanediol.
4-01-00-02556 (Beilstein Handbook Reference); AI3-03307; BRN 1633461; 1,6-Dihydroxyhexane; EINECS 211-074-0; HDO;

320

Hexamethylene glycol; Hexamethylenediol; 1,6-Hexanediol; α,ω-Hexanediol; HSDB 6488; NSC 508. Crystals; mp = 42.8°; bp = 208°; insoluble in C_6H_6, slightly soluble in Et_2O, soluble in H_2O, EtOH, Me_2CO.

1930 Hexamethylene glycol diacrylate
13048-33-4   H

$C_{12}H_{18}O_4$
2-Propenoic acid, 1,6-hexanediyl ester.
Acrylic acid, hexamethylene ester; Ageflex HDDA; BRN 1870540; C 716; CCRIS 4823; EINECS 235-921-9; HDDA; HDODA; Hexamethylene acrylate; Hexa-methylene diacrylate; Hexamethylene glycol diacrylate; 1,6-Hexamethylene diacrylate; 1,6-Hexanediol di-2-propenoate; 1,6-Hexanediol diacrylate; HSDB 6109; Kayarad HDDA; NK Ester A HD; Photomer 4017; Sartomer 238; Sartomer SR 238; Setalux UV 2243; SR 238; Viscoat 230. Fast curing monomer providing adhesion to metal and glass, flexibility in inks and coatings, water resistant, good weatherability; reactive diluent for radiation-curable oligomers. Liquid; d_{20}^{20} = 1.01; mp <-20°; flash pt. >93°. *Rit-Chem.*

1931 Hexamethylenediamine
124-09-4   4713   H

$C_6H_{16}N_2$
1,6-Hexamethylenediamine.
4-04-00-01320 (Beilstein Handbook Reference); AI3-37283; BRN 1098307; CCRIS 6224; 1,6-Diaminohexane; EINECS 204-679-6; Hexamethylene diamine; 1,6-Hexamethylenediamine; Hexane, 1,6-diamino-; 1,6-Hexanediamine; 1,6-Hexylenediamine; HMDA; HSDB 189; NCI-C61405; NSC 9257; UN1783; UN2280. Crystals; mp = 41.5°; bp = 205°; very soluble in EtOH, C_6H_6, freely soluble in H_2O.

1932 Hexamethylenediamine monoadipate
3323-53-3   H

$C_6H_{16}N_2.C_6H_{10}O_4$
Adipic acid, compd. with 1,6-hexanediamine (1:1).
Adipan hexamethylendiaminu; Adipic acid, compd. with 1,6-hexanediamine; EINECS 222-037-3; Hexamethyl-enediamine adipate (1:1); Hexamethylenediamine monoadipate; Nylon 66 salt.

1933 Hexamethylenediamine sebacate
6422-99-7   H

$C_{10}H_{18}O_4.C_6H_{16}N_2$
Decanedioic acid, compd. with 1,6-hexanediamine.
Decanedioic acid, compd. with 1,6-hexanediamine (1:1); EINECS 229-177-4; Hexamethylenediamine sebacate; Hexamethylenediammonium sebacate; Nylon 6-10 salt; Sebacic acid, compd. with 1,6-hexanediamine (1:1); Sebakan hexamethylendiaminu.

1934 β,1,1,2,3,3-Hexamethylindan-5-ethanol
1217-08-9   H

$C_{17}H_{26}O$
2,3-Dihydro-β,1,1,2,3,3-hexamethyl 1H-indene-5-ethan-ol.
EINECS 214-934-3; β,1,1,2,3,3-Hexamethylindan-5-eth-anol; 1H-Indene-5-ethanol, 2,3-dihydro-β,1,1,2,3,3-hexa-methyl-; 5-Indanethanol, β,1,1,2,3,3-hexamethyl-.

1935 Hexamethylsilanediamine
3768-58-9   G

$C_6H_{18}N_2Si$
Bis(dimethylamino)dimethylsilane.
Bis(dimethylamino)dimethylsilane; CB2100; EINECS 223-200-1;. Coupling agent, chemical Intermediate, blocking agent, release agent, lubricant, primer, reducing agent. Liquid; mp = -98°; bp = 128-129°; d = 0.8090. *Degussa-Hüls Corp.*

1936 Hexamethylsilazane
999-97-3   4708   H

$C_6H_{19}NSi_2$
1,1,1-Trimethyl-N-(trimethylsilyl)silanamine.
AI3-51467; Bis(trimethylsilyl)amine; CCRIS 2456; Disilazane, 1,1,1,3,3,3-hexamethyl-; EINECS 213-668-5; Hexamethyldisilazane; 1,1,1,3,3,3-Hexamethyldisila-zane; Hexamethylsilazane; HMDS; NSC 93895; Prosil® HMDS; OAP; Silanamine, 1,1,1-trimethyl-N-(trimethyl-silyl)-; SZ 6079; 1,1,1-Trimethyl-N-(trimethylsilyl)-silanamine; Tri-Sil; TSL 8802. A silica reaction product; chemical intermediate, used in chromatographic packings and as a silylating agent. Used for the preparation of volatile derivatives of a wide range of biologically active compounds for glc analysis. Coupling agent; silica treatment for silicone elastomers, novolac photoresist adhesion promoter. Liquid; bp = 125°; d^{25} = 0.7741; LD_{50} (rat orl) = 850 mg/kg. *Dow Corning; Hüls Am.; Janssen Chimica; PCR; Sigma-*

1937 Hexane

110-54-3 4712 G H

C6H14

n-Hexane.
Al3-24253; CCRIS 6247; Dipropyl; EINECS 203-777-6; Esani; Exxsol® Hexane; Gettysolve-B; Heksan; Hexane; n-Hexane; Hexyl hydride; HSDB 91; NCI-C60571; NSC 68472; Skellysolve B; UN1208. Used as a solvent, alcohol denaturant, paint diluent, polymerization reaction medium; filling for thermometers. Liquid; mp = -95.3°; bp = 68.7°; d^{25} = 0.6548; insoluble in water; soluble in Et2O, CHCl3, very soluble in EtOH; LD50 (rat orl) = 32 mg/kg. *Ashland; BP Chem.; Exxon; Humphrey; Mitsui Petroleum; Phillips 66; Texaco.*

1938 Hexaneperoxoic acid, 2-ethyl-, 1,1,4,4-tetramethyl-1,4-butanediyl ester

13052-09-0 G

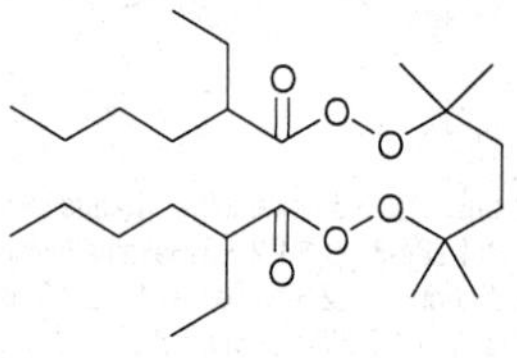

C24H46O6

2,5-dimethyl-2,5-di(2-ethyl hexanoyl peroxy) hexane.
EINECS 235-935-5; Hexaneperoxoic acid, 2-ethyl-, 1,1,4,4-tetramethyl-1,4-butanediyl ester; 1,1,4,4-Tetramethylbutane-1,4-diyl bis(2-ethylperoxyhexanoate); USP®-245; 2,5-Dimethyl-2,5-bis(2-ethyl-1-hexanoyl-peroxy)hexane; 2,5-Dimethyl-2,5-di(2-ethylhexanoyl-peroxy)hexane; 2,5-Dimethyl-2,5-hexanediol bis(2-ethylperoxyhexanoate); 2,5-Dimethylhexane 2,5-diper-2-ethylhexanoate; Lupersol 256; USP 245. Catalyst for heated curing of polyester resin systems; features rapid cures and outstanding surface finishes. *CK Witco Corp.*

1939 Hexanoic acid, 2-ethyl-, diester with tetraethylene glycol

18268-70-7 H

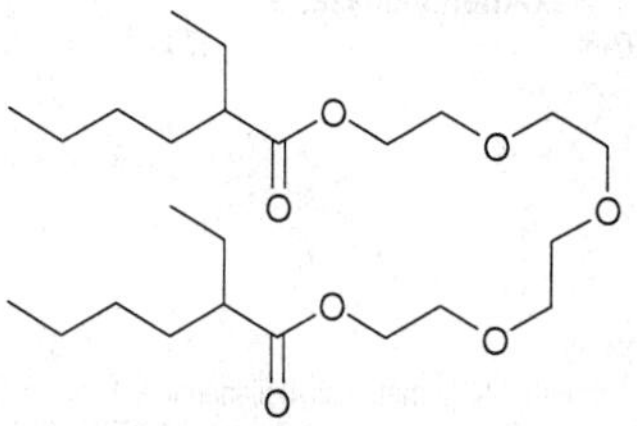

C24H46O7

3,6,9-Trioxaundecamethylene bis(2-ethylhexanoate).
EINECS 242-149-6; Flexol 4GO; Hexanoic acid, 2-ethyl-, oxybis(2,1-ethyldiyloxy-2,1-ethanediyl)ester; Hexanoic acid, 2-ethyl-, diester with tetraethylene glycol; Plasticizer 4GO; Polyethylene glycol 200 di(2-ethylhexoate); TegMeR 804; Tetraethylene glycol di(2-ethylhexoate); 3,6,9-Trioxaundecamethylene bis(2-ethylhexanoate).

1940 Hexanoic acid, 2-ethyl-, zirconium salt

22464-99-9 H

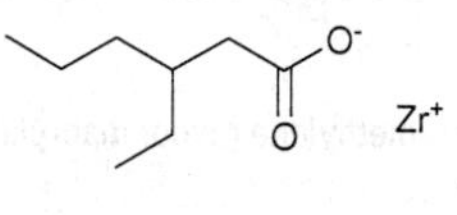

C8H15O2Zr

2-Ethylhexanoic acid, zirconium salt.
EINECS 245-018-1; 2-Ethylhexanoic acid, zirconium salt; Hexanoic acid, 2-ethyl-, zirconium salt.

1941 Hexazinone

51235-04-2 4734 G P

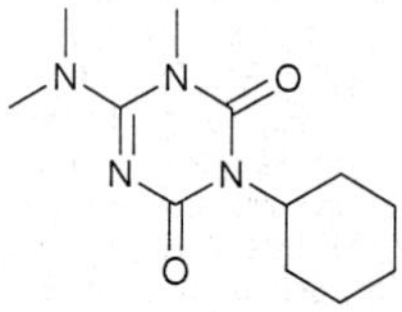

C12H20N4O2

3-cyclohexy-6-(dimethylamino)-1-methyl-1,3,5-triazine-2,4(1H,3H)-dione.
5-26-10-00171 (Beilstein Handbook Reference); BRN 0618801; Brushkiller; Caswell No. 271AA; CCRIS 5273; 3-Cyclohexyl-1-methyl-6-(dimethylamino)-s-trazine-2,4(1H,3H)-dione; 3-Cyclohexyl-6-(dimethylamino)-1-methyl-1,3,5-triazine-2,4(1H,3H)-dione; 3-Cyclohexyl-6-dimethylamino-1-methyl-1,2,3,4-tetrahydro-1,3,5-triazine-2,4-dione; DPX 3674; EINECS 257-074-4; EPA Pesticide Chemical Code 107201; Gridball; Hexazinone; HSDB 6670; s-Triazine-2,4(1H,3H)-dione, 3-cyclohexyl-6-(dimethylamino)-1-methyl-; 1,3,5-Triazine-2,4(1H,3H)-dione, 3-cyclohexyl-6-(dimethylamino)-1-methyl-; SHA 107201; Velpar; Velpar L; Velpar weed killer. Non-selective contact herbicide. Inhibits photosynthesis. Used for control of annual, biennial, perennial weeds and woody plants in non-crop areas and coniferous plantations. mp = 115-117°; d = 1.25; soluble in H2O (33 g/l), more soluble in organic solvents; LD50 (rat orl) = 1590 mg/kg. *DuPont UK; Selectokil Ltd (Velpar).*

1942 Hexedine

5980-31-4 4718 D G

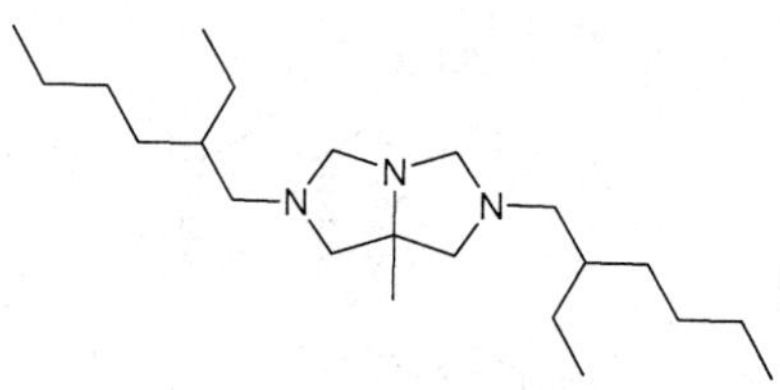

C22H45N3

2,6-Bis(2-ethylhexyl)-hexahydro-7a-methyl-1H-imidazo[1,5-c]imidazole.
2,6-Bis(2-ethylhexyl)-hexahydro-7a-methyl-1H-imidazo-(1,5-c)imidazole; Esedina; Hexedina; Hexedine; Hexedinum; 1H-Imidazo(1,5-c)imidazole, 2,6-bis(2-ethylhexyl)hexahydro-7a-methyl-; Sterisol; W 4701. Antibacterial. Liquid; bp0.025 = 131°. *Parke-Davis.*

1943 Hexene

25264-93-1 H

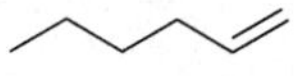

C6H12 ʹ
Hexene, isomers.
EINECS 246-768-2; Hexene; Hexene, isomer; Hexene, isomers; Hexylene; HSDB 5143. Mixture of isomers.

1944   Hex-1-ene
592-41-6   G H

C6H12
Butyl ethylene.
Al3-28797; Butyl ethylene; Butylethylene; Dialene 6; EINECS 209-753-1; Hexene; 1-Hexene; 1-n-Hexene; Hexene-1; Hexylene; HSDB 1079; Neodene® 6; NSC 74121; UN2370. Intermediate for biodegradable surfactants and specialty industrial chemicals (flavors, perfumes, dyes, resins); polymer modifier. Liquid; mp = -139.7°; bp = 63.4°; d^{20} = 0.6731; very soluble in C6H6, EtOH, Et2O, petroleum ether. *Degussa-Hüls Corp.; Monsanto (Solaris); Phillips 66; Shell.*

1945   Hexyl acrylate
2499-95-8   G

C9H16O2
n-Hexyl 2-propenoate.
4-02-00-01466 (Beilstein Handbook Reference); Acrylic acid, hexyl ester; Ageflex FA-6; Ageflex n-HA; Al3-15732; BRN 1757327; CCRIS 7038; EINECS 219-698-5; Hexyl 2-propenoate; Hexyl acrylate; n-Hexyl acrylate; HSDB 5463; NSC 11786; 2-Propenoic acid, hexyl ester. Monomer for UV-cured inks and coatings, glass coating, viscosity Index improver for functional oils, polymer cements and sealants; polymer modifier. Liquid; mp = -45°; bp24 = 88-90°, bp1 = 40°; d^{20}_{20} = 0.8780; very soluble in EtOH, Et2O; LD50 (rat orl) = 26 mg/kg. *Rit-Chem.*

1946   Hexyl alcohol
111-27-3   4715   G P

C6H14O
n-Hexyl alcohol.
4-01-00-01694 (Beilstein Handbook Reference); Al3-08157; Alcohol C6; Amylcarbinol; BRN 0969167; C6 alcohol; Caproic alcohol; Caproyl alcohol; Caswell No. 482E; EINECS 203-852-3; EPA Pesticide Chemical Code 079047; EPAL 6; Fatty alcohol(C6); FEMA Number 2567; Hexanol; Hexan-1-ol; 1-Hexanol; n-Hexanol; n-Hexan-1-ol; Hexyl alcohol; 1-Hexyl alcohol; n-Hexyl alcohol; HSDB 565; 1-Hydroxyhexane; Nacol® 6-98; NSC 9254; Pentylcarbinol. Detergent and emulsifier intermediate, used in pharmaceuticals (antiseptics, perfume esters), and as a solvent and plasticizer. Registered by EPA as a herbicide and insecticide (cancelled). Liquid; mp = -44.6°; bp = 157.6°; d^{20} = 0.8136; slightly soluble in H2O, CCl4, soluble in EtOH, Me2CO, CHCl3, freely soluble in C6H6, Et2O; LD50 (rat orl) = 4.59 g/kg. *Amoco Lubricants; Ashland; Condea Vista Co.; Penta Mfg.; Sigma-Aldrich Fine Chem.*

1947   Hexylamine
111-26-2   H

C6H15N
1-Hexylamine.
4-04-00-00709 (Beilstein Handbook Reference); Al3-16554; 1-Aminohexane; BRN 1731298; EINECS 203-851-8; ENT 16554; 1-Hexanamine; Hexylamine; 1-Hexylamine; n-Hexylamine; Mono-n-hexylamine; NSC 2590. Liquid; mp = -22.9°; bp = 132.8; d^{20} = 0.7660; slightly soluble in H2O, soluble in CHCl3, freely soluble in EtOH, Et2O.

1948   Hexyl cellosolve
112-25-4   H

C8H18O2
Ethylene glycol n-hexyl ether.
4-01-00-02383 (Beilstein Handbook Reference); BRN 1734691; Cellosolve, n-hexyl-; EINECS 203-951-1; Ethanol, 2-(hexyloxy)-; Ethylene glycol monohexyl ether; Ethylene glycol-n-monohexyl ether; Ethylene glycol n-hexyl ether; Glycol monohexyl ether; Hexyl cellosolve; n-Hexyl cellosolve; HSDB 5569. Liquid; mp = -45.1°; bp = 208°; d^{20} = 0.8878; slightly soluble in H2O, very soluble in EtOH, Et2O.

1949   α-Hexylcinnamaldehyde
101-86-0   H

C15H20O
2-Hexylcinnamaldehyde.
Al3-05096; Cinnamaldehyde, α-hexyl-; EINECS 202-983-3; FEMA No. 2569; Hexyl cinnamic aldehyde (VAN); 2-Hexylcinnamaldehyde; α-Hexylcinnamaldehyde; α-Hexylcinnamyl aldehyde; α-n-Hexyl-β-phenylacrolein; 2-Hexyl-3-phenyl-2-propenal; NSC 406799; Octanal, 2-(phenylmethylene)-.

1950   Hexylene glycol
107-41-5   4730   G H P

C6H14O2
2-Methyl-2,4-pentanediol.
4-01-00-02565 (Beilstein Handbook Reference); 4-Methyl-2,4-pentanediol; Al3-00919; BRN 1098298; Caswell No. 574; Diolane; EINECS 203-489-0; EPA Pesticide Chemical Code 068601; Hexylene glycol; HSDB 1126; Isol; (±)-2-Methyl-2,4-pentanediol; NSC 8098; Pinakon; α,α,α'-Trimethyltrimethylene glycol. Used in hydraulic brake fluids, printing inks, as a coupling agent and penetrant for textiles, cosmetics; ice inhibitor in carburetors; fuel, lubricant additive. Registered by EPA (cancelled). Liquid; mp = -50°; bp = 197.1°; d^{15} = 0.923; λ_m = 302 nm (ε = 3631, H2SO4); soluble in H2O, EtOH, Et2O, CCl4; LD50 (rat orl)= 4.7 g/kg. *Allchem Ind.; Ashland; BP Chem.; Coyne; Elf Atochem N. Am.; Great Western; McIntyre; Mitsui Petroleum; Penta Mfg.; Ruger; Shell; Sigma-Aldrich Fine Chem.; Spectrum Chem. Manufacturing; Union Carbide Corp.*

1951   Hexyltrichlorosilane
928-65-4   G

323

C6H13Cl3Si

Trichlorohexylsilane.

CH7332; EINECS 213-178-1; Hexyltrichlorosilane; HSDB 2005; NSC 139843; Silane, hexyltrichloro-; Silane, trichlorohexyl-; UN1784. Coupling agent, chemical intermediate, blocking agent, release agent, lubricant, primer, and reducing agent. Liquid; bp = 190°; d^{20} = 1.1100; freely soluble in H2O. *Degussa-Hüls Corp.*

1952   Hippuric acid
495-69-2   4735   G

C9H9NO3

Benzamidoacetic acid.

4-09-00-00778 (Beilstein Handbook Reference); Acetic acid, (benzoylamino)-; Acido ippurico; AI3-01062; Benzoylglycine; N-Benzoylglycine; BRN 1073987; EINECS 207-806-3; Glycine, N-benzoyl-; NSC 9982; Phenylcarbonylaminoacetic acid. Used in organic synthesis and medicine. Prisms; mp = 191.5°; d^{20} = 1.371; λ_m = 226 nm (ε = 12700, MeOH); insoluble in petroleum ether, CS2, slightly soluble in Et2O, C6H6, CHCl3, soluble in H2O, EtOH, EtOAc. *Lancaster Synthesis Co.; Penta Mfg.; U.S. BioChem.*

1953   Holmium
7440-60-0   4747   G

Ho

Ho

Holmium.

EINECS 231-169-0; Holmium. Metallic element; getter in vacuum tubes, research in electrochemistry, spectroscopy. Metal; mp = 1474°; bp = 2700°. *Atomergic Chemetals; Cerac; Noah Chem.; Rhône-Poulenc.*

1954   Holmium oxide
12055-62-8   G

Ho2O3

Holmia.

EINECS 235-015-3; Holmium oxide; Holmium oxide (Ho2O3). Refractories, special catalyst. *Atomergic Chemetals; Cerac; Noah Chem.; Rhône-Poulenc.*

1955   Homosalate
118-56-9   4759   D G

C16H22O3

3,3,5-Trimethylcyclohexyl salicylate.

Benzoic acid, 2-hydroxy-, 3,3,5-trimethylcyclohexyl ester; Caswell No. 482B; CCRIS 4885; Coppertone; EINECS 204-260-8; EPA

Pesticide Chemical Code 076603; Filtersol "A"; Heliopan; Homomenthyl salicylate; m-Homomenthyl salicylate; Homosalate; Homosalato; Homosalatum; Kemester® HMS; Metahomomenthyl salicylate; NSC 164918; Salicylic acid, m-homomenthyl ester; Salicylic acid, 3,3,5-trimethylcyclohexyl ester; 3,3,5-Trimethyl-cyclohexyl 2-hydroxybenzoate; 3,3,5-Trimethyl-cyclohexyl salicylate. UV absorber; sunscreen agent; used in cosmetic skin preparations. Used as an ultraviolet screen in Coppertone® products. Oil; bpss4 = 161-165°; d_{25}^{25} = 1.045.; insoluble in H2O, soluble in organic solvents. *Greeff R.W. & Co.; Witco/Humko.*

1956   Hydrargaphen
14235-86-0   D G

C33H24Hg2O6S2

Phenylmercuric dinaphthylmethane disulfonate.

Ba 13155; 3,3'-Bis(naphthalene-2-sulfonic acid) phenyl mercury salt; Bis(phenylmercuri)methylenedinaphthalene-sulfonate; Conotrane; Diphenylmercuridinaphthyl-meth-anedisulfonate; EINECS 238-107-1; Fibrotan; Hidrarg-afeno; Hydraphen; Hydrargaphen; Hydrarga-phene; Hyd-rargafenum; Idrargafene; Mercury, (μ-(3,3'-methyl-enedi-2-naphthalenesulfonato))diphenyldi-; Methylenedi-naphthalenesulfonic acid bisphenylmercuri salt; 2-Naphthalenesulfonic acid, 3,3'-((phenyl-mercuri)methyl-ene)di-; 2-Naphthalenesulfonic acid, 3,3'-methylenedi-phenyl-mercury; P.M.F.; Penotrane; Phenyl mercuric fixtan; Phenylmercuric methylenebis(2-naphthyl-3-sulfonic acid); Phenylmercuric 3,3'-methylenebis(2-naphthalenesulfonate); Phenylmercury methylenedi-naphthalenesulfonate; Phenylmercuric dinaphthylmeth-anedisulfonate; Phenylmercury 2,2'-dinaphthylmethane-3,3'-disulphonate; Septotan; Versa-trane; Versotrane. Bactericide and fungicide for the treatment of wool, hides, leather, textiles, timber and wood-pulp, paints and adhesives. Antiseptic; anti-infective (topical). Amorphous powder, Practically insoluble in H2O; forms colloidal solutions in alkali methal dinaphthylmethane sulfonates; colloid tends to absorb at interfaces and form charged hydrated aggregates; LD50 (mus orl) = 80 mg/kg. *Ward Blenkinsop.*

1957   Hydrazine
302-01-2   4789   G

H2N—NH2

H4N2

Nitrogen hydride.

CCRIS 335; Diamine; EINECS 206-114-9; HSDB 544; Hydrazine; Hydrazine base; Hydrazines; Hydrazyna; Levoxine; Oxytreat 35; RCRA waste number U133; Ultra Pure; UN2029; UN3293; Zerox. Chemical intermediate and reducing agent. A propellant for mono and bi-propellant systems. Hydrazine solutions (35% in water); used for water treatment. Liquid; mp = 1°; bp = 114°; miscible with H2O; LD50 (rat orl) = 60 mg/kg. *Akzo Chemie; Fairmount; Miles Inc.; Olin Corp; Schering AG; Sigma-Aldrich Fine Chem.; Spectrum Chem. Manufacturing.*

1958   Hydrazine hydrobromide
13775-80-9   G

$$H_2N-NH_2 \quad H-Br$$

BrH5N2
Hydrazine monohydrobromide.
EINECS 237-412-7; Hydrazine, monohydrobromide; Hydrazinium bromide; Hyflux M. Reducing agent. Crystals; mp = 87-92°. *ABM Chemicals Ltd.*

1959   Hydrazine sulfate
10034-93-2   4791   G

$$H_2N-NH_2 \quad O=\overset{OH}{\underset{O}{\overset{|}{\underset{||}{S}}}}-OH$$

H4N2.H2SO4
Diamine sulfate.
AI3-18433; CCRIS 336; Diamidogen sulfate; Diamine sulfate; EINECS 233-110-4; HS; HSDB 5086; Hydrazine dihydrogen sulfate salt; Hydrazine hydrogen sulfate; Hydrazine monosulfate; Hydrazine sulfate; Hydrazine, sulfate (1:1); Hydrazine sulphate; Hydrazinium sulfate; Hydrazinium(2+) sulfate; Hydrazinium(2+) sulphate; Hydrazonium sulfate; Idrazina solfato; NSC-150014; NSC 215190; Siran hydrazinu. Chemical intermediate, condensation reactions, catalyst for making acetate fibers; analysis of minerals, slags, fluxes; determination of arsenic in metals; separation of polonium from tellurium; fungicide, germicide. Plates or prisms; mp = 254°; d = 1.378; soluble in cold H2O (3 g/100 ml), freely soluble in hot H2O, insoluble in EtOH. *Fairmount; Janssen Chimica; Mallinckrodt Inc.; Otsuka Pharm. Co. Ltd.; Spectrum Chem. Manufacturing.*

1960   Hydriodic acid
10034-85-2   4797   G

$$H-I$$

HI
Hydroiodic acid.
Acide iodhydrique; Acido yodhidrico; Anhydrous hydriodic acid; Caswell No. 482C; EINECS 233-109-9; EPA Pesticide Chemical Code 046912; HSDB 2155; Hydriodic acid; Hydrogen iodide; Hydrogen iodide (HI); Hydrogen iodide, anhydrous; Hydroiodic acid; Iodure d'hydrogene anhydre; UN1787; UN2197; Yoduro de hidrogeno anhidro. Used in preparation of iodine salts, organic preparations, analytical reagent, disinfectant, pharmaceuticals. Liquid; d = 1.70; bp = 127°; pH = 1.0 (0.1 molar solution). *Janssen Chimica.*

1961   Hydrobromic acid
10035-10-6   4799, 4814   D G

$$H-Br$$

HBr
Hydrogen bromide.
Acide bromhydrique; Acido bromhidrico; Acido bromidrico; Anhydrous hydrobromic acid; Bromowodor; Bromure d'hydrogene anhydre; Bromuro de hidrogeno anhidro; Bromwasserstoff; Broomwaterstof; EINECS 233-113-0; HSDB 570; Hydrobromic acid; Hydrogen bromide; Hydrogen bromide (HBr); UN1048; UN1788. Used in analytical chemistry, as a solvent for ore minerals, in manufacture of inorganic and some alkyl bromides, alkylation catalyst. Used medicinally as a sedative-hypnotic. Colorless gas; mp = -86.9°; bp = -66.8°; d = 2.71 (air = 1); freely soluble in H2O, soluble in EtOH, organic solvents; LC50 (rat ihl) = 2858 ppm, (mus ihl) = 814 ppm. Aqueous solution, marketed in various concentrations: 50%

HBr (d = 1.517), 40% HBr (d = 1.38), 34% HBr (d = 1.31), 1-% HBr (d = 1,08); solution contains up to 68.85% HBr in H2O; forms azeotropes with H2O, bp100 = 74.12° (49.80%), bp400 = 107.00° (48.47%), bp700 = 122.0° (47.74%), bp800 = 125.79° (47.56%). *Allchem Ind.; EM Ind. Inc.; Ethyl Corp.; Great Lakes Fine Chem.; Octel Chem. Ltd.*

1962   Hydrochloric acid
7647-01-0   4801   G P

$$H-Cl$$

HCl
Hydrogen chloride.
4-D Bowl Sanitizer; Acide chlorhydrique; Acido clorhidrico; Acido cloridrico; Anhydrous hydrochloric acid; Aqueous hydrogen chloride; Bowl Cleaner; Caswell No. 486; Chloorwaterstof; Chlorhydric acid; Chlorowodor; Chlorure d'hydrogene; Chlorure d'hydrogene anhydre; Chloruro de hidrogeno; Chlorwasserstoff; Cloruro de hidrogeno anhidro; EINECS 231-595-7; Emulsion Bowl Cleaner; EPA Pesticide Chemical Code 045901; HSDB 545; Hydrochloric acid; Hydrogen chloride; Hygeia Creme Magic Bowl Cleaner; Marine acid; Muriatic acid; NSC 77365; Percleen Bowl and Urinal Cleaner; Soldering acid; Spirits of salt; UN 1050 (anhydrous); UN 1789 (solution); UN 2186 (refrigerated liquefied gas); Varley Poly-Pak Bowl Creme; Varley's Ocean Blue Scented Toilet Bowl Cleaner; White Emulsion Bowl Cleaner; Wuest Bowl Cleaner Super Concentrated. Used as a chemical intermediate and reactant. Registered by EPA as an antimicrobial, fungicide and herbicide (cancelled). FDA GRAS, Japan restricted, Europe listed, UK approved, FDA approved for injectables, parenterals, inhalants, intravenous, ophthalmics, orals, otics, topicals. USP/NF, BP. Ph.Eur., JP compliant. In the FDA list of inactive ingredients, (injectants, inhalants, ophthalmics, orals, otics, rectals and topicals). Acidifier, buffer, neutralizing agent. Preservative in soft lens products. Used in injectables, parenterals, inhalants, intravenous, ophthalmics, orals, otics and topicals. Gas; mp = -114°; bp = -85°; azeotrope with H2O (20% HCl): bp = 109°; d²⁵ = 1.096.

1963   Hydrochlorothiazide
58-93-5   4802   D

C7H8ClN3O4S2
6-Chloro-3,4-dihydro-2H-1,2,4-benzothiadiazine-7-sulfonamide 1,1-dioxide.
Accuretic; Acesistem; Acuilix; Aldactazide; Aldazida; Aldectazide 50/50; Aldoril; Apo-Hydro; Apresazide; Aquarills; Aquarius; Aquazide H; Bremil; Briazide; BRN 0625101; Caplaril; Capozide; Carozide; Catiazida; CCRIS 2082; Chlorizide; Chlorosulthiadil; Chlorzide; Chlothia; Cidrex; Clorana; Concor Plus; Condiuren; Diaqua; Dichlorosal; Dichlotiazid; Dichlotride; Diclotride; Didral; Dihydran; Dihydrochlorothiazid; Dihydrochlorothiazide; Dihydrochlorothiazidum; Dihydroxychlorothiazidum; Direma; Disalunil; Disothiazid; Diu 25 Vigt; Diu-melusin; Diurogen; Dixidrasi; Drenol; Dyazide; EINECS 200-403-3; Esidrex; Esidrix; Esimil; Esoidrina; Fluvin; H.H. 25/25; H.H. 50/50; HCT-Isis; HCTZ; HCZ; Hidril; Hidro-Niagrin; Hidrochlortiazid; Hidroclorotiazida; Hidroronol; Hidro-saluretil; Hidrotiazida; HSDB 3096; Hyclosid; Hydril; Hydro-Aquil; Hydro-D; Hydro-Diuril; Hydro Par; Hydro-Saluric; Hydro-T; Hydrochlorothiazid; Hydrochloro-thiazidum; Hydrochlorthiazide; Hydrochlorthiazidum; Hydrocot; Hydrodiuretic; Hydrodiuril; Hydropres; Hydrosaluric; Hydrothide; Hydrozide; Hypothiazid; Hypothiazide; Hytrid; Hyzaar; Idroclorotiazide; Idrotiazide; Inderide;

325

Indroclor; Ivaugan; Jen-Diril; Lopressor HCT; Lotensin HCT; Manuril; Maschitt; Maxzide; Mazide 25 mg; Medozide; Megadiuril; Microzide; Mictrin; Mikorten; Modurcen; Moduretic; Natrinax; NCI-C55925; Nefrix; Nefrol; Neo-codema; Neo-Flumen; Neo-Minzil; Newtolide; Novodiurex; NSC 53477; Oretic; Pantemon; Panurin; Prinzide; Raunova Plus; Ro-Hydrazide; Roxane; Saldiuril; Selozide; Ser-Ap-Es; Servithiazid; Spironazide; SU 5879; Tandiur; Thiuretic; Timolide; Unazid; Unipres; Urodiazin; Urozide; Vaseretic; Vasoretic; Vetidrex; Ziac; Zide.; component of: Micardis (with telmisartan). A thiazide diuretic used in the treatment of hypertension. Solid; mp= 273-275°; λ_m = 317, 271, 226 nm ($A^{1\%}_{1cm}$ 130, 654, 1280 MeOH/HCl); soluble in MeOH, EtOH, Me2CO; insoluble in H2O; LD50 (mus iv) = 590 mg/kg, (mus orl) > 8000 mg/kg. *Abbott Labs Inc.; Ciba-Geigy Corp.; Lederle Labs.; Lemmon Co.; Merck & Co.Inc.; Parke-Davis; Searle G.D. & Co.; Solvay Pharm. Inc.; Squibb E.R. & Sons; Wallace Labs; Wyeth Labs.*

1964   Hydrocinnamic acid
501-52-0   4805   G

C9H10O2
3-Phenylpropionic acid.
4-09-00-01752 (Beilstein Handbook Reference); AI3-00892; Benzenepropanoic acid; Benzenepropionic acid; Benzylacetic acid; BRN 0907515; CCRIS 3199; Dihydrocinnamic acid; EINECS 207-924-5; FEMA No. 2889; Hydrocinnamic acid; NSC 9272; Phenylpropanoic acid; 3-Phenylpropanoic acid; β-Phenylpropionic acid; 3-Phenylpropionic acid. Fixative for perfumes, flavoring. Solid; mp = 47-48°; bp = 280°; λ_m = 247, 252, 257, 260, 263, 267 nm (MeOH); soluble in H2O (5.8 g/l), EtOH, Et2O, CCl4, CHCl3, CS2, very soluble in C6H6. *Greeff R.W. & Co.; Janssen Chimica; Penta Mfg.*

1965   Hydrocodone
125-29-1   4806   D

- C18H21NO3
4,5α-Epoxy-3-methoxy-17-methylmorphinan-6-one.
dihydrocodeinone; Bekadid; Dicodid. Analgesic, narcotic; antitussive. Federally controlled substance (opiate). CAUTION: may be habit forming. mp = 198°; soluble in alcohol, dilute acids; insoluble in H2O; λ_m = 280 nm (ε 1310); LD50 (mus sc) = 85.7 mg/kg. *Merck & Co.Inc.*

1966   Hydrofluosilicic acid
16961-83-4   4209   G

F6H2Si
Fluorosilicic acid.

Acide fluorosilicique; Acide fluosilicique; Acido fluosilicico; Caswell No. 463; CCRIS 2296; Dihydrogen hexafluorosilicate; Dihydrogen hexafluorosilicate (2-); EINECS 241-034-8; EPA Pesticide Chemical Code 075305; FKS; Fluorosilicic acid; Fluorosilicic acid (H2SiF6); Fluosilicic acid; Hexafluorokieselsaeure; Hexafluorokieselsäure; Hexafluorokiezelzuur; Hexa-luorosilicic acid; Hexafluosilicic acid; HSDB 2018; Hydrofluorosilicic acid; Hydrofluosilicic acid; Hydrogen hexafluorosilicate; Hydrosilicofluoric acid; Keramyl; Kiezelfluorwaterstofzuur; NSC 16894; Sand acid; Silicate (2-), hexafluoro-, dihydrogen; Silicofluoric acid; Silicofluoride; Silicon hexafluoride dihydride; UN1778. Stable in 60-70% aqueous solution. Etches glass. Used as 2% solution to sterilise equipment and in electroplating. Liquid; $d^{17.5}_{17.5}$ (30% soln) = 1.2742; 60-70% solution forms crystaline dihydrate around 19°.

1967   Hydrogenated Castor Oil
8001-78-3   1909   E P

C57H110O9
Glyceryl tri-(12-hydroxystearate).
Castor oil, hydrogenated; Castorwax®; Castorwax MP 70®; Castorwax MP 80®; Castorwax NF®; Caswell No. 486A; EINECS 232-292-2; EPA Pesticide Chemical Code 031604; Glyceryl tri(12-hydroxystearate); Hydrogenated castor oil; Octadecanoic acid, 12-hydroxy-, 1,2,3-propanetriyl ester; Olio di ricino idrogenato; Opalwax®; Rice syn wax; Simulsol®; Thixcin E; Trihydroxystearin; Unitina HR. Registered by EPA as an antimicrobial, fungicide, herbicide, insecticide and rodenticicide (cancelled). FDA listed, USP/NF compliance. Used as a stiffening agent, wax for ointments, lubricant, tableting aid, in tablet coatings and as an extended release agent, solvent for intramuscular injections, has laxative effect. Used in orals and topicals. Hard white wax; mp = 86-88°; d = 0.98 - 1.1; insoluble in H2O, soluble in Me2CO, CHCl3. LD50 (rat orl) > 5000 mg/kg. *Akzo Nobel; Amber Syn.; Hoechst AG; Southern Clay Products.*

1968   Hydrogenated rosin, pentaerythritol ester
64365-17-9   H

Resin acids and Rosin acids, hydrogenated, esters with pentaerythritol.
EINECS 264-848-5; Hydrogenated rosin, pentaerythritol ester; Resin acids and Rosin acids, hydrogenated, esters with pentaerythritol; Resin and rosin acids, hydrogenated, esters with pentaerythritol.

1969   Hydrogenated Soybean Oil
8016-70-4   E H

Soybean oil hydrogenated.
Akolizer S; Clarity; EINECS 232-410-2; Hydrogenated soybean oil; Neustrene® 064; Oils, soybean, hydrogenated; Partially hydrogenated soybean oil; Soya bean oil fatty acids, hydrogenated; Soybean oil, hardened; Soybean oil, hydrogenated; Soybean oil, partially hydrogenated; Sterotex® HM. Pharmaceutical intermediate, crystallization promoter, mp modifier, bodying agent, lubricant, moisturizer, diluent.

1970   Hydrogenated Tallow
8030-12-4   E H

Hydrogenated Tallow.
Beef tallow, fully hydrogenated; EINECS 232-442-7; Hydrogenated tallow; Tallow, hydrogenated.; Special Fat 168T. FDA listed. Hydrophobing agent, used as a raw material for pharmaceuticals.

1971   Hydrogenated tallow acid
61790-38-3   H

Fatty acids, tallow, hydrogenated.
Acids, tallow, hydrogenated; EINECS 263-130-9; Fatty acids, tallow,

hydrogenated; Hydrogenated tallow acid; Tallow acid, hydrogenated; Tallow fatty acids, hydrogenated.

1972 Hydrogenated tallow glyceride
61789-09-1   H

Hydrogenated tallow fatty acids, glycerol monoester.
EINECS 263-031-0; Glycerides, hydrogenated tallow mono-; Glycerides, tallow mono-, hydrogenated; Hydrogenated tallow fatty acids, glycerol monoester; Hydrogenated tallow glyceride; Monoglycerides, hydrogenated tallow.

1973 Hydrogenated tallow nitrile
61790-29-2   H

Nitriles, tallow, hydrogenated.
EINECS 263-122-5; Nitriles, tallow, hydrogenated.

1974 Hydrogenated tallowamine
61788-45-2   H

Amines, hydrogenated tallow alkyl.
Amines, hydrogenated tallow alkyl; EINECS 262-976-6; Hydrogenated tallowamine; Tallow amine, hydrogenated.

1975 Hydrogenated terphenyls
61788-32-7   H

Terphenyl, partially hydrogenated.
EINECS 262-967-7; Hydrogenated terphenyls; Terphenyl, hydrogenated; Terphenyl, partially hydrogenated; Terphenyls, hydrogenated.

1976 Hydrogen fluoride
7664-39-3   4811   G

$$H—F$$

FH
Hydrofluoric acid gas.
Acide fluorhydrique; Acido fluorhidrico; Acido fluoridrico; Antisal 2B; Caswell No. 484; EINECS 231-634-8; EPA Pesticide Chemical Code 045601; Fluorhydric acid; Fluorowodor; Fluorure d'hydrogene anhydre; Fluoruro de hidrogeno anhidro; Fluorwasser-stoff; Fluorwaterstof; HSDB 546; Hydrofluoric acid; Hydrofluoride; Hydrogen fluoride; RCRA waste number U134; Rubigine; UN 1052 (anhydrous); UN 1790 (solution). Catalyst; used in aluminum production, fluorocarbons, pickling stainless steel, etching glass, acidizing oil wells, gasoline production, processing uranium. Liquid; mp = -83.55°; LD$_{50}$ (rat ihl 1h) = 1278 ppm; soluble in water, alcohol, and many organic solvents, slightly soluble in ether. *Allied Signal; DuPont; Farleyway Chem. Ltd.; General Chem; Hoechst Celanese; Seimi Chem.*

1977 Hydrogen peroxide
7722-84-1   4819   D G P

$$HO—OH$$

H$_2$O$_2$
Dihydrogen dioxide.
Albone; Albone 35; Albone 35CG; Albone 50; Albone 50CG; Albone 70; Albone 70CG; Albone DS; Caswell No. 486AAA; CCRIS 1060; Dihydrogen dioxide; EINECS 231-765-0; Elawox; EPA Pesticide Chemical Code 000595; Hioxyl; HSDB 547; Hydrogen dioxide; Hydrogen peroxide; Hydrogen peroxide (H$_2$O$_2$); Hydroperoxide; Inhibine; Interox; Kastone; Lensept; NSC 19892; Oxydol; Perhydrol; Perone; Perone 30; Perone 35; Perone 50; Perossido di idrogeno; Peroxaan; Peroxal; Peroxan; Peroxide; Peroxol; Peroxyde d'hydrogene; Proxy; Puresept; Superoxol; T-Stuff; UN2014; UN2015; UN2984; Wasserstoffperoxid; Waterstofperoxyde.

Bleaching and oxidizer used for deodorizing textiles, wood pulp, hair, fur, etc.; plasticizers; refining and cleaning metals; viscosity control for starch and cellulose derivatives. Registered by EPA as an antimicrobial, fungicide, herbicide and rodenticide. FDA GRAS (3 ppm maximum in wine, 35% solution maximum, 200 ppm maximum in distilling materials), Japan restricted, FDA approved for topicals, BP compliance. Used as a topical antiseptic to cleanse wounds, skin ulcers, local infections, in oral rinse products; in treatment of inflammatory conditions of the external ear canal; foaming agent in dentrifices, used in topicals and mouthwash gargles. Liquid; mp = -11°; bp = 150°; d = 1.4067. *Degussa AG; DuPont; Elf Atochem N. Am.; Farleyway Chem. Ltd.; FMC Corp.; Mallinckrodt Inc.; Mitsubishi Gas.*

1978 Hydrogen sulfide
7783-06-4   4823   G

$$H\overset{S}{\diagup}{}^{\diagdown}H$$

H$_2$S
Hydrogen monosulfide.
Acide sulfhydrique; Acide sulphhydrique; Dihydrogen monosulfide; Dihydrogen sulfide; EINECS 231-977-3; FEMA No. 3779; HSDB 576; Hydrogen sulfide; Hydrogen sulfide (H$_2$S); Hydrogen sulfure; Hydrogen sulfuric acid; Hydrogen sulphide; Hydrogen sulphide; Hydrogene sulphure; Hydrosulfuric acid; Idrogeno solforato; RCRA waste number U135; Schwefelwasserstoff; Sewer gas; Siarkowodor; Stink DAMP; Sulfur hydride; Sulfureted hydrogen; UN 1053; UN1053; Zwavelwaterstof. Used as a chemiucal feedstock and as an analytical reagent. Gas; mp = -85°; bp = -60°; soluble in H$_2$O (4.1 g/l), similarly soluble in organic solvents; LC$_{50}$ (rat inh, 4 hr) = 444 ppm.

1979 Hydrogen telluride
7783-09-7   4824   G

$$H\overset{Te}{\diagup}{}^{\diagdown}H$$

H$_2$Te
Hydrogen monotelluride.
Dihydrogen telluride; EINECS 231-981-5; Hydrogen telluride. Gas; mp = -49°; bp = -2°; d$_{4}^{12}$ = 2.68; soluble in H$_2$O.

1980 Hydroquinone
123-31-9   4833   D G H

C$_6$H$_6$O$_2$
p-Dihydroxybenzene.
AI3-00072; Arctuvin; Artra; Benzene, p-dihydroxy-; 1,4-Benzenediol; Benzohydroquinone; Benzoquinol; Black and White Bleaching Cream; CCRIS 714; Derma-Blanch; Diak 5; Dihydroxybenzene; EINECS 204-617-8; Eldopaque; Eldopaque Forte; Eldoquin; Eldoquin Forte; HE 5; Hidroquinone; HSDB 577; Hydrochinon; Hydroquinol; Hydroquinole; Hydroquinone; Idrochinone; NCI-C55834; NSC 9247; Phiaquin; Pyrogentistic acid; Quinol; Solaquin Forte; Tecquinol; Tenox HQ; Tequinol; UN2662; USAF EK-356. Photographic developer (not for color film); dye intermediate, inhibitor; stabilizer in paints and varnishes; motor fuels and oils; antioxidant for fats and oils. Antioxidant for synthetic latexes, fats, oils, monomers, polyester resins. Used medically as a depigmentor. Prisms or needles; mp = 172.3°; bp = 287°; d^{15} = 1.328; λm = 225, 294 nm (ε = 5180, 2810, MeOH); insoluble in C$_6$H$_6$, solulbe in H$_2$O,

Et2O, very soluble in EtOH, Me2CO, freely soluble in CCl4; LD50 (rat orl) = 320 mg/kg. *Allchem Ind.; Eastman Chem. Co.; Goodyear Tire & Rubber; ICN Pharm. Inc.; Kraeber GmbH; Schering-Plough HealthCare Products; Schering-Plough Pharm.; Sigma-Aldrich Fine Chem.*

1981   Hydroquinone diethylol ether
104-38-1   H

C10H14O4
Hydroquinone bis(2-hydroxyethyl) ether.
EINECS 203-197-3; Hydroquinone bis(β-hydroxyethyl) ether; Hydroquinone bis(2-hydroxyethyl) ether; Hydroquinone diethylol ether; NSC 1862; Vernatzer 30/10.

1982   o-Hydroxyacetophenone
118-93-4   G

C8H8O2
2'-hydroxyacetophenone.
4-08-00-00320 (Beilstein Handbook Reference); Acetophenone, 2'-hydroxy-; Acetophenone, o-hydroxy-; 2-Acetylphenol; o-Acetylphenol; AI3-12134; BRN 0386123; EINECS 204-288-0; Ethanone, 1-(2-hydroxyphenyl)-; FEMA No. 3548; 2'-Hydroxyacetophenone; o-Hydroxyacetophenone; 1-(2-Hydroxyphenyl)ethanone; 2-Hydroxyphenyl methyl ketone; o-Hydroxyphenyl methyl ketone; Methyl 2-hydroxyphenyl ketone; NSC 16933; USAF KE-20. Used in organic synthesis. mp = 2.5°; bp = 218°, bp717 = 213°; d^{20} = 1.1307; λ_m = 217, 250, 309 nm (ε = 19500, 8080, 2290, MeOH); insoluble in ligroin, slightly soluble in H2O; very soluble in EtOH, Et2O, C6H6, CHCl3. *Greeff R.W. & Co.; Hoechst Celanese; Janssen Chimica; Penta Mfg.*

1983   p-Hydroxyacetophenone
99-93-4   G

C8H8O2
4-hydroxyacetophenone.
Acetophenone, 4'-hydroxy-; Acetophenone, p-hydroxy-; 4-Acetylphenol; p-Acetylphenol; AI3-12133; EINECS 202-802-8; Ethanone, 1-(4-hydroxyphenyl)-; 4-Hydroksy-acetofenol; 4-Hydroxyacetophenone; p-Hydroxyaceto-phenone; 4'-Hydroxyacetophenone; 1-(4-Hydroxy-phenyl)ethanone; p-Hydroxyphenyl methyl ketone; Methyl p-hydroxyphenyl ketone; NSC 3698; p-Oxyacetophenone; Phenol, p-acetyl-; Piceol; USAF KF-15. Used in organic synthesis. Needles; mp = 109.5°; bp3 = 147-148°; d^{109} = 1.1090; λ_m = 220, 275 nm (MeOH); slightly soluble in H2O, DMSO, very soluble in EtOH, Et2O. *Greeff R.W. & Co.; Hoechst Celanese; Janssen Chimica; Lancaster Synthesis Co.*

1984   p-Hydroxybenzoic acid
99-96-7   4837   H

C7H6O3
Benzoic acid, 4-hydroxy-.
Acido p-idrossibenzoico; AI3-01003; Benzoic acid, 4-hydroxy-; Benzoic acid, p-hydroxy-; EINECS 202-804-9; Kyselina 4-hydroxybenzoova; NSC 4961; p-Hydroxy-benzoic acid; 4-Hydroxybenzoic acid; p-Oxybenzosaeure; p-Oxybenzoesäure; p-Salicylic acid. White crystals; mp = 214.5°; bp = 270-280°; d^{25} = 1.46; λ_m = 254 nm (ε = 15100, MeOH); soluble in H2O (0.25 g/100 ml at 25°, 2 g/100 ml at 50°, 3.3 g/100 ml at 80°), EtOH (50 g/100 ml), 95% EtOH (33 g/100 ml), 50% EtOH (16.6 g/100 ml), Et2O (10 g/100 ml), glycerin (1.6 g/100 ml), propylene glycol (5 g/100 ml), peanut oil (0.5 g/100 ml). LD50 (mus sc) = 1200 mg/kg, (mus ip) = 960 mg/kg, (dog orl) = 3000 mg/kg.

1985   1-Hydroxybenzotriazole
2592-95-2   G

C6H5N3O.3H2O
1-Hydroxy-1H-benzotriazole.
4-26-00-00095 (Beilstein Handbook Reference); Benzazimidol hydrate; 1H-Benzotriazole, 1-hydroxy-; 1H-Benzotriazole, 1-hydroxy-, hydrate; BRN 0004515; CCRIS 2605; EINECS 219-989-7; HOBt; 1-Hydroxybenzotriazole; 1-Hydroxybenzotriazole hydrate; 1-Hydroxy-1H-benzotriazole hydrate; N-Hydroxy-benzotriazole hydrate. Widely used additive to decrease racemization during dicyclohexylcarbodiimide-catalyzed peptide coupling. Crystals; mp = 157-158°. *Janssen Chimica; Lancaster Synthesis Co.; Sigma-Aldrich Fine Chem.*

1986   Hydroxybutylmethylphenylchlorobenzotriazole
3896-11-5   G

C17H18ClN3O
2-(2'-Hydroxy-3'-t-butyl-5'-methylphenyl)-5-chloro-benzotriazole.
Bumetrizol; Bumetrizole; Bumetrizolum; 2-t-Butyl-6-(5-chloro-2H-benzotriazol-2-yl)-p-cresol; 2-(5-Chloro-2H-benzotriazol-2-yl)-6-(1,1-dimethylethyl)-4-methylphenol; EINECS 223-445-4; 2-(2-Hydroxy-3-t-butyl-5-methyl-phenyl)-5-chloro-2H-benzotriazole; 2-(2'-Hydroxy-3'-t-butyl-5'-methylphenyl)-5-chlorobenzo-triazole; Lowitite® 26; Phenol, 2-(5-chloro-2H-benzotriazol-2-yl)-6-(1,1-dimethylethyl)-4-methyl-; Tinuvin; UV Absorber-6; Uvazol 236. UV stabilizer for polyolefins, polyester resins and coatings. Used to protect polyolefins, styrenics, acrylics, PVC, unsaturated polyesters, elastomers, PC, and PU. *EniChem SpA; Lowi.*

1987   4-Hydroxy-3,5-di-t-butylphenylpropionic   acid thioglycolate
41484-35-9   H

C38H58O6S

3,5-Bis(1,1-dimethylethyl)-4-hydroxybenzenepropanoic acid thiodi-2,1-ethanediyl ester.

Benzenepropanoic acid, 3,5-bis(1,1-dimethylethyl)-4-hydroxy-, thiodi-2,1-ethanediyl ester; 3,5-Bis(1,1-dimethylethyl)-4-hydroxybenzenepropanoic acid thiodi-2,1-ethanediyl ester; BRN 2407120; EINECS 255-392-8; Fenozan 30; 4-Hydroxy-3,5-di-t-butylphenylpropionic acid thioglycolate; Irganox 1035; Thiodiethylene bis(3-(3,5-di-t-butyl-4-hydroxyphenyl)propionate); Thiodiethyl-eneglycolbis(3,5-di-t-butyl-4-hydroxyhydrocinnamate).

1988   4-Hydroxycoumarin
1076-38-6   G

C9H6O3

4-Hydroxy-2H-1-benzopyran-2-one.

5-18-01-00378 (Beilstein Handbook Reference); AI3-52393; 2H-1-Benzopyran-2-one, 4-hydroxy-; Benzo-tetronic acid; BRN 0129768; Coumarin, 4-hydroxy-; 4-Coumarinol; EINECS 214-060-2; 4-Hydroxycoumarin; NSC 11889. Chemical intermediate. Crystals; mp = 213-214°. Greeff R.W. & Co.; Janssen Chimica; Penta Mfg.

1989   2-Hydroxy-3,3-dimethylbutyric acid
4026-20-4   H

C6H12O3

3,3-Dimethyl-2-hydroxybutyric acid.

Butanoic acid, 2-hydroxy-3,3-dimethyl-; Butyric acid, 2-hydroxy-3,3-dimethyl-; 3,3-Dimethyl-2-hydroxybutyric acid; EINECS 223-698-0; 2-Hydroxy-3,3-dimethylbutyric acid.

1990   Hydroxyethyl acrylate
818-61-1   H

C5H8O3

2-Propenoic acid, 2-hydroxyethyl ester.

4-02-00-01469 (Beilstein Handbook Reference); Acrylic acid, 2-hydroxyethyl ester; 2-(Acryloyloxy)ethanol; Bisomer 2HEA; BRN 0969853; CCRIS 3431; EINECS 212-454-9; Ethylene glycol, acrylate; Ethylene glycol, monoacrylate; HEA; HSDB 1123; Hydroxyethyl acrylate; 2-Hydroxyethyl acrylate; 2-Hydroxyethylester kyseliny akrylove; 2-Propenoic acid, 2-hydroxyethyl ester.

1991   Hydroxyethyl methacrylate
868-77-9   G H

C6H10O3

2-Propenoic acid, 2-methyl-, 2-hydroxyethyl ester.

4-02-00-01530 (Beilstein Handbook Reference); Bisomer 2HEMA; BRN 1071583; CCRIS 6879; EINECS 212-782-2; Ethylene glycol methacrylate; Ethylene glycol, monomethacrylate; Glycol methacrylate; Glycol monomethacrylate; HSDB 5442; Hydroxyethyl methacrylate; 2-Hydroxyethyl methacrylate; Methacrylic acid, 2-hydroxyethyl ester; 2-(Methacryloyloxy)ethanol; Mhoromer; Monomer MG-1; Monomethacrylic ether of ethylene glycol; NSC 24180; 2-Propenoic acid, 2-methyl-, 2-hydroxyethyl ester; Sipomer® HEM-D. A monomer which permits the production of polymers with side chain hydroxyl groups suitable for crosslinking and the production of thermosetting acrylic surface coating adhesives. Use to manufacture acrylic resins and as a binder for nonwoven fabrics, enamels. Liquid; mp = -12°; bp13 = 103°, bp3 = 67°; d^{20} = 1.079; soluble in H2O, organic solvents; LD50 (rat orl) = 5050 mg/kg. BP Chem.; Mitsubishi Gas; Rhône Poulenc Surfactants; Rohm & Haas Co.

1992   Hydroxyethylidene diphosphonic acid, potassium salt
67953-76-8   H

C2H4K4O7P2

Phosphonic acid, (1-hydroxyethylidene) bis-, potassium salt.

EINECS 267-956-0; Hydroxyethylidene diphosphonic acid, potassium salt; (1-Hydroxyethylidene)bisphosphonic acid, potassium salt; 1-Hydroxylethanediphosphonic acid, potassium salt; Phosphonic acid, (1-hydroxyethylidene) bis-, potassium salt.

1993   Hydroxyethylidene diphosphonic acid, sodium salt
29329-71-3   H

C2H7NaO7P2

Phosphonic acid, (1-hydroxyethylidene)bis-, sodium salt.

EINECS 249-559-4; (1-Hydroxyethylidene)bisphosphonic acid, sodium salt; 1-Hydroxyethanediphosphonic acid, sodium salt; Hydroxyethylidene diphosphonic acid, sodium salt; Phosphonic acid, (1-hydroxyethylidene)di-, sodium salt.

1994   Hydroxyethylpiperazine
103-76-4   H

C6H14N2O

4-(2-Hydroxyethyl)piperazine.

5-23-01-00406 (Beilstein Handbook Reference); AI3-25357; BRN 0104361; CCRIS 6687; EINECS 203-142-3; Ethanol, 2-(1-

piperazinyl)-; Hydroxyethylpiperazine; N-(β-Hydroxyethyl)piperazine; N-(2-Hydroxyethyl)piper-azine; NSC 26884; 1-Piperazinethanol; 2-(1-Piperaz-inyl)ethanol; USAF DO-22. Liquid; bp = 246°; d^{25} = 1.061.

1995 (Hydroxyimino)cyclohexane
100-64-1   H

C6H11NO
Cyclohexanone, oxime.
4-07-00-00021 (Beilstein Handbook Reference); AI3-07288; Antioxidant D; BRN 1616769; CCRIS 1383; Cyclohexanone oxime; EINECS 202-874-0; HSDB 5337; (Hydroxyimino)cyclohexane; NSC 6300. Prisms; mp = 90°; bp = 206°; λ_m = 192, 310 nm (ε = 6918, 15, isooctane); slightly soluble in CHCl3, soluble in H2O, EtOH, MeOH, Et2O.

1996 Hydroxylamine
7803-49-8   4853   G

$$H_2N-OH$$

H3NO
Hydroxylamine.
EINECS 232-259-2; HSDB 579; Hydroxylamine; Oxammonium. White needles; mp = 33°; bp22 = 58°; d^0 = 1.2255; very soluble in H2O, EtOH; LD50 (mus ip) = 60 mg/kg.

1997 Hydroxymethylacrylamide
924-42-5   G H

C4H7NO2
N-(Hydroxymethyl)-2-propenamide.
4-02-00-01472 (Beilstein Handbook Reference); Acrylamide, N-(hydroxymethyl)-; AI3-25447; BRN 0506646; CCRIS 2380; EINECS 213-103-2; HSDB 4361; Hydroxymethylacrylamide; Methylolacrylamide; Monomethylolacrylamide; NCI-C60333; NM-AMD; NSC 553; 2-Propenamide, N-(hydroxymethyl)-; Uramine T 80. A cross-linking agent used in adhesives, binders for paper, crease-resistant textiles, resins, latex film, and sizing agents. Solid; mp= 74-75°; insoluble in H2O, soluble in organic solvents. Am. Cyanamid.

1998 Hydroxymethyl-1-aza-3,7-dioxabicyclo-[3.3.0]octane
6542-37-6   G P

C6H11NO3
1-Aza-3,7-dioxa-5-hydroxymethylbicyclo(3.3.0)octane.
4-27-00-06389 (Beilstein Handbook Reference); 1-Aza-3,7-dioxabicyclo(3.3.0)oct-5-ylmethanol; 1-Aza-5-meth-ylol-3,7-dioxabicyclo(3.3.0)octane; 1-Aza-5-hydroxy-methyl-3,7-dioxabicyclo(3.3.0)octane; M3; Zoldine® ZT-55; Bonding agent M 3; BRN 0107344; Caswell No. 495AB; EPA Pesticide Chemical Code 107002; GDUE; Hydroxymethyl dioxoazabicyclooctane; 5-(Hydroxy-methyl)-1-aza-3,7-dioxabicyclo(3.3.0)octane; 7-Hydroxy-methyl-1,5-dioxo-3-aza-bicyclooctane; M 3; M 3 (curing agent); M 3 (heterocycle); 5-Methylol-1-aza-3,7-dioxa-bicyclo(3.3.0)octane; NSC 270787; Nuosept 95; Oxazolidine T; Oxazolo(3,4-c)oxazole-7a(7H)-methanol; 1H,3H,5H-Oxazolo(3,4-c)oxazole-7a,(7H)-methanol; Zoldine ZT 100; Zoldine ZT 40; Zoldine ZT 55; Zoldine ZT 65. Cross-linking agent for resorcinol phenol-formaldehyde or protein-based resin systems; raw material for synthesis; used in hair care products. Used as an antibacterial preservative in coatings including those that contact food. Registered by EPA as an antimicrobial and disinfectant. Angus; Creanova Inc.

1999 2-Hydroxy-5-methyl-2'-nitroazobenzene
1435-71-8   H

C13H11N3O3
2-((o-Nitrophenyl)azo)-p-cresol.
EINECS 215-863-0; p-Cresol, 2-((o-nitrophenyl)azo)-; 2-Hydroxy-5-methyl-2'-nitroazobenzene; 2-((o-Nitro-phen-yl)azo)-p-cresol; Phenol, 4-methyl-2-((2-nitro-phenyl)-azo)-.

2000 Hydroxymethyl-2-nitro-1,3-propanediol
126-11-4   9823   G P

C4H9NO5
1,1,1-Tris(hydroxymethyl)nitromethane.
AI3-02259; Caswell No. 495; Cimcool wafers; EINECS 204-769-5; EPA Pesticide Chemical Code 083902; Isobutylglycerol, nitro-; Methane, trimethylolnitro-; Nitroisobutylglycerol; Nitromethylidynetrimethanol; Nitr-otris(hydroxymethyl)methane; NSC 17675; S.S.T® Sump Saver Tablets; Trihydroxymethylnitromethane; Trimethyl-olnitromethane; Tris-Nitro®; Tris(hydroxy-methyl)nitro-methane. Antibacterial agent, preservative for water treatment, metalworking fluids, oil production, de-odorizing; formaldehyde releaser. Used as a bactericide in pesticide products and consumer products, building materials or furnishings. Registered by EPA as an antimicrobial and disinfectant. Crystals; mp = 165°; bp dec; very soluble in H2O (220 g/100 ml), EtOH, Et2O. Dow Chem. U.S.A.; Hess & Clark Inc.; Whittaker Clark & Daniels.

2001 Hydroxymethylphthalimide
118-29-6   H

C9H7NO3

N-(Hydroxymethyl)phthalimide.

5-21-10-00366 (Beilstein Handbook Reference); AI3-28943; BRN 0140946; EINECS 204-241-4; Hydroxy-methylphthalimide; Methanol, phthalimido-; N-(Hydroxymethyl)phthalimide; N-Methylolphthalimide; NSC 27350; Oxymethyl phthalimide; Phthalimide, N-(hydroxymethyl)-; Phthalimidomethyl alcohol.

2002   (Hydroxymethyl)urea
1000-82-4   H

C2H6N2O2

N-(Hydroxymethyl)urea.

EINECS 213-674-8; HSDB 5776; 1-(Hydroxymethyl)urea; Methylol urea; Methylolurea; Methylolureas; Mono-(hydroxymethyl)urea; Monomethylolurea; N-(Hydroxy-methyl)urea; N-Methylolurea; NSC 13181; Urea, (hydroxymethyl)-. Used in manufacture of urea-formaldehyde resins. Colorless crystals; mp = 111°; soluble H2O, EtOH, MeOH, AcOH, insoluble in Et2O.

2003   3-Hydroxy-2-naphthoic acid
92-70-6   4860   H

C11H8O3

2-Naphthalenecarboxylic acid, 3-hydroxy-.

4-10-00-01184 (Beilstein Handbook Reference); AI3-00894; B.o.n. acid; BON; BON acid; BONA; BRN 0744100; CI Developer 8; CI Developer 20; CCRIS 2298; Developer 8; Developer BON; EINECS 202-180-8; HSDB 5278; 2-Hydroxy-3-naphthalenecarboxylic acid; 2-Hydroxy-3-naphthoic acid; Kyselina 3-hydroxy-2-naftoova; Miketazol Developer ONS; β-Naphthoic acid, 3-hydroxy-; Naphthol B.O.N.; Naphthol bon; NSC 3719; β-Oxynaphthoic acid. Needles; mp = 222.5°; λm = 235, 284 nm (MeOH); slightly soluble in H2O, readily soluble in EtOH, Et2O, C6H6, CHCl3, C7H8.

2004   Hydroxypivalaldehyde
597-31-9   H

C5H10O2

3-Hydroxy-2,2-dimethylpropionaldehyde.

2,2-Dimethyl-3-hydroxypropanal; EINECS 209-895-4; HSDB 5711; Hydracrylaldehyde, 2,2-dimethyl; 3-Hydroxy-2,2-dimethyl-propionaldehyde; 3-Hydroxy-2,2-dimethylpropanal; Hydroxypivalaldehyde; 3-Hydroxy-pivalaldehyde; Pent-aldol; Propanal, 2,2-dimethyl-3-hydroxy-; Propanal, 3-hydroxy-2,2-dimethyl-; Propion-aldehyde, 3-hydroxy-2,2-dimethyl-.

2005   Hydroxypivalyl hydroxypivalate
1115-20-4   H

C10H20O4

2,2-dimethyl-, 3-hydroxy-2,2-dimethylpropyl hydracryl-ate.

2,2-Dimethyl-, 3-hydroxy-2,2-dimethylpropyl hydracryl-ate; EINECS 214-222-2; Esterdiol 204; HSDB 5783; Hydracrylic acid, 2,2-dimethyl-, 3-hydroxy-2,2-dimethyl-propyl ester; Hydroxyneopentyl hydroxypivalate; Hydroxypivalic acid neopentyl glycol ester; Hydroxypivalyl hydroxypivalate; Neopentyl glycol monohydroxypivalate.

2006   Hydroxypropyl acrylate
25584-83-2   H

C6H10O3

Acrylic acid, monoester with propane-1,2-diol.

Acrylic acid, hydroxypropyl ester; Acrylic acid 2-hydroxypropyl ester; Acrylic acid, monoester with 1,2-propanediol; CCRIS 7745; EINECS 247-118-0; HSDB 596; Hydroxypropyl acrylate; 2-Propenoic acid, monoester with 1,2-propanediol; Propylene glycol acrylate; Propylene glycol monoacrylate.

2007   Hydroxypropyl methacrylate
27813-02-1   G H

C7H12O3

2-Propenoic acid, 2-methyl-, 2-hydroxymethylethyl ester.

Bisomer 2HPMA; EINECS 248-666-3; HSDB 597; Hydroxypropyl methacrylate; 2-Hydroxypropyl meth-acrylate; Methacrylic acid, ester with 1,2-propanediol; Methacrylic acid, monoester with propane-1,2-diol; 1,2-Propanediol, monomethacrylate; 2-Prop-enoic acid, 2-methyl-, monoester with 1,2-propanediol; 2-Propenoic acid, 2-methyl-, 2-hydroxymethylethyl ester; Propylene glycol monomethacrylate; Rocryl 410. Mixture of isomers. Monomer for cross-linkable acrylic resins, nonwoven fabric binders, detergent lube oil additives. BP Chem.; Dajac Labs.; Rohm & Haas Co.

2008   1-Hydroxy-2-pyridine
142-08-5   G

C5H5NO

2-Hydroxypyridine.

AI3-19236; EINECS 205-520-3; 2-Hydroxypyridine; NSC 172522; 2-Oxopyridine; 2-Pyridinol; α-Pyridone; 2-Pyridinone; 2(1H)-Pyridinone; Pyridone-2; 2-Pyridone; 2(1H)-Pyridone. Used in the preparation of 1-alkenyl-2-pyridones; useful in Diels-Alder reactions. Needles; mp = 107.8°; bp = 280°; d20 = 1.3910; λm = 227, 298 nm (MeOH); slightly soluble in Et2O, DMSO, ligroin, soluble in H2O,

EtOH, C6H6, CHCl3.

2009 Hydroxystearic acid
106-14-9 H

C18H36O3
12-Hydroxyoctadecanoic acid.
4-03-00-00942 (Beilstein Handbook Reference); AI3-19730; Barolub FTO; BRN 1726730; Cerit Fac 3; Ceroxin GL; DL-12-Hydroxystearic acid; EINECS 203-366-1; EINECS 253-004-1; Harwax A; HSDB 5368; Hydrofol acid 200; 12-Hydroxyoctadecanoic acid; Hydroxystearic acid; 12-Hydroxystearic acid; KOW; Loxiol G 21; NSC 2385. A lubricant used in plastics processing.

2010 Hydroxyzine
68-88-2 4875 D

C21H277ClN2O2
2-[2-[4-(p-Chloroα-phenylbenzyl)-1-piperazinyl]ethoxy]-ethanol.
5-23-01-00462 (Beilstein Handbook Reference); Atara; Atarax; Atarax base; Ataraxoid; Atarazoid; Atarox; Atazina; Aterax; BRN 0321392; Deinait; EINECS 200-693-1; Equipoise; Fenarol; Hidroxizina; HSDB 3098; Hychotine; Hydroksyzyny; Hydroxine; Hydroxizine; Hydroxizinum; Hydroxycine; Hydroxyzin; Hydroxyzine; Hydroxyzinum; Hydroxyzyne; Idrossizina; Neo-calma; Neurozina; Nevrolaks; NP 212; NSC 169188; Pamazone; Parenteral; Paxistil; Placidol; Plaxidol; Tran-Q; Tranquizine; Traquizine; UCB-4492; Vesparaz-wirkstoff; Vistaril. Anxiolytic; antihistaminic. Has been used as a minor tranquillizer with doses of 75-150 mg/day. Pfizer Inc.

2011 Hydroxyzyne Hydrochloride
2192-20-3 4875 D

C21H29Cl3N2O2
2-[2-[4-(p-Chloroα-phenylbenzyl)-1-piperazinyl]ethoxy]ethanol dihydrochloride.
AI3-50162; Alamon; Atarax; Atarax dihydrochloride; Ataraxoid dihydrochloride; Aterax; Aterax dihydro-chloride; Durrax; EINECS 218-586-3; Hydroxizine dihydrochloride; Hydroxizine hydrochloride; Marax; Neurolax; Orgatrax; Quiess; QYS; Tran-Q dihydro-chloride; Tranquizine dihydrochloride; Vistaril Parenteral; Vistaril steraject. A H1 receptor antagonist. Anxiolytic; antihistaminic. Has been used as a minor tranquillizer. Crystals; mp = 193°; soluble in H2O (<70 g/100 ml), CHCl3 (6 g/100 ml), Me2CO (0.2 g/100 ml), Et2O (< 0.01 g/100

ml); LD50 (rat ip) = 126 mg/kg, (rat orl) = 950 mg/kg. Elkins-Sinn; Forest Pharm. Inc.; KV Pharm.; Pfizer Inc.; Roerig Div. Pfizer Pharm.

2012 Hymexazol
10004-44-1 4882 G P

C4H5NO2
5-Methyl-3(2H)-isoxazolone.
Bucid; Bucide; Butsid; EINECS 233-000-6; F-319; Hydroxyisoxazole; Hymexazol; Hymexazole; Isoxazole, 3-hydroxy-5-methyl-; Itachigarden; NSC 217971; RTY-319; SF-6505; Tachigaren; Tachigaren 70. Fungicide for pelleting sugar beet seed. Registered by EPA as a fungicide and herbicide. White needles; mp = 84 - 85°, 86 - 87°; soluble in H2O (8.5 g/100 ml at 25°), soluble in polar organic solvents; LD50 (mmus orl) = 2148 mg/kg, (fmus orl) = 1968 mg/kg, (mrat orl) = 4678 mg/kg, (frat orl) = 3909 mg/kg, (mmus sc) = 1297, (fmus sc) = 1167 mg/kg, (mrat sc) = 1924 mg/kg, (frat sc) = 1884 mg/kg, (mmus iv) = 445 mg/kg, (fmus iv) = 514 mg/kg, (mrat iv) >1000 mg/kg, (frat iv) > 1000 mg/kg, (rat der) > 10000 mg/kg, (rbt der) > 2000 mg/kg, (m Japanese quail orl) = 1698 mg/kg, (ckn orl) > 1000 mg/kg; LC50 (carp 48 hr.) = 165 mg/l, (Japanese killifish) > 40 mg/l; non-toxic to bees.. Sankyo Co. Ltd.

2013 Hypochlorous acid
7790-92-3 4891 G P

$$HO-Cl$$

ClHO
Hypochlorous acid.
HyPure A. Used for bleaching, in organic synthesis; disinfectant In aqueous solution, decomposes slowly to chlorine, oxygen and perchloric acid. Olin Corp.

2014 Hypoxanthine
68-94-0 4895 G

C5H4N4O
1,7-Dihydro-6H-purin-6-one.
AI3-52242; 1,7-Dihydro-6H-purin-6-one; EINECS 200-697-3; HX; 6-Hydroxypurine; 6-Hydroxy-1H-purine; Hypoxanthine; Hypoxanthine enol; NSC 14665; 6-Oxopurine; 3H-Purin-6-ol; Purin-6-ol; 9H-Purin-6-ol; 6(1H)-Purinone; 9H-Purin-6(1H)-one; Purin-6(1H)-one; Purin-6(3H)-one; Sarcine; Sarkin; Sarkine. Used in biochemical research. Crystals; dec 150°; λm = 249 nm (ε 11000 MeOH); slightly soluble in H2O, more soluble in acid or alkali, poorly soluble in organic solvents. Lancaster Synthesis Co.; Sigma-Aldrich Fine Chem.

2015 Ibuprofen
15687-27-1 4906 D

C13H18O2

α-2-(p-Isobutylphenyl)propionic acid.

Act-3; Adex 200; Adran; Advil; Alaxan; Algofen; Am-Fam 400; Amersol; Anafen; Anco; Andran; Anflagen; Antagil; Antalfene; Antarene; Antiflam; Apo-Ibuprofen; Apsifen; Artofen; Artril; Atril 300; Balkaprofen; Bayer Select Pain Relief; Betaprofen; Bloom; Bluton; BRN 2049713; Brofen; Brufanic; Brufen; Brufen Retard; Bruflam; Brufort; Buburone; Bufeno; Bufigen; Bukrefen; Bupron; Buracaps; Burana; Butacortelone; Butylenin; Carol; CCRIS 3223; Cesra; Citalgan; Cobo; Codral Period Pain; Combiflam; Cunil; Daiprophen; Dalsy; Dansida; Deep Relief; Dentigoa; Dibufen; Dignoflex; Dolgit; Dolibu; Dolmaral; DOLO PUREN; Dolocyl; Dolofen; Dolofin; Dolofort; Dologel; Dolomax; Doloren; Dolormin; Doltibil; Dolven; Donjust B; Dorival; DRG-0069; Drin; Dristan Sinus; Duafen; Dularbuprofen; Duobrus; Dura-Ibu; Dysdolen; Easifon; Ebufac; EINECS 239-784-6; Emflam; Emflam-200; Epobron; Eputex; Ergix; Esprenit; Exneural; Faspic; Femafen; Femapirin; Femidol; Fenbid; Fendol; Fenspan; Fibraflex; Gelufene; Gofen; Grefen; Gynofug; Haltran; Hemagene Tailleur; HSDB 3099; IB-100; Ibol; Ibren; Ibu; Ibu-slo; Ibu-slow; Ibu-Tab; Ibubeta; Ibubest; Ibucasen; Ibudol; Ibudolor; Ibufen; Ibuflamar; Ibufug; Ibugel; Ibugen; Ibugesic; Ibuhexal; Ibulagic; Ibular; Ibulav; Ibuleve; Ibulgan; Ibumed; Ibumerck; Ibumetin; Ibupirac; Ibuprin; Ibuprocin; Ibuprofen; (-)-Ibuprofen; (RS)-Ibuprofen; (±)-Ibuprofen; Ibuprofene; Ibuprofeno; Ibuprofenum; Ibuprohm; Ibusal; Ifen; Inabrin; Inflam; IP-82; Ipren; Irfen; Isodol; Jenaprofen; Junifen; Kesan; Kontagripp Mono; Kratalgin; Lamidon; Librofem; Lidifen; Liptan; Lopane; Malafene; Manypren; Medipren; Melfen; Mensoton; Midol 200; Midol IB Cramp Relief; Moment; Motrin; Mynosedin; Nagifen-D; Napacetin; Narfen; Neo-Helvagit; Neo-Mindol; Neobrufen; Nerofen; Noalgil; Nobafon; Nobfelon; Nobfen; Nobgen; Noritis; Norton; Novadol; Novo Dioxadol; Novo-Profen; Novogent; Novoprofen; NSC 256857; Nuprilan; Nuprin; Nurofen; Optifen; Opturem; Oralfene; Ostarin; Ostofen; Ozonol; Paduden; Panafen; Pantrop; Paxofen; PediaProfen; Perofen; Proartinal; Profen; Proflex; Provon; Quadrax; Rafen; Ranofen; RD 13621; Rebugen; Relcofen; Rhinadvil; Rofen; Roidenin; Rufen; Rufin; Rupan; Sadefen; Salivia; Schmerz-Dolgit; Sednafen; Seklodin; Seskafen; Sine-Aid IB Caplets; Siyafen; Solpaflex; Solufen; Stelar; Sugafen; Suprafen; Suspren; Syntofene; Tabalon; Tabalon 400; Tatanal; Tempil; Tofen; Togal N; Tonal; Trendar; U-18,573; UCB 79171; Unipron; Upfen; Uprofen; Urem; VUFB 9649; Zafen; Zofen. Anti-inflammatory. White crystals; mp = 75-77°; fairly insoluble in H2O; soluble in most organic solvents; LD50 (mus ip) = 495 mg/kg, (mus orl) = 1255 mg/kg. *Boots Pharmaceuticals Inc.; Bristol-Myers Squibb Co.; Ciba-Geigy Corp.; McNeil Pharm.; Sterling Health U.S.A.; Upjohn Ltd.; Whitehall Labs. Inc.*

2016   Ibuprofen aluminum
61054-06-6   D

C26H35AlO5

(±)-Hydroxybis(p-isobutylhydratropato)aluminum.
Ibuprofen aluminiumhydroxid; Ibuprofen aluminum; Motrin-A; U 18,573G. Anti-inflammatory. *Upjohn Ltd.*

2017   Ibuprofen lysinate
57469-77-9   D

C19H32N2O4

(±)-2-(p-Isobutylphenyl)propionate lysine salt.
EINECS 260-751-7; MK-223; Saren; Solprofen; Solufen; Solufenum; Soluphene. Anti-inflammatory. *Boots Pharmaceuticals Inc.*

2018   Ibuprofen methylglucamine Salt
135861-34-6   D

C13H18O2

(±)-2-(p-Isobutylphenyl)propionate methylglucamine salt.
Artrene. Anti-inflammatory. *Boots Pharmaceuticals Inc.*

2019   Ibuprofen piconol
64622-45-3   D

C19H23NO2

2-Pyridylmethyl (±)-p-isobutylhydratropate.
Be-100; EINECS 264-979-8; Ibuprofen piconol; Pimeprofen; Staderm; U-18573G; U 75630. Anti-inflammatory (topical). *Hisamitsu Pharm. Co. Ltd.; Upjohn Ltd.*

2020   Idoxuridine
54-42-2   4916   G

C9H11IN2O5

2'-Deoxy-5-iodouridine.
4-24-00-01235 (Beilstein Handbook Reference); AI3-50861; Allergan 211; BRN 0030397; CCRIS 2827; Dendrid; 1β-D-2'-Deoxyribofuranosyl-5-iodouracil; 2'-Deoxy-5-iodouridine; EINECS 200-207-8; Emanil; Herpes-Gel; Herpesil; Herpidu; Herplex; Herplex liquifilm; Idexur; Idossuridina; Idoxene; Idoxuridin; Idoxuridina; Idoxuridine; Idoxuridinum; Idu Oculos; IDU; Iducher; Idulea; Iduoculos; IDUR; Iduridin; Iduviran; Iododeoxyridine; Iododuridine; 5-

Iodo-2'-deoxyuridine; 5-Iododeoxyuridine; 5-Iodouracil deoxyriboside; 5-IUDR; 5IUDR; IUDR; Joddeoxiuridin; Kerecid; NSC 39661; Ophthalmadine; SK&F 14287; Spectanefran; Stoxil; Synmiol; Uracil, 5-iodo-1-(2-deoxy-β-D-ribofuranosyl)-; Uridine, 2'-deoxy-5-iodo-; Uridine, 5-iodo-2'-deoxy-; Virudox. A proprietary preparation of idoxuridine; an ocular antiseptic and antiviral agent. Crystals; dec 160°(dec) also 190-195°, 240° and > 175°; λ_m = 288nm (log ε 3.87); $[\alpha]_D^{25}$=.4° (c = 0.108 in water). *SmithKline Beecham Pharm.*

2021 Imazalil

35554-44-0   3616   G P V

C14H14Cl2N2O

(±)-1-[2-(2,4-Dichlorophenyl)-2-(2-propenyloxy)ethyl]-1H-imidazole. 5-23-04-00320 (Beilstein Handbook Reference); Allyl-1-(2,4-dichlorophenyl)-2-imidazol-1-ylethyl ether; (±)-1-(-(Allyloxy)-2,4-dichlorophenethyl)imidazole; 1-(2-(Allyl-oxy)-2-(2,4-dichlorophenyl)ethyl)-1H-imidazole; BRN 0545683; Bromazil; Caswell No. 497AB; CGA 41333; Chloramizol; Clinafarm; Deccozil; Deccozil S 75; (±)-1-(2-(2,4-dichlorophenylethyl)-2-(2-propenyl-oxy)ethyl)-1H-imidazole; 1-(2-(2,4-Dichlorophenyl)-2-(2-propenyloxy)-ethyl)-1H-imidazole; 1-(2-(2,4-Dichlorphenyl)-2-(2-prop-enyloxy) äthyl)-1H-imidazole; 1-(2-(2,4-Dichlor-phenyl)-2-(2-prop-enyloxy)aethyl)-1H-imidazol; EINECS 252-615-0; Enilconazole; Eniloconazol; EPA Pesticide Chemical Code 111901; Fecundal; Florasan; Freshgard; Freshguard; Fungaflor; Fungazil; HSDB 6672; Imaverol; Imazalil; 1H-Imidazole, 1-(2-(2,4-dichloro-phenyl)-2-(2-propenyloxy)-ethyl)-, (±)-; R-23979. Ergosterol biosynthesis inhibitor, used as a systemic fungicide with protective and curative action. Used to control fungal diseases in fruit, vegetables and ornamentals. Used to treat veterinary dermato-mycoses; disinfectant; agricultural fungicide. Registered by EPA as a fungicide. Yellow-brown solid; mp = 50°; sg23 = 1.243; soluble in H2O (0.14 g/100 ml), EtOH, MeOH, xylene, C6H6 (all > 50 g/100 ml), i-PrOH, C7H8, C7H16, C6H14, petroleum ether; LD50 (mrat orl) = 320 mg/kg, (dog orl) > 640 mg/kg, (rat der) = 4200 - 4880 mg/kg; LC50 (bobwhite quail 8 day) = 6290 mg/kg, (mallard duck 8 day) = 5620 mg/kg, (rainbow trout) 2.5 mg/l, (bluegill sunfish) = 3.2 mg/l; non-toxic to bees. *DowElanco Ltd.; Gustafson LLC; Hortichem Ltd.; Janssen Chimica; Janssen Pharmaceutical, Belgium; Makhteshim Chemical Works Ltd.; Makhteshim-Agan; Schering-Plough Animal Health; Uniroyal; Wilbur Ellis Co.*

2022 Imazapyr

81334-34-1   4929   G P

C13H15N3O3

2-[4,5-Dihydro-4-methyl-4-(1-methylethyl)-5-oxo-1H-imidazol-2-yl]-3-pyridinecarboxylic acid.

AC 243,997; Arsenal; BRN 5442754; Caswell No. 003F; CL

263,284; 2-(4,5-Dihydro-4-methyl-4-(1-methylethyl)-5-oxo-1H-imidazol-2-yl)-3-pyridinecarboxylic acid; EPA Pesticide Chemical Code 128821; Imazapyr; Imazapyr acid; 2-(4-Isopropyl-4-methyl-5-oxo-2-imidazolin-2-yl)-nicotinic acid; Nicotinic acid, 2-(4-isopropyl-4-methyl-5-oxo-2-imidazolin-2-yl); 3-Pyridinecarboxylic acid,2-(4,5-dihydro-4-methyl-4-(1-methylethyl)-5-oxo-1H-imidazol-2-yl)-. Non-selective systemic herbicide. Inhibits acetohydroxy acid synthase. Used for control of annual and perennial grass and broad-leaved weeds in non-crop areas. Registered by EPA as a herbicide. White solid; mp = 169-173°; soluble in H2O (10-15 g/l), organic solvents; LD50 (rat orl) >5000 mg/kg. *Am. Cyanamid; American Cyanamid, Divn. AHP Corp.; Chipman Ltd.; Cyanamid of Great Britain Ltd.; DowElanco Ltd.*

2023 Imazaquin

81335-37-7   4930   P

C17H17N3O3

2-(4,5-Dihydro-4-methyl-4-(1-methylethyl)-5-oxo-1H-imidazol-2-yl)-3-quinolinecarboxylic acid.

AC 252214; BRN 5450078; Caswell No. 003C; CL 252214; EPA Pesticide Chemical Code 128848; HSDB 6677; Image 1.5LC; Imazaquin; Imazaquin acid; Imazaquine; Scepter; Scepter 1.5L; Scepter Herbicide. Registered by EPA as a herbicide. Crystals; mp = 219 - 224° (dec); soluble in H2O (0.006 - 0.012 g/100 ml), C7H8 (0.04 g/100 ml), DMF (6.8 mg/kg), DMSO (15.9 g/100 ml), CH2Cl2 (1.4 g/100 ml); LD50 (rat orl) > 5000 mg/kg, (fmus orl) = 2363 mg/kg, (rbt der) > 2000 mg/kg, (bobwhite quail, mallard duck orl) > 3160 mg/kg, (bee contact) > 0.1 mg/bee; LC50 (rat ihl) > 5 mg/l air, (bobwhite quail, mallard duck 8 day dietary) > 5000 mg/kg diet, (rainbow trout, bluegill sunfish, channel catfish 96 hr.) > 100 mg/l. *BASF AG; BASF Corp.; BASF Specialty Products.*

2024 Imazaquin ammonium salt

81335-47-9   4930   P

C17H20N4O3

Imazaquin, ammonium salt.

Ammonium salt of imazaquin; 3-(4,5-Dihydro-4-methyl-4-(1-methylethyl)-5-oxo-1H-imidazol-2-yl)-3-quinolinecarboxylic; acid monoammonium salt; Image; Imazaquin ammonium; Imazaquin, monoammonium salt; imazquin; 3-Quinolinecarboxylic; acid, 2-(4,5-dihydro-4-methyl-4-(1-methylethyl)-5-oxo-1H-imidazol-2-yl)-, mono-noammonium salt; Scepter; Scepter 1.5. Registered by EPA as a herbicide. Solid; soluble in H2O. *Ambrands; BASF Corp.*

2025 Imazaquin sodium salt

81335-46-8   4930   P

C17H16N3NaO3
Imazaquin, sodium salt.
Registered by EPA as a herbicide. *BASF Corp.*

2026 Imidacloprid
105827-78-9   4933   G P

C9H10ClN5O2
1-((6-Chloro-3-pyridinyl)methyl)-4,5-dihydro-N-nitro-1H-imidazol-2-amine.
[138261-41-3]; Admire; BAY-NTN 33893; 1-((6-chloro-3-pyridinyl)methyl)-4,5-dihydro-N-nitro-1H-imidazol-2-amine; 1-((6-Chloro-3-pyridyl)methyl)-N-nitro-2-imidazo-lidinimine; Confidor; Confidor 200 SL; CP 1; Gaucho; Imazethapyr; Imidacloprid; 1H-Imidazol-2-amine, 4,5-dihydro-1-((6-chloro-3-pyridinyl)methyl)-N-nitro-; 2-Imid-azolidinimine, 1-((6-chloro-3-pyridinyl)-methyl)-N-nitro-; Merit; Merit (insecticide); NTN 33823; NTN 33893; Provado. Systemic insecticide with contact and stomach action; applied as a foliar or soil treatment especially against sucking pests (virus vectors), e.g., aphids, thrips, whiteflies and leafhoppers on rice, potatoes, vegetables, cotton, tobacco, citrus, pome, stone fruit and other crops; also against some biting insects such as rice water weevil, Colorado potato beetle, wireworm, frit fly, beet fly, onion fly and citrus and apple leaf miners. Insecticide and termiticide. Registered by EPA as an insecticide and rodenticide. Crystals; mp = 136.4°, 143.8°; soluble in H2O (0.05 g/100 ml); LD50 (rat orl) = 450 mg/kg, (rat der) > 5000 mg/kg, (Japanese quail orl) = 31 mg/kg; LC50 (golden orfe 96 hr.) = 237 mg/l; EC50 (*Daphnia* 24, 48 hr.) > 32 mg/l. *Bayer Corp.; Gustafson LLC.*

2027 Imidazole
288-32-4   4935   G

C3H4N2
1,3-diaza-2,4-cyclopentadiene.
5-23-04-00191 (Beilstein Handbook Reference); AI3-24703; BRN 0103853; CCRIS 3345; 1,3-Diaza-2,4-cyclopentadiene; 1,3-Diazole; EINECS 206-019-2; Formamidine, N,N'-vinylene-; Glioksal; Glyoxaline; IMD; Imidazol; Imidazole; 1H-Imidazole; Imutex; Methanimidamide, N,N'-1,2-ethenediyl-; Miazole; NSC 60522; Pyrro(b)monazole; USAF EK-4733. Used for biological control of pests, especially fabric-feeding insects; contact insecticide in an oil spray. Crystals; mp = 90-91°; soluble in H2O, organic solvents; LD50 (mus orl) = 1880 mg/kg. *BASF Corp.; Janssen Chimica; Organon Inc.; Sigma-Aldrich Fine Chem.*

2028 1H-Imidazole-1-ethanol, 4,5-dihydro-, 2-nortall oil
61791-39-7   H

4,5-Dihydro-7-nortall oil-1H-imidazole-1-ethanol.
4,5-Dihydro-7-nortall oil-1H-imidazole-1-ethanol; EINECS 263-171-2; 1-(2-Hydroxyethyl)-2-(tall oil alkyl)-2-imidazoline; 1H-Imidazole-1-ethanol, 4,5-dihydro-, 2-nortall-oil alkyl derivs.; 2-Imidazoline, 1-(2-hydroxyethyl)-2-(tall oil alkyl)-; Miramine TOC; 2-Nortall oil-1H-imidazole-1-ethanol, 4,5-dihydro-; 2-(Tall oil alkyl)-1-(2-hydroxyethyl)-2-imidazoline; Tall oil hydroxy-ethyl imidazoline.

2029 Iminodiacetic acid
142-73-4   4940   H

C4H7NO4
N-(Carboxymethyl)glycine.
4-04-00-02428 (Beilstein Handbook Reference); Acetic acid, 2,2'-iminobis-; Acetic acid, iminodi-; Aminodiacetic acid; BRN 0878499; Diglycin; Diglycine; Diglycocoll; Diglykokoll; EINECS 205-555-4; Glycine, N-(carboxymethyl)-; Hampshire; HSDB 2852; IDA; IDA (chelating agent); IMDA; NSC 18467; USAF DO-55. Prisms; mp = 247.5°; insoluble in EtOH, Et2O, slightly soluble in H2O.

2030 Imperatorin
482-44-0   4949   G

C16H14O4
9-[(3-Methylbut-2-enyl)oxy]-7H-furo[3,2-g][1]benzo-pyran-7-one.
AI3-61725; Ammidin; 5-Benzofuranacrylic acid, 6-hydroxy-7-((3-methyl-2-butenyl)oxy)-, δ-lactone; CCRIS 4346; EINECS 207-581-1; 7H-Furo(3,2-g)(1)benzopyran-7-one, 9-((3-methyl-2-butenyl)oxy)-; HSDB 3497; 6-Hydroxy-7-(3-methyl-2-butenyloxy)-5-benzofuranacrylic acid ω-lactone; 8-Isoamylenoxypsoralen; 8-Isopentenyl-oxypsoralene; Imperatorin; Marmelosin; 9-((3-methyl-2-butenyl)oxy)-9-(3-Methylbut-2-enyloxy)-7H-furo(3,2-g)-chromen-7-one; 9-((3-Methyl-2-butenyl)oxy)-7H-furo(3,2-g)(1)benzopyran-7-one; 9-(3-Methylbut-2-enyloxy)furo-(3,2-g)chromen-7-one; NSC 402949; Pentosalen. Chemical intermediate. mp = 102°; λm = 215, 247, 300 nm (ε = 26303, 22387, 12023, EtOH); slightly soluble in H2O, soluble in EtOH, Et2O, C6H6, petroleum ether, very soluble in CHCl3.

2031 2-Indanone
615-13-4   G

C9H8O
1,3-Dihydro-2H-inden-2-one.
4-07-00-01002 (Beilstein Handbook Reference); AI3-39163; BRN 0636550; EINECS 210-410-3; Indan-2-one; 2-Indanone; 2H-Inden-2-one, 1,3-dihydro-. Chemical intermediate. Needles; mp = 59°; bp =

218°; d[69] = 1.0712; λ_m = 206 nm (ε = 10, cyclohexane); insoluble in H2O, very soluble in EtOH, Et2O, Me2CO, CHCl3. *Lancaster Synthesis Co.; Penta Mfg.; Sigma-Aldrich Fine Chem.*

2032   Indene, 3a,4,7,7a-tetrahydro-
3048-65-5   H

C9H12
3a,4,7,7a-Tetrahydro-°1H-indene.
Bicyclo(4,3,0)nona-3,7-diene; BRN 1902859; EINECS 221-260-3; Indene, 3a,4,7,7a-tetrahydro-; 1H-Indene, 3a,4,7,7a-tetrahydro-; Tetrahydroindene; 3a,4,7,7a-Tetrahydro-1H-indene; 3a,4,7,7a-Tetrahydroindene; 4,7,8,9-Tetrahydroindene.

2033   Indican
487-94-5   4966   G

C14H17NO6
Indol-3-yl potassium sulfate.
1H-Indol-3-ol, hydrogen sulfate (ester); 3-Indoxyl sulfate; 3-Indoxylsulfuric acid; Indican; Indican (metabolic indican); Indican (metabolic indolyl sulfate). Chemical intermediate. Crystals; mp = 179-180° (dec); soluble in H2O, insoluble in organic solvents.

2034   Indigo
482-89-3   4968   G H

C16H10N2O2
2-(1,3-Dihydro-3-oxo-2H-indol-2-ylidene)-1,2-dihydro-3H-indol-3-one.
11669 Blue; 5-24-08-00503 (Beilstein Handbook Reference); AI3-09080; (2,2'-Biindoline)-3,3'-dione; Blue No. 201; BRN 0088275; CCRIS 4379; CI 73000; CI Pigment Blue 66; CI Vat Blue 1; Cystoceva; D & C Blue No. 6; Diindogen; EINECS 207-586-9; HSDB 4372; Indigo; Indigo Blue; Indigo Ciba; Indigo Ciba SL; Indigo J; Indigo N; Indigo NAC; Indigo nacco; Indigo P; Indigo PLN; Indigo powder W; Indigo Pure BASF; Indigo Pure BASF Powder K; Indigo synthetic; Indigo VS; Indigotin; Lithosol Deep Blue B; Lithosol Deep Blue V; Mitsui indigo paste; Mitsui Indigo Pure; Modr Kypova 1; Monolite Fast Navy Blue BV; NCI-C61392; NSC 8645; Pigment Blue 66; Synthetic indigo; Synthetic indigo TS; Vat Blue 1; Vulcafix Blue R; Vulcafor Blue A; Vulcanosine Dark Blue L; Vulcol Fast Blue GL; Vynamon Blue A. Natural indigo is obtained by steeping the leaves of indigo bearing plants in water then oxidizing the extract. Synthetic indigo is prepared by several methods; used for cotton, wool, silk, by steeping the material in a vat containing the leuco compound. Dark blue powder, mp = 290° (dec); sublimes 300°; λ_m = 600 nm (CHCl3), 607 nm (EtOH); insoluble in water, EtOH, Et2O, acids, soluble in nonpolar solvents with red and in polar solvents with blue color.

2035   Indigo Carmine
860-22-0   4969   G

C16H8N2Na2O8S2
Indigotin disulfonate sodium.
1311 Blue; 12070 Blue; 1H-Indole-5-sulfonic acid, 2-(1,3-dihydro-3-oxo-5-sulfo-2H-indol-2-; A.F. Blue No. 2; Acid Blue 74; Acid Blue W; Acid Leather Blue IC; AI3-09087; Airedale Blue IN; Amacid Brilliant Blue; Aniline Carmine powder; Atul Indigo Carmine; $\Delta^{2,2'}$-Biindoline)-5,5'-disulfonic acid, 3,3'-dioxo-, disodium salt; Bucacid indigotine B; C.I. 73015; C.I. 75781; C.I. Acid Blue 74; C.I. Food Blue 1, disodium salt; C.I. Food Blue 1; C.I. Natural Blue 2; Canacert Indigo Carmine; Carmine Blue (biological stain); CCRIS 1865; CI 73015; Cilefa Blue R; Dinatrium 3,3'-dioxo-(delta2,2'-biindolin)-5,5'-disulfonat; Dinatrium indigolin-5,5'-disulfonat; Disodium 3,3'-dioxo-($\Delta^{2,2'}$-biindoline)-5,5'-disulfonate; Disodium 5,5'-disulfoindigo; Disodium 5,5'-indigotin disulfonate; Disodium indigo-5,5-disulfonate; Disodium salt of 1-indigotin-S,S'-disulfonic acid; Disodium salt of 1-indigotin-S,S'-disulphonic acid; Dolkwal Indigo Carmine; E 132; Edicol Supra Blue X; EINECS 212-728-8; FD and C Blue No. 2; FD&C Blue No. 2; Food Blue 1; Food Blue 2; Grape Blue A; Grape Blue A Geigy; HD Indigo Carmine; HD Indigo Carmine Supra; Hexacert Blue No. 2; Hexacol Indigo Carmine Supra; Indigo carmine; Indigo Carmine; Indigo Carmine (biological stain); Indigo Carmine A; Indigo Carmine AC; Indigo Carmine BP; Indigo Carmine Conc. FQ; Indigo Carmine disodium salt; Indigo Carmine powder; Indigo Carmine X; Indigo disulfonate (biological stain); Indigo extract; Indigo-karmin; Indigocarmin; Indigotin-5,5'-disulfonic acid disodium salt; Indigotin I; Indigotindisulfonate Sodium; Indigotine; Indigotine B; Indigotine Blue LZ; Indigotine conc. powder; Indigotine disodium salt; Indigotine Extra Pure A; Indigotine I; Indigotine IA; Indigotine Lake; Indigotine N; Indocarmine F; Intense Blue; L-Blau 2; L Blue Z 5010; Maple Indigo Carmine; Mitsui Indigo Carmine; Modr Kysela 74; Modr Pigment 63; Modr Potravinarska 1; Murabba; NSC 8646; Sachsischblau; San-ei Indigo Carmine; Schultz Nr. 1309; Sodium 5,5'-indigotindisulfonate; Sodium indigo-5,5'-bisulfonate; Sodium indigotindisulfonate; Soluble indigo; Sumitomo Wool Blue SBC; Usacert Blue No. 2. Aid in diagnosis of kidney function. Also used as a dyestuff. Crystals; Soluble in H2O (1 g/100 ml); slightly soluble in EtOH, insoluble in organic solvents. *Hynson Westcott & Dunning; Mitsui Toatsu.*

2036   Indium
7440-74-6   4971   G

In

In
Indium.
EINECS 231-180-0; HSDB 6972; Indium. Metallic element,; automobile bearings, electronic and semiconductor devices, brazing and soldering alloys, reactor control rods, electroplated coatings on aircraft bearings. Metal; mp = 155°; bp = 2000°; d = 7.3; specific heat = 0.0568 cal/g/°. *Atomergic Chemetals; Cerac; Noah Chem.; Sigma-Aldrich Fine Chem.*

2037   Indium oxide
1312-43-2   4975   G

In2O3
Indium(III) oxide.
Diindium trioxide; EINECS 215-193-9; India; Indium (3+) oxide; Indium (III) oxide; Indium oxide; Indium oxide (In2O3); Indium

sesquioxide; Indium trioxide. Used in the manufacture of special types of glass. Solid; d = 7.18; volatilizes at 850°; insoluble in water; soluble in hot mineral acids. *Atomergic Chemetals; Cerac; Noah Chem.*

2038   Indol-3-ylacetic acid
87-51-4   4986   P

C10H9NO2
1H-Indole-3-acetic acid.
Acetic acid, indolyl-; AI3-24131; 3-(Carboxymethyl)-indole; CCRIS 1014; EINECS 201-748-2; EPA Pesticide Chemical Code 128915; Heteroauxin; IAA; 3-Iaa; Indole-3-acetic acid; 1H-Indole-3-acetic acid; 3-Indoleacetic acid; Indoleacetic acid; -Indoleacetic acid; Indol-3-ylacetic acid; -Indolylacetic acid; Kyselina 3-indolyloctova; NSC 3787; Rhizopin; Rhizopon A; -Skatole carboxylic acid. A root growth promoter and plant growth regulator. Pale, salmon-colored crystals; mp = 168.5°; pK = 4.75; poorly soluble in H_2O (0.15 g/100 ml), soluble in EtOH (10 - 100 g/100 ml), Me_2CO (3 - 10 g/100 ml), Et_2O (3 - 10 g/100 ml), $CHCl_3$ (1 - 3 g/100 ml); LD_{50} (mus der) = 1000 mg/kg; non-toxic to bees. *ACF; Fargro Ltd.; Generic.*

2039   4-Indol-3-ylbutyric acid
133-32-4   4987   G P

C12H13NO2
4-(3-indole)butyric acid.
5-22-03-00140 (Beilstein Handbook Reference); AI3-17434; BRN 0171120; Butyric acid, 4-(indolyl)-; Caswell No. 499; CCRIS 1020; Chryzoplus, Chryzopon, Chryzosan, Chryzotek; EINECS 205-101-5; EPA Pesticide Chemical Code 046701; Hormex rooting powder; Hormodin; IBA; Indolbutyric acid; Indole-3-butanoic acid; 1H-Indole-3-butanoic acid; Indolebutyric acid; Indole butyric acid; Indole-3-butyric acid; 1H-Indole-3-butyric acid; β-Indolebutyric acid; 3-Indolebutyric acid; 4-(Indol-3-yl)butyric acid; 4-Indol-3-ylbutyric acid; 3-Indolyl-γ-butyric acid; Indol-3,4'-yl butyric acid; Jiffy grow; Kyselina 4-indol-3-ylmaselina; NSC 3130; Rhizopon AA; Seradix; Seradix 2; Seradix 3; Seradix B 2; Seradix B 3; hormodin; IBA. A root growth promoter. Used for promotion and acceleration of root formation in plant clippings. mp = 124.5°; λ_m = 222, 282, 290 nm (ε = 55900, 9140, 7740, MeOH); insoluble in H_2O, petroleum ether, soluble in DMSO, very soluble in c_6H_6; LD_{50} (mus ip) = 100 mg/kg. *Embetec Crop Protection Ltd.; Fargro Ltd.; Penta Mfg.; Pfalz & Bauer; Spectrum Chem. Manufacturing.*

2040   Inositol
87-89-8   5001   D G

C6H12O6
Hexahydroxycyclohexane.
AI3-16111; Bios I; CCRIS 6745; Cyclohexanehexol; Cyclohexitol; Dambose; EINECS 201-781-2; i-Inositol; Inositene; Inositina; Inositol; Insitolum; Isoinositol; Meat sugar; meso-Inositol; Mesoinosit; Mesoinosite; Mesoinositol; Mesol; Mesovit; mouse antialopecia factor; Muscle sugar; myo-Inositol; Myoinosite; Myoinositol; NSC 404118; Nucite; Phaseomannite; Phaseomannitol; rat antispectacled eye factor; Scyllite. A lipotropic agent, used as a nutrient. mp = 225-227°; d = 1.752; soluble in H_2O (14 g/100 ml at 25°, 28 g/100 ml at 60°, slightly soluble in EtOH, insoluble on other organic solvents.

2041   Insulin, human
11061-68-0   5003   D

C257H383N65O77S6
Human insulin.
EINECS 234-279-7; Human insulin; Humulin; Humuline; Insulin; Insulin (Cercopithecus aethiops); Insulin (Macaca fascicularis); Insulin (Macaca mulatta); Insulin (Pan troglodytes); Insulin, human synthetic; Insulina humana; Insuline humaine; Insulinum humanum; Novolin; Novolin R; Penfil R; Ultraphane; Velosulin. The natural antidiabetic principle produced by the human pancreas. Used as an antidiabetic agent. *Christiaens S.A.; Novo Nordisk Pharm. Inc.*

2042   Insulin, isophane
9004-17-5   5003   D

Isophane insulin.
Depo-insulin; Deposulin; Humulin N-U 100; Insulin, depo; Insulin, nph; Insulin, protamine zinc; Insulin retard RI; Insulina isofana; Insulina-zinco protaminato iniettabile; Insulini zinci protaminati injectio; Insulini zinco protamini aquosuspensa; Insulinum isophanum; Insulyl-retard; Inyectable de insulina cinc protamina; Isophane insulin; Isophane insuline; Isophanum insulinum; neutral protein Hagedorn insulin; NPH humulin insulin; NPH Iletin; NPH insulin; Prep. injectable d'insuline zinc protam.; Protamine zinc insulin; Protaphane HM; PZI insulin; Suspension injectable d'insuline protamine zinc; Zinc protamine insulin; Zinc-protamininsulinum. Antidiabetic agent. Crystallized form consisting of protamine, insulin, and zinc. Activity begins 3-4 hours after sc injection; lasts 18-28 hours. pH = 7.1-7.4. *Christiaens S.A.; Novo Nordisk Pharm. Inc.*

2043   Insulin, Lispro
133107-64-9   5008   D

28B-L-Lysine-29B-L-prolineinsulin (human).
Humalog; Insulin lispro; LY 275585. Antidiabetic agent. *Eli Lilly & Co.*

2044   Iodamide meglumine
18656-21-8   5033   G

$C_{19}H_{28}I_3N_3O_9$

3-(Acetylamino)-5-[(acetylamino)methyl]-2,4,6-triiodo-benzoic acid N-methyl-D-glucamine salt.

Conraxin H; 1-Deoxy-1-(methylamino)-D-glucitol α,5-diacetamido-2,4,6-triiodo-m-toluate; α,5-Diacetamido-2,4,6-triiodo-m-toluic acid compound + 1-deoxy-1-(methylamino)-D-glucitol; EINECS 242-480-6; Glucitol, 1-deoxy-1-(methylamino)-,α,5-diacetamido-2,4,6-triiodo-m-toluate (salt), D-; Iodamide 300; Iodamide methylglucamine; Iodamide N-methyl-D-glucamine salt; Isteropac E.R.; Jodamid methylglucaminsalz; Meglumine iodamide; Meglumine Sodium Iodamide Injection; Opacist E.R.; Renovue-65; Renovue-DIP. Radiopaque medium used as a diagnostic aid. Sparingly soluble in H_2O, alcohol, insoluble in other organic solvents; LD50 (rat iv) = 11.4 g/kg. Bristol-Myers Squibb Co.

2045   Iodine
7553-56-2   5036   D G P

I_2

Actomar; AI3-08544; Caswell No. 501; Diiodine; EINECS 231-442-4; EPA Pesticide Chemical Code 046905; Eranol; Ethanolic solution of iodine; HSDB 34; IODE; Iodine; Iodine (resublimed); Iodine-127; Iodine colloidal; Iodine crystals; Iodine solution; Iodine sublimed; Iodine Tincture USP; Iodio; Iosan superdip; Jod; Jood; Molecular iodine; NSC 42355; Tincture iodine; Vistarin. Nonmetallic halogen element; dyes, alkylation and condensation catalyst, iodides, iodates, antiseptics, germicides, x-ray contrast media, food and feed additive, stabilizers, photographic film, water treatment, pharmaceuticals, medicinal soaps. Registered by EPA as an antimicrobial, fungicide and herbicide. BP, Ph.Eur. compliance. Used in medicinal soaps, x-ray contrast media, radioisotopes used in diagnosis and treatment, used in expectorants and thinners, in cough medicines for treatment of asthma, in topical antiseptics and anesthetics. Antihyperthyroid and topical anti-infective. Also used, particularly in veterinary medicine, to treat goiter. CAUTION: Overexposure could cause irritation of eyes and nose, lacrimation, headache, tight chest, skin burns or rash, cutaneous hypersensitivity. Highly corrosive on the GI tract. Metallic crystals; mp = 113.60°; bp = 185.24°; soluble in water, organic solvents. Andeno; Atomergic Chemetals; Cerac; Mallinckrodt Inc.; Sigma-Aldrich Fine Chem.

2046   Iodobenzene
591-50-4   5051   G

C_6H_5I

Phenyl iodide.

AI3-16898; Benzene iodide; Benzene, iodo-; EINECS 209-719-6; Iodinebenzol; Iodobenzene; NSC 9244; Phenyl iodide. Chemical intermediate. Liquid; mp = -31.3°; bp = 188.4°; d^{20} = 1.8308; λ_m = 226, 250 nm (ε = 11700, 732, MeOH); insoluble in H_2O; slightly

soluble in EtOH, freely soluble in Et_2O, Me_2CO, C_6H_6, CCl_4, ligroin. Greeff R.W. & Co.

2047   Iodoform
75-47-8   5055   G

CHI_3

Triiodomethane.

AI3-52396; Carbon triiodide; CCRIS 346; Dezinfekt V; EINECS 200-874-5; HSDB 4099; Iodoform; Jodoform; Methane, triiodo-; NCI-C04568; NSC 26251; Triiodomethane; Trijodmethane. Used as a topical anti-infective and disinfectant. Yellow crystals; mp = 119°; bp = 218°; d^{25} = 4.008; insoluble in H_2O, more soluble in organic solvents; LD50 (rat orl) = 355 mg/kg. Alfa Aesar; Allchem Ind.; Clariant Corp.; Lancaster Synthesis Co.; Molekula Fine Chemicals; Walton Pharm.

2048   Iodol
87-58-1   5063   G

C_4HI_4N

Tetraiodopyrrole.

EINECS 201-754-5; Iodol; Iodopyrrole; 2,3,4,5-Tetraiodopyrrole. Used externally as a disinfectant for superficial lesions. Yellow needles; mp = 150° (dec); soluble in H_2O (2.04 g/100 ml), more soluble in Me_2CO, Et_2O, $CHCl_3$.

2049   Iodopanoic acid
96-83-3   5074   G

$C_{11}H_{12}I_3NO_2$

3-Amino-α-ethyl-2,4,6-triiodobenzenepropanoic acid.

4-14-00-01741 (Beilstein Handbook Reference); Acide iopanoique; Acido iopanoico; Acidum iopanoicum; 3-Amino-α-ethyl-2,4,6-triiodohydrocinnamic acid; 3-Amino-α-ethyl-2,4,6-triiodobenzenepropanoic acid; 2-(3-Amino-2,4,6-triiodobenzyl)butyric acid; β-(3-Amino-2,4,6-triiodophenyl)-α-ethylpropionic acid; 3-(3-Amino-2,4,6-triiodophenyl)-2-ethylpropanoic acid; Benzene-propanoic acid, 3-amino-α-ethyl-2,4,6-triiodo-; Bilijodon; BRN 2220381; Choladine; Cholevid; Cistobil; Colepax; Copanoic; EINECS 202-539-9; 2-Ethyl-3-(3-amino-2,4,6-triiodophenyl)propionic acid; HSDB 3345; Hydro-cinnamic acid, 3-amino-α-ethyl-2,4,6-triiodo-; Iopagnost; Iopanoic acid; Iopanoicum; Jopagnost; Jopanoic acid; NSC 41706; Polognost; Telepaque; Teletrast; WIN 2011. Radiopaque agent, used as a diagnostic aid. (dl) form; Crystals; mp = 155-157°; soluble in H_2O, organic solvents; LD50 (rat orl) = 3870

2050 Iodophthalein
386-17-4   5060   G

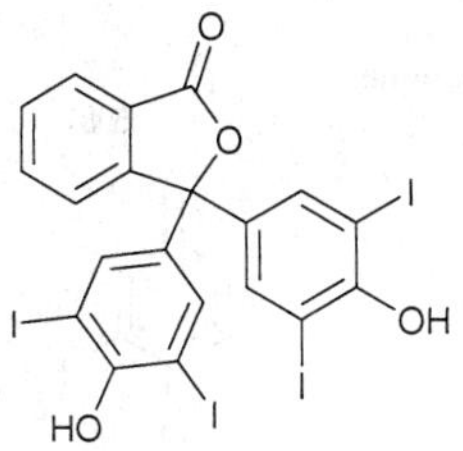

$C_{20}H_{10}I_4O_4$
 Tetraiodophenolphthalein.
 4-18-00-01949 (Beilstein Handbook Reference); 3,3-Bis(4-hydroxy-3,5-diiodophenyl)phthalide; 3,3-Bis(4-hydroxy-3,5-diiodophenyl)-1(3H)-isobenzofuranone; BRN 0351654; EINECS 206-857-9; Iodophene; Iodophthalein sodium free acid; 1(3H)-Isobenzofuranone, 3,3-bis(4-hydroxy-3,5-diiodophenyl)-; Jodphthaleinum; Kerasol; Phenolphthalein, 3',3",5',5"-tetraiodo-; 3',3",5',5"-Tetra-iodophenolphthalein. Amorphous solid; mp = 308°; d^{22} = 2.0201; slightly soluble in $CHCl_3$, insoluble in H_2O, EtOH, Et_2O.

2051 Iodoquinol
83-73-8   5064   D G

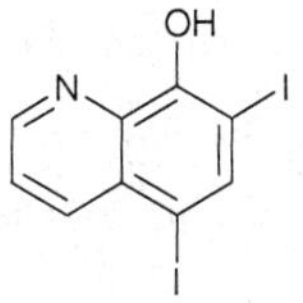

$C_9H_5I_2NO$
 5,7-diiodo-8-quinolinol.
 5-21-03-00296 (Beilstein Handbook Reference); AI3-16443; BRN 0153639; Caswell No. 354; Cor-Tar-Quin; Di-quinol; Diamoebin; Diiodohidroxiquinoleina; Diiodo-hydroxyquin; Diiodohydroxyquinoleine; Diiodohydroxy-quinoline; 5,7-Diiodo-8-hydroxyquinoline; 5,7-Diiodo-oxine; 5,7-Diiodo-8-quinolinol; Diiodohydroxy-quino-linum; Diiodoidrossichinolina; Di-iodoquin; Dijodoxi-chinoline; Dijodoxichinolinum; Di-noleine; Diodohydroxyquin; Diodoquin; Diodoquine; Diodoxylin; Diodoxy-quinoleine; Direxiode; Disoquin; Dyodin; EINECS 201-497-9; Embequin; Enterodiamoebin; Enterosept; EPA Pesticide Chemical Code 024003; Floraquin; Fluoraquin; HSDB 3224; 8-Hydroxy-5,7-diiodoquinoline; Iodoquinol; Ioquin; Ioquin suspension; Lanodoxin; Moebiquin; NSC 8704; Quinadome; 8-Quinolinol, 5,7-diiodo-; Rafamebin; Searlequin; Sebaquin; SS 578; Stanquinate; Vytone; Yodoxin; Zoaquin. Anti-amebic. Yellow needles; mp = 210° (dec); λ_m = 251, 313, 349 nm (ε = 19055, 4467, 4786, i-PrOH); insoluble in H_2O, Et_2O, C_6H_6, $CHCl_3$, ACOH, soluble in EtOH, hot C_5H_5N, dioxane. *Glenwood Inc.; Searle G.D. & Co.*

2052 Iodothymol
552-22-7   G

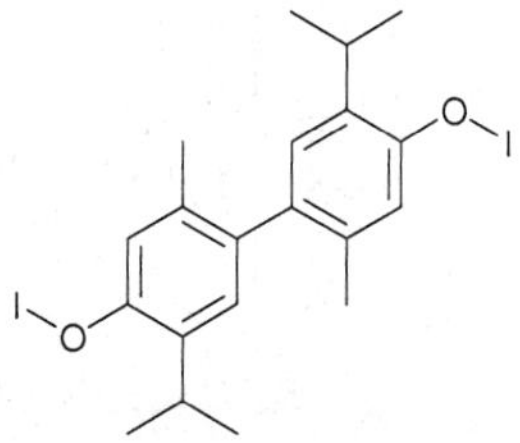

$C_{20}H_{24}I_2O_2$
 Hypoiodous acid, 2,2'-dimethyl-5,5'-bis(1-methylethyl)-[1,1'-biphenyl]-4,4'-diyl ester.
 Annidalin; Aristol; 4,4'-Bis(iodooxy)-2,2'-dimethyl-5,5'-bis(1-methylethyl-1,1'-biphenyl; Bithymol diiodide; Diiododithymol; 5,5'-Diisopropyl-2,2'-dimethylbiphenyl-4,4'-diyl dihypoiodite; Dithymol diiodide; EINECS 209-007-5; HSDB 867; Hypoiodous acid, 2,2'-dimethyl-5,5'-bis(1-methylethyl)(1,1'-biphenyl)-4,4'-diyl ester; Iodistol; Iodohydromol; Iodosol; Iosol; Iothymol; Lothymol; NSC 2222; Thymiode; Thymiodol; Thymodin; Thymol iodide; Thymotol.

2053 Iodotope I-125
7790-26-3   8706   D G

Na—I

^{125}INa
 Sodium iodide ^{125}I.
 Iodotope; Oriodide; Radiocaps-125; Theriodide-125. Diagnostic aid (thyroid function); radioactive agent. *Bristol-Myers Squibb Co.*

2054 Iodotope I-131
7790-26-3   8706   D G

Na—I

^{131}INa
 Sodium iodide ^{131}I.
 Iodotope; Oriodide; Radiocaps-131; Theriodide-131. Antineoplastic; diagnostic aid; radioactive agent. *Bristol-Myers Squibb Co.*

2055 Iohydrin
534-08-7   5083   G

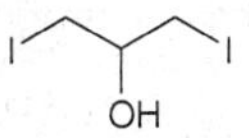

$C_3H_6I_2O$
 1,3-Diiodoisopropyl alcohol.
 3-01-00-01476 (Beilstein Handbook Reference); Agojodo; AI3-61824; BRN 1732080; 1,3-Diiodoisopropyl alcohol; 1,3-Diiodo-2-propanol; 1,3-Diiodopropan-2-ol; Dijodan; EINECS 208-586-1; Glycerol-α,γ-diiodohydrin; Ioprop-ane; Iothion; Iotone; Jothion; 2-Propanol, 1,3-diiodo-. Chemical intermediate. Solid; d = 2.4-2.5; soluble in H_2O (1.25 g/100 ml); more soluble in organic solvents.

2056 Iopamidol
60166-93-0   5073   G

C17H22I3N3O8

(S)-N,N'-Bis[2-hydroxy-1-(hydroxymethyl)ethyl]-5-[(2-hydroxy-1-oxopropyl)amino]-2,4,6-triiodo-1,3-benzene-dicarboxamide.

B 15000; 1,3-Benzenedicarboxamide, N,N'-bis(2-hydroxy-1-(hydroxymethyl)ethyl)-5-((2-hydroxy-1-oxo-propyl)amino)-2,4,6-triiodo-, (S)-; L-(+)-N,N'-Bis(2-hydr-oxy-1-hydroxymethylethyl)-2,4,6-triiodo-5-lactamide iso-phthalamide; (S)-N,N'-bis(2-Hydroxy-1-(hydroxy-meth-yl)ethyl)-2,4,6-triiodo-5-lactamidoisophthalamide; (S)-N, N'-Bis(2-hydroxy-1-(hydroxymethyl)ethyl)-5-[(2-hydroxy-1-oxopropyl)amino)-2,4,6-triiodo-1,3-benzene-dicarbox-diamide; BRN 6250226; EINECS 262-093-6; L-5α-Hydroxypropionyl-amino-2,4,6-triiodoisophthalic acid di(1,3-dihydroxy-2-propylamide); L-5α-Idrossipropionil-amino-2,4,6-triiodo-isoftal-di(1,3-diidrossi-2-propilamide); Iopamidol; Iopam-idol 300; Iopamidolum; Iopamiro; Iopamiro 370; Iopam-iron; Iopamiron 300; Iopamiron 370; Isovue; Isovue-370; Jopamiron 200; Niopam; Niopam 300; Solutrast; Solutrast 370; SQ 13396. Radiopaque medium used as a diagnostic aid. Crystals; dec 300°; $[\alpha]_D^{20}$= -2.01°; soluble in H2O, methanol, insoluble in chloroform; LD50 (rat iv) = 28.2 g/kg. Bristol-Myers Squibb Co.

2057   lophendylate
99-79-6   5076   G

C19H29IO2

Ethyl 10-(p-iodophenyl)undecanoate.

3-09-00-02627 (Beilstein Handbook Reference); Benzenedecanoic acid, 4-iodo-ι-methyl-, ethyl ester; BRN 2661373; EINECS 202-787-8; Ethiodan; Ethyl 10-(p-iodophenyl)undecylate; Ethyl 10-(p-iodophenyl)-hendecanoate; Ethyl iodophenylundecylate; HSDB 3346; Iofendilato; Iofendylate; Iofendylatum; Ioglunide; Iophendylate; Jofendylatum; Mulsopaque; Myodil; Neurotrast; Pantopaque; Undecanoic acid, 10-(p-iodophenyl)-, ethyl ester. Diagnostic aid - X-ray contrast medium for myelography. d_{20}^{20} = 1.240-1.263; insoluble in H2O, soluble in organic solvetns; LD50 (rat ip) = 19 g/kg. Alcon Labs; British Drug Houses.

2058   lopronic acid
41473-08-9   5079   G

C15H18I3NO5

2-[[2-[3-(acetylamino)-2,4,6-triiodophenoxy]ethoxy]-methyl]butanoic acid.

(±)-2-[[2-(3-acetamido)-2,4,6-triiodophenoxy]ethoxy]-methyl]butyric acid; B-11420; Bilimiro; Bilmiron; Iopronic acid; Oravue; SQ-21983; Videobil. Radio-opaque medium used as a diagnostic aid. Solid; LD50 (mus iv) = 32 g/kg, (rat iv) = 23 g/kg. Bracco Diagnostics Inc.; Bristol-Myers Squibb Co.

2059   lopydone
5579-93-1   5081   G

C5H3I2NO

3,5-Diiodo-4(1H)-pyridinone.

5-21-07-00164 (Beilstein Handbook Reference); BRN 1447538; 3,5-Diiodo-4-pyridone; 3,5-Diiodo-4(1H)-pyridone; DJP; EINECS 226-969-1; Hytrast; Iopidona; Iopydon; Iopydone; Iopydonum; Jopydonum; NSC 135284; 4(1H)-Pyridinone, 3,5-diiodo-; 4(1H)-Pyridone, 3,5-diiodo-. Diagnostic aid. A radiopaque medium used in bronchography. Solid; mp = 321° (dec); insoluble in H2O, organic solvents.

2060   lotrolan
79770-24-4   5084   G

C37H48I6N6O18

5,5'-[(1,3-Dioxo-1,3-propanediyl)bis(methylimino)]bis-[N,N'-bis[2,3-dihydroxy-1-(hydroxymethyl)propyl]-2,4,6-triiodo-1,3-benzenedicarboxamide.

1,3-Benzenedicarboxamide, 5,5'-((1,3-dioxo-1,3-prop-anediyl)bis(methylimino)bis(N,N'-bis(2,3-dihydroxy-1-(hydroxymethyl)propyl)-2,4,6-triiodo-; DL 3-117; DL-3117; Iotrol; Iotrolan; Iotrolanum; Iotrolum; Iotrovist; Isovist; 5,5'-(Malonylbis(methylimino))bis(N,N'-bis(2,3-dihydroxy-1-(hydroxymethyl)propyl)-2,4,6-triiodoiso-phthalamide); SH 437; ZK 39482. Radiopaque medium used as a diagnostic aid. Solid; soluble in H2O (> 850 g/l); LD50 (rat iv) = 27.1 g/kg.

2061   loxynil
1689-83-4   H P

C7H3I2NO

4-Hydroxy-3,5-diiodobenzonitrile.

15380 RP; 3-10-00-00373 (Beilstein Handbook Reference); ACP 63303; Actril; Actrilawn; Bantrol; Bentrol; BRN 2364041; CA 69-15; Caswell No. 353A; Certrol; Cipotril; EINECS 216-881-1; EPA Pesticide Chemical Code 353200; HSDB 1584; Iotox; Iotril; Ioxynil;

Joxynil; Loxynil; M&B 11641; Mate; Topper; Totril; Trevespan. Contact herbicide for use in turf, onion crops. Colorless crystals; mp = 212 - 213°; soluble in H_2O (0.005 g/100 ml), Me_2CO (7 g/100 ml), EtOH (2 g/100 ml), MeOH (2 g/100 ml), cyclohexanone (14 g/100 ml), THF (34 g/100 ml), DMF (74 g/100 ml), $CHCl_3$ (1 g/100 ml), CCl_4 (< 0.1 g/100 ml); LD_{50} (rat orl) = 110 mg/kg, (mus orl) = 230 mg/kg, (rat der) > 2000 mg/kg, (pheasant orl) = 75 mg/kg; LC_{50} (harlequin fish 48 hr.) = 3.3 mg/l (sodium salt); non-toxic to bees. *Aventis Crop Science; CFPI Agro SA; Makhteshim-Agan; Rhône-Poulenc.*

2062   Iphaneine
102-13-6   G

$C_{12}H_{16}O_2$
2-Methyl propyl phenyl acetate.
Acetic acid, phenyl-, isobutyl ester; AI3-01969; Benzeneacetic acid, 2-methylpropyl ester; CCRIS 7324; EINECS 203-007-9; FEMA No. 2210; Isobutyl α-toluate; Isobutyl phenylacetate; Isobutyl phenylethanoate; 2-Methylpropyl benzeneacetate; 2-Methylpropyl phenylacetate; NSC 6602; Phenylacetic acid, isobutyl ester. Used in perfumery as a sweet musk chocolate amber scent. Solid; bp = 247°; d^{18} = 0.999; λ_m = 257, 264 nm (MeOH); insoluble in H_2O, soluble in EtOH, Et_2O. *Bush Boake Allen; IFF.*

2063   Ipratropium Bromide
66985-17-9   D

$C_{20}H_{30}BrNO_3 \cdot H_2O$
(8R)-3α-Hydroxy-8-isopropyl-1αH,5αH-tropanium   bromide monohydrate.
Atem; Atrovent; Bitrop; Combivent; Ipratropium bromide; Itrop; Narilet; Rinatec; Sch-1000-Br-monohydrate.; component of: Combivent. An anticholinergic bronchodilator and antiarrhythmic. White solid; mp = 230-232°; soluble in H_2O, MeOH, EtOH; insoluble in Et_2O, $CHCl_3$, fluorohydrocarbons; LD_{50} (mmus orl) = 1001 mg/kg, (mmus iv) = 12.29 mg/kg, (mmus sc) = 300 mg/kg, (fmus orl) = 1083 mg/kg, (fmus iv) = 14.97 mg/kg, (fmus sc) = 340 mg/kg, (mrat orl) = 1663 mg/kg, (mrat iv) = 15.89 mg/kg, (frat orl) = 1779 mg/kg, (frat iv) = 15.70 mg/kg. *Boehringer Ingelheim Ltd.; Schering AG.*

2064   Iprodione
36734-19-7   5096   P

$C_{13}H_{13}Cl_2N_3O_3$
3-(3,5-Dichlorophenyl)-N-(1-methylethyl)-2,4-dioxo-1-imidazolidinecarboxamide.
26019 rp; 330.16; 5-24-05-00201 (Beilstein Handbook Reference); Anfor; BRN 0895003; Caswell No. 470A; Chipco 26019; EINECS 253-178-9; EPA Pesticide Chemical Code 109801; FA 2071; FRP 26019; Glycophen; Glycophene; HSDB 6855; Iprodial; Iprodine; Iprodione; Kidan; LFA 2043; MRC 910; NRC 910; Promidione; ROP 500; ROP 500 F; Rovral; Rovral 50WP; Rovral Flo; Rovral HN; Rovral PM; Rovrol; RP 26019; Verisan. Registered by EPA as a fungicide. Colorless crystals; mp = 136°; soluble in H_2O (0.0013 g/100 ml at 20°), EtOH (2.5 g/100 ml), MeOH (2.5 g/100 ml), Me_2CO (30 g/100 ml), CH_2Cl_2 (50 g/100 ml), DMF (50 g/100 ml); LD_{50} (rat orl) = 4000 mg/kg, (mus orl) = 3500 mg/kg. *Andersons Lawn Fertilizer Div. Inc.; Aventis Crop Science; Micro-Flo Co. LLC; Scotts Co.*

2065   Irbesartan
138402-11-6   5100   D

$C_{25}H_{28}N_6O$
2-Butyl-3-[p-(o-1H-tetrazol-5-ylphenyl)benzyl]-1,3-diazaspiro[4.4]non-1-en-4-one.
Avapro; BMS 186295; Irbesartan; SR 47436. Non-peptidic angiotensin II type-1 receptor antagonist. Used as an antihypertensive. Crystals; mp = 180-181°. *Bristol-Myers Squibb Pharm. R&D; Sanofi Winthrop.*

2066   Irigenin
548-76-5   5107   G

$C_{18}H_{16}O_8$
3',5,7-trihydroxy-4',5',6,-trimethoxyisoflavone.
5,7-Dihydroxy-3-(3-hydroxy-4,5-dimethoxyphenyl)-6-methoxy-4-benzopyrone; EINECS 208-958-3; Irigenin; 3',5,7-Trihydroxy-4',5',6,-trimethoxyisoflavone. The aglycone of iridin. A pigment. Crystals; mp = 185°; λ_m = 267 nm; insoluble in H_2O, soluble in organic solvents.

2067   Iron
7439-89-6   5109   G

341

Fe
Iron.
3ZhP; A 227; Ancor B; Ancor en 80/150; Armco iron; Atomel 28; Atomel 95; Atomel 300M200; Atomel 500M; Atomiron 44MR; Atomiron 5M; Atomiron AFP 5; Atomiron AFP 25; ATW 230; ATW 432; Carbonyl iron; CCRIS 1580; Copy Powder CS 105-175; DSP 1000; DSP 128B; DSP 135; DSP 135C; DSP 138; EF 250; EF 1000; EFV 200/300; EFV 250; EFV 250/400; EINECS 231-096-4; EO 5A; F 60 (metal); Ferrous iron; Ferrovac E; Ferrum; FT 3 (element); GS 6; HF 2 (element); HL (iron); Hoeganaes ATW 230; Hoeganaes EH; HQ (metal); HS (iron); HS 4849; HSDB 604; Iron; Iron, elemental; LOHA; Malleable iron; NC 100; PZh-1M3; PZh-2; PZh1M1; PZh2M; PZh2M1; PZh2M2; PZh3; PZh3M; PZh4M; PZhO; Remko; SUY-B 2; Wrought iron. Used to form steels by alloying with other elements such as C, Ni, Mn, Cr. Radioisotopes of iron are used in biological research and in medicine. Metallic solid; mp = 1535°; bp = 3000°; d = 7.8600; electrical resistivity = 9.71 microhm-cm.

2068 Isatin
91-56-5   5119   G

$C_8H_5NO_2$
1H-Indole-2,3-dione.
5-21-10-00221 (Beilstein Handbook Reference); AI3-03111; o-Aminobenzoylformic anhydride; BRN 0383659; 2,3-Diketoindoline; 2,3-Dioxo-2,3-dihydroindole; 2,3-Dioxoindoline; EINECS 202-077-8; Indole-2,3-dione; Indoline-2,3-dione; 2,3-Indolinedione; 1H-Indole-2,3-dione; Isatic acid lactam; Isatin; Isatine; Isatinic acid anhydride; Isotin; 2,3-Ketoindoline; NSC 9262; Pseudoisatin; Tribulin. Intermediate for production of pharmaceuticals and dyes, stabilizer used in the plastics industry. Analytical reagent for mercaptans, thiophene, indican and Cu^{2+} ions. Orange prisms; mp = 203° (dec); λ_m = 212, 295 nm (ε = 9800, 10600, MeOH); poorly soluble in H_2O, more soluble in EtOH, less soluble in Et_2O, Me_2CO, C_6H_6, soluble in organic solvents. BASF Corp.

2069 Isatoic anhydride
118-48-9   H

$C_8H_5NO_3$
N-Carboxyanthranilic acid cyclic anhydride.
4-27-00-03330 (Beilstein Handbook Reference); AI3-24983; Benzoic acid, 2-(carboxyamino)-, cyclic anhydride; BRN 0136786; N-Carboxyanthranilic acid cyclic anhydride; N-Carboxyanthranilic anhydride; EINECS 204-255-0; HSDB 5017; Isatoic acid anhydride; Isatoic anhydride; NSC 104662.

2070 Isoamyl acetate
123-92-2   5127   G

$C_7H_{14}O_2$
Isopentyl Acetate.
4-02-00-00157 (Beilstein Handbook Reference); Acetic acid, 3-methylbutyl ester; Acetic acid, isopentyl ester; AI3-00576; Amyl acetate; i-Amyl acetate; Amylacetic ester; Banana oil; BRN 1744750; 1-Butanol, 3-methyl-, acetate; CCRIS 6051; EINECS 204-662-3; FEMA Number 2055; HSDB 1818; Isoamyl acetate; Isoamyl ethanoate; Isoamylester kyseliny octove; Isopentyl acetate; Isopentyl alcohol, acetate; Isopentyl ethanoate; Jargonelle pear essence; 3-Methyl-1-butanol acetate; β-Methylbutyl acetate; 2-Methylbutyl ethanoate; 3-Methylbutyl acetate; 3-Methyl-1-butyl acetate; 3-Methylbutyl ethanoate; NSC 9260; Pear oil; Pentyl acetate, all isomers; β-methyl butyl acetate; Amyl acetate, common. Provides a pear flavor. Used as a flavorant in foods, beverages and confectionery and as a solvent. Also used in the manufacture of fruit essences. mp = -78.5°; bp = 142.5°; d^{15} = 0.876; slightly soluble in H_2O (0.25 g/100 ml), soluble in Me_2CO, $CHCl_3$, AmOH, freely soluble in EtOH, Et_2O; LD_{50} (rat orl) = 16600 mg/kg. Penta Mfg.; Sigma-Aldrich Fine Chem.; Spectrum Chem. Manufacturing.

2071 Isoascorbic acid
89-65-6   5143   G H

$C_6H_8O_6$
D-erythro-Hex-2-enonic acid, γ-lactone.
5-18-05-00026 (Beilstein Handbook Reference); Araboascorbic acid; D-Araboascorbic acid; BRN 0084271; CCRIS 6568; EINECS 201-928-0; Erycorbin; Erythorbic acid; D-Erythorbic acid; Erythroascorbic acid, D-; FEMA Number: 2410; Glucosaccharonic acid; HSDB 584; Isoascorbic acid; D-Isoascorbic acid; Isovitamin C; Mercate 5; Neo-cebicure; NSC 8117; Saccharosonic acid. Used as an antioxidant and antimicrobial agent (industrial, food, brewing), and as a reducing agent in photography. Isomer of ascorbic acid. Crystals; mp = 174° (dec); $[\alpha]_D^{20}$ = -16.6°; soluble in H_2O, EtOH, C_5H_5N, moderately soluble in Me_2CO, slightly soluble in glycerol. Ashland; Spice King.

2072 Isobornyl acrylate
5888-33-5   G

$C_{13}H_{20}O_2$
exo-1,7,7-Trimethylbicyclo(2.2.1)hept-2-yl acrylate.
Acrylic acid, isobornyl ester; Ageflex IBOA; Al-co-cure IBA; Ebecryl IBOA; EINECS 227-561-6; IBOA; Isobornyl acrylate; Light Acrylate IB-XA; 2-Propenoic acid, (1R,2R,4R)-1,7,7-trimethylbicyclo(2.2.1)hept-2-ylester, rel-; 2-Propenoic acid, 1,7,7-trimethylbicyclo(2.2.1)hept-2-yl ester, exo-; QM 589; Sartomer 506; SR 506; SR 506 (acrylate). Intermediate in the preparation of monomers and for acrylic polymers which, when cured provide hardness, low shrinkage, abrasion resistant, heat and water resistant, good weatherability in automotive coatings, electronics, adhesives. Liquid; mp = -60°; d_{20}^{20} = 0.986; m.p. -60°; f.p. <-20°; flash

pt. 84°; irritant. *Rit-Chem; Sigma-Aldrich Fine Chem.*

2073 Isobornyl methacrylate
7534-94-3   G H

C14H22O2

2-Propenoic acid, 2-methyl-, 1,7,7-trimethylbicyclo-(2.2.1)hept-2-yl ester, exo-.

Ageflex IBOMA; EINECS 231-403-1; HSDB 6088; Isobornyl methacrylate; Methacrylic acid, isobornyl ester; 2-Propenoic acid, 2-methyl-, 1,7,7-trimethylbicyclo-(2.2.1)hept-2-yl ester, exo-; 2-Propenoic acid, 2-methyl-, (1R,2R,4R)-1,7,7-trimethylbicyclo(2.2.1)hept-2-yl ester, rel-; Sipomer® IBOMA; exo-1,7,7-Trimethylbicyclo-(2.2.1)hept-2-yl methacrylate. Monomers when cured providing hardness, low shrinkage, abrasion resistance, heat and water resistance, good weatherability in automotive coatings, electronics, adhesives, and other acrylic polymers. Liquid; mp = -60°; bp = 245°; d_{20}^{20} = 0.983. *Rhône Poulenc Surfactants; Rit-Chem.*

2074 Isobutane
75-28-5   G H

C4H10

2-Methylpropane.

A 31 (hydrocarbon); Caswell No. 503A; 1,1-Dimethyl-ethane; EINECS 200-857-2; EPA Pesticide Chemical Code 097101; HSDB 608; Isobutane; 2-Methylpropane; Propane, 2-methyl-; R 600a; Trimethylmethane. Hydro-carbon gas; used as a fuel and an Aerosol® propellant. Colorless gas; mp = -138.3°; bp = -11.7°; d^{25} = 0.5510; insoluble in H_2O, soluble in organic solvents. *Air Prods & Chems; Phillips 66.*

2075 Isobutanol
78-83-1   5148   G H

C4H10O

2-Methyl-1-propanol.

4-01-00-01588 (Beilstein Handbook Reference); Al3-01777; Alcool isobutylique; BRN 1730878; CCRIS 2300; EINECS 201-148-0; FEMA Number 2179; Fermentation butyl alcohol; HSDB 49; Iso-butyl alcohol; Isobutanol; Isobutyl alcohol; Isobutylalkohol; Isopropylcarbinol; NSC 5708; RCRA waste number U140; UN1212. Used as a solvent and chemical feedstock. Clear liquid; mp = -107.8°; bp = 108°; d^{20} = 0.8018; soluble in H = 72O (5 g/100 ml), more soluble in organic solvents; LD50 (rat orl) = 2.46 g/kg. *BASF Corp.; Eastman Chem. Co.; Hoechst Celanese; Neste UK; Shell UK; Shell; Usines de Melle.*

2076 Isobutene
115-11-7   5157   H

C4H8

2-Methylpropene.

CCRIS 2281; 1,1-Dimethylethene; 1,1-Dimethylethylene; EINECS 204-066-3; γ-Butylene; HSDB 613; Isobutene; Isobutylene; Isopropylidenemethylene; Methylpropene; 2-Methyl-1-propene; 2-Methylpropene; 2-Methylpropylene; Propene, 2-methyl-; 1-Propene, 2-methyl-; Propene, 2-methyl-; UN 1055. Gas; mp = -140.4°; bp = -6.9°; d^{25} = 0.589; λ_m = 159, 184, 188, 192, 200 nm (ε = 7943, 12589, 12589, 7943, 7943, gas); insoluble in H_2O, soluble in C_6H_6, H_2SO_4, petroleum ether, very soluble in EtOH, Et2O.

2077 N-Isobutoxymethylacrylamide
16669-59-3   G

C8H15NO2

N-((2-Methylpropoxy)methyl)acrylamide.

Acrylamide, N-(isobutoxymethyl)-; EINECS 240-715-7; Isobu-M-AMD; N-((2-Methylpropoxy)methyl)-2-propen-amide; 2-Propenamide, N-((2-methylpropoxy)methyl)-; Synocure 3165. Monomer. Liquid; bp0.3 = 99-100°. *Am. Cyanamid.*

2078 Isobutyl acetate
110-19-0   5147   G H

C6H12O2

2-Methylpropyl acetate.

4-02-00-00149 (Beilstein Handbook Reference); Acetate d'isobutyle; Acetic acid, 2-methylpropyl ester; Acetic acid, isobutyl ester; Al3-15305; BRN 1741909; EINECS 203-745-1; FEMA Number 2175; HSDB 609; Isobutyl acetate; Isobutyl ethanoate; Isobutylester kyseliny octove; 2-Methyl-1-propyl acetate; 2-Methylpropyl acetate; 2-Methylpropyl ethanoate; β-Methylpropyl ethanoate; NSC 8035; UN1213. Solvent for nitrocellulose; in thinners, sealants, topcoat lacquers; perfumery; flavoring agent. Liquid; mp = -98.8°; bp = 116.5°; d^{20} = 0.8712; λ_m = 208 nm (ε = 58, MeOH); slightly soluble in H_2O, CCl4, soluble in Me2CO, freely soluble in EtOH, Et2O. *BASF Corp.; Eastman Chem. Co.; Hoechst Celanese; Janssen Chimica; Union Carbide Corp.*

2079 Isobutyl acrylate
106-63-8   H

C7H12O2

Isobutyl 2-propenoate.

4-02-00-01465 (Beilstein Handbook Reference); Acrylic acid, isobutyl ester; Al3-15696; BRN 1749388; CCRIS 4828; EINECS 203-417-8; HSDB 610; Isobutyl 2-propenoate; Isobutyl acrylate; Isobutyl propenoate; Isobutylester kyseliny akrylove; NSC 20949; Propenoic acid, isobutyl ester; UN2527. Liquid; mp = -61°; BP = 132°; d^{20} = 0.8896; slightly soluble in H_2O, soluble in EtOH, Et2O, MeOH.

2080   Isobutylbenzene
538-93-2   5151   H

C10H14
(2-Methylpropyl)benzene.
4-05-00-01042 (Beilstein Handbook Reference); Benzene, (2-methylpropyl)-; Benzene, isobutyl-; BRN 1852218; EINECS 208-706-2; Isobutylbenzene; (2-Methylpropyl)benzene; 2-Methyl-1-phenylpropane; NSC 24848; 1-Phenyl-2-methylpropane; 2-Phenyl-2-methyl-propane. Liquid; mp = -51.4°; bp = 172.7; d^{20} =0.8532; λ_m = 259, 261, 264, 268 nm (ϵ = 221, 216, 170, 174, MeOH); insoluble in H2O, freely soluble in EtOH, Et2O, Me2CO, C6H6, CCl4, petroleum ether.

2081   O-Isobutylbenzoin
22499-12-3   G

C18H20O2
2-Isobutoxy-2-phenylacetophenone.
EINECS 245-039-6; Ethanone, 2-(2-methylpropoxy)-1,2-diphenyl-; Vicure®. Photosensitizer for UV curable systems, e.g., coatings, inks, graphic arts. bp0.5 = 133°; d = 0.9850. Akzo.

2082   Isobutyl formate
542-55-2   5159   G

C5H10O2
2-Methyl-1-propyl formate.
4-02-00-00029 (Beilstein Handbook Reference); AI3-24240; BRN 1738888; EINECS 208-818-1; FEMA No. 2197; Formic acid, 2-methylpropyl ester; Formic acid, isobutyl ester; Iso-butyl formate; Isobutyl formate; Isobutyl methanote; Isobutylester kyseliny mravenci; 2-Methylpropyl formate; 2-Methyl-1-propyl formate; NSC 6968; Tetryl formate; UN2393. Industrial solvent. Liquid; mp = -95.8°; bp = 98.2°; d^{20} = 0.8776; slightly soluble in H2O (1 g/100 ml), CHCl3, very soluble in Me2CO, freely soluble in EtOH, Et2O.

2083   Isobutyl heptyl ketone
123-18-2   H

C12H24O
2,6,8-Trimethylnonan-4-one.
4-01-00-03385 (Beilstein Handbook Reference); EINECS 204-607-3; Isobutyl heptyl ketone; 4-Nonanone, 2,6,8-trimethyl-; NSC 66186; 2,6,8-Trimethyl-4-nonanone; 2,6,8-Trimethylnonan-4-one.

2084   Isobutylidene diurea
6104-30-9   H

C6H14N4O2
N,N''-(2-Methylpropylidene)bisurea.
BRN 1908981; Diureidoisobutane; 1,1-Diureidisobutane; EINECS 228-055-8; IBDU; 1,1'-Isobutylidenebisurea; Isobutyldiurea; Isobutylenediurea; Isobutylidene diurea; N,N''-(Isobutylidene)diurea; Isodur; N,N''-(2-Methyl-propylidene)bisurea; Urea, 1,1'-isobutylidenedi-; Urea, N,N''-(2-methylpropylidene)bis-.

2085   Isobutyl isobutyrate
97-85-8   5161   H

C8H16O2
Propanoic acid, 2-methyl-, 2-methylpropyl ester.
4-02-00-00847 (Beilstein Handbook Reference); AI3-06122; BRN 1701355; EINECS 202-612-5; FEMA Number 2189; HSDB 5311; Isobutyl 2-methyl-propanoate; Isobutyl isobutanoate; Isobutyl isobutyrate; Isobutylester kyseliny isomaselne; Isobutyric acid, isobutyl ester; NSC 6538; Propanoic acid, 2-methyl-, 2-methylpropyl ester; UN2528. Liquid; mp = -80.7°; bp = 148.6°; d^{20} = 0.8542; poorly soluble in H2O, CCl4, soluble in EtOH, Me2CO, freely soluble in Et2O.

2086   Isobutylmethylcarbinol
108-11-2   G H

C6H14O
4-Methylpentan-2-ol.
4-01-00-01717 (Beilstein Handbook Reference); 4-Methylpentan-2-ol; 4-Metilpentan-2-olo; Alcohol methyl amylique; Alcool methyl amylique; BRN 1098268; CCRIS 2304; EINECS 203-551-7; HSDB 1154; Isobutyl-methylcarbinol; Isobutylmethylmethanol; M.I.B.C.; MAOH; Methyl amyl alcohol; Methyl isobutyl carbinol; dl-Methylisobutylcarbinol; Methyl-isobutyl-karbinol; Metilamil alcohol; MIBC; MIC; 3-MIC; NSC 9384; UN 2053. Solvent for dyestuffs, oils, gums, resins, waxes, nitrocellulose, ethylcellulose; organic synthesis; froth flotation; brake fluids. Liquid; mp = -90°; bp = 131.6°; d^{20} = 0.8075; slightly soluble in H2O, CCl4, Soluble in EtOH, Et2O; LD50 (rat orl) = 2590 mg/kg. *Allchem Ind.; Ashland; Shell; Union Carbide Corp.*

2087   Isobutyl salicylate
87-19-4   G

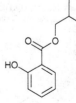

344

C11H14O3

2-Methylpropyl 2-hydroxybenzoate.
3-10-00-00121 (Beilstein Handbook Reference); Al3-24370; Benzoic acid, 2-hydroxy-, 2-methylpropyl ester; BRN 2615955; EINECS 201-729-9; FEMA No. 2213; 2-Isobutoxycarbonylphenol; Isobutyl o-hydroxybenzoate; Isobutyl salicylate; 2-Methyl-1-propyl salicylate; 2-Methylpropyl o-hydroxybenzoate; NSC 62140; Salicylic acid, isobutyl ester. Used in perfumery. Liquid; mp = 5.9°; bp = 261°; d^{20} = 1.0639; λ_m = 238 nm (cyclohexane); insoluble in H2O, soluble in EtOH, Et2O, CCl4.

2088   Isobutyl stearate

646-13-9   5167   G H

C22H44O2

Octadecanoic acid, 2-methylpropyl ester.
3-02-00-01017 (Beilstein Handbook Reference); BRN 1792857; EINECS 211-466-1; Emerest® 2324; HSDB 2177; Isobutyl stearate; Kemester® 5415; Kessco® IBS; 2-Methylpropyl octadecanoate; Octadecanoic acid, 2-methylpropyl ester; Stearic acid, 2-methylpropyl ester; Stearic acid, isobutyl ester; Uniflex® IBYS. Emollient, lubricant for textiles, metalworking compounds; slip aid, wetting agent for pigmented lipsticks, bath oils, nail polish and removers, skin cleaners, creams, lotions. Waxy solid; mp = 28.9°; bp15 = 223°; d^{20} = 0.8498; very soluble in Et2O. CK Witco (Europe) S.A.; CK Witco Chem. Corp.; Henkel/Emery; Henkel; Stepan; Union Camp.

2089   Isobutyltrimethoxysilane

18395-30-7   G

C7H18O3Si

Trimethoxy(2-methylpropyl)silane.
CI7810; Dynasylan® IBTMO; EINECS 242-272-5; Prosil® 178; Silane, trimethoxy(2-methylpropyl)-; Trimethoxy(2-methylpropyl)silane. Coupling agent for hydrophobic treatment, chemical intermediate, blocking agent, release agent, lubricant, primer, and reducing agent. Degussa-Hüls Corp.; PCR.

2090   Isobutyraldehyde

78-84-2   5171   H

C4H8O

2-Methylpropionaldehyde.
Al3-15311; CCRIS 1101; EINECS 201-149-6; FEMA No. 2220; HSDB 614; Isobutaldehyde; Isobutanal; Isobutylaldehyde; Isobutyral; Isobutyraldehyd; Isobutyraldehyde; Isobutyric aldehyde; Isobutyryl aldehyde; Isopropyl aldehyde; Isopropyl formaldehyde; Isopropylformaldehyde; Methylpropanal; NCI-C60968; NSC 6739; Propanal, 2-methyl-; Propionaldehyde, 2-methyl-; UN2045; Valine aldehyde. Used in chemical synthesis. Liquid; mp = -65.9°; bp = 64.5°; d^{20} = 0.7891; moderately soluble in H2O, organic solvents; LD50 (rat orl) = 3700 mg/kg. Eastman Chem. Co.; Lancaster Synthesis Co.; Mallinckrodt Inc.; Sigma-Aldrich Fine Chem.

2091   Isobutyric acid

79-31-2   5172   G H

C4H8O2

2-Methylpropionic acid.
4-02-00-00843 (Beilstein Handbook Reference); Acetic acid, dimethyl-; Al3-24260; BRN 0635770; Caswell No. 503AA; Cenex RP b2; Dimethylacetic acid; EINECS 201-195-7; EPA Pesticide Chemical Code 101502; FEMA No. 2222; HSDB 5228; Iso-butyric acid; Isobutanoic acid; Isobutyric acid; Isopropylformic acid; Kyselina isomaselna; Methylpropanoic acid, 2-; NSC 62780; Propanoic acid, 2-methyl-; Propionic acid, 2-methyl-; Tenox IBP-2; Tenox IBP-2 Grain Pr.; UN2529. Used in manufacture of esters for solvents, flavors, perfume bases, disinfecting agent, varnish, deliming hides, tanning agent. Liquid; mp = -46°; bp = 154.4°; d^{20} = 0.9681; soluble in H2O, EtOH, less soluble in Et2O, CCl4. Degussa-Hüls Corp.; Eastman Chem. Co.; Hoechst Celanese.

2092   Isobutyric   acid,   1-isopropyl-2,2-dimethyltrimethylene ester

6846-50-0   G H

C16H30O4

2,2,4-Trimethyl-1,3-pentanediol diisobutyrate.
BRN 1878083; EINECS 229-934-9; 1-Isopropyl-2,2-dimethyltrimethylene diisobutyrate; Kodaflex® TXIB; 1,3-Pentanediol, 2,2,4-trimethyl-, diisobutyrate (ester); Propanoic acid, 2-methyl-, 2,2-dimethyl-1-(1-methyl-ethyl)-1,3-propanediyl ester; TXIB. Primary plasticizer used in surface coatings, vinyl floorings, molding, and vinyl products; compatible with film-forming vehicles used in lacquers for wood, paper, and metals; primary plasticizer for PVC plastisols for rotocasting and slush molding; used in PNC organosols processed by extrusion and injection molding. Liquid; mp = -70°; bp = 280°; d = 0.941. Eastman Chem. Co.

2093   Isobutyronitrile

78-82-0   5173   H

C4H7N

2-Methylpropanenitrile.
4-02-00-00853 (Beilstein Handbook Reference); Al3-28525; BRN 1340512; Dimethylacetonitrile; EINECS 201-147-5; HSDB 5221; Isobutyronitrile; Isopropyl cyanide; Isopropyl nitrile; Isopropylkyanid; Isopropylnitrile; NSC 60536; Propanenitrile, 2-methyl-; Propanoic acid, 2-methyl-, nitrile; UN2284. Used as a solvent. Liquid; mp = -71.5°; bp = 103.9°; d^{20} = 0.7704; slightly soluble in H2O, more soluble in organic solvents; LD50 (rat orl) = 200 mg/kg, (mmus orl) = 25 mg/kg, (mus ip) = 25 mg/kg. Lancaster Synthesis Co.; Mallinckrodt Inc.; Sigma-Aldrich Fine Chem.

2094   Isocetyl alcohol

36311-34-9   G

345

C16H34O
Isohexadecyl alcohol.
EINECS 252-964-9; Isocetyl alcohol; Isohexadecanol; Isohexadecyl alcohol. Used in organic synthesis. *M. Michel.*

2095 Isocyanobenzotrifluoride
329-01-1   H

C8H4F3NO
Isocyanic acid, (m-trifluoromethylphenyl) ester.
4-12-00-01848 (Beilstein Handbook Reference); Benzene, 1-isocyanato-3-(trifluoromethyl)-; BRN 0744880; EINECS 206-341-3; 1-Isocyanato-3-(trifluoro-methyl)benzene; Isocyanic acid, (m-trifluoromethyl-phenyl) ester; Isocyanic acid, α,α,α-trifluoro-m-tolyl ester; Isocyanobenzotrifluoride; (α,α,α-Trifluoro-m-tolyl) iso-cyanate; α,α,α-Trifluoro-3-tolyl isocyanate.

2096 Isodecanol
25339-17-7   H

C10H22O
Isodecyl alcohol.
EINECS 246-869-1; HSDB 616; Isodecanol; Isodecyl alcohol.

2097 Isodecyl acrylate
1330-61-6   G H

C13H24O2
Isodecyl prop-2-enoate.
Acrylic acid, isodecyl ester; Ageflex FA-10; EINECS 215-542-5; HSDB 615; Isodecyl acrylate; Isodecyl alcohol, acrylate; Isodecyl propenoate; 2-Propenoic acid, isodecyl ester; Sipomer® IDA. Used in adhesives, coatings, UV-curable reactive diluent in inks and coatings, viscosity index improver. Liquid; mp = -100°; bp = 304°; bp3 = 161-163°; d$_{20}^{20}$ = 0.864. *Rhône Poulenc Surfactants; Rit-Chem.*

2098 Isodecyl diphenyl phosphate
29761-21-5   G H

C22H31O4P
Phosphoric acid, isodecyl diphenyl ester.
CCRIS 4830; Diphenyl isodecyl phosphate; EINECS 249-828-6; HSDB 6797; Isodecyl diphenyl phosphate; Phosflex 390; Phosphoric acid, isodecyl diphenyl ester; Santicizer 148; Syn-O-Ad® 8479.

Antiwear and extreme pressure agents in non-crankcase lubricants; ferrous metal passivator; fluid base stock where inhibition of flame propagation is desired. Liquid; mp = -35°; insoluble in H2O, soluble in organic solvents. *Akzo Chemie.*

2099 Isodecyl methacrylate
29964-84-9   G H

C14H26O2
2-Propenoic acid, 2-methyl-, isodecyl ester.
Ageflex FM-10; CCRIS 4660; EINECS 249-978-2; HSDB 5065; Isodecyl 2-methyl-2-propenoate; Isodecyl 2-methylpropenoate; Isodecyl methacrylate; Methacrylic acid, isodecyl ester; 2-Methyl-2-propenoic acid, isodecyl ester; 2-Propenoic acid, 2-methyl-, isodecyl ester. Pressure-sensitive adhesives, coatings for leather, textiles, paper, nonwovens, polymer modifier/stabilizer, viscosity index improver, dispersion for plastic and rubber, floor waxes, potting compounds, sealants, adhesives. Liquid; d$_{20}^{20}$ = 0.878; bp = 126°; flash pt. 121 °; insoluble in H2O, soluble in organic solvents; LD50 (rat ip) = 2467 mg/kg. *Rohm & Haas Co.*

2100 Isodecyl oleate
59231-34-4   G

C28H54O2
9-Octadecenoic acid (Z)-, isodecyl ester.
Ceraphyl 140-A; EINECS 261-673-6; Isodecyl oleate; 9-Octadecenoic acid, isodecyl ester; 9-Octadecenoic acid (Z)-, isodecyl ester; 9-Octadecenoic acid (9Z)-, isodecyl ester; Wickenol® 144. Emollient, wetting agent and pigment binder for cosmetics. *CasChem.*

2101 Isodecyloxypropylamine
30113-45-2   H

C13H29NO
1-Propanamine, 3-(isodecyloxy)-.
EINECS 250-056-7; Isodecyloxypropylamine; 3-(Isodecyl-oxy)propylamine; 1-Propanamine, 3-(isodecyloxy)-; Prop-ylamine, 3-isodecoxy-. *Rit-Chem.*

2102 Isodecyl phosphite
25448-25-3   H

$C_{30}H_{63}O_3P$
Phosphorous acid, tris(isodecyl) ester.
EINECS 246-998-3; Isodecyl phosphite; Phosclere T 310; Phosphorous acid, triisodecyl ester; Phosphorous acid, tris(isodecyl) ester; Triisodecyl phosphite.

2103   Isoflupredone Acetate
338-98-7   5192   D G V

$C_{23}H_{29}FO_6$
(11β)-9-Fluoro-11,17,21-trihydroxypregna-1,4-diene-3,20-dione 21-acetate.
Biorinil; EINECS 206-423-9; Isoflupredone acetate; NSC 12600; Predef; U 6013. Anti-inflammatory used in veterinary medicine. Crystals; mp = 244-246° (dec); $[\alpha]_D^{23}$ = +108° (c = 0.735 dioxane); λ_m = 240 nm (ε 16250 EtOH). *CIBA plc; Schering-Plough HealthCare Products; Upjohn Ltd.*

2104   Isoflurane
26675-46-7   5193   D G

$C_3H_2ClF_5O$
1-Chloro-2,2,2-trifluoroethyl difluoromethyl ether.
Aerrane; BRN 1852087; CCRIS 3043; Compound 469; EINECS 247-897-7; Forane; Forene; Isoflurane; Isoflurano; Isofluranum; R-E 235dal. An inhalation anesthetic; solvent and dispersant for fluorinated materials. Gas; bp = 48.5°; sg = 1.45; non-flammable; soluble in most organic solvents. *Anaquest; Elf Atochem N. Am.*

2105   Isohexadecane
4390-04-9   G

$C_{16}H_{34}$
2,2,4,4,6,8,8-Heptamethylnonane.
EINECS 224-506-8; 2,2,4,4,6,8,8-Heptamethylnonane; HMN; Nonane, 2,2,4,4,6,8,8-heptamethyl-; NSC 77129; Permethyl 101A. A cosolubilizer for nonhydrocarbon materials; in eyeliners, mascaras,

sun care and skin products; cleanser for eye and face makeup. Liquid; bp = 240°; d = 0.7930. *Presperse.*

2106   Isohexyl alcohol
105-30-6   H

$C_6H_{14}O$
2-Methylpentan-1-ol.
3-01-00-01665 (Beilstein Handbook Reference); AI3-21997; sec-Amyl carbinol; BRN 1718974; EINECS 203-285-1; HSDB 2890; Isohexyl alcohol; 2-Methyl-1-pentanol; 2-Methylpentan-1-ol; 2-Methylpentanol-1; (±)-2-Methylpentanol; (±)-2-Methyl-1-pentanol; 2-MPOH; NSC 6250; 1-Pentanol, 2-methyl-. Liquid; bp = 149°; d^{20} = 0.8263; soluble in EtOH, Et2O, Me2CO, CCl4.

2107   Iso-Iodeikon
18265-54-8   7347   G

$C_{20}H_8I_4Na_2O_4$
4,5,6,7-Tetraiodophenolphthalein sodium.
Phentetiothalien sodium; Phenoltetraiodothalein sodium. Used in cholecystography. Bronze-purple granules, soluble in H2O, EtOH.

2108   Isonicotinic acid
55-22-1   5206   G

$C_6H_5NO_2$
Pyridine-4-carboxylic acid.
Acide iso-nicotinique; AI3-19239; 4-Carboxypyridine; 1,4-Dihydroisonicotinic acid; EINECS 200-228-2; γ-Picolinic acid; γ-Pyridinecarboxylic acid; INA; Isonicotinic acid; NSC 1483; 4-Picolinic acid; 4-Pyridinecarboxylic acid; p-Pyridinecarboxylic acid. Chemical intermediate in the manufacture of its hydrazide, isoniazid, an antituberculosis drug. Colorless crystals; mp = 319°; soluble in H2O (0.52 g/100 ml), less soluble in organic solvents. *Raschig GmbH; Reilly Ind.*

2109   Isononylic acid
26896-18-4   H

$C_9H_{18}O_2$
Isononanoic acid.

2110   Isonox® 132
17540-75-9   G

C18H30O
2,6-Di-*tert-butyl*-4-*sec*-butylphenol.
4-sec-Butyl-2,6-di-t-butylphenol;   2,6-Di-t-butyl-4-sec-butylphenol;
EINECS 241-533-0; Isonox® 132; NSC 14460; Phenol, 2,6-bis(1,1-dimethylethyl)-4-(1-methyl-propyl)-;   Phenol,   4-sec-butyl-2,6-di-t-butyl-; Vanox® 1320. Antioxidant used in polyols, rubber systems and PU foam; oxidation inhibitor and scorch preventer for manufacture and storage of bun stock; stabilizer for PU foam. Crystals; mp = 25°; bp10 = 141-142°; d = 0.902; $[\alpha]^{25}$ = 0° (c = 1, CHCl3). *Schenectady; Vanderbilt R.T. Co. Inc.*

2111   Isooctadecanoic acid, monoamide with 2-[(2-aminoethyl)amino]ethanol
68443-85-6   H

C22H46N2O2
Isooctadecanoic   acid,   monoamide   with   2-((2-aminoethyl)amino)ethanol.
EINECS 270-561-6; Isooctadecanoic acid, monoamide with 2-((2-aminoethyl)amino)ethanol.

2112   Isooctane
540-84-1   5212   H

C8H18
2,2,4-Trimethylpentane.
4-01-00-00439 (Beilstein Handbook Reference); AI3-23976; BRN 1696876;   EINECS   208-759-1;   HSDB   5682;
Isobutyltrimethylmethane; Isooctane; NSC 39117; Pentane, 2,2,4-trimethyl-; 2,2,4-Trimethylpentane. Liquid; mp = -107.3; bp = 99.2°; d^{25} = 0.6877; insoluble in H2O, soluble in Et2O, CCl4, CHCl3; freely soluble in EtOH, Me2CO, C6H6, CHCl3, C7H16.

2113   Isooctyl acrylate
29590-42-9   G H

C11H20O2
2-Propenoic acid, isooctyl ester.
Acrylic acid, isooctyl ester; Ageflex FA-8; AI3-28217; CCRIS 4352; EINECS 249-707-8; Isooctyl acrylate; Isooctyl acrylate monomer; 2-Propenoic acid, isooctyl ester; SR-440. Curing agent. Used in pressure-sensitive adhesives, coatings, caulks and sealants. Liquid; d_{20}^{20} = 0.880; flash point = 79°. *Rit-Chem; Sartomer.*

2114   Isooctyl alcohol
26952-21-6   5213   G

C8H18O
Isooctanol.
EINECS 248-133-5; Exxal 8; HSDB 6486; Isooctanol; Isooctan-1-ol; Isooctyl alcohol. Mixture of isomers, mostly with a methyl branch at C3, C4 or C5. Intermediate in manufacture of plasticizers; intermediate for nonionic detergents and surfactants, hydraulic fluids; resin, solvent, emulsifier, antifoaming agent.

2115   Isooctyl diphenyl phosphite
26401-27-4   G H

C20H27O3P
Phosphorous acid, isooctyl diphenyl ester.
Doverphos® DPIOP; DPIOP; EINECS 247-658-7; Isooctyl diphenyl phosphite; Phosphorous acid, isooctyl diphenyl ester; Weston® ODPP. Color stabilizer for ABC, PC; stabilizer for PVC; antioxidant

Liquid; bp = 190°; d = 1.040-1.047. *Dover; Sigma-Aldrich Fine Chem.*

2116   Isooctyl mercaptoacetate
25103-09-7   G H

C10H20O2S
Mercaptoacetic acid, isooctyl ester.
Acetic acid, mercapto- isooctyl ester; AI3-26088; EINECS 246-613-9; HSDB 2704; Isooctyl mercaptoacetate; Isooctyl thioglycolate; Mercaptoacetic acid, isooctyl ester; NSC 9590. Antioxidants, fungicides, oil additives, plasticizers, insecticides, stabilizers, polymerization modifiers, stabilizer for tin-sulfur compounds, stripping agent for polysulfide rubber. *Bock, Bruno Chemische Fabrik KG.*

2117   Isopentane
78-78-4   G H

C5H12
2-Methylbutane.
AI3-28787; Butane, 2-methyl-; Dimethylethylmethane; EINECS 201-142-8; Ethyldimethylmethane; Exxsol® Isopentane; HSDB 618; Isoamylhydride; Isopentane; NSC 119476; Pentane, all isomers. Used as a non-aromatic solvent. Volatile liquid; mp = -159.9°; bp = 27.8°; d^{20} = 0.6201; insoluble in H_2O, soluble in organic solvents. *ExxonMobil Chem. Co.; Phillips 66.*

2118   Isopentyl salicylate
87-20-7   5141   G

C12H16O3
3-Methylbutyl 2-hydroxybenzoate.
4-10-00-00153 (Beilstein Handbook Reference); AI3-00378; Benzoic acid, 2-hydroxy-, 3-methylbutyl ester; BRN 2580465; EINECS 201-730-4; FEMA No. 2084; 2-Hydroxybenzoic acid, 3-methylbutyl ester; Isoamyl o-hydroxybenzoate; Isoamyl salicylate; Isoamylester kyseliny salicylove; Isopentyl-2-hydroxyphenyl methanoate; Isopentyl o-hydroxybenzoate; Isopentyl salicylate; 3-Methylbutyl o-hydroxybenzoate; 3-Methylbutyl salicylate; NSC 7952; Salicylic acid, isopentyl ester. Used in perfumery. Liquid; bp = 278°, bp15 = 151°; d^{20} = 1.0535; λ_m = 238, 306 nm (ε = 8420, 3870, MeOH); insoluble in H_2O, soluble in organic solvents.

2119   Isophorone
78-59-1   5215   G H

C9H14O
3,5,5-Trimethyl-2-cyclohexen-1-one.
4-07-00-00165 (Beilstein Handbook Reference); AI3-00046; BRN 1280721; Caswell No. 506; CCRIS 1353; EINECS 201-126-0; EPA Pesticide Chemical Code 047401; FEMA No. 3553; HSDB 619; Isoacetophorone; Isoforon; Isoforone; Isooctopherone; Isophorone; Izoforon; NCI-C55618; NSC 403657. Used in solvent mixtures for printing inks and finishes, for polyvinyl and nitrocellulose resins, pesticides, stoving lacquers. Liquid; mp = -8.1°; bp = 215.2°; d^{20} = 0.9255; λ_m = 225, 337 nm (ε 14125, 32 isooctane); soluble in H_2O (1.2 g/100 ml), organic solvents; LD50 (rat orl)= 2700 mg/kg. *Albright & Wilson Americas Inc.; BP Chem.; FMC Corp.; Union Carbide Corp.*

2120   Isophorone diamine
2855-13-2   G H

C10H22N2
5-Amino-1,3,3-trimethylcyclohexanemethylamine.

CCRIS 6680; Cyclohexanemethanamine, 5-amino-1,3,3-trimethyl-; EINECS 220-666-8; HSDB 4058; Isophorone diamine; UN2289; Vestamin® TMD. Epoxy curative for flexible coatings, adhesives, castings and composites. *Degussa-Hüls Corp.*

2121   Isophorone diisocyanate
4098-71-9   G H

C12H18N2O2
3-Isocyanatomethyl-3,5,5-trimethylcyclohexylisocyanate.
BRN 2726467; CCRIS 6252; Cyclohexane, 5-isocyanato-1-(isocyanatomethyl)-1,3,3-trimethyl-; EINECS 223-861-6; HSDB 6337; IPDI; Isophorone diamine diisocyanate; Isophorone diisocyanate; UN2290; Vestanat® IPDI. Raw material for mfg. of light stable polyurethanes. *Degussa-Hüls Corp.*

2122   Isophthalic acid
121-91-5   5216   H

C8H6O4
Benzene-1,3-dicarboxylic acid.
4-09-00-03292 (Beilstein Handbook Reference); Acide isophtalique; AI3-16107; Benzene-1,3-dicarboxylic acid; BRN 1909332; EINECS 204-506-4; HSDB 2090; IPA; Isophthalate; Isophthalic acid; Kyselina isoftalova; m-Phthalic acid; NSC 15310. Needles; mp = 347°; bp sublimes; λ_m = 279 nm (MeOH); insoluble in Et_2O, C_6H_6, ligroin, slightly soluble in H_2O, soluble in EtOH, AcOH.

2123   Isophthalodinitrile
626-17-5   H

C8H4N2
3-Cyanobenzonitrile.
AI3-25034; 1,3-Benzendikarbonitril; 1,3-Benzene-dicarbonitrile; 1,3-Benzodinitrile; m-Benzenedinitrile; CCRIS 4132; 3-Cyanobenzonitrile; 1,3-Dicyanobenzene; m-Dicyanobenzene; Dinitrile of isophthalic acid; EINECS 210-933-7; HSDB 5724; IPN; Isoftalonitril; Isoftalonitril; Isophthalodinitrile; Isophthalonitrile; Nitril kyseliny isoftalove; NSC 87880; m-Phthalodinitrile; m-Phthalodinitrile. Needles; mp = 162°; bp sublimes; d^{40} = 0.992; λ_m = 221, 227, 231, 279, 287 nm (ε = 11500, 12000, 11100, 532, 570, MeOH); insoluble in petroleum ether, slightly soluble in H_2O, soluble in Et_2O, C_6H_6, $CHCl_3$, very soluble in EtOH.

2124   Isophthaloyl dichloride
99-63-8   H

C8H4Cl2O2
1,3-Benzenedicarbonyl dichloride.
4-09-00-03295 (Beilstein Handbook Reference); m-Benzenedicarbonyl chloride; BRN 0638342; Dichlorid kyseliny isoftalove; EINECS 202-774-7; HSDB 5326; Isophthalic acid chloride; Isophthalic acid dichloride; Isophthalic chloride; Isophthaloyl chloride; Isophthaloyl dichloride; Isophthalyl chloride; m-Phthalic dichloride; m-Phthaloyl chloride; m-Phthalyl dichloride; meta-Phthalyl dichloride; NSC 41884. Solid; mp = 43.5°; bp = 276°; d^{17} = 1.3880; slightly soluble in H2O, EtOH, soluble in Et2O.

2125   Isoprene
78-79-5   5220   H

C5H8
2-Methyl-1,3-butadiene.
CCRIS 6253; EINECS 201-143-3; HSDB 620; Isopentadiene; Isoprene; β-Methylbivinyl; 2-Methyl-divinyl; 3-Methyl-1,3-butadiene; NSC 9237; UN1218. Used in rubbers, sealants, latex paints and as a solvent. Volatile liquid; mp = -145.9°; bp = 34°; d^{20} = 0.679; insoluble in H2O, soluble in organic solvents. *Equistar Chemical Products; Lancaster Synthesis Co.; Mallinckrodt Inc.; Sigma-Aldrich Fine Chem.*

2126   Isoprocarb
2631-40-5   G P

C11H15NO2
2-(1-Methylethyl)phenyl methylcarbamate.
4-06-00-03212 (Beilstein Handbook Reference); AI3-25670; BAY 105807; BAY 39731; Bayer 39731; BRN 1875020; Carbamic acid, methyl-, o-cumenyl ester; Carbamic acid, methyl-, 2-(1-methylethyl)phenyl ester; Carbamic acid, methyl-, o-isopropylphenyl ester; Caswell No. 512B; o-Cumenyl methylcarbamate; o-Cumenyl N-methylcarbamate; EINECS 220-114-6; ENT 25670; EPA Pesticide Chemical Code 512300; Etrofolan; Hytox; Isoprocarb; Isoprocarbe; 2-Isopropylphenyl methylcarba-mate; 2-Isopropylphenyl N-methylcarbamate; Isopropyl-phenol methyl-carbamate; o-Isopropylphenyl methyl-carbamate; o-Isopropylphenyl N-methylcarbamate; KHE 0145; 2-(1-Methylethyl)phenyl methylcarbamate; MIPC; Mipcin; Mipcine; Mipsin; NSC 191479; OMS-32; Phenol, 2-(1-methylethyl)-, methylcarbamate; Phenol, o-iso-propyl-, methylcarbamate; PPC 3; Ro 7-5050. Fast-acting insecticide with contact and stomach action. Effective against leafhoppers, aphids, codling moths, capsids and bugs in rice, cocoa, vegetables, cereals, hops, coffee, potatoes, sugarcane, deciduous fruits and other crops. Colorless crystals; mp = 88 - 93°; insoluble in H2O, soluble in organic solvents, Me2CO, MeOH; LD50 (mrat orl) = 485 mg/kg; (mus orl) = 487 - 512 mg/kg, (rbt orl) ≅500 mg/kg, (mrat der) > 500 mg/kg; LC50 (carp 48 hr.) = 4.2 mg/l; toxic to bees. *Bayer*

Corp.; Jin Hung; Mitsubishi Chem. Corp.; Planters Products; Shinung; Sundat; Sunko; Taiwan Tainan Giant.

2127   Isopropanol
67-63-0   5228   P

C3H8O
Isopropyl alcohol.
4-01-00-01461 (Beilstein Handbook Reference); AI3-01636; Alcojel; Alcolo; Alcool isopropilico; Alcool isopropylique; Alcosolve; Alcosolve 2; Alkolave; Arquad DMCB; Avantin; Avantine; BRN 0635639; Caswell No. 507; CCRIS 2308; Chromar; Combi-schutz; Dimethylcarbinol; EINECS 200-661-7; EPA Pesticide Chemical Code 047601; FEMA Number 2929; Hartosol; Hibistat; HSDB 116; Imsol A; IPA; Iso-propyl alcohol; Iso-propylalkohol; Isohol; Isopropanol; Isopropyl alcohol; Isopropyl alcohol, rubbing; Lavacol; Lutosol; NSC 135801; Petrohol; Propan-2-ol; i-Propanol; sec-propanol; n-Propan-2-ol; i-Propyl alcohol; i-Propylalkohol; sec-Propyl alcohol; Propol; Rubbing Alcohol; Spectrar; Sterisol hand disinfectant; Surfactants; Takineocol; UN1219; Visco 1152. Registered by EPA as an antimicrobial, fungicide, herbicide and insecticide. Used as a hard surface treatment disinfectant, sanitizer, sterilant, virucide, fungicide, and mildewcide. Also is used in combination with other pesticide active ingredients to kill fleas, ticks, and other household insects. Used in antifreeze, as a solvent and as an antiseptic. Colorless liquid; mp = -88°; bp = 82°; d = 0.7850; soluble in H2O, all organic solvents; viscosity - 2.43 mPa at 20°.
LD50 (rat orl) = 5050 mg/kg, (rat iv) = 1090 mg/kg, (rat ip) = 2740 mg/kg, (rbt skn) = 12800 mg/kg, (rbt iv) = 1180 mg/kg, (rbt ip) = 670 mg/kg, (mus iv) = 1510 mg/kg, (mus ip) = 4480 mg/kg, (mus orl) = 3600 mg/kg, (hmtr ip) = 3440 mg/kg, (gpg ip) = 2560 mg/kg, (dog orl) = 4800 mg/kg. *Apollo Industries; Burnishine Products; Champion Technologies Inc; Chardon Laboratories, Inc; Chase Products Co.; Chemisco; Garratt-Callahan Co; Midco Products; Noble Pine Products; Palmero Health Care; PDI Res. Labs; Quest Chemical Corp.; Stepan.*

2128   Isopropanolamine
78-91-1   467   G

C3H9NO
2-amino-1-propanol.
2-Amino-2-methylethanol; 2-Aminopropanol; 2-Amino-1-propanol; 2-Aminopropan-1-ol; EINECS 201-156-4; EINECS 228-207-3; 1-Hydroxy-2-aminopropane; 1-Methyl-2-hydroxyethylamine; NSC 1360; β-Propanol-amine; 1-Propanol, 2-amino-; MIPA; Aliphatic amine. Solubilizer, neutralizer, emulsifying agent; plasticizers, insecticides. Liquid; bp = 173-176°; soluble in H2O, organic solvents. *Ashland; BASF Corp.; Mitsui Toatsu.*

2129   Isopropoxypropanol
3944-37-4   G

C6H14O2
1-Isopropoxy-2-propanol.
EINECS 223-534-8; 2-Isopropoxypropanol; Solvenon® IPP. Solvent for resins and dyes, surface coatings, cleaning agent for printing

plates. *BASF Corp.*

2130 Isopropyl acetate
108-21-4   5225   H

C5H10O2
1-Methylethyl acetate.
4-02-00-00141 (Beilstein Handbook Reference); Acetate d'isopropyle; Acetic acid, 1-methylethyl ester; Acetic acid, isopropyl ester; BRN 1740761; CCRIS 6053; EINECS 203-561-1; FEMA No. 2926; HSDB 159; Isopropile (acetato di); Isopropyl (acetate d'); Isopropyl acetate; Isopropyl ethanoate; Isopropylacetaat; Isopropylacetat; Isopropyle (acetate d'); Isopropylester kyseliny octove; NSC 9295; Paracetat; UN1220. Liquid; mp = -73.4°; bp = 88.6°; d^{20} = 0.8718; λ_m = 209 nm (ϵ = 59, MeOH); soluble in H2O, EtOH, Me2CO, CHCl3, freely soluble in Et2O.

2131 Isopropylamine
75-31-0   5229   H

C3H9N
2-Propanamine.
AI3-15636; 2-Amino-propaan; 2-Amino-propano; 2-Aminopropan; 2-Aminopropane; CCRIS 4318; EINECS 200-860-9; HSDB 804; Isopropilamina; Isopropylamine; 1-Methylethylamine; Monoisopropylamine; NSC 62775; Propanal, 2-amino-; 2-Propanamine; Propane, 2-amino-; 2-Propaneamine; 2-Propylamine; sec-Propylamine; UN1221. Used as a solvent in manufacture of intermediates for pesticides, medicines and dyes, a raw material for making rubber accelerants, as a cleaning agent, emulsifing agent, surfactant, detergent and water treatment chemical. Liquid; mp = -95.1°; bp = 31.7°; d^{20} = 0.6891; very soluble in H2O and organic solvents; LD50 (rat orl) = 820 mg/kg. *Atofina; Hoechst Celanese; Lancaster Synthesis Co.; Sigma-Aldrich Fine Chem.*

2132 Isopropylamine dodecylbenzenesulfonate
26264-05-1   H

C18H30O3S.C3H9N
Dodecylbenzenesulfonic acid monoisopropanolamine salt.
Arylan PWS; Atlas G 3300; Atlas G 711; Benzenesulfonic acid, dodecyl-, compd. with isopropylamine; Dodecylbenzenesulfonic acid, isopropylamine salt; Dodecylbenzenesulfonic acid monoisopropanolamine salt; Dodecylbenzenesulphonic acid, compound with isopropylamine; EINECS 247-556-2; G 3300; G 711; Isopropylamine dodecylbenzenesulfonate; Nansa YS 94; P 10-59; Polystep A 11; Rhodacal 330; Siponate 330; Witconate P 10-59.

2133 Isopropylamine Glyphosate
38641-94-0   4525   G P

C6H17N2O5P
N-(Phosphonomethyl) glycine, isopropylamine salt.
Azural AT; Buggy; Caswell No. 471AAB; CCRIS 6431; CP 70139; EPA Pesticide Chemical Code 103601; Fosulen; Glifosato estrella; Glycel; Glycine, N-(phosphonomethyl)-, compound with 2-propanamine (1:1); Glyphosate isopropylamine salt; HSDB 3433; Isopropylamine glyphosate; Rodeo; Landmaster; MON 139; MON 39; Mono-isopropylammoniova sul; N-(Phosphonomethyl)-glycine isopropylamine salt; Nitosorg; 2-Propanamine, compd. with N-(phosphonomethyl)glycine (1:1); Rattler; Rodeo; Ron-do; Roundup; Utal; Utal (herbicide); Vision (herbicide). Aquatic herbicide. Registered by EPA as a herbicide and plant growth regulator. White solid; very soluble in H2O. *Albaugh Inc.; Monsanto Co.*

2134 4-Isopropylaminodiphenylamine
101-72-4   G H

C15H18N2
N-Isopropyl-N'-phenyl-p-phenylenediamine.
4-13-00-00115 (Beilstein Handbook Reference); 4010 NA; Antigene 3C; Antioxidant 4010 NA; Antioxidant 40NA; Antioxidant IP; ASM 4010ma; Benzenediamine, N-(1-methylethyl)-N'-phenyl-, 1,4-; BRN 2213195; CCRIS 4833; Cyzone; Cyzone IP; Diafen FP; Diaphen FP; EINECS 202-969-7; Elastozone 34; Flexzone 3C; HSDB 5342; Ipognox 44; IPPD; NA 4010; NCI-C56304; Nocrac 810NA; Nocrack 810NA; Nonox ZA; NSC 41029; Orflex PP; Ozonon 3C; Permanax 115; Permanax IPPD; S-IP; Santoflex 36; Santoflex IP; Vulkanox® 4010 NA. Staining and discoloring antioxidant/antiozonant for protection of rubber from ozone attack, oxidation, heat aging, flexcracking, and rubber poisons; suitable for natural and synthetic rubbers; used for dynamically stressed goods, technical goods, spring components, conveyor and transmission belting, seals, insulation. An antidegradant in natural rubber, styrene-butadiene, nitrile-butadiene, butadiene, and chloroprene rubber. Solid; mp = 72-76°; bp1 = 161°; insoluble in H2O, soluble in organic solvents; LD50 (rat orl) = 555 mg/kg. *Akrochem Chem. Co.; Akzo Chemie; Bayer AG; Polysar; Uniroyal.*

2135 N-Isopropylaniline
768-52-5   H

C9H13N
N-Phenylisopropylamine.
4-12-00-00255 (Beilstein Handbook Reference); Aniline, N-isopropyl-; Benzenamine, N-(1-methylethyl)-; BRN 2205871; CCRIS 4831; EINECS 212-196-7; HSDB 6133; Isopropylaniline; N-Isopropylaniline; N-Phenylisopropyl-amine. Liquid; bp = 203°; d^{25} = 0.9526; λ_m = 248, 295 nm (ϵ = 11000, 1640, MeOH); soluble in EtOH, Et2O, Me2CO, C6H6.

351

2136 Isopropylbenzene

98-82-8   2644   G H

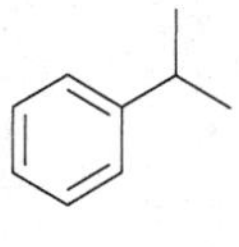

C_9H_{12}
Benzene, (1-methylethyl)-.
AI3-04630; Benzene, (1-methylethyl)-; Benzene, isopropyl; Cumeen; Cumene; Cumol; EINECS 202-704-5; HSDB 172; Isopropilbenzene; Isopropyl-benzol; Isopropylbenzeen; Isopropylbenzene; Isopropylbenzol; NSC 8776; Propane, 2-phenyl; RCRA waste number U055; UN1918. Used in production of phenol, acetone and α-methylstyrene and as a solvent. Oil; mp = -96°; bp = 152-154°; d^{20} = 0.8618; λ_m = 258 nm (cyclohexane); insoluble in H_2O, freely soluble in EtOH, Et_2O, Me_2CO, C_6H_6, CCl_4, petroleum ether; LD_{50} (rat orl) = 2.91 g/kg. *Ashland; Chevron; Degussa-Hüls Corp.; Georgia Gulf; Mitsubishi Chem. Corp.*

2137 4-Isopropylbiphenyl

25640-78-2   G H

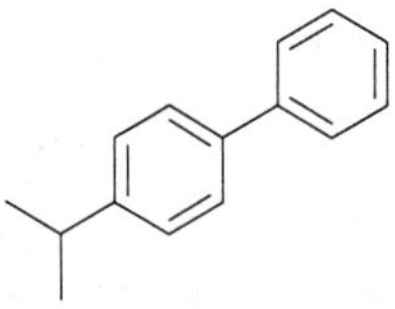

$C_{15}H_{16}$
(1-Methylethyl)-1,1'-biphenyl.
Biphenyl, isopropyl-; 1,1'-Biphenyl, (1-methylethyl)-; EINECS 247-156-8; HSDB 6160; Isopropylbiphenyl; 4-Isopropylbiphenyl; (1-Methylethyl)-1,1'-biphenyl; Mono-isopropylbiphenyl; Nusolv ABP-62; Wemcol. Alkyl biphenyl mixture; solvent possessing excellent solvency, chemical and thermal stability, nonvolatility; for Aerosols, adhesives, electronic parts manufacturing and cleaning, industrial cleaning, inks and paper coatings, metal cleaning, paints, plastics, sealants, tile manufacturing. *Ridge Tech.*

2138 Isopropyl chloroformate

108-23-6   G H

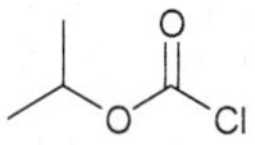

$C_4H_7ClO_2$
Chloroformic acid isopropyl ester.
4-03-00-00024 (Beilstein Handbook Reference); AI3-26269; BRN 0506416; Carbonochloridic acid, 1-methylethyl ester; Carbonochloridic acid, 1-methylethyl ester; Chloroformic acid isopropyl ester; EINECS 203-563-2; Formic acid, chloro-, isopropyl ester; HSDB 2848; Isopropyl chlorocarbonate; Isopropyl chloroformate; Isopropyl chloromethanoate; Isopropylester kyseliny chlormravenci; UN2407. Chemical intermediate for free-radical polymerization initiators, also used in organic synthesis. Liquid; bp = 105°; soluble in Et_2O. *BASF Corp.; Elf Atochem N. Am.; PPG Ind.*

2139 Isopropyl ethylthiocarbamate

141-98-0   H

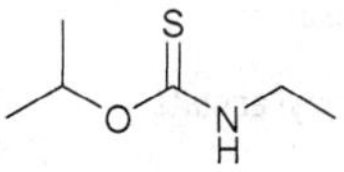

$C_6H_{13}NOS$
Ethylcarbamothioic acid, O-(1-methylethyl) ester.
Carbamic acid, ethylthio-, O-isopropyl ester; Carbamothioic acid, ethyl-, O-(1-methylethyl) ester; EINECS 205-517-7; Ethylcarbamothioic acid, O-(1-methylethyl) ester; O-Isopropyl ethylthiocarbamate; Z-200.

2140 Isopropyl formate

625-55-8   H

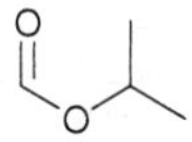

$C_4H_8O_2$
Formic acid, isopropyl ester.
4-02-00-00027 (Beilstein Handbook Reference); AI3-15407; BRN 1735844; EINECS 210-901-2; FEMA No. 2944; Formic acid, 1-methylethyl ester; Formic acid, isopropyl ester; HSDB 6401; Isopropyl formate; Isopropyl methanoate; 1-Methylethyl formate; NCI-C60106. Liquid; bp = 68.2°; d^{20} = 0.8728; λ_m = 213 nm (ε = 65, MeOH); slightly soluble in H_2O, CCl_4, soluble in $CHCl_3$, very soluble in Me_2CO, freely soluble in EtOH, Et_2O.

2141 4,4'-Isopropylidenebis(o-t-butylphenol)

79-96-9   H

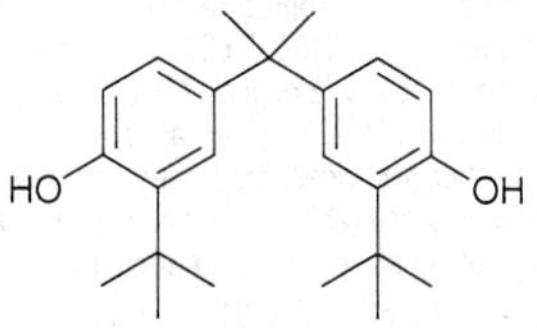

$C_{23}H_{32}O_2$
4,4'-(1-methylethylidene)bis(2-(1,1-dimethylethyl)phenol.
BRN 2294370; EINECS 201-239-5; 4,4'-Isopropylidene-bis(2-t-butylphenol); 4,4'-Isopropylidenebis(o-t-butyl-phenol); Phenol, (2,2'-di-t-butyl-4,4'-isopropylene)di-; Phenol, 4,4'-(1-methylethylidene)bis(2-(1,1-dimethyl-ethyl)-; Phenol, 4,4'-isopropylidenebis(2-t-butyl-; TBD.

2142 4,4'-((Isopropylidene)bis(p-phenyleneoxy))bis(N-methylphthalimide)

54395-52-7   H

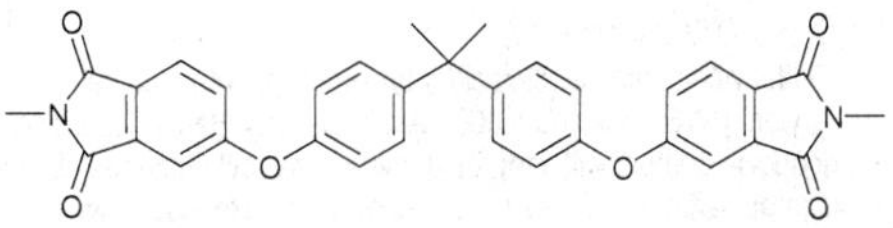

$C_{33}H_{26}N_2O_6$
N,N'-Dimethyl-2,2-bis(4-(3,4-dicarboxyphenoxy)-phenyl)propane diimide.
N,N'-Dimethyl-2,2-bis(4-(3,4-dicarboxyphenoxy)-phenyl)propane diimide; EINECS 259-143-4; 1H-Iso-indole-1,3(2H)-dione, 5,5'-((1-methylethylidene)bis(4,1-phenyleneoxy))bis(2-methyl-; 4,4'-((Isopropylidene)bis(p-phenyleneoxy))bis(N-methylphthalimide).

2143 Isopropyl mercaptan
75-33-2   G

SH (structure)

C_3H_8S
2-propanethiol.
AI3-22988; EINECS 200-861-4; HSDB 625; Isopropane-thiol; Isopropyl mercaptan; Isopropylthiol; 1-Methylethanethiol; NSC 87537; 2-Mercaptopropane; 2-Propanethiol; 2-Propylmercaptan. Unpleasant stench oder. Used as a leak detecting additive to natural gas. Liquid; mp = -130.5°; bp = 52.6°; d^{20} = 0.8143; poorly soluble in H_2O, very soluble in organic solvents. *Sigma-Aldrich Fine Chem.; U.S. BioChem.*

2144 Isopropyl methyl ketone
563-80-4   6113   H

(structure)

$C_5H_{10}O$
2-Methyl-3-butanone.
2-Acetyl propane; 2-Acetylpropane; AI3-24194; Caswell No. 555A; EINECS 209-264-3; EPA Pesticide Chemical Code 044104; Isopropyl methyl ketone; Ketone, isopropyl methyl; Methyl butanone-2; 3-Methylbutanone; Methyl isopropyl ketone; Methylbutanone; MIPK; NSC 9379; UN2397. Liquid; mp = -92°; bp = 94.3°; d^{20} = 0.8051; λ_m = 280 nm (ϵ = 21, EtOH); slightly soluble in H_2O, soluble in CCl_4, very soluble in Me_2CO, freely soluble in EtOH, Et_2O.

2145 Isopropyl myristate
110-27-0   5235   E G H

(structure) $C_{13}H_{27}$

$C_{17}H_{34}O_2$
1-Methylethyl tetradecanoate.
4-02-00-01132 (Beilstein Handbook Reference); Bisomel; BRN 1781127; Caswell No. 511E; Crodamol I.P.M.; Crodamol IPM; Deltyl Extra; Deltylextra; EINECS 203-751-4; Emcol-IM; Emerest 2314; EPA Pesticide Chemical Code 000207; Estergel; FEMA No. 3556; HSDB 626; IPM; Isomyst; Isopropyl myristate; IPM; Isopropyl tetradecanoate; JA-FA IPM; Kessco® IPM; Kessco isopropyl myristate; Kesscomir; Myristic acid, isopropyl ester; NSC 406280; Plymouth IPM; Promyr; Radia® 7190 Sinnoester MIP; Starfol IPM; Stepan D-50; Tegester; Tetradecanoic acid, 1-methylethyl ester; Tetradecanoic acid, isopropyl; Tetradecanoic acid, isopropyl ester; 1-Tridecanecarboxylic acid, isopropyl ester; Unimate® IPM; Wickenol® 101; Wickenol® 105; myristic acid isopropyl ester. Biodegradable replacement for mineral oil; used as lubricants in textile spin finish, coning oils, carding, dye bath; emollient, solubilizer, vehicle for makeup, shaving preparations, bath oils, hair preparations. Chemical intermediate, chemical synthesis; lubricant in mineral, cutting, lamination, textile oils, and rust inhibitors; textile and leather application; also as emollient, plasticizer, solubilizer of active components in cosmetics and pharmaceuticals. Cosmetic creams, topical medicinals. Liquid; bp_{20} = 193°, bp_2 = 140°; d^{20} = 0.8532; insoluble in H_2O, soluble in EtOH, Et_2O, $CHCl_3$, very soluble in Me_2CO, C_6H_6. *Amerchol; CasChem; Fina Chemicals; Goldschmidt; Henkel/Emery; Inolex; Lanaetex; Stepan; Unichema; Union Camp.*

2146 Isopropyl palmitate
142-91-6   G H

(structure) $C_{15}H_{31}$

$C_{19}H_{38}O_2$
1-Methylethyl hexadecanoate.
4-02-00-01167 (Beilstein Handbook Reference); AI3-05733; BRN 1786567; Crodamol IPP; Deltyl; Deltyl prime; EINECS 205-571-1; Emcol-IP; Emerest® 2316; Estol 103; Hexadecanoic acid, 1-methylethyl ester; Hexadecanoic acid, isopropyl ester; HSDB 2647; Isopal; Isopalm; Isopropyl hexadecanoate; Isopropyl n-hexadecanoate; Isopropyl palmitate; Ja-fa ippkessco; JA-FA Ipp; Kessco® IPP; Kessco isopropyl palmitate; Nikkol IPP; NSC 69169; Palmitic acid, isopropyl ester; Plymouth IPP; Propal; Radia® 7200; Sinnoester PIT; Starfol IPP; Stepan D-70; Tegester isopalm; Unimate® IPP; USAF KE-5; Wickenol® 111. Emollient and solvent for cosmetics, toiletries, makeups; nongreasing rub in. Colorless liquid with low viscosity; good penetration and spreading properties; non-oily feel, low viscosity, good penetration and spreading properties. Chemical intermediate, chemical synthesis; lubricant in mineral, cutting, lamination, textile oils, and rust inhibitors; textile and leather application; Lubricant used for synthetic fibers in applications where low friction is essential; emollient in cosmetic formulations; high purity. Liquid; mp = 13.5°; d = 0.852, bp_2 = 160°; d^{38} = 0.8404; insoluble in H_2O, very soluble in EtOH, Me_2CO, Et_2O, C_6H_6. *Amerchol; CasChem; Fina Chemicals; Goldschmidt; Henkel/Cospha; Henkel/Emery; Inolex; Lipo; Stepan; Unichema; Union Camp.*

2147 Isopropylphenol
25168-06-3   H

(structure) OH

$C_9H_{12}O$
Phenol, (1-methylethyl)-.
EINECS 246-699-8; Isopropylphenol; Phenol, (1-methylethyl)-; Phenol, isopropyl-.

2148 Isopropylphenyl diphenyl phosphate
28108-99-8   G

(structure)

$C_{21}H_{21}O_4P$
Phosphoric acid (1-methylethyl)phenyl diphenyl ester.
CCRIS 4661; EINECS 248-848-2; HSDB 6795; Isopropylphenyl diphenyl phosphate; Kronitex 100; Phosflex 41P; Phosphoric acid, (1-methylethyl)phenyl diphenyl ester; Syn-O-Ad® 8480. Antiwear and EP agents in non-crankcase lubricants; ferrous metal passivator; fluid base stock where inhibition of flame propagation is desired. Insoluble in H_2O, soluble in organic solvents. *Akzo Chemie.*

2149　Isopropyl stearate
112-10-7　　　　　　　　　　　　　　　　　　　G

$C_{21}H_{42}O_2$
1-Methylethyl octadecanoate.
4-02-00-01219 (Beilstein Handbook Reference); BRN 1791443; EINECS 203-934-9; Isopropyl stearate; Octadecanoic acid, 1-methylethyl ester; Stearic acid, isopropyl ester; Wickenol® 127. Emollient, cosolvent and lubricant. Oily solid; mp = 28°; bp6 = 207°; d^{34} = 0.8403; very soluble in Me_2CO, Et_2O, EtOH, $CHCl_3$. *CasChem.*

2150　p-Isopropyltoluene
99-87-6　　　　　2792　　　　　　　　　　H

$C_{10}H_{14}$
Benzene, 1-methyl-4-(1-methylethyl)-.
AI3-02272; Benzene, 1-isopropyl-4-methyl-; Benzene, 1-methyl-4-(1-methylethyl)-; Camphogen; Cumene, p-methyl-; Cymene, p-; p-Cymene; p-Cymol; Dolcymene; EINECS 202-796-7; FEMA No. 2356; HSDB 5128; p-Isopropylmethylbenzene; p-Isopropyltoluene; p-Methyl-cumene; p-Methylisopropylbenzene; NSC 4162; Paracymene; Paracymene; Paracymol; UN2046. Liquid; mp = -68.9°; bp = 177.1°; d^{20} = 0.8573; λ_m = 279 nm (cyclohexane); insoluble in H_2O, freely soluble in EtOH, Et_2O, Me_2CO, C_6H_6, CCl_4, petroleum ether.

2151　Isoproturon
34123-59-6　　　　　5238　　　　　　　H P

$C_{12}H_{18}N_2O$
N,N-Dimethyl-N'-[4-(1-methylethyl)phenyl]urea.
35689 R.P.; Alon; Arelon; Arelon R; Augur; Avanon; Belgran; BRN 2214033; CGA-18731; CL 12150; 3-p-Cumenyl-1,1-dimethylurea; DPX 6774; Graminon; HOE 16410; Hytane 500L; IP 50; IP-Flo; IPU Stefes; Isotop; Isoproturon; N-(4-Isopropylphenyl)-N',N'-dimethylharn-stoff; N-(Isopropyl-4-phenyl)-N',N'-dimethyluree; N-4-Isopropylphenyl-N,N-dimethylurea; 3-(4-Isopropyl-phenyl)-1,1-dimethylurea; N,N-Dimethyl-N'-(4-(1-methylethyl)phenyl)urea; Nocilon; Sabre; Swing; Tolkan; Urea, 1,1-dimethyl-3-(p-isopropylphenyl)-; Urea, 3-p-cumenyl-1,1-dimethyl-. Herbicide. Crystals; mp = 155 - 156°; soluble in H_2O (0.0072 g/100 ml), MeOH (5.6 g/100 ml), CH_2Cl_2 (7.3 g/100 ml), C_6H_6 (0.5 g/100 ml), C_6H_{14} (0.01 g/100 ml); LD_{50} (mrat orl) = 1826 mg/kg, (frat orl) = 2417 mg/kg, (mus orl) = 3350 mg/kg, (rat der) > 2000 mg/kg, (Japanese quail) > 3000 mg/kg, (pgn orl) > 5000 mg/kg; LC_{50} (rat ihl 4 hr.) > 0.67 mg/l air, (goldfish 96 hr.) = 100 mg/l, (bluegill sunfish 96 hr.) = 100 mg/l, (guppy 96 hr.) = 90 mg/l, (rainbow trout 96 hr.) = 240 mg/l; non-toxic to bees. *Aventis Crop Science; Ciba-Geigy Corp.; Gharda USA Inc.; Hoechst AG; Rhône-Poulenc; Syngenta Crop Protection.*

2152　Isosorbide Mononitrate
16051-77-7　　　　　5246　　　　　　　D

$C_6H_9NO_6$
1,4:3,6-Dianhydro-D-glucitol-5-mononitrate.
AHR-4698; AHR 4698; BM-22.145; BRN 5851319; CCRIS 1911; Conpin; Corangin; Duride; Edistol; EINECS 240-197-2; Elan; Elantan; Epicordin; Etimonis; Fem-Mono; IHD; Imazin; Imdur; Imdur Durules; Imodur; Imtrate; 5-Ismn; IS 5-MN; IS 5MN; Ismexin; ISMN; ISMO; Ismo-20; Ismox; Isomon; Isomonat; Isomonit; Isopen-20; Iturol; Medocor; Momo Mack; Monicor; Monis; Monit; Monizid; Mono-Cedocard; Mono Corax; Mono Mack; Mono-Sanorania; Monoclair; Monocord 20; Monocord 40; Monocord 50 SR; Monodur Durules; Monoket; Monolong; Mononit; Mononitrate d'isosorbide; Mononitrato de isosorbida; Monopront; MonoSigma; Monosorb; Monosorbitrate; Monosordil; Monotrate; Multitab; Nitex; Nitramin; Olicard; Olicardin; Orasorbil; Pentacard; Percorina; Pertil; Plodin; Promocard; Sigacora; Sorbimon; Titarane; Turimonit; UN3251; Uniket; Vasdilat; Vasotrate. A metabolite of isosorbide dinitrate. Coronary vasodilator used as an antianginal agent. White crystals; mp = 88-91°. *Boehringer Mannheim GmbH; Key Pharm.; Schwarz Pharma Kremers Urban Co.; Wyeth-Ayerst Labs.*

2153　Isostearic acid
2724-58-5　　　　　　　　　　　　　　　　　G

$C_{18}H_{36}O_2$
16-Methylheptadecanoic acid.
EINECS 220-336-3; Isooctadecanoic acid; Isostearic acid; 16-Methylheptadecanoic acid; 16-methylheptadecanoic acid; Emersol® 871; Proto-Lan IP; Prisorine 3508; Imwitor® 780K. Mixture of branched chain 18 carbon aliphatic acids; used in cosmetics, chemicals, dispersant, softener in rubber compounds, food packaging, suppositories and ointments. Liquid; d = 0.96-0.98; viscosity = 700-900 mPa·s; HLB=3.7; acid no = 3 max; iodine no = 10 max; sapon no = 240-260; LD_{50} (rat orl) >5 g/kg. *Degussa-Hüls Corp.; Henkel/Emery; Nissan Chem. Ind.; Unichema; Union Camp.*

2154　Isostearic acid
30399-84-9　　　　　　　　　　　　　　　　H

$C_{18}H_{36}O_2$
Isooctadecanoic acid.
875D; EINECS 250-178-0; Emersol 871; Emersol 875; Emery 871; Emery 875D; Haimaric MKH(R); Isooctadecanoic acid; Isostearic acid EX; Prisorine 3501; Prisorine 3502; Prisorine 3508; Unimac 5680.

2155　Isotridecyl isononanoate
59231-37-7　　　　　　　　　　　　G

$C_{22}H_{44}O_2$
Isotridecyl 3,5,5-trimethylhexanoate.
EINECS 261-675-7; Hexanoic acid, 3,5,5-trimethyl-, isotridecyl ester; Isomeric tetramethyl-1-nonyl 3,5,5-trimethyl hexanoate; Isotridecyl isononanoate; Isotridecyl 3,5,5-trimethylhexanoate; Wickeno® 153. Silky emollience and solvent characteristics for skin and hair care products; pigment wetter, moisturizer. *CasChem.*

2156　Isovalinonitrile
4475-95-0　　　　　　　　　　　　H

$C_5H_{10}N_2$
2-Amino-2-methylbutanenitrile.
2-Amino-2-methylbutyronitrile; Butanenitrile, 2-amino-2-methyl-; Butyronitrile, 2-amino-2-methyl-; EINECS 224-752-6; Isovalinonitrile; Vazo 67 aminonitrile.

2157　Isoxaben
82558-50-7　　　　5257　　　　G P

$C_{18}H_{24}N_2O_4$
N-[3-(1-Ethyl-1-methylpropyl)-5-isoxazolyl]-2,6-dimethoxybenzamide.
Benzamide, 2,6-dimethoxy-N-(3-(1-ethyl-1-methyl-propyl)-5-isoxazolyl)-; Benzamizole; Caswell No. 419F; Cent 7; Compound 121607; 2,6-Dimethoxy-N-(3-(1-ethyl-1-methylpropyl)-5-isoxazolyl)benzamide; EL 107; EPA Pesticide Chemical Code 125851; Flexidor; Gallery; Isoxaben; Knot-out; N-(3-(1-Ethyl-methylpropyl)-5-isox-azolyl)-2,6-dimethoxybenzamide; NA 8318; Prolan; Prolan (herbicide); Ratio; X-Pand; EL-107; EPA Pesticide Chemical Code 125851; Flexidor; Gallery; Isoxaben; N-(3-(1-Ethyl-methylpropyl)-5-isoxazolyl)-2,6-dimethoxy-benzamide; NA 8318. A residual herbicide for the control of broad-leaved weeds in winter and spring cereals, grass leys and herbage seed crops, used for control of annual dicotyledons in cereals, grass and fruit. Registered by EPA as a herbicide. Colorless crystals; mp = 176-179°; poorly soluble in H_2O (0.0001 g/100 ml), more soluble in MeOH (5 - 10 g/100 ml), EtOAc (5 -10 g/100 ml), CH_3CN (3 - 5 g/100 ml), C_7H_8 (0.4 - 0.5 g/100 ml), C_6H_{14} (0.007 - 0.008 g/100 ml); LD_{50} (rat mus orl) > 10000 mg/kg, (dog orl) > 5000 mg/kg, (rbt der) > 200 mg/kg, (rat ip) > 2000 mg/kg, (mus ip) > 5000 mg/kg, (bobwhite quail orl) > 2000 mg/kg; LC_{50} (rat ihl) > 1.99 mg/l air, (mallard duck, bobwhite quail 5 day dietary) > 5000 mg/kg, (bluegill sunfish, rainbow trout 96 hr.) > 1.1 mg/l, (*Daphnia 48 hr.) > 1.3 mg/l(Selenastrum capricornutum* 14 day) > 1.4 mg/l, non-toxic to bees. *Dow AgroSciences; Synchemicals Ltd; Tripart Farm Chemicals Ltd.*

2158　Itaconic acid
97-65-4　　　　　　5264　　　　H

$C_5H_6O_4$
Butanedioic acid, methylene-.
AI3-16901; Butanedioic acid, methylene-; EINECS 202-599-6; HSDB 5308; Itaconic acid; Methylenebutanedioic acid; Methylenesuccinic acid; NSC 3357; Propylenedicarboxylic acid; Succinic acid, methylene-. Crystals; mp = 175°; bp dec; d^{25} = 1.632; soluble in H_2O, EtOH, Me_2CO, $CHCl_3$, poorly soluble in Et_2O, C_6H_6, petroleum ether, CS_2.

2159　Jasmacyclat
61699-38-5　　　　　　　　　　　　G

$C_{10}H_{18}O_3$
Methylcyclooctylcarbonate.
BRN 2254088; Carbonic acid, cyclooctyl methyl ester; Cyclooctyl methyl carbonate; EINECS 262-912-7; Jasmacyclat; Methyl cyclooctyl carbonate. Raw material for fragrance in floral notes. Henkel/Cospha.

2160　Jasmacyclene
2500-83-6　　　　　　　　　　　　G

$C_{12}H_{16}O_2$
Tricyclodecenyl acetate.
Cyclacet; EINECS 219-700-4; Greenyl acetate; Herbyl acetate; Hexahydro-4,7-methanoinden-5(or 6)-yl acetate; 3a,4,5,6,7,7a-Hexahydro-4,7-methano-1H-inden-5-yl acetate; 4,7-Methano-1H-inden-5-ol, 3a,4,5,6,7,7a-hexahydro-, acetate; 4,7-Methanoinden-5-ol, 3a,4,5,6,7,7a-hexahydro-, acetate; Jasmacyclene; NSC 142428; Tricyclo(5.2.1.0^{2,6})dec-3-en-9-yl acetate; Tricyclodecenyl acetate; Verdyl acetate. Used in perfumery. *Quest Intl.*

2161　Jasmopyrane
18871-14-2　　　　　　　　　　　　G

$C_{12}H_{22}O_3$
4-Acetoxy-3-pentyltetrahydropyran.
4-Acetoxy-3-pentyltetrahydropyran; 3-Amyl-4-acetoxy-tetrahydropyran; EINECS 242-640-5; Jasmal; Jasmophyll; Jasmopyrane; 2H-Pyran-4-ol, tetrahydro-3-pentyl-, acetate; Tetrahydro-3-pentyl-2H-pyran-4-yl acetate. Used in perfumery as a fresh, sweet, oily, jasmin, watery fragrance. *Quest UK.*

2162 Jodfenphos

18181-70-9   5054   G P

$C_8H_8Cl_2IO_3PS$

O-(2,5-Dichloro-4-iodophenyl) O,O-dimethyl phosphoro-thioate.

Al3-27408; Alfacron; BRN 1885795; C 9491; Caswell No. 309B; Ciba C-9491; Ciba-Geigy C-9491; Compound C-9491; O-(2,5-Dichloro-4-iodophenyl) O,O-dimethyl phosphorothioate; O,O-Dimethyl-O-(2,5-dichlor-4-jod-phenyl)-thionophosphat; O,O-Dimethyl-O-(2,5-dichlor-4-jodphenyl)-monothiophosphat; O,O-Dimethyl-O-2,5-di-chloro-4-iodophenyl thiophosphate; EINECS 242-069-1; ENT 27408; EPA Pesticide Chemical Code 309700; HSDB 1568; Iodofenfos; Iodofenphos; Iodophenphos; Iodophos; Jodfenphos; Monocron 9491; NSC 190998; Nuvanol N; OMS-1211; Phosphorothioic acid, O-(2,5-dichloro-4-iodophenyl) O,O-dimethyl ester; Trix; O,O-Dimethyl-O-(2,5-dichlor-4-jodphenyl)-monothiophosphat [; ENT 27408; EPA Pesticide Chemical Code 309700; HSDB 1568; Iodofenfos; Iodofenphos; Iodophenphos; Iodophos; Jodfenphos; Monocron 9491; NSC 190998; Nuvanol N; OMS-1211; Phenol, 3,4-dichloro-, O-ester with O-methyl methylphosphoramidothioate; Phosphoro-thioic acid, O-(2,5-dichloro-4-iodophenyl) O,O-dimethyl ester; Trix. An organophosphorus insecticide and acaricide. Colorless crystals; mp = 76°; sg^{20} = 2.0; poorly soluble in H_2O (< 0.0002 g/100 ml), soluble in Me_2CO (48 g/100 ml), CH_2Cl_2 (86 g/100 ml), C_6H_6 (61 g/100 ml), i-PrOH (23 g/100 ml), C_6H_{14} (3.3 g/100 ml); LD_{50} (rat orl) = 2100 mg/kg, (mus orl) = 3000 mg/kg, (rbt orl) = 2000 mg/kg, (dog orl) = 3000 mg/kg, (rat der) >2000 mg/kg, (rbt der) > 500 mg/kg; LC_{50} (rat ihl 6 hr.) > 0.246 mg/l air; (rainbow trout 96 hr.) = 0.06 - 0.10 mg/l, (goldfish 96 hr.) = 1.00 - 1.33 mg/l, (bluegill sunfish 96 hr.) = 0.42 - 0.75 mg/l; toxic to bees. *Ciba Geigy Agrochemicals.*

2163 Juglone

481-39-0   5288   G

$C_{10}H_6O_3$

5-Hydroxynaphthoquinone.

4-08-00-02368 (Beilstein Handbook Reference); Akhnot; BRN 1909764; C.I. 75500; C.I. Natural Brown 7; Caswell No. 515AA; CCRIS 5423; EINECS 207-567-5; 5-Hydroxy-1,4-naftochinon; 5-Hydroxy-1;4-naphthalenedione; 5-Hydroxynaphthoquinone; 5-Hydroxy-1,4-naphtho-quinone; 8-Hydroxy-1,4-naphthoquinone; 5-Hydroxy-1,4-naphthosemiquinone; Juglon; Juglone; 1,4-Naph-thalenedione, 5-hydroxy-; 1,4-Naphthoquinone, 5-hydroxy-; 1,4-Naphthoquinone, 8-hydroxy-; NSC 153189; Nucin; Regianin; Walnut extract; Yuglon. Brown pigment from the walnut. Used as a chemical intermediate. mp = 155°; λ_m = 420 nm (log ε 3.56 MeOH), 248 nm (ε = 12900, MeOH); insoluble in cold H_2O, slightly soluble in ligroin, soluble EtOH, Et_2O, C_6H_6, very soluble in $CHCl_3$, AcOH; LD_{50} (rat orl) = 112 mg/kg.

2164 Kaempferol

520-18-3   5293   G

$C_{15}H_{10}O_6$

3,5,7-Trihydroxy-2-(4-hydroxyphenyl)-4H-1-benzopyran-4-one.

5-18-05-00251 (Beilstein Handbook Reference); Al3-36096; 4H-1-Benzopyran-4-one, 3,5,7-trihydroxy-2-(4-hydroxyphenyl)-; BRN 0304401; C.I. 75640; Campherol; CCRIS 41; EINECS 208-287-6; Flavone, 3,4',5,7-tetra-hydroxy-; Indigo Yellow; Kaemferol; Kaempferol; Kaem-pherol; Kampferol; Kampherol; Kempferol; Nimbecetin; NSC 407289; Pelargidenolon; Pelargidenolon 1497; Populnetin; Rhamnolutein; Rhamnolutin; Robigenin; Swartziol; Trifolitin; 3,4',5,7-Tetrahydroxyflavone; 5,7,4'-Trihydroxyflavonol; 3,5,7-Trihydroxy-2-(4-hydroxyphen-yl)-4H-1-benzopyran-4-one. The coloring matter of the blue flowers of *Delphinium consolida.* Yellow needles; mp = 276-278°; λ_m = 266, 368 nm (ε = 18197, 11220, EtOH); insoluble in C_6H_6, slightly soluble in H_2O, $CHCl_3$, soluble in alkali, very soluble in EtOH, Et_2O, Me_2CO.

2165 Kainite

1318-72-5   G

$ClKMgO_4S.3H_2O$

Kainite.

Kainite; Potassium magnesiom sulfate-magnesium chloride. A salt found in the Stassfurt deposits, consisting mainly of potassium magnesium sulfate and magnesium chloride. The crude material consists of a mixture of kainite and rock salt, and contains 23% of pot; used in chemicals and fertilizers.

2166 Kamillosan

23089-26-1   1241   G

$C_{15}H_{26}O$

α-(-)-Bisabolol.

-α-Bisabolol; α-Bisabolol (-)-form; α-Bisabolol, l-; α-Bisabolol, L-; Bisbalol; 3-Cyclohexene-1-methanol, α,4-dimethyl-α-(4-methyl-3-pentenyl)-, (S)-; 1-methyl-4-(1,5-dimethyl-1-hydroxyhex-4(5)-enyl)cyclohexen-1; 6-meth-yl-2-(4-methyl-3-cyclohexene). Pharmaceutical prep-aration containing the active principle of *Matricaria chamomilla*; used as an anti-inflammatory. Liquid; bp_{12} = 153°; d^{20} = 0.9211; $[α]_D$ = -55.7°. *Degussa AG.*

2167 Kanamycin A Sulfate

25389-94-0   5299   D G

356

C18H36N4O11.H2SO4

O-3-Amino-3-deoxy-α-D-glucopyranosyl-(1→6)-O-[6-amino-6-deoxy-α-D-glucopyranosyl-(1→4)]-2-deoxy-D-streptamine sulfate.

Amforol; Cantrex; Cristalomicina; EINECS 246-933-9; Enterokanacin; Kamycin; Kanabristol; Kanacedin; Kanamycin A sulfate; Kanamycin monosulfate; Kanamycin sulfate; Kanamytrex; Kanaqua; Kanasig; Kanatrol; Kanescin; Kanicin; Kannasyn; Kano; Kantrex; Kantrexil; Kantrim; Kantrox; Kasmynex; Klebcil; Ophtalmokalixan; Otokalixin; Resistomycin.; component of: Amforol. Aminoglycoside antibiotic from *Streptomyces kanamyceticus*. Powder; dec > 250°; freely soluble in H2O; insoluble in most organic solvents; LD50 (mus orl) = 20700 mg/kg, (mus ip) = 1450 mg/kg. *Apothecon; Bristol-Myers Squibb Pharm. R&D; Merck & Co.Inc.; SmithKline Beecham Pharm.*

2168   Kaolin
1332-58-7   G

Al2O3.2SiO2.2H2O
Hydrated aluminum silicate.

AA Kaolin; Acidic white clays; Airflo V 8; Alfaplate; Alphacoat; Alphagloss; Altowhite; Altowhites; Amazon 88; Amazon 90; Amazon Kaolin 855D; Apsilex; Arcilla blanca; Argiflex; Argilla; Argilla alba; Argirec B 22; Argirec KN 15; ASP (mineral); ASP Ultrafine; Astra-Glaze; Bentone; Beta Coat; Bilt Plate 156; BOL Blanc; Bolus alba; Burgess 10; CB 1 (clay); CB 2 (clay); Century HC; China clay; Clay (kaolin); Clay 347; Clay slurry; Clays, white, acidic; Comalco; Comalco Kaolin; Donnagel; Electros; Emathlite; EPA Pesticide Chemical Code 100104; Fitrol; Fitrol desiccite 25; Glomax; HSDB 630; Hydrite; Kao-gel; Kaolin; Kaolin clay slurry; Kaolin colloidal; Kaopaous; Kaopectate; Kaophills-2; Kayphobe-ABO; Langford; Light kaolin; Mcnamee; Osmo kaolin; Parclay; Porcelain clay; Snow tex; Vanclay; White bole. Filler and coatings for paper, rubber, refractories, ceramics; in anticaking preparations, paint; adsorbent for clarification of liquids. *Burgess Pigment; Dry Branch Kaolin; ECC Intl.; Huber J. M.; Kaopolite; Southern Clay Products; Vanderbilt R.T. Co. Inc.*

2169   Karate
68085-85-8   2787   G P

C23H19ClF3NO3
λ-cyhalothrin.

3-(2-Chloro-3,3,3-trifluoro-1-propenyl)-2,2-dimethyl-cyclopropanecarboxylic acid cyano(3-phenoxyphenyl)-methyl ester; Grenade. Pyrethroid insecticide. White solid; mp = 49.2°; poorly soluble in H2O (0.5 g/100 ml), soluble in organic solvents; LD50 (mrat orl) = 79 mg/kg, (frat orl) = 56 mg/kg, (mrat der) = 632 mg/kg, (frat der) = 696 mg/kg. *ICI Chem. & Polymers Ltd.*

2170   Kemamide® W-20
110-31-6   G

C38H72N2O2
N,N'-Ethylenebisoleamide.

N,N'-Dioleoylethylenediamine; EINECS 203-756-1; N,N'-1,2-Ethanediylbis-9-octadecenamide; N,N'-Ethane-1,2-diylbisoleamide; N,N'-Ethylenebisoleamide; N,N'-Ethylenedioleamide; Ethylene bis(oleamide); Ethylene dioleamide; NSC 131419; 9-Octadecenamide, N,N'-1,2-ethanediylbis-, (Z,Z)-; 9-Octadecenamide, N,N'-1,2-ethanediylbis-, (9Z,9'Z)-; Oleamide, N,N'-ethylenebis- (8CI); Oleic acid-ethylenediamine condensate. Lubricant, slip, antiblock, and mold release agent for plastics, crayons, petrol. products, asphalts, inks, metals, textiles; mold release agent for thermoplastic resins in injection molding; defoamer and water repellent in industrial/household Solid; mp = 115-118°. *CK Witco Chem. Corp.*

2171   Kerosene
8008-20-6   5311   G H P

Petroleum base oil.

AF 100 (pesticide); Avtur; Avtur (pesticide); Bayol 35; Bitumen Cutter; Caswell No. 517; CCRIS 1359; Coal oil; Deodorized kerosene; Distillate fuel oils, light; EINECS 232-366-4; EPA Pesticide Chemical Code 063501; Escaid 100; Escaid 110; Exxsol D 200/240; Fuel No. 1 [Oil, fuel]; Fuel oil No. 1; Fuels, kerosine; HSDB 632; Ink oil; Jet fuels, JP-5; Jp-5 navy fuel/marine diesel fuel; JP5 Jet fuel; Kerosene; Kerosene (deodorized); Kerosene, straight run; Kerosene [Flammable liquid]; Kerosine; Kerosine (petroleum); KO 30 (solvent); Navy fuels JP-5; Neochiozol; Nysolvin 75A; Odorless Solvent 3440; Parasol; Pegasol 3040; Range-oil; Shell 140; Shellsol 2046; Straight-run kerosene; SX 7; SX 12; UN1223. Registered by EPA as an insecticide and herbicide. Colorless mobile liquid; d = 0.8, bp = 175 -325°; insoluble in H2O, soluble in most non-polar organic solvents; LD50 (rbt orl) = 22400 mg/kg. *Drexel Chemical Co.; Triangle; UAP - Platte Chemical.*

2172   α-Ketoglutaric acid
328-50-7   5320   G

C5H6O5
2-Oxopentanedioic acid.

AI3-26938; Bis(L-arginin)-2-oxoglutarat; EINECS 206-330-3; Glutaric acid, α-keto; Glutaric acid, 2-oxo-; α-Ketoglutaric acid; α-Oxoglutaric acid; NSC 17391; 2-Oxoglutaric acid; 2-Oxopentanedioic acid; 2-Oxo-1,5-pentanedioic acid; Pentanedioic acid, 2-oxo-. Crystals; mp = 115.5°; λ_m = 312 nm (MeOH); very soluble in H2O, EtOH, Et2O, Me2CO.. *Am. Biorganics; Lancaster Synthesis Co.; Penta Mfg.; U.S. BioChem.*

2173   Krystallazurin
14283-05-7   9256   G

CuH12N4O4S
Tetraamminecopper sulfate.

Copper ammine sulfate; Copper tetraammine sulfate; Copper(2+), tetraammine-, sulfate (1:1); Cupric sulfate ammoniate; Cupric sulfate-tetraammonia complex; EINECS 238-177-3; NSC 305303; Tetraaminecopper sulfate; Tetraamminecopper sulfate; Tetraamminecopper sulphate; Tetraamminecopper(II) sulfate. A fungicide consisting of ammoniacal copper sulfate. [monohydrate]; large dark blue crystals; d_4^{20} 1.81; solulbe in H_2O (18.5 g/100 ml).

2174 Lactic acid

50-21-5   5351   E G

$C_3H_6O_3$
2-Hydroxypropanoic acid.
Acetonic acid; Acidum lacticum; Aethylidenmilchsaeure; AI3-03130; BRN 5238667; CCRIS 2951; E270; Eco-Lac®; EINECS 200-018-0; EINECS 209-954-6; EPA Pesticide Chemical Code 128929; Ethylidenelactic acid; FEMA Number 2611; HSDB 800; Kyselina 2-hydroxypropanova; Kyselina mlecna; L18; Lactic acid; DL-Lactic acid; Lactovagan; Milchsäure; Milchsaeure; DL-Milchsaeure; Milchsaure; Milk acid; NSC 367919; Patlac LAS®; Propanoic acid, 2-hydroxy-; Propel; Propionic acid, 2-hydroxy-; Purac 88 PH; Purac®; Racemic lactic acid; SY-83; Tonsillosan. Organic acid; cultured dairy products, as acidulant, chemicals (salts, plasticizers, adhesives, pharmaceuticals), mordant in wool dyeing, food additive, manufacture of lactates. Registered by EPA as a plant growth regulator. FDA, USDA, FEMA GRAS (not to be used in infant foods); Japan, UK approved, Europe listed, FDA approved for injectables, parenterals, orals, topicals, vaginals. USP/NF, BP, Ph.Eur., JP compliance. In the FDA list of inactive ingredients, (injectables, oral syrups and tablets, topicals and vaginals). Crystals; mp = 17°; bp15 = 122°, bp0.5-1 = 82 - 85°; d = 1.249; λ_m = 264.5 nm ($E_{1cm}^{1\%}$ 459, hexane); LD50 (rat orl) = 3730 mg/kg; soluble with water, glycerol, furfural, insoluble in CHCl3, CS2, petroleum ether. *AB Technology Ltd.; ADM; Ashland; Chr Hansen's Lab.; Dinoval; Ellis & Everard; Honeywell & Stein; Integra; Jungbunzlauer; Lohmann; Mallinckrodt Inc.; Mitsubishi Corp.; Penta Mfg.; Pfanstiehl Labs Inc.; Pointing; Purac Biochem. BV; Ruger; Sibner-Hegner; Spectrum Chem. Manufacturing; Todd's; Van Waters & Rogers.*

2175 D-lactic acid

10326-41-7   5350   G

$C_3H_6O_3$
(R)-2-hydroxypropanoic acid.
EINECS 233-713-2; (R)-α-Hydroxypropionic acid; (R)-2-Hydroxypropanoic acid; (R)-2-Hydroxypropionsaeure; (R)-2-Hydroxypropionsäure; Lactic acid (D); 1-Lactic acid; (-)-Lactic acid; D-Lactic acid; (D)-(-)-Lactic acid; (R)-Lactic acid; (R)-(-)-Lactic acid; (R)-Milchsaeure; D-Milchsaeure; (R)-Milchsäure; D-Milchsäure; Propanoic acid, 2-hydroxy-, (2R)-; Propanoic acid, 2-hydroxy-, (R)-. Crystals; mp = 53°; $[\alpha]_{546}^{22}$ = -2.6° (c = 8); soluble in H_2O.

2176 DL-lactic acid

598-82-3   5351   G P V

$C_3H_6O_3$
2-hydroxypropanoic acid.
Acidum lacticum; Aethylidenmilchsaeure; Äthylidenmilchsäure; AI3-03130; BRN 5238667; CCRIS 2951; EINECS 200-018-0; EINECS 209-954-4; EPA Pesticide Chemical Code 128929; Ethylidenelactic acid; FEMA Number 2611; HSDB 800; 1-Hydroxyethane-carboxylic acid; 2-Hydroxypropanoic acid; (±)-2-Hydroxypropanoic acid; α-Hydroxypropionic acid; 2-Hydroxypropionic acid; (RS)-2-Hydroxypropionsaeure; Kyselina 2-hydroxypropanova; Kyselina mlecna; Lactate; Lactic acid; DL-Lactic acid; Lactovagan; Milchsaeure; DL-Milchsaeure; Milchsäure; DL-Milchsäure; Milk acid; NSC 367919; Ordinary lactic acid; Propanoic acid, 2-hydroxy-; Propel; Propionic acid, 2-hydroxy-; Racemic lactic acid; SY-83; Tonsillosan. Occurs in sour milk. Used as an acidulant and, in veterinary medicine, as an internal antiseptic. Also used as a plant growth regulator. Registered by EPA as a plant growth regulator. Yellow liquid; mp = 18°; bp15 = 122°; d^{21} = 1.2060; very soluble in H_2O, EtOH, soluble in Et2O; LD50 (rat orl) = 3.73 g/kg; LD50 (rat orl) = 3730 mg/kg, (mus orl) = 4880 mg/kg, (mus sc) = 4500 mg/kg, (gpg orl) = 1810 mg/kg. *AB Technology Ltd.; ADM; Ashland; Chr Hansen's Lab.; Ellis & Everard; Honeywell & Stein; Integra; Jungbunzlauer; Mallinckrodt Inc.; Mitsubishi Corp.; Penta Mfg.; Pfanstiehl Labs Inc.; Pointing; Purac Biochem. BV; Ruger; Sibner-Hegner; Spectrum Chem. Manufacturing; Todd's; Van Waters & Rogers.*

2177 L-lactic acid

79-33-4   5352   G

$C_3H_6O_3$
(S)-2-hydroxypropanoic acid.
d-lactic acid; sarcolactic acid; paralactic acid. Crystals; mp = 53°; $[\alpha]_{546}^{22}$ = 2.6° (c = 2.5); soluble in H_2O.

2178 Lactofen

77501-63-4   P

$C_{19}H_{15}ClF_3NO_7$
5-(2-Chloro-4-(trifluoromethyl)phenoxy)-2-nitrobenzoic acid2-ethoxy-1-methyl-2-oxoethyl ester.
Benzoic acid, 5-(2-chloro-4-(trifluoromethyl)phenoxy)-2-nitro-, 2-ethoxy-1-methyl-2-oxoethyl ester; 1'-(Carbo-ethoxy)ethyl5-(2-chloro-4-(trifluoromethyl)phenoxy)-2-nitrobenzoate; 5-(2-Chloro-4-(trifluoromethyl)phenoxy)-2-nitrobenzoic acid 2-ethoxy-1-methyl-2-oxoethyl ester; Cobra; Cobra 2E; Cobra herbicide; EPA Pesticide Chemical Code 128888; HSDB 6991; Lactofen; PPG 844; Cobra; Cobra 2E; Cobra herbicide; (±)-2-Ethoxy-1-methyl-2-oxoethyl5-(2-chloro-4-(trifluoromethyl)phenoxy)-2 nitrobenzoate; Lactofen; PPG 844; [83513-60-4].
Herbicide. Dark brown oil; almost insoluble in H_2O (< 0.0001 g/100 ml at 20°); LD50 (rat orl) > 5000 mg/kg, (rat der) = 2000 mg/kg. *BASF Corp.; PPG Ind.; Valent USA Corp.*

2179   Lactonitrile
78-97-7   H

HO
  \
   ∕—CN

C3H5NO

2-Hydroxypropionitrile.
3-03-00-00451 (Beilstein Handbook Reference); Acetaldehyde, cyanohydrin; Acetocyanohydrin; AI3-24276; BRN 0605366; CCRIS 6048; EINECS 201-163-2; EINECS 255-852-8; Ethylidene cyanohydrin; HSDB 5225; Lactonitrile; Laktonitril; NSC 7764; Propanenitrile, 2-hydroxy-; Propionitrile, 2-hydroxy-. Used as a solvent and an intermediate in production of ethyl lactate and lactic acid. Liquid; mp = -40°; bp = 183°; d^{20} = 0.9877; freely soluble in H_2O, organic solvents; LD_{50} (mrat orl) = 31 mg/kg; (frat orl) = 41 mg/kg; LCLo (rat ihl) = 124 ppm, 4 hr. *Lancaster Synthesis Co.; Mallinckrodt Inc.; Sigma-Aldrich Fine Chem.; Supernutrition.*

2180   Lactose
63-42-3   5358   G P

C12H22O11

4-O-β-D-Galactopyranosyl-D-glucose.
4-17-00-03066 (Beilstein Handbook Reference); AHL; AI3-08876; Aletobiose; BRN 0093796; CCRIS 7078; EINECS 200-559-2; Fast-flo; Fast-flo Lactose; Galactinum; 4-O-β-D-Galactopyranosyl-D-glucose; 4-(β-D-Galactos-ido)-D-glucose; D-Glucose, 4-O-β-D-galactopyranosyl-; Lactin; Lactin (carbohydrate); Lactobiose; Lactose; D-Lactose; D-(+)-Lactose; (+)-Lactose; Lactose; Lactose, anhydrous; Lactose Fast-flo; Milk sugar; Pharmatose 21; Pharmatose 450M; Saccharum lactin; Tablettose; Zeparox EP. Used to supplement animal feeds. Two crystalline forms. α form: mp = 201 - 202°$[\alpha]_D^{20}$= +92.6°→+83.5°. β form: $[\alpha]_D^{25}$= +39°→+52.3°.

2181   α-D-Lactose
5989-81-1   G

C12H22O11

4-O-β-D-Galactopyranosyl-D-glucose.
α-D-Glucopyranose, 4-O-β-D-galactopyranosyl-, monohydrate; α-Lactose monohydrate; Milk sugar; d-glucose; 4-(β-D-galactosido)-D-glucose; Saccharum lactis; Disaccharide. Pharmacy; infant foods; baking and confectionary; manufacture of penicillin, yeast; adsorbent in chromatography. White solid; d = 1.23; mp = 201-202°; $[\alpha]_D^{20}$= +96°; soluble in H_2O, EtOH, insoluble in CHCl3, Et2O. *Dajac Labs.;*

Lancaster Synthesis Co.; Penta Mfg.; Simonis BV.

2182   Lanolin
8006-54-0   5373   H P

Lanolin.
Adeps lanae; Agnin; Agnolin; alapurin; Anhydrous Lanolin Grade P.95; Anhydrous Lanolin Grade 2; Anhydrous Lanolin Grade 1; Anhydrous Lanolin P.80; Anhydrous Lanolin Superfine; Anhydrous Lanolin USP Pharmaceutical Grade; Anhydrous Lanolin USP Deodorized AAA; Anhydrous Lanolin USP Cosmetic Grade; Anhydrous Lanolin USP; Anhydrous Lanolin USP Cosmetic AA; Anhydrous Lanolin USP Cosmetic; Anhydrous Lanolin USP Pharmaceutical; Clearlin; Corona®; Glossylan; Lanain; Lanalin; Lanesin; Lanichol; Laniol; Lanolin anhydrous; Lanum; Oesipos; Purified lanolin; Refined wool fat; Wool alcohol; Wool alcohols; Wool fat; Wool grease; Wool wax; Woolwax ester. Emulsifying agent, ointment base. Japan approved, FDA approved for use in ophthalmics and topicals, USP/NF, BP, Ph.Eur. compliance. In the FDA list of inactive ingredients, (ophthalmics, otics, topicals and vaginals). Used as an emollient, ointment base, filler, emulsifier, vehicle. Used in ophthalmics, topicals, suppositories, as a protectant against diaper rash, in hemorrhoidal and antibiotic ointments. Solid; mp = 38-40°; d^{15} = 0.932 - 0.945; insoluble in H_2O, soluble in most organic solvents. Non-toxic, possible allergenic. *Amerchol; Croda; Henkel/Emery; Integra; Lanaetex; Rita; Ruger; Sigma-Aldrich Fine Chem.; Spectrum Chem. Manufacturing; Stevenson Cooper; Westbrook Lanolin.*

2183   Lansoprazole
103577-45-3   5378   D

C16H14F3N3O2S

2-[[[3-Methyl-4-(2,2,2-trifluoroethoxy)-2-pyridyl]methyl]-sulfinyl]benzimidazole.
A-65006; AG 1749; Agopton; Amarin; Aprazol; Bamalite; Blason; BRN 4333393; Compraz; Dakar; Ilsatec; Ketian; Lancid; Lanproton; Lansopep; Lansoprazol; Lansoprazole; Lansoprazolum; Lanston; Lanz; Lanzol-30; Lanzopral; Lanzor; Lasoprol; Limpidex; Mesactol; Monolitum; Ogast; Ogastro; Opiren; Prevacid; Prezal; Pro Ulco; Promp; Prosogan; Suprecid; Takepron; Ulpax; Zoprol; Zoton. Gastric proton pump inhibitor. Antiulcerative. Crystals; mp = 178-182° (dec). *Takeda Chem. Ind. Ltd.*

2184   Lanthanum
7439-91-0   5379   G

La

La
Lanthanum.
EINECS 231-099-0; Lanthanum. Lanthanum salts, electronic devices, pyrophoric alloys, rocket propellants, reducing agent catalyst for conversion of nitrogen oxides to nitrogen in exhaust gases, phosphors in x-ray screens. Metallic solid; mp = 920°; bp = 3464°; d = 6.162; reacts slowly with cold water, more readily on heating. *Atomergic Chemetals; Cerac; Rhône-Poulenc; Sigma-Aldrich Fine Chem.*

2185   Lanthanum chloride
10099-58-8   G

Cl₃La

Lanthanum(III) chloride anhydrous.

CCRIS 6887; EINECS 233-237-5; Lanthanum chloride; Lanthanum chloride (LaCl₃); Lanthanum chloride, anhydrous; Lanthanum trichloride; Lanthanum(III) chloride. Anhydrous trichloride used to prepare the rare-earth metal. Solid; mp = 860°; bp = 1000°. *Atomergic Chemetals; Cerac; Spectrum Chem. Manufacturing.*

2186   Lanthanum nitrate
10099-59-9   G

LaN₃O₉.6H₂O

Lanthanum(III) nitrate hexahydrate.

Lanthanum nitrate (La(NO₃)₃); Lanthanum trinitrate; Lanthanum(III) nitrate; Nitric acid, lanthanum (3+) salt. Antiseptic, gas mantles. *Atomergic Chemetals; Cerac; Rhône-Poulenc.*

2187   Lanthanum oxide
1312-81-8   G

La₂O₃

Lanthanum(III) oxide.

Dilanthanum oxide; Dilanthanum trioxide; EINECS 215-200-5; Lanthana; Lanthania (La₂O₃); Lanthanum oxide; Lanthanum sesquioxide; Lanthanum trioxide; Lanthanum(3+) oxide. Used in calcium lights, optical glass, technical ceramics, cores for carbon-arc electrodes, fluorescent phosphors, refractories. Solid; mp = 2315°; bp = 4200°; d = 6.5100. *Atomergic Chemetals; Cerac; Rhône-Poulenc; Seimi Chem.*

2188   Lapyrium chloride
6272-74-8   5385   G

C₂₁H₃₅ClN₂O₃

1-[[2-Oxo-2-[(1-oxododecyl)oxy]ethyl]amino]ethyl]pyridinium chloride.

N-(Acylcolaminoformylmethyl)pyridinum chloride; Caswell No. 519; Chlorure de lapirium; Cloruro de lapirio; N-(Colaminoformylmethyl)pyridinium chloride laurate; N-((N-(2-Dodecanoyloxyethyl)carbamoyl)-methyl)pyridinium chloride; EINECS 228-464-1; Emcol E-607; EPA Pesticide Chemical Code 069131; 1-(((2-Hydroxyethyl)carbamoyl)methyl)pyridinium chloride laurate; 1-(2-Hydroxyethyl)carbamoyl methyl pyridinium chloride laurate; Lapirii chloridum; Lapirium chloride; Lapyrium chloride; N-(Lauroyl colamino formylmethyl)-pyridinium chloride; N-(Lauroylcolaminoformylmethyl)-pyridinium chloride; NSC-33659; 1-(2-Oxo-2-((2-((1-oxododecyl)oxy)ethyl)amino)ethyl)pyridinium chloride; Pyridinium, 1-(((2-hydroxyethyl)carbamoyl)methyl)-, chloride, laurate; Pyridinium, 1-(2-oxo-2-((2-((1-oxo-dodecyl)oxy)ethyl)-amino)ethyl)-, chloride; Pyridinium, 1-(2-hydroxyethylcarbamoylmethyl)-, chloride, dodecan-

oate; 1-[[[(2-hydroxyethyl)carbamoyl]methyl]pyridinium chloride laurate (ester); N-(lauroylcolaminoformyl-methyl)pyridinium chloride; N-(acylcolaminoformyl-methyl)pyridinium chloride; emulsepr (obsolete); E-607; Emcol E-607. Emollient, emulsifier, conditioner, foamer, cleanser, substantive agent, deodorant for cosmetics, toiletries, industrial applications; hair conditioner; antistat; detergent-germicide. White powder. *CK Witco Chem. Corp.; CK Witco Corp.*

2189   Lard
61789-99-9   H

Lard.

EINECS 263-100-5; Fat, lard; Lard.

2190   Lard oil
8016-28-2   H

Lard oil.

EINECS 232-405-5; HSDB 5150; Lard oil; Lard oil [Oil, edible]; Oil, lard; Oils, lard.

2191   Latanoprost
130209-82-4   5391   D

C₂₆H₄₀O₅

Isopropyl (Z)-7-[(1R,2R,3R,5S)-3,5-dihydroxy-2-[(3R)-3-hydroxy-5-phenylpentyl]-cyclopentyl]-6-heptenoate.

Latanoprost; PhXA 41; XA 41; Xalatan. Prostaglandin used as an antiglaucoma agent. [α]$_D^{20}$ = 31.57° (c = 0.91 CH₃CN). *Kabi Pharmacia Diagnostics; Pharmacia & Upjohn.*

2192   Laureth 4
9002-92-0   7641   D E G

C₁₄H₃₀O₂

Polidocanol.

40L (polyether); Actinol L 7; Actinol L3; Adeka Carpol M 2; Adeka Carpol MBF 100; Adekatol LA 1275; Aethoxysklerol; Akyporox RLM 22; Akyporox RLM 40; Akyporox RLM 160; Akyporox RLM 230; Aldosperse L 9; Alkasurf LAN 1; Alkasurf LAN 3; Arapol 0712; Atlas G 2133; Atlas G 3705; Atlas G 3707; Atlas G 4829; B 205; Base LP 12; BASE LP 12; BL 9; BL 9 (polyglycol); Brij 22; Brij 23; Brij 30; Brij 30ICI; Brij 30SP; Brij 35; Brij 35L; Brij 36T; Calgene 40L; Carsonol L 2; Carsonol L 3; CCRIS 3397; Chemal LA 23; Chimipal AE 3; Cimagel; Conion 275-100; Conion 275-20; Conion 275-30; Conion 275-80; Conion 2P80; Dodecanol, ethoxylate; Dodecyl alcohol, ethoxylated; Dodecyl alcohol polyoxyethylene ether; α-Dodecyl-ω-hydroxypoly(oxyethylene); α-Dodecyl-ω-hydroxypoly(oxy-1,2-ethanediyl); Du Pont WK; Ethosperse LA 12; Ethosperse LA 23; Ethoxylated lauryl alcohol; G 3707; Glycols, polyethylene, monododecyl ether; HSDB 4351; Hydroxypoly-ethoxydodecane; LA (Alcohol); LA 7; Laureth; Laureth 4; Laureth 9; Laureth 11; Lauromacrogol 400; Lipal 4LA; Lubrol 12A9; Lubrol PX; Marlipal 1217; Mergital LM 11; NCI-C54875; Newcol 1203; Nikkol BL; Noigen ET 160; Noigen ET 170; Noigen YX 500; 3,6,9,12,15,18,21, 24,27-Nonaoxanonatriacontan-1-ol; Noniolite AL 20; PEG-11 Lauryl ether; Pegnol L 12; Polidocanol; Poly(oxy-1,2-ethanediyl), α-dodecyl-

360

omega-hydroxy-; Polyethylene glycol monododecyl ether; Polyoxyethylene dodecyl mono ether(8Cl); Polyoxyethylene lauryl ether; Rokanol L; Romopal LN; Simulsol P 23; Simulsol P 4; Siponic L; Slovasol S; Standamul LA 2; Stmer 135; Surfactant WK; Texofor B 9; Thesat; Thesit. Emulsifier for emulsion polymerization. FDA approved for topicals, used as an emollient, emulsifier, thickener, and stabilizer. Used in topicals; anti-irritant in deodorant and anti-perspirants. White solid; mp = 40-42°; soluble in H2O, EtOH, C7H8; miscible with hot mineral, natural and synthetic oils, fats and fatty alcohols; LD50 (mus orl) = 1170 mg/kg, (mus iv) = 125 mg/kg. *Chem-Y GmbH.*

2193   Lauric acid
143-07-7   5400   E G H

C12H24O2
n-Dodecanoic acid.
4-02-00-01082 (Beilstein Handbook Reference); ABL; AI3-00112; Aliphat No. 4; BRN 1099477; C-1297; C12 fatty acid; CCRIS 669; Dodecanoic acid; n-Dodecanoic acid; Dodecoic acid; Dodecylic acid; Duodecylic acid; EINECS 205-582-1; FEMA No. 2614; HSDB 6814; Hydrofol acid 1255; Hydrofol acid 1295; Hystrene 9512; Lauric acid; Laurostearic acid; Neo-fat 12; Neo-fat 12-43; Ninol AA62 Extra; NSC-5026; Philacid 1200; Univol U-314; Vulvic acid; Wecoline 1295. Used in manufacture of alkyd resins, wetting agents, soaps, detergents, cosmetics, insecticides, food additives. Intermediate for manufacturing of toilet soaps, synthetic detergents, cosmetics and pharmaceuticals. Needles; mp = 43.2°; bp100 = 225°; d20 = 0.8679; insoluble in H2O, slightly soluble in CHCl3, soluble in Me2CO, petroleum ether, very soluble in EtOH, MeOH, Et2O, freely soluble in C6H6; LD50 (rat orl) = 12 g/kg. *Akzo Chemie; Henkel/Emery; Mirachem Srl.; Unichema; Witco/Humko.*

2194   Lauroyl chloride
112-16-3   G

C12H32OCl
n-Dodecanoyl chloride.
AI3-52409; Dodecanoic acid, chloride; Dodecanoyl chloride; n-Dodecanoyl chloride; EINECS 203-941-7; HSDB 5567; Lauric acid chloride; Lauroyl chloride. Surfactant, polymerization initiator, antienzyme agent, foamer; synthesis of lauroyl peroxide, sodium lauroyl sarcosinate, other sarcosinates. Liquid; mp = -17°; bp18 = 145°; d25 = 0.9169; very soluble in Et2O. *Degussa-Hüls Corp.; Elf Atochem N. Am.; PPG Ind.*

2195   Lauroyl peroxide
105-74-8   G

C24H46O4
Dodecanoyl peroxide.
4-02-00-01102 (Beilstein Handbook Reference); Alperox C; Alperox F; BRN 1804936; CCRIS 2455; Dilauroyl peroxide; Dilauryl peroxide; Dodecanoyl peroxide; Dyp-97F; EINECS 203-326-3; HSDB 352;

Laurox; Lauroyl peroxide; Laurydol; LYP 97; LYP 97F; NSC 670; Peroxide, bis(1-oxododecyl)-; Peroxide, didodecanoyl; Peroxyde de lauroyle. Initiator for bulk, solution, and suspension polymerization, high-temperature curing of polyester resins, and cure of acrylic syrup. White plates; mp = 49°; insoluble in H2O, soluble in CHCl3. *Elf Atochem N. Am.*

2196   Lauroyl sarcosine
97-78-9   G

C15H29NO3
N-Methyl-N-(1-oxododecyl)glycine.
Crodasinic L; EINECS 202-608-3; Glycine, N-methyl-N-(1-oxododecyl)-; Hamposyl L; Lauroyl sarcosine; Maprosyl L; N-Lauroyl-N-methylaminoacetic acid; N-Lauroyl sarcosinate; N-Lauroylsarcosine; N-Lauryl-sarcosine; N-Methyl-N-(1-oxododecyl)glycine; NSC 96994; Sarcosine, N-lauroyl-; Sarcosyl L; Sarkosyl; Crodasinic L; Vanseal® LS; Nikkol Sarcosinate LH; Oramix L; N-Methyl-N-(1-oxododecyl) glycine; Sarcosyl® L. Detergent whose properties include mild detergency, high foaming, bacteriostatic activity, enzyme inhibition, corrosion inhibition, hard water tolerance and stability in mildly acid formulations; anionic surfactant used in cosmetics and toiletries, dentrifices and shampoos, carpet shampoos, emulsion polymerization, metal treatment, food and food packaging, textile and fine fabric detergents. Detergent, wetting and foaming agent, foam stabilizer, emulsifier, anticorrosive agent, conditioner for hair and rug shampoos, cosmetics, skin cleansers; biodegradable. White solid. *Chemplex; Ciba Geigy Agrochemicals; Ciba-Geigy Corp.; Croda Surfactants; Grace W.R. & Co.; Hampshire; Nikko Chem. Co. Ltd.; Vanderbilt R.T. Co. Inc.*

2197   Laurtrimonium chloride
112-00-5   G H

C15H34ClN
N-Dodecyl-N,N,N-trimethylammonium chloride.
Adogen® 412; Alicop; Aliquat 4; Ammonium, dodecyltrimethyl-, chloride; Arquad® 12; Arquad® 12-23; Arquad® 12-33; Arquad® 12-37W; Arquad® 12/50; Arquad® 12D; Arquad® MC 50; Catinal LTC 35A; Catiogen L; Cation BB; Cation FB; Dehyquart LT; Dodecyltrimethylammonium chloride; n-Dodecyl-trimethylammonium chloride; N-Dodecyl-N,N,N-tri-methylammonium chloride; DTAC; EINECS 203-927-0; Laurtrimonium chloride; Lauryl trimethyl ammonium chloride; Monolauryltrimethylammonium chloride; Nissan Cation BB; Nissan Cation BB 300; Nissan Cation FB; NSC 6931; Octosol 562; Quartamin 24P; Quartamin 24W; Radiaquat 6465; Redicote E 5; Rewoquat B 18; Rhodaquat M 242C29; Swanol CA 2150; Trimethyldodecylammonium chloride; N,N,N-Trimethyl-1-dodecanaminium chloride; Trimethyllaurylammonium chloride; Varisoft® LAC. A biodegradable emulsifier, foaming, wetting, dispersing agents, corrosion inhibitor, softener, dyeing aid, antistat for textiles, paper, cosmetics; industrial, agriculture, plastics, petrol. industry, acid pickling baths; bactericide, algicide; gel sensitizer for latex foam. Gel sensitizer for latex foam rubber. Base for hair conditioners and cream rinses. Softener for textile, laundry and paper. Solid; mp = 236°. *Akzo Chemie; CK Witco Corp.; Sherex.*

2198   Lauryl acrylate
2156-97-0   G

$C_{15}H_{28}O_2$

n-Dodecyl acrylate.

Acrylic acid, dodecyl ester; AI3-03198; Dodecyl acrylate; n-Dodecyl acrylate; EINECS 218-463-4; Lauryl acrylate; n-Lauryl acrylate; NSC 24177; 2-Propenoic acid, dodecyl ester. UV-curable reactive diluent in inks and coatings, adhesives, viscosity index improver, finishing aid for leather. Solid; d_{20}^{20} = 0.884. *Rit-Chem; Sartomer.*

2199 Lauryl alcohol
112-53-8   3439   G H

$C_{12}H_{26}O_2$

1-Dodecanol.

4-01-00-01844 (Beilstein Handbook Reference); Adol 10; Adol 11; Adol 12; AI3-00309; Alcohol C12; Alfol 12; BRN 1738860; C12 alcohol; Cachalot® L-50; Cachalot® L-90; CCRIS 662; CO-1214; CO-1214N; CO-1214S; Conol 20P; Conol 20PP; Dodecanol; Dodecan-1-ol; Duodecyl alcohol; Dytol J-68; EINECS 203-982-0; Emery® 3326; Emery® 3332; EPA Pesticide Chemical Code 001509; EPAL 12; Fatty alcohol (C12); FEMA Number 2617; Hainol 12SS; HSDB 1075; Hydroxydodecane; 1-Hydroxydodecane; Karukoru 20; Lauric alcohol; Laurinic alcohol; Lauroyl alcohol; Lauryl 24; Lauryl alcohol; Lorol 5; Lorol 7; Lorol 11; Lorol C 12; Lorol C 12/98; MA-1214; NAA 42; Nacol® 12-96; Nacol® 12-99; NSC 3724; Pisol; S 1298; Siponol 25; Siponol L2; Siponol L5; Sipol L12; Undecyl carbinol; Unihydag Wax 12. Chemical intermediate for detergent manufacture. Used in the manufacture of sulfuric acid esters which are used as wetting agents; synthetic detergents, lube additives, pharmaceuticals, rubber, textiles, perfumes, flavoring agents. Registered by EPA as an insecticide. White solid; mp = 24°; bp = 259°; d24 = 0.8309; soluble in EtOH, Et2O, c6h6, insoluble in H2O. LD50 (rat orl) = 12800 mg/kg. *Amoco Lubricants; Condea Vista Co.; Henkel/Emery; Michel M.; Penta Mfg.; Procter & Gamble Co.; Procter & Gamble Pharm. Inc.; Schweizerhall Inc.; Sigma-Aldrich Fine Chem.; Spectrum Chem. Manufacturing.*

2200 Lauryl chloride
112-52-7   H

$C_{12}H_{25}Cl$

n-Dodecyl chloride.

1-Chlorododecane; CCRIS 5810; Dodecane, 1-chloro-; Dodecyl chloride; n-Dodecyl chloride; EINECS 203-981-5; Lauryl chloride; NSC 57107. Liquid; mp = -9.6°; bp = 216.3°; d^{20} = 0.7487; insoluble in H2O, soluble in C6H6; very soluble in EtOH, freely soluble in Me2CO, CCl4, ligroin.

2201 Lauryldimethylamine
112-18-5   G H

$C_{14}H_{31}N$

N,N-Dimethyl-n-dodecylamine.

ADMA 2; AI3-16726; Antioxidant DDA; Armeen DM-12D; Barlene 125; DDA; DDA (antioxidant); DDA (corrosion inhibitor); Dimethyl lauramine; Dimethyl laurylamine; Dimethyl-n-dodecylamine; 1-Dodecan-amine, N,N-dimethyl-; Dodecylamine, N,N-dimethyl-;

Dodecyldimethylamine; EINECS 203-943-8; Empigen AB; Farmin DM 20; Farmin DM 2098; Genamin LA 302D; HSDB 5568; IPL; Lauryldimethylamine; Monolauryl dimethylamine; NSC 7332; Onamine 12; RC 5629. Liquid cationic detergent; corrosion inhibitor; acid stable emulsifier. Used as an intermediate in the synthesis of surfactants, antioxidants, oil and grease additives. Liquid; mp = -20°; d = 0.7870. *Albemarle Corp.; Asahi Denka Kogyo; CK Witco Corp.; Ethicon Inc.; Fluka; Lonzagroup; Mason; Sigma-Aldrich Fine Chem.*

2202 Laurylethanolamide
142-78-9   G H

$C_{14}H_{29}NO_2$

N-(2-hydroxyethyl)dodecanamide.

Ablumide LME; Amisol LDE; Amisol LME; CCRIS 4834; Comperlan LM; Copramyl; Crillon LME; Cyclomide LM; Dodecanamide, N-(2-hydroxyethyl)-; EINECS 205-560-1; HSDB 5644; Lauramide MEA; Lauric acid ethanolamide; Lauric acid monoethanolamide; Lauric N-(2-hydroxy-ethyl)amide; Lauridit LM; N-Lauroylethanolamine; Monoethanolamine lauric acid amide; Rewomid® L 203; Rolamid CM; Stabilor CMH; Steinamid L 203; Ultrapole H; Vistalan. Foam stabilizer, thickener for shampoos, bubble baths, liquid detergents, toiletries. Detergent, thickener, foam booster/stabilizer, superfatting agent for detergent preparations; fixation of perfumes; stabilizer of emulsions. *Rewo; Taiwan Surfactants.*

2203 Lauryl isoquinolinium bromide
93-23-2   G

$C_{21}H_{32}BrN$

2-Dodecylisoquinolinium bromide.

Caswell No. 520; Dodecyl isoquinolinium bromide; n-Dodecylisoquinolinium bromide; EINECS 202-230-9; EPA Pesticide Chemical Code 069130; Intexsan LQ75; Isoquinolinium, 2-dodecyl-, bromide; Isothan; Isothan Q-15; Isothan Q-75; Isothan Q-90; Lauryl isoquinolinium bromide; 2-Laurylisoquinolinium bromide; N-Laurylisoquinolinium bromide; NSC 20909. Anti-infective. Use as an antibacterial agent. *MLPC International.*

2204 Laurylpyridinium chloride
104-74-5   G

$C_{17}H_{30}ClN$

1-Dodecylpyridinium chloride.

AI3-02741; C 2; Dehyquart C; Dehyquart C Crystals; 1-Dodecylpyridinium chloride; N-Dodecylpyridinium chloride; DPC (onium compound); EINECS 203-232-2; Eltren; Laurylpyridinium chloride; 1-Laurylpyridinium chloride; N-Laurylpyridinium chloride; Ledmin LPC; LPC; Newkalgen B 251; NSC 35027; Pyridinium, 1-dodecyl-, chloride; Quaternario LPC; Dehyquart C Crystals; Ledmin LPC. Quaternary ammonium compound; Cationic detergent, dispersing and wetting agent; ingredient of fungicides and bactericides. Used as an emulsifier in creams and lotions; hair conditioners, skin creams; antistat for hair and fiber; bactericide, fungicide, corrosion inhibitor, sequestrant; conditioner used in personal care products. Crystals; mp = 66-70°. *Henkel Surface*

2205 N-Laurylpyrrolidinone
2687-96-9

G

$C_{16}H_{31}NO$
Lauryl pyrrolidone.
5-21-06-00330 (Beilstein Handbook Reference); Agsol Ex 12; BRN 0155011; 1-Dodecyl-2-pyrrolidone; 1-Dodecyl-2-pyrrolidinone; N-Dodecylpyrrolidinone; Lauryl pyrrolidone; 1-Lauryl-2-pyrrolidone; 2-Pyrrolidinone, 1-dodecyl-. Conditioner, foam stabilizer, wetting agent; special solvent for commercial cleaning, textile processing, water-borne coatings, inks; replacement for volatile organic compounds. ISP; Sigma-Aldrich Fine Chem.

2206 Lawsone
83-72-7   5410   D G

$C_{10}H_6O_3$
2-Hydroxy-1,4-naphthalenedione.
4-08-00-02360 (Beilstein Handbook Reference); AI3-12099; BRN 1565260; C.I. 75480; C.I. Natural Orange 6; CCRIS 6248; EINECS 201-496-3; Flower of paradise; HANA; Henna; Henna leaves; 2-Hydroxy-1,4-naphthalenedione; 2-Hydroxynaphthoquinone; 2-Hydroxy-1,4-naphthoquinone; 2-Hydroxy-1,4-naptho-quinone; Lawson; Lawsone; Lawsonia alba; Mehendi; Mendi; 1,4-Naphthalenedione, 2-hydroxy-; 1,4-Naphthoquinone, 2-hydroxy-; NSC 27285; henna. Egyptian privet; flower of paradise, derived from the leaves and roots of Lawsonia inermis or L. Alba; used as a dye, and for staining the hair, in medicine as an antifungal agent and an ultra-violet screen. Yellow prisms; dec 195-196°; λ_m = 243, 249, 275, 332 nm (ϵ = 19500, 21500,17600, 3200 MeOH); very soluble in EtOH, soluble in AcOH, insoluble in Et_2O, C_6H_6, $CHCl_3$.

2207 Lead
7439-92-1   5414   G

Pb

Pb
Lead.
C.I. 77575; C.I. Pigment Metal 4; CCRIS 1581; EINECS 231-100-4; Glover; Haro® Mix CE-701, Haro® Mix CK-711, Haro® Mix MH-204; HSDB 231; KS-4; Lead; Lead element; Lead, elemental; Lead flake; Lead, inorganic; Lead metal; Lead S 2; Olow; Omaha & Grant; Pb-S 100; Plumbum; Rough lead bullion; SSO 1; Haro® Mix CE-701, Haro® Mix CK-711, Haro® Mix MH-204. Lead one-pack system; one-pack lead heat stabilizer systems for rigid PVC applications, UPVC pressure and nonpressure pipes; construction material; radiation protection; alloys; manufacture of pigments. Soft, grey metal; mp = 327°; bp = 1740°. Harcros; National Lead Co.

2208 Lead acetate
301-04-2   5415   D G

$C_4H_6O_4Pb$
Lead(II) acetate.
Acetate de plomb; Acetic acid, lead salt; Acetic acid, lead(2+) salt; Bleiacetat; Caswell No. 523A; CCRIS 356; Dibasic lead acetate; EINECS 206-104-4; EPA Pesticide Chemical Code 048001; HSDB 1404; Lead acetate; Lead acetate (anhydrous); Lead acetate ($Pb(Ac)_2$); Lead acetate ($Pb(O_2C_2H_3)_2$); Lead di(acetate); Lead diacetate; Lead dibasic acetate; Lead(2+) acetate; Lead(II) acetate; Normal lead acetate; NSC 75797; Plumbous acetate; RCRA waste number U144; sal Saturni; Salt of saturn; Sugar of lead; UN1616; Unichem PBA. Used as a mordant in dyeing of textiles, waterproofing varnishes, insecticides, lead driers, chrome pigments, hair dye, weighting silks. Used as an astringent. [trihydrate]; colorless crystals; mp = 75°; dec. 200°; d = 2.55; soluble in H_2O (62.5 g/100 ml, 25°, 200 g/100 ml, 100°), EtOH (3.3 g/100 ml), freely soluble in glycerol; LD_{50} (rat ip) = 24 mg/kg. Am. Biorganics; Cerac; Chemetall Chem. products; Hoechst Celanese; Mallinckrodt Inc.

2209 Lead antimonate
13510-89-9   5416   G

$O_8Pb_3Sb_2$
Lead antimonate(V).
Antimony lead oxide ($Sb_2Pb_3O_8$); Diantimony trilead octaoxide; EINECS 236-845-9; Lead antimonate; Lead antimonate(V); Naples yellow; Paris yellow. An orange-yellow pigment. A mixture of this material with carbonate and chromate of lead is also sold under this name. Cadmium sulfide, CPS, and a pale yellow ocher have been identified by this term; used to stain glass, crockery and porcelain. Insoluble in H_2O.

2210 Lead, [1,2-benzenedicarboxylato(2-)]dioxo-tri-
69011-06-9   G H

$C_8H_4O_6Pb_3$
1,2-Benzenedicarboxylic acid, lead complex.
1,2-Benzenedicarboxylic acid, lead complex; EINECS 273-688-5; Halthal; Lead, (1,2-benzenedicarboxylato(2-))dioxotri-; Lead, (1,2-benzenedicarboxylato(2-))dioxotri-; (Phthalato(2-))dioxotrilead. Halstab.

2211 Lead carbonate
1319-46-6   G

CO_3Pb
Basic lead carbonate.
Basic lead carbonate; Basic lead carbonate ($2PbCO_3.Pb(OH)_2$); Berlin white; Carbonic acid, lead salt, basic; Ceruse; Cerussa; C.I. 77597; C.I. pigment white 1; EINECS 215-290-6; Flake lead; Halcarb 20; HSDB 5701; Kremser white; Lead, bis(carbonato(2-))dihydroxytri-; Lead, bis(carbonato)dihydroxytri-; Lead carbonate hydroxide ($Pb_3(OH)_2(CO_3)_2$); Lead hydroxide carbonate; Lead subcarbonate; Silver white; Trilead bis(carbonate) dihydroxide; White lead; White lead, hydrocerussite. Used as a PVC stabilizer. White solid; mp = 400° (dec). Halstab.

2212 Lead chloride
7758-95-4   5422   G

Cl—Pb—Cl

Cl2Pb
Hydrochloric acid lead salt.
CCRIS 7565; EINECS 231-845-5; HSDB 6309; Lead chloride; Lead (II) chloride; Lead chloride (PbCl2); Lead dichloride; Lead(2+) chloride; Leclo; Plumbous chloride. Used in pigments, solders and flux. White crystalline powder; d = 5.85; mp = 501°; bp = 950°; soluble in H2O, NH4Cl, NH4NO3; MLD (guinea pig orl) = 1.5-2.0 g/kg. *Mechema Chemicals Ltd.*

2213   Lead chromate
7758-97-6   5423   G

$$O=Cr-O^- \quad Pb^{2+}$$

CrO4Pb
Lead chromate(VI).
CCRIS 357; Chromate de plomb; Chrome yellow; Chromic acid, lead(2+) salt (1:1); C.I. Pigment Yellow 34; C.I. 77600; Cologne Yellow; EINECS 231-846-0; HSDB 1650; King's yellow; Lead chromate; Lead chromate (PbCrO4); Lead chromate(VI); Leipzig yellow; Paris yellow; Phoenicochroite; Plumbous chromate. Is known as chrome green. It dyes chromed wool green, also used in cotton printing. The name has also been used for various other pigments. Orange-yellow powder; mp = 844°; d = 6.3; generally insoluble; LD75 (gpg ip) = 156 mg/kg.

2214   Lead chromate silicate
11113-70-5   H

$$Pb^{2+}$$

$$2Cr^{3+}$$

Silicic acid, chromium lead salt.
EINECS 234-347-6; Lead chromate silicate; Silicic acid, chromium lead salt.

2215   Lead diacetate
6080-56-4   G

$$Pb^{2+}$$

$$3H_2O$$

C4H6O4Pb.3H2O
Lead(II) acetate trihydrate.
Acetic acid, lead(+2) salt trihydrate; Bis(acetato)-trihydroxytrilead; Bleiazetat; Lead acetate (II), trihydrate; Lead acetate trihydrate; Lead diacetate trihydrate; Lead(II) acetate trihydrate. Anhydrous

form [301-04-2]. Solid; mp = 75°, d = 2.55; soluble in H2O (630 mg/ml), poorly soluble in organic solvents; LD50 (rat ip) = 150 mg/kg.

2216   Lead fluoborate
13814-96-5   G

$$Pb^{2+}$$

B2F8Pb
Borate(1-), tetrafluoro-, lead (2+); Borate(1-), tetrafluoro-, lead (2+) (2:1); EINECS 237-486-0; HSDB 1991; Lead bis(tetrafluoroborate); Lead borofluoride; Lead boron fluoride; Lead fluoborate; Lead fluoroborate; Lead fluoroborate (Pb(BF4)2); Lead fluoroborate solution; Lead tetrafluoroborate; Lead tetrafluoroborate (Pb(BF4)2); Lead(II) tetrafluoroborate. Salt for electroplating lead. *Atomergic Chemetals; Elf Atochem N. Am.; Hoechst Celanese; M&T Harshaw.*

2217   Lead formate
811-54-1   5427   G

$$H \quad Pb^{2+} \quad H$$

C2H2O4Pb
Formic acid lead salt.
EINECS 212-371-8; Formic acid, lead(2+) salt; Lead diformate; Lead formate; Lead formate (Pb(HCO2)2); Lead(2+) formate; NSC 112233; Ledfo. Solid; d = 4.63; dec. at 190°; soluble in water; insoluble in alcohol. *Mechema Chemicals Ltd.*

2218   Lead molybdate
10190-55-3   5430   G

$$O=Mo-O^- \quad Pb^{2+}$$

MoO4Pb
Molybdic acid lead salt.
EINECS 233-459-2; Lead molybdate; Lead molybdate(VI); Lead molybdenum oxide (PbMoO4). Used in analytical chemistry and in pigments. Yellow powder; insoluble in H2O; soluble in HNO3, NaOH. *AAA Molybdenum; Atomergic Chemetals; Cerac.*

2219   Lead monoxide
1317-36-8   5431   G

$$Pb=O$$

OPb
Lead(II) oxide.
C.I. 77577; C.I. Pigment Yellow 46; EINECS 215-267-0; HSDB 638; Lead monooxide; Lead monoxide; Lead oxide; Lead oxide (PbO); Lead Oxide Yellow; Lead protoxide; Lead(2+) oxide; Lead(II) oxide; Litharge; Litharge Pure; Litharge Yellow L-28; Massicot; Massicotite; NSC 57634; Plumbous oxide; Vulcaid; Yellow Lead Ocher; Massicot; Lead Oxide yellow; plumbous oxide; Lead protoxide. Obtained as a by-product in silver refining; has a more reddish color than Massicot, which is made by roasting lead. Used in pigments and as a vulcanizing agent. Red-yellow crystals; mp = 886°; bp = 1470°; d = 9.5300; LD40 (rat ip) = 40 mg Pb/100 g). *National Lead Co.*

2220   Lead nitrate
10099-74-8   5432   G

N$_2$O$_6$Pb
 Lead(II) nitrate.
 CCRIS 1945; EINECS 233-245-9; HSDB 637; Lead dinitrate; Lead nitrate (Pb(NO$_3$)$_2$); Lead(2+) nitrate; Lead(II) nitrate; Ledni; Nitrate de plomb; Nitric acid, lead(2+) salt; Plumbous nitrate; UN1469. Used as a mordant for dyeing and printing textiles and staining mother of pearl, matches, oxidizer in dye industry, sensitizer for photography, explosives, tanning, process engraving, lithography. White translucent crystals; mp = 470°; d = 4.5300; soluble in H$_2$O, EtOH, MeOH, insoluble in concentrated HNO$_3$; toxic. *Blythe, Williams Ltd.; Cerac; Mallinckrodt Inc.; Noah Chem.; Spectrum Chem. Manufacturing.*

2221   Lead peroxide
1309-60-0   5425   G

PbO$_2$
 Lead(IV) oxide.
 Bioxyde de plomb; C.I. 77580; CCRIS 6254; EINECS 215-174-5; HSDB 4335; Lead Brown; Lead dioxide; Lead oxide; Lead oxide (PbO$_2$); Lead Oxide Brown; Lead peroxide; Lead peroxide (PbO$_2$); Lead superoxide; Lead(IV) oxide; Lepro; LP-100; Peroxyde de plomb; UN1872. Catalyst/curing agent for polysulfide, low molecular weight butyl and polyisoprene rubber; oxidizer in manufacture of dyes and to control burning rate of incendiary fuses or pyrotechnics. Dark brown powder; d = 9.38; insoluble in H$_2$O, LD$_{50}$ (gpg ip) = 200 mg/kg. *Eagle-Picher; Mechema Chemicals Ltd.; National Lead Co.*

2222   Lead phosphite
12141-20-7   G

HO$_3$PPb
 Lead phosphite dibasic.
 Dibasic lead phosphite; EINECS 235-252-2; Halphos; Lead oxide phosphite; Lead oxide phosphonate (Pb$_3$O$_2$(HPO$_3$)); Trilead dioxide phosphonate. A heat/light stabilizer used with PVC. *Halstab.*

2223   Lead silicate
10099-76-0   G

O$_3$PbSi
 Lead(II) metasilicate.
 BSWL 202; EINECS 233-246-4; Lead metasilicate (PbSiO$_3$); Lead monosilicate; Lead silicate (PbSiO$_3$); Lead silicon oxide (PbSiO$_3$); Lead silicon trioxide; Lead(2+) silicate. Basic silicate white lead; white pigment acting as heat stabilizer for chlorinated polyethylene, chlorosulfonated polyethylene, PVC, and polyepi-chlorohydrin; rust-

inhibitive pigment in the automobile industry; used industrial or maintenance paints. Used in ceramics, fireproofing fabrics, paints, electrode position process in the automotive industry. Solid; mp = 680-730°; d = 6.50. *Eagle-Picher; Hammond Lead Products.*

2224   Lead silicate
11120-22-2   H

O$_3$PbSi
 Silicic acid, lead salt.
 EINECS 234-363-3; EP 202 (silicate); Lead silicate; Silicic acid, lead salt; Stabinex S.

2225   Lead stearate
1072-35-1   5437   G

C$_{36}$H$_{70}$O$_4$Pb
 Lead(II) n-octadecanoate.
 5002G; EINECS 214-005-2; Lead distearate; Lead stearate; Lead(2+) octadecanoate; Lead(2+) stearate; Lead(II) octadecaoate; Lead(II) stearate; Listab 28ND; Octadecanoic acid, lead(2+) salt; P-51; SL 1000 (stabilizer); Stabinex NC18; Stearic acid, lead(2+) salt. Used in extreme pressure lubricants, as a dryer in varnishes and as a heat stabilizer for PVC. Solid; mp = 125°; insoluble in H$_2$O, slightly soluble in EtOH. *Goodrich B.F. Co. Europe; Goodrich B.F. Co.; Ore & Chem.; Syn Prods.; Vanderbilt R.T. Co. Inc.*

2226   Lead subacetate
1335-32-6   5438   G
C$_4$H$_{10}$O$_8$Pb$_3$
 Monobasic lead acetate.
 Basic lead acetate; Bis(acetato)dihydroxytrilead; Bis-(acetato)tetrahydroxytrilead; BLA; CCRIS 68; EINECS 215-630-3; HSDB 1651; Lead acetate, basic; Lead, bis(acetato-O)tetrahydroxytri-; Lead, bis(acetato)tetra-hydroxytri-; Lead monosubacetate; Lead subacetate; Monobasic lead acetate; RCRA waste number U146; Subacetate lead. Used as a decolorizing agent. Crystals; soluble in H$_2$O (6.2-25.0 g/100 ml).

2227   Lead subcarbonate
598-63-0   5442   G

2PbCO$_3$·Pb(OH)$_2$
 Carbonic acid, lead(2+) salt (1:1).
 Carbonic acid, lead(2+) salt (1:1); Cerussete; Cerussite; Dibasic lead carbonate; EINECS 209-943-4; HSDB 1649; Lead carbonate; Lead carbonate (PbCO$_3$); Lead(2+) carbonate; Plumbous carbonate; White lead. Exterior paint pigment, ceramic glazes. White solid; mp = 125°; insoluble in H$_2$O, slightly soluble in EtOH. *Halstab.*

2228   Lecithin
8002-43-5   5447   D E G

L-α-phosphatidyl choline.

Alcolec® Granules; Asol; Capcithin™; Capsulec 51-SB; Capsulec 51-UB; Capsulec 56-SB; Capsulec 56-UB; Capsulec 62-SB; Capsulec 62-UB. Yolk powder. Mixture of the diglycerides of stearic, palmitic and oleic acids linked to the choline ester of phosphoric acid; found in plants and animals; R and R' are fatty acid groups; edible surfactant and emulsifier for food use, pharmaceuticals, cosmetics, leather treatment, textiles. FDA GRAS; USDA 0.5% maximum in oleomargarine, Japan, UK approved, Europe listed, FDA approved for orals, topicals, USP/NF compliance. In the FDA list of inactive ingredients, (inhalations, im and iv injectables, oral capsules, suspensions and tablets, rectals, topicals and vaginals). Used as an edible surfactant, emulsifier, stabilizer, solubilizer, wetting agent, emollient for pharmaceuticals; used in liposome technology, orals, topicals; nutritional in gel capsules and tablet form; binding agent in tableting; choline source in dementia. Solid; d_4^{24}= 1.0305; insoluble in H_2O, soluble in $CHCl_3$, Et_2O, petroleum ether, mineral oils and fatty acids. Generally non-toxic. *ADM Lecithin; Am. Lecithin; Am. Roland; Anstead D. F.; CanAmera Foods; Central Soya; Cleary W. A.; Duphar BV; Great Western; Greef K & K; Grunau; Hatrick A. C.; Landers-Segal Color; Meyer Lucas; Penta Mfg.; Quest UK; Reichold; Rhône-Poulenc; Riken; Ruger; Sibner-Hegner; Solvay Duphar Labs Ltd.; Spectrum Chem. Manufacturing; Spice King; Stern-France; U.S. BioChem.; Varno Mills; Westin.*

2229   Lenacil
2164-08-1   5455   G P

C13H18N2O2

3-Cyclohexyl-6,7-dihydro-1H-cyclopentapyrimidine-2,4(3H,5H)-dione.

5-24-07-00375 (Beilstein Handbook Reference); Adol; Adol (pesticide); BRN 0751331; Buracyl; Caswell No. 525A; CCRIS 1936; 3-Cyclohexyl-1,5,6,7-tetrahydro-cyclopentapyrimidine-2,4(3H)-dione; 3-Cyclohexyl-5,6-trimethylenuracil; 3-Cyclohexyl-5,6-trimethyleneuracil; 3-Cyclohexyl-6,7-dihydro-1H-cyclopentapyrimidine-2,4(3H,5H)-dione; 31H-Cyclopentapyrimidine-2,4(3H, 5H)-dione, 3-cyclohexyl-6,7-dihydro-; 1H-Cyclopenta-pyrimidine-2,4(3H,5H)-dione, 6,7-dihydro-3-cyclohexyl-; 6,7-Dihydro-3-cyclohexyl-1H-cyclopentapyrimidine-2,4(3H,5H)-dione; Du Pont 634; EINECS 218-499-0; Elbatan; EPA Pesticide Chemical Code 525200; Experimental herbicide 634; Herbicide 634; Hexilure; Lenacil; Lenacile; Uracil 634; Venzar; Vizor. Selective herbicide, inhibits photosynthesis. Used for control of annual grass and broad-leaved weeds in a variety of crops. Wettable powder containing 80% lenacil; used for control of annual dicotyledons and meadow grass in beet, fruit herbaceous perennials. Crystals; mp = 315-317°; sg = 1.32; soluble in H_2O (6 mg/l), more soluble in organic solvents; LD50 (rat orl) > 11000 mg/kg. *Chemolimpex; DuPont UK; DuPont; Fahlberg-List; Farm Protection Ltd.; ICI Chem. & Polymers Ltd.*

2230   Leucine
61-90-5   5470   G

C6H13NO2

l-Leucine.

AI3-08899; α-aminoisocaproic acid; (S)-2-Amino-4-methylpentanoic acid; 2-Amino-4-methylpentanoic acid (L); 2-Amino-4-methylpentanoic acid, (S)-; α-amino-γ-methylvaleric acid; (S)-2-Amino-4-methylvaleric acid; 2-amino-4-methyl-valeric acid; 2-amino-4-methylvaleric acid (L); EINECS 200-522-0; FEMA No. 3297; L; Leu; Leucin; Leucina; Leucine; L-Leucine; Leucine, L-; (S)-Leucine; Leucinum; L-Norvaline, 4-methyl-; NSC 46709; Pentanoic acid, 2-amino-4-methyl-, (S)-; Valeric acid, 2-amino-4-methyl-, (S)-. Nutrient and dietary supplement, biochemical research. Sublimes at 145-148°; mp 293-295° (dec); d = 1.293; $[\alpha]_D^{25}$= -10.8°; pKa (25°) = 9.6; soluble in water, alcohol, acetic acid, insoluble in ether. *Degussa AG; Greeff R.W. & Co.; Nippon Rikagakuyakuhin; Tanabe U.S.A. Inc.; U.S. BioChem.*

2231   Levamisole
14769-73-4   5480   D G

C11H12N2S

(-)-2,3,5,6-Tetrahydro-6-phenylimidazol[2,1-b]thiazole.

EINECS 238-836-5; Ergamisol; Ketrax; Lepuron; Levamisol; Levamisole; Levamisolum; Levomysol; Levovermax; Totalon; Vermisol 150; Wormicid.; tetramisole [as DL-form]; tetramizole [as DL-form]; dexamisole [as L(+)-form]. Immunomodulator with anthelmintic activity (targets nematodes). Crystals; mp = 60-61.5°; $[\alpha]_D^{25}$= -85.1° (c = 10 CHCl3); [DL-form]: mp = 87-89°; [D-(+)-form]: mp = 60-61.5°; $[\alpha]_D^{25}$= 85.1° (c = 10 CHCl3). *Am. Cyanamid; Bayer Corp. Pharm. Div.; ICI; ICI Chem. & Polymers Ltd.; Janssen Pharm. Inc.; McNeil Pharm.*

2232   Levamisole hydrochloride
16595-80-5   5480   D G V

C11H12N2SCl

(S)-2,3,5,6-tetrahydro-6-phenylimidazo[2,1-b]thiazole hydrochloride.

Ascaridil; Citarin L; Decaris; Dekaris; DRG-0245; EINECS 240-654-6; Ergamisol; Imidazo(2,1-b)thiazole, 2,3,5,6-tetrahydro-6-phenyl-, monohydrochloride, (-)-; Imidazo-(2,1-b)thiazole, 2,3,5,6-tetrahydro-6-phenyl-, mono-hydrochloride, L-(-)-; Imidazo(2,1-b)thiazole, 2,3,5,6-tetrahydro-6-phenyl-, monohydrochloride, (S)-; KW-2-LE-T; KW 2LE-T; L-(-)-2,3,5,6-Tetrahydro-6-phenyl-imidazo-(2,1-b)thiazole hydrochloride; Levacide; Levadin; Leva-misole hydrochloride; Levasole; Levomysol hydro-chloride; Meglum; Nemicide; Nilverm forte; Niratic hydrochloride; Niratic-puron hydrochloride; NSC-177023; R 12,564; Ripercol-L; Solaskil; Spartakon; Stimamizol hydrochloride; (-)-2,3,5,6-Tetrahydro-6-phen-ylimidazo(2,1-b)thiazole monohydrochloride; Tetramisole hydrochloride; Tramisol®; Tramisole; Worm-chek. Immunomodulator with anthelmintic activity (targets nematodes). Used as a veterinary anthelmintic. Crystals; mp = 227-229°; $[\alpha]_D^{20}$= -124° ± 2° (c = 0.9 H_2O); soluble in H_2O; [DL-form]: mp = 264-265°; soluble in H_2O (21 g/100 ml), MeOH, propylene glycol; sparingly soluble in EtOH, CHCl3, C6H14, Me2CO;

LD$_{50}$ (mus iv) = 22 mg/kg, (mus sc) = 84 mg/kg, (mus orl) = 210 mg/kg, (rat iv) = 24 mg/kg, (rat sc) = 130 mg/kg, (rat orl) = 480 mg/kg; [D-(+)-form]: mp = 227-227.5°; $[\alpha]_D^{20}$= 125° (c = 0.7 H$_2$O). *Am. Cyanamid; Bayer Corp. Pharm. Div.; ICI; Janssen Pharm. Inc.; McNeil Pharm.*

2233 Levodopa
59-92-7   5485   D G

C$_9$H$_{11}$NO$_4$
L-3-(3,4-Dihydroxyphenyl)-L-alanine.
Alanine, 3-(3,4-dihydroxyphenyl)-, (-)-; Alanine, 3-(3,4-dihydroxyphenyl)-, L-; 2-Amino-3-(3,4-dihydroxy-phenyl)propanoic acid; Bendopa; Biodopa; Brocadopa; CCRIS 3766; Cerepap; Cerepar; Cidandopa; DA; Deadopa; Dihydroxy-L-phenylalanine; (-)-3-(3,4-Dihydroxyphenyl)-L-alanine; β-(3,4-Dihydroxyphenyl)-α-L-alanine; β-(3,4-Dihydroxyphenyl)-L-alanine; L-(o-Dihydroxyphenyl)alanine; L-β-(3,4-Dihydroxyphenyl)-α-alanine; L-β-(3,4-Dihydroxyphenyl)alanine; L-3-(3,4-Dihydroxyphenyl)alanine; L-O-Dihydroxyphenylalanine; 3,4-Dihydroxy-L-phenylalanine; 3,4-Dihydroxyphenyl-L-alanine; 3,4-Dihydroxyphenylalanine; (-)-Dopa; L-(-)-Dopa; L-Dopa; Dopaflex; Dopaidan; Dopal; Dopal-fher; Dopalina; Dopar; Doparkine; Doparl; Dopasol; Dopaston; Dopaston SE; Dopastral; Doprin; EINECS 200-445-2; Eldopal; Eldopar; Eldopatec; Eurodopa; Helfo-dopa; HSDB 3348; 3-Hydroxy-L-tyrosine; Insulamina; L-O-Hydroxytyrosine; 3-Hydroxy-L-tyrosine; Larodopa; Levodopa; Levodopum; Levopa; Madopa; Maipedopa; NSC 118381; Pardopa; Prodopa; Ro 4-6316; Sinemet; Sobiodopa; Syndopa; L-Tyrosine, 3-hydroxy-; Veldopa.; component of: Madopa, Sinemet. Naturally occurring isomer of dopa, the biological precursor of catecholamine. Used to treat Parkinson's disease. Crystals; mp = 276-278° (dec), 284-286°; $[\alpha]_D^{13}$= -13.1° (c = 5.12 in 1N HCl); λ_m = 220.5, 280 nm (log ε 3.79, 3.42 0.001N HCl); soluble in H$_2$O (0.165 g/100 ml); insoluble in EtOH, C$_6$H$_6$, EtOAc, CHCl$_3$; LD$_{50}$ (mus orl) = 3650 mg/kg, (rat orl) = 4000 mg/kg, (rbt orl) = 609 mg/kg. *DuPont-Merck Pharm.; Hoffmann-LaRoche Inc.; ICN Pharm. Inc.; Lemmon Co.; Roberts Pharm. Corp.; SmithKline Beecham Pharm.*

2234 Levothyroxine sodium
55-03-8   9491   D

C$_{15}$H$_{10}$I$_4$NNaO$_4$
O-(4-Hydroxy-3,5-diiodophenyl)-3,5-diiodo-L-tyrosine monosodium salt.
Dathroid; EINECS 200-221-4; Eltroxin; Euthyrox; Laevoxin; Letter; Levaxin; Levo-T; Levoroxine; Levo-throid; Levothyrox; Levothyroxin-natrium; Levothyroxine sodique; Levothyroxine sodium; Levothyroxinum natricum; Levotiroxina sodica; Levoxyl; NSC 259940; Oroxine; Ro-thyroxine; Roxstan; Sodium l-thyroxin; Sodium-L-thyroxine; Sodium levothyroxine; Sodium thyroxin; Sodium thyroxinate; Sodium thyroxine; Synthroid sodium; Tetroid; Thyradin S; Thyro-Tabs; Thyronamin; Thyroxevan; Thyroxin sodium;

Thyroxine sodique; Thyroxine sodium; Thyroxinum natricum. Thyroid hormone. Sodium salt of the amino acid L-thyroxine. Obtained from thyroid gland of domesticated animals or synthesized. Used in treatment of hyperthyroidism. Cream-colored powder; d^{20}= 2.381; $[\alpha]^{20}$= -4.4° (c = 3 70% EtOH); soluble in H$_2$O (0.015 g/100 ml); soluble in aq. acid and alkaline solutions, EtOH, slightly soluble in CHCl$_3$, Et$_2$O, pH of satd. aq. solution = 8.35 - 9.35. *Astra USA Inc.; Forest Pharm. Inc.; Knoll Pharm. Co.*

2235 Levulinic acid
123-76-2   5492   G

C$_5$H$_8$O$_3$
γ-Ketovaleric acid.
4-03-00-01560 (Beilstein Handbook Reference); Acetopropionic acid; β-Acetylpropionic acid; 3-Acetylpropionic acid; 3-Acetylpropionsaeure; 3-Acetylpropionsäure; Acidum laevulinicum; AI3-03377; BRN 0506796; EINECS 204-649-2; FEMA No. 2627; 3-Ketobutane-1-carboxylic acid; γ-Ketovaleric acid; 4-Ketovaleric acid; Laevulic acid; Laevulinic acid; LEVA; Levulic acid; Levulinic acid; Levulinsaeure; Levulinsäure; NSC 3716; 4-Oxopentanoic acid; 4-Oxopentansaeure; 4-Oxopentansäure; 4-Oxovaleric acid; Pentanoic acid, 4-oxo-; Propionic acid, 3-acetyl-; USAF CZ-1; Valeric acid, 4-oxo-; 4-oxopentanoic acid; levulic acid. Intermediate for plasticizers, solvents, reins, flavors, pharmaceuticals, acidulant, and preservative; chrome plating; solder flux; stabilizer for calcium greases; control of lime deposits. Plates; mp = 33°; bp = 245° (dec); d^{20} = 1.1335; λ_m = 256 nm (ε 32); soluble in CHCl$_3$, freely soluble in H$_2$O, EtOH, Et$_2$O. *Chemie Linz N. Am.; Lancaster Synthesis Co.; Otsuka Pharm. Co. Ltd.; Penta Mfg.*

2236 Lichen sugar
149-32-6   3708   G

C$_4$H$_{10}$O$_4$
Erythritol.
Antierythrite; 1,2,3,4-Butanetetrol, (2R,3S)-rel-; 1,2,3,4-Butanetetrol, (R*,S*)-; C*Eridex; CCRIS 7901; EINECS 205-737-3; Erythrit; Erythritol; L-Erythritol; Erythritol, meso-; meso-Erythritol; Erythrol; NIK 242; NSC 8099; Paycite; Phycitol. Crystals; mp = 121.5°; bp = 330.5°; d^{20} = 1.451; $[\alpha]_D$= +11.1° (c = 5, EtOH), -4.4° (C = 5, H$_2$O); soluble in H$_2$O, slightly soluble in Et$_2$O, C$_6$H$_6$; LD$_{50}$ (dog iv) = 5.0 g/kg.

2237 Lidocaine
137-58-6   5503   D G

C$_{14}$H$_{22}$ON$_2$
α-Diethylaminoaceto-2,6-xylidide.
4-12-00-02538 (Beilstein Handbook Reference); Anestacon; BRN

2215784; Cappicaine; Cito optadren; Cracked Heel Relief Cream; Cuivasil; Dalcaine; Duncaine; EINECS 205-302-8; Emla Cream; Esracaine; Gravocain; HSDB 3350; Isicaina; Isicaine; L-Caine; Leostesin; Lida-Mantle; Lidaform HC; Lidamantle-HC; Lidocaina; Lidocaine; Lidocainum; Lidothesin; Lignocaine; Lignocainum; Maricaine; Neosporin Plus; NSC 40030; Rucaina; Solarcaine; Solcain; Xilina; Xilocaina; Xycaine; Xylestesin; Xylocain; Xylocaine; Xylocitin; Xylocitin; Xyloneural; Xylotox. Used medically as a local anesthetic and antiarrhythmic (class IB). Crystals; mp = 68-69°; bp$_4$ = 180-182°, bp$_2$ = 159-160°; soluble in EtOH, Et$_2$O, C$_6$H$_6$, CHCl$_3$; insoluble in H$_2$O. *Alcoa Ind. Chem.; Astra USA Inc.; Bayer Corp.; Glaxo Wellcome Inc.; Greeff R.W. & Co.; Schering-Plough Animal Health; Schering-Plough Pharm.*

2238   Lignoceryl alcohol
506-51-4   H

C$_{24}$H$_{50}$O
1-Tetracosanol.
EINECS 208-043-9; HSDB 5674; Lignoceric alcohol; Lignocerol; Lignoceryl alcohol; NSC 93768; Tetracosanol; 1-Tetracosanol; n-Tetracosanol; Tetracosyl alcohol.

2239   Ligroin
8032-32-4   5510   H P

Ligroin.
Benzine; Benzine (light petroleum distillate); Benzoline; BP 2; BP 2 (solvent); Canadol; EINECS 232-453-7; Isoparaffinic hydrocarbons; Ligroin; Ligroine; Mineral spirits; Naphtha, ligroine; Naphtha, varnish makers' and painters'; Painters' naphtha; Petroleum ether; Refined solvent naphtha; Skellysolve F; Skellysolve G; V.M. and P. naphtha; Varnish makers' and painters naphtha; Varnish makers' naphtha; Varnish makers' naphtha and painters' naphtha; Varnish marker's naphtha; VM & P naphtha; VM and P Naphtha. Registered by EPA as an insecticide and herbicide (cancelled). Mobile, flammable liquid; bp = ca. 145°; d$^{15.6}_{15.6}$ = 0.850 - 0.870. *Sigma-Aldrich Co.*

2240   Lilial®
80-54-6   H

C$_{14}$H$_{20}$O
p-t-Butyl-α-methylhydrocinnamaldehyde.
4-07-00-00802 (Beilstein Handbook Reference); AI3-36196; Benzenepropanal, 4-(1,1-dimethylethyl)-α-methyl-; BRN 0880140; EINECS 201-289-8; Hydrocinnamaldehyde, p-t-butyl-α-methyl-; Lilial; Lilyal; NSC 22275. Used as a fragrance. *Givaudan.*

2241   Lime nitrogen
156-62-7   1663   G

$$N\equiv C-N^{2-}\quad Ca^{2+}$$

CaCN$_2$
Calcium cyanamide.
Aero-cyanamid; Aero cyanamid granular; Aero cyanamid special grade; AI3-28780; Alzodef; Calcii carbimidum; Calcium carbimide;

Calcium cyanamide; Calciumcarbimidum; Carbimida calcica; Carbimide calcique; Carbodimid calcium; Caswell No. 140; CCRIS 118; CY-L 500; Cyanamide calcique; Cyanamide, calcium salt (1:1); Dormex; EINECS 205-861-8; EPA Pesticide Chemical Code 014001; HSDB 1328; Lime nitrogen; NCI-C02937; Nitrogen lime; Nitrolime; NSC 7078; Perlka; UN1403; USAF CY-2. Used as a fertilizer, nitrogen products, pesticide, hardening iron and steel. Solid; mp approx. 1340°; d^{20} = 2.29; insoluble in H$_2$O (reacts). *Am. Cyanamid.*

2242   d-Limonene
5989-27-5   G H

C$_{10}$H$_{16}$
1-Methyl-4-(1-methylethenyl)cyclohexene, (R)-.
AI3-15191; Biogenic SE 374; Carvene; CCRIS 671; Citrene; Cyclohexene, 1-methyl-4-(1-methylethenyl)-, (4R)-; EC 7; EINECS 227-813-5; FEMA No. 2633; Glidesafe; Glidsafe; HSDB 4186; Kautschiin; Limonene, (+)-; Limonene, D-; d-(+)-Limonene; d-Limonene; d-Limoneno; D-(+)-Limonene; D-Limonene; NCI-C55572; p-Mentha-1,8-diene, (R)-(+)-; d-p-Mentha-1,8-diene; 1-Methyl-4-(1-methylethenyl)cyclohexene, (R)-; Refchole. Terpene; used in flavoring, as a fragrance, and in perfume materials, solvents, as a wetting agent and in resin manufacture. Liquid; mp = -74.3°; bp = 178°; d^{20} = 0.8411; [α]$^{20}_D$= +125.6°; λ$_m$ = 220, 250 nm (ε = 257, 23, isooctane); insoluble in H$_2$O, soluble in CCl$_4$, freely soluble in EtOH, Et$_2$O. *Allchem Ind.; IFF; Langley Smith Ltd.; Penta Mfg.; SCM Glidco Organics.*

2243   Linalool
78-70-6   5517   G H

C$_{10}$H$_{18}$O
3,7-Dimethyl-1,6-octadien-3-ol.
0-01-00-00462 (Beilstein Handbook Reference); AI3-00942; allo-Ocimenol; BRN 1721488; Caswell No. 526A; CCRIS 3726; EINECS 201-134-4; EINECS 245-083-6; EPA Pesticide Chemical Code 128838; FEMA Number 2635; HSDB 645; Linalol; Linalool; (±)-Linalool; β-Linalool; p-Linalool; Linalool 95; Linalool ex bois de rose oil; Linalool ex ho oil; Linalool ex orange oil; Linalyl alcohol; NSC 3789; Phantol; One of the principal components of bergamot or french lavender. Pure linalool. Used in perfumery. Registered by EPA as an insecticide. Colorless liquid; bp$_{720}$ = 194-197°, bp$_{14}$ = 89-91°; d = 0.868; soluble in EtOH, Et$_2$O, fixed oils, propylene glycol, insoluble in glycerin, H$_2$O. LD$_{50}$ (rat orl) = 2790 mg/kg. [l form]: bp = 198°. bp$_{25}$ = 98-98.3°, bp$_{14}$ = 86-87°; d^{20} = 0.8622; [α]$^{20}_D$= -20.1°. [d form]: bp = 198-200°, bp$_{26}$ = 114-114.5°, bp$_{15.5}$ = 93-94°, bp$_{12}$ = 86°; d^{20}= 0.8733; [α]$^{20}_D$= +19.3°. *BASF Corp.; Bush Boake Allen; IFF; Florida Treatt; Sigma-Aldrich Fine Chem.*

2244   Linalyl acetate
115-95-7   5518   G H

C12H20O2

3,7-Dimethyl-1,6-octadien-3-yl acetate.

Acetic acid linalool ester; AI3-00941; Bergamiol; Bergamol; Bergamot mint oil; Dehydrolinalool, acetate; 3,7-Dimethyl-1,6-octadien-3-yl acetate; EINECS 204-116-4; EINECS 254-806-4; Ex bois de rose (synthetic); FEMA No. 2636; HSDB 644; Licareol acetate; Linalol acetate; Linalool acetate; Lynalyl acetate; NSC 2138; 1,6-Octadien-3-ol, 3,7-dimethyl-, acetate; Phanteine. Found in volatile oils such as bergamot and lavendar oils. Used in perfumery. Liquid; bp = 220°; d$_4^{20}$= 0.895; insoluble in H$_2$O, soluble in organic solvents. Bush Boake Allen; IFF.

2245   Lincomycin hydrochloride monohydrate
7179-49-9   5522   D G

C18H34N2O6S.HCl.H2O

Methyl   6,8-dideoxy-6-trans-(1-methyl-4-propyl-L-2-pyrrolidinecarboxamido)-1-thio-D-erythro-α-D-galactooctopyranoside hydrochloride monohydrate.

Albiotic; Cillimycin; Frademicina; Lincocin [as hemihydrate]; Lincomix; Lincomycin hydrochloride; Lincomycin, hydrochloride hydrate; Lincomycin hydrochloride monohydrate; Mycivin; NSC 70731; Waynecomycin. Antibacterial. White powder; mp = 145-147°; [α]^{25}= 137° (H$_2$O); freely soluble in H$_2$O, MeOH, EtOH; sparingly soluble in most organic solvents; LD$_{50}$ (mus, rat orl) = 4000 mg/kg, (mus, rat ip) = 1000 mg/kg. Pharmacia & Upjohn.

2246   Lindane
58-89-9   5523   H P

C6H6Cl6

(1α,2α,3β,4α,5α,6β)-1,2,3,4,5,6-Hexachlorocyclo-hexane.

4-05-00-00058 (Beilstein Handbook Reference); Aalindan; Aficide; Agrisol g-20; Agrocide; Agrocide 2; Agrocide 6g; Agrocide 6G; Agrocide 7; Agrocide III; Agrocide WP; Agronexit; AI3-07796; Ameisenatod; Ameisenmittel merck; Ameisentod; Aparasin; Aphtitria; Aplidal; Arbitex; Arcotal S; BBH; Ben-Hex; Benhexachlor; Benhexol; Bentox 10; Benzene hexachloride; Benzene-1,2,3,4,5,6-hexachloride; Benzex; Bexol; BHC; BHC (α-, β-, γ-); Borer Spray; BRN 1907337; Caswell No. 079; CCRIS 329; Celanex; Chloresene; Codechine; Cyclohexane, 1α,2α,3β,4α,5α,6β-hexachloro-; DBH; Detmol Extract; Detox 25; Devoran; Dol Granule; Drilltox-Spezial Aglukon; EINECS 200-401-2; ENT 7,796; Entomoxan; EPA Pesticide Chemical Code 009001; Esoderm; Exagama; Fenoform forte; Forlin; Forst-Nexen; Gallogama; Gamacarbatox; Gamacid; Gamacide; Gamacide 20; Gamaphex; Gamene; Gamiso; γ-BHC; γ-COL; γ-HCH; γ-Hexachlorocyclohexane; γ-Hexachloro-cyclohexane [Lindane and other hexachlorocyclohexane isomers]; γ-lindane;

Gamma-mean 400; Gammahexa; Gammahexane; Gammalin; Gammalin 20; Gammalin 20; Gammaterr; Gammex; Gammexane; Gammopaz; Geobilan; Geolin G 3; Gexane; HCC; HCCH; HCH; Heclotox; HEXA; Hexachloran; Hexachlorane; Hexachlorocyclohexane; Hexachlorocyclohexane, γ-; Hexatox; Hexaverm; Hexcidum; Hexicide; Hexyclan; HGI; Hilbeech; Hortex HSDB 646; Hungaria L7; Inexit; Isotox; Jacutin; Kokotine; Kwell; Kwell-R; Lacco HI lin; Lasochron; Latka 666; Lendine; Lentox; Lidenal; Lindafor; Lindagam; Lindagrain; Lindagranox; Lindalo; Lindamul; Lindane; Lindano; Lindanum; Lindapoudre; Lindaterra; Lindatox; Lindex; Lindosep; Lintox; Linvur; Lorexane; Mglawik L; Milbol 49; Mszycol; NCI-C00204; Neo-scabicidol; Nexen FB; Nexit; Nexit-stark; Nexol-E; Nicochloran; Novigam; Novigan; Omnitox; Ovadziak; Owadziak; Pedraczak; Pflanzol; PLK; Quellada; RCRA waste number U129; Sang gamma; Scabene; Silvanol; Spritz-Rapidin; Spritzlindane; Spruehpflanzol; Streunex; TAP 85; Tri-6; Verindal Ultra; Viton. Registered by EPA as an insecticide. Banned by EPA from use as a pesticide and restricted by FDA in use as a treatment for lice and scabies in humans. Crystals; mp = 112.5°; slightly soluble in H$_2$O (0.00073 g/100 ml), soluble in Me$_2$CO (43.5 g/100 ml), C$_6$H$_6$ (28.9 g/100 ml), CHCl$_3$ (24.0 g/100 ml), Et$_2$O (20.8 g/100 ml), EtOH (6.4 g/100 ml), C$_7$H$_8$ (2.76 g/100 ml), xylene (2.47 g/100 ml), CCl$_4$ (0.67 g/100 ml), cyclohexanonone (3,67 g/100 ml), dioxane (3.14 g/100 ml), AcOH (1.28 g/100 ml); LD$_{50}$ (mrat orl) = 88 - 125 mg/kg, (frat orl) = 91 mg/kg, (mus orl) = 59 - 246 mg/kg, (rat der) = 900 - 1000 mg/kg, (bobwhite quail orl) = 120 - 130 mg/kg; LC$_{50}$ (guppy 48 hr.) = 0.16 - 0.3 mg/l; toxic to bees. Drexel Chemical Co.; Earth Care; Gustafson LLC; UAP - Platte Chemical; Wilbur Ellis Co.; Zeneca Pharm.

2247   Linoleic acid
60-33-3   5527   G H

C18H32O2

9,12-Octadecadienoic acid (9Z,12Z)-.

AI3-11132; CCRIS 650; EINECS 200-470-9; Emersol 310; Emersol 315; Extra Linoleic 90; FEMA No. 3380; Grape seed oil; HSDB 5200; Linoleic acid; α-Linoleic acid; 9Z,12Z-Linoleic acid; cis,cis-Linoleic acid; 9-cis,12-cis-Linoleic acid; Linolic acid; NSC 281243; 9,12-Octadecadienoic acid; 9,12-Octadecadienoic acid (9Z,12Z)-; 9,12-Octadecadienoic acid, (Z,Z)-; (Z,Z)-9,12-Octadecadienoic acid; Oils, grape; Oils, grape seed; Polylin 515; Polylin No. 515; Telfairic acid; Unifac 6550. Unsaturated fatty acid; used in manufacture of paints, coatings, emulsifiers, vitamins. Oil; bp$_1$ = 230-232°; d$_4^8$= 0.914; insoluble in H$_2$O, soluble in organic solvents. Arizona; CasChem; Henkel/Emery; Hercules Inc.; Langley Smith Ltd.

2248   Linolenic acid
463-40-1   5528   G

C18H30O2

(Z,Z,Z)-9,12,15-Octadecatrienoic acid.

AI3-23986; CCRIS 656; EINECS 207-334-8; FEMA No. 3380; Industrene® 120; Linolenate; Linolenic acid; α-Linolenic acid; Linolenic acid; Linolenic acid, crude; NSC 2042; 9,12,15-Octadecatrienoic acid, (9Z,12Z,15Z)-; 9,12,15-Octadecatrienoic acid, (Z,Z,Z)-; 9-cis,12-cis,15-cis-Octadecatrienoic acid; cis,cis,cis-9,12,15-Octadeca-trienoic acid; cis-Δ^{9,12,15}-Octadecatrienoic acid; all-cis-9,12,15-Octadecatrienoic acid; (Z,Z,Z)-Octadeca-9,12,15-trienoic acid. Chemical intermediate. Liquid; mp = -16.5°; bp$_{17}$ = 213°, bp$_{0.05}$ = 129°; d^{20} = 0.9164; λ$_m$ = 233 nm (ε = 126); insoluble in water;

slightly soluble in C_6H_6, soluble in EtOH, Et_2O. *Witco/Humko.*

2249   **Linseed Oil**
8001-26-1   5530   E P V

Linseed oil.
Aceite de Linaza; Acid refined linseed oil; Acidulated linseed soapstock; Bodied linseed oil; Caswell No. 527A; EINECS 232-278-6; EPA Pesticide Chemical Code 031603; Fats and Glyceridic oils, linseed; Fats and Glyceridic oils, flaxseed; Flaxseed oil; Groco; HSDB 5155; Huile de Lin; L-310; Leinol; Linseed absolute; Linseed fatty acids, glycerin ester; Linseed oil; Linseed oil absolute; Linseed oil, alkali refined; Linseed oil, bleached; Linseed oil extract; Linseed oil fatty acids, glycerol triester; Linseed oil, wash recovered; Oils, glyceridic, flaxseed or linseed; Oils, linseed; Oleum Lini; Sunflower oil. Registered by EPA as an antimicrobial, fungicide, herbicide, insecticide and rodenticide (cancelled). FDA, BP compliance. Used as a demulcent and emollient in medicinal soaps, soothing to skin, used in cough medicines and as a purgative in veterinary medicine. Amber drying oil; d = 0.921 - 0.936; mp = -19°; bp = 343°; soluble in Et_2O, $CHCl_3$, CS_2, turpentine, slightly soluble in EtOH. Allergen and skin irritant. *Arista Ind.; Ferro/Bedford; Penta Mfg.; Ruger; Spectrum Chem. Manufacturing.*

2250   **Lisinopril**
83915-83-7   5538   D

$C_{21}H_{31}N_3O_5 \cdot 2H_2O$
1-[N^2[(S)-1-Carboxy-3-phenylpropyl]-L-lysyl]-L-proline dihydrate.
Acerbon; Alapril; Carace; Coric; HSDB 6852; Lisinopril; Lisinopril dihydrate; MK-521; Novatec; Prinil; Prinivil; Prinzide; Renacor; RS-10029; Tensopril; Vivatec; Zestril.; component of: Prinzide. Angiotensin-converting enzyme inhibitor. Orally active peptidyldipeptide hydrolase inhibitor. Antihypertensive. Crystals; λ_m = 246, 254, 258, 261, 267 nm (A$^{1\%}_{1cm}$ 4.0, 4.5, 5.1, 5.1, 3.7 0.1N NaOH), 246 253, 258, 264, 267 nm (A$^{1\%}_{1cm}$ 3.2, 3.9, 4.5, 3.0, 2.8 0.1NHCl); $[\alpha]^{25}_{405}$ = -120° (c = 1 0.25M Zn(OAc)$_2$ pH 6.4). *Merck & Co.Inc.*

2251   **Lithium**
7439-93-2   5542   G

Li

Li
Lithium.
EINECS 231-102-5; HSDB 647; Lithium; Lithium, elemental; Lithium, metallic; UN1415. Metallic element. Used as a scavenger and degasifier for stainless and mild steels in molten state; deoxidizer in copper and alloys; rocket propellants; pharmaceuticals. Brittle metal; mp = 180.54°; bp = 1347°; reacts violently with inorganic acids; soluble in liquid ammonia; does not react with oxygen at room temperature. *Atomergic Chemetals; Cerac; FMC*

Corp.

2252   **Lithium 12-hydroxystearate**
7620-77-1   H

$C_{18}H_{35}LiO_3$
Octadecanoic acid, 12-hydroxy-, monolithium salt.
AI3-19768; EINECS 231-536-5.

2253   **Lithium hypochlorite**
13840-33-0   G

Cl—O$^-$ Li$^+$

ClLiO
Hypochlorous acid lithium salt.
Caswell No. 528A; CCRIS 4023; EINECS 237-558-1; EPA Pesticide Chemical Code 014702; Hypochlorous acid, lithium salt; HyPure L; Lithium chloride oxide; Lithium hypochlorite; Lithium oxychloride; UN1471. Laundry bleach, swimming pool chlorination. *FMC Corp.; Olin Corp.*

2254   **Lithium iodide**
10377-51-2   5557   G

Li—I

Ili
Iodic acid lithium salt.
EINECS 233-822-5; Lithium iodide; Lithium iodide (LiI); Tenephrol. Used in photography. Crystals; mp = 446°; bp = 1171°; d = 3.4900; soluble in H_2O (2 g/ml), EtOH, Me_2CO.

2255   **Lithium methyl 12-oxidooctadecanoate**
53422-16-5   H

$C_{19}H_{37}LiO_3$
n-Octadecanoic acid, 12-hydroxy-, methyl ester, lithium salt.
EINECS 258-547-8; Lithium methyl 12-oxidoocta-decanoate; Methyl 12-hydroxy stearate, lithium salt; Octadecanoic acid, 12-hydroxy-, methyl ester, lithium salt.

2256   **Lithium molybdate**
13568-40-6   G

Li_2MoO_4
Dilithium molybdate.
Dilithium molybdate; EINECS 236-977-7; Molybdate (MoO_4^{2-}), dilithium, (T-4)-. Use in manufacture of steel coating and as a petroleum cracking catalyst. Solid; mp = 705°. *AAA Molybdenum; Atomergic Chemetals; Cerac.*

2257   **Lithium nitrate**
7790-69-4   5558   G

LiNO3
Nitric acid lithium salt.
EINECS 232-218-9; Lithium nitrate; Nitric acid, lithium salt; UN2722. Ceramics, pyrotechnics, salt baths, heat exchange media, refrigeration systems, rocket propellant. Crystals; mp = 251°; d = 2.3800. *Atomergic Chemetals; Cerac; Mallinckrodt Inc.*

2258   Lithium stearate
4485-12-5   G

$C_{18}H_{35}LiO_2$
Lithium octadecanoate.
Afco-Chem LIS; Al3-19767; EINECS 224-772-5; Lithalure; Lithium octadecanoate; Lithium stearate; Litholite; Octadecanoic acid, lithium salt; Stavinor; Stearic acid, lithium salt. Lubricant for metal sintering. Lubricant and stabilizer for resins. Pigment dispersant, mold releasing agent, waterproofing agent and lubricant additive. Used in cosmetics, plastics, waxes, greases, a lubricant in powder metallurgy, corrosive inhibitor, flatting agent, high-temperature lubricant. White solid; mp= 220°; LD50 (rat orl) = 15 gm/kg. *Adeka Fine Chem.; Chemetall Chem. products; CK Witco Corp.; Lancaster Synthesis Co.; Syn Prods.*

2259   Lithopone
1345-05-7   5566   G

$ZnS.BaSO_4$
Zinc sulfide/barium sulfate.
Beckton white; C.I. Pigment White 5; Duresco; EINECS 215-715-5; Enamel white; Fulton white; Griffith's zinc white; Jersey lily white; Knights patent zinc white; Lithopone; Marbon white; Nevin; Oleum white; Orrs white; Pinolith; Porcelain white; Ross white; Zinc baryta white; Zincolith. A mixture of zinc sulfide, barium sulfate, and some zinc oxide. White pigment for paints and coatings. Used in water and oil paints to provide thixotropy, improve gloss and flow. *Ore & Chem.; Sachtleben Chemie GmbH.*

2260   Lofepramine Hydrochloride
26786-32-3   5578   G D

$C_{26}H_{28}Cl_2N_2O$
4'-Chloro-2-[[3-(10,11-dihydro-5H-dibenz[b,f]azepin-5-yl)propyl]methylamino]acetophenone monohydro-chloride.
Amplit; Clopepramine hydrochloride; DB-2182; EINECS 248-002-2; Gamanil; Gamonil; Iopramine hydrochloride; Leo-640; Leo 640 hydrochloride; Lofepramine hydrochloride; Timelit; Tymelyt; WHR 2908A. Tricyclic antidepressant. Crystals; mp = 152-154°; soluble in alcohol, CHCl3; practically insoluble in H2O; LD50 (mus orl) > 2500 mg/kg, (rat orl) > 1000 mg/kg, (mus ip) > 920 mg/kg; (rat ip) > 1000 mg/kg, (mus/rat sc) > 1000 mg/kg. *E. Merck.*

2261   Lofetamine ^{123}I hydrochloride
85068-76-4   5067   G

$C_{12}H_{18}$[123]IN
(±)-4-(Iodo-^{123}I)-α-methyl-N-(1-methylethyl) benzene ethanamine hydrochloride.
Benzeneethanamine, 4-(iodo-^{123}I)-α-methyl-N-(1-methyl-ethyl)-, hydrochloride, (±)-; (±)-p-Iodo-^{123}I-N-isopropyl-α-methylphenethylamine hydrochloride; lofetamine hydro-chloride I[123]; lofetamine I[123] hydrochloride; (^{123}I)(±)-N-Isopropyl-p-iodoamphetamine hydrochloride; Perfus-amine; Spectamine. Lipid-soluble radioactive brain imaging agent. Crystals; mp = 156-158°.

2262   Loratadine
79794-75-5   5599   D

$C_{22}H_{23}ClN_2O_2$
Ethyl 4-(8-chloro-5,6-dihydro-11H-benzo[5,6]cyclo-hepta[1,2-b]pyridin-11-ylidine)-1-piperidinecarboxylate.
Aerotina; Alarin; Alerpriv; Allertidin; Anhissen; Bedix Loratadina; Biloina; Bonalerg; Civeran; Claratyne; Clarinase; Clarinase Reperabs; Claritin; Claritine; Clarityn; Clarityne; Fristamin; Histaloran; HSDB 3578; Lergy; Lertamine; Lesidas; Lisino; Loracert; Loradex; Loradif; Loranox; Lorantis; Lorastine; Loratadina; Loratadine; Loratadinum; Loratyne; Loraver; Lorfast; Loritine; Lowadina; Nularef; Optimin; Polaratyne; Pylor; Restamine; Rhinase; Rinomex; Sanelor; Sch 29851; Sensibit; Sinhistan Dy; Sohotin; Tadine; Talorat Dy; Velodan; Versal; Zeos. Antihistaminic. Crystals; mp = 134-136°. *Schering AG.*

2263   Lorazepam
846-49-1   5600   D

$C_{15}H_{10}Cl_2N_2O_2$
7-Chloro-5-(o-chlorophenyl)-1,3-dihydro-3-hydroxy-2H-1,4-benzodiazepin-2-one.
5-25-02-00248 (Beilstein Handbook Reference); Almazine; Anxiedin; Anxira; Anzepam; Aplacasse; Aplacassee; Apo-Lorazepam; Aripax; Ativan; Azurogen; Bonatranquan; Bonton; BRN

0759084; DEA No. 2885; Delormetazepam; Demethyllormetazepam; Donix; Duralozam; Efasedan; EINECS 212-687-6; Emotion; Emotival; Equitam; Idalprem; Kalmalin; Larpose; Laubeel; Lomesta; Lopam; Lorabenz; Lorafen; Loram; Lorans; Lorapam; Lorat; Lorax; Lorazene; Lorazep; Lorazepam; Lorazepam-Efeka; Lorazepam Fabra; Lorazepam Genericon; Lorazepam Lannacher; Lorazepam Medical; Lorazepamum; Lorazepan Chobet; Lorazepan Richet; Lorazin; Lorazon; Lorenin; Loridem; Lorivan; Lorsedal; Lorsilan; Lorzem; Lozepam; Max Pax; Merlit; Nervistop L; NIC; Norlormetazepam; Novhepar; Novolorazem; NSC 289758; Nu Loraz; Orfidal; Pro Dorm; Psicopax; Punktyl; Quait; Renaquil; Rocosgen; Securit; Sedatival; Sedazin; Sedizepan; Sidenar; Silence; Sinestron; Somagerol; Stapam; Tavor; Temesta; Titus; Tolid; Tranqipam; Trapax; Upan; Vigiten; WY 4036; Wypax. Pharmaceutical preparation for the treatment of anxiety. Crystals; mp = 166-168°; λ_m = 229 nm (1N NaOH), 233 nm (1N HCl); soluble in H_2O (0.008 g/100 ml), $CHCl_3$ (0.33 g/100 ml), EtOH (1.4 g/100 ml), propylene glycol (1.6 g/100 ml), EtOAc (3.0 g/100 ml); LD_{50} (mus orl) = 3178 mg/kg, (rat orl) > 5000 mg/kg. *Wyeth-Ayerst Labs.*

2264   Losartan
114798-26-4   5604   D

$C_{22}H_{23}ClN_6O$
2-Butyl-4-chloro-1-[p-(o-1H-tetrazol-5-ylphenyl)benzyl]-imidazole-5-methanol.
DUP 89; Losartan. Non-peptide angiotensin II receptor antagonist used as an antihypertensive agent. Crystals; mp = 183.5-184.5°. *DuPont-Merck Pharm.; Merck & Co.Inc.*

2265   Losartan monopotassium salt
124750-99-8   5604   D

$C_{22}H_{22}ClKN_6O$
2-Butyl-4-chloro-1-[p-(o-1H-tetrazol-5-ylphenyl)benzyl]-imidazole-5-methanol monopotassium salt.
Cozaar; Du Pont 753; DuP-753; HSDB 7043; Hyzaar; L-158086; Lortaan; Lorzaan; Lorzaar; Losacar; Losacor; Losaprex; Losartan monopotassium salt; Losartan potassium; Lotim; MK-0954; MK-954; Neo Lotan; Niten; Ocsaar; Tenopres; component of: Hyzaar. Nonpeptide angiotensin II receptor antagonist. Used as an antihypertensive. *DuPont-Merck Pharm.; Merck & Co.Inc.*

2266   LSD
50-37-3   5654   G

$C_{20}H_{25}N_3O$
Lysergic acid diethylamide.
4-25-00-00939 (Beilstein Handbook Reference); ACID; BRN 0094179; Cubes; DEA No. 7315; Delysid; 9,10-Didehydro-N,N-diethyl-6-methyl-ergoline-8-β-carbox-amide; Diethylamid kyseliny lysergove; N,N-Diethyl-D-lysergamide; N,N-Diethyllysergamide; EINECS 200-033-2; Ergoline-8β-carboxamide, 9,10-didehydro-N,N-diethyl-6-methyl-; Ergoline-8-carboxamide, 9,10-didehydro-N,N-diethyl-6-methyl-, (8β)-; Heavenly Blue; HSDB 3920; Liserigide; Lisergido; LSD; D-LSD; LSD-25; D-Lysergic acid diethylamide; Lysergamid; Lysergamide, N,N-diethyl-; Lysergsäure diethylamid; Lysergic acid diethylamide; Lysergic acid diethylamide-25; Lysergide; Lysergidum; Lysergsaeurediaethylamid; Lysergsäurediäthylamid; Pearly gates; Royal Blue; Wedding bells. Hallucinogen. Not sold in the US.

2267   2,6-Lutidine
108-48-5   5636   G

C_7H_9N
2,6-Dimethylpyridine.
AI3-24282; α,α'-Dimethylpyridine; 2,6-Dimethyl-pyridine; EINECS 203-587-3; FEMA No. 3540; HSDB 79; α,α'-Lutidine; 2,6-Lutidine; NSC 2155; Pyridine, 2,6-dimethyl-; α,α'-lutidin. Used in the manufacture of pharmaceuticals, resins, dyestuffs, rubber accelerators, insecticides. Liquid; mp = -6.1°; bp = 144.1°; d^{20} = 0.9226; λ_m = 268 nm (ε = 12589, H_2SO_4); freely soluble in H_2O, soluble in Et_2O, Me_2CO, $CHCl_3$, slightly soluble in EtOH. *Janssen Chimica; Raschig GmbH; Reilly Ind.; Sigma-Aldrich Fine Chem.*

2268   Lyral
31906-04-4   H

$C_{13}H_{22}O_2$
4-(4-Hydroxy-4-methylpentyl)cyclohex-3-enecarbalde-hyde.
BRN 2046455; 3-Cyclohexene-1-carboxaldehyde, 4-(4-hydroxy-4-methylpentyl)-; EINECS 250-863-4; 4-(4-Hydroxy-4-methylpentyl)cyclohex-3-enecarbaldehyde; 4-(4-Hydroxy-4-methylpentyl)-3-cyclohexene-1-carbox-aldehyde; Lyral.

2269   Lysidine
534-26-9   5655   G

C4H8N2
4,5-Dihydro-2-methyl-1H-imidazole.
5-23-03-00385 (Beilstein Handbook Reference); AI3-16866; BRN 0104225; 4,5-Dihydro-2-methyl-1H-imidazole; EINECS 208-596-6; 1H-Imidazole, 4,5-dihydro-2-methyl-; 2-Imidazoline, 2-methyl-; Lysidine; Methylglyoxalidine; 2-Methylimidazoline. Crystals; mp = 107°; bp = 196.5°; insoluble in Et$_2$O, soluble in CHCl$_3$, very soluble in H$_2$O, EtOH.

2270   Lysine
56-87-1   5656   G

C$_6$H$_{14}$N$_2$O$_2$
(S)-2,6-Diaminohexanoic acid.
4-04-00-02717 (Beilstein Handbook Reference); AI3-26523; Aminutrin; BRN 1722531; (S)-α,ε-Diaminocaproic acid; EINECS 200-294-2; h-Lys-oh; Hexanoic acid, 2,6-diamino-, (S)-; HSDB 2108; Lisina; LYS; Lysine; α-Lysine; L-Lysine; (S)-Lysine; Lysine acid; Lysine, L-; Lysinum; L-Norleucine, 6-amino-. Biochemical and nutritional research pharmaceuticals, culture media, fortification of foods and feeds, nutrient supplement, animal feed additive. White solid; dec 224.5°; [α]$_D^{20}$ = 14.6°; pK$_1$ = 2.20; soluble in water, insoluble in common neutral solvents. Degussa AG; Greeff R.W. & Co.; Indofine Chem. Co. Inc.; Sigma-Aldrich Fine Chem.; Walton Pharm.

2271   Lysine monohydrochloride
70-53-1   5656   G

C$_6$H$_{15}$ClN$_2$O$_2$
DL-Lysine hydrochloride.
AI3-18306; EINECS 200-739-0; Lysine, monohydro-chloride; Lysine monohydrochloride, dl-; NSC 46705; [657-27-2]. Used to enrich cereals and feeds. Crystals; mp = 263 - 264°; [α]$_D^{25}$ = +14.6° (c = 2 0.6N HCl). Degussa AG; Greeff R.W. & Co.; Spectrum Chem. Manufacturing; Tanabe U.S.A. Inc.

2272   L-(+)-Lysine hydrochloride
657-27-2   5656   G

C$_6$H$_{14}$N$_2$O$_2$.HCl
L-Lysine monohydrochloride.
AI3-52405; Darvyl; EINECS 211-519-9; Enisyl; L-Gen; Lyamine; Lysine hydrochloride; L-Lysine hydrochloride; Lysine monohydrochloride; Lysine, monohydrochloride, L-; L-Lysine monohydrochloride; Lysion; NSC 9253; Lysine hydrochloride; K; (S)-2,6-diaminohexanoic acid. Used in the enrichment of cereals and feeds Crystals; dec 210°; [α]$_D^{20}$ = +14.6°; soluble in H$_2$O; insoluble in

non-polar organic solvents. Person & Covey Inc.

2273   Lysochlor
59-50-7   2152   G P

C$_7$H$_7$ClO
4-Chloro-m-cresol.
4-06-00-02064 (Beilstein Handbook Reference); AI3-00075; Aptal; Baktol; Baktolan; BRN 1237629; Candaseptic; Caswell No. 185A; CCRIS 1938; Chlorcresolum; Chlorkresolum; Chlorocresol; m-Cresol, 4-chloro-; 4-Chloro-3-cresol; 6-Chloro-m-cresol; Chlorocresol; Chloro-3-cresol; Chlorocresolo; Chlorocresolum; 2-Chloro-5-hydroxytoluene; 2-Chloro-hydroxytoluene; 6-Chloro-3-hydroxytoluene; 4-Chloro-1-hydroxy-3-methylbenzene; 4-Chloro-3-methylphenol; Chlorokresolum; Clorocresolo; CMK; EINECS 200-431-6; EPA Pesticide Chemical Code 064206; HSDB 5198; 3-Methyl-4-chlorophenol; NSC 4166; Ottafact; Parachlorometacresol; Parmetol; Parol; PCMC; Peritonan; Phenol, 4-chloro-3-methyl-; Preventol CMK; Raschit; Raschit K; Rasen-Anicon; RCRA waste number U039. A phenol-based disinfectant. Has been used medically as a topical antiseptic and as a preservative for latex. Crystals; mp = 55.5°, 66°; bp = 235°; soluble in H$_2$O (0.38 g/100 ml), soluble in organic solvents. Kalle BV; Lancaster Synthesis Co.; Sigma-Aldrich Fine Chem.

2274   Maclurin
519-34-6   5665   G

C$_{13}$H$_{10}$O$_6$
(3,4-Dihydroxyphenyl)(2,4,6-trihydroxyphenyl)methan-one.
Benzophenone, 2,3',4,4',6-pentahydroxy-; C.I. 75240; C.I. Natural Yellow 11; (3,4-Dihydroxyphenyl)(2,4,6-trihydroxyphenyl)methanone; EINECS 208-268-2; Fustic extract; Kino-yellow; Laguncurin; Maclurin; Maklurin; Methanone, (3,4-dihydroxyphenyl)(2,4,6-trihydroxy-phenyl)-; Morintannic acid; Moritannic acid; NSC 83240; Patent Fustin; 2,3',4,4',6-Pentahydroxybenzophenone. Used to dye fabrics. Crystals; mp = 222-223°; soluble in H$_2$O (0.48 g/100 ml), more soluble in organic solvents.

2275   Magnesia alba
39409-82-0   5682   G

C$_4$H$_2$Mg$_5$O$_{14}$.5H$_2$O
Magnesium carbonate hydroxide.
Carbonic acid, magnesium salt (1:1), mixture with magnesium hydroxide (Mg(OH)$_2$), hydrate; CCRIS 7884; Magnesium carbonate basic; Magnesium carbonate hydroxide; Marinco C. A basic magnesium carbonate of variable composition. Used in fireproofing, as an antacid and laxative and in manufacture of magnesium compounds.

2276   Magnesium
7439-95-4   5675   G

Mg

Mg
Magnesium.
EINECS 231-104-6; HSDB 654; Magnesio; Magnesium; Magnesium powdered; Magnesium sheet; Rieke's active magnesium; RMC; UN1418; UN1869; UN2950. Metallic element; aluminum alloys for structural parts; used in pyrotechnics; photography; production of iron, nickel, zinc, titanium, zirconium, steel; gasoline additive; magnesium compounds; cathodic protection; reducing agent; precision instruments and optical mirrors. Metal; mp = 651°; bp = 1100°; d^{20} = 1.738. *Norsk Hydro AS; Pechiney Electrométallurgie; Sigma-Aldrich Fine Chem.*

2277   Magnesium acetate
142-72-3   5676   G H

C4H6MgO4
Acetic acid magnesium salt.
Acetic acid magnesium salt; CCRIS 7883; Cromosan; EINECS 205-554-9; Magnesium acetate; Magnesium diacetate; Mg Acetate; NSC 75798; Nu-Mag. Dye fixative in textile printing, deodorant, disinfectant, antiseptic. Solid; mp = 80°; d = 1.45; LD50 (mus iv) = 18 mg/kg. *EM Ind. Inc.; Hoechst AG; Verdugt BV.*

2278   Magnesium aluminum hydroxide carbonate
11097-59-9   H

CH16Al2Mg6O19
(Carbonato(2-))hexadecahydroxybis(aluminium)hexa-magnesium.
(Carbonato(2-))hexadecahydroxybis(aluminium)hexa-magnesium; Aluminate (Al(OH)63-), (OC-6-11)-, magnesium carbonate hydroxide (2:6:1:4); EINECS 234-319-3; Magnesium, (carbonato(2-))hexadecahydroxy-bis(aluminum)hexa-; Magnesium, (carbonato)hexadeca-hydroxydialuminumhexa-; Magnesium aluminum hydroxide carbonate.

2279   Magnesium borate
13703-82-7   5680   G

B2MgO4
Magnesium metaborate.
Antifungin; EINECS 237-235-5; Magnesium borate; Magnesium metaborate. An antiseptic and fungicide. Occurs in a variety of minerals. White powder, slightly soluble in H2O.

2280   Magnesium carbonate
546-93-0   D G

CMgO3
Magnesium(II) carbonate.
AI3-00768; Apolda; Carbonate magnesium; Carbonic acid,

magnesium salt (1:1); Caswell No. 530; CI 77713; DCI light magnesium carbonate; Destab; EINECS 208-915-9; Elastocarb Tech Light, Heavy; EPA Pesticide Chemical Code 073503; Gold Star (carbonate); GP 20 (carbonate); HSDB 211; Hydromagnesite; Kimboshi; MA 70 (carbonate); Magfy; Magmaster; Magnesite; Magnesite dust; Magnesium carbonate; Magnesium carbonate (1:1); Magnesium carbonate (MgCO3); Magnesium carbonate anhydrous; Magnesium carbonate basic; Magnesium carbonate hydroxide; Magnesium(II) carbonate (1:1); Magocarb-33; NSC 83511; Stan-mag magnesium carbonate. Inorganic filler providing flame retardancy and smoke suppression to elastomers, plastics, and thermosets incl. EPDM, PP, PE, PVC; used as a flame retardant in wire and cable compounds, conduit/tubing, film and sheet. *Kaopolite; Morton Intn'l.*

2281   Magnesium chloride
7786-30-3   5684   D G P

Cl2Mg
Magnesium chloride anhydrous.
Aerotex Accelerator MX; Caswell No. 531; CCRIS 3961; DUS-top; EINECS 232-094-6; EPA Pesticide Chemical Code 013902; HSDB 657; Magnesium chloride; Magnesium chloride (MgCl2); Magnesium chloride anhydrous; Magnesium dichloride; Magnogene; NSC 529832; TMT 2. Source of magnesium; disinfectants, fire extinguishers, fireproofing wood, cement, refrigerating brines, ceramics, cooling drilling tools, textile sizes and lubricants, paper manufacture, dust control on roads, flocculating agent, catalyst. Solid; mp = 712°; d = 2.41; soluble in H2O; LD50 (rat orl) = 8.1 g/kg. *Magnesia GmbH; Mallinckrodt Inc.; Schaeffer Salt & Chem.; Sigma-Aldrich Fine Chem.*

2282   Magnesium ethoxyethoxide
14064-03-0   H

C8H18MgO4
Ethanol, 2-ethoxy-, magnesium salt.
EINECS 237-912-5; Ethanol, 2-ethoxy-, magnesium salt; Magnesium bis(2-ethoxyethanolate); Magnesium ethoxyethoxide.

2283   Magnesium fluoride
7783-40-6   5688   G

F2Mg
Hydrofluoric acid magnesium salt.
Afluon; EINECS 231-995-1; Irtran 1; Magnesium fluoride; Magnesium fluoride (MgF2); Magnesium fluorure; Magtran; Sellaite. Used in manufacture of ceramics and glass. Solid; mp = 1248°; bp = 2260°; d = 3.148; soluble in H2O (8.7 g/100 ml), insoluble in organic solvents; LD50 (gpg orl) = 1 g/kg. *BDH Laboratory Supplies.*

2284   Magnesium gluconate
3632-91-5   4469   D G

374

C12H22MgO14.2H2O
Magnesium D-gluconate.
Almora; EINECS 222-848-2; Glucomag; Gluconic acid, magnesium salt (2:1), D-; D-Gluconic acid, magnesium salt (2:1); Glucosium; GYN; Magnesium D-gluconate (1:2); Magnesium digluconate; Magnesium gluconate; Magnesium gluconate anhydrous; Menesia. Mineral source for pharmaceutical and food products. Used medicinally as a magnesium replenisher. Crystals; soluble in H_2O. *Akzo Chemie; Atomergic Chemetals; Forest Pharm. Inc.; Spectrum Chem. Manufacturing.*

2285   Magnesium hydroxide
1309-42-8   5693   D G

H2MgO2
Magnesium(II) hydroxide.
200-06H; Alcanex NHC 25; Arthritis Pain Formula Maximum Strength; Asahi Glass 200-06; Ascriptin; Baschem 12; Calcitrel; Camalox; CCRIS 3342; Combustrol 500; Di-Gel; DP 393; DSB 100; Duhor; Duhor N; Ebson RF; EINECS 215-170-3; FloMag H; FloMag HUS; Gelusil; Haley's MO; HSDB 659; Hydro-mag MA; Hydrofy G 1.5; Hydrofy G 2.5; Hydrofy N; Ki 22-5B; Kisuma 4AF; Kisuma 5; Kisuma 5A; Kisuma 5B; Kisuma 5B-N; Kisuma 5BG; Kisuma 5E; Kisuma 78; Kisuma S 4; Kudrox; KX 8S(A); KX 8S(B); Kyowamag F; Lycal 96 HSE; Maalox; Maalox Plus; Mag Chem MH 10; Magnesia hydrate; MagneClear 58; Magnesia magma; Magnesia, [Milk of]; Magnesiamaito; Magnesium dihydroxide; Magnesium hydroxide; Magnesium hydroxide (Mg(OH)2); Magnesium hydroxide gel; Magnesium hydroxide slurry; Magnifin H 10; Magoh-S; Magox; Marinco H; Marinco H 1241; Martinal VPF 8812; Milk of magnesia; Milmag; Mint-O-Mag; Mylanta; Nemalite; Phillips Magnesia Tablets; Phillips Milk of Magnesia Liquid; Reachim; S/G 84; Simeco Suspension; Star 200; Versamag® DC; Wingel; Zerogen® 10, Zerogen® 60. Fire retardant filler for plastics; extender pigment for flame retardant coatings. White solid; insoluble in H_2O, aqueous suspension has pH 9.5-10.5. *Kaopolite.*

2286   Magnesium laureth sulfate
62755-21-9   G

C24H52MgO8S2.(C2H4O)n
Magnesium lauryl ether sulfate.
Elfan® NS 243 S Mg; Empicol® EGB; Empicol® EGC; Empicol® EGC70; Magnesium laureth sulfate; Magnesium lauryl ether sulfate; Poly(oxy-1,2-ethanediyl), α-sulfo-ω-(dodecyloxy)-, magnesium salt; Sulfochem® MgLES; Zoharpon MgES. Detergent and toiletry raw material. *Albright & Wilson UK Ltd.*

2287   Magnesium nitrate
10377-60-3   5697   G

MgN2O6
Magnesium(II) nitrate.
EINECS 233-826-7; HSDB 660; Magnesium dinitrate; Magnesium nitrate; Magnesium(II) Nitrate (1:2); Magniosan; Magnisal; Nitric acid, magnesium salt; UN1474. Used for agricultural (for curing magnesium deficiency via irrigation system, direct soil application or foliar spray) and technical (in metal, textile, ceramic and other industries) applications. Used in pyrotechnics and in the manufacture of nitric acid. [hexahydrate]; solid; mp = 95°; soluble in H_2O (1.25 g/ml), more soluble in EtOH. *Blythe, Williams Ltd.; EM Ind. Inc.; Haifa Chemicals Ltd; Hoechst Celanese; Mallinckrodt Inc.*

2288   Magnesium oxide
1309-48-4   5700   D G

$$Mg=O$$

MgO
Magnesium(II) oxide.
100A (oxide); Akro-mag; AM 2 (cement additive); Animag; Anscor P; BayMag; Calcined brucite; Calcined magnesia; Causmag; Caustic magnesite; CCRIS 3659; Corox; EINECS 215-171-9; Elastomag 100; Elastomag 170; Fert-O-Mag; FloMag HP; FloMag HP-ER; FMR-PC; Granmag; Hamag LP; Heavy calcined magnesia; Heavy magnesia; Heavy magnesium oxide; HP 10 (oxide); HSDB 1652; Ken-Mag®; KM 3 (oxide); KM 40; KMACH-F; KMB 100-200; Kyowaad 100; Kyowamag 20; Kyowamag 30; Kyowamag 40; Kyowamag 60; Kyowamag 100; Kyowamag 150; Kyowamag 150B; Kyowaway 150; Light magnesia; Liquimag A; Liquimag B; Luvatol MK 35; Mag Chem 10; Mag Chem 10-40; Mag Chem 10-200; Mag Chem 10-325; Mag Chem 35; Mag Chem 200AD; Mag Chem 200D; Magcal; Magchem 100; Maglite; Maglite D; Maglite de; Maglite K; Maglite S; Maglite Y; Magnesa preprata; Magnesia; Magnesia monoxide; Magnesia USTA; Magnesium oxide; Magnesium oxide, fume; Magnezu tlenek; Magotex; Magox; Magox 85; Magox 90; Magox 95; Magox 98; Magox OP; Magrods; Marmag; Oxymag; Periclase; Seasorb; Seawater magnesia; SLO 369; SLO 469. A proprietary insulating material having a great thermal conductivity and high electrical insulating power; consists essentially of magnesium oxide. White solid; mp = 2800°. *Kaopolite.*

2289   Magnesium perchlorate
10034-81-8   5702   G

Cl2MgO8.3H2O
Magnesium perchlorate trihydrate.
Anhydrone; Anhydrous magnesium perchlorate; Dehydrite®; EINECS 233-108-3; HSDB 661; Magnesium perchlorate; Perchlorate de magnesium; Perchloric acid, magnesium salt; UN1475. Used as a drying agent for gases. Solid; dec > 250°; d = 2.6000; soluble in H_2O, (exothermic), EtOH.

2290   Magnesium peroxide
1335-26-8   D G

MgO2
Magnesium dioxide.
EINECS 215-627-7; Magnesium peroxide; Novozone®. Prepared for medical purposes; an antiseptic used internally, and externally as an ointment, for wounds and gatherings.

2291 Magnesium stearate

557-04-0   5714   G H

$C_{17}H_{35}$—C(=O)—O⁻

Mg^{2+}

$C_{17}H_{35}$—C(=O)—O⁻

C36H70MgO4

Octadecanoic acid, magnesium salt.

Afco-Chem MGS; AI3-01638; Dibasic magnesium stearate; EINECS 209-150-3; HSDB 713; Magnesium distearate; Magnesium octadecanoate; Magnesium stearate; NP 1500; NS-M (salt); Octadecanoic acid, magnesium salt; Petrac MG 20NF; SM 1000; SM-P; Stearic acid, magnesium salt; Synpro 90; Synpro Magnesium Stearate 90. Lubricant for metal sintering. Lubricant and stabilizer for resins. Pigment dispersant, mold releasing agent, waterproofing agent and lubricant additive. Used in baby dusting powder; lubricants in making tablets; as a drier in paints and varnishes; stabilizer and lubricant for plastics; emulsifying agent for cosmetics.

White solid; mp = 130-140°; insoluble in H_2O.

2292 Magnesium sulfate

7487-88-9   5715   D G P

Mg^{2+}

O=S(O⁻)(O⁻)=O

MgO4S

Magnesium sulfate.

Caswell No. 534; EINECS 231-298-2; EPA Pesticide Chemical Code 050503; Epsom salt; Epsom salts; HSDB 664; Kieserite (monohydrate); Magnesium sulfate; Magnesium sulfate (1:1); Magnesium sulfate anhydrous; Magnesium sulphate; NSC 146179; OT-S; OT-S (drying agent); Sal Amarum; Sal Angalis; Sal Anglicum; Sal Catharticum; Sal de Seidlitz; Sal Seidlitense; Sel d'Angleterre; Sulfuric acid magnesium salt; Sulfuric acid magnesium salt (1:1); Tomix OT (monohydrate); Salts of England; Hair salt; Bitter salt. Used in fireproofing, textiles (warp sizing, dyeing, etc.), mineral waters, catalyst carrier, paper (sizing), cosmetic lotions. Used in horticulture as a magnesium source. Registered by EPA as a herbicide and insecticide (cancelled). Used medicinally as a magnesium replenisher. The heptahydrate is an anticonvulsant and cathartic. Also effective in termination of refractory ventricular tachyarrhythmias. Loss of deep tendon reflex is a sign of overdose.

Heptahydrate: d = 1.67; soluble in H_2O. Blythe, Williams Ltd.; Heico; Mallinckrodt Inc.; PQ Corp.

2293 Magnevist

86050-77-3   4348   G

Gd^{3+}

C28H54GdN5O20

Dimeglumine gadopentetate.

Diethylenetriaminepentaacetic acid dimeglumine salt gadolinium chelate; Dihydrogen (N,N-bis(2-(bis(carboxy-methyl)amino)ethyl)glycinato(5-))gadolinate(2-), compound with 1-deoxy-1-(methylamino)-D-glucitol; Dimeglumine-gadolinium-dtpa; Gadolinate(2-), (N,N-bis(2-(bis(carboxymethyl)amino)ethyl)glycinato(5-))-, dihydrogen, compd. with 1-deoxy-1-(methylamino)-D-glucitol (1:2); Gadopentetate Dimeglumine; Gadopentetic acid dimeglumine salt; Magnevist; Meglumine gadopentetate; Resovist; SHL 451A; ZK 93035. MRI agent. Soluble in H_2O; LD50 (rat iv) = 9.37 g/kg. Schering Health Care Ltd.

2294 Malathion

121-75-5   5723   G H P

C10H19O6PS2

Diethyl [(dimethoxyphosphinothioyl)thio]butanedioate.

4-03-00-01136 (Beilstein Handbook Reference); AC 4049; AC 26691; AI3-17034; American Cyanamid 4,049; BRN 1804525; Butanedioic acid, ((dimethoxyphosphino-thioyl)thio)-, diethyl ester; Calmathion; Camathion; Carbethoxy Malathion; Carbetovur; Carbetox; Carbofos; Carbophos; CCRIS 368; Celthion; Chemathion; Cimexan; Cleensheen; Compound 4049; Cython; Detmol MA; Dorthion; EINECS 204-497-7; EL 4049; Emmatos; Emmatos Extra; ENT 17034; Ethiolacar; Etiol; Extermathion; Flair; Formal; Forthion; Fosfothion; Fosfotion; Fosfotion 550; Fyfanon; Hilthion; HSDB 665; IFO 13140; Insecticide No. 4049; Karbofos; Kill-A-Mite; Kop-thion; Kypfos; Latka 4049; Lice Rid; Malacide; Malafor; Malagran; Malakill; Malamar; Malamar 50; Malaphos; Malasol; Malaspray; Malataf; Malathiazol; Malathion; Malathion E50; Malathion LV concentrate; Malathon; Malathyl; Malation; Malatol; Malatox; Maldison; Malmed; Malphos; Mercaptothion; Mercaptotion; MLT; Moscarda; NCI-C00215; NSC 6524; Oleophosphothion; Ortho malathion; Paladin; Phosphothion; Prioderm; Sadofos; Sadophos; SF 60; Siptox I; Sumitox; TAK; Taskil; TM-4049; UN 2783. A minimally toxic insecticide effective against insect pests on livestock, stored crops, agriculture, home and garden. Cholinesterase inhibitor, behaves as a non-systemic acaricide and insecticide with contact, stomach and respiratory action. Used for broad spectrum control of sucking and chewing insects and mites in a wide variety of crops. Particularly useful on Mediterranean fruit fly. Registered by EPA as an insecticide. Clear, amber liquid; mp = 1.4°; bp$_{0.7}$ = 156°; d$_{20}$ = 1.2076; soluble in H$_2$O (0.0145 g/100 ml), more soluble in organic solvents; LD$_{50}$ (rat orl) = 1375 - 2800 mg/kg, (mus orl) = 775 - 3320 mg/kg, rbt der) = 4100 mg/kg, (bee topical) = 0.071 mg/bee; LC$_{50}$ (bobwhite quail 5 day dietary) = 3500 mg/kg diet, (ringneck pheasant 5 day dietary) = 4320 mg/kg diet, (bluegill sunfish 96 hr.) = 0.1 mg/l. *A/S Cheminova; Agriliance LLC; Allchem Ind.; Am. Cyanamid; Amvac Chemical Corp.; Aventis Environmental Science USA LP; Bonide Products, Inc.; Cape Fear Chemicals Inc; Drexel Chemical Co.; Earth Care; Helena Chemical Co.; Prentiss Inc.; Riverdale Chemical Co; Rockland Corp.; Sariaf SpA; Tomic Insecticide Co., Inc.; UAP - Platte Chemical; Value Gardens Supply LLC.*

2295   Maleated rosin

8050-28-0   H

Maleated rosin.
EINECS 232-480-4; Gum rosin, maleic anhydride resin; Rosin, maleated; Rosin, maleic acid polymer; Rosin-maleic anhydride reaction product; Rosin, maleic anhydride adduct.

2296   Maleic acid

110-16-7   5726   G H

```
        COOH


        COOH
```

C$_4$H$_4$O$_4$
2-Butenedioic acid (Z)-.
4-02-00-02199 (Beilstein Handbook Reference); AI3-01002; BRN 0605762; Butenedioic acid, (Z)-; 2-Butenedioic acid (Z)-; cis-Butenedioic acid; CCRIS 1115; EINECS 203-742-5; 1,2-Ethylenedicarboxylic acid, (Z); cis-1,2-Ethylenedicarboxylic acid; HSDB 666; Kyselina maleinova; Maleic acid; Maleinic acid; Malenic acid; NA2215; NSC 25940; Toxilic acid. Used to retard rancidity in fats and oils; dyeing and finishing textiles; as intermediate in synthesis. Monoclinic prisms; mp = 130.5°; d^{20} = 1.590; λ_m = 203 nm (ε = 14791, EtOH); insoluble in C$_6$H$_6$, CHCl$_3$, slightly soluble in DMSO, soluble in Et$_2$O, AcOH, conc. H$_2$SO$_4$; very soluble in EtOH, Me$_2$CO. *General Chem; Penta Mfg.; Thor.*

2297   Maleic anhydride

108-31-6   5727   G H

```
        O



        O


        O
```

C$_4$H$_2$O$_3$
cis-Butenedioic acid anhydride.
5-17-11-00055 (Beilstein Handbook Reference); AI3-24283; Anhydrid kyseliny maleinove; BRN 0106909; cis-Butenedioic acid anhydride; CCRIS 2941; Dihydro-2,5-dioxofuran; EINECS 203-571-6; 2,5-Furandione; HSDB 183; Maleic acid anhydride; Maleic anhydride; Maleinanhydrid; NSC 137651; RCRA waste number U147; Toxilic anhydride; UN2215. Crystals; mp = 52.8°; bp = 202°; d^{60} = 1.314; slightly soluble in ligroin, soluble in H$_2$O (reacts), Et$_2$O, Me$_2$CO, CHCl$_3$.

2298   Maleic anhydride-methyl vinyl ether polymer

9011-16-9   G

(C$_4$H$_2$O$_3$.C$_3$H$_6$O)x
2,5-Furandione, polymer with methoxyethylene.
Agrimer VEMA-H-240; Ethene, methoxy-, polymer with 2,5-furandione; 2,5-Furandione, polymer with methoxy-ethene; 2,5-Furandione, polymer with methoxyethylene; Gantrez; Gantrez 39; Gantrez 149; Gantrez 169; Gantrez 903; Gantrez AN; Gantrez AN 119; Gantrez AN 139; Gantrez AN 149; Gantrez AN 169; Gantrez AN 179; Gantrez AN-1195; Gantrez S 95; Maleic anhydride-methoxyethylene copolymer; Maleic anhydride methyl vinyl ether, copolymer; Methoxyethylene-maleic anhydride copolymer; Methyl vinyl ether-maleic anhydride polymer; Methyl vinyl oxide-maleic anhydride polymer; NSC 79367; NSC 79368; NSC 79369; NSC 130566; NSC 130567; NSC 130568; NSC 130569; PVM/MA Copolymer; Vinyl methyl ether-maleic anhydride polymer; Viscofas; Viscofas X 100000. Dispersant, coupling, stabilizer, thickener, emulsifier, solubilizer, corrosion inhibitor, film former, antistat, used in paper and textile industries, chemical processing, industrial products, detergents, cosmetics, emulsion polymerization; sequestrant. *ISP; Sigma-Aldrich Fine Chem.*

2299   Maleic hydrazide

123-33-1   5728   G P

```
   HO                    OH

          N   N
```

C$_4$H$_4$N$_2$O$_2$
1,2-Dihydro-3,6-pyridazinedione.
AI3-18870; Antergon; Antyrost; Burtolin; Caswell No. 352; CCRIS 1879; Chemform; Chiltern Fazor; De-cut; De-sprout; 3,6-Dihydroxypyridazine; 1,2-Dihydro-3,6-pyridazinedione; 1,2-Dihydropyridazine-3,6-dione; 3,6-Dioxopyridazine; Drexel-super P; EINECS 204-619-9; ENT 18,870; EPA Pesticide Chemical Code 051501; Fair-2; Fair 30; Fair plus; Fair PS; Fazor; Gotax; HSDB 1162; Hydrazid kyseliny maleinove; Hydrazide maleique; 6-Hydroxy-2H-pyridazin-3-one; 6-Hydroxy-3(2H)-pyrid-azinone; KMH; MAH; Maintain 3; Malazide; Maleic acid cyclic hydrazide; Maleic acid hydrazide; Maleic hydrazide; Maleic hydrazine; Malein 30; Malein-saeurehydrazid; Maleinsäurehydrazid; N,N-Maleoyl-hydrazine; Malzid; Mazide; Mazide 25; MG-T; MH; MH 30; MH 36 Bayer; MH-40; Milurit; MSS MH18; NSC 13892; 3,6-Pyridazinediol; Pyridazine-3,6-diol; 3(2H)-Pyridazinone, 6-hydroxy-; 3,6-Pyridazinedione, 1,2-dihydro-; RCRA waste number U148; Regulox; Regulox 36; Regulox 50 W; Regulox K; Regulox W; Retard; Royal MH-30; Royal Slo-Gro; Slo-Gro; Sprout/off; Sprout-stop; Stuntman; Sucker-stuff; Super-de-sprout; Super sprout stop; Super sucker-stuff HC; Super sucker-stuff; 1,2,3,6-Tetrahydro-3,6-dioxopyridazine; Unriprim; Vondalhyde; Vondrax; 6-Hydroxy-2H-pyridazin-3-one; 1,2-dihydro-pyridazine-3,6-dione; 6-hydroxy-3(2H)-pyridazinone;; Burtolin; De-Cut; Fair-2; Fair-Plus; Malzid; Mazide; MH-30; Slo-Gro; Super-De-Sprout; Vondalhyd. Plant growth regulator, absorbed by leaves and roots. Used for suppression of growth of grass, sprouts on potatoes and other vegetables. Used with 2,4-D as a herbicide. As a tree growth inhibitor containing 185 g/liter maleic hydrazide (as the potassium salt); used to control shoots on the trunk and suckers around the base of street trees; it also inhibits the development of buds on the trunk which remain dormant following treatment. mp = 292-298°; d^{25} = 1.60; soluble in H$_2$O (6 g/kg), more soluble in polar organic solvents; LD$_{50}$ (rat orl) > 5000 mg/kg. *Bos Chemicals Ltd.; Chiltern Farm Chemicals Ltd; Fair Products; Fisons plc, Horticultural Div.; Mirfield Sales Services Ltd; Pennwalt Holland; Rhône Poulenc Crop Protection Ltd.; Rhône-Poulenc Environmental Prods. Ltd; Synchemicals Ltd; Uniroyal.*

2300 Malic acid

617-48-1   5730   G

$C_4H_6O_5$

(±)-1-Hydroxy-1,2-ethanedicarboxylic acid.

AI3-06292; Butanedioic acid, hydroxy-; Caswell No. 537; CCRIS 2950; CCRIS 6567; Deoxytetraric acid; EINECS 210-514-9; EINECS 230-022-8; EPA Pesticide Chemical Code 051101; FDA 2018; FEMA Number 2655; HSDB 1202; Hydroxybutanedioic acid; Hydroxybutandisaeure; Hydroxybutandisäure; 2-Hydroxyethane-1,2-dicarboxylic acid; Hydroxysuccinic acid; α-Hydroxysuccinic acid; Kyselina hydroxybutandiova; Kyselina jablecna; Malic acid; dl-Malic acid; DL-Malic acid; Malic acid, dl-; R,S(±)-Malic acid; Monohydroxybernsteinsaeure; Mono-hydroxybernsteinsäure; Musashi-no-Ringosan; NSC 25941; Pomalus acid; Succinic acid, hydroxy-. Chelating, buffering and flavoring agent. Crystals; mp = 13 -132°; insoluble in C_6H_6, slightly soluble in Et_2O, very soluble in EtOH, MeOH, Me_2CO, H_2O, dioxane.

2301 Malic acid

6915-15-7   5730   G H P

$C_4H_6O_5$

(±)-1-Hydroxy-1,2-ethanedicarboxylic acid.

AI3-06292; Apple acid; Butanedioic acid, hydroxy-; Caswell No. 537; CCRIS 2950; CCRIS 6567; Deoxytetraric acid; dl-Malic acid; DL-Malic acid; EINECS 210-514-9; EINECS 230-022-8; EPA Pesticide Chemical Code 051101; FDA 2018; FEMA Number 2655; HSDB 1202; Hydroxybutandisäure; Hydroxybutanedioic acid; 2-Hydroxyethane-1,2-dicarboxylic acid; Hydroxysuccinic acid; α-Hydroxysuccinic acid; Kyselina hydroxy-butandiova; Kyselina jablecna; Malic acid; Malic acid, dl-; R,S(±)-Malic acid; Monohydroxybernsteinsäure; Musashi-no-Ringosan; NSC 25941; Pomalus acid; Succinic acid, hydroxy-. Used in manufacture of esters and salts, wines; chelating agent, food acidulant, flavoring. Registered by EPA (cancelled). [DL-form: 617-48-1]: crystals, mp = 131 - 132°; soluble in MeOH (83 g/100 ml), Et_2O (0.84 g/100 ml), EtOH (46 g/100 ml), Me_2CO (18 g/100 ml), dioxane (23 g/100 ml), H_2O (56 g/100 ml), insoluble in C_6H_6; [D-(+)-form: 636-61-3]: crystals, mp = 101°; [L-(-)-form: 97-67-6]: crystals, mp = 100°; dec. 140°; $[\alpha]_D$ = -2.3° (c = 8.5); soluble in MeOH (197 g/100 ml), Et_2O (3 g/100 ml), EtOH (87 g/100 ml), Me_2CO (61 g/100 ml), dioxane (74 g/100 ml), H_2O (36 g/100 ml), insoluble in C_6H_6. Allchem Ind.; Fluka; Haarmann & Reimer GmbH; Janssen Chimica; Lancaster Synthesis Co.

2302 Malladrite

16893-85-9   8698   G P

F_6Na_2Si

Silicate(2-), hexafluoro-, disodium.

AI3-01501; Caswell No. 771; Destruxol applex; Disodium hexafluorosilicate; Disodium hexafluorosilicate (2-); Disodium silicofluoride; EINECS 240-934-8; Ens-zem weevil bait; ENT 1,501; EPA Pesticide Chemical Code 075306; Fluorosilicate de sodium; Fluosilicate de sodium; HSDB 770; Natriumhexafluorosilicat; Natriumsilico-fluorid; Ortho earwig bait; Ortho weevil bait; Prodan; PSC Co-Op weevil bait; Safsan; Salufer; Silicate(2-), hexafluoro-, disodium; Silicon sodium fluoride; Sodium fluorosilicate; Sodium fluosilicate; Sodium hexafluorosilicate; Sodium hexafluosilicate; Sodium silica fluoride; Sodium silicofluoride; Sodium silicon fluoride; SSF; Super prodan; UN2674. Fluoridation, laundry soaps, opalescent glass, vitreous enamel frits, metallurgy (aluminum and beryllium), insecticides, rodenticides, chemical intermediate, glue, leather and wood preservative, moth repellent, manufacture of pure silicon. Used in enamels and as a moth repellent, insecticide and rodenticide. Registered by EPA as an insecticide (cancelled). White granules; d = 2.68; soluble in H_2O (0.66 g/100 ml at 25°, 2.5 g/100 ml at 100°), very slightly soluble in EtOH; LD$_{50}$ (rat orl) = 125 mg/kg. Alfa Aesar; Faezy & Besthoff; ICN Biomedical Res. Products; LaRoche Ind.; Mitsui Toatsu; Pfalz & Bauer; Riedel de Haen (Chinosolfabrik); Whiting Peter Ltd.

2303 Malonic acid

141-82-2   5732   G

$C_3H_4O_4$

Methanedicarboxylic acid.

4-02-00-01874 (Beilstein Handbook Reference); AI3-15375; BRN 1751370; Carboxyacetic acid; Dicarboxy-methane; EINECS 205-503-0; Kyselina malonova; Malonic acid; Methanedicarboxylic acid; NSC 8124; Propanedioic acid; USAF EK-695. Intermediate in manufacture of barbiturates and other pharmaceuticals. Crystals; mp = 135° (dec); d^{10} = 1.619; λ_m = 228 nm (ε = 219, MeOH); very soluble in H_2O (153 g/100 ml), C_5H_5N, soluble in EtOH, Et_2O, insoluble in C_6H_6. Greeff R.W. & Co.; Lonzagroup; Penta Mfg.; Sigma-Aldrich Fine Chem.

2304 Maltodextrin

9050-36-6   E H

$[C_6H_{10}O_5]n$

Maltodextrin.

C*Dry MD®; Dextrin, malto; Glucidex®; Lycatab DSH®; Maltagran®; Maltodextrin; Maltodextrin 24DE; Maltodextrin I; Maltrin®; Mar Rex 1918; Paselli®; Snowflake; Star-Dri®; Wickenol® 550.; Absorbent for lipophilic materials for powder bath applications; foodgrade carrier for flavors. Coating agent, tablet and capsule diluent; tablet binder and viscosity-increasing agent. FDA GRAS, In the FDA list of inactive ingredients, (oral tablets and granules). Nonsweet nutritive polymer, carrier, bulking agent, adsorbent for pharmaceuticals. Used in non-parenteral medicines. White powder; d = 1.334 - 1.425; freely soluble in H_2O, slightly soluble in anhydrous EtOH; pH = 4-7; viscosity of 20% aqueous solution = 3.45 mPa. Non-toxic. Am. Maize Products; Avebe UK; Chemcentral; Grain Processing; Kingfood Australia; National Starch & Chem. UK; Roquette UK; Sigma-Aldrich Fine Chem.; Sweeteners Plus; Welding; Westin; Zumbro.

2305 Maltol

118-71-8   5735   G

C6H6O3

2-Methyl-3-hydroxy-4-pyrone.

5-18-01-00114 (Beilstein Handbook Reference); AI3-18547; BRN 0112169; CCRIS 3467; Corps praline; EINECS 204-271-8; FEMA No. 2656; 3-Hydroxy-2-methyl-4-pyrone; 3-Hydroxy-2-methyl-γ-pyrone; 3-Hydroxy-2-methyl-4H-pyran-4-one; Larixic acid; Larixinic acid; Maltol; 2-Methyl-3-hydroxypyrone; 2-Methyl-3-oxy-γ-pyrone; 2-Methyl pyromeconic acid; NSC 2829; Palatone; 4H-Pyran-4-one, 3-hydroxy-2-methyl-; Talmon; Vetol. Flavoring agent Prisms; mp = 161.6°; bp sublimes at 93°; λ_m = 214, 277 nm (ε = 10900, 8360, MeOH); insoluble in petroleum ether, slightly soluble in H_2O, Et_2O, C_6H_6; soluble in Me_2CO, very soluble in Me_2CO. *Pfizer Group Ltd.; Pfizer Inc.*

2306   Maltose
69-79-4   5736   G

C12H22O11·H2O

4-O-α-D-glucopyranosyl-D-glucose.

Malt sugar; Maltobiose; Maltos; Martos-10. Malt sugar, an isomer of cellobiose; nutrient, sweetener, culture media, stabilizer for polysulfides, brewing. [Monohydrate]: White crystals; mp = 102 - 103°; $[\alpha]_D^{20}$ = +112°→+130.4° (c = 4); soluble in H_2O, EtOH, insoluble in organic solvents. *Am. Biorganics; Avebe BV; Penta Mfg.; Pfanstiehl Labs Inc.*

2307   Mancozeb
8018-01-7   5738   P

C4H6MnN2S4.C4H6N2S4Zn

(Ethylenebis(dithiocarbamato))manganese mixture with (ethylenebis(dithiocarbamato))zinc.

Acarie M; Agrox 16D; Blecar MN; Carbamic acid, ethylenebis(dithio-, manganese zinc complex; Carm-azine; Caswell No. 913A; CCRIS 2495; Crittox MZ; Dithane 945; Dithane DG; Dithane LF; Dithane M 45; Dithane M 45 Poudrage; Dithane S 60; Dithane SPC; Dithane ultra; EPA Pesticide Chemical Code 014504; ((1,2-Ethanediylbis(carbamodithioato))(2-)) manganese mixture; (Ethylenebis(dithiocarbamato))manganese mix-ture with(ethylenebis(dithiocarbamato))zinc; Ethylene-bis-(dithiocarbamic acid) manganese zinc complex; F 2966; Fore; Green-daisen M; HSDB 6792; Indofil M 45; Karamate; Karamate N; Kascade; Liro manzeb; Mancofol; Mancomix; Mancozeb; Mancozebe; Mancozi; Mancozin; Maneb-zinc; Maneb-zineb-komplex; Maneb-zineb-mischkomplex; Mangan-zink-äthylendiamin-bis-dithio-carbamat; Mangan-zink-aethylendiamin-bis-dithio-carb-amat; Manganese ethylenebis(dithiocarbamate) (poly-meric) complex with zinc salt; Manoseb; Manzate 200; Manzeb; Manzin; Manzin 80; Marzidan; Marzin; Milcozebe; Nemispor; Novozir MN 80; Othane M 45; Pace; Penncozeb; Policar MZ; Policar S; Sandozebe; Tanzeo M45; Thane M45; Tritogol MZ; Vondozeb; Vondozeb plus; Ziman-Dithane; Zimanat; Zimaneb; Zinc manganese ethylenebisdithiocarbamate; Zineb-maneb mixture. A protectant fungicide for fruit, field crops, and roses. Used for control of potato blight and rust, blight and mildew in winter wheat. Registered by EPA as a fungicide. Grey-yellow powder; dec. 192-194°; insoluble in H_2O and most organic solvents; generally low toxicity, LD50 (rat orl) > 5000 mg/kg, (rat sc) > 10000 mg/kg, (rbt) > 5000 mg/kg. *Agrimont; Akzo Chemie; All-India Medical; Crystal; Diachem; DuPont; Ercros; Pan Britannica Industries Ltd.; Pennwalt Holland; Rohm & Haas Co.; Rohm & Haas*

UK; Sanachem; Shell UK.

2308   Mandelic acid
90-64-2   5739   G

C8H8O3

Phenylglycolic acid.

Acido mandelico; AI3-06293; Almond acid; Amygdalic acid; Benzeneacetic acid, α-hydroxy-; EINECS 202-007-6; EINECS 210-277-1; Glycolic acid, phenyl-; α-Hydroxybenzeneacetic acid; α-Hydroxybenzeneacetic acid, (±)-; (±)-α-Hydroxybenzeneacetic acid; (±)-α-Hydroxyphenylacetic acid; DL-2-Hydroxy-2-phenylacetic acid; α-Hydroxyphenylacetic acid; DL-Hydroxy(phenyl)-acetic acid; (±)-2-Hydroxy-2-phenylethanoic acid; α-Hydroxy-α-toluic acid; α-Toluic acid, α-hydroxy-; 2-Phenyl-2-hydroxyacetic acid; 2-Phenylglycolic acid; Kyselina 2-fenyl-2-hydroxyethanova; Kyselina mandlova; Mandelic acid; (±)-Mandelic acid; dl-Mandelic acid; p-Mandelic acid; (RS)-Mandelic acid; NSC 7925; Paramandelic acid; Phenylglycolic acid; Phenylhydroxy-acetic acid; Racemic mandelic acid; Uromaline. Used in organic synthesis and in medicine as a urinary antiseptic. Alternate CAS RN [611-72-3]. Crystals; mp = 119°; d = 1.30; soluble in H_2O (15.9 g/100 ml), more soluble in organic solvents. *Grace W.R. & Co.; Greeff R.W. & Co.*

2309   Maneb
12427-38-2   5743   G P

C4H6MnN2S4

Manganese, [[1,2-ethanediylbis[carbamodithioato]](2-)]-.

AAmangan; Agrox flowable; Agrox N-M; AI3-14875; Akzo chemie maneb; BASF-maneb spritzpulver; Carbamic acid, ethylenebis(dithio-, manganese salt; Caswell No. 539; CCRIS 1107; Chem neb; Chloroble M; CR 3029; Curzate M; Delsene M; Diphar; Dithane M 22; Dithane M 22 special; EINECS 235-654-8; ENT 14,875; EPA Pesticide Chemical Code 014505; Ethylenebis(dithiocarbamato), manganese; Ethylenebis-dithiocarbamate manganese; F 10 (Pesticide); Farmaneb; Granol NM; Griffin manex; HSDB 4063; Kypman 80; Lonocol M; Luxan maneb 80; M-Diphar; Maneb; Maneb 80; Maneb-R; Maneb ZL4; Maneba; Manebe; Manebe 80; Manebgan; Manesan; Manex; Manganese ethylene-1,2-bis-dithiocarbamate; Manzate; Manzate D; Manzin; MEB; MnEBD; Nereb; Nespor; Plantifog 160M; Polyram M; Remasan chloroble M; Rhodianebe; Sopranebe; Sup'r flo; Tersan LSR; Trimangol; Trimangol 80; Tubothane; UN2210; UN2968; Unicrop maneb; Vancide Maneb 80; Vancide Maneb 90. Fungicide with protective action. Used in control of many fungal diseases, e.g. blight, leaf spot, rust, downy mildew, scab. Registered by EPA as a fungicide. Yellow powder; dec. 192-204°; insoluble in H_2O and common solvents; LD50 (rat orl) > 5000 mg/kg, (rat, rbt der) > 5000 mg/kg; LC50 (rat ihl 4hr.) = 3.8 mg/l air, (mallard duck, bobwhite quail 8 day dietary) > 10000 mg/kg diet, (carp 48 hr.) = 1.8 mg/l; non-toxic to bees. *BASF Corp.; Chiltern Farm Chemicals Ltd; Crystal; Cumberland; Drexel Chemical Co.; DuPont; MTM AgroChemicals Ltd.; Pennwalt Holland; Rohm & Haas Co.; Schering AG; Universal Crop Protection Ltd.; WBC Technology Ltd.*

2310   Manganese
7439-96-5   5745   G

Mn

Mn
 Manganese.
 CCRIS 1579; Colloidal manganese; Cutaval; EINECS 231-105-1; HSDB 550; Magnacat; Mangan; Mangan nitridovany; Manganese, elemental; Manganese fume; Manganese metal alloy; Tronamang. Used in ferroalloys and steel manufacture; improves corrosion resistance and hardness in nonferrous alloys; purifying and scavenging agent in metal production; manufacture of aluminum. Metal; mp = 1244°; bp = 2095°. *Atomergic Chemetals; Cerac; Kerr-McGee.*

2311   Manganese borate
12228-91-0   5747   G

BH_3 Mn^{2+} O^{2-}

$MnB_4O_7.8H_2O$
 Tetraboron manganese heptaoxide.
 Boron manganese oxide (B_4MnO_7); EINECS 235-446-7; Manganese borate; Siccative; Tetraboron manganese heptaoxide. Mixed with linseed oil and resin, used as a drying agent for impregnating leather. Insoluble in H_2O, organic solvents.

2312   Manganese carbonate
598-62-9   5749   G

$CMnO_3$
 Carbonic acid manganese salt.
 Carbonic acid, manganese(2+) salt (1:1); CCRIS 3660; EINECS 209-942-9; HSDB 790; Manganese carbonate; Manganese carbonate (1:1); Manganese carbonate ($MnCO_3$); Manganese(2+) carbonate; Manganese(2+) carbonate (1:1); Manganese(II) carbonate; Manganese white; Manganous carbonate; Natural rhodochrosite; NSC 83512; Elastocarb Tech Light, Tech Heavy; Magocarb-33. Inorganic filler providing flame retardancy and smoke suppression to elastomers, plastics, and thermosets such as EPDM, PP, PE and PVC; used in wire and cable compounds, conduit/tubing, film and sheet. A pigment, also used as a drier for varnishes and in animal feeds. Solid; d = 3.1; soluble in acids; insoluble in alcohol, H_2O; incompatible with formaldehyde; heated to decomposition emits acrid smoke and irritating fumes. *Allchem Ind.; Chemisphere; Fluka; Giulini; Lohmann; Lonzagroup; Magnesia GmbH; Mallinckrodt Inc.; Marine Magnesium; Martin Marietta; Morton Intn'l.*

2313   Manganese chloride
7773-01-5   5751   G

$Cl_2Mn.4H_2O$
 Manganese(II) chloride tetrahydrate.
 CCRIS 6882; EINECS 231-869-6; HSDB 2154; Manganese bichloride; Manganese chloride; Manganese chloride ($MnCl_2$); Manganese chloride anhydrous; Manganese dichloride; Manganese(II) chloride; Manganous chloride; Mangatrace; NSC 9879; Scacchite. Mineral supplement. Red monoclinic crystals; mp = 58°; d = 2.01; soluble in H_2O (1.43 g/ml), EtOH, insoluble in organic

solvents; LD50 (mus sc) = 180-250 mg/kg. *Armour Pharm. Co. Ltd.*

2314   Manganese dioxide
1313-13-9   5753   G

MnO_2
 Manganese(IV) oxide.
 AI3-52833; Black manganese oxide; BOG manganese; Braunstein; Bruinsteen; C.I. 77728; C.I. Pigment Brown 8; C.I. Pigment Black 14; Cement Black; EINECS 215-202-6; Glassmaker's soap; KM Manganese Dioxide; Mangaanbioxyde; Mangaandioxyde; Mangandioxid; Manganese (biossido di); Manganese (bioxyd de); Manganese (diossido di); Manganese (dioxyde de); Manganese binoxide; Manganese Black; Manganese dioxide; Manganese oxide (MnO_2); Manganese peroxide; Manganese superoxide; Pyrolusite Brown. Oxidizing agent, used in pyrotechnics, matches, catalyst, laboratory reagent, scavenger and decolorizer, textile dyeing, source of metallic manganese (as pyrolusite). Used in AB and SB battery active grades 90% minimum MnO_2 for use in Leclanché, alkaline and zinc chloride dry cell batteries. Black powder; mp = 535° (dec); d = 5.0260; insoluble in cold water, nitric or cold sulfuric acid; LD50 (rbt iv) = 45 mg/kg. *Atomergic Chemetals; Eagle-Picher; Hoechst Celanese; Kerr-McGee; Nichia Kagaku Kogyo; Sigma-Aldrich Fine Chem.*

2315   Manganese sesquioxide
1317-34-6   5761   G

$Mn_2O_3 \cdot H_2O$
 Manganese(III) oxide.
 Brown manganese ore; Dimanganese trioxide; EINECS 215-264-4; Manganese manganate; Manganese oxide (Mn_2O_3); Manganese sesquioxide; Manganese sisquioxide; Manganese trioxide; Manganese(3+) oxide; Manganese(III) oxide; Manganic oxide. A hydrated oxide of manganese. Black powder; d = 4.50; insoluble in H_2O, soluble in HCl.

2316   Manganese sulfate
7785-87-7   5763   G

$MnO_4S.4H_2O$
 Manganese(II) sulfate.
 CCRIS 6916; EINECS 232-089-9; HSDB 2187; Man-Gro; Manganese monosulfate; Manganese sulfate; Manganese sulfate ($MnSO_4$); Manganese sulfate anhydrous; Manganese sulphate; Manganese(2+) sulfate (1:1); Manganese(II) sulfate; Manganous sulfate; NCI C61143; Sorba-spray Mn; Sorba-Spray Manganese; Sulfuric acid, manganese (II) salt (1:1); Sulfuric acid, manganese(2+) salt (1:1). Used in fertilizers, feed additive, paints and varnishes, ceramics, textile dyes, medicine, fungicide, ore flotation, catalyst in viscose process, synthetic manganese dioxide. Solid; mp = 700°; bp = 850°; d = 2.9500. *Chemetall Chem. products; Mallinckrodt Inc.; Nihon Kagaku Sangyo; Sigma-Aldrich Fine Chem.*

2317   Manganous acetate
638-38-0   5746   G H

380

C4H6MnO4
Acetic acid, manganese(II) salt (2:1).
Acetic acid, manganese(2+) salt; Acetic acid, manganese(II) salt (2:1); Diacetylmanganese; EINECS 211-334-3; HSDB 5734; Manal; Manganese acetate; Manganese acetate (Mn(OAc)2); Manganese(2+) acetate; Manganese(II) acetate; Manganese diacetate; Manganous acetate; Octan manganaty. Used as a dye mordant and as a drier for paints and varnishes. Used in textile dyeing, oxidation catalyst, paint and varnish drier, fertilizer, food packaging, feed additive. Crystals; d = 1.59; soluble in H_2O, EtOH; LD_{50} (rat orl) = 3.73 g/kg. *Atomergic Chemetals; Hoechst Celanese; Mechema Chemicals Ltd.; Nihon Kagaku Sangyo; Spectrum Chem. Manufacturing; Verdugt BV.*

2318 Mannitol
69-65-8   5769   D

C6H14O6
D-Mannitol.
4-01-00-02841 (Beilstein Handbook Reference); BRN 1721898; CCRIS 369; Cordycepic acid; Diosmol; EINECS 200-711-8; HSDB 714; Invenex; Isotol; Maniton-S; Manna sugar; Mannazucker; Mannidex; Mannigen; Mannistol; Mannit; Mannite; Mannitol; D-Mannitol; Mannitol, D-; Mannogem 2080; Marine Crystal; Mushroom sugar; NCI-C50362; NSC 407017; Osmitrol; Osmofundin; Osmosal; Resectisol; SDM No. 35. Used in organic synthesis and as a base for dietetic foods, diluent, determination of boron, pharmaceutical products, medicine, thickener, and stabilizer in food products. White crystals; mp = 168°; $bp^{3.5}$ = 295; d^{20} = 1.489; $[\alpha]_D^{20}$= +23°→+24°; very soluble in H_2O, less soluble in EtOH, Et_2O, C_5H_5N. *Lancaster Synthesis Co.; Sigma-Aldrich Fine Chem.*

2319 Mannose
3458-28-4   5772   G

C6H12O6
D-Mannose.
AI3-18442; Carubinose; Mannose; Mannose, D-; (+)-Mannose; D-Mannose; D(+)-Mannose; EINECS 222-392-4; NSC 26247; Seminose. Orthorhombic prisms; mp = 132° (dec); d^{20} = 1.539; $[\alpha]_D^{20}$= -17° → +14.6° (c = 3, H_2O); insoluble in Et_2O, C_6H_6, slightly soluble in EtOH, MeOH, very soluble in H_2O.

2320 Manzanate
39255-32-8   G

C8H16O2
Ethyl-2-methyl pentanoate.
AI3-33618; EINECS 254-384-1; Ethyl 2-methylpentanoate; Ethyl 2-methylvalerate; Ethyl α-methylvalerate; FEMA No. 3488; Pentanoic acid, 2-methyl-, ethyl ester. Oil with a natural, fruity, pineapple odor. Used in specialty perfumes and fruit flavorings. Oil; bp_{15} = 60°; d^{25} = 0.861-0.865. *Quest Intl.*

2321 MCPA
94-74-6   5787   G H P

C9H9ClO3
(4-Chloro-o-tolyloxy)acetic acid.
2,4-MCPA; 2M-4Ch; 2M-4Kh; 4K-2M; 4-06-00-01991 (Beilstein Handbook Reference); Acetic acid, ((4-chloro-o-tolyl)oxy)-; Acme MCPA Amine 4; Agroxon; Agroxone; Anicon kombi; Anicon M; B-Selektonon M; Banvel M; BH MCPA; Bordermaster; BRN 2051752; Brominal M & plus; Caswell No. 557C; CCRIS 1022; Cekherbex; Chiptox; Chwastox; Chwastox 30; Chwastox Extra; CMP acetate; Cornox-M; Ded-weed; Dicopur-M; Dikotex; EINECS 202-360-6; Emcepan; Empal; EPA Pesticide Chemical Code 030501; FLUID 4; Hedapur M 52; Hedarex M; Hedonal; Hedonal M; Herbicide M; Hormotuho; Hornotuho; HSDB 1127; Kilsem; Krezone; Kwas 4-chloro-2-metylofenoksyoctowy; Kyselina 4-chlor-2-methylfenoxyoctova; Legumex DB; Leuna M; Leyspray; Linormone; M 40; MCP; MCP ester; MCPA; MCPA Concentrate Weedone MCPA ester; MCPA Ester; MCPA Weedar; MCPA [Chlorophenoxy herbicides]; Mephanac; Metaxon; Methoxone; Methylchloro-phenoxyacetic acid; Netazol; NSC 2351; Okultin M; Phenoxylene 50; Phenoxylene plus; Phenoxylene super; Raphone; Razol dock killer; Rhomenc; Rhomene; Rhonox; Selektonon M; Seppic MMD; Shamrox; Soviet technical herbicide 2M-4C; Trasan; U 46 M-Fluid; Ustinex; Vacate; Vesakontuho; Vesakontuho MCPA; Weed-rhap; Weedar; Weedar MCPA Concentrate; Weedar Sodium MCPA; Weedone; Weedone MCPA ester; Zelan. Selective, systemic, hormone-like herbicide used for post-emergence control of annual and perennial broad-leaved weeds in cereal crops and grassland. Registered by EPA as a herbicide. Colorless crystals; mp = 118-119°; λ_m = 227, 278, 285 nm (MeOH); soluble in H_2O (0.0825 g/100 ml), EtOH (153 g/100 ml), Et_2O (77 g/100 ml), C_7H_8 (6.2 g/100 ml), xylene (4.9 g/100 ml), C_7H_{16} (0.5 g/100 ml); LD_{50} (rat orl) = 700 mg/kg, (mus orl) = 550 mg/kg, (rat der) > 1000 mg/kg, (bobwhite quail orl) = 377 mg/kg; LC_{50} (rainbow trout 96 hr.) = 232 mg/l; non-toxic to bees. *Agrichem (International) Ltd.; Albaugh Inc.; Atlas Interlates Ltd; BritAg Ind. Ltd.; Farmers Crop Chemicals Ltd.; Fisons plc, Horticultural Div.; ICI Agrochemicals; Makhteshim Chemical Works Ltd.; Mirfield Sales Services Ltd; Murphy Chemical Co Ltd.; Rhône Poulenc Crop Protection Ltd.; Rhône-Poulenc Environmental Prods. Ltd; Schering Agrochemicals Ltd.; Star Agrochem Ltd; Syngenta Crop Protection; Universal Crop Protection Ltd.*

2322 MCPB
94-81-5   G P

C11H13ClO3
4-(4-Chloro-2-methylphenoxy) butyric acid.
4-06-00-01996 (Beilstein Handbook Reference); 4MCPB; Belmac Straight; Bexane; Bexone; BRN 2215202; Butanoic acid, 4-(4-chloro-2-methylphenoxy)-; Butyric acid, 4-((4-chloro-o-tolyl)oxy)-; Can-Trol; Caswell No. 558; CCRIS 1463; 4-(4-Chlor-2-methylphenoxy)-buttersaeure; 4-(4-Chlor-2-methylphenoxy)-buttersäure; 4-(4-Chloro-2-methylphenoxy)butanoate; 4-(4-Chloro-2-methylphenoxy)butanoic acid; 4-(4-Chloro-2-methyl-phenoxy)butyric acid; (4-Chloro-o-tolyloxy)butyric acid; 4-((4-Chloro-o-tolyl)oxy)butyric acid; EINECS 202-365-3; EPA Pesticide Chemical Code 019201; Fisons 18-15, MCPB; γ-(4-Chloro-2-methylphenoxy)butyric acid; γ-MCPB; HSDB 1737; Kyselina 4-(4-chlor-2-methyl-fenoxy)maselna; Legumex; 2M 4KhM; 2,4-MCPB; MCP-butyric; 2-Methyl-4-chlorophenoxy γ-butyric acid; 2-Methyl-4-chlorophenoxybutyric acid; 4-(2-Methyl-4-chlorphenoxy)-buttersaeure; 4-(2-Methyl-4-chlor-phen-oxy)-buttersäure; 4-(2-Methyl-4-chlorophenoxy)-butyric acid; MCPB; MCPB - acid; NSC 102796; PDQ; Thitrol; Tropotox; Trotox; U46 MCPB. Selective systemic hormone-like herbicide. Used for post-emergence control of annual and perennial broad-leaved weeds in cereal and grassland. Used as a systemic fungicide active against chocolate spot disease in broad beans. White solid; mp = 99-100°; λ_m = 228, 279, 287 nm (ϵ = 9950, 1690, 1490 MeOH); soluble in H2O (4.4 g/100 ml), more soluble in organic solvents; LD50 (rat orl) = 680 mg/kg. *Fisons plc, Horticultural Div.; MTM AgroChemicals Ltd.*

2323   MCPP
93-65-2   H P

C10H11ClO3
(±)-2-(4-Chloro-2-methylphenoxy)propanoic acid.
3-06-00-01266 (Beilstein Handbook Reference); Acide 2-(4-chloro-2-methyl-phenoxy)propionique; Acido 2-(4-cloro-2-metil-fenossi)-propionico; Anicon B; Anicon P; Astix; BRN 2212752; Caswell No. 559; CCRIS 1464; Celatox CMPP; Chipco turf herbicide mcpp; CMPP; Compitox; Duplosan New System CMPP; Duplosan® CMPP; EINECS 202-264-4; EINECS 230-386-8; EPA Pesticide Chemical Code 031501; FBC CMPP; HSDB 1738; Iso-Cornox; Isocarnox; Kilprop; Kwas 4-chloro-2-metylofenoksypropionowy; Kyselina 2-(4-chlor-2-methylfenoxy)propionova; Liranox; MCPP; 2-MCPP; 2M-4CP; 2M 4KhP; Mechlorprop; Mecomec; Mecopar; Mecopeop; Mecoper; Mecopex; Mecoprop; Mecoturf; Mecprop; Mepro; Methoxone; Morogal; N.b. mecoprop; NSC 60282; Okultin MP; Propanoic acid, 2-(4-chloro-2-methylphenoxy)-; Propionic acid, 2-(4-chloro-2-methylphenoxy); Proponex-plus; Rankotex; RD 4593; Runcatex; SYS 67 Mecmin; U 46 KV Fluid; Vi-Par; Vi-Pex. Selective herbicide for control of broadleaf weeds in cereals, meadows, pastures. Registered by EPA as a herbicide. Colorless crystals; mp = 94-95°; soluble in H2O (0.062 g/100 ml), Me2CO (> 100 g/100 ml), CHCl3 (33.9 g/100 ml), Et2O (> 100 g/100 ml), EtOH (> 100 g/100 ml), EtOAc (82.5 g/100 ml); LD50 (rat orl) = 930-1166 mg/kg, 1050 mg/kg, (rat der) > 4000 mg/kg, (quail orl) = 500 mg/kg; LC50 (trout 96 hr.) = 150 - 220 mg/l, non-toxic to bees. *Agrichem (International) Ltd.; Akzo Chemie; Aventis Crop Science; BASF Corp.; Bayer AG; BritAg Ind. Ltd.; Cleanacres Ltd.; Farmers Crop Chemicals Ltd.; Fermenta; Marks; Rhône Poulenc Crop Protection Ltd.; Rhône-Poulenc Environmental Prods. Ltd; Rhône-Poulenc; Rigby Taylor*

Ltd.; SBC Technology Ltd.; Schering Agrochemicals Ltd.; Schering; Universal Crop Protection Ltd.

2324   Mebendazole
31431-39-7   5791   D G V

C16H13N3O3
Methyl 5-benzoyl-2-benzimidazolecarbamate.
Bantenol; (5-Benzoyl-1H-benzimidazol-2-yl)carbamic acid methyl ester; 5-Benzoyl-2-benzimidazolecarbamic acid methyl ester; Besantin; CCRIS 4479; EINECS 250-635-4; Equi-Vurm Plus; Equivurm Plus; HSDB 3232; Lomper; MBDZ; Mebendazol; Mebendazole; Mebendazolum; Mebenvet; Mebutar; Methyl 5-benzoyl-2-benzimidazolecarbamate; Noverme; NSC 184849; Ovitelmin; Pantelmin; R 17635; Telmin; Vermex; Vermicidin; Vermirax; Vermox; Verpanyl. Anthelmintic. Targets nematodes. Used as a veterinary anthelmintic, especially with horses. Crystals; mp = 288.5°; insoluble in H2O, EtOH, Et2O, CHCl3; soluble in formic acid; LD50 (sheep orl) > 80 mg/kg, (mus, rat, chk) > 40 mg/kg. *Janssen Chimica; Janssen Pharm. Ltd.*

2325   Meclizine
569-65-3   5800   D

C25H27ClN2
1-(p-Chloro-α-phenylbenzyl)-4-(m-methylbenzyl)-piperazine.
5-23-01-00235 (Beilstein Handbook Reference); Ancolan; Bonadettes; Bonine; BRN 0332002; Calmonal; Chiclida; EINECS 209-323-3; Histamethine; Histamethizine; Histametizine; Histametizyne; HSDB 3113; Itinerol; Marex; Meclizine; Meclozina; Meclozine; Meclozinum; Monamine; Navicalm; Neo-istafene; Neo-suprimal; Neo-suprimel; NSC 169189; Parachloramine; Peremesin; Postafene; Ravelon; Sabari; Sea-Legs; Siguran; Suprimal; Travelon; UCB 170; UCB 5062; Vibazine; Vomisseis. Antiemetic. bp2= 230°; soluble in CS2. *KV Pharm.; Pfizer Intl.; Roerig Div. Pfizer Pharm.*

2326   Meclizine Hydrochloride
31884-77-2   5800   D

C25H29Cl3N2.2H2O
1-(p-Chloro-α-phenylbenzyl)-4-(m-methylbenzyl)piperazine dihydrochloride monohydrate.
UCB-5062; Ancolan; Antivert; Bonamine; Bonine; Calmonal; Diadril; Histametizine; Navicalm; Neo-Istafene; Peremesin; Postafene;

Sabari; Sea-Legs; Veritab. Antiemetic. Solid; insoluble in H_2O (0.1 g/100 ml); freely soluble in $CHCl_3$, C_5H_5N. *KV Pharm.; Pfizer Intl.; Roerig Div. Pfizer Pharm.*

2327 Mecoprop
7085-19-0   5807   G P

$C_{10}H_{11}ClO_3$
(±)-2-(4-Chloro-2-methylphenoxy)propanoic acid.
3-06-00-01266 (Beilstein Handbook Reference); Acide 2-(4-chloro-2-methyl-phenoxy)propanique; Acido 2-(4-cloro-2-metil-fenossi)-propionico; Anicon B; Anicon P; BRN 2212752; Caswell No. 559; CCRIS 1464; Celatox CMPP; Chipco turf herbicide mcpp; 2-(4-Chloor-2-methyl-fenoxy)-propionzuur; 2-(4-Chlor-2-methyl-phenoxy)-propionsaeure; 2-(4-Chlor-2-methyl-phenoxy)-prop-ionsäure; 2-(4-Chloro-2-methylphenoxy)propanoic acid; 2-(4-Chloro-2-methylphenoxy)propionic acid; 4-Chloro-2-methylphenoxy-α-propionic acid; 2-(4-Chlorophenoxy-2-methyl)propionic acid; 2-(4-Chloro-o-tolyl)oxylprop-ionic acid; 2-(4-Chloro-2-tolyloxy)propionic acid; 2-(p-Chloro-o-tolyloxy)propionic acid; (+)-α-(4-Chloro-2-methylphenoxy) propionic acid; (±)-2-(((4-Chloro-o-tolyl)oxy)propionic acid (8Cl); CMPP; Compitox; EINECS 202-264-4; EINECS 230-386-8; EPA Pesticide Chemical Code 031501; FBC CMPP; HSDB 1738; Iso-Cornox; Isocarnox; Kilprop; Kwas 4-chloro-2-metylofenoksy-propionowy; Kyselina 2-(4-chlor-2-methylfenoxy)-propionova; Liranox; 2M-4CP; 2M 4KhP; MCPP; 2-Mcpp; Mechlorprop; Mecomec; Mecopar; Mecoper; Mecopex; Mecoprop; Mecoturf; Mecprop; Mepro; Methoxone; 2-(2-Methyl-4-chlorophenoxy)propanoic acid; 2-(2-Methyl-4-chlorophenoxy)propionic acid; α-(2-Methyl-4-chloro-phenoxy)propionic acid; 2-(2-Methyl-4-chlorophenoxy)-propionsaeure; 2-(2-Methyl-4-chlorophenoxy)-propion-säure; Morogal; N.b. mecoprop; NSC 60282; Okultin MP; Propanoic acid, 2-(4-chloro-2-methylphenoxy)-; Prop-ionic acid, 2-(2-methyl-4-chlorophenoxy)-; Propionic acid, 2-((4-chloro-o-tolyl)oxy)-; Proponex-plus; Rankotex; RD 4593; Runcatex; SYS 67 Mecmin; U 46 KV Fluid; Vi-Par; Vi-Pex; 2-(4-Chlor-2-methyl-phenoxy)-propion-saeure; 2-(4-Chloro-2-methylphenoxy)propionic acid; (+)-α-(4-Chloro-2-methylphenoxy) propionic acid; 2-(4-Chloro-2-methylphenoxy)propanoic acid; 4-Chloro-2-methylphenoxy-α-propionic acid; 2-(4-Chlorophenoxy-2-methyl)propionic acid; 2-(4-Chloro-2-tolyloxy)propionic acid; 2-(4-Chloro-o-tolyl)oxylpropionic acid; (±)-2-((4-Chloro-o-tolyl)oxy)propionic acid (8Cl); 2-(p-Chloro-o-tolyloxy)propionic acid; CMPP; Compitox; EINECS 202-264-4; EINECS 230-386-8; EPA Pesticide Chemical Code 031501; FBC CMPPHSDB 1738; Iso-Cornox; Isocarnox; Kilprop; Kwas 4-chloro-2-metylofenoksypropionowy; Kyselina 2-(4-chlor-2-methylfenoxy)propionova; Liranox; MCPP; 2-Mcpp; 2M-4CP; 2M 4KhP; 2M4KhP; Mechlorprop; Mecomec; Mecopar; Mecopeop; Mecoper; Mecopex; Mecoprop; Mecoturf; Mecprop; Mepro; Methoxone; 2-(2-Methyl-4-chlorophenoxy)propionic acid; 2-(2-Methyl-4-chlorophenoxy)propanoic acid; 2-(2-Methyl-4-chlorphenoxy)-propionsäure; Morogal; N.b. mecoprop; NSC 60282; Okultin MP; Propanoic acid, 2-(4-chloro-2-methylphenoxy)-; Propionic acid, 2-((4-chloro-o-tolyl)oxy)-; Proponex-plus; Rankotex; RD 4593; Runcatex; SYS 67 Mecmin; U 46 KV Fluid; Vi-Par; Vi-Pex. Selective, systemic hormone-type herbicide, absorbed by leaves, translocated to roots. Post emergence control of broad-leaf weeds such as clovers, chickweeds, plantains, cleavers and weeds in cereals and grassland. Registered by EPA as a herbicide. Colorless crystals; mp = 94-95°; soluble in H_2O (0.062 g/100 ml), Me_2CO (> 100 g/100 ml), $CHCl_3$ (33.9 g/100 ml), Et_2O (> 100 g/100 ml), EtOH (> 100 g/100 ml), EtOAc (82.5 g/100 ml); LD_{50} (rat orl) = 930-1166 mg/kg, 1050

mg/kg, (rat der) > 4000 mg/kg, (quail orl) = 500 mg/kg; LC_{50} (trout 96 hr.) = 150 - 220 mg/l, non-toxic to bees. *Agrichem (International) Ltd.; Akzo Chemie; Aventis Crop Science; BASF Corp.; Bayer AG; BritAg Ind. Ltd.; Cleanacres Ltd.; Farmers Crop Chemicals Ltd.; Fermenta; Marks; Rhône Poulenc Crop Protection Ltd.; Rhône-Poulenc Environmental Prods. Ltd; Rhône-Poulenc; Rigby Taylor Ltd.; SBC Technology Ltd.; Schering Agrochemicals Ltd.; Schering; Universal Crop Protection Ltd.*

2328 Mecoprop-P
16484-77-8   H P

$C_{10}H_{11}ClO_3$
(R)-2-(4-Chloro-2-methylphenoxy)propanoic acid.
2M-4XP; Duplosan KV; EINECS 240-539-0; (+)-Mcpp; Mecoprop, D-; Mecoprop-P. Herbicide. Colorless crystals; mp = 95°; soluble in H_2O (0.086g/100 ml at 20°, pH 7), Me_2CO (> 79 g/100 ml), Et_2O (> 70 g/100 ml), EtOH (> 79 g/100 ml), CH_2Cl_2 (97 g/100 ml), C_6H_{14} (0.6 g/100 ml), C_7H_8 (28 g/100 ml); LD_{50} (rat orl) = 1050 mg/kg, (rat der) > 4000 mg/kg, (quail orl) ≅500 mg/kg; LC_{50} (rat ihl 4 hr.) > 5.6 mg/l air; (trout 96 hr.) = 1250 - 220 mg/l; non-toxic to bees. *BASF Corp.*

2329 Medroxyprogesterone
520-85-4   5817   D V

$C_{22}H_{32}O_3$
17α-Hydroxy-6α-methylprogesterone.
4-08-00-02211 (Beilstein Handbook Reference); Amen; BRN 2510965; Curretab; EINECS 208-298-6; HSDB 3114; Medrossiprogesterone; Medroxiprogesterona; Medroxiprogesteronum; Medroxyprogesteron; Medroxy-progesterone; Medroxyprogesteronum; NSC 27408; U 8840. Orally active progestogen used with estrogens (eg. ethinyl estradiol) in oral contraceptives. Used in veterinary medicine for estrus regulation. Crystals; mp = 214.5°; $[\alpha]_D^{25}$= 75° ($CHCl_3$); λ_m = 241 nm (ε 16000 EtOH); very soluble in $CHCl_3$. *Farmitalia Carlo Erba Ltd.; Pharmacia & Upjohn.*

2330 Medroxyprogesterone acetate
71-58-9   5817   D V

383

C24H34O4

17α-Hydroxy-6α-methylprogesterone 17-acetate.

4-08-00-02212 (Beilstein Handbook Reference); AI3-60127; Amen; Aragest; Aragest 5; BRN 2066112; CCRIS 371; Clinofem; Clinovir; Curretab; Cycrin; Cykrina; Depcorlutin; Depo-Clinovir; Depo-Map; Depo-Prodasone; Depo-Progestin; Depo-Progevera; Depo-Promone; Depo-Provera; Depo-Provera Contraceptive; Depo-Ralovera; Depocon; Deporone; DMPA; DP150; Dugen; EINECS 200-757-9; Farlutal; Farlutin; G-Farlutal; Gestapuran; Hysron; Indivina; Lutopolar; Lutoral; MAP; Med-Pro; Medrosterona; Medroxyacetate progesterone; Medroxyprogesterone acetate; Mepastat; Meprate; Methylacetoxyprogesterone; Metigestrona; MPA-β; MPA GYM; MPA Hexal; MPA-Noury; Nadigest; Nidaxin; NSC 21171; NSC-26386; Oragest; Perlutex; Perlutex Leo; Prodasone; Progestalfa; Progeston; Progevera; Promone-E; Provera; Provera dosepak; Proverone; Ralovera; Repromap; Repromix; Sirprogen; Sodelut G; Sumiferm; Supprestral; Suprestral; U 8839; Veramix; Veraplex.; component of: Provest. Orally active progestogen once used with estrogens in oral contraceptives. Used in veterinary medicine for estrus regulation. White powder; mp = 220 - 223.5°; $[\alpha]_D^{25} = 5°$ (CHCl3); λm = 241 nm (ε 16000 EtOH). *Farmitalia Carlo Erba Ltd.; Pharmacia & Upjohn; Solvay Pharm. Inc.; Wyeth-Ayerst Labs.*

2331 Mefenacet
73250-68-7   5820   G P

C16H14N2O2S

2-(2-Benzothiazolyloxy)-N-methyl-N-phenylacetamide.

Acetamide, 2-(2-benzothiazolyloxy)-N-methyl-N-phenyl-; 2-(2-Benzothiazolyloxy)-N-methyl-N-phenylacetamide; 2-(1,3-Benzothiazol-2-yloxy)-N-methylacetanilide; 2-(Benzothiazol-2-yloxy)-N-methyl-N-phenylacetamide; BRN 1143987; FOE 1976; Hinochloa; Mefenacet; NTN 801; Rancho. Selective herbicide, inhibits cell division. Used for control of grass weeds, especially *Echinochloa cur-galli* and cyperaceous weeds, pre- and early post-emergence in rice. Solid; mp = 135°; slightly soluble in H2O (0.0004 g/100 ml at 20°), CH2Cl2 (> 20 g/100 ml), C6H14 (0.01 - 0.10 g/100 ml), C7H8 (2 - 5 g/100 ml), i-PrOH (0.5 - 1.0 g/100 ml); LD50 (rat, mus, dog orl) > 5000 mg/kg, (rat, mus der) > 5000 mg/kg; LC50 (rat ihl 4 hr.) = 0.02 mg/l air; (carp 96 hr.) = 8.0 mg/l, (trout 96 hr.) = 6.8 mg/l. *Bayer AG; Bayer Corp., Agriculture.*

2332 Mefluidide
53780-34-0   5825   G P

C11H13F3N2O3S

N-(2,4-Dimethyl-5-(((trifluoromethyl)sulfonyl)amino)-phenyl)acetamide.

5'-Acetamido-2',4'-dimethyltrifluoromethanesulfonanili-de; Acetamide, N-(2,4-dimethyl-5-(((trifluoromethyl)-sulf-onyl)amino)phenyl)-; 5-Acetamido-2,4-dimethyl-trifluoro-

methanesulfonanilide; Acetanilide, 2',4'-di-methyl-5'-((tri-fluoromethyl)sulfonamido)-; BRN 2819120; Echo; Em-bark; MBR12325; Mowchem; Trimcut; 2',4'-Dimethyl-5'-(trifluoromethanesulphonamido)acetanilide; Embark; Em-bark 2S; Embark plant growth regulator; MBR 12325; Mefluidide; Methafluoridamid; N-(2,4-Dimethyl-5-(((tri-fluoromethyl)sulfonyl)amino)phenyl)acetamide; 5'-(1,1,1-Trifluoromethanesulphonamido)acet-2',4'-xylidide; Trim-Cut; VEL 3973; Vistar; Vistar herbicide. Plant growth regulator and herbicide which inhibits growth and development of grasses. Used in lieu of grass cutting, e.g. on road verges and embankments. Registered by EPA as a herbicide. Crystalline solid; mp = 183-185°; soluble in H2O (0.018 g/100 ml), C6H6 (0.031 g/100 ml), CH2Cl2 (0.21 g/100 ml), CH3CN (6.4 g/100 ml), EtOAc (5 g/100 ml), n-octanol (1.7 g/100 ml), MeOH (31 g/100 ml), Me2CO (35 g/100 ml), Et2O (0.39 g/100 ml), xylene (0.12 g/100 ml); LD50 (rat orl) > 4000 mg/kg, (mus orl) = 1920 mg/kg, (rbt der) > 4000mg/kg, (mallard duck, bobwhite quail orl) > 4620 mg/kg; LC50 (mallard duck, bobwhite quail 8 day dietary) > 10000 mg/kg diet; (rainbow trout, bluegill sunfish 96 hr.) > 100 mg/l; non-toxic to bees. *Generic; Scotts Co.*

2333 Meglumine
6284-40-8   6102   G

C7H17NO5

1-Deoxy-1-(methylamino)-D-glucitol.

1-Deoxy-1-methylaminosorbitol; 1-Deoxy-1-(methyl-amino)-D-glucitol; EINECS 228-506-9; Glucitol, 1-deoxy-1-(methylamino)-, D-; D-Glucitol, 1-deoxy-1-(methyl-amino)-; Meglumin; Meglumina; Meglumine; Meglum-inum; Methylglucamin; N-Methylglucamine; N-Methyl-D-glucamine; N-Methyl-D(-)-glucamine; D-(-)-N-Methyl-glucamine; N-Methylsorbitylamine; NSC 7391; NSC 52907; Sorbitol, 1-deoxy-1-methylamino-. Used in the synthesis of surfactants, pharmaceuticals and dyestuffs. Crystals; mp = 128.5°; $[\alpha]_D^{25} = -16.4°$ (c = 10 H2O); soluble in H2O (100 g/100 ml), less soluble in organic solvents.

2334 Melamine
108-78-1   5834   H

C3H6N6

1,3,5-Triazine-2,4,6-triamine.

4-26-00-01253 (Beilstein Handbook Reference); ADK Stab ZS 27; Aero; AI3-14883; BRN 0124341; CCRIS 373; Cyanuramide; Cyanuric triamide; Cyanurotriamide; Cyanurotriamine; DG 002; DG 002 (amine); EINECS 203-615-4; Hicophor PR; HSDB 2648; Isomelamine; Mark ZS 27; Melamine; NCI-C50715; NSC 2130; Pluragard; Pluragard C 133; Spinflam ML 94M; Teoharn; Theoharn; Triaminotriazine; 1,3,5-Triazine-2,4,6-triamine; 2,4,6-Triaminotriazine; 2,4,6-Triamino-1,3,5-triazine; Virset 656-4; Yukamelamine; ZS 27. Crystals; mp = 345° (dec); sublimes; d16 = 1.573; insoluble in Et2O, slightly soluble in H2O, EtOH.

2335 Melilot
122-00-9   G

384

C9H10O

p-Methylacetophenone.

Acetophenone, 4'-methyl-; p-Acetotoluene; p-Acetyltoluene; 1-Acetyl-4-methylbenzene; AI3-00734; EINECS 204-514-8; Ethanone, 1-(4-methylphenyl)-; FEMA No. 2677; Melilotal; 4'-Methylacetophenone; p-Methyl aceto-phenone; p-Methylacetophenone; 1-Methyl-4-acetyl-benzene; 1-(4-Methylphenyl)ethanone; 4-Methylphenyl methyl ketone; Methyl p-tolyl ketone; NSC 9401; 1-p-Tolylethanone; Tolyl methyl ketone, p-. Extracted from the dried leaves and flowering tops of *Melilotus officinalis*. It imparts the honey-like fragrance of sweet clover, and is used for perfuming soap. Needles; mp = 28°; bp = 226°, bp7 = 93.5°; d^{20} = 1.0051; λ_m = 252 nm (ε = 12300, MeOH); very soluble in EtOH, Et2O, C6H6, CHCl3.

2336   Menadiol Sodium Diphosphate

131-13-5   5852   D G

C11H8Na4O8P2.6H2O

2-Methyl-1,4-naphthalenediol   bis(dihydrogen   phosphate) tetrasodium salt hexahydrate.

EINECS 205-012-1; Kappadione; Kipca water soluble; Menadiol sodium diphosphate; Menadiol tetrasodium diphosphate; Menadione diphosphate tetrasodium salt; Menadione sodium phosphate; Procoagulo; Sodium menadione diphosphate; Synkavit; Synkavite; Synkayvite; Thylokay. Vitamin, vitamin source. Prothrombogenic. Very soluble in H2O; insoluble in MeOH, EtOH, Et2O, Me2CO. *Eli Lilly & Co.; Hoffmann-LaRoche Inc.*

2337   Menhaden Oil

8002-50-4   5859   E H

Oils, menhaden.
Oil obtained from the small Atlantic fish, *Brevoortia tyrannus*. FDA listed. Used in nutritional supplements, used therapeutically and in soaps and creams. Yellow-brown oil; mp = 38.5-47.2°; d = 0.925-0.933; soluble in Et2O, petroleum ether, C6H6, CS2, naphtha, kerosene. *Abitec Corp.; Arista Ind.*

2338   Menthol

89-78-1   5861   G P

C10H20O

(±)-(1α,2β,5α)-5-Methyl-2-(1-methylethyl)cyclohexanol.

2-06-00-00052 (Beilstein Handbook Reference); 4-06-00-00151 (Beilstein Handbook Reference); 5-Methyl-2-(1-methylethyl)cyclohexanol, (1α,2beta,5α)-; BRN 1902288; BRN 3194263; CCRIS 375; CCRIS 4666; Cyclohexanol, 5-methyl-2-(1-methylethyl)-, (1R,2S,5R)-rel-; dl-Menthol; EINECS 239-388-3; FEMA No. 2665; Headache crystals; Hexahydrothymol; HSDB 593; Menthacamphor; p-Menthan-3-ol; Menthol; (±)-(1R*,3R*,4S*)-Menthol; (±)-Menthol; Menthol racemic; Menthol racemique; Menthomenthol; NCI-C50000; NSC 2603; Peppermint camphor; rac-Menthol; Racementhol; Racementholum; Racementol; Racemic menthol; Tra-kill tracheal mite killer. Insecticide. Registered by EPA as an insecticide and rodenticide. Also used in perfumery, cigarettes, liqueurs, flavoring agent, chewing gum, chest rubs, cough drops and nasal inhalers. Crystals with peppermint taste and odor; mp = 41 - 43°; bp = 212°; d = 0.89; $[\alpha]_D^{18}$ = -50° (c = 10 EtOH); slightly soluble in H2O, very soluble in EtOH, CHCl3, Et2O, petroleum ether, AcOH; LD50 (rat orl) = 3180 mg/kg. *Janssen Chimica; Penta Mfg.; Quest Chemical Corp.; Robeco; Sigma-Aldrich Fine Chem.*

2339   Mephaneine

101-41-7   G

C9H10O2

Methyl phenylacetate.

Acetic acid, phenyl-, methyl ester; AI3-01971; Benzeneacetic acid, methyl ester; EINECS 202-940-9; FEMA No. 2733; Methyl α-toluate; Methyl 2-phenylacetate; Methyl benzeneacetate; Methyl benzeneethanoate; Methyl phenylacetate; Methyl phenylethanoate; NSC 401667; Phenylacetic acid, methyl ester. Used in perfumery: Sweet Floral Fruity Honey Spice. Liquid; bp = 216.5°, d^{16} = 1.0622; λ_m = 247, 252, 258, 264 nm (ε = 234, 268, 289, 214, MeOH); insoluble in H2O, soluble in Me2CO, CCl4, freely soluble in EtOH, Et2O; LD50 (rat orl) = 2550 mg/kg. *Bush Boake Allen; IFF.*

2340   Mephosfolan

950-10-7   5880   G P

C8H16NO3PS2

Diethyl (4-methyl-1,3-dithiolan-2-ylidene)phosphor-amidate.

AC 47470; AI3-25991; American Cyanamid CL-47470; CL-47,470; Cyclic propylene (diethoxyphosphinyl)-dithioimidocarbonate; Cyclic propylene P,P-diethyl phosphonodithioimidocarbonate; Cytrolane; (Diethoxy-phosphinyl)dithioimidocarbonic acid cyclic propylene ester; 2-(Diethoxyphosphinylimino)-4-methyl-1,3-dithio-lane; O,O-Diethyl(4-methyl-1,3-dithiolan-2-ylidene)phos-phoramidate; 1,3-Dithiolane, 2-(diethoxyphosphinyl-imino)-4-methyl-; EI-47470; ENT-25,991; HSDB 6411; Imidocarbonic acid, phosphonodithio-, cyclic propylene P,P-diethyl ester; Mephosfolan; Mephospholan; Phosphonodithioimidocarbonic acid cyclic propylene P,P-diethyl ester; Phosphoramidic acid, (4-methyl-1,3-dithiolan-2-ylidene)-,

385

diethyl ester; P,P-Diethyl cyclic propylene ester of phosphonodithioimidocarbonic acid; 1,2-Propanedithiol, cyclic ester with P,P-diethyl-phosphonodithioimidocarbonate. Insecticide and acaricide. Used for control of damsonhop aphid in hops. Amber liquid; bp$_{0.001}$ = 120°; soluble in H_2O (0.0057 g/100 ml at 25°), EtOH, Me_2CO, C_6H_6; LD$_{50}$ (mrat orl) = 8.9 mg/kg, (mus orl) = 11.3 mg/kg, (mrbt der) = 28.7 mg/kg, (Japanese quail orl) = 12.8 mg/kg, (bee topical) = 0.0035 mg/bee; LC$_{50}$ (rainbow trout 96 hr.) = 2.1 mg/l, (carp 96 hr.) = 54.5 mg/l; toxic to bees. *American Cyanamid, Divn. AHP Corp.*

2341 Mepiquat
15302-91-7   H P

(C$_7$H$_{16}$N)$^+$
1,1-Dimethylpiperidinium.
1,1-Dimethylpiperidinium; Mepiquat; Piperidinium, 1,1-dimethyl-. Plant growth regulator. *BASF Corp.*

2342 Mepiquat Chloride
24307-26-4   5882   G H

C$_7$H$_{16}$ClN
1,1-Dimethylpiperidinium chloride.
BAS-083; BAS-08300W; BAS 08301W; BAS 08305 W; BAS 08306 W; BAS 08307 W; BAS 083W; Bas85559X; Caswell No. 380AB; 1,1-Dimethylpiperidinium chloride; EPA Pesticide Chemical Code 109101; Mepiquat chloride; Methylpiperidine hydrochloride; N,N-Dimethylpiperidinium chloride; Piperidinium, 1,1-dimethyl-, chloride; PIX. Plant growth regulator for reduction of undesired vegetative growth of cotton, better boil retention, earlier maturity; improves yield and market quality of garlic and onions. Used in combination with ethephon to control growth of cereal crops. Registered by EPA as a herbicide. Colorless crystals; mp = 223°; freely soluble in H_2O (> 100 g/100 ml), EtOH (12.8 g/100 ml), Me_2CO (> 0.1 g/100 ml), CHCl$_3$ (1.48 g/100 ml), sparingly soluble in C_6H_6, EtOAc, Et$_2O$; LD$_{50}$ (rat orl) = 1420 mg/kg, (rbt orl) = 1780 mg/kg, (rat der) > 7800 mg/kg; LC$_{50}$ (rat ihl 7 hr.) = 3.2 mg/l air, (trout 96 hr.) = 4300 mg/l; non-toxic to hens and wildfowl; non-toxic to bees. *BASF Corp.; Griffin LLC; Micro-Flo Co. LLC.*

2343 Mequinol
150-76-5   D G H

C$_7$H$_8$O$_2$
p-Hydroxymethoxybenzene.
AI3-00841; BMS 181158; CCRIS 5531; Eastman® HQMME; EINECS 205-769-8; p-Guaiacol; HQMME; HSDB 4258; Hydroquinone methyl ether; Hydroquinone monomethyl ether; Leucobasal; Leucodine B; Mechinolo; Mechinolum; MEHQ; Mequinol; Mequinolum; Monomethyl ether hydroquinone; Novo-Dermoquinona; NSC 4960; Phenol, 4-methoxy-; PMF (antioxidant); USAF AN-7. Antioxidant for monomers. Used in manufacture of

pharmaceuticals, plasticizers, dyestuffs; stabilizer for chlorinated hydrocarbons and ethylcellulose; UV inhibitor; inhibitor for arcylic and vinyl monomers and acrylonitrile. Studied for its utility in treatment of malignant melanomas. Crystals; mp = 57°; bp = 243°; d^{20} = 1.55; λ_m = 225, 292 nm (MeOH); soluble in H_2O, C_6H_6, CCl$_4$, very soluble in EtOH, Et$_2O$. *Alemark; Alfa; Arenol; Chemdesign Corp.; Eastman Chem. Co.; Fluka; Kincaid Enterprises; Penta Mfg.; SpecialtyChem. Prods.*

2344 Meralluride
8069-64-5   5891   D G

C$_{16}$H$_{23}$HgN$_6$NaO$_8$
[3-[[[(3-Carboxylato-1-oxopropyl)amino]carbonyl]amino]-2-methoxypropyl]hydroxymercurate(1-) sodium compound with 3,7-dihydro-1,3-dimethyl-1H-purine-2,6-dione.
Butanoic acid, 4-(((((2-methoxypropyl)amino)carbonyl)-amino)-4-oxo-, mercury complex; Dilurgen; N-((3-(Hydroxymercuri)-2-methoxypropyl)carbamoyl)-succinamic acid and theophylline; Meralluride; Meralluridum; Meralurida; Mercardan; Mercuhydrin; Mercurate(1-), (3-((((3-carboxylato-1-oxopropyl)amino)-carbonyl)amino)-2-methoxypropyl) hydroxy-, sodium, mixed with 3,7-dihydro-1,3-dimethyl-1H-purine-2,6-dione; Mercuretin; Mercury, (3-((((3-carboxy-1-oxopropyl)amino)carbonyl)amino)-2-methoxypropyl)-hydroxy-, monosodium salt, with 3,7-dihydro-1,3-dimethyl-1H-purine-2,6-dione; Muralluride. Used as a diuretic. Slightly soluble in H_2O, insoluble in organic solvents; LD$_{50}$ (rat sc) = 28 mg/kg. *Marion Merrell Dow Inc.*

2345 2-Mercaptoacetic acid
68-11-1   9410   G H

C$_2$H$_4$O$_2$S
Acetic acid, mercapto-.
4-03-00-00600 (Beilstein Handbook Reference); Acetic acid, mercapto-; Acide thioglycolique; AI3-24151; BRN 0506166; CCRIS 4873; EINECS 200-677-4; Glycolic acid, thio-; Glycolic acid, 2-thio-; HSDB 2702; Kyselina merkaptooctova; Kyselina thioglykolova; Mercaptoacetic acid; NSC 1894; Thioglycolate; Thioglycolic acid; Thiovanic acid; UN1940; USAF CB-35. In chemical analysis for the spectrophotometric determination of palladium; cosmetics (intermediates for hairwaving, depilatories), vinyl stabilizer intermediate, reaction intermediate for radiation-cured plastics; reagent for iron; manufacture of thioglycolates. Liquid; mp = -16.5°; bp^{20} = 120°; d^{20} = 1.3253; very soluble in H_2OEtOH, Et$_2O$, less soluble in CHCl$_3$; LD$_{50}$ (rat orl) = 198 mg/kg. *Elf Atochem N. Am.; EM Ind. Inc.; Evans Chemetics; Kreussler Chemische-Fabrik; Lancaster Synthesis Co.; Sigma-Aldrich Fine Chem.; Witco/Humko.*

2346 Mercaptoacetic acid 2-ethylhexyl ester
7659-86-1   H

C10H20O2S

2-Ethylhexyl mercaptoacetate.

Acetic acid, mercapto-, 2-ethylhexyl ester; EINECS 231-626-4; 2-Ethylhexyl mercaptoacetate; 2-Ethylhexyl thioglycolate; 2-Ethylhexylthioglycolate; Mercaptoacetic acid 2-ethylhexyl ester; Thioglycolic acid 2-ethylhexyl ester; Thioglykolsaeure-2-aethylhexyl ester; Thioglykol-säure-2-äthylhexyl ester. Liquid; bp = 133.5°; d^{20} = 0.97.

2347   2-Mercaptobenzimidazole
583-39-1   1083   G

C7H6N2S

2-Benzimidazolethiol.

AI3-18633; Antiegene MB; Antigen MB; Antigene MB; Antioxidant MB; AOMB; ASM MB; Benzimidazolethiol; Benzimidazole-2-thiol; 1H-Benzimidazole-2-thiol; 2-Benzimidazolethiol; 2H-Benzimidazole-2-thione, 1,3-dihydro-; Benzimidazoline-2-thione; 2-Benzimidazoline-thione; 2-Benzimidazolinthion; CCRIS 4837; EINECS 209-502-6; 2-Merkaptobenzimidazol; 2-Mercaptobenz-imidazole; 2-Mercaptobenzoimidazole; Mercaptobenz-imidazole; Mercaptobenzoimidazole; Merkaptobenz-imidazol; NCI-C56268; NCI-C60980; NSC 21414; NSC 186246; o-Phenylenethiourea; Permanax 21; 2-Thiobenzimidazole; 2-Thiol benzimidazole; USAF EK-6540; USAF XF-21. An antioxidant. Plates; mp = 298°; slightly soluble in H2O, more soluble in EtOH; LD50 (rat orl) = 1230 mg/kg. Aceto Corp.; Bayer AG.

2348   2-Mercaptobenzothiazole
149-30-4   5893   G P

C7H5NS2

2-Benzothiazolinethione.

2-MBT; Accel M; Accelerator M; Accelerator Mercapto; AG 63; AI3-00985; Benzothiazolethiol; Captax; Caswell No. 541; CCRIS 891; Dermacid; EINECS 205-736-8; Ekagom G; EPA Pesticide Chemical Code 051701; HSDB 4025; Kaptaks; Kaptax; MBT; Mebetizole; Mebithizol; Mercaptobenzothiazol; Mercaptobenzothiazole; Mertax; NCI-C56519; NSC 2041; Nuodeb 84; Nuodex 84; Pennac MBT powder; Pneumax MBT; Rokon; Rotax; Royal MBT; Soxinol M; Sulfadene; Thiotax; USAF GY-3; USAF XR-29; Vulkacit M; Vulkacit mercapto; Vulkacit mercapto/C. Curing accelerator for rubber. Used with 1,3-diphenylguanidine. Mineral oil coated; semi-ultra accelerator for NR, IR, BR, SBR, NBR, IIR, and EPDM; used alone in bulky goods or in combination for molded and extruded goods, hoses, conveyor belts, tires, footwear, cables, expanded rubber goods. Zinc and sodium salts used as a fungicide. Registered by EPA as an antimicrobial and fungicide (cancelled). Pale yellow needles; mp = 181°; d^{20} = 1.42; λ_m = 206, 238, 325 nm (MeOH); insoluble in H2O, soluble in EtOH (2 g/100 ml at 25°), Et2O (1 g/100 ml), Me2CO (10 g/100 ml), C6H6 (1 g/100 ml), CCl4 (< 0.2 g/100 ml), naphtha (< 0.5 g/100 ml), moderately soluble in AcOH. Bayer AG; Fluka; Monsanto

2349   2-Mercaptoethanol
60-24-2   5894   G H

C2H6OS

1-Ethanol-2-thiol.

4-01-00-02428 (Beilstein Handbook Reference); AI3-07710; BRN 0773648; CCRIS 2097; EINECS 200-464-6; Emery 5791; Ethanol, 2-mercapto-; Ethylene glycol, monothio-; HSDB 5199; Hydroxyethyl mercaptan; 2-Hydroxyethanethiol; β-Hydroxyethanethiol; 2-Hydroxy-1-ethanethiol; 2-Hydroxyethyl mercaptan; β-Hydroxy-ethylmercaptan; 1-Hydroxy-2-mercaptoethane; 2-ME; 2-Mercaptoethanol; 2-Mercapto-1-ethanol; 2-Mercapto-ethyl alcohol; Mercaptoetanol; Mercaptoethanol; 1-Mercapto-2-hydroxyethane; Monothioethylene glycol; Monothioglycol; NSC 3723; 2-Thioethanol; Thioethylene glycol; Thioglycol; Thiomonoglycol; UN2966; USAF EK-4196. Solvent for dyestuffs, intermediate for producing dyestuffs, pharmaceuticals, rubber chemicals, flotation agents, insecticides, plasticizers, reducing agent, biochemical reagent, PVC stabilizers, agricultural chemicals, textile auxiliary. BASF AG; Morton Intn'l.; Rhône Poulenc Surfactants.

2350   2-Mercaptoethyl oleate
59118-78-4   H

C20H38O2S

9-Octadecenoic acid (Z)-, 2-mercaptoethyl ester.

EINECS 261-609-7; 2-Mercaptoethyl oleate; 9-Octadecenoic acid (Z)-, 2-mercaptoethyl ester; 9-Octadecenoic acid (9Z)-, 2-mercaptoethyl ester.

2351   3-Mercaptopropyltrimethoxysilane;
4420-74-0   G

C6H16O3SSi

γ-Mercaptopropyltrimethoxy silane.

A 189 (silicone); Aktisil MM; AZ 6129; BRN 2038119; EINECS 224-588-5; GF 70; KBE 803; KBM 803; M 8500; M 8500 (coupling agent); (3-Mercaptopropyl)trimethoxy-silane; γ-Mercaptopropyltrimethoxysilane; (γ-Mercapto-propyl)trimethoxysilane; MPS; MPS-M; NUCA 189; 1-Propanethiol, 3-(trimethoxysilyl)-; Prosil® 196; SH 6062; Sila-Ace S 810; Silane, 3-mercaptopropyltrimethoxy-; Silane A 189; Silquest A 189; 3-(Sulfanylpropyl)-trimethoxysilane; (3-Thiopropyl)trimethoxysilane; 3-(Tri-methoxysilyl)propanethiol; 3-Trimethoxysilylpropane-1-thiol; 3-(Trimethoxysilyl)propyl mercaptan; TSL 8380; TSL8380E; Union carbide® A-189; Z 6062. Coupling agent with both organic and inorganic reactivity; for acrylic, epichlorohydrin, nitrile, polysulfone, PS, PVC, urethane thermoplastics; thermoset acrylic, epoxy, nitrile/phenolic, phenolic, polybutadiene; and elasto-merics. Crosslinking agent adhesion promoter for coatings; provides durability; features active hydrogen reaction, chain transfer, end blocking. Filler for sulfur and metal oxide-cured systems. Liquid; bp = 198°, 215°; d = 1.0390; LD50 (rat orl) = 2940 mg/kg. Hoffmann Mining; PCR; Union Carbide Corp.

387

2352 Mercufenol Chloride

90-03-9   D G

C_6H_5ClHgO

Chloro(o-hydroxyphenyl)mercury.

4-16-00-01736 (Beilstein Handbook Reference); AI3-23201; BRN 3662387; Caswell No. 194A; Chloro(o-hydroxyphenyl)mercury; EINECS 201-962-6; EPA Pesticide Chemical Code 294100; Hydroxychloro-mercuribenzene; Mercresin; Mercufenol chloride; Myringacaine Drops; NP-27; NSC 5579; Salicresin Fluid; U 7743.; component of: Mercresin. Anti-infective (topical); disinfectant. CAUTION: poisonous. Prepared from phenol and mercuric acetate. Solid; mp = 150-152°; slightly soluble in cold H_2O; moderately soluble in boiling H_2O; freely soluble in alcohol, hot C_6H_6; sparingly soluble in $CHCl_3$. *Upjohn Ltd.*

2353 Mercuric acetate

1600-27-7   5898   G

$C_4H_6HgO_4$

Mercury(II) acetate.

Acetic acid, mercuridi-; Acetic acid, mercury(2+) salt; AI3-04458; Anthracene, 1,4-dihydro-, compd. with mercury diacetate (1:1); Bis(acetyloxy)mercury; Caswell No. 543A; CCRIS 7488; Diacetoxymercury; EINECS 216-491-1; EPA Pesticide Chemical Code 052104; HSDB 1244; Mercuriacetate; Mercuric acetate; Mercuric diacetate; Mercury acetate; Mercury di(acetate); Mercury diacetate; Mercury(2+) acetate; Mercury(II) acetate; Mercuryl acetate; NSC 215199; UN1629. Catalyst for organic synthesis, pharmaceuticals. Used for mercuration of organic compounds. Crystals; mp = 178-180°; soluble in H_2O (40 g/100 ml), EtOH; LD_{50} (rat orl) = 4 mg/kg. *Atomergic Chemetals; Cerac; Noah Chem.; Thor.*

2354 Mercuric chloride

7487-94-7   5901   D G P

Cl_2Hg

Mercury(II) chloride.

Abavit B; Bichloride of mercury; Bichlorure de mercure; Calo-Clor; Calochlor; Calocure; Caswell No. 544; CCRIS 4838; Chlorid rtutnaty; Chlorure mercurique; Cloruro di mercurio; Corrosive mercury chloride; Corrosive sublimate; Dichloromercury; EINECS 231-299-8; EPA Pesticide Chemical Code 052001; Fungchex; HSDB 33; Hydraargyrum bichloratum; Mercuric bichloride; Mercuric chloride; Mercury bichloride; Mercury chloride (HgCl2); Mercury chloride (2); Mercury dichloride; Mercury perchloride; Mercury(2+) chloride; Mercury(II) chloride; NCI-C60173; NSC 353255; Perchloride of mercury; Quecksilber chlorid; Sublimat; Sublimate; Sulem; Sulema; UN1624; Mercury Chloride; dichloromercury; perchloride of mercury; sublimate; TL 898. Used in manufacture of calomel (mercurous chloride), other mercury compounds; disinfectant, organic synthesis,

analytical reagent, metallurgy, tanning, catalyst, sterilant, fungicide, insecticide, wood preservative (kyanizing), embalming fluids, textile printing, photography and dry batteries.Used as a topical antiseptic and disinfectant. Highly toxic. Antiseptic (topical); disinfectant. Registered by EPA as a fungicide (cancelled). Solid; mp = 277°; bp = 302°; d = 5.4; soluble in H_2O (74 mg/ml), less soluble in organic solvents; highly toxic, LD_{50} (rat orl) = 1 mg/kg. *Atomergic Chemetals; Sigma-Aldrich Fine Chem.; Spectrum Chem. Manufacturing; Thor.*

2355 Mercuric sulfate

7783-35-9   5914   G

HgO_4S

Mercury(II) sulfate.

EINECS 231-992-5; HSDB 1247; Mercuric bisulphate; Mercuric sulfate; Mercuric sulphate; Mercurous bisulphate; Mercury bisulfate; Mercury disulfate; Mercury persulfate; Mercury sulfate (HgSO4); Mercury sulphate; Mercury(II) sulfate; Mercury(II) sulfate (1:1); Sulfate mercurique; Sulfuric acid, mercury(2+) salt (1:1). Catalyst in the conversion of acetylene to acetaldehyde, extracting gold and silver from roasted pyrites, battery electrolyte. Solid; dec 450°; d = 6.47; reacts with H_2O; LD_{50} (rat orl) = 57 mg/kg. *Atomergic Chemetals; Lancaster Synthesis Co.; Noah Chem.*

2356 Mercuric sulfide, red and black

1344-48-5   5915   G

$Hg=S$

HgS

Mercury(II) sulfide.

Almaden; C.I. 77766; C.I. Pigment Red 106; Chinese vermilion; EINECS 215-696-3; Ethiops mineral; Hydrargyrum sulfuratum rubrum; Mercuric sulfide; β-Mercuric sulfide; Mercuric sulfide, black; Mercuric sulfide, red; Mercuric sulphide; Mercury monosulfide; Mercury sulfide; Mercury sulfide (HgS); Mercury sulphide; Mercury sulphide, natural; Mercury(2+) sulfide; Mercury(II) sulfide; Monomercury sulfide; Orange vermilion; Pigment Red 106; Pure English (Quicksilver) Vermilion; Quecksilber(II)-sulfid, rotes; Red cinnabar; Red mercury sulphide; Rotes Quecksilbersulfid; Scarlet vermilion; Vermilion; Vermilion (HgS). Red pigment. Black amorphous powder; mp = 583°; d = 8.1; insoluble in H_2O, EtOH, acids. *Atomergic Chemetals; Cerac; Noah Chem.*

2357 Mercurous chloride

10112-91-1   5921   D G V

Cl_2Hg_2

Mercury(II) chloride.

Calogreen; Calomel; Calotab; Chlorure mercureux; Dimercury dichloride; EINECS 233-307-5; Mercurous chloride; Mercurous chloride (Hg2Cl2); Mercury chloride (Hg2Cl2); Mercury subchloride; Mercury(I) chloride. Fungicide. Used for control of clubroot in brassicas and white rot in onions. Antiseptic; cathartic; diuretic; antisyphilitic. Used as a cathartic, local antiseptic and desiccant in veterinary medicine. Sublimes at 400-500°; d = 7.15; practically insoluble in H_2O; insoluble in alcohol, Et_2O; incompatible with bromides, iodides, alkali chlorides, sulfates, sulfites, carbonates, hydroxides, lime H_2O, acacia, ammonia, golden antimony sulfide, cocain. *Hortichem Ltd.*

2358   Mercurous oxide
15829-53-5   G

$$Hg-O-Hg$$

Hg2O
Mercury(I) oxide.
EINECS 239-934-0; Hahnmann's mercury; HSDB 4487; Mercurous oxide; Mercurous oxide (Hg2O); Mercury oxide; Mercury oxide (Hg2O); Mercury oxide black; Mercury(I) oxide; Quecksilberoxid. Black oxide of mercury.

2359   Mercury
7439-97-6   5925   D G

$$Hg$$

Hg
Mercury.
Caswell No. 546; CCRIS 1578; Colloidal mercury; EINECS 231-106-7; EPA Pesticide Chemical Code 052301; HSDB 1208; Hydrargyrum; KWIK; Liquid silver; Mercure; Mercurio; Mercury; Mercury, elemental; Mercury, elemental and inorganic forms; Mercury, metallic; Mercury vapor; Metallic mercury; NCI-C60399; Quecksilber; Quicksilver; RCRA waste number U151; RTEC; UN 2024 (liquid compounds); UN2809. Metallic element; Hg; Amalgam, catalyst, electrical apparatus, cathodes for production of chlorine and caustic soda, thermometers, barometers. Has been used medicinally as a laxative-cathartic. Metallic liquid; mp = -39°; bp = 357°; insoluble in H2O, organic solvents. *Atomergic Chemetals; Cerac; Cox Chemical Co.; Sigma-Aldrich Fine Chem.; Spectrum Chem. Manufacturing.*

2360   Mercury oxide
21908-53-2   5908   G

$$Hg=O$$

HgO
Mercuric oxide.
AI3-02738; C.I. 77760; Caswell No. 544A; EINECS 244-654-7; EPA Pesticide Chemical Code 052102; Gelbes quecksilberoxyd; HSDB 1265; Hydrargyrum oxid flav; Hydrargyrum oxydatum rubrum; Kankerex; Mercuric oxide; Mercuric oxide (HgO); Mercuric oxide, red; Mercuric oxide, yellow; Mercury monoxide; Mercury oxide; Mercury oxide (HgO); Mercury(2+) oxide; Mercury(II) oxide; Natural montroydite; Oxide mercurique jaune; Oxido amarillo de mercurio; Oxyde de mercure; Oxyde mercurique; Red mercuric oxide; Red oxide of mercury; Red Precipitate; Santar; Santar M; UN1641; Yellow mercuric oxide; Yellow oxide of mercury; Yellow precipitate. Red and yellow forms represent different physical states of an orthorhombic crystal form. Yellow is more finely divided and more reactive. The two forms can be interconverted. For control of canker in apples and pears. Used in marine bottom paints, ceramic paints; in dry batteries and as a reagent. Red: chemicals, paint pigment, perfumery, cosmetics, pharmaceuticals, ceramics, dry batteries, polishes, analytical reagent, antifouling paints, fungicide, antiseptic; Yellow: antiseptic, mercury compounds. Both forms used as topical anti-infectives. Amorphous solid; mp = 500° (dec); d = 11.14; LD50 (rat orl) = 18 mg/kg. *Cerac; Noah Chem.; Spectrum Chem. Manufacturing; Thor Chemicals (UK) Ltd.; Universal Crop Protection Ltd.*

2361   Merphos
150-50-5   H P

C12H27PS3
Tributyl phosphorotrithioite.
4-01-00-01564 (Beilstein Handbook Reference); AI3-25783; BRN 1703869; Butyl phosphorotrithioite; Butyl phosphorotrithioite ((BuS)3P); Caswell No. 865; Chemagro B-1776; Deleaf defoliant; Easy off-D; EINECS 205-761-4; EPA Pesticide Chemical Code 074901; Folex; Folex 6EC; HSDB 1777; Merphos; NSC 27720; Tributyl phosphoro-trithioite; Tributyl trithiophosphite; Tributylthiofosfin. Herbicide. Registered by EPA as a herbicide and fungicide (cancelled). Crystals; mp = 100°; bp15 = 176°, bp0.7 = 137°; d^{20} = 1.02; slightly soluble in H2O, soluble in organic solvents.

2362   Mesidine
88-05-1   G

C9H13N
2,4,6-trimethylaniline.
Aminomesitylene; 2-Aminomesitylene; 1-Amino-2,4,6-trimethylbenzen; 2-Amino-1,3,5-trimethylbenzene; Aniline, 2,4,6-trimethyl-; Benzenamine, 2,4,6-trimethyl-; CCRIS 2871; EINECS 201-794-3; HSDB 2694; Mesidin; Mesidine; Mesitylamine; Mesitylene, 2-amino-; 2,4,6-Trimethylaniline; 2,4,6-Trimethylbenzenamine. Used as an intermediate in the manufacture of dyestuffs. mp = -2.5°; bp = 232.5°; d^{25} = 0.96330; λ_m = 234, 28 nm (ε = 7770, 2050 MeOH); slightly soluble in CCl4; LD50 (rat orl) = 743 mg/kg.

2363   Mesitol
527-60-6   H

C9H12O
2,4,6-Trimethylphenol.
4-06-00-03253 (Beilstein Handbook Reference); Benzene, 2-hydroxy-1,3,5-trimethyl-; BRN 1859675; EINECS 208-419-2; HSDB 5677; 2-Hydroxymesitylene; 1-Hydroxy-2,4,6-trimethylbenzene; Mesitol; \Mesityl alcohol; NSC 5353; Phenol, 2,4,6-trimethyl-; 1,3,5-Trimethylphenol; 2,4,6-Trimethylofenol; 2,4,6-Trimethylphenol; 2,4,6-Trimetylofenol. Needles; mp = 73°; bp = 220°; very soluble in EtOH, Et2O.

2364   Mesitylene
108-67-8   5934   H

389

C9H12

1,3,5-Trimethylbenzene.

AI3-23973; Benzene, 1,3,5-trimethyl-; 3,5-Dimethyl-toluene; EINECS 203-604-4; Fleet-X; HSDB 92; Mesitylene; NSC 9273; TMB; 1,3,5-Trimethylbenzene; sym-Trimethylbenzene; Trimethylbenzol; UN2325. Liquid; mp = -44.7°; bp = 164.7°; d^{20} = 0.8652; λ_m = 207, 213, 217, 282 nm (ε = 11900, 12100, 12700, 17400, MeOH); insoluble in H2O, freely soluble in EtOH, Et2O, Me2CO, C6H6, CCl4, petroleum ether.

2365   Mesotartaric acid

147-73-9   9159   G

C4H6O6

Meso-tartaric acid monohydrate.

Antiweinsäure; Butanedioic acid, 2,3-dihydroxy-, (2R,3S)-rel-; (R*,S*)-2,3-Dihydroxybutanedioic acid; L-(+)-Dihydroxysuccinic acid; EINECS 205-696-1; internally compensated tartaric acid; meso-Tartaric acid; unresolvable tartaric acid. Prepared by racemization of L-tartaric acid. Crystals; mp = 140°; d^{20} 1.66; soluble in H2O (125 g/100 ml).

2366   Mesulfen

135-58-0   5943   G V

C14H12S2

2,7-Dimethylthianthrene.

2,7-Dimethylthianthrene; EINECS 205-202-4; Mesulfene; Mesulfeno; Mesulfenum; Mesulphen; Mitigal; Odylen®Soufrol; Thiantholum; Thiotal. Scabicide and anti-pruritic veterinary preparation; for external use against ectoparasites. mp = 123°; bp14= 228-231°, bp3 = 184°; insoluble in H2O, very soluble in Me2CO, Et2O, petroleum ether, CHCl3. Bayer AG.

2367   Metal Deactivator S

94-91-7
                                                  G

C17H18N2O2

N,N'-Disalicylidene-1,2-propane diamine.

N,N'-Bis(salicylidene)-1,2-diaminopropane; Carlisle metal deactivator; Copper inhibitor 50; o-Cresol, α,α'-(propylenedinitrilo)di-; Cuvan 80; α,α'-Dipropyl-enedinitrilodi-o-cresol; Disalicylalpropylenediimine; N,N'-Disalicylidene-1,2-diaminopropane; N,N'-Disali-cyclidene-1,2-propanediamine; N,N'-Disalicylidene-1,2-propylenediamine; DMD; Du Pont metal deactivator; EINECS 202-

374-2; N,N'-(2-Hydroxybenzylidene)-1,2-propandiamine; Keromet MD; 2,2'-[(1-Methyl-1,2-ethanediyl)bis(nitrilomethylidyne)]bis-phenol; NSC 67004; Phenol, 2,2'-((1-methyl-1,2-ethanediyl)bis-(nitrilomethylidyne))bis-; N,N'-Propylenebis(salicylidene-imine); α,α'-Propylenedinitrilodi-o-cresol; Tenamene 60; UOP copper deactivator. Copper chelating agent used in the refinery industry. Hart Prod.

2368   Metalaxyl

57837-19-1   5947   G P

C15H21NO4

N-(2,6-Dimethylphenyl)-N-(methoxyacetyl)-DL-alanine methyl ester.

DL-Alanine, N-(2,6-dimethylphenyl)-N-(methoxy-acetyl)-, methyl ester; Alanine, N-(2,6-dimethylphenyl)-N-(2,6-xylyl)-, methyl ester, DL-; Allegiance; Apron; Apron 2E; Apron FL; Apron SD 35; BRN 2947777; Caswell No. 375AA; CG 117; CGA 48988; (±)-N-(2,6-Dimethylphenyl)-N-(2'-methoxyacetyl)alaninate de methyle; D,L-N-(2,6-Di-methylphenyl)-N-(2'-methoxyacetyl)alaninate de methyle; EINECS 260-979-7; EPA Pesticide Chemical Code 113501; HSDB 7061; IPO-FS; Jiashuangling; Metalaxil; Metalaxyl; (±)-Metalaxyl; Metanaxin; Metasyl; Metaxanin; Methyl N-(2-methoxyacetyl)-N-(2,6-xylyl)-DL-alaninate; Methyl N-(2,6-dimethylphenyl)-N-(methoxyacetyl)-DL-alaninate; N-(2,6-Dimethylphenyl)-N-(methoxyacetyl)-alanine methyl ester; Ridomil; Ridomil 2E; Ridomil 72WP; Ridomil Vino; Subdue; Subdue 2E; Subdue 5SP.Fungicide. Effective against *Phytophthora infestans*, the cause of potato blight. Registered by EPA as a fungicide. White crystals; mp = 71 - 72°; soluble in H2O (0.71 g/ 100 ml), CH2Cl2 (75 g/100 ml), MeOH (85 g/100 ml), C6H6 (55 g/100 ml), i-PrOH (27 g/100 ml), n-octanol (13 g/100 ml), C6H14 (0.91 g/100 ml); LD50 (rat orl) = 669 mg/kg, (rat der) > 3100 mg/kg; LC50 (rainbow trout, carp, bluegill sunfish 96 hr.) > 100 mg/l; non-toxic to birds; non-toxic to bees. *Agriliance LLC; Drexel Chemical Co.; Gustafson LLC; Micro-Flo Co. LLC; Nation's AG LLC; Syngenta Crop Protection; Wilbur-Ellis Co.*

2369   Metaldehyde

108-62-3
                                                  G H P

C8H16O4

2,4,6,8-Tetramethyl-1,3,5,7-tetraoxacyclooctane.

Acetaldehyde, tetramer; AI3-15376; Antimilace; Ariotox; Caswell No. 548; Cekumeta; Corry's Slug Death; EINECS 203-600-2; EPA Pesticide Chemical Code 053001; Halizan; Metacetaldehyde; Metaldehyd; Metaldehyde; Metaldeide; Metason; Namekil; Slug-Tox; Tetramethyl-1,3,5,7-tetroxocane; 2,4,6,8-Tetramethyl-1,3,5,7-tetra-oxacyclooctane; R-2,C-4,C-6,C-8-Tetramethyl-1,3,5,7-tetroxocane; 1,3,5,7-Tetroxocane, 2,4,6,8-tetramethyl-; UN1332; See also: [9002-91-9]. Molluscicide with contact and stomach action. Contact with the foot makes the mollusk torpid and induces an increased secretion of mucus, leading to dehydration and death. Used for control of slugs and snails. Registered by EPA as an insecticide and molluscicide. Colorless crystals; mp = 246°; sublimes 112 - 115°; soluble in H2O (0.02 g/100 ml at 17°, 0.026 g/100 ml at 30°), freely soluble in C_6H_6, $CHCl_3$, less soluble in EtOH, Et_2O; LD_{50} (rat orl) = 630 mg/kg, (dog orl) = 600 - 1000 mg/kg, (gpg orl) = 600 mg/kg), non-toxic to fish. *Amvac Chemical Corp.; E.M. Matson Jr., Co.; Farm Protection Ltd.; Fisons plc, Horticultural Div.; Generic; Hi-Yield Chemical Co.; Metro Biological Lab.; Scotts Co.; Value Gardens Supply LLC.*

2370   Metamitron
41394-05-2   5949   G H P

$C_{10}H_{10}N_4O$
  4-Amino-3-methyl-6-phenyl-1,2,4-triazin-5(4H)-one.
  5-26-04-00395 (Beilstein Handbook Reference); 4-Amino-3-methyl-6-phenyl-1,2,4-triazin-5(4H)-one; 4-Amino-3-methyl-6-phenyl-1,2,4-triazin-5-one; 4-Amino-4,5-dihydro-3-methyl-6-phenyl-1,2,4-triazin-5-one; BAY-DRW 1139; BRN 0613129; DRW 1139; EINECS 255-349-3; Goltix; Goltix®; Herbrak; Metamiton; Metamitron; Metamitrone; Methiamitron; 3-Methyl-4-amino-6-phenyl-1,2,4-triazin(4H)-on; as-Triazin-5(4H)-one, 4-amino-3-methyl-6-phenyl-. A water dispersible granular form-ulation containing 70% w/w metamitron; used to control annual weeds in sugar beet grown on mineral and organic soils and red beet, fodder beet and mangolds grown on mineral soils. Crystals; mp = 169°; LD_{50}(rat, orl) = 3343 mg/kg. *Bayer AG.*

2371   Metam-sodium
137-42-8   5976   G H P

$C_2H_4NNaS_2$
  Sodium N-methyldithiocarbamate.
  A7 Vapam; Basamid-fluid; BAY 5590; Carbam; Carbamic acid, methyldithio-, monosodium salt; Carbamodithioic acid, methyl-, monosodium salt; Carbathion; Carbathione; Carbation; Carbothion; Diethylamino-2,6-acetoxylidide; EINECS 205-293-0; Geort; HSDB 1767; Karbation; Mapasol; Masposol; Metam-fluid Basf; Metam-sodium; Metham; Metham sodium; Methylcarbamo-dithioic acid sodium salt; Methyldithiocarbamic acid, sodium salt; Methyldithiokarbaman sodny; Monam; Monosodium methylcarbamodithioate; N-869; Natrium-N-methyl-dithiocarbamaat; Natrium-N-methyl-dithiocarb-amat; Nematin; NSC 3515; Sectagon II; Sepivam; Sistan; SMDC; Sodium metam; Sodium metham; Sodium methylcarbamodithioate; Sodium methyldithiocarbamate; Soil-Prep; Solasan 500; Sometam; Trapex; Trimaton; Vapam; VPM; VPM (fungicide). Soil-fumigant which acts by decomposition to methyl isothiocyanate. Used as a soil sterilant, controlling soil fungi, nematodes, weed seeds and soil insects. Also used as a herbicide. Registered by EPA as an antimicrobial, herbicide, nematicide and fungicide. [Dihydrate]: crystals with unpleasant odor; loses $2H_2O$ at 130°; soluble in H_2O (72.2 g/100 ml at 20°), moderately soluble in MeOH, EtOH, insoluble in other organic solvents; LD_{50} (mrat orl) = 1800 mg/kg, (frat orl) = 1700 mg/kg, (mus orl) = 285 mg/kg, (rbt der) = 1300 mg/kg; LC_{50} (mallard duck, Japanese quail 5 day dietary) > 5000 mg/kg diet, (guppy 96 hr.) = 4.2 mg/l, (bluegill sunfish 96 hr.) = 0.39 mg/l, (rainbow trout 96 hr.) = 0.079 mg/l; non-toxic to bees. *Amvac Chemical Corp.; Buckman Laboratories Inc.; Micro-Flo Co. LLC; Osmose Inc.; Sewer Sciences Inc.; Tessenderlo Kerley, Inc.;*

UAP - Platte Chemical; UCB SA; Wilbur-Ellis Co.

2372   Metanilic acid
121-47-1   5952   G

$C_6H_7NO_3S$
  3-aminobenzenesulfonic acid.
  1-aminobenzene-3-sulfonic acid; m-aniline sulfonic acid; m-sulfanilic acid; aniline-m-sulfonic acid. Used as a chemcial intermediate. Needles or prisms; mp dec.; d = 1.69; λ_m = 235, 294 nm (ε = 6310, 1585, H_2O); insoluble in EtOH, Et_2O, slightly soluble in H_2O (<1 g/100ml).

2373   Metaxalone
1665-48-1   D

$C_{12}H_{15}NO_3$
  5-[(3,5-Xylyloxy)methyl]-2-oxazolidinone.
  AHR-438; BRN 0884592; CL 39,148; 5-((3,5-Dimethylphenoxy)methyl)-2-oxazolidinone; EINECS 216-777-6; HSDB 3236; Metassalone; Metaxalon; Metaxalona; Metaxalone; Metaxalonum; Methaxalonum; Methoxolone; NSC 170959; 2-Oxazolidinone, 5-((3,5-dimethylphenoxy)methyl)-; 2-Oxazolidinone, 5-((3,5-xylyloxy)methyl)-; Skelaxin; Zorane; 5-((3,5-Xylyloxy)-methyl)-2-oxazolidinone; 5-(3,5-Xyloloxymethyl)oxa-zolidin-2-one. Skeletal muscle relaxant. Crystals; mp = 121.5-123°. *Robins, A. H. Co.,.*

2374   Metazachlor
67129-08-2   G P

$C_{14}H_{16}ClN_3O$
  2-Chloro-N-(2,6-dimethylphenyl)-N-(1H-pyrazol-1-ylmethyl)acetamide.
  5-23-04-00126 (Beilstein Handbook Reference); Acetamide, 2-chloro-N-(2,6-dimethylphenyl)-N-(1H-pyr-azol-1-ylmethyl)-; BAS 479H; BRN 0621550; Butisan S; 2-Chloro-N-(2,6-dimethylphenyl)-N-(1H-pyrazol-1-yl-methyl)-acetamide; 2-Chloro-N-(pyrazol-1-ylmethyl)acet-2',6'-xylidide; Metazachlor; Metazachlore. Selective herbicide, inhibits germination. Used for control of annual grasses and broad-leaved weeds in fruit and vegetable crops, weed control in brassicas and ornamental crops. Metazachlor is a selective herbicide, absorbed by the hypocityls and roots and inhibiting germination. Yellow crystals; mp = 85°; soluble in H_2O (0.0017 g/100

ml at 20°), Me2CO (> 79 g/100 ml), CHCl3 (>148 g/100 ml), EtOAc (53 g/100 ml), EtOH (15.8 g/100 ml); LD50 (rat orl) = 2150 mg/kg, (rat der) > 6810 mg/kg; toxic to trout, moderately toxic to carp; non-toxic to bees. *BASF Corp.; Kommer-Brookwick Ltd.*

2375 Metepa
57-39-6   5961   G

C9H18N3OP
Tris(2-methyl-1-aziridinyl)phosphine oxide.
AI3-50003; Aziridine, 1,1',1''-phosphinylidynetris(2-methyl-; BRN 1345447; C 3172; CCRIS 1435; EINECS 200-326-5; ENT 50,003; APO; Mapo®; Metapoxide; Metepa; Methaphoxide; Methyl aphoxide; NSC 54054; Phosphine oxide, tris(1-(2-methyl)aziridinyl)-; 1,1',1''-Phosphinylidynetris(2-methyl)aziridine; Trimethyl-aziridinylphosphine oxide; Tris(1-methylethylene)-phosphoric triamide; Tris(1,2-propylene)phosphoramide; N,N',N''-Tris(1-methylethylene)phosphoramide; Tris(2-methylaziridinyl)phosphine oxide; Tris(methylaziridinyl)-phosphine oxide. Used for creaseproofing and flameproofing textiles; resin raw material, crosslinker, adhesion promoter and chemosterilant. Oil; bp0.15 = 90-92°; LD50 (mrat orl) = 136 mg/kg, (frat orl) = 213 mg/kg. *Aceto.*

2376 Metformin
657-24-9   5963   D

C4H11N5
N,N-Dimethylbiguanide.
Diabetosan; Diabex; Dimethylbiguanide; DMGG; EINECS 211-517-8; Fluamine; Flumamine; Glifage; Gliguanid; Glucinan; Haurymelin; Islotin; LA-6023; Melbin; Metformin; Metformina; Metformine; Metforminum; Metiguanide; NNDG; Siofor; Stagid.; Glucinan [as p-chlorophenoxyacetate]; metformin pamoate [as emboate]; Stagid [as emboate]. Antidiabetic agent. *Bristol-Myers Squibb Co.*

2377 Metformin hydrochloride
1115-70-4   5963   D

C4H12ClN5
N,N-Dimethylimidocarbonimidic diamide monohydro-chloride.
AI3-51264; Apo-Metformin; Benofomin; Biocos; D 15095; D.B.I.; Dabex; Dextin; Diabefagos; Diaberit; Diabesin; Diabetex; Diabetmin; Diabetosan; Diabex; Diaformin; Diamin; Diaphage; Diformin; Diformin Retard; Dimethylbiguanide hydrochloride; DMGG hydrochloride; EINECS 214-230-6; Fornidd; Geamet; Glibomet; Glucofago; Glucoform; Glucoliz; Glucomet; Glucomine; Gluconil; Glucophage; Gluformin; Glupermin; Glyciphage; Glycoran; Glyformin; Haurymellin hydrochloride; HSDB 7080; Islotin retard; LA 6023; Meglucon; Meguan; Mescorit; Metaglip; Metbay; Metforal; Metiguanide; Metiguanide monohydrochloride; Metolmin; Metomin; Miformin; Novo-Metformin; NSC 91485; Riomet; Risidon; Siamformet; Thiabet; Walaphage. Antidiabetic agent. Crystals; mp =

218-220°, 232°; soluble in H2O, EtOH; insoluble in Et2O, CHCl3; LD50 (rat orl) = 1000 mg/kg, (rat sc) = 300 mg/kg. *Bristol-Myers Squibb Co.*

2378 Methabenzthiazuron
18691-97-9   5964   G P

C10H11N3OS
N-2-Benzothiazolyl-N,N'-dimethylurea.
4-27-00-04842 (Beilstein Handbook Reference); BAY 74283; Bayer 5633; 1-(2-Benzothiazolyl)-1,3-dimethyl-urea; N-2-Benzothiazolyl-N,N'-dimethylurea; 1-Benzo-thiazol-2-yl-1,1'-dimethylurea; BRN 0196633; Caswell No. 081B; CCRIS 6766; 1,3-Dimethyl-3-(2-benz-thiazolyl)-harnstoff; 1,3-Dimethyl-3-(2-benzothiazolyl)-urea; EPA Pesticide Chemical Code 281300; Metabenzthiazuron; Methabenzthiazuron; Methbenz-thiazuron; N-Methyl-N'-methyl-N'-(2-benzothiazolyl)-urea; Preparation 5633; S 25128; Tribunil; Urea, 1-(2-benzothiazolyl)-1,3-dimethyl-. Selective herbicide, acts as a photosynthesis inhibitor and used for pre- and post-emergence control of annual grasses and broad-leaved weeds in garlic, onions, chives, leeks, peas, field beans, cereals, grass seed crops, lucerne, maize, potatoes, artichokes, stonefruit and tree nurseries. Crystals; mp = 119 - 121°; soluble in H2O (0.0059 g/100 ml at 20°), Me2CO (12 g/100 ml), MeOH (7 g/100 ml), DMF (≅10 g/100 ml), CH2Cl2 (> 20 g/100 ml), i-PrOH (2 - 5 g/100 ml), C7H8 (5 - 10 g/100 ml), C6H14 (0.1 - 0.2 g/100 ml); LD50 (mrat, gpg orl) > 2500 mg/kg, (fmus, rbt, cat, dog orl) > 1000 mg/kg, (rat der) > 5000 mg/kg; LC50 (rat ihl 4 hr.) > 0.5 mg/l air; (rainbow trout 96 hr.) = 15.9, (golden orfe 96 hr.) = 29 mg/l; non-toxic to bees. *Bayer AG; Bayer Corp., Agriculture; DuPont.*

2379 Methacetin
51-66-1   52   H

C9H11NO2
N-(4-Methoxyphenyl)acetamide.
4-13-00-01092 (Beilstein Handbook Reference); Acetamide, N-(4-methoxyphenyl)-; Acetanilide, 4'-methoxy-; p-Acetanisidide; p-Acetanisidine; Aceto-p-anisidide; Acetyl-p-anisidide; N-Acetyl-p-anisidine; N-Acetyl-p-methoxyaniline; AI3-00798; BRN 0387887; EINECS 200-114-2; Metacetin; 4-Methoxyacetanilide; p-Methoxyacetanilide; N-(4-Methoxyphenyl)acetamide; NSC 4687. Crystals; mp = 130 - 132°; soluble in Me2CO, EtOH, CHCl3, slightly soluble in H2O.

2380 Methacrylic acid
79-41-4   5967   H

C4H6O2
2-Methyl-2-propenoic acid.
4-02-00-01518 (Beilstein Handbook Reference); Acide

methacrylique; Acido metacrilico; Acrylic acid, 2-methyl-; AI3-15724; BRN 1719937; CCRIS 5925; EINECS 201-204-4; HSDB 2649; Kyselina methakrylova; Methacrylic acid; NSC 7393; Propenoic acid, 2-methyl-; Propionic acid, 2-methylene-; UN2531. Used in the manufacture of methacrylate polymers. Liquid; mp = 16°; bp = 162.5°; d^{2-} = 1.-153; soluble in EtOH, less soluble in H_2O, Et_2O, $CHCl_3$.

2381 Methacrylic acid, 2-(dimethylamino)ethyl ester
2867-47-2   G H

$C_8H_{15}NO_2$
2-(Dimethylamino)ethyl 2-methyl-2-propenoate.
4-04-00-01432 (Beilstein Handbook Reference); Ageflex FM-1; BRN 1757048; Dimethylaminoethyl methacrylate; 2-(Dimethylamino)ethyl methacrylate; N,N-Dimethyl-aminoethyl methacrylate; N,N-Dimethylethanolamine methacrylate; EINECS 220-688-8; Ethanol, 2-(dimethyl-amino)-, methacrylate; HSDB 5464; Methacrylic acid, 2-(dimethylamino)ethyl ester; NSC 20959; 2-Propenoic acid, 2-methyl-, 2-(dimethylamino)ethyl ester; Sipomer® 2M1M; UN2522; USAF RH-3. Detergent and sludge dispersant in lubricants; viscosity index improver; flocculant for waste water treatment; retention aid for paper manufacturing; acid scavenger in PU foams; corrosion inhibitor; resin and rubber modifier; used in acrylic polishes and paints, hair preparation copolymers, sugar and water clarification and adhesives. Monomer to produce polymers with pendant amino function. Liquid; mp = -30°; bp = 182°, bp6 = 62-65°; d = 0.933; very soluble in H_2O, soluble in organic solvents; LD_{50} (rat orl) = 1751 mg/kg. Rhône Poulenc Surfactants; Rit-Chem.

2382 Methacrylic acid, 3-(trichlorosilyl)propyl ester
7351-61-3   H

$C_7H_{11}Cl_3O_2Si$
2-Propenoic acid, 2-methyl-, 3-(trichlorosilyl)propyl ester.
EINECS 230-878-2; 2-Propenoic acid, 2-methyl-, 3-(trichlorosilyl)propyl ester; 3-(Trichlorosilyl)propyl methacrylate.

2383 (2-(Methacryloyloxy)ethyl)trimethyl-ammonium chloride
5039-78-1   G H

$C_9H_{18}ClNO_2$
Ethanaminium, N,N,N-trimethyl-2-((2-methyl-1-oxo-2-propenyl)oxy)-, chloride.
Ageflex FM-1Q80MC; EINECS 225-733-5; Ethanaminium, N,N,N-trimethyl-2-((2-methyl-1-oxo-2-propenyl)oxy)-, chloride; Madquat Q-6; (2-(Methacryloyloxy)ethyl)-trimethylammonium chloride. Quaternary antistatic finish for polyester fibers, flocculant and

coagulant for industrial process water treatment, flocculant for mineral recovery, ion exchange resins, adhesives and acid dye receptivity. Rhône Poulenc Surfactants; Rit-Chem.

2384 2-(Methacryloyloxy)ethyltrimethyl-ammonium methyl sulfate
6891-44-7   G

$C_{10}H_{21}NO_6S$
N,N,N-Trimethyl-2-(1-oxo-2-methyl-2-propenyloxy)ethanaminium methylsulfate.
Ageflex FM-1Q80DMS; EINECS 229-995-1; Ethan-aminium, N,N,N-trimethyl-2-((2-methyl-1-oxo-2-prop-enyl)oxy)-, methyl sulfate; Ethanaminium, N,N,N-trimethyl-2-((2-methyl-1-oxo-2-propenyl)oxy)-, methyl sulfate; (2-(Methacryloyloxy)ethyl)trimethylammonium methyl sulphate. Antistatic finish for polyester fibers, flocculant and coagulant for industrial process water treatment, flocculant for mineral recovery, ion exchange resins, adhesives, acid dye receptivity, electrostatic coatings on wood, retention aids for paper. Solid; soluble in H_2O, d_{20}^{20} = 1.183; cationic. Rit-Chem.

2385 Methacryloxypropyltrimethoxysilane
2530-85-0   G

$C_{10}H_{20}O_5Si$
γ-Methacryloxypropyltrimethoxysilane.
A 174; BRN 1952435; Dow Corning® Z-6030; Dynasylan MEMO; EINECS 219-785-8; HSDB 5468; KBM 503; KDM 503; KH 570; M 8550; Methacrylic acid, 3-(trimethoxysilyl)propyl ester; Methacryloyloxypropyl-trimethoxysilane; 3-Methacryloxypropyl-trimethoxy-silane; (3-(Methacryloxy)propyl)trimethoxysilane; α-Methylacryloxypropyltrimethoxysilane; γ-Methacryloxy-propyltrimethoxysilane; Mops-M; NSC 93591; NUCA 174; 1-Propanol, 3-(trimethoxysilyl)-, methacrylate; 2-Propenoic acid, 2-methyl-3-(trimethoxysilyl)propyl ester; Prosil® 248; Q 174; Silane, (3-hydroxypropyl)trimethoxy-, methacrylate; Silane, 3-methacryloxypropyltrimethoxy-; Silane adduct, allyl methacrylate trimethoxy-; Silicone A-174; 3-(Trimethoxysilyl)-1-propanol methacrylate; Trimethoxysilyl-3-propylester kyseliny methakrylove; Trimethoxysilylpropyl methacrylate; 3-Trimethoxysilyl-propyl methacrylate; 3-(Trimethoxysilyl)propyl meth-acrylate; Union carbide® A-174; Z 6030. Coupling agent, crosslinking agent providing gloss, durability, hiding power, adhesion promotion to coatings. A coupling agent with reactive methacrylate and trimethoxysilyl groups; improves adhesion of organic thermoset resins to inorganic materials such as fiberglass, clay, quartz, and other siliceous surfaces; for ABS, acrylic, polyethylene, polyimide, polymethacrylate, PP, PS, silicone, SAN, urethane thermoplastics, alkyd, DAP, epoxy, polybutadiene, polyester, cross-linked polyethylene thermosets and elastomers. Liquid; bp = 190°; d = 1.0450. Dow Chem. U.S.A.; PCR; Union Carbide Corp.

2386 ⁺Methallyl chloride
563-47-3   2167   H

C4H7Cl
3-Chloro-2-methylpropene.
4-01-00-00803 (Beilstein Handbook Reference); AI3-14901; BRN 0878160; CCRIS 869; Chlorure de methallyle; Cloruro di metallile; EINECS 209-251-2; γ-Chloroisobutylene; HSDB 1149; Isobutenyl chloride; MAC; Methallyl chloride; Methyl allyl chloride; NCI-C54820; NSC 7303; Propene, 3-chloro-2-methyl-; UN2554. Liquid; bp = 71.5°; d^{20} = 0.9165; soluble in Me2CO, very soluble in CHCl3, freely soluble in EtOH, Et2O.

2387   Methamidophos
10265-92-6   5974   G P

C2H8NO2PS
O,S-Dimethyl phosphoramidothioate.
Acephate-met; AI3-27396; Amidor; Bayer 5546; BRN 1098870; Caswell No. 378A; Chevron 9006; Chevron ortho 9006; CKB 1220; EINECS 233-606-0; ENT 27,396; EPA Pesticide Chemical Code 101201; Filitox; Hamidop; HSDB 1593; Metamidofos estrella; Methamidophos; Methyl phosphoramidothioate ((MeO)(MeS)P(O)(NH2)); Monitor; Monitor (insecticide); MTD; NSC 190987; Ortho 9006; Ortho Monitor; O,S-Dimethyl phosphoramido-thioate; Patrole; Phosphoramidothioic acid, O,S-dimethyl ester; Pillaron; RE 9006; Sniper; SRA 5172; Tahmabon; Tamanox; Tamaron; Tamaron®; Thiophosphorsäure-O,S-dimethylesteramid; Thiophosphorsaeure-O,S-dimethyl-esteramid.Broad spectrum systemic insecticide and acaricide with stomach and contact action, used for treatment of sucking and biting insects and spider mites on a wide range of crops including cotton, tobacco, vegetables, potatoes, sugar beets. Registered by EPA as an insecticide and acaricide. Crystals; mp = 46.1°, 54°; sg^{20} = 1.31; very soluble in H2O (> 200 g/100 ml), soluble in i-PrOH (140 g/100 ml), C6H6 (< 10 g/100 ml), xylene (> 10 g/100 ml), CH2Cl2 (> 2.5 g/100 ml), Et2O (> 2.5 g/100 ml), kerosene (> 1 g/100 ml), C6H14 (> 1 g/100 ml); LD50 (mrat orl) = 25 mg/kg, (frat orl) = 27 mg/kg, (gpg orl) = 30 - 5- mg/kg, (rbt orl) = 10 - 30 mg/kg, (hen orl) = 25 mg/kg, bobwhite quail orl) = 57 mg/kg, toxic to bees; LC50 (rat ihl 4 hr.) = 0.2 mg/l air, (guppy 96 hr.) = 46 mg/l, (rainbow trout 96 hr.) = 51 mg/l, (carp, goldfish 96 hr.) = 100 mg/l. Bayer Corp.; Valent USA Corp.

2388   Methane
74-82-8   5979   G H

CH4
Methane.
4-01-00-00003 (Beilstein Handbook Reference); Biogas; BRN 1718732; EINECS 200-812-7; Fire Damp; HSDB 167; Marsh gas; Methane; Methyl hydride; R 50 (refrigerant); UN1971; UN1972. Light carburetted hydrogen. Major constituent of American natural gas. Used as a fuel and in chemical manufacturing. Colorless, odorless gas; mp = -183°; bp = -161°; soluble in H2O (3.5 ml/100 ml), more soluble in organic solvents. BOC Gases; Mesa Specialty Gases & Equipment.

2389   Methanesulfonamide, N-(2-(4-amino-N-ethyl-m-toluidino)ethyl)-, sulfate (2:3)
25646-71-3   H

C24H44N6O10S3
N-(2-(4-Amino-N-ethyl-m-toluidino)ethyl)methane-sulfonamide sulfate (2:3).
CD 003; CD 3; CD III; EINECS 247-161-5; FCD 03; Kodak CD-3; Methanesulfonamide, N-(2-((4-amino-3-methylphenyl)ethylamino)ethyl)-, sulfate (2:3); N-(2-(4-Amino-N-ethyl-m-toluidino)ethyl)methanesulfonamide sulfate (2:3); NSC 164932.

2390   Methanesulfonic acid
75-75-2   5981   G

CH4O3S
Methanesulfonicate.
4-04-00-00010 (Beilstein Handbook Reference); AI3-28532; BRN 1446024; CCRIS 2783; EINECS 200-898-6; HSDB 5004; Kyselina methansulfonova; Methane-sulfonate; Methanesulfonic acid; Methanesulphonic acid; Methylsulfonic acid; MSA; NSC 3718. Used as a polymerization catalyst, as a solvent, and in chemical synthesis. Solid; mp = 20°; bp10 = 167°; d^{18} = 1.4812; soluble in H2O, poorly soluble in organic solvents. Lancaster Synthesis Co.; Mallinckrodt Inc.; Sigma-Aldrich Fine Chem.

2391   Methanesulfonyl chloride
124-63-0   5982   H

CH3ClO2S
Methanesulfonic acid chloride.
AI3-52234; Chloromethyl sulfone; EINECS 204-706-1; HSDB 5605; Mesyl chloride; Methanesulfonic acid chloride; Methanesulfonyl chloride; Methanesulfuryl chloride; Methanesulphonyl chloride; Methyl sulfochloride; Methylsulfonyl chloride; NSC 15039; UN3246. Liquid; bp = 162°, bp11 = 55°; d^{18} = 1.4805; insoluble in H2O, soluble in EtOH, Et2O.

2392   Methanethiol
74-93-1   5983   H

CH4S

Methyl mercaptan.

4-01-00-01273 (Beilstein Handbook Reference); BRN 1696840; EINECS 200-822-1; FEMA No. 2716; HSDB 813; Mercaptan methylique; Mercaptomethane; Methaanthiol; Methanethiol; Methanthiol; Methvtiolo; Methyl mercaptan; Methyl sulfhydrate; Methyl-mercaptaan; Methylmercaptan; Metilmercaptano; RCRA waste number U153; Thiomethanol; Thiomethyl alcohol; UN 1064. Used in manufacture of pesticides, plastics and jet fuels. Odiferous gas; mp = -123°; bp = -5.9°; d^{20} = 0.8665; soluble in organic solvents, slightly soluble in H_2O.

2393   Methazole

20354-26-1   5990   G P

C9H6Cl2N2O3

2-(3,4-Dichlorophenyl)-4-methyl-1,2,4-oxadiazolidine-3,5-dione.

Bioxone; BRN 0533604; Caswell No. 549AA; Chlor-methazole; 2-(3,4-Dichlorophenyl)-4-methyl-1,2,4-oxa-diazolidinedione; 2-(3,4-Dichlorophenyl)-4-methyl-1,2,4-oxadiazolidine-3,5-dione; EINECS 243-761-6; EPA Pesticide Chemical Code 106001; HSDB 3436; Metazol; Metazole; Methazole; Mezopur; 1,2,4-Oxadiazolidine-3,5-dione, 2-(3,4-dichlorophenyl)-4-methyl-; Oxydiazol; Paxilon; Probe; Probe 75; Tunic; VCS 438. Broad spectrum herbicide. Registered by EPA as a herbicide (cancelled). Tan crystals; mp = 123 - 124°; sg^{25} = 1.24; soluble in H_2O (0.00015 g/100 ml at 25°), DMF (32 g/100 ml), CH_2Cl_2 (26 g/100 ml), cyclohexanone (17 g/100 ml), Me_2CO (4 g/100 ml), xylene (5.5 g/100 ml), MeOH (0.65 g/100 ml); LD50 (rat orl) = 2500 mg/kg, (rbt der) > 12500 mg/kg; LC50 (rat ihl 4hr.) > 200 mg/l, (mallard duck 8 day dietary) = 11200 mg/kg diet, (bobwhite quail 8 day dietary) = 1825 mg/kg diet, (bluegill sunfish 96 hr.) = 4.47 mg/l, (rainbow trout 96 hr.) = 4.09 mg/l; toxic to bees. *ICI Chem. & Polymers Ltd.*

2394   Methenamine

100-97-0   5994   D G H

C6H12N4

1,3,5,7-Tetraazatricyclo[3.3.1.1^{3,7}]decane.

Aceto HMT; Al3-09611; Aminoform; Aminoform-aldehyde; Ammoform; Ammonioformaldehyde; Anti-hydral; Caswell No. 482; CCRIS 2297; Cystamin; Cystogen; Duirexol; EINECS 202-905-8; Ekagom H; EPA Pesticide Chemical Code 045501; Esametilentetramina; Formamine; Formin; Grasselerator 102; Herax UTS; Heterin; Hexa (vulcanization accelerator); Hexa-Flo-Pulver; Hexaform; Hexaloids; Hexamethylenamine; Hexamethylene tetramine; Hexamethyleneamine; Hexa-methylenetetraamine; Hexamethylenetetramine; Hexa-methylenetetraminum; Hexamine; Hexaminum; Hexasan; Hexilmethylenamine; HMT; HSDB 563; Metenamina; Methenamin; Methenamine; Methenaminum; Metramine; Nocceler H; NSC 26346; Preparation AF; Resotropin; S 4 (heterocycle); Sanceler H; UN1328; Uramin; Uratrine; Uritone; Uro-Phosphate; Urodeine; Urotropin; Urotropine; Vulkacit H 30; Xametrin.; component of: Uro-Phosphate. Used in the curing of phenol formaldehyde and resorcinol formaldehyde, resins, rubber to textile adhesives, protein modifier, organic synthesis,

pharmaceuticals. Also used as an antibacterial agent. Rhombohedral crystals; mp >250°, sublimes; d^{-5} = 1.331; very soluble in H_2O, soluble in EtOH, Me_2CO, $CHCl_3$, slightly soluble in Et_2O, C_6H_6. *Allchem Ind.; Dajac Labs.; ECR Pharm.; Greeff R.W. & Co.; Mitsubishi Gas; OxyChem; Parke-Davis.*

2395   Methidathion

950-37-8   5999   G P

C6H11N2O4PS3

O,O-Dimethyl S-((5-methoxy-2-oxo-1,3,4-thiadiazol-3(2H)-yl)methyl) phosphorodithioate.

Al3-27193; BRN 0619915; Caswell No. 378B; CCRIS 7085; Ciba-Geigy GS 13005; S-(2,3-Dihydro-5-methoxy-2-oxo-1,3,4-thiadiazol-3-methyl)dimethylphosphorothio-lothionate; 3-Dimethoxyphosphinothioylthiomethyl-5-methoxy-1,3,4-thiadiazol-2(3H)-one; DMTP (insectic-ide); ENT 27193; EPA Pesticide Chemical Code 100301; Fisons NC 2964; Geigy 13005; Geigy GS-13005; GS 13005; HSDB 1594; Methidathion; Metidation; OMS 844; (O,O-Dimethyl)-S-(-2-methoxy-delta(sup2)-1,3,4-thiadiazolin-5-on-4-ylmethyl)dithio-phosphate; O,O-Di-methyl-S-(2-methoxy-1,3,4-thiadiazol-5(4H)-onyl-(4)-methyl) dithiophosphat; O,O-Dimethyl-S-(2-methoxy-1,3,4-thiadiazol-5(4H)-onyl-(4)-methyl)-phosphorodithi-oate; O,O-Dimethyl-S-((2-methoxy-1,3,4(4H)thiodiazol-5-on-4-yl)-methyl)-dithiofosfaat; O,O-Dimethyl S-(2,3-dihydro-5-methoxy-2-oxo-1,3,4-thiadiazol-3-ylmethyl)-phosphorodithioate; O,O-Dimethyl S-(5-methoxy-1,3,4-thiadiazolinyl-3-methyl)dithio-phosphate; O,O-Dimetil-S-((2-metossi-1,3,4-(4H)-tia-diazol-5-on-4-il)-metil)-ditiofos-fato; O,O'-Dimethyl-S-((2-methoxy-1,3,4-thiadiazole-5 (4H)-one-4-yl)methyl)-dithiophosphate; S-((2-Methoxy-5-oxo-1,3,4-thia-diazolin-4-yl)methyl)dimethylphosphoro-thiolothionate; S-((5-Methoxy-2-oxo-1,3,4-thiadiazol-3(2H)-yl)methyl) O,O-dimethylphosphorodithioate; S-2-Metoksy-1,3,4-tiadiazolo-5-on-N-metylo-O,O-dwumetyl-owy; Somonil; Supracid; Supracide; Supracide Ulvair; Suprathion; Ultracid; Ultracid 40; Ultracid EC 40; Ultracide; Ulvair 250. Used as an insecticide and acaricide. Registered by EPA as an acaricide and insecticide. Crystals; mp = 39 - 40°; soluble in H_2O (0.024 g/100g), cyclohexanone (80 g/100 ml), Me_2CO (62 g/100 ml), xylene (52 g/100 ml), EtOH (20.5 g/100 ml), octanol (4.4 g/100 ml); LD50 (rat orl) = 25 - 54 mg/kg, (mus orl) = 25 - 70 mg/kg, (rbt orl) = 63 - 80 mg/kg, (gpg orl) = 25 mg/kg, (rbt der) = 200 mg/kg, (rat der) = 1546 mg/kg, (hen orl) = 80 mg/kg; LC50 (rat ihl 4 hr.) = 3.6 mg/l air, (rainbow trout 96 hr.) = 0.01 mg/l, (bluegill sunfish 96 hr.) = 0.002 mg/l; slightly toxic to bees. *Gowan Co.; Novartis; Syngenta Crop Protection.*

2396   Methiocarb

2032-65-7   6050   G P

C11H15NO2S

Phenyl-3,5-dimethyl-4-(methylthio)-, methylcarbamate.

Carbamic acid, methyl-, 4-(methylthio)-3,5-xylyl ester; B 37344; Bay 9026; Bayer 37344; BAY 37344; BAY 5024; BAY 9026; Club; Draza; DCR 736; Esurol; ENT 25,726; mercaptodimethur; Mesurol®;

Mesurol phenol; methyl carbamic acid, 4-(methylthio)-3,5-xylyl ester; Metmercapturan; Metmercapturon; Methiocarbe. Versatile product formulated for different uses; especially as a moluscicide against slugs and snails as well as a seed dressing for repelling depredating birds; also as insecticide/acaricide against foliar-feeding caterpillars and sucking pests on various crops. Molluscicide with neurotoxic action. Used for control of slugs and snails in a wide variety of agricultural situations. Pellets containing 4% w/w methiocarb; snail and slug bait. Crystals; mp = 121.5°; soluble in H_2O (27 mg/l), more soluble in organic solvents; LD50 (rat orl) = 20 mg/kg. *Bayer AG; ICI Agrochemicals; Mobay.*

2397   **Methiodal sodium**
126-31-8   6002   G

CH2INaO3S
Iodomethanesulfonic acid sodium salt.
Abrodan; Abroden; Abrodil; Conturex; Diagnorenol; EINECS 204-782-6; HSDB 4377; Iodomethanesulfonic acid sodium salt; Kontrast-U; Methanesulfonic acid, iodo-, sodium salt; Methiodal; Methiodal s; Methiodal sodique; Methiodal sodium; Methiodalnatrium; Methiodalum natricum; Methoidal sodium; Metiodal sodico; Monoiodomethanesulfonic acid, sodium salt; Myelotrast; NCI-C03849; Neo-sombraven; NSC 510652; Radiographol; Sergosin; Sergosinum; Sergozin; Skiodan; Skiodan Sodium; Sodium iodomethanesulfonate; Sodium methiodal; Sodium monoiodomethanesulfonate. Radiopaque medium used primarily in urology as a diagnostic aid, radio-opaque medium, urographic. Soluble in H_2O (70 g/100 ml), EtOH (2.5 g/100 ml), slightly soluble in C6H6, Et2O, Me2CO. *Schering.*

2398   **Methional**
3268-49-3   H

C4H8OS
3-(Methylmercapto)propionaldehyde.
4-01-00-03974 (Beilstein Handbook Reference); AI3-36656; BRN 1739289; EINECS 221-882-5; FEMA No. 2747; Methional; Methylmercaptopropionic   aldehyde;   β-(Methylmercapto)propionaldehyde;   3-(Methylmercapto)-propionaldehyde;   3-Methylmercaptopropyl   aldehyde;   3-(Methylthio)propanal;   β-(Methylthio)propionaldehyde;   3-(Methylthio)propionaldehyde; NSC 15874; Propanal, 3-(methylthio)-; Propionaldehyde, 3-(methylthio)-; 4-Thiapentanal; UN2785. Liquid; bp11 = 62°.

2399   **Methomyl**
16752-77-5   6012   G P

C6H12N2O3S
S-Methyl-N[(methylcarbamoyl) oxy]thioacetimidate.
Nudrin; Lannate; Lannate(R); Lanox; Methomyl 5G; Lannabait; Lannate LB; Insecticide 1179; Lanox 216; LANOX 90; Mesomile; Nu-bait II; Thiobutan-2-one, O-(methylcarbamoyl)oxime; Flytek; Kipsin; Dupont 1179; Memilene; Methavin; Methomex; Nudrin; N-[(methylcarbamoyl)oxy]thioacetamidic acid, methyl ester; SD 14999; WL

18236. A cholinesterase inhibitor used as a systemic insecticide and acaricide. Used for control of a wide range of insects and spider mites in many fruits and vegetables. Also for control of flies in animal and poultry houses and in dairies. Registered by EPA as an insecticide. Used as an acaricide. Colorless crystals; mp = 78-79°; d_4^{24}= 1.2946; soluble in H_2O (5.8 g/100 ml), MeOH (80 g/100 ml), Me2CO (57.5 g/100 ml), i-PrOH (17 g/100 ml), C7H8 (2.6 g/100 ml); LD50 (rat orl) = 17 mg/kg, (frat orl) = 24 mg/kg, (rbt der) > 5000 mg/kg, (mallard duck orl) = 15.9 mg/kg, (pheasant orl) = 15.4 mg/kg; LC50 (rat ihl 4 hr.) = 0.3 mg/l air, (Pekin duck 8 day dietary) = 1890 mg/kg diet, (bobwhite quail 8 day dietary) = 3680 mg/kg diet; toxic to bees. *Burlington Scientific Corp.; Denka International; E. I. DuPont de Nemours Inc.; Farnam Companies Inc.; Glades Formulating Corp.; Makhteshim Chemical Works Ltd.; Troy Biosciences Inc.; Wellmark International.*

2400   **Methomyl oxime**
13749-94-5   H

C3H7NOS
Methyl N-hydroxyacetimidothioate.
Acetohydroximic acid, thio-, methyl ester; EINECS 237-332-2; Ethanimidothioic acid, N-hydroxy-, methyl ester; Methyl N-hydroxyacetimidothioate;   Methyl   N-hydroxy-ethanimidothioate; Methyl N-hydroxythio-imidoacetate; Methyl thioacetohydroxamate; 1-(Methylthio)-acetaldehyde oxime; 1-(Methylthio)acetaldoxime.

2401   **Methoxsalen**
298-81-7   6018   D G

C12H8O4
8-Methoxy-2',3',6,7-furocoumarin.
5-19-06-00015 (Beilstein Handbook Reference); 6-Hydroxy-7-methoxy-5-benzofuranacrylic   acid   delta-lactone;   7H-Furo(3,2-G)(1)benzopyran-7-one, 9-methoxy-; Ammodin; Ammoidin; 5-Benzofuranacrylic acid, 6-hydroxy-7-methoxy-, δ-lactone; BRN 0196453; CCRIS 2083; DRG-0088; EINECS 206-066-9; Geroxalen; HSDB 2505; Meladinin; Meladinine; Meladoxen; Meloxine; Methoxa-Dome; Methoxasalen; Methoxsalen; Methox-salen plus ultraviolet radiation; Methoxsalen with ultraviolet A therapy; 8-Methoxy-(furano-3'.2':6.7-coumarin);   8-Methoxy-2',3',6,7-furocoumarin;   8-Methoxypsoralen with ultraviolet A therapy; 8-MOP; 8-MOP Capsules;   9-Methoxy-7H-furo(3,2-g)(1)benzopyran-7-one;   9-Methoxyfuro(3,2-g)chromen-7-one;   9-Methoxypsoralen;   NCI-C55903; New-meladinin; NSC 45923; Oxsoralen Lotion; Oxsoralen Ultra; Oxypsoralen; Psoralon-MOP; Puvalen; Uvadex; Xanthotoxin; Xanthotoxine;   6-hydroxy-7-methoxy-5-benzofuranacrylic   acid   δ-lactone; meladinin; methoxalen; oxypsoralen; proralone-mop; 9-methoxy-7H-furo(3,2-g)benzopyran-7-one;   xanthoxin;   7-furocoumarin;   zanthotoxin;   psoralen-mop;   8-Methoxyfuranocoumarin. Pigmentation agent. Prisms or needles; mp = 148°; λm = 219, 249, 300 nm (ε = 39811, 6026, 4898, EtOH), 238, 270, 312 nm (ε = 35481, 6918, 5495, cyclohexane); slightly soluble in H_2O, EtOH, very soluble in CHCl3; LD50 (rat ip) = 470 mg/kg. *Bayer Corp. Pharm. Div.; Hoffmann-LaRoche Inc.; ICN Pharm. Inc.*

2402   **Methoxychlor**
72-43-5   6020   G P

C16H15Cl3O2

1,1'-(2,2,2-Trichloroethylidene)bis(4-methoxybenzene).

4-06-00-06691 (Beilstein Handbook Reference); 4,4-(2,2,2-trichloroethylidene)dianisole; 4,4-(2,2,2-Trichloro-ethylidene)dianisole; Benzene, 1,1'-(2,2,2-trichloroethyl-idenebis(4-methoxy-; Bis(p-methoxyphenyl)-1,1,1-tri-chloroethane; 2,2-bis(p-methoxyphenyl)-1,1,1-trichloroethane; 2,2-Di-(p-methoxyphenyl)-1,1,1-trichloroethane; Bis(p-anisyl)-1,1,1-trichloroethane; 2,2-Bis(p-anisyl)-1,1,1-trichloro-ethane; 2,2-Di-p-anisyl-1,1,1-trichloro-ethane; BRN 2057367; Caswell No. 550; CCRIS 380; Chemform methoxychlor; Chemform methoxychlor; Dianisyl trichloroethane; Dimethoxy-DDT; Dimethoxydiphenyl-trichloroethane; di(p-methoxyphenyl) trichloromethyl methane;; p,p'-Dimethoxydiphenyltrichloroethane; p,p'-dmdt; DMDT; double-m ec; p,p'-Dwumetoksydwu-fenylotrojchloroetan; ENT 1,716; EPA Pesticide Chemical Code 034001; Ethane, 1,1,1-trichloro-2,2-bis(p-methoxy-phenyl)-; Ethane, 2,2-bis(p-anisyl)-1,1,1-trichloro-; flo pro; Higalmetox; HSDB 1173; Maralate; Marlate; Maxie; mcseed protectant; Methoxcide; methoxo; Methoxy-DDT; Methoxychlor; Methoxychlor 2 EC; Methoxychlore; Metoksychlor; Metox; Mezox K; Moxie; NCI-C00497; NSC 8945; OMS 466; RCRA waste number U247; 1,1,1-Trichlor-2,2-bis(4-methoxy-phenyl)-äthan; 1,1,1-trichloro-2,2-bis(p-methoxyphenyl)ethane; 1,1,1-Trichloro-2,2-bis(p-anisyl)ethane; 1,1,1-Trichloro-2,2-bis(4-methoxy-phenyl)ethane; 2,2,2-trichloro-1,1-bis(4-methoxyphenyl)-ethane; 2,2,2-Trichloro-1,1-bis(4-methoxyphenyl)ethane. Registered by EPA as an insecticide and acaricide. Insecticide effective against mosquito larvae and houseflies; recommended for use in dairy barns. Gray powder - colorless crystals; mp = 89°; sg^{25} = 1.41; poorly soluble in H2O (0.00001 g/100 ml), soluble in CHCl3 (65 g/100 ml), xylene (38 g/100 ml), MeOH (4 g/100 ml), soluble in aromatic, chlorinated, ketonic solvents, vegetable oils; LD50 (rat orl) = 6000 mg/kjg, (rbt der) > 2000 mg/kg, (mallard ducks orl) > 2000 mg/kg; LC50 (quail, pheasant, dietary 8 day) > 5000 mg/kg diet; (rainbow trout 24 hr.) = 0.052 mg/l, (bluegill sunfish 24 hr.) = 0.067 mg/l, *Daphnia* 48 hr.) = 0.00078 mg/l. *Bonide Products, Inc.; Kincaid Enterprises; Novartis; Prentiss Inc.; Riverdale Chemical Co.*

2403   Methoxydiglycol
111-77-3   6064   G H

C5H12O3

2-(2-Methoxyethoxy)ethanol.

4-01-00-02392 (Beilstein Handbook Reference); AI3-18364; BRN 1697812; Caswell No. 338B; DEGME; Diethylene glycol methyl ether; Diethyleneglycol monomethyl ether; Diglycol monomethyl ether; Dowanol 16; Dowanol DM; EGME, di-; EINECS 203-906-6; Ektasolve® DM; EPA Pesticide Chemical Code 042204; 2-(2-Methoxyethoxy)ethanol; Ethanol, 2-(2-methoxy-ethoxy)-; Ethylene diglycol monomethyl ether; HSDB 96; Jeffersol DM; MECB; Methoxydiglycol; Methyl digol; Methyl dioxitol; Methyl karbitol; NSC 2261; Poly-Solv® DM. Evaporating solvent used in brushing lacquers and dye stains; useful in wood stains, printing inks, and dye pastes for textiles; coalescing aid for PVAc latex paints; used in stamp pad and stencil inks; diluent for hydraulic brake fluids. Used in hard-surface cleaners, leather dyeing, paints, coatings, printing inks, textile vat dyeing and printing, adhesives, antifreeze, floor waxes/polishes, insect repellents; solubilizer for dyes; plasticizer. Liquid; mp < -84°; bp = 193°; d^{20} = 1.035; soluble in EtOH, Et2O,

freely soluble in H2O, Me2CO; LD50 (rat orl) = 9.21 g/kg. *Eastman Chem. Co.; Olin Corp.*

2404   Methoxyethanol
109-86-4   6066   G P

C3H8O2

2-Methoxyethanol.

4-01-00-02375 (Beilstein Handbook Reference); Äthylenglykol-monomethylaether; Aethylenglykol-mono-methylaether; AI3-18363; Amsco-Solv EE; BRN 1731074; Caswell No. 551; CCRIS 5826; Cellosolve; Dowanol 7; Dowanol EM; EGM; EGME; EINECS 203-713-7; Ektasolve; EPA Pesticide Chemical Code 042202; Ethanol, 2-methoxy-; Ether monomethylique de l'ethylene-glycol; Ethylene glycol methyl ether; Ethylene glycol monomethyl ether; Glycol ether EM; Glycol monomethyl ether; Glycomethyl ether; HSDB 97; 1-Hydroxy-2-methoxyethane; Jeffersol EM; MECS; 2ME; Methoxyethanol; Methoxyethanol, 2-; 2-Methoxy-1-ethanol; 2-Methoxy-äthanol; 2-Methoxy-aethanol; 2-Methoxyethyl alcohol; 2-Metossietanolo; Methyl Cellosolve®; Methyl ethoxol; Methyl Glycol; Methoxyhydroxyethane; Methyl oxitol; Methylcelosolv; Methylglykol; Metil cellosolve; Metoksyetylowy alkohol; Monoethylene glycol methyl ether; Monomethyl ether of ethylene glycol; Monomethyl ethylene glycol ether; Monomethyl glycol; NSC 1258; 3-Oxa-1-butanol; Poly-Solv EM; Prist; UN1188. A colorless and nearly odorless liquid boiling at 124.5°C; has the lowest boiling-point and greatest rate of evaporation of all available glycol ethers; it is a solvent for cellulose acetate. Used as a solvent for brake fluids, hard-surface cleaners, leather dyeing, paints, coatings, printing inks, textile vat dyeing and printing, adhesives antifreeze, floor waxes/polishes, insect repellents; solubilizer for dyes; plasticizer. Registered by EPA as an antimicrobial and fungicide (cancelled). Liquid; mp = -85.1°; bp = 124.1°, bp20 - 34 - 41°; d^{20} = 0.9647; freely soluble in H2O, Et2O, C6H6, very soluble in EtOH, soluble in Me2CO, slightly soluble in CHCl3; LD50 (rat orl) = 2460 mg/kg, (gpg orl) = 950 mg/kg; MLC (mus 7 hr. in air) = 4.6 mg/l. *Arco; Ashland; Fluka; Olin Corp; OxyChem; Union Carbide Corp.*

2405   Methoxyethyl acrylate
3121-61-7   G

C6H10O3

Methoxyethyl acrylate.

4-02-00-01469 (Beilstein Handbook Reference); Acrylic acid, 2-methoxyethyl ester; Acrylic acid, 2-methoxyethoxy ester; Ageflex MEA; AI3-15726; BRN 1754333; EINECS 221-499-3; Ethanol, 2-methoxy-, acrylate; Ethylene glycol monomethyl ether acrylate; Glycol monomethyl ether acrylate; 2-Methoxyethanol, acrylate; 2-Methoxyethoxy acrylate; 2-Methoxyethyl acrylate; Methyl cellosolve acrylate; NSC 24153; 2-Propenoic acid, 2-methoxyethyl ester; Sipomer MCA. Solvent-resistant elastomer, polyacrylate rubber, UV-curable reactive diluent, soft contact lenses, PVC impact modifier, fabric coatings, barrier coatings for polyethylene, textile coatings. Liquid; bp17 = 61°; d^{20} = 1.012; soluble in H2O, organic solvents; moderately toxic by ingestion and inhalation; LD50 (rat orl) = 810 mg/kg. *Rit-Chem.*

2406   Methoxyisopropanol
107-98-2   H

$C_4H_{10}O_2$
1-Methoxy-2-propanol.
 3-01-00-02146 (Beilstein Handbook Reference); Arcosolv® PM; AI3-15573; BRN 1731270; Closol; Dowanol® 33B; Dowtherm 209; EINECS 203-539-1; Icinol PM; HSDB 1016; Methoxyisopropanol; 1-Methoxy-2-hydroxypropane; 1-Methoxy-2-propanol; 2-Methoxy-1-methylethanol; Methyl proxitol; NSC 2409; PGME; Poly-Solv® MPM; 2-Propanol, 1-methoxy-; Propasol solvent M; Propylene glycol methyl ether; Propylene glycol monomethyl ether; Solvenon® PM; Solvent PM; Ucar Solvent LM; UN3092. Solvent for coatings, cleaners, inks, agricurtural products, cosmetics, chemical intermediate applications. Coupling agent providing improved surface wetting, soil penetration in household, commercial and industrial cleaning products. Solvent used in brake fluids, hard-surface cleaners, leather dyeing, paints, coatings, printing inks, textile vat dyeing and printing, adhesives, antifreeze, floor waxes/polishes, insect repellents; solubilizer for dyes; plasticizer. Liquid; mp = -97°; bp = 119°; d^{20} = 0.9620; soluble in H_2O, organic solvents; LD50 (rat orl) = 5660 mg/kg. *Arco; BASF Corp.; ICI Spec.; Olin Corp.*

2407 Methoxymethanol
4461-52-3   H

$C_2H_6O_2$
 Methoxymethanol.
 EINECS 224-722-2; Methanol, methoxy-; Methoxy-methanol.

2408 p-Methoxyphenylacetic acid
104-01-8   G

$C_9H_{10}O_3$
 4-Methoxyphenylacetic acid.
 4-10-00-00544 (Beilstein Handbook Reference); 2-(p-Anisyl)acetic acid; Acetic acid, p-methoxyphenyl-; Benzeneacetic acid, 4-methoxy-; BRN 1101737; EINECS 203-166-4; Homoanisic acid; 4-Methoxybenzeneacetic acid; 4-Methoxyphenylacetic acid; p-Methoxy-phenylacetic acid; NSC 27799. Used in manufacture of pharmaceuticals and other organic compounds. Plates; mp = 87°; bp2 = 138°; λ_m = 225, 277, 283 nm (ε = 9640, 1700, 1460, MeOH); slightly soluble in H_2O (0.6 g/100 ml), CHCl3, ligroin, soluble in Et2O, C6H6, very soluble in EtOH; LD50 (rat orl) = 1550 mg/kg. *Lancaster Synthesis Co.; Penta Mfg.*

2409 Methoxy-1-propanol
28677-93-2   H

$C_4H_{10}O_2$
 Methoxy-1-propanol.
 EINECS 249-146-9; Methoxy-1-propanol.

2410 3-Methoxypropionitrile
110-67-8   H

C_4H_7NO
 1-Cyano-2-methoxyethane.
 4-03-00-00708 (Beilstein Handbook Reference); AI3-25449; BRN 1739284; 1-Cyano-2-methoxyethane; 2-Cyanoethyl methyl ether; EINECS 203-790-7; β-Methoxypropionitrile; β-Methoxypropionitrile; 3-Methoxy-propanenitrile; 3-Methoxypropannitril; 3-Methoxy-propionitrile; 3-Methoxypropylnitrile; Methyl β-cyano-ethyl ether; NSC 4090; Propanenitrile, 3-methoxy-. Liquid; bp = 163°; d^{20} = 0.9379; soluble in EtOH, Et2O, CHCl3.

2411 Methyl acetate
79-20-9   6038   G H

$C_3H_6O_2$
 Acetic acid, methyl ester.
 Acetate de methyle; Acetic acid, methyl ester; CCRIS 5846; Devoton; EINECS 201-185-2; Ethyl ester of monoacetic acid; FEMA Number 2676; HSDB 95; Methyl acetate; Methyl acetic ester; Methyl ethanoate; Methylacetaat; Methylacetat; Methyle (acetate de); Methylester kiseliny octove; Metile (acetato di); NSC 405071; Octan metylu; Tereton; UN1231. Used in paint remover compounds, as a lacquer solvent, intermediate, synthetic flavoring; solvent for nitrocellulose, acetylcellulose; manufacture of artificial leather. Liquid; mp = -98°; bp = 56.8°; d^{20} = 0.9342; insoluble in H_2O, soluble in organic solvents; LD50 (rbt orl) = 3705 mg/kg. *Akzo Chemie; Hoechst Celanese; Penta Mfg.*

2412 Methyl acetoacetate
105-45-3   6039   H

$C_5H_8O_3$
 Methyl 3-oxobutyrate.
 Acetoacetic acid, methyl ester; Acetoacetic methyl ester; AI3-06000; Butanoic acid, 3-oxo-, methyl ester; CCRIS 2302; EINECS 203-299-8; HSDB 1083; Methyl 3-oxobutanoate; Methyl 3-oxobutyrate; Methyl aceto-acetate; Methyl acetylacetate; Methyl acetylacetonate; Methylacetoacetate; Methylester kyseliny acetoctove. Solid; mp = 27.5°; bp = 171.7°; d^{20} = 1.0762; λ_m = 240 nm (ε = 1514, EtOH); soluble in CCl4, very soluble in H_2O, freely soluble in EtOH, Et2O.

2413 Methyl acetyl ricinoleate
140-03-4   G

$C_{21}H_{38}O_4$
 Methyl 12-acetoxyoleate.
 0-03-00-00387 (Beilstein Handbook Reference); 12-(Acetyloxy)-9-octadecenoic acid, methyl ester; BRN 1729461; CCRIS 7333;

EINECS 205-392-9; Flexricin® P-4; Flexricin® P-4; Methyl 12-acetoxy-9-octadecenoate; Methyl 12-acetoxyoleate; Methyl acetyl ricinoleate; Methyl O-acetylricinoleate; Methyl ricinoleate, acetate; Methylester kyseliny acetylricinolejove; NSC 2398; 9-Octadecenoic acid, 12-(acetyloxy)-, methyl ester; 9-Octadecenoic acid, 12-(acetyloxy)-, methyl ester, (9Z,12R)-; 9-Octadecenoic acid, 12-(acetyloxy)-, methyl ester, (R-(Z)- (9CI); Ricinoleic acid, methyl ester, acetate; Ricinoleic acid, methyl ester, acetate; Methyl 12-acetoxy-9-octadecenoate. All purpose plasticizer, lubricant for vinyls and lacquers. Liquid; mp = -15°; d = 0.938; soluble in most organic solvents; insoluble in water; LD_{50} (mus orl) = 34,900 mg/kg. CasChem.

2414 Methyl acrylate
96-33-3   6041   G H

$C_4H_6O_2$
2-Propenoic acid, methyl ester.
4-02-00-01457 (Beilstein Handbook Reference); Acrylate de methyle; Acrylic acid, methyl ester; Acrylsaeure-methylester; Acrylsäuremethylester; AI3-15715; BRN 0605396; CCRIS 1839; Curithane 103; EINECS 202-500-6; HSDB 194; Methoxycarbonylethylene; Methyl 2-propenoate; Methyl-acrylat; Methyl acrylate; Methyl propenoate; Methylacrylaat; Methylacrylate; Methylester kyseliny akrylove; Metilacrilato; NSC 24146; UN1919. Used in manufacture of acrylic polymers, amphoteric surfactants, as a chemical intermediate, in leather finish resins, textile and paper coatings and plastic films. Liquid; mp < -75°; bp = 80.7°; d^{20} = 0.9535; slightly soluble in H2O (6 g/100 ml) , soluble in EtOH, Et2O, Me2CO, C6H6, CHCl3; LD_{50} (rat orl) = 277 mg/kg. BASF Corp.; Hoechst Celanese.

2415 Methyl alcohol
67-56-1   5984   G P

CH_4O
Monohydroxymethane.
AI3-00409; Alcohol, methyl; Alcool methylique; Alcool metilico; Bieleski's solution; Carbinol; Caswell No. 552; CCRIS 2301; Coat-B1400; Colonial spirits; Columbian Spirit; EINECS 200-659-6; EPA Pesticide Chemical Code 053801; Eureka Products, Criosine; Eureka Products Criosine Disinfectant; Freers Elm Arrester; HSDB 93; Ideal Concentrated Wood Preservative; Metanol; Metanolo; Methanol; Methyl alcohol; Methyl hydrate; Methyl hydroxide; Methylalkohol; Methylol; Metylowy alkohol; Monohydroxymethane; NSC 85232; Pyroxylic Spirit; RCRA waste number U154; Surflo-B17; UN1230; Wilbur-Ellis Smut-Guard; Wood alcohol; Wood naphtha; Wood Spirit; X-Cide 402 Industrial Bactericide. Registered by EPA as an antimicrobial, fungicide, herbicide and insecticide (cancelled). FDA approved for orals, USP/NF, BP compliance. Used in orals as a solvent and excipient. Industrial solvent; feedstock in organic synthesis. Additive in antifreeze, gasoline. Chemical intermediate and solvent. Used in the manufacture of formaldehyde, acetic acid, as a solvent, in chemical synthesis of methyl amines, methyl chloride and methyl methacrylate. Liquid; mp = -97.6°; bp = 64.6°; soluble in H2O, EtOH, Et2O, C6H6, ketones; d^{20} = 0.7914. LD_{50} (rat orl) = 5628 mg/kg; TLV = 200 ppm in air. Albright & Wilson Americas Inc.; Ashland; Brown; Coyne; DuPont; Eastman Chem. Co.; General Chem; Hoechst Celanese; Integra; Mallinckrodt Inc.; Mitsui Toatsu; Nissan Chem. Ind.; Norsk Hydro AS; Quantum/USI; Rit-Chem; Ruger; Sigma-Aldrich Fine Chem.; Spectrum Chem. Manufacturing;

Veckridge.

2416 Methylamine
74-89-5   6044   H

CH_5N
Methanamine.
AI3-15637; Aminomethane; Carbinamine; CCRIS 2508; EINECS 200-820-0; HSDB 810; Mercurialin; Methanamine; Methylamine; N-Methylamine; Methylaminen; Metilamine; Metyloamina; MMA; Monomethylamine; UN1061; UN1235. Used in tanning and in organic synthesis. Colorless gas; mp = -93.4°; bp = -6.3°; d^{25} = 0.656, soluble in H2O, organic solvents; LD_{50} (rat orl) = 100-200 mg/kg. Lancaster Synthesis Co.; Mallinckrodt Inc.; Sigma-Aldrich Fine Chem.

2417 p-Methylanisole
104-93-8   H

$C_8H_{10}O$
Methyl 4-methylphenyl ether.
AI3-07621; Anisole, p-methyl-; Benzene, 1-methoxy-4-methyl-; EINECS 203-253-7; FEMA Number 2681; HSDB 5363; Methyl 4-methylphenyl ether; Methyl p-cresol; Methyl p-cresyl ether; Methyl p-methylphenyl ether; Methyl p-tolyl ether; Methyl-para-cresol; NSC 6254; Toluene, 4-methoxy-. Oil; mp = -32°; bp = 175.5°; d^{25} = 0.969; λ_m = 278, 284 nm (MeOH); insoluble in H2O, soluble in EtOH, Et2O, CHCl3.

2418 Methyl anthranilate
134-20-3   6049   G P

$C_8H_9NO_2$
o-Aminobenzoic acid methyl ester.
4-14-00-01008 (Beilstein Handbook Reference); AI3-01022; Amino methyl benzoate, o-; Anthranilic acid, methyl ester; Antranilato de metilo; artificial Neroli oil; Benzoic acid, 2-amino-, methyl ester; BRN 0606965; Carbomethoxyaniline; CCRIS 1349; EINECS 205-132-4; EPA Pesticide Chemical Code 128725; FEMA No. 2682; HSDB 1008; Methyl 2-aminobenzoate; Methyl o-aminobenzoate; Methyl anthranilate; Methylester kyseliny anthranilove; Neroli oil, artifical; Nevoli oil; NSC 3109. Flavoring, fragrance, perfume, cosmetics, pomades; intermediate for pharmaceuticals and dyes. FDA, FEMA GRAS, Japan approved as a flavoring. Used as a grape and berry flavoring agent, as an orange scent in ointments and suntan oils. Registered by EPA as a bird repellent. Used as a bird repellent. Crystals; mp = 24.5°; bp = 259°; bp_{15} = 135.5°; d^{15} = 1.167 - 1.175; λ_m = 218, 247, 337 nm (ε = 27900, 7110, 4910, MeOH); soluble in fixed oils, propylene glycol, slightly soluble in H2O, insoluble in glycerol; LD_{50} (rat orl) = 2910 mg/kg, (mus orl) = 3900 mg/kg. BASF Corp.; Becker Underwood, Inc.; Bell F & F; Generic; Haarmann &

Reimer GmbH; PMC; Sigma-Aldrich Fine Chem.; Spectrum Chem. Manufacturing.

2419 **Methyl behenate**
929-77-1   G

C23H46O2
Methyl docosanoate.
AI3-36456; Behenic acid, methyl ester; Docosanoic acid, methyl ester; EINECS 213-207-8; HSDB 2724; Kemester® 9022; Methyl docosanoate; NSC 158426. Intermediate in production of superamides, in metalworking lubricants, as solv. Needles; mp = 54°; very soluble in EtOH, Et2O. CK Witco Corp.

2420 **Methylbenzenesulfonic acid, sodium salt**
12068-03-0   H

C7H7NaO3S
ar-Toluenesulfonic acid, sodium salt.
Benzenesulfonic acid, methyl-, sodium salt; Cyclophil STS 70; EINECS 235-088-1; Eltesol ST 34; Eltesol ST 90; Methylbenzenesulfonic acid, sodium salt; Pilot STS 32; Sodium methylbenzenesulfonate; Sodium toluene-sulfonate; Sodium toluenesulphonate; Toluenesulfonic acid sodium salt; ar-Toluenesulfonic acid, sodium salt. Mixture of o, m and p-isomers.

2421 **Methyl benzoate**
93-58-3   6052   G H

C8H8O2
Benzoic acid, methyl ester.
AI3-00525; Benzoic acid, methyl ester; CCRIS 5851; Clorius; EINECS 202-259-7; Essence of niobe; FEMA No. 2683; HSDB 5283; Methyl benzenecarboxylate; Methyl benzoate; Methylester kyseliny benzoove; Niobe oil; NSC 9394; Oil of niobe; UN2938. Used in perfumery, as a solvent for cellulose esters and ethers, resins, rubber; flavoring. Liquid; mp = -15°; bp = 199°; d^{15} = 1.0933; insoluble in H2O, soluble in organic solvents; λ_m = 228, 272, 280 nm (ε = 11000, 830, 686, MeOH); LD$_{50}$ (rat orl) = 1177 mg/kg. Lancaster Synthesis Co.; Morflex; Penta Mfg.; Pentagon Chems. Ltd.; Sybron.

2422 **2-Methylbenzophenone**
131-58-8   G

C14H12O
Phenyl tolyl ketone.
AI3-11216; Benzophenone, 2-methyl-; EINECS 205-032-0; Methanone, (2-methylphenyl)phenyl-; 2-Methylbenzophenone; o-Methylbenzophenone; NSC 67362; Phenyl o-tolyl ketone. Perfume additive (fixative). Liquid; mp <-18°; bp = 308°; bp$_{12}$ = 128°, bp$_{0.3}$ = 125-127°; d^{20} = 1.1098; λ_m = 252, 332 nm (ε = 15136, 151, EtOH); insoluble in H2O, soluble in organic solvents, very soluble in EtOH. Janssen Chimica; Spectrum Chem. Manufacturing.

2423 **Methyl benzoquate**
13997-19-8   6498   G V

C22H23NO4
6-Butyl-1,4-dihydro-4-oxo-7-(phenylmethoxy)-3-quinolinecarboxylic acid methyl ester.
AI3-52600; AY 20385; EINECS 237-796-6; ICI 55,052; Methyl 7-(benzyloxy)-6-butyl-1,4-dihydro-4-oxo-3-quino-linecarboxylate; Methyl benzoquate; Nequinate; Nequinato; Nequinatum; 3-Quinolinecarboxylic acid, 6-butyl-1,4-dihydro-4-oxo-7-(phenylmethoxy)-, methyl ester; Statyl. Coccidiostat used in poultry feedstuffs. Crystals; mp = 287-288°. ICI Agrochemicals; Purina Mills Inc.

2424 **5-Methylbenzotriazole**
136-85-6   G

C7H7N3
5-methyl-1H-benzo-1,2,3-triazole.
4-26-00-00144 (Beilstein Handbook Reference); 1H-Benzotriazole, 5-methyl-; BRN 0116658; CCRIS 6780; EINECS 205-265-8; 5-Methyl-1,2,3-benzotriazole; 6-Methyl-1,2,3-benzotriazole; 5-Methyl-1H-benzotriazole; 6-Methylbenzotriazole; NSC 122012; Retrocure® G; Tolutriazole; Vulkalent® TM. Blend of 4- and 5-methylbenzotriazole; prevulcanization retarder for sulfur-modified CR and halobutyl rubbers; also effective in NBR when Vulkacit DM/sulfur curing systems are used; applications including conveyor belting, hose, and molded goods. Solid; mp = 80-82°; bp$_{12}$ = 210-212°. Akrochem Chem. Co.; Bayer AG; Polysar.

2425 **Methyl-1H-benzotriazole, sodium salt**
64665-57-2   H

C7H6N3Na
1H-Benzotriazole, 4(or 5)-methyl-, sodium salt.
1H-Benzotriazole, 4(or 5)-methyl-, sodium salt; EINECS 265-004-9;

Methyl-1H-benzotriazole, sodium salt; Sodium 4(or 5)-methyl-1H-benzotriazolide; Tolyltriazole, sodium salt.

2426   Methyl benzoylformate
15206-55-0   G

C9H8O3
Methyl phenylglyoxalate.
AI3-07037; Benzeneacetic acid, α-oxo-, methyl ester; EINECS 239-263-3; Glyoxylic acid, phenyl-, methyl ester; Methyl oxophenylacetate; Methyl phenylglyoxylate; NSC 171206; Phenylglyoxylic acid, methyl ester; Vicure® 55. High purity photoinitiator for UV curable systems, especially acrylate-based formulations. Liquid; bp = 246-248°; d = 1.1550. *Akzo Chemie.*

2427   Methyl bromide
74-83-9   6056   H

CH3Br
Bromomethane.
AI3-01916; Bercema; Brom-methan; Brom-O-Gas; Brom-O-Gas Methyl Bromide Soil Fumigant; Brom-O-Gaz; Brom-O-Sol; Bromometano; Bromomethane; Bromur di metile; Bromure de methyle; Bromuro di metile; Broommethaan; Caswell No. 555; CCRIS 385; Celfume; Curafume; Dawson 100; Detia gas EX-M; Dowfume MC-2; Dowfume MC-2 soil fumigant; Dowfume MC-2 Fumigant; Dowfume MC-2; Dowfume MC-2R; Dowfume MC-33; Drexel Plant Bed Gas; Edco; EINECS 200-813-2; Embafume; EPA Pesticide Chemical Code 053201; F 40B1; Fumigant-1 (Obs.); Halon 1001; Haltox; HSDB 779; Iscobrome; Kayafume; M-B-C Fumigant; M-B-R 98; MB; MBC-33 Soil Fumigant; MBC Soil Fumigant; MBX; MC-33; MeBr; Metafume; Meth-O-Gas; Methane, bromo-; Methogas; Methyl Bromide Rodent Fumigant (with chloropicrin); Methyl fume; Methylbromid; Metylu bromek; Monobromomethane; Pestmaster; Pestmaster Soil Fumigant-1; Profume; R 40B1; RCRA waste number U029; Rotox; Superior Methyl Bromide-2; Terabol; Terr-O-Cide II; Terr-O-Gas; Terr-O-Gas 67; Terr-O-Gas 100; Tri-Brom; UN 1062; Zytox. Registered by EPA as an insecticide and bactericide. Fumigant used as an insecticide, acaricide and rodenticide in grain storage facilities. Colorless gas; mp = -93°; bp = 4.5°; soluble in H2O 1.34 g/100 ml), readily soluble in lower alcohols, ethers, esters, ketones, aromatic and halogenated hydrocarbons, CS2; LC100 (rat inhl 6 hr.) = 0.63 mg/l air, not dangerous to bees. *AmeriBrom Inc.; Great Lakes Chemical Corp; La-Co Industries; Soil Chemicals Corp.*

2428   2-Methylbutanol
137-32-6   6057   H

C5H12O
2-Methyl-1-butanol.
4-01-00-01666 (Beilstein Handbook Reference); Active amyl alcohol; Active primary amyl alcohol; AI3-24190; BRN 1718810; sec-Butylcarbinol; dl-sec-Butyl carbinol; EINECS 205-289-9; EINECS 252-163-4; HSDB 5626; NSC 8431; Primary active amyl alcohol. Liquid; bp = 128°; d^{25} = 0.8150; slightly soluble in H2O, very soluble

in Me2CO, freely soluble in EtOH, Et2O.

2429   2-Methyl-2-butenenitrile
30574-97-1   H

C5H7N
2-Butenenitrile, 2-methyl-, (2E)-.
2-Butenenitrile, 2-methyl-, (2E)-; 2-Butenenitrile, 2-methyl-, (E)-; EINECS 250-247-5; (E)-2-Methyl-2-butenenitrile.

2430   2-Methyl-3-butenenitrile
16529-56-9   H

C5H7N
3-Cyanobut-1-ene.
AI3-30534; 3-Butenenitrile, 2-methyl-; CCRIS 6056; 3-Cyanobut-1-ene; EINECS 240-596-1; 2-Methyl-3-butenenitrile.

2431   Methyl t-butyl ether
1634-04-4   6059   H

C5H12O
2-Methoxy-2-methylpropane.
4-01-00-01615 (Beilstein Handbook Reference); BRN 1730942; CCRIS 7596; EINECS 216-653-1; Ether, tert-butyl methyl; HSDB 5847; MTBE; Propane, 2-methoxy-2-methyl-; t-Butyl methyl ether; tert-Butyl methyl ether; UN2398. Liquid; mp = -108.6°; bp = 55.2°; d^{20} = 0.7405; soluble in H2O, very soluble in EtOH, Et2O.

2432   Methyl caprylate
111-11-5   H

C9H18O2
Methyl n-octanoate.
4-02-00-00986 (Beilstein Handbook Reference); AI3-01979; BRN 1752270; Caprylic acid methyl ester; EINECS 203-835-0; FEMA No. 2728; HSDB 5544; Methyl caprylate; Methyl octanoate; Methyl n-octanoate; Methyl octylate; NSC 3710; Octanoic acid, methyl ester; Uniphat A20. Oil; mp = -40°; bp = 192.9°; d^{20} = 0.8775; insoluble in H2O, slightly soluble in CCl4, very soluble in EtOH, Et2O.

2433   Methyl m-chlorobenzoate
2905-65-9   H

C8H7ClO2
Benzoic acid, 3-chloro-, methyl ester.
Benzoic acid, 3-chloro-, methyl ester; Benzoic acid, m-chloro-, methyl ester; m-Chlorobenzoic acid methyl ester; EINECS 220-814-1; HSDB 5465; Methyl 3-chlorobenzoate.

2434   Methyl chloroform
71-55-6   9710   G H

C2H3Cl3
1,1,1-Trichloroethane.
4-01-00-00138 (Beilstein Handbook Reference); Aerothene TT; AI3-02061; Algylen; Baltana; BRN 1731614; Caswell No. 875; CCRIS 1290; CF 2; Chloroethene (inhibited); Chloroform, methyl-; Chlorotene; Chlorothene; Chlorothene NU; Chlorothene SM; Chlorothene VG; Chlorten; Chlorylen; Cleanite; Dowclene LS; EINECS 200-756-3; EPA Pesticide Chemical Code 081201; Ethana; Ethana NU; Ethane, 1,1,1-trichloro-; F 140a; Gemalgene; Genklene; Genklene LB; HCC 140a; HCC 140a; HSDB 157; ICI-CF 2; Inhibisol; Methyl chloroform; Methylchloroform; Methyltrichloromethane; NCI-C04626; NSC 9367; NUEthane; RCRA waste number U226; Solvent 111; α-T; Tafclean; Tcea; Three One A; Three One S; Trichloran; Trichloro-1,1,1-ethane; Trichloroethane; α-Trichloro-ethane; Trichloroethane; Trichloroethane, 1,1,1-; Tri-chloromethylmethane; Trielene; UN 2831. Registered by EPA as an insecticide (cancelled). Used as a solvent for cleaning precision instruments, metal degreasing, pesticide, textile processing. Liquid; mp = -30.4°; bp = 74.0°; d20 = 1.3390; insoluble in H2O, soluble in Me2CO, C6H6, CCl4, MeOH, Et2O. Sigma-Aldrich Co.

2435   Methyl chloroformate
79-22-1   6071

C2H3ClO2
Chloroformic acid methyl ester.
4-03-00-00023 (Beilstein Handbook Reference); BRN 0605437; Carbonochloridic acid, methyl ester; Chlorameisensaeure methylester; Chlorameisensäure methylester; Chlorocarbonate de methyle; Chlorocarbonic acid methyl ester; Chloroformiate de methyle; Chloroformic acid methyl ester; EINECS 201-187-3; Formic acid, chloro-, methyl ester; HSDB 1116; K-Stoff; MCF; Methoxycarbonyl chloride; Methyl chloro-carbonate; Methyl chloroformate; Methylchloorformiaat; Methylester kyseliny chlormravenci; Methylester kyseliny chloruhlicite; Metilcloroformiato; RCRA waste number U156; TL 438; UN1238. Extremely lachrymatory. Active chemical intermediate, used in pharmaceutical, agricultural and pesticide manufacturing. Liquid; bp = 70.5°; d20 = 1.2231; very soluble in EtOH, Et2O, less soluble in C6H6, CHCl3. Esprit Chem. Co.

2436   Methyl cocoate
61788-59-8   H

Coconut oil acid, methyl esters.
Coconut fatty acid methyl ester; Coconut oil acid, methyl esters; EINECS 262-988-1; Fatty acids, coco, Me esters; Methyl cocoate.

2437   Methyl cyanoacetate
105-34-0   G

C4H5NO2
Methyl 2-cyanoacetate.
4-02-00-01889 (Beilstein Handbook Reference); Acetic acid, cyano-, methyl ester; AI3-05599; BRN 0773945; Cyanoacetic acid methyl ester; EINECS 203-288-8; Methyl cyanoacetate; Methyl cyanoethanoate; Methylester kyseliny kyanoctove; NSC 3113; USAF KF-22. Used in organic synthesis and manufacture of pharmaceuticals and dyes. Liquid; mp = -22.5°; bp = 200.5°; d25 = 1.1225; insoluble in H2O, very soluble in EtOH, Et2O. Degussa-Hüls Corp.; Greeff R.W. & Co.; Lonzagroup.

2438   Methyl cyanocarbamate
21729-98-6   H

C3H4N2O2
Carbamic acid, cyano-, methyl ester.
Carbamic acid, cyano-, methyl ester; EINECS 244-548-0; Methyl cyanocarbamate.

2439   Methyl cyclohexane
108-87-2   6074   G

C7H14
Hexahydrotoluene.
AI3-18132; Cyclohexane, methyl-; Cyclohexylmethane; EINECS 203-624-3; Hexahydrotoluene; HSDB 98; Methylcyclohexane; Metylocykloheksan; NSC 9391; Sextone B; Toluene hexahydride; Toluene, hexahydro-; UN2296. Solvent for cellulose ethers, organic synthesis. Liquid; mp = -126.6°; bp = 100.9°; d20 = 0.7694; insoluble in H2O, soluble EtOH, Et2O, freely soluble in Me2CO, C6H6, CCl4, ligroin. LD50 (mus orl) = 2250 mg/kg. Janssen Chimica; Penta Mfg.; Phillips 66.

2440   Methylcyclopentadiene dimer
26472-00-4   H

C12H16
3a,4,7,7a-Tetrahydrodimethyl-4,7-methano-1H-indene.
Bis(methylcyclopentadiene); Dimethyldicyclopentadiene; EINECS 247-724-5; 4,7-Methano-1H-indene, 3a,4,7,7a-tetrahydrodimethyl-; 4,7-Methanoindene, 3a,4,7,7a-tetrahydrodimethyl-; Methylcyclopentadiene dimer; 3a,4,7,7a-Tetrahydrodimethyl-4,7-methanoindene.

2441   Methylcyclopentadienyl manganese tricarbonyl
12108-13-3   6244   H

C9H7MnO3
2-Methylcyclopentadienylmanganese tricarbonyl.

AI3-61450; AK-33X; Antiknock-33; CI-2; Combustion improver -2; EINECS 235-166-5; HSDB 2014; Manganese, (methylcyclopentadienyl)tricarbonyl-; Man-ganese, tricarbonyl methylcyclopentadienyl; Methyl-cyclopentadienyl manganese tricarbonyl; Methyl-cyklopentadientrikarbonylmanganium; Methylcym-antrene; MMT; NSC 22316; Tricarbonyl(methylcyclopentadienyl)manganese; Tricarbonyl(2-methylcyclopentadienyl)manganese. Used as a fuel additive.

2442   Methyl decanoate
110-42-9   G H

C11H22O2
Methyl n-decanoate.
4-02-00-01044 (Beilstein Handbook Reference); AI3-26168; BRN 1759170; Capric acid methyl ester; CCRIS 673; n-Decanoic acid, methyl ester; EINECS 203-766-6; HSDB 5399; Metholene 2095; Methyl caprate; Methyl n-caprate; Methyl caprinate; Methyl decanoate; Methyl n-decanoate; NSC 3713; Uniphat A30. Intermediate in manufacture of detergents, emulsifiers, wetting agents, stabilizer, resins, lubricants, plasticizer. Liquid; mp = -18°; bp = 224°; d^{20} = 0.8730; insoluble in H2O, slightly soluble in CCl4; very soluble in EtOH, Et2O, freely soluble in CHCl3. *Penta Mfg.; Procter & Gamble Co.*

2443   Methyldichlorosilane
75-54-7   H

CH4Cl2Si
Dichloromethylsilane.
4-04-00-04096 (Beilstein Handbook Reference); BRN 1071194; CCRIS 849; Dichloro(methyl)silane; Dichlorohydridomethylsilicon; Dichloromethylsilane; EINECS 200-877-1; HSDB 1167; Methyl-dichlorsilan; Methyldichlorosilane; Monomethyldichlorosilane; Silane, dichloromethyl-; UN1242. Used as a silylating agent. Liquid; mp = -93°; bp = 41°; d^{25} = 1.105. *Lancaster Synthesis Co.; Sigma-Aldrich Fine Chem.; United Chemical Corp.*

2444   Methyl   3-(2,2-dichlorovinyl)-2,2-dimethyl-cyclopropanecarboxylate
61898-95-1   H

C9H12Cl2O2
Cyclopropanecarboxylic acid, 3-(2,2-dichloroethenyl)-2,2-dimethyl-, methyl ester.
Cyclopropanecarboxylic acid, 3-(2,2-dichloroethenyl)-2,2-dimethyl-, methyl ester; EINECS 263-308-6; Methyl 3-(2,2-dichlorovinyl)-2,2-dimethylcyclopropanecarboxyl-ate.

2445   Methyl   3-(3,5-di-t-butyl-4-hydroxyphenyl)-propionate
6386-38-5   H

C18H28O3
Hydrocinnamic acid, 3,5-di-t-butyl-4-hydroxy-, methyl ester.
Benzenepropanoic acid, 3,5-bis(1,1-dimethylethyl)-4-hydroxy-, methyl ester; EINECS 228-985-4; Hydro-cinnamic acid, 3,5-di-t-butyl-4-hydroxy-, methyl ester; Methyl 3-(3,5-di-t-butyl-4-hydroxyphenyl)propionate.

2446   Methyl 3,3-dimethylpent-4-enoate
63721-05-1   H

C8H14O2
3,3-Dimethyl-4-pentenoic acid, methyl ester.
3,3-Dimethyl-4-pentenoic acid, methyl ester; EINECS 264-431-8; Methyl 3,3-dimethyl-4-pentenoate; Methyl 3,3-dimethylpent-4-enoate; 4-Pentenoic acid, 3,3-dimethyl-, methyl ester; Penten-4-oic acid, 3,3-dimethyl, methyl ester.

2447   N-Methyldioctadecylamine
4088-22-6   H

C37H77N
1-Octadecanamine, N-methyl-N-octadecyl-.
Dioctadecylamine, N-methyl-; EINECS 223-819-7; N-Methyldioctadecylamine; 1-Octadecanamine, N-methyl-N-octadecyl-.

2448   Methyl disulfide
624-92-0   G H

C2H6S2
Methyldithiomethane.
AI3-25305; CCRIS 2939; Dimethyl disulfide; Dimethyl disulphide; Dimethyldisulfide; Disulfide, dimethyl; 2,3-Dithiabutane; EINECS 210-871-0; FEMA No. 3536; HSDB 6400; Methyl disulfide; Methyldisulfide; Methyl-dithiomethane; NSC 9370; Sulfa-Hitech® 0382; UN2381. Solvent used to dissolve sulfur in the production of sour gas wells, in sour-gas pipelines, and in refinery and chemical plant flowlines. Liquid; mp = -85°; bp = 109.8°; d^{20} = 1.0625; λ_m = 256 nm (ε = 347, MeOH); insoluble in H2O, freely soluble in EtOH, Et2O. *Elf Atochem N. Am.*

2449   Methyl eicosenate
1120-28-1   G

$C_{21}H_{42}O_2$
Eicosanoic acid, methyl ester.
AI3-36455; EINECS 214-304-8; Kemester® 2050; Methyl aracidate; Methyl icosanoate. Intermediate in production of alkanolamides, in metalworking lubricants, as specialized solvents; foam depressant and nutrient in fermentation. Leaflets; mp = 54.5°; bp_{10} = 215°; very soluble in C_6H_6, Et_2O, EtOH, $CHCl_3$. *CK Witco Corp.*

2450   Methylene bis(thiocyanate)
6317-18-6   G P

$C_3H_2N_2S_2$
Thiocyanic acid, methylene ester.
3-03-00-00288 (Beilstein Handbook Reference); Amerstat® 282; Antiblu 3737; BRN 1743370; Busan 110; Caswell No. 565; CCRIS 636; CP 17879; Cytox; Dithiocyanatomethane; EINECS 228-652-3; EPA Pesticide Chemical Code 068102; HSDB 4497; Methane, dithiocyanato-; Methylendirhodanid; Methylendithio-kyanat; Methylene bis(thiocyanate); Methylene dithio-cyanate; Methylene thiocyanate; Methylenedirhodanid; Methylenedirhodanide; N-948® Biocide; Nalco D-1994; Nalfloc N 206; NSC 40464; Proxel MB; Slimicide MC; Tolcide MBTV 709. Antimicrobial in industrial water systems; preservative in water-containing systems. Industrial biocide controlling algae, bacteria, yeast and fungi. Biocide and slimicide for use in water treatment, paper, antifoulant paint, leather, timber preservation. Registered by EPA as an antimicrobial and slimicide. Used as a slimicide. Crystals; mp = 104-106°; reacts with H_2O, soluble in organic solvents; LD_{50} (rat orl) = 161 mg/kg. *Akzo Chemie; Albright & Wilson Americas Inc.; Drew Industrial Div.*

2451   Methylene bisacrylamide
110-26-9   G

$C_7H_{10}N_2O_2$
N,N'-Methylenebis(2-propenamide).
4-02-00-01472 (Beilstein Handbook Reference); Acrylamide, N,N'-methylenebis-; AI3-08643; BRN 1706297; CCRIS 4672; EINECS 203-750-9; MBA; Methylenebisacrylamide; Methylenediacrylamide; N,N'-Methylidenebisacrylamide; N,N'-Methylenebis(2-propen-amide); N,N'-Methylenediacrylamide; NAPP; NSC 406836; 2-Propenamide, N,N'-methylenebis-. Cross-linking agent for preparation of polyacrylamides. N,N-Methylene-bis-acrylamide is an acrylimide compound cross-reacting with unidentified primary sensitizers in NAPP and Nyloprint UV-cured printing plates. Used as an organic intermediate and crosslinking agent. Crystals; mp > 300°; d = 1.235; soluble in H_2O (0.1 - 1.0 mg/ml), more soluble in organic solvents; LD_{50} (rat orl) = 390 mg/kg. *Am. Cyanamid; Bio-Rad Labs.; Fluka; Lancaster Synthesis Co.*

2452   Methylene chloride
75-09-2   6088   G H

CH_2Cl_2
Dichloromethane.
4-01-00-00035 (Beilstein Handbook Reference); Aerothene MM; AI3-01773; BRN 1730800; Caswell No. 568; CCRIS 392; Chlorure de methylene; DCM; Dichloromethane; EINECS 200-838-9; EPA Pesticide Chemical Code 042004; F 30; F 30 (chlorocarbon); Freon 30; HCC 30; HSDB 66; Khladon 30; Metaclen; Methane dichloride; Methane, dichloro-; Methylene bichloride; Methylene chloride; Methylene dichloride; Methylenum chloratum; Metylenu chlorek; Narkotil; NCI-C50102; NSC 406122; R 30; R30 (refrigerant); RCRA waste number U080; Salesthin; Solaesthin; Soleana VDA; Solmethine; UN 1593. Registered by EPA as an insecticide (cancelled). Used as a solvent, propellant, degreaser and cleaner, insecticide, vapor pressure depressant and carrier solvent. Colorless liquid; mp = -95.1°; bp = 40°; d^0_4 = 1.36174, d^{15}_4= 1.33479, d^{20}_4= 1.3266, d^{20}_4= 1.30777; LD_{50} (rat orl) = 1.6 ml/kg (2.11 mg/kg). *Ashland; Dow Chem. U.S.A.; Elf Atochem N. Am.; Farleyway Chem. Ltd.; ICI Spec.; Mallinckrodt Inc.; Mitsui Toatsu; OxyChem.*

2453   Methylene dibromide
74-95-3   6086   H

CH_2Br_2
Methane, dibromo-.
4-01-00-00078 (Beilstein Handbook Reference); AI3-52311; BRN 0969143; CCRIS 939; Dibromomethane; EINECS 200-824-2; HSDB 1334; Methane, dibromo-; Methylene bromide; Methylene dibromide; NSC 7293; RCRA waste number U068; UN2664. Used in the production of leaded gasoline, as a fumigant for stored products and as a nematocide. Also used in the manufacture of fire retardent chemicals. Dense, mobile liquid; mp = -52.5°; bp = 97°; d^{20} = 2.4969; soluble in organic solvents, slightly soluble in H_2O.

2454   Methylene glutaronitrile
1572-52-7   H

$C_6H_6N_2$
2-Methylenepentanedinitrile.
EINECS 216-391-8; Glutaronitrile, 2-methylene-; Methylene glutaronitrile; 2-Methyleneglutaronitrile; 2-Methylenepentanedinitrile.

2455   Methylene iodide
75-11-6   6091   G

CH_2I_2
Diiodomethane.

Diiodomethane; Dijodmethan; EINECS 200-841-5; Methane, diiodo-; Methylene diiodide; Methylene iodide; Methylenjodid; MI-Gee; NSC 35804. Used to separate mixtures of minerals, in determination of specific gravity and in the manufacture of x-ray contrast materials. Liquid; mp = 6.1°; bp = 182°; d^{20} = 3.3212; soluble in H_2O (14 g/l), more soluble in organic solvents. *Sigma-Aldrich Fine Chem.; U.S. BioChem.*

2456   Methylenebis(chloroaniline)
101-14-4   6084   H

C13H12Cl2N2
Benzenamine, 4,4'-methylenebis[2-chloro-.
1-13-00-00074 (Beilstein Handbook Reference); Aniline, 4,4'-methylenebis(2-chloro-; Benzenamine, 4,4'-methyl-enebis(2-chloro-; Bis amine; Bis-amine A; Bis(3-chloro-4-aminophenyl)methane; Bis(4-amino-3-chlorophenyl)-methane; Bisamine; Bisamine S; BRN 1882318; CCRIS 389; CL-Mda; Cuamine M; Cuamine MT; Curalin M; Curene 442; Cyanaset; Dacpm; Di-(4-amino-3-clorofenil)metano; Di(-4-amino-3-chlorophenyl)methane; Diamet Kh; EINECS 202-918-9; HSDB 2629; LD 813; MBOCA; Methylene 4,4'-bis(o-chloroaniline); Methyl-enebis(2-chloroaniline); Methylenebis(3-chloro-4-amino-benzene); Methylenebis(chloroaniline); Millionate M; MOCA (curing agent); NSC 52954; Quodorole; RCRA waste number U158. Solid; λ_m = 247, 298 nm (MeOH); soluble in CCl4.

2457   Methylenediphenyl diisocyanate
26447-40-5   H

C15H10N2O2
Isocyanic acid, methylenediphenylene ester.
Benzene, 1,1'-methylenebis(isocyanato-; Crude MDI; 4,4'-,2,4'-,2,2'-Diisocyanatodiphenylmethane; EINECS 247-714-0; Generic MDI; Isocyanic acid, methyl-enediphenylene ester; Methylenediphenyl diisocyanate; Non-isomeric-specific MDI; PMDI; Polymeric 4,4-methylenediphenyl diisocyanate; Polymeric MDI.

2458   Methyl esters of fatty acids (C6-C12)
67762-39-4   H

(C6-C12) Alkylcarboxylic acid methyl ester.
(C6-C12) Alkylcarboxylic acid methyl ester; Caswell No. 568C; EINECS 267-017-5; EPA Pesticide Chemical Code 079034; Fatty acids, C6-12, Me esters; Fatty acids, methyl esters; Methyl esters of fatty acids (C8-C12); Methyl octanoate and methyl decanoate; Octanoic acid, methyl ester mixed with methyl decanoate; Off-Shoot O.

2459   Methylethanolamine
109-83-1   6045   H

C3H9NO
2-(N-Methylamino)ethanol.
4-04-00-01422 (Beilstein Handbook Reference); Amietol M 11; BRN 1071196; Caswell No. 489A; CCRIS 4845; EINECS 203-710-0; EPA

Pesticide Chemical Code 489200; HSDB 1128; Methyl(β-hydroxyethyl)amine; Methyl(2-hydroxyethyl)amine; Methylethanolamine; N-Monomethylaminoethanol; NSC 62776; USAF DO-50. Liquid; bp = 158°; d^{20} = 0.937; freely soluble in H_2O, EtOH, Et2O.

2460   Methylethylacetic acid
116-53-0   H

C5H10O2
2-Methylbutyric acid.
4-02-00-00889 (Beilstein Handbook Reference); Active valeric acid; AI3-24202; BRN 1098537; Butanoic acid, 2-methyl-; Butyric acid, 2-methyl-; EINECS 204-145-2; EINECS 209-982-7; Ethylmethylacetic acid; FEMA No. 2695; (1)-2-Methylbutyric acid; α-Methylbutyric acid; 2-Methybutyric acid; 2-Methylbutanoic acid; Methylethyl-acetic acid; NSC 7304; Valeric acid, active.

2461   Methyl ethyl ketone
78-93-3   6097   G H

C4H8O
n-Butan-2-one.
Äthylmethylketon; Acetone, methyl-; Aethylmethylketon; AI3-07540; Butan-2-one; Butanone; Butanone 2; Caswell No. 569; CCRIS 2051; EINECS 201-159-0; EPA Pesticide Chemical Code 044103; Ethyl methyl cetone; Ethyl methyl ketone; Ethylmethylcetone; Ethylmethylketon; FEMA No. 2170; HSDB 99; Ketone, ethyl methyl; Meetco; MEK; Methyl acetone; Methyl ethyl ketone; Metiletilcetona; Metiletilchetone; Metyl ethyl ketone; Metyloetyloketon; Oxobutane; RCRA waste number U159; UN1193. Solvent and chemical intermediate. Solvent in nitrocellulose coatings and vinyl films, paint removers, cements, adhesives, organic synthesis; manufacture of smokeless powder, cleaning fluids, priming, catalyst carrier, acrylic coatings, in paint and lacquer thinners, natural and synthetic resins, gums and rubbers, printing inks, PVC cloth manufacture, cleaning agent for metal surfaces, adhesives and cements; refining and dewaxing of mineral and lubricating oils. Liquid; mp = -86.6°; bp = 79.5°; d^{20} = 0.8054; soluble in H_2O (27 g/100 ml), organic solvents; LD50 (rat orl) = 5.5 g/kg. *BP Chem.; Elf Atochem N. Am.; Exxon; Fluka; Hoechst Celanese; Mallinckrodt Inc.; Sasolchem; Shell UK; Texaco; Union Carbide Corp.*

2462   Methyl ethyl ketone peroxide
1338-23-4   G H

C8H18O6
MEK peroxide.
2-Butanone, peroxide; Butanox LPT; Butanox M 105; Butanox M 50; CCRIS 6216; Chaloxyd MEKP HA 1; EINECS 215-661-2; Esperfoam FR; Ethyl methyl ketone peroxide; FR 222; HI-Point 90; HI-Point 180; HI-Point PD-1; HSDB 4181; Kayamek A; Kayamek M; Ketonox; Lucidol DDM 9; Lucidol delta X; Lupersol DDA 30; Lupersol DDM; Lupersol DEL; Lupersol DNF; Lupersol DSW; MEK peroxide; MEKP; MEKPO; Mepox; Methy ethyl ketone peroxide; Methyl ethyl ketone hydroperoxide; NCI-C55447; Permek G; Permek N; Quickset Extra; Quickset super; RCRA waste number U160; Sprayset MEKP; Superox 46-710; Thermacure; Trigonox M 50. Initiator and catalyst for cure of unsaturated polyester resins. Liquid; bp = 110°; insoluble in H_2O (<0.5 g/100 ml). *Akzo Chemie; CK Witco*

405

Corp.; Elf Atochem N. Am.; Norac.

2463   Methyl ethyl sulfide
624-89-5   G

C_3H_8S
Methylthioethane.
AI3-18786; EINECS 210-868-4; Ethane, (methylthio)-; Ethyl methyl sulfide; Ethyl methyl sulphide; Methyl ethyl sulphide. Unpleasant stench oder. Used as a leak detecting additive to natural gas. Liquid; mp = -105.9°; bp = 66.7°; d^{20} = 0.8422; λ_m = 214 nm (ε = 977 isooctane); insoluble in H_2O, soluble in Et_2O, $CHCl_3$, freely soluble in EtOH.

2464   Methyl eugenol
93-15-2   G

$C_{11}H_{14}O_2$
4-allyl-1,2-dimethyoxybenzene.
4-06-00-06337 (Beilstein Handbook Reference); 1-Allyl-3,4-dimethoxybenzene; 4-Allyl-1,2-dimethoxybenzene; 4-Allylveratrole; AI3-21040; Benzene, 1,2-dimethoxy-4-(2-propenyl)-; Benzene, 4-allyl-1,2-dimethoxy-; BRN 1910871; Caswell No. 579AB; CCRIS 746; 1,2-Dimethoxy-4-allylbenzene; 1-(3,4-Dimethoxyphenyl)-2-propene; 1,2-Dimethoxy-4-(2-propenyl)benzene; 3,4-Dimethoxyallylbenzene; EINECS 202-223-0; ENT 21040; EPA Pesticide Chemical Code 203900; 1,3,4-Eugenol methyl ether; Eugenyl methyl ether; FEMA Number 2475; HSDB 4504; Methyl eugenol; Methyl eugenol ether; Methyl eugenyl ether; Methyleugenol; NSC 209528; O-Methyl eugenol; Veratrole methyl ether. Insect attractant, flavoring. Oil; mp = -2°; bp = 254.7°; d^{20} = 1.0396; λ_m = 230, 280 nm (MeOH); insoluble in H_2O, soluble in organic solvents; LD_{50} (rat orl) = 1179 mg/kg. *Firmenich; Lancaster Synthesis Co.; Penta Mfg.*

2465   Methyl formamide
123-39-7   6100   H

C_2H_5NO
N-Methyl formamide.
4-04-00-00170 (Beilstein Handbook Reference); AI3-26076; BRN 1098352; EINECS 204-624-6; EK 7011; Formamide, N-methyl-; Formic acid amide, N-methyl-; HSDB 100; Methylformamide; Monomethylformamide; N-Methyl formamide; NSC 3051; X 188. Liquid; mp = -3.8°; bp = 199.5°; d^{19} = 1.011; very soluble in H_2O, Me_2CO, EtOH.

2466   Methyl formate
107-31-3   6101   H

$C_2H_4O_2$
Formic acid, methyl ester.
AI3-00408; Caswell No. 570; CCRIS 6062; EINECS 203-481-7; EPA Pesticide Chemical Code 053701; Formiate de methyle; Formic acid, methyl ester; HSDB 232; Methyl formate; Methyl methanoate;

Methyle (formiate de); Methylester kyseliny mravenci; Methylformiaat; Methylformiat; Metil (formiato di); Mravencan methylnaty; UN1243. Liquid; mp = -99°; bp = 31.7°; d^{20} = 0.9742; λ_m = 207 nm (ε = 63, H_2O); freely soluble in EtOH, very soluble in H_2O, soluble in Et_2O, $CHCl_3$, MeOH.

2467   Methyl p-formylbenzoate
1571-08-0   H

$C_9H_8O_3$
Methyl benzaldehyde-4-carboxylate.
Benzoic acid, 4-formyl-, methyl ester; p-Carbomethoxybenzaldehyde; CCRIS 6063; EINECS 216-385-5; p-Formylbenzoic acid methyl ester; HSDB 5842; Methyl 4-formylbenzoate; Methyl p-formylbenzoate; Methyl benzaldehyde-4-carboxylate; Methyl terephth-aldehydate; NSC 28459; Terephthalaldehydic acid, methyl ester.

2468   Methyl D-glucopyranoside
3149-68-6   H

$C_7H_{14}O_6$
1-O-Methyl-D-glucopyranose.
EINECS 221-581-9; GEO-MEG 365; Glucopyranoside, methyl, D-; D-Glucopyranoside, methyl; Glucosidizer 100; Methyl D-glucopyranoside; Methyl glucoside; Sta-Meg 104; Sta-Meg 106; Sta-Meg 200.

2469   Methyl-α-D-glucopyranoside
97-30-3   6104   H

$C_7H_{14}O_6$
α-D-Glucopyranoside, methyl.
AI3-18790; EINECS 202-571-3; Glucopyranoside, methyl, α-D-; Methyl α-D-glucopyranoside; Methyl α-D-glucoside; NSC 102101. Needles; mp = 168°; $bp_{0.2}$ = 200°; d^{30} = 1.46; $[\alpha]_D^{20}$ = +158.9° (H=72O); Very soluble in H_2O.

2470   Methyl glutaronitrile
4553-62-2   H

$C_6H_8N_2$
2-Methyl-1,5-valerodinitrile.
4-02-00-01990 (Beilstein Handbook Reference); BRN 1741955; CCRIS 4673; Diacrylonitrile; 1,3-Dicyan-obutane; 2,4-Dicyanobutane; EINECS 224-923-5; Glutanonitrile, α-methyl-;

406

Glutaronitrile, 2-methyl-; HSDB 6502; Methyl glutaronitrile; 2-Methylglutaronitrile; α-Methylglutarsäuredinitril; 2-Methylpentanedinitrile; 2-Methyl-1,5-valerodinitrile; Pentanedinitrile, 2-methyl-; 1,5-Valerodinitrile, 2-methyl-.

2471   Methyl green
14855-76-6   6106   G

C27H35BrClN3
 4-[[4-(Dimethylamino)phenyl][4-(dimethylimino)-2,5-cyclohexadien-1-ylidene]methyl]-N-ethyl-N,N-dimethyl-benzeneaminium bromide chloride.
 Ammonium, (α-(p-(dimethylamino)phenyl)-α-(4-(di-meth-ylimino)-2,5-cyclohexadien-1-ylidene)-p-tolyl)-ethyldi-methyl-, bromide, chloride; Benzenaminium, 4-((4-(dimethylamino)phenyl)(4-(dimethyliminio)-2,5-cyclo-hexadien-1-ylidene)methyl)-N-ethyl-N,N-dimethyl-, bromide chloride; C.I. 42590; (α-(p-(Dimethylamino)-phenyl)-alpha-(4-(methylimino)-2,5-cyclohexadien-1-ylid-ene)-p-tolyl)ethyldimethylammonium bromide metho-chloride; 4-((4-(Dimethylamino)phenyl)(4-(dimethyl-imino)cyclohexa-2,5-dien-1-ylidene)methyl)-N-ethyl-N,N-dimethylanilinium bromide chloride; EINECS 238-920-1; Ethyl green; Iodin green (Griesbach); NSC 3091. Green powder, soluble in H2O. Used as a biological stain and in dyeing and printing of textiles. Green powder; soluble in H2O.

2472   Methylheptenone
409-02-9   H

C8H14O
 2-Methylhept-2-en-6-one.
 EINECS 206-990-2; Heptenone, methyl-; HSDB 5565; Methylheptenone.

2473   Methyl hydrogenated rosinate
8050-15-5   H

 Methyl hydrogenated rosinate.
 EINECS 232-476-2; Hercolyn D; Hercolyn D 55W; Hercolyn D-E; Hercolyn H; Hydrogenated resin acid Me esters; Hydrogral M; Methyl ester of hydrogenated rosin; Methyl ester of rosin, partially hydrogenated; Methyl hydrogenated rosinate; Resin acids and Rosin acids, hydrogenated, Me esters; Rosin, hydrogenated, methyl ester. Hercules Inc.

2474   Methyl hydroxystearate
141-23-1   H

C19H38O3
 Methyl 12-hydroxyoctadecanoate.

AI3-19731; Cenwax ME; EINECS 205-471-8; HSDB 5635; Kemester 1288; Methyl 12-hydroxyoctadecanoate; Methyl-12-hydroxystearate; Methyl hydroxystearate; NSC 2392.

2475   Methyl iodide
74-88-4   6110   G

CH3I
 Iodomethane.
 CCRIS 395; EINECS 200-819-5; Halon 10001; HSDB 1336; Iodometano; Iodomethane; Iodure de methyle; Jod-methan; Joodmethaan; Methane, iodo-; Methyl iodide; Methyl iodide; Methyliodide; Methyljodid; Methyljodide; Metylu jodek; Monoiodomethane; Monoioduro di metile; NSC 9366; RCRA waste number U138; UN2644. Used in organic synthesis, microscopy, testing for pyridine. Colorless liquid; mp = -66.4°; bp = 42.5°; d^{20} = 2.279; insoluble in H2O ,soluble in organic solvents; LC50 (rat orl) = 1300 mg/l/4H. Akzo Chemie; Andeno; Burlington Scientific Corp.; Fairmount; Greeff R.W. & Co.

2476   2-Methyl imidazole
693-98-1   G

C4H6N2
 2-Methyl-1H-imidazole.
 2MZ; AI3-50033; CCRIS 2459; EINECS 211-765-7; Imidazole, 2-methyl-; 1H-Imidazole, 2-methyl-; 2-Methylimidazole; NSC 21394. Dyeing auxiliary for acrylic fibers, plastic foams. Crystals; mp = 144°; bp = 267°; very soluble in H2O, EtOH; LD50 (mus orl) = 1400 mg/kg. Allchem Ind.; BASF Corp.; Janssen Chimica.

2477   Methyl isoamyl ketone
110-12-3   H

C7H14O
 5-Methyl-2-hexanone.
 4-01-00-03329 (Beilstein Handbook Reference); BRN 0506163; EINECS 203-737-8; 2-Hexanone, 5-methyl-; HSDB 2885; Isoamyl methyl ketone; Isopentyl methyl ketone; Ketone, methyl isoamyl; 2-Methyl-5-hexanone; 5-Methyl-2-hexanone; 5-Methylhexan-2-one; Methyl isoamyl ketone; Methyl isopentyl ketone; MIAK; UN2302. Liquid; bp = 144°; d^{20} = 0.888; λm = 176, 188, 192, 196 nm (gas); slightly soluble in H2O, soluble in CCl4, very soluble in Me2CO, C6H6, freely soluble in EtOH, Et2O.

2478   Methyl isobutyl ketone
108-10-1   5227   H P

C6H12O

Methyl-2-pentanone.

4-01-00-03305 (Beilstein Handbook Reference); AI3-01229; BRN 0605399; Caswell No. 574AA; CCRIS 2052; EINECS 203-550-1; EPA Pesticide Chemical Code 044105; FEMA Number 2731; Hexanone; Hexon; Hexone; HSDB 148; Isobutyl methyl ketone; Isobutyl-methylketon; Isohexanone; Isopropyl acetone; Ketone, isobutyl methyl; Methyl-2-pentanone; Methyl-isobutyl-cetone; Methyl isobutyl ketone [Flammable liquid]; Methylisobutylketon; 4-Methyl-2-pentanone; 4-Metil-pentan-2-one; Metilisobutilchetone; Metyloizobutylo-keton; MIBK; MIK; NSC .5712; 2-Pentanone, 4-methyl-; RCRA waste number U161; Shell MIBK; UN1245. Registered by EPA (cancelled). Colorless liquid; bp = 117 - 118°; d_4^{20} = 0.801; soluble in H_2O (1.91 g/100 ml), soluble in EtOH, C_6H_6, Et_2O; LD50 (rat orl) = 2080mg/kg. *Fluka*.

2479   Methyl isocyanate
624-83-9   6112   H

O=C=N

C3H3NO

Isocyanic acid, methyl ester.

4-04-00-00247 (Beilstein Handbook Reference); AI3-28280; BRN 0605318; CCRIS 1385; EINECS 210-866-3; HSDB 1165; Isocyanatomethane; Isocyanate de methyle; Isocyanate, methyl-; Isocyanic acid, methyl ester; Methane, isocyanato-; Methyl carbonimide; Methyl isocyanat; Methyl isocyanate; Methyl isocyanide; Methylcarbylamine; Methylisocyanaat; Methylisokyanat; Metil isocianato; NSC 64323; RCRA waste number P064; TL 1450; UN 2480. Liquid; mp = -45°; bp = 39.5°; d^{27} = 0.9230; very soluble in H_2O.

2480   Methyl-α-isoionone
127-51-5   H

C14H22O

3-Methyl-4-(2,6,6-trimethyl-2-cyclohexen-1-yl)-3-buten-2-one.

AI3-36074; 3-Buten-2-one, 3-methyl-4-(2,6,6-trimethyl-2-cyclohexen-1-yl)-; Cetone α; α-Cetone; EINECS 204-846-3; FEMA No. 2714; α-Ionone, isomethyl-; Isomethyl-α-ionone; Isomethylionone, α-; Methyl-α-isoionone; 3-Methyl-4-(2,6,6-trimethyl-2-cyclohexen-1-yl)-3-buten-2-one; NSC 66432; 4-(2,6,6-Trimethyl-2-cyclohexen-1-yl)-3-methyl-3-buten-2-one.

2481   Methyl isothiocyanate
556-61-6   6114   G H P

N=C=S

C2H3NS

Isothiocyanatomethane.

AI3-28257; Biomet 33; Caswell No. 573; EINECS 209-132-5; EPA Pesticide Chemical Code 068103; HSDB 6396; Isothiocyanate de methyle; Isothiocyanatomethane; Isothiocyanic acid, methyl ester; Isotiocianato de metilo; Isotiocianato di metile; MeNCS; Methane, isothiocyanato-; Methyl-isothiocyanat; Methyl isothiocyanate; Methyl mustard; Methyl mustard oil; Methylisothiocyanaat; Methylisothiocyanaat; Methylisothiocyanate; Methyliso-thiokyanat; Methylsenföl; Methylsenfoel; Metile iso-tiocianato; MIT; Mitc; Morton EP-161E; Trapex; Trapex-40; Trapexide; UN2477; Vorlex (Nor-Am);

Vorlex 201 (Nor-Am); Vortex; WN 12. Used as a soil fumigant for control of nematodes, fungi, insects and weed seeds. Registered by EPA as a fungicide, herbicide and insecticide. Used as a fungicide and nematicide. Colorless crystals; mp = 35 - 36°; bp = 118 - 119°; soluble in H_2O (0.82 g/100 ml), freely soluble in EtOH, Et_2O; LD50 (rat orl) = 220 mg/kg, (mus orl) = 110 mg/kg, (mrat orl) = 175 mg/kg, (mmus orl) = 90 mg/kg, (rat der) = 2780 mg/kg, (rbt der) = 263 mg/kg, (mmus der) = 1870 mg/kg, (mallard duck orl) = 136 mg/kg; LC50 (rat ihl 1 hr.) = 1.9 mg/l air, (mallard duck 5 day dietary) = 10936 mg/kg diet, (pheasant 5 day dietary) > 5000 mg/kg diet, (bluegill sunfish 96 hr.) = 0.13 mg/l, (rainbow trout 96 hr.) = 0.37 mg/l, (mirror carp 96 hr.) = 0.37 - 0.57 mg/l; non-toxic to bees. *AgrEvo; Aventis Crop Science; MLPC International; Osmose Inc.*

2482   Methyl lactate
547-64-8   6116   H

C4H8O3

(±)-Methyl 2-hydroxypropanoate.

AI3-00584; DL-Methyl lactate; EINECS 208-930-0; EINECS 218-449-8; HSDB 5687; 2-Hydroxypropanoic acid methyl ester; Lactic acid, methyl ester; Methyl 2-hydroxypropanoate; Methyl α-hydroxypropionate; (±)-Methyl 2-hydroxypropanoate; Methyl lactate; (±)-Methyl lactate; NSC 406248; Propanoic acid, 2-hydroxy-, methyl ester. Colorless liquid; bp = 144-145°; d^{19} = 1.09; soluble in EtOH, Et_2O.

2483   Methyl laurate
111-82-0   G H

C13H26O2

Methyl n-dodecanoate.

4-02-00-01090 (Beilstein Handbook Reference); AI3-00669; BRN 1767780; Dodecanoic acid, methyl ester; EINECS 203-911-3; FEMA number 2715; HSDB 5550; Lauric acid, methyl ester; Metholene 2296; Methyl dodecanoate; Methyl dodecylate; Methyl laurate; Methyl laurinate; Methyl n-dodecanoate; NSC 5027; Stepan C40; Uniphat A40. Used as an intermediate in the manufacture of detergents, emulsifiers, wetting agent, stabilizers, lubricants, plasticizers, textiles, flavoring. Liquid; mp = 5.2°; bp = 267°; d^{20} = 0.8702; insoluble in H_2O, soluble in MeOH, $CHCl_3$, CCl_4, EtOAc, freely soluble in EtOH, Et_2O, Me_2CO, C_6H_6. *Henkel/Emery; Procter & Gamble Co.; Stepan.*

2484   Methyl ledate
19010-66-3   G

Pb²⁺

C6H12N2PbS4

Lead dimethyldithiocarbamate.

Bis(dimethylcarbamodithioato-S,S')lead; Bis(dimethyldi-thiocarbamato)lead; Carbamic acid, dimethyldithio-, lead salt; CCRIS 359; EINECS 242-748-2; HSDB 2886; Lead bis(dimethyldithiocarbamate); Lead, bis(dimethyldithio-carbamato)-; Lead dimethyl dithiocarb-amate; Ledate®; Methyl ledate; NCI-C02891. Ultra accelerator used with NR, SBR, IIR, IR, and BR rubbers; used for ultra acceleration, high speed, high temperature vulcanization. Recommended for improved dynamic properties in natural and polyisoprene rubbers. Solid; mp = 310°; d = 3.43; insoluble in H_2O, soluble in organic solvents. *Vanderbilt R.T. Co. Inc.*

2485   Methyl 3-mercaptopropionate
2935-90-2   H

$C_4H_8O_2S$
Propionic acid, 3-mercapto-, methyl ester.
EINECS 220-912-4; HSDB 5905; 3-Mercaptopropionic acid methyl ester; Methyl mercaptopropionate; Methyl 3-mercaptopropionate; NSC 137814; Propanoic acid, 3-mercapto-, methyl ester. Liquid; bp_{14} = 54-55°; d^{25} = 1.085.

2486   Methyl methacrylate
80-62-6   5967   G H

$C_5H_8O_2$
Methyl 2-methylpropenoate.
4-02-00-01519 (Beilstein Handbook Reference); Acryester M; Acrylic acid, 2-methyl-, methyl ester; AI3-24946; BRN 0605459; CCRIS 1364; Diakon; EINECS 201-297-1; HSDB 195; Metakrylan metylu; Methacrylate de methyle; Methacrylic acid, methyl ester; Methacrylsaeuremethyl ester; Methacrylsäuremethyl ester; Methyl 2-methyl-2-propenoate; Methyl 2-methylpropenoate; Methylmethacrylat; Methyl methacrylate; Methyl methacrylate monomer; Methyl methylacrylate; Methylester kyseliny methakrylove; Methylmethacrylaat; Methylmethacrylate; Metil met-acrilat; MMA; Monocite Methacrylate monomer; NCI-C50680; NSC 4769; Pegalan; RCRA waste number U162; TEB 3K; UN1247. Monomer for polymethacrylate resins, impregnation of concrete. Oil; mp = -48°; bp = 100.5°; d^{20} = 0.9440; soluble in EtOH, Me_2CO, Et_2O, $CHCl_3$, ethylene glycol, slightly soluble in H_2O; TLV = 100 ppm. *Allchem Ind.; Degussa AG; Mitsubishi Gas; Rohm & Haas Co.; Sigma-Aldrich Fine Chem.; Transol Chem. UK Ltd.*

2487   p-Methyl morpholine
109-02-4   G

$C_5H_{11}NO$
4-Methyl morpholine.
AI3-24289; CCRIS 6691; EINECS 203-640-0; Methylmorpholine; Morpholine, 4-methyl-; Morpholine, N-methyl-; 4-Methylmorfolin; 1-Methylmorpholine; 4-Methylmorpholine; N-Methylmorpholine; NSC 9382; Texacat® NMM; UN2535. Catalyst for polyester urethane flexible foam, high rise rigid foam panels, extraction solvent; stabilizer for chlorinated hydrocarbons; self-polishing waxes; corrosion inhibitor; pharmaceuticals. Oil; mp = -65°; bp = 116°; d^{20} = 0.9051; soluble in H_2O, EtOH, Et_2O; LD_{50} (rat orl) = 1960 mg/kg. *Texaco.*

2488   Methyl myristate
124-10-7   G H

$C_{15}H_{30}O_2$
Methyl n-tetradecanoate.
AI3-01980; EINECS 204-680-1; Emery® 2214; FEMA No. 2722; HSDB 5602; Metholeneat 2495; Methyl myristate; Methyl tetradecanoate; Methyl n-tetradecanoate; Myristic acid, methyl ester; NSC 5029; Tetradecanoic acid, methyl ester; Uniphat A50. Intermediate for myristic acid detergents; emulsifiers, wetting agents, stabilizers, resins, lubricants, plasticizers, textiles, animal feeds; standard for gas chromatography; flavoring. Used as a solvent for pesticides and herbicides. Liquid; mp = 19°; bp = 295°; bp_7 = 155°; d^{20} = 0.8671; λ_m = 212 nm (ε = 71, C_7H_{16}); insoluble in H_2O, freely soluble in EtOH, Et_2O, Me_2CO, C_6H_6, CCl_4, $CHCl_3$, MeOH, EtOAc. *Henkel/Emery; Stepan.*

2489   Methyl nadic anhydride
25134-21-8   H

$C_{10}H_{10}O_3$
Methylnorbornene-2,3-dicarboxylic anhydride.
5-17-11-00199 (Beilstein Handbook Reference); 5-Norbornene-2,3-dicarboxylic anhydride, methyl-; BRN 0162395; EINECS 246-644-8; Epicure NMA; Hardener HY906; HSDB 6093; Kayahard MCD; MEA 610; 4,7-Methanoisobenzofuran-1,3-dione, 3a,4,7,7a-tetrahydro-methyl-,; Methendic anhydride; Methyl-5-norbornene-2,3-dicarboxylic anhydride; Methyl nadic anhydride; Methylbicyclo(2.2.1)heptene-2,3-dicarboxylic anhydride isomers; Methylendic anhydride; Methylnorbornene-2,3-dicarboxylic anhydride; Nadic methyl anhydride; NMA; TK 10 524; XMNA.

2490   Methylnaphthalene
1321-94-4   H P

$C_{11}H_{10}$
Methylnaphthalenes.
CCRIS 7916; EINECS 215-329-7; HSDB 1143; Methyl naphthalene; Methyl naphthalene (molten); Methylated naphthalenes; Methylnaftalen; Methylnaphthalene; 2-Methylnaphthalene; Methylnaphthalenes, liquid; Methyl-naphthalenes, solid; Methylnapthalene; Naphthalene, methyl-. Mixture of α- and β-isomers. Has been used as an insecticide. Mixture of isomers of methylnaphthalene.

2491   1-Methylnaphthalene
90-12-0   G

409

C11H10
α-methylnaphthalene.
AI3-15378; CCRIS 6151; EINECS 201-966-8; FEMA No. 3193; HSDB 5268; α-Methylnaphthalene; 1-Methyl-naphthalene; Naphthalene, α-methyl-; Naphthalene, 1-methyl-; NSC 3574. Carrier for polyester/wool blended fabrics. Liquid; mp = -30.4°; bp = 244.7°; d^{20} = 1.0202; insoluble in H_2O, insoluble in H_2O, very soluble in EtOH, Et2O, less soluble in C6H6; LD50 (rat orl) = 1840 mg/kg. *Allchem Ind.; Crowley Tar Prods.*

2492   β-Methylnaphthalene
91-57-6   G

C11H10
2-Methylnaphthalene.
AI3-17554; EINECS 202-078-3; HSDB 5274; β-Methylnaphthalene; 2-Methylnaphthalene; Naphthalene, β-methyl-; Naphthalene, 2-methyl-; NSC 3575. Solvent; has been used as an insecticide. Solid; mp = 34.4°; bp = 241.1°; d20 = 1.0058; λ_m = 223, 275, 304, 311, 318 nm (ε = 16800, 5370, 495, 314, 515, MeOH); insoluble in H_2O, soluble in organic solvents; LD50 (rat orl)= 1630 mg/kg. *Allchem Ind.; Crowley Tar Prods.*

2493   1-Methyl-2-nitro-benzene
88-72-2   6684   H

C7H7NO2
Benzene, 1-methyl-2-nitro-.
AI3-16311; Benzene, 1-methyl-2-nitro-; CCRIS 1224; EINECS 201-853-3; HSDB 2189; o-Methylnitrobenzene; Nitrotoluene, o-; o-Nitrotoluene; ortho-Nitrotoluol; NSC 9577; ONT; Toluene, o-nitro-; UN1664. Liquid; mp = -10°; bp = 222°; d^{19} = 1.1611; λ_m = 254 nm (ε= 4330, MeOH); insoluble in H_2O, very soluble in EtOH, Et2O, less soluble in CCl4.

2494   N-Methyl-4-nitrophthalimide
41663-84-7   H

C9H6N2O4
4-Nitro-N-methylphthalimide.
AI3-28673; EINECS 255-483-2; 1H-Isoindole-1,3(2H)-dione, 2-methyl-5-nitro-; N-Methyl-4-nitrophthalimide; 4-Nitro-N-methylphthalimide.

2495   Methyl oleate
112-62-9   G H

C19H36O2
Methyl (Z)-9-octadecenoate.
ADJ 100; AI3-00651; CCRIS 675; Edenor Me 90/95V; Edenor MeTiO5; EINECS 203-992-5; Emerest 2801; Emery® 2301; Emery oleic acid ester 2301; Esterol 112; Exceparl M-OL; HSDB 5572; Kemester® 104; Methyl (Z)-9-octadecenoate; Methyl 9-octadecenoate; Methyl cis-9-octadecenoate; Methyl oleate; Nissan Unister M 182A; NSC 406282; 9-Octadecenoic acid, methyl ester; 9-Octadecenoic acid (Z)-, methyl ester; 9-Octadecenoic acid (9Z)-, methyl ester; (Z)-9-Octadecenoic acid, methyl ester; Oleic acid, methyl ester; Phytorob 926-67; Priolube 1403; Unister M 182A; Witconol 2301. Emulsifier, emollient for cosmetics; lubricant for leather; carrier for agricultural spray products. Defoamer, lubricant, moisture barrier. Solvent for pesticides and herbicides. Liquid; mp = -19.9°; bp^{20} = 218.5°; d^{20} = 0.8739λ_m = 230 nm (ε = 3162, EtOH); insoluble in H_2O, soluble in CHCl3, freely soluble in EtOH, Et2O. *CK Witco Corp.; Henkel/Emery.*

2496   Methylolmethacrylamide
923-02-4   H

C5H9NO2
N-(Hydroxymethyl)-2-methyl-2-propenamide.
4-02-00-01538 (Beilstein Handbook Reference); Acryl-amide, N-(hydroxymethyl)-2-methyl-; BRN 0506510; EINECS 213-086-1; Methylolmethacrylamide; N-(Hydr-oxymethyl)-2-methyl-2-propenamide; N-(Hydroxymeth-yl)-methacrylamide; N-Methylolmethacryl-amide; NSC 2691; 2-Propenamide, N-(hydroxymethyl)-2-methyl-.

2497   Methyl orthoformate
149-73-5   H

C4H10O3
Orthoformic acid, trimethyl ester.
4-02-00-00022 (Beilstein Handbook Reference); AI3-23842; BRN 0969215; EINECS 205-745-7; HSDB 1006; Methane, trimethoxy-; Methyl orthoformate; Methylester kyseliny orthomravenci; NSC 147479; Orthoformic acid methyl ester; Orthoformic acid, trimethyl ester; Orthomravencan methylnaty; Trimethoxymethane; Trimethyl orthoformate; Trimethylester kyseliny orthomravenci. Liquid; mp = 15°; bp = 104°, d^{20} = 0.9676; soluble in EtOH, Et2O.

2498   Methyl 2-oxogluconate
21063-40-1   H

410

C7H12O7

D-arabino-Hexulosonic acid, methyl ester.

D-arabino-Hexulosonic acid, methyl ester; D-arabino-2-Hexulosonic acid, methyl ester; EINECS 244-187-9; Methyl 2-keto-D-gluconate; Methyl 2-oxogluconate.

2499   Methyl palmitate
112-39-0   G

C17H34O2

Methyl hexadecanoate.

AI3-03509; EINECS 203-966-3; Emery® 2216; Hexadecanoic acid, methyl ester; n-Hexadecanoic acid methyl ester; HSDB 5570; Metholene 2216; Methyl hexadecanoate; Methyl n-hexadecanoate; Methyl palmitate; NSC 4197; Palmitic acid, methyl ester; Radia® 7120; Uniphat A60; Emery® 2216; Radia® 7120. Chemical intermediate, chemical synthesis; lubricant in mineral, cutting, lamination, textile oils, and rust inhibitors; textile and leather application. Used as a chemical intermediate, chemical synthesis; lubricant in mineral, cutting, lamination, textile oils, and rust inhibitors; textile and leather application. Solvent for pesticides and herbicides. Solid; mp = 30°; bp = 417°, bp30 = 211°, bp2 = 148°; d20 = 0.8520, d75 = 0.8247; insoluble in H2O, soluble in Et2O, very soluble in EtOH, Me2CO, C6H6, CHCl3. *Fina Chemicals; Henkel/Emery; Penta Mfg.; Stepan.*

2500   Methylparaben
99-76-3   6129   G P

C8H8O3

Methyl 4-hydroxybenzoate.

4-10-00-00360 (Beilstein Handbook Reference); Abiol; AI3-01336; Aseptoform; Benzoic acid, 4-hydroxy-, methyl ester; Benzoic acid, p-hydroxy-, methyl ester; BRN 0509801; Caswell No. 573PP; CCRIS 3946; E218; EINECS 202-785-7; EPA Pesticide Chemical Code 061201; FEMA Number 2710; Germaben® II; HSDB 1184; 4-Hydroxybenzoic acid methyl ester; Lexgrad® M; Maseptol; Metaben; Methaben; Methyl butex; Methyl Chemosept®; Methyl p-hydroxybenzoate; Methyl 4-hydroxybenzoate; Methyl p-oxybenzoate; Methyl paraben; Methyl parahydroxybenzoate; Methyl Parasept®; Methyl Parasept® NF/FCC®; Methylben; Methylester kyseliny p-hydroxybenzoove; Methyl-paraben; Methylparaben®; Metoxyde; Moldex; Nipagin; Nipagin®; Nipagin M; Nipagin M®; Nipaguard® BPX; Nipasta; NSC 3827; Paridol; Phenonip; Preserval; Preserval M; Septos; Solbrol; Solbrol M; Solbrol M®; Tegosept M; Tegosept M®; Uniphen P-23; Antimicrobial, antifungal and antibacterial preservative. Used as a preservative in foods, beverages and cosmetics, in parenterals, injectables, inhalants, intravenous, ophthalmics, orals, rectals, topicals, dental anesthetics and insulin preparations. Disinfectant, possibly antibacterial. Registered by EPA (cancelled). Used as a preservative in foods, beverages and cosmetics. White crystals; mp = 125-128°; bp = 270-280°; d = 1.352; soluble in H2O (0.25 g/100 ml at 25°, 2 g/100 ml at 50°, 3.3 g/100 ml at 80°), EtOH (50 g/100 ml), 95% EtOH (33 g/100 ml), 50% EtOH (16.6 g/100 ml), Et2O (10 g/100 ml), glycerin (1.6 g/100 ml), propylene glycol (5 g/100 ml), peanut oil (0.5 g/100 ml). LD50 (mus sc) = 1200 mg/kg, (mus ip) = 960 mg/kg, (dog orl) = 3000 mg/kg. *Aceto; Ashland; Charkit; E. Merck; Greeff R.W. & Co.; Kraft; Napp Labs Ltd; Nipa; Ruger; Sigma-Aldrich Fine Chem.; Spectrum*

Chem. Manufacturing; Sutton Labs.; Tri-K Ind.

2501   2-Methylpentanediamine
15520-10-2   G H

C6H16N2

2-Methyl-1,5-pentanediamine.

3-04-00-00609 (Beilstein Handbook Reference); BRN 1732701; Dytek® A; EINECS 239-556-6; 2-Methyl-1,5-pentanediamine; 2-Methylpentamethylenediamine; 2-Methylpentane-1,5-diamine; 2-Methylpentanediamine; 1,5-Pentanediamine, 2-methyl-. Epoxy curing agent; also used in polyurethanes, wet strength resins, scale and corrosion inhibitors, motor oil and gasoline additives, polyamide plastics, films, adhesives, and inks. Liquid; mp = -50 - -60°; bp = 193°; d25 = 0.86; LD50 (rat orl) = 1690 mg/kg. *DuPont.*

2502   Methylphenidate
113-45-1   6132   D

C14H19NO2

Methyl α-phenyl-2-piperidine acetate.

4311/B Ciba; C 4311; Calocain; Centedein; DEA No. 1724; EINECS 204-028-6; HSDB 3126; Meridil; Methyl phenidate; Methyl phenidylacetate; Methylofenidan; Methylphenidan; Methylphenidate; Methylphenidatum; Metilfenidato; NCI-C56280; Phenidylate; Plimasine; Ritalin; Ritaline. Psychomotor stimulant. Used as a CNS stimulant. Liquid; bp0.6 = 135-137°; insoluble in H2O, petroleum ether; soluble in EtOH, EtOAc, Et2O. *Ciba-Geigy Corp.*

2503   Methylphenidate Hydrochloride
298-59-9   6132   D

C14H20ClNO2

Methyl α-phenyl-2-piperidine acetate hydrochloride.

CCRIS 6258; Centedrin; Centedrine; EINECS 206-065-3; Meridil hydrochloride; Metadate CD; Methylin; Methylphenidate hydrochloride; Methylphenidylacetate hydrochloride; Metilfenidat hydrochloride; NSC 169868; Ritalin; Ritalin hydrochloride. CNS, psychomotor stimulant. Crystals; mp = 224-226°; soluble in H2O, EtOH, CHCl3; LD50 (mus orl) = 190 mg/kg. *Ciba-Geigy Corp.*

2504   4-Methylphthalic anhydride
19438-61-0   H

411

$C_9H_6O_3$
1,3-Isobenzofurandione, 5-methyl-.
EINECS 243-073-6; 1,3-Isobenzofurandione, 5-methyl-; 4-Methylphthalic anhydride; Phthalic anhydride, 4-methyl-.

2505 N-Methylphthalimide
550-44-7 H

$C_9H_7NO_2$
2-Methyl-1H-isoindole-1,3(2H)-dione.
5-21-10-00273 (Beilstein Handbook Reference); AI3-01393; BRN 0124428; EINECS 208-982-4; 1H-Isoindole-1,3(2H)-dione, 2-methyl-; 2-Methyl-1H-isoindole-1,3(2H)-dione; N-Methylphthalimide; NSC 44059; Phthalimide, N-methyl-. Needles; mp = 134°; bp = 286°; insoluble in H_2O; slightly soluble in EtOH.

2506 Methyl pivalate
598-98-1 H

$C_6H_{12}O_2$
Methyl 2,2-dimethylpropionate.
EINECS 209-959-1; Methyl 2,2-dimethylpropionate; Methyl pivalate; Pivalic acid, methyl ester; Propanoic acid, 2,2-dimethyl-, methyl ester. Liquid; bp = 101.1°; $d^0 = 0.891$; very soluble in EtOH, Et_2O.

2507 Methylprednisolone
83-43-2 6133 D

$C_{22}H_{30}O_5$
(6α,11β)-11,17,21-Trihydroxy-6-methylpregna-1,4-diene-3,20-dione.
4-08-00-03498 (Beilstein Handbook Reference); A-Methapred; Artisone-Wyeth; Besonia; BRN 2340300; Depo-Medrol (acetate); Dopomedrol; DRG-0050; EINECS 201-476-4; Esametone; Firmacort; HSDB 3127; Lemod; Medesone; Medixon; Medlone 21; Medrate; Medrol; Medrone; MEPRDL; Mesopren; Metastab; Methyleneprednisolone; Methylprednisolone; Methyl-prednisolonum;

Metilbetasone; Metilprednisolona; Metilprednisolone; Metrisone; Metrocort; Metysolon; Moderin; Nirypan; Noretona; NSC-19987; Predni N Tablinen; Prednol- L; Promacortine; Reactenol; Sieropresol; Solomet; Solu-medrol; Summicort; Suprametil; U-67,590A; U 7532; Urbason; Urbasone; Wyacort. Glucocorticoid. mp = 228-237°; $[α]_D^{20} = 83°$ (dioxane); $λ_m = 243$ nm (α_M 14875 95% EtOH). *Schering-Plough HealthCare Products; Upjohn Ltd.*

2508 2-Methylpropanoic anhydride
97-72-3 H

$C_8H_{14}O_3$
Propanoic acid, 2-methyl-, anhydride.
AI3-28521; EINECS 202-603-6; HSDB 5309; Isobutyric acid anhydride; Isobutyric anhydride; Isobutyryl anhydride; Propanoic acid, 2-methyl-, anhydride; UN2530. Liquid; mp = -53.5°; bp = 183°, $bp^{32} = 89°$; $d^{20} = 0.9535$; freely soluble in Et_2O, soluble in $CHCl_3$.

2509 Methyl propionate
554-12-1 6135 H

$C_4H_8O_2$
Propionic acid, methyl ester.
4-02-00-00704 (Beilstein Handbook Reference); AI3-10621; BRN 1737628; EINECS 209-060-4; FEMA Number 2742; HSDB 5688; Methyl propanoate; Methyl propionate; Methyl propylate; Methylester kyseliny propionove; NSC 9375; Propanoic acid, methyl ester; Propionate de methyle; Propionic acid, methyl ester; UN1248. Liquid; mp = -87.5°; bp = 79.8°; $d^{20} = 0.9150$; $λ_m = 211$ nm (ε = 62, isooctane), 204 nm (ε = 50, H_2O); slightly soluble in H_2O, soluble in Me_2CO, CCl_4, freely soluble in EtOH, Et_2O.

2510 2-Methylpropyl methacrylate
97-86-9 H

$C_8H_{14}O_2$
2-Propenoic acid, 2-methyl-, 2-methylpropyl ester.
4-02-00-01526 (Beilstein Handbook Reference); BRN 1747595; CCRIS 4829; EINECS 202-613-0; HSDB 5312; Isobutyl α-methacrylate; Isobutyl 2-methyl-2-propenoate; Isobutyl methacrylate; Isobutylester kyseliny methakrylove; Methacrylic acid, isobutyl ester; NSC 18607; Propenoic acid, 2-methyl, isobutyl ester; UN2283. Liquid; bp = 155°; $d^{20} = 0.8858$; insoluble in H_2O, freely soluble in EtOH, Et_2O.

2511 4-(1-Methylpropyl)phenol
99-71-8 H

C10H14O

Phenol, 4-(1-methylpropyl)-.

4-06-00-03279 (Beilstein Handbook Reference); AI3-18887; BRN 1364714; 4-sec-Butylphenol; p-(sec-Butyl)phenol; EINECS 202-781-5; NSC 2210; Phenol, 4-(1-methylpropyl)-; Phenol, p-(sec-butyl)-. Solid; mp = 61.5°; bp = 241°; d^{20} = 0.986; $[\alpha]_D^{20}$ = +13.3° (m-xylene); insoluble in H_2O, soluble in EtOH, very soluble in Et_2O.

2512   N-Methylpyrrolidone
872-50-4   6140   G H

C5H9NO

N-Methyl-2-pyrrolidone.

AI3-23116; CCRIS 1633; EINECS 212-828-1; HSDB 5022; M-Pyrol; Methylpyrrolidone; Methylpyrrolidone, N-; 1-Methyl-2-pyrrolidone; 1-Methyl-2-pyrrolidinone; 1-Methylazacyclopentan-2-one; 1-Methylpyrrolidinone; 1-Methylpyrrolidone; 2-Pyrrolidinone, 1-methyl-.; N-Methyl-2-pyrrolidinone; N-Methyl-2-pyrrolidone; NMP; Norleucine, 5-oxo-, DL-; NSC 4594.Solvent for resins, acetylene, pigment dispersant; petroleum processing; spinning agent for PVC; intermediate. Used as coatings solvent, in stripping and cleaning of paints and varnishes, industrial cleaning, mold cleaning, petrochemical processing, agricultural cleaning, polymer solvent. Liquid; mp = -24°; bp = 202°; d = 1.033; λ_m = 205 nm (ϵ = 2884, MeOH); very soluble in H_2O, soluble in Et_2O, Me_2CO, $CHCl_3$; LD_{50} (rat orl) = 3.9 g/kg. *Arco; Ashland; BASF Corp.; ISP; Janssen Chimica; Sigma-Aldrich Fine Chem.*

2513   Methyl ricinoleate
141-24-2   G

C19H36O3

12-Hydroxy-9-octadecenoic acid methyl ester.

AI3-10523; Castor oil acid, methyl ester; EINECS 205-472-3; Flexricin P-1; HSDB 5636; 12-Hydroxy-9-octadecenoic acid, methyl ester; Methyl 12-hydroxy-9-octadecenoate; Methyl ricinoleate; NSC 1254; 9-Octadecenoic acid, 12-hydroxy-, methyl ester; 9-Octadecenoic acid, 12-hydroxy-, methyl ester, (9Z,12R)-; 9-Octadecenoic acid, 12-hydroxy-, methyl ester, (R-(Z))- (9Cl); Ricinoleic acid methyl ester. Plasticizer, lubricant, cutting oil additive, wetting agent. *Penta Mfg.; Reilly-Whiteman.*

2514   Methyl Salicylate
119-36-8   6143   D G P

C8H8O3

2-Hydroxybenzoic acid methyl ester.

4-10-00-00143 (Beilstein Handbook Reference); AI3-00090; Analgit; Benzoic acid, 2-hydroxy-, methyl ester; Betula; Betula Lenta; Betula oil; Birch oil, sweet; BRN 0971516; Caswell No. 577; CCRIS 6259; EINECS 204-317-7; EPA Pesticide Chemical Code 076601; Exagien; FEMA Number 2745; Flucarmit; Gaultheria oil; Gaultheriaöl; Gaultheriaoel; Heet; HSDB 1935; o-Hydroxybenzoic acid, methyl ester; Methyl 2-hydroxybenzoate; Methyl o-hydroxybenzoate; Methyl salicylate; Methylester kyseliny salicylove; Metylester kyseliny salicylove; Natural wintergreen oil; NSC 8204; Oil of wintergreen; Panalgesic; Salicylic acid, methyl ester; Spicewood Oil; Sweet birch oil; Synthetic wintergreen oil; Teaberry oil; Theragesic; Wintergreen oil; Wintergruenöl; Wintergruenoel. Used as a flavor in foods and beverages, pharmaceuticals, and in perfumery. Registered by EPA as an antimicrobial, fungicide and insecticide. Used medically as an analgesic. Used in food packaging as an insect repellant and possibly an insecticide. Yellow-red oil; mp = -8.6°; bp = 220.9°; d_{25} = 1.181; d = 1.180; λ_m = 237, 305 nm (ϵ = 11200, 5300, MeOH); soluble in H_2O (0.06 g/100 ml), $CHCl_3$, Et_2O, freely soluble in EtOH, AcOH; LD_{50} (rat orl) = 887 mg/kg. *Caraustar Industries Inc.; Plantabbs Corp.*

2515   Methyl selenac
144-34-3   G

C12H24N4S8Se

Selenium dimethyldithiocarbamate.

Carbamodithioic acid, dimethyl-, tetraanhydrosulfide with; EINECS 205-624-9; HSDB 2929; Methyl selenac; Orthothioselenious acid; Selenium dimethyldithio-carbamate; Selenium, tetrakis(dimethyldithiocarbamato)-; Tetrakis(di-methyl carbamodithioato-S,S')selenium. Rubber accelerator for NR, SBR, and IIR, vulcanizing agents; effective in low sulfur and sulfurless heat resistant compounds. *Vanderbilt R.T. Co. Inc.*

2516   Methyl stearate
112-61-8   G

C19H38O2

Methyl octadecanoate.

4-02-00-01216 (Beilstein Handbook Reference); AI3-07960; BRN 1786213; EINECS 203-990-4; Emery® 2218; HSDB 2901; Kemester® 4516; Kemester® 9018; Kemester® 9718; Metholene 2218; Methyl octadecanoate; Methyl n-octadecanoate; Methyl stearate; NSC 9418; Octadecanoic acid, methyl ester; n-Octadecanoic acid methyl ester; Stearic acid, methyl ester; Emery® 2218; Kemester® 4516. Intermediate for stearic acid detergents, emulsifiers, wetting agents, stabilizers, resins, lubricants, plasticizers. Intermediate in production of alkanolamides, in metalworking lubricants, as specialized solvents; foam depressant and nutrient in fermentation. Solvent for pesticides and herbicides. Solid; mp = 39.1°; bp = 443°; bp_{15} = 215°; d^{40} = 0.8498; insoluble in H_2O, soluble in Et_2O, $CHCl_3$. *CK Witco Corp.; Ferro/Keil;*

413

2517   α-Methylstyrene
98-83-9   G H

C9H10

Benzene, (1-methylethenyl)-.
AI3-18133; as-Methylphenylethylene; Benzene, (1-methylethenyl)-; CCRIS 6067; EINECS 202-705-0; HSDB 196; Isopropenil-benzolo; Isopropenyl-benzeen; Iso-propenyl-benzol; Isopropenylbenzene; α-Methyl-styrol; α-Methylstyreen; α-Methylstyrene; NSC 9400; 2-Phenylpropene; 2-Phenyl-1-propene; Styrene, α-methyl-; UN2303. Liquid; mp = -23.2°; bp = 165.4°; d^{20} = 0.9106; λ_m = 420 nm (ε = 22387, liq. NH3); insoluble in H2O, solubel in EtOH, Et2O, freely soluble in Me2CO, C6H6, CCl4, petroleum ether.

2518   p-Methylstyrene
622-97-9   H

C9H10

1-Ethenyl-4-methylbenzene.
4-05-00-01369 (Beilstein Handbook Reference); Benzene, 1-ethenyl-4-methyl-; BRN 1209317; CCRIS 3488; EINECS 210-762-8; 1-Ethenyl-4-methylbenzene; HSDB 6503; 4-Methylstyrene; 1-Methyl-4-vinylbenzene; p-Methyl styrene; para-Methylstyrene; Styrene, p-methyl-; 1-p-Tolylethene; p-Vinyltoluene; 4-Vinyltoluene. Liquid; mp = -34.1°; bp = 172.8°; d^{25} = 0.9173; λ_m = 210, 216, 251, 284, 294 nm (ε = 19500, 11200, 15200, 1070, 770, MeOH); insoluble in H2O, soluble in C6H6.

2519   Methyl sulfide
75-18-3   6146   G H

C2H6S

Dimethyl sulfide.
4-01-00-01275 (Beilstein Handbook Reference); AI3-25274; BRN 1696847; Dimethyl monosulfide; Dimethyl sulfide; Dimethyl sulphide; Dimethyl thioether; Dimethylsulfid; DMS; EINECS 200-846-2; Exact-S; FEMA No. 2746; HSDB 356; Methane, thiobis-; Methanethiomethane; Methyl monosulfide; Methyl sulfide; Methyl sulphide; Methyl thioether; Methylthiomethane; Sulfure de methyle; 2-Thiapropane; Thiobis(methane); UN1164. Coking suppressor for ethylene production and for steel mill furnace walls; odorant for natural gas; presulfiding agent for catalysts in refinery processes. Used with t-butyl mercaptan [75-66-1] as an additive to natural gas for leak detection. mp = -98.3°; bp = 37.3°; d^{20} = 0.8483; insoluble in H2O, soluble in organic solvents. *Gaylord.*

2520   Methyl tallowate
61788-61-2   H

Tallow, methyl ester.
EINECS 262-989-7; Fatty acids, tallow, Me esters; Methyl esters of tallow; Methyl tallowate; Tallow, methyl ester.

2521   Methyl   4,6,6,6-tetrachloro-3,3-dimethyl-hexanoate
64667-33-0   H

C9H14Cl4O2

Hexanoic acid, 4,6,6,6-tetrachloro-3,3-dimethyl-, methyl ester. EINECS 265-005-4; Hexanoic acid, 3,3-dimethyl-4,6,6,6-tetrachloro, methyl ester; Hexanoic acid, 4,6,6,6-tetrachloro-3,3-dimethyl-, methyl ester; Methyl 4,6,6,6-tetrachloro-3,3-dimethylhexanoate; 4,6,6,6-Tetrachloro-3,3-dimethylhexanoic acid, methyl ester.

2522   4-Methylthiazole
693-95-8   H

C4H5NS

4-Methylthiazole.
4-27-00-00969 (Beilstein Handbook Reference); BRN 0105228; EINECS 211-764-1; FEMA No. 3716; HSDB 5747; 4-Methylthiazole; NSC 42976; Thiazole, 4-methyl-. Liquid; bp = 133.3°; d^{25} = 1.112; λ_m = 250 nm (ε = 3467, EtOH); soluble in H2O, EtOH, Et2O.

2523   (Methylthio)acetaldehyde oxime
10533-67-2   H

C3H7NOS

Acetaldehyde, (methylthio)-, oxime.
(Methylthio)acetaldoxime; (Methylthio)acetaldehyde oxime; Acetaldehyde, (methylthio)-, oxime; EINECS 234-096-2.

2524   Methyltin trichloride
993-16-8   H

CH3Cl3Sn

Trichloromethylstannane.
CCRIS 6327; EINECS 213-608-8; Methyltin trichloride; Methyltrichlorostannane; Methyltrichlorotin; Mono-methyltin trichloride; Stannane, methyltrichloro-; Stannane, trichloromethyl-; Tin, methyl-, trichloride; Trichloromethylstannane; Trichloromethyltin.

2525   Methyltin tris(isooctyl mercaptoacetate)
54849-38-6   H

$C_{31}H_{60}O_6S_3Sn$

Stannane, tris(((isooctylthio)acetyl)oxy)methyl-.

Acetic acid, 2,2',2''-((methylstannylidyne)tris(thio))tris-,triisooctyl ester; CCRIS 6070; EINECS 259-374-0; HSDB 6101; Methyltin S,S',S''-tris(isooctyl mercaptoacetate); Methyltin tris(isooctyl mercaptoacetate); Methyltin tris(isooctyl thioglycolate); Monomethyltin tris(isooctyl mercaptoacetate); Monomethyltin tris(isooctyl thioglycolate); Stannane methyltris((carboxymethyl)-thio)tris isooctyl ester; Stannane, tris(((isooctylthio)-acetyl)oxy)methyl-; Triisooctyl 2,2',2''-((methylstannyl-idyne)tris(thio))triacetate.

2526 Methyl p-toluate
99-75-2   H

$C_9H_{10}O_2$

Benzoic acid, 4-methyl-, methyl ester.

4-09-00-01726 (Beilstein Handbook Reference); Al3-04243; Benzoic acid, 4-methyl-, methyl ester; BRN 1100609; p-Carbomethoxytoluene; CCRIS 6071; EINECS 202-784-1; HSDB 5327; Methyl 4-methylbenzoate; Methyl p-methylbenzoate; Methyl 4-toluate; Methyl p-toluate; Methyl p-toluenecarboxylate; NSC 24761; p-Toluic acid, methyl ester. Crystals; mp = 33.2°; bp = 220°; λ_m = 238, 281 nm (ϵ = 14200, 557, MeOH); insoluble in H_2O, very soluble in EtOH, Et_2O.

2527 Methyltriacetoxysilane
4253-34-3   H

$C_7H_{12}O_6Si$

Methylsilanetriol triacetate.

4-04-00-04208 (Beilstein Handbook Reference); APK 1 (silane derivative); BRN 1788668; EINECS 224-221-9; K 10S; Methylsilanetriol triacetate; Methylsilanetriyl triacetate; Methyltriacetoxysilane; Methyltrihydroxysilane triacetate; NSC 139845; Silane, methyltriacetoxy-; Silanetriol, methyl-, triacetate; Triacetoxymethylsilane.

2528 Methyltrichlorosilane
75-79-6   H

CH_3Cl_3Si

Trichloromethylsilane.

4-04-00-04212 (Beilstein Handbook Reference); Al3-51465; BRN 1361381; CCRIS 1322; EINECS 200-902-6; HSDB 840; Methyltrichlorsilan; Methylsilicochloroform; Methylsilyl trichloride; Methyltrichlorosilane; Mono-methyltrichlorosilane; NSC 77069; Silane, methyl-trichloro-; Silane, trichloromethyl-; Trichlormethylsilan; Trichloro(methyl)silane; Trichloromethylsilane; Trichloromethylsilicon; UN1250. Coupling agent, chemical intermediate, blocking agent, release agent, lubricant, primer, reducing agent. Liquid; mp = -90°; bp = 65.6°; d^{20} = 1.273; soluble in H_2O and organic solvents. *Degussa-Hüls Corp.*

2529 Methyltridecylsilane
18769-78-3   G

$C_{31}H_{66}Si$

M9030.

Tridecylmethylsilane. Thermally stable fluid with superior metal-on-metal lubrication and wear characteristics. *Degussa-Hüls Corp.*

2530 Methyltrimethoxysilane
1185-55-3   G H

$C_4H_{12}O_3Si$

Trimethoxy(methyl)silane.

Dynasylan® MTMS; EINECS 214-685-0; Methyl-trimethoxysilane; NSC 93883; Silane A-163; Silane, methyltrimethoxy-; Silane, trimethoxymethyl-; Tri-methoxy(methyl)silane; Trimethoxymethylsilane; Union carbide A-163; Union Carbide® A-163; Z 6070. Coupling agent, chemical intermediate, blocking agent, release agent, lubricant, primer, reducing agent. Crosslinking agent providing durability, gloss, hiding power to coatings. Liquid; bp = 102°; d;ss20 = 0.9548; soluble in $CHCl_3$; LD_{50} (rat orl) = 12500 mg/kg. *Degussa-Hüls Corp.; Union Carbide Corp.*

2531 Methyltrimethylolmethane
77-85-0   H

$C_5H_{12}O_3$

1,1,1-Tris(hydroxymethyl)ethane.

EINECS 201-063-9; Ethane, 1,1,1-tris(hydroxymethyl)-; Ethylidynetrimethanol; HSDB 5217; Methriol; Methyltrimethanolmethane; Methyltrimethylolmethane; Metriol; NSC 65581; Pentaglycerine; Pentaglycerol; Trimet; Trimethylolethane; Tris(hydroxymethyl)ethane. Oil; bp22 = 114°, bp = 221°; d^{20} = 0.989; slightly soluble in H_2O, more soluble in organic solvents.

2532 Methyl vinyl ether

107-25-5   H

C3H6O

1-Methoxyethylene.

Agrisynth MVE; EINECS 203-475-4; Ethene, methoxy-; Ether, ethenyl methyl; Ether, methyl vinyl; HSDB 1033; Methoxyethene; 1-Methoxyethylene; Methyl vinyl ether; UN1087; Vinyl methyl ether. Gas; mp = -122°; bp = 5.5°; d^0 = 0.7725; slightly soluble in H_2O, very soluble in EtOH, Et_2O, Me_2CO, C_6H_6.

2533 Metiram

9006-42-2   P

$C_{16}H_{33}N_{11}S_{16}Zn_3$

Tris(amine)(ethylenebis(dithiocarbamato))zinc(2+)(tetrahydro-1,2,4,7-dithiadiazocene-3,8-dithione), polymer.

Amarex; Carbatene; Caswell No. 041A; CCRIS 2500; EPA Pesticide Chemical Code 014601; FMC-9102; HSDB 6705; Metiram; Metiram-complex; Metirame zinc; NIA 9102; Polikarbatsin; Polycarbacin; Polycarbacine; Polycarbazin; Polycarbazine; Polymarcin; Polymarcine; Polymarsin; Polymarzin; Polymarzine; Polymat; Polyram; Polyram 80; Polyram 80WP; Polyram combi; Polyram DF; Zinc ammoniate ethylenebis(dithiocarbamate)-poly-(ethylenethiuram disulfide); Zinc metiram; Zineb-ethylene thiuram disulfide adduct. Used to prevent crop damage in the field, during storage, or transport. Effective against a broad spectrum of fungi, and is used to protect fruits, vegetables, field crops, and ornamentals from foliar diseases and damping off. Practically non-toxic. Registered by EPA as a fungicide. Yellow powder; dec. 140°; insoluble in H_2O, common organic solvents, sol-uble in C_5H_5N; LD_{50} (rat orl) > 10000 mg/kg, (mus orl) > 5400 mg/kg, (gpg orl) = 2400 - 4800 mg/kg, (rat der) > 2000 mg/kg, (bee orl) > 0.04 mg/bee, (bee contact) > 0.016 mg/bee; LC_{50} (rat ihl 4 hr.) > 5.7 mg/l air, (carp 96 hr.) = 85 mg/l, (rainbow trout 96 hr.) = 1.1 mg/l, (harlewuim fish 48 hr.) = 17 mg/l. Amvac Chemical Corp.; BASF AG; UAP - Platte Chemical.

2534 Metobromuron

3060-89-7   6163   G H

C9H11BrN2O2

N'-(4-Bromophenyl)-N-methoxy-N-methylurea.

3-(4-Bromophenyl)-1-methoxy-1-methylurea; 3-(4-Brom-phenyl)-1-methoxyharnstoff; 3-(p-Bromophenyl)-1-meth-yl-1-methoxyurea; BRN 2103964; C-3126; Caswell No. 579A; CCRIS 6765; CIBA-3126; EPA Pesticide Chemical Code 035901; HSDB 1741; Metbromuron; Metobrom-uron; Metobromurone; Monobromuron; N-(4-Bromo-phenyl)-N'-methyl-N'-methoxy-harnstoff; N'-(4-Bromo-phenyl)-N-methoxy-N-methylurea; Patoran; Patt-onex; Urea, 3-(p-bromophenyl)-1-methoxy-1-methyl-; Urea, N'-(4-bromophenyl)-N-methoxy-N-methyl-. A substituted urea which inhibits photosynthesis and is used for pre-emergence control of annual broad-leaved weeds and grasses in vegetable crops. Used as a herbicide. Colorless crystals; mp = 95 - 96°; sg^{20} = 1.6; soluble in H_2O (0.033 g/100 ml at 20°), Me_2CO (50 g/100 ml), CH_2Cl_2 (55 g/100 ml), MeOH (24 g/100 ml), C_7H_8 (10 g/100 ml), n-octanol (7 g/100 ml), $CHCl_3$ (6.25 g/100 ml), C_6H_{14} (0.26 g/100 ml); LD_{50} (rat orl) = 2603 mg/kg, (rat der) > 3000 mg/kg, (rbt der) > 10200 mg/kg; LC_{50} (rat ihl 4 hr.) > 1.1 mg/l air, (rainbow trout 96 hr.) = 36 mg/l, (bluegill sunfish

96 hr.) = 40 mg/l, (crucian carp 96 hr.) = 40 mg/l; non-toxic to bees. BASF Corp.; CIBA plc; Ciba-Geigy Corp.; Makhteshim-Agan; Syngenta Crop Protection.

2535 Metoclopramide

364-62-5   6164   D

$C_{14}H_{22}ClN_3O_2$

4-Amino-5-chloro-N-[2-(diethylamino)ethyl]-2-methoxy-benzamide.

AHR-3070-C; BRN 1884366; Cerucal; Clopromate; DEL 1267; Draclamid; EINECS 206-662-9; Emperal; Eucil; Gastrese; Gastro-tablinen; Gastro-Timelets; Gastrobid; Gastromax; Gastrosil; Gastrotem; Maxeran; Maxolon; MCP-ratiopharm; Meclopran; Metamide; Metho-chloropramide; Metochloropramide; Metoclol; Meto-clopramida; Metoclopramide; Metoclopramidum; Metocobil; Metramid; Moriperan; Parmid; Paspertin; Plasil; Primperan; Regla; Reliveran. Antiemetic with neuroleptic activity. Crystals; mp = 146-148°; soluble in H_2O (0.2 g/100 ml), more soluble in organic solvents. Abbott Labs Inc.; Apothecon; Lemmon Co.; SmithKline Beecham Pharm.; Whitehall-Robins.

2536 Metoclopramide Hydrochloride

54143-57-6   6164   D

$C_{14}H_{23}Cl_2N_3O_2.H_2O$

4-Amino-5-chloro-N-[2-(diethylamino)ethyl]-2-methoxybenzamide monohydrochloride monohydrate.

AHR-3070-C; Cerucal; Clopromate; Draclamid; Emperal; Eucil; Gastrese; Gastro-tablinen; Gastro-Timelets; Gastrobid; Gastromax; Gastrosil; Gastrotem; Maxeran; Maxolon; MCP-ratiopharm; Meclopran; Metamide; Metoclol; Metoclopramide hydrochloride; Meto-clopramide monohydrochloride monohydrate; Metocobil; Metramid; Moriperan; Mygdalon; Parmid; Paspertin; Peraprin; Plasil; Pramiel; Reglan. Antiemetic with neuroleptic activity. Crystals; mp = 182.5-184°; slightly soluble in H_2O, more soluble in organic solvents. Abbott Labs Inc.; Apothecon; Lemmon Co.; SmithKline Beecham Pharm.; Whitehall-Robins.

2537 Metol

55-55-0   6046   G

$C_{14}H_{20}N_2O_6S$

p-Methylaminophenol sulfate.

AI3-15404; Armol; Bis(4-hydroxy-N-methylanilinium) sulphate; CCRIS 4842; EINECS 200-237-1; Elon; Elon (developer); Genol;

416

Graphol; Metatyl; Methyl-p-aminophenol sulfate; Metol; N-Methyl-4-hydroxyaniline hemisulfate; NSC 148345; p-Methylaminophenol sulfate; p-Methylaminophenol sulphate; Photo-Rex; Photol; Pictol; Planetol; Rhodol; Verol. Used as a developing agent in photography. Also used for dyeing furs. Crystals; mp = 260° (dec); soluble in H_2O (5-15 g/100 ml); less soluble in organic solvents. *Fabrichem.*

2538   Metolachlor
51218-45-2   6167   G H P

C15H22ClNO2
 2-Chloro-N-(2-ethyl-6-methylphenyl)-   N-(2-methoxy-1-methylethyl)acetamide.
 2-Äthyl-6-methyl-N-(1-methyl-2-methoxyäthyl)-chloracet-anilid; 2-Ethyl-6-methyl-1-N-(2-methoxy-1-methylethyl)-chloroacetanilide; 2-Etylo-6-metylo-N-(1'-metylo-2'-metoksyetylo)chloroacetanilid; Bicep; Bicep 6L; BRN 2743537; Caswell No. 188DD; CGA-24705; α-Chlor-6'-äthyl-N-(2-methoxy-1-methyläthyl)-aceto-toluidin;   α-Chloro-2'-ethyl-6'-methyl-N-(1-methyl-2-methoxyethyl)-acetanilide; 2-Chloro-2'-ethyl-6'-methyl-N-(2-methoxy-1-methyl-ethyl)-6'-methylacetanilide; 2-Chloro-6'-ethyl-N-(2-meth-oxy-1-methylethyl)-o-acetotoluidide;   2-Chloro-6'-ethyl-N-(2-methoxy-1-methylethyl)acet-o-toluide; 2-Chloro-6'-ethyl-N-(2-methoxy-1-methyl-ethyl)-o-acetoluidide; 2-Chloro-6'-ethyl-N-(2-methoxy-1-methyl-ethyl)-o-aceto-toluidine;   2-Chloro-N-(2-ethyl-6-methyl-phenyl)-N-(2-methoxy-1-methylethyl)acetamide;   Codal; Cotoran multi; Dual; Dual 8E; Dual 25G; Dual 720EC; Dual 960 EC; Dual II; Dual Magnum; Dual Triple; EINECS 257-060-8; EPA Pesticide Chemical Code 108801; HSDB 6706; Humextra; Metelilachlor;   N-(2'-Methoxy-1'-methylethyl)-2'-ethyl-6'-methyl-2-chloro-acetanilide;   N-(1-Methyl-2-methoxyethyl)-N-chloro-acetyl-2-ethyl-6-methylaniline;   Metolachlor;   Metola-chlore;   Metolaclor; Ontrack 8E; Pace 6L; Pennan; Pennant; Primagram; Primextra; Turbo; Yibingjiacaoan. Selective herbicide used for control of annual grasses and some broad-leaved weeds. Registered by EPA as a herbicide. Colorless liquid; bp0.001 = 100°; d = 1.12; soluble in H_2O (0.0530 g/100 ml), more soluble in organic solvents, insoluble in ethylene glycol, propylene glycol, petroleum ether; LD50 (rat orl) = 2780 mg/kg, (rat der) > 3170 mg/kg; LC50 (rat ihl 6 hr.) = 1.75 mg/l air, (bobwhite quail, mallard duck 8 day dietary) > 10000 mg/kg, (rainbow trout 96 hr.) = 2 mg/l, (carp 96 hr.) = 4.9 mg/l, (bluegill sunfish 96 hr.) = 15 mg/l; non-toxic to bees. *BASF Corp.; Bayer Corp.; Drexel Chemical Co.; Syngenta Crop Protection.*

2539   Metoprolol
51384-51-1   6172   D

C15H25NO3
 1-(Isopropylamino)-3-[p-(2-methoxyethyl)phenoxy]-2-propanol. CGP-2175; EINECS 257-166-4; H-93/26. Antianginal; class II antiarrhythmic; antihypertensive. A β-adrenergic blocker. Lacks inherent sympathomimetic activity. *Apothecon; Astra Chem. Ltd.; Ciba-Geigy Corp.; Lemmon Co.*

2540   Metoprolol succinate
98418-47-4   6172   D

C34H56N2O10
 1-(Isopropylamino)-3-[p-(2-methoxyethyl)phenoxy]-2-propanol succinate (2:1) (salt).
 Beloc-Zok; H 93/26 succinate; Metoprolol succinate; Seloken ZOC; Selozok; Spesicor Dos; Toprol XL. A β-adrenergic blocker. Lacks inherent sympathomimetic activity. Antianginal; class II antiarrhythmic; anti-hypertensive. *Apothecon; Astra Chem. Ltd.; Astra Sweden; Ciba-Geigy Corp.; Lemmon Co.*

2541   Metoprolol tartrate
56392-17-7   6172   D

C34H56N2O12
 1-(Isopropylamino)-3-[p-(2-methoxyethyl)phenoxy]-2-propanol (2:1) dextro-tartrate salt.
 Apo-Metoprolol; Arbralene; Azumetop; Beloc; Beprolo; Betaloc; Bloksan; Bloxan; Cardoxone; CGP 2175E; EINECS 260-148-9; H 93/26; HCTCGP 2175E; HSDB 6531; Jeprolol; Kapodine; Lanoc; Lopresor; Lopressor; Lopressor HCT; Mepolol; Meto AbZ; Meto-Isis; Metoberag; Metoberta; Metocard; Metolol; Metomerck; Metop; Metoproferm; Metoprolin; Metoprolol acis; Metoprolol AL; Metoprolol Apogepha; Metoprolol Atid; Metoprolol Basics; Metoprolol-GRY; Metoprolol hemitartrate; Metoprolol Heumann; Metoprolol PB; Metoprolol-radiopharm;   Metoprolol   Stada; Metoprolol tartrate; Metoprolol Verla; Metoprolol-Wolf; Minax; PMS-Metroprolol-B; Prelis; Proken M; Prolaken; Ritmolol; Selectadril; Selo-Zok; Selokeen; Seloken; Selopral; Sigaprolol; Slow-Lopresor; Spesicor; Topromel; Vasocardin.; component of: Lopressor HCT. A β-adrenergic blocker. Lacks inherent sympathomimetic activity. Antianginal; class II antiarrhythmic; anti-hypertensive. Crystals; soluble in MeOH (50 g/100 ml), H_2O (> 100 g/100 ml), CHCl3 (49.6 g/100 ml), Me2CO (0.11 g/100 ml), CH3CN (0.089 g/100 ml); insoluble in C6H14; LD50 (fmus iv) = 118 mg/kg, (fmus orl) = 3090 mg/kg, (mrat iv) ≅ 90 mg/kg, (mrat orl) = 3090 mg/kg, (mrat iv) ≅ 90 mg/kg, (mrat orl) = 3090 mg/kg. *Apothecon; Astra Chem. Ltd.; Ciba-*

2542 Metribuzin
21087-64-9   6175   G P

$C_8H_{14}N_4OS$

4-Amino-6-(1,1-dimethylethyl)-3-(methylthio)-1,2,4-tri-azin-5(4H)-one.

5-26-06-00432 (Beilstein Handbook Reference); 4-Amino-6-(1,1-dimethylethyl)-3-(methylthio)-1,2,4-triazin-5(4H)-one; 4-Amino-6-t-butyl-3-(methylthio)-1,2,4-tri-azin-5-one; 4-Amino-6-t-butyl-4,5-dihydro-3-methyl-thio-1,2,4-triazin-5-one; 4-Amino-6-t-butyl-3-methylthio-1,2, 4-triazin-5(4H)-one; 4-Amino-6-t-butyl-3-(methyl-thio)-as-triazin-5(4H)-one; Bay 94337; BAY 61597; BAY dic 1468; Bayer 6159H; Bayer 6443H; Bayer 94337; BRN 0746650; Caswell No. 033D; DIC 1468; EINECS 244-209-7; EPA Pesticide Chemical Code 101101; HSDB 6844; Lexone; Lexone 4L; Lexone DF; 3-Methylthio-4-amino-6-t-butyl-1,2,4-triazin-5-one; Metribuzin; Metribuzine; NTN 70; Sencor; Sencor 4F; Sencor DF; Sencorex; Sencorex L.F.; Senkor; as-Triazin-5(4H)-one, 4-amino-6-t-butyl-3-(methylthio)-; 1,2,4-Triazin-5-one, 4-amino-6-t-butyl-3-(methylthio)-; 1,2,4-Triazin-5(4H)-one, 4-amino-6-(1,1-dimethylethyl)-3-(methylthio)-; Zenkor. Systemic herbicide used for control of many important grasses and broad-leaved weeds in soybeans, potatoes, tomatoes, sugarcane, alfalfa and asparagus; suitable for pre-and in some cases post emergence application. Registered by EPA as a herbicide. Colorless crystals; mp = 125.5 - 126.5°; sg^{20} = 1.28; d$^{20}_4$ = 1.28; slightly soluble in H_2O (0.105 g/100 ml), soluble in DMF (167 g/100 ml), cyclohexanone (94 g/100 ml), $CHCl_3$ (126 g/100 ml), Me_2CO (65 g/100 ml)), MeOH (36 g/100 ml), CH_2Cl_2 (44.3 g/100 ml), C_6H_6 (19 g/100 ml), n-BuOH (12 g/100 ml), EtOH (15 g/100 ml), C_7H_8 (10 g/100 ml), xylene (8 g/100 ml), i-PrOH (4 - 8 g/100 ml), C_6H_{14} (0.13 g/100 ml); LD_{50} (rat orl) = 2200 mg/kg, (mus orl) = 698 - 711 mg/kg, (gpg orl) = 250 mg/kg, (cat orl) > 500 mg/kg, (rat, rbt der) > 2000 mg/kg, (bobwhite quail orl) = 164 mg/kg; LC_{50} (bobwhite quail, mallard duck 5 day dietary) > 4000 mg/kg diet; (bluegill sunfish 96 hr.) = 80 mg/l, (rainbow trout 96 hr.) = 76 mg/l; non-toxic to bees. Bayer AG; Bayer Corp.; E. I. DuPont de Nemours Inc.; FMC Corp.; Syngenta Crop Protection; UAP - Platte Chemical.

2543 Metronidazole
443-48-1   6178   D P

$C_6H_9N_3O_3$

2-Methyl-5-nitroimidazole-1-ethanol.

5-23-05-00063 (Beilstein Handbook Reference); Acromona; Anagiardil; Arilin; Atrivyl; Bayer 5360; Bexon; BRN 0611683; Caswell No. 579AA; CCRIS 410; Clont; Cont; Danizol; Deflamon; Deflamonwirkstoff; Efloran; EINECS 207-136-1; Elyzol; Entizol; EPA Pesticide Chemical Code 120401; Eumin; Flagemona; Flagesol; Flagil; Flagyl; Fossyol; Giatricol; Gineflavir; HSDB 3129; Klion; Klont; Meronidal; Metro Cream & Gel; Metro Gel; Metro I.V.; Metrolag; Metrolyl; Metronidaz; Metro-nidazol; Metronidazole; Metronidazolo;

Metronidazolum; Metrotop; Mexibol; Mexibol 'silanes'; Mixture Name; Monagyl; Monasin; Nalox; Neo-tric; NIDA; Noritate; Novonidazol; NSC-50364; Orvagil; Protostat; Rathimed; RP 8823; Sanatrichom; Satric; SC 10295; Takimetol; Trichazol; Trichex; Tricho Cordes; Tricho-Gynaedron; Trichocide; Trichomol; Trichomonacid 'pharmachim'; Trichopol; Tricocet; Tricom; Tricowas B; Trikacide; Trikamon; Trikojol; Trikozol; Trimeks; Trivazol; Vagilen; Vagimid; Vertisal; Wagitran; Zadstat.; component of: Flagyl I.V. RTU, Metro I.V. Antiamebic, antibacterial and antiprotozoal against Trichomonas. Crystals; mp = 160.5°; λ_m = 227, 314 nm (ε = 3467, 8913, EtOH); soluble in H_2O (1.0 g/100 ml), EtOH (0.5 g/100 ml), Et_2O (< 0.05 g/100 ml), $CHCl_3$ (< 0.05 g/100 ml); sparingly soluble in DMF. Bayer Corp. Pharm. Div.; Elkins-Sinn; Galderma Labs Inc.; Lemmon Co.; McGaw Inc.; Ortho Pharm. Corp.; Savage Labs; SCS Pharm.

2544 Metronidazole hydrochloride
69198-10-3   6178   D

$C_6H_{10}ClN_3O_3$

2-Methyl-5-nitroimidazole-1-ethanol.

Flagyl I.V.; Imidazole-1-ethanol, 2-methyl-5-nitro-, hydro-chloride; 1H-Imidazole-1-ethanol, 2-methyl-5-nitro-, hydrochloride; 2-Methyl-5-nitroimidazole-1-ethanol monohydrochloride; 2-Methyl-5-nitroimidazole-1-ethanol hydrochloride; Metronidazole hydrochloride; SC 32642. Antiprotozoal against Trichomonas. Antiamebic and antibacterial. SCS Pharm.

2545 Metsulfuron
79510-48-8   P

$C_{13}H_{13}N_5O_6S$

2-[[[[(4-Methoxy-6-methyl-1,3,5-triazin-2-yl)amino]-carbonyl]amino]sulfonyl]benzoic acid.

Benzoic acid, 2-(((((4-methoxy-6-methyl-1,3,5-triazin-2-yl)amino)carbonyl)amino)sulfonyl)- 2-(((((4-Methoxy-6-methyl-1,3,5-triazin-2-yl)amino)carbonyl)amino)sulfonyl)-benzoic acid; 2-(3-(4-Methoxy-6-methyl-1,3,5-triazin-2-yl)ureidosulfonyl)benzoic acid; Metsu 453; Metsulfuron; 2-(3-(4-Methoxy-6-methyl-1,3,5-triazin-2-yl)ureidosulph-onyl)benzoic acid; 2-(4-Methoxy-6-methyl-1,3,5-triazin-2-ylcarbamoylsulfamoyl)benzoic acid; Metsu 453; Metsulfuron. Herbicide. E. I. DuPont de Nemours Inc.

2546 Metsulfuron-methyl
74223-64-6   6180   G P

C14H15N5O6S

Methyl 2-[[[[(4-methoxy-6-methyl-1,3,5-triazin-2-yl)-amino]carbonyl]amino]sulfonyl]benzoate.

Allie; Ally; Ally 20DF; BRN 0587472; Brush-off; Caswell No. 419H; DPD 63760H; DPD 63760M; DPX 6376; DPX-T 6376; EPA Pesticide Chemical Code 122010; Escort; Escort (pesticide); Finesse; Granstar; Gropper; HCHA 92HA; HSDB 6849; Metsulfuron Me; Metsulfuron methyl; Metsulfuron methyl ester; T 6376. Plant growth regulator, used for control of annual dicotyledons and pre- or post-emergence control of annual and perennial broad-leaved weeds in wheat, barley and oats. Registered by EPA as a herbicide. Colorless crystals; mp = 158°; slightly soluble in H2O (0.00011 g/100 ml at pH 5, 0.00095 g/100 ml at pH 7), soluble in xylene (0.058 g/100 ml), C6H14 (0.079 g/100 ml), EtOH (0.23 g/100 ml), MeOH (0.73 g/100 ml), Me2CO (3.6 g/100 ml), CH2Cl2 (12.1 g/100 ml); LD50 (rat orl) > 5000 mg/kg, (rbt der) > 2000 mg/kg, (mallard duck orl) > 5000 mg/kg; LC50 (rat ihl 4 hr.) > 5 mg/l air, (rainbow trout, bluegill sunfish 96 hr.) > 12.5 mg/l, (mallard duck, bobwhite quail 8 day dietary) > 5620 mg/kg diet; non-toxic to bees. DuPont UK; E. I. DuPont de Nemours Inc.; Makhteshim-Agan; Riverdale Chemical Co; Scotts Co.

2547   Metyrapone
54-36-4   6181   G

C14H14N2O

2-Methyl-1,3-di-3-pyridiyl-1-propanone.

1,2-Di-3-pyridyl-2-methyl-1-propanone; EINECS 200-206-2; HSDB 2500; Mepyrapone; Methapyrapone; Methbipyranone, SU-4885; Methopirapone; Metho-pyrapone; Methopyrinine; Methopyrone; 2-Methyl-1,2-di-3-pyridinyl-1-propanone; 2-Methyl-1,2-di-3-pyridyl-1-propanone; Metirapona; Metopiron; Metopirone; Metopyrone; Metroprione; Metyrapon; Metyrapone; Metyraponum; NSC 25265; 1-Propanone, 1,2-di-3-pyridyl-2-methyl-; 1-Propanone, 2-methyl-1,2-di-3-pyridinyl-; SU 4885. Diagnostic aid used to measure pituitary function. Solid: mp = 50-51°. Ciba-Geigy Corp.

2548   Metyridine
114-91-0   6182   G V

C8H11NO

2-(β-methoxyethyl)pyridine.

Al3-26615; Dekelmin; EINECS 204-060-0; Emthyridine; Farmintic; 2-(2-Methoxyethyl)pyridine; Methydrine; Metiridin; Metiridina; Metyridine; Metyridinum; Minthic; Mintic; NSC 34071; Prominthic; Promintic; Pyridine, 2-(2-methoxyethyl)-; 2-(2-Pyridyl)ethyl methyl ether; 2-Pyridyl-2-methoxyethane. Anthelmintic, used in veterinary medicine. Liquid; bp = 203°, bp17 = 96°; d20 = 0.988; very soluble in H2O, EtOH.

2549   Mica
12001-26-2   G

SiO2, Al2O3, K2O,Fe2O3, Na2O, CaO, TiO2, MnO2, P, S
Silicates: mica.

Abhrak; C 1000; CI 77019; Davenite P 12; HSDB 2539; HX 610; Mica; Mica, fluorian; Mica-group minerals; Mica, respirable fraction; Micacoat®; Micatex; Micromica W 1; Muscovite; Muscovite mica; P 80P; Silicate, mica; Silicates (<1% crystalline silica):MICA; Suzorite; Suzorite 60S; Suzorite mica. Chemically coupled muscovite mica, coarse and fine grinds: used for polymer composites and high performance coatings. NYCO® Minerals Inc..

2550   Michler's base
101-61-1   6200   G

C17H22N2

4,4'-Methylene bis(N,N'-dimethylaniline).

Al3-09165; Aniline, 4,4'-methylenebis(N,N-dimethyl-; Arnold's Base; BAZE michlerova; CCRIS 390; 4,4'-(Dimethylamino)diphenylmethane; Diphenylmethane, tetramethyldiamino-; EINECS 202-959-2; HSDB 2856; Methane, bis(p-(dimethylamino)phenyl)-; Methane-diamine, tetramethyl-N,N'-diphenyl-; Methylene base; Michler's hydride; Michler's ketone, reduced; Michler's methane; NCI-C01990; N,N'-Tetramethy-diamino-diphenylmethane; NSC 36782; Reduced Michler's ketone; Tetra-base; Tetramethyldiaminodiphenylmethane; methylene base; Michler's hydride; reduced Michler's ketone;; 4,4'-Methylene bis(N,N-dimethyl)benzenamine. Used in manufacture of dyes and as a reagent for lead. Crystals; mp = 91.5°; bp = 390° (dec), bp3 = 183°; insoluble in H2O, slightly soluble in EtOH, very soluble in Et2O, C6H6, CS2.

2551   Michler's hydrol
119-58-4   G

C17H22N2O

4,4'-Bis(dimethylamino)benzhydrol.

4-13-00-02148 (Beilstein Handbook Reference); Al3-17495; Benzenemethanol, 4-(dimethylamino)-α-(4-(dimethylamino)phenyl)-; Benzhydrol, 4,4'-bis-(dimethylamino)-; α,α-Bis(p-dimethylaminophenyl)-methanol; Bis(4-(dimethylamino)phenyl)methanol; BRN 2131843; EINECS 204-335-5; Michler's hydrol; p,p'-Michler's hydrol; NSC 3563; Tetramethyldiamino-benzhydrol. Used as a dye intermediate and in organic synthesis. Crystals; mp = 100-102°.

2552   Michler's ketone
90-94-8   6201   G

419

$C_{17}H_{20}N_2O$

Tetramethyldiaminobenzophenone.
AI3-22412; Benzophenone, 4,4'-bis(dimethylamino)-; CCRIS 412; EINECS 202-027-5; HSDB 2865; Methanone, bis(4-(dimethylamino)phenyl)-; Michler's ketone; p,p'-Michler's ketone; NCI-C02006; NSC 9602; Tetramethyldiaminobenzophenone. Used for making pigments and dyestuffs, especially auramine derivatives. Crystalline solid; mp = 179°; bp = 360° (dec); λ_m = 245 nm (MeOH); slightly soluble in H_2O, more soluble in organic solvents. *Hill Brothers Chemical Co.*

2553   Microcrystalline wax
63231-60-7   H

Petroleum wax, microcrystalline.
EINECS 264-038-1; Microcrystalline wax; Paraffin waxes and hydrocarbon waxes, microcryst.; Petroleum wax, microcrystalline.

2554   Minocycline
10118-90-8   6224   D

$C_{23}H_{27}N_3O_7$

4,7-Bis(dimethylamino)-1,4,4a,5,5a,6,11,12a-octahydro-3,10,12,12a-tetrahydroxy-1,1-dioxo-2-naphthacenecarb-oxamide.
BRN 3077644; CL 59806; HSDB 3130; Minociclina; Minocyclin; Minocycline; Minocyclinum; Minocyn. Tetracycline antibiotic. Semi-synthetic. Effective against tetracycline-resistant staphylococci. $[\alpha]_D^{25}$ = -166° (c = 0.524); λ_m 352 263 nm (4.16 4.23 0.1N HCl), 380 243 nm (log ε 4.30 4.38 0.1N NaOH). *Lederle Labs.; Parke-Davis.*

2555   Minocycline hydrochloride
13614-98-7   6224   D

$C_{23}H_{28}ClN_3O_7$

4,7-Bis(dimethylamino)-1,4,4a,5,5a,6,11,12a-octahydro-3,10,12,12a-tetrahydroxy-1,1-dioxo-2-naphthacenecarb-oxamide hydrochloride.
Arestin; EINECS 237-099-7; Klinomycin; Minocin; Minocycline chloride; Minocycline hydrochloride; Minomax; Minomycin; Mynocine hydrochloride; NSC 141993; Tri-mino; Tri-minocycline; Vectrin. Tetracycline antibiotic. Yellow crystalline powder; sensitive to light and to oxidation. *Lederle Labs.; Parke-Davis.*

2556   Mirex
2385-85-5   6229   P

$C_{10}Cl_{12}$

Dodecachlorooctahydro-1,3,4-metheno-1H-cyclobuta-(cd)pentalene.
AI3-25719; Bichlorendo; Caswell No. 411; CCRIS 413; CG-1283; Cyclopentadiene, hexachloro-, dimer; 1,3-Cyclopentadiene, 1,2,3,4,5,5-hexachloro-, dimer; Dechl-orane; 1,1a,2,2,3,3a,4,5,5,5a,5b,6-Dodecachloro-octa-hydro-1,3,4-metheno-1H-cyclobuta(cd)pentalene; Do-decachloropentacyclodecane; Dodecachloropentacyclo-$(5.2.1.0^{2,6}.0^{3,9}.0^{5,8})$decane; Dodecachloropenta-cyclo-$(3.2.2.0^{2,6}.0^{3,9}.0^{5,10})$decane; Dodecaclor; EINECS 219-196-6; ENT 25,719; EPA Pesticide Chemical Code 039201; Ferriamicide; Fire Ant Bait; GC 1283; Hexachlorocyclopentadiene dimer; 1,2,3,4,5,5-Hexa-chloro-1,3-cyclopentadiene dimer; HRS 1276; HSDB 1659; 1,3,4-Metheno-1H-cyclobuta(cd)pentalene, 1,1a,2,2,3,3a,4,5,5,5a,5b,6-dodecachlorooctahydro-; 1,3,4-Metheno-1H-cyclobuta(cd)pentalene, dodecachloroocta-hydro-; Mirex; NCI-C06428; NSC 26107; NSC 37656; NSC 124102; Paramex; Pentacyclodecane, dodeca-chloro-; Perchlordecone; Perchlorodihomo-cubane; Per-chloropentacyclo(5.2.1.02,6.0^{3,9}.0^{5,8})decane; Perchloropentacyclodecane. Used as a fire retardant and formerly, as an insecticide for control of fire ants. Designated as a Persistent Organic Pollutant (POP) under the Stockholm convention. White crystals; dec. 485°; insoluble in H_2O, soluble in dioxane (15 g/100 ml), xylene (14 g/100 ml), C_6H_6 (12 g/100 ml), CCl_4 (7 g/100 ml), MEK (6 g/100 ml), LD_{50} (frat orl) = 600 mg/kg.

2557   Mirtazapine
61337-67-5   6230   D

$C_{17}H_{19}N_3$

1,2,3,4,10,14b-Hexahydro-2-methylpyrazino[2,1-a]pyrido[2,3-c]benzazepine.
EINECS 288-060-6; Mepirzepine; Mirtazepine; Org-3770; Remeron. An α_2-adrenergic blocker; analog of mianserin. Tetracyclic antidepressant. Crystals; mp = 114-116°. *Chinoin; Organon Inc.*

2558   MNT
1321-12-6   G

C7H7NO2
  Mononitrotoluenes.
  Benzene, methylnitro-; EINECS 215-311-9; HSDB 6301; Methylnitrobenzene; MNT; Mononitrotoluene; Nitrophenylmethane; Nitrotoluene; Nitrotoluene (all isomers); Nitrotoluenes; Nitrotoluidines (mono); Nitrotoluol; UN2660. Mixture of o, m and p-isomers. Used as intermediates in the preparation of trinitrotoluene.

2559 Molybdenum
  7439-98-7   6257   G

  Mo
  Molybdenum.
  Amperit 105.054; Amperit 106.2; EINECS 231-107-2; HSDB 5032; MChVL; Metco 63; Molybdenum; Molybdenum, elemental; Molybdenum, metallic; TsM1. Metallic element; alloying agent in steels and cast iron; pigments for printing inks, paints, ceramics; catalyst; solid lubricants; missile and aircraft parts; reactor vessels; cermets; die-casting copper-base alloys; special batteries. Metal; mp = 2622°; bp = 4825°; d = 10.28. *AAA Molybdenum; Atomergic Chemetals; Cerac; Climax Molybdenum Co.*

2560 Molybdenum dioxide
  18868-43-4   G

  MoO2
  Molybdenum(IV) oxide.
  EINECS 242-637-9; Molybdenum dioxide; Molybdenum oxide (MoO2). *AAA Molybdenum; Atomergic Chemetals; Climax Molybdenum Co.*

2561 Molybdenum disulfide
  1317-33-5   6258   G

  MoS2
  Molybdenum(IV) sulfide.
  C.I. 77770; C.I. Pigment Black 34; DAG 206; DAG 325; DAG-V 657; DM 1; EINECS 215-263-9; HSDB 1660; Liqui-Moly LM 2; Liqui-Moly LM 11; LM 13; LM 13 (lubricant); M 5; M 5 (lubricant); MD 40; MD 40 (lubricant); Moly Powder B; Moly Powder C; Moly Powder PA; Moly Powder PS; Molybdenite; Molybdenum bisulfide; Molybdenum disulfide; Molybdenum sulfide (MoS2); Molybdenum(IV) sulfide; Molycolloid CF 626; Molyke R; Molykote; Molykote Microsize Powder; Molykote Z; Molysulfide; Mopol M; Mopol S; Motimol; Natural molybdenite; Nichimoly C; Pigment Black 34; Solvest 390A; Sumipowder PA; T-Powder. Used as a dry lubricant and hydrogenation catalyst. Lead-gray powder; mp = 2375°; sublimes at 450°; d15 = 5.06; insoluble in H2O. *AAA Molybdenum; Climax Molybdenum Co.; Dow Corning.*

2562 Molybdenum trioxide
  1313-27-5   6261   G

  MoO3
  Molybdenum(VI) oxide.
  CCRIS 1163; EINECS 215-204-7; HSDB 1661; Mo-1202T; Molybdena; Molybdenum oxide; Molybdenum peroxide; Molybdenum trioxide; Molybdenum(VI) oxide; Molybdenumperoxide;

Molybdic acid anhydride; Molybdic anhydride; Molybdic oxide; Molybdic trioxide; Natural molybdite; NSC 215191. Source of Mo; reagent for analytical chemistry; agriculture; manufacture of metallic molybdenum; corrosion inhibitor; ceramic glazes; enamels; pigments; catalyst. Yellow-blue powder; mp = 795°; bp = 1155°; d4^26 = 4.696; soluble in H2O (0.49 g/l), insoluble in organic solvents; LD50 (rat orl) = 2689 mg/kg. *AAA Molybdenum; Atomergic Chemetals; Cerac; Climax Molybdenum Co.*

2563 Mometasone furoate
  83919-23-7   6264   D

  C27H30Cl2O6
  (11β,16α)-9,21-Dichloro-17-[(2-furanylcarbonyl)oxy]-11-hydroxy-16-methylpregna-1,4-diene-3,20-dione.
  BRN 4340538; Danitin; Ecural; Elocon; Elocone; Flumeta; Mometasone furoate; Nasonex; Nosorex; Rimelon; Sch 32088. Topical corticosteroid. Anti-inflammatory. Crystals; mp = 218-220°; [α]D^26 = + 58.3° (in dioxane); λm = 247 nm (ε 26300 in MeOH). *Schering-Plough HealthCare Products.*

2564 Monacetin
  26446-35-5   6265   G

  C5H10O4
  1,2,3-propanetriol monoacetate.
  Acetin; Acetin, mono-; Acetoglyceride; Acetyl monoglyceride; AI3-24158; CCRIS 5881; EINECS 247-704-6; Glycerin monoacetate; Glycerine monoacetate; Glycerol acetate; Glycerol monoacetate; Glyceryl acetate; Glyceryl monoacetate; HSDB 4285; Monacetin; Mono-acetin; Monoacetin; 1,2,3-Propanetriol, monoacetate. Mixture of the 1- and 2-monoacetates. Used for tanning; solvent for dyes; food additive; gelatinizing agent in explosives. Liquid; bp17 = 158°; d4^20 = 1.206; soluble in H2O, EtOH, insoluble in other organic solvents; LD50 (rat sc) = 6.6 g/kg.

2565 Monacrin
  134-50-9   406   G

  C13H11ClN2
  Aminacrine hydrochloride.
  Acramine Yellow; 9-Acridinamine, monohydrochloride; Acridin-9-ylamine hydrochloride monohydrate; Acridine, 9-amino-,

421

hydrochloride; Acridine, 9-amino-, mono-hydrochloride; AI3-16947; Aminacrine hydrochloride; Aminoacridine hydrochloride; 5-Aminoacridine hydrochloride; 9-Aminoacridine hydrochloride; 9-Aminoacridinium chloride; Aminoakridin; Caswell No. 033A; CCRIS 3802; EINECS 205-145-5; EPA Pesticide Chemical Code 055503; HSDB 4505; Monacrin; Monacrin hydrochloride; Mycosert; NSC-7571; Quench; Acridinamine, monohydrochloride; Aminoacridine monohydrochloride; Aminoacridinium chloride; NSC-7571. Anti-infective, topical. Highly fluorescent. Crystals; mp = 241°; insoluble in H_2O, soluble in organic solvents; LD_{50} (mus orl) = 78 mg/kg. *Sterling Res. Labs.*

2566   Monensin
17090-79-8   6270   D G

$C_{36}H_{62}O_{11}$
2-[2-Ethyloctahydro-3'-methyl-5'-[tetrahydro-6-hydroxy-6-(hydroxymethyl)-3,5-dimethyl-2H-pyran-2-yl][2,2'-bifuran-5-yl]]-9-hydroxy-β-methoxy-α,γ,2,8-tetramethyl-1,6-dioxapsiro[4.5]decan-7-butanoic acid.
63714; A 3823A; ATCC 15413; Coban (as sodium salt); EINECS 241-154-0; Elancoban; HSDB 7031; Lilly 673140; Monelan; Monensic acid; Monensin; Monensin A; Monensina; Monensinum; Rumensin (as sodium salt). Antibiotic produced by *Streptomyces cinnamonensis.* Antifungal, antibiotic and antiprotozoal. Solid; mp = 103-105°; [α]D = 47.7°; slightly soluble in H_2O, more soluble in hydrocarbons, very soluble in organic solvents; LD_{50} (mus orl) = 43.8 ± 5.2 mg/kg, (chicks orl) = 284 ± 47 mg/kg. *Eli Lilly & Co.*

2567   Monobutyl maleate
925-21-3   H

$C_8H_{12}O_4$
2-Butenedioic acid (Z)-, monobutyl ester.
2-Butenedioic acid (Z)-, monobutyl ester; Butyl hydrogen maleate; EINECS 213-116-3; Maleic acid, monobutyl ester.

2568   Monochlorodifluoroethane
75-68-3   H

$C_2H_3ClF_2$
1-Chloro-1,1-difluoroethane.
4-01-00-00127 (Beilstein Handbook Reference); BRN 1731584; CFC 142b; Chlorodifluoroethane; Chloroethyl-idene fluoride; Chlorofluorocarbon 142b; Difluoro-1-chloroethane; EINECS 200-891-8; Ethane, 1-chloro-1,1-difluoro-; FC142b; Fluorocarbon 142b; Fluorocarbon FC142b; Freon 142; Freon 142b; Genetron 101; Genetron 142b; Gentron 142b; HCFC-142b; HSDB 2881; Hydrochlorofluorocarbon 142b; Monochlorodifluoroethane; Propellant 142b; R-142b; UN2517. Used as a refrigerant. Gas; mp = -130.8°; bp = -9.7°; d^{25} = 1.107; insoluble in H_2O, soluble in C_6H_6. *Lancaster Synthesis Co.; Mallinckrodt Inc.; Sigma-Aldrich Fine Chem.*

2569   Mono(2-ethylhexyl)phosphate   tert-dodecyl-amine salt
67763-14-5   H

$C_{12}H_{27}N.xC_8H_{19}O_4P$
Phosphoric acid, mono(2-ethylhexyl) ester, compd. with tert-dodecanamine.
EINECS 267-037-4; 2-Ethylhexyl dihydrogen phosphate, compound with tert-dodecylamine; Mono(2-ethyl-hexyl)phosphate tert-dodecylamine salt; Phosphoric acid, mono(2-ethylhexyl) ester, compound with tert-dodecanamine.

2570   Mono(2-hydroxyethyl)   3-(hydroxymethyl-phosphinyl)propionate
68334-62-3   H

$C_6H_{13}O_5P$
Propanoic acid, 3-(hydroxymethylphosphinyl)-, mono(2-hydroxyethyl) ester.
EINECS 269-837-9; Mono(2-hydroxyethyl) 3-(hydroxy-methylphosphinyl)propionate; Propanoic acid, 3-(hydroxymethylphosphinyl)-, mono(2-hydroxyethyl) ester.

2571   Monoisopropanolamine
78-96-6   H

C_3H_9NO
1-Amino-2-propanol.
4-04-00-01665 (Beilstein Handbook Reference); AI3-14653; BRN 0605275; CCRIS 2284; EINECS 201-162-7; HSDB 5224; Isopropanolamine; Mipa; Mono-iso-propanolamine; Monoisopropanolamine; NSC 3188; Threamine. Used in personal care products and in manufacture of cement and concrete. Colorless liquid; mp = -2°; bp = 159.5°; d = 0.96; soluble in H_2O, organic solvents. *Dow Chem. U.S.A.*

2572   Monoisopropanolamine oleic acid amide
111-05-7   G

C21H41NO2

9-Octadecenamide, N-(2-hydroxypropyl)-, (Z)-.
EINECS 203-828-2; N-(2-Hydroxypropyl)-9-octadecen-amide; N-(2-Hydroxypropyl)oleamide; Monoisopropanol-amine oleic acid amide; 9-Octadecenamide, N-(2-hydroxypropyl)-; 9-Octadecenamide, N-(2-hydroxy-propyl)-, (Z)-; 9-Octadecenamide, N-(2-hydroxypropyl)-, (9Z)-; Oleamide MIPA; Oleic monoisopropanolamide; Steinamid IPE 280; Witcamide® 61. Used as a hair conditioner, emulsifier, lubricant; cosmetics and toiletries. *CK Witco Corp.*

2573　Monomethyl terephthalate
1679-64-7　　　　　　　　　　　　　　　　H

C9H8O4

1,4-Benzenedicarboxylic acid, monomethyl ester.
Acetic acid, (4-carboxyphenyl)-; 1,4-Benzenedicarboxylic acid, monomethyl ester; 4-(Carbomethoxy)benzoic acid; p-Carboxy-α-toluic acid; EINECS 216-849-7; Homoterephthalic acid; HSDB 5849; Hydrogen methyl terephthalate; 4-Methoxycarbonylbenzoic acid; Methyl hydrogen terephthalate; Methyl terephthalate; Monomethyl terephthalate; NSC 210838; Terephthalic acid, monomethyl ester. Used as a plasticizer.

2574　1-Monoolein
111-03-5　　　　　　　　　　　　　　　　G

C21H40O4

Glyceryl oleate.
Ablunol GMO; Aldo HMO; Aldo MO; Danisco MO 90; 2,3-Dihydroxypropyl oleate; EINECS 203-827-7; FEMA No. 2526; Glyceryl monooleate; Glycerin 1-monooleate; Glycerol α-monooleate; Glycerol α-cis-9-octadecenate; Glyceryl oleate; 1-Glyceryl oleate; 1-Mono(cis-9-octacenoyl)glycerol; Monoolein; α-Monoolein; 1-Monoolein; 1-Monooleoylglycerol; 1-Monooleoyl-rac-glycerol; 1-Oleoylglycerol; NSC 406285; 9-Octadecenoic acid, 2,3-dihydroxypropyl ester; 9-Octadecenenoic acid (Z)-, 2,3-dihydroxypropyl ester; 9-Octadecenoic acid (9Z)-, 2,3-dihydroxypropyl ester; 9-Octadecenoic acid, monoester with 1,2,3-propanetriol; Olein, 1-mono-; rac-1-Monoolein; rac-1-Monooleoylglycerol; Radiasurf® 7150; Witconol 2421; 1-Glyceryl oleate; Monolein; 1-Monoolein.alpha.-Monoolein; 1-Monooleoylglycerol; 1-Oleoylglycerol; 1-Oleylglycerol. Internal lubricant for PVC; biodegradable surfactant, wetting agent, emulsifier for cosmetics, pharmaceuticals, agriculture, chemical synthesis, explosives, polymers, glass fibers, surface coatings, textiles and leather. Used as a defoamer, oil-water emulsifier, lubricant and moisture barrier. Oil; mp = 35°; bp3 = 238-240°; d^{20} = 0.9420; insoluble in H2O, soluble in EtOH, Et2O, CHCl3. *CK Witco Corp.*;

Fina Chemicals; Taiwan Surfactants.

2575　Monosodium glutamate
142-47-2　　　　　　　6278　　　　　　　G

C5H8NNaO4.H2O

Glutamic acid monosodium salt.
Accent; Accent (food additive); Al3-18393; Ajinomoto; Ancoma; CCRIS 3625; Chinese seasoning; EINECS 205-538-1; FEMA No. 2756; Glutacyl; Glutamat sodny; Glutamate monosodium salt; Glutamate Sodium; Glutamic acid, monosodium salt; Glutamic acid, sodium salt; Glutamic acid, L-, sodium salt; L-Glutamic acid sodium salt; Glutammato monosodico; Glutavene; HSDB 580; Monosodioglutammato; Monosodium glutamate; Monosodium L-glutamate; MSG; Natrium L-hydrogenglutamat; Natriumglutaminat; NSC 135529; RL-50; Sodium glutamate; Sodium hydrogen glutamate; Sodium L-glutamate; Vetsin; Zest. Flavor enhancer for foods. Crystals; mp = 232° (dec); $[\alpha]_D^{20}$= 25° (c = 5 1N HCl); soluble in H2O. *Ajinomoto Co. Inc.; Allchem Ind.; Asahi Chem. Industry; Lancaster Synthesis Co.; Penta Mfg.*

2576　Montelukast
158966-92-8　　　　　　6281　　　　　　　D

C35H36ClNO3S

1-[[[(R)-m-[(E)-2-(7-Chloro-2-quinolyl)vinyl]-α-[o-(1-hydroxy-1-methylethyl)phenethyl]benzyl]thio]methyl]-cyclopropane acetic acid.
Montelukast. Selective leukotriene D4 receptor antagonist. Used as an antiasthmatic. *Merck & Co.Inc.*

2577　Montelukast sodium
151767-02-1　　　　　　6281　　　　　　　D

C35H35ClNNaO3S

Sodium 1-[[[(R)-m-[(E)-2-(7-chloro-2-quinolyl)vinyl]-α-[o-(1-hydroxy-1-methylethyl)phenethyl]benzyl]thio]methyl]-cyclopropaneacetate.
MK 476; Montelukast monosodium salt; Montelukast sodium; Singulair. Selective leukotriene D4 receptor antagonist. Used as an antiasthmatic. *Merck & Co.Inc.*

2578 Montmorillonite
1318-93-0   6283   G

$HAlSi_2O_6$
White montmorillonite.
Alabama Blue Clay; Albagen 4439; AMS (mineral); Arcillite; Bedelix; Ben-A-Gel; Ben-A-Gel EW; Benclay MK 101; Bentolite; Bentolite L 3; BPW 009; BPW 009-3; BPW 009-10; BPW 015-10; Brock; Deriton; DH 1; DH 1 (catalyst); DH 2; Dis-Thix Extra; EINECS 215-288-5; Envirobent; Flygtol GA; Furonaito 101; Furonaito 113; Galleonite 136; Gelwhite GP; Gelwhite H; Gelwhite L; Hydrocol 2D1; Imvite E; Imvite K; K 10 (clay); K 129; KM 1 (mineral); Kunipia G; Kunipia G 4; Kunipia TO; Lavioplast C; Metaloid; Mineral Colloid BP; Montmorillonite; Montmorillonite (AlH(SiO3)2); Mont-morillonite (HAlSi2O6); Montmorillonite K 10; Montmorillonite KSF; Neokunibond; Optigel CL; Osmos N. Major constituent of Bentonite and Fuller's earth. Used as a binder and plasticizer for ceramic formulations; ion exchange builder in detergents; thixotropic agent for liq. soaps; flocculant for water treatment. Industrial chromatographic techniques White powder; bulk d = 300-370 g/l. *Kaopolite.*

2579 Morantel tartrate
26155-31-7   6289   G V

$C_{12}H_{16}N_2S \cdot C_4H_6O_6$
(E)-1,4,5,6-tetrahydro-1-methyl-2-[2-(3-methyl-2-thienyl)ethenyl]pyrimidine tartrate.
AI3-29747; Banminth II; CP 12009-18; EINECS 247-481-5; Morantel tartrate; Paratect; Pyrimidine, 1,4,5,6-tetrahydro-1-methyl-2-(2-(3-methyl-2-thienyl)ethenyl)-, (E)-, (R-(R*,R*))-2,3-dihydroxybutanedioate (1:1); Pyr-imidine, 1,4,5,6-tetrahydro-1-methyl-2-(2-(3-methyl-2-thi-enyl)vinyl)-, (E)-, tartrate (1:1); (E)-1,4,5,6-Tetrahydro-1-methyl-2-(2-(3-methyl-2-thienyl)vinyl)pyrimidine tartrate (1:1); Thelmesan; Suiminth. Anthelmintic.

2580 Morpholine
110-91-8   6303   G H

C_4H_9NO
Tetrahydro-1,4-isoxazine.
4-27-00-00015 (Beilstein Handbook Reference); AI3-01231; BASF 238; BRN 0102549; Caswell No. 584; CCRIS 2482; Diethylene imidoxide; Diethylene oximide; Diethylenimide oxide; Drewamine; EINECS 203-815-1; EPA Pesticide Chemical Code 054701; HSDB 102; p-Isoxazine, tetrahydro-; Morpholine; NSC 9376; Tetrahydro-1,4-isoxazine; Tetrahydro-p-oxazine; UN2054. Used as a rubber accelerator; solvent; additive to boiler water; optical brightener for detergents; corrosion inhibitor; organic intermediate. Liquid; mp = -4.9°; bp = 128°; d^{20} = 1.0005; very soluble in H_2O, soluble in EtOH, Et_2O, Me_2CO; LD50 (rat orl) = 1.05 g/kg. *Air Products & Chemicals Inc.; BASF Corp.; Nippon Nyukazai; PMC; Texaco.*

2581 2-(4-Morpholinyldithio)benzothiazole
95-32-9   G H

$C_{11}H_{12}N_2OS_3$
Benzothiazole, 2-(4-morpholinyldithio)-.
4-27-00-01864 (Beilstein Handbook Reference); Accel DS; AI3-27133; Benzothiazole 2-(4-morpholinyl); Benzothiazole, 2-(4-morpholinyldithio); BRN 0222552; Disulfal MG; EINECS 202-410-7; HSDB 5289; Morfax; Morpholino 2-benzothiazolyl disulfide; N-Morpholinyl-2-benzothiazolyl disulfide; Nocceler MDB; NSC 519695; N-Oxydiethyl-2-benzthiazolsulfenamid; Sulfenax mob; Vulcuren 2. An accelerator for natural rubber, isoprene butadiene, styrene-butadiene, and nitrile-butadiene rubber products; provides good curing activity; suggested for tires and mechanical goods requiring maximum strength and quality. *Goodyear Tire & Rubber.*

2582 4-((4-Morpholinylthio)thioxomethyl)-morpholine
13752-51-7   H

$C_9H_{16}N_2O_2S_2$
Morpholine, 4-((4-morpholinylthio)thioxomethyl)-.
Accelerator otos; BRN 1214828; Cure-Rite 18; EINECS 237-335-9; Morpholine, 4-((4-morpholinylthio)thioxo-methyl)-; Morpholine, 4-((morpholinothiocarbonyl)thio)-; 4-((Morpholinothio)thioxomethyl)morpholine; 4-((Morph-olinothiocarbonyl)thio)morpholine; OTOS; N-Oxydi-ethylene thiocarbamyl-N'-oxydiethylene sulfenamide.

2583 MP Diol Glycol
2163-42-0   G

$C_4H_{10}O_2$
2-Methyl-1,3-propanediol.
Propane-1,3-diol, 2-methyl-. Used in unsaturated polyesters, gel coats, saturated polyester and alkyd coatings, plasticizers. *Arco.*

2584 MSMA
2163-80-6   6020   G H P

CH_4AsNaO_3
Monosodium methylarsonate.
Ansar; Ansar 170; Ansar 170L; Ansar 529; Ansar 529 HC; Ansar 6.6; Arsonate Liquid; Arsonic acid, methyl-, monosodium salt; Asazol; Bueno; Bueno 6; Caswell No. 582; CCRIS 4676; Daconate; Daconate 6; Dal-E-Rad; Dal-E-Rad 120; Drexar; Drexar 530; EINECS 218-495-9; EPA Pesticide Chemical Code 013803; Gepiron; Herb-All; Herban M; HSDB 754; Merge; Merge 823; Mesamate; Mesamate-400; Mesamate-600; Mesamate concentrate; Mesamate H.C.; Methanearsonic acid, monosodium salt; Methylarsenic acid, sodium salt; Methylarsonat monosodny;

Methylarsonic acid, sodium salt; Monate; Monex; Monoban; Monomethylarsonic acid sodium salt; Monosodium acid methanearsonate; Monosodium acid metharsonate; Monosodium methanearsenate; Monosodium methanearsonic acid; Monosodium methylarsonate; MSMA; NCI-C60071; Neoarsycodyl; Phyban; Phyban H.C.; Silvisar 550; Sodium acid methanearsonate; Sodium hydrogen methylarsonate; Sodium methanearsonate; Sodium methylarsonate; Super Arsonate; Target MSMA; Trans-vert; Versar; Weed 108; Weed-E-Rad; Weed-Hoe-108; Weed-Hoe; Neoarsycodyl; monosodium methylarsonate; monosodium methanearsonate. Selective contact herbicide with some systemic properties. Used in post-emergence control of grass weeds in cotton, sugar cane and under trees, as a herbicide on turf to control established crabgrass, dollis grass, and nutselse. Colorless crystals; mp = 113 - 116° (sesquihydrate), mp = 132 - 139° (hexahydrate); soluble in H_2O (140 g/100 ml (anhydrous); soluble in MeOH, insoluble in most organic solvents; LD50 (rat orl) = 900 mg/kg; LC50 (bluegill sunfish 48 hr.) > 1000 mg/l. *American Brand Chemical Co.; Bayer Corp.; Drexel Chemical Co.; Earth Care; Fermenta; Inter-Ag; KMG-Bernuth Inc.; Lawn & Garden Products Inc.; Luxembourg-Pamol, Inc.; PBI/Gordon Corp; Shinung; UAP - Platte Chemical; Vertac; Vineland Lab.*

2585 Mucochloric acid
87-56-9   6323   H

$C_4H_2Cl_2O_3$
 2-Butenoic acid, 2,3-dichloro-4-oxo-, (2Z)-.
 4-03-00-01720 (Beilstein Handbook Reference); Acrylic acid, 2,3-dichloro-3-formyl-; AI3-26601; Aldehydodi-chloromaleic acid; BRN 1705641; CCRIS 6597; Dichloromalealdehydic acid; EINECS 201-752-4; Kyselina mukochlorova; Malealdehydic acid, dichloro-4-oxo-, (Z)-; Malealdehydic acid, dichloro-; Mucochloric acid; NSC 15905. Prisms; mp = 127°; slightly soluble in H_2O, organic solvents; LD50 (rat orl) = 500 - 1000 mg/kg, (rat ip) = 10 - 25 mg/kg.

2586 Mupirocin
12650-69-0   6327   D

$C_{26}H_{44}O_9$
 (E)-(2S,3R,4R,5S)-5-([[(2S,3S,4S,5S)-2,3-Epoxy-5-hydroxy-4-methylhexyl]tetrahydro-3,4-dihydroxy-β-methyl-2H-pyran-2-crotonic acid ester with 9-hydroxynonanoic acid.
 Bactoderm; Bactroban; BRL 4910A; Mupirocin; Mupirocina; Mupirocine; Mupirocinum; Plasimine; Pseudomonic acid; Pseudomonic acid A; Turixin. Antibacterial. Crystals; mp = 77-78°; $[\alpha]_D^{20}$= -19.3° (c = 1 MeOH); λ_m 222 nm (ε 14500 EtOH). *SmithKline Beecham Pharm.*

2587 Murexide
3051-09-0   6330   G

NH4+

$C_8H_8N_6O_6$
 5-[(Hexahydro-2,4,6-trioxo-5-pyrimidinyl)imino]-2,4,6(1H,3H,5H)-pyrimidinetrione monoammonium salt.
 Ammonium purpurate; Ammonium salt purpuric acid; Ammonium 5-(2,4,6-trioxoperhydropyrimid-5-ylidene-amino)barbiturate; Barbituric acid, 5,5'-nitrilodi-, monoammonium salt; EINECS 221-266-6; Murexide; Naples red; NSC 215208; 2,4,6(1H,3H,5H)-Pyrimidinetrione, 5-((hexahydro-2,4,6-trioxo-5-pyrimid-inyl)imino)-, monoammonium salt. A red basic dyestuff, now obsolete, obtained by the action of nitric acid upon guano, and subsequently treating the product with ammonia. Used as an indicator in complexometric titrations. Crystals; slightly soluble in H_2O, insoluble in organic solvents; λ_m = 520 nm (H_2O).

2588 Muscarine
300-54-9   6334   D G

$[C_7H_{20}NO_2]^+$
 [2S-(2α,4β,5α)]-tetrahydro-4-hydroxy-N,N,N,5-tetramethyl-2-furanmethanaminium.
 Ammonium, trimethyl(tetrahydro-4-hydroxy-5-methyl-furfuryl)-; EINECS 206-094-1; 2-Furanmethanaminium, tetrahydro-4-hydroxy-N,N,N,5-tetramethyl-, (2S-2α,4β,5α))-; D-ribo-Hexitol, 2,5-anhydro-1,4,6-trideoxy-6-(trimethyl)ammonio)-; Muscarin; Muscarine; (+)-Muscarine; L-(+)-Muscarine; (+)-(2S,4R,5S)-Muscarine; Muskarin; (2S-(2α,4β,5α))-(Tetrahydro-4-hydroxy-5-methylfurfuryl)trimethylammonium. Toxic principle of the mushroom *Amanita muscaria*. Cholinergic agent. [chloride]: mp = 180-181°; $[\alpha]_D^{25}$= 8.1° (c = 3.5 EtOH); soluble in H_2O, EtOH; less soluble in Et_2O, $CHCl_3$, Me_2CO; LD50 (mus iv) = 0.23 mg/kg.

2589 Muscone
541-91-3   6337   G

$C_{16}H_{30}O$
 3-Methylcyclopentadecanone.
 AI3-38746; Cyclopentadecanone, 3-methyl-; EINECS 208-795-8; FEMA No. 3434; HSDB 1219; 3-Methyl-1-cyclopentadecanone; 3-Methylcyclopentadecanone; 3-Methylcyclopentadecan-1-one; 3-Methylcyclopenta-decanone, dl-; 5-Methyl-1-cyclopentadecanone; Methyl-exaltone; Moschus ketone; Muscone; Muskone. Odiferous secretion of the musk. No longer used in perfumery. Liquid; bp = 328°; bp0.5 = 130°; $[\alpha]_D^{17}$= -13°; insoluble in H_2O, very soluble in EtOH, Et_2O, Me_2CO.

2590 Musk ambrette
83-66-9

G

C12H16N2O5
2,6-Dinitro-3-methoxy-1-methyl-4-t-butylbenzene.
4-06-00-03402 (Beilstein Handbook Reference); AI3-02439; Amber musk; Ambrette musk; Anisole, 6-t-butyl-3-methyl-2,4-dinitro-; Artificial musk ambrette; Benzene, 1-(1,1-dimethylethyl)-2-methoxy-4-methyl-3,5-dinitro-; BRN 1889437; 4-t-Butyl-3-methoxy-2,6-dinitrotoluene; 4-t-Butyl-3-methoxy-1-methyl-2,6-dinitrobenzene; 6-t-But-yl-3-methyl-2,4-dinitroanisole; CCRIS 2390; 1-(1,1-Di-methylethyl)-2-methoxy-4-methyl-3,5-dinitrobenzene; 2,4-Dinitro-3-methyl-6-t-butylanisole; 2,6-Dinitro-3-methoxy-4-t-butyltoluene; EINECS 201-493-7; 4-Methoxy-1,3-dinitro-2-methyl-5-t-butylbenzene; Musk ambrette; NSC 46122; Synthetic musk ambrette. Used in perfumery. Yellow leaflets; mp = 85°; bp[16] = 185°, bp = 135°; λ_m = 265 nm (MeOH); insoluble in H2O, poorly soluble in EtOH, soluble in Et2O, CHCl3.

2591 Musk ketone
81-14-1

G

C14H18N2O5
4-t-Butyl-2,6-dimethyl-3,5-dinitroacetophenone.
4-07-00-00808 (Beilstein Handbook Reference); Aceto-phenone, 4'-t-butyl-2',6'-dimethyl-3',5'-dinitro-; 2-Acetyl-5-t-butyl-4,6-dinitroxylene; AI3-02440; BRN 2062638; 4-t-Butyl-3,5-dinitro-2,6-dimethylaceto-phenone; 4'-t-Butyl-2',6'-dimethyl-3',5'-dinitroaceto-phenone; CCRIS 4677; 1-(4-(1,1-Dimethylethyl)-2,6-dimethyl-3,5-dinitrophenyl)-ethanone; 2,6-Dinitro-3,5-di-methyl-4-acetyl-t-butyl-benzene; 3,5-Dinitro-2,6-di-methyl-4-t-butylacetophen-one; EINECS 201-328-9; Ethanone, 1-(4-(1,1-dimethyl-ethyl)-2,6-dimethyl-3,5-dinitrophenyl)-; Musk ketone; NSC 15339; Acetophenone, 4'-t-butyl-2',6'-dimethyl-3',5'-dinitro-; 1-Acetyl-4-t-butyl-2,6-dimethyl-3,5-dinitro-benzene; 4-t-Butyl-2,6-dimethyl-3,5-dinitro-acetophen-one; 4'-t-Butyl-2',6'-dimethyl-3',5'-dinitro-acetophenone; 1-[4-(1,1-Dimethylethyl)-2,6-dimethyl-3,5-dinitrophenyl]-ethanone. An artificial musk perfume. Solid; mp = 138-140°. AM Aromatics Pvt, Ltd.; Zhejiang Winsun Imp. And Exp. Co., Ltd.

2592 Mustard gas
505-60-2

6340

G

C4H8Cl2S
1,1'-Thiobis[2-chloroethane].
4-01-00-01407 (Beilstein Handbook Reference); Bis(β-chloroethyl)sulfide; Bis(β-chloroethyl)sulphide; Bis(2-chloroethyl) sulfide; Bis(2-chloroethyl)sulphide; BRN 1733595; CCRIS 570; 1-Chloro-2-(β-chloroethylthio)-ethane; β,β-Dichlor-ethyl-sulphide; β,β'-Dichloroethyl sulphide; 2,2'-Dichlorodiethyl sulfide; 2,2'-Dichlorodiethyl sulphide; 2,2'-Dichloroethyl sulfide; 2,2'-Dichloroethyl sulphide; Di-2-chloroethyl sulfide; Di-2-chloroethyl sulphide; Diethyl sulfide, 2,2'-dichloro; Distilled mustard; Ethane, 1,1'-thiobis(2-chloro-; Gelbkreuz; HD; HS; HSDB 336; Iprit; Kampstoff lost; Lost; Mustard gas; Mustard HD; Mustard, sulfur; Mustard vapor; S-lost; S mustard; S-Yperite; Schwefel-lost; Senfgas; Sulfide, bis(2-chloroethyl); Sulfur mustard; Sulfur mustard gas; Sulphur mustard; Sulphur mustard gas; 1,1'-Thiobis(2-chloroethane); UN 2927; Yellow cross liquid; Yellow Cross Gas; Yperite. Used as a military poison gas. Inactivated by sodium or calcium hypochlorite. Gas; mp = 13-14°; bp = 216°, bp10 = 98°; d4[20] 1.2741; LD50 (rat iv) = 3.3 mg/kg.

2593 Myclobutanil
88671-89-0

6346

G P

C15H17ClN4
α-Butyl-α-(4-chlorophenyl)-1H-1,2,4-triazole-1-propane-nitrile.
2-(4-Chlorophenyl)-2-(1H-1,2,4-triazol-1-ylmethyl)hex-anenitrile; 2-p-Chlorophenyl-2-(1H-1,2,4-triazol-1-yl-methyl)hexanenitrile; Caswell No. 723K; EPA Pesticide Chemical Code 128857; HOE 39304F; HSDB 6708; Myclobutanil; Nova (pesticide); Nova W; Nu-Flow M; Rally; RH 3866; RH-53,866; Systhane; Systhane 6 Flo; 1H-1,2,4-Triazole-1-propanenitrile, α-butyl-α-(4-chloro-phenyl)-. Systemic fungicide with protective and curative action. Used in control of Ascomycetes, Fungi Imperfecti and Basidomycetes in a wide variety of crops. Registered by EPA as a fungicide. Yellow crystals; mp = 63-68°; bp1 = 202-208°; soluble in H2O (0.0142 g/100ml), polar organic solvents; LD50 (mrat orl) = 1600 mg/kg, (frat orl) = 2229 mg/kg, (rbt der) = 7500 mg/kg, (bobwhite quail orl) = 510 mg/kg, (grey partridge orl) = 635 mg/kg; LC50 (bluegill sunfish 96 hr.) = 2.4 mg/l, (rainbow trout 96 hr.) = 4.2 mg/l, (Daphnia 48 hr.) = 11 mg/l; non-toxic to bees. Andersons Lawn Fertilizer Div. Inc.; Chemisco; Dow AgroSciences; Hoechst UK Ltd.; Pan Britannica Industries Ltd.; Rohm & Haas Co.

2594 Mycose
99-20-7

9655

G

C12H22O11
Trehalose.
'α-D-Glucopyranoside, α-D-glucopyranosyl; EINECS 202-739-6; Ergot sugar; Mycose; Natural trehalose; NSC 2093; Trehalose; α-D-Trehalose; Trehalose dihydrate. Used in biochemical research. Crystals; mp = 203°; d24 = 1.58; [α]D[20] 199° (H2O, c = 6); soluble in H2O, EtOH, less soluble in Et2O, C6H6.

2595 Myrcene
123-35-3

6356

H

C10H16

7-Methyl-3-methylene-1,6-octadiene.
4-01-00-01108 (Beilstein Handbook Reference); Al3-00738; BRN 1719990; CCRIS 3725; EINECS 204-622-5; FEMA No. 2762; HSDB 1258; 7-Methyl-3-methyl-eneocta-1,6-diene; Myrcene; β-Myrcene; NSC 406264. Liquid; bp = 167°; d^{15} = 0.8013; λ_m = 225 nm (ϵ = 17378, EtOH); insoluble in H_2O, soluble in EtOH, Et_2O, C_6H_6, $CHCl_3$, AcOH.

2596   Myrcenol
543-39-5   H

C10H18O

3-Methylene-7-methyl-1-octen-7-ol.
4-01-00-02280 (Beilstein Handbook Reference); 7-Octen-2-ol, 2-methyl-6-methylene-; 7-Octen-2-ol, 2-methyl-6-methylene-; BRN 1744474; EINECS 208-843-8; 3-Methylene-7-methyl-1-octen-7-ol; 2-Methyl-6-methyl-eneoct-7-en-2-ol; Myrcenol.

2597   Myristalkonium chloride
139-08-2   G

C23H42ClN

Benzyldimethylmyristylammonium chloride.
Ammonium, benzyldimethyltetradecyl-, chloride; Arquad DM14B-90; Barquat MB 50; Barquat MS 100; Benzenemethanaminium, N,N-dimethyl-N-tetradecyl-, chloride; Benzyldimethyltetradecylammonium chloride; N-Benzyl-N-tetradecyldimethylammonium chloride; BTC 824P100; Chlorur de miristalkonium; Cloruro de miristalkonio; Dibactol; Dimethylbenzylmyristyl-amm-onium chloride; Dimethylbenzyltetradecyl-ammonium chloride; N,N-Dimethyl-N-tetradecylbenzenemethan-aminium chloride; EINECS 205-352-0; Faringets; HSDB 5627; JAQ Powdered Quaternary; Miristalkonii chloridum; Miristalkonium chloride; Myristalkonium chloride; Myristyldimethylbenzyl-ammonium chloride; Nissan cation M2-100; Quarton 14 BCL; Quaternario 14B; Sanibond 200lg; Tetradecylbenzyldimethyl-ammonium chloride; Tetra-decyl-dimethyl-benzyl-amm-onium chloride; Trimethyl tetradecylphenyl ammonium chloride; Zephiramine; Zephiramine chloride; Zephir-amine; Benzenemethan-aminium, N,N-dimethyl-N-tetra-decyl-, chloride; C14 benzyl dimethyl ammonium chloride; C14 dimethyl benzyl ammonium chloride; Benz-enemethaminium, N-tetradecyl-N,N-dimethyl, chloride; Benzyl dimethyl tetradecyl ammonium chloride; C14-alkylbenzyl-dimethylammonium chloride; Roccal MC-14; Tetradecyl dimethyl benzyl ammonium chloride. Used in formulation of disinfectants, sanitizers, and swimming pool algicides. White solid; Soluble in H_2O. Huntington Professional Products.

2598   Myristamine
2016-42-4   G

C14H31N

1-Tetradecylamine.
4-04-00-00812 (Beilstein Handbook Reference); Al3-15076; Alamine 5D; Amine 14D; 1-Aminotetradecane; Armeen 14; BRN 1633740; EINECS 217-950-9; Monotetradecylamine; Myristylamine; NSC 66437; Tetradecanamine; 1-Tetradecanamine; Tetradecylamine; n-Tetradecylamine. Emulsifier; chemical intermediate; quaternized end products used as bactericides. Solid; mp = 83.1°; bp = 291.2°; d^{20} = 0.8079; insoluble in H_2O, soluble in Me_2CO, very soluble in EtOH, Et_2O, C_6H_6, $CHCl_3$. Berol Nobel AB.

2599   Myristic acid
544-63-8   6359   G H

C14H28O2

n-Tetradecanoic acid.
4-02-00-01126 (Beilstein Handbook Reference); Al3-15381; BRN 0508624; C14 fatty acid; CCRIS 4724; Crodacid; EINECS 208-875-2; Emery 655; FEMA No. 2764; HSDB 5686; Hydrofol acid 1495; Hystrene 9014; Myristic acid; Neo-fat 14; NSC 5028; Philacid 1400; Tetradecanoic acid; n-Tetradecanoic acid; n-Tetradecan-1-oic acid; Tetradecoic acid; n-Tetradecoic acid; Univol U 316S. Detergent surfactant. Organic acid used in the manufacture of soaps, cosmetics, synthesis of esters for flavors and perfumes; component of food-grade additives. Leaflets; mp = 53.9°; bp_{100} = 250°; λ_m = 210 nm (ϵ = 71, C_7H_{16}); insoluble in H_2O, slightly soluble in Et_2O, soluble in EtOH, MeOH, Me_2CO, $CHCl_3$; very soluble in C_6H_6; LD50 (mus iv) = 432.6 mg/kg. Akzo Chemie; Henkel/Emery; Mirachem Srl.; Sigma-Aldrich Fine Chem.; Unichema; Witco/Humko.

2600   Myristyl Alcohol
112-72-1   6361   G H

C14H30O

1-Tetradecanol.
4-01-00-01864 (Beilstein Handbook Reference); Al3-00943; Alcohol C14; Alfol 14; BRN 1742652; C14 alcohol; Dehydag® Wax; Dytol R-52; EINECS 204-000-3; EPA Pesticide Chemical Code 001510; Fatty alcohol(C14); HSDB 5168; Lanette® K; Lanette® Wax KS; Lanette® 14; Loxanol V; Myristic alcohol; Nacol® 14-95; Nacol® 14-98; NSC 8549; Tetradecanol; Tetradecan-1-ol; 1-Tetradecanol; n-Tetradecanol; n-Tetradecanol-1; n-Tetradecyl alcohol; Tetradecyl alcohol; Unihydag Wax 14. Surfactant intermediate; organic synthesis; antifoam agent; perfume fixative for soaps and cosmetics; specialty cleaning preparations; emollient for cold creams.Registered by EPA as an insecticide. Leaflets; mp = 39.5°; bp = 289°; d^{38} = 0.8236; insoluble in H_2O, very soluble in EtOH, Et_2O, Me_2CO, C_6H_6, $CHCl_3$. Amoco Lubricants; Condea Vista Co.; Condor; Greeff R.W. & Co.; Michel M.; Ruger; Schweizerhall Inc.; Sigma-Aldrich Fine Chem.; Spectrum Chem. Manufacturing.

2601   Myristyl lactate
1323-03-1   G

C17H34O3

Tetradecyl 2-hydroxypropanoate.

AI3-14482; EINECS 215-350-1; 2-Hydroxypropanoic acid, tetradecyl ester; Myristyl lactate; Propanoic acid, 2-hydroxy-, tetradecyl ester; Tetradecyl 2-hydroxy-propanoate; Tetradecyl lactate; Wickenol® 506. Mixture of o, m and p-isomers. Emollient with smooth, satiny afterfeel. Imparts lubricity and sheen to cosmetic and pharmaceutical preparations. CasChem; Dinoval.

2602   Myristyl myristate
3234-85-3   G

C28H56O2

Tetradecyl tetradecanoate.

3-10-00-01641 (Beilstein Handbook Reference); BRN 1801576; Ceraphyl 424; Cyclochem MM; EINECS 221-787-9; Myristic acid, tetradecyl ester; Myristyl myristate; Tetradecanoic acid, tetradecyl ester; Tetradecyl myristate; Waxenol® 810. Emollient with unusual afterfeel; especially for stick preparations. CasChem.

2603   Myrtrimonium bromide
1119-97-7   6362   G

C17H38BrN

N,N,N-Trimethyl-1-tetradecanaminium bromide.

Ammonium, tetradecyltrimethyl-, bromide; Ammonium, trimethyltetradecyl-, bromide; Bromure de tetradonium; Bromuro de tetradonio; EINECS 214-291-9; Morpan T; Myristyl trimethyl ammonium bromide; Myristyltrimethyl-ammonium bromide; Myrtrimonium bromide; Mytab; Quaternium 13; Rhodaquat® M214B/99; Sumquat® 6110; 1-Tetradecanaminium, N,N,N-trimethyl-, bromide; Tetradecyltrimethylammonium bromide; Tetradonii bromidum; Tetradonio bromuro; Tetradonium bromide; Trimethylmyristylammonium bromide; Trimethyltetra-decylammonium bromide; N,N,N-Trimethyl-1-tetradecanaminium bromide. A cationic germicidal detergent, disinfectant and deodorant. A conditioner with superior antistatic properties and light feel. Solid; mp = 245-250°; soluble in H2O (20 g/100 ml); LD50 (rat iv) = 15 mg/kg. Hexcel; Rhône Poulenc Surfactants.

2604   Naftifine
65473-14-5   6381   G

H—Cl

C21H22ClN

(E)-N-Methyl-N-(3-phenyl-2-propenyl)-1-naphthalene-methanamine. AW 105-843; AW 105843; (E)-N-Cinnamyl-N-methyl-1-naphthalenemethylamine hydrochloride; Exoderil; (E)-N-Methyl-N-(1-naphthylmethyl)-3-phenyl-2-propen-1-amine-hydrochloride; Naftifine hydrochloride; Naftin; 1-Naphthalenemethanamine, N-methyl-N-(3-phenyl-2-propenyl)-, hydrochloride, (E)-; SN 105843; N-methyl-N-(1-naphthylmethyl)-3-phenylpropen-1-amine; naftifung-in(E)-N-cinnamyl-N-methyl-1-naphthalenemethyl-amine.   Squalene epoxidase inhibitor. Blocks ergosterol biosynthesis and is used as a topical antifungal. Colorless oil; bp11 = 162-167°. Sandoz Pharm. Corp.

2605   Naganol
129-46-4   9096   D G V

C51H34N6Na6O23S6

8,8'-[carbonylbis[imino-3,1-phenylenecarbonylimino(4-methyl-3,1-phenylene)carbonylimino]]bis-1,3,5-naph-thalenetrisulfonic acid hexasodium salt.

309 F; Antrypol; BAY 205; Bayer 205; 1-(3-Benzamido-4-methylbenzamido)naphthalene-4,6,8-trisulfonic acid; EINECS 204-949-3; Fourneau 309; Germanin; Moranyl; Naganin; Naganine; Naganinum; Naganol; Naphuride sodium; 1,3,5-Naphthalenetrisulfonic acid, 8,8'-(carbonylbis(imino-3,1-phenylenecarbonylimino(4-methyl-3,1-phenylene)carbonylimino))bis-, hexasodium salt; NF060; NSC 34936; SK 24728; Sodium suramin; Suramin Hexasodium; Suramin sodium; Suramina sodica; Suramine sodique; Suramine sodium; Suraminum natricum; sym-3"-urea sodium salt. Veterinary preparation; for the control of trypanosomiasis of domestic animals. Solid; Soluble in H2O, insoluble in organic solvents; LD50 (mus iv) = 620 mg/kg. Bayer AG.

2606   Naphtha
8030-30-6   7265   H

Naphtha.

Amsco H-J; Amsco H-SB; Aromatic solvent; Benzin; Benzin B70; Benzine; Benzyna DO lakierow C; Coal tar; EINECS 232-443-2; Herbitox; Hi-flash naphthayethylen; HI-Flash naphtha; HSDB 2892; Hydrotreated naphtha; Light ligroin; Mineral spirits; Mineral spirits No. 10; Mineral thinner; Mineral turpentine; Naphtha; Naphtha 49 degree be-coal tar type; Naphtha, hydrotreated; Naphtha, petroleum; Naphtha VM & P; Naphtha VM & P, high flash; Naphtha VM & P, regular; Naphtha VM & P, 50 degree flash; Petroleum benzin; Petroleum-derived naphtha; Petroleum distillates (naphtha); Petroleum ether; Petroleum, light; Petroleum naphtha; Rubber solvent; Rubber solvent (Naphtha); Skelly-solve H; Skelly-solve R; Skelly-solve S; Skelly-solve S-66; Solvent naphtha; Super VMP; Unleaded gasoline; Varsol; VM & P Naphtha; White spirit. Used as a fuel and solvent. Has been used medicinally as a counter-irritant. Colorless liquid; bp = 35-80°; d = 0.625-0.660; insoluble in H2O, soluble in organic solvents.

2607   Naphtha
8052-41-3   6223   H

Naphtha, solvent.

Caswell No. 802; EINECS 232-489-3; EPA Pesticide Chemical Code 063504; Mineral spirits Type I; Naphtha, solvent; Naphtha,

Stoddard solvent; Organic solvents, Stoddard solvent; Petroleum distillates; Solvents, naphthas; Stoddard solvent; White spirit; White spirits. Used as a solvent and pain thinner. Liquid; bp = 149°; $d^{15.6}$ = 0.754-0.820.

2608   Naphthalene

91-20-3   6396   D G H

$C_{10}H_8$
Naphthalene.
AI3-00278; Albocarbon; Camphor tar; Caswell No. 587; CCRIS 1838; Dezodorator; EINECS 202-049-5; EPA Pesticide Chemical Code 055801; HSDB 184; Mighty 150; Mighty RD1; Moth balls; Moth flakes; Naftalen; Naphtalene; Naphthalene; Naphthalin; Naphthaline; Naphthene; NCI-C52904; NSC 37565; RCRA waste number U165; Tar camphor; UN1334; UN2304; White tar. Major feedstock in the polymer, dyestuffs and pharmaceutical industries. Used in lubricants and fuels. Anthelmintic. Targets Cestodes. Crystals; mp = 80.2°; bp = 217.9°; d^{20} = 1.0253; d^{100} = 0.9628; λ_m = 221, 258, 266, 275, 279, 283, 286, 301, 304, 311 nm (ε = 10600, 3470, 4990, 5530, 313, 3760, 294, 224, 239, MeOH); 3710, insoluble in H_2O; soluble in EtOH or MeOH (7.7 g/100 ml), C_6H_6 or C_7H_8 (28.6 g/100 ml), CHCl3 or CCl4 (50 g/100 ml), CS2 (83.3 g/100 ml). *Allied Signal; Aristech; Crowley Tar Prods.; Dragon Chemical Corp.; Farnam Companies Inc.; Stanchem; Texaco; Willert Home Products Inc.*

2609   1-Naphthaleneacetic acid

86-87-3   6397   G P

$C_{12}H_{10}O_2$
2-(1-Naphthyl)acetic acid.
Alphaspra; Acide naphthylacetique; Acide naphtylacetique; Agronaa; AI3-16113; Alman; ANU; Appl-set; Biokor; Caswell No. 589; Celmone; EPA Pesticide Chemical Code 056002; Etifix; Floramon; Fruit Fix; Fruitofix; Fruitone; Fruitone N; Hormofix; HSDB 2038; Klingtite; Kyselina 1-naftyloctova; Liqui-stik; N 10; NAA; NAA 800; 1-NAA; -NAA; Naphthaleneacetic acid; 1-Naphthalene acetic acid; Naphthalene-1-acetic acid; 1-Naphthylacetic acid; α-Naphthylacetic acid; α-Naphthyleneacetic acid; α-Naphthylessigsäure; 2-(α-Naphthyl)-ethanoic aid; Nafusaku; Naphthyl-1-essigsäure; Naphthylacetic acid; Naphyl-1-essigsäure; Niagara-stik; NSC 15772; NU-Tone; Phyomone; Pimacol-sol; Planofix; Planofixe; Plucker; Pomoxon; Primacol; Rasin; Rhizopon B; Rhodofix; Stafast; STIK; Stop-Drop; Tekkam; Tip-Off; Tipoff; Transplantone; Tre-Hold; Vardhak. A plant growth regulator used to control suckering in fruit trees. Registered by EPA as a plant growth regulator. Needles; mp = 135°; bp dec.; λ_m = 223, 280, 313 nm (MeOH); soluble in H_2O (0.038 g/100 ml at 17°), xylene (5.5 g/100 ml), CCl4 (1.06 g/100 ml), EtOH (3.3 g/100 ml), Me2CO, Et2O, CHCl3; LD50 (rat orl) = 1000 - 5900 mg/kg, (mus orl) = 700 mg/kg (sodium salt), (rbt der) > 5000 mg/kg; LC50 (mallard duck 8 day dietary) > 10000 mg/kg diet; non-toxic to bees. *Amvac Chemical Corp.; Aventis Crop Science; Fargro Ltd.; ICI Agrochemicals; Syngenta Crop Protection; UAP - Platte Chemical; Value Gardens*

Supply LLC.

2610   1,6-Naphthalenediol

575-44-0   G

$C_{10}H_8O_2$
1,6-dihydroxynaphthalene.
C.I. 76630; CCRIS 7894; 1,6-Dihydroxynaphthalene; 2,5-Dihydroxynaphthalene; EINECS 209-386-7; 6-Hydroxy-1-naphthol; Naphthalene, 1,6-dihydroxy-; 1,6-Naphth-alenediol; 2,5-Naphthalenediol; NSC 7201. Used in chemical synthesis. Prisms; mp = 138°; bp sublimes; $\lambda mLks$ = 223, 243, 286, 298, 321, 332 nm (MeOH); insoluble in ligroin; slightly soluble in H_2O, EtOH, CHCl3, soluble in Et2O, Me2CO, C_6H_6. *Aceto Corp.*

2611   2,3-Naphthalenediol

92-44-4   G

$C_{10}H_8O_2$
2,3-Dihydroxynaphthalene.
4-06-00-06564 (Beilstein Handbook Reference); AI3-18148; BRN 0742375; 2,3-Dihydroxynaphthalene; 2,3-Dihydroxynapthalene; EINECS 202-156-7; 2,3-Naphthalenediol; Naphthalene-2,3-diol; NSC 8707. Used as a complexing reagent. Crystals; mp = 162-164°.

2612   2,7-naphthalenediol

582-17-2   G

$C_{10}H_8O_2$
2,7-dihydroxynaphthalene.
4-06-00-06570 (Beilstein Handbook Reference); BRN 2042383; CI 76645; 2,7-Dihydroxynaphthalene; EINECS 209-478-7; 2,7-Naphthalenediol; Naphthalene-2,7-diol; NSC 407541. Reagent. Used in synthesis. Needles or plates; mp = 193°; bp sublimes; λ_m = 231, 285, 313, 320, 328 nm (ε = 75800, 3360, 2400, 1940, 3080, MeOH); insoluble in ligroin, CS2, slightly soluble in Me2CO, soluble in H_2O, EtOH, Et2O, C_6H_6, CHCl3, C7H8. *Aceto Corp.*

2613   Naphthalene-2,6-disulfonic acid

581-75-9   6400   G

$C_{10}H_8O_6S_2$
2,6-Naphthalenedisulfonic acid.
Ebert-Merz β-acid; EINECS 209-471-9; 2,6-Naphth-alenedisulfonic acid; Naphthalene-2,6-disulphonic acid; NSC 37041. Chemical intermediate. Crystals; very soluble in H_2O, EtOH; practically

insoluble in Et$_2$O.

2614   **Naphthalene-2,7-disulfonic acid;**
92-41-1   6401   G

C$_{10}$H$_8$O$_6$S$_2$
2,7-Naphthalenedisulfonic acid.
Ebert and Merz's α-acid; EINECS 202-154-6; 2,7-Naphthalenedisulfonic acid; 2,7-Naphthalenedisulphonic acid; Naphthalene-2,7-disulfonic acid; Naphthalene-2,7-disulphonic acid; NSC 9589. Used as an intermediate for dyes. Needles; mp = 199°; λ$_m$ = 233, 266, 308, 315, 323 ε = 87096, 3890, 257, 240, 178, EtOH); soluble in H$_2$O.

2615   **2-Naphthalene sulfonic acid**
120-18-3   6403   G

C$_{10}$H$_8$O$_3$S
Naphthalene-2-sulfonic acid.
4-11-00-00527 (Beilstein Handbook Reference); Al3-18435; BRN 1955756; EINECS 204-375-3; Kyselina 2-naftalensulfonova; Naphthalene-2-sulphonic acid; β-Naphthalenesulfonic acid; 2-Naphthalenesulfonic acid; 2-Naphthalenesulfonic acid, monohydrate; β-Naphthylsulfonic acid. Used in manufacture of β-naphthol and to solubilize phenols in water. Platelets; mp = 91°; d^{25} = 1.441; λ$_m$ = 227, 265, 274, 305, 312, 320 nm (MeOH); slightly soluble in C$_6$H$_6$, soluble in Et$_2$O, very soluble in H$_2$O, EtOH; LD$_{50}$ (rat orl) = 4440 mg/kg.

2616   **Naphthalol**
613-78-5   6441   D G

C$_{17}$H$_{12}$O$_3$
2-Naphthyl salicylate.
Al3-00890; Benzoic acid, 2-hydroxy-, 2-naphthalenyl ester; Berol; Betol; EINECS 210-355-5; Naphthalol; Naphthosalol; β-Naphthol salicylate; β-Naphthyl salicylate; 2-Naphthyl salicylate; NSC 5538; Salicylic acid, 2-naphthyl ester; Salinaphthol. Antiseptic; anti-inflammatory. Solid; mp = 95.5°; d^{116} = 1.11; insoluble in H$_2$O, glycerol; slightly soluble in cold alcohol; soluble in C$_6$H$_6$, Et$_2$O, boiling alcohol.

2617   **Naphthazarin**
475-38-7   G

C$_{10}$H$_6$O$_4$
5,8-Dihydroxy-1,4 naphthoquinone.
CCRIS 6670; 5,8-Dihydroxynaphthoquinone; 5,8-Dihydroxy-1,4-naphthoquinone; 5,8-Dihydroxy-1,4-naphthalenedione; 5,8-Dihydroxy-1,4-naphthosemi-quinone; EINECS 207-495-4; 1,4-Naphthalenedione, 5,8-dihydroxy-; 1,4-Naphthoquinone, 5,8-dihydroxy-; Naphthazarin; Naphthazarine; Naphthazarone; NSC 26647. Used in manufacture of dyestuffs. Dark red crystals; mp = 232°; bp sublimes; λ$_m$ = 268, 516, 556 nm (ε = 10000, 6310, 3162, EtOH), 520 nm (ε = 3981, DMF); slightly soluble in H$_2$O, EtOH, Et$_2$O, soluble in AcOH.

2618   **Naphthenic acid**
1338-24-5   H

C$_{11}$H$_8$O$_2$
Carboxylic acids, naphthenic.
Acide naphtenique; Agenap; Carboxylic acids, naphthenic; EINECS 215-662-8; HSDB 1178; Naphid; Naphthenic acid; Naphthenic acids; Sunaptic acid B; Sunaptic acid C.

2619   **Naphthionic acid**
84-86-6   6429   G

C$_{10}$H$_9$NO$_3$S
α-Naphthylamine-4-sulfonic acid.
Al3-52278; 1-Aminonaphthalene-4-sulfonic acid; 4-Amino-1-naphthalenesulfonic acid; 4-Amino-1-naphth-alinsulfonsaeure; 4-Aminonaphthalene-1-sulphonic acid; 1-Amino-4-sulfonaphthalene; EINECS 201-567-9; Kyselina 1-naftylamin-4-sulfonova; Kyselina nafthionova; 1-Naphthalenesulfonic acid, 4-amino-; 1-Naphthylamin-4-sulfonsaeure; α-Naphthylamine-4-sulfonic acid; α-Naphthylamine-p-sulfonic acid; 1-Naphthylamine-4-sulfonic acid; Naphthionic acid; 1,4-Naphthionic acid; Naphthionsaeure; Naphthionsäure; NSC 4155; Piria's acid; Sulfonaphtin; USAF M-5. Intermediate in manufacture of dyestuffs such as Congo Red, Fast Red A and Azo Rubine. The sodium salt is used as a hemostatic. Solid, decomposes without melting; d$_4^{25}$= 1.6703; slightly soluble in H$_2$O (0.28 g/l), sparingly soluble in organic solvents.

2620   **Naphthol**
135-19-3   6410   G H

C$_{10}$H$_8$O
2-Naphthol.
Al3-00081; Antioxygene BN; Azogen developer A; C.I. 37500; C.I. Azoic Coupling Component 1; C.I. Developer 5; Caswell No. 590;

Developer A; Developer AMS; Developer BN; Developer sodium; EINECS 205-182-7; EPA Pesticide Chemical Code 010301; HSDB 6812; Isonaphthol; β-Monoxynaphthalene; β-Naftol; β-Naftolo; β-Naphthol; Naphthol, β; Naphthol B; β-Naphtol; β-Naphthyl alcohol; β-Naphthyl hydroxide; NSC 2044. An antioxidant and chemical intermediate. Registered by EPA as a fungicide and insecticide (cancelled). Has been used as a nematicide. Crystals; mp = 123°; bp = 285°; d^{20} = 1.28; λ_m = 225, 263, 273, 284, 318, 329 nm (ε 91194, 3911, 4559, 3301, 1861, 2163, 95% EtOH); soluble in H_2O (0.1 g/100 ml at 25°, 1.25 g/100 ml at 100°, EtOH 125 g/100 ml), $CHCl_3$ (5.9 g/100 ml), Et_2O (77 g/100 ml), glycerol, olive oil. *Allied Colloids; Riedel de Haen (Chinosolfabrik).*

2621   1-Naphthol
90-15-3   6409   H

$C_{10}H_8O$
1-Naphthalenol.
AI3-00106; Basf Ursol ERN; CCRIS 1172; CI 76605; CI Oxidation Base 33; Durafur developer D; EINECS 201-969-4; Fouramine ERN; Fourrine 99; Fourrine ERN; Furro ER; HSDB 2650; Nako TRB; NSC 9586; Tertral ERN; Ursol ERN; Zoba ERN. Used in the manufacture of dyestuffs and synthetic perfumes. Crystalline solid; mp = 95°; bp = 288°, bp_{40} = 184°; d^{99} = 1.0989; λ_m =233, 295, 309, 323 nm (MeOH); soluble in organic solvents, insoluble in H_2O; LD_{50} (rat orl) = 2590 mg/kg.

2622   Naphthoresorcin
132-86-5   6422   G

$C_{10}H_8O_2$
1,3-Dihydroxy-naphthalene.
4-06-00-06543 (Beilstein Handbook Reference); AI3-08780; BRN 2044002; CCRIS 7896; EINECS 205-079-7; 3-Hydroxybenzocyclohexadien-1-one; 1,3-Naphthalene-diol; Naphthoresorcinol; NSC 115890. Inhibitor of prostaglandin synthase. Used to assay sugars, oils and glucuronic acid in urine. White leaflets; mp = 123.5°; λ_m = 214, 235, 287, 336 nm (MeOH); slightly soluble in Me_2CO, C_6H_6, ligroin, soluble in H_2O, EtOH, Et_2O, AcOH.

2623   α-Naphthylamine
134-32-7   6424   G

$C_{10}H_9N$
1-Naphthalenamine.
4-12-00-03009 (Beilstein Handbook Reference); AI3-00085; alfa-Naftyloamina; Alfanaftilamina; 1-Amino-naftalen; α-Aminonaphthalene; 1-Aminonaphthalene; Aminogen I; BRN 0386133; CCRIS 423; CI 37265; CI Azoic Diazo Component 114;

EINECS 205-138-7; Fast Garnet B Base; Fast Garnet Base B; HSDB 1080; α-Naftalamin; 1-Naftilamina; α-Naftylamin; 1-Naftylamin; 1-Naftylamine; 1-Naphthalamine; 1-Naphthalenamine; Naphthalidam; Naphthalidine; 1-Naphthylamin; α-Naphthylamine; 1-Naphthylamine; RCRA waste number U167; UN2077. Used in manufacture of dyestuffs, as a rubber vulcanization accelerator and as a chemical feedstock. mp = 49.2°; bp = 300.8°; d^{20} = 1.0228; λ_m = 243, 318, 328 nm (cyclohexane); soluble in H_2O (1.7 mg/l), soluble in $CHCl_3$; LD_{50} (rat orl) = 779 mg/kg.

2624   2-Naphthylphenylamine
135-88-6   G

$C_{16}H_{13}N$
β-Naphthylphenylamine.
4-12-00-03128 (Beilstein Handbook Reference); Aceto PBN; AgeRite Powder; AI3-00068; AK 1; AK 1 (stabilizer); Anilinonaphthalene; 2-Anilinonaphthalene; Antioxidant 116; Antioxidant D; Antioxidant PBN; BRN 2211188; CCRIS 853; EINECS 205-223-9; Fenyl-β-naftylamin; N-Fenyl-2-aminonaftalen; HSDB 2888; 2-Naphthalenamine, N-phenyl-; 2-Naphthylamine, N-phenyl-; N-2-Naphthyl-aniline; N-(2-Naphthyl)aniline; β-naphthylphenylamine; 2-naphthylphenylamine; N-(2-naphthyl)-N-phenylamine; N-β-naphthyl-N-phenylamine; Naftam 2; NCI-C02915; Neosone D; Neozon D; Neozone; Neozone D; Nilox PbNa; Nilox PBNA; Noclizer D; Nocrac D; Nonox D; Nonox DN; NSC 37151; PBNA; 2-Phenylamino-naphthalene; phenyl-2-naphthylamine; Phenyl-β-naph-thylamine; Phenyl-β-naphtilamine; Phenyl-2-naphthyl-amine; N-Phenyl-β-naphthylamine; N-Phenyl-2-naph-thylamine; Stabilator A.R; Stabilizer AR; Stabilizer AR; Vulkanox PBN. An antioxidant. Crystals; mp = 108°; bp = 395.5°; d = 1.24, λ_m = 221, 271 nm (cyclohexane); insoluble in H_2O, slightly soluble in $CHCl_3$, soluble in EtOH, Et_2O, C_6H_6, AcOH; LD_{50} (rat orl) = 8730 mg/kg.

2625   Naproxen
22204-53-1   6443   D

$C_{14}H_{14}O_3$
(+)-6-Methoxy-α-methyl-2-napthaleneacetic acid.
Acusprain; Anax; Anexopen; Apo-Napro-NA; Apo-Naproxen; Apronax; Artagen; Arthrisil; Artrixen; Artroxen; Atiflan; Axer; Bipronyl; Bonyl; Calosen; CCRIS 5265; CG 3117; Clinosyn; Congex; Danaprox; Daprox; Diocodal; Duk; Dysmenalgit; Dysmenalgit N; Ec-Naprosyn; EINECS 244-838-7; Equiproxen; Flanax Forte; Flexen; Flexipen; Floginax; Fuxen; Genoxen; Headlon; HSDB 3369; Laraflex; Laser; Lefaine; Leniartil; MNPA; Nafasol; Naixan; Nalyxan; Napflam; Napmel; Naposin; Napratec; Napren; Napren E; Naprium; Naprius; Naprontag; Naprosyn; Naprosyn LLE; Naprosyn LLE Forte; Naprosyne; Naproxen; Naproxene; Naproxeno; Naproxenum; Naproxi 250; Naproxi 500; Naprux; Napxen; Narma; Narocin; Naxen; Naxen F; Naxid; Naxopren; Naxyn; Naxyn 250; Naxyn 500; Noflam; Novonaprox; Nycopren; Panoxen; Patxen; Prafena; Prexan; Priaxen; Pronaxen; Proxen; Proxen LE; Proxen LLE; Proxine; Rahsen; Reuxen; Rheumaflex; Roxen; RS 3540; Saritilron; Sinartrin; Sinton; Soproxen; Sutolin; Sutony; Tohexen; Traumox; U-Ritis; Velsay; Veradol; Vinsen; Xenar. Anti-inflammatory; analgesic; antipyretic. Crystals; mp = 152-154°; $[\alpha]_D$ = +66° (c = 1 in $CHCl_3$); nearly

insoluble in H$_2$O; soluble in organic solvents; LD$_{50}$ (mus orl) = 1234 mg/kg, (rat orl) = 534 mg/kg, (rat ip) = 575 mg/kg. *Syntex Labs. Inc.*

2626 Naproxen Piperazine
70981-66-7   6443   D

C$_{32}$H$_{38}$N$_2$O$_6$

(+)-6-Methoxy-α-methyl-2-napthaleneacetate piperazine salt (2:1). EINECS 275-083-1; Naproxen piperazine salt; Numidan; Numide; Piproxen. Anti-inflammatory; analgesic; antipyretic. *Syntex Labs. Inc.*

2627 Naproxen Sodium
26159-34-2   6443   D

C$_{14}$H$_{13}$NaO$_3$

(-)-Sodium 6-methoxy-α-methyl-2-napthaleneacetate. A-Nox; Aleve; Alfa; Anapran; Anaprotab; Anaprox; Apo-Napro-Na; Apranax; Aprol; Aprowell; Axer; Axer Alfa; Causalon Pro; Diocodal; Dysmenalgit; EINECS 247-486-2; Flanax; Flogen; Floginex; Flogogin; Floneks; Floxalin; Gibinap; Gibixen; Gynestrel; Kapnax; Karoksen; Laser; Leniartril; Miranax; Monarit; Naixan; Naprium; Naprodil; Naprodol; Naprosyn; Naprovite; Naproxen natrium; Naproxen sodium; Naproxen sodium salt; Naprux; Naprux Gesic; Natrioxen; Nixal; Opraks; Pactens; Prexan; Primeral; Proxen; RS 3650; Sodimax; Synflex; Tandax; Veradol; Xenar. Anti-inflammatory; analgesic; antipyretic. White crystals; mp = 244-246°; [α]$_D$ = -11° (in MeOH). *Syntex Labs. Inc.*

2628 Narasin
55134-13-9   6445   G V

C$_{43}$H$_{72}$O$_{11}$

4S-Methylsalinomycin. A 28086A; AI3-29798; Antibiotic A-28086 factor A; Antibiotic A 28086A; Antibiotic C 7819B; BRN 1678311; C 7819B; Compound 79891; Lilly 79891; 4-Methylsalinomycin; Monteban; Narasin; Narasin A; Narasine; Narasino; Narasinum; Salinomycin, 4-methyl-, (4S). Main component of an antibiotic complex produced by *Streptomyces aureofaciens* NRRL 5758 and NRRL 8092. A veterinary medicine used as a coccidiostat and growth stimulant. Crystals; mp = 98-100°; λ_m = 285 nm (ϵ 58); [α]$_D^{25}$ = -54° (c = 0.2 MeOH); insoluble in H$_2$O, soluble in organic solvents; LD$_{50}$ (mus ip) = 7.15 mg/kg.

2629 Narceol
140-39-6   G

C$_9$H$_{10}$O$_2$

p-Tolylacetate. 4-06-00-02112 (Beilstein Handbook Reference); Acetic acid, 4-methylphenyl ester; Acetic acid, p-tolyl ester; 4-Acetoxytoluene; p-Acetoxytoluene; AI3-01266; BRN 1908125; p-Cresol acetate; p-Cresyl acetate; Cresyl acetate; Cresyl acetate, p-; Cresylic acetate, p-; EINECS 205-413-1; FEMA No. 3073; 4-Methylbenzoic acid methyl ester; 4-Methylphenyl acetate; p-Methylphenyl acetate; Narceol; NSC 43244; Paracresyl acetate; p-Tolyl acetate; 4-Tolyl acetate; p-Tolyl ethanoate. A synthetic perfume. Liquid; bp = 212.5°; d^{17} = 1.0512; λ_m = 265, 271 nm (ϵ = 446, 406, MeOH); slightly soluble in H$_2$O, soluble in EtOH, Et$_2$O, CCl$_4$, CHCl$_3$.

2630 Natural gas
8006-14-2   H

Natural gas. EINECS 232-343-9; Gas, natural; Liquified natural gas; Natural gas; Synthetic natural gas.

2631 Neanthine
123-88-6   G

C$_3$H$_7$ClHgO

Methoxyethyl mercury chloride. Agallo forte; Agallol; Agallolat; Aretan; Aretan 6; Atiran; BRN 4123443; CCRIS 5835; Ceranit 6; Ceresan; Ceresan Universal Wet; Chloro(2-methoxyethyl)mercury; EINECS 204-659-7; Emisan; Emisan 6; Gramisan; Higosan; MEMC; Merchlorate; Mercury, chloro(2-methoxyethyl)-; Methoxyethyl mercuric chloride; 2-Methoxyethyl-mercuric chloride; Methoxyethylmercury chloride; β-Methoxyethylmercury chloride; 2-Methoxyethyl-merk-urichlorid; Methoxyäthylquecksilberchlorid; Meth-oxyaethylquecksilberchlorid; Neanthine; Sedresan; Tafasan; Tafasan 6W; Tayssato; Triadimenol; Vegoll. Seed dressing for control of fungal diseases on cereals, rice, cotton, vegetables, flower bulbs and sugar cane cuttings. A potato fungicide. Crystals; mp = 65°; insoluble in H$_2$O, organic solvents; LD$_{50}$ (rat orl) = 22 mg/kg. *Bayer AG; ICI Plant Protection.*

2632 Neburon
555-37-3   6463   G P

C$_{12}$H$_{16}$Cl$_2$N$_2$O

N-butyl-N'-(3,4-dichlorophenyl)-N-methylurea. BRN 2733280; Butyl-3-(3,4-dichlorophenyl)-1-methyl-urea; Butyl-N'-(3,4-dichlorophenyl)-N-methylurea; 1-Butyl-3-(3,4-dichlorophenyl)-1-methylurea; 1-n-Butyl-3-(3,4-dichlorophenyl)-1-methylurea; N-Butyl-N'-(3,4-di-chlorophenyl)-N-methylurea; Caswell No. 594; 3-(3,4-Dichlorophenyl)-1-methyl-1-butylurea; 3-(3,4-Dichlor-phenyl)-1-n-butyl-harnstoff; EINECS 209-096-0; EPA Pesticide Chemical Code 012001; Granurex; Herbalt; Kloben; Kloben neburon; Neburea; Neburex; Neburon; Urea, 1-butyl-3-(3,4-dichlorophenyl)-1-methyl-;

Urea, N-butyl-N'-(3,4-dichlorophenyl)-N-methyl-. Selective herbicide, absorbed through the roots, inhibits photosynthesis. Used for pre-emergence control of annual broad-leaved weeds and grasses in beans, peas, lucerne, garlic, cereals, beets, strawberries and ornamentals, and in forestry. Crystals; mp = 102-103°; soluble in H_2O (5 mg/l), sparingly soluble in hydrocarbon solvents; LD_{50} (rat orl) >11000 mg/kg. *DuPont UK.*

2633   Neo Heliopan® 303
6197-30-4   D G

C24H27NO2
2-Ethylhexyl-2-cyano-3,3-diphenylacrylate.
CCRIS 4814; EINECS 228-250-8; 2-Ethylhexyl 2-cyano-3,3-diphenylacrylate; 2-Ethylhexyl 2-cyano-3,3-diphenyl-2-prop-enoate; 2-Ethylhexyl 2-cyano-3,3-diphenyl-acryl-ate; Neo Heliopan® 303; Octocrilene; Octocrileno; Octocrilenum; Octocrylene; 2-Propenoic acid, 2-cyano-3,3-diphenyl-, 2-ethylhexyl ester; UV Absorber-3; Uvinul® N-539. UV-B absorber for cosmetics, waterproof sunscreens. Used in flexible and rigid PVC; in NC lacquers, varnishes, vinyl flooring, and oil-based paints; in Aerosol® and oil-based suntan lotions; nonreactive with metallic driers. Liquid; mp = -10°; bp1.5 = 218°; d = 1.051; insoluble in H_2O, soluble in organic solvents. *BASF Corp.; Haarmann & Reimer GmbH.*

2634   Neodecaneperoxoic   acid,   1,1-dimethyl-propyl ester
68299-16-1   G

C15H30O3
t-Amyl peroxyneodecanoate.
EINECS 269-597-5; Neodecaneperoxoic acid, 1,1-dimethylpropyl ester; tert-Amyl peroxyneodecanoate; Trigonox® 123-C75. Supplied in odorless mineral spirits; initiator for PVC. *Akzo Chemie.*

2635   Neodecanoic acid
26896-20-8   H

C10H20O2
2,5-Dimethyl-2-ethylhexanoic acid.
Caswell No. 365A; 2,4-Dimethyl-2-isopropylpentanoic acid; 2,5-Dimethyl-2-ethylhexanoic acid; EINECS 248-093-9; EPA Pesticide Chemical Code 097501; Neodecanoic acid; 2,2,3,5-Tetramethylhexanoic acid; Topper 5E; Wiltz 65.

2636   Neodecanoyl chloride
40292-82-8   H

C10H19ClO
7,7-Dimethyloctanoyl chloride.
EINECS 254-875-0; Neodecanoic chloride; Neodecanoyl chloride.

2637   Neoflex® 9
68527-05-9   G

C9H20O
Isononyl alcohol.
Octene, hydroformylation products. *Shell.*

2638   Neopentanetetrayl 3-(dodecylthio)prop-ionate
29598-76-3   H

C65H124O8S4
Propionic acid, 3-(dodecylthio)-, neopentanetetrayl ester.
2,2-Bis((3-(dodecylthio)-1-oxopropoxy)methyl)propane-1,3-diyl bis(3-(dodecylthio)propionate); EINECS 249-720-9; Propionic acid, 3-(dodecylthio)-, neopentanetetrayl ester.

2639   Neopentyl glycol
126-30-7   6486   G H

C5H12O2
2,2-Dimethylpropane-1,3-diol.
4-01-00-02551 (Beilstein Handbook Reference); AI3-05739; BRN 0605291; Dimethylolpropane; Dimethyl-trimethylene glycol; EINECS 204-781-0; Hydroxypivalyl alcohol; Neol; Neopentanediol; Neopentyl glycol; Neopentylene glycol; NPG®; NPG® Glycol; NSC 55836; Propanediol, 2,2-dimethyl-, 1,3-. Resin intermediate; insect repellent. Used in manufacture of plasticizers and polyesters. Needles; mp = 130°; bp = 208°; soluble in H_2O (65 g/ml), C_6H_6, $CHCl_3$, very soluble in EtOH, Et2O. *BASF Corp.; Degussa-Hüls Corp.; Eastman Chem. Co.; Mitsubishi Gas.*

2640   Neopentyl glycol dioleate
42222-50-4   H

433

C41H76O4

2,2-Dimethyl-1,3-propanediyl dioleate.

2,2-Dimethyl-1,3-propanediyl dioleate; EINECS 255-713-1; Neopentyl glycol dioleate; NSC 65468; 9-Octadecenoic acid (9Z)-, 2,2-dimethyl-1,3-propanediyl ester; 9-Octadecenoic acid (Z)-, 2,2-dimethyl-1,3-propanediyl ester.

2641 Neopentyl glycol oleate

67989-24-6   H

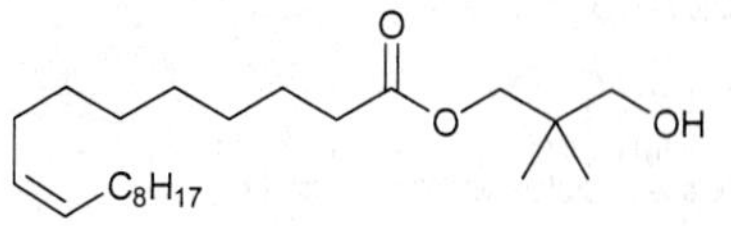

C23H44O3

9-Octadecenoic acid (9Z)-, ester with 2,2-dimethyl-1,3-propanediol. EINECS 268-008-9; Neopentyl glycol oleate; 9-Octadecenoic acid (9Z)-, ester with 2,2-dimethyl-1,3-propanediol; 9-Octadecenoic acid (Z)-, ester with 2,2-dimethyl-1,3-propanediol.

2642 Neophyl chloride

515-40-2   6487   H

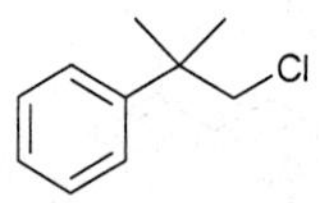

C10H13Cl

2-Methyl-2-phenylpropyl chloride.

Benzene, (2-chloro-1,1-dimethylethyl)-; (β-Chloro-α,α-dimethyl)ethylbenzene; (β-Chloro-t-butyl)benzene; (2-Chloro-1,1-dimethylethyl)benzene; β,β-Dimethylphen-ethyl chloride; EINECS 208-197-7; 2-Methyl-2-phenyl-propyl chloride; Neophyl chloride; NSC 54159. Liquid; bp = 223°, bp18 = 105°; very soluble in EtOH, Et2O, Me2CO, C6H6.

2643 Neopine

467-14-1   6488   G

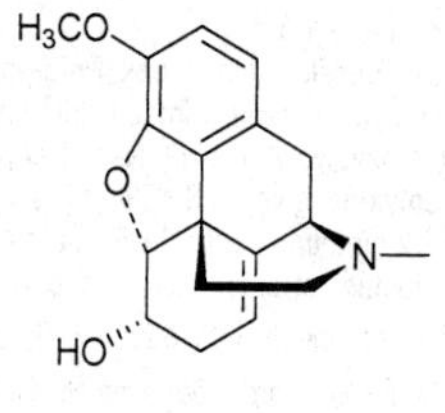

C18H21NO3

8,14-didehydro-4,5α-epoxy-3-methoxy-17-methylmorph-inan-6α-ol.

β-codeine; (5α,6α)-8,14-Didehydro-4,5-epoxy-3-meth-oxy-17-methylmorphinan-6-ol; EINECS 207-387-7; Neopine. Alkaloid from opium, isomeric with codeine. Needles; mp = 127.5°; $[\alpha]_D^{23}$ = -28° (c = 7.5 CHCl3); λ_m = 210, 284 nm (ε = 39811, 1585, aq HBr); slightly soluble in ligroin, soluble in H2O. EtOH, Et2O, C6H6, very soluble in MeOH, CHCl3.

2644 Neoscan

41183-64-6   4367   G

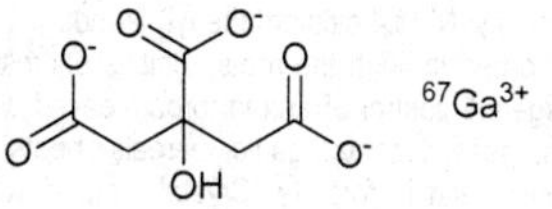

C6H5^{67}GaO7

Gallium citrate-^{67}Ga.

Citrate de gallium (67 Ga); Citrato de galio (^{67}Ga); EINECS 255-248-4; Gallii (^{67}Ga) citras; Gallium (^{67}Ga) citrate; (^{67}Ga) Gallium citrate; Gallium-(^{67}Ga) citrate (1:1); Gallium-67 citrate; Gallium citrate (^{67}Ga)); Neoscan; 1,2,3-Propanetricarboxylic acid, 2-hydroxy-, gallium-(^{67}Ga) (1:1) salt. Diagnostic aid; radioactive agent. *Medi-Physics Inc.*

2645 Nerol

106-25-2   6500   G H

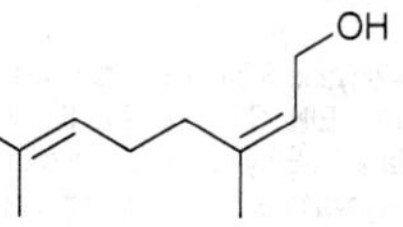

C10H18O

(Z)-3,7-Dimethyl-2,6-octadien-1-ol.

AI3-28202; 2-cis-3,7-Dimethyl-2,6-octadien-1-ol; 3,7-Dimethyl-2,6-octadien-1-ol, cis-; EINECS 203-378-7; FEMA No. 2770; cis-Geraniol; (Z)-Geraniol; Nerol; Neryl alcohol; NSC-46105; 2,6-Octadien-1-ol, 3,7-dimethyl-, (Z)-; Vernol®. A sweet, fresh, citrus-rose odor used in fragrances. Oil; bp745 = 224-225°; bp25 = 125°; d^{15} = 0.8813. *Bush Boake Allen; IFF.*

2646 Nerolidol

7212-44-4   6501   G P

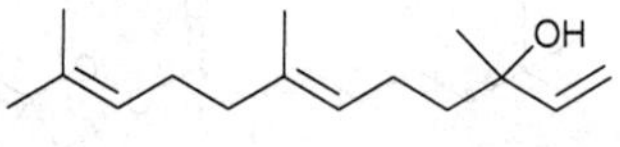

C15H26O

3,7,11-Trimethyl-1,6,10-dodecatrien-3-ol.

3-01-00-02042 (Beilstein Handbook Reference); 3,7,11-Trimethyldodeca-1,6,10-trien-3-ol; 3,7,11-Trimethyldo-deca-1,6,10-trien-3-ol,mixed isomers; AI3-10519; BRN 1724135; CCRIS 7678; 1,6,10-Dodecatrien-3-ol, 3,7,11-trimethyl-; FEMA No. 2772; Methylvinylhomogeranyl carbinol; Nerolidol. Fragrance and flavoring; sweetly floral, green, woody, lilly-like. Occurs in *cis and trans* forms. Has been used as an insecticide. [trans-form, 40716-66-3]: liquid; bp0.15 = 78°; [cis-form, 3790-78-1]: liquid; bp0.10 = 70°; [cis/trans-form]: liquid; bp3 = 122°; d_4^{25}= 0.8720; soluble in 70% EtOH (30 g/100 ml). *BASF Corp.; Troy Biosciences Inc.*

2647 Nerolin

93-04-9   6026   G

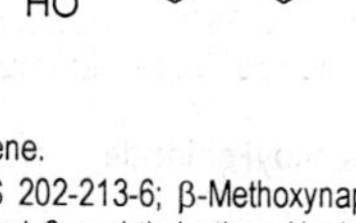

C12H12O

2-Methoxynaphthalene.

AI3-21213; EINECS 202-213-6; β-Methoxynaphthalene; Methyl β-naphthyl ether; Methyl 2-naphthyl ether; Naphthalene, 2-methoxy-; β-Naphthol methyl ether; β-Naphthyl methyl ether; 2-Naphthyl methyl ether; Nerolin; NSC 4171; Yara-Yara; Yura Yara. Used in perfumery. Leaflets; mp = 73.5°; bp = 274°; d_{20}^{20} = 1.0640; λ_m = 227,

261, 271, 327 nm (MeOH); soluble in C_6H_6, Et_2O, $CHCl_3$. .

2648   Nevile and Winther's acid
84-87-7   6417   G

$C_{10}H_8O_4S$
'α-Naphthol-4-sulfonic acid.
EINECS 201-568-4; 1-Hydroxy-4-naphthalenesulfonic acid; 4-Hydroxy-1-naphthalenesulfonic acid; 4-Hydroxynaphthalene-1-sulphonic acid; 1-Naphthalene-sulfonic acid, 4-hydroxy-; 1-Naphtho-4-sulfonic acid; 1-Naphthol-4-sulfonic acid; Neville and Winther's acid; Neville-winther acid; NSC 9587; NW Acid; 1,4-Oxy Acid. Used in manufacture of azo dyes. Crystals; dec 170°; λm = 234, 300, 320 nm (ε = 31623, 7079, 4365, EtOH - 5% HCl); very soluble in H_2O.

2649   Niacin
59-67-6   6552   D G

$C_6H_5NO_2$
3-Pyridinecarboxylic acid.
5-22-02-00057 (Beilstein Handbook Reference); Acide nicotinique; Acido nicotinico; Acidum nicotinicum; AI3-18994; Akotin; Apelagrin; Bionic; BRN 0109591; Caswell No. 598; CCRIS 1902; Daskil; Davitamon PP; Diacin; Direktan; Efacin; EINECS 200-441-0; Enduracin; EPA Pesticide Chemical Code 056701; HSDB 3134; Kyselina nikotinova; Linic; NAH; Naotin; Niac; Niacin; Niacor; Niaspan; Nicacid; Nicagin; Nicamin; Nicangin; Nico-400; Nico-span; NICO; Nicobid; Nicocap; Nicocidin; Nicocrisina; Nicodan; Nicodelmine; Nicodon; Nicolar; Niconacid; Niconat; Niconazid; Nicorol; Nicosan 3; Nicoside; NicoSpan; Nicosyl; Nicotamin; Nicotene; Nicotil; Nicotine acid; Nicotinic acid; Nicotinipca; Nicotinsaure; Nicovasan; Nicovasen; Nicyl; Nipellen; NSC 169454; Nyclin; P.P. Factor; Pellagramin; Pellagrin; Pelonin; Peviton; S115; SK-Niacin; Slo-niacin; SR 4390; Tega-Span; Tinic; Vitaplex N; Wampocap. Vitamin (enzyme cofactor). mp = 236.6°; λm = 263 nm; soluble in H_2O (1.67 g/100 ml); freely soluble in H_2O at 100°, EtOH at 76°; insoluble in Et_2O; LD50 (raty sc) = 5000 mg/kg. *Abbott Labs Inc.; Apothecon; Forest Pharm. Inc.; Marion Merrell Dow Inc.; Rhône-Poulenc Rorer Pharm. Inc.; Wallace Labs.*

2650   Niacinamide
98-92-0   6550   D H

$C_6H_5NO_2$
3-Pyridinecarboxylic acid.
5-22-02-00057 (Beilstein Handbook Reference); Acide nicotinique;

Acido nicotinico; Acidum nicotinicum; AI3-18994; Akotin; Apelagrin; Bionic; BRN 0109591; Caswell No. 598; CCRIS 1902; Daskil; Davitamon PP; Diacin; Direktan; Efacin; EINECS 200-441-0; Enduracin; EPA Pesticide Chemical Code 056701; HSDB 3134; Kyselina nikotinova; Linic; NAH; Naotin; Niac; Niacin; Niacor; Niaspan; Nicacid; Nicagin; Nicamin; Nicangin; Nico-400; Nico-span; NICO; Nicobid; Nicocap; Nicocidin; Nicocrisina; Nicodan; Nicodelmine; Nicodon; Nicolar; Niconacid; Niconat; Niconazid; Nicorol; Nicosan 3; Nicoside; NicoSpan; Nicosyl; Nicotamin; Nicotene; Nicotil; Nicotine acid; Nicotinic acid; Nicotinipca; Nicotinsaure; Nicovasan; Nicovasen; Nicyl; Nipellen; NSC 169454; Nyclin; P.P. Factor; Pellagramin; Pellagrin; Pelonin; Peviton; S115; SK-Niacin; Slo-niacin; SR 4390; Tega-Span; Tinic; Vitaplex N; Wampocap.; component of: Medriatric. Vitamin (enzyme cofactor).
Crystals; mp = 128-131°; bp0.0005 = 150-160°; λm 261 nm ($A_{1 cm}^{1\%}$ 451); soluble in H_2O (100 g/100 ml), EtOH (66.6 g/100 ml), glycerol (10 g/100 ml); LD50 (rat sc) = 1680 mg/kg. *Wyeth-Ayerst Labs.*

2651   Niagara blue
72-57-1   9862   G

$C_{34}H_{24}N_6Na_4O_{14}S_4$
Sodium ditolyldisazobis-8-amino-1-naphthol-3,6-disulph-onate.
AI3-26698; Amanil Sky Blue R; Azidinblau 3B; Azidine Blue 3B; Azurro diretto 3B; Bencidal Blue 3B; Benzaminblau 3B; Benzamine Blue; Benzamine Blue 3B; Benzanil Blue 3BN; Benzanil Blue R; Benzo Blue; Benzo Blue 3B; Benzo Blue 3BS; Benzoblau 3B; Bleu diamine; Bleu diazole N 3B; Bleu directe 3B; Bleu trypane N; Bleue diretto 3B; Blue 3B; Blue EMB; Brasilamina Blue 3B; Brasilazina Blue 3B; C.I. 23850; C.I. Direct Blue 14; C.I. Direct Blue 14, tetrasodium salt; CCRIS 616; Centraline Blue 3B; Chloramiblau 3B; Chloramine Blue; Chloramine Blue 3B; Chlorazol Direct Blue 3B; Chrome Leather Blue 3B; Congo Blue; Congoblau 3B; Cresotine Blue 3B; Diaminblau 3B; Diamine Blue 3B; Diamineblue; Diaminine Blue; Dianil Blue; Dianil Blue H3G; Dianilblau; Dianilblau H3G; Diaphtamine Blue TH; Diazine Blue 3B; Diazol Blue 3B; Diphenyl Blue 3B; Direct Blue 14; Direct Blue 3B; Direct Blue 3BX; Direct Blue D3B; Direct Blue H3G; Direct Blue M3B; Directakol Blue 3BL; Directblau 3B; EINECS 200-786-7; Hispamin Blue 3BX; HSDB 2945; Modr Prima 14; Modr Trypanova; Naphthaminblau 3BX; Naphthamine Blue 3BX; Naphthylamine Blue; NCI C61289; Niagara Blue; Niagara Blue 3B; NSC 11247; Orion Blue 3B; Paramine Blue 3B; Parkibleu; Parkipan; Pontamine Blue 3BX; Pyrazol Blue 3B; Pyrotropblau; RCRA waste number U236; Renolblau 3B; Trianol Direct Blue 3B; Triazolblau 3BX; Tripan Blue; Trypan (Congo) Blue; Trypan blue; Trypan blue (commercial grade); Trypan Blue BPC; Trypanblau; Trypane Blue; Trypan blue. General use dyestuff, Trypan blue. Also used as a biological stain. Blue-gray powder; soluble in H_2O, insoluble in organic solvents; LD100 (rat iv) = 300 mg/kg; LD50 (rat orl) = 6200 mg/kg, (rat iv) = 300 mg/kg; LDLO (rat ip) = 300 mg/kg, (gpg sc) = 300 mg/kg, (mus iv) = 100 mg/kg. *Baker J. T.; Cambrex.*

2652 Nicfo
3349-06-2   6533   G

$C_2H_2NiO_4$
Nickel(II) formate.
EINECS 222-101-0; Formic acid, nickel(2+) salt; Nickel diformate; Nickel formate. Used in manufacture of nickel compounds, including catalysts. Green crystals; dec 180-200°; $d^{20.2}$ = 2.154; soluble in H_2O, insoluble in organic solvents. *Mechema Chemicals Ltd.*

2653 Nickel
7440-02-0   6523   G

Ni

Ni
Nickel.
Alcan 756; C.I. 77775; Carbonyl nickel powder; CCRIS 427; EINECS 231-111-4; EL12; Fibrex; Fibrex P; HSDB 1096; Ni 0901-S; Ni 270; Ni 4303T; Nichel; Nickel 200; Nickel 201; Nickel 205; Nickel 207; Nickel 270; Nickel catalyst; Nickel compounds; Nickel, elemental; Nickel, elemental/metal; Nickel particles; Nickel sponge; NP 2; Raney alloy; Raney nickel; RCH 55/5. Metallic element, used in electroplating, as a hydrogenation catalyst and in iron- and copper-based alloys. Metal; mp = 1453°; bp (calc) = 2732°; d= 8.908. *Atomergic Chemetals; Inco, Europe; Lancaster Synthesis Co.; Sigma-Aldrich Fine Chem.*

2654 Nickel acetate
373-02-4   6524   G

$C_4H_6NiO_4$
Nickel(II) acetate.
Acetic acid, nickel(2+) salt; AI3-26110; CCRIS 6206; EINECS 239-086-1; HSDB 1029; Nickel acetate; Nickel di(acetate); Nickel diacetate; Nickel(2+) acetate; Nickel(II) acetate; Nickel(II) acetate (1:2); Nickelous acetate; NSC 112210. Textile mordant, catalyst. Crystals; d = 1.744; soluble in H_2O (16.7 g/100 ml), EtOH. *Alcoa Ind. Chem.; Ashland; Mallinckrodt Inc.; Nihon Kagaku Sangyo.*

2655 Nickel carbonate
3333-67-3   G

$CNiO_3$
Nickel (II) carbonate.
Carbonic acid, nickel(2+) salt (1:1); EINECS 222-068-2; HSDB 1662; Nickel carbonate; Nickel carbonate ($NiCO_3$); Nickel monocarbonate; Nickel(2+) carbonate ($NiCO_3$); Nickel(II) carbonate; Nickelous carbonate. Chemical intermediate in manufacture of nickel oxide, nickel powder, and nickel catalysts. Used in vacuum tubes and transistor cans, as a catalyst to remove organic contaminants from wastewater or potable water in the preparation of colored glass, of nickel pigments, as a neutralizing compound in nickel electroplating solution, and in the preparation of many specialty nickel compounds. mp= 57°.

2656 Nickel carbonate, basic
12607-70-4   6527   G

$CH_4Ni_3O_7$
Nickel carbonate hydroxide.
Basic nickel(II) carbonate; (Carbonato(2-))tetrahydroxy-trinickel; Carbonic acid, nickel salt, basic; EINECS 235-715-9; HSDB 6154; Nickel, (carbonato(2-)tetra-hydroxytri-; Nickel carbonate hydroxide ($Ni_3(CO_3)(OH)_4$); Nickel carbonate hydroxide. Electroplating, preparation of nickel catalysts, ceramic colors and glazes. [tetrahydrate]; Yellow-green crystals; insoluble in H_2O, soluble in NH_3, dilute acids. *Ashland; Elf Atochem N. Am.; Farleyway Chem. Ltd.; M&T Harshaw; Nihon Kagaku Sangyo.*

2657 Nickel chloride
7718-54-9   6529   G

$NiCl_2$
Nickel(II) chloride.
CCRIS 1788; EINECS 231-743-0; EINECS 253-399-0; HSDB 860; Nickel chloride; Nickel chloride ($NiCl_2$); Nickel dichloride; Nickel(2+) chloride; Nickel(II) chloride; Nickelous chloride; NSC 254532; Hexahydrate [7791-20-0]. Electroplated nickel coatings, chemical reagent. Yellow crystals; bp = 987°; soluble in H_2O (1 g/ml), EtOH; LD (dog iv) = 40-80 mg/kg. *Ashland; Atomergic Chemetals; Elf Atochem N. Am.; Mallinckrodt Inc.; Nihon Kagaku Sangyo.*

2658 Nickel dibutyldithiocarbamate
13927-77-0   G H

$C_{18}H_{36}N_2NiS_4$
Carbamic acid, dibutyldithio-, nickel(II) salt.
AI3-26152; Bis(dibutyldithiocarbamato)nickel; Carbamo-dithioic acid, dibutyl-, nickel(2+) salt; Dibutyldithio-carbamic acid, nickel salt; (Dibutyldithiocarbamato)-nickel(II); EINECS 237-696-2; HSDB 2950; Naugard® NBC; NBC; Nickel bis(dibutyldithiocarbamate); Nickel(II) dibutyldithiocarbamate; Nocrac NBC; NSC 4797; Perkacit® NDBC; Rylex NBC; UV Chek AM 104; Vanguard N. Nickel chelating UV stabilizer for polyolefins and accelerator for rubber. *Akzo Chemie; DuPont UK; Uniroyal.*

2659 Nickel dimethyldithiocarbamate
15521-65-0   G P

$C_6H_{12}N_2NiS_4$
Nickel dimethylcarbamodithioate.
AI3-50831; Bis(dimethylcarbamodithioato-S,S')nickel; Bis(dimethyldithiocarbamato)nickel; Dimethyldithio-carb-amatonickel; Methyl niclate®; Nickel, bis(dimethyl-carb-amodithioato-kappaS,kappaS')-, (SP-4-1)-; Nickel, bis(di-methylcarbamodithioato-S,S')-; Nickel, bis(dimethyl-carb-amodithioato-S,S')-, (SP-4-1)-; Nocrac NMC; Robac Ni D.D.; Sankel. Antioxidant for epichlorohydrin and peroxide vulcanized elastomers. Fungicide and bactericide. Green powder; dec. > 250°; insoluble in

H2O, soluble in CHCl3 (0.029 g/100 ml), DMF (0.056 g/100 ml), THF (0.048 g/100 ml); LD50 (rat orl) > 36000 mg/kg, (mus orl) > 30000 mg/kg, (rat der) > 5000 mg/kg; LC50 (carp 48 hr.) = 360 mg/l. *Misaka; Vanderbilt R.T. Co. Inc.*

2660 Nickel nitrate
13478-00-7   G

N2NiO6.6H2O
 Nickel(II) nitrate.
 Nickel dinitrate hexahydrate; Nickel(2+) dinitrate hexahydrate; Nickel(2+) nitrate, hexahydrate; Nickel(II) nitrate, hexahydrate (1:2:6); Nickelous nitrate hexahydrate; Nitric acid, nickel(2+) salt, hexahydrate. Used in nickel plating, preparation of nickel catalysts, manufacture of brown ceramic colors. Crystals; mp = 57°; bp = 137°; d = 2.05; soluble in H2O (2.5 g/ml), EtOH; LD50 (rat orl) = 1.62 g/kg. *Mallinckrodt Inc.; Nihon Kagaku Sangyo; Noah Chem.*

2661 Nickel oxide
1313-99-1   6536   G

NiO
 Nickel (II) oxide.
 Black nickel oxide; Bunsenite; C.I. 77777; CCRIS 431; CI 77777; EINECS 215-215-7; Green nickel oxide; HSDB 1664; Mononickel oxide; Nickel (II) oxide; Nickel monoxide; Nickel oxide; Nickel oxide (NiO); Nickel oxide sinter 75; Nickel protoxide; Nickel(2+) oxide; Nickel(II) oxide; Nickelous oxide. Used in preparation of nickel salts, in porcelain painting and fuel cell electrodes. Green powder; mp = 1960°; d = 6.6700; insoluble in H2O; soluble in acids. *Atomergic Chemetals; Cerac; Nihon Kagaku Sangyo; Noah Chem.*

2662 Nickel rutile yellow
8007-18-9   G

NiSbTi
 C.I. Pigment Yellow 53.
 C.I. Pigment Yellow 53; EINECS 232-353-3; Nickel antimony titanium yellow rutile; Nickel Rutile Yellow; V-9415. Pigment for thermoplastic and thermoset resins, especially high temperature engineering resins, PVC siding and profile, and industrial finishes. *Ferro.*

2663 Nickel sulfamate
13770-89-3   G

H4N2NiO6S2
 Nickel (II) sulfamate dihydrate.
 Aeronikl 250; Aeronikl 400; Aeronikl 575; AI3-18003; EINECS 237-

396-1; Nickel sulfamate; Nickel (II) sulfamate; Nickel bis(sulphamidate); Nimate; NSC 78888; Sulfamic acid, nickel(2+) salt (2:1). For nickel electroforming and plating; aerospace and electronics application. *Albright & Wilson Americas Inc.; Atomergic Chemetals; CK Witco Corp.; Elf Atochem N. Am.; M&T Harshaw.*

2664 Nickel sulfate
7786-81-4   6541   G

NiO4S
 Nickel(II) sulfate.
 EINECS 232-104-9; HSDB 1114; NCI-C60344; Nickel sulfate; Nickel sulfate (NiSO4); Nickel sulfate(1:1); Nickel sulphate; Nickel(2+) sulfate; Nickel(II) sulfate; Nickelous sulfate; NSC 51152; Sulfuric acid, nickel(2+) salt (1:1); Sulphuric acid, nickel(II) salt. Also occurs as the hexahydrate [10101-97-0; 232-104-9] and the heptahydrate [10101-98-1; 205-788-1]. Used in the electroplating trade. Hexahydrate loses 5H2O at 100° and becomes anhydrous at 280°. Soluble in H2O (0.7 g/ml); LD50 (gpg sc) = 62 mg/kg.

2665 Nickel, tetrakis(tritolyl phosphite)-
35884-66-3   H

C84H84NiO12P4
 Nickel, tetrakis(tris(methylphenyl) phosphite-P)-.
 EINECS 252-777-2; Nickel, tetrakis(tris(methylphenyl) phosphite-P)-; Nickel, tetrakis(tritolyl phosphite)-; Tetrakis(tritolyl phosphite)nickel.

2666 Niclosamide
50-65-7   6543   G V

C13H8Cl2N2O4
 5-chloro-N-(2-chloro-4-nitrophenyl)-2-hydroxybenz-amide.
 AI3-25823; Atenase; B 2353; BAY 2353; Bayer 2353; Bayer 73; Bayluscid; BRN 2820605; CCRIS 3437; Cestocid; Chemagro 2353; Devermin; Devermine; Dichlosale; EINECS 200-056-8; ENT 25823; Fedal-Telmin; Fenasal; Helmiantin; HL 2447; HSDB 1572; Iomesan; Iomezan; Lintex; Mansonil; Mato; Nasemo; Niclocide; Niclosamida; Niclosamide; Niclosamidum; Nitrophenyl chlorsalicylamide; NSC 178296; Phenasal; Radeverm; Sagimid; Salicylanilide, 2',5-dichloro-4'-nitro-; SR 73; Sulqui; Tredemine; Vermitid; Vermitin; WR 46234; Yomesan; Zestocarp. Used in medicine and veterinary medicine as an anthelmintic. Veterinary preparation; anthelmintic (Cestodes); used against tapeworm infestation in ruminants, dogs, and cats. The ethanolamine salt is used as a molluscicide. mp = 225-230°; insoluble in H2O, sparingly soluble in organic solvents. *Bayer AG.*

2667 Nicotine
54-11-5   6551   D G

C10H14N2

3-(1-methyl-2-pyrollidinyl)pyridine.

AI3-03424; Black leaf; Black Leaf 40; Caswell No. 597; CCRIS 1637; Destruxol orchid spray; EINECS 200-193-3; Emo-nik; ENT 3,424; EPA Pesticide Chemical Code 056702; Flux MAAG; Fumetobac; Habitrol; HSDB 1107; Mach-Nic; L-3-(1-Methyl-2-pyrrolidyl)pyridine; Micotine; Niagara P.A. dust; Nic-Sal; Nico-dust; Nico-fume; Nicocide; Nicotin; Nicotina; Nicotine; l-Nicotine; L-Nicotine; Nicotine alkaloid; Nicotine and salts; Nikotin; Nikotyna; NSC 5065; Ortho N-4 dust; Ortho N-4 and N-5 dusts; Ortho N-5 dust; Prostep; Pyridine, 3-((2S)-1-methyl-2-pyrrolidinyl)-; Pyridine, 3-(1-methyl-2-pyrrolidinyl)-, (S)-; Pyrrolidine, 1-methyl-2-(3-pyridal)-; RCRA waste number P075; Tendust; Tetrahydronicotyrine, dl-; UN1654; XL-All Insecticide. Component of leaves of *Nicotiana tabacum and N. rustica*. An alkaloid insecticide which is the addictive principal in tobacco. Oil; bp = 243-248°; d = 1.017; $[\alpha]^{20}_D$ = -168±5° (c = 5, H_2O); miscible with H_2O, soluble in organic solvents; LD_{50} (mus iv) = 0.3 mg/kg, (mus ip) = 9.5 mg/kg, (mus orl) = 230 mg/kg. *Synchemicals Ltd.*

2668   Nifedipine

21829-25-4   6555   D

C17H18N2O6

Dimethyl   1,4-dihydro-2,6-dimethyl-4-(o-nitrophenyl)-3,5-pyridinedicarboxylate.

5-22-04-00268 (Beilstein Handbook Reference); Adalat; Adalat CC; Adalate; Adapress; Aldipin; Alfadat; Anifed; Aprical; Bay a 1040; Bonacid; BRN 0497773; Camont; CCRIS 6074; Chronadalate; Citilat; Coracten; Cordicant; Cordilan; Cordipin; Corinfar; Corotrend; Duranifin; Ecodipi; EINECS 244-598-3; Fenihidin; Fenihidine; Hexadilat; Introcar; Kordafen; Nifedicor; Nifedin; Nifedipine; Nifedipino; Nifedipinum; Nifelan; Nifelat; Nifensar XL; Orix; Oxcord; Pidilat; Procardia; Procardia XL; Sepamit; Tibricol; Zenusin. Dihydropyridine calcium channel blocker. Coronary vasodilator used as an antianginal agent. mp = 172-174°; λ_m = 340, 235 nm (ε 5010, 21590 MeOH), 338, 238 nm (ε 5740, 20600 0.1N HCl), 340, 238 nm (5740, 20510 0.1N NaOH); soluble in Me_2CO (25.0 g/100 ml), CH_2Cl_2 (16 g/100ml), $CHCl_3$ (14 g/100 ml), EtOAc (5 g/100 ml), MeOH (2.6 g/100 ml), EtOH (1.7 g/100 ml); LD_{50} (mus orl) = 494 mg/kg, (mus iv) = 4.2 mg/kg, (rat orl) = 1022 mg/kg, (rat iv) = 15.5 mg/kg. *Bayer AG; Miles Inc.; Pfizer Inc.; Pratt Pharm.*

2669   Nifuraldezone

3270-71-1   6558   D G

C7H6N4O5

5-Nitro-2-furaldehyde semioxamazone.

4-17-00-04465 (Beilstein Handbook Reference); Aldefur Bolus Veterinary; BRN 0239535; CBC 503239; EINECS 221-890-9; Entefur Bolus Veterinary; Furamazone; NF-84; Nifuraldezon; Nifuraldezona; Nifuraldezone; Nifural-dezonum; NSC-3184. Antibacterial. Crystals; mp = 270° (dec). *Eaton Labs.; Norwich*

Eaton.

2670   Nifurpirinol

13411-16-0   6564   D G

C12H10N2O4

6-[2-(5-Nitro-2-furyl)vinyl]-2-pyridinemethanol.

BRN 1216943; CCRIS 1046; EINECS 236-503-9; Furanace; Furanace-10; Furpirinol; Furpyrinol; NF 323; Nifurpirinol; Nifurpirinolum; P-7138. Used to treat bacterial diseases in fish. Antibacterial agent. Yellow needles; mp = 170-171°; LD_{50} (eel orl) = 1780 mg/kg. *Daiichi Seiyaku; Yamanouchi U.S.A. Inc.*

2671   Nimidane

50435-25-1   6578   G P

C9H8ClNS2

4-Chloro-N-1,3-dithietan-2-ylidene-2-methylbenzene-amine.

84633; Abequito; AC 84633; AI3-29106; Benzenamine, 4-chloro-N-1,3-dithietan-2-ylidene-2-methyl-; Benzene-amine, 4-chloro-N-1,3-dithietan-2-ylidene-2-methyl-; (4-Chloro-2-tolyl)-dithio-imidocarbonic acid cyclic methyl-ene ester; CL 84,633; Cyclic methylene (4-chloro-o-tolyl)-dithioimidocarbonate; ENT 29106; Nimidane; Nimidano; Nimidanum. Acaricide. Crystals; mp = 43-46°. *Am. Cyanamid.*

2672   Ninhydrin

485-47-2   6582   G

C9H6O4

2,2-Dihydroxy-1H-indene-1,3(2H)-dione.

4-07-00-02786 (Beilstein Handbook Reference); AI3-04464; BRN 1910963; CCRIS 4849; 2,2-Dihydroxy-1,3-indandione; 2,2-Dihydroxy-1H-indene-1,3(2H)-dione; EINECS 207-618-1; Indan-1,2,3-trione; 1,2,3-Indantrione, 2-hydrate; 1,2,3-Indantrione monohydrate; 1,3-Indandione, 2,2-dihydroxy-; 1H-Indene-1,3(2H)-dione, 2,2-dihydroxy-; Ninhydrin; Ninhydrin hydrate; Triketohydrindene hydrate. A colorimetric reagent used for aminoacids. Pale yellow crystals; mp = 241-243° (dec); λ_m = 485 nm (ε 1259, pH 7.8), 250, 294 nm (ε = 7943, 1000, pH 4.2); slightly soluble in Et_2O, soluble in EtOH, alkali, very soluble in H_2O; LD_{50} (mus ip) = 78 mg/kg.

2673   Niobium

7440-03-1   6584   G

Nb

Nb
Niobium.

Columbium; EINECS 231-113-5; Niobium; Niobium-93; Niobium element; VN 1. Metallic element; used in ferrous metallurgy; for superconducting and magnetic alloys, cermets, missiles and rockets, cryogenic equipment, ferroniobium for alloy steels. Metal; mp = 2468°; bp = 4927°; d = 8.57. *Atomergic Chemetals; Cabot Carbon Ltd.; Cerac; Noah Chem.*

2674 Niobium oxide
1313-96-8   6587   G

Nb2O5
Niobium(V) pentoxide.
  Diniobium pentaoxide; Diniobium pentoxide; EINECS 215-213-6; Niobia; Niobium oxide; Niobium oxide (Nb2O5); Niobium pentaoxide; Niobium pentoxide; Niobium(5+) oxide; Niobium(V) oxide. Used as a chemical intermediate, in electronics fabrication. White crystals; mp = 1520°; d= 4.6; insoluble in H2O, soluble in HF or H2SO4. *Atomergic Chemetals; Cabot Carbon Ltd.; Cerac; Noah Chem.*

2675 Niter
7757-79-1   7733   D G P

KNO3
Potassium nitrate.
  AI3-51245; Caswell No. 697; CCRIS 3667; Collo-Bo; EINECS 231-818-8; EPA Pesticide Chemical Code 076103; HSDB 1227; Kalii nitras; Kaliumnitrat; Niter; Nitrate of potash; Nitre; Nitric acid, potassium salt; NSC 57632; Petre; Potassium nitrate; Prismatic niter; Sal niter; Sal prunella; Salt peter; Saltpeter; Saltpeter flour; UN1486; Vicknite; saltpeter. Used in pyrotechnics, explosives, matches, fertilizer, reagent, glass manufacture, tempering steel, curing foods, oxidizer in solid rocket propellants. Registered by EPA as an antimicrobial and rodenticide. Used as a fertilizer, but leads to water runoff pollution. Used medicinally as a diuretic. Colorless prisms; mp = 333°; dec 400°; d = 2.11; soluble in H2O (35.7 g/100 ml 25°, 200 g/100 ml 100°), EtOH (0.16 g/100 ml); LD50 (rbt orl) = 1.16 g/kg. *Am. Biorganics; EM Ind. Inc.; Mallinckrodt Inc.; San Yuan Chem. Co. Ltd.; Whiting Peter Ltd.*

2676 Nitralin
4726-14-1   6600   G P

C13H19N3O6S
4-(methylsulfonyl)-2,6-dinitro-N,N-dipropylbenzene-amine.
  SD 11831; Planavin. Herbicide. mp = 150-151°; slighlty soluble in H2O (0.6 mg/l), more soluble in polar organic solvents; LD50 (rat orl) >2 g/kg.

2677 Nitrapyrin
1929-82-4   6604   P

C6H3Cl4N
α,α,α-6-Tetrachloro-2-picoline.
  5-20-05-00500 (Beilstein Handbook Reference); BRN 1618997; Caswell No. 217; CCRIS 4599; 2-Chloro-6-(trichloromethyl)pyridine; Dowco-163; EINECS 217-682-2; EPA Pesticide Chemical Code 069203; N-Serve; N-Serve nitrogen stabilizer; Nitrapyrin; Nitrapyrine; Pyridine, 2-chloro-6-(trichloromethyl)-. Used to control *Nitrosomonas* bacteria in soils. Registered by EPA as an antimicrobial and disinfectant. Crystals; mp = 62.5 - 62.9°; bp11 = 136 - 137.5°, soluble in H2O (0.004 g/100 ml), anhydrous NH3 (54 g/100 ml), EtOH (23.7 g/100 ml), Me2CO (156 g/100 ml), CH2Cl2 (246 g/100 ml), xylene (89.4 g/100 ml); LD50 (rat orl) = 1072 - 1231 mg/kg, (rbt der) = 2830 mg/kg, (ckn orl) = 235 mg/kg; LC50 (mallard duck 8 day dietary) = 1466 mg/kg, (Japanese quail 8 day dietary) = 820 mg/kg, (channel catfish) 5.8 mg/l; non toxic to Ramshorn snails or to *Daphnia*. *Dow AgroSciences; DowElanco Ltd.; UAP - Platte Chemical.*

2678 Nitric acid
7697-37-2   6608   G

HNO3
Hydrogen nitrate.
  Acide nitrique; Acido nitrico; Acidum nitricum; Aqua fortis; Azotic acid; Azotowy kwas; EINECS 231-714-2; Engraver's acid; HSDB 1665; Kyselina dusicne; Nital; Nitric acid; Nitric acid, anhydrous; Nitric acid, red fuming; Nitrous fumes; Nitryl hydroxide; Red fuming nitric acid; RFNA; Salpetersaeure; Salpetersäure; Salpetersaure; Salpeterzuuroplossingen; UN2031; UN2032. Used in manufacture of ammonium nitrate for fertilizer and explosives, organic synthesis (dyes, drugs, explosives, cellulose nitrate, nitrate salts), metallurgy, photo-engraving, etching steel, ore flotation, urethanes, rubber chemicals, nuclear fuel. Liquid; mp = -42°; bp = 83°; d4^25 = 1.50269. *Aceto Corp.; Air Prods & Chems; Am. Cyanamid; Angus; Asahi Chem. Industry; DuPont; Miles Inc.; Monsanto Co.; Nissan Chem. Ind.; Norsk Hydro AS.*

2679 Nitrilotriacetic acid
139-13-9   6612   G

C6H9NO6
Triglycollamic acid.
  4-04-00-02441 (Beilstein Handbook Reference); Acetic acid, nitrilotri-; AI3-52483; Aminotriacetic acid; Aminotriethanoic acid; N,N-Bis(carboxymethyl)glycine; BRN 1710776; CCRIS 436; CHEL 300; Complexon I; EINECS 205-355-7; Glycine, N,N-bis(carboxymethyl)-; Hampshire NTA acid; HSDB 2853; Komplexon I; Kyselina nitrilotrioctova; NCI-C02766; Nitrilo-2,2',2''-triacetic acid; Nitriloacetate; Nitrilotriacetate; Nitrilo-triacetic acid; Nitrilotriessigsaeure; Nitrilo-triessigsäure; NSC 2121; NTA; TG Buffer; Titriplex I; Tri(carboxy-methyl)amine; Triglycine; Trilon A;

α,α',α''-Trimethyl-aminetricarboxylic acid; Versene NTA acid; A chelating, sequestering and metal complexing agent; builder in synthetic detergents. Cream-colored crystals; mp = 242° (dec); slightly soluble in H2O (1.28 g/l), DMSO, soluble in EtOH.. *Grace W.R. & Co.; Greeff R.W. & Co.; Hampshire.*

2680   Nitrilotriacetic acid, trisodium salt
5064-31-3   G P

$C_6H_6NNa_3O_6$
Trisodium N,N-bis(carboxymethyl)glycinate.
Acetic acid, nitrilotri-, trisodium salt; CCRIS 1404; Cheelox NTA-14 - Na3; Chemcolox 365 powder; EINECS 225-768-6; Glycine, N,N-bis(carboxymethyl)-, trisodium salt; Hampshire NTA; Hampshire NTA 150; HSDB 1013; Masquol np 140; Nitrilotriacetic acid sodium salt; Nitrilotriacetic acid, trisodium salt; NTA Trisodium salt; Ntana3; Sodium nitrilotriacetate; Syntron A; Trilon A; Trisodium 2,2',2''-nitrilotriacetate; Trisodium amino-triacetate; Trisodium nitrilotriacetate; Trisodium NTA; Versene NTA 150; Versene NTA 335. A chelating agent used as a sequestrant. Used primarily as a chelating agent and as a binder in synthetic detergents.The compound sequesters magnesium and calcium ions present in hard water which would normally inhibit the activity of detergent surfactants. Also used as a chelator in water treatment, textile manufacture and analytical chemistry, as a boiler feedwater additive and in metal plating, metal cleaning and pulp and paper processing. Registered by EPA as an antimicrobial and disinfectant. *Biosentry, Inc.; Rhône Poulenc Surfactants.*

2681   Nitrilotrimethanephosphonic acid
6419-19-8   G H

$C_3H_{12}NO_9P_3$
Aminotri(methylene phosphonic acid).
4-01-00-03070 (Beilstein Handbook Reference); AI3-51572; BRN 1715724; Dequest 2000; Dowell L 37; EINECS 229-146-5; Ferrofos 509; Fostex AMP; Phosphoric acid, (nitrilotris-(methylene))tris-; Tris(phos-phonomethyl)amine. Scale inhibitor and sequestrant used in water treatment. Liquid; mp = -14°; bp = 105°; d = 1.33; LD50 (rat orl) = 2100 mg/kg. *Henkel.*

2682   2,2',2''-Nitrilotrisethanol formate
24794-58-9   H

$C_6H_{15}NO_3.CH_2O_2$
Formic acid, compd. with 2,2',2''-nitrilotriethanol (1:1).
EINECS 246-463-4; Formic acid, compd. with 2,2',2''-nitrilotris(ethanol) (1:1); 2,2',2''-Nitrilotrisethanol formate.

2683   4-Nitroaniline
100-01-6   6617   G H

$C_6H_6N_2O_2$
p-Nitroaniline.
AI3-08926; para-Aminonitrobenzene; Aniline, 4-nitro-; Aniline, p-nitro-; Azoamine Red ZH; Azofix Red GG Salt; Azoic Diazo Component 37; Benzenamine, 4-nitro-; CCRIS 1184; CI 37035; CI Azoic Diazo Component 37; CI Developer 17; Developer P; Devol Red GG; Diazo Fast Red GG; EINECS 202-810-1; Fast Red 2G Salt; Fast Red 2G Base; Fast Red Base GG; Fast Red Base 2J; Fast Red GG Salt; Fast Red GG Base; Fast Red MP Base; Fast Red P Base; Fast Red P Salt; Fast Red Salt 2J; Fast Red Salt GG; HSDB 1156; Naphtoelan Fast Red Base; NCI-C60786; Nitrazol CF Extra; para-Nitroaniline; NSC 9797; PNA; RCRA waste number P077; Red 2G Base; Shinnippon Fast Red GG Base; UN1661. Intermediate for dyes, antioxidants; gasoline gum inhibitors; corrosion inhibitor. Pale yellow needles; mp = 147°; bp = 332°; d^{20} = 1.424; λ_m = 228 nm (MeOH); insoluble in H2O, slightly soluble in C6H6, DMSO, soluble in EtOH, Et2O, Me2CO, CHCl3, C7H8, very soluble in MeOH. *EniChem Am.; Hoechst AG; Monsanto Co.*

2684   o-Nitroaniline
88-74-4   6616   H

$C_6H_6N_2O_2$
Benzenamine, 2-nitro-.
AI3-02916; Aniline, o-nitro-; Azoene Fast Orange GR Salt; Azoene Fast Orange GR Base; Azofix Orange GR Salt; Azogene Fast Orange GR; Azoic Diazo Component 6; Benzenamine, 2-nitro-; Brentamine Fast Orange GR Base; Brentamine Fast Orange GR Salt; CCRIS 2317; CI 37025; CI Azoic Diazo Component 6; Devol Orange B; Devol Orange Salt B; Diazo Fast Orange GR; EINECS 201-855-4; Fast Orange Base JR; Fast Orange Base GR; Fast Orange GR Base; Fast Orange GR Salt; Fast Orange O Salt; Fast Orange O Base; Fast Orange Salt GR; Fast Orange Salt JR; Hiltonil Fast Orange GR Base; Hiltosal Fast Orange GR Salt; Hindasol Orange GR Salt; HSDB 1132; Natasol Fast Orange GR Salt; NSC 9796; ONA; Orange Base Ciba II; Orange Base IRGA II; Orange GRS Salt; Orange Salt Ciba II; Orange Salt IRGA II; Orthonitroaniline; UN1661. Used in manufacture of dyestuffs. Crystals; mp = 71.2°; bp = 284°; d^{15} = 1.442; λ_m = 232, 277 nm (MeOH); poorly soluble in H2O, more soluble in EtOH, very soluble in Et2O, Me2CO, C6H6, CHCl3.

2685   o-Nitrobenzaldehyde
552-89-6   G

C7H5NO3
2-Nitrobenzaldehyde.
AI3-02415; Benzaldehyde, 2-nitro-; Benzaldehyde, o-nitro-; CCRIS 2322; EINECS 209-025-3; 2-Nitro-benzaldehyde; o-Nitrobenzaldehyde; NSC 5713. Used in the synthesis of dyes, pharmaceuticals, surface active agents. Yellow needles; mp = 43.5°; bp_{23} = 153°; d^{20} = 1.1111; λ_m = 252 nm (ε - 4400, MeOH); slightly soluble in H2O, CHCl3, very soluble in EtOH, Et2O, Me2CO, C6H6. *Lancaster Synthesis Co.; Penta Mfg.*

2686   p-Nitrobenzaldehyde
555-16-8   G

C7H5NO3
4-Nitrobenzaldehyde.
AI3-52475; Benzaldehyde, 4-nitro-; Benzaldehyde, p-nitro-; CCRIS 1675; EINECS 209-084-5; p-Formylnitrobenzene; NSC 6103; 4-Nitrobenzaldehyde; p-Nitrobenzaldehyde. Used in the synthesis of dyes, pharmaceuticals, surface active agents. Prisms; mp = 107°; bp sublimes; d^{25} = 1.496; λ_m = 265 nm (ε = 10600, MeOH); slightly soluble in H2O, Et2O, ligroin, soluble in C6H6, CCl3, AcOH, very soluble in EtOH. *Lancaster Synthesis Co.; Penta Mfg.*

2687   p-Nitrobenzoic acid
62-23-7   G H

C7H5NO4
Benzoic acid, 4-nitro-.
AI3-00149; Benzoic acid, 4-nitro-; Benzoic acid, p-nitro-; CCRIS 1185; EINECS 200-526-2; HSDB 2140; Kyselina p-nitrobenzoova; Nitrodracylic acid; 4-Nitrodracylic acid; NSC 7707; p-Nitrobenzenecarboxylic acid; p-Nitrobenzoic acid. Organic synthesis. mp = 242°; d^{20} = 1.610; λ_m = 259 nm (MeOH); soluble in H2O (4.2 g/100 ml), more soluble in organic solvents. *Akzo Nobel; DuPont; Hüls Am.; Lancaster Synthesis Co.; Mallinckrodt Inc.; Schweizerhall Inc.; Sigma-Aldrich Fine Chem.*

2688   Nitro Blue tetrazolium
298-83-9   G

C40H30Cl2N10O6
2H-tetrazolium, 3,3'-(3,3'-dimethoxy[1,1'-biphenyl]-4,4'-diyl)bis[2-(4-nitrophenyl)-5-phenyl-, dichloride.
EINECS 206-067-4; NBT; NBT (dye); Nitro Blue tetrazolium salt; Nitro Blue tetrazolium chloride; Nitro BT; Nitro tetrazolium BT; Nitroblue tetrazolium; Nitroblue tetrazolium salt; (4-Nitrophenyl)-5-phenyl-,dichloride; Nitrotetrazolium Blue; Nitrotetrazolium Chloride Blue; NSC 27622; p-Nitro Blue tetrazolium; p-Nitro Blue tetrazolium chloride; p-Nitrotetrazolium Blue; 2H-Tetrazolium, 3,3'-(3,3'-dimethoxy(1,1'-biphenyl)-4,4'-diyl)bis(2-4,4'-ylene)ditetrazolium dichloride; Tetraz-olium nitro BT; Tetrazolium Nitro Blue. Activity stain for electrophoresis and histochemistry. Substrate for dehydrogenase. Crystals; mp = 200°. *Monomer-Polymer & Dajac.*

2689   Nitrobenzene
98-95-3   6621   G H

C6H5NO2
Benzene, nitro-.
AI3-01239; Benzene, nitro-; Caswell No. 600; CCRIS 2841; EINECS 202-716-0; EPA Pesticide Chemical Code 056501; Essence of mirbane; HSDB 104; Mirbane oil; Mononitrobenzene; NCI-C60082; Nitrobenzeen; Nitro-benzen; Nitrobenzene; p-Nitrobenzene; Nitrobenzol; NSC 9573; Oil of mirbane; Oil of myrbane; RCRA waste number U169; UN1662. Used as a chemical intermediate and as a solvent. Used in the manufacture of aniline, solvent for cellulose ether, modifying esterfication of cellulose acetate, ingredient of metal polishes; manufacture of benzoline, quinoline, and azobenzene. Liquid; mp = 5.7°; bp = 250°, bp_{16} = 137°; d^{24} = 1.126; LD50 (rat orl) = 780 mg/kg.

2690   Nitrobutylmorpholine
2224-44-4   G P

C8H16N2O3
4-(2-Nitrobutyl)morpholine.
4-27-00-00027 (Beilstein Handbook Reference); Bioban P 1487; BRN 0149001; Caswell No. 601A; EINECS 218-748-3; EPA Pesticide Chemical Code 100801; Morpholine, 4-(2-nitrobutyl)-; 4-(2-Nitrobutyl)morph-oline; N-(2-Nitrobutyl)morpholine; Vancide 40;

Vancide F 5386. A mixture of 4,4-(2-Ethyl-2-nitro-trimethylene) dimorpholine (30%) and 4-(2-Nitrobutyl)morpholine (70%), known as Bioban P 1487, is used as a preservative. Registered by EPA as an antimicrobial and disinfectant. *Dow Chem. U.S.A.*

2691   2-Nitro-p-cresol
119-33-5   H

C7H7NO3
4-Hydroxy-3-nitrotoluene.
4-06-00-02149 (Beilstein Handbook Reference); 4-Hydroxy-3-nitrotoluene; AI3-15389; BRN 1868022; EINECS 204-315-6; NSC 5387; 2-Nitro-4-cresol; 2-Nitro-p-cresol; 2-Nitro-4-methylphenol; o-Nitro-p-cresol; p-Cresol, 2-nitro-. Yellow needles; mp = 36.5°; bp^{22} = 125°; d^{20} = 1.2399; very soluble in Me_2CO, C_6H_6, Et_2O, EtOH.

2692   p-Nitrodiphenylamine
836-30-6   H

C12H10N2O2
4-Nitrodiphenylamine.
4-12-00-01619 (Beilstein Handbook Reference); AI3-02915; Amine, diphenyl, 4-nitro-; Benzenamine, 4-nitro-N-phenyl-; BRN 2051910; CCRIS 5174; Diphenylamine, 4-nitro-; EINECS 212-646-2; HSDB 5763; 4-Ndpa; 4-Nitro-N-phenylaniline; 4-Nitrodifenylamin; 4-Nitro-diphenylamine; p-Nitrodiphenylamine; N-(4-Nitrophenyl)benzenamine; p-Nitrophenylphenylamine; NSC 33836. Crystals; mp = 135.3°; bp_{30} = 211°; λ_m = 258, 390 nm (ε = 8330, 16900, MeOH); insoluble in H_2O, slightly soluble in Me_2CO, soluble in conc. H_2SO_4; very soluble in EtOH, AcOH.

2693   Nitroethane
79-24-3   6630   G H

C2H5NO2
Nitroethane.
AI3-00110; CCRIS 3088; EINECS 201-188-9; Ethane, nitro-; HSDB 105; NE; Nitroetan; Nitroethane; NSC 8800; UN2842. Intermediate, stabilizer for halogenated solvents, additive in fuels and explosives, solvent for coatings or industrial processes. Liquid; mp = -89.5°; bp = 114°; d^{25} = 1.0448; soluble in H_2O (4.5 g/100 ml), very soluble in EtOH, moderately soluble in Et_2O, Me_2CO, $CHCl_3$. *Grace W.R. & Co.; Spectrum Chem. Manu-facturing; Whittaker Clark & Daniels.*

2694   Nitrofurantoin
67-20-9   6632   D

C8H6N4O5
1-[(5-Nitrofurfurylidene)amino]hydantoin.
AI3-26388; Alfuran; Benkfuran; Berkfuran; Berkfurin; CCRIS 1192; Ceduran; Chemiofuran; Cistofuran; Cyantin; Cystit; Dantafur; EINECS 200-646-5; Fua Med; FuaMed; Fur-ren; Furabid; Furachel; Furadantin; Furadantin Retard; Furadantina MC; Furadantine; Furadantine MC; Furadantoin; Furadoin; Furadoine; Furadonin; Furadonine; Furadoninum; Furadontin; Furagin; Furalan; Furaloid; Furantoin; Furantoina; Furatoin; Furazidin; Furedan; Furina; Furobactina; Furophen T; Furophen T-Caps; Gerofuran; HSDB 3135; Ituran; Macpac; Macrobid; Macrodantin; Macrodantina; Macrofuran; Macrofurin; NCI-C55196; Nierofu; Nifurantin; Nifuretten; Nitoin; Nitrex; Nitrofur-C; Nitrofuradantin; Nitrofurantoin; Nitrofurantoina; Nitrofurantoine; Nitrofurantoinum; Novofuran; NSC 2107; Orafuran; Parfuran; Phenurin; PiyEloseptyl; Ro-Antoin; Siraliden; Trantoin; Uerineks; Urantoin; Urizept; Uro-Selz; Uro-Tablinen; Urodin; Urofuran; Urofurin; Urolisa; Urolong; USAF EA-2; Welfurin; Zoofurin. Nitrofuran antibiotic. Used to treat mastitis in cows. Solid; dec 270-272°; λ_m = 370 nm ($E_{1cm}^{1\%}$ 776); soluble in H_2O (0.019 g/100 ml at pH 7), EtOH 0.051 g/1oo ml), Me_2CO (0.51o g/100 ml), dmg (8 g/100 ml), glycerol (0.06 g/100 ml), polyethylene glycol (1.5 g/100 ml). *Eaton Labs.; Norwich Eaton; Parke-Davis; Procter & Gamble Pharm. Inc.*

2695   Nitrogen dioxide
10102-44-0   6636   G

N2O4
Dinitrogen tetroxide.
Azote; Azoto; CCRIS 4040; Dioxido de nitrogeno; EINECS 233-272-6; HSDB 718; Nitrito; Nitro; Nitrogen dioxide; Nitrogen dioxide (NO_2); Nitrogen oxide (NO_2); Nitrogen peroxide; Nitrogen peroxide, liquid; Peroxyde d' azote; RCRA waste number P078; Stickstoffdioxid; Stikstofdioxyde. Used as a chemical intermediate, oxidizing agent and propellant. Brown gas; mp = -9°; bp = 21°; d_4^{20} 1.448.

2696   Nitroglycerin
55-63-0   6640   D H

C3H5N3O9
1,2,3-Propanetriol trinitrate.
4-01-00-02762 (Beilstein Handbook Reference); Adesitrin; Aldonitrin; Angibid; Anginine; Angiolingual; Angiplex; Anglix; Angonist; Angorin; Aquo-Trimitrosan; Blasting gelatin; Blasting oil; BRN 1802063; Buccard; Cardabid; Cardamist; Cardinit; Cardiodisco; CCRIS 4089; Chitamite; Colenitral; Corangin Nitrokapseln; Cordipatch; Corditrine; Coro-Nitro; Dauxona; Deponit; Deponit 5; Deponit TTS 5; Deponit TTS 10; Diafusor; Discotrine; EINECS 200-240-8; Epinitril; Gepan Nitroglicerin; Gilucor nitro; Gilustenon; Glonoin; Glycerintrinitrate; Glycerol (trinitrate de); Glycerol, nitric acid triester; Glycerol trinitrate; Glycerol(trinitrate de); Glyceroltrinitraat;

Glyceryl nitrate; Glyceryl trinitrate; Glycerylnitrat; Glytrin; GTN; GTN-Pohl; Herwicard; Herzer; HSDB 30; Klavikordal; Lenitral; Lentonitrina; Mi-Trates; Millisrol; Minitram; Minitran; Mionitrat; Myocon; Myoglycerin; Myovin; Natispray; Neos nitro OPT; NG; Niglin; Niglycon; Niong; Niong Retard; Nirmin; Nit-Ret; Nitora; Nitradisc; Nitradisc Pad; Nitradisc TTS; Nitrek; Nitric acid triester of gylcerol; Nitriderm TTS; Nitrin; Nitrine; Nitrine-TDC; Nitro-Bid; Nitro-Dur; Nitro-Dur 10; Nitro-Dur 5; Nitro Dur TTS; Nitro-Gesanit Retard; Nitro-lent; Nitro-M-Bid; Nitro Mack Retard; Nitro-Par; Nitro-Pflaster; Nitro Retard; Nitro Rorer; Nitro-Span; Nitro-Time; Nitroard; Nitrobaat; Nitrobid Oint; NitroBid; Nitrobukal; Nitrocap T.D.; Nitrocerin; Nitrocine; Nitrocine 5; Nitroclyn; Nitrocontin; Nitrocontin Continus; Nitrocor; NitroCor; Nitrocot; Nitroderm; Nitroderm TTS; Nitroderm TTS-5; Nitroderm TTS Ext; Nitrodisc; Nitrodyl; Nitrodyl TTS; Nitrogard; Nitrogard-SR; Nitroglicerina; Nitrogliceryna; Nitroglin; Nitroglycerin; Nitroglycerin-ACC; Nitro-glycerin, spirits of; Nitroglycerine; Nitroglycerol; Nitroglyn; Nitrol; Nitrol Ointment; Nitrolan; Nitroletten; Nitrolin; Nitrolingual; Nitrolingual Spray; Nitrolowe; Nitromack Retard; Nitromel; Nitromex; Nitromint; Nitromint Aerosol; Nitromint Retard; Nitronal Aqueous; Nitronet; Nitrong; Nitrong Retard; Nitrong-SR; Nitropatch; Nitropen; Nitropercuten; Nitroperlinit; Nitroplast; Nitroprol; Nitropront; Nitroprontan; Nitrorectal; Nitroretard; Nitrorex; Nitrospan; Nitrostabilin; Nitrostat; Nitrovis; Nitrozell retard; NK-843; NTG; NTS; Nysconitrine; Percutol; Percutol Oint.; Perganit; Perglottal; Perlinganit; Plastranit; Polnitrin; Propanetriol trinitrate; Ratiopharm; RCRA waste no. P081; RCRA waste number P081; SDM No. 17; SDM No. 27; SK-106N; Solution glyceryl trinitrate; Soup; Spirit of glonoin; Spirit of glyceryl trinitrate; Spirit of trinitroglycerin; Susadrin; Suscard; Sustonit; Temponitrin; Top-Nitro; Transderm-N TTS; Transderm-Nitro TTS; Transiderm-Nitro; Tridil; Trinalgon; Trinipatch; Triniplas; Trinitrin; Trinitrina Erba; Trinitrine Simple Laleuf; Trinitroglicerina Fabra; Trinitroglycerin; Trinitroglycerol; Trinitrol; Trinitron; Trinitrosan; Turicard; UN0143; UN0144; UN1204; UN3064; UN3319; Vasoglyn; Vasolator; Vernies; Willong. Antianginal. A coronary vasodilator used in the treatment of angina pectoris. Also used in manufacture of dynamite. CAUTION: Acute poisoning can cause nausea, vomiting abdominal cramps, headache, mental confusion, delirium, bradypnea, bradycardia, paralysis, convulsions, methemoglobinemia, cyanosis, circulatory collapse, death. Chronic poisoning can cause severe headaches, hallucinations, skin rashes. Alcohol aggravates symptoms. Toxic effects may occur by ingestion, inhalation, absorption. [labile form]: liquid; mp = 2.8°; [stable form]: liquid; mp = 13.5°; begins to decompose at $\cong$ 50°; d$_{15}^{15}$ = 1.599l; n$_D^{15}$ = 1.474; heat of combustion = 1580 cal/g; slightly soluble in H$_2$O (0.125 g/100 ml), EtOH (0.5 g/100 ml); more soluble in MeOH (2.25 g/100 ml), CS$_2$ (15 g/100 ml); miscible with Et$_2$O, Me$_2$CO, glacial AcOH, EtOAc, C$_6$H$_6$, nitrobenzene, C$_5$H$_5$N, CHCl$_3$, ethylene bromide, dichloroethylene; sparingly soluble in petroleum Et$_2$O, liquid petrolatum, glycerol. *3M Pharm.; Ciba-Geigy Corp.; Hoechst AG (USA); ICI Americas Inc.; Key Pharm.; KV Pharm.; Marion Merrell Dow Inc.; Parke-Davis; Rhône-Poulenc Rorer Pharm. Inc.; Schwarz Pharma Kremers Urban Co.; Searle G.D. & Co.; U.S. Ethicals Inc.; Zeneca Pharm.*

2697   2-Nitro-2'-hydroxy-3',5'-bis(α,α-dimethyl-benzyl)azobenzene
70693-50-4   H

2698   5-Nitroisophthalic acid
618-88-2   G

C$_8$H$_5$NO$_6$
5-Nitrobenzene-1,3-dicarboxylic acid.
1,3-Benzenedicarboxylic acid, 5-nitro-; EINECS 210-568-3; Isophthalic acid, 5-nitro-; 1-Nitrobenzene-3,5-dicarboxylic acid; 5-Nitro-1,3-benzenedicarboxylic acid; 5-Nitroisophthalic acid; 5-Nitro-m-phthalic acid; NSC 66545. Crystals; mp = 260-261°. *Lancaster Synthesis Co.; Pfister.*

C$_{30}$H$_{29}$N$_3$O$_3$
Phenol, 2,4-bis(1-methyl-1-phenylethyl)-6-[(2-nitro-phenyl)azo]-.
2,4-Bis(1-methyl-1-phenylethyl)-6-((2-nitrophenyl)azo)-phenol;
EINECS 274-768-2; Phenol, 2,4-bis(1-methyl-1-phenylethyl)-6-((2-nitrophenyl)azo)-; 2-Nitro-2'-hydroxy-3',5'-bis(α,α-dimethylbenzyl)azobenzene.

2699   Nitroisopropane
79-46-9   6661   G H

C$_3$H$_7$NO$_2$
2-Nitropropane.
AI3-00109; CCRIS 453; Dimethylnitromethane; EINECS 201-209-1; HSDB 1134; Isonitropropane; Nipar S-20; Nipar S-20 solvent; Nipar S-30 solvent; Nitroisopropane; NSC 5369; Propane, 2-nitro-; RCRA waste number U171. Used as a solvent in inks, adhesives, paints and varnishes and polymers. Also used as a chemical intermediate and as a rocket propellant, gasoline additive. Liquid; mp = -91.3°; bp = 120.2°; d^{25} = 0.9821; soluble in CHCl$_3$, less soluble in H$_2$O. *Ashland; Grace W.R. & Co.; Whittaker Clark & Daniels.*

2700   Nitromethane
75-52-5   6644   G H

CH$_3$NO$_2$
Nitromethane.
AI3-00111; CCRIS 1205; EINECS 200-876-6; HSDB 106; Methane, nitro-; NM; Nitrocarbol; Nitrometan; Nitromethane; NSC 428; UN1261. Stabilizer for chlorinated solvents; chemical intermediate; solvent for cellulosic compounds, polymers, waxes; rocket fuel, gasoline additive; in coatings industry. Liquid; mp = -28.5°; bp = 101.1°; d^{20} = 1.1371; soluble in H$_2$O (10 g/100 ml), soluble in organic solvents; LD$_{50}$ (rat orl) = 1.44 g/kg. *Grace W.R. & Co.; Whittaker Clark & Daniels.*

2701   2-Nitro-2-methylpropanol
76-39-1   G H

443

C4H9NO3

2-Methyl-2-nitro-1-propanol.

AI3-02258; EINECS 200-957-6; HSDB 5213; 2-Methyl-2-nitropropanol; 2-Methyl-2-nitropropan-1-ol; 2-Methyl-2-nitro-1-propanol; 2-Nitro-2-methyl-1-propanol; 2-Nitro-2-methylpropanol; NMP; NMP Conc.; NSC 17676; 1-Propanol, 2-methyl-2-nitro-. Chemical and pharm-aceutical intermediate, in tire cord adhesives, as formaldehyde release agents, deodorants, antimicrobials. Chemical intermediate, formaldehyde donor, textile reactant; reduces formaldehyde on finished cloth. Solid; mp = 89.5°; bp^{10} = 94°; λ_m = 278 nm (MeOH); slightly soluble in H_2O, soluble in organic solvents. *Sigma-Aldrich Fine Chem.; Whittaker Clark & Daniels.*

2702   Nitron

2218-94-2   6646   G

C20H16N4

1,4-Diphenyl-3-(phenylamino)-1H-1,2,4-triazolium inner salt.

1,4-Diphenyl-3-(phenylammonio)-1H-1,2,4-triazolium; EINECS 218-724-2; Nitron; NSC 5038; 1H-1,2,4-Triazolium, 1,4-diphenyl-3-(phenylamino)-, hydroxide, inner; 1H-1,2,4-Triazolium, 1,4-diphenyl-3-(phenyl-amino)-, inner salt; 2,3,5,6-Tetraazabicyclo(2.1.1)hex-1-ene, 3,5,6-triphenyl-. A base which forms a nitrate almost insoluble in water; used for the determination of nitric acid; also used as a rubber vulcanization accelerator; also a proprietary trade name for a cellulose nitrate plastic. Yellow leaflets; mp = 189°, insoluble in H_2O, soluble in organic solvents.

2703   4-Nitrophenetole

100-29-8   H

C8H9NO3

Benzene, 1-ethoxy-4-nitro-.

4-06-00-01283 (Beilstein Handbook Reference); Benzene, 1-ethoxy-4-nitro-; BRN 0972473; CCRIS 2333; EINECS 202-837-9; p-Ethoxynitrobenzene; Ethyl p-nitrophenyl ether; HSDB 5333; p-Nitrophenetol; p-Nitrophenetole; NSC 9812; Phenetole, p-nitro-. Prisms; mp = 60°; bp = 283°; λ_m = 227, 307 nm (MeOH); slightly soluble in H_2O, EtOH, soluble in petroleum ether, freely soluble in Et2O, Me2CO, C6H6.

2704   4-Nitrophenol

100-02-7   6654   G P

C6H5NO3

p-Nitrophenol.

AI3-04856; Caswell No. 603; CCRIS 2316; EINECS 202-811-7; EPA Pesticide Chemical Code 056301; HSDB 1157; NCI-C55992; Niphen; NSC 1317; p-Hydroxynitrobenzene; p-Nitrofenol; p-Nitrophenol; Paranitrofenol; Paranitrofenolo; Paranitrophenol; PNP; RCRA waste number U170; UN1663. Important chemical intermediate; also used as an indicator, colorless below pH 5.6, yellow above pH 7.6. Registered by EPA as a fungicide (cancelled). Slightly yellow crystals; mp =113.8°; d^{20} = 1.479, d_4^{120} = 1.270; λ_m = 228, 311 nm (ε = 7510, 11000, MeOH); freely soluble in EtOH, CHCl3, Et2O, soluble in H_2O; LD50 (mus orl) = 467 mg/kg, (rat orl) = 616 mg/kg. *Fluka.*

2705   6-((2-Nitrophenyl)azo)-2,4-di-t-pentylphenol

52184-19-7   H

C22H29N3O3

Phenol, 2,4-bis(1,1-dimethylpropyl)-6-((2-nitrophenyl)-azo)-.

EINECS 257-716-3; 6-((2-Nitrophenyl)azo)-2,4-di-t-pentylphenol; Phenol, 2,4-bis(1,1-dimethylpropyl)-6-((2-nitrophenyl)azo)-.

2706   1-Nitropropane

108-03-2   6660   G H

C3H7NO2

Propane, 1-nitro-.

AI3-02264; CCRIS 1329; EINECS 203-544-9; HSDB 2526; 1-Nitropropane; 1-NP; NSC 6363; Propane, 1-nitro-. Used as a solvent for cellulose acetate and vinyl resins; also as an intermediate and propellant. Liquid; mp = -108°; bp = 131.1°; d^{25} = 0.9961; λ_m = 280 nm (cyclohexane); slightly soluble in H_2O, soluble in CHCl3, freely soluble in EtOH, Et2O. *Ashland; Grace W.R. & Co.; Whittaker Clark & Daniels.*

2707   p-Nitrosophenol

104-91-6   6676   H

C6H5NO2

4-Nitrosophenol.

4-07-00-02073 (Beilstein Handbook Reference); AI3-19026; BRN 1856695; p-Chinonmonoxim; CCRIS 4710; EINECS 203-251-6; HSDB 5362; Nitrosophenol; NSC 3124; p-Nitrosophenol; 4-Nitrosofenol; 4-Nitrosophenol; para-Nitrosophenol; Phenol, 4-nitroso-

; Phenol, p-nitroso-; Quinone monoxime; p-Quinone monooxime; Quinone oxime. Yellow crystals; mp = 144° (dec); d^{20} = 1.479; λ_m = 228, 311 nm (ε = 7510, 11000, MeOH); slightly soluble in H_2O, C_6H_6, CS_2; soluble in $CHCl_3$, C_7H_8, C_5H_5N, very soluble in EtOH, Et_2O, Me_2CO.

2708   Nitro-starch
9056-38-6   G

$C_{12}H_{12}N_8O_{26}$
Xyloidin.
Pyroxylin; Starch nitrate; UN0146; UN1337. A nitric ester of starch, probably the octonitrate used for blasting explosives, either alone, or by mixing 10% of it with a mixture of sodium nitrate and carbonaceous material. See pyroxylin [9004-70-0].

2709   Nitrosylsulfuric acid
7782-78-7   6682   G

HNO_5S
Nitrosyl sulfate.
EINECS 231-964-2; Lead chamber crystals; Nitro acid sulfite; Nitrosulfonic acid; Nitrose; Nitrososulfuric acid; Nitrosulfonic acid; Nitrosylsulfuric acid; Nitrosylsulphuric acid; Nitroxylsulfuric acid; Sulfuric acid, monoanhydride with nitrous acid; UN2308; Weber's acid. Used to bleach flour. Crystals; dec 73°.

2710   Nitrothal-isopropyl
10552-74-6   G P

$C_{14}H_{17}NO_6$
5-Nitro-1,3-Benzenedicarboxylic acid, bis(1-methylethyl) ester.
Bis(1-methylethyl) 5-nitro-1,3-benzenedicarboxylate; Di-isopropyl 5-nitroisophthalate; Isophthalic acid, 5-nitro-, diisopropyl ester; Nitrothal-isopropyl; Nitrothale-isopropyl. Non-systemic contact fungicide with curative action. Used in combination with other fungicides to control powdery mildews on apples, vines, hops, vegetables and ornamentals. Yellow crystals; mp = 65°; soluble in H_2O (0.00004 g/100 ml at 20°), Me_2CO, C_6H_6, $CHCl_3$, EtOAc (all > 80 g/100 ml), Et_2O (61 g/100 ml), EtOH (5.5 g/100 ml); LD_{50} (rat orl) > 6400 mg/kg, (rat der) > 2500 mg/kg, (rbt der) > 4000 mg/kg; non-toxic to birds, toxic to fish; non-toxic to bees. *BASF Corp.*

2711   m-Nitrotoluene
99-08-1   6684   H

$C_7H_7NO_2$
Benzene, 1-methyl-3-nitro-.
Benzene, 1-methyl-3-nitro-; CCRIS 2312; EINECS 202-728-6; HSDB 2937; m-Methylnitrobenzene; m-Nitrotoluene; m-Nitrotoluol; meta-Nitrotoluol; Nitro-toluene, m-; NSC 9578; Toluene, m-nitro-; UN1664. Liquid; mp = 15.5°, bp = 232°; d^{20} = 1.1581; λ_m = 264 nm (ε 7700, MeOH); insoluble in H_2O, soluble in EtOH, C_6H_6, CCl_4, very soluble in Et_2O.

2712   p-Nitrotoluene
99-99-0   6684   H

$C_7H_7NO_2$
Benzene, 1-methyl-4-nitro-.
AI3-15394; Benzene, 1-methyl-4-nitro-; CCRIS 2313; EINECS 202-808-0; HSDB 1158; p-Methylnitrobenzene; NCI 9579; NCI-C60537; Nitrotoluene, p-; p-Nitrotoluene; Nitrotoluenos; para-Nitrotoluol; NSC 9579; Toluene, p-nitro-; UN1664. Orthorhombic crystals; mp = 51.6°, bp = 238.3°; d^{75} = 1.1038; λ_m = 215, 274 nm (ε = 9440, 11700, MeOH); insoluble in H_2O, soluble in EtOH, very soluble in Et_2O, Me_2CO, C_6H_6, CCl_4, $CHCl_3$, C_5H_5N, C_7H_8, CS_2.

2713   Nitrous Oxide
10024-97-2   6687   D G

$^-N{\equiv}N^+{=}O$

N_2O
Nitrogen oxide.
CCRIS 1225; Dinitrogen monoxide; Dinitrogen oxide; EINECS 233-032-0; Factitious air; FEMA No. 2779; HSDB 504; Hyponitrous acid anhydride; Laughing gas; Nitral; Nitrogen hypoxide; Nitrogen monoxide; Nitrogen oxide; Nitrogen protoxide; Nitrous oxide; Oxido nitroso; Protoxyde d'azote; Stickdioxyd; UN1070; UN2201. Anesthetic (inhaled); analgesic. Asphyxiant and narcotic in high concentrations. Less irritating than other oxides of nitrogen. Gas; dissociates > 300°; mp = -90.91°; bp_{760} = -88.46°; d^{-89}(liquid) = 1.226; d(gas) = 1.53; soluble in sulfuric acid, alcohol, Et_2O, oils. *DuPont-Merck Pharm.*

2714   Nitrovin
804-36-4   6688   G

$C_{14}H_{12}N_6O_6$
2-[3-[(5-Nitro-2-furanyl)-1-[2-(5-nitro-2-furanyl)ethenyl]-2-propenylidene]hydrazinecarboximidamide.
4-19-00-01824 (Beilstein Handbook Reference); 1,5-Bis(5-nitro-2-furanyl)-1,4-pentadien-3-one, (aminoimino-methyl)hydrazone; sym-Bis(5-nitro-2-furfurylidene) acet-one guanylhydrazone; Bis(2-(5-nitro-2-furyl)vinyl)methyl-enehydrazinoformamidine; 1,5-Bis(5-nitro-2-furyl)-3-pen-tadienone guanylhydrazone; 1,5-Bis(5-nitro-2-furyl)-3-pentadienone amidinonhydrazone; Bis(5-nitro-furfuryl-idene)acetone guanylhydrazone; BRN 4827520; Difur-azone; EINECS 212-358-7; Guanidine, ((3-(5-nitro-2-fur-yl)-1-(2-(5-nitro-2-furyl)vinyl)allylidene)amino)-; Hydraz-inecarboximidamide, 2-(3-(5-

nitro-2-furanyl)-1-(2-(5-nitro-2-furanyl)ethenyl)-2-propenylidene)-; ((3-(5-Nitro-2-furyl)-1-(2-(5-nitro-2-furyl)vinyl)allylidene)amino)-guanid-ine; 1-(3-(5-Nitro-2-furyl)-1-(2-(5-nitro-2-furyl)-vinyl)allyl-ideneamino)guanidine hydrochloride; Nitrovin; Panazon; Payzone. Growth promoter and antibacterial. mp = 217° (dec).

2715   3-Nitro-o-xylene
83-41-0

H

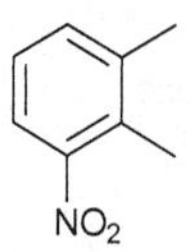

C8H9NO2
1,2-Dimethyl-3-nitrobenzene.
4-05-00-00930 (Beilstein Handbook Reference); AI3-29558; Benzene, 1,2-dimethyl-3-nitro-; BRN 2045105; CCRIS 3117; EINECS 201-474-3; NSC 5402; o-Xylene, 3-nitro-. Used as a chemical intermediate. Needles; mp = 15°; bp = 240°; d^{20} = 1.1402; λ_m = 207, 258 nm (MeOH); insoluble in H2O, soluble in EtOH, CCl4. *Albright & Wilson Americas Inc.; Deepak Chemicals Group; Shanxi Kangyuan Chemical Co. Ltd.*

2716   4-Nitro-m-xylene
89-87-2

H

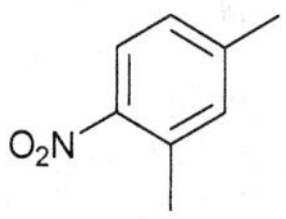

C8H9NO2
Benzene, 2,4-dimethyl-1-nitro-.
Benzene, 2,4-dimethyl-1-nitro-; CCRIS 3120; EINECS 201-947-4; NSC 50661; m-Xylene, 4-nitro-. Liquid; mp = 9°; bp = 247°, bp18 = 122°; d^{15} = 1.135; insoluble in H2O, soluble in EtOH, Me2CO, C6H6, CHCl3.

2717   Nitroxynil
1689-89-0   6690   G V

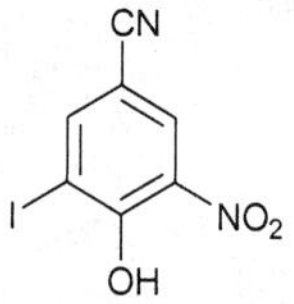

C7H3IN2O3
4-Hydroxy-3-iodo-5-nitrobenzonitrile.
Benzonitrile, 4-hydroxy-3-iodo-5-nitro-; BRN 2213717; CCRIS 5743; Dovenix; EINECS 216-884-8; 4-Hydroxy-3-iodo-5-nitrobenzonitrile; Nitroxinil; Nitroxinilo; Nitroxinilum; Nitroxynil; Trodax. Used in veterinary medicine as an anthelmintic. Crystals; mp = 137-138°; sparingly soluble in H2O, soluble in organic solvents.

2718   Nonanal
124-19-6

H

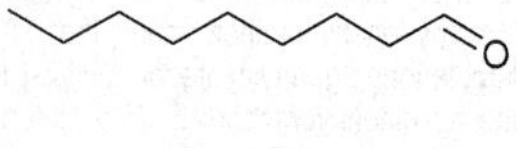

C9H18O
n-Nonaldehyde.

4-01-00-03352 (Beilstein Handbook Reference); AI3-04859; Aldehyde C9; BRN 1236701; C9 aldehyde; CCRIS 664; EINECS 204-688-5; FEMA No. 2782; NCI-C61018; Nonaldehyde; 1-Nonaldehyde; n-Nonaldehyde; Nonanal; 1-Nonanal; n-Nonanal; Nonanaldehyde; Nonanoic aldehyde; Nonoic aldehyde; Nonyl aldehyde; 1-Nonyl aldehyde; Nonylic aldehyde; NSC 5518; Pelargonaldehyde; Pelargonic aldehyde. Liquid; bp = 191°; d^{22} = 0.8264; soluble in Et2O, CHCl3.

2719   Nonane
111-84-2

H

C9H20
n-Nonane.
CCRIS 6081; EINECS 203-913-4; HSDB 107; Nonane; n-Nonane; Nonyl hydride; NSC 72430; Shellsol 140. Liquid; mp = -53.5°; bp = 150.8°; d^{20} = 0.7176; insoluble in H2O, very soluble in EtOH, Et2O, freely soluble in Me2CO, C6H6, CHCl3, C7H16.

2720   Nonanoic acid, 4-sulfophenyl ester, sodium salt
89740-11-4

H

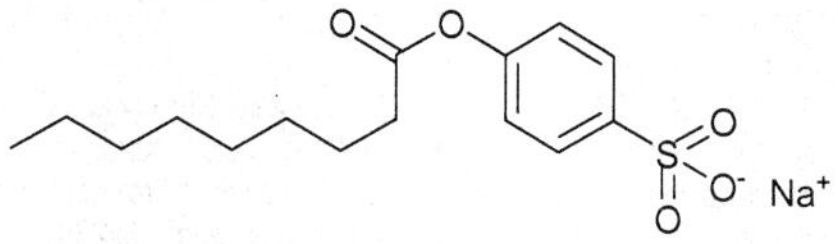

C15H21NaO5S
Nonanoic acid, 4-sulfophenyl ester, sodium salt.

2721   Nonanol
3452-97-9

G

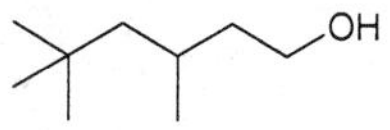

C9H20O
3,5,5-Trimethylhexan-1-ol.
AI3-22142; Caswell No. 892A; EINECS 222-376-7; EPA Pesticide Chemical Code 492200; FEMA No. 3324; 1-Hexanol, 3,5,5-trimethyl-; Nonylol; NSC 83151; NSC 97226; 3,5,5-Trimethyl-1-hexanol; 3,5,5-Trimethylhexan-1-ol; 3,5,5-Trimethylhexanol; Trimethylhexyl alcohol; 3,5,5-Trimethylhexyl alcohol. Plasticizer, primarily 3,5,5-trimethylhexanol. Liquid; bp = 194°; d^{25} = 0.8236. *ICI Chem. & Polymers Ltd.*

2722   Nonene
27215-95-8

H

C9H18
Nonene (all isomers).
EINECS 248-339-5; HSDB 756; Nonene; Nonene (all isomers); Nonene (mixed isomers); Nonene (non-linear); Nonene mixture; Propylene trimer; Tripropylene.

2723   Nonyl alcohol
28473-21-4

H

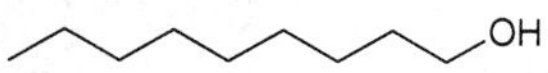

C9H20O
Nonan-1-ol.
EINECS 249-048-6; Nonan-1-ol; Nonanol; Nonyl alcohol.

2724   **sec-Nonyl alcohol**
108-82-7   H

C9H20O
2,6-Dimethyl-4-heptanol.
4-01-00-01810 (Beilstein Handbook Reference); AI3-14496; BRN 1733804; Diisobutyl carbinol; 2,6-Dimethyl-4-heptanol; 2,6-Dimethyl heptanol-4; EINECS 203-619-6; FEMA No. 3140; 4-Heptanol, 2,6-dimethyl-; HSDB 5140; 4-Hydroxy-2,6-dimethyl heptane; Nonyl alcohol, secondary; sec-Nonyl alcohol; NSC 62683. Liquid; bp = 174.5°; d^{20} = 0.8114; insoluble in H_2O, slightly soluble in CCl_4, soluble in EtOH, Et_2O.

2725   **tert-Nonyl mercaptan**
25360-10-5   H

C9H20S
1,1-Dimethylheptanethiol.
1,1-Dimethylheptanethiol; EINECS 246-896-9; tert-Nonanethiol; tert-Nonyl mercaptan.

2726   **Nonyl phenol**
25154-52-3   6715   G H

C15H24O
2,6-Dimethyl-4-heptylphenol, (O and P).
3-06-00-02067 (Beilstein Handbook Reference); AI3-14638; BRN 2047450;; EINECS 246-672-0; HSDB 1032; Hydroxyl No. 253; Mononononylphenol; Nonyl phenol; Nonyl phenol (mixed isomers); n-Nonylphenol; NSC 71410; Phenol, nonyl-; Prevostsel VON-100. A mixture of isomeric monoalkyl phenols; nonionic surfactant (nonbiodegradable), lube oil additives, stabilizers, petroleum demulsifiers, fungicides, antioxidants for plastics and rubber. Liquid; bp = 293-297°. *Allchem Ind.; Ashland; Berol Nobel AB; Degussa-Hüls Corp.; GE Specialities; Mitsui Toatsu; Texaco.*

2727   **(p-Nonylphenoxy)acetic acid**
3115-49-9   6716   G

C17H26O3
(4-Nonylphenoxy)-acetic acid.
Acetic acid, (4-nonylphenoxy)-; Akypo NP 70; EINECS 221-486-2; Nonoxynol-8 carboxylic acid; (4-Nonylphenoxy)acetic acid; (p-Nonylphenoxy)acetic acid. Emulsifier. *Chem-Y GmbH.*

2728   **Nonylphenoxypolyethoxyethanol**
9016-45-9   G H P

$(C_2H_4O)_n.C_{15}H_{24}O$
Nonyl phenyl polyethylene glycol.
Alcosist PN; Alfenol; Alfenol 8; Alfenol 10; Alfenol 18; Alfenol 22; Alfenol 28; Alfenol 710; Alfenol N 8; Alkasurf NP; Alkasurf NP 11; Alkasurf NP 15; Alkasurf NP 8; Antarox 897; Antarox CO; Antarox CO 430; Antarox CO 530; Antarox CO 630; Antarox CO 730; Antarox CO 850; Arkopal N-090; Carsonon N-9; Caswell No. 605; Chemax NP series; Conco NI-90; Dowfax 9N20; Emulgen - 913; EPA Pesticide Chemical Code 079005; Ethoxylated nonylphenol; Glycols, polyethylene, mono(nonylphenyl) ether; Glycols, polyethylene mono(nonylphenyl) ether (nonionic); HSDB 6825; Igepal CO-630; Lissapol NX; Neutronyx 600; Nonoxinolum; Nonoxynol-3; Nonoxyol-4; Nonoxynol-30; Nonoxynol-44; Nonyl phenol, ethox-ylated; Nonyl phenyl polyethylene glycol ether; Nonyl phenyl polyethylene glycol; Nonylphenol, poly-oxyethylene ether; Nonylphenoxypoly(ethyleneoxy)-ethanol, branched; Nonylphenoxypolyethoxyethanol; Nonylphenoxypolyethoxyethanol; (Nonylphenoxy)poly-ethylene oxide; α-(Nonylphenyl)-ω-hydroxypolyoxy-ethylene; α-(Nonylphenyl)-ω-hydroxypoly(oxy-1,2-eth-anediyl); ω-Hydroxy-α-(nonylphenyl)poly(oxy-1,2-eth-anediyl); PEG-3 Nonyl phenyl ether; PEG-4 Nonyl phenyl ether; PEG-9 nonyl phenyl ether; PEG-13 Nonyl phenyl ether; PEG-15 Nonyl phenyl ether; PEG-30 Nonyl phenyl ether; PEG-44 Nonyl phenyl ether; Poly(oxy-1,2-ethanediyl), α-(nonylphenyl)- ω-hydroxy-; Poly(oxy-1,2-ethanediyl), α-(nonylphenyl)- ω-hydroxy-; Polyethylene glycol (13) nonyl phenyl ether; Polyethylene glycol (15) nonyl phenyl ether; Polyethylene glycol (3) nonyl phenyl ether; Polyethylene glycol (30) nonyl phenyl ether; Polyethylene glycol (44) nonyl phenyl ether; Polyethylene glycol 200 nonyl phenyl ether; Polyethylene glycol 450 nonyl phenyl ether; Polyethylene glycol nonylphenyl ether; Polyoxyethylene (3) nonyl phenyl ether; Polyoxyethylene (4) nonyl phenyl ether; Polyoxyethylene (9) nonyl phenyl ether; Polyoxyethylene (13) nonyl phenyl ether; Polyoxyethylene (15) nonyl phenyl ether; Polyoxyethylene (30) nonyl phenyl ether; Polyoxy-ethylene (44) nonyl phenyl ether; Polyoxyethylene nonylphenol; Prevocel 12; Protachem 630; Rewopol HV-9; Synperonic NX; Tergetol NP; Tergitol NP-10; Tergitol NP-14; Tergitol NP-27; Tergitol NP-33 (nonionic); Tergitol NP-35 (nonionic); Tergitol NP-40 (nonionic); Tergitol NPX; Tergitol TP-9 (nonionic); Triton N-100; Trycol NP; EPA Pesticide Chemical Code 079005; Ethoxylated nonylphenol; Glycols, polyethylene mono(nonylphenyl) ether (nonionic); HSDB 6825; Igepal CO-630; Lissapol NX; Neutronyx 600; Nonoxinolum; Nonoxynol-3; Nonoxynol-4; Nonoxynol-30; Nonoxynol-44; Nonyl phenol, ethoxylated; Nonyl phenyl polyethylene glycol ether; Nonylphenol, polyoxyethylene ether; Nonylphenoxypoly(ethyleneoxy)ethanol, branched; Nonylphenoxypolyethoxyethanol; (Nonylphenoxy)poly-ethylene oxide; α-(Nonylphenyl)-ω-hydroxypolyoxy-ethylene; α-(Nonylphenyl)-ω-hydroxypoly(oxy-1,2-eth-anediyl); ω-Hydroxy-α-(nonylphenyl)poly(oxy-1,2-eth-anediyl); PEG-3 Nonyl phenyl ether; PEG-4 Nonyl phenyl ether; PEG-9 nonyl phenyl ether; PEG-13 Nonyl phenyl ether; PEG-15 Nonyl phenyl ether; PEG-30 Nonyl phenyl ether; PEG-44 Nonyl phenyl ether; Polyethylene glycol (n) nonyl phenyl ether; Polyoxyethylene nonylphenol; Polyoxyethylene (n) nonyl phenyl ether; Prevocel 12; Protachem 630; Rewopol HV-9; Synperonic NX; Tergetol NP; Tergitol NPX; Tergitol TP-9 (nonionic); Triton N-100; Trycol NP-1. Detergent and dispersant used with petroleum oils; intermediate in manufacture of surfactants and antistats; co-emulsifier for fats, oils and waxes. Registered by EPA as a bactericide and insecticide (cancelled). Liquid; d_4^{20} = 1.06. *Fluka; Taiwan Surfactants.*

2729   **Norelgestromin**
53016-31-2   D

C21H29NO2

13-ethyl-17α-hydroxy-18,19-dinorpregn-4-en-20-yn-3-one, 3-oxime. BRN 4202099; 17-Deacetylnorgestimate; Levonorgestrel oxime; 18-Methylnorethindrone oxime; Norelgestromin; D-Norgestrel 3-oxime; Norplant 3-oxime; Ortho-Evra; RWJ 10553. FDA-approved 20 November, 2001 for use with ethinyl estradiol in a patch formulation to be used as a contraceptive. *Johnson R. W. Pharm. Res. Institute; Organon Inc.*

2730   Norethindrone
68-22-4   6729   D

C20H26O2

17-Hydroxy-19-nor-17α-pregn-4-en-20-yn-3-one. 4-08-00-01221 (Beilstein Handbook Reference); AI3-26422; Anhydrohydroxynorprogesterone; Anovulatorio; Anovule; Brevicon; Brevinor-1 21; Brevinor-1 28; Brevinor 21; Brevinor 28; BRN 1915671; CCRIS 484; Ciclovulan; Conludaf; Conludag; EINECS 200-681-6; Estrinor; Ethinylnortestosterone; Ethynylnortestosterone; Gencept; Genora; Gestest; HSDB 3370; Jenest-28; Microneth; Micronett; Micronor; Micronovum; Milli; Mini-pe; Mini-pill; Minovlar; Modicon; N.E.E.; Nelova; Nodiol; Nor-Q.D.; Noraethisteronum; Noralutin; Norcept-E; Norcolut; Noresthisterone; Norethadrone; Norethin; Norethin 1/35 E; Norethin 1/50 M; Nor-ethindrone; Norethisteron; Norethisterone; Nor-ethisteronum; Norethyndron; Norethynodron; Norethyn-odrone; Noretisterona; Noretisterone; Norfor; Norgestin; Noriday; Noriday 28; Norinyl; Norluten; Norlutin; Norluton; Norpregneninlone; Norpregneninotone; NSC-9564; Orlest; Ortho 1 35; Ortho 7 7 7; Ortho-Novum; Ortho-Novum 1 50; Ortho-Novum 1 35; Ortho-Novum 7 7 7; Ovcon; Ovysmen 0.5 35; Ovysmen 1 35; Primolut-N; Proluteasi; SC 4640; Synphase; Synphasic 28; Tri-Norinyl; Triella; Trinovum 21; Utovlan; Utovlar.; Conludag; Menzol; Micronor; Micronovum; Mini-Pe; mini-pill; Norcolut; Noriday; Norluten; Norlutin; Nor-QD; Primolut N; Utovlan; component of: Binovum, Brevicon, Brevinor, Conceplan, Modicon, Neocon 1/35, Norimin, Norinyl 1+35, Norquentiel, Ortho-Novum 1/35, Ortho-Novum 7/7/7, Ovcon, Ovysmen, Synphase, Tri-Norinyl, Trinovum, Norinyl-1, Ortho-Novin 1/50, Ortho-Novum 1/50. Progestogen. Used in combination with estrogens as an oral contraceptive. White crystals; mp = 203-204°; $[\alpha]_D^{20}$ = -31.7° (CHCl3); λ_m = 240 nm (log ε 4.24). *Syntex Labs. Inc.*

2731   Norethindrone Acetate
51-98-9   6729   D

C22H28O3

17β-Hydroxy-19-norpregn-4-en-20-yn-3-one acetate. 4-08-00-01221 (Beilstein Handbook Reference); Aygestin; BRN 2064104; CCRIS 485; EINECS 200-132-0; ENTA; Ethinyl-nortestosterone acetate; Gestakadin; Loestrin; Milli-Anovlar; Milligynon; Miniphase; Monogest; Norethindrone acetate; Norethisteron acetate; Norethisterone acetate; Norethynyltestosterone acetate; Norethysterone acetate; Norlestrin; Norlutate; Norlutin A; Norlutin acetate; Norlutine acetate; NSC 22844; Orlutate; Primolut-Nor; Progylut; SH 420.; component of: Anovlar, Estrostep, Etalontin, Gynovlar, Loestrin, Minovlar, Norlestrin, Primosiston. Progestogen. Used in combination with estrogens as an oral contraceptive. Crystals; mp = 161-162°; $[\alpha]_D^{20}$ = -31.7° (CHCl3); λ_m = 240 nm (ε 18690). *Bristol-Myers Squibb Pharm. R&D; Parke-Davis; Schering AG; Syntex Labs. Inc.; Wyeth-Ayerst Labs.*

2732   Norflurazon
27314-13-2   6733   H P

C12H9ClF3N3O

4-Chloro-5-(methylamino)-2-(α,α,α-trifluoro-m-tolyl)-3(2H)-pyridazinone. 5-25-14-00201 (Beilstein Handbook Reference); BRN 0757115; Caswell No. 195AA; 4-chloro-5-(methyl-amino)-2-(3-(trifluoromethyl)phenyl)-3(2H)-Pyridazinone; EPA Pesticide Chemical Code 105801; Evital; Evitol; H-52,143; H 52143; H 9789; HSDB 6845; Monomet-flurazon; Monometflurazone; Norflurazon; Norflurazone; SAN 9789; SAN 9789 H; San 97895; Solicam; Solicam Rapid; Telok; 1-(3-Trifluoromethylphenyl)-4-methyl-amino-5-chloropyridazone; Zorial. Registered by EPA as a herbicide. Crystals; mp = 183 - 185°, 174 - 180°; soluble in H2O (0.0028 g/100 ml), EtOH (14.2 g/100 ml), Me2CO (5 g/100 ml), xylene (0.25 g/100 ml), sparingly soluble in hydrocarbon solvents; LD50 (rat orl) = 9400 mg/kg, (rat der) > 5000 mg/kg, (rbt der) > 20000 mg/kg, (bobwhite quail, mallard duck orl) > 1250 mg/kg, (bee) >0.235 mg/bee; LC50 (catfish, goldfish) > 200 mg/l; non-toxic to bees. *Syngenta Crop Protection.*

2733   Nortriptyline
72-69-5   6749   D

C19H21N

3-(10,11-Dihydro-5H-dibenzo[a,d]cyclohepten-5-ylid-ene)-N-methyl-1-propanamine. Ateben; Avantyl; Aventyl; BRN 2216786; Demethyl-amitriptyline; Desitriptilina; Desmethylamitriptyline; EINECS 200-788-8; HSDB 3371; Lumbeck; Noramitriptyline; Noritren; Nortriptilina; Nortriptyline; Nortriptylinum; Nortryptiline; Nortryptyline; Psychostyl; Sensaval; Sesaval. Tricyclic antidepressant. *Merck & Co.Inc.*

2734 Nortriptyline Hydrochloride
894-71-3   6749   D

C19H22ClN

3-(10,11-Dihydro-5H-dibenzo[a,d]cyclohepten-5-ylid-ene)-N-methyl-1-propanamine hydrochloride.

Acetexa; Allegron; Altilev; Ateben hydrochloride; Aventyl allegron; Aventyl Hydrochloride; Desmethylamitriptyline hydrochloride; EINECS 212-973-0; Lilly 38489; N 7048; Noramitriptyline hydrochloride; Noritren; Nortab hydrochloride; Nortrilen; Nortriptylin hydrochloride; Nortriptyline hydrochloride; Norzepine; NSC 78248; NSC 169453; Pamelor; Psychostyl; Sensival; Vividyl. A tricyclic antidepressant. mp = 213-215°; λ_m = 240 nm (ϵ 13900); soluble in H2O, EtOH, CHCl3, insoluble in other organic solvents. *Eli Lilly & Co.; Merck & Co.Inc.; Sandoz Pharm. Corp.*

2735 Nuarimol
63284-71-9   G P

C17H12ClFN2O

5-(2-Chloro-4'-fluorobenzhydryl)-4-hydroxypyrimidine.

BRN 6223667; Caswell No. 419G; 5-(2-Chloro-4'-fluorobenzhydryl)-4-hydroxypyrimidine; (±)-2-Chloro-4'-fluoro-α-(pyrimidin-5-yl)benzhydryl alcohol; 2-Chloro-4'-fluoro-α-(pyrimidin-5-yl)diphenylmethanol; α-(2-Chloro-phenyl)-α-(4-fluorophenyl)-5-pyrimidinemethanol; EINECS 264-071-1; EL 228; EL 2289; EPA Pesticide Chemical Code 224100; Guantlet; Murox; Nuarimol; 5-Pyrimidinemethanol, α-(2-chlorophenyl)-α-(4-fluoro-phenyl)-; Trimidal; Trimifruit SC; Triminol; EL 228; EL 2289; EPA Pesticide Chemical Code 224100; Guantlet; Murox; Nuarimol; 5-Pyrimidinemethanol, α-(2-chlorophenyl)-α-(4-fluorophenyl)-; Trimidal; Trimifruit SC; Triminol. Systemic foliar fungicide wit curative and protective action. Ergosterol biosynthesis inhibitor. Used for control of a wide range of pathogenic fungi. Used as a foliar spray or seed treatment. Crystals; mp = 126 - 127°; slightly soluble in H2O (0.0026 g/100 ml at 25°, pH 7), Me2CO (17 g/100 ml), MeOH (5.5 g/100 ml), xylene (2 g/100 ml), freely soluble in CH3CN, C6H6, CHCl3; LD50 (mrat orl) = 1250 mg/kg, (frat orl) = 2500 mg/kg, (mmus orl) = 2500 mg/kg, (fmus orl) = 3000 mg/kg, (beagle dog orl) = 500 mg/kg, (rbt der) > 2000 mg/kg, (bobwhite quail orl) = 200 mg/kg; LC50 (bluegill sunfish 96 hr.) ≅12.1 mg/l; non-toxic to bees. *Dow AgroSciences; DowElanco Ltd.*

2736 Nylon 6
25038-54-4   6768   G

(C6H13NO2)n

Poly[imino(1-oxo-1,6-hexanediyl)].

A 1030; A 1030N0; Akulon; Akulon M 2W; Alkamid; Amilan CM 1001; Amilan CM 1001C; Amilan CM 1001G; Amilan CM 1011; Amilan CM 1031; 6-Aminohexanoic acid homopolymer; ATM 2 (nylon); Aviamide-6; B-35; B-203; B-216; B-300; B-350; Bonamid; Cabelec® 1015; Capran 77C; Capran 80; Caproamide polymer; Caprolactam oligomer; Caprolactam polymer; Caprolon B; Caprolon V; Capron; Capron 8250; Capron 8252; Capron 8253; Capron 8256; Capron B; Capron GR 8256; Capron GR 8257; Capron GR 8258; Capron PK 4; Chemlon; CM 1001; CM 1011; CM 1031; CM 1041; Danamid; DULL 704; Durethan BK; Durethan BK 30S; Durethan BKV 30H; Durethan BKV 55H; ε-Caprolactam polymer; ε-Caprolactam polymere; Ertalon 6SA; Extron 6N; Grilon; Hexahydro-2H-azepin-2-one homopolymer; Itamid; Itamide 25; Itamide 35; Itamide 250; Itamide 250G; Itamide 350; Itamide S; Kaprolit; Kaprolit B; Kaprolon; Kaprolon B; Kapromin; Kapron; Kapron A; Kapron B; KS 30P; Maranyl F 114; Maranyl F 124; Maranyl F 500; Metamid; Miramid H 2; Miramid WM 55; Nylon; Nylon 6; Nylon A1035SF; Nylon CM 1031; Nylon X 1051; Orgamid rmnocd; Orgamide; P 6 (Polyamide); PA 6 (Polymer); PK 4; PKA; Plaskin 8200; Plaskon 201; Plaskon 8201; Plaskon 8201HS; Plaskon 8202C; Plaskon 8205; Plaskon 8207; Plaskon 8252; Plaskon XP 607; Policapram; Policapramum; Policapran; Poly(ε-aminocaproic acid); Poly(hexahydro-2H-azepin-2-one); Poly(imino(1-oxo-1,6-hexanediyl)); Poly(iminocarbonylpentamethylene); Poly-amide 6; Polycaproamide; Polycaprolactam; Relon P; Renyl MV; Sipas 60; Spencer 401; Spencer 601; Tarlon X-A; Tarlon XB; Tarnamid T; Tarnamid T 2; Tarpamid T 27; TNK 2G5; Torayca N 6; UBE 1022B; Ultramid B 3; Ultramid B 4; Ultramid B 5; Ultramid BMK; Vidlon; Widlon; Zytel 211. Nylon 6/carbon black compound; conductive compound for injection molding applications; gives rigidity with permanent electrical conductivity; suggested for packaging and electronic production handling applications; e.g., for handling explosives, electronic components. Used in tire cord; fishing lines; tow ropes; hose manufacture; woven fabrics. Amorphous solid; mp = 223°; d = 1.084. *Cabot Carbon Ltd.; Snia UK.*

2737 Nylon 6-12 salt
13188-60-8   H

C12H22O4.C6H16N2

Dodecanedioic acid, compd. with 1,6-hexanediamine (1:1).

Dodecanedioic acid, compound with hexane-1,6-diamine (1:1); EINECS 236-143-2; Nylon 6-12 salt.

2738 Nylon-66
32131-17-2   G H

(C12H22N2O2)n

Poly(imino(1,6-dioxo-1,6-hexanediyl)imino-1,6-hexane-diyl).

126AM30P; A 100 (polyamide); A 142 (nylon); A 146 (nylon); A 153 (nylon); A 153P0118; A 175 (nylon); A 203 (nylon); A 205 (polyamide); A 216V35; A 216V50; 1,8-Diazacyclotetradecane-2,7-dione homopolymer, sru; Nylon-66; Poly(imino(1,6-dioxo-1,6-hexanediyl)imino-1,6-hexanediyl); Ultramid A3 times G7. Polymeric amide formed by the reaction of adipic acid with hexylenediamine. Thermoplastic resin for injection molding, extrusion. *Asahi Chem. Industry; Snia UK.*

2739 Nystatin
1400-61-9   6770   D

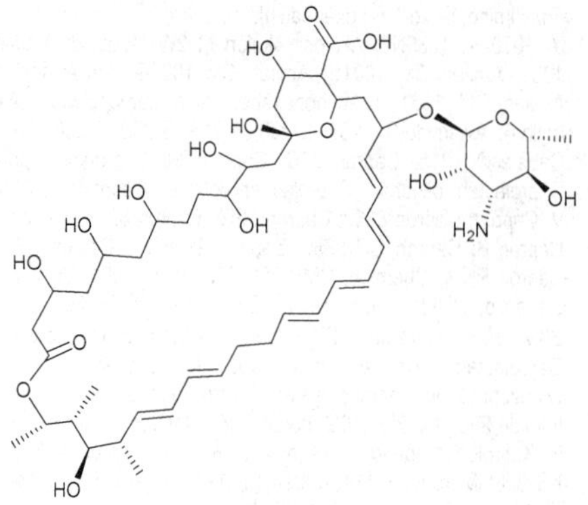

C47H75NO17

Nystatin.

Biofanal; Candex; Candio-Hermal; Comycin; Diastatin; Fungicidin; Moronal; Myco-Triacet II; Mycolog II; Mycostatin; Mycostatin Pastilles; Mytrex; Nystaform; Nystaform HC; Nystatin; Nystavescent; Nystex; O-V Statin; Panolog Cream; Terrastatin.; component of: Mycolog II, Myco-Triacet II, Mytrex, Nystaform, Nystaform HC, Panolog Cream, Terrastatin. Mixture of Nystatin A_1, A_3 and A_3, biologically active polyene antibiotics. Used as an antifungal agent. Dec 250°; $[\alpha]_D^{25}$ = -10° (AcOH), 21° (C_5H_5N), 12° (DMF), -7° (0.1N HCl/EtOH); λ_m = 290, 307, 322 nm; soluble in H_2O (0.4 g/100 ml), MeOH (1.12 g/100 ml), EtOH (0.12 g/100 ml), CCl_4 (0.123 g/100 ml), $CHCl_3$ (0.048 g/100 ml), C_6H_6 (0.028 g/100 ml), ethylene glycol (0.875 g/100 ml); LD_{50} (mus ip) ≅ 200 mg/kg. *Apothecon; Bayer AG; Bristol-Myers Oncology; Bristol-Myers Squibb Co.; Lederle Labs.; Lemmon Co.; Pfizer Inc.; Savage Labs; Solvay Animal Health Inc.*

2740   Nystatin A₁
34786-70-4   6770   D

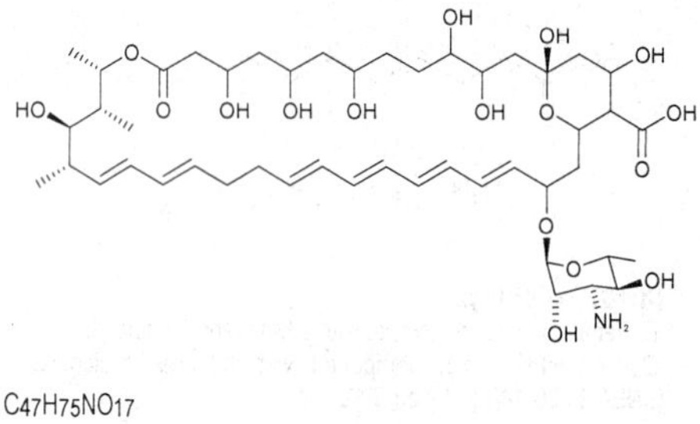

C47H75NO17

Nystatin A1.

Major component of Nystatin. Antifungal. *Apothecon; Bayer AG; Bristol-Myers Oncology; Bristol-Myers Squibb Co.; Lederle Labs.; Lemmon Co.; Pfizer Inc.; Savage Labs; Solvay Animal Health Inc.*

2741   Octabromobiphenyl ether
32536-52-0   G H

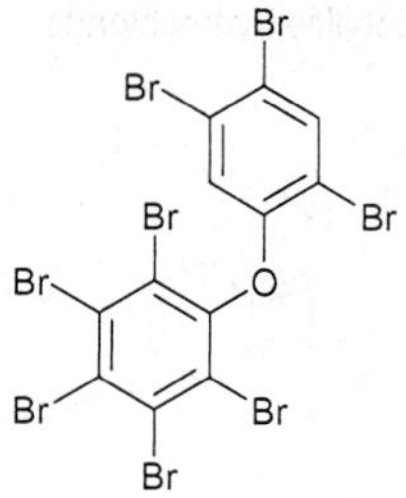

C12H2Br8O

1,1'-Oxybisbenzene octabromo deriv.

Benzene, 1,1'-oxybis-, octabromo deriv.; Bromkal 79-8DE; CD 79; DE 79; Diphenyl ether, octabromo derivative; EB 8; EINECS 251-087-9; FR 1208; FR 143; OBDPO; OCTA; Octabromobiphenyl ether; Octa-bromodiphenyl oxide; 1,1'-Oxybisbenzene octabromo deriv.; Phenyl ether, octabromo deriv.; Saytex® 111; Tardex 80. Flame retardant for thermoplastics, e.g., ABS, HIPS, LDPE, PP random copolymers, polyamides, elastomers, adhesives, and coatings; semiplasticizing additive for styrenic polymers and copolymers such as ABS; recommended for injection moldings.   White to off-white powder; sg = 2.9; mp = 125-165; LD_{50} (rat orl) > 5000 mg/kg. *Albemarle Corp.; Allchem Ind.; Amerihaas; Dead Sea Bromine; Ethyl Corp.; Great Lakes Fine Chem.*

2742   Octacosanol
557-61-9   6776   H

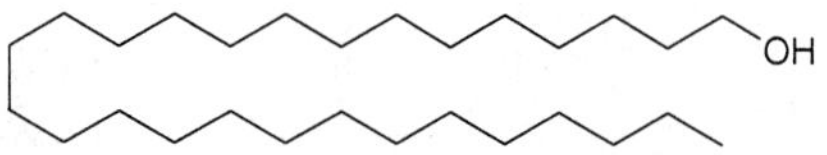

C28H58O

1-Octacosanol.

Cluytyl alcohol; EINECS 209-181-2; Montanyl alcohol; NSC 10770; Octacosanol; 1-Octacosanol; Octacosyl alcohol. Solid; mp = 83.3°; bp1 = 200°.

2743   Octadecadienoic acid
121250-47-3   H

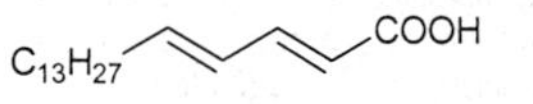

C18H32O2

9,11(or 10,12)-Octadecadienoic acid.

CCRIS 7064; Conjugated linoleic acid; Octadecadienoic acid; 9,11(or 10,12)-Octadecadienoic acid; Octadeca-dienoic acid, conjugated.

2744   Octadecanonitrile
638-65-3   H

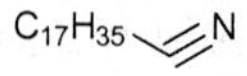

C18H35N

n-Heptadecyl cyanide.

4-02-00-01242 (Beilstein Handbook Reference); AI3-14660; BRN 0972464; 1-Cyanoheptadecane; EINECS 211-345-3; Heptadecyl cyanide; HSDB 5735; Nitril kyseliny stearove; NSC 5541; Octadecanenitrile; Oktadekannitril; Stearinsäurenitril; Stearonitrile. Crystals; mp = 41°; bp = 362°; d^{20} = 0.8325; insoluble in H_2O, soluble in EtOH, very soluble in Et_2O, Me_2CO, $CHCl_3$.

2745   1-Octadecene
112-88-9   G H

$C_{18}H_{36}$
n-Octadec-1-ene.
AI3-06521; EINECS 204-012-9; Neodene® 18; NSC 66460; α-Octadecene; 1-Octadecene; n-Octadec-1-ene; Octadecylene α-. Intermediate for biodegradable surfactants and specialty industrial chemicals. Solid; mp = 17.5°; bp$_{15}$ = 179°, bp$_8$ = 145°; d^{20} = 0.7891; insoluble in H$_2$O, soluble in Me$_2$CO, CCl$_4$. *Shell.*

2746   9-Octadecenoic acid (9Z)-, epoxidized, ester with propylene glycol
68609-92-7   H

$C_{21}H_{42}O_4$
Oleic acid, 1,2-propylene glycol epoxidized ester.
EINECS 271-842-6; 9-Octadecenoic acid (Z)-, epoxidized, ester with propylene glycol; Oleic acid, 1,2-propylene glycol epoxidized ester; Vikoflex 4964.

2747   N-(9-Octadecenyl)-3-aminopropionitrile
26351-32-6   H

$C_{21}H_{40}N_2$
Propanenitrile, 3-((9Z)-9-octadecenylamino)-.
EINECS 247-628-3; N-(9-Octadecenyl)-3-aminopropio-nitrile; (Z)-3-(9-Octadecenylamino)propiononitrile; Prop-anenitrile, 3-((9Z)-9-octadecenylamino)-; Propionitrile, 3-(9-octadecenylamino)-, (Z)-.

2748   Octadecenylsuccinic anhydride
28777-98-2   H

$C_{22}H_{38}O_3$
Dihydro-3-(octadecenyl)furan-2,5-dione.
Dihydro-3-(octadecenyl)-2,5-furandione; Dihydro-3-(octadecenyl)furan-2,5-dione; EINECS 249-210-6; 2,5-Furandione, dihydro-3-(octadecenyl)-; Octadecenyl-succinic anhydride; Succinic anhydride, octadecenyl-.

2749   Octadecyl   3,5-di-t-butyl-4-hydroxyhydro-cinnamate
2082-79-3   G H

$C_{35}H_{62}O_3$
3,5-Bis(1,1-dimethylethyl)-4-hydroxybenzenepropanoic acid.
ADK Stab AO 50; Anox PP 18; Antioxidant 1076; AO 4; E 376; EINECS 218-216-0; HSDB 5865; Hydrocinnamic acid, 3,5-di-t-butyl-4-hydroxy-, octadecyl ester; I 1076; IR 1076; Irganox 1076; Irganox 1906; Irganox 1976; Irganox I 1076; Irganox L 107; Mark AO 50; Naugard® 76; Octadecyl 3-(3',5'-di-t-butyl-4'-hydroxyphenyl)propion-ate; Octadecyl 3,5-di-t-butyl-4-hydroxyhydrocinnamate; Ralox 530; Stearyl 3,5-di-t-butyl-4-hydroxyhydro-cinnamate; Sumilizer BP 76; Tominokusu SS; U 276; Ultranox® 276. High molecular weight antioxidant/ stabilizer for styrenics, polyolefins, PVC, urethane and acrylic coatings, adhesives, and elastomers; effective replacement for BHT in polyolefins. Used to stabilize polymeric substances such as polyolefins, styrenics, EPDM, and PVC; provides good thermal and color stability. Solid; mp = 50-52°. *GE Specialities; Uniroyal.*

2750   Octadecyl chloride
3386-33-2   H

$C_{18}H_{37}Cl$
n-Octadecyl chloride.
AI3-28591; 1-Chlorooctadecane; EINECS 222-207-7; NSC 5543; Octadecane, 1-chloro-; Octadecyl chloride; n-Octadecyl chloride. Solid; mp = 28.6°; bp = 348°; d^{20} = 0.8641; insoluble in H$_2$O, slightly soluble in CCl$_4$.

2751   Octadecyl methacrylate
32360-05-7   G H

$C_{22}H_{42}O_2$
2-Propenoic acid, 2-methyl-, octadecyl ester.
Ageflex FM-68; Ageflex FM-1620; AI3-25418; EINECS 251-013-5; Methacrylic acid, octadecyl ester; Methacrylic acid, stearyl ester; Octadecyl methacrylate; 2-Propenoic acid, 2-methyl-, octadecyl ester. Natural, C$_{16-18}$ methacrylates, with 100 ppm hydroquinone inhibitor; lube oil additive, pour point depressant, paper coatings, textile finishes, paints, varnishes, pressure-sensitive adhesives. Lube oil additive, pour point depressant. Used in paper coatings, textile finishes, paints, varnishes, pressure-sensitive adhesives. Liquid; bp$_5$ = 181°, d^{20} = 0.868; f.p. -20°; flash pt. 110°. *Rit-Chem; Rohm & Haas Co.; Sartomer.*

2752   Octamethylcyclotetrasiloxane
556-67-2   6779   G H

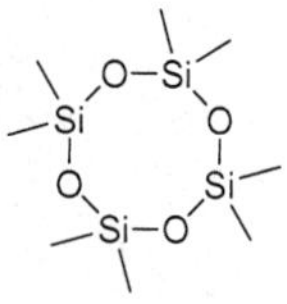

C8H24O4Si4

Cyclic dimethylsiloxane tetramer.

4-04-00-04125 (Beilstein Handbook Reference); BRN 1787074; CCRIS 1327; Cyclic dimethylsiloxane tetramer; Cyclotetrasiloxane, octamethyl-; EINECS 209-136-7; HSDB 6131; KF 994; NSC 345674; NUC silicone VS 7207; O9810; Octamethylcyclotetrasiloxane; Okta-methylcyklotetrasilraxilan; SF 1173; Silicone SF 1173; UC 7207; Union carbide 7207; VS 7207. As cleaning, polishing and damping media; offers low toxicity, inertness. Used to deactivate glass chromatography columns. Liquid; mp = 17.5°; bp = 175.8°; d^{20} = 0.9561; insoluble in H2O, soluble in CCl4. *Degussa-Hüls Corp.*

2753   Octamethyltrisiloxane
107-51-7   6780   G

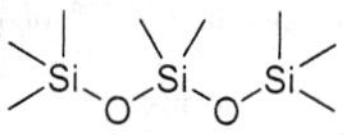

C8H24O2Si3

O9816.

CCRIS 3198; EINECS 203-497-4; Octamethyltrisiloxane; Trisiloxane, octamethyl-. As cleaning, polishing and damping media; offers low toxicity, inertness. Also used as a foam suppressant. Liquid; mp= -80°; bp = 153°, bp17 = 50-52°; d^{20} = 0.8200; soluble in C6H6, petroleum ether, slightly soluble in EtOH. *Hüls Am.*

2754   Octane
111-65-9   6782   H

C8H18

n-Octane.

Al3-28789; EINECS 203-892-1; HSDB 108; NSC 9822; Octane; n-Octane; Oktan; Oktanen; Ottani. Mobile liquid; mp = -56.8°; bp = 125.6°; d^{25} = 0.6986; insoluble in H2O, freely soluble in EtOH, Et2O.

2755   Octane-1-thiol
111-88-6   H

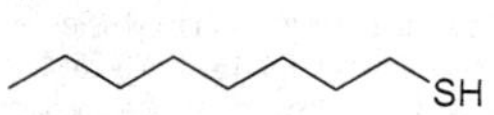

C8H18S

1-Mercaptooctane.

Al3-06557; EINECS 203-918-1; HSDB 5552; 1-Mercaptooctane; 1-Octanethiol; n-Octyl mercaptan; 1-Octylthiol; NSC 41903; Octane-1-thiol; Octyl mercaptan; Octylthiol. Liquid; mp = -49.2°, bp = 199.1°; d^{20} = 0.8433; slightly soluble in CCl4, soluble in EtOH.

2756   Octanesulfonyl chloride
7795-95-1   H

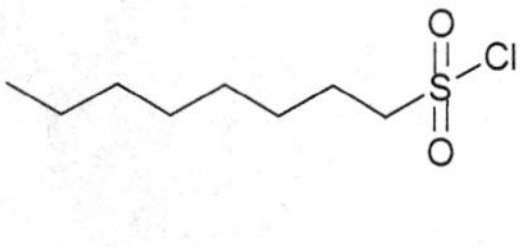

C8H17ClO2S

1-Octanesulfonyl chloride.

EINECS 232-249-8; 1-Octanesulfonyl chloride; 1-Octanesulphonyl chloride.

2757   Octanesulfonyl fluoride
40630-63-5   H

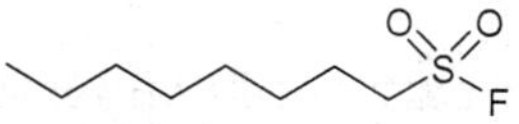

C8H17FO2S

1-Octanesulfonyl fluoride.

EINECS 255-012-0; 1-Octanesulfonyl fluoride; 1-Octanesulphonyl fluoride.

2758   Octanoic acid
124-07-2   1771   G H

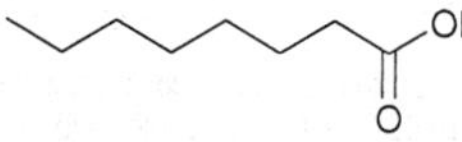

C8H16O2

n-Octane carboxylic acid.

4-02-00-00982 (Beilstein Handbook Reference); Acide octanoique; Acido octanoico; Acidum octanocium; Al3-04162; BRN 1747180; C8 acid; Caprylic acid; Caprylsäure; CCRIS 4689; EINECS 204-677-5; Enantic acid; FEMA No. 2799; Hexacid 898; HSDB 821; Kyselina kaprylova; neo-Fat 8; NSC 5024; Octanoic acid; Octic acid; Octoic acid; Octylic acid. Fatty acid; Used in manufacture of dyes, drugs, perfumes. Liquid; mp = 16.3°; bp = 239°; d^{20} = 0.9106; slightly soluble in H2O (68 mg/100 g), freely soluble in EtOH, CHCl3, CH3CN; LD50 (rat orl) = 10,080 mg/kg. *Akzo Chemie; Henkel/Emery; Procter & Gamble Co.; Sigma-Aldrich Fine Chem.; Unichema.*

2759   Octanoic acid, decanoic acid, trimethylolpropane ester
11138-60-6   H

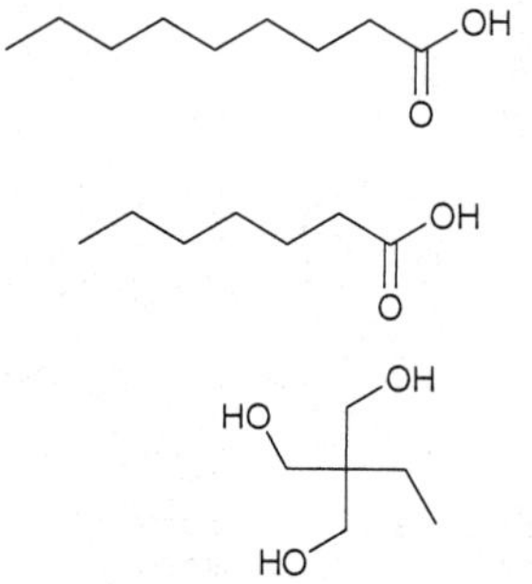

C10H20O2.C8H16O2.C6H14O3

Caprylic acid, capric acid, trimethylolpropane ester.

Caprylic acid, capric acid, trimethylolpropane ester; Decanoic acid, ester with 2-ethyl-2-(hydroxymethyl)-1,3-propanediol octanoate; EINECS 234-392-1; Octanoic acid, decanoic acid, trimethylolpropane ester; Trimethylolpropane, caprylate caprate triester.

2760   Octanoic acid, diethanolamine salt
16530-72-6   H

452

$C_8H_{16}O_2 \cdot C_4H_{11}NO_2$
Octanoic acid, compd. with 2,2'-iminodiethanol (1:1).
Caprylic acid, diethanolamine salt; EINECS 240-600-1; Octanoic acid, compd. with 2,2'-iminobis(ethanol) (1:1).

2761 Octanol
111-87-5   6784   G P

$C_8H_{18}O$
1-n-Octyl alcohol.
AI3-02169; Alcohol C8; Alfol 8; C8 alcohol; Capryl alcohol; n-Capryl Alcohol; Caprylic alcohol; n-Caprylic alcohol; Caswell No. 611A; Dytol M-83; EINECS 203-917-6; EPA Pesticide Chemical Code 079037; EPAL 8; Fatty alcohol(C8); FEMA Number 2800; Heptyl carbinol; HSDB 700; Lorol 20; NSC 9823; Octanol; n-Octanol; n-Octan-1-ol; Octilin; Octyl alcohol; n-Octyl alcohol; Primary octyl alcohol; Sipol L8. Solvent and chemical intermediate. Used as a basis for silicone oils, also as a foam suppressant. Registered by EPA as a herbicide and insecticide. Colorless liquid; mp = -15.5°; bp = 195.1°; d^{25} = 0.8262; insoluble in H_2O, slightly soluble in CCl_4; freely soluble in EtOH, Et_2O. *Coastal; Ethicon Inc.; Lancaster Synthesis Co.; Michel M.; Penta Mfg.*

2762 Oct-1-ene
111-66-0   1770   G H

C_8H_{16}
n-1-Octene.
AI3-28403; Caprylene; 1-Caprylene; EINECS 203-893-7; HSDB 1084; Octene; Octene-1; α-Octene; 1-Octene; Octylene; 1-Octylene; n-1-Octene; NSC 8457. Intermediate for surfactants and specialty industrial chemicals. Liquid; mp = -101.7°; bp = 121.2°. d_4^{20} 0.7149; insoluble in H_2O, slightly soluble in CCl_4, soluble in Et_2O, Me_2CO, C_6H_6, very soluble in $CHCl_3$. *Air Products & Chemicals Inc.; Chevron; Ethicon Inc.; Shell; Sigma-Aldrich Fine Chem.; Texaco.*

2763 Oct-2-ene
25377-83-7   H

C_8H_{16}
2-Octene (mixed cis and trans isomers).
EINECS 246-920-8; HSDB 5146; Octene; 2-Octene (mixed cis and trans isomers); Octene (all isomers); Octylene.

2764 Octenylsuccinic anhydride
26680-54-6   H

$C_{12}H_{18}O_3$
Dihydro-3-(octenyl)furan-2,5-dione.

Dihydro-3-(octenyl)furan-2,5-dione; EINECS 247-899-8; 2,5-Furandione, dihydro-3-(octenyl)-; Succinic anhydride, octenyl-.

2765 Octhilinone
26530-20-1   6788   G P

$C_{11}H_{19}NOS$
2-Octyl-4-isothiazolin-3-one.
BRN 1211137; Caswell No. 613C; CCRIS 6082; EPA Pesticide Chemical Code 099901; 3(2H)-Isothiazolone, 2-octyl-; 4-Isothiazolin-3-one, 2-octyl-; Kathon®; Kathon® 893; Kathon® 893F; Kathon® 4200; Micro-Chek® 11; Kathon LM; Kathon LP preservative; Kathon SP 70; Micro-chek 11; Micro-chek 11D; Micro-chek skane; Microbicide M-8; Octhilinone; 2-Octyl-3-isothiazolone; 2-n-Octyl-4-isothiazolin-3-one; 2-Octyl-2H-isothiazol-3-one; 2-Octyl-3(2H)-isothiazolone; 2-Octyl-4-isothiazolin-3-one; Pancil; Pancil-T; RH 893; Skane 8; Skane M8; Vinylzene IT 3000DIDP. Antimicrobial, mildewcide for PVC, polyurethane, other polymers for use in roofing membranes, exterior automotive trims, awnings, tarpaulins, pond liners, marine upholstery, shower curtains, outdoor furniture, leather, textiles, paper, plastics floor polish and adhesives. Mildewcide for paints. Fungicide, biocide in cooling towers, cutting oils, cosmetics and shampoos. Leather preservative. A paste containing 1% w/w octhilinone is used as a fruit tree canker paint. Registered by EPA as a fungicide. Liquid; $bp_{0.01}$ = 120°; λ_m = 280 nm (log ε 3.88 MeOH); LD_5- (rat orl) = 1470 mg/kg, (rbt der) = 4500 mg/kg.; toxic to fish. *Buckman Laboratories Inc.; Ferro; Rohm & Haas Co.; Sanitized Inc; Thomson Research Associates; Thor; Troy Chemical Corp.*

2766 Octogen
2691-41-0   6790   H

$C_4H_8N_8O_8$
1,3,5,7-Tetranitro-1,3,5,7-tetraazacyclooctane.
4-26-00-01645 (Beilstein Handbook Reference); BRN 0361386; Cyclotetramethylenetetranitramine; EINECS 220-260-0; HMX; β-Hmy; HSDB 5893; HW 4; LX 14-0; Octahydro-1,3,5,7-tetranitro-1,3,5,7-tetrazocine; Octo-gen; Oktogen; Tetramethylenetetranitramine; UN0226; UN0484. White crystals; mp = 281°; insoluble in H_2O.

2767 Octrizole
3147-75-9   D G H

$C_{20}H_{25}N_3O$
2-(2-Hydroxy-5-t-octylphenyl)benzotriazole.
2-(2H-Benzotriazol-2-yl)-4-(1,1,3,3-tetramethylbutyl)-phenol; Cyasorb® UV 5411; EINECS 221-573-5; 2-(2-Hydroxy-5-t-

octylphenyl)benzotriazole; Octrizol; Octri-zole; Octrizolum; Phenol, 2-(2H-benzotriazol-2-yl)-4-(1,1,3,3-tetramethylbutyl)-; Spectra-Sorb UV 5411; UV Absorber-5; Uvazol 311. Light stabilizer and UV absorber for polymeric systems including polyester, PVC, styrenics, acrylics, PC, polyvinyl butyral, PNMA, PS, ABS and unsaturated polyesters. End-uses including molding, sheet, and glazing materials for window, marine, and auto applications; also in coatings, photoproducts, sealants and elastomeric materials. *Am. Cyanamid; EniChem SpA.*

2768   Octyl acetate
103-09-3   6795   G H

$C_{10}H_{20}O_2$
2-Ethylhexyl acetate.
4-02-00-00166 (Beilstein Handbook Reference); Acetic acid, 2-ethylhexyl ester; AI3-07924; BRN 1758321; EINECS 203-079-1; 2-Ethylhexyl acetate; 2-Ethyl-1-hexyl acetate; 22-Ethylhexylester kyseliny octove; FEMA Number 2806; HSDB 2668; NSC 8897; Octyl acetate. Solvent for nitrocellulose, resins, lacquers, baking finishes. Liquid; mp = -80°; bp = 199°; d^{20} = 0.8718; insoluble in H_2O, soluble in EtOH, Et2O; LD50 (rat orl) = 3 g/kg. *Degussa-Hüls Corp.; Eastman Chem. Co.; MTM Spec. Chem. Ltd.; Penta Mfg.*

2769   tert-Octylacrylamide
4223-03-4   H

$C_{11}H_{21}NO$
2-Propenamide, N-(1,1,3,3-tetramethylbutyl)-.
4-04-00-00772 (Beilstein Handbook Reference); Acrylamide, N-(1,1,3,3-tetramethylbutyl)-; AI3-26325; BRN 1759951; EINECS 224-169-7; N-(1,1,3,3-Tetra-methylbutyl)-2-propenamide; N-(1,1,3,3-Tetramethyl-butyl)acrylamide; NSC 9035; tert-Octylacrylamide; N-tert-Octylacrylamide; 2-Propenamide, N-(1,1,3,3-tetramethylbutyl)-.

2770   Octyl acrylate
103-11-7   G H

$C_{11}H_{20}O_2$
2-Ethylhexyl acrylate.
4-02-00-01467 (Beilstein Handbook Reference); Acrylic acid, 2-ethylhexyl ester; AI3-03833; BRN 1765828; CCRIS 3430; EINECS 203-080-7; 2-Ethyl-1-hexyl acrylate; 2-Ethylhexyl 2-propenoate; 2-Ethylhexyl acrylate; 2-Ethylhexylester kyseliny akrylove; HSDB 1121; Mono(2-ethylhexyl) acrylate; NSC 4803; Octyl acrylate; 2-Propenoic acid, 2-ethylhexyl ester. Liquid; mp = -90°; bp60 = 123-127°; d^{25} = 0.880; insoluble in H_2O (< 1 mg/ml). *BASF Corp.; Hoechst Celanese; Sartomer; Union Carbide Corp.*

2771   Octyl alcohol
104-76-7   3842   G H

$C_8H_{18}O$
2-Ethylhexan-1-ol.
4-01-00-01783 (Beilstein Handbook Reference); 2-Aethylhexanol; 2-Äthylhexanol; Aerofroth 88; AI3-00940; Alcohol, 2-ethylhexyl; BRN 1719280; CCRIS 2292; EINECS 203-234-3; Ethylhexanol; Ethylhexanol, 2-; 1-Hexanol, 2-ethyl-; 2-Ethyl-1-hexanol; 2-Ethylhexan-1-ol; 2-Ethylhexanol; 2-Ethylhexyl alcohol; FEMA No. 3151; HSDB 1118; NSC 9300; Octyl alcohol. Plasticizer for PVC resins; defoaming agent, wetting agent, organic synthesis, solvent mix for nitrocellulose; penetrant for plasticizing inks. Used in the mining industry; mercerizing textiles; solvent for dyes, resin, oils; antifoaming properties. Liquid; mp = -70°; bp = 184.6°; slightly soluble in H_2O (0.14 g/100 ml), soluble in EtOH, Et2O, Me2CO, C6H6, CHCl3; LD50 (rat orl) = 12.46 mg/kg. *Am. Cyanamid; Aristech; Ashland; BASF Corp.; BP Chem.; Eastman Chem. Co.; Shell.*

2772   Octyl chloride
111-85-3   H

$C_8H_{17}Cl$
1-Chlorooctane.
Capryl chloride; 1-Chlorooctane; EINECS 203-915-5; HSDB 5551; NSC 5406; Octane, 1-chloro-; Octyl chloride; 1-Octyl chloride; n-Octyl chloride. Liquid; mp = -57.8°; bp = 181.5°; d^{20} = 0.8738; insoluble in H_2O, slightly soluble in CCl4, very soluble in EtOH, Et2O.

2773   sec-Octyl epoxytallate
61789-01-3   H

2-Ethylhexyl epoxy tallate.
EINECS 263-024-2; 2-Ethylhexanol, tall oil fatty acids epoxidized ester; 2-Ethylhexyl epoxy tallate; Expoxidized 2-ethylhexyl ester of tall oil fatty acid; Fatty acids, tall-oil, epoxidized, 2-ethylhexyl esters; Flexol EP-8; Flexol plasticizer EP-8; sec-Octyl epoxytallate; Tall oil, epoxidized, 2-ethylhexyl esters.

2774   Octyl hydroxystearate
29710-25-6   G

$C_{26}H_{52}O_3$
2-Ethylhexyl 12-hydroxy octadecanoate.
EINECS 249-793-7; 2-Ethylhexyl 12-hydroxy octa-decanoate; 2-Ethylhexyl 12-hydroxystearate; Octa-decanoic acid, 12-hydroxy-, 2-ethylhexyl ester; Wickenol® 171. Emollient, moisturizer, pigment wetter/dispersant; increases water vapor porosity of fatty components used in cosmetic and topical pharmaceutical preparations; refatting agent, counterirritant, cosolvent, solubilizer. *CasChem.*

2775 Octyl methoxycinnamate
5466-77-3   6799   G

$C_{18}H_{26}O_3$
3-(4-Methoxyphenyl)-2-propenoic acid 2-ethylhexyl ester.
AI3-05710; CCRIS 6200; EINECS 226-775-7; Escalol; Escalol®
557; 2-Ethylhexyl methoxycinnamate; 2-Ethylhexyl-4-
methoxycinnamate; 2-Ethylhexyl p-meth-oxycinnamate; 3-(4-
Methoxyphenyl)-2-propenoic acid, 2-ethylhexyl ester; Neo
Heliopan®; Neo Heliopan® AV; NSC 26466; Octinoxate; Octyl
methoxycinnamate; Parsol® MOX; 2-Propenoic acid, 3-(4-
methoxyphenyl)-, 2-ethylhexyl ester; 2-ethylhexyl p-methoxy-
cinnamate; Octyl methoxy cinnamate; Parsol MCX; Parsol MOX;
Neo Heliopan, Type AV; Beclovent Inhaler. UV-B absorber,
sunscreening agent in the wavelength range of 2900-3200 A where
causes sunburn and skin damage result, stimulates tanning. Used in
cosmetic applications, waterproof sunscreens. *Bernel; Givaudan;
Glaxo Labs.; Haarmann & Reimer GmbH; Van Dyk.*

2776 Octyl pelargonate
59587-44-9   G

$C_{17}H_{34}O_2$
2-Ethylhexyl pelargonate.
EINECS 261-819-9; 2-Ethylhexyl nonanoate; 2-Ethylhexyl
pelargonate; Nonanoic acid, 2-ethylhexyl ester; Octyl pelargonate;
Wickenol® 160. Emollient, moisturizer, pigment wetter/dispersant;
increases water vapor porosity of fatty components used in cosmetic
and topical pharmaceutical preparations; improves stick
formulations. *CasChem.*

2777 Octyl phenol
27193-28-8   H

$C_{14}H_{22}O$
(1,1,3,3-Tetramethylbutyl)phenol.
Caswell No. 613D; EINECS 248-310-7; EPA Pesticide Chemical
Code 064118; Octyl phenol; Octylphenol; tert-Octylphenol; Phenol,
(1,1,3,3-tetramethylbutyl)-; Phenol, octyl-; (1,1,3,3-
Tetramethylbutyl)phenol; USAF RH-6. Mixture of o, m and p-
isomers.

2778 p-Octylphenol
140-66-9   H

$C_{14}H_{22}O$
p-(1,1,3,3-Tetramethylbutyl)phenol.
4-06-00-03484 (Beilstein Handbook Reference); AI3-10011; BRN
0513992; EINECS 205-426-2; HSDB 5411; NSC 5427; p-
Octylphenol; 4-t-Octylphenol; 4-t-Octylphenol; p-terc.Oktylfenol;
para-t-Octylphenol; Phenol, 4-(1,1,3,3-tetramethylbutyl)-.

2779 p-Octylphenol
1806-26-4   H

$C_{14}H_{22}O$
1-(p-Hydroxyphenyl)-n-octane.
EINECS 217-302-5; HSDB 5857; 1-(p-Hydroxyphenyl)-octane;
Phenol, 4-octyl-; Phenol, p-octyl-; 4-Octylphenol; p-Octylphenol;
para-Octylphenol.

2780 Octyl salicylate
118-60-5   G

$C_{15}H_{22}O_3$
Octyl salicylate.
3-10-00-00124 (Beilstein Handbook Reference); Benzoic acid, 2-
hydroxy-, 2-ethylhexyl ester; BRN 2730664; Dermoblock OS;
EINECS 204-263-4; Ethyl hexyl salicylate; 2-Ethylhexyl salicylate;
Ethylhexyl salicylate; 2-Ethylhexyl 2-hydroxybenzoate; Neo
Heliopan® OS; NSC 46151; Octisalate; Octyl salicylate; Salicylic
acid, 2-ethylhexyl ester; Sunarome O; Sunarome WMO; USAF DO-
11; Uvinul® O-18; WMO. UV-B absorber for cosmetic applications,
waterproof sunscreens; solubilizer for oxybenzone. Solid; bp_{21} =
189-190°; d = 1.014. *Akzo Chemie; BASF Corp.; Haarmann &
Reimer GmbH.*

2781 Octyl stearate
22047-49-0   G

$C_{26}H_{52}O_2$
2-Ethylhexyl octadecanoate.
Cetiol 868; EINECS 244-754-0; 2-Ethylhexyl stearate; Octadecanoic
acid, 2-ethylhexyl ester; Octyl stearate; Stearic acid, 2-ethylhexyl
ester; Wickenol® 156. Emollient, moisturizer, pigment
wetter/dispersant; increases water vapor porosity of fatty
components used in cosmetic and topical pharmaceutical
preparations. *CasChem.*

2782 Octylaldehyde
123-05-7   H

C8H16O
2-Ethylhexaldehyde.
4-01-00-03345 (Beilstein Handbook Reference); AI3-26805; BRN 1700556; Butylethylacetaldehyde; CCRIS 4643; EINECS 204-596-5; Ethylbutylacetaldehyde; Ethyl-hexaldehyde; α-Ethylcaproaldehyde; 2-Ethylcaproalde-hyde; 2-Ethylhexaldehyde; 2-Ethylhexanal; 2-Ethyl-hexylaldehyde; Hexanal, 2-ethyl-; HSDB 5142; NSC 42871; Octyl aldehyde. Liquid; mp <-100°; bp = 163°; d20ks = 0.8540; insoluble in H2O; slightly soluble in CCl4, soluble in EtOH, Et2O.

2783   Octylaldehyde
124-13-0   1772   H

C8H16O
n-Octyl aldehyde.
4-01-00-03337 (Beilstein Handbook Reference); AI3-03961; Aldehyde C8; Antifoam-LF; BRN 1744086; C8 aldehyde; Caprylaldehyde; Caprylic aldehyde; EINECS 204-683-8; FEMA No. 2797; HSDB 5147; NSC 1508; Octaldehyde; n-Octaldehyde; Octanal; n-Octanal; Octanaldehyde; Octanoic aldehyde; Octyl aldehyde; Octylaldehyde. Solid; mp = 171°; d20 = 0.8211; λm = 295 nm (ε = 13, C6H14); very soluble in Me2CO, C6H6, Et2O, EtOH.

2784   Octylamine
111-86-4   G H

C8H19N
n-Octylamine.
4-04-00-00751 (Beilstein Handbook Reference); AI3-11522; Amine-8D; 1-Aminooctane; Armeen 8; Armeen 8D; BRN 1679227; Caprylamine; Caprylylamine; EINECS 203-916-0; NSC 9824; 1-Octanamine; Octylamine; 1-Octylamine; n-Octylamine; n-Octylamine, mono-. Chemical intermediate. Liquid; mp = 0°; bp = 179.6°; d20 = 0.7826; slightly soluble in H2O; soluble in CCl4, very soluble in EtOH, Et2O.soluble in organic solvents.

2785   Octylated diphenylamines
68921-45-9   G

C28H43N
Diphenylamine reaction product with styrene and diisobutylene.
Agerite® Stalite; Benzenamine, N-phenyl-, reaction products with styrene and 2,4,4-trimethylpentene; Diphenylamine reaction product with styrene and diisobutylene; EINECS 272-940-1; Reaction product of N-phenylbenzenamine, ethenylbenzene, and diisobutylene. A proprietary antioxidant. Insoluble in H2O, soluble in organic solvents. Vanderbilt R.T. Co. Inc.

2786   Octyldodecanol
5333-42-6   G

C20H42O
2-Octyldodecan-1-ol.
3-01-00-01844 (Beilstein Handbook Reference); AI3-19966; BRN 1763479; 1-Dodecanol, 2-octyl-; EINECS 226-242-9; Eutanol G; Exxal 20; Isofol 20; Kalcohl 200G; Kalcohl 200GD; Michel XO-150-20; NSC 2405; 2-Octyldodecanol; 2-Octyl dodecanol; 2-Octyl-1-dodecanol; 2-Octyldodecyl alcohol; Rilanit G 20; Standamul G.

Lubricant, emollient for cosmetics and pharmaceuticals; for oil-soluble active ingredients; pigment dispersant. Used in organic synthesis. Henkel/Cospha; Michel M.

2787   Octyltriethoxysilane
2943-75-1   G

C14H32O3Si
n-Octyltriethoxysilane.
A 137; A 137 (coupling agent); Dynasylan® OCTEO; EINECS 220-941-2; NSC 42964; Octyl(triethoxy)silane; Prosil® 9202; Prosil® 9234; Silane, triethoxyoctyl-; Silquest A 137; Triethoxyoctylsilane; Union Carbide® A-137; Y 9187. Coupling agent for hydrophobic treatment, crosslinking agent providing durability, gloss, hiding power to coatings. Chemical intermediate, blocking agent, release agent, lubricant, primer, reducing agent. Liquid; bp2 = 98-99°; d = 0.878. Degussa-Hüls Corp.; PCR; Union Carbide Corp.

2788   4-Octyne-3,6-diol, 3,6-dimethyl-
78-66-0   G

C10H18O2
3,6-Dimethyl-4-octyne-3,6-diol.
4-01-00-02706 (Beilstein Handbook Reference); AI3-07312; BRN 1722390; 3,6-Dimethyl-4-octyne-3,6-diol; 3,6-Dimethyl-octin-4-diol-(3,6); 3,6-Dimethyloct-4-yne-3,6-diol; EINECS 201-131-8; NSC 1025; 4-Octyne-3,6-diol, 3,6-dimethyl-; Surfynol 82; Surfynol 85. Surfactant, defoamer, wetting agent, visocsity reducer used with aqueous systems, pesticide concentrates, shampoos, vinyl plastisols, starch solutions, flexographic inks, electroplating baths. Detergent for radiator cleaners, Solid; mp = 53-55°; bp680 = 208-214°. Air Products & Chemicals Inc.

2789   (S)-(-)-Ofloxacin
100986-85-4   6800   D

C18H20FN3O4
(S)-(-)-9-Fluoro-2,3-dihydro-3-methyl-10-(4-methyl-1-piperazinyl)-7-oxo-7H-pyrido[1,2,3-de]-1,4-benzoxazine-6-carboxylic acid.
CCRIS 4074; Cravit; DL-8280; DR-3355; DR3355; DRG-0129; Elequine; Exocin; Floxin; HOE-280; HR 355; Levaquin; Levofloxacin; Levofloxacine; Levofloxacino; Levofloxacinum; Ocuflox; (-)-Ofloxacin; (S)-Ofloxacin; (S)-(-)-Ofloxacin; Ofloxacin S-(-)-form; RWJ 25213-097; Synonyms. Quinolone antibiotic. Crystals; mp = 225-227° (dec); [α]D23 = -76.9° (c = 0.385 0.5N NaOH); LD50 (mmus orl) = 1881 mg/kg, (fmus orl) = 1803 mg/kg, (mrat orl) = 1478 mg/kg, (frat orl) = 1507 mg/kg. Daiichi Seiyaku; Hoechst AG.

2790   Ofurace
58810-48-3   F

$C_{14}H_{16}ClNO_3$

2-Chloro-N-(2,6-dimethylphenyl)-N-(tetrahydro-2-oxo-3-furyl)acetamide.

5-18-11-00316 (Beilstein Handbook Reference); Acet-amide, 2-chloro-N-(2,6-dimethylphenyl)-N-(tetra-hydro-2-oxo-3-furanyl)-; Acetanilide, 2-chloro-2',6'-di-methyl-N-(2-oxotetrahydro-3-furyl)-; BRN 1653655; Chevron 20615; (±)- -2-Chloro-N-2,6-xylylacetamido- -butyro-lactone; 2-Chloro-2',6'-dimethyl-N-(2-oxotetrahydro-3-furyl)acetanilide; 2-Chloro-N-(2,6-di-methylphenyl)-N-(tetrahydro-2-oxo-3-furanyl)acetamide; 2-Chloro-N-(2,6-dimethylphenyl)-N-(tetrahydro-2-oxo-3-furyl)acetamide; EINECS 261-451-9; Milfuram; Ofurace; Ortho 20615; RE 20615. Fungicide used for the control of potato blight. Colorless crystals; mp = 145 - 146°; $sg^{20} = 1.366$; soluble in H_2O ((0.014 g/100 ml at 20°), $CHCl_3$ (37 g/100 ml), Me_2CO (135 g/100 ml), DMF (34 g/100 ml), cyclohexanone (14 g/100 ml), EtOAc (4.4 g/100 ml), i-PrOH (0.6 g/100 ml), readily soluble in CH_2Cl_2, N-methylpyrrolidone, chlorobenzene, insoluble in C_6H_{14}, kerosene; LD_{50} (mrat orl) = 3500 mg/kg, (frat orl) = 2600 mg/kg, (rbt der) > 5000 mg/kg; non-toxic to bees. *Aventis Crop Science; DowElanco Ltd.*

2791   Oil of Pennyroyal
8007-44-1   6863   G P

$C_{10}H_{18}O$

Pennyroyal oil.

American pennyroyal oil; Caswell No. 638B; EPA Pesticide Chemical Code 040509; Hedeoma oil; Oil of pennyroyal; Oils, pennyroyal, Hedeoma pulegioides; Pennyroyal; Pennyroyal oil; Pennyroyal oil, American; Pennyroyal oil, European. Registered by EPA as an insecticide (cancelled). Pale yellow liquid; $d_{25}^{25} = 0.920 - 0.935$; $[\alpha]_D^{20} = +18° - +22°$; slightly soluble in H_2O, soluble in 70% EtOH (33 g/100 ml), very soluble in $CHCl_3$, Et_2O.

2792   Oil of turpentine, α-pinene fraction
65996-96-5   H

Turpentine-oil, α-pinene fraction.
EINECS 266-031-9; Oil of turpentine, α-pinene fraction; Turpentine-oil, α-pinene fraction.

2793   Oil of turpentine, β-pinene fraction
65996-97-6   H

Turpentine-oil, β-pinene fraction.
EINECS 266-032-4; Oil of turpentine, β-pinene fraction; Turpentine-oil, β-pinene fraction.

2794   Oil of turpentine, limonene fraction
65996-99-8   H

Turpentine-oil, limonene fraction.
EINECS 266-035-0; Oil of turpentine, 30% or greater limonene fraction; Turpentine-oil, limonene fraction.

2795   Olanzapine
132539-06-1   6892   D

$C_{17}H_{20}N_4S$

2-Methyl-4-(4-methyl-1-piperazinyl)-10H-thieno[2,3-b][1,5]benzodiazepine.

Lanzac; LY 170053; Olansek; Olanzapine; Zyprexa. A serotonin and dopamine receptor antagonist with anticholinergic activity; a tricyclic antipsychotic. Antipsychotic. Crystals; mp = 195°. *Eli Lilly & Co.*

2796   Oleamine
112-90-3   G H

$C_{18}H_{37}N$

9-Octadecen-1-amine, (Z)-.

4-04-00-01132 (Beilstein Handbook Reference); Alamine 11; Armeen O; BRN 1723960;; EINECS 204-015-5; HSDB 5579; Jet Amine PO; Kemamine® P 989; Noram O; OA; 9-Octadecen-1-amine, (Z)-; 9-Octadecenylamine, (Z)-; cis-9-Octadecenylamine; Oleamine; Oleinamine; Oleyl amine; Radiamine 6172. Emulsifier for herbicides, ore flotation, pigment dispersion; auxiliary for textiles, leather, rubber, plastics, and metal industries. Corrosion inhibitor, flotation agent, emulsifier, mold release agent, lube oil additive, fertilizer anticaking agent. Used as a dispersing and flushing agent, as a chemical intermediate, in metalworking oils, as fuel oil additive; mold release for rubber and plastics; lubricant and spinning aid in metalworking oils. Mineral flotation, corrosion inhibitor, pigment dispersant; cosmetics; lubricant and mold release for hard rubber, textile chemical, chemical synthesis; antistat and antifog additive for plastic foils. Solid; mp = 15-22°; bp - 349°; d = 0.8130; insoluble in H_2O, soluble in organic solvents. *CK Witco Corp.; Croda; Fina Chemicals.*

2797   Oleic acid
112-80-1   6898   G P

$C_{18}H_{34}O_2$

cis -9-Octadecenoic acid.

4-02-00-01641 (Beilstein Handbook Reference); Acide oleique; l'Acide oleique; Al3-01291; BRN 1726542; Caswell No. 619; CCRIS 682; Century cd fatty acid; Crodolene®; D 100; D 100 (fatty acid); EINECS 204-007-1; Elaic acid; Elaidoic acid; Elainic acid; Emersol®; Emersol® 205; Emersol® 210; Emersol® 211; Emersol® 213; Emersol® 220 white oleic acid; Emersol® 221 low titer white oleic acid; Emersol® 233LL; Emersol® 6313NF; Emersol® 6321; EPA Pesticide Chemical Code 031702; Extra Oleic 80R; Extra Oleic 90; Extra Oleic 99; Extra Olein 80; Extra Olein 90; Extra Olein 90R; FEMA Number 2815; Glycon®; Glycon® RO; Glycon® WO; Groco® 5l; Groco® 6; Groco®; HSDB 1240; Hy-phi® 1055; Hy-phi® 1088;

Hy-phi® 2066; Hy-phi® 2088; Hy-phi® 2102; Hy-Phi®; HY-Phi® 1055; HY-Phi® 1088; HY-Phi® 2066; HY-Phi® 2088; HY-Phi® 2102; Industrene® 105; Industrene® 205; Industrene® 206; Industrene®;; K 52; Lunac O-CA; Lunac O-LL; Lunac O-P; Metaupon; NAA 35; Neo-Fat 92-04; NSC 9856; 9-Octadecenoic acid (9Z)-; Z-9-Octadecenoic acid; cis-Δ⁹-Octadecenoic acid; cis-Octadec-9-enoic acid; cis-Δ⁹-cis-Oleic acid; Ölsäuere; Oelsauere; Oleic acid; Oleine 7503; Oleinic acid; Pamolyn; Pamolyn 100; Priolene 6906; Priolene 6907; Priolene 6928; Priolene 6930; Priolene 6933; Tego-oleic 130; Vopcolene 27; Wecoline OO; White oleic acid; Wochem No. 320. Detergent intermediate for personal care, emollient, household and industrial applications. Registered by EPA as an antimicrobial, fungicide, herbicide, insecticide and rodenticicide (cancelled). Yellowish, oily liquid; mp = 13.4°; bp₁₀₀ = 286°; insoluble in H₂O, freely soluble in EtOH, Et₂O, C₆H₆, CHCl₃; LD₅₀ (rat orl) = 74000 mg/kg, (rat iv) = 2.4 mg/kg, (mus iv) = 230 mg/kg. *Akzo Nobel; Arizona; Brown; Fluka; Henkel/Emery; Hercules Inc.; Ruger; Schweizerhall Inc.; Spectrum Chem. Manufacturing; Unichema; Union Derivan SA; Witco/Oleo-Surf.*

2798   Oleic acid diethanolamide
93-83-4   G H

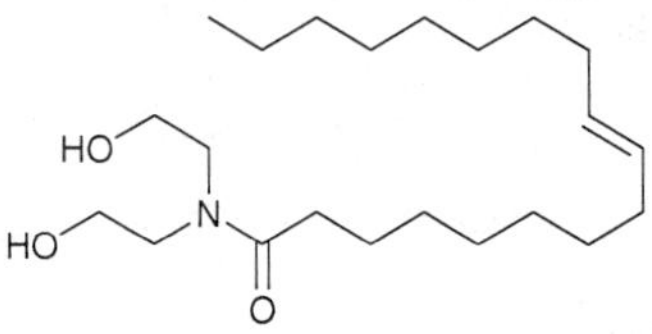

C₂₂H₄₃NO₃
9-Octadecenamide, N,N-bis(2-hydroxyethyl)-, (9Z)-.
Alkamide® DO-280; Alkamide® WRS 1-66; Alrosol O; Amisol ode; BRN 1985420; Clindrol 2000; Clindrol 2020; Comperlan OD; Diethanolamine oleic acid amide; Diethanololeamide; N,N-Diethanololeamide; EINECS 202-281-7; EMID 6545; Emulsifier WHC; Lauridit OD; Mackamide O; Marlamid D 1885; Nitrene NO; Oleamide DEA; Oleamide, N,N-bis(2-hydroxyethyl)-; Oleic acid diethanolamide; Oleic diethanolamide; Rewomid® DO 280; Schercomid ODA; Stafoam DO; Steinamid DO 280SE; Witcamide 511C. Emulsifier for highly nonpolar aliphatic hydrocarbons and chlorinated aliphatic hydrocarbons; rust inhibitor. Used with diethanolamine as a detergent and in the cosmetic, cleaning, and detergent industries, as a water-oil emulsifier, pigment dispersant, conditioner, corrosion inhibitor, and viscosity builder; emulsifier for aromatic and aliphatic hydrocarbon solvents; used in gel-type pine cleaners, shampoo formulations, hair conditioning agent. *Rohm & Haas Co.; Scher*; 1261.

2799   Oleic acid nitrile
112-91-4   H

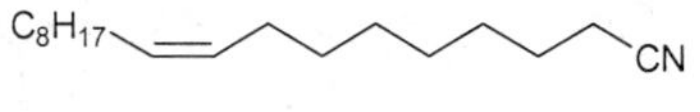

C₁₈H₃₃N
9-Octadecenenitrile, (Z)-.
4-02-00-01668 (Beilstein Handbook Reference); BRN 1102081; EINECS 204-016-0; HSDB 5580; 9-Octadecenenitrile, (9Z)-; 9-Octadecenenitrile, (Z)-; 9-Octadecenoic acid (cis), nitrile (cis); Oleic acid nitrile; Oleic nitrile; Oleonitrile; Oleoylnitrile; Oleyl nitrile; Oleylonitrile; Olsaeurenitril. Liquid; mp = -1°; bp = 332° (dec); d¹⁷ = 0.847; very soluble in EtOH.

2800   Oleic acid, propylene ester
105-62-4   H

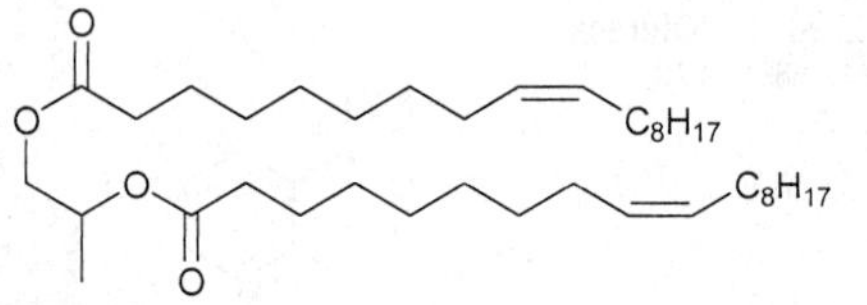

C₃₉H₇₂O₄
1-Methyl-1,2-ethanediyl 9-octadecenoate.
AI3-09503; EINECS 203-315-3; 1-Methyl-1,2-ethanediyl 9-octadecenoate; 1-Methyl-1,2-ethanediyl dioleate; 9-Octadecenoic acid, 1,3-propanediyl ester; 9-Octadecenoic acid (9Z)-, 1-methyl-1,2-ethanediyl ester; Oleic acid, propylene ester; Propylene glycol dioleate.

2801   Oleth-12
9004-98-2   E G

(C₂H₄O)+nC₁₈H₃₆O
Poly(oxy-1,2-ethanediyl),α-9-octadecenyl-ω-hydroxy-, (Z)-.
Ablunol OA-6; Ahco 3998; Amerox OE-20; Ameroxol OE 10; Ameroxol OE 2; Atlas G 3915; Atlas G 3920; Atmer 137; Blaunon EN 905; Blaunon EN 909; Blaunon EN 1504; Blaunon EN 1540; Brij 96; Brij 97; BRIJ 92; BRIJ 92((2)-oleyl); BRIJ 96((10) oleyl); BRIJ 98; BRIJ 98((20) oleyl); BRIJ 99; Chemal OA 9; EL-620; EL-719; Emalex 505; Emalex 510; Emalex 515; Emalex 520; Emery 6802; Emulgen 3200; Emulgen 404; Emulgen 408; Emulgen 409P; Emulgen 420; Emulgen 430; Emulphor; Emulphor O; Emulphor ON-870; Emulphor ON 870; Emulphor surfactants; Emulsogen MS 12; EO 20; Ethal OA 23; Ethosperse OA 2; Ethoxol 20; Ethoxylated oleyl alcohol; Ethylene oxide - oleyl alcohol adduct; Eumulgin EP 4; G 3910; G 3915; G 3920; Genapol O; Genapol O 020; Genapol O 050; Glycols, polyethylene, mono-9-octadecenyl ether, (Z)-; Lipal 20-OA; Lipocol O-20; Novol Poe 20; Oleth-12; Oleth-15; Oleth-20; Oleth-23; Oleth-25; Oleth-30; Oleth-4; Oleth-40; Oleth-44; Oleth-50; Oleth-6; Oleth-7; Oleth-8; Oleth-9; Oleyl alcohol EO (10); Oleyl alcohol EO (20); Oleyl alcohol EO (2); Oleylpolyoxyethylene glycol ether; PEG-12 Oleyl ether; PEG-15 Oleyl ether; PEG-20 oleyl ether; PEG-20 Oleyl ether; PEG-23 Oleyl ether; PEG-25 Oleyl ether; PEG-4 Oleyl ether; PEG-40 Oleyl ether; PEG-44 Oleyl ether; PEG-50 Oleyl ether; PEG-6 Oleyl ether; PEG-7 Oleyl ether; PEG-8 Oleyl ether; PEG-9 Oleyl ether; Polyoxyethylene monooleyl ether; Polyoxyl 10 oleyl ether; Polyoxyl 10 oleyl ether; Procol OA-20; Procol OA-25; Siponic Y-501; Standamul O20; Trycol HCS; Volpo 20.; Glycols, polyethylene, mono-9-octadecenyl ether, (Z)-; Amerox oe-20; Ameroxol EO 2; Ameroxol OE 10; Ameroxol OE-20; Atlas G-3915; Atlas G-3920; BO 2; BO 7; Brij 92; Brij 93; Brij 96; Brij 97; Brij 98; Brij 99; Brij 92((2)-oleyl); Brij 96((10) oleyl); Decaethoxy oleyl ether; Dehydol 100; EL-620; EL-719; Emalex 515; Emery 6802; Emulgen 408; Emulgen 420; Emulgen 430; Emulgin 010; Emulgin 05; Emulphor; Emulphor ON-870; Emulphor O. Emollient, lubricant, emulsifier, solubilizer for cosmetic application; emulsifier in astringent creams and lotions; clear gel formation; superfatting in shampoos, foaming bath preparations, and Carbopol gels; used in cold waves, depilatories, and hair straighteners; solubilizer for bromoacids in lipsticks and liquid rouge; spreading agent for bath oils. *Taiwan Surfactants.*

2802   Oleyl alcohol
143-28-2   6900   G H

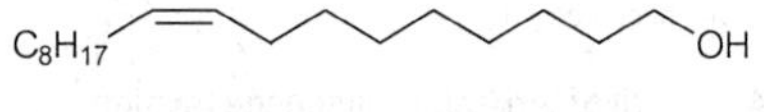

C₁₈H₃₆O
9-Octadecen-1-ol, (Z)-.
Adol® 34; Adol® 80; Adol® 85; Adol® 90; Adol® 320; Adol® 330; Adol® 340; AI3-07620; Atalco O; Cachalot O-1; Cachalot O-3; Cachalot O-8; Cachalot O-15; Conditioner 1; Crodacol A.10; Crodacol-O; Dermaffine; EINECS 205-597-3; H.D. eutanol; HD-Ocenol 90/95; HD-Ocenol K; HD oleyl alcohol 70/75; HD oleyl

alcohol 80/85; HD oleyl alcohol 90/95; HD oleyl alcohol CG; HSDB 6484; Lancol; Loxanol 95; Loxanol M; Novol; NSC 10999; Ocenol; Oceol; Octadecenol; Oleic alcohol; Oleo alcohol; Oleol; Oleoyl alcohol; Oleyl alcohol; Olive alcohol; Satol; Sipol O; Siponol OC; Unjecol 50; Unjecol 70; Unjecol 90; Unjecol 110; Witcohol 85; Witcohol 90. Super refined oleyl alcohollient, emulsion stabilizer, superfatting agent, pigment suspending aid, used in cosmetics, personal care products; lipsticks, sunscreens, antiperspirants, bath oils. Used as an emulsifier, lubricant, foam control agent, cosolvent, plasticizer and emollient. Oil; mp = 6.5°; bp13 = 207°; d^{20} = 0.8489; insoluble in H_2O, slightly sluble in CCl_4, soluble in EtOH, Et_2O. *Croda; Lanaetex; Michel M.; Ronsheim & Moore; Sherex.*

2803 Oleyl hydroxyethyl imidazoline;
95-38-5   407   G

$C_{22}H_{42}N_2O$
2-(8-Heptadecenyl)-4,5-dihydro-1H-imidazole-1-ethanol.
5-23-05-00319 (Beilstein Handbook Reference); Amine 220®; BRN 0025816; EINECS 202-414-9; 2-(2-Heptadec-8-enyl-2-imidazolin-1-yl)ethanol; 2-(8-Heptadecenyl)-4, 5-dihydro-1H-imidazole-1-ethanol; 2-(8-Heptadec-enyl)-2-imidazoline-1-ethanol; 2-(8-Heptadecenyl)-4,5-dihyd-ro-1H-imidazole-1-ethanol; 1-(2-Hydroxyethyl)-2-hepta-decenylglyoxalidine; 1-(2-Hydroxyethyl)-2-n-hepta-dec-enyl-2-imidazoline; 1-(2-Hydroxyethyl)-2-(8-hepta-decen-yl)-2-imidazoline; 1-(2-Hydroxyethyl)-2-heptadec-enyl-2-imidazoline; 1-Hydroxyethyl-2-heptadecenyl-glyoxalid-ine; 1H-Imidazole-1-ethanol, 2-(8-heptadec-enyl)-4,5-di-hydro-; 2-Imidazoline-1-ethanol, 2-(8-hepta-decenyl)-; Marlowet® 5440; Nalcamine G-13; NSC 231649; Oleyl imidazoline; Sovatex IM17H; UCL 5410. Used as an emulsifier for mineral oils, corrosion protection, car wash rinses. Also used as a fungicide and soil stabilizer, corrosion inhibitor; lubricant; antistatic agent; as a base for cationic surface active agents. See USP 2,987,515; 3,020,276. *Baker Petrolite Corp.; Hüls Am.*

2804 Oleyl oleate
3687-45-4   G

$C_{36}H_{68}O_2$
cis 9,10 octadecenyl-cis-9,10-octadecenoate.
EINECS 222-980-0; (Z)-Octadec-9-enyl oleate; 9-Octadecenoic acid, 9-octadecenyl ester; 9-Octadecenoic acid (Z)-, 9-octadecenyl ester, (Z)-; 9-Octadecenoic acid (9Z)-, (9Z)-9-octadecenyl ester; Cis 9,10 octadecenyl cis 9,10 octadecanoate; Oleyl oleate; Wickenol® 143. Lubricant used in cosmetic and toilet preparations; replacement for sperm oil in addition to functioning as cosmetic additive. *CasChem.*

2805 Oleyl palmitamide
16260-09-6   G

$C_{34}H_{67}NO$
Hexadecanamide, N-9-octadecenyl-.
EINECS 240-367-6; Hexadecanamide, N-(9Z)-9-octadecenyl-; Hexadecanamide, N-9-octadecenyl-, (Z)-; N-9-Octadecenyl hexadecanamide; (Z)-N-Octadec-9-enylhexadecan-1-amide; Oleyl palmitamide. Release agent providing slip, antiblocking to thermoplastics incl. PP film, nylon. *Croda; Witco/Humko.*

2806 N-Oleyl-1,3-propanediamine
7173-62-8   H

$C_{21}H_{44}N_2$
1,3-Propanediamine, N-9-octadecenyl-, (Z)-.
EINECS 230-528-9; (Z)-N-9-Octadecenylpropane-1,3-diamine; 1,3-Propanediamine, N-(9Z)-9-octadecenyl-.

2807 Oleylamide
301-02-0   G H

$C_{18}H_{35}NO$
9-Octadecenamide, (Z)-.
Adogen 73; AI3-36742; Armoslip CP; Crodamide O; Crodamide OR; EINECS 206-103-9; HSDB 5560; NSC 26987; Oleamide; Oleic acid amide; Oleyl amide; Polydis® TR 121; Slip-eze; Unislip 1759. Lubricant, slip and antiblock agent for extrusion of polyethylene; wax additive; ink additive. A brain lipid that induces physiological sleep when injected into rats. This lipid may represent a new class of biological signaling molecules. *Akzo Chemie; Chemax; Chemron; Croda; Henkel/Emery; Mona; Syn Prods.; Unichema; Witco/Humko.*

2808 Oleylamine, ethoxylated
26635-93-8   H

$(C_2H_4O)_x.(C_2H_4O)_y$-$C_{22}H_{45}NO_2$
Glycols, polyethylene, (9-octadecenylimino)diethylene ether, (Z)-.
N,N'-Bis(polyoxyethylene)oleylamine; Ethoxylated oleyl-amine; Glycols, polyethylene, (9-octadecenylimino)-diethylene ether, (Z)-; α,α'-((9-Octadecenylimino)di-2,1-ethanediyl)bis(omega-hydroxypoly(oxy-1,2-ethanediyl)-,(Z); N-Polyoxyethylated-N-oleylamine hydrochloride; Oleylamine, ethoxylated.

2809 Omeprazole
73590-58-6   6913   D

$C_{17}H_{19}N_3O_3S$
5-Methoxy-2-[[(4-methoxy-3,5-dimethyl-2-pyridyl)-methyl]sulfinyl]benzimidazole.
Antra; Audazol; Aulcer; Belmazol; CCRIS 7099; Ceprandal; Demeprazol; Desec; Dudencer; Elgam; Emeproton; Epirazole; Erbolin; Exter; Gasec; Gastrimut; Gastroloc; Gibancer; H 168/68; HSDB 3575; Indurgan; Inhibitron; Inhipump; Lensor; Logastric;

Lomac; Losec; Mepral; Miol; Miracid; Mopral; Morecon; Nilsec; Nopramin; Ocid; Olexin; Omapren; Omed; Omegast; OMEP; Omepral; Omeprazen; Omeprazol; Omeprazole; Omeprazolum; Omeprazon; Omeprol; Omesek; Omezol; Omezolan; Omid; Omisec; Omizac; OMP; Ompanyt; OMZ; Ortanol; Osiren; Ozoken; Paprazol; Parizac; Pepticum; Pepticus; Peptilcer; Prazentol; Prazidec; Prazolit; Prilosec; Procelac; Proclor; Prysma; Ramezol; Regulacid; Sanamidol; Secrepina; Tedec Ulceral; Ulceral; Ulcesep; Ulcometion; Ulcozol; Ulcsep; Ulsen; Ultop; Ulzol; Victrix; Zefxon; Zepral; Zimor; Zoltum. Gastric antisecretory agent (proton pump inhibitor). Used in treatment of Zollinger-Ellison syndrome. Antiulcerative. Crystals; mp = 156°; LD$_{50}$ (mus iv) = 80 mg/kg, (mus orl) > 4000 mg/kg, (rat iv) > 50 mg/kg, (rat orl) > 4000 mg/kg. *Astra Hassle AB; Astra Sweden; Merck & Co.Inc.*

2810 Omeprazole Sodium
95510-70-6   D

C$_{17}$H$_{18}$N$_3$NaO$_3$S
5-Methoxy-2-[[(4-methoxy-3,5-dimethyl-2-pyridyl)methyl]sulfinyl]benzimidazole sodium salt.
Andra; H 168/68 sodium; H 168/68; Losec Sodium; Omeprazole sodium. Gastric antisecretory agent (proton pump inhibitor). Antiulcerative. *Astra Hassle AB; Astra Sweden; Merck & Co.Inc.*

2811 Omethoate
1113-02-6   G P

C$_5$H$_{12}$NO$_4$PS
O,O-Dimethyl S-[2-(methylamino)-2-oxoethyl] phospho-rothioate.
Dimethoate-met; Bay 45432; Folimat®; S-6876. Systemic insecticide and acaricide with contact and stomach action. Used for control of spider mites, aphids, beetles, caterpillars, scale insects, thrips, suckers, fruit flies etc. in fruit and vegetable crops and in forestry. Registered as an insecticide by EPA (cancelled). Yellow oil; dec. 135°; d^{20} = 1.32; readily soluble in H$_2$O and most organic solvents; LD$_{50}$ (rat, rbt, cat orl) = 50 mg/kg, (gpg orl) = 100 mg/kg, (rat der) = 700 mg/kg, (hen orl) = 125 mg/kg; LC$_{50}$ (mrat ihl 1 hr.) > 15 mg/l air, (goldfish 96 hr.) = 10 - 100 mg/l; toxic to bees. *Bayer AG; Bayer Corp.; Bayer plc.*

2812 Orange Oil
8008-57-9   E D H

sweet orange oil.
Absolue orange flower from water; Absolue orange flower decoloree; Absolue orange flower; Absolute orange flowers; Absolute petitgrain; Caswell No. 425A; CCRIS 683; Citrus sinensis Oil; Citrus sinensis peel oil; EPA Pesticide Chemical Code 040517; EPA Pesticide Chemical Code 040501; FEMA No. 2822; HSDB 1934; Neat oil of sweet orange; Neroli oil; Neroli oil, pommade; Oil of orange; Oil of sweet orange; Oils, orange; Oils, orange, sweet; Orange flower absolute; Orange flower oil; Orange flower water absolute; Orange leaf oil; Orange oil; Orange oil concentrate; Orange oil, sweet, expressed; Orange oil, sweet; Orange oil, terpeneless (Citrus sinensis (L.) Osbeck); Orange peel oil, sweet; Orange sweet oil, expressed; Petitgrain bigarade sur fleurs d'oranger; Petitgrain citronnier oil; Sweet orange oil; Sweet orange oil, terpeneless. FDA, FEMA GRAS, Europe listed, no restrictions, FDA approved for orals, BP compliance. Used as a flavoring agent, a carminative, treatment for stomach disorders, expectorant and tonic. Used in orals. Orange liquid; d$_{25}^{25}$ = 0.842 - 0.846; [α]$_D^{25}$= +84° - +99°; slightly soluble in H$_2$O, soluble in EtOH, CS$_2$, AcOH. *Caminiti Foti & Co. Srl; Chr Hansen's Lab.; Florida Treatt; Ibrahim N. I.; Ruger.*

2813 Orcinol
504-15-4   6932   G

C$_7$H$_8$O$_2$
1,3-Dihydroxy-5-methylbenzene.
4-06-00-05892 (Beilstein Handbook Reference); Al3-23954; 1,3-Benzenediol, 5-methyl-; BRN 1071903; 1,3-Dihydroxy-5-methylbenzene; 3,5-Dihydroxytoluene; EINECS 207-984-2; 3-Hydroxy-5-methylphenol; 5-Methyl-1,3-benzenediol; 5-Methylresorcin; 5-Methyl-resorcinol; NSC 12441; Orcin; Orcinol; Resorcinol, 5-methyl-. Reagent for beet sugar, lignin, and pentoses. Prisms or leaflets; mp = 107.5°; bp = 289.5°; λ_m = 275, 281 nm (ε = 1510, 1490, MeOH); slightly soluble in ligroin, petroleum ether, soluble in H$_2$O, EtOH, Et$_2$O, C$_6$H$_6$; LD$_{50}$ (rat orl) = 844 mg/kg.

2814 Orpiment
1303-33-9   815   G

As$_2$S$_3$.
Arsenic trisulfide.
Al3-01006; Arsenic Red; Arsenic sesquisulfide; Arsenic sesquisulphide; Arsenic sulfide; Arsenic sulfide (As$_2$S$_3$), (yellow); Arsenic sulfide (As$_2$S$_3$); Arsenic sulfide Yellow; Arsenic sulphide; Arsenic tersulfide; Arsenic tersulphide; Arsenic trisulfide; Arsenic Trisulphide; Arsenic Yellow; Arsenious sulfide; Arsenious sulphide; Arsenous sulfide; Auripigment; Auripigmentum; C.I. 77086; C.I. Pigment Yellow 39; Cl 77086; Diarsenic trisulfide; Diarsenic trisulphide; EINECS 215-117-4; HSDB 428; King's Gold; NA1557; Pigment Yellow 39; Sekio; Shio; Yellow arsenic sulfide. Used in manufacture of glass, oilcloth and linoleum, in semi- and photoconductors, as a pigment and depilatory. Orange powder; mp = 300-325°; insoluble in H$_2$O. *Atomergic Chemetals.*

2815 Ortholate
88-41-5   G

C$_{12}$H$_{22}$O$_2$
2-t-Butylcyclohexanol acetate.
1-Acetoxy-2-t-butylcyclohexane; 2-t-Butylcyclohexyl acetate; o-t-Butylcyclohexyl acetate; Cyclohexanol, 2-(1,1-dimethylethyl)-, acetate; 2-(1,1-Dimethylethyl)-cyclo-hexanol acetate; 2-(1,1-Dimethylethyl)cyclohexyl acetate; EINECS 201-828-7; Green acetate;; Grumex; Ortholate; Verdox; Ylanat ortho. Used in

perfumery. Clear yellow oil with a fresh fruity woody green apple odor. Solid; mp = 23 - 27°; bp1.5: = 98°; d^{25} = 0.937; LD$_{50}$ (rbt der) >5000 mg/kg, (rat orl) = 4600 mg/kg. *Privi Organics Ltd.; Quest Intl.; Quest UK; Quest USA.*

2816   Oryzalin
19044-88-3   6953   G H P

C$_{12}$H$_8$N$_4$O$_6$S
  4-(Dipropylamino)-3,5-dinitrobenzenesulfonamide.
  Benzenesulfonamide, 4-(dipropylamino)-3,5-dinitro-; BRN 2177305; Caswell No. 623A; Compound 67019; DF; Dinitrodipropylsulfanilamide; 3,5-Dinitro-N4,N4-dipropylsulfanilamide; 3,5-Dinitro-N4,N4-dipropylsulph-anilamide; 3,5-Dinitro-N^4,N^4-dipropylsulfanilamide; 3,5-Dinitro-N',N'-dipropylsulfanilamide; 4-(Dipropylamino)-3,5-dinitrobenzenesulfonamide; Dirimal; Dirimal Extra; EL-119; EPA Pesticide Chemical Code 104201; HSDB 6858; N4,N4-dipropyl-3,5-dinitrosulfanilamide; Oryzalin; Rycelan; Ryzelan; Sulfanilamide, 3,5-dinitro-N(sup 4),N(sup 4)-dipropyl-; Surflan; Surflan 4AS; Surflan AS; Surflan. Selective herbicide; inhibits cell division and germination. Used for pre-emergence control of annual grasses. Used for pre-emergence control of annual grasses and used on ornamentals, trees, roses, flower beds, bulbs, and warm season turf; controls annual grasses any many broadleaf weeds; may be tank-mixed with Roundup. Registered by EPA as a herbicide. Yellow-orange crystals; mp = 141-142°; slightly soluble in H$_2$O (0.00025 g/100 ml), soluble in Me$_2$CO (> 50 g/100 ml), methyl Cellosolve (> 50 g/100 ml), CH$_3$CN (> 15 g/100 ml), MeOH (5 g/100 ml), CH$_2$Cl$_2$ (> 3 g/100 ml), C$_6$H$_6$ (0.4 g/100 ml), xylene (0.2 g/100 ml), insoluble in C$_6$H$_{14}$; LD$_{50}$ (rat, mus orl) > 10000 mg/kg, (cat, dog orl) > 1000 mg/kg, (rbt der) > 2000 mg/kg, (ckn orl) > 1000 mg/kg, (bee orl) = 0.011 mg.bee; (bobwhite quail, mallard duck orl) > 500 mg/kg, LC$_{50}$ (bluegill sunfish) = 2.88 mg/l, (rainbow trout) = 3.26 mg/l. *Dow AgroSciences; Lawn & Garden Products Inc.; Lebanon Seaboard Corporation.*

2817   Osmium tetroxide
20816-12-0   6962   G

OsO$_4$
  Osmic acid.
  EINECS 244-058-7; HSDB 719; Osmic acid; Osmic acid anhydride; Osmium oxide (OsO4), (β-4)-; Osmium tetraoxide; Osmium tetroxide; Perosmic acid anhydride; Perosmic oxide; RCRA waste number P087; UN2471; perosmic oxide. Used as a microscopic stain, in photography, as a oxidation catalyst in organic synthesis. Pale yellow crystals; mp = 40°; bp = 130°; soluble in H$_2$O (7.24 g/100 g), more soluble in organic solvents; highly toxic. *Atomergic Chemetals; Degussa AG; Janssen Chimica; Sigma-Aldrich Fine Chem.; Spectrum Chem. Manufacturing.*

2818   Osyrol
41890-92-0   G

C$_{11}$H$_{24}$O$_2$
  3,7-Dimethyl-7-methoxyoctan-2-ol.
  BRN 1921024; Dihydromethoxyelgenol; 3,7-Dimethyl-7-methoxy-2-octanol; EINECS 255-574-7; Elesant; 7-Methoxy-3,7-dimethyloctan-2-ol; Methoxyelgenol; 2-Octanol, 3,7-dimethyl-7-methoxy-; 2-Octanol, 7-meth-oxy-3,7-dimethyl-; Osirol. Used in perfumery. *Bush Boake Allen* (IFF bought 2000).

2819   Oxadiazon
19666-30-9   6975   G P

C$_{15}$H$_{18}$Cl$_2$N$_2$O$_3$
  2-t-Butyl-4-(2,4-dichloro-5-isopropoxyphenyl)-Δ^2-1,3,4-oxadiazoline-5-one.
  BRN 0558070; 2-t-Butyl-4-(2,4-dichloro-5-isopropoxy-phenyl)- Δ^2-1,3,4-oxadiazoline-5-one; 2-t-Butyl-4-(2,4-di-chloro-5-isopropyloxyphenyl)-1,3,4-oxadiazolin-5-one; 5-t-Butyl-3-(2,4-dichloro-5-isopropoxyphenyl)-1,3,4-oxa-diazol-2(3H)-one; Caswell No. 624A; 3-(2,4-Dichloro-5-(1-methylethoxy)-phenyl)-5-(1,1-dimethylethyl)-1,3,4-oxadiazol-2(3H)-one; 3-(2,4-Dichloro-5-isopropyloxy-phenyl)-Δ^4-5-(tert-butyl)-1,3,4-oxa-diazoline-2-one; EINECS 243-215-7; EPA Pesticide Chemical Code 109001; G 315; HSDB6936; 1,3,4-Oxadiazol-2(3H)-one,3-(2,4-dichloro-5-(1-methylethoxy)phenyl)-5-(1,1-di-methylethyl)-; Oxadiazon; Oxadiazone; Oxy-diazon; Ronstan; Ronstar; Ronstar 2G; Ronstar 50W; RP 17623; RP-17623; Scotts OH I; Δ^2-1,3,4-Oxadiazolin-5-one, 2-t-butyl-4-(2,4-dichloro-5-isopropyloxyphenyl)-. Selective systemic herbicide. Registered by EPA as a herbicide. White crystals; mp = 88 - 90°; insoluble in H$_2$O (0.00007 g/100 ml), soluble in EtOH (10 g/100 ml), MeOH (10 g/100 ml), cyclohexanone (20 g/100 ml), Me$_2$CO, acetophenone, anisole, isophorone, MEK (60 g/100 ml), CCl$_4$, C$_6$H$_6$, CHCl$_3$, CH$_2$Cl$_2$, C$_7$H$_8$ (100 g/100 ml); LD$_{50}$ (rat orl) > 8000 mg/kg, (rat der) > 8000 mg/kg, (mallard duck orl) > 1000 mg/kg, (bobwhite quail orl) = 6000 mg/kg; LC$_{50}$ (rat ihl) > 200 mg/l air, (carp 96 hr.) = 9 - 15.4 mg/l; toxic to bees. *Aventis Crop Science; Aventis Environmental Science USA LP; Aventis ES Turf & Ornamental; Earth Care; Regal Chemical Co.; Rhône Poulenc Crop Protection Ltd.; Rhône-Poulenc Environmental Prods. Ltd; Scotts Co.; UAP - Platte Chemical; Wilbro Inc.*

2820   Oxadixyl
77732-09-3   6976   G P

C14H18N2O4

N-(2,6-Dimethylphenyl)-2-methoxy-N-(2-oxo-3-oxazolid-inyl)acetamide.

Acetamide, N-(2,6-dimethylphenyl)-2-methoxy-N-(2-oxo-3-oxazolidinyl)-; N-(2,6-Dimethylphenyl)-2-methoxy-N-(2-oxo-3-oxazolidinyl)acetamide; M 10797; 2-Methoxy-N-(2-oxo-1,3-oxazolidin-3-yl)acet-2',6'-xylidide; 2-Meth-oxy-N-(2-oxo-1,3-oxazolidine-3-yl)-acet-2,6-xylidine; Oxadixyl; Recoil; Ripost; SAN 371; SAN 371F; Sandofan; Wakil. Systemic fungicide with protective and curative action. Used in combination with contact fungicides (e.g. mancozeb; captofol etc.) for control of downy mildews, late blights and rusts in vines, potatoes, maize, tobacco, hops, sunflowers, citrus, fruits and vegetables. Registered by EPA as a fungicide (cancelled). Colorless crystals; mp = 104 - 105°; soluble in H2O (0.34 g/100 ml at 25°), Me2CO (27.1 g/100 ml), DMSO (42.9 g/100 ml), EtOH (3.9 g/100 ml), xylene (1.53 g/100 ml), Et2O (0.42 g/100 ml); LD50 (mrat orl) = 3480 mg/kg, (frat orl) = 12860 mg/kg, (mmus orl) = 1860 mg/kg, (fmus orl) = 2150 mg/kg, (rat der) > 2000 mg/kg, (rbt der) > 2000 mg/kg, (mrat ip) = 490 mg/kg, (frat ip) = 550 mg/kg, (mallard duck orl) > 2510 mg/kg, (bee contact) > 0.1 mg/bee, (bee orl) > 0.2 mg/bee; LC50 (mallard duck, Japanese quail 8 day dietary) > 5620 mg/kg diet, (carp 96 hr.) > 300 mg/l, (rainbow trout 48 hr.) > 320 mg/l, (bluegill sunfish 48 hr.) = 360 mg/l, (*Daphnia* 48 hr.) = 530 mg/l. *Syngenta Crop Protection.*

2821   Oxalic acid
144-62-7   6980   G P

C2H2O4

Ethane-1,2-dioic acid.

4-02-00-01819 (Beilstein Handbook Reference); Acide oxalique; Acido ossalico; Acidum oxalicum; AI3-26463; Aktisal; Aquisal; BRN 0385686; Caswell No. 625; CCRIS 1454; EINECS 205-634-3; EPA Pesticide Chemical Code 009601; Ethanedioic acid; HSDB 1100; Kyselina stavelova; NCI-C55209; NSC 62774; Oxaalzuur; Oxalate; Oxalic acid; Oxalsäure; Oxalsaeure; Oxiric acid. Automobile radiator cleanser, metal and equipment cleaning, purifying agent, intermediate, leather tanning, catalyst, laboratory reagent, stripping agent, textile bleaching, rare-earth processing, printing and dyeing auxiliary. Used for control of varroa in honeybees and as a rust preventative, wood preservative and cleaner. Registered by EPA as an antimicrobial and disinfectant (cancelled). Orthorhombic crystals; mp = 189.5°; sublimes 157°; d^{17} = 1.900; soluble in H2O (6.7 g/100 ml at 15°, 8.3 g/100 ml at 20°, 9.8 g/100 ml at 25°. [dihydrate]: monoclinic prisms; mp = 101 - 102°; $d^{18.5}$ 1.653; pK1 = 1.27, pK2 = 4.238; pH of 0.1M solution = 1.3; soluble in H2) (14.3 g/100 ml at 25°, 50 g/100 ml at 100°), EtOH (40 g/100 ml at 20°, 55.5 g/100 ml at 76°), Et2O (1 g/100 ml), glycerol (18.2 g/100 ml), insoluble in C6H6, CHCl3, petroleum ether; LD50 (rat orl) = 0.475 mg/kg. *Ashland; Fluka; General Chem; Hoechst Celanese; La Littorale; Mallinckrodt Inc.; Mitsubishi Gas.*

2822   Oxalic acid dihydrate
6153-56-6   G

2H2O

C2H2O4.2H2O

Ethanedioic acid, dihydrate.

Ethandionic acid, dihydrate; Ethanedionic acid dihydrate. Crystals;

mp = 101°; d^{18} = 1.653; slightly soluble in Et2O, soluble in H2O, EtOH..

2823   Oxamniquine
21738-42-1   6988   D G

C14H21N3O3

1,2,3,4-Tetrahydro-2-[[(1-methylethyl)amino]methyl]-7-nitro-6-quinolinemethanol.

5-22-11-00475 (Beilstein Handbook Reference); BRN 0485597; CCRIS 4113; EINECS 244-556-4; HSDB 6510; Mansil; NSC 352888; Oxaminiquine; Oxamniquina; Oxamniquine; Oxamniquinum; UK 4271; Vansil. Anthelmintic. Targets schistosoma. Crystals; mp = 147-149°; soluble in Me2CO, CHCl3, MeOH, H2O (0.03 g/100 ml); λ_m = 205.5, 249.5, 389.5 (A $^{1\%}_{1cm}$ 486, 695, 62.5 MeOH); LD50 (mus im) > 2000 mg/kg, (mus orl) = 1300 mg/kg, (rbt im) > 1000 mg/kg, (rbt orl) = 800 mg/kg. *Pfizer Inc.*

2824   Oxamyl
23135-22-0   6989   G H P

C7H13N3O3S

N',N'-Dimethyl-N-[(methylcarbamoyl)oxy]-1-thiooxam-imidic   acid methyl ester.

BRN 2050910; Caswell No. 561A; CCRIS 1963; D-1410; 2-(Dimethylamino)-N-(((methylamino)carbonyl)oxy)-2-oxoethanimidothioic acid methyl ester; 2-Dimethyl-amino-1-(methylthio)glyoxal O-methylcarb-amoylmon-oxime; N',N'-Dimethyl-N-((methylcarbamoyl)-oxy)-1-thiooxamimidic acid methyl ester; N',N'-Dimethylcarbamoyl(methylthio)methylenamine N-methylcarbamate; N,N-Dimethyl-α-methylcarbamoyl-oxyimino-α-(methylthio)acetamide; N,N-Dimethyl-2-methylcarbamoyloxyimino-2-(methylthio)acetamide; Dioxamyl; DPX 1410; DPX 1410L; Du Pont 1410; DuPont 1410; EINECS 245-445-3; EPA Pesticide Chemical Code 103801; Ethanimidothioic acid, 2-(dimethylamino)-N-(((methylamino)carbonyl)oxy)-, methyl ester; Ethanimidothioic acid, 2-(dimethylamino)-N-(((methylamino)carbonyl)oxy)-2-oxo-, methyl ester; Formidic acid, 1-(dimethylcarbamoyl)-N-((methylcarb-amoyl)oxy)thio-, methyl ester; HSDB 6453; Insecticide-nematicide 1410; Methyl 2-(dimethylamino)-N-(((methylamino)carbonyl)oxy)-2-oxoethanimidothioate; Methyl 1-(dimethylcarbamoyl)-N-(methylcarbamoyloxy)-thioformimidate; S-Methyl 1-(dimethylcarbamoyl)-N-((methylcarbamoyl)oxy)thioformimidate; S-Methyl N',N'-dimethyl-N-(methylcarbamoyloxy)-1-thio-oxamimidate; Methyl N',N'-dimethyl-N-((methylcarbamoyl)oxy)-1-thio-oxamimidate; NSC 379588; Oxamil; Oxamimidic acid, N',N'-dimethyl-N-((methylcarbamoyl)oxy)-1-thio-, methyl ester; Oxamimidic acid, N',N'-dimethyl-N-((methyl-carbamoyl)oxy)-1-methylthio-; Oxamyl; RCRA waste no. P194; Thioxamyl; Vydate; Vydate-G; Vydate K; Vydate L; Vydate L insecticide/nematicide; Vydate L oxamyl insecticide/nematicide. Contact and systemic insecticide, acaricide and nematicide. Cholinesterase inhibitor. Absorbed by roots and foliage with translocation. Used for control of nematodes in ornamentals, fruit trees, vegetables, curcubits, beet, bananas, pineapples, groundnuts, cotton, soybeans, tobacco, potatoes and other crops. Registered by EPA as an acaricide and insecticide.

Used as a nematicide. Crystals; mp = 108-110°; sg^{25} = 0.97; soluble in H_2O (28 g/100 ml), MeOH (115 g/100 ml), EtOH (26 g/100 ml), i-PrOH (8.6 g/100 ml), Me$_2$CO (52.8 g/100 ml), C$_7$H$_8$ (0.86 g/100 ml); LD$_{50}$ (rat orl) = 5.4 mg/kg, (rbt der) = 2960 mg/kg, (bobwhite quail orl) = 4.18 mg/kg; LC$_{50}$ (bluegill sunfish 96 hr.) = 5.6 mg/l, (goldfish 96 hr.) = 27.5 mg/l, (rainbow trout 96 hr.) = 4.2 mg/l; toxic to bees. *DuPont.*

2825   Oxantel Pamoate
68813-55-8   6991   D G

C$_{36}$H$_{32}$N$_2$O$_7$
(E)-m-[2-(1,4,5,6-Tetrahydro-1-methyl-2-pyrimidinyl)-vinyl]phenol
4,4'-methylenebis[3-hydroxy-2-naphthoate] (1:1) (salt).
CP 14,445-16; EINECS 272-332-6; Oxanel pamoate; Oxantel ebonate; Telopar. Anthelmintic. Targets nematodes. *Pfizer Inc.*

2826   Oxazolidine
497-25-6   G

C$_3$H$_5$NO$_2$
2-oxazolidine.
4-27-00-02516 (Beilstein Handbook Reference); AI3-38980; BRN 0106251; Carbamic acid, (2-hydroxyethyl)-, gamma-lactone; EINECS 207-840-9; NSC 35382; 2-Oxazolidinone; 1,3-Oxazolidin-2-one; Oxazolidone; 2-Oxazolidone. Used as a preservative and antibacterial agent in water-based paints, latexes, emulsions, metalworking fluids; for oilfield water-flooding operations; corrosion inhibitor; crosslinking agent, catalyst for resin systems. Crystals; mp = 86-89°; bp$_{48}$ = 220°. *Whittaker Clark & Daniels.*

2827   Oxfendazole
53716-50-0   7004   G V

C$_{15}$H$_{13}$N$_3$O$_3$S
[5-(Phenylsulfinyl)-1H-benzimidazol-2-yl]carbamic acid methyl ester.
2-Benzimidazolecarbamic acid, 5-(phenylsulfinyl)-, methyl ester; BRN 0761290; Carbamic acid, (5-(phenylsulfinyl)-1H-benzimidazol-2-yl)-, methyl; EINECS 258-714-5; Fenbendazole sulfoxide; HOE 8105; Methyl 5-(phenylsulfinyl)-1H-benzimidazol-2-yl carbamate; Methyl 5-(phenylsulfinyl)-2-benzimidazolecarbamate; OFDZ; Oxfendazol; Oxfendazole; Oxfendazolum; (5-(Phenylsulfinyl)-1H-benzimidazol-2-yl)carbamic acid methyl ester; 5-(Phenylsulfinyl)-2-

benzimidazolecarbamic acid methyl ester; 5-Phenylsulfinyl-2-carbomethoxy-aminobenzimidazole; Repidose; RS 8858; Synanthic; Synanthic (Veterinary); Systamex; Systemax. Veterinary anthelmintic. Crystals; mp = 253° (dec); LD$_{50}$ (rat) > 6400 mg/kg.

2828   2-Oxogluconic acid
669-90-9   H

C$_6$H$_{10}$O$_7$
D-arabino-2-Hexulosonic acid.
D-arabino-Hexulosonic acid; D-arabino-2-Hexulosonic acid; EINECS 211-574-9; 2-Oxogluconic acid.

2829   5-Oxoproline
98-79-3   8091   G

C$_5$H$_7$NO$_3$
L-Pyroglutamic acid.
Acide pidolique; Acido pidolico; Acidum pidolicum; Ajidew A-100; 2-Benzothiazolesulfenemorpholide; 2-Benzothiazolesulfenic acid morpholide; 5-Carboxy-2-pyrrolidinone; L-5-Carboxy-2-pyrrolidinone; EINECS 202-700-3; L-Glutamic Acid Gamma-lactam; Glutimic acid; L-Glutimic Acid; L-Glutiminic acid; L-Glutiminic Acid; NSC 143034; Oxoproline; 5-Oxo-l-proline; L-5-Oxoproline; (5S)-2-Oxopyrrolidine-5-carboxylic Acid; (S)-5-Oxo-2-pyrrolidinecarboxylic Acid; 2-Oxopyrrolidine-5-carbox-ylic Acid; 5-Oxo-2-pyrrolidinecarboxylic Acid; L-5-Oxo-2-pyrrolidinecarboxylic Acid; PCA; Pidolic acid; Proline, 5-oxo-; Proline, 5-oxo-, l-; L-Proline, 5-oxo-; Pyro-glutamic acid; L-Pyroglutamic acid; (S)-Pyroglutamic Acid; Pyrrolidinonecarboxylic acid; Pyrrolidone-carboxylic acid; Pyrrolidone-5-carboxylic acid; (-)-2-Pyrrolidone-5-carboxylic Acid; (S)-(-)-2-Pyrrolidone-5-carboxylic Acid; 2-Pyrrolidinone-5-carboxylic Acid; 2-l-Pyrrolidone-5-carboxylic Acid; 2-Pyrrolidone-5-carbox-ylate; 2-Pyrrolidone-5-carboxylic Acid; 5-Pyrrolidinone-2-carboxylic Acid; L-Pyrrolidinonecarboxylic acid; L-2-Pyrrolidone-5-carboxylic Acid; L-Pyrrolidonecarboxylic acid. A natural humectant used in cosmetics, soaps, dentifrices, medicinal supplies, tobacco, cellulose film, paper products, fiber products, paints; additive to dyeing agent, softening agent, finishing agent, and antistatic agent; Crystals; mp = 184.7°; soluble in H_2O (10 g/100 ml), DMSO; $[\alpha]_D^{20}$ = -8.7° (c = 13, H_2O); pH 1.8-2.2. *Ajinomoto Co. Inc.; Ajinomoto USA Inc.*

2830   Oxyanthracene
529-86-2   691   G

C$_{14}$H$_{10}$O
9-Anthranol.
4-06-00-04930 (Beilstein Handbook Reference); Anthracene, 9-hydroxy-; 9-Anthracenol; Anthranol; 9-Anthrol; BRN 1869102; 9-Hydroxy-anthracene; NSC 39886. Used in the manufacture of dyestuffs. Yellow-red leaflets; mp = 152°.

2831   Oxybenzone

131-57-7   7023   D G H

C14H12O3
2-Hydroxy-4-methoxybenzophenone.
4-08-00-02442 (Beilstein Handbook Reference); Advastab 45; AI3-23644; Anuvex; Benzophenone, 2-hydroxy-4-methoxy-; Benzophenone-3; BRN 1913145; CCRIS 1078; Chimassorb 90; Cyasorb® UV 9; Cyasorb UV 9 Light Absorber; DuraScreen; EINECS 205-031-5; Escalol 567; HMBP; HSDB 4503; Methanone, (2-hydroxy-4-methoxyphenyl)phenyl-; MOB; MOD; Neo Heliopan® BB; NCI-C60957; NSC-7778; Ongrostab HMB; Oxibenzona; Oxibenzonum; Oxybenzon; Oxybenzone; Oxybenzonum; PreSun 15; PreSun 46; Rhodialux A; Solaquin; Spectra-sorb UV 9; Sunscreen UV-15; Syntase 62; UF 3; USAF CY-9; UV 9; Uvinul® 9; Uvinul® M-40; Uvistat 24. UV absorber; similar to Uvinul 400 except for higher solubility in aromatic solvents; good weather resistance in resins and plastics; stabilizes PVC and polyesters against UV-light degradation; used in NC lacquers, varnishes, and oil-based paints. Light stabilizer and UV absorber for plastics and coatings; especially for flexible and rigid PVC, unsaturated polyesters, and acrylics; used in outdoor sheeting and glazing applications, molded products, adhesives. UV-A and UV-B broad spectrum absorber for sunscreen formulations. mp = 65.5°; bp5 = 150-160°; λ_m = 326 nm (ε = 9772, EtOH), 242, 284, 324 nm (ε = 10233, 15136, 10233, MeOH); insoluble in H2O (<0.1 g/100 ml), soluble in CCl4; LD50 (rat orl) > 12800 mg/kg. *Am. Cyanamid; BASF Corp.; Carrington Labs Inc.; Haarmann & Reimer GmbH; ICN Pharm. Inc.; Rhône-Poulenc; Schwarz Pharma Kremers Urban Co.; Westwood-Squibb Pharm. Inc.*

2832   Oxybisbenzenesulfonic acid dihydrazide

80-51-3   H

C12H14N4O5S2
4,4'-Oxybisbenzenesulfonic acid dihydrazide.
4-11-00-00597 (Beilstein Handbook Reference); Benzenesulfonic acid, 4,4'-oxybis-, dihydrazide; Benzenesulfonic acid, oxybis-, dihydrazide; BRN 2954604; Cellmic S; Celmike S; Celogen; Celogen OT; Cenitron OB; Diphenyl ether 4,4'-disulfohydrazide; Diphenyloxide-4,4'-disulfohydrazide; EINECS 201-286-1; Genitron ob; HSDB 5237; NSC 5318; OBSH; Oxybis(benzenesulfonylhydrazide); Oxybisbenzene-sulf-onic acid dihydrazide; Serogen; Zhenitron OV. A low-temperature sulfonylhydrazide used in low process temperature plastics, such as LDPE, EVA, and soft vinyl compounds. It decomposes at 320°F (160°C). Blowing agent for sponge rubber and expanded plastics (LDPE wire/cable, structural foam injection moldings, and rotational casting, flexible PVC structural foam injection molding). Dec 160°. *Purecha Group; Uniroyal.*

2833   Oxybutynin

5633-20-5   7024   D

C22H31NO3
4-(Diethylamino)-2-butynyl α-phenylcyclohexaneglycol-ate.
Benzeneacetic acid, α-cyclohexyl-α-hydroxy-, 4-(diethyl-amino)-2-butynyl ester; CCRIS 1923; Cyclohexane-glycolic acid, α-phenyl-, 4-(diethylamino)-2-butynyl ester; 4-Diethylamino-2-butynyl α-phenylcyclohexaneglycol-ate; 4-Diethylamino-2-butinyl α-cyclohexylmandelat; HSDB 3270; Oxibutinina; Oxibutyninum; Oxybutynin; Oxybutynine; Oxybutyninum; Oxytrol. Anticholinergic. Solid; soluble in H2O (7 g/100 ml pH 1, 0.08 g/100 ml pH 6, 0.0012 g/100 ml pH 9.6). *Hoechst AG (USA).*

2834   Oxybutynin Chloride

1508-65-2   7024   D

C22H32ClNO3
4-(Diethylamino)-2-butynyl α-phenylcyclohexaneglycol-ate hydrochloride.
5058; Cystrin; Ditropan; Dridase; EINECS 216-139-7; MJ-4309-1; Oxibutinina hydrochloride; Oxybutynin chloride; Oxybutynin hydrochloride; Pollakisu; Tropax. Anti-cholinergic. Crystals; mp = 129-130°; LD50 (rat orl) = 1220 mg/kg. *Hoechst AG (USA).*

2835   Oxycarboxin

5259-88-1   G P

C12H13NO4S
5,6-Dihydro-2-methyl-N-phenyl-1,4-Oxathiin-3-carbox-amide 4,4-dioxide.
BRN 1432554; Carboxin sulfone; Caswell No. 627A; Dcmod; 2,3-Dihydro-5-carboxanilido-6-methyl-1,4-oxa-thiin, 4,4-dioxide; 2,3-Dihydro-6-methyl-5-phenylcarb-amoyl-1,4-oxathiin 4,4-dioxide; 5,6-Dihydro-2-methyl-N-phenyl-1,4-oxathiin-3-carboxamide 4,4-dioxide; 5,6-Dihydro-2-methyl-3-carboxanilido-1,4-oxathiin-4,4-di-oxid; 5,6-Dihydro-2-methyl-1,4-oxathiin-3-carboxanilide 4,4-dioxide; Dioxide of vitavax; EINECS 226-066-2; EPA Pesticide Chemical Code 090202; F 461; F 461 (Pesticide); HSDB 1747; Methyl-6 phenylcarbamoyl-5 dihydro-2,3 oxathiine-1,4-dioxyde-4,4; NSC 232673; 1,4-Oxathiin,

2,3-dihydro-5-carboxanilido-6-methyl-, 4,4-di-oxide; 1,4-Oxathiin-3-carboxamide, 5,6-dihydro-2-meth-yl-N-phenyl-, 4,4-dioxide; 1,4-Oxathiin-3-carboxanilide, 5,6-dihydro-2-methyl-, 4,4-dioxide; Oxicarboxin; Oxy-carboxin; Oxycarboxine; Plant wax; Plantvax; Plantvax 20; Ringmaster; Vitavax sulfone; Vitavex. A systemic fungicide with curative action. Used for the control of rust diseases in ornamentals, nursery trees, wheat and fairy rings in grass. Registered by EPA as a fungicide. Colorless crystals; mp = 127-130°; soluble in H_2O (0.1 g/100 ml), DMSO (246 g/100 ml), Me_2CO (28.4 g/100 ml), MeOH (5.6 g/100 ml), C_6H_6 (3.0 g/100 ml), EtOH (2.4 g/100 ml); LD$_{50}$ (rat orl) = 2000 mg/kg, (rbt der) > 16000 mg/kg; LC$_{50}$ (mallard duck 8 day dietary) > 4640 mg/kg diet, (bobwhite quail dietary) > 10000 mg/kg diet; (rainbow trout 96 hr.) = 19.9 mg/l, (bluegill sunfish 96 hr.) = 28.1 mg/l; non-toxic to bees. *Crompton Corp.; Fargro Ltd.; ICI Agrochemicals; Rhône-Poulenc Environmental Prods. Ltd; Rhône-Poulenc; Uniroyal.*

2836 Oxyclozanide
2277-92-1   7027   G V

$C_{13}H_6Cl_5NO_3$
2,3,5-trichloro-N-(3,5-dichloro-2-hydroxyphenyl)-6-hydroxybenzamide.
Benzamide, 2,3,5-trichloro-N-(3,5-dichloro-2-hydroxy-phenyl)-6-hydroxy-; Benzanilide, 2,2'-dihydroxy-3,3',5,5',6-pentachloro-; BRN 2014120; CCRIS 5744; 2,2'-Dihydroxy-3,3',5,5',6-pentachlorobenzanilide; Diplin; EINECS 218-904-0; ICI 46638; Oxiclozanida; Oxiclozanidum; Oxyclozanid; Oxyclozanide; Oxyclozanidum; 3,5,6,3',5'-Pentachloro-2,2'-dihydroxybenz-anilide; 3,3',5,5',6-Pentachloro-2'-hydroxysalicylanilide; Salicylanilide, 3,3',5,5',6-pentachloro-2'-hydroxy-; 2,3,5-Trichloro-N-(3,5-dichloro-2-hydroxyphenyl)-6-hydroxy-benzamide; Zanil; Zanilox. A vetrinary anthelmintic (Trematodes). Used in combination with Haloxon. Solid; mp = 209-211°.

2837 Oxycodone
76-42-6   7028   D

$C_{18}H_{21}NO_4$
4,5α-Epoxy-14-hydroxy-3-methoxy-17-methylmorphinan-6-one.
4-27-00-03681 (Beilstein Handbook Reference); BRN 0043446; DEA No. 9143; Dihydrohydroxycodeinone; Dihydrone; Dihydroxycodeinone; Diphydrone; EINECS 200-960-2; Endine; Endone; Eubine; Eucodalum; HSDB 3142; NSC-19043; Ossicodone; Oxanest; Oxicodona; Oxicon; Oxycodeinone; Oxycodon; Oxycodone; Oxy-codonum; Pancodone retard; Percobarb; Percocet; Percodan; Prodalone [as pectinate]; Roxicet; Roxicodone; Supendol; Tylox. Analgesic, narcotic. Federally controlled substance (opiate). mp = 218-220°; insoluble in H_2O, Et_2O, base; soluble in alcohol, CHCl$_3$; [tautomeric form]: mp = 219-220°; tautomeric form is more

soluble in alcohol than other form. *DuPont-Merck Pharm.*

2838 Oxycodone Hydrochloride
124-90-3   7028   D

$C_{18}H_{22}ClNO_4$
4,5α-Epoxy-14-hydroxy-3-methoxy-17-methylmorphinan-6-one hydrochloride.
Dihydrone hydrochloride; Dihydrooxycodeinone hydro-chloride; Dihydroxycodeinone hydrochloride; Dinarkon; EINECS 204-717-1; Eubine; Eukodal; Eutagen; Oxikon; Oxycodon hydrochloride; Oxycon; Oxycontin; Oxygesic; Oxykodal; Oxykon; Pancodine; Percocet; Percodan; Percodan Demi; Percodan hydrochloride; Stupenone; Tecodin; Tecodine; Tekodin; Thecodin; Thecodine; Thecodinum; Thekodin; Tylox.; component of: Percodan. Analgesic, narcotic. Federally controlled substance (opiate). Dec 270-272°; $[\alpha]_D^{20}$= -125° (c = 2.5); soluble in H_2O (10 g/100 ml); slightly soluble in alcohol. *DuPont-Merck Pharm.*

2839 Oxyfluorfen
42874-03-3   7032   P

$C_{15}H_{11}ClF_3NO_4$
2-Chloro-1-(3-ethoxy-4-nitrophenoxy)-4-(trifluoromethyl)-benzene.
Benzene, 2-chloro-1-(3-ethoxy-4-nitrophenoxy)-4-(trifluoromethyl)-; Boxer (Obs.); BRN 2065259; Caswell No. 188AAA; 2-Chloro-1-(3-ethoxy-4-nitrophenoxy)-4-(trifluoromethyl)benzene; 2-Chloro-4-trifluoromethyl 3'-ethoxy-4'-nitrodiphenyl ether; 2-Chloro-α,α,α-trifluoro-p-tolyl-3-ethoxy-4-nitrophenyl ether; EINECS 255-983-0; EPA Pesticide Chemical Code 111601; Ether, 2-chloro-α,α,α-trifluoro-p-tolyl 3-ethoxy-4-nitrophenyl; Goal; Goal 1.6E; Goldate; Hada F; Koltar; Oxyfluorfen; Oxyfluorfene; Oxyfluorfen; Oxygold; RH-2915; RH 32915; RH 915; Zoomer. Registered by EPA as a herbicide. Orange crystals; mp = 83 - 84°; bp = 358°; almost insoluble in H_2O (0.00001 g/100 ml), soluble in Me_2CO (5.7 g/100 ml), cyclohexanone (5.8 g/100 ml), isophorone (5.7 g/100 ml), DMF (> 4.7 g/100 ml), CHCl$_3$ (7.4 - 8.1 g/100 ml), mesityl oxide (3.4 - 4.3 g/100 ml); LD$_{50}$ (mrat orl) > 5000 mg/kg, (rat. dog orl) > 5000 mg/kg, (rbt der) > 10000 mg/kg, (mallard orl) > 4000 mg/kg, (bobwhite quail orl) > 4000 mg/kg; LC$_{50}$ (bluegill sunfish, trout 96 hr.) < 1 mg/l; non-toxic to bees at 0.025 mg/bee. *Dow AgroSciences; Monsanto Co.; Rohm & Haas Co.; Scotts Co.*

2840 Oxygen
7782-44-7   7033   G

O=O

O₂

Oxygen.

CCRIS 1228; EINECS 231-956-9; HSDB 5054; Hyperoxia; Liquid oxygen; LOX; Molecular oxygen; Oxigeno; Oxygen; Oxygen (liquid); Oxygen-16; Oxygene; Oxygenium; Oxygenium medicinale; Pure oxygen; Sauerstoff; UN1072; UN1073. Gaseous element; used in copper smelting, steel production; in manufacture of ammonia, methyl alcohol, acetylene; oxidizer for rocket propellants; resuscitation, heart stimulant; decompression chambers; spacecraft; chemical intermediate; in oxidation of municipal and industrial wastes. Colorless, odorless gas; mp = -218°; bp = -183°; d⁰ (gas) = 1.429 g/l. *Air Products & Chemicals Inc.; Norsk Hydro AS; Showa Denko.*

2841 Oxynone
537-65-5   2999   G

C₁₂H₁₃N₃

1,4-Benzenediamine, N-(4-aminophenyl)-.

AI3-12116; N-(4-Aminophenyl)-1,4-benzenediamine; Aniline, 4,4'-iminodi-; Benzenamine, 4,4'-iminobis-; 1,4-Benzenediamine, N-(4-aminophenyl)-; Bis(p-amino-phenyl)amine; CI 76120; 4,4'-Diaminodiphenylamine; p,p'-Diaminodiphenylamine; Di(p-aminophenyl)amine; Diazol Black C; Diphenylamine, 4,4'-diamino-; EINECS 208-673-4; 4,4'-Iminodianiline; Indamine; NSC 33417; p-Phenylenediamine, N-(p-aminophenyl)-. A proprietary trade name for a rubber antioxidant. Also used to dye fur and to detect hydrogen cyanide. Leaflets; mp = 158°; bp dec; very soluble in EtOH, Et₂O.

2842 Oxythioquinox
2439-01-2   7048   G P

6-Methyl-2,3-quinoxalinedithiol cyclic S,S-dithiocarbon-ate.

Bay 36205; Morestan; Forstan; Quinomethionate. Selective non-systemic contact fungicide and acaricide. Used for control of powdery mildews and spider mites on fruits, ornamentals, curcubits, coffee and tea and various other crops. Registered by EPA as a fungicide and insecticide (cancelled). *Bayer Corp.*

2843 Paclobutrazol
76738-62-0   7053   G P

C₁₅H₂₀ClN₃O

(R*,R*)-(±)-β-[(4-Chlorophenyl)methyl]-α-(1,1-dimethyl-ethyl-1H-1,2,4-triazole-1-ethanol.

(2RS,3RS)-1-(4-Chlorophenyl)-4,4-dimethyl-2-(1H-1,2,4-triazol-1-yl)pentan-3-ol;   (±)-R*,R*-β-((4-chlorophenyl)-methyl)-   α-(1,1-dimethylethyl)-1H-1,2,4-triazol-1-ethanol;   ((R*,R*)-(±)-   β-((4-chlorophenyl)methyl)-   α-(1,1-dimethyl-ethyl)-1H-1,2,4-triazole-1-ethanol Clipper;   (R*,R*)-(±)-   β-((4-Chlorophenyl)methyl)-   α-(1,1-dimethylethyl)-1H-1,2,

4-triazole-1-ethanol;   α-t-Butyl-β-((4-chlorophenyl)methyl)-1H-triazol-1-ethanol;   β-((4-Chlorophenyl)methyl)-   α-(1,1-dimethylethyl)-1H-1,2,4-triazole-1-ethanol; Bonsai; Bonzi; Caswell No. 628C; Cultar; Duo Xiao Zuo; EINECS 266-325-7; EPA Pesticide Chemical Code 125601; Friazole; ICI-PP 333; Paclobutrazol; Parlay; PP 333; 1H-1,2,4-Triazole-1-ethanol,   β-((4-chlorophenyl)methyl)-   α-(1,1-dimethyl-ethyl)-, (R*,R*)-(±)- Trimmit; R*,R*)-(±)- 1H-1,2,4-Triazole-1-ethanol,   β-((4-chlorophenyl)methyl)-α-(1,1-dimethylethyl)-; Trimmit. Plant growth regulator, gibberellin biosynthesis inhibitor. Used on fruit trees, flowers, turf and rice. Registered by EPA as a herbicide. Crystals; mp = 165°; d = 1.22; soluble in H₂O (35 mg/l), more soluble in organic solvents; LD₅₀ (rat orl) = 2000 mg/kg. *ICI Chem. & Polymers Ltd.; Syngenta Crop Protection; Syngenta Professional Prod.; Uniroyal.*

2844 Palite
22128-62-7   G

C₂H₂Cl₂O₂

Chloromethyl chloroformate.

Carbonochloridic acid, chloromethyl ester; Chloromethyl chloroformate; EINECS 244-793-3; UN2745. A military poison gas.

2845 Palladium chloride
7647-10-1   7058   G

Cl₂Pd

Palladium(II) chloride.

CCRIS 6263; Dichloropalladium; EINECS 231-596-2; Enplate activator 440; HSDB 4362; NCI-C60184; NSC 146183; Palladium chloride; Palladium chloride (PdCl₂); Palladium dichloride; Palladium(2+) chloride; Palladium(II) chloride; Palladous chloride. Analytical chemistry, electrodeless coatings for metals, photography, leak detection in gas lines, indelible inks, catalyst. Solid; mp = 678-680°; soluble in H₂O, EtOH, Me₂CO; MLD (rbt iv) = 18.6 mg/kg. *Atomergic Chemetals; Dajac Labs.; Degussa AG; Sigma-Aldrich Fine Chem.; Spectrum Chem. Manufacturing.*

2846 Palladium nitrate
10102-05-3   7060   G

N₂O₆Pd

Palladium(II) nitrate.

EINECS 233-265-8; Hydrogen tetranitropalladate (II); Nitric acid, palladium(2+) salt; Palladium dinitrate; Palladium nitrate. Analytical reagent, used to separate Cl₂ and I₂, catalyst. Brown crystals; poorly soluble in H₂O. *Atomergic Chemetals; Degussa AG.*

2847 Palladium oxide
1314-08-5   7061   G

OPd

Palladium(II) monoxide.

EINECS 215-218-3; Palladium monoxide; Palladium oxide; Palladium oxide (PdO). Reduction and hydrogenation catalyst in organic synthesis. Black powder; d = 8.3; insoluble in H₂O.

2848   Palm kernel oil
8023-79-8   7066   H

Palm kernel oil.
EINECS 232-425-4; HSDB 1977; Oils, glyceridic, palm kernel; Oils, palm kernel; Palm kernel oil; Palm kernel oil, fatty acid; Palm kernel oil [Oil, edible]; Palm nut oil (from seed). Used in the manufacture of soap, has been used in oliniments and ointments. Yellow-white oil; mp = 26-30°; d = 0.952.

2849   Palm Oil
8002-75-3   7065   H

Palm Oil.
CCRIS 7935; EINECS 232-316-1; HSDB 2188; Oils, palm; Palm acidulated soapstock; Palm butter; Palm oil; Palm oil (from fruit); Palm oil [Oil, edible]. Yellow-white oil; mp = 26-30°; d = 0.952.

2850   Palm oil fatty acids, sodium salts
61790-79-2   H

Fatty acids, palm-oil, sodium salts.
EINECS 263-162-3; Fatty acids, palm-oil, sodium salts; Fatty acids, palm-oil, sodium salts.

2851   Palmitamine
143-27-1   G H

$C_{16}H_{35}N$
n-Hexadecylamine.
4-04-00-00818 (Beilstein Handbook Reference); AI3-16867; Alamine 6; Alamine 6D; 1-Amine 16D; Aminohexadecane; Armeen® 16D; BRN 1634065; CCRIS 4654; Cetylamin; Cetylamine; n-Cetylamine; Crodamine 1.16D; EINECS 205-596-8; Hetaflur hexadecylamine; Hexadecylamine; 1-Hexadecylamine; n-Hexadecylamine; Kemamine® P-880D; Nissan amine PB; NSC 8489; Palmitamine; Palmitylamine. Emulsifier; chemical intermediate; end products such as quaternary ammonium compounds used as bactericides and in shampoo formulations. Emulsifier for herbicides, ore flotation, pigment dispersion; auxiliary for textiles, leather, rubber, plastic, and metal industries. Leaflets; mp = 46.8°; bp = 322.5°; d^{20} = 0.8129; insoluble in H_2O; soluble in $CHCl_3$, Me_2CO, very soluble in EtOH, Et_2O, C_6H_6. *Akzo Chemie; Croda; Witco/Humko.*

2852   Palmitic acid
57-10-3   7064   G H

$C_{16}H_{32}O_2$
n-Hexadecanoic acid.
4-02-00-01157 (Beilstein Handbook Reference); AI3-01594; BRN 0607489; C_{16} fatty acid; Cetylic acid; EINECS 200-312-9; Emersol 140; Emersol 143; FEMA No. 2832; Hexadecanoic acid; n-Hexadecanoic acid; n-Hexadecoic acid; Hexadecylic acid; HSDB 5001; Hydrofol; Hystrene 8016; Hystrene 9016; Industrene 4516; NSC 5030; Palmitate; Palmitic acid; Pentadecanecarboxylic acid; 1-Pentadecanecarboxylic acid. A saturated fatty acid, used in the manufacture of metallic palmitates, soaps, lube oils, waterproofing, food-grade additives. A detergent intermediate; opacifier in cosmetics, soaps, emulsifiers, chemical specialties. White scales; mp = 63-64°; bp_{15} = 215°; d_4^{62}= 0.853; n_D^{80}= 1.4273; insoluble in H_2O,

soluble in organic solvents; LD50 (mus iv) = 57 mg/kg. *Akzo Nobel Chemicals Inc; Ashland; Henkel/Emery; Lancaster Synthesis Ltd.; Lonza Ltd.; Sigma-Aldrich Fine Chem.; Unichema; UOP; Witco/Humko.*

2853   Palmitonitrile
629-79-8   H

$C_{16}H_{31}N$
Hexadecanenitrile.
AI3-11234; 1-Cyanopentadecane; EINECS 211-110-5; Hexadecanenitrile; NSC 2137; Palmitic acid, nitrile; Palmitonitrile. Solid; mp = 31°; bp = 333°; d^{20} = 0.8303; insoluble in H_2O, freely soluble in EtOH, Et_2O, Me_2CO, C_6H_6, $CHCl_3$.

2854   Panogen M
151-38-2   G

$C_5H_{10}HgO_3$
2-Methoxyethylmercury acetate.
(Acetato-O)(2-methoxyethyl)mercury; Acetato(2-methoxy-ethyl)mercury; Ba 2743; Caswell No. 551B; Cekusil universal A; EINECS 205-790-2; EPA Pesticide Chemical Code 041508; HSDB 6387; Landisan; Mema; Mema RM; Mercuran; Mercury, (acetato-O)(2-methoxyethyl)-; Mercury, acetoxy(2-methoxyethyl)-; Merkuran; Methoxy-ethyl mercuric acetate; Methoxyethylmercuric acetate; Methoxyethylmercury acetate; (2-Methoxyethyl)mercury acetate; 2-Methoxyethylmercury acetate; 2-Methoxy-ethylmerkuriacetat; NSC 202875; Panogen; Panogen M; Panogen Metox; Radosan. Cereal seed treatment. Formulated as aqueous solutions or dusts and used chiefly as a seed protectant. Use of alkyl mercury fungicides in the United States has been virtually prohibited for several years. Phenyl mercuric acetate is still used to control diseases of turf, but other applications have been sharply restricted. *Embetec Crop Protection Ltd.*

2855   Pantoprazole
102625-70-7   7084   D

$C_{16}H_{15}F_2N_3O_4S$
5-(Difluoromethoxy)-2-[[(3,4-dimethoxy-2-pyridyl)methyl]sulfinyl]benzimidazole.
BY 1023; Pantoprazol; Pantoprazole; Pantoprazolum; SK&F 96022. Gastric proton pump inhibitor. Antiulcerative. Crystals; mp = 139-140° (dec); [sodium salt]: mp >130° (dec); λ_m = 289 (ε 16400). *Byk Gulden Lomberg GmbH; SmithKline Beecham Pharm.*

2856   Pantoprazole sodium salt
138786-67-1   7084   D

467

C16H14F2N3NaO34S
5-(Difluoromethoxy)-2-[[(3,4-dimethoxy-2-pyridyl)methyl]sulfinyl]benzimidazole sodium salt.
Controloc; Pantoloc; Pantoprazole sodium; Protonix; Protonix I.V. Gastric proton pump inhibitor. Used as an antiulcerative. Crystals; mp > 130° (dec); λ_m 289 nm (ε 16400 MeOH). *Byk Gulden Lomberg GmbH; SmithKline Beecham Pharm.*

2857 Paraffin
8002-74-2   G H

Paraffin wax.
905 (mineral hydrocarbons); Cream E45; Derma-Oil; Duratears; Fischer-Tropsch wax; Granugen; Hard wax; Koster Keunen Paraffin Wax; Parachoc; Paraffin wax (petroleum); Paraffin wax fume; Paraffin Wax; Paraffin Wax, granular; Paraffin waxes; Paraffinum durum; Paraffinum solidum; Poly(methylene)wax; Replens; Wax extract.; petroleum wax; crystalline; Hydrocarbon. Solid mixture of hydrocarbons obtained from petroleum; characterized by relatively large crystals; Candles; paper coating, protective sealant for food products; lubricants; hot-melt carpet backing, floor polishes, cosmetics, chewing gum base; raising melting points of ointments. Ointment base and stiffening agent. FDA, FEMA GRAS, Canada, Japan approved, FDA approved for implants, orals, topicals, USP/NF, BP compliance. In the FDA list of inactive ingredients, (oral capsules and tablets, topical emulsions and ointments). Used as a stiffening agent, tablet coating agent, used in implants, orals, topicals, ointments. White solid; mp = 50-57°; d = 0.9; insoluble in H2O, EtOH, soluble in C6H6, CHCl3, Et2O, CS2, oils. Non-toxic. *Astor Wax; Condea Vista Co.; EM Ind. Inc.; ExxonMobil L&PS; Humphrey; Ibrahim N. I.; IGI; Jonk BV; Koster Keunen; Mobil; N. Am. Philips; N.V. Philips; Penreco; Ross Frank B.; Ruger; Shell; Sigma-Aldrich Fine Chem.; Spectrum Chem. Manufacturing; Stevenson Cooper; Texaco.*

2858 Paraformaldehyde
110-88-3   9804   H

C3H6O3
1,3,5-Trioxane.
AI3-01363; Aldeform; CCRIS 4732; EINECS 203-812-5; Formagene; Formaldehyde, trimer; HSDB 3416; Marvosan; Metaformaldehyde; NSC 26347; Para-formaldehyde; Polymerized formaldehyde; Polyoxy-methylene; Triformol; Triossimetilene; 1,3,5-Trioxa-cyclohexane; Trioxan; 1,3,5-Trioxan; Trioxane; s-Tri-oxane; 1,3,5-Trioxane; Trioxin; Trioxymethylen; Trioxy-methylen; Trioxymethylene; s-Trixane. Needles; mp = 60.2°; bp = 114.5°; d^{65} = 1.17; very soluble in H2O, soluble in EtOH, Et2O, C6H6, CHCl3, CCl4, CS2, insoluble in petroleum ether.

2859 Paraformaldehyde
30525-89-4   7096   G P

HO(CH2O)nH
Para-formaldehyde.
Aldacide; Caswell No. 633; EPA Pesticide Chemical Code 043002; Flo-Mor; Formagene; Formaldehyde polymer; HSDB 4070; Oilstop,

Halowax; Paraform; Paraform; Paraformaldehyde; Paraformaldehydum; Paraformic aldehyde; Poly(oxymethylene); Polyformaldehyde; Polymerised formaldehyde; Polyoxymethylene; Polyoxymethylene glycol; UN2213. A polymer of formaldehyde in which n = 8-100; used in fungicides, bactericides, disinfectants, adhesives, hardener and waterproofing agent for gelatin contraceptive creams. Registered by EPA as an antimicrobial and fungicide. Solid; mp = 132-136°; slowly soluble in H2O, insoluble in EtOH, Et2O. *Degussa-Hüls AG; Hoechst Celanese; Mitsubishi Gas; Mitsui Toatsu.*

2860 Paraldehyde
123-63-7   7098   D G

C6H12O3
2,4,6-trimethyl-1,3,5-trioxane.
5-19-09-00112 (Beilstein Handbook Reference); Acetaldehyde, trimer; p-Acetaldehyde; AI3-03115; BRN 0080142; DEA No. 2585; EINECS 204-639-8; Elaldehyde; HSDB 3375; NSC 9799; Paracetaldehyde; Paral; Paraldehyd; Paraldehyde; Paraldehyde Draught (BPC 1973); Paraldehyde Enema (BPC 1973); Paraldeide; RCRA waste number U182; 1,3,5-Trimethyl-2,4,6-trioxane; 2,4,6-Trimethyl-1,3,5-trioxane; 2,4,6-Trimethyl-1,3,5-trioxacyclohexane; 2,4,6-Trimethyl-1,3,5-trioxaan; 2,4,6-Trimethyl-s-trioxane; 2,4,6-Trimetil-1,3,5-triossano; s-Trimethyltrioxymethylene; 1,3,5-Trioxane, 2,4,6-trimethyl-; s-Trioxane, 2,4,6-trimethyl-; Triacetaldehyde; UN1264. Polymer of acetaldehyde. Use in the manufacture of organic chemicals and also as a sedative and hypnotic. Liquid; mp = 12.6°; bp = 124.3°; d^{20} = 0.9943; slightly soluble in H2O, freely soluble in EtOH, Et2O, CHCl3; LD50 (rat orl) = 1650 mg/kg. *Forest Pharm. Inc.; Phillips 66.*

2861 Paraoxon
311-45-5   7012   G

C10H14NO6P
Phosphoric acid diethyl 4-nitrophenyl ester.
4-06-00-01327 (Beilstein Handbook Reference); AI3-16087; BRN 1915526; CCRIS 7780; Chinorta; Chinorto; Diaethyl-p-nitrophenylphosphorsaeureester; Diäthyl-p-nitrophenylphosphorsäureester; Diethyl 4-nitrophenyl phosphate; Diethyl-p-nitrofenyl ester kyseliny fosforecne; Diethyl p-nitrophenyl phosphate; Diethyl paraoxon; Diethylparaoxon; E 600; E 600 (pesticide); EINECS 206-221-0; ENT 16,087; Ester 25; Ethyl p-nitrophenyl ethylphosphate; Ethyl paraoxan; Eticol; Fosfakol; HC 2072; HSDB 6044; Mintaco; Mintacol; Mintisal; Miotisal A; NSC 404110; O,O-Dietyl-O-p-nitrofenyl-fosfat; O,O-Diethyl O-p-nitrophenyl phosphate; O,O'Diethyl-p-nitrophenylphosphat; O,O-Diethyl phos-phoric acid O-p-nitrophenyl ester; Oxyparathion; p-Nitrophenyl diethyl phosphate; Paraoxon; Paraoxone; Paroxan; Pestox 101; Phenol, p-nitro-, ester with diethyl phosphate; Phosphachole; Phosphacol; Phosphakol; Phosphonothioic acid, diethylparanitrophenyl ester; Phosphoric acid, diethyl p-nitrophenyl

468

ester; Phosphoric acid, diethyl 4-nitrophenyl ester; RCRA waste number P041; Soluglacit; Soluglaucit; TS 219. Cholinesterase inhibitor. Has been used as an insecticide. Oil; $bp_{1.0}$ = 169-170°; d_4^{25} = 1.2683; λ_m = 264 nm (ε 8900); soluble in H_2O (0.24 g/100 ml), more soluble in organic solvents; LD_{50} (rat orl) = 1.8 mg/kg.

2862 Paraquat
4685-14-7   7103   G H P

C12H14N2
1,1'-Dimethyl-4,4'-bipyridinium.
  4,4'-Bipyridinium, 1,1'-dimethyl-; CCRIS 7731; Dextrone; Dextrone X; 1,1'-Dimethyl-4,4'-bipyridinium; 1,1'-Dimethyl-4,4'-bipyridinium cation; 1,1'-Dimethyl-4,4'-bipyridinium salt; 1,1'-Dimethyl-4,4'-bipyridyldiylium; N,N'-Dimethyl-4,4'-bipyridinium dication; N,N'-Dimethyl-4,4'-bipyridinium; N,N'-Dimethyl-γ,γ'-dipyr-idylium; N,N'-Dimethyl-gamma,gamma'-dipyridylium; Dimethyl viologen; EINECS 225-141-7; HSDB 1668; Methyl viologen (2+); Methyl viologen ion(2+); Paraquat; Paraquat dication; Paraquat ion; Spraytop-graze; Starfire; Weedol. Non-selective contact herbicide. [dichloride]; colorless crystals; dec. 300°; d^{20} = 1.25; very soluble in H_2O (70 g/100 ml), slightly soluble in EtOH, MeOH, insoluble in other organic solvents; LD_{50} (rat orl) = 157 mg/kg, (mus orl) = 104 mg/kg, (gpg orl) = 22-42 mg/kg, (dog orl) = 25-50 mg/kg, (cow, sheep orl) = 50-75 mg/kg, (hen orl) = 262-280 mg/kg, (mallard duck orl) = 4048 mg/kg, (Japanese quail orl) = 970 mg/kg; LC_{50} (rainbow trout 96hr.) = 32 mg/l; non-toxic to bees. *Cia-Shen; Comlets; Crystal; ICI Agrochemicals; Pillar; Productos Quimicos y Alimenticios Osku SA; Sanex Agro Inc.; Shinung.*

2863 Paraquat bis(methylsulfate)
2074-50-2   7103   H P

C14H20N2O8S2
4,4'-Bipyridinium, 1,1'-dimethyl-, bis(methyl sulfate).
  Caswell No. 635; EINECS 218-196-3; EPA Pesticide Chemical Code 061602; Gramoxone methyl sulfate; Paraquat; Paraquat bis(methyl sulfate); Paraquat dimethyl sulfate; Paraquat dimethyl sulphate; Paraquat I; Paraquat methosulfate; Paraquat methylsulfate; PP 910. Registered by EPA as a herbicide (cancelled). Yellow solid; LD_{50} (mrat orl) = 100 mg/kg.

2864 Paraquat dichloride
1910-42-5   7103   H P

C12H14Cl2N2
1,1'-Dimethyl-4,4'-bipyridinium dichloride.
  AH 501; AI3-61943; Caswell No. 634; Cekuquat; Crisquat; Dextrone; Dextrone-X; Dexuron; Dimethyl viologen chloride; Dwuchlorek 1,1'-dwumetylo-4,4'-dwupirydyniowy; EINECS 217-615-7; EPA Pesticide Chemical Code 061601; Esgram; Galokson; Goldquat 276; Gramixel; Gramoxone; Gramoxone D; Gramoxone dichloride; Gramoxone S; Gramoxone W; Gramuron; Herbaxon; Herboxone; Herboxone; Methyl viologen; Methyl viologen chloride; Methyl viologen dichloride; NSC 88126; NSC 263500; OK 622; Ortho paraquat CL; Parakwat; Paraquat; Paraquat chloride; Paraquat CL; Paraquat dichloride; Pathclear; Pillarquat; Pillarxone; Toxer total; Viologen, methyl-. A pre-emergence bipyridinium herbicide to control weeds in field crops and ornamentals. Registered by EPA as a herbicide. Colorless crystals; mp = 300° (dec); very soluble in H_2O (70 g/100 ml), slightly soluble in MeOH, EtOH, insoluble in most other organic solvents; LD_{50} (rat orl) = 125 mg/kg, 157 mg/kg, (mus orl) = 104 mg/kg, (gpg orl) = 22 - 42 mg/kg, (dog orl) = 25 - 50 mg/kg, (cat orl) 40 - 50 mg/kg, (cow, sheep orl) = 50 75 mg/kg, (rbt der) = 236 - 500 mg/kg, (hen orl) = 282 - 380 mg/kg, (mallard duck orl) = 4048 mg/kg, (Japanese quail orl) = 960 mg/kg; LC_{50} (rainbow trout 96 hr.) = 32 mg/l; non-toxic to bees. *Ashlade Formulations Ltd.; Cyanamid of Great Britain Ltd.; ICI Agrochemicals; ICI Chem. & Polymers Ltd.; Schering AG; Schering Agrochemicals Ltd.; Syngenta Crop Protection.*

2865 Pararosaniline
569-61-9   G

C19H18ClN3
4-((4-Aminophenyl)(4-imino-2,5-cyclohexadien-1-ylidene)methyl)benzenamine monohydrochloride.
  Basic parafuchsine; Basic Red 9; Benzenamine, 4-((4-aminophenyl)(4-imino-2,5-cyclohexadien-1-ylidene)-methyl)-, monohydrochloride; C.I. 42500; C.I. Basic Red 9; C.I. Basic Red 9, monohydrochloride; Calcozine magenta N; CCRIS 1350; Cerven zasadita 9; CI 42500; EINECS 209-321-2; p-Fuchsin; Fuchsine DR-001; Fuchsine SP; Fuchsine SPC; HSDB 2952; 4,4'-((4-Imino-2,5-cyclohexadien-1-ylidene)methylene)dianiline mono-hydrochloride; 4,4'-(4-Iminocyclohexa-2,5-dienylidene-methylene)dianiline hydrochloride; NCI-C54739; NSC 10460; Orient Para Magenta Base; Parafuchsin; Parafuchsine; Parafuksin; Pararosaniline chloride; p-Rosaniline hydrochloride; Schultz-tab No. 779; 4-Toluidine, α-(p-aminophenyl)-α-(4-imino-2,5-cyclohexa-dien-1-ylidene)-monohydrochloride; 4,4',4-Triaminotri-phenylmethane hydrochloride; Parafuchsine; Calcozine Magenta N; Pararosaniline Chloride; Basic Red 9; Basic Red 9, monohydrochloride. Dyes wool and silk, purple-red, and cotton with mordants. Red crystals; mp = 268-270; insoluble in H_2O.

2866 Pararosaniline base, N,N',N''-triphenyl-
23681-60-9   H

C37H31N3O
4,4',4''-Trianilinotrityl alcohol.
Benzenemethanol, 4-(phenylamino)-α,α-bis(4-(phenyl-amino)phenyl)-; EINECS 245-822-2; Methanol, tris(p-anilinophenyl)-; Pararosaniline base, N,N',N''-triphenyl-; 4,4',4''-Trianilinotrityl alcohol.

2867　Parathion
56-38-2　　　　　　　　　7105　　　　　　H

C10H14NO5PS
O,O-Diethyl O-p-nitrophenyl phosphorothioate.
3422; 4-06-00-01337 (Beilstein Handbook Reference); AAT; AATP; AC 3422; ACC 3422; AI3-15108; Alkron; Alleron; American Cyanamid 3422; Aphamite; Aqua 9-Parathion; Aralo; B 404; BAY E-605; Bayer E-605; Bladan; Bladan F; Bladen; BRN 2059093; Caswell No. 637; CCRIS 493; Compound 3422; Corothion; Corthione; Danthion; Diethyl parathion; Dietil tiofosfato de p-nitrofenila; DNTP; Drexel parathion 8E; Durathion; E 605; E 605 F; E 605 forte; Ecatox 20; Ecatox 20; EINECS 200-271-7; Ekatin WF & WF ULV; Ekatox; Ekatox 20; ENT 15,108; EPA Pesticide Chemical Code 057501; EPA Shaughnessy Code: 057701; Ethlon; Ethyl parathion; Ethyl parathion 50 EC; Etilon; Etylparation; Foliclal; Folidol; Folidol E; Folidol E; Folidol E & E 605; Folidol E605; Folidol oil; Fosferno; Fosferno 50; Fosfive; Fosova; Fostox; Gearphos; Genithion; HSDB 197; Kolphos; Kypthion; Lethalaire G-54; Lirothion; Murfos; NA1967; NA2783; NCI-C00226; Niran; Niran E-4; Nitrostigmin; Nitrostigmine; Nitrostygmine; NIUIF-100; Nourithion; NSC 8933; Oleofos 20; Oleoparaphene; Oleoparathion; OMS 19; Orthophos; PAC; Pacol; Panthion; Paradust; Paramar; Paramar 50; Paraphos; Parathene; Parathion; Parathion 8E; Parathion 20 wp; Parathion-äthyl; Parathion-aethyl; Parathion-E; Parathion-ethyl; Parawet; Penncap E; Pestox plus; Pethion; Phoskil; Phosphemol; RB; RCRA waste number P089; Rhodiasol; Rhodiatox; Selephos; Sixty-three special EC insecticide; SNP; Soprathion; Soprothion; Stabilized ethyl parathion; Stathion; Strathion; Sulfos; Sulphos; Super Rodiatox; T-47; Thiofos; Thiomex; Thionspray No.84; Thiophos; Thiophos 3422; Thiophosphate de O,O-diethyle et de O-(4-nitrophenyle); Tiofos; TOX 47; UN 2783; Vapophos; Viran; Vitrex. A powerful insecticide and acaricide. Registered by EPA as an insecticide. Yellow-brown liquid; mp = 6°; bp = 375°, bp0.6 = 150°; d_4^{25}= 1.26; insoluble in H2O (0.0024 g/100 ml), soluble in organic solvents; LD50 (rat orl) = 3.6 - 13 mg/kg, (mus orl) = 12 mg/kg, (gpg orl) = 16 - 32 mg/kg, (rat der) = 6.8 mg/kg, (mallard duck orl) = 2 mg/kg, (pheasant orl) = 12.4 mg/kg; LC50 (rat ihl 4 hr.) = 0.05 mg/l air, (rainbow trout 96 hr.) = 1.5 mg/l, (golden orfe 96 hr.) = 0.57 mg/l; toxic to bees. *A/S Cheminova; Bayer Corp.; Drexel Chemical Co.*

2868　Paroxetine
61869-08-7　　　　　　　　7115　　　　　　D

C19H20FNO3
(-)-(3S,4R)-4-[(p-Fluorophenyl)-3-[(3,4-methylenedioxy)-phenoxy]methyl]piperidine.
Aropax; BRL 29060; FG 7051; Paroxetina; Paroxetine; Paroxetinum; Paxil; Paxil CR; Seroxat. Serotonin uptake inhibitor. Antidepressant. [hydrochloride (C19H21ClFN O3)]: mp = 118°; [hydrochloride hemihydrate (C19H21Cl FNO3.0.5H2O)]: mp = mp = 129-131°; [maleate]: mp = 136-138°; [α]D = -87° (c = 5 EtOH); LD50 (mus sc) = 845 mg/kg, (mus orl) = 500 mg/k. *Beecham Res. Labs. UK; Ferrosan A/S; SmithKline Beecham Pharm.*

2869　Parsley camphor
523-80-8　　　　　　　　741　　　　　　G

C12H14O4
4,7-dimethoxy-5-(2-propenyl)-1,3-benzodioxole.
5-19-03-00307 (Beilstein Handbook Reference); 1-Allyl-2,5-dimethoxy-3,4-methylenedioxybenzene; 5-Allyl-4,7-dimethoxy-1,3-benzodioxol; 5-Allyl-4,7-dimethoxy-1,3-benzodioxole; AI3-14843; Apiol; Apiole; Apiole (parsley); Apioline; Benzene, 1-allyl-2,5-dimethoxy-3,4-(methyl-enedioxy)-; 1,3-Benzodioxole, 4,7-dimethoxy-5-(2-propenyl)-; BRN 0195747; EINECS 208-349-2; NSC 9070; Parsley apiol; Parsley apiole; Parsley camphor; Petersiliencampher. Is synergistic with insecticides. Needles; mp = 29.5°; bp = 294°, bp35 = 179°; insoluble in H2O, very soluble in C6H6, EtOH, Me2CO, ligroin.

2870　Pavlin
524-30-1　　　　　　　　4289　　　　　　G

C16H18O10
8-(`D-glucopyranosyloxy)-7-hydroxy-6-methoxy-2H-1-benzopyran-2-one.
2H-1-Benzopyran-2-one, 8-(β-D-glucopyranosyloxy)-7-hydroxy-6-methoxy-; EINECS 208-355-5; Fraxin; 8-(β-D-Glucopyranosyloxy)-7-

hydroxy-6-methoxy-2H-1-benzo-pyran-2-one. A substance which occurs in the bark of the common ash. Yellow needles; mp = 205°; sparingly soluble in H_2O, soluble in EtOH, insoluble in Et_2O.

2871 Peanut oil
8002-03-7   7132   E H

Oil of peanut.
Arachis oil; Calchem IVO-112®; Earthnut oil; Groundnut oil; Katchung oil; Lipex 101®; Nut oil; Peanut oil; Pecan shell powder; Solvent crude peanut oil; Super Refined® Peanut Oil.; Oleaginous vehicle and solvent. FDA GRAS, FDA approved for injectables, orals, vaginals, USP/NF, BP compliance. In the FDA list of inactive ingredients, (im injections, topicals, oral capsules and vaginal emulsions). Used as a solvent for ointments, liniments, oleaginous vehicle for medicine, nutritive emulsions, softener for ear wax, enemas, in emollient creams, gall bladder evacuant; used in injectables, orals, vaginals and sunburn preparations. Yellow liquid; mp = -5°; d = 0.916 - 0.922; insoluble in H_2O, soluble in C_6H_6, EtOH, Et_2O, $CHCl_3$; viscosity at 37° = 35.2 mPa. Non-toxic. Possibly tumorigenic and mutagenic. *Abitec Corp.; Arista Ind.; Black S.; Bunge Foods; C&T Refinery; CanAmera Foods; Cargill; Charkit; Croda; Daniel H. E.; Flavors of N. America; Gittens Edwd.; MLG Enterprises; Penta Mfg.; Ruger; SKW Chem.; Spectrum Chem. Manufacturing; Tri-K Ind.; Vamo-Fuji Specialties; Van Den Bergh.*

2872 Pecilocin
19504-77-9   7134   D G

$C_{17}H_{25}NO_3$
[R-(E,E,E)]-1-(8-Hydroxy-6-methyl-1-oxo-2,4,6-dodeca-trienyl)-2-pyrrolidone.
EINECS 243-116-9; Leofungine; NSC 291839; Pecilocin; Pecilocina; Pecilocine; Pecilocinum; Supral; Variotin. Antifungal antibiotic obtained from cultures of *Paecyllomyces varioti Banier var. antibioticus.* Used in the treatment of fungal skin infections. Oil; $[\alpha]^{28}$ = -5.68° (MeOH); freely soluble in MeOH, EtOH, Me_2CO, EtOAc, C_6H_6, Et_2O, $CHCl_3$, C_5H_5N, dioxane, AcOH, slightly soluble in H_2O, petroleum ether; λ_m = 318, 324 nm ($E_1^{1\%}cm$ 1198 MeOH); [monohydrate]; mp = 41.5-42.5°; λ_m = 320 nm (ε 46000). *LeoAB; Nippon Kayaku Co. Ltd.*

2873 PEG-8 dioleate
9005-07-6   E H

POE (8) dioleate.
Glycols, polyethylene, dioleate; 9-Octadecenoic acid, oxybis(2,1-ethanediyloxy-2,1-ethanediyl)ester; Oleic acid, diester with polyethylene glycol; α-(1-Oxo-9-octadecenyl)-ω-((1-oxo-9-octadecenyl)oxy)poly(oxy-1,2-ethanediyl), (Z,Z)-; PEG Dioleate; PEG-4 Dioleate; PEG-8 Dioleate; PEG-20 Dioleate; Poly(oxy-1,2-ethanediyl), α-(1-oxo-9-octadecenyl)-ω-((1-oxo-9-octadecenyl)oxy)-, (Z,Z)-; Poly(oxy-1,2-ethanediyl), α-((9Z)-1-oxo-9-octadec-enyl)-ω-(((9Z)-1-oxo-9-octadecenyl)oxy)-; Polyethylene glycol dioleate; Polyethylene glycol 200 dioleate; Polyethylene glycol 400 dioleate; Polyethylene glycol 1000 dioleate; Polyoxyethylene (4) dioleate; Polyoxyethylene (8) dioleate; Polyoxyethylene (20) dioleate; Polyoxyethylene dioleate. Emulsifier, lubricant, solubilizer, wetting agent, dispersant for pharmaceuticals.

2874 PEG-2 laurate
141-20-8   3147   G

$C_{16}H_{32}O_4$
Dodecanoic acid 2-(2-hydroxyethoxy)ethyl ester.
AI3-00969; Atlas G-2124; Diethylene glycol laurate; Diethylene glycol monolaurate; Diethylene glycol sesquilaurate; Diglycol laurate; Diglycol monolaurate; 2,2'-Dihydroxyethyl ether monododecanoate; Dodeca-noic acid, 2-(2-hydroxyethoxy)ethyl ester; EINECS 205-468-1; Emcol RDC-D; Ethanol, 2-(2-hydroxyethoxy)-, laurate; G 2124; Glaurin; 2-(2-Hydroxyethoxy)ethyl laurate; Lauric acid, 2-(2-hydroxyethoxy)ethyl ester; Lauro-Sebum; Nonex 413; NSC 3868; PEG-2 Laurate; Pegosperse 100L; Pegosperse 100 LN; Polyethylene glycol 100 monolaurate; Polyoxyethylene (2) monolaurate. Water-oil emulsifier, dispersant, antistat, defoamer and plasticizer for textile, paper processing, cutting oils, polishes, emulsion cleaners, rubber latex, wool lubricants, paints. Light yellow oil; mp = 17.5°; bp > 270° (dec); d^{25} = 0.96; soluble in C_6H_6, C_7H_8, freely soluble in $EtOHEt_2O$, Me_2CO. *Henkel/Emery; Karlshamns; Lonzagroup; Mona; Stepan; Witco/Humko.*

2875 PEG-6 sorbitan stearate
9005-67-8   8796   E G V

$C_{64}H_{126}O_{26}$
POE (6) sorbitan monostearate.
CCRIS 702; FEMA No. 2916; Glycosperse S-20; Peg-20 sorbitan stearate; Peg sorbitan stearate; PEG-3 Sorbitan stearate; PEG-6 Sorbitan stearate; PEG-40 Sorbitan stearate; PEG-60 Sorbitan stearate; Polyethylene glycol (3) sorbitan monostearate; Polyethylene glycol 300 sorbitan monostearate; Polyethylene glycol 2000 sorbitan stearate; Polyethylene glycol 3000 sorbitan monostearate; Polyoxyethylene (3) sorbitan monostearate; Polyoxy-ethylene (4) sorbitan monostearate; Polyoxyethylene (6) sorbitan monostearate; Polyoxyethylene (20) sorbitan monostearate; Polyoxyethylene (40) sorbitan stearate; Polyoxyethylene (60) sorbitan monostearate; Poly-oxyethylene sorbitan monostearate; Polysorbate 60; Polysorbate 61; Sorbimacrogol stearate 300; Sorbitan, monooctadecanoate, poly(oxy-1,2-ethanediyl) derivs.; Sorbitan, monostearate polyoxyethylene deriv.; Tween 60. Mixture of stearate esters of sorbitol and sorbitol anhydrides, with approximately 20 moles ethylene oxide; (generic); industrial chemicals, solvent, emulsifier, pharmaceuticals, veterinary drug. Emulsifier and solubilizer. *Spectrum Chem. Manufacturing.*

2876 PEG-12 stearate
9004-99-3   E G H

$(C_2H_4O)_nC_{18}H_{36}O_2$
POE (12) stearate.
Crodet S12; Diethylene glycol stearate; Hodag 60-S; Kessco® PEG 600 MS; Mapeg® 600 MS; PEG 400 monostearate; PEG 600 monostearate; Polyoxyethylene (8) stearic acid (monoester); Polyoxyethylene (40) stearic acid (monester); Polyoxyethylene (50) stearic acid (monoester). Emulsifier, wetting agent, solubilizer, dispersant, emollient, spreading agent for pharmaceuticals. Toxic by ip, iv routes. *Rhône Poulenc Surfactants; Taiwan Surfactants.*

2877 Pelargonic acid
112-05-0   7141   G H P

C9H18O2

n-Nonanoic acid.

4-02-00-01018 (Beilstein Handbook Reference); AI3-04164; BRN 1752351; Cirrasol 185A; EINECS 203-931-2; Emery® 1202; Emfac 1202; EPA Pesticide Chemical Code 217500; FEMA No. 2784; Hexacid C9; HSDB 5554; Nonanoic acid; n-Nonanoic acid; Nonoic acid; n-Nonoic acid; n-Nonylic acid; Nonylic acid; n-Nonylic acid; NSC 62787; Pelargic acid; Pelargon; Pelargonic acid. Intermediate in manufacture of detergents. Herbicide and plant growth regulator. Oily liquid; mp = 12.3°; bp = 254.5°; bp14 = 143 - 145°, bp6.3 = 132 - 133°; insoluble in H_2O, soluble in EtOH, CHCl3, Et2O; LD50 (mus iv) = 224 mg/kg. *Dow AgroSciences; Henkel/Emery; Japan Tobacco.*

2878   Penconazole
66246-88-6   G P

C13H15Cl2N3

1-[2-(2,4-Dichlorophenyl)pentyl]-1H-1,2,4-triazole.

5-26-01-00149 (Beilstein Handbook Reference); Award; BRN 0541488; CGA 71818; 1-(2-(2,4-Dichloro-phenyl)pentyl)-1H-1,2,4-triazole; EINECS 266-275-6; Onmex; Penconazole; Topas 100; Topas C; Topas MZ; Topaz; Topaze; Topaze C; 1H-1,2,4-Triazole, 1-(2-(2,4-dichlorophenyl)pentyl)-. Systemic fungicide with protective and curative action. Inhibits biosynthesis of ergosterol in the cell membrane. Used for control of powdery mildew. Crystals; mp = 60°; soluble in H_2O (0.007 g/100 ml at 20°), Me2CO, cyclohexanone (both 58 g/100 ml), CH2Cl2 (106 g/100 ml), MeOH (64 g/100 ml), C6H14 (1.7 g/100 ml), xylene, i-PrOH (50 g/100 ml); LD50 (rat orl) = 2125 mg/kg, (mus orl) = 2444 mg/kg, (rat der) > 3000 mg/kg, (Japanese quail orl) = 2424 mg/kg, (Peking duck orl) > 3000 mg/kg, (mallard duck orl) > 1590 mg/kg; LC50 (rainbow trout 96 hr.) = 1.7 - 4.3 mg/l, (carp 96 hr.) = 3.8 - 4.6 mg/l, (bluegill sunfish 96 hr.) = 2.1 - 2.8 mg/l; non-toxic to bees. *Ciba-Geigy Corp.; Syngenta Crop Protection.*

2879   Pencycuron
66063-05-6   G P

C19H21ClN2O

N-[(4-Chlorophenyl)methyl]-N-cyclopentyl-N'-phenyl-urea.

BAY NTN 19701; BRN 2154416; Caswell No. 638A; 1-((4-Chlorophenyl)methyl)-1-cyclopentyl-3-phenylurea; 1-(p-Chlorobenzyl)-1-cyclopentyl-3-phenylurea; N-((4-Chlorophenyl)methyl)-N-cyclopentyl-N'-phenylurea; EINECS 266-096-3; EPA Pesticide Chemical Code 128823; Monceren; NTN 19701; Pencycuron; Urea, N-((4-chlorophenyl)methyl)-N-cyclopentyl-N'-phenyl-. Non-systemic fungicide used fro control of *Rhizoctonia and Pellicularia* spp. in potatoes, rice, cotton and vegetables. In

particular, control of black scurf of potatoes, sheath blight of rice, and damping-off of ornamentals. Colorless crystals; mp = 129°; almost insoluble in H_2O (0.00004 g/100 ml at 20°), soluble in CH2Cl2 (10 - 100 g/100 ml), C7H8 (1 - 3 g/100 ml), i-PrOH (0.1 - 1.0 g/100 ml), C6H14 (< 0.1 g/100 ml); LD50 (rat, mus, dog orl) > 5000 mg/kg, (cat orl) > 1000 mg/kg, (rat, mus der) > 2000 mg/kg, (Japanese quail orl) > 2500 mg/kg, (hen orl) > 2500 mg/kg, (canary orl) > 1000 mg/kg; LC50 (rat ihl 1 hr.) > 0.625 mg/l air, (carp 96 hr.) = 8.8 mg/l, (guppy 96 hr.) = 5 - 10 mg/l; non-toxic to bees. *Bayer AG; Bayer Corp., Agriculture; Mobay.*

2880   Pendimethalin
40487-42-1   7155   G H P

C13H19N3O4

N-(1-Ethylpropyl)-3,4-dimethyl-2,6-dinitrobenzenamine.

AC 92553; Accotab; Aniline, 3,4-dimethyl-2,6-dinitro-N-(1-ethylpropyl)-; Benzenamine, 3,4-dimethyl-2,6-dinitro-N-(1-ethylpropyl)-; Benzenamine, N-(1-ethylpropyl)-3,4-dimethyl-2,6-dinitro-; BRN 2157711; Caswell No. 454BB; Dirimal; EPA Pesticide Chemical Code 108501; Go-Go-San; Herbadox; Herbodox; N-(1-Äthylpropyl)-2,6-dinitro-3,4-xylidin; N-(1-Äthylpropyl)-3,4-dimethyl-2,6-dinitroanilin; N-(1-Ethylpropyl)-2,6-dinitro-3,4-xylidine; Pay-off; Penoxalin; Pentagon; Pre-M 60DG; Prowl; Prowl 3.3E; Prowl 4E; Sipaxol; Stomp; Way Up; N-(1-Ethylpropyl)-3,4-dimethyl-2,6-dinitroaniline; N-(1-Ethyl-propyl)-3,4-dimethyl-2,6-dinitrobenzenamine; Horbadox; HSDB 6721; Pendimethalin; Pendimethaline; Penoxalin; Penoxaline; Penoxyn; Phenoxalin; Prowl; Sipaxol; Stomp; Stomp 330D; Stomp 330E; Tendimethalin; Wax Up; Way Up; Wayup; 3,4-Xylidine, 2,6-dinitro-N-(1-ethylpropyl)-. Selective herbicide, absorbed by roots and leaves. Used for control of most annual grasses and many broad-leaved weeds in cereal, vegetable and fruit crops. Emulsifiable or suspension concentrate containing pendimethalin; a dinitroaniline herbicide for cereals and bush fruit. Registered by EPA as a herbicide. Orange crystals; mp = 56-57°; sg25 = 1.19; soluble in H_2O (0.00003 g/100 ml), soluble in Me2CO (70 g/100 ml), xylene (62.8 g/100 ml), corn oil (14.8 mg/kg), C7H16 (13.8 g/100 ml), i-PrOH (7.7 g/100 ml), readily soluble in C6H6, C7H8, CHCl3, CH2Cl2, slightly soluble in petroleum ether; LD50 (mrat orl) = 1250 mg/kg, (frat orl) = 1050 mg/kg, (mmus orl) = 1620 mg/kg, (fmus orl) = 1340 mg/kg, (rbt orl) > 5000 mg/kg, (beagle dog orl) > 5000 mg/kg, (rbt der) > 5000 mg/kg, (bee topical) > 0.05 mg/bee; LC50 (bobwhite quail 8 day dietary) = 4187 mg/kg diet, (mallard duck 8 day dietary) = 10388 mg/kg diet; (rainbow trout 96 hr.) = 0.14 mg/l, (bluegill sunfish 96 hjr.) = 0.2 mg/l; non-toxic to bees. *BASF Corp.; Bayer AG; Cyanamid of Great Britain Ltd.; Hortichem Ltd.; UAP - Platte Chemical.*

2881   Penicillin G potassium
113-98-4   7165   D G

472

C16H17KN2O4S

[2S-(2α,5α,6β)]-3,3-Dimethyl-7-oxo-6-[(phenylacetyl)-amino]-4-thia-1-azabicyclo[3.2.0]heptane-2-carboxylic acid monopotassium salt.

Benzylpenicillin potassium; Benzylpenicillin potassium salt; Benzylpenicilline potassique; Benzylpenicillinic acid potassium salt; C-Cillin; Capicillin; Cillin; Cilloral potassium salt; Cintrisul; Cosmopen; Cosmopen potassium salt; Cristapen; Crytapen; DRG-0128; EINECS 204-038-0; Eskacillin; Falapen; Forpen; Hipercilina; Hyasorb; Hylenta; Lemopen; Liquapen; M-Cillin; Megacillin tablets; Monopen; Notaral; Novocillin vet.; NSC 131815; Paclin G; Penalev; Penicillin G K; Penicillin G potassium; Penisem; Pentid; Pentids; Pfizerpen; Potassium benzylpenicillinate; Potassium penicillin G; Potassium 6-(phenylacetamido)penicillanate; Potassium benzylpenicillin; Potassium benzylpenicillin G; Potassium benzylpenicillinate; Potassium penicillin G; Potassium salt of benzylpenicillin; Qidpen G; Scotcil; SK-Penicillin G; Sugracillin; Tabilin; Tu Cillin; Van-Pen-G. Antibacterial Solid; mp = 214-217° (dec); [α]$_D^{22}$= 285-310° (c = 0.7); freely soluble in H2O. Bristol-Myers Squibb Co.

2882   Penicillin V
87-08-1   7171   D G

C16H18N2O5S

[2S-(2α,5α,6β)]-3,3-Dimethyl-7-oxo-6-[(phenoxyacetyl)-amino]-4-thia-1-azabicyclo[3.2.0]heptane-2-carboxylic acid.

4-27-00-05884 (Beilstein Handbook Reference); Acipen V; Apopen; Beromycin; BRN 0096259; Calcipen; CCRIS 752; Compocillin V; Crystapen V; Distaquaine V; EINECS 201-722-0; Eskacillin V; Fenacilin; Fenospen; Fenossi-metilpenicillina; Fenoxypen; HSDB 6314; Meropenin; Oracillin; Oracilline; Oratren; Ospen; P-Mega-Tablinen; Pen-Oral; Pen-V; Pen-Vee; Penicillin phenoxymethyl; Penicillin V; Phenocillin; Phenomycilline; Phenopeni-cillin; Phenoximethyl-penicillinum; Phenoxymethyl peni-cillin; Phenoxymethyl-penicilline; Phenoxy-methylpenicillin; Phenoxy-methylpenicilline; Phenoxy-methylpenicillinum; Phenoxy-methylpenicillinic acid; Rocilin; Stabicillin; V-Cil; V-Cillin; V-Cylina; V-Cyline; V-Tablopen; Vebecillin. Anibacterial agent. Monoclinic hemimorphic crystals; Dec 120-128°; λm = 268, 274 nm (ε 1330, 1100); soluble in H2O at pH 1.8 (25 mg/100 ml); soluble in polar organic solvents; practically insoluble in vegetable oils, liquid petrolatum; LD50 (mus sc) = 2300 mg/kg. Eli Lilly & Co.; Wyeth-Ayerst Labs.

2883   Penicillin V potassium
132-98-9   7171   D

C16H17KN2O5S

[2S-(2α,5α,6β)]-3,3-Dimethyl-7-oxo-6-[(phenoxyacetyl)-amino]-4-thia-1-azabicyclo[3.2.0]heptane-2-carboxylic acid monopotassium salt.

Antibiocin; Apsin VK; Arcacil; Arcasin; Aspin VK; Beepen-VK; Beromycin; Beromycin 400; Betapen V; Betapen-VK; Calciopen K; CCRIS 750; Cliacil; Compocillin-VK; Distakaps V-K; Distaquaine V-K; Dov-k; Dowpen V-K; DQV-K; EINECS 205-086-5; Fenocin; Fenocin Forte; Fenoxypen; HSDB 6315; Icipen; Isocillin; Ispenoral; Kavepenin; Ledercillin VK; Megacillin Oral; Oracil-VK; Orapen; Ospeneff; Pedipen; Pen-V-K; Pen-Vee K; Penagen; Penapar VK; Pencompren; Penicillin Potassium Phenoxymethyl; Penicillin V potassium salt; Penicillin V Potassium; Penicillin VK; Penvikal; Pfizerpen VK; Phenoxymethylpenicillin potassium; Potassium penicillin V salt; Potassium penicillin V; Potassium phenoxymethyl penicillin; Potassium phenoxy-methylpenicillin; Primcillin; PVK; Qidpen VK; Robicillin VK; Rocillin-VK; Roscopenin; SK-Penicillin VK; Stabillin VK syrup; Sumapen VK; Suspen; Uticillin VK; V-CIL-K; V-Cillin K; Veetids; Vepen. Antibacterial agent. Solid; mp = 120-128° (dec); λm = 268, 274 nm (ε 1330 1100); soluble in H2O (0.025 g/100 ml), polar organic solvents. Apothecon; Eli Lilly & Co.; Lederle Labs.; Parke-Davis; Robins, A. H. Co.,; SmithKline Beecham Pharm.; Wyeth-Ayerst Labs.

2884   Pentabromodiphenyl ether
32534-81-9   G H

C12H5Br5O

Diphenyl ether, pentabromo derivative.

Benzene, 1,1'-oxybis-, pentabromo deriv.; Bromkal G 1; CCRIS 4851; DE 71; Diphenyl ether, pentabromo derivative; EINECS 251-084-2; FR-1205; Fyrol® PBR; Pentabromodiphenyl ether; Pentabromodiphenyl oxide; Pentabromophenoxybenzene; Planelon PB 501; Saytex 125. Flame retardant for use in laminates (both epoxy and phenolic), unsaturated polyesters, synthetic fibers, and flexible PU foams; suitable for textiles. Flame retardant additive for flexible polyurethane foams; low propensity for discoloration caused by high exotherm in the processing of flexible polyether urethane foam. Liquid at 50-60°; d = 2.25; soluble in CCl4; CH3Cl2, C6H6, Me2CO; insoluble in water; slightly soluble in methanol; LD50 (rat orl) = 5200 mg/kg, (rbt der) > 2000 mg/kg. Akzo Chemie; Albemarle Corp.; Amerihaas; Dead Sea Bromine; Ethyl Corp.; Great Lakes Fine Chem.

2885   Pentabromotoluene
87-83-2   G

C7H3Br5

2,3,4,5,6-Pentabromotoluene.
Benzene, pentabromomethyl-; CCRIS 4854; EINECS 201-774-4; Flammex 5bt; FR 705; HSDB 5253; Penta-bromomethylbenzene; Pentabromotoluene; 2,3,4,5,6-Pentabromotoluene; Toluene, 2,3,4,5,6-pentabromo-. Flame retardant for use with unsaturated polyesters, polyethylene, PP, PS, SBR latex, textiles and rubbers. Crystals; mp = 288°; d^{17} = 2.97; λ_m = 222, 270 nm (ε = 41700, 925, MeOH); insoluble in H_2O, soluble in C_6H_6, less soluble in EtOH, AcOH; LD_{50}=rat oral>5000 mg/kg; irritant. *Dead Sea Bromine; Great Lakes Fine Chem.; Sigma-Aldrich Fine Chem.*

2886   **Pentacarbonyliron**
13463-40-6   5114   H

C5FeO5

Iron carbonyl (Fe(CO)5), (TB-5-11)-.
EINECS 236-670-8; FER pentacarbonyle; HSDB 6347; Iron carbonyl; Iron carbonyl (Fe(CO)5); Iron carbonyl (Fe(CO)5), (TB-5-11)-; Iron carbonyl, (Fe(CO)5); Iron carbonyl, (TB-5-11)-; Iron carbonyl compounds; Iron pentacarbonyl; Pentacarbonyliron; UN1994.

2887   **Pentachlorethane**
76-01-7   7179   G

C2HCl5

Ethane pentachloride.
4-01-00-00147 (Beilstein Handbook Reference); BRN 1736845; Caswell No. 639B; CCRIS 494; EINECS 200-925-1; EPA Pesticide Chemical Code 598300; Ethane pentachloride; Ethane, pentachloro-; HSDB 2034; NCI-C53894; Pentachloorethaan; Pentachloraethan; Pentachloräthan; Pentachlorethane; Pentachloroethane; Pentacloroetano; Pentalin; RCRA waste number U184; UN1669. Solvent, degreaser, cleaning agent. Liquid; mp = -29°; bp =159.8°; d^{20} = 1.6796; insoluble in H_2O, very soluble in EtOH, Et_2O, other organic solvents; LC_{50} (rat ihl) = 4238 ppm/2 hr. *Lancaster Synthesis Co.*

2888   **Pentachlorophenol**
87-86-5   7180

C6HCl5O

2,3,4,5,6-Pentachlorophenol.
4-06-00-01025 (Beilstein Handbook Reference); AD 73; AI3-00134; BRN 1285380; Caswell No. 641; CCRIS 1663; Chem-Penta; Chem-Tol; Chlon; Chlorophen; CM 613; CP 1309; Cryptogil oil; Dow pentachlorophenol; Dow pentachlorophenol dp-2; Dowicide 7; Dowicide EC-7; DP-2 antimicrobial; Dura-Treet; Dura Treet II; Durotox; EP 30; EP 30 (pesticide); EPA Pesticide Chemical Code 063001; Forepen; Forpen-50 Wood Preservative; Fungifen; Glazd penta; Grundier arbezol; HSDB 894; 1-Hydroxy-2,3,4,5,6-pentachlorobenzene; 1-Hydroxypentachlorobenzene; Lauxtol; Lauxtol a; Liroprem; MB 333; NCI-C54933; NCI-C55378; NCI-C56655; NSC 263497; Ontrack WE Herbicide; Ortho Triox Liquid Vegetation Killer; Osmoplastic; Osmose Wood Preserving Compound; Oz-88; PCP; Penchlorol; Penta; Penta Concentrate; Penta-kil; Penta ready; Penta WR; Pentachloorfenol; Pentachlorofenol; Pentachloro-phenate; Pentachlorophenol; 2,3,4,5,6-Pentachloro-phenol; Pentachlorophenol [UN3155] [Poison]; Penta-clorofenolo; Pentacon; Penwar; Peratox; Permacide; Permagard; Permasan; Permatox DP-2; Permatox penta; Permite; Phenol, pentachloro-; Pol Nu; Pole topper fluid; Preventol P; Priltox; RCRA waste number U242; santobrite; Santophen; Santophen 20; Sinituho; Term-i-trol; Thompson's wood fix; UN3155; Watershed Wood Preservative; Weed and Brush Killer; Weedone; Witophen P; Woodtreat A. Used as a preharvest defoliant on selected crops. Has been used as a herbicide, algacide, defoliant, wood preservative, germicide, fungicide, and molluscicide. Insecticide used in termite control, as a defoliant and general herbicide. Also used for wood preservation. Registered by EPA as an antimicrobial, fungicide and insecticide. Colorless crystals; mp = 174°, 191°; bp = 310° (dec); sg^{22} = 1.978; λ_m = 224, 273, 280 nm (ε = 8220, 1050, 911, MeOH); soluble in H_2O (0.008 g/100 ml at 20°), Me_2CO (21.5 g/100 ml), most other organic solvents; LD_{50} (rat orl) = 210 mg/kg; LC_{50} (rainbow trout 48 hr.) = 0.17 mg/l (sodium salt). *Dow AgroSciences; KMG-Bernuth Inc.; Nyco Minerals; Penta Mfg.; Vulcan; Wood Protection Products.*

2889   **Pentachloropyridine**
2176-62-7   H

C5Cl5N

2,3,4,5,6-Pentachloropyridine.
5-20-05-00422 (Beilstein Handbook Reference); BRN 0155197; EINECS 218-535-5; HSDB 5867; NSC 26286; Pentachloropyridine; 2,3,4,5,6-Pentachloropyridine; Per-chloropyridine; Pyridine, 2,3,4,5,6-pentachloro-; Pyrid-ine, pentachloro-. Crystals; mp = 125.5°; bp = 280°; λ_m = 261 nm (ε = 372, C_6H_{14}), 260 nm (ε = 339, EtOH); very soluble in EtOH, C_6H_6, ligroin.

2890   **Pentachlorothiophenol**
133-49-3   G

C6HCl5S
Pentachlorobenzenethiol.
4-06-00-01642 (Beilstein Handbook Reference); Al3-23118; Akrochem® Peptizer PTP; Benzenethiol, pentachloro-; BRN 1108638; EINECS 205-107-8; HSDB 6124; NSC 5578; PCTP; Pentachloro-benzenethiol; Pentachlorothiophenol; Pentachlorthiofenol; Renacit 7; RPA 6; USAF B-51. Peptizer for natural rubber, polyisoprene, styrene/butadiene rubber, polybutadiene, NBR, butyl, chloroprene and blends; absorbed on clay, used as a peptizing agent facilitating open mill and internal mixer mastication in rubber industry. Mildly toxic by ingestion; severe eye irritant. *Akrochem Chem. Co.; Bayer AG; Polysar.*

2891 Pentadecane
629-62-9   H

C15H32
n-Pentadecane.
4-01-00-00529 (Beilstein Handbook Reference); BRN 1698194; EINECS 211-098-1; HSDB 5729; NSC 172781; Pentadecane; n-Pentadecane. Liquid; mp = 9.9°; bp = 270.6°; d^{20} = 0.7685; insoluble in H_2O, very soluble in EtOH, Et_2O.

2892 Pentadecanol
629-76-5   H

C15H32O
n-Pentadecan-1-ol.
Al3-33881; EINECS 211-107-9; Neodol® 5; NSC 66446; Pentadecanol; 1-Pentadecanol; n-Pentadecanol; n-1-Pentadecanol; Pentadecyl alcohol. Detergent intermediate. Solid; mp = 45-46°; bp = 269-271°. *Shell.*

2893 Pentadiene, 1,3-
504-60-9   H

C5H8
1-Methylbutadiene.
4-01-00-00995 (Beilstein Handbook Reference); BRN 1523657; EINECS 207-995-2; HSDB 6063; 1-Methylbutadiene; Pentadiene, 1,3-; 1,3-Pentadiene; Piperylene; RCRA waste number U186.

2894 Pentaerythritol
115-77-5   7182   G H

C5H12O4
Tetrakis(hydroxymethyl)methane.
4-01-00-02812 (Beilstein Handbook Reference); Al3-19571; Auxinutril; BRN 1679274; CCRIS 2306; EINECS 204-104-9; Hercules® P6; Hercules® Mono-PE; HSDB 872; Maxinutril; Metab-Auxil; Methane tetramethylol; Monopentaerythritol; Monopentek; NSC 8100; PE 200; Penetek; Pentaerythrital; Pentaerythrite; Pentaerythritol; Pentek; PETP; Tetrahydroxymethylmethane; Tetrakis-(hydroxymethyl)methane; Tetramethylolmethane; THME;. Used in production of alkyd resins, rosin esters, urethane resins, drying oils, synthetic lubricants, plasticizer, intumescent paints, plastics, stabilizers for plastics, explosives and as a chemical intermediate. Crystals; mp = 260°; bp30 = 276°, sublimes; d = 1.396; soluble in H_2O (55 mg/ml), alcohols, slightly soluble in Et_2O, C_6H_6. *Aqualon; Degussa AG; Hercules Inc.; Hoechst Celanese; Mitsubishi Gas; Mitsui Toatsu; Penta Mfg.; Perstorp AB.*

2895 Pentaerythritol tetrabenzoate
4196-86-5   G

C33H28O8
2,2-Bis((benzoyloxy)methyl)dibenzoate propanediol.
3-09-00-00688 (Beilstein Handbook Reference); Benzoflex S-552; Benzoic acid, tetraester with pentaerythritol; 2,2-Bis((benzoyloxy)methyl)-1,3-propane-diol dibenzoate; BRN 3513249; CCRIS 5971; EINECS 224-079-8; NSC 166502; Pentaerythritol tetrabenzoate; 1,3-Propanediol, 2,2-bis((benzoyloxy)methyl)-, dibenz-oate; Uniplex 552. Plasticizer for adhesives intended for heat seal applications Solid; mp = 102-104°. *Unitex.*

2896 Pentaerythritol tetrakis(3,5-di-t-butyl-4-hydroxyhydrocinnamate)
6683-19-8   G H

C73H108O12

Hydrocinnamic acid, 3,5-di-t-butyl-4-hydroxy-, neo-pentanetetrayl ester.

ADK Stab AO 60; Anox 20AM; ANOX 20; AO 60; AO3; BP 101; BRN 2035465; Dovernox 10; EINECS 229-722-6; Fenozan 22; Fenozan 23; IR 1010; Irganox 1010; Irganox 1010; Irganox 1010FF; Irganox 1010FP; Irganox 1040; MARK AO 60; Naugard® 10; Phenosane 23; RA 1010; Ralox 630; Sumilizer BP 101; Tetraalkofen BPE. Antioxidant effective against thermal oxidative degradation during long term heat aging; for polyolefins, styrenics, elastomers, adhesives, lubricants, and oils. Crystals; mp = 110-125°. *Uniroyal.*

2897   Pentaerythrityl tetrapelargonate
14450-05-6   H

C41H76O8

Nonanoic acid, 2,2-bis(((1-oxononyl)oxy)methyl)-1,3-propanediyl ester.

AI3-14798; 2,2-Bis(((1-oxononyl)oxy)methyl)propane-1,3-diyl dinonan-1-oate; EINECS 238-430-8; Nonanoic acid, 2,2-bis(((1-oxononyl)oxy)methyl)-1,3-propanediyl ester; Nonanoic acid, neopentanetetrayl ester; Pentaerythrityl tetrapelargonate.

2898   Pentaerythrityl tetrastearate
115-83-3   G H

C77H148O8

2,2-Bis(octadecanoyloxymethyl)-1,3-propanediyl diocta-decanoate. EINECS 204-110-1; Octadecanoic acid, 2,2-bis(((1-oxooctadecyl)oxy)methyl)-1,3-propanediyl ester; Penta-erythrityl tetrastearate; Radia® 7176, Radiasurf® 7175; Stearic acid, neopentanetetrayl ester. Lubricant, chemical intermediate; used in formulation of cutting, lamination, textile oils and corrosion inhibitors, in chemical synthesis, polishes, coatings and textile finishes. Solid; mp = 71°; d = 0.940. *Fina Chemicals; Hercules Inc.; Lipo; Lonzagroup.*

2899   Pentaerythrityl stearate
8045-34-9   H

C23H46O5

Octadecanoic acid, ester with 2,2-bis(hydroxymethyl)-1,3-propanediol.

EINECS 232-457-9; Octadecanoic acid, ester with 2,2-bis(hydroxymethyl)-1,3-propanediol; Stearic acid, ester with pentaerythritol.

2900   Pentaerythrityl triacrylate
3524-68-3   G

C14H18O7

(2-Propenoic acid-2-(hydroxymethyl)-2-(((1-oxo-2-propenyl)oxy)methyl)-1,3-propanediol ester).

Acrylic acid, triester with pentaerythritol; Aronix M 305; CCRIS 3436; EINECS 222-540-8; Gafgard 233; 2-(Hydroxymethyl)-2-(((1-oxoallyl)oxy)methyl)-1,3-propane-diyl diacrylate; Kayarad PET 30; Light Acrylate PE 3A; NK Ester A-TMM 3; NK Ester A-TMM 3L; NK Ester TMM 50T; P 300 (acrylate); Pentaerythritol triacrylate; PETA; 2-Propenoic acid, 2-(hydroxymethyl)-2-(((1-oxo-2-propenyl)oxy)methyl)-1,3-propanediyl ester; Sartomer SR 444; Setalux UV 2242; SR 444; SR 444C; Tetra-methylolmethane triacrylate. Crosslinking agent used in adhesives, coatings, inks, textile products, photoresists, castings, modifiers for polyester, fiberglass, or polymers. A trifunctional cross-linking acrylic monomer cured by UV light used in the production of polyfunctional aziridine, added to paint primer and floor top coatings as a self-curing cross-linker or hardener. Often admixed with 300-400 ppm MEHQ inhibitor. Liquid; bp = 205-215° (detonates); insoluble in H2O. *Sartomer.*

2901   Pentaerythritol tristearate
28188-24-1   H

C59H114O7

Stearic acid, triester with pentaerythritol.

EINECS 248-889-6; 2-(Hydroxymethyl)-2-(((1-oxooctadecyl)oxy)methyl)propane-1,3-diyl distearate; Octadeca-noic acid, 2-(hydroxymethyl)-2-(((1-oxooctadecyl)oxy)-methyl)-1,3-propanediyl ester; Pentaerythritol tristearate; Stearic acid, triester with pentaerythritol.

2902   Pentaethylene glycol
4792-15-8   H

C10H22O6

3,6,9,12-Tetraoxatetradecane-1,14-diol.

4-01-00-02405 (Beilstein Handbook Reference); BRN 1635593; EINECS 225-341-4; Pentaethylene glycol.

476

2903 Pentafluoroethane
354-33-6   G

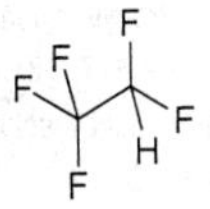

C_2HF_5

1,1,2,2,2-Pentafluoroethane.

CCRIS 7630; EINECS 206-557-8; Ethane, pentafluoro-; F 125; FC 125; Fron 125; Genetron® HFC 125; HCFC 125; HFC-125; HSDB 6755; Khladon 125; Pentafluoroethane; R 125; Refrigerant gas R 125; UN3220. For use in low temperature refrigeration applications Gas; bp = -48°. *Allied Signal.*

2904 Pentagastrin
5534-95-2   7188   D G

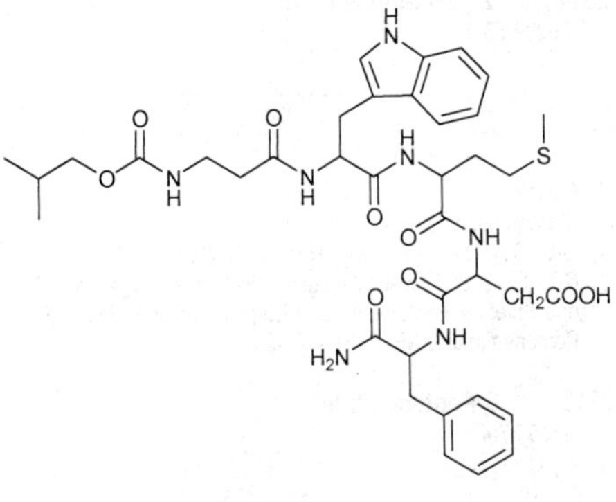

$C_{37}H_{49}N_7O_9S$

N-[(1,1-Dimethylethoxy)carbonyl]-β-alanyl-L-tryptophan-yl-L-methionyl-L-α-aspartyl--L-phenylalaninamide.

Alaninamide, N-carboxy-β-alanyl-L-tryptophyl-L-meth-ionyl-L-aspartylphenyl-, N-t-butyl ester, L-; AY 6608; Boc-β-ala-try-met-asp-phe(NH2); BRN 5472892; EINECS 226-889-7; Gastrodiagnost; HSDB 3247; ICI 50123; L-Phenyl-alaninamide, N-((1,1-dimethylethoxy)carbonyl)-β-alanyl-L-tryptophyl-L-methionyl-L-α-aspartyl-; N-(N-(N-(N-(N-t-Butoxycarbonyl-β-alanyl)-L-tryptophanyl)-L-meth-ionyl)-L-aspartyl)-L-phenylalaninamide; N-Carboxy-β-alanyl-L-tryptophyl-L-methionyl-L-aspartylphenyl-L-alaninamide N-t-butyl ester; N-t-Butyloxycarbonyl-β-alanyl-L-trypto-phyl-L-methionyl-L-aspartyl-L-phenylalan-ine amide; NSC 367746; Pentagastrin; Pentagastrina; Pentagastrine; Pentagastrinum; Peptavlon; Petogasrin. Gastric secretion stimulant used as a diagnostic aid. Solid; mp = 229-230°; $[\alpha]_D^{22}$ = -29° (DMF); λ_m = 280, 289 nm (ε 5340, 4590, 2N NH4OH); almost insoluble in H2O, organic solvents. *ICI.*

2905 Pentamethylheptane
13475-82-6   G

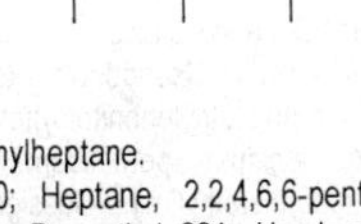

$C_{12}H_{26}$

2,2,4,6,6-Pentamethylheptane.

EINECS 236-757-0; Heptane, 2,2,4,6,6-pentamethyl-; 2,2,4,6,6-Pentamethylheptane; Permethyl 99A. Used as a cosolubilizer for nonhydrocarbon materials; for mascara, eyeliner, antiperspirant and where residual film is not desirable; solvent for debris on skin. *Presperse.*

2906 Pentamethylindane
1203-17-4   H

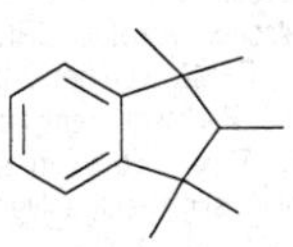

$C_{14}H_{20}$

1,1,2,3,3-Pentamethylindane.

1H-Indene, 2,3-dihydro-1,1,2,3,3-pentamethyl-; EINECS 214-868-5; HSDB 5793; Indan, 1,1,2,3,3-pentamethyl-; 1,1,2,3,3-Pentamethylindan; 1,1,2,3,3-Pentamethyl-indane.

2907 Pentane
109-66-0   7193   G H

C_5H_{12}

n-Pentane.

AI3-28785; Amyl hydride; Caswell No. 642AA; EINECS 203-692-4; EPA Pesticide Chemical Code 098001; HSDB 109; NSC 72415; Pentan; Pentane; n-Pentane; Pentanen; Pentani; Skellysolve A; Tetrafume; Tetrakil; Tetraspot; UN1265. Solvent; artificial ice manufacture, low-temperature thermometers, solvent extract processes, blowing agent in plastics (expandable polystyrene), pesticide. Liquid; mp = -129.7°; bp = 36.0°; d^{20} = 0.6262; slightly soluble in H2O, soluble in CCl4; freely soluble in EtOH, Et2O, Me2CO, CHCl3, C6H6, C7H16; LC100 (mus ihl) = 128200 ppm. *Ashland; Phillips 66.*

2908 Pentanochlor
2307-68-8   8851   G P

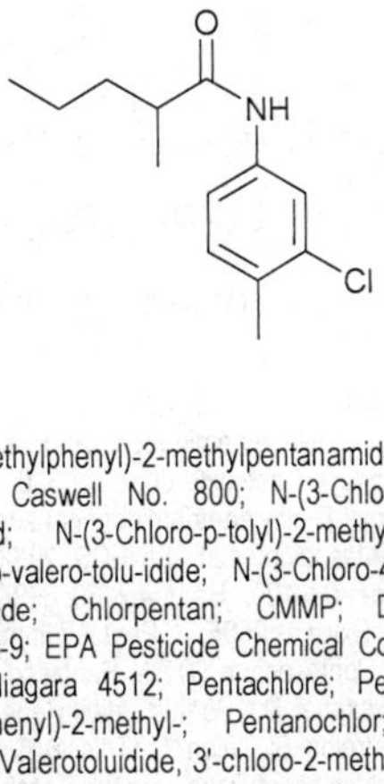

$C_{13}H_{18}ClNO$

N-(3-Chloro-4-methylphenyl)-2-methylpentanamide.

BRN 2722779; Caswell No. 800; N-(3-Chlor-methylphenyl)-2-methylpentanamid; N-(3-Chloro-p-tolyl)-2-methylvaleramide; 3'-Chloro-2-methyl-p-valero-tolu-idide; N-(3-Chloro-4-methylphenyl)-2-methylpentan-amide; Chlorpentan; CMMP; Dakuron; Dutom; EINECS 218-988-9; EPA Pesticide Chemical Code 020901; FMC 4512; Hortox; Niagara 4512; Pentachlore; Pentanamide, N-(3-chloro-4-methylphenyl)-2-methyl-; Pentanochlor; Pentanochlore; Solan; Solane; p-Valerotoluidide, 3'-chloro-2-methyl-. Pre- and post-emergence herbicide used in carrots, celery, fennel, parsley, parsnip, tomatoes and some flowers. Used as a contact spray with carnations, roses, fruit and ornamental trees and shrubs. Pale yellow powder; mp = 85-86°; d^{20} = 1.106; soluble in H2O (0.085 g/100 ml); freely soluble in organic solvents; LD50 (rat orl) > 10000 mg/kg; not very toxic to bees. *Atlas Interlates Ltd; Hortichem Ltd.*

2909 n-Pentanol
71-41-0   7195   G H

C5H12O
1-Pentanol.

4-01-00-01640 (Beilstein Handbook Reference); AI3-01293; Alcool amylique; Amyl alcohol; Amylol; BRN 1730975; Butyl carbinol; EINECS 200-752-1; FEMA Number 2056; HSDB 111; NSC 5707; Pentanol; Pentanol-1; Pentasol; Pentyl alcohol; Pentan-1-ol; Primary amyl alcohol; UN1105. Raw material for pharmaceutical preparations, organic synthesis solvent; flotation agent, organic synthesis, medicine (sedative). Liquid; mp= -78.9°; bp = 137.9°; d^{20} = 0.8144; slightly soluble in H_2O, freely soluble in organic solvents; LD_{50} (rat orl) = 2200 mg/kg. *Ashland; Condea Vista Co.; Hoechst Celanese; MTM Spec. Chem. Ltd.; Union Carbide Corp.*

2910   2-Pentanone
107-87-9   6137   G H

C5H10O
Methyl propyl ketone.

4-01-00-03271 (Beilstein Handbook Reference); AI3-32118; BRN 0506058; EINECS 203-528-1; Ethyl acetone; FEMA Number 2842; HSDB 158; Methyl n-propyl ketone; Methyl-propyl-cetone; Methyl propyl ketone; Metylopropyloketon; NSC 5350; 2-Pentanone; Propyl methyl ketone; UN 1249. Solvent, substitute for diethyl ketone, flavoring. Mobile liquid; mp = -76.9°; bp = 102.2°; d^{20} = 0.809; λ_m = 280 nm (cyclohexanone); slightly soluble in H_2O (4 g/100 ml), Cl4, freely soluble in EtOH, Et2O; LD_{50} (rat orl) = 1600 mg/kg. *Ashland; CasChem; Janssen Chimica; Penta Mfg.; Sigma-Aldrich Fine Chem.*

2911   Pentasodium pentetate
140-01-2
                                                          G H

C14H18N3Na5O10
Pentasodium diethylenetriaminepentaacetate.

Caswell No. 642B; Cheelox® 80; Chel 330; Chelest P; Clewat DP 80; Detarex py; Diethylenetriaminepenta-acetic acid pentasodium salt; Dtpa pentasodium salt; EINECS 205-391-3; EPA Pesticide Chemical Code 039120; Hamp-ex 80; HSDB 5629; Kiresuto P; Pentaquest Extra 0685; Pentasodium diethylenetriamine-pentaacetate; Pentasodium DTPA; Pentasodium pent-etate; Perma kleer 140; Plexene D; Sodium diethylene-triaminepentaacetate; Syntron C; Tetralon B; Trilon C; Trilon® C Liq; Versenex 80. All purpose chelating agent used in pulp bleaching applications using hydrogen peroxide. Chelating agent for iron chelation up to pH 11.5, used when higher metal chelate stability is required; used in peroxide bleach systems. Solid; mp = -40°; bp = 106°; d = 1.299; soluble in H_2O. *BASF Corp.; Clough; Rhône Poulenc Surfactants.*

2912   1-Pentene
109-67-1   7199   H

C5H10
n-Pent-1-ene.

'α-Amylene; α-n-Amylene; EINECS 203-694-5; HSDB 1086; 1-Pentene; 1-Pentylene; Propylethylene. Liquid; mp = -165.2°, bp = 29.9°; d^{20} = 0.6405; λ_m = 177, 181, 187 nm (16596, 13804, 4467, gas); insoluble in H_2O, slightly soluble in CCl4, soluble in C6H6, freely soluble in EtOH, Et2O.

2913   2-(E)-Pentenenitrile
26294-98-4   H

C5H7N
(E)-Pent-2-enenitrile.

EINECS 247-593-4; (E)-Pent-2-enenitrile; 2-Pentenenitrile, (2E)-; 2-Pentenenitrile, (E)-.

2914   2-(Z)-Pentenenitrile
25899-50-7   H

C5H7N
2-Pentenenitrile, (Z)-.

4-02-00-01546 (Beilstein Handbook Reference); BRN 1719855; CCRIS 1229; 1-Cyano-1-butene; cis-1-Butenyl cyanide; (Z)-Pent-2-enenitrile; 2-Pentenenitrile, (2Z)-; 2-Pentenenitrile, (Z)-; cis-2-Pentenenitrile; EINECS 247-323-5.

2915   3-Pentenenitrile
4635-87-4   H

C5H7N
3-Pentenenitrile.

CCRIS 6090; EINECS 225-060-7; HSDB 6779; 3-Pentenenitrile. Liquid; bp = 144°; d^4 = 0.83.

2916   Pentetic acid
67-43-6   7202   D G

C14H23N3O10
Diethylenetriaminepentaacetic acid.

4-04-00-02454 (Beilstein Handbook Reference); Acetic acid, ((carboxymethylimino)bis(ethylenenitrilo))tetra-; Acide pentetique; Acido pentetico; Acidum penteticum; BRN 1810219; 2,2'-(Carboxymethylimino)bis(ethylimino-diessigsaeure); 2,2'-(Carboxymethylimino)bis(ethylimino-diessigsäure); Chel DTPA; CHEL 330 acid; Complexon V; Dabeersen 503; Detapac; Detarex; Diethylenetriamine-N,N,N',N'',N''-pentaacetic acid; (Diethylenetrinitrilo)-pentaacetic acid; 1,1,4,7,7-

Diethylenetriaminepenta-acetic acid; Dissolvine D; DTPA; EINECS 200-652-8; Glycine, N,N-bis(2-(bis(carboxymethyl)amino)ethyl)-; Hamp-Ex Acid; Monaquest CAI; N,N-Bis(2-(bis(carboxymethyl)amino)ethyl)glycine; NSC 7340; Pentacarboxymethyldiethylenetriamine; Pentetic acid; Penthamil; Perma kleer; Titriplex V; 3,6,9-Triaza-undecanedioic acid, 3,6,9-tris(carboxymethyl)-. Chelating agent with affinity for iron; scale and corrosion inhibitor for aqueous systems; stable to hydrogen peroxide. Sequestrant used in peroxide bleach baths. Crystals; mp = 220° (dec); soluble in H_2O (5 g/l). *Cia-Shen; Clough.*

2917   Pentyl acetate
624-41-9   H

C7H14O2
 2-Methylbutyl acetate.
 1-Butanol, 2-methyl-, acetate; EINECS 210-843-8; FEMA No. 3644; 2-Methyl-1-butyl acetate; 2-Methylbutyl acetate; Pentyl acetate. Liquid; bp = 140°; d^{20} = 0.8740; very soluble in Me_2CO, EtOH, Et_2O.

2918   Perchloric acid
7601-90-3   7232   G

ClHO4
 Perchloric acid.
 EINECS 231-512-4; Fraude's reagent; HSDB 1140; Perchloric acid; UN1802; UN1873. Used in analytical chemistry, as a catalyst, in manufacture of esters, as an ingredient of electrolytic bath in deposition of lead, electropolishing, explosives. Oxidizing agent and bleach. Liquid; mp = -112°; bp_{11} = 19°; d^{22} = 1.768. *Spectrum Chem. Manufacturing.*

2919   Perfluidone
37924-13-3   7236   G P

C14H12F3NO4S2
 Trifluoro-N-(2-methyl-4-(phenylsulfonyl)phenyl)methane-sulfonamide.
 BRN 2180629; Caswell No. 903D; Destun; EPA Pesticide Chemical Code 108001; Flamprop-m-methyl; Lancer; MBR 8251; Methanesulfonamide, 1,1,1-trifluoro-N-(2-methyl-4-(phenylsulfonyl)phenyl)-; Methanesulfonamide, N-(4-phenylsulfonyl-o-tolyl)-1,1,1-trifluoro-; Methyl-4-(phenylsulfonyl)trifluoromethanesulfonanilide; Methyl-N-benzoyl-N-(3-chloro-4-fluorophenyl)alaninate, (R)-; (R)-Methyl-N-benzoyl-N-(3-chloro-4-fluorophenyl)alaninate; 2-Methyl-4-(phenylsulfonyl)trifluoromethanesulfonanil-ide; Perfluidone; N-(4-

Phenylsulfonyl-o-tolyl)-1,1,1-tri-fluoromethanesulfonamide; SB 1528; 1,1,1-Trifluoro-2'-methyl-4'-(phenylsulphonyl)methane-sulphonanilide; 1,1,1-Trifluoro-4'-(phenylsulfonyl)meth-anesulfono-o-toluidide; 1,1,1-Trifluoro-N-(2-methyl-4-(phenylsulfonyl)-phenyl)methanesulfonamide; 1,1,1-Trifluoro-N-(4-phenyl-sulphonyl-o-tolyl)methanesulphonamide; 1,1,1-Trifluoro-N-(4-phenylsulfonyl-o-tolyl)methanesulf-onamide; WL 43423. Herbicide for wild oat control. Registered by EPA as a herbicide (cancelled). Crystalline solid; mp = 142 - 144°; pKa = 2.5; soluble in H_2O (0.006 g/100 ml), Me_2CO (75 g/100 ml), C_6H_6 (1.1 g/100 ml), CH_2Cl_2 (16.2 g/100 ml), MeOH (60 g/100 ml). *ICI Chem. & Polymers Ltd.*

2920   Perfluoro(1-methyldecalin)
306-92-3   G

C11F20
 Perfluoro-1-methyldecalin.
 3-05-00-00269 (Beilstein Handbook Reference); BRN 2030021; EINECS 206-191-9; Flutec PP9; 1,1,2,2,3,3,4,4,4a,5,5,6,6,7,7,8,8a-Heptadecafluorodeca-hydro-8-(trifluoromethyl)naphthalene; Heptadecafluoro-decahydro-1-(trifluoromethyl)naphthalene; Naphthalene, heptadecafluorodecahydro-1-(trifluoromethyl)-; PP 9. A fluorinated hydrocarbon.

2921   Perfluoro-1,3-dimethylcyclohexane
335-27-3   G

C8F16
 Decafluoro-1,3-bis(trifluoromethyl)cyclohexane.
 Cyclohexane, decafluoro-1,3-bis(trifluoromethyl)-; Cyclo-hexane, 1,1,2,2,3,3,4,5,5,6-decafluoro-4,6-bis(trifluoro-methyl)-; Cyclohexane, perfluoro-1,3-dimethyl-; 1,1,2,2,3,3,4,5,5,6-Decafluoro-4,6-bis(trifluoromethyl)-cyclohexane; EINECS 206-386-9; Flutec PP3; Flutec PP3102; NSC 4782; Perfluoro(1,3-dimethylcyclohexane). A fluorinated hydrocarbon used in heat transfer. Liquid; mp = -55°; bp = 101-102°; d = 1.8280.

2922   Perfluorohexane
355-42-0   G

$$C_6F_{14}$$

C6F14
 Perfluoro-n-hexane.
 AF0150; EINECS 206-585-0; Fluorinert FC72; Flutec PP1; Hexane, tetradecafluoro-; Perflexane; Perfluorohexane; Perfluoro-n-hexane; Tetradecafluorohexane; n-Tetra-decafluorohexane. A fluorinated hydrocarbon, used in heat transfer. Liquid; mp = -87.1°; bp = 56.6°; d^{20} = 1.6995; insoluble in H_2O, soluble in Et_2O, C_6H_6, $CHCl_3$.

2923   Perfluoromethylcyclohexane
355-02-2   G

C7F14

Tetradecafluoromethylcyclohexane.

Cyclohexane, 1-trifluoromethyl-1,2,2,3,3,4,4,5,5,6,6-undecafluoro-; Cyclohexane, undecafluoro(trifluoro-methyl)-; EINECS 206-573-5; Flutec PP2; NSC 4779; Perfluoro(methylcyclohexane); Tetradecafluoromethyl-cyclohexane; Undecafluoro(trifluoromethyl)cyclohexane. A fluorinated hydrocarbon used in heat transfer. Liquid; mp = -44.7°; bp = 76.3°; d;ss25 = 1.7878; soluble in Me2CO, C6H6, CCl4, C7H8, EtOAc.

2924   Perfluorooctanoic acid

335-67-1   G

C8HF15O2

Pentadecafluorooctanoic acid.

4-02-00-00994 (Beilstein Handbook Reference); AI3-19341; BRN 1809678; CCRIS 4386; EINECS 206-397-9; Fluorad® FC-26; Hexanoyl fluoride, 3,3,4,4,5,5,6,6,6-nonafluoro-2-oxo-; NSC 95114; Octanoic acid, penta-decafluoro-; Pentadecafluoro-1-octanoic acid; Pentadeca-fluoro-n-octanoic acid; Perfluorocaprylic acid; Perfluor-octanoic acid; Perfluoroheptanecarboxylic acid; Per-fluorooctanoic acid; PFOA. Intermediate for preparation of monomers and surfactants. Solid; mp = 59-60°; bp = 192°; soluble in H2O (3.4 g/l), organic solvents. 3M Company.

2925   Perfluorooctylsulfonyl fluoride

307-35-7   G H

C8F18O2S

Heptadecafluoro-1-octanesulfonyl fluoride.

EINECS 206-200-6; Fluorad® FX-8; Heptadecafluoro-octanesulphonyl fluoride; 1,1,2,2,3,3,4,4,5,5,6,6,7,7,8, 8,8-Heptadecafluoro-1-octanesulfonyl fluoride; HSDB 5561; 1-Octanesulfonyl fluoride, 1,1,2,2,3,3,4,4,5,5,6,6, 7,7,8,8,8-heptadecafluoro-; N-Perfluorooctanesulfonyl fluoride; Perfluorooctanesulfonyl fluoride. Intermediate for preparation of monomers and surfactants for textile treatment, paper sizes, inert fluids. Liquid; mp = -1°; bp = 154-155°; d = 1.8240. 3M.

2926   Periodic acid

10450-60-9   7248   G

H5IO6

Periodic Acid Dihydrate.

EINECS 233-937-0; Orthoperiodic acid; Periodic acid; Periodic acid (H5IO6). Oxidizing agent, increases wet strength of paper, photographic paper. Used in organic synthesis. Monoclinic crystals; mp = 122°; soluble in H2O (30 g/ml), less soluble in organic solvents.

Atomergic Chemetals; Blythe, Williams Ltd.; EM Ind. Inc.; Janssen Chimica; Spectrum Chem. Manufacturing.

2927   Perkadox® 26-fl

53220-22-7   G

C30H58O6

Dimyristyl peroxydicarbonate.

Ditetradecyl peroxydicarbonate; EINECS 258-436-4; Peroxydicarbonic acid, dimyristyl ester; Peroxydicarbonic acid, ditetradecyl ester. Catalyst, initiator. Akzo Chemie.

2928   Permethrin

52645-53-1   7257   G P

C21H20Cl2O3

(±)-(cis,trans)-3-Phenoxybenzyl   3-(2,2-dichlorovinyl)-2,2-dimethylcyclopropanecarboxylate.

3768; Acticin; Adion; AI3-29158; Ambush; Ambushfog; Anomethrin N; Antiborer; Antiborer 3768; Atroban; Bematin 987; Bio Flydown; Biomist; BRN 2063148; BW-21-Z; Caswell No. 652BB; CCRIS 2001; Chinetrin; Cooper; Coopex; Corsair; Cosair; Cyclopropane-carboxylic acid, 3-(2,2-dichloroethenyl)-2,2-dimethyl-,(3-phenoxyphenyl)methyl ester; Cyclopropanecarboxylic acid, 3-(2,2-dichlorovinyl)-2,2-dimethyl-, 3-phenoxy-benzyl ester, (±)-, (cis,trans)-; (±)-cis,trans-3-(2,2-dichloroethenyl)-2,2-dimethylcyclopropanecarboxylate m-phenoxybenzyl; 3-(2,2-Dichloroethenyl)-2,2-dimethyl-cyclopropane carboxylic acid, (3-phenoxyphenyl) methyl ester; Diffusil H; Dragnet; Dragnet FT; Dragon; Ecsumin; Ectiban; Efmethrin; EINECS 258-067-9; Eksmin; Elimite; EPA Pesticide Chemical Code 109701; Epigon; Exmin; Expar; Exsmin; Flee; FMC 33297; FMC 41655; HSDB 6790; ICI-PP 557; Imperator; Indothrin; Insorbcid MP; Ipitox; JF 7065; Jureong; Kafil; Kaleait; Kavil; Kestrel; Kestrel (pesticide); Ketokil; Kudos; LE 79-519; Lyclear; m-Phenoxybenzyl (+1)-cis,trans-3-(2,2-dichlorovinyl)-2,2-dimethylcyclopropanecarboxylate; m-Phenoxybenzyl (±)-3-(2,2-dichlorovinyl)-2,2-dimethylcyclopropanecarb-oxylate; m-Phenoxybenzyl 3-(2,2-dichlorovinyl)-2,2-dimethylcyclopropanecarboxylate; (±)-3-Phenoxybenzyl 3-(2,2-dichlorovinyl)-2,2-dimethylcyclopropanecarboxyl-ate; 3-Phenoxy-benzyl(±)-cis, trans-3-(2,2-dichlorovinyl)-2,2-dimethyl-cyclopropane-1-carboxylate; 3-Phenoxy-benzyl (1RS)-cis-trans-3-(2,2-dichlorovinyl)-2,2-dimethyl-cyclopropanecarboxylate; 3-Phenoxybenzyl 2,2-dimethyl-3-(2,2-dichlorovinyl)cyclopropanecarboxylate; (3-Phenoxyphenyl)-methyl 3-(2,2-dichlorethenyl)-2,2-dimethylcyclopropane-carboxylate; Matadan; Mitin BC; MP79; NIA 33297; Nix; NRDC 143; OMS 1821; Outflank; Outflank-stockade; Perigen; Perigen W; Permanone; Permanone 10; Permanone 40; Permanone 80; Permasect; Permasect-25EC; Permethrin; Permethrin, racemic; Permethrine; Permethrinum; Permetrin; Permetrina; Permit; Permitrene; Persect; Perthrine; Pertox; Picket; Picket G; Pounce; PP 557; Pramex; Pynosect; Quamlin; Ridect Pour-On; S 3151; SBP-1513; SBP-1513TEC; SBP 15131TEC; Spartan; Stomoxi; Stomoxin; Stomoxin P; Stomozan;

480

Talcord; WL 43479. Non-systemic insecticide with contact and stomach action. Used for control of the larvae of lepidopterous and coleopterous insect pests in fruit and vegetable crops. Registered by EPA as an insecticide. Colorless crystals - pale yellow liquid; mp = 35°; bp$_{0.05}$ = 220°; d^{20}= 1.190 - 1.272; almost insoluble in H$_2$O (< 0.0001 g/100 ml), insoluble in ethylene glycol, soluble in xylene (> 86 g/100 ml), C$_6$H$_{14}$ (> 66 g/100 ml), MeOH (20.6 g/100 ml), soluble in most other organic solvents; LD$_{50}$ (frat orl) = 3801 mg/kg, (mrat orl) = 1500 mg/kg, (mrat 8-days old orl) = 340.5 mg/kg, (ckn orl) > 3000 mg/kg, (Japanese quail orl) > 13,500 mg/kg; LC$_{50}$ (rat ihl) > 23.5 mg/l air, (rainbow trout 48 hr.) = 0.0054 mg/l, (bluegill sunfish) = 0.0018 mg/l; toxic to bees. *Am. Cyanamid; Bengal Products Inc.; Boehringer Ingelheim Vetmedica Inc.; Bonide Products, Inc.; Chemisco; Chem-Tech Ltd.; CLL Custom Manufacturing Inc.; Dragon Chemical Corp.; DVM Pharmaceuticals Inc.; Farnam Companies Inc.; FMC Corp.; Mclaughlin Gormley King Co; PBI/Gordon Corp; PET Chemicals; S.C. Johnson & Son, Co; Schering-Plough Veterinary Operations, Inc.; Speer Products Inc.; Sumitomo Corp.; Syngenta Crop Protection; Unicorn Laboratories; Valent Biosciences Corp.; Value Gardens Supply LLC; Virbac AH Inc.; Zeneca Pharm.*

2929   Permethyl 102A
93685-79-1   G

C$_{20}$H$_{42}$
Isoeicosane.
EINECS 297-627-7; Hydrocarbons, C4, 1,3-butadiene-free, polymd, pentaisobutylene fraction, hydrogenated. Isoeicosane, cosolubilizer for nonhydrocarbon materials; for skin care and sun care products; plasticizer for mascara. *Presperse.*

2930   Peroxyacetic acid
79-21-0   7229   G P

C$_2$H$_4$O$_3$
Monoperacetic acid.
4-02-00-00390 (Beilstein Handbook Reference); Acetic Peroxide; Acetyl Hydroperoxide; Acide peracetique; Acide peroxyacetique; Acido peroxiacetico; BRN 1098464; Caswell No. 644; CCRIS 686; Desoxon 1; EINECS 201-186-8; EPA Pesticide Chemical Code 063201; Estosteril; Ethaneperoxic acid; Ethaneperoxoic acid; HSDB 1106; Hydroperoxide, acetyl; Kyselina peroxyoctova; Monoperacetic acid; Osbon AC; Oxymaster; PAA; Peracetic acid; Peroxoacetic acid; Proxitane 4002. Registered by EPA as an antimicrobial, fungicide, herbicide and rodenticicide. A strong oxidizing agent, used in sewage treatment. Liquid with acrid odor; mp = -0.2°; bp = 110°; decomposes explosively at 110°; d^{15} = 1.226; freely soluble in H$_2$O, EtOH, Et$_2$O, H$_2$SO$_4$. *Degussa Ltd.; Diverseylever; Ecolab Inc.; Enviro Tech Chemical Services; FMC Corp.; Minntech Corp; Solvay Interox Inc.; Steris Corp.*

2931   Perstoff
503-38-8   3368   G

C$_2$Cl$_4$O$_2$
Trichloromethyl-chloroformate.
4-03-00-00033 (Beilstein Handbook Reference); BRN 0970225; Carbonochloridic acid, trichloromethyl ester; Difosgen; Diphosgen; Diphosgene; EINECS 207-965-9; Formic acid, chloro-, trichloromethyl ester; HSDB 371; Methanol, trichloro-, chloroformate; Perchloromethyl formate; Superpalite; Trichlormethylester kyseliny chlormravenci; Trichloromethyl chloroformate. A poison gas. Liquid; mp = -57°; bp = 128°; d^{14} = 1.6525; very soluble in EtOH, Et$_2$O.

2932   Perthane
72-56-0   G

C$_{18}$H$_{20}$Cl$_2$
Diethyldiphenyldichloroethane.
4-05-00-01985 (Beilstein Handbook Reference); AI3-17082; Benzene, 1,1'-(2,2-dichloroethylidene)bis(4-ethyl-; 1,1-Bis(ethylphenyl)-2,2-dichloroethane; 1,1-Bis(p-ethylphenyl)-2,2-dichloroethane; 2,2-Bis(p-ethylphenyl)-1,1-dichloroethane; BRN 2054366; Caswell No. 337; CCRIS 292; DDD; α,α-Dichloro-2,2-bis(p-ethylphenyl)-ethane; 1,1-Dichloro-2,2-bis(p-ethylphenyl)ethane; 1,1-Dichloro-2,2-bis(4-ethylphenyl)ethane; 2,2-Dichloro-1,1-bis(p-ethylphenyl)ethane; Di(p-ethylphenyl)dichloro-ethane; Diethyl-diphenyl dichloroethane; EINECS 200-785-1; ENT 17,082; EPA Pesticide Chemical Code 032101; Ethane, 1,1-dichloro-2,2-bis(p-ethylphenyl)-; Ethane, 2,2-bis(p-ethylphenyl)-1,1-dichloro-; Ethylan; HSDB 4009; NCI-C02868; Perthane; p,p'-Ethyl DDD; Q-137. An agricultural insecticide supplied as a wettable powder or emulsifiable concentrate; controls insects on plants and livestock; has a lower toxicity to both insects and mammals than its structural analogs, DDT and DDD, and is of moderate persistence in the environment; also used as a moth protection for textiles. First marketed in 1950 for use against houseflies and cloth moths, it has since been used on vegetables, pears, and livestock. Not a carcinogen. Solid; mp = 60-61°; insoluble in H$_2$O, soluble in organic solvents; LD$_{50}$ (rat orl) = 6600 mg/kg. *Pillar.*

2933   Pertscan-99m
23288-60-0   8730   G

NaO$_4$^{99m}Tc
Sodium pertechnitate ^{99m}Tc.
MPI Tc 99m Generator; NeoTect; Pertechnetic acid (H^{99m}TcO$_4$), sodium salt; Pertscan-99m; Sodium pertechnetate (Na^{99m}TcO$_4$); Sodium pertechnetate (Tc-99M); Sodium pertechnetate ^{99m}Tc; Sodium pertechnetate Tc 99m; Ultra-Technekow FM. A radioactive agent. Used in brain scans, thyroid function tests. *Abbott Labs Inc.*

2934   PETN
78-11-5   7186   H

C5H8N4O12
Pentaerythritol tetranitrate.
4-01-00-02816 (Beilstein Handbook Reference); Angicap; Angitet; Antora; Arcotrate; Baritrate; BRN 1716886; C 2; Cardiacap; CCRIS 2387; CHot; CHOT; Deltrate 20; Dilcoran 80; Dipentrate; Duotrate; EINECS 201-084-3; El PETN; Erinit; Erynitum; Extex; Hasethrol; HSDB 6313; Kaytrate; Lentrat; Lowetrate; LX 02-1; LX 08-0; LX 13; LX 16 (explosive); Martrate 45; Metranil; Mikardol; Miltrate; Mycardol; Myotrate 10; NA0150; NCI-C55743; Neo-corovas; Neopentanetetrayl nitrate; Niperyt; Niperyth; Nitrin; Nitrinal; Nitrine; Nitrinol; Nitro-Riletten; Nitrolong; Nitropent; Nitropenta; Nitropenta 7W; Nitropentaerythrite; Nitropentaerythritol; Nitropenton; Nitrotalans; PBXN 301; Pen-Tetra; Pencard; Penta; Pentaerithrityl tetranitrate; Pentaerithrityli tetranitras; Pentaeritrile tetranitrato; Pentaerythrite tetranitrate; Pentaerythritol nitrate; Pentaerythritol tetranitrate; Pentaerythritylium tetranitricum; Pentafin; Pentanitrine; Pentanitrol; Pentanitrolum; Pentarit; Pentestan-80; Pentetrate unicelles; Penthrit; Penthrite; Pentitrate; Pentral 80; Pentrate; Pentriol; Pentrite; Pentritol Tempules; Pentryate; Pentryate 80; Pergitral; Peridex; Peridex-LA; Peritrate; Perityl; PETN; Prevangor; Quintrate; Rythritol; SDM No. 23; SDM No. 35; Subicard; Tanipent; Tentrate-20; Terpate; Tetranitrate de pentaerithrityle; Tetranitrato de pentaeritritilo; Tetranitropentaerythrite; Tetrasule; Tranite D-lay; UN0411; Vasitol; Vaso-80; Vaso-80 unicelies; Vasodiatol; Vasolat; XTX 8003. Explosive used in Primacord® detonating fuse. Diluted with, for example, lactose, used medically as a vasodilator to treat angina pectoris. Solid; mp = 140.5°; d^{20} = 1.773; slightly soluble in H_2O, MeOH, EtOH, Et_2O, more soluble in Me_2CO, C_6H_6, C_7H_8, C_5H_5N. *EMS-Dottikon AG.*

2935   Petrolatum, liquid
8009-03-8   D H

Petrolatum.
A+D Original Ointment; Adepsine oil; Alboline; Carrafoam Incontinence Skin Care Kit; Carraklenz Incontinence Skin Care Kit; Caswell No. 645A; Clearteck; Cosmoline; Cream white; Drakeol; EINECS 232-373-2; EPA Pesticide Chemical Code 598400; Extra amber; Glymol; Hevyteck; HSDB 1138; Kaydol; Kremol; liquid paraffin; mineral oil; Mineral fat; Mineral grease (petrolatum); Mineral jelly; Mineral wax; Moisture Barrier Cream with Zinc; Moisture Barrier Cream; Moisture Guard; Nujol; Paraffin jelly; Paraffin oil; Paroleine; Pennsoline soft yellow; Penreco white; Perfecta; Petrolatum; Petrolatum amber; Petrolatum USP; Petrolatum white; Petrolatum, yellow; Petroleum jelly; Preparation H Cream; Preparation H Ointment; Protopet, alba; Protopet, white 1S; Protopet, white 2L; Saxol; Saxoline; Snow white; Stanolind; Ultima white; Vaseline; Vasoliment; white mineral oil; White petrolatum USP; White petroleum jelly; White protopet; White vaseline; Yellow petrolatum. Used as a laxative-cathartic. d (light oil) = 0.83 - 0.86; d (heavy oil) = 0.875 - 0.905; insoluble in H_2O, EtOH; soluble in C_6H_6, $CHCl_3$, Et_2O, CS_2, petroleum ether, oils.

2936   Petroleum distillates
8002-05-9   H

Petroleum distillates Naphtha, Rubber Solvent.
BASE oil; Caswell No. 632A; Coal oil [Oil, misc.]; Crankcase oil, used mineral-based; Crankcase Oil, Used; Crude oil; Crude oils; Crude petroleum; EINECS 232-298-5; EPA Pesticide Chemical Code 063503; Mineral oils, highly-refined oils [Mineral oils]; Naphtha;

Oil, crude; Paraffinic oil; Petroleum; Petroleum crude; Petroleum crude oil [Flammable liquid]; Petroleum distillate; Petroleum distillates; Petroleum distillates, n.o.s. or petroleum products, n.o.s.; Rock oil; Rubber solvent; Seneca oil; UN1267; UN1268; Used Mineral-based Crankcase Oil; Virol.

2937   Petroleum sulfonates
61789-85-3   H

Sulfonic acids, petroleum.
Caswell No. 647B; EINECS 263-092-3; EPA Pesticide Chemical Code 598500; Petrolatum acid sulfonate; Petroleum sulfonates; Sulfonic acids, petroleum.

2938   Petroleum sulfonic acids, calcium salts
61789-86-4   H

Sulfonic acid, petroleum, calcium salt.
Calcium petroleum sulfonate; EINECS 263-093-9; Petroleum sulfonic acids, calcium salts; Sulfonic acid, petroleum, calcium salt.

2939   Petroleum sulfonic acids, magnesium salts
61789-87-5   H

Sulfonic acid, petroleum, magnesium salt.
EINECS 263-094-4; Sulfonic acids, petroleum, magnesium salts.

2940   Phenarsazine oxide
58-36-6   7030   G P

C24H16As2O3
10-10' Oxybisphenoxyarsine.
2-27-00-00948 (Beilstein Handbook Reference); Bis(10-phenoxarsinyl)oxide; 10,10'-Bis(phenoxyarsinyl) oxide; Bis(phenoxarsin-10-yl) ether; BRN 0055641; DID 47; Diphenoxarsin-10-yl oxide; Durotex; EINECS 200-377-3; HSDB 6375; OBPA; 10,10'-Oxidiphenoxarsine; 10,10'-Oxybis-10H-phenoxarsine; 10H-Phenoxarsine, 10,10'-oxybis-; Phenarsazine oxide; Phenoxaksine oxide; Phenoxarsine, 10,10'-oxydi-; SA 546; Vinadine; Vinyzene; Vinyzene (pesticide); Vinyzene BP 5; Vinyzene BP 5-2; Vinyzene SB 1. Condensation product of epoxidized soy bean oil and 10, 10'-oxybisphenoxyarsine; used as a fungicide and bactericide, particularly in consumer plastics such as shower curtains, wall coverings and upholstery. Solid; mp= 184-185°; d = 1.41; insoluble in H_2O, soluble in organic solvents; LD_{50} (rat orl) = 35. mg/kg. *Morton Intn'l.; Rohm & Haas Co.; Rohm & Haas UK.*

2941   Phenetidine
156-43-4   7307   G H

C8H11NO
p-Ethoxyaniline.
4-13-00-01017 (Beilstein Handbook Reference); AI3-09042; 4-

Aminoethoxybenzene; 4-Aminophenetole; Aniline, p-ethoxy-; Benzenamine, 4-ethoxy-; BRN 0606666; CCRIS 2878; CP 5685; EINECS 205-855-5; 4-Ethoxyaniline; 4-Ethoxybenzenamine; Ethyl p-aminophenol; p-Fenetidin; NSC 3116; p-Phenetidin; p-Phenetidine; Phenetidine. An antipyretic. Used in the manufacture of acetophenetidin. Liquid; mp = 3°; bp = 253-255°; d_4^{16} = 1.0652; insoluble in H_2O, soluble in organic solvents; LD_{50} (rat orl) = 540 mg/kg.

2942   Phenetole
103-73-1   7308   G

C8H10O
Ethoxy benzene.
AI3-05616; Benzene, ethoxy; EINECS 203-139-7; Ether, ethyl phenyl; Ethoxybenzene; Ethyl phenyl ether; HSDB 112; NSC 406706; Phenetol; Phenetole; Phenyl ethyl ether. Liquid; mp = -29.5°; bp = 169.8°; d^{w20} = 0.9651; λ_m = 220, 271, 278 nm (ε = 6400, 1400, 1190, MeOH); insoluble in H_2O, soluble in EtOH, Et2O, CCl4; LD_{50} (mus orl) = 2200 mg/kg.

2943   Phenmedipham
13684-63-4   7317   G P

C16H16N2O4
3-[(Methoxycarbonyl)amino]phenyl (3-methylphenyl)-carbamate.
Beta; Betaflow; Betalion; Betamix; Betanal; Betanal E; Alegro; Beetomax; Beetup; Betosip; BRN 2395027; Carbamic acid, (3-methylphenyl)-, 3-((methoxycarbonyl)-amino)phenyl ester; Carbanilic acid, m-hydroxy-, methyl ester, m-methylcarbanilate; 3-(Carbomethoxyamino)-phenyl 3-methylcarbanilate; Caswell No. 648B; CCRIS 6091; EINECS 237-199-0; EP-452; EPA Pesticide Chemical Code 098701; Fender; Fenmedifam; Goliath; Gusto; HSDB 1402; Kemifam; m-Hydroxycarbanilic acid methyl ester m-methylcarbanilate; Medipham; Methyl 3-(3-methylcarbaniloyloxy)carbanilate; 3-(Methylphenyl)-carbamic acid 3-phenyl ester; Morton EP 452; Phenmedipham; Phenmediphame; Pistol; Protrum K; Protrum K SN 38584; S 4075; Schering 4072; Schering 38584; SN-38584; SN 4075; Spin-Aid; Spn-aid; Suplex; Synbetan P; Vangard. Selective systemic herbicide, absorbed through the leaves, inhibits photosynthesis. Used for post-emergence control of annual broad-leaved weeds in sugar beet, fodder beet, beetroot, mangels, spinach and strawberries. Selective systemic herbicide, absorbed through the leaves, inhibits photosynthesis. Used for post-emergence control of annual broad-leaved weeds in sugar beet, fodder beet, beetroot, mangels, spinach and strawberries. Registered by EPA as a herbicide. Colorless crystals; mp = 143 - 144°; soluble in H_2O (< 0.00047 g/100 ml at 25°), Me2CO (20 g/100 ml), cyclohexanone (20 g/100 ml), MeOH (5 g/100 ml), CHCl3 (2 g/100 ml), C6H6 (0.25 g/100 ml), C6H14 (0.05 g/100 ml); LD_{50} (rat, mus orl) > 8000 mg/kg, (gpg, dog orl) > 4000 mg/kg, (rat ip) > 5000 mg/kg, (ckn orl) > 3000 mg/kg, (mallard duck orl) = 2100 mg/kg; LC_{50} (mallard duck, bobwhite quail 8 day dietary)> 10000 mg/kg diet, (harlequin fish 96 hr.) = 16.5 mg/l (15.9% formulation); non-toxic to bees. *ABM Chemicals Ltd.; Atlas Interlates Ltd; Aventis Crop Science; Farmers Crop Chemicals Ltd.; Fine Agrochemicals Ltd.; MTM AgroChemicals Ltd.; Rhone-Poulenc UK; SBC Technology Ltd.; Schering Agrochemicals Ltd.; Universal*

Crop Protection Ltd.

2944   Phenobarbital sodium
57-30-7   7319   D G

C12H11N2NaO3
5-Ethyl-5-phenylbarbituric acid sodium salt.
 Barbituric acid, 5-ethyl-5-phenyl-, sodium salt; CCRIS 503; EINECS 200-322-3; 5-Ethyl-5-phenyl-2,4,6-(1H,3H,5H)pyrimidinetrione monosodium salt; Fenobarbital natrium; Fenobarbital sodico; Gardenal sodium; Linasen; Luminal sodium; Phenemalnatrium; Phenemalum; Phenobal sodium; Phenobarbital elixir; Phenobarbital Na; Phenobarbital sodique; Phenobarbital sodium; Phenobarbital sodium salt; Pheno-barbitalnatrium; Phenobarbitalum natricum; Pheno-barbiton-natrium; Phenobarbitone sodium; Pheno-barbitone sodium salt; Phenyl-aethyl-barbitursaeure natrium; Phenyl-äthyl-barbitursäure natrium; Phenyl-ethylbarbituric acid, sodium salt; 2,4,6(1H,3H,5H)-Pyrimidinetrione, 5-ethyl-5-phenyl-, monosodium salt; Sodium 5-ethyl-5-phenylbarbiturate; Sodium luminal; Sodium phenobarbital; Sodium phenobarbitone; Sodium phenobarbiturate; Sodium phenylethylbarbiturate; Sodium phenylethylmalonylurea; Sol phenobarbital; Sol phenobarbitone; Soluble phenobarbital; Soluble phenobarbitone. A sedative, hypnotic and anticonvulsant. Used as a hypnotic, may be habit forming; a controlled substance. Hygroscopic crystals or white powder; soluble in H_2O (100g/100 ml), EtOH (10g/100ml), insoluble in Et2O, CHCl3; pH9.3; LD_{50} (rat orl) = 660mg/kg. *Rhône Poulenc Rorer.*

2945   Phenoctide
78-05-7   7322   G

C27H42ClNO
 Benzyldiethyl-2-(p-(1,1,3,3-tetramethylbutyl)phenoxy)-ethylammonium chloride.
 Benzyldiethyl(2-(4-(1,1,3,3-tetramethylbutyl)phenoxy)-ethyl)ammonium chloride; EINECS 201-078-0; Octafonii chloridum; Octafonium chloride; Octaphen; Octa-phonium chloride; Phenoctidiumchlorid. Used as a topical anti-infective and as a source of its orthophosphate, a lubricant. Solid; mp = 112-114°.

2946   Phenol
108-95-2   7323   G P

C6H6O
Carbolic acid.

483

Acide carbolique; AI3-01814; Anbesol; Baker's P & S liquid & Ointment; Baker's P and S Liquid and Ointment; Benzene, hydroxy-; Benzenol; Campho-Phenique Cold Sore Gel; Campho-Phenique Gel; Campho-Phenique Liquid; Carbolic acid; Carbolic oil; Carbolsaure; Caswell No. 649; CCRIS 504; EINECS 203-632-7; EPA Pesticide Chemical Code 064001; FEMA No. 3223; Fenol; Fenolo; HSDB 113; Hydroxybenzene; Izal; Liquid phenol; Monohydroxybenzene; Monophenol; NCI-C50124; NSC 36808; Oxybenzene; Paoscle; Phenic acid; Phenol; Phenole; Phenyl alcohol; Phenyl hydrate; Phenyl hydroxide; Phenylic acid; Phenylic alcohol; PhOH; RCRA waste number U188; UN 2812 (solution); UN 1671 (solid); UN 2312 (molten). Solvent and chemical intermediate. Used in rubber manufacture, solvents, in phenolic resins. Registered by EPA as an antimicrobial and disinfectant. In trade, the term carbolic acid is used for pure phenol, the cresols and their mixtures with phenol, and also for crude tar oils. Crystals; mp = 40.9°; bp = 181.8°; d^{45} = 1.0545; λ_m = 264, 270, 277 nm (MeOH); soluble in H_2O (6.6 g/100 ml), C_6H_6 (8.3 g/100 ml); very soluble in EtOH, $CHCl_3$, Et_2O, CS_2; insoluble in petroleum ether; LD_{50} (rat orl) = 530 mg/kg. *Robins, A. H. Co.,; Sterling Health U.S.A.; Whitehall Labs. Inc.*

2947   Phenol Carbonate
102-09-0   7363   G

C13H10O3
Carbonic acid diphenyl ester.
4-06-00-00629 (Beilstein Handbook Reference); AI3-00063; BRN 1074863; Carbonic acid, diphenyl ester; Diphenyl carbonate; EINECS 203-005-8; HSDB 5346; NSC 37087; Phenyl carbonate; Phenyl carbonate ((PhO)2CO). Solvent for nitrocellulose, resins, lacquers, baking finishes. Lustrous needles; mp = 83°; bp = 306°; bp15 = 168°; insoluble in water; soluble in EtOH, Et2O, CCl4, AcOH. *Degussa-Hüls Corp.; Eastman Chem. Co.; MTM Spec. Chem. Ltd.; Penta Mfg.*

2948   Phenol red
143-74-8   7329   G

C19H14O5S
Phenol-sulfonphthalein.
5-19-03-00457 (Beilstein Handbook Reference); 3H-2,1-Benzoxathiole, 3,3-bis(4-hydroxyphenyl)-, 1,1-dioxide; 4,4'-(3H-2,1-Benzoxathiol-3-ylidene)diphenol S,S-diox-ide; BRN 0326470; EINECS 205-609-7; Fenolipuna; NSC 10459; Phenol, 4,4'-(3H-2,1-benzoxathiol-3-ylidene)di-, S-S-dioxide; Phenol, 4,4'-(1,1-dioxido-3H-2,1-benzoxa-thiol-3-ylidene)bis-; Phenol Red; Phenolsulfonphthalein; Phenolsulphonphthalein; PSP; PSP (indicator); Sulfonphthal; Sulphental; Sulphonthal. Used as an acid-base indicator; diagnostic reagent medicine and laboratory reagent. Crystals; mp >300°; λ_m = 236, 268, 416, 509 nm (ϵ = 19500, 6830, 5330, 2990, MeOH); soluble in KOH solution, slightly soluble in H_2O (0.08 g/100 ml), EtOH, Me2CO, C6H6, insoluble in Et2O, CHCl3;

LD_{50} (rat orl) >600 mg/kg.

2949   Phenol, 2-(2H-benzotriazol-2-yl)-4,6-di-t-pentyl-
25973-55-1   H

C22H29N3O
Phenol, 2-(2H-benzotriazol-2-yl)-4,6-bis(1,1-dimethyl-propyl)-.
2-(2H-Benzotriazol-2-yl)-4,6-ditertpentylphenol; EINECS 247-384-8; Phenol,   2-(2H-benzotriazol-2-yl)-4,6-bis(1,1-dimethylpropyl)-; Phenol, 2-(2H-benzotriazol-2-yl)-4,6-di-t-pentyl-.

2950   Phenol, 2-ethyl-
90-00-6   3871   G H

C8H10O
o-Ethylphenol.
4-06-00-03011 (Beilstein Handbook Reference); Benzene, 1-ethyl-2-hydroxy-; BRN 1099397; CCRIS 6038; EINECS 201-958-4; o-Ethylphenol; Florol; HSDB 5267; NSC 10112; Phenol, 2-ethyl-; Phenol, o-ethyl-; Phlorol. Organic intermediate. Liquid; mp = 18°; bp = 204.5°; d^{25} = 1.0146; soluble in EtOH, Me2CO, Et2O, C6H6; LD_{50} (mus orl) = 600 mg/kg.

2951   Phenol, 2,2'-methylenebis(4-methyl-6-nonyl-
7786-17-6   H

C33H52O2
p-Cresol, 2,2'-methylenebis(6-nonyl-.
EINECS 232-092-5; 2,2'-Methylenebis(6-nonyl-p-cresol); 2,2'-Methylenebis(6-nonyl-4-methylphenol); Nauga White; Naugawhite; Noclizer NS 90; NSC 111145.

2952   Phenol, (1-methyl-1-phenylethyl)-
27576-86-9   H

C15H16O
(1-Methyl-1-phenylethyl)phenol.
EINECS 248-539-2; (1-Methyl-1-phenylethyl)phenol; Phenol, (1-methyl-1-phenylethyl)-.

2953 Phenol, 2-((2-methyl-2-propenyl)oxy)-
4790-71-0   H

C10H12O2
2-((2-Methyl-2-propenyl)oxy)phenol.
2-((2-Methyl-2-propenyl)oxy)phenol; Phenol, 2-((2-methyl-2-propenyl)oxy)-.

2954 Phenolsulfonic acid
1333-39-7   G H

C6H6O4S
Hydroxy benzenesulfonic acid.
Benzenesulfonic acid, hydroxy-; Caswell No. 650; EINECS 215-587-0; Eltesol® PSA 65; EPA Pesticide Chemical Code 064102; HSDB 5818; Hydroxy-benzenesulfonic acid; Hydroxybenzenesulphonic acid; Phenolsulfonic acid; Phenolsulphonic acid; Sulfocarbolic acid; UN1803. Mixture of o, m and p-isomers. Catalyst for foundry resins; descaling agent for metal cleaning; anti-stress additive and plating aid in electroplating bath; curing aid in the plastics industry; raw material in the manufacture of dyes and pigments; detergents industry; pharmaceutical chemicals. *Albright & Wilson Americas Inc.; Albright & Wilson UK Ltd.*

2955 Phenol, thiobis[tetrapropylene-
68815-67-8   H

C36H58O2S
Thiobis(tetrapropylenephenol).
EINECS 272-388-1; Phenol, thiobis(tetrapropylene-; Thiobis((tetrapropylene)phenol); Thiobis(tetrapropylene-phenol).

Phenothiazine
92-84-2   7335   G V

C12H9NS
Dibenzothiazine.

4-27-00-01214 (Beilstein Handbook Reference); AFI-Tiazin; Agrazine; AI3-00038; Antiverm; Biverm; BRN 0143237; Caswell No. 652; CCRIS 5877; Contaverm; Contavern; contraverm; Dibenzo-1,4-thiazine; Dibenzo-p-thiazine; Dibenzoparathiazine; Dibenzothiazine; Early bird wormer; EINECS 202-196-5; ENT 38; EPA Pesticide Chemical Code 064501; Feeno; Fenothiazine; Fenotiazina; Fenoverm; Fentiazin; Fentiazine; Helmetina; Helmetine; HSDB 5279; Lethelmin; Nemazene; Nemazine; Nexarbol; NSC 2037; Orimon; Padophene; Penthazine; Phenegic; Phenosan; Phenothiazine; 10H-Phenothiazine; Phenothiazinum; phenothiazone; Phenovarm; Phenoverm; Phenovis; Phenoxur; Phenthiazine; Phenthiazinum; Phenzeen; Reconox; Souframine; Thiodifenylamine; Thiodiphenylamin; Thiodiphenylamine; Tiodifenilamina; Vermitin; Wurm-thional; XL-50. Insecticide and intermediate in chemical and pharmaceutical synthesis. Used in veterinary medicine as an anthelmintic. Yellow prisms; mp = 187.5°; bp = 371°; bp26.6 = 235°; λ_m = 254, 318 nm (MeOH); insoluble in H2O, soluble in Me2CO, C6H6, Et2O, EtOH; LD50 (mus orl) = 5 g/kg.

2957 Phenothrin
26002-80-2   7336   G P

C23H26O3
(±)-3-Phenoxyphenyl)methyl 2,2-dimethyl-3-(2-methyl-1-propenyl)cyclopropanecarboxylate.
AI3-29062; Anchimanaito 20S; Anvil; Anvil 2+2 ULV; Anvil 10+10 ULV; Benzyl alcohol, m-phenoxy-, 2,2-dimethyl-3-(2-methylpropenyl)cyclopropanecarboxylate; Caswell No. 652B; CCRIS 2502; Cyclopropanecarboxylic acid, 2,2-dimethyl-3-(2-methyl-1-propenyl)-, 3-(phenoxy-phenyl)methyl ester, cis,trans-(±)-; Cyclopropane-carboxylic acid, 2,2-dimethyl-3-(2-methyl-1-propenyl)-, 3-(phenoxyphenyl)methyl ester; Cyclopropanecarboxylic acid, 2,2-dimethyl-3-(2-methylpropenyl)-, m-phenoxy-benzyl ester; 2,2-Dimethyl-3-(2-methyl-1-propenyl)cyclo-propanecarboxylic acid (3-phenoxyphenyl)methyl ester; 2,2-Dimethyl-3-(2-methylpropenyl)cyclopropane-carboxylic acid m-phenoxybenzyl ester; Duet; EINECS 247-404-5; ENT 27 972; EPA Pesticide Chemical Code 069005; Fenothrin; Fenotrina; Forte; HSDB 3922; Multicide 2154; Multicide Concentrate F-2271; OMS 1809; OMS 1810; Phenothrin; d-Phenothrin; Phenothrine; Phenothrinum; 3-Phenoxybenzyl chrys-anthemate; 3-Phenoxybenzyl (±)-cis,trans-chrysanth-emate; 3-Phenoxybenzyl (1R)-cis/trans chrysanthemate; 3-Phenoxybenzyl (1RS)-cis,trans-chrysanthemate; 3-Phenoxybenzyl (1RS)-cis,trans-2,2-dimethyl-3-(2-methyl-prop-1-enyl)cyclopropanecarboxylate; m-Phenoxybenzyl (±)-cis,trans-2,2-dimethyl-3-(2-methylpropenyl)-cyclopropanecarboxylate; 3-Phenoxybenzyl 2-dimethyl-3-(methylpropenyl)cyclopropanecarboxylate; 3-Phenoxy-benzyl 2,2-dimethyl-3-(2-methylprop-1-enyl)cycloprop-anecarboxylate; 3-Phenoxybenzyl ester of DL-cis-trans-chrysanthemum monocarboxylic acid; 3-Phenoxybenzyl (1RS,3RS;1RS,3SR)-2,2-dimethyl-3-(2-methylprop-1-enyl)cyclopropanecarboxylate; (3-Phenoxyphenyl)methyl 2,2-dimethyl-3-(2-methyl-1-propenyl)cyclopropane-carboxylate; Phenoxythrin; Pibutin; PT 515; S-2539; Solo; Solo (insecticide); Sumithrin; Sumitrin; Wellcide. Non-systemic insecticide with rapid knockdown action. Used for control of insect pests, including mosquitoes in public health and in stored grain. Mixture of isomers. Registered by EPA as an insecticide. Supplied as a mixture of isomers; pale yellow liquid; bp > 290°; d25 =

485

1.06; insoluble in H_2O (< 0.0002 g/100 ml), soluble in MeOH, Me_2CO, C_6H_{14}, xylene (all > 80 g/100 ml); LD_{50} (rat orl) = 300 - 400 mg/kg, (mus orl) = 350 - 400 mg/kg, (dog orl) > 500 mg/kg, (gpg orl) = 400 mg/kg, (rbt orl) = 210 mg/kg, (rat der) = 2100 mg/kg, (mus der) > 5000 mg/kg, (pheasant orl) = 218 mg/kg, (quail orl) = 300 mg/kg, (bee) = 0.00012 mg/bee; LC_{50} (guppy 96 hr.) = 0.69 mg/l, (goldfish 96 hr.) = 2.9 mg/l; toxic to bees. *Hartz Mountain Products Corp.; Mclaughlin Gormley King Co; S.C. Johnson & Son, Co; Sergeant's Pet Products; Speer Products Inc.; Sumitomo Corp.; Summit Chemical Co.; Value Gardens Supply LLC; Whitmire Micro-Gen Research Laboratories Inc; ZEP Manufacturing Co.*

2958   Phenoxyacetic acid
122-59-8   7338   G

$C_8H_8O_3$
 Phenoxyethanoic acid.
 4-06-00-00634 (Beilstein Handbook Reference); Acetic acid, phenoxy-; Acide phenoxyacetique; AI3-06295; BRN 0907949; CCRIS 7275; EINECS 204-556-7; FEMA No. 2872; Glycol acid phenyl ether; Glycolic acid phenyl ether; Glycollic acid phenyl ether; NSC 9810; Phenoxy-acetic acid; Phenoxyethanoic acid; Phenylglycolic acid, O-; o-Phenylglycolic acid; POA. Intermediate for dyes, pharmaceuticals, pesticides, other organics, fungicides, flavoring, laboratory reagent, precursor in antibiotic fermentations especially penicillin V. Needles or plates; mp = 98.5°; bp = 285° (dec); soluble in H_2O (13 mg/ml), very soluble in EtOH, Et_2O, C_6H_6, CS_2, AcOH; LD_{50} (rat orl) = 1500 mg/kg. *Am. Xyrofin; Chemie Linz N. Am.; Great Lakes Fine Chem.; Lancaster Synthesis Co.; Penta Mfg.*

2959   m-Phenoxybenzaldehyde
39515-51-0   H

$C_{13}H_{10}O_2$
 3-Phenoxybenzaldehyde.
 4-08-00-00242 (Beilstein Handbook Reference); Benzaldehyde, 3-phenoxy-; BRN 0511662; EINECS 254-487-1; m-(Phenyloxy)benzaldehyde; 3-Phenoxybenz-aldehyde; m-Phenoxybenzaldehyde.

2960   m-Phenoxybenzyl alcohol
13826-35-2   H

$C_{13}H_{12}O_2$
 3-(Hydroxymethyl)diphenyl ether.
 3-06-00-04545 (Beilstein Handbook Reference); Benzenemethanol,

3-phenoxy-; Benzyl alcohol, m-phenoxy-; BRN 0475312; EINECS 237-525-1; 3-(Hydroxymethyl)diphenyl ether; 3-Phenoxybenzene-methanol; 3-Phenoxybenzyl alcohol; m-Phenoxybenzyl alcohol; 3-Phenoxybenzylic alcohol; (3-Phenoxy-phenyl)methanol.

2961   Phenoxyethyl acrylate
48145-04-6   G

$C_{11}H_{12}O_3$
 2-Phenoxyethyl acrylate.
 3-06-00-00572 (Beilstein Handbook Reference); Acrylic acid, 2-phenoxyethyl ester; Ageflex PEA; AI3-03194; BRN 2102773; Chemlink 160; Ebecryl 110; EINECS 256-360-6; Ethanol, 2-phenoxy-, acrylate; Light ester PO-A; Melcril 4087; 2-Phenoxyethanol acrylate; Phenoxyethyl acrylate; 2-Phenoxyethyl acrylate; Phenyl cellosolve acrylate; POA; 2-Propenoic acid, 2-phenoxyethyl ester; R 561; SR 339. UV-curable reactive diluent in inks and coatings, adhesives, viscosity index improver, tile coating. Supplied with 100 ppm HQ inhibitor; high boiling monomeric ester of acrylic acid; polymerization initiated by heat, catalysis, and/or radiation; copolymerization with other acrylic-type monomers is easily achieved; curing agent. Liquid; bp_2 = 110°; d^{20} = 1.090; soluble in Me_2CO, Et_2O, $CHCl_3$; flash pt. 90°. *CPS Chemical Co. Inc.*

2962   m-Phenoxytoluene
3586-14-9   H

$C_{13}H_{12}O$
 1-Methyl-3-phenoxybenzene.
 4-06-00-02041 (Beilstein Handbook Reference); Benz-ene, 1-methyl-3-phenoxy-; BRN 2045714; EINECS 222-716-4; Ether, phenyl m-tolyl-; 1-Methyl-3-phenoxy-benzene; 3-Methyldiphenyl ether; 3-Methylphenyl phenyl ether; m-Methylphenyl phenyl ether; 3-Phenoxytoluene; m-Phenoxytoluene; Phenyl m-tolyl ether. Liquid; = 272°; d^{25} = 1.051.

2963   Phenprocoumon
435-97-2   7344   D G

$C_{18}H_{16}O_3$
 3-(α-Ethylbenzyl)-4-hydroxycoumarin.
 5-18-02-00343 (Beilstein Handbook Reference); BRN 1291115; EINECS 207-108-9; Falithrom; Fencumar; Fenprocoumona; Fenprocoumone; HSDB 3248; Liquamar; Marcoumar; Marcumar; Phenprocoumarol; Phenpro-coumarole; Phenprocoumon; Phenprocoumone; Phen-procoumonum; Phenprocumonum; Ro 1-4849. Used as an anticoagulant. Crystals; mp = 179-180°.

2964   Phenyl cellosolve
122-99-6   7341   D G H

C8H10O2
2-Phenoxyethyl alcohol.
4-06-00-00571 (Beilstein Handbook Reference); AI3-00752; Arosol; BRN 1364011; Dowanol EP; Dowanol EPH; EGMPE; EINECS 204-589-7; Emeressence® 1160; Emeressence® 1160 Rose Ether; Emery 6705; Ethanol, 2-phenoxy-; Ethylene glycol phenyl ether; Fenyl-cellosolve; Fenylcelosolv; Glycol monophenyl ether; HSDB 5595; Igepal® Cephene Distilled; NSC 1864; Phenoxethol; Phenoxetol; Phenoxyethanol; 2-Phenoxyethyl alcohol; 2-Phenoxyethyl alcohol; Phenoxytol; Phenyl cellosolve; Phenylmonoglycol ether; Plastiazan-41; Rewopal® MPG 10; Rose ether. Solvent, surfactant and solubilizer for preservatives. Cosmetic preservative; effective against gram negative microorganisms. Liquid; mp = 14°; bp = 245°; d^{22} = 1.102; λ_m = 220, 271, 277 nm (ε = 7740, 1690, 1420, MeOH); insoluble in H_2O, soluble in EtOH, Et_2O, $CHCl_3$, alkali; LD_{50} (rat orl) = 1260 mg/kg, 960 mg/kg. *Henkel/Emery; Rewo; Rhône Poulenc Surfactants.*

2965   Phenyl dimethicone
2116-84-9   G

C15H32O3Si4
Methyl phenyl polysiloxane.
Abil® AV 8853, 20-1000; EINECS 218-320-6; Emalex MTS-30E; 1,1,5,5,5-Hexamethyl-3-phenyl-3-((trimethyl-silyl)oxy)trisiloxane; Methyl phenyl polysiloxane; Phenyl trimethicone; Phenyl trimethyl siloxane; Phenyl-tris(trimethylsiloxyl)silane; Polyphenylmethyl siloxane; Trisiloxane, 1,1,1,5,5,5-hexamethyl-3-phenyl-3-((tri-methylsilyl)oxy)-. Emollient providing skin protection; barrier against aqueous media; perfumery ingredient and fixative; provides improved rub and spreadability; faster penetration; non-sticky; prevents Aerosol® clogging. Oil-phase cosmetic ingredient for alcoholic milky-white lotions. *Goldschmidt; Nihon Nohyaku Co. Ltd.*

2966   N-Phenyl-1-naphthylamine
90-30-2   G H

C16H13N
1-Naphthalenamine, N-phenyl-.
4-12-00-03015 (Beilstein Handbook Reference); Aceto PAN; Additin 30; AI3-00528; Akrochem® Antioxidant PANA; Amoco 32; Antigene PAN; Antioxidant PAN; BRN 2211174; C.I. 44050; CCRIS 4701; EINECS 201-983-0; Fenyl-α-naftylamin; Naugard® PANA; Neozone A; Nonox A; NSC 2622; PANA; Phenyl-1-naphthylamine; N-Phenyl-1-naphthylamine; Phenylnaphthylamine; Vulka-nox PAN. Antioxidants for the mineral oil industries; gum inhibitor and staining antioxidant for rubber technical goods and heavily stressed goods, antiflexcracking under dynamic stress; antiflexing agent for NR and IR; storage stabilizer for petroleum products. mp = 60-62°; bp_{15} = 226°; d = 1.2; insoluble in H_2O, soluble in C_6H_6, Me_2CO, CCl_4, EtOAc, EtOH, CH_2Cl_2; LD_{50} (rat orl) = 1625 mg/kg. *Akrochem Chem. Co.; Bayer AG; Uniroyal.*

2967   N-Phenylaminopropyltrimethoxysilane
3068-76-6   G

C12H21NO3Si
N-(3-(Trimethoxysilyl)propyl)aniline.
Benzenamine, N-(3-(trimethoxysilyl)propyl)-; CP0156; EINECS 221-328-2; N-(γ-(Trimethoxysilyl)propyl)aniline. Coupling agent, chemical intermediate, blocking agent, release agent, lubricant, primer, reducing agent. *Degussa-Hüls Corp.*

2968   o-Phenylphenol
90-43-7   7388   G P

C12H10O
Hydroxy-2-phenylbenzene.
4-06-00-04579 (Beilstein Handbook Reference); AI3-00062; Anthrapole 73; Biphenyl, 2-hydroxy-; Biphenyl-2-o1; Biphenylol; o-Biphenylol; BRN 0606907; Caswell No. 623AA; CCRIS 1388; o-Diphenylol; Dowicide 1; Dowicide 1 antimicrobial; Dowicide A; EINECS 201-993-5; EPA Pesticide Chemical Code 064103; HSDB 1753; Hydroxdiphenyl; o-Hydroxybiphenyl; Invalon OP; Kiwi lustr 277; NCI-C50351; Nectryl; Nipacide® OPP; NSC 1548; OPP; Orthohydroxydiphenyl; Orthophenylphenol; Orthoxenol; Phenylphenol; o-Phenylphenol; Phenol, o-phenyl-; Phenyl-2 phenol; Phenylphenol; Preventol O extra; Remol TRF; Tetrosin OE; Topane; Torsite; Tumescal OPE; USAF EK-2219; Xenol; o-Xenol; o-Xonal. Disinfectant, preservative for detergents, cooling lubricants, textile/leather finishing, adhesives, paper, citrus fruit, polishes, wax emulsions, ceramic glazes, soap solutions. Fungicide. Registered by EPA as an antimicrobial and fungicide. Intermediate in manufacture of dyes, germicides, fungicides, rubber chemicals, laboratory reagents, food packaging and disinfectants. Colorless crystals; mp = 59°; bp = 286°; d^{25} = 1.213; λ_m = 246, 287 nm (ε = 5560, 2470, MeOH); soluble in H_2O (0.07 g/100 ml), most organic solvents; LD_{50} (mrat orl) = 2700 mg/kg, (frat orl) = 2000 mg/kg; toxic to fish. *Bayer Corp.; Caltech Industries Inc.; Cerexagri, Inc.; Chase Products Co.; Clariant Corp.; Coalite Chem. Div.; Colcide Inc.; Dow Chem. U.S.A.; Ecolab Inc.; Generic; Huntington Professional Products; Illinois Tool Works, Inc.; King Research Inc; Lonza Ltd.; National Chemical Laboratories, Inc.; Nipa; Quest Chemical Corp.; Reckitt Benckiser Inc.; S.C. Johnson & Son, Co; Spartan Chemical Co.; Steris Corp.*

2969   Phenyl salicylate
118-55-8   7394   D G

$C_{13}H_{10}O_3$

2-Hydroxybenzoic acid phenyl ester.

4-10-00-00154 (Beilstein Handbook Reference); AI3-00195; BRN 0393969; CCRIS 4859; EINECS 204-259-2; Fenylester kyseliny salicylove; Musol; NSC 33406; Phenol salicylate; Phenyl 2-hydroxybenzoate; Phenyl salicylate; Salol; Salphenyl. Analgesic; anti-inflammatory; antipyretic. Used as a UV absorber in plastics and as an analgesic, antipyretic and anti-inflammatory. Also used in the manufacture of plastics, laquers, adhesives, waxes, polishes; used in suntan oils and creams. Has some light absorbing and plasticizer properties. Crystals; mp = 130.5°; bp12 = 173°; d^{30} = 1.2614; λ_m = 241, 308 nm (ε = 12400, 4970, MeOH); insoluble in H_2O (0.014g/100 ml); soluble in Et_2O, AcOH, very soluble in alcohol (17 g/100 ml), C_6H_6 (67 g/100 ml), amyl alcohol (20 g/100 ml), liquid paraffin (10 g/100 ml), almond oil (25 g/100 ml); soluble in Me_2CO.

2970   Phenyl xylyl ethane

6196-95-8   H

$C_{16}H_{18}$

4-(1-Phenylethyl)-o-xylene.

Benzene, 1,2-dimethyl-4-(1-phenylethyl)-; 1-(3,4-Dimethylphenyl)-1-phenylethane; EINECS 228-249-2; Ethane, 1-phenyl-1-(3,4-xylyl)-; Phenyl-1-(3,4-dimethyl)-phenylethane; 1-Phenyl-1-(3,4-dimethylphenyl)ethane; 4-(1-Phenylethyl)-o-xylene; Phenyl xylyl ethane.

2971   Phenylacetic acid

103-82-2   7352   G

$C_8H_8O_2$

α-toluic acid.

4-09-00-01614 (Beilstein Handbook Reference); Acetic acid, phenyl-; AI3-08920; Benzenacetic acid; Benzene-acetic acid; Benzylcarboxylic acid; BRN 1099647; EINECS 203-148-6; FEMA No. 2878; HSDB 5010; Kyselina fenyloctova; NSC 125718; Phenylacetic acid; 2-Phenylacetic acid; Phenylethanoic acid; α-Toluic acid. Perfume, precursor in manufacture of penicillin G, fungicide, flavoring, laboratory reagent. Synthetic intermediate in perfumery industry. Solid; mp = 76.7°; bp = 265.5°; d^6 = 1.228, λ_m = 258 nm (MeOH); insoluble in ligroin, slightly soluble in H_2O, $CHCl_3$, soluble in Me_2CO, very soluble in EtOH, Et_2O. *Calaire Chimie S.A.; Lancaster Synthesis Co.; Penta Mfg.*

2972   4-(N-Phenylamidino)thiazole hydrochloride

13631-64-6   H

$C_{10}H_{10}ClN_3S$

4-Thiazolecarboxamidine, N-phenyl-, hydrochloride.

EINECS 237-117-3; 4-(N-Phenylamidino)thiazole hydro-chloride; 4-Thiazolecarboxamidine, N-phenyl-, hydro-chloride.

2973   Phenylbenzimidazol-5-sulfonic acid

27503-81-7   G

$C_{13}H_{10}N_2O_3S$

2-Phenylbenzimidazole-5-sulfonic acid.

1H-Benzimidazole-5-sulfonic acid, 2-phenyl-; EINECS 248-502-0; Ensulizole; Eusolex 232; Neo Heliopan® Hydro; Novantisol; Phenylbenzimidazole sulfonic acid; 2-Phenylbenzimidazole-5-sulfonic acid; 2-Phenyl-1H-benzimidazole-5-sulfonic acid; 2-Phenyl-1H-benz-imidazole-5-sulphonic acid. UV-B filter for sunscreen formulations. Used in creams, lotions and subscreens. Solid; mp >300°. *Haarmann & Reimer GmbH.*

2974   Phenyldimethylchlorosilane

768-33-2   G

$C_8H_{11}ClSi$

Dimethylphenylchlorosilane.

Chlorodimethylphenylsilane; CP0160; EINECS 212-193-0; NSC 95425; Silane, chlorodimethylphenyl-. Coupling agent, chemical intermediate, blocking agent, release agent, lubricant, primer, reducing agent. Liquid; bp = 195°, bp16 = 82°; d^{20} = 1.032. *Degussa-Hüls Corp.*

2975   1,3-Phenylenebismaleimide

3006-93-7   G

$C_{14}H_8N_2O_4$

m-Phenylenedimaleimide.

4-21-00-04640 (Beilstein Handbook Reference); 1,3-Bismaleimidobenzene; BRN 0249503; 1,3-Dimale-imidobenzene; m-Dimaleimidobenzene; EINECS 221-112-8; HVA 2; HVA-2 curing agent; Maleimide, N,N'-(m-phenylene)di-; m-PHDM; m-Phenylenebismaleimide; N,N'-(m-Phenylene)bismaleimide; 1,1'-(1,3-

Phenylene)-bis-1H-pyrrole-2,5-dione; 1,1'-(m-Phenylene)bis-1H-pyrrole-2,5-dione; m-Phenylenedimaleimide; N,N'-m-Phenylenemaleimide; N,N'-(m-Phenylenedimaleimide); N,N'-1,3-Phenylenedimaleimide; NSC 19639; 1H-Pyrrole-2,5-dione, 1,1'-(1,3-phenylene)bis-; 1H-Pyrrole-2,5-dione, 1,1'-(phenylene)bis-; Vanax® MBM. Accelerator; free radical regulator and coagent in peroxide-cured polymers. Also used for cross-linking proteins. Solid; mp = 197-200°. DuPont UK; Vanderbilt R.T. Co. Inc.

2976 (1,3(or 1,4)-Phenylenebis(1-methylethyl-idene))bis(1,1-dimethylethyl)peroxide
25155-25-3 G H

C20H34O4
(Phenylenediisopropylidene)bis(tert-butylperoxide).
Bisbutyl peroxy diisopropyl benzene; Bis(tert-butyldioxy-isopropyl)benzene; α,α'-Bis(tert-butylperoxy)diisopropyl-benzene; CCRIS 4588; EINECS 246-678-3; Luperox 802; Perkadox® 14; Peroxide, (1,3(or 1,4)-phenylenebis(1-methylethylidene))bis((1,1-dimethylethyl); Peroxide, (phenylenebis(1-methylethylidene))bis(1,1-dimethylethyl)-; Peroxide, (phenylenediisopropylidene)bis(tert-butyl-; (Phenylenediisopropylidene)bis(tert-butylperoxide); Retilox® F 40 MG; Vul-Cup; Vul-Cup 40KE; Vul-Cup R. Low reactivity peroxide useful as a finishing initiator at high temperatures for styrenics; synergist for some halogen-containing flame retardants; also for cross-linking of EPM, EPDM, polyethylene, silicone rubbers, NBR, EVA copolymers, SBR, chlorosulfonated polyethylene, PVC, PU rubbers, polybutadiene rubbers and neoprene. Crosslinking agent for thermoplastic modification, curing elastomers, and high temperature cure of polyesters; initiator for vinyl polymerization. Solid; mp = 44-48°; insoluble in H2O (< 1 mg/ml). Akrochem Chem. Co.; Akzo Chemie; Elf Atochem N. Am.; Hercules Inc.

2977 m-Phenylenediamine
108-45-2 7367 G H

C6H8N2
1,3-Diaminobenzene.
4-13-00-00079 (Beilstein Handbook Reference); Al3-52607; APCO 2330; Benzene, 1,3-diamino-; BRN 0471357; CCRIS 1236; CI Developer 11; Developer 11; Developer C; Developer H; Developer M; Direct Brown BR; Direct Brown GG; EINECS 203-584-7; HSDB 5384; m-Phenylenediamine; Metaphenylenediamine; NSC 4776; UN1673. Used in the manufacture of dyes and hair dyes; as a rubber curing agent; in photography; as an analytical reagent for gold and bromine. Rhombohedral crystals; mp = 63.5°, bp = 285°; $d^{58} = 1.0096$; $\lambda_m = 293$ nm (MeOH); very soluble in H2O, soluble in EtOH, Et2O, C6H6. DuPont.

2978 p-Phenylenediamine
106-50-3 7369 G H

C6H8N2
1,4-Diaminobenzene.
6PPD; Al3-00710; 4-Aminoaniline; Aminogen II; BASF ursol D; 1,4-Benzenediamine; Benzofur D; CCRIS 509; CI 76060; CI Developer 13; CI Oxidation Base 10; Developer 13; Developer PF; 1,4-Diaminobenzene; Durafur Black R; EINECS 203-404-7; Fenylenodwuamina; Fouramine D; Fourrine 1; Fourrine D; Fur Black 41866; Fur Black 41867; Fur Yellow; FUR Brown 41866; Furro D; Futramine D; HSDB 2518; Nako H; NSC 4777; Orsin; Oxidation Base 10; 1,4-Phenylenediamine; 4-Phenyl-enediamine; PARA; Paraphenylen-diamine; Paraphenyl-enediamine; Pelagol D; Pelagol DR; Pelagol Grey D; Peltol D; Phenylenediamine; Phenylenediamine, para-; Renal PF; Rodol D; Santoflex IC; Santoflex LC; Tertral D; UN1673; Ursol D; USAF EK-394; Vulkanox 4020; Zoba Black D. Used as an azo dye intermediate; photographic developing agent; intermediate in manufacture of antioxidants, accelerators for rubber, synthetic fibers; dyeing hair and fur. White platelets; mp= 146°; bp = 267°; $\lambda_m = 244$ nm ($\varepsilon = 10965$, EtOH); slightly soluble in H2O, soluble in EtOH; Et2O, C6H6, CHCl3; LD50 (rat orl) = 80 mg/kg. BASF Corp.; DuPont; Hoechst Celanese; Janssen Chimica.

2979 p-Phenylenediamine, N,N'-bis(1,4-dimethylpentyl)-
3081-14-9 H

C20H36N2
N,N'-Bis(1,4-dimethylpentyl)-p-phenylenediamine.
77PPD; Antioxidant 4030; BRN 2739028; CCRIS 4750; Eastozone; Eastozone 33; EINECS 221-375-9; Elastozone 33; Flexzone 4L; NCI-C56337; p-Phenylenediamine, N,N'-bis(1,4-dimethylpentyl)-; Santoflex 77; Tenamene 4; UOP 788. An antioxidant and antiozonant.

2980 p-Phenylenediamine, N-(1,3-dimethyl-butyl)-N'-phenyl-
793-24-8 H

C18H24N2
N-(1,3-Dimethylbutyl)-N'-phenyl-1,4-phenylenediamine.
4-13-00-00115 (Beilstein Handbook Reference); 6PPD; Antage 6C; Antioxidant 4020; Antioxidant cd; Antioxidant CD 13; Antozite 67; Antozite 67F; 1,4-Benzenediamine, N-(1,3-dimethylbutyl)-N'-phenyl-; BRN 2215491; CCRIS 2352; CCRIS 4801; CD 13; DBDA; Diafen 13; Diafen FDMB; DMBPD; Dusantox 6PPD; EINECS 212-344-0; Flexzone 7F; Flexzone 7L; Forte 6C; HSDB 5755; N-(1,3-Dimethylbutyl)-N'-phenyl-1,4-phenylenediamine; N-(1,3-Dimethylbutyl)-N'-phenyl-p-phenylenediamine; NCI-C56315; Nocrac 6C; Nocrane 6C; Nocrane 7 L; Ozonon 6C; p-Phenylenediamine, N-(1,3-dimethylbutyl)-N'-phenyl-; Permanax 120; Permanax 6PPD; Santoflex 13; Santoflex 13F; Santoflex 6PPD; UOP 562; UOP 588;

Vulkanox 4020; Wingstay 300.

2981 p-Phenylenediamine, N-(1,4-dimethyl-pentyl)-N'-phenyl-
3081-01-4   H

C19H26N2
N-(1,4-Dimethylpentyl)-N'-phenylbenzene-1,4-diamine.
1,4-Benzenediamine, N-(1,4-dimethylpentyl)-N'-phenyl-; N-(1,4-Dimethylpentyl)-N'-phenylbenzene-1,4-diamine; N-(1,4-Dimethylpentyl)-N'-phenyl-1,4-benzenediamine; EINECS 221-374-3; p-Phenylenediamine, N-(1,4-dimethylpentyl)-N'-phenyl-; Santoflex 14; Santoflex 14 antiozonant. An antioxidant and antiozonant.

2982 p-Phenylenediamine, N-(1-methylheptyl)-N'-phenyl-
15233-47-3   H

C20H28N2
N-(1-Methylheptyl)-N'-phenyl-p-phenylenediamine.
1,4-Benzenediamine, EINECS 239-281-1; N-(1-methyl-heptyl)-N'-phenyl-; N-(1-Methylheptyl)-N'-phenyl-p-phenylenediamine; N-(1-Methylheptyl)-N'-phenyl-1,4-benzenediamine; p-Phenylenediamine, N-(1-methylhep-tyl)-N'-phenyl-; Uop 688.

2983 Phenylethyl alcohol
60-12-8   7304   H

C8H10O
Benzeneethanol.
4-06-00-03067 (Beilstein Handbook Reference); AI3-00744; Benzeneethanol; Benzylcarbinol; Benzyl-methanol; BRN 1905732; Caswell No. 655C; EINECS 200-456-2; EPA Pesticide Chemical Code 001503; Ethanol, 2-phenyl-; FEMA No. 2858; FEMA Number 2858; β-Fenethylalkohol; β-Fenylethanol; HSDB 5002; β-Hydroxyethylbenzene; Methanol, benzyl-; NSC 406252; PEA; Phenethanol; β-Phenylethanol; Phenethyl alcohol; Phenylethyl alcohol. Has the smell of roses. Used in perfumery and as an antimicrobial agent. Clear liquid; mp= -27°; bp750 = 219-221°; d$_4^{25}$ = 1.017; n$_D^{20}$= 1.530; soluble in H2O (2 g/100 ml), more soluble in organic solvents; LD50 (rat orl)= 1790 mg/kg. Lancaster Synthesis Co.; Sigma-Aldrich Fine Chem.

2984 2-Phenyl imidazole
670-96-2   G

C9H8N2
2-phenylimidazole.
AI3-50034; EINECS 211-581-7; Imidazole, 2-phenyl-; 1H-Imidazole, 2-phenyl-; NSC 255226; Phenylimidazole; 2-Phenylimidazole; 2-Phenyl-1H-imidazole. Epoxy curing agent for printed circuit boards, molding compounds, potting; accelerator for dicyandiamide and anhydrides. Leaflets; mp = 1469.3°; bp = 340°; insoluble in H2O, very soluble in EtOH. BASF Corp.; Janssen Chimica; Lancaster Synthesis Co.

2985 Phenylmercuric oleate
104-60-9   G P

C24H38HgO2
Mercury, (Z)-(9-octadecenoato-O)phenyl-.
Caswell No. 657J; EPA Pesticide Chemical Code 066022; Intercide PMO 11; Mercury, (9-octadecenoato-O)phenyl-, (Z)-; Mercury, (9-octadecenoato-kappaO)phenyl-; Merc-ury, (oleato)phenyl-; Mercury, (oleoyloxy)phenyl-; (Oleato)phenylmercury; (Z)-(9-Octadecenoato-O)phenyl-mercury; Phenylmercuric oleate; PMO 10. Mildew-proofing agent for paints; fungicide, germicide. Registered by EPA as an antimicrobial, fungicide, herbicide and insecticide (cancelled). Noah Chem.; Thor.

2986 Phenylmercury acetate
62-38-4   7383   G P

C8H8HgO2
Phenylmercuric acetate.
(Acetato-O)phenylmercury; (Acetato)phenylmercury; (Acetoxymercuri)benzene; Acetate de phenylmercure; Acetate phenylmercurique; Acetic acid, phenylmercury deriv.; Acetic acid, phenylmercury(II) salt; Acetoxy-phenylmercury; Agrosan D; Agrosan GN 5; AI3-14668; Algimycin 200; Anticon; Antimucin WBR; Antimucin WDR; Benzene, (acetoxymercuri)-; Benzene, (acetoxymercurio)-; Bufen; Bufen 30; Caswell No. 656; CCRIS 4858; Cekusil; Celmer; Ceresan slaked lime; Ceresol; Contra Creme; Dyanacide; EINECS 200-532-5; EPA Pesticide Chemical Code 066003; Femma; Fenylmercuriacetat; Fenylmerkuriacetat; Fungicide R; Fungitox OR; Gallotox; Hexasan; Hexasan (fungicide); HI-331; Hong nien; Hostaquick; Hostaquik; HSDB 1670; Intercide 60; Intercide PMA 18; Kwiksan; Liquiphene; Lorophyn; Meracen; Mercron; Mercuriphenyl acetate; Mercuron; Mercury, (acetato-O)phenyl-; Mercury, (acetato)phenyl-; Mercury(II) acetate, phenyl-; Mergal A 25; Mersolite; Mersolite 8; Mersolite D; Metasol 30; Neantina; Norforms; NSC 35670; Nuodex PMA 18; Nylmerate; Octan fenylrtutnaty; Pamisan; Panomatic; Parasan; Parasan (bactericide); Phenmad; Phenomercuric acetate; Phenyl mercuric acetate; Phenylquecksilber-acetate; Phenylquecksilberacetat; Phix; PMA; PMA (fungicide); PMAC; PMAL; PMAS; Programin; Purasan-SC-10; Puraturf 10; Quicksan; Quicksan 20; RCRA waste number P092; Riogen; Ruberon; Samtol; Sanitized SPG; Sanitol; Sanmicron; Sc-110; Scutl; Seed Dressing R; Seedtox; Setrete; Shimmerex; Single Purpose; Spor-Kil; Spruce Seal; Tag; Tag 331; Tag Fungicide; Tag

HL-331; Trigosan; Troysan 30; Troysan PMA 30; UN1674; Verdasan; Volpar; Zaprawa Nasienna R; Ziarnik. Organomercury compound used as a fungicide seed dressing for cereals and fodder beet. mp = 149°; soluble in H_2O (1.6 mg/ml), more soluble in organic solvents; LD_{50} (rat orl) = 22 mg/kg. *Allchem Ind.; DowElanco Ltd.; EM Ind. Inc.*; 907;1191.

2987　Phenylphosphine dichloride

644-97-3　　　　　　　　　　　　　　　　　　　　H

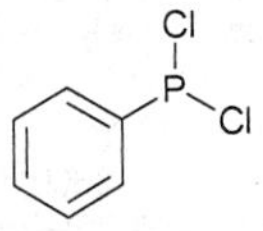

$C_6H_5Cl_2P$

Phenyl phosphorus dichloride.

4-16-00-00972 (Beilstein Handbook Reference); AI3-15063; Benzene phosphorus dichloride; BRN 0508189; Dichlorophenylphosphine; EINECS 211-425-8; HSDB 2729; NSC 66478; Phenyl phosphorus dichloride; Phenyldichlorophosphine; Phenylphosphine dichloride; Phenylphosphonous dichloride; Phenylphosphorus dichloride; Phosphine, dichlorophenyl-; Phosphonous dichloride, phenyl-; UN2798. Liquid; mp = -51°; bp = 225°, bp_{57} = 142°; d^{20} = 1.356; very soluble in C_6H_6.

2988　Phenylphosphinic acid

1779-48-2　　　　　　　　　　　　　　　　　　　H

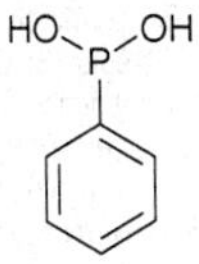

$C_6H_7O_2P$

Benzenephosphonous acid.

4-16-00-01033 (Beilstein Handbook Reference); AI3-15040; Benzenephosphinic acid; Benzenephosphonous acid; BRN 2802244; EINECS 217-217-3; Hydroxy-phenylphosphine oxide; NSC 2670; Phenylphosphinic acid; Phenylphosphonous acid; Phosphinic acid, phenyl-.

2989　Phenyltetrazole

18039-42-4　　　　　　　　　　　　　　　　　　G

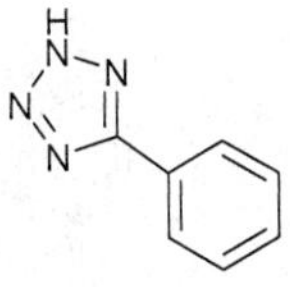

$C_7H_6N_4$

5-Phenyl-1H-tetrazole.

EINECS 241-950-8; Expandex 5PT; Expandex OX 5PT; Kempore 50XPT; MA 1623; NSC 11138; Phenyltetrazole; 5-Phenyltetrazole; 5-Phenyl tetrazole; 5-Phenyl-2H-tetrazole; Tetrazole, 5-phenyl-; 1H-Tetrazole, 5-phenyl-; 2H-Tetrazole, 5-phenyl-. Blowing agent for foaming plastics and elastomers at elevated temperatures. *Degussa-Hüls Corp.*

2990　Phenyltrichlorosilane

98-13-5　　　　　　　　　　　　　　　　　　　G H

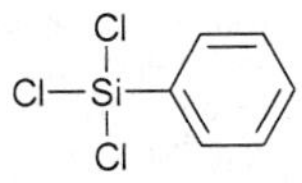

$C_6H_5Cl_3Si$

Silane, trichlorophenyl-.

4-16-00-01560 (Beilstein Handbook Reference); AI3-51469; BRN 0508730; CP0280; EINECS 202-640-8; HSDB 1039; NSC 77080; Phenylsilicon trichloride; Phenyltrichlorosilane; Silane, phenyltrichloro-; Silane, trichlorophenyl-; Silicon phenyl trichloride; Trichloro-phenylsilane; UN1804. Coupling agent, chemical intermediate, blocking agent, release agent, lubricant, primer, reducing agent. Intermediate for silicones; laboratory reagent. bp = 201°; d^{20} = 1.3210; soluble in $CHCl_3$, CCl_4, CS_2; LD_{50} (rat orl) = 2390 mg/kg. *Degussa-Hüls Corp.; PCR; Sigma-Aldrich Fine Chem.*

2991　Phenyltriethoxysilane

780-69-8　　　　　　　　　　　　　　　　　　G

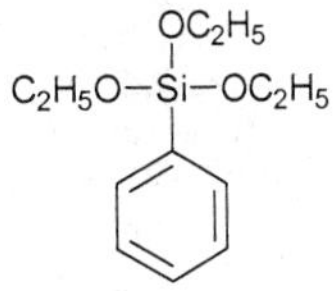

$C_{12}H_{20}O_3Si$

Silane, phenyltriethoxy-.

Benzeneorthosiliconic acid, triethyl ester; CP0320; EINECS 212-305-8; NSC 77115; Phenyltriethoxysilane; Silane, phenyltriethoxy-; Silane, triethoxyphenyl-; Triethoxy(phenyl)silane; Triethoxyfenylsilan; Triethoxy-phenylsilane. Coupling agent, chemical intermediate, blocking agent, release agent, lubricant, primer, reducing agent. Liquid; bp = 232°, bp_{10} = 113°; d^{25} = 0.9960; λ_m = 210, 253, 259, 265, 272 nm (ε = 8318, 251, 363, 417, 302, EtOH); LD_{50} (rat orl) = 2818 mg/kg. *Degussa-Hüls Corp.*

2992　Phenyltrimethoxysilane

2996-92-1　　　　　　　　　　　　　　　　　　G

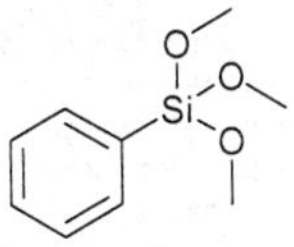

$C_9H_{14}O_3Si$

(Trimethoxysilyl)benzene.

A 153; AI3-60043; CP0330; EINECS 221-066-9; NSC 93925; Phenyltrimethoxysilane; Silane, phenyltrimeth-oxy-; Silane, trimethoxyphenyl-; Trimethoxyphenylsilane. Coupling agent, chemical intermediate, blocking agent, release agent, lubricant, primer, reducing agent. Liquid; bp_{45} = 130°, bp_{20} = 110°; d^{20} = 1.064; soluble in CCl_4, CS_2. *Degussa-Hüls Corp.; Janssen Chimica; PCR.*

2993　Phenylurethane

101-99-5　　　　　　　　　　　7404　　　　　　G

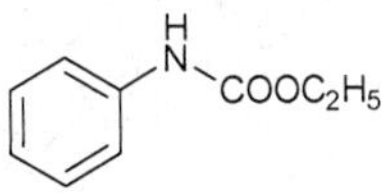

$C_9H_{11}NO_2$

Phenylcarbamic acid ethyl ester.

4-12-00-00619 (Beilstein Handbook Reference); AI3-15353; BRN 1942785; Carbamic acid, phenyl-, ethyl ester; Carbanilic acid, ethyl ester; EINECS 202-995-9; EPC (the plant regulator); Ethanol, carbanilate; N-(Ethoxycarbonyl)aniline; Ethyl carbanilate; Ethyl-N-phenylcarbamate; Ethyl N-phenylurethan; Ethyl N-phenylurethane; Ethyl phenylcarbamate; Ethylester kyseliny karbanilove; Euphorin; HSDB 5344; Keimstop; NSC 3245; Phenylethyl carbamate; Phenylurethan; Phenylurethane; N-Phenylurethane; Urethan, phenyl-; Urethane, phenyl-. Chemical intermediate made by reaction of ethyl chloroformate and aniline. Needles or plates; mp = 53°; bp = 237° (dec); d^{30} = 1.1064; λ_m = 235, 273, 281 nm (ε = 8770, 512, 393, MeOH); insoluble in H_2O, soluble in C_6H_6, CCl_4, very soluble in EtOH, Et_2O.

2994 Phenytoin
57-41-0   7406   D

$C_{15}H_{12}N_2O_2$
5,5-Diphenylhydantoin.
AI3-52498; Aleviatin; Auranile; Causoin; CCRIS 515; Comitoina; Convul; Danten; Dantinal; Dantoinal; Dantoinal klinos; Dantoine; Denyl; Di-Hydan; Di-Lan; Di-Phetine; Didan TDC 250; Difenilhidantoina; Difenin; Difetoin; Diffhydan; Dihycon; Dihydantoin; Dilabid; Dilantin; Dilantin acid; Dilantine; Dillantin; Dintoin; Dintoina; Diphantoin; Diphedal; Diphedan; Diphenin; Diphenine; Diphentyn; Diphenylan; Diphenylhydantoin; Diphenylhydantoine; Diphenylhydatanoin; Ditoinate; DPH; EINECS 200-328-6; Ekko; Elepsindon; Enkelfel; Epamin; Epasmir 5; Epdantoin Simple; Epelin; Epifenyl; Epihydan; Epilan-D; Epilantin; Epinat; Epised; Eptal; Fenantoin Mn Pharma; Fenidantoin s; Fenitoina; Fentoin; Fenylepsin; Fenytoin Dak; Fenytoine; Gerot-epilan-D; Hidan; Hidantal; Hidantilo; Hidantina; Hidantina senosian; Hidantina vitoria; Hidantomin; Hydantal; Hydantin; Hydantoinal; Hydantol; Ictalis simple; Idantoil; Idantoin; Iphenylhydantoin; Kessodanten; Labopal; Lehydan; Lepitoin; Lepsin; Mebroin; Minetoin; NCI-C55765; Neoshidantoina; Neosidantoina; Novantoina; Novophenytoin; NSC 8722; OM hidantoina simple; OM-Hydantoine; Oxylan; Phanantin; Phanatine; Phenatine; Phenatoine; Phenhydan; Phenhydanin; Phentoin; Phenytoin; Phenytoine; Phenytoinum; Ritmenal; Saceril; Sanepil; Silantin; Sinergina; Sodanthon; Sodantoin; Sodanton; Solantin; Sylantoic; Thilophenyl; Toin unicelles; TOIN; Zentronal; Zentropil.; component of: Mebroin. Anticonvulsant; class IB antiarrhythmic. Crystals; mp = 295-298°; insoluble in H_2O; soluble in EtOH (1.67 g/100 ml), Me_2CO (3.32 g/100 ml); LD_{50} (mus iv) = 92 mg/kg, (mus sc) = 110 mg/kg. Parke-Davis; Sterling Winthrop Inc.

2995 Phenytoin Sodium
630-93-3   7406   D

$C_{15}H_{11}N_2NaO_2$
5,5-Diphenylhydantoin sodium salt.
Aladdin; Alepsin; Aleviatin sodium; Antilepsin; Antisacer; Auranile;

Beuthanasia-D; Citrullamon; Cumatil; Danten; Dantoin; Decatona; Denyl; Denyl sodium; Derizene; Di-len; Difenin; Difetoin; Diffhydan; Dihydan; Dilantin; Dilantin sodium; Diphantoine; Diphedan; Diphenin; Diphenine; Diphenine sodium; Diphentoin; Diphenylan sodium; Diphenylhydantoin sodium; Ditoin; Ditomed; EINECS 211-148-2; Enkefal; Epanutin; Epdantoin; Epelin; Epilan-D; Epilantin-E; Epileptin; Epinat; Episar; Epsolin; Eptoin; Eptolin; Fenidantoin; Fenigramon; Fenitoin sodium; Fenitoina Sodica; Fenitron; Fenytoin; Hidantal sodium; Hidantin; HSDB 3160; Hydantin; Hydantoinal; Lepitoin sodium; Metinal Idantoina; Minetoin; Muldis; Novodiphenyl; OM-Hydantoine sodium; Phenilep; Phenytoin soluble; Phenytoin sodium; SDPH; Sodium phenytoin; Solantoin; Solantyl; Soluble phenytoin; Tacosal.; component of: Beuthanasia-D. Anticonvulsant; class IB antiarrhythmic. Soluble in EtOH (9.5 g/100 ml), H_2O (1.5 g/100 ml); insoluble in Et_2O, $CHCl_3$; LD_{50} (mus orl) = 490 mg/kg. Elkins-Sinn; Schering-Plough Animal Health.

2996 Phloroglucinol
108-73-6   7413   G

$C_6H_6O_3$
1,3,5-Benzenetriol.
4-06-00-07361 (Beilstein Handbook Reference); AI3-08848; Benzene, trihydroxy; Benzene, 1,3,5-trihydroxy-; Benzene-1,3,5-triol; Benzene-s-triol; 1,3,5-Benzenetriol; BRN 1341907; CCRIS 4147; 3,5-Dihydroxyphenol; Dilo-span S; EINECS 203-611-2; Floroglucin; Floroglucinol; 5-Hydroxyresorcinol; NSC 1572; 5-Oxyresorcinol; Phloro-glucin; Phloroglucine; Phloro-glucinol; Spasfon-Lyoc; s-Trihydroxybenzene; sym-Trihydroxybenzene; 1,3,5-Trihydroxybenzene; 1,3,5-Trihydroxycyclohexatriene. Used in analytical chemistry, as a decalcifying agent for bones, in preparation of pharmaceuticals and dyes, resins, as a preservative for cut flowers, in textile dyeing and printing. Available in anhydrous and dihydrate forms. Leaflets or plates; mp = 218.5°; λ_m = 268 nm (ε = 398, MeOH); slightly soluble in H_2O (10 g/l), soluble in Me_2CO, very soluble in EtOH, Et_2O, C_6H_6, C_5H_5N; LD_{50} (rat orl) = 4 g/kg. Goldschmidt; Lancaster Synthesis Co.; Schering AG; Spectrum Chem. Manufacturing.

2997 Phlorone
137-18-8   G

$C_8H_8O_2$
p-Xyloquinone.
AI3-61044; p-Benzoquinone, 2,5-dimethyl-; CCRIS 7150; 2,5-Cyclohexadiene-1,4-dione, 2,5-dimethyl-; 2,5-Di-methyl-1,4-benzoquinone; 2,5-Dimethyl-1,4-benzo-chinon; 2,5-Dimethyl-2,5-cyclohexadiene-1,4-dione; 2,5-Dimethyl-4-benzoquinone; 2,5-Dimethyl-p-benzo-quinone; 2,5-Dimetilbenzochinone (1:4); 3,6-Dimethyl-p-benzoquinone; EINECS 205-283-6; Floron; Florone; NSC 15309; 2,5-Xyloquinone; p-Xyloquinone; Phlorone. Organic intermediate. Crystals; mp = 124-126°; LD_{50} (mus orl) = 290 mg/kg.

2998 Phorate

298-02-2   7416   G

C7H17O2PS3
O,O-Diethyl S-[(ethylthio)methyl]phosphorodithioate.
4-01-00-03090 (Beilstein Handbook Reference); AC 3911; Agrimet; AI3-24042; American Cyanamid 3,911; BRN 1708517; Caswell No. 660; CCRIS 2747; Chim; O,O-Diaethyl-S-(aethylthio-methyl)-dithiophosphat; O,O-Diethyl ethylthiomethyl phosphorodithioate; O,O-Diethyl S-(ethylthio)methyl phosphorodithioate; O,O-Diethyl S-ethylmercaptomethyl dithiophosphate; O,O-Diethyl S-ethylmercaptomethyl dithiophosphonate; O,O-Diethyl-S-(ethylthio-methyl)-dithiofosfaat; O,O-Diethyl S-ethylthiomethyl thiothionophosphate; O,O-Diethyl S-ethylthiomethyl dithiophosphonate; O,O-Diethyl-S-((ethylthio)methyl)phosphorodithioate; O,O-Dietil-S-(etiltio-metil)-ditiofosfato; Dithiophosphate de O,O-diethyle et d'ethylthiomethyle; EI3911; EINECS 206-052-2; ENT 24,042; EPA Pesticide Chemical Code 057201; Experimental insecticide 3911; Foraat; Forate; Geomet; Granutox; HSDB 1183; L 11/6; Methanethiol, (ethylthio)-, S-ester with O,O-diethylphosphorodithioate; Phorat; Phorate; Phorate-10G; Phosphorodithioic acid, O,O-diethyl S-(ethylthio)methyl ester; Phosphorodithioic acid, O,O-diethyl S-((ethylthio)methyl) ester; Rampart; RCRA waste number P094; Terrathion; Thimenox; Thimet; Timet; Vegfru; Vegfru Foratox; Volphor; VUAgT 182. Systemic insecticide and acaricide with contact and stomach action. A cholinesterase inhibitor with some nematicidal activity. Terrathion consists of granules containing 10% w/w phorate; an organophosphorus insecticide. Oil; mp <-15°; bp0.8= 118-120°; d25 = 1.167; nD25= 1.5349; soluble in H2O (50 mg/l), more soluble in organic solvents; LD59 (rat orl) = 3.7 mg/kg. *Am. Cyanamid; Farmers Crop Chemicals Ltd.; MTM AgroChemicals Ltd.*

2999 Phosalone

2310-17-0   7489   G P

C12H15ClNO4PS2
S-[(6-chloro-2-oxo-3(2H)-benzoxazolyl)methyl] O,O-diethylphosphorodithioate.
11 974 rp; Agria 1060; Agria 1060 A; AI3-27163; Azofene; Benzophosphate; S-((3-Benzoxazolinyl-6-chloro-2-oxo)methyl) O,O-diethyl phosphorodithioate; Benzphos; BRN 0694650; Caswell No. 660A; CCRIS 2000; Chipman 11974; 6-Chloro-3-diethoxyphosphino-thioylthiomethyl-1,3-benzoxazol-2(3H)-one; 6-Chloro-3-(O,O-diethyldithiophosphorylmethyl)benzoxazolone; S-6-Chloro-2,3-dihydro-2-oxo-benzoxazol-3-ylmethyl O,O-diethyl phosphorodithioate; S-6-Chloro-2,3-dihydro-2-oxo-1,3-benzoxazol-3-ylmethyl O,O-diethyl phosphoro-dithioate; S-(6-Chloro-3-(mercaptomethyl)-2-benzoxa-zolinone) O,O-diethyl phosphorodithioate; 3-(6-Chloro-2-oxobenzoxazolin-3-yl)methyl-O,O-diethyl phosphoro-thiolothionate; S-((6-Chloro-2-oxo-3(2H)-benzoxazolyl)-methyl) O,O-diethyl phosphorodithioate; S-((6-Chloro-2-oxo-3H)-benzoxazolylmethyl)O,O-diethyl phosphoro-dithioate; S-(6-Chloro-2-oxobenzoxazolin-3-yl)methyl   diethyl

phosphorothiolothionate; O,O-Diethyl-S-((6-chloor-2-oxo-benzoxazolin-3-yl)-methyl)-dithiofosfaat; O,O-Diethyl S-(6-chlorobenzoxazolinyl-3-methyl) dithio-phosphate; O,O-Diaethyl-S-(6-chlor-2-oxo-ben(b)-1,3-oxalin-3-yl)-methyl-dithiophosphat; O,O-Diäthyl-S-(6-chlor-2-oxo-ben(b)-1,3-oxalin-3-yl)-methyl-dithio-phosphat; O,O-Diethyl-S-(6-chloro-2-oxobenzoxazolin-3-yl-methyl)-phosphorodithioate; O,O-Diethyl-S-(6-chloro-2-oxo-benzoxazolin-3-yl)methyl-phosphorothiolo-thionate; O,O-Dietil-S-((6-cloro-2-oxo-benzossazolin-3-il)-metil)-ditiofosfato; 3-Diethyldithiophosphorylmethyl-6-chlorobenzoxazolone-2; 3-(O,O-Diethyldithiophos-phorylmethyl)-6-chlorobenzoxazolone; O,O-Diethyl phosphorodithioate, S-ester with 6-chloro-3-(mercaptomethyl)-2-benzoxazolinone; EINECS 218-996-2; ENT 27,163; EPA Pesticide Chemical Code 097701; Fosalon; Fozalon; HSDB 4050; NIA-9241; Niagara 9241; NPH-1091; P-974; Phosalon; Phosalone; Phosalone 35 EC; Phosphorodithioic acid, O,O-diethyl ester, S-ester + 6-Cl-3-(mercaptomethyl)-2-benzoxazolinone; Phosphoro-dithioic acid-S-ester of 6-chloro-3-mercaptomethylbenz-oxazoyl-2-one; Phosphorodithioic acid, S-((6-chloro-2-oxo-3(2H)-benzoxazolyl)methyl) O,O-diethyl ester; Phozalon; Rhodia RP 11974; RP 11,974; Rubitox; Zolone; Zolone DT; Zolone PM; Zone;. Non-systemic insecticide and acaricide with contact and stomach action. Used for control of sucking and chewing insects, spider mites, Colorado beetles, aphids, bollworms, stem borers, insects and spider mites on fruit and vegetables. Colorless crystals; mp = 45-48°; soluble in H2O (10 mg/l), more soluble in organic solvents; LD50 (rat orl) = 120-175 mg/kg. *Hortichem Ltd.; Rhône Poulenc Crop Protection Ltd.; Voltas.*

3000 Phosgene

75-44-5   7421   H

CCl2O
Carbonic acid dichloride.
4-03-00-00031 (Beilstein Handbook Reference); BRN 1098367; Carbon dichloride oxide; Carbon oxychloride; Carbone (oxychlorure de); Carbonic acid dichloride; Carbonic chloride; Carbonic dichloride; Carbonio (ossicloruro di); Carbonyl chloride; Carbonyl dichloride; Carbonylchlorid; Chloroformyl chloride; Combat gas; EINECS 200-870-3; Fosgeen; Fosgen; Fosgene; HSDB 796; Koolstofoxychloride; NCI-C60219; Phosgen; Phosgene; RCRA waste number P095; UN 1076. Used in the synthesis of isocyanates, polyurethane, and polycarbonate resins; organic carbonates and chloroformates; pesticides and herbicides; dye manufacture. Highly toxic; has been used as a chemical warfare agent. Gas; mp = -127.9°; bp = 8°; d25 = 1.3719; soluble in organic solvents, slightly soluble in H2O. *BASF Corp.; Bayer AG; Dow Chem. U.S.A.; DuPont; GE Silicones; PPG Ind.; Syngenta Professional Prod.*

3001 Phosphomolybdic acid

11104-88-4   7428   G

24MoO3.P2O5.H2O
Molybdenum hydroxide oxide phosphate.
EINECS 234-336-6; Molybdenum phosphorus hydroxide oxide; Molybdophosphoric acid; Phosphomolybdic acid; Phosphoric acid, anhydride with molybdic acid. Reagent for alkaloids; pigments; catalyst; fixing agent in photography; additive in plating processes; imparts water resistance to plastics, adhesives, and cement. Bright yellow crystals; soluble in H2O (>250 g/100 ml). *AAA Molybdenum; Atomergic Chemetals; Noah Chem.*

3002 2-Phosphonobutane-1,2,4-tricarboxylic acid

37971-36-1   H

C7H11O9P
1,2,4-Butanetricarboxylic acid, 2-phosphono-.
Butanetricarboxylic acid, 2-phosphono-1,2,4-; 1,2,4-Butanetricarboxylic acid, 2-phosphono-; EINECS 253-733-5; 2-Phosphono-1,2,4-butanetricarboxylic acid; 2-Phosphonobutane-1,2,4-tricarboxylic acid.

3003 Phosphonomethyliminodiacetic acid
5994-61-6   H

C5H10NO7P
N-(Carboxymethyl)-N-(phosphonomethyl)glycine.
EINECS 227-824-5; N-(Carboxymethyl)-N-(phosphono-methyl)glycine; Glycine, N-(carboxymethyl)-N-(phos-phonomethyl)-; Phosphonomethyliminodiacetic acid.

3004 Phosphoramidothioic acid, O,O-dimethyl ester
17321-47-0   H P

C2H8NO2PS
O,O-Dimethyl thiophosphoramide.
Amidate; Amidate (pesticide); O,O-Dimethyl phosphoramidothioate; O,O-Dimethyl thiophosphor-amidate; O,O-Dimethyl thiophosphoramide; E-118 amide; EINECS 241-342-2; NSC 133028; Phosphoramidothioic acid, O,O-dimethyl ester.

3005 Phosphoric acid
7664-38-2   7430   E G P

H3O4P
Orthophosphoric acid.
Acide phosphorique; Acido fosforico; Acidum phosphoricum; Caswell No. 662; CCRIS 2949; EINECS 231-633-2; EPA Pesticide Chemical Code 076001; Evits; FEMA No. 2900; Fosforzuuroplossingen; HSDB 1187; Hydrogen phosphate; NSC 80804; Orthophosphoric acid; Phosphoric acid; Phosphorsaeure; Phosphorsäure; Phosphorsaeureloesungen; Phosphorsäurelösungen; Son-ac; UN1805; Wc-reiniger; White phosphoric acid. Inorganic acid; in manufacture of inorganic phosphates, fertilizers, detergents, chemical polishing, priming metals, petroleum refining; acid catalyst; as acidulant and flavoring agent, antioxidant and sequestrant in food; pharmaceutic acid; in dental cements; as an analytical reagent. Registered by EPA as an antimicrobial, fungicide and herbicide. FDA GRAS, approved for parenterals, intramuscular injectables, orals, topicals, vaginals.

USP/NF, BP, Ph.Eur.-compliant. Acidulant and buffering agent. Used in parenterals, intramuscular injectables, orals, topicals, vaginals, in dental cements and etchants. Diluted, as a tonic and for management of nausea and vomiting. Used in fuel cells and as a masonry and tile cleaner. Solid; mp = 42°; bp = 260°; d = 1.685; soluble in H2O. Albright & Wilson Americas Inc.; Ashland; Farleyway Chem. Ltd.; FMC Corp.; Mallinckrodt Inc.; Mitsui Toatsu; Monsanto Co.; Rasa; Rhône-Poulenc.

3006 Phosphoric acid, 2,2-bis(chloromethyl)-1,3-propanediyl tetrakis(2-chloroethyl) ester
38051-10-4   H

C13H24Cl6O8P2
2,2-Bis(chloromethyl)trimethylene bis(bis(2-chloroethyl)-phosphate). EINECS 253-760-2; Phosphoric acid, 2,2-bis(chloro-methyl)-1,3-propanediyl tetrakis(2-chloroethyl) ester.

3007 Phosphoric acid, isooctyl ester, potassium salt
68647-19-8   H

C8H17K2O4P
Isooctyl alcohol phosphate, potassium salt.
EINECS 271-944-0; Isooctyl alcohol phosphate, potassium salt; Isooctyl alcohol, phosphoric anhydride condensation product, potassium salt; Isooctyl phosphate, potassium salt; Phosphoric acid, isooctyl ester, potassium salt.

3008 Phosphoric acid, trixylyl ester
25155-23-1   G H

C24H27O4P
Dimethylphenol phosphate (3:1).
CCRIS 4891; Coalite NTP; Dimethylphenol phosphate (3:1); EINECS 246-677-8; HSDB 6094; Ivviol-3; Phenol, dimethyl-, phosphate (3:1); Phosflex 179; Phosphate, trixylyl; Phosphoric acid, trixylyl ester; Reofos 95; Syn-O-Ad® 8475M; Tri-dimethyl phenyl phosphate; Tri-xylenyl phosphate; Trixylenyl phosphate; Trixylyl phosphate; Xylenol, phosphate (3:1); Xylyl phosphate. Trixylenyl (mixed) phosphate; antiwear and extreme pressure agents in non-crankcase lubricants; ferrous metal passivator; fluid base stock where inhibition of flame propagation is desired. Liquid; bp10 = 243-265°; d = 1.155; insoluble in H2O, soluble in organic solvents. Akzo Chemie.

3009 Phosphorous acid, 2-(1,1-dimethylethyl)-4-(1-(3-(1,1-dimethylethyl)-4-hydroxyphenyl)-1-methylethyl)phenyl bis(4-nonylphenyl) ester
20227-53-6   H

C53H77O4P

2-(tert-Butyl)-4-(1-(3-(tert-butyl)-4-hydroxyphenyl)-1-methylethyl)phenyl bis(4-nonylphenyl) phosphite.

2-(tert-Butyl)-4-(1-(3-(tert-butyl)-4-hydroxyphenyl)-1-methylethyl)phenyl bis(4-nonylphenyl) phosphite; EINECS 243-610-4; Phosphorous acid, 2-(1,1-dimethylethyl)-4-(1-(3-(1,1-dimethylethyl)-4-hydroxyphenyl)-1-methylethyl)-phenyl bis(4-nonylphenyl) ester; Phosphorous acid, 2-t-butyl-alpha-(3-t-butyl-4-hydroxyphenyl)-p-cumenyl bis(p-nonylphenyl) ester.

3010 Phosphorus
7723-14-0   7433   G P

P

P

Violet phosphorus.
Amgard CPC; Amgard CPC 405; Black phosphorus; Bonide blue death rat killer; Caswell No. 663; Common sense cockroach and rat preparations; EINECS 231-768-7; EPA Pesticide Chemical Code 066502; Exolit 385; Exolit 405; Exolit LPKN; Exolit LPKN 275; Exolit RP 605; Exolit RP 650; Exolit RP 652; Exolit RP 654; Exolit VPK-n 361; FR-T 2 (element); Gelber phosphor; Hishigado; Hishigado AP; Hishigado CP; Hishigado NP 10; Hishigado PL; Hittorf's phosphorus; Hostaflam RP 602; Hostaflam RP 614; Hostaflam RP 622; Hostaflam RP 654; HSDB 1169; Masteret 70450; Nova Sol R 20; Novaexcel 140; Novaexcel 150; Novaexcel F 5; Novaexcel ST 100; Novaexcel ST 140; Novaexcel ST 300; Novared 120UF; Novared 120UFA; Novared 120VFA; Novared 140; Novared 280; Novared C 120; Novared F 5; NVE 140; Phosphorus; Phosphorus (red); Phosphorus-31; Phosphorus, amorphous; Phosphorus, white; Rat-Nip; Red phosphorus; Violet phosphorus; White Phosphorus. The coarse-grained red allotrope of phosphorus is metallic or violet phosphorus. Registered by EPA as a rodenticide (cancelled). Red-violet powder; sublimes 416°; d = 2.34; insoluble in organic solvents.

3011 Phosphorus acid
13598-36-2   7432   G

H3O3P

Orthophosphorous acid.
Dihydroxyphosphine oxide; EINECS 237-066-7; Orthophosphorus acid; Phosphonic acid; Phosphorous acid; Phosphorus trihydroxide; Trihydroxyphosphine. Intermediate for manufacture of diphosphonic acids and phosphite salts used as pesticides, chelates, and plastic additives; restricts color formation in esterification and condensation reactions (in small quantities); chemical reducing agent. Usually marketed as a 20% aqueous solution. White solid; mp = 73°; d_4^{21}= 1.65; soluble in H2O, EtOH. *Albright & Wilson Americas Inc.; Astra Hassle AB; CK Witco Corp.; Janssen Chimica; Lonzagroup; Rasa.*

3012 Phosphorus chloride
7719-12-2   7444   G

Cl3P
Phosphorus trichloride.
EINECS 231-749-3; Fosforo(tricloruro di); Fosfor-trichloride; HSDB 1031; Phosphine, trichloro-; Phosphore(trichlorure de); Phosphorous chloride; Phosphorous trichloride; Phosphortrichlorid; Phosphorus chloride; Phosphorus chloride (PCl3); Phosphorus trichloride; Trichlorophosphine; Trojchlorek fosforu; UN1809. Used in manufacture of phosphite and phosphonate esters as the source of phosphorus; chlorinating agent in manufacture of organic acid chlorides; analysis; catalyst. Liquid; mp = -112°; bp = 76°; d_4^{21} 1.574; reacts with H2O, EtOH; soluble in organic solvents; LD50 (rat orl) = 18 mg/kg. *Albright & Wilson Americas Inc.; Atomergic Chemetals; FMC Corp.; Rhône-Poulenc; Sigma-Aldrich Fine Chem.*

3013 Phosphorus oxychloride
10025-87-3   7435   G

Cl3OP
Phosphoryl chloride.
EINECS 233-046-7; Fosforoxychlorid; HSDB 784; Oxychlorid fosforecny; Phosphoric trichloride; Phosphoroxychloride; Phosphorus chloride oxide (POCl3); Phosphorus oxide trichloride; Phosphorus oxychloride; Phosphorus oxytrichloride; Phosphoryl chloride; Phosphoryl trichloride; Trichlorophosphine oxide; Trichlorophosphorus oxide; UN1810. Chlorinating agent, used in the manufacture of insecticides, phosphate esters, pharmaceuticals; gasoline additives; dopant for semiconductor grade silicon, tricresyl phosphate, and fire-retarding agents. Liquid; mp = 1°; bp = 105°; d = 1.6450. *Albright & Wilson Americas Inc.; Cerac; FMC Corp.; Hoechst Celanese; Rhône-Poulenc; Sigma-Aldrich Fine Chem.*

3014 Phosphorus pentachloride
10026-13-8   7437   G

Cl5P
Phosphorus(V) chloride.
EINECS 233-060-3; Fosforo(pentacloruro di); Fosfor-pentachloride; HSDB 1205; Pentachlorophosphorane; Phosphorane, pentachloro-; Phosphore(pentachlorure de); Phosphoric chloride; Phosphoric perchloride; Phosphorous pentachloride; Phosphorpentachlorid; Phosphorus chloride (PCl5); Phosphorus pentachloride; Phosphorus perchloride; Phosphorus(V) chloride; Pieciochlorek fosforu; UN1806. Used as a chlorinating and dehydrating agent and catalyst. Solid; mp = 148°; bp = 160°; d = 1.6000. *Atomergic Chemetals; Cerac; Hoechst Celanese; Sigma-Aldrich Fine Chem.*

3015 Phosphorus pentoxide
1314-56-3   7441   G P

O5P2

Phosphorus(V) oxide.

Diphosphorus pentaoxide; Diphosphorus pentoxide; EINECS 215-236-1; HSDB 847; Phosphoric acid anhydride; Phosphoric acid, anhydrous; Phosphoric anhydride; Phosphoric oxide; Phosphoric pentoxide; Phosphorus oxide; Phosphorus oxide (P2O5); Phosphorus, oxide, pent-; Phosphorus pentaoxide; Phosphorus pentoxide; UN1807. Used in manufacture of chemicals, dessicant (drying agent), surfactants, condensing agent in organic synthesis, sugar refining, lab reagent, fire extinguishing, special glasses. Deliquescent hexagonal crystals; mp = 340°; d = 2.390; very soluble in H2O, reacts to form phosphoric acid. *Albright & Wilson UK Ltd.; Cerac; Elf Atochem N. Am.; Hoechst Celanese; Rasa; Rhône-Poulenc; Sigma-Aldrich Fine Chem.*

3016   Phosphotungstic acid
12067-99-1   7450   G

H6O56P2W24.24H2O

Phosphotungstic acid hydrate.

EINECS 235-087-6; Phosphotungstic acid; Tungsten hydroxide oxide phosphate; Tungstophosphoric acid. Reagent in analytical chemistry to detect alkaloids and other nitrogenous bases and in biology; manufacture of organic pigments; additive in plating industry; imparts water resistance to plastics, adhesives, cement; catalyst for organic reactions; photographic fixative; textile antistat. Yellow-green crystals; soluble in H2O (200 g/100 ml), EtOH, Et2O. *Atomergic Chemetals; Noah Chem.; Spectrum Chem. Manufacturing.*

3017   Phostoxin
20859-73-8   358   G P

$$Al\equiv P$$

AlP

Aluminum Monophosphide.

AL-Phos; AlP; Aluminium fosfide; Aluminium phosphide; Aluminum monophosphide; Aluminum phosphide; Aluminum phosphide (AlP); Caswell No. 031; Celphide; Celphine; Celphos; Celphos (indian); Delicia; Delicia gastoxin; Detia; Detia-Ex-B; Detia gas Ex-B; EINECS 244-088-0; EPA Pesticide Chemical Code 066501; Fosfuri di alluminio; L-Fume; Fumitoxin; Gastion; Gastoxin; HSDB 6035; Phos-Kill; Phosphures d'alumium; Phostoxin; Phostoxin-A; Quick-Phos;; Quickphos; Quik-Fume; RCRA waste number P006; UN1397. Reacts with water to give phosphine (PH3). Used for gassing of rabbits and moles. Registered by EPA as an insecticide. Dark gray-yellow crystals; stable below 1000°; d^{15}_4 = 2.85. *Bernardo Chemicals Inc.; D&D Holdings Inc.; Inventa Corp.; Kommer-Brookwick Ltd.; Midland Fumigant Inc.*

3018   Photine
2606-93-1   1309   G

C28H22N4Na2O8S2

Disodium 4,4'-bis(3-phenylureido)stilbene-2,2'-disulfon-ate.

Amar White RWS; Benzensulfonic acid, 2,2'-(1,2-ethenediyl)bis(5-; Blancol C; Blancophor R; Blankophor R; C.I. 40600; C.I. Fluorescent Brightener 30; EINECS 220-021-0; Leucophor R; Lumisol RV; (((Phenyl-amino)carbonyl)amino)-, disodium salt; Phorwite RN; Photine R; Pontamine White BR; Tintophen X. Fluorescent whitening agents used with paper textiles and detergents. *Hickson & Welch Ltd.*

3019   Photocure 51
24650-42-8   G

C16H16O3

2,2-Dimethoxy-2-phenylacetophenone.

Benzil dimethyl acetal; Benzil dimethyl ketal; Benzil mono(dimethyl acetal); Benzil mono(dimethyl ketal); α,α-Dimethoxy-α-phenylacetophenone; α,α-Dimethoxy-deoxybenzoin; 2,2-Dimethoxy-1,2-diphenylethan-1-one; 2,2-Dimethoxy-1,2-diphenylethanone; 2,2-Dimethoxy-2-phenylacetophenone; 2,2-Dimethoxyphenylaceto-phen-one; ω,ω-Dimethoxy-ω-phenylacetophenone; 1,2-Diphenyl-2,2-dimethoxyethanone; 1,2-Diphenylethane-1,2-dione, dimethyl ketal; DMPA; EINECS 246-386-6; Esacure KB 1; Ethanone, 2,2-dimethoxy-1,2-diphenyl-; IR 651; IRG 651; Irgacure 621; Irgacure 641; Irgacure 651; Irgacure 951; Irgacure E 651; Irgacure I 651; Kayacure BDMK; KB 1; Lucirin BDK; 2-Phenyl-2,2-dimethoxy-acetophenone; Photomer 51. UV photoinitiator. Crystals; mp = 67-70°. *Aceto Corp.*

3020   Phoxim
14816-18-3   7452   G P

C12H15N2O3PS

4-Ethoxy-7-phenyl-3,5-dioxa-6-aza-5-phosphaoct-6-ene-8-nitrile 4-sulfide.

AI3-27448; B 77488; BAY 5621; BAY 77488; BAY sra 7502; Bayer 77488; Baythion; Benzeneacetonitrile, -(((diethoxyphosphinothioyl)oxy)imino)-; Benzoyl cyanide O-(diethoxyphosphinothioyl)oxime; Caswell No. 902L; O,O-Diaethyl-O-(α -cyano-benzyliden-amino)-mono-thiophosphat; O,O-Diaethyl-O-(α-cyan-benzyliden-amino)thionphosphat; O,O-Diäthyl-O-(α-cyano-benz-ylidenamino)-mono-thiophosphat; O,O-Diäthyl-O-(α-cyano-cyanbenzyliden-amino)-thionphosphat; Diethoxyphos-phinothioyloxyimino(phenyl)acetonitrile; α -(((Diethoxy-phosphinothioyl)oxy)imino)benzeneaceto-nitrile; (Diethoxythiophosphoryloxyimino)-phenyl acetonitrile; 2-(Diethoxyphosphinothioyloxyimino)-2-phenyl-acetonitrile; O,O-Diethyl phosphorothioate, O-ester with phenylglyoxylonitrile oxime; O,O-Diethyl α-cyano-benzylideneamino-oxyphosphono-thioate; O,O-Diethyl-α-cyanobenzylidineaminooxyphosphonothiate; 3,5-Di-oxa-6-aza-4-phosphaoct-6-ene-8-nitrile, 4-ethoxy-7-phenyl-, 4-sulfide; 4-Ethoxy-7-phenyl-3,5-dioxa-6-aza-4-phosphaoct-6-ene-8-nitrile 4-sulfide; EINECS 238-887-3; ENT 27488; EPA Pesticide Chemical Code 598800; Foxima; Glyoxylonitrile, phenyl-, oxime,

O,O-diethyl phosphorothioate; OMS 1170; Phenylglyoxylnitrile oxime O,O-diethyl phosphorothioate; Phoxim; Phoxime; Phoximum; Phoxin; Sebacil; Sebacil®; SRA 7502; Valexon; Valexone; Volaton; Volthion®. Foliage- and soil-applied insecticide for control of lepidopterous larvae, beetles and their larvae and locusts on a wide range of crops including cotton maize and vegetables. For control of all ectoparasites, especially mange mites of domestic animals; veterinary medicine. Yellow liquid; mp = 5 - 6°; dec. on distillation; sg^{20} = 1.176; slightly soluble in H_2O (0.0007 g/100 ml at 20°), soluble in C_7H_8, CH_2Cl_2, i-PrOH (all > 20 g/100 ml), slightly soluble in hydrocarbons and vegetable and mineral oils; LD_{50} (rat orl) = 1976 - 2170 mg/kg, (mus orl) = 1935 - 2340 mg/kg, (rbt orl) = 250 - 375 mg/kg, (dog, cat orl) = 250 - 500 mg/kg, (gpg orl) $\cong$600 mg/kg, (rat der) > 1000 mg/kg, (ckn orl) = 40 mg/kg; LC_{50} (rat ihl 1 hr.) = 3.2 mg/l air, (rainbow trout 48 hr.) = 0.1 - 1.0 mg/l, (mirror carp 48 hr.) = 0.1 - 1.0 mg/l, (goldfish 48 hr.) = 1 - 10 mg/l; toxic to bees. *Bayer AG; Bayer Corp., Agriculture; Bayer Corp.*

3021   Phthalic anhydride
85-44-9   7457   H

C8H4O3
o-Phthalic acid anhydride.
AI3-04869; Anhydrid kyseliny ftalove; Anhydride phtalique; Anidride ftalica; CCRIS 519; EINECS 201-607-5; ESEN; Ftaalzuuranhydride; Ftalanhydrid; Ftalowy bezwodnik; HSDB 4012; Isobenzofuran, 1,3-dihydro-1,3-dioxo-; NCI-C03601; NSC 10431; Pan; Phthalandione; Phthalic acid anhydride; o-Phthalic acid anhydride; Phthalic anhydride; Phthalsaeureanhydrid; Phthalsäure-anhydrid; RCRA waste number U190; Retarder AK; Retarder esen; Retarder PD; TGL 6525; UN2214; Vulkalent B; Vulkalent B/C; Wiltrol P. Modified phthalic anhydride; nondiscoloring retarding agent used to reduce scorching of rubber compounds at processing temps; also acts as an activator for certain blowing agents. Used in alkyd resins and plasticizers, as a hardener for resins, polyester, in insecticides, as a laboratory reagent, in the manufacture of phthaleins, phthalates, benzoic acid. White needles, mp = 130.8°; bp = 295°; d^4 = 1.527; λ_m = 224, 274 nm (ε = 8770, 1290, MeOH); soluble in H_2O (6 mg/ml), more soluble in organic solvents. *Akrochem Chem. Co.; Aristech; BASF Corp.; Elf Atochem N. Am.; ExxonMobil Chem. Co.; Mitsubishi Gas; Mitsui Toatsu; OxyChem; Sigma-Aldrich Fine Chem.; Stepan; UCB Res. Inc.*

3022   Phthalic anhydride, polyester with ethylene glycol, maleic anhydride and 1,2-propanediol
26588-55-6   H

$(C_8H_4O_3.C_4H_2O_3.C_3H_8O_2.C_2H_6O_2)_x$
Phthalic anhydride, maleic anhydride, ethylene glycol, propylene glycol polymer.
Ethylene glycol, phthalic anhydride, propylene glycol, maleic anhydride polymer; 1,3-Isobenzofurandione, 1,2-ethanediol, 2,5-furandione, 1,2-propanediol polymer; 1,3-Isobenzofurandione, polymer with 1,2-ethanediol, 2,5-furandione and 1,2-propanediol; Maleic anhydride, phthalic anhydride, ethylene glycol, propylene glycol polymer; Phthalic anhydride, maleic anhydride, ethylene glycol, propylene glycol polymer; Phthalic anhydride, polyester with ethylene glycol, maleic anhydride and 1,2-propanediol; Phthalic anhydride, propylene glycol, maleic anhydride, ethylene glycol polymer; Polymer of propylene glycol, phthalic anhydride, maleic anhydride, ethylene glycol; 1,2-Propanediol, 1,2-ethanediol, cis-butenedioic anhydride, phthalandione polymer; Propylene glycol,

ethylene glycol, maleic anhydride, phthalic anhydride polymer.

3023   Phthalimide
85-41-6   7458   H

C8H5NO2
1,2-Benzenedicarboximide.
5-21-10-00270 (Beilstein Handbook Reference); AI3-07565; Benzoimide; BRN 0118522; EINECS 201-603-3; Ftalimmide; HSDB 5007; Isoindole-1,3-dione; NSC 3108; Phenylimide; o-Phthalic imide; Phthalimid; Phthalimide. Used in manufacture of fungicides such as Folpet and Phosmet. Crystals; mp = 238°; λ_m = 216, 229, 237, 291 (ε = 30500, 12900, 8960, 1390, MeOH); very soluble in C_6H_6; soluble in EtOH (5g/100 ml at 76°), slightly soluble in H_2O. *Allchem Ind.*

3024   Phthalocyanine green
1328-53-6   H

Phthalocyanine Green.
Accosperse Cyan Green G; Brilliant Green Phthalocyanine; Calcotone Green G; CCRIS 4702; Ceres Green 3B; Chromatex Green G; CI 74260; CI Pigment Green 7; CI Pigment Green 42; Colanyl Green GG; Copper Phthalocyanine Green; Cromophtal Green GF; Cromophthal Green GF; Cyan Green 15-3100; Cyanine Green GP; Cyanine Green NB; Cyanine Green T; Cyanine Green Toner; Dainichi Cyanine Green FGH; Dainichi Cyanine Green FG; Daltolite Fast Green GN; Duratint Green 1001; EINECS 215-524-7; Fastogen Green 5005; Fastogen Green B; Fastolux Green; Fenalac Green G Disp; Fenalac Green G; Granada Green Lake GL; Graphtol Green 2GLS; Heliogen Green 8680; Heliogen Green 8681K; Heliogen Green 8682T; Heliogen Green 8730; Heliogen Green A; Heliogen Green G; Heliogen Green GA; Heliogen Green GN; Heliogen Green GNA; Heliogen Green GTA; Heliogen Green GV; Heliogen Green GWS; Hostaperm Green GG; HSDB 4198; Irgalite Fast Brilliant Green GL; Irgalite Fast Brilliant Green 3GL; Irgalite Green GLN; Klondike Yellow X-2261; Lutetia Fast Emerald J; Microlit Green GK; Microlith Green G-FP; Monarch Green WD; Monastral Fast Green GF; Monastral Fast Green GD; Monastral Fast Green GN; Monastral Fast Green GNA; Monastral Fast Green GTP; Monastral Fast Green GV; Monastral Fast Green GWD; Monastral Fast Green GX; Monastral Fast Green GXB; Monastral Fast Green GYH; Monastral Fast Green LGNA; Monastral Fast Green BGNA; Monastral Fast Green G; Monastral Fast Green 2GWD; Monastral Green GFNP; Monastral Green B; Monastral Green B Pigment; Monastral Green G; Monastral Green GFN; Monastral Green GH; Monastral Green GN; Monolite Fast Green GVSA; NCI-C54637; Non-flocculating Green G 25; Opaline Green G 1; Permanent Toner GT-376; Phthalocyanine Brilliant Green; Phthalocyanine Green; Phthalocyanine Green LX; Phthalocyanine Green V; Phthalocyanine Green VFT 1080; Phthalocyanine Green WDG 47; Pigment Fast Green G; Pigment Fast Green GN; Pigment green 7; Pigment Green 7; Pigment Green Phthalocyanine; Pigment Green Phthalocyanine V; Polychloro copper phthalocyanine; Polymo Green FBH; Polymo Green FGH; Polymon Green 6G; Polymon Green G; Polymon Green GN; Pv-Fast Green G; Ramapo; Resinated Phthalocyanine Green G-5025; Sanyo Cyanine Green; Sanyo Phthalocyanine Green FB Pure; Sanyo Phthalocyanine Green F6G; Segnale Light Green G; Sherwood Green A 4436; Siegle Fast Green G; SolFast Green; SolFast Green 63102; Synthaline Green; Termosolido Green FG Supra; Thalo Green No. 1; Versal Green G; Vulcal Fast Green F2G; Vulcanosine Fast Green G; Vulcol Fast Green F2G; Vynamon Green BE; Vynamon Green

BES; Vynamon Green GNA.

3025 Phytic acid
83-86-3   7471   G

$C_6H_{18}O_{24}P_6$

Hexakis(dihydrogen phosphate) myo-inositol.
Acide fytique; Acido fitico; Acidum fyticum; Alkalovert; Alkovert; CCRIS 4513; EINECS 201-506-6; Fytic acid; Inosithexaphosphorsaeure; Inosithexaphosphorsäure; Ino-sitol, hexakis(dihydrogen phosphate), myo-; Inositol hexaphosphate; myo-Inosistol hexakisphosphate; myo-Inositol, hexakis(dihydrogen phosphate); myo-Inositol hexaphosphate; NSC 269896; Phytate; Phytic acid; Saeure des phytins; Säure des phytins. Used for chelation of heavy metals in processing of animal fats and vegetable oils; corrosion inhibitor, metal treating, in the treatment of hard water. The sodium salt is used medically as a hypocalcemic agent. Straw-colored viscous liquid; d= 1.2850; miscible with H_2O, insoluble in organic solvents. *A.G. Scientific, Inc.; Ciba-Geigy Corp.; Corn Products; Staley, A. E. Mfg.*

3026 Phytol
150-86-7   7474   G

$C_{20}H_{40}O$

3,7,11,15-Tetramethyl-2-hexadecen-1-ol.
4-01-00-02208 (Beilstein Handbook Reference); AI3-24344; BRN 1726098; EINECS 205-776-6; 2-Hexadecen-1-ol, 3,7,11,15-tetramethyl-, (2E,7R,11R)-; 2-Hexadecen-1-ol, 3,7,11,15-tetramethyl-, (R-(R*,R*-(E)))-; Phytol; trans-Phytol; 3,7,11,15-Tetramethyl-2-hexadecen-1-ol; 3,7,11, 15-Tetramethylhexadec-2-en-1-ol. Formed in the decomposition of chlorophyll, the coloring matter of plants. Used in preparation of vitamins E and K. bp_{10} = 203°; d^{25} = 0.8497; LD_{50} (rat orl) >5 g/kg.

3027 Picloram
1918-02-1   7482   G H P

$C_6H_3Cl_3N_2O_2$

4-Amino-3,5,6-trichloro-2-pyridine carboxylic acid.
5-22-13-00585 (Beilstein Handbook Reference); Amdon; Amdon grazon; Borolin; BRN 0479075; Caswell No. 039; CCRIS 520; EINECS 217-636-1; EPA Pesticide Chemical Code 005101; Grazon; HSDB 1151; K-Pin; NCI-C00237; NSC 233899; Picloram; Piclorame; Tordon; Tordon 101 mixture; Tordon 10K; Tordon 22K; Tordon 22Kk-Pin. Herbicide; broadleaf and brush killer for forestry, grain and corn. Registered by EPA as a herbicide. Colorless crystals; dec.

215°; soluble in H_2O (0.0430 g/100 ml), Me_2CO (1.98 g/100 ml), EtOH (1.05 g/100 ml), i-PrOH (0.55 g/100 ml), CH_3CN (0.16 g/100 ml), Et_2O (0.12 g/100 ml), CH_2Cl_2 (0.16 g/100 ml), C_6H_6 (0.02 g/100 ml), CS_2 (< 0.005 g/100 ml); LD_{50} (rat orl) = 8200 mg/kg, (mus orl) = 2000 - 4000 mg/kg, (rbt orl) = 2000 mg/kg, (gpg orl) = 3000 mg/kg, (sheep orl > 1000 mg/kg, (cattle orl) > 750 mg/kg, (rbt der) > 4000 mg/kg, (ckn orl) = 6000 mg/kg; LC_{50} (mallard duck, bobwhite quail, Japanese quail 8 day dietary) > 5000 mg/kg, (rainbow trout 96 hr.) = 19.3 mg/l, (bee) = 1000 mg/kg; non-toxic to bees. *Chipman Ltd.; Dow AgroSciences; Dow UK.*

3028 α-Picoline
109-06-8   7484   G H

C_6H_7N

2-Methylpyridine.
AI3-2409; AI3-24109; CCRIS 1721; EINECS 203-643-7; HSDB 101; α-Methylpyridine; 2-Methylpyridine; NSC 3409; Picoline; α-Picoline; 2-Picoline; o-Picoline; Pyridine, 2-methyl-; RCRA waste number U191; UN2313. Organic intermediate for pharmaceuticals, dyes, rubber chemicals, solvent, source for vinyl pyridine, laboratory reagent. Liquid; mp = -66.7°; bp = 129.3°; d^{20} = 0.9443; λ_m = 256, 261, 268 nm (MeOH); soluble in CCl_4, very soluble in H_2O, Me_2CO, freely soluble in EtOH, Et_2O; LD_{50} (rat orl) = 790 mg/kg. *Lancaster Synthesis Co.; Lonzagroup; Nepera.*

3029 β-Picoline
108-99-6   7485   G H

C_6H_7N

3-Methylpyridine.
AI3-24110; CCRIS 1722; EINECS 203-636-9; HSDB 4254; m-Methylpyridine; β-Methylpyridine; 3-Methyl-pyridine; NSC 18251; β-Picoline; 3-Picoline; m-Picoline; Pyridine, 3-methyl-; UN2313. Used as a solvent in synthesis of pharmaceuticals, resins, dyestuffs, rubber accelerators, insecticides; preparation of nicotinic acid, nicotinic acid amide, waterproofing agents; laboratory reagent. Liquid; mp = -18.1°; bp = 144.1°; d^{20} = 0.9566; λ_m = 257, 262, 269 nm (MeOH); soluble in CCl_4, very soluble in Me_2CO, freely soluble in H_2O, EtOH, Et_2O; LD_{50} (rat orl) = 400 mg/kg. *Lancaster Synthesis Co.; Nepera.*

3030 γ-Picoline
108-89-4   7486   G H

C_6H_7N

4-Methylpyridine.
AI3-24111; Ba 35846; CCRIS 1723; EINECS 203-626-4; HSDB 5386; 4-Methylpyridine; p-Methylpyridine; NSC 18252; 4-Picoline; p-Picoline; γ-Picoline; Pyridine, 4-methyl-; UN2313. Solvent used in

the synthesis of pharmaceuticals, resins, dyestuffs, rubber accelerators, pesticides, waterproofing agents; laboratory reagent; manufacture of isoniazid; catalyst; curing agent. Liquid; mp = 3.66°; bp = 145.3°; d^{20} =).9548; λ_m = 255, 262 nm (MeOH); soluble in Me$_2$CO, CCl$_4$, freely soluble in H$_2$O, EtOH, Et$_2$O; LD$_{50}$ (rat orl) = 1290 mg/kg. *Lancaster Synthesis Co.; Lonzagroup.*

3031 Picolinonitrile
100-70-9   H

C$_6$H$_4$N$_2$
2-Pyridinecarbonitrile.
2-Cyanopyridine; EINECS 202-880-3; HSDB 5338; NSC 59697; Picolinic acid nitrile; Picolinonitrile; 2-Pyridinecarbonitrile; 2-Pyridinecarboxylic acid, nitrile; 2-Pyridyl nitrile. Needles; mp = 29°; bp = 224.5°; d^{25} = 1.0810; λ_m = 216, 219, 260, 265, 273 nm (MeOH); slightly soluble in CCl$_4$, ligroin, soluble in H$_2$O, CHCl$_3$, very soluble in EtOH, Et$_2$O, C$_6$H$_6$.

3032 Picric acid
88-89-1   7492   G

C$_6$H$_3$N$_3$O$_7$
2,4,6-Trinitrophenol.
Acide picrique; Acido picrico; Acidum picrinicum; AI3-15403; C.I. 10305; Carbazotic acid; CCRIS 3106; CI 10305; EINECS 201-865-9; HSDB 2040; 2-Hydroxy-1,3,5-trinitrobenzene; Kyselina pikrova; Melinite; NA1344; Nitroxanthic acid; NSC 36947; Phenol, 2,4,6-trinitro-; Phenol trinitrate; Picral; Picric acid; Picronitric acid; Pikrinezuur; Pikrinsaeure; Pikrinsäure; Pikrynowy kwas; Reflorit; Trinitrophenol; 1,3,5-Trinitrophenol; 2,4,6-Trinitrophenol; 2,4,6-Trinitrophenol; 2,4,6-Trinitrophenol; UN0154; UN1344. Explosive, used in matches, for etching copper and production of colored glass. Also used in the disinfection of seed-corn. Yellow crystals; mp = 122.5°; bp > 300° (explosive); d = 1.763; λ_m = 350 nm (ε = 6740 MeOH); soluble in H$_2$O (1.28 g/100 ml), more soluble in organic solvents.

3033 Picrite
556-88-7   6641   H

CH$_4$N$_4$O$_2$
1-Nitroguanidine.
AI3-16306; EINECS 209-143-5; Guanidine, nitro-; Guanidine, 1-nitro-; HSDB 5693; Nitroguanidine; 1-Nitroguanidine; 2-Nitroguanidine; NSC 41036; Picrite; Picrite (the explosive); UN0282; UN1336. Needles or prisms; mp = 239° (dec); λ_m = 218, 266 nm (ε = 6310, 15849, EtOH); insoluble in Et$_2$O, slightly soluble in H$_2$O, EtOH, very soluble in alkali.

3034 Picrocrocin
138-55-6   7493   G

C$_{16}$H$_{26}$O$_7$
(R)-4-(β-D-glucopyranosyloxy)-2,6,6-trimethyl-1-cyclo-hexene-1-carboxaldehyde.
1-Cyclohexene-1-carboxaldehyde, 4-(β-D-glucopyran-osyloxy)-2,6,6-trimethyl-, (R)-; (R)-4-(β-D-Glucopyran-osyloxy)-2,6,6-trimethyl-1-cyclohexene-1-carboxaldeh-yde; 4-(β-D-Glucopyranosyloxy)-2,6,6-trimethyl-1-cyclo-hexene-1-carboxaldehyde; Picrocrocin. From the stigmas of *Crocus Sativus L. Iridaceae.* Important in growth control of the plant. Crystals; mp = 154-156°; $[\alpha]_D^{20}$ = -58° (c = 0.6); soluble in H$_2$O, EtOH, less soluble in organic solvents.

3035 Pigment yellow 83
5567-15-7   H

C$_{36}$H$_{32}$Cl$_4$N$_6$O$_8$
2,2'-((3,3'-Dichloro-1,1'-biphenyl)-4,4'-diyl)bis(azo)bis(N-(4-chloro-2,5-dimethoxyphenyl)-3-oxobutanamide).
Acetoacetanilide, 2,2''-((3,3'-dichloro-4,4'-biphenylyl-ene)bis(azo))bis(4'-chloro-2',5'-dimethoxy-; CI 21108; Pigment yellow 83; EINECS 226-939-8.

3036 Pinacolone
75-97-8   7522   G H

C$_6$H$_{12}$O
Methyl tert-butyl ketone.
4-01-00-03310 (Beilstein Handbook Reference); AI3-03075; BRN 1209331; t-Butyl methyl ketone; tert-Butyl methyl ketone; EINECS 200-920-4; HSDB 5210; Ketone, t-butyl methyl; Ketone, tert-butyl methyl; Methyl t-butyl ketone; Methyl tert-butyl ketone; NSC 935; Pinacolin; Pinacoline; Pinacolone; Pinakolin. Used in chemical synthesis, manufacture. Liquid; mp = -52.5°; bp = 106.1°; d^{25} = 0.7229; slightly soluble in H$_2$O (2.5 g/100 ml), more soluble in organic solvents; LD$_{50}$ (rat orl) = 610 mg/kg. *Lancaster Synthesis Co.; Mallinckrodt Inc.; Sigma-Aldrich Fine Chem.*

3037   cis-Pinane
6876-13-7   H

C10H18
Bicyclo(3.1.1)heptane, 2,6,6-trimethyl-, (1α,2β,5α)-.
AI3-26466; Bicyclo(3.1.1)heptane, 2,6,6-trimethyl-, (1R,2S,5R)-rel-; EINECS 229-978-9; Pinane, stereoisomer; (1α,2β,5α)-2,6,6-Trimethylbicyclo(3.1.1)heptane.

3038   Pinan-2α-ol
4948-28-1   H

C10H18O
Bicyclo(3.1.1)heptan-2-ol, 2,6,6-trimethyl-, (1α,2α,5α)-.
Bicyclo(3.1.1)heptan-2-ol, 2,6,6-trimethyl-, (1R,2S,5S)-rel-; Bicyclo(3.1.1)heptan-2-ol, 2,6,6-trimethyl-, (1α,2α,5α)-; Caswell No. 663L; EINECS 225-591-4; Pinan-2-α-ol; 2-Pinanol, cis-; (1α,2α,5α)-2,2,6-Tri-methylbicyclo(3.1.1)heptan-2-ol; 2,6,6-Trimethylbicyclo-(3.1.1)heptan-2-ol, (1α,2α,5α)-.

3039   Pinan-2β-ol
4948-29-2   H

C10H18O
2-Pinanol, trans-.
Bicyclo(3.1.1)heptan-2-ol, 2,6,6-trimethyl-, (1α,2β,5α)-; Bicyclo(3.1.1)heptan-2-ol, 2,6,6-trimethyl-, (1R,2R,5S)-rel-; NSC 2326; Pinan-2β-ol; 2-Pinanol, trans-; trans-2-Pinanol; 2,6,6-Trimethylbicyclo(3.1.1)heptan-2-ol, (1α,2β,5α)-.

3040   Pinane hydroperoxide
28324-52-9   H

C10H18O2
2,6,6-Trimethylbicyclo(3.1.1)hept-2-yl hydroperoxide.
AI3-19188; EINECS 248-969-0; Hydroperoxide, 2,6,6-trimethylbicyclo(3.1.1)heptyl; Pinane hydroperoxide; Pinanyl hydroperoxide.

3041   (±)-α-Pinene
80-56-8   7527   G H

C10H16
2,6,6-Trimethylbicyclo(3.1.1)-2-hept-2-ene.
4-05-00-00456 (Beilstein Handbook Reference); Acintene A; AI3-24594; Bicyclo(3.1.1)hept-2-ene, 2,6,6-trimethyl-; BRN 3194807; CCRIS 697; DL-Pin-2(3)-ene; EINECS 201-291-9; EINECS 219-445-9; FEMA Number 2902; HSDB 720; NSC 7727; UN2368. Solvent for protective coating, synthesis of camphene, pine oil, odorant, lube oil additives, flavoring, insecticides. Oil; mp = -55°; bp = 156.2°; d20 = 0.8582; insoluble in H2O, soluble in organic solvents. *Arizona; Hercules Inc.; SCM Glidco Organics; Sigma-Aldrich Fine Chem.; Veitsiluoto Oy.*

3042   β-Pinene
127-91-3   7528   G H

C10H16
2,2,6-Trimethylbicyclo(3.1.1)hept-2-ene.
AI3-24483; Bicyclo(3.1.1)heptane, 6,6-dimethyl-2-methylene-; (1)-6,6-Dimethyl-2-methylenebicyclo(3.1.1)-heptane; 6,6-Dimethyl-2-methylenebicyclo(3.1.1)hept-ane; 6,6-Dimethyl-2-methylenenorpinane; EINECS 204-872-5; EINECS 245-424-9; FEMA No. 2903; HSDB 5615; Nopinen; Nopinene; NSC 21447; β-Pinene; 2(10)-Pinene; Pin-2(10)-ene; Pseudopinen; Pseudopinene; Rosemarel; Terbenthene; Terebenthene. Found in many essential oils; used as a chemical intermediate and an intermediate for perfumes and flavorings. Oil; mp = -61.5°; bp = 166°; d15 = 0.874; [α]D = -22°; insoluble in H2O, soluble in C6H6, EtOH, Et2O, CHCl3. *Arizona; Penta Mfg.; SCM Glidco Organics.*

3043   Pine Tar Oil
8002-09-3   P

Pine needle oil.
Arizole; C 30 (pine oil); Caswell No. 665; Dertol 90; EPA Pesticide Chemical Code 067002; Essential oils, pine; Essential pine oil; Glico 150; Glidsol 150; Oil of fir - Siberian; Oil of pine; Oils, essential, pine; Oils, pine; Oils, pine, synthetic; Oils, pine wood; Oleum abietis; Oulo 02; Pine oil; Pine oil absolute; Pine oil, synthetic; Pine oil, white (*Pinus* spp.); Pine oil [Flammable liquid]; Pine oil [Oil, misc.]; Pine Oil C 30; Pine tar oil; Pine wood oil; Polyiff 272; RT 1712; Terpentinöl; Terpentinoel; UN1272; Unipine; Unipine 80; Unipine 85; Yarmor; Yarmor 60; Yarmor 302; Yarmor F; Yarmor pine oil; Yarmor pine oil (*Pinus palustris* MilL.). Disinfectant, possibly antibacterial. Used as a fragrance and as a disinfectant in household cleaners and laundries. Registered by EPA as a disinfectant. Pale yellow oil, d = ca. 0.9; bp = 200 - 220°; insoluble in H2O, soluble in most organic solvents. *ABC Compounding Co.; Bullen Co; Bush Boake Allen (IFF bought 2000); Carroll Co.; Clorox Co.; Hercules Inc.; James Austin Co; JF Daley International Ltd; Lonza Ltd.; Mason; Millennium; National Chemical Laboratories, Inc.; Reckitt Benckiser Inc.; T&R Chemicals.*

3044   Pioglitazone
111025-46-8   7533   D

hydroxyethylpiperazine; Piperazineethanol.

C19H20N2O3S
(±)-5-[p-[2-[(5-Ethyl-2-pyridyl)ethoxy]benzyl]-2,4-thiazolidinedione.
Actos; AD-4833; Pioglitazona; Pioglitazone; Pioglitazon-um. Antidiabetic agent. Insulin sensitizer. Crystals; mp = 183-184°. *Takeda Chem. Ind. Ltd.*

3045   Pioglitazone hydrochloride
112529-15-4   7533   D

C19H21ClN2O3S
(±)-5-[p-[2-[(5-Ethyl-2-pyridyl)ethoxy]benzyl]-2,4-thiazol-idinedione monohydrochloride.
Actos; Pioglitazone hydrochloride; U-72107A. Antidia-betic agent. Insulin sensitizer. Crystals; mp = 193-194°. *Takeda Chem. Ind. Ltd.*

3046   Piperazine
110-85-0   7545   G H

C4H10N2
Hexahydropyrazine.
5-23-01-00030 (Beilstein Handbook Reference); Antiren; Asca-Trol No. 3; BRN 0102555; CCRIS 5950; 1,4-Diazacyclohexane; 1,4-Diethylenediamine; Diethylene-diamine; Diethyleneimine; Dispermine; EINECS 203-808-3; Entacyl; Eraverm; Hexahydro-1,4-diazine; Hexahydro-pyrazine; HSDB 1093; Lumbrical; NSC 474; Piper-azidine; Piperazin; Piperazine; 1,4-Piperazine; Pipersol; Pyrazine hexahydride; Pyrazine, hexahydro-; UN2579; Upixon; Uvilon; Vermex; Worm-A-Ton; Wurmirazin. Corrosion inhibitor, anthelmintic, insecticide and accelerator for curing polychloroprene. Plates or leaflets; mp = 106°; bp = 146°; λ_m = 196 nm (ϵ = 5012, gas); very soluble in H_2O, ethylene glycol, soluble in EtOH, CHCl3, insoluble in Et2O. *Allchem Ind.; BASF Corp.; Bayer AG; Janssen Chimica.*

3047   Piperazineethanol
25154-38-5   H

C6H14N2O
Monohydroxyethylpiperazine.
EINECS 246-671-5; Hydroxyethylpiperazine; Mono-

3048   Piperidine
110-89-4   7549   G H

C5H11N
Hexahydropyridine.
AI3-24114; Azacyclohexane; CCRIS 967; Cyclo-pentimine; Cypentil; EINECS 203-813-0; FEMA No. 2908; Hexahydropyridine; Hexazane; HSDB 114; Pentamethyleneimine; Pentamethylenimine; Perhydro-pyridine; Piperidin; Piperidine; Pyridine, hexahydro-; UN2401. Solvent and intermediate, curing agent for rubber and epoxy resins, catalyst for condensation reactions, ingredient in oils and fuels, complexing agent. Liquid; mp = -11.03°; bp = 106.2; d^{20} = 0.8606; λ_m = 190 nm (ϵ = 4677, cyclohexane); very soluble in H_2O, EtOH, soluble in Et2O, Me2CO, CHCl3, C6H6; LD50 (rat orl) = 400 mg/kg. *Janssen Chimica; Lancaster Synthesis Co.; Nepera; Sigma-Aldrich Fine Chem.*

3049   Piperidinium pentamethylene dithiocarb-amate
98-77-1   G

C11H22N2S2
Piperidinecarbodithioic acid, piperidinium salt.
522 Rubber accelerator; Accelerator 552; AI3-03659; Akrochem® P.P.D; EINECS 202-698-4; NSC 1906; Pentamethyleneammonium pentamethylenedithiocarb-amate; PIP; Pip-Pip; 1-Piperidinecarbodithioic acid, compound with piperidine; 1-Piperidinecarbodithioic acid, piperidine salt; Piperidine pentamethylene-dithiocarbamate; Piperidinium N,N-pentamethylene-dithiocarbamate; Piperidinium pentamethylene dithio-carbamate; Piperidinium piperidine-1-carbodithioate; Piperidinium piperidinedithiocarbamate; PMP (accel-erator); Vanax® 552. Accelerator for latex; in rubber cements, compounds containing white factice and for tank linings; peptizing agent; nondiscoloring and nonstaining. Crystals; mp = 167° (dec); insoluble in H_2O, soluble in CHCl3, Me2co; C7H8, EtOH; poisonous by ingestion; eye irritant. *Akrochem Chem. Co.; Vanderbilt R.T. Co. Inc.*

3050   Piperonyl butoxide
51-03-6   7557   G

C19H30O5
4,5-methylenedioxy-2-propylbenzyldiethylene glycol butyl ether.
4-19-00-00779 (Beilstein Handbook Reference); AI3-14250; Anvil 10+10 ULV; Anvil 2+2 ULV; BRN 0288063; Butacide; Butocide; Butoxide; Butoxide (synergist); Butyl carbitol 6-propylpiperonyl ether; (Butylcarbityl)(6-propylpiperonyl)ether; Caswell No. 670; CCRIS 522; EINECS 200-076-7; ENT 14,250; EPA Pesticide Chemical Code 067501; Ethanol butoxide; FMC 5273; HSDB 1755; NCI-C02813; NIA 5273; NSC 8401; Nusyn-noxfish; PB; Piperonyl butoxide; Piperonyl butoxyde; Pyrenone 606; Scourge. Synergist in insecticides, especially pyethrins and rotenone; used in combinations with pyrethrins (Derringer, Duracide, Grovex, Prentox;

Scourge). $bp_{1.0} = 180°$; d = 1.04-1.07; $n_D^{20} = 1.50$; insoluble in H_2O, soluble in organic solvents; LD_{50} (rat orl) = 7500 mg/kg. *Burlington Biomedical; Wellcome Foundation Ltd.*

3051 Pirimicarb
23103-98-2   7579   G H P

$C_{11}H_{18}N_4O_2$
  2-(Dimethylamino)-5,6-dimethyl-4-pyrimidinyl   N,N-dimethylcarbamate.
  5-25-13-00063 (Beilstein Handbook Reference); ABOL; Aficida; AI3-27766; Aphox; BRN 0663442; Carbamic acid, dimethyl-, 2-(dimethylamino)-5,6-dimethyl-4-pyrimidinyl ester; Caswell No. 359C; 2-(Dimethylamino)-5,6-dimethyl-4-pyrimidinyldimethylcarbamate; Dimethyl-carbamic acid 2-(dimethylamino)-5,6-dimethyl-4-pyr-imidinyl ester; 5,6-Dimethyl-2-dimethylamino-4-pyrimid-inyldimethylcarbamate; 5,6-Dwumetylo-2-dwumetylo-amino-4-pirimidynylodwukarbaminian; EINECS 245-430-1; ENT-27766; EPA Pesticide Chemical Code 106101; Fernos; HSDB 7005; Pirimicarb; Pirimicarbe; Pirimor; Pirimor 50 DP; Pirimor G; Pirimor granulate; PP 062; Primicarbe; Pyrimor; RapidGarden insecticide. Registered by EPA as an insecticide. Crystals; mp = 90.5°; soluble in H_2O (0.27 g/100 ml), Me_2CO (0.4 g/100 ml), EtOH ().25 g/100 ml), xylene (0.29 g/100 ml), $CHCl_3$ (0.33 g/10 ml); LD_{50} (rat orl) = 147 mg/kg, (mus orl) = 107 mg/kg, (dog orl) = 100 - 200 mg/kg, (rat der) > 500 mg/kg, (poultry orl) = 25 - 50 mg/kg, (mallard duck orl) = 17.2 mg/kg, (bobwhite quail orl) = 8.2 mg/kg; LC_{50} (rainbow trout 96 hr.) = 29 mg/l, (bluegill sunfish (96 hr.) = 55 mg/l; non-toxic to bees. *ICI Chem. & Polymers Ltd.; Syngenta Crop Protection; Zeneca Pharm.*

3052 Pivalic acid
75-98-9   7594   H

$C_5H_{10}O_2$
  2,2-Dimethylpropanoic acid.
  4-02-00-00908 (Beilstein Handbook Reference); Acetic acid, trimethyl-; AI3-04165; BRN 0969480; EINECS 200-922-5; HSDB 5211; Kyselina 2,2-dimethylpropionova; Kyselina pivalova; Neopentanoic acid; NSC 65449; tert-Pentanoic acid; Pivalic acid; Propionic acid, 2,2-dimethyl-; Trimethylacetic acid; Versatic 5. Solid; mp = 35°; bp = 164°; $d^{50} = 0.905$; soluble in H_2O (2.5 g/100 ml), more soluble in EtOH, Et_2O. *Lancaster Synthesis Co.; Mallinckrodt Inc.; Shiva Pharmachem; Sigma-Aldrich Fine Chem.*

3053 Pivaloyl chloride
3282-30-2   H

C_5H_9ClO
  2,2-Dimethylpropanoyl chloride.
  2,2-Dimethylpropanoyl chloride; EINECS 221-921-6; Pivaloyl chloride; Propanoyl chloride, 2,2-dimethyl-; Trimethylacetyl chloride;

UN2438. Liquid; bp = 107°; $d^{20} = 1.003$; very soluble in Et_2O.

3054 Plant indican
487-60-5   4967   G

$C_{14}H_{17}NO_6$
  1H-Indol-3-yl-β-D-glucopyranoside.
  'β-D-Glucopyranoside, 1H-indol-3-yl; 3-(Glucosyloxy)-indole; Indican (glucoside); Indican (plant indican); Indican, plant; Indikan; Indole, 3-(β-D-glucopyranosyl-oxy)-; 1H-Indol-3-yl β-glucopyranoside; Indoxyl-β-D-glucoside; NSC 87517; Uroxanthin. Found in leaves of *Indigofera tinctoria*. Crystals; mp = 57-58°; $[\alpha]_D^{19}$ = -66°; soluble in H_2O, organic solvents.

3055 Platinum
7440-06-4   7612   G

Pt

Pt
  Platinum.
  C.I. 77795; EINECS 231-116-1; HSDB 6479; Liquid bright platinum; Platin; Platinum; Platinum Black; Platinum, elemental; Platinum, metal; Platinum sponge. Pt; Metallic element; catalyst, laboratory ware, rayon and glass fiber manufacture, jewelry, dentistry, electrical contacts, thermocouples, surgical wire, bushings, electroplating, electric furnace windings, chemical reaction vessels, permanent magnets. Metal; mp = 1773°; bp = 3827°; d = 21.447. *Degussa AG; Handy & Harman; Noah Chem.; Sigma-Aldrich Fine Chem.*

3056 Poly(acrylonitrile-co-butadiene)
9003-18-3   G

$[-CH_2CH(CN)]_x-(CH_2CH=CHCH_2)_y$

  Acrylonitrile-butadiene copolymer.
  Butadiene/acrylonitrile copolymer; 1,3-Butadiene, polymer with 2-propenenitrile; Krynac® 19.65; Krynac® 20H35; Krynac® 34.140; Krynac® 34.35; Krynac® 34.50; Krynac® 34.80; Krynac® 823X2; Krynac® 843; Krynac® PXL 34.17; 2-Propenenitrile, polymer with 1,3-butadiene. A nonstaining, cold polymerized copolymer; used for low temperature oil well specialties, belt covers, and idler rolls for low-temperature service, o-rings, seals, gaskets, hydraulic hose; vulcanized with sulfur, sulfur donor, or peroxide. White solid; d = 0.980. *Bayer AG; Polysar.*

3057 Poly(oxyethylene)-p-t-octylphenyl ether
9002-93-1   6793   G

$[C_{16}H_{26}O_2]_n$

  Octoxynol-1.
  Alfenol 3; Alfenol 9; Antarox A-200; CCRIS 985; Conco nix-100; Ethoxylated p-tert-octylphenol; Glycols, polyethylene, mono(p-(1,1,3,3-tetramethylbutyl)phenyl) ether; Hydrol SW; Hyonic pe-250; Igepal CA-630; Marlophen 820; Neutronyx 605; NSC 406472; 3,6,9,12,15,18,21,24-Octaoxahexacosan-1-ol, 26-(octyl-phenoxy)-; Octoxinol; Octoxinolum; Octoxynol; Octoxynol-3; Octoxynol-9; Octoxynol-11; Octoxynol-12; Octoxynol-25; Octoxynol-33; Octoxynol-40; Octyl phenol condensed with 12-13 moles ethylene oxide; Octylphenoxypolyethoxyethanol; 26-(4-Octylphenoxy)-3,6,9,12,15,18,21,24-octaoxahexacosan-1-ol; 26-(Octyl-phenoxy)-

3,6,9,12,15,18,21,24-octaoxahexacosan-1-ol; p-tert-(Octylphenoxypolyethoxy)ethanol; 4-tert-Octyl-phenyl peg ether; OPE 30; Ortho-gynol; p-(1,1,3,3-Tetramethylbutyl)phenol ethoxylate; Peg (P-(1,1,3,3-tetramethylbutyl)phenyl) ether; Peg 4-isooctylphenyl ether; Peg 4-tert-octylphenyl ether; Peg-9 octyl phenyl ether; Peg P-tert-octylphenyl ether; PEG-9 Octyl phenyl ether; PEG-11 Octyl phenyl ether; PEG-12 Octyl phenyl ether; PEG-25 Octyl phenyl ether; PEG-33 Octyl phenyl ether; PEG-40 Octyl phenyl ether; Phenol, P-(1,1,3,3-tetramethylbutyl)-, monoether with polyethylene glycol; Poletoxol; Poly(oxy-1,2-ethanediyl), α-(octylphenyl)-ω-hydroxy-; Poly(oxy-1,2-ethanediyl), α-(4-(1,1,3,3-tetra-methylbutyl)phenyl)-ω-hydroxy-; Polyethylene glycol mono(4-octylphenyl) ether; Polyethylene glycol mono(octylphenyl) ether; Polyethylene glycol mono(p-(1,1,3,3-tetramethylbutyl)phenyl) ether; Polyethylene glycol mono(p-tert-octylphenyl) ether; Polyethylene glycol monoether with p-tert-octylphenyl; Polyethylene glycol octylphenol ether; Polyethylene glycol octylphenol ether; Polyethylene glycol p-octylphenyl ether (VAN); Polyethylene glycol p-tert-octylphenyl ether; Poly-ethyleneglycol 4-(tert-octyl)phenyl ether; Poly-ethylene glycol p-1,1,3,3-tetramethylbutylphenyl ether; Poly-oxyethylene (9) octyl phenyl ether; Polyoxyethylene (11) octyl phenyl ether; Polyoxyethylene (12) octyl phenyl ether; Polyoxyethylene (13) octylphenyl ether; Polyoxyethylene (25) octyl phenyl ether; Polyoxyethylene (33) octyl phenyl ether; Polyoxyethylene (40) octyl phenyl ether; Polyoxyethylene 4-(1,1,3,3-tetramethylbutyl)phenyl ether; Polyoxyethylene mono(octylphenyl) ether (VAN); Polyoxyethylene octyl phenyl ether; Preceptin; 4-(1,1,3,3-Tetramethylbutyl)phenol, ethoxylated; 4-(1,1,3,3-Tetra-methylbutyl)phenyl hydroxypoly(oxyethylene); α-(4-(1,1,3,3-Tetramethylbutyl)phenyl)-ω-hydroxypoly(oxy-1,2-eth-anediyl); α-(P-(1,1,3,3-Tetramethylbutyl)phenyl)-ω-hydro-xypoly(oxyethylene); Texofor FP 300; Triton x-45; Triton X; Triton X 35; Triton X 45; Triton X 100; Triton X 101; Triton X 102; Triton X 165; Triton X 305; Triton X 405; Triton X 705; TX 100; Ethylene glycol octyl phenyl ether; Polyoxyethylene octyl phenyl ether; Octylphenoxy-polyethoxyethanol; Polyethylene glycol mono [4-(1,1,3,3-tetramethylbutyl)phenyl] ether; Poly(oxyethylene)-p-tert-octylphenyl ether; POE octylphenol; polyoxyethylene (10) octylphenol; POE (10) octylphenol. Surfactant, coupling agent, emulsifier for industrial/household cleaners, emulsion polymerization, agriculture, latex stabilizer. *Union Carbide Corp.*

3058   Poly(vinylidene fluoride)
24937-79-9   G

(-CH2CF2-)n

Ethene, 1,1-difluoro-, homopolymer.
 Kynar® 301 F; Kynar® 460; Kynar® 700 Series; Kynar® Flex® 2800, 2801; Kynar® Flex® 2900; Polyvinylidene fluoride; PVDF. A crystalline high molecular weight polymer used for solvent-based coatings; produces films with high resistance to gamma radiation and transparency to UV radiation. A tough engineering resin with high abrasion resistance and stability in harsh thermal, chemical, and UV environments; readily melt processable in molding and extrusion; used for coatings incl. corrosion-resistance coatings for chemical process equipment, long life decorative finishes on building panes, film, filter cloth, intrumentation, control equipment linings, membranes, static mixers, pipes and fittings, pumps, stock shapes, valves and electronic and electrical jacketing. Used in wire and cable construction and other uses requiring high flexibility, improved impact resistance, extremely low smoke emission and low flame spread. Tm = 165-172º. *Elf Atochem N. Am.*

3059   Polyacrylamide
9003-05-8   G

[CH2CHCONH2]x

2-Propenamide Homopolymer.
 Acrylamide homopolymer; Acrylamide polymer; Aerofloc 3453;

American Cyanamid KPAM; American Cyanamid P-250; Aminogen PA; AP 273; Bio-Gel P 2; BioGel P-100; Cyanamer P 35; Cyanamer P 250; Cytame 5; Diaclear MA 3000H; Dow 164; Dow ET 597; Dow J 100; ET 597; Flokonit E; Flygtol GB; Gelamide 250; Himoloc OK 507; Himoloc SS 200; HSDB 1062; J 100; K 4 (acrylic polymer); K-PAM; Magnafloc R 292; Nacolyte 673; NSC 116573; NSC 116574; NSC 116575; NSC 118185; P 250; P 300; PAA; PAA-1; PAA 70L; PAARK 123sh; PAM; PAM (polymer); PAM-50; Pamid; Percol 720; Poly(2-propenamide); Polyacrylamide; Polyacrylamide resin; Polyhall 27; Polyhall 402; Polystolon; Polystoron; Porisutoron; Praestol 2800; 2-Propenamide, homo-polymer; Q 41F; Reten 420; Sanpoly A 520; Solvitose 433; Stipix AD; Stokopol D 2624; Sumirez A 17; Sumirez A 27; Sumitex A 1; Superfloc 84; Superfloc 900; Sursolan P 5; Taloflote; Versicol W 11. Used as a thickener, dispersant, antiprecipitant, solubilizer, binder, sizing, flocculating, suspending, crosslinking agent, filtering aid, lubricant; used in adhesives, agriculture, cement, coatings, cosmetics, detergents, latex manufacture, printing ink. *Allied Colloids; Am. Cyanamid; Calgon Carbon; Rhône Poulenc Surfactants; Sigma-Aldrich Fine Chem.*

3060   Polyacrylic acid
9003-01-4   G P

[C3H4O2]n

Acumer 1000.
 Acrylic acid homopolymer; Acrylic acid polymer; Acrylic acid resin; Acrylic polymer; Acrylic polymer resins; Acrylic resin; Acrysol A 1; Acrysol A 3; Acrysol A 5; Acrysol AC 5; Acrysol ase-75; Acrysol lmw-20X; Acrysol WS-24; Antiprex 461; Antiprex A; Arasorb 750; Arasorb S 100F; Arolon; Aron; Aron A 10H; Atactic poly(acrylic acid); Carbomer; Carbomer 910; Carbomer 934; Carbomer 934p; Carbomer 940; Carbomer 941; Carbomer 1342; Carbopol 910; Carbopol 934; Carbopol 934P; Carbopol 940; Carbopol 941; Carbopol 960; Carbopol 961; Carbopol 971P; Carbopol 974P; Carbopol 980; Carbopol 981; Carbopol 1342; Carboset 515; Carboset Resin No. 515; Carboxy vinyl polymer; Carboxypolymethylene; Carpolene; CCRIS 3234; Colloids 119/50; Cyguard 266; Dispex C40; Dow Latex 354; G-Cure; Good-rite K 37; Good-rite K 700; Good-rite K 702; Good-rite K 727; Good-rite K 732; Good-rite WS 801; Haloflex 202; Haloflex 208; Joncryl 678; Junlon 110; Jurimer AC 10H; Jurimer AC 10P; Nalfloc 636; Neocryl A-1038; NSC 106034; NSC 106035; NSC 106036; NSC 106037; NSC 112122; NSC 112123; NSC 114472; NSC 165257; OLD 01; P 11H; PA 11M; PAA-25; Pemulen TR-1; Pemulen TR-2; Poly(acrylic acid); Polyacrylate; Polyacrylate elastomers; Polyacrylic acid; Polyacrylic acid; Polymer, carboxy vinyl; Polymer of 2-propenoic acid, cross-linked with allyl ethers of sucrose or pentaerythritol; Polymer of 2-propenoic acid, cross-linked with allyl ethers of sucrose; Polymerized acrylic acid; Polytex 973; Primal ASE 60; Propenoic acid polymer; 2-Propenoic acid, homopolymer; R968; Racryl; Revacryl A 191; Rohagit SD 15; Sokalan PAS; Solidokoll N; Synthemul 90-588; TB 1131; Tecpol; Texcryl; Versicol E 7; Versicol E15; Versicol E9; Versicol K 11; Versicol S 25; Viscalex HV 30; Viscon 103; WS 24; WS 801; XPA. Used as a scale inhibitor in industrial water treatment and oil production. Powder; irritant; the viscosity of a neutralized 1.0 percent aqueous dispersion of Carbomer1342 is between 9,500 and 26,500 centipoises. *Rohm & Haas Co.*

3061   Polybutadiene
9003-17-2   G

(C4H6)n

Butadiene rubber.
 Alfine; Atactic butadiene polymer; B 7; B 11; B 3000; Budium RK 622; Butadiene homopolymer; 1,3-Butadiene, homopolymer; Butadiene oligomer; 1,3-Butadiene, polymers; Butadiene polymer; Butadiene resin; Butarez 15; CB 10; Diene 35 NF; Dienite 556; Dienite 643; Dienite X 555; Dienite X 644; FCR 126; FCR 1261; FCR 1261pd; HSDB 1196; Hystl; Hystl B 300; Hystl B 1000; Hystl B 2000;

Hystl B 3000; LCB 150; Nisso BN 1000; Nisso PB 100; Nisso PB 3000; Nisso PB 4000; Nisso PB-B 4000; Nisso PB-GQ 3000; Nisso PR 2000; Pbc200; Poly-1,3-butadiene; Polybutadiene; Polyoil 110; Polyoil 130; Quintol B 1000; S 820; XPDR-A 288. BR; elastomer for tire industry, footwear, molded goods; blending ingredient in SBR; additive for plastics; coating resin in liquid form. *Asahi Chem. Industry; BASF Corp.; Goodyear Tire & Rubber; Phillips 66; Reichold.*

3062   Polybutene
9003-28-5   G

[C4H8]n

Polybutylene.
1-Butene, homopolymer; PIB; Polybutene; Poly-α-butylene; U.S. National L. Polymer formed by polymerization of a mixture of iso and normal butenes. Thermoplastic resin; used as tackifier, strengthener, and extender in adhesives, as plasticizer for rubber, as vehicle and fugitive binder for coatings. *Amoco Lubricants; Ashland; BP Chem.; Harcros.*

3063   Polydimethylsiloxane
9006-65-9   3241   G

(C2H6OSi)xC4H12Si

Dimethyl polysiloxane, dimethyl-terminated.
AF 10 FG; Belsil DM 1000; CCRIS 3957; DC 1664; Dimethicone; Dimethicone 350; Dimethicones; Dimethyl polysiloxane, bis(trimethylsilyl)-terminated; Dimethyl-polysiloxane; Dimeticona; Dimeticone; Dimeticonum; Dow Corning 1664; HSDB 1808; Mirasil DM 20; Poly(oxy(dimethylsilylene)), α-(trimethylsilyl)-omega-methyl-; Polysilane; Sentry Dimethicone; α-(Trimethyl-silyl)-omega-methylpoly(oxy(dimethylsilylene)); Viscasil 5M; Durkex 100DS; SF18-350;; AF 9020; AF 30 FG. Silicone antifoam agent used for general food, poultry, and meat processing applications; anticaking agent. Soluble in hydrocarbon solvents, CHCl3, Et2O, H2O; suspected carcinogen. *Harcros.*

3064   Polyethylene
9002-88-4   7650   G P

[C2H4]x

Ethylene homopolymer.
Alathon 7140; Alathon 7511; Alcowax 6; Aldyl A; Alithon 7050; Alkathene; Allied PE 617; Alphex FIT 221; Ambythene; Amoco 610A4; Bakelite DFD 330; Bakelite DHDA 4080; Bakelite DYNH; Bareco polywax 2000; Bareco wax C 7500; Bicolene C; BPE-I; Bralen KB 2-11; Bulen A; Carlona PXB; CCRIS 3553; Courlene-X3; Epolene C; Epolene E; Epolene N; Ethene, homopolymer; Etherin; Ethylene polymer; Grex; HI-Fax; Hizex; Plastipore; Polyethylene. Polymer of ethylene monomers. Used in laboratory tubing; prostheses; electrical insulation; packaging materials; kitchenware; tank and pipe linings; paper coatings; textile stiffeners. Polyethylene wax; processing lubricant, melt index modifier, pigment dispersant, mold release aid; external lubricant PVC, color concentrates, polyolefin flow modifiers; thickener for cosmetic and pharmaceutical gels. Rodenticide. Registered by EPA as a rodenticide (cancelled). FDA approved for dentals, ophthalmics, orals, topicals, vaginals. Used as an excipient, gel thickener, emollient, moisture controller, film former, non-irritating abrasive, used in dentals, ophthalmics, orals, topicals and vaginals. White solid; mp = 85-110°, 130-145°; d$_4^{20}$= 0.922; soluble in hot C6H6, insoluble in H2O. *Allied Signal; Asahi Chem. Industry; Ashland; Cabot Carbon Ltd.; Eastman Chem. Co.; Elf Atochem N. Am.; EniChem Am.; ExxonMobil Chem. Co.; LNP; Mitsubishi Corp.; Quantum/USI.*

3065   Polyethylene glycol
25322-68-3   7651   G H

(C2H4O)n.H2O

Ethanol, 2,2'-(oxybis(2,1-ethanediyloxy)bis-.
Alcox E 30; Alkapol PEG 300; Alkox E 45; Alkox E 60; Alkox E 75; Alkox E 100; Alkox E 130; Alkox E 160; Alkox E 240; Alkox R 15; Alkox R 150; Alkox R 400; Alkox R 1000; Alkox SR; Antarox E 4000; Aquacide III; Aquaffin; Atpeg 300; Badimol; BDH 301; Bradsyn PEG; Breox 20M; Breox 550; Breox 2000; Breox 4000; Breox PEG 300; CAFO 154; Carbowax; Carbowax 20; Carbowax 100; Carbowax 200; Carbowax 300; Carbowax 400; Carbowax 600; Carbowax 1000; Carbowax 1350; Carbowax 1500; Carbowax 3350; Carbowax 4000; Carbowax 4500; Carbowax 4600; Carbowax 14000; Carbowax 20000; Carbowax 25000; Carbowax Sentry; CCRIS 979; Colyte; DD 3002; Deactivator H; Emkapol 4200; Ethanol, 2,2'-(oxybis(2,1-ethanediyloxy)bis-; Ethoxylated 1,2-ethanediol; Ethylene glycol homopolymer; Ethylene glycol polymer; Gafanol E 200; Glycols, polyethylene; HM 500; HSDB 5159; Lutrol; Macrogol; Merpol OJ; Modopeg; Nosilen; NSC 152324; Nycoline; Oxide Wax AN; Oxyethylene polymer; PEG; PEG; PEG 400; PEG 6000DS; Pluracol E; Pluracol E 400, E 600, E 1450; Pluriol E 200; Poly-G; Poly-G600; Polyethylene glycol; Polyox wsr-N 60; WSR-301. Intermediate in manufacture of surfactants; binder/ lubricant in pharmaceuticals; plasticizer; paper softener; humectant; solvent; antistat; for cosmetics, textile, plastics processing, dyes and inks. *Rhône Poulenc Surfactants.*

3066   Polyethylene glycol diester of tall oil acids
61791-01-3   H

Polyoxyethylene (12) ditallate.
Fatty acids, tall-oil, diesters with polyethylene glycol; PEG-8 Ditallate; PEG-12 Ditallate; Polyethylene glycol 400 ditallate; Polyethylene glycol 600 ditallate; Polyethylene glycol diester of tall oil acids; Polyoxyethylene (8) ditallate; Polyoxyethylene (12) ditallate; Tall oil fatty acid diester of polyethylene glycol.

3067   Polyethylene glycol phenyl ether phosphate
39464-70-5   H

(C2H4O)x(C6H6O)y.zH3PO4

Poly(oxy-1,2-ethanediyl), α-phenyl-ω-hydroxy-, phosph-ate.
Phenol, ethoxylate, phosphate; Phenol, ethoxylated, phosphated; Poly(oxy-1,2-ethanediyl), α-phenyl-ω-hydroxy-, phosphate; Polyethylene glycol phenyl ether phosphate.

3068   Polyethylene glycol, tridecyl ether, phosphate, potassium salt
68186-36-7   H

(C2H4O)y.C13H28O.xH3KO4P

Tridecyl alcohol, ethoxylated, phosphate, potassium salt.
Poly(oxy-1,2-ethanediyl)α-tridecyl-omega-hydroxy-, phosphate, potassium salt; Polyethylene glycol, tridecyl ether, phosphate, potassium salt; Tridecyl alcohol, ethoxylate, phosphate, potassium salt; Tridecylalcohol, ethoxylated, phosphated, potassium salt.

3069   Polyethylene terephthalate
25038-59-9   G

(-OCH2CH2O2CC6H4-4-CO-)n
Poly(oxyethyleneoxyterephthaloyl).
  Amilar; Arnite A; Arnite A-049000; Arnite A 200; Arnite FP 800; Arnite G; Arnite G 600; Cassappret SR; Celanar; Cleartuf; Clertuf; Crastin S 330; Crastin S 350; Crastin S 440; Daiya foil; Dowlex; E 20 [Japanese polyester]; Estrofol; Estrofol B; Estrofol Ow; Ethylene terephthalate oligomer; Ethylene terephthalate polymer; Fiber V; Hostadur; Hostadur A; Hostadur K; Hostadur K-VP 4022; Hostaphan; Hostaphan BNH; Hostaphan RN; Iambolen; KLT 40; Kodapak® 5214A; Kodapak® PET Copolyester 13339; Kodar® PETG Copolyester 6763 Lavsan; Lawsonite; Lumilar 100; Lumirror; Lumirror 38S; Meliform; Melinex; Melinex O; Mylar; Mylar A; Mylar C; Mylar C-25; Mylar HS; Mylar T; Nitron (polyester); Nitron lavsan; Pegoterate; Pegoterato; Pegoteratum; Poly(ethylene terephthalate); Poly(ethylene terephthalate) glycol; Poly(oxy-1,2-ethanediyloxycarbonyl-1,4-phenylenecarbonyl); Polyethylene terephthalate film; Scotch par; Superfloc; Terephtahlic acid-ethylene glycol polyester; Terfan; Tergal; Terom; Terphan; VFR 3801; Vituf. Amorphous glycol-modified PET; offers sparkling clarity in film and sheet form, easy thermoformability, toughness, sterilizability with ethylene oxide or gamma rays for medical applications; used for medical containers, thermoformed food containers and lids, blister packaging for cosmetics, pharmaceuticals, heavy hardware and electronic parts. FDA compliant. PET polyester hydropolymer; a light weight material resistant to breaking, bursting and shattering; improved barrier properties; for bottles of carbonated soft drinks fruit juices and foods. *Eastman Chem. Co.*

3070   **Polyisoprene**
9003-31-0   G

[C5H8]x

Poly(2-methyl-1,3-butadiene).
  Betaprene H; 1,3-Butadiene, 2-methyl-, homopolymer; HSDB 747; IR; Isoprene D; Isoprene oligomer; Isoprene polymer; Isoprene, polymers; Isoprene rubber; 2-Methyl-1,3-butadiene, homopolymer; Natsyn® 2200; Polyiso-prene; Poly(isoprene), cis; cis-1,4-Polyisoprene; cis-1,4-Polyisoprene rubber; Poly(2-methyl-1,3-butadiene); Poly-1-methylbutenylene; Rubber; Rubber (all-cis). Polyiso-prene, solution polymerized; nonstaining synthetic elasto-mer for light colored goods, adhesives, footwear, sponge products, tires, pharmaceutical goods, rubber bands, molded and mechanical goods. *A. Schulman; Goodyear Tire & Rubber.*

3071   **Polymethylmethacrylate**
9011-14-7   G

(C5H8O2)n

Methyl methacrylate polymer.
  Methacrylic acid methyl ester polymers; Methyl methacrylate homopolymer; 2-Methyl-2-propenoic acid methyl ester homopolymer; PMMA; Poly(methyl methacrylate); Polymethyl methacrylate; 2-Propenoic acid, 2-methyl-, methyl ester, homopolymer. Polymer of methyl methacrylate; thermoplastic used as main constituent of acrylic sheet, molding and extrusion compounds. Used in coatings, barrier coatings for PS, vinyl topcoats, product finishes, printing inks. *Aristech; Cyro Industries; Elf Atochem N. Am.; Shuman Plastics; Sybron.*

3072   **Polyoxyethylene monobutyl ether**
9004-77-7   E H

Polyoxyethylene monobutyl ether.
  Poly(oxy-1,2-ethanediyl), α-butyl-ω-hydroxy-; Polyethyl-ene glycol butyl ether; Polyethylene glycol, monobutyl ester; Polyoxyethylene monobutyl ether.

3073   **Polyoxyethylene monolaurate**
9004-81-3   E G

(C2H4O)n.C12H24O2

PEG 200 laurate.
  Ablunol 200ML; Dodecanoic acid, 2-(2-(2-(2-hydroxyethoxy)ethoxy)ethoxy)ethyl ester; Dodecanoic acid, 23-hydroxy-3,6,9,12,15,18,21-heptaoxatricos-1-yl; G-2129; Glycols, polyethylene, monolaurate; 2-(2-(2-(2-Hydroxyethoxy)ethoxy)ethoxy)ethyl dodecanoate; 23-Hydroxy-3,6,9,12,15,18,21-heptaoxatricos-1-yl dodecan-oate; Lauric acid, ethylene oxide adduct; Laurox 9; Nopalcol 6-L; PEG-4 Laurate; PEG-8 Laurate; PEG-10 Laurate; PEG-12 Laurate; PEG-14 Laurate; PEG-20 Laurate; PEG-32 Laurate; PEG-75 Laurate; PEG-150 Laurate; Polyethylene glycol 200 monolaurate; Polyethylene glycol 400 monolaurate; Polyethylene glycol 500 monolaurate; Polyethylene glycol 600 monolaurate; Polyethylene glycol 1000 monolaurate; Polyethylene glycol 1540 monolaurate; Polyethylene glycol 4000 monolaurate; Polyethylene glycol 6000 monolaurate; Polyglycol laurate; Polyoxyethylene (4) monolaurate; Polyoxyethylene (8) monolaurate; Polyoxyethylene (10) monolaurate; Polyoxyethylene (12) monolaurate; Polyoxyethylene (14) monolaurate; Polyoxyethylene (20) monolaurate; Polyoxyethylene (32) monolaurate; Polyoxyethylene (75) monolaurate; Polyoxyethylene (150) monolaurate; Glycols, polyethylene, monolaurate; Aquafil I; Aquafil II; Atlas G-2127; Atlas G-2129; Cirrasol TCS; Emanon 1112; Empilan AP-100; Ester 14; Ethylan L; Ethylan L 3; G 2129; Hallco CPH 43; Ionet ML-400; Lauric; Lauric acid, ethylene oxide adduct; Lipo-Peg 4-L; Lonzest PEG-4L; Macrogol laurate 600; Newcol 150; Nissan Nonion L 4; Nonex 139; Nonex 27; Nonex 31; Nonex 39; Nonex 55; Nonex 56; Nonex 99; Nonion L 4; Nopalcol 1-L; Nopalcol 6-L; Nopalcol 10-L. Emulsifier, lubricant, dispersing and leveling agent; defoamer used in cosmetics, textiles, paint, dyestuffs and other industrial uses. *Taiwan Surfactants.*

3074   **Polyoxypropylene cetyl ether**
9035-85-2   G

(C3H6O)nC16H34O

Poly[oxy(methyl-1,2-ethanediyl)], α-hexadecyl-ω-hydroxy-.
  Poly(oxy(methyl-1,2-ethanediyl)), α-hexadecyl-ω-hydroxy-; Polyoxypropylene (10) cetyl ether; Poly-oxypropylene (28) cetyl ether; Polyoxypropylene (30) cetyl ether; Polyoxypropylene (50) cetyl ether; Polyoxypropylene cetyl ether; Polyoxypropylene hexadecyl ether; Polypropylene glycol (10) cetyl ether; Polypropylene glycol (28) cetyl ether; Polypropylene glycol (30) cetyl ether; Polypropylene glycol (50) cetyl ether; PPG-10 Cetyl ether; PPG-28 Cetyl ether; PPG-30 Cetyl ether; PPG-50 Cetyl ether; Wickenol® 707; Polypropylene glycol monocetyl ether; Polypropylene glycol monohexadecyl ether; PPG-10 cetyl ether; Procetyl; Procetyl alcohol 30; Procetyl AWS; Propoxylated hexadecyl alcohol; Wickenol 707. PPG-30 cetyl ether; all-purpose fluid, nongreasy emollient with hydroalcoholic compatibility; provides coupling and emulsion stability; foam modifier; enhances sheen and manageability. *CasChem.*

3075   **Polyphosphoric acid**
8017-16-1   G

Superphosphoric acid.
  Condensed phosphoric acid; EINECS 232-417-0; HSDB 1176; Phospholeum; Polyphosphoric acid; Polyphosphoric acids;

Superphosphoric acid; Tetraphosphoric acid. Acid used in the manufacture of phosphates, phosphate esters, catalysts, fuel cell electrolytes, metal cleaning and brightening, organic reactions. Viscous liquid; d = 2.1000. *Albright & Wilson Americas Inc.*

3076 Polypropylene

9003-07-0   7663   G

[C3H6]x

Propene, homopolymer.
413S; A-Fax; Admer PB 02; AMCO; Amerfil; Amoco 1010; Ampol C 60; AT 36; Atactic polypropylene; Avisun; Avisun 101; Avisun 12-270A; Avisun 12-407A; Azdel; Beamette; Bicolene P; Carlona K 571; Carlona KM 61; Carlona P; Carlona PM 61 naturel; Carlona PPLZ 074; CD 419; Celgard; Celgard 2400W; Celgard 2500; Celgard 3501; Celgard KKX 2; Chisso 507B; Chisso polypro 1014; Clysar; Coathylene PF 0548; Courlene PY; CPP 25S; D 151; Daplen AD; Daplen APP; Daplen AS 50; Daplen AT 10; Daplen ATK 92; Daplen DM 55U; Dexon E 117; DLP; DS 8620; Eastobond L 8080-270A; Eastobond M 3; Eastobond M 5; Eastobond M 5H; EL Rexene PP 115; Elpon; EM 490; Enjay CD 392; Enjay CD 460; Enjay CD 490; Enjay E 117; Enjay E 11S; Epolene M 5H; Epolene M 5K; Epolene M 5W; Escon 622; Escon CD 44A; Escon EX 375; F 080PP; Gerfil; GPCD 398; Hercoflat 135; Hercotuf 110A; Hercotuf PB 681; Hercules 6523; Herculon; HF 20; HO 50; Hostalen N 1060; Hostalen PP; Hostalen PP-U; Hostalen PPH 1050; Hostalen PPN; Hostalen PPN 1060; Hostalen PPN 1075 F; Hostalen PPN 1076 F; Hostalen PPR 1042; Hostalen PPT VP 7090A; HSDB 1069; HULS P 6500; ICI 543; Isotactic polypropylene; J 400; J 700; JGD 1800; JMD 4500; K 300; Lambeth; Lanco Wax PP 1362D; Lupareen; LYM 42; Marlex 9400; Marlex HGH 050-01; Maurylene; Meraklon; MFR 4; MH 4; Mitsui polypro B 220; MM2A; Moplen; Moplen AD 50N; Moplen AD 5ON; Moplen AS 50; Moplen Q 51C; Moplen T 30G; Mosten; Nablen S 50; Noblen; Noblen 2VH501; Noblen BC 8; Noblen D 101; Noblen D 501; Noblen EBG; Noblen FA 3; Noblen FL; Noblen FL 4; Noblen FP; Noblen FS 101; Noblen FS 2011; Noblen H; Noblen H 101; Noblen H 501; Noblen HS; Noblen JHHG; Noblen JK-M; Noblen MA 4; Noblen MH 6; Noblen MM 2A; Noblen S 101; Noblen SHG; Noblen W 101; Noblen W 501; Noblen W 502; Noblen WF 464; Noplen FL 6314; Novamont 2030; Novolen; Novolen 1300ZX; Novolen KR 1300P; Oletac 100; P 6500; Paisley 750; Paisley polymer; Pellon 2505; Pellon 2506; Pellon FT 2140; Polipropene 25; Poly(5+)propylene; Polypropylene; Propene polymers; W 101. Polymer of propylene monomers; three forms; isotactic (fiber-forming), syndiotactic, atactic (amorphous). (nucleated); Isotactic; fishing gear, ropes, filter cloths, laundry bags, protective clothing, blankets, fabrics, carpets, yarns. *Amoco Lubricants; Aristech; Ashland; Chisso Corp.; Degussa-Hüls Corp.; Eastman Chem. Co.; ExxonMobil Chem. Co.; Fina Chemicals; LNP; Mitsubishi Corp.; Mitsui Toatsu; Neste UK; Quantum/USI; Shell; Solvay Polymers.*

3077 Polypropylene glycol

25322-69-4   H

(C3H6O)n.H2O

Polyoxypropylene glycol.
Actocol 51-530; Alkapal PPG-1200; Alkapal PPG-2000; Alkapal PPG-4000; Bloat guard; Caswell No. 680; Desmophen 360C; Emkapyl; EPA Pesticide Chemical Code 068602; Glycols, polypropylene; HSDB 1266; Jeffox PPG 400; Laprol 2002; Laprol 702; Lineartop E; Methyloxirane homopolymer; Napter E 8075; Niax 1025; Niax 11-27; Niax 61-582; Niax polyol ppg 4025; Niax ppg; Niax ppg 1025; Niax ppg 3025; Niax ppg 425; Oopg 1000; Oxirane, methyl-, homopolymer; P 400; P 4000 (polymer); P.P.G. 150; P.P.G. 400; P.P.G. 425; P.P.G. 750; P.P.G. 1000; P.P.G. 1025; P.P.G. 1200; P.P.G. 1800; P.P.G 2025; P.P.G 3025; P.P.G 4025; Pluracol 1010; Pluracol 2010; Pluracol P 2010; Pluracol P 410; Poly(propylene oxide); Polyglycol P 400; Polyglycol P-2000; Polyglycol P-4000; Polyglycol type P250; Polyglycol type P400; Polyglycol type P750; Polyglycol type P1200; Polyglycol type P2000;

Polyglycol type P3000; Polymer 2; Polyoxypropylene; Polyoxypropylene (9); Polyoxypropylene (12); Polyoxy-propylene (15); Polyoxypropylene (17); Polyoxypropylene (20); Polyoxypropylene (26); Polyoxypropylene (30); Polyoxypropylene (34); Polyoxypropylene glycol; Poly-propylene glycol; Polypropylene glycol (9); Poly-propylene glycol (12); Polypropylene glycol (15); Poly-propylene glycol (17); Polypropylene glycol (20); Poly-propylene glycol (26); Polypropylene glycol (30); Poly-propylene glycol (34); Polypropylene glycol 400; Poly-propylene glycol 425; Polypropylene glycol 750; Poly-propylene glycol 1000; Polypropylene glycol 1200; Poly-propylene glycol 1800; Polypropylene glycol 150; Poly-propylene glycol 1025; Polypropylene glycol 2000; Poly-propylene glycol 2025; Polypropylene glycol 3025; Poly-propylene glycol 4000; Polypropylene glycol 4025; Poly-propylenglykol; Ppg-15; PPG; PPG-9; PPG-12; PPG-15; PPG-17; PPG-20; PPG-26; PPG-30; PPG-34; Propylan 8123; Propylene oxide homopolymer; Propylene oxide, propylene glycol polymer; SKF 18667; Voranol P 1010; Voranol P 2000; Voranol P 4000.

3078 Polystyrene

9003-53-6   G

(C8H8)x

Styrene polymer.
3A; 31N (styrene polymer); 168N15; 454H; 475U; 550P (styrene polymer); 666U; 686E; 825TV; A 180 (vinyl polymer); A 3-80; A 75 (vinyl polymer); Adion H; Afcolene; Afcolene 492; Afcolene 666; Afcolene S 100; Atactic polystyrene; Bactolatex; Bakelite SMD 3500; BASF III; BDH 29-790; Benzene, ethenyl-, homopolymer; Bextrene XL 750; Bicolastic A 75; Bicolene H; Bio-Beads S-S 2; BP-Klp; BSB-S 40; BSB-S-E; Bustren; Bustren K 500; Bustren K 525-19; Bustren U 825; Bustren U 825E11; Bustren Y 3532; Bustren Y 825; Cadco 0115; Carinex GP; Carinex HR; Carinex HRM; Carinex SB 59; Carinex SB 61; Carinex SL 273; Carinex tgx/MF; Copal Z; Cosden 550; Cosden 945E; Denka QP3; Diarex 43G; Diarex HF 55-247; Diarex HF 55; Diarex HF 77; Diarex HS 77; Diarex HT 190; Diarex HT 500; Diarex HT 88; Diarex HT 88A; Diarex HT 90; Diarex YH 476; Dorvon; Dorvon FR 100; Dow 360; Dow 456; Dow 665; Dow 860; Dow 1683; Dow MX 5514; Dow MX 5516; Dylark 250; Dylene; Dylene 8; Dylene 8G; Dylite F 40; Dylite F 40L; Edistir RB; Esbrite; Esbrite 2; Esbrite 4; Esbrite 4-62; Esbrite 500HM; Esbrite 8; Esbrite G 10; Esbrite G-P 2; Esbrite LBL; Escorez 7404; Estyrene 4-62; Estyrene 500SH; Estyrene G 15; Estyrene G 20; Estyrene G-P 4; Estyrene H 61; Ethenylbenzene, homopolymer; FC-MY 5450; FG 834; Foster grant 834; Gedex; HF 10; HF 55; HF 77; HH 102; HHI 11; HI-Styrol; Hostyren N; Hostyren N 4000; Hostyren N 4000V; Hostyren N 7001; Hostyren S; HT 88; HT 88A; HT 91-1; HT-F 76; IT 40; K 525; KB (Polymer); KM (Polymer); Koplen 2; KR 2537; Krasten 052; Krasten 1.4; Krasten SB; Lacqren 506; Lacqren 550; Latex; LS 061A; LS 1028E; Lustrex; Lustrex H 77; Lustrex HH 101; Lustrex HP 77; Lustrex HT 88; MX 4500; MX 5514; MX 5516; MX 5517-02; N 4000V; NaPst; NBS 706; Owispol GF; Pelaspan 333; Pelaspan Esp 109s; Piccolastic; Piccolastic A; Piccolastic A 5; Piccolastic A 25; Piccolastic A 50; Piccolastic A 75; Piccolastic C 125; Piccolastic D; Piccolastic D-100; Polystyrene; Polystyrene resin; R 3; R 3612; S 173; Styrene polymer; Styrofoam; U625; X 600. Polymer; grades: crystal, impact, expandable. thermoplastic resin for injection molding, extrusion of egg carton foam, pill bottles, packaging, appliances, electronics, toys, recreation and construction, expendable polystyrene for insulation and protective packaging. *Asahi Chem. Industry; Ashland; BASF Corp.; Chevron; Degussa-Hüls Corp.; Elf Atochem N. Am.; Fina Chemicals; LNP; Mitsubishi Chem. Corp.; Mitsui Toatsu; Westlake Plastics.*

3079 Polytetrafluoroethylene

9002-84-0   7665   G

(C2F4)n (n > 20,000)

Ethene, tetrafluoro-, homopolymer.

PTFE; Teflon; Politef; Tetrafluoroethylene Resin; Fluon; Tetran. Versatile, chemically inert thermoplastic homopolymer. Used as tubing or sheeting for chemical laboratory and process work; gaskets and pump packings; as electrical insulators especially in high frequency applications, filtration fabrics, protective clothing, prosthetic aids. White solid; usable between -270° and 265°. *DuPont; Janssen Chimica.*

3080 Polyvinyl acetate
9003-20-7   G

[C4H6O3]x

Polyvinyl acetate homopolymer.
76 Res; Acetic acid, ethenyl ester, homopolymer; Acetic acid, vinyl ester, polymer; Asahisol 1527; ASB 516; AYAA; AYAF; AYJV; Bakelite AYAA; Bakelite AYAF; Bakelite AYAT; Bakelite LP 90; Bond CH 3; Bond CH 18; Bond CH 1200; Booksaver; Borden 2123; Cascorez; Cemedine 196; Cevian 380; Cevian A 678; D 50; D 50 M; Danfirm; Daratak; DCA 70; Duvilax; Duvilax BD 20; Duvilax HN; Duvilax LM 52; Elmer's Glue All; Elvacet 81-900; Emultex F; En-cor; EP 1208; EP 1436; EP 1437; EP 1463; Esnil P 18; Ethenyl acetate homopolymer; Everflex B; Formvar 1285; Gelva; Gelva 25; Gelva CSV 16; Gelva GP 702; Gelva S 55H; Gelva TS 22; Gelva TS 23; Gelva TS 30; Gelva TS 85; Gelva V 100; Gelva V 15; Gelva V 25; Gelva V 800; Gohensil E 50Y; Gohsenyl E 50 Y; HSDB 1250; Kurare OM 100; Lemac; Lemac 1000; Meikatex 5000NG60; Merckogel OR; Merckogen 6000; Mokotex D 2602; Movinyl; Movinyl 114; Movinyl 50M; Movinyl 801; Mowilith 30; Mowilith 50; Mowilith 70; Mowilith 90; Mowilith D; Mowilith DV; Mowilith M70; National 120-1207; National starch 1014; NS 2842; OM 100; OR 1500; P-170; Pioloform F; Plyamul 40-155; Plyamul 40-350; Polisol S-3; Poly(vinyl acetate); Polyco 953; Polyco 117FR; Polyco 2116; Polyco 2134; Polyfox P 20; Polyfox PO; Polysol 1000; Polysol 1000AX; Polysol 1200; Polysol PS 10; Polysol S 5; Polysol S 6; Polyvinyl acetate; Protex (polymer); PS 3h; PVAE; R 10688; Raviflex 43; Resyn 25-1014; Resyn 25-1025; Rhodopas; Rhodopas 010; Rhodopas 5000SMR; Rhodopas 5425; Rhodopas A 10; Rhodopas AM 041; Rhodopas B; Rhodopas BB; Rhodopas HV 2; Rhodopas M; RV225-5B; S-Nyl-P 42; Sakunol SN 08; Soloid; Soviol; SP 60; Toabond 2; Toabond 40H; Toabond 6; TS2; Ucar 15; Ucar 130; UK 131; V 501; VA 0112; Vinac; Vinac ASB 10; Vinac B 7; Vinac RP251; Vinacet D; Vinalite D 50N; Vinalite DS 41/11; Vinamul 9300; Vinapol A 16; Vinipaint 555; Vinnapas B; Vinnapas B 100; Vinnapas B 17; Vinnapas UW 50. Resin with weathering resistance; used for paints; adhesives for food packaging, paper, wood, glass, and metals; primer sealers, dry wall cement; intermediate for conversion to polyvinyl alcohol and acetals; paper coating, component of lacquers and inks. *Asahi Chem. Industry; Fuller H. B.; Monsanto Co.; National Starch & Chem. UK; Sigma-Aldrich Fine Chem.; Union Carbide Corp.; Wacker Chemie GmbH.*

3081 Polyvinyl alcohol
9002-89-5   7667   G

[C2H4O]x

Ethenol, homopolymer.
Alcotex 17F-H; Alcotex 88/05; Alcotex 88/10; Alcotex 99/10; Alkotex; Alvyl; Aracet APV; Cipoviol W 72; Covol; Covol 971; Elvanol; Elvanol 50-42; Elvanol 51-05G; Elvanol 5105; Elvanol 52-22; Elvanol 52-22G; Elvanol 522-22; Elvanol 70-05; Elvanol 71-30; Elvanol 73125G; Elvanol 90-50; Elvanol T 25; Enbra OV; EP 160; Ethenol, homopolymer; FH 1500; Galvatol 1-60; Gelvatol; Gelvatol 1-30; Gelvatol 1-60; Gelvatol 1-90; Gelvatol 20-30; Gelvatol 2060; Gelvatol 2090; Gelvatol 3-91; GH 20; GL 02; GL 03; GLO 5; GM 14; Gohsenol; Gohsenol AH 22; Gohsenol GH; Gohsenol GH 17; Gohsenol GH 20; Gohsenol GH 23; Gohsenol GL 03; Gohsenol GL 05; Gohsenol GL 08; Gohsenol GM 14; Gohsenol GM 14L; Gohsenol GM 94; Gohsenol KH 17; Gohsenol MG 14; Gohsenol N 300; Gohsenol NH 05; Gohsenol NH 17; Gohsenol NH 18; Gohsenol NH 20; Gohsenol NH 26; Gohsenol NK 114; Gohsenol NL 05;

Gohsenol NM 114; Gohsenol NM 14; Gosenol KH-17; Gtohsenol GL 05; HSDB 1038; HypoTears; Kuralon VP; Kurare 217; Kurare Poval 120; Kurare Poval 1700; Kurare PVA 205; Kurate Poval 120; Lamephil OJ; Lamicel; Lamicel; Lemol; Lemol 5-88; Lemol 5-98; Lemol 12-88; Lemol 16-98; Lemol 24-98; Lemol 30-98; Lemol 51-98; Lemol 60-98; Lemol 75-98; Lemol GF-60; Liquifilm Forte; Liquifilm Tears; M 13/20; Mowiol; Mowiol 26-88; Mowiol 4-88; Mowiol N 30-88; Mowiol N 50-98; Mowiol N 50/88; Mowiol N 70-98; NH 18; NM 11; NM 14; NSC 108129; Poly(1-hydroxyethylene); Polydesis; Polyethenol; Polysizer 173; Polyvinol; Polyvinyl alcohol; Polyvinyl alcohol 18/11; Polyviol; Polyviol M 13/140; Polyviol MO 5/140; Polyviol W 25/140; Polyviol W 28/20; Polyviol W 40/140; Poval; Poval 117; Poval 120; Poval 1700; Poval 203; Poval 205; Poval 205S; Poval 217; Poval 217S; Poval 420; Poval C 17; Prefrin Liquifilm; PVA; PVA (VAN); PVA 008; PVAL 45/02; PVAL 55/12; PVS 4; Refresh; Relief; Resistoflex; Rhodoviol; Rhodoviol 16/200; Rhodoviol 4/125; Rhodoviol 4-125P; Rhodoviol R 16/20; Sloviol R; Solvar; Sumitex H 10; Tears Plus; VasoClear; VasoClear A; Vibatex S; Vinacol DT; Vinacol MH; Vinalak; Vinarol; Vinarol DT; Vinarol ST; Vinarole; Vinavilol 2-98; Vinyl alcohol, polymers. Used in plastics industry in molding compounds, surface coatings, films resistant to gasoline, textile sizes; for elastomers (artificial sponges, fuel hoses); printing inks; pharmaceutical finishing; cosmetics; film and sheeting; ophthalmic lubricant. *British Traders & Shippers; Dajac Labs.; Honeywell & Stein.*

3082 Polyvinyl chloride
9002-86-2   7668   G

(-CH2CHCl-)n

Ethene, chloro-, homopolymer.
101EP; 1032X; Armodour; Aron TS 700; Atactic poly(vinyl chloride); Bakelite; Boltaron; Carina; CCRIS 634; Chloroethene homopolymer; Chloroethylene polymer; Corvic 55/9; Dacovin; Darvic 110; Dynadur; Ekavyl SD 2; Ethylene, chloro-, polymer; Exon 605; Expanded polyvinyl chloride; FC 4648; Flocor; Geno-therm; GEON 51; Halvic 223; Hostalit; HSDB 1213; Korogel; Koroplate;L 5; Poly(vinyl chloride); Polyvinyl chloride; PVC; Resinite 90; U 1; Vestolit®. Rubber substitutes; electrical wire and cable coverings; pliable thin sheeting; film finishes for textiles; nonflammable upholstery; raincoats; tubing; belting; gaskets; shoe soles. White solid; d = 1.406. *Ashland; Chisso Corp.; Degussa-Hüls Corp.; Elf Atochem N. Am.; Georgia Gulf; Goodrich B.F. Co.; Goodyear Tire & Rubber; Mitsui Toatsu; Norsk Hydro AS; OxyChem; Sigma-Aldrich Fine Chem.; Vista; Wacker Chemie GmbH.*

3083 Polyvinylpyrrolidone - vinyl acetate co-polymer
25086-89-9   G

(C6H9NO.C4H6O2)x

Vinyl pyrrolidone/vinyl acetate copolymer.
Acetic acid, ethenyl ester, polymer with 1-ethenyl-2-pyrrolidinone; Acetic acid vinyl ester, polymer with 1-vinyl-2-pyrrolidinone; Agrimer VA 6; Copolyvidon; E 335; E 535; Ethenyl acetate, polymer with 1-ethenyl-2-pyrrolidinone; 1-Ethenyl-2-pyrrolidinone, polymer with acetic acid ethenyl ester; 1-Ethenyl-2-pyrrolidinone, polymer with ethenyl acetate; GAF-S 630; Ganex E 535; Gantron PVP; Gantron S 630; Gantron S 860; I 535; I 635; I 735; Kolima 10; Kolima 35; Kolima 75; Kollidon VA 64; Luviskol VA 64; Luviskol VA 28I; Luviskol VA 37E; Luviskol VA 37I; Luviskol VA 55E; Luviskol VA 55I; Luviskol VA 73E; NSC 114023; Polectron 845; PVP-VA; PVP/VA copolymer; PVP-VA-E 735; PVP/VA-S 630; 2-Pyrrolidinone, 1-ethenyl-, polymer with ethenyl acetate; S 630; Vinyl acetate-1-vinyl-2-pyrrolidinone polymer; Vinyl acetate-N-vinyl-2-pyrrolidinone copolymer; Vinyl acetate-N-vinylpyrrolidinone copolymer; Vinyl acetate-N-vinylpyrrolidinone polymer; Vinyl acetate-vinyl-pyrrolidinone polymer; Vinylpyrrolidinone-vinyl acetate polymer; 1-Vinyl-2-pyrrolidone-vinyl acetate copolymer; N-Vinylpyrrolidone-vinyl acetate polymer; PVP/VA copolymer. Film-former used in hairsprays, gels, hair thickeners, tints, and dyes; suspending agent, dispersant, thickener,

stabilizer, adhesion promoter, coatings. ISP.

3084 Potash Soap
8046-74-0   H

Fatty acids, potassium salts; Potassium fatty acid soap; Potassium salts of fatty acids; Potassium soap; Soap, potassium; Soaps, potassium.

3085 Potassium
7440-09-7   7686   G

K

K
Potassium.
EINECS 231-119-8; HSDB 698; Kalium; Potassium; Potassium, (liquid alloy); Potassium, metal; UN2257. Metallic element; K; intermediate for potassium peroxide, heatexchange alloys; laboratory reagent; component of fertilizers (as potassium chloride). Used, with sodium, as a heat transfer agent in nuclear reactors. mp = 63°; bp = 765°; d^{20} = 0.856; reacts with H_2O, soluble in liquid NH_3, ethylenediamine, aniline. *Atomergic Chemetals; Sigma-Aldrich Fine Chem.*

3086 Potassium acetate
127-08-2   7687   D G H V

C2H3KO2
Acetic acid, potassium salt.
Acetic acid, potassium salt; Diuretic salt; EINECS 204-822-2; FEMA No. 2920; Octan draselny; Potassium acetate; Potassium ethanoate; Sal diureticum. Dehydrating agent, textile conditioner, analytical reagent, medicine, cacodylic derivatives, crystal glass, synthetic flavors. An alkalizer; has been used to treat veterinary cardiac arrhythmias. Crystals; mp = 292°; d = 1.8000; soluble in H_2O (200 g/100 ml), poorly soluble in organic solvents; LD_{50} (rat orl) = 3.25 g/kg. *Am. Intl. Chem.; EM Ind. Inc.; General Chem; Heico; Honeywell & Stein; Niacet.*

3087 Potassium arsenate
7784-41-0   7689   G

AsH2KO4
Potassium dihydrogen arsenate.
Arsenic acid, monopotassium salt; EINECS 232-065-8; HSDB 1235; Macquer's salt; Monopotassium arsenate; Monopotassium dihydrogen arsenate; Potassium acid arsenate; Potassium arsenate; Potassium arsenate, monobasic; Potassium dihydrogen arsenate; Potassium dihydrogen arsenate (KH2AsO4); Potassium hydrogen arsenate (KH2AsO4); UN1677. Used in textile, tanning and paper industries. Also in insecticide formulations. Crystals; mp = 288°; d = 2.8; slightly soluble in H_2O (20 g/100 ml), insoluble in EtOH.

3088 Potassium benzoate
582-25-2   G H

C7H5KO2
Benzoic acid potassium salt.
Benzoic acid, potassium salt; EINECS 209-481-3; Potassium benzoate. Anti-corrosive, preservative, fermentation-inhibitor, anti-fungal agent for tobacco production, pyrotechnical additive. White solid; soluble in H_2O, EtOH. *Am. Biorganics; Lancaster Synthesis Co.; Mallinckrodt Inc.; Pentagon Chems. Ltd.; Verdugt BV.*

3089 Potassium bicarbonate
298-14-6   7691   D G

CHKO3
Carbonic acid, monopotassium salt.
CCRIS 3510; EINECS 206-059-0; EPA Pesticide Chemical Code 073508; K-Lyte; K-Lyte/Cl; K-Lyte DS; Monopotassium carbonate; Potassium acid carbonate; Potassium bicarbonate; Potassium hydrogen carbonate. Electrolyte replenisher. Used in pharmaceutical formulation. White crystals; soluble in H_2O (35.7 g/100 ml at 20°, 50 g/100 ml at 50°), insoluble in EtOH. *Bristol Laboratories; Mead Johnson Labs.*

3090 Potassium bisulfate
7646-93-7   7695   D G P

HKO4S
Potassium hydrogen sulfate.
Acid potassium sulfate; AI3-04459; Caswell No. 682C; EINECS 231-594-1; EPA Pesticide Chemical Code 005605; Monopotassium hydrogen sulfate; Monopotassium sulfate; NSC 409996; Potassium acid sulfate; Potassium bisulfate; Potassium bisulphate; Potassium hydrogen sulfate; Potassium hydrogen sulfate, solid; Potassium hydrogensulphate; Potassium hydrosulfate (KHSO4); Potassium sulfate; Potassium sulfate (KHSO4); Sal enixum; Sulfuric acid, monopotassium salt; Sulfuric acid potassium salt (1:1); Tartarline; UN2509. Used as a substitute for tartaric acid for industrial purposes; conversion of wine lees and tartrates into potassium bitartrate; in the manufacture of fertilizers. Used as a leavening and pH control agent. Registered by EPA as an antimicrobial, fungicide and herbicide (cancelled). Used medicinally as a laxative-cathartic. mp = 197°; d = 2.24; soluble in H_2O (0.55 g/ml).

3091 Potassium bitartrate
868-14-4   7697   G

C4H5KO6
Monopotassium 2,3-dihydroxybutanedioate, (R-(R*,R*))-.
Acid potassium tartrate; Butanedioic acid, 2,3-dihydroxy-, (R-(R*,R*))-, monopotassium salt; Butanedioic acid, 2,3-dihydroxy-(2R,3R)-, monopotassium salt; CCRIS 7329; Cream of tartar;

Cremor tartari; EINECS 212-769-1; Faccla; Faccula; Faecla; Faecula; HSDB 1264; Monopotassium tartrate; NSC 155080; Potassium acid salt of L-(+)-tartaric acid; Potassium acid tartrate; Potassium bitartrate; Potassium hydrogen tartrate; Potassium L-bitartrate; Tartar; Tartar cream; Tartaric acid, monopotassium salt. Baking powder, preparation of other tartrates, galvanic tinning of metals, food additive. Colorless crystals; soluble in H_2O (0.61 g/100 ml 20°, 6.25 g/100 ml 100°), EtOH (0.011 g/100 ml). *Penta Mfg.; Spectrum Chem. Manufacturing.*

3092   Potassium bromate
7758-01-2   7700   G

BrKO3
Bromic acid, potassium salt.
Bromic acid, potassium salt; CCRIS 529; EEC No. E924; EINECS 231-829-8; HSDB 1253; NSC 215200; Potassium bromate; UN1484. Laboratory reagent, oxidizing agent, permanent wave compounds, dough conditioner, food additive. White crystals; mp = 350°; d = 3.2700; soluble in H_2O (8.0 g/100 ml), insoluble in EtOH. *Allchem Ind.; Gist-Brocades Intl.*

3093   Potassium bromide
7758-02-3   7701   D G P

K—Br

BrK
Hydrobromic acid potassium salt.
Bromide salt of potassium; Bromure de potassium; Caswell No. 684; CCRIS 6095; EINECS 231-830-3; EPA Pesticide Chemical Code 013903; HSDB 5044; Kalii bromidum; NSC 77367; Potassium bromide; Potassium bromide (KBr); Tripotassium tribromide. Used in photography, process engraving and lithography, special soaps, spectroscopy, infrared transmission, lab reagent. Registered by EPA as an antimicrobial and disinfectant (cancelled). Anticonvulsant; sedative. CAUTION: Large doses may cause CNS suppression. Has been used as a disinfectant on food contact surfaces, food handling equipment and eating utensils in food processing establishments. Solid; mp = 730°; d = 2.75; soluble in H_2O (0.67 g/ml), less soluble in organic solvents. *Great Lakes Fine Chem.; Mallinckrodt Inc.; Morton Intn'l.; Sigma-Aldrich Fine Chem.*

3094   Potassium butanolate
3999-70-0   H

C4H9KO
n-Butyl alcohol, potassium salt.
1-Butanol, potassium salt; Butyl alcohol, potassium salt; EINECS 223-646-7; Potassium butanolate.

3095   Potassium carbonate
584-08-7   7702   G P

CK2O3
Carbonic acid potassium salt.
Carbonate of potash; Carbonic acid, dipotassium salt; Caswell No. 685; CCRIS 7320; Dipotassium carbonate; EINECS 209-529-3; EPA

Pesticide Chemical Code 073504; HSDB 1262; K-Gran; Kalium carbonicum; Kaliumcarbonat; Pearl ash; Potash; Potassium carbonate; Potassium carbonate (2:1); Potassium carbonate (K2CO3); Potassium carbonate, anhydrous; Sal absinthii; Salt of tartar; Salt of wormwood. Used in chemical manufacturing and in medicine as an alkalizer and diuretic. Used in manufacture of special glasses (optical, TV tubes), potassium silicate, fertilizer manufacture, dehydrating agent, pigments, printing inks, lab reagent, general purpose food additive, textile, dyeing. White solid; mp = 891°; d = 2.9; soluble in H_2O (100 g/100 ml), insoluble in organic solvents; LD50 (rat orl) = 1.87 g/kg. *Degussa-Hüls Corp.; Mallinckrodt Inc.; OxyChem.*

3096   Potassium chlorate
3811-04-9   7703   G

ClKO3
Chloric acid, potassium salt.
AI3-02907; Anforstan; Berthollet salt; Berthollet's salt; Chlorate de potassium; Chlorate of potash; Chlorate of potassium; Chloric acid, potassium salt; EINECS 223-289-7; Fekabit; HSDB 1110; NSC 68505; Oxymuriate of potash; Potash chlorate (DOT); Potassio (clorato di); Potassium (chlorate de); Potassium chlorate; Potassium oxymuriate; Potcrate; Salt of tarter; UN1485; UN2427. Oxidizing agent. Used in explosives, fireworks and matches. Also in printing and dyeing of cotton and wool. Crystals; mp = 368°; bp = 400°; d = 2.32; soluble in water (7 g/100 ml), insoluble in organic solvents.

3097   Potassium chloride
7447-40-7   7704   D G V

K—Cl

ClK
Potassium Chloride.
Acronitol; Addi-K; Apo-K; Caswell No. 686; CCRIS 1962; Celeka; Cena-K; Chlorid draselny; Chloride of potash; Chloropotassuril; Chlorvescent; Clor-K-Zaf; Colyte; Diffu-K; Dipotassium dichloride; Durekal; Durules; Durules-K; EINECS 231-211-8; Emplets Potassium Chloride; Enpott; Enseal; EPA Pesticide Chemical Code 013904; HSDB 1252; Infalyte; K-10; K-Care; K-Contin; K-Dur; K-Grad; K-Lease; K-Lor; K-Lyte/Cl; K-Norm; K-Predne-Dome; K-Sol; K-SR; K-Tab; Kadalex; Kalcorid; Kaleorid; Kaliduron; Kaliglutol; Kalilente; Kalinor-Retard P; Kalinorm; Kalinorm Depottab; Kaliolite; Kalipor; Kalipoz; Kalitabs; Kalitrans Retard; Kalium-Durettes; Kalium-Duriles; Kalium Duriles; Kalium-R; Kalium Retard; Kalium S.R.; Kalium SR; Kaochlor; Kaon-Cl; Kaon-Cl 10; Kaon-Cl TABS; Kaskay; Kato; Kay-Cee-L; Kay Ciel; Kay-EM; Kayback; Kayciel; Kaysay; KCl-retard Zyma; KCL Retard; Kelp salt; Keylyte; Klor-Con; Klor-Lyte; Kloren; Klorvess; Klotrix; Kolyum; KSR; Lento-K; Lento-Kalium; Leo-K; Micro-K; Micro-K Extentcaps; Micro-Kalium Retard; Miopotasio; Monopotassium chloride; Muriate of potash; Natural sylvite; Neobakasal; NSC 77368; Nu-K; Peter-Kal; Pfiklor; Plus Kalium Retard; Potasion; Potasol; Potassium chloride; Potassium chloride (K3Cl3); Potassium chloride (KCl); Potassium monochloride; Potassium muriate; Potavescent; Rekawan; Rekawan Retard; Repone-K; Rum-K; Slow-K; Span-K; Steropotassium; Super K; Ten-K; Tripotassium trichloride; Ultra-K-Chlor.; component of: Colyte, Infalyte, K-Lyte/Cl, Kolyum, K-Predne-Dom. Electrolyte replenisher. Used in veterinary medicine as a potassium supplement. Large doses may cause gastrointestinal irritation, weakness, circulatory disturbances. Used in fertilizers; foods, pharmaceuticals; in photography; in buffer solutions, electrode cells; used medicinally as an electrolyte replenisher and as a potassium supplement. Crystals; mp = 773°; d = 1.98; pH 7; soluble in H_2O (35.7 g/100 ml at 20°,

55.5 g/100 ml at 100°), glycerol (7.1 g/100 ml), EtOH (0.4 g/100 ml). *Abbott Labs Inc.; Apothecon; Bayer Corp.; Bristol Laboratories; Ciba-Geigy Corp.; Fisons Pharm. Div.; Forest Pharm. Inc.; ICN Pharm. Inc.; Key Pharm.; McGaw Inc.; Parke-Davis; Robins, A. H. Co.,; Sandoz Pharm. Corp.; Savage Labs; Schwarz Pharma Kremers Urban Co.*

3098   Potassium 4-chloro-3,5-dinitrobenzene-sulfonate
38185-06-7   H

C6H2ClKN2O7S
4-Chloro-3,5-dinitrobenzenesulfonic acid potassium salt.
4-Chloro-3,5-dinitrobenzenesulfonic acid potassium salt; Benzenesulfonic acid, 4-chloro-3,5-dinitro-, potassium salt; EINECS 253-816-6; NSC 123956; Potassium 4-chloro-3,5-dinitrobenzenesulphonate.

3099   Potassium   3-(2-chloro-4-(trifluoromethyl)-phenoxy)benzoate
72252-48-3   H

C14H7ClF3KO3
Benzoic acid, 3-[2-chloro-4-(trifluoromethyl)phenoxy]-, potassium salt.
Potassium 3-(2-chloro-4-(trifluoromethyl)phenoxy)benz-oate.

3100   Potassium citrate
866-84-2   7706   G H V

C6H5K3O7
1,2,3-Propanetricarboxylic acid, 2-hydroxy-, tripotassium salt.
CCRIS 6566; Citric acid, tripotassium salt; EINECS 212-755-5; HSDB 1248; Kajos; Kaliksir; Litocit; Polycitra K; Porekal; Potassium citrate; Potassium citrate anhydrous; Potassium tribasic citrate; Seltz-K; Tripotassium citrate; Urocit K. Food additive; antacid, antiurolithic. Used in veterinary medicine as a diuretic. White solid; soluble in H2O (150 g/100 ml), less soluble in alcohols, insoluble in organic solvents. *Lancaster Synthesis Co.; Mallinckrodt Inc.; Pfizer Inc.*

3101   Potassium cobalt nitrite
13782-01-9   2457   G

CoK3N6O12
Potassium cobaltic nitrite.
Aureolin; C.I. 77357; C.I. Pigment Yellow 40; Cobalt Yellow; Cobaltate(3-), hexakis(nitrito-N)-, tripotassium, (OC-6-11)-; Cobaltate(3-), hexanitro-, tripotassium; Cobaltic potassium nitrite; EINECS 237-435-2; Fischer's salt; Fischer's yellow; Hexanitrocobaltate(3-) tripotassium; Potassium hexanitrocobaltate(III); Potassium nitro-cobaltate(III); Tripotassium hexanitritocobaltate. Used for the detection and determination of potassium. Also used as a pigment in painting glass, porcelain and rubber. [sesquihydrate]; yellow cubic crystals; slightly soluble in H2O, soluble in EtOH.

3102   Potassium cocoate
61789-30-8   H

Potassium coconut oil soap.
Cocoa fatty acids, potassium salts; Coconut fatty acid, potassium salt; Coconut oil acids, potassium salt; Coconut oil fatty acid, potassium salt; Coconut oil, potassium salts; Coconut Oil potassium soap; EINECS 263-049-9; Fatty acids, coco, potassium salts; Fatty acids, coconut oil, potassium salts; Potassium cocoate; Potassium coconut oil soap; Potassium coconut soap; Soap, potassium coconut.

3103   Potassium cyanide
151-50-8   7709   G

N≡C—K

CKN
Hydrocyanic acid, potassium salt.
4-02-00-00050 (Beilstein Handbook Reference); AI3-28749; BRN 4652394; Caswell No. 688A; Cyanide of potassium; Cyanure de potassium; EINECS 205-792-3; EPA Pesticide Chemical Code 599600; HSDB 1245; Hydrocyanic acid, potassium salt; Kalium cyanid; M-44 capsules (potassium cyanide); Potassium cyanide; RCRA waste number P098; UN1680. Used in gold and silver extraction, as an analytical reagent, insecticide, fumigant, in electroplating. Crystals; mp = 634°; d = 1.52; soluble in H2O (0.5 g/ml), less soluble in organic solvents; LD50 (rat orl) = 10 mg/kg. *Degussa AG; DuPont; Elf Atochem N. Am.; Grace W.R. & Co.; Mallinckrodt Inc.*

3104   Potassium dichloroisocyanurate
2244-21-5   9836   D G P

C3Cl2KN3O3
1,3-Dichloro-s-triazine-2,4,6(1H,3H,5H)trione potassium salt.
ACL-56; ACL-59; ACL-60; ACL-66; AI3-26480; Caswell No. 689; CCRIS 4878; CP17251; Dichloroisocyanuric acid potassium salt; EINECS 218-828-8; EPA Pesticide Chemical Code 081403; Fluonon; HSDB 5871; Isocyanuric acid, dichloro-, potassium salt; Laitonon; Neochlor 59; NSC 251767; Potassium dichloro isocyanurate; Potassium dichlorocyanurate; Potassium troclosene; Troclosene potassique; Troclosene potassium; Troclosen potasico.

Bleaching compound, sanitizer, disinfectant, oxidizer in dishwashing compositions. Active ingredient in dry bleaches, water and sewage treatment. Used as a topical anti-infective. Anti-infective (topical). Halogen containing compound, the source of Cl in solid bleach and detergent formulations. Used for swimming pool maintenance. Registered by EPA as an antimicrobial and disinfectant (cancelled). Crystals; mp = 250°; soluble in H₂O (10-50 mg/ml). *3V; Biachem; ICI Spec.; Monsanto Co.; Nissan Chem. Ind.*

3105 Potassium 2,5-dichlorophenolate

68938-81-8   H

$C_6H_3Cl_2KO$
 Phenol, 2,5-dichloro-, potassium salt.
 EINECS 273-147-3; Phenol, 2,5-dichloro-, potassium salt; Potassium 2,5-dichlorophenolate.

3106 Potassium dichromate

7778-50-9   7710   G P

$Cr_2K_2O_7$
 Potassium dichromate(VI).
 Bichromate of potash; Caswell No. 690; CCRIS 2409; Chromic acid, dipotassium salt; Chromium potassium oxide (K₂Cr₂O₇); Dichromic acid dipotassium salt; Dipotassium bichromate; Dipotassium dichromate; Dipotassium dichromium heptaoxide; EINECS 231-906-6; EPA Pesticide Chemical Code 068302; HSDB 1238; Iopezite; Kaliumdichromat; NSC 77372; Potassium bichromate; Potassium chromate (K₂Cr₂O₇); Potassium dichromate; Potassium dichromate(VI); SRM 935a. Oxidizing agent, analytical reagent, brass pickling, electroplating, pyrotechnics, explosives, matches, textile dyeing and printing, adhesives, tanning leather, wood stains, lithography, synthetic perfumes, pigments, alloys, ceramics, batteries. Registered by EPA as a fungicide (cancelled). Orange crystals; mp = 398°; d = 2.6760; soluble in H₂O (4 g/100 ml 20°, 50 g/100 ml 100°). *Hoechst Celanese; Mallinckrodt Inc.*

3107 Potassium dihydrogen phosphate

7778-77-0   7744   G P

H_2KO_4P
 Monobasic potassium phosphate.
 EINECS 231-913-4; EPA Pesticide Chemical Code 076413; HSDB 5046; KDP; MKP; Monobasic potassium phosphate; Monopotassium dihydrogen phosphate; Monopotassium monophosphate; Monopotassium ortho-phosphate; Monopotassium phosphate; Orthophosphoric acid, monopotassium salt; Phosphoric acid, monopotassium salt; Potassium acid phosphate; Potassium dihydrogen orthophosphate; Potassium dihydrogen phosphate; Potassium dihydrogenortho-phosphate; Potassium hydrogen phosphate (KH₂PO₄); Potassium orthophosphate, dihydrogen; Potassium phosphate; Potassium phosphate, monobasic; Sorensen's potassium phosphate Used in food products, baking powder, nutrient solutions, water treatment, buffer and sequestrant,

laboratory reagent. Registered by EPA as a fungicide. Used in turf, ornamentals and bedding plants. Colorless crystals; loses H₂O at 400°; d = 2.34; soluble in H₂O (22 g/100 ml), insoluble in EtOH; pH of aqueous solution = 4.4 - 4.7. *Albright & Wilson Americas Inc.; FMC Corp.; Heico; Monsanto Co.; Sigma-Aldrich Fine Chem.*

3108 Potassium dodecanoate

10124-65-9   P

$C_{12}H_{23}KO_2$
 Dodecanoic acid, potassium salt.
 Caswell No. 694A; Dodecanoic acid, potassium salt; EINECS 233-344-7; EPA Pesticide Chemical Code 079021; Lauric acid, potassium salt; Potassium dodecanoate; Potassium laurate. Registered by EPA as an antimicrobial, fungicide, herbicide, insecticide and rodenticide. *Micro-Flo Co. LLC; Organica Inc; Safer Inc; W Neudorff GMBH KG; Woodstream Corp.*

3109 Potassium 2-ethylhexanoate

3164-85-0   H

$C_8H_{15}KO_2$
 Hexanoic acid, 2-ethyl-, potassium salt.
 EINECS 221-625-7; 2-Ethylhexanoic acid, potassium salt; Hexanoic acid, 2-ethyl-, potassium salt; Potassium 2-ethylhexanoate.

3110 Potassium gluconate

299-27-4   7716   D G V

$C_6H_{11}KO_7$
 Gluconic acid potassium salt.
 EINECS 206-074-2; Gluconal® K; Gluconic acid, monopotassium salt; D-Gluconic acid, monopotassium salt; Gluconic acid potassium salt; Gluconsan K; HSDB 3165; K-Iao; Kalium-beta; Kalium Gluconate; Kaon; Kaon elixir; Katorin; KOK; Kolyum; Monopotassium D-gluconate; Potalium; Potasoral; Potassium D-gluconate; Potassium gluconate; Potassuril; Sirokal; Twin-K; Glucosan K; Kalimozan; Kaon; Potasoral; Potassuril; K-IAO; Tumil-K; Jaon; Katorin. Electrolyte replenisher. Pharmaceutical/food grade mineral source for human and veterinary pharmaceutical preparations, dietary supplements, fortified foods and animal feed. Dec 180°; freely soluble in H₂O; insoluble in EtOH, Et₂O, C₆H₆, CHCl₃. *Adria Labs.; Akzo Chemie; Boots Co.; Fisons plc; Greeff R.W. & Co.; Knoll Pharm. Co.; Savage Labs.*

3111 Potassium hydroxide

1310-58-3   7724   G P

K—OH

HKO
 Potash.
 Caswell No. 693; Caustic potash; Caustic potash solution; CCRIS

511

6569; Cyantek CC 723; EINECS 215-181-3; EPA Pesticide Chemical Code 075602; HSDB 1234; Hydroxyde de potassium; Potash; Potash lye; Potassa; Potasse caustique; Potassio (idrossido di); Potassium (hydroxyde de); Potassium hydrate; Potassium hydroxyde; Potassium hydroxide (K(OH)); UN1813; UN1814; Vegetable alkali. Used to manufacture soap; absorbs CO_2; removes paint and varnish; used in organic syntheses. Registered by EPA as an antimicrobial, fungicide and herbicide (cancelled). FDA GRAS, Europe listed, UK approved; FDA approved for intravenous, parenterals, orals, topicals, USP/NF, BP, JP compliant. Alkalizing agent, used in intravenous, parenterals, orals, topicals, caustic for wart removal, cuticle solvent, used in escharotic preparations. White solid; mp = 405°, bp = 1320°; d = 2.044; soluble in H_2O (111 g/100 ml at 20°, 167 g/100 ml at 100°), EtOH (33 g/100 ml), glycerol (40 g/100 ml); pH of 0.1M aqueous solution = 13.5. LD_{50} (rat orl) = 1230 mg/kg, 365 mg/kg, TLV = 2 mg/m³ (air). *Avrachem; Hüls Am.; ICI Spec.; Integra; Mallinckrodt Inc.; Occidental Chem. Corp.; Olin Res. Ctr.; Sigma-Aldrich Fine Chem.; Spectrum Chem. Manufacturing.*

3112   Potassium 4-hydroxyphenyl sulfate
37067-27-9   G

$C_6H_5KO_5S$
Potassium 4-hydroxyphenyl sulfate.
EINECS 253-332-5; Potassium p-hydroxyphenyl sulphate; Potassium 4-hydroxyphenyl sulfate. *Rhone-Poulenc UK.*

3113   Potassium hypochlorite
7778-66-7   G

$$Cl—O^- \quad K^+$$

ClKO
Hypochlorous acid, potassium salt.
EINECS 231-909-2; HyPure K; Potassium chloride oxide; Potassium hypochlorite. For liquid bleach, water treatment, hard surface cleaners. *Olin Corp.*

3114   Potassium Iodide
7681-11-0   7727   D G P V

$$K—I$$

IK
Iodic acid potassium salt.
AI3-52931; Asmofug E; Caswell No. 694; Dipotassium diiodide; EINECS 231-659-4; Embamix; EPA Pesticide Chemical Code 075701; HSDB 5040; Iodure de potassium; Jodid; Joptone; K1-N; Kali iodide; Kalii iodidum; Kisol; Knollide; Mixture Name; Mudrane Tablets; Mudrane-2 Tablets; NSC 77362; Pima; Potassium diiodide; Potassium iodide (KI); Potassium monoiodide; Potassium salt of hydriodic acid; Potide; Quadrinal; Thyro-Block; Thyrojod; Iripotassium triiodide; Jodid; Thyroblock; Thyrojod. Reagent in analytical chemistry; photographic emulsions; animal feed additive; dietary supplement; in table salt; nylon stabilizer. Antifungal; expectorant; iodine supplement. Used in veterinary medicine (orally) to treat goiter, actinobacillosis, actinomycosis, iodine deficiency, lead or mercury poisoning. Registered by EPA as an antimicrobial, fungicide and herbicide. FDA GRAS, BP, Ph.Eur. compliance. Used as a source of dietary iodine, a dye remover and an antiseptic. Crystals; mp = 680°; d = 3.12; soluble in H_2O (140 g/100 ml), EtOH, MeOH, Me_2CO, glycerol, ethylene glycol; LD_{50} (rat iv) = 285 mg/kg. *Atomergic Chemetals; Greeff R.W. & Co.; May &*

Baker Ltd.; Mitsui Toatsu; Sigma-Aldrich Fine Chem.

3115   Potassium metabisulfite
16731-55-8   7729   G

$K_2O_5S_2$
Disulfurous acid, dipotassium salt.
CCRIS 1427; Dipotassium disulfite; Dipotassium disulphite; Dipotassium metabisulfite; Dipotassium pyrosulfite; Disulfurous acid, dipotassium salt; EINECS 240-795-3; HSDB 5062; Potassium disulfite; Potassium disulfite ($K_2S_2O_5$); Potassium metabisulfite; Potassium pyrosulfite; Pyrosulfurous acid, dipotassium salt. Inorganic salt; antiseptic; reagent; source of sulfurous acid; photographic developing agent; brewing, wine making; food preservative; bleaching agent. White crystals; freely soluble in H_2O, insoluble in organic solvents. *Allchem Ind.; Farleyway Chem. Ltd.; Mallinckrodt Inc.*

3116   Potassium N-phenylglycinate
19525-59-8   H

$C_8H_8KNO_2$
Glycine, N-phenyl-, monopotassium salt.
AI3-15398; EINECS 243-133-1; Glycine, N-phenyl-, monopotassium salt; N-Phenylglycine potassium salt; NSC 405072; Potassium N-phenylglycinate.

3117   Potassium oleate
143-18-0   7735   G P

$C_{18}H_{33}KO_2$
Potassium 9-(Z)-octadecenoate.
Caswell No. 698B; EINECS 205-590-5; EPA Pesticide Chemical Code 079095; HSDB 5643; Octosol 449; Oleic acid, potassium salt; Potassium 9-octadecenoate; Potassium 9-octadecenoate, (Z)-; Potassium cis-9-octadecenoic acid; Potassium oleate; Trenamine D-200; Trenamine D-201. Foaming agent, stabilizer, emulsifier, dispersant; primary frothing aid in gelled latex foam compounds. Liquid soap for hand cleaners, tire mounting lubricant; emulsifier and corrosion control in paint strippers. Biodegradable. Registered by EPA as an antimicrobial, fungicide, herbicide, insecticide and rodenticicide (cancelled). Yellow-brown solid; soluble in H_2O (> 100 mg/ml), EtOH. *Emkay; Fluka; Norman-Fox.*

3118   Potassium oxalate
6487-48-5   G

$2K^+ \quad 2H_2O$

512

C2K2O4.2H2O
Potassium oxalate monohydrate.
Oxalic acid, dipotassium salt monohydrate. Analytical reagent, source of oxalic acid, bleaching and cleaning, removing stains from textiles, photography, anticoagulant. Used as an anti-coagulant, in photography and for cleaning and bleaching straw. Crystals; d = 2.1270; soluble in H2O (30 g/100 ml). *Am. Intl. Chem.; General Chem; Heico; Verdugt BV.*

3119 Potassium perchlorate
7778-74-7   7738   G

$$O=\overset{\overset{O}{\|}}{\underset{\underset{O}{\|}}{C}l}-O^-\quad K^+$$

ClKO4
Perchloric acid, potassium salt.
Astrumal; EINECS 231-912-9; HSDB 1222; Irenal; Irenat; KM Potassium perchlorate; Perchloric acid, potassium salt; Perchloric acid, potassium salt (1:1); Peroidin; Potassium hyperchloride; Potassium perchlorate; Potassium perchlorate (KClO4); Spectrex Fire Extinguishant Formulation A; UN1489. Used in explosives, as an oxidizing agent, in photography, pyrotechnics and flares, reagent, oxidizer in solid rocket propellants. Crystals; mp = 400° (dec); d = 2.52; soluble in H2O (0.75 g/100 ml), insoluble in organic solvents. *Am. Intl. Chem.; Eka Nobel Ltd.; Kerr-McGee; Mallinckrodt Inc.; San Yuan Chem. Co. Ltd.*

3120 Potassium periodate
7790-21-8   7739   G

$$O=\overset{\overset{O^-}{|}}{\underset{\underset{O}{\|}}{I}}=O\quad K^+$$

IKO4
Potassium metaperiodate.
EINECS 232-196-0; Periodic acid (HIO4), potassium salt; Potassium periodate. Analysis, oxidizing agent. Solid; mp = 582°; d = 3.6180. *Atomergic Chemetals; Cerac; Spectrum Chem. Manufacturing.*

3121 Potassium permanganate
7722-64-7   7740   D G P

$$^-O-\overset{\overset{O}{\|}}{\underset{\underset{O}{\|}}{M}n}=O\quad K^+$$

KMnO4
Permanganic acid potassium salt.
Al3-52835; Algae-K; Argucide; C.I. 77755; Cairox; Caswell No. 699; CCRIS 5561; Chameleon mineral; CI 77755; Condy's crystals; Diversey Diversol CX with Arodyne; Diversey Diversol CXU; EINECS 231-760-3; EPA Pesticide Chemical Code 068501; Hilco 88; HSDB 1218; Icc 237 Disinfectant, Sanitizer, Destainer, and Deodorizer; Insta-perm; Kaliumpermanganat; Manganese potassium oxide (KMnO4); NSC 146182; Permanganate de potassium; Permanganate of potash; Permanganato potasico; Permanganic acid (HMnO4), potassium salt; Permanganic acid, potassium salt; Potassio (permanganato di); Potassium (permanganate de); Potassium permanganate; Pure Light E 2; Sin red; Solo San Soo; UN1490; Walko Tablets. Powerful oxidizing agent. Used in bleaches, dyeing wood brown, printing fabrics, washing CO2, in photography, tanning, purifying water and as a chemical reagent. Disinfectant, deodorizer, dye, tanning, decontamination of radioactive skin, analytical reagent, medicine (antiseptic), manufacture of organic chemicals, air and water purification. Anti-infective (topical). Registered by EPA as an antimicrobial and disinfectant. Purple crystals; dec 240°; d = 2.7; soluble in H2O (70-280 mg/ml); LD50 (rat orl) = 1.09 g/kg. *Am. Biorganics; Am. Intl. Chem.; Blythe, Williams Ltd.; Lancaster Synthesis Co.; Mallinckrodt Inc.*

3122 Potassium persulfate
7727-21-1   7741   G

$$^-O-\overset{\overset{O}{\|}}{\underset{\underset{O}{\|}}{S}}-O-O-\overset{\overset{O}{\|}}{\underset{\underset{O}{\|}}{S}}-O^-\quad 2K^+$$

K2O8S2
Peroxydisulfuric acid dipotassium salt.
Anthion; Caswell No. 700; Dipotassium peroxodisulphate; Dipotassium peroxydisulfate; Di-potassium persulfate; EINECS 231-781-8; EPA Pesticide Chemical Code 063602; HSDB 2638; Peroxydisulfuric acid, dipotassium salt; Peroxydisulfuric acid (((HO)S(O)2)2O2), dipotassium salt; Potassium peroxydi-sulfate; Potassium peroxydisulfate (K2(S2O8)); Potassium peroxydisulphate; Potassium persulfate; UN1492. Bleaching, oxidizing agent, reducing agent in photography, antiseptic, soap manufacture, analytical reagent, polymerization promoter, pharmaceuticals, starch modifier, flour-maturing agent, textile desizing. Used as a hypo eliminator in photography. Colorless crystals; dec 100°; soluble in H2O (2 g/100 ml 25°, 4 g/100 ml 40°), insoluble in EtOH. *Allchem Ind.; DuPont; FMC Corp.; Mallinckrodt Inc.; San Yuan Chem. Co. Ltd.; Transol Chem. UK Ltd.*

3123 Potassium phosphate, dibasic
7758-11-4   7743   D G

$$O=\overset{\overset{OH}{|}}{\underset{\underset{O^-}{|}}{P}}-O^-\quad 2K^+$$

K2HPO4
Potassium phosphate, dibasic.
CCRIS 6544; Dibasic potassium phosphate; Dipotassium hydrogen phosphate; Dipotassium hydrogenortho-phosphate; Dipotassium monohydrogen phosphate; Dipotassium monophosphate; Dipotassium ortho-phosphate; Dipotassium phosphate; DKP; EINECS 231-834-5; HSDB 935; Hydrogen dipotassium phosphate; Isolyte; Phosphoric acid, dipotassium salt; Potassium dibasic phosphate; Potassium monohydrogen phosphate; Potassium monophosphate; Potassium phosphate (dibasic). DKP; food and automotive industry; buffer in antifreezes; nutrient; humectant; pharmaceuticals. Medicinally, a calcium regulator and cathartic. Also used as a buffer. White crystals, very soluble in H2O (150 g/100 ml). *Albright & Wilson Americas Inc.; FMC Corp.; Heico; Monsanto Co.; Sigma-Aldrich Fine Chem.; U.S. BioChem.*

3124 Potassium phosphate, tribasic
7778-53-2   7745   G P

$$O=\overset{\overset{O^-}{|}}{\underset{\underset{O^-}{|}}{P}}-O^-\quad 3K^+$$

K3O4P
Potassium phosphate tribasic monohydrate.
Caswell No. 700A; CCRIS 7321; EPA Pesticide Chemical Code

076407; Phosphoric acid, tripotassium salt; Potassium orthophosphate; Potassium phosphate; Potassium phosphate (K3PO4); Potassium phosphate, tribasic; TKP; Tripotassium orthophosphate; Tripotassium phosphate. Detergent, water treatment, automotive products, fertilizer, foods as emulsifier. Used as a food additive. Registered by EPA as an antimicrobial, fungicide and herbicide (cancelled). Deliquescent crystals; mp = 1340°; d$_4^{17}$= 2.564; soluble in H2O (44 g/100 ml at 0°, 51 g/100 ml at 25°, 58 g/100 ml at 45°), aqeous solutions are strongly alkaline, insoluble in EtOH; [octahydrate]: plates; mp = 45.1°. *Albright & Wilson Americas Inc.; Ashland; FMC Corp.; Monsanto Co.; Sigma-Aldrich Co.*

3125   Potassium phthalimide
1074-82-4   H

C8H4KNO2
1H-Isoindole-1,3(2H)-dione, potassium salt.
EINECS 214-046-6; HSDB 5781; N-Potassiophthalimide; N-Potassium phthalimide; NSC 167070; Phthalimide, potassium salt; Potassium phthalimidate; Potassium phthalimide.

3126   Potassium polyacrylate
25608-12-2   G

(C3H4O2)x ·xK
Polyacrylic acid, potassium salt.
2-Propenoic acid, homopolymer, potassium salt; Polyacrylic acid, potassium salt; Potassium polyacrylate. Used as a dispersant for latex paints and coatings, pigments.

3127   Potassium pyrophosphate
7320-34-5   7748   G

K4O7P2.3H2O
Tetrapotassium pyrophosphate.
Diphosphoric acid, tetrapotassium salt; EINECS 230-785-7; Potassium diphosphate; Potassium pyrophosphate; Pyrophosphoric acid, tetrapotassium salt; Tetrakal; Tetrapotassium diphosphate; Tetrapotassium diphos-phorate; Tetrapotassium pyrophosphate; TKPP. Soap and detergent builder, sequestering agent, peptizing and dispersing agent. Crystals; soluble in H2O, insoluble in EtOH, organic solvents. *Albright & Wilson Americas Inc.; Elf Atochem N. Am.; FMC Corp.; Monsanto Co.*

3128   Potassium ricinoleate
7492-30-0   G P

C18H33KO3
12-Hydroxy-9-octadecenoic acid, monopotassium salt.
Caswell No. 701A; EINECS 231-314-8; EPA Pesticide Chemical Code 079023; HSDB 2167; 12-Hydroxy-9-octadecenoic acid, monopotassium salt; 9-Octadecenoic acid, 12-hydroxy-, monopotassium salt; 9-Octadecenoic acid, 12-hydroxy-, monopotassium salt, (R-(Z))-; 9-Octadecenoic acid, 12-hydroxy-, monopotassium salt, (9Z,12R)-; Ricinoleic acid, monopotassium salt; Solricin® 135. Detergent, emulsifier, mild germicide, glycerized rubber lubricant, foam stabilizer in foamed rubber; making of cutting and solvent oils, household and cosmetic products. Has been used as an antimicrobial in catfish farms. Registered by EPA as an antimicrobial, fungicide, herbicide, insecticide and rodenticicide (cancelled). *CasChem.*

3129   Potassium rosinate
61790-50-9   H

Rosin acids, potassium salts.
Disproportionated rosin acid, potassium salt; Disproportionated rosin, potassium soap; EINECS 263-142-4; Potassium rosinate; Potassium salt of wood rosin acids; Potassium soap of rosin; Resin acids and Rosin acids, potassium salts; Rosin acids, potassium salts; Rosin, disproportionated, potassium salt; Rosin, potassium salt.

3130   Potassium silicate
1312-76-1   7753   G

K2Si2O5 to K2Si3O7
Potassium polysilicate.
Caswell No. 701B; EINECS 215-199-1; EPA Pesticide Chemical Code 072606; HSDB 5798; Kasil; Kasil 6; Potassium metasilicate; Potassium silicate; Potassium silicate solution; Potassium water glass; PS 7; Pyramid 120; Silicic acid, potassium salt; Soluble potash glass; Soluble potash water glass. Alkaline builder for liquid detergents and cleaners, binder for phosphors in cathode ray tubes. Used as a binder in carbon electrodes, lead pencils etc., also in protective coatings, detergents and glass. White powder; poorly soluble in H2O, insoluble in organic solvents. *PQ Corp.*

3131   Potassium sodium 2,6-dichlorosalicylate
68938-79-4   H

C7H2Cl2KNaO3
Benzoic acid, 3,6-dichloro-2-hydroxy-, potassium sodium salt.
Benzoic acid, 3,6-dichloro-2-hydroxy-, potassium sodium salt; 3,6-Dichlorosalicylate, sodium, potassium salt; EINECS 273-145-2; Potassium sodium 2,6-dichloro-salicylate.

3132   Potassium sorbate
24634-61-5   7756   G H

C6H7KO2
2,4-Hexadienoic acid, (E,E)-, potassium salt.
AI3-26043; CCRIS 1894; E 202; EINECS 246-376-1; 2,4-Hexadienoic acid, (E,E)-, potassium salt; 2,4-Hexadienoic acid, potassium salt, (E,E)-; 2,4-Hexadienoic acid, potassium salt, (2E,4E)-; Potassium (E,E)-hexa-2,4-dienoate; Potassium 2,4-hexadienoate, (E,E)-; Potassium (E,E)-2,4-hexadienoate; Potassium sorbate; Potassium sorbate (E); Potassium (E,E)-sorbate; Sorbic acid, potassium salt, (E,E)-; Sorbic acid, potassium salt; Sorbistat potassium. Used as a mold and yeast inhibitor. Crystals; mp = 270° (dec); d_{20}^{25} = 1.363; soluble in H2O (58 g/100 ml), less soluble in organic solvents. *Chisso Corp.; Gist-Brocades Intl.; Hoechst Celanese; Pfizer Inc.; Protameen.*

3133   Potassium stannate
12142-33-5   7757   G

K2O3Sn·3H2O
Potassium stannate(IV).
Dipotassium tin trioxide; EINECS 235-255-9; Potassium stannate; Potassium stannate(IV); Stannate (SnO3^2-), dipotassium. Textile dyeing and printing, alkaline tinplating bath. [trihydrate]; d = 3.197; soluble in H2O (100 g/100 ml), insoluble in EtOH. *Allchem Ind.; Blythe, Williams Ltd.; Elf Atochem N. Am.; M&T Harshaw; Nihon Kagaku Sangyo.*

3134   Potassium stearate
593-29-3   7758   G

C18H35KO2
Octadecanoic acid, potassium salt.
EINECS 209-786-1; Octadecanoic acid, potassium salt; Potassium n-octadecanoate; Potassium octadecanoate; Potassium stearate; Steadan 300; Stearates; Stearic acid, potassium salt. Used in manufacture of textile softeners. White waxy solid; slightly soluble in H2O. *CK Witco Corp.; Original Bradford Soapworks.*

3135   Potassium sulfate
7778-80-5   7759   D G P

K2O4S
Sulfuric acid dipotassium salt.
Arcanum duplicatum; Caswell No. 702; Dipotassium sulfate; EINECS 233-558-0; EPA Pesticide Chemical Code 005603; Glazier's salt; HSDB 5047; Kalium sulphuricum; Potassium sulfate; Potassium sulphate; Sal Polychrestum; Sulfuric acid, dipotassium salt; Sulphuric acid, potassium salt; Tartarus vitriolatus. Analytical reagent, medicine (cathartic), gypsum cements, fertilizer, manufacture of alum and glass, food additive. Registered by EPA as a herbicide and insecticide. Colorless, hard crystals; mp = 1067°; d = 2.66; soluble in H2O (12 g/100 ml at 25°, 25 g/100 ml at 100°),

glycerol 1.3 g/100 ml), insoluble in EtOH; pH of aqueous solution = 7. *Chisso Corp.; General Chem; Heico; Mallinckrodt Inc.; Sigma-Aldrich Fine Chem.*

3136   Potassium sulfite
10117-38-1   7761   D G

K2O3S
Sulfurous acid potassium salt.
Dipotassium sulfite; EINECS 233-321-1; HSDB 5052; Potassium sulfite; Potassium sulfite (K2SO3); Potassium sulphite; Stahl's sulfur salt; Sulfurous acid, dipotassium salt; Sulfurous acid, potassium salt. Used as a photographic developer, medicine, in food and wine as a preservative. Laxative-cathartic. Soluble in H2O (28.5 g/100 ml), slightly soluble in EtOH.

3137   Potassium tallate
61790-44-1   H

Tall oil acid, potassium salt.
EINECS 263-136-1; Fatty acids, tall-oil, potassium salts; Potassium soap of tall oil fatty acids (C18); Potassium tallate; Tall oil acids, potassium salt; Tall oil fatty acids, potassium salt; Tall oil, potassium salt; Tall oil soaps, potassium.

3138   Potassium titanium fluoride
16919-27-0   G

F6K2Ti
Titanium(IV) potassium Fluoride.
Aflammit TI; Dipotassium hexafluorotitanate; Dipotass-ium hexafluorotitanate(2-); Dipotassium monotitanium hexafluoride; Dipotassium titanium hexafluoride; EINECS 240-969-9; Fluotitanate de potassium; NSC 187663; Potassium fluorotitanate (K2TiF6); Potassium hexa-fluorotitanate; Potassium titanium fluoride; Titanate(2-), hexafluoro-, dipotassium; Titanium potassium fluoride. Flameproofing agent used in wool processing. *Thor Chemicals (UK) Ltd.*

3139   Potassium toluene sulfonate
30526-22-8   G

C7H7KO3S
Benzenesulfonic acid, methyl-, potassium salt.
Caswell No. 703B; EINECS 250-228-1; EPA Pesticide Chemical Code 079031; Hartotrope KTS 44; Naxonate® 4KT; Potassium toluene sulfonate; Potassium toluene sulphonate; Toluenesulfonic acid, potassium salt. Mixture of o, m and p-isomers. Hydrotrope for high activity or built liquid detergent systems; for use where sodium ion undesirable. Hydrotrope, stabilizer, solubilizer used in formulating detergents, inks, electroplating baths, dyestuffs, polymers. *Hart Prod.; Ruetgers-Nease.*

3140   Potassium tripolyphosphate
13845-36-8   G

515

5K+

$K_5O_{10}P_3$

Pentapotassium tripolyphosphate.

EINECS 237-574-9; Pentapotassium triphosphate; Potass-ium triphosphate ($K_5P_3O_{10}$); Potassium tripoly-phosphate; Triphosphoric acid, pentapotassium salt. Used in detergents, paints, cleaners, specialty fertilizers, sequestrant. *Albright & Wilson Americas Inc.; FMC Corp.*

3141 Potassium vanadate
13769-43-2   G

KO_3V

Potassium metavanadate.

EINECS 237-388-8; K-Van; Potassium m-vanadate; Potassium metavanadate; Potassium trioxovanadate; Potassium vanadate; Potassium vanadate(V) (KVO_3); Potassium vanadium trioxide; UN2864; Vanadate (VO_3^{1-}), potassium; Vanadic acid, potassium salt. Crystals; mp = $520°$; d = 2.840. *Kerr-McGee.*

3142 Potassium zirconium fluoride
16923-95-8   7723   G

F_6K_2Zr

Zirconium(IV) potassium fluoride.

Aflammit ZR; Dipotassium hexafluorozirconate(2-); Dipotassium zirconium hexafluoride; EINECS 240-985-6; HSDB 2019; NSC 310011; Potassium fluorozirconate; Potassium fluorozirconate (K_2ZrF_6); Potassium hexafluorozirconate; Potassium zirconifluoride; Potassium zirconium fluoride; Potassium zirconium hexafluoride; Zirconate(2-), hexafluoro-, dipotassium; Zirconium potassium fluoride. Flameproofing agent used in wool processing. Also used in manufacture of zirconium. Solid; slightly soluble in cold H_2O. *Thor Chemicals (UK) Ltd.*

3143 Pravastatin
81093-37-0   D

$C_{23}H_{36}O_7$

(+)-(3R,5R)-3,5-Dihydroxy-7-[(1S,2S,6R,8S,8aR)-6-hydroxy-2-methyl-8-[(S)-2-methylbutyryloxy]-1,2,6,7,8,   8a-hexahydro-1-naphthyl]heptanoic acid.

CCRIS 7557; 3β-Hydroxycompactin; Eptastatin; Prava-statin; Pravastatina; Pravastatine; Pravastatinum. Antihyperlipoproteinemic.

HMG-CoA reductase inhibitor; used as an antihypercholesterolemic agent. Active metabolite of mevastatin. *Sankyo Co. Ltd.*

3144 Pravastatin sodium
81131-70-6   7800   D

$C_{23}H_{35}NaO_7$

Sodium   (+)-(3R,5R)-3,5-dihydroxy-7-[(1S,2S,6R,8S,8aR)-6-hydroxy-2-methyl-8-[(S)-2-methylbutyryloxy]-1,2,6,7,8,   8a-hexahydro-1-naphthyl]heptanoate.

Aplactin; Bristacol; CS-514; DRG-0319; Elisor; Epastatin sodium; 3-β-Hydroxycompactin sodium salt; Lipemol; Lipidal; Lipostat; Liprevil; Mevalotin; Oliprevin; Pralidon; Prareduct; Pravachol; Pravacol; Pravaselect; Pravasin; Pravasine; Pravastatin Natrium Mayrho Fer; Pravastatin sodium; Pravigard PAC; Sanaprav; Selectin; Selektine; Selipran; SQ-31000; Vasen; Vasten. HMG-CoA reductase inhibitor; used as an antihypercholesterolemic agent. Active metabolite of mevastatin. Antihyperlipoprotein-emic. λ_m = 230, 237, 245 nm. *Sankyo Co. Ltd.*

3145 Prednisone
53-03-2   7810   D

$C_{21}H_{26}O_5$

17,21-Dihydroxypregna-1,4-diene-3,11,20-trione.

Adasone; AI3-52939; Ancortone; Apo-Prednisone; Bicortone; Cartancyl; CCRIS 2646; Colisone; Cortan; Cortancyl; Cortidelt; Cotone; Dacorten; Dacortin; Decortancyl; Decortin; Decortisyl; Dekortin; Delcortin; Dellacort; Dellacort A; delta-Cortelan; delta-Cortisone; delta-Dome; delta E; Deltacortene; Deltacortisone; Deltacortone; Deltasone; Deltison; Deltisona; Deltra; Di-Adreson; DRG-0227; Econosone; EINECS 200-160-3; Encorton; Encortone; Enkorton; Fiasone; Hostacortin; HSDB 3168; IN-Sone; Incocortyl; Juvason; Liquid Pred; Lisacort; Me-Korti; Metacortandracin; Meticorten; Metrevet; NCI-C04897; Nisona; Nizon; Novoprednisone; NSC-10023; Nurison; Orasone; Origen Prednisone; Panafcort; Panasol; Paracort; Parmenison; Pehacort; Predeltin; Prednicen-M; Prednicorm; Prednicort; Prednicot; Prednidib; Prednilonga; Prednison; Prednisona; Prednisone; Prednisonum; Prednitone; Prednizon; Prednovister; Presone; Pronison; Rectodelt; Retrocortine; Servisone; SK-Prednisone; Sone; Sterapred; Supercortil; U 6020; Ultracorten; Ultracortene; Winpred; Wojtab; Zenadrid. Glucocorticoid. Used as an antiinflammatory agent. Crystals; Dec 233-235°; $[\alpha]_D^{25}$= +172° (in dioxane); λ_m = 238 nm (ε 15500); slightly soluble in H_2O; more soluble in organic solvents.

3146   Prednisone 21-acetate
125-10-0   7810   D

$C_{23}H_{28}O_6$

21-(Acetyloxy)-17-hydroxypregna-1,4-diene-3,11,20-trione.
4-08-00-03532 (Beilstein Handbook Reference); BRN 2342061; Cortancyl; Delcortin; Delta-corlin; Delta-Cortelan; Deltalone; EINECS 204-726-0; Ferrosan; Hostacortin; Nisone; NSC 10965; Prednisone acetate; Prednisone 21-acetate; U 6167. Glucocorticoid. Dec 226-232°; $[\alpha]_D^{25}$ = +186° (dioxane); λ_m = 238 nm (ϵ 16100 EtOH). Schering-Plough HealthCare Products.

3147   Prenol
556-82-1   G H

$C_5H_{10}O$

3-Methyl-2-butenyl alcohol.
4-01-00-02129 (Beilstein Handbook Reference); BRN 1633479; 2-Buten-1-ol, 3-methyl-; 3,3-Dimethylallyl alcohol; Dimethylallyl alcohol; EINECS 209-141-4; FEMA No. 3647; 3-Methyl-2-buten-1-ol; 3-Methylcrotyl alcohol; NSC 158709; Prenol; Prenyl alcohol. Fragrance and flavoring (fresh, herbal, green, fruity, slightly lavender-like). Liquid; mp = 0°; bp = 140°; d^{25} = 0.8480; soluble in H_2O (170 g/l), organic solvents; LD_{50} (rat orl) = 810 mg/kg. BASF Corp.

3148   Pretilachlor
51218-49-6   7829   H P

$C_{17}H_{26}ClNO_2$

2-Chloro-N-(2,6-diethylphenyl)-N-(2-propoxyethyl)acet-amide.
Acetamide, 2-chloro-N-(2,6-diethylphenyl)-N-(2-propoxy-ethyl)-; BRN 2754162; CG 113; CGA 26423; 2-Chloro-2',6'-diethyl-N-(2-propoxyethyl)acetanilide; 2-Chloro-N-(2,6-diethylphenyl)-N-(2-propoxyethyl)acetamide; Pretil-achlor; Pretilachlore; Rifit; Solnet. Herbicide. Colorless liquid; bp$_{0.001}$ = 135°; sg^{20} = 1.076; soluble in H_2O (0.005 g/100 ml at 20°), very soluble in C_6H_6, C_6H_{14}, MeOH, CH_2Cl_2; LD_{50} (rat orl) = 6099 mg/kg, (rat der) > 3100 mg/kg; LC_{50} (rat ihl 4 hr.) = 2.8 mg/l air, (rainbow trout 96 hr.) = 0.9 mg/l, (catfish 96 hr.) = 2.7 mg/l, (crucian carp 96 hr.) = 2.3 mg/l; slightly toxic to

birds, slightly toxic to bees. Ciba-Geigy Corp.; Syngenta Crop Protection.

3149   Printer's acetate
8006-13-1   322   D G

$AlC_6H_9O_6$
Burow's solution.
Aluminum acetate solution; Burow's solution; Buro-Sol Concentrate; Domeboro; component of: Otic Domeboro; See [139-12-8]. Aluminum acetate. Antiseptic (topical); astringent. Colorless liquid; d 1.002; pH (1:20 aqueous solution) = 4.2. Doak Pharmacal Co. Inc.; Miles Inc.

3150   Pristane
1921-70-6   7841   G

$C_{19}H_{40}$
2,6,10,14-tetramethylpentadecane.
3-01-00-00570 (Beilstein Handbook Reference); BRN 1720538; Bute hydrocarbon; EINECS 217-650-8; Norphytan; Norphytane; NSC 114852; Pentadecane, 2,6,10,14-tetramethyl-; Pristane; 2,6,10,14-Tetramethyl-pentadecane. Isolated from Shark liver oil. A chemically inert oil used as a lubricant and heat transfer oil in transformers. Oil; bp = 296°; d_4^{20} 0.78267; n_D^{20} 1.43848; insoluble in H_2O, soluble in organic solvents.

3151   Prochloraz
67747-09-5   7849   P

$C_{15}H_{16}Cl_3N_3O_2$
N-Propyl-N-[2-(2,4,6-trichlorophenoxy)ethyl]-1H-imidaz-ole-1-carboxamide.
Ascurit; BTS 40542; BTS 40542-7877; Caswell No. 704E; Dibavit; DMI; EINECS 266-994-5; EPA Pesticide Chemical Code 128851; KI 835; Mirage; Octave; Omega; Prelude; Prochloraz; Sporgon; Sportak; Sportak PF; Sprint. Broad spectrum fungicide wth protective and eradicative action. Applied as a foliar spray to control Rhynchosporium, Helminthosporium, Septoria, Fusarium, Pseudocercosporella, Erysiphe and Pyrenophora species in cereal crops. Crystals; mp = 36 - 41°; sg^{20} = 1.42; soluble in H_2O (0.0055 g/100 ml at 25°), CHCl$_3$ (250 g/100 ml), Et$_2$O (250 g/100 ml), C$_7$H$_8$ (250 g/100 ml), xylene (250 g/100 ml), Me$_2$CO (350 g/100 ml); LD_{50} (rat orl) = 1600 mg/kg, (mus orl) =2400 mg/kg, (rat der) > 5000 mg/kg, (rbt der) > 3000 mg/kg, (mallard duck orl) = 3132 mg/kg, (bee orl) = 0.06 mg/bee, (bee topical) = 0.05 mg/bee; LC_{50} (rat ihl 4 hr.) > 2.16 mg/l air, (rainbow trout 96 hr.) = 1 mg/l, (bluegill sunfish 96 hr.) = 2.2 mg/l; non-toxic to bees. Aventis Crop Science; Schering Agrochemicals Ltd.

3152   Procymidone
32809-16-8   7854   P

C13H11Cl2NO2
3-(3,5-Dichlorophenyl)-1,5-dimethyl-3-azabicyclo[3.1.0]-hexane-2,4-dione.
5-21-10-00069 (Beilstein Handbook Reference); 3-Azabicyclo(3.1.0)hexane-2,4-dione, 3-(3,5-dichlorophen-yl)-1,5-dimethyl-; BRN 1539058; 1,2-Cycloprop-anedicarboximide, N-(3,5-dichlorophenyl)-1,2-dimethyl-; 3-(3,5-Dichlorophenyl)-1,5-dimethyl-3-azabicyclo(3.1.0)-hexane-2,4-dione; Dicyclidine; N-(3',5'-Dichlorophenyl)-1,2-dimethylcyclopropane-1,2-dicarboximide; Procymid-one; S 7131; SP 751011; Sumilex; Sumisclex. Fungicide. Light brown crystals; mp = 165 - 167°; sg^{25} = 1.452; soluble in H2O (0.00045 g/100 ml at 25°), Me2CO (18 g/100 ml), xylene (18 g/100 ml), CHCl3 (21 g/100 ml), DMF (23 g/100 ml); slightly soluble in alcohols; LD50 (mrat orl) = 6800 mg/kg, (frat orl) = 7700 mg/kg, (mus orl) = 7800 - 9100 mg/kg, (mmus der) = 7800 mg/lkg, (fmus der) = 9100 mg/kg, (rat der) > 2500 mg/kg; LC50 (bluegill sunfish 96 hr.) = 22.9 mg/l, (rainbow trout 96 hr.) = 3.6 mg/l, (carp 48 hr.) = 10 mg/l; non-toxic to bees. *Sumitomo Corp.*

3153 Profenofos

41198-08-7   7860   H P

C11H15BrClO3PS
O-(4-Bromo-2-chlorophenyl)-O-ethyl-S-propyl phosph-orothioate.
AI3-29236; BRN 2150258; Caswell No. 266AA; CGA 15324; Curacron; EPA Pesticide Chemical Code 111401; O-(4-Bromo-2-chlorophenyl) O-ethyl S-propyl phosph-orothioate; Phosphorothioic acid, O-(4-bromo-2-chlorophenyl) O-ethyl S-propyl ester; Polycron; Profenofos; Profenophos; Selecron. Used as an insecticide and acaricide. Registered by EPA as an insecticide. Yellow liquid; bp0.001 = 110°; d^{20} = 1.455; soluble in H2O (0.002 g/100 ml at 20°), freely soluble in MeOH, CH2Cl2, C6H6, C6H14; LD50 (rat orl) = 358 - 400 mg/kg, (rbt orl) = 700 mg/kg, (rat der) = 472 mg/kg, 3300 mg/kg, LC50 (rat ihl 4 hr.) = 3.0 mg/l air, LC50 (rainbow trout 96 hr.) = 0.08 mg/l, (crucian carp 96 hr.) = 0.09 mg/l, (bluegill sunfish 96 hr.) = 0.3 mg/l, toxic to birds, toxic to bees. *Novartis; Syngenta Crop Protection.*

3154 Progallin LA

1166-52-5   G

C19H30O5
Dodecyl gallate.

4-10-00-02006 (Beilstein Handbook Reference); Benzoic acid, 3,4,5-trihydroxy-, dodecyl ester; BRN 2701981; CCRIS 5568; Dodecyl 3,4,5-trihydroxybenzoate; Dodecyl gallate; Dodecylester kyseliny gallove; EINECS 214-620-6; Gallic acid, dodecyl ester; Gallic acid, lauryl ester; Lauryl 3,4,5-trihydroxybenzoate; Lauryl gallate; Nipagallin LA; NSC 133463; Progallin LA; 3,4,5-Trihydroxybenzoic acid, dodecyl ester. Antioxidant used in cosmetics. *Nipa.*

3155 Proglumide

6620-60-6   7866   D G

C18H26N2O4
(±)-4-Benzamido-N,N-dipropylglutaramic acid.
242 DL; Binoside; BRN 4151696; CR-242; EINECS 229-567-4; Gastrotopic; KXM; Midelid; Milid; Milide; Nulsa; Proglumida; Proglumide; Proglumidum; Promid; Promide (parasympatholytic); Ulcutin; W-5219; Xyde; Xylamide. Cholecystokinin inhibitor. Used as an anticholinergic agent. Crystals; mp = 142-145°; LD50 (mus iv) = 2211-2649 mg/kg, (mus orl) = 7350-8861 mg/kg. *Wallace Labs.*

3156 Proline

147-85-3   7871   G

C5H9NO2
L-Proline.
AI3-26710; CB 1707; EINECS 205-702-2; FEMA Number 3319; HSDB 1210; NSC 46703; PRO (IUPAC abbreviation); Prolina; Proline; (-)-(S)-Proline; (-)-Proline; (L)-Proline; L-(-)-Proline; L-Proline; Prolinum; 2-Pyrrolidinecarboxylic acid, (S)-; (-)-2-Pyrrolidine-carboxylic acid; L-Pyrrolidine-2-carboxylic acid; L-α-Pyrrolidinecarboxylic acid; (S)-2-Pyrrolidinecarboxylic acid. A nonessential amino acid; used in moisturizers, biochemical and nutritional research, microbiological tests, culture media, dietary supplements, lab reagent. Crystals; dec 220-222°; [α]$_D^{20}$= -52.6° (c = 0.57 in 0.5N HCl); soluble in H2O (127 g/100 ml), less soluble in organic solvents. *Am. Biorganics; Degussa AG; Greeff R.W. & Co.; Nippon Rikagakuyakuhin; Penta Mfg.*

3157 Promethazine

60-87-7   7878   D

C17H20N2S
10-[2-(Dimethylamino)propyl]phenothiazine.

3277 RP; 4-27-00-01253 (Beilstein Handbook Reference); A-91033; Antiallersin; Aprobit; Atosil; Avomine; BRN 0088554; Camergan; CCRIS 7056; Dimapp; Dimethylamino-isopropyl-phenthiazin; Diphergan; Dipra-zine; Diprozin; EINECS 200-489-2; Fargan; Fenazil; Fenetazina; Fenetazine; Hiberna; Histargan; HSDB 3173; Isophenergan; Lilly 1516; Metaryl; NCI-C60673; NSC 30321; Pelpica; Phargan; Phenargan; Phenerzine; Phenoject-50; Phensedyl; Pilothia; Pipolphene; Pro-50; Proazamine; Procit; Promacot; Promazinamide; Pro-mergan; Promesan; Prometazina; Prometh; Promethacon; Promethazin; Promethazine; Promethazinum; Pro-methegan; Promezathine; Prorex; Protazine; Provigan; Pyrethiazine; RP-3277; SKF 1498; Tanidil; Thiergan; Vallergine; WY 509. Antihistaminic with antiemetic and CNS depressant properties. Crystals; mp = 60°; bp3 = 190-192°; moderatley soluble in H2O, soluble in dil. HCl. *Rhône-Poulenc; Wyeth Labs.*

3158 Promethazine hydrochloride
58-33-3   7878   D

$C_{17}H_{21}ClN_2S$
10-[2-(Dimethylamino)propyl]phenothiazine hydro-chloride.
Allerfen; Anergan 25; Anergan 50; Atosil; Bonnox; Closin; Dorme; Duplamin; Eusedon Mono; Farganesse; Fellozine; Fenazil; Frinova; Genphen; Goodnight; Hibechin; Hiberna; Histantil; Lergigan; Mepergan; Phanergan D; Phencen; Phenergan; Phenergan-D; Phenergan VC; PMS Promethazine; Prome; Prometazina; Promethazine hydrochloride; Proneurin; Prorex; Prothazin; Prothazine; Provigan; Remsed; RP-3389; Soporil. Antihistaminic with antiemetic and CNS depressant properties. Crystals; mp = 230-232°; λ_m = 249, 297 nm (ε 28770, 3400 H2O); soluble in H2O, EtOH, CHCl3; insoluble in Me2CO, EtOAc, Et2O; LD50 (mus iv) = 55 mg/kg. *Rhône-Poulenc; Wyeth Labs.*

3159 Prometon
1610-18-0   7880   H P

$C_{10}H_{19}N_5O$
2,4-Bis(isopropylamino)-6-methoxy-s-triazine.
Al3-60364; 2,4-Bis(isopropylamino)-6-methoxy-s-triazine; 2,4-Bis(isopropylamino)-6-methoxy-1,3,5-triazine; 4,6-Bis(isopropylamino)-2-methoxy-s-triazine; BRN 0613574; Caswell No. 096; 2,6-Diisopropylamino-4-methoxy-triazine; N²,N⁴-Di-isopropyl-6-methoxy-1,3,5-triazine-2,4-diamine; N,N'-Diisopropyl-6-methoxy-1,3,5-triazine-2,4-diyldiamine; N,N'-Diisopropyl-6-methoxy-1,3,5-tri-azine-2,4-diamine; EINECS 216-548-0; EPA Pesticide Chemical Code 080804; G-31435; Gesafram; Gesafram 50; HSDB 1519; 2-Methoxy-4,6-bis(isopropylamino)-1,3,5-triazine; 2-Methoxy-4,6-bis(isopropylamino)-s-tri-azine; 6-Methoxy-N,N'-bis(1-methylethyl)-1,3,5-triazine-2,4-diamine; Methoxypropazine; NSC 163048; Ontracic 800; Ontrack; Ontrack-we-2; Pramitol; Pramitol 5P; Primatol 25e; Prometon; Prometone; 2,4-Prometone; s-Triazine, 2,4-bis(isopropylamino)-6-methoxy-; 1,3,5-Triazine-2,4-diamine, 6-methoxy-N,N'-bis(1-methyl-ethyl)- Herbicide. Registered by EPA as

an antimicrobial and herbicide. Crystalline solid; mp = 91 - 92°; soluble in H2O (0.075 g/100 ml at 20°), C6H6 (> 25 g/100 ml), MeOH (> 50 g/100 ml), Me2CO (> 50 g/100 ml), CH2Cl2 (35 g/100 ml), C7H8 (25 g/100 ml); LD50 (rat orl) = 2980 mg/kg, (mus orl) = 2160 mg/kg, (rbt der) > 2000 mg/kg; LC50 (bluegill sunfish 96 hr.) > 32 ppm, (rainbow trout 96 hr.) = 20 ppm, (goldfish 96 hr.) = 8.6 mg/l; non-toxic to bees. *ABC Compounding Co.; Agriliance LLC; Athea Laboratories, Inc.; Makhteshim-Agan; Riverdale Chemical Co; Sungro Chemicals Inc; Syngenta Crop Protection; UAP - Platte Chemical.*

3160 Prometryn
7287-19-6   7881   G H P

$C_{10}H_{19}N_5S$
1,3,5-Triazine-2,4-diamine, N,N'-bis(1-methylethyl)-6-(methylthio)-.
A 1114; Al3-60366; BRN 0613575; Caparol; Caparol 80W; Caswell No. 097; N,N-Di-isopropyl-6-methylthio-1,3,5-triazine-2,4-diamine N²,N⁴-Di-isopropyl-6-methyl-thio-1,3,5-triazine-2,4-diamine; N,N'-Diisopropyl-6-methylthio-1,3,5-triazine-2,4-diyldiamine; EINECS 230-711-3; EPA Pesticide Chemical Code 080805; G 34161; Gesagard; Gesagard 50; Gesagarde 50 Wp; HSDB 4060; Mercasin; Merkazin; 2-Methylmercapto-4,6-bis(isopropyl-amino)-s-triazine; 2-Methylthio-4,6-bis(isopropylamino)-s-triazine; NSC 163049; Polisin; Primatol; Primatol Q; Promepin; Promethryn; Prometrene; Prometrex; Prometrin; Prometryn; Prometryne; Selectin; Selectin 50; Selektin; Sesagard; s-Triazine, 2,4-bis(isopropylamino)-6-(methylthio)-; s-Triazine, 4,6-bis(isopropylamino)-2-methylmercapto-; 1,3,5-Triazine-2,4-diamine, N,N'-bis(1-methylethyl)-6-(methylthio)-; Uvon. A selective pre- and post-emergence herbicide for the control of broadleaf and grass weeds in a variety of crops. Absorbed by roots and foliage. Used for pre- and post-emergence control of most annual grasses and broad-leaved weeds. Used for selective weed control in cotton and celery crops. Registered by EPA as a herbicide. Colorless crystals; mp = 118-120°; d20 = 1.157; soluble in H2O (0.0048 g/100 ml), Me2CO (24 g/100 ml), MeOH (16 g/100 ml), CH2Cl2 (30 g/100 ml), C6H14 (0.55 g/100 ml), C7H8 (17 g/100 ml); LD50 (rat orl) = 5235 mg/kg, (rbt der) > 3100 mg/kg; LC50 (bobwhite quail 5-7 day dietary) = 16140 mg/kg, (mallard duck 5-7 day dietary) = 38736 mg/kg (rainbow trout 96 hr.) = 2.5 mg/l, (bluegill sunfish 96 hr.) = 10.0 mg/l, (goldfish 96 hr.) = 3.5 mg/l, (carp 96 hr.) = 8 mg/l. *Agan Chemical Manufacturers; Agriliance LLC; Ciba Geigy Agrochemicals; Griffin LLC; Syngenta Crop Protection; UAP-West.*

3161 Propachlor
1918-16-7   7885   G P

$C_{11}H_{14}ClNO$
2-Chloro-N-(1-methylethyl)-N-phenylacetamide.

Acetamide, 2-chloro-N-(1-methylethyl)-N-phenyl-; Acet-anilide, 2-chloro-N-isopropyl-; Acilid; AI3-51503; Bexton; Bexton 4L; BRN 2103903; Caswell No. 194; Chloressigsaeure-N-isopropylanilid; Chloressigsäure-N-isopropylanilid; α-Chloro-N-isopropylacetanilide; 2-Chloro-N-(1-methylethyl)-N-phenylacetamide; 2-Chloro-N-isopropyl-N-phenylacetamide; 2-Chloro-N-isopropyl-acetanilide; CP 31393; EINECS 217-638-2; EPA Pesticide Chemical Code 019101; HSDB 1200; Kartex A; N-Isopropyl-α-chloroacetanilide; N-Isopropyl-2-chloroacet-anilide; Nitacid; Niticid; Prolex; Propachlor; Propachlore; Ramrod; Ramrod 65; Satecid; Ramrod 20G; Ramrod Flowable; propachlor + atrazine; Chloro-N-isopropyl-acetanilide; Isopropyl-2-chloroacetanilide; Albrass; Cp 31393; Croptex; Amber; Niticid; Orange; Prolex; Ramrod; Satecid; Sentinel. A pre-emergence herbicide for various horticultural crops. Selective herbicide, absorbed by seedling shoots and roots. Used for control of annual grasses and some broad-leaved weeds in vegetable crops. Crystals; mp = 77°; bp:0.03 = 110°; soluble in H2O (613 mg/l), more soluble in organic slvents; LD50 (rat orl) = 1800 mg/kg. *Agan Chemical Manufacturers; Atlas Interlates Ltd; Dow UK; Hortichem Ltd.; ICI Chem. & Polymers Ltd.; ICI Plant Protection; Monsanto plc; Portman Agrochemicals Ltd.; Tripart Farm Chemicals Ltd.*

3162   1-Propanaminium, 3-amino-N-(carboxy-methyl)-N,N-dimethyl-, N-coco acyl derivs, chlorides, sodium salts
61789-39-7   H

1-Propanaminium, 3-amino-N-(carboxymethyl)-N,N-dimethyl-, N-coco acyl derivs, chlorides, sodium salts.
EINECS 263-057-2; 1-Propanaminium, 3-amino-N-(carboxymethyl)-N,N-dimethyl-, N-coco acyl derivs, chlorides, sodium salts.

3163   Propane
74-98-6   7891   G H

C_3H_8
Propane.
A-108; Dimethylmethane; EINECS 200-827-9; HC 290; HSDB 1672; Hydrocarbon Propellant A-108; Liquefied petroleum gas; LPG; Petroleum gas, liquefied; Propane; Propyl hydride; Propyldihydride; R 290; UN1978. Used as a propellant and in organic synthesis, as a household and industrial fuel, in the manufacture of ethylene, as an extractant, solvent, refrigerant and gas enricher. Colorless gas; mp = -187.6°; bp = -42.1°; d^{25} = 0.493; slightly soluble in H2O, soluble in organic solvents. *Air Prods & Chems; Fina Chemicals; Phillips 66.*

3164   Propanenitrile, 3-(isodecyloxy)-
64354-92-3   H

$C_{13}H_{25}NO$
3-(Isodecyloxy)propiononitrile.
EINECS 264-840-1; 3-(Isodecyloxy)propylnitrile; 3-Isodecoxypropionitrile; Propanenitrile, 3-(isodecyloxy)-.

3165   Propanethiol
107-03-9   G H

C_3H_8S
1-Propyl mercaptan.
4-01-00-01449 (Beilstein Handbook Reference); BRN 1696860; CCRIS 1246; EINECS 203-455-5; FEMA No. 3521; HSDB 1037;

Propanethiol; Propane-1-thiol; Propyl mercaptan; n-Propylmercaptan; Propylthiol; n-Propylthiol; Thiopropyl alcohol; n-Thiopropyl alcohol. Unpleasant stench oder. Used as a leak detecting additive to natural gas and as a chemical intermediate and herbicide. Liquid; mp = -113.3°; bp = 67.8°; d^{20} = 0.8411; λ_m = 240 nm; slightly soluble in H2O, soluble in EtOH, Et2O, Me2CO, C6H6. *Elf Atochem N. Am.; Phillips 66.*

3166   Propanil
709-98-8   7896   G H P

$C_9H_9Cl_2NO$
N-(3,4-Dichlorophenyl)propanamide.
AI3-31382; B-30,130; BAY 30130; Bayer 30 130; BRN 2365645; Caswell No. 325; CCRIS 3009; Cekupropanil; Chem-Rice; Crystal propanil-4; DCPA; Dichloro-propionanilide; Dipram; DPA; Drexel Prop-Job; EINECS 211-914-6; EPA Pesticide Chemical Code 028201; Erban; Erbanil; Farmco propanil; FW 734; Grascide; Herbax; Herbax 3E; Herbax 4E; Herbax LV-30; Herbax technical; HSDB 1226; Montrose propanil; NSC 31312; Prop-Job; Propanamide, N-(3,4-dichlorophenyl)-; Propanex; Prop-anid; Propanide; Propanil; Propanilo; Propionanilide, 3',4'-dichloro-; Propionic acid 3,4-dichloroanilide; Riselect; Rogue; Rosanil; S 10165; Stam; Stam 80EDF; Stam F 34; Stam LV 10; Stam M-4; Stam supernox; Stampede; Stampede 360; Stampede 3E; Strel; Supernox; Surcopur; Synpran N; Vertac Propanil 3; Vertac Propanil 4; Wham EZ. Post-emergence applied herbicide with no residual effect for control of numerous grasses and broad-leaved weeds in rice crops. Also used as a nematocide. Registered by EPA as a herbicide. Crystals; mp = 92°; d^{25} = 1.25; soluble in H2O (22.5 g/100 ml), C6H6 (7 g/100 ml), Me2CO (170 g/100 ml); EtOH (110 g/100 ml); LD50 (rat orl) = 1384 mg/kg. *Agriliance LLC; Bayer AG; Dow AgroSciences; Drexel Chemical Co.; Helena Chemical Co.; RICECO; Rohm & Haas Co.; Syngenta Crop Protection.*

3167   Propanol, (2-methoxymethylethoxy)-, acetate
88917-22-0   H

$C_9H_{18}O_4$
Propanol, 1(or 2)-(2-methoxymethylethoxy)-, acetate..
Propanol, (2-methoxymethylethoxy)-, acetate; Propanol, 1(or 2)-(2-methoxymethylethoxy)-, acetate. Contains different dihydroxypropane isomers.

3168   Propanolamine methyl ether
5332-73-0   H

$C_4H_{11}NO$
3-Methoxy-1-propanamine.
4-04-00-01623 (Beilstein Handbook Reference); AI3-25438; 1-Amino-3-methoxypropane; 3-Aminopropyl methyl ether; BRN 0878144; CCRIS 6178; EINECS 226-241-3; 3-Methoxy-1-propanamine; 3-Methoxy-n-propylamine; 3-Methoxypropylamine; 3-MPA; γ-Methoxypropylamine; NSC 552; 1-Propanamine, 3-methoxy-

; Propanolamine methyl ether; Propylamine, 3-methoxy-. Liquid; bp = 117.5°; d^{20} = 0.8727; soluble in H_2O, Me_2CO, C_6H_6, CCl_4, $CHCl_3$, MeOH.

3169   n-Propanol
71-23-8   7934   G P

OH

C_3H_8O

n-Propyl alcohol.
4-01-00-01413 (Beilstein Handbook Reference); AI3-16115; Albacol; Alcohol, propyl; Alcool propilico; Alcool propylique; BRN 1098242; Caswell No. 709A; CCRIS 3202; EINECS 200-746-9; EPA Pesticide Chemical Code 047502; Ethyl carbinol; FEMA Number 2928; HSDB 115; NSC 30300; Optal; Osmosol Extra; Propanol; Propanol-1; n-Propan-1-ol; n-Propanol; n-Propyl alcohol; Propanole; Propanolen; Propanoli; Propyl alcohol; Propylic alcohol; Propylowy alkohol; UN1274. Registered by EPA as an antimicrobial, fungicide, herbicide and insecticide (cancelled). Used as a chemical intermediate and as a solvent, especially for cellulose esters, waxes, vegetable oils, natural and synthetic resins. Liquid with alcoholic odor; mp = -127°; bp = 97.2°; d_4^{20} = 0.8053; d_4^{25} = 0.8016; soluble in H_2O, EtOH, Et_2O. LD50 (rat orl) = 1870 mg/kg. *Arco; Eastman Chem. Co.; Fluka; Hoechst Celanese; Mallinckrodt Inc.; Ruger; Spectrum Chem. Manufacturing; Union Carbide Corp.*

3170   2-Propanol, 1,3-dichloro-, phosphate
13674-87-8   4341   G H

$C_9H_{15}Cl_6O_4P$

Phosphoric acid, tris(1,3-dichloro-2-propyl) ester.
3-01-00-01473 (Beilstein Handbook Reference); BRN 1715458; CCRIS 6284; CRP; CRP (fireproofing agent); EINECS 237-159-2; Emulsion 212; Fosforan troj-(1,3-dwuchloroizopropylowy); Fyrol® FR 2; HSDB 4364; PF 38; PF 38/3; Phosphoric acid, tris(1,3-dichloro-2-propyl) ester; TCP; TDCPP; Tri(β,β'-dichloroisopropyl)phosphate; Tris-(1,3-dichloro-2-propyl)-phosphate; Tris(2-chloro-1-(chloromethyl)ethyl) phosphate. Flame retardant for flexible urethane foams Liquid; bp5 = 236-237°; n_D^{20} = 1.5022; soluble in H_2O (100 mg/l); LD50 (rat orl) = 1.85 g/kg. *Akzo Chemie.*

3171   2-Propanol, 1-(2-methoxy-1-methyl-ethoxy)-
20324-32-7   H

$C_7H_{16}O_3$

1-(2-Methoxy-1-methylethoxy)propan-2-ol.
EINECS 243-733-3; 1-(2-Methoxy-1-methylethoxy)-2-propanol; 1-(2-Methoxy-1-methylethoxy)propan-2-ol; 2-Propanol, 1-(2-methoxy-1-methylethoxy)-.

3172   Propantheline bromide
50-34-0   7897   G

$C_{23}H_{30}BrNO_3$

N-methyl-N-(1-methylethyl)-N-[2-[(9H-xanthen-9-ylcarbonyl)oxy]ethyl]-2-propanaminium bromide.
Ammonium, diisopropyl(2-hydroxyethyl)methyl-, brom-ide, xanthene-9-carboxylate; Bromure de propantheline; Bromuro de proanteline; CCRIS 6271; Diisopropyl(2-hydroxyethyl)methylammonium bromide xanthene-9-carboxylate; EINECS 200-030-6; Ercorax; Ercotina; (2-Hydroxyethyl)diisopropylmethylammonium bromide xanthene-9-carboxylate; Ketaman; Kivatin; NCI-C56257; Neometantyl; Neopepulsan; Pantas; Pantheline; Pervagal; Pro-Banthine; Pro-Gastron; Probantine; Prodixamon; 2-Propanaminium, N-methyl-N-(1-methylethyl)-N-(2-((9H-xanthen-9-ylcarbonyl)oxy)ethyl)-, Br; Propantel; Prop-antelina bromuro; Propantheline bromide; Propanthelini bromidum; SC-3171; Xanthene-9-carboxylic acid, ester with (2-hydroxyethyl)diisopropylmethylammonium brom-ide. Anticholinergic agent. mp = 159-161°; very soluble in H_2O, EtOH, $CHCl_3$, insoluble in Et_2O, C_6H_6. *Ellis & Everard.*

3173   Propargyl alcohol
107-19-7   7901   H

OH

C_3H_4O

1-Propyn-3-ol.
4-01-00-02214 (Beilstein Handbook Reference); Acetylene carbinol; Agrisynth PA; AI3-24359; BRN 0506003; CCRIS 6781; EINECS 203-471-2; Ethynyl carbinol; HSDB 6054; 1-Hydroxy-2-propyne; 3-Hydroxy-1-propyne; Methanol, ethynyl-; NA1986; NSC 8804; Propargyl alcohol; Propiolic alcohol; Propynyl alcohol; 1-Propyn-3-ol; 2-Propyn-1-ol; 3-Propynol; RCRA waste number P102. Liquid; mp = -51.8°, bp = 113.6°; d^{20} = 0.9478; freely soluble in EtOH, Et_2O, soluble in H_2O, $CHCl_3$.

3174   Propazine
139-40-2   7904   G H P

$C_9H_{16}ClN_5$

6-Chloro-N,N'-bis(1-methylethyl)-1,3,5-triazine-2,4-di-amine.
AI3-60348; BRN 0747081; Caswell No. 184; CCRIS 1026; EINECS 205-359-9; EPA Pesticide Chemical Code 080808; G-30028; Geigy 30,028; Gesamil; HSDB 1400; Maxx 90; Milo-pro; Milocep; Milogard; NSC 26002; Plantulin; Primatol P; Propasin; Propazin; Propazine; Prozinex. Pre-emergent selective systemic herbicide used for control of annual grasses and broad-leaved weeds in sorghum and crops such as carrots, chervil and parsley. Registered by EPA as a herbicide. Used in greenhouses. Colorless crystals; mp = 213°; soluble in H_2O (0.0005 g/100 ml), soluble in C_6H_6 (0.55 g/100 ml), C_7H_8 (0.53 g/100 ml), Et_2O (0.35 g/100 ml), CCl_4 (0.40 g/100 ml); LD50 (rat orl) >7000 mg/kg, (rat der) > 3100 mg/kg, (rbt der) > 10200

mg/kg; LC$_{50}$ (rbt ihl 4 hr.) = 2.04 mg/l air, (bobwhite quail, mallard duck 8 day dietary) > 10000 mg/kg, (rainbow trout 96 hr.) = 17.5 mg/l, (bluegill sunfish 96 hr.) > 100 mg/l, (goldfish 96 hr.) > 32 mg/l; non-toxic to bees. *Agan Chemical Manufacturers; Griffin LLC; Syngenta Crop Protection.*

3175 Propetamphos
31218-83-4   7907   G P

C$_{10}$H$_{20}$NO$_4$PS
1-Methylethyl 3-(((ethylamino)methoxyphosphinothioyl)-oxy)-2-butenoate.
1-methylethyl 3-(((ethylamino)methoxyphosphinothioyl)oxy)-2-butenoate, (E)-; Butenoic acid, 3-(((ethylamino)-methoxyphosphinothioyl)oxy)-, 1-methylethyl ester; Methylethyl (E)-3-(((ethylamino)methoxyphosphino-thioyl)oxy)-2-butenoate; Phosphoramidothioic acid, N-ethyl-, (E)-O-(2-isopropoxycarbonyl-1-methylvinyl) O-methyl ester; Safrotin; Seraphos; TSAR; Zoecon. Used as an ectoparasiticide. Registered by EPA as an insecticide. Yellow liquid; bp$_{0.005}$ = 87 - 89°; d$_4^{20}$ 1.1294; soluble in H$_2$O (0.011 g/100 ml at 24°), soluble in most organic solvents; LD$_{50}$ (mrat orl) = 82 mg/kg, 119 mg/kg, (mrat der) = 2300 mg/kg, (mallard duck orl) = 197 mg/kg; LC$_{50}$ (carp 96 hr.) = 8.8 mg/l. *Novartis; Wellmark International.*

3176 Propham
122-42-9   7908   G H P

C$_{10}$H$_{13}$NO$_2$
Phenylcarbamic acid 1-methylethyl ester.
4-12-00-00620 (Beilstein Handbook Reference); Agermin; AI3-14879; Ban-Hoe; Beet-Kleen; Birgin; BRN 2209666; Carbamic acid, phenyl-, 1-methylethyl ester; Carbanilic acid, isopropyl ester; Caswell No. 510; Chem-Hoe; Collavin; EINECS 204-542-0; EPA Pesticide Chemical Code 047601; HSDB 602; IFC; IFK; INPC; IPPC; Iso.ppc.; Isopropil-N-fenil-carbammato; Isopropyl carbanilate; Isopropyl carbanilic acid ester; Isopropyl-N-fenyl-carbamaat; Isopropyl-N-phenyl-carbamat; Isopropyl-N-phenylcarbamate; O-Isopropyl N-phenyl carbamate; Isopropyl-N-phenylurethan; Isopropyl phenyl urethane; Isopropyl phenylcarbamate; Isopropylester kyseliny karbanilove; 1-Methylethyl phenylcarbamate; NSC 2105; Ortho grass killer; N-Phenylcarbamate d'isopropyle; Phenylcarbamic acid, 1-methylethyl ester; N-Phenylcarbamic acid, isopropyl ester; N-Phenyl isopropyl carbamate; Premalox; Profam; Propham; Prophame; Prophos; RCRA waste no. U373; Tixit; Triherbide; Triherbide-ipc; Tuberit; Tuberite; USAF D-9; Y 2. Selective sytemic herbicide and growth regulator. Used to control annual grasses and some broad-leaf weeds. Plant growth regulator for control of sprouting in stored potatoes and in some cases as herbicide against weeds in vegetables. Registered by EPA as a herbicide (cancelled). Colorless crystals; mp = 90°; sg^{20} = 1.09; poorly soluble in H$_2$O (0.025 g/100 ml at 20°), soluble in organic solvents; LD$_{50}$ (rat orl) = 5000 mg/kg, (mrat orl) = 3724 mg/kg, (frat orl) = 4315 mg/kg), (mus

orl) = 3000 mg/kg, (rat ip) = 600 mg/kg, (mus ip) = 1000 mg/kg, (mallard duck orl) > 2000 mg/kg; LC$_{50}$ (bluegill sunfish 48 hr.) = 32 mg/l, (guppy 48 hr.) = 35 mg/l; non-toxic to bees. *Bayer AG; Syngenta Crop Protection.*

3177 Propiconazole
60207-90-1   7910   G P

C$_{15}$H$_{17}$Cl$_2$N$_3$O$_2$
1-(2-(2',4'-Dichlorophenyl)-4-propyl-1,3-dioxolan-2-yl-methyl)-1H-1,2,4-triazole.
5-26-01-00205 (Beilstein Handbook Reference); Banner; BRN 0841361; Caswell No. 323EE; CGA-64250; CGD 92710F; Desmel; EINECS 262-104-4; EPA Pesticide Chemical Code 122101; HSDB 6731; Orbit; Proconazole; Propiconazole; Radar; Tilt; Wocosin. A systemic triazole fungicide for control of powdery mildew and rust in wheat and barley. Registered by EPA as a fungicide. Yellow liquid; bp$_{0.01}$ = 180°; sg^{20} = 1.27; soluble in H$_2$O (0.011 g/100 ml at 20°), C$_6$H$_{14}$ (3.96 g/100 ml), soluble in most organic solvents; LD$_{50}$ (rat orl) = 1517 mg/kg, (rat der) > 4000 mg/kg; LC$_{50}$ (carp 96 hr.) > 100 mg/l, (brown trout 96 hr.) = 20 mg/l; non-toxic to birds, non-toxic to bees. *Ciba Geigy Agrochemicals; Contechem Inc.; Diacon Technologies Ltd.; Dow AgroSciences; Farm Protection Ltd.; ICI Chem. & Polymers Ltd.; Janssen Pharmaceutical, Belgium; Kop-Coat Inc.; Syngenta Crop Protection.*

3178 Propiolactone
57-57-8   7912   G

C$_3$H$_4$O$_2$
3-Hydroxypropionic acid β-lactone.
5-17-09-00003 (Beilstein Handbook Reference); AI3-24257; Betaprone; BPL; BRN 0001360; Caswell No. 709; CCRIS 536; EINECS 200-340-1; EPA Pesticide Chemical Code 010901; HSDB 811; Hydracrylic acid β-lactone; 3-Hydroxypropionic acid β-lactone; 3-Hydroxypropionic acid lactone; NSC-21626; Oxetanone; Oxetan-2-one; 2-Oxetanone; Propanilide; Propanoic acid, 3-hydroxy-, β-lactone; Propanolide; 3-Propanolide; β-Propanoic acid lactone; β-Propiolactone; β-Propiolakton; β-Propiono-lactone; β-Proprolactone; 1,3-Propiolactone; 3-Propio-lactone; Propiolactona; Propiolactone; Propiolactone, β-; Propiolactonum; Propiolattone; Propionic acid, 3-hydroxy-, β-lactone; Propionolactone. Chemical inter-mediate; used as a disinfectant. Liquid; mp = -34.4°; bp = 162° (dec); d$_4^{20}$ 1.1460; soluble in H$_2$O (35 g/100 ml); LC$_{50}$ (rat ihl) = 25 ppm/6H. *O'Neal Jones & Feldman Pharm.*

3179 Propiolic acid
471-25-0   7913   G

C3H2O2
Propargylic acid.
Acetylenecarboxylic acid; Carboxyacetylene; EINECS 207-437-8; NSC 16152; Propargylic acid; Propiolic acid; Propynoic acid; 2-Propynoic acid. Used in chemical synthesis. mp = 9°; bp = 144° (dec), bp$_{50}$ = 70-75; d$^{20}_4$ 1.1380; very soluble in H_2O, EtOH, Et_2O, $CHCl_3$; LD$_{50}$ (rat orl) = 100 mg/kg.

3180   Propionaldehyde
123-38-6   7915   G H

C3H6O
Methylacetaldehyde.
AI3-16114; Aldehyde propionique; CCRIS 2917; EINECS 204-623-0; FEMA Number 2923; HSDB 1193; Methylacetaldehyde; NCI-C61029; NSC 6493; Propaldehyde; Propanal; n-Propanal; Propional; Propionaldehyde; Propionic aldehyde; Propyl aldehyde; Propylic aldehyde; UN1275. Used in the manufacture of propionic acid, polyvinyl and other plastics, synthesis of rubber chemicals, and preservatives. Liquid; mp = -80°; bp = 48°; d^{25} = 0.8657; λ$_m$ = 282 nm (ε = 8, H_2O); soluble in H_2O, freely soluble in EtOH, Et_2O; LD$_{50}$ (rat orl) = 1410 mg/kg.

3181   Propionic acid
79-09-4   7917   G H

C3H6O2
Propanoic Acid.
4-02-00-00695 (Beilstein Handbook Reference); Acide propionique; AI3-04167; BRN 0506071; C3 acid; Carboxyethane; Caswell No. 707; CCRIS 6096; EINECS 201-176-3; EPA Pesticide Chemical Code 077702; Ethanecarboxylic acid; Ethylformic acid; FEMA Number 2924; HSDB 1192; Kyselina propionova; Luprosil; Metacetonic acid; Methyl acetic acid; Monoprop; Propanoic acid; Propionic acid; Propionic acid grain preserver; Prozoin; Pseudoacetic acid; Sentry grain preserver; Tenox P grain preservative; UN1848. Registered by EPA as a herbicide. Esterifying agent; in production of cellulose propionates, etc.; as mold inhibitors and preservatives; in manufacture of ester solvents, fruit flavors, perfume bases; antifungal. Liquid; mp = -21.5°; bp = 141.1°, bp$_{400}$ = 122°; bp$_{100}$ = 85.8°, bp$_1$ = 4.6°; viscosity = 1.175 cp at 15°, 1.020 cp at 25°, 0.956 cp at 30°, 0.668 at 60°, 0.495 cp at 90°; soluble in H_2O, EtOH, Et_2O, $CHCl_3$. LD$_{50}$ (rat orl) = 4290 mg/kg, TLV:TWA = 10 ppm. *Alltech Inc.; Eastman Chem. Co.; Kemin Industries Inc.; West Agro Inc.*

3182   Propionic anhydride
123-62-6   7918   H

C6H10O3
Propionic acid anhydride.
4-02-00-00722 (Beilstein Handbook Reference); AI3-26975;

Anhydrid kyseliny propionove; BRN 0507066; Caswell No. 708; EINECS 204-638-2; EPA Pesticide Chemical Code 077704; HSDB 1215; Methylacetic anhydride; Propanoic acid, anhydride; Propanoic anhydride; Propionic acid anhydride; Propionic anhydride; Propionyl oxide; UN2496. Liquid; mp = -45°; bp = 170°, bp$_{18}$ = 67.5; d^{20} = 1.0110; slightly soluble in CCl_4, freely soluble in Et_2O.

3183   Propionitrile
107-12-0   7919   H

C3H5N
Cyanoethane.
4-02-00-00728 (Beilstein Handbook Reference); AI3-08777; BRN 0773680; CCRIS 4706; Cyanoethane; EINECS 203-464-4; Ether cyanatus; Ethyl cyanide; Ethylkyanid; HSDB 117; Hydrocyanic ether; NSC 7966; Propanenitrile; Propannitril; Propionic nitrile; Propionitrile; Propiononitrile; Propylnitrile; RCRA waste number P101; UN2404. Liquid; mp = -92.8°; bp = 97.1°; d^{20} = 0.7818; very soluble in H_2O, soluble in EtOH, Et_2O, Me_2CO, C_6H_6, CCl_4.

3184   Propoxur
114-26-1   7929   G H P

C11H15NO3
o-Isopropoxyphenyl Methylcarbamate.
58 12 315; AI3-25671; Aprocarb; BAY 5122; BAY 9010; BAY 39007; Bayer 39007; Bayer B 5122; Baygon; Bifex; Blattanex; Blattosep; Bolfo; Boruho; Boruho 50; BRN 1879891; Brygou; Carbamic acid, methyl-, o-isopropoxyphenyl ester; Caswell No. 508; CCRIS 1392; Chemagro 9010; Dalf dust; DMS 33; EINECS 204-043-8; ENT 25,671; EPA Pesticide Chemical Code 047802; HSDB 603; o-IMPC; Invisi-Gard; IPMC; Isocarb; Isopropoxyphenyl methylcarbamate; 2-Isopropoxyphenyl N-methylcarbamate; Methyl-2-isopropoxyphenylcarb-amate; Methylcarbamic acid, o-isopropoxyphenol ester; Mrowkozol; NSC 379584; OMS-33; PHC; PHC (carbamate); PHC 7; Phenol, 2-(1-methylethoxy)-, methylcarbamate; Pillargon; Propoksuru; Propotox; Propoxur; Propoxure; Propoxylor; Propyon; Rhoden; Sendran; Suncide; Tendex; Tugon fliegenkugel; Unden; Unden (pesticide); Unden 50PM. Non-systemic insecticide with contact and stomach action; used for treatment of sucking and biting insects, e.g., aphids, mealybugs, scales, leafhoppers, caterpillars on vegetables, pome and stone fruit, cocoa, rice, oil palms and other crops. Used for protection of flowers and fruits. Used for control of cockroaches, flies, fleas, mosquitoes, bugs, ants, millepedes and other insect pests in houses and food storage areas. Registered by EPA as an insecticide. Crystals; mp = 84 - 87°, 91.5°; sg^{20} = 1.12; soluble in H_2O (0.2 g/100 ml at 20°), MeOH, Me_2CO, CH_2Cl_2 (> 20 g/100 ml), C_7H_8 (10 g/100 ml), $CHCl_3$ (10 g/100 ml); LD$_{50}$ (mrat orl) = 95 mg/kg, (frat orl) = 104 mg/kg, (mmus orl) = 100 - 109 mg/kg, (mrat der) = 800 - 1000 mg/kg, (mrbt der) > 500 mg/kg, (bobwhite quail orl) = 25.9 mg/kg; LC$_{50}$ (rat ihl 1 hr.) = 1.44 mg/l air, (bobwhite quail 5 day dietary) = 2828 mg/kg diet, (mallard duck 5 day dietary) > 5000 mg/kg diet, (bluegill sunfish 96 hr.) = 6.6 mg/l, (rainbow trout 96 hr.) = 3.7 mg/l; highly toxic to bees. *Amrep Inc.; Bayer Corp.; Buhl Products Co; MBL Industries; Sergeant's Pet Products; Theochem Laboratories Inc.; Value Gardens Supply LLC; Whitmire Micro-Gen*

3185   2-(2-Propoxyethoxy)ethanol
6881-94-3

G H

C7H16O3

2-(2-Propoxyethoxy)ethanol.
EINECS 229-985-7; Ethanol, 2-(2-propoxyethoxy)-. Evaporating, water-miscible solvent used in solution and water-dilutable coatings; active for many coating materials including NC, acrylic copolymers, epoxy resins, chlorinated rubber, and alkyd resins; strong coupling agent with some resin systems in water-dilutable coatings. Liquid; bp = 202°; soluble in H_2O, organic solvents; d = 0.963; causes eye irritation. *Ashland; Eastman Chem. Co.; Great Western.*

3186   Propoxyphene
469-62-5   7931

D

C22H29NO2

(2S,3R)-(+)-4-(Dimethylamino)-3-methyl-1,2-diphenyl-2-butanol propionate (ester).
Algafan; Antalvic; Darvocet; Depromic; Destroprop-ossifene; Dextropropoxifeno; Dextropropoxyphen; Dextropropoxyphene; Dextropropoxyphenum; Dextro-proxifeno; Diméprotane (α-dl-form); EINECS 207-420-5; Femadol; HSDB 3175; Levopropoxyphene (α-l-form); Phenylbenzeneethanol propanoate; Propoxyphene; d-Propoxyphene; Proxagesic; SK 65. Analgesic, narcotic. Bulk dextropropoxyphene (non-dosage form) is a federally controlled substance (opiate); dextropropoxy-phene is a controlled substance (narcotic). *See* levo-propoxyphene (α-l-form). Crystals; mp = 75-76°; $[\alpha]_D^{25}$= +67.3° (c = 0.06 in CHCl3); [β-dl-form]: mp = 187-188°; more soluble than α-form. *Eli Lilly & Co.*

3187   Propoxyphene hydrochloride
1639-60-7   7931

D

C22H30ClNO2

(2S,3R)-(+)-4-(Dimethylamino)-3-methyl-1,2 diphenyl 2 butanol propionate (ester) hydrochloride.
Abalgin; Anatalvic; Antalvic; Darvon; Darvon with ASA; Deprancol; Depronal retard; Develin; Dextropropofixen Dak; Dolene; Dolocap; Dolotard; Erantin; Femadol; Harmar; Kesso-Gesic; Liberen; Mardon; Margesic Improved; Novopropoxyn; Paljin; Prophene; Propoxy-chel; Propoxyphene Compound 65; Propoxyphene hydrochloride; Proxagesic; Romidon; SK-65; SK-65 Apap; SK-65 Compound; Wygesic; Zideron.; component of: Darvon with ASA, SK-65 Apap, SK-65 Compound, Wygesic. Analgesic, narcotic. Federally

controlled substance (opiate). Crystals; mp = 163-165°; $[\alpha]_D^{25}$= + 59.8° (c = 0.06 in H_2O); soluble in H_2O, alcohol, CHCl3, Me2CO; practically insoluble in C6H6, Et2O; LD50 (mus iv) = 28 mg/kg, (mus ip) 111 mg/kg, (mus sc) = 211 mg/kg, (mus orl) = 282 mg/kg. *Eli Lilly & Co.*

3188   Propranolol
525-66-6   7932

D

C16H21NO2

1-(Isopropylamino)-3-(1-naphthyloxy)-2-propanol.
Avlocardyl; Betalong; CCRIS 3082; Corpendol; EINECS 208-378-0; Euprovasin; HSDB 3176; Inderal; Propanix; Propanolol; Propranolol; β-Propranolol; Propranololo; Propranololum; Proprasylyt; Reducor; Sawatal; Sumial. A β-adrenergic blocker. Antianginal; antihypertensive; antiarrhythmic (class II) agent. Crystals; mp = 96°. *ICI; Parke-Davis; Quimicobiol; Wyeth-Ayerst Labs.*

3189   Propranolol hydrochloride
318-98-9   7932

D

C16H22ClNO2

1-(Isopropylamino)-3-(1-naphthyloxy)-2-propanol hydrochloride.
Acifol; Anaprilin; Anapriline; Angilol; Apsolol; Arcablock; Artensol; Avlocardyl; AY-64043; Bedranol; Beprane; Beprane; Berkolol; Biocard; Blocaryl; Cardinol; Caridolol; CCRIS 1105; Cinlol; Ciplar; Corbeta; Deralin; Detensol; Dibudinate; Dociton; Dumopranol; Duranol; Duraprox; Efectolol; EINECS 206-268-7; Elbol; Elbrol; Emforal; Farprolol; Frekven; Frekven; Half-Inderal; Hemipralon; Herzbase; Herzul; I 2065; ICI-45520; Ikopal; Inderal; Inderalici; InderalLA; Inderex; Inderide; Indermigran; Indobloc; InnoPran XL; Intermigran; Kemi S; KEMI; Kidoral; Naprilin; Nedis; β-Neg; Nelderal; Noloten; Novopranol; NSC-91523; Obsidan; Oposim; Panolol; Prandol; Pranix; Prano-Puren; Proberta LA; Procor; Pronovan; Propabloc; Propadex; Propalong; Propayerst; Prophylux; Propra vt ct; Proprahexal; Propral; Propranolol chloride; Propranolol hydrochloride; Propranovitan; Propranur; Prosin; Pur-Bloka; Pylapron; Rapynogen; Sagittol; Sawatol; Scandrug; Servanolol; Sinal; Sloprolol; Sudenol; Sumial; β-Tablinen; Tensiflex; Tesnol; β-Timelets; Tiperal; Tonum. A β-adrenergic blocker. Antianginal; antihypertensive; antiarrhythmic (class II) agent. Crystals; mp = 163-164°; soluble in H_2O, alcohol; insoluble in Et2O, C6H6, EtOAc; LD50 (mus orl) = 565 mg/kg, (mus iv) = 22 mg/kg, (mus ip) = 107 mg/kg. *ICI; Parke-Davis; Quimicobiol; Wyeth Labs.*

3190   Propyl acetate
109-60-4   7933

G H

C5H10O2
n-Propyl acetate.

4-02-00-00138 (Beilstein Handbook Reference); Acetate de propyle normal; Acetic acid n-propyl ester; Acetic acid, propyl ester; 1-Acetoxypropane; AI3-24156; BRN 1740764; EINECS 203-686-1; FEMA No. 2925; HSDB 161; NSC 72025; Octan propylu; Propyl acetate; 1-Propyl acetate; n-Propyl acetate; Propyl ethanoate; n-Propyl ethanoate; Propylester kyseliny octove; UN1276. Flavoring agent, perfumery, solvent for nitrocellulose and other cellulose derivatives, natural and synthetic resins, lacquers, plastics, organic synthesis, laboratory reagent. Liquid; mp = -93°; bp = 101.5°; d^{20} = 0.8878; λ_m = 203, 209 nm (ε = 51, 55, MeOH); slightly soluble in CCl4, freely soluble in EtOH, Et2O; LD50 (rat orl) = 9370 mg/kg. *BASF Corp.; BP Chem.; Eastman Chem. Co.; Hoechst Celanese; Union Carbide Corp.*

3191   Propyl chlorothioformate
13889-92-4   H

C4H7ClOS
Carbonochloridothioic acid, S-propyl ester.

4-03-00-00279 (Beilstein Handbook Reference); AI3-52665; BRN 1633747; Carbonochloridothioic acid, S-propyl ester; CCRIS 6097; EINECS 237-656-4; Formic acid, chlorothio-, S-propyl ester; HSDB 6155; NSC 72086; Propyl chlorothioformate; n-Propyl chloro-thioformate; S-Propyl chlorothioformate; n-Propyl thiochloroformate; S-Propyl thiochloroformate.

3192   Propylene
115-07-1   7941   G H

C3H6
1-Propene.

CCRIS 1356; EINECS 204-062-1; HSDB 175; Methylethene; Methylethylene; NCI-C50077; Propene; 1-Propene; Propylene; 1-Propylene; R 1270; UN1077. Chemical intermediate for manufacture of isopropyl alcohol, polypropylene, synthetic glycerol, acrylonitrile, propylene oxide, heptene, cumene, polymer gasoline, acrylic acid, vinyl resins, oxo chemicals. Gas; mp = -185.2°; bp = -47.6°; d^{25} = 0.505; very soluble in H2O (23 g/100 ml), EtOH, AcOH. *Air Prods & Chems; Amoco Lubricants; BP Chem.; Chevron; Exxon; Fina Chemicals; Mobil; OxyChem; Phillips 66; Shell; Texaco.*

3193   Propylene carbonate
108-32-7   G H

C4H6O3
4-Methyl-1,3-dioxolan-2-one.

5-19-04-00021 (Beilstein Handbook Reference); AI3-19724; Arconate® 5000; Arconate® Propylene Carbonate; BRN 0107913; Carbonic acid, cyclic propylene ether; Carbonic acid, propylene ester; Cyclic 1,2-propylene carbonate; Cyclic methylethylene carbonate; Cyclic propylene carbonate; Dipropylene carbonate; EINECS 203-572-1; HSDB 6806; 1-Methylethylene carbonate; 4-Methyl-1,3-dioxolan-2-one; NSC 11784; 1,2-Propanediol cyclic carbonate; Propylene carbonate; 1,2-Propylene carbonate; Propylene glycol cyclic carbonate; Propylenester kyseliny uhlicite; Solvenon® PC; Texacar® PC. Solvent with high boiling point, low

toxicity, broad range of applications; reactive diluent for woodbinders, urethane foams and coatings, foundry sand binders, in textile and synthetic fiber industry, natural gas treating. Solvent for pigments and dyes, in screen printing dyes; extracting agent; washing liquid for natural and synthetic gases; intermediate for organic syntheses. Solvent for organic and inorganic materials; Rule 66 exempt; also used as reactant and plasticizer in fibers and textiles, hydraulic fluids, plastics and resins, gas treating, aromatic hydrocarbon extraction, metal extraction, Used as a chemical intermediate and in the production of resins and coatings. Liquid; mp = -48.8°; bp = 242°; d^{20} = 1.2047; very soluble in H2O, EtOH, Et2O, Me2CO, C6H6; LD55 (rat orl) = 29 g/kg. *Arco; BASF Corp.; Texaco.*

3194   Propylene dichloride
78-87-5   7946   G H

C3H6Cl2
1,2-Dichloropropane.

3-01-00-00225 (Beilstein Handbook Reference); AI3-15406; Bichlorure de propylene; BRN 1718880; Caswell No. 324; CCRIS 951; Dichloro-1,2 propane; 1,2-dichloropropane; Dwuchloropropan; EINECS 201-152-2; ENT 15,406; EPA Pesticide Chemical Code 029002; HSDB 1102; NCI-C55141; NSC 1237; Propane, 1,2-dichloro-; Propylene dichloride; RCRA waste number U083; UN1279. Intermediate for perchloroethylene, CCl4; lead scavenger for antiknock fluids; solvents for fats, oils, waxes, gums, resins, cellulose esters and ethers; scouring compounds; metal degreasers; soil fumigant for nematodes. Clear mobile liquid; bp = 95 - 96°; d^{25} = 1.159; LD50 (rat orl) = 1379 mg/kg. *Lancaster Synthesis Co.; Mallinckrodt Inc.; Sigma-Aldrich Fine Chem.*

3195   Propylene glycol
57-55-6   7947   P H

C3H8O2
1,2-Propanediol.

3-01-00-02142 (Beilstein Handbook Reference); Acid Proof; Adeka Propylene Glycol (P); AI3-01898; Aloe-Moist™; Aloe-Moist™ A; Annatto Liquid 3968; Annatto OS 2894; Annatto OS 2922; Annatto OS 2923; Arlacel® 186; AZP-908PG; AZZ-908PG; Biophytex®; BRN 1340498; Carmine AS; Caswell No. 713; CCRIS 5929; Cremophor® RH 455; Crodarom Nut A; Dehymuls® F; Dowfrost; EINECS 200-338-0; EPA Pesticide Chemical Code 068603; FEMA No. 2940; Germaben® II; Grillocin® AT Basis; HSDB 174; Isopropylene glycol; Lanotein AWS 30; Lipo PE Base GP-55; Lubrajel® CG; Lubrajel® DV; Lubrajel® MS; Lubrajel® TW; Methylethyl glycol; Methylethylene glycol; Monawet MO-70R; Monopropylene glycol; NAT-50-PGReach®; Natipide® II PG; NSC 69860; Oxynex® 2004; PG 12; Phosal®; Phosal® 50 PG; Phosal® 60 PG; Propane-1,2-diol; Propylene glycol; Reach®; Rezal® 36GPG; SDM No. 27; Sentry Propylene Glycol; Sirlene®; Solar Winter Ban®; Solargard P; Tagat® R63; Tepescohuite AMI Watersoluble; Tepescohuite HG; Tepescohuite HS; Trimethyl glycol; Ucar 35; Unibix AP (Acid Proof); Westchlor A2Z 8106. Registered by EPA as an antimicrobial, fungicide and insecticide. Used as a coolant in the manufacture of beer, wine, milk and other liquids; also used to freeze poultry and fish. Also as a solvent, emulsifier, in production of paints, resins, foods, drugs, as an antifreeze in breweries and dairies; as a substitute for ethylene glycol and glycerol; as a mold growth and fermentation inhibitor. FDA GRAS, USDA, EPA, BATF reg., approved for some drugs, Japan approved with limitations, Europe listed, FEMA GRAS, FDA approved for orals, parenterals, topicals,

USP/NF, BP, Ph. Eur., Japan compliance. Used as a solvent, vehicle, humectant, preservative, plasticizer, stabilizer in vitamin preparations, protectant in hemorrhoidal products, used in orals, otics, parenterals, vaginals and topical products. Clear viscous liquid; mp = -59°; bp = 188.2°; d = 1.0362; [(R)-form]: $[\alpha]_D^{20}$ = -15.0° (neat); [(S)-form]: $[\alpha]_D^{20}$ = +15.8° (neat); insoluble in oils, soluble in H_2O, Me_2CO, $CHCl_3$, EtOH, glycerol, Et_2O 16.7 g/100 ml); dynamic viscosity = 58.1 mPa at 20°. LD_{50} (rat orl) = 25900 mg/kg, (rat sc) = 21700- 29000 mg/kg, (rat iv) = 6200 - 12700 mg/kg, (rat ip) = 13000 - 16800 mg/kg, (rat im) = 13000 - 20700 mg/kg, (rbt orl) = 15700 - 19200 mg/kg, (rbt iv) = 5000 - 6500 mg/kg, (rbt im) = 6000 mg/kg, (mus sc) = 15500 - 19200 mg/kg, (mus orl) = 23900 mg/kg, (mus iv) = 7600 - 8300 mg/kg, (mus ip) = 6800 - 13600 mg/kg, (gpg sc) = 13000 - 15500 mg/kg, (gpg orl) = 18400 - 19600, (dog orl) = 10000 - 20000 mg/kg, (dog iv) = 25900 mg/kg. Amrep Inc.; Arco; Asahi Denka Kogyo; Ashland; BP Chem.; Chemical Packaging Corp.; Crompton & Knowles; Eastman Chem. Co.; Ellis & Everard; Greef K & K; Hays; Honeywell & Stein; Hüls Am.; Ibrahim N. I.; Integra; Mallinckrodt Inc.; McIntyre; Meer; Nisso Petrochem. Ind.; Norman-Fox; Olin Res. Ctr.; Penta Mfg.; PET Chemicals; Primachem; Pronova; Ruger; Showa Denko; Sigma-Aldrich Fine Chem.; Spectrum Chem. Manufacturing; Speer Products Inc.; Texaco; Todd's; Treatt R. C.; Van Waters & Rogers; Veckridge; Westco.

3196   Propylene glycol Alginate
9005-37-2
E G

(C9H14O7)8

Propane-1,2-diol alginate.
Alginic acid, ester with 1,2-propanediol; Alginic acid propylene glycol ester; CCRIS 3654; FEMA Number 2941; HSDB 1907; Hydroxypropyl alginate; Kelcoloid; Kelcoloid HVF; Manucol ester E/REP; 1,2-Propanediol alginate; Propylene alginate; Propylene glycol alginate; 1,2-Propylene glycol alginate; hydroxypropyl alginate; KELCOLOID®; KELCOLOID® D; KELCOLOID® DH; KELCOLOID® DSF; KELCOLOID® HVF; KELCOLOID® LVF; KELCOLOID® S; Kimiloid HV; Kimiloid NLS-K; Manucol ester®; Pronova®; propane 1,2-diol alginate; Protanal®; component of: Ches® 500. Mixture of propylene glycol esters of alginic acid. food additive (human). FDA, FEMA GRAS, UK, Japan approved (1% maximum), Europe listed, FDA approved for orals, USP/NF compliance. Used as a flavoring adjuvant, suspending agent, thickener, gellant, film former, emulsifier, formulation aid, stabilizer, solvent and defoamer. Used in orals. While granular or fibrous powder; soluble in H_2O, dilute organic acids; dynamic viscosity of 1% aqueous solution = 20 - 400 mPa. Non-toxic; LD_{50} (rat orl) = 7200 mg/kg, (rbt orl) = 7600 mg/kg, (mus orl) = 7800 mg/kg, (hmtr orl) = 7000 mg/kg. Frutarom; Kelco Intl.; Meer; Pronova; Spectrum Chem. Manufacturing.

3197   Propylene glycol dibenzoate
19224-26-1
H

C17H16O4
Propane-1,2-diyl dibenzoate.
EINECS 242-894-7; FEMA No. 3419; Propane-1,2-diyl dibenzoate; 1,2-Propanediol, dibenzoate; Propylene glycol dibenzoate; Propyleneglycol dibenzoate.

3198   Propylene glycol dinitrate
6423-43-4
H

C3H6N2O6
1,2-Propylene glycol dinitrate.
4-01-00-02497 (Beilstein Handbook Reference); BRN 1709968; EINECS 229-180-0; Isopropylene nitrate: NSC 62614; PGDN; 1,2-Propanediol, dinitrate; Propane-1,2-diyl dinitrate; Propylene dinitrate; Propylene glycol 1,2-dinitrate; Propylene glycol dinitrate; Propylene nitrate.

3199   Propylene glycol monolaurate
10108-22-2
G

C15H30O3
Dodecanoic acid, 3-hydroxypropyl ester.
Atlas G 917; Atlas G 3851; EINECS 233-292-5; 3-Hydroxypropyl laurate; Lauric acid, 3-hydroxypropyl ester; NSC 406283; Propylene glycol monolaurate. Surfactant.

3200   Propylene glycol monomethyl ether acetate
108-65-6
G H

C6H12O3
Acetic acid, 2-methoxy-1-methylethyl ester.
4-02-00-00220 (Beilstein Handbook Reference); Acetic acid, 2-methoxy-1-methylethyl ester; AI3-18548; Arcosolv® PMA; BRN 1751656; Dowanol® PMA glycol ether acetate; EINECS 203-603-9; NSC 2207; PGMEA; Propylene glycol monomethyl ether acetate. Slow-evaporating solvent with good solvency for many commonly used coating resins, e.g., acrylics, NC, and urethanes; used in lacquers, water-based paints. Coupling agent providing improved surface wetting, soil penetration in household, commercial and industrial cleaning products Liquid; mp = -67°; bp = 150°; d = 0.9690; soluble in H_2O (19 g/100 ml), organic solvents; LD_{50} (rat orl) = 8532 mg/kg. Arco; Dow AgroSciences; Dow Chem. U.S.A.

3201   Propylene glycol monostearate
1323-39-3
G H

C21H42O3
1,2-Propanediol mono-n-octadecanoate.
AI3-00975; Aldo® PGHMS KFG; Atlas G 924; Cerasynt PA; Cerasynt PN; Crill 26; Dragil-P; EINECS 215-354-3; Emcol PS-50 RHP; Emerest 2381; Monosteol; Monosteol TG; Noca; Nonex 32; NSC 4841; Octadecanoic acid, monoester with 1,2-propanediol; Pegosperse PS; Promodan SP; 1,2-Propanediol monooctadecanoate; 1,2-Propanediol monostearate; Propylene glycol monooctadecanoate; Propylene glycol monostearate; 1,2-

Propylene glycol monostearate; Propylene glycol octadecanoate; Propylene glycol stearate; Propylene glycol stearic acid ester; Prostearin; Stearic acid, monoester with propane-1,2-diol; Tegin P; USAF KE-13; Witconol 2380. Oil-water emulsifier, lubricant, and opacifier. Food emulsifier, kosher food grade emulsifier, whipping agent. CK Witco Corp.; Grindsted UK; Lonzagroup.

3202   Propylene glycol ricinoleate
26402-31-3   G

$C_{21}H_{40}O_4$
12-Hydroxy-9-octadecenoic acid, monoester with 1,2propanediol.
Cithrol PGMR N/E; EINECS 247-669-7; Flexricin® 9; 12-Hydroxy-9-octadecenoic acid, monoester with 1,2-propanediol; (R)-12-Hydroxyoleic acid, monoester with propane-1,2-diol; Naturechem® PGR; 9-Octadecenoic acid, 12-hydroxy-, monoester with 1,2-propanediol; Propylene glycol monoricinoleate; Propylene glycol ricinoleate. Wetting agent, dye solv., wax plasticizer, stabilizer for textile, household, and cosmetic applications, rewetting dried skins. Liquid; mp < -16°; d = 0.960. CasChem.

3203   Propylene oxide
75-56-9   7948   G H

C_3H_6O
1,2-Propylene oxide.
5-17-01-00017 (Beilstein Handbook Reference); AD 6; AI3-07541; BRN 0079763; Caswell No. 713A; CCRIS 540; EINECS 200-879-2; EPA Pesticide Chemical Code 042501; Epoxypropane; 1,2-Epoxypropane; 2,3-Epoxypropane; Ethylene oxide, methyl-; HSDB 173; Methyl Ethylene Oxide; Methyl oxirane; Methyloxacyclopropane; Methyloxirane; NCI-C50099; Oxirane, methyl-; Oxyde de propylene; Propane, epoxy-; Propane, 1,2-epoxy-; Propene oxide; Propylene epoxide; Propylene oxide; 1,2-Propylene oxide; UN1280. Intermediate; polyols for urethane foams, propylene glycols, surfactants, detergents, isopropanolamines, fumigant, synthetic lubricants, synthetic elastomers, solvent. Also used as an antifreeze, food emulsifier and preservative. Colorless liquid; mp = -112°; bp = 34°; soluble in H_2O (40 g/100 ml at 20°), EtOH, Et2O, LD50 (rat orl) = 1140 mg/kg. Aberco Inc.; Arco; Ashland; Degussa-Hüls Corp.; Texaco.

3204   Propylene phenoxetol
770-35-4   G H

$C_9H_{12}O_2$
Phenoxyisopropanol.
AI3-14682; EINECS 212-222-7; NSC 24015; Phenoxyisopropanol; 1-Phenoxy-2-propanol;   1-Phenoxypropan-2-ol;   2-Phenoxy-1-methylethanol; 2-Propanol, 1-phenoxy-; Propylene phenoxetol. Liquid; bp = 233°, d20 = 134°; d^{20} = 1.0622.

3205   Propylene tetramer
6842-15-5   H

$C_{12}H_{16}$
1-Propene, tetramer.
Propene, tetramer; Propylene tetramer; Tetrapropylene; UN2850.

3206   Propyl gallate
121-79-9   7951   G

$C_{10}H_{12}O_5$
3,4,5-Trihydroxybenzoic acid propyl ester.
AI3-17136; Benzoic acid, 3,4,5-trihydroxy-, propyl ester; CCRIS 541; EINECS 204-498-2; FEMA No. 2947; Gallic acid, propyl ester; HSDB 591; n-Propyl 3,4,5-trihydroxybenzoate; n-Propyl ester of 3,4,5-trihydroxybenzoic acid; n-Propyl gallate; NCI-C505888; NIPA 49; Nipagallin P; NSC 2626; Progallin P; Propyl 3,4,5-trihydroxybenzoate; Propyl gallate; Propylester kyseliny gallove; Sustane® PG; Tenox PG; 3,4,5-Tri-hydroxybenzene-1-propylcarboxylate; 3,4,5-Trihydroxy-benzoic acid, propyl ester. Food and feed antioxidant, flavor and packaging material. Used as a preservative and antioxidant for fats and oils. Solid; mp = 130°; λ_m = 275 nm (EtOH); slightly soluble in H_2O (0.35 g/100 ml), more soluble in organic solvents; LD50 (rat orl) = 2.1-7.0 g/kg. Aceto; Eastman Chem. Co.; Nipa; UOP.

3207   Propyl p-hydroxybenzoate, sodium salt
35285-69-9   G

$C_{10}H_{11}NaO_3$
Sodium propylparaben.
Benzoic acid, 4-hydroxy-, propyl ester, sodium salt; Caswell No. 714A; EINECS 252-488-1; EPA Pesticide Chemical Code 061204; 4-Hydroxybenzoic acid, propyl ester, sodium salt; Nipasol M Sodium; Parasept; Propyl p-hydroxybenzoate, sodium salt; Propylparaben sodium; Propylparaben, sodium salt; Sodium 4-propoxy-carbonylphenoxide; Sodium propylparaben. Preservative, bactericide, fungicide for pharmaceuticals, cosmetics, foods, medicinal preparations, industrial applications. Laboratories; Ltd; Nipa.

3208   Propyliodone
587-61-1   7955   G

527

C10H11I2NO3

N-Propyl 3,5-di-iodo-4-pyridone-N-acetate.

5-21-07-00164 (Beilstein Handbook Reference); A.G. 33-107; BRN 0206829; Bronchodiagnostin; Bronkho-diagnostin; Brosombra; 3,5-Diiodo-4-oxo-1(4H)-pyridine-acetic acid propyl ester; 3,5-Diiodo-4-pyridone-N-acetic acid propyl ester; Dionosil; Diostril; EINECS 209-603-5; NSC 97103; Propiliodona; Propiodone; Propyl 3,5-diiodo-4-oxo-1(4H)-pyridineacetate; Propyl 3,5-diiodo-4-pyridone-N-acetate; Propyliodone; Propyliodonum; 1(4H)-Pyridineacetic acid, 3,5-diiodo-4-oxo-, propyl ester. Radiopaque medium. Used in medicine as a diagnostic aid. Crystals; 186° (dec); λ_m = 217, 238, 281 nm (ε = 18200, 16300, 12500, MeOH); soluble in H_2O (14 mg/100 ml), more soluble in organic solvents; LD_{50} (mus iv) = 300 mg/kg.

3209 Propylparaben
94-13-3   8051   E G P

C10H12O3

Propyl p-hydroxybenzoate.

4-10-00-00374 (Beilstein Handbook Reference); AI3-01341; Aseptoform P; Benzoic acid, 4-hydroxy-, propyl ester; Benzoic acid, p-hydroxy-, propyl ester; Betacide P; Betacine P; Bonomold Op; BRN 1103245; Caswell No. 714; Chemacide pk; Chemocide pk; Chemoside PK; EINECS 202-307-7; EPA Pesticide Chemical Code 061203; FEMA Number 2951; HSDB 203; 4-Hydroxybenzoic acid, propyl ester; p-Hydroxybenzoic acid propyl ester; p-Hydroxybenzoic propyl ester; p-Hydroxypropyl benzoate; Nipagin P; Nipasol; Nipasol M; Nipasol P; Nipazol; NSC 23515; p-Oxybenz-oesaurepropylester; Paraben; Parabens; Parasept; Pasep-tol; Preserval P; Propagin; Propyl 4-hydroxybenzoate; Propyl aseptoform; Propyl butex; Propyl Chemosept; Propyl p-hydroxybenzoate; n-Propyl p-hydroxybenzoate; Propyl paraben; Propyl parahydroxybenzoate; Propyl Parasept; Propylester kyseliny p-hydroxybenzoove; Propylparaben; Propylparasept; Protaben P; Pulvis conservans; Solbrol P; Tegosept P. Preservative, bactericide, fungicide for pharmaceuticals, cosmetics, foods, medicinal preparations, industrial applications. Used for mold control in sausage casings and as a pharmaceutical aid. Crystals; mp = 97°; bp = 133°; d^{102} = 1.0630; λ_m = 256 nm (ε = 17100, MeOH); slightly soluble in H_2O (0.05 g/100 ml), more soluble in EtOH, Et2O. *Allchem Ind.; Greeff R.W. & Co.; Nipa; Penta Mfg.*

3210 n-Propyltrichlorosilane
141-57-1
                                            G

C3H7Cl3Si

Trichloropropylsilane.

AI3-60044; CP0800; EINECS 205-489-6; HSDB 889; NSC 93878; Propyl trichlorosilane; n-Propyltrichlorosilane; Silane, propyltrichloro-; Silane, trichloropropyl-; Tri-chloro-n-propylsilane; Trichloro(propyl)silane; Trichloro-propylsilane; UN1816. Coupling agent, chemical intermediate, blocking agent, release agent, lubricant, primer, reducing agent. Intermediate in preparation of silicones. Liquid; bp = 123.5°; d^{20} = 1.195. *Degussa-Hüls Corp.; PCR.*

3211 Propyne
74-99-7   H

C3H4

1-Propyne.

4-01-00-00958 (Beilstein Handbook Reference); Acetylene, methyl-; Allylene; BRN 0878138; CCRIS 6830; EINECS 200-828-4; HSDB 2508; Methyl acetylene; Methylacetylene; Propine; Propyne. Used in chemical manufacturing. Colorless gas: mp = -102-7°; bp = -23.2°; d^{25} = 0.607; poorly soluble in H_2O, soluble in organic solvents.

3212 Propyzamide
23950-58-5   7964   G P

C12H11Cl2NO

3,5-Dichloro-N-(1,1-dimethyl-2-propynyl)benzamide.

Benzamide, 3,5-dichloro-N-(1,1-dimethyl-2-propynyl)-; BRN 0882391; Caswell No. 306A; Clanex; 3,5-Dichloro-N-(1,1-dimethyl-2-propynyl)benzamide; 3,5-Dichloro-N-(1,1-dimethylprop-2-ynyl)benzamide; 3,5-Dichloro-N-(1,1-dimethylpropynyl)benzamide; N-(1,1-Dimethylpropynyl)-3,5-dichlorobenzamide; EINECS 245-951-4; EPA Pesticide Chemical Code 101701; HSDB 5118; Kerb Propyzamide 50; KERB; KERB 50W; Pronamid; Pronamide; Propyzamide; Rapier; RCRA waste number U192; RH 315. A residual herbicide in a wettable powder form for a wide range of agricultural crops such as oil seed rape. Registered by EPA as a herbicide. Colorless powder; mp = 155-156°; soluble in H_2O (0.0015 g/100 ml), MeOH (15 g/100 ml), i-PrOH (15 g/100 ml), cyclohexanonoe (20 g/100 ml), MEK (20 g/100 ml), DMSO (33 g/100 ml), moderately soluble in petroleum ether; LD_{50} (mrat orl) = 8350 mg/kg, (frat orl) = 5620 mg/kg, (dog orl) > 10000 mg/kg, (rbt der) > 3160 mg/kg, (dck orl) > 14000 mg/kg; LC_{50} (rat ihl) = 5.0 mg/l air, (goldfish 96 hr.) = 350 mg/l, (guppy 96 hr.) = 150 mg/l, (rainbow trout 96 hr.) = 72 mg/l; non-toxic to bees. *Dow AgroSciences; Earth Care; Farmers Crop Chemicals Ltd.; Kommer-Brookwick Ltd.; MTM AgroChemicals Ltd.; Pan Britannica Industries Ltd.; Rohm & Haas Co.; Rohm & Haas UK.*

3213 Protect
81-84-5   H

C12H6O3

1,8-Naphthalenedicarboxylic acid anhydride.

5-17-11-00492 (Beilstein Handbook Reference); AI3-09071; BRN 0153190; EINECS 201-380-2; Naphthalene-1,8-dicarboxylic anhydride; Naphthalic anhydride; NSC 5747; Pakarli; Protect; Protect (agrochemical). Herbicide safener. *Anshan HIFI Chemical Co., Ltd.; Mil-Spec Industries Inc.*

3214 Prothiofos
34643-46-4
                                            G P

C11H15Cl2O2PS2

O-(2,4-Dichlorophenyl) O-ethyl S-propyl phosphorodi-thioate.

AI3-29305; BAY-NTN 8629; Bideron; BRN 1998314; Caswell No. 714H; Dichlorpropaphos; O-(2,4-Dichlorophenyl) O-ethyl S-propyl dithiophosphate; EPA Pesticide Chemical Code 128858; O-Ethyl-O-(2,4-dichlorophenyl)-S-n-propyl-dithiophosphate; NTN-8629; NTN 8629,4541; Phosphorodithioic acid, O-(2,4-dichlorophenyl) O-ethyl S-propyl ester; Prothiofos; Prothiophos; Tokuthion; Toyodan; Toyothion. Non-systemic insecticide with contact and stomach action; cholinesterase inhibitor, especially effective against leaf-eating caterpillars. Colorless liquid; $bp_{0.1}$ = 125 - 128°; sg^{20}= 1.3; slightly soluble in H_2O (0.00017 g/100 ml at 20°), freely soluble in cyclohexanone, i-PrOH, C7H8, CH2Cl2, ligroin (all > 90 g/100 ml); LD50 (mrat orl) ≅1500 mg/kg, (mrat der) > 5000 mg/kg; LC50 (rat ihl 4 hr.) > 2.7 mg/l air, (golden orfe 96 hr.) = 4 - 8 mg/l, (goldfish 96 hr.) = 6 - 20 mg/l, (rainbow trout 96 hr.) 0.5 - 1.0 mg/l; non-toxic to bees. *Bayer Corp., Agriculture; Bayer Corp.; Mobay.*

3215　Protocatechuic acid
99-50-3　　　　　　　7986　　　　　　　G

C7H6O4

3,4-Dihydroxybenzoic acid.

4-10-00-01459 (Beilstein Handbook Reference); Benzoic acid, 3,4-dihydroxy-; BRN 1448841; 4-Carboxy-1,2-dihydroxybenzene; CCRIS 6291; 3,4-Dihydroxybenzoic acid; 4,5-Dihydroxybenzoic acid; EINECS 202-760-0; NSC 16631; Protocatechuic acid. Used in chemical synthesis. mp = 200-202°; d^4 = 1.524; λ_m = 270, 290 nm (ε = 2754, 3890, H_2O); insoluble in C6H6, slightly soluble in H_2O (2 g/100 ml), soluble in Et2O, very soluble in EtOH. *Dinoval.*

3216　Prussic acid
74-90-8　　　　　　　4816　　　　　　　G

$$H-C\equiv N$$

CHN

Hydrocyanic acid.

4-02-00-00050 (Beilstein Handbook Reference); Acide cyanhydrique; Acido cianidrico; Aero Liquid HCN; AI3-31100-X; Blausaeure; Blauwzuur; BRN 1718793; Carbon hydride nitride (CHN); Caswell No. 483; Cyaanwaterstof; Cyanwasserstoff; Cyclon; Cyclone B; Cyjanowodor; EINECS 200-821-6; EPA Pesticide Chemical Code 045801; Evercyn; Formic anammonide; Formonitrile; HCN; HSDB 165; Hydrocyanic acid; Hydrogen cyanide; NA1613; Prussic acid; RCRA waste number P063; UN 1051; UN 1613; UN 1614; UN 3294; Zaclondiscoids; Zootic acid; Zyklon B. Present in apricot and peach pits in low concentrations, extremely toxic, used to exterminate rodents and insects. Used in the manufacture of acrylonitrile, acrylates, adiponitrile, chelates, dyes, and pesticides. Colorless or pale blue liquid or gas with a bitter almond odor detectable at 1 to 5 ppm; mp = -13°; bp = 26°; soluble in H_2O, LC50 (rat ihl 5 min) = 544 ppm. *Am. Cyanamid; BP Chem.; Ciba Geigy Agrochemicals; Ciba-Geigy Corp.; Degussa AG; Dow Chem. U.S.A.;*

DuPont; Monsanto Co.; Rohm & Haas Co.

3217　Pseudocumene
95-63-6　　　　　　　8006　　　　　　　H

C9H12

1,2,4-Trimethylbenzene.

AI3-03976; Asymmetrical trimethylbenzene; Benzene, 1,2,4-trimethyl-; Benzene, 1,2,5-trimethyl-; EINECS 202-436-9; HSDB 5293; NSC 65600; Pseudocumene; Pseudocumol; as-Trimethylbenzene; Uns-trimethylbenzene. Used as a solvent in chemical manufacturing and in scintillation counters. Oil; mp = -43.8°; bp = 169.3°; d^{20} = 0.8578; λ_m = 267, 277 nm (MeOH); insoluble in H_2O, freely soluble in EtOH, Et2O, Me2CO, C6H6, CCl4, petroleum ether.

3218　Pulegone
89-82-7　　　　　　　8028　　　　　　　G

C10H16O

1-Methyl-4-isopropylidene-3-cyclohexanone.

AI3-11218; CCRIS 5746; Cyclohexanone, 5-methyl-2-(1-methylethylidene)-, (5R)-; EINECS 201-943-2; FEMA No. 2963; 1-Isopropylidene-4-methyl-2-cyclohexanone; 4(8)-p-Menthen-3-one, delta-; (1R)-(+)-p-Menth-4(8)-en-3-one; p-Menth-4(8)-en-3-one; p-Menth-4(8)-en-3-one, (R)-(+)-; 1-Methyl-4-isopropylidene-3-cyclohexanone; 3-Methyl-6-isopropylidenecyclohexanone; 5-Methyl-2-(1-methyl-ethylidene)cyclohexanone, (R)-; NSC 15334; Pulegon; Pulegone; d-Pulegone; (+)-(R)-Pulegone; (+)-Pulegone; (R)-(+)-Pulegone; (R)-Pulegone. Found in natural oils such as pennyroyal oil. Oil; bp = 224°, bp12 = 97°; $[\alpha]_D^{20}$= 23.4° (neat); d^{45} = 0.9346; λ_m = 253 nm (ε = 6340, MeOH); insoluble in H_2O, soluble in organic solvents.

3219　Purpurin
81-54-9　　　　　　　8036　　　　　　　G

C14H8O5

1,2,4-Trihydroxy-9,10-anthraquinone.

4-08-00-03568 (Beilstein Handbook Reference); 9,10-Anthracenedione, 1,2,4-trihydroxy-; Anthraquinone, 1,2,4-trihydroxy-; BRN 1887127; C.I. 1037; C.I. 58205; C.I. 75410; CCRIS 3527; EINECS 201-359-8; Hydroxylizaric acid; NSC 10447; Purpurin; Purpurine; Smoke Brown G; 1,2,4-Trihydroxy-9,10-anthraquinone; 1,2,4-Trihydroxyanthrachinon; 1,2,4-Trihydroxyanthra-quinone; Verantin. An anthraquinone-based dyestuff. Orange-red crystals; mp = 259°; insoluble in H_2O, soluble in organic solvents.

3220　Putrescine
110-60-1　　　　　　　8038　　　　　　　G

$C_4H_{12}N_2$

1,4-Butanediamine.

4-04-00-01283 (Beilstein Handbook Reference); AI3-25444; BRN 0605282; 1,4-Butanediamine; Butylene-diamine; 1,4-Butylenediamine; CCRIS 6751; 1,4-Diaminobutane; EINECS 203-782-3; NSC 60545; Putrescin; Putrescine; Tetramethyldiamine; Tetramethyl-enediamine; 1,4-Tetramethylenediamine. Ubiquitous biological chemical formed by decarboxylation in tissue of ornithine or arginine. Used in biochemical and biological research. Solid; mp = 27.5°; bp = 158.5°; d^{25} = 0.8770; soluble in H_2O (40 g/l), organic solvents.

3221   PVP

9003-39-8   7783   E G

$C_6H_9NO_x$

Polyvinylpyrrolidone.

143 RP; Agent AT 717; Agrimer; Albigen A; Aldacol Q; Antaron P 804; AT 717; Bolinan; Caswell No. 681; CCRIS 3611; Crospovidone; Crospovidonum; EPA Pesticide Chemical Code 079033; 1-Ethenyl-2-pyrrolidinone homopolymer; Ganex P 804; Hemodesis; Hemodez; HSDB 205; α-Hydro-ω-(p-iodobenzyl)poly(1-(2-oxo-1-pyrrolidinyl)ethylene)-^{131}I; K 15; K 25 (polymer); K 30 (polymer); K 60 (polymer); K 90; K115; K 115 (polyamide); Kollidin CLM; Kollidon; Kollidon 17; Kollidon 25; Kollidon 30; Kollidon CL; Luviskol; Luviskol K30; Luviskol K90; MPK 90; NCI-C60582; Neocompensan; NSC 114022; NSC 142693; Peragal ST; Peregal ST; Periston; Periston-N; Peviston; Plasdone; Plasdone 4; Plasdone K-26/28; Plasdone K 29-32; Plasdone No. 4; Plasdone XL; Plasmosan; Polividona; Polividone; Poly-N-vinyl pyrrolidone; Poly(1-(2-oxo-1-pyrrolidinyl)ethylene); Poly(1-(2-oxo-1-pyrrolidinyl)-1,2-ethanediyl), α-hydro-ω-((4-(iodo-^{131}I)phenyl)methyl)-; Poly(1-ethenyl-2-pyrrolidinone); Poly(N-vinylbutyro-lac-tam); Poly(vinylpyrrolidinone); Poly(1-vinylpyrrolidin-one); Poly(N-vinylpyrrolidinone); Poly(N-vinyl-2-pyrro-lidinone); Poly(1-vinyl-2-pyrrolidinone); Poly(1-vinyl-2-pyrrolidinone) Hueper's Polymer No.1; Poly(1-vinyl-2-pyrrolidinone) Hueper's Polymer No.2; Poly(1-vinyl-2-pyrrolidinone) Hueper's Polymer No.3; Poly(1-vinyl-2-pyrrolidinone) Hueper's Polymer No.4; Poly(1-vinyl-2-pyrrolidinone) Hueper's Polymer No.5; Poly(1-vinyl-2-pyrrolidinone) Hueper's Polymer No.6; Poly(1-vinyl-2-pyrrolidinone) Hueper's Polymer No.7; Poly(N-vinylpyrrolidone); Poly(N-vinyl-2-pyrrolid-one); Polyclar AT; Polyclar H; Polyclar L; Polygyl; Poly-plasdone; Polyplasdone XL; Polyvidone; Poly-vidone; Polyvinylpyrrolidone; Povidone; Protagent; PVP; PVP 1; PVP 2; PVP 3; PVP 4; PVP 5; PVP 6; PVP 7; PVP 40; PVP K 3; PVP-K 15; PVP-K 30; PVP-K 60; PVP-K 90; PVPP; 2-Pyrrolidinone, 1-ethenyl-, homopolymer; 2-Pyrrolidinone, 1-vinyl-, polymers; Refresh; Sauflon; Soothe; Subtosan; Tears Plus; Tolpovidone I-131; Toxobin; Vinisil; Vinylpyrrolidone polymer; N-Vinylbutyrolactam polymer; Vinylpyrrolidinone polymer; 1-Vinyl-2-pyrrolidinone homopolymer; 1-Vinyl-2-pyrrolidinone polymer; N-Vinylpyrrolidinone polymer; N-Vinylpyrrolidone polymer; 1-Vinyl-2-pyrrolidone poly-mer; N-Vinyl-2-pyrrolidone polymer. Polymer of 1-vinyl-2-pyrrolidone monomers; a film-forming agent, hair fixative, thickener, protective colloid, suspending agent, and dispersant for cosmetics industry, technical applications, drug vehicle and retardant; tablet binder, pharmaceutical excipient; used in adhesives and detergents. FDA listed (must be removed by filtration), BATF listed (limitation 6 lb/1000 gals. wine). Used as a binder and disintegrant in pharmaceutical tablets and as a suspension aid.

White hygroscopic powder; insoluble in H_2O. *Abbott Labs Inc.; Allchem Ind.; BASF Corp.; ISP.*

3222   Pyrazophos

13457-18-6   8052   G P

$C_{14}H_{20}N_3O_5PS$

Ethyl 2-[(diethoxyphosphinothioyl)oxy]-5-methylpyr-azolo[1,5-a]pyrimidine-6-carboxylate.

Afugan; BRN 0577209; Caswell No. 714D; Curamil; 2-(O,O-Diethyl-thionophosphoryl)-5-methyl-6-carbethoxy-pyrazolo-(1,5a)pyrimidine; EINECS 236-656-1; EPA Pesticide Chemical Code 447500; Ethyl 2-diethoxythiophosphoryloxy-5-methylpyrazolo(1,5-a)pyr-imidine-6-carboxylate; Hoechst 2873; Missile; NSC 232671; O-6-Ethoxycarbonyl-5-methylpyrazolo(1,5-a)-pyrimidin-2-yl O,O-diethyl phosphorothioate; O,O-Diethyl-O-(5-methyl-6-ethoxy-carbonyl-pyrazolo(1.5-a)pyrimid-2-yl)-thionophosphate; Phosphorothioic acid, O,O-diethyl ester, O-ester with (6-ethoxycarbonyl-5-methyl)pyrazolo(1,5-a)pyrimidin-2-ol; Pyrazolo-(1,5a)pyr-imidine, 2-(O,O-diethyl-thionophosphoryl)-5-methyl-6-ethoxycarbonyl-; Pyrazophos; Pyrazolo(1,5-a)pyrimidine-6-carboxylic acid, 2-((diethoxyphosphinothioyl)oxy)-5-methyl-, ethyl ester; Pyrazolo(1,5-a)pyrimidine-6-carboxylic acid, 2-hydroxy-5-methyl-, ethyl ester, O-ester with O,O-diethyl phosphorothioate. Systemic organophosphorus fungicide. Crystals; mp - 50 - 51°; dec. on distillation; sg^{25} = 1.348; slightly soluble in H_2O (0.00042 g/100 ml) at 20°), soluble in Me_2CO (121 g/100 ml), C_7H_8 (> 98 g/100 ml), EtOAc (90 g/100 ml), EtOH (9.5 g/100 ml), C_6H_{14} (1.1 g/100 ml), soluble in most organic solvents, such as xylene, C_6H_6, CCl_4, CH_2Cl_2, trichlorethylene; LD_{50} (rat orl) = 140 - 632 mg/kg, (rat der) > 2000 mg/kg, (quail orl) = 118 - 480 mg/kg; LC_{50} (carp 96 hr.) = 6.1 mg/l, (rainbow trout 96 hr.) = 0.48 mg/l; non-toxic to bees up to 1 g/l (in a spray). *Aventis Crop Science; Hoechst AG; Hoechst UK Ltd.*

3223   Pyrethrin I

121-21-1   8054   H P

$C_{21}H_{28}O_3$

(1R-(1α(S*(Z)),3β))- 2,2-Dimethyl-3-(2-methyl-1-prop-enyl)-2-methyl-4-oxo-3-(2,4-pentadienyl)-2-cyclopenten-1-ylcyclopropanecarboxylate.

3-09-00-00215 (Beilstein Handbook Reference); BRN 2004306; Caswell No. 715; Chrysanthemum mono-carboxylic acid pyrethrolone ester; EINECS 204-455-8; EPA Pesticide Chemical Code 069001; HSDB 6302; Piretrina 1; Pyrethrin I; Pyrethrine I; Pyrethrins; Pyrethrolone, chrysanthemum monocarboxylic acid ester; Pyrethronyl (+)-trans-chrysanthemate; Pyrethrum; RCRA waste number P008. Registered by EPA as an insecticide. Viscous liquid; bp0.1 = 170°; d^{18} = 1.5192; $[\alpha]_D^{20}$= -14° (isooctane); λ_m = 225 nm (ε 36400 95% EtOH); insoluble in H_2O, soluble in EtOH, petroleum ether, kerosene, CCl_4, ethylene dichloride, nitromethane; LD_{50} (rat orl) = 584 - 900 mg/kg, (mus orl) = 273 - 796 mg/kg, (rat der) > 1500 mg/kg, (rbt der) = 5000 mg/kg; highly toxic to fish; toxic to bees, with repellant effect.

3224 Pyridaben
96489-71-3   8057   P

C19H25ClN2OS
4-Chloro-2-(1,1-dimethylethyl)-5-(((4-(1,1-dimethylethyl)phenyl)methyl)thio)-3(2H)-pyridazinone.
HSDB 7052; NC 129; NCI 129; Nester; Pyridaben; Sanmite. Registered by EPA as an miticide and insecticide. White crystals; mp = 111-112°; d_4^{20} = 1.2; soluble in Me2CO (46 g/100 ml), corn oil (4.2 g/100 ml), EtOH (5.7 g/100 ml), methyl cellosolve (111 g/100 ml), xylene (39 g/100 ml), C6H6 (11 g/100 ml), cyclohexane (32 g/100 ml), C6H14 (1.0 g/100 ml), n-octanol (6.3 g/100 ml), H2O (0.0012 mg/100 ml); LD50 (mrat orl) = 435 mg/kg, (frat orl) = 358 mg/kg, (Bobwhite quail orl) > 2250 mg/kg, (Mallard duck orl) > 2500 mg/kg, (mrbt der) > 2000 mg/kg, (frbt der) > 2000 mg/kg. *BASF Corp.; Nissan Chem. Ind.*

3225 Pyridine
110-86-1   8060   G H

C5H5N
Pyridine.
AI3-01240; Azabenzene; Azine; Caswell No. 717; CCRIS 2926; CP 32; EINECS 203-809-9; EPA Pesticide Chemical Code 069202; FEMA Number 2966; HSDB 118; NCI-C55301; NSC 406123; Piridina; Pirydyna; Pyridin; Pyridine; RCRA waste number U196; UN1282. Used as a solvent and in synthesis of agrochemicals, pharm-aceuticals, photographic materials, coatings, curing agents, rubber chemicals, plastics, antidandruff shampoos, textiles, dyestuffs. Mobile liquid; mp = -41.6°; bp = 115.2°; d^{20} = 0.9819; λ_m =250, 256, 261 nm (MeOH); very soluble in H2O, EtOH, Et2O, Me2CO, C6H6, CHCl3; LD50 (rat orl) = 1.58 g/kg. *Lancaster Synthesis Co.; Nepera; Penta Mfg.; Whitecourt.*

3226 2-Pyridinecarbonitrile, 4-amino-3,5,6-tri-chloro
14143-60-3   H

C6H2Cl3N3
Picolinonitrile, 4-amino-3,5,6-trichloro-.
4-Amino-3,5,6-trichloropyridine-2-carbonitrile; EINECS 237-992-1; Picolinonitrile, 4-amino-3,5,6-trichloro-; 2-Pyridinecarbonitrile, 4-amino-3,5,6-trichloro.

3227 Pyridine, 3,6-dichloro-2-(trichloromethyl)-
1817-13-6   H

C6H2Cl5N
2,5-Dichloro-6-(trichloromethyl)pyridine.
2,5-Dichloro-6-(trichloromethyl)pyridine; Pyridine, 3,6-dichloro-2-(trichloromethyl)-.

3228 Pyridine hydrochloride
628-13-7   H

C5H6ClN
Pyridine hydrochloride.
AI3-30571; EINECS 211-027-4; Pyridine hydrochloride; Pyridinium chloride; Pyridinium monochloride. Platelets; mp = 146°; bp = 222°; λ_m = 266 nm (ε = 5012, EtOH), 255 nm (ε = 5248, EtOH-HCl); very soluble in H2O, EtOH, CHCl3.

3229 Pyrimidine
289-95-2   8076   G

C4H4N2
Metadiazine.
1,3-Diazabenzene; 1,3-Diazine; m-Diazine; EINECS 206-026-0; Metadiazine; Miazine; NSC 89305; Pyrimidine. Used in biochemical research. Crystals; mp = 22°; bp = 123.8°; d = 1.0160; λ_m = 238, 243, 280 nm (ε = 2220, 2360, 298, MeOH); soluble in EtOH, very soluble in H2O.

3230 6-Pyrimidinol, 2-isopropyl-4-methyl
2814-20-2   H

C8H12N2O
2-Isopropyl-4-methyl-6-hydroxypyrimidine.
2-Isopropyl-4-methyl-6-hydroxypyrimidine; 2-Isopropyl-6-methyl-1H-pyrimidin-4-one; 4(3H)-Pyrimidinone, 2-isopropyl-6-methyl-; 6-Pyrimidinol, 2-isopropyl-4-methyl; EINECS 220-561-7; G 27550; HSDB 5899; IMHP.

3231 Pyrimithate
5221-49-8   8078   G P

C11H20N3O3PS

Phosphorothioic acid O-[2-(dimethylamino)-6-methyl-4-pyrimidinyl] O,O-diethyl ester.
ICI-29661; Diothyl; pyrimitate. Used as an acaricide and an insecticide. Liquid; $bp_{0.04}$ = 128-132°; sg = 1.165; insoluble in H_2O, soluble in organic solvents. *ICI Agrochemicals.*

3232   Pyrithione Zinc
13463-41-7   8084   D G

C10H8N2O2S2Zn

Bis[1-hydroxy-2(1H)-pyridinethionato]zinc.
AI3-62421; BC-J; Biocut ZP; Breck One Dandruff Shampoo; Caswell No. 923; CCRIS 4894; Danex; Desquaman; EINECS 236-671-3; EPA Pesticide Chemical Code 088002; Evafine P 50; Finecide ZPT; FSB 8332; Head and Shoulders; Hokucide ZPT; HSDB 4498; Niccanon SKT; NSC 290409; OM-1563; Omadine Zinc; Piritionato cincico; Pyrithione zinc; Pyrithione zincique; Pyrithionum zincicum; Sebulon Shampoo; Tomicide Z 50; Tomicide ZPT 50; Top Brass; Vancide P; Vancide ZP; Wella Crisan; Zinc Omadine; Zinc PT; Zinc pyrethion; Zinc pyridine-2-thiol-1-oxide; Zinc pyridinethione; Zinc - pyrion; Zinc pyrithione; Zinci pyrithionum; Zincon Dandruff Shampoo; Zincpolyanemine; Zn - pyrion; ZNP Bar; ZnPT; ZPT.; component of: Head and Shoulders. Antibacterial; antifungal; antiseborrheic. Ingredient in Head and Shoulders. *Allergan Herbert; Herbert; Lederle Labs.; Olin Res. Ctr.; Stiefel Labs Inc.; Westwood-Squibb Pharm. Inc.*

3233   Pyrogallol
87-66-1   8090   D G

C6H6O3

1,2,3-Benzenetriol.
215; 4-06-00-07327 (Beilstein Handbook Reference); AI3-00709; BRN 0907431; C.I. 76515; C.I. Oxidation Base 32; CCRIS 1940; CI 76515; CI Oxidation Base 32; EINECS 201-762-9; Fouramine Brown AP; Fourrine 85; Fourrine PG; HSDB 794; NSC 5035; Pyro; Pyrogallic acid; Pyrogallol. Protective colloid in preparation of metallic colloidal solutions, used in photography, in preparation of dyes, intermediates, in medicine, as a reducing agent and antioxidant. Absorbs oxygen, used in gas analysis. Chemical intermediate. Used as a photographic developer, dyeing wool, staining leather, in metallurgy and analytical chemistry. Used in manufacture of dyes, intermediate, synthetic drugs, laboratory reagent, reducing agent, antioxidant in lubricating oils. A catechol-O-methyltransferase inhibitor with antipsoriatic activity. Crystals; mp = 133°; bp = 309°; d^4 = 1.453; λ_m = 267 nm (EtOH); soluble in H_2O (58.8 g/100 ml), EtOH (76.9 g/100 ml), Et2O (62.5 g/100 ml); slightly soluble in C6H6, CHCl3, CS2; LD50 (rbt orl) = 1600 mg/kg. *Burlington*

Biomedical; Fuji Chem. Ind.; Hoechst Celanese; Mallinckrodt Inc.; Schering AG.

3234   Pyrogallol dimethylether
91-10-1   G

C8H10O3

2,6-Dimethoxy phenol.
Aldrich; 1,3-Di-o-methylpyrogallol; 1,3-Dimethoxy-2-hydroxybenzene; 2,6-Dimethoxyphenol; 1,3-Dimethyl pyrogallate; 2,6-Dwumetoksyfenol; EINECS 202-041-1; FEMA No. 3137; 2-Hydroxy-1,3-dimethoxybenzene; Phenol, 2,6-dimethoxy-; Pyrogallol 1,3-dimethyl ether; Syringol. Crystals; mp = 56.5°; bp = 261°; λ_m = 268 nm (ε 1096, MeOH), 270 nm (ε = 1072, EtOH); soluble in H_2O (2 g/100 ml), more soluble in organic solvents; LD50 (rat orl) = 550 mg/kg. *Janssen Chimica; Lancaster Synthesis Co.; Penta Mfg.; Spectrum Chem. Manufacturing.*

3235   Pyromellitic dianhydride
89-32-7   H

C10H2O6

1,2,4,5 Benzenetetracarboxylic 1,2:4,5 dianhydride.
EINECS 201-898-9; NSC 4798; Pyromellitic 1,2:4,5-dianhydride; Pyromellitic acid anhydride; Pyromellitic acid dianhydride; Pyromellitic anhydride; Pyromellitic dianhydride.

3236   Pyronine B
2150-48-3   8096   G

C21H27ClN2O

3,6-Bis(diethylamino)xanthylium chloride.
AI3-52461; Ammonium, (6-(diethylamino)-3H-xanthen-3-ylidene)diethyl-, chloride; Ammonium, diethyl(6-(diethyl-amino)-3H-xanthen-3-ylidene)-, chloride; 3,6-Bis(diethyl-amino)xanthylium chloride; C.I. 45010; (6-(Diethyl-amino)-3H-xanthen-3-ylidine)diethylammonium chloride; N-(6-(Diethylamino)-3H-xanthen-3-ylidine)-N-ethyl-ethanaminium chloride; E tetraethylpyronin; EINECS 218-429-9; Ethanaminium, N-(6-(diethylamino)-3H-xanthen-3-ylidene)-N-ethyl-, chloride (9CI); NSC 44690; Pyronin B; Pyronine B; Xanthylium, 3,6-bis(diethylamino)-, chloride. With ferric chloride, forms a green complex which is used as a stain for bacteria, molds and RNA.

3237   Pyronine Y
92-32-0   8097   G

C17H19ClN2O

Methanaminium, N-(6-(dimethylamino)-3H-xanthen-3-ylidene)-N-methyl-, chloride.

AI3-52862; Ammonium, (6-(dimethylamino)-3H-xanten-3-ylidene)dimethyl-, chloride; 3,6-Bis(dimethylamino)-xanthylium chloride; C.I. 45005; EINECS 202-147-8; Methyl pyronin; NSC 10454; Pyronin G; Pyronin Y; Pyronin Yellow; Pyronine; Pyronine G; Pyronine Y; Pyronine ZH; Schultz No. 853; Tetramethyl pyronin; Xanthylium, 3,6-bis(dimethylamino)-, chloride. Used in the dyestuffs industry. With ferric chloride, gives a green complex which is used as a bacteriological and biological stain. Soluble in H2O (9 g/100 ml), EtOH; λ_m = 576 nm. *Allchem Ind.; Enrique Silvestrini y Cia; Oliver Y Batlle.*

3238 Pyroxylin
9004-70-0   8101   D E G

C12H16(ONO2)4O6

Cellulose nitrate.

BK2-W; BK2-Z; C 2018; CA 80-15; Celex; Celloidin; Cellulose, nitrate; Cellulose tetranitrate; CN 85; Collodion; Collodion cotton; Collodion, flexible; Collodion wool; Colloxylin; Colloxylin VNV; Corial EM finish F; Daicel RS 1; E 1440; Flexible collodion; FM-Nts; Fulmicoton; Guncotton; H 1/2; HSDB 1973; HX 3/5; Kodak LR 115; LR 115; Nitrocel S; Nitrocellulose; Nitrocotton; Nitron; Nitron (nitrocellulose); Nixon N/C; NP 11; NTs 218; NTs 222; NTs 539; NTs 542; NTs 62; Parlodion; Pirossilina; Piroxilina; Pyralin; Pyroxylin; Pyroxyline; Pyroxylinum; R.S.Nitrocellulose; RF 10; RS; RS 1/2; RS Nitrocellulose; Shadolac MT; Soluble gun cotton; Synpor; Tsapolak 964; UN0340; UN0341; UN0342; UN0343; UN2059; UN2555; UN2556; UN2557; Xyloidin. Cellulose derivative used in lacquers, high explosives, rocket propellant, printing ink base, leather finishing, and as a topical protectant. White filaments; d_{25}^{25} = 0.765 - 0.775; soluble in EtOH, Me2CO, AcOH. *Allchem Ind.; Aqualon; Asahi Chem. Industry; Hercules Inc.; SNPE North America.*

3239 Pyrrole
109-97-7   8104   G

C4H5N
1H-Pyrrole.
AI3-18817; 1-Aza-2,4-cyclopentadiene; Azole; CCRIS 2933; Divinyleneimine; Divinylenimine; EINECS 203-724-7; FEMA No. 3386; HSDB 119; Imidole; Monopyrrole; NSC 62777; Parzate; Pyrrol; Pyrrole; 1H-Pyrrole. Chemical intermediate. Used in the manufacture of pharmaceuticals. Oil; mp = -23.4°; bp = 129.7°; d^{20} = 0.9698; λ_m = 209 nm (ε = 6730, MeOH); slightly soluble in H2O (60 g/l), soluble in C6H6, EtOH, Et2O, Me2CO, CHCl3..

3240 Pyruvic acid
127-17-3   8110   G

C3H4O3
2-Oxopropionic acid.
4-03-00-01505 (Beilstein Handbook Reference); Acetylformic acid; AI3-11220; BRN 0506211; BTS; EINECS 204-824-3; FEMA No. 2970; α-Ketopropionic acid; 2-Ketopropionic acid; NSC 179; 2-Oxopropanoic acid; 2-Oxopropionic acid; Propanoic acid, 2-oxo-; Pyroracemic acid; Pyruvic acid. Chemical intermediate and used in biochemical research. Liquid; mp = 13.8°; bp = 165° (dec), bp_{10} = 54°; d^{20} = 1.2272; λ_m = 285m nm (ε = 8, H2O); soluble in Me2CO, freely soluble in H2O, EtOH, Et2O. *Lancaster Synthesis Co.; Penta Mfg.; U.S. BioChem.*

3241 Quartz
14808-60-7   8567   G

SiO2
Crystallized silicon dioxide.
Agate; Amethyst; CCRIS 2475; Chalcedony; Cherts; Crystallized silicon dioxide; D & D; DQ12; EINECS 238-878-4; Flint; Flintshot; Gold bond R; Imsil; Min-U-Sil; MIN-U-sil α quartz; Novaculite; Onyx; Quartz; α-Quartz; Quartz (SiO2); Quartz dust; Quartz silica; Quazo puro; Rock crystal; Rose quartz; Sand; SF 35; Sicron F 300; Siderite (SiO2); Sikron F 100; Sil-Co-Sil; Silica, Silica, crystalline, quartz; Silica dust; Silica flour (powdered crystalline silica); Silicates (<1% crystalline silica):Graphite, natural; Silicon oxide, di- (sand); Silver bond B; Snowit; TGL 16319; Tiger-eye; W 12; W 12 (Filler). Silicon dioxide; silica; crystallized silicon dioxide; there are three types. a) crystalline, such as tridymite and cristobalite. b) crypto-crystalline, such as chalcedony, and c) hydrated silica or opal; electronic components; TV components. Transparent crystals; d = 2.65; insoluble in H2O, organic solvents. *U.S. Silica Inc.; Unimin; Westco.*

3242 Quaternium-24
32426-11-2   G

C20H44ClN
Decyl dimethyl octyl ammonium chloride.
Ammonium, decyldimethyloctyl, chloride; BTC® 818; Caswell No. 613A; 1-Decaminium, N-octyl-N,N-di-methyl-, chloride; 1-Decanaminium, N,N-dimethyl-N-octyl-, chloride; Decylocytldimethylammonium chloride; EINECS 251-035-5; EPA Pesticide Chemical Code 069165; N,N-Dimethyl-N-octyl-1-decanaminium chloride; Octyl decyl dimethyl ammonium chloride; Quaternium-24. Disinfectant, sanitizer, and fungicide for hard surfaces; excellent sanitizer in hard water to 800 ppm as CaCO3. Liquid; soluble in H2O; d = 0.93; flash point = 86°F; cationic. *Stepan Canada; Stepan.*

3243 Quaternium-27
86088-85-9   G

R2C8H15N3O5S, (R represents tallow alkyl)

Methyl-1-tallow amido ethyl-2-tallow imidazolinium methyl sulfate.
ACCOSOFT® 808-90; Empigen® FRC90S;; Incrosoft S-75; Incrosoft S-90; Incrosoft S-90M; Varisoft® TIMS; Ditallow imidazoline methyl sulfate; Tallow imidazolinium methosulfate. Softener base, lubricant, antistat and rewetting agent for fabrics and synthetics. Pale amber soft paste; d = 0.95; set point = 20°; pH =

3244   Quercetin
117-39-5   8122   G

C15H10O7
2-(3,4-Dihydroxyphenyl)-3,5,7-trihydroxy-4H-1-benzo-pyran-4-one.
5-18-05-00494 (Beilstein Handbook Reference); AI3-26018; 4H-1-Benzopyran-4-one, 2-(3,4-dihydroxy-phenyl)-3,5,7-trihydroxy-; BRN 0317313; C.I. 75670; C.I. natural red 1; C.I. natural yellow 10 & 13; CCRIS 1639; CI Natural Yellow 10; Cyanidelonon 1522; EINECS 204-187-1; Flavin meletin; Flavone, 3,3',4',5,7-pentahydroxy-; HSDB 3529; Kvercetin; Meletin; Natural Yellow 10; NCI-C60106; NSC 9219; 3,5,7,3',4'-Pentahydroxyflavone; Quercetin; Quercetine; Quercetol; Quercitin; Quertine; Sophoretin; T-Gelb bzw. grun 1; 3',4',5,7-Tetrahydroxyflavan-3-ol; 3,3',4',5,7-Pentahydroxyflavone; 3,5,7-Trihydroxy-2-(3,4-dihydroxy-phenyl)-4H-chromen-4-on; Xanthaurine. The aglycone of quercitrin. Has been used in medicine as a capillary protectant; used in manufacture of epoxy resins. Yellow needles; mp = 316.5° (dec); λ_m = 257, 374 nm (ε =21878, 21878, EtOH); slightly soluble in H2O, Et2O, MeOH, soluble in EtOH, Me2CO, C5H5N, AcOH; LD50 (mus orl) = 160 mg/kg. *EM Ind. Inc.; U.S. BioChem.*

3245   Quercitol
488-73-3   8124   G

C6H12O5
1,2,3,4,5-Cyclopentanepentol.
D-chiro-Inositol, 2-deoxy-; 2-Deoxy-D-chiro-inositol; D-1-Deoxy-muco-inositol; (+)-Protoquercitol; (+)-Quercitol; d-Quercitol. Sugar found in acorns. Used as a chemical intermediate. Crystals; mp = 234-235°; $[\alpha]_D^{20}$= 24-26°; soluble in H2O, slightly soluble in hot EtOH, insoluble in cold EtOH, Et2O.

3246   Quinacridone
1047-16-1   H

C20H12N2O2
5,12-Dihydroquino(2,3-b)acridine-7,14-dione.
CI 46500; CI 73900; CI Pigment Red 122; CI Pigment Violet 19; Cinquasia B-RT 796D; Cinquasia Red; Cinquasia Red B; Cinquasia Red Y; Cinquasia Red Y-RT 759D; Cinquasia Violet; Cinquasia

Violet R; Cinquasia Violet R-RT 791D; Dark violet; E 3B Red; EINECS 213-879-2; Fastogen Super Red YE; Fastogen Super Red BN; Hostaperm Red E 5B; Hostaperm Red Violet ER; Hostaperm Red Violet ER 02; HSDB 6136; Linear quinacridone; Linear trans quinacridone; Monastral Red; Monastral Red B; Monastral Red Y; Monastral Violet 4R; Monastral Violet R; Monastrol Red Y; NSC 316165; Paliogen Red BG; Permanent Magenta; Permanent Red E 3B; Permanent Red E 5B; Pigment Pink Quinacridone S; Pigment Quinacridone Red; Pigment Violet 19; Pigment Violet Quinacridone; PV Fast Red E 5B; PV Fast Red E 3B; Quinacridone; Quinacridone Red; Quinacridone Red MC; Quinacridone Violet; Quinacridone Violet MC; Red E 3B; Sunfast Red 19; Sunfast Violet. Red pigment used by artists. *Daikin Kogyo; Fahlberg-List.*

3247   Quinaldine
91-63-4   8135   G

C10H9N
2-Methylquinoline.
5-20-07-00375 (Beilstein Handbook Reference); AI3-11528; BRN 0110309; CCRIS 1155; Chinaldine; EINECS 202-085-1; Khinaldin; 2-Methylchinolin; 2-Methyl-quinoline; NSC 3397; Quinaldine; Quinoline, 2-methyl-. Manufacture of dyes, pharmaceuticals, fine organic chemicals, acid-base indicators. Liquid; mp = -0.8°; bp = 246.5°, d^{25} = 1.06; λ_m = 207, 226, 229, 233, 274, 290, 297, 303, 309, 319 nm (MeOH); insoluble in H2O, soluble in organic solvents; LD50 (rat orl) = 1.23 g/kg. *Allchem Ind.*

3248   Quinalphos
13593-03-8   G P

C12H15N2O3PS
O,O-Diethyl O-2-quinoxalinyl phoshporothioate.
5-23-11-00455 (Beilstein Handbook Reference); AI3-27394; BAY 5821; BAY 77049; Bayer 77049; Bayrusil; BRN 0754823; CCRIS 3398; Chinalphos; Diethquinalphion; Diethquinalphione; O,O-Diäthyl-O-(chinoxalyl-(2))-monothiophosphat; O,O-Diethyl O-(2-chinoxalyl)-phosphorothioate; O,O-Diethyl O-2-quinoxalinyl phosphorothioic acid ester; O,O-Diethyl O-quinoxalin-2-yl phosphorothioate; Ekalux; Ekalux 25EC; ENT 27394; NSC 190986; Phosphorothioic acid, O,O-diethyl O-(2-quinoxalinyl) ester; Quinalphos; Quinaltaf; S-6538; SAN 6538 I; SAN 6626 I; Sandoz 6538; Savall; Spencer S-6538; SRA 7312; Wie oben. Insecticide and acaricide. Crystals; mp = 31 - 32°; bp0.0003 = 142° (dec); sg20 = 1.235; soluble in H2O (0.0022 g/100 ml at 24°), C6H14 (25 g/100 ml), freely soluble in C7H8, xylene, Et2O, EtOAc, Me2CO, CH3CN, EtOH, MeOH, slightly soluble in petroleum ether; LD50 (rat orl) = 71 mg/kg, (rat der) = 1750 mg/kg; LC50 (quail 8 day dietary) = 150 mg/kg, (mallard duck 8 day dietary) = 220 mg/kg, (carp 96 hr.) = 2.8 mg/l, (goldfish 96 hr.) = 1 - 10 mg/l; highly toxic to bees. *All-India Medical; Bayer Corp., Agriculture; Bayer Corp.; Novartis; Sandoz; United Phosphorus Inc.*

3249   Quinapril
85441-61-8   8139   D

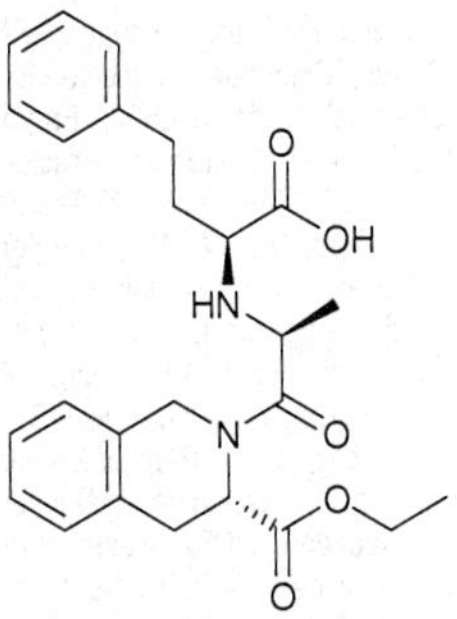

C25H30N2O5
(S)-2-[(S)-N-[(S)-1-Carboxy-3-phenylpropyl]alanyl]-1,2,3,4-tetrahydro-3-isoquinolinecarboxylic acid 1-ethyl ester.
Accuretic; Acequide; Koretic; Quinapril; Quinaprilum.; component of: Accuretic; Acequide; Koretic. Angiotensin-converting enzyme inhibitor. Orally active peptidyl-dipeptide hydrolase inhibitor. Antihypertensive. *Parke-Davis.*

3250   Quinapril Hydrochloride
82586-55-8   8139   D

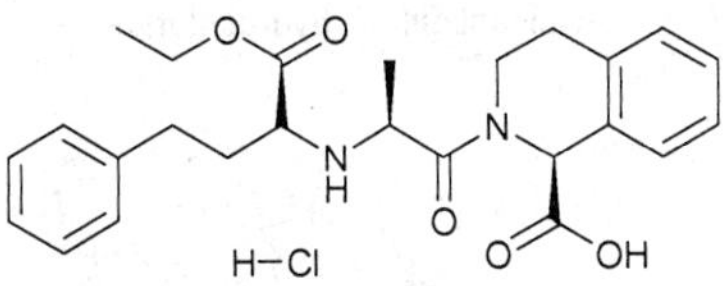

C25H31ClN2O5
(S)-2-[(S)-N-[(S)-1-Carboxy-3-phenylpropyl]alanyl]-1,2,3,4-tetrahydro-3-isoquinolinecarboxylic acid 1-ethyl ester monohydrochloride.
Accupril; Accuprin; Accupro; Accupron; Accuretic; Acequide; Acequin; Acuitel; Acuprel; Asig; CI-906; Conan; Continucor; Ectren; Hemokvin; HSDB 7046; Korec; Korectic; Koretic; Lidaltrin; PD-109452-2; Quinapril hydrochloride; Quinazil; component of: Accuretic; Acequide; Korectic. Angiotensin-converting enzyme inhibitor. Antihypertensive. Crystals; mp = 120-130°, 119-121.5°; $[\alpha]_D^{23}$ = 14.5° (c = 1.2 EtOH), $[\alpha]_D^{25}$ = 15.4° (c = 2.0 MeOH); LD50 (mmus orl) = 1739 mg/kg, (mmus iv) = 504 mg/kg, (fmus orl) = 1840 mg/kg, (fmus iv) = 523 mg/kg, (mrat orl) = 4280 mg/kg, (mrat iv) = 158 mg/kg, (frat orl) = 3541 mg/kg, (frat iv) = 107 mg/kg. *Parke-Davis.*

3251   Quindoxin
2423-66-7   D G V

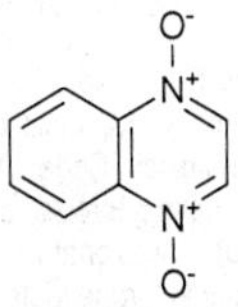

C8H6N2O2
Quinoxaline-1,4-dioxide copper(II) salt.
BAY-Va 9391; Bayo N-Ox; CCRIS 1570; Celbar; Chindoxin; Chinoxalin-1,4-dioxid; EINECS 219-352-3; Grofas; ICI 8173; NSC 21653; NSC 193508; Quindoxin; Quindoxina; Quindoxine; Quindoxinum; Quinoxaline di-N-oxide; Quinoxaline 1,4-dioxide; Quinoxaline 1,4-di-N-oxide; Quinoxaline dioxide; USAF H-1. An antibacterial and growth promoter used in animal husbandry. CAUTION: high degree of phototoxicity and mutagenicity. *Bayer AG.*

3252   Quinine iodo-sulfate
7631-46-1   8152   G

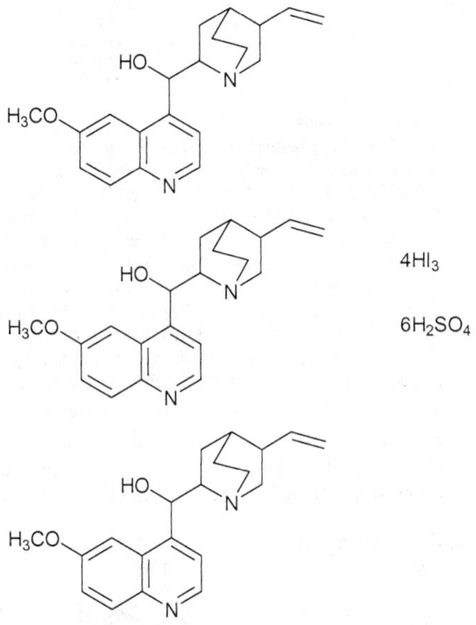

4HI3

6H2SO4

C60H96I12N6O30S6
Tetraquinine octahydrogen hexaiodide tris(sulphate).
Cinchonan-9-ol, 6'-methoxy-, (8α,9R)-, (hydrogen triiod-ide); EINECS 231-544-9; Herapath's salt; Quinine iodosulfate sulfate (salt) (4:2:3). Used in the manufacture of polarizing glasses and plastics. Olive-green crystals; dec >100°; soluble in H2O (0.12 g/100 ml 25°, 0.10 g/100 ml 100°), EtOH (2 g/100 ml 76°), AcOH (1.6 g/100 ml 80°); crystals polarize light strongly.

3253   Quinizarin
81-64-1   8156   G

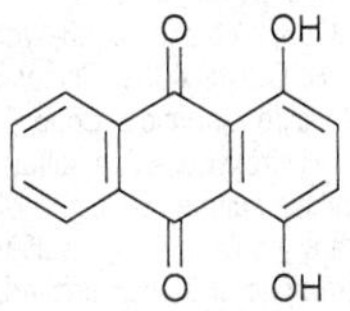

C14H8O4
1,4-Dihydroxyanthraquinone.
AI3-17616; 9,10-Anthracenedione, 1,4-dihydroxy-; An-thraquinone, 1,4-dihydroxy-; C.I. 58050; CCRIS 3524; Chinizarin; CI 58050; 1,4-Dihydroxy-9,10-anthracene-dione; 1,4-Dihydroxy-9,10-anthraquinone; 1,4-Di-hydr-oxyanthrachinon; 1,4-Dioxyanthraquinone; 1,4-Doa; EINECS 201-368-7; HSDB 5242; Macrolex Orange GG; NSC 15367; Quinizarin; Quinizarine; Smoke Orange R; Solvent Orange 86. Antioxidant in synthetic lubricants, dyes. Yellow-red crystals; mp = 200°; λ_m = 225, 249, 279325, 450 (ε = 21878, 30903, 10233, 2455, 8511 EtOH); insoluble in H2O, soluble in organic solvents. *BASF Corp.; Sandoz.*

3254   8-Quinolinol
148-24-3   4869   G

535

C9H7NO

8-Hydroxyquinoline.

5-21-03-00252 (Beilstein Handbook Reference); AI3-00483; 1-Azanaphthalene-8-ol; Bioquin; BRN 0114512; Caswell No. 719; CCRIS 340; 8-Chinolinol; EINECS 205-711-1; EPA Pesticide Chemical Code 059803; Fennosan; Fennosan H 30; Fennosan HF-15; HSDB 4073; 8-Hydroxy-chinolin; 8-Hydroxychinolin; 8-Hydroxyquinoline; Hydroxy quinoline; Hydroxybenzopyridine; NCI-C55298; NSC 2039; 8-OQ; o-Oxychinolin; 8-Oxyquinoline; Oxin; Oxine; Oxybenzopyridine; Oxychinolin; Oxyquinoline; Phenopyridine; 8-Quinol; Quinoline, 8-hydroxy-; 8-Quinolinol; Quinophenol; Tumex; USAF EK-794. Use for precipitating and separating metals, preparation of fungicides, chelating agent, disinfectant. Needles; mp = 75.5°; bp = 267°; λ_m = 243, 318 nm (cyclohexane); insoluble in H$_2$O, Et$_2$O, soluble in Me$_2$CO, alkali, acid, very soluble in EtOH, C$_6$H$_6$, CHCl$_3$; LD$_{50}$ (mus ip) = 48 mg/kg. Penta Mfg.; Spectrum Chem. Manufacturing; Tanabe Seiyaku Co. Ltd.; Tanabe U.S.A. Inc.

3255   Quinosol
134-31-6   4869   G

C18H14N2O2.H2SO4.H2O

8-Quinolinol sulfate monohydrate.

Aci-Jel; AI3-03968; Albisal; Bis(8-hydroxyquinolinium) sulphate; Caswell No. 719B; CCRIS 4659; Chinosol; Cryptonol; EINECS 205-137-1; EPA Pesticide Chemical Code 059804; Happy; 8-Hydroxy-chinolin-sulfat; 8-Hydroxyquinoline sulfate; Khinozol; Octofen; Oxine sulfate; Oxyquinoline sulfate; 8-Quinolinol, hydrogen sulfate (2:1); 8-Quinolinol sulfate; 8-Quinolinol, sulfate (2:1) (salt); Quinosol; Quinosol Extra; Solfato di 8-ossichinolina; Solvochin-Extra; Sunoxol; Superol. The potassium salt of oxyquinoline sulfate; used as an antiseptic. mp = 177.5°; λ_m = 242, 253, 309, 318 nm (ε = 62800, 43900, 3810, 3730, MeOH); very soluble in H$_2$O, soluble in EtOH, insoluble in Et$_2$O.

3256   Quintozene
82-68-8   8172   G P

C6Cl5NO2

Pentachloronitrobenzene.

101 brand PCNB 75 Wettable; 4-05-00-00728 (Beilstein Handbook Reference); AI-23024; AI3-23024; Avicol; Avicol (pesticide); Batrilex; Benzene, nitropentachloro-; Benzene, pentachloronitro-; Botrilex;

Brassicol; Brassicol 75; Brassicol super; BRN 1914324; Caswell No. 640; CCRIS 495; Chinozan; Earthcide; EINECS 201-435-0; EPA Pesticide Chemical Code 056502; Fartox; Folosan; Fomac 2; Fungiclor; GC 3944-3-4; Gustafson Terraclor 80% Dust Concentrate; HOE 026014; HSDB 1749; KOBU; KOBUKP 2; Kobutol; KP 2; Liro-PCNB; Marisan forte; NCI-C00419; Nitropentachlorobenzene; NSC 58427; Olin Terraclor 75% Wettable Powder; Olin Terraclor 90% Dust Concentrate; Olin Terraclor Technical Grade PCNB 99% Soil Fungicide; Olpisan; Pcnb 100; Pcnb Technical Material for Manufacturing Purposes Only; PCNB; Pentachlornitrobenzol; Pentachloronitrobenzene; Pentachloronitrobenzol; Pentagen; Phomasan; PKhNb; Quinosan; Quintobenzene; Quintocene; Quintozen; Quintozene; RCRA waste number U185; RTU 1010; Saniclor 30; Technical Grade PCNB 95%; Terrachlor; Terraclor; Terraclor 30 G; Terrafun; Terrazan; Tilcarex; Tri-pcnb; Tritisan. Registered by EPA as a fungicide. Seed and soil-fungicide used for control of fungal diseases in fruits and vegetables. Pale yellow crystals; mp = 144°; bp = 328° (dec); d^{25} = 1.718; insoluble in H$_2$O (0.000044 g/100 ml), soluble in EtOH (1.6 g/100 ml) and other organic solvents; LD$_{50}$ (rat orl) > 12000 mg/kg aqueous suspension, (mrat orl) = 1710, (frat orl) = 1650 mg/kg in maize oil, (rat ip) = 5000 mg/kg, nonlethal to golden orfe or rainbow trout at 1.2 mg/l over 96 hr.; non-toxic to bees. Amvac Chemical Corp.; Bos Chemicals Ltd.; ICI Chem. & Polymers Ltd.; Rhône-Poulenc; Scotts Co.; Uniroyal; Wilbur Ellis Co.

3257   3-Quinuclidinone hydrochloride
1193-65-3   G

C7H12ClNO

1-Azabicyclo [2.2.2] octan-3-one hydrochloride.

1-Azabicyclo(2.2.2)octan-3-one, hydrochloride; EINECS 214-776-5; NSC 91498; Quinuclidin-3-one hydro-chloride; 3-Quinuclidinone hydrochloride. Janssen Chimica; Lancaster Synthesis Co.

3258   Quizalofop-ethyl
76578-14-8   8178   G P

C19H17ClN2O4

2-[4-[(6-Chloro-2-quinoxalinyl)oxy]phenoxy]propanoic acid ethyl ester.

Assure; Caswell No. 215D; 2-(4-((6-Chloro-2-quinoxalinyl)oxy)phenoxy)propanoic acid ethyl ester; DPX-Y 6202; EPA Pesticide Chemical Code 128201; Ethyl 2-(4-(6-chloro-quinoxalin-2-yl-oxy)phenoxy)propan-oate; Ethyl 2-(4-(6-chloro-2-quinoxalyloxy)phenoxy)-propionate; Ethyl 2-(4-(6-chloro-2-quinoxalinyloxy)-phenoxy)propanoate; EXP-3864; FBC-32197; INY-6202; NC 302; NCI 96683; Pilot; Propanoic acid, 2-(4-((6-chloro-2-quinoxalinyl)oxy)phenoxy)-, ethyl ester; Quino-fop-ethyl; Quizalofop ethyl ester; Quizalofop-Et; Quiz-alofop-ethyl; Targa; Xylofop-ethyl. Post-emergence herbicide used for control of grasses in broad-leaved crops. Selective herbicide for control of grasses in mustard, rape and beet crops. Registered by EPA as a herbicide. Crystals; mp = 92-93°; bp$_{0.2}$ = 220°; poorly soluble in H$_2$O (0.00003 g/100 ml at 20°), soluble in C$_6$H$_6$ (29 g/100 ml), xylene (12 g/100 ml), Me$_2$CO (11 g/100 ml), EtOH (0.9 g/100 ml), C$_6$H$_{14}$ (0.26 g/100 ml); LD$_{50}$ (mrat orl) = 1670 mg/kg, (frat orl) = 1480 mg/kg, (mmus orl) = 2350 mg/kg, (fmus orl) = 2360 mg/kg, (rat, mus der) > 10000 mg/kg; (mallard duck orl) = 200 mg/kg; LC_{50} (rainbow trout 96 hr.) = 10.7 mg/l, (bluegill sunfish 96 hr.) = 0.46 - 2.8 mg/l; non-toxic to bees. *E. I. DuPont de Nemours Inc.; FMC Corp.; Monsanto Co.; Nissan Chem. Ind.; Schering Agrochemicals Ltd.*

3259 Rabeprazole
117976-89-3   8181   D

$C_{18}H_{21}N_3O_3S$

2-[[[4-(3-Methoxypropoxy)-3-methyl-2-pyridinyl]methyl]-sulfinyl]benzimidazole.

1H-Benzimidazole, 2-(((4-(3-methoxypropoxy)-3-methyl-2-pyridinyl)methyl)sulfinyl)-; 2-(((4-(3-Methoxypropoxy)-3-methyl-2-pyridinyl)methyl)sulfinyl)-1H-benzimidazole; Pariprazole; Rabeprazole. Gastric proton pump inhibitor. Antiulcerative. White crystals; mp = 99-100°. *Eisai Co. Ltd.; Eli Lilly & Co.*

3260 Rabeprazole Sodium
117976-90-6   8181   D

$C_{18}H_{20}N_3NaO_3S$

2-[[[4-(3-Methoxypropoxy)-3-methyl-2-pyridinyl]methyl]-sulfinyl]benzimidazole sodium salt.

Aciphex; E 3810; LY 307640 sodium; Rabeprazole sodium; Rabeprazole sodium salt. Gastric proton pump inhibitor. Antiulcerative. Crystals; mp = 140-141° (dec). *Eisai Co. Ltd.; Eli Lilly & Co.*

3261 R-Acid
148-75-4   6413   G

$C_{10}H_8O_7S_2$

2-Naphthol-3,6-disulfonic acid.

EINECS 205-724-2; 3-Hydroxynaphthalene-2,7-disulph-onic acid; 2-Naphthol-3,6-disulfonic acid; 2,7-Naph-thalenedisulfonic acid, 3-hydroxy-. Used as an azo dye intermediate. The disodium salt is used as a reagent to detect nitrogen dioxide in the air. Solid; mp dec; very soluble in H_2O, EtOH, insoluble in Et_2O.

3262 Racumin®
5836-29-3   G P

$C_{19}H_{16}O_3$

2H-1-Benzopyran-2-one, 4-hydroxy-3-(1,2,3,4-tetra-hydro-1-naphthalenyl)-.

Coumetralyl; 2H-1-Benzopyran-2-one, 4-hydroxy-3-(1,2,3,4-tetrahydro-1-naphthalenyl)-; Coumarin, 4-hydroxy-3-(1,2,3,4-tetrahydro-1-naphthyl)-; BAY 25634; Bay ene 11183 B; Bayer 25 634; Coumatetralyl; Cumatetralyl; Endox; Endrocid; Endrocide; Ene 11183 B; 4-Hydroxy-3-(1,2,3,4-tetrahydro-1-naftyl)-cumarine; 4-Hydroxy-3-(1,2,3,4-tetrahydro-1-naphthalenyl)-2H-1-benzopyran-2-one; 4-Hydroxy-3-(1,2,3,4-tetrahydro-1-naphthyl)coumarine; 4-Hydroxy-3-(1,2,3,4-tetrahydro-1-naphthyl)cumarin; 3-(1,2,3,4-Tetrahydro-1-naphtyl)-4-hydroxycumarine; 3-(α-Tetralinyl)-4-hydroxycoumarin; 3-(α-Tetral)-4-oxycoumarin; 3-(D-Tetralyl)-4-hydroxy-coumarin; 3-(α-Tetralyl)-4-hydroxycoumarin; Rodentin; 3-(1,2,3,4-Tetrahydro-1-naphthyl)-4-hydroxycumarin. A ready-to-use anticoagulant rodenticide. Registered by EPA as a rodenticide. Colorless crystals; mp = 172 - 176°; soluble in H_2O (0.0004 g/100 ml at pH 4.2, 0.002 g/100 ml at pH 5, 0.0425 g/100 ml at pH 7, 0.01 - 0.02 g/100 ml at pH 9), soluble in DMF, alcohols, Me_2CO, slightly soluble in C_6H_6, C_7H_8, Et_2O, CH_2Cl_2 (5 - 10 g/100 ml), i-PrOH (2 - 5 g/100 ml); LD_{50} (rat orl) = 6.5 mg/kg, (mus orl) > 1000 mg/kg, (rat der) = 25 - 50 mg/kg; LC_{50} (rat orl 5 days) = 0.3 mg/kg/day, (hen 8 day dietary) > 50 mg/kg/day, (guppy 96 hr.) = 1000 mg/l. *Bayer Corp., Agriculture.*

3263 Radia® 7060
67762-38-3   G

$C_{18}H_{34}O_2$

(Z)-9-octadecenoic acid methyl ester.

(C16-C18) and (C18) Unsaturated alkylcarboxylic acid methyl ester; EINECS 267-015-4; Fatty acids, C16-18 and C18-unsatd, Me esters; SDA 11-010-00. Chemical intermediate, lubricant; chemical synthesis; lubricity improvers in mineral oils; formulation of cutting, lamination, and textile oils; rust inhibitors; textile and leather industry. bp_2 = 168-170°; d_4^{18} = 0.879; n_D^{26} = 1.4510; soluble in EtOH, Et_2O. *Fina Chemicals.*

3264 Radia® 7110
85586-21-6   G

$C_{19}H_{38}O_2$

Methyl stearate.

EINECS 287-824-6; Fatty acids, C16-18, Me esters. Chemical intermediate, chemical synthesis; lubricant in mineral, cutting, lamination, textile oils, and rust inhibitors; textile and leather application. Liquid; mp = 38-39°; bp_{15} = 215°; insoluble in H_2O, soluble in organic solvents. *Fina Chemicals.*

3265 Radia® 7187
85049-36-1   G

$C_{20}H_{38}O_2$

(Z)-9-octadecenoic acid ethyl ester.

EINECS 285-206-0; Fatty acids, C16-18 and C18-unsatd., Et esters. Chemical intermediate, chemical synthesis; lubricant in mineral, cutting, lamination, textile oils, and rust inhibitors; textile and leather application; also as emollient, plasticizer, solubilizer of active components in cosmetics and pharmaceuticals. Oil; bp = 205-208°; d

= 0.87; insoluble in H$_2$O, soluble in organic solvents. *Fina Chemicals.*

3266 **Radiasurf® 7125**
68154-36-9   G

C$_{18}$H$_{34}$O$_6$
Sorbitan laurate.
Alkamuls SML; Anhydrosorbitol monococoate; Arlacel 20; Coconut fatty acid, sorbitan ester (1:1); Coconut oil fatty acids, sorbitan monoester; EINECS 268-910-2; Emsorb 2515; Fatty acids, coco, monoesters with sorbitan; Glycomul L; Sorbitan monolaurate; Sorbitan cocoate; Sorbitan monococoate; Span 20. Emulsifier, descouring aid, antistat; anticorrosive agent for pipelines; cleaner for metallic surfaces; superfatting, bodying and antifog aid; pigment dispersant; detergent; emulsion of solvents; Insoluble in H$_2$O, soluble in organic solvents. *Fina Chemicals.*

3267 **Raffinose**
512-69-6   8188   G

C$_{18}$H$_{32}$O$_{16}$
β-D-fructofuranosyl-O-α-D-galactopyranosyl-(1→6)-α-D-glucopyranoside.
AI3-19427; d-(+)-Raffinose; D-Raffinose; EINECS 208-146-9; α-D-Glucopyranoside, β-D-fructofuranosyl O-α-D-galactopyranosyl-(1→6)-; α-D-Glucopyranoside, β-D-fructofuranosyl O-α-D-galactopyranosyl (1 → 6)-, hydrate; Gossypose; Melitose; Melitriose; NSC 170228; Raffinose. A trisaccharide used in bacteriology, in the preparation of other saccharides. The pentahydrate [17629-30-0] has EINECS Number 208-146-9. Crystals; mp = 80°; d^{25} = 1.465; [α]$_D^{20}$= 105° (c = 4); insoluble in Et$_2$O, slightly soluble in EtOH, soluble in C$_5$H$_5$N, very soluble in MeOH, H$_2$O (0.14 g/ml).

3268 **Rafoxanide**
22662-39-1   8189   G

C$_{19}$H$_{11}$Cl$_2$I$_2$NO$_3$
N-[3-chloro-4-(4-chlorophenoxy)phenyl]-2-hydroxy-3,5-diiodobenzamide.
AI3-29020; Benzamide, N-(3-chloro-4-(4-chlorophen-oxy)phenyl)-2-hydroxy-3,5-diiodo-; Bovanide; BRN 2228187; 3'-Chlor-4'-(4-chlorphenoxy)-3,5-diiodsalicyl-anilid; 3'-Chloro-4'-(p-chlorophenoxy)-3,5-diiodosalicyl-anilide; Disalan; Duofas; EINECS 245-148-9; Flukanide; MK-990; NSC 355278; Rafoxanida; Rafoxanide; Rafoxanidum; Ranide; Ranide, veterinary; Salicylanilide, 3'-chloro-4'-

(p-chlorophenoxy)-3,5-diiodo-. Fasciolicide, anthelmintic. Crystals; mp = 168-170°; insoluble in H$_2$O, soluble in Me$_2$CO, CH$_3$CN. *Merck & Co.Inc.*

3269 **Raloxifene**
84449-90-1   8190   D

C$_{28}$H$_{27}$NO$_4$S
6-Hydroxy-2-(p-hydroxyphenyl)benzo[b]thien-3-yl-p-(2-piperidinoethoxy)phenyl ketone.
CCRIS 7129; Keoxifene; LY-139481; Raloxifene; Raloxifeno; Raloxifenum. Antiosteoporotic; antiestro-genic. Used to treat osteoporosis. Crystals; mp = 143-147°; λ$_m$ = 290 nm (ε 34000 EtOH). *Eli Lilly & Co.*

3270 **Raloxifene hydrochloride**
82640-04-8   8190   D

C$_{28}$H$_{28}$ClNO$_4$S
6-Hydroxy-2-(p-hydroxyphenyl)benzo[b]thien-3-yl-p-(2-piperidinoethoxy)phenyl ketone hydrochloride.
Evista; Keoxifene hydrochloride; LY 156758; Raloxifene hydrochloride. Nonsteroidal estrogen receptor mixed agonist-antagonist. Antiestrogen; used to treat osteo-porosis. Solid; mp = 258°; λ$_m$ = 286 nm (ε 32800 EtOH). *Eli Lilly & Co.*

3271 **Rametin**
1491-41-4   6380   G V

C$_{16}$H$_{16}$NO$_6$P
2-[(Diethoxyphosphinyl)oxy]-1H-benz[de]isoquinoline-1,3(2H)-dione.
AI3-25567; B-9002; BAY 9002; Bayer 9002; Bayer 25820; Bayer S940; 1H-Benz(de)isoquinoline-1,3(2H)-dione, 2-((diethoxyphosphinyl)oxy)-; BRN 1551429; Caswell No. 495AA; Chemagro B-9002; N-(Diethoxy-phosphinoyloxy)-1,8-naphthalindicarboximid; O,O-Di-ethyl N-hydroxynaphthalimide phosphate; E 9002; EINECS 216-078-6; ENT 25,567; EPA Pesticide Chemical Code 204100; N-Hydroxynaphthalimide diethyl phos-

phate; Maretin; Naftalofos; Naftalofosum; Naphthalimide, N-hydroxy-, diethyl phosphate; Naphthalophos; NSC 229795; Phosphoric acid, diethyl ester, N-naphthalimide deriv.; Phtalophos; Rametin; Rawetin; S-940. A proprietary preparation of naphthalophos; a veterinary anthelmintic. Brown crystals; mp = 174-179°; insoluble in H$_2$O, sligthly soluble in EtOH, Et$_2$O, C$_6$H$_6$, Me$_2$CO. *Bayer AG.*

3272 Ramipril
87333-19-5   8139   D

C$_{23}$H$_{32}$N$_2$O$_5$
(2S,3aS,6aS)-1-[(S)-N-[(S)-1-Carboxy-3-phenylpropyl]-alanyl]octahydrocyclopenta[b]-pyrrole-2-carboxylic acid.
Acovil; Altace; Carasel; Cardace; Delix; HOE-498; Hytren; Lostapres; Pramace; Quark; Ramace; Ramipril; Ramiprilum; Triatec; Tritace; Unipril; Vesdil. Angiotensin-converting enzyme inhibitor. Used to treat hypertension. Orally active peptidyldipeptide hydrolase inhibitor. Crystals; mp = 109°; [α]$_D^{24}$= 33.2° (c= 1 0.1N ethanolic HCl); LD$_{50}$ (mmus iv) = 1194 mg/kg, (mmus orl) = 10933 mg/kg, (fmus iv) = 1158 mg/kg, (fmus orl) = 10048 mg/kg, (mrat iv) = 687 mg/kg, (mrat orl) > 10000 mg/kg, (frat iv) = 608 mg/kg, (frat orl) > 10000 mg/kg. *Hoechst Roussel Pharm. Inc.*

3273 Ranitidine
66357-35-5   8198   D

C$_{13}$H$_{22}$N$_4$O$_3$S
N-[2-[[5-[(Dimethylamino)methyl]furfuryl]thio]ethyl]-N'-methyl-2-nitro-1,1-ethenediamine.
Achedos; Acidex; Atural; Axoban; Coralen; Curan; Duractin; EINECS 266-332-5; Ezopta; Gastrial; Gastrosedol; HSDB 3925; Istomar; Logast; Mauran; Microtid; Ptinolin; Quantor; Quicran; Radinat; Randin; Ranidine; Ranin; Raniogas; Ranisen; Raniter; Ranitidina; Ranitidine; Ranitidinum; Ranitiget; Rantacid; Ratic; Raticina; RND; Sampep; Taural; Ul-Pep; Ulceranin; Urantac; Verlost; Vesyca; Vizerul; Weichilin; Weidos; Xanidine; Zantab; Zantadin. Histamine H$_2$ receptor antagonist; inhibits gastric secretion. Antiulcerative. Crystals; mp = 69-70°. *Glaxo Labs.*

3274 Ranitidine hydrochloride
66357-59-3   8198   D

C$_{13}$H$_{23}$ClN$_4$O$_3$S
N-[2-[[5-[(Dimethylamino)methyl]furfuryl]thio]ethyl]-N'-methyl-2-nitro-1,1-ethenediamine hydrochloride.
AH-19065; Alquen; Alter-H2; Alvidina; Apo-Ranitidin; Artomil; Azantac; Azuranit; CCRIS 5268; Coralen; Digestosan; DRG-0114; EINECS 266-333-0; Ergan; Esofex; Fendibina; Gastridina; Gastrolav; Kuracid; Label; Lake; Logat; Melfax; Mideran; Neugal; Noctone; Noktome; Normon; Novo-Radinine; Nu-Ranit; Pep-Rani; Quadrin; Quantor; Radin; Ran H2; Ran Lich; RAN; Rani 2; Rani AbZ; Rani-BASF; Rani-nerton; Rani-Puren; Rani-Q; Rani-Sanorania; Raniben; Raniberl; Raniberta; Ranibloc; Ranic; Ranicux; Ranidil; Ranidin; Ranidura; Ranifur; Ranigasan; Ranigast; Ranigen; Ranilonga; Ranimerck; Raniplex; Ranisan; Ranitab; Ranitic; Ranitidin; Ranitidin 1A Pharma; Ranitidin AL; Ranitidin Arcana; Ranitidin Atid; Ranitidin AWD; Ranitidin Basics; Ranitidin-Cophar; Ranitidin Duncan; Ranitidin Dyna; Ranitidin Helvepharm; Ranitidin Heumann; Ranitidin Hexal; Ranitidin-Isis; Ranitidin Merck; Ranitidin Millet; Ranitidin NM; Ranitidin Normon; Ranitidin PB; Ranitidin-ratiopharm; Ranitidin Stada; Ranitidin von ct; Ranitidina predilu Grif; Ranitidina Tamarang; Ranitidine hydrochloride; Ranitin; Ranitine; Ranobel; Ranuber; Regalil; Renatac; Rozon; Rubiulcer; Santanol; Serviradine; Sostril; Tanidina; Taural; Terposen; Toriol; Trigger; Ulcecur; Ulcex; Ulcirex; Ulcodin; Ulcolind Rani; Ulcosan; Ulsaven; Ultidine; Viserul; Zandid; Zantac; Zantarac; Zantic. Histamine H$_2$ receptor antagonist; inhibits gastric secretion. Antiulcerative. Crystals; mp = 133-134°; soluble in H$_2$O, AcOH; less soluble in EtOH, MeOH; insoluble in organic solvents. *Glaxo Labs.*

3275 Realgar
1303-32-8   804   G

As$_2$S$_2$
Arsenic disulfide.
Arsenic disulfide; Arsenic monosulfide; Arsenic orange; Arsenic sulfide (As$_2$S$_2$); Arsenic Sulfide Red; Arsino, thioxo-; C.I. 77085; Caswell No. 058; EPA Pesticide Chemical Code 006901; Red algar; Red arsenic; Red arsenic glass; Ruby arsenic; Red orpiment; Ruby sulfur; Thioxoarsino. Red pigment used in the leather industry, paint, pyrotechnics, and taxidermy. Red solid; mp = 320°; bp = 565°; insoluble in H$_2$O. *Atomergic Chemetals.*

3276 Red lead
1314-41-6   5444   G

O$_4$Pb$_3$
Lead tetroxide.
C.I. 77578; C.I. Pigment Red 105; EINECS 215-235-6; Gold satinobre; Heuconin 5; HSDB 5801; Lead orthoplumbate; Lead oxide; Lead oxide (3:4); Lead oxide (Pb$_3$O$_4$); Lead Oxide Red; Lead tetraoxide; Lead tetroxide; Mennige; Mineral Orange; Mineral Red; Minium; Minium Non-Setting RL 95; Minium Red; Orange Lead; Paris Red; Pigment Red 105; Plumboplumbic oxide; Red Lead; Red Lead Oxide; Red Lead Oxide (Pb$_3$O$_4$); Sandix; Saturn Red; Trilead tetraoxide; Trilead tetroxide. A pigment, made by heating litharge, PbO. There are several kinds on the market distinguished by their color and amount of lead dioxide they contain; used in storage batteries. Bright red powder; mp = 500° (dec); d = 9.1; insoluble in H$_2$O, EtOH; LD$_{50}$ (gpg ip) = 220 mg/kg. *National Lead Co.*

3277 Reinecke's salt
13573-16-5   8213   G

539

C4H10CrN7S4

Chromate(1-), diaminetetrakis(thiocyanato-N)-, ammon-ium, (OC-6-11)-.

AI3-28741; Ammonium diamminetetrakis(thiocyanato-N)chromate(1-); Chromate(1-), diamminetetrakis(thio-cyanato-N)-, ammonium; Chromate(1-), diamminetetra-kis(thiocyanato-N)-, ammonium, (OC-6-11)-; EINECS 237-003-3; Reinecke salt. Produced when ammonium cyanate is melted and ammonium bichromate added; used as a precipitating agent for organic bases and amino acids and as a reagent for mercury. mp = 270° (dec).

3278   Remazol black B
17095-24-8   H

C26H21N5Na4O19S6

4-Amino-5-hydroxy-3,6-bis((4-((2-(sulfooxy)ethyl)-sulfon-yl)phenyl)azo)-2,7-naphthalenedisulfonic acid tetra-sodium salt.

4-Amino-5-hydroxy-3,6-bis((4-((2-(sulfooxy)ethyl)-sulfon-yl)phenyl)azo)-; C.I. 20505; C.I. Reactive Black 5; Cav-alite Black B; Celmazol Black B; Diamira Black B; Dri-marene Black R/K 3B; EINECS 241-164-5; Intracron Black VS-B; Levafix Black E-B; Primazin Black BN; Reactive Black 5; Remazol Black B; Remazol Black GF; Sumifix Black B.

3279   Reprodin®
73523-00-9   5631   D G V

C21H29ClO6S

[1S-[1α(Z),2β(R*), 3α,5α]]-7-[2-[[3-(3-chlorophenoxy)-2-hydroxypropyl]thio]-3,5-dihydroxycyclopentyl]-5-hepten-oic acid.

3,5-Dihydroxycyclopentyl)-, (1α(Z),2β(R*),3α,5α)-; EMD-34946; 5-Heptenoic acid, 7-(2-((3-(3-chlorophenoxy)-2-hydroxypropyl)thio)-; Luprostenol; Luprostiol; Pronilin; Prosolvin. A luteolytic prostaglandin for use with cattle, horses, pigs and sheep. Affects regulation of estrus cycle. *E. Merck.*

3280   Resacetophenone
89-84-9   8225   G

C8H8O3

2,4-Dihydroxyacetophenone.

4-08-00-01792 (Beilstein Handbook Reference); 4-Acetylresorcinol; Acetophenone, 2',4'-dihydroxy-; AI3-00866; BRN 1282505; 1-(2,4-Dihydroxyphenyl)ethan-one; 2',4'-Dihydroxyacetophenone; EINECS 201-945-3; Ethanone, 1-(2,4-dihydroxyphenyl)-; NSC 10883; Res-acetophenone; Resoacetophenone; Resorcinol, 4-acetyl-. Reagent specific for iron. Solid; mp = 146°; d141 = 1.1800; insoluble in H2O, soluble in EtOH, TFA, C5H5N, AcOH, less soluble in Et2O, C6H6.

3281   Resazurin
550-82-3   8226   G

C12H7NO4

7-Hydroxy-3H-phenoxazin-3-one 10-oxide.

4-27-00-03594 (Beilstein Handbook Reference); Azoresorcin; BRN 0221382; Diazoresorcinol; EINECS 208-987-1; 7-Hydroxy-3H-fenoxazin-3-on-10-oxid; 7-Hydroxy-3H-phenoxazin-3-one 10-oxide; NSC 10438; 3H-Phenoxazin-3-one, 7-hydroxy-, 10-oxide; Resazoin; Resazurin; Resazurine. An acid-base indicator; pH 3.8 orange, pH 6.5 dark violet. Also used in detection of hyposulfite, in food research and in enzymology - as a reductase detector. Dark red crystals; bp sublimes; λ_m = 355, 520 nm (ε = 3981, 7943, pH 4.0), 280, 375, 590 nm (ε = 6310, 3981, 19953, pH 6.0), 283, 375, 585 nm (ε = 6310, 5012, 25119, pH 8.0); insoluble in H2O, Et2O, slightly soluble in EtOH, AcOH, soluble in alkali.

3282   Resmethrin
10453-86-8   G P

C22H26O3

(5-(Phenylmethyl)-3-furanyl)methyl 2,2-dimethyl-3-(2-methyl-1-propenyl)cyclopropanecarboxylate 2,2-di-meth-yl-3-(2-methylpropenyl)-, (5-benzyl-3-furyl)methyl ester.

AI3-27474; ARI-B; Benzofuroline; (5-Benzyl-3-furyl)methyl-2,2-dimethyl-3-(2-methylpropenyl)-cyclo-propanecarboxylate; 5-Benzyl-3-furyl-methyl-(1RS)-cis-trans-2,2-dimethyl-3-(2-methylprop-1-enyl)cyclopropane-carboxylate; 5-Benzyl-3-furylmethyl (±)-cis, trans-chrysanthemate; 5-Benzyl-3-furyl-methyl (1RS)-cis,trans-chrysanthemate; 5-Benzylfurfuryl chrysanthemate; Bioresmethrin (d trans isomer); Bioresmethrin (D-trans isomer); Caswell No. 083E; CCRIS 2501; Chryson; Chrysron; Crossfire; Enforcer; ENT 27474; EPA Pesticide Chemical Code 097801; FMC 17370; For-syn; HSDB 1516; Isathrine; NIA-17370; NRDC 104; NSC 195022; OMS-1206; Penick 1382; Penncapthrin; [5-(phenylmethyl)-3-furanyl]-methyl ester)oxy)imino)-benzeneaceto-nitrile; Premgard; Pynosect; Pyresthrin; Pyretherm; Resmethrin; Resmethrine; Resmetrina; S.B. Penick 1382; SBP-1382; Scourge; Synthrin. Non-systemic insecticide with contact action. Used as a household and garden insecticide and in agricultural premises. Registered by EPA as an insecticide. Colorless crystals; mp = 43 - 48°; sg20 = 0.958 - 0.968; insoluble in H2O (< 0.0001 g/100 ml), soluble in MeOH (6.5 g/100 ml), C6H14 (14.5 g/100 ml), xylene (86 g/100 ml); LD50 (rat orl) > 2500 mg/kg, (rbt der) = 2500 mg/kg, (rat der) > 3000 mg/kg; toxic to fish; toxic to bees. *Bonide Products, Inc.; Chase Products Co.; Huntington Professional Products; Quest Chemical Corp.; S.C. Johnson & Son, Co; Scotts Co.; Speer Products Inc.; Sumitomo Corp.; Unicorn Laboratories; Valent Biosciences Corp.; W.F. Young Inc.*

3283 Resorcinol
108-46-3   8240   D G H

C6H6O2
3-Hydroxyphenol.
4-06-00-05658 (Beilstein Handbook Reference); Acnomel; AI3-03996; Benzene, m-dihydroxy-; BRN 0906905; C.I. 76505; C.I. Developer 4; C.I. Oxidation Base 31; Caswell No. 723; CCRIS 4052; Developer O; Developer R; Developer RS; Dihydroxybenzol; Durafur developer G; EINECS 203-585-2; EPA Pesticide Chemical Code 071401; FEMA No. 3589; Fouramine RS; Fourrine 79; Fourrine EW; HSDB 722; 3-Hydroxyphenol; Nako TGG; NCI-C05970; NSC 1571; Pelagol Grey RS; Pelagol RS; Phenol, m-hydroxy-; RCRA waste number U201; Resorcin; Resorcine; Resorcinil; Resorcinolum; Resorzin; Rezamid; Sulforcin; UN2876. Used in the manufacture of resorcinol-formaldehyde resins, dyes, cosmetics, pharm-aceuticals and as a crosslinking agent in the manufacture of neoprene. Used in tanning, manufacture of resins and resin adhesives, explosives, dyes and cosmetics, in dyeing and printing of textiles, as a reagent for zinc and medically, as a keratolytic and antiseborrheic. Crystals; mp = 111°; bp = 280°, bp16 = 178°; d^{25} = 1.2717; λm = 275 nm (MeOH); soluble in H2O (111.1 g/100 ml at 20°, 500 g/100 ml at 80°), EtOH (111.1 g/100 ml); freely soluble in CCl4, glycerol; slightly soluble in CHCl3, C6H6. *Dermik Labs. Inc.; Galderma Labs Inc.; Menley & James Labs Inc.*

3284 Resorcinol benzoate
136-36-7   G

C13H10O3
Resorcinol monobenzoate.
4-09-00-00372 (Beilstein Handbook Reference); 1,3-Benzenediol, monobenzoate; Benzoic acid, m-hydroxyphenyl ester; BRN 1873897; Eastman® Inhibitor RMB; EINECS 205-241-7; 3-Hydroxyphenyl benzoate; NSC 4807; Resorcinol monobenzoate. Industrial grade UV absorber/stabilizer for cellulosic plastics and PVC formulations. White crystalline solid; mp = 133-135°; insoluble in H2O, C6H6, soluble in EtOH, Me2CO; LD50 (rat orl) = 1600 mg/kg. *Eastman Chem. Co.; Monomer -Polymer & Dajac.*

3285 Resorufin
635-78-9   G

C12H7NO3
Hydroxyphenazone.
4-27-00-02263 (Beilstein Handbook Reference); BRN 0174850; EINECS 211-241-8; 7-Hydroxyphenoxazin-3-one; 7-Hydroxy-3H-phenoxazin-3-one; NSC 12097; 3H-Phenoxazin-3-one, 7-hydroxy-; Resorufin; Resorufine.

3286 Retene
483-65-8   8245   G

C18H18
1-methyl-7-(1-methylethyl)phenanthrene.
AI3-00840; CCRIS 3180; EINECS 207-597-9; 7-Isopropyl-1-methylphenanthrene; 1-Methyl-7-(1-methylethyl)phen-anthrene; 1-Methyl-7-isopropylphenanthrene; NSC 26317; Phenanthrene, 1-methyl-7-(1-methylethyl)-; Phenanthrene, 7-isopropyl-1-methyl-; Reten; Retene. Crystals; mp = 101°; bp = 390°; λm = 258, 279, 288, 300, 318 nm (ε = 68700, 14700, 11600, 14800, 325, MeOH); insoluble in H2O (> 1 mg/ml), soluble in EtOH, Et2O, C6H6, CS2, ligroin, AcOH.

3287 Rhamnose
3615-41-6   8256   G

C6H12O5H2O
L-(+)-Rhamnose monohydrate.
6-Deoxy-L-mannose; EINECS 222-793-4; FEMA No. 3730; Isodulcit; Isodulcitol; Locaose; Mannomethylose; Mannomethylose, L-; L-Mannose, 6-deoxy-; NSC 2056; Rhamnose; L-Rhamnose; Rhamnose, L-. Food sweetener. Crystals; mp = 93-95°; d$_4^{20}$ = 1.4708; [α]$_D^{20}$= 8° (c = 10 H2O). *Penta Mfg.*

3288 Rhapontin
155-58-8   8258   G

C21H24O9
4'-Methoxy-3,3',5-stilbenetriol-3-glucoside.
EINECS 205-845-0; Glucopyranoside, rhapontigenin-3, β-D-; β-D-Glucopyranoside, 3-hydroxy-5-(2-(3-hydroxy-4-methoxyphenyl)ethenyl)phenyl; 3-Hydroxy-5-(2-(3-hydr-oxy-4-methoxyphenyl)ethenyl)phenyl-β-D-glucopyranos-ide; 3-Hydroxy-5-(2-(3-hydroxy-4-methoxyphenyl)-vinyl)-phenyl-β-D-glucopyranoside; NSC 43321; Ponticin; Rhaponticine; Rhapontigenin, 3-β-D-glucopyranoside; Rhapontin. The crystalline substance from the

541

common English rhubarb. Crystals; mp = 236° (dec); $[\alpha]_D^{32}$ = -59° (Me2CO); soluble in EtOH, H2O, Me2CO, less soluble in other organic solvents.

3289   Rhein
478-43-3   2860   G

C15H8O6

9,10-Dihydro-4,5-dihydroxy-9,10-dioxo-2-anthracene-carboxylic acid.
4-10-00-04088 (Beilstein Handbook Reference); 2-Anthracenecarboxylic acid, 9,10-dihydro-4,5-dihydroxy-9,10-dioxo-; 2-Anthroic acid, 9,10-dihydro-4,5-di-hydr-oxy-9,10-dioxo-; BRN 2222155; Cassic acid; CCRIS 5129; Chrysazin-3-carboxylic acid; 9,10-Dihydro-4,5-dihydroxy-9,10-dioxoanthracene-2-carboxylic acid; 9,10-Dihydro-4,5-dihydroxy-9,10-dioxo-2-anthracenecarbox-ylic acid; 1,8-Dihydroxyanthraquinone-3-carboxylic acid; EINECS 207-521-4; Monorhein; NSC 38629; Rheic acid; Rhein; Rhubarb Yellow; 4,5-dihydroxy-2-anthraquinone-carboxylic acid; 1,8-di-hydr-oxy-3-carboxyanthra-quinone. A yellow pigment found in rhubarb root. Yellow needles; mp = 321°; bp suplimes; λ_m = 229, 258, 435 nm (ε 36800, 20100, 11100, MeOH); slightly soluble in H2O (< 0.1 g/100 ml), EtOH, Et2O, Me2CO, C6H6, CHCl3, soluble in H2SO4, alkali, very soluble in C5H5N.

3290   Rhodamine B
81-88-9   8266   G

C28H31ClN2O3
9-(2-carboxyphenyl)-3,6-bis(diethylamino)xanthylium chloride.
11411 Red; Acid Brilliant Pink B; ADC Rhodamine B; Aizen Rhodamine BH; Aizen Rhodamine BHC; Akiriku Rhodamine B; Ammonium, (9-(o-carboxyphenyl)-6-(di-ethylamino)-3H-xanthen-3-ylidene)diethyl-, chloride; Basazol Red 71P; Basic Rose Extract; Basic Rose Red; Basic Violet 10; Basonyl Red 540; Basonyl Red 545; Basonyl Red 545FL; Brilliant Pink B; 9-(2-Carboxy-phenyl)-3,6-bis(diethylamino)xanthylium chloride; 9-o-Carboxyphenyl-6-diethylamino-3-ethylimino-3-isoxanth-ene, 3-; Calcozine Red BX; Calcozine Rhodamine BXP; CCRIS 3985; Cerise Toner X 1127; Certiqual Rhodamine; CI 45170; CI Basic Violet 10; CI Food Red 15; Cogilor Red 321.10; Cosmetic Brilliant Pink Bluish D Conc; D&C Red 19; D&C Red No. 19; Diabasic Rhodamine B; Edicol Supra Rose BS; Edicol Supra Rose B; EINECS 201-383-9; Elcozine Rhodamine B; Eriosin Rhodamine B; Ethanaminium, N-(9-(2-carboxyphenyl)-6-(diethylamino)-3H-xanthen-3-ylidene)-N-ethyl-, Cl; Ethanaminium, N-(9-(2-carboxyphenyl)-6-(diethylamino)-3H-xanthen-3-ylid-ene)-N-e- thyl-, cl; FD&C Red No. 19; Food Red 15; Geranium Lake N; Hexacol Rhodamine B Extra; HSDB 5244; Ikada Rhodamine B; Iragen Red L-U; Mitsui Rhodamine BX; N-(9-(2-Carboxyphenyl)-6-(diethyl-amino)-3H-xanthen-3-ylidene)-N-e-

thylethanaminium cl; NSC 10475; Red No. 213; Rheonine B; Rhodamine; Rhodamine B; Rhodamine B 20-7470; Rhodamine B chloride; Rhodamine B Extra S; Rhodamine B Extra; Rhodamine B Extra M 310; Rhodamine B500; Rhodamine B500 hydrochloride; Rhodamine BA; Rhodamine BA Export; Rhodamine BF; Rhodamine BL; Rhodamine, Blue Shade; Rhodamine BN; Rhodamine BS; Rhodamine BX; Rhodamine BXL; Rhodamine BXP; Rhodamine FB; Rhodamine FB CL; Rhodamine Lake Red B; Rhodamine O; Rhodamine S; Rhodamine, tetraethyl-; Sicilian Cerise Toner A-7127; Symulex Magenta F; Symulex Pink F; Symulex Rhodamine B Toner F; Takaoka Rhodamine B; Tetraethyldiamino-o-carboxyphenyl xanthenyl chloride; Tetraethylrhodamine; Violet zasadita 10; Xanthylium, 9-(2-carboxyphenyl)-3,6-bis(diethylamino)-, chloride. A dyestuff which colors wool and silk bluish-red with fluorescence also tannined cotton, violet-red. mp = 165°; λ_m = 546 nm (ε 72443 EtOH); soluble in H2O, EtOH; LD50 (rat iv) = 89.5 mg/kg.

3291   Rhodamine G
989-38-8   G

C28H31ClN2O3
Xanthylium, 9-[2-(ethoxycarbonyl)phenyl]-3,6-bis(ethyl-amino)-2,7-dimethyl-, chloride.
Aizen Rhodamine 6GCP; Basic Red 1; Basic Rhodamine Yellow; Benzoic acid, 2-(6-(ethylamino)-3-(ethylimino)-2,7-dimethyl-3H-xanthen-9-yl)-, ethyl ester, HCl Benzoic acid, o-(6-(ethylamino)-3-(ethylimino)-2,7-dimethyl-3H-xanthen-9-yl)-, ethyl ester, HCl C.I. 45160; C.I. Basic Red 1; C.I. Basic Red 1, monohydrochloride; Calcozine Red 6G; Calcozine Rhodamine 6GX; CCRIS 2388; Cerven zasadita 1; CI 45160; CI Basic Red 1; CI Basic Red 1, monohydrochloride; EINECS 213-584-9; Elcozine Rhodamine 6GDN; Eljon Pink Toner; 9-(2-(Ethoxy-carbonyl)phenyl)-3,6-bis(ethylamino)-2,7-dimethylxanth-ylium chloride; Fanal Pink B; Fanal Pink GFK; Fanal Red 25532; Flexo Red 482; Heliostable Brilliant Pink B Extra; HSDB 4179; Mitsui Rhodamine; Mitsui Rhodamine 6GCP; NCI-C56122; NSC 36345; Nyco Liquid Red GF; Red 169; Rh 6G; Rhodamin 6G; Rhodamine 4GD; Rhodamine 4GH; Rhodamine 590 chloride; Rhodamine 5GDN; Rhodamine 5GL; Rhodamine 6 G extra; Rhod-amine 6 GDN Extra; Rhodamine 6 GDN; Rhodamine 69DN Extra; Rhodamine 6G; Rhodamine 6G; Rhodamine 6G; Rhodamine 6G (biological stain); Rhodamine 6G chloride; Rhodamine 6G Extra; Rhodamine 6G Extra Base; Rhodamine 6G hydrochloride; Rhodamine 6G Lake; Rhodamine 6GB; Rhodamine 6GBN; Rhodamine 6GCP; Rhodamine 6GD; Rhodamine 6GDN; Rhodamine 6GDN Extra; Rhodamine 6GEX ethyl ester; Rhodamine 6GH; Rhodamine 6GO; Rhodamine 6GX; Rhodamine 6JH; Rhodamine 6Zh-DN; Rhodamine 6ZH; Rhodamine 6ZH-DN; Rhodamine 7JH; Rhodamine F 5GL; Rhod-amine F4G; Rhodamine F5G; Rhodamine F5G chloride; Rhodamine GDN; Rhodamine J; Rhodamine Lake Red 6G; Rhodamine Y 20-7425; Rhodamine ZH; Rhodanine 6GDN; Silosuper Pink B; Vali Fast Red 1308; Xanthylium, 9-(2-(ethoxycarbonyl)phenyl)-3,6-bis(ethylamino)-2,7-dimethyl-, chloride; C.I. basic red 1, monohydrochloride; rhodamine 6gex ethyl ester; silosuper pink b; fanal pink gfk; rhodamine f4g; rhodamine 6gb; rhodamine gdn; rhodamine 6 gdn; rhodamine 4gh; rhodamine 6gx; rhodamine lake red 6g; rhodamine y 20-7425; rhodamine zh; rhodamine 6zh. Dyestuffs, consisting chiefly of triethyl-rhodamine; dyes wool, silk, and tannined cotton,

red. Crystals; insoluble in H_2O (<0.1 g/100 ml). *Mitsui Toatsu.*

3292   Rhodapex® 674/C
25446-78-0   G

$C_{19}H_{39}NaO_7S$
2-[2-[2-(tridecyloxy)ethoxy]ethoxy]ethanol hydrogen sulf-ate sodium salt.
Sodium trideceth sulfate. Surfactant. *Rhône-Poulenc; Surf.*

3293   Rhodinol
6812-78-8   8269   G

$C_{10}H_{20}O$
S-(-)-3,7-Dimethyl-7-octen-1-ol.
(S)-3,7-Dimethyl-7-octen-1-ol; EINECS 229-887-4; FEMA No. 2980; 7-Octen-1-ol, 3,7-dimethyl-, (3S)-; 7-Octen-1-ol, 3,7-dimethyl-, (S)-; Rhodinol. A terpene alcohol prepared from the oils of rose, geranium, and citronella. It is essentially pure geraniol and is used in perfume manufacture and as a flavoring agent. Oil; bp12 = 114-115°; d^{20}= 0.8549; $[\alpha]_D^{20}$ = -2.9°; λ_m = 186-189 nm (ϵ 9000); slightly soluble in H_2O, very soluble in EtOH, Et_2O.

3294   Rhodium
7440-16-6   8270   G

Rh

Rh
Rhodium.
EINECS 231-125-0; HSDB 2534; Rh; Rhodium; Rhodium, elemental; Rhodium fume; Rhodium, metal. Metallic element in the platinum group; alloy with platinum for high temperature thermocouples, furnace windings, laborabory crucibles, spinerets for rayon, electrical contacts, jewelry, catalyst, mp = 1966°; d^{20} = 12.41. *Atomergic Chemetals; Degussa AG; Noah Chem.; Sigma-Aldrich Fine Chem.*

3295   Rhodium chloride
10049-07-7   8272   G

Cl_3Rh
Rhodium(III) chloride.
EINECS 233-165-4; Rhodium chloride; Rhodium chloride (RhCl3); Rhodium trichloride; Rhodium(III) chloride (1:3). Used in manufacture of rhodium trifluoride. Solid; insoluble in H_2O; LD50 (rat iv) = 198 mg/kg. *Atomergic Chemetals; Degussa AG; Noah Chem.; Sigma-Aldrich Fine Chem.*

3296   Ricinoleic acid
141-22-0   8295   G

$C_{18}H_{34}O_3$
9-Octadecenoic acid, 12-hydroxy-, [R-(Z)]-.
Acide ricinoleique; l'Acide ricinoleique; AI3-17956; Castor oil acid; D-12-Hydroxyoleic acid; EINECS 205-470-2; HSDB 5634; (9Z,12R)-12-Hydroxy-9-octadecen-saeure; 12-Hydroxy-9-octadecenoic acid; 12-Hydroxy-cis-9-octadecenoic acid; (R-(Z))-12-Hydroxy-9-octadec-enoic acid; Kyselina 12-hydroxy-9-oktadecenova; Kysel-ina ricinolova; NSC 281242; 9-Octadecenoic acid, 12-hydroxy-; 9-Octadecenoic acid, 12-hydroxy-, (cis)-; 9-Octadecenoic acid, 12-hydroxy-, (R-(Z))-; 9-Octadecenoic acid, 12-hydroxy-, (Z)-; 9-Octadecenoic acid, 12-hydroxy-, (9Z,12R)-; Oleic acid, 12-hydroxy-; Ricinic acid; Ricinoleic acid; Ricinolic acid. Chemical intermediate; imparts lubricity and rust-proofing to soluble cutting oils; basis for grease, soaps, resin plasticizers, and ethoxylated derivatives. Liquid; mp = 5°; bp120 = 245°, bp10 = 227°; d^{221} = 0.9450; $[\alpha]_D^{02}$= +5.1° (c = 5 Me2CO); insoluble in H_2O, very soluble in EtOH, Et_2O. *CasChem.*

3297   Rintal®
58306-30-2   3973   G V

$C_{20}H_{22}N_4O_6S$
[[2-[(methoxyacetyl)amino]-4-(phenylthio)phenyl]carbon-imidoyl]biscarbamic acid dimethyl ester.
BAY Vh5757; BRN 2195764; Carbamic acid, ((2-((methoxyacetyl)amino)-4-(phenylthio)phenyl)carbon-imidoyl)bis-, dimethyl ester; Combotel; Dimethyl ((2-(2-methoxyacetamido)-4-(phenylthio)phenyl)imido-carbon-yl)dicarbamate; Drontal; Drontal Plus; EINECS 261-205-0; Febantel; Febantelum; ((2-((Methoxyacetyl)-amino)-4-(phenylthio)phenyl)carbonimidoyl)bis(carbamic acid), di-methyl ester; N-(2-(2,3-Bis-(methoxycarbonyl)-guanid-ino)-5-(phenylthio)-phenyl)-2-methoxyacetamide; Nega-bot Plus Paste; Oratel; Rintal; Vercom. Broad-spectrum anthelmintic for use in sheep, goat, camels, cattle, horses; veterinary medicine. Used to control strongyles, ascarids and pinworms in horses and intestinal parasites in dogs. Crystals; mp = 129-130°. *Bayer AG.*

3298   Risedronic acid
105462-24-6   8315   D

$C_7H_{11}NO_7P_2$
[1-Hydroxy-2-(3-pyridyl)ethylidene]diphosphonic acid.
Acide risedronique; Acido risedronico; Acidum risedronicum; NE 58019; Risedronate; Risedronic acid. Calcium regulator, used as a bone resorption inhibitor. *Merck & Co.Inc.; Procter & Gamble Pharm. Inc.*

3299   Risedronic acid monosodium salt
115436-72-1   8315   D

C7H10NNaO7P2

Sodium   trihydrogen   [1-hydroxy-2-(3-pyridyl)ethylidene]-
diphosphonate.
Actonel; NE-58095; Risedronate sodium; Risedronic acid
monosodium salt. Calcium regulator, used as a bone resorption
inhibitor in treatment of osteoporosis. *Merck & Co.Inc.; Procter &
Gamble Pharm. Inc.*

3300   Risolex
57018-04-9   9587   G P

C9H11Cl2O2PS

O-(2,6-dichloro-4-methylphenyl) O,O-dimethyl phosph-orothioate.
BRN 2136521; O-(2,6-Dichloro-p-tolyl) O,O-dimethyl ester of
phosphorothioic   acid;   O-(2,6-Dichloro-p-tolyl)   O,O-dimethyl
thiophosphate;   O-(2,6-Dichloro-4-methylphenyl)   O,O-dimethyl
phosphorothioate;   O-2,6-Dichloro-p-tolyl   O,O-dimethyl
phosphorothioate;   O,O-Dimethyl   O-(2,6-dichloro-4-
methylphenyl)phosphoro-thioate;   EINECS   260-515-3;
Phosphorothioic acid, O-(2,6-dichloro-4-methylphenyl) O,O-dimethyl
ester; Phosphorothioic acid, O-(2,6-dichloro-p-tolyl) O,O-dimethyl
ester; Risolex; Rizolex; S-3349; Toclofos-methyl; Tolclofos-methyl.
An organophosphorus fungicide which gives protection against soil-
borne diseases. Colorless crystals; mp = 78 - 80°; poorly soluble in
H2O (0.00003 - 0.00004 g/100 ml at 23°), soluble in Me2CO (39.5
g/100 ml), CHCl3 (72.5 g/100 ml), cyclohexanone (50.8 g/100 ml),
xylene (46.4 g/100 ml), MeOH (4.8 g/100 ml), C6H14 (2.6 g/100 ml);
LD50 (rat orl) = 5000 mg/kg, (mus orl) = 23500 - 3600 mg/kg, (rat
der) > 5000 mg/kg, (rat ip) = 4900 - 5000 mg/kg, (mus ip) = 1070 -
1260 mg/kg, (rat, mus sc) > 5000 mg/kg, (mallard duck, bobwhite
quail orl) > 5000 mg/kg; LC50 (carp 96 hr.) = 2.13 mg/l. *Schering
Agrochemicals Ltd.; Sumitomo Corp.*

3301   Risperidone
106266-06-2   8316   D

C23H27FN4O2

3-[2-[4-(6-Fluoro-1,2-benzisoxazol-3-yl)piperidino]ethyl]-6,7,8,9-
tetrahydro-2-methyl-4H-pyrido[1,2-a]pyridimin-4-one.
BRN 4891881; R-64766; Risperdal; Risperdal Consta; Risperidona;
Risperidone; Risperidonum; Risperin; Rispolept; Rispolin; Sequinan.
A combined serotonin and dopamine receptor antagonist, a
benzisoxazole anti-psychotic. Antipsychotic. Crystals; mp = 170°;
LD50 (rat iv) = 29.7 mg/kg, (dog iv) = 14.1 mg/kg, (rat orl), 82.1=
mg/kg, (dog orl) = 18.3 mg/kg. *Janssen Pharm. Ltd.*

3302   Rofecoxib
162011-90-7   8330   D

C17H14O4S

4-[4-(Methylsulfonyl)phenyl]-3-phenyl-2(5H)-furanone.
MK 966; Rofecoxib; Vioxx. A COX-2 specific inhibitor used in the
treatment of osteoarthritis. Analgesic; anti-inflammatory. Crystals;
Sparingly soluble in Me2CO; slightly soluble in MeOH, isopropyl
acetate; very slightly soluble in EtOH; practically insoluble in octanol;
insoluble in H2O. *Merck & Co.Inc.*

3303   Ronnel
299-84-3   8415   G P

C8H8Cl3O3PS

Dimethyl trichlorophenyl thiophosphate.
4-06-00-00994 (Beilstein Handbook Reference); Al3-23284; Blitex;
BRN 1885571; Caswell No. 724; Dermafos; Dermafosu; Dermaphos;
Dimethyl (2,4,5-trichlorophenyl) phosphorothionate; O,O-Dimethyl O-
(2,4,5-trichlorophenyl)   phosphorothioate;   O,O-Dimethyl-O-(2,4,5-
trichlorphenyl)-thionophosphat;   O,O-Dimethyl   O-(2,4,5-
trichlorophenyl)thiophosphate; Dow ET 14; Dow ET 57; Ectoral;
EINECS 206-082-6; ENT 23,284; EPA Pesticide Chemical Code
058301; ET 14; ET 57; Etrolene; Fenchloorfos; Fenchlorfos;
Fenchlorfosu; Fenchlorphos; Fenclofos; Fenclofosum; Gesektin K;
HSDB 667; Korlan; Korlane; Moorman's Medicated Rid-Ezy;
Nanchor; Nanker; Nankor; NSC 8926; OMS 123; Phenchlorfos;
Phenol,   2,4,5-trichloro-,   O-ester   with   O,O-dimethyl
phosphorothioate; Phosphorothioic acid, O,O-dimethyl O-(2,4,5-
trichlorophenyl)   ester;   Remelt;   Ronnel;   Rovan;   Smear;
Thiophosphate de O,O-dimethyle et de O-(2,4,5-trichlorophenyl);
Trichlormetaphos;   O-(2,4,5-Trichloor-fenyl)-O,O-dimethyl-
monothiofosfaat;   Trichlorometafos;   O-(2,4,5-Trichlor-phenyl)-O,O-
dimethyl-monothiophos-phat;   O-(2,4,5-Tricloro-fenil)-O,O-dimetil-
monotiofos-fato; Trolene; Trolene 20L; Viozene. Insecticide with
systemic action. Used on cattle for the control of ticks, files,
maggots, and lice. Solid; mp = 41°; bp0.4 = 151-1564°; d32 = 1.44;
insoluble in H2O, soluble in organic solvents; LD50 (rat orl) = 2630
mg/kg. *Dow UK.*

3304 Rose bengal sodium
632-69-9   8343   G

$C_{20}H_2Cl_4I_4Na_2O_5$

4,5,6,7-Tetrachloro-3',6'-dihydroxy-2',4',5',7'-tetraiodo-spiro[isobenzofuran-1(3H),9'[9H]xanthen]-3-one dipotas-sium salt.

Disodium 3,4,5,6-tetrachloro-2-(2,4,5,7-tetraiodo-6-ox-ido-3-oxoxanthen-9-yl)benzoate; EINECS 211-183-3; Flu-orescein, 4,5,6,7-tetrachloro-2',4',5',7'-tetraiodo-, di-sodium salt; Food Red Color No. 105, sodium salt; Food Red No. 105, sodium salt; R105 sodium; Rose bengal disodium salt; Rose bengal sodium; Rose bengale; Spiro(isobenzofuran-1(3H),9'-(9H)xanthen)-3-one, 4,5,6, 7-tetrachloro-3',6'-dihydroxy-2',4',5',7'-tetraiodo-, di-sodium salt; 2,4,5,7-Tetraido(m,p,o',m')tetrachloro-flu-orescein, disodium salt; 4,5,6,7-Tetrachloro-2',4',5',7'-tetraiodofluorescein disodium salt; 9-(3',4',5',6'-Tetra-chloro-o-carboxyphenyl)-6-hydroxy-2,4,5,7-tetraiodo-3-isoxanthone 2Na;.

3305 Rose ester
90-17-5   G

$C_{10}H_9Cl_3O_2$

2,2,2-Trichloro-1-phenylethyl acetate.

4-06-00-03049 (Beilstein Handbook Reference); Acetic acid, α-(trichloromethyl)benzyl ester; AI3-02454; Benzenemethanol, α-(trichloromethyl)-, acetate; Benzyl alcohol, α-(trichloromethyl)-, acetate; BRN 2270144; EINECS 201-972-0; NSC 165582; Rosacetol; Rose crystals; Trichlor phenyl methyl carbinyl acetate; Trichloromethyl phenyl carbinyl acetate; α-(Trichloro-methyl)benzyl acetate; α-(Trichloromethyl)benzenemeth-anol acetate; α-(Trichloromethyl)benzyl alcohol acetate;. A rose scent used in the perfumery industry. Liquid; mp = 85-87°. *Bedoukian Research Inc.; Frutarom; Nipa.*

3306 Rosiglitazone
122320-73-4   8346   D

$C_{18}H_{19}N_3O_3S$

(±)-5-[p-[2-(Methyl-2-pyridylamino)ethoxy]benzyl]-2,4-thiazolidinedione.

Avandia; BRL 49653; Rosiglitazone. Antidiabetic; safe if used in combination with metformin hydrochloride. *Bristol-Myers Squibb*

Pharm. R&D; SmithKline Beecham Pharm.

3307 Rosiglitazone maleate
155141-29-0   8346   D

$C_{22}H_{23}N_3O_7S$

(±)-5-[p-[2-(Methyl-2-pyridylamino)ethoxy]benzyl]-2,4-thiazolidinedione maleate (1:1).

Avandia; BRL 49653-C; Rosiglitazone Maleate. Anti-diabetic; safe if used in combination with metformin hydrochloride. *Bristol-Myers Squibb Pharm. R&D; SmithKline Beecham Pharm.*

3308 Rosin

8050-09-7   948, 8347   E H P

Colophony.

BALS 3A; Bandis G100; Caswell No. 667; Colophony; Disproportionated rosin; EINECS 232-475-7; EM 3; EPA Pesticide Chemical Code 067205; Gum rosin; Highrosin; Hongkong rosin WW; KE 709; Rondis R; Rosin; Rosin gum; Rosin WW; Shiragiku rosin; Wood rosin; WW Wood rosin; Yellow resin. Registered by EPA as a fungicide and insecticide. FDA listed, Japan approved, FDA approved for orals. Used as a stiffening agent in orals, ointments and coated and sustained action tablets. Also used in varnishes and paints, printing inks, cements and soaps. Amber translucent solid; mp = 100-150°; d = 0.987 - 0.994; $[\alpha]_D^{20}$ = +1° - +4°; insoluble in H_2O, soluble in EtOH, C_6H_6, Et_2O, AcOH, CS_2, oils. *Akzo Chemie; Arakawa Chem. Ind.; Arizona; Browning; Chemcentral; Cytec; Fluka; Focus; Georgia Pacific; Hercules Inc.; Meer; Natrochem; Punda Mercantile; Spectrum Chem. Manufacturing; Veitsiluoto Oy; Westvaco.*

3309 Rosin pentaerythritol ester
8050-26-8   H

Pentaerythritol rosinate.

EINECS 232-479-9; Pentaerythritol ester of rosin; Pentaerythritol rosinate; Resin acids and Rosin acids, esters with pentaerythritol; Rosin, pentaerythritol ester; Rosin, pentaerythritol polymer; Rosin, pentaerythritol resin.

3310 Rotenone
83-79-4   8350   G H P

$C_{23}H_{22}O_6$

(2R-(2α,6aα,12aα))-1,2,12,12a-Tetrahydro-8,9-dimeth-oxy-2-(1-methylethyl)benzopyrano(3,4-b)furo(2,3-h)(1)-benzopyran-6(6aH)-one.

AI3-00133; Barbasco; Canex; Caswell No. 725; CCRIS 895; Cenol garden dust; Chem-mite; Chem-Fish; Chem-Fish Synergized; Cube extract; Cube-Pulver; Cube root; Cubor; Curex flea duster; Dactinol; Deril; Derrin; Derris; Derris (insecticide); Derris root; Dri-Kil; ENT 133; EPA Pesticide Chemical Code 071003; Extrax; Fish-Tox; Foliafume; Foliafume E.C.; Gerane; Green cross warble powder; Haiari; HSDB 1762; Hydrogenated rotenone; Liquid derris; Mexide; NCI-C55210; Nekoe; Nicouline; Noxfire; Noxfish; NSC 26258; Nusyn; Nusyn-Noxfish; Paraderil; PB-Nox; Powder and root; Prenfish; Prentox; Prentox Synpren-Fish; Pro-Nox fish; Protax; RO-KO; Ronone; Rotacide; Rotacide E.C.; Rotefive; Rotefour; Rotenon; Rotenona [Spanish]; Rotenone; (-)-cis-Rotenone; (-)-Rotenone; 5'-β-Rotenone; Rotenox; Rotenox 5EC; Rotessenol; Rotocide; Synpren; Tubatoxin; tubotoxin. A crystalline material found in the roots of derris, a plant grown in the rubber plantations of the Malay Peninsula; also found in the South American cube plant. It occurs up to 5.5% in derris and up to 7% in cube. Registered by EPA as an insecticide and acaricide. Orthorhombic crystals (dimorphic); mp = 165-166°, 185-186°; $[\alpha]_D^{20}$ = -228° (c = 2.22 C_6H_6); slightly soluble in H_2O (0.0015 g/100 ml), readily soluble in Me_2CO, CS_2, EtOAc, $CHCl_3$, moderately soluble in Et_2O, alcohols, CCl_4, petroleum ether; LD_{50} (mus ip) = 2.8 mg/kg, (rat orl) = 132 - 1500 mg/kg, (rat iv) = 6 mg/kg, (mus orl) = 350 mg/kg, (hmn orl) = 300 - 500 mg/kg, very toxic to pigs; LC_{50} (rainbow trout 96 hr.) = 0.031 mg/l, (bluegill sunfish 96 hr.) = 0.023 mg/l; non-toxic to bees. AgrEvo; Amvac Chemical Corp.; Bonide Products, Inc.; Chem-Tech Ltd.; Dragon Chemical Corp.; Earth Care; Farmland Industries Inc.; Foreign Domestic Chemicals Corp.; Green Light Co.; ICI Garden Products; Prentiss Inc.; Rockland Corp.; Speer Products Inc.; Tifa Ltd.; Value Gardens Supply LLC.

3311   Rubidium carbonate
584-09-8   G

CO_3Rb_2
Carbonic acid rubidium salt.
Carbonic acid dirubidium salt; Dirubidium carbonate; Dirubidium monocarbonate; EINECS 209-530-9; NSC 112222; Rubidium carbonate. Used in the manufacture of special glass formulations. Solid; mp = 837°. Atomergic Chemetals; Cabot Carbon Ltd.; Cerac; Noah Chem.

3312   Rubidium chloride
7791-11-9   8365   G

Rb—Cl

RbCl
Hydrochloric acid rubidium salt.
EINECS 232-240-9; NSC 84273; Rubidium-chlorid; Rubidium chloride; Rubidium chloride (RbCl); Rubidium monochloride. Analysis (testing for perchloric acid), source of rubidium metal. Also used as a gasoline additive and as an antidepressant. Solid; mp = 718°; bp = 1390°; d = 2.8000. Atomergic Chemetals; Cabot Carbon Ltd.; Cerac; Noah Chem.

3313   Rubidium iodide
7790-29-6   8367   D G

Rb—I

IRb
Iodic acid rubidium salt.
EINECS 232-198-1; Rubidium-iodid; Rubidium iodide; Rubidium iodide (RbI); Rubidium jodatum. Used as a source of iodine. Crystals; mp = 642°, bp = 1300°; d = 3.55; soluble in H_2O (150 g/100

ml), EtOH. Atomergic Chemetals; Cabot Carbon Ltd.; Cerac; Noah Chem.

3314   Rubidium sulfate
7488-54-2   G

O_4Rb_2S
Sulfuric acid, rubidium salt.
Dirubidium sulfate; EINECS 231-301-7; NSC 84274; Rubidium sulfate; Rubidium sulfate ($Rb_2(SO_4)$); Rubidium sulphate; Sulfuric acid, dirubidium salt. Atomergic Chemetals; Cabot Carbon Ltd.; Cerac; Noah Chem.

3315   Rubierythric acid
152-84-1   8361   G

$C_{25}H_{26}O_{13}$
1-Hydroxy-2-[(6-O-β-D-xylopyranosyl-β-D-glucopyran-osyl)oxy]-9,10-anthracenedione.
4-17-00-03462 (Beilstein Handbook Reference); Alizarin primeveroside; BRN 0071586; CCRIS 4531; EINECS 205-808-9; Glucopyranoside, 1-hydroxy-2-anthraquinonyl 6-O-β-D-xylopyranosyl-, β-D-; β-D-1-Hydroxy-2-anthra-quinonyl 6-O-β-D-xylopyranosylglucopyranoside; 1-Hyd-roxy-2-((6-O-β-D-xylopyranosyl-β-D-glucopyran-osyl)-oxy)anthraquinone; Ruberythric acid; Rubianic acid. Crystals; mp = 259-261°; soluble in H_2O, less soluble in organic solvents.

3316   Rutin trihydrate
153-18-4   8383   G

$C_{27}H_{30}O_{16}$
3-[[6-O-(6-Deoxy-α-L-mannopyranosyl)-β-D-glucopyran-osyl]oxy]-2-(3,4-dihydroxyphnyl)-5,7-dihydroxy-4H-1-benzopyran-4-one.
5-18-05-00519 (Beilstein Handbook Reference); AI3-19098; Bioflavonoid; Birutan; Birutan Forte; BRN 0075455; C.I. 75730; CCRIS 7564; EINECS 205-814-1; Eldrin; Flavone, 3,3',4',5,7-pentahydroxy-, 3-(O-rham-nosylglucoside); Globulariacitrin; Globularicitrin; Gluco-pyranoside, quercetin-3 6-O-α-L-rhamnopyranosyl-, β-D; Glucopyranoside, quercetin-3 6-O-(6-deoxy-α-L-manno-pyranosyl)-, β-D-; 3,3',4',5,5',7-Hexahydroxy-flavone (6-O-α-L-rhamnosyl-β-D-glucoside); Ilixanthin; Melin; Myrticolorin; NSC 9220; Osyritin; Oxyritin; Paliuroside; 3,3',4',5,7-Pentahydroxyflavone-3-rutinoside; Phytomel-in; Quercetin, 3-(6-O-(6-deoxy-α-L-mannopyran-osyl)-β-D-glucopyranoside); Quercetin, 3-

546

(6-O-α-L-rhamno-pyranosyl-β-D-glucopyranoside); Quercetin rhamno-glucosine; Quercetol 3-rhamnoglucoside; Quercetin 3-rutinoside; Quercetin-3β-rutinoside; Quercetin 3-O-rutinoside; Quercetin 3-O-β-D-rutinoside; 3-Rhamno-glucoside of 3,3',4',5,7-pentahydroxyflavone; 3-Rhamno-glucosylquercetin; 3-Rutinosyl quercetin; Rutabion; Rutin; Rutin trihydrate; Rutine; Rutinic acid; Rutinion acid; Rutinoside, 2-(3,4-dihydroxyphenyl)-5,7-dihydroxy-4-oxo-4H-1-; Rutinoside, quercetin-3, β-; Rutinum; Rutosid; Rutoside; Rutosido; Rutosidum; Rutozyd; Sophorin; Tanrutin; Troxerutin; USAF CF-5; Venoruton; Viola-quercitrin; Vitamin P. Protects capillary structure. Crystals; mp = 214-215° (dec); $[\alpha]_D^{23}$= 14° (EtOH); insoluble in H_2O, more soluble in polar organic solvents; LD50 (mus iv) = 950 mg/kg. *Lederle Labs.*

3317 Saccharin
81-07-2   8390   G H

C7H5NO3S
1,2-Benzisothiazolin-3-one 1,1-dioxide.
AI3-38107; Anhydro-o-sulfaminebenzoic acid; Benzo-2-sulphimide; Benzoic acid sulfimide; Benzoic sulfimide; Benzoic sulphimide; Benzosulfimide; Benzosulfinide; Benzosulfimide, O-; Benzosulphimide; Benzoylsulfonic Imide; CCRIS 3707; EINECS 201-321-0; Garantose; Glucid; Gluside; Hermesetas; HSDB 669; Insoluble saccharin; Kandiset; Natreen; NSC 5349; RCRA waste number U202; Sacarina; Saccharimide; Saccharin; Saccharin acid; Saccharin and salts; Saccharin insoluble; Saccharina; Saccharine; Saccharinol; Saccharinose; Saccharol; Sacharin; Saxin; Sucre edulcor; Sucrette; Sulfobenzimide, O-; Sweeta; Sykose; Zaharina. Non-nutritive sweetener (500 times sweeter than cane sugar), organic compound; noncaloric, pharmaceutic aid; in formulations for electroplating bath brighteners. Crystals; mp = 228° (dec); d^{25} = 0.828; soluble in H_2O, EtOH, Et2O, Me2CO, C6H6, CHCl3, DMSO. *Aisan Chem. Co. Ltd.; Dinoval; Greeff R.W. & Co.; Maruzen Fine Chem.; PMC; Rit-Chem; Spice King.*

3318 Saccharin ammonium
6381-61-9   H

C7H8N2O3S
Ammonium 1,2-benzisothiazolin-3-one 1,1-dioxide.
Ammonium 1,2-benzisothiazolin-3-one 1,1-dioxide; Ammonium O-benzosulfimide; Ammonium saccharin; 1,2-Benzisothiazolin-3-one, 1,1-dioxide, ammonium salt; Daramin; EINECS 228-971-8; Saccharin ammonium; Saccharinate ammonium.

3319 Safrole
94-59-7   8395   G

C10H10O2
5-(2-propenyl)-1,3-benzodioxole.
5-19-01-00553 (Beilstein Handbook Reference); 5-Allyl-1,3-benzodioxole; Allylcatechol methylene ether; Allyldioxybenzene methylene ether; 1-Allyl-3,4-methylenedioxybenzene; m-Allylpyrocatechin methylene ether; Allylpyrocatechol methylene ether; 4-Allyl-pyrocatechol; 4-Allylpyrocatechol formaldehyde acetal; AI3-00514; Benzene, 1,2-methylenedioxy-4-allyl-; Benz-ene, 4-allyl-1,2-(methylenedioxy)-; 1,3-Benzo-dioxole, 5-(2-propenyl)-; 1,3-Benzodioxole, 5-allyl-; BRN 0136380; Caswell No. 729; CCRIS 553; EINECS 202-345-4; EPA Pesticide Chemical Code 097901; HSDB 2653; 1,2-Methylenedioxy-4-allylbenzene; 3,4-(Methylenedioxy)-allylbenzene; NSC 11831; 5-(2-Propenyl)-1,3-benzo-dioxole; RCRA waste no. U203; RCRA waste number U203; Rhyuno oil; Safrene; Safrol; Safrole; Sassafras; Shikimole; Shikomol. The methylene ether of allyl-pyrocatechol, It is the chief constituent of oil of sassafras, and is obtained from red oil of camphor; used in the place of oil of sassafras, in perfumes, soaps, and medicine. Oil; mp = 11°; bp = 234.5°; d^{20} = 1.1000; λ_m = 236, 285 nm (ε = 4169, 3802, EtOH); insoluble in H_2O, soluble in EtOH, Et2O, CHCl3; LD50 (rat orl) = 1750 mg/kg. *Dr. Ehrenstorfer GmbH; Protocol Analytical Supplies; SanhuiPerfume Chemical Co. Ltd.*

3320 Salicylaldehyde
90-02-8   8405   G H

C7H6O2
o-Hydroxybenzaldehyde.
4-08-00-00176 (Beilstein Handbook Reference); AI3-02174; Benzaldehyde, 2-hydroxy-; Benzaldehyde, o-hydroxy-; BRN 0471388; CCRIS 7451; EINECS 201-961-0; FEMA No. 3004; HSDB 721; NSC 49178; o-Formylphenol; o-Hydroxybenzaldehyde; SAH; Salicylal; Salicylaldehyde; Salicylic aldehyde. Used in analytical chemistry, perfumery, synthesis of coumarin, as an auxiliary fumigant and in flavoring. Oil; mp = -7°; bp = 197°; d^{20} = 1.1674; λ_m = 259, 328 nm (cyclohexane); very soluble in EtOH, Et2O, Me2CO, C6H6, less soluble in CHCl3, H2O; MLD (rat sc) = 900 - 1000 mg/kg. *Janssen Chimica; Penta Mfg.; Seimi Chem.*

3321 Salicylic acid
69-72-7   8411   D G P

C7H6O3
2-Hydroxybenzoic acid.
4-10-00-00125 (Beilstein Handbook Reference); Acido o-idrossibenzoico; Acido salicilico; Acidum salicylicum; Advanced Pain Relief Callus Removers; Advanced Pain Relief Corn Removers; AI3-02407; Benzoic acid, 2-hydroxy-; BRN 0774890; Caswell No. 731; CCRIS 6714; Clear away Wart Remover; Compound W; Domerine; Dr. Scholl's Callus Removers; Dr. Scholl's Corn Removers; Dr. Scholl's Wart Remover Kit; Duofil Wart Remover; Duoplant; EINECS 200-712-3; EPA Pesticide Chemical Code 076602; Fostex; Fostex Medicated Bar and Cream; Freezone; HSDB 672; Ionil; Ionil-Plus; Keralyt; Kyselina 2-hydroxybenzoova; Kyselina salicylova; NSC 180; Orthohydroxybenzoic acid; Pernox; Phenol-2-carboxylic acid; Psoriacid-S-stift; Retarder W; Rutranex; Salicylic acid; Salicylic Acid & Sulfur Soap; Salicylic Acid collodion; Salicylic Acid Soap; Saligel; Salonil; Sebucare; Sebulex; Solarcaine First Aid Spray; Stri-Dex; Verrugon; 51; 171; 311; 334; 337; 380; k537; k557. Registered

by EPA as an antimicrobial, fungicide and insecticide (cancelled). Aromatic acid; preservative for foods; used in manufacture of methyl salicylate, acetylsalicylic acid, dyestuffs, etc., as a reagent in analytical chemistry; a topical keratolytic agent. Used as a synthetic intermediate. Crystals; mp = 157-159°; bp$_{20}$ = 211°; d^{20} = 1.443; soluble in H$_2$O (0.22 g/100 ml), boiling H$_2$O (6.7 g/100 ml), EtOH (37 g/100 ml), Me$_2$CO (33.3 g/100 ml), CHCl$_3$ (2.4 g/100 ml), Et$_2$O (33.3 g/100 ml), C$_6$H$_6$ (0.74 g/100 ml), oil turpentine (1.9 g/100 ml), glycerol (1.67 g/100 ml); LD$_{50}$ (mus iv) = 500 mg/kg. *Bayer AG; Galderma Labs Inc.; Nippon Shinyaku Japan; Schering-Plough Pharm.; Sterling Health U.S.A.; Stiefel Labs Inc.; Westwood-Squibb Pharm. Inc.; Whitehall-Robins.*

3322 Salinomycin
53003-10-4   8415   G V

C$_{42}$H$_{70}$O$_{11}$
Salinomycin.
AHR 3096; Bio-cox; Coxistac; EINECS 258-290-1; HSDB 7032; K 364; Salinomicina; Salinomycin; Salinomycine; Salinomycinum. An ionophorous antibiotic used as an anticoccidial agent in poultry. Used in veterinary medicine as an anticoccidial agent. Crystals; mp = 112.5-113.5°; [α]^{25}$_D$ = -63° (c = 1, EtOH); λ$_m$ = 284 nm (ε = 126, EtOH-H$_2$O, 2:1); LD$_{50}$ (mus orl) = 50 mg/kg. *Pfizer Intl.*

3323 Salmeterol
89365-50-4   8416   D

C$_{25}$H$_{37}$NO$_4$
(±)-4-Hydroxy-α'-[[[6-(4-phenylbutoxy)hexyl]amino]-methyl]-m-xylene-α,α'-diol.
Aeromax; Astmerole; Benzenedimethanol; GR 33343X; Salmeterol; Salmeterolum; Xylene-α,α'-diol. Structural analog of albuterol. β2-Adrenergic agonist. Ephedrine derivative. Used as a bronchodilator. Crystals; mp = 75.5-76.5°. *Glaxo Labs.*

3324 Salmeterol xinafoate
94749-08-3   8416   D

C$_{36}$H$_{45}$NO$_7$
(±)-4-Hydroxy-α'-[[[6-(4-phenylbutoxy)hexyl]amino]-methyl]-m-xylene-α,α'-diol 1-hydroxy-2-napthoate (salt).
Arial; Asmerole; Beglan; Betamican; Dilamax; GR-3343G; Inaspir; Salmetedur; Salmeterol xinafoate; Serevent; Serevent Inhaler and Disks; Ultrabeta. Bronchodilator. Crystals; mp = 137-138°; soluble in MeOH; slightly soluble in EtOH, CHCl$_3$, iPrOH; sparingly soluble in

H$_2$O. *Glaxo Labs.*

3325 Salt, Amido-R
92-28-4   401   G

C$_{10}$H$_9$NO$_6$S$_2$
3-Amino-2,7-naphthalenedisulfonic acid.
Amido-R-acid; 2-Amino-3,6-disulfonaphthalene; 2-Aminonaphthalene-3,6-disulfonic acid; 3-Amino-2,7-naphthalenedisulfonic acid; 3-Aminonaphthalene-2,7-disulph-onic acid; Amino-R-acid; EINECS 202-143-6; 2-Naph-thylamine-3,6-disulfonic acid; 2,7-Naphthalenedisulfonic acid, 3-amino-; NSC 5525. Used in manufacture of dyes. Crystals; soluble in H$_2$O.

3326 Samarium
7440-19-9   8425   G

Sm

Sm
Samarium.
EINECS 231-128-7; Samarium. A rare-earth metallic element; neutron absorber, dopant for laser crystals, metallurgical research, permanent magnets. mp = 1074°; bp = 1794°; d = 7.536. *Atomergic Chemetals; Cerac; Rhône-Poulenc; Sigma-Aldrich Fine Chem.*

3327 Samarium oxide
12060-58-1   G

O$_3$Sm$_2$
Samaria.
Disamarium trioxide; EINECS 235-043-6; Samaria; Samarium (III) oxide; Samarium oxide; Samarium oxide (Sm$_2$O$_3$); Samarium sesquioxide; Samarium trioxide; Samarium(3+) oxide; Samarium(III) oxide. Catalyst in the dehydrogenation of ethanol, infrared-absorbing glass, neutron absorber, preparation of samarium salts. Solid; d = 8.347. *Noah Chem.; Rhône-Poulenc.*

3328 Santowhite
85-60-9   G

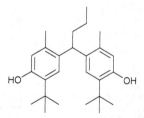

C$_{26}$H$_{38}$O$_2$
4,4'-Butylidenebis(6-t-butyl-m-cresol).
4-06-00-06807 (Beilstein Handbook Reference); Annulex PBA 15; Antage W300; Anullex PBA 15; BBM; 1,1-Bis(2-methyl-4-hydroxy-5-tert-butylphenyl)butane; 6,6'-Di-tert-butyl-4,4'-butylidenedi-m-cresol; BRN 2016567; 4,4'-Butylidene bis(6-tert-butyl-3-methylphenol); 4,4'-Butyl-idenebis(3-methyl-6-tert-butylphenol); 4,4'-Butylidene-bis(6-tert-butyl-m-cresol); CCRIS 4916; m-Cresol, 4,4'-butylidenebis(6-tert-butyl-; EINECS 201-618-5; HSDB 5250; Noclizer NS 30; NSC 67485; Phenol, 4,4'-butylidenebis(2-(1,1-dimethylethyl)-5-methyl-; Santo-white; Santowhite powder; Sumilit bbm; Sumilizer BBM;

Sumilizer BBM-S; SWP; SWP (antioxidant); Yoshinox BB. Antioxidant for polypropylene, polyethylene, nylon molding powders, and other polymer resins. *DuPont; Great Lakes Fine Chem.; Monsanto Co.*

3329   Sarcosine
107-97-1   8450   G

C3H7NO2
 N-Methylglycine.
 4-04-00-02363 (Beilstein Handbook Reference); Acetic acid, (methylamino)-; Al3-15410; BRN 1699442; EINECS 203-538-6; Glycine, N-methyl-; (Methylamino)acetic acid; N-Methylaminoacetic acid; (Methylamino)ethanoic acid; Methylglycine; N-Methylglycine; Methylglycocoll; Sarcosin; Sarcosine; Sarcosinic acid; Sargosine hydrochloride. Used in the synthesis of foaming antienzyme compounds for toothpaste, cosmetics, and pharmaceuticals. mp = 212° (dec); soluble in H2O (2.8 g/ml). *BASF Corp.; Grace W.R. & Co.; Hampshire; Lancaster Synthesis Co.*

3330   Sarcosine, monosodium salt
4316-73-8   H

C3H6NaNO2
 Sodium N-methylglycinate.
 EINECS 224-338-5; Glycine, N-methyl-, sodium salt; Hamposyl sodium salt; N-Methylglycine, sodium salt; NSC 10480; Sarcosine, monosodium salt.; Sarcosine sodium salt; SARCOSINE; Sodium (methylamino)acetate; Sodium N-methylaminoacetate; Sodium N-methylglycin-ate; Sodium sarcosinate.

3331   Scandium
7440-20-2   8468   G

Sc

Sc
 Scandium.
 EINECS 231-129-2; Scandium. Metallic element; no major industrial use; some application in semiconductor field; an artificial radioactive isotope has been used in tracer studies and leak detection. mp = 1541; bp = 2836°; d = 2.9890. *Cerac; Rhône-Poulenc;* 3973Atomergic; Cerac; Chemetals; Rhone-Poulenc.

3332   Scandium oxide
12060-08-1   G

O3Sc2
 Scandium(III) oxide.
 EINECS 235-042-0; Scandia; Scandium oxide; Scandium oxide (Sc2O3). Used in preparation of scandium fluoride. *Atomergic Chemetals; Cerac; Noah Chem.; Sigma-Aldrich Fine Chem.*

3333   Schaeffer's acid
93-01-6   6418   G

C10H8O4S
 2-Naphthol-6-sulfonic acid.
 4-11-00-00622 (Beilstein Handbook Reference); Armstrong acid; Baum's acid; 2-Hydroxy-6-naph-thalenesulfonic acid; 6-Hydroxy-2-naphthalenesulfonic acid; 6-Hydroxynaphthalene-2-sulphonic acid; BRN 0521151; EINECS 202-209-4; Kyselina 2-naftol-6-sulfonova; Kyselina schaferova; β-Naphthol-6-sulfonic acid; β-Naphtholsulfonic acid S; 2-Naphthol-6-sulfonic acid; 2-Naphthalenesulfonic acid, 6-hydroxy-; 2-Naphtol-6-sulfosaeure; 2-Naphtol-6-sulfosäure; NSC 5871; Schaeffer's β-naphtholsulfonic acid; Schaeffer's acid. Used as an azo dye intermediate. mp = 125°; λ_m = 275, 280, 330 nm (ϵ = 5012, 5012, 1000, EtOH); soluble in H2O, EtOH, insoluble in non-polar organic solvents. *Allchem Ind.; Taixing Chemical Co. Ltd.; Uni Impex India.*

3334   Scheibler's reagent
51312-42-6   8739   G

Na4O2.P2O5.W12O36.H36O18
 Sodium phosphotungstate.
 EINECS 257-132-9; Phosphotungstic acid, sodium salt; Sodium phosphotungstate; Sodium tungsten hydroxide oxide phosphate; Sodium tungstophosphate; Tungstophosphoric acid, sodium salt. Obtained by dissolving 100 g sodium tungstate and 70 g sodium phosphate in 500 ml water, and acidifying with nitric acid; used as a testing reagent for alkaloids.

3335   Schollkopf's acid
82-75-7   6431   G

C10H9NO3S
 1-Naphthylamine-8-sulfonic acid.
 1-Amino-8-naphthalene sulfonate; 1-Aminonaphthalene-8-sulfonic acid; 8-Aminonaphthalene-1-sulphonic acid; 8-Amino-1-naphthalenesulfonic acid; EINECS 201-437-1; Naphthylaminemonosulfonic acid S; α-Naphthylamine-8-sulfonic acid; 1-Naphthalenesulfonic acid, 8-amino-; 1-Naphthylamine-8-sulfonic acid; NSC 7798; peri Acid. Used in the manufacture of dyestuffs as an azo dye intermediate. Soluble in H2O (20 mg/100 ml at 20°, 416 mg/100 ml at 100°), freely soluble in AcOH. *Kalpana Chemicals; M/s. Textile Colour and Chemical (Exports).*

3336   Schradan
152-16-9   8471   G

C8H24N4O3P2
 Octamethyldiphosphoramide.
 4-04-00-00288 (Beilstein Handbook Reference); Al3-17291; Bis(bisdimethylaminophosphonous)anhydride;

Bis(dimethylamino)phosphonous anhydride; Bis-N,N,N',N'-tetramethylphosphorodiamidic anhydride; BRN 1801375; Caswell No. 610; Diphosphoramide, octamethyl-; EINECS 205-801-0; ENT 17,291; EPA Pesticide Chemical Code 058601; HSDB 1744; Lethalaire G-59; NSC 8929; Octamethyl-difosforzuur-tetramide; Octamethyl-diphosphorsaeure-tetramid; Octamethyl pyrophosphoramide; Octamethyl pyrophosphortetramide; Octamethyl tetramido pyrophosphate; Octamethyl-diphosphoramide; Octamethylpyrophosphoramide; Octa-methylpyrophosphoric acid amide; Octamethylpyro-phosphoric acid tetramide; Octamidophos; Oktamethyl; Oktamidofos; OMPA; Ompacide; Ompatox; Ompax; Ottometil-pirofosforammide; Pestox; Pestox 3; Pestox 66; Pestox III; Pyrophosphoramide, octamethyl-; Pyrophos-phoric acid octamethyltetraamide; Pyrophosphoryl-tetrakisdimethylamide; RCRA waste number P085; Schradan; Schradane; Systam; Systophos; Sytam; Tetrakisdimethylaminophosphonous anhydride.

Insecticide. Oil; mp = 14-20°; bp2 = 154°; d_4^{25} = 1.09; soluble in H_2O, polar organic solvents; LD_{50} (rat orl) = 9.1 mg/kg.

3337   Sebacic acid
111-20-6   8490   G H

C10H18O4
Decanedioic acid.
4-02-00-02078 (Beilstein Handbook Reference); Acide sebacique; AI3-09127; BRN 1210591; CCRIS 2290; Decanedicarboxylic acid; Decanedioic acid; n-Decanedioic acid; 1,10-Decanedioic acid; EINECS 203-845-5; Ipomic acid; NSC 19492; 1,8-Octanedicarboxylic acid; Sebacic acid; Sebacinsaeure; Sebacinsäure; USAF HC-1. Stabilizer; raw material in manufacture of alkyd resins, maleic and other polyesters, plasticizers, polyester rubbers, synthetic polyamide fibers. Leaflets; mp = 130.8°; bp100 = 295°, bp10 = 232°; d^{20} = 1.2705; soluble in EtOH, Et2O, slightly soluble in H_2O (0.1 g/100 ml), insoluble in C6H6. *Janssen Chimica; Penta Mfg.; Union Camp.*

3338   Selenium
7782-49-2   8505   G

Se

Se
Colloidal selenium.
C.I. 77805; Caswell No. 732; CCRIS 4250; CI 77805; Colloidal selenium; EINECS 231-957-4; Elemental selenium; EPA Pesticide Chemical Code 072001; Gray selenium; HSDB 4493; Selen; Selenate; Selenium; Selenium alloy; Selenium base; Selenium, colloidal; Selenium dust; Selenium elemental; Selenium homopolymer; Selenium, metallic; Selenium powder [Keep away from food]; UN2658; Vandex. A nonmetallic element; electronics, colorant for glass (ceramics), rectifiers, relays, solar batteries. Also used as a vulcanizing agent for rubber. Has three allotropic forms, black, grey and red selenium. [Black selenium]; brown-red liquid; bp = 685°; [red selenium] mp < 200°; d = 4.46; [gray selenium]; mp = 217°; d = 4.81,. *Appleby Group Ltd;; Asarco; Atomergic Chemetals; Cerac; Shinko Chemical; Vanderbilt R.T. Co. Inc.*

3339   Selenium dioxide
7446-08-4   8508   G

O2Se
Selenous acid anhydride.
CCRIS 5820; EINECS 231-194-7; HSDB 677; NSC 56753; RCRA waste number U204; Selenious anhydride; Selenium dioxide; Selenium dioxide dimer; Selenium oxide; Selenium oxide (1:2); Selenium oxide (Se2O4); Selenium oxide (SeO2); Selenium(IV) dioxide; Selenium(IV) dioxide (1:2); Selenous acid anhydride. Used in analysis (testing for alkaloids), oxidizing agent, antioxidant in lubricating oils, catalyst. Powder; mp = 340°; d_4^{15} = 3.954; soluble in H_2O (38 g/100 ml), less soluble in organic solvents. Aldrich; Atomergic; Cerac; Chem.; Chemetals; Shinko.

3340   L-Serine
56-45-1   8534   G

C3H7NO3
α-amino-β-hydroxy-propionic acid.
L-serine; (S)-(-)-serine; Ser; S; L-2-amino-3-hydroxypropionic acid; L-(-)-serine; 2-amino-3-hydroxypropionic acid; 3-hydroxyalanine. Naturally occurring enantiomer of serine. Used in biochemical research, as a dietary supplement, in culture media, microbiological tests, feed additive. mp = 228° (dec); $[\alpha]_D^{20}$ = -7° (c = 10.41 H_2O); soluble in H_2O, insoluble in organic solvents. *Degussa AG; Janssen Chimica; Mitsui Toatsu; Nippon Rikagakuyakuhin; U.S. BioChem.*

3341   Sertraline hydrochloride
79559-97-0   8541   D

C17H18Cl3N
(1S,4S)-4-(3,4-Dichlorophenyl)-1,2,3,4-tetrahydro-N-methyl-1-naphthalenamine hydrochloride.
Atruline; Cp-51974-1; Gladem; Lustral; Serad; Sertraline hydrochloride; Tatig; Tresleen; Zoloft. Serotonin uptake inhibitor. Antidepressant. Crystals; mp = 243-245°; $[\alpha]_D^{23}$ = 38° (c = 2 MeOH). *Pfizer Inc.; Roerig Div. Pfizer Pharm.*

3342   Sethotope
1187-56-0   G

C5H11NO2^{75}Se
2-Amino-4-(methyl-^{75}Se-seleno)butanoic acid.
(S)-2-Amino-4-(methylselenyl-^{75}Se)butyric acid; Butanoic acid, 2-amino-4-(methylseleno-^{75}Se)-, (S)-; L-Seleno-methionine (^{75}Se); L-Selenomethioninum (^{75}Se); Seleno-methionine ^{75}Se-injection; Selenomethionine ^{75}Se; Selenomethioninum (^{75}Se); Selenometionina (^{75}Se); L-(^{75}Se)-Selenothionine; Sethotope. Selenomethionine ^{75}Se is used as a diagnostic aid and radioactive

imaging agent. Crystals; mp = 265° (dec). Bristol-Myers; Co; Inc; Squibb.

d^{20} = 0.876. *Hüls Am.*

3343 Sethoxydim
74051-80-2   8549   G P

C$_{17}$H$_{29}$NO$_3$S

(±)-(EZ)-2-(1-Ethoxyiminobutyl)-5-[2-(ethylthio)propyl]-3-hydroxycyclohex-2-enone.

Aljaden; BAS 9052; BAS 9052H; BAS 90520H; BRN 5955471; Caswell No. 072A; Checkmate; 2-Cyclohexen-1-one, 2-(1-(ethoxyimino)butyl)-5-(2-(ethylthio)propyl)-3-hydroxy-; Cyethoxydim; EINECS 277-682-3; EPA Pesticide Chemical Code 121001; (±)-(ZE)-2-(1-Eth-oxyiminobutyl)-5-(2-(ethylthio)propyl)-3-hydroxycyclo-hex-2-enone; 2-(1-(Ethoxyimino)butyl)-5-(2-(ethylthio)-propyl)-3-hydroxy-2-cyclohexen-1-one; 2-(1-(Ethoxy-imino)butyl)-5-(2-ethylthiopropyl)-3-hydroxycyclohex-2-en-1-one; 2-(N-ethoxybutyrimidoyl)-5-(2-ethylthiopropyl)-3-hydroxy-2-cyclohexen-1-one; Expand; Fervinal; Grasidim; Nabu; NABU; NP 55; Poast; Poast 1.5E; Poast plus; Sethoxydim; Sethoxydime; SN 81742; Tritex-extra. Selective systemic herbicide absorbed by foliage, used in control of annual and perennial grasses in broad-leaved crops. Post-emergence graminicide against annual and perennial grasses. Registered by EPA as a herbicide. Odorless oil; bp$_{0.00003}$ > 90°; d^{25} = 1.043; soluble in H$_2$O (0.0025 g/100 ml), soluble in Me$_2$CO, C$_6$H$_6$, EtOAc, C$_6$H$_{14}$, MeOH (all > 80 g/100 ml); LD$_{50}$ (mrat orl) = 3200 mg/kg, (frat orl) = 2676 mg/kg, (mmus orl) = 5600 mg/kg, (fmus orl) = 6300 mg/kg, (rat der) > 5000 mg/kg, (mus der) > 5000 mg/kg, (Japanese quail orl) > 5000 mg/kg; LC$_{50}$ (rat ihl 4 hr.) > 6.28 mg/l air, (trout 96 hr.) = 30 mg/l, (carp 96 hr.) = 1.6 mg/l, (*Daphnia* 3 hr.) = 1.5 mg/l; non-toxic to bees. *BASF Corp.; Bayer AG; Embetec Crop Protection Ltd.; Ewos; Micro Flo Co. LLC; Nippon Soda Co. Ltd.; Nippon Soda; Rhône-Poulenc; Schering; Sipcam.*

3344 Silane
7803-62-5   8562   G

SiH$_4$

Silicon hydride.

CCRIS 6831; EINECS 232-263-4; Flots 100SCO; HSDB 6351; Monosilane; Silane; Silicane; Silicon tetrahydride; UN2203. Used as a doping agent, source of ultra-pure silicon for electronics fabrication. Gas; mp = -185°; bp = -112°; d^{-185} = 0.68; dec. in H$_2$O, insoluble in organic solvents.

3345 Silane, butylchlorodimethyl-
1000-50-6   G

C$_6$H$_{15}$ClSi

n-Butyldimethylchlorosilane.

CB2785. Coupling agent, chemical intermediate, blocking agent, release agent, lubricant, primer, reducing agent. Liquid; bp = 139°;

3346 Sildenafil citrate
171599-83-0   8563   D

C$_{28}$H$_{38}$N$_6$O$_{11}$S

1-[[3-(6,7-dihydro-1-methyl-7-oxo-3-propyl-1H-pyrazolo[4,3-d]pyrimidin-5-yl)-4-ethoxyphenyl]sulfonyl]-4-methylpiperazine citrate.

Sildenafil citrate; UK 92480-10; Viagra. A selective inhibitor of cyclic guanosine monophosphate (cGMP)-specific phosphodiesterase type 5 (PDE5), which allows smooth muscle relaxation and inflow of blood to the corpus cavernosum. An oral therapy for erectile dysfunction. Crystals; Soluble in H$_2$O (3.5 mg/ml). *Pfizer Intl.*

3347 Silica
7631-86-9   8567   G P

SiO$_2$

Silicon dioxide.

Acticel; Aerogel 200; Aerosil®; Aerosil® 130; Aerosil® 150; Aerosil® 200; Aerosil® 300; Aerosil® 380; Aerosil® A 300; Aerosil® bs-50; Aerosil®-Degussa; Aerosil® E 300; Aerosil® K 7; Aerosil® M-300; Al3-25549; Amorphous silica; Amorphous silica dust; Amorphous silica gel; Cab-o-sil® L-90; Cab-o-sil® M-5; Cab-o-sil® N-70TS; Cabosil® N 5; Cabosil® st-1; Carplex; Carplex 30; Carplex 80; Caswell No. 734A; CCRIS 3699; Celite superfloss; Chalcedony; CI 7811; Clarcell; Colloidal silicon dioxide; Corasil II; Cristobalite; Diatomaceous earth; Diatomaceous earth, calcined; Diatomaceous silica; Diatomite; Dimethyl siloxanes and silicones; Dri-Die; EINECS 231-545-4; ENT 25,550; EPA Pesticide Chemical Code 072605; Extrusil; Flint; Fossil flour; Fused silica; Glass; Hi-Sil; HK 400; HSDB 682; Hydrophobic silica 2482; Infusorial earth; J Slip NS-77; Krystallos; Ludox hs 40; Manosil vn 3; Micro-cel; Min-U-sil; N1030; Nalco 1050; Nalfloc N 1050; Neosil; Neosyl; Opal; Pigment White 27; Porasil; Positive sol 130M; Positive sol 232; Quartz; Quso 51; Quso G 30; Santocel; Sg-67; Silanox 101; Silica; Silica (SiO2); Silica 2482, hydrophobic; Silica, amorphous; Silica, amorphous, fumed; Silica, amorphous fused; Silica, colloidal; Silica particles; Silica slurry; Siliceous earth; Siliceous earth, purified; Silicic anhydride; Silicon dioxide; Silicon dioxide (amorphous); Silicon dioxide, chemically prepared; Silicon dioxide, fumed; Silicon oxide; Silikil; Sillikolloid; Siloxid; Sipernat; Snowtex 30; Snowtex O; Superfloss; Syton 2X; Tamsil 8; Tamsil Gold Bond; Tokusil TPLM; Tridymite; U 333; Ultrasil VH 3; Ultrasil VN 3; Vitasil 220; Vulkasil S; Wessalon; White carbon; Zeofree 80; Zipax; Zorbax sil. Highly active filler for natural and synthetic rubber; thickening agent; tabletting and dragee production auxiliary; thixotropizing agent for polyester resins; antisetting agent. Anti-caking and free flow agent with high absorption capacity; for adhesives, food, cosmetics, paint, paper, film, pesticides, pharmaceuticals, plastics, silicone rubber, inks, sealants; a thixotrope for greases and mineral oils. Dispersant,

anticaking agent for foods, agricultural products, and powders for cosmetics and coatings industries. Filler for finishes, enamels, maintenance paints, plastic film antiblocks, urethane rubber, polishes, traffic paint, interior/exterior coatings, maintenance and marine coatings, mastics and adhesives, electrical epoxy compounds and buffing compounds. Registered by EPA as an antimicrobial, fungicide and insecticide. Used in grain protectant insecticides. Refractory white solid; d^0 = 2.2 (for quartz, 2.65; melts to a glass; insoluble in all solvents except HF. *BYK-Chemie; Cabot Carbon Ltd.; Catalysts & Chem. Ind.; Chisso Corp.; Degussa AG; DuPont; Elf Atochem N. Am.; Geltech; Huber J. M.; Nippon Silica Ind.; Nissan Chem. Ind.; PPG Ind.; PQ Corp.; Sybron; Unimin.*

3348   Silica gel
1343-98-2   8564   G

$SiO_2.xH_2O$, x varies with method of precipitation and extent of drying Silica, hydrated.
Akrochem® Rubbersil RS-150/RS-200; Caswell No. 734; Cubosic; EINECS 215-683-2; EPA Pesticide Chemical Code 072602; G 952; H-Ilerit; Hydrosilisic acid; K 320DS; K 60; K 60 (silicate); Mikronisil; Neoxyl ET; Polyorthosilicic acid; Polysilicic acid; Silica acid; Silica gel; Silicic acid; Silicic acid hydrate; Silicon hydroxide; Silton TF 06; Sipernat 17; Sipernat 50; Sipernat 50S; Sipernat D 10; Sipernat S; Sizol 030; Vulcasil S/GR; Zeosil 45. Highly reinforcing filler for use in synthetic and natural rubber compounding, for mechanical rubber goods, tires, adhesives, footwear, soling. Insoluble in H_2O or acids except HF; eye irritant; poison by intravenous route; TLV: TWA 10 mg/m^3. *Akrochem Chem. Co.*

3349   Silicic acid, ethyl ester
11099-06-2   H

$C_8H_{20}O_4Si$
Silicic acid, tetraethylester polymer.
EINECS 234-324-0; Polysilicic acid, ethyl ester; Silicic acid (H_4SiO_4), tetraethyl ester, homopolymer; Silicic acid, ethyl ester; Silicic acid, tetraethylester polymer; Tetraethyl orthosilicate polymer.

3350   Silicon
7440-21-3   8565   G

Si

Si
Silicon.
CCRIS 6599; Defoamer S-10; EINECS 231-130-8; HSDB 5033; Polyeristalline silicon powder; Silicon; Silicon dust; Silicon metal; Silicon powder, amorphous; Silicone; UN1346. Nonmetallic element; Si; semiconductor in solid-state devices; organosilicon compounds; silicon carbide; alloying agent in steels, aluminum, copper, bronze, iron; cermets, refractories; halogenated silanes; deoxidizer in steel manufacture. mp = 1410°; d$^{25}_4$ = 2.33; insoluble in H_2O, organic solvents. *Atomergic Chemetals; Cerac; Dow Corning; Eagle-Picher; Pechiney Electrométallurgie; Shin-Etsu Chem.*

3351   Silicon monoxide
10097-28-6   8570   G

$^-Si\equiv O^+$

OSi
Silicon(II) oxide.
EINECS 233-232-8; Monox; Silicon monoxide; Silylene, oxo-. A product containing mainly silicon monoxide, with some silicon, silicon dioxide, and small quantities of silicon carbide. It is obtained by heating sand with silicon, carborundum, or coke, in the electric furnace; good thermal and electrical insulator. Brown-black scales; d = 2.18; electrical non-conductor.

3352   Silicon nitride
12033-89-5   8571   G

N_4Si_3
Trisilicon tetranitride.
EINECS 234-796-8; Roydazide; Silicon nitride; Silicon nitride (Si_3N_4). A proprietary trade name for silicon nitride; used for the manufacture of turbine blades and generally where a temperature of up to 1650°C is required. Two crystalline forms, α and β; mp = 1900°; d = 3.2. *Doulton.*

3353   Silicon tetrachloride
10026-04-7   8574   G

Cl_4Si
Tetrachlorosilane.
CCRIS 1324; Chlorid kremicity; CT1800; EINECS 233-054-0; HSDB 683; Silane, tetrachloro-; Silicio(tetracloruro di); Silicium(tetrachlorure de); Siliciumtetrachlorid; Siliciumtetrachloride; Silicon chloride; Silicon chloride (SiCl4); Silicon tetrachloride; Tetrachlorosilane; Tetra-chlorosilicon; Tetrachlorure de silicium; UN1818. Coupling agent, chemical intermediate, blocking agent, release agent, lubricant, primer, reducing agent. Used in generation of smoke screens; manufacture of ethyl silicate, silicones, highpurity silica, fused silica glass; source of silicon, silica, and hydrogen chloride; laboratory reagent. Liquid; mp = -70°; bp = 59°; d$^0_4$ = 1.52; decomposed by H_2O, soluble in organic solvents. *Air Products & Chemicals Inc.; Atomergic Chemetals; Chisso Corp.; Degussa AG; Degussa-Hüls Corp.; Dow Corning; PCR; Union Carbide Corp.*

3354   Silicon tetrafluoride
7783-61-1   8575   G

F_4Si
Tetrafluorosilane.
EINECS 232-015-5; HSDB 684; Perfluorosilane; Silane, tetrafluoro-; Silicon fluoride; Silicon fluoride (SiF4); Silicon tetrafluoride; Tetrafluorosilane; UN1859. Used in manufacture of fluosilicic acid, intermediate in manufacture of pure silicon, to seal water out of oil wells during drilling. Colorless gas; mp = -90°; d^{-80} = 1.590. *Air Products & Chemicals Inc.; Degussa-Hüls Corp.; La Littorale; PCR.*

3355 Silver
7440-22-4   8577   D G

Ag

Ag
Silver.
Silver, colloidal. Metallic element; used in manufacture of silver nitrate, etc.; sterilant; water purification; for coinage; manufacture of tableware, jewelry, ornaments; for electroplating; as catalyst; in dental alloys. Registered by EPA as an antimicrobial, fungicide and herbicide. White metal; mp = 960.5°; bp ca. 2000°; d^{15} = 10.49. *Asarco; Cerac; Degussa AG; Handy & Harman; Mariovilla SpA; Sigma-Aldrich Fine Chem.*

3356 Silver acetate
563-63-3   8578   G

$C_2H_3AgO_2$
Acetic acid silver salt.
Acetic acid, silver(1+) salt; EINECS 209-254-9; NSC 112212; Silver acetate; Silver monoacetate; Silver(1+) acetate; Silver(I) acetate. Laboratory reagent, oxidizing agent. Crystals; d = 3.26; soluble in H_2O (1-3 g/100ml). *Spectrum Chem. Manufacturing.*

3357 Silver carbonate
534-16-7   8580   G

CAg_2O_3
Silver(I) carbonate.
Carbonic acid, disilver(1+) salt; Disilver carbonate; EINECS 208-590-3; Fetizon's reagent; NSC 112223; Silver carbonate; Silver carbonate (2:1); Silver(1) carbonate. Laboratory reagent. Solid; dec 220°; d = 6.08; slightly soluble in H_2O (0.03 g/100 ml), readily soluble in acids, ammonia. *Sigma-Aldrich Fine Chem.; Spectrum Chem. Manufacturing.*

3358 Silver chloride
7783-90-6   8582   G P

Ag—Cl

AgCl
Silver(I) Chloride.
Caswell No. 735A; EPA Pesticide Chemical Code 072506; Silver chloride; Silver chloride (AgCl); Silver monochloride. Used in photography, photometry and optics, batteries, photochromic glass, silver plating, production of pure silver, antiseptic. Used as a bactericide. Registered by EPA as an antimicrobial, fungicide and herbicide (cancelled). Light-sensitive white powder; mp = 455°; bp = 1550°; d = 5.56; almost insoluble in H_2O (0.000193 g/100 ml), soluble in conc. HCl (0.4 g/100 ml), 10% NH_4OH (7.7 g/100 ml). *Degussa AG; La Littorale; Noah Chem.; Sigma-Aldrich Co.*

3359 Silver citrate
126-45-4   8584   G

$C_6H_5Ag_3O_7$
2-hydroxy-1,2,3-propanetricarboxylic acid silver salt.
EINECS 204-786-8; Itrol; Silver citrate; 1,2,3-Propanetricarboxylic acid, 2-hydroxy-, trisilver(1+) salt; Trisilver citrate. Anti-infective dusting powder. Soluble in H_2O (0.29 mg/ml).

3360 Silver fluoride
7775-41-9   8587   D G P

Ag—F

AgF
Silver monofluoride.
Argentous fluoride; Caswell No. 736; EINECS 231-895-8; EPA Pesticide Chemical Code 072502; Lachiol; Silver fluoride; Silver fluoride (AgF); Tachiol. Fluorinating agent, also used as a topical anti-infective. Registered by EPA as an antimicrobial, fungicide and herbicide (cancelled). Used in dentistry as a fluoridation agent. Cubic crystals; mp = 435°; bp = 1150°; d = 5.8520; soluble in H_2O (182 g/100 ml).

3361 Silver iodide
7783-96-2   8589   G

Ag—I

AgI
Silver(I) iodide.
Colloidal silver iodide; EINECS 232-038-0; HSDB 2930; Neosiluol; Neosilvol; Silver iodide; Silver iodide (AgI); Silver monoiodide; Silver(1+) iodide. Used in photography, cloud seeding for artificial rainmaking, laboratory reagent, antiseptic. Solid; mp = 552°; d = 5.67; poorly soluble in H_2O (0.03 mg/l), soluble in ammonia. *Atomergic Chemetals; Cerac; Noah Chem.; Sigma-Aldrich Fine Chem.*

3362 Silver nitrate
7761-88-8   8591   D G P

AgNO3
Nitric acid silver salt.
Argenti nitras; Argerol; Caswell No. 737; EINECS 231-853-9; EPA Pesticide Chemical Code 072503; HSDB 685; Lunar caustic; Nitrate d'argent; Nitrato de plata; Nitric acid, silver(1+) salt; Nitric acid silver(I) salt; NSC 135800; Silbernitrat; Silver mononitrate; Silver nitrate; Silver(I) nitrate (1:1); Silver saltpeter; UN1493; UTS Silverator Water Treatment Unit. Photographic film, catalyst for ethylene oxide, indelible inks, silver plating, silver salts, silvering mirrors, germicide, hair dyeing, antiseptic, laboratory reagent. Registered by EPA as an antimicrobial, fungicide and herbicide (cancelled). Has been used to keep cut flowers fresh, but has heavy metal toxicity. An antiseptic and energetic caustic for wounds and sores. White crystals; mp = 212°; d = 4.35; soluble in H_2O (2.5 - 10 g/ml), EtOH, Me_2CO. *Accurate Chem. & Sci. Corp.; Degussa AG; Johnson Matthey; Sigma-Aldrich Fine Chem.; Spectrum Chem. Manufacturing.*

3363 Silver oxide
20667-12-3   8595   G V

Ag—O—Ag

Ag2O
Silver(l) oxide.
Argentous oxide; Disilver oxide; EINECS 243-957-1; Silver oxide; Silver oxide (Ag2O); Silver(1+) oxide. Used in polishing glass, coloring glass yellow, catalyst, purifying drinking water, laboratory reagent. Used in veterinary medicine as a germicide and parasiticide. Brown-black powder; dec. 200-300°; d_4^{25} = 7.22; soluble in H2O (25 mg/l), freely soluble in HNO3, ammonia; LD50 (rat orl) = 2.82 g/kg. *Atomergic Chemetals; Degussa AG; Elf Atochem N. Am.; Spectrum Chem. Manufacturing.*

3364 Silver sulfate
10294-26-5   8603   G

O=S(—O⁻)(=O)—O⁻   2Ag⁺

Ag2O4S
Sulfuric acid silver salt.
Disilver monosulfate; Disilver sulfate; Disilver(1+) sulfate; Silver sulfate (Ag2SO4); Sulfuric acid, disilver(1+) salt; Sulfuric acid silver salt (1:2). Used as a laboratory reagent. Solid; mp = 657°; dec 1085°; d = 5.45; soluble in H2O (8 mg/ml), HNO3 and ammonia. *Atomergic Chemetals; Noah Chem.*

3365 Silver sulfide
21548-73-2   8604   G

Ag—S—Ag

Ag2S
Silver(l) sulfide.
Disilver sulphide; EINECS 244-438-2; Silver sulfide; Silver sulfide (Ag2S). Used for inlaying in niello metal work, ceramics. Grey-black powder; mp = 845°; d_4^{20} = 7.234; insoluble in H2O. *Atomergic Chemetals; Cerac.*

3366 Silvex
93-72-1   8606   G P

C9H7Cl3O3
2-(2,4,5-Trichlorophenoxy)propionic acid.
3-06-00-00721 (Beilstein Handbook Reference); Acide 2-(2,4,5-trichloro-phenoxy) propionique; Acido 2-(2,4,5-tricloro-fenossi)-propionico; Amchem 2,4,5 TP; Aqua-vex; BRN 1985768; Caswell No. 739; CCRIS 1467; Color-set; Double strength; EINECS 202-271-2; EPA Pesticide Chemical Code 082501; Fenoprop; (±)-Fenoprop; Fenormone; Fruitone T; Herbicides, silvex; HSDB 686; Kuron; Kurosal G; Kwas 2,4,5-trojchlorofenoksypropionowy; Kyselina 2-(2,4,5-trichlorfenoxy)propionova; Miller Nu Set; Propanoic acid, 2-(2,4,5-trichlorophenoxy)-; Propionic acid, 2-(2,4,5-trichlorophenoxy)-; Propon; RCRA waste number U233; Silvex; (±)-Silvex; Silvi-Rhap; Sta-Fast; 2-(2,4,5-Trichloor-fenoxy)-propionzuur; 2-(2,4,5-Trichlorophenoxy)propion-ic acid; (±)-2-(2,4,5-Trichlorophenoxy)propanoic acid (9Cl); α-(2,4,5-Trichlorophenoxy)propionic acid; 2-(2,4,5-Trichlor-phenoxy)-

propionsaeure; 2,4,5-TCPPA; 2,4,5-TP; 2,4,5-TP acid. Herbicide containing silvex as the active ingredient; herbicide used in ponds and other still water for the control of aquatic weeds, as well as control of brush on rangeland; also used industrially on railroads or under power lines for the control of Crystals; mp = 181.6°, also 181.6°; soluble in H2O (14 g/100 ml), Me2CO (15 g/100 ml), C6H6 (1.6 g/100 ml), CCl4 (2.4 g/100 ml), C7H16 (1.7 g/100 ml), MeOH (10.5 g/100 ml). *Dow AgroSciences; Dow Chem. U.S.A.; Dow Europe; Dow UK.*

3367 Simazine
122-34-9   8608   G H P

C7H12ClN5
2,4-Bis(ethylamino)-6-chloro-s-triazine.
27692G; 4-26-00-01208 (Beilstein Handbook Reference); A 2079; AI3-51142; Aktinit S; Amizine; Aquazine; Azotop; Batazina; Batazine FLO; Bitemol; Bitemol S 50; BRN 0010895; Caswell No. 740; CAT; CCRIS 1469; CDT; Cekusan; Cekusima; Cekuzina-S; CET; EINECS 204-535-2; EPA Pesticide Chemical Code 080807G; G-27692; Geigy 27,692; Gesapun; Gesaran; Gesatop; Gesatop 50; H 1803; Herbatoxol S; Herbazin; Herbazin 50; Herbex; Herboxy; HSDB 1765; Hungazin DT; NSC 25999; Premazine; Primatel S; Primatol S; Princep; Printop; Radocon; Radokor; Simadex; Simanex; Simatsin-neste; Simazin; Simazine; Simazine 80W; Symazine; Tafazine; Tafazine 50-W; Taphazine; Triazine A 384; W 6658; Yrodazin; Zeapur. Selective systemic herbicide, absorbed through roots; used to control most germinating grasses and broad-leaved weeds in fruit, vegetables, trees. Registered by EPA as a herbicide. Crystals; mp = 226 - 227°; insoluble in H2O (0.0005 g/100 ml), soluble in MeOH (0.04 g/100 ml), CHCl3 (0.09 g/100 ml), Et2O (0.03 g/100 ml), slightly soluble in dioxane, ethyl Cellosolve; LD50 (rat, mus, rbt orl) > 5000 mg/kg, (rat der) > 3100 mg/kg, (rbt der) > 10200 mg/kg, (hen, pgn orl) > 5000 mg/kg; LC50 (rat ihl 4 hr.) > 2 mg/l air, (bobwhite quail 8 day dietary) =8800 mg/kg, (8 day dietary mallard duck) = 51200 mg/kg, (bluegill sunfish 96 hr.) = 90 mg/l, (rainbow trout 96 hr.) > 100 mg/l, (crucian carp 96 hr.) > 100 mg/l, (guppy 96 hr.) = 49 mg/l; non-toxic to bees. *Agriliance LLC; Drexel Chemical Co.; Syngenta Crop Protection; UAP - Platte Chemical.*

3368 Simvastatin
79902-63-9   8613   D

C25H38O5
2,2-Dimethylbutyric acid 8-ester with (4R,6R)-6-[2-[(1S,2S,6R,8S,8aR)-1,2,6,7,8,8a-hexahydro-8-hydroxy-2,6-dimethyl-1-naphthyl]ethyl]tetrahydro-4-hydroxy-2H-pryan-2-one.
BRN 4768037; CCRIS 7558; Cholestat; Coledis; Colemin; Corolin; Denan; DRG-0320; L 644128-000U; Labistatin; Lipex; Liponorm; Lodalès; Lodales; Medipo; MK-0733; MK-733; Nivelipol; Pantok;

Rendapid; Simovil; Simvastatin; Simvastatina; Simvastatine; Simvastatinum; Sinvacor; Sivastin; Synvinolin; Vasotenal; Zocor; Zocord. Antihyperlipoproteinemic. HMG-CoA reductase inhibitor; used as an antihypercholesterolemic agent. Synthetic analog of lovastatin. Crystals; mp = 135-138°. *Merck & Co.Inc.*

3369 Sinapoline
1801-72-5   G

C7H12N2O
Diallyl-urea.
1,3-Diallylurea; N,N'-Diallylurea; DAU; EINECS 217-291-7; NSC 102722; Urea, 1,3-diallyl-; Urea, N,N'-di-2-propenyl-. mp = 90-93°.

3370 Sinigrin
3952-98-5   8619   G

C10H16KNO9S2
1-Thio-β-D-glucopyranose   1-[N-(sulfoxy)-3-butenimidate] monopotassium salt.
Allyl glucosinolate; CCRIS 3715; EINECS 223-545-8; Glucopyranose, 1-thio-, 1-(3-butenohydroximate) NO-(hydrogen sulfate), monopotassium salt, β-D-; β-D-Glucopyranose, 1-thio-, 1-(N-(sulfooxy)-3-butenimidate), monopotassium salt; Glucoside of allyl isothiocyanate; NSC 90774; NSC 407279; Potassium 1-(β-D-gluco-pyranosylthio)but-3-enylideneaminooxysulphonate; 2-Propenyl glucosinolate; Sinigrin. A constituent of black mustard seed and horseradish root. Substrate for thioglucosidase. The monohydrate is [64550-88-5; 223-545-8]. [monohydrate]; mp = 127-129°; $[\alpha]_D^{18}$= -16° (c = 1 H2O); soluble in H2O, EtOH, insoluble in organic solvents.

3371 Sodium
7440-23-5   8641   G

Na

Na
Natrium.
EINECS 231-132-9; HSDB 687; Natrium; Sodio; Sodium; Sodium (liquid alloy); Sodium-23; Sodium ion; Sodium metal; UN1428. Metallic element; polymerization catalyst for synthetic rubber, laboratory reagent, coolant in nuclear reactors, heat transfer agent, manufacture of tetraethyl and tetramethyl lead, sodium peroxide, sodium hydride; radioactive isotopes in tracer studies and medicine. mp = 98°; bp = 892°; SG = 0.9; reacts with H2O. *Foseco (FS) Ltd; Octel Chem. Ltd.*

3372 Sodium acetaldehyde bisulfite
918-04-7   42   H

C2H5NaO4S
1-Hydroxyethanesulfonic acid, monosodium salt.

Acetaldehyde sodium bisulfite; Azetaldehydschweflig-sauren natriums; EINECS 213-037-4; Ethanesulfonic acid, 1-hydroxy-, monosodium salt; 1-Hydroxyethanesulfonic acid, monosodium salt; Monosodium 1-hydroxy-ethanesulfonate; Sodium 1-hydroxyethanesulfonate; Sodium 1-hydroxyethanesulphonate. Crystals; freely soluble in H2O, insoluble in EtOH.

3373 Sodium acetate
127-09-3   8642   D G H

C2H3NaO2
Acetic acid, sodium salt.
Acetic acid, sodium salt; Anhydrous sodium acetate; Caswell No. 741A; EINECS 204-823-8; EPA Pesticide Chemical Code 044006; FEMA Number 3024; HSDB 688; Natrium aceticum; Natriumacetat; NSC 77459; Octan sodny; Sodii acetas; Sodium acetate; Sodium ethanoate. Dye and color intermediate, pharmaceuticals, cinnamic acid, soaps, photography, purification of glucose, meat preservation, medicine, electroplating, tanning, dehydrating agent, buffer, laboratory reagent, food additive. Trihydrate: mp = 58°; d = 1.45; soluble in H2O (1.25 g/ml), less soluble in organic solvents. *EM Ind. Inc.; General Chem; Heico; Honeywell & Stein; Lonzagroup; Niacet; Sigma-Aldrich Fine Chem.; Verdugt BV.*

3374 Sodium acifluorfen
62476-59-9   H

C14H6ClF3NNaO5
5-(2-Chloro-4-(trifluoromethyl)phenoxy)-2-nitrobenzoic acid, sodium salt.
Acifluorfen sodium; Acifluorfen sodium salt; Blazer; Blazer 2L; Blazer 2S; Carbofluorfen; Caswell No. 755D; EINECS 263-560-7; EPA Pesticide Chemical Code 114402; HSDB 6551; LS 80.1213; MC 10978; RH 6201; RH 6205; Scifluorfen; Sodium Acifluorfen; Sodium salt of acifluorfen; Tackle; Tackle 2AS. Used for post-emergence control of broad-leaved weeds and suppression of some annual grasses in soybeans. Registered by EPA as a herbicide. Solid; soluble in H2O (25 g/100 ml at 25°); LC50 (rainbow trout 96 hr.) = 31 mg/l, (bluegill sunfish 96 hr.) = 54 mg/l; non-toxic to bees. *Aventis Crop Science; BASF Corp.; Bayer AG; Monsanto Co.*

3375 Sodium acrylate
7446-81-3   H

C3H3NaO2
2-Propenoic acid, sodium salt.
Acrylic acid, sodium salt; EINECS 231-209-7; HSDB 6087.

3376 Sodium 3-(allyloxy)-2-hydroxypropane-sulfonate
52556-42-0   H

C6H11NaO5S
1-Propanesulfonic acid, 2-hydroxy-3-(2-propenyloxy)-, monosodium salt.
EINECS 258-004-5; 1-Propanesulfonic acid, 2-hydroxy-3-(2-propenyloxy)-, monosodium salt; Sodium 3-(allyloxy)-2-hydroxypropanesulfonate; Sodium 3-(allyloxy)-2-hydroxypropanesulphonate.

3377   Sodium aluminate
1302-42-7   8645   G

AlNaO2
Aluminum sodium oxide.
Aluminate (AlO2^{1-}), sodium; Aluminate, sodium; Aluminium sodium dioxide; Aluminum sodium oxide (Al2Na2O4); EINECS 215-100-1; HSDB 5023; Sodium aluminate; Sodium aluminate (Na2Al2O4); Sodium aluminate (NaAlO2); Sodium aluminate solution; Sodium aluminum dioxide; Sodium aluminum oxide (NaAlO2); Sodium metaaluminate; Sodium metaaluminate (NaAlO2). Used as a mordant, in zeolites, water purification, sizing paper, manufacture of milk glass, soap and cleaning compounds. White solid; mp = 1650°; very soluble in H2O, insoluble in EtOH. *Asada Chem. Ind. Ltd.; Degussa-Hüls Corp.; Laporte Fine Chem.; Nalco Diversified Technologies, Inc.*

3378   Sodium aluminum phosphate
7785-88-8   G

Al3Na3O16P4
Sodium aluminum phosphate acidic.
Aluminum sodium phosphate; EINECS 232-090-4; HSDB 691; Phosphoric acid, aluminium sodium salt; Sodium aluminum phosphate. Food additive used in baked products. *Monsanto Co.; Rhône-Poulenc; Whiting Peter Ltd.*

3379   Sodium aminotris(methylenephosphonate)
20592-85-2   H

C3H11NNaO9P3
Nitrilotris(methylene phosphonic acid), sodium salt.
Aminotri(methylene phosphonic acid), sodium salt; EINECS 243-900-0; Nitrilotris(methylene phosphonic acid), sodium salt; (Nitrilotris(methylene))trisphosphonic acid, sodium salt; Phosphonic acid, (nitrilotris-(methylene))tri-, sodium salt; Sodium (nitrilotris-(methylene))triphosphonate; Sodium aminotris(methylenephosphonate).

3380   Sodium ascorbate
134-03-2   837   G

C6H7NaO6
L-Ascorbic acid sodium salt.
Ascorbate de sodium; Ascorbato sodico; Ascorbic acid sodium salt; L-Ascorbic acid sodium salt; L-Ascorbic acid, monosodium salt; Ascorbicin; Ascorbin; CCRIS 3291; Cebitate; Cenolate; Cevalin; EINECS 205-126-1; HBL 508; HSDB 694; Iskia-C; monosodium ascorbate; Monosodium L-ascorbate; Natrascorb; Natri-C; Natrii ascorbas; 3-Oxo-L-gulofuranolactone sodium; Soda-scorbate; Sodium ascorbate; Sodium L-ascorbate; Vitamin C, sodium salt; Vitamin C sodium. Antioxidant in food products. Source of vitamin C.
White solid; mp = 218° (dec); [α]$_D^2$= +104°; soluble in H2O (62 g/100 ml). *BASF Corp.; EM Ind. Inc.; Hoffmann-LaRoche Inc.; Pfizer Inc.; Spice King; Takeda Chem. Ind. Ltd.*

3381   Sodium benzoate
532-32-1   8654   G H

C7H5NaO2
Benzoic acid sodium salt.
AI3-07835; Antimol; Benzoan sodny; Benzoate of soda; Benzoate sodium; Benzoesäure (Na-salz); Benzoic acid, sodium salt; Caswell No. 746; CCRIS 3921; EINECS 208-534-8; EPA Pesticide Chemical Code 009103; FEMA Number 3025; HSDB 696; Natrium benzoicum; Sobenate; Sodium benzoate; Ucephan. Used as a fungicide; preservative in pharmaceuticals and foods, especially in slightly acidic media; as a clinical reagent (bilirubin assay). Solid; mp >300°;
λm = 223, 269 nm (H2O); soluble in H2O; LD50 (rat orl) = 4.07 g/kg. *Aceto Corp.; Dinoval; DSM Spec. Prods.; Haarmann & Reimer GmbH; Mallinckrodt Inc.; Pentagon Chems. Ltd.*

3382   Sodium bicarbonate
144-55-8   8655   D G P

CHNaO3
Sodium hydrogen carbonate.
Acidosan; Baking soda; Baros; Bicarbonate of soda; Carbonic acid monosodium salt; Caswell No. 747; CCRIS 3064; Col-evac; Colyte; EINECS 205-633-8; EPA Pesticide Chemical Code 073505; HSDB 697; Jusonin; Meylon; Monosodium carbonate; Monosodium hydrogen carbonate; Natrii hydrogencarbonas; Natrium bicarbonicum; Natrium hydrogencarbonicum; Natrium-hydrogenkarbonat; Natron; Neut; NSC 134031; Saleratus; Sel De vichy; Soda; Soda Mint; Sodium acid carbonate; Sodium bicarbonate; Sodium carbonate (Na(HCO3)); Sodium hydrocarbonate; Sodium hydrogen carbonate; Soludal; Tronacarb Sodium Bicarbonate. Used in baking, effervescent salts, beverages, fire extinguishers and cleaning compounds. Used in manufacture of many sodium salts, as a baking additive, in extinguishers, cleaning agents and medically, as an antacid. White solid; mp = 270° (decomposes to sodium carbonate and CO2 when heated); d = 2.1590; soluble in H2O (10 g/100 ml at 25°, 8.3 g/100 ml at 18°), insoluble in EtOH; pH (aqueous solution) = 8.3. *Kerr-McGee; O'Neal Jones & Feldman Pharm.*

3383   Sodium   3,5-bis(methoxycarbonyl)benzene-sulfonate
3965-55-7   H

C10H9NaO7S
 Isophthalic acid, 5-sulfo-, 1,3-dimethyl ester, sodium salt.
 Benzenesulfonic acid, 3,5-bis(methoxycarbonyl)-, sodium salt; 1,3-Benzenedicarboxylic acid, 5-sulfo-, 1,3-dimethyl ester, sodium salt; 3,5-Bis-(methoxykarbonyl)benzen-sulfonan sodny; 3,5-Bis-methylkarboxy-benzensulfonan sodny; EINECS 223-578-8; Isophthalic acid, 5-sulfo-, 1,3-dimethyl ester, sodium salt; Sodium 3,5-Bis(methoxy-carbonyl)benzenesulfonate; Sodium dimethyl 5-sulph-onatoisophthalate; 5-Sulfoisophthalic acid, dimethyl ester, sodium salt.

3384   Sodium bisulfide
    16721-80-5   8659   G

                    Na—SH

HNaS
 Sodium hydrosulfide.
 AI3-14915; EINECS 240-778-0; Hidrosulfuro sodics; HSDB 5165; Hydrogen sodium sulfide; Hydrogenosulfure de sodium; NA2922; NSC 158264; Sodium bisulfide; Sodium hydrogen sulfide; Sodium hydrogen sulfide (NaHS); Sodium hydrogensulphide; Sodium hydrosulfide; Sodium hydrosulphide; Sodium mercaptan; Sodium mercaptide; Sodium sulfhydrate; Sodium sulfide; Sodium sulfide (Na(SH)); UN2318; UN2949. Used in paper pulping, dyestuffs processing, rayon and cellophane desulfurizing, dehairing hides, bleaching reagent. [dihydrate]; mp = 55°; d = 1.79; soluble in H2O, organic solvents; [trihydrate]; rhombic crystals; mp = 22°. *Nissan Chem. Ind.*

3385   Sodium bisulfite
    7631-90-5   8660   G P

                O
                ||
          O=S—O⁻   Na⁺
                |
                H

HNaO3S
 Sodium acid sulfite.
 Abisol; AI3-08582; Bisulfite de sodium; Caswell No. 750; CCRIS 3950; EINECS 231-548-0; EPA Pesticide Chemical Code 078201; Fr-62; HSDB 724; Hydrogen sodium sulfite; Hydrogen sulfite sodium; Leucogen; Monosodium sulfite; NSC 60680; Sodium acid sulfite; Sodium bisulfite; Sodium bisulfite (NaHSO3); Sodium bisulphite; Sodium hydrogen sulfite; Sodium hydrogen sulfite, solution; Sodium hydrosulfite(DOT); Sodium metabisulfite; Sodium sulfite (NaHSO3); Sodium sulhydrate; Sulfurous acid, monosodium salt; Uantax SBS. Fusion of minerals to make solutions for analysis; pickling metals; carbonizing wool. Used as a disinfectant and bleach, as a reducing agent and in bleaching and paper making. Registered by EPA as a fungicide. Crystals; d = 1.48; soluble in 3.5 parts cold H2O, 2 parts boiling H2O, 70 parts EtOH; LD50 (rat iv) = 115 mg/kg. *BASF Corp.; Degussa-Hüls Corp.; DuPont; Hoechst Celanese; Penreco; Sigma-Aldrich Fine Chem.*

3386   Sodium borate
    1330-43-4   8662   D G P

B4Na2O7
 Sodium tetraborate.
 Anhydrous borax; Borates, tetrasodium salts, anhydrous; Borax,

anhydrous; Borax, dehydrated; Borax, fused; Borax glass; Boric acid (H2B4O7) , sodium salt; Boric acid (H2B4O7) , disodium salt; Boric acid, disodium salt; Boron sodium oxide; Boron sodium oxide (B4Na2O7); Disodium tetraborate; Disodium tetraborate, anhydrous; EINECS 215-540-4; FR 28; Fused borax; Fused sodium borate; HSDB 5025; Rasorite 65; Sodium biborate; Sodium borate; Sodium borate, anhydrous; Sodium boron oxide; Sodium pyroborate; Sodium tetraborate. Used in heat-resistant glass; porcelian enamel; detergents, herbicides, fertilizers, rust inhibitors, pharmaceuticals, leather, photography, bleaches, paint, boron compounds, flame retardant fungicide for wood, soldering flux, cleaners and lab. reagents. A highly soluble form of sodium borate; used to correct boron deficiency in plants by applying either as a foliar spray, in nutrient feeds or with herbicides. Has been used as an antiseptic, detergent and astringent for mucous membranes. White solid; mp = 741°; d = 2.3670. *Borax Europe Ltd; Lancaster Synthesis Co.; U.S. Borax.*

3387   Sodium borohydride
    16940-66-2   8664   G

                H
                |
          H—B⁻—H   Na⁺
                |
                H

BH4Na
 Sodium tetrahydroborate.
 Borate(1-), tetrahydro-, sodium; Borohydrure de sodium; Borol; EINECS 241-004-4; Hidkitex DF; HSDB 699; Sodium borohydrate; Sodium borohydride; Sodium borohydride (Na(BH4)); Sodium hydroborate; Sodium tetrahydridoborate(1-); Sodium tetrahydroborate; Sodium tetrahydroborate(1-); UN1426. Source of H2 and other borohydrides; bleaching wood pulp; blowing agent for plastics; decolorizer for plasticizers. White solid; mp = 36°; d = 1.074. *Atomergic Chemetals; Morton Intn'l.; Tennant Trading Ltd.*

3388   Sodium bromide
    7647-15-6   8666   D G P V

                    Na—Br

BrNa
 Hydrobromic acid sodium salt.
 Bromide salt of sodium; Bromnatrium; Caswell No. 750A; EINECS 231-599-9; EPA Pesticide Chemical Code 013907; HSDB 5039; NSC 77384; Sedoneural; Sodium bromide; Sodium bromide (NaBr); Trisodium tribromide. Used for preparation of bromides and in photography. Registered by EPA as an antimicrobial, fungicide, herbicide, insecticide and molluscicide. Used to control bacterial slimes in paper and pulp mills. Sedative/hypnotic; anticonvulsant. Used in veterinary medicine as a sedative and to control convulsions. Crystals; mp = 755°; d = 3.21; soluble in H2O (0.9 g/ml), less soluble in organic solvents; crystallizes as a dihydrate; LD50 (rat orl) = 3.5 g/kg. *Ethyl Corp.; Great Lakes Fine Chem.; Hawks Chem Co Ltd; Morton Intn'l.*

3389   Sodium 2-(2-butoxyethoxy)ethanolate
    38321-18-5   H

C8H17NaO3
 Ethanol, 2-(2-butoxyethoxy)-, sodium salt.
 EINECS 253-880-5; Ethanol, 2-(2-butoxyethoxy)-, sodium salt; Sodium 2-(2-butoxyethoxy)ethanolate.

3390   Sodium 2-butoxyethoxide
52663-57-7   H

$C_6H_{13}NaO_2$
Ethanol, 2-butoxy-, sodium salt.
EINECS 258-079-4; Ethanol, 2-butoxy-, sodium salt; Sodium 2-butoxyethanolate; Sodium 2-butoxyethoxide.

3391   Sodium butylate
2372-45-4   H

C_4H_9NaO
Butanol, sodium salt.
Butanol, sodium salt; 1-Butanol, sodium salt; Butyl alcohol, sodium salt; EINECS 219-144-2; HSDB 5872; Sodium butanolate; Sodium butoxide; Sodium 1-butoxide; Sodium butylate; Sodium n-butylate.

3392   Sodium butyrate
156-54-7   G

$C_4H_7NaO_2$
Butyric acid sodium salt.
Butanoic acid, sodium salt; Butyrate sodium; Butyric acid, sodium salt; CCRIS 7068; EINECS 205-857-6; HSDB 5655; NSC 174280; Sodium butanoate; Sodium butyrate; Sodium n-butyrate; Sodium propanecarboxylate. Used in chemical laboratory as agent for alteration of chromatin structure; enhances transfection efficiency and improved cell survival. Crystals; mp = 250-253°. *Penta Mfg.*

3393   Sodium carbonate
497-19-8   8668   G P

CNa_2O_3
Carbonic acid disodium salt.
Bisodium carbonate; Calcined soda; Carbonic acid, disodium salt; Carbonic acid sodium salt; Carbonic acid sodium salt (1:2); Caswell N0 752; CCRIS 7319; Crystol carbonate; Disodium carbonate; Disodium carbonate (Na_2CO_3); Dynamar L 13890; EINECS 207-838-8; EINECS 231-420-4; EPA Pesticide Chemical Code 073506; HSDB 5018; Light Ash; Mild alkali; Na-X; Natrium Carbonicum Calcinatum; Natrium Carbonicum Siccatum; NSC 156204; Sal soda; Salt of soda; Scotch soda; Snowlite 1; Snowlite I; Soda; Soda ash; Soda Ash Light 4P; Soda, calcined; Sodium carbonate; Sodium carbonate (2:1); Sodium carbonate ($Na_2(CO_3)$); Sodium carbonate, anhydrous; Sodium Carbonate, Anhydrous ASTM D458; Sodium Carbonate, Anhydrous GE Materials D4D5; Solvay soda; Suprapur 6395; Trona; Trona soda ash; V 20N; V Soda; Washing soda. Registered by EPA as an antimicrobial, fungicide and herbicide. FDA GRAS, Japan approved, Europe listed, FDA approved for injectables, parenterals, ophthalmics, orals, rectals, USP/NF, BP, Ph.Eur. compliant. Alkalizing agent, antacid, reagent. Used in injectables, parenterals, ophthalmics, orals, rectals, mouthwashes, foot preparations and vaginal douches. Used in food additives as a preservative. White solid; mp = 851°; d = 2.53; soluble in H_2O (28.6

g/100 ml at 20°, 45.5 g/100 ml at 35°), insoluble in EtOH. LD_{50} (mus ip 30 day) = 116.6 mg/kg. *ABC Compounding Co.; Albright & Wilson Americas Inc.; Chardon Laboratories, Inc; EM Ind. Inc.; FMC Corp. Pharm. Div.; General Chem; Huntington Professional Products; Integra; Lohmann; Mallinckrodt Inc.; Medical Chemical Corp; Norsk Hydro AS; Rhône-Poulenc; Ruger; Sigma-Aldrich Fine Chem.; Solvay Pharm. SA; Spectrum Chem. Manufacturing; Speer Products Inc.; Texasgulf; Unit Chemical Corp.*

3394   Sodium carbonate
5968-11-6   G

$CNa_2O_3.H_2O$
Sodium carbonate monohydrate.
Carbonic acid disodium salt, monohydrate; Disodium carbonate monohydrate; Getrocknetes natriumkcarbonat; Natrii carbonas monohydricus; Natrium carbonicum monohydricum; Soda ash; Sodium carbonate monohydrate; Sodium Carbonate. White solid; bp = 851°; d = 2.25; soluble in 3 parts H_2O, insoluble in organic solvents. *Church & Dwight.*

3395   Sodium chlorate
7775-09-9   8670   G P

$ClNaO_3$
Chloric acid sodium salt.
Agrosan; Asex; Atlacide; B-Herbatox; Caswell No. 753; Chlorate de sodium; Chlorate salt of sodium; Chloric acid, sodium salt; Chlorsaure; Chlorsäure; De-Fol-Ate; Defol; Dervan; Desolet; Drexel defol; Drop-Leaf; EINECS 231-887-4; EPA Pesticide Chemical Code 073301; Evau-super; Fall; Grain sorghum harvest-aid; Granex O; Harvest-aid; Hibar C; HSDB 732; KM; KM Sodium Chlorate; Kusatohru; Kusatol; Leafex 2; Leafex 3; Natrium chloraat; Natrium chlorat; Ortho C-1 defoliant & weed killer; Ortho-C-1-Defoliant; Oxycil; Rasikal; Shed-A-leaf; Shed-A-Leaf L; Soda chlorate; Sodakem; Sodio (clorato di); Sodium (chlorate de); Sodium chlorate; Sodium chlorate ($NaClO_3$); Sodium(chlorate de); Travex; Tumbleaf; Tumbleleaf; UN1495; UN2428; United Chemical Defoliant No. 1; VAL-DROP; Weed Killer. Oxidizing agent, pulp bleaching, defoliant and herbicide; leather tanning and finishing; textile mordant. Also used in pyrotechnics. Herbicide. Registered by EPA as a herbicide and fungicide. FDA approved for injectables. Used in injectables and as an astringent. White granules; mp = 248°; d = 2.5; soluble in H_2O, less soluble in organic solvents; LD_{50} (rat orl) = 12000 mg/kg. *Albright & Wilson Americas Inc.; Degussa-Hüls Corp.; Eka Nobel Ltd.; Georgia Gulf; Kerr-McGee; OxyChem; PPG Ind.*

3396   Sodium chloride
7647-14-5   8671   D G P

Na—Cl

$ClNa$
Hydrochloric acid sodium salt.
Adsorbanac; Alberger® Natural Flake; Arm-A-Vial; Ayr; Betrox; Caswell No. 754; CCRIS 982; Colyte; Common salt; Dendritis; EINECS 231-598-3; EPA Pesticide Chemical Code 013905; Extra Fine 200 Salt; Extra Fine 325 Salt; Flexivial; Gingivyl; Halite; H.G. blending; Halite; HSDB 6368; Hypersal; Hyposaline; Iodized salt; Marine salt; Muriate of salt; Natriumchlorid; NSC 77364; Purex; Rock salt; Sal culinaris; Sal commune; Saline; Salt; Sea salt; Slow Sodium; Sodium chloric; Sodium chloride; Sodium chloride brine, purified; Sodium monochloride; SS salt; Table salt; Top flake; Trisodium trichloride; White crystal. Used as a chemical intermediate and reagent. Also used as a fertilizer. Registered by EPA as an

antimicrobial, fungicide, herbicide, insecticide and molluscicide. FDA listed, BP, Ph. Eur compliance. Used as a flavoring agent and preservative and as a tonicity agent, (injectables, nasals, dentals, parenterals, inhalants, ophthalmics, orals, rectals and topicals). Used as an aquaculture feed additive and to improve fruit quality, also as a de-icer. Electrolyte replenisher; emetic; topical anti-inflammatory. Cubic crystals; mp = 804°; soluble in H_2O (0.36 g/ml), insoluble in organic solvents; LD_{50} (rat orl) = 3.75 g/kg. *Akzo Chemie; ICI Chem. & Polymers Ltd.*

3397 Sodium chlorite
7758-19-2   8672   G P

$ClNaO_2$
Chlorous acid sodium salt.
Adox 3125; Alcide LD; Caswell No. 755; CCRIS 1426; Chlorite sodium; Chlorous acid, sodium salt; EINECS 231-836-6; EPA Pesticide Chemical Code 020502; HSDB 733; Neo Silox D; Sodium chlorite; Textile; Textone; UN1496. Antimicrobial for water and waste water treatment. Improves taste and odor of potable water; oxidizing agent, bleaching agent for textiles, paper pulp, disinfecting. Registered by EPA as an antimicrobial, fungicide, herbicide and insecticide. Used in germicidal teat-dip preparations. Crystals; dec 180-200°; soluble in H_2O (34 g/100 ml). *Albright & Wilson Americas Inc.; Elf Atochem N. Am.; International Dioxcide Inc.; Olin Corp.*

3398 Sodium chloroacetate
3926-62-3   2129   G H P

$C_2H_2ClNaO_2$
Chloroacetic acid sodium salt.
Acetic acid, chloro-, sod; Acetic acid, chloro-, sodium salt; Caswell No. 755A; Chloroctan sodny; Dow defoliant; EPA Pesticide Chemical Code 355200; Monoxone; SMA; SMCA; Sodium chloroacetate; Sodium monochloroacetate; UN2659. Used as a herbicide. White crystals; soluble in H_2O (85 g/100 ml at 20°); LD_{50} (rat orl) = 76 mg/kg, (mus orl) = 255 mg/kg, (gpg orl) = 80 mg/kg. *Atlas Interlates Ltd; Rhône-Poulenc; Syngenta Crop Protection.*

3399 Sodium chromate
7775-11-3   8674   G P

Cr_2NaO_4
Sodium chromate(VI).
Caswell No. 757; CCRIS 883; Chromate of soda; Chromic acid (H_2CrO_4), disodium salt; Chromic acid, disodium salt; Chromium disodium oxide; Chromium sodium oxide ($CrNa_2O_4$); Disodium chromate; EINECS 231-889-5; EPA Pesticide Chemical Code 068303; HSDB 2962; Neutral sodium chromate; Rachromate; Sodium chromate; Sodium chromate(VI); chromium disodium oxide; chromium sodium oxide ($CrNa_2O_4$). Inks, dyeing, paint pigment, leather tanning, other chromates, protection of iron against corrosion, wood preservative. Registered by EPA as a fungicide (cancelled). Has tetrahydrate and decahydrate (mp = 20°); soluble in

H_2O (1 g/ml). *Elf Atochem N. Am.; OxyChem.*

3400 Sodium citrate
68-04-2   8675   H

$C_6H_8Na_3O_7$
1,2,3-Propanetricarboxylic acid, 2-hydroxy-, trisodium salt, anhydrous.
CCRIS 3293; Citnatin; Citreme; Citric acid, trisodium salt; Citrosodina; Citrosodine; Citrosodna; EINECS 200-675-3; FEMA No. 3026; HSDB 5201; Natrocitral; Sodium citrate; Sodium citrate ($Na_3C_6H_5O_7$); Sodium citrate anhydrous; Trisodium 2-hydroxy-1,2,3-propanetricarboxylate; Tri-sodium citrate. [Dihydrate]: White crystals; soluble in H_2O (77 g/100 ml at25°, 166 g/100 ml at 100°), insoluble in EtOH; LD_{50} (rat ip) = 1764 mg/kg. [Pentahydrate]: crystalline solid; less stable than the dihydrate, loses water easily.

3401 Sodium citrate
994-36-5   H

$C_6H_7NaO_7$
1,2,3-Propanetricarboxylic acid, 2-hydroxy-, sodium salt.
Bicitra; Citric acid, sodium salt; EINECS 213-618-2; Pneucid; 1,2,3-Propanetricarboxylic acid, 2-hydroxy-, sodium salt; Sodium 2-hydroxy-1,2,3-propanetricarb-oxylate; Sodium citrate.

3402 Sodium cocoate
61789-31-9   H

Sodium coconut oil soap.
Coconut fatty acid, sodium salt; Coconut oil fatty acids, sodium salt; EINECS 263-050-4; Fatty acids, coco, sodium salts; Fatty acids, coconut oil, sodium salts; Soap, coconut oil; Sodium cocoate; Sodium coconate.

3403 Sodium coconut monoglyceride sulfate
61789-04-6   H

Glycerides, coco mono-, sulfated, sodium salts.
EINECS 263-026-3; Glycerides, coco mono-, sulfated, sodium salts; Glycerides, coconut oil mono-, sulfated, sodium salts; Sodium cocomonoglyceride sulfate; Sodium coconut monoglyceride sulfate.

3404 Sodium cocoyl isethionate
61789-32-0   H

Fatty acids, coco, 2-sulfoethyl esters, sodium salts.
Coconut fatty acid, 2-sulfoethyl ester, sodium salt; EINECS 263-052-5; Fatty acids, coco, 2-sulfoethyl esters, sodium salts; Fatty acids, coconut oil, sulfoethyl esters, sodium salts; Igepon AC-78; Jordapon CI; Sodium cocoyl isethionate.

559

3405 Sodium cresylate
34689-46-8   H

C7H7NaO
 Methylphenol sodium salt.
 Caswell No. 757A; Cresol sodium salt; Cresylic acid sodium salt; EINECS 252-154-5; EPA Pesticide Chemical Code 357200; HSDB 5139; Methylphenol sodium salt; Phenol, methyl-, sodium salt; Sodium cresolate; Sodium cresoxide; Sodium cresylate. Mixture of o, m and p-isomers.

3406 Sodium cumenesulfonate
28348-53-0   H

C9H11NaO3S
 Benzenesulfonic acid, (1-methylethyl)-, sodium salt.
 Benzenesulfonic acid, (1-methylethyl)-, sodium salt; ar-Cumenesulfonic acid, sodium salt; EINECS 248-983-7; Sodium cumenesulfonate; Sodium cumenesulphonate.

3407 Sodium cumenesulfonate
32073-22-6   G H

C9H11NaO3S
 (1-Methylethyl)benzene, monosulfo deriv., sodium salt.
 Benzene, (1-methylethyl)-, monosulfo deriv., sodium salt; Cumene, monosulpho derivative, sodium salt; EINECS 250-913-5; (1-Methylethyl)benzene, monosulfo deriv., sodium salt; Sodium cumenesulfonate; Witconate SCS 45%. Hydrotrope, solubilizer, coupler and processing aid in detergent manufacturing and industrial processes; antiblocking and anticaking agent in powder products; formulates shampoos, Aerosols®s, cutting oils, glue; textile finishing. CK Witco Corp.

3408 Sodium cyanide
143-33-9   8679   G

$$N\equiv C-Na$$

CNNa
 Hydrocyanic acid sodium salt.
 4-02-00-00050 (Beilstein Handbook Reference); BRN 3587243; Caswell No. 758; CCRIS 7712; Cianuro di sodio; Cyanasalt H; Cyanasalt S; Cyanide of sodium; Cyanide salts; Cyanobrik; Cyanogran; Cyanure de sodium; Cymag; EINECS 205-599-4; EPA Pesticide Chemical Code 074002; HSDB 734; Hydrocyanic acid, sodium salt; Kyanid sodny; M-44 cyanide capsules; NSC 77379; RCRA waste number P106; Sodium cyanide; Sodium cyanide solution; UN1689. Used in extraction of gold and silver from ores, electroplating, heat treatment of metals, making hydrogen cyanide, insecticide, metal cleaning, fumigation, dyes and pigments, nylon intermediates, chelating compounds, ore flotation. Crystals; mp = 563°; soluble in H2O, poorly soluble in organic solvents; LD50 (rat orl) = 15 mg/kg. Degussa AG; DSM Spec. Prods.; DuPont; Elf Atochem N. Am.; FMC Corp.; Grace W.R. & Co.; ICI Spec.; Mitsui Toatsu.

3409 Sodium cyclamate
139-05-9   G

C6H12NNaO3S
 Cyclohexylsulfamic acid, monosodium salt.
 AI3-24213; Assugrin; Assurgrin Feinsuss; Assurgrin Vollsuss; Asugryn; CCRIS 187; Ciclamato sodico; Cyclamate de sodium; Cyclamate sodium; Cyclamate sodium/saccharin sodium; Cyclamate, sodium salt; Cyclamic acid sodium salt; Cyclohexanesulfamic acid, monosodium salt; Cyclohexanesulphamic acid, mono-sodium salt; Cyclohexylsulfamate sodium; Cyclohexyl-sulfamic acid, sodium salt; Cyclohexylsulphamate sodium; Cyclohexylsulphamic acid, monosodium salt; N-Cyklohexylsulfamat sodny; Dulzor-Etas; EINECS 205-348-9; Hachi-Sugar; Ibiosuc; Monosodium cyclohexyl-sulfamate; Natraiumcyclohexyl-amidosulfat; Natreen; Natrii cyclamas; Natrium cyclamicum; Natrium cyclohexylsulfamat; Natrium-cyclohexylamidosulfat; Natriumzyklamate; NSC 42195; Sodium cyclamate; Sodium cyclohexanesulfamate; Sodium cyclohexane-sulphamate; Sodium cyclohexyl amidosulphate; Sodium cyclohexyl sulfamate; Sodium cyclohexyl sulphamate; Sodium cyclohexylsulphamidate; Sodium N-cyclohexylsulfamate; Sodium sucaryl; Sucaryl sodium; Sucrosa; Sucrum 7; Sucrun 7; Suessette; Suestamin; Sugarin; Sugaron; Sulfamic acid, cyclohexyl-, monosodium salt. Artificial sweetener; about 30x as sweet as cane sugar. Crystals; mp = 265°; soluble in H2O, insoluble in organic solvents; LD50 (rat orl) = 15.25 g/kg.

3410 Sodium decanoate
1002-62-6   H

C10H19NaO2
 n-Decanoic acid, sodium salt.
 Capric acid, sodium salt; Caprinic acid sodium salt; Decanoic acid, sodium salt; EINECS 213-688-4; Sodium caprate; Sodium caprinate; Sodium decanoate; Sodium-n-decanoate.

3411 Sodium decyl sulfate
142-87-0   G

C10H21NaO4S
 Sulfuric acid, monodecyl ester, sodium salt.
 Atlasol 103; EINECS 205-568-5; Sodium decyl sulfate; Sodium decyl sulphate; Sulfuric acid, decyl ester, sodium salt. Emulsifier, wetting agent, dispersant, fiber lubricant, synthetic fatliquor; for textile, leather, and general industrial applications. Atlas Chemical Corp.

3412   Sodium dibutyldithiocarbamate
136-30-1   G

C9H18NNaS2
 Dibutyldithiocarbamic acid sodium salt.
 Accel TP; AI3-08293; Butyl namate; Carbamic acid, dibutyldithio-,
sodium salt; Carbamodithioic acid, dibutyl-, sodium salt;
Dibutyldithiokarbaman sodny; EINECS 205-238-0; HSDB 2900;
Octopol NB-47; Pennac; Pennac SDB; Sodium DBDT; Sodium
dibutylcarbamodithioate; Sodium dibutyldithiocarbamate; Sodium
N,N-dibutyldithiocarbamate; Tepidone; Tepidone rubber accelerator;
USAF B-35; Vulcacure; Vulcacure NB. Ultra accelerator for SBR and
natural rubber latex compounds; in polymerization of chloroprene
rubber, especially for latex compounds where copper staining of zinc
salt dithiocarbamates is a problem.

3413   Sodium dicamba
1982-69-0   H P

C8H5Cl2NaO3
 Sodium 2-methoxy-3,6-dichlorobenzoate.
 o-Anisic acid, 3,6-dichloro-, sodium salt; Banvel II; Benzoic acid,
3,6-dichloro-2-methoxy-, sodium salt; Dicamba-sodium; Dicamba
sodium salt; 3,6-Dichloro-2-methoxybenzoic acid, sodium salt; 3,6-
Dichloro-o-anisic acid, sodium salt; EINECS 217-846-3; HSDB 5861;
2-Methoxy-3,6-dichlorobenzoic acid sodium salt; Sodium 2-methoxy-
3,6-dichlorobenzoate; Sodium 3,6-dichloro-2-methoxybenzoate;
Sodium 3,6-dichloro-o-anisate; Sod-ium dicamba. Registered by
EPA as a herbicide. Solid; soluble in H2O (39.7 g /100 ml); LD50 (rat
orl) = 6764 mg/kg. BASF Corp.; Micro-Flo Co. LLC; Nissan Chem.
Ind.; Syngenta Crop Protection.

3414   Sodium dichloroisocyanurate
2893-78-9   G P

C3Cl2N3NaO3
 1,3-Dichloro-1,3,5-triazine-2,4,6(1H,3H,5H)-trione sodium salt.
 ACL 60; BX 94; Caswell No. 759; CCRIS 4788; CDB 63; CP 17254;
Dichloroisocyanurate sodium; Dikonit; Dimanin C; EINECS 220-767-
7; EPA Pesticide Chemical Code 081404; Fi Clor 60S; HSDB 4235;
Isocyanuric acid, dichloro-, sodium salt; Monosodium 1,3-
dichloroiso-cyanurate; NCI-C55732; OCl 56; SDIC; Simpla; Sodium
dichlorisocyanurate; Sodium dichlorocyanurate; Tro-closene sodium.
Active ingredient in dry bleaches, water and sewage treatment.
Topical anti-infective. Used as a water disinfectant, bleach and
media constituent. Registered by EPA as an antimicrobial and
disinfectant. White, granular powder; mp = 250°; soluble in H2O;
LD50 (hmn orl) = 3570 mg/kg, (rat orl) = 1420 mg/kg. Alden Leeds
Inc.; Aqua Clear Industries Inc; Atofina; Biolabs Inc; CFR Packaging
Inc; Clearon Corp; Ecolab Inc.; Hasa Inc; ICI Americas Inc.; Nissan
Chem. Ind.; Occidental Chem. Corp.; PR Pharmaceuticals; Pro

Packaging Inc.

3415   Sodium 2,4-dichlorophenolate
3757-76-4   H

C6H3Cl2NaO
 Phenol, 2,4-dichloro-, sodium salt.
 EINECS 223-163-1; Phenol, 2,4-dichloro-, sodium salt; Sodium 2,4-
dichlorophenolate.

3416   Sodium 2,5-dichlorophenolate
52166-72-0   H

C6H3Cl2NaO
 Phenol, 2,5-dichloro-, sodium salt.
 EINECS 257-697-1; Phenol, 2,5-dichloro-, sodium salt; Sodium 2,5-
dichlorophenolate.

3417   Sodium dichromate
7789-12-0   G P

Cr2Na2O7
 Sodium bichromate.
 Chromic acid (H2Cr2O7), disodium salt, dihydrate; CRIS 6344;
Dichromic acid (H2Cr2O7), disodium salt, dihydrate; Sodium
dichromate dihydrate. Used in colorimetry (copper determination), as
a complexing agent, oxidation inhibitor in ethyl ether. Registered by
EPA as a fungicide. Crystals; d = 2.350; suspected carcinogen.
British Chrome & Chemical; Chemisphere; Occidental Chem. Corp.;
OxyChem; Rit-Chem.

3418   Sodium dicyclohexyl sulfosuccinate
23386-52-9   H

C16H25NaO7S
 Sulfosuccinic acid, 1,4-dicyclohexyl ester, sodium salt.
 Aerosol® A 196; Bis-cyclohexyl sodium sulfosuccinate (80%);
Butanedioic acid, sulfo-, 1,4-dicyclohexyl ester, sodium salt;
Dicyclohexyl   sodium   sulfosuccinate;   1,4-Dicyclohexyl

sulfobutanedioate sodium salt; EINECS 245-629-3; Pelex CS; Sodium 1,4-dicyclohexyl sulphonato-succinate; Sodium 1,4-dicyclohexyl sulfobutanedioic acid; Sodium dicyclohexyl sulfosuccinate; Succinic acid, sulfo-, 1,4-dicyclohexyl ester, sodium salt; Sulfosuccinic acid, dicyclohexyl ester, sodium salt; Sulfosuccinic acid, 1,4-dicyclohexyl ester, sodium salt.

3419 Sodium diethyl oxaloacetate
40876-98-0   H

C8H11NaO5
Diethyl oxaloacetate, monosodium salt.
AI3-04820; Butanedioic acid, oxo-, diethyl ester, ion(1-), sodium; Diethyl oxalacetate sodium salt; Diethyl oxaloacetate, monosodium salt; Diethyl oxobutanedioate ion(1-) sodium; Diethyl sodiooxalacetate; Diethyl sodium oxalacetate; EINECS 255-122-9; NSC 126906; Oxalacetic acid diethyl ester sodium salt; Sodium diethyl oxaloacetate; Sodium diethyl oxobutanedioate.

3420 Sodium diethyl phosphorodithioate
3338-24-7   H

C4H10NaO2PS2
Phosphorodithioic acid, O,O-diethyl ester, sodium salt.
O,O-Diethyldithiofosforecnan sodny; EINECS 222-079-2; Ethyl sodium phosphorodithioate; Hostaflot LET; HSDB 2606; Phosphorodithioic acid, O,O-diethyl ester, sodium salt; Sodium aerofloat; Sodium diethyl dithiophosphate; Sodium diethyl phosphorodithioate; Sodium O,O-diethyl dithiophosphate.

3421 Sodium dihexyl sulfosuccinate
3006-15-3   G

C16H29NaO7S
Sodium 1,4-dihexyl sulfobutanedioate.
Aerosol MA80; AI3-18858; Bis(1-methylamyl) sodium sulfosuccinate; Butanedioic acid, sulfo-, 1,4-dihexyl ester, sodium salt; Dihexyl sodiosulfosuccinate; Dihexyl sodium sulfosuccinate; Dihexyl sulfosuccinate sodium salt; 1,4-Dihexyl sulfosuccinate sodium salt; EINECS 221-109-1; Monawet MM 80; Octosol HA-80; Sodium 1,4-dihexyl sulphonatosuccinate; Sodium 1,4-dihexyl sulfosuccinate; Sodium dihexyl sulfosuccinate; Succinic acid, sulfo-, dihexyl ester, sodium salt; Succinic acid, sulfo-, 1,4-dihexyl ester, sodium salt; Sulfosuccinic acid, dihexyl ester, sodium salt; SV 1017. Emulsifier used in latex emulsion polymerization. *Tiarco.*

3422 Sodium dihydroxysuccinate
51307-92-7   H

C4H4Na2O6
Butanedioic acid, 2,3-dihydroxy-, disodium salt.
Butanedioic acid, 2,3-dihydroxy-, disodium salt, (R*,R*)-(±)-; Butanedioic acid, 2,3-dihydroxy-, disodium salt, (2R,3R)-rel-.

3423 Sodium dimethyl dithiophosphate
26377-29-7   H

C2H6NaO2PS2
Phosphorodithioic acid, O,O-dimethyl ester, sodium salt.
O,O-Dimethyldithiofosforecnan sodny; EINECS 247-636-7; Methyl sodium phosphorodithioate; Phosphorodithioic acid, O,O-dimethyl ester, sodium salt; Sodium dimethyl dithiophosphate; Sodium dimethyl phosphorodithioate; Sodium O,O-dimethyl phosphorodithioate; Sodium O,O-dimethyl dithiophosphate.

3424 Sodium dimethyldithiocarbamate
128-04-1   G P

C3H6NNaS2
Sodium dimethylcarbamodithioate.
Aceto SDD 40; AI3-14673; Alcobam NM; Amersep MP 3R; Aquatreat SDM; Brogdex 555; Carbam-S; Carbamodithioic acid, dimethyl-, sodium salt; Carbon s; Caswell No. 762; CCRIS 5535; DDC; Diaprosim AB 13; Dibam; Dibam A; Dimethyldithiocarbamic acid, sodium salt; Diram; DMDK; EINECS 204-876-7; EPA Pesticide Chemical Code 034804; Freshgard 40; MetalPlex 143; Methyl namate®; MSL; MSL (carbamate); Nalmet A 1; Nocceler S; NSC 85566; Octopol SDM-40; Perkacit® SDMC; Sanceler S; SDDC; SDMDTC; Sharstop 204; Sodam; Sodium dimethyl dithiocarbamate; Sta-Fresh 615; Steriseal liquid 40; Thiostop N; Vinditat; Vinstop; Vulnopol NM; Wing Stop B. A water treatment chemical; readily forms water-insoluble salts with heavy metals such as cadmium, copper, chromium, and nickel; clarification agent for wastewater from plating, photo finishing. Polymerization shortstop in SBR rubber; precipitant for heavy metals in waste water treatment. Accelerator for rubber industry; shortstopper for polymer production. Fungicide. Registered by EPA as an antimicrobial and fungicide. Crystals; mp = 120-122°; soluble in H2O, polar organic solvents; λm = 257, 290 nm (ε 1200, 13000 EtOH); LD50 (rat orl) = 1000 mg/kg, (rat ip) = 1000 mg/kg, (mus orl) = 1500 mg/kg, (mus ip) = 573 mg/kg, (rat orl) = 300 mg/kg. *Akzo Chemie; Alco Chem. Corp.; IBC Manufacturing Co.; Quest Chemical Corp.; Uniroyal; Vanderbilt R.T. Co. Inc.; Vinings Industries Inc.*

3425 Sodium dodecylbenzenesulfonate
25155-30-0   8686   G P

C18H29NaO3S

Sodium laurylbenzenesulfonate.

35SL; AA-9; AA-10; Abeson nam; Arylan SBC; Benzenesulfonic acid, dodecyl-, sodium salt; Bio-soft D-35x; Bio-soft D-40; Bio-soft D-60; Bio-soft D-62; C 550; Calsoft F-90; Calsoft L-40; Calsoft L-60; Caswell No. 765; Conco aas-35; Conco AAS-35; Conco AAS 35H; Conco AAS-40; Conco AAS-65; Conco AAS-90; Conco C-50; Conoco SD 40; Detergent HD-90; Dodecyl benzene sodium sulfonate; DS 60; EINECS 246-680-4; EPA Pesticide Chemical Code 079010; HS 85S; HSDB 740; KB; Marlon 375a; Mercol 25; Mercol 30; Naccanol NR; Naccanol SW; Nacconol; Nacconol 35SL; Nacconol 40f; Nacconol 90f; Nansa HF 80; Nansa HS 85S; Neopelex 05; Pelopon A; Pilot hd-90; Pilot sf-40; Pilot sf-40b; Pilot sf-40fg; Pilot sf-60; Pilot sf-96; Richonate 1850; Richonate 40b; Richonate 45b; Richonate 45B; Richonate 60b; Sandet 60; Santomerse; Santomerse 3; Santomerse No. 1; Santomerse No. 85; SDBS; Sinnozon; Sodium dodecylphenylsulfonate; Sodium lauryl benzene sulfonate; Sol sodowa kwasu laurylobenzenesulfono-wego; Solar 40; Solar 90; Steinaryl NKS 50; Steinaryl NKS 100; Stepan DS 60; Stepantan DS 40; Sulfapol; Sulfapolu; Sulfaril paste; Sulframin 1238; Sulframin 1238 slurry; Sulframin 1240; Sulframin 1250 slurry; Sulframin 40; Sulframin 40RA; Sulframin 85; Sulframin 90; Sulfuril; Trepolate F 40; Ultrawet 60K; Ultrawet K; Ultrawet KX; Ultrawet SK; Ultrawet XK; Vista C 550; Witconate 1238; Witconate 1250; Witconate 60 B; X 2073. Registered by EPA as an antimicrobial, fungicide and insecticide. FDA, USDA listed, FDA approved for topicals. Used in topicals as a surfactant. Used in rug and upholstery cleaners. Detergent. White to light yellow flakes, granules or powder; mp > 300°; soluble in H2O (0.5 - 1.0 g/100 ml at 19°). LD50 (mus orl) = 2000 mg/kg, (mus iv) = 105 mg/kg. *Albright & Wilson Americas Inc.; DuPont; Emkay; Fluka; Norman-Fox; Pilot; Rhône-Poulenc; Spectrum Chem. Manufacturing; Stepan; Tokyo Kasei Kogyo; Unger Fabrik Chem. Ind.; Witco/Oleo-Surf.*

3426   Sodium erythorbate
6381-77-7   G H

C6H7NaO6

Monosodium D-erythro-hex-2-enonic acid γ-lactone.

Araboascorbic acid, monosodium salt, D-; CCRIS 2517; EINECS 228-973-9; Eribate; Hex-2-enonic acid gamma-lactone, D-erythro-, monosodium salt; D-erythro-Hex-2-enonic acid, γ-lactone, monosodium salt; HSDB 741; Isoascorbic acid sodium salt; Isona; Mercate 20; Neo-cebitate®; Sodium isoascorbate. Antioxidant and antimicrobial agent used in meat, poultry, and seafood industries. Crystals; mp = 168-170°; [α]$_D^{25}$ = 95° (c = 10 H2O); very soluble in H2O 916 g/100 ml). *Lancaster Synthesis Co.; PMP Fermentation Prods.; Rhône-Poulenc Food Ingredients.*

3427   Sodium ethoxide
141-52-6   8687   H

C2H5NaO

Sodium ethanolate.

Al3-52660; Caustic alcohol; EINECS 205-487-5; Ethanol, sodium salt; Ethoxysodium; Ethyl alcohol, sodium salt; HSDB 5637; Sodium ethanolate; Sodium ethoxide; Sodium ethylate.

3428   Sodium ethylenesulfonate
3039-83-6   H

C2H3NaO3S

Ethylenesulfonic acid, sodium salt.

EINECS 221-242-5; Ethenesulfonic acid, sodium salt; NSC 8957; Sodium ethenesulfonate; Sodium ethylenesulfon-ate; Sodium ethylenesulphonate; Sodium vinylsulfonate.

3429   Sodium 2-ethylhexyl sulfate
126-92-1   G

C8H17NaO4S

Sodium 2-ethylhexyl-sulfate.

08-Union carbide; CCRIS 2461; EINECS 204-812-8; Emcol D 5-10; Emersal 6465; Etasulfate de sodium; Etasulfato sodico; Ethasulfate sodium; 2-ethyl-1-hexanol sulfate sodium salt; 2-Ethyl-1-hexanol sodium sulfate; 2-Ethyl-1-hexanol hydrogen sulfate sodium salt; 2-Ethylhexyl sodium sulfate; 2-Ethylhexylsiran sodny; 2-Ethylhexylsulfate sodium; Hexanol, 2-ethyl-, hydrogen sulfate, sodium salt; 1-Hexanol, 2-ethyl-, sulfate, sodium salt; 1-Hexanol, 2-ethyl-, hydrogen sulfate, sodium salt; HSDB 1314; Mono(2-ethylhexyl) sulfate sodium salt; Natrii etasulfas; NCI-C50204; NIA proof 08; Niaproof® Anionic Surfactant 08; NSC 4744; Pentrone ON; Propaste 6708; Rhodapon® BOS; Sipex bos; Sodium (2-ethylhexyl)alcohol sulfate; Sodium 2-ethylhexyl sulfate; Sodium etasulfate; Sodium ethasulfate; Sodium mono(2-ethylhexyl) sulfate; Sodium octyl sulfate, iso; Sodium octyl sulfate; Sodium(2-ethylhexyl)alcohol sulfate; Sole Tege TS-25; Sulfirol 8; Sulfuric acid, mono(2-ethylhexyl) ester, sodium salt; Tergemist; tergimist; Tergitol 08; Tergitol anionic 08. Biodegradable detergent, wetting agent, penetrant, emulsifier used in textile mercerizing, metal cleaning, electroplating, photo chemicals, adhesives, emulsion polymerization, household and industrial cleaners, agricultural, pharmaceuticals; stable to high concentrations of electrolytes. Used as a latex stabilizer, nickel brightener and in metal treatment. Liquid; bp = 96-104°; soluble in H2O (>10 g/100 ml); d21.7 = 1.114; insoluble in organic solvents; LD50 (rat orl) = 4 g/kg. *Niacet; Rhône Poulenc Surfactants.*

3430   Sodium ferrocyanide
13601-19-9   8689   G

C6FeN6Na4

Sodium ferrocyanide.

Al3-28762;   EINECS   237-081-9;   Ferrate(4-),   hexacyano-,

563

tetrasodium; Ferrate(4-), hexakis(cyano-C)-, tetrasodium; Ferrate(4-), hexakis(cyano-C)-, tetrasodium, (OC-6-11)-; HSDB 742; Sodium ferrocyanide; Sodium ferrocyanide (Na4(Fe(CN)6)); Sodium hexacyanoferrate(II); Sodium prussiate yellow; Tetrasodium ferrocyanide; Tetrasodium hexacyanoferrate; Tetrasodium hexacyanoferrate(4-); Yellow prussiate of soda. Manufacture of sodium ferricyanide, blue pigments, blueprint paper, anticaking agent for salt, ore flotation, pickling metals, polymerization catalyst, photographic fixative. [decahydrate]; pale yellow crystals; mp = 81.5°, dec 435°; soluble in H2O (14%), insoluble in organic solvents. *Atomergic Chemetals; Degussa AG; Rit-Chem.*

3431   Sodium fluoride
7681-49-4   8691   D G P

H—F

FNa
Hydrofluoric acid sodium salt.
AI3-01500; Alcoa sodium fluoride; Antibulit; Caswell No. 769; Cavitrol; CCRIS 1573; Chemifluor; Credo; Disodium difluoride; EINECS 231-667-8; EPA Pesticide Chemical Code 075202; F1-Tabs; FDA 0101; Floridine; Florocid; Flozenges; Fluonatril; Fluor-O-kote; Fluoraday; Fluoral; Fluorid sodny; Fluoride, sodium; Fluorident; Fluorigard; Fluorineed; Fluorinse; Fluoritab; Fluorocid; Fluorol; fluoros; Fluorure de sodium; Flura; Flura drops; Flura-gel; Flura-loz; Flurcare; Flurexal; Flursol; Flux; Fungol B; Gel II; GEL II; Gelution; Gleem; HSDB 1766; Iradicav; Kari-rinse; Karidium; Karigel; Koreberon; Lea-Cov; Lemoflur; Les-Cav; Liquiflur; Luride; Luride lozi-tabs; Luride SF; Minute-Gel; Na frinse; Nafeen; NaFpak; Natrium fluoride; NCI-C55221; Neutra Care; NSC 77385; Nufluor; Ossalin; Ossin; osteofluor; Osteopor-F; Pediaflor; Pedident; Pennwhite; Pergantene; Phos-Flur; Point two; Predent; Raflour; Rescue squad; Roach salt; So-flo; SO-Flo; Sodium fluoride; Sodium fluoride (NaF); Sodium fluoride cyclic dimer; Sodium fluoride, solid (DOT); Sodium fluorure; Sodium hydrofluoride; Sodium monofluoride; Stay-Flo; Studafluor; Super-dent; T-Fluoride; Theraflur; Thera-flur-N; Trisodium trifluoride; UN1690; Villiaumite; Xaridium; Zymafluor. Fluoridation of municipal water, degassing steel, wood preservative, insecticide, fungicide, rodenticide, chemical cleaning, electroplating, glass manufacture, vitreous enamels, preservative for adhesives, toothpastes, disinfectants, dental prophylaxis. Registered by EPA as an insecticide and fungicide. Crystals; mp = 993°; bp = 1704°; d = 2.78; soluble in H2O (4 g/100 ml); insoluble in organic solvents; LD50 (rat orl) = 0.18 g/kg. *Cerac; EM Ind. Inc.; General Chem; Hoechst Celanese; Solvay Deutschland GmbH; Whiting Peter Ltd.*

3432   Sodium fluoroborate
13755-29-8   8690   G

F—B⁻—F Na+

BF4Na
Sodium tetrafluoroborate.
Apreton R; Borate(1-), tetrafluoro-, sodium; Boron sodium fluoride (BNaF4); EINECS 237-340-6; NSC 77386; Pyricit; Sodium borofluoride; Sodium boron fluoride; Sodium boron tetrafluoride; Sodium fluoborate; Sodium fluoborate (Na(BF4)); Sodium fluoroborate (NaBF4); Sodium tetrafluoroborate; Sodium tetrafluoroborate(1-). Used as a fluorinating agent. Rectangular prisms; mp = 384°; d²⁰ = 2.4700; soluble in H2O (108 g/100 ml), less soluble in organic solvents.

3433   Sodium fluorophosphate
10163-15-2
                                       G

O=P—O⁻   2Na⁺

FNa2O3P
Disodium monofluorophosphate.
AI3-16931; Albaphos Dental Na 211; Disodium fluorophosphate; Disodium monofluorophosphate; Disodium phosphorofluoridate; EINECS 233-433-0; NSC 248; Phosphorofluoridic acid, disodium salt; Sodium fluorophosphate (Na2PO3F); Sodium monofluorophosphate; Sodium phosphorofluoridate. A fluorine-containing component for toothpastes, the toxic effects of which are only 1/3 of those of sodium fluoride. *Hoechst UK Ltd.*

3434   Sodium formaldehyde bisulfite
870-72-4   4260   G H

HO   SO₃Na

CH3NaO4S
Hydroxymethanesulfonic acid, monosodium salt.
AI3-23356; EINECS 212-800-9; Formaldehyde sodium bisulfite; Formbis; HSDB 5766; Hydroxymethanesulfonic acid, monosodium salt; Methylolsulfonic acid sodium salt; Monosodium hydroxymethanesulfonate; NSC 2441; Rongalit® C; Sodium formaldehyde bisulfite; Sodium hydroxymethanesulfonate; Sodium hydroxymethane-sulphonate. Reducing and discharge agent for textile printing. Needles; mp = 200°. *BASF Corp.*

3435   Sodium formate
141-53-7   8694   G H

H   O⁻ Na⁺

CHNaO2
Formic acid, sodium salt.
CCRIS 1037; EINECS 205-488-0; Formax; Formic acid, sodium salt; HSDB 744; Mravencan sodny; NSC 77457; Salachlor; Sodium formate. A reducing agent, used in the manufacture of formic acid, oxalic acid and sodium dithionite, organic chemicals, mordant, complexing agent, analytical reagent (noble metal precipitant), buffering agent. Crystals; mp = 253°; bp = 360°; d = 1.92; soluble in H2O (77 g/100 ml), slightly soluble in EtOH, glycerol; LD50 (mus orl) = 11200 mg/kg. *Aqualon; Heico; Hoechst Celanese; May & Baker Ltd.; Perstorp AB; Spectrum Chem. Manufacturing.*

3436   Sodium glucoheptonate
31138-65-5   H

C7H13NaO8
Monosodium D-glucoheptonate.
EINECS 250-480-2; D-Gluco-heptonic acid, monosodium salt, (2.xi)-; Monosodium D-glucoheptonate; Sodium glucoheptonate.

3437   Sodium gluconate
527-07-1   8695   G H

564

C6H11NaO7

Monosodium D-gluconate.

EINECS 208-407-7; Glonsen; Gluconato di sodio; Gluconic acid, monosodium salt, D-; D-Gluconic acid, monosodium salt; Gluconic acid, sodium salt; Monosodium D-gluconate; Monosodium gluconate; Pasexon 100T; Sodium D-gluconate; Sodium gluconate. As sequestering agent; in metal plating, tanning of hides, mordants for fabrics, paints; rust remover; mineral source in foods and pharmaceuticals. Soluble in H_2O (59 g/100 ml), less soluble in organic solvents. *Akzo Chemie; Albright & Wilson Americas Inc.; Pfizer Inc.; PMP Fermentation Prods.; Rit-Chem.*

3438   Sodium hexametaphosphate
10124-56-8   G

Na6O18P6

Metaphosphoric acid, hexasodium salt.

Calgon; Calgon (old); Calgon S; Caswell No. 772; Chemi-charl; EINECS 233-343-1; EPA Pesticide Chemical Code 076402; FEMA 3027; Graham's salt; Hexameta-phosphate, sodium salt; Hexasodium hexametaphos-phate; Hexasodium metaphosphate; HMP; HSDB 5053; Kalex; Medi-Calgon; Metaphosphoric acid ($H_6P_6O_{18}$), hexasodium salt; Metaphosphoric acid, hexasodium salt; Natrium hexametaphosphat; Phosphate, sodium hexameta; Polyphos; SHMP; Sodaphos; Sodium hexa-metaphosphate; Sodium hexametaphosphate ($Na_6P_6O_{18}$); Sodium metaphosphate; Sodium phosphate, tribasic; Sodium polymetaphosphate. Sequestering agent for calcium and magnesium ions for certain sensitive textile application. Used in water softeners and detergents such as Calgon, Giltex; Quadrafos, Hagan Phosphate and Micromet. Also used in leather tanning, dyeing, laundries and textile processing. Crystals; mp = 628°; soluble in H_2O. *FMC Corp.; Hart Prod.*

3439   Sodium hyaluronate
9067-32-7   4776   G V

(C14H20NNaO11)x

Hyaluronic acid, sodium salt.

ARTZ; CCRIS 4127; Connettivina; Equron (Veterinary); Healon; Healonid; Hyacid; Hyalgan; Hyalovet; Hyaluronate Sodium; Hyalurone sodium; Hyaluronic acid, sodium salt; Hyaluronic acid sodium; Hyonate; Ial; Legend; NRD101; Opegan; Provisc; SI-4402; SL-1010; Sodium hyaluronate; SPH; Synacid (veterinary). Protein for use in skin and hair care preparations. Used as an aid in ophthalmology. In veterinary medicine, used to treat non-infectious synovitis and osteoarthritis in dogs and horses. Crystals; $[\alpha]_D^{25}$= -74° (c = 0.35 H_2O). *Rita.*

3440   Sodium   hydrogen   4-amino-5-hydroxy-naphthalene-2,7-disulfonate
5460-09-3   H

C10H8NNaO7S2

2,7-Naphthalenedisulfonic acid, 4-amino-5-hydroxy-, monosodium salt.

4-Amino-5-hydroxy-2,7-naphthalenedisulfonic acid, monosodium; 8-Amino-1-naphthol-3,6-disulfonic acid monosodium salt; Ash acid; EINECS 226-736-4; H acid monosodium salt; 2,7-Naphthalenedisulfonic acid, 4-amino-5-hydroxy-, monosodium salt; NSC 8632; Sodium hydrogen 4-amino-5-hydroxynaphthalene-2,7-disulphon-ate.

3441   Sodium hydrosulfite
7775-14-6   8700   G

Na2O4S2

Disodium dithionite.

Blankit; Blankit IN; Burmol; Caswell No. 774; CCRIS 1428; combustible]; Disodium dithionate; Disodium dithionite; Disodium hydrosulfite; Dithionous acid, disodium salt; EINECS 231-890-0; EPA Pesticide Chemical Code 078202; HSDB 746; Hydros; Hydrosulfite R Conc; Sodium dithionate; Sodium dithionite; Sodium hydrosulfite; Sodium hydrosulfite ($Na_2S_2O_4$); Sodium hydrosulphite; Sodium hyposulfite; Sodium sulfoxylate; UN1384; V-Brite B; Vatrolite. Used as a chemical reagent for the reduction of aldehydes and ketones to alcohols; vat dying of fibers and textiles; bleaching sugar, soap, oils; oxygen scavenger for synthetic rubbers. Gray-white crystals; mp = 52°; soluble in H_2O, less soluble in organic solvents. *BASF Corp.; Farleyway Chem. Ltd.; Henkel/Organic Prods.; Hoechst Celanese; Mitsubishi Gas; Morton Intn'l.; Nissan Chem. Ind.; Olin Corp.*

3442   Sodium hydroxide
1310-73-2   8701   G

Na—OH

HNaO

Caustic soda.

Aetznatron; Ascarite; Caswell No. 773; Caustic soda; Collo-Grillrein; Collo-Tapetta; EINECS 215-185-5; EPA Pesticide Chemical Code 075603; Fuers Rohr; HSDB 229; Hydroxyde de sodium; Lewis-red devil lye; Liquid-plumr; Natrium causticum; Natrium-hydroxid, reinstes; Natriumhydroxid; Natriumhydroxyde; NSC 135799; Plung; Rohrputz; Rohrreiniger Rofix; Soda, caustic; Soda, hydrate; Soda, kaustische; Soda lye; Sodio(idrossido di); Sodium hydrate; Sodium hydroxide; Sodium(hydroxyde de); UN 1823 (solid); UN 1824 (solution); White caustic; White caustic solution. Used in chemical manufacturing, rayon, cellophane, neutralizing agent in petroleum refining; textile processing; pulp, paper, soaps; vegetable oil refining; reclaiming rubber; food additive; etching and electroplating. White flakes; mp = 318°; bp = 1390°; SG = 1.045; soluble in H_2O (0.5 g/ml); LD50 (rbt orl) = 500 mg/kg (10% solution). *Akzo Chemie; Asahi Chem. Industry; Asahi Denka Kogyo; Elf Atochem N. Am.; Georgia Gulf; Georgia Pacific; ICI Spec.; Nissan Chem. Ind.; Norsk Hydro AS; Olin Corp; OxyChem; Rasa.*

3443　Sodium p-hydroxybenzenesulfonate
825-90-1　　　　　　　　　　　　　　　　H

C6H5NaO4S
Sodium p-hydroxybenzenesulfonate.
Benzenesulfonic acid, 4-hydroxy-, monosodium salt; Benzenesulfonic acid, p-hydroxy-, monosodium salt; EINECS 212-550-0; 4-Hydroxybenzenesulfonic acid, monosodium salt; p-Hydroxybenzenesulfonic acid, monosodium salt; NSC 147483; Sodium 4-hydroxy-benzenesulphonate; Sodium p-hydroxybenzenesulfon-ate; Sodium p-hydroxyphenylsulphonate.

3444　Sodium hydroxymethylamino aAcetate
70161-44-3　　　　　　　　　　　　　　G P

C3H6NNaO3
Sodium hydroxymethyl glycinate.
Glycine, N-(hydroxymethyl)-, monosodium salt; Hydroxy-methylaminoacetic acid, sodium salt; N-(Hydroxy-methyl)glycine, monosodium salt; Sodium hydroxy-methylamino acetate; Sodium N-(hydroxymethyl)-glycinate; Suttocide® A. Antimicrobial preservative for cosmetics. Used as an industrial preservative for coatings, adhesives, printing inks, pastes, and polishes. Registered by EPA as an antimicrobial and disinfectant. ISP; Sutton Labs.

3445　Sodium hydroxytriphenylborate
12113-07-4　　　　　　　　　　　　　　H

C18H16BNaO
Sodium hydroxytriphenylborate(1-).
Borate(1-), hydroxytriphenyl-, sodium, (T-4)-; EINECS 235-171-2; Sodium hydroxytriphenylborate; Sodium hydroxytriphenylborate(1-).

3446　Sodium hypochlorite
7681-52-9　　　　　　　8702　　　　　D G P V

Na⁺　⁻O—Cl　5H2O

CINaO.5H2O
Sodium hypochlorite pentahydrate.
AD Gel; Adeka Hypote; Antiformin; B-K liquid; Carrel-Dakin solution; Caswell No. 776; CCRIS 708; Chlorinated water (sodium hypochlorite); Chloros; Chlorox; Cloralex; Cloropool; Clorox; Clorox liquid bleach; Dakin's solution; Deosan; Deosan Green Label Steriliser; Dispatch; EINECS 231-668-3; EPA Pesticide Chemical Code 014703; Hospital Milton; Household bleach; HSDB 748; Hyclorite; Hypochlorite sodium; Hypochlorous acid, sodium salt; Hyposan and Voxsan; Hypure; Hypure N; Javel water; Javelle water; Javex; Klorocin; Milton; Milton Crystals; Modified Dakin's solution; Neo-cleaner; Neoseptal CL; Parozone; Purin B; Sodium hypochlorite; Sodium hypochlorite (NaOCl); Sodium oxychloride; Solutions, Dakin's; Sunnysol 150; Surchlor; UN 1791; XY 12; Youxiaolin; Hychlorite; Javelle water; Mera industries 2MOM3B; Milton; Modified dakin's solution; Piochlor. Bleaching paper, pulp and textiles; intermediate; organic chemicals; water purification; medicine; fungicide; swimming pool disinfectant; laundering; reagent; germicide. Widely used as a bleach and water disinfectant. Registered by EPA as an antimicrobial and fungicide. Used in veterinary medicine as an anti-infective and topical antiseptic - a

diluted soda solution used as an antiseptic for wound irrigation. Crystals; mp = 18°; d = 1.250; soluble in H2O (29 g/100 ml). Asahi Denka Kogyo; Mann George; Norsk Hydro AS; Olin Corp; OxyChem; Showa Denko.

3447　Sodium iodate
7681-55-2　　　　　　8704　　　　　D G V

INaO3
Iodic acid sodium salt.
EINECS 231-672-5; HSDB 749; Iodic acid, sodium salt; Iodic acid (HIO3), sodium salt; Natriumjodat; NSC 77387; Sodium iodate. Antiseptic, disinfectant, feed additive reagent. Used medicinally as an antiseptic, particularly in mucous membranes. Solid; d = 4.28; soluble in H2O (9 g/100 ml), insoluble in organic solvents; LD (dog iv) = 200 mg/kg. Atomergic Chemetals; Mallinckrodt Inc.

3448　Sodium iodide
7681-82-5　　　　　　8705　　　　　D G

Na—I

INa
Hydriodic acid sodium salt.
Caswell No. 777; EINECS 231-679-3; EPA Pesticide Chemical Code 075702; HSDB 750; Iodure de sodium; Ioduril; Jodid sodny; Natrii iodidum; Natriumjodid; NSC 77388; Sodium iodide; Sodium iodide (NaI); Sodium monoiodide; Soiodin. Photography, solvent for iodine, organic chemicals, reagent, feed additive, cloud seeding, scintillation. Used in medicine as an iodine supplement and expectorant. White crystals; mp = 651°; bp = 1300°; d = 3.67; soluble in H2O (200 g/100 ml), less soluble in organic solvents; MLD (rat iv) = 1.3 g/kg. EM Ind. Inc.; Mallinckrodt Inc.; Sigma-Aldrich Fine Chem.

3449　Sodium iodohippurate
133-17-5　　　　　　5056　　　　　　G

C9H7¹²³INNaO3
N-(2-Iodobenzoyl)glycine monosodium salt.
EINECS 205-097-5; Glycine, N-(2-iodobenzoyl)-, mono-sodium salt; Hippodin; Hippuran; Hippuran-¹³¹I; Hippuric acid, o-iodo-, monosodium salt; Iodairal; Iodohippura; Iodohippuran; Iodohippurate sodium; o-Iodohippurate sodium; Jodairol; o-Jodhippursaeure natrium; Medopaque; Nephroflow; NSC 63351; Orthoiodin; Renumbral; Sodium 2-iodohippurate; Sodium iodohippurate; Sodium o-iodohippurate; Sodium orthoiodohippurate; Urocontrast. Diagnostic aids; The unlabelled form provides a radiopaque medium, the radioactive (¹³¹I) form serves as a radioactive imaging agent. Soluble in H2O. Mallinckrodt Inc.; Medi-Physics Inc.

3450　Sodium isethionate
1562-00-1　　　　　　　　　　　　　　G H

C2H5NaO4S

Sodium 1-hydroxy-2-ethanesulfonate.

2-Hydroxyethanesulfonic acid sodium salt; EINECS 216-343-6; Ethanesulfonic acid, 2-hydroxy-, monosodium salt; Ethanesulfonic acid, 2-hydroxy-, sodium salt; HSDB 5838; Isethionic acid sodium salt; NSC 124283; Sodium β-hydroxyethanesulfonate; Sodium 1-hydroxy-2-ethane-sulfonate; Sodium 2-hydroxy-1-ethanesulfonate; Sodium 2-hydroxyethanesulfonate; Sodium 2-hydroxyethane-sulphonate; Sodium 2-hydroxyethyl sulfonate; Sodium 2-hydroxyethylsulfonate; Sodium hydroxyethylsulfonate; Sodium isethionate; Witconate NIS. Detergent, foaming agent, and wetting agent. CK Witco Corp.

3451   Sodium isoascorbate
7378-23-6   H

C6H7NaO6

D-erythro-Hex-2-enonic acid, γ-lactone, sodium salt.
EINECS 230-938-8; D-erythro-Hex-2-enonic acid, γ-lactone, sodium salt; Isoascorbic acid, sodium salt.

3452   Sodium isopropyl xanthate
140-93-2   8708   G H P

C4H7NaOS2

Sodium O-isopropyl dithiocarbonate.
Aeroxanthate 343; AI3-18899; Caswell No. 788; EINECS 205-443-5; EPA Pesticide Chemical Code 076301; Good-rite NIX; HSDB 5633; O-Isopropyl sodium dithio-carbonate; Isopropylxanthic acid, sodium salt; Isopropylxanthogenan sodny; Natrium-O-isopropyl-dithiokarbonat; NIX; NSC 35596; Proxan sodium; Proxan-sodium; Sodium isopropyl xanthate; Xanthic acid, isopropyl-, sodium salt; Z 11. Used in the mining industry and as a weed control in bean and pea crops. Am. Cyanamid.

3453   Sodium lactate
72-17-3   8709   G

C3H5NaO3

2-Hydroxypropanoic acid, monosodium salt.
AI3-03131; CCRIS 7316; EINECS 200-772-0; EINECS 206-231-5; Lacolin; Lactic acid, monosodium salt; Lactic acid sodium salt; Monosodium 2-hydroxypropanoate; Monosodium lactate; NSC 31718; Per-glycerin; Propan-oic acid, 2-hydroxy-, monosodium salt; Sodium (dl)-lactate; Sodium α-hydroxypropionate; Sodium lactate. Hygroscopic agent; glycerol substitute; plasticizer for casein; corrosion inhibitor in alcoholic antifreeze; electrolyte replenisher.
Colorless, viscous, odorless liquid; d = 1.326; soluble in H2O. Am. Biorganics; EM Ind. Inc.; Greeff R.W. & Co.; Patco; Rita; Verdugt BV.

3454   Sodium laureth sulfate
3088-31-1   H

C16H33NaO6S

Ethanol, 2-(2-(dodecyloxy)ethoxy)-, hydrogen sulfate, sodium salt.
Diethylene glycol monododecyl ether sodium sulfate; EINECS 221-416-0; HSDB 6143; Lauryl diethylene glycol ether sulfonate sodium; PEG-(1-4) Lauryl ether sulfate, sodium salt; Poly(oxy-1,2-ethanediyl), α-sulfo-ω(dodecyl-oxy)-, sodium salt; Polyethylene glycol (1-4) lauryl ether sulfate, sodium salt; Polyoxyethylene (1-4) lauryl ether sulfate, sodium salt; Sodium 2-(2-dodecyloxyethoxy)ethyl sulphate; Sodium diethylene glycol dodecyl ether sulfate; Sodium dioxyethylenedodecyl ether sulfate; Sodium laureth sulfate; Sodium lauryl alcohol diglycol ether sulfate; Sodium lauryl di(oxyethyl) sulfate; Sodium lauryl ether sulfate; Sodium lauryloxyethoxyethyl sulfate; Sodium polyoxyethylene lauryl sulfate; Tergentol.

3455   Sodium laureth sulfate
9004-82-4   E G

$$CH_3(CH_2)_{10}CH_2(OCH_2CH_2)_nOSO_3Na$$

Sodium lauryl ether sulfate (n=1-4).
α-Sulfo-ω-(dodecyloxy)poly(oxy-1,2-ethanediyl) sodium salt; Avirol 100E; Conco Sulfate WE; Cycloryl NA; Dodecanol, ethoxylated, monoether with sulfuric acid, sodium salt; Elfan 242; Elfan NS 242; Elfan NS 243; Empicol ESB 3; Empicol ESB 30; Empimin KSN; Empimin KSN 27; Empimin KSN 60; Empimin KSN 70; Etoxon EPA; Glycols, polyethylene, mono(hydrogen sulfate), dodecyl ether, sodium salt; HSDB 752; Laureth-8 carboxylic acid, sodium salt; Maprofix 60S; Maprofix ES; PEG-5 Lauryl ether sulfate, sodium salt; PEG-7 Lauryl ether sulfate, sodium salt; PEG-8 Lauryl ether sulfate, sodium salt; PEG-12 Lauryl ether sulfate, sodium salt; Poly(oxy-1,2-ethanediyl), α-sulfo-ω-(dodecyloxy)-, sodium salt; Polyethylene glycol (5) lauryl ether sulfate, sodium salt; Polyethylene glycol (7) lauryl ether sulfate, sodium salt; Polyethylene glycol 400 lauryl ether sulfate, sodium salt; Polyethylene glycol 600 lauryl ether sulfate, sodium salt; Polyethylene glycol sulfate monododecyl ether sodium salt; Polyoxyethylene (5) lauryl ether sulfate, sodium salt; Polyoxyethylene (7) lauryl ether sulfate, sodium salt; Polyoxyethylene (8) lauryl ether sulfate, sodium salt; Polyoxyethylene (12) lauryl ether sulfate, sodium salt; Polyethylene glycol, mono(hydrogen sulfate), dodecyl ether, sodium salt; Rewopol NL-2; Rhodapex ESY; Sipon ES; Sipon ESY; Sipon LES 25; Sodium (lauryloxypolyethoxy)-ethyl sulfate; Sodium dodecylpoly(oxyethylene) sulfate; Sodium laureth sulfate; Sodium laureth-5 sulfate; Sodium laureth-7 sulfate; Sodium laureth-8 sulfate; Sodium laureth-12 sulfate; Sodium lauryl ether sulfate; Sodium lauryl sulfate ethoxylate; Sodium laurylpoly(oxyethylene) sulfate; Sodium polyoxyethylene (5) lauryl ether sulfate; Sodium polyoxyethylene (7) lauryl ether sulfate; Sodium polyoxyethylene (8) sulfate; Sodium polyoxyethylene (12) lauryl ether sulfate; Standapol ES 2; Standapol ES-3; Texapon N40; Witcolate ES-2, Witcolate LES-60C; Zetesol LES 2; poly(oxyethylene) lauryl ether sulfate sodium salt; polyethylene glycol, mono(hydrogen sulfate), dodecyl ether, sodium salt. Foam stabilizer, detergent, flash foamer, wetter for detergent systems, personal care products; emulsion polymerization; shampoo base. Chemron; CK Witco Chem. Corp.; Lonzagroup; Norman-Fox; Pilot; Sandoz; Stepan; Unger Fabrik Chem. Ind.; Vista.

3456   Sodium lauroyl sarcosinate
137-16-6   4383   G H

C15H28NNaO3

Sodium N-methyl-N-(1-oxododecyl)glycinate.

Caswell No. 778B; Compound 105; Crodasinic LS30; EINECS 205-281-5; EPA Pesticide Chemical Code 000174; Gardol; Gardol; Hamposyl L-30; Lauroyl-sarcosine sodium salt; Maprosyl 30; Medialan LL-99; NSC 117874; Sarcosine, N-lauroyl-, sodium salt; Sarcosyl NL; Sarcosyl NL 30; Sarkosyl NL; Sarkosyl NL 30; Sarkosyl NL 35; Sarkosyl NL 97; Sarkosyl NL 100; Sodium lauroyl sarcosinate; Sodium N-lauroylsarcosine; Sodium N-methyl-N-(1-oxododecyl)glycinate; Vanseal® NALS-30; Zoharsyl L-30. Foaming, wetting agent and detergent for acidic conditions; corrosion inhibitor; bacteriastat and inhibitor; used in dental care preparations, pharmaceuticals, personal care products, household and industrial applications. Biodegradable surfactant, foaming and wetting agent, detergent, foam booster for soaps, bath gels, shampoos, shaving creams, dentifrices, rug shampoos, oven cleaners, dishwash, textile/leather processing; offers tolerance to hard water, mildness. Raw material for manufacturing of hair shampoos, con-ditioners, toothpastes, carpet and upholstery shampoos; anticorrosive properties. Solid; slightly souble in H2O. *Croda; Grace W.R. & Co.; Vanderbilt R.T. Co. Inc.*

3457   Sodium lauryl sulfate

151-21-3   8710   G P

C12H25NaO4S

Dodecyl alcohol, hydrogen sulfate, sodium salt.

AI3-00356; Akyposal SDS; Anticerumen; Aquarex ME; Aquarex methyl; Avirol 101; Avirol 118 conc; Berol 452; Carsonol SLS; Carsonol SLS Paste B; Carsonol SLS Special; Caswell No. 779; CCRIS 6272; Conco sulfate; Conco sulfate WA; Conco sulfate WA-1200; Conco sulfate WA-1245; Conco sulfate WN; Conco Sulfate WAG; Conco Sulfate WAN; Conco Sulfate WAS; CP 75424; Cycloryl; Cycloryl 21; Cycloryl 31; Cycloryl 580; Cycloryl 585N; Dehydag sulfate GL emulsion; Dehydag sulphate GL emulsion; Dehydrag sulfate gl emulsion; Detergent 66; Dodecyl sodium sulfate; Dreft; Dupanol WAQ; Duponal; Duponal WAQE; Duponol; Duponol C; Duponol ME; Duponol methyl; Duponol QC; Duponol QX; Duponol WA; Duponol WA Dry; Duponol WAQ; Duponol WAQA; Duponol WAQE; Duponol WAQM; EINECS 205-788-1; EMAL 10; Emersal 6400; Empicol; Empicol LPZ; Empicol LS 30; Empicol LX 28; Emulsifier No. 104; EPA Pesticide Chemical Code 079011; Finasol osr2; Gardinol; Hexamol SLS; HSDB 1315; Incronol SLS; Irium; Jordanol SL-300; Lanette wax-S; Lauryl sodium sulfate; Lauryl sulfate, sodium salt; Laurylsiran sodny; Maprobix NEU; Maprofix; Maprofix 563; Maprofix LK; Maprofix NEU; Maprofix WAC; Melanol; Melanol CL; Melanol CL 30; Monagen Y 100; Monododecyl sodium sulfate; Monogen Y 100; Montopol LA paste; NALS; Natrium lauryl-sulfuricum; Natriumlaurylsulfat; NCI-C50191; Neutra-zyme; Nikkol SLS; NSC 402488; Odoripon AL 95; Orvus WA paste; P and G emulsifier 104; Perklankrol ESD 60; Perlandrol L; Perlankrol L; Product No. 75; Product No. 161; Quolac EX-UB; Rewopol NLS 30; Richonol; Richonol A; Richonol AF; Richonol C; SDS; Sinnopon; Sinnopon LS 100; Sinnopon LS 95; Sintapon L; Sipex; Sipex OP; Sipex SB; Sipex SD; Sipex SP; Sipex UB; Sipon; Sipon LS; Sipon LS 100; Sipon LSB; Sipon PD; Sipon WD; SLS; Sodium dodecyl sulfate; Sodium dodecyl sulphate; Sodium lauryl sulfate; Sodium lauryl sulphate; Sodium monododecyl sulfate; Sodium monolauryl sulfate; Solsol needles; Standapol; Standapol 112 conc; Standapol WA-AC; Standapol WAQ; Standapol WAQ Special; Standapol WAS 100; Steinapol NLS 90; Steinapol NLS 90; Stepanol; Stepanol ME; Stepanol ME Dry; Stepanol ME Dry AW; Stepanol methyl; Stepanol methyl dry AW; Stepanol T 28; Stepanol WA; Stepanol WA-100; Stepanol WA paste; Stepanol WAC; Stepanol WAQ; Sterling WA paste; Sterling WAQ-CH. Detergent used in shampoos and other cosmetic products. Registered by EPA as an antimicrobial, fungicide, insecticide and rodenticide. Cream-colored crystals; mp = 204-207°; soluble in H2O (10 g/100 ml); LD50 (rat orl) = 1288 mg/kg. *Growth Plus Laboratories.*

3458   Sodium magnesium silicate

53320-86-8   G

Sodium magnesium silicate.

EINECS 258-476-2; Laponite® XLG, D; Lithium magnesium sodium silicate; Silicic acid, lithium magnesium sodium salt; Sodium magnesium silicate; Synthetic magnesium lithium silicate; Synthetic magnesium silicateLaponite® XLG, D. Inert base/carrier for active ingredients; suspending agent; promotes thixotropy giving stable suspensions; thickens cosmetic, toiletry creams, lotions, toothpaste products. *Laporte Fine Chem.*

3459   Sodium 2-mercaptobenzothiazolate

2492-26-4   G P

C7H4NNaS2

Sodium benzothiazolethiolate.

AI3-17229; Benzothiazolethiol, sodium salt; Caswell No. 541C; Duodex; EINECS 219-660-8; EPA Pesticide Chemical Code 051704; HSDB 6141; Mercapto-benzothiazole sodium salt; Nacap®; NaMBT; NSC 191935; Nuodex 84; Sodium MBT; Sodium mercapto-benzothiazolate; Vancide 51. Metal deactivator; corrosion inhibitor for water, alcohol, and glycol systems; used in antifreeze; chemical intermediate. Antimicrobial preservative for fibrous substrates, adhesives and aqueous emulsions. Fungicide. Registered by EPA as an antimicrobial and fungicide. Liquid; mp = -6°; bp = 103°; d = 1.25; soluble in H2O (>10 g/100 ml). *Degussa-Hüls Corp.; Vanderbilt R.T. Co. Inc.*

3460   Sodium metabisulfite

7681-57-4   8712   G

Na2O5S2

Disulfurous acid, disodium salt.

AI3-51684; Campden Tablets; CCRIS 3951; Disodium disulfite; Disodium disulphite; Disodium metabisulfite; Disodium pyrosulfite; Disulfurous acid, disodium salt; EINECS 231-673-0; Fertisilo; HSDB 378; Natrii disulfis; Natrium metabisulfurosum; Natrium pyrosulfit; Natrium-bisulfit; Natriummetabisulfit; Pyrosulfurous acid, disod-ium salt; Sodium bisulfite anhydrous; Sodium disulfite; Sodium metabisulfite; Sodium metabisulphite; Sodium pyrosulfite; Sodium pyrosulfite (Na2S2O5). Used in foods, as a preservative, antioxidant, laboratory reagent. Antioxidant used in pharmaceuticals. Solid; soluble in H2O, slightly soluble in EtOH. *BASF Corp.; Blythe; Williams Ltd.; EM Ind. Inc.; General Chem; Mallinckrodt Inc.*

3461   Sodium metaborate

7775-19-1   8713   G P

BNaO2
 Monosodium metaborate.
 Boric acid (HBO2), sodium salt; Boric acid, monosodium salt; Borosoap; Caswell No. 779AA; EINECS 231-891-6; EPA Pesticide Chemical Code 011104; HSDB 5045; Kodalk; Rasorite; Sodium borate; Sodium borate (NaBO2); Sodium metaborate; Sodium metaborate, anhydrous. Used in herbicides. Registered by EPA as a fungicide, herbicide and insecticide. White powder; mp = 966°; soluble in H2O. *Ashland; Borax Consolidated Ltd; Borax Europe Ltd; U.S. Borax.*

3462 Sodium metaperiodate
7790-28-5   8714   G

INaO4
 Sodium periodate.
 EINECS 232-197-6; Metaperiodate; Periodic acid (HIO4), sodium salt; Periodic acid, sodium salt; Sodium metaperiodate; Sodium periodate. Source of periodic acid, analytical reagent, oxidizing agent. Crystals; mp = 300°; d_4^{16} = 3.865; soluble in H2O; trihydrate: dec 175°; d_4^{18} = 3.219; soluble in H2O (0.12 g/ml). *Atomergic Chemetals; Noah Chem.*

3463 Sodium metaphosphate
10361-03-2   5668   G

(NaPO3)n, n = 3-10
 Sodium metaphosphate.
 EINECS 233-782-9; Graham's salt; HSDB 5055; Insoluble metaphosphate; Insoluble sodium metaphosphate; Kurrol's salt; Maddrell's salt; Metafos; Metaphosphoric acid (HPO3), sodium salt; Metaphosphoric acid, sodium salt; Poly(sodium metaphosphate); Polymeric sodium metaphosphate; Sodium Kurrol's salt; Sodium meta-phosphate; Sodium metaphosphate (NaPO3); Sodium metaphosphate, insoluble; Sodium phosphate (NaPO3); Sporix. (Cyclic) or larger (polymers); dental polishing agents, detergent builders, water softening, sequestrants, emulsifiers, food additives, textile processing, laundering. *Atomergic Chemetals.*

3464 Sodium metasilicate
6834-92-0   8716   G

Na2O3Si
 Silicic acid, disodium slat.
 B-W; Crystarnet; Disodium metasilicate; Disodium monosilicate; EINECS 229-912-9; HSDB 753; Metso 20; Metso beads 2048; Metso beads, drymet; Metso pentabead 20; Orthosil; Silicic acid (H2SiO3), disodium salt; Silicic acid, disodium salt; Simet A; Sodium metasilicate; Sodium metasilicate (Na2SiO3); Sodium metasilicate, anhydrous; Sodium silicate; Water glass. Inorganic salt used in cosmetics, laundry, dairy, and metal cleaning as soap builder and detergent, bleaching agent; as flocculant and dispersant in metallurgy and mining. Solid; mp = 1089°; d = 2.614; soluble in cold H2O, insoluble in organic solvents. *Eka Nobel Ltd.; PQ Corp.; Rhône-Poulenc.*

3465 Sodium metasilicate pentahydrate
10213-79-3   G

Na2O3Si.5H2O
 Sodium metasilicate pentahydrate.
 Silicic acid, disodium salt, pentahydrate; Sodium metasilicate pentahydrate. Ingredient in detergents for food processing equipment and general cleaning in dairies, bakeries, packing houses, in laundry, textile, paper, oil, metal industries. *OxyChem; PQ Corp.*

3466 Sodium methallylsulfonate
1561-92-8   G

C4H7NaO3S
 Sodium 2-methylprop-2-ene-1-sulfonate.
 AI3-16415; EINECS 216-341-5; Geropon® MLS/A; Methallyl sulfonate; Methallylsulfonic acid, sodium salt; 2-Methyl-2-propene-1-sulfonic acid, sodium salt; NSC 2253; 2-Propene-1-sulfonic acid, 2-methyl-, sodium salt; Sodium 2-methyl-2-propenesulfonate; Sodium 2-methylprop-2-ene-1-sulphonate. Dye improver reactive co-monomer for acrylic fibers polymerization; reactive emulsifier or coemulsifier in latex emulsion polymerization. *Rhône-Poulenc.*

3467 Sodium methanethiolate
5188-07-8   H

CH3NaS
 Methyl mercaptan sodium salt.
 EINECS 225-969-9; Methanethiol, sodium salt; Methyl mercaptan sodium salt; Sodium methanethiolate; Sodium thiomethylate.

3468 Sodium methoxide
124-41-4   8717   H

CH3NaO
 Sodium methanolate.
 AI3-52659; EINECS 204-699-5; Feldalat NM; HSDB 755; Methanol, sodium salt; Methoxysodium; Methylate de sodium; Metilato sodico; Sodium methanolate; Sodium methoxide; Sodium methylate; UN1289; UN1431.

3469 Sodium 2-methyl-2-((1-oxoallyl)amino)-propanesulfonate
5165-97-9   H

C7H12NaNO4S
1-Propanesulfonic acid, 2-acrylamido-2-methyl-, sodium salt.
2-Acrylamido-2-methylpropanesulfonic acid sodium salt; 2-Methyl-2-((1-oxo-2-propenyl)amino)-1-propanesulfonic acid, sodiumsalt; EINECS 225-948-4; 1-Propanesulfonic acid, 2-acrylamido-2-methyl-, sodium salt; Sodium 2-methyl-2-((1-oxoallyl)amino)propanesulphonate.

3470 Sodium molybdate
10102-40-6   8718   G

MoNa2O4.2H2O
Sodium molybdate(VI) dihydrate.
CCRIS 3682; Disodium molybdate dihydrate; Molybdic acid, disodium salt, dihydrate; Molyhibit 100; Sodium molybdate; Sodium molybdate dihydrate; Sodium molybdate(VI) dihydrate; Anhydrous sodium molybdate [77631-95-0]. Reagent in analytical chemistry, used in manufacture of paint pigment, corrosion inhibitor, catalyst in dye and pigment production, additive for fertilizers and feeds, micronutrient. Crystalline powder; dec 100°; soluble in H2O (60 g/100 ml). *AAA Molybdenum; Cerac; Climax Molybdenum Co.; Mallinckrodt Inc.; PMC.*

3471 Sodium monofluorophosphate
7631-97-2   G

FNa2O3P
Monofluorophosphoric acid sodium salt.
EINECS 231-552-2; Fluorophosphoric acid, sodium salt; MFP; Phosphorofluoridic acid, sodium salt; SMFP; Sodium monofluorophosphate. Used in manufacture of toothpaste. *Albright & Wilson Americas Inc.; Elf Atochem N. Am.*

3472 Sodium N,N'-1,3-propanediylbis(N-(carboxymethyl))glycinate
18719-03-4   H

C11H14N2Na4O8
Glycine, N,N'-1,3-propanediylbis(N-(carboxymethyl))-, tetrasodium salt.
Sodium N,N'-1,3-propanediylbis(N-(carboxymethyl))-gly-cinate; Glycine, N,N'-1,3-propanediylbis(N-(carboxy-methyl))-, tetrasodium salt.

3473 Sodium naphthenate
61790-13-4   H

Naphthenic acids, sodium salts.
Caswell No. 589E; EINECS 263-108-9; EPA Pesticide Chemical Code 589600; Naphthathenic soap; Naphthenic acid, sodium salt solution; Naphthenic acids, sodium salts; Sodium naphthenate

solution.

3474 Sodium α-naphthylsulfonate
130-14-3   G

C10H7NaO3S
Sodium naphthalene sulfonate.
Aerosol® NS; 1-Naphthalenesulfonic acid, sodium salt; EINECS 204-976-0; α-Naphthalenesulfonic acid sodium salt; NSC 37036; α salt; Sodium α-naphthalenesulfonate; Sodium α-naphthalenesulfonic acid; Sodium 1-naphthalenesulfonate; Sodium naphthalene-1-sulphonate. Dispersant for pigments, extenders and fillers. Usable in aqueous media over a wide pH range. *Am. Cyanamid.*

3475 Sodium nitrate
7631-99-4   8720   G P

NNaO3
Nitric acid sodium salt.
Caswell No. 781; CCRIS 558; Chile saltpeter; Cubic niter; EINECS 231-554-3; EPA Pesticide Chemical Code 076104; HSDB 726; Niter; Nitrate de sodium; Nitrate of soda; Nitratine; Nitric acid monosodium salt; Nitric acid, sodium salt; NSC 77390; Peri Saltpeter; Saliter; Saltpeter (Chile); Soda niter; Sodium nitrate; Sodium saltpeter; Sodium(I) nitrate (1:1); UN1498. Oxidizing agent; solid rocket propellants; fertilizer, flux, glass manufacture, pyrotechnics, dynamites; color fixative and preservative in cured meats, fish, enamel for pottery; modifying burning properties of tobacco. Registered by EPA as an antimicrobial and rodenticide. Used as a nitrogen fertilizer. Crystals; mp = 308°; d = 2.26; soluble in H2O (0.9 g/ml), less soluble in organic solvents; LD50 (rbt orl) = 1.96 g/kg. *BASF Corp.; Faezy & Besthoff; Farleyway Chem. Ltd.; Mallinckrodt Inc.; Nissan Chem. Ind.; Spice King.*

3476 Sodium nitrite
7632-00-0   8721   D G P

NNaO2
Nitrous acid sodium salt.
Anti-rust; Azotyn sodowy; Caswell No. 782; CCRIS 559; Diazotizing salts; Dusitan sodny; EINECS 231-555-9; EPA Pesticide Chemical Code 076204; Erinitrit; Filmerine; HSDB 757; Natrium nitrit; NCI-C02084; Nitrite de sodium; Nitrito sodico; Nitrous acid, sodium salt; NSC 77391; Sodium nitrite; Synfat 1004; UN1500. Diazo-tization, rubber accelerators, color fixative and pre-servative in cured meats, meat products, fish; pharm-aceuticals, photographic and analytical reagent, dye manufacture, Vasodilator and antidote for cyanide poisoning. Registered by EPA as an antimicrobial and rodenticide (cancelled). Has been used as a preservative in foods, it also imparts a fresh color to many foods. It is however carcinogenic and its use has been banned in Germany and Holland. Antidote to cyanide poisoning; vasodilator. Also used as a reagent in

manufacture of inorganic and organic compounds, as well as in textile dyeing and printing, photography, meat curing and preserving. Crystals; mp = 271°; dec > 320°; d = 2.17; soluble in H_2O (0.7 g/ml), less soluble in organic solvents; LD50 (rat orl) = 180 mg/kg. *BASF Corp.; DuPont; EM Ind. Inc.; Farleyway Chem. Ltd.; General Chem; ICI Spec.; PMC.*

3477 Sodium m-nitrobenzenesulfonate
127-68-4   H

$C_6H_4NNaO_5S$
3-Nitrobenzenesulfonic acid sodium salt.
Benzenesulfonic acid, 3-nitro-, sodium salt; Benzene-sulfonic acid, m-nitro-, sodium salt; EINECS 204-857-3; HSDB 5614; Ludigol; Ludigol F,60; Nacan; m-Nitrobenzenesulfonic acid sodium salt; Nitrobenzen-m-sulfonan sodny; Nitrol S; NSC 9795; Sodium 3-nitrobenzenesulfonate; Sodium 3-nitrobenzenesulphon-ate; Tiskan.

3478 Sodium p-nitrophenolate
824-78-2   H P

$C_6H_5NaNO_3$
Sodium 4-nitrophenolate.
AI3-09021; Atonik; EINECS 212-536-4; EPA Pesticide Chemical Code 129077; HSDB 2592; Phenol, 4-nitro-, sodium salt; Phenol, p-nitro-, sodium salt; PNSP; Sodium 4-nitrophenolate; Sodium 4-nitrophenoxide; Sodium nitrophenate; Sodium p-nitrophenate; Sodium p-nitro-phenol; Sodium p-nitrophenolate; Sodium p-nitro-phenoxide. Plant growth regulator. Registered by EPA as a plant growth regulator. *Asahi Chem. Industry.*

3479 Sodium octanoate
1984-06-1   G H

$C_8H_{15}NaO_2$
n-Octanoic acid, sodium salt.
AI3-50473; Caprylic acid sodium salt; EINECS 217-850-5; HSDB 5862; Natrium octanoat; Octanoic acid, sodium salt; Sodium caprylate; Sodium n-octanoate; Sodium octanoate; Sodium octoate. Source of caprylic acid, an intermediate in manufacture of dyestuffs and surfactants. *Hart Prod.; Penta Mfg.; Sigma-Aldrich Fine Chem.*

3480 Sodium octyl sulfate
142-31-4   G

$C_8H_{17}NaO_4S$
Sulfuric acid, monooctyl ester, sodium salt.
Cycloryl OS; Duponol 80; EINECS 205-535-5; Octyl sodium sulfate; Octyl sulfate, sodium salt; Rhodapon® OLS; Sipex ols; Sodium capryl sulfate; Sodium octyl sulphate; SOS. Used as a wetting agent; rinse aid for industrial, institutional and household cleaners; mercerizing agent for cotton goods; surfactant in electrolyte baths for metal cleaning; hard surface cleaning; neoprene dispersant; emulsifier for emulsion polymerization. *Rhône Poulenc Surfactants.*

3481 Sodium octylsulfonate
5324-84-5   H

$C_8H_{17}NaO_3S$
1-Octanesulfonic acid, sodium salt.
EINECS 226-195-4; NSC 2738; 1-Octanesulfonic acid, sodium salt; Sodium 1-octanesulfonate; Sodium octane-1-sulphonate monohydrate; Sodium octanesulfonate; Sodium octylsulfonate.

3482 Sodium oleate
143-19-1   6898   P G

$C_{18}H_{33}NaO_2$
Sodium 9-octadecenoate.
AI3-19806; CCRIS 1964; EINECS 205-591-0; Eunatrol; HSDB 758; 9-Octadecenoic acid, sodium salt; 9-Octadecenoic acid, sodium salt, (Z)-; 9-Octadecenoic acid (Z)-, sodium salt; 9-Octadecenoic acid (9Z)-, sodium salt; Olate flakes; Oleic acid sodium salt; Osteum; Potassium oleate; Sodium 9-octadecenoate; Sodium 9-octadecenoate, (Z)-; Sodium oleate. Used for ore flotation, waterproofing textiles and emulsification of oil/water systems. Solid; mp = 232-235°; soluble in H_2O (10 g/100 ml), less soluble in EtOH, organic solvents. *Hart Prod.; Lancaster Synthesis Co.; Norman-Fox; Witco/Humko.*

3483 Sodium oleylmethyltauride
137-20-2   G H

$C_{21}H_{40}NNaO_4S$
Sodium 2-(N-methyloleamido)ethane-1-sulfonate.
Adinol OT; Adinol T; Alkyl sodium N-methyltaurate; Arkopon T; Concogel 2 conc.; EINECS 205-285-7; Hostapon T; HSDB 5624; Igepon T; Igepon T 33; Igepon T-43; Igepon T 51; Igepon T-71; Igepon T-73; Igepon T 77; Igepon TE; Metaupon paste; N-Methyl-N-oleoyl-taurine sodium salt; Nissan diapion S; Nissan diapon T; Oleoylmethyltaurine sodium salt; OMT; Sodium methyl oleoyl taurate; Sodium oleylmethyl-tauride; Taurine, N-methyl-N-oleoyl-, sodium salt. A biodegradable anionic surfactant with detergent, wetting, emulsifying and foaming properties. Used in the textile and dyeing industries and in the manufacture of leather and paper. A dispersing and wetting agent used in wettable powders, plant protection and pest control. *Croda; Hoechst AG.*

3484 Sodium N-phenylglycinate
10265-69-7   H

$C_8H_8NNaO_2$
Glycine, N-phenyl-, monosodium salt.
EINECS 233-605-5; Glycine, N-phenyl-, monosodium salt; Sodium N-phenylglycinate.

3485 Sodium o-phenylphenate
132-27-4   G P

$C_{12}H_9NaO$
Sodium o-phenylphenoxide.
AI3-09076; Bactrol; Caswell No. 787; CCRIS 693; D.C.S.; Dowicide; Dowicide A; Dowicide A & A flakes; Dowicide A Flakes; Dowizid A; EINECS 205-055-6; EPA Pesticide Chemical Code 064104; Mil-Du-Rid; Mystox WFA; Natriphene; NSC 1547; OPP-sodium; Orphenol; Phenol, o-phenyl-, sodium deriv.; Phenylphenol, sodium salt; Preventol-ON; Preventol ON extra; Sodium 2-biphenylate; Sodium 2-hydroxydiphenyl; Sopp; Stopmold B; Topane. Fungicide, antiseptic and germicide. Registered by EPA as an antimicrobial and fungicide. Crystals; mp = 55-57°; bp = 275°; d_4^{25} = 1.2130; soluble in H_2O (>10 g/100ml), organic solvents; LD_{50} (rat orl) = 2000 mg/kg. *Bayer Corp.; Brogdex Co.; Central Solutions, Inc.; Cerexagri, Inc.; Decco; Dow Chem. U.S.A.; FMC Corp.; IBC Manufacturing Co.; Penn Champ; West Chemical Products, Inc.; Wexford Labs, Inc.*

3486 Sodium perborate
7632-04-4   8725   G

$BNaO_3$
Sodium perborate anhydrous.
Dexol; EINECS 231-556-4; HSDB 1676; Perborax; Perborin; Perboric acid ($HBO(O_2)$), sodium salt; Perboric acid (HBO_3), sodium salt; Perboric acid, sodium salt; Per-oxydol; Sodium perborate ($BaBO_3$); Sodium peroxo-borate; Sodium peroxoborate, anhydrous; Sodium per-oxometaborate; UN3247. Topical antiseptic, denture cleaner, oxygen source. A deodorant and washing and bleaching agent, used in washing powders. [tetrahydrate]; White crystals; mp = 271°; dec 320°; d = 2.17; soluble in H_2O (65 g/100 ml), slightly soluble in EtOH; LD_{50} (rat orl) = 180 mg/kg. *Degussa AG; DuPont; Eka Nobel Ltd.; ICI Spec.; Mitsubishi Gas.*

3487 Sodium percarbonate
15630-89-4   G

$C_2H_6Na_4O_{12}$
Carbonic acid disodium salt, compd. with hydrogen peroxide (H_2O_2) (2:3).
Carbonic acid, disodium salt, compd. with hydrogen peroxide (2:3); Disodium carbonate, compound with hydrogen peroxide (2:3); Disodium carbonate, hydrogen peroxide (2:3); EINECS 239-707-6; FB Sodium percarbonate; Oxyper; Perdox; Peroxy sodium carbonate; Sodium carbonate peroxide; Sodium percarbonate. Bleaching agent for domestic and industrial use, denture cleaner, mild antiseptic. *Chemoxal; Degussa AG; Interox Am.*

3488 Sodium perchlorate
7601-89-0   8726   D G

$ClNaO_4$
Perchloric acid sodium salt.
CPL 46; EINECS 231-511-9; HSDB 5038; Irenat; KM Sodium Perchlorate; Natriumperchloraat; Natriumper-chlorat; Perchlorate de sodium; Perchloric acid, sodium salt; Sodio (perclorato di); Sodium (perchlorate de); Sodium perchlorate; UN1502. Used in explosives; has been used medicinally as a thyroid inhibitor. Crystals.dec about 130°; d = 2.02. *Kerr-McGee.*

3489 Sodium peroxide
1313-60-6   8728   G

O_2Na_2
Disodium peroxide.
Disodium dioxide; Disodium peroxide; EINECS 215-209-4; Flocool 180; HSDB 763; Natrium peroxydatum; Natriumsuperoxid; Oxolin; Oxone; Oxygen powder; Sodium dioxide; Sodium oxide (Na_2O_2); Sodium oxide, per-; Sodium peroxide; Sodium peroxide ($Na_2(O_2)$); Sodium superoxide; Solozone; UN1504. Compressed sodium peroxide, used in washing powders, scouring powders, metal cleaners, hair wave neutralizers, general oxidizing reactions. Yellow-white granules; mp = 460° (dec); d = 2.8050. *DuPont.*

3490 Sodium persulfate
7775-27-1   8729   G

$Na_2O_8S_2$
Sodium peroxydisulfate.
Disodium peroxodisulphate; EINECS 231-892-1; Peroxy-disulfuric acid, disodium salt; Persulfate de sodium; Sodium peroxydisulfate; Sodium persulfate; UN1505. Bleaching agent (fats, oils, fabrics, soaps), battery depolarizers, emulsion polymerization. Soluble in H_2O (549 g/l; MLD (rbt iv) = 178 mg/kg. *Degussa AG; FMC Corp.;*

3491 Sodium phenate
139-02-6   7323   P

C6H5NaO
 Phenol, sodium salt.
 Caswell No. 786A; EINECS 205-347-3; EPA Pesticide Chemical Code 064002; HSDB 5623; Phenol sodium; Phenol, sodium salt; Phenolate sodium; Sodium carbolate; Sodium phenate; Sodium phenolate; Sodium phenoxide; 10, 36. Registered by EPA as an antimicrobial and disinfectant. *Sporicidin International.*

3492 Sodium phosphate
7558-80-7   8734   G P

H2NaO4P.2H2O
 Monosodium dihydrogen phosphate dihydrate.
 Acid sodium phosphate; EINECS 231-449-2; HSDB 738; Monobasic sodium phosphate; Monosodium dihydrogen orthophosphate; Monosodium dihydrogen phosphate; Monosodium hydrogen phosphate; Monosodium mono-phosphate; Monosodium orthophosphate; Monosodium phosphate; Monosodium phosphate, anhydrous; Mono-sorb XP-4; MSP; Phosphoric acid, monosodium salt; Primary sodium phosphate; Recresal; Sodium acid phosphate; Sodium biphosphate anhydrous; Sodium dihydrogen monophosphate; Sodium dihydrogen orthophosphate; Sodium dihydrogen phosphate (1:2:1); Sodium dihydrogen phosphate (NaH2PO4); Sodium orthophosphate, primary; Sodium phosphate; Sodium phosphate (Na(H2PO4)); Sodium phosphate, monobasic; Sodium primary phosphate; Uro-phosphate. Controls pH in mildly acidic solutions; food products, water treatment, metal treatment. Registered by EPA as an antimicrobial, fungicide and herbicide (cancelled). FDA GRAS, Japan approved, FDA approved for buccals, parenterals, ophthalmics, orals, topicals, vaginals, USP/NF, BP compliant. Buffering agent, precipitation agent, laxative and effervescent. Used in buccals, parenterals, ophthalmics, orals, topicals, vaginals. Crystals; mp = 60°; d = 1.915. *Albright & Wilson Americas Inc.; BritAg Ind. Ltd.; Sigma-Aldrich Fine Chem.*

3493 Sodium phosphate
10101-89-0   G

H24Na3O16P
 Trisodium phosphate, dodecahydrate.
 CCRIS 7322; Phosphoric acid, trisodium salt, dodeca-hydrate; Sodium phosphate dodecahydrate; Sodium phosphate tribasic dodecahydrate; Sodium phosphate, tribasic; TSP-12; Trisodium phosphate; Phosphoric acid, trisodium, 12-hydrate; Sodium Phosphate 12-Water; Sodium Phosphate Tribasic Dodecahydrate; TSP dodecahydrate. Used in food products, cleaning compounds, water treatment. Crystals; mp = 75°; d = 1.6. *Albright & Wilson Americas Inc.; Rhône-Poulenc.*

3494 Sodium phosphinate
10039-56-2   G

NaO2P.H2O
 Sodium hypophosphite monohydrate.
 Fosfornan sodny; Phosphinic acid, sodium salt, monohydrate; Sodium phosphinate hydrate; Sofibex. Used for surface treatment, electrodeless nickel plating. Crystals; mp = 90°; soluble in H2O (1 g/ml), alcohols, glycerol, insoluble in other organic solvents. *Elf Atochem N. Am.*

3495 Sodium phytate
7205-52-9   7471   G

C6H9Na9O24P6
 myo-Inositol hexakis(dihydrogen phosphate) sodium salt.
 Inositol hexaphosphoric acid hexasodium salt; myo-Inositol hexakis(dihydrogen phosphate) sodium salt; Phytat D.B.; Phytate sodium; Rencal; Sodium phytate. Chelating agent. Used to remove traces of heavy metal ions and also as a fermentation nutrient. Solid; soluble in H2O. *Bristol-Myers Squibb Co.*

3496 Sodium polynaphthalenesulfonate
9084-06-4   G

(C10H8O3S.CH2O)x.xNa
 Sodium naphthalene sulfonate formaldehyde condensate.
 Ablusol ML; Atlox 4862; Barra Super; Bevaloid 35; Blancol; Blancol dispersant; Darvan 1; Darvan No. 1; Daxad 11; Daxad 15; Daxad 18; Daxad No. 11; Dispergator NF; Disperser NF; Dispersing agent NF; Dispersol ACA; Flube; Humifen NBL 85; Leukanol NF; Lissatan AC; Lomar D; Lomar LS; Lomar PW; Na-Cemmix; Naphthalenesulfonic acid-formaldehyde condensate sodium salt; Naphthalenesulfonic acid, polymer with formaldehyde, sodium salt; NF; NF (dispersant); NF-A; Pozzolith 400N; QR 819; Sodium salt of sulfonated naphthaleneformaldehyde condensate; Surfactant NF; Tamol L; Tamol SN. Plasticizer and water reducing agent used in pourable and high-strength concrete. *Taiwan Surfactants.*

3497 Sodium propionate
137-40-6   8743   D G P V

C4H5NaO2
 Propionic acid, sodium salt.
 Bioban-S; Caswell No. 707A; CCRIS 1896; Deketon; E281; EINECS 205-290-4; EPA Pesticide Chemical Code 077703; Ethylformic acid, sodium salt; HSDB 766; Impedex; Keenate; Methylacetic acid, sodium salt; Mycoban; Napropion; Natriumpropionat; Ocuseptine; Propanoic acid, sodium salt; Propi-ophtal; Propiofar; Propion; Propionan sodny; Propionic acid, sodium salt; Propisol; Sodium propanoate; Sodium propionate; Whit-pro. Sodium salt of propionic acid; food additive (preservative); fungicide. Has been used in veterinary medicine as a ketosis treatment. An antifungal agent. Registered by EPA as a fungicide (cancelled). FDA

573

GRAS, Japan approved with limitations, Europe listed, FDA approved for orals; USP/NF compliance. In the FDA List of inactive ingredients, (oral capsules, powder, suspensions and syrups). Used as an antimicrobial preservative in orals and in treatment of skin fungal infections. Crystals; mp = 287-289°; soluble in H_2O (1 g/ml), slightly soluble in EtOH, less soluble in organic solvents. LD_{50} (rat orl) = 3920 mg/kg, (mus orl) = 2350 mg/kg, (rbt skn) = 1640 mg/kg. *Browning; Gist-Brocades Intl.; Lohmann; Niacet; Ruger; Rystan Co. Inc.; Sigma-Aldrich Fine Chem.; Spectrum Chem. Manufacturing; Verdugt BV.*

3498   Sodium pyrithione
3811-73-2   D G P

C5H4NNaOS

N-Hydroxy-2-pyridinethione, sodium salt.
AI3-22596; EINECS 223-296-5; Fonderma; (1-Hydroxy-2-pyridinethione), sodium salt; 2-Mercaptopyridine 1-oxide sodium salt; 2-Mercaptopyridine-N-oxide sodium salt; 2-Mercaptopyridine oxide sodium salt; NSC 4483; Omadine sodium; 1-Oxo-2-pyridinethiol sodium salt; 2-Pyridinethiol, 1-oxide, sodium salt; 2-Pyridinethiol N-oxide sodium salt; Sodium 2-mercaptopyridine 1-oxide; Sodium 2-pyridinethiol 1-oxide; Sodium 2-pyridinethiol N-oxide; Sodium 2-pyridinethiolate 1-oxide; Sodium, (2-pyridinylthio)-, N-oxide; Sodium (2-pyridylthio)-N-oxide; Sodium omadine; Sodium pyrithione; Thione (reagent). Industrial microbiostat; chelating agent; used in aqueous metal coolant and cutting fluids, latex emulsion, inks, fiber lubricants. Used medicinally as an antibacterial and antifungal agent. Solid; mp = -25-30° (dec); bp 109°; d = 1.2200. *Olin Corp.*

3499   Sodium pyrophosphate
7722-88-5   9312   G P

Na4O7P2

Tetrasodium pyrophosphate.
Anhydrous tetrasodium pyrophosphate; Caswell No. 847; Diphosphoric acid, tetrasodium salt; EINECS 231-767-1; EPA Pesticide Chemical Code 076405; HSDB 854; Natrium pyrophosphat; NSC 56751; Phosphotex; Pyrophosphoric acid, tetrasodium salt; Sodium diphosphate (Na4P2O7); Sodium diphosphate, anhydrous; Sodium phosphate (Na4P2O7); Sodium pyrophosphate; Sodium pyrophosphate, tetrabasic; Tetranatriumpyro-phosphat; Tetrasodium diphosphate; Tetrasodium pyrophosphate; TSPP; Victor TSPP; tetrasodium salt; Sodium diphosphate (Na4P2O7); Sodium phosphate (Na4P2O7); Sodium pyrophosphate (4:1); Sodium pyrophosphate (Na4P2O7); Tetrasodium diphosphate; Victor TSPP. Water softener, synthetic detergent builder, dispersant, emulsifier, metal cleaner, boiler water treatment, viscosifier for drilling muds, deinking news-print, synthetic rubber, textile dyeing, wool scouring, buffer, sequestrating agent. Registered by EPA as an antimicrobial, fungicide and herbicide (cancelled). Corrosion inhibitor, used as a water softener, dispersing and emulsifying agent, metal cleaner, sequestrant, and a nutritional supplement. Also used in drilling muds, boiler water treatment, soaps and detergents, dyes, and in wool scouring. Crystals; mp = 988°; d = 2.534; soluble in H_2O (6.23 g/100 ml), insoluble in organic solvents. *Farleyway Chem. Ltd.; FMC Corp.; Mitsubishi Gas; Mitsui Toatsu; Monsanto Co.; Spectrum Chem. Manufacturing.*

3500   Sodium pyrophosphate
7758-16-9   8643   G

2Na+

6H2O

H2Na2O7P2.6H2O
Disodium dihydrogen pyrophosphate hexahydrate.
Dinatriumpyrophosphat; Diphosphoric acid, disodium salt; Disodium acid pyrophosphate; Disodium dihydrogen diphosphate; Disodium dihydrogen pyrophosphate; Di-sodium diphosphate; Disodium pyrophosphate; EINECS 231-835-0; HSDB 377; Pyrophosphoric acid, disodium salt; SAPP; Sodium acid pyrophosphate; Sodium pyro-phosphate (Na2H2P2O7); Taterfos®; Victor Cream®. Leavening agent for baking, cereals. Used in baking powder. Solid; mp = 220° (dec); d = 1.86; soluble in H_2O. *Rhône-Poulenc Food Ingredients.*

3501   Sodium ricinoleate
5323-95-5   8295   D G

C18H33NaO3
[R-(Z)]-12-hydroxy-9-octadecenoic acid sodium salt.
AI3-52679; CCRIS 5930; Colidosan; EINECS 226-191-2; HSDB 4280; 12-Hydroxy-9-octadecenoic acid, sodium salt; NSC 2835; 9-Octadecenoic acid, 12-hydroxy-, sodium salt; 9-Octadecenoic acid, 12-hydroxy-, monosodium salt, (R-(Z))-; 9-Octadecenoic acid, 12-hydroxy-, monosodium salt, (9Z,12R)-; Ricinoleic acid; Ricinoleic acid, monosodium salt, (+)-; Ricinoleic acid sodium salt; Sodium (R)-12-hydroxyoleate; Sodium ricinate; Sodium ricinolate; Sodium ricinoleate; Soricin; Soricinol 40. Mold release. Solid, soluble in H_2O, alcohols. *Actrachem.*

3502   Sodium rosinate
61790-51-0   H

Rosin acids, sodium salts.
Disproportionated rosin acid, sodium salt; EINECS 263-144-5; Resin acids and Rosin acids, sodium salts; Rosin acid, monosodium salt; Rosin acids, sodium salt; Rosin acids, sodium salts; Rosin, disproportionated, sodium salt; Rosin, sodium salt; Rosin, sodium soap; Sodium soap of disproportationed rosin; Wood rosin, sodium salt.

3503   Sodium saccharin
128-44-9   G H

574

C7H4NNaO3S

1,2-Benzisothiazolin-3-one, 1,1-dioxide, sodium salt.
Artificial sweetening substanz gendorf 450; Benzoic acid sulfimide, sodium; CCRIS 706; Cristallose; Crystallose; Dagutan; EINECS 204-886-1; FEMA No. 2997; Kristallose; Madhurin; NSC 4867; ODA; Saccharin sodium; Saccharinnatrium; Saccharoidum natricum; Saxin; Sodium benzosulphimide; Sodium saccharinate; Soluble gluside; Soluble saccharin; Sucra; Sucromat; Sulphobenzoic imide, sodium salt; Sweeta; Sykose; Willosetten. Non-caloric sweetener, pharmaceutic aid (flavorant). Solid; soluble in H_2O (83 g/100 ml), less soluble in organic solvents; LD_{50} (rat orl) = 17.0 g/kg.

3504   Sodium salicylate
54-21-7   8411   H

C7H6NaO3

Sodium o-hydroxybenzoate.
Alysine; Aroall; Benzoic acid, 2-hydroxy-, monosodium salt; CCRIS 6715; Clin; Corilin; Diuretin; EINECS 200-198-0; Enterosalicyl; Enterosalil; Entrosalyl; Glutosalyl; 2-Hydroxybenzoic acid monosodium salt; o-Hydroxy-benzoic acid monosodium salt; Idocyl novum; Kera-salicyl; Kerosal; Magsalyl; Monosodium 2-hydroxy-benz-oate; Monosodium salicylate; Nadisal; Natium salic-ylicum; Natrium salicylat; Neo-salicyl; NSC 202167; Pabalate; Parbocyl-rev; Salicylic acid, monosodium salt; Salicylic acid, sodium salt; Salisod; Salsonin; Sodium 2-hydroxybenzoate; Sodium o-hydroxybenzoate; Sodium salicylate. Used to treat rheumatoid arthritis and in manufacture of salicylate drugs. Also used in synthesis of azo dyes and as a food preservative. *Spectrum Chem. Manufacturing; Thiele Kaolin.*

3505   Sodium selenite
26970-82-1   G

Na2O3Se.5H2O

Sodium selenite pentahydrate.
CCRIS 1378; Selenious acid, disodium salt, pentahydrate; Sodium selenite pentahydrate. Glass manufacture (removes green color during manufacture), reagent in bacteriology, testing germination of seeds, decorating porcelain. Crystals; soluble in H_2O, insoluble in organic solvents; LD_{50} (rat orl) = 7 mg/kg. Atomergic; Chem.; Chemetals; Degussa; Noah.

3506   Sodium sesquicarbonate
533-96-0   8749   G

C2HNa3O6

Carbonic acid, sodium salt (2:3).
Carbonic acid, sodium salt (2:3); EINECS 208-580-9; HSDB 769; Magadi Soda; Snowflake crystals; Sodium carbonate (3:2); Sodium sesquicarbonate; SQ 810; Trisodium hydrogendicarbonate; Trona; Urao. Inorganic salt used as a detergent and soap builder; mild alkaline agent for general cleaning and water softening; bath crystal; alkaline agent in leather tanning; food additive. White solid; d =

1.2112; soluble in H_2O (13 g/100 ml). *FMC Corp.*

3507   Sodium silicate
1344-09-8   8750   G

Na4O4Si

Silicic acid, sodium salt.
49FG; Acsil; Agrosil LR; Agrosil S; Britesil; Britesil H 20; Britesil H 24; Carsil; Carsil (silicate); Caswell No. 792; DP 222; Dryseq; Dupont 26; EINECS 215-687-4; EPA Pesticide Chemical Code 072603; HK 30 (van); HSDB 5028; L 96 (salt); Metso 99; N 38; Portil A; Pyramid 1; Pyramid 8; Q 70; Sikalon; Silica E; Silica K; Silica N; Silica R; Silican; Silicic acid, sodium salt; Sodium polysilicate; Sodium sesquisilicate; Sodium silicate; Sodium silicate glass; Sodium silicate solution; Sodium siliconate; Sodium water glass; Soluble glass; Star; Waterglass. Used as a preservative agent, for lining reactors and as a binder in abrasive wheels. Used for lining Bessemer converters, acid concentrators; manufacture of grindstones, abrasive wheels; as solution, for preserving eggs; fireproofing fabrics; detergent in soaps; as adhesive, for waterproofing walls, a builder for laundry detergents and cleaners, an adhesive for corrugated, spiral wound and laminated paper products, a binder for foundry cores and molds, in the manufacture of silica gels, zeolites and catalysts, in cements and paints. Solid; bp = 102°; insoluble in H_2O. *Aichi Silicate Chem Co Ltd; Asahi Denka Kogyo; Crosfield; Lancaster Synthesis Co.; OxyChem; PQ Corp.*

3508   Sodium stannate
12058-66-1   8752   G

H6Na2O6Sn

Sodium stannate(IV).
Disodium stannate; Disodium tin trioxide; EINECS 235-030-5; Preparing salt; Sodium stannate; Sodium stannate(IV); Sodium tin oxide; Stannate (SnO_3^{2-}), disodium; Tin sodium oxide. Dyeing mordant, ceramics, glass, source of tin for electroplating, textile fireproofing, stabilizer for hydrogen peroxide, blueprint paper, laboratory reagent. [trihydrate]; white crystals; soluble in H_2O (59 g/100 ml), insoluble in EtOH. *Blythe, Williams Ltd.; Elf Atochem N. Am.; M&T Harshaw; Spectrum Chem. Manufacturing.*

3509   Sodium stearate
822-16-2   8753   G H

C18H35NaO2

n-Octadecanoic acid, sodium salt.
AI3-19808; Bonderlube 235; EINECS 212-490-5; Flexichem B; HSDB 5759; Octadecanoic acid, sodium salt; Prodhygine; Sodium octadecanoate; Sodium stearate; Stearic acid, sodium salt. Waterproofing and gelling agent; toothpaste, cosmetics; stabilizer in plastics; emulsifier and stiffener in pharmaceuticals; in glycerol suppositories. *CK Witco Corp.; Elf Atochem N. Am.; Magnesia GmbH; Norman-Fox; Original Bradford Soapworks.*

3510   Sodium stearoyl-2-lactylate
25383-99-7   H

C24H43NaO6

Sodium 2-(1-carboxyethoxy)-1-methyl-2-oxoethyl octa-decanoate.
EINECS 246-929-7; Octadecanoic acid, 2-(1-carboxyethoxy)-1-methyl-2-oxoethyl ester, sodium salt; Sodium 2-(1-carboxyethoxy)-1-methyl-2-oxoethyl octa-decanoate; Sodium stearoyl lactylate; Sodium 2-stearoyllactate; Sodium stearoyl-2-lactylate; Sodium stearyl-2-lactylate; Stearic acid, ester with lactic acid bimol. ester, sodium salt.

3511 Sodium sulfate
7727-73-3   8755   D G

Na2O4S ·10H2O
Sodium sulfate decahydrate.
Colyte; Disodium sulfate decahydrate; Disodium sulfate decahydrate (Na2SO4.10H2O); Glauber's salt; Mirabilite; Sodium sulfate; Sodium sulfate (Na2SO4) decahydrate; Sodium sulfate decahydrate; Sodium sulfate decahydrate (Na2SO4.10H2O); Sulfuric acid disodium salt, deca-hydrate. Used in solar heat storage, air conditioning and in freezing mixtures. Used medicinally as a calcium regulator and a laxative-cathartic. Colorless crystals; mp = 32.4°; d = 1.46; soluble in H2O (66 g/100 ml 25°, 33 g/100 ml 15°), insoluble in EtOH.

3512 Sodium sulfate
7757-82-6   8755   D G

Na2O4S
Sodium sulfate, anhydrous.
Bisodium sulfate; Dibasic sodium sulfate; Disodium monosulfate; Disodium sulfate; Disodium sulphate; HSDB 5042; Kemsol; Natriumsulfat; Sal Mirabil; Salt cake; Sodium sulfate; Sodium sulfate (2:1); Sodium sulfate (Na2SO4); Sodium sulfate anhydrous; Sulfuric acid disodium salt. Manufacture of Kraft paper, etc.; filler in synthetic detergents; processing textile fibers, dyes, tanning, pharmaceuticals, lab reagent, food additive. The decahydrate is known as Glauber's salt. Used as a calcium regulator and a laxative-cathartic. Crystals; mp = 884°; SG = 2.68; soluble in H2O (0.28 g/ml), insoluble in organic solvents. Akzo Chemie; Elf Atochem N. Am.; Kemira Kemi UK; Lenzing AG; Occidental Chem. Corp.

3513 Sodium sulfide
1313-82-2   8756   G

Na2S
Sodium monosulfide.
AI3-09342; Disodium monosulfide; Disodium sulfide; Disodium sulphide; EINECS 215-211-5; HSDB 772; NSC 41874; Sodium monosulfide; Sodium sulfide; Sodium sulfide (anhydrous); Sodium sulfide (Na2S); Sodium sulfide solution; Sodium sulfuret; Sodium sulphide; UN1385; UN1849; ,. Used in manufacture of organic chemicals, sulfur dyes, intermediates, viscose rayon, leather depilatory, paper pulp, hydrometallurgy of gold ores, sulfiding oxidized lead and copper ores, sheep dips, photographic reagent, engraving and lithography, analytical reagent. Nonahydrate [16721-80-5]. Cubic crystals; mp = 1180°; d4 = 1.856; soluble in H2O (18.6 g/100 ml), less soluble in organic solvents. Cerac; Farleyway Chem.

Ltd.; PPG Ind.; Sigma-Aldrich Fine Chem.

3514 Sodium sulfite
7757-83-7   8757   G P

Na2O3S
Sulfurous acid, sodium salt (1:2).
Anhydrous sodium sulfite; CCRIS 1429; Disodium sulfite; Disodium sulfite (Na2SO3); EINECS 231-821-4; Exsiccated sodium sulfite; HSDB 5043; Natrii sulphis; Natrium sulfurosum; Natriumsulfid; Natriumsulfit; S-WAT; Sodium sulfite; Sodium sulfite (2:1); Sodium sulfite (Na2SO3); Sodium sulfite anhydrous; Sodium sulfite, exsiccated; Sodium sulphite; Sulftech; Sulfurous acid, disodium salt. Used in the paper industry; reducing agent (dyes); food preservative and antioxidant; textile bleaching; in photographic developers. Registered by EPA as a fungicide (cancelled). Solid; soluble in H2O (31 g/100 ml), less soluble in organic solvents; LD50 (mus iv) = 175 mg/kg. BASF Corp.; Blythe, Williams Ltd.; EM Ind. Inc.; Ferro/Grant; Indspec; Nissan Chem. Ind.; Rhône-Poulenc.

3515 Sodium sulphooctadecanoate
67998-94-1   H

C18H35NaO5S
Octadecanoic acid, sulfo-, sodium salt.
EINECS 268-062-3; Octadecanoic acid, sulfo-, sodium salt; Oleic acid, sulfonated, sodium salt; Sodium sulphooctadecanoate.

3516 Sodium tallate
61790-45-2   H

Tall oil acid, sodium salt.
EINECS 263-137-7; Fatty acids, tall-oil, sodium salts; Sodium tallate; Tall oil acids, sodium salt; Tall oil fatty acids, sodium salt; Tall oil fatty acids, sodium soap.

3517 Sodium tallowate
8052-48-0   H

Sodium tallowate.
EINECS 232-491-4; Fatty acids, tallow, sodium salts; Sodium tallow soap; Sodium tallowate; Tallow fatty acids, sodium salt; Tallow, sodium salt.

3518 Sodium tartrate
868-18-8   8759   G

C4H4Na2O6
Disodium tartrate.
Bisodium tartrate; Butanedioic acid, 2,3-dihydroxy- (R-(R*,R*))-, disodium salt; CCRIS 7318; Disodium 2,3-dihydroxybutanedioate,

(R-(R*,R*))-; Disodium L-(+)-tartrate; Disodium salt of L-(+)-tartaric acid; Disodium tartrate; EINECS 212-773-3; Natrium (RR)-tartrat; Sal tartar; Sodium tartrate; Sodium L-tartrate; Sodium L-(+)-tartrate; Tartaric acid, disodium salt. Reagent, food additive, sequestrant, stabilizer. Used as a cathartic. Crystals; d = 1.82; soluble in H_2O (30 g/100 ml), insoluble in organic solvents. *Lancaster Synthesis Co.; Sigma-Aldrich Fine Chem.*

3519   Sodium tetradecyl sulfate
139-88-8   8764   G

$C_{14}H_{29}NaO_4S$
7-Ethyl-2-methyl-4-undecanol hydrogen sulfate sodium salt.
EINECS 205-380-3; 4-Ethyl-1-isobutyloktylsiran sodny; 7-Ethyl-2-methyl-4-undecanol sulfate sodium salt; 7-Ethyl-2-methyl-4-hendecanol sulfate sodium salt; Natrii tetracylis sulfas; Natrii tetradecylis sulfas; Natrii tetradecylsulfas; Natrium 4-ethyl-1-isobutyloctylsulfat; Niaproof® Anionic Surfactant 4; Obliterol; Sodium 2-methyl-7-ethylundecyl sulfate-4; Sodium 2-methyl-7-ethylundecanol-4-sulfate; Sodium 7-ethyl-2-methyl-4-undecanol sulfate; Sodium 7-ethyl-2-methylundecyl-4-sulfate; Sodium sotradecol; Sotradecol; Tergitol; Tergitol 4; Tergitol anionic 4; Tergitol penetrant 4; Tetradecil-sulfato sodico; Tetradecyl sulfate de sodium; Trombovar; 4-Undecanol, 7-ethyl-2-methyl-, hydrogen sulfate, sod-ium salt; Varicol. Detergent, wetting agent, penetrant, emulsifier used in adhesives and sealants, coatings, photo chemicals, emulsion polymerization, metal processing, electrolytic cleaning, pickling baths, plating, pharm-aceuticals, leather, textiles. Solid; soluble in H_2O and organic solvents; LD_{50} (rat orl) = 4.95 g/kg. *Niacet.*

3520   Sodium tetrathiocarbonate
7345-69-9   H P

CNa_2S_4
Tetrathioperoxycarbonic acid, disodium salt.
Carbono(dithioperoxo)dithioic acid, disodium salt; Disodium tetrathioperoxycarbonate; EINECS 230-865-1; Enzone; EPA Pesticide Chemical Code 128904; Sodium perthiocarbonate; Sodium tetrathiocarbonate; Sodium thioperoxycarbonate (Na_2CS_4); STTC; Tetrathioperoxy-carbonic acid disodium salt. Registered by EPA as a nematicide. *Entek Corp.*

3521   Sodium thioantimonate(V)
13776-84-6   G

$Na_3S_4Sb \cdot 9H_2O$
Trisodium tetrathioantimonate.
Antimonate(3-), tetrathioxo-, trisodium, (T-4)-; EINECS 237-414-8; Schlippe's salt; Sodium thiantimonate(V); Trisodium tetrathioantimonate. Soluble in H_2O (300 mg/ml), insoluble in EtOH.

Source of antimony sulfide.

3522   Sodium thiophenate
930-69-8   H

C_6H_5NaS
Benzenethiol, sodium salt.
Benzenethiol, sodium salt; EINECS 213-224-0; HSDB 5771; Sodium benzenethiolate; Sodium phenylmercap-tide; Sodium phenylsulfide; Sodium phenylthiolate; Sodium thiophenate; Sodium thiophenolate; Sodium thio-phenoxide; Sodium thiophenylate; Thiophenyl sodium salt.

3523   Sodium thiosulfate
7772-98-7   8769   D G

$Na_2O_3S_2$
Sodium thiosulfate.
AI3-01237; Chlorine Control; Chlorine Cure; Declor-It; Disodium thiosulfate; EINECS 231-867-5; HSDB 592; S-Hydril; Hypo; NSC 45624; Sodium hyposulfite; Sodium oxide sulfide; Sodium thiosulfate; Sodium thiosulfate ($Na_2S_2O_3$); Sodium thiosulfate anhydrous; Sodium thio-sulphate; Sodothiol; Thiosulfuric acid, disodium salt. Photographic fixative, chrome tanning, chlorine removal in bleaching and papermaking, extraction of silver, dechlorination of water, mordant, reagent, bleaching, reducing agent in chrome dyeing, sequestrant in salt, antidote for cyanide poisoning. Crystals; mp = 48°; d = 1.69; soluble in H_2O, insoluble in organic solvents; LD_{50} (rat iv) >2.5 g/kg. *Blythe, Williams Ltd.; Ferro/Grant; General Chem; May & Baker Ltd.; Nissan Chem. Ind.; Sigma-Aldrich Fine Chem.*

3524   Sodium p-toluenesulfonate
657-84-1   9611   G

$C_7H_7NaO_3S$
Methylbenzenesulfonic acid, sodium salt.
AI3-50010; Benzenesulfonic acid, 4-methyl-, sodium salt; EINECS 211-522-5; Eltesol® ST 40; HSDB 5738; Naxonat® 4ST; Naxonate hydrotrope; NSC 203318; Sodium 4-methylbenzenesulfonate; Sodium p-methylbenzenesulfonate; Sodium p-toluenesulfonate; Sodium p-tolyl sulfonate; Sodium paratoluene sulphonate; Sodium toluenesulfonate; Sodium toluene-4-sulphonate; Sodium toluene-p-sulphonate; Sodium tosylate; 4-Toluenesulfonic acid sodium salt; p-Toluenesulfonic acid, sodium salt; para-Toluenesulfonic acid, sodium salt; Tosylate, sodium. Hydrotrope, solubilizer, coupling agent, and viscosity modifier in liq. formulations; cloud point depressant in detergent formulations. Orthorhombic plates, very soluble in H_2O. *Albright & Wilson Americas Inc.; Albright & Wilson UK Ltd.; Degussa-Hüls Corp.; Ruetgers-Nease.*

3525 Sodium trichloroacetate
650-51-1   9700   G P

C2Cl3NaO2

Trichloracetic acid, sodium salt.

Acetic acid, trichloro-, sodium salt; ACP grass killer; Allied Arcadian Sodium TCA; Antiperz; Antyperz; Caswell No. 797; Dow sodium TCA inhibited; EPA Pesticide Chemical Code 081001; Green cross couch grass killer; HSDB 6737; Natriumtrichlooracetaat; Natriumtrichloracetat; Sodio(trichloroacetato di); Sodium (trichloracetate de); Sodium TCA; Sodium TCA inhibited; Sodium trichloroacetate; Sodium trichloroacetic acid; STCA; TCA; TCA sodium; Trichloressigsäures natrium; Trichloroacetic acid sodium salt; Trichloroctan sodny; Varitox; Weedmaster grass killer. Registered by EPA as a herbicide (cancelled). Yellow powder; mp > 300°; dec 165 - 200°; soluble in H2O (120 g/100 ml), MeOH (23.2 g/100 ml), Me2CO (0.76 g/100 ml), Et2O (0.02 g/100 ml), C6H6 (0.007 g/100 ml), CCl4 (0.004 g/100 ml), C7H16 (0.002 g/100 ml); LD50 (rat orl) = 3200 - 5000 mg/kg, (mus orl) = 3600 - 5600 mg/kg, (ckn orl) = 4280 mg/kg, (rat der) > 2000 mg/kg, LC50 (rat ihl 4 hr.) = 365 mg/l air; non-toxic to fish below 11000 mg/l, non-toxic to bees. *Dow AgroSciences; May & Baker Ltd.; Sigma-Aldrich Co.*

3526 Sodium 3,5,6-trichloropyridin-2-olate
37439-34-2   H

C5HCl3NNaO

3,5,6-Trichloro-2-pyridinol, sodium salt.

EINECS 253-506-0; 2(1H)-Pyridinone, 3,5,6-trichloro-, sodium salt; Sodium 3,5,6-trichloropyridin-2-olate; 3,5,6-Trichloro-2-pyridinol, sodium salt.

3527 Sodium tridecylbenzene sulfonate
26248-24-8   H

C19N31NaO3S

Tridecylbenzenesulfonic acid, sodium salt.

Benzenesulfonic acid, tridecyl-, sodium salt; Conoco C 650; EINECS 247-536-3; HSDB 775; Sodium tridecyl-benzene sulfonate; Sodium n-tridecylbenzenesulfonate; Sodium tridecylbenzenesulphonate; Tridecylbenzene-sulfonic acid, sodium salt.

3528 Sodium tungstate
13472-45-2   8772   G

Na2O4W

Sodium tungstate(VI) dihydrate.

CCRIS 5814; Disodium tetraoxatungstate (2-); Disodium tungstate; Disodium tungstate (Na2WO4); Disodium wolframate; EINECS 236-743-4; HSDB 5057; Sodium tungstate; Sodium tungstate (Na2WO4); Sodium tungstate dihydrate; Sodium tungstate(VI); Sodium tungsten oxide; Sodium wolframate; Tungstic acid, disodium salt. Intermediate for tungsten compounds (e.g., phosphotungstate), reagent, fireproofing textiles, alkaloid precipitant. [dihydrate]; soluble in H2O (90 g/100 ml), insoluble in organic solvents; LD50 (gpg orl) = 990 mg/kg. *Cerac; La Littorale; Noah Chem.*

3529 Sodium undecylenate
3398-33-2   G

C11H19NaO2

10-Undecenoic acid, sodium salt.

AI3-19721; EINECS 222-264-8; Sodium undec-10-enoate; Sodium undecylenate; 10-Undecenoic, sodium salt. Used as a bacteriostat and fungistat in cosmetics and pharmaceuticals. *Elf Atochem N. Am.*

3530 Sodium vanadate
13718-26-8   8774   G

NaO3V

Sodium metavanadate.

AI3-51890; CCRIS 6881; EINECS 237-272-7; Meta-wanadan sodowy; Monosodium trioxovanadate(1-); Northovan; NSC 79535; Sodium metavanadate; Sodium trioxovanadate(1-); Sodium vanadate; Sodium vanadate (meta); Sodium vanadate(V); Sodium vanadate(V) (NaVO3); Sodium vanadium oxide; Vanadate (VO3^{1-}), sodium; Vanadic acid (HVO3), sodium salt; Vanadic acid, monosodium salt. Used in manufacture of inks and in photography. mp = 630°; insoluble in H2O; LD75 (rat ip) = 4-5 mg V/kg.

3531 Sodium xylenesulfonate
1300-72-7   G H P

C8H9NaO3S

Dimethylbenzenesulfonic acid, sodium salt.

Benzenesulfonic acid, 3,4-dimethyl-, sodium salt; Caswell No. 799A; CCRIS 4893; Conco SXS; Cyclophil sxs30; 3,4-Dimethylbenzenesulfonic acid, sodium salt; EINECS 215-090-9; Eltesol® SX 30; EPA Pesticide Chemical Code 079019; HSDB 776; Hydrotrope; Naxonate®; Naxonate® 4L; Naxonate® G; NCI-C55403; Pilot SXS-40; Richonate SXS; Sodium dimethylbenzene-sulfonate; Sodium xylenesulfonate; Sodium xylene-sulphonate; Stepanate X; Surco SXS; Ultrawet 40SX; Witconate SXS 40%; Xylenesulfonic acid, sodium salt. Hydrotrope, solubilizer, coupler and processing aid in detergent manufacturing and industrial processes; antiblocking and anticaking agent in powder products; formulates shampoos, Aerosols®, cutting oils, glue; textile finishing. Coupling agent, hydrotrope, solubilizer, solvent, stabilizer; used in liquid cleaners, organic polymers and dyestuffs, petroleum industry, pulping, animal glues. Solid; mp = 27°; bp = 157°; d = 1.023; soluble in H_2O (>10 g/100 ml), less soluble in organic solvents; LD50 (rat orl) = 1000 mg/kg. *Albright & Wilson Americas Inc.; ICI; Pilot; Ruetgers-Nease.*

3532   Soluble tartar
921-53-9   7762   G

C4H4K2O6
Potassium tartrate.
Butanedioic acid, 2,3-dihydroxy- (2R,3R)-, dipotassium salt; Dipotassium tartrate; EINECS 213-067-8; Potassium tartrate. [hemihydrate]; white crystals; d = 1.98; soluble in H_2O (200 g/100 ml); insoluble in EtOH.

3533   Sorbic acid
110-44-1   8793   G H

C6H8O2
2,4-Hexadienoic acid, (E,E).
Acetic acid, (2-butenylidene)-; Acetic acid, crotylidene-; AI3-14851; Caswell No. 801; CCRIS 5748; E 200; EINECS 203-768-7; EPA Pesticide Chemical Code 075901; Hexadienoic acid; Hexa-2,4-dienoic acid; 2,4-Hexa-dienoic acid, (E,E)-; trans-trans-2,4-Hexadienoic acid; HSDB 590; Kyselina 1,3-pentadien-1-karboxylova; Kyselina sorbova; Panosorb; 2-Propenylacrylic acid; Sorbic acid; α-trans-γ-trans-Sorbic acid; trans,trans-Sorbic acid; Sorbistat. Organic acid; used as an animal feed preservative. Solid; mp = 134°; bp = 228° (dec); soluble in H_2O (0.25 g/100 ml), more soluble in organic solvents; LD50 (rat orl) = 7.36 g/kg. *Allchem Ind.; Chisso Corp.; Hoechst Celanese; Honeywell & Stein; Penta Mfg.; Spice King.*

3534   Sorbitan diisostearate
68238-87-9   G

C42H80O7
Sorbitan, diisooctadecanoate.
Anhydrohexitol diisostearate; EINECS 269-410-7; Emsorb® 2518;

Isostearic acid, sorbitan ester (2:1); Sorbitan, diisooctadecanoate; Sorbitan diisostearate. Auxiliary emulsifier, solubilizer, corrosion inhibitor in lubricants, metal protectants and cleaners, emulsion polymerization. *Henkel/Emery.*

3535   Sorbitan monolaurate
1338-39-2   8796   G H

C18H34O6
1,4-Anhydro-D-glucitol, 6-dodecanoate.
Ablunol S-20; Alkamuls® S 20; Alkamuls® SML; Anhydrosorbitol monolaurate; Arlacel 20; Armotan ML; Atmer 100; CCRIS 709; Dehymuls SML; Disponil SML 100; EINECS 215-663-3; Emasol 110; Emasol L 10; Emasol L 10(F); Emasol Super L 10(F); Emsorb 2515; Glycomul L; Glycomul LC; Ionet S 20; Kemester® S20; Kemotan S 20; L 250; L 250 (ester); Laurate de sorbitan; Laurato de sorbitano; Lauric acid sorbitan ester; Lonzest SML; ML 33F; Montane 20; Nissan Nonion LP 20R; Nonion LP 20R; Nonion LR 20R; NRF 201; Radiasurf 7125; Rheodol SP-L 10; Rheodol Super SP-L 10; S-Maz® 20; Sorbitan laurate; Sorbitan, monododecanoate; Sorbitan monolaurate; Sorbitani lauras; Sorbon S 20; Sorgen 90; SP-L 10; Span 20; T 20; Texnol SPT; Value SP 20. Lipophilic emulsifier, emulsion stabilizer; thickener for cosmetic, pharmaceutical, food applications; textile fiber lubricant, softener, anti-fogging agent. emulsifier for oils and fats in cosmetic, metalworking and industrial oil products; corrosion inhibitor; antistat for PVC. Lubricant, antistat, textile softener, process defoamer, opacifier, coemulsifier, solubilizer, dispersant, suspending agent, coupler; prepares excellent w/o emulsions; with T-Maz Series used as oil-water emulsifiers in cosmetics, food formulations, industrial oils and household products. Liquid; d= 1.032; insoluble in H_2O, soluble in organic solvents. *CK Witco Corp.; PPG Ind.; Rhône Poulenc Surfactants; Taiwan Surfactants; Toho Chem. Co.*

3536   Sorbitan monooleate
1338-43-8   8796   G H

C24H44O6
1,4-Anhydro-D-glucitol, 6-(9-octadecenoate).
Ablunol S-80; Alkamuls® SMO; Anhydrosorbitol monooleate; Arlacel 80; Armotan MO; Atmer 05; CCRIS 710; Crill 4; Dehymuls SMO; Disponil 100; EINECS 215-665-4; Emasol 410; Emasol O 10; Emasol O 10F; Emsorb 2500; G 946; Glycomul O; HSDB 5822; Ionet S-80; Kemester® S80; Kemmat S 80; Kosteran O 1; Lonzest SMO; ML 33F; ML 55F; MO 33F; Monodehydrosorbitol monooleate; Montan 80; Montane 80 VGA; Newcol 80; Nikkol SO 10; Nikkol SO-15; Nissan Nonion OP 80R; Nonion OP80R; NSC 406239; O 250; Oleate de sorbitan; Oleato de sorbitano; Radiasurf® 7155; Rheodol AO 10; Rheodol SP-O 10; Rikemal O 250; S 270; S 271 (surfactant); S 80; S-MAX 80; Sorbester P 17; Sorbitan, mono-(9Z)-9-octadecenoate; Sorbitan O; Sorbitan oleate; Sorbitani oleas; Sorbon S 80; Sorgen 40; Sorgen 40A; Span 80; Witconol 2500. Emulsifier, emulsion stabilizer, thickener for cosmetics, pharmaceutical and food applications; textile fiber lubricant, softener, antifogging agent, wet processing of synthetic PU leather. Coupling agent, wetting agent for medicants, petroleum oils, fats, and waxes in the industrial, textile, metalworking, and cosmetic industries; textile and leather

lubricant and softener; corrosion inhibitor. Descouring aid, antistat; anticorrosive agent for pipelines; cleaner for metallic surfaces; superfatting, bodying and antifog aid; pigment dispersant; detergent; emulsion of solvent; cutting oils; textile lubricant additive. Solid; d = 0.986; insoluble in H_2O, propylene glycol, soluble in organic solvents. *CK Witco Corp.; Fina Chemicals; Rhône Poulenc Surfactants; Taiwan Surfactants; Toho Chem. Co.*

3537 Sorbitan monostearate
1338-41-6   8796   G H

$C_{24}H_{46}O_6$
1,4-Anhydro-D-glucitol, 6-octadecanoate.
Ablunol S-60; Alkamuls® S-60; Anhydrosorbitol monostearate; Anhydrosorbitol stearate; Arlacel 60; Armotan MS; Crill 3; Crill K 3; Drewsorb 60; Durtan 60; EINECS 215-664-9; Emsorb 2505; Estearato de sorbitano; FEMA No. 3028; Glycomul S; Hodag SMS; HSDB 778; Ionet S 60; Kemester® S60; Liposorb S; Liposorb S-20; Montane 60; MS 33; MS 33F; Newcol 60; Nikkol SS 30; Nissan nonion SP 60; Nonion SP 60; Nonion SP 60R; Polyoxyethylene sorbitan monooleate; Radiamuls® Sorb 2145; Radiamuls® Sorb 2161 Radiamuls® Sorb 2166; Rikemal S 250; Sorbitan 0; Sorbitan C; Sorbitan, esters, monooctadecanoate; Sorbitan, monooctadecanoate; Sorbitan, monostearate; Sorbitan stearate; Sorbitani stearas; Sorbon S 60; Sorgen 50; Span 55; Span 60; Stearate de sorbitan; Stearic acid, monoester with sorbitan. Emulsifier, emulsion stabilizer, thickener for cosmetics, pharmaceutical and food applications; textile fiber lubricant, softener, antifogging agent, silicone defoamer emulsions. Water-oil emulsifier, lubricant and softener for the textile industry; secondary suspending agent, porosity modifier in PVC suspensions. Used in cosmetics, pharmaceuticals and foods. Waxy solid; mp = 49-65°; insoluble in H_2O, Me_2CO, soluble in EtOH, CCl_4, C_7H_8; acid value = 5-11; saponification value = 140-157; hydroxyl value = 230-260. *CK Witco Corp.; Fina Chemicals; Rhône Poulenc Surfactants; Taiwan Surfactants; Toho Chem. Co.*

3538 Sorbitan palmitate
26266-57-9   G

$C_{22}H_{42}O_6$
Sorbitan monohexadecanoate.
Ablunol S-40; AI3-03901; 1,4-Anhydro-D-glucitol, 6-hexadecanoate; Arlacel 40; Crill 2; EINECS 247-568-8; Emsorb 2510; D-Glucitol, 1,4-anhydro-, 6-hexadecanoate; Glycomul P; Liposorb P; Montane 40; Nikkol SP10; Nissan nonion PP 40R; Nissan nonion PP40; Nonion PP40; Palmitate de sorbitan; Palmitato de sorbitano; Protachem SMP; Rheodol SP-P 10; Sorbitan, esters, monohexadecanoate; Sorbitan monopalmitate; Sorbitan palmitas; Sorbitan palmitate; Sorbitani palmitas; Sorbon S-40; Sorgen 70; Span 40. Emulsifier, emulsion stabilizer, thickener for cosmetics, pharmaceutical, food applications; textile fiber lubricant, softener, antifogging agent. *Taiwan Surfactants; Toho Chem. Co.*

3539 Sorbitan sesquioleate
8007-43-0   8796   E H

$C_{33}H_{60}O_{6.5}$
Sorbitan, 9-octadecenoate (2:3).
Anhydrohexitol sesquioleate; Anhydrosesquioleate; Arlacel C®; Arlacel® 83; Crill 43; Crill 43®; Dehymuls® SSO; Glycomul® SOC; Hodag SSO®; Liposorb SQO®; Montane 83®; Nikkol SO-15®; Nissan Nonion OP-83RAT; Protachem SOC®; Rheodol AO-15; Sorbirol SQ; Sorbitan, 9-octadecenoate (2:3); Sorgen 30; Sorgen 30®; Sorgen S-30-H®; Sporbirol SQ;; component of: Alcolan®, Alcolan® 36W, Alcolan® 40, Dehymuls® E, Dehymuls® K. FDA approved for topicals. Used in topicals as water-oil emulsifier, solubilizer and antifoamer. Yellow oily liquid; soluble in EtOH, iPrOH; mineral oil, cottonseed oil, insoluble in H_2O, propylene glycol. Non-toxic. *Sigma-Aldrich Fine Chem.; Spectrum Chem. Manufacturing.*

3540 Sorbitan trioleate
26266-58-0   H

$C_{60}H_{108}O_8$
Sorbitan, tri-9-octadecenoate.
Ablunol S-85; AI3-03902; Anhydro-D-glucitol trioleate; Anhydrosorbitol trioleate; Arlacel 85; Crill 5; EINECS 247-569-3; Emasol 430; Emsorb 2503; Glycomul TO; Ionet S 85; Liposorb TO; Nissan nonion OP 85; Nissan nonion OP 85R; OP 85R; Protachem STO; Rheodol SP 030; Sorbitan, tri-9-octadecenoate; Sorbitan, tri-(9Z)-9-octadecenoate; Sorbitan, tri-9-octadecenoate, (Z,Z,Z)-; Sorbitan trioleate; Sorbitan, tris(9-octadecenoate), (Z)-; Sorbitani trioleas; Span 85; TE 33; Trioleate de sorbitan; Trioleato de sorbitano; Witconol 2503. Water-oil emulsifier, emulsion stabilizer, thickener for cosmetics, pharmaceutical and food applications; textile fiber lubricant, softener, antifogging agent. *CK Witco Corp.; Taiwan Surfactants.*

3541 Sorbitan tristearate
26658-19-5   G

$C_{60}H_{114}O_8$
Anhydrosorbitol tristearate.
Anhydrosorbitol tristearate; EINECS 247-891-4; Radia-muls® Sorb 2345; Sorbitan, esters, trioctadecanoate; Sorbitan, trioctadecanoate; Sorbitan tristearate; Sorbitani tristearas; Span 65; Triestearato de sorbitano; Tristearate de sorbitan; sorbitan trioctadecanoate. Non-ionic surfactant; emulsifier for foods, cosmetics, household products, industrial applications. *Fina Chemicals; Henkel/Emery; ICI Spec.; Lonzagroup.*

3542 Sorbitol
50-70-4   8797   G

C6H14O6
D-(-)-Sorbitol.
AI3-19424; CCRIS 1898; Cholaxine; Diakarmon; EINECS 200-061-5; Esasorb; FEMA No. 3029; Foodol D 70; D-Galactitol; Glucarine; Glucitol; D-Glucitol; Glucitol, D-; Gulitol; Hexahydric alcohol; 1,2,3,4,5,6-Hexanehexol; D-1,2,3,4,5,6-Hexanehexol; HSDB 801; Karion; Karion instant; L-Gulitol; Multitol; Neosorb; Neosorb 20/60DC; Neosorb 70/02; Neosorb 70/70; Neosorb P 20/60; Neosorb P 60; Nivitin; NSC 25944; Probilagol; Sionit; Sionit K; Sionite; Sionon; Siosan; Sorbex M; Sorbex R; Sorbex RP; Sorbex S; Sorbex X; Sorbicolan; Sorbilande; D-Sorbit; Sorbite; D-Sorbite; Sorbitol; Sorbitol F; Sorbitol solutions; Sorbitol syrup C; (-)-Sorbitol; Sorbo; Sorbol; D-Sorbol; Sorbostyl; Sorvilande. Naturally occurring, synthesized by hydrogenation of glucose; used in manufacture of sorbose, ascobic acid, propylene glycol, synthetic plasticizers and resins, as humectant and softener. Nutrient and dietary supplement, food additive; bodying agent for paper, textile, liquid pharmaceuticals; in manufacture of sorbose, ascorbic acid, propylene glycol, synthetic plasticizers, resins; as humectant, sequestrant. mp = 93-97°, 110-112°; bp = 105°; n_D^{20} 1.4600; $[\alpha]_D^{20}$ = -2.0°; soluble in H2O (83%); moderately soluble in EtOH, alcohols, phenol, Me2CO, AcOH, DMF, pyridine, acetamide, insoluble in most other solvents. *Cerestar Intl.; EM Ind. Inc.; Fanning; ICI Americas Inc.; ICI Spec.; Lipo; Lonzagroup; Sigma-Aldrich Fine Chem.*

3543 L-Sorbose
87-79-6   8798   G

C6H12O6
L-xylo-hexulose.
AI3-19425; EINECS 201-771-8; Esorben; HSDB 780; NSC 97195; L-1,2,3,4,5,6-Pentahydroxyhexan-2-one; Sorbin; Sorbinose; L-Sorbinose; Sorbose; Sorbose, L-; L-Sorbose; l-Sorbose; L-xylo-2-Hexulose. Made from sorbitol (itself made by reduction of glucose) by fermentation and used in the manufacture of ascorbic acid (vitamin C). mp = 165°; $[\alpha]_D^{20}$ = -43.2° (c = 5); soluble in H2O, insoluble in organic solvents.

3544 Soyamine
61790-18-9   H

Amines, soya alkyl.
Amines, soya alkyl; EINECS 263-112-0; Soyamine.

3545 Soybean Oil
8001-22-7   8802   E P

Soybean oil.
Acidulated soybean soapstock; Calchem IVO-114®; Caswell No. 801B; Crude soybean oil, solvent extracted; Degummed soybean oil; EINECS 232-274-4; EPA Pesticide Chemical Code 031605; Extract of soy; Lipex 107®; Lipex 200®; Oils, soybean; Refined soybean oil; Refined undeodorized soybean oil; Soja bean oil; Soy germ extract; Soy oil; Soya-bean oil; Soya bean oil [Oil, edible]; Soya oil; Soybean acidulated soapstock; Soybean deodorizer distillate; Soybean oil; Soybean oil, bleached; Soybean oil bleaching; Soybean oil, degummed; Soybean oil deodorization; Soybean oil, deodorized; Soybean oil fatty acids, glycerol triester; Soybean oil, refined; Soybean vegetable oil, winter fraction; VT 18. Registered by EPA as an antimicrobial, fungicide, herbicide, insecticide and rodenticicide. FDA GRAS, FDA approved for orals, topicals, USP/NF, BP compliance, (iv injectables, oral capsules and topicals). Used as an oleaginous vehicle, used in parenteral pharmaceuticals, orals, topicals. Oil; mp = 22-31°; d^{25} = 0.916 - 0.922; insoluble in H2O, soluble in EtOH, Et2O, CHCl3, CS2; d = 0.924 - 0.929; dynamic viscosity = 172.9 mPa at o°, 99.7 mPa at 10°, 50.09 mPa at 25°, 28.86 mPa at 40°. Non-toxic, LD50 (rat iv) = 16500 mg/kg, (mus iv) = 22100 mg/kg. *Aarhus Oliefabrik A/S; Abitec Corp.; Am. Roland; Amcan France; Arista Ind.; Black S.; C&T Refinery; Calgene; CanAmera Foods; Cargill; Central Soya; Charkit; Pammark Farms Ltd.; Penta Mfg.; Riceland Foods; Ruger; Sigma-Aldrich Fine Chem.; Soya Mainz; Spectrum Chem. Manufacturing; Spectrum Naturals; Stoller Enterprises Inc.; Tri-K Ind.; Vamo-Fuji Specialties.*

3546 Sozoiodolic acid
554-71-2   8805   G

C6H4I2O4S
4-Hydroxy-3,5-diiodo-benzenesulfonic acid.
Acidum sozojodolicum; 2,6-Diiodphenol-4-sulfonsaeure; 3,5-Dijod-4-hydroxybenzolsulfonsaeure; EINECS 209-069-3; 4-Hydroxy-3,5-diiodobenzenesulfonic acid; 4-Hydroxy-3,5-diiodobenzenesulphonic acid; Iodozol. Used in a test for albumin. Also a diagnostic aid; used as a radiopaque medium in retrograde pyelography.

3547 Spermaceti
8002-23-1   E G

C32H64O2
Cetyl Esters.
EINECS 232-302-5; Spermaceti; Spermaceti wax; Spermaceti wax, refined. A wax obtained from the head of the sperm whale. The crude product is obtained by chilling the head and blubber oils. It consists principally of cetyl palmitate. FDA approved for orals, topicals, USP/NF compliance. Used as an emollient, stiffening agent and viscosity builder for pharmaceutical preparations. Used in orals, topicals. Waxy solid; mp = 43-47°; d = 0.820; soluble in Et2O, CHCl3, boiling EtOH, insoluble in H2O. Non-toxic. *Koster Keunen; Robeco; Smith Werner G.; Spectrum Chem. Manufacturing.*

3548 Spirit Blue
2152-64-9   H

C37H30ClN3

N-Phenyl-4-((4-(phenylamino)phenyl)(4-(phenylimino)-2,5-cyclohexadien-1-ylidene)methyl)benzenamine mono-hydrochloride. Benzenamine, N-phenyl-4-((4-(phenylamino)phenyl)(4-(phenylimino)-2,5-cyclohexadien-1-ylidene)methyl)-, monohydrochloride; CI 42760; CI Solvent Blue 23; CI Solvent Blue 23, monohydrochloride; EINECS 218-441-4; HSDB 466; Opal Blue SS; N-Phenyl-4-((4-(phenyl-amino)phenyl)(4-(phenylimino)cyclohexa-2,5-dien-1-ylidene)methyl)aniline monohydrochloride; Spirit Blue.

3549   Spiro(isobenzofuran-1(3H),9'-(9H)xanth-en)-3-one, 6'-(diethylamino)-2'-((2,4-dimethyl-phenyl)amino)-3'-methyl-
36431-22-8   H

C33H32N2O3

Spiro(isobenzofuran-1(3H),9'-(9H)xanthen)-3-one,   6'-(di-ethylamino)-2'-((2,4-dimethylphenyl)amino)-3'-methyl-.

3550   Spironolactone
52-01-7   8839   D

C24H32O4S

17α-Pregn-4-ene-21-carboxylic acid, 17-hydroxy-7α-mercapto-3-oxo-, γ-lactone acetate.
4-18-00-01601 (Beilstein Handbook Reference); Acelat; 7α-Acetylthio-3-oxo-17α-pregn-4-ene-21,17β-carbo-lactone; 7α-(Acetylthio)-17α-hydroxy-3-oxopregn-4-ene-21-carboxylic acid, γ-lactone; Aldactazide; Aldactone; Aldactone A; Alderon; BRN 0057767; Dira; Duraspiron; EINECS 200-133-6; Espironolactona; Euteberol; HSDB 3184; Lacalmin; Lacdene; Laractone; Melarcon; Nefuro-fan; NSC 150399; Osiren; Osyrol; Sagisal; SC 15983; SC 9420; Sincomen; Spiresis; Spiretic; Spiridon; Spiro-Tablinen; Spiroctan; Spiroderm; Spirolactone; Spirolang; Spirolone; Spirone; Spironocompren; Spironolactone; Spironolactone A; Spironolactonum; Spironolattone; Supra-Puren; Suracton; Uractone; Urusonin; Verospiron; Verospirone; Xenalon. Aldosterone antagonist. Used as a potassium sparing diuretic. Used for edema in cirrhosis of the liver, nephrotic syndrome, congestive heart failure, potentiation of thiazide and loop diuretics, hypertension and Conn's syndrome. Crystals; mp = 134-135°, 201-202°; $[\alpha]_D^{20}$ = -33.5° (CHCl3); λ_m = 238 nm (ε 20200); insoluble in H2O, soluble in most organic solvents. Abbott Labs Inc.; Parke-Davis; Searle G.D. & Co.

3551   Squalane
111-01-3   8846   G

C30H62

Dodecahydrosqualene.
4-01-00-00593 (Beilstein Handbook Reference); AI3-36494; BRN 0776019; Caswell No. 482A; Cosbiol®; Dodecahydrosqualene; EINECS 203-825-6; EPA Pesticide Chemical Code 045503; Hexamethyltetracosane; 2,6,10, 15,19,23-Hexamethyltetracosane; NSC 6851; Perhydro-squalene; Robane®; Spinacane; Squalane; Tetracosane, 2,6,10,15,19,23-hexamethyl-; Vitabiosol. Saturated branched chain hydrocarbon obtained by hydrogenation of shark liver oil or other natural oils; high-grade lubricating oil, perfume fixative, gas chromatographic analysis, transformer oil; in cosmetics and pharmaceutals. A high quality oil for cosmetics and pharmaceuticals used as a moisturizer, emollient, lubricant, humectant; aids spread of topical agents over the skin, increases skin respiration, prevents insensible water loss, imparts suppleness to skin without greasy feel; cosmetics and pharmaceuticals. Oil; mp = -38°; bp = 350°, bp10 = 263°; d_4^{15} 0.8115; insoluble in H2O, slightly soluble in EtOH, Me2CO, AcOH, soluble in Et2O, CHCl3, MeOH, very soluble in petroleum ether, freely soluble in C6H6. Arista Ind.; Robeco.

3552   Squalene
111-02-4   8847   G

C30H50

(all-E)-2,6,10,15,19,23-hexamethyl-2,6,10,14,18,22-tetracosahexaene.
2,6,10,15,19,23-Hexamethyl-2,6,10,14,18,22-tetracosa-hexaene; 2,6,10,15,19,23-Hexamethyltetracosa-2,6,10,   14,18,22-hexaene; Spinacene; Squalene; (E,E,E,E)-Squal-ene; Supraene; 2,6,10,14,18,22-Tetracosahexaene, 2,6, 10,15,19,23-hexamethyl-, EINECS 203-826-1. Used in biochemical and pharmaceutical research; a precursor of cholesterol in biosynthesis; chemical intermediate for manufacture of pharmaceuticals, organic colorants, rubber chemicals, aromatics, surfactants; bactericide. Oil; mp <-20°; bp4 = 241°; bp17 = 280°, bp25 = 285°; d_4^{20} 0.8584; insoluble in H2O, slightly soluble in EtOH, soluble in Et2O, Me2CO, CCl4. Arista Ind.; Robeco.

3553   Stabinol
3567-08-6   G

C13H17N3O3S2

N-(5-Isobutyl-1,3,4-thiadiazol-2-yl)-*p*-methoxybenzene-sulfonamide. 4-27-00-08084 (Beilstein Handbook Reference); 8002 C. B.; 8002 CB; 2-(p-Anisylsulfonamido)-5-isobutyl-1,3,4-thiadiazole; Benzenesulfonamide, 4-methoxy-N-(5-(2-methylpropyl)-1,3,4-thiadiazol-2-yl)-; Benzenesulfon-amide, N-(5-isobutyl-1,3,4-thiadiazol-2-yl)-p-methoxy-; BRN 0295174; FWH 114; Glisobuzole; Glysobuzole; Gly-sobuzol; 5-Isobutyl-2-p-methoxybenzenesulfon-ami-do-1,3,4-thiadiazole; N-(5-Isobutyl-1,3,4-thiadiazol-2-yl)-p-methoxybenzenesulfonamide; Isobuzol; Isobuz-ole; 2-(p-Methoxybenzenesulfonamido)-5-isobutyl-1,3,4-thia-diazole; NSC 34014; Stabinol; 1,3,4-Thiadiazole, 5-isobutyl-2-(p-methoxybenzenesulfonamido)-.

A proprietary trade name for isobuzole, an oral hypoglycemic agent for the treatment of diabetes mellitus. *Sankyo Co. Ltd.*

3554   Standapol® MG
3097-08-3   G

C24H50MgO6S2

Dodecyl magnesium sulphate.

Caswell No. 532A; Dodecyl sulfate, magnesium salt; EINECS 221-450-6; EPA Pesticide Chemical Code 079017; Magnesium lauryl sulfate; Magnesium monododecyl sulfate; Standapol® MG; Sulfuric acid, monododecyl ester, magnesium salt. Foamer with good dermatological properties; used in personal care products. *Henkel/Cospha.*

3555   Stannic chloride
7646-78-8   8852   G

Cl4Sn

Tin(IV) tetrachloride.

CCRIS 6886; EINECS 231-588-9; Etain (tetrachlorure d'); HSDB 781; Libavius fuming spirit; NSC 209802; Stagno (tetracloruro di); Stannane, tetrachloro-; Stannic chloride; Stannic chloride, anhydrous; Stannic tetrachloride; Stannous chloride; Tetrachlorostannane; Tetrachlorotin; Tin chloride; Tin chloride (1:4); Tin chloride (SnCl4); Tin chloride, fuming (DOT); Tin perchloride; Tin tetra-chloride; Tin Tetrachloride, anhydrous; Tin(IV) chloride; Tin(IV) tetrachloride; Tintetrachloride; UN1827; Zinn-tetrachlorid. Coatings that are electroconductive and electroluminescent, textile dye

mordant, perfume stabilizer, manufacture of fuchsin, blueprint paper, color lakes, ceramic coatings, bleaching agent for sugar, stabilizer for resins, tin salts, soap bactericide and fungicide. Liquid; mp = -33°; bp770 = 114°; d = 2.2260; soluble in H2O, organic solvents. *Chemisphere; CK Witco Corp.; Elf Atochem N. Am.; M&T Harshaw; Nihon Kagaku Sangyo; Sigma-Aldrich Fine Chem.*

3556   Stannic chloride pentahydrate
10026-06-9   G

Cl4Sn

Tin(IV) chloride pentahydrate.

CCRIS 6330; Stannic chloride, pentahydrate; Tetra-chlorostannane pentahydrate; Tin chloride pentahydrate; Tin(IV) chloride, pentahydrate (1:4:5); UN2440.

3557   Stannic sulfide
1315-01-1   8858   G

S2Sn

Tin(IV) sulfide.

EINECS 215-252-9; Mosaic gold; Stannic sulfide; Tin bronze; Tin disulphide; Tin sulfide (SnS2)tin disulfide;. Used for gilding and bronzing metals and other surfaces. Golden leaflets; d = 4.5; insoluble in H2O, soluble in alkali, aqua regia.

3558   Stannous 2-ethylhexoate
301-10-0   G H

C16H32O4Sn

2-Ethylhexanoic acid tin(2+) salt.

EINECS 206-108-6; 2-Ethylhexanoic acid tin(2+) salt; Metacure® T-9; NSC 75857; Nuocure 28; Stannous 2-ethylhexanoate; Stannous 2-ethylhexoate; Stannous octoate; Tin 2-ethylhexanoate; Tin bis(2-ethylhexanoate); Tin dioctoate; Tin ethylhexanoate; Tin octoate; Tin(2+) 2-ethylhexanoate; Tin(II) 2-ethylhexanoate; Tin(II) 2-ethylhexylate; Tin(II) bis(2-ethylhexanoate). Catalyst used in production of PU coatings, adhesives, and sealants; uniform activity and excellent stability. *Air Products & Chemicals Inc.*

3559   Stannous chloride
7772-99-8   8861   G

Cl2Sn

Tin (II) chloride anhydrous.

AI3-51686; Anhydrous stannous chloride; C.I. 77864; CCRIS 560; CI 77864; Dichlorotin; EINECS 231-868-0; HSDB 582; NCI-C02722; Stannous chloride; Stannous chloride anhydrous; Stannous dichloride; Tin chloride (SnCl2); Tin dichloride; Tin Protochloride; Tin(II) chloride; Uniston CR-HT 200; tin salt. Reducing agent for intermediates, dyes, polymers, phosphors; manufacture of lakes; textile dyeing and printing; tin galvanizing; analytical reagent; silvering mirrors; antisludge for lubricants; food preservative; perfume stabilizer, soldering flux. Used as a wool mordant for dyeing cochineal scarlet, for dyeing blacks on silk, for weighting silk, for calico printing, in stabilizers and soldering flux. Orthorhombic crystals; mp = 37-38°, bp = 247°; d = 3.95; soluble in H_2O, polar organic solvents; LD_{50} (mus ip) = 66 mg/kg. *Blythe, Williams Ltd.; Cerac; Elf Atochem N. Am.; M&T Harshaw; Noah Chem.; Sigma-Aldrich Fine Chem.*

3560   Stannous chloride dihydrate
10025-69-1   8861   G

$Cl_2Sn.2H_2O$
Tin(II) chloride.
CCRIS 3953; Dihydrated stannous chloride; Stannochlor; Stannous chloride; Stannous dichloride dihydrate; Tin chloride ($SnCl_2$) dihydrate; Tin Chloride, dihydrate; Tin(II) chloride, dihydrate (1:2:2).

3561   Stannous chromate
38455-77-5   8853   G

$Cr_2H_2O_8Sn$
Stannic chromate(VI).
Chromic acid (H_2CrO_4), tin(2+) salt (2:1); EINECS 253-946-3; Stannous chromate; Tin chromate; Tin(II) chromate. Rose-violet pigment used to color paper, porcelain and china. Soluble in H_2O.

3562   Starch, acid hydrolyzed
65996-63-6   H

Acid hydrolyzed starch.
Acid modified, corn starch; Acid-treated starch; Corn flour, hydrochloric acid modified; Starch, acid hydrolyzed; Starch, acid-modified; Starch, acid-treated; Starch, thin-boiling; Wheat starch, acid modified.

3563   Starch, enzyme hydrolyzed
65996-64-7   H

Enzyme hydrolyzed starch.
Enzyme converted starch; Starch, converted; Starch enzyme; Starch, enzyme converted; Starch, enzyme-hydrolyzed.

3564   Stearalkonium chloride
122-19-0   G H

$C_{27}H_{50}ClN$
Benzyldimethyloctadecylammonium chloride.
2B; Albumine 280; AI3-09119; Algene SC 25; Alkaquat DMB-ST; Ammonium, benzyldimethyloctadecyl-, chlor-ide; Ammonyx 4; Ammonyx 4002; Ammonyx 485; Ammonyx 490; Ammonyx CA special; Arquad DM18B-90; Barquat SB-25; Benzenemethanaminium, N,N-dimethyl-N-octadecyl-, chloride; Carsoquat sdq-25; Carsoquat sdq-85; Cation S (surfactant); CCRIS 6022; Cycloton® SCS; EINECS 204-527-9; HSDB 6123; Incroquat SDQ 25; Intexan SB-85; Intexsan SB-85; J Soft C 4; Larostat HTS 905; Mackernium SDC 25; Mackernium SDC 85; Maquat SC 18; Nissan cation S2-100; NSC 20199; Octadecylbenzyldimethylammonium chloride; Orthosan MB; Quaternol 1; SDQ 25; SDQ 85; Stearalkonium chloride;

Stearyldimethylbenzylammon-ium chloride; Stebac; Stedbac; Sumquat® 6210; Swanol CA 1485; Tallow benzyl dimethyl ammonium chloride; Triton CG 400; Triton CG 500; Triton X-40; Triton X-400; Varisoft SDC 85. Surfactant used in cosmetics and antimicrobials. Antistat, hair conditioner and germicidal detergent. Solid; insoluble in H_2O (< 1 mg/ml), soluble in EtOH, organic solvents; LD_{50} (rat orl) = 1250 mg/kg. *Ferrosan A/S; Hexcel; Lonzagroup; Mason; McIntyre; Rhône Poulenc Surfactants; Sherex; Taiwan Surfactants.*

3565   Stearamide
124-26-5   G H

$C_{18}H_{37}NO$
n-Octadecylamide.
4-02-00-01240, 4-04-00-00794 (Beilstein Handbook Reference); Adogen 42; AI3-10003; AI3-15083; Alamine 4; Amine BB; 1-Aminododecane; Armeen 12D; BRN 0909006; BRN 1633576; Dodecanamine; 1-Dodecan-amine; Dodecylamine; 1-Dodecylamine; EINECS 204-690-6; EINECS 204-693-2; Farmin 20D; HSDB 723; HSDB 2645; Kemamide® S; Kemamine P690; Lauramine; Laurinamine; Laurylamine; n-Laurylamine; Monododecyl-amine; Nissan amine BB; NSC 66462; Octadecamide; Octadecanamide; Octadecylamide; Stearamide; Stearic acid amide; Stearic amide; Stearoylamide; Stearylamide; Uniwax 1750. Lubricant, slip, antiblock, and mold release agent for plastics, crayons, petrol. products, asphalts, inks, metals, textiles; mold release agent for thermoplastic resins in injection molding; defoamer and water repellent in industrial/household application. Slip/antiblock agent for LDPE, HDPE, PP. Antiblock additive for polyolefin films; lubricant, release agent for rubber compounding, and PVC processing.
Leaflets; mp = 109°; bp$_{12}$ = 250°; very soluble in Et_2O; $CHCl_3$. *Akzo Chemie; Astor Wax; Chemax; CK Witco Corp.; Croda; Henkel/Emery; Syn Prods.; Unichema; Witco/Humko.*

3566   Stearamide MEA-stearate
14351-40-7   G

$C_{38}H_{75}NO_3$
2-((1-Oxooctadecyl)amino)ethyl stearate.
EINECS 238-310-5; Octadecanoic acid, 2-((1-oxoocta-decyl)amino)ethyl ester; 2-((1-Oxooctadecyl)amino)ethyl octadecanoate; 2-((1-Oxooctadecyl)amino)ethyl stearate; Stearamide MEA-stearate; Stearic monoethanolamide stearate; Witcamide® MAS. Opacifier, conditioner, lubricant, gelling agent for personal care, household, and institutional liquid soaps; used as partial or total replacement for vegetable waxes in polishes; coating agent for paper and textiles; mold release agent for industrial processing; additive for raising melting points of pertoleum or glyceride waxes and fats; ingredient of insulating coatings or barriers and of water-repellent compounds. *CK Witco Chem. Corp.*

3567   Steareth-20
9005-00-9   E G

$C_{58}H_{118}O_{21}$
POE (20) stearyl ether.
Ablunol SA-7; Ethanol, 2-(2-(octadecyloxy)ethoxy)-; 3,6,9,12,15,18,21-Heptaoxanonatriacontan-1-ol; 2-(2-(Octadecyloxy)ethoxy)ethanol; Octadecyl polyoxyethyl-ene ether;

PEG-2 Stearyl ether; PEG-7 Stearyl ether; PEG-11 Stearyl ether; PEG-13 Stearyl ether; PEG-14 Stearyl ether; PEG-15 Stearyl ether; PEG-16 Stearyl ether; PEG-20 Stearyl ether; PEG-21 Stearyl ether; PEG-25 Stearyl ether; PEG-27 Stearyl ether; PEG-30 Stearyl ether; PEG-40 Stearyl ether; PEG-50 Stearyl ether; PEG-100 Stearyl ether; Poly(oxy-1,2-ethanediyl), α-octadecyl-ω-hydroxy-; Poly-ethylene glycol (7) stearyl ether; Polyethylene glycol (11) stearyl ether; Polyethylene glycol (13) stearyl ether; Polyethylene glycol (14) stearyl ether; Polyethylene glycol (15) stearyl ether; Polyethylene glycol (16) stearyl ether; Polyethylene glycol (21) stearyl ether; Polyethylene glycol (25) stearyl ether; Polyethylene glycol (27) stearyl ether; Polyethylene glycol (30) stearyl ether; Polyethylene glycol (50) stearyl ether; Polyethylene glycol (100) stearyl ether; Polyethylene glycol 1000 stearyl ether; Polyethylene glycol 2000 stearyl ether; Polyoxyethylated stearyl alcohol; Polyoxyethylene (7) stearyl ether; Polyoxyethylene (11) stearyl ether; Polyoxyethylene (13) stearyl ether; Polyoxyethylene (14) stearyl ether; Polyoxyethylene (15) stearyl ether; Polyoxyethylene (16) stearyl ether; Polyoxyethylene (2) stearyl ether; Polyoxyethylene (20) stearyl ether; Polyoxyethylene (21) stearyl ether; Poly-oxyethylene (25) stearyl ether; Polyoxyethylene (27) stearyl ether; Polyoxyethylene (30) stearyl ether; Poly-oxyethylene (40) stearyl ether; Polyoxyethylene (50) stearyl ether; Polyoxyethylene (100) stearyl ether; Poly-oxyethylene monooctadecyl ether; Steareth-7; Steareth-11; Steareth-13; Steareth-14; Steareth-15; Steareth-16; Steareth-2; Steareth-20; Steareth-21; Stear-eth-25; Steareth-27; Steareth-30; Steareth-40; Steareth-50; Steareth-100; Stearyl alcohol EO (10); Stearyl alcohol EO (20); Stearyl alcohol, ethoxylated; Stearyl alcohol ethylene oxide (2); Stereal alcohol EO (2). Emulsifier for waxes and cosmetics. Used as an emulsifier, gellant, stabilizer, wetting agent, solubilizer for pharmaceuticals. *Taiwan Surfactants.*

3568 Stearic acid

57-11-4   8882   G

C18H36O2
 n-Octadecanoic acid.
 4-02-00-01206 (Beilstein Handbook Reference); AI3-00909; Barolub fta; BRN 0608585; Caswell No. 801D; CCRIS 2305; Century 1210; Century 1220; Century 1230; Century 1240; Cetylacetic acid; Darchem 14; EINECS 200-313-4; Emersol 120; Emersol 132; Emersol 150; Emersol 153; Emersol 6349; EPA Pesticide Chemical Code 079082; FEMA No. 3035; Formula 300; Glycon DP; Glycon S-70; Glycon S-80; Glycon S-90; Glycon TP; Groco 54; Groco 55; Groco 55l; Groco 55L; Groco 58; Groco 59; 1-Heptadecanecarboxylic acid; HSDB 2000; Humko Industrene R; Hy-phi 1199; Hy-phi 1205; Hy-phi 1303; Hy-phi 1401; Hydrofol 1895; Hydrofol acid 150; Hydrofol acid 1655; Hydrofol acid 1855; Hystrene 80; Hystrene 4516; Hystrene 5016; Hystrene 7018; Hystrene 9718; Hystrene S-97; Hystrene T-70; Industrene 5016; Industrene 8718; Industrene 9018; Industrene R; KAM 1000; KAM 2000; KAM 3000; Loxiol g 20; Lunac s 20; NAA 173; Neo-fat 18; Neo-fat 18-55; Neo-fat 18-53; Neo-fat 18-54; Neo-fat 18-59; Neo-fat 18-61; Neo-fat 18-S; NSC 25956; Octadecanoic acid; n-Octadecanoic acid; n-Octadecylic acid; PD 185; Pearl stearic; Promulsin; Proviscol wax; Stearex beads; Stearic acid; Stearophanic acid; Tegostearic 254; Tegostearic 255; Tegostearic 272; Vanicol. Used in cosmetics, chemicals, as a dispersant and softener in rubber compounds, in food packaging; suppositories and ointments. White solid; mp = 67-69°; bp = 361°; d = 0.8450; slightly soluble in H2O, more soluble in organic solvents; LD50 (rat iv) = 21 mg/kg. *Akrochem Chem. Co.; Akzo Chemie; Henkel/Emery; Lonzagroup; Sherex; Syn Prods.; Unichema; UOP; Witco/Humko.*

3569 Stearic acid monoethanolamide

111-57-9   G

C20H41NO2
 Stearamide MEA.
 Ablumide SME; Clindrol 200-MS; Comperlan HS; Cycloamide SM; EINECS 203-883-2; N-(2-Hydroxy-ethyl)octadecanamide; N-(2-Hydroxyethyl)-stearamide; N-(Hydroxyethyl)stearamide; Loramine S 280; Marlamid M 18; Monoethanolamine stearic acid amide; Ninol 1301; NSC 3377; Octadecanamide, N-(2-hydroxyethyl)-; Onyx Wax EL; Stearamide MEA; Stearamyl; Stearic acid monoethanolamide; Stearic ethanolamide; Stearic ethylolamide; Stearic mono-ethanolamide; Stearic monoethanolamine; N-Stearoylethanolamine; Stearoyl monoethanolamide; Stearoylethanolamine; Stearoyl-ethanolamine; Teric CME7; Witcamide® 70. Opacifier, conditioner, lubricant, thickener, gelling agent, mold release agent, binder; cosmetics and toiletries; base for antiperspirant and makeup sticks. Used as a thickener for shampoo, cream rinse and bubble bath. *CK Witco Corp.; Taiwan Surfactants.*

3570 Steartrimonium chloride

112-03-8   G H

C21H46ClN
 Trimethyloctadecylammonium chloride.
 Aliquat 7; Ammonium, trimethyloctadecyl-, chloride; Arquad® 18; Arquad® 18-50; Cation AB; EINECS 203-929-1; Monostearyl trimethyl ammonium chloride; Nissan cation AB; Octadecyltrimethylammonium chloride; Octosol 474; Quaternium-10; Stac; Stear-trimonium chloride; Stearyltrimethylammonium chloride; Trimethyloctadecylammonium chloride; Trimethylstearyl-ammonium chloride; N,N,N-Trimethyl-1-octadecanam-inium chloride. Biodegradable emulsifier for cationic or cationic/anionic emulsion systems. Used in foaming, wetting, dispersing agents, corrosion inhibitor, softener, dyeing aid, antistat for textiles, paper, cosmetics; industrial, agriculture, plastics, petroleum industry, acid pickling baths; bactericide, algicide; dye leveling agent, viscosity stabilizer in lubricant compounding. *Akzo Chemie.*

3571 Stearyl alcohol

112-92-5   8883   G H

C18H38O
 n-Octadecyl alcohol.
 4-01-00-01888 (Beilstein Handbook Reference); Adol® 62 NF; Adol® 68; AI3-01330; Alcohol stearylicus; Alcohol(C18); Aldol 62; Alfol 18; Atalco S; BRN 1362907; C18 alcohol; Cachalot S-43; CCRIS 3960; CO-1895; CO-1897; Crodacol-S; Decyl octyl alcohol; Dytol E-46; EINECS 204-017-6; Fatty alcohol(C18); HSDB 1082; 1-Hydroxyoctadecane; Lanol S; Lorol 28; Mixture Name; NSC 5379; Octadecanol; 1-Octadecanol; n-Octadecanol; n-1-Octadecanol; Octadecanol, 1-; Octadecyl alcohol; n-Octadecyl alcohol; Octadecylalkohol; Polaax; Sipol S; Siponol S; Siponol SC; SSD AF; Stearic alcohol; Stearol; Stearyl alcohol; Stenol; Steraffine; USP XIII stearyl alcohol; Varonic® BG. Emollient, auxiliary emulsifier, texturizer; nonoily, velvety feel; higher viscosity in emulsions. Used in perfumery, cosmetics, intermediate, surfactants, lubricants, resins, antifoam agent, glass frit binders, waxes, emulsion stabilizers, esters, tertiary amines, surfactants, polymers, as a chemical

intermediate and in cosmetic formulations. A self-emulsifying wax and viscosity modifier for hair conditioners, creams and lotions; an emulsifier and solubilizer. Leaflets; mp = 59.5°; bp = 340-355°, bp_{15} = 210.5°; d^{59} = 0.8124; insoluble in H_2O, slightly soluble in Me_2CO, C_6H_6, soluble in EtOH, Et_2O, $CHCl_3$. *Aarhus Oliefabrik A/S; Amerchol; Chemron; Croda; Ethicon Inc.; Lipo; Lonzagroup; Michel M.; Procter & Gamble Co.*

3572 Stearyl erucamide
10094-45-8   G

$C_{40}H_{79}NO$

N-Octadecyl-13-docosenamide.

(13-Docosenamide, N-octadecyl-; 13-Docosenamide, N-octadecyl-, (13Z)-; 13-Docosenamide, N-octadecyl-, (Z)-; EINECS 233-226-5; (Z)-N-Octadecyldocos-13-enamide; Stearyl erucamide. Surfactant and release agent providing slip, antiblocking to thermoplastics incl. PP film, nylon. *Croda; Hexcel; Witco/Humko; Zeeland Chem. Inc.*

3573 Stearyl hydroxyethyl imidazoline
95-19-2   G

$C_{22}H_{44}N_2O$

1-(2-Hydroxyethyl)-2-heptadecyl-2-imidazoline.

5-23-04-00119 (Beilstein Handbook Reference); Amine 225; Atlasol KAD; BRN 0236487; Calgene C-100-S; Casamine SH; CCRIS 1053; Crodazoline; Crodazoline O; Crodazoline S; EINECS 202-397-8; Fungacide 337; Fungicide 337; 2-(2-Heptadecyl-2-imidazolin-1-yl)-ethanol; 2-Heptadecyl-1-(2-hydroxyethyl)-2-imidazoline; 2-Heptadecyl-2-imidazoline-1-ethanol;; 2-Heptadecyl-3-hydroxyethylimidazoline; 2-Heptadecyl-4,5-dihydro-1H-imidazole; Hodag C-100-S; 1-(β-Hydroxethyl)-2-hepta-decylimidazoline; 1-(2-Hydroxyethyl)-2-(hepta-decyl)-imidazoline; 1-(2-Hydroxyethyl)-2-heptadecyl-2-imid-azoline; 1-(Hydroxyethyl)-2-(heptadecyl)imidazoline; 1-Hydroxyethyl-2-heptadecyl glyoxalidine; 1H-Imidazole-1-ethanol, 2-heptadecyl-4,5-dihydro-; Imid-azoline SOH; 2-Imidazoline-1-ethanol, 2-heptadecyl-; 2-Imidazoline, 2-heptadecyl-1-hydroxyethyl-; Miramine® GS; Monazoline S; NSC 22372; Schercozoline S; Stearic acid, aminoethylethanolamine amide-imidazoline; Stearyl hydroxyethyl imidazoline; Stearyl Imidazoline; Unamine® S. Intermediate for quaternary ammonium compounds; strongly absorbed on textiles, paper and many metal surfs.; for agric., asphalt, cleaners, corrosion inhibitors, demulsifiers, flotation, metalworking, paints, pigment grinding, inks, textiles, wax emulsions Moderately toxic by ingestion. *Calgene; Croda Surfactants; Lonzagroup; Mona; Scher.*

3574 Stearyl stearate
2778-96-3   G

$C_{36}H_{72}O_2$

n-Octadecyl stearate.

EINECS 220-476-5; Octadecanoic acid, octadecyl ester; Octadecyl stearate; Radia® 7501; Stearic acid, stearyl ester; Stearyl stearate. Chemical intermediate, chemical synthesis; lubricant in mineral, cutting, lamination, textile oils, and rust inhibitors; textile and leather application; also used for wax formulation due to its high melting point. *Fina Chemicals.*

3575 Stearylamine
124-30-1   G H

$C_{18}H_{39}N$

n-Octadecylamine.

4-04-00-00825 (Beilstein Handbook Reference); Ado-genen 142; AI3-14661; Alamine 7; Alamine 7D; Amine 18-90; Amine AB; 1-Aminooctadecane; Armeen® 118D; Armeen® 18; Armeen® 18D; Armid® HTD; Armofilm; BRN 0636111; CCRIS 4688; Crodamine 1.18D; EINECS 204-695-3; Farmin 80; HSDB 1194; Kemamine® P990; Monooctadecylamine; n-Stearylamine; Nissan amine AB; Noram SH; NSC 9857; Octadecanamine; 1-Octa-decanamine; Octadecylamine; 1-Octadecylamine; n-Octadecylamine; Oktadecylamin; Steamfilm FG; Stear-amine; Stearylamine. Emulsifier, flotation agent, dis-persing and flushing agent, intermediate, used in metalworking oils, as fuel oil additive; mold release for rubber and plastics; lubricant and spinning aid in metalworking oils. Corrosion inhibitor, anticaking agent; hard rubber mold release agent. Processing aid for high viscosity rubber compounds; facilitating flow behavior and improving mold release. Emulsifier for herbicides, ore flotation, pigment dispersion; auxiliary for textiles, leather, rubber, plastics, and metal industries. Gorrosion inhibitor for control of corrosion caused by carbon dioxide and oxygen in the afterboiler section of steam-generating systems by forming a nonwettable, monomolecular film on metal surfaces. Solid; mp = 52.9°; bp = 346.8°, bp_{32} = 232°; d^{20} = 0.8618; insoluble in H_2O, slightly soluble in EtOH, soluble in Et_2O, Me_2CO, $CHCl_3$, ligroin; LD_{50} (rat orl) = 2395 mg/kg. *Akzo Chemie; Croda Surfactants; 2161;3244.*

3576 Stearylamine acetate
2190-04-7   G

$C_{20}H_{43}NO_2$

1-Octadedecylamine acetate.

Acetamin 86; Acetamin T; Alamac 7D; Armac 18D; Armac OD; EINECS 218-583-7; NSC 81938; Octadecanamine acetate; 1-Octadecanamine, acetate; Octadecylamine, acetate; n-Octadecylamine acetate; Octadecylammonium acetate; Stearamine acetate. Surface coating agent for pigments, anticaking agent for fertilizer; emulsifier, dispersant, and softening agent for textiles; mineral flotation reagent. Used for flotation of minerals, in anticaking agents and as an emulsifier bactericide. *Kao Corp. SA.*

3577 Stilbene
588-59-0   8895   G

C14H12
1,1'-(1,2-Ethenediyl)bis[benzene].
Benzene, 1,1'-(1,2-ethenediyl)bis-; Bibenzal; Bibenzyl-idene; BRN 1904445; α,β-Diphenylethylene; 1,2-Di-phenylethylene; Eccobrite RB; 1,1'-(1,2-Ethenediyl)-bisbenzene; EINECS 209-621-3; Stilben; Stilbene; bi-benzal; bibenzylidene. Whitening agent for cotton and acetates. Exists as cis-isomer [645-49-8] or trans-isomer [103-30-0]. *Eastern Color & Chem.*

3578   cis-Stilbene
645-49-8   8895   G

C14H12
1,1'-(1,2-Ethenediyl)bis[benzene].
Benzene, 1,1'-(1,2-ethenediyl)bis-, (Z)-; CCRIS 5932; cis-Diphenylethene; cis-1,2-Diphenylethylene; cis-Stilbene; EINECS 211-445-7; HSDB 4270; Isostilbene; NSC 66424; Stilbene, (Z)-; (Z)-Stilbene. Used in chemical synthesis. Liquid; mp = -5°; bp$_{10}$ = 135°; d^{20} = 1.0143; λ_m = 223, 276 nm (ε = 20600, 10900, MeOH); insoluble in H$_2$O, soluble in EtOH, Et$_2$O, Me$_2$CO, C$_6$H$_6$, CHCl$_3$, petroleum ether.

3579   Stronscan-85
10476-85-4   8920   G

Cl$_2$^{85}Sr
Strontium chloride-^{85}Sr.
strontium chloride-(^{85}Sr). Radioactive agent. Solid; mp = 868°; soluble in H$_2$O (120 g/100 ml); LD$_{50}$ (mus iv) = 148 mg/kg; [hexahydrate]; crystals; mp = 61°; d = 1.96; loses 5H$_2$O at 100°, loses last H$_2$O at 150°. *Abbott Labs Inc.*

3580   Strontium
7440-24-6   8915   G

Sr

Sr
Strontium.
EINECS 231-133-4; HSDB 2545; Strontium; Strontium, elemental. Metallic element; alloys of strontium used in electron tubes as a 'getter' to combine chemically with active gases and to hold inactive gases by adsorption. Used in fireworks and tracer bullets. mp = 757°; bp = 1366°; d = 2.6; reacts with oxygen. *Atomergic Chemetals; Degussa AG; Noah Chem.*

3581   Strontium carbonate
1633-05-2   8918   G

CO$_3$Sr
Strontium(II) carbonate.
C.I. 77837; Carbonic acid strontium salt (1:1); CCRIS 3203; CI 77837; EINECS 216-643-7; HSDB 5845; NSC 112224; Strontianite;

Strontium carbonate; Strontium carbonate (SrCO$_3$). Catalyst, used in radiation-resistant glass for color TV tubes, ceramic ferrites, pyrotechnics. Also used in sugar refining. White solid; dec 1100°; d = 3.5; soluble in H$_2$O (0.001 g/100 ml). *Atomergic Chemetals; Cerac; Mallinckrodt Inc.; Solvay Deutschland GmbH.*

3582   Strontium chloride
10476-85-4   8920   G

Cl$_2$Sr
Hydrochloric acid strontium salt.
EINECS 233-971-6; Strontium chloride; Strontium chloride (SrCl$_2$). Used in pyrotechnics, electron tubes. Solid; mp = 868°; soluble in H$_2$O (120 g/100 ml); LD$_{50}$ (mus iv) = 148 mg/kg; [hexahydrate]; crystals; mp = 61°; d = 1.96; loses 5H$_2$O at 100°, loses last H$_2$O at 150°. *Atomergic Chemetals; Hoechst Celanese; Sigma-Aldrich Fine Chem.*

3583   Strontium nitrate
10042-76-9   8926   G

N$_2$O$_6$Sr
Strontium(II) nitrate.
EINECS 233-131-9; HSDB 787; Nitrate de strontium; Nitric acid, strontium salt; Strontium dinitrate; Strontium nitrate; Strontium nitrate (Sr(NO$_3$)$_2$); Strontium(II) nitrate (1:2); UN1507. Pyrotechnics, marine signals, railroad flares, matches. Solid; mp = 570°; d = 2.99; soluble in H$_2$O (0.66 g/ml), slightly soluble in EtOH, Me$_2$CO; LD$_{50}$ (rat ip) = 540 mg/kg. *Hoechst Celanese; Noah Chem.; Solvay Deutschland GmbH.*

3584   Strontium oxide
1314-11-0   8928   G

OSr
Strontium(II) oxide.
EINECS 215-219-9; Strontium oxide; Strontium oxide (SrO). Used in the manufacture of strontium salts, pyrotechnics, greases, and soaps. Grey-white solid; mp = 2430°; d = 4.7; reacts with H$_2$O.

3585   Strontium titanate
12060-59-2   G

O$_3$SrTi
Strontium titanium oxide.
EINECS 235-044-1; Strontium titanate; Strontium titanium oxide (SrTiO$_3$); Strontium titanium trioxide; Titanate (TiO$_3$$^{2-}$), strontium (1:1). Used in electronic devices and in electrical insulation. *Atomergic Chemetals; Ferro/ Transelco; Tam Ceramics.*

587

3586 Strotope
10042-76-9   8926   G

$N_2O_6{}^{85}Sr$

Strontium nitrate-^{85Sr}.

EINECS 233-131-9; HSDB 787; Nitrate de strontium; Nitric acid, strontium salt; Strontium dinitrate; Strontium nitrate; Strontium nitrate ($Sr(NO_3)_2$); Strontium(II) nitrate (1:2); UN1507. Radioactive agent. Solid; mp = 570°; d = 2.99; soluble in H_2O (0.66 g/ml), slightly soluble in EtOH, Me_2CO; LD_{50} (rat ip) = 540 mg/kg. *Bristol-Myers Squibb Co.*

3587 Styphnic acid
82-71-3   8943

$C_6H_3N_3O_8$

2,4,6-Trinitro-1,3-benzenediol.

Al3-08927; 1,3-Benzenediol, 2,4,6-trinitro-; CCRIS 3107; 2,4-Dihydroxy-1,3,5-trinitrobenzene; EINECS 201-436-6; NSC 36932; Resorcinol, 2,4,6-trinitro-; Styphnic acid; 2,4,6-Trinitro-1,3-benzenediol; 2,4,6-Trinitroresorcinol; UN0219; UN0394. Used in explosives as a priming agent. Yellow crystals; mp = 175.5°; λ_m = 208, 265, 336, 392 nm (ε = 14700, 12500, 8820, 9780 MeOH); soluble in H_2O (0.6-1.2 g/100 ml), soluble in organic solvents. *Dynamit Nobel GmbH.*

3588 Styracin
122-69-0   2327   G

$C_{18}H_{16}O_2$

3-Phenyl-2-propen-1-yl 3-phenyl propenoate.

Al3-02445; CCRIS 904; Cinnamic acid, cinnamyl ester; Cinnamyl β-phenylacrylate; Cinnamyl alcohol, cinnamate; Cinnamyl cinnamate; Cinnamylester kyseliny skoricove; EINECS 204-566-1; FEMA No. 2298; NSC 46161; Phenylallyl cinnamate; 3-Phenylallyl cinnamate; 3-Phenyl-2-propen-1-yl 3-phenylpropenoate; 3-Phenyl-2-propen-1-yl cinnamate; 3-Phenyl-2-propenyl 3-phenyl-2-propenoate; 2-Propenoic acid, 3-phenyl-, 3-phenyl-2-propenyl ester; Styracin. Used in chemical synthesis. Needles; mp 44°; d^4 = 1.1565; λ_m = 272 nm (MeOH); insoluble in H_2O, soluble in EtOH, $CHCl_3$, very soluble in Et_2O.

3589 Styrenated phenol
61788-44-1   H

Phenol, styrenated.
CCRIS 4915; EINECS 262-975-0; Phenol, styrenated; Styrenated phenol.

3590 Styrene
100-42-5   8944   G H

C_8H_8

Benzene, ethenyl-.

Al3-24374; Benzene, ethenyl-; Benzene, vinyl-; Bulstren K-525-19; CCRIS 564; Cinnamene; EINECS 202-851-5; Ethenylbenzene; Ethylene, phenyl-; FEMA Number 3234; HSDB 171; NCI-C02200; NSC 62785; Phenethylene; Phenylethene; Phenylethylene; Stirolo; Styreen; Styren; Styrene; Styrol; Styrole; Styrolene; Styropol SO; UN2055; Vinyl benzene; Vinylbenzen; Vinylbenzene; Vinylbenzol. Used in the manufacture of polystyrene, SBR, ABS, and SAN resins. Oil; mp = -31°; bp = 145°; d^{20} = 0.9060; λ_m = 246, 272, 281, 289 nm (cyclohexane); insoluble in H_2O, slightly soluble in CCl_4, soluble in EtOH, Et_2O, Me_2CO, MeOH, CS_2; freely soluble in C_6H_6, petroleum ether.

3591 Styrene oxide
96-09-3   H

C_8H_8O

Oxirane, phenyl-.

5-17-01-00577 (Beilstein Handbook Reference); Al3-18151; Benzene, (1,2-epoxyethyl)-; Benzene, (epoxy-ethyl)-; BRN 0108582; CCRIS 1268; EINECS 202-476-7; EP-182; Epoxyethyl benzene; Epoxystyrene; Fenyloxiran; HSDB 2646; NCI-C54977; NSC 637; Oxirane, phenyl-; Phenethylene oxide; Phenyl oxirane; Phenylethylene oxide; Phenyloxirane; Styrene-7,8-oxide; Styrene epoxide; Styrene oxide; Styryl oxide. Liquid; mp = -35.6°; bp = 194.1°; d^{16} = 1.0523; λ_m = 261 nm (cyclohexane); insoluble in H_2O, soluble in EtOH, Et_2O, $CHCl_3$.

3592 Styrene-butadiene copolymer
9003-55-8   G

$$[-CH_2CH(C_6H_5)]_x-(CH_2CH=CHCH_2)_y$$

Poly(styrene-co-butadiene).

8000A; Afcolac B 101; Andrez; BASE 661; Benzene, ethenyl-, polymer with 1,3-butadiene; 1,3-Butadiene-ethenylbenzene copolymer; 1,3-Butadiene, polymer with styrene; 1,3-Butadiene-styrene copolymer; Butadiene-styrene copolymer; Butadiene-styrene resin; Butadiene-styrene rubber; Butadiene-styrol copolymer; Butakon 85-71; Diarex 600; Dienol S; Dow 209; Dow 234; Dow 460; Dow 620; Dow 680; Dow 816; Dow latex 612; Dowtex TL 612; DST 50; DST 75; Duranit; Duranit 40; Edistir RB 268; Ethenylbenzene polymer with 1,3-butadiene; Goodrite 1800X73; Histyrene S 6F; Hycar LX 407; K 55E; K-Resin Polymer KR01; K-Resin Polymer KR03; K-Resin Polymer KR04; K-Resin Polymer KR05; K-Resin Polymer KR10; Kopolymer butadien styrenovy; KR01 K-Resin Polymer; KR04 K-Resin Polymer; Kraton®; Kraton® D 1101; Kraton® D 1116; Kraton® D 2103; KRO 1; KRO 2; Krylene® 606; Krylene® 608; Litex CA; Lytron 5202; Marbon 8000A; Marbon 9200; Nipol 407; Pharos 100.1; Plioflex; Pliolite 151; Pliolite 160; Pliolite

491; Pliolite 55B; Pliolite S 50; Pliolite S 5A; Pliolite S-5B; Pliolite S-5C; Pliolite S 5D; Pliolite S-5E; Pliolite S5; Poly(butadiene-co-styrene); Polybutadiene-polystyrene copolymer; Polyco 2410; Polyco 2415; Polystyrene-polybutadiene; PS-SU2; R-104; Ricon 100; Ricon 109; Ricon 181; Ricon 182; Ricon 183; Ricon A-84; Rubber, butadiene-styrene; S6F histyrene resin; SBS; SBS (block polymer); SBS copolymer; SD 345 (Polymer); SD 354; SKS 85; Soil stabilizer 661; Solprene 300; Solprene 303; Styrene, 1,3-butadiene polymer; Styrene-1,3-butadiene copolymer; Styrene block polymer with 1,3-butadiene; Styrene-butadiene block copolymer; Styrene-butadiene copolymer; Styrene polymer with 1,3-butadiene; Synpol 1500; Thermoplastic 125; TR 201; UP 1E; Vestyron HI. Linear styrene-butadiene-styrene transparent and shatter resistant block copolymer containing at least 70 weight percent polymerized styrene; thermoplastic rubber requiring no vulcanization; for formulating adhesives for the building/construction trade, hot-melt adhesives. A copolymer with clarity, toughness, rigidity in extruded parts, and detail on fast production cycles; for blending with general purpose PS; Used for injection molding and extrusion processing; for crystal clear and warp resistant parts; in housings, blister packs, extruded tubes, molded boxes with integral hinges, toys, and where impact PS, oriented PS sheet, cellulosics and rigid PVC have been used. Used to manufacture thermoplastic rubbers, footwear and adhesives, thermoformed blister packs, disposable containers, and in applications where impact PS, oriented PS sheet, polyesters, cellulosics and rigid PVC have been used. Resin can be tinted or colored in a variety of transparent and opaque shades.and is FDA compliant. Used in solvent-based construction mastic adhesives, for medical/food, sporting goods, and misc. molded items, dust covers, point-of-purchase displays, molded boxes, and containers, lids, office supplies, toys, medical devices and tool handles. Other uses include blister packages, bottles, jars, medical devices and extruded tubes and profiles, medical packaging, shrink wrap, overwrap, skin packaging, produce wrap, windows for envelopes and boxes and twist wrap. White solid; d = 0.965. *Phillips 66; Shell.*

3593   Styrolyl alcohol
93-56-1   8945   G

$C_8H_{10}O_2$
1-Phenyl-1,2-ethanediol.
4-06-00-05939 (Beilstein Handbook Reference); AI3-03789; BRN 1306723; α,β-Dihydroxyethylbenzene; 1,2-Dihydroxyethylbenzene; 1,2-Dihydroxy-1-phenylethane; EINECS 202-258-1; 1,2-Ethanediol, 1-phenyl-; 1,2-Ethanediol, phenyl-; 1-Fenyl-1,2-ethandiol; Fenylglycol; NSC 406601; Phenyl-1,2-ethanediol; Phenyl glycol; Phenyl glycol ether; Phenylethanediol; Phenylethane-1,2-diol; Phenylethylene glycol; 1-Phenylethylene glycol; Styrene glycol; Styrolyl alcohol. Used in perfumery. Its esters used as plasticizers. White needles; mp = 67.5°; bp = 273°; soluble in H_2O, organic solvents.

3594   Suberic acid
505-48-6   8946   G

$C_8H_{14}O_4$
Octane-1,8-dioic acid.
AI3-52672; EINECS 208-010-9; Hexamethylenedicarb-oxylic acid;

1,6-Hexanedicarboxylic acid; NSC 25952; Octanedioic acid; 1,8-Octanedioic acid; Suberic acid. Used in plastics manufacture. Needles or plates; mp = 144°; bp100 = 279°, bp 20 = 219°; insoluble in H_2O (0.16 g/100 ml), slightly soluble in DMSO, freely soluble in Et_2O, C_6H_6. *BASF Corp.; Degussa-Hüls Corp.; Penta Mfg.*

3595   Succinic acid
110-15-6   8953   G H

$C_4H_6O_4$
1,4-Butanedioic acid.
4-02-00-01908 (Beilstein Handbook Reference); AI3-06297; Amber acid; Asuccin; Bernsteinsaure; BRN 1754069; Butanedioic acid; 1,4-Butanedioic acid; Dihydrofumaric acid; EINECS 203-740-4; 1,2-Ethane-dicarboxylic acid; Ethylene dicarboxylic acid; Ethylene-succinic acid; HSDB 791; Katasuccin; Kyselina jantarova; NSC 106449; Sal succini; Salt of amber; Succinic acid; Wormwood; Wormwood acid. Used in organic synthesis; manufacture of lacquers, dyes, esters for perfumes, photography, in foods as a sequestrant, buffer, neut-ralizing agent. Triclinic or monoclinic prisms; mp = 188°; bp = 235° (dec); d^{25} = 1.572; λ_m = 222 nm (ε = 4, 0.01N NaOH); insoluble in C_6H_6, C_7H_8; slightly soluble in H_2O, DMSO, soluble in EtOH, MeOH, Et_2O, Me_2CO. *Am. Biorganics; Degussa-Hüls Corp.; DuPont; General Chem; Mallinckrodt Inc.; Pentagon Chems. Ltd.*

3596   Succinic acid, sulfo-, 1,4-bis(1,3-dimethyl-butyl) ester, sodium salt
2373-38-8   H

$C_{16}H_{29}NaO_7S$
Butanedioic acid, sulfo-, 1,4-bis(1,3-dimethylbutyl) ester, sodium salt.
Butanedioic acid, sulfo-, 1,4-bis(1,3-dimethylbutyl) ester, sodium salt; EINECS 219-147-9; Sodium 1,4-bis(1,3-dimethylbutyl) sulphonatosuccinate; Succinic acid, sulfo-, 1,4-bis(1,3-dimethylbutyl) ester, sodium salt.

3597   Succinic anhydride
108-30-5   8954   G

$C_4H_6O_3$
Dihydro-2,5-furandione.
5-17-11-00006 (Beilstein Handbook Reference); AI3-52664; Bernsteinsäure-anhydrid; BRN 0108441; Butanedioic anhydride; CCRIS 2386; Dihydro-2,5-furandione; 2,5-Diketotetrahydrofuran; EINECS 203-570-0; 2,5-Furandione, dihydro-; HSDB 792; NCI-C55696; NSC 8518; Rikacid SA; Succinic acid anhydride; Succinic

anhydride; Succinyl oxide; Tetrahydro-2,5-dioxofuran; Tetrahydro-2,5-furandione. Manufacture of chemicals, pharmaceuticals, esters; hardner for resins, starch modifier in foods. Needles; mp = 119°; bp = 261°; d^{20} = 1.2; insoluble in H_2O, slightly soluble in Et_2O, soluble in EtOH, $CHCl_3$. *Degussa-Hüls Corp.; Humphrey; Lancaster Synthesis Co.; Penta Mfg.*

3598   Succinimide
123-56-8   8955   G

$C_4H_5NO_2$
2,5-Diketopyrrolidine.
5-21-09-00438 (Beilstein Handbook Reference); AI3-08539; BRN 0108440; Butanimide; Dihydro-3-pyrroline-2,5-dione; 3,4-Dihydropyrrole-2,5-dione; 2,5-Diketo-pyrrolidine; 2,5-Dioxopyrrolidine; EINECS 204-635-6; Lubrizol® 2153; NSC 11204; 2,5-Pyrrolidinedione; Succinic acid imide; Succinic imide; Succinimide; Succinimide-sauba. Pigment dispersant and wetting agent for color concs., inks, plastisols and organosols. Growth stimulants for plants, organic synthesis. Plates; mp = 126.5°; bp = 287° (dec); d^{25} = 1.418; λ_m = 276 nm (MeOH); soluble in H_2O (0.3-1.4 g/ml), slightly soluble in EtOH, Et_2O, Me_2CO; LD_{50} (rat orl) = 14 g/kg. *Chemie Linz N. Am.; Chemie Linz UK.*

3599   Sucrose
57-50-1   8966   D H

$C_{12}H_{22}O_{11}$
β-D-Fructofuranosyl-α-D-glucopyranoside.
AI3-09085; Amerfand; Amerfond; Beet sugar; Cane sugar; CCRIS 2120; Confectioner's sugar; EINECS 200-334-9; Fructofuranoside, α-D-glucopyranosyl, β-D; β-D-Fructo-furanosyl-α-D-glucopyranoside; Glucopyranoside, β-D-fructofuranosyl, α-D; α-D-Glucopyranoside, β-D-fructo-furanosyl-; α-D-Glucopyranosyl β-D-fructofuranoside; (α-D-Glucosido)-β-D-fructofuranoside; Granulated sugar; HSDB 500; Microse; NCI-C56597; NSC 406942; Rock candy; Rohrzucker; Saccharose; Saccharum; Sucraloxum; Sucrose; D-Sucrose; Sucrose, dust; Sucrose, pure; Sugar; Table sugar; White sugar. Disaccharide; sweetening agent in food, pharmaceuticals; in fermentation; as flavoring agent, preservative, antioxidant (in form of invert sugar); granulation agent and excipient for tablets; in plastics and cellulose industry, rigid polyurethane foams and manufacture of ink. White crystals; dec 160-186°; d_4^{25} = 1.587; $[\alpha]_D^{25}$ = 66.5°; soluble in H_2O (2 g/ml), less soluble in organic solvents. *Am. Biorganics; Mallinckrodt Inc.; Mendell Edward; Pfanstiehl Labs Inc.*

3600   Sucrose acetate isobutyrate
126-13-6
                                                                        H

$C_{40}H_{62}O_{19}$
6-O-Acetyl-1,3,4-tris-O-(2-methyl-1-oxopropyl)-β-D-fructofuranosyl, 6-acetate 2,3,4-tris(2-methylpropanoate).
AI3-25354; EINECS 204-771-6; HSDB 5657; Isobutyric acid, hexaester with sucrose diacetate; Saccharose acetate isobutyrate; SAIB; SAIB 100S; Sucrose acetate isobutyrate; Sucrose acetoisobutyrate; Sucrose di(acetate) hexaisobutyrate.

3601   Sucrose octa-acetate
126-14-7   8967   G

$C_{28}H_{38}O_{19}$
D-(+)-sucrose octa-acetate.
5-17-08-00410 (Beilstein Handbook Reference); AI3-00071; BRN 0079290; EINECS 204-772-1; FEMA No. 3038; Fructofuranosyl, tetraacetate; α-D-Glucopyran-oside, 1,3,4,6-tetra-O-acetyl-β-D-; NSC 1695; Octa-acetylsucrose; Octa-O-acetylsucrose; 2,3,4,6,1',3',4',6'-Octa-O-acetylsucrose; Sucrose octaacetate; 1,3,4,6-Tetra-O-acetyl-β-D-fructofuranosyl-α-D-glucopyranoside.
Used as an adhesive and for impregnating and insulating paper. Needles; mp = 86.5°; bp_1 = 250°; d^{16} = 1.27; $[\alpha]_D^{20}$ = 59.6° (c = 1 $CHCl_3$); slightly soluble in H_2O (0.9 g/100 ml), soluble in EtOH, Et_2O, Me_2CO, C_6H_6, $CHCl_3$.

3602   Sudan I
842-07-9
                                                                        G

$C_{16}H_{12}N_2O$
1-phenylazo-2-naphthol.
4-16-00-00228 (Beilstein Handbook Reference); Atul Orange R; Benzene-1-azo-2-naphthol; Benzeneazo-β-naphthol; Brasilazina Oil Orange; Brilliant Oil Orange R; BRN 0651992; C. I. Solvent Yellow 14; C.I. 12055; C.I. Disperse Yellow 97; Calco Oil Orange 7078-Y; Calco Oil Orange 7078; Calco Oil Orange Z-7078; Calcogas M;

Calcogas Orange NC; Campbelline Oil Orange; Carminaph; CCRIS 174; Ceres Orange R; Cerotinorange G; CI 12055; Disperse Yellow 97; Dispersol Yellow PP; Dunkelgelb; EINECS 212-668-2; Enial Orange I; Fast Oil Orange; Fast Oil Orange I; Fast Orange; Fat Orange 4A; Fat Orange G; Fat Orange I; Fat Orange R; Fat Orange RS; Fat Soluble Orange; Fettorange 4a; Fettorange 4A; Fettorange IG; Fettorange Ig; Fettorange R; Grasal Orange; Grasan Orange R; Hidaco Oil Orange; HSDB 4132; 2-Hydroxy-1-phenylazonaphthalene; 2-Hydroxy-naphthyl-1-azobenzene; Lacquer Orange VG; Morton Orange Y; Motiorange R; 2-Naphthalenol, 1-(phenylazo)-; 2-Naphthol, 1-(phenylazo)-; 2-Naphtholazobenzene; NCI-C53929; NSC 11227; NSC 51524; Oil Orange; Oil Orange 2B; Oil Orange 31; Oil Orange 2311; Oil Orange 7078-V; Oil Orange E; Oil Orange EP; Oil Orange PEL; Oil Orange PS; Oil Orange R; Oil Orange R-14; Oil Orange Z-7078; Oil Soluble Orange; Oleal Orange R; orange 3RA soluble in grease; orange a l'huile; orange r fat soluble; orange resenole no. 3; orange soluble a l'huile; Orange 2 insoluble; Orange 3RA soluble in grease; Orange á l'huile; Orange Insoluble OLG; Orange pel; Orange R Fat Soluble; Orange Resenole 3; Orange Resenole No. 3; Orange soluble á l'huile; Organol Orange; Orient Oil Orange PS; Petrol Orange Y; α-Phenylazo-β-naphthol; 1-(Phenylazo)-2-naphthalenol; 1-Benzoazo-2-naphthol; 1-Phenylazo-β-naphthol; 1-Phenylazo-2-naphthol; Plastoresin Orange F4A; Pyronalorange; Resinol Orange R; Resoform Orange G; Sansel Orange G; Scharlach B; Silotras Orange TR; Solvent Yellow 14; Somalia Orange I; Soudan I; Spirit Orange; Spirit Yellow I; Stearix Orange; Sudan 1; Sudan I; Sudan J; Sudan Orange R; Sudan Orange RA New; Sudan Orange RA; Sudan Yellow; Tertrogras Orange SV; Toyo Oil Orange; Waxakol Orange GL; Waxoline Yellow I; Waxoline Yellow IM; Waxoline Yellow IP; Waxoline Yellow IS; Zlut rozpoustedlova 14. A dyestuff, used for coloring oils and varnishes. Crystals; mp = 131-133°; λ_m = 230, 280, 304 nm (ε = 38905, 6607, 6918, cycl;ohexane), 229, 256, 280, 313, 408, 421 nm (ε = 35481, 11220, 5129, 7079, 15488, 10715, EtOH); insoluble in H_2O.

3603   Sudan III
85-86-9   8969   G

$C_{22}H_{16}N_4O$

1-[4-(phenylazo)phenylazo]-2-naphthol.
111440 Red; 4-16-00-00248 (Beilstein Handbook Reference); AI3-02854; Atul Oil Red G; Benzeneazo-benzeneazo-β-naphthol; Brasilazina Oil Scarlet; BRN 0931185; C.I. 23; C.I. 26100; C.I. Solvent Red 23; CCRIS 7074; Cerasin Red; Cerasinrot; Cerotinscharlach R; Certiqual Oil Red; Cerven rozpoustedlova 23; CI 26100; D & C Red No. 17; EINECS 201-638-4; Fast Oil Scarlet III; Fat Red (bluish); Fat Red HRR; Fat Red R; Fat Red RS; Fat Scarlet LB; Fat Soluble Red ZH; FD and C Red No. 17; Fettponceau G; Fettrot; Fettscharlach; Fettscharlach LB; Grasal Brilliant Red G; Grasan Brilliant Red G; Japan Red 225; Motirot 2R; 2-Naphthalenol, 1-((4-(phenylazo)-phenyl)azo)-; 2-Naphthol, 1-((p-(phenylazo)phenyl)azo)-; NSC 65825; Oil Red; Oil Red 3G; Oil Red 6566; Oil Red AS; Oil Red G; Oil Scarlet; Oil Scarlet AS; Oil Scarlet G; Organol Red BS; Organol Scarlet; 1-(4-(Phenylazo)-phenylazo)-2-naphthol; 1-(Phenylazophenylazo)-2-hydroxynaphthalene; Ponceau, insoluble, OLG; Pyro-nalrot B; Red No. 225; Red ZH; Rouge cerasine; Scarlet B Fat Soluble; Schultz No. 31; Silotras Scarlet TB; Solvent red 23; Somalia Red III; Soudan III; Stearix Scarlet; Sudan 3; Sudan G III; Sudan III; Sudan III (G); Sudan P III; Sudan Red III; Tetrazobenzene-β-naphthol; Toney Red. Used for coloring oils and varnishes. Approved by FDA for external use.

Brown leaflets with a green lustre; mp = 195°; λ_m = 345, 505 nm (ε = 16596, 30903, cellosolve); insoluble in H_2O, soluble in organic solvents. *Medical Chemical Corp.*

3604   Sudan IV
85-83-6   8469   G

$C_{24}H_{20}N_4O$

1-[[2-methyl-4-[(2-methylphenyl)azo]phenyl]azo]-2-naphthalenol.
4-16-00-00249 (Beilstein Handbook Reference); AI3-02855; Biebrich Scarlet BPC; Biebrich Scarlet R medicinal; Biebrich Scarlet Red; Brasilazina Oil Red B; BRN 0709018; C.I. 258; C.I. 26105; C.I. Solvent Red 24; Calco Oil Red D; Candle Scarlet 2B; Candle Scarlet B; Candle Scarlet G; CCRIS 2430; Ceres Red BB; Cerotine Ponceau 3B; Cerven rozpoustedlova 24; 2',3-Dimethyl-4-(2-hydroxynaphthylazo)azobenzene; Dispersol Red PP; EINECS 201-635-8; Enial Red IV; Fast Oil Red B; Fast Red BB; Fat Ponceau R; Fat Red 2B; Fat Red B; Fat Red BB; Fat Red BS; Fat Red TS; Fat Soluble Dark Red; Grasal Brilliant Red B; Grasan Brilliant Red B; Hidaco Oil Red; Lacquer Red V; Lacquer Red VS; Lipid Crimson; 1-((2-Methyl-4-((2-methylphenyl)azo)phenyl)-azo)-2-naphthalenol; 1-(2-Methyl-4-(2-methylphenylazo)-phenylazo)-2-naphthol; 2-Naphthalenol, 1-((2-methyl-4-((2-methylphenyl)azo)-phenyl)azo)-; 2-Naphthol, 1-((4-(o-tolylazo)-o-tolyl)azo)-; NSC 10472; Oil Red D; Oil Red APT; Oil Red TAX; Oil Red S; Oil Red RR; Oil Red RC; Oil Red PEL; Oil Red A; Oil Red IV; Oil Red BB; Oil Red BS; Oil Red ZD; Oil Red ED; Oil Red F; Oil Red GO; Oil Red 2B; Oil Red 3; Oil Red 3B; Oil Red 4B; Oil Red 7; Oil Red 47; Oil Red 282; Oil Scarlet; Oil Scarlet 48; Oil Soluble Dark Red; Oleal Red BB; Organol Red B; Orient Oil Red RR; Phenoplaste Organol Red B; Plastoresin Red F; Red 3R soluble IN grease; Resinol Red 2B; Resoform Red G; Rubrum scarlatinum; Sarlach R; Scarlet oil; Scarlet R (michaelis); Scarlet red; Scarlet red, biebrich; Scarlet Red; Scharlachrot; Schultz No. 541; Silotras Red T3B; Solvent red 24; Somalia Red IV; Stearix Red 4B; Stearix Red 4S; Sudan IV; Sudan P; Sudan Red 4BA; Sudan Red BB; Sudan Red BBA; Sudan Red IV; Tertrogras Red N; o-Tolueneazo-o-tolueneazo-β-naphthol; o-Tolueneazo-o-toluene-β-naphthol; o-Tolyl-azo-o-tolylazo-β-naphthol; o-Tolylazo-o-tolylazo-2-naph-thol; α-Tolylazo-o-tolylazo-β-naphthol; 1-(4-o-Tolylazo-o-tolylazo)-2-naphthol; Toyo Oil Red BB; Waxakol Red BL; Waxoline Red O; Waxoline Red OM; Waxoline Red OS. Used as a dyestuff to stain fats. Dark brown powder; mp = 186° (dec); bp = 260° (dec); insoluble in H_2O, soluble in organic solvents. *Medical Chemical Corp.*

3605   Sulbenox
58095-31-1   8976   G D V

$C_9H_{10}N_2O_2S$

(4,5,6,7-Tetrahydro-7-oxobenzo[b]thien-4-yl)urea.
BRN 1643663; CL 206576; Sulbenox; Sulbenoxum; (4,5,6,7-Tetrahydro-7-oxobenzo(b)thien-4-yl)urea; Urea, (4,5,6,7-tetrahydro-7-oxobenzo(b)thien-4-yl)-; Vigazoo. Veterinary growth stimulant.

Crystals; mp = 245-246°; LD$_{50}$ (rat orl) > 5000 mg/kg. *Am. Cyanamid.*

3606   Sulconazole
61318-90-9   8978   G D

C$_{18}$H$_{15}$Cl$_3$N$_2$S
(±)-1-[2,4-Dichloro-β-[(p-chlorobenzyl)yhio]phenethyl]-imidazole.
(-)-1-(2,4-Dichlor-β-((4-chlorbenzyl)thio)phenethyl)-imidazol;   (-)-1-
(2,4-Dichloro-β-((p-chlorobenzyl)thio)-phenethyl)imidazole;
Exelderm; (±)-1H-Imidazole, 1-(2-(((4-chlorophenyl)methyl)thio)-2-
(2,4-dichlorophenyl)-ethyl)-;   Sulconazol;   Sulconazole;
Sulconazolum. Topical antifungal agent. *C.M. Ind.; ICI Chem. & Polymers Ltd.; Syntex Labs. Inc.*

3607   Sulfalene
152-47-6   8995   D G

C$_{11}$H$_{12}$N$_4$O$_3$S
3-Methoxy-2-sulfapyrazine.
5-25-12-00574   (Beilstein   Handbook   Reference);   2-(p-
Aminobenzenesulfonamido)-3-methoxypyrazine;   4-Amino-N-(3-
methoxypyrazinyl)benzenesulfonamide;   AS   18908;
Benzenesulfonamide,   4-amino-N-(3-methoxy-pyrazinyl)-;   BRN
0622512; Dalysep; EINECS 205-804-7; F.I. 5978; Farmitalia
204/122; Kelfizin; Kelfizina; Kelfizine; Kelfizine W; Longum; N^1-(3-
Methoxy-2-pyrazinyl)sulfanilamide;   N^1-(3-Methoxypyrazinyl)-sulf-
anilamide; 2-Methoxy-3-sulfanilamidopyrazine; 3-Meth-oxypyrazine
sulfanilamide; NSC-110433; Policydal; Polycidal; Pyrazine, 2-
sulfanilamido-3-methoxy-; SMP; SMP2; Solfametopirazina; Sulfalen;
Sulfalene;   Sulfaleno;   Sulfalenum;   Sulfamethopyrazine;
Sulfamethoxypyrazine; Sulfametopyrazine; Sulfanilamide, N^1-(3-
methoxy-2-pyrazinyl)-; Sulfanilamide, N'-(3-methoxypyrazinyl)-; 2-
Sulfanilamide   3-methoxypyrazine;   2-Sulfanilamido-3-
methoxypyrazine;   Sulfapyrazinemethoxyine;   Sulfapyr-
azinemethoxine; Sulfapyrazinemethoxyne; Vetkelfizina; WR 4629. A
sulfonamide antibiotic. Crystals; mp = 176°; LD$_{50}$ (mus orl) = 2164
mg/kg, (mus ip) = 1410 mg/kg. *Abbott Labs Inc.; Farmitalia Societa Farmaceutici.*

3608   Sulfamethazine
57-68-1   9000   D

C$_{12}$H$_{14}$N$_4$O$_2$S
N^1-(4,6-Dimethyl-2-pyrimidinyl)sulfanilamide.
5-25-10-00250 (Beilstein Handbook Reference); A-502; AI3-26817;
Azolmetazin; BN 2409; BRN 0261304; Calfspan Tablets; CCRIS
3701; Cremomethazine; Diazil; DiazilSulfadine; Diazyl; Dimezathine;
Dimidin-R; S-Dimidine; EINECS 200-346-4; Hava-Span; HSDB
4157; Intradine; Kelametazine; Mermeth; Metazin; NCI-C56600;
Neasina; Neazina; NSC 67457; Pirmazin; SA III; Solfadimidina;
Spanbolet; Sulfa-Isodimerazine; Sulfa-dimerazine; Sulfadimesin;
Sulfadimesine;   Sulfadimethyl-diazine;   Sulfadimethylpyrimidine;
Sulfadimezin;   Sulfa-dimezine;   Sulfadimezinum;   Sulfadimidin;
Sulfadimidina;   Sulfadimidine;   Sulfadimidinum;   Sulfadine;
Sulfametazina; Sulfametazyny; Sulfamethazine; Sulfamethiazine;
Sulfa-mezathine;   SulfaSURE   SR   Bolus;   Sulfodimesin;
Sulfodimezine; Sulka K Boluses; Sulka S Boluses; Sulmet;
Sulphadimethylpyrimidine; Sulphadimidine; Sulpha-methasine;
Sulphamethazine;   Sulphamezathine;   Sulpha-midine;
Sulphodimezine; Superseptil; Superseptyl; Verto-lan. Sulfonamide
antibiotic. Used as a bacteriostatic agent. Crystals; mp = 170-176°,
178-179°, 198-199°, 205-207°; λ$_m$ 241 nm (E$^{1\%}_{1cm}$ 670 H$_2$O, pH 6.6),
243 257 nm (E$^{1\%}_{1cm}$ 765 776 0.01N NaOH), 241 297 nm (E$^{1\%}_{1cm}$ 561
266 0.01N HCl); soluble in H$_2$O (0.15 g/100 ml at 29°, 0.192 g/100
ml at 37°, pH 7.00); LD$_{50}$ (mus ip) = 1060 mg/kg. *Fermenta Animal Health Co.; ICI; Inst. Chemioter.; Merck & Co.Inc.; Solvay Animal Health Inc.*

3609   Sulfamic acid
5329-14-6   9006   G P

H$_3$NO$_3$S
Amidosulfonic acid.
AI3-15024; Amidosulfonic acid; Amidosulfuric acid; Aminosulfonic
acid; Aminosulfuric acid; Caswell No. 809; EINECS 226-218-8; EPA
Pesticide Chemical Code 078101; HSDB 795; Imidosulfonic acid;
Jumbo; Kyselina amidosulfonova; Kyselina sulfaminova; NSC 1871;
Sulfamic acid; Sulfamidic acid; Sulfaminic acid; Sulphamic acid;
Sulphamidic acid; UN2967. Used in metal/ceramic cleaning, nitrite
removal in azo-dyeing, gas-liberating compositions, organic
synthesis, analytical standard, chlorine stabilizer, bleaching paper
pulp and textiles, catalyst for ureaformaldehyde resins, sulfonating
agent, used for pH control and as a weedkiller. Registered by EPA
as an antimicrobial and disinfectant (cancelled). Orthorhombic
crystals; mp = 205° (dec); d = 2.15; soluble in H$_2$O (15.3 g/100 ml at
0°, 50 g/100 ml at 80°), sparingly soluble in EtOH, MeOH, slightly
soluble in Me$_2$CO, insoluble in Et$_2$O; pH of 1% aqueous solution =
1.18; MLD (rat orl) = 1600 mg/kg. *General Chem; Lancaster Synthesis Co.; PMC; Sigma-Aldrich Co.; Transol Chem. UK Ltd.*

3610   Sulfanilic acid
121-57-3   9012   D G H

C$_6$H$_7$NO$_3$S
p-Aminobenzenesulfonic acid.
AI3-15414; Aniline-4-sulfonic acid; Aniline-p-sulfonic acid; Aniline-p-
sulphonic acid; Benzenesulfonic acid, 4-amino-; CCRIS 4576;
EINECS 204-482-5; HSDB 5590; Kyselina sulfanilova; NSC 7170; p-
Aminobenzenesulfonic acid; Sulfanilic acid; Sulfanilsäure;
Sulphanilic acid. Used in dyestuffs, organic synthesis, medicine,

reagent. Medically as a sulfone antibiotic. Crystals; mp = 288°, d^{25} = 1.485; λ_m = 252 nm (ε = 3060, MeOH); slightly soluble in H_2O (1.0 g/100 ml at 20°, 1.45 g/100 ml at 30°, 1.94 g/100 ml at 40°); insoluble in EtOH, C_6H_6, Et_2O; slightly soluble in hot MeOH. *3V; Am. Cyanamid; Penta Mfg.; U.S. BioChem.*

3611　Sulfaquinoxaline

59-40-5　　　　　　　　9025　　　　　　　D G

$C_{14}H_{12}N_4O_2S$

N^1-2-Quinoxalinylsulfanilamide.
5-25-11-00125 (Beilstein Handbook Reference); AI3-17254; Anti-K; Avicocid; BRN 0290026; Caswell No. 721; Compound 3-120; EINECS 200-423-2; EPA Pesticide Chemical Code 077901; Italquina; Kokozigal; NSC 41805; SQ 40; SQX; Sulfa-Q 20; Sulfabenzpyrazine; Sulfacox; Sulfaline; Sulfaquinoxalina; Sulfaquinoxaline; Sulfaquinoxalinum; Sulphaquinoxaline; Sulquin; Sulquin 6-50 Concentrate; Ursokoxaline.; component of: Sulquin 6-50 Concentrate. Sulfonamide antibiotic. Crystals; mp = 247-248°; λ_m = 252, 360 nm ($E^{1\%}_{cm}$ 1110, 275 H_2O pH 6.6), soluble in H_2O (0.00075 g/100 ml pH 7), 95% EtOH (0.073 g/100 ml), Me_2CO (0.43 g/100 ml), soluble in aq. NaOH, Na_2CO_3 solutions. *Solvay Animal Health Inc.*

3612　Sulfasan R

103-34-4　　　　　　　3408　　　　　　　G H

$C_8H_{16}N_2O_2S_2$

Di(morpholin-4-yl) disulphide.
4-27-00-00613 (Beilstein Handbook Reference); Accel R; AI3-08625; Akrochem® Accelerator R; Bismorpholino disulfide; BRN 0126214; Deovulc M; Di(morpholin-4-yl) disulphide; Dimorpholine disulfide; Dimorpholine N,N'-disulfide; Dimorpholino disulfide; Disulfide, dimorphol-ino-; Dithiobismorpholine; 4,4'-Dithiobismorpholine; 4,4'-Dithiodimorpholine; EINECS 203-103-0; HSDB 5351; Morpholine disulfide; Naugex SD-1; NSC 65239; Sanfel R; Sulfasan; Sulfasan R; USAF B-17; USAF EK-T-6645; Vanax® A; Vulnoc. A sulfur donor used as a partial or total replacement of sulfur for resistance to heat and ageing in NR, SBR, NBR, and EPDM. Functions as primary accelerator for NR, IR, SBR, NBR, IIR elastomers, and as primary and secondary accelerator in EPDM. Accelerator for uses where a nonblooming or nonstaining sulfur donor is required; used for EV or semi-EV compounds, in synthetic and natural rubbers. Vulcanizing agent for natural and synthetic rubbers. Has also been used as a fungicide. Crystals; mp = 124-125°; d^{25} = 1.36; λ_m = 251 nm (ε = 5129, cyclohexane); soluble in Me_2CO, C_6H_6, EtOAc.

3613　Sulfated fish oil, sodium salt

61788-64-5　　　　　　　　　　　　　　　　H

Fish oil, sulfated, sodium salt.

EINECS 262-992-3; Fats and Glyceridic oils, fish, sulfated, sodium salts; Fish oil, sulfated, sodium salt; Mixed fish oils, sulfated, sodium salt; Oils, fish, sulfated, sodium salts; Sulfated fish oil, sodium salt.

3614　Sulfated tall oil, sodium salt

61790-35-0　　　　　　　　　　　　　　　　H

Tall oil, sulfated, sodium salt.
EINECS 263-127-2; Tall oil sulfonate, sodium salt; Sulfated tall oil, sodium salt.

3615　Sulfazamet

852-19-7　　　　　　　　9033　　　　　　　G V

$C_{16}H_{16}N_4O_2S$

4-Amino-N-(3-methyl-1-phenyl-1H-pyrazol-5-yl)benz-enesulfonamide.
Ba 18605; Benzenesulfonamide, 4-amino-N-(3-methyl-1H-pyrazol-5-yl)-; EINECS 212-707-3; N1-(3-Methyl-1-phenylpyrazol-5-yl)sulfanilamide; 1-Phenyl-3-methyl-5-sulfanilamidopyrazole; Solfapirazolo; Sulfapira-zol; Sulfapyrazole; Sulfapyrazolum; Sulphapyrazole; Ves-ulong veterinary. Antibacterial agent used in veterinary medicine. Crystals; mp = 195°, 181-182°. *CIBA plc.*

3616　Sulfolane

126-33-0　　　　　　　　9044　　　　　　　H

$C_4H_8O_2S$

2,3,4,5-Tetrahydrothiophene-1,1-dioxide.
5-17-01-00039 (Beilstein Handbook Reference); AI3-09541; Bondelane A; Bondolane A; BRN 0107765; CCRIS 2310; Cyclic tetramethylene sulfone; Cyclo-tetramethylene sulfone; Dihydrobutadiene sulfone; Dihydrobutadiene sulphone; Dioxothiolan; EINECS 204-783-1; HSDB 122; NSC 46443; Sulfalone; Sulfolan; Sulfolane; Sulpholane; Sulphoxaline; Tetrahydrothiofen-1,1-dioxid; Tetrahydrothiophene dioxide; Tetrahydro-thiophene 1,1-dioxide; Tetramethylene sulfone; Thia-cyclopentane dioxide; Thiocyclopentane-1,1-dioxide; Thiolane-1,1-dioxide; Thiophan sulfone; Thiophane dioxide; Thiophene, tetrahydro-, 1,1-dioxide. Solid; mp = 27,6°; bp = 287.3°; d^{18} = 1,2723; soluble in $CHCl_3$.

3617　3-Sulfolene

77-79-2　　　　　　　　9045　　　　　　　H

$C_4H_6O_2S$

2,5-Dihydrothiophene 1,1-dioxide.
5-17-01-00177 (Beilstein Handbook Reference); AI3-23457; BRN 0107004; Butadiene sulfone; CCRIS 569; 2,5-Dihydrothiophene 1,1-dioxide; 2,5-Dihydrothio-phene dioxide; 2,5-Dihydrothiophene sulfone; EINECS 201-059-7; HSDB 2903; NCI-C04557; NSC 48532; β-Sulfolene; 1-Thia-3-cyclopentene 1,1-dioxide; Thio-phene, 2,5-

dihydro-, 1,1-dioxide. Solid; mp = 64.5°; soluble in H_2O, $CHCl_3$. *Dow Chem. U.S.A.; Shell UK; Shell.*

3618   Sulfonic acids, petroleum, barium salts
61790-48-5   H

Sulfonic acids, petroleum, barium salts.
EINECS 263-140-3; Sulfonic acids, petroleum, barium salts; Sulfonic acids, petroleum, barium salts, overbased.

3619   4,4'-Sulfonyldiphenol
80-09-1   H

$C_{12}H_{10}O_4S$
Bis(p-hydroxyphenyl) sulfone.
4-06-00-05809 (Beilstein Handbook Reference); Al3-08667; Bis(4-hydroxyphenyl) sulfone; Bis(p-hydroxy-phenyl)sulfone; Bisphenol S; 4,4'-Bisphenol S; BPS 1; BRN 2052954; CCRIS 2647; 4,4'-Dihydroxydiphenyl sulfone; Diphone C; EINECS 201-250-5; 4-Hydroxy-phenyl sulfone; NSC 8712; Phenol, 4,4'-sulfonylbis-; Phenol, 4,4'-sulfonyldi-; p,p'-Dihydroxydiphenyl sulfone; 4,4'-Sulfonylbisphenol; 4,4'-Sulfonyldiphenol; 4,4'-Sulphonyldiphenol. Used in manufacture of fire retardant chemicals, polyether sulfone resins, polycarbonate resins, polyester resins, epoxy resins, formaldehyde condens-ation products, speciality resins. FDA GRAS. Needles; mp = 240.5°; d^{15} = 1.3663; insoluble in H_2O, soluble in EtOH, Et_2O, C_6H_6. *Hindustan Monomers; Zhangjiagang Gangda Chemical Company Ltd.*

3620   4-Sulfophthalic acid
89-08-7   H

$C_8H_6O_7S$
1,2-Benzenedicarboxylic acid, 4-sulfo-.
EINECS 201-881-6; HSDB 5264; NSC 100615; Phthalic acid, 4-sulfo-; 4-Sulfophthalic acid; 4-Sulphophthalic acid.

3621   Sulfur
7704-34-9   9059   D G P

S_8
Sulfur, pharmaceutical.
Acnomel; Agri-Sul; AN-Sulfur Colloid Kit; Aquilite; Asulfa-Supra; Atomic sulfur; Bensulfoid; Biosulphur powder; Brimstone; Caswell No. 812; Colloidal-S; Colloidal sulfur; Collokit; Colsul; Corosul D and S; Cosan; Cosan 80; Crystex; Devisulphur; EINECS 231-722-6; Elosal; EPA Pesticide Chemical Code 077501; Flour sulfur; Flour sulphur; Flowers of sulfur; Flowers of sulphur; Fostril; Gofrativ; Ground vocle sulfur; Ground vocle sulphur; Hexasul; HSDB 5166; Kolloidschwefel 95; Kolo 100; Kolofog; Kolospray; Kristex; Kumulus®; Kumulus® FL; Liquamat; Magnetic 6; Magnetic 70; Magnetic 90; Magnetic 95; Micowetsulf; Microflotox; Microthiol; NA1350; NA2448; Netzschwefel; NSC 403664; Octocure 456;

Pernox; Polsulkol extra; Precipitated sulfur; Rc-schwefel extra; Rezamid; Roll sulfur; Salicylic acid & sulfur soap; Sastid; Schwefel, feinverteilter; Sebulex; Shreesul; Sofril; Solfa; Soufre; Sperlox-S; Spersul; Spersul thiovit; Sublimed sulfur; Sublimed sulphur; Suffa; Sufran; Sufran D; Sulfex; Sulfidal; Sulforcin; Sulforon; Sulfospor; Sulfoxyl; Sulfur; Sulfur, monoclinic; Sulfur ointment; Sulfur, pharmaceutical; Sulfur, precipitated; Sulfur, rhombic; Sulfur Soap; Sulfur, solid; Sulfur, sublimed; Sulfur vapor; Sulikol; Sulkol; Sulphur; Sulsol; Sultaf; Super cosan; Super Six; Svovl; TechneColl; Tesuloid; Thiolux; Thion; Thiovit; Thiovit S; Thiozol; Transact; Ultra Sulfur; UN1350; UN2448; Vassgro Flowable Sulphur; Wettasul; Zolvis. Scabicide; ectoparasiticide. Used for control of diseases and spider mites in fruit, vines, vegetable, ornamentals, and agricultural crops. Registered by EPA as a fungicide. Rubber accelerator; also used for vulcanization processes in aqueous latex compounds. Required for sulfuric acid manufacture, petroleum refining, dyes and chemicals, fungicide, insecticides, explosives, detergents. Used medicinally as a scabicide and ectoparasiticide, incorporated into products for impure skin, oily hair and dandruff. Yellow powder; mp = 115.21°; bp = 444°; d = 2.06; insoluble in H_2O, sparingly soluble in alcohol and ether; non-toxic to humans and animals. *Agriliance LLC; BASF Corp.; Bayer Corp.; Cleary W. A.; Dermik Labs. Inc.; Drexel Chemical Co.; ECR Pharm.; Galderma Labs Inc.; Generic; Helena Chemical Co.; Henkel/Cospha; ICI Chem. & Polymers Ltd.; L W Vass (Agricultural) Ltd; Menley & James Labs Inc.; Micro-Flo Co. LLC; Norsk Hydro AS; Pan Britannica Industries Ltd.; Richter; Shell; Stiefel Labs Inc.; Texaco; Tiarco; Westwood-Squibb Pharm. Inc.; Wilbur Ellis Co.*

3622   Sulfur dioxide
7446-09-5   9061   G P

SO_2
Sulfurous oxide.
Caswell No. 813; EINECS 231-195-2; EPA Pesticide Chemical Code 077601; FEMA No. 3039; Fermenicide liquid; Fermenicide powder; HSDB 228; Schwefeldioxid; Schwefeldioxyd; Siarki dwutlenek; Sulfur dioxide; Sulfur dioxide (SO_2); Sulfur superoxide; Sulfurous acid anhyd-ride; Sulfurous anhydride; Sulfurous oxide; Sulphur dioxide; Surfur dioxide (anhydrous); UN 1079. Used to process sulfite paper pulp, ore and metal refining, soybean protein, intermediates, solvent extraction, bleaching oils and starch, sulfonation of oils, disinfectant, fumigant, food additive, reducing agent, antioxidant. Registered by EPA as a fungicide. FDA, FEMA GRAS, Japan approved (0.03 - 5.0 g/kg), Europe listed, UK approved, USP/NF, BP compliance. Used as an antioxidant preservative for pharmaceuticals. Gas; mp = -72°; bp = -10°; soluble in H_2O (10 g/100 ml), more soluble in organic solvents. *Air Prods & Chems; Boliden Intertrade; Hoechst Celanese; Outokumpu Oy; Rhône-Poulenc.*

3623   Sulfur fluoride
2551-62-4   9063   G

F_6S
Sulfur hexafluoride.
EINECS 219-854-2; Elegas; Esaflon; Hexafluorure de soufre; HSDB 825; Sulfur fluoride, (OC-6-11)-; Sulfur fluoride (SF_6), (OC-6-11)-; Sulfur hexafluoride; Sulphur hexafluoride; UN1080. Used in electrical circuit breakers and in electronic ultra-high frequency piping. Gas; mp = -51°; soluble in oil. *Montedison UK.*

3624 Sulfuric acid

7664-93-9   9064   G P

H2O4S

Hydrogen sulfate.

Acide sulfurique; Acido solforico; Acido sulfurico; Acidum sulfuricum; Battery acid; BOV; Caswell No. 815; Dihydrogen sulfate; Dipping acid; EINECS 231-639-5; Electrolyte acid; EPA Pesticide Chemical Code 078001; HSDB 1811; Hydrogen sulfate; Mattling acid; Oil of vitreol; Oil of vitriol; Schwefelsaeure; Schwefelsäure; Schwefelsaeureloesungen; Schwefelsäurelösungen; Spirit of alum; Spirit of vitriol; Sulfuric acid; Sulfuric acid, spent; Sulphuric acid; UN1830; UN1832; UN2796; Vitriol Brown Oil; Zwavelzuuroplossingen. Fertilizers, chem-icals, dyes and pigments, laboratory reagents, electro-plating baths. Registered by EPA as an antimicrobial, fungicide and herbicide. FDA GRAS, Japan restricted, Europe listed, UK approved, FDA approved for injectables, parenterals, inhalants, ophthalmics, orals. USP/NF, BP. compliant. Acidifying agent used in parenterals, inhalants, intramuscular injectables, ophthalmic and orals. Astringent in diarrhea, in mixtures to stimulate appetites. Oily liquid; mp = 10°; bp = 290°, dec 340°; d = 1.84; miscible with H2O, EtOH; LD50 (rat orl) = 2140 mg/kg. *Akzo Chemie; Am. Cyanamid; Amax; Boliden Intertrade; DuPont; Metallgesellschaft GmbH; Nissan Chem. Ind.; OxyChem; Pasminco Europe; Rasa; Rhône-Poulenc.*

3625 Sulfurized lard oil

61790-49-6   H

Oils, lard, sulfurized.

EINECS 263-141-9; Lard oil, sulfurized; Oils, lard, sulfurized.

3626 Sulfuryl Fluoride

2699-79-8   9071   G P

F2O2S

Sulfur difluoride dioxide.

Caswell No. 816A; EINECS 220-281-5; EPA Pesticide Chemical Code 078003; Fluoro de sulfurilo; Fluorure de sulfuryle; HSDB 828; Sulfonyl fluoride; Sulfur difluoride dioxide; Sulfuric oxyfluoride; Sulfuryl difluoride; Sulfuryl fluoride; Sulphuryl difluoride; Sulphuryl fluoride; UN2191; Vikane; Vikane fumigant. A restricted use pesticide, a gas that is used in structures, vehicles, and wood products for control of drywood termites, wood-infesting beetles, and certain other insects and rodents. There are no registered uses for sulfuryl fluoride on food or feed crops. Registered by EPA as an antimicrobial and disinfectant. Odorless, colorless gas; mp = -136°; bp =-55°; sg = 1.349; soluble in H2O (0.075 g/100 ml), EtOH (24 - 27 ml/100 ml), C7H8 (0.21 - 0.22 l/100 ml), CCl4 (0.136 - 0.138 l/100 ml); stable below 400°; not hydrolyzed by H2O; LD50 (rat orl) = 100 mg/kg; LC50 (mrat ihl 4 hr.) = 5.11 mg/l air, (frat ihl 4 hr.) = 4.92 mg/l air, (mrat ihl 1 hr.) = 15.0 mg/l air, (frat ihl 1 hr.) = 12.2 mg/l air. *Dow AgroSciences; Dow UK; Integrated Environments International Inc.*

3627 Sulisobenzone

4065-45-6   9074   G

C14H12O6S

2-Benzoyl-5-methoxy-1-phenol-4-sulfonic acid.

Benzenesulfonic acid, 5-benzoyl-4-hydroxy-2-methoxy-; Benzophenone 4; Benzophenone-4; 5-Benzoyl-4-hydr-oxy-2-methoxybenzolsulfonsäure; 5-Benzoyl-4-hydroxy-2-methoxybenzolsulfonsaeure; 5-Benzoyl-4-hydroxy-2-methoxybenzene sulfonic acid; BRN 2889165; Cyasorb UV 284; EINECS 223-772-2; 2-Hydroxy-4-methoxy-benzophenone-5-sulfonic acid; MS 40; NSC 60584; 1-Phenol-4-sulfonic acid, 2-benzoyl-5-methoxy-; Rhodialux S; Seesorb 101S; Spectra-Sorb UV 284; Sungard; Sulisobenzona; Sulisobenzone; Sulisobenzonum; Sungard; Syntase 230; Uval; Uvinuc ms 40; Uvinul; Uvinul MS 40; Uvinul MS-40 substanz; Uvistat 1121. UV absorber, used as a UV screen. Tan-colored powder; mp = 145°; soluble in H2O (25.0 g/100 ml), less soluble in organic solvents. *Dorsey Pharmaceuticals; Rhône-Poulenc.*

3628 Sulprofos

35400-43-2   9081   H P

C12H19O2PS3

O-Ethyl O-(4-(methylthio)phenyl) S-propyl phosphoro-dithioate.

AI3-29149; BAY-NTN-9306; Bayer NTN 9306; Bolstar; Bolstar 6; Bolstar(R); BRN 1990231; Caswell No. 453AA; EPA Pesticide Chemical Code 111501; Ethyl O-(4-(methylthio)phenyl) S-propyl phosphorodithioate; O-Ethyl O-[4-(methylthio)phenyl]phosphorodithioic acid S-propyl ester; O-Ethyl O-(4-methylthiophenyl) S-propyl dithiophosphate; O-Ethyl O-(4-(methylmercapto)phenyl)-S-n-propylphosphorothionothiolate; Heliothion; Merca-profos; Mercaprophos; Merpafos; Morpafos; NTN 9306; Phosphorodithioic acid O-ethyl O-[4-(methylthio)phenyl] S-propyl ester; Sulprofos; Sulprophos. Registered by EPA as an insecticide (cancelled). Tan liquid; bp0.1 = 155 - 158°; d20/20 = 1.20; almost insoluble in H2O (0.000031 g/100 ml), soluble in cyclohexanone (12 g/100 ml), i-PrOH (40 - 60 g/100 ml), CH2Cl2 (120 g/100 ml), C6H14 (120 g/100 ml), C7H8 (> 120 g/100 ml); LD50 (rat orl) = 150 mg/kg, 227 mg/kg, (mus orl) = 1700 mg/kg, (mrbt der) = 820 mg/kg, (frbt der) = 994 mg/kg, (rat der) > 1000 mg/kg, (bobwhite quail orl) = 47 mg/kg; LC50 (rat ihl 4 hr.) = 0.66 mg/l, (bobwhite quail 5 day dietary) = 99 mg/kg diet, (bluegill sunfish 96 hr.) = 11 mg/l, (rainbow trout 96 hr.) = 23 mg/l, (carp 96 hr.) = 5.2 mg/l. *Bayer Corp.*

3629 Sumatriptan

103628-46-2   9088   D

C14H21N3O2S

3-[2-(Dimethylamino)ethyl]-N-methylindole-5-methane-sulfonamide. BRN 6930870; GR 43175; GR 43175X; Imitrex; Sumatran; Sumatriptan; Sumatriptanum; Sumax. Sero-tonin 5HT1 receptor agonist. Antimigraine. Crystals; mp = 169-171°. *Glaxo Wellcome Inc.*

3630 Sumatriptan succinate
103628-48-4   9088   D

C18H27N3O6S

3-[2-(Dimethylamino)ethyl]-N-methylindole-5-methanesulfonamide succinate (1:1).
Antibet; Arcoiran; Diletan; GR-43175C; Imigran; Imijekt; Imitrex; Micralgin; Migmax; Migratran; Migratriptan; Novelian; Permicran; Sumadol; Sumatrin; Sumatriptan succinate; Sumigrene. Serotonin 5HT1 receptor agonist. Antimigraine. Crystals; mp = 165-166°. *Glaxo Wellcome Inc.*

3631 Sunflower oil
8001-21-6   9090   E H

Sunflower seed oil [Oil, edible].
EINECS 232-273-9; Extract of sunflower; Extract of sunflower seeds; GTO 80; GTO 90; GTO 90E; Helianthus annuus extract; NS-20; Oils, sunflower seed; Oleum helianthi; Solvent sunflower oil; Sunflower extract; Sunflower oil; Sunflower seed extract; Sunflower seed oil; Sunyl® 80; Sunyl® 80 RBD; Sunyl® 80 RBWD; Sunyl® 80 RBWD ES; Sunyl® 80 RDB ES; Sunyl® 90; Sunyl® 90 RBD; Sunyl® 90 RBWD ES 1016; Sunyl® 90 RBWD; Sunyl® 90E RBWD; Sunyl® HS 500. Soluble in EtOH, Et2O, CHCl3, CS2; d = 0.924 - 0.926. Non-toxic. *Abitec Corp.; Arista Ind.; Charkit; Holme & Clark; Lipo; Penta Mfg.; Tri-K Ind.; Welch.*

3632 Sylvan
534-22-5   G

C5H6O
2-Methylfuran.
5-17-01-00322 (Beilstein Handbook Reference); 5-Methylfuran; Al3-24245; BRN 0103733; CCRIS 2920; EINECS 208-594-5; Furan, 2-methyl-; Methylfuran; α-Methylfuran; 2-Methylfuran; NSC 3707; Silvan; Sylvan; UN2301. A constituent of wood tar. Used in perfumery. Crystals; mp = -87.5°; bp = 65°; d20 = 0.9132; λm = 212 nm (MeOH); slightly soluble in H2O, CCl4, soluble in EtOH, Et2O.

3633 Symclosene
87-90-1   9103   G H P

C3Cl3N3O3
1,3,5-Trichloro-s-triazine-2,4,6(1H,3H,5H)-trione.
4-26-00-00642 (Beilstein Handbook Reference); ACL 85; ACL 90 Plus; Al3-17193; BRN 0202022; Caswell No. 876B; CBD 90; CCRIS 2311; CDB 90; Chloreal; EPA Pesticide Chemical Code 081405; Fi Clor 91; Fichlor 91; HSDB 885; Isocyanuric chloride; Kyselina trichloiso-kyanurova; Queschlor; Neochlor 90; NSC-405124; Sin-closeno; Symclosen; Symclosene; Symclosenum; 1,3,5-Triazine-2,4,6(1H,3H,5H)-trione, 1,3,5-trichloro-; s-Tri-azine-2,4,6(1H,3H,5H)-trione, 1,3,5-trichloro-; Trichloro-isocyanic acid; 1,3,5-Trichloroisocyanuric acid; Tri-chlorocyanuric acid; Trichloroisocyanuric acid, dry [UN2468] [Oxidizer]; N,N',N-Trichloroisocyanuric acid; 1,3,5-Trichloro-s-triazinetrione; 1,3,5-Trichloro-1,3,5-tri-azine-2,4,6(1H,3H,5H)-trione; 1,3,5-Trichloro-2,4,6-tri-oxohexahydro-s-triazine; Trichlorinated isocyanuric acid; Trichloro-s-triazinetrione; Trichloro-s-triazine-2,4,6(1H, 3H,5H)-trione; Trichloroisocyanurate; UN2468. A chlorinated isocyanurate containing 90% available chlorine; a high performance chlorine microbiocide used to control algae, bacteria, and fungi growth in recirculating water and cooling towers. A topical anti-infective agent. A chlorinating agent, disinfectant, and industrial deodorant, component of household cleaners. Bleaching compound, used in sanitizers, disinfectants, detergents, dishwashing compounds, swimming pool disinfectants, bactericides, algicides, deodorants; active ingredient in dry household cleaners. Registered by EPA as an antimicrobial disinfectant. CAUTION: Irritating to eyes, skin, mucous membranes. Needles; mp = 246.7° (dec); pH 4.4; slightly soluble in H2O (2 g/100 ml); insoluble in non-polar organic solvents, soluble in chlorinated and highly polar solvents. *3V; Alden Leeds Inc.; Allchem Ind.; Arch Chemicals Inc; Atofina; Bio-Lab, Inc.; Colgate-Palmolive; Hasa Inc; ICI Americas Inc.; Monsanto Co.; Nissan Chem. Ind.; Occidental Chem. Corp.; Qualco Inc; Quest Intl.; Shikoku Chemicals Corp.; Stellar Technology Co.; Suncoast Chemicals Co.*

3634 Talc
14807-96-6   9127   G

H2Mg3O12Si4
Hydrated magnesium silicate.
Agalite; Alpine talc USP, bc 127; Asbestine; B 9; B 13; B 13 (mineral); Beaver White 200; CCRIS 3656; CI 77718; Cosmetic talc; CP 10-40; CP 38-33; Crystalite CRS 6002; Desertalc 57; EINECS 238-877-9; Emtal 500; Emtal 549; Emtal 596; Emtal 599; EX-IT; Fibrene C 400; Finntalc C10; Finntalc M05; Finntalc M15; Finntalc P40; Finntalc PF; French chalk; FW-XO; HSDB 830; Hydrous magnesium silicate; IT Extra; LMR 100; Lo Micron talc USP, bc 2755; Magnesium silicate, hydrous; Magnesium silicate talc; Micro Ace K1; Micro Ace L1; Micron White 5000A; Micron White 5000P; Micron White 5000S; Microtalco IT Extra; Mistron 139; Mistron 2SC; Mistron frost P; Mistron RCS; Mistron Star; Mistron super frost; Mistron vapor; MP 12-50; MP 25-38; MP 40-27; MP 45-26; MST; Mussolinite; NCI-C06008; Nonasbestiform talc; Nonfibrous talc;

Nytal 200; Nytal 400; P 3; P 3 (Mineral); PK-C; PK-N; Polytal 4641; Polytal 4725; Silicates (<1% quartz):talc (not containing asbestos); Silicates: talc (containing no asbestos); Snowgoose; Soapstone; Steatite; Steatite talc; Steawhite; Supreme; Supreme dense; Talc; Talc (Mg3H2(SiO3)4); Talc (powder); Talc (powder), containing no asbestos fibers; Talc, containing no asbestos fibers; Talc, non-asbestos form; Talc, not containing asbestiform fibers; Talcan PK-P; Talcron CP 44-31; Talcum; TY 80; Vantalc®; Vertal 92; Vertal 200; soapstone; steatite; talcum. Native, hydrous magnesium silicate somethimes containing small portion of aluminum silicate; ceramics, cosmetics, pharmaceuticals; as filler and pigment in rubber, paints, soaps. Dusting agent; lubricant; electrical insulation, reinforcement in black PP compounds; dusting/parting agent for rubbers; filler in autobody compounds, caulks, putties, and sealants; as filler and pigment in rubber, paints, soaps. Dusting agent; lubricant; electrical insulation. White fibrous solid; mp = 800°; insoluble in H2O, organic solvents. *Cyprus Industrial Min.; Luzenac Am.; Pfizer Inc.; Salomon L. A.; Vanderbilt R.T. Co. Inc.; Whittaker Clark & Daniels.*

3635   Tall oil
8002-26-4   E H

Tallol.
Acintol C; Acintol D 29 LR; EINECS 232-304-6; HSDB 5049; Lignin liquor; Liquid rosin; Rosin, liquid; Tall oil; Tall oil (crude and distilled); Tall oil rosin; Tall oil rosin and fatty acids; Tall oil [Oil, misc.]; Talleol; Tallol; Unitol CX; Unitol DT 40; Yatall MA. By-product of sulfate pulp manufacture; contains 2.2% material soluble in petroleum ether, 12.4% unsaponifiable matter, 30.4% resin acid, and 54.9% fatty acids. The resin acid consists of abietic acid, and the fatty acids contain oleic, linoleic, and linolenic acids; used as paint vehicles, in oil drilling muds, lubricants and greases, asphalt derivatives; in rubber reclaiming and as chemical intermediates. FDA GRAS as indirect food additive; FDA approved for topicals. Used as a fungicide in topicals. Mild allergen. *Spectrum Chem. Manufacturing.*

3636   Tall oil fatty acids
61790-12-3   H

Fatty acids, tall-oil.
Acids, tall oil; Disproportionated tall oil fatty acid; EINECS 263-107-3; Fatty acids, tall oil; Tall oil acid; Tall oil acids; Tall oil fatty acids.

3637   Tall oil rosin
8052-10-6   H

Tall oil rosin.
EINECS 232-484-6; Tall oil rosin.

3638   Tall oil, sodium salt
65997-01-5   H

Tall oil, sodium salt.
EINECS 266-037-1; Tall oil, sodium salt.

3639   Tall-oil pitch
8016-81-7   H

Tall-oil pitch.
EINECS 232-414-4; Tall-oil pitch.

3640   Tallow acid
61790-37-2   H

Fatty acids, tallow.
Acids, tallow; EINECS 263-129-3; Fatty acids, tallow; Tallow acid; Tallow fatty acid.

3641   Tallow nitrile
61790-28-1   H

Nitriles, tallow.
EINECS 263-120-4; Nitriles, tallow; Tallow nitrile.

3642   Tallow oil
61789-97-7   H

Tallow.
Beef tallow; EINECS 263-099-1; HSDB 5167; Mutton tallow; Tallow; Tallow, beef; Tallow, mutton; Tallow oil.

3643   N-Tallow-1,3-propylenediamine
61791-55-7   H

Amines, N-tallow alkyltrimethylenedi-.
Amines, N-tallow alkyltrimethylenedi-; EINECS 263-189-0; N-(Tallow alkyl)-1,3-propanediamine; N-(Tallowalkyl)-trimethylenediamine; N-Tallow-1,3-propylenediamine.

3644   Tallow trimonium chloride
8030-78-2   H

Tallow trimethylammonium chloride.
Adogen 471; Ammonium, trimethyltallow alkyl-, chlorides; Arquad T; Arquad T-50; EINECS 232-447-4; Noramium MS 50; Quaternary ammonium compounds, trimethyltallow alkyl-, chloride; Quaternary ammonium compounds, tallow alkyl trimethyl, chlorides; Tallow trimethyl ammonium chloride; N-Tallow-trimethylammonium surfactant; Tallowtrimonium chloride; Trimethyl(tallowalkyl)ammonium chloride. Emulsifier.

3645   Tallowamine
61790-33-8   H

Amines, tallow alkyl.
Amines, tallow alkyl; EINECS 263-125-1; (Tallow alkyl)amine; Tallow amine; Tallowamine.

3646   Tamoxifen
10540-29-1   9137   D

C26H29NO
(Z)-2-[p-(1,2-Diphenyl-1-butenyl)phenoxy]-N,N-dimethylethylamine. Citofen; Crisafeno; Diemon; Istubol; Oncomox; Retaxim; Tamizam; Tamoxen; Tamoxifen; Valodex. Antiestrogen. Antineoplastic. Nonsteroidal estrogen antagonist used in prevention and palliative treatment of breast cancer. Crystals; mp = 96-98°; [cis-form]: mp = 72-74°. *Zeneca Pharm.*

3647   Tamoxifen citrate
54965-24-1   9137   D

C32H37NO8

(Z)-2-[p-(1,2-Diphenyl-1-butenyl)phenoxy]-N,N-dimethylethylamine citrate.

Apo-Tamox; Emblon; Farmifeno; Genox; Ginarsan; ICI-46474 citrate; ICI-47699; Jenoxifen; Kessar; Ledertam; Nolgen; Noltam; Nolvadex; Nourytam; Nourytan; Nox-item; Oncotam; Tafoxen; Tamax; Tamofen; Tamoxifen citrate; Taxus; TMX; Tomaxasta; Zemide; Zitazonium.; ICI-47699 [as cis-form citrate]. A nonsteroidal estrogen antagonist used in prevention and palliative treatment of breast cancer. Antiestrogen. Antineoplastic. Crystals; mp = 140-142°; slightly soluble in H_2O; more soluble in EtOH, MeOH, Me2CO; LD_{50} (mus ip) = 200 mg/kg, (mus iv) = 62.5 mg/kg, (mus orl) = 3000-6000 mg/kg, (rat ip) = 600 mg/kg, (rat iv) = 62.5 mg/kg, (rat orl) = 1200-2500 mg/kg; [cis-form citrate]: mp = 126-128°. ICI; Zeneca Pharm.

3648   Tamsulosin
106133-20-4   9138   D

C20H28N2O5S

(-)-(R)-5-[2-[[2-(o-Ethoxyphenoxy)ethyl]amino]propyl]-2-methoxybenzenesulfonamide.

Amsulosin; Tamsulosin; Tamsulosina; Tamsulosine; Tamsulosinum. Specific α_1-adrenoceptor antagonist. Used in treatment of benign prostatic hypertrophy. Boehringer Ingelheim Pharm. Inc.; Yamanouchi U.S.A. Inc.

3649   (R)-Tamsulosin hydrochloride
106463-17-6   9138   D

HCl

C20H29ClN2O5S

(-)-(R)-5-[2-[[2-(o-Ethoxyphenoxy)ethyl]amino]propyl]-2-methoxybenzenesulfonamide hydrochloride.

Alna;   Amsulosin   hydrochloride;   R-(-)-5-(2-((2-(2-Ethoxy-phenoxy)ethyl)amino)propyl)-2-methoxybenzenesulfon-amide Expros; Flomax; Josir; LY253351; Omic; Omix; Omnic; Pradif; hydrochloride; Secotex; Tamsulosin (R)-form hydrochloride; Tamsulosin hydrochloride; Urolosin; YM 617; YM617; R-(-)-YM-

12617. Specific α_1-adrenoceptor antagonist. Used in treatment of benign prostatic hypertrophy. Crystals; mp = 228-230°; $[\alpha]_D^{24}$ = -4.0° (c = 0.35 MeOH). Boehringer Ingelheim Pharm. Inc.; Yamanouchi U.S.A. Inc.

3650   (S)-Tamsulosin hydrochloride
106463-19-8   9138   D

H

C20H29ClN2O5S

(+)-(S)-5-[2-[[2-(o-Ethoxyphenoxy)ethyl]amino]propyl]-2-methoxybenzenesulfonamide hydrochloride.

YM-12617-2. Used in treatment of benign prostatic hypertrophy. Specific α_1-adrenoceptor antagonist. Crystals; mp = 228-230°; $[\alpha]_D^{24}$ = +4.2° (c = 0.36 MeOH). Boehringer Ingelheim Pharm. Inc.; Yamanouchi U.S.A. Inc.

3651   dl-Tamsulosin hydrochloride
106463-17-6   9138   D

H

C20H29ClN2O5S

(±)-(R)-5-[2-[[2-(o-Ethoxyphenoxy)ethyl]amino]propyl]-2-methoxybenzenesulfonamide hydrochloride.

Alna; Amsulosin; Amsulosin hydrochloride; Expros; Flomax; Josir; LY-253351; Omic; Omix; Omnic; Pradif; R-(-)-YM-12617; Secotex; Urolosin; YM-617. Specific α_1-adrenoceptor antagonist. Used in treatment of benign prostatic hypertrophy. Crystals; mp = 254-256°. Boehringer Ingelheim Pharm. Inc.; Yamanouchi U.S.A. Inc.

3652   Tannic acid
1401-55-4   D G H

C76H52O46

Hydrolyzable gallotannin.

Acacia mollissima tannin; Acid, tannic; Acide tannique; d'Acide tannique; Castanea sativa Mill tannin; Caswell No. 819; CCRIS 571; Chestnut tannin; EINECS 215-753-2; EPA Pesticide Chemical Code 078502; FEMA No. 3042; Gallotannic acid; Gallotannin; Glycerite; HSDB 831; Hydrolyzable gallotannin; Liquidambar styraciflua; Mimosa tannin; Quebracho extract; Quebracho tannin; Quebracho wood extract; Schinopsis lorentzii tannin; Sweet gum; Tannic acid; Tannic acid (Quercus spp.); Tannic acid and tannins; Tannin; Tannin from chestnut; Tannin from mimosa; Tannin from quebracho; Tannin from sweet gum; Tannins. A mixture of organic acids occurring in the bark and fruit of many plants, e.g., oak species, sumac; mordant in dyeing; manufacture of ink, imitation tortoise shell; sizing paper; printing fabrics; tanning; clarifying beer; in photography; as coagulant in rubber; analytical chemistry reagent; astringent. Burlington Biomedical; Crompton & Knowles; Fuji Chem. Ind.; Mallinckrodt Inc.; Thiem; Ulrich GmbH.

3653   Tantalum oxide
1314-61-0   9146   G

O5Ta2

Tantalum(V) oxide.

Ditantalum pentaoxide; EINECS 215-238-2; Tantalic acid anhydride; Tantalum oxide; Tantalum oxide (Ta2O5); Tantalum oxide dusts; Tantalum penta oxide; Tantalum pentoxide. Used in production of tantalum, tantalum carbide; optical glass; piezoelectric and laser applications; dielectric layers in electronic circuits. White powder; insoluble in H_2O, soluble in HF; LD_{50} (rat orl) = 8000 mg/kg. *Atomergic Chemetals; Cabot Carbon Ltd.; Cerac; Sigma-Aldrich Fine Chem.*

3654   Tartaric acid

133-37-9   9157   G

C4H6O6

DL-2,3-dihydroxybutanedioic acid.

Butanedioic acid, 2,3-dihydroxy-, (2R,3R)-rel-; Butane-dioic acid, 2,3-dihydroxy-, (R*,R*)-; Butanedioic acid, 2,3-dihydroxy-, (R*,R*)-(±)-; EINECS 205-105-7; FEMA No. 3044; NSC 148314; Paratartaric aicd; Racemic acid; Racemic tartaric acid; Resolvable tartaric acid; (±)-Tartaric acid; dl-Tartaric acid; DL-Tartaric acid; Tartaric acid D,L; (2RS,3RS)-Tartaric acid; DL-Tartrate; Traubensaure; Uvic acid. Used the manufacture of cream of tartar, tartar emetic, acetaldehyde; sequestrant, tanning, effervescent beverages, baking powder, fruit esters, ceramics, galvanoplastics, photography, textiles, silvering mirrors, coloring metals, acidulant in foods. Prisms; mp = 206°; d^{25} = 1.788; insoluble in C6H6, soluble in H_2O, EtOH. *Bromhead & Denison Ltd.; Greeff R.W. & Co.; Mallinckrodt Inc.; Penta Mfg.; Rit-Chem.*

3655   Taurine

107-35-7   9163   G

C2H7NO3S

2-Aminoethanesulfonic acid.

AI3-18307; Aminoethanesulfonic acid; 2-Aminoethane-sulfonic acid; β-Aminoethylsulfonic acid; 2-Aminoethyl-sulfonic acid; CCRIS 4721; EINECS 203-483-8; Ethane-sulfonic acid, 2-amino-; FEMA No. 3813; NCI-C60606; NSC 32428; O-Due; 2-Sulfoethylamine; Tauphon; Taurina; Taurine; L-Taurine; Taurinum. Used in biochemical research, pharmaceuticals, wetting agents. Crystals; mp = 328° (dec); very soluble in H_2O (6.5 g/100 ml), insoluble in organic solvents. *Chemisphere; Lancaster Synthesis Co.; Mitsui Toatsu; Penta Mfg.; Tanabe U.S.A. Inc.*

3656   Tebuconazole

107534-96-3   9176   P

C16H22ClN3O

(RS)-1-(4-Chlorophenyl)-4,4-dimethyl-3-(1H-1,2,4-triazol-1-ylmethyl)pentan-3-ol.

BAY-HWG 1608; Elite; Ethyltrianol; Etiltrianol; Fenetr-azole; Folicur; Folicur 1.2 EC; GWG 1609; HWG 1608; Lynx; Lynx 1.2; Preventol A 8; Raxil; Tebuconazole; Tebuconazole (±); Terbuconazole; Terbutrazole. Fung-icide with systemic properties/broad spectrum activity against rusts, leaf spot diseases, e.g., *Septoria spp., powdery mildew and several Fusarium species on cereals; whitemold, Phoma* and various leaf spot diseases on oilseed. Cereal seed dressing with systemic properties for control of seed-borne diseases such as stinking smut, loose smuts and covered smut; highly effective at low dosage rates. Registered by EPA as a fungicide. Colorless crystals; mp = 102.4°; soluble in H_2O (0.0032 g/100 ml), CH2Cl2 (> 20 g/100 ml), C6H14 (0.02 - 0.05 g/100 ml), i-PrOH (5 - 10 g/100 ml), C7H8 (5 - 10 g/100 ml); LD_{50} (rat orl) > 4000 mg/kg, (mus orl) > 2000 mg/kg, (rat der) > 5000 mg/kg, (m Japanese quail orl) = 4438 mg/kg, (f Japanese quail orl) = 2912 mg/kg, (ckn orl) = 4488 mg/kg; LC_{50} (rat ihl 4 hr.) > g/l air, (rainbow trout 96 hr.) = 6.4 mg/kg, (golden orfe 96 hr.) = 8.7 mg/l, (*Daphnia* 48 hr.) = 10 - 12 mg/l; non-toxic to bees. *Bayer AG; Bayer Corp.; Gustafson LLC.*

3657   Tebutam

35256-85-0   G P

C15H23NO

2,2-Dimethyl-N-(1-methylethyl)-N-(phenylmethyl)propan-amide.

N-Benzyl-N-isopropyl-2,2-dimethylpropionamide; N-Benzyl-N-isopropyltrimethylacetamide; N-Benzyl-N-iso-propylpivalamide; BRN 2838127; Butam; Caswell No. 083EE; Comodor; Comodor 600; 2,2-Dimethyl-N-(1-methylethyl)-N-(phenylmethyl)propanamide; EINECS 252-470-3; EPA Pesticide Chemical Code 219500; GCP-5544; Propanamide, 2,2-dimethyl-N-(1-methylethyl)-N-(phenylmethyl)-; Propionamide, N-benzyl-2,2-dimethyl-N-isopropyl-; S-15544; Tebutam; Tebutame. Selective herbicide, acting by inhibition of weed germination. Used for pre-emergence control of annual grasses and broad-leaved weeds. Colorless oil; bp0.1 = 95 - 97°; sg25 = 0.975; poorly soluble in H_2O (0.000079 g/100 ml at 25° and pH 7), soluble in Me2CO, C6H14, MeOH, C7H8, CHCl3 (all > 50 g/100 ml); LD_{50} (rat orl) = 6210 mg/kg, (gpg orl) = 2025 mg/kg, (rbt der) > 2000 mg/kg, (bee orl) = 0.1 mg/bee; LC_{50} (rat ihl) > 2.18 mg/l air, (mallard duck, bobwhite quail 8 day dietary) > 5000 mg/kg, (rainbow trout 96 hr.) = 23 mg/l, (bluegill sunfish 96 hr.) = 19 mg/l, (moonfish 96 hr.) = 18.7 mg/l; toxic to bees. *Farm Protection Ltd.; ICI Agrochemicals; ICI Plant Protection; La Quinoleine S.A.; LaRoche Ind.; Maag; Syngenta Crop Protection.*

3658   Tebuthiuron

34014-18-1   9178   H P

C9H16N4OS

N-[5-(1,1-Dimethylethyl)-1,3,4-thiadiazol-2-yl]-N,N'-dimethylurea.

BRN 0527479; Brulan; Brush bullet; 1-(5-tert-Butyl-1,3,4-thiadiazol-2-yl)-3-dimethylharnstoff; 1-(5-tert-Butyl-1,3,4-thiadiazol-2-yl)-1,3-dimethylurea; Caswell No. 366AA; EL-103; Pesticide Chemical Code

105501; Graslan; Graslan 40P; Graslan 250 brush bullets; HSDB 6863; N-(5-(1,1-Dimethyläthyl)-1,3,4-thiadiazol-2-yl)-N,N'-dimethylharnstoff; N-(5-(1,1-Dimethylethyl)-1,3,4-thiadia-zol-2-yl)-N,N'-dimethylurea; Perflan; Prefmid;; SHA 105501; Spike; Spike 20P; Spike 40P; Spike 40W; Spike 5G; Spike 80W; Spike DF; Tebulan; Tebuthiuron; Tiurolan; Urea, 2-(5-tert-butyl-1,3,4-thiadiazol-2-yl)-1,3-dimethyl-; Urea, N-(5-(1,1-di-methylethyl)-1,3,4-thiadi-azol-2-yl)-N,N'-dimethyl-. Wettable powder or granules used for total weed control in noncrop areas. Registered by EPA as a herbicide. Solid; mp = 160 - 163°; soluble in H_2O (0.25 g/100 ml), C_6H_6 (0.378 g/100 ml), C_6H_{14} (0.61 g/100 ml), 2-methoxyethanol (6 g/100 ml), CH_3CN (6 g/100 ml), Me_2CO (7 g/100 ml), MeOH (17 g/100 ml), $CHCl_3$ (25 g/100 ml); LD_{50} (rat orl) = 644 mg/kg, (mus orl) = 579 mg/kg, (rbt orl) = 286 mg/kg, (dog orl) > 500 mg/kg, (cat orl) > 200 mg/kg, (ckn, mallad duck, bobwhite quail orl) > 500 mg/kg; LC_{50} (ckn orl 1 month) > 1000 mg/kg, (rainbow trout 96 hr.) = 144 mg/l, (goldfish 96 hr.) > 160 mg/l, (bluegill sunfish 96 hr.) = 112 mg/l. Dow AgroSciences; Rhône-Poulenc Environmental Prods. Ltd.

3659 Tecnazene
117-18-0

G P

$C_6HCl_4NO_2$
1,2,4,5-tetrachloro-3-nitrobenzene.
4-05-00-00728 (Beilstein Handbook Reference); AI3-22329; Altritan; Arena; Benzene, 1,2,4,5-tetrachloro-3-nitro-; Benzene, 3-nitro-1,2,4,5-tetrachloro-; BRN 1973805; Bygran; Caswell No. 831; CCRIS 5939; Chipman 3,142; Easytec; EINECS 204-178-2; EPA Pesticide Chemical Code 055201; Folosan; Folosan DB 905; Fumite; Fumite Techalin; Fusarex; Fusarex G; Fusarex T; Hickstor; HSDB 1772; Hystor; Hystore; Hytec; Myfusan; Nebulin; New Hystore; NSC 10235; Storite SS; TCNB; 2,3,5,6-TCNB; Tecnazen; Tecnazene; Teknazen; Terraclor; 2,3,5,6-Tetrachlor-1-nitrobenzol; 2,3,5,6-Tetra-chlor-3-nitrobenzol; Tetrachloronitrobenzene; 1,2,4,5-Tetrachloro-3-nitrobenzene; 2,3,5,6-Tetrachloro-1-nitro-benzene; 2,3,5,6-Tetrachloronitrobenzene; Tubodust; Tubostore; Turbostore. Fungicide with protective and curative action. Tecgran is provided as granules or dispersible powder containing tecnazene; protectant fungicide and potato sprout suppressant. Used to control dry rot in both ware and seed potatoes, and sprouting in ware potatoes. Registered by EPA as a fungicide (cancelled). Solid; mp = 99°; bp = 304° (dec); soluble in H_2O (0.000044 g/100 ml), EtOH (4 g/100 ml), readily soluble in C_6H_6, CS_2, $CHCl_3$, ketones; LD_{50} (mrat orl) = 2047 mg/kg, (frat orl) = 1256 mg/kg; non-toxic to bees. Atlas Interlates Ltd; Bayer Corp., Agriculture; ICI Agrochemicals; ICI Plant Protection; Sigma-Aldrich Co.; Tripart Farm Chemicals Ltd.

3660 Tellurium
13494-80-9

9201

G

Te

Te
Tellurium.
Aurum paradoxum; EINECS 236-813-4; HSDB 2532; Metallum problematum; NCI-C60117; Telloy®; Tellur; Tellurium; Tellurium element; Tellurium, elemental; Tellurium, metallic. Used in alloys, as a secondary rubber vulcanizing agent, in manufacture of iron and steel casting; as a coloring agent for glass and ceramics; in thermoelectric devices. Grey-white powder; mp = 450°; bp = 990°; d = 6.11-6.27. Atomergic Chemetals; Cabot Carbon Ltd.; Cerac;

Vanderbilt R.T. Co. Inc.

3661 Tellurium oxide
7446-07-3

9203

G

O_2Te
Tellurium dioxide.
EINECS 231-193-1; Tellurium dioxide; Tellurium oxide; Tellurium oxide (TeO_2). Solid; mp= 733°; d = 5.75 or 6.04; slightly soluble in H_2O. Asarco; Atomergic Chemetals; Cerac; Noah Chem.

3662 Temazepam
846-50-4

9213

D

$C_{16}H_{13}ClN_2O_2$
7-Chloro-1,3-dihydro-3-hydroxy-1-methyl-5-phenyl-2H-1,4-benzodiazepin-2-one.
Dasuen; ER-115; Euhypnos; Euipnos; Gelthix; K-3917; Lenal; Levanxene; Levanxol; Neodorm SP; Nocturne; Nomapam; Norkotral Tema; Normison; Normitab; Nortem; Oxydiazepam; Perdorm; Planum; Pronervon T; Remestan; Restoril; Ro-5-5354; Temador; Temazep von ct; Temazepam; Temtabs; Tenox; Uvamin Retard; Wy-3917. Used medically as a sedative/hypnotic and anxiolytic. Crystals; mp = 119-121°. Hoffmann-LaRoche Inc.; Sandoz Pharm. Corp.

3663 Temephos
3383-96-8

9214

D G P

$C_{16}H_{20}O_6P_2S_3$
Tetramethyl O,O'-thiodi-p-phenylene phosphorothioate.
Abaphos; Abate; Abate 1-SG; Abate 2-CG; Abate 4-E; Abate 5CG; Abathion; AC 52160; AI3-27165; American Cyanamid AC 52,160; American Cyanamid CL-52160; American Cyanamid E.I. 52,160; Biothion; Bis-p-(O,O-dimethyl O-phenylphosphorothioate)sulfide; Bithion; BRN 1896901; Caswell No. 845; CL 52160; Difenphos; Difos; O,O-Dimethyl phosphorothioate O,O-diester with 4,4'-thiodiphenol; Diphos; Diphos (pesticide); Ecopro 1707; EI 52160; ENT 27,165; EPA Pesticide Chemical Code 059001; Experimental insecticide 52160; HSDB 956; Lypor; Nephis; Nephis 1G; Nimitex; Nimitox; OMS 786; Phenol, 4,4'-thiodi-, O,O-diester with O,O-dimethyl phosphorothioate; phosphorothioate; Phosphorothioic acid, O,O-dimethyl ester, O,O-diester with 4,4'-thiodiphenol; Procida; Swebat; Swebate; Temefos; Temefosum; Temephos; Temephosn; Tempephos; Tetrafenphos; O,O,O',O'-tetramethyl O,O'-(thiodi-4,1-phenylene) phosphorothioate; O,O,O',O'-tetramethyl O,O'-thiodi-p-phenylene; O,O'-(Thiodi-4,1-phenylene)-bis(O,O-dimethyl phosphorothioate); O,O'-(Thiodi-4,1-phenylene)phosphorothioic acid O,O,O',O'-tetramethyl ester;; O,O'-(Thiodi-4,1-phenylene) O,O,O',O'-tetra-methyl di(phosphorothioate); O,O'-(Thiodi-p-phenylene) O,O,O',O'-tetramethyl bis(phos-phorothioate); O,O'-thiodi-p-

600

phenylene phosphoro-thioate. Granular and emulsifiable concentrated herbicide. Registered by EPA as an insecticide. Has been used medically as an ectoparasiticide. Crystals; mp = 30 - 30.5°; almost insoluble in H_2O (< 0.000003 g/100 ml), soluble in C_6H_{14}, CH_3CN, CCl_4, Et_2O, C_7H_8, dichloro-ethane; LD_{50} (mrat orl) = 8600 mg/kg, (frat orl) = 13000 mg/kg, (rbt der 24 hr.) = 2181 mg/kg, (rat der) > 4000 mg/kg, (bee topical) = 0.00155 mg/bee); LC_{50} (mallard duck 5 day dietary) = 1200 mg/kg diet, (ring-necked pheasant 5 day dietary) = 170 mg/kg diet, (rainbow trout) = 31.8 mg/l; highly toxic to bees. *Am. Cyanamid; Cyanamid Agri. De Puerto Rico Inc.; Riverdale Chemical Co; Sanex Agro Inc.; Value Gardens Supply LLC.*

3664 Terazosin [anhydrous]
63590-64-7   9229   D

$C_{19}H_{25}N_5O_4$
1-(4-Amino-6,7-dimethoxy-2-quinazolinyl)-4-(tetrahydro-2-furoyl)piperazine.
Blavin; Flumarc; Fosfomic; Terazosin; Terazosina; Terazosine; Terazosinum; Vasomet.; Anhydrous Hydro-chloride salt [63074-08-8]. An α_1-adrenergic blocker related to prazosin. Used as an antihypertensive agent and in treatment of BPH. White crystals; mp = 272.6-274°; λ_m = 212, 245, 330 nm (a 65.7, 127.5, 24.0 H_2O); soluble in MeOH (3.37 g/100 ml), H_2O (2.97 g/100 ml), EtOH (0.41 g/100 ml), $CHCl_3$ (0.12 g/100 ml), Me_2CO (.1 mg/100 ml); insoluble in C_6H_{14}. *Abbott Labs Inc.*

3665 Terazosin hydrochloride dihydrate
70024-40-7   9229   D

$C_{19}H_{26}ClN_5O_4$
1-(4-Amino-6,7-dimethoxy-2-quinazolinyl)-4-(tetrahydro-2-furoyl)piperazine hydrochloride dihydrate.
Abbott-45975; Adecur; Deflox; Dysalfa; Flotrin; Heitrin; Hitrin; Hydracin; Hytracin; Hytrin; Hytrine; Hytrinex; Isontyn; Itrin; Magnurol; Sinalfa; Teralfa; Teraprost; Terazosin Abbot; Terazosin hydrochloride dihydrate; Terazosin Hydrochloride; Terazosin monohydrochloride dihydrate; Unoprost; Urodie; Uroflo; Vasocard; Vasomet; Vicard; Anhydrous Hydrochloride salt [63074-08-8]. An α_1-adrenergic blocker related to prazosin. Used to treat hypertension and benign antiprostatic hypertrophy. Crystals; mp = 271-274°; soluble in H_2O (2.42 g/100 ml); LD_{50} (mrat iv) = 277 mg/kg, (frat iv) = 293 mg/kg; [anhydrous hydrochloride]: mp = 278-279°; soluble in H_2O (76.12 g/100 ml); LD_{50} (mus iv) = 259.3 mg/kg. *Abbott Labs Inc.*

3666 Terbacil
5902-51-2   9230   G H P

$C_9H_{13}ClN_2O_2$
5-Chloro-3-(1,1-dimethylethyl)-6-methyl-2,4-(1H,3H)-pyrimidinedione.
5-24-07-00036 (Beilstein Handbook Reference); BRN 0643054; 3-t-Butyl-5-chloro-6-methyluracil; 3-tert-Butyl-5-chloro-6-methyluracil; 3-tert.Butyl-5-chlor-6-methyl-uracil; Caswell No. 821A; 5-Chloro-3-(1,1-dimethyl-ethyl)-6-methyl-2,4(1H,3H)-pyrimidinedione; 5-Chloro-3-tert-butyl-6-methyluracil; Compound 732; DPX-D732; Du Pont 732; Du Pont herbicide 732; EPA Pesticide Chemical Code 012701; Experimental herbicide 732; Geonter; HSDB 1418; 2,4(1H,3H)-Pyrimidinedione, 5-chloro-3-(1,1-dimethylethyl)-6-methyl-; Sinbar; Sinbar 80W; Terbacil; Turbacil; Uracil, 3-tert-butyl-5-chloro-6-methyl-. Selective general-use herbicide effective in the control of annual weeds and perennial grasses in sugarcane, apples, alfalfa, peaches, mints, and pecans. Used for control of annual broad-leaved weeds, most annual grasses and some perennial weeds. Registered by EPA as a herbicide. Crystals; mp = 175-177°; sg = 1.34; soluble in H_2O (0.0710 g/100 ml), DMF (31.7 g/100 ml), cyclohexanone (20.7 g/100 ml), methyl isobutyl ketone (9.7 g/100 ml), butyl acetate (7.9 g/100 ml), xylene (5.6 g/100 ml), sparingly soluble in mineral oil and petroleum ether; LD_{50} (rat orl) > 5000 mg/kg, (rbt der) > 5000 mg/kg; LC_{50} (Pekin duckling 8 day dietary) > 56000 mg/kg diet, (pheasant chick 8 day dietary) > 31450 mg/kg diet, (fiddler crab 48 hr.) > 1000 mg/l, (pumpkinseed sunfish 48 hr.) = 86 mg/l; non-toxic to bees. *DuPont UK; E. I. DuPont de Nemours Inc.*

3667 Terbium
7440-27-9   9232   G

Tb

Tb
Terbium.
EINECS 231-137-6; Terbium. A lanthanide element; phosphor activator, dope for solid-state devices. mp = 1356°; bp= 3230°; d = 8.27. *Atomergic Chemetals; Cerac; Rhône-Poulenc; Sigma-Aldrich Fine Chem.*

3668 Terbium oxide
12037-01-3   G

O_3Tb_2
Terbia.
EINECS 234-856-3; Terbium oxide; Terbium oxide (Tb_4O_7); Tetraterbium heptaoxide. *Atomergic Chemetals; Cerac; Rhône-Poulenc.*

3669 Terbufos
13071-79-9   9233   G P

$C_9H_{21}O_2PS_3$
S-t-Butylthiomethyl O,O-diethylphosphorodithioate.
4-01-00-03092 (Beilstein Handbook Reference); AC 92100; AI3-27920; BRN 1710115; Caswell No. 131A; CCRIS 4772; Contraven;

Counter; Counter 15G soil insecticide; Counter 15G soil insecticide-nematicide; ENT 27920; EPA Pesticide Chemical Code 105001; HSDB 6444; Phosphorodithioic acid, O,O-diethyl S-(((1,1-dimethylethyl)thio)methyl) ester; Phosphorodithioic acid, S-(((1,1-dimethylethyl)thio)methyl) O,O-diethyl ester; Phosphorodithioic acid S-((tert-butylthio)methyl) O,O-diethyl ester; S-(((1,1-Dimethylethyl)thio)methyl) O,O-diethyl phosphorodithioate; St-100; Terbufos. A soil insecticide and nematicide with stomach and contact action. A cholinesterase inhibitor. Used for control of soil insects in vegetable and fruit crops. Registered by EPA as an insecticide and fungicide. Used as a nematicide. Colorless liquid; mp = -29.2°, bp0.01 = 69°; d24 = 1.105; soluble in organic solvents, H_2O (0.45 - 1.5 mg/100 ml); LD50 (rat orl) = 1.6 mg/kg, (mus orl) = 5 mg/kg, (rat der) = 9.8 mg/kg, (rbt der) = 1.0 mg/kg, (bee topical) = 0.0041 mg/bee, (quail orl) = 15 mg/kg; LC50 (mallard duck 8 day dietary) = 185 mg/kg diet, (ring-necked pheasant 8 day dietary) = 145 mg/kg diet, (rainbow trout 96 hr.) = 0.01 mg/l, (bluegill sunfish 96 hr.) = 0.004 mg/l. *Aceto Agriculture Chemicals Corp.; Agriliance LLC; Am. Cyanamid; BASF Corp.; UAP - Platte Chemical; Uniroyal.*

3670   Terbuthylazine
5915-41-3   G P

C9H16ClN5
2-(t-Butylamino)-4-chloro-6-(ethylamino)-s-triazine.
Caswell No. 125B; ChlorCaragard; 6-Chloro-N-(1,1-dimethylethyl)-N'-ethyl-1,3,5-triazine-2,4-diamine; EINECS 227-637-9; EPA Pesticide Chemical Code 080814; G 13529; Gardeprim A 1862; Gardoprim; GS 13529; HSDB 6148; N2-tert-Butyl-6-chloro-N4-ethyl-1,3,5-triazine-2,4-diamine; Primatol M; Primatol-M80; s-Triazine, 2-(tert-butylamino)-4-chloro-6-(ethylamino)-; 1,3,5-Triazine-2,4-diamine, 6-chloro-N-(1,1-dimethyl-ethyl)-N'-ethyl-; Sorgoprim; Terbutazine; Terburthylazine; Terbutylazine; Terbutylethylazine; Turbulethylazin. Herb-icide, absorbed mainly by roots. Used for broad spectrum weed control. Registered by EPA as an algicide and herbicide. Colorless crystals; mp = 177 - 179°; sg20 = 1.188; soluble in H_2O (0.00085 g/100 ml), DMF (10 g/100 ml), EtOAc (4 g/100 ml), i-PrOH (1 g/100 ml), tetralin (1 g/100 ml), xylene (1 g/100 ml), n-octanol (1.43 g/100 ml); LD50 (rat orl) = 2160 mg/kg, (rbt der) > 3000 mg/kg; LC50 (rat ihl 4 hr.) > 3.51 mg/l air, (rainbow trout 96 hr.) = 4.6 mg/l, (goldfish 96 hr.) = 9.4 mg/l, (bluegill sunfish 96hr.) = 66 mg/l, (crucian carp 96 hr.) = 52 mg/l; non-toxic to bees. *Bio-Lab, Inc.; Syngenta Crop Protection; Vinings Industries Inc.*

3671   Terbutryn
886-50-0   G H P

C10H19N5S
1,3,5-Triazine-2,4-diamine,   N(1,1-dimethylethyl)-N'-ethyl-6-(methylthio)-.
4-Äthylamino-2-tert-butylamino-6-methylthio-s-triazin; A 1866; BRN

0611817; 2-tert-Butylamino-4-ethylamino-6-methylthio-s-triazine; 2-tert-Butylamino-4-ethylamino-6-methylmercapto-s-triazine; 2-tert-Butylamino-4-ethyl-amino-6-methylthio-1,3,5-triazine; 2-tert.Butylamino-4-äthylamino-6-methylthio-1,3,5-triazin; Caswell No. 125D; Clarosan; EPA Pesticide Chemical Code 080813; GS 14260; HS-14260; HSDB 1525; Igran; Igran 50; Igran 500; 2-Methylthio-4-ethylamino-6-tert-butylamino-s-tri-azine; N-(1,1-Dimethylethyl)-N'-ethyl-6-(methylthio)-1,3, 5-triazine-2,4-diamine; N2-tert-Butyl-N4-ethyl-6-methyl-thio-1,3,5-triazine-2,4-diamine; N-tert-Butyl-N-ethyl-6-methylthio-1,3,5-triazine-2,4-diamine; Prebane; Saterb; Short-stop E; Shortstop; Terbutrex; Terbutryn; Terbutryne. Selective herbicide absorbed by roots and foliage and used for pre-emergence and post-emergence weed control of most grasses in winter cerals, vegetables and citrus fruit, and aquatic weed control. Registered by EPA as a herbicide (cancelled). Colorless crystals; mp = 104-105°; bp0.06 = 154-160°; d20 = 1.115; soluble in H_2O (0.0025 g/100 ml at 20°), Me2CO (28 g/100 ml), C6H14 (0.9 g/100 ml), CH2Cl2 (30 g/100 ml), octanol (13 g/100 ml), MeOH (28 g/100 ml), C7H8 (4.5 g/100 ml), dioxane, Et2O, xylene, CHCl3, CCl4, DMF, sparingly soluble in petroleum ether; LD50 (rat orl) = 2045 mg/kg, (mus orl) = 500 mg/kg, (rat der) > 2000 mg/kg, (rbt der) > 10200 mg/kg; LC50 (rat ihl 4 hr.) = 8 mg/l, (bobwhite quail 8 day dietary) > 20000 mg/kg, (mallard duck 8 day dietary) > 4640 mg/kg, (rainbow trout 96 hr.) = 3 mg/l, (bluegill sunfish, carp, perch 96 hr.) = 4 mg/l; non-toxic to bees. *Agan Chemical Manufacturers; Chemolimpex; Ciba Geigy Agrochemicals; Ciba-Geigy Corp.; Makhteshim-Agan; Probelte; Syngenta Crop Protection.*

3672   Terephthalaldehydic acid
619-66-9   H

C8H6O3
4-Formylbenzoic acid.
4-10-00-02752 (Beilstein Handbook Reference); Benzoic acid, 4-formyl-; BRN 0471734; 4-Carboxybenzaldehyde; EINECS 210-607-4; 4-Formylbenzoic acid; HSDB 5719; NSC 15797; p-Carboxybenzaldehyde; p-Formylbenzoic acid; Terephthalaldehydic acid.

3673   Terephthalic acid
100-21-0   9238   H

C8H6O4
1,4-Benzenedicarboxylic acid.
4-09-00-03301 (Beilstein Handbook Reference); Acide terephtalique; Al3-16108; p-Benzenedicarboxylic acid; BRN 1909333; p-Carboxybenzoic acid; CCRIS 2786; p-Dicarboxybenzene; EINECS 202-830-0; HSDB 834; Kyselina tereftalova; NSC 36973; p-Phthalic acid; para-Phthalic acid; TA 12; TA-33MP; Tephthol; Terephthalic acid; WR 16262. Crystals; sublimes at 300°; λ_m = 241, 286 nm (ϵ = 17500, 1960, dioxane); insoluble in H_2O, EtOH, Et2O, CHCl3, CCl4, AcOH.

3674   Terephthaloyl chloride
100-20-9   H

C8H4Cl2O2
1,4-Benzenedicarbonyl dichloride.
4-09-00-03318 (Beilstein Handbook Reference); BRN 0607796; EINECS 202-829-5; HSDB 5332; NSC 41885; p-Phenylenedicarbonyl dichloride; p-Phthaloyl dichloride; p-Phthaloyl chloride; p-Phthalyl dichloride; Tere-phthalic acid chloride; Terephthalic acid dichloride; Terephthalic dichloride; Terephthaloyl chloride; Tere-phthalyl dichloride. Needles, mp = 83.5°; bp = 258, bp9 = 125°; soluble in Et2O.

3675   Terphenyl
26140-60-3   H

C18H14
Diphenylbenzene.
AI3-01405; Delowax OM; Delowax S; Diphenylbenzene; EINECS 247-477-3; Gilotherm OM 2; Santowax OM; Santowax R; Terbenzene; Terphenyl; Terphenyls; Triphenyl.

3676   α-Terpineol
98-55-5   9248   D G H

C10H18O
3-Cyclohexene-1-methanol, α,α,4-trimethyl-.
2-06-00-00067 (Beilstein Handbook Reference); AI3-00275; BRN 1906604; Carvomenthenol; CCRIS 3204; EINECS 202-680-6; EINECS 219-448-5; FEMA Number 3045; HSDB 5316; Lily of valley, artificial; p-Menth-1-en-8-ol; NSC 21449; PC 593; Terpenol; Terpilenol, α-; Terpineol; α-Terpineol; dl-α-Terpineol; Terpineol 350; Terpineol schlechthin. One of three isomers of terpineol. Has been used as an antiseptic. Liquid; [dl-form]: bp3 = 85°; d15 = 0.9386.

3677   α-Terpineol, acetate
80-26-2   H

C12H20O2
p-Menth-1-en-8-yl acetate.
4-06-00-00253 (Beilstein Handbook Reference); AI3-00522; BRN 3198769; EINECS 201-265-7; EINECS 234-183-5; FEMA No. 3047; p-Menth-1-en-8-ol, acetate; p-Menth-1-en-8-yl acetate; Ravensara; Terpinyl acetate. Used in aromatherapy. Liquid; bp40 = 140°, bp11 =

105°; d21 = 0.9659; [α]D = +52.5°; insoluble in H2O, soluble in EtOH, Et2O, C6H6. *Essex County Ltd.; Ravenwood.*

3678   Terpinolene
586-62-9   H

C10H16
1-Methyl-4-isopropylidene-1-cyclohexene.
AI3-24378; Cyclohexene, 1-methyl-4-(1-methylethyl-idene)-; EINECS 209-578-0; FEMA Number 3046; HSDB 5702; 4-Isopropylidene-1-methylcyclohexene; Isoterp-inene; Nofmer TP; p-Mentha-1,4(8)-diene; Tereben; 1,4(8)-Terpadiene; Terpinolen; Terpinolene; UN2541. Oil; bp = 186°; d15 = 0.8632; insoluble in H2O, soluble in C6H6, CCl4, freely soluble in EtOH, Et2O.

3679   Tetraammonium ethylenediaminetetra-acetate
22473-78-5   H

C10H28N6O8
Glycine, N,N'-1,2-ethanediylbis(N-(carboxymethyl)-, tetraammonium salt.
Acetic acid, (ethylenedinitrilo)tetra-, tetraammonium salt; EINECS 245-022-3; Glycine, N,N'-1,2-ethanediylbis(N-(carboxymethyl)-, tetraammonium salt; Tetraammonium ethylenediaminetetraacetate.

3680   Tetrabromobisphenol A
79-94-7   G H

C15H12Br4O2
4,4'-Isopropylidenebis(2,6-dibromophenol).
BA 59; Bromdian; CCRIS 6274; EINECS 201-236-9; FG 2000; Fire Guard 2000; Firemaster BP 4A; Great Lakes BA-59P; HSDB 5232; NSC 59775; Phenol, 4,4'-(1-methylethylidene)bis(2,6-dibromo-; Phenol, 4,4'-iso-propylidenebis(2,6-dibromo-; Saytex RB 100PC; Tetra-bromobisphenol A; Tetrabromodian; Tetrabromodi-phenylopropane. Reactive or additive source of bromine for flame retardancy; reactive intermediate for preparation of brominated epoxy resins, polycarbonates, and unsaturated polyesters; additive for ABS, PS, and phenolic resins. Reactive flame retardant used in the manufacture of epoxy, PC, ABS, phenolic, PS, and polyester resins, rubber; flame retardant intermediate. White powder; mp = 180 - 184°; d = 2.17; poorly soluble in H2O, soluble in Me2CO, C6H6, EtOH; acetone, benzene, alcohol; LD50 (rat orl) >50,000 mg/kg, (rbt der) >2000 mg/kg. *Dead Sea Bromine; Ethicon Inc.; Rohm & Haas Co.*

3681   Tetrabromodipentaerythritol
109678-33-3

G

$C_{10}H_{18}Br_4O_3$
1-Propanol, 3,3'-oxybis(2,2-bis(bromomethyl)-.
FR-1034; 1-Propanol, 3,3'-oxybis(2,2-bis(bromomethyl)-; TBDPE.
Flame retardant for PP extruded fibers; processing aid for ABS and HIPS. Powder; mp = 75-82°; LD50 (rat orl) > 5000mg/kg. AmeriHaas; Bromine; Dead; Sea.

3682   1,1,2,2-Tetrabromoethane
79-27-6   9261

G

$C_2H_2Br_4$
sym-Tetrabromoethane.
4-01-00-00162 (Beilstein Handbook Reference); Acetylene tetrabromide; Al3-08850; BRN 1098321; CCRIS 1272; EINECS 201-191-5; Ethane, 1,1,2,2-tetrabromo-; HSDB 1600; Muthmann's liquid; NSC 406889; s-Tetrabromoethane; TBE; Tetrabromoacetylene; 1,1,2,2-Tetrabromoaethan; 1,1,2,2-Tetrabromäthan; 1,1, 2,2-Tetrabromoetano; 1,1,2,2-Tetrabromoethane; 1,1,2,2-Tetrabroomethaan. Used as a solvent and in flotation of minerals for separation. mp = 0°; bp54 = 151°, bp = 243.5°; d^{20} = 2.9655; insoluble in H_2O, soluble in organic solvents; LD50 (mus ip) = 443 mg/kg.

3683   Tetrabromophenolphthalein
76-62-0   9262

$C_{20}H_{10}Br_4O_4$
3,3-Bis(3,5-dibromo-4-hydroxyphenyl)phthalide.
3,3-Bis(3,5-dibromo-4-hydroxyphenyl)-1(3H)-isobenzo-furanone; EINECS 200-974-9; 1(3H)-Isobenzofuranone, 3,3-bis(3,5-dibromo-4-hydroxyphenyl)-; NSC 21261; Phenolphthalein, 3',3",5',5"-tetrabromo-; 3',3",5',5"-Tetrabromophenolphthalein. Disodium salt used as a radiopaque medium in diagnostic work. Needles; mp = 296°; insoluble in H_2O, slightly soluble in EtOH, AcOH, soluble in Et_2O, alkali (violet).

3684   Tetrabromophthalic anhydride
632-79-1

G H

$C_8Br_4O_3$
3,4,5,6-Tetrabromophthalic anhydride.
5-17-11-00265 (Beilstein Handbook Reference); BRN 0018908; Bromphthal; Bromphthal; CCRIS 6201; Dion 6692; EINECS 211-

185-4; FG 400; FG 4000; FireMaster PHT 4; Great Lakes PHT4; HSDB 5438; NSC 4874; PHT 4; Phthalic acid, tetrabromo-, anhydride; Phthalic anhydride, tetrabromo-; Saytex® RB 49; Tetrabromo-phthalic acid anhydride; Tetrabromophthalic anhydride; 3,4,5,6-Tetrabromophthalic anhydride. Flame retardant in production of unsaturated polyester resins and rigid PU polyols; co-hardener for epoxy resins; cost efficient additive for latex emulsions; derivatives used as flame retardants in diverse applications (wire coating, and wool, etc.). Crystals; mp = 279-281°; insoluble in H_2O. *Ethyl Corp.; Great Lakes Fine Chem.*

3685   Tetrabutoxysilane
4766-57-8

G

$C_{16}H_{36}O_4Si$
Tetra-n-butoxysilane.
Butyl silicate; Butyl silicate (($BuO)_4Si$); CT1750; EINECS 225-305-8; NSC 89762; Silane, tetrabutoxy-; Silicic acid (H_4SiO_4), tetrabutyl ester; Silicon tetrabutoxide; T1750; Tetrabutoxysilane; Tetra-n-butoxysilane; Tetrabutyl ortho-silicate; Tetrabutyl silicate. Coupling agent, chem-ical intermediate, blocking agent, release agent, lubricant, primer, reducing agent. Used in heat exchange applications, as dielectric fluids; as a lubricant in airborne radar. Liquid; bp = 256°, bp3ks = 120°; d^{20} = 0.8990. *Degussa-Hüls Corp.*

3686   Tetrabutyl ammonium bromide
1643-19-2

G

$C_{16}H_{36}BrN$
Tetra-N-butylammonium bromide.
1-Butanaminium, N,N,N-tributyl-, bromide; EINECS 216-699-2; TBAB; Tetrabutylammonium bromide. Quaternary ammonium salt. Solid; mp = 103-104°. *Hawks Chem Co Ltd; Lancaster Synthesis Co.; Pentagon Chems. Ltd.; Sigma-Aldrich Fine Chem.; Zeeland Chem. Inc.*

3687   Tetrabutyldiethylhexamethylenediammon-ium ethylsulfate
68052-49-3

H

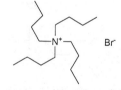

$C_{30}H_{68}N_2O_8S_2$
1,6-Hexanediaminium, bis(ethylsulfate).

N,N,N',N'-tetrabutyl-N,N'-di-ethyl-,

S,S'-Diethyl N,N'-hexane-1,6-diylbis(dibutylethyl-amm-onium) disulphate; EINECS 268-327-3; 1,6-Hexane-di-aminium, N,N,N',N'-tetrabutyl-N,N'-diethyl-, bis(ethyl-sulfate); Tetrabutyldiethylhexamethylenediammonium ethylsulfate.

3688   N,N,N',N'-Tetrabutylhexane-1,6-diamine
27090-63-7   H

$C_{22}H_{48}N_2$
1,6-Hexanediamine, N,N,N',N'-tetrabutyl-.
EINECS 248-219-2; 1,6-Hexanediamine, N,N,N',N'-tetrabutyl-; N,N,N',N'-Tetrabutylhexane-1,6-diamine.

3689   Tetrabutylthiuram disulfide
1634-02-2   G

$C_{18}H_{36}N_2S_4$
Tetrabutylthioperoxydicarbonic diamide.
4-04-00-00595 (Beilstein Handbook Reference); Akro-chem® TBUT; BRN 1715575; Disulfide, bis(dibutyl-thiocarbamoyl); EINECS 216-652-6; Tetrabutylthioperoxy-dicarbamic acid; Tetrabutylthiuram disulfide; Tetrabutyl-thiuram disulphide; Thioperoxydicarbonic diamide (((H_2N)C(S))_2S_2), tetrabutyl-; Thiuram disulfide tetrabutyl; Thiuram, tetrabutyl-, disulfide. Rubber accelerator, sulfur donor, accelerator, vulcanizing agent. Combustible liquid; d$_{20}^{20}$ = 1.03-1.06; soluble in CS_2, C_6H_6, $CHCl_3$, ligroin, gasoline, insoluble in H_2O. *Akrochem Chem. Co.*

3690   Tetrabutyltin
1461-25-2   H

$C_{16}H_{36}Sn$
Tetrabutylstannane.
4-04-00-04312 (Beilstein Handbook Reference); BRN 3648237; CCRIS 6322; EINECS 215-960-8; HSDB 6074; NSC 22330; NSC 28131; NSC 65524; Stannane, tetrabutyl-; Tetra-n-butylcin; Tetra-n-butyltin; Tetrabutyl-stannane; Tetrabutyltin; Tin, tetrabutyl-.

3691   Tetrachlorobenzene
634-66-2   H

$C_6H_2Cl_4$
1,2,3,4-Tetrachlorobenzene.
4-05-00-00667 (Beilstein Handbook Reference); AI3-01834; Benzene, 1,2,3,4-tetrachloro-; BRN 1910025; Caswell No. 825; CCRIS 5935; EINECS 211-214-0; EPA Pesticide Chemical Code 061101; HSDB 4268; NSC 50729; 1,2,3,4-Tetrachlorobenzene. Needles; mp = 47.5°; bp = 254°; λ_m = 282 nm (cyclohexane); insoluble in H_2O, slightly soluble in EtOH, very soluble in Et_2O, CS_2, AcOH, ligroin.

3692   1,2,4,5-Tetrachlorobenzene
95-94-3   H

$C_6H_2Cl_4$
Benzene, 1,2,4,5-tetrachloro-.
4-05-00-00668 (Beilstein Handbook Reference); AI3-01835; Benzene, 1,2,4,5-tetrachloro-; Benzene tetra-chloride; BRN 1618315; CCRIS 766; EINECS 202-466-2; HSDB 2733; NSC 27003; RCRA waste number U207; s-Tetrachlorobenzene. Crystals; mp = 139.5°; bp = 244.5°; λ_m = 232, 295 nm (cyclohexane); insoluble in H_2O, poorly soluble in EtOH, soluble in Et_2O, C_6H_6, $CHCl_3$, CS_2.

3693   Tetrachloroethane
79-34-5   9264   G H

$C_2H_2Cl_4$
1,1,2,2-Tetrachloroethane.
4-01-00-00144 (Beilstein Handbook Reference); Acetosal; Acetylene tetrachloride; AI3-04597; Bonoform; BRN 0969206; Caswell No. 826; CCRIS 578; Cellon; Di-chloro-2,2-dichloroethane; EINECS 201-197-8; EPA Pesticide Chemical Code 078601; Ethane, 1,1,2,2-tetrachloro-; HSDB 123; NCI-C03554; NSC 60912; RCRA waste number U209; s-Tetrachloroethane; TCE Tetra-chlorethane; Tetrachloroethane; Tetrachloroethane, 1,1,2,2-; Tetrachlorure d'acetylene; Westron. Solvent for fats and oils, used in dry cleaning, and manufacture of paint and varnish. Used in soil sterilization and in formulations of herbicides and insecticides. Liquid; mp = -43.8°; bp = 146.5°; d^{20}= 1.5953; sparingly soluble in H_2O (0.29 g/100 ml), soluble in organic solvents; LD$_{50}$ (rat orl) = 0.20 ml/kg.

3694   Tetrachloroethane
630-20-6   H

$C_2H_2Cl_4$
1,1,1,2-Tetrachlorethane.
4-01-00-00143 (Beilstein Handbook Reference); BRN 1733216; CCRIS 577; EINECS 211-135-1; Ethane, 1,1,1,2-tetrachloro-; HSDB

4148; NCI-C52459; RCRA waste no. U208; 1,1,1,2-Tetrachlorethane. Liquid; mp = -70.2°; bp = 130.5°; d^{20} = 1.5406; slightly soluble in H_2O, soluble in Me_2CO, C_6H_6, $CHCl_3$, freely soluble in EtOH, Et_2O.

3695 Tetrachloroethene
127-18-4   9265   D G H

C_2Cl_4
1,1,2,2-Tetrachloroethylene.
4-01-00-00715 (Beilstein Handbook Reference); AI3-01860; Ankilostin; Antisol 1; BRN 1361721; Carbon bichloride; Carbon dichloride; Caswell No. 827; CCRIS 579; ChemIDplus; Czterochloroetylen; Didakene; Dilatin PT; Dow-per; Dowper; EINECS 204-825-9; ENT 1,860; EPA Pesticide Chemical Code 078501; Ethene, tetrachloro-; Ethylene tetrachloride; Ethylene, tetrachloro-; Fedal-UN; HSDB 124; NCI-C04580; Nema; NSC 9777; PCE; PER; Perawin; Perc; Perchloorethyleen, per; Perchlor; Perchloräthylen, per; Perchloraethylen, per; Perchlorethylene; Perclene; Perclene D; Percloroetilene; Percosolv; Percosolve; Perk; Perklone; Persec; RCRA waste number U210; Tetlen; Tetracap; Tetrachlooretheen; Tetrachloräthen; Tetrachloraethen; Tetrachlorethylene; Tetrachloroethene; Tetrachloroethylene; Tetracloroetene; Tetraguer; Tetraleno; Tetralex; Tetravec; Tetrochloro-ethane; Tetroguer; Tetropil; UN1897. Solvent, degreaser. Dry-cleaning solvent; vermifuge; drying agent; degreasing metals. EPA Registered as a nematicide (cancelled). Colorless liquid; mp = -22°; bp = 121°; d_4^{15}= 1,6311, d_4^{20}= 1.6230; soluble in EtOH, Et_2O, $CHCl_3$, C_6H_6, slightly soluble in H_2O (0.01 g/100 ml); LD_{50} (mus orl) = 8850 mg/kg, LC_{50} (mus ihl) = 5925 ppm. *Asahi Chem. Industry; Ashland; Elf Atochem N. Am.; General Chem; ICI Spec.; OxyChem; PPG Ind.*

3696 Tetrachlorophthalic anhydride
117-08-8   G H

$C_8Cl_4O_3$
4,5,6,7-Tetrachloro-1,3-isobenzofurandione.
5-17-11-00260 (Beilstein Handbook Reference); AI3-09048; BRN 0211560; CCRIS 6202; CP 626; EINECS 204-171-4; HSDB 2922; NCI-C61585; Niagathal; NSC 1484; Phthalic anhydride, tetrachloro-; Tetrachloro-phthalic anhydride; Tetrathal. Flame retardant used with polyester resins and polyols. Crystals; mp = 254.5°; bp = 349-354° sublimes; d^{275} = 1.49; λ_m = 298 nm (MeOH); slightly soluble in Et_2O. *Monsanto Co.*

3697 Tetrachloropicolinonitrile
17824-83-8   H

$C_6Cl_4N_2$
Tetrachloro-2-cyanopyridine.
EINECS 241-784-6; Picolinonitrile, tetrachloro-; 2-Pyridinecarbonitrile, 3,4,5,6-tetrachloro-; 3,4,5,6-Tetra-chloropyridine-2-carbonitrile; Tetrachloro-2-cyanopyr-idine; Tetrachloropicolinonitrile.

3698 Tetrachloropyridine
2402-79-1   H

C_5HCl_4N
2,3,5,6-Tetrachloropyridine.
5-20-05-00421 (Beilstein Handbook Reference); BRN 0129639; EINECS 219-283-9; HSDB 5874; NSC 2009; Pyridine, 2,3,5,6-tetrachloro-; 2,3,5,6-Tetrachloropyr-idine; Tetrachloropyridine, 2,3,5,6-. Crystals; mp = 90.5°; bp = 250.5°; very soluble in EtOH, Et_2O, petroleum ether.

3699 Tetrachloroterephthalic dichloride
719-32-4   H

$C_8Cl_2O_2$
2,3,5,6-Tetrachloroterephthaloyl dichloride.
1,4-Benzenedicarbonyl dichloride, 2,3,5,6-tetrachloro-; EINECS 211-947-6; HSDB 5749; Perchloroterephthaloyl chloride; Terephthaloyl chloride, tetrachloro-; Tetra-chloroterephthaloyl chloride; 2,3,5,6-Tetrachlorotere-phthaloyl chloride.

3700 Tetracycline
60-54-8   9271   D G

$C_{22}H_{24}N_2O_8$
(4S,4aS,5aS,6S,12aS)-4-(Dimethylamino)-1,4,4a,5,5a,6, 11,12a-octahydro-3,6,10,12,12a-pentahydroxy-6-methyl-1,11-dioxo-2-naphthacenecarboxamide.
Abramycin; Abricycline; Achromycin; Agromicina; Am-bramicina; Ambramycin; Amycin; Bio-Tetra; Biocycline; Cefracycline; Cefracycline suspension; Centet (base); Ciclibion; Copharlan; Criseociclina; Cyclomycin; Cyclo-par; Democracin; Deschlorobiomycin; Dumocyclin; EINECS 200-481-9; Hostacyclin; HSDB 3188; Lexacycline; Limecycline; Liquamycin; Mericycline; Micycline; Mysteclin-F; Neocycline; NSC 108579; Omegamycin; Orlycycline; Panmycin; Piracaps (base); Polycycline; Polyotic; Purocyclina; Robitet; Roviciclina; SK-Tetracycline; Solvocin; Sumycin; Sumycin syrup; T-125; Talsutin; Tetra-co; Tetrabon; Tetraciclina; Tetracycl; Tetracyline; Tetracycline I; Tetracycline II; Tetra-cyclinum; Tetracyn; Tetradecin; Tetrafil; Tetraverine; Tsiklomistsin; Tsiklomitsin; Veracin; Vetacyclinum; Vet-quamycin-324 (free base).; component of: Mysteclin-F. Tetracycline antibiotic

606

produced in *Streptomyces* species. Antiamebic; antibacterial; antirickettsial. [trihydrate]: Dec 170-175°; $[\alpha]_D^{25}$ = -257.9° (0.1 N HCl), -239° (MeOH); λ_m = 220, 268, 355 nm (ϵ 13000, 18040, 13320 0.1N HCl); soluble in H_2O (1.7 mg/ml), MeOH (> 20 mg/ml); LD_{50} (rat orl) = 707mg/kg, (mus orl) = 808 mg/kg. *Bristol-Myers Squibb Co.; Pfizer Inc.*

3701 Tetradecane
629-59-4   H

C14H30
n-Tetradecane.
4-01-00-00520 (Beilstein Handbook Reference); AI3-04240; BRN 1733859; CCRIS 715; EINECS 211-096-0; HSDB 5728; NSC 72440; Tetradecane; n-Tetradecane. Liquid; mp = 5.8°; bp = 253,5°; d^{20} = 0.7628; insoluble in H_2O, soluble in CCl_4, very soluble in EtOH, Et_2O.

3702 Tetradec-1-ene
1120-36-1   G H

C14H28
n-Tetradec-1-ene.
AI3-10509; CCRIS 3785; EINECS 214-306-9; HSDB 1087; Neodene® 14; α-Tetradecene; 1-Tetradecene; 1-Tetradecylene; n-Tetradec-1-ene; NSC 66434. An intermediate for biodegradable surfactants and specialty industrial chemicals. Liquid; mp =-12°; bp = 233°; d^{25} = 0.7745; insoluble in H_2O, slightly soluble in CCl_4, soluble in C_6H_6, very soluble in EtOH, Et_2O. *Shell.*

3703 Tetradecyl methacrylate
2549-53-3   H

C18H34O2
2-Propenoic acid, 2-methyl-, n-tetradecyl ester.
EINECS 219-835-9; HSDB 5885; Methacrylic acid, tetradecyl ester; Myristyl methacrylate; Tetradecyl 2-methyl-2-propenoate; Tetradecyl methacrylate.

3704 Tetradifon
116-29-0   9273   G P

C12H6Cl4O2S
4-Chlorophenyl 2,4 5-trichlorophenylsulfone.
4-06-00-01636 (Beilstein Handbook Reference); AI3-23737;

Akaritox; Aracnol K; Aredion; Benzene, 1,2,4-trichloro-5-((4-chlorophenyl)sulfonyl)-; BRN 2292528; Caswell No. 836; CCRIS 4026; Childion; 4-Chlorophenyl 2,4,5-trichlorophenyl sulfone; 4-Chlorophenyl 2,4,5-trichlorophenyl sulphone; Dorvert; Duphar; EINECS 204-134-2; ENT 23,737; EPA Pesticide Chemical Code 079202; FMC 5488; HSDB 1773; Mition; NIA 5488; p-Chlorophenyl 2,4,5-trichlorophenyl sulfone; p-Chloro-phenyl 2,4,5-trichlorophenyl sulphone; Polacaritox; Roztoczol; Roztoczol extra; Roztozol; Sulfone, 2,4,4',5-tetrachlorodiphenyl; Sulfone, p-chlorophenyl 2,4,5-trichlorophenyl; Tedane; Tedane Combi; Tedion; Tedion V-18; 2,4,4',5-Tetrachloor-difenyl-sulfon; 2,4,4',5-Tetrachlor-diphenyl-sulfon; 2,4,4',5-Tetrachlorodiphenyl sulfone; 2,4,4',5-Tetrachlorodiphenyl sulphone; 2,4,4',5-Tetracloro-difenil-solfone; 3,4,6,4'-Tetrachlor-diphenyl-sulfon; Tetradichlone; Tetradifon; Tetradiphon; Tetrafidon; 1,2,4-Trichloro-5-((4-chlorophenyl)sulfonyl)-benzene; Turbair Acaricide; V-18. A selective acaricide for use against mite infestation in orchards, citrus fruit plantations, hop fields, groundnut plantations, vegetable plots,cotton fields and on ornamental plants; red spider mite control in horticultural crops. A long-acting, non-systemic acaricide used to control eggs of phytophagous mites on fruit trees. Registered by EPA (cancelled). Crystals; mp = 146°; d_{20} = 1.151; soluble in H_2O (0.000005 g/100 ml at 10°, 0.000008 g/100 ml at 20°, 0.000034 g/100 ml at 50°), Me_2CO (8.2 g/100 ml), C_6H_6 (14.8 g/100 ml), $CHCl_3$ (25.5 g/100 ml), cyclohexanone (20 g/100 ml), dioxane (22.3 g/100 ml), kerosene (1 g/100 ml), MeOH (1 g/100 ml), C_7H_8 (13.5 g/100 ml), xylene (11.5 g/100 ml); LD_{50} (mrat orl) > 14700 mg/kg, (rbt der) = 10000 mg/kg; LC_{50} (bobwhite quail, Japanese quail, pheasant, mallard duck 8 day dietary) > 5000 mg/kg diet, (carp 3 hr.) > 10 mg/l; non-toxic to bees. *Diachem; Duphar BV; Hortichem Ltd.; Uniroyal.*

3705 4,4,15,15-Tetraethoxy-3,16-dioxa-8,9,10, 11-tetrathia-4,15-disilaoctadecane
40372-72-3   G

C18H42O6S4Si2
3,16-Dioxa-8,9,10,11-tetrathia-4,15-disilaoctacane.
Aktisil PF 216; EINECS 254-896-5; 3,16-Dioxa-8,9,10,11-tetrathia-4,15-disilaoctadecane, 4,4,15,15-tetraethoxy-; 4, 4,15,15-Tetraethoxy-3,16-dioxa-8,9,10,11-tetrathia-4,15-disilaoctadecane. Filler for sulfur-cured systems. *Hoffmann Mining.*

3706 Tetraethoxysilane
78-10-4   3882   H

C8H20O4Si
Silicon tetraethoxide.
4-01-00-01360 (Beilstein Handbook Reference); AI3-15098; BRN 1422225; Dynasil A; EINECS 201-083-8; ES 100; ES 28; Ethyl orthosilicate; Ethyl silicate; Ethyl silicate, ((EtO)4Si); Etylu krzemian; HSDB 534; NSC 4790; Orthosilicic acid, tetraethyl ester; Silane, tetraethoxy-; Silester; Silicate d'ethyle; Silicate tetraethylique; Silicic acid, tetraethyl ester; Silicic acid (H4SiO4), tetraethyl ester; Silicon ethoxide; Silicon tetraethoxide; Silikan L; TEOS; Tetraethoxysilane;

Tetraethoxysilicon; Tetraethyl-O-silicate; Tetraethyl orthosilicate; Tetraethyl silicate; Tetraethylsilikat; UN1292. Intermediate for manufacture of ethyl silicate products; produces binders; chemical and heat-resistant paints, cements, weatherproofing; protective coatings. Used to weatherproof and harden stone and in the manufacture of weather- and acid-proof mortars and cements. Liquid; mp = -77°; bp = 168°; d= 0.94. *Akzo Chemie; Degussa-Hüls Corp.; Greeff R.W. & Co.; Monsanto Co.; PCR; Sigma-Aldrich Fine Chem.; Wacker Silicones.*

3707　　Tetraethyl lead
78-00-2　　　　　　　　9277　　　　　　　　H

$C_8H_{20}Pb$
Lead, tetraethyl.
4-04-00-04349 (Beilstein Handbook Reference); BRN 3903146; CCRIS 1565; Czteroetylek olowiu; EINECS 201-075-4; HSDB 841; Lead tetraethide; Lead tetraethyl; NA1649; NCI-C54988; NSC 22314; Piombo tetra-etile; Plumbane, tetraethyl-; RCRA waste number P110; Tetra(methylethyl)lead; Tetraethyl lead; Tetraethylolovo; Tetraethylplumbane. Has been used as a gasoline additive as an anti-knock agent, improving the octane rating of fuels. No longer used in the US and Europe. Colorless liquid; bp = 200°; d^{20} = 1.653; almost insoluble in H_2O, slightly soluble in EtOH, more soluble in C_6H_6; LD_{50} (rat orl) = 12.3 mg/kg. *Great Lakes Chemical Corp.*

3708　　Tetraethylene glycol
112-60-7　　　　　　　　　　　　　　　　　G H

$C_8H_{18}O_5$
3,6,9-Trioxaundecane-1,11-diol.
4-01-00-02403 (Beilstein Handbook Reference); AI3-01838; BRN 1634320; Carbitol, diethyl; EINECS 203-989-9; Ethanol, 2,2'-(oxybis(2,1-ethanediyloxy))bis-; HI-Dry; HSDB 843; NSC 1262; PEG-4; Tetraethylene glycol; 3,6,9-Trioxaundecane-1,11-diol. Lubricant for rubber molds, textile fibers, metalworking; in food and food pkg.; in cosmetics and hair preparations; pharmaceutic aid; in gas chromatography; in paints, paper coatings, polishes, ceramics. Oil; mp = -6.2°; bp = 328°; d^{15} = 1.1285; soluble in EtOH, Et_2O, CCl_4, dioxane, very soluble in H_2O; LD_{50} (rat orl) = 29 g/kg.

3709　　Tetraethylene glycol diacrylate
17831-71-9　　　　　　　　　　　　　　　　G

$C_{14}H_{22}O_7$
2-Propenoic acid oxybis(2,1-ethanediyloxy-2,1-ethane-diyl) ester. Acrylic acid, diester with tetraethylene glycol; Acrylic acid, oxybis(ethyleneoxyethylene) ester; Ageflex T4EGDA; Aronix M 240; CCRIS 3434; EINECS 241-789-3; Oxybis(2,1-ethanediyloxy-2,1-ethanediyl)diacrylate; PEG-4 diacrylate; PEG 200 diacrylate; Photomer™4013; Polyethylene glycol 1000 diacrylate; 2-Propenoic acid oxybis(2,1-ethanediyloxy-2,1-ethanediyl) ester; SR 268; TTEGDA; Viscoat 335HP. Fast curing monomer providing good adhesion and flexibility, low shrinkage, and good impact strength in inks, coatings, adhesives, photo resists, and rubber products. Liquid;

bp$_{0.3}$ = 120°; d$_{20}^{20}$ =1.11; slightly soluble in H_2O; f.p. <20°; flash pt. >93°. *Rit-Chem.*

3710　　Tetraethylene glycol-di-n-heptanoate
70729-68-9　　　　　　　　　　　　　　　　H

$C_{22}H_{42}O_7$
Heptanoic acid, oxybis(2,1-ethanediyloxy-2,1-ethanediyl) ester. BRN 2011337; EINECS 274-829-3; Heptanoic acid, oxy-bis(2,1-ethanediyloxy-2,1-ethanediyl) ester; Oxybis-(eth-ane-2,1-diyloxyethane-2,1-diyl) bisheptanoate; Tegdh; Tetraethylene glycol diheptanoate; Tetraethylene glycol di-n-heptanoate.

3711　　Tetraethylene glycol monomethyl ether
23783-42-8　　　　　　　　　　　　　　　　H

$C_9H_{20}O_5$
3,6,9,12-Tetraoxotridecanol.
EINECS 245-883-5; NSC 345692; Tetraethylene glycol monomethyl ether; 2,5,8,11-Tetraoxatridecan-13-ol; 3,6,9,12-Tetraoxotridecanol; 3,6,9,12-Tetraoxatridecan-1-ol.

3712　　Tetraethylene glycol, monobutyl ether
1559-34-8　　　　　　　　　　　　　　　　H

$C_{12}H_{26}O_5$
3,6,9,12-Tetraoxahexadecan-1-ol.
EINECS 216-322-1; Tetraethylene glycol, monobutyl ether; 3,6,9,12-Tetraoxahexadecan-1-ol.

3713　　Tetraethylenepentamine
112-57-2　　　　　　　　　　　　　　　　G H

$C_8H_{23}N_5$
3,6,9-Triazaundecane-1,11-diamine.
4-04-00-01244 (Beilstein Handbook Reference); AI3-10049; BRN 0506966; CCRIS 6275; DEH 26; EINECS 203-986-2; HSDB 5171; NSC 88603; TEPA; Tetra-ethylene pentamine; Tetraethylenepentamine; Tetraethyl-pentylamine; Tetren; Texlin® 400; UN2320. Lube oil additive; intermediate in asphalt additives, corrosion inhibitors, epoxy curing agents, surfactants, and in the paper industry. Solvent for sulfur, acid gases, various resins and dyes; saponifying agent for acidic materials; manufacture of synthetic rubber; dispersant in motor oils; intermediate for oil additives. Liquid; mp = -40°; bp =341.5°; d = 0.9980; soluble in H_2O. *Texaco; Tosoh; Union Carbide Corp.*

3714   Tetrafluoroethylene

116-14-3   G H

C2F4

1,1,2,2-Tetrafluoroethylene.

CCRIS 7738; EINECS 204-126-9; Ethene, tetrafluoro-; Ethylene, tetrafluoro-; Fluoroplast 4; HSDB 844; MS-122; Perfluoroethene; Perfluoroethylene; Tetrafluoroethene; Tetrafluoroethylene; 1,1,2,2-Tetrafluoroethylene; TFE; UN1081. Release agent, dry lubricant for use on cold molds, esp. for epoxy potting/encapsulating, PU, nylon, acrylics, PP, PC phenolics, PS, foams, rubber molding. Gas; mp = -142.5°; bp = -75.9°; d^{-76} = 1.519; insoluble in H2O.

3715   Tetraglyme

143-24-8   9282   G

C10H22O5

Tetraethylene glycol dimethyl ether.

4-01-00-02404 (Beilstein Handbook Reference); AI3-01596; Ansol E-181; Ansul ether 181AT; Bis(2-(2-methoxyethoxy)ethyl) ether; BRN 1760005; Dimethoxy-tetraethylene glycol; Dimethoxytetraglycol; E181 (Ether); EINECS 205-594-7; Ether, bis(2-(2-methoxyethoxy)ethyl); Glyme 5; Methyltetraglyme200; Nissan uniox MM 200; NSC 65624; 2,5,8,11,14-Pentaoxapentadecane; Tetra-ethylene glycol dimethyl ether. A trademark for tetraethylene glycol dimethyl ether (tetraglyme), a solvent. Liquid; bp = 275.3°; d^{20} = 1.0114; soluble in EtOH, Et2O, CCl4, freely soluble in H2O.

3716   Tetrahydroabietyl alcohol

13393-93-6   H

C20H36O

1-Phenanthrenemethanol, tetradecahydro-1,4a-dimethyl-7-(1-methylethyl)-.

Abietyl alcohol, tetrahydro-; AI3-04505; EINECS 236-476-3; 1-Phenanthrenemethanol, tetradecahydro-1,4a-dimethyl-7-(1-methylethyl)-, (1R,4aR,4bS,10aR)-; Tetra-decahydro-7-isopropyl-1,4a-dimethylphenanthren-1-methanol; Tetrahydroabietyl alcohol.

3717   Tetrahydro-3,5-dimethylol-4-pyrone

67845-26-5   H

C7H12O4

4H-Pyran-4-one, tetrahydro-3,5-bis(hydroxymethyl)-.

EINECS 267-305-0; 4H-Pyran-4-one, tetrahydro-3,5-bis(hydroxymethyl)-; Tetrahydro-3,5-bis(hydroxymethyl)-4H-pyran-4-one; Tetrahydro-3,5-dimethylol-4-pyrone.

3718   Tetrahydrofuran

.109-99-9   9285   G H

C4H8O

1,4-Epoxybutane.

Agrisynth THF; AI3-07570; Butane, α,δ-oxide; Butane, 1,4-epoxy-; Butylene oxide; CCRIS 6276; Cyclotetra-methylene oxide; Diethylene oxide; EINECS 203-726-8; 1,4-Epoxybutane; Furan, tetrahydro-; Furanidine; HSDB 125; Hydrofuran; NCI-C60560; NSC 57858; Oxacyclo-pentane; Oxolane; RCRA waste number U213; Tetra-hydrofuraan; Tetrahydrofuranne; Tetraidrofurano; Tetra-methylene oxide; THF; UN2056. Solvent used in Grignard reactions, reductions, and polymerizations; chemical intermediate and monomer. Clear, mobile liquid; mp = -108.3°; bp = 65°; d^{20} = 0.8892; soluble in H2O, CHCl3, very soluble in EtOH, Et2O, C6H6. *Arco; Ashland; BASF Corp.; Degussa-Hüls Corp.; Great Lakes Fine Chem.; Janssen Chimica; QO.*

3719   Tetrahydrofurfuryl alcohol

97-99-4   9287   G

C5H10O2

Tetrahydrofuryl carbinol.

5-17-03-00115 (Beilstein Handbook Reference); AI3-00104; BRN 0102723; CCRIS 2923; EINECS 202-625-6; FEMA No. 3056; 2-Furanmethanol, tetrahydro-; Furfuryl alcohol, tetrahydro-; HSDB 5314; 2-(Hydroxymethyl)-tetrahydrofuran; NSC 15434; QO Thfa; Tetrahydro-2-furancarbinol; Tetrahydro-2-furanmethanol; Tetrahydro-2-furanylmethanol; Tetrahydro-2-furfuryl alcohol; Tetra-hydro-2-furylmethanol; Tetrahydrofurfuryl alcohol; Tetra-hydrofurfurylalkohol; Tetrahydrofuryl carbinol; Tetra-hydrofurylalkohol; THFA. Solvent for vinyl resins, dyes for leather, chlorinated rubber, cellulose esters, coupling agent, solvent-softener for nylon. Liquid; mp <-80°; bp = 178°; d^{20} = 1.0524; miscible with H2O, freely soluble in EtOH, Me2CO. *Lancaster Synthesis Co.; Penta Mfg.; QO.*

3720   Tetrahydrofurfuryl methacrylate

2455-24-5   G

C9H14O3

Methacrylic acid tetrahydrofurfuryl ester.

Ageflex THFMA; AI3-08497; EINECS 219-529-5; HSDB 5461; Methacrylic acid tetrahydrofurfuryl ester; NSC 32634; 2-Propenoic acid, 2-methyl-, (tetrahydro-2-furanyl)methyl ester; Sartomer SR 203; SR 203; Tetrahydrofurfuryl methacrylate; THFMA. Used to produce anaerobic adhesives and sealants, printed circuit boards, artificial finger nails, modifier for hard rubber rolls, wire and cable coatings, screen printing inks, emulsion polymerization, plastic modifier, EB-curable coatings. Often stabilized with 4-methoxyphenol/-hydroquinone. Liquid; bp = 265°, bp4 = 83-84°; d= 1.040. *Rit-Chem; Sartomer.*

3721 Tetrahydromethylphthalic anhydride
11070-44-3   H

C9H10O3
Tetrahydromethyl-1,3-isobenzofurandione.
EINECS 234-290-7; 1,3-Isobenzofurandione, tetrahydromethyl-; Tetrahydromethylphthalic anhydride.

3722 Tetrahydro-1-naphthol
529-33-9   H

C10H12O
1,2,3,4-Tetrahydro-1-naphthol.
3-06-00-02457 (Beilstein Handbook Reference); AI3-07039; BRN 2046227; EINECS 208-459-0; 1-Hydroxy-tetralin; 1-Naphthalenol, 1,2,3,4-tetrahydro-; 1-Naphthol, 1,2,3,4-tetrahydro-; NSC 5172; Tetrahydro-1-naphthol; 1,2,3,4-Tetrahydro-α-naphthol; 1,2,3,4-Tetrahydro-1-naphthol; 1,2,3,4-Tetrahydronaphthalen-1-ol; α-Tetralol; Tetralin-1-ol; 1-Tetralol. Crystals; mp = 34.5°; bp = 255°, bp2 = 103°; d20 = 1.0996; λm = 265, 272 nm (ε = 464, 488, MeOH).

3723 Tetrahydrophthalic anhydride
85-43-8   H

C8H8O3
4-Cyclohexene-1,2-dicarboxylic anhydride.
5-17-11-00134 (Beilstein Handbook Reference); AI3-22626; Anhydrid kyseliny tetrahydroftalove; BRN 0082340; Butadiene-maleic anhydride adduct; EINECS 201-605-4; HSDB 846; Maleic anhydride adduct of butadiene; NSC 82642; Phthalic anhydride, 1,2,3,6-tetrahydro-; Rikacid TH; Tetrahydroftalanhydrid; Tetrahydrophthalic acid anhydride; Tetrahydrophthalic anhydride; THPA.

3724 Tetrahydrophthalimide
85-40-5   H

C8H9NO2
4-Cyclohexene-1,2-dicarboximide.
5-21-10-00130 (Beilstein Handbook Reference); BRN 0128764; CCRIS 3648; EINECS 201-602-8; HSDB 5006; Isoindole-1,3-dione, 3a,4,7,7a-tetrahydro-; NSC 59011; Tetrahydrophthalic acid imide; Tetrahydrophthalimide. Used in manufacture of pesticides Captan

and Captafol. *Allchem Ind.*

3725 Tetrahydrothiophene
110-01-0   9291   G

C4H8S
Thiophane.
AI3-30989; EINECS 203-728-9; HSDB 6122; NSC 5272; Pennodorant 1013; Pennodorant 1073; Tetrahydro-thiofen; Tetrahydrothiophen; Tetrahydrothiophene; Tetra-methylene sulfide; Thiacyclopentane; Thilane; Thio-fan; Thiolane; Thiophane; Thiophene, tetrahydro-; THT; UN2412. Solvent, intermediate and odorant used in fuel gases. Liquid; mp = -96.1°; bp = 121°; d20 = 0.9987; λm = 210 nm (ε = 1000, EtOH); insoluble in H2O, soluble in CHCl3, freely soluble in EtOH, Et2O, Me2CO, C6H6; LC50 (mus inh) = 27 mg/l. *Elf Atochem N. Am.*

3726 Tetralin
119-64-2   9294   G H

C10H12
1,2,3,4-Tetrahydronaphthalene.
AI3-01257; Bacticin; Benzocyclohexane; Caswell No. 842A; CCRIS 3564; EINECS 204-340-2; EPA Pesticide Chemical Code 055901; HSDB 127; Naphthalene, 1,2,3,4-tetrahydro-; NSC 77451; Tetrahydronaphthalene; 1,2,3,4-Tetrahydronaphthalene; Tetralin; Tetralina; Tetra-line; Tetranap. Chemical intermediate, solvent for greases, fats, oils, waxes; substitute for turpentine. Liquid; mp = -35.7°; bp = 207.6°; d25 = 0.9660; λm = 267, 274 nm (ε = 550, 603, EtOH); insoluble in H2O, soluble in CHCl3, very soluble in EtOH, Et2O. *Degussa-Hüls Corp.; DuPont.*

3727 Tetralol
530-91-6   9295   G

C10H12O
1,2,3,4-Tetrahydro-2-naphthalenol.
3-06-00-02460 (Beilstein Handbook Reference); Ac-tetrahydro-β-naphthol; AI3-05563; BRN 2046968; EINECS 208-497-8; HSDB 5679; 2-Hydroxytetralin; 2-Hydroxytetraline; 2-Naphthalenol, 1,2,3,4-tetrahydro-; 2-Naphthol, 1,2,3,4-tetrahydro-; NSC 44875; Tetrahydronaphthol-2; 1,2,3,4-Tetrahydro-2-naphthol; 1,2,3,4-Tetrahydronaphthalen-2-ol; 2-Tetralinol; Tetralol; β-Tetralol. Used as an antiseptic. Crystals; mp = 15.5°; bp12 = 140°; LD50 (rat orl) = 1.0 ml/kg.

3728 1-Tetralone
529-34-0   H

C10H10O

1,2,3,4-Tetrahydronaphthalen-1-one.
4-07-00-01015 (Beilstein Handbook Reference); AI3-19569; BRN 0607374; 3,4-Dihydro-1(2H)-naphthal-enone; 3,4-Dihydro-2H-naphthalen-1-one; EINECS 208-460-6; HSDB 5678; 1(2H)-Naphthalenone, 3,4-dihydro-; NSC 5171; 1-Oxotetralin; 1,2,3,4-Tetrahydronaphthalen-1-one; α-Tetralone; 1-Tetralone. Liquid; mp = 8°; bp6 = 115°; d^{16} = 1.0988; λm = 206, 249, 292 nm (ε = 26000, 12000, 1800, EtOH).

3729   Tetramethoxysilane
681-84-5   H

$$H_3CO-\underset{\underset{OCH_3}{|}}{\overset{\overset{OCH_3}{|}}{Si}}-OCH_3$$

C4H12O4Si
Silicic acid, tetramethyl ester.
AI3-11596; EINECS 211-656-4; HSDB 5511; Methyl orthosilicate; Methyl silicate ((CH3)4SiO4); Methyl silicate, ((MeO)4Si); Methyl silicate 28; Methyl silicate 39; MSP 150; NSC 67383; Silane, tetramethoxy-; Silicic acid (H4SiO4), tetramethyl ester; Silicic acid, methyl ester of ortho-; Silicic acid, tetramethyl ester; SIT 7510.0; Tetra-methyl orthosilicate; Tetramethoxysilane; Tetramethyl-silikat; TL 190; TMOS; TSL 8114; UN2606. Liquid; mp = -1.0°; bp = 121°; d^{20} = 1.0232; very soluble in EtOH.

3730   Tetramethrin
7696-12-0   9296   G P

C19H25NO4
2,2-Dimethyl-3-(2-methyl-1-propenyl)cyclopropanecarb-oxylic acid (1,3,4,5,6,7-hexa-hydro-1,3-dioxo-2H-iso-indol-2-yl)methyl ester.
AI3-27339; Bioneopynamin; Caswell No. 844; CCRIS 3284; (1-Cyclohexane-1,2-dicarboximido)methyl chrys-anthemumate; Cyclohex-1-ene-1,2-dicarboximidomethyl (±)-cis-trans-chrysanthemate; Cyclohex-1-ene-1,2-dicarb-oximidomethyl (1RS)-cis,trans-2,2-dimethyl-3-(2-methyl-prop-1-enyl)cyclopropane carboxylate; (1-Cyclohexene-1,2-dicarboximido)methyl-2,2-dimethyl-3-(2-methyl-propenyl)cyclopropanecarboxylate; Cyclopropanecarb-oxylic acid, 2,2-dimethyl-3-(2-methyl-1-propenyl)-,(1,3,4,5,6,7-hexahydro-1,3-dioxo-2H-isoindol-2-yl)-methyl ester; 2,2-Dimethyl-3-(2-methyl-1-propenyl)cyclo-propanecarboxylic acid (1,3,4,5,6,7-hexahydro-1,3-di-oxo-2H-isoindol-2-yl)methyl ester; 2,2-Dimethyl-3-(2-methylpropenyl)cyclopropanecarboxylic acid ester with N-(hydroxymethyl)-1-cyclohexene-1,2-dicarboximide; EINECS 231-711-6; ENT 27339; EPA Pesticide Chemical Code 069003; FMC-9260; (1,3,4,5,6,7-Hexahydro-1,3-dioxo-2H-isoindol-2-yl) methyl 2,2-dimethyl-3-(2-methyl-1-propenyl)cyclopropanecarboxylate; HSDB 6738; Insec-tol; Killgerm® Py-Kill W; Multicide; N-(3,4,5,6-Tetra-hydrophthalimido)methyl-cis,trans-chrysanthemate; N-(3,4,5,6-Tetrahydrophthalimido)-methyl dl-cis-trans-chrysanthemate; N-(Chrysanthemoxymethyl)-1-cyclohex-ene-1,2-dicarboximide; Neopynamin; Neopinamin; Neopynamin forte; NIA-9260; Niagara nia-9260; NSC 190939; Phthalthrin; d-Phthalthrin; Py-kill; SP-1103; SP 1103 forte; Sumitomo SP-1103; Tetralate; 2,3,4,5-Tetrahydrophthalimidomethylchrysanthemate; 3,4,5,6-Tetrahydro-phthalimidomethylester der dl-cis-trans-chrysanthemumsaeure; 3,4,5,6-Tetrahydrophthalimido-methyl (±)-cis-trans-chrysanthemate;

3,4,5,6-Tetrahydro-phthalimidomethyl cis and trans dl chrysanth-emummonocarboxylic acid; Tetramethrin; d-Tetra-methrin; Tetramethrine; Tetramethrinum; Tetrametrina; Weo-Pynamin;. Non-systemic insecticide with rapid knockdown. Used in combination with synergists such as piperonyl butoxide for control of flies, cockroaches, mosquitoes, wasps and other insect pests. Used for control of flies in livestock houses. Registered by EPA as an insecticide. Liquid; mp = 65-80°; d$^{20}_{20}$ = 1.108; poorly soluble in H2O (0.00046 g/100 ml), very soluble in organic solvents; LD50 (mus orl) = 1000 mg/kg. Killgerm Chemicals Ltd.

3731   Tetramethylammonium hydroxide
75-59-2   9297   G

$$-\overset{\overset{|}{}}{\underset{\underset{|}{}}{N^+}}-\quad OH^-$$

C4H13NO
N,N,N-trimethylmethanaminium hydroxide.
Ammonium, tetramethyl-, hydroxide; EINECS 200-882-9; Hydroxyde de tetramethylammonium; Methanaminium, N,N,N-trimethyl-, hydroxide; NMD 3; NMW-W; Tetramethylammonium hydroxide; TMAH; UN1835. Usually marketed as a 10% aqueous solution. d$^{25}_{4}$ = 1.00. [Pentahydrate]; crystals; mp = 63°. Fluka; Janssen Chimica; Sigma-Aldrich Fine Chem.

3732   Tetramethyl decynediol
126-86-3   G H

C14H26O2
2,4,7,9-Tetramethyl-5-decyne-4,7-diol.
AI3-07159; EINECS 204-809-1; HSDB 5612; NSC 5630; Surfynol® 104; Surfynol® 104A; Surfynol® 104E; Syrfynol® 104; Tetramethyl decynediol. Defoamer and dye dispersant in paints, inks, dyestuffs, pesticides; surfactant in rinse aids; substrate pigment wetting agent for industrial coatings and adhesives; wetting agent for industrial cleaners; viscosity reducer for vinyl dispersions. Used with ethylene glycol as a wetting agent, defoamer, dispersant, viscosity stabilizer. Solid; mp = 40-42°; bp = 254-255°. Air Products & Chemicals Inc.

3733   Tetramethyldisiloxane
3277-26-7   G

C4H14OSi2
Bis(dimethylsilyl) ether.
Bis(dimethylsilyl) oxide; CT2030; 1,3-Dihydrotetra-methyldisiloxane; Dimethylsilyl ether; Disiloxane, 1,1,3,3-tetramethyl-; EINECS 221-906-4; NSC 155369; Tetramethyldisiloxane; 1,1,3,3-Tetramethyldisiloxane. Coupling agent, chemical intermediate, blocking agent, release agent, lubricant, primer, reducing agent. Liquid; bp = 71°; d = 0.7600; LD50 (mus orl) = 3 g/kg. Degussa-Hüls Corp.

3734   Tetramethyl-1,2-ethanediamine
110-18-9   H

611

C6H16N2

N,N,N',N'-Tetramethyl-1,2-ethanediamine.

Al3-26631; 1,2-Bis-(dimethylamino)ethane; CCRIS 4870; Dimethyl(2-(dimethylamino)ethyl)amine; EINECS 203-744-6; 1,2-Ethanediamine, N,N,N',N'-tetramethyl-; Ethylenediamine, N,N,N',N'-tetramethyl-; HSDB 5396; Propamine D; Temed; Tetrameen; Tetramethyldiamino-ethane; N,N,N',N'-Tetramethyl-1,2-ethanediamine; Tetra-methyl ethylene diamine; N,N,N',N'-Tetramethyl-ethylenediamine; TMEDA; UN2372. A liquid catalyst miscible with both water and organic liquids. Liquid; mp = -55°; bp = 121°; d^{25} = 0.77; soluble in H_2O, organic solvents; LD_{50} (rat orl) = 1020 mg/kg. *Harcros.*

3735   Tetramethylsilane
75-76-3   G H

C4H12Si

Silicon, tetramethyl-.

CT2050; EINECS 200-899-1; NSC 5210; Silane, tetramethyl-; Silicon, tetramethyl-; Tetramethyl silane; Tetramethylsilane; Tetramethylsilicane; UN2749. Coup-ling agent, chemical intermediate, blocking agent, release agent, lubricant, primer, reducing agent. Liquid; mp = -99°; bp = 26.6°; d^{19} = 0.648; insoluble in H_2O, soluble in organic solvents. *Degussa-Hüls Corp.*

3736   Tetramethylthiuram monosulfide
97-74-5   G H

C6H12N2S3

Thiodicarbonic diamide ([(H2N)C(S)]2S), tetramethyl-.

4-04-00-00238 (Beilstein Handbook Reference); Aceto TMTM; Al3-00984; Ancazide IS; Bis(dimethylthiocarb-amoyl)sulfide; BRN 1775650; CP 2113; Cyuram MS; EINECS 202-605-7; Ekagom TM; Formamide, 1,1'-thiobis(N,N-dimethylthio-; HSDB 2902; Monex; Mono-thiurad; Monosulfure de tetramethylthiurame; Mono-thiuram; NSC 3400; Pennac MS; Perkacit® TMTM; Sulfide, bis(dimethylthiocarbamoyl); Tetramethylthiuram monosulfide; Tetramethylthiuram monosulphide; Thiodi-carbonic diamide (((H2N)C(S))2S), tetramethyl-; Thiuram MM; Thiuram monosulfide, tetramethyl-; TMTM; TMTMS; Unads®; USAF B-32; USAF EK-P-6255; Vulkacit MS; Vulkacit thiuram MS/C; Vulkacit Thiuram MS; Vulcaid 222. An accelerator and activator for natural rubber nitrile-butadiene, and butyl rubber. A nonstaining and nondiscoloring delayed action accelerator; has a short sharp curing range with normal to high sulfur in natural rubber; used in natural, SBR, butyl, nitrile and neoprene rubbers for wire insulation, druggist sundries, mechanicals, sponges. Used alone or in combination in NR, SBR, NBR, butyl rubber, neoprene, and reclaim rubber. Crystals; mp = 109.5°; d^{25} = 1.37; λ_m = 279 nm (ε = 16218, MeOH); insoluble in H_2O, poorly soluble in Et_2O, soluble in EtOH, Me_2CO, C_6H_6, $CHCl_3$. *Akrochem Chem. Co.; Akzo Chemie; Uniroyal; Vanderbilt R.T. Co. Inc.*

3737   m-Tetramethylxylene diisocyanate
2778-42-9   H

C14H16N2O2

1,3-Bis(1-isocyanato-1-methylethyl)benzene.

Benzene, 1,3-bis(1-isocyanato-1-methylethyl)-; BRN 2811946; EINECS 220-474-4; Isocyanic acid, α,α,α',α'-tetramethyl-m-xylylene; Isocyanic acid, m-phenylene-diisopropylidene ester; m-Tetramethylxylene diisocyan-ate; m-TMXDI; Tetramethyl-m-xylylene diisocyanate.

3738   Tetraoctylstannane
3590-84-9   H

C32H68Sn

Tetra-n-octylstannane.

4-04-00-04314 (Beilstein Handbook Reference); BRN 3907209; EINECS 222-733-7; NSC 65527; Stannane, tetraoctyl-; Tetraoctylstannane; Tetra-n-octylstannane; Tetraoctyltin; Tetra-n-octyltin; Tin, tetraoctyl-.

3739   Tetrapropenylsuccinic acid
27859-58-1   H

C16H28O4

Butanedioic acid, (tetrapropenyl)-.

Butanedioic acid, (tetrapropenyl)-; EINECS 248-698-8; Succinic acid, (tetrapropenyl)-; (Tetrapropenyl)butane-dioic acid; (Tetrapropenyl)succinic acid.

3740   Tetrapropenylsuccinic anhydride
26544-38-7   H

C16H26O3

Dihydro-3-(tetrapropenyl)furan-2,5-dione.

Al3-28007; Dihydro-3-(tetrapropenyl)-2,5-furandione; Di-hydro-3-(tetrapropenyl)furan-2,5-dione; DSA; DSA (cross-linking agent); EINECS 247-781-6; 2,5-Furandione, di-hydro-3,3,4,4-tetra-1-propenyl-; 2,5-Furandione, dihydro-3-(tetrapropenyl)-; RD 174; Succinic anhydride, (tetra-propenyl)-; Tetrapropenylsuccinic anhydride.

3741   Tetrapropoxysilane
682-01-9   G

$$C_3H_7O-\underset{\underset{OC_3H_7}{|}}{\overset{\overset{OC_3H_7}{|}}{Si}}-OC_3H_7$$

C12H28O4Si
Tetra-n-propoxysilane.

CT2090; EINECS 211-659-0; Silicic acid (H4SiO4), tetrapropyl ester; Tetrapropyl orthosilicate; Tetrapropyl silicate. Coupling agent, chemical intermediate, blocking agent, release agent, lubricant, primer, reducing agent. Liquid; bp = 226°, bp5 = 94°; d^{20} = 0.9158; soluble in CCl4, CS2. *Degussa-Hüls Corp.*

3742 Tetrapropylenephenyl phenyl ether
68938-96-5   H

C24H34O
Benzene, phenoxytetrapropylene-.
Benzene, phenoxytetrapropylene-; EINECS 273-153-6; Phenoxytetrapropylenebenzene; Tetrapropylenephenyl phenyl ether.

3743 Tetrasodium edetate
64-02-8   3544   G P

C10H12N2Na4O8
Tetrasodium ethylenediaminetetraacetate.
AI3-17182; Aquamoline BC; Aquamollin; Calsol; Caswell No. 846; CCRIS 6797; Celon E; Celon H; Celon IS; Cheelox BF; Cheelox BF-12; Cheelox BF-13; Cheelox BF-78; Cheelox BR-33; Chelest 400; Chelon 100; Chemcolox 200; Chemcolox 240 powder; Clewat S 2; Clewat T; Complexone; Conigon BC; Distol; Distol 8; Edathanil tetrasodium; Edetate Sodium; Edetic acid tetrasodium salt; EDTA sodium; EDTA, tetrasodium; EINECS 200-573-9; Endrate tetrasodium; EPA Pesticide Chemical Code 039107; Ergon; Ergon B; Hamp-ene 100; Hamp-ene 100S; Hamp-ene 215; Hamp-ene 220; Hamp-ene Na4; HSDB 5003; Irgalon; Kalex; Kemplex 100; Komplexon; Kutrilon CS; Metaquest C; Na4EDTA; Natrii edetas; Natrium aedeticum; Nervanaid B; Nervanaid B liquid; Nullapon; Nullapon B; Nullapon BF-12; Nullapon BF-78; Nullapon BFC; Nullapon BFC Liquid; Nullapon BFC Xonc Beads; Nullapon BFC Xonc; Perma-kleer 100; Perma kleer 50 crystals; Perma kleer tetra CP; Questex; Questex 4; Sequestrene; Sequestrene 30A; Sequestrene Na 4; Sequestrene ST; Sodium (edetate de); Sodium edetate; Sodium EDTA; Syntes 12A; Syntron B; Tetracemate tetrasodium; Tetracemin; Tetranatrium ethylendiamin-tetraacetat; Tetrasodium (ethylenedinitrilo)tetraacetate; Tetrasodium edetate; Tetrasodium EDTA; Tetrine; Trilon B; TST; Tyclarosol; Versene; Versene 67; Versene 100; Versene 220; Versene beads; Versene FE 3; Versene flake; Versene powder; Versene powder tetra sodium; Warkeelate PS-42; Warkeelate PS-43; Warkeelate PS-47; Warkeelate S-42. A general purpose chelating agent; complexes Ca, Mg and other common metals over wide pH range; used in pulp/paper processing. Complexes most common metals over pH range, iron at acidic pH; chelating agents used in soaps, detergents, water treatment, metal finishing and plating, synthesis of polymers and photographic products. Registered by EPA as an antimicrobial, fungicide, herbicide and insecticide. FDA approved for IM, IV injectables, inhalants, ophthalmics, orals, topicals. Used in iv, im injectables as a chelating agent, used in inhalants, ophthalmics, orals, topicals and drug stabilization, treatment of heavy metal poisoning. Solid; mp > 300°; soluble in H2O (103 g/100 ml), pH = 11.3; slightly soluble in EtOH. LD50 (mus ip) = 330 mg/kg. *ABC Compounding Co.; Akzo Nobel; Biosentry, Inc.; Champion Chemical Co.; Chardon Laboratories, Inc; Chemplex; Complex Quimica SA; Edsan Chemical Co.; GFS; Great Western; Hampshire; J.I. Holcomb Manufacturing Co.; PET Chemicals; Pioneer Chemical Co.; Rhône-Poulenc; Spectrum Chem. Manufacturing; Unit Chemical Corp; Universal Laboratories.*

3744 Tetrazolium chloride
1871-22-3   9316   G

C40H36Cl2N8O2
2,4,5-Triphenyltetrazolium chloride.
AI3-50892; 4-Anisyltetrazolium blue; Blue tetrazolium; Blue tetrazolium chloride; BT (VAN); 3,3'-Dianisole-bis(4,4'-(3,5-diphenyl)tetrazoliumchloride); 3,3'-(3,3'-Di-methoxy-4,4'-biphenylene)bis(2,5-diphenyl-2H-tetrazolium) chloride; Dimethoxy neotetrazolium; Ditetra-zolium chloride; EINECS 217-488-8; NSC 27623; Tetra-zolium blue; 2H-Tetrazolium, 2,2'-(3,3'-dimethoxy(1,1'-biphenyl)-4,4'-diyl)bis(3,5-diphenyl-, dichloride; TTC; 3,3'-(3,3'-Dimethoxy-4,4'-biphenylene)bis(2,5-diphenyl-2H-tetrazolium chloride); BT; Blue Tetrazolium chloride. Used in germination and viability testing of seeds. Also used as a stain for bacteria and molds and to detect redox enzymes in cells. Yellow crystals; mp = 242-245° (dec); slightly soluble in H2O, soluble in MeOH, EtOH, CHCl3, insoluble in non-polar organic solvents. *Dajac Labs.; U.S. BioChem.*

3745 Tetryl
479-45-8   6602   G

C7H5N5O8

N-Methyl-N,2,4,6-tetranitrobenzenamine.

4-16-00-00895 (Beilstein Handbook Reference); Aniline, N-methyl-N,2,4,6-tetranitro-; Benzenamine, N-methyl-N,2,4,6-tetranitro-; BRN 0964788; CCRIS 3143; CE; EINECS 207-531-9; HSDB 2857; N-Methyl-N-picryl-nitramine; N-Methyl-N,2,4,6-tetranitroaniline; N-Methyl-N-2,4,6-tetranitrobenzenamine; Nitramine; NSC 2166; Picrylmethylnitramine; N-Picryl-N-methylnitramine; Picrylnitromethylamine; Tetralite; Tetril; Tetryl; 2,4,6-Tetryl; Trinitrophenylmethylnitramine; 2,4,6-Trinitrophenyl-methylnitramine; 2,4,6-(Trinitrophenyl)methylnitroamine; 2,4,6-Trinitrophenyl-N-methylnitramine; UN0208. Used in explosives; a detonator known as tetryl contains 0.4 gram tetranitrophenylmethylnitramine, and 0.3 gram of a mixture of 87.5% mercury fulminate and 12.5% potassium chlorate. Also used as an acid-base indicator. mp = 131.5°, explodes at 180°; d^{10} = 1.57; λ_m = 224 nm (ϵ = 22800, MeOH); insoluble in H_2O, CS_2, slightly soluble in EtOH, $CHCl_3$, Et_2O, soluble in Me_2CO, C_6H_6, C_5H_5N.

3746 Thallium
7440-28-0   9327   G

Tl

Tl
Thallium.

EINECS 231-138-1; HSDB 4496; Ramor; Thallium; Thallium, elemental; Thallium, metallic. Metallic elem-ent; thallium salts, mercury alloys, low-melting glasses, rodenticides, photoelectric applications, electrodes in dissolved oxygen analyzers. mp = 303°; bp = 1457°; d = 11.85. *Atomergic Chemetals; Cerac; Noah Chem.; Sigma-Aldrich Fine Chem.*

3747 Thallium chloride
7791-12-0   9331   G

Tl—Cl

CITl

Thallium(I) chloride-^{201}Tl.
EINECS 232-241-4; HSDB 6066; NSC 15197; RCRA waste number U216; Thallium chloride; Thallium chloride (TlCl); Thallium monochloride; Thallium(1+) chloride; Thallium(I) chloride; Thallous chloride. Radioactive agent used as a diagnostic aid. Crystals; mp = 430°; bp = 720°; d = 7.0; soluble in H_2O (4 mg/ml), insoluble in EtOH. *Amersham Corp.*

3748 Thenium closylate
4304-40-9   9349   G V

C21H24ClNO4S2

N,N-Dimethyl-N-(2-phenoxyethyl)-2-thiophenmethanaminium salt with 4-chlorobenzene-sulfonic acid (1:1).

611 C 65; Ammonium, dimethyl(2-phenoxyethyl)-2-the-nyl-, salt with p-chlorobenzenesulfonic acid (1:1); Bancaris; Canopar; Closilate de thenium; Closilato de tenio; Dimethyl(2-phenoxyethyl)-2-thenylammonium p-chlorobenzenesulfonate; Dimethyl(2-phenoxyethyl)-2-thenylammonium closylate; N,N-Dimethyl-N-(2-phen-oxyethyl)-N-(2-thenyl)ammonium 4-chlorbenzol-sulfonat; EINECS 224-318-6; NSC 106569; Theni closylas; Thenii closilas; Thenium closilate; Thenium closylate; Thenium p-chlorobenzenesulfonate; 2-Thiophenemethanaminium, N,N-dimethyl-N-(2-phenoxyethyl)-, 4-chlorobenzenesulf-onic acid. Used in veterinary medicine as an anthelmintic. Crystals; mp = 159-160°; soluble in H_2O (0.6 g/100 ml at 20°). *Burroughs Wellcome Inc.*

3749 Thiabendazole
148-79-8   9360   D G

C10H7N3S

1H-Benzimidazole, 2-(4-thiazolyl)-.

AI3-50598; APL-luster; Benzimidazole, 2-(4-thiazolyl)-; 1H-Benzimidazole, 2-(4-thiazolyl)-; 4-(2-Benzimid-azolyl)thiazole; Biogard; Bioguard; Bovizole; BRN 0611403; Captan T; Caswell No. 849A; CCRIS 4510; Chemviron TK 100; Cropasal; E-Z-Ex; EINECS 205-725-8; EPA Pesticide Chemical Code 060101; Eprofil; Equizole; Equizole A; G 491; Helmindrax octelmin; Hokustar HP; HSDB 2027; Lombristop; Mertec; Mertect; Mertect 160; Metasol TK-100; Mintesol; Mintezol; Mintezole; Minzolum; MK 360; Mycozol; Nemapan; NSC 525040; Omnizole; Ormogal; Pitrizet; Polival; Rival; RPH; RTU Flowable Fungicide; Sanaizol 100; Sistesan; Storite; Syntol M100; Tbdz; TBZ; TBZ-6; TBZ 60W; Tebuzate; Tecto; Tecto 10P; Tecto 40F; Tecto 60; Tecto RPH; Testo; Thiaben; Thiabendazol; Thiabendazole; Thiabendazo-lum; Thiabenzole; 2-Thiazole-4-ylbenzimidazole; 2-(1,3-Thiazol-4-yl)benzimidazole; 2-(4-Thiazolyl)-1H-benz-imidazole; 2-(4-Thiazolyl)benzimidazole; Thibenzol; Thibenzole; Thibenzole 200; Thibenzole ATT; Thiprazole; Tiabenda; Tiabendazol; Tiabendazole; Tiabendazolum; Tobaz; Top form wormer; Tresaderm; Triasox. Systemic fungicide with protective and curative action. Absorbed by leaves and roots; used for control of fungus in fruits, vegetables and cereals. Crystals; mp = 300° (dec); sublimes 305°; soluble in H_2O (250 mg/l at pH 2-5), more soluble in organic solvents; LD$_{50}$ (rat orl) = 2080 mg/kg, 3300 mg/kg. *Agrichem (International) Ltd.; BASF Corp.; Ciba Geigy Agrochemicals; DowElanco Ltd.; MSD Agvet; Pennwalt Corp.*

3750 Thiambutosine
500-89-0   G

C19H25N3OS

Thiourea, N-(4-butoxyphenyl)-N'-[4-(dimethylamino)-phenyl]-.

4-13-00-01179 (Beilstein Handbook Reference); BRN 2819159; 4-Butoxy-4'-(dimethylamino)thiocarbanilide; 1-(p-Butoxyphenyl)-3-(p-dimethylaminophenyl)-2-thiourea; Carbanilide, 4-butoxy-4'-(dimethylamino)thio-; Ciba 1906; EINECS 207-914-0; NSC 682; SU 1906; Summit 1906; Thiambutosine; Thiambutosinum; Thiourea, N-(4-butoxyphenyl)-N'-(4-(dimethylamino)phenyl)-; Tiambutosina. A proprietary preparation containing thiambutosine, a leprostatic agent.

3751 Thifensulfuron methyl
79277-27-3   9387   G P

$C_{12}H_{13}N_5O_6S_2$

3-[[[[(4-Methoxy-6-methyl-1,3,5-triazin-2-yl)amino]carb-onyl]amino]sulfonyl]-2-thiophenecarboxylic acid methyl ester.

Caswell No. 573S; DPX-M 6316; DPX-M6316; EPA Pesticide Chemical Code 128845; Harmony; Harmony Extra; INM 6316; 3-(((((4-methoxy-6-methyl-1,3,5-triazin-2-yl)amino)carbonyl)amino)sulfonyl)-, methyl ester; 3-[[[[(4-methoxy-6-methyl-1,3,5-triazin-2-yl)amino]-carb-onyl]amino]sulfonyl]-2-thiophenecarboxylic acid methyl ester; Methyl 3-(4-methoxy-6-methyl-1,3,5-triazin-2-ylcarbamoylsulfamoyl)-2-thenoate; methyl 3-(((((4-meth-oxy-6-methyl-1,3,5-triazin-2-yl)amino)carbonyl)-amino)-sulfonyl)-2-thiophenecarboxylate; Pinnacle; Refine; Thifensulfuron Me; Thiameturon-methyl; Thia-meturon methyl ester; Thiophenecarboxylic acid; 2-Thio-phenecarboxylic acid,3-(((((4-methoxy-6-methyl-1,3,5-tri-azin-2-yl)amino)carbonyl)amino)sulfonyl)-, methyl ester. Used to control annual dicotyledons in cereals.Registered by EPA as a herbicide. Colorless, odorless crystals; mp = 186°; sg = 1.49; soluble in H_2O (0.0024 g/100 ml at pH 4, 0.026 g/100 at pH5, 0.24 g/100 ml at pH 6), C_6H_{14} (< 0.01 g/100 ml), EtOH (0.09 g/100 ml), xylene (0.02 g/100 ml), MeOH (0.26 g/100 ml), EtOAc (0.26 g/100 ml), CH_3CN (0.73 g/100 ml), Me_2CO (1.19 g/100 ml), CH_2Cl_2 (2.25 g/100 ml); LD_{50} (rat orl) > 5000 mg/kg, (rbt der) > 2000 mg/kg, (mallard duck orl) > 2510 mg/kg, (bee topical) = 0.0125 mg/bee; LC_{50} (rat ihl) > 7.9 mg/l air, (rat dietary 90 day) > 100 mg/kg diet, (mallard duck, Japanese quail 8 day dietary) > 5620 mg/kg diet, (rainbow trout, bluegill sunfish 96 hr.) > 100 mg/l, (*Daphnia* 48 hr.) = 1000 mg/l. *DuPont; E. I. DuPont de Nemours Inc.*

3752 Thioacetamide
62-55-5   9391   G

C_2H_5NS

Ethanethioamide.

Acetamide, thio-; Acetic acid, thiono-, amide; Acetimidic acid, thio-; Acetothioamide; AI3-17220; CCRIS 584; EINECS 200-541-4; HSDB 1318; NSC 2120; RCRA waste number U218; TAA; Thiacetamide; Thioacetamide; Thioacetimidic acid; USAF CB-21; USAF EK-1719. Replacement for gaseous hydrogen sulfide in qualitative analysis. Crystals; mp = 115.5°; λ_m = 210 nm (ϵ 3700), 265 nm (ϵ 11000); very soluble in H_2O (16.3 g/100 ml), EtOH (26.4 g/100 ml), less soluble in Et_2O, C_6H_6. DMSO; MLD (rat orl) = 200 mg/kg. *Aceto Corp.; Burlington Scientific Corp.; Lancaster Synthesis Co.; Lombart Lenses Ltd. Inc.; Mallinckrodt Inc.; Penta Mfg.*

3753 4,4'-Thiobis(6-tert-butyl-3-cresol)
96-69-5   G H

$C_{22}H_{30}O_2S$

Phenol, 4,4'-thiobis[2-(1,1-dimethylethyl)-5-methyl-.

4-06-00-06043 (Beilstein Handbook Reference); Antage Crystal; Antioxidant AO; Antioxidant TMB 6; Bis(3-tert-butyl-4-hydroxy-6-methylphenyl) sulfide; BRN 1147776; CCRIS 4917; Disperse MB-61; EINECS 202-525-2; HSDB 5304; Nocrac 300; Nonflex BPS; NSC 35388; Rutenol; Santonox; Santonox BM; Santonox R; Santowhite crystals; Santox; Sumilizer WX; Sumilizer WX-R; Thioalkofen BM 4; Thioalkofen BMCH; Thioalkofen MBCH; Thioalko-phene BM-4; Ultranox® 236; USAF B-15; Yoshinox S; Yoshinox SR. Antioxidant for use in adhesives, rubber articles for repeated use, polymers include polyolefins, PVC, acrylic ethyl cellulose; antioxidant for lubricants, cutting oils, water-sol. oils, hydraulic oils. Solid; mp = 161-164°; insoluble in H_2O. *GE Silicones.*

3754 Thiocarbohydrazide
2231-57-4   H

CH_6N_4S

1,3-Diamino-2-thiourea.

4-03-00-00388 (Beilstein Handbook Reference); AI3-52269; BRN 0506657; Carbohydrazide, thio-; Carbono-thioic dihydrazide; EINECS 218-769-8; HSDB 5869; Hydrazinecarbohydrazonothioic acid; NSC 689; TCH; Thiocarbazide; Thiocarbohydrazide; Thiocarbonic di-hydrazide; Thiocarbonohydrazide; USAF EK-7372. Needles or plates; mp = 170° (dec): very soluble in H_2O.

3755 Thiocyanuric acid
638-16-4   G

$C_3H_3N_3S_3$

2,4,6-Trimercapto-s-triazine.

AI3-61105; Cyanuric acid, trithio-; EINECS 211-322-8; NSC 62071; NSC 65480; 1,3,5-Triazine-2,4,6-trimercaptan; 2,4,6-Triazinetrithiol; s-Triazine-2,4,6-trithiol; 1,3,5-Triazine-2,4,6(1H,3H,5H)-trithione; s-Triazine-2,4,6(1H,3H,5H)-trithione; 1,3,5-Trimercaptotri-azine; 2,4,6-Trimercapto-1,3,5-triazine; 2,4,6-Trimercap-to-s-triazine; Trimercaptocyanuric acid; Trismercapto-triazine; Trithiocyanuric acid; USAF TH-3; Zisnet F-PT. Curing agent for epichlorohydrin rubber; used in place of ethylene thiourea and red lead; gives improved heat resistance, less mold fouling, reduced toxicity; oil treated to reduce dusting. Solid; mp > 300°. *Nippon Zeon.*

3756 Thiodicarb
59669-26-0   9403   P

C10H18N4O4S3

Dimethyl N,N'-(thiobis((methylimino)carbonyloxy))bis-(ethanimidothioate).

Al3-29311; Bismethomyl thioether; Bissulfide; BRN 2015026; Caswell No. 900AA; CGA 45156; Dicarbasulf; Dicarbosulf; Dimethyl N,N'-(thio-bis((methylimino)carb-onyloxy))bis(thioimidoacetate); Dimethyl-N,N'-(thiobis-(((methylimino)carbonyl)oxy))bis(ethanimidothioate); EINECS 261-848-7; EPA Pesticide Chemical Code 114501; Ethanimidothioic acid, N,N'-(thiobis-((methylimino)carbonyloxy))bis-, dimethyl ester; HSDB 6940; Larvin; Larvin thio dicarb insecticide ovicide; Lepicron; Liushuanwei; Nivral; N,N'-(Thiobis-((methyl-imino)carbonyloxy))bisethanimidothioic acid dimethyl ester; N,N'-Bis(1-methylthioacetaldehyde O-(N-methyl-carbamoyl)oxime)sulfide; RCRA waste no. U410; Semevin; 3,7,9,13-Tetramethyl-5,11-dioxa-2,8,14-trithia-4,7,9,12-tetra-azapentadeca-3,12-diene-6,10-dione; Thiodicarb; UC 51762; UC 51769; UC 80502. Registered by EPA as an insecticide. Crystals; mp = 173 - 174°; sg^{20} = 1.4; soluble in H$_2$O (0.0035 g/100 ml), CH$_2$Cl$_2$ (20 g/100 ml), Me$_2$CO (0.63 g/100 ml), MeOH (0.4 g/100 ml), xylene (0.26 g/100 ml); LD$_{50}$ (rat orl) = 66 mg/kg (in H$_2$O), 120 mg/kg (in corn oil), (dog orl) > 800 mg/kg, (mky orl) > 467 mg/kg, (rbt der) > 2000 mg/kg, (rat der) > 1600 mg/kg, (Japanese quail orl) = 2023 mg/kg; LC$_{50}$ (rat ihl 4 hr.) = 0.0015 - 0.0022 mg/l air, (bluegill sunfish 96 hr.) = 1.21 mg/l, (rainbow trout 96 hr.) = 2.55 mg/l, (Daphnia magna 48 hr.) = 0.053 mg/l; moderately toxic to bees. Agriliance LLC; Aventis Crop Science; Rhône-Poulenc.

3757   Thiodiglycol
111-48-8   9404   G H

C4H10O2S

Thiodiethylene glycol.

4-01-00-02437 (Beilstein Handbook Reference); Al3-05541; Bis(β-hydroxyethyl) sulfide; Bis(2-hydroxyethyl)-sulfide; Bis(2-hydroxyethyl) thioether; BRN 1236325; β,β'-Dihydroxydiethyl sulfide; Di(2-hydroxyethyl) sulfide; β,β'-Dihydroxyethyl sulfide; EINECS 203-874-3; Ethanol, 2,2'-thiobis-; Ethanol, 2,2'-thiodi-; Glycine A; β-Hydroxyethyl sulfide; Kromfax solvent; NSC 6289; Sulfide, bis(2-hydroxyethyl); Tedegyl; 2,2'-Thiobisethanol; Thiodiethanol; 2,2'-Thiodiethanol; Thiodiethylene glycol; Thiodiglycol; 2,2'-Thiodiglycol; Thiodiglycolum; Tiodi-glicol; Tiodiglicolo. Intermediate for elastomers and antioxidants, solvent for dyes in textile printing. Liquid; mp = -10.2°; bp = 282°; d^{25} = 1.1793; slightly soluble in C$_6$H$_6$; soluble in Et$_2$O, freely soluble in H$_2$O, EtOH, CHCl$_3$, EtOAc. Morton Intn'l.

3758   Thiodiglycolic acid
123-93-3   9405   G

C4H6O4S

2,2'-Thiodiacetic acid.

4-03-00-00612 (Beilstein Handbook Reference); A 6402; Acetic acid, 2,2'-thiobis-; Acetic acid, thiodi-; Al3-25085; BRN 1764392; (Carboxymethylthio)acetic acid; Dicarb-oxymethyl sulfide; Dimethylsulfide-α-α'-dicarboxylic acid; EINECS 204-663-9; HSDB 2707; Mercaptodiacetic acid; NSC 28743; TDGA; 2,2'-Thiobisacetic acid; Thiodi(acetic acid); Thiodiacetic acid; 2,2'-Thio-diethanoic acid; Thiodiglycolic acid; 2,2'-Thiodiglycolic acid; Thiodiglycollic acid; USAF CB-36; USAF E-2. Analytical reagent, used for detection of metals such as copper, lead, mercury and silver. White solid; mp = 129°; slightly soluble in H$_2$O, soluble in C$_6$H$_6$, very soluble in EtOH. CK Witco Corp.

3759   3,3'-Thiodipropionic acid
111-17-1   9406   G

C6H10O4S

3,3'-thiobis[propanoic acid].

Al3-25276; Bis(2-carboxyethyl) sulfide; CCRIS 3288; Diethyl sulfide 2,2'-dicarboxylic acid; EINECS 203-841-3; HSDB 858; Kyselina β,β'-thiodipropionova; Kyselina 3,3-thiodipropionova; NSC 8166; Propanoic acid, 3,3'-thiobis-; Propionic acid, 3,3'-thiodi-; Sulfide, bis(2-carboxyethyl); TDPA; 4-Thiaheptanedioic acid; Thia-hydracrylic acid; 3,3'-Thiobis(propanoic acid); 3,3'-Thio-dipropionic acid; Thiodihydracrylic acid; Thiodipropion-ic acid; Tyox A. Used as an antioxidant in food packaging, soaps, plasticizers, lubricants, fats, and oils. Crystalline white powder; mp = 129°; soluble in H$_2$O (37 g/l), very soluble in EtOH, Me$_2$CO. CK Witco Corp.; Evans Chemetics; Janssen Chimica.

3760   Thioglycerol
96-27-5   9409   G

C3H8O2S

Thioglycerin.

3-01-00-02339 (Beilstein Handbook Reference); Al3-25462; BRN 1732046; 2,3-Dihydroxypropanethiol; EINECS 202-495-0; Glycerol, 1-thio-; Glycerol-1-thiol; HSDB 2184; 1-Mercaptoglycerol; 1-Mercapto-2,3-prop-anediol; 3-Mercapto-1,2-propanediol; 3-Mercaptoprop-ane-1,2-diol; Monothioglycerin; Monothioglycerol; NSC 5370; 1,2-Propanediol, 3-mercapto-; Thioglycerin; Thio-glycerine; 1-Thioglycerol; α-Thiolglycerol; 1-Thio-2,3-propanediol; Thiovanol®; USAF B-40; USAF CB-37. Stabilizer for acrylonitrile polymers; crosslinking agent for hard highgloss coatings; accelerator for epoxy-amine condensation reactions; reducing agent; used in hair waving and straightening, hair dyes, depilatories, textiles, furs, pharmaceuticals, Liquid; bp = 100-101°; bp$_5$ = 118°; d^{20} = 1.2455; very soluble in EtOH, Me$_2$CO, slightly soluble in H$_2$O, Et$_2$O, C$_6$H$_6$, CHCl$_3$, insoluble in Et$_2$O. Evans Chemetics.

3761   Thiohexam
95-33-0   G H

C13H16N2S2

2-Benzothiazolesulfenamide, N-cyclohexyl-.

4-27-00-01867 (Beilstein Handbook Reference); Accelerator CZ; Accicure HBS; Al3-16782; Benzothiazyl-2-cyclohexylsulfenamide;

BRN 0192376; CBS; CBTS; CCRIS 4910; Conac A; Conac H; Conac S; Curax; Cyclohexyl benzothiazolesulfenamide; Delac S; Durax®; EINECS 202-411-2; Ekagom CBS; HSDB 2868; Nocceler CZ; NSC 4809; Pennac CBS; Perkacit® CBS; Rhodifax 16; Royal CBTS; Sanceler CM-PO; Santocure; Santocure Pellets; Santocure Powder; Santocure vulcanization accelerator; Soxinol CZ; Sulfenamide TS; Sulfenax; Sulfenax CB; Sulfenax CB 30; Sulfenax CB/K; Sulfenax TsB; Thiohexam; Thiohexam; Vulcafor CBS; Vulcafor HBS; Vulkacit® C; Vulkacit® CZ; Vulkacit® CZ/C; Vulkacit® CZ/EGC; Vulkacit® CZ/K; Vulkacit® DZ/EGC; Vulkacite CZ. Accelerator used in natural rubber and styrene-butadienethiazyl sulfenamide rubber. An all-purpose delayed action accelerator which combines superior scorch safety with shorter curing cycles; used in tire tread, carcass, camelback and mechanical goods. Used for tires, dynamically stressed technical goods, technical moldings and extrudates. Solid; mp = 93-100°; d = 1.27; insoluble in H_2O, soluble in C_6H_6, $CHCl_3$. *Akrochem Chem. Co.; Akzo Chemie; Bayer AG; Goodyear Tire & Rubber; Monsanto Co.; Uniroyal; Vanderbilt R.T. Co. Inc.*

3762   Thiometon
640-15-3   G P

$C_6H_{15}O_2PS_3$
S-[2-(ethylthio)ethyl] O,O-dimethyl phosphorodithioate.
S-2-ethylthioethyl O,O-dimethyl phosphorodithioate; dithiometon; M-81; Bay 23129; Ekatin; Medrin; nimeton. Systemic insecticide and acaricide with contact and stomach action. Cholinesterase inhibitor, used for control of sucking insects in fruit and vegetable crops. Oil; $bp_{0.1}$ = 110°; d_{20} = 1.209; soluble in H_2O (0.02 g/100 ml), more soluble in organic solvents; LD50 (rat orl)= 125 mg/kg. *Sandoz.*

3763   Thionine
581-64-6   G

$C_{12}H_9ClN_3S$
3,7-Diaminophenazathionium chloride.
C.I. 52000; CCRIS 5616; Cyanine; 3,7-Diaminophen-azathionium chloride; 3,7-Diaminophenothiazin-5-ium chloride; EINECS 209-470-3; Katalysin; Lauth's Violet; Lauthsches violett; NSC 9591; NSC 56342; Phenothiazin-5-ium, 3,7-diamino-, chloride; Thionin; Thionine. Used as a microscopic stain.

3764   Thiophanate-methyl
23564-05-8   9427   G P

$C_{12}H_{14}N_4O_4S_2$
Dimethyl ester 4,4'-o-Phenylenebis(3-Thioallophanic) Acid.
3336WP; AI3-27905; Allophanic acid, 4,4'-o-phenyl-enebis[3-thio-, dimethyl ester; BAS 32500F; Bis(3-(methoxycarbonyl)-2-thioureido)benzene; 1,2-Bis(3-(methoxycarbonyl)-2-thioureido)benzene; 1,2-Bis(meth-oxycarbonylthioureido)benzene; BRN 0937942; Caligran; Carbamic acid, (1,2-phenylenebis(imino-carbonothioyl))-bis-, dimethyl ester; Caswell No. 375A; CCRIS 6101; Cercobin M; Cercobin M 70; Cercobin methyl; Cycosin; 1,2-Di-(3-methoxycarbonyl-2-thio-ureido)benzene; Dimethyl ((1,2-phenylene)bis-(iminocarbonothioyl))bis-(carbamate); Dimethyl 4,4'-o-phenylenebis(3-thio-allophanate); Dimethylbis; Ditek; Dragon Systemic Fungicide; Easout; EINECS 245-740-7; Enovit M; Enovit M 70; Enovit methyl; Enovit Super; Enovit-Supper; EPA Pesticide Chemical Code 102001; F 6385; Ferti-lome Halt Systemic; frumidor; Fungo; Fungo 50; Green Light Systemic Fungicide; HSDB 6937; Labilite; Methyl thiophanate; Methyl topsin; Methylthiofanate; Methylthiophanate; Metoben; Mildothane; Neotopsin; NF 44; NSC 170811; o-Bis(3-methoxycarbonyl-2-thioureido)benzene; PEI 190; PELT 14; PELT-44; (1,2-phenylenebis(iminocarbonothioyl))bis-(carbamate); 4,4'-o-phenylenebis(3-thioallophanic acid) dimethyl ester; RCRA waste no. U409; Rilon; SA Thiomyl Systemic Fungicide;; Sigma; Sipcaplant; Sipcasan; Sipcavit; Spot Kleen; TD 1771; Thiopan; Thiophanate; Thiophanate M; Thiophanate-Me; Thiophanate-methyl; THIOPHANNATE-METHYL; Tiofanate metile; Tops; Topsin M; Topsin M 70; Topsin Methyl; Topsin NF-44; Topsin turf and ornamentals; Topsin WP methyl; Trevin; Zyban. A broad-spectrum systemic fungicide and wound protectant which controls a variety of diseases for use on turf and ornamentals. Registered by EPA as a fungicide. Colorless prisms; mp = 172° (dec); poorly soluble in H_2O (0.0026 g/100 ml), soluble in MeOH (2.3 g/100 ml), Me_2CO (4.6 g/100 ml), $CHCl_3$ (3.8 g/100 ml), CH_3CN (1.9 g/100 ml), cyclohexanonoe (4.04 g/100 ml), EtOAc (1.08 g/100 ml); LD50 (mrat orl) = 7500 mg/kg, (frat orl) = 6640 mg/kg, (rbt orl) = 2270 mg/kg, (mus orl) = 3510 mg/kg, (gpg orl) = 3640 mg/kg, (Japanese quail orl) > 5000 mg/kg; LC50 (rainbow trout 48 hr.) = 7.8 mg/l, (carp 48 hr.) = 11 mg/l; non-toxic to bees. *Cerexagri, Inc.; Cleary Chemical Corp.; Micro-Flo Co. LLC; Nippon Soda; Regal Chemical Co.; Scotts Co.*

3765   Thiophene
110-02-1   9428   G

C_4H_4S
Thiophene.
5-17-01-00297 (Beilstein Handbook Reference); AI3-15417; BRN 0103222; CCRIS 2935; CP 34; Divinylene sulfide; EINECS 203-729-4; Furan, thio-; HSDB 130; Huile H50; Huile H5O; NSC 405073; Thiacyclo-pentadiene; Thiaphene; Thiofen; Thiofuram; Thiofuran; Thiofurfuran; Thiole; Thiophen; Thiophene; Thiotetrole; UN2414; USAF EK-1860. Used in organic synthesis (condenses with phenol and formaldehyde, copolymerizes with maleic anhydride), and as a solvent, dye, and in pharmaceutical manufacturing. Liquid; mp = -39.4°; bp = 84°; d^{20} = 1.0649; λ_m = 233, 237 nm (MeOH); insoluble in H_2O, slightly soluble in $CHCl_3$, freely soluble in EtOH, Et_2O, Me_2CO, C_6H_6, CCl_4; C_7H_{16}; C_5H_5N, dioxane, C_7H_8. *Elf Atochem N. Am.; Penta Mfg.*

3766   Thiophenol
108-98-5   9430   G H

C6H6S

Phenyl mercaptan.

4-06-00-01463 (Beilstein Handbook Reference); AI3-15418; Benzene, mercapto-; Benzenethiol; BRN 0506523; EINECS 203-635-3; FEMA No. 3616; HSDB 5387; Mercaptobenzene; NSC 6953; Phenol, thio-; Phenyl mercaptan; Phenylthiol; RCRA waste number P014; Thiofenol; Thiophenol; UN2337; USAF XR-19. Intermediate in synthesis of pesticides and pharmaceuticals. Oil; mp = -14.8°; bp = 169.1°; d^{20} = 1.0775; λ_m = 240 nm (ϵ = 7244, EtOH); insoluble in H_2O, slightly soluble in CCl_4, soluble in EtOH, Et_2O, C_6H_6. *ICI Americas Inc.; Janssen Chimica; Lancaster Synthesis Co.; Sigma-Aldrich Fine Chem.*

3767   Thiophosgene
463-71-8   G

CCl2S

Thiocarbonyl chloride.

4-03-00-00281 (Beilstein Handbook Reference); BRN 1633495; Carbon chlorosulfide; Carbonic dichloride, thio-; Carbonothioic dichloride; Carbonyl chloride, thio-; Dichlorothiocarbonyl; EINECS 207-341-6; HSDB 861; Phosgene, thio-; Thiocarbonic dichloride; Thiocarbonyl chloride; Thiocarbonyl dichloride; Thiofosgen; Thiokarbonylchlorid; Thiophosgene; UN2474. Used as an intermediate in organic synthesis. Red solid; bp = 73°; d^{15} = 1.5080; insoluble in H_2O, soluble in EtOH, Et_2O. *Fine Chem. Corp.; Fluka; Pfalz & Bauer; Sigma-Aldrich Fine Chem.*

3768   3-Thiopropionic acid
107-96-0   H

C3H6O2S

2-Mercaptoethanecarboxylic acid.

3MPA; 4-03-00-00726 (Beilstein Handbook Reference); AI3-26090; BRN 0773807; EINECS 203-537-0; HSDB 5381; Hydracrylic acid, 3-thio-; β-Mercaptopropionic acid; 2-Mercaptoethanecarboxylic acid; 3-Mercapto-propanoic acid; 3-Mercaptopropionic acid; NSC 437; Propanoic acid, 3-mercapto-; Propionic acid, 3-merc-apto-; 3-Thiopropanoic acid; 3-Thiopropionic acid; β-Thiopropionic acid. Solid; mp = 18°; bp15 = 111°, bp3 = 86°; d^{21} = 1.218; soluble in H_2O, EtOH, Et_2O, CCl_4.

3769   Thiostop E, N
20624-25-3   G

C5H10NNaS2.3H2O

Diethyldithiocarbamic acid sodium salt trihydrate.

Carbamic acid, diethyldithio-, sodium salt, trihydrate; Diethyldithiocarbamic acid sodium salt trihydrate; Diethyldithiocarbamate sodium salt trihydrate; Diethyl-dithiocarbamate sodium trihydrate; Diethyldithio-karbaman sodny trihydrat; Dithiocarb trihydrate; Sodium diethyldithiocarbamate trihydrate. An ultra-accelerator for NR and SBR latexes; an activator

for guanidine type accelerators. Also used in analytical chemistry to assay copper. Solid; mp = 95-99°. *Uniroyal.*

3770   Thiourea
62-56-6   9443   G

CH4N2S

Thiocarbamide.

AI3-03582; Caswell No. 855; CCRIS 588; EINECS 200-543-5; EPA Pesticide Chemical Code 080201; HSDB 1401; Isothiourea; NSC 5033; Pseudothiourea; Pseudo-urea, 2-thio-; RCRA waste number U219; Sulfourea; Thiocarbonic acid diamide; Thiomocovina; β-Thio-pseudourea; 2-Thiopseudourea; 2-Thiourea; Thiourea; Thiuronium; THU; Tsizp 34; Urea, 2-thio-; Urea, thio-; USAF EK-497. Photography, photocopy papers, organic synthesis (intermediate, dyes, drugs, hair preparations), rubber accelerator, analytical reagent, amino resins, mold inhibitor. Crystals; mp = 174-177°; d = 1.4100; soluble in H_2O (9.1 g/100 ml), soluble in EtOH, poorly soluble in other organic solvents; LD50 (rat orl) = 1830 mg/kg. *Allchem Ind.; Dajac Labs.; Fairmount; Greeff R.W. & Co.; Sigma-Aldrich Fine Chem.*

3771   Thiram
137-26-8   9448   D G P

C6H12N2S74

Tetramethylthiuram disulfide.

4-04-00-00242 (Beilstein Handbook Reference); Aapirol; AAtack; Aatiram; Accel TMT; Accelerator T; Accelerator thiuram; Aceto TETD; AI3-00987; Akrochem® TMTD; Anles; Arasan; Arasan 42-S; Arasan 70; Arasan 70-S Red; Arasan 75; Arasan-M; Arasan-SF; Arasan-SF-X; Atiram; Attack; Aules; Bis((dimethylamino)carbonothioyl) disulf-ide; Bis(dimethyl-thiocarbamoyl)-disulfid; Bis(dimethyl thiocarbamoyl)disulfide; Bis(dimethylthiocarbamoyl) di-sulphide; Bis(dimethylthiocarbamoyl) disulphide; BRN 1725821; Caswell No. 856; CCRIS 1282; Cunitex; Cyuram DS; Delsan; Disolfuro di tetrametiltiourame; Disulfide, bis(dimethylthiocarbamoyl); Disulfure de tetramethylthiourame; Disulfuro di tetrametiltiourame; EINECS 205-286-2; Ekagom TB; ENT 987; EPA Pesticide Chemical Code 079801; Falitiram; Fermide; Fermide 850; Fernacol; Fernasan; Fernasan A; Fernide; Flo Pro T Seed Protectant; FMC 2070; Formalsol; Hermal; Hermat TMT; Heryl; Hexathir; HSDB 863; Hy-Vic; Kregasan; Mercuram; Mercuran; Methyl thiram; Methyl thiuramdisulfide; Methyl tuads; Metiur; Metiurac; Micropearls; Nobecutan; Nocceler TT; Nomersam; Nomersan; Normersan; NSC 1771; Orac TMTD; Panoram 75; Perkacit® TMTD; Pol-Thiuram; Polyram-Ultra; Pomarsol; Pomarsol forte; Pomasol; Puralin; Radothiram; RCRA waste number U244; Rezifilm; Royal TMTD; Sadoplon; Sadoplon 75; Spotrete; Spotrete-F; Spotrete WP 75; SQ 1489; Sranan-sf-X; Teramethylthiuram disulfide; Tersan; Tersan 75; Tersantetramethyldiurane sulfide; Tetrapom; Tetrasipton; Tetrathiuram disulfide; Tetrathiuram disulphide; Thianosan; Thillate; Thimar; Thimer; Thioknock; Thioperoxydicarbonic diamide, tetramethyl-; Thiosan; Thioscabin; Thiotex; Thiotox; Thiram; Thiram 75; Thiram 80; Thiram B; Thiramad; Thirame; Thirampa; Thiramum; Thirasan; Thiulin; Thiulix; Thiurad; Thiuram; Thiuram d; Thiuram m; Thiuramin; Thiuramyl; Thylate; Tirampa; Tiuramyl; TMTD; TMTDS; Trametan; Tridipam; Tripomol; Tuads; TUEX; Tulisan; Vancide TM; Vancide TM-95;

Vuagt-i-4; vulcafor tmtd; vulkacit mtic; vulkacit thiuram. Vulcanizer, bacteriostat, antifungal and animal repellent. Very active, sulfur-bearing, nondiscoloring organic accelerator and activator; for curing systems requiring very low or no sulfur and for butyl and ethylene-propylene-diene rubbers compounds. Fungicide with animal repellent properties. Also used in vulcanization and as a bacteriostat. Registered by EPA as a fungicide. White-yellow crystals; mp = 155.6°; bp20 = 129°; λ_m = 243, 282 nm (ε = 12589, 11482, CHCl3); poorly soluble in H_2O (0.003 g/100 ml at 25°), soluble in EtOH (> 1 g/100 ml), Me2CO (8 g/100 ml), CHCl3 (23 g/100 ml); LD50 (rat orl)= 780 - 865 mg/kg, (mus orl) = 1500 - 2000 mg/kg, (rbt orl) = 210 mg/kg, (rat der) > 1000 mg/kg, (redwing blackbird orl) > 100 mg/kg; LC50 (rat ihl 4 hr.) = 0.5 mg/l air, (bluegill sunfish (48 hr.) = 0.23 mg/l, (trout 48 hr.) = 0.13 mg/l, (carp 48 hr.) = 4 mg/l; non-toxic to bees. *Agrichem (International) Ltd.; Akrochem Chem. Co.; Akzo Chemie; Avon Packers Ltd; Bayer AG; Bristol Myers Squibb Pharm. Ltd.; Cleary W. A.; DuPont; Ford Smith & Co Ltd.; ICI Chem. & Polymers Ltd.; Monsanto Co.; Universal Crop Protection Ltd.*

3772 Thonzide
553-08-2   9450   G

C32H55BrN4O
N-[2-[[(4-methoxyphenyl)methyl]-2-pyrimidinylamino]-ethyl]-N,N-dimethyl-1-hexadecanaminium bromide.
Ammonium,   hexadecyl(2-((p-methoxybenzyl)-2-pyrim-idinylamino)ethyl)dimethyl-, bromide; Bromure de tonz-onium; Bromuro de tonzonio; Coly-Mycin S Otic; EINECS 209-032-1; 1-Hexadecanaminium,   N-(2-(((4-methoxy-phenyl)methyl)-2-pyrimidinylamino)ethyl)-N,N-dimethyl-, bromide; Hexadecyl(2-((p-methoxybenzyl)-2-pyrimidinyl-amino)ethyl)dimethylammonium bromide; NC 1264; Nebair; NSC 5648; Thonzide; Thonzonium bromide; Tonzonii bromidum; Tonzonium bromide. Detergent. White solid; mp= 91-92°. *Parke-Davis.*

3773 Thorium dioxide
1314-20-1   9455   G

O2Th
Thorium(IV) dioxide.
EINECS 215-225-1; HSDB 6364; Thoria; Thorianite; Thorium anhydride; Thorium dioxide; Thorium oxide; Thorium oxide (ThO2); Thorotrast; Thortrast; Umbrathor. Used in ceramics, gas mantles, nuclear fuel, medicine and non-silica optical glass. White crystalline powder; mp = 3390°; d = 10.0; insoluble in H_2O; carcinogen.

3774 Thujone
546-80-5   9469   G

C10H16O
4-Methyl-1-(1-methylethyl)bicyclo[3.1.0]hexan-3-one.
Bicyclo(3.1.0)hexan-3-one, 4-methyl-1-(1-methylethyl)-, (1S,4R,5R)-; BRN 4660369; EINECS 208-912-2; 1-Iso-propyl-4-methylbicyclo(3.1.0)hexan-3-one; Isothujone, (-)-; (-)-3-Thujone; 6-Ketosabinane; NSC 93742; Tan-acetone; 3-Thujanone, (-)-; 3-Thujanone, (1S,4R,5R)-(-)-; Thujon; Thujone; l-Thujone. A monoterpene ketone; used as a solvent. Oil; d^{25} = 0.9109; λ_m = 300 nm (ε 23, isooctane); insoluble in H_2O, soluble in organic solvents; LD50 (mus sc)= 134.2 mg/kg.

3775 Thulium oxide
12036-44-1   G

O3Tm2
Thulium(III) oxide.
EINECS 234-851-6; Thulia; Thulium oxide; Thulium oxide (Tm2O3). Source of thulium metal. *Atomergic Chemetals; Cerac; Rhône-Poulenc.*

3776 Thymine
65-71-4   9475   G

C5H6N2O2
2,4-Dihydroxy-5-methylpyrimidine.
AI3-25479; CCRIS 5584; EINECS 200-616-1; 5-Methyl-uracil; NSC 14705; 2,4(1H,3H)-Pyrimidinedione, 5-methyl-; Thymin; Thymin (purine base); Thymine; Thym-ine anhydrate. Obtained by the hydrolysis of nucleic acids; used in biochemical research. Crystalline solid; mp = 316°, dec 335 - 337°; λ_m = 205, 264.5 nm (ε 9500, 7900 pH 7); λ_m = 206, 262 nm (ε 9400, 7860 MeOH), 292 nm (ε 8900 MeOH/KOH); slightly soluble in H_2O (4 g/l), poorly soluble in organic solvents. *Lancaster Synthesis Co.; Mallinckrodt Inc.; Sigma-Aldrich Fine Chem.; U.S. BioChem.*

3777 Thymol
89-83-8   9476   D G

C10H14O
5-Methyl-2-(1-methylethyl)phenol.
AI3-00708; Caswell No. 856A; CCRIS 7299; Cymo-phenol, α-; EINECS 201-944-8; EPA Pesticide Chemical Code 080402; FEMA Number 3066; HSDB 866; Isopropyl cresol; NSC 11215; Thyme camphor; Thymic acid; Thymol; m-Thymol; Thymol (natural). An antibacterial and antifungal agent used in perfumery, microscopy, preservative, antioxidant, flavoring, as a laboratory reagent, in the manufacture of menthol. Used in the prevention of mold and mildew, in flavoring and perfumery, as a preservative and antioxidant and a topical antiseptic. Crystalline solid; mp = 51.5°; bp = 232.5°; d^{25}= 0.9699; λ_m = 276 nm (ε = 2320, MeOH); soluble in H_2O (0.1 g/100 ml), EtOH (100 g/100 ml), CHCl3 (143 g/100 ml); Et2O (66.7 g/100

3778 Timolol
91524-16-2   9521   D

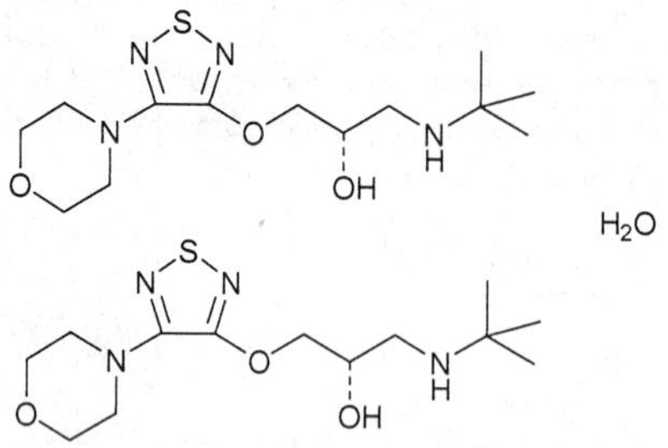

H$_2$O

C$_{13}$H$_{24}$N$_4$O$_3$S.1/2H$_2$O
(S)-1-(tert-Butylamino)-3-[(4-morpholino-1,2,5-thiadiazol-3-yl)oxy]-2-propanol hemihydrate.
Blocadren; EINECS 248-032-6; HSDB 6533; Istalol; Timolol; Timololum; Timopic.; See also [26839-75-8]. Antianginal agent with antiarrhythmic (class II), antihypertensive and antiglaucoma properties. A β-adrenergic blocker. [(±) form]: mp = 71.5-72.5°. Merck & Co.Inc.

3779 Timolol maleate
26921-17-5   9521   D

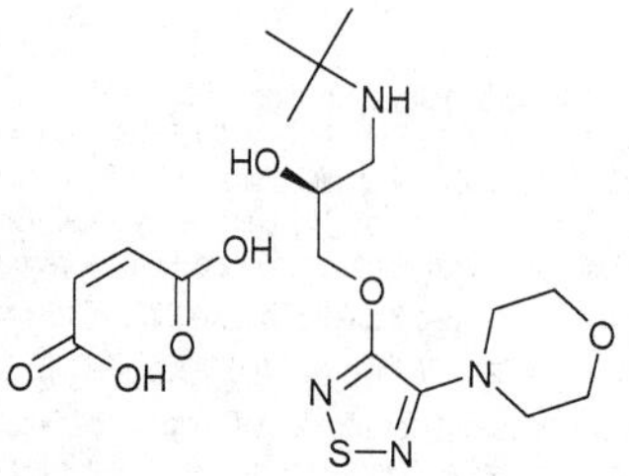

C$_{17}$H$_{28}$N$_4$O$_7$S
(S)-1-(tert-Butylamino)-3-[(4-morpholino-1,2,5-thiadiazol-3-yl)oxy]-2-propanol maleate (1:1).
Aquanil; Betim; Betime; Blocadren; CCRIS 1057; Cosopt; EINECS 248-111-5; MK 950; Proflax; Temserin; Tenopt; Timacar; Timacor; Timolide; Timolol hydrogen maleate salt; Timolol hydrogen maleate; Timolol maleate; Tim-optic; Timoptol.; component of: Cosopt, Timolide. A β-adrenergic blocker. Antianginal agent with anti-arrhythmic (class II), antihypertensive and antiglaucoma properties. Crystals; mp = 201.5-202.5°; [α]$_{405}^{24}$ = -12.0° (c = 5 1N HCl); [α]$_{D}^{25}$ = -4.2°; λ$_m$ = 294 nm (A$_{cm}^{1\%}$ 200 0.1N HCl); soluble in EtOH, MeOH; poorly soluble in CHCl$_3$, cyclohexane; insoluble in isooctane, Et$_2$O. Merck & Co.Inc.

3780 Tin
7440-31-5   9523   G

Sn

Sn
Tin.
CI 77860; CI Pigment metal 5; EINECS 231-141-8; HSDB 5035; Metallic tin; Silver matt powder; Tin; Tin, elemental; Tin flake; Tin, inorganic compounds (except oxides); Tin, metal; Tin powder; Wang; Zinn. Element; Used in manufacture of tin plate, anodes, corrosion-resistant coatings, manufacture of chemicals. mp = 232°; bp= 2507°; d = 7.31. Atomergic Chemetals; Cerac; M&T Harshaw; Noah Chem.;

3781 Tin dioxide
18282-10-5   8856   G

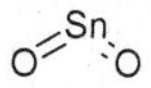

O$_2$Sn
Tin(IV) oxide.
C.I. 77861; EINECS 242-159-0; Flowers of tin; HSDB 5064; Stannic anhydride; Stannic dioxide; Stannic oxide; Stannic oxide (SnO$_2$); T 10 (oxide); Tin ash; Tin dioxide; Tin oxide; Tin oxide (SnO$_2$); Tin Peroxide; Tin(IV) oxide; White tin oxide. Used to polish glass and metals; tin salts, catalyst, ceramic glazes and colors, putty, perfume, cosmetic preparations (fingernail polish), manufacture of special glasses. White-grey powder; mp = 1127°; d = 6.9500; insoluble in H$_2$O. Atomergic Chemetals; Cerac; Goldschmidt.

3782 Tin(II) oxalate
814-94-8   8865   G

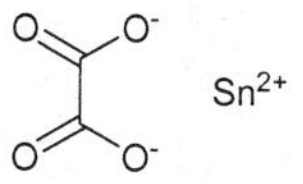

Sn^{2+}

C$_2$O$_4$Sn
Stannous oxalate.
EINECS 212-414-0; Ethanedioic acid, tin(2+) salt (1:1); Oxalic acid, tin(2+) salt (1:1); Stannous oxalate; Stavelan cinaty; Tin oxalate; Tin(2+) oxalate; Tin(II) oxalate. Catalyst for stannous esterification; dying and printing textiles. Solid; d = 3.56; insoluble in H$_2$O. Elf Atochem N. Am.

3783 Tinopal 5BM
13863-31-5   H

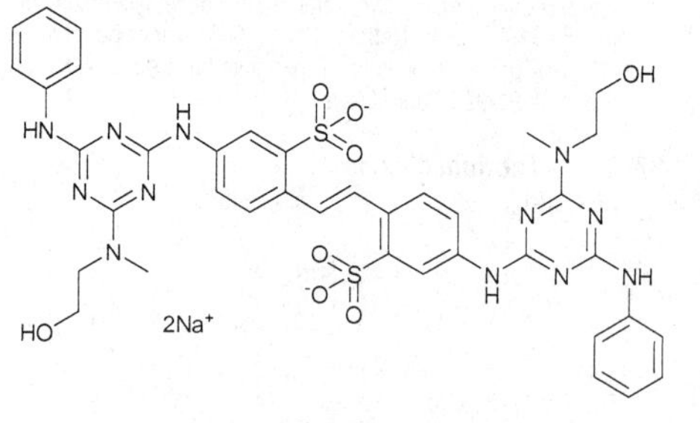

2Na$^+$

C$_{38}$H$_{38}$N$_{12}$Na$_2$O$_8$S$_2$
4,4'-Bis((4-anilino-6-((2-hydroxyethyl)methylamino)-s-triazin-2-yl)amino)2,2'-stilbenedisulfonate.
Disodium 4,4'-bis-(2-sulfostyryl)biphenyl; EINECS 237-600-9; HSDB 5059; Phorwhite RKH; Tinopal 5bm; Tinopal 5BM-XC.

3784 Tiox
61570-90-9   9534   G V

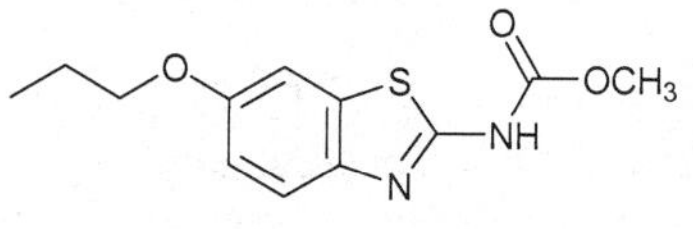

C$_{12}$H$_{14}$N$_2$O$_3$S
(6-propoxy-2-benzothiazolyl)carbamic acid methyl ester.
Carbamic acid, (6-propoxy-2-benzothiazolyl)-, methyl ester;

EINECS 262-854-2; Methyl 6-propoxy-2-benzothiazolecarbamate; Tiox (Veterinary); Tioxidazol; Tioxidazole; Tioxidazolum. Anthelmintic, used in horses. Crystals; mp - 178-180°; insoluble in H2O, soluble in organic solvents. *Schering.*

3785   Titanium dioxide
13463-67-7   9549   D G

$$O{=}Ti{=}O$$

O2Ti

Titanium Dioxide.
1385RN 59; 1700 White; 234DA; 500HD; 63B1 White; A 200 (pigment); A 330 (pigment); A-Fil; A-Fil Cream; A-FN 3; Aerolyst 7710; Aerosil P 25; Aerosil P 25S6; Aerosil P 27; Aerosil T 805; AI3-01334; AK 15 (pigment); Amperit 780.0; AMT 100; AMT 600; Atlas white titanium dioxide; AUF 0015S; Austiox R-CR 3; B 101 (pigment); Bayer R-FD 1; Bayertitan A; Bayertitan AN 3; Bayertitan R-FD 1; Bayertitan R-FK 21; Bayertitan R-FK-D; Bayertitan R-KB; Bayertitan R-U-F; Bayertitan R-U 2; Bayertitan R-V-SE 20; Bayertitan T; Bistrater L-NSC 200C; Blend White 9202; BR 29-7-2; Brookite; C 97 (oxide); C-Weiss 7; C.I. 77891; C.I. Pigment White 6; Cab-O-Ti; Calcotone White T; CCRIS 590; CG-T; CI 77891; CI Pigment white 6; CL 310; Cosmetic Hydrophobic TiO2 9428; Cosmetic Micro Blend TiO2 9228; Cosmetic White C47-5175; Cosmetic White C47-9623; Covermark; E 171; EINECS 236-675-5; Flamenco; Hombitan; Hombitan R 101D; Hombitan R 610K; Horse Head A-410; Horse Head A-420; Horse Head R-710; HSDB 869; KH 360; Kronos®; Kronos® 1000; Kronos® 2020; Kronos® 2073; Kronos® 3020; Kronos® cl 220; Kronos® RN 40P; Kronos® RN 56; Kronos® titanium dioxide; Levanox White RKB; NCI-C04240; NSC 15204; Orgasol 1002D White 10 Extra Cos; P 25 (oxide); Pigment White 6; R 680; Rayox; Ro 2; Runa ARH 20; Runa ARH 200; Runa RH20; Rutile; Rutiox CR; Ti-pure R 900; Ti-Pure; Ti-Pure R 901; Tichlor; Tin dioxide dust; Tiofine; Tiona T.D.; Tiona td; Tioxide; Tioxide; Tipaque; Tipaque R 820; Titafrance; Titan White; Titandioxid; Titandioxid; Titania; Titanic anhydride; Titanic oxide; Titanium dioxide; Titanium oxide; Titanium peroxide; Titanium White; Titanium(IV) oxide; Titanox; Titanox 2010; Titanox ranc; Trioxide(s); Tronox; Unitane; Uniwhite AO; Uniwhite KO; Uniwhite OR 450; Uniwhite OR 650; Zopaque; Zopaque LDC. Used to impart opacity and brightness; for cosmetics, paints, inks, plastics, paper, glass, rubber, ceramics, glazes, vitreous enamels, welding rods, fibers and coatings.Used with different pigments in a cream base as a skin masking cream and topical protectant. Used for decontamination of radioactive skin. White powder, widely used as a pigment; mp = 1855°; d = 4.23, 3.90 or 4.13; insoluble in H2O or mineral acids. *Allergan Herbert; Bayer NV; British Traders & Shippers; Degussa AG; DuPont; Ferro/Transelco; Herbert; Kerr-McGee; Kronos; Lederle Labs.; Miles Inc.; Olin Res. Ctr.; Rheox Inc.; SCM Glidco Organics; Stiefel Labs Inc.; Tioxide Am.; Westwood-Squibb Pharm. Inc.*

3786   Titanium isopropoxide
546-68-9   9551   H

C12H32O4Ti
Isopropyl alcohol, titanium(4+) salt.
A 1 (titanate); EINECS 208-909-6; HSDB 848; Isopropyl alcohol, titanium(4+) salt; Isopropyl orthotitanate; Isopropyl titanate(IV);

Isopropyl titanate(IV) ((C3H7O)4Ti); NSC 60576; Orgatix TA 10; TA 10; Tetraisopropanol-atotitanium; Tetraisopropoxide titanium; Tetraisopropoxy-titanium; Tetraisopropoxytitanium(IV); Ti Isopropylate; Tilcom TIPT; Titanic acid isopropyl ester; Titanium isopropylate; Titanium tetraisopropoxide; Titanium(4+) isopropoxide; Titanium(IV) isopropoxide; Tyzor TPt.

3787   Tobias acid
81-16-3   6432   G H

C10H9NO3S
2-Naphthylamine-1-sulfonic acid.
4-14-00-02792 (Beilstein Handbook Reference); BRN 0613084; EINECS 201-331-5; HSDB 5239; Kyselina 2-naftylamin-1-sulfonova; Kyselina tobiasova; NSC 5523; Tobias acid. Used as an azo dye intermediate and as an optical brightener. Crystalline solid; λm = 225, 244, 283, 313, 318 nm (ε 40900, 18600, 5900, 714, 669 MeOH); slightly soluble in H2O, EtOH, Et2O. *DuPont.*

3788   Tolan
501-65-5   9581   G

C14H10
1,1'-(1,2-ethanediyl)bisbenzene.
Acetylene, diphenyl-; AI3-04360; Benzene, 1,1'-(1,2-ethynediyl)bis-; Biphenylacetylene; Diphenylacetylene; 1,2-Diphenylacetylene; Diphenylethyne; EINECS 207-926-6; Ethyne, diphenyl-; NSC 5185; Tolan; Tolane. Used in organic synthesis. Prisms or plates; mp = 62.5°; bp = 300°; d^{100} = 0.9657; λm = 216, 221, 269, 272, 279, 288, 297 nm (ε 20600, 20300, 23450, 25200, 33000, 23250, 29400); insoluble in H2O, slightly soluble in EtOH, CCl4, very soluble in Et2O.

3789   Tolterodine
124937-51-5   9603   D

C22H31NO
(+)-(R)-2-[α-[2-(Diisopropylamino)ethyl]benzyl]-p-cresol.
Detrol; Kabi 2234; PNU 200583; Tolterodina; Toltero-dine; (+)-Tolterodine; Tolterodinum. Antimuscarinic. Used to treat urinary incontinence. *Pharmacia & Upjohn.*

3790   Toluene
108-88-3   9607   G H

621

C7H8
Methylbenzene.
Al3-02261; Antisal 1a; Benzene, methyl-; Caswell No. 859; CCRIS 2366; CP 25; EINECS 203-625-9; EPA Pesticide Chemical Code 080601; HSDB 131; Methacide; Methane, phenyl-; Methylbenzene; Methylbenzol; NCI-C07272; NSC 406333; Phenylmethane; RCRA waste number U220; Tolu-Sol; Tolueen; Toluen; Toluene; Tolueno; Toluol; Toluolo; UN 1294. Used as an aviation gasoline additive and high octane blending stock; as a solvent for paint; diluent and thinner in nitrocellulose lacquers; adhesive solvent in plastic toys; in the manufacture of benzoic acid, benzaldehyde, explosives and dyes; in extraction of various principles from plants. Liquid; mp = -94.9°; bp = 110.6°; d^{20} = 0.8669; λ_m = 254, 255, 260, 261, 264, 268 nm (ε = 163, 175, 198, 238, 167, 222, MeOH); insoluble in H_2O; soluble in Me_2CO, $CHCl_3$, CS_2, ligroin; freely soluble in EtOH, Et_2O; LD50 (rat orl) = 636 mg/kg. *Ashland; Chevron; Exxon; ExxonMobil Chem. Co.; Fina Chemicals; Mitsubishi Corp.; Phillips 66; Texaco.*

3791   Toluenediamine
496-72-0   H

C7H10N2
4-Methyl-1,2-benzenediamine.
4-13-00-00260 (Beilstein Handbook Reference); 4-Methyl-1,2-diaminobenzene; 4-Methyl-1,2-phenylene-diamine; 4-Methyl-o-phenylenediamine; BRN 0507965; CCRIS 4611; EINECS 207-826-2; HSDB 6070; NSC 1495; RCRA waste no. U221; Toluene-3,4-diamine; Toluene-diamine. Plates; mp = 89.5°; bp = 265°; λ_m = 297 nm (MeOH); soluble in ligroin, very soluble in H_2O.

3792   Toluenediamine
823-40-5   H

C7H10N2
2-Methyl-1,3-phenylenediamine.
3-13-00-00291 (Beilstein Handbook Reference); Benzenediamine, ar-methyl-; 1,3-Benzenediamine, 2-methyl-; BRN 2079476; CCRIS 3031; 2,6-Diamino-1-methylbenzene; 2,6-Diaminotoluene; EINECS 212-513-9; HSDB 4131; 2-Methyl-1,3-phenylenediamine; 2-Methyl-1,3-benzenediamine; 2-Methyl-m-phenylenediamine; NCI-C50317; NSC 147490; RCRA waste no. U221; Toluenediamine; Toluene-2,6-diamine; o-Toluene di-amine; 2,6-Toluylenediamine. Prisms; mp = 106°; soluble in H_2O, EtOH, C_6H_6.

3793   Toluenediamine
25376-45-8   H

C7H10N2
Diaminotoluene (mixed isomers).
Benzenediamine, ar-methyl-; Diaminotoluene; Diamino-toluene (mixed isomers); EINECS 246-910-3; HSDB 6059; Methylphenylenediamine; Phenylenediamine, ar-methyl-; RCRA waste number U221; TDA 8020; Toluene-ar,ar'-diamine; Toluenediamine; Tolylenediamine.

3794   Toluene-2,3-diamine
2687-25-4   H

C7H10N2
3-Methyl-1,2-phenylenediamine.
3-13-00-00277 (Beilstein Handbook Reference); 1,2-Benzenediamine, 3-methyl-; BRN 0907184; CCRIS 7692; 1,2-Diamino-3-methylbenzene; 2,3-Diaminotoluene; EINECS 220-248-5; HSDB 6077; 3-Methyl-o-phenylene-diamine; 1-Methyl-2,3-phenylenediamine; Toluene-2,3-diamine; 2,3-Tolylenediamine. Crystals; mp = 63.5°; bp = 255°; very soluble in Me_2CO, C_6H_6, EtOH.

3795   Toluene-2,4-dicarbamic acid,   2-(N,N-dimethylaminoethyl) 4-(2-ethylhexyl) ester, lactate
68227-46-3   H

C21H35N3O4.C3H6O3
Lactic acid, compound with 3-(2-(dimethylamino)ethyl) 1-(2-ethylhexyl) toluene-2,4-dicarbamate (1:1).
EINECS 269-358-5; Toluene-2,4-dicarbamic acid, 2-(N,N-dimethylaminoethyl) 4-(2-ethylhexyl) ester, lactate; Lactic acid, compound with 3-(2-(dimethylamino)ethyl) 1-(2-ethylhexyl) toluene-2,4-dicarbamate (1:1); Propanoic acid, 2-hydroxy-, compd. with 3-(2-(dimethylamino)ethyl) 1-(2-ethylhexyl) (4-methyl-1,3-phenylene)bis(carbamate) (1:1); Propanoic acid, 2-hydroxy-, compd. with 3-(2-(dimethylamino)ethyl) 1-(2-ethylhexyl) (4-methyl-1,3-phenylene)bis(carbamate) (1:1).

3796   Toluene-2,6-diisocyanate
91-08-7   H

C9H6N2O2
Benzene, 1,3-diisocyanato-2-methyl-.
4-13-00-00259 (Beilstein Handbook Reference); Benzene, 1,3-diisocyanato-2-methyl-; Benzene, 2,6-diisocyanato-1-methyl-; BRN 2211546; CCRIS 3741; EINECS 202-039-0; HSDB 5272; Isocyanic

acid, 2-methyl-m-phenylene ester; RCRA waste no. U223; Toluene diisocyanate; Toluene 2,6-diisocyanate; Tolylene 2,6-diisocyanate; m-Tolylene diisocyanate; UN 2078. Used as the monomer in polyurethane coatings. Solid; mp = 18.3°; freely soluble in H_2O, less soluble in Me_2CO, C_6H_6. *Air Prods & Chems; Dow Chem. U.S.A.*

3797 Toluene diisocyanate
584-84-9   9608   G H

$C_9H_6N_2O_2$
2,4-Toluene diisocyanate.
4-13-00-00243 (Beilstein Handbook Reference); AI3-15101; Benzene, 1,3-diisocyanatomethyl-; Benzene, 2,4-diisocyanato-1-methyl-; BRN 0744602; CCRIS 3742; Cresorcinol diisocyanate; Desmodur T80; Di-iso-cyanatoluene; Di-isocyanate de toluylene; Diisocyanat-toluol; EINECS 209-544-5; HSDB 874; Hylene T; Hylene tlc; Isocyanic acid, 4-methyl-m-phenylene ester; Mondur TD; Mondur TD-80; Mondur TDS; Nacconate IOO; NCI-C50533; NSC 4791; NSC 56759; RCRA waste number U223; TDI; 2,4-TDI; TDI; TDI-80; Tolueen-diisocyanaat; Toluen-disocianato; Toluene diisocyanate; Toluene 2,4-diisocyanate; 2,4-Toluene diisocyanate; Toluilenodwu-izocyjanian; Toluylene-2,4-diisocyanate; Tolyene 2,4-diisocyanate; UN 2078; Voranate T-80, Type I; Voranate T-80, Type II. In manufacture of polyurethane foams, elastomers, and coatings. Used by the PU industry for manufacture of flexible slabstock foam, molded flexible foam. Liquid; mp = 20.5°; bp = 251°; d^{20}_4 = 1.2244; reacts with H_2O, very soluble in Me_2CO, C_6H_6, Et_2O. *Bayer AG; Bayer Corp.; Dow Chem. U.S.A.; ICI Chem. & Polymers Ltd.; Nippon Chem. Industrial Co. Ltd.; Olin Corp.*

3798 Toluene diisocyanate
26471-62-5   H

$C_9H_6N_2O_2$
Isocyanic acid, methyl-m-phenylene ester.
Benzene, 1,3-diisocyanatomethyl-; CCRIS 596; Desmodur T 80; Desmodur T100; Diisocyanatotoluene; 1,3-Diisocyanatomethylbenzene; EINECS 247-722-4; HSDB 6003; Isocyanic acid, methyl-m-phenylene ester; Methyl-m-phenylene isocyanate; Methylphenylene isocyanate; Mondur TD-80; Nacconate-100; Niax iso-cyanate TDI; RCRA waste number U223; Rubinate TDI; Rubinate TDI 80/20; TDI; TDI 80-20; Toluene-2,4-diisocyanate (mixt. with Toluene-2,6-diisocyanate); Toluene diisocyanate; 2,4/2,6-Toluene diisocyanate iso-meric mixture; 2,4-Toluene diisocyanate; 2,4- and 2,6-Toluene diisocyanate; Toluenediisocyanate (mixed isomers); Tolylene diisocyanate; Tolylene isocyanate; m-Tolylidene diisocyanate; UN 2078. Mixture of 2,4- and 2.6-diisocyanates. A polyurethane monomer, used in paints and varnishes.

3799 Toluenesulfonamide
1333-07-9   H

$C_7H_9NO_2S$
Benzenesulfonamide, ar-methyl-.
ar-Toluenesulfonamide; Benzenesulfonamide, ar-methyl-; EINECS 215-578-1; HSDB 5816; Toluenesulfonamide; Toluenesulphonamide. Mixture of o, m and p-isomers.

3800 o-Toluenesulfonamide
88-19-7   G

$C_7H_9NO_2S$
2-Methylbenzenesulfonamide.
4-11-00-00229 (Beilstein Handbook Reference); AI3-23216; Benzenesulfonamide, 2-methyl-; BRN 1102362; EINECS 201-808-8; HSDB 5256; NSC 2185; o-Methylbenzenesulfonamide; o-Toluenesulfonamide; 2-Toluenesulfonamide; ortho-Toluenesulfonamide; ortho-Toluenesulphonamide; ortho-Toluolsulfonamid; Tolu-ene-2-sulfonamide; Toluene-2-sulphonamide; 2-Tolyl-sulfonamide; Uniplex 171. Plasticizer for thermoplastic and thermoset resins; imparts gloss and wetting to melamine, urea and phenolic resins. Component (with p-isomer) of Uniplex 171. Crystals; mp = 156-158°. *Unitex.*

3801 p-Toluenesulfonamide
70-55-3   G

$C_7H_9NO_2S$
4-Toluenesulfonic acid, amide.
4-11-00-00376 (Beilstein Handbook Reference); AI3-19503; Benzenesulfonamide, 4-methyl-; BRN 0472689; EINECS 200-741-1; HSDB 5203; 4-Methylbenzene-sulfonamide; p-Methylbenzenesulfonamide; NSC 9908; para-Toluenesulfonamide; p-Toluenesulfamide; 4-Tolu-enesulfanamide; p-Toluenesulfonamide; p-Toluenesulf-onylamide; p-Tolylsulfonamide; p-Tosylamide; Toluene-4-sulfonamide; Toluene-4-sulphonamide; Toluene-p-sulphonamide; Tolylsulfonamide; Tosylamide. Used in organic synthesis; in manufacture of plasticizers, resins; fungicide and mildewcide in paints and coatings. Crystals; mp = 135-137°. *Allchem Ind.; Honeywell & Stein; ICI Spec.; Unitex.*

3802 o-Toluenesulfonic acid, 4-amino-5-methoxy-
6471-78-9   H

$C_8H_{11}NO_4S$
4-Amino-5-methoxy-2-methylbenzenesulfonic acid.

Benzenesulfonic acid, 4-amino-5-methoxy-2-methyl-; EINECS 229-319-5; 5-Methoxy-2-methylsulphanilic acid.

3803   p-Toluene sulfonic acid
104-15-4   9611   G P

C7H8O3S
4-Methylbenzenesulfonic acid.
4-11-00-00241 (Beilstein Handbook Reference); ar-Toluenesulfonic acid; AI3-26478; Benzenesulfonic acid, 4-methyl-; BRN 0472690; Cyclophil P T S A; Cyzac 4040; EINECS 203-180-0; Eltesol®; Eltesol® TSX; Eltesol® TSX/A; Eltesol® TSX/SF; HSDB 2026; K-Cure 040; K-Cure 1040; Kyselina p-toluenesulfonova; Kyselina p-toluensulfonova; Manro PTSA 65 E; Manro PTSA 65 H; Manro PTSA 65 LS; Methylbenzenesulfonic acid; 4-Methylbenzenesulfonic acid; p-Methylbenzenesulfonic acid; Nacure 1040; NSC 167068; PTSA; p-Toluene sulfonate; 4-Toluenesulfonic acid; p-Toluenesulfonic acid; p-Toluenesulphonic acid; Tosic acid; TSA-HP; TSA-MH. Catalyst for organic synthesis, synthetic resins, manufacture of p-cresol, toluene derivatives, pharm-aceutical products, dyestuffs; chemical intermediate. Registered by EPA as an antimicrobial and fungicide (cancelled). Crystals; mp = 104.5°; bp_{20} = 140°, $bp_{0.1}$ = 185-187°; λ_m = 221 nm (MeOH); very soluble in H2O (67 g/100 ml), soluble in EtOH, Et2O. *Albright & Wilson Americas Inc.; Albright & Wilson UK Ltd.; Bayer Corp.; Boliden Intertrade; BYK-Chemie; Eastman Chem. Co.; Ferro/Grant; Nissan Chem. Ind.; PMC; Ruetgers-Nease; Sigma-Aldrich Fine Chem.; Spectrum Chem. Manufacturing; Witco/Oleo-Surf.*

3804   p-Toluene sulfonylhydrazide
877-66-7   G

C7H10N2O2S
4-(Methylsulphonyl)phenylhydrazine.
Celogen® TSH; EINECS 212-895-7. Chemical blowing agent for thermoset polyester. Crystals; mp = 202°. *Uniroyal.*

3805   p-Toluenesulfonyl isocyanate
4083-64-1   H

C8H7NO3S
Benzenesulfonyl isocyanate, 4-methyl-.
Benzenesulfonyl isocyanate, 4-methyl-; EINECS 223-810-8; p-Toluenesulfonic acid, anhydride with isocyanic acid; p-Toluenesulfonyl isocyanate; p-Toluenesulphonyl isocyanate.

3806   m-Toluic acid
99-04-7   G H

C8H8O2
Benzoic acid, 3-methyl-.
4-27-00-07537 (Beilstein Handbook Reference); AI3-15626; Benzoic acid, 3-methyl-; BRN 0970526; EINECS 202-723-9; m-Methylbenzoic acid; m-Toluic acid; m-Toluylic acid; meta-Toluic acid; NSC 2214. Used in manufacture of N,N-diethyl-m-toluamide, a broad-spectrum insect repellent. Crystals; mp = 108.7°; bp = 263°; d^{112} = 1.054slightly soluble in H2O, CHCl3, very soluble in EtOH, Et2O;; slightly soluble in H2O, soluble in organic solvents. *CK Witco Chem. Corp.; CK Witco Chem. Ltd.; Mitsubishi Gas.*

3807   o-Toluic acid
118-90-1   G H

C8H8O2
o-Methylbenzoic acid.
4-27-00-07537 (Beilstein Handbook Reference); AI3-15625; Benzoic acid, 2-methyl-; BRN 1072103; EINECS 204-284-9; 2-Methylbenzoic acid; NSC 2193; o-Methylbenzoic acid; Orthotoluic acid; Toluic acid; 2-Toluic acid; o-Toluic acid; o-Toluylic acid. Bacteriostat. Prisms or needles; mp = 103.7°; bp = 259°; d^{115} = 1.062; λ_m = 229, 278 nm (ε = 8460, 1250, MeOH); insoluble in H2O, soluble in CHCl3, very soluble in EtOH, Et2O. *Mitsubishi Gas.*

3808   p-Toluic acid
99-94-5   G H

C8H8O2
Benzoic acid, 4-methyl-.
4-27-00-07537 (Beilstein Handbook Reference); AI3-15627; Benzoic acid, 4-methyl-; BRN 0507600; p-Carboxytoluene; Crithminic acid; EINECS 202-803-3; 4-Methylbenzoic acid; NSC 2215; p-Methylbenzoic acid; p-Toluic acid; 4-Toluic acid; p-Toluylic acid. An animal feed supplement used in agricultural chemicals. Crystals; mp = 179.6°; bp = 274-275°; λ_m = 236, 280 nm (ε = 13500, 614, MeOH); insoluble in H2O, slightly soluble in TFA, very soluble in EtOH, Et2O, MeOH. *Degussa-Hüls Corp.; National Starch & Chem. UK; Penta Mfg.*

3809   p-Toluidine
106-49-0   9614

C7H9N
4-Toluidine.
AI3-19858; 1-Amino-4-methylbenzene; 4-Amino-1-methylbenzene; 4-Aminotoluen; 4-Aminotoluene; Anil-ine, p-methyl-; Benzenamine, 4-methyl-; CCRIS 598; CI 37107; CI Azoic Coupling Component 107; EINECS 203-403-1; HSDB 2044; Naphtol AS-KG; Naphtol AS-

KGLL; NSC 15350; RCRA waste number U353; Toluidine, p-; UN1708. Leaflets; mp = 43.7; bp = 200.4°; d^{20} = 0.9619; λ_m = 237, 288, 291, 294 nm (cyclohexane); slightly soluble in H_2O, soluble in Et_2O. Me_2CO, CCl_4, very soluble in EtOH, C_5H_5N.

3810   p-Toluoyl chloride
874-60-2   G

C_8H_7ClO
4-Methylbenzoyl chloride.
4-09-00-01733 (Beilstein Handbook Reference); Benzoyl chloride, 4-methyl-; BRN 0471492; EINECS 212-864-8; 4-Methylbenzoic acid chloride; 4-Methylbenzoyl chloride; p-Methylbenzoyl chloride; p-Toluic acid chloride; 4-Toluoyl chloride; p-Toluoyl chloride; p-Toluyl chloride. Used in chemical synthesis. Liquid; mp = -1.5°; bp = 226°; d^{20} = 1.1686; soluble in CCl_4. *James River; Sigma-Aldrich Fine Chem.*

3811   p-Tolyl aldehyde
104-87-0   G

C_8H_8O
4-Methylbenzaldehyde.
AI3-24380; Benzaldehyde, 4-methyl-; CCRIS 2942; EINECS 203-246-9; FEMA No. 3068; p-Formyltoluene; HSDB 5361; 4-Methylbenzaldehyde; p-Methylbenz-aldehyde; para-Methylbenzaldehyde; NSC 2224; 4-Tolu-aldehyde; p-Tolualdehyde; p-Toluylaldehyde; p-Tolyl-aldehyde; para-Toluladehyde; para-Toluyl aldehyde. Intermediate in the perfumes, pharmaceuticals and dyestuffs industries. Additive for resins, also used as a flavoring agent. Liquid; mp = -6°; bp = 205°, bp_{10} =106°; d^{17} = 1.0194; λ_m = 255 nm (ε = 21300, MeOH); slightly soluble in H_2O, very soluble in $CHCl_3$, freely soluble in EtOH, Et_2O, Me_2CO. *BASF Corp.; Mallinckrodt Inc.; Mitsubishi Gas.*

3812   Toluhydroquinone
95-71-6   G

$C_7H_8O_2$
2,5-dihydroxytoluene.
methylhydroquinone. Used as an antioxidant and polymerization inhibitor. Solid; mp = 128°; bp_{11} = 163°; λ_m = 292 nm (MeOH); very soluble in H_2O, EtOH, Et_2O, less soluble in Me_2CO, C_6H_6. *Eastman Chem. Co.*

3813   m-Toluidine
108-44-1   9614   H

C_7H_9N
·3-Aminotoluene.
4-12-00-01813 (Beilstein Handbook Reference); AI3-19859; m-Aminotoluene; Aniline, 3-methyl-; Benzen-amine, 3-methyl-; BRN 0635944; CCRIS 4325; EINECS 203-583-1; HSDB 2043; m-Methylaniline; m-Methyl-benzenamine; NSC 15349; m-Toluidin; m-Toluidine; m-Toluidyna; m-Tolylamine; UN1708. Liquid; mp = -31.2°; bp= 203.3°; d^{20} = 0.9889; λ_m = 236, 286 nm (ε = 8900, 1460, MeOH); very soluble in Me_2CO, EtOH, Et_2O, C_6H_6.

3814   o-Toluidine
95-53-4   9614   H

C_7H_9N
Benzenamine, 2-methyl-.
AI3-24383; o-Aminotoluene; Aniline, 2-methyl-; Benzen-amine, 2-methyl-; CI 37077; CCRIS 597; EINECS 202-429-0; HSDB 2042; o-Methylaniline; o-Methylbenzen-amine; NSC 15348; RCRA waste number U328; o-Toluidin; o-Toluidine; o-Toluidyna; o-Tolylamine; UN 1708. Oil; mp = -16.3°; bp = 200.3°; d^{20} = 0.9984; λ_m = 233, 285 nm (MeOH); poorly soluble in H_2O, freely soluble in EtOH, Et_2O, CCl_4.

3815   o-Toluidine, 6-ethyl-
24549-06-2   H

$C_9H_{13}N$
Benzenamine, 2-ethyl-6-methyl-.
4-12-00-02638 (Beilstein Handbook Reference); Aniline, 2-methyl-6-ethyl-; Benzenamine, 2-ethyl-6-methyl-; BRN 2079468; C 25702; EINECS 246-309-6; 2-Ethyl-6-methylaniline; 6-Ethyl-2-toluidine; 6-Ethyl-o-toluidine; 2-Methyl-6-ethylaniline; o-Toluidine, 6-ethyl-.

3816   m-Tolunitrile
620-22-4   H

C_8H_7N
1-Methyl-3-cyanobenzene.
AI3-17129; Benzonitrile, 3-methyl-; CCRIS 4736; 3-Cyanotoluene; m-Cyanotoluene; EINECS 210-631-5; 1-Methyl-3-cyanobenzene; 3-Methylbenzonitrile; m-Methylbenzonitrile; MTN; Nitril kyseliny m-toluylove; NSC 75453; 3-Toluenkarbonitril; m-Toluenenitrile; 3-Tolunitrile; m-Tolunitrile; m-Toluonitrile; m-Tolylnitrile. Liquid; mp = -23°; bp = 213°; d^{20} = 1.0316; λ_m = 227, 275, 283 nm (ε = 11000,

1130, 1130, MeOH); insoluble in H_2O, slightly soluble in CCl_4, freely soluble in EtOH, Et_2O.

3817 Toluylene
103-30-0   8895   G

$C_{14}H_{12}$
1,1'-(1,2-Ethanediyl)bis[benzene].
4-05-00-02160 (Beilstein Handbook Reference); Al3-52677; Benzene, 1,1'-(1,2-ethenediyl)bis-, (E)-; Benzene, 1,1'-(1E)-1,2-ethenediylbis-; BRN 1616740; CCRIS 5933; (E)-1,2-Diphenylethylene; trans-Diphenylethene; trans-α,β-Diphenylethylene; trans-1,2-Diphenylethene; EINECS 203-098-5; (E)-1,1'-(1,2-Ethenediyl)bisbenzene; HSDB 5020; NSC 2069; Stilbene, (E)-; (E)-Stilbene; trans-Stilbene; Toluylene. Used in the manufacture of optical bleaches and dyes. Crystals; mp = 123°; bp = 307°; bp_{12} = 166°; λ_m = 227, 294, 307 nm (ϵ = 21000, 33200, 32100, MeOH); λ_m = 296, 305 nm (ϵ 28100, 26700, 95% EtOH); insoluble in H_2O, slightly soluble in EtOH, $CHCl_3$, very soluble in Et_2O, C_6H_6.

3818 o-Tolyldiguanide
93-69-6   G

$C_9H_{13}N_5$
N-(2-Methylphenyl)imidodicarbonimidic diamide.
4-12-00-01764 (Beilstein Handbook Reference); Aliant; Biguanide, 1-o-tolyl-; BRN 0612193; EINECS 202-268-6; Eponoc B; Imidodicarbonimidic diamide, N-(2-methylphenyl)-; Nocceler BG; NSC 164906; o-Tolyl-biguanide; 1-o-Tolylbiguanide; 2-Tolylbiguanide; Sop-anox; Vulkacit 1000. A rubber vulcanization accelerator. Bayer AG.

3819 Tolyltriazole
29385-43-1   G H

$C_7H_7N_3$
1H-Benzotriazole, methyl-.
1H-Benzotriazole, 4(5)-methyl-; 1H-Benzotriazole, 4(or 5)-methyl-; 1H-Benzotriazole, methyl-; CCRIS 4738; Cobratec TT 100; EINECS 249-596-6; Methyl-1H-benzotriazole; Olin 53734; Preventol® Cl7-100; Tolu-triazole; Tolyl triazole. Corrosion inhibitor for copper, copper alloys and other metals; particularly suitable for antifreezes, coolants, cutting fluids and hydraulic fluids. Crystals; mp = 76-87°; bp_2 = 160°; d = 1.24; insoluble in H_2O, soluble in organic solvents; LD_{50} (rat orl) = 1600 mg/kg. Bayer AG; Dinoval; PMC; Sandoz.

3820 Tonophosphan
575-75-7   9589   G V

$C_9H_{13}NNaO_2P$
(4-Dimethylamino-o-tolyl)phosphonous acid sodium salt.
EINECS 209-391-4; Foston; Phosphinic acid, (4-(di-methylamino)-2-methylphenyl)-, sodium salt; Sodium (4-(dimethylamino)-2-methylphenyl)phosphinate; Toldimfos sodium; Tonofosfan. A proprietary preparation of tol-dimfos sodium; a source of phosphorus used for vet-erinary purposes. Soluble in H_2O and EtOH.

3821 Topiramate
97240-79-4   9625   D

$C_{12}H_{21}NO_8S$
2,3:4,5-Di-O-isopropylidene-β-D-fructopyranose sulfam-ate.
BRN 5988957; McN-4853; RWJ-17021; RWJ-17021-000; Tipiramate; Tipiramato; Topamax; Topiramate; Topir-amato; Topiramatum. An anticonvulsant. Crystals; mp = 125-126°; $[\alpha]_D^{23}$ = -34.0° (c = 0.4 MeOH). Johnson & Johnson Med. Inc.; McNeil Pharm.

3822 Toxaphene
8001-35-2   9633   H P

$C_{10}H_{10}Cl_8$
Technical chlorinated camphene.
Agricide; Agricide Maggot Killer; Agricide Maggot Killer (F); Agro-Chem Brand Toxaphene 6E; Agro-Chem Brand Torbidan 28; Agsco toxaphene; Agway toxaphene 6E; Alltex; Alltox; Anatox; Attac 4-2; Attac 4-4; Attac 6; Attac 6-3; Attac 8; Camphechlor; Camphechlore; Camphene, chlorinated; Camphene, octachloro-; Camphochlor; Camphofene huileux; Caswell No. 861; CCRIS 600; Chem-Phene; Chlorinated Camphenes; Chlorocamphene; Clor Chem T-590; Clor Chem T-590 Insecticide; Compound 3956; Coopertox; Cotton-Tox MP 82; Crestoxo; Cristoxo; Cristoxo 90; Dr Roger's TOX-ENE; EINECS 232-283-3; ENT 9,735; EPA Pesticide Chemical Code 080501; Estonox; Fasco-terpene; Felco/Land O'Lakes Toxaphene; Geniphene; Grower Service Toxaphene 6E; Grower Service Toxaphene MP; Gy-phene; Hercules 3956; Hercules Toxaphene Emulsifiable Concentrate; HSDB 1616; Kamfochlor; Latka 3956; M 5055; Melipax; Motox; NCI-C00259; Octachloro-camphene; PchK; Penphene; Phenacide; Phenatox; Poly-chlorcamphene; Polychlorinated camphenes; Polychloro-camphene; RCRA waste number P123; Red Top Toxaphene 8 Spray; Rigo Toxaphene 8; Royal Brand Bean Tox 82; Security Motox 63 cotton spray; Security Tox-MP cotton spray; Security Tox-Sol-6; Strobane-T; Strobane T-90; Synthetic 3956; Technical chlorinated camphene; Toxadust; Toxafeen; Toxakil; Toxaphen; Toxaphene; Toxaphene 8 EC; Toxaphene 8 Emulsifiable Insecticide; Toxaphene 90-10; Toxaphene E-8; Toxon 63; Toxyphen; Vertac 90%; Vertac toxaphene 90. Registered by EPA as an insecticide (cancelled). Designated as a Persistent Organic Pollutant (POP) under the Stockholm con-vention. Yellow solid; mp = 65-90°; d^{25} = 1.630; log P = 6.44; slightly

soluble in H_2O (0.0003 g/100 ml), readily soluble in aromatic hydrocarbons; LD_{50} (mrat orl) = 90 mg/kg, (frat orl) = 80 mg/kg, (mrat der) = 1075 mg/kg, (frat der) = 780 mg/kg. *Sigma-Aldrich Co.*

3823   Tramazoline
1082-57-1   9643   D

C13H17N3
2-[(5,6,7,8-Tetrahydro-1-naphthyl)amino]-2-imidazoline.
EINECS 214-105-6; Tramazolina; Tramazoline; Tramaz-olinum. An α-adrenergic agonist. Used as a nasal de-congestant. Crystals; mp = 142-143°. *Boehringer Ingelheim GmbH; Thomae GmbH Dr. Karl.*

3824   Tramazoline hydrochloride monohydrate
3715-90-0   9643   D

C13H20ClN3O
2-[(5,6,7,8-Tetrahydro-1-naphthyl)amino]-2-imidazoline monohydrochloride monohydrate.
Adolonta; Amadol; Biciron; Calmador; Contramol; Dolol; Dolsic; Dolzam; EINECS 223-064-3; Ellatun; Exopen; Forgesic; Fortradol; Jutadol; KB-227; Lumidol; Mandolgin; Nobligan; Nycodol; Paxilfar; Prontofort; Rhinaspray; Rhinogutt; Rhinospray; Rinogutt; Sylator; Tiparol; Towk; Tradol; Tradolan; Tralgiol; Trama 1A Pharma; Trama AbZ; Trama-BASF; Trama-Dorsch; Trama KD; Trama-Sonorania; Tramaβ; Tramabene; Tramadol acis; Tramadol AL; Tramadol-Dolgit; Tramadol Helvepharm; Tramadol Heumann; Tramadol-Mepha; Tramadol PB; Tramadol Stada; Tramadolhydrochlorid Fresenius; Tramadol-hydrochlorid Gerot; Tramadolor; Tramagit; Tramake; Tramandol AWD; Tramazoline hydrochloride; Tram-azoline hydrochloride monohydrate; Tramdolar; Tram-undin; Trodon; Trunal DX; Zamadol; Zydol; Zyndol. An α-adrenergic agonist (vasoconstrictor). Used as a nasal decongestant. mp = 172-174°; soluble in H_2O; LD_{50} (mus orl) = 195 mg/kg. *Boehringer Ingelheim GmbH; Thomae GmbH Dr. Karl.*

3825   trans-1,2-dichloroethylene
156-60-5   H

C2H2Cl2
(E)-1,2-Dichloroethylene.
4-01-00-00709 (Beilstein Handbook Reference); AI3-28786; BRN 1420761; CCRIS 2505; trans-1,2-Dichloroethene; trans-1,2-Dichloroethylene; EINECS 205-860-2; Ethene, 1,2-dichloro-, (E)-; Ethylene, 1,2-dichloro-, (E)-; Ethylene, 1,2-dichloro-, trans-; HCC 1130t; HSDB 6361; NSC 60512; R 1130t; RCRA waste number U079. Liquid; mp = -49.8°, bp = 48.7°; d^{20} = 1.2565; slightly soluble in H_2O, soluble in CCl4, very soluble in C6H6, CHCl3, very soluble in

EtOH, Et2O, Me2CO.

3826   trans-But-2-ene
624-64-6   H

C4H8
1,2-Dimethylethylene (E).
(E)-But-2-ene; 2-Butene, (E)-; 2-Butene, (2E)-; 2-Butene-trans; 2-trans-Butene; trans-2-Butene; β-trans-Butylene; EINECS 210-855-3; HSDB 5723; Low-boiling butene-2; trans-1,2-Dimethylethylene. Gas; mp = -105.5°; bp = 0.8°; d^{25} = 0.599; λ_m = 163, 177, 187, 202 nm (ε = 7943, 12589, 6310, 501, gas); soluble in C6H6.

3827   Trazodone
19794-93-5   9654   D

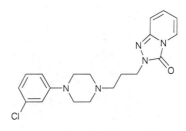

C19H24ClN5O
2-[3-[4-m-Chlorophenyl)-1-piperazinyl]propyl]-s-triazolo-[4,3-a]pyridin-3(2H)-one.
Beneficat; BRN 0628010; Desirel; EINECS 243-317-1; Sideril; Trazalon; Trazodil; Trazodon; Trazodona; Trazodone; Trazodonum; Trazonil. Antidepressant. mp = 86-87°; pKa (50% EtOH) = 6.14. *Angelini Francesco.*

3828   Trazodone hydrochloride
25332-39-2   9654   D

C19H25Cl2N5O
2-[3-[4-m-Chlorophenyl)-1-piperazinyl]propyl]-s-triazolo-[4,3-a]pyridin-3(2H)-one monohydrochloride.
AF 1161; Apo-Trazodone; Azona; Bimaran; Desyrel; Devidon; EINECS 246-855-5; HSDB 7048; KB-831; Molipaxin; NSC 292811; Pragmazone; Thombran; Tombran; Trazodone hydrochloride; Trazolan; Tritico; Triticum; Trittico. Antidepressant. Crystals; mp = 233°; soluble in CHCl3, less soluble in polar solvents, practically insoluble in common organic solvents; λ_m = 211, 246, 274, 312 nm (ε 50100, 11730, 3840, 3820); LD_{50} (mus iv) = 96 mg/kg. *Angelini Francesco; Lemmon Co.; Mead Johnson Pharmaceuticals.*

3829   Trenbolone acetate
10161-34-9   9659   D G

C20H24O3

17β-Hydroxyestra-4,9,11-trien-3-one acetate.
BRN 2012395; Component T-S; EINECS 233-432-5; Finaplix; RU-1697; Trenbolone acetate. An anabolic agent. Crystals; mp = 96-97°; $[\alpha]_D^{20} = 36.8°$ (c = 0.37 MeOH). *Roussel-UCLAF*.

3830   Tri-2-ethylhexyl trimellitate
3319-31-1   G H

C33H54O6

1,2,4-Benzenetricarboxylic acid, tris(2-ethylhexyl) ester.
1,2,4-Benzenetricarboxylic acid, tris(2-ethylhexyl) ester; BRN 2683525; CCRIS 4733; EINECS 222-020-0; Hatcol 200; HSDB 6145; Kodaflex® TOTM; Monosizer W710L; Morflex 510; PX-388; Staflex TOTM; TOTM; Tri-(2-ethylhexyl)trimellitate; Trimex T 08; Tris(2-ethylhexyl) trimellitate. Primary plasticizer used in vinyl film and vinyl-coated fabrics. Oil; bp = 414°; d = 0.9890; LD50 (mus orl) >60 g/kg. *Aristech; Eastman Chem. Co.*

3831   Triacetin
102-76-1   9664   G H

C9H14O6

Glyceryl triacetate.
4-02-00-00253 (Beilstein Handbook Reference); AI3-00661; BRN 1792353; EINECS 203-051-9; Enzactin; FEMA Number 2007; Fungacetin; Glycerin triacetate; Glycerol triacetate; Glyceryl triacetate; Glyped; HSDB 585; Kesscoflex TRA; Kodaflex® triacetin; NSC 4796; Triacetin; Triacetina; Triacetine; Triacetinum; Triacetyl glycerine; Triacetylglycerol; Vanay. Low-toxicity plasticizer for vinyl compounds; used in adhesives, resinous and polymeric coatings, paper, and paperboard for food contact; water-insol. hydroxyethyl cellulose films, as a plasticizer; fixative in perfumery; manufacture of cosmetics; specialty solvents, manufacture of celluloid, photographic films; used medically as a topical antifungal agent. Crystals; mp = -78°; bp = 259°; d20 = 1.1583; slightly soluble in H2O, CS2, ligroin;

very soluble in Me2CO, freely soluble in EtOH, Et2O, C6H6, CHCl3; LD50 (mus iv) = 1600 ± 81 mg/kg. *Bayer AG; Eastman Chem. Co.; MTM Spec. Chem. Ltd.; Penta Mfg.; Spectrum Chem. Manufacturing; Unichema; Whitehall-Robins*.

3832   Triacetoxyvinylsilane
4130-08-9   G

C8H12O6Si

Vinyltriacetoxysilane.
CV4800; Dow Corning® Z-6075; EINECS 223-943-1; NSC 93913; SH 6075; Silanetriol, ethenyl-, triacetate; Silanetriol, vinyl-, triacetate; Triacetoxyvinylsilane; Vinyltriacetoxysilane; Z 6075. Coupling agent, chemical intermediate, blocking agent, release agent, lubricant, primer, reducing agent. Used for polyesters, polyolefins, EPDM, EPM (peroxide cured). Liquid; mp = 7°; bp13 = 112°, bp10 = 115°; d20 = 1.169. *Degussa-Hüls Corp.; Dow Corning*.

3833   Triadimefon
43121-43-3   9723   G P

C14H16ClN3O2

1-(4-Chlorophenoxy)-3,3-dimethyl-1-(1H-1,2,4-triazol-1-yl) butan-2-one.
5-26-01-00123 (Beilstein Handbook Reference); Acizol; Adifon; Amiral; Azocene; Bay 6681 f; Bay MEB 6447; Bayleton; Bayleton 250 ec; Bayleton BM; Bayleton BM gel; Bayleton CF; Bayleton Total; Bayleton triple; Bonide Bayleton Systemic Fungicide; 2-Butanone, 1-(4-chlorophenoxy)-3,3-dimethyl-1-(1,2,4-triazol-1-yl)-; 2-Butanone, 1-(4-chlorophenoxy)-3,3-dimethyl-1-(1H-1,   2,4-triazol-1-yl)-; BRN 0619231; Caswell No. 862AA; 1-(4-Chlorophenoxy)-3,3-dimethyl-1-(1,2,4-triazol-1-yl)but-anone; 1-(4-Chlorophenoxy)-3,3-dimethyl-1-(1H-1,2,4-triazol-1-yl)-2-butanone; Diametom B; EINECS 256-103-8; EPA Pesticide Chemical Code 109901; Green Light Fung-Away Fungicide; Haleton; HSDB 6857; MEB 6447; Mighty; Miltek; Monterey Bayleton; NSC 303303; Nurex; Otria 25; Reach; Rofon; SA Systemic Fungicide for Turf & Ornamentals; Strike; Tenor; Tidifon; Traidimefon; Tria-dimefone; Triadimephon; Triamefon; 1-(1,2,4-Triazoyl-1)-1-(4-chloro-phenoxy)-3,3-dimethylbutanone; 1H-1,2,4-Tri-azole, 1-((tert-butylcarbonyl-4-chlorophenoxy)-methyl)-; Tridemifone; Typhon. Systemic fungicide with protective and curative action. Used for control of powdery mildews in cereals, apples, hops, raspberries, strawberries and other cane fruits plus American gooseberry mildew on all varieties of blackcurrants 'and gooseberries. Also used for the control of diseases in turf and the control of powdery mildew, rusts and blight of ornamental plants. Crystals; mp = 82°; sg20 = 1.22; soluble in H2O (0.026 g/100 ml), moderately soluble in organic solvents, CH2Cl2 (> 20 g/100 ml), C7H8 (> 20 g/100 ml), i-PrOH (10 - 20 g/100 ml), C6H14 (1 - 2 g/100 ml); LD50 (mrat orl) = 568 mg/kg, (frat orl) = 313 mg/kg, (mus orl) = 989 - 1071 mg/kg, (rbt orl) = 500 mg/kg, (mdog orl) > 500 mg/kg(rat der) > 5000 mg/kg, (rbt der) > 2000 mg/kg, (canary orl) > 1000 mg/kg, (Japanese quail 1750 - 2500 mg/kg), (mallard duck orl) >

628

4000 mg/kg, (ckn orl) = 5000 mg/kg; LC_{50} (rat ihl 4 hr.) = 0.48 mg/l air, (mallard duck 8 day dietary) > 10000 mg/kg diet, (bobwhite quail 8 day dietary) > 4640 mg/kg diet, (bluegill sunfish 96 hr.) = 11 mg/l, (goldfish 96 hr.) = 10 - 50 mg/l, (orfe 96 hr.) = 13.8 mg/l, (rainbow trout 96 hr.) = 14 mg/l, (carp 48 hr) = 7.6 mg/l; non-toxic to bees. *Bayer AG; Lawn & Garden Products Inc.*

3834   Triadimenol
55219-65-3   9667   G P

C14H18ClN3O2

β-(4-Chlorophenoxy)-α-(1,1-dimethylethyl)-1,2,4-triazole-1-ethanol. BAY KWG 0519; Bayfidan; Bayfrdan EW; Baytan; Baytan 15; Baytan TF 3479B; Caswell No. 074A; EINECS 259-537-6; EPA Pesticide Chemical Code 127201; KWG 0519; Spinnaker; Summit; Triadimenol; Triafol; Triaphol; Tridan Fungicide; UK 199. Fungicide used to control powdery mildew, rusts, and rhychosporium in winter and spring crops of wheat, barley, oats, and rye.Inhibits ergosterol biosynthesis in fungi. Emulsifiable concentrate containing 250 g/l triadimenol; used to control powdery mildew, rusts and rhychosporium in winter and spring crops of cereals, beet and brassicas. Registered by EPA as a fungicide. Colorless crystals; mp = 112 - 117°; soluble in H2O (0.0095 g/100 ml), soluble in C6H14 (0.01 - 0.1 g/100 ml), CH2Cl2 (10 - 20 g/100 ml), i-PrOH (10 - 20 g/100 ml), C7H8 (2 - 5 g/100 ml), also soluble in EtOH, ketones; LD_{50} (rat orl) = 700 - 1200 mg/kg, (mrat orl) = 1161 mg/kg, (frat orl) = 1105 mg/kg, (rat der 24 hours) > 5000 mg/kg, (Japanese quail orl) = 1750 - 2500 mg/kg, (hen orl) > 2000 mg/kg, (canary orl) > 1000 mg/kg; LC_{50} (rat ihl 4 hr.) = 0.45 mg/l air, (goldfish 96 hr.) = 10 - 50 mg/l, (golden orfe 96 hr.) = 17.4 mg/l, (rainbow trout 96 hr.) = 17 - 25 mg/l, (bluegill sunfish 96 mr.) = 15 mg/l; non-toxic to bees. *Bayer AG; Gustafson LLC; Shell UK; Wilbur Ellis Co.*

3835   Triallate
2303-17-5   9726   G H P

C10H16Cl3NOS

S-(2,3,3-trichloro-2-propenyl) bis(1-methylethyl)carbamo-thioate. Avadex BW; Bis(1-methylethyl)carbamothioic acid, S-(2,3,3-trichloro-2-propenyl)ester; BRN 1875853; Carb-amic acid, diisopropylthio-, S-(2,3,3-trichloroallyl) ester; Carbamothioic acid, bis(1-methylethyl)-, S-(2,3,3-tri-chloro-2-propenyl) ester; Caswell No. 870A; CCRIS 5383; CP 23426; Diisopropylthiocarbamic acid S-(2,3,3-trichloroallyl) ester; N-Diisopropylthiocarbamic acid S-2,3,3-trichloro-2-propenyl ester; Diisopropyltrichloro-allylthiocarbamate; N,N-Diisopropyl-2,3,3-trichlorallyl-thiolcarbamat; Dipthal; EINECS 218-962-7; EPA Pesticide Chemical Code 078802; Far-Go; HSDB 1780; NSC 379698; 2-Propene-1-thiol, 2,3,3-trichloro-, diisopropyl-carbamate; RCRA waste no. U389; Thiocarbamic acid, N-diisopropyl-, S-2,3,3-trichloroallyl ester; Tri-allate; Triallat; Triallate; Triamyl; 2,3,3-Trichloroallyl diisopropylthiocarbamate; 2,3,3-Trichlorallyl-N,N-(diiso-propyl)-thiocarbamat; S-(2,3,3-Trichloroallyl) diisopropyl-thiocarbamate; S-2,3,3-Trichloroallyl N,N-diisopropyl-thiocarbamate; 2,3,3-Trichloro-2-propene-1-thiol diiso-propylcarbamate; S-(2,3,3-Trichloro-2-propenyl) bis(1-

methylethyl)carbamothiote. Selective herbicide for control of wild oats, slender foxtail and bent grass in sugar beet and feed turnips, summer and winter barley, winter rye. Registered by EPA as a herbicide. Amber oil - crystals; mp = 29-30°; $bp_{0.0003}$ = 117°; sg^{25} = 1.273; slightly soluble in H2O (0.0004 g/100 ml), more soluble in organic solvents; LD_{50} (rat orl) = 1100 mg/kg, (rbt der) = 8200 mg/kg, (bobwhite quail orl) > 2251 mg/kg; LC_{50} (mallard duck, bobwhite quail 8 day dietary) > 5000 mg/kg diet, (rainbow trout 96 hr.) = 1.2 mg/l, (bluegill sunfish 96 hr.) = 1.3 mg/l, (*Daphnia* 48 hr.) = 0.43 mg/l; non-toxic to bees. *BASF Corp.; Bayer AG; Monsanto Co.*

3836   Triallyl cyanurate
101-37-1   G

C12H15N3O3

1,3,5-Triazine, 2,4,6-tris(2-propenyloxy)-. 5-26-03-00532 (Beilstein Handbook Reference); Activator OC; AI3-25448; BRN 0235560; Cyanuric acid, tri-2-propenyl ester; Cyanuric acid triallyl ester; EINECS 202-936-7; NSC 4804; Perkalink® 300; Perkalink® 300-50D; Rhenofit TAC; TAC; Triallyl cyanurate; Tripropargyl cyanurate; 2,4,6-Tri(allyloxy)-s-triazine; 2,4,6-Triallyloxy-1,3,5-triazine; 1,3,5-Triazine, 2,4,6-tris(2-propenyloxy)-; s-Triazine, 2,4,6-tris(allyloxy)-; 2,4,6-Triprop-2-ynyloxy-s-triazine; 2,4,6-Tris(allyloxy)triazine; 2,4,6-Tris(allyloxy)-1,3,5-triazine; 2,4,6-Tris(allyloxy)s-triazine. Co-agent to improve efficiency of peroxide-induced crosslinking of rubber; sensitizer for radiation-cured compounds. Used in polymers as a monomer and modifier; also used as an organic intermediate. *Akzo Chemie; Am. Cyanamid; Degussa AG; National Starch & Chem. UK.*

3837   Triallyl isocyanurate
1025-15-6   G

C12H15N3O3

Triallyl-s-triazine-2,4,6(1H,3H,5H)-trione. 4-26-00-00637 (Beilstein Handbook Reference); AI3-60290; BRN 0225482; CCRIS 6105; DIAK 7; EINECS 213-834-7; Isocyanuric acid triallyl ester; NSC 11692; TAIC; Triallyl isocyanurate; 1,3,5-Triallyl-1,3,5-triazine-2,4,6(1H,3H,5H)-trione; 1,3,5-Triallyl isocyanurate; 1,3,5-Triallylisocyanuric acid; 1,3,5-Triazine-2,4,6(1H,3H, 5H)-trione, 1,3,5-tri-2-propenyl-; s-Triazine-2,4,6(1H,3H, 5H)-trione, 1,3,5-triallyl-.; isocyanuric acid triallyl ester; 1,3,5-triallyl-s-triazine-2,4,6(1H,3H,5H)-trione; DIAK 7; TAIC. Co-agent improving peroxide-induced crosslinking of rubber. Stabilized with 100 ppm BHT. Liquid; mp = 20.5°; bp_4 = 149-152°, $bp_{0.5}$ = 105°; d^{20} = 1.1590. *Akzo Chemie.*

3838   Triallyl trimellitate
2694-54-4   G

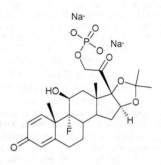

C18H18O6

Tri-2-propenyl 1,2,4-benzenetricarboxylate.

1,2,4-Benzenetricarboxylic acid, tri-2-propenyl ester; 1,2,4-Benzenetricarboxylic acid, triallyl ester; EINECS 220-264-2; Sipomer® TATM; TATM; Triallyl benzene-1,2,4-tricarboxylate; Triallyl trimellitate; 1,2,4-Triallyl trimellitate; Triam 705; Trimellitic acid triallyl ester. Liquid; mp = -30°; bp4 = 210°; d = 1.160. *Rhône Poulenc Surfactants.*

3839   Triamcinolone
124-94-7   9670   D

C21H27FO6

(11β,16α)-9-Fluoro-11,16,17,21-tetrahydroxypregna-1,4-diene-3,20-dione.

4-08-00-03629 (Beilstein Handbook Reference); Adcortyl; Aristocort; BRN 2341955; Celeste; Cinolone; Cinolone-T; CL-19823; Delphicort; EINECS 204-718-7; Fluoxy-prednisolone; HSDB 3194; Kenacort; Kenacort-AG; Ledercort; NSC 13397; Omcilon; Omicilon; Orion; Polcortolon; Rodinolone; SK-Triamcinolone; Tiamcinol-onum; Triam-Tablinen; Triamcet; Triamcinlon; Triam-cinolon; Triamcinolona; Triamcinolone; Triamcinolonum; Tricortale; Volon. Glucocorticoid. mp = 269-271°; $[\alpha]_D^{25}$= 5° (Me2OH); λ_m = 238 nm (ε 15800). *Am. Cyanamid; Bristol-Myers Squibb Co.; Fujisawa Pharm. USA Inc.*

3840   Triamcinolone acetonide
76-25-5   9671   D

C24H31FO6

(11β,16α)-9-Fluoro-11,21-dihydroxy-16,17-[(1-methylethylidene)bis(oxy)]pregna-1,4-diene-3,20-dione.

5-19-06-00568 (Beilstein Handbook Reference); Acetospan; Adcortyl; Adcortyl A; Aristocort; Aristocort A; Aristocort acetonide; Aristoderm; Aristogel; Azmacort; BRN 0060069; CCRIS 5231; Coupe-A; Delphicort; EINECS 200-948-7; Extracort; Flutone;

Ftorocort; Kenacort-A; Kenalog; Kenalone; Ledercort Cream; Myco-Triacet II; Mycolog II; Mytrex; Nasacort; Nasacort AQ; NSC 21916; Omcilon A; Panolog; Polcortolon; Respicort; Rineton; Solodelf; TAC-3; TAC-40; Tramacin; Triacet; Triaceton; Triam; Triam-Injekt; Triamcinolone acetonide; Triamonide 40; Triamsinolone acetonide; Tricinolon; Trymex; Vetalog; Volon A; Volon A 40; Volonimat.; component of: Mycolog II, Myco-Triacet II, Mytrex, Panolog. Glucocorticoid; antiasthmatic (inhalant); antiallergic (nasal). mp = 292-294°; $[\alpha]_D^{23}$= +109° (c = 0.75 CHCl3); λ_m = 238nm (ε 14600 EtOH); sparingly soluble in MeOH, Me2OH, EtOAc. *Am. Cyanamid; Baker Petrolite Corp.; Bristol-Myers Squibb Co.; Forest Pharm. Inc.; Herbert; Johnson & Johnson Med. Inc.; Lemmon Co.; Olin Mathieson; Rhône-Poulenc Rorer Pharm. Inc.; Savage Labs.*

3841   Triamcinolone acetonide 21-hemisuccinate
                    9671   D

C28H35FO9

(11β,16α)-9-Fluoro-11-hydroxy-16,17-[(1-methylethyl-idene)-bis(oxy)]pregna-1,4-diene-3,20-dione 21-hemi-succinate. Solutedarol. Glucocorticoid; antiasthmatic (inhalant); antiallergic (nasal). *Am. Cyanamid; Olin Mathieson.*

3842   Triamcinolone acetonide sodium phos-phate
1997-15-5   9671   D

C24H30FNa2O9P

(11β,16α)-9-Fluoro-11-hydroxy-16,17-[1-methylethyl-idinebis(oxy)]-21-(phosphonooxy)pregna-1,4-diene3,20-dione disodium salt.

Aristosol; CL 61965; CL 106359; EINECS 217-878-8; Triamcinolone acetonide 21-disodium phosphate; Triam-cinolone acetonide sodium Phosphate. Glucocorticoid; antiasthmatic (inhalant); antiallergic (nasal). *Am. Cyanamid; Olin Mathieson.*

3843   Triamterene
396-01-0   9674   D

C12H11N7

2,4,7-Triamino-6-phenylpteridine.

Ademin; Ademine; Amteren; Anjal; Apo-triazide; Dazid; Diarol; Dinazide; Diucelpin; Diutensat; Diuteren; Dyazide; Dyberzide; Dyrenium; Dytenzide; Esiteren; Fluss 40; Hidiurese; Hydrene; Hypertorr; Isobar; Jatropur; Jenateren; Kalspare; Masuharmin; Maxzide; Nephral; NSC-77625; Pterofen; Pterophene; Renezide; Reviten; Sali-Puren; SK&F-8542; Teriam; Thiazid Wolff; Triamizide; Triamterene; Triamthiazid; Triazide; Tricilone; Trispan; Triteren; Triurene; Trizid; Turfa; Uretren; Urocaudal. Aldosterone antagonist. Potassium sparing diuretic. mp= 316°, 327°; λ_m = 356 nm (ε =

21000, 4.5% HCOOH), 288, 361 nm (ε = 7079, 19953, EtOH). *SmithKline Beecham Pharm.*

3844   s-Triazine-2,4,6(1H,3H,5H)-trione,   1,3,5-tris(3,5-di-tert-butyl-4-hydroxybenzyl)-
27676-62-6   H

C48H69N3O6
1,3,5-Tris(3,5-di-tert-butyl-4-hydroxybenzyl)-1,3,5-tri-azine-2,4,6(1H,3H,5H)-trione.
EINECS 248-597-9; s-Triazine-2,4,6(1H,3H,5H)-trione, 1,3,5-tris(3,5-di-tert-butyl-4-hydroxybenzyl)-; 1,3,5-Tris-(3,5-di-tert-butyl-4-hydroxybenzyl)-1,3,5-triazine-2,4,6(1H,3H,5H)-trione.

3845   1,3,5-Triazine-2,4,6(1H,3H,5H)-trione, 1,3,5-tris(6-isocyanatohexyl)-
3779-63-3   H

C24H36N6O6
(2,4,6-Trioxotriazine-1,3,5(2H,4H,6H)-triyl)tris(hexa-methylene)isocyanate.
EINECS 223-242-0; Isocyanic acid, (2,4,6-trioxo-s-tri-azine-1,3,5(2H,4H,6H)-triyl)tris(hexamethylene) ester; (2, 4,6-Trioxotriazine-1,3,5(2H,4H,6H)-triyl)tris(hexamethyl-ene)isocyanate; 1,3,5-Triazine-2,4,6(1H,3H,5H)-trione, 1,3,5-tris(6-isocyanatohexyl)-.

3846   s-Triazole
288-88-0   9679   G H

C2H3N3
1H-1,2,4-Triazole.
AI3-51031; CGA-71019; EINECS 206-022-9; NSC 83128; Pyrrodiazole; 1,2,4-Triazole; 1H-1,2,4-Triazole; s-Triaz-ole; TA. Chemical intermediate. Crystals mp = 119-121°; bp = 260°; soluble

in H2O, organic solvents; LD50 (rat orl) = 1750 mg/kg.

3847   Triazophos
24017-47-8   9680   G P

C12H16N3O3PS
O,O-Diethyl O-(1-phenyl-1H-1,2,4-triazol-3-yl) phosph-orothioate.
AI3-27764; BRN 0682554; O,O-Diethyl O-(1-phenyl-1H-1,2,4-triazol-3-yl)phosphorothioate; EINECS 245-986-5; HOE 2960; HOE 2960 OJ; Hostathion; HSDB 6455; Methoxone; 1-Phenyl-3-(O,O-diethyl-thionophosphoryl)-1,2,4-triazole; 1-Phenyl-1,2,4-triazolyl-3-(O,O-diethyl-thionophosphate); Phosphorothioic acid, O,O-diethyl O-(1-phenyl-1,2,4-triazolyl) ester; Phosphorothioic acid, O,O-diethyl O-(1-phenyl-1H-1,2,4-triazol-3-yl) ester; Triazofos; Triazofosz; Triazophos. Broad spectrum insecticide and acaricide. Used for control of aphids and insects in a wide variety of crops. Pale yellow oil; mp = 2 - 5°; dec. < bp; d^{20} = 1.247; soluble in H2O (0.003 - 0.004 g/100 ml), EtOAc (> 90 g/100 ml), Me2CO (> 79 g/100 ml), EtOH (26 g/100 ml), C7H8 (28 g/100 ml), C6H14 (0.59 g/100 ml); LD50 (rat orl) = 57 - 68 mg/kg, (dog orl) > 320 mg/kg, (rat der) = 1100 mg/kg, (Japanese quail orl) = 4.2 - 27.1 mg/kg; LC50 (carp 96 hr.) = 5.6 mg/l, (goldfish 96 hr.) = 8.4 mg/l), (golden orfe 96 hr.) = 11 mg/l; toxic to bees. *AgrEvo; ICI Chem. & Polymers Ltd.*

3848   Triazoxide
72459-58-6   9681   P

C10H6ClN5O
7-Chloro-3-(1H-imidazol-1-yl)-1,2,4-benzotriazine-1-oxide.
EINECS 276-668-4; SAS 9244; Triazoxide. Fungicide. Light yellow crystals; mp = 182°; soluble in H2O (0.003 g/100 ml at 20°), C6H14 (< 0.1 g/100 ml), CH2Cl2 (5 - 10 g/100 ml), i-PrOH (0.2 - 0.5 g/100 ml), C7H8 (2 - 5 g/100 ml); LD50 (rat orl) = 100 - 200 mg/kg, (rat der) > 5000 mg/kg; LC50 (rat ihl 4 hr.) = 0.8 - 3.2 mg/l air. *Bayer Corp., Agriculture.*

3849   Tribenzoin
614-33-5   G

C24H20O6
Glyceryl tribenzoate.
AI3-08196; EINECS 210-379-6; FEMA No. 3398; Glycerol tribenzoate; NSC 2230; 1,2,3-Propanetriol, tribenzoate; Tribenzoin; Uniplex 260. Polymer modifier; plasticizer; for heat seal applications, lacquers, films, in PVAc-based adhesives, cellophane coatings, nitrocellulose coatings, nail lacquer formulations, printing inks, polishes; extrusion and injection molding processing aid. Needles; mp = 76°; d^{12} = 1.228; insoluble in H_2O, soluble in EtOH, very soluble in Et_2O, Me_2CO, C_6H_6, $CHCl_3$. Unitex.

3850   Tribromoneopentyl alcohol
36483-57-5   G

C5H9Br3O
2,2-Bis(bromomethyl)-3-bromo-1-propanol.
2,2-Dimethyl-1-propanol tribromo deriv.; 2,2-Dimethyl-propan-1-ol, tribromo derivative; EINECS 253-057-0; FR 1360; 1-Propanol, 2,2-dimethyl-, tribromo deriv.; TBNPA; Tribromoneopentyl alcohol; FR-513. Flame retardant for flexible and rigid PU; flame retardant intermediate. White to off white flakes; mp = 62-67°; d = 2.28; soluble in alchols; LD50 (rat orl) = 2823 mg/kg; irritant to eyes, mild irritant to skin. Amerihaas; Dead Sea Bromine.

3851   Tribromophenol
118-79-6   9687   G H

C6H3Br3O
2,4,6-Tribromophenol.
AI3-14896; Bromkal pur 3; Bromol; CCRIS 1658; EINECS 204-278-6; Flammex 3BP; HSDB 5584; NSC 2136; Phenol, 2,4,6-tribromo-; TA 10; Tribromophenol; 2,4,6-Tribromophenol. Reactive flame retardant used mainly as an intermediate for polymeric flame retardants. Also used as a caustic and disinfectant. White needles or prisms; mp = 95.5°; bp = 286°; d^{20} = 2.55; λ_m = 289, 297 nm (cyclohexane); slightly soluble in H_2O, CCl_4, soluble in Et_2O, C_6H_6, $CHCl_3$, AcOH, very soluble in EtOH; LD50 (rat orl) = 2000 mg/kg; hazard by ingestion, inhalation, skin. AmeriBrom Inc.; Dead Sea Bromine; Fluka; Great Lakes Fine Chem.; Sigma-Aldrich Fine Chem.

3852   Tribromophenyl allyl ether
3278-89-5   G

C9H7Br3O
Allyl 2,4,6-Trobromophenyl ether.
2-(Allyloxy)-1,3,5-tribromobenzene; Allyl 2,4,6-tribromophenyl ether; Benzene, 1,3,5-tribromo-2-(2-prop-enyloxy)-; EINECS 221-913-2; Ether, allyl 2,4,6-tribromo-phenyl; NSC 35767; TBP-AE; 2,4,6-

Tribromophenylallyl ether; FR-913. Aromatic flame retardant for expandable PS; synergist with hexabromocyclododecane. White to off-white crystal powder; mp = 74-76°; d = 2.20; LD50 (rat orl) > 5000 mg/kg; (rbt der) > 2000 mg/kg. AmeriBrom Inc.; Amerihaas; Dead Sea Bromine.

3853   Tribromsalan
87-10-5   9690   G P

C13H8Br3NO2
3,4',5-Tribromosalicylanilide.
Agramed; AI3-25516; ASC-4; Benzamide, 3,5-dibromo-N-(4-bromophenyl)-2-hydroxy-; BRN 2146888; Brom-salans; Caswell No. 863; CCRIS 5745; Diaphene; 3,5-Dibromo-N-(4-bromophenyl)-2-hydroxybenzamide; 3,5-Dibromosalicylic acid p-bromoanilide; ENT 25516; EPA Pesticide Chemical Code 077404; ET-394; Lamar L-300; Temaspet IV; Tuasol 100; NSC-20526; Polybrominated salicylanilide; Salicylanilide, 3,4',5-tribromo-; Sherstat TBS; Stecker asc-4; TBS; TBS 95; Temasept; Temasept II; Temasept IV; Tempasept II; Tribromosalicylanilide; Tribromosalicylanilide, 3,4',5-; 3,4',5-Tribromosalicyl-anilide; Tribromsalan; Tribromsalanum; Trisanil; Trisanyl; Tuasal; Tuasol 100; Vancide TBS; WR 34912. Antimicrobial used in resins, latex emulsions, plastics. Registered by EPA as an antimicrobial and disinfectant (cancelled). Crystals; mp = 227°; λ_m = 220, 283, 330 nm (ε = 33884, 13804, 9772 EtOH); practically insoluble in H_2O; soluble in hot Me_2CO, DMF. Colgate-Palmolive; Dow Chem. U.S.A.; Hexcel; Merrell Dow Pharm. Inc.; Sogeras.

3854   Tributylaluminum
1116-70-7   H

C12H27Al
Tri-n-butylaluminum.
Aluminum, tributyl-; EINECS 214-240-0; HSDB 5784; Tri-n-butylaluminum; Tributylaluminum. Pyrophoric liquid.

3855   Tributylamine
102-82-9   9691   H

C12H27N
Tri-n-butylamine.
4-11-00-00122 (Beilstein Handbook Reference); AI3-15424; Amine, tributyl-; BRN 1698872; CCRIS 4879; EINECS 203-058-7; HSDB 877; N,N-Dibutyl-1-butan-amine; Tri-n-butylamine; Tributilamina; Tributylamine; Tris-n-butylamine; UN2542; UN3254. Clear liquid; mp = -70°; bp = 216.5°; d^{20} = 0.7770; λ_m = 242 nm (ε = 1000, $CHCl_3$); slightly soluble in H_2O, CCl_4, soluble in Me_2CO, C_6H_6, very soluble

in EtOH, Et2O. *Union Carbide Corp.*

3856 2,4,6-Tri(3,5-di-tert-butyl-4-hydroxybenzyl)mesitylene
1709-70-2   H

C54H78O3
α,α',α-(2,4,6-Trimethyl-s-phenenyl)tris(2,6-di-tert-butyl-p-cresol.
Agidol 40; Ahydol; Antioxidant 40; Antioxidant 330; AO-40; BRN
2034522; EINECS 216-971-0; Ethanox 330; Ethyl 330; Ethyl
Antioxidant 330; Ionox 330; Irganox 330; Methylene bis ethyl butyl
phenol; NSC 85846; Santoquin emulsion; Santoquin mixture 6;
α,α',α-(2,4,6-Trimethyl-s-phenenyl)tris(2,6-di-tert-butyl-p-cresol;
1,3,5-Trimethyl-2,4,6-tris(3,5-di-tert-butyl-4-hydroxybenzyl)benzene;
1,3,5-Tris(3,5-di-tert-butyl-4-hydroxybenzyl)mesitylene.

3857 Tributyl phosphate
126-73-8   9692   G H

C12H27O4P
Phosphoric acid, tributyl ester.
4-01-00-01531 (Beilstein Handbook Reference); AI3-00399; BRN
1710584; Butyl phosphate, ((BuO)3PO); But-yl phosphate, tri-;
CCRIS 6106; Celluphos 4; Disflamoll TB; EINECS 204-800-2; HSDB
1678; Kronitex® TBP; MCS 2495; NSC 8484; Phosphoric acid
tributyl ester; TBP; Tri-n-butyl phosphate; Tributilfosfato; Tributoxyphosphine oxide; Tributyl phosphate; Tributyl phosph-ateol; Tributyle
(phosphate de); Tributylfosfaat; Tributyl-fosfat; Tributylphosphat;
Tributylphosphate. Heat-ex-change medium; solvent extraction of
metal ions; plasticizer for cellulose esters, lacquers, plastics, vinyl
resins; antifoam agent; dielectric. Used as an antifoam for paints and
as a pigment dispersant. Liquid; mp = -79°; bp = 289°; d25 = 0.9727;
soluble in H2O, Et2O, C6H6, CS2, freely soluble in EtOH. *Akzo
Chemie; Chemron; FMC Corp.*

3858 Tributyl phosphite
102-85-2   G

C12H27O3P
tri-n-butyl phosphite.
4-01-00-01527 (Beilstein Handbook Reference); AI3-15022; BRN
1703866; 1-Butanol, 1,1',1''-phosph-inidynetri-; Butyl phosphite
((C4H9O)3P); EINECS 203-061-3; JP 304; NSC 2675; Phosphorous
acid, tributyl ester; Phosphorus tributoxide (P(OBu)3); Syn-O-Ad® P-
312; Tri-n-butyl phosphite; Tributoxyphosphine; Tributyl phosphite;
Tributylfosfit. Antioxidant and antiwear agent in gear and
transmission oils. Used as an additive for greases and extreme
pressure lubricants; stabilizer for fuel oils and polyamides; gasoline
additive. Liquid; mp = -80°; bp26 = 137°; bp12 = 122°; bp7 = 118-
125°; d20 = 0.9259; slightly soluble in CCl4, soluble in EtOH, very
soluble in Et2O. *Akzo Chemie; Albright & Wilson Americas Inc.;
Albright & Wilson UK Ltd.; Janssen Chimica.*

3859 Tributylphosphorotrithioate
78-48-8   G H

C12H27OPS3
S,S,S-Tributyl phosphorotrithioate.
AI3-25812; B-1,776; B 1776; BRN 1910992; Butifos; Butiphos; Butyl
phosphorotrithioate; Butyl phosphoro-trithioate ((BuS)3PO); Caswell
No. 864; Chemagro B-1776; De-Green; DEF; DEF defoliant; E-Z-Off
D; EINECS 201-120-8; EPA Pesticide Chemical Code 074801; Fos-
Fall A; Fossfall; HSDB 668; Ortho phosphate defoliant;
Phosphorotrithioic acid, S,S,S-tributyl ester; TBTP; Tribufos;
Tribuphos; S,S,S-Tributyl phosphorotrithioate; S,S,S-Tributyl
trithiophosphate; S,S,S-Tributylphosphoro-trithioate; S,S,S-
Tributyltrithiofosfat; S,S,S-Tributyltrithio-phosphate; Folex 6EC;
FOS-FALL "A". Plant growth regulator which acts as a defoliant.
Used on cotton to facilitate harvesting. Registered by EPA as a
herbicide. mp < -25°; bp0.3 = 150°; d20 = 1.057; nD25 = 1.532; poorly
soluble in H2O (2.3 mg/l), more soluble in organic solvents; LD50 (rat
orl) = 233 mg/kg. *Aventis Crop Science; Bayer AG; Bayer Corp.;
Micro-Flo Co. LLC.*

3860 Tributyltin Chloride
1461-22-9   P

C12H27ClSn
Tri-n-butyltin chloride.
Caswell No. 867A; CCRIS 6319; Chlorid tri-n-butylcinicity; Chlorotri-
n-butylstannane; Chlorotri-n-butyltin; Chlorotributylstannane;
Chlorotributyltin; EINECS 215-958-7; EPA Pesticide Chemical Code
083107; HSDB 6816; Monochlorotributyltin; NSC 22323; Stannane,
chlorotributyl-; Stannane, tributylchloro-; Tin, tri-n-butyl-, chloride;
Tri-n-butylchlorotin; Tri-n-butyltin chloride; Tri-N-butylzinn-chlorid;
Tributylchlorostannane; Tributylchlorotin; Tributylstannium chloride;
Tributyl-stannyl chloride; Tributyltin chloride; WR 3396. Registered
by EPA as a fungicide (cancelled). Oil; bp25 = 171-173°; d = 1.200.
Fluka.

3861   Tricaprylin
538-23-8   G H

C27H50O6
Octanoic acid, 1,2,3-propanetriyl ester.
4-02-00-00991 (Beilstein Handbook Reference); BRN 1717202; Caprylic acid, 1,2,3-propanetriyl ester; Caprylic acid triglyceride; Caprylin; Captex® 8000; EINECS 208-686-5; Glycerol tricaprylate; Glycerol trioctanoate; Maceight; Miglyol® 808; NSC 4059; Octanoic acid, 1,2,3-propanetriyl ester; Octanoic acid triglyceride; Octanoin, tri-; Panacete 800; 1,2,3-Propanetriol trioctanoate; Propane-1,2,3-triyl trioctanoate; Radiamuls® MCT 2108; Rato; Sefsol 800; Tricaprilin; Tricaprylic glyceride; Tricaprylin; Trioctanoin; Trioctanoin oil; Trioctanoylglycerol. Nonoily lubricant imparting rich feel to the skin; for cosmetics and pharmaceuticals; carrier for essential oils, flavors; vehicle for vitamins, medicinals, nutritional products. Good skin spreading/penetrating properties. Edible oil; forms very thin films for coating confectionery and dried fruits; mold release aid for bakery, confectionery; lubricant for food processing equipment; viscosity depressant, carrier of actives in oleoresins; lipid for dietetic foods. Oil; mp = 10°; bp = 233°; d^{20} = 0.9540; insoluble in H2O, very soluble in Et2O, C6H6, CHCl3, petroleum ether, ligroin, freely soluble in EtOH; LD50 (rat orl) = 33300 mg/kg. *Degussa-Hüls Corp.; Fina Chemicals; Karlshamns.*

3862   Trichlorethylene
79-01-6   9713   H

C2HCl3
Ethylene trichloride.
4-01-00-00712 (Beilstein Handbook Reference); Acetylene trichloride; AI3-00052; Algylen; Anamenth; Benzinol; Blacosolv; Blancosolv; BRN 1736782; Caswell No. 876; CCRIS 603; Cecolene; Chlorilen; Chlorylea; Chlorylen; Chorylen; Circosolv; CirCosolv; Clor; Crawhaspol; Densinfluat; Disparit B; Dow-Tri; Dukeron; EINECS 201-167-4; EPA Pesticide Chemical Code 081202; Ethene, trichloro-; Ethinyl trichloride; Ethylene trichloride; Ethylene, trichloro-; F 1120; Fleck-flip; Flock FLIP; Fluate; Germalgene; HSDB 133; Lanadin; Lethurin; Narcogen; Narkosoid; NCI-C04546; Nialk; NSC 389; Per-A-; Perm-A-chlor; Petzinol; Philex; R 1120; RCRA waste number U228; TCE; Threthylene; Threthylene; Trethylene; Tri; Tri-clene; Tri-plus; Tri-plus M; Triad; Trial; Triasol; Trichlooretheen; Trichloorethyleen, tri; Trichloraethen; Trichloräthen; Trichloraethylen, tri; Trichloräthylen, tri; Trichloraethylenum; Trichloran; Trichloren; Trichlor-ethylene; Trichlorethylenum; Trichloroethene; Trichloro-ethylenum; Tricloretene; Tricloroetilene; Tricloroetileno; Trielene; Trielin; Trielina; Trieline; Triklone N; Trilen; Trilene; Trilene TE-141; Trimar; UN 1710; Vestrol; Vitran; Westrosol. Metal degreasing; extraction solvent for oils, fats, waxes; solvent for cellulose esters and ethers; drycleaning; solvent dyeing; manufacture of organic chemicals, pharmaceuticals. Also used as a disinfecting cleaning compound. Liquid; mp = -84.7°; bp = 87.2°; d^{20} = 1.4642; slightly soluble in H2O (0.11 g/100 ml), soluble in organic solvents; LD50 (rat orl)= 4.92 ml/kg. *Asahi Chem. Industry; Ashland; Elf Atochem N. Am.; General Chem; PPG Ind.*

3863   Trichlorfon
52-68-6   9696   G

C4H8Cl3O4P
Dimethyl (2,2,2-trichloro-1-hydroxyethyl)phosphonate.
4-01-00-03147 (Beilstein Handbook Reference); Aerol 1; Aerol 1 (pesticide); Agroforotox; AI3-19763; Anthon; Bay 15922; Bay L13/59; BAY 15922; BAY-a 9826; BAY-L 1359; Bayer 15922; Bayer L 13/59; Bayer L 1359; Bilarcil; Bovinox; Briten; Briton; Britten; BRN 1709434; Caswell No. 385; CCRIS 1289; Cekufon; Chlorak; Chlorfos; Chlorofos; Chloroftalm; Chlorophos; Chloro-phosciclosom; Chlorophthalm; Chloroxyphos; Chlorphos; Ciclosom; Clorofos; Combot; Combot equine; Danex; Denkaphon; DEP; DEP (pesticide); Depthon; DETF; Dicontal Fort; Dimetox; Dioxaphos; Dipterex; Dipterex 50; Dipterex SL; Dipterex WP 80; Diptevur; Ditrifon; Ditriphon 50; Dylox; Dyrex; Dyvon; EINECS 200-149-3; ENT 19,763; EPA Pesticide Chemical Code 057901; Equino-aid; Ertefon; Flibol E; Fliegenteller; Forotox; Foschlor; Foschlor 25; Foschlor R; Foschlor R-50; Foschlorem; HSDB 881; Hypodermacid; Khloroftalm; Leivasom; Loisol; Masoten; Mazoten; Methyl chlorophos; Metrifonate; Metrifonato; Metrifonatum; Metriphonate; Metriphonatum; NCI-C54831; Neguvon; Nevugon; NSC 8923; OMS 800; Phoschlor; Phoschlor R50; Polfoschlor; Proxol; Ricifon; Ritsifon; Satox 20WSC; Soldep; Sotipox; Totalene; Trichloorfon; Trichlorfon; (±)-Trichlorfon; Trichlorophon; Trichlorphon; Trichlorphon FN; Trinex; Tugon; Tugon fly bait; Tugon stable spray; Vermicide bayer 2349; Volfartol; Votexit; WEC 50; Wotexit; Zeltivar. Non-systemic insecticide with stomach and contact action. Used for insect control in agriculture and horticulture. Crystals; mp = 75-79°; bp0.1 = 100°; SG = 1.73; n$_D^{20}$ 1.3439; soluble in H2O (120 g/l), more soluble in organic solvents; LD50 (rat orl) = 560 mg/kg. *Agrolinz; Bayer AG; Cequisa; Chemolimpex; Denka International; Diachem; ICI Agrochemicals; Jin Hung; Makhteshim-Agan.*

3864   Trichloroacetic acid
76-03-9   9700   G P

C2HCl3O2
Trichloroacetic acid.
4-02-00-00508 (Beilstein Handbook Reference); Acetic acid, trichloro-; Acetic acid, trichloro- (solid); Aceto-caustin; Acide trichloracetique; Acido tricloroacetico; AI3-24157; Amchem grass killer; BRN 0970119; Caswell No. 870; CCRIS 4015; EPA Pesticide Chemical Code 081002; Na Ta; Farmon TCA; HSDB 1779; Konesta; Kyselina trichloroctova; NSC 215204; TCA; Tecane; Trichloorazijnzuur; Trichloracetic acid; Trichloressig-säure; Trichloroacetate; Trichloroacetic acid; Trichloro-acetic acid; Trichloroacetic acid [UN1839] [Corrosive]; Trichloroethanoic acid; Trichloromethanecarboxylic acid; UN1839; UN2564. Registered by EPA as a herbicide (cancelled). Used for control of weeds in field crops. Also used as a decalcifier and fixative in microscopy and a protein precipitant. mp = 57.5°; bp = 196.5°; d^{64} = 1.6126; soluble in H2O (10 g/ml), less soluble in organic solvents; LD50 (rat orl) = 5000 mg/kg. *Hoechst UK Ltd.*

3865   Trichloroacetyl chloride
76-02-8   G

634

C2Cl4O
Trichloromethyl chloroformate.
4-02-00-00519 (Beilstein Handbook Reference); Acetyl chloride, trichloro-; BRN 0774120; CCRIS 6764; Diphosgene; EINECS 200-926-7; HSDB 6321; NSC 190466; Superpalite; Trichloroacetic acid chloride; Trichloroacetochloride; UN2442. Has been used as a military poison gas. mp = -57°; bp = 117.9°; d^{20} = 1.6202; very soluble in organic solvents; reacts with H_2O. *Sigma-Aldrich Fine Chem.*

3866   Trichlorobenzene
120-82-1   9704   G H

C6H3Cl3
1,2,4-Trichlorobenzene.
4-05-00-00664 (Beilstein Handbook Reference); AI3-07775; Benzene, 1,2,4-trichloro-; BRN 0956819; CCRIS 5945; EINECS 204-428-0; Hostetex L-pec; HSDB 1105; NSC 406697; 1,2,4-Trichlorobenzene; unsym-Trichloro-benzene; 1,2,4-Trichlorobenzol; Trojchlorobenzen. Solv-ent in chemical manufacture, dyes, intermediates, dielec-tric fluid, synthetic transformer oils, lubricants, heat-transfer media, insecticides. Rhombohedral crystals; mp = 17°; bp = 213.5°; d^{25} = 1.459; λ_m = 220, 227, 269 277, 286 nm (ε = 9340, 9200, 323, 482, 438, MeOH); insoluble in H_2O, slightly soluble in EtOH, CCl4, very soluble in Et2O. *Ashland.*

3867   1,2,3-Trichlorobenzene
87-61-6   9703   H

C6H3Cl3
Benzene, 1,2,3-trichloro-.
4-05-00-00664 (Beilstein Handbook Reference); AI3-15516; Benzene, 1,2,3-trichloro-; BRN 0956882; CCRIS 5944; EINECS 201-757-1; HSDB 1502; NSC 43432; vic-Trichlorobenzene. Plates; mp = 53.5°; bp = 218.5°; λ_m = 224, 264, 271, 279 nm (ε = 8080, 144, 178, 137, MeOH); insoluble in H_2O, poorly soluble in EtOH, CHCl3, more soluble in Et2O, C6H6.

3868   2,3,4-Trichloro-1-butene
2431-50-7   H

C4H5Cl3
2,3,4-Trichloro-1-butene.
3-01-00-00725 (Beilstein Handbook Reference); BRN 1745897; Butene, 2,3,4-trichloro-, 1-; 1-Butene, 2,3,4-trichloro-; EINECS 219-397-9; HSDB 5878; 2,3,4-Tri-chloro-1-butene; 2,3,4-Trichlorobut-1-ene; 2,3,4-Tri-chlorobutene-1. Liquid; bp_{20} = 60°, bp_{10} = 40°; d^{20} = 1.3430; very soluble in Me2CO, CHCl3.

3869   Trichlorododecylsilane
4484-72-4   G

C12H25Cl3Si
Dodecyltrichlorosilane.
CD6220; Dodecyl trichlorosilane; EINECS 224-769-9; HSDB 387; NSC 139836; Silane, dodecyltrichloro-; Silane, trichlorododecyl-; Trichloro(dodecyl)silane; Tri-chlorododecylsilane; UN1771. Coupling agent, chemical intermediate, blocking agent, release agent, lubricant, primer, reducing agent. Liquid; bp_{16} = 125°; d^{20} = 1.0602. *Degussa-Hüls Corp.*

3870   Trichloroethyl phosphate
115-96-8   G H

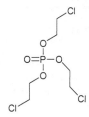

C6H12Cl3O4P
Tris(2-chloroethyl) phosphate.
4-01-00-01379 (Beilstein Handbook Reference); AI3-15023; Antiblaze 100; BRN 1710938; CCRIS 1302; Celluflex; Celluflex CEF; Disflamoll TCA; EINECS 204-118-5; Ethanol, 2-chloro-, phosphate (3:1); Fyrol CEF; Genomoll P; HSDB 2577; NCI-C60128; Niax 3CF; Niax Flame Retardant 3CF; NSC 3213; Phosphoric acid, tris(2-chloroethyl)ester; Tri-β-chloroethyl phosphate; Tri(2-chloroethyl) phosphate; Trichlorethyl phosphate; Tris-(2-chloroethyl)fosfat; Tris-(2-chloroethyl)fosfat; Tris(β-chloro-ethyl) phosphate; Tris(2-chloroethyl) orthophosphate; Tris(2-chloroethyl) phosphate. A plasticizer. Also used as a flame retardant. Liquid; mp = -15°; bp = 330°, bp_{10} = 194°; d^{25} = 1.39; soluble in H_2O (7 g/l), CCl4; LD50 (rat orl) = 1230 mg/kg. *Hoechst UK Ltd.*

3871   Trichloroethylsilane
115-21-9   G H

C2H5Cl3Si
Ethyl trichlorosilane.
4-04-00-04227 (Beilstein Handbook Reference); BRN 1361384; CE6350; EINECS 204-072-6; Ethyl silicon trichloride; Ethyl trichlorosilane; HSDB 884; Silane, ethyl(trichloro)-; Silane, trichloroethyl-; Silicane, trichloroethyl-; Trichloro(ethyl)silane; Trichloroethyl-silane; UN1196. Coupling agent, chemical intermediate, blocking agent, release agent, lubricant, primer, reducing agent. Intermediate in manufacture of silicones. Liquid; mp = -105.6°; bp = 100.5°; d^{20} = 1.2373; soluble in CCl4. *Degussa-Hüls*

3872   Trichlorofluoromethane
75-69-4   9714   H

CCl3F

Fluorotrichloromethane.
4-01-00-00054 (Beilstein Handbook Reference); Algofrene type 1; Arcton 9; Arcton 11; BRN 1732469; Caswell No. 878; CCRIS 604; CFC 11; Chladone 11; Daiflon 11; Daiflon S 1; Dymel 11; EINECS 200-892-3; Electro-CF 11; EPA Pesticide Chemical Code 000013; Eskimon 11; F-11; F 11 (halocarbon); F 11B; FC 11; FC 11 (halocarbon); FKW 11; Fluon 11; Fluorocarbon 11; Fluorochloroform; Fluorotrichloro-methane; Fluorotroj-chlorometan; Flurotrichloromethane; Freon 11; Freon 11A; Freon 11B; Freon HE; Freon MF; Frigen 11; Frigen 11A; Frigen S 11; Genetron 11; Genetron 11SBA; Halocarbon 11; Halon 11; HSDB 138; Isceon 131; Isotron 11; Kaltron 11; Khaladon 11; Khladon 11; Ledon 11; Methane, fluorotrichloro-; Methane, trichlorofluoro-; Monofluorotrichloromethane; NCI-C04637; Propellant 11; R 11; R 11 (refrigerant); RCRA waste no. U121; Refrigerant 11; Refrigerant R 11; Trichlorofluorocarbon; Trichlorofluoromethane; Tri-chloromethyl fluoride; Trichloromonofluoromethane; Ucon flurocarbon 11; Ucon refrigerant 11. Registered by EPA as an insecticide (cancelled). Colorless liquid below 24°; mp = -111.1; bp = 23.7°, bp400 = 6.8°, bp200 = -9.1°, bp100 = -23°, bp60 = -32.3°, bp400 = 6.8°, bp40 = -39°, bp20 = -49.7°, bp10 = -59°, bp5 = -67.6°, bp1.0 = -84.3°; d$_4^{17.2}$ 1.494, d$_{gas}^{25}$ = 5.04 (air = 1). Sigma-Aldrich Co.

3873   Trichloromethanethiol
594-42-3   H

CCl4S

Trichloromethanethiol.
4-03-00-00290 (Beilstein Handbook Reference); BRN 0506034; Clairsit; EINECS 209-840-4; HSDB 6052; HSDB 886; Mercaptan methylique perchlore; Methane-sulfenic acid, trichloro-, chloride; Methanesulfenyl chloride, trichloro-; Methanethiol, trichloro-; NSC 66404; PCM; Perchlorinemethylmercaptan; Perchlormethyl-merkaptan; Perchloro-methyl-mercaptan; Perchloro-methanethiol; Perchloromethyl mercaptan; PMM; RCRA waste number P118; Thiocarbonyl tetrachloride; Trichlormethyl sulfur chloride; Trichlorofluoromethane; Trichloromethane sulfenyl chloride; Trichloromethane sulphuryl chloride; Trichloromethanethiol; Trichloro-methyl mercaptan; Trichloromethyl sulfochloride; Trichloromethyl sulphochloride; Trichloromethylsulfenyl chloride; Trichloromethylsulphenyl chloride; UN1670. Yellow oil; bp = 147.5°; d^{20} = 1.6947; λm = 324 nm (ε = 12, CHCl3), 322 nm (ε = 10, petroleum ether); soluble in Et2O.

3874   Trichloronaphthalene
1321-65-9   G

C10H5Cl3

1,2,8-Trichloronaphthalene.
EINECS 215-321-3; Halowax; HSDB 4325; Naphthalene, trichloro-; Nibren; Nibren wax; Seekay wax; Trichloronaphthalene. Chlorinated naphthalene. Crystals; mp = 93°; bp = 304-354°; insoluble in H2O.

3875   Trichloropentylsilane
107-72-2   G

C5H11Cl3Si

Amyltrichlorosilane.
4-04-00-04249 (Beilstein Handbook Reference); Amyl trichlorosilane; BRN 1737947; EINECS 203-515-0; HSDB 887; Pentylsilicon trichloride; Pentyltrichlorosilane; Silane, pentyltrichloro-; Silane, trichloropentyl-; Tri-chloroamylsilane; Trichloropentylsilane; UN1728. Coup-ling agent, chemical intermediate, blocking agent, release agent, lubricant, primer, reducing agent. Liquid; bp = 172°, bp15 = 60.5; d^{20} = 1.1330. Degussa-Hüls Corp.

3876   Trichlorophenethylsilane
940-41-0   G

C8H9Cl3Si

2-Phenethyltrichlorosilane.
4-16-00-01568 (Beilstein Handbook Reference); BRN 2936059; CP0110; EINECS 213-371-0; NSC 139848; Phenethyltrichlorosilane; (2-Phenylethyl)trichlorosilane; 1-Phenyl-2-(trichlorosilyl)ethane; Silane, trichloro(2-phenylethyl)-; Silane, trichlorophenethyl-; Trichlor-2-fenylethylsilan; Trichloro-2-phenylethylsilane; Trichloro-(2-phenylethyl)silane. Coupling agent, chemical inter-mediate, blocking agent, release agent, lubricant, primer, reducing agent. Liquid; bp = 242°, bp5 = 98-99°; d^{20} = 1,2397. Degussa-Hüls Corp.

3877   2,3,6-Trichlorophenol
933-75-5   G

C6H3Cl3O

2,3,6-Trichlorophenol.
4-06-00-00962 (Beilstein Handbook Reference); BRN 1867596; CCRIS 1937; EINECS 213-271-7; HSDB 5773; Phenol, 2,3,6-trichloro-; 2,3,6-Trichlorophenol. Used as a disinfectant and fungicide. Needles; mp = 58°; bp = 253°; bp0.4 = 88°; λm = 280, 289 nm (ε = 1905, 1950, 0.1N HCl), 304 nm (ε = 5623, 0.1N NaOH); slightly soluble in H2O, soluble in AcOH, ligroin, very soluble in EtOH, Et2O, C6H6. LD50 (rat ip) = 308 mg.kg.

3878   2,4,5-Trichlorophenol
95-95-4   9716   G P

C6H3Cl3O
2,4,5-Trichlorophenol.
4-06-00-00962 (Beilstein Handbook Reference); BRN 0607569; CCRIS 718; Collunosol; Dowicide 2; HSDB 4067; NCI-C61187; NSC 2266; Nurelle; Phenol, 2,4,5-trichloro-; Preventol; Preventol I; RCRA waste number U230; TCP; 2,4,5-TCP; Trichlorophenol, 2,4,5-; 2,4,5-Trichlorophenol. Fungicide. Registered by EPA as an antimicrobial, fungicide and herbicide (cancelled). Used as a fungicide and bactericide. Sodium salt is Dowicide B, soluble in H_2O and organic solvents. Needles; mp = 67°; bp = 253°, bp_{746} = 248°; pK = 7.37; slightly soluble in H_2O (< 0.2 g/100 ml), soluble in Me_2CO (615 g/100 ml), C_6H_6 (163 g/100 ml), CCl_4 (51 g/100 ml), Et_2O (525 g/100 ml), EtOH (525 g/100 ml), MeOH (615 g/100 ml), soybean oil (79 g/100 ml), C_7H_8 (122 g/100 ml); LD_{50} (rat orl) = 820 mg/kg. *Fluka; Sigma-Aldrich Co.*

3879   **2,4,5-Trichlorophenol, sodium salt**
136-32-3   G P

C6H2Cl3NaO
Sodium 2,4,5-trichlorphenate.
Caswell No. 797A; Dowicide B; EINECS 205-239-6; EPA Pesticide Chemical Code 064217; HSDB 5620; Phenol, 2,4,5-trichloro-, sodium salt; Preventol 1; Sodium, (2,4,5-trichlorophenoxy)-; Sodium 2,4,5-trichlorophenate; Sodium 2,4,5-trichlorophenolate; Sodium 2,4,5-trichlorophenoxide; Sodium salt of 2,4,5-trichlorophenol; 2,4,5-Trichlorophenol, sodium salt. A proprietary trade name for sodium 2,4,5-trichlorphenate; an antiseptic and germicide. *Chemical; Dow.*

3880   **2,4,6-Trichlorophenol**
88-06-2   9717   G

C6H3Cl3O
2,4,6-trichlorophenol.
4-06-00-01005 (Beilstein Handbook Reference); AI3-00142; BRN 0776729; CCRIS 605; Dowicide 2S; EINECS 201-795-9; HSDB 4013; NCI-C02904; NSC 2165; OMAL; Phenachlor; Phenaclor; Phenol, 2,4,6-trichloro-; RCRA waste number U231; 2,4,6-T; TCP; 2,4,6-Trichlorfenol; Trichloro-2-hydroxybenzene; 2,4,6-Trichlorophenol. Used as a fungicide, bactericide and preservative, Commonly used as disinfectant and topical anti-infective. mp = 69°; bp = 246°; d^{75} = 1.4901; λ_m = 289, 296 nm (ε = 2470, 2550, MeOH); poorly soluble in H_2O (< 1 mg/ml), soluble in organic solvents.

3881   **1,2,3-Trichloropropane**
96-18-4   H

C3H5Cl3
Propane, 1,2,3-trichloro-.
4-01-00-00199 (Beilstein Handbook Reference); AI3-26040; Allyl trichloride; BRN 1732068; CCRIS 5874; EINECS 202-486-1; HSDB 1340; NCI-C60220; NSC 35403; Propane, 1,2,3-trichloro-; Trichloropropane. Liquid; mp = -14.7°; bp = 157°; d^{20} = 1.3889; slightly soluble in H_2O, soluble in EtOH, Et_2O, $CHCl_3$, CCl_4.

3882   **Trichlorosilane**
10025-78-2   9719   G

Cl3HSi
Trichloromonosilane.
EINECS 233-042-5; HSDB 890; Silane A-19; Silane, trichloro-; Silici-chloroforme; Siliciumchloroform; Silico-chloroform; Silicon chloride hydride (SiHCl3); Trichloorsilaan; Trichloromonosilane; Trichlorosilane; Trichlorsilan; Triclorosilano; UN1295. Used in organic synthesis, purification of silicon; as an intermediate. Liquid; mp = -127°; bp = 32°; d_4^{20} 1.3417; soluble in organic solvents; LD_{50} (rat orl) = 1.03 g/kg. *Chisso Corp.; Degussa-Hüls Corp.; PCR.*

3883   **Trichlorotrifluoroethane**
76-13-1   G H

C2Cl3F3
1,1,2-Trichlorotrifluoroethane.
4-01-00-00142 (Beilstein Handbook Reference); AI3-62874; Arcton 63; Arklone P; Asahifron 113; BRN 1740335; CFC-113; Chlorofluorocarbon 113; Daiflon S 3; EINECS 200-936-1; F 113; Ethane, 1,1,2-trichloro-1,2,2-trifluoro-; FC 113; Flugene 113; Fluorocarbon 113; Forane 113; Freon 113; Freon 113TR-T; Freon F113; Freon R 113; Freon TF; Frigen 113; Frigen 113 TR; Frigen 113 TR-T; Frigen 113A; Frigen 113tr-T; Frigen 113TR; Frigen 113TR-N; Genesolv D; Genetron 113; Halocarbon 113; HSDB 145; Isceon 113; Kaiser chemicals 11; Kaltron 113MDR; Khladon 113; Ledon 113; MS-180 Freon® TF Solv; Propellant 113; R 113; Racon 113; Refrigerant 113; Refrigerant R 113; Trichlorotrifluoroethane; 1,1,2-Trichloro-1,2,2-trifluoroethane; 1,1,2-Trichlorotrifluoro-ethane; 1,1,2-Trifluoro-1,2,2-trichloroethane; 1,1,2-Trifluorotrichloroethane; Ucon fluorocarbon 113. Liquid; mp = -35°; bp = 47.7°; d^{25} = 1.5635; insoluble in H_2O, soluble in organic solvents. *Air Products & Chemicals Inc.; Allied Signal; Elf Atochem N. Am.; PCR.*

3884   **Trichlorotrifluoroethane**
354-58-5   H

C2Cl3F3
1,1,1-Trichloro-2,2,2-trifluoroethane.
4-01-00-00142 (Beilstein Handbook Reference); BRN 1699455; CFC 113a; EINECS 206-564-6; Ethane, 1,1,1-trichloro-2,2,2-trifluoro-; F 113a; FC 113a; FC 133A; Fluorocarbon 113; Freon FT; HSDB 6501; Precision cleaning agent; T-WD602; Trichlorotrifluoroethane; 1,1,1-Trichloro-2,2,2-trifluoroethane; 1,1,1-Trifluorotri-chloroethane. Liquid; mp = 14.2°; bp = 46.1°; d^{20} =

637

1.5790; insoluble in H2O, soluble in EtOH, Et2O, CHCl3.

3885 1,1,2-Trichloro-1,2,2-trimethyldisilane
13528-88-6   H

Cl
|
Cl—Si—Si—
| |
| Cl

C3H9Cl3Si2
Disilane, 1,1,2-trichloro-1,2,2-trimethyl-.
Disilane, 1,1,2-trichloro-1,2,2-trimethyl-; EINECS 236-870-5; 1,1,2-Trichloro-1,2,2-trimethyldisilane; 1,1,2-Tri-methyltrichlorodisilane.

3886 Trichlorovinylsilane
75-94-5   G H

Cl
|
Cl—Si—
|
Cl

C2H3Cl3Si
Trichloroethenylsilane.
4-04-00-04258 (Beilstein Handbook Reference); A 150; BRN 1743440; CV-4900; EINECS 200-917-8; HSDB 891; NSC 93872; Silane, trichloroethenyl-; Silane, trichloro-vinyl-; Silane, vinyl trichloro A-150; Trichloroethenyl-silane; Trichloro(vinyl)silane; Trichlorovinyl silicane; Tri-chlorovinylsilane; Trichlorovinylsilicon; UN1305; Union carbide A-150; Vinyl trichlorosilane; Vinylsilicon tri-chloride; Vinyltrichlorosilane; Vtcs. Coupling agent, chemical intermediate, blocking agent, release agent, lubricant, primer, reducing agent. Intermediate for silicones; coupling agent in adhesives and bonds.
Liquid; mp = -95°; bp = 91.5°; d^{20} = 1.2426; soluble in CHCl3; LD50 (rat orl) = 1280 mg/kg. *Degussa-Hüls Corp.; Lancaster Synthesis Co.; PCR; Union Carbide Corp.*

3887 Triclocarban
101-20-2   9727   G H

Cl   H   H
| | |
Cl—   —N—C—N—   —Cl
| ||
Cl   O

C13H9Cl3N2O
Urea, N-(4-chlorophenyl)-N'-(3,4-dichlorophenyl)-.
4-12-00-01265 (Beilstein Handbook Reference); AI3-26925; BRN 2814890; Carbanilide, 3,4,4'-trichloro-; Caswell No. 874; CCRIS 4880; N-(4-Chlorophenyl)-N'-(3,4-dichlorophenyl)urea; CP 78416; Cusiter; Cutisan; N-(3,4-Dichlorophenyl)-N'-(4-chlorophenyl)urea; EINECS 202-924-1; ENT 26925; EPA Pesticide Chemical Code 027901; Genoface; HSDB 5009; NSC-72005; Procutene; Solubacter; TCC; TCC (soap bacteriostat); TCC Soap; Trichlocarban; Triclocarban; Triclocarbanum; Trilocarb-an; Urea, N-(4-chlorophenyl)-N'-(3,4-dichlorophenyl)-. Bacteriostatic agent for bar soaps. Crystals; mp = 255-256°. *Monsanto Co.*

3888 Triclopyr
55335-06-3   9730   G P

HO   Cl
| |
O—   —Cl
| |
| N
| H
Cl

C7H4Cl3NO3
[(3,5,6-Trichloro-2-pyridinyl)oxy]acetic acid.
4-21-00-00362 (Beilstein Handbook Reference); Acetic acid, (3,5,6-trichloro-2-pyridyloxy)-; BRN 0225301; Caswell No. 882I; Confront; Crossbow Turflon; Curtail; Dowco 233; EINECS 259-597-3; EPA Pesticide Chemical Code 116001; Garlon; Garlon 2; Garlon 4; Garlon 250; Grazon ET; HSDB 7060; NSC 190671; Redeem; Release; Remedy; Timbrel; Triclopyr; Turflon. Herbicide to control perennial and woody weeds. Registered by EPA as a herbicide.
Solid; mp = 148-150°; bp= 290° (dec); soluble in H2O (0.0440 g/100 ml), Me2CO (77.9 g/100 ml), n-octanol (25.5 g/100 ml), CH3CN (9.8 g/100 ml), xylene (2.4 g/100 ml), C6H6 (2.4 g/100 ml), CHCl3 (4.04 g/100 ml), C6H14 (0.03 g/100 ml); LD50 (rat orl) = 713 mg/kg, (gpg orl) = 310 mg/kg, (rbt orl) = 550 mg/kg, (rbt der) > 2000 mg/kg, (mallard duck orl) = 1698 mg/kg, (bee) > 0.0604 mg/bee; LC50 (mallard duck 8 day dietary) > 5000 mg/kg, (Japanese quail 8 day dietary) = 3278 mg/kg, (bobwhite quail 8 day dietary) = 2935 mg/kg, (rainbow trout 96 hr.) = 117 mg/l, (bluegill sunfish 96 hr.) = 148 mg/l; non-toxic to bees. *Dow AgroSciences; DowElanco Ltd.; Riverdale Chemical Co.*

3889 Triclopyr triethylammonium salt
57213-69-1   P

C13H19Cl3N2O3
3,5,6-Trichloro-2-pyridinyloxyacetic acid, TEA salt.
Acetic acid, ((3,5,6-trichloro-2-pyridinyl)oxy)-, compd. with N,N-diethylethanamine (1:1); Caswell No. 882J; EPA Pesticide Chemical Code 116002; Garlon 3A; M 3724; N,N-Diethylethanamine compd. with ((3,5,6-trichloro-2-pyridinyl)oxy)aceticacid (1:1); ((3,5,6-Trichloro-2-pyridin-yl)oxy)acetic acid compd. with N,N-diethylethanamine (1:1); (3,5,6-Trichloro-2-pyridinyl)oxyacetic acid, triethyl-amine salt; ((3,5,6-Trichloro-2-pyridyl)oxy)acetic acid, compound with triethylamine(1:1); Triclopyr triethyl-amine; Triclopyr, triethylamine salt; Triclopyr triethyl-ammonium salt; Triethylamine triclopyr; Triethyl-ammonium triclopyr. Registered by EPA as a herbicide. *Dow AgroSciences; Riverdale Chemical Co; Scotts Co.*

3890 Triclopyr butoxyethyl ester
64700-56-7   P

C13H16Cl3NO4
2-Butoxyethyl [(3,5,6-trichloropyridin-2-yl)oxy]acetate.
2-Butoxyethyl [(3,5,6-trichloropyridin-2-yl)oxy]acetate; Acetic acid, ((3,5,6-trichloro-2-pyridinyl)oxy)-, 2-butoxy-ethyl ester; Butoxyethyl triclopyr; Caswell No. 882K; Crossbow; Crossbow 3L; EPA Pesticide Chemical Code 116004; Garlon 4; Garlon 4E; M 4021; ((3,5,6-Trichloro-2-pyridinyl)oxy)acetic acid 2-butoxyethyl ester; Turflon D. Registered by EPA as a herbicide. *Dow AgroSciences.*

3891 Tricresyl phosphate
78-30-8   9833   G

638

$C_{21}H_{21}O_4P$
o-Tolyl phosphate.
4-06-00-01979 (Beilstein Handbook Reference); AI3-00520; BRN 1892885; CCRIS 6421; o-Cresyl phosphate; EINECS 201-103-5; HSDB 4084; NSC 438; Phosflex 179C; Phosphoric acid, tri-o-cresyl ester; Phosphoric acid, tri-o-tolyl ester; Phosphoric acid, tri(2-tolyl)ester; Phosphoric acid, tris(2-methylphenyl) ester; Plastic X; TOCP; TOTP; o-Trikresylphosphate; Tri 2-methylphenyl phosphate; Tri-2-tolyl phosphate; Tri-o-cresyl phosphate; Tri-o-tolyl phosphate; Tri-ortho-cresylphosphate; Triortho-cresyl phosphate; Tris(o-cresyl)-phosphate; Tris(o-methyl-phenyl)phosphate; Tris(o-tolyl) phosphate; Trojkrezylu fosforan. A proprietary plasticizer with specific gravity from 1.177-1.20. Generally used in lacquers and varnishes. Usd as a plasticizer, fire retardant for plastics, air filter medium, waterproofing, additive to extreme pressure lubricants. Pale yellow liquid; mp = 11°; bp = 410°, bp_{20} = 265°; d^{20} = 1.1955; insoluble in H_2O, soluble in organic solvents. *Akzo Chemie; Chemron; Daihachi Chem. Ind. Co. Ltd; FMC Corp.*

3892   Tridecane
629-50-5   H

$C_{13}H_{28}$
n-Tridecane.
EINECS 211-093-4; HSDB 5727; NSC 66205; Tridecane; n-Tridecane. Liquid; mp = -5.3°; bp = 235.4°; d^{20} = 0.7564; insoluble in H_2O, soluble in CCl_4, very soluble in EtOH, Et_2O.

3893   Tridecanol
112-70-9   G H

$C_{13}H_{28}O$
n-Tridecan-1-ol.
4-01-00-01860 (Beilstein Handbook Reference); AI3-35264; BRN 1739991; EINECS 203-998-8; HSDB 5574; NSC 5252; Tridecanol; 1-Tridecanol; Tridecyl alcohol. Commercial material is usually a mixture of isomers; used in manufacture of esters for synthetic lubricants, detergents, antifoam agents, other tridecyl compounds; in perfumery. Crystals; mp - 32.5°; bp_{14} = 152°; d^{31} = 0.8223; insoluble in H_2O, soluble in EtOH, Et_2O. *Allchem Ind.; Penta Mfg.*

3894   Tridecylaluminum
1726-66-5   H

$C_{30}H_{63}Al$
Tri-n-decylaluminum.
Aluminum, tris(decyl)-; EINECS 217-040-1; HSDB 5852; Tri-1-decylaluminum; Tri-n-decylaluminum; Tridecyl-aluminium; Tris(decyl)aluminum; Tris(n-decyl)aluminum.

3895   Tridecyl alcohol, ethoxylated, phosphated.
9046-01-9   H

PEG-3 Tridecyl ether phosphate.
PEG-3 Tridecyl ether phosphate; PEG-6 Tridecyl ether phosphate; PEG-10 Tridecyl ether phosphate; Phosphoric acid, (ethoxylated tridecyl alcohol) esters; Poly(oxy-1,2-ethanediyl), α-tridecyl-ω-hydroxy-, phosphate; Poly-ethylene glycol (3) tridecyl ether phosphate; Polyethylene glycol 300 tridecyl ether phosphate; Polyethylene glycol 500 tridecyl ether phosphate; Polyoxyethylene (3) tridecyl ether phosphate; Polyoxyethylene (6) tridecyl ether phosphate; Polyoxyethylene (10) tridecyl ether phosphate; Trideceth-3 phosphate; Trideceth-6 phosphate; Trideceth-10 phosphate.

3896   Tridecylbenzenesulfonic acid
25496-01-9   G H

$C_{19}H_{32}O_3S$
Tridecylbenzenesulfonic acid.
Benzenesulfonic acid, tridecyl-; EINECS 247-036-5; Nansa® TDB; Tridecylbenzenesulfonic acid; Tridecyl-benzenesulphonic acid. Used in manufacture of detergents. *Albright & Wilson UK Ltd.*

3897   Tridecyl methacrylate
2495-25-2   G

$C_{17}H_{32}O_2$
Tridecyl-2-methyl-2-propanoate.
Ageflex FM-25; EINECS 219-671-8; HSDB 5882; Methacrylic acid, tridecyl ester; 2-Propenoic acid, 2-methyl-, tridecyl ester; Tridecyl methacrylate. Liquid; d = 0.88. *Rit-Chem.*

3898   3-(Tridecyloxy)propionitrile
68239-19-0   H

$C_{16}H_{31}NO$
Propanenitrile, 3-(tridecyloxy)-.
EINECS 269-431-1; Propanenitrile, 3-(tridecyloxy)-; 3-(Tridecyloxy)propiononitrile; 3-Tridecoxypropanenitrile.

3899   Tridecyl stearate
31556-45-3   H

$C_{31}H_{62}O_2$
n-Octadecanoic acid, tridecyl ester.
Cirrasol LN-GS; EINECS 250-696-7; NSC 152080; Octadecanoic

acid, tridecyl ester; Stearic acid, tridecyl ester; Tridecanol stearate; Tridecyl stearate. Used as an emollient in creams and lotions.

3900   Tridemorph
81412-43-3   P

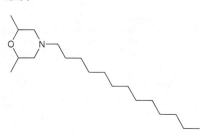

$C_{19}H_{39}NO$
Dimethyl-2,6 tridecyl-4 morpholine.
BAS 2203F; Calixin; Dimethyl-2,6 tridecyl-4 morpholine; Tridemorf; Tridemorph; Tridemorphe. Fungicide. Color-less oily liquid; sg = 0.86, bp$_{0.5}$ = 134°; soluble in H_2O (0.00117 g/100 ml at 20°), EtOH, Me$_2$CO, EtOAc, Et$_2$O, cyclohexane, C$_6$H$_6$, CHCl$_3$, olive oil; LD$_{50}$ (rat orl) = 480 mg/kg, (rat der) > 4000 mg/kg; LC$_{50}$ (rat ihl 4 hr.) = 4.5 mg/l air, (guppy 96 hr.) = 3.5 mg/l; non-toxic to bees. *BASF Corp.*

3901   Tridiphane
58138-08-2   9736   G P

$C_{10}H_7Cl_5O$
(±)-2-(3,5-Dichlorophenyl)-2-(2,2,2-trichloroethyl)-oxirane.
2-(3,5-dichlorophenyl)-2-(2,2,2-trichloroethyl)-oxirane;   (RS)-2-(3,5-Dichlorophenyl)-2-(2,2,2-trichloroethyl)-oxirane; Dowco 356; Nelpon; Oxirane, 2-(3,5-dichloro-phenyl)-2-(2,2,2-trichloroethyl)-; Tandem; Tridiphane. A selective non-systemic herbicide. Used for control of annual grass seedlings and broad-leaved weeds in maize. Registered by EPA as a herbicide. Colorless crystals; mp = 43°; soluble in H_2O (0.00018 g/100 ml), soluble in MeOH (0.078 g/100 ml), Me$_2$CO (0.72 g/100 ml), xylene (0.40 g/100 ml), chlorobenzene (0.62 g/100 ml), CH$_2$Cl$_2$ (0.94 g/100 ml); LD$_{50}$ (rat orl) = 1743-1918 mg/kg, (rbt der) = 3536 mg/kg, (mallard duck orl) > 2510 mg/kg; LC$_{50}$ (mallard duck, bobwhite quail 8 day dietary) = 5620 mg/kg diet; (rainbow trout 96 hr.) = 0.53 mg/l, (bluegill sunfish 96 hr.) = 0.37 mg/l. *Dow AgroSciences; Dow UK.*

3902   Tridocosylaluminum
6651-25-8   H

$C_{66}H_{135}Al$
Tri(docosyl) aluminum.
Aluminum, tridocosyl-; EINECS 229-672-5; Tridocosyl-aluminium.

3903   Tridodecylaluminum
1529-59-5   H

$C_{36}H_{75}Al$
Aluminum, tridodecyl-.
Aluminum, tridodecyl-; EINECS 216-219-1; HSDB 5837; Tri(dodecyl) aluminum; Tridodecylaluminium; Trido-decylaluminum; Tris(dodecyl)aluminum.

3904   Trieicosylaluminum
1529-57-3   H

$C_{60}H_{123}Al$
Aluminum, trieicosyl-.
Aluminum, trieicosyl-; EINECS 216-217-0; HSDB 5835; Trieicosylaluminum; Triicosylaluminium.

3905   Trientine
112-24-3   9737   D G H

$C_6H_{18}N_4$
N,N'-Bis(2-aminoethyl)-1,2-ethanediamine.
4-04-00-01242 (Beilstein Handbook Reference); Al3-24384; Araldite hardener HY 951; Araldite HY 951; BRN 0605448; CCRIS 6279; DEH 24; EINECS 203-950-6; Ethylenediamine, N,N'-bis(2-aminoethyl)-; HSDB 1002; HY 951; N,N-Bis(2-aminoethyl)-1,2-diaminoethane; NSC 443; Tecza; TETA; Texlin® 300; Trien; Trientina; Trientine; Trientinum; Triethylene tetramine; Triethylene-tetramine; UN2259. Chelating agent used in detergents and softening agents, synthesis of dyestuffs, pharm-aceuticals, rubber accelerator; as thermosetting resin; epoxy curing agent; lubricating oil additive; analytical reagent for Cu, Ni. Improves yields and reduces processing time for many applications; intermediate in asphalt additives, corrosion inhibitors, epoxy curing agents, surfactants in fabric softener and textile additives; in paper industry, petrol. products. Chelating agent used as a drug for the treatment of Wilson's disease. Liquid; mp = 12°; bp = 266.5°; d^{15} = 0.9817; soluble in H_2O, EtOH; LD$_{50}$ (rat orl) = 2500 mg/kg. *Merck & Co.Inc.; Rit-Chem; Texaco; Tosoh; Union Carbide Corp.*

3906   Trietazine
1912-26-1   9797   G P
$C_9H_{16}ClN_5$
6-Chloro-N,N,N'-triethyl-1,3,5-triazine-2,4-diamine.
2-chloro-4-diethylamino-6-ethylamino-s-triazine;   2-chloro-4,6-diethylamino-s-triazine; G-27901; NC-1667; Aventox; Gesafloc. Herbicide. mp = 100-102°; soluble in H_2O (20 mg/l), more soluble in organic solvents; LD$_{50}$ (rat orl) = 1750 mg/kg. *Schering Agrochemicals Ltd.*

3907   Triethanolamine
102-71-6   9739   G P

640

C6H15NO3

Trihydroxytriethylamine.

Al3-01140; Alkanolamine 244; Caswell No. 886; CCRIS 606; Daltogen; EINECS 203-049-8; EPA Pesticide Chemical Code 004208; Ethanol, 2,2',2"-nitrilotri-; HSDB 893; Nitrilotriethanol; Nitrilo-2,2',2"-Triethanol; 2,2',2"-Nitrilotriethanol; NSC 36718; Sodium ISA; Sterolamide; Sting-Kill; T-35; TEA; TEA (amino alcohol); Thiofaco T-35; Tri(hydroxyethyl)amine; Triethanolamine; Triethylolamine; Trihydroxytriethylamine; Tris(β-hydroxy-ethyl)amine; Tris(2-hydroxyethyl)amine; Trolamine. Inter-mediate in manufacture of surfactants, textile specialties, waxes, polishes, toiletries, cutting oils, fatty acid soaps (drycleaning), cosmetics, household detergents, emulsions; solvent for casein, shellac, dyes. Registered by EPA as an antimicrobial and insecticide (cancelled). Leaflets; mp = 20.5°; bp = 335.4°; d_4^{20} 1.1242; very soluble in H_2O, EtOH, soluble in $CHCl_3$, slightly soluble in Et_2O (1.6 g/100 ml), C_6H_6 (4.2 g/100 ml), $n-C_7H_{14}$ (< 0.1 g/100 ml), ligroin; pH of 0.1N aqueous solution = 10.5; LD_{50} (rat orl) = 8000 mg/kg. *Hüls Am.; Mitsui Toatsu; Nippon Shokubai; Occidental Chem. Corp.; Schweizerhall Inc.; Sigma-Aldrich Fine Chem.; Spectrum Chem. Manufacturing; Texaco; Union Carbide Corp.*

3908 Triethanolamine dodecylbenzene sulfonate
27323-41-7   H

C18H30O3S.C6H15NO3

Dodecylbenzenesulfonic acid, compd. with 2,2',2"-nitrilotris(ethanol) (1:1).

Al3-26730-X; Benzenesulfonic acid, dodecyl-, compd. with 2,2',2"-nitrilotris(ethanol) (1:1); Caswell No. 887AA; Dodecylbenzenesulfonic acid, compd. with 2,2',2"-nitrilotris(ethanol) (1:1); Dodecylbenzenesulfonic acid, triethanolamine salt; EINECS 248-406-9; EPA Pesticide Chemical Code 079020; TEA-Dodecylbenzenesulfonate; Triethanolamine dodecylbenzene sulfonate; Witconate 5725; Witconate 60L; Witconate 60T; Witconate 79S; Witconate S-1280; Witconate Tab.

3909 Triethanolamine hydrochloride
637-39-8   G

C6H16ClNO3

2,2',2-Nitrilotris(ethanol) hydrochloride.

EINECS 211-284-2; Ethanol, 2,2',2"-nitrilotri-, hydro-chloride; Ethanol, 2,2',2"-nitrilotris-, hydrochloride; 2,2',2"-Nitrilotriethanol hydrochloride; 2,2',2"-Nitrilo-trisethanol hydrochloride; TEA-

Hydrochloride; Triethan-olamine hydrochloride; Triethanolammonium chloride; Tris(2-hydroxyethyl)ammonium chloride. Crystals; mp = 171°. *Cia-Shen; Janssen Chimica; Spectrum Chem. Manufacturing; U.S. BioChem.*

3910 Triethanolamine lauryl sulfate
139-96-8   G P

C18H41NO7S

Triethanolamine n-dodecyl sulfate.

Akyposal TLS; Cycloryl TAWF; Cycloryl WAT; Dodecyl sulfate triethanolamine salt; Drene; EINECS 205-388-7; Elfan 4240 T; EMAL T; Emersal 6434; HSDB 2899; Lauryl sulfate triethanolamine salt; Laurylsulfuric acid triethanolamine salt; Maprofix TLS; Maprofix TLS 65; Maprofix TLS 500; Melanol LP20T; Propaste T; Rewopol TLS 40; Richonol T; Sipon LT; Sipon LT-6; Sipon LT-40; Standapol T; Standapol TLS 40; Steinapol TLS 40; Stepanol WAT; Sterling wat; Sulfetal lt; Sulfuric acid, dodecyl ester, triethanolamine salt; TEA-Lauryl sulfate; Texapon T-35; Texapon T-42; Texapon TH; Triethanolamine dodecyl sulfate; Triethanolamine lauryl sulfate; Tylorol LT 50; Witcolate TLS-500. Emulsifier, detergent, wetting agent, dispersant, foaming agent used for household cleaning products, cosmetics, emulsion polymerization. Registered by EPA as an antimicrobial, fungicide, insecticide and rodenticide (cancelled). *Chemron; CK Witco Corp.; Lonzagroup; Norman-Fox; Sandoz; Stepan.*

3911 Triethoxy(3-aminopropyl)silane
919-30-2   G H

C9H23NO3Si

3-(Triethoxysilyl)-1-propanamine.

4-04-00-04273 (Beilstein Handbook Reference); A 1100; A 1112; AGM-9; Aktisil AM; 3-Aminopropyltriethoxy-silane; γ-Aminopropyltriethoxysilane; APTES; BRN 1754988; CA0750; EINECS 213-048-4; HSDB 5767; NSC 95428; Nuca 1100; Propylamine, 3-(triethoxysilyl)-; Prosil® 220; Silane 1100; Silane amg-9; Silane, γ-amino-propyltriethoxy-; Silicone A-1100; Triethoxy(3-amino-propyl)silane; 3-(Triethoxysilyl)-1-propanamine; Uc-A 1100; Union Carbide® A-1100. Filler for thermosets, thermoplastics. Reacts with porous glass of, for example, chromatography columns, to form the aminopropyl derivative of glass, an adsorbent for affinity chromatography. Coupling agent, chemical intermediate, blocking agent, release agent, lubricant, primer, reducing agent. Technical grade coupling agent enhancing and promoting chemical bonding between inorganic and organic molecules; for acetal, acrylic, epichlorohydrin, nitrile, NC, polyamide, PC, polyethylene, polyimide, polymethacrylate, PP, polysulfone, PS, PVC, urethane and vinyl thermoplastics; thermoset acrylic (thermoset, latex), alkyd, epoxy, furan, melamine, nitrile/phenolic, phenolic, polyester, vinyl butyral/phenolic and elastomers. Liquid; mp < -70°; bp = 217°, bp29 = 119°; d^{20} = 0.9506; LD_{50} (rat orl) = 1780 mg/kg. *Degussa-Hüls Corp.; Gelest; PCR; Union Carbide Corp.*

3912   Triethoxy(3-isocyanatopropyl)silane
24801-88-5   G

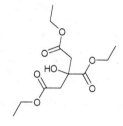

C10H21NO4Si
Isocyanatopropyltriethoxy silane.
CI7840; EINECS 246-467-6; I 7840; 3-Isocyanatopropyl-triethoxysilane; Isocyanatopropyltriethoxysilane; Iso-cyanic acid, 3-(triethoxysilyl)propylester; Silane, tri-ethoxy(3-isocyanatopropyl)-; Y 9030; YH 9030. Coupling agent, chemical intermediate, blocking agent, release agent, lubricant, primer, and reducing agent. *Degussa-Hüls Corp.*

3913   Triethoxymethylsilane
2031-67-6   G

C7H18O3Si
Methyltriethoxysilane.
4-04-00-04204 (Beilstein Handbook Reference); A 162; BRN 1742453; CCRIS 1323; Dynasylan® MTES; EINECS 217-983-9; ICI-EP 5850; Methaneorthosiliconic acid, triethyl ester; Methyl triethoxysilane; Methyltriethoxy-silane; NSC 5226; Silane, methyltriethoxy-; Silane, triethoxymethyl-; Triethoxy(methyl)silane; Triethoxysilyl-methane; Union carbide® A-162. Coupling agent, chemical intermediate, blocking agent, release agent, lubricant, primer, reducing agent, crosslinking agent providing durability, gloss, hiding power to coatings. Liquid; bp = 141-143°; d = 0.8950; LD_{50} (rat orl) = 7.67 g/kg. *Degussa-Hüls Corp.; Union Carbide Corp.*

3914   Triethoxysilane
998-30-1   G

$$C_2H_5O-Si-H$$ with OC_2H_5 above and OC_2H_5 below

C6H16O3Si
Silane, triethoxy-.
4-01-00-01359 (Beilstein Handbook Reference); BRN 1738989; CT2500; EINECS 213-650-7; HSDB 6332; NSC 124134; Triethoxysilane. Coupling agent, chemical intermediate, blocking agent, release agent, lubricant, primer, reducing agent. Liquid; bp = 133.5°; d^{20} = 0.8745. *Degussa-Hüls Corp.*

3915   Triethylaluminum
97-93-8   H

C6H15Al
Aluminum, triethyl-.
4-04-00-04398 (Beilstein Handbook Reference); Aluminum, triethyl-; ATE; BRN 3587229; EINECS 202-619-3; HSDB 4016; TEA; Triethylalane; Triethylalum-inium; Triethylaluminum.

3916   Triethylbenzene
25340-18-5   H

C12H18
Benzene, triethyl- (mixed isomers).
Benzene, triethyl-; Benzene, triethyl- (mixed isomers); EINECS 246-875-4; HSDB 5174; Triethylbenzene.

3917   Triethyl citrate
77-93-0   G

C12H20O7
Triethyl 2-hydroxy-1,2,3-propanetricarboxylate.
4-03-00-01276 (Beilstein Handbook Reference); AI3-00659; BRN 1801199; Citric acid, triethyl ester; Citroflex 2; EINECS 201-070-7; Ethyl citrate; Eudraflex; FEMA No. 3083; HSDB 729; Hydragen CAT; 2-Hydroxy-1,2,3-propanetricarboxylic acid, triethyl ester; NSC 8907; 1,2,3-Propanetricarboxylic acid, 2-hydroxy-, triethyl ester; Triaethylcitrat; Triäthylcitrat; Triethyl citrate; Triethylester kyseliny citronove. Plasticizer for food packaging. Active agent in non-microbiocidal deodorants. Oil; bp = 294°; d^{20}= 1.137; n^{20}_D = 1.4455; soluble in H_2O (7g/100 ml), more soluble in organic solvents. *Henkel/Cospha; Henkel; Yamanouchi Pharma.*

3918   Triethylenediamine
280-57-9   9742   G H

C6H12N2
1,4-Diazabicyclo(2.2.2)octane.
AI3-24809; Bicyclo(2,2,2)-1,4-diazaoctane; CCRIS 6692; D 33LV; Dabco; Dabco 33LV; Dabco crystal; Dabco EG; Dabco R-8020; Dabco S-25; EINECS 205-999-9; N,N'-endo-Ethylenepiperazine; HSDB 5556; NSC 56362; TED; TEDA; TEDA-L33; Texacat TD 100; Thancat TD 33; Triethylenediamine. In solution in dipropylene glycol, used as a gelling catalyst for PU slabstock and molded flexible, rigid, and elastomer shoe shoe applications. mp = 159°; bp = 174°; soluble in H_2O (45 g/100g), soluble in $CHCl_3$. *Tosoh.*

3919   Triethylene glycol
112-27-6   9743   P

C6H14O4

2,2'-(1,2-Ethanediylbis(oxy))bisethanol.

4-01-00-02400 (Beilstein Handbook Reference); AI3-01453; Bis(2-hydroxyethoxyethane); BRN 0969357; Caswell No. 888; Di-β-hydroxyethoxyethane; EINECS 203-953-2; EPA Pesticide Chemical Code 083501; Ethanol, 2,2'-(1,2-ethanediylbis(oxy))bis-; Ethanol, 2,2'-(ethylenedioxy)di-; Ethylene glycol-bis-(2-hydroxyethyl ether); Glycol Bis(Hydroxyethyl) Ether; HSDB 898; NSC 60758; TEG; Triethylene glycol; Triethylenglykol; Trigen; Triglycol; Trigol. Registered by EPA as an antimicrobial and disinfectant. Colorless liquid; mp = -7°; bp = 285°, bp14 = 165°; d$^{15}_4$ = 1.1274; freely soluble in H_2O, EtOH, C6H6, C7H8, slightly soluble in Et2O, CHCl3, insoluble in petroleum ether; LD50 (rat orl) = 1500 - 2200 mg/kg, (mus orl) = 2100 mg/kg, (rat iv) = 11700 mg/kg, (mus iv) = 730 - 950 mg/kg. *Amrep Inc.; Chemical Packaging Corp.; Quest Chemical Corp.; S.C. Johnson & Son, Co; Waterbury Companies Inc.*

3920   Triethylene glycol diacetate

111-21-7   H

C10H18O6

2,2'-(Ethylenedioxy)di(ethyl acetate).

4-02-00-00215 (Beilstein Handbook Reference); Acetic acid, triethylene glycol diester; AI3-02502; BRN 1789453; EINECS 203-846-0; 2,2'-(Ethylenedioxy)di(ethyl acetate); Ethanol, 2,2'-ethylenedioxydi-, diacetate; HSDB 5547; NSC 2591; TDAC; Triethylene glycol diacetate; Triglycol, diacetate. Liquid; mp = -50°; bp = 286°; d^{20} = 1.1153; very soluble in H_2O, EtOH, Et2O.

3921   Triethylene glycol di(2-ethylhexanoate)

94-28-0   H

C22H42O6

Hexanoic acid, 2-ethyl-, 1,2-ethanediylbis(oxy-2,1-ethanediyl) ester.

4-02-00-01005 (Beilstein Handbook Reference); AI3-01451; BRN 1806809; EINECS 202-319-2; Flexol 3GO; Flexol plasticizer 3GO; Hexanoic acid, 2-ethyl-, diester with triethylene glycol; Hexanoic acid, 2-ethyl-, 1,2-ethanediylbis(oxy-2,1-; Triethylene glycol, bis(2-ethyl-hexanoate); Triethylene glycol di(2-ethylhexoate). *Capital Resin Corp.*

3922   Triethylene glycol monoethyl ether

112-50-5   H

C8H18O4

2-(2-(2-Ethoxyethoxy)ethoxy)ethanol.

4-01-00-02401 (Beilstein Handbook Reference); AI3-14498; BRN

1700466; Dowanol TE; EINECS 203-978-9; Ethanol, 2-(2-(2-ethoxyethoxy)ethoxy)-; 2-(2-(2-Ethoxy-ethoxy)ethoxy)ethanol; Ethoxy triglycol; Ethoxytriethylene glycol; Ethoxytriglycol; Ethyltriglycol; HSDB 899; Poly-Solv TE; Triethylene glycol ethyl ether; Triethylene glycol, monoethyl ether; Triglycol monoethyl ether; 3,6,9-Trioxaundecan-1-ol.

3923   Triethylene glycol monomethyl ether

112-35-6   H

C7H16O4

2-(2-(2-Methoxyethoxy)ethoxy)ethanol.

4-01-00-02401 (Beilstein Handbook Reference); BRN 1700198; Dowanol TMAT; EINECS 203-962-1; Ethanol, 2-(2-(2-methoxyethoxy)ethoxy)-; HSDB 1001; 2-(2-(2-Methoxyethoxy)ethanol; Methoxytriethylene glyc-ol; Methoxytriglycol; Methyltrioxitol; NSC 97395; Poly-Solv TM; Triethylene glycol methyl ether; Triethylene glycol monomethyl ether; Triglycol monomethyl ether; 3,6,9-Trioxa-1-decanol.

3924   Triethyl phosphate

78-40-0   9747   H

C6H15O4P

Phosphoric acid, triethyl ester.

4-01-00-01339 (Beilstein Handbook Reference); AI3-00653; BRN 1705772; CCRIS 4882; EINECS 201-114-5; Ethyl phosphate ((EtO)3PO); HSDB 2561; NSC 2677; Phosphoric acid, triethyl ester; TEP; Triethoxyphosphine oxide; Triethyl phosphate; Triethylfosfat; Triethylphos-phate; Tris(ethyl) phosphate. Intermediate in manufacture of agricultural insecticides; in floor polishes, unsaturated polyesters, lubricants; as plasticizer, solvent, and catalyst. Liquid; mp = -56.4°; bp = 215.5°; d^{20} = 1.0695; soluble in H_2O, EtOH, Et2O. *Eastman Chem. Co.; Miles Inc.*

3925   Triethyl phosphite

122-52-1   G H

C6H15O3P

Phosphorous acid, triethyl ester.

4-01-00-01333 (Beilstein Handbook Reference); AI3-15624; BRN 0956578; EINECS 204-552-5; Ethyl phosphite, (EtO)3P; Fosforyn trojetylowy; HSDB 895; NSC 5284; Phosphorous acid, triethyl ester; Triethoxy-phosphine; Triethyl phosphite; UN2323. Used as a plasticizer and a reducing agent. Used in stabilizers, lube and grease additives. Liquid; bp = 157.9°; d^{20} = 0.9629; λ_m = 260 nm (ε = 1, EtOH); insoluble in H_2O, very soluble in EtOH, Et2O; LD50 (rat orl) = 1840 mg/kg. *Akzo Chemie; Albright & Wilson Americas Inc.; Albright & Wilson UK Ltd.; ICI Americas Inc.; Janssen Chimica.*

3926   Triethylsilane

617-86-7   G

C6H16Si

Triethylsilicon hydride.
CT2523; EINECS 210-535-3; NSC 93579; Silane, triethyl-; Triethylsilane. Coupling agent, chemical intermediate, blocking agent, release agent, lubricant, primer, reducing agent. Liquid; mp = -156.9°; bp = 109°; d^{20} = 0.7302; insoluble in H_2O, H_2SO_4. *Degussa-Hüls Corp.*

3927   Triflic acid
1493-13-6   G

CHF3O3S

Trifluoromethanesulfonic acid.
AI3-62912; EINECS 216-087-5; Fluorad® FC-24; Methanesulfonic acid, trifluoro-; Perfluoromethane-sulfonic acid; Trifluoromethanesulfonic acid; Trifluoro-methanesulphonic acid. Catalyst and reactant increasing yields in polymerization of epoxies, styrenes, THF, in alkylation and acylation reactions; improves octane rating; used with nitric acid for higher yields of pharmaceuticals, explosives, dyes, and intermediates. Liquid; mp = -40°; bp = 162°; d = 1.6960. *3M; Lancaster Synthesis Co.*

3928   Triflumuron
64628-44-0   9751   G P

C15H10ClF3N2O3

2-Chloro-N-[[[4-(trifluoromethoxy)phenyl]amino]carbonyl]benzamide.
AI3-29368; Alsystin; Alsystine; Bay-Sir 8514; BAY-Vi 7533; Benzamide, 2-chloro-N-(((4-(trifluoromethoxy)-phenyl)amino)carbonyl)-; BRN 2776684; Caswell No. 217C; 1-(o-Chlorobenzoyl)-3-(p-(trifluoromethoxy)-phen-yl)urea; 2-Chloro-N-(((4-(trifluoromethoxy)phenyl)-am-ino)carbonyl)benzamide; EPA Pesticide Chemical Code 118201; Mascot; OMS 2015; SIR 8514; Triflumuron; Trifluron; Trifumuron. Insecticide. Chitin synthesis inhibitor, used for control of Lepidoptera, Psyllidae, Dipter and Coleoptera on fruit, soy beans, forest trees and cotton. Colorless powder; mp = 195°; almost insoluble in H_2O (0.0000025 g/100 ml at 20°), soluble in CH_2Cl_2 (2 - 5 g/100 ml), i-PrOH (0.1 - 0.2 g/100 ml), C_7H_8 (0.2 - 0.5 g/100 ml); LD50 (rat, mus orl) > 5000 mg/kg, (dog orl) > 1000 mg/kg, (rat der) > 5000 mg/kg, (Japanese quail orl) > 5000 mg/kg, (fcanary orl) > 1000 mg/kg; LC50 (rat ihl 4 hr.) > 0.12 mg/l air, (carp, golden orfe 96 hr.) >100 mg/l. *Bayer Corp., Agriculture; Bayer Corp.*

3929   Trifluoroacetic acid
76-05-1   9754   G

C2HF3O2

Trifluoroacetic acid.
4-02-00-00458 (Beilstein Handbook Reference); Acetic acid, trifluoro-; AI3-28549; BRN 0742035; EINECS 200-929-3; Kyselina trifluoroctova; NSC 77366; Perfluoro-acetic acid; Trifluoracetic acid; Trifluoroacetic acid; Trifluoroethanoic acid; UN2699. Strong nonoxidizing acid, laboratory reagent, solvent, catalyst. mp = -15.2°; bp = 73°; d^{25} = 1.5351; soluble in H_2O, organic solvents; LD50 (mus iv) = 1200 mg/kg. *Alcoa Ind. Chem.; Allied Signal; Janssen Chimica; Sigma-Aldrich Fine Chem.; Solvay America Inc.*

3930   Trifluoroethanol
75-89-8   G

C2H3F3O

2,2,2-Trifluoroethanol.
4-01-00-01370 (Beilstein Handbook Reference); AI3-25486; BRN 1733203; EINECS 200-913-6; Ethanol, 2,2,2-trifluoro-; Fluorinol 85; NSC 451; TFE; Trifluoro-ethanol; β,β,β-Trifluoroethyl alcohol; 2,2,2-Trifluoroethyl alcohol. Used in synthetic chemistry and as a coating for metals in the automobile industry. Liquid; mp = -43.5°; bp = 74°; d^{20} = 1.3842; soluble in EtOH, less soluble in non-polar solvents. *American Nickeloid Co.; Lancaster Synthesis Co.; Sigma-Aldrich Fine Chem.; Solvay America Inc.*

3931   Trifluoromethyl-3,5-dinitro-4-chloro-benzene
393-75-9   H

C7H2ClF3N2O4

1-Chloro-2,6-dinitro-4-(trifluoromethyl)benzene.
1-Chloro-2,6-dinitro-4-(trifluoromethyl)benzene; 2,6-Di-nitro-4-trifluoromethylchlorobenzene; 3,5-Dinitro-4-chloro-α,α,α-trifluorotoluene; 3,5-Dinitro-4-chloro-benzotrifluoride; 4-Chloro-3,5-dinitrobenzotrifluoride; Benzene, 2-chloro-5-(trifluoromethyl)-1,3-dinitro-; Benzo-trifluoride, 4-chloro-3,5-dinitro-; BRN 1220937; CCRIS 2818; EINECS 206-889-3; HSDB 4262; NSC 88274; Toluene, 4-chloro-3,5-dinitro-α,α,α-trifluoro-; Trifluoro-methyl-3,5-dinitro-4-chlorobenzene.

3932   Trifluralin
1582-09-8   9757   G H P

C13H16F3N3O4

2,6-Dinitro-4-trifluormethyl-N,N-dipropylaniline.
Agreflan; Agriflan 24; Agriphan 24; AI3-28203; BRN 1893555; Caswell No. 889; CCRIS 607; Crisalin; Crisalina; Digermin; EINECS

216-428-8; Elancolan; EPA Pesticide Chemical Code 036101; HSDB 1003; Ipersan; L-36352; Lilly 36,352; NCI-C00442; Nitran; Nitran K; Olitref; Su seguro carpidor; Super-Treflan; Synfloran; Trefanocide; Treficon; Treflan; Treflan EC; Treflanocide elancolan; TRI-4; Trifluralin; Trifluralina 600; Trifluraline; Triflurex; Trifurex; Trikepin; Trim; Tristar. A herbicide against annual weeds such as cottonweeds, bean, tomato, pear, garlic and sunflower weeds; very efficient against amaranth, bristle-grass, knapweed etc. Used for pre-emergence control of annual grasses and broad-leaved weeds. Registered by EPA as a herbicide and plant growth regulator. Yellow crystals; mp = 46 - 47°; bp$_{4.2}$ = 139 - 140°, slightly soluble in H_2O (0.0024 g/100 ml), freely soluble in Me_2CO (40 g/100 ml), xylene (58 g/100 ml); LD$_{50}$ (rat orl) = 500 mg/kg, (rat orl) > 10000 mg/kg, (mus orl) = 500 mg/kg, (dog, rbt orl) > 2000 mg/kg, (ckn orl) > 2000 mg/kg, (bee orl) = 0.011 mg/bee; LC$_{50}$ (rainbow trout 96 hr.) = 0.01 - 0.04 mg/l, (bluegill sunfish 96 hr.) 0.02 - 0.09 mg/l. *Agan Chemical Manufacturers; Agrimont; Am. Cyanamid; Chemical Combine; Dow AgroSciences; DowElanco Ltd.; Drexel Chemical Co.; Earth Care; Griffin LLC; Helena Chemical Co.; ICI; Knox Fertilizer Co. Inc.; Riverdale Chemical Co; Rockland Corp.; Shell; UAP - Platte Chemical.*

3933 Triforine
26644-46-2   9762   G P

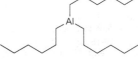

C$_{10}$H$_{14}$Cl$_6$N$_4$O$_2$
N,N'-(1,4-   Piperazinediyl-bis(2,2,2-trichloroethylidine))-bis(formamide).
5-23-01-00042 (Beilstein Handbook Reference); Biformylchlorazin; N,N'-Bis(1-formamido-2,2,2-trichloro-ethyl)piperazine; 1,4-Bis(2,2,2-trichloro-1-formamido-ethyl)piperazine; BRN 0626358; CA 70203; CA 73021; Caswell No. 890AA; Cela W-524; CELA 50; CELA W 524; CME 74770; Compound W; CW 524; Denarin; EINECS 247-872-0; EPA Pesticide Chemical Code 107901; Formamide, N,N'-(1,4-piperazinediylbis(2,2,2-trichloroethylidene))bis-; NSC 263493; Ortho Rose Pride Funginex Rose and Shrub Disease Control; Piperazine, 1,4-bis(1-formamido-2,2,2-trichloroethyl)-; Funginex; 1,1'-Piperazine-1,4-diyldi-(N-(2,2,2-trichloroethyl)-formamide); N,N'-(1,4-Piperazinediyl-bis(2,2,2-trichloro-ethylidene))bis(formamide); N,N'-(1,4-Piperazinediylbis-(2,2,2-trichloroethylidene))bisformamide; N,N'-(Piper-azine-1,4-diylbis((trichloromethyl)methylene))diform-amide; Saprol; Triforin; Triforine; W 524. Fungicide. Registered by EPA as a fungicide. White crystals; mp = 155°; soluble in H_2O (0.0027 - 0.0029 g/100 ml), insoluble in most organic solvents, soluble in DMF (33 g/100 ml), N-methylpyrrolidone (47.6 g/100 ml), THF; Me_2CO (1.1 g/100 ml), CH_2Cl_2 (0.1 g/100 ml), MeOH 1 g/100 ml), dioxane, cyclohexanone; half-life in soil ca. 3 weeks; LD$_{50}$ (rat orl) > 16000 mg/kg, (mus orl) > 6000 mg/kg, (dog orl) > 2000 mg/kg, (rbt, rat der) > 10000 mg/kg, (rat ip) > 4000 mg/kg, (bobwhite quail orl) > 5000 mg/kg; LC$_{50}$ (rat ihl 1 hr.) > 4.5 mg/l air; (bluegill sunfish, rainbow trout 96 hr.) > 1000 mg/l; non-toxic to bees. *BASF Corp.; Scotts Co.; Synchemicals Ltd.*

3934 Trihexacosylaluminum
10449-71-5   H

C$_{26}$H$_{53}$—Al—(C$_{26}$H$_{53}$)$_2$

C$_{78}$H$_{159}$Al
Tri(n-hexacosyl) aluminum.
Aluminum, trihexacosyl-; EINECS 233-934-4; Tri(hexa-cosyl) aluminum; Trihexacosylaluminium; Trihexacosyl-aluminum.

3935 Trihexadecyaluminum
1726-65-4   H

C$_{16}$H$_{33}$—Al—(C$_{16}$H$_{33}$)$_2$

C$_{48}$H$_{99}$Al
Tri-n-hexadecyaluminum.
Aluminum, trihexadecyl-; EINECS 217-039-6; HSDB 5851; Tri(hexadecyl) aluminum; Trihexadecylaluminium.

3936 Trihexylaluminum
1116-73-0   H

C$_{18}$H$_{39}$Al
Tri-n-hexylaluminum.
Aluminum, trihexyl-; EINECS 214-241-6; HSDB 5785; Tri-n-hexylaluminium; Trihexylaluminium; Trihexylalum-inum; Tri-n-hexylaluminum.

3937 Triisobutylaluminum
100-99-2   H

C$_{12}$H$_{27}$Al
Tris(2-methylpropyl)aluminum.
4-04-00-04400 (Beilstein Handbook Reference); Aluminum, triisobutyl-; Aluminum, tris(2-methylpropyl)-; BRN 3587328; EINECS 202-906-3; HSDB 1004; Tibal; Triisobutylalane; Triisobutylaluminium; Triisobutyl-aluminum; Tris(2-methylpropyl)aluminum; Tris(isobutyl)-alane; Tris(isobutyl)aluminum.

3938 Triisononyl benzene-1,2,4-tricarboxylate
53894-23-8   H

645

C36H60O6

1,2,4-Benzenetricarboxylic acid, triisononyl ester.
1,2,4-Benzenetricarboxylic acid, triisononyl ester; EINECS 258-847-9; Triisononyl benzene-1,2,4-tricarboxylate.

3939 Triisooctyl (stibylidynetrithio)triacetate
27288-44-4   H

C30H60O6S3Sb

Antimony tris(isooctyloxycarbonylmethylmercaptide).
Acetic acid, (stibylidynetrithio)tri-, triisooctyl ester; Acetic acid, mercapto-, isooctyl ester, antimony(3+) salt; Antimony tris(isooctyloxycarbonylmethylmercaptide); Antimony(3+) tris(2-(isooctyloxy)-2-oxoethanethiolate); EINECS 248-388-2; (Stibylidynetrithio)triacetic acid, triisooctyl ester; Triisooctyl (stibylidynetrithio)triacetate.

3940 Triisooctyl phosphite
25103-12-2   G

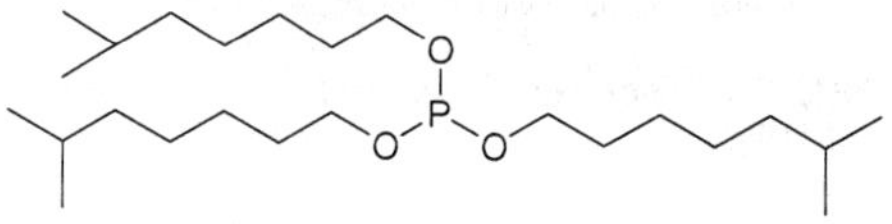

C24H51O3P

Phosphorous acid tri-isooctyl ester.
EINECS 246-614-4; Isooctyl phosphite ((C8H17O)3P); Phosphorous acid, triisooctyl ester; Triisooctyl phosphite; Weston® TIOP. Intermediate; insecticides; lubricant additive; specialty solvents; stabilizer for acrylics, nylon, unsaturated polyester, PVC; improves antiwear and antifriction properties. Stabilizer for hot-melt adhesives, PU, polyesters; used in molding, extrusion, and film applications in PP, HDPE, LDPE, PVC, and polyesters; also useful for PP fiber applications and calendering of PVC. Oil; d = 0.891. *Albright & Wilson Americas Inc.; GE Specialities; Stave.*

3941 Triisooctyl trimellitate
27251-75-8   H

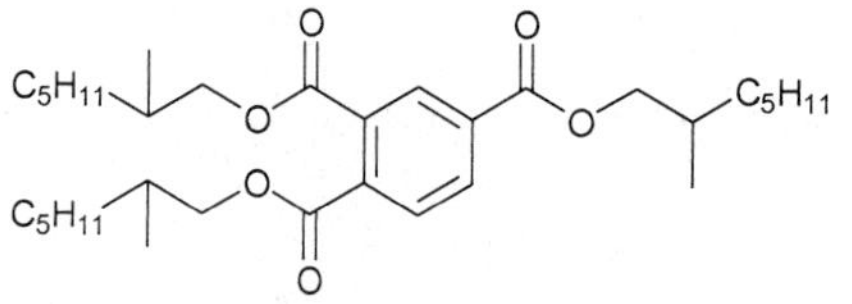

C33H54O6

1,2,4-Benzenetricarboxylic acid, triisooctyl ester.
1,2,4-Benzenetricarboxylic acid, triisooctyl ester; EINECS 248-365-7; Triisooctyl benzene-1,2,4-tricarboxylate; Tri-isooctyl trimellitate.

3942 Triisopropanolamine
122-20-3   G H

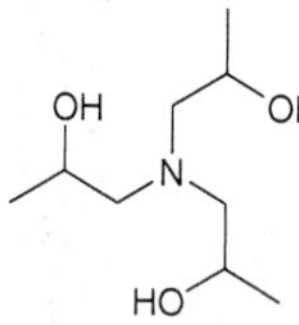

C9H21NO3

1,1',1''-Nitrilotri-2-propanol.
4-04-00-01680 (Beilstein Handbook Reference); AI3-01450; BRN

1071570; Caswell No. 891; CCRIS 4884; EINECS 204-528-4; EPA Pesticide Chemical Code 004209; HSDB 5593; 1,1',1''-Nitrilotri-2-propanol; NSC 4010; 2-Propanol, 1,1',1''-nitrilotris-; TIPA; Tri-2-propanolamine; Tri-iso-propanolamine; Triisopropanol-amine. Emulsifying agent. Solid; mp = 45°; bp32 = 190°, bp10 = 170-180°; d^{20} = 1.0; slightly soluble in CHCl3, soluble in H2O, EtOH. *Ashland; Sigma-Aldrich Fine Chem.*

3943 Triisopropyl borate
5419-55-6   H

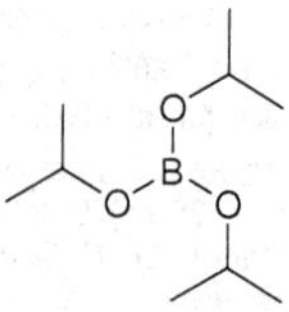

C9H21BO3

Boric acid (H3BO3), triisopropyl ester.
4-01-00-01488 (Beilstein Handbook Reference); AI3-61082; Boric acid (H3BO3), tris(1-methylethyl) ester; Boric acid, (H3BO3), triisopropyl ester; Boric acid (H3BO3), triisopropyl ester; Boric acid, triisopropyl ester; Boric acid, tris(1-methylethyl) ester; Boron isopropoxide; Boron triisopropoxide; BRN 1701469; EINECS 226-529-9; Isopropyl borate; NSC 9779; Triisopropoxy borane; Triisopropoxy boron; Triisopropyl borate; Triisopropyl orthoborate; Trisisopropoxyborane; UN2616. Liquid; bp = 140°, bp76 = 75°; d^{20} = 0.8251; very soluble in EtOH, Et2O, C6H6, n-PrOH.

3944 Trilaurin
538-24-9   G

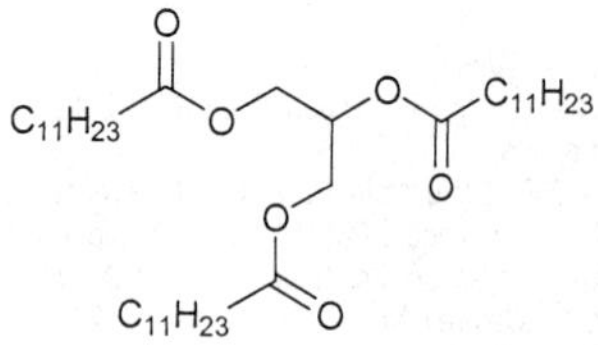

C39H74O6

Glyceryl tri-laurate.
AI3-11124; CCRIS 6991; Dodecanoic acid, 1,2,3-propanetriyl ester; EINECS 208-687-0; Glycerin trilaurate; Glyceryl tridodecanoate; Glyceryl trilaurate; Lauric acid triglyceride; Lauric acid triglycerin ester; Laurin, tri-; NSC 4061; 1,2,3-Propanetriol tridodecanoate; Trilaurin. Needles; mp = 45-47°; d^{20} = 0.8986; insoluble in H2O, soluble in EtOH, Et2O, CHCl3, petroleum ether, very soluble in Me2CO, C6H6.

3945 Trilauryl phosphite
3076-63-9   G

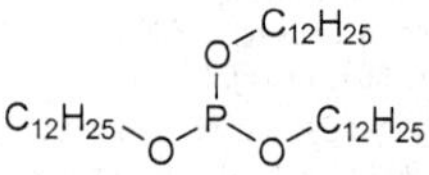

C36H75O3P

Dodecyl phosphite.
AI3-51074; Dodecyl phosphite, (C12H25O)3P; Dover-phos® 53; Doverphos® TLP; EINECS 221-356-5; NSC 44603; P 2; P 2 (antioxidant); Phosclere T 312; Phos-phorous acid, tridodecyl ester; Tridodecyl phosphite; Tri-n-dodecyl phosphite; Trilauryl phosphite.

Antioxidant and EP additive for lubricants. *Dover.*

3946   Trimellitic acid
528-44-9   9775   G H

C9H6O6
1,2,4-Benzenetricarboxylic acid.
4-09-00-03746 (Beilstein Handbook Reference); 1,2,4-Benzenetricarboxylic acid; 1,3,4-Benzenetricarboxylic acid; 1,4,5-Benzenetricarboxylic acid; 4-Carboxyphthalic acid; BRN 2214815; EINECS 208-432-3; NSC 72986; 1,2,4-Tricarboxybenzene; Trimellitic acid. Used in organic synthesis. Needles; mp = 219°; λmLks = 296 nm (ε = 1585, H2O); very soluble in H2O (2.1 g/100 ml), EtOH, Et2O.

3947   Trimellitic anhydride
552-30-7   9776   H

C9H4O5
1,2,4-Benzenetricarboxylic acid 1,2-anhydride.
5-18-08-00562 (Beilstein Handbook Reference); Anhyd-ride-ethomid ht polymer; Anhydrotrimellic acid; Anhyd-rotrimellitic acid; Benzene-1,2,4-tricarboxylic acid 1,2-anhydride; BRN 0009394; CCRIS 6282; EINECS 209-008-0; Epon 9150; HSDB 4299; 5-Isobenzofurancarboxylic acid, 1,3-dihydro-1,3-dioxo-; NCI-C56633; NSC 60252; 5-Phthalanacarboxylic acid, 1,3-dioxo-; TMA; Trimellic acid anhydride; Trimellic acid 1,2-anhydride. Crystals; mp = 162°; bp14 = 240-243°.

3948   Trimesic acid
554-95-0   G

C9H6O6
1,3,5-Benzenetricarboxylic acid.
AI3-06468; 5-Carboxyisophthalic acid; EINECS 209-077-7; NSC 3998; 1,3,5-Tricarboxybenzene; Trimesic acid; Trimesinic acid; Trimesitinic acid. Chemical intermediate. White solid; mp > 330°.

3949   Trimethoprim
738-70-5   9782   D

C14H18N4O3
5-[(3,4,5-Trimethoxyphenyl)methyl]-2,4-pyrimidinedi-amine.
Abaprim; Acuco; AI3-52594; Alcorim-F; Alprim; Anitrim; Antrima; Antrimox; Apo-Sulfatrim; Bacdan; Bacidal; Bacide; Bacin; Bacta; Bacterial; Bacticel; Bactifor; Bactoprim; Bactramin; Bactrim; Bencole; Bethaprim; Biosulten; Briscotrim; BRN 0625127; BW 56-72; CCRIS 2410; Centrim; Chemotrin; Cidal; Co-Trimoxizole; Colizole; Conprim; Cotrim; Cotrimel; Deprim; Diseptyl; Dosulfin; DRG-0030; Duocide; EINECS 212-006-2; Esbesul; Espectrin; Euctrim; Exbesul; Fermagex; Fortrim; Futin; HSDB 6781; Idotrim; Ikaprim; Instalac; Kombinax; Lagatrim; Lagatrim Forte; Lastrim; Lescot; Lidaprim; Methoprim; Metoprim; Monoprim; Monotrim; Monotrimin; NIH 204; Novotrimel; NSC-106568; Omstat; Pancidim; Primosept; Primsol; Proloprim; Protrin; Purbal; Resprim; Roubac; Roubal; Salvatrim; Septra; Septrin DS; Septrin Forte; Septrin S; Setprin; Sinotrim; Smz-Tmp; SMZ/TMP; Stopan; Streptoplus; Sugaprim; Sulfamar; Sulfamethoprim; Sulfamethoxazole Trimetho-prim; Sulfatrim; Sulfoxaprim; Sulmeprim; Sulthrim; Sultrex; Syraprim; Tiempe; Tmp Smx; Toprim; Trimanyl; Trimeth/Sulfa; Trimethoprim; Trimethoprime; Trimetho-primum; Trimetoprim; Trimetoprima; Trimexol; Trimez-IFSA; Trimezol; Trimogal; Trimono; Trimopan; Trimpex; Triprim; Trisul; Trisulcom; Trisulfam; Trisural; U-Prin; Uretrim; Uro-D S; Urobactrim; Utetrin; Velaten; Veltrim; Wellcoprim; Wellcoprin; WR 5949; Xeroprim; Zamboprim. A 2,4-diaminopyrimidine antibiotic. Cryst-als; mp = 199-203°; soluble in dimethylacetamide (13.86 g/100 ml), benzyl alcohol (7.29 g/100 ml), propylene glycol (2.57 g/100 ml), CHCl3 (1.82 g/100 ml), MeOH (1.21 g/100 ml), H2O (0.04 g/100 ml), Et2O (0.003 g/100 ml), C6H6 (0.002 g/100 ml); LD50 (mus orl) = 7000 mg/kg. *Glaxo Wellcome Inc.; Hoffmann-LaRoche Inc.*

3950   Trimethoxy(3-chloropropyl)silane
2530-87-2   G H

C6H15ClO3Si
3-(Trimethoxysilyl)propyl chloride.
A 143; CC3300; 3-Chloropropyltrimethoxysilane; CPS-M; CPTMO; Dynasylan® CPTMO; EINECS 219-787-9; KBM 703; NSC 83878; SH 6076; Sila-Ace S 620; Silane, (3-chloropropyl)trimethoxy-; Trimethoxy(3-chloropropyl)-silane; 3-(Trimethoxysilyl)propyl chloride; Z 6076. Coupling agent, chemical intermediate, blocking agent, release agent, lubricant, primer and reducing agent. Used with epoxies, nylons and urethanes. Liquid; bp = 195-196°; d = 1.081; irritating to eyes, respiratory sytem, skin. *Degussa-Hüls Corp.; Dow Corning; Howard Hall; OSI Specialties; PCR; Union Carbide Corp.*

3951   Trimethoxysilane
2487-90-3   H

C3H10O3Si
Trimethoxysilane.
4-01-00-01266 (Beilstein Handbook Reference); BRN 1697990; EINECS 219-637-2; HSDB 6320; NA9269; Silane, trimethoxy-;

Trimethoxysilane.

3952   Trimethoxysilylpropane
1067-25-0   G

C6H16O3Si
n-Propyltrimethoxysilane.
CP0810; Dynasylan PTMO; EINECS 213-926-7; n-Propyltrimethoxysilane; Silane, trimethoxypropyl-; Tri-methoxypropylsilane. Coupling agent, chemical intermediate, blocking agent, release agent, lubricant, primer, reducing agent. *Hüls Am.*

3953   3-(Trimethoxysilyl)propyl 2-propenoate
4369-14-6   G

C9H18O5Si
3-Acryloxypropyltrimethoxysilane.
3-Acryloxypropyltrimethoxysilane; CA0397; 2-Propenoic acid, 3-(trimethoxysilyl)propyl ester. Coupling agent, chemical intermediate, blocking agent, release agent, lubricant, primer, reducing agent. Liquid; bp3 = 97°. *Degussa-Hüls Corp.*

3954   Trimethylamine
75-50-3   9783   H

C3H9N
N,N-Dimethylmethanamine.
Al3-15639; CCRIS 6283; Dimethylmethaneamine; EINECS 200-875-0; FEMA Number 3241; HSDB 808; Methanamine, N,N-dimethyl-; N-Trimethylamine; N,N-Dimethylmethanamine; Trimethylamine; UN1083; UN1297. Used to impart an odor to natural gas, as an insect attractant and in chemical synthesis. Gas; mp = -117.1°; bp = 2.8°; d^{25} = 0.0627; soluble in H_2O, organic solvents. *Alfa Aesar; Alfa Chem. Ltd.; DuPont; Fluka; Lancaster Synthesis Co.; Sigma-Aldrich Fine Chem.*

3955   Trimethylammonium chloride
593-81-7   H

C3H10ClN
Trimethylamine monohydrochloride.
EINECS 209-810-0; HSDB 5710; Methanamine, N,N-dimethyl-, hydrochloride; NSC 91484; Trimethylamine, hydrochloride; Trimethylamine monohydrochloride; Trimethylammonium chloride. Crystals; mp = 277.5°; bp sublimes; very soluble in H_2O, EtOH, $CHCl_3$.

3956   2,6,6-Trimethylbicyclo(3.1.1)heptanol
98510-89-5   H

C10H18O
Bicyclo[3.1.1]heptanol, 2,6,6-trimethyl-.
2,6,6-Trimethyl-bicyclo(3.1.1)heptanol.

3957   2,2,5-Trimethyl-3-(dichloroacetyl)-1,3-oxazolidine
52836-31-4   H

C8H13Cl2NO2
3-(Dichloroacetyl)-2,2,5-trimethyloxazolidine.
3-(Dichloroacetyl)-2,2,5-trimethyloxazolidine; EINECS 258-214-7; Oxazolidine, 3-(dichloroacetyl)-2,2,5-tri-methyl-; R 29148; 2,2,5-Trimethyl-3-dichloroacetyl-1,3-oxazolidine.

3958   Trimethyl-1,2-dihydroquinoline
147-47-7   H

C12H15N
2,2,4-Trimethyl-1,2-dihydroquinone.
Acetone anil; Acetone anil (quinoline deriv.); Agerite resin D; Al3-17714; CCRIS 4795; EINECS 205-688-8; Flectol A; Flectol H; Flectol pastilles; HSDB 1103; NCI-C60902; NSC 4175; Polnox R; Quinoline, 1,2-dihydro-2,2,4-trimethyl-; Trimethyl dihydroquinoline; Trimethyl-1,2-dihydroquinoline; Vulkanox HS/LG; Vulkanox HS/powder.

3959   Trimethyloctadecylammonium bromide
1120-02-1   G

C21H46BrN
Trimethyl octadecyl ammonium bromide.
EINECS 214-294-5; 1-Octadecanaminium, N,N,N-trimethyl-, bromide; Zeonet B; Stearyltrimethyl-ammonium bromide; N,N,N-Trimethyl-1-octadecan-aminium bromide; Trimethyloctadecylammonium bromide. Accelerator for Nipol AR and HyTemp 4050 series polyacrylate elastomers. *Nippon Zeon.*

3960   Trimethylolpropane
77-99-6   G H

C6H14O3

1,1,1-Tris(hydroxymethyl)propane.

4-01-00-02786 (Beilstein Handbook Reference); AI3-24124; BRN 1698309; EINECS 201-074-9; Ethriol; Ethyltrimethylolmethane; Etriol; Ettriol; Hexaglycerine; Hexaglycerol; HSDB 5218; Methanol, (propanetriyl)tris-; NSC 3576; Propane, 1,1,1-tris(hydroxymethyl)-; Propyl-idynetrimethanol; TMP; Tri(hydroxymethyl)-propane; Tri-methylolpropane; Tris(hydroxymethyl)propane. Conditi-oner, manufacture of varnishes, alkyd resins, synthetic drying oils, urethane foams and coatings, silicone lube oils, lactone plasticizers, textile finishes, surfactants, epoxidation products. White crystals; mp = 58°; bp5 = 160°; freely soluble in H2O, EtOH. Hoechst Celanese; Mitsubishi Gas; Perstorp AB.

3961   Trimethylolpropane ethoxytriacrylate
28961-43-5   H

$$(C_2H_4O)_x.(C_2H_4O)_y.(C_2H_4O)_z.C_{15}H_{20}O_6$$

Ethanol, 2,2',2"-(propylidynetris(methyleneoxy))tri-, tri-acrylate. Ethanol, 2,2',2"-(propylidynetris(methyleneoxy))tri-, tri-acrylate; Ethoxylated trimethylolpropane triacrylate; Trimethylolpropane ethoxylated, triacrylate; Trimethyl-olpropane ethoxytriacrylate; Trimethylolpropane poly-oxyethylene triacrylate.

3962   Trimethylolpropane oleate
70024-57-6   H

C18H34O2.xC6H14O3

9-Octadecenoic acid (9Z)-, ester with 2-ethyl-2-(hydroxy-methyl)-1,3-propanediol.

EINECS 274-260-0; 9-Octadecenoic acid (Z)-, ester with 2-ethyl-2-(hydroxymethyl)-1,3-propanediol; Trimethylolpropane oleate.

3963   Trimethylolpropane trimethacrylate
3290-92-4   G H

C18H26O6

2-Ethyl-2(hydroxymethyl)-1,3-propanediol trimethacryl-ate.

Acryester TMP; Ageflex TMP 402; Ageflex TM 403; Ageflex TM 404; Ageflex TM 410; Ageflex TM 421; Ageflex TM 423; Ageflex TM 451; Ageflex TM 461; Ageflex TM 462; Ageflex TMPTMA; Blemmer PTT; BRN 5961805; CCRIS 530; Chemlink 30; Chemlink 3080; EINECS 221-950-4; Hi-Cross M; Light Ester TMP; Methacrylic acid, 1,1,1-trihydroxymethyl propane triester; Monocizer TD 1500; NK Ester TMPT; NK Ester M TMPT; NSC 84261; Perkalink® 400; Perkalink® 400-50D; Propylidynetri-methyl trimethacrylate; PTMA; Saret 515; Sartomer 350; Sartomer SR 350; SR 350; TD 1500; TD 1500 S; TMPT (crosslinking agent); Trimethylolpropane trimethacrylate. Used as a processing aid for extrusion and molding of plastisols and rubber compounds (improves abrasion resistance, adhesion in PVC plastisols, scorch and chemical resistance, elevated temperature stability). Co-agent to improve efficiency of peroxide-induced cross-linking of rubber and sensitizer for radiation-cured compounds; used with wire and cable, hard rubber rolls, polybutadiene and polyethylene, moisture barrier films and coatings, plastisols and vinyl acetate latexes, adhesives, molding compounds, textile products. Liquid; bp = 185°; d = 1.060. Akzo Chemie; Rit-Chem; Sartomer; Sigma-Aldrich Fine Chem.; US Chemical Corp.

3964   Trimethylolpropane trioleate
57675-44-2   H

C60H110O6

2-Ethyl-2-(((1-oxooleyl)oxy)methyl)-1,3-propanediyl dioleate.

EINECS 260-895-0; 2-Ethyl-2-(((1-oxooleyl)oxy)methyl)-1,3-propanediyl dioleate; 9-Octadecenoic acid (9Z)-, 2-ethyl-2-((((9Z)-1-oxo-9-octadecenyl)oxy)methyl)-1,3-prop-anediyl ester; Trimethylolpropane trioleate.

3965   Trimethylolpropane tripelargonate
126-57-8   H

C33H62O6

2-Ethyl-2-(((1-oxononyl)oxy)methyl)propane-1,3-diyl dinonan-1-oate.

EINECS 204-793-6; 2-Ethyl-2-(((1-oxononyl)oxy)methyl)-propane-1,3-diyl dinonan-1-oate; Nonanoic acid, 2-ethyl-2-(((1-oxononyl)oxy)methyl)-1,3-propanediyl ester; Non-anoic acid, triester with 2-ethyl-2-(hydroxymethyl)-1,3-propanediol; Trimethylolpropane tripelargonate.

3966   Trimethyl orthoacetate
1445-45-0   H

$C_5H_{12}O_3$
Ethane, 1,1,1-trimethoxy-.
AI3-24332; EINECS 215-892-9; Ethane, 1,1,1-trimethoxy-; Orthoacetic acid, trimethyl ester; 1,1,1-Trimethoxyethane; Trimethyl orthoacetate. Liquid; bp = 108°; d^{25} = 0.9438; very soluble in EtOH, Et_2O.

3967   Trimethylpentanediol
144-19-4   G H

$C_8H_{18}O_2$
2,2,4-Trimethyl-1,3-pentanediol.
4-01-00-02604 (Beilstein Handbook Reference); AI3-02706; BRN 1698098; Caswell No. 893; EINECS 205-619-1; EPA Pesticide Chemical Code 041002; HSDB 1136; NSC 6368; 1,3-Pentanediol, 2,2,4-trimethyl-; TMPD®; TMPD® (alcohol); TMPD® Glycol; 2,2,4-Trimethyl-1,3-pentanediol. Resin intermediate. Plates; mp = 51.5°; bp = 235°, bp1 = 81-82°; d^{15} = 0.936; slightly soluble in H_2O, soluble in C_6H_6, $CHCl_3$, very soluble in EtOH, Et_2O. *Eastman Chem. Co.*

3968   2,2,4-Trimethyl-1,3-pentanediol monoisobutyrate
25265-77-4   G H

$C_{12}H_{24}O_4$
Isobutyric acid, ester with 2,2,4-trimethyl-1,3-pentane-diol.
Chissocizer CS 12; CS 12; EINECS 246-771-9; Isobutyraldehyde Tishchenko trimer; Isobutyric acid, ester with 2,2,4-trimethyl-1,3-pentanediol; 1,3-Pentanediol, 2,2,4-trimethyl-, monoisobutyrate; Propanoic acid, 2-methyl-, monoester with 2,2,4-trimethyl-1,3-pentanediol; Texanol; Texanol® Ester-Alcohol; 2,2,4-Trimethyl-1,3-pentanediol monoisobutyrate. Liquid; bp = 244°; d = 0.9500. *Chisso Corp.; Eastman Chem. Co.*

3969   2,4,4-Trimethyl-2-pentene
107-40-4   H

C_8H_{16}
2,4,4-Trimethylpent-2-ene.
AI3-16047; EINECS 203-488-5; HSDB 5377; 2-Pentene, 2,4,4-trimethyl-; 2,2,4-Trimethyl-3-pentene; 2,4,4-Tri-methyl-2-pentene; 2,4,4-Trimethylpent-2-ene; 2,4,4-Tri-methylpentene-2; Propene, 1-tert-butyl-2-methyl-. Liquid; mp = -106.3°; bp = 104.9°; d^{20} = 0.7218; λ_m = 194 nm (ε = 7943, C_6H_{14}).

3970   2,3,6-Trimethylphenol
2416-94-6   H

$C_9H_{12}O$
1-Hydroxy-2,3,6-trimethylbenzene.
EINECS 219-330-3; HSDB 5876; 1-Hydroxy-2,3,6-trimethylbenzene; 3-Hydroxypseudocumene; NSC 91509; Phenol, 2,3,6-trimethyl-; 2,3,6-Trimethylphenol.

3971   Trimethyl phosphite
121-45-9   H

$C_3H_9O_3P$
Phosphorous acid, trimethyl ester.
AI3-60394; EINECS 204-471-5; Fosforyn trojmetylowy; HSDB 1007; NSC 6513; Phosphorous acid, trimethyl ester; Trimethoxyfosfin; Trimethoxyphosphine; Trimethyl phosphite; Trimethylfosfit; Trimethylphosphite; UN2329. Liquid; bp = 111.5°; d^{20} = 1.0518; λ_m = 251, 257, 263 nm (ε = 575, 631, 457, EtOH); slightly soluble in CCl_4, very soluble in EtOH, Et_2O.

3972   Trimethylpyruvic acid
815-17-8   H

$C_6H_{10}O_3$
3,3-Dimethyl-2-oxobutyric acid.
AI3-11509; Butanoic acid, 3,3-dimethyl-2-oxo-; tert-But-ylglyoxylic acid; Butyric acid, 3,3-dimethyl-2-oxo-; 3,3-Dimethyl-2-oxobutyric acid; 3,3-Dimethyl-2-oxobutan-oic acid; EINECS 212-418-2; Glyoxylic acid, tert-butyl-; HSDB 5757; NSC 16648; Pyruvic acid, trimethyl-; Trimethylpyruvic acid. Crystals; mp = 90.5°; bp = 189, bp15 = 80°; λ_m = 312 nm (ε = 30, etOH); slightly soluble in H_2O, soluble in Et_2O, C_6H_6, $CHCl_3$, CS_2.

3973   Trimethylsilylacetamide
13435-12-6   G

$C_5H_{13}NOSi$
N-Trimethylsilylacetamide.
4-04-00-04011 (Beilstein Handbook Reference); Acetamide, N-(trimethylsilyl)-; (Acetylamino)trimethyl-silane; BRN 0741928; CT3250; EINECS 236-565-7; NSC 139859; Trimethylsilylacetamide; N-(Trimethylsilyl)acet-amide; N-Trimethylsilylacetamide. Coupling agent, chemical intermediate, blocking agent, release agent, lubricant, primer, reducing agent. Solid; mp = 40-46°; bp18 = 84°. *Degussa-Hüls Corp.*

3974 Trimethylsilyl chloride
75-77-4   G H

C3H9ClSi
Trimethylchlorosilane.
4-04-00-04007 (Beilstein Handbook Reference); BRN 1209232; CCRIS 790; Chlorotrimethylsilane; EINECS 200-900-5; HSDB 1009; Monochlorotrimethylsilicon; NSC 15750; Silane, chlorotrimethyl-; Silane, trimethyl-chloro-; Silicane, chlorotrimethyl-; Silylium, trimethyl-, chloride; TL 1163; Trimethyl chlorosilane; Trimethylchlorosilane; Trimethylsilyl chloride; UN1298. Coupling agent, chemical intermediate, blocking agent, release agent, lubricant, primer, reducing agent. Liquid; mp = -40°; bp = 60°; d^{25} = 0.856; LD$_{50}$ (rat orl) = 4.8 g/kg. *Degussa-Hüls Corp.*

3975 Trimethylsilyl imidazole
18156-74-6   G

C6H12N2Si
N-(Trimethylsilyl)-imidazole.
5-23-04-00456 (Beilstein Handbook Reference); AI3-52901; BRN 0606148; CT3600; EINECS 242-040-3; Imidazole, 1-(trimethylsilyl)-; Imidazole, N-(trimethyl-silyl)-; 1H-Imidazole, 1-(trimethylsilyl)-; (Trimethylsilyl)-imidazole; 1-(Trimethylsilyl)imidazole; 1-(Trimethylsilyl)-1H-imidazole; N-(Trimethylsilyl)-imidazole; N-(Trimethylsilyl)imidazole; N-(Trimethylsilyl)imidazol; NSC 139860; TSIM. Coupling agent, chemical intermediate, blocking agent, release agent, lubricant, primer, reducing agent. Liquid; bp$_{14}$ = 93-94°; d = 0.9560; soluble in CHCl$_3$. *Degussa-Hüls Corp.*

3976 Trimethylsilyl iodide
16029-98-4   G

C3H9ISi
Iodotrimethylsilane.
CT3610; EINECS 240-171-0; Iodotrimethylsilane; Silane, iodotrimethyl-; Trimethyliodosilane. Coupling agent, chemical intermediate, blocking agent, release agent, lubricant, primer, reducing agent. bp = 106°; d = 1.4060; n$_D^{20}$= 1.4770. Am; Hüls.

3977 Trimethylsilyl triflate
27607-77-8   9790   G

C4H9F3O3SSi
Trimethylsilyl trifluoromethane sulfonate.
CT3795; EINECS 248-565-4; Methanesulfonic acid, trifluoro-, trimethylsilyl ester; Trifluoromethanesulfonic acid trimethylsilyl ester; Trimethylsilyl triflate; Trimethylsilyl trifluoromethanesulfonate; Trimethylsilyl trifluoromethanesulphonate; Trimethylsilyl trifluoromethylsulfonate. Coupling agent, chemical intermediate, blocking agent, release agent, lubricant, primer, reducing agent. Liquid; bp$_{80}$ = 77°; d = 1.1500. *Degussa-Hüls Corp.*

3978 Trimethyl thiourea
2489-77-2   G

C4H10N2S
N,N,N'-Trimethylthiourea.
AI3-62303; CCRIS 611; EINECS 219-644-0; HSDB 4095; NCI-C02186; NSC 153385; Thiate E; Thiourea, trimethyl-; Trimethylthiourea; Trimethyl-2-thiourea; 1,1,3-Tri-methylthiourea; 1,1,3-Trimethyl-2-thiourea; Urea, 1,1,3-trimethyl-2-thio-. A proprietary accelerator. *Greef K & K.*

3979 Trimethylvinylmethane
558-37-2   H

C6H12
3,3-Dimethyl-1-butene.
1-Butene, 3,3-dimethyl-; 2,2-Dimethyl-3-butene; 3,3-Dimethylbutene; 3,3-Dimethyl-1-butene; EINECS 209-195-9; Neohexene; NSC 74119; tert-Butylethene; tert-Butylethylene; tert-Hexene; Trimethylvinylmethane. Liquid; mp = -115.2°; bp = 41.2°; d^{20} = 0.6529; λ$_m$ = 177 nm (ε = 12589); insoluble in H$_2$O, soluble in EtOH, Et$_2$O, CHCl$_3$, CCl$_4$.

3980 Trinitrocresol
28905-71-7   G

C7H5N3O7
4-Hydroxy-2,3,5-trinitrobenzene.
Methyltrinitrophenol; Trinitrocresolate.

3981 Trinitromethane
517-25-9   9800   G

CHN3O6
Nitroform.
4-01-00-00107 (Beilstein Handbook Reference); BRN 1708361; EINECS 208-236-8; Methane, trinitro-; Nitroform; Trinitromethane. Used in the manufacture of explsives and propellants. Liquid; mp = 15°; bp explodes; d$^{20.}$ = 1.479; very soluble in EtOH, Me$_2$CO.

3982 Trinitrotoluene
118-96-7   9801   G

$C_7H_5N_3O_6$
2-Methyl-1,3,5-trinitrobenzene.
Benzene, 2-methyl-1,3,5-trinitro-; CCRIS 1299; EINECS 204-289-6; Gradetol; HSDB 1146; 1-Methyl-2,4,6-trinitrobenzene; NCI-C56155; NSC 36949; TNT; α-TNT; TNT-tolite; 2,4,6-Trinitrotolueen; Trinitrotoluene; 2,4,6-Trinitrotoluene; s-Trinitrotoluene; s-Trinitrotoluol; sym-Trinitrotoluol; 2,4,6-trinitrotoluol; Tolit; Tolite; Toluene, 2,4,6-trinitro-; Tritol; Trojnitrotoluen; Trotyl; Trotyl oil; UN0209; UN1356. An explosive constituent. Orthorhombic crystals; mp = 80.1°; bp explodes at 240°; d^{25} = 1.654; $λ_m$ = 227 nm (ε = 19055, MeOH); insoluble in H_2O, slightly soluble in EtOH, soluble in Et_2O, very soluble in Me_2CO, C_6H_6, C_7H_8, C_5H_5N.

3983 Trioctacosylaluminum
6651-27-0   H

$C_{84}H_{171}Al$
Tri(octacosyl)aluminum.
Aluminum, trioctacosyl-; EINECS 229-674-6; Triocta-cosylaluminium.

3984 Trioctadecylaluminum
3041-23-4   H

$C_{54}H_{111}Al$
Tri-n-octadecyl aluminum.
Aluminum, trioctadecyl-; EINECS 221-246-7; Tri(octa-decyl) aluminum; Trioctadecylaluminium.

3985 Trioctyl phosphate
78-42-2   G H

$C_{24}H_{51}O_4P$
Tris(2-ethylhexyl)phosphate.
4-01-00-01786 (Beilstein Handbook Reference); AI3-07852; BRN 1715839; CCRIS 615; Disflamoll TOF; EINECS 201-116-6; Flexol plasticizer TOF; Flexol TOF; HSDB 2562; Kronitex TOF; NCI-C54751; NSC 407921; Phosphoric acid, tris(2-ethylhexyl)ester; TOF; Tri(2-ethylhexyl) phosphate; Triethylhexyl phosphate; Trioctyl phosphate; Tris-(2-ethylhexyl)fosfat; Tris-2(2-ethylhexyl)-fosfat; Tris(2-ethylhexy)phosphate; Tris(ethyl-hexyl) phos-phate. Paint and varnish driers (metallic salts); esters as plasticizers. Solvent, antifoaming agent; flame retardant. Liquid; bp8 = 220-230°; d = 0.924; soluble in EtOH, Me_2CO, Et_2O. Akzo Chemie; Albright & Wilson Americas Inc.; Ashland; BASF Corp.; Eastman Chem. Co.;

Neste UK; Sigma-Aldrich Fine Chem.; Union Carbide Corp.

3986 Trioctyl phosphite
301-13-3   G

$C_{24}H_{51}O_3P$
Tris(2-ethylhexyl) phosphite.
4-01-00-01785 (Beilstein Handbook Reference); AI3-18581; BRN 1712332; CCRIS 6203; EINECS 206-111-2; JP 308; 1-Hexanol, 2-ethyl-, phosphite (3:1); NSC 3233; Phosphorous acid, tris(2-ethylhexyl) ester; Syn-O-Ad® P-374; Tri-(2-ethylhexyl)phosphite; Trioctyl phosphite; Tris(2-ethylhexyl) phosphite. Antioxidant and antiwear agent in gear and transmission oils Oil; bp0.3 = 163-164°; d = 0.902; insoluble in H_2O, soluble in organic solvents. Akzo Chemie.

3987 Trioctyl trimellitate
89-04-3   G

$C_{33}H_{54}O_6$
Trioctyl benzene-1,2,4-tricarboxylate.
ADK CIZER C-8; 1,2,4-Benzenetricarboxylic acid, trioctyl ester; Benzene tricarboxylic acid, trioctyl ester; Diplast® TM; Diplast® TM8; EINECS 201-877-4; HSDB 5263; Jayflex® TOTM; Kodaflex® TOTM; Nuoplaz® 6959; PX 338; TOTM; Tri (2-ethylhexyl) trimellitate; Tri-n-octyl trimellitate; Trimex N 08; Trioctyl trimellitate; Nuoplaz® TOTM; Palatinol® TOTM; Plasthall® TOTM; PX-338; Staflex TOTM; Uniflex® TOTM. Plasticizer Solid; d_{20}^{20} = 0.989; bp = 414°. Exxon; ExxonMobil Chem. Co.

3988 Trioctylaluminum
1070-00-4   H

$C_{24}H_{51}Al$
Tri-n-octylaluminum.
Aluminum, trioctyl-; EINECS 213-964-4; HSDB 5779; Tri-n-octylaluminum; Trioctylaluminium; Trioctylaluminum.

3989 Triolein
122-32-7   9802   G H

C57H104O6

Glycerol, tri(cis-9-octadecenoate).

Aldo TO; EINECS 204-534-7; Emery 2423; Emery oleic acid ester 2230; Glycerin trioleate; Glycerol, tri(cis-9-octadecenoate); Glycerol trioleate; Glycerol triolein; Glyceryl trioleate; Glyceryl-1,2,3-trioleate; HSDB 5594; Kemester® 1000; Oleic acid triglyceride; Oleic triglyceride; Olein; Olein, tri-; Oleyl triglyceride; Radia® 7363; Raoline; Triolein; Trioleoylglyceride; Trioleoyl-glycerol. Emollient used in cosmetics, textiles, leather, metalworking lubricants; base for sulfation. Lubricant, chemical intermediate; used in formulation of cutting, lamination, and textile oils; corrosion inhibitors; chemical synthesis; as carbon source in antibiotic culture broths. Solid; mp = -32°; bp18 = 235-240°; d40 = 0.8988; insoluble in H2O, slightly soluble in EtOH, soluble in CHCl3, petroleum ether, very soluble in Et2O. *CK Witco Corp.; Fina Chemicals.*

3990 Trioxsalen

3902-71-4   9805   D G

C14H12O3

2,5,9-Trimethyl-7H-Furo[3,2-g][1]benzopyran-7-one.

5-19-04-00472 (Beilstein Handbook Reference); 6-Hydroxy-β,2,7-trimethyl-5-benzofuranacrylic acid γ-lac-tone; 7H-Furo(3,2-g)(1)benzopyran-7-one, 2,5,9-tri-methyl-; BRN 0221723; EINECS 223-459-0; Elder 8011; NSC-71047; 2',4,8-Trimethylpsoralen; 2,5,9-Trimethyl-7H-furo(3,2-g)(1)benzopyran-7-one; 4,5',8-Trimethyl-psoralen; 4,5',8-Trimethylpsoralene; Trioxisaleno; Triox-salen; Trioxysalen; Trioxysalene; Trioxysalenum; Trisor-alen. Photosensitizer and pigmentation agent. Solid; mp = 234.5-235°; λm 250 295 335 nm (log ε 4.35, 3.99, 3.80 MeOH); slightly soluble in EtOH, CHCl3; soluble in CH2Cl2; insoluble in H2O. *ICN Pharm. Inc.*

3991 Triphenyl phosphate

115-86-6   9813   G H

C8H15O4P

Phosphoric acid, triphenyl ester.

4-06-00-00720 (Beilstein Handbook Reference); AI3-04491; BRN 1888236; CCRIS 4888; Celluflex TPP; Disflamoll TP; EINECS 204-112-2; HSDB 2536; Kronitex® TPP; NSC 57868; Phenyl phosphate ((PhO)3PO); Phosflex TPP; Phosphoric acid, triphenyl ester; TP;

TPP; Trifenylfosfat; Triphenoxyphosphine oxide; Triphenyl phosphate. Fire-retarding agent; noncombustible substitute for camphor in celluloid; plasticizer for cellulose acetate and nitrocellulose; in lacquers and varnishes; impregnating roofing paper. Flame retardant plasticizer for engineering resins, cellulosics. Prisms or needles; mp = 50.5°; bp11 = 245°; d50 = 1.2055; λm = 256, 262, 268 nm (ε = 955, 1175, 912, C6H14); insoluble in H2O; soluble in EtOH; very soluble in Et2O, C6H6, CCl4, CHCl3. *FMC Corp.; Monsanto Co.*

3992 Triphenyl phosphite

101-02-0   H

C18H15O3P

Phosphorous acid, triphenyl ester.

4-06-00-00695 (Beilstein Handbook Reference); ADK Stab TPP; Advance TPP; AI3-07866; BRN 1079456; CCRIS 4890; Doverphos® 10; EFED; EINECS 202-908-4; HSDB 2571; JP 360; Mellite 310; NSC 43789; P 36 (stabilizer); Phenyl phosphite; Phenyl phosphite ((C6H5O)3P); Phosclere T 36; Phosphorous acid, triphenyl ester; Stabilizer P 36; Sumilizer TPP-R; Sumilizer TTP-R; Syn-O-Ad® P-399; TP 1 (plasticizer); TPP (plasticizer); Trifenoxyfosfin; Trifenylfosfit; Triphenoxyphosphine; Triphenyl phosphite; Tris(phenoxy)phosphine; Weston® EGTPP; Weston® TPP. Antioxidant and antiwear agent in gear and transmission oils. Improves color stability in polyesters, polyurethanes, and alkyd resins; also aids curing and hardening in epoxies. Used as a chemical intermediate, stabilizer systems for resins, metal scavenger, diluent for epoxy resins, antioxidant and antiwear agent in gear and transmission oils. Often used with 0.5% w/w triisopropanolamine; reactive diluent for epoxy applications including adhesives, coatings, laminates, potting and soldering compounds, tooling; viscosity reducer. A stabilizer used in epoxies, hot-melt adhesives, PU, polyester, SBR, PP; in extrusion of PP, HDPE, LDPE, HIPS, PC, ABS, PVC, polyesters, in calendering of ABS, PVC; in film applications of PP, PE, PVC; fiber applications of PP, polyesters. Crystals; mp = 25°; bp = 360°; d20 = 1.1842; λm = 264 nm (MeOH); insoluble in H2O, very soluble in EtOH, organic solvents. *Akzo Chemie; Dover; GE Silicones.*

3993 Triphenylborane

960-71-4   H

C18H15B

Triphenylborane.

AI3-60391; Borane, triphenyl-; Borine, triphenyl EINECS 213-504-2; Triphenylborane; Triphenylborine; Triphenyl-boron.

3994   Triphenylmethane triisocyanate
2422-91-5   G

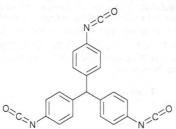

C22H13N3O3
Triphenylmethane-4,4',4-triisocyanate.
Benzene, 1,1',1''-methylidynetris(4-isocyanato-; Desmo-dur® RE; EINECS 219-351-8; Methylidynetri-p-phenylene triisocyanate; Triphenylmethane triisocyanate; Triphenyl-methane-4,4',4-triisocyanate. A room temperature cross-linking agent for adhesives based on Desmocoll and Baycoll polymers, natural rubber, and synthetic rubbers. Also used as a plasticizer. *BASF Corp.; Polysar.*

3995   Triphenylphosphine
603-35-0   9814   G H

C18H15P
Triphenylphosphorus.
4-16-00-00951 (Beilstein Handbook Reference); BRN 0610776; CCRIS 4889; EINECS 210-036-0; HSDB 4266; NSC 10; NSC 215203; Phosphine, triphenyl-; PP 360; Trifenylfosfin; Triphenylphosphane; Triphenylphosphide; Triphenylphosphine; Triphenylphosphorus. Used in organic synthesis, in the manufacture of phosphonium salts and other phosphorus compounds, as a polymerization initiator. White solid; mp = 81°; bp > 360°; d$_4^{25}$ 1.194; insoluble in H2O, soluble in organic solvents. *BASF Corp.; Elf Atochem N. Am.; Janssen Chimica; Morton Intn'l.*

3996   Tripropylene glycol
24800-44-0   G H

C9H20O4
2-(2-(2-Hydroxypropoxy)propoxy)-1-propanol.
AI3-14657; BRN 2235421; EINECS 246-466-0; HSDB 2655; 2-(2-(2-Hydroxypropoxy)propoxy)-1-propanol; ((Methylethylene)bis(oxy))dipropanol; Propanol, ((1-methyl-1,2-ethanediyl)bis(oxy))bis-; Tripropylene glycol. Intermediate in resins, plasticizers, pharmaceuticals, insecticides, dyestuffs, and mold lubricants. *Union Carbide Corp.*

3997   Tripropylene glycol diacrylate
42978-66-5   H

C15H24O6
Acrylic acid, propylenebis(oxypropylene) ester.
Acrylic acid, propylenebis(oxypropylene) ester; EINECS 256-032-2; EINECS 272-647-9; (1-Methyl-1,2-ethane-diyl)bis(oxy(methyl-2,1-ethanediyl)) diacrylate; 2-Propen-oic acid, 1,3-propanediylbis(oxy-3,1-propanediyl) ester; 2-Propenoic acid, (1-methyl-1,2-ethanediyl)bis(oxy-(methyl-2,1-ethanediyl)) ester; Propane-1,3-diylbis(oxy-propane-1,3-diyl) diacrylate; Tripropylene glycol diacrylate. Mixture of isomers of propylene glycol.

3998   Tripropylene glycol diacrylate
68901-05-3   G

C15H24O6
2-Propenoic acid, (1-methyl-1,2-ethanediyl)bis(oxy-(methyl-2,1-ethanediyl)) ester.
Acrylic acid, propylenebis(oxypropylene) ester; Ageflex TPGDA; EINECS 256-032-2; EINECS 272-647-9; (1-Methyl-1,2-ethanediyl)bis(oxy(methyl-2,1-ethanediyl)) diacrylate; PPG-3 diacrylate; Propane-1,3-diylbis(oxy-propane-1,3-diyl) diacrylate; 2-Propenoic acid, 1,3-propanediylbis(oxy-3,1-propanediyl) ester; 2-Propenoic acid, (1-methyl-1,2-ethanediyl)bis(oxy(methyl-2,1-ethane-diyl)) ester; Tripropylene glycol diacrylate; TRPGDA. Crosslinking monomer for UV-curable inks/coatings, floor tiles, wood coatings and fillers, adhesives, textile finishes, and rubber compounds; thinner for radiation curing systems, inks, etc. Liquid; bp >120°; flash pt. >100°; d$_{20}^{20}$ = 1.039; irritant. *CPS Chemical Co.*

3999   Tripropylene glycol methyl ether
20324-33-8   H

C10H22O4
2-Propanol, 1-(2-(2-methoxy-1-methylethoxy)-1-methyl-ethoxy)-.
4-01-00-02475 (Beilstein Handbook Reference); BRN 1701854; Dowanol 62B; Dowanol TPM glycol ether; EINECS 243-734-9; 1-(2-(2-Methoxy-1-methylethoxy)-1-methylethoxy)propan-2-ol; 2-Propanol, 1-(2-(2-methoxy-1-methylethoxy)-1-methylethoxy)-; Propasol solvent TM; Tripropylene glycol methyl ether; Tripropylene glycol monomethyl ether.

4000   Tris(butoxyethyl) phosphate
78-51-3   G H

C18H39O7P
Tris(2-butoxyethyl) phosphate.
4-01-00-02422 (Beilstein Handbook Reference); AI3-04596; BRN

654

1716010; CCRIS 5942; EINECS 201-122-9; Ethanol, 2-butoxy-, phosphate (3:1); HSDB 2564; KP 140; Kronitex KP-140; NSC 4839; Phosflex T-bep; Phosphoric acid, tributoxyethyl ester; Phosphoric acid, tris(2-butoxyethyl) ester; TBEP; Tri(2-butoxyethanol) phosphate; Tributoxyethyl phosphate; Tri(2-butoxyethyl) phosphate; Tributyl cellosolve phosphate; Tris-(2-butoxyethyl)fosfat; Tris(butoxyethyl) phosphate. Surfactant and primary plasticizer for most resins and elastomers; floor finishes and waxes; flame-retarding agent; latex paints; as defoamer. Nonfoaming emulsifier, wetting agent, dispersant, leveling agent; acid and electrolyte stable. Plasticizer; leveling agent in floor polish formulations allowing uniform coverage, eliminates high and low spots in gloss, and preventing streaking, crazing, powder, and film contracting; flame retardant for plastics or synthetic rubber. Liquid; bp4 = 215-228°; d = 1.0060; LD50 (rat orl) = 3000 mg/kg. *Akzo Chemie; Albright & Wilson Americas Inc.; FMC Corp.; Rhône Poulenc Surfactants; Rhône-Poulenc.*

4001   1,3,5-Tris((4-tert-butyl-3-hydroxy-2,6-xylyl)methyl)-1,3,5-triazine-2,4,6(1H,3H,5H)-trione
40601-76-1   H

C39H51N3O6
 EINECS 254-996-9; 1,3,5-Triazine-2,4,6(1H,3H,5H)-trione, 1,3,5-tris((4-(1,1-dimethylethyl)-3-hydroxy-2,6-di-methylphenyl)methyl)-; 1,3,5-Tris((4-tert-butyl-3-hydroxy-2,6-xylyl)methyl)-1,3,5-triazine-2,4,6(1H,3H,5H)-trione.

4002   Tris(2,4-di-tert-butylphenyl) phosphite
31570-04-4   G H

C42H63O3P
 Phenol, 2,4-bis(1,1-dimethylethyl)-, phosphite .
 EINECS 250-709-6; Hostanox® PAR 24; Irgafos 168; Naugard® 524; Phenol, 2,4-bis(1,1-dimethylethyl)-, phosphite (3:1); Phenol, 2,4-di-tert-butyl-, phosphite (3:1); Tris(2,4-di-tert-butylphenyl) phosphite. Antioxidant used in thermoplastic and thermoset polymers (LLDPE, HDPE, PP) where color and processing stability

are critical. *Hoechst Celanese; Uniroyal.*

4003   Tris(2-chloroethyl) phosphite
140-08-9   H

C6H12Cl3O3P
 Phosphorous acid, tris(2-chloroethyl) ester.
 4-01-00-01378 (Beilstein Handbook Reference); BRN 1704979; CCRIS 5949; EINECS 205-397-6; Ethanol, 2-chloro-, phosphite (3:1); HSDB 2584; NSC 6514; Phosphorous acid, tris(2-chloroethyl) ester; Tris(β-chloroethyl) phosphite; Tris(2-chloroethyl) phosphite . Liquid; bp3 = 120°; d^{26} = 1.3443.

4004   Tris(2-chloropropyl) phosphate
6145-73-9   H

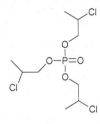

C9H18Cl3O4P
 2-Chloro-1-propanol phosphate (3:1).
 Antiblaze 80; AP 33; 2-Chloro-1-propanol phosphate (3:1); EINECS 228-150-4; Fyrol PCF; NSC 524664; 1-Propanol, 2-chloro-, phosphate (3:1); Tris(β-chloropropyl) phosphate; Tris(2-chloropropyl) phosphate. Flame retardant.

4005   Tris(1-chloro-2-propyl)phosphate
13674-84-5   H

C9H18Cl3O4P
 Phosphoric acid, tris(2-chloro-1-methylethyl) ester.
 Amgard TMCP; BRN 1842347; CCRIS 6111; EINECS 237-158-7; Hostaflam OP 820; Phosphoric acid, tris(2-chloro-1-methylethyl) ester; 2-Propanol, 1-chloro-, phosphate (3:1); Tri-(2-chloroisopropyl)phosphate; Tris(1-chloro-2-propyl)phosphate; Tris(2-chloro-1-methylethyl) phos-phate. *ABIC.*

4006   Tris(2,3-dibromopropyl) phosphate
126-72-7   9821   G

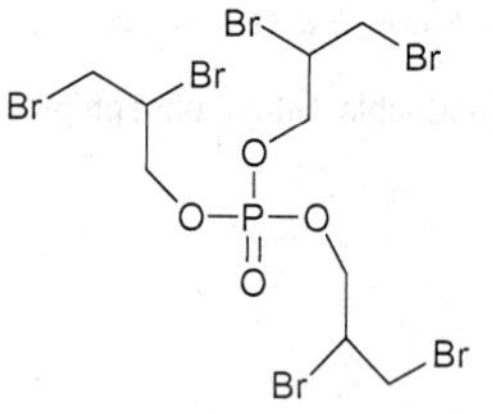

C9H15Br6O4P

1-Propanol, 2,3-dibromo-, phosphate.

3PBR; AI3-19170; Anfram 3PB; APEX 462-5; BRN 1915153; Bromkal P 67-6HP; CCRIS 614; 2,3-Dibromo-1-propanol phosphate; (2,3-Dibromopropyl) phosphate; EINECS 204-799-9; ES685; Firemaster LV-T 23P; FireMaster T 23; Firemaster T23P; Firemaster T23p-lv; Flacavon R; Flamex T 23P; Flammex AP; Flammex LV-T 23P; Flammex T 23P; Fyrol HB32; HSDB 2581; NCI-C03270; NSC 3240; Phoscon PE 60; Phoscon UF-S; Phosphoric acid, tris(2,3-dibromopropyl) ester; 1-Propanol, 2,3-dibromo-, phosphate (3:1); RCRA waste number U235; T 23P; TDBP; TDBPP; Tris-(2,3-dibrompropyl)fosfat; Tris (flame retardant); Tris-2,3-dibrompropyl ester kyseliny fosforecne; Tris-BP; TRIS; Tris(2,3-dibromo-1-propyl) phosphate; Tris(2,3-dibromo-propyl) phosphate; Tris(2,3-dibromopropyl) phosphoric acid ester; Tris(dibromopropyl)phosphate; Tris(dibromo-propyl) phosphate; USAF DO-41; Zetifex ZN; Zetofex ZN; FR-2406. A proprietary fire-retardant additive used in acrylics, epoxies, latices, phenolics, polyesters, polystyrenes, polyvinyl chloride, rayon celluloses and polyurethanes. Pale yellow viscous liquid; soluble in CHCl3; LD50 (rat orl) = 2830 mg/kg. Dow; UK.

4007   Tris(1,1-dimethylethyl)phenol
732-26-3   H

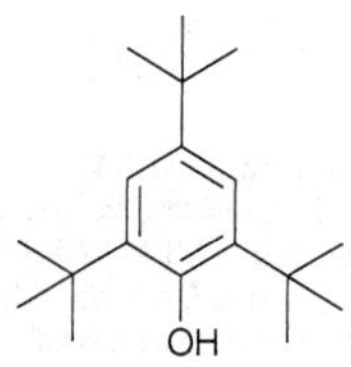

C18H30O

2,4,6-Tri-tert-butylphenol.

Alkofen B; 2,4,6-Tri-t-butylphenol; 2,4,6-Tri-tert-butyl-1-hydroxybenzene; 2,4,6-Tri-tert-butylphenol; CCRIS 5845; 2,4,6-Tris(1,1-dimethylethyl)phenol; EINECS 211-989-5; NSC 14459; P 23; Phenol, 2,4,6-tri-tert-butyl-; Phenol, 2,4,6-tris(1,1-dimethylethyl)-; TM02; Tris(1,1-dimethyl-ethyl)phenol; Voidox. Solid; mp = 131°; bp = 278°; d^{27} = 0.864; λ_m = 280 nm (cyclohexane); insoluble in H2O, alkali, soluble in EtOH, Me2CO, CCl4.

4008   Tris(2-hydroxyethyl) isocyanurate
839-90-7   H

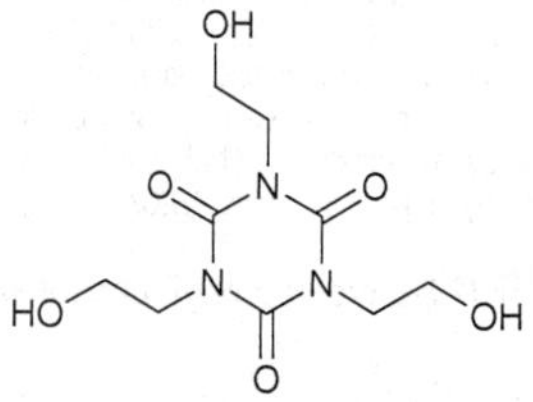

C9H15N3O6

Isocyanuric acid tris(2-hydroxyethyl) ester.

AI3-60291; CCRIS 6113; EINECS 212-660-9; HSDB 6135; Isocyanuric acid tris(2-hydroxyethyl) ester; NSC 11680; Theic; s-Triazine-2,4,6(1H,3H,5H)-trione, 1,3,5-tris(2-hydroxyethyl)-; Tris(β-hydroxyethyl) isocyanurate; Tris(2-hydroxyethyl)-1,3,5-triazinetrione; Tris(2-hydroxy-ethyl) isocyanurate; N,N',N''-Tris(2-hydroxyethyl) iso-cyanurate; Tris(2-hydroxyethyl)-s-triazine-2,4,6-trione; Tris(2-hydroxyethyl)cyanurate; Tris(hydroxyethyl) cyanur-ate; 1,3,5-Tris(2-hydroxyethyl) isocyanurate; 1,3,5-Tris(2-hydroxyethyl) isocyanuric acid; 1,3,5-Tris(2-hydroxy-ethyl)triazine-2,4,6-trione.

4009   Tris(hydroxymethyl) aminomethane hydro-chloride
1185-53-1   G

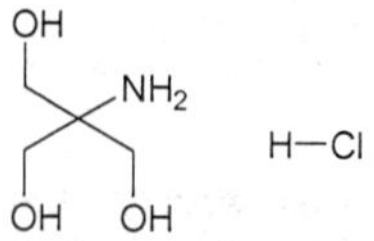

C4H12ClNO3

Tris hydrochloride.

2-Amino-2-(hydroxymethyl)propane-1,3-diol hydrochloride; 2-Amino-2-hydroxymethyl-1,3-propane-diol hydrochloride; EINECS 214-684-5; 1,3-Propanediol, 2-amino-2-(hydroxymethyl)-, hydrochloride; Tris hydro-chloride; Tris hydrochloride buffer. A buffer used exten-sively in biochemical research. Crystals; mp = 150-152° (dec). Am. Biorganics; Janssen Chimica; Sigma-Aldrich Fine Chem.; U.S. BioChem.

4010   Tris(isocyanatohexyl)biuret
4035-89-6   H

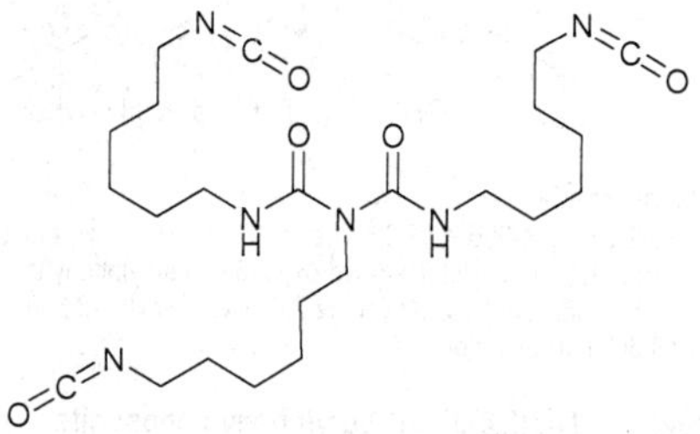

C23H38N6O5

N,N',2-Tris(6-isocyanatohexyl)-imidodicarbonic diamide.

BRN 2192715; EINECS 223-718-8; Hexamethylene diisocyanate biuret; Imidodicarbonic diamide, N,N',2-tris(6-isocyanatohexyl)-; Isocyanic acid, triester with 1,3,5-tris(6-hydroxyhexyl)biuret; N,N',2-Tris(6-isocyan-atohexyl)-imidodicarbonic diamide; Tris(6-isocyanato-hexyl) biuret; 1,3,5-Tris(6-isocyanatohexyl)biuret.

4011   Tris(isooctyl thioglycollate) methyl tin
57583-34-3   H

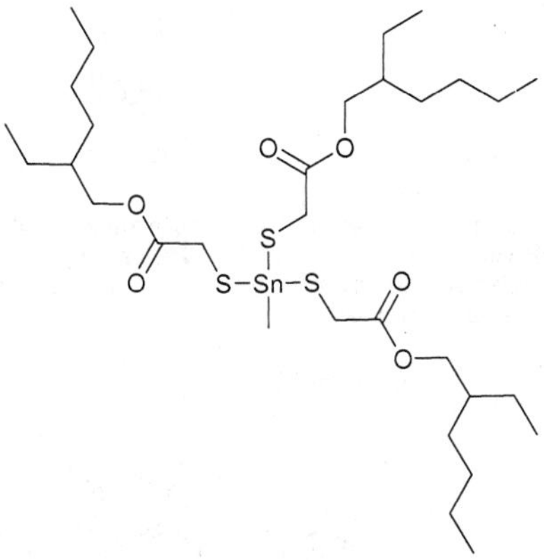

C$_{31}$H$_{60}$O$_6$S$_3$Sn

Methyltris(2-ethylhexyloxycarbonylmethylthio)stannane.
EINECS 260-828-5; Methyltris(2-ethylhexyloxycarbonyl-methylthio)stannane; Stannane, methyltris(2-ethylhexyl-oxycarbonylmethylthio)-; Tin, methyl-, tris(isooctyl thio-glycollate); Tris(isooctyl thioglycollate) methyl tin.

4012　Tris(2-methoxyethoxy)vinylsilane
1067-53-4　　　　　　　　　　　　　　　G H

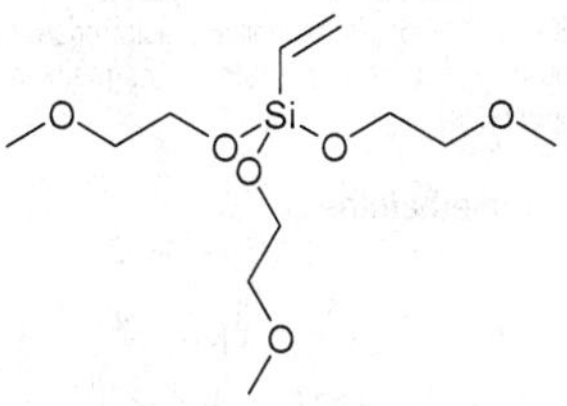

C$_{11}$H$_{24}$O$_6$Si

Vinyltris(2-methoxyethoxy)silane.
4-04-00-04257 (Beilstein Handbook Reference); A 172; Aktisil VM; BRN 1795316; CV-5000; EINECS 213-934-0; GF 58; NSC 78465; NUCA 172; Prosil 248; Q 174; SH 6030; Silane, tris(2-methoxyethoxy)vinyl-; Silicone A-172; Tris(2-methoxyethoxy)vinylsilane; Union Carbide® A-172; Z 6030. Coupling agent, chemical intermediate, blocking agent, release agent, lubricant, primer, reducing agent. As a crosslinking agent provides gloss, hides power to coatings. Filler for peroxide-cured systems. Liquid; mp = -30°; bp = 285°; d = 1.0300; reacts with H$_2$O; LD$_{50}$ (rat orl) = 2960 mg/kg. Degussa-Hüls Corp.; Union Carbide Corp.

4013　Trisodium　　　(2-Hydroxyethyl)ethylenedi-aminetriacetate
139-89-9　　　　　　　　　10029　　　　　G P

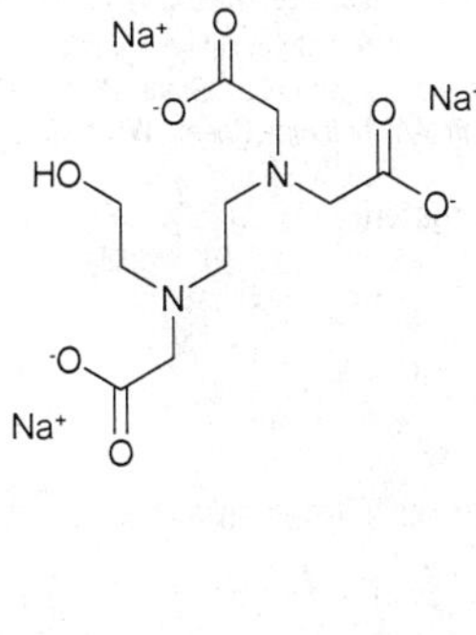

C$_{10}$H$_{18}$N$_2$Na$_3$O$_7$

Trisodium (2-Hydroxyethyl)ethylenediaminetriacetate.
Caswell No. 487C; Chel dm 41; Chemcolox 800; Detarol trisodium salt; EINECS 205-381-9; Emkasene 800; EPA Pesticide Chemical Code 039109; Hamp-ol 120; Hamp-ol crystals; HSDB 5628; Kalex OH; Monaquest ica-120; NSC 148338; Perma kleer 80; Trilon® D Liq; Trisodium (2-hydroxyethyl)ethylenediaminetriacetate; Trisodium HEDTA; Versen-01; Versen-ol; Versenol 120 . General purpose chelate for control of iron in pH range of 6.5-9.5, Ca, Mg; used in soaps, detergents, water treatment, metal finishing and plating, pulp and paper manufacturing, synthesis of polymers, photographic products, textiles, chemical cleaning. Chelating agent for sequestering calcium, magnesium, and ferric ions in solution, sequesters Ca, Mg, iron at pH 8.0-10.5. Registered by EPA as an antimicrobial, fungicide and herbicide (cancelled). Crystals; mp = 288-290° (dec). BASF Corp.; Emkay; Fluka; Hart Prod.

4014　Trisodium　　bis(3-hydroxy-4-((2-hydroxy-1-naphthyl)azo)-7-nitronaphthalene-1-sulphonato(3-))chromate(3-)
57693-14-8　　　　　　　　　　　　　　　H

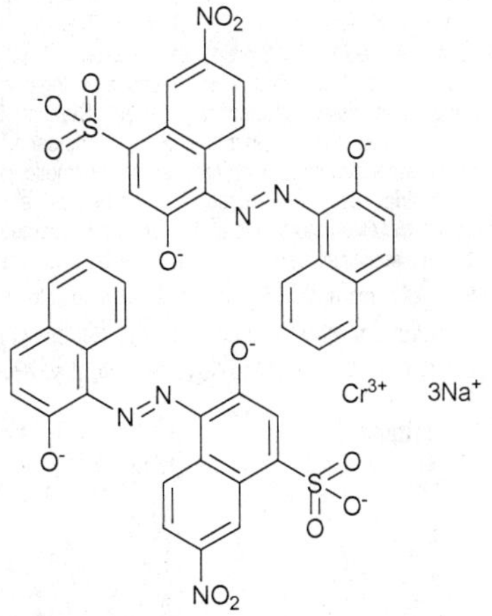

C$_{40}$H$_{14}$CrN$_6$Na$_3$O$_{14}$S$_2$

Chromate(3-), bis(3-hydroxy-4-((2-hydroxy-1-naphthal-eneyl)azo)-7-nitro-1-naphthalenesulfonato(3-))-, tri-sodium; EINECS 260-906-9; Trisodium bis(3-hydroxy-4-((2-hydroxy-1-naphthyl)azo)-7-nitronaphthalene-1-sulphonato(3-))chromate(3-).

4015　Trisodium phosphate
7601-54-9　　　　　　　　8736　　　　　　G P

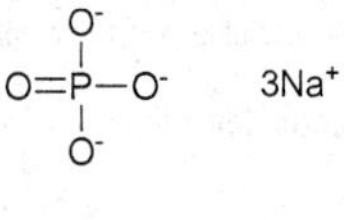

Na$_3$O$_4$P

Sodium phosphate, tribasic, anhydrous.
Antisal 4; Caswell No. 898; CCRIS 7086; Dri-Tri; EINECS 231-509-8; Emulsiphos 440/660; EPA Pesticide Chemical Code 076406; HSDB 583; NSC 215202; Nutrifos STP; Oakite; Phosphoric acid, trisodium salt; Sodium orthophosphate; Sodium orthophosphate, tertiary; Sodium orthophosphate, tribasic; Sodium phosphate; Sodium phosphate (Na$_3$PO$_4$); Sodium phosphate, tribasic; Sodium tertiary phosphate; Tertiary sodium phosphate; Tribasic sodium orthophosphate; Tribasic sodium phosphate; Trinatriumphosphat; Trisodium orthophosphate; Tri-sodium phosphate; Tromete; TSP; TSP-O. Inorganic salt; For pH adjustment in food systems; cleaning compounds, water treatment, textiles. Water softening agent subject

to restrictions on phosphates and thus seeing diminishing use. Registered by EPA as an antimicrobial, fungicide and herbicide. Crystals; soluble in H_2O (0.29 g/ml), insoluble in organic solvents; LD_{50} (rat orl)= 7.40 g/kg. *Albright & Wilson Americas Inc.; Kemira Kemi UK; Monsanto Co.*

4016   Tristearin
555-43-1   9825   G

$C_{57}H_{110}O_6$

Glyceryl tristearate.
AI3-01633; Dynasan 118; EINECS 209-097-6; Glycerol, trioctadecanoate; Glycerol tristearate; Glycowax S 932; Hardened oil; HSDB 5690; Kemester® 5500; Kemester® 6000; Neobee® 62; Octadecanoic acid, 1,2,3-propanetriyl ester; 1,2,3-Propanetriol trioctadecanoate; Spezialfett 118; Stearic acid triglyceride; Stearic acid triglycerin ester; Stearic triglyceride; Stearin; Stearin, tri-; Stearoyl triglyceride; Trioctadecanoin; Tristearin. Emollient, emulsifier; stabilizer, plasticizer, lubricant for cosmetic, paper, textile, and industrial uses. Cosmetic and industrial emulsifier and plasticizer for elastomers. Industrial lubricant; cosmetic emollient. Solubilizer, stabilizer used in food applications. Used to make candles. Solid; mp about 55°; d^{80} = 0.8559; insoluble in H_2O, EtOH, ligroin, petroleum ether, slightly soluble in C_6H_6, CCl_4, soluble in Me_2CO, $CHCl_3$, CS_2. *CK Witco Corp.; Stepan; Witco/Humko.*

4017   Tritane
519-73-3   9812   G

$C_{19}H_{16}$

Triphenylmethane.
AI3-02337; Benzene, 1,1',1''-methylidynetris-; CCRIS 5194; EINECS 208-275-0; Methane, triphenyl-; NSC 4049; Triphenylmethane; Tritane. Used as a dye. Solid; mp = 93°; bp = 360°; d^{100}_{4} 1.0134; insoluble in H_2O, soluble in organic solvents.

4018   Tritetracosylaluminum
6651-26-9   H

$C_{72}H_{147}Al$
Tri(tetracosyl)aluminum.
Aluminum, tritetracosyl-; EINECS 229-673-0; Tritetracosylaluminium.

4019   Tritetradecylaluminum
1529-58-4   H

$C_{42}H_{87}Al$
Aluminum, tris(tetradecyl)-.
Aluminum, tris(tetradecyl)-; Aluminum, tritetradecyl-; EINECS 216-218-6; HSDB 5836; Tri(tetradecyl) aluminum; Tris(tetradecyl)aluminum; Tritetradecyl-aluminum; Tritetradecylaluminium.

4020   Tritolyl phosphite
25586-42-9   H

$C_{21}H_{21}O_3P$
Phosphorous acid, tris(methylphenyl) ester.
EINECS 247-119-6; Phosphorous acid, tris(methylphenyl) ester; Phosphorous acid, tritolyl ester; Tris(methylphenyl) phosphite; Tritolyl phosphite.

4021   Tromethamine
77-86-1   9842   H

$C_4H_{11}NO_3$
1,1,1-Tris(hydroxymethyl)methanamine .
Addex-tham; AI3-03948; Aminotrimethylolmethane; Aminotris(hydroxymethyl)methane; Apiroserum Tham; Caswell No. 036; EINECS 201-064-4; EPA Pesticide Chemical Code 083901; HSDB 3408; Methanamine, 1,1,1-tris(hydroxymethyl)-; NSC 6365; Pehanorm; Talatrol; Tham; THAM; THAM-E; Trimethylolaminomethane; Tris; Tris (buffering agent); Tris-base; Tris buffer; Tris, free base; Tris-hydroxymethyl-aminomethan; Tris-hydroxymethylaminomethane; Tris-steril; Tris(hydroxymethyl)aminomethane; Tris(hydroxymethyl)methylamine; Tris(hydroxymethyl)methanamine; Trisamine; Trisaminol; Trispuffer; Trizma; Trometamol; Trometamolum; Tro-methamine; Tromethamolum; Tromethane; Tutofusin tris. Pigment dispersant, neutralizing amine, corrosion inhibitor, acid-salt catalyst, pH buffer, chemical and pharmaceutical intermediate, solubilizer. Emulsifying agent for oils, fats, waxes; absorbent for acidic gases, medicine; chemical intermediate. Used medically as an alkalizer. Solid; mp = 171.5°; bp_{10} = 219-220°; soluble in H_2O and organic solvents. *Abbott Labs Inc.; Dajac Labs.; Grace W.R. & Co.; Heico; Janssen Chimica; Sigma-Aldrich Fine Chem.; Whittaker Clark & Daniels.*

4022   Tungsten
7440-33-7   9884   G

W

W
Tungsten.
EINECS 231-143-9; HSDB 5036; Tungsten; Tungsten, elemental;

VA (tungsten); Wolfram. Metallic element; high-speed tool steel, alloys, filaments for electric light bulbs, contact points, x-ray and electron tubes, welding electrodes, heating elements, rocket nozzles, sheet steel, chemical apparatus, high-speed rotors, solar energy devices. mp = 3410°; bp = 5660°; d_4^{20}= 18.7-19.3; LD50 (rat ip) = 5 g/kg. *Atomergic Chemetals; Cerac; Noah Chem.; Sigma-Aldrich Fine Chem.*

4023   Tungsten hexafluoride
7783-82-6   9885   G

F6W
Tungsten (VI) fluoride.
EINECS 232-029-1; Tungsten fluoride; Tungsten fluoride (WF6); Tungsten hexafluoride; UN2196. Used in vapor-phase deposition of tungsten, as a fluorinating agent. Gas; mp= 2°; bp = 17°; d^{15} = 3.441. *Air Products & Chemicals Inc.; Akzo Chemie; Atomergic Chemetals; Elf Atochem N. Am.*

4024   Tungsten trioxide
1314-35-8   9886   G

O3W
Tungsten(VI) oxide.
C.I. 77901; EINECS 215-231-4; HSDB 5800; Tungsten Blue; Tungsten oxide; Tungsten oxide (WO3); Tungsten trioxide; Tungstic acid; Tungstic acid anhydride; Tungstic anhydride; Tungstic oxide; Wolframic acid, anhydride . Forms metals by reduction, alloys, preparation of tungstates for xray screens, fireproofing fabrics, yellow pigment in ceramics. Yellow powder; insoluble in H2O, slightly soluble in alkali, soluble in acids. *Atomergic Chemetals; Cerac; Climax Molybdenum Co.*

4025   Tungstic acid
7783-03-1   9887   G

H2O4W
Tungstic(VI) acid.
Dihydrogen wolframate; EINECS 231-975-2; Ortho-tungstic acid; Tungstate (WO4^{2-}), dihydrogen, (β-4)-; Tungstic acid (H2WO4); Tungstic(VI) acid. Used in textiles as a mordant and color resist, plastics, tungsten metal, wire, etc. Poorly soluble in H2O. *Am. Intl. Chem.; Atomergic Chemetals; Noah Chem.*

4026   Turpentine
9005-90-7   9894   H

Skipidar.
EINECS 232-688-5; FEMA No. 3088; Galipot; Gum thus; Gum turpentine; petropine; Pine gum; Pine resin; Skipidar; Turpentine; Turpentine gum (Pinus spp.); Turpentine gum-1. Used as an insecticide, solvent for waxes, in polishes. Yellow gum; insoluble in H2O, soluble in CHCl3, Et2O, AcOH.

4027   Turpentine oil
8006-64-2   H

C10H16
Turpentine Oil.
Caswell No. 900; EINECS 232-350-7; EPA Pesticide Chemical Code 084501; FEMA No. 3089; Gum spirits of turpentine; Gum turpentine; HSDB 204; Oil of turpentine, rectified; Oil of turpentine; Oil of turpentine, distillation residue; Purified gum spirits; Purified turpentine; Rectified turpentine; Spirit of turpentine; Spirits of turpentine; Sulfate turpentine; Terebenthine; Terpentin oel; Terpentine; Turpentine; Turpentine oil, rectified; Turpentine spirits; Turpentine, steam-distilled (Pinus spp.); Turpentine steam distilled; Turpentine substitute [Flammable liquid]; Turpentine [Flammable liquid]; UN1299; UN1300; Wood turpentine. FDA approved for inhalants, BP compliance. Used as a solvent, rubifacient, diuretic, used in inhalants, liniments and in preparations for respiratory tract disorders. Colorless liquid; insoluble in H2O; d^{15} = 0.860 - 0.875. TLV = 100 ppm in air. *Spectrum Chem. Manufacturing.*

4028   Turpeth mineral
1312-03-4   G

HgSO4·2H2O
Mercuric subsulfate.
Basic mercuric sulfate; EINECS 215-191-8; HSDB 1188; Mercuric basic sulfate; Mercuric subsulfate; Mercury oxide sulfate; Mercury oxide sulfate (Hg3O2(SO4)); Mercury oxonium sulfate; Mercury sulfate, basic; Queen's yellow; Trimercury dioxide sulphate; Turpeth mineral. Mercuric subsulfate, a yellow basic sulfate of mercury. Solid; insoluble in H2O.

4029   Tylosin
1401-69-0   9900   G V

C46H77NO17
Tylosine.
AI3-29799; EINECS 215-754-8; Fradizine; HSDB 7022; Tilosina; Tylan; Tylocine; Tylosin; Tylosin A; Tylosine; Tylosinum; Vubityl 200. A veterinary antibiotic derived from an actinomycete resembling *Streptomyces fradioe* . White powder; mp = 128-132°; $[\alpha]_D^{25}$ = -46° (c = 2 MeOH); λm = 282 nm (E$_{1cm}^{1\%}$ 245); soluble in H2O (0.5 g/100 ml), more soluble in organic solvents. *ADM Animal Health & Nutrition Divn.; Boehring-Ingelheim Veterinary; Elanco Animal Health.*

4030   Tyloxapol
25301-02-4   9901   D G

C8H11NO
p-(1,1,3,3-Tetramethylbutyl)phenol polymer with ethyl-ene oxide and formaldehyde.
Alevaire; Enuclene; Exosurf; Formaldehyde, polymer with oxirane and 4-(1,1,3,3-tetramethylbutyl)phenol; Macro-cyclon; NSC 90255; Oxyethylated tertiary octyl-phenol-formaldehyde polymer; p-Isooctylpolyoxyethylenephenol formaldehyde polymer; Phenol, 4-(1,1,3,3-tetramethyl-butyl)-, polymer with formaldehyde and oxirane; Phenol, p-(1,1,3,3-tetramethylbutyl)-, polymer with ethylene oxide and formaldehyde; Superinone; 4-(1,1,3,3-Tetramethylbutyl)phenol, polymer with formaldehyde & oxirane; p-(1,1,3,3-Tetramethylbutyl)phenol polymer with ethylene oxide and formaldehyde; Tiloxapol; Triton A-20; Triton W.R.1339; Triton WR

1339; Tyloxapol; Tyloxapol-um; Tyloxypal. A nonionic detergent with surface-tens-ion-reducing properties, used in pulmonary surfactants. Used as a mucolytic. Freely soluble in H_2O; soluble in C_6H_6; C_7H_8, $CHCl_3$, CCl_4, CS_2, AcOH. *Alcon Labs; Glaxo Wellcome Inc.; Rohm & Haas Co.; Sterling Winthrop Inc.*

4031   Ultra-DMC
598-64-1   G

C5H14N2S2
Dimethylamine dimethyldithiocarbamate.
AI3-62285; Carbamic acid, dimethyldithio-, dimethyl-amine salt (1:1); Carbamodithioic acid, dimethyl-, compound with N-methylmethanamine; Dimethyl dithiocarbamate dimethylammonium salt; Dimethylamine dimethyldithiocarbamate; Dimethylammonium dimethyl-dithiocarbamate; Dimethyldithiocarbamic acid compd. with dimethylamine (1:1); Dimethyldithiocarbamic acid dimethyl amine salt; Dimethyldithiocarbamic acid dimethylamine salt; Dimethyldithiocarbamic acid dimethylammonium salt; EINECS 209-945-5; NSC 100885; NSC 6205. A vulcanization accelerator.

4032   Umbelliferone
93-35-6   9915   G

C9H6O3
7-hydroxy-2H-1-benzopyran-2-one.
5-18-01-00386 (Beilstein Handbook Reference); AI3-38054; BRN 0127683; CCRIS 3591; Coumarin, 7-hydroxy-; EINECS 202-240-3; 7 HC; Hydrangin; Hydr-angine; 7-Hydroxycoumarin; NSC 19790; 7-Oxycoum-arin; Skimmetin; Skimmetine; Umbelliferon; Umbelliferone. Aglucon of skimmin, obtained by distillation of *umbelliferae* resins. Formed by metabolism in coumarin in humans. Used in sunscreens and as medical reagent. Needles; mp = 230.5°; d^{25} = 1.25; λ_m = 216, 254, 324 nm (ε = 10000, 2030, 14600, MeOH); soluble in H_2O (10 g/l); more soluble in organic solvents. *Caledon Laboratories Limited; Dr. Ehrenstorfer GmbH; Penn Bio-organics, Inc.; Showa Denko.*

4033   Undecane
1120-21-4   H

C11H24
n-Undecane.
4-01-00-00487 (Beilstein Handbook Reference); AI3-21126; BRN 1697099; CCRIS 3796; EINECS 214-300-6; Hendecane; HSDB 5791; NSC 66159; UN2330; Undecane; n-Undecane. Liquid; mp = -25.6°; bp = 195.9°; d^{20} = 0.7402; insoluble in H_2O, freely soluble in EtOH, Et_2O.

4034   Undecanol
30207-98-8   H

C11H24O
n-Undecanol.

EINECS 250-092-3; HSDB 5175; Undecanol.

4035   Undecyl alcohol
112-42-5   G H

C11H24O
1-Undecyl alcohol.
4-01-00-01835 (Beilstein Handbook Reference); AI3-00330; Alcohol C11; Alcohol, undecyl; BRN 1698334; C11 alcohol; Decyl carbinol; EINECS 203-970-5; Fatty alcohol(C11); FEMA No. 3097; Hendecanoic alcohol; 1-Hendecanol; Hendecyl alcohol; n-Hendecylenic alcohol; HSDB 1089; Neodol® 1; Neodol® 11; NSC 403667; Tip-Nip; Undecanol; 1-Undecanol; Undecyl alcohol; 1-Undecyl alcohol. Intermediate in manufacture of detergents. Solid; mp = 19°; bp = 243°; d^{20} = 0.8298; insoluble in H_2O, soluble in EtOH, very soluble in Et_2O. *Shell.*

4036   Undecyl dodecyl phthalate
68515-47-9   G

C34H58O4
1,2-Benzenedicarboxylic acid, di-C11-14-branched alkyl esters, C13-rich.
1,2-Benzenedicarboxylic acid, di-C11-14-branched alkyl esters, C13-rich; EINECS 271-089-3; Jayflex® UDP. Plasticizer Liquid; d = 0.959; pour point = -40°; flash point = 437°F (TCC). *Exxon.*

4037   Undecylenic acid
112-38-9   9916   G

C11H20O2
Undecenoic acid.
4-02-00-01612 (Beilstein Handbook Reference); AI3-02065; BRN 1762631; Caswell No. 901; Declid; Desenex; Desenex solution; EINECS 203-965-8; EPA Pesticide Chemical Code 085501; FEMA No. 3247; Fulvidex (Veterinary); Hendecenoic acid, omega-; 10-Hendecenoic acid; Kyselina 9-decen-1-karboxylova; Kyselina undecylenova; Mixture Name; NSC 2013; Renselin; Sevinon; Undecenoic acid; Undecenoic acid, ω-; 10-Undecenoic acid; Undecylenic acid; Undecyl-10-enic acid; 9-Undecylenic acid . Used in perfumery, flavoring, plastics, modifying agent, medicine (antifungal agent); intermediate in chemical synthesis, polyamide plastics and fibers, synthetic floors. Crystals; mp = 24.5°; bp = 275° (dec); d^{24} = 0.9072; insoluble in H_2O, slightly soluble in CCl_4, soluble in EtOH, $CHCl_3$, Et_2O; LD_{50} (mus orl) = 8.15 g/kg. *CasChem; Elf Atochem N. Am.; Lancaster Synthesis Co.*

4038   Union Carbide® A-1106
58160-99-9   G

C3H11NO3Si
Silanetriol, (3-aminopropyl)-.
(3-Aminopropyl)silanetriol; EINECS 261-145-5; Silane-triol, (3-aminopropyl)-. Aminoalkyl silicone solution; finish for woven fiberglass; coupling agent for glass fiber sizes; filler treatments; additive to water-soluble/dispersion resins include vinyl and acrylic

latex, epoxies, and phenolic binder dispersions. Used for coatings, adhesives. *Union Carbide Corp.*

4039 Uracil
66-22-8   9918   G

C4H4N2O2
2,4-Pyrimidinedione.
AI3-25470; BMS 205603-01; 2,4-Dihydroxypyrimidine; 2,4-Dioxopyrimidine; CCRIS 3077; EINECS 200-621-9; Hybar X; NSC 3970; Pirod; 2,4-Pyrimidinediol; 2,4(1H,3H)-Pyrimidinedione; Pyrod; RU 12709; SQ 6201; SQ 7726; SQ 8493; Ura; Uracil. Biochemical research. Solid; mp = 338°; λm = 202.5, 259.5 (ε 9200, 8200, pH 7), 283 (0.01N NaOH); soluble in H2O, insoluble in organic solvents. *PCR; Penta Mfg.; Schweizerhall Inc.; Sigma-Aldrich Fine Chem.; U.S. BioChem.*

4040 Uramil
118-78-5   9920   G

C4H5N3O3
5-amino-2,4,6(1H,3H,5H)-pyrimidinetrione.
AI3-52683; 5-Amino-2,4,6-pyrimidinetriol; 5-Amino-barbituric acid; Barbituric acid, 5-amino-; Dialuramide; EINECS 204-277-0; Murexan; NSC 264287; 2,4,6(1H,3H,5H)-Pyrimidinetrione, 5-amino-; Uramil. mp > 400°; soluble in H2O, CHCl3, conc. H2SO4; insoluble in Et2O, C6H6.

4041 Urea
57-13-6   9935   H P

CH4N2O
Carbonyl diamine.
AI3-01202; Alphadrate; Aqua Care; Aquadrate; B-I-K; Calmurid; Carbaderm; Carbamide; Carbamide resin; Carbamimidic acid; Carbonyl diamide; Carbonyldiamine; Caswell No. 902; CCRIS 989; EINECS 200-315-5; EPA Pesticide Chemical Code 085702; Harnstoff; HSDB 163; Isourea; Keratinamin; Mocovina; NCI-C02119; NSC 34375; Nutraplus; Panafil; Pastaron; Pre-spersion, 75 urea; Pseudourea; Supercel 3000; Ultra Mide; UR; Urea; Urea-13C; Urea ammonium nitrate solution; Urea solution; Ureacin-10 lotion; Ureacin-20; Ureacin-40 Creme; Ureaphil; Ureophil; Urepearl; Urevert; Varioform II. Registered by EPA as an antimicrobial and disinfectant. FDA GRAS, approved for injectables, orals, BP, Ph.Eur. compliance. Used as a diuretic, antiseptic, keratin softener for dry skin products, in ammoniated dentrifices, in injectables and orals. Used as a nitrogen fertilizer. Crystals; mp = 132.7°; soluble in H2O (0.1 g/100 ml), EtOH (5 g/100 ml), MeOH (16 g/100 ml); insoluble in CHCl3, Et2O. LD50 (rat orl) = 14,300 mg/kg. *Air Prods & Chems; Bio-Rad Labs.; Chisso Corp.; Elf Atochem N. Am.; EM Ind. Inc.;*

Entek Corp.; Heico; Mallinckrodt Inc.; Mitsui Toatsu; Nissan Chem. Ind.; Norsk Hydro AS; Occidental Chem. Corp.; Ruger; Showa Denko; Sigma-Aldrich Fine Chem.; Spectrum Chem. Manufacturing; Tech Chems. & Prods.

4042 Urea sulfate
17103-31-0   H

C2H10N4O6S
Dicarbamide dihydrogensulfate.
Dicarbamide dihydrogensulfate; Diuronium sulphate; EINECS 241-175-5; Urea, sulfate (2:1).

4043 Ureidopropyltriethoxysilane
23779-32-0   G

C10H24N2O4Si
N-(Triethoxysilylpropyl) urea.
CT2507; Dynasylan 2201; EINECS 245-876-7; N-(Triethoxysilylpropyl)urea; (3-(Triethoxysilyl))propyl)urea; Urea, (3-(triethoxysilyl)propyl)-. Coupling agent, chemical intermediate, blocking agent, release agent, lubricant, primer, reducing agent. *Degussa-Hüls Corp.*

4044 Urethane
51-79-6   9942   G

C3H7NO2
Carbamic acid ethyl ester.
A 11032; Aethylcarbamat; Äthylcarbamat; Aethylurethan; Äthylurethan; AI3-00553; Carbamic acid, ethyl ester; Carbamidsaeure-aethylester; Carbamidsäure-ähylester; CCRIS 619; EINECS 200-123-1; Estane 5703; Ethyl carbamate; Ethyl urethan; Ethyl urethane; O-Ethylurethane; Ethylcarbamate; Ethylester kyseliny karbaminove; Ethylurethan; Ethylurethane; HSDB 2555; Leucethane; NSC 746; Pracarbamin; Pracarbamine; RCRA waste number U238; U-Compound; Uretan etylowy; Uretano; Urethan; Urethane; Urethanum; X 41. Used as an intermediate for pharmaceuticals and pesticides; in biochemical research and medicine. Crystals; mp = 48-50°; soluble in H2O (2 g/ml), less soluble in organic solvents; MLD (mus ip) = 2.1 g/kg. *Agro-Kanesho; Hokko; Ihara; Shinung.*

4045 Uric acid
69-93-2   9943   G

C5H4N4O3

Purine-2,6,8(1H,3H,9H)-trione.

AI3-15432; 7,9-Dihydro-1H-purine-2,6,8(3H)-trione; EINECS 200-720-7; Lithic acid; NSC 3975; 1H-Purine-2,6,8-triol; Purine-2,6,8(1H,3H,9H)-trione; 1H-Purine-2,6,8(3H)-trione, 7,9-dihydro-; 2,6,8-Trihydroxypurine; 2,6,8-Trioxopurine; Uric acid. Used in organic synthesis. Solid; d^{25} = 1.89; λ_m = 286 nm (MeOH); poorly soluble in H2O, organic solvents. *Burlington Biomedical; ICI Spec.; Spectrum Chem. Manufacturing.*

4046   Uridine
58-96-8   9945   D G

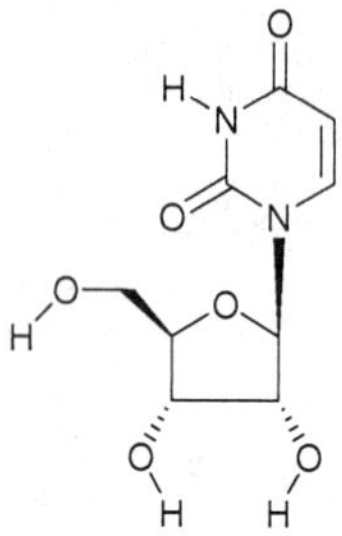

C9H12N2O6

1-β-D-ribofuranosyluracil.

AI3-52690; EINECS 200-407-5; NSC 20256; 1-β-D-Ribofuranosyluracil; Uracil-1-β-D-ribofuranoside; Uracil, 1-β-D-ribofuranosyl-; D-Ribosyl uracil; Uracil riboside; Urd; Uridin; Uridine. Used in biochemical research. Crystalline solid; mp = 165°; $[\alpha]_D^{25}$= 4° (c = 2); λ_m = 261, 205 nm (ε 10100, 9800); soluble in H2O. *Am. Biorganics; Greeff R.W. & Co.; Penta Mfg.; Sigma-Aldrich Fine Chem.*

4047   Valacyclovir
124832-26-4   9966   D

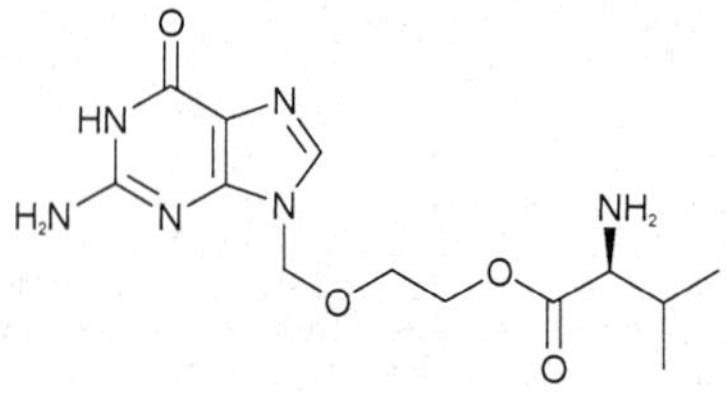

C17H21NO3

2-[(2-Amino-1,6-dihydro-6-oxo-9H-purin-9-yl)methoxy ethyl ester of L-valine.

256U87; Valaciclovir; ValACV; Valacyclovir. Antiviral agent. Prodrug of acyclovir. *Glaxo Wellcome Inc.*

4048   Valacyclovir hydrochloride
124832-27-5   9966   D

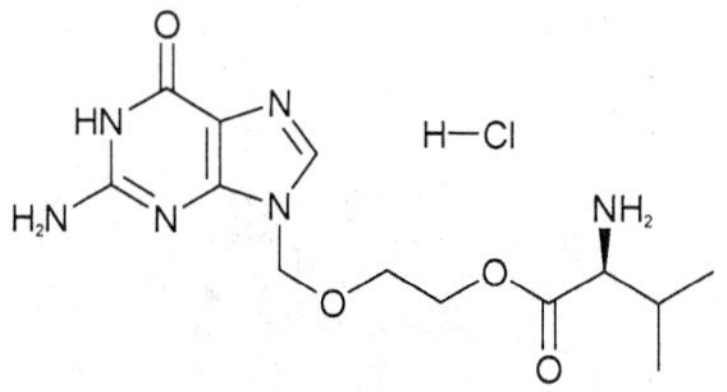

C17H22ClNO3

2-[(2-Amino-1,6-dihydro-6-oxo-9H-purin-9-yl)methoxy ethyl ester of L-valine monohydrochloride.

256U; 256U87 hydrochloride; BW-256; BW-256U87; DRG-0119; Valacyclovir hydrochloride; Valtrex. Antiviral agent. Crystals; λ_m = 252.8 nm (ε 8250, H2O); soluble in H2O (17.4 g/100 ml). *Glaxo Wellcome Inc.*

4049   Valdecoxib
181695-72-7   D

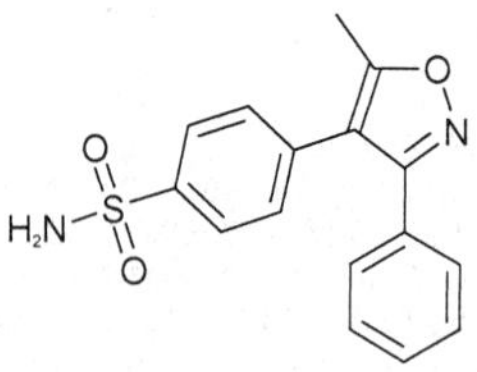

C16H14N2O3S

4-(5-Methyl-3-phenyl-4-isoxazolyl)-benzenesulfonamide .

Benzenesulfonamide, 4-(5-methyl-3-phenyl-4-isoxazolyl)-; Bextra; SC 65872; Valdecoxib. USP 6,090,834 (18 July, 2000) to G. D. Searle. FDA-approved 16 November, 2001 for the relief of the signs and symptoms of osteoarthritis and adult rheumatoid arthritis. *G.D. Searle & Co.*

4050   Valeraldehyde
110-62-3   9968   H

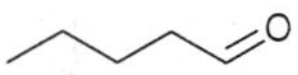

C5H10O

n-Pentyl aldehyde.

4-01-00-03268 (Beilstein Handbook Reference); AI3-16105; Amyl aldehyde; BRN 1616304; Butyl formal; CCRIS 3220; EINECS 203-784-4; FEMA Number 3098; HSDB 851; NSC 35404; Pentanal; n-Pentanal; Pentyl aldehyde; UN2058; Valeral; Valeraldehyde; n-Valeraldehyde; Valerianic aldehyde; Valeric acid aldehyde; Valeric aldehyde; n-Valeric aldehyde; Valeryl aldehyde; Valerylaldehyde. Liquid; mp = -91.5°; bp = 103°; d^{20} = 0.8095; λ_m = 178, 182, 184 nm (gas); slightly soluble in H2O, soluble in EtOH, Et2O.

4051   Valeric acid
109-52-4   9970   G H

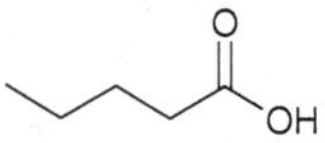

C5H10O2

Butanecarboxylic acid.

4-02-00-00868 (Beilstein Handbook Reference); AI3-08657; BRN 0969454; Butanecarboxylic acid; 1-Butanecarboxylic acid; EINECS 203-677-2; FEMA No. 3101; HSDB 5390; Kyselina valerova; NSC 406833; Pentanoic acid; n-Pentanoic acid; Propylacetic acid; Valerianic acid; Valeric acid; n-Valeric acid. Intermediate for flavors and perfumes, ester-type lubricants, plasticizers, pharmaceuticals, and vinyl stabilizers. Crystals; mp = -34°; bp = 186.1°; d^{20} = 0.9391; soluble in H2O, EtOH, Et2O, CCl4. *BASF Corp.; Sigma-Aldrich Fine Chem.; Union Carbide Corp.*

4052   Valsartan
137862-53-4   9982   D

C24H29N5O3
N-[p-(o-1H-Tetrazol-5-ylphenyl)benzyl]-N-valeryl-L-valine.
CGP-48933; Diovan; Valsartan. A non-peptide angioten-sin II AT1-receptor antagonist. Used as an anti-hypertensive agent. Crystals; mp = 116-117°. *Ciba-Geigy Corp.*

4053   Vanadium
7440-62-2   9984   G

V

V
Vanadium.
EINECS 231-171-1; HSDB 1022; Vanadium; Vanadium (fume or dust); Vanadium dust; Vanadium, elemental. Metallic element; target material for x-rays, manufacture of alloy steels. Metal; mp = 1917°; $d^{18.7}$ = 6.11. *Atomergic Chemetals; Cerac; Noah Chem.*

4054   Vanadium oxychloride
10213-09-9   9991   G

Cl2OV
Vanadium oxydichloride.
AI3-52349; Dichlorooxovanadium; EINECS 233-517-7; NSC 79526; Vanadium chloride oxide; Vanadium dichloride oxide; Vanadium, dichlorooxo-; Vanadium oxychloride; Vanadyl chloride; Vanadyl dichloride. Strong reducing agent, used in purification of hydrogen chloride from arsenic and as a mordant in printing fabrics. Green crystals; d = 2.88; disproportionates at 384°; soluble in EtOH, AcOH, decomposes with H2O. *Atomergic Chemetals; Noah Chem.*

4055   Vanadium pentoxide
1314-62-1   9987   G

V2O5
Vanadium (V) oxide.
AI3-52159; Anhydride vanadique; C.I. 77938; CCRIS 3206; CI 77938; Divanadium pentaoxide; Divanadium pentoxide; EINECS 215-239-8; HSDB 1024; KM Vanadium Pentoxide; RCRA waste number P120; UN2862; Vanadic acid anhydride; Vanadic anhydride; Vanadin(V) oxide; Vanadio, pentossido di; Vanadium oxide; Vanadium oxide (V2O5); Vanadium pentaoxide; Vanadium pentoxide; Vanadium, pentoxyde de; Vanadiumpentoxid; Vanadiumpentoxyde; Wanadu pieciotlenek. Catalyst for oxidation of sulfur dioxide in sulfuric acid manufacture, ferrovanadium, catalyst for organic reactions, ceramic coloring material, vanadium salts, inhibiting UV transmission in glass, photographic developer, textile dyeing. Yellow-brown crystals; mp = 690°; d = 3.35; soluble in H2O

(0.8 g/100 ml), acids, alkalis. *Atomergic Chemetals; Cerac; Kerr-McGee; Sigma-Aldrich Fine Chem.*

4056   Vancocin hydrochloride
1404-93-9   9995   D G

C66H76Cl3N9O24
Vancomycin hydrochloride.
Lyphocin; Vancor. Peptide antibiotic. Antibacterial agent. White solid; λm 282 nm (E$^{1\%}_{1cm}$ 40 H2O); soluble in H2O (> 10 g/100 ml); moderately soluble in dilute MeOH; less soluble in higher alcohols, Me2CO, Et2O; LD50 (mus iv) = 489 mg/kg, (mus ip) = 1734 mg/kg, (mus sc, orl) = 5000 mg/kg. *Eli Lilly & Co.*

4057   Vanillin
121-33-5   9998   G H

C8H8O3
p-Hydroxy-m-methoxybenzaldehyde.
4-08-00-01763 (Beilstein Handbook Reference); AI3-00093; Benzaldehyde, 4-hydroxy-3-methoxy-; BRN 0472792; CCRIS 2687; EINECS 204-465-2; FEMA No. 3107; HSDB 1027; 4-Hydroxy-3-methoxybenzaldehyde; 4-Hydroxy-m-anisaldehyde; Lioxin; Methylprotocatechu-ic aldehyde; NSC 15351; Protocatechualdehyde 3-methyl ether; Protocatechualdehyde, methyl-; Vanilla; Vanill-aldehyde; Vanillic aldehyde; Vanillin; Vanilline; Zimco. Used in perfumes, flavorings and pharmaceuticals, as a lab reagent, source of L-dopa. Pharmaceutic aid (flavor). Tetragonal crystals; mp = 81.5°; bp = 285°; d^{25} = 1.056; λm = 208, 232, 279, 309 nm (ε = 12023, 14454, 10233, 10471, EtOH); slightly soluble in H2O, soluble in C6H6, ligroin, very soluble in EtOH, Et2O, Me2O, CHCl3, CS2; LD50 (rat orl)= 1580 mg/kg. *Lancaster Synthesis Co.; Penta Mfg.; Trafford Chem. Ltd.*

4058   Variamine
101-64-4   6028   G

C13H14N2O
N-(p-Methoxyphenyl)-p-phenylenediamine.
N-(4-Aminophenyl)-p-anisidine; 1,4-Benzenediamine, N-(4-methoxyphenyl)-; EINECS 202-962-9; N-(p-Methoxy-phenyl)-p-phenylenediamine; Variamine Blue Base. Used in manufacture of azoic dyestuffs. Crystals; mp = 102°; bp12 = 238°. *Hoechst AG.*

4059   Vegetable gelatin
9002-18-0   184   D E G

(C12H18O9)x
Agar.
Agar; Agar (Gelidium spp.); Agar-agar; Agar agar flake; Agar-agar gum; Agaropectin, mixt with agarose; Agarose, mixt with agaropectin; Bengal; Bengal gelatin; Bengal isinglass; CCRIS 16; Ceylon; Ceylon isinglass; Chinese gelatin; Chinese isinglass; Digenea simplex mucilage; EINECS 232-658-1; FEMA No. 2012; Gelose; HSDB 1901; Isinglass, japanese; Japan agar; Japan isinglass; Japanese gelatin; Japanese isinglass; Layor Carang; Macassar gelatin; NCI-C50475; Vegetable gelatin. Suspending agent, used as a pharmaceutical aid. Used in biochemical research. Used medicinally as a laxative-cathartic. FDA, FEMA GRAS, Europe listed, UK approved, USP/NF, BP, Ph.Eur. compliance. Used as a protective colloid, stabilizer, carrier, emulsifier, emollient for orals, slow release capsules, suppositories, surgical lubricants, emulsions; excipient, disintegrant in tablets, laxative, bulking agent, suspending agent for barium sulfate. Transparent, odorless, fine powder; insoluble in cold H2O; slowly soluble in hot H2O to a viscid solution.

A 1% solution forms a stiff jelly on cooling. LD50 (rat orl) = 11,000 mg/kg. *Alcoa Ind. Chem.; Am. Roland; Brown & Dureau Intl.; Calaga Food Ingredients; Charkit; Chart; Diamalt; Frutarom; Gumix Intl.; Hatrick A. C.; Hercules Inc.; Honeywell & Stein; Meer; MLG Enterprises; Quest USA; Roeper C. E. GmbH; Ruger; Schweizerhall Inc.; Sigma-Aldrich Fine Chem.; Spectrum Chem. Manufacturing; Spice King; TIC Gums; U.S. BioChem.*

4060   Vegetable oil fatty acids
61788-66-7
                                                          H

Fatty acids, vegetable-oil.
Acidulated vegetable oil soapstock; EINECS 262-994-4; Fatty acids, vegetable-oil; Mixed vegetable oil acids; Vegetable oil fatty acids.

4061   Vegetable rouge
36338-96-2   1882   G

C43H42O22
Carthamin.
Carthamin; Carthamine; EINECS 252-981-1; 6-β-D-Glucopyranosyl-2-[[3-β-D-glucopyranosyl-2,3,4-tri-hydroxy-5-[3-(4-hydroxyphenyl)-1-oxo-2-propenyl]-6-oxo-1,4-cyclohexadien-1-yl]methylene]-5,6-dihydroxy-4-[3-(4-hydroxyphenyl)-1-oxo-2-propenyl]-4-cyclohexene-1,3-dione; cathamic acid; safflor carmine; safflor red; C.I. Natural Red 26; C.I. 75140. Carthamin, the coloring matter of *Carthamus tinctorius* mixed with French chalk; used as a cosmetic.

4062   Venlafaxine
93413-69-5   10008   D

C17H27NO2
(±)-1-[α-[(Dimethylamino)methyl]-p-methoxybenzyl]-cyclohexanol.

Elafax; Venlafaxina; Venlafaxine; Venlafaxinum; Venla-fexine. A serotonin noradrenaline reuptake inhibitor. Used as an antidepressant. Crystals; mp = 102-104°; [(-)-form)]: $[\alpha]_D^{25}$ = -27.1° (c = 1.04 in 95% EtOH); [(+)-form]: $[\alpha]_D^{25}$ = +27.6° (c = 1.07 in 95% EtOH). *Am. Home Products; Wyeth-Ayerst Labs.*

4063   Venlafaxine hydrochloride
99300-78-4   10008   D

C17H28ClNO2
(±)-1-[α-[(Dimethylamino)methyl]-p-methoxybenzyl]-cyclohexanol hydrochloride.
Effexor; Effexor XR; HSDB 6699; Venlafaxine hydro-chloride; Wy-45,030; Wy-45,651; Wy-45,655. A serotonin noradrenaline reuptake inhibitor. Antidepress-ant. Crystals; mp = 215-217°; soluble in H2O (57.2 g/100 ml); [(+) or (-) form]: mp = 240-240.5°; [(-)-form]: $[\alpha]_D^{25}$ = +4.6° (c = 1.0 in EtOH); [(+)-form]: $[\alpha]_D^{25}$ = -4.7° (c = 0.945 in EtOH). *Am. Home Products; Wyeth-Ayerst Labs.*

4064   Verapamil
52-53-9   10012   D

C27H38N2O4
5-[(3,4-Dimethoxyphenethyl)methylamino]-2-(3,4-di-methoxyphenyl)-2-isopropylvaleronitrile.
CCRIS 6749; CP-16533-1; D-365; Dilacoran; EINECS 200-145-1; Iproveratril; Isoptimo; Vasolan; Verapamil; Verapamilo; Verapamilum. Antianginal. class IV antiarrhythmic agent. Coronary vasodilator with calcium channel blocking activity. Oil; bp0.001 = 243-246°; insoluble in H2O; slightly soluble in C6H6, C6H16, Et2O; soluble in EtOH, MeOH, Me2CO, EtOAc, CHCl3. *Bristol-Myers Squibb Co.*

4065   Verapamil hydrochloride
152-11-4   10012   D

C27H39ClN2O4
5-[(3,4-Dimethoxyphenethyl)methylamino]-2-(3,4-dimethoxyphenyl)-2-isopropylvaleronitrile hydrochloride.
Akilen; Anpec; Apo-Verap; Arapamyl; Arpamyl LP; Berkatens;

Calan; Calan SR; Calaptin; Calaptin 240 SR; Calcan hydrochloride; Cardiabeltin; Cardiagutt; Cardibeltin; Cardioprotect; Caveril; Civicor; Civicor Retard; Coraver; Cordilox; Cordilox SR; Corpamil; Covera-HS; D-365 hydrochloride; Dignover; Dilacoran HTA; Drosteakard; Durasoptin; EINECS 205-800-5; Elthon; Falicard; Finoptin; Flamon; Geangin; Harteze; Hexasoptin; Hexasoptin Retard; Hormitol; HSDB 3928; Ikacor; Ikapress; Inselon; Iproveratril hydrochloride; Isoptin; Isoptin Retard; Isoptin SR; Isoptine; Isoptino; Izoptin; Jenapamil; Lekoptin; Lodixal; LU 20175; Magotiron; Manidon; Manidon Retard; Novapamyl LP; Novo-Veramil; NSC 272366; Nu-Verap; Ormil; Praecicor; Quasar; Rapam; Robatelan; Securon; Tarka; Univer; Univex; Vasolan; Vasomil; Vasopten; Vera-Sanorania; Verabeta; Veracaps SR; Veracim; Vera-cor; Verahexal; Veraloc; Veramex; Veramil; Verapamil-AbZ; Verapamil Acis; Verapamil AL; Verapamil Atid; Verapamil Basics; Verapamil chloridrate; Verapamil Ebewe; Verapamil Henning; Verapamil hydrochloride; Verapamil Injection; Verapamil MSD; Verapamil NM; Verapamil NM Pharma; Verapamil Nordic; Verapamil PB; Verapamil Riker; Verapamil SR; Verapamil Verla; Verapin; Verapress 240 SR; Veraptin; Verasal; Verasifar; Veratensin; Verdilac; Verelan; Verelan PM; Verelan SR; Verexamil; Veroptinstada; Verpamil; Vetrimil; Vortac. Antianginal; class IV antiarrhythmic agent. Coronary vasodilator with calcium channel blocking activity. Crystals; mp = 138.5-140.5°; soluble in H_2O (7 g/100 g, 83 mg/ml), EtOH (26 mg/ml), propylene glycol (93 mg/ml), MeOH (> 100 mg/ml), iPrOH (4.6 mg/ml), EtOAc (1.0 mg/ml), DMF (> 100 mg/ml), CH_2Cl_2 (> 100 mg/ml), C_6H_{14} (0.001 mg/ml); LD_{50} (rat iv) = 16 mg/kg, (mus iv) = 8 mg/kg. *Knoll Pharm. Co.; Lederle Labs.; Parke-Davis; Searle G.D. & Co.*

4066 Veratrole

91-16-7   10018   G

$C_8H_{10}O_2$
1,3-dimethoxybenzene.
pyrocatechol   dimethylether;   1,2-dimethoxybenzene;   o-dimethoxybenzene. Used medically as an antiseptic. Solid; mp = 22.5°; bp = 206°; d^{25} = 1.0810; λ_m = 225, 275 nm (ε = 11700, 3920, MeOH); soluble in EtOH, Et_2O, CCl_4, poorly soluble in H_2O; LD_{50} (rat orl) = 1360 mg/kg, (mus orl) = 2020 mg/kg. *Alfa Aesar; Garuda Chemicals; Krupa Scientific.*

4067 Videne

25655-41-8   7784   G

$(C_6H_9NO)_n.xI$
1-ethenyl-2-pyrrolidinone polymers complexed with iodine.
Argentyne; Betadine; Betaisodona; Braunol; Braunosan H; Bridine; Disadine D.P.; Disphex; Efo-dine; 1-Ethenyl-2-pyrrolidinone homopolymer compound with iodine; HSDB 6831; Inadine; Iodinated poly(vinylpyrrolidone); Iodine-poly(vinylpyrrolidinone); Iodopoly(vinyl pyrrolid-inone); Isobetadyne; Isodine; NSC 26245; Poly(1-(2-oxo-1-pyrrolidinyl)ethylene)iodine complex; Poly(vinylpyrrol-idinone) iodide; Povadyne; Povidone-iodine; Proviodine; Pvp-1; PVP-I; PVP iodine; PVP-Iodine, 30-06; 1-Vinyl-2-pyrrolidinone polymer, compound with iodine; 1-Vinyl-2-pyrrolidinone polymers, iodine complex; 2-Pyrrol-idinone, 1-ethenyl-, homopolymer, compd. with iodine; 2-Pyrrolidinone, 1-vinyl-, polymers, compd. with iodine (8CI); Traumasept; Ultradine; Videne Disinfectant Solution, Videne Disinfectant Tincture, Videne Powder. Pre-operative skin antiseptics containing povidone-iodine USP. Soluble in H_2O, EtOH, $CHCl_3$, insoluble in Et_2O. *3M Pharm.*

4068 Vinaconic acid

598-10-7   G

$C_5H_6O_4$
Propene-3,3-dicarboxylic acid.
Cyclopropane-1,1-dicarboxylic   acid;   1,1-Cyclopropane-dicarboxylate; EINECS 209-917-2. Crystals; mp = 135-140°.

4069 Vinclozolin

50471-44-8   10046   G P

$C_{12}H_9Cl_2NO_3$
3-(3,5-Dichlorophenyl)-5-ethenyl-5-methyl-2,4-oxazolid-inedione.
BAS 352 F; BAS 35204F; BRN 1080192; Caswell No. 323C; Dichlorophenyl)-5-ethenyl-5-methyl-2,4-oxazol-idinedione;   (±)-3-(3,5-dichlorophenyl)-5-ethenyl-5-methyl-2,4-oxazolidinedione; 3-(3,5-Dichlorophenyl)-5-methyl-5-vinyl-2,4-oxazolidinedione;   3-(3,5-Dichloro-phenyl)-5-methyl-5-vinyl-1,3-oxazolidine-2,4-dione;   3-(3,5-Dichlorphenyl)-5-methyl-5-vinyl-1,3-oxazolidin-2,4-dion;   N-3,5-Dichlorophenyl-5-methyl-5-vinyl-1,3-oxaz-olidine-2,4-dione;   EPA Pesticide   Chemical   Code   113201;   HSDB   6747;   Ornalin; Oxazolidinedione, 3-(3,5-dichlorophenyl)-5-ethenyl-5-methyl-;   2,4-Oxazolidine-dione, 3-(3,5-dichlorophenyl)-5-ethenyl-5-methyl-;   2,4-Oxazolidinedione, 3-(3,5-dichlorophenyl)-5-methyl-5-vinyl-; Ronilan; Vinchlozoline;   Vinclozalin;   Vinclozolin;   Vinclozoline;   Vorlan. Registered by EPA as a fungicide. Crystals; mp = 108°; bp$_{0.05}$ = 131°; soluble in H_2O (0.1 g/100 ml at 20°); Me_2CO (34.3 g/100 ml), EtOH (1.1 g/100 ml), C_6H_6 (12.8 g/100 ml), $CHCl_3$ (47.2 g/100 ml), EtOAc (22.8 g/100 ml), cyclohexane (0.7 g/100 ml), Et_2O (4.2 g/100 ml), xylene (9.5 g/100 ml), cyclohexanone (50.8 g/100 ml); LD_{50} (rat orl) > 10000 mg/kg, (gpg orl) = 8000 mg/kg, (rat der) > 2500 mg/kg; LC_{50} (rat ihl 4 hr.) > 29.1 mg/l air, (trout 96 hr.) = 52.5 mg/l, (guppy 96 hr.) = 130 mg/l; non-toxic to bees. *BASF Corp.; Mallinckrodt Inc.*

4070 Vinyl acetate

108-05-4   10053   G H

$C_4H_6O_2$
1-Acetoxyethylene.
Acetate de vinyle; Acetic acid ethenyl ester; Acetic acid, vinyl ester; Acetoxyethene; 1-Acetoxyethylene; AI3-18437; CCRIS 1306; EINECS 203-545-4; Ethanoic acid, ethenyl ester; Ethenyl acetate; Ethenyl ethanoate; Everflex® 81L; HSDB 190; NSC 8404; Octan winylu; UN1301; Vinile (acetato di); Vinyl A monomer; Vinyl acetate; Vinyl ethanoate; Vinylacetaat; Vinylacetat; Vinylacetate; Vinyle (acetate de); Vinylester kyseliny octove; Zeset T. A paint and coating emulsion used as a binder for clay coating of paper and paperboard; factory finishes, ceiling tile, wall board, and textile treatments. Intermediate for industrial and consumer products including polyvinyl acetate, used in paints, adhesives and coatings, polyvinyl alcohol,

used in adhesives, coatings and packaging films, polyvinyl acetals, used in insulation and as a safety glass interlayer and ethylene vinyl acetate copolymers, used in films, coatings and adhesives. Liquid; mp = -93.2°; bp = 72.5°; d= 0.932; λ_m = 258 nm (C_6H_{14}); insoluble in H_2O, soluble in Et_2O, Me_2CO, C_6H_6, CCl_4, $CHCl_3$, freely soluble in EtOH; LD_{50} (rat orl) = 2.92 g/kg. *Exxon; Grace W.R. & Co.; Hoechst Celanese; Quantum/USI; Reichold; Union Carbide Corp.; Wacker Chemie GmbH.*

4071   Vinyl amyl carbinol
3391-86-4   G P

$C_8H_{16}O$
1-Octene-3-ol.
AI3-28627; Amyl vinyl carbinol; Amylvinylcarbinol; BRN 1744110; EINECS 222-226-0; EPA Pesticide Chemical Code 069037; FEMA No. 2805; 3-Hydroxy-1-octene; Matsuica alcohol; Matsutake alcohol; Morillol®; NSC 87563; Oct-1-ene-3-ol; 1-Octen-3-ol; 3-Octenol; 1-Okten-3-ol; Pentyl vinyl carbinol; Pentylvinylcarbinol; 1-Vinylhexanol. Oil with herbaceous and mushroom odor. Used as a fragrance and flavoring. Registered by EPA as an insecticide. Liquid; bp = 174°; d = 0.8300; insoluble in H_2O, soluble in organic solvents. *Aberdeen Road Co.; Armatron International, Inc.; BASF Corp.; Biosensory, Inc.*

4072   Vinyl bromide
593-60-2   G

C_2H_3Br
Monobromoethylene.
4-01-00-00718 (Beilstein Handbook Reference); BRN 1361370; Bromoethene; Bromoethylene; Bromure de vinyle; CCRIS 620; EINECS 209-800-6; Ethene, bromo-; Ethylene, bromo-; HSDB 1030; NCI-C50373; Saytex® VBR; UN1085; Vinile (bromuro di); Vinyl bromide; Vinylbromid; Vinyle (bromure de). Flame retardant; intermediate in organic synthesis in the manufacturing. of flame retardants, polymers, copolymers, pharmaceuticals, fumigants, and other chemicals; also used in textiles, adhesives, coating, photographic plates and films. Gas; mp = -137.8°; bp = 15.8°; d^{20} = 1.4933; insoluble in H_2O, soluble in EtOH, Et_2O, Me_2CO, C_6H_6, $CHCl_3$. *Ethyl Corp.*

4073   Vinyl chloride
75-01-4   10055   H

C_2H_3Cl
Ethene, chloro- .
4-01-00-00700 (Beilstein Handbook Reference); BRN 1731576; CCRIS 621; Chlorethene; Chlorethylene; Chloroethene; Chloroethylene; Chlorure de vinyle; Cloruro di vinile; EINECS 200-831-0; Ethene, chloro-; Ethylene, chloro-; Ethylene monochloride; HSDB 169; Monochloroethene; Monochloroethylene; Monovinyl chloride; RCRA waste number U043; Trovidur; UN 1086; VC; VCM; Vinile (cloruro di); Vinyl C monomer; Vinyl chloride; Vinylchlorid; Vinylchloride; Vinyle(chlorure de); Winylu chlorek. Used as a monomer, refrigerant and in chemical synthesis. Gas; mp = -153.7°, bp = -13.3°; poorly soluble in H_2O, more soluble in organic solvents; LD_{50} (mus ihl) = 293.75 mg/l, (rat ihl) = 390 mg/l, (gpg ihl) = 595 mg/l, (rbt ihl) = 295 mg/l. *BOC Gases; Dow Chem. U.S.A.; Dow UK;*

EniChem Am.

4074   Vinylcyclohexene tetrabromide
3322-93-8   G

$C_8H_{12}Br_4$
1,2-Dibromo-4-(1,2-dibromoethyl)cyclohexane.
3-05-00-00093 (Beilstein Handbook Reference); BRN 1927455; CCRIS 3743; Citex BCL 462; Cyclohexane, 1,2-dibromo-4-(1,2-dibromoethyl)-; 1,2-Dibromo-4-(1,2-di-bromoethyl)cyclohexane; 1-(1,2-Dibromoethyl)-3,4-di-bromocyclohexane; 4-(1,2-Dibromoethyl)-1,2-dibromo-cyclohexane; EINECS 222-036-8; HSDB 6146; Saytex® BCL 462. Flame retardant for expandable, crystalline and high-impact PS, SAN resins, adhesives, coatings, textile treatment, PU. Solid; mp = 68-90°; insoluble in H_2O (<0.1 g/100 ml). *Ethyl Corp.*

4075   4-Vinylcyclohexene
100-40-3   H

C_8H_{12}
Cyclohexene, 4-ethenyl-.
4-05-00-00406 (Beilstein Handbook Reference); AI3-08499; BRN 1901553; Butadiene dimer; CCRIS 1422; Cyclohexene, 4-ethenyl-; Cyclohexene, 4-vinyl-; Cyclohexenylethylene; EINECS 202-848-9; Ethenyl-1-cyclohexene; HSDB 2872; NCI-C54999; NSC 15760; 1,2,3,4-Tetrahydrostyrene; Vinylcyclohexene; 4-Vinyl-cyclohexene; 4-Vinylcyclohexene-1. Liquid; mp = -108.9°; bp = 128°; d^{20} = 0.8299; insoluble in H_2O, soluble in Et_2O, C_6H_6, petroleum ether.

4076   Vinyl fluoride
75-02-5   H

C_2H_5F
Ethene, fluoro- .
4-01-00-00694 (Beilstein Handbook Reference); BRN 1731574; CCRIS 7213; EINECS 200-832-6; Ethene, fluoro-; Ethylene, fluoro-; Fluoroethene; Fluoroethylene; HSDB 807; Monofluoroethene; Monofluoroethylene; UN1860; Vinyl fluoride . Used as a monomer and in chemical synthesis. Gas; mp = -160.5°; bp = -72°; slightly soluble in H_2O, soluble in organic solvents. *E. I. DuPont de Nemours Inc.; Lancaster Synthesis Co.; Sigma-Aldrich Fine Chem.*

4077   Vinyl pyridine
1337-81-1   G

C_7H_7N
Pyridine, ethenyl-.
Ethenylpyridine; Pyridine, ethenyl; Pyridine, vinyl-; Vinylpyridine. Mixture of α-, β- and γ-isomers. Used for adhesives, tire cord, and

industrial goods dips. *Penta Mfg.; Raschig GmbH; Reilly Ind.*

4078   2-Vinylpyridine
100-69-6   H

C7H7N
Pyridine, 2-ethenyl-.
Al3-24116; CCRIS 5238; EINECS 202-879-8; 2-Ethenylpyridine; HSDB 1508; NSC 18255; Pyridine, 2-ethenyl-; Pyridine, 2-vinyl-; 2-Vinylpyridine; 2VP. Oil; bp = 159.5°; d^{20} = 0.9983; λ_m = 234, 278 nm (MeOH); slightly soluble in H2O, very soluble in EtOH, Et2O, Me2CO, CHCl3.

4079   N-Vinyl-2-pyrrolidinone
88-12-0   7783   H

C6H9NO
2-Pyrrolidinone, 1-ethenyl- .
5-21-06-00330 (Beilstein Handbook Reference); BRN 0110513; EINECS 201-800-4; NSC 10222; V-Pyrol; Vinyl-2-pyrrolidone; Vinylbutyrolactam; Vinylpyrrolidin-one; N-Vinyl-2-pyrrolidinone; Vinylpyrrolidone; N-Vinyl-2-pyrrolidone; N-Vinyl pyrrolidone; N-Vinylpyrrolidin-one. Liquid; mp = 13.5°; bp_{400} = 193°, bp_{11} = 93°; d^{20} = 1.04.

4080   Vinyl toluene
25013-15-4   H

C9H10
3- and 4-Vinyl toluene (mixed isomers).
4-05-00-01369 (Beilstein Handbook Reference); ar-Methylstyrene; Benzene, ethenylmethyl-; BRN 1209317; CCRIS 2369; EINECS 246-562-2; Ethenylmethylbenzene; HSDB 1035; Methylethenylbenzene; Methylstyrene; Methylvinylbenzene; NCI-C56406; NSC 4832; Styrene, ar-methyl-; Styrene, methyl-; Styrene, methyl- (mixed isomers); α,β-Styrene; Toluene, vinyl- (mixed isomers); Tolylethylene; UN2618; Vinyl toluene; 3- and 4-Vinyl toluene (mixed isomers); Vinyl toluene; Vinyltoluene; Vinyltoluene, industrial; Vinyltoluenes, inhibited.

4081   Vinyl trichloride
79-00-5   9711   H

C2H3Cl3
1,1,2-Trichloroethane.
4-01-00-00139 (Beilstein Handbook Reference); BRN 1731726; Caswell No. 875A; CCRIS 602; EINECS 201-166-9; EPA Pesticide Chemical Code 081203; Ethane, 1,1,2-trichloro-; HSDB 1412; NCI-C04579; NSC 405074; RCRA waste number U359; RCRA waste number U227; Trojchloroetan(1,1,2); Vinyl trichloride. Used as a solvent for fats and waxes. Liquid; mp = -36.6°; bp = 113.8°; d^{20} = 1.4397; insoluble in H2O, moderately soluble in organic solvents; LD50 (rat orl) = 835 mg/kg. *Mallinckrodt Inc.; Sigma-Aldrich Fine Chem.*

4082   Vinyltriethoxysilane
78-08-0   G

C8H18O3Si
(Triethoxysilyl)ethylene.
4-04-00-04256 (Beilstein Handbook Reference); Al3-51468; BRN 1767229; CCRIS 2645; CV-4910; EINECS 201-081-7; KBE 1003; NV 1107; Polyscience VTES; Silane A 151; Silane, ethenyltriethoxy-; Silane, triethoxy-vinyl-; Silane, vinyl triethoxy 1-151; Triethoxy(vinyl)sil-ane; Triethoxyvinyl silane; Triethoxyvinylsilicane; Union Carbide A-151; Vinyltriethoxysilane; VTES; VTS-E. Intermediate, especially when acidic by-products are undesirable; filler. bp = 160-161°, bp_{20} = 62°; d^{20} = 0.901; soluble in CHCl3. *Degussa-Hüls Corp.; PCR; Union Carbide Corp.*

4083   Vinyltrimethoxysilane
2768-02-7   G

C5H12O3Si
(Trimethoxysilyl)ethene.
4-04-00-04256 (Beilstein Handbook Reference); A 171; BRN 1099136; CV-4917; EINECS 220-449-8; Ethenyl-trimethoxysilane; KBM 1003; Silane, ethenyltrimethoxy-; Silane, trimethoxyvinyl-; SZ 6300; Trimethoxyvinylsilane; Union Carbide® A-171; V 4917; Vinyl trimethoxy silane; Vinyltrimethoxysilane; VTS-M; Y 4302. Coupling agent, chemical intermediate, blocking agent, release agent, lubricant, primer, reducing agent. Used as a crosslinking agent providing gloss, hiding power to coatings. Liquid; bp = 123°; d = 1.1300; LD50 (rat orl) = 11300 mg/kg. *Degussa-Hüls Corp.; Union Carbide Corp.*

4084   Vinyl tris(trimethylsiloxy) silane
5356-84-3   G

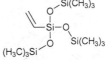

C11H30O3Si4
1,1,1,5,5,5-Hexamethyl-3-((trimethylsilyl)oxy)-3-vinyltrisiloxane.
CV5100; EINECS 226-342-2; Trisiloxane, 3-ethenyl-1,1,1,5,5,5-hexamethyl-3-((trimethylsilyl)oxy)-. Coupling agent, chemical intermediate, blocking agent, release agent, lubricant, primer, reducing agent. *Degussa-Hüls Corp.*

4085 Violuric acid

87-39-8

G

C4H3N3O4

2,4,5,6(1H,3H)-Pyrimidinetetrone-5-oxime.
Alloxan 5-oxime; EINECS 201-741-4; 5-Hydroxyimino-barbituric acid; 5-Isonitrosobarbituric acid; NSC 56338; 2,4,5,6(1H,3H)-Pyrimidinetetrone, 5-oxime; Violuric acid. Crystals; mp = 240-241° (dec); insoluble in H2O, soluble in EtOH.

4086 Vitamin A

68-26-8

10073

D G

C20H30O

3,7-Dimethyl-9-(2,6,6,-trimethyl-1-cyclohexen-1-yl)-2,4,   6,8-nonatetraen-1-ol.
4-06-00-04133 (Beilstein Handbook Reference); A-Mulsal; A-Sol; A-Vi-Pel; A-Vitan; Acon; Afaxin; Agiolan; Agoncal; Alcovit A; Alphalin; Alphasterol; Anatola; Anat-ola A; Anti-infective vitamin; Antixerophthalmic vitamin; Antixerophthalmisches Vitamin; Aoral; Apexol; Apostavit; Aquasol A; Aquasynth; Atars; Atav; Avibon; Avita; Avitol; Axerol; Axerophthol; Axerophtholum; Bentavit A; Biosterol; BRN 0403040; CCRIS 5444; Chocola A; Del-VI-A; Disatabs Tabs; Dofsol; Dohyfral A; EINECS 200-683-7; Epiteliol; HI-A-Vita; Homagenets Aoral; HSDB 815; Lard factor; Myvpack; Nio-A-let; Nonatetraen-1-ol; NSC 122759; Oleovitamin A; Ophthalamin; Plivit A; Prepalin; Retinol; Retinol, all trans-; all-trans-Retinol; Retinolo; Retinolum; all-trans-Retinyl alcohol; Retro-vitamin A; Sehkraft A; Solu-A; Super A; Testavol; Testavol S; Vaflol; Vafol; Veroftal; Vi-α; Vi-Dom-A; VI-alpha; Vio-A; Vitamin A; all-trans-Vitamin A; all-trans-Vitamin A alcohol; Vitamin A alcohol, all-trans-; Vitamin A1; Vit-amin A1 alcohol; Vitamine A; Vitaminum A; Vitavel-A; Vitpex; Vogan; Vogan-Neu; Wachstumsvitamin. Vitamin, vitamin source. An antixerophthalmic agent with hematopoietic properties. Crystals; mp = 62-64°; bp0.005 = 120-125°; λm 324-325 nm (E$_{1cm}^{1\%}$ 1835; insoluble in H2O, glycerol; soluble in EtOH, MeOH, CHCl3, fats and oils; LD50 (10 day) (mus ip) = 1510 mg/kg, (mus orl) = 2570 mg/kg. Abbott Laboratories Inc.; Astra Chem. Ltd.; BASF Corp.; Eli Lilly & Co.; Parke-Davis; Purdue Pharma L.P.

4087 Vitamin B1

67-03-8

9366

D G

C12H17N4OSCI·HCl

Thiamine hydrochloride .
Betabion; Betabion hydrochloride; Betalin S; Betaxin; AI3-18993; Aneurine hydrochloride; Apate drops; Beat-ine; Bedome; Begiolan; Benerva; Bequin; Berin; Bethia-zine; Beuion; Bevitex; Bevitine; Bewon; Biamine; Bithia-min; Biuno; Bivatin; Bivita; CCRIS 1906;

Chloride-hydro-chloride salt of thiamine; Clotiamina; EINECS 200-641-8; Eskapen; Eskaphen; FEMA No. 3322; Hybee; Lixa-β; Metabolin; NSC 36226; Slowten; THD; Thiadoxine; Thiamin chloride; Thiamin dichloride; Thiamin hydro-chloride; Thiaminal; Thiamine chloride hydrochloride; Thiamine dichloride; Thiamine monohydrochloride; Thia-minium chloride; Thiaminium chloride hydrochloride; Thiamol; Thiavit; Tiamidon; Tiaminal; Troph-Iron; Troph-ite; Trophite + Iron; USAF CB-20; Vetalin S; Vinothiam; Vitamin B hydrochloride; Vitamin B1 hydrochloride; Vita-neuron. Essential vitamin. Used in medicine, as a nutrient, in enriched flours. Crystals; mp = 248° (dec); λm = 240, 265 nm (MeOH); LD50 (mus iv) = 89.2 mg/kg, (mus orl) = 8224 mg/kg. BASF Corp.; EM Ind. Inc.; Hoffmann-LaRoche Inc.; Roche Vitamins Inc.; Sigma-Aldrich Fine Chem.; Takeda Chem. Ind. Ltd.

4088 Vitamin B1

59-43-8

9366

D G

C12H17N4OS

3-[(4-Amino-2-methyl-5-pyrimidinyl)methyl]-5-(2-hydroxyethyl)-4-methylthiazolium chloride monohydro-chloride.
Aneurin; Aneurine; Apatate drape; B-Amin; Beivon; Beta-bion; Bethiamin; CCRIS 5823; EINECS 200-425-3; HSDB 220; Oryzanin; Oryzanine; Thiamin; Thiamine; Thia-mine, chloride; Thiaminium chloride; Tiamina; Vitaneur-in. Vitamin (enzyme cofactor). Used in medicine, as a nutrient, in enriched flours; available as thiamine hydrochloride and thiamine mononitrate. Hoffmann-LaRoche Inc.; Hoffmann-LaRoche Ltd.; Honeywell & Stein; Parke-Davis; Roche Vitamins Inc.

4089 Vitamin B12

68-19-9

10074

D G

C63H88CoN14O14P

Cyanocobalamin.
Anacobin; B-12; B-Twelve; B-Twelve Ora; Berubigen; Betalin-12; Betolvex; Bevatine-12; Bevidox; Biocobal-amine; BRN 4122889; Byladoce; Cabadon M; CCRIS 3955; Chromagen; Cianocobalamina; CN-B12; Cobadoce forte; Cobalamin, cyanide; Cobalin; Cobamin; Cobavite; Cobione; Coobalamed; Copharvit 5000; Cosmo-Rey; Cotel; Covit; Crystamin; Crystamine; Crystimin; Crystimin-1000; Crystwel; Cyano-B12; Cyanocobalamin; Cyanocobalamine; Cyanocobalaminum; Cycobemin; Cycolamin; Cykobemin; Cykobeminet; Cynobal; Cyomin; Cyredin; Cytacon; Cytamen; Cytobion; Depinar; Dicopac Kit; Distivit (B12 peptide); Dobetin; Docémine; Docibin; Docigram; Docivit; Dodecabee; Dodecavite; Dodex; Ducobee; Duodecibin; EINECS 200-680-0; Embiol; Emociclina; Eritrone; Erycytol; Erythrotin; Euhaemon; Extrinsic factor; Factor II (vitamin B12); Fresmin; Geriplex; Hemo-B-doze;

Hemomin; Hepacon-B12; Hepagon; Hepavis; Hepcovite; HSDB 2850; Hylugel plus; Lactobacillus lactis dorner factor; LLD factor; Macrabin; Magravon; Megalovel; Milbedoce; Millevit; Nagravon; Nascobal; Normocytin; Novidroxin; NSC 80365; Pernaemon; Pernaevit; Pernipuron; Plecyamin; Poyamin; Rebramin; Redamina; Redisol; Rhodacryst; Rubesol; Rubramin; Rubramin PC; Rubripca; Rubrocitol; Sytobex; Tolfrinic; Troph-Iron; Trophite; Trophite + Iron; Vi-Twel; Vibalt; Vibisone; Virubra; Vita-rubra; Vitamin B₁₂; Vitamin B12; Vitarubin; Vitral; component of: Chromagen, Geriplex, Spondylonal, Tolfrinic, Troph-Iron, Trophite, Trophite+Iron. Vitamin, vitamin source. Used as a hematopoietic agent. Dietary supplement; deficiency in man causes pernicious anemia and neural degeneration. Red crystals; mp > 300°; $[\alpha]_{656}^{23}$ = -59° ± 9° (dilute aqueous solution); λ_m = 278, 361, 550 nm (A$_{1cm}^{1\%}$ 115, 204, 64 H₂O). *Amersham Corp.; Ascher B.F. & Co.; Eli Lilly & Co.; Elkins-Sinn; Forest Pharm. Inc.; Lederle Labs.; Marion Merrell Dow Inc.; Menley & James Labs Inc.; Merck & Co.Inc.; Parke-Davis; Roerig Div. Pfizer Pharm.; Savage Labs; Wyeth-Ayerst Labs.*

4090 Vitamin B₂
83-88-5   8284   G

C₁₇H₂₀N₄O₆
7,8-dimethyl-10-(D-*ribo*-2,3,4,5-tetrahydroxypentyl)iso-alloxazine.
AI3-14697; Aqua-Flave; Beflavin; Beflavine; CCRIS 1904; Dermadram; 6,7-Dimethyl-9-D-ribitylisoalloxazine; 7,8-Dimethyl-10-ribitylisoalloxazine; 7,8-Dimethyl-10-(D-ribo-2,3,4,5-tetrahydroxypentyl)isoalloxazine; EINECS 201-507-1; Fiboflavin; Flavaxin; Flavin BB; Flaxain; HSDB 817; Hyflavin; HYRE; Isoalloxazine, 7,8-dimethyl-10-(D-ribo-2,3,4,5-tetrahydroxypentyl)-; Isoalloxazine, 7,8-dimethyl-10-D-ribityl-; Lactobene; Lactoflavin; Lacto-flavine; NSC 33298; Ribipca; D-Ribitol, 1-deoxy-1-(3,4-dihydro-7,8-dimethyl-2,4-; Ribocrisina; Riboderm; Ribo-flavin; Riboflavina; Riboflavine; Riboflavinequinone; Riboflavinum; Ribosyn; Ribotone; Ribovel; Russupteridine Yellow III; Vitaflavine; Vitamin B2; Vitamin Bi; Vitamin G; Vitasan B2. Organic compound; crystalline pigment; dietary supplement; principal growth-promoting factor of vitamin B2 complex (functions as flavor protein in tissue respiration). An essential dietary factor. Yellow crystals; mp = 280° (dec); $[\alpha]_D^{25}$ = -117° (H₂O); λ_m = 223, 268, 373, 445 nm (ε = 31623, 31623, 10000, 10000 H₂O); slightly soluble in H₂O, organic solvents; LD₅₀ (rat orl > 10 g/kg. *Am. Biorganics; BASF Corp.; Bio-Rad Labs.; EM Ind. Inc.; Hoffmann-LaRoche Inc.; Honeywell & Stein; Roche Vitamins Inc.; Takeda Chem. Ind. Ltd.*

4091 Vitamin B₆ hydrochloride
58-56-0   8072   D G

C₈H₁₁NO₃.HCl
3-Hydroxy-4,5-dihydroxymethyl-2-methylpyridine hydrochloride.
Adermin hydrochloride; Aderomine hydrochloride; Ader-oxin; Aderoxine; AI3-19016; Alestrol; Becilan; Bécilan; Beesix; Benadon; Bendectin; 4,5-Bis(hydroxymethyl)-2-methylpyridin-3-ol hydrochloride; Bonadoxin; Bonasanit; Campoviton 6; CCRIS 1903; DRG-0125; EINECS 200-386-2; Godabion; Gravidox; Hexa-betalin; Hexabione hydrochloride; Hexavibex; Hexermin; Hexobion; HSDB 1212; Hydoxin; 5-Hydroxy-6-methyl-3,4-pyridine-dicarbinol hydrochloride; NSC 36225; Paxa-don; Pydox; Pyridipca; 3,4-Pyridinedimethanol, 5-hydr-oxy-6-methyl-, hydrochloride; Pyridox; Pyridoxin hydrochloride; Pyrid-oxine chloride; Pyridoxine hydrochloride; Pyridoxinium chloride; Pyridoxinum hydrochloricum; Pyridoxol, hydrochloride; Rodex; Rodex (R); Spondylonal. Vitamin, nutritional factor. An essential dietary factor. Used in medicine and nutrition. Crystals; mp = 214° (dec); λ_m = 290 nm (ε 8400 0.1N HCl); soluble in H₂O (0.22 g/ml), less soluble in organic solvents. *Napp Labs Ltd; Sigma-Aldrich Fine Chem.*

4092 Vitamin C
50-81-7   856   D G H

C₆H₈O₆
L-Ascorbic acid.
Acide ascorbique; Acido ascorbico; Acidum ascorbicum; Acidum ascorbinicum; Adenex; Allercorb; Antiscorbic vitamin; Antiscorbutic vitamin; Arco-cee; Ascoltin; Ascor-B.I.D.; Ascorb; Ascorbajen; Ascorbate; L-Ascorbate; Ascorbic acid; L-Ascorbic acid; Ascorbicap; Ascorbutina; Ascorin; Ascorteal; Ascorvit; C-Level; C-Long; C-Quin; C-Span; C-Vimin; Cantan; Cantaxin; Caswell No. 061B; Catavin C; CCRIS 57; ce lent; Ce-Mi-Lin; Ce-Vi-Sol; CE-VI-Sol; Cebicure; Cebid; Cebion; Cebione; Cecon; Cee-caps TD; Cee-vite; Cegiolan; Ceglion; Celaskon; Celin; Cemagyl; Cemill; Cenetone; Cenolate; Cereon; Cergona; Cescorbat; Cetamid; Cetane; Cetane-caps TD; Cetane-Caps TC; Cetebe; Cetemican; Cevalin; Cevatine; Cevex; Cevi-bid; Cevimin; Cevital; Cevitamic acid; Cevitamin; Cevitan; Cevitex; Cewin; Chromagen; Ciamin; Cipca; Citriscorb; Colascor; Concemin; Cortalex; Davitamon C; Dora-C-500; Duoscorb; EINECS 200-066-2; FEMA No. 2109; Ferancee; Hicee; HSDB 818; Hybrin; IDO-C; Kyselina askorbova; Laroscorbine; Lemascorb; Liqui-Cee; L-Lyxoascorbic acid; Mediatric; Meri-C; Natrascorb; Natrascorb injectable; NCI-C54808; NSC 33832; Planavit C; Proscorbin; Redoxon; Ribena; Roscorbic; Scorbacid; Scorbu-C; Secorbate; Stuartinic; Testascorbic; Tolfrinic; Vicelat; Vicomin C; Viforcit; Viscorin; Vitace; Vitacee; Vitacimin; Vitacin; Vitamin C; Vitamisin; Vitascorbol; Xitix; component of: Chromagen, Freancee, Mediatric, Stuartinic, Tolfrinic, Veliten. Vitamin, vitamin source. Deficiency in man causes scurvy. It probably plays a part in the production of collagen in the tissues; used in nutrition, color fixing, flavoring, dietary supplement. mp = 190-192°; d =1.65; $[\alpha]_D^{25}$= 20.5 - 21.5° (c = 1), $[\alpha]_D^{23}$ =48° (c = 1 MeOH); λ_m = 245 nm (acid solution), 265 nm (neutral solution); soluble in H₂O (33 g/100 ml, 40 g/100 ml at 45°, 80 g/100 ml at 100°), EtOH (3.3 g/100 ml), absolute EtOH (2 g/100 ml); glycerol (1 g/100 ml), propylene glycol (5 g/100 ml), insoluble in Et₂O, CHCl₃, C₆H₆, petroleum ether. *Abbott Labs Inc.; Ascher B.F. & Co.; BASF Corp.; Eli Lilly & Co.; Forest Pharm. Inc.; ICN Pharm. Inc.; Johnson & Johnson-Merck Consumer Pharm.; Lederle Labs.; Marion Merrell Dow Inc.; Savage Labs; Wyeth-Ayerst Labs.*

4093 Vitamin D₃
67-97-0   10079   D G V

C27H44O

9,10-Secocholesta-5,7,10(19)-trien-3-ol.

Activated 7-dehydrocholesterol; Arachitol; Calciol; CC; CCRIS 6286; Cholecalciferol; Cholecalciferol, D3; Chole-calciferolum; Colecalciferol; Colecalciferolo; Colecal-ciferolum; Colecalcipherol; D3-Vicotrat; D3-Vigantol; Delsterol; Deparal; Duphafral D3 1000; Ebivit; EINECS 200-673-2; EPA Pesticide Chemical Code 202901; HSDB 820; Irradiated 7-dehydrocholesterol; NEO Dohyfral D3; NSC 375571; Oleovitamin D3; Quintox; Rampage; Rick-eton; Trivitan; Vi-de-3-hydrosol; Vigantol; Vigorsan; Vita-min D3; Vitinc dan-dee-3. Essential vitamin with anti-rachitic properties. In an oily solution, used as a veterinary preparation to prevent rickets and osteomalacia. Solid; mp = 84-85°; $[\alpha]_D^{20}$ = 84.8° (c = 1.6 Me2CO), 51.9° (c = 1.6 CHCl3); λ_m = 264.5 nm (E$_{1cm}^{1\%}$ 450-490 EtOH or C6H14); soluble in most organic solvents, insoluble in H2O. Bayer AG.

4094   Vitamin D4

511-28-4   10080   G

C28H46O

(3β,5Z,7E)-9,10-secoergosta5,7,10(19)-trien-3-ol.

EINECS 208-127-5; 9,10-Secocholesta-5,7,10(9)-trien-3-ol, 24-methyl-; 9,10-Secoergosta-5(Z),7(E),10(19)-trien-3β-ol; Vitamin D4. A synthetic vitamin derived from 22-dihydroergosterol by irridation with UV light. It has less activity than vitamins D2 and D3. Solid; mp = 96-98°; $[\alpha]_D^{18}$ = 89° (c = 0.47, Me2CO); λ_m= 265 nm; insoluble in H2O, organic solvents.

4095   Vitamin E

59-02-9   9571   D H

C29H50O2

(2R)-3,4-Dihydro-2,5,7,8-tetramethyl-2-[(4R,8R)-4,8,12-

trimethyltridecyl]-2H-1-benzopyran-6-ol .

Almefrol; Antisterility vitamin; Aquasol E; CCRIS 3588; Covi-Ox; Covitol F 1000; Denamone; E 307; E 307 (tocopherol); E-Oil 1000; E Prolin; E-Vimin; EINECS 200-412-2; Emipherol; Endo E; Eprolin; Eprolin S; Epsilan; Esorb; Etamican; Etavit; Evion; Evitaminum; HSDB 2556; Ilitia; Lan-E; Med-E; Mixed tocopherols; NSC 20812; Phytogermine; Profecundin; Rhenogran Ronotec 50; Spavit E; Syntopherol; Tenox GT 1; α-Tocopherol; α-Tocopherol acid; (+)-α-Tocopherol; d-α-Tocopherol; D-α-Tocopherol; (2R,4'R,8'R)-α-Tocopherol; (all-R)-α-Toco-pherol; (R,R,R)-α-Tocopherol; α-Tokoferol; Tokopharm; Vascuals; Verrol; Vi-E; Viprimol; Vita E; Vitamin E; Vitamin E α; Vitaplex E; Vitayonon; Viteolin; Viterra E . A fat soluble vitamin; used as an antioxidant, in meat curing, nutrient. Occurs largely in plants. Used as an antioxidant in vegetable oils and shortening. Antioxidant and moisturizer for sun protection and skin care products Oil; mp = 3°; bp 0.1 = 200-220°; d_4^{25} = 0.950; n_D^{25} = 1.5045; λ_m = 294 nm (E$_{1cm}^{1\%}$ 71); $[\alpha]_{546}^{25}$ = -3° (C6H6); insoluble in H2O, soluble in organic solvents. Henkel/Cospha; Henkel/Organic Prods.; U.S. Vitamin.

4096   Vitamin K1

84-80-0   7465   D G

C31H46O2

[R-[R*,R*-(E)]]-2-Methyl-3-(3,7,11,15-tetramethyl-2-hexa-decenyl)-1,4-naphthalenedione.

Antihemorrhagic vitamin; Aqua mephyton; Aquamephyt-on; Combinal K1; EINECS 201-564-2; Fitomenadiona; Fitomenadione; HSDB 3162; K-Ject; Kativ N; Kephton; Kinadion; Konakion; Mephyton; Mono-Kay; Monodion; NSC 270681; Phyllochinon; Phyllochinonum; Phyllo-quinone; Phytomenadione; Phytomenadionum; Phytona-dione; Phytonadionum; Phytylmenadione; Synthex P; Veda-K1; Veta-K1; Vitamin K1; Vitamin K1. Vitamin, vita-min source. A fat soluble vitamin; deficiency gives rise to hemorrhage. Oil; mp = -20°; bp0.001 = 140 - 145°; d^{20} = 0.964; $[\alpha]_D^{25}$ = -0.28° (dioxane); λ_m = 242, 248, 260, 269, 325 nm (E$_{1cm}^{1\%}$ 396, 4319, 383, 387, 68 petroleum ether); insoluble in H2O; sparingly soluble in MeOH; soluble in EtOH, Me2CO, C6H6, petroleum ether, C6H14, dioxane, CHCl3, Et2O, fats and oils; [dihydro form (phytonadiol; dihydrovitamin K1)]: insoluble in H2O, sparingly soluble in petroleum ether, freely soluble in Et2O; [dihydro form sodium diphosphate (C31H48Na2O8P2; phytonadiol sod-ium diphosphate; Kayhydrin)]: mp = 138°; soluble in H2O, MeOH. Abbott Labs Inc.; Hoffmann-LaRoche Inc.; Merck & Co.Inc.; Roche Labs; Roche Products Ltd.

4097   Vitamin K2

84-81-1   5855   D G

C41H56O2

(all-E)-2-(3,7,11,15,19,23,27-Heptamethyl-2,6,10,14,18,   22,26-

octacosaheptaenyl)-3-methyl-1,4-naphthalene-dione.

Menaquinone 6; MK 6; Vitamin K2(30). Vitamin, vitamin source. A fat soluble vitamin. 2-methyl-3-difarnesyl-1,4-napthaquinone; synthesized in the gut by bacteria. mp = 50°; λ_m 243, 248, 261, 270, 325-328 nm (E$^{1\%}_{1cm}$ 304, 320, 290, 292, 53 petroleum ether).

4098   Vitreosil
60676-86-0   G

SiO2
Silicon dioxide glass.
Accusand; Admafine SO 25H; Admafine SO 25R; Admafine SO 32H; Admafine SO-C 2; Admafine SO-C 3; AF-SO 25R; Amorphous quartz; Amorphous silica; As 1 (silica); BF 100; Borsil P; Corning 7940; CP-SilicaPLOT; CRS 1102RD8; Cryptocrystalline quartz; Denka F 90; Denka FB 30; Denka FB 44; Denka FB 74; Denka FS 30; ED-C (silica); EF 10; EINECS 262-373-8; Elsil 100; Elsil BF 100; EQ 912; F 44; F 44 (filler); F 125; F 125 (silica); F 160 (silica); FB 5 (silica); FB 20 (silica); FS 74; Fused quartz; Fused silica; Fuselex; Fuselex RD 40-60; Fuselex RD 120; Fuselex ZA 30; GP 7I; GP 11I; Microcrystalline quartz; MR 84; Nalcast; Optocil; Optocil (quartz); QG 100; Quartz glass; Quartz sand; Quarzsand; Rancosil; RD 8; RD 120; S-Col; SGA; Silica; Silica, amorphous, fused; Silica, crystalline - fused; Silica, fused; Silica, fused, respirable dust; Silica, fused [Silica, amorphous]; Silica, vitreous; Silicon dioxide; Silicon dioxide (vitreous); Silicone dioxide; Siltex; Spectrosil; Suprasil; Suprasil W; Vitreosil IR; Vitreous quartz; Vitreous silica; Vitrified silica; Y 40. Used as an ablative material in rocket engines, as fibers in reinforced plastics; special camera lenses.

4099   Vitride®
22722-98-1   8635   G

C6H16AlNaO4
Dihydrobis(2-methoxyethanolato-O,O')aluminate(1-) sodium.
Aluminate(1-), dihydrobis(2-methoxyethanolato-O,O')-, sodium; Dihydrobis(2-methoxyethoxy)aluminate; EINECS 245-178-2; Red-Al; SMEAH; Sodium dihydridobis(2-methoxyethanolato)aluminate(1-); Vitride. Reducing agent. Viscous liquid; d = 1.0360. Zeeland Chem. Inc.

4100   Warfarin
81-81-2   10097   D G P

C19H16O4
3-(α-Acetonylbenzyl)-4-hydroxycoumarin.
5-18-04-00162 (Beilstein Handbook Reference); Arab Rat Death; Athrombine-K; BRN 1293536; Brumolin; Caswell No. 903; Co-Rax; Compound 42; Coumadin; Coumafen; Coumafene; Coumaphene; Coumarins; Coumefene; Cov-R-Tox; d-Con; Dethmor; Dethnel; Dicusat E; Eastern states duocide; EINECS 201-377-6; EPA Pesticide Chemical Code 086002; Fasco fascrat powder; Frass-ratron; HSDB 1786; Killgerm® Sewarin P; Kumader; Kumadu; Kumatox; Kypfarin; Latka 42; Liqua-tox; Maag rattentod cum; Mar-

frin; Martin's mar-frin; Maveran; Mouse pak; NSC 59813; Panwarfin; Place-Pax; Prothromadin; Rat & mice bait; Rat-A-way; Rat-B-gon; Rat-Gard; Rat-Kill; Rat-Mix; Rat-O-cide 2; Rat-ola; Rat-Trol; Ratorex; Ratox; Ratoxin; Ratron; Ratron G; Rats-No-More; Ratten-koederrohr; Rattenstreupulver Neu Schacht; Rattentraenke; Rattunal; Rax; RCRA waste number P001; Ro-Deth; Rodafarin; Rodafarin C; Rodex; Rodex blox; Rosex; Rough & Ready mouse mix; Solfarin; Spray-trol brand roden-trol; Temus W; Tox-Hid; Twin light rat away; Vampirinip II; Vampirinip III; WARF compound 42; Warfarat; Warfarin; rac-Warfarin; Warfarin Q; Warfarina; Warfarine; Warfarinum; Zoocoumarin. Used as a rodent-icide and an anticoagulant. Orally effective, fat soluble derivative of 4-hydroxycoumarin that induce hypo-coagulability only in vivo by inducing the formation of structurally incomplete clotting factors. Crystals; mp = 161°; λ_m = 308 nm (ϵ 13610 H2O); soluble in Me2CO, dioxane; slightly soluble in MeOH, EtOH, iPrOH, some oils; insoluble in H2O, C6H6, C6H12. Killgerm Chemicals Ltd.; Mechema Chemicals Ltd.

4101   Warfarin sodium
129-06-6   10097   D P

C19H15NaO4
3-(α-Acetonylbenzyl)-4-hydroxycoumarin sodium salt.
3-(α-Acetonylbenzyl)-4-hydroxycoumarin sodium salt; Aldocumar; Athrombin; 2H-1-Benzopyran-2-one, 4-hydroxy-3-(3-oxo-1-phenylbutyl)-, sodium salt; Caswell No. 903A; Coumadan Sodico; Coumadin sodium; Coum-adine; Coumafene sodium; Coumarin, 3-(α-acetonyl-benzyl)-4-hydroxy-, sodium salt; Dicusat; EINECS 204-929-4; EPA Pesticide Chemical Code 086003; Jantoven; Marevam; Marevan; Orfarin; Panwarfin; Prothromadin; Ratsul soluble; Simarc; Sodium, ((3-(α-acetonylbenzyl)-2-oxo-2H-1-benzopyran-4-yl)oxy)-; Sodium coumadin; Sodium warfarin; Tintorane; UniWarfin; Varfine; Waran; Warcoumin; Warfarin sodium; (±)-Warfarin sodium; Warfilone; Zoocoumarin sodium salt. Used as a rodenticide and an anticoagulant. Crystals; Very soluble in H2O, EtOH; slightly soluble in CHCl3, Et2O; LD50 (mrat orl) = 323 mg/kg, (frat orl) = 58 mg/kg, (mus orl) = 374 mg/kg, (rbt orl) $\cong$ 800 mg/kg, (mrat orl) = 100.3 mg/kg, (frat orl) = 8.7 mg/kg. Abbott Labs Inc.; DuPont-Merck Pharm.

4102   White mineral oil
8042-47-5   H

White mineral oil.
EINECS 232-455-8; Mineral oil, white; Slab oil (Obs.); White mineral oil; White mineral oil, petroleum.

4103   Wickenol® 131
68171-33-5   G

C21H42O2
Isopropyl isostearate.
EINECS 250-651-1; Isooctadecanoic acid, 1-methylethyl ester;

Isopropyl isostearate; Isostearic acid, isopropyl ester; 1-Methylethyl isooctadecanoate; Nikkol IPIS; 2-Propyl isooctadecanoate; Wickenol 131. Lubricant, emollient, solubilizer. *CasChem.*

4104   **Wickenol® 159**
2915-57-3   G

C20H38O4
Dioctyl succinate.
AI3-00998; Bis(2-ethylhexyl) succinate; Butanedioic acid, bis(2-ethylhexyl) ester; Di-(2-ethylhexyl) succinate; Di-2-ethylhexyl succinate; Di(2-ethylhexyl)butanedioate; EINECS 220-836-1; Succinic acid, bis(2-ethylhexyl) ester; Wickenol 159. Emollient, moisturizer, pigment wetter/ dispersant; increases water vapor porosity of fatty components used in cosmetic and topical pharmaceutical preparations. *CasChem.*

4105   **Wurster's blue**
100-22-1   9301   G

C10H16N2
Tetramethyl-p-phenylenediamine.
AI3-51106; Benzene, 1,4-bis(dimethylamino)-; 1,4-Benzenediamine, N,N,N',N'-tetramethyl-; 1,4-Bis-(di-methylamino)benzene; p-Bis(dimethylamino)benzene; CCRIS 4728; EINECS 202-831-6; p-Phenylenediamine, N,N,N',N'-tetramethyl-; Tetramethyl-p-phenylenedi-am-ine; N,N,N',N'-Tetramethyl-p-fenylendiamin; N,N,N',N'-Tetramethyl-p-phenylenediamine; N,N,N,N-Tetramethyl-1,4-benzenediamine; TL 85; Wurster's Blue; Wurster's reagent. An oxidation product of tetramethyl-p-phenylenediamine; used as an acid-base indicator. Leaflets; mp = 51°; bp = 260°; λm = 228, 264, 304 nm (ε = 4677, 16218, 1861, cyclohexane); slightly soluble in H2O, soluble in ligroin, freely soluble in EtOH, Et2O, C6H6, CHCl3.

4106   **Wytox® 312**
26523-78-4   G H

C45H69O3P
Tris (nonylphenyl) phosphite.
EINECS 247-759-6; Irgafos TNPP; JP 351; Mark 829; Mark 1178; Nonylphenyl phosphite; P 3; P 3 (anti-oxidant); Phenol, nonyl-,

phosphite (3:1); Phosphorous acid, tris(nonylphenyl) ester; Tri(mononnonylphenyl) phosphite; Tris(nonylphenyl) phosphite; Wytox® 312; Trinonylphenol phosphite; Tri(nonylphenyl) phosphite; Tris(nonyl phenyl) phosphite; UVI-NOX 3100. Antioxidant for PE, PP, PS, PVC, ABS, nylon, food packaging; processing, and color stabilizer. A non staining, nondiscoloring, low volatility antioxidant for polyolefins, vinyl chloride polymers, high impact polystyrenes. *National Polychemicals; Uniroyal.*

4107   **Xanthene**
92-83-1   G

C13H10O
9H-Xanthene.
5-17-02-00252 (Beilstein Handbook Reference); AI3-01544; BRN 0133939; EINECS 202-194-4; NSC 46931; 10H-9-Oxaanthracene; Xanthene. Used in organic synthesis and as a fungicide. Yellow leaflets; mp= 100.5°; bp = 311°; λm = 246, 282, 291 nm (ε = 7410, 2150, 1730, MeOH); insoluble in H2O, poorly soluble in EtOH, soluble in Et2O, C6H6, CCl4, CHCl3, AcOH, ligroin. *Lancaster Synthesis Co.; Sigma-Aldrich Fine Chem.*

4108   **Xanthine**
69-89-6   10116   G

C5H4N4O2
3,7-dihydro-1H-purine-2,6-dione.
AI3-52268; CCRIS 994; 3,7-Dihydro-1H-purine-2,6-dione; 2,6-Dioxo-1,2,3,6-tetrahydropurine; 2,6-Dioxo-purine; EINECS 200-718-6; EPA Pesticide Chemical Code 116900; Isoxanthine; NSC 14664; Pseudoxanthine; Purine-2(3H),6(1H)-dione; Purine-2,6-diol; 1H-Purine-2,6-diol; 9H-Purine-2,6-diol; 1H-Purine-2,6-dione, 3,7-dihydro-; 2,6(1,3)-Purinedion; Purine-2,6-(1H,3H)-dione; 9H-Purine-2,6-(1H,3H)-dione; USAF CB-17; XAN; Xanth-ic oxide; Xanthin; Xanthine. Used in organic synthesis and medicine. Yellowish powder; soluble in H2O (0.007 g/100 ml at 25°, 0.07 g/100 ml at 100°), less soluble in EtOH. *Am. Biorganics; Dajac Labs.; U.S. BioChem.*

4109   **Xanthophyll**
127-40-2   10120   G

C40H56O2
β,ε-carotene-3,3'-diol.
Bo-Xan; β,ε-Carotene-3,3'-diol; β,ε-Carotene-3,3'-diol, (3R,3'R,6'R)-; EINECS 204-840-0; Lutein; all-trans-Lutein; Lutein, all-trans-; Lutein ester; Luteine; NSC 59193; Vegetable lutein; Vegetable luteol; Xanthophyll; all-trans-(+)-Xanthophyll; Xanthophyll, all-trans-(+)-. The yellow pigment occurring in green vegetation and some animal products. Yellow or violet prisms; mp = 196°; [α]18D = 160° (c = 0.7, CHCl3); λm = 340, 457, 488 nm (ε = 6607, 123027, 112202, C6H6); insoluble in H2O, very soluble in C6H6, EtOH, Et2O,

petroleum ether.

4110 Xylazine
7361-61-7   10135   D G V

C12H16N2S
5,6-Dihydro-2-(2,6-xylidino)-4H-1,3-thiazine.
Bay 1470; Bay Va 1470; EINECS 230-902-1; Narcoxyl; Rompun; Wh-7286; Xilazina; Xylapan; Xylazine; Xylazinum. Used in veterinary medicine as a sedative, analgesic and muscle relaxant. [free base]; crystals; mp = 140-142°, 136-139°; soluble in dilute acids, C6H6, Me2CO, CHCl3; slightly soluble in petroleum ether; nearly insoluble in H2O, alkali; LD50 (mus iv) = 42 mg/kg, (mus orl) = 240 mg/kg, (rat orl) = 130 mg/kg. *Bayer Corp. Pharm. Div.*

4111 Xylene
1330-20-7   10136   G H P

C24H30
Dimethylbenzene.
4-05-00-00951 (Beilstein Handbook Reference); AI3-02209-X; Benzene, dimethyl-; BRN 1901563; Caswell No. 906; CCRIS 903; Dimethylbenzene; EINECS 215-535-7; EPA Pesticide Chemical Code 086802; HSDB 4500; Ksylen; Methyltoluene; NCI-C55232; RCRA waste number U239; Socal aquatic solvent 3501; UN 1307; Violet 3; Xiloli; Xylene; Xylene (mixed isomers); Xylenen; Xylenes; Xylol; Xylole. Aromatic compound.; commercial mixture of 3 isomers: o-, m-, p-xylene; solvent; raw material for production of benzoic acid, phthalic anhydride, dyes, other organics; aviation gasoline; protective coating; solvent for alkyd resins, lacquers, enamels andrubber cements. Registered by EPA as an insecticide (cancelled). Mobile liquid; d = 0.86; bp = 137 - 140°; insoluble in H2O, miscible with EtOH, Et2O and most other organic solvents. *Ashland; Crowley Tar Prods.; ExxonMobil Chem. Co.; Fina Chemicals; Mallinckrodt Inc.; Mitsubishi Corp.; Mitsui Petroleum; Riedel de Haen (Chinosolfabrik); Shell; Sigma-Aldrich Co.; Texaco.*

4112 m-Xylene
108-38-3   10136   G H

C8H10
1,3-Dimethylbenzene.
AI3-08916; Benzene, 1,3-dimethyl-; Benzene, m-dimethyl-; CCRIS 907; 1,3-Dimethylbenzene EINECS 203-576-3; m-Dimethylbenzene; HSDB 135; m-Methyltoluene; NSC 61769; m-Xylene; 1,3-Xylene; 3-Xylene; m-Xylol; UN1307; Xylene, m-. Solvent and raw material for preparation of organic compounds. Mobile liquid; mp = -47.8°; bp = 139.1°; d^{20} = 0.8642; λ_m = 265 nm (cyclohexane); insoluble in H2O, soluble in CHCl3, freely soluble in EtOH, Et2O, Me2CO, C6H6, petroleum ether.

4113 o-Xylene
95-47-6   10136   G H

C8H10
Benzene, 1,2-dimethyl-.
AI3-08197; Benzene, 1,2-dimethyl-; Benzene, o-di-methyl-; CCRIS 905; o-Dimethylbenzene; EINECS 202-422-2; HSDB 134; o-Methyltoluene; NSC 60920; UN1307; Xylene, o-; o-Xylene; o-Xylol. Solvent and raw material for preparation of organic compounds. mp = -25.2°; bp= 144.5°; d;ss10 = 0.8802; λ_m = 262, 270 nm (ε = 254, 211, MeOH); insoluble in H2O, freely soluble in EtOH, Et2O, Me2CO, C6H6, CCl4, petroleum ether.

4114 p-Xylene
106-42-3   10136   G H

C8H10
1,4-Dimethylbenzene.
AI3-52255; Benzene, 1,4-dimethyl-; Benzene, p-di-methyl-; CCRIS 910; Chromar; 1,4-Dimethylbenzene; EINECS 203-396-5; HSDB 136; p-Methyltoluene; NSC 72419; Scintillar; UN1307; Xylene, p-; p-Xylene; 1,4-Xylene; 4-Xylene; p-Xylol. Solvent and raw material for preparation of organic compounds. Mobile liquid; mp = 13.2°, bp = 138.3°; d^{20} = 0.8611; λ_m = 212, 275 nm (cyclohexane); insoluble in H2O, slightly soluble in CHCl3, freely soluble in EtOH; Et2O, Me2CO, C6H6, petroleum ether.

4115 Xylenol
105-67-9   P

C8H10O
2,4-Dimethylphenol.
4-06-00-03126 (Beilstein Handbook Reference); AI3-17612; Bacticin; Benzene, 2,4-dimethyl-1-hydroxy-; BRN 0636244; Bulk Lysol Brand Disinfectant; Caswell No. 907A; CCRIS 721; Dimethylphenol, 2,4-; 2,4-Dimethyl-phenol; 2,4-DMP; 4,6-Dimethylphenol; Du Cor Concen-trated Fly Insecticide; EPA Pesticide Chemical Code 086804; Gable-Tite Light Creosote (Creola); Gable-Tite Dark Creosote (Creola); Gallex; HSDB 4253; 1-Hydroxy-2,4-Dimethylbenzene; 4-Hydroxy-1,3-dimethylbenzene; 4-hydroxy-m-xylene; Lysol Brand disinfectant; NSC 3829; Phenol, 2,4-dimethyl-; RCRA waste number U101; asym-o-xylenol; m-Xylenol; 2,4-Xylenol . Registered by EPA as an antimicrobial and fungicide. When formulated with m-cresol, xylenol has bacteriostatic activity against the causal agents of crown gall and olive knot on fruit, ornamental and shade trees and ornamental woody shrubs and vines and control of the genetic/physiological disorder, burr knot, on apples. The pesticide product that contains these two active ingredients, Gallex, is a ready-to-use liquid that is brushed or painted onto the infected areas of trees and ornamentals. Treatments may be made every 4 to 6 months, or about twice a year. Although usage data are not available, EPA assumes that the volume of use is relatively low. Crystals; mp = 24.5°; bp= 210.9°; d^{20} = 0.9650; λ_m = 292 nm (MeOH); slightly soluble in H2O, soluble in CCl4, freely soluble in EtOH, Et2O; LD50 (rat orl) = 3200 mg/kg, LD50 (mus orl) =

809 mg/kg.

4116 o-Xylenol
526-75-0   H

$C_8H_{10}O$
2,3-Dimethylphenol.
4-06-00-03096 (Beilstein Handbook Reference); Benz-ene, 1,2-dimethyl-3-hydroxy-; BRN 1906267; 2,3-Di-methylphenol; EINECS 208-395-3; HSDB 5676; 1-Hydroxy-2,3-dimethylbenzene; NSC 62011; Phenol, 2,3-dimethyl-; 2,3-Xylenol; o-3-Xylenol; o-Xylenol; vic-o-Xylenol; Xyellenol 100. Liquid; mp = 72.8°; bp = 216.9°; λ_m = 273 nm (ε = 1445, MeOH), 269, 273, 278 nm (ε = 1380, 1413, 1549, C_6H_{14}), 280 nm (ε = 1549, cyclohexane); slightly soluble in H_2O, soluble in EtOH, Et_2O.

4117 2,5-Xylenol
95-87-4   H

$C_8H_{10}O$
Phenol, 2,5-dimethyl-.
4-06-00-03164 (Beilstein Handbook Reference); AI3-01551; BRN 1099260; CCRIS 722; EINECS 202-461-5; FEMA No. 3595; HSDB 5296; 6-Methyl-m-cresol; NSC 2599; Phenol, 2,5-dimethyl-; p-Xylenol. Needles or prisms; mp = 74.8°; bp = 211.1; λ_m = 276 nm (ε = 1810, MeOH); very soluble in Et_2O, soluble in H_2O, EtOH, poorly soluble in $CHCl_3$.

4118 2,6-Xylenol
576-26-1   H

$C_8H_{10}O$
2,6-Dimethylphenol.
AI3-08524; Dimethylphenol, 2,6-; 2,6-Dimethylphenol; 2,6-Dmp; EINECS 209-400-1; FEMA No. 3249; HSDB 5697; 1-Hydroxy-2,6-dimethylbenzene; 2-Hydroxy-1,3-dimethylbenzene; NSC 2123; Phenol, 2,6-dimethyl-; Vic-m-xylenol; Xylenol 235; 2,6-Xylenol. Leaflets or needles; mp = 45.7°; bp = 201°, λ_m = 272 nm (ε = 1480, MeOH), 240, 278 nm (ε = 1800, 1560, MeOH-KOH); soluble in H_2O, EtOH, Et_2O, CCl_4.

4119 3,5-Xylenol
108-68-9   H

$C_8H_{10}O$
1-Hydroxy-3,5-dimethylbenzene.
AI3-01553; Benzene, 1,3-dimethyl-5-hydroxy-; CCRIS 724; 3,5-Dimethylphenol; 3,5-Dmp; EINECS 203-606-5; HSDB 5385; 1-Hydroxy-3,5-dimethylbenzene; NSC 9268; Phenol, 3,5-dimethyl-; Sym-m-xylenol; Xylenol 200; 1,3,5-Xylenol; 3,5-Xylenol. Needles; mp = 63.6°; bp = 221.7°; d^{20} = 0.9680; λ_m = 203, 276, 282 nm (MeOH); soluble in H_2O, EtOH, CCl_4.

4120 Xylidine
1300-73-8   10139   H

$C_8H_{11}N$
Dimethylaminobenzene.
AI3-24178-X (USDA); Aminodimethylbenzene; Dimethyl-aminobenzene; Dimethylaniline; Dimethylphenylamine; EINECS 215-091-4; HSDB 6464; UN1711; Xilidine; Xylidine; Xylidine (mixed isomers); Xylidine isomers; Xylidine mixed ortho-meta-para isomers; Xylidinen; Xylidines. Liquid; d = 0.97-0.99; bp = 213-226°; slightly soluble in H_2O, soluble in EtOH, acids.

4121 Xylose
58-86-6   10142   D G

$C_5H_{10}O_5$
D-(+)-xylose.
4-01-00-04223 (Beilstein Handbook Reference); AI3-19010; BRN 1562108; CCRIS 1899; EINECS 200-400-7; FEMA No. 3606; HSDB 3273; NSC 122762; 2,3,4,5-Tetrahydroxypentanal; Wood sugar; Xylo-Med; Xylo-Pfan; Xylose; (+)-Xylose; D-Xylose; Xylose, D-; Xylose, pure. Used in dyeing, tanning, diabetic food, source of ethanol. Crystals; mp = 144-145°; d^{20}_4 = 1.525; $[\alpha]_D^{20}$ = 92° → 18.6° (c = 10, 16 hours); soluble in H_2O, EtOH. Et_2O, poorly soluble in non-polar organic solvents. Am. Biorganics; Penta Mfg.; Sigma-Aldrich Fine Chem.

4122 1-Xylylazo-2-naphthol
3118-97-6   G

$C_{18}H_{16}N_2O$
1-(2,4-Dimethylphenylazo)-2-naphthol.
A.F. Red No. 5; AI3-02853; Aizen Food Red No. 5; Brasilazina Oil Scarlet 6G; Brilliant Oil Scarlet B; C.I. 12140; C.I. Solvent Orange 7; Calco Oil Scarlet BL; CCRIS 906; Ceres Orange RR; Cerisol Scarlet G; Cerotinscharlach G; CI 12140; Color Index No: 12140; 1-(2,4-Dimethylphenylazo)-2-naphthol; 1-((2,4-Dimethyl-phenyl)azo)-2-naphthalenol; EINECS 221-490-4; Ext D & C Red No. 14; Extract D and C Red No. 14; Fast Oil Orange II; Fat Red (yellowish); Fat Scarlet 2G; FD & C No. 32; Fettorange B; Grasan Orange 3R; HSDB

6365; Japan Red 5; Japan Red 505; Lacquer Orange VR; Motirot G; 2-Naphthalenol, 1-((2,4-dimethyl-phenyl)azo)-; 2-Naphthol, 1-(2,4-xylylazo)-; NSC 10457; Oil Orange 2R; Oil Orange KB; Oil Orange N Extra; Oil Orange R; Oil Orange X; Oil Orange XO; Oil Red GRO; Oil Red RO; Oil Red XO; Oil Scarlet; Oil Scarlet 371; Oil Scarlet 6G; Oil Scarlet APYO; Oil Scarlet BL; Oil Scarlet L; Oil Scarlet Y; Oil Scarlet YS; Orange Insoluble RR; Orange oil KB; Oranz rozpoustedlova 7; Ponceau insoluble olg; Pyronalrot R; Red B; Red No. 5; Resin Scarlet 2R; Resoform Orange R; Rot B; Rot GG fettloeslich; Solvent Orange 7; Somalia Orange 2R; Somalia Orange A2R; Sudan ax; Sudan II; Sudan Orange; Sudan Orange RPA; Sudan Red; Sudan Scarlet 6G; Sudan X; Waxakol vermilion; Waxakol vermilion L; 1-(2,4-Xylylazo)-2-naphthol; 1-(o-Xylylazo)-2-naphthol. Used for coloring oils and varnishes. Red needles; mp = 166°; very soluble in EtOH, Et2O.

4123 Yellow prussiate of potash
14459-95-1   G

$$C_6FeK_4N_6 \cdot 3H_2O$$

Potassium ferrocyanide trihydrate.
Ferrate(4-), hexakis(cyano-C)-, tetrapotassium, trihydrate, (OC-6-11)-; Ferro prussiate of potassium; Potassium ferricyanide trihydrate; Potassium ferrocyanide; potassium hexkis(cyano-C)ferrate(4-) trihydrate; potassium hexa-cyanoferrate(II); Anhydrous form, RN 13943-58-3. Used for tempering steel, engraving, and as a laboratory reagent. Yellow crystals; d = 1.85.

4124 Ytterbium
7440-64-4   10160   G

Yb

Yb
Ytterbium.
EINECS 231-173-2; Ytterbium. Metallic element, valencies 2 and 3; used in lasers, dopant for garnets, portable x-ray source, chemical research. Metal; d = 6.977; has face-centered or body-centered cubic structure. Atomergic Chemetals; Cerac; Rhône-Poulenc; Sigma-Aldrich Fine Chem.

4125 Ytterbium oxide
1314-37-0   G

O_3Yb_2
Ytterbium(III) oxide.
EINECS 215-234-0; Ytterbium (III) oxide; Ytterbium oxide; Ytterbium oxide (Yb2O3). Used in special alloys, dielectric ceramics, carbon rods for industrial lighting, catalyst, and special glasses. Solid; soluble in dilute acids. Atomergic Chemetals; Noah Chem.; Rhône-Poulenc.

4126 Yttrium
7440-65-5   10161   G

Y

Y
Yttrium.
EINECS 231-174-8; Yttrium; Yttrium-89. Metallic element; nuclear technology, iron and other alloys, deoxidizer for vanadium and other nonferrous metals, microwave ferrites, coating on high-temperature

alloys, and special semiconductors. Metal; mp = 1509°; bp ≅ 3000°; d = 4.472. Atomergic Chemetals; Cerac; Noah Chem.; Rhône-Poulenc.

4127 Yttrium oxide
1314-36-9   G

O_3Y_2
Yttrium(III) oxide.
Diyttrium trioxide; EINECS 215-233-5; YO 3-245; Yttria; Yttrium oxide; Yttrium oxide (Y2O3); Yttrium sesquioxide; Yttrium trioxide; Yttrium(3+) oxide. Phosphors for color TV tubes, yttrium-iron garnets for microwave filters, stabilizer for high-temperature service materials. Powder; d = 5.03; LD50 (rat ip) = 500 mg/kg. Atomergic Chemetals; New Metals & Chems. Ltd; Noah Chem.; Rhône-Poulenc; Shin-Etsu Chem.

4128 Zentralin
611-92-7   G

$C_{15}H_{16}N_2O$
1,3-Dimethyl-1,3-diphenylurea.
4-12-00-00839 (Beilstein Handbook Reference); BRN 2126077; Carbanilide, N,N'-dimethyl-; Centralit II; Centralite-2; Centralite II; 1,3-Dimethyl-1,3-diphenylurea; EINECS 210-283-4; Methyl centralite; N,N'-Dimethyl carbanilide; N,N'-Dimethylcarbanilide; N,N'-Dimethyl-N,N'-diphenylurea; NSC 59781; Urea, N,N'-dimethyl-N,N'-diphenyl-; Zentralit I; Zentralit II. Used for explosives. Plates; mp = 122°; bp = 350°; λm = 243 nm (ε = 9270, MeOH); slightly soluble in Et2O, C6H6, CS2, very soluble in H2O, EtOH, Me2CO, CHCl3.

4129 Zinc
7440-66-6   10180   G P

Zn

Zn
Zinc.
Asarco L 15; Blue powder; CCRIS 1582; EINECS 231-175-3; Emanay zinc dust; Granular zinc; HSDB 1344; Jasad; Lead refinery vacuum zinc; LS 2; LS 6; Merrillite; Rheinzink; UN1435; UN1436; Zinc; Zinc (dust or fume); Zinc (fume or dust); Zinc (metallic); Zinc ashes; Zinc dust; Zinc, elemental; Zinc powder; Zinc powder or zinc dust. Metallic element; Zn; alloys, galvanizing iron and other metals, and fungicides. Registered by EPA as an antimicrobial, fungicide and herbicide. Used as a micronutrient fertilizer. Metal; mp = 419.5°; bp = 908°; d^{25} = 7.14. Cerac; Cuproquim Corp.; Ferro/Bedford; Pasminco Europe; Sigma-Aldrich Fine Chem.; Zinc Corp. of Am.

4130 Zinc ammonium chloride
14639-98-6   543   G

$Cl_5H_{12}N_3Zn$
Ammonium pentachlorozincate.
Ammonium pentachlorozincate; EINECS 238-688-1; Triammonium pentachlorozincate(3-); Zinc ammonium chloride; Zincate(3-), pentachloro-, triammonium. Used in welding and soldering flux, in dry batteries and galvanizing. Bipyramidal crystals; sublimes 340°; d

= 1.81; very soluble in H_2O. *Blythe, Williams Ltd.; DuPont.*

4131 Zinc O,O-bis(1,3-dimethylbutyl)dithio-phosphate
2215-35-2   H

$C_{24}H_{52}O_4P_2S_4Zn$
 Phosphorodithioic acid, O,O-bis(1,3-dimethylbutyl) ester, zinc salt.
 EINECS 218-679-9; Phosphorodithioic acid, O,O-bis(1,3-dimethylbutyl) ester, zinc salt; Zinc O,O-Bis(1,3-dimethylbutyl)dithiophosphate; Zinc O,O,O',O'-tetrakis-(1,3-dimethylbutyl) bis(phosphorodithioate).

4132 Zinc bis(bis(dodecylphenyl)) bis(dithio-phosphate)
54261-67-5   H
 $C_{72}H_{116}O_4P_2S_4Zn$
 Phenol, dodecyl-, hydrogen phosphorodithioate, zinc salt.
 EINECS 259-048-8; Phenol, dodecyl-, hydrogen phosphorodithioate, zinc salt; Phosphorodithioic acid, O,O-bis(dodecylphenyl) ester, zinc salt; Zinc bis(bis-(dodecylphenyl)) bis(dithiophosphate); Zinc, bis(O,O-bis(dodecylphenyl) phosphorodithioato-S,S')-.

4133 Zinc bis(O-(2-ethylhexyl)) bis(O-(isobutyl)) bis(dithiophosphate)
26566-95-0   H

$C_{24}H_{52}O_4P_2S_4Zn$
 Phosphorodithioic acid, O-(2-ethylhexyl) O-isobutyl ester, zinc salt.
 EINECS 247-810-2; Phosphorodithioic acid, O-(2-ethylhexyl) O-isobutyl ester, zinc salt; Zinc, bis(O-(2-ethylhexyl) O-(2-methylpropyl) phosphorodithioato-kappaS,kappaS')-, (T-4)-; Zinc bis(O-(2-ethylhexyl)) bis(O-(isobutyl)) bis(dithiophosphate); Zinc, bis(O-(2-ethylhexyl) O-(2-methylpropyl) phosphorodithioato-S,S')-, (β-4)-.

4134 Zinc bis(O,O-bis(2-ethylhexyl)) bis(dithiophosphate)
4259-15-8   H

$C_{32}H_{68}O_4P_2S_4Zn$
 Phosphorodithioic acid, O,O-bis(2-ethylhexyl) ester, zinc salt.
 EINECS 224-235-5; Phosphorodithioic acid, O,O-bis(2-ethylhexyl) ester, zinc salt; Zinc bis(O,O-bis(2-ethyl-hexyl)) bis(dithiophosphate).

4135 Zinc borate
1332-07-6   D G P

 Zinc(II) borate.
 Alcanex FR 100; Alcanex FRC 600; Bonrex FC; Borax 2335; Boric acid, zinc salt; Climax ZB 467; EINECS 215-566-6; EPA Pesticide Chemical Code 128859; Firebrake ZB; Firebrake ZB 500; Flamtard Z 10; FRC 600; HSDB 1046; JS 9502; SZB 2335; XPI 187; ZB 112; ZB 237; ZB 467 Lite; Zinc borate; ZN 100; ZSB 2335; ZT; ZT (fire retardant). Medicine, fireproofing textiles, fungistat, mildew inhibitor. Zinc Borate 2335 (Borax Consolidated Ltd) is a specialty flame retardant additive to plasticized PVC and other polymers to reduce afterglow and smoke. Solid; mp = 980°; d = 3.64; insoluble in H_2O, slightly soluble in dilute acids. *BA Chem. Ltd.; Borax Consolidated Ltd; Borax Europe Ltd; Climax Performance; U.S. Borax.*

4136 Zinc borate
12447-61-9
                                               G

$$2BH_3 \quad 2O^{2-} \quad 2Zn^{2+} \quad 15H_2O$$

 $B_6O_{11}Zn_2.xH_2O$
 Boron zinc oxide.
 Boric acid ($H_4B_6O_{11}$), zinc salt (1:2), hydrate; Boron zinc oxide ($B_6Zn_2O_{11}$), hydrate (2:15); Caswell No. 909B; EPA Pesticide Chemical Code 128859; Firebrake ZB; ZB2335; Zinc borate. Zinc borate for flame retardancy. *Borax Europe Ltd.*

4137 Zinc bromide
7699-45-8   10182   G

 Br_2Zn
 Zinc bromide anhydrous.
 EINECS 231-718-4; HSDB 1047; Zinc bromide; Zinc bromide ($ZnBr_2$); Zinc dibromide. Used in photographic emulsions, manufacture of rayon, radiation viewing shields. Crystals; mp = 394°; bp = 697°; d = 4.22; soluble in H_2O, organic solvents. *Atomergic Chemetals; Cerac; Ethyl Corp.; Great Lakes Fine Chem.; Hoechst Celanese; Ryvan Ltd.*

4138 Zinc butyl xanthate
150-88-9
                                               G

 $C_{10}H_{18}O_2S_4Zn$
 Carbonodithioic acid, O-butyl ester, zinc salt.
 O-Butyl hydrogen dithiocarbonate, zinc salt; Butylxanth-ate, zinc salt; Butylxanthogenic acid zinc salt; Carbonic acid, dithio-, O-butyl ester, zinc salt; EINECS 205-777-1; Nocceler ZBX; NSC 402837; Vulcaid 27; Zinc, bis(O-butyl carbonodithioato-S,S')-, (T-4)-; Zinc butyl xanthate; Zinc butylxanthate; Zinc, bis(o-butyl carbonodithioato-S,S')-, (t-4)-; Zinc O-butyldithio-carbonate. A rubber

vulcanization accelerator.

4139　Zinc carbonate
3486-35-9　　　　　　10184　　　　　D G

CO3Zn

Carbonic acid, zinc salt (1:1).

Akrochem® 9930 Zinc Oxide Transparent; Carbonic acid, zinc salt (1:1); CI 77950; EINECS 222-477-6; HSDB 1048; Natural smithsonite; Zinc carbonate; Zinc carbonate (1:1); Zinc monocarbonate; Zincspar. Accelerator-activator for transparent natural and synthetic rubber goods, adhesives; as pigment; fireproofing filler for rubber and plastics; topical antiseptics, cosmetics, and lotions. Used medically as an astringent. Solid; mp = 168°; d = 4.3980; poorly soluble in H_2O (0.001 g/100 ml at 15°), soluble in dilute mineral acid, alkali. *Akrochem Chem. Co.; Allchem Ind.; Harcros Durham; Nihon Kagaku Sangyo; Spectrum Chem. Manufacturing.*

4140　Zinc chloride
7646-85-7　　　　　　10185　　　　　D G P

Cl2Zn

Hydrochloric acid zinc salt.

AI3-04470; Butter of zinc; Caswell No. 910; CCRIS 3509; Chlorure de zinc; EINECS 231-592-0; EPA Pesticide Chemical Code 087801; HSDB 1050; NSC 529648; UN1840; UN2331; Zinc (chlorure de); Zinc butter; Zinc chloride; Zinc chloride (ZnCl2); Zinc chloride, anhydrous; Zinc chloride fume; Zinc chloride, solution; Zinc dichloride; Zinc(II) chloride; Zinco (cloruro di); Zine dichloride; Zinkchlorid; Zinkchloride; Zintrace. Catalyst, dehydrating and condensing agent in organic synthesis, fireproofing, preserving food, electroplating, antiseptic denaturant for alcohol. An astringent and dentin desens-itizer. Used with chelating agents to form solutions of zinc that are biologically available to plants and animals. Also used in specialty corrosion inhibitors in cooling towers, potable water, and in gas and oil wells. Registered by EPA as an antimicrobial, fungicide and herbicide. White solid; mp =290°; bp = 732°; d^{25} = 2.907; soluble in H_2O (432 g/100 g), less soluble in organic solvents; LD50 (rat iv) = 60-90 mg/kg. *Allied Signal; Armour Pharm. Co. Ltd.; Blythe, Williams Ltd.; Elf Atochem N. Am.; EM Ind. Inc.; Mallinckrodt Inc.*

4141　Zinc chromate
13530-65-9　　　　　　10186　　　　　G

O4CrZn2.H2O

Zinc chromate(VI) hydroxide.

Basic zinc chromate; Buttercup Yellow; C.I. 77955; CCRIS 7569; Chromic acid (H_2CrO_4), zinc salt (1:1); Chromic acid, zinc salt; Chromic acid, zinc salt (1:1); Chromium zinc oxide (ZrCrO4); CI 77955; EINECS 236-878-9; HSDB 6188; Pigment Yellow 36; Zinc chromate; Zinc chromate AM; Zinc chromate C; Zinc chromate O; Zinc chromate T; Zinc chromate Z; Zinc chromate(VI) hydroxide; Zinc chrome; Zinc chrome (anti-corrosion); Zinc Chrome Yellow;

Zinc chromium oxide (ZrCrO4); Zinc hydroxychromate; Zinc tetraoxychromate; Zinc tetraoxychromate 76A; Zinc tetraoxychromate 780B; Zinc tetroxychromate; Zinc yellow; Zincro ZTO. Used as a yellow pigment in paints, varnishes, oil colors, linoleum, rubber etc. [hydrate]; fine yellow powder; slightly soluble in H_2O. *Landers-Segal Color.*

4142　Zinc diacrylate
14643-87-9　　　　　　　　　　　　　H

C6H6O4Zn

2-Propenoic acid, zinc salt.

Acrylic acid, zinc salt; Ageflex ZDA; EINECS 238-692-3; 2-Propenoic acid, zinc salt; Zinc acrylate; Zinc diacrylate. Crosslinker for molded polybutadiene compounds, conductive and protective coatings, coagent for SBR compounds and reactive pigments; activator for rubber compounding; scorch retarder. *Rit-Chem.*

4143　Zinc dibenzyldithiocarbamate
14726-36-4　　　　　　　　　　　　　G

C30H28N2S4Zn

Zinc bis(dibenzyldithiocarbamate).

Akrochem® Z.B.E.D; Arazate®; (T-4)-Bis(bis(phenyl-methyl)carbamodithioato-S,S')zinc; Dibenzyldithiocarb-amic acid, zinc salt; EINECS 238-778-0; Naftocit® ZBEC; Octocure ZBZ-50; Perkacit® ZBEC; Vulkacit ZBEC; ZBeDC; Zinc, bis(bis(phenylmethyl)carbamodithioato-S,S')-, (β-4)-; Zinc bis(dibenzyldithiocarbamate); Zinc dibenzyldithiocarbamate. Accelerator for rubber, latex dispersions, cements; nondiscoloring and nonstaining. Solid; mp = 186°; insoluble in H_2O, Me2CO, gasoline, soluble in C6H6, ethylene dichloride. *Akrochem Chem. Co.; Akzo Chemie; Uniroyal.*

4144　Zinc O,O-diisodecyl dithiophosphate
25103-54-2　　　　　　　　　　　　　H

C40H84O4P2S4Zn

Phosphorodithioic acid, O,O-diisodecyl ester, zinc salt.

EINECS 246-618-6; Elco 106; HSDB 6158; Isodecanol, hydrogen phosphorodithioate, zinc salt; Isodecyl ZDDP; Phosphorodithioic acid, O,O-diisodecyl ester, zinc salt; Zdtp; Zinc, bis(O,O-diisodecyl phosphorodithioato-S,S')-; Zinc isodecyl phosphorodithioate; Zinc O,O-bisisodecyl dithiophosphate; Zinc O,O-Diisodecyl dithiophosphate.

4145 Zinc diisooctyl dithiophosphate
28629-66-5   H

$C_{32}H_{68}O_4P_2S_4Zn$
Phosphorodithioic acid, O,O-diisooctyl ester, zinc salt.
Dithiophosphoric acid, O,O'-diisooctyl ester, zinc salt; EINECS 249-109-7; HSDB 6161; Isooctyl ZDDP; Oronite; Phosphorodithioic acid, O,O-diisooctyl ester, zinc salt; Zinc bis(diisooctyl dithiophosphate); Zinc, bis(O,O-diiso-octyl phosphorodithioato-S,S')-; Zinc bis(O,O-diisooctyl) bis(dithiophosphate); Zinc diisooctyl dithiophosphate; Zinc O,O-diisooctyl dithiophosphate.

4146 Zinc 2-ethylhexanoate
136-53-8   G H

$C_{16}H_{32}O_4Zn$
Zinc bis(2-ethylhexanoate).
EINECS 205-251-1; Hexanoic acid, 2-ethyl-, zinc salt; Octoate Z; Zinc 2-ethylhexoate; Zinc bis(2-ethyl-hexanoate). Rubber activator used in soluble cure systems in place of stearic acid and partial replacement of zinc oxide for natural and synthetic rubbers. *Vanderbilt R.T. Co. Inc.*

4147 Zinc fluoride
7783-49-5   10189   G

F_2Zn
Hydrofluoric acid zinc salt.
EINECS 232-001-9; HSDB 1052; Zinc difluoride; Zinc fluoride; Zinc fluoride (ZnF_2); Zinc fluorure. Used in phosphors, ceramic glazes, as a wood preservative, in electroplating and organic fluorination. Solid; mp = 872°; bp = 1500°; d^{25} = 5.00; anhydrous form is poorly soluble in H_2O. *Atomergic Chemetals; Cerac; Elf Atochem N. Am.; Hoechst Celanese; Noah Chem.*

4148 Zinc hydroxide
20427-58-1   G

H_2O_2Zn
Zinc dihydroxide.
Caswell No. 916; Collozine; Colloidal zinc hydroxide; EINECS 243-814-3; EPA Pesticide Chemical Code 088501; Zinc dihydroxide; Zinc hydroxide; Zinc hydroxide $(Zn(OH)_2)$. White solid; d = 3.050.

4149 Zinc hydroxystannate
12027-96-2   G

$ZnSn (OH)_6$
Stannate $(Sn(OH)_6^{2-})$, zinc (1:1), (OC-6-11)-.
Flamtard H. Flame retardant for PVC, polychloroprene, chlorosulfonated polyethylene, other halopolymers. White powder; d = 3.4; dec 180°; soluble in strong acids and bases; LD_{50} (rat orl) > 5000 mg/kg, (rat der) > 2466 mg/kg. *Alcan Chem.; Atomergic Chemetals; Blythe, Williams Ltd.; Joseph Storey.*

4150 Zinc lactate
16039-53-5   10195   G

$C_6H_{10}O_6Zn$
Lactic acid zinc salt.
EINECS 240-178-9; Zinc dilactate; Zinc lactate. White solid; soluble in H_2O (2%). *Patco.*

4151 Zinc 2-mercaptobenzothiazolate
155-04-4   G P

$C_{14}H_8N_2S_4Zn$
2(3H)-Benzothiazolethione, zinc salt .
Bantex; Bis(2-benzothiazolylthio)zinc; Caswell No. 917; EINECS 205-840-3; EPA Pesticide Chemical Code 051705; Hermat Zn-mbt; HSDB 5419; Mercaptobenzo-thiazole zinc salt; Octocure ZMBT-50; OXAF; Pennac ZT; Perkacit® ZMBT; Tisperse MB-58; USAF GY-7; Vulkacit® Merkapto/MGC; Vulkacit® ZM; Zenite; Zenite special; Zetax®; Zinc 2-benzothiazolethiolate; Zinc 2-mercapto-benzothiazole; ZMBT; ZnMB. A medium temperature accelerator widely used in latex compounding; also used in proofing, wire and druggist sundries where fast curing and a minimum of odor are required. Also used as a fungicide. Registered by EPA as an antimicrobial and fungicide (cancelled). Used in the form of a soluble concentrate or liquid and wettable powder to control mold, mildew, bacteria and fungi which degrade aqueous industrial products, fabrics, and yarns; and slime-forming bacteria and fungi in industrial water systems. There are no registered food uses for either the sodium or zinc salts of 2-mercaptobenzothiazole. Solid; d^{25} = 1.70.

4152 Zinc naphthenate
12001-85-3   G P

Naphthenic acid, zinc salt .
CCRIS 1171; EINECS 234-409-2; Naphtenate de zinc; Naphthenic acids, zinc salts; Zinc naphthenate; Zinc uversol; Zinc Uversol Fungicide . Drier and wetting agent in paints, varnishes, resins; insecticide, fungicide, mildew preventive; wood preservative; waterproofing textiles; and insulating materials. Fungicide. Registered by EPA as a fungicide, herbicide and insecticide. Solid; d = 0.962; flash point = 135°F. *Generic; Green Products Co.; Jasco Chemical Corp.; Lanco Manufacturing Corp.; Mobile Paint Manufacturing Co. Inc.; OMG Americas Inc.; Sherwin-Williams Co.*

4153 Zinc nitrate
10196-18-6   G

$N_2O_6Zn.6H_2O$
Zinc nitrate hexahydrate.

Dusicnan zinecnaty; Nitric acid, zinc salt, hexahydrate; Zinc nitrate hexahydrate; Zinc(II) nitrate, hexahydrate (1:2:6). Acidic catalyst, latex coagulant, reagent, intermediate, mordant. Crystals; mp = 36°; d= 2.065; soluble in water (50%), alcohol. *Blythe, Williams Ltd.; Mallinckrodt Inc.; Nihon Kagaku Sangyo.*

4154 Zinc oxide
1314-13-2   10200   G P

Ozn
Zinc(II) oxide.

A&D Medicated Ointment; Activox; Activox B; Actox 14; Actox 16; Actox 216; AI3-00277; Akro-zinc bar 90; Akro-zinc bar 85; Amalox; Azo 22; Azo-33; Azo-55; Azo-55TT; Azo-66; Azo-66TT; Azo-77; Azo-77TT; Azodox; Azodox-55; Azodox-55TT; Blanc de Zinc; C-Weiss 8; C.I. 77947; C.I. Pigment White 4; Cadox XX 78; Caswell No. 920; CCRIS 1309; Chinese White; CI 77947; CI Pigment white 4; Cynku tlenek; Dome Paste Bandage; EINECS 215-222-5; Electox 2500; Electrox 2500; Emanay zinc oxide; EMAR; EPA Pesticide Chemical Code 088502; Felling zinc oxide; Flores de zinci; Flowers of zinc; GIAP 10; Green seal-8; HSDB 5024; Hubbuck's White; K-Zinc; Kadox 15; Kadox-25; Kadox 72; Ken-Zinc®; No-Genol; Nogenol; Outmine; Ozide; Ozlo; Permanent White; Philosopher's wool; Pigment white 4; Powder base 900; Protox type 166; Protox type 167; Protox type 168; Protox type 169; Protox type 267; Protox type 268; Red Seal 9; RVPaque; Snow White; Supertah; Unichem ZO; Vandem VAC; Vandem VOC; Vandem VPC; White seal-7; XX 78; XX 203; XX 601; Zinc gelatin; Zinc monoxide; Zinc oxide; Zinc oxide (ZnO); Zinc White; Zinca 20; Zinci Oxicum; Zinci Oxydum; Zincite; Zincoid; Zincum Oxydatum; Zinkoxyd Activ®; Zinox; Ziradryl; Zn 0701T; Chinese white; zinc white; Tertiary zinc oxide; Zinc Oxide No. 185; Zinc Oxide No. 318. UV absorber; accelerator activator in elastomer compounding; flame retardant; pigment in white paints, cosmetics, driers, dental cements; mold inhibitor in paints; in manufacture of opaque glass, enamels, tires, printing inks, porcelains; reagent in analytical chemistry. Used as a flowable nutrient for rice seed dressing. Highly disperse precipitated zinc oxide is used as a vulcanization accelerator activator for rubber goods based on natural and synthetic elastomers and latex applications; suitable at high levels for dynamically stressed articles, e.g., buffers and rollers, and in low concentrations in transparent and translucent goods. Amorphous powder or crystals; odorless, bitter taste; mp = 1975°; d = 5.67; soluble in dilute acetic or mineral acids, alkalis, insoluble in H_2O, EtOH; pH 6.95 (American process); LD_{50} (rat orl) = 240 mg/kg. *Am. Chemet; Asarco; Bayer AG; Durham Chemicals Ltd.; Eagle Zinc; General Chem; Griffin LLC; Harcros Durham; Harcros; Kenrich Petrochem.; Mallinckrodt Inc.; Miles Inc.; Polysar; Zinc Corp. of Am.*

4155 Zinc phosphate
7779-90-0   10205   G

$O_8P_2Zn_3$
Phosphoric acid zinc salt.

704TVM; Bonderite 40; Bonderite 181; Bonderite 880; C.I. 77964; C.I. Pigment White 32; Delaphos; Delaphos 2M; EINECS 231-944-3; Fleck's Extraordinary Cement; Granodine 16NC; Granodine 80; Heucophos™ ZP 10; J 0852; LF Bowsei PW 2; LF-PW 2; Man-Gill 51339; Man-Gill 51355; Microphos 90; Neutral zinc phosphate; Phosphinox PZ 06; Phosphoric acid, zinc salt (2:3); Pigment White 32; Sicor ZNP/M; Sicor ZNP/S; Tribasic zinc phosphate; Trizinc bis(orthophosphate); Trizinc diphosphate; Virchem 931; Weather coat 1000; Zinc acid phosphate; Zinc orthophosphate; Zinc phosphate (3:2); Zinc phosphate cement; ZP-DL; ZP-SB; ZPF. Anticorrosive pigment for paints, particularly primer formulations; in flame retardant for plastics. Used in dental cements, phosphors. White powder; mp = 900°; d^{15} = 3.998; insoluble in H_2O, EtOH; soluble in dilute mineral acids, AcOH, ammonium hydroxide, alkali hydroxide solutions. *BASF Corp.; Calgon Carbon; CK Witco Chem. Corp.; Elf Atochem N. Am.; Halox Pigments; Hammonds Fuel Additives Inc.; Heucotech Ltd.; Landers-Segal Color; Lohmann; McGean-Rohco; Min. Pigments; National Chem.; Pasminco Europe; Sino-Am. Pigment Systems; Whitfield Chem. Ltd.*

4156 Zinc silicate
13597-65-4   10211   G

O_4SiZn_2
Zinc orthosilicate.

Dizinc orthosilicate; EINECS 237-057-8; Silicic acid (H_4SiO_4), zinc salt (1:2); Tornusil; Willemite; Zinc silicate. A proprietary two-pack moisture cured inorganic zinc silicate primer. Insoluble in H_2O. Coatings; *Sigma.*

4157 Zinc stannate
12036-37-2   G

O_3SnZn
Zinc stannate.

Flamtard S; Tin zinc oxide ($SnZnO_3$). Flame retardant for PVC, polychloroprene, chlorosulfonated polyethylene, polypropylene, other halopolymers. White powder; d = 3.4; dec 180°; soluble in strong acids and bases; LD_{50} (rat orl) > 5000 mg/kg, (rat der) > 2466 mg/kg. *Alcan Chem.; Atomergic Chemetals; Blythe, Williams Ltd.; Joseph Storey.*

4158 Zinc stearate
557-05-1   10212   G H

C36H70O4Zn
Octadecanoic acid, zinc salt.
AI3-00388; Caswell No. 926; Coad; Dermarone; Dibasic zinc stearate; EINECS 209-151-9; EPA Pesticide Chemical Code 077002; HSDB 212; Hydense; Hytech; Mathe; Metallac; Metasap 576; NSC 25957; Octadecanoic acid, zinc salt; Petrac ZN-41; Stavinor ZN-E; Stearates; Stearic acid, zinc salt; Synpro stearate; Synpro stearate (VAN); Talculin Z; Unichem ZS; Witco Zinc Stearate USP; Zinc distearate; Zinc distearate, pure; Zinc octadecanoate; Zinc stearate; Zinc stearate W. S; Zn Stearate. Used in cosmetics, pharmaceuticals, lacquers, ointments, tablet manufacture; as a mold release agent for plastic; filler, antifoamer; as a flatting agent in lacquers; as a drying lubricant and dusting agent for rubber; waterproofing agent for concrete, paper and textiles. Solid; mp = 130°; insoluble in H_2O, EtOH, Et_2O, soluble in C_6H_6. *Akzo Chemie; Ferro/Grant; Magnesia GmbH; Mallinckrodt Inc.; Norac; Stave; Syn Prods.*

4159   Zinc sulfate
7446-19-7   10213   G P

$ZnSO_4.2O$
Zinc sulfate, monohydrate.
Sulfuric acid, zinc salt (1:1), monohydrate; Zinc sulfate monohydrate. Used in rayon manufacture, agricultural sprays, chemical intermediate, dyestuffs, electroplating. Registered by EPA as an antimicrobial, fungicide and herbicide. FDA GRAS, Japan approved with limitations, BP. Ph. Eur. compliance. Used in orals, eye lotions, astringents, styptics, gargle sprays and skin tonics. Used medicinally as an emetic. Crystals; soluble in H_2O, insoluble in EtOH. *Allchem Ind.; Blythe, Williams Ltd.; EM Ind. Inc.*

4160   Zinc sulfate
7446-20-0   10213   D G

$O_4SZn.7H_2O$
Zinc sulfate heptahydrate.
CCRIS 5563; Op-Thal-Zin; Sulfuric acid, zinc salt (1:1), heptahydrate; VasoClear A; Verazinc; White vitriol (heptahydrate); Zinc sulfate; Zinc sulfate heptahydrate; Zinc sulfate (1:1) heptahydrate; Zinc sulfate (ZnSO4) heptahydrate; Zinc vitriol (heptahydrate); Zincfrin; white copperas; zinc vitriol. Used in rayon, manufacture, dietary supplement, animal feeds, mordant, wood preservative, analytical reagent. Used as an astringent in ophthalmology. Electrolyte replenisher. Zinc supplement. Solid; mp = 40°; d = 1.9570. *Alcon Labs; Mallinckrodt Inc.; Sigma-Aldrich Fine Chem.*

4161   Zinc sulfate
7733-02-0   10213   D G P

O4SZn
Zinc sulfate anhydrous .
AI3-03967; Bonazen; Bufopto zinc sulfate; Caswell No. 927; CCRIS 3664; Complexonat; EINECS 231-793-3; EPA Pesticide Chemical Code 089001; HSDB 1063; Medizinc; Neozin; NSC 32677; NSC 135806; NU-Z; OP-Thal-zin; Optised; Optraex; Orazinc; Phenylzin; Prefrin-Z; Solvezinc; Solvezink; Sulfate de zinc; Sulfuric acid, zinc salt; Sulfuric acid, zinc salt (1:1); Verazinc; Visine-ac; White copperas; White vitriol; Zinc-200; Zinc sulfate; Zinc sulfate (1:1); Zinc sulfate (ZnSO4); Zinc sulphate; Zinc vitriol; Zincate; Zincfrin; Zinci Sulfas; Zincomed; Zincum Sulfuricum; Zink-Gro; Zinklet; Zinkosite. Used in rayon manufacture, in animal feeds, as a wood preservative and analytical reagent. Registered by EPA as an antimicrobial, fungicide and herbicide (cancelled). Electrolyte replenisher. Zinc supplement. Astringent. FDA GRAS, Japan approved with limitations, BP. Ph. Eur. Compliance. Used in orals, eye lotions, astringents, styptics, gargle sprays and skin tonics. Used medicinally as an emetic. Zinc Sulfate is the optimum source of zinc in animal feeds as the nutritional source of zinc due to its biological availability. Zinc is important in maintaining normal health and increased yields, especially in swine and poultry. Zinc Sulfate is also used in production of pigments, preservation and clarification of glue, in flame proofing compounds, as a mining flotation agent, in wood preservatives, in electro metallurgy and electro galvanizing. White solid.

4162   Zinc sulfide
1314-98-3   10214   G

$$Zn=S$$

SZn
Zinc(II) sulfide.
Albalith; C.I. Pigment White 7; Cleartran; EINECS 215-251-3; HSDB 5802; Irtran 2; Pigment White 7; Sachtolith®; Sachtolith® HD-S; Spalerite; Sphalerite; Wurtzite; Zinc blende; Zinc monosulfide; Zinc sulfide; Zinc sulfide (ZnS); Zinc sulphide. White pigment; used in white and opaque glass, plastics, dyeing, paints, linoleum, leather, dental rubber; fungicide; anhydrous in x-ray screens, and TV screens. White-grey powder; insoluble in H_2O. *Cerac; Chemson GmbH; Noah Chem.; Ore & Chem.; Sachtleben Chemie GmbH.*

4163   Zinc undecenoate
557-08-4   9916   D E G

C22H38O4Zn
zinc 10-undecenoate.
EINECS 209-155-0; Mycoseptin; NSC 402438; Tineafax; 10-Undecenoic acid, zinc salt; Zinc diundec-10-enoate; Zinc undecenoate; Zinc 10-undecenoate; Zinc undecyl-enate. Used as a base for ointments. Has antifungal activity. Solid; mp= 115-116°.

4164   Zineb
12122-67-7   10220   G P

Zn^{+2}

(C4H6N2S4Zn)x
[[1,2-Ethanediylbis[carbamodithioato]](2-)]zinc.
  Aaphytora; Acuprex; Aphytora; Aspor; Asporum; Bercema; Blightox; Blizene; Bombardier; Carbadine; Carbamodithioic acid, 1,2-ethanediylbis-, zinc salt; Caswell No. 930; CCRIS 2503; CHEM zineb; Cineb; Clortocaffaro; Crittox; Crystal Zineb; Cynkotox; Daisen; Deikusol; Devizeb; Diiner; Dipher; Discon; Discon-Z; Dithane 65; Dithane Z; Dithane Z-78; Ditiamina; Ditiozin; EINECS 235-180-1; Enozin; ENT 14,874; EPA Pesticide Chemical Code 014506; Ethylenebis(dithio-carbamato)zinc; Fitodith 80; Fungo-pulvit; Funjeb; Hexaphane; Hexathane; HSDB 1787; Kupratsin; Kypzin; Lipotan; Lirotan; Lonacol; Micide; Micide 55; Novosir N; Novozin N 50; Novozir; Novozir N; Novozir N 50; NSC 49513; Pamosol 2 forte; Parzate; Parzate C; Parzate zineb; Permilan; Perosin; Perosin 75B; Perozin; Perozine; Perozine 75B; Phytox; Pilzol SZ; Polyram-Z; Sepineb; Shaughnessy Number 014506; Sperlox-Z; Taloberg; Tanazon; Thiodow; Thionic M; Tiazin; Tiezene; Tritoftorol; Tsineb; Unizeb; Zebenide; Zebtox; Zidan; Zidanit; Zinc, (ethylenebis(dithiocarbamato))-; Zincethyl-enebisdithiocarbamate; Zineb; Zineb 75; Zineb 75 WP; Zineb 80; Zineb-R; Zinebe; Zinosan; Zinugec; Zipar . A fungicide and insecticide. Foliar fungicide with protective action, insecticide. Repellent to birds and rodents. Used for control of fungi in fruits, vines, vegetables and ornamentals. Controls scab in apples and pears. Registered by EPA as a fungicide (cancelled). Pale yellow crystalline powder; dec 157°; sparingly soluble in H2O (0.001 g/100 ml), soluble in C5H5N and CS2, insoluble in other organic solvents; insoluble in organic solvents; LD50 (rat orl) > 5200 mg/kg, (rat der) > 6000 mg/kg; LC50 (perch) = 2 mg/l, (roach (6 - 8 mg/l); non-toxic to bees. *Agrimont; Bayer AG; Diachem; DuPont; Makhteshim Chemical Works Ltd.; Pennwalt Holland; Rhône-Poulenc; Rohm & Haas Co.; Visplant.*

4165   Zirconium boride
12045-64-6   G

B2Zr
Zirconium diboride.
  EINECS 234-963-5; Zirconium boride; Zirconium boride (ZrB2); Zirconium diboride; Zirconium diboride (ZrB2). Refractory for aircraft and rocket applications, thermocouple protection tubes, high-temp. electrical conductor, cutting-tool component, coating tantalum, cathode in high-temp. electrochemical systems; oxidation-resistant composites. *Atomergic Chemetals; Cerac; Noah Chem.*

4166   Zirconium fluoride
7783-64-4   10228   G

F4Zr
Zirconium tetrafluoride.
  EINECS 232-018-1; Zirconium fluoride; Zirconium fluoride (ZrF4), (β-4)-; Zirconium tetrafluoride. Component of molten salts used in nuclear reactors. Solid; mp = 640°; bp = 905°; d^{16} = 4.6; LD50 (mus orl) = 98 mg/kg. *Atomergic Chemetals; Cerac; Elf Atochem N. Am.*

4167   Zirconium hydride
7704-99-6   10229   G

H2Zr
Zirconium dihydride.
  EINECS 231-727-3; UN1437; Zirconium hydride; Zirconium hydride (ZrH2). Vacuum tube getter, powder metallurgy, source of hydrogen, metal-foaming agent, nuclear moderator, reducing agent, and hydrogenation catalyst. Stable, grey-black powder. *Atomergic Chemetals; Cerac; Degussa AG; Morton Intn'l.*

4168   Zirconium oxide
1314-23-4   10233   G P

O2Zr
Zirconium(IV) oxide.
  AI3-29087; C.I. 77990; C.I. Pigment White 12; CAP (oxide); CC 10; CCRIS 6601; E 101; EINECS 215-227-2; Kontrastin; Nissan Zirconia Sol NZS 20A; Norton 9839; NSC 12958; Nyacol Zr (acetate); NZS 30A; PCS (filler); Pigment White 12; Rhuligel; S945; S975; S987; S992; S994; Torayceram Sol ZS-OA; TZ 3YTSK; ZD 100; Zedox; Zircoa 5027; Zirconia; Zirconic anhydride; Zirconium dioxide; Zirconium oxide (ZrO2); Zirconium White; Zirox Zt 35; zirconic anhydride. Monoclinic zirconium dioxide with a zirconia content (including hafnia) of 94.5%, 97.5% or 98.7%, 99.2% or 99.4%; used for manufacture of ceramic pigments, welding fluxes and insulating material, in production of piezoelectric crystals, high-frequency induction coils, ceramic glazes, glasses, heat-resistant fibers; hydrous: odor absorbent, and poison ivy treatment. White monoclinic crystals; mp = 2680°; bp = 4300°; d = 5.85; insoluble in H2O, soluble in mineral acids. *Atomergic Chemetals; Degussa-Hüls Corp.; Ferro/Transelco; Magnesium Elektron; Tam Ceramics; Zircar.*

4169   Zirconium silicate
10101-52-7   10234   G

O4SiZr
Zirconium orthosilicate.
  A-PAX-SA; EINECS 233-252-7; Excelopax; Micro-Pax; Micro-Pax 20A; Micro-Pax SP; MZ 1000B; Oscal 1224; Silicic acid (H4SiO4), zirconium(4+) salt (1:1); Silicon zirconium oxide (SiZrO4); Tam 418; Zircon; Zircon 30MY; Zirconium orthosilicate; Zirconium orthosilicate (ZrSiO4); Zirconium silicate; Zirconium silicon oxide (ZrSiO4); Zircosil; Zircosil 1. Used in refractories, cer-amics, cements and coatings for casting molds. Also in jewelry. Tetragonal crystals; mp = 2550°. *Anzon.*

4170   Zirconium silicate
14940-68-2   G

O4SiZr
Silicic acid, zirconium salt (1:1).

681

A-PAX 45M; EINECS 239-019-6; Hyacinth; Silicic acid, zirconium(4+) salt (1:1); Standard SF 200; Ultrox 500W; Zircon; Zircon (Zr(SiO4)); Zirconite; Zirconium(IV) silicate (1:1); Zircosil 15. Glaze opacifier; stabilizes color shades; used in white and colored glazes for sanitary ware, wall tile, glazed brick, structural tile, stoneware, dinnerware, special porcelains, refractory compositions, epoxy formulations, and encapsulating resins. White solid; mp = 2550°. *Atomergic Chemetals; DuPont; Elf Atochem N. Am.; Tam Ceramics.*

4171   Zirconium tetrachloride
10026-11-6   10227   G

Cl4Zr
Zirconium (IV) chloride.
EINECS 233-058-2; HSDB 2531; Tetrachlorozirconium; UN2503; Zirconium chloride; Zirconium chloride (ZrCl4); Zirconium chloride, tetra-; Zirconium tetrachloride; Zirconium(IV) chloride (1:4). Source of the pure metal, analytical chemistry, water repellents for textiles, tanning agent, zirconium compounds, special catalysts (Friedel-Crafts, Ziegler). Solid; mp = 437°; d = 2.8030; LD50 (rat orl) = 1588 mg/kg. *Cerac;* 3973*Atomergic; Chem; Chemetals; Noah.*

4172   Zoldine® MS-52
137796-06-6   G

C11H23NO
4-Ethyl-2-methyl-2-(3-methylbutyl)-1,3-oxazolidine.
Oxazolidine,   4-ethyl-2-methyl-2-(3-methylbutyl)-.   Moist-ure scavenger; urethane crosslinker. *Angus.*

4173   Zolpidem
82626-48-0   10242   D

C19H21N3O
N,N,6-Trimethyl-2-p-tolylimidazo[1,2-a]pyridine-3-acetamide.
DEA No. 2783; Lorex; SL-800750; Zolpidem. Sedative/ hypnotic. Crystals; mp = 196°. *Synthelabo Pharmacie.*

4174   Zolpidem Tartrate
99294-93-6   10242   D

C42H48N6O8
N,N,6-Trimethyl-2-p-tolylimidazo[1,2-a]pyridine-3-acetamide   L-(+)-tartrate (2:1).
Ambien; Bilcam; Cymerion; Dalparan; DEA 2783; Durnit; Eudorm; Ivadal; Myslee; Niotal; SL-800750-23N; Somit; Stilnoct; Stimox; Sumenan;   Zolpidem   L-(+)-hemitartrate;   Zolpidem   tartrate. Sedative/hypnotic. Crystals; Soluble in H2O (2.3 g/100 ml). *Synthelabo Pharmacie.*

INDEXES

INDEXES

EINECS Numbers

EINECS Number Index

200-001-8	Formaldehyde	1786	200-386-2	Vitamin B₆ hydrochloride	4091
200-001-8	Paraformaldehyde	2859	200-400-7	Xylose	4121
200-014-9	Ergocalciferol	1582	200-401-2	Lindane	2246
200-018-0	Lactic acid	2174	200-403-3	Hydrochlorothiazide	1963
200-023-8	Estradiol	1588	200-407-5	Uridine	4046
200-024-3	DDT	1056	200-412-2	Vitamin E	4095
200-030-6	Propantheline bromide	3172	200-414-3	Ethopabate	1602
200-033-2	LSD	2226	200-419-0	Folic acid	1782
200-041-6	Amitriptyline	166	200-420-6	Carbostyril	691
200-056-8	Niclosamide	26266	200-423-2	Sulfaquinoxaline	3611
200-061-5	Sorbitol	3542	200-425-3	Vitamin B₁	4088
200-062-0	Alloxan	87	200-431-6	Lysochlor	2273
200-064-1	Aspirin	265	200-441-0	Niacin	2649
200-066-2	Vitamin C	4092	200-445-2	Levodopa	2233
200-075-1	Glucose	1832	200-449-4	Edetic acid	1560
200-076-7	Piperonyl butoxide	3050	200-455-7	Butter yellow	518
200-081-4	Benzimidazole	332	200-456-2	Phenylethyl alcohol	2983
200-087-7	2,4-Dinitrophenol	1434	200-464-6	2-Mercaptoethanol	2349
200-114-2	Methacetin	2379	200-467-2	Diethyl ether	1267
200-123-1	Urethane	4044	200-470-9	Linoleic acid	2247
200-132-0	Norethindrone Acetate	2731	200-480-3	Dimethoate	1344
200-133-6	Spironolactone	3550	200-481-9	Tetracycline	3700
200-143-0	Bronopol	486	200-484-5	Dieldrin	1233
200-145-1	Verapamil	4064	200-489-2	Promethazine	3157
200-149-3	Trichlorfon	3863	200-500-0	5'-Adenylic acid	61
200-157-7	Cysteine hydrochloride	1041	200-522-0	Leucine	2230
200-158-2	Cysteine	1040	200-526-2	p-Nitrobenzoic acid	2687
200-160-3	Prednisone	3145	200-529-9	Calcium disodium edetate	618
200-193-3	Nicotine	2667	200-532-5	Phenylmercury acetate	2986
200-198-0	Sodium salicylate	3504	200-539-3	Aniline	223
200-206-2	Metyrapone	2547	200-540-9	Calcium acetate	607
200-207-8	Idoxuridine	2020	200-541-4	Thioacetamide	3752
200-221-4	Levothyroxine sodium	2234	200-543-5	Thiourea	3770
200-228-2	Isonicotinic acid	2108	200-547-7	Dichlorvos	1211
200-231-9	Fenthion	1731	200-552-4	Diphemanil methylsulfate	1464
200-237-1	Metol	2537	200-555-0	Carbaryl	673
200-240-8	Nitroglycerin	2696	200-559-2	Lactose	2180
200-251-8	Biguanide	381	200-559-2	α-D-Lactose	2181
200-262-8	Carbon tetrachloride	688	200-565-5	Arecoline	249
200-268-0	Bis(tributyltin) Oxide	425	200-567-6	Colfosceril palmitate	935
200-270-1	Benzyltriethyl ammonium chloride	370	200-573-9	Tetrasodium edetate	3743
200-271-7	Parathion	2867	200-578-6	Ethanol	1593
200-272-2	Glycine	1856	200-579-1	Formic acid	1790
200-273-8	Alanine	67	200-580-7	Acetic acid	17
200-274-3	L-Serine	3340	200-589-6	Diethyl sulfate	1273
200-283-2	Adenosine triphosphate	60	200-616-1	Thymine	3776
200-285-3	Coumaphos	953	200-618-2	Benzoic acid	335
200-289-5	Glycerol	1838	200-621-9	Uracil	4039
200-294-2	Lysine	2270	200-641-8	Vitamin B₁	4087
200-296-3	Cystine	1042	200-646-5	Nitrofurantoin	2694
200-300-3	Benzyl trimethyl ammonium chloride	369	200-652-8	Pentetic acid	2916
200-302-4	Chlorhexidine acetate	760	200-655-4	Choline chloride	842
200-311-3	Cetrimonium bromide	734	200-659-6	Methyl alcohol	2415
200-312-9	Palmitic acid	2852	200-661-7	Isopropanol	2127
200-313-4	Stearic acid	3568	200-662-2	Acetone	23
200-315-5	Urea	4041	200-663-8	Chloroform	790
200-322-3	Phenobarbital sodium	2944	200-664-3	Dimethyl sulfoxide	1419
200-326-5	Metepa	2375	200-666-4	Hexachloroethane	1907
200-328-6	Phenytoin	2994	200-673-2	Vitamin D₃	4093
200-333-3	Fructose	1796	200-675-3	Sodium citrate	3400
200-334-9	Sucrose	3599	200-677-4	2-Mercaptoacetic acid	2345
200-338-0	Propylene glycol	3195	200-679-5	Dimethylformamide	1388
200-340-1	Propiolactone	3178	200-680-0	Vitamin B₁₂	4089
200-346-4	Sulfamethazine	3608	200-681-6	Norethindrone	2730
200-349-0	Chlordane	755	200-683-7	Vitamin A	4086
200-353-2	Cholesterol	841	200-686-3	Hexachloro-p-xylene	1909
200-362-1	Caffeine	606	200-693-1	Hydroxyzine	2010
200-375-2	Promethazine hydrochloride	3158	200-697-3	Hypoxanthine	2014
200-377-3	Phenarsazine oxide	2940	200-709-7	Ampicillin	210

EINECS Number Index

EINECS Number Index

EINECS Number Index

EINECS Number Index

EINECS Number Index

EINECS Number Index

EINECS Number Index

EINECS Number Index

EINECS Number Index

EINECS Number Index

EINECS Number Index

EINECS Number Index

INDEXES

INDEXES

CAS Registry Numbers

CAS Registry Number Index

CAS Registry Number Index

CAS Registry Number Index

CAS Registry Number Index

CAS Registry Number Index

134-50-9	Monacrin	2565	141-53-7	Sodium formate	3435
134-62-3	Diethyl toluamide	1275	141-57-1	n-Propyltrichlorosilane	3210
135-19-3	Naphthol	2620	141-62-8	Decamethyltetrasiloxane	1061
135-20-6	Cupferron	972	141-78-6	Ethyl Acetate	1609
135-58-0	Mesulfen	2366	141-82-2	Malonic acid	2303
135-88-6	2-Naphthylphenylamine	2624	141-86-6	2,6-Diaminopyridine	1109
136-23-2	Butyl zimate	583	141-97-9	Ethyl acetoacetate	1610
136-26-5	Capramide DEA	660	141-98-0	Isopropyl ethylthiocarbamate	2139
136-30-1	Sodium dibutyldithiocarbamate	3412	142-08-5	1-Hydroxy-2-pyridine	2008
136-32-3	2,4,5-Trichlorophenol, sodium salt	3879	142-16-5	Dioctyl maleate	1452
136-36-7	Resorcinol benzoate	3284	142-17-6	Calcium oleate	632
136-51-6	Calcium 2-ethylhexanoate	620	142-18-7	Ablunol GML	4
136-52-7	Cobalt 2-ethylhexanoate	908	142-22-3	Diallyl glycol carbonate	1104
136-53-8	Zinc 2-ethylhexanoate	4146	142-26-7	Acetamide MEA	13
136-60-7	n-Butyl benzoate	530	142-30-3	Dimethylhexynediol	1393
136-85-6	5-Methylbenzotriazole	2424	142-31-4	Sodium octyl sulfate	3480
137-08-6	Calcium d-pantothenate	636	142-47-2	Monosodium glutamate	2575
137-16-6	Sodium lauroyl sarcosinate	3456	142-62-1	Caproic acid	661
137-18-8	Phlorone	2997	142-71-2	Cupric acetate	973
137-20-2	Sodium oleylmethyltauride	3483	142-72-3	Magnesium acetate	2277
137-26-8	Thiram	3771	142-73-4	Iminodiacetic acid	2029
137-29-1	Copper dimethyldithiocarbamate	940	142-77-8	Butyl oleate	565
137-30-4	Bis(dimethyldithiocarbamato)zinc	401	142-78-9	Laurylethanolamide	2202
137-32-6	2-Methylbutanol	2428	142-82-5	Heptane	1891
137-40-6	Sodium propionate	3497	142-84-7	Dipropylamine	1484
137-42-8	Metam-sodium	2371	142-87-0	Sodium decyl sulfate	3411
137-58-6	Lidocaine	2237	142-90-5	Dodecyl methacrylate	1538
137-97-3	Di-o-tolylthiourea	1513	142-91-6	Isopropyl palmitate	2146
138-22-7	Butyl lactate	556	142-96-1	Dibutyl ether	1142
138-25-0	Dimethyl 5-sulfoisophthalate	1418	143-07-7	Lauric acid	2193
138-32-9	Cetrimonium tosylate	736	143-18-0	Potassium oleate	3117
138-55-6	Picrocrocin	3034	143-19-1	Sodium oleate	3482
138-86-3	Dipentene	1462	143-22-6	Butoxytriethylene glycol	517
138-89-6	p-(Dimethylamino)nitrosobenzene	1369	143-23-7	Bis(6-aminohexyl)amine	384
138-92-1	Betazole hydrochloride	375	143-24-8	Tetraglyme	3715
139-02-6	Sodium phenate	3491	143-27-1	Palmitamine	2851
139-05-9	Sodium cyclamate	3409	143-28-2	Oleyl alcohol	2802
139-07-1	Dehyquart LDB	1085	143-29-3	Bis(butoxyethoxyethoxy)methane	389
139-08-2	Myristalkonium chloride	2597	143-33-9	Sodium cyanide	3408
139-13-9	Nitrilotriacetic acid	2679	143-74-8	Phenol red	2948
139-33-3	Ethylenediamine Tetraacetic acid, Disodium Salt	1635	144-19-4	Trimethylpentanediol	3967
139-40-2	Propazine	3174	144-21-8	Disodium methanearsonate	1499
139-88-8	Sodium tetradecyl sulfate	3519	144-34-3	Methyl selenac	2515
139-89-9	Trisodium (2-Hydroxyethyl)ethylenediaminetriacetate	4013	144-35-4	Dipentite	1463
			144-55-8	Sodium bicarbonate	3382
139-96-8	Triethanolamine lauryl sulfate	3910	144-62-7	Oxalic acid	2821
140-01-2	Pentasodium pentetate	2911	144-79-6	Chloromethyldiphenylsilane	796
140-03-4	Methyl acetyl ricinoleate	2413	144-87-6	Glycarsamide	1836
140-04-5	Butyl acetyl ricinoleate	521	147-14-8	Copper phthalocyanine	946
140-08-9	Tris(2-chloroethyl) phosphite	4003	147-47-7	Trimethyl-1,2-dihydroquinoline	3958
140-11-4	Benzyl acetate	356	147-73-9	Mesotartaric acid	2365
140-31-8	Aminoethylpiperazine	150	147-85-3	Proline	3156
140-39-6	Narceol	2629	148-01-6	3,5-Dinitro-o-toluamide	1435
140-40-9	Aminitrazole	135	148-18-5	Diethyldithiocarbamate sodium	1253
140-66-9	p-Octylphenol	2778	148-24-3	8-Quinolinol	3254
140-67-0	Estragole	1589	148-25-4	Chromotrope acid	856
140-72-7	Cetylpyridinium bromide	742	148-75-4	R-Acid	3261
140-80-5	Ethyl acrylate	1611	148-79-8	Thiabendazole	3749
140-93-2	Sodium isopropyl xanthate	3452	149-30-4	2-Mercaptobenzothiazole	2348
141-04-8	Diisobutyl adipate	1300	149-32-6	Lichen sugar	2236
141-08-2	Glyceryl monoricinoleate	1845	149-44-0	Formaldehyde sodium sulfoxylate	1787
141-17-3	Dibutoxyethoxyethyl adipate	1135	149-57-5	Ethyl hexanoic acid	1654
141-20-8	PEG-2 laurate	2874	149-73-5	Methyl orthoformate	2497
141-22-0	Ricinoleic acid	3296	149-91-7	Gallic acid	1812
141-23-1	Methyl hydroxystearate	2474	150-38-9	Edetate Trisodium	1559
141-24-2	Methyl ricinoleate	2513	150-50-5	Merphos	2361
141-32-2	Butyl acrylate	522	150-69-6	Dulcin	1556
141-43-5	Ethanolamine	1596	150-76-5	Mequinol	2343
141-52-6	Sodium ethoxide	3427	150-84-5	Citronellyl acetate	886
			150-86-7	Phytol	3026

CAS Registry Number Index

506-68-3	Cyanogen bromide	994	541-91-3	Muscone	2589
506-87-6	Baker's salt	284	542-05-2	Acetone dicarboxylic acid	25
506-93-4	Guanidine nitrate	1875	542-10-9	Ethylidene acetate	1666
507-70-0	Borneol	450	542-18-7	Cyclohexyl chloride	1021
511-28-4	Vitamin D$_4$	4094	542-42-7	Calcium palmitate	635
512-69-6	Raffinose	3267	542-55-2	Isobutyl formate	2082
513-35-9	Amylene	215	542-75-6	1,3-Dichloropropene	1205
513-74-6	Ammonium dithiocarbamate	182	542-85-8	Ethyl isothiocyanate	1669
513-77-9	Barium carbonate	289	542-92-7	Cyclopentadiene	1029
513-79-1	Cobaltous carbonate	920	543-39-5	Myrcenol	2596
513-86-0	Acetyl methyl carbinol	37	543-80-6	Barium acetate	286
514-10-3	Abietic acid	2	544-17-2	Calcium formate	622
514-78-3	Canthaxanthin	659	544-19-4	Cupric formate	979
515-40-2	Neophyl chloride	2642	544-63-8	Myristic acid	2599
515-69-5	α-Bisabolol	382	544-76-3	Hexadecane	1911
517-25-9	Trinitromethane	3981	546-68-9	Titanium isopropoxide	3786
517-88-4	Alkanet	82	546-80-5	Thujone	3774
518-47-8	Fluorescein disodium salt	1763	546-93-0	Magnesium carbonate	2280
519-34-6	Maclurin	2274	547-64-8	Methyl lactate	2482
519-73-3	Tritane	4017	547-91-1	Ferron	1741
520-18-3	Kaempferol	2164	548-62-9	Gentian violet	1817
520-45-6	Dehydroacetic acid	1083	548-76-5	Irigenin	2066
520-85-4	Medroxyprogesterone	2329	549-18-8	Amitriptyline Hydrochloride	167
522-51-0	Dequalinium Chloride	1088	550-44-7	N-Methylphthalimide	2505
523-80-8	Parsley camphor	2869	550-82-3	Resazurin	3281
524-30-1	Pavlin	2870	551-92-8	Dimetridazole	1427
525-37-1	Eiver-Pick acid	1565	552-22-7	Iodothymol	2052
525-66-6	Propranolol	3188	552-30-7	Trimellitic anhydride	3947
525-82-6	Flavone	1754	552-89-6	o-Nitrobenzaldehyde	2685
526-75-0	o-Xylenol	4116	553-08-2	Thonzide	3772
526-95-4	D-Gluconic acid	1830	553-26-4	Bipyridyl	380
527-07-1	Sodium gluconate	3437	554-12-1	Methyl propionate	2509
527-09-3	Copper gluconate	941	554-71-2	Sozoiodolic acid	3546
527-60-6	Mesitol	2363	554-95-0	Trimesic acid	3948
528-29-0	Dinitrobenzene	1429	555-16-8	p-Nitrobenzaldehyde	2686
528-44-9	Trimellitic acid	3946	555-31-7	Aluminum isopropoxide	111
529-33-9	Tetrahydro-1-naphthol	3722	555-37-3	Neburon	2632
529-34-0	1-Tetralone	3728	555-43-1	Tristearin	4016
529-86-2	Oxyanthracene	2830	555-75-9	Aluminum ethoxide	104
530-91-6	Tetralol	3727	556-50-3	N-Glycylglycine	1867
532-05-8	1,3-Di-6-quinolylurea	1492	556-61-6	Methyl isothiocyanate	2481
532-27-4	Chloroacetophenone	769	556-67-2	Octamethylcyclotetrasiloxane	2752
532-32-1	Sodium benzoate	3381	556-82-1	Prenol	3147
532-82-1	Chrysoidine	860	556-88-7	Picrite	3033
533-74-4	Dazomet	1055	557-04-0	Magnesium stearate	2291
533-96-0	Sodium sesquicarbonate	3506	557-05-1	Zinc stearate	4158
534-08-7	Iohydrin	2055	557-08-4	Zinc undecenoate	4163
534-15-6	Dimethylacetal	1355	557-09-5	Bärostab® L 230	304
534-16-7	Silver carbonate	3357	557-61-9	Octacosanol	2742
534-17-8	Cesium carbonate	723	558-37-2	Trimethylvinylmethane	3979
534-22-5	Sylvan	3632	560-95-2	Bromonitroform	478
534-26-9	Lysidine	2269	563-12-2	Ethion	1600
534-52-1	4,6-Dinitrocresol	1433	563-43-9	Dichloroethylaluminum	1193
535-89-7	Crimidine	965	563-47-1	Methallyl chloride	2386
536-90-3	m-Anisidine	226	563-63-3	Silver acetate	3356
537-00-8	Cerous acetate	720	563-72-4	Calcium oxalate	633
537-65-5	Oxynone	2841	563-80-4	Isopropyl methyl ketone	2144
538-23-8	Tricaprylin	3861	564-25-0	Doxycycline	1551
538-24-9	Trilaurin	3944	569-61-9	Pararosaniline	2865
538-75-0	Dicyclohexyl carbodiimide	1225	569-65-3	Meclizine	2325
538-93-2	Isobutylbenzene	2080	575-44-0	1,6-Naphthalenediol	2610
540-10-3	Cetyl palmitate	741	575-75-7	Tonophosphan	3820
540-59-0	Acetylene dichloride	36	576-26-1	2,6-Xylenol	4118
540-84-1	Isooctane	2112	577-11-7	Docusate sodium	1525
540-88-5	tert-Butyl Acetate	520	578-94-9	Adamsite	58
540-97-6	Dodecamethylcyclohexasiloxane	1526	579-66-8	2,6-Diethylaniline	1250
541-02-6	Decamethylcyclopentasiloxane	1060	581-64-6	Thionine	3763
541-05-9	Hexamethylcyclotrisiloxane	1924	581-75-9	Naphthalene-2,6-disulfonic acid	2613
541-41-3	Ethyl chloroformate	1620	582-17-2	2,7-naphthalenediol	2612
541-73-1	m-Dichlorobenzene	1172	582-25-2	Potassium benzoate	3088

CAS Registry Number Index

CAS Registry Number Index

CAS Registry Number Index

CAS Registry Number Index

CAS Registry Number Index

CAS Registry Number Index

CAS Registry Number Index

CAS Registry Number Index

CAS Registry Number Index

CAS Registry Number Index

INDEXES

Names and Synonyms

INDEXES

Names and subjects

Name and Synonym Index

Name and Synonym Index

Name and Synonym Index

Name and Synonym Index

Name and Synonym Index

Name and Synonym Index

Name and Synonym Index

Name and Synonym Index

Name and Synonym Index

Name and Synonym Index

Name and Synonym Index

Name and Synonym Index

Name and Synonym Index

AI3-02065	4037	AI3-03111	2068
AI3-02066	1890	AI3-03112	841
AI3-02073	1892	AI3-03115	2860
AI3-02169	2761	AI3-03130	2174, 2176
AI3-02173	1065	AI3-03131	3453
AI3-02174	3320	AI3-03194	2961
AI3-02209-X	4111	AI3-03198	2198
AI3-02239	614	AI3-03254	1653
AI3-02246	1420	AI3-03266	1659
AI3-02247	1398	AI3-03268	1198
AI3-02254	1680	AI3-03271	1547
AI3-02257	1682	AI3-03294	1020
AI3-02258	2701	AI3-03307	1929
AI3-02259	2000	AI3-03311	1388
AI3-02261	3790	AI3-03313	1101
AI3-02264	2706	AI3-03314	37
AI3-02266	33	AI3-03340	1374
AI3-02272	2150	AI3-03346	1523
AI3-02293	1257	AI3-03357	141
AI3-02330	648	AI3-03358	151
AI3-02337	4017	AI3-03377	2235
AI3-02360	957	AI3-03386	1228
AI3-02370	1199	AI3-03424	2667
AI3-02372	1908	AI3-03438	702
AI3-02392	228	AI3-03464	730
AI3-02394	17	AI3-03485	1858
AI3-02406	448	AI3-03509	2499
AI3-02407	3321	AI3-03545	1572
AI3-02408	233	AI3-03574	40
AI3-02411	604	AI3-03582	3770
AI3-02415	2685	AI3-03611	302
AI3-02439	2590	AI3-03659	3049
AI3-02440	2591	AI3-03700	62
AI3-02445	3588	AI3-03710	335
AI3-02454	3305	AI3-03717	1110
AI3-02460	579, 582	AI3-03718	1513
AI3-02480	1415	AI3-03737	332
AI3-02502	3920	AI3-03776	1167
AI3-02503	539	AI3-03789	3593
AI3-02531	1106	AI3-03797	754
AI3-02574	1107	AI3-03833	2770
AI3-02583	350	AI3-03836	357
AI3-02587	1909	AI3-03840	1244
AI3-02602	625	AI3-03873	817
AI3-02615	1116	AI3-03901	3538
AI3-02692	1232	AI3-03902	3540
AI3-02706	3967	AI3-03947	156
AI3-02729	213	AI3-03948	4021
AI3-02738	2360	AI3-03949	155
AI3-02741	2204	AI3-03961	2783
AI3-02822	779	AI3-03967	4161
AI3-02836	13	AI3-03968	3255
AI3-02853	4122	AI3-03976	3217
AI3-02854	3603	AI3-03995	708
AI3-02855	3604	AI3-03996	3283
AI3-02901	1190	AI3-04009	438
AI3-02903	607	AI3-04085	1856
AI3-02904	96	AI3-04119	48
AI3-02907	3096	AI3-04162	2758
AI3-02911	1470	AI3-04164	2877
AI3-02915	2602	AI3-04165	3052
AI3-02916	2684	AI3-04167	3181
AI3-02920	1428	AI3-04169	963
AI3-02930	566	AI3-04240	3701
AI3-02955	367	AI3-04243	2526
AI3-02956	265	AI3-04257	24
AI3-03050	1644	AI3-04273	406
AI3-03053	223	AI3-04278	1306
AI3-03066	331	AI3-04360	3788
AI3-03075	3036	AI3-04453	1064
AI3-0310	335	AI3-04458	2353
AI3-03101	611	AI3-04459	3090

AI3-04462	474	AI3-07835	3381
AI3-04463	244	AI3-07852	3985
AI3-04464	2672	AI3-07853	1468
AI3-04466	1803	AI3-07866	3992
AI3-04470	4140	AI3-07870	1452
AI3-04491	3991	AI3-07924	2768
AI3-04492	324	AI3-07958	562
AI3-04494	514	AI3-07960	2516
AI3-04501	1804	AI3-07963	1287
AI3-04504	394	AI3-07965	1451
AI3-04505	3716	AI3-08004	1702
AI3-04596	4000	AI3-08011	326
AI3-04597	3693	AI3-08087	1054
AI3-04630	2136	AI3-08105	1690
AI3-04705	688	AI3-08157	1946
AI3-04707	138	AI3-08191	1225
AI3-04820	3419	AI3-08196	3849
AI3-04822	20	AI3-08197	4113
AI3-04856	2704	AI3-08222	1014
AI3-04859	2718	AI3-08223	1857
AI3-04869	3021	AI3-08224	1326
AI3-05096	1949	AI3-08269	971
AI3-05532	1408	AI3-08293	3412
AI3-05541	3757	AI3-08416	1255
AI3-05563	3727	AI3-08497	3720
AI3-05598	664	AI3-08499	4075
AI3-05599	2437	AI3-08515	1634
AI3-05612	590	AI3-08524	4118
AI3-05613	1268	AI3-08538	1202
AI3-05616	2942	AI3-08539	3598
AI3-05710	2775	AI3-08542	203
AI3-05733	2146	AI3-08544	2045
AI3-05739	2639	AI3-08582	3385
AI3-06000	2412	AI3-08584	227
AI3-06026	1389	AI3-08621	1156
AI3-06066	1323	AI3-08625	3612
AI3-06080	1375	AI3-08632	835
AI3-06122	2085	AI3-08643	2451
AI3-06286	883	AI3-08657	4051
AI3-06292	2300, 2301	AI3-08667	3619
AI3-06293	2308	AI3-08678	1399
AI3-06295	2958	AI3-08751	1365
AI3-06297	3595	AI3-08765	1538
AI3-06299	273	AI3-08777	3183
AI3-06468	3948	AI3-08778	594
AI3-06521	2745	AI3-08780	2622
AI3-06522	1911	AI3-08831	417
AI3-06556	731	AI3-08848	2996
AI3-06557	2755	AI3-08850	3682
AI3-06625	353	AI3-08870	1214
AI3-07037	2426	AI3-08876	2180
AI3-07039	3722	AI3-08877	704
AI3-07159	3732	AI3-08899	2230
AI3-07288	1995	AI3-08903	518
AI3-07312	2788	AI3-08916	4112
AI3-07422	139	AI3-08920	2971
AI3-07498	1104	AI3-08926	2683
AI3-07540	2461	AI3-08927	3587
AI3-07541	3203	AI3-08931	1556
AI3-07551	511	AI3-08935	686
AI3-07553	502	AI3-08937	179
AI3-07565	3023	AI3-09021	3478
AI3-07570	3718	AI3-09042	2941
AI3-07577	1537	AI3-09048	3696
AI3-07620	2802	AI3-09057	1615
AI3-07621	2417	AI3-09064	1042
AI3-07662	388	AI3-09065	160
AI3-07701	661	AI3-09071	3213
AI3-07710	2349	AI3-09073	235
AI3-07775	3866	AI3-09076	3485
AI3-07776	778	AI3-09080	2034
AI3-07796	2246	AI3-09085	3599

Name and Synonym Index

Name and Synonym Index

Name and Synonym Index

Name and Synonym Index

Name and Synonym Index

Name and Synonym Index

Name and Synonym Index

Name and Synonym Index

Name and Synonym Index

Name and Synonym Index

Amdon grazon	3027	Amid kyseliny mravenci	1788
Amdon	3027	Amidate (pesticide)	3004
Amdry 6410	850	Amidate	3004
Amea 100	1607	Amide, dodecyl-,	409
Ameisenatod	2246	Amidex AME	13
Ameisenmittel merck	2246	Amidex CP	660
Ameisensäure	1790	Amidinoquanidine	381
Ameisensaeure	1790	Amido betaine C, C-45	923
Ameisentod	2246	Amido-G acid	132
Amen	2329, 2330	Amido-G-acid	132
Amephyt	130	Amidor	2387
Amercide	668	Amido-R-acid	3325
Amerfand	3599	Amidosulfonic acid	3609
Amerfil	3076	Amidosulfuric acid	3609
Amerfond	3599	Amidotrizoate de sodium	1118
American Cyanamid 3422	2867	Amidotrizoate meglumine	1117
American Cyanamid 3,911	2998	Amidotrizoato sodico	1118
American Cyanamid 4,049	2294	Amidox	1202
American Cyanamid 5223	1544	Amietol M 11	2459
American Cyanamid 12880	1344	Amietol M 21	1057
American Cyanamid AC 52,160	3663	Amilan CM 1001	2736
American Cyanamid CL-47,300	1722	Amilan CM 1001C	2736
American Cyanamid CL-47470	2340	Amilan CM 1001G	2736
American Cyanamid CL-52160	3663	Amilan CM 1011	2736
American Cyanamid E.I. 52,160	3663	Amilan CM 1031	2736
American Cyanamid KPAM	3059	Amilar	3069
American Cyanamid P-250	3059	Amilfenol	217
American pennyroyal oil	2791	Amilit	167
American vermilion	845	Amilmetacresol	214
Amerox OE-20	2801	Amilperoxy pivalate	133
Amerox oe-20	2801	Amilphenol	217
Ameroxol EO 2	2801	Aminacrine hydrochloride	2565
Ameroxol OE 10	2801	Amincene C 140	1275
Ameroxol OE 2	2801	Amincene C-EM	1275
Ameroxol OE-20	2801	Amine 4 2,4-D Weed Killer	1202
Amersep MP 3R	3424	Amine 10	1071
Amersol	2015	Amine 12	1533
Amerstat 274	1633	Amine 12-98D	1071
Amerstat® 233	1055	Amine 14D	2598
Amerstat® 282	2450	Amine 18-90	3575
Amerstat® 294	442	Amine 220®	2803
Amerstat® 300	1130	Amine 225	3573
Ametazole dihydrochloride	375	Amine AB	3575
A-Methapred	2507	Amine BB	1533, 3565
Amethyst	3241	Amine CS-1135®	1404
Ametoterina	135	Amine D	1082
Ametrex	130	Amine M210D	1231
Ametrine	130	Amine, diisobutyl-	1301
Ametryn	130	Amine, dipentyl	1113
Ametryne	130	Amine, diphenyl, 4-nitro-	2692
Am-Fam 400	2015	Amine, tributyl-	3855
Amfamox	1714	Amine-8D	2784
Amfebutamon hydrochlorid	493	Amine-C$_4$	525
Amfebutamona	492	Amines, coco alkyl	928
Amfebutamone	492	Amines, soya alkyl	3544
Amfebutamonum	492	Amines, tallow alkyl	3645
Amfipen V	210	Amineurin	167
Amfipen	210	Aminic acid	1790
AM-Fol	170	Aminitrazol	135
Amforol (Veterinary)	433	Aminitrazole	135
Amforol	2167	Aminitrozol	135
Amgard CPC 405	3010	Aminitrozole	135
Amgard CPC	3010	Aminitrozolum	135
Amgard TMCP	4005	Aminoacetic acid	1856
Ami-Anelun	167	Aminoacridine hydrochloride	2565
Amical 48	1298	Aminoacridine monohydrochloride	2565
Amicarbalida	131	Aminoacridinium chloride	2565
Amicarbalide	131	Aminoakridin	2565
Amicarbalidum	131	Aminoazijnzuur	1856
Amichlor	498	Aminobenzamide	232
Amicide	201	Aminobenzene	223
Amid kyseliny akrylove	48	p-Aminobenzenesulfonic acid	3610

783

Name and Synonym Index

Name and Synonym Index

Araldite HT	1054	Arecolin	249
Araldite HT 907	1921	Arecoline	249
Araldite HT 976	1054	Arecoline-acetarsol	1554
Araldite HT 986	995	Aredion	3704
Araldite HY 951	3905	Areecop	945
Araldite HY 960	1523	Arekolin	249
Araldite PT 810	1852	Arelon R	2151
Araldite XB 2879B	995	Arelon	2151
Araldite XB 2979B	995	Arena	3659
Aralo	2867	Arendal	76
Arancio cromo	845	Aresenid	253
Arapamyl	4065	Aresin	250
Arapol 0712	2192	Arestin	2555
Arasan	3771	Aretan	2631
Arasan 42-S	3771	Aretan 6	2631
Arasan 70	3771	Arezin	250
Arasan 70-S Red	3771	Arezine	250
Arasan 75	3771	Arfen	15
Arasan-M	3771	Argenti nitras	3362
Arasan-SF	3771	Argentous fluoride	3360
Arasan-SF-X	3771	Argentous oxide	3363
Arasemide	1807	Argentyne	4067
Arasorb 750	3060	Argerol	3362
Arasorb S 100F	3060	Argezin	269
Arathane	1443	Argiflex	2168
Arazate®	4143	Argilla alba	2168
Arbestab DSTDP	1505	Argilla	2168
Arbinex 30TN	1887	Argirec B 22	2168
Arbitex	2246	Argirec KN 15	2168
Arbitol E	1288	Argucide	3121
Arbocel BC 200	714	Argus DLTDP	1337
Arbocel	714	Arial	3324
Arbocell B 600/30	714	ARI-B	3282
Arbogal	1722	Aricept	1546
Arborseal	666	Aries Antox	1607
Arbortrine	313	Arilat	673
Arbralene	2541	Arilate	313, 673
Arcablock	3189	Arilin	2543
Arcacil	2883	A-Rin	1567
Arcanum duplicatum	3135	Ariotox	2369
Arcasin	2883	Aripax	2263
Arcilla blanca	2168	Aristocort A	3840
Arcillite	2578	Aristocort acetonide	3840
Arco-cee	4092	Aristocort	3839, 3840
Arcoiran	3630	Aristoderm	3840
Arconate® 5000	3193	Aristogel	3840
Arconate® Propylene Carbonate	3193	Aristol	2052
Arconol	523	Aristosol	3842
Arcosolv DPM	1488	Aritromicina	277
Arcosolv® PM	2406	Arizole	3043
Arcosolv® PMA	3200	Arklone P	3883
Arcotal S	2246	Arkofix NG	1403
Arcotrate	2934	Arkofix	1403
Arcton 1	1768	Arkopal N-090	2728
Arcton 3	829	Arkopon T	3483
Arcton 4	784	Arkotine	1056
Arcton 6	1185	Arlacel C®	3539
Arcton 9	3872	Arlacel 20	3535
Arcton 11	3872	Arlacel 20	3266
Arcton 12	1185	Arlacel 40	3538
Arcton 22	784	Arlacel 60	3537
Arcton 33	1206	Arlacel 80	3536
Arcton 63	3883	Arlacel® 83	3539
Arcton 114	1206	Arlacel 85	3540
Arcton	829	Arlacel 129	1840
Arctuvin	1980	Arlacel 161	1841
Ardap	1038	Arlacel 165	1846
Arecaidine methyl ester	249	Arlacel 169	1841
Arecaidine	248	Arlacel® 186	3195
Arecaine	248	Arlacide A	760
Arecaline	249	Arlacide G	761
Arecholine	249	Arlamol DOA	1449

Name and Synonym Index

Name and Synonym Index

Atcardil	266	Atlantic Sky Blue A	861
ATCC 15413	2566	Atlantic	684
ATE	3915	Atlas G 711	2132
Ateben hydrochloride	2734	Atlas G 917	3199
Ateben	2733	Atlas G 924	3201
ATEC	40	Atlas G 2124	2874
Atecard	266	Atlas G 2127	3073
Atehexal	266	Atlas G 2129	3073
Atem	2063	Atlas G 2133	2192
Atemi C	1039	Atlas G 2142	1864
Atemi	1039	Atlas G 2144	1864
Atemur	1780	Atlas G 2146	1263
Atenase	2666	Atlas G 3300	2132
Atenblock	266	Atlas G 3705	2192
Atendol	266	Atlas G 3707	2192
Atenet	266	Atlas G 3802	732
Ateni	266	Atlas G 3816	732
Atenil	266	Atlas G 3851	3199
Atenol 1A pharma	266	Atlas G 3915	2801
Atenol acis	266	Atlas G 3915	2801
Atenol AL	266	Atlas G 3920	2801
Atenol Atid	266	Atlas G 3920	2801
Atenol Cophar	266	Atlas G 4829	2192
Atenol ct	266	Atlas Linuron	267
Atenol Fecofar	266	Atlas white titanium dioxide	3785
Atenol Gador	266	Atlasol 103	3411
Atenol Genericon	266	Atlasol KAD	3573
Atenol GNR	266	Atlox 4862	3496
Atenol Heumann	266	ATM 2 (nylon)	2736
Atenol MSD	266	Atmer 05	3536
Atenol NM Pharma	266	Atmer 100	3535
Atenol Nordic	266	Atmer 137	2801
Atenol PB	266	Atmer 1007	1840
Atenol Quesada	266	Atmonil	1817
Atenol Stada	266	Atmos 150	1841
Atenol Tika	266	Atmul 67	1841
Atenol Trom	266	Atmul 84	1841
Atenol von ct	266	Atmul 124	1841
Atenolin	266	Atocin	874
Atenol-Mepha	266	Atofan	874
Atenolol	266	Atomel 28	2067
Atenololum	266	Atomel 300M200	2067
Atenol-ratiopharm	266	Atomel 500M	2067
Atenol-Wolff	266	Atomel 95	2067
Atenomel	266	Atomic Red	869
Atens	1567	Atomic sulfur	3621
Atensina	896	Atomiron 44MR	2067
Atensine	1121	Atomiron 5M	2067
Aterax dihydrochloride	2011	Atomiron AFP 25	2067
Aterax	2010, 2011	Atomiron AFP 5	2067
Atereal	266	Atomit	612
Aterol	266	Atomite	612
Atgard C	1211	Atonalyt	696
Atgard V	1211	Atonik	3478
Atgard	1211	Atophan	874
Athrombin	4101	Atorvastatin Calcium	268
Athrombine-K	4100	Atosil	3157, 3158
Athylen	1627	Atox B	245
Athyl-gusathion	274	Atox F	245
Atiflan	2625	Atox R	245
Atigoa	874	Atox S	245
Atilen	1121	Atoxan	673
Atipi	60	ATP	60
Atiram	3771	ATP (nucleotide)	60
Atiran	2631	Atpeg 300	3065
Atisuril	86	Atral	1492
Ativan	2263	Atralidon	15
ATJ-S graphite	1873	Atranex	269
ATJ-S	1873	Atrasine	269
Atlacide	201	Atrataf	269
Atlacide	3395	Atratol A	269
Atlantic Chrysoidine Y	860	Atratol	269

Name and Synonym Index

Name and Synonym Index

Name and Synonym Index

Name and Synonym Index

Name and Synonym Index

Name and Synonym Index

4-01-00-02437	3757	4-02-00-00157	2070
4-01-00-02471	516	4-02-00-00166	2768
4-01-00-02473	411	4-02-00-00022	2497
4-01-00-02474	1486	4-02-00-00204	1821
4-01-00-02475	3999	4-02-00-00214	1605
4-01-00-02497	3198	4-02-00-00215	3920
4-01-00-02501	486	4-02-00-00215	534
4-01-00-02515	502	4-02-00-00217	1857
4-01-00-02550	1682	4-02-00-00220	3200
4-01-00-02551	2639	4-02-00-00253	3831
4-01-00-02554	1131	4-02-00-00386	18
4-01-00-02556	1929	4-02-00-00390	2930
4-01-00-02565	1950	4-02-00-00391	567
4-01-00-02597	1664	4-02-00-00395	34
4-01-00-02604	3967	4-02-00-00458	3929
4-01-00-02687	584	4-02-00-00474	767
4-01-00-02701	1381	4-02-00-00481	1619
4-01-00-02706	2788	4-02-00-00488	771
4-01-00-02751	1838	4-02-00-00504	1169
4-01-00-02762	2696	4-02-00-00508	3864
4-01-00-02786	3960	4-02-00-00519	3865
4-01-00-02812	2894	4-02-00-00695	3181
4-01-00-02816	2934	4-02-00-00704	2509
4-01-00-02841	2318	4-02-00-00708	574
4-01-00-03028	1195	4-02-00-00722	3182
4-01-00-03046	785	4-02-00-00728	3183
4-01-00-03070	2681	4-02-00-00753	1050
4-01-00-03090	2998	4-02-00-00779	588
4-01-00-03092	3669	4-02-00-00802	589
4-01-00-03103	1355	4-02-00-00806	594
4-01-00-03103	9	4-02-00-00819	1888
4-01-00-03121	12	4-02-00-00843	2091
4-01-00-03142	750	4-02-00-00847	2085
4-01-00-03143	751	4-02-00-00853	2093
4-01-00-03147	3863	4-02-00-00868	4051
4-01-00-03215	768	4-02-00-00889	2460
4-01-00-03229	586	4-02-00-00908	3052
4-01-00-03250	506	4-02-00-00912	569
4-01-00-03268	4050	4-02-00-00917	661
4-01-00-03271	2910	4-02-00-00958	1892
4-01-00-03305	2478	4-02-00-00982	2758
4-01-00-03310	3036	4-02-00-00986	2432
4-01-00-03313	1203	4-02-00-00991	3861
4-01-00-03314	1890	4-02-00-00994	2924
4-01-00-03318	1894	4-02-00-01005	3921
4-01-00-03323	593	4-02-00-01018	2877
4-01-00-03329	2477	4-02-00-01041	1064
4-01-00-03337	2783	4-02-00-01044	2442
4-01-00-03345	2782	4-02-00-01082	2193
4-01-00-03352	2718	4-02-00-01090	2483
4-01-00-03360	1304	4-02-00-01102	2195
4-01-00-03385	2083	4-02-00-01104	1529
4-01-00-03625	1868	4-02-00-01126	2599
4-01-00-03644	1101	4-02-00-01132	2145
4-01-00-03659	1833	4-02-00-01132	562
4-01-00-03662	33	4-02-00-01157	2852
4-01-00-03974	2398	4-02-00-01167	2146
4-01-00-04023	1099	4-02-00-01168	741
4-01-00-04119	1295	4-02-00-01206	3568
4-01-00-04223	4121	4-02-00-01216	2516
4-01-00-04302	1832	4-02-00-01218	1695
4-02-00-00171	737	4-02-00-01219	2149
4-02-00-00027	2140	4-02-00-01219	577
4-02-00-00029	2082	4-02-00-01222	1263, 1649
4-02-00-00050	3103, 3216, 3408	4-02-00-01240	3565
4-02-00-00094	17	4-02-00-01990	2470
4-02-00-00138	3190	4-02-00-01242	2744
4-02-00-00141	2130	4-02-00-01455	50
4-02-00-00143	519	4-02-00-01457	2414
4-02-00-00149	2078	4-02-00-01460	1611
4-02-00-00151	520	4-02-00-01463	522
4-02-00-00152	213	4-02-00-01465	2079

Name and Synonym Index

Name and Synonym Index

4-04-00-01685	1561	4-05-00-00866	1439, 1440
4-04-00-01705	141	4-05-00-00930	2715
4-04-00-01740	1368	4-05-00-00951	4111
4-04-00-01740	156	4-05-00-00968	1909
4-04-00-01881	155	4-05-00-01042	2080
4-04-00-02363	3329	4-05-00-01067	1251
4-04-00-02369	373	4-05-00-01125	1326
4-04-00-02387	1916	4-05-00-01126	1327
4-04-00-02428	2029	4-05-00-01130	579, 582
4-04-00-02441	2679	4-05-00-01200	1534
4-04-00-02449	1560	4-05-00-01369	2518
4-04-00-02453	1638	4-05-00-01369	4080
4-04-00-02454	2916	4-05-00-01542	1234
4-04-00-02533	1370	4-05-00-00048	1021
4-04-00-02717	2270	4-05-00-01885	1056
4-04-00-02823	164	4-05-00-01985	2932
4-04-00-02998	263	4-05-00-02160	3817
4-04-00-03155	1042	4-06-00-00831	836
4-04-00-03304	1276	4-06-00-00036	1020
4-04-00-03499	1399	4-06-00-00041	1221
4-04-00-03991	1604	4-06-00-00072	1017
4-04-00-03994	1608	4-06-00-00151	2338
4-04-00-04007	3974	4-06-00-00253	3677
4-04-00-04011	3973	4-06-00-00568	1471
4-04-00-04096	2443	4-06-00-00571	2964
4-04-00-04101	1382	4-06-00-00629	2947
4-04-00-04111	1187	4-06-00-00634	2958
4-04-00-04114	1923	4-06-00-00695	3992
4-04-00-04125	2752	4-06-00-00710	1152
4-04-00-04128	1060	4-06-00-00718	1472
4-04-00-04169	1196	4-06-00-00720	3991
4-04-00-04184	1197	4-06-00-00773	1770
4-04-00-04201	1240	4-06-00-00020	1018
4-04-00-04204	3913	4-06-00-00885	1200
4-04-00-04208	2527	4-06-00-00908	1202
4-04-00-04212	2528	4-06-00-00927	1046
4-04-00-04227	3871	4-06-00-00942	1201
4-04-00-04249	3875	4-06-00-00962	3877, 3878
4-04-00-04256	4082, 4083	4-06-00-00994	3303
4-04-00-04257	4012	4-06-00-01005	3880
4-04-00-04258	3886	4-06-00-01025	2888
4-04-00-04271	993	4-06-00-01061	1132
4-04-00-04272	997	4-06-00-01283	2703
4-04-00-04273	3911	4-06-00-01290	417
4-04-00-04312	3690	4-06-00-01327	2861
4-04-00-04314	3738	4-06-00-01337	2867
4-04-00-04349	3707	4-06-00-01463	3766
4-04-00-04359	447	4-06-00-01587	1190
4-04-00-04398	3915	4-06-00-05878	955
4-04-00-04400	3937	4-06-00-01642	2890
4-05-00-01783	1070	4-06-00-01979	3891
4-05-00-00058	2246	4-06-00-01991	2321
4-05-00-00169	1008	4-06-00-01996	2322
4-05-00-00381	1905	4-06-00-02035	957
4-05-00-00406	4075	4-06-00-02041	2962
4-05-00-00456	3041	4-06-00-02064	2273
4-05-00-00664	3866, 3867	4-06-00-02112	2629
4-05-00-00667	3691	4-06-00-02149	2691
4-05-00-00668	3692	4-06-00-02152	1432
4-05-00-00670	1904	4-06-00-02222	358
4-05-00-00726	1198	4-06-00-02597	1177
4-05-00-00728	3256	4-06-00-02632	366
4-05-00-00728	3659	4-06-00-02651	1128
4-05-00-00805	823	4-06-00-03011	2950
4-05-00-00809	793	4-06-00-03020	1686
4-05-00-00815	832	4-06-00-03049	3305
4-05-00-00816	780	4-06-00-03067	2983
4-05-00-00817	316	4-06-00-03096	4116
4-05-00-00820	350	4-06-00-03099	1407
4-05-00-00823	779	4-06-00-03126	4115
4-05-00-00855	804	4-06-00-03152	835
4-05-00-00865	1437	4-06-00-03164	4117

4-06-00-03212	2126	4-08-00-02360	2206
4-06-00-03219	1408	4-08-00-02368	2163
4-06-00-03221	969	4-08-00-02442	2831
4-06-00-03225	1219	4-08-00-02442	340
4-06-00-03253	2363	4-08-00-03163	1459
4-06-00-03279	1447	4-08-00-03256	81
4-06-00-03279	2511	4-08-00-03268	236
4-06-00-03292	572	4-08-00-03272	231
4-06-00-03397	537	4-08-00-03498	2507
4-06-00-03400	561	4-08-00-03505	341, 342
4-06-00-03402	2590	4-08-00-03532	3146
4-06-00-03484	2778	4-08-00-03568	3219
4-06-00-03493	1150	4-08-00-03629	3839
4-06-00-03796	219	4-09-00-03188	1107
4-06-00-03817	1589	4-09-00-00307	359
4-06-00-04133	4086	4-09-00-00356	1257
4-06-00-04219	673	4-09-00-00372	3284
4-06-00-04579	2968	4-09-00-00715	353
4-06-00-04648	331	4-09-00-00721	351
4-06-00-04722	1217	4-09-00-00778	1952
4-06-00-04761	971	4-09-00-01614	2971
4-06-00-04930	2830	4-09-00-01716	1275
4-06-00-05557	708	4-09-00-01726	2526
4-06-00-05658	3283	4-09-00-01733	3810
4-06-00-05684	817	4-09-00-01752	1964
4-06-00-05772	801	4-09-00-02176	1
4-06-00-05809	3619	4-09-00-02425	1680
4-06-00-01636	3704	4-09-00-02505	1465
4-06-00-05892	2813	4-09-00-03172	1689
4-06-00-05939	3593	4-09-00-03175	1153
4-06-00-06013	553	4-09-00-03177	1306
4-06-00-06043	3753	4-09-00-03180	1455
4-06-00-06074	1145	4-09-00-03181	1165
4-06-00-06337	2464	4-09-00-03181	406
4-06-00-06543	2622	4-09-00-03183	1445
4-06-00-06564	2611	4-09-00-03186	1075, 1232
4-06-00-06570	2612	4-09-00-00290	530
4-06-00-06658	1199	4-09-00-03189	1227
4-06-00-06659	1908	4-09-00-03218	532
4-06-00-06691	2402	4-09-00-03241	1349
4-06-00-06717	438	4-09-00-03242	1136
4-06-00-06801	408	4-09-00-03256	512
4-06-00-06806	1000	4-09-00-03292	2122
4-06-00-06807	3328	4-09-00-03293	1398
4-06-00-06811	1296	4-09-00-03294	1270
4-06-00-07327	3233	4-09-00-03295	1105
4-06-00-07361	2996	4-09-00-03295	2124
4-06-00-07602	234	4-09-00-03301	3673
4-07-00-00802	2240	4-09-00-03306	404
4-07-00-00165	2119	4-09-00-03318	3674
4-07-00-00213	656	4-09-00-03640	1626
4-07-00-00230	1292	4-09-00-03746	3946
4-07-00-00316	704	4-10-00-00138	265
4-07-00-00641	769	4-10-00-00143	2514
4-07-00-00723	970	4-10-00-00149	1693
4-07-00-00021	1995	4-10-00-00153	2118
4-07-00-00808	2591	4-10-00-00153	576
4-07-00-01002	2031	4-10-00-00154	2969
4-07-00-01015	3728	4-10-00-00157	368
4-07-00-02073	2707	4-10-00-00125	3321
4-07-00-02680	659	4-10-00-00360	2500
4-07-00-02786	2672	4-10-00-00367	1665
4-08-00-00176	3320	4-10-00-00374	3209
4-08-00-00242	2959	4-10-00-00375	566
4-08-00-00320	1982	4-10-00-00378	1897
4-08-00-01221	2730, 2731	4-10-00-00475	482
4-08-00-01279	337	4-10-00-00544	2408
4-08-00-01763	4057	4-10-00-01184	2003
4-08-00-01765	1706	4-10-00-01459	3215
4-08-00-01792	3280	4-10-00-01998	1813
4-08-00-02211	2329	4-10-00-02006	3154
4-08-00-02212	2330	4-10-00-02752	3672

Name and Synonym Index

4-10-00-04088	3289	4-13-00-01092	2379
4-11-00-00248	1702	4-13-00-01154	1556
4-11-00-00049	328	4-13-00-01179	3750
4-11-00-00050	324	4-13-00-01306	1054
4-11-00-00051	326	4-13-00-00038	137
4-11-00-00122	1057, 1137, 1301, 1484	4-13-00-02148	2551
4-11-00-00122	3855	4-14-00-02792	3787
4-11-00-00229	3800	4-14-00-01010	232
4-11-00-00241	3803	4-14-00-01016	1374
4-11-00-00027	325	4-14-00-01092	138
4-11-00-00376	3801	4-14-00-01741	2049
4-11-00-00527	2615	4-14-00-01008	2418
4-11-00-00597	2832	4-14-00-02804	890
4-11-00-00622	3333	4-14-00-02811	132
4-12-00-00018	1379	4-14-00-02823	152, 157
4-12-00-00022	1222	4-16-00-00228	3602
4-12-00-00072	1225	4-16-00-00248	3603
4-12-00-00008	1024	4-16-00-00249	3604
4-12-00-00102	1003	4-16-00-00895	3745
4-12-00-00250	1614	4-16-00-00951	3995
4-12-00-00255	2135	4-16-00-00972	2987
4-12-00-00619	2993	4-16-00-01033	2988
4-12-00-00620	3176	4-16-00-01523	1479
4-12-00-00741	672	4-16-00-01526	1189
4-12-00-00839	4128	4-16-00-01560	2990
4-12-00-01141	1236	4-16-00-01568	3876
4-12-00-01149	837	4-16-00-01736	2352
4-12-00-01257	1170	4-17-00-00993	1854
4-12-00-01263	1518	4-17-00-01005	1851
4-12-00-01265	3887	4-17-00-03066	2180
4-12-00-01504	465	4-17-00-03462	3315
4-12-00-01558	1369	4-17-00-04465	2669
4-12-00-01619	2692	4-17-00-06070	758
4-12-00-01681	1214	4-17-00-06699	1083
4-12-00-01689	1428	4-18-00-01601	3550
4-12-00-01734	473	4-18-00-01949	2050
4-12-00-01737	1481	4-18-00-00344	953
4-12-00-01764	1514	4-19-00-00779	3050
4-12-00-01764	3818	4-19-00-00641	1345
4-12-00-01777	21	4-19-00-01824	2714
4-12-00-01813	3813	4-19-00-01935	251
4-12-00-01816	1703	4-19-00-01936	1631
4-12-00-01843	139	4-19-00-02917	1571
4-12-00-01848	2095	4-20-00-01016	1461
4-12-00-01985	825	4-21-00-00362	3888
4-12-00-01986	826	4-21-00-04640	2975
4-12-00-02538	2237	4-24-00-01235	2020
4-12-00-02638	3815	4-25-00-02184	965
4-12-00-02841	1250	4-25-00-03033	893
4-12-00-03005	1082	4-25-00-00939	2266
4-12-00-03009	2623	4-26-00-00093	348
4-12-00-03015	2966	4-26-00-00095	1985
4-12-00-03128	2624	4-26-00-00144	2424
4-13-00-01946	1523	4-26-00-00022	1028
4-13-00-00079	2977	4-26-00-00460	274, 275
4-13-00-00111	1151	4-26-00-00637	3837
4-13-00-00115	2134, 2980	4-26-00-00642	3633
4-13-00-00116	1477	4-26-00-01208	3367
4-13-00-00119	65	4-26-00-01244	334
4-13-00-00235	1110	4-26-00-01253	2334
4-13-00-00243	3797	4-26-00-01274	1922
4-13-00-00259	3796	4-26-00-01645	2766
4-13-00-00260	3791	4-26-00-03615	61
4-13-00-00390	1116	4-27-00-01214	2956
4-13-00-00396	1475	4-27-00-00023	1679
4-13-00-00806	227	4-27-00-00027	2690
4-13-00-00865	20	4-27-00-00613	3612
4-13-00-00953	226	4-27-00-00969	2522
4-13-00-00969	1246	4-27-00-00015	2580
4-13-00-01009	1108	4-27-00-01253	3157
4-13-00-01017	2941	4-27-00-01435	1476
4-13-00-01038	386	4-27-00-01862	388

Name and Synonym Index

4-27-00-01864	2581	5-19-03-00457	2948
4-27-00-01866	1328	5-19-03-00460	471
4-27-00-01866	531	5-19-03-00307	2869
4-27-00-01867	3761	5-19-04-00472	3990
4-27-00-01868	347	5-19-04-00021	3193
4-27-00-02263	3285	5-19-06-00015	2401
4-27-00-02516	2826	5-19-06-00456	1762
4-27-00-03330	2069	5-19-06-00568	3840
4-27-00-03594	3281	5-19-07-00251	693
4-27-00-03681	2837	5-19-08-00189	1509
4-27-00-04675	159	5-19-09-00112	2860
4-27-00-04676	135	5-20-02-00071	1727
4-27-00-04842	2378	5-20-04-00003	1920
4-27-00-05884	2882	5-20-05-00422	2889
4-27-00-06389	1998	5-20-05-00500	2677
4-27-00-07436	1055	5-20-05-00421	3698
4-27-00-07537	1299	5-20-06-00093	936
4-27-00-07537	3806, 3807, 3808	5-20-07-00375	3247
4-27-00-07875	233	5-20-08-00247	671
4-27-00-08084	3553	5-21-02-00055	1489
4-27-00-09796	58	5-21-03-00296	2051
5-17-01-00169	1917	5-21-03-00252	3254
5-17-01-00020	1572	5-21-06-00330	2205
5-17-01-00039	3616	5-21-06-00328	1691
5-17-01-00056	1684	5-21-06-00330	4079
5-17-01-00017	3203	5-21-06-00331	1023
5-17-01-00177	3617	5-21-06-00444	662
5-17-01-00297	3765	5-21-06-00566	1007
5-17-01-00322	3632	5-21-07-00164	2059
5-17-01-00577	3591	5-21-07-00164	3208
5-17-02-00252	4107	5-21-09-00438	3598
5-17-03-00012	90	5-21-09-00543	818
5-17-03-00035	1853	5-21-10-00069	3152
5-17-03-00115	3719	5-21-10-00130	3724
5-17-03-00196	1293	5-21-10-00136	666, 668
5-17-03-00338	1805	5-21-10-00221	2068
5-17-03-00011	550, 551	5-21-10-00270	3023
5-17-04-00048	681	5-21-10-00273	2505
5-17-08-00410	3601	5-21-10-00366	2001
5-17-09-00003	3178	5-21-11-00117	1027
5-17-09-00034	663	5-21-11-00118	1783
5-17-09-00115	1335	5-21-11-00141	1507
5-17-09-00292	1803	5-22-01-00322	1880
5-17-09-00306	1804	5-22-01-00322	248, 249
5-17-10-00552	1754	5-22-02-00046	900
5-17-10-00143	954	5-22-02-00057	2649, 2650
5-17-11-00055	2297	5-22-02-00115	998
5-17-11-00134	3723	5-22-03-00140	2039
5-17-11-00199	2489	5-22-03-00484	874
5-17-11-00260	3696	5-22-04-00268	2668
5-17-11-00006	3597	5-22-07-00582	1741
5-17-11-00265	3684	5-22-11-00255	1109
5-17-11-00492	3213	5-22-11-00475	2823
5-18-01-00378	1988	5-22-13-00585	3027
5-18-01-00386	4032	5-23-01-00030	3046
5-18-01-00114	2305	5-23-01-00042	3933
5-18-02-00343	2963	5-23-01-00235	2325
5-18-03-00211	1053	5-23-01-00257	150
5-18-04-00162	4100	5-23-01-00406	1994
5-18-04-00381	1286	5-23-01-00462	2010
5-18-04-00497	1818	5-23-03-00041	272
5-18-05-00251	2164	5-23-03-00385	2269
5-18-05-00494	3244	5-23-04-00119	3573
5-18-05-00519	3316	5-23-04-00126	2374
5-18-05-00026	2071	5-23-04-00191	2027
5-18-06-00102	1806	5-23-04-00320	2021
5-18-08-00562	3947	5-23-04-00456	3975
5-18-09-00555	1807	5-23-05-00058	1427
5-18-09-00269	1825	5-23-05-00062	699
5-18-11-00316	2790	5-23-05-00063	2543
5-19-01-00553	3319	5-23-05-00319	2803
5-19-01-00016	1458	5-23-06-00196	332

Name and Synonym Index

Name and Synonym Index

Bentolite L 3	2578	Benzidene Yellow YB-1	871
Bentone 660	315	Benzidine Lacquer Yellow G	871
Bentone	2168	Benzidine Yellow 1178	871
Bentonit T	315	Benzidine Yellow 45-2650	871
Bentonite	315	Benzidine Yellow 45-2680	871
Bentonite 2073	315	Benzidine Yellow 45-2685	871
Bentonite magma	315	Benzidine Yellow AAOT	872
Bentonite, calcian-sodian	315	Benzidine Yellow ABZ 249	872
Bentosolon 82	315	Benzidine Yellow ABZ-245	871
Bentox 10	2246	Benzidine Yellow E	871
Bentrol	2061	Benzidine Yellow G	871, 872
Benur	1550	Benzidine Yellow GF	871
Ben-u-ron	15	Benzidine Yellow GGT	872
Benylate	359	Benzidine Yellow GR	871
Benzac W	1125	Benzidine Yellow GT	871
Benzac	1125	Benzidine Yellow GTR	871
Benzagel 10	1125	Benzidine Yellow HG PLV	871
Benzagel	1125	Benzidine Yellow HG	871
Benzaknen	1125	Benzidine Yellow L	872
Benzal alcohol	358	Benzidine Yellow OT (6CI)	872
Benzal chloride	316	Benzidine Yellow OT	872
Benzalaceton	363	Benzidine Yellow OTYA 8055	872
Benzalacetone	363	Benzidine Yellow Toner YA-8081	871
Benzalacetophenone	745	Benzidine Yellow Toner YT-378	871
Benzalconio cloruro	83	Benzidine Yellow Toner	871
Benzaldehyde	317	Benzidine Yellow WD-266 (water dispersible)	871
Benzaletas	730	Benzidine Yellow YB 5722	871
Benzalkonii chloridum	83	Benzidine Yellow YB-1	871
Benzalkonium A	83	Benzidine Yellow	871
Benzalkonium chloride	83	Benzile (cloruro di)	361
Benzalox	900	Benzilideneacetone	363
Benzaminblau 3B	2651	Benzimidazole	332
Benzamine Blue 3B	2651	Benziminazole	332
Benzamine Blue	2651	Benzin (Obs.)	321
Benzamizole	2157	Benzin B70	2606
Benzamycin	1125	Benzin	321, 1815, 2239, 2606
Benzanil Blue 3BN	2651	Benzine (light petroleum distillate)	2239
Benzanil Blue R	2651	Benzinoform	688
Benzanil Sky Blue	861	Benzinol	3862
Benzanilide, 2-iodo-	312	Benzisothiazolin-3-one	333
Benzar	309	Benzisotriazole	348
Benzashave	1125	Benzo Blue	2651
Benzatropine methanesulfonate	355	Benzo Blue 3B	2651
Benzazimide	318	Benzo Blue 3BS	2651
Benzazimidol hydrate	1985	Benzo Flex 2-45	1257
Benzazimidone	318	Benzo Sky Blue A-CF	861
Benzeen	321	Benzo Sky Blue S	861
Benzen	321	Benzo(a)phenanthrene	859
Benzenamine	223	Benzoan sodny	3381
Benzene hexachloride	2246	Benzoate of soda	3381
Benzene iodide	2046	Benzoate sodium	3381
Benzene	321	Benzoate	335
Benzenesulfonic acid	325	Benzoblau 3B	2651
Benzenesulfonyl chloride	328	Benzochloryl	1056
Benzenethiol	3766	Benzocyclohexane	3726
Benzenol	2946	Benzoepin	1569
Benzethoni chloridum	330	Benzoesäure	335
Benzethonii chloridum	330	Benzoesäure (Na-salz)	3381
Benzethonium chloride	330	Benzoesäure GK	335
Benzethonium	330	Benzoesäure GV	335
Benzetonio cloruro	330	Benzoesäurebenzylester	359
Benzetonium chloride	330	Benzoesaeure GK	335
Benzex	2246	Benzoesaeure GV	335
Benzhydrol	331	Benzoesaeure	335
Benzhydryl alcohol	331	Benzoesaeurebenzylester	359
Benzidam	223	Benzoflex 9-88	1487
Benzidene Yellow	871	Benzoflex S-552	2895
Benzidene Yellow ABZ-245	871	Benzofur D	2978
Benzidene Yellow ABZ-249	872	Benzofur GG	160
Benzidene Yellow G	872	Benzofur MT	1110
Benzidine Yellow (Grease proof)	872	Benzofur P	161
Benzidene Yellow WD-266 (Water Dispersible)	871	Benzofuroline	3282

Name and Synonym Index

Name and Synonym Index

Name and Synonym Index

BRN 0028784	1804	BRN 0112169	2305
BRN 0030397	2020	BRN 0112366	248
BRN 0043446	2837	BRN 0113176	380
BRN 0054346	1825	BRN 0113915	818
BRN 0054612	61	BRN 0114512	3254
BRN 0055641	2940	BRN 0116039	1055
BRN 0057767	3550	BRN 0116658	2424
BRN 0060069	3840	BRN 0118522	3023
BRN 0060420	893	BRN 0121832	1023
BRN 0063410	1571	BRN 0122031	1007
BRN 0071586	3315	BRN 0123045	249
BRN 0075455	3316	BRN 0124246	1002
BRN 0077011	1286	BRN 0124341	2334
BRN 0078195	260	BRN 0124428	2505
BRN 0079290	3601	BRN 0124982	1874
BRN 0079763	3203	BRN 0126214	3612
BRN 0079785	1572	BRN 0126797	159
BRN 0080142	2860	BRN 0127683	4032
BRN 0082340	3723	BRN 0127995	965
BRN 0083882	1043	BRN 0128710	1345
BRN 0084271	2071	BRN 0128764	3724
BRN 0088275	2034	BRN 0129639	3698
BRN 0088554	3157	BRN 0129768	1988
BRN 0092693	758	BRN 0130665	1427
BRN 0093796	2180	BRN 0133939	4107
BRN 0094179	2266	BRN 0136380	3319
BRN 0094324	1762	BRN 0136786	2069
BRN 0096259	2882	BRN 0140946	2001
BRN 0102364	84	BRN 0143237	2956
BRN 0102411	1684	BRN 0146013	1186
BRN 0102549	2580	BRN 0149001	2690
BRN 0102551	1458	BRN 0153190	3213
BRN 0102555	3046	BRN 0153223	334
BRN 0102723	3719	BRN 0153639	2051
BRN 0102969	1679	BRN 0155011	2205
BRN 0103222	3765	BRN 0155197	2889
BRN 0103483	551	BRN 0157021	1476
BRN 0103668	550	BRN 0157598	1754
BRN 0103733	3632	BRN 0158370	531
BRN 0103853	2027	BRN 0162395	2489
BRN 0104225	2269	BRN 0167361	135
BRN 0104361	1994	BRN 0171120	2039
BRN 0104363	150	BRN 0174850	3285
BRN 0104541	1335	BRN 0177776	1328
BRN 0105228	2522	BRN 0178698	58
BRN 0105755	1803	BRN 0191684	347
BRN 0105871	90	BRN 0192376	3761
BRN 0106251	2826	BRN 0192803	874
BRN 0106291	1805	BRN 0193373	1783
BRN 0106909	2297	BRN 0195747	2869
BRN 0106919	663	BRN 0196453	2401
BRN 0106934	662	BRN 0196633	2378
BRN 0107004	3617	BRN 0199405	251
BRN 0107274	1293	BRN 0202022	3633
BRN 0107283	936	BRN 0206829	3208
BRN 0107344	1998	BRN 0211560	3696
BRN 0107711	998	BRN 0212344	1631
BRN 0107765	3616	BRN 0217725	1489
BRN 0107913	3193	BRN 0221382	3281
BRN 0107971	1691	BRN 0221723	3990
BRN 0108440	3598	BRN 0222552	2581
BRN 0108441	3597	BRN 0223133	222
BRN 0108513	1109	BRN 0223832	1741
BRN 0108582	3591	BRN 0225301	3888
BRN 0109092	1851	BRN 0225482	3837
BRN 0109591	2649, 2650	BRN 0235560	3836
BRN 0109682	332	BRN 0236487	3573
BRN 0110149	1806	BRN 0239535	2669
BRN 0110309	3247	BRN 0249503	2975
BRN 0110428	1917	BRN 0258464	1119
BRN 0110513	4079	BRN 0261304	3608
BRN 0112133	348	BRN 0273790	1122

BRN 0280476	275	BRN 0506142	1012
BRN 0285796	388	BRN 0506163	2477
BRN 0288063	3050	BRN 0506166	2345
BRN 0288466	1028	BRN 0506211	3240
BRN 0290026	3611	BRN 0506325	992
BRN 0295174	3553	BRN 0506416	2138
BRN 0297468	274	BRN 0506422	750
BRN 0298051	1461	BRN 0506455	1619
BRN 0304401	2164	BRN 0506510	2496
BRN 0313390	1084	BRN 0506523	3766
BRN 0317313	3244	BRN 0506646	1997
BRN 0321392	2010	BRN 0506657	3754
BRN 0326470	2948	BRN 0506719	957
BRN 0327083	953	BRN 0506731	1390
BRN 0332002	2325	BRN 0506796	2235
BRN 0351654	2050	BRN 0506966	3713
BRN 0356813	1922	BRN 0507066	3182
BRN 0361386	2766	BRN 0507407	1254
BRN 0372527	471	BRN 0507434	1156
BRN 0383644	954	BRN 0507468	1614
BRN 0383659	2068	BRN 0507600	3808
BRN 0384788	1818	BRN 0507757	876
BRN 0385636	37	BRN 0507950	769
BRN 0385653	1620	BRN 0507965	3791
BRN 0385686	2821	BRN 0508152	350
BRN 0385737	18	BRN 0508189	2987
BRN 0386119	226	BRN 0508509	232
BRN 0386123	1982	BRN 0508624	2599
BRN 0386133	2623	BRN 0508730	2990
BRN 0386210	227	BRN 0509801	2500
BRN 0387672	139	BRN 0510203	832
BRN 0387887	2379	BRN 0511662	2959
BRN 0391839	337	BRN 0513992	2778
BRN 0393969	2969	BRN 0521151	3333
BRN 0397241	762	BRN 0527479	3658
BRN 0403040	4086	BRN 0530220	314
BRN 0471175	1024	BRN 0533604	2393
BRN 0471201	1672	BRN 0541488	2878
BRN 0471300	1409	BRN 0545683	2021
BRN 0471308	793	BRN 0558070	2819
BRN 0471357	2977	BRN 0565491	272
BRN 0471388	3320	BRN 0577209	3222
BRN 0471389	351	BRN 0587472	2546
BRN 0471401	708	BRN 0605263	1633
BRN 0471434	1241	BRN 0605266	1640
BRN 0471492	3810	BRN 0605275	2571
BRN 0471603	138	BRN 0605282	3220
BRN 0471700	780	BRN 0605287	1187
BRN 0471734	3672	BRN 0605291	2639
BRN 0471803	233	BRN 0605293	1371
BRN 0472689	3801	BRN 0605303	34
BRN 0472690	3803	BRN 0605310	52
BRN 0472792	4057	BRN 0605314	1264
BRN 0473755	900	BRN 0605315	1235
BRN 0474706	1116	BRN 0605317	394
BRN 0475312	2960	BRN 0605318	2479
BRN 0475735	386	BRN 0605328	1862
BRN 0479075	3027	BRN 0605349	48
BRN 0485597	2823	BRN 0605363	1322
BRN 0497773	2668	BRN 0605366	2179
BRN 0505943	785	BRN 0605369	768
BRN 0505947	558	BRN 0605390	1833
BRN 0505974	1484	BRN 0605391	24
BRN 0505979	156	BRN 0605396	2414
BRN 0506001	1137	BRN 0605398	1101
BRN 0506003	3173	BRN 0605399	2478
BRN 0506007	17	BRN 0605437	2435
BRN 0506012	148	BRN 0605438	767
BRN 0506034	3873	BRN 0605439	771
BRN 0506058	2910	BRN 0605448	3905
BRN 0506061	586	BRN 0605459	2486
BRN 0506071	3181	BRN 0605737	875

Name and Synonym Index

BRN 1875862	673	BRN 1913062	402
BRN 1878083	2092	BRN 1913145	2831
BRN 1879891	3184	BRN 1914037	81
BRN 1880877	1107	BRN 1914064	1153
BRN 1881200	1816	BRN 1914324	3256
BRN 1881718	236	BRN 1914746	341
BRN 1882318	2456	BRN 1915153	4006
BRN 1882657	1056	BRN 1915198	344
BRN 1884366	2535	BRN 1915474	755
BRN 1884514	1199	BRN 1915526	2861
BRN 1885571	3303	BRN 1915671	2730
BRN 1885795	2162	BRN 1915994	1455
BRN 1885803	1626	BRN 1916263	1445
BRN 1886299	1217	BRN 1916919	1296
BRN 1887087	342	BRN 1919922	1379
BRN 1887127	3219	BRN 1921024	2818
BRN 1887367	1722	BRN 1927455	4074
BRN 1888236	3991	BRN 1936365	6
BRN 1888840	1047	BRN 1942785	2993
BRN 1889288	1227	BRN 1948305	1859
BRN 1889437	2590	BRN 1952435	2385
BRN 1890696	406	BRN 1952749	801
BRN 1892885	3891	BRN 1955756	2615
BRN 1893077	1232	BRN 1973805	3659
BRN 1893555	3932	BRN 1974129	1731
BRN 1894045	1517	BRN 1976809	1046
BRN 1896901	3663	BRN 1978326	820
BRN 1898520	659	BRN 1978448	1896
BRN 1900796	1021	BRN 1978786	1432
BRN 1901008	1008	BRN 1985420	2798
BRN 1901553	4075	BRN 1985768	3366
BRN 1901563	4111	BRN 1988797	1562
BRN 1902271	1015	BRN 1990231	3628
BRN 1902277	539	BRN 1990738	1435
BRN 1902288	2338	BRN 1997888	1221
BRN 1902859	2032	BRN 1998314	3214
BRN 1903396	1251	BRN 2004306	3223
BRN 1903999	878	BRN 2005093	1165
BRN 1904092	1228	BRN 2006754	1136
BRN 1904175	823	BRN 2007363	512
BRN 1904445	3577	BRN 2009141	1075
BRN 1905012	1408	BRN 2011337	3710
BRN 1905732	2983	BRN 2012395	3829
BRN 1905828	1326	BRN 2014120	2836
BRN 1906042	1380	BRN 2015026	3756
BRN 1906225	515	BRN 2016207	1000
BRN 1906267	4116	BRN 2016567	3328
BRN 1906471	1292	BRN 2023076	1515
BRN 1906543	1020	BRN 2025982	1734
BRN 1906604	3676	BRN 2030021	2920
BRN 1907120	572	BRN 2034522	3856
BRN 1907337	2246	BRN 2035465	2896
BRN 1907611	656	BRN 2038119	2351
BRN 1907692	1201	BRN 2039935	1668
BRN 1908117	969	BRN 2042383	2612
BRN 1908125	2629	BRN 2042864	817
BRN 1908224	217	BRN 2044002	2622
BRN 1908225	561	BRN 2045054	1870
BRN 1908981	2084	BRN 2045105	2715
BRN 1909107	1534	BRN 2045714	2962
BRN 1909332	2122	BRN 2046227	3722
BRN 1909333	3673	BRN 2046455	2268
BRN 1909764	2163	BRN 2046711	1275
BRN 1910025	3691	BRN 2046968	3727
BRN 1910383	1150	BRN 2047450	2726
BRN 1910871	2464	BRN 2049262	1128
BRN 1910963	2672	BRN 2049280	359
BRN 1910992	3859	BRN 2049542	1145
BRN 1911160	1465	BRN 2049713	2015
BRN 1912251	1398	BRN 2050274	1812
BRN 1912500	1689	BRN 2050910	2824
BRN 1912585	1904	BRN 2051282	1909

Bromkal 79-8DE	2741	Bromure d'hydrogene anhydre	1961
Bromkal 82-0DE	1058	Bromuro de cetrimonio	734
Bromkal 83-10DE	1058	Bromuro de hidrogeno anhidro	1961
Bromkal G 1	2884	Bromuro de proantelina	3172
Bromkal P 67-6HP	4006	Bromuro de tetradonio	2603
Bromkal pur 3	3851	Bromuro de tonzonio	3772
Brom-methan	2427	Bromuro di etile	1640
Bromnatrium	3388	Bromuro di metile	2427
Bromo DNA	473	Bromwasserstoff	1961
Bromo	464	Broncho Inhalat	69
Bromoacetanilide	465	Bronchodiagnostin	3208
Bromoanilide	465	Bronchoselectan	31
Bromoantifebrin	465	Broncho-Spray	68
Bromobutanol	463	Bronchospray	69
Bromocet	742	Broncodil	69
Bromochloromethane	469	Broncovaleas	68
Bromociclen	470	Bronidiol	486
Bromociclene	470	Bronkhodiagnostin	3208
Bromocicleno	470	Bronocot	486
Bromociclenum	470	Bronopol	486
Bromocresol green	471	Bronopolu	486
Bromocresol purple	461	Bronopolum	486
Bromocyan	994	Bronosol	486
Bromocyane	994	Bronotak	486
Bromocyanide	994	Bronter	68
Bromocyanogen	994	Brontex	933
Bromocyclen	470	Bronze powder	937
Bromocyclene	470	Bronze Red 16913 Yellowish	869
Bromodan	470	Bronze Red RO	869
Bromoeosin	1571	Bronze Scarlet CA	869
Bromoethane	474	Bronze Scarlet CBA	869
Bromoethene	4072	Bronze Scarlet CT	869
Bromoethylene	4072	Bronze Scarlet CTA	869
Bromoflor	1599	Bronze Scarlet Toner	869
Bromofluoroform	476	Brookite	3785
Bromoform	477	Broom	464
Bromoforme	477	Broommethaan	2427
Bromoformio	477	Broomwaterstof	1961
Bromofume	1640	Broprodifacoum	460
Brom-O-Gas	2427	Brosombra	3208
Brom-O-Gaz	2427	Brown Acetate of Lime	607
Bromol	3851	Brown acetate	607
Bromometano	2427	Brown copper oxide	988
Bromomethane	2427	Brown for Fur T	1110
Bromone	460	Brown manganese ore	2315
Bromonitroform	478	Broxynil	482
Bromophthal	3684	Brozil	1816
Bromopicrin	479	Brufanic	2015
Bromore	460	Brufen Retard	2015
Brom-O-Sol	2427	Brufen	2015
Bromotrifluoromethane	476	Bruflam	2015
Bromotril	482	Brufort	2015
Bromotrimethylsilane	481	Bruinsteen	2314
Bromotrinitromethane	478	Brulan	3658
Bromowodor	1961	Brumolin	4100
Bromoxynil butyrate	483	Brunol	576
Bromoxynil heptanoate	484	Brush bullet	3658
Bromoxynil octanoate	485	Brush buster	1161
Bromoxynil	482	Brushkiller	1941
Bromphthal	3684	Brush-off	2546
Bromsalans	3853	Brush-rhap	1202
Brom-tetragnost	462	Bruzem	1340
Bromur di metile	2427	Brygou	3184
Bromure de cetrimonium	734	BS 8T	1158
Bromure de cyanogen	994	BS	577
Bromure de methyle	2427	BSB-S 40	3078
Bromure de potassium	3093	BSB-S-E	3078
Bromure de propantheline	3172	BSC-refine D	328
Bromure de tetradonium	2603	B-Selektonon	1202
Bromure de tonzonium	3772	B-Selektonon M	2321
Bromure de vinyle	4072	BSWL 202	2223
Bromure d'ethyle	474	BT	3744

Name and Synonym Index

Name and Synonym Index

Name and Synonym Index

Name and Synonym Index

Name and Synonym Index

Name and Synonym Index

Camphechlor	3822
Camphechlore	3822
Camphene, chlorinated	3822
Camphene, octachloro-	3822
Camphene	655
Campherol	2164
Camphochlor	3822
Camphofene huileux	3822
Camphogen	2150
Camphol	450
Campho-Phenique Cold Sore Gel	656, 2946
Campho-Phenique Gel	656, 2946
Campho-Phenique Liquid	656, 2946
Camphor tar	2608
Camphor, synthetic	656
Camphor	656
Camphostyl	657
Campicillin	210
Campilit	994
Camposan	1599
Campoviton 6	4091
Camsylate	657
Canacert erythrosine BS	1585
Canacert Indigo Carmine	2035
Canacert Sunset Yellow FCF	865
Canacert tartrazine	42
Canadien 2000	460
Canadol	2239
Canamex Glicepol 182	1840
Cancarb	684
Candaseptic	2273
Candeptin	658
Canderel	262
Candex	269
Candex	2739
Candicidin	658
Candimon	658
Candio-Hermal	2739
Candle Scarlet 2B	3604
Candle Scarlet B	3604
Candle Scarlet G	3604
Cane sugar	3599
Canesorb	683
Canex	3310
Canguard® 327	1404
Canlub	1873
Cannon	66
Canogard	1211
Canopar	3748
Cantan	4092
Cantaxin	4092
Canthaxanthine	659
Cantrex	2167
Can-Trol	2322
Canzol	1758
CaO	537, 634
CAO 1	1147
CAO 3	1147
CAO 5	408
CAO 14	408
Caocobre	988
CAP (oxide)	4168
Caparol 80W	3160
Caparol	3160
Capcithin™	2228
Capcure EH 30	1523
Capicillin	2881
Capital with Codeine	15
Capitol	83
Caplaril	1963
Caplenal	86
Capmul MCM-C 8	1843

Caporit	627
Capoten	669
Capozide	1963
Capozide	669
Capozide	669
Cappicaine	2237
Capramide DEA	660
Capran 77C	2736
Capran 80	2736
Caprane	1443
Capresin	896
Capric acid	1064
Capric acid chloride	1067
Capric acid diethanolamide	660
Capric acid methyl ester	2442
Capric acid, sodium salt	3410
Capric alcohol	1065
Caprinic acid sodium salt	3410
Caprinic acid	1064
Caprinic alcohol	1065
Caprinitrile	1063
Caprinoyl chloride	1067
Caproamide polymer	2736
Caprodat	696
Caproic acid	661
Caproic alcohol	1946
Caprolactam oligomer	2736
Caprolactam polymer	2736
Caprolactam	662
Caprolactone	663
Caprolattame	662
Caprolin	673
Caprolon B	2736
Caprolon V	2736
Capron 8250	2736
Capron 8252	2736
Capron 8253	2736
Capron 8256	2736
Capron B	2736
Capron GR 8256	2736
Capron GR 8257	2736
Capron GR 8258	2736
Capron PK 4	662, 2736
Capron	2736
Capronic acid	661
Caproyl alcohol	1946
Capryl alcohol	664, 2761
Capryl chloride	2772
Capryl o-phthalate	1165
Capryl peroxide	1448
Caprylaldehyde	2783
Caprylamine	2784
Capryldinitrophenyl crotonate	1443
Caprylene	2762
Caprylic acid	2758
Caprylic acid methyl ester	2432
Caprylic acid sodium salt	3479
Caprylic acid triglyceride	3861
Caprylic acid, 1,2,3-propanetriyl ester	3861
Caprylic alcohol	664, 2761
Caprylic alcohol, secondary	664
Caprylic aldehyde	2783
Caprylic/capric triglyceride	665
Caprylin	1837
Caprylin	3861
Caprylsäure	2758
Caprylyl peroxide	1448
Caprylylamine	2784
Caprynic acid	1064
Caprysin	896
Capsulec 51-SB	2228
Capsulec 51-UB	2228

Name and Synonym Index

Name and Synonym Index

Casamine SH	3573	Caswell No. 055	747
Cascorez	3080	Caswell No. 056	253
Cashew nut oil	705	Caswell No. 057	254
Cashew nut shell liquid	705	Caswell No. 058	3275
Cashew nut shell oil (untreated)	705	Caswell No. 059	258
Cashew, nutshell liq.	705	Caswell No. 061B	4092
Casiflux	646	Caswell No. 062	264
Casiflux VP 413-004	646	Caswell No. 063	269
Casing head gasoline	1815	Caswell No. 063B	276
Cassadan	95	Caswell No. 068A	286
Cassappret SR	3069	Caswell No. 069	289
Cassella's acid	706, 889	Caswell No. 070	292
Cassia aldehyde	875	Caswell No. 071B	302
Cassic acid	3289	Caswell No. 072A	3343
Cassurit LR	1403	Caswell No. 073A	1085
Castanea sativa Mill tannin	3652	Caswell No. 074A	3834
Castor Oil	707	Caswell No. 075A	313
Castor oil (Ricinus communis L.)	707	Caswell No. 077	321
Castor oil [Oil, edible]	707	Caswell No. 077A	1734
Castor oil acid	3296	Caswell No. 079	2246
Castor oil acid, methyl ester	2513	Caswell No. 079A	333
Castor oil aromatic	707	Caswell No. 081	335
Castor oil, dehydrated	1081	Caswell No. 081B	2378
Castor oil, hydrogenated	1967	Caswell No. 081EA	356
Castorwax MP 70®	1967	Caswell No. 081F	358
Castorwax MP 80®	1967	Caswell No. 081G	339
Castorwax NF®	1967	Caswell No. 082	359
Castorwax®	1967	Caswell No. 083E	3282
Castrix Grains	965	Caswell No. 083EE	3657
Castrix	965	Caswell No. 087	378
Casul 70HF	619	Caswell No. 093	1217
Caswell No. 002A	6	Caswell No. 096	3159
Caswell No. 003	17	Caswell No. 097	3160
Caswell No. 003A	18	Caswell No. 101	425
Caswell No. 003B	22	Caswell No. 104	442
Caswell No. 003C	2023	Caswell No. 106	264
Caswell No. 003F	2022	Caswell No. 106A	445
Caswell No. 004	23	Caswell No. 109	448
Caswell No. 005AB	39	Caswell No. 109B	449
Caswell No. 009	47	Caswell No. 111	458
Caswell No. 009A	50	Caswell No. 112	464
Caswell No. 010	52	Caswell No. 114AAA	457
Caswell No. 011	66	Caswell No. 116	470
Caswell No. 011A	73	Caswell No. 116A	486
Caswell No. 012	75	Caswell No. 119	482
Caswell No. 024	84	Caswell No. 119A	485
Caswell No. 026	88	Caswell No. 119B	498
Caswell No. 028A	97	Caswell No. 119C	505
Caswell No. 029	100	Caswell No. 119D	503
Caswell No. 031	3017	Caswell No. 119E	483
Caswell No. 031A	123	Caswell No. 120	511
Caswell No. 033A	2565	Caswell No. 121	513
Caswell No. 033D	2542	Caswell No. 121B	547
Caswell No. 033G	233	Caswell No. 124A	523
Caswell No. 036	4021	Caswell No. 125	140
Caswell No. 037	156	Caswell No. 125B	3670
Caswell No. 039	3027	Caswell No. 125D	3671
Caswell No. 041	170	Caswell No. 125G	532
Caswell No. 041A	2533	Caswell No. 125H	547
Caswell No. 041B	173	Caswell No. 128A	1343
Caswell No. 042	284	Caswell No. 128F	1442
Caswell No. 044B	187	Caswell No. 128GG	544
Caswell No. 045	190	Caswell No. 130	311
Caswell No. 047	201	Caswell No. 130A	566
Caswell No. 048	202	Caswell No. 130BB	552
Caswell No. 048A	205	Caswell No. 130E	570
Caswell No. 049A	213	Caswell No. 131A	3669
Caswell No. 050	217	Caswell No. 131B	512
Caswell No. 051B	218	Caswell No. 132A	589
Caswell No. 051C	223	Caswell No. 132B	592
Caswell No. 051E	225	Caswell No. 135	600
Caswell No. 052A	235	Caswell No. 136AA	605

Name and Synonym Index

Name and Synonym Index

Name and Synonym Index

CCRIS 162		CCRIS 390	2550
CCRIS 170	860	CCRIS 392	2452
CCRIS 172	865	CCRIS 395	2475
CCRIS 174	869	CCRIS 410	2543
CCRIS 178	3602	CCRIS 412	2552
CCRIS 179	897	CCRIS 413	2556
CCRIS 180	1642	CCRIS 423	2623
CCRIS 181	953	CCRIS 427	2653
CCRIS 183	954	CCRIS 431	2661
CCRIS 184	956	CCRIS 436	2679
CCRIS 187	972	CCRIS 453	2699
CCRIS 191	3409	CCRIS 475	1485
CCRIS 192	1052	CCRIS 484	2730
CCRIS 194	1054	CCRIS 485	2731
CCRIS 202	1056	CCRIS 491	386
CCRIS 203	1110	CCRIS 493	2867
CCRIS 204	871	CCRIS 494	2887
CCRIS 218	1122	CCRIS 495	3256
CCRIS 224	1157	CCRIS 503	2944
CCRIS 225	1191	CCRIS 504	2946
CCRIS 230	1641	CCRIS 508	137
CCRIS 231	1211	CCRIS 509	2978
CCRIS 233	1217	CCRIS 513	147
CCRIS 235	1233	CCRIS 515	2994
CCRIS 236	1253	CCRIS 519	3021
CCRIS 237	1449	CCRIS 520	3027
CCRIS 242	406	CCRIS 522	3050
CCRIS 243	1273	CCRIS 529	3092
CCRIS 245	1277	CCRIS 530	3963
CCRIS 246	1344	CCRIS 536	3178
CCRIS 248	1345	CCRIS 540	3203
CCRIS 251	1611	CCRIS 541	3206
CCRIS 265	518	CCRIS 553	3319
CCRIS 266	1417	CCRIS 558	3475
CCRIS 268	1420	CCRIS 559	3476
CCRIS 269	1438	CCRIS 560	3559
CCRIS 275	1458	CCRIS 564	3590
CCRIS 277	1569	CCRIS 569	3617
CCRIS 280	1572	CCRIS 570	2592
CCRIS 292	1588	CCRIS 571	3652
CCRIS 293	2932	CCRIS 577	3694
CCRIS 294	1632	CCRIS 578	3693
CCRIS 295	1559	CCRIS 579	3695
CCRIS 297	1640	CCRIS 582	1506
CCRIS 298	1650	CCRIS 584	3752
CCRIS 304	1651	CCRIS 588	3770
CCRIS 307	1698	CCRIS 590	3785
CCRIS 310	1174	CCRIS 596	3798
CCRIS 311	1731	CCRIS 597	3814
CCRIS 314	1734	CCRIS 598	3809
CCRIS 315	1760	CCRIS 600	3822
CCRIS 318	1786	CCRIS 602	4081
CCRIS 324	1816	CCRIS 603	3862
CCRIS 325	1887	CCRIS 604	3872
CCRIS 329	1904	CCRIS 605	3880
CCRIS 330	2246	CCRIS 606	3907
CCRIS 331	1907	CCRIS 607	3932
CCRIS 335	1908	CCRIS 611	3978
CCRIS 336	1957	CCRIS 612	1733
CCRIS 340	1959	CCRIS 614	4006
CCRIS 346	3254	CCRIS 615	3985
CCRIS 356	2047	CCRIS 616	2651
CCRIS 357	2208	CCRIS 619	4044
CCRIS 359	2213	CCRIS 620	4072
CCRIS 368	2484	CCRIS 621	4073
CCRIS 369	2294	CCRIS 622	1192
CCRIS 371	2318	CCRIS 625	401
CCRIS 373	2330	CCRIS 626	86
CCRIS 375	2334	CCRIS 628	234
CCRIS 380	2338	CCRIS 629	339
CCRIS 385	2402	CCRIS 630	1125
CCRIS 389	2427	CCRIS 634	3082
	2456		

Name and Synonym Index

Name and Synonym Index

CCRIS 2924	592	CCRIS 3415	1259
CCRIS 2926	3225	CCRIS 3430	2770
CCRIS 2933	3239	CCRIS 3431	1990
CCRIS 2935	3765	CCRIS 3434	3709
CCRIS 2939	2448	CCRIS 3436	2900
CCRIS 2941	2297	CCRIS 3437	2666
CCRIS 2942	3811	CCRIS 3466	33
CCRIS 2949	3005	CCRIS 3467	2305
CCRIS 2950	2300, 2301	CCRIS 3478	995
CCRIS 2951	2174, 2176	CCRIS 3488	2518
CCRIS 3002	670	CCRIS 3491	43
CCRIS 3009	3166	CCRIS 3493	864
CCRIS 3031	3792	CCRIS 3500	1477
CCRIS 3043	2104	CCRIS 3501	1185
CCRIS 3044	1850	CCRIS 3509	4140
CCRIS 3047	1576	CCRIS 3510	3089
CCRIS 3057	1369	CCRIS 3524	3253
CCRIS 3064	3382	CCRIS 3527	3219
CCRIS 3077	4039	CCRIS 3530	81
CCRIS 3082	3188	CCRIS 3553	3064
CCRIS 3088	2693	CCRIS 3564	3726
CCRIS 3091	1429	CCRIS 3588	4095
CCRIS 3097	1198	CCRIS 3591	4032
CCRIS 3102	1434	CCRIS 3611	3221
CCRIS 3106	3032	CCRIS 3613	1041
CCRIS 3107	3587	CCRIS 3617	1510
CCRIS 3111	1214	CCRIS 3625	2575
CCRIS 3117	2715	CCRIS 3645	1024
CCRIS 3118	1401	CCRIS 3647	1166
CCRIS 3120	2716	CCRIS 3648	3724
CCRIS 3129	1430	CCRIS 3652	941
CCRIS 3143	3745	CCRIS 3653	694
CCRIS 3150	236	CCRIS 3654	3196
CCRIS 3151	1818	CCRIS 3656	3634
CCRIS 3155	66	CCRIS 3657	618
CCRIS 3180	3286	CCRIS 3658	1635
CCRIS 3182	850	CCRIS 3659	2288
CCRIS 3183	2	CCRIS 3660	2312
CCRIS 3184	338	CCRIS 3663	315
CCRIS 3187	863	CCRIS 3664	4161
CCRIS 3191	878	CCRIS 3665	983
CCRIS 3198	2753	CCRIS 3666	648
CCRIS 3199	1964	CCRIS 3667	2675
CCRIS 3202	3169	CCRIS 3668	641
CCRIS 3203	3581	CCRIS 3682	3470
CCRIS 3204	3676	CCRIS 3697	425
CCRIS 3206	4055	CCRIS 3699	3347
CCRIS 3208	746	CCRIS 3701	3608
CCRIS 3220	4050	CCRIS 3704	1086
CCRIS 3221	586	CCRIS 3707	3317
CCRIS 3223	2015	CCRIS 3715	3370
CCRIS 3234	3060	CCRIS 3716	842
CCRIS 3245	701	CCRIS 3725	2595
CCRIS 3276	659	CCRIS 3726	2243
CCRIS 3278	47	CCRIS 3741	3796
CCRIS 3284	3730	CCRIS 3742	3797
CCRIS 3288	3759	CCRIS 3743	4074
CCRIS 3291	3380	CCRIS 3744	1644
CCRIS 3292	883	CCRIS 3766	2233
CCRIS 3293	3400	CCRIS 3783	655
CCRIS 3335	1796	CCRIS 3784	1119
CCRIS 3342	2285	CCRIS 3785	3702
CCRIS 3345	2027	CCRIS 3796	4033
CCRIS 3348	1654	CCRIS 3800	1833
CCRIS 3349	788	CCRIS 3801	969
CCRIS 3397	2192	CCRIS 3802	2565
CCRIS 3398	3248	CCRIS 3921	3381
CCRIS 3401	522	CCRIS 3928	1014
CCRIS 3402	774	CCRIS 3936	1337
CCRIS 3406	143	CCRIS 3946	2500
CCRIS 3409	39	CCRIS 3947	1095
CCRIS 3410	1059	CCRIS 3950	3385

Name and Synonym Index

Name and Synonym Index

Name and Synonym Index

CCRIS 5977	479	CCRIS 6222	875
CCRIS 5984	502	CCRIS 6224	1931
CCRIS 5985	513	CCRIS 6225	1113
CCRIS 5996	801	CCRIS 6227	36
CCRIS 6000	824	CCRIS 6228	1222
CCRIS 6005	903	CCRIS 6231	1459
CCRIS 6006	957	CCRIS 6232	1301
CCRIS 6009	1121	CCRIS 6233	1304
CCRIS 6010	1142	CCRIS 6234	1322
CCRIS 6011	1169	CCRIS 6235	1324
CCRIS 6018	1275	CCRIS 6239	1763
CCRIS 6022	3564	CCRIS 6240	1788
CCRIS 6026	65	CCRIS 6246	1874
CCRIS 6029	421	CCRIS 6247	1937
CCRIS 6030	1530	CCRIS 6248	2206
CCRIS 6035	1579	CCRIS 6252	2121
CCRIS 6036	1609	CCRIS 6253	2125
CCRIS 6037	1705	CCRIS 6254	2221
CCRIS 6038	2950	CCRIS 6258	2503
CCRIS 6039	1790	CCRIS 6259	2514
CCRIS 6041	1890	CCRIS 6260	1596
CCRIS 6042	1892	CCRIS 6261	1613
CCRIS 6048	2179	CCRIS 6263	2845
CCRIS 6051	2070	CCRIS 6271	3172
CCRIS 6053	2130	CCRIS 6272	3457
CCRIS 6056	2430	CCRIS 6274	3680
CCRIS 6060	1199	CCRIS 6275	3713
CCRIS 6062	2466	CCRIS 6276	3718
CCRIS 6063	2467	CCRIS 6279	3905
CCRIS 6067	2517	CCRIS 6282	3947
CCRIS 6070	2525	CCRIS 6283	3954
CCRIS 6071	2526	CCRIS 6284	3170
CCRIS 6074	2668	CCRIS 6286	4093
CCRIS 6081	2719	CCRIS 6291	3215
CCRIS 6082	2765	CCRIS 6319	3860
CCRIS 6090	2915	CCRIS 6320	1188
CCRIS 6091	2943	CCRIS 6321	1140
CCRIS 6095	3093	CCRIS 6322	3690
CCRIS 6096	3181	CCRIS 6325	833
CCRIS 6097	3191	CCRIS 6327	2524
CCRIS 6101	3764	CCRIS 6330	3556
CCRIS 6105	3837	CCRIS 6369	333
CCRIS 6106	3857	CCRIS 6421	3891
CCRIS 6111	4005	CCRIS 6431	2133
CCRIS 6112	1852	CCRIS 6544	3123
CCRIS 6113	4008	CCRIS 6552	588
CCRIS 6150	1773	CCRIS 6560	1497
CCRIS 6151	2491	CCRIS 6566	3100
CCRIS 6177	1099	CCRIS 6567	2300, 2301
CCRIS 6178	3168	CCRIS 6568	2071
CCRIS 6186	854	CCRIS 6569	3111
CCRIS 6187	845	CCRIS 6597	2585
CCRIS 6188	1630	CCRIS 6599	3350
CCRIS 6190	1227	CCRIS 6600	714
CCRIS 6191	1457	CCRIS 6601	4168
CCRIS 6193	1306	CCRIS 6605	96
CCRIS 6194	1312	CCRIS 6667	1167
CCRIS 6195	1316	CCRIS 6670	2617
CCRIS 6196	1455	CCRIS 6678	150
CCRIS 6197	1515	CCRIS 6680	2120
CCRIS 6198	1517	CCRIS 6682	1109
CCRIS 6199	1472	CCRIS 6687	1994
CCRIS 6200	2775	CCRIS 6690	1412
CCRIS 6201	3684	CCRIS 6691	2487
CCRIS 6202	3696	CCRIS 6692	3918
CCRIS 6203	3986	CCRIS 6693	362
CCRIS 6206	2654	CCRIS 6696	337
CCRIS 6211	218	CCRIS 6714	3321
CCRIS 6212	1285	CCRIS 6715	3504
CCRIS 6216	2462	CCRIS 6744	1282
CCRIS 6217	353	CCRIS 6745	2040
CCRIS 6218	571	CCRIS 6749	4064

Name and Synonym Index

Chinaldine	3247	Chloral	750
Chinalphos	3248	Chloraldural	751
Chinchen	696	Chloraldurat	751
Chindoxin	3251	Chloralex	751
Chinese gelatin	4059	Chlorali Hydras	751
Chinese isinglass	4059	Chlorallylene	89
Chinese red	845	Chloralone	753
Chinese seasoning	2575	Chloralvan	751
Chinese vermilion	2356	Chlorameisensäure methylester	2435
Chinese White	4154	Chlorameisensäureäthylester	1620
Chinese white	4154	Chlorameisensaeure methylester	2435
Chinetrin	2928	Chloramiblau 3B	2651
Chinizarin	3253	Chloramin B	752
Chinmix	1038	Chloramin Dr. Fahlberg	753
Chinoin-fundazol	313	Chloramin Heyden	753
Chinorta	2861	Chloramine B	752
Chinorto	2861	Chloramine Blue	2651
Chinosol	3255	Chloramine Blue 3B	2651
Chinoxalin-1,4-dioxid	3251	Chloramine Sky Blue A	861
Chinozan	3256	Chloramine Sky Blue 4B	861
Chinufur	681	Chloramine T	753
Chip-Cal Granular	608	Chloramine	753
Chipco 26019	2064	Chloramine-Heyden, Pyrgos	753
Chipco buctril	482	Chloramizol	2021
Chipco Crab Kleen	482, 1499	Chlorammonic	179
Chipco Florel PRO	1599	Chloramon	179
Chipco turf herbicide D	1202	Chloran 542	758
Chipco turf herbicide MCPP	2323, 2327	Chloranil	754
Chipman 3,142	3659	Chloranile	754
Chipman 11974	2999	Chloraprep	761
Chiptox	2321	Chlorasan	753
Chisso 507B	3076	Chlorasept 2000	760
Chisso polypro 1014	3076	Chloraseptine	753
Chissocizer CS 12	3968	Chlorate de potassium	3096
Chissonox 221 monomer	1577	Chlorate de sodium	3395
Chitamite	2696	Chlorate of potash	3096
Chitan, N-acetyl-	749	Chlorate of potassium	3096
Chitin Tc-L	749	Chlorate salt of sodium	3395
Chitin	749	Chlorazan	753
Chitina	749	Chlorazene	753
ChKhz 21	282	Chlorazol Blue 3B	2651
ChKhz 21R	282	Chlorazone	753
ChKhz 21r	282	Chlorbenzene	778
ChKhz 57	279	Chlorbenzol	778
Chladone 11	3872	Chlorbismol	432
Chlofazimine	893	ChlorCaragard	3670
Chlofenotan	1056	Chlorcholinchlorid	766
Chloksil	1909	Chlorcholine chloride	766
Chlolincocin	891, 892	Chlorcosane	763
Chlon	2888	Chlorcresolum	2273
Chlonazepam	894	Chlordan(e)	755
Chloor	764	Chlordane, α & γ isomers	755
Chloorbenzeen	778	Chlordantoin	213
Chloordaan	755	Chlordene	756
Chloorethaan	788	Chlordene 50	756
Chloorfacinon	807	Chlordetal	752
Chloor-methaan	792	Chlordiethylsulfid	821
Chloorpikrine	810	Chlordimethylether	785
Chloorwaterstof	1962	Chlore	764
Chlophazolin	896	Chloreal	3633
Chlor IFK	837	Chlorendic acid	757
Chlor IPC	837	Chlorendic anhydride	758
Chlor Kil	755	Chlorene	788
Chlor	764	Chloresene	2246
Chloracetic acid	767	Chlorethene	4073
Chloracetone	768	Chlorethephon	1599
Chloracetyl chloride	771	Chlorethyl	788
Chloradorm	751	Chlorethylene	4073
Chlorak	3863	Chlorex	394
Chloral Hydrate	751	Chlorez 700	763
Chloral monohydrate	751	Chlorez 700hmp	763
Chloral, anhydrous, inhibited	750	Chlorfacinon	807

Name and Synonym Index

Name and Synonym Index

Name and Synonym Index

Name and Synonym Index

Name and Synonym Index

Name and Synonym Index

Name and Synonym Index

Cystein	1040	D&C Red No. 12	293
Cysteine	1040	D&C Red No. 19	3290
Cysteine chlorhydrate	1041	D&C Red No. 21	1571
Cysteine disulfide	1042	D&C Yellow No. 7	1762
Cysteine hydrochloride	1041	D&C Yellow No. 8	1763
Cysteine, L-	1040	D.B.I.	2377
Cysteine, L-, hydrochloride	1041	D.C.P.	638
Cysteinum	1040	D.D.T. technique	1056
Cystin	1042	D.E.H. 20	1255
Cystine	1042	D.E.H. 52	1255
Cystine (L)-	1042	D.E.R. 332	439
Cystine acid	1042	D.T.S.	1084
Cystisine	1043	D	263
Cystit	2694	D-1410	2824
Cystoceva	2034	D-365	4064
Cystogen	2394	D-365 hydrochloride	4065
Cystografin	1117	D3770	1060
Cystographin Dilute	1117	D3780	1061
Cystokon	31	D3-Vicotrat	4093
Cystrin	2834	D3-Vigantol	4093
Cytacon	4089	D-735	693
Cytagon	1835	D-9998	1774
Cytame 5	3059	DA	2233
Cytamen	4089	Dab	518, 1045
Cytel	1722	Dab (Carcinogen)	518
Cython	2294	Dabco	3918
Cytisine	1043	Dabco 33LV	3918
Cytiton	1043	Dabco® B-16	362
Cytitone	1043	Dabco® BDO	502
Cytizin	1043	Dabco BL 11	397
Cytobion	4089	Dabco BL 19	397
Cytofol	1782	Dabco BL 19I	397
Cytox	2450	Dabco crystal	3918
Cytrolane	2340	Dabco DEOA-LF	1235
Cyuram DS	3771	Dabco® DMEA	1057
Cyuram MS	3736	Dabco EG	3918
Cyzac 4040	3803	Dabco R-8020	3918
Cyzine premix	135	Dabco S-25	3918
Cyzone IP	2134	Dabco TMR 30	1523
Cyzone	2134	Dabeersen 503	2916
Czterochlorek wegla	688	Dabex	2377
Czterochloroetylen	3695	Dab-O-lite P 4	646
Czteroetylek olowiu	3707	Dabonal	1567
		Dabrosin	86
D & C Blue No. 6	2034	Dabroson	86
D & C Red No. 17	3603	DAC 2787	820
D & D	3241	DAC 4	1047
D 1 (antioxidant)	1337	DAC 893	1047
D 2EHPA	1269	d'Acide tannique	3652
D 10P	1230	Dacobre	820
D 33LV	3918	Daconate	2584
D 50 M	3080	Daconate 6	2584
D 50	3080	Daconil	820
D 65MT	95	Daconil 2787	820
D 35	1055	Daconil Flowable	820
D 43	1393	Daconil M	820
D 100	2797	Daconit	1730
D 100 (fatty acid)	2797	Dacorten	3145
D 151	3076	Dacortin	3145
D 1221	681	Dacosoil	820
D 1991	313	Dacovin	3082
D 15095	2377	Dacpm	2456
D and C Orange No. 15	81	Dacthal	1047
D and C Orange Number 15	81	Dacthalor	1047
D and C Orange Number 15D	81	Dactin	1186
D and C Red No. 21	1571	Dactinol	3310
D and C Yellow 6	865	Dacutox	1122
D and C Yellow No. 5	42	Dadpe	386
D&C Red 9	869	Dadpm	1116
D&C Red 19	3290	DADPS	1054
D&C Red No. 3	1585	DAG 206	2561
D&C Red No. 7	870	DAG 325	2561

Name and Synonym Index

Name and Synonym Index

Name and Synonym Index

Name and Synonym Index

Name and Synonym Index

Name and Synonym Index

Name and Synonym Index

Name and Synonym Index

Name and Synonym Index

Name and Synonym Index

Dimethylacetic acid	2091	Dinitolmida	1435
Dimethylacetone	1271	Dinitolmide	1435
Dimethylacetonitrile	2093	Dinitolmidum	1435
Dimethylacetylenecarbinol	1387	Dinitra	1434
Dimethylacetylenylcarbinol	1387	Dinitrall	1447
Dimethyladipate	1358	Dinitrate de diethylene-glycol	1259
Dimethylaethanolamin	1057	Dinitrate d'ethylene glycol	1860
Dimethylamid kyseliny acetoctove	1357	Dinitrato de etilenglicol	1860
Dimethylamid kyseliny mravenci	1388	Dinitrile of isophthalic acid	2123
Dimethylamid kyseliny octove	1356	Dinitrobenzene	1429
Dimethylamide acetate	1356	Dinitrobenzoic acid	1430
Dimethylamine hydrochloride	1360	Dinitrobutylphenol	1447
Dimethylamine	1359	Dinitrocaprylphenyl crotonate	1443
Dimethylaminoaethanol	1057	Dinitrochlorobenzene	1431
Dimethylaminobenzene	4120	Dinitrochlorobenzol	1431
Dimethylaniline	1373	Dinitrocresol	1433
Dimethylaniline	4120	Dinitrodiglicol	1259
Dimethylbenzene	4111	Dinitrodipropylsulfanilamide	2816
Dimethylbenzylamine	362	Dinitrofenolo	1434
Dimethylchloroether	785	Dinitrogen monoxide	2713
Dimethylchlorosilane	787	Dinitrogen oxide	2713
Dimethylglyoxal	1101	Dinitroglicol	1860
Dimethylglyoxime	1390	Dinitroglycol	1860
Dimethylketal	23	Dinitrol	1433
Dimethylketol	37	Dinitrophenylmethane	1436
Dimethylmethane	3163	Dinitrosol	1433
Dimethylmethaneamine	3954	Dinitrotoluene (mixed isomers)	1436
Dimethylsulfaat	1417	Dinitrotoluene	1436, 1438
Dimethylsulfat	1417	Dinitrotoluenes	1436
Dimethylsulfoxide	1419	Dinitrotoluol	1436
Dimethylsulfoxyde	1419	Dinobuton	1442
Dimethyrimol	1343	Dinocap	1443
Dimeticona	3063	Dinofan	1434
Dimeticone	3063	Dinofen	1442
Dimeticonum	3063	Dinokap	1443
Dimetil sulfoxido	1419	Dinokapu	1443
Dimetilformamide	1388	Dinoleine	2051
Dimetilsolfato	1417	Dinonyl phenol	1444
Dimetilsolfossido	1419	Dinonyl phthalate	1445
Dimeton	1344	Dinonylnaphthalenesulphonic acid	1446
Dimetox	3863	Dinonylphenol	1444
Dimetridazol	1427	Dinopol 235	1075
Dimetridazole	1427	Dinopol NOP	1455
Dimetridazolo	1427	Dinoseb	1447
Dimetridazolum	1427	Dinosebe	1447
Dimetylformamidu	1388	DINP	1316
Dimevur	1344	DINP2	1316
Dimexide	1419	DINP3	1316
Dimezathine	3608	Dintoin	2994
Dimidin-R	3608	Dintoina	2994
S-Dimidine	3608	Dioa	1318
Dimilin G1	1279	Diocimex	1552
Dimilin G4	1279	Diocodal	2625, 2627
Dimilin ODC-45	1279	Dioctadecyl thiodipropionate	1505
Dimilin WP-25	1279	Dioctanoyl peroxide	1448
Dimilin	1279	Dioctlyn	1525
Dimitone	703	Dioctyl adipate	1449, 1450
Dimodan GMO 90	1840	Dioctyl azelate	1451
Dimodan LSQK	1840	Dioctyl maleate	1452, 1453, 1454
Dimorpholine disulfide	3612	Dioctyl nonanedioate	1451
Dimorpholino disulfide	3612	Dioctyl phosphite	1456
Dimpilato	1122	Dioctyl phthalate	1455
Dimpylat	1122	Dioctyl phthalate	406
Dimpylate	1122	Dioctyl sebacate	1457
Dimpylatum	1122	Dioctyl sodium sulfosuccinate	1525
Dimycon	1758	Dioctyl sodium sulphosuccinate	1525
Dinarkon	2838	Dioctylal	1525
Dinate	1499	Dioctyl-medo forte	1525
Dinatrium ethylendiamintetraacetat	1635	Diodoquin	2051
Dinatriumpyrophosphat	3500	Diodoquine	2051
Dinazide	3843	Diodoxylin	2051
Diniobium pentoxide	2674	Diodoxyquinoleine	2051

Name and Synonym Index

Name and Synonym Index

Name and Synonym Index

DSS	1054, 1525	Duo Xiao Zuo	2843
DST 50	3592	Duobrus	2015
DST 75	3592	Duocide	3949
DSTDP	1505	Duodecibin	4089
DSTP	1505	Duodecyl alcohol	2199
DTA	1509	Duodecylic acid	2193
DTAC	2197	Duodex	3459
DTBHQ	1145	Duofas	3268
DTBP	1148	Duofil Wart Remover	3321
DTC	1253	Duofilm	3321
DTDP	1515	Duoplant	3321
DTG	1514	Duoscorb	4092
DTMC	1217	Duosol	1525
Dtpa pentasodium salt	2911	Duotrate	2934
DTPA	2916	Duovel	1714
DU 112307	1279	DUP 89	2264
Du Cor Concentrated Fly Insecticide	4115	DuP-753	2265
Du Pont 634	2229	Dupanol WAQ	3457
Du Pont 732	3666	Duphacid	1279
Du Pont 753	2265	Duphacillin	210
Du Pont 1410	2824	Duphafral D3 1000	4093
Du Pont Gasoline Antioxidant No. 22	1151	Duphar	3704
Du Pont herbicide 732	3666	Dupin	1121
Du Pont herbicide 976	458	Duplamin	3158
Du Pont metal deactivator	2367	Duplosan KV	2328
Du Pont WK	2192	Duplosan New System CMPP	2323
DU-A 1	690	Duplosan® CMPP	2323
DU-A 2	690	Dupon 4472	1506
DU-A 3	690	Duponal WAQE	3457
DU-A 3C	690	Duponal	3457
DU-A 4	690	Duponol	3457
Duafen	2015	Duponol 80	3480
Dual	2538	Duponol C	3457
Dual 8E	2538	Duponol ME	3457
Dual 25G	2538	Duponol methyl	3457
Dual 720EC	2538	Duponol QC	3457
Dual 960 EC	2538	Duponol QX	3457
Dual II	2538	Duponol WA Dry	3457
Dual Magnum	2538	Duponol WA	3457
Dual Murganic RPB	693	Duponol WAQ	3457
Dual Triple	2538	Duponol WAQA	3457
Dublofix	788	Duponol WAQE	3457
Dubronax	1054	Duponol WAQM	3457
Ducene	1121	DuPont 26	3507
Duckalgin	79	Dupont 1179	2399
Ducobee	4089	DuPont 1410	2824
Dudencer	2809	DuPont 1991	313
Duet	2957	DuPont asana SP	1586
Dugen	2330	DuPont fungicide 4472	1506
Duhor N	2285	Dura Al	86
Duhor	2285	Dura AX	209
Duirexol	2394	Dura Treet II	2888
Duk	2625	Duraatenolol	266
Dukeron	3862	Durabetason	374
Duksen	1121	Duracef	710
Dularbuprofen	2015	Duraclon	895, 896
Dulcin	1556	Duractin	3273
Dulcine	1556	Durad	964
Dulcite	1810	Duradoxal	1552
Dulcitol	1810	Durafur Black R	2978
D-Dulcitol	1810	Durafur Brown RB	161
Dulcose	1810	Durafur developer C	708
Dulein	1556	Durafur developer D	2621
DULL 704	2736	Durafur developer G	3283
Dulsivac	1525	Durafurid	1807
Dulzor-Etas	3409	Duraglucon	1835
Dumitone	1054	Dura-Ibu	2015
Dumitone	1054	Dural	96
Dumocyclin	3700	Duralozam	2263
Dumopranol	3189	Duran	1518
Duncaine	2237	Duranifin	2668
Dunkelgelb	3602	Duranit	3592

Name and Synonym Index

Name and Synonym Index

Elfan NS 243	3455	Emalex 115	732
Elfan WA sulphonic acid	1535	Emalex 120	732
Elfan® NS 243 S Mg	2286	Emalex 505	2801
Elfonal	1567	Emalex 510	2801
Elftex	684	Emalex 515	2801
Elgam	2809	Emalex 520	2801
Elgetol	1433	Emalex EG-di-L	1858
Elgetol 30	1433	Emalex MTS-30E	2965
Elgetol 318	1447	Emalex O.T.G	1837
Elgetox	1433	Emalsy MO	1840
Elimite	2928	Emalsy OL	1840
Elisor	3144	Emanay atomized aluminum powder	97
Elite	3656	Emanay zinc dust	4129
Eliten	1793	Emanay zinc oxide	4154
Eliur	1807	Emanil	2020
Eljon Lake Red C	869	Emanon 1112	3073
Eljon Lithol Red MS	293	Emanon 4115	1864
Eljon Pink Toner	3291	EMAR	4154
Eljon Rubine BS	868	Emasol 110	3535
Ellatun	3824	Emasol 410	3536
Elliott's Lawn Sand	1740	Emasol 430	3540
Elliott's Moss Killer	1740	Emasol L 10(F)	3535
Elmer's Glue All	3080	Emasol L 10	3535
Elmogan	1816	Emasol MO 50	1840
Elocon	2563	Emasol O 10	3536
Elocone	2563	Emasol O 10F	3536
Elon	2537	Emasol Super L 10(F)	3535
Elon (developer)	2537	Emathlite	2168
Elosal	3621	Embafume	2427
Eloxyl	1125	Embamix	3114
Elpon	3076	Embanox	528
Elsil 100	4098	Embarin	86
Elsil BF 100	4098	Embark 2S	2332
Eltesol®	3803	Embark plant growth regulator	2332
Eltesol ST 34	2420	Embark	2332
Eltesol ST 90	2420	Embephen	1199
Eltesol® AC60	180	Embequin	2051
Eltesol® AX 40	207	Embiol	4089
Eltesol® PSA 65	2954	Emblem	311
Eltesol® ST 40	3524	Emblon	3647
Eltesol® SX 30	3531	Embutone	1046
Eltesol® TSX/A	3803	Embutox	1046
Eltesol® TSX/SF	3803	Emcepan	2321
Eltesol® TSX	3803	Emcol CA	1841
Elthon	4065	Emcol CAD	1263
Eltren	2204	Emcol D 5-10	3429
Eltroxin	2234	Emcol DS-50 CAD	1263
Elvacet 81-900	3080	Emcol E-607	2188
Elvanol	3081	Emcol ETS	1263
Elvanol 50-42	3081	Emcol H 31A	1864
Elvanol 5105	3081	Emcol H-2A	1864
Elvanol 51-05G	3081	Emcol MSK	1841
Elvanol 52-22	3081	Emcol O	1840
Elvanol 522-22	3081	Emcol PS-50 RHP	3201
Elvanol 52-22G	3081	Emcol RDC-D	2874
Elvanol 70-05	3081	Emcol® 6748, Coco betaine	923
Elvanol 71-30	3081	Emcol® DG, NA30	923
Elvanol 73125G	3081	Emcol-IM	2145
Elvanol 90-50	3081	Emcol-IP	2146
Elvanol T 25	3081	Emcor	437
Elvaron	1166	EMD 33 512	436
Elyzol	2543	EMD 33512	437
EM 3	3308	EMD-34946	3279
EM 490	3076	Emdecassol	260
EM-379	1879	Emeproton	2809
EMA2O	1678	Emeressence 1150	1631
EMAL 10	3457	Emeressence® 1160 Rose Ether	2964
EMAL O	3457	Emeressence® 1160	2964
EMAL T	3910	Emerest 2314	2145
Emalex 103	732	Emerest 2325	577
Emalex 105	732	Emerest 2350	1649
Emalex 110	732	Emerest 2355	1646

Name and Synonym Index

Name and Synonym Index

Name and Synonym Index

Name and Synonym Index

Name and Synonym Index

Etylu chlorek	788	E-Vimin	4095
Etylu krzemian	3706	Evion	4095
Eubine	2837, 2838	Eviplast 80	406
Eucalmyl	272	Eviplast 81	406
Eucanine GB	1110	Evista	3270
Eucardic	703	Evital	2732
Eucil	2535, 2536	Evitaminum	4095
Euclamin	1835	Evitocor	266
Euclorina	753	Evitol	2732
Eucodalum	2837	Evits	3005
Euctrim	3949	Evofenac	1213
Eudextran	1096	Evola	1174
Eudigox	1286	Evorel	1588
Eudorm	4174	Eweiss	302
Eudraflex	3917	EX 0030 (clay)	315
Eufor	1772	EX 0276	315
Eugenyl methyl ether	2464	EX 10781	309
Euglucan	1835	Ex bois de rose (synthetic)	2244
Euglucon	1835	Exact-S	2519
Euglucon 5	1835	Exagama	2246
Euglykon	1835	Exagien	2514
Euhaemon	4089	Exbesul	3949
Euhypnos	3662	Excalibur	1035
Euipnos	3662	Excel O 95F	1840
Eukodal	2838	Excel O 95N	1840
Eulimen	1462	Excel O 95R	1840
Eumin	2543	Excelopax	4169
Eumulgin EP 4	2801	Excelsior	684
Eunasin	309	Exceparl M-OL	2495
Eunatrol	3482	Exchem GO-1	1660
Euonymit	1810	Excis	1038
Euparen M	1712	Exelderm	3606
Euparen	1166	Exhaust gas	687
Euparen® M	1712	Exhorran	1506
Euparene	1166	EX-IT	3634
Euphorin	2993	Exitelite	239, 245
Euphorin P	1121	EX-M 703	315
Eupressin	1567	Exmin	2928
Euprovasin	3188	Exneural	2015
Euradal	437	Exocin	2789
Eureka Products, Criosine	2415	Exoderil	2604
Eureka Products Criosine Disinfectant	2415	Exodin	1122
Eurocert Orange FCF	865	Exofene	1908
Eurocert tartrazine	42	Exolit 385	3010
Eurodopa	2233	Exolit 405	3010
Euroquat C45, CPB, K, LA, LAC	923	Exolit LPKN 275	3010
Eurosan	1121	Exolit LPKN	3010
Eurtadal	437	Exolit RP 605	3010
Eusedon Mono	3158	Exolit RP 650	3010
Eusolex 232	2973	Exolit RP 652	3010
Eusolvan	1670	Exolit RP 654	3010
Eustidil	1884	Exolit VPK-n 361	3010
Eutagen	2838	Exolon XW 60	96
Eutanol G	2786	Exon 605	3082
Euteberol	3550	Exopen	3824
Eutensin	1807	Exosurf	935, 4030
Euthyrox	2234	Exotherm	820
Euxyl K 100	358	Exotherm Termil	820
Evacalm	1121	EXP-49	1382
Evade	820	EXP-51	1608
Evafine P 50	3232	EXP 5598	1038
Evansite	117	EXP-3864	3258
Evanstab® 18	1505	Expand	3343
Evatin	1714	Expanded polyvinyl chloride	3082
Evau-super	3395	Expandex 5PT	2989
Evazol	1088	Expandex OX 5PT	2989
Eveite DOTG	1514	Expar	2928
Evercide 2362	1734	Experimental fungicide 5223	1544
Evercide fenvalerate	1734	Experimental herbicide 634	2229
Evercyn	3216	Experimental herbicide 732	3666
Everflex B	3080	Experimental insecticide 3911	2998
Everflex® 81L	4070	Experimental Insecticide 7744	673

Name and Synonym Index

FEMA No. 2221	588	FEMA No. 2676	2411
FEMA No. 2222	2091	FEMA No. 2677	2335
FEMA No. 2224	606	FEMA No. 2681	2417
FEMA No. 2228	607	FEMA No. 2682	2418
FEMA No. 2229	655	FEMA No. 2683	2421
FEMA No. 2242	697	FEMA No. 2695	2460
FEMA No. 2245	702	FEMA No. 2710	2500
FEMA No. 2249	704	FEMA No. 2714	2480
FEMA No. 2263	707	FEMA No. 2715	2483
FEMA No. 2286	875	FEMA No. 2716	2392
FEMA No. 2286	875	FEMA No. 2718	1374
FEMA No. 2288	876	FEMA No. 2722	2488
FEMA No. 2294	878	FEMA No. 2728	2432
FEMA No. 2298	3588	FEMA No. 2731	2478
FEMA No. 2303	882	FEMA No. 2733	2339
FEMA No. 2306	883	FEMA No. 2742	2509
FEMA No. 2307	884	FEMA No. 2745	2514
FEMA No. 2309	885	FEMA No. 2746	2519
FEMA No. 2311	886	FEMA No. 2747	2398
FEMA No. 2330	697	FEMA No. 2756	2575
FEMA No. 2337	958, 960	FEMA No. 2762	2595
FEMA No. 2341	970	FEMA No. 2764	2599
FEMA No. 2349	1020	FEMA No. 2770	2645
FEMA No. 2356	2150	FEMA No. 2772	2646
FEMA No. 2364	1064	FEMA No. 2779	2713
FEMA No. 2365	1065	FEMA No. 2782	2718
FEMA No. 2370	1101	FEMA No. 2784	2877
FEMA No. 2373	1154	FEMA No. 2797	2783
FEMA No. 2396	1415	FEMA No. 2799	2758
FEMA No. 2398	1501	FEMA No. 2800	2761
FEMA No. 2411	1589	FEMA No. 2801	664
FEMA No. 2414	1609	FEMA No. 2805	4071
FEMA No. 2415	1610	FEMA No. 2806	2768
FEMA No. 2418	1611	FEMA No. 2815	2797
FEMA No. 2419	1593	FEMA No. 2822	2812
FEMA No. 2433	1650	FEMA No. 2832	2852
FEMA No. 2440	1670	FEMA No. 2842	2910
FEMA No. 2458	1693	FEMA No. 2858	2983
FEMA No. 2464	1706	FEMA No. 2858	2983
FEMA No. 2475	2464	FEMA No. 2872	2958
FEMA No. 2478	1715	FEMA No. 2878	2971
FEMA No. 2487	1790	FEMA No. 2881	363
FEMA No. 2488	1799	FEMA No. 2889	1964
FEMA No. 2489	1803	FEMA No. 2900	3005
FEMA No. 2491	1805	FEMA No. 2903	3042
FEMA No. 2497	1808	FEMA No. 2908	3048
FEMA No. 2507	1819	FEMA No. 2916	2875
FEMA No. 2509	1821	FEMA No. 2920	3086
FEMA No. 2525	1838	FEMA No. 2925	3190
FEMA No. 2526	2574	FEMA No. 2926	2130
FEMA No. 2527	1846	FEMA No. 2929	2127
FEMA No. 2541	1890	FEMA No. 2940	3195
FEMA No. 2544	1894	FEMA No. 2944	2140
FEMA No. 2546	593	FEMA No. 2947	3206
FEMA No. 2548	1893	FEMA No. 2963	3218
FEMA No. 2554	1912	FEMA No. 2970	3240
FEMA No. 2559	661	FEMA No. 2980	3293
FEMA No. 2567	1946	FEMA No. 2981	886
FEMA No. 2569	1949	FEMA No. 2997	3503
FEMA No. 2587	590	FEMA No. 3004	3320
FEMA No. 2611	2174, 2176	FEMA No. 3026	3400
FEMA No. 2614	2193	FEMA No. 3027	3438
FEMA No. 2617	2199	FEMA No. 3028	3537
FEMA No. 2627	2235	FEMA No. 3029	3542
FEMA No. 2633	2242	FEMA No. 3035	3568
FEMA No. 2635	2243	FEMA No. 3038	3601
FEMA No. 2636	2244	FEMA No. 3039	3622
FEMA No. 2655	2300, 2301	FEMA No. 3042	3652
FEMA No. 2656	2305	FEMA No. 3044	3654
FEMA No. 2665	2338	FEMA No. 3047	3677
FEMA No. 2670	225	FEMA No. 3056	3719
FEMA No. 2671	955	FEMA No. 3068	3811

Name and Synonym Index

|---|---|
Fenkem	1734
Fenkill	1734
Fenmedifam	2943
Fennosan B 100	1055
Fennosan H 30	3254
Fennosan	3254
Fennosan HF-15	3254
Fenobarbital natrium	2944
Fenobarbital sodico	2944
Fenobrate	1723
Fenocin	2883
Fenocin Forte	2883
Fenofibrate	1723
Fenoform forte	2246
Fenol	2946
Fenolipuna	2948
Fenolo	2946
Fenolovo	1733
Fenolovo acetate	1732
Fenom	1038
Fenom (pesticide)	1038
Fenopon CO 436	191
Fenopon EP 110	191
Fenopon EP 120	191
Fenoprop	3366
Fenormone	3366
Fenospen	2882
Fenossimetilpenicillina	2882
Fenotard	1723
Fenothiazine	2956
Fenothrin	2957
Fenotiazina	2956
Fenotrina	2957
Fenoverm	2956
Fenoxaprop-P	1724
Fenoxaprop-P-ethyl	1725
Fenoxin	1734
Fenoxyl	1434
Fenoxyl Carbon N	1434
Fenoxypen	2882, 2883
Fenozan 22	2896
Fenozan 23	2896
Fenozan 30	1987
Fenprocoumona	2963
Fenprocumone	2963
Fenpropanate	1726
Fenpropathrin	1726
Fenpropathrine	1726
Fenpropidin	1727
Fenpropidine	1727
Fenpropimorph	1728
Fenpropimorphe	1728
Fenpyroximate	1729
Fenspan	2015
Fensulfothion	1730
Fenthion	1731
Fenthion 4E	1731
Fenthione	1731
Fenthion-methyl	1731
Fentiazin	2956
Fentiazine	2956
Fentin	1733
Fentin acetaat	1732
Fentin acetat	1732
Fentin acetate	1732
Fentin azetat	1732
Fentin chloride	833
Fentin hydroxide	1733
Fentinacetat	1732
Fentine	1733
Fentine acetate	1732
Fentoin	2994
Fentrothione	1722
Fenval	1734
Fenvalerate	1734
Fenvalerate α	1586
Fenvalerate A α	1586
Fenvalerate (S,S)-isomer	1586
N-Fenyl-2-aminonaftalen	2624
Fenyl-α-naftylamin	2966
Fenyl-β-naftylamin	2624
Fenylbutatin oxide	1721
Fenylbutylstannium oxide	1721
Fenyl-cellosolve	2964
Fenylcelosolv	2964
Fenylenodwuamina	2978
Fenylepsin	2994
Fenylester kyseliny salicylove	2969
Fenylglycol	3593
Fenylkyanid	338
Fenylmercuriacetat	2986
Fenylmerkuriacetat	2986
Fenyloxiran	3591
Fenytoin Dak	2994
Fenytoin	2995
Fenytoine	2994
Fenzen	321
Feosol	1746
Feospan	1746
FER pentacarbonyle	2886
Ferancee	4092
Feredato sodico	1739
Feredetate de sodium	1739
Fer-In-Sol	1746
Ferisan	1739
Ferkethion	1344
Fermagex	3949
Fermenicide liquid	3622
Fermenicide powder	3622
Fermentation alcohol	1593
Fermentation butyl alcohol	2075
Fermide	3771
Fermide 850	3771
Fermine	1411
Fernacol	3771
Fernasan	3771
Fernasan A	3771
Fernesta	1202
Fernide	3771
Fernimine	1202
Fernos	3051
Fernoxene	1044
Fernoxone	1044
Ferobuff	1746
Ferolix	1746
Ferotine	1714
Ferralyn	1746
Ferriamicide	2556
Ferric chloride	1735
Ferric hydroxide	1736
Ferric nitrate	1737
Ferric oxide	1738
Ferric oxide (colloidal)	1738
Ferric oxide [Haematite and ferric oxide]	1738
Ferric persulfate	1740
Ferric sesquioxide	1738
Ferric sesquisulfate	1740
Ferric sodium edetate	1739
Ferric sodium EDTA	1739
Ferric sulfate	1740
Ferric tersulfate	1740
Ferrihemate	1886
Ferriheme hydroxide	1886
Ferriheme	1886

Firemaster BP 4A	3680
FireMaster FF 680	424
Firemaster LV-T 23P	4006
FireMaster PHT 4	3684
FireMaster T 23	4006
Firemaster T23P	4006
Firemaster T23p-lv	4006
Fireshield FSPO 405	245
FireShield H	245
FireShield LS-FR	245
Fireshield® H	245
Fireshield® HPM	245
Fireshield® HPM-UF	245
Fireshield® L	245
Firmacort	2507
Firmatex RK	1403
Fischer's salt	3101
Fischer's yellow	3101
Fischer-Tropsch wax	2857
Fish oil, sulfated, sodium salt	3613
Fish-Tox	3310
Fisons 18-15, MCPB	2322
Fisons NC 2964	2395
Fisons NC 6897	310
Fitodith 80	4164
Fitomenadiona	4096
Fitomenadione	4096
Fitrol desiccite 25	2168
Fitrol	2168
Fixanol C	742, 743
Fixapret CP 40	1403
Fixapret CP	1403
Fixapret CPK	1403
Fixapret CPN	1403
Fixapret CPNS	1403
FKS	1966
FKW 11	3872
FKW 12	1185
FKW 22	784
FKW 114	1206
FKW 115	806
Flacavon R	4006
Flagemona	2543
Flagesol	2543
Flagil	2543
Flagyl	2543
Flagyl I.V. RTU, Metro I.V.	2543
Flagyl I.V.	2544
Flair	2294
Flake lead	2211
Flake White	432, 434
Flame	1752
Flame Cut 110R	1058
Flame Cut 610	245
Flame Cut 610R	245
Flame Cut Br 100	1058
Flameguard VF 59	245
Flamenco	3785
Flamex T 23P	4006
Flammex 3BP	3851
Flammex 5bt	2885
Flammex AP	4006
Flammex LV-T 23P	4006
Flammex T 23P	4006
Flamon	4065
Flamprop-M-isopropyl	1752
Flamprop-M-methyl	2919
Flamruss	684
Flamtard H	4149
Flamtard S	4157
Flamtard Z 10	4135
Flanax	2627
Flanax Forte	2625
Flavaxin	4090
Flavin BB	4090
Flavin meletin	3244
Flavone	1754
Flavone, 3,4',5,7-tetrahydroxy-	2164
Flavor orange	1462
flavors	358
Flavyl	166
Flaxain	4090
Flaxseed oil	2249
Flea Prufe	448
FLEABOR	1500
Fleck-flip	3862
Fleck's Extraordinary Cement	4155
Flectol A	3958
Flectol H	3958
Flectol pastilles	3958
Flectron	1038
Flee	2928
Fleet-X	2364
Flemoxine	209
Flexagilt	696
Flexagit	696
Flexal	696
Flexamine G	1477
Flexartal	696
Flexchlor	763
Flexen	2625
Flexeril	1005
Flexeril hydrochloride	1005
Flexiban	1005
Flexible collodion	3238
Flexichem	647
Flexichem B	3509
Flexichem CS	647
Flexidon	696
Flexidor	2157
Fleximel	406
Flexipen	2625
Flexivial	3396
Flexo Red 482	3291
Flexol 3GO	3921
Flexol 4GO	1939
Flexol A 26	1449
Flexol DOP	406
Flexol EP-8	2773
Flexol EPO	1575
Flexol plasticizer 3GO	3921
Flexol plasticizer 10-A	1449
Flexol plasticizer A-26	1449
Flexol Plasticizer DIOP	1321
Flexol Plasticizer DOP	406
Flexol plasticizer EP-8	2773
Flexol Plasticizer TCP	964
Flexol plasticizer TOF	3985
Flexol TOF	3985
Flexricin P 6	521
Flexricin P-1	2513
Flexricin® 9	3202
Flexricin® 15	1863
Flexricin® P-4	2413
Flexzone 3C	2134
Flexzone 4L	2979
Flexzone 7F	2980
Flexzone 7L	2980
Fliibol E	3863
Fliegenteller	3863
Flint	3241, 3347
Flintshot	3241
Flit 406	668
Flit	668

FMC 17370	3282	Fonganil	1802
FMC 18739	377	Fongarid	1802
FMC 30980	1038	Fongitar	902
FMC 33297	2928	Fontex	1773
FMC 35001	692	Food orange 5	701
FMC 41655	2928	Food orange 8	659
FMC 45497	1038	Food Red 17	864
FMC 45498	1086	Food Red Color No. 105, sodium salt	3304
FMC 45806	1038	Food Red No. 40	864
FM-Nts	3238	Food Yellow 3	865
FMR-PC	2288	Food Yellow 4	42
Foamid AME-70	13	Food Yellow 5	42
Foamid AME-75	13	Food Yellow 6	865
Foamid AME-100	13	Food Yellow No. 5	865
Foamtaine CAB, CAB-G	923	Foodol D 70	3542
Focus	1033	Foraat	2998
Focus Ultra	1033	Foral 85	1784
FOE 1976	2331	Forane	2104
Folacid	1782	Forane 12	1185
Folacin	1782	Forane 22 B	784
Folaemin	1782	Forane 22	784
Folan	1782	Forane 113	3883
Folasic	1782	Forate	2998
Folate	1782	Forbel	1728
Folbal	1782	Forcan	1758
Folcid	666	Fore	2307
Folcidin	1782	Foredex 75	1202
Folcord	1038	Forene	2104
Folcysteine	1782	Forepen	2888
Foldine	1782	Forgesic	3824
Folettes	1782	Foriod	247
Folex 6EC	2361	Forlin	2246
Folex 6EC	3859	Formagene	2858, 2859
Folex	2361	Formal	1350, 2294
Foliafume	3310	Formal glycol	1785
Foliafume E.C.	3310	Formaldehyd	1786
Foliamin	1782	Formaldehyde	1786
Folic acid	1782	Formaldehyde polymer	2859
Folicet	1782	Formaldehyde sodium bisulfite	3434
Foliclal	2867	Formaldehyde, trimer	2858
Folico	1782	Formalin	1786
Folicur	3656	Formalin 40	1786
Folicur 1.2 EC	3656	Formalina	1786
Folidol	2867	Formaline	1786
Folidol E & E 605	2867	Formalin-lösungen	1786
Folidol E	2867	Formalin-loesungen	1786
Folidol E605	2867	Formalith	1786
Folidol oil	2867	Formalsol	3771
Foligan	86	Formamide	1788
Folimat®	2811	Formamide, N-methyl-	2465
Folina	1782	Formamine	2394
Folipac	1782	Formapon	1787
Folithion	1722	Format	900
Folithion EC 50	1722	Formax	3435
Follicyclin	1588	Formbis	3434
Folnit	1783	Formiate de methyle	2466
Folosan DB 905	3659	Formic acid	1790
Folosan	3256, 3659	Formic acid amide, N-methyl-	2465
Folpan	1783	Formic acid, amide	1788
Folpel	1783	Formic acid, sodium salt	3435
Folpet	1783	Formic aldehyde	1786
Folsan	1782	Formic anammonide	3216
Folsaure	1782	Formimidic acid	1788
Folsav	1782	Formin	2394
Foltaf	666	Formira	1790
Folvite	1782	Formisoton	1790
Fomac	1908	Formistin	733
Fomac 2	3256	Formol	1786
Fomrez sul-3	1157	Formonitrile	3216
Fomrez sul-4	585	Formopan	1787
Fonderma	3498	Formu	1202
Fondril	437	Formula 40	1202

Name and Synonym Index

Formula 144	330
Formula 300	3568
Formvar 1285	3080
p-Formylbenzoic acid	3672
Formylformic acid	1869
p-Formyltoluene	3811
Formylic acid	1790
Fornidd	2377
Forotox	3863
Forpen	2881
Forpen-50 Wood Preservative	2888
Forstan	2842
Forst-Nexen	2246
For-syn	3282
Fortafil 5Y	1873
Forte 6C	2980
Forte	2957
Forthion	2294
Fortigro	670
Fortion NM	1344
Fortodyl	1582
Fortombrine-N	31
Fortradol	3824
Fortrim	3949
Fortrol	991
Forturf	820
Fosalan	77
Fosamax	77
Foschlor	3863
Foschlor 25	3863
Foschlor R	3863
Foschlor R-50	3863
Foschlorem	3863
Fosenopril	1792
Fosfakol	2861
Fos-Fall A	3859
FOS-FALL "A"	3859
Fosfamid	1344
Fosfato de adenosina	61
Fosfatox R	1344
Fosferno	2867
Fosferno 50	2867
Fosfive	2867
Fosfomic	3664
Fosfornan sodny	3494
Fosforo(pentacloruro di)	3014
Fosforo(tricloruro di)	3012
Fosforoxychlorid	3013
Fosforpentachloride	3014
Fosfortrichloride	3012
Fosforyn trojetylowy	3925
Fosforyn trojmetylowy	3971
Fosforzuuroplossingen	3005
Fosfothion	2294
Fosfotion	2294
Fosfotion 550	2294
Fosfotox	1344
Fosfotox R 35	1344
Fosfotox R	1344
Fosfuri di alluminio	3017
Fosgeen	3000
Fosgen	3000
Fosgene	3000
foshagite	645
foshallasite	645
Fosinil	1793
Fosinopril	1792
Fosinopril sodium	1793
Fosinorm	1793
Fosipres	1793
Fositen	1793
Fositens	1793
Fosova	2867
Fospirat	1794
Fospirate	1794
Fospirate methyl	1794
Fospirato	1794
Fospiratum	1794
Fossfall	3859
Fossil flour	3347
Fossyol	2543
Fostex	1125, 3321
Fostex AMP	2681
Fostex BPO	1125
Fostex Medicated Bar and Cream	3321
Fostion MM	1344
Foston	3820
Fostox	2867
Fostril	1908, 3621
Fosulen	2133
Foundrox	1738
Fouramine Brown AP	3233
Fouramine D	2978
Fouramine ERN	2621
Fouramine J	1110
Fouramine OP	160
Fouramine P	161
Fouramine PCH	708
Fouramine RS	3283
Fourneau 309	2605
Fourrine 1	2978
Fourrine 68	708
Fourrine 79	3283
Fourrine 84	161
Fourrine 85	3233
Fourrine 94	1110
Fourrine 99	2621
Fourrine D	2978
Fourrine ERN	2621
Fourrine EW	3283
Fourrine M	1110
Fourrine P Base	161
Fourrine PG	3233
Fowler's Solution	252
Foxetin	1773
Foxima	3020
Foziretic	1793
FPW 400	646
FPW 800	646
FR-1	177
FR 10	1058
FR 10 (ether)	1058
FR 28	3386
Fr-62	3385
FR 143	2741
FR 222	2462
FR 300	1058
FR 300BA	1058
FR-513	3850
FR-521	1131
FR-522	1131
FR-612	1132
FR 705	2885
FR-913	3852
FR-1034	3681
FR 1138	1131
FR-1205	2884
FR-1206	1902
FR 1208	2741
FR 1360	3850
FR-2406	4006
Frademicina	2245
Fradizine	4029
Fraissite	364

Name and Synonym Index

Name and Synonym Index

Furobeta	1807	FWH 114	3553
Furocot	1807	FW-XO	3634
Furodiurol	1807	Fycol 8	945
Furodrix	1807	Fycop 40A	945
Furol	1803	Fycop	945
Furole	1803	FYDE	1786
Furomen	1807	Fyfanon	2294
Furomex	1807	Fyran 200 K	201
Furomide M.D.	1807	Fyran J 3	201
Furonaito 101	2578	Fyraway	245
Furonaito 113	2578	Fyrebloc	245
Furophen T	2694	Fyrex	196
Furophen T-Caps	2694	Fyrol CEF	3870
Furo-Puren	1807	Fyrol® DMMP	1399
Furorese	1807	Fyrol® FR 2	3170
Furosan	1807	Fyrol HB32	4006
Furose	1807	Fyrol® PBR	2884
Furosedon	1807	Fyrol PCF	4004
Furosemid	1807	Fyrquel 150	964
Furosemida	1807	Fytic acid	3025
Furosemide	1807	Fytolan	945
Furosemide mita	1807		
Furosemidu	1807	G 0	1920
Furosemidum	1807	G 0 (Oxide)	96
Furosemix	1807	G 2 (Oxide)	96
Furoside	1807	G-4	1199
Furosifar	1807	G-11	1908
Furosix	1807	G 20	1125
Furoter	1807	G 25	810
Furovite	1807	G 100 (clay)	315
Furpirinol	2670	G 301	1122
Furpyrinol	2670	G 315	2819
Fur-ren	2694	G 339S	647
Furro D	2978	G-444E	754
Furro ER	2621	G 491	3749
Furro P Base	161	G 665	674
Fursemid	1807	G 711	2132
Fursemida	1807	G 946	3536
Fursemide	1807	G 952	3348
Fursol	1807	G 996	1599
Furucton	1796	G 2124	2874
Fury	1038	G 2129	3073
Furyl-methanal	1803	G 3300	2132
Fusarex	3659	G 3707	2192
Fusarex G	3659	G 3802	732
Fusarex T	3659	G 3802POE	732
Fused borax	3386	G 3804POE	732
Fused boric acid	449	G 3816	732
Fused quartz	4098	G 3820	732
Fused silica	3347, 4098	G 3910	2801
Fused sodium borate	3386	G 3915	2801
Fusel oil	1808	G 3920	2801
Fusel oil, refined	1808	G 13529	3670
Fusel oil, refined (mixed amyl alcohols)	1808	G 27550	3230
Fusel oil, sugar beet	1808	G 30027	269
Fuselex RD 40-60	4098	G 34161	3160
Fuselex RD 120	4098	G 34360	1091
Fuselex ZA 30	4098	G1V Gard DXN	1345
Fuselex	4098	G-2129	3073
Fuselöl	1808	G-24480	1122
Fusid	1807	G-25804	754
Fusilade	1755	G-27692	3367
Fustic extract	2274	G-27901	3906
Futin	3949	G-30028	3174
Futramine D	2978	G-30320	893
Fuwalip	646	G-31435	3159
Fuxen	2625	G-32883	671
FW 50	646	G-704,650	77
FW 200 (mineral)	646	GA	1825
FW 293	1217	GA3	1825
FW 325	646	Gabapentin	1809
FW 734	3166	Gabapentine	1809

Gabapentino	1809
Gabapentinum	1809
Gabapetin	1809
Gable-Tite Dark Creosote (Creola)	4115
Gable-Tite Light Creosote (Creola)	4115
Gadopril	1566
Gafanol E 200	3065
Gafcol EB	513
Gafgard 233	2900
GAF-S 630	3083
Gagro	1599
Galactinum	2180
Galactitol	1810
Galaxolide	1811
Galaxy	314
Galenamox	209
Galesan	1122
Galipan	309
Galipot	4026
Galisol	247
Galleonite 136	2578
Gallery	2157
Gallex	958, 4115
Gallic acid	1812
Gallic acid methyl ester	1813
Gallic acid, dodecyl ester	3154
Gallic acid, lauryl ester	3154
Gallic acid, propyl ester	3206
Gallicin	1813
Gallii (^{67}Ga) citras	2644
Gallium	1814
Gallium, elemental	1814
Gallium (^{67}Ga) citrate	2644
Gallium citrate (^{67}Ga))	2644
Gallium-(^{67}Ga) citrate (1:1)	2644
Gallium-67 citrate	2644
Gallogama	2246
Gallotannic acid	3652
Gallotannin	3652
Gallotox	2986
Galloxon	1884
Galokson	2864
Galoxone	1884
Galtak	309
Galvatol 1-60	3081
Gamacarbatox	2246
Gamacid	2246
Gamacide	2246
Gamacide 20	2246
Gamanil	2260
Gamaphex	2246
Gamasol 90	1419
Gamene	2246
Gamiso	2246
Gamma-mean 4	2246
Gamonil	673, 2260
Gamophen	1908
Gamophene	1908
Ganex E 535	3083
Ganex P 804	3221
Ganocide	1553
Ganor	1714
Gansil	753
Gantrez	2298
Gantrez 39	2298
Gantrez 149	2298
Gantrez 169	2298
Gantrez 903	2298
Gantrez AN	2298
Gantrez AN 119	2298
Gantrez AN 139	2298
Gantrez AN 149	2298
Gantrez AN 169	2298
Gantrez AN 179	2298
Gantrez AN-1195	2298
Gantrez S 95	2298
Gantron PVP	3083
Gantron S 630	3083
Gantron S 860	3083
Garantose	3317
Garbenda	674
Garden Tox	1122
Garden	1640
Gardenal sodium	2944
Gardeprim A 1862	3670
Gardinol	3457
Gardol	3456
Gardoprim	3670
Garlon	3888
Garlon 2	3888
Garlon 3A	3889
Garlon 4	3888, 3890
Garlon 4E	3890
Garlon 250	3888
Garox	1125
Garranil	669
Garvox	310
Gas Black	684
Gas, natural	2630
Gas, producer	1798
Gasec	2809
Gas-furnace black	684
Gasoline (casinghead)	1815
Gasoline	1815
Gasoline, natural	1815
Gaster	1714
Gastex	684
Gastion	3017
Gastoxin	3017
Gastramine	375
Gastrese	2535, 2536
Gastrial	3273
Gastridan	1714
Gastridin	1714
Gastridina	3274
Gastrimut	2809
Gastrion	1714
Gastro	1714
Gastrobid	2535, 2536
Gastrodiagnost	2904
Gastrodomina	1714
Gastrofam	1714
Gastrografin Oral (Veterinary)	1117
Gastrografin	1117, 1118
Gastrolav	3274
Gastroloc	2809
Gastromax	2535, 2536
Gastropen	1714
Gastrosedol	3273
Gastrosidin	1714
Gastrosil	2535, 2536
Gastro-tablinen	2535, 2536
Gastrotem	2535, 2536
Gastro-Timelets	2535, 2536
Gastrotopic	3155
Gaucho	2026
Gaultheria oil	2514
Gaultheriaöl	2514
Gaultheriaoel	2514
GBL	593
GBN 5	1835
GC 1283	2556
GC 3944-3-4	3256
GC 6936	1732

Name and Synonym Index

Name and Synonym Index

Gold Star (carbonate)	2280
Gold trichloride	1872
Gold, colloidal	1871
Goldate	2839
Golden antimony sulfide	242
Goldenleaf tobacco spray	1569
Goldflush II	1462
Goldquat 276	2864
Goldsugar	1832
Goliath	2943
Goltix	2370
Goltix®	2370
Goodnight	3158
Goodrite 1800X73	3592
Good-rite GP 264	406
Good-rite K 37	3060
Good-rite K 700	3060
Good-rite K 702	3060
Good-rite K 727	3060
Good-rite K 732	3060
Good-rite NIX	3452
Good-rite WS 801	3060
Good-rite® GP-223	1449
Good-rite® GP-265	1075
Gosenol KH-17	3081
Gossypose	3267
Gotax 86,	2299
Gozid	1816
GP 7I	4098
GP 11I	4098
GP 20 (carbonate)	2280
GP 60	1873
GP 60S	1873
GP 63	1873
GP-137	1238
GP-45840	1213
GPCD 398	3076
GPKh	1887
GR-3343G	3324
GR 33343X	3323
GR 43175	3629
GR-43175C	3630
GR 43175X	3629
Gradetol	3982
Grafalin	68
Grafoil GTA	1873
Grafoil	1873
Graham's salt	3438, 3463
Grahamsches salz	1827
Grain alcohol	1593
Grain sorghum harvest-aid	3395
Gramevin	1051
Gramidil	209
Graminon	2151
Graminon-plus	314
Gramisan	2631
Gramixel	2864
grammite	645
Gramoxone	2864
Gramoxone D	2864
Gramoxone dichloride	2864
Gramoxone methyl sulfate	2863
Gramoxone S	2864
Gramoxone W	2864
Grampenil	210
Gramuron	2864
Granada Green Lake GL	3024
Granex O	3395
Granmag	2288
Granodine 16NC	4155
Granodine 80	4155
Granol NM	2309
Granox	1904
Granox NM	1904
Granox PFM	668
Granstar	2546
Granudoxy	1552
Granugen	2857
Granular zinc	4129
Granulated sugar	3599
Granulin	1097
Granurex	2632
Granutox	2998
Grape seed oil	2247
Grape sugar	1832
Graphite	1873
Graphite, natural	1873
Graphite, synthetic	683, 1873
Graphitic acid	1873
Graphlox	1905
Graphnol N 3M	1873
Graphol	2537
Graphtol Green 2GLS	3024
Graphtol Yellow GXS	872
Grasal Brilliant Red B	3604
Grasal Brilliant Red G	3603
Grasal Brilliant Yellow	518
Grasal Orange	3602
Grasan Brilliant Red B	3604
Grasan Brilliant Red G	3603
Grasan Orange 3R	4122
Grasan Orange R	3602
Grascide	3166
Grasex	750
Grasidim	3343
Graslan	3658
Graslan 40P	3658
Graslan 250 brush bullets	3658
Grasselerator 102	2394
Grassland weedkiller	309
Gravidox	4091
Gravocain	2237
Gray acetate	607
Gray Acetate of Lime	607
Gray arsenic	252
Gray selenium	3338
Grazon ET	3888
Grazon	3027
Great Lakes BA-59P	3680
Great Lakes CD-75P™	1902
Great Lakes PHT4	3684
Great Lakes SP-75™	1902
Green acetate	2815
Green Charm Multi-Purpose Fungicide	820
Green Chrome Oxide	850
Green chromic oxide	850
Green chromium oxide	850
Green cinnabar	850
Green cross couch grass killer	3525
Green cross warble powder	3310
Green densic GC 800	690
Green densic	690
Green Light Fung-Away Fungicide	3833
Green Light Systemic Fungicide	3764
Green nickel oxide	2661
Green Oil	230
Green Oxide of Chromium	850
Green rouge	850
Green Salts	1746
Green seal-8	4154
Green spar	939
Green Thumb Lawn & Garden Fungicide	820
Green verditer	939
Green Vitriol	1746

Name and Synonym Index

Name and Synonym Index

HSDB 1120	1690	HSDB 1258	2595
HSDB 1121	2770	HSDB 1262	3095
HSDB 1122	1894	HSDB 1264	3091
HSDB 1123	1990	HSDB 1265	2360
HSDB 1126	1950	HSDB 1266	3077
HSDB 1127	2321	HSDB 1314	3429
HSDB 1128	2459	HSDB 1315	3457
HSDB 1132	2684	HSDB 1316	386
HSDB 1134	2699	HSDB 1318	3752
HSDB 1136	3967	HSDB 1319	1170
HSDB 1137	388	HSDB 1320	1369
HSDB 1138	2935	HSDB 1321	233
HSDB 1139	1200	HSDB 1322	802
HSDB 1140	2918	HSDB 1326	1148
HSDB 1142	1428	HSDB 1328	2241
HSDB 1143	2490	HSDB 1329	1057
HSDB 1144	1438	HSDB 1332	1672
HSDB 1146	3982	HSDB 1333	1195
HSDB 1147	1147	HSDB 1334	2453
HSDB 1149	2386	HSDB 1336	2475
HSDB 1151	3027	HSDB 1340	3881
HSDB 1152	1099	HSDB 1343	824
HSDB 1154	2086	HSDB 1344	4129
HSDB 1156	2683	HSDB 1345	1455
HSDB 1157	2704	HSDB 1400	3174
HSDB 1158	2712	HSDB 1401	3770
HSDB 1160	1668	HSDB 1402	2943
HSDB 1162	2299	HSDB 1403	1731
HSDB 1165	2479	HSDB 1404	2208
HSDB 1167	2443	HSDB 1412	4081
HSDB 1169	3010	HSDB 1418	3666
HSDB 1171	275	HSDB 1421	50
HSDB 1173	2402	HSDB 1427	218
HSDB 1175	153	HSDB 1432	448
HSDB 1176	3075	HSDB 1433	608
HSDB 1178	2618	HSDB 1434	611
HSDB 1179	1373	HSDB 1436	708
HSDB 1183	2998	HSDB 1441	640
HSDB 1184	2500	HSDB 1442	1302
HSDB 1187	3005	HSDB 1445	1447
HSDB 1188	4028	HSDB 1500	1183
HSDB 1192	3181	HSDB 1501	1181
HSDB 1193	3180	HSDB 1502	3867
HSDB 1194	3575	HSDB 1508	4078
HSDB 1196	3061	HSDB 1510	73
HSDB 1200	3161	HSDB 1516	3282
HSDB 1202	2300, 2301	HSDB 1519	3159
HSDB 1205	3014	HSDB 1522	458
HSDB 1208	2359	HSDB 1523	482
HSDB 1210	3156	HSDB 1525	3671
HSDB 1212	4091	HSDB 1527	1442
HSDB 1213	3082	HSDB 1530	681
HSDB 1215	3182	HSDB 1532	693
HSDB 1218	3121	HSDB 1533	754
HSDB 1219	2589	HSDB 1541	766
HSDB 1222	3119	HSDB 1542	801
HSDB 1226	3166	HSDB 1546	820
HSDB 1227	2675	HSDB 1548	968
HSDB 1229	197	HSDB 1549	988
HSDB 1234	3111	HSDB 1557	1234
HSDB 1235	3087	HSDB 1565	1166
HSDB 1236	1800	HSDB 1567	222
HSDB 1238	3106	HSDB 1568	2162
HSDB 1240	2797	HSDB 1570	1214
HSDB 1242	1075	HSDB 1572	2666
HSDB 1244	2353	HSDB 1580	1730
HSDB 1245	3103	HSDB 1582	1275
HSDB 1247	2355	HSDB 1583	1509
HSDB 1248	3100	HSDB 1584	2061
HSDB 1250	3080	HSDB 1586	1344
HSDB 1252	3097	HSDB 1590	1722
HSDB 1253	3092	HSDB 1593	2387

Name and Synonym Index

HSDB 1594	2395	HSDB 1787	4164
HSDB 1597	1443	HSDB 1788	401
HSDB 1600	3682	HSDB 1801	67
HSDB 1603	228	HSDB 1802	189
HSDB 1604	246	HSDB 1808	3063
HSDB 1605	709	HSDB 1809	1462
HSDB 1607	371	HSDB 1811	3624
HSDB 1608	434	HSDB 1813	959
HSDB 1609	449	HSDB 1814	960
HSDB 1611	558	HSDB 1815	958
HSDB 1613	605	HSDB 1818	2070
HSDB 1615	634	HSDB 1820	1350
HSDB 1616	3822	HSDB 1824	974
HSDB 1618	812	HSDB 1901	4059
HSDB 1619	850	HSDB 1907	3196
HSDB 1622	937	HSDB 1909	79
HSDB 1623	954	HSDB 1910	172
HSDB 1635	9	HSDB 1933	707
HSDB 1636	1273	HSDB 1934	2812
HSDB 1639	1249	HSDB 1935	2514
HSDB 1641	1411	HSDB 1973	3238
HSDB 1642	1055	HSDB 1977	2848
HSDB 1643	1651	HSDB 1991	2216
HSDB 1644	1679	HSDB 1995	1192
HSDB 1646	1790	HSDB 2000	3568
HSDB 1649	2227	HSDB 2002	1782
HSDB 1650	2213	HSDB 2004	584
HSDB 1651	2226	HSDB 2005	1951
HSDB 1652	2288	HSDB 2013	1612
HSDB 1655	313	HSDB 2014	2441
HSDB 1659	2556	HSDB 2017	55
HSDB 1660	2561	HSDB 2018	1966
HSDB 1661	2562	HSDB 2019	3142
HSDB 1662	2655	HSDB 2026	3803
HSDB 1664	2661	HSDB 2027	3749
HSDB 1665	2678	HSDB 2033	1907
HSDB 1666	803	HSDB 2034	2887
HSDB 1668	2862	HSDB 2038	2609
HSDB 1670	2986	HSDB 2040	3032
HSDB 1672	3163	HSDB 2042	3814
HSDB 1676	3486	HSDB 2043	3813
HSDB 1678	3857	HSDB 2044	3809
HSDB 1701	1499	HSDB 2045	775
HSDB 1705	1544	HSDB 2046	774
HSDB 1709	1710	HSDB 2047	776
HSDB 1716	1664	HSDB 2060	825
HSDB 1721	1760	HSDB 2063	1335
HSDB 1724	1904	HSDB 2064	33
HSDB 1729	1091	HSDB 2067	148
HSDB 1737	2322	HSDB 2070	198
HSDB 1738	2323, 2327	HSDB 2072	215
HSDB 1741	2534	HSDB 2073	227
HSDB 1744	3336	HSDB 2074	235
HSDB 1747	2835	HSDB 2076	350
HSDB 1749	3256	HSDB 2078	428
HSDB 1753	2968	HSDB 2079	1028
HSDB 1755	3050	HSDB 2085	874
HSDB 1759	762	HSDB 2089	530
HSDB 1762	3310	HSDB 2090	2122
HSDB 1765	3367	HSDB 2101	187
HSDB 1766	3431	HSDB 2105	366
HSDB 1767	2371	HSDB 2107	532
HSDB 1769	1052	HSDB 2108	2270
HSDB 1772	3659	HSDB 2109	1040
HSDB 1773	3704	HSDB 2117	1812
HSDB 1777	2361	HSDB 2118	1823
HSDB 1779	3864	HSDB 2123	1862
HSDB 1780	3835	HSDB 2125	1871
HSDB 1782	1036	HSDB 2126	995
HSDB 1783	1732	HSDB 2127	1877
HSDB 1784	1733	HSDB 2128	1762
HSDB 1786	4100	HSDB 2131	1272

Name and Synonym Index

Name and Synonym Index

Name and Synonym Index

Name and Synonym Index

Name and Synonym Index

Name and Synonym Index

Name and Synonym Index

Name and Synonym Index

Name and Synonym Index

Name and Synonym Index

Name and Synonym Index

Name and Synonym Index

Name and Synonym Index

Name and Synonym Index

Name and Synonym Index

Name and Synonym Index

Name and Synonym Index

Name and Synonym Index

Name and Synonym Index

Name and Synonym Index

Name and Synonym Index

Name and Synonym Index

Morpan T	2603	Mravencan sodny	3435
Morphine monomethyl ether	934	Mravencan vapenaty	622
Morphine-3-methyl ether	934	MRC 910	2064
Morpholine	2580	Mrowkozol	3184
Morpholine disulfide	3612	MS 33	3537
Morpholine, 4-methyl-	2487	MS 33F	3537
Morton EP 452	2943	MS 40	3627
Morton EP-161E	2481	MS-122	3714
Morton Orange Y	3602	MS-180 Freon® TF Solv	3883
Mosaic gold	3557	MS 193	1887
Mosanon	79	MSA	2390
Mosatil	618	MSG	2575
Moscarda	2294	MSL	3424
Moschus ketone	2589	MSL (carbamate)	3424
Moss Green	974	MSMA	2584
Mosul	1714	MSP	3492
Moth balls	2608	MSP 150	3729
Moth flakes	2608	MSS MH18	2299
Moth Snub D	1233	MST	3634
Motiax	1714	MTBE	2431
Motilyn	1095	MTBHQ	553
Motimol	2561	MTD	2387
Motiorange R	3602	MTN	3816
Motirot 2R	3603	Muclox	1714
Motirot G	4122	Mucochloric acid	2585
Motor benzol	321	Mudrane Tablets	3114
Motox	3822	Mudrane-2 Tablets	3114
Motrin-A	2016	Mugan	673
Mottenhexe	1907	Muldis	2995
Mountain Green	939, 974	Mulsiferol	1582
Mouse antialopecia factor	2040	Mulsopaque	2057
Mouse pak	4100	Mult	3730
Movinyl	3080	Multamat	310
Movinyl 50M	3080	Multichlor	753
Movinyl 114	3080	Multicide 2154	2957
Movinyl 801	3080	Multicide Concentrate F-2271	2957
Mowchem	2332	Multimet	310
Mowilith 30	3080	Multitab	2152
Mowilith 50	3080	Multitol	3542
Mowilith 70	3080	Mupirocin	2586
Mowilith 90	3080	Mupirocina	2586
Mowilith D	3080	Mupirocine	2586
Mowilith DV	3080	Mupirocinum	2586
Mowilith M70	3080	Muralluride	2344
Moxacef Kapseln	710	Murexan	4040
Moxal	209	Murexide	2587
Moxaline	209	Murfos	2867
Moxie	2402	Murganic	693
Mozal	68	Muriacite	648
MP-2	1409	Muriate of Ammonia	179
MP 12-50	3634	Muriate of sdoa	3396
MP 25-38	3634	Muriatic acid	1962
MP 40-27	3634	Muriatic ether	788
MP 45-26	3634	Muriol	807
MP 79	2928	Murox	2735
MP 620	840	Murphy's rose fungicide	1783
MP 1023	31	Murvin	673
MP Diol Glycol	2583	Musashi-no-Ringosan	2300, 2301
MPA GYM	2330	Muscarin	2588
MPA Hexal	2330	Muscarine	2588
MPA-β	2330	Muscatox	953
MPA-Noury	2330	Muscle adenylic acid	61
MPG 6	1873	Muscle sugar	2040
MPI Tc 99m Generator	2933	Muscol	1275
MPK 90	3221	Muscone	2589
MPP	1731	Muscovite	2549
MPP (pesticide)	1731	Muscovite mica	2549
MPS	2351	Mushroom sugar	2318
MPS-M	2351	Musk ambrette	2590
M-Pyrol	2512	Musk ketone	2591
MR 84	4098	Musk T	1631
Mravencan methylnaty	2466	Muskarin	2588

Name and Synonym Index

NAA 42	2199	Naldetuss	15
NAA 173	3568	Nalfloc 636	3060
NAA 800	2609	Nalfloc N 206	2450
Nabac	1908	Nalfloc N 1050	3347
NABU	3343	Nalkil	458
NAC	673	Nalkylene 500	1534
NAC (insecticide)	673	Nalmet A 1	3424
Nacan	3477	Nalox	2543
Nacap®	3459	Nalyxan	2625
Naccanol NR	3425	Namate	1499
Naccanol SW	3425	NaMBT	3459
Nacconol 35SL	3425	Namekil	2369
Nacconol 40f	3425	Nanchor	3303
Nacconol 90f	3425	Nanker	3303
Nacconol 98SA	1535	Nankor	3303
Nacconate-100	3798	Nansa 1042P	1535
Nacconate 300	1475	Nansa HF 80	3425
Nacconate H 12.	1226	Nansa HS 85S	3425
Nacconate IOO	3797	Nansa SSA	1535
Nacconol	3425	Nansa YS 94	2132
Nacconol LAL	1539	Nansa® AS 40	184
Na-Cemmix	3496	Nansa® TDB	3896
Nacol® 6-98	1946	Naotin	2649, 2650
Nacol 10-99	1065	Napagin A	1665
Nacol® 12-96	2199	Napflam	2625
Nacol® 12-99	2199	Naphid	2618
Nacol® 14-95	2600	Naphtalene	2608
Nacol® 14-98	2600	Naphtenate de cobalt	909
Nacolyte 673	3059	Naphtenate de cuivre	943
Nacure 1040	3803	Naphtenate de zinc	4152
Na-ddtc	1253	Naphtha	2606, 2607, 2936
Nadib	1835	Naphtha 49 degree be-coal tar type	2606
Nadic methyl anhydride	2489	Naphtha VM & P	2606
Nadigest	2330	Naphtha VM & P, 50 degree flash	2606
Nadisal	3504	Naphtha VM & P, high flash	2606
Nadone	1019	Naphtha VM & P, regular	2606
Nafasol	2625	Naphtha, hydrotreated	2606
Nafeen	3431	Naphtha, ligroine	2239
NaFpak	3431	Naphtha, petroleum	2606
Naftalen	2608	Naphtha, solvent	2607
Naftalofos	3271	Naphtha, Stoddard solvent	2607
Naftalofosum	3271	Naphtha, varnish makers' and painters'	2239
Naftam 2	2624	Naphthalane	1059
Naftifine	2604	Naphthalene	2608
Naftifine hydrochloride	2604	Naphthalene oil	903
Naftin	2604	Naphthalene, decahydro-	1059
Naftocit® ZBEC	4143	Naphthalene, 1,6-dihydroxy-	2610
Naftolite	909	Naphthalene, diisononyl-	1315
Nafusaku	2609	Naphthalene, α-methyl-	2491
Naganin	2605	Naphthalene, β-methyl-	2492
Naganine	2605	Naphthalene, 1-methyl-	2491
Naganinum	2605	Naphthalidam	2623
Naganol	2605	Naphthalidine	2623
Nagravon	4089	Naphthalin	2608
NAH	2649, 2650	Naphthaline	2608
Naixan	2625, 2627	Naphthalol	2616
Nako Brown R	161	Naphthalophos	3271
Nako H	2978	Naphthan	1059
Nako TGG	3283	Naphthathenic soap	3473
Nako TMT	1110	Naphthazarin	2617
Nako TRB	2621	Naphthazarine	2617
Nako Yellow 3GA	160	Naphthazarone	2617
Nako Yellow ga	160	Naphthene	2608
Nalcamine G-13	2803	Naphthenic acid, cobalt salt	909
Nalcast	4098	Naphthenic acids	2618
Nalco 1050	3347	Naphthenic acids, copper salts	943
Nalco 7046	666	Naphthenic acids, sodium salts	3473
Nalco 7530	123	Naphthenic acids, zinc salts	4152
Nalco 8676	110	Naphthionsäure	2619
Nalco D-1994	2450	Naphthionsaeure	2619
Nalcon 243	1055	Naphthoelan Navy Blue	147
Naldegesic	15	Naphthol	2620

Name and Synonym Index

Naphthol B.O.N.	2003
Naphthol B	2620
Naphthol bon	2003
Naphthol, β	2620
Naphthoresorcin	2622
Naphthoresorcinol	2622
Naphthosalol	2616
Naphthylaminemonosulfonic acid S	3335
Naphtoelan Red GG Base	2683
Naphtol AS-KG	3809
Naphtol AS-KGLL	3809
Naphuride sodium	2605
Napizide	1828
Naples red	2587
Naples yellow	2209
Napmel	2625
Naposin	2625
NAPP	2451
Napratec	2625
Napren E	2625
Napren	2625
Naprilene	1567
Naprilin	3189
Naprium	2625, 2627
Naprius	2625
Naprodil	2627
Naprodol	2627
Naprontag	2625
Napropion	3497
Naprosyn	2625, 2627
Naprosyn LLE	2625
Naprosyn LLE Forte	2625
Naprosyne	2625
Naprovite	2627
Naproxen	2625
Naproxen natrium	2627
Naproxen Piperazine	2626
Naproxen Sodium	2627
Naproxene	2625
Naproxeno	2625
Naproxenum	2625
Naproxi 250	2625
Naproxi 500	2625
Naprux Gesic	2627
Naprux	2625, 2627
Napter E 8075	3077
Napxen	2625
Narasin	2628
Narasin A	2628
Narasine	2628
Narasino	2628
Narasinum	2628
Narceol	2629
Narcogen	3862
Narcotile	788
Narcoxyl	4110
Narcylen	35
Narilet	2063
Naritec	1567
Narkosoid	3862
Narkotil	2452
Narma	2625
Narocin	2625
Narol	496
Nasacort AQ	3840
Nasacort	3840
Nascobal	4089
Nasemo	2666
Nasonex	2563
NAT-50-PGReach®	3195
Natacillin	1900
Natasol Fast Orange GR Salt	2684

Natigoxin	1286
National 120-1207	3080
National starch 10	3080
Natipide® II PG	3195
Natispray	2696
Natium salicylicum	3504
Native calcium sulfate	649
Natraiumcyclohexylamidosulfat	3409
Natrascorb	3380
Natreen	3317, 3409
Natri-C	3380
Natrii acetrizoas	31
Natrii amidotrizoas	1118
Natrii ascorbas	3380
Natrii carbonas monohydricus	3394
Natrii cyclamas	3409
Natrii d	1525
Natrii disulfis	3460
Natrii edetas	3743
Natrii etasulfas	3429
Natrii feredetas	1739
Natrii hydrogencarbonas	3382
Natrii iodidum	3448
Natrii sulphis	3514
Natrii tetracylis sulfas	3519
Natrii tetradecylis sulfas	3519
Natrii tetradecylsulfas	3519
Natrioxen	2627
Natriphene	3485
Natriumglutaminat	2575
Natriumhexafluorosilicat	2302
Natriumphosphat	1501
Natriumsilicofluorid	2302
Natrocitral	3400
Natron	3382
Natsyn® 2200	3070
Natural	1738
Natural anhydrite	648
Natural bromellite	371
Natural calcium carbonate	613
Natural fluorite	621
Natural gas	2630
Natural gasoline	1815
Natural gasoline (natural gas)	1815
Natural hematite	1738
Natural iron oxides	1738
Natural lawrencite	1744
Natural marshite	987
Natural molybdenite	2561
Natural molybdite	2562
Natural molysite	1735
Natural montroydite	2360
Natural Pearl Essence	1877
Natural red 4	698
Natural rhodinol, acetylated	886
Natural rhodochrosite	2312
Natural roesslerite	1618
Natural smithsonite	4139
Natural tenorite	982
Natural trehalose	2594
Natural White 1	1877
Natural whitlockite	641
Natural wintergreen oil	2514
Natural Yellow 10	3244
Natural Yellow 26	701
Naturechem® PGR	3202
Naturon	1877
Nauga White	2951
Naugalube® 403	1151
Naugalube 428L	1466
Naugalube® 438	421
Naugard DSTDP	1505

Naugard® 10	2896	NCI-C00533	810
Naugard® 76	2749	NCI-C00566	1569
Naugard 451	1145	NCI-C00920	1644
Naugard® 524	4002	NCI-C01730	233
Naugard® NBC	2658	NCI-C01821	1369
Naugard® PANA	2966	NCI-C01865	1438
Naugatuck DET	1275	NCI-C01990	2550
Naugawhite	2951	NCI-C02006	2552
Naugex SD-1	3612	NCI-C02028	1157
Nauli gum	218	NCI-C02039	776
Navicalm	2325, 2326	NCI-C02040	825
Navicote 2000	425	NCI-C02073	1556
Navy fuels JP-5	2171	NCI-C02084	3476
Na-X	3393	NCI-C02119	4041
Naxaine® C, CO, Cocbetaine	923	NCI-C02186	3978
Naxen	2625	NCI-C02200	3590
Naxen F	2625	NCI-C02233	147
Naxid	2625	NCI-C02302	1110
Naxol	1018	NCI-C02551	605
Naxonat® 4ST	3524	NCI-C02686	790
Naxonate®	3531	NCI-C02722	3559
Naxonate® 4AX	207	NCI-C02733	606
Naxonate® 4KT	3139	NCI-C02766	2679
Naxonate® 4L	3531	NCI-C02799	1786
Naxonate® G	3531	NCI-C02813	3050
Naxonate hydrotrope	3524	NCI-C02857	1698
Naxopren	2625	NCI-C02868	2932
Naxy	888	NCI-C02891	2484
Naxyn	2625	NCI-C02904	3880
Naxyn 250	2625	NCI-C02915	2624
Naxyn 500	2625	NCI-C02937	2241
NBC	2658	NCI-C02959	1506
NBC wormer	535	NCI-C02960	766
NBT	2688	NCI-C02982	956
NBT (dye)	2688	NCI-C03054	1186
NC 100	2067	NCI-C03065	159
NC 129	3224	NCI-C03134	1593
NC 302	3258	NCI-C03258	972
NC 1264	3772	NCI-C03270	4006
NC-1667	3906	NCI-C03372	1651
NC 6897	310	NCI-C03521	348
NC 8438	1601	NCI-C03554	3693
NCI 129	3224	NCI-C03598	1147
NCI 9579	2712	NCI-C03601	3021
NCI 96683	3258	NCI-C03689	1458
NCI-554813	474	NCI-C03736	223
NCI-C505888	3206	NCI-C03816	1277
NCI C61198	654	NCI-C03827	1052
NCI C61143	2316	NCI-C03849	2397
NCI C07103	954	NCI-C03974	1559
NCI CO2835	1253	NCI-C03985	1205
NCI-C00044	75	NCI-C04535	1191
NCI-C00066	275	NCI-C04546	3862
NCI-C00077	668	NCI-C04557	3617
NCI-C00099	755	NCI-C04568	2047
NCI-C00102	820	NCI-C04579	4081
NCI-C00113	1211	NCI-C04580	3695
NCI-C00124	1233	NCI-C04591	686
NCI-C00135	1344	NCI-C04604	1907
NCI-C00180	1887	NCI-C04615	89
NCI-C00215	2294	NCI-C04626	2434
NCI-C00226	2867	NCI-C04637	3872
NCI-C00237	3027	NCI-C04897	3145
NCI-C00259	3822	NCI-C05970	3283
NCI-C00260	1733	NCI-C06008	3634
NCI-C00419	3256	NCI-C06111	358
NCI-C00431	897	NCI-C06155	535
NCI-C00442	3932	NCI-C06224	788
NCI-C00486	1217	NCI-C06360	361
NCI-C00497	2402	NCI-C06428	2556
NCI-C00511	1641	NCI-C06508	356
NCI-C00522	1640	NCI-C07001	1572

Name and Synonym Index

Name and Synonym Index

Name and Synonym Index

Name and Synonym Index

Name and Synonym Index

Name and Synonym Index

Name and Synonym Index

Name and Synonym Index

Nortryptyline	2733	Novonaprox	2625
Norvalamine	525	Novonidazol	2543
Norvasc	168	Novopranol	3189
Norway saltpeter	190, 631	Novoprednisone	3145
Norwegian saltpeter	190, 631	Novopropoxyn	3187
Norzepine	2734	Novoprotect	167
No-Scald DPA 283	1466	Novoquin	880
No-Scald	1466	Novo-Radinine	3274
Nosilen	3065	Novosalmol	68
Nosophene sodium	247	Novoscabin	359
Nosorex	2563	Novosir N	4164
Nospasm	696	Novotox	1211
Notaral	2881	Novotrimel	3949
Noten	266	Novotriptyn	167
Notidin	1714	Novo-Veramil	4065
Nouralgine	79	Novovitamin-D	1582
Nourithion	2867	Novox	353
Nouryset 200	1104	Novozin N 50	4164
Nourytam	3647	Novozir	4164
Nourytan	3647	Novozir MN 80	2307
Nova (pesticide)	2593	Novozir N 50	4164
Nova Chem AB	303	Novozir N	4164
Nova Sol R 20	3010	Novozone®	2290
Nova W	2593	Noxaben	1215
Novabritine	209	Noxamine CA 30	925
Novacorn	482	Noxfire	3310
Novaculite	3241	Noxfish	3310
Novaexcel 140	3010	Noxitem	3647
Novaexcel 150	3010	NP 2	2653
Novaexcel F 5	3010	NP 11	3238
Novaexcel ST 100	3010	NP-27	2352
Novaexcel ST 140	3010	NP 55	3343
Novaexcel ST 300	3010	NP 212	2010
Novantisol	2973	NP 1500	2291
Novapamyl LP	4065	NPG®	2639
Novapirina	1213	NPG® Glycol	2639
Novared 120UF	3010	NPH 1320	485
Novared 120UFA	3010	NPH humulin insulin	2042
Novared 120VFA	3010	NPH Iletin	2042
Novared 140	3010	NPH insulin	2042
Novared 280	3010	NRC 910	2064
Novared C 120	3010	NRD	2928
Novared F 5	3010	NRD101	3439
Novatec	2250	NRDC 104	3282
Novathion	1722	NRDC 107	377
Novazam	1121	NRDC 143	2928
Novelian	3630	NRDC 149	1038
Noverme	2324	NRDC 161	1086
Novhepar	2263	NRF 201	3535
Novidat	880	NS 02	313
Novismuth	434	NS 02 (fungicide)	313
Novo-Alprazol	95	NS 11	1403
Novo-ampicillin	210	NS-20	3631
Novocal	623	NSC 2	388
Novochlorhydrate	751	NSC 4	159
Novocillin vet.	2881	NSC 9	1390
Novo-Clonidine	896	NSC 10	3995
Novo-Cycloprine	1005	NSC 11	1145
Novo-Dermoquinona	2343	NSC 55	1128
Novodigal [inj.]	1286	NSC 140	1143
Novodiphenyl	2995	NSC 142	767
Novo-Famotidine	1714	NSC 144	233
Novofolacid	1782	NSC 149	335
Novofuran	2694	NSC 166	1861
Novogel® ST	129	NSC 172	480
Novol	2802	NSC 179	3240
Novol Poe 20	2801	NSC 180	3321
Novolexin	715	NSC 218	1214
Novolin	2041	NSC 232	1370
Novolin R	2041	NSC 247	1170
Novolorazem	2263	NSC 248	3433
Novo-Metformin	2377	NSC 389	3862

NSC 4807	3284	NSC 5541	2744
NSC 4809	3761	NSC 5543	2750
NSC 4811	1127	NSC 5571	992
NSC 4814	562	NSC 5575	33
NSC 4820	577	NSC 5578	2890
NSC 4823	1461	NSC 5579	2352
NSC 4832	4080	NSC 5590	225
NSC 4833	1520	NSC 5595	1392
NSC 4839	4000	NSC 5605	363
NSC 4840	1136	NSC 5630	3732
NSC 4841	3201	NSC 5648	3772
NSC 4847	1787	NSC 5670	1416
NSC 4867	3503	NSC 5707	2909
NSC 4874	3684	NSC 5708	2075
NSC 4886	970	NSC 5711	1019
NSC 4893	886	NSC 5712	2478
NSC 4959	1235	NSC 5713	2685
NSC 4960	2343	NSC 5747	3213
NSC 4961	1984	NSC 5761	1477
NSC 4963	1322	NSC 5776	149
NSC 4969	955	NSC 5871	3333
NSC 4971	835	NSC 5999	13
NSC 4972	553	NSC 6085	1063
NSC 4974	12	NSC 6088	1073
NSC 4983	889	NSC 6098	1105
NSC 5024	2758	NSC 6101	1227
NSC 5025	1064	NSC 6103	2686
NSC 5026	2193	NSC 6110	1679
NSC 5027	2483	NSC 6170	882
NSC 5028	2599	NSC 6175	859
NSC 5029	2488	NSC 6183	775
NSC 5030	2852	NSC 6188	702
NSC 5033	3770	NSC 6197	415
NSC 5035	3233	NSC 6202	1205
NSC 5036	606	NSC 6203	1133
NSC 5038	2702	NSC 6205	4031
NSC 5065	2667	NSC 6213	1132
NSC 5163	522	NSC 6236	518
NSC 5171	3728	NSC 6237	971
NSC 5172	3722	NSC 6250	2106
NSC 5175	541	NSC 6254	2417
NSC 5185	3788	NSC 6275	704
NSC 5188	1538	NSC 6284	1001
NSC 5210	3735	NSC 6289	3757
NSC 5212	1195	NSC 6292	1431
NSC 5226	3913	NSC 6295	1198
NSC 5237	1739	NSC 6296	1201
NSC 5246	1104	NSC 6300	1995
NSC 5252	3893	NSC 6303	1299
NSC 5270	1499	NSC 6329	1113
NSC 5272	3725	NSC 6343	1300
NSC 5284	3925	NSC 6347	1147
NSC 5318	2832	NSC 6362	52
NSC 5341	324	NSC 6363	2706
NSC 5342	362	NSC 6364	155
NSC 5349	3317	NSC 6365	4021
NSC 5350	2910	NSC 6368	3967
NSC 5353	2363	NSC 6370	1153
NSC 5354	137	NSC 6373	1453
NSC 5356	1388	NSC 6381	1321
NSC 5369	2699	NSC 6401	836
NSC 5370	3760	NSC 6493	3180
NSC 5379	3571	NSC 6509	1097
NSC 5387	2691	NSC 6513	3971
NSC 5402	2715	NSC 6514	4003
NSC 5406	2772	NSC 6524	2294
NSC 5427	2778	NSC 6526	88
NSC 5517	1361	NSC 6533	556
NSC 5518	2718	NSC 6538	2085
NSC 5523	3787	NSC 6602	2062
NSC 5525	3325	NSC 6647	368
NSC 5538	2616	NSC 6700	565

NSC 10295	1770	NSC 15319	1232
NSC 10309	832	NSC 15334	3218
NSC 10431	3021	NSC 15339	2591
NSC 10438	3281	NSC 15348	3814
NSC 10447	3219	NSC 15349	3813
NSC 10454	3237	NSC 15350	3809
NSC 10457	4122	NSC 15351	4057
NSC 10459	2948	NSC 15367	3253
NSC 10460	2865	NSC 15398	1411
NSC 10472	3604	NSC 15434	3719
NSC 10480	3330	NSC 15635	1177
NSC 10770	2742	NSC 15750	3974
NSC 10883	3280	NSC 15760	4075
NSC 10965	3146	NSC 15772	2609
NSC 10977	821	NSC 15797	3672
NSC 10999	2802	NSC 15874	2398
NSC 11016	579, 582	NSC 15905	2585
NSC 11138	2989	NSC 16152	3179
NSC 11204	3598	NSC 16201	1450
NSC 11213	1358	NSC 16236	1920
NSC 11215	3777	NSC 16241	1791
NSC 11571	1854	NSC 16572	473
NSC 11680	4008	NSC 16596	1362
NSC 11687	1928	NSC 16631	3215
NSC 11690	1237	NSC 16648	3972
NSC 11692	3837	NSC 16894	1966
NSC 11784	3193	NSC 16933	1982
NSC 11786	1945	NSC 16935	875
NSC 11801	1632	NSC 17069	406
NSC 11831	3319	NSC 17391	2172
NSC 11889	1988	NSC 17558	998
NSC 12016	1531	NSC 17581	774
NSC 12097	3285	NSC 17675	2000
NSC 12441	2813	NSC 17676	2701
NSC 12561	1479	NSC 17706	1368
NSC 12600	2103	NSC 18251	3029
NSC 12958	4168	NSC 18252	3030
NSC 13181	2002	NSC 18255	4078
NSC 13203	1042	NSC 18467	2029
NSC 13392	1833	NSC 18589	812
NSC 13397	3839	NSC 18596	90
NSC 13563	318	NSC 18597	91
NSC 13892	2299	NSC 18607	2510
NSC 13979	41	NSC 19028	1754
NSC 14190	1825	NSC 19043	2837
NSC 14459	4007	NSC 19311	1471
NSC 14460	2110	NSC 19492	3337
NSC 14490	1245	NSC 19493	273
NSC 14663	350	NSC 19639	2975
NSC 14664	4108	NSC 19790	4032
NSC 14665	2014	NSC 19892	1977
NSC 14666	59	NSC 19987	2507
NSC 14685	14	NSC 20103	1812
NSC 14705	3776	NSC 20199	3564
NSC 14759	664	NSC 20200	330
NSC 14864	743	NSC 20256	4046
NSC 14910	1425	NSC 20263	1497
NSC 14916	378	NSC 20264	61
NSC 15012	138	NSC 20291	1854
NSC 15039	2391	NSC 20293	1588
NSC 15043	38	NSC 20526	3853
NSC 15087	872	NSC 20610	1108
NSC 15136	1304	NSC 20812	4095
NSC 15197	3747	NSC 20909	2203
NSC 15198	724	NSC 20939	89
NSC 15199	727	NSC 20949	2079
NSC 15258	1333	NSC 20952	1365
NSC 15309	2997	NSC 20954	1268
NSC 15310	2122	NSC 20956	559
NSC 15313	1398	NSC 20959	2381
NSC 15316	1306	NSC 20964	357
NSC 15318	1455	NSC 20968	1022

Name and Synonym Index

NSC 158709	3147	NSC 203387	1084
NSC 159120	1439	NSC 206315	67
NSC 159432	680	NSC 209528	2464
NSC 163046	269	NSC 209529	219
NSC 163048	3159	NSC 209802	3555
NSC 163049	3160	NSC 210838	2573
NSC 163103	540	NSC 212537	1408
NSC 163175	1493	NSC 215190	1959
NSC 163400	344	NSC 215191	2562
NSC 163904	1379	NSC 215196	188
NSC 164906	3818	NSC 215199	2353
NSC 164915	517	NSC 215200	3092
NSC 164918	1955	NSC 215202	4015
NSC 164932	2389	NSC 215203	3995
NSC 165582	3305	NSC 215204	3864
NSC 165706	978	NSC 215208	2587
NSC 166062	260	NSC 215210	1466
NSC 166351	587	NSC 217971	2012
NSC 166454	1088	NSC 218451	1166
NSC 166502	2895	NSC 218452	1509
NSC 166511	373	NSC 220327	1003
NSC 167068	3803	NSC 221683	498
NSC 167070	3125	NSC 226253	1427
NSC 167822	681	NSC 226979	1789
NSC 169188	2010	NSC 227401	672
NSC 169189	2325	NSC 227995	1708
NSC 169453	2734	NSC 229795	3271
NSC 169454	2649, 2650	NSC 231649	2803
NSC 169864	671	NSC 232671	3222
NSC 169868	2503	NSC 232673	2835
NSC 169900	1005	NSC 233899	3027
NSC 169913	355	NSC 240503	164
NSC 170228	3267	NSC 241716	1489
NSC 170811	3764	NSC 244436	1571
NSC 170959	2373	NSC 249815	1270
NSC 170976	272	NSC 251767	3104
NSC 171184	1254	NSC 252156	815
NSC 171206	2426	NSC 252159	1238
NSC 172124	696	NSC 254532	2657
NSC 172522	2008	NSC 255226	2984
NSC 172781	2891	NSC 259940	2234
NSC 172971	773	NSC 263489	313
NSC 173379	1005	NSC 263490	1343
NSC 174140	777	NSC 263492	693
NSC 174280	3392	NSC 263493	3933
NSC 174502	1150	NSC 263497	2888
NSC 177023	2232	NSC 263500	2864
NSC 177699	1707	NSC 264287	4040
NSC 178296	2666	NSC 269896	3025
NSC 179737	1771	NSC 270681	4096
NSC 179742	1036	NSC 270787	1998
NSC 179913	894	NSC 272366	4065
NSC 184849	2324	NSC 281242	3296
NSC 186246	2347	NSC 281243	2247
NSC 187663	3138	NSC 289758	2263
NSC 190466	3865	NSC 289928	69
NSC 190671	3888	NSC 290195	746
NSC 190986	3248	NSC 290409	3232
NSC 190987	2387	NSC 291839	2872
NSC 190998	2162	NSC 292811	3828
NSC 191479	2126	NSC 293873	699
NSC 191935	3459	NSC 296964	1852
NSC 193508	3251	NSC 298885	1786
NSC 195022	3282	NSC 302962	1095
NSC 195058	1794	NSC 303303	3833
NSC 195106	1718	NSC 305303	2173
NSC 195319	1771	NSC 309818	1897
NSC 202167	3504	NSC 310011	3142
NSC 202753	1447	NSC 312447	1028
NSC 202875	2854	NSC 313680	1756
NSC 202959	182	NSC 315851	493
NSC 203318	3524	NSC 316165	3246

Name and Synonym Index

Name and Synonym Index

Name and Synonym Index

Name and Synonym Index

Name and Synonym Index

Name and Synonym Index

Name and Synonym Index

Name and Synonym Index

Name and Synonym Index

Name and Synonym Index

Name and Synonym Index

Name and Synonym Index

Name and Synonym Index

Name and Synonym Index

Name and Synonym Index

Name and Synonym Index

Name and Synonym Index

Name and Synonym Index

Name and Synonym Index

Name and Synonym Index

Name and Synonym Index

Name and Synonym Index

Name and Synonym Index

Name and Synonym Index

Name and Synonym Index

Rat-O-cide	4100	RCRA waste no. U007	48
Rat-O-cide 2	4100	RCRA waste no. U008	50
Rat-ola	4100	RCRA waste no. U009	52
Ratorex	4100	RCRA waste no. U012	223
Ratox	4100	RCRA waste no. U017	316
Ratoxin	4100	RCRA waste no. U019	321
Ratron G	4100	RCRA waste no. U020	328
Ratron	4100	RCRA waste no. U023	350
Rats-No-More	4100	RCRA waste no. U024	1195
Ratsul soluble	4101	RCRA waste no. U025	394
Ratten-koederrohr	4100	RCRA waste no. U028	406
Rattenstreupulver Neu Schacht	4100	RCRA waste no. U031	504
Rattentraenke	4100	RCRA waste no. U034	750
Rattex	954	RCRA waste no. U036	755
Rattler	2133	RCRA waste no. U037	778
Rat-Trol	4100	RCRA waste no. U041	1572
Rattunal	4100	RCRA waste no. U044	790
RAV 7N	1104	RCRA waste no. U045	792
Ravelon	2325	RCRA waste no. U046	785
Ravensara	3677	RCRA waste no. U050	859
Raviac	807	RCRA waste no. U051	903
Ravotril	894	RCRA waste no. U052	957, 958, 959, 960
Rawetin	3271	RCRA waste no. U053	966
Rax	4100	RCRA waste no. U056	1014
Raxil	3656	RCRA waste no. U057	1019
Rayon	714	RCRA waste no. U067	1640
Rayon flock	714	RCRA waste no. U069	1153
Rayophane	714	RCRA waste no. U070	1173
Rayweb Q	714	RCRA waste no. U070	1174
Razol dock killer	2321	RCRA waste no. U071.	1172
RBA 777	835	RCRA waste no. U071	1174
RC 4 (peroxide)	1155	RCRA waste no. U072	1174
RC 17	652	RCRA waste no. U074	543
RC 5629	2201	RCRA waste no. U075	1185
RC Comonomer DBF	1143	RCRA waste no. U076	1191
RC Comonomer DBM	1146	RCRA waste no. U077	1641
RC Comonomer DIOM	1453	RCRA waste no. U078	1192
RC Comonomer DOM	1452	RCRA waste no. U081	1200
RC Plasticizer B-17	577	RCRA waste no. U084	1205
RC Plasticizer DBP	1153	RCRA waste no. U088	1689
RC Plasticizer DOP	406	RCRA waste no. U092	1359
RCH 55/5	2653	RCRA waste no. U096	969
RCRA waste	3822	RCRA waste no. U102	1411
RCRA waste no. P003	47	RCRA waste no. U103	1417
RCRA waste no. P004	75	RCRA waste no. U105	1438
RCRA waste no. P005	88	RCRA waste no. U106	1440
RCRA waste no. P009	198	RCRA waste no. U107	1455
RCRA waste no. P010	253	RCRA waste no. U108	1458
RCRA waste no. P011	254	RCRA waste no. U110	1484
RCRA waste no. P012	258	RCRA waste no. U112	1609
RCRA waste no. P015	372	RCRA waste no. U113	1611
RCRA waste no. P020	1447	RCRA waste no. U115	1650
RCRA waste no. P021	616	RCRA waste no. U116	1651
RCRA waste no. P022	686	RCRA waste no. U117	1267
RCRA waste no. P024	776	RCRA waste no. U118	1672
RCRA waste no. P028	361	RCRA waste no. U120	1761
RCRA waste no. P037	1233	RCRA waste no. U121	3872
RCRA waste no. P044	1344	RCRA waste no. U122	1786
RCRA waste no. P048	1434	RCRA waste no. U123	1790
RCRA waste no. P050	1569	RCRA waste no. U125	1803
RCRA waste no. P056	1766	RCRA waste no. U127	1904
RCRA waste no. P059	1887	RCRA waste no. U130	1905
RCRA waste no. P069	24	RCRA waste no. U131	1907
RCRA waste no. P070	73	RCRA waste no. U133	1957
RCRA waste no. P119	188	RCRA waste no. U134	1976
RCRA waste no. P189	692	RCRA waste no. U135	1978
RCRA waste no. P205	401	RCRA waste no. U203	3319
RCRA waste no. U001	10	RCRA waste no. U208	3694
RCRA waste no. U002	23	RCRA waste no. U211	688
RCRA waste no. U003	27	RCRA waste no. U221	1110
RCRA waste no. U004	28	RCRA waste no. U221	3791, 3792
RCRA waste no. U006	34	RCRA waste no. U223	3796

Name and Synonym Index

Name and Synonym Index

Name and Synonym Index

Name and Synonym Index

Name and Synonym Index

Name and Synonym Index

Saccharina	3317	Salbusian	68
Saccharinate ammonium	3318	Salbutalan	68
Saccharine	3317	Salbutamol hemisulfate	69
Saccharinnatrium	3503	Salbutamol sulfate	69
Saccharinol	3317	Salbutamol	68
Saccharinose	3317	Salbutamolum	68
Saccharoidum natricum	3503	Salbutan	68
Saccharol	3317	Salbutol	68
Saccharose	3599	Salbuven	68
Saccharose acetate isobutyrate	3600	Salbuvent	68
Saccharosonic acid	2071	Saleratus	3382
Saccharum	3599	Salesthin	2452
Saccharum lactin	2180	Salicresin Fluid	2352
Saccharum lactis	2181	Salicylal	3320
Sacharin	3317	Salicylaldehyde	3320
Sachtolith® HD-S	4162	Salicylic acid	3321
Sachtolith®	4162	Salicylic Acid & Sulfur Soap	3321
SADH	1052	Salicylic Acid collodion	3321
Saeure des phytins	3025	Salicylic Acid Soap	3321
Saeure fluoride	1766	Salicylic acid, methyl ester	2514
safflor carmine	4061	Salicylic acid, monosodium salt	3504
safflor red	4061	Salicylic acid, sodium salt	3504
Safrene	3319	Salicylic aldehyde	3320
Safrol	3319	Salicylic ether	1693
Safrole	3319	Salicylic ethyl ester	1693
Safrotin	3175	Saligel	3321
Safsan	2302	Salinaphthol	2616
Sagimid	2666	Saline	3396
Sagisal	3550	Salinomicina	3322
Sagittol	3189	Salinomycin	3322
SAH	3320	Salinomycine	3322
SAIB	3600	Salinomycinum	3322
SAIB 100S	3600	Sali-Puren	3843
Saisan	1553	Salisod	3504
Sakarat Special	807	Saliter	3475
Sakresote 100	903	Sallbupp	68
Sal absinthii	3095	Salmaplon	68
Sal Amarum	2292	Salmetedur	3324
Sal ammonia	179	Salmeterol	3323
Sal ammoniac	179	Salmeterol xinafoate	3324
Sal Angalis	2292	Salmeterolum	3323
Sal Anglicum	2292	Salmiac	179
Sal Catharticum	2292	Salol	2969
Sal Chalybis	1746	Salomol	68
Sal commune	3396	Salonil	3321
Sal culinaris	3396	Salotan	1693
Sal de Seidlitz	2292	Salpetersäure	2678
Sal diureticum	3086	Salpetersaeure	2678
Sal enixum	3090	Salpetersaure	2678
Sal ether	1693	Salpeterzuuroplossingen	2678
Sal ethyl	1693	Salphenyl	2969
Sal Martis	1747	Salpix	31
Sal Mirabil	3512	Salsonin	3504
Sal niter	2675	Salt cake	3512
Sal Polychrestum	3135	Salt of amber	3595
Sal prunella	2675	Salt of Hartshorn	284
Sal Saturni	2208	Salt of saturn	2208
Sal Sedativus	448	Salt of soda	3393
Sal Seidlitense	2292	Salt of steel	1747
Sal soda	3393	Salt of tartar	3095
Sal succini	3595	Salt of tarter	3096
Sal tartar	3518	Salt of wormwood	3095
Salachlor	3435	Salt Perlate	1501
Salammonite	179	Salt peter	2675
Salamol	68	Salt, Amido-R	3325
Salbetol	68	Salt	3396
Salbron	68	Saltpeter	631, 2675
Salbu-BASF	68	Saltpeter (Chile)	3475
Salbu-Fatol	68	Saltpeter flour	2675
Salbuhexal	68	Salts of England	2292
Salbulin	68	Salufer	2302
Salbupur	68	Salut	1344

Name and Synonym Index

Name and Synonym Index

Name and Synonym Index

Name and Synonym Index

Name and Synonym Index

Name and Synonym Index

Name and Synonym Index

Name and Synonym Index

Name and Synonym Index

Name and Synonym Index

Name and Synonym Index

Name and Synonym Index

TA-33MP	3673	Tall oil soaps, potassium	3137
TA-4708	437	Tall oil sulfonate, sodium salt	3614
TAA	3752	Tall oil, potassium salt	3137
Tabax	1043	Tall oil, sodium salt	3638
Tabex	1043	Talleol	3635
Tabfine 097(HS)	1832	Tall-oil pitch	3639
Tabilin	2881	Tallol	3635
Table salt	3396	Tallow	3642
Table spate	645	Tallow acid	3640
Table sugar	3599	Tallow acid, hydrogenated	1971
Tablettose	2180	Tallow amine	3645
Taborcil	1816	Tallow amine, hydrogenated	1974
Tabular spar	646	Tallow benzyl dimethyl ammonium chloride	3564
Tac dessicant	1570	Tallow fatty acid	3640
TAC	3836	Tallow fatty acids, hydrogenated.	1971
TAC-3	3840	Tallow fatty acids, sodium salt	3517
TAC-40	3840	Tallow imidazolinium methosulfate	3243
Tachigaren	2012	Tallow nitrile	3641
Tachigaren 70	2012	Tallow oil	3642
Tachiol	3360	Tallow trimethyl ammonium chloride	3644
Tackle	43, 3374	Tallow trimonium chloride	3644
Tackle 2AS	3374	Tallow, beef	3642
Tacosal	2995	Tallow, hydrogenated	1970
Tadenan	1524	Tallow, methyl ester	2520
Tadine	2262	Tallow, mutton	3642
Taeniatol	1199	Tallow, sodium salt	3517
Tafapon	1051	Tallowamine	3645
Tafasan	2631	Tallowtrimonium chloride	3644
Tafasan 6W	2631	Talmon	2305
Tafazine	3367	Taloberg	820, 4164
Tafazine 50-W	3367	Talodex	1731
Tafclean	2434	Talofloc	1383
Tafil	95	Taloflote	3059
Tafoxen	3647	Talon Rat Bait	457
Tagat	79	Talon rodenticide	457
Tagat® R63	3195	Talon	457, 1278
Tahmabon	2387	Talorat Dy	2262
Tai-Ace K 20	118	Talpran	1464
Tai-Ace K 150	118	Talsutin	3700
Tai-Ace S 100	123	Tam 418	4169
Tai-Ace S 150	123	Tamanox	2387
TAIC	3837	Tamaron	2387
Takanarumin	86	Tamaron®	2387
Takatol	161	Tamax	3647
Takepron	2183	Tame	1726
Takimetol	2543	Tamizam	3646
Takineocol	2127	Tamofen	3647
Taktic	165	Tamogam	460
Talan	1442	Tamol L	3496
Talatrol	4021	Tamol SN	3496
Talc	3634	Tamoxen	3646
Talcord	2928	Tamoxifen	3646
Talculin Z	4158	Tamoxifen citrate	3647
talcum	3634	Tampules	753
Talisman	826	Tamraghol	945
Tall oil	3635	Tamsulosin	3648
Tall oil (crude and distilled)	3635	Tamsulosin hydrochloride	3649
Tall oil [Oil, misc.]	3635	Tamsulosina	3648
Tall oil acid	3636	Tamsulosine	3648
Tall oil acids	3636	Tamsulosinum	3648
Tall oil acids, potassium salt	3137	Tamtron X 7R262L	303
Tall oil acids, sodium salt	3516	Tamtron X 7R302H	303
Tall oil fatty acid diester of polyethylene glycol	3066	Tanacetone	3774
Tall oil fatty acids	3636	Tanazon	4164
Tall oil fatty acids, cobalt salts	916	Tandax	2627
Tall oil fatty acids, potassium salt	3137	Tandem	3901
Tall oil fatty acids, reaction product with diethylenetriamine	1266	Tangantangan oil	707
Tall oil fatty acids, sodium salt	3516	Tanidil	3157
Tall oil fatty acids, sodium soap	3516	Tannic acid	3652
Tall oil hydroxyethyl imidazoline	2028	Tannic acid (*Quercus* spp.)	3652
Tall oil rosin	3635, 3637	Tannic acid and tannins	3652
Tall oil rosin and fatty acids	3635	Tannin	3652

Name and Synonym Index

Tannin from chestnut	3652	TBS 95	3853
Tannin from mimosa	3652	TBS	3853
Tannin from quebracho	3652	TBT	579
Tannin from sweet gum	3652	TBTO	425
Tannins	3652	TBTP	3859
Tantalic acid anhydride	3653	TBZ 60W	3749
Tantalum oxide (Ta$_2$O$_5$)	3653	TBZ	3749
Tantalum oxide	3653	TBZ-6	3749
Tantalum oxide dusts	3653	TC 8 (catalyst)	1155
Tantalum penta oxide	3653	TCA	3525, 3864
Tantalum pentoxide	3653	TCA sodium	3525
Tanzeo M45	2307	TCC	3887
Taphazine	3367	TCE	3862
Tar	902	TCE Tetrachlorethane	3693
Tar camphor	2608	Tcea	2434
Tar oil	903	TCH	3754
Tar, coal, colorless purified	902	TCIN	820
Tar, coal, purified colorless	902	TCLP extraction fluid 2	17
Tar, coal	902	TCNB	3659
Tar, coking	902	TCP	964, 3170, 3878, 3880
Tarcron 180	902	TCTH	1036
Tarcron 180L	902	TCTP	1047
Tarcron 230	902	TD 1500 S	3963
Tardex 80	2741	TD 1500	3963
Tardex 100	1058	TD 1771	3764
Tardiol D	1128	TDA	1110
Targa	3258	TDA 8020	3793
Target MSMA	2584	TDAC	3920
Tarimyl	1054	TDBP	4006
Tarka	4065	TDBPP	4006
Tarragon	1589	TDCPP	3170
Tartar	3091	T-Det	1540
Tartar cream	3091	TDGA	3758
Tartaric acid, disodium salt	3518	TDI 80-20	3798
Tartaric acid, monopotassium salt	3091	TDI	3797, 3798
Tartaric acid	3654	TDI-80	3797
Tartarline	3090	TDPA	3759
Tartarus vitriolatus	3135	TE 33	3540
Task	1728	TE-031	888
Tasteless salts	1501	TEA	3907, 3915
TAT chlor 4	755	TEA (amino alcohol)	3907
Taterfos®	3500	Teaberry oil	2514
Taterpex	837	TEAC	370
Tatig	3341	TEA-Dodecylbenzenesulfonate	3908
TATM	3838	TEA-Hydrochloride	3909
Tatoo	310	TEA-Lauryl sulfate	3910
Tattoo	310	TEB 3K	2486
Tatuzinho	75	TEBA	370
Tauphon	3655	TEBAC	370
Taural	3273	Tebol 88	504
Taurina	3655	Tebol 89	504
Taurine	3655	Tebuconazole	3656
Taurinum	3655	Tebuconazole (±)	3656
Tauriscite mineral	1747	Tebulan	3658
Tavor	2263	Tebutam	3657
Taxus	3647	Tebutame	3657
Taylors Lawn Sand	1740	Tebuthiuron	3658
Tayssato	2631	Tebuzate	3749
TB 1	303	Tecane	3864
TBA	523	Technical heptachlor	1887
TBAB	3686	Technical Hydrox	942
TBD	2141	Tecnazen	3659
TBDPE	3681	Tecnazene	3659
Tbdz	3749	Tecodin	2838
TBE	3682	Tecodine	2838
TBEP	4000	Tecquinol	1980
TBHP-70	552	Tecsol	1593
TBHQ	553	Tecsol C	1593
TBNPA	3850	Tecto	3749
TBOT	425	Tecto 10P	3749
TBP	3857	Tecto 40F	3749
TBP-AE	3852	Tecto 60	3749

Name and Synonym Index

Name and Synonym Index

Name and Synonym Index

Name and Synonym Index

Name and Synonym Index

Name and Synonym Index

Name and Synonym Index

Name and Synonym Index

Tolane	3788
Tolcide MBTV 709	2450
Tolclofos-methyl	3300
Toldimfos sodium	3820
Tolfrinic	4089, 4092
Tolid	2263
Tolit	3982
Tolite	3982
Tolkan	2151
Tolterodina	3789
Tolterodine	3789
Tolterodinum	3789
Tolueen	3790
Tolueen-diisocyanaat	3797
Toluen	3790
Toluen-disocianato	3797
Toluene	3790
Toluene diisocyanate	3796, 3797, 3798
Toluene trichloride	350
Toluene, vinyl- (mixed isomers)	4080
Toluenediamine	3791, 3792, 3793
Toluenediisocyanate (mixed isomers)	3798
Toluenesulfonamide	3799
Toluenesulfonic acid, potassium salt	3139
Toluenesulphonamide	3799
Toluene-p-sulphonamide	3801
Tolueno	3790
Toluhydroquinone	3812
Toluic acid	3807
meta-Toluic acid	3806
Toluilenodwuizocyjanian	3797
Toluol	3790
Toluolo	3790
Tolurane	826
Tolurex	826
Tolu-Sol	3790
Tolutriazole	2424, 3819
Toluylene	3817
Tolyfluanid	1712
Tolyfluanide	1712
Tolyl chloride	361
Tolyl triazole	3819
Tolylenediamine	3793
Tolylene-2,4-diamine	1110
Tolylene diisocyanate	3798
Tolylene isocyanate	3798
Tolylethylene	4080
Tolylfluanid	1712
Tolylfluanide	1712
Tolylsulfonamide	3801
Tolyltriazole	3819
Tomathrel	1599
Tomaxasta	3647
Tombran	3828
Tomicide Z 50	3232
Tomicide ZPT 50	3232
Tominokusu SS	2749
Tomix OT (monohydrate)	2292
Tomofan	714
Tonarol	1147
Tonarsan	1499
Tonarsin	1499
Tonite	768
Tonka bean camphor	954
Tonocardin	1550
Tonofosfan	3820
Tonophosphan	3820
Tonox	1116
Tonsillosan	2174, 2176
Tonzonii bromidum	3772
Tonzonium bromide	3772
Top Brass	3232
Top flake	3396
Top Hand	22
Topamax	3821
Topane	3485, 2968
Topanex 100BT	1555
Topanol	1147
Topanol O	1147
Topanol OC	1147
Topanol® M	1151
Topas 100	2878
Topas C	2878
Topas MZ	2878
Topaz	2878
Topaze C	2878
Topaze	2878
Topclip	1038
Tophol	874
Tophosan	874
Topichlor 20	755
Topiclor	755
Topiclor 20	755
Topiramate	3821
Topiramato	3821
Topiramatum	3821
Topitox	807
Toplan	482
Topnotch	22
Toppel	1038
Topper	2061
Topper 5E	2635
Topri	3949
Toprol XL	2540
Topromel	2541
Tops	3764
Topsin M	3764
Topsin M 70	3764
Topsin Methyl	3764
Topsin NF-44	3764
Topsin turf and ornamentals	3764
Topsin WP methyl	3764
Topsym	1419
Topusyn	1091
Torayceram Sol ZS-OA	4168
Torbin	1579
Tordon	3027
Tordon 101 mixture	3027
Tordon 10K	3027
Tordon 22K	3027
Tordon 22Kk-Pin	3027
Torelle	1794
Torion	135
Tornusil	4156
Toro	826
Torque	1721
Torsite	2968
Tosic acid	3803
Tosilcloramida sodica	753
Tosyl	751
Tosylamide	3801
Tosylate, sodium	3524
Tosylchloramid-natriu	753
Totalon	2231
TOTM	3830, 3987
TOTP	3891
Totril	2061
Towk	3824
Toxaphene	3822
Toxer total	2864
Tox-Hid	4100
Toxichlor	755
Toxilic acid	2296
Toxilic anhydride	2297

Name and Synonym Index

Name and Synonym Index

Name and Synonym Index

Name and Synonym Index

Name and Synonym Index

Name and Synonym Index

Name and Synonym Index

Name and Synonym Index

Name and Synonym Index

UN2038	1436	UN2381	2448
UN2045	2090	UN2383	1484
UN2046	2150	UN2387	1767
UN2049	1251	UN2393	2082
UN2050	1303	UN2397	2144
UN2051	1057	UN2398	2431
UN2052	1462	UN2401	3048
UN2054	2580	UN2404	3183
UN2055	3590	UN2407	2138
UN2056	3718	UN2412	3725
UN2058	4050	UN2414	3765
UN2059	3238	UN2427	3096
UN2077	2623	UN2428	3395
UN2079	1264	UN2432	1249
UN2186	1962	UN2438	3053
UN2191	3626	UN2440	3556
UN2192	1824	UN2442	3865
UN2196	4023	UN2468	3633
UN2197	1960	UN2471	2817
UN2201	2713	UN2474	3767
UN2203	3344	UN2477	2481
UN2209	1786	UN2488	1026
UN2210	2309	UN2489	1475
UN2213	2859	UN2491	1596
UN2214	3021	UN2493	1920
UN2215	2297	UN2496	3182
UN2243	1020	UN2498	1791
UN2247	1062	UN2503	4171
UN2248	1137	UN2509	3090
UN2252	1378	UN2517	2568
UN2253	1373	UN2518	1011
UN2257	3085	UN2521	1335
UN2259	3905	UN2522	2381
UN2261	1406	UN2525	1272
UN2264	1379	UN2527	2079
UN2265	1388	UN2528	2085
UN2270	1613	UN2529	2091
UN2272	1614	UN2530	2508
UN2276	1663	UN2531	2380
UN2277	1672	UN2535	2487
UN2280	1931	UN2541	3678
UN2281	1928	UN2542	3855
UN2283	2510	UN2554	2386
UN2284	2093	UN2555	3238
UN2289	2120	UN2556	3238
UN2290	2121	UN2557	3238
UN2296	2439	UN2564	3864
UN2301	3632	UN2565	1222
UN2302	2477	UN2579	3046
UN2303	2517	UN2582	1735
UN2304	2608	UN2606	3729
UN2308	2709	UN2616	3943
UN2313	3028, 3029, 3030	UN2618	4080
UN2318	3384	UN2644	2475
UN2320	3713	UN2646	1905
UN2323	3925	UN2651	1116
UN2325	2364	UN2658	3338
UN2329	3971	UN2659	3398
UN2330	4033	UN2660	2558
UN2331	4140	UN2662	1980
UN2337	3766	UN2664	2453
UN2346	1101	UN2666	1621
UN2357	1024	UN2670	1002
UN2361	1301	UN2674	2302
UN2362	1191	UN2686	1243
UN2363	1671	UN2687	1223
UN2368	3041	UN2699	3929
UN2370	1944	UN2722	2257
UN2372	3734	UN2729	1904
UN2375	1274	UN2745	2844
UN2377	1355	UN2748	1657
UN2380	1382	UN2749	3735

Name and Synonym Index

Name and Synonym Index

Name and Synonym Index

Velsicol heptachlor	1887	Vermitid	2666
Velvetex® BA-35	923	Vermitin	2666, 2956
Vencedor	948	Vermoestricid	688
Venceweed	1046	Vermox	2324
Vencronyl	68	Vernatzer 30/10	1981
Vendex	1721	Vernol®	2645
Venetlin	69	Veroftal	4086
Venlafaxina	4062	Verol	2537
Venlafaxine	4062	Verolan KAF	791
Venlafaxine hydrochloride	4063	Verospiron	3550
Venlafaxinum	4062	Verospirone	3550
Venlafexine	4062	Verpanyl	2324
Ventamol	68	Verrol	4095
Ventilan	68	Verrugon	3321
Ventiloboi	68	Versal	2262
Ventolin HFA	69	Versamag® DC	2285
Ventolin Inhaler	68	Versapen	1899
Ventolin Rotacaps	68	Versa r	2584
Ventolin	68	Versar DSMA LQ	1499
Ventoline	68	Versatic 5	3052
Ventox	52	Versatill	900
Venturol	1544	Versatrane	1956
Venus copper	973	Versatrex	1899
Venzar	2229	Versen-01	4013
Vepen	2883	Versene	1560
Verabeta	4065	Versene acid	1560
Veracaps SR	4065	Versene Diammonium EDTA	1112
Veracillin	1215	Versene Na2	1635
Veracim	4065	Versene NTA 150	2680
Veracin	3700	Versene NTA 335	2680
Veracor	4065	Versene NTA acid	2679
Veradol	2625, 2627	Versene-9	1559
Verahexal	4065	Versenex 80	2911
Veraloc	4065	Versenol 120	4013
Veramex	4065	Versen-ol	4013
Veramil	4065	Versicol W 11	3059
Veramix	2330	Versneller NL 49	908
Verantin	3219	Versneller NL 63/10	1373
Verapam	4065	Versonol 120	1635
Verapamil	4064	Versotrane	1956
Verapamil Acis	4065	Vertac Dinitro Weed Killer	1447
Verapamil AL	4065	Vertac General Weed Killer	1447
Verapamil Atid	4065	Vertac Propanil 3	3166
Verapamil Basics	4065	Vertac Propanil 4	3166
Verapamil chloridrate	4065	Vertac Selective Weed Killer	1447
Verapamil hydrochloride	4065	Vertenex	540
Verapamil-AbZ	4065	Verthion	1722
Verapamilo	4064	Vertisal	2543
Verapamilum	4064	Vertolan	3608
Veraplex	2330	Veruca-sep	1833
Verapret DH	1403	Verv®	653
Verapret DKh	1403	Vesamin	31
Vera-Sanorania	4065	Vesdil	3272
Veratrole methyl ether	2464	Vesparaz-wirkstoff	2010
Veratrole	4066	Vesta PP	634
Verazinc	4160, 4161	Vestamin® TMD	2120
Vercom	3297	Vestanat® IPDI	2121
Verditer blue	939	Vestinol 9	1316
Verdox	2815	Vestinol DZ	1312
Verdyl acetate	2160	Vestinol NN	1316
Verdyol Super	80	Vestinol OA	1449
Veresene disodium salt	1635	Vestolit®	3082
Verisan	2064	Vesulong veterinary	3615
Veritab	2326	Vesyca	3273
Verlost	3273	Vetacyclinum	3700
Vermex	2324, 3046	Veta-K$_1$	4096
Vermicidin	2324	Vetalin S	4087
Vermilion	2356	Vetalog	3840
Vermilion (HgS)	2356	Vetkelfizina	3607
Vermirax	2324	Vetol	2305
Vermisol 150	2231	Vetquamycin-324 (free base)	3700
Vermithana	1199	Vetsin	2575

Name and Synonym Index

Name and Synonym Index

Name and Synonym Index

Name and Synonym Index

Name and Synonym Index

Name and Synonym Index

Name and Synonym Index

Name and Synonym Index

INDEXES

Manufacturers and Suppliers

3M Company, Industrial Chemicals
3M Center
St. Paul, MN 55144
(612)-733-1110, 800-541-6752
www.3m.com

3M Pharmaceuticals
3M Center 2751
St. Paul, MN 55144
(612)-733-1110
www.3m.com

3V Inc.
9140 Arrowpoint Blvd., Suite 120,
Charlotte, NC, 28273-8120, USA
(704) 523-5252, Fax (704) 522-1763

AAA Molybdenum Products Inc.
7233 W. 116th Place,
Broomfield, CO 88020 USA
303-460-0844, 800-443-6812,
Fax 303-460-0851

Aarhus Oliefabrik A/S
M.P. Bruuns Gade 27
DK-8100 Aarhus, C, Denmark
+45-8730-6000, Fax +45-8730-6012
www.aarhus.com

Abbott Laboratories Inc.
100 Abbott Park Rd.
Abbott Park, IL 60064,USA
(847) 936-6100, Fax (847) 937-1511

ABC Compounding Co. Inc.
PO Drawer 585,
Morrow, GA, 30260
(800) 795-9222, (770) 968-9222
www.abccompounding.com

ABCR-Gelest (UK) Ltd.
Division of ABCR GmbH
Levenshulme, UK

ABCR GmbH & Co. KG
Im Schlehert 10
D-76187 Karlsruhe, Germany
+49 (0)7 21 / 950 61-0
Fax +49 (0)7 21 950 61-80
www.abcr.de

ABM Chemicals Ltd.
Poleacre Lane, Woodley,
Stockport, Ches, SK6 1PQ,UK

AC Ind. Inc.
Pechiney Chemicals Division
North Tower
Stamford Harbour Park
Stamford, CT 06902

Accurate Chem. & Sci. Corp.
300 Shames Drive
Westbury, NY 11590-1725
800-645-6264; 516-333-2221,
Fax 516-997-4948
www.accuratechemical.com

ACE Chemicals (PVT) Ltd.
Sunshine Court, 65
Gothami Lane, Colombo 8,
Sri Lanka
(941) 699937, Fax (941) 698531
acechem@slt.lk

Aceto Agriculture Chemicals Corp.
One Hollow Lane
Lake Success, NY, 11042,USA
(516) 627-6000

Aceto Corp.
One Hollow Lane
Lake Success, NY, 11042,USA
(516) 627-6000
www.aceto@aceto.com

ACF Chemiefarma NV
Postbus 5, Straatweg 2
Maarsen, 3600 AA, Netherlands

Acme-Hardesty Co.
1787 Sentry Parkway West
Suite 18-460
Blue Bell, PA 19422
(215) 591-3610, (800) 223-7054,
Fax (215) 591-3620
www.acme-hardesty.com

Acros Organics B.V.B.A.
Janssen Pharmaceuticalaan 3A
2440 Geel, Belgium
+32 14 57 52 11, Fax +32 145934 34

Acros Organics - USA
See: Fisher Scientific International

Adeka Fine Chemicals
Yoko Bldg., 4-5, 1-chome,
Bunkyo-ku, Tokyo, 113, Japan
81 3 5869-8681, Fax 81 3 5689-8680

ADM Alliance Nutrition Inc.
PO Box 492, 24th Street Terrace,
Higginsville, MO, 64037

ADM Ethanol
Archer Daniels Midland Company
4666 Faries Parkway
Decatur, IL62526
1-800-637-5843
www.admworld.com

ADM Tronics Unlimited,
224-S Pegasus Ave,
Northvale, NJ 07647, USA
201-767-6040, Fax 201-784-0620
www.admtronics.com

AGA Gas
Linde Gas LLC, P.O. Box 94737
6055 Rockside Woods Blvd.
Cleveland, OH 44101-4737
216-642-6600
www.us.lindegas.com

Agan Chemical Manufacturers
551 5th. Ave., Suite 1100,
New York, NY, 10175,USA
(212) 661-9800, Fax (212) 661-9043,
(212) 661-9048
www.agan.co.il

Agbiochem Inc.
166 N.W. Fritz Place,
Orina, CA, 94563,USA

AgrEvo Inc.,
see Aventis Crop Science

Agrichem (International) Ltd.
Station Rd.
Whittlesey,Cambridgeshire, PE7 2EY,
UK
(01733) 204019, (01733) 204162
info@agrichem.co.uk

Agriliance Crop Nutrient
PO Box 64089
St. Paul, MN, 55164-0089,USA
(651) 451-5151
www.agriliance.com

Agriliance LLC
PO Box 64089
St. Paul, MN, 55164-0089,USA
(651) 451-5151
www.agriliance.com

Agrimont,Gruppo Montedison
Agrimont Foro Bonaparto 31
Milan, 20121,Italy

Agro Logistic Systems Inc.
9812 Odessa Avenue
North Hills, CA, 91343,USA
(818) 892-1610
Agro19@worldnet.att.net

Agrolinz Melamin GmbH
St.-Peter-Strasse 25
Linz, 4021,Austria
0732 6914-0, Fax 0732 6914-3581
www.agrolinz.com

Agsco Inc.
2600 Mill Road,
Grand Forks, ND, 58203-1506,USA
(800) 859-3047, (701) 746-0934,
Fax (701) 775-9587
info@agscoinc.com

Manufacturer and Supplier Directory

Agtrol International
7322 Southwest Fwy., Suite 1400
Houston, TX, 77074-2019,USA
(713) 995-0111

Agvalue Inc.
1124 North Chinowth Street
Visalia CA, 93291, USA
866 511-3171, Fax 559 627-6962
contact@agvalue.net

Air Liquide America Corp.
2700 Post Oak Boulevard, Suite
1800,
Houston, TX, 77056
(800) 820-2522
www.us.airliquide.com

Air Products/Polyurethane,
7201 Hamilton Blvd.,
Allentown, PA 18195-1501, USA
610-481-6799, 800-345-3148, Fax
610-481-4381
www.airproducts.com

**Air Products and Chemicals
Germany,**
Postfach 5108, Robert-Koch-Str. 27,
D-22821 Norderstedt, Germany.
49-40-529009-0, Fax 49-40-
52900999
www.airproducts.com

Air Products Nederland BV,
Kanaalweg 15, PO Box 3193,
3502 GD Utrecht, The Netherlands
31-30-857100, Fax 31-30-857111
www.airproducts.com

Air Products and Chemicals Mexico,
Rio Guadiana 23, Piso 5,
Colonia Cuauhtemoc,
Mexico D.F., 06500, Mexico
525-591-0800, Fax 525-592-3018
www.airproducts.com

Air Products and Chemicals Inc.
7201 Hamilton Blvd.,
Allentown, PA, 18195-1501
(610) 481-4911; Fax: (610) 481-5900
www.airproducts.com

Ajinomoto Co. Inc.
1-15-1 Kyobashi
Chuo-ku Tokyo 104
Japan
+81 (3) 5250-8111
www.ajinomoto.com

Ajinomoto Europe Sal,
Stubbenhuk 3,
20459 Hamburg, Germany
40-3749-3650, Fax 40-372087
www.ajinomoto.com

Ajinomoto USA Inc.,
Glenpointe Centre West,
500 Frank W. Burr Blvd.,
Teaneck, NJ 07666-6894,USA
201-907-3250, Fax 201-907-3252
www.ajinomoto.com

Akcros Chemicas America
500 Jersey Avenue,
New Brunswick, NJ, 08903
(732) 247-2202; Fax: (732) 247-2287
www.akzonobel.com

Akrochem Chemical Co,
255 Fountain St.,
Akron, OH 44304, USA
216-535-2108, 800-321-2260, Fax
216-535-8947

Akrochem Chemical Co,
255 Fountain St.,
Akron, OH 44304, USA
216-535-2108, 800-321-2260
Fax 216-535-8947

Akros Chemicals Ltd (Azko Nobel),
PO Box 1, Eccles,
Manchester M3O 0BH, UK
44-161-7851111, Fax 44-161-
7887886
www.akzonobel.com/

Akzo Nobel
Terhulpsesteenweg 166
Chee de la Hulpe 166
Brussels, Belgium

Akzo Nobel Base Chemicals,
4 Stationsplien/3818 LE
PO Box 247
3800 AE Amersfoort, The
Netherlands
31-33-4676270, Fax 31-33-4676110
www.akzonobel.com/hc/home.htm

Akzo Nobel Chemicals,
1 City Center Dr., Suite 320,
Mississauga,Ontario, L5B IM2,
Canada
905-273-5959, Fax 905-273-7339
www.akzonobel.com

Akzo Nobel Chemicals,
300 S. Riverside Plaza,
Chicago, IL 60606-6697, USA
312-906-7500, 800-227-7070, Fax
312-906-7811
www.akzonobel.com

Alba International Inc.
508 Clearwater Dr.,
N. Aurora, IL, 60542 USA
708-897-4200, 800-669-9333, Fax
708-377-5330

Albaugh Inc./Agri Star
121 NE 18th Street,
Ankeny, IA, 50021
(800) 247-8013, Fax (515) 964-7813
www.albaughinc.com

Albemarle Corp.
451 Florida Street
Baton Rouge LA 70801-1785
+1 (225) 388-7402, Fax +1 (225)
388-7848

Albright & Wilson Americas Inc.
4851 Lake Brook Drive
PO Box 4439
Glen Allen, VA 23060
+1 (804) 968-6300, +1 (804) 968
6385
www.albright-wilson.com

Albright & Wilson UK Ltd.
PO Box 3, 210-222 Hagley Road
West Oldbury
West Midlands B68 ONN, UK
+44 (0121) 429-4942, +44 (0121)
429-5151
www.albright-wilson.com

Alcan Chemicals,
3690 Orange Place, Suite 400,
Cleveland, OH 44122-4438,USA
216-765-2550, 800-321-3864,
Fax 216-765-2570

Alcan Chemicals Europe,
Ditton Rd.,
Widnes, Ches., WA8 0PH, UK
44-1592-411000, Fax 44-151-802
2999

Alcan Chemicals Ltd.,
Chalfont Park, Gerrards Cross,
Bucks, SL9 0QB, UK
44-1753-887373, Fax 44-1753-
881556

Manufacturer and Supplier Directory

Alco Chemical Corp.
909 Mueller Dr.,
PO Box 5401,
Chattanooga, TN 37406-0401,USA
423-629-1405, 800-251-1080,
Fax 423-698-9367

Alcoa
201 Isabella St./7th. St. Bridge
Pittsburgh, PA 15212-5858
412-553-4545
www.alcoa.com

Alcoa Industrial Chemicals
4701 Alcoa Rd.
PO Box 300
Bauxite, AR, 72011, USA
501-776-4717, 800-643-8771,
Fax 501-776-4904
www.alcoa.com

Alcoa Industrial Chemicals/Asia
2 Havelock RD,
#07-5 Apollo Center, 0105,
Singapore
65-538-0070
Fax 65-538-3237
www.alcoa.com

Alcoa Industrial Chemicals/Europe
Im Atzelnest 3
D-6380 Bad Homburg, Germany
49-06172-4068-0, Fax 49-06172-
4068-13
www.alcoa.com

Alcoa Inter-America
396 Alhambra Circle, Suite 200
Coral Gables, FL 33114, USA
305-445-8544, Fax 305-444-8924
www.alcoa.com

Alcoa Kasei Ltd.
Toranomom 4-chrome,
Minato-ku, Tokyo 105, Japan
81-3-5472-3201, Fax 81-3-5472-
3209
www.alcoa.com

Alcon Laboratories
6201 South Freeway
Ft Worth, TX 76134, USA
817-293-0450
www/alconlabs/com

Aldrich Chemical Co.
See: Sigma-Aldrich Co.
www.sigma-aldrich.com

Alfa Aesar
26 Parkridge Road
Ward Hill, MA 01835 USA
800-343-0660 or 978-521-6300
www.alfa.com

Allan Chem. Corp.   **Amerchol, Ik**
Fort Lee, NJ 07024
USA
(201) 592-8122
www.allanchem.com

Allchem Industries
4001 Newberry Rd, Suite E-3,
Gainesville, FL 32607, USA
904-378-9696, Fax 904-338-0400

Allen & Hanbury
See: GlaxoSmithKline

Allergan Inc.
P. O. Box 19534
Irvine, CA 92623-9534
714-246-4500 800-347-4500,
Fax 714-246-6987
www.allergan.com

Allergan Herbert
Allergan Inc.
2525 Dupont Drive
Irvine, CA 92612
(714) 246-4500, Fax (714) 246-4971
www.allergan.com

Alliance Packaging Inc.
109 Northpark Blvd.,
Covington, LA, 70433-5001
(985) 892-5521

Allied Colloids
2301 Wilroy Rd.,
Suffolk, VA, 23434
(757) 538-3700, Fax (757) 538-0204

Allied Colloids
P. O. Box 38, Cleckheaton Rd.,
Low Moor, Bradford, BD12 0JZ, U.K.
44-1274-41-70-00, Fax 44-1274-60-
64-99

Allied Signal Inc.
PO Box 2245, 101 Columbia Rd.
Morristown, NJ 07962, USA
201-455-2000, Fax 973-455-5445
800-522-8001, 800-810-4340
www.alliedsignal.com

All-India Medical
8th. Road, Akhand Jyoti,
Santacruz (East),
Bombay, India, 400055

Alpharma Inc.
400 State Street
Chicago Heights, IL 60411
(708) 758-0111, Fax (708) 757-2510
www.alpharma.com

Amerchol, Ikeda Corp.
New Tokyo Bldg., No. 3-1,
Marunouchi 3-Chome,
Chiyoda-Ku, Tokyo,100, Japan
81-3-3212-8791, Fax 813-3215-5069

Amersham International,
Amersham Place, Little Chalfont,
Amersham, Bucks, HP7 9NA, UK
44 1494 544000, Fax 44 1494
542266

Amoco Chemical Asia,
16th Floor, Great Eagle Centre,
23 Harbour Rd.
Hong Kong
852-2586-8899
www.amoco.com

Amoco Performance Products,
10th Floor, Tonichi Bldg.,
2-31 Roppongi 6-Chome,
Mianto Ku, Tokyo, 106, Japan
www.amoco.com

Amax Inc.
See: Phelps Dodge Corp.

Amber Synthetics
1011 High Ridge Road
Stamford, CT 06905
(203) 329-6500, Fax (203) 329-6600
www.amsyn.com

Ambrands Inc.
2255 Cumberland Pky., Bldg. 500,
Suite 200,
Atlanta, GA, 30339-4515
(770) 333-8999; Fax: (770) 333-1887
www.amdro.com

Amerchol Corp.
PO Box 4051
136 Talmadge Rd.,
Edison, NJ USA
732-248-6000, 800-367-3534,Fax
732-287-4186

Amerchol Europe
Havenstraat 86
B-1800 Vilvoorde, Belgium
32-2-252-4012, Fax 32-2-252-4909

Ameribrom Inc.
2115 Linwood Ave,
Fort Lee , NJ, 07024-5004
(770) 333-8999; Fax: (201) 242-6560
www.ameribrom.com

Ameribrom Inc.
2115 Linwood Ave,
Fort Lee , NJ, 07024-5004
(770) 333-8999; Fax: (201) 242-6560
www.ameribrom.com

Manufacturer and Supplier Directory

American Biorganics Inc.
2236 Liberty Drive
Niagara Falls NY 14304-3796

American Bio-Synthetics Corp.
Bell Aromatics
710 W. National Avenue
Milwaukee, WI 53204-1715
414-384-7017, Fax 414-384-1369

American Camellia Society
1 Massee Lane,
Fort Valley, GA, 31030
(478) 967-2358, (478) 967-2358
www.camellias-acs.com

American Chemet Corp.
400 Lake Cook Rd.,
Deerfield, IL 60015, USA
708-948-0800, Fax 708-948-0811

American Colloid Co.,
1500 W. Shure Dr.,
Arlington Hts., IL 60004-1434, USA
708-394-8730, Fax 708-506-6199

American Colloid Co.
Highway 212 West,
PO Box 160,
Belle Fourche, SD 57717, USA
605-892-2591, 800-535-1935, Fax
605-892-4880

American Cyanamid Co.
#1 Campus Dr.,
Parsippany, NJ 7054, USA
973-683-2000, Fax 973-683-4041

American Cyanamid, Divn. AHP
American Cyanamid Co.
Pearl River, NY, 10965
(201) 242-6560

American Maize Products Co.
See: Cerestar USA Inc.,

American Roland Chemical Corp.
See: Amerol Corp.,

Amerihaas Inc.

Amerol Corp.
71 Carolyn Blvd,
Farmingdale NY 11735 USA
631-694-9090, Fax 631-694-9177
www.amerolcorp.com

Amersham Corp.,
2636 S Clearbrook Dr.,
Arlington Heights, IL 60005, USA
708-593-6300

Amoco Chemicals
200 East Randolph Dr.,
Mail code 7802,
Chicago, IL 60601, USA
312-856-4729, 800-621-4567,
Fax 312-856-6225
www.amoco.com

Amoco Chemical (Europe)
15, Rue Rothschild,
1211 Geneva 21, Switzerland
41-22-715-0701
www.amoco.com

Amspec Chem.

AMVAC Chemical Corp.
4100 E. Washington Blvd.,
Los Angeles, CA, 90023
(888) 462-6822, (888) 462-6822,
(323) 264-3910
www.amvac-chemical.com

Andeno BV
Grubbenvorsterweg 8,
5928 NX, Venlo-Holland, The
Netherlands
31-77-899-555, Fax 31-77-299300

Andersen Sterilizers Inc.

Anderson Chemical Co. Inc.
PO Box 4507,
Macon, GA, 31213
(478) 745-0455, Fax (478) 742-6332
www.anderonchemical.com

Andersons Lawn Fertilizer Div. Inc.
PO Box 119,
Maumee, OH, 43537

Andrulex Trading Ltd.

Angus
Zeppelinstrasse 30
49479 Ibbenbüren, Germany
49-5459-560, Fax 49-5459-56-267
www.angus.de

Anzon Ltd.
Cookson House, Willington Quay,
Wallsend, Tyne & Wear,
NE28 6UQ, UK

Apothecon
See: Bristol-Myers Squibb Co.

Aquacide Corp.
5600 Cenex Dr,
Inver Grove Hts, MN 55077
(651) 451-5151; (800) 328-9350;
Fax: (651) 429-0563
www.killlakeweeds.com

Aqualon Canada Inc.,
5407 Eglinton Ave. West,
Etobicoke, Ont., M9C 5K6, Canada
416-620-5400

Aqualon Co.,
Hercul

Aqualon Co.
Hercules Aqualon Division
1313 North Market St.,
PO Box 8740,
Wilmington, DE 19899-8740, USA
302-594-6600, 800-345-8104, Fax
302-594-6660

Aqualon France,
3 Rue Eugene & Armand Peugeot,
92500 Rueil-Malmaison, France
33-1-4751-2919, Fax 33-1-4777-
0614

Aquarium Pharmaceuticals Inc.
PO Box 218,
Chalfont, PA, 18914-0218
(800) 847-0659
www.acquariumpharm.com

Aqua-Serv Engineers Inc.
13560 Colombard Court,
Fontana, CA, 92337-7600
(800) 637-0787; (909) 681-9696
Fax: (909) 681-9698
www.aqua-serv.com

Aquashade Inc.
See: Applied Biochemists,

Aquatec Quimica SA
Av. Paulista 37 - 12th. Andar,
Sao Paulo-SP, Brazil, 01311-000
55 11 284-4188; Fax: 55 11 288-
4431

Aquatronics Inc.
1212 NE 5th. St.,
Redmond, OR, 97756
(541) 548-2110; Fax: (541) 548-2117
www.aquatronics.com

ARC Specialty Products
See: ARCO Chemical Co.

Arch Chemicals Inc.
501 Merritt 7,
Norwalk, CT, 6851
(203) 229-2900
www.archchemicals.com

Arch Wood Protection Inc.
1955 Lake Park Drive, Suite 250,
Smyrna, GA, 30080
(770) 801-6600; (404) 362-3970,
Fax: (404) 363-8585
www.dricon.com

ARCO Chemical Co.,
3801 West Chester Pike
Newtown Square, PA 19073-2387
610-359-200, Fax 610-359-2722
www.arcochem.com

ARCO Chemical Europe Inc.,
ARCO Chemical House, Bridge Ave.,
Maidenhead, Berks, SL6 1YP, UK
44-1628-77-5000, Fax 44-1628-77-
5180
www.arcochem.com

ARCO Chemical Asia Pacific Ltd.,
41st. Floor, The Lee Gardens
33 Hysan Ave., Causeway Bay
Hong Kong
852-2-822-2668, Fax 852-2-840-
1690
www.arcochem.com

Manufacturer and Supplier Directory

Arco Chemical Europe,
Weenahuis,Weena 141,
NL-30 13 CK Rotterdam, The
Netherlands
31-10-401 0400, Fax 31-10-411
4849
www.arco.com

Arista Industries Inc.
1082 Post Rd.,
Darien, CT 06820, USA
203-655-0881, 800-255-6457, Fax
203-656-0328

Aristech Chemical Co.,
600 Grant St., Room 1170,
Pittsburgh, PA 15230-0250, USA
412-433-7800, Fax 412-433-7721

Arizona Chemical Co.,
1001 E. Business Hwy. 98,
Panama City, FL 32401-3633, USA
904-785-6700, 800-526-5294
Fax 904-785-2203

Armour Pharmaceuticals,
500 Arcola Road,
P.O. Box 1200,
Collegeville, PA 19422, USA
215-454-8000, 800-72-RORER,
Fax 215-454-8940

Arvesta Corp.
100 First St., Suite 1700,
San Francisco, CA, 94105
(415) 536-3480
www.arvesta.com

Asahi Chemical Industries,
Hibiya Mitsui Bldg., 1-2, Yura,
Chiyoda-ku, Tokyo 100, Japan
81-3-3507-2730, Fax 81-3-3507-
2495
www.asahi.com

Asahi Denka Kogyo Kogyo
Furukawa Bldg. 2-8,
Nihonbashi Muro-machi 2-chome
Ch,
Tokyo 103, Japan,
81-3-5255-9002, Fax 81-3-3270-
2463
www.asahi.com

Asarco
180 Maiden Lane, Fl. 22
New York, NY 10038
(212) 510-2000, Fax (212) 510-2122
www.asarco.com

Ascher B.F. & Co.

Ashe Chemicals

Ashlade Formulations Ltd.
see: Nufarm-Whyte Agriculture Ltd.

Ashland Chemical Co.
PO Box 2219,
Columbus, OH 43216, USA
614-790-3333, 800-526-4032, Fax
614-889-3465
www.ashchem.com

Ashland Chemical Co.
2463 Royal Windsor Drive,
Mississauga, Ont., L5J 1K9, Canada
905-823-7975
www.ashland.com

Ashland-Chemie
Reisholzstrasse 16,
40721 Hilden, Germany
49-21-711030, Fax 49-21-711-0335
www.ashland.com

Astra Chemicals GmbH
Postfach 249
Tinsdaler Weg 18
D-22876 Wedel, Germany
49-4103-70 80, Fax 49-4103-70-82-
93
www.astra.com

Astra Pharmaceutical
50 Otis St.
Westboro, MA, 01581-4500, USA
508-366-1100, 800-225-6333, Fax
508-366-7406
www.astra.com

**Atabay Agrochemicals & Vet.
Products SA**
Acibadem, Koftunku Sok., Kadikoy,
Istanbul, Turkey, 81010

Athea Laboratories, Inc.
PO Box 240014,
Milwaukee, WI, 53224
(800) 743-6417, (800) 743-6417
www.athea.com

Atlas Interlates Ltd.
See: Allied Colloids

Atofina Chemicals
2000 Market St.,
Philadelphia, PA, 19103-3222
(215) 419-7000
www.atofinachemicals.com

Atomergic Chemetals
222 Sherwood Ave.,
Farmingdale, NY 11735-1718, USA
516-694-9090, Fax 516-694-9177

Avebe BV
Avebeweg 1
9607 PT Foxhol
Netherlands
+31 598 662670, Fax +31 598
662122
www.avebe.com

Avecia Inc.
PO Box 15457, 1405 Foulk Road
Wilmington, DE, 19850-5457
(302) 477-8000, Fax (302) 477-8250
www.avecia.com

Aventis Crop Science
PO Box 12014
2 T. W. Alexander Drive
Research Triangle Park, NC, 27709
(800) 334-9745, Fax (919) 549-2000'
www.rp.rpna.com

**Aventis Environmental Science USA
LP**
95 Chestnut Ridge Rd.,
Montvale, NJ, 07645
(201) 307-9700
www.rp.rpna.com

Avrachem AG
Weststrasse 119, P.O. Box 9865
8036 Zurich, Switzerland
+41 14513132, Fax +41 14513222

BA Chemicals Ltd.
Chalfont Park, Gerrards Cross
Bucks, SL9 0QB, UK
+44 1753-887373, Fax +44 1753-
889602

Baiye

Bakelite GmbH
Postfach 7154,
Gennaer Strasse 2-4,
58642 Iserlohn, Germany
49-2374-510, Fax 49-2374-51409

Bakelite Polymers (UK)
Syer House, Stafford Court,
Stafford Park, Telford, Shropshire,
TF3 3BD, UK

Baker, J.T. Inc.
Corporate Headquarters, USA,
222 Red School Lane,
Phillipsburgh, NJ, 08865
(908) 859-2151, (908) 859-6905
www.jtbaker.com

Baker Petrolite Corp.
12645 West Airport Blvd,
Sugar Land, Houston, TX, 77478
(281) 276-5400; Fax: (281) 275-7218
www.bakerpetrolite.com

Balchem Corp.
ARC Specialty Products,
2007 Route 284,
Slate Hill, NY, 10973-4220
(845) 355-5300
www.balchem.com

BASF AG
PO Box 13528, 26 Davis Drive,
Research Triangle Park, NC, 27709-
3528
(919) 547-2000
www.basf.com

Manufacturer and Supplier Directory

BASF Corp.
26 Davis Drive,
Research Triangle Park, NC, 27709
(919) 547-2000; (800) 669-2273,
(800) 962-7830;
(800) 669-1770; (800) 874-0081
www.basf.com

BASF Microcheck Ltd.
Ruddington Fields, Business Park
Mere Way
Ruddington, Nottingham, UK, NG11
6JS
(44) 115 912-4586; Fax: (44) 115
912-4592
www.basf.de

BASF plc
Carl-Bosch-Strasse 38,
Ludwigshafen, Germany 67056
(49) 621 60-0; Fax: (49) 621 60-
42525
www.basf-ag.de

BASF Specialty Products
26 Davis Drive,
Research Triangle Park, NC, 27709
(800) 545-9525
www.basf.com

Battle Hayward & Bower Ltd.
Crofton Drive
Allenby Road Ind. Estate
Lincoln LN3 4NP, UK
+44 01522 529206, Fax 01522
538960
bhb@battles.co.uk

Baxter Health Care Systems
One Baxter Pkwy,
Deerfield, IL 60015, USA
708-948-2000
www.baxter.com

Bayer Corp.
PO Box 4913,
Kansas City, MO, 64120
(816) 242-2000; (800) 842-8020
www.bayer.com

Bayer Corp.
PO Box 390,
Shawnee Mission, KS, 66201
www.bayer.com

Bayer plc
BayerWerk,
Leverkusen, Germany, 'D-5090
(49) 214-301, (49) 214-306-5136
www.bayer.com

Bayer Corp., Agriculture
Carl-Bosch-Strasse 38,
Ludwigshafen, Germany, D-67056
(49) 621-6060, (49) 621-604-2525
www.bayer.com

BDH Laboratory Supplies
See: EMD Chemicals Inc.

Becker Underwood Inc.
801 Dayton Avenue,
Ames, IA, 50010
(800) 232-5907; (515) 232-5907
www.bucolor.com

Becton Dickinson
1 Becton Drive
Franklin Lakes, NJ 07417
(201) 847-6800

Bell Flavors & Fragrances, Inc.
500 Academy Dr.,
Northbrook, IL, 60062-2497
(847) 291-8300, (847) 291-1217
www.bellff.com

Bell Laboratories/Motomco Ltd.
3699 Kinsman Blvd.,
Madison, WI, 53704
(800) 323-6628, (608) 241-0202

Berje
5 Lawrence St.
Bloomfield, NJ 07003-4604
973-748-8980, Fax 973-680-9618

Berk Pharmaceuticals
St Leonards House, St Leonards,
Eastbourne, Sussex, BN21 3YG, UK
44-1323-501111

Berlex Laboratories Inc.
300 Fairfield Road,
Wayne, NJ 07470, USA
201-694-4100, 800-221-1756
Fax 201-305-5365

Bernel Chemical Co.
174 Grand Ave.,
Englewood, NJ 07631, USA
201-569-8934, Fax 201-569-1741

Berol Nobel AB
S-44485 Stenungsund,
Sweden
46-303-85000
Fax 46-303-84659

Berol Nobel Ltd.
23 Grosvenor Road,
St. Albans, Herts, AL1 3AW, UK
44-1727-841421, Fax 44-1727-
841529

Betzdearborn
4636 Somerton Road,
Trevose, PA, 19053
(215) 355-3300; Fax: (215) 953-5524
www.betzdearborn.com

Bio-Cide International Inc.
PO Box 722170, 2845 Broce Drive,
Norman, OK, 73072-2448
(405) 329-5556
www.bio-cide.com

Bio-Rad Labs.
1000 Alfred Nobel Drive
Hercules, CA 94547
510-724-7000, Fax 510-741-5817
www.biorad.com

Biosensory, Inc.
322 Main Street, Bldg. 1, 2nd. Fl.,
Willimantic, CT, 06226-3149
(860) 423-3009
www.nomorebites.com

Biosentry, Inc.
1481 Rock Mountain Road,
Stone Mountain, GA, 30083
(770) 723-9211
www.biosentry.com

Biosynth AG
(800) 270-2436
www.biosynth.com

Biosys
Certis USA LLC
9145 Guilford Road, Suite 175
Columbia, MD, 21046
(800) 847-5620

Biotex Laboratories International Ltd.
www.biotex.com

Blythe, William Ltd.
Church St.
Accrington, Lancs
BB5 4PD, UK
+44 1254 87872, Fax +44 1254
872000
www.wm-blyth.co.uk

BOC Gases
See: BOC Group plc.

BOC Group plc,
Chertsey Road,
Windlesham, Surrey, GU20 6HJ, UK
44-1276-77222
Fax 44-1276-71333
www.boc.com

Boehringer Ingelheim GmbH
Binger Strasse 173,
D-55216 Ingelheim, Germany
49-6132-773-666
Fax 49-6132-773-755
www.boehringer-ingelheim.com

Boehringer Ingelheim Ltd.
Ellesfield Avenue,
Bracknell, Berks., RG12 8YS, UK
44-1344-424-600
Fax 44-1344-424-600
www.boehringer-ingelheim.com

Manufacturer and Supplier Directory

Boehringer Ingelheim Vetmedica Inc.
2621 N. Belt Hwy.,
St. Joseph, MO, 64506-2046
(816) 233-2571; (800) 325-9167,
Fax: (816) 236-2717
www.bi-vetmedica.com

Boehringer Ingelheim Pharmaceuticals
900 Ridgebury Road,
P.O. Box 368,
Ridgefield, CT 06877-0368, USA
www.boehringer-ingelheim.com

Boehringer Mannheim GmbH
See: Roche Diagnostics Corp.

Bofors

Boliden Intertrade
3379 Peachtree Rd. NE, Suite 3,
Atlanta, GA 30326, USA
404-239-6700, 800-241-1912
Fax 404-239-6701

Bonide Products, Inc.
6301 Sutliff Rd.,
Oriskany, NY, 13424
(800) 536-8231, (315) 736-8231
www.bonideproducts.com

Boots Co. plc,
Head Office,
Nottingham, Notts, NG2 3AA, UK
44-115-950-6111, Fax 44-115-959-2727
www.boots.co.uk

Boots Pharmaceuticals Inc.
300 Tri-State International Center,
Suite 200,
Lincolnshire, IL, 60069-4415
(708) 405-7400, (708) 405-7505

Borax Consolidated Ltd
London, UK

Borax Europe Ltd.,
170 Priestley Rd.,
Guildford, Surrey, GU2 5RQ, UK
44-1483-242035
Fax 44-1483-242097
www.borax.com

Borax Ltd.,
Gorsey Lane,
Widnes,Cheshire, WA8 0RP, UK
44-1483-734000
Fax 44-1483-457676
www.borax.com

Borderland Products Inc.
PO Box 73
Toldeo, OH 43566

Bos Chemicals Ltd.
Paget Hall, Tydd St. Giles,
Wisbech Cambs., UK, PE13 5LF
(44) 1945 870 014

BP Chemicals Ltd.
Britannic House, 6th Floor,
1 Finsbury Circus, London, EC2M
7BA, UK
44-171-496-4867
Fax 44-171-496-4898
www.bp.com

Bracco Diagnostics Inc.
107 College Road East
Princeton, NJ 08540
(609) 514-2200, (800) 631-5245,
Fax (609) 514-2424

Bracco Industria Chimicas
Casella Postale 12064,
Via E Folli 50,
I-20134 Milan, Italy,
39-2-2 1771, Fax 39-2-21773

Bretagne Chimie Fine SA
Boisel
F-56140 Pleucadeuc, France
+33 97 26 91 21, Fax +33 97 26 90 46

Bristol Laboratories
PO Box 4500755,
Princeton, NJ 08543-4500, USA
609-897-2000
Fax 315-432-4804

Bristol-Myers Co. Ltd.
Swakeleys House, Milton Rd.,
Ickenham, Uxbridge, Middlesex,
UB10 8NS, UK
44-1895-639-911
Fax 44-1895-636-975
www.bristolmyers.com

Bristol-Myers Squibb Co.
345 Park Avenue,
New York, NY 10154-0037, USA
212-546-4000
Fax 212-546-5664
www.bristolmyers.com

Bristol-Myers Squibb Pharmaceuticals
BMS House,
141-149 Staines Road,
Hounslow, Middlesex, TW3 3JA, UK
44-181-572-7422
Fax 44-181-577-1756
www.bristolmyers.com

British Arkady Co. Ltd.
Arkady Soya Mills
Skerton Road, Old Trafford
Manchester M16 0NJ, UK
+44 161 872 7161, Fax +44 161 873 8083

British Arkady Group
62/70 rue Ivan Tourgueneff
78380 Bougival
France
+33 1 3969 7070, Fax +33 1 3918 4610

British Chrome & Chemicals
Urlay Nook, Eaglescliffe,
Stockton-on-Tees, Cleveland, TS16 0QG, UK
44-1642-787-755, Fax 44-1642-781-935

British Drug Houses
See: EMD Chemicals Inc.

British Traders & Shippers
429-431 Rainham Rd. South
Dagenham, Essex RM9 6SL, UK
+44 2085967500, Fax +44 2085967509

Britz Fertilizers Inc.
PO Box 6001 3265 West Figarden,
Fresno, CA, 93715
(559) 448-8000 Fax, (559) 448-8020

Bromhead & Denison Ltd.

Bromine & Chems. Ltd.
See: Solaris Chemtech Ltd.

Brotherton Specialty Products Ltd.
Calder Vale Road
Wakefield, W. Yorks, WF1 5PH, UK
+44 1924 371919, Fax +44 1924 290108

Brown Chemical Co., Inc.
302 W. Oakland Ave.
Oakland, NJ 07436-1381
(201) 337-0900, Fax, (201) 337-9026

Brown Butlin Ltd.
Brook House, Ruskington
Sleaford NG34 9EP, Lincs, UK
+44 01526-831000, Fax +44 01526-832967

Browning Chemical Corp.
707 Westchester Ave.,
White Plains, NY, 10604

Buckman Laboratories Inc.
1256 North Mclean Blvd.,
Memphis, TN, 38108-0305
(901) 278-0330; Fax: (901) 276-5343
www.buckman.com

Buckton Scott Commodities Ltd.
Black Horse House, Bentalls
Pipps Hill Estate
Basildon, Essex, SS14 3BX, UK
+44 1268 531308 Fax, +44 1268 531316

Buffalo Color Corp.
100 Lee St.
Buffalo, NY 14240-7027

Burroughs Wellcome Inc.
3030 Cornwallis Rd,
Research Triangle Park, NC 27709,
USA
919-248-3000, 800-722-9292
Fax 919-248-8375

Burts & Harvey

Bush Boake Allen
See: International F&F
Blackhorse Lane,
Walthamstow, London, E17 5QP, UK
44-181-531-4211, Fax 44-181-527-2360
www.bushboakeallen.com

C.M. Ind.

Cabot Carbon Ltd.
Barry Site, Sully Moors Rd.,
Sully, S. Glamorgan, CF6 2XP, UK
44-1446-736999, Fax 44-1446-737123
www.cabot-corp.com

Cabot Carbon Ltd.
Lees Lane,
Stanlow-Ellesmere Port,
S. Wirral, Cheshire, L65 4HT, UK
44-151-355-3677, Fax 44-151-356-0712
www.cabot-corp.com

Caffaro SpA
Via Friuli, 55,
20031 Cesano Maderno, Italy
39-362-51-4266, Fax 39-362-51-4405

Calgene
See: Lambent Technologies Corp.,

Calgon Carbon
400 Calgon Drive
Pittsburgh, PA 15205-1131
(800) 422-7266, (412) 787-6700, Fax
412-787-6676

Callery Chem. Co.
See: BASF Corp

Callery Chem. Co. UK
See: BASF Corp

Camphor and Allied Products

Cape Fear Chemicals Inc
PO Box 695, 4271 US Highway 701
S,
Elizabethtown, NC, 28337
(910) 862-3139

Carborundum Corp.
168 Creekside Dr,
Amherst, NY 14228, USA
716-691-2051

Carbo-Tech Composites GmbH
Eugen Müller Str. 16
A-5020 Salzburg, Austria
+43 662 850382 0, Fax +43 662
850382 77
www.carbotech.at

Cargill
Cargill Office Center
PO Box 9300
Minneapolis, MN 55440-9300
(952) 742-7575
www.cargill.com

Cargill Ltd.
Fairmile Lane, Cobham, Surrey
KT11 2PD, UK
+44 1932 861000
www.cargill.com

Carter-Wallace Ltd.
Wear Bay Road,
Folkestone, Kent, CT19 6PG, UK
44-1303-850661

CasChem Inc.
40 Ave. A,
Bayonne, NJ 07002, USA
201-858-7900, 800-CASCHEM, Fax
201-437-2728

Catalyst Resources Inc.

Catalysts & Chem. Ind.
Nippon Building
6-2, Ohte-machi 2-chrome
Chiyoda-ku
Tokyo 100, Japan
+81 3270-6086, Fax +81 3246-0617

Catomance Ltd.
96 Bridge Road East
Welwyn Garden City
Herts, AL7 1JW, UK
+44 1707 324373, Fax +44 1707
372191

Ceca S.A.
22, place de l'Iris,
Cedex 54, 92062 Paris-La Defense,
France
33-1-47-96-9090, Fax 33-1-47-96-9234

Cedar Chemical Corp.
5100 Poplar Avenue, Suite 2414,
Memphis, TN, 38137
(901) 685-5348
www.cedarchem.com

Celaflor GmbH
Konrad-Adenauer-Str. 30
Ingelheim, Germany, 55218
49 6132-7803-0, 49 6132-7803-142
www.cefaclor.de

Celtic Chem. Ltd.

Celite Corp.
PO Box 519,
Lompoc, CA 93438-0519, USA
805-735-7791, 800-342-8667
Fax 805-735-5699

Celite Corp. (Canada)
295 The West Mall,
Etobicoke, Ont., M9C 4Z7, Canada
416-626-8175
Fax 416-626-8235

Celite France
9 rue du Colonel-de-Rochebrune,B.P.
240,
92504 Rueil-Malmaison, France
33-47-49-0560
Fax 33-47-08-3025

**Celite Mexico S.A. d'Alejandro
Dumas**
No. 103, 3er, Col. Polanco, C.P.
11560,
Mexico
52-5-203-5611
Fax 52-5-255-1835

Celite Pacific
Suite 284, Sui On Centre,
8 Harbor Road, Hong Kong
852 582 5609
Fax 852 827 9392

Celite (UK) Ltd.,
Livingston Rd.,
Hessle, North Humberside, HU13
OEG, UK
44-1482-64-5265
Fax 44-1482-64-1176

Central Garden & Pet Company
3697 Mt. Diablo Blvd., Suite 310,
Lafayette, CA, 94549
(925) 283-4573
www.centralgardenandpet.com

Cequisa SA
Rua Luis de Freitas Branco, 42
Corpo, A-2A,
Lisbon, Portugal, 1600-491
21 758 86 62

Cerac Inc.
PO Box 1178, 407 N. 13th St.
Milwaukee, WI 53201-1178
(414) 289-9800, Fax (414) 289-9805
www.cerac.com

Cerestar U.S.A. Inc.
1100 Indianapolis Boulevard,
Hammond, IN, 46320-1094
219-473-5887, 219-473-6601
www.cerestar.com

Cerexagri Decco Inc.
see: Cerexagri, Inc.

Manufacturer and Supplier Directory

Cerexagri, Inc.
630 Freedom Business Ctr., Suite 402,
King of Prussia, PA, 19406
(800) 438-6071
www.cerexagri.com

Certis USA LLC
9145 Guilford Rd., Suite 175,
Columbia, MD, 21046
(800) 847-5620
www.certisusa.com

CFPI Agro SA
Lobeco Products Inc.
PO Box 630,
Lobeco, SC, 29931-0630
(843) 846-8171

Champion Technologies Inc.
3355 W. Alabama, Ste. 400,
Houston, TX, 77098
(713) 627-3303; Fax: (713) 623-8083
www.champ-tech.com

Charkit Chemical Corp.
9 Old Kings Highway South
P.O. Box 1725
Darien, CT 06820-1725
(203) 655-3400, Fax (203) 655-8643
www.charkit.com

Chase Products Co.
19th and Gardner Road,
Broadview, IL, 60155
(800) 242-7326; (708) 865-1000,
(708) 865-0923
www.chaseproducts.com

Cheltec Inc.
2215 Industrial Blvd.,
Sarasota, FL, 34234
(800) 536-8667; (941) 335-1045
www.cheltec.com

Chemax International
155 N. Main St.,
New City, NY 10956, USA
914-634-0451
Fax 914-634-0937
www.chemax.com

Chemax, Inc.
PO Box 6067,
Greenville, SC 29606, USA
864-277-7000, 800-334-6234,
Fax 864-277-7807
www.chemax.com

Chemcentral Corp.
7050 W. 71 St.,
PO Box 730,
Bedford Park, IL 60499-0730, USA
708-594-7000, 800-331-6174,
Fax 708-594-6328

Chemical Cealin Co. Inc.
961 Blaine Ave.,
Salt Lake City, UT, 84105
(801) 484-1145

Chemical Packaging Corp.
2700 SW 14th. St.,
Pompano Beach, FL, 33069,
(954) 974-5440

Chemical Specialties Inc.
One Woodlawn Green, Suite 250,
Charlotte, NC, 28217
(704) 522-0825; (800) 421-8661;
Fax: (704) 527-8232
www.treatedwood.com

Chemie Linz N. Am.
See: DSM Chemie Linz GmbH

Cheminova A/S
1700 State Route 23 # 210,
Wayne, NJ, 07470-7536
(800) 548-6113, (973)305-6600
www.cheminova.dk

Chemisco
8494 Chapin Industrial Drive,
St. Louis, MO, 63114
(800) 332-5553

Chemisphere
38 King Street
Chester, Ches., CH1 2AH, UK
+44 (0) 1244 320878, Fax +44 (0)
1244 320858
www.chemisphere.co.uk

Chemplex Chemicals Inc.
201 Route 17, Suite 300,
Rutherford, NJ 07070
(201) 935-8903, Fax (201) 935-9051

Chemron
3115 Propeller Dr,
Paso Robles, CA, 93446-8524
(805) 239-1550
www.chemron.com

Chem-Serv Inc.
3205 Maverick Drive,
Kilgore, TX, 75662
(903) 988-2215; Fax: (903) 988-1136
www.chem-serv.com

Chemtreat Inc.
4301 Dominion Boulevard,
Glen Allen, VA, 23060
(804) 935-2000; Fax: (804) 965-0154
www.chemtreat.com

Chem-Y GmbH
See: Kao Corp.

Chevron Chemical Co.
1301 McKinney,
PO Box 3766,
Houston, TX 77253-3766, USA
713-754-2000, 800-231-3260
www.chevron.com

Chevron Chemical Co.
6001 Bollinger Canyon Rd.,
San Ramon, CA 94583, USA
925-842-5764, Fax 925-842-0378
www.chevron.com

Chevron International
PO Box 7146,
San Francisco, CA 94120-7146, USA
415-894-5341, 800-344-5650,
Fax 415-894-1083
www.chevron.com

Chiltern Farm Chemicals
11 High Street,
Thornborough, Buckingham
Bucks, MK18 2DF,UK
44-1280-822400, Fax 44-1280-
822082

Chinoin Co. Ltd.
Tc utca 1-5, PO Box 110
Budapest 1045, Hungary
+36 1 1690900, Fax +36 1 1690293

Chipman Ltd.
Nomix-Chipman Ltd,
Portland Bldg., Portland St., Staple
Hill,
Bristol, UK, BS16 4PS
44 1423 568-658; Fax: 44 1423 504-
654
www.nomix-chipman.co.uk

Chisso Corp.
Tokyo Building 7-3, Marunouchi
2-Chome, Chiyoda-ku
Tokyo 100-8333, Japan
+81 3-3284-8749, Fax +81 3-3284-
8750
www.chisso.co.jp

Chisso Corp., USA
1185 Avenue of the Americas, 32
Fl.,
New York, NY, 10036
(212) 302-0500; Fax: (212)302-0643
www.chisso.co.jp

Christiaens S.A.
See: Nycomed Christiaens S.A.

Cia-Shen

Ciba Geigy Agrochemicals
Whittlesford,
Cambridge, Cambridgeshire, CB2
4QT, UK
44-1223-833621, Fax 44-1223-
835211
www.ciba.com/

Ciba Pharmaceuticals Co.
556 Morris Ave.,
Summit, NJ 07901, USA
800-742-2422
www.ciba.com/

Manufacturer and Supplier Directory

CIBA plc
540 White Plains Road
Tarrytown, NY 10591
914-785-2000
www.ciba.com/

CIBA Vision Corp.
11460 Johns Creek Parkway
Duluth, GA 30097
(678) 415-4255, Fax (678) 415-4260

CIBA Vision UK Ltd.
Park West, Royal London Park
Flanders Road, Hedge End
Southampton, Hants, S030 2LG, UK
+44 1 489-786-580, Fax +44 1 489-786-803

Ciba-Geigy (Japan) L,
10-66, Miyuki-cho,
Takarazuka-shi, Hyogo, 665, Japan
81-797-74-2472
Fax 81-797-74-2515
www.ciba.com

Ciba-Geigy Corp.
CH-4002, Basel, Switzerland
41-061- 696-4534, Fax 41-061-697-1111
www.ciba.com

CID Lines NV
19, Rue Delezenne,
Lille, France, 59000
(33) 3 202 30111; Fax: (33) 3 202 32648
www.cidlines.be

Cilag-Chemie Ltd.

CK Witco (Europe) S.A.
See: Witco (Europe)

CK Witco Chem. Corp.
See: Witco Corp.

CK Witco Corp.
See: Witco Corp.

Clariant Corp.
4000 Monroe Rd.,
Charlotte, NC, 28205
(908) 903-9170; Fax: (908) 903-9168
www.clariant.com

Clearwater Inc.
5605 Grand Avenue,
Pittsburgh, PA, 15225
(412) 269-9363; Fax: (412) 264-1616
www.cwichem.com

Cleary Chemical Corp.
178 Ridge Road, Dayton, NJ, 8810
(800) 524-1662; Fax: 908-274-0894
www.clearychemical.com

Cleary W. A.
See: Cleary Chemical Corp.

Climax Molybdenum Co.
2598 Highway 61,
Fort Madison, IA, 52627
(319) 463-7151
www.climaxmolybdenum.com

Climax Performance Materials,
PO Box 22015,
Tempe, AZ 85285-2015, USA

Clough
251 St. Georges Terrace, Level 6
Perth, W. Australia 6000

CNM Technologies
94 Gardiners Ave, Suite 242
Levittown, NY, 11756

Coalite Chemicals Division
PO Box 152,
Buttermilk Lane,
Bolsover, Chesterfield, Derbyshire,
S44 6AZ, UK
44-1246-826816
Fax 44-1246-240309

Coastal Chemical Corp.
PO Box 1287, 8305 Otto Rd.,
Cheyenne, WY, 82003-9502
(800) 443-2754; (307) 637-2200,
(800) 832-7601

Colcide Inc.
11500 W. Hill Dr.,
Rockville, MD, 20852
(301) 588-8454

Columbus Foods Co.
730 N. Albany Avenue
Chicago, IL 60612-1006
(773) 265-6500, Fax (773) 265-6985

Comlets Chemical Industrial Co. Ltd.
No. 196, Shinping Rd., Sec. 1
Taiping, Taichung Hsien, Taiwan,
886-2-23213850, 4-22702121
Fax: 886-2-23963125, 4-22706335

Compania Quimica
Sarmiento 329
Buenos Aires, Argentina 1041

Condea Vista Co.
900 Threadneedle,
Houston, TX, 77224-9029
(800) 231-8216; (281) 588-3219;
Fax: (281) 588-3107
www.condeavista.com

Conseal International Inc.
728 Industry Rd.,
Longwood, FL, 32750-3632
(407) 834-8728

Continental Sulfur Co. LLC
5100 Poplar Avenue, Suite 2700
Memphis, TN, 38137
(901) 763-4017

Controlled Solutions
5903 Genoa-Red Bluff,
Pasadena, TX, 77507
(800) 242-5562; (713) 473-3345

Coopers Creek Chemical Corp.
884 River Rd,
W Conshohocken, PA, 19428-2699
(610) 828-0375

Corn Products International Inc.
6500 S, Archer Road,
Bedford Park, IL 60501-1933
(708) 563-2400, Fax (708) 563-6878

Courtaulds Water Soluble Polymers
PO Box 5,
Spondon, Derbyshire, DE21 7BP, UK
44-1332- 661422, Fax 44-1332-661078
www.courtaulds.com

Coyne
3015 State Road
Croydon, PA 19021-6997
(215) 785-3000, Fax (215) 785-1585
www.coynechemical.com

Creanova Inc.
220 Davidson Ave.,
Somerset, NJ, 8873
(732) 560-6326, (732) 560-6356
www.creanovainc.com

Creative Sales Inc.
222 N Park Ave,
Fremont, NE, 68025-4964
(402) 727-4800

Croda Inc.
7 Century Dr,
Parsippany, NJ 07054-4698, USA
973-644-4900
www.croda.com

Croda Surfactants Ltd.
Cowick Hall, Snaith,
Goole, North Humberside, DN14
9AA, UK
44-1405-860551, Fax 44-1405-860205
www.croda.com

Crompton Corp.
1 American Lane,
Greenwich, CT, 06831-2559
(203) 353-5400
www.cromptoncorp.com

Crookes Healthcare Ltd.
See: Boots Co. Plc

Crosfield Chemical Inc.
101 Ingalls Avenue
Joliet, IL 60435-4373
(800) 727-3651, (815) 727-3651, Fax
(815) 727-5312

Manufacturer and Supplier Directory

Crowley Tar Prods.
261 Madison Avenue, Fl. 14
New York, NY 10016-2303
(212) 682-1200

Crown Metro

Crystal Inc./H & S Chemical
970 E. Tipton St.,
Huntington, IN, 46750
(219) 358-9154, Fax (219) 358-9154

CTX-Cenol Inc.
25801 Solon Rd.,
Bedford Heights, OH, 44146
(888) 281-7025

Cuproquim Corp.
6075 Poplar, Suite 500,
Memphis, TN, 38119
(901) 761-0050

Cyanamid of Great Britain Ltd.
Fareham Rd.,
Gosport, Hants, UK, PO13 0AS

Cytec
5 Garret Mountain Plaza,
W. Patterson, NJ, 07424
(800) 652-6013
www.cytec.com

Cyprus Industrial Min.
2600 N. Central Avenue
Phoenix, AZ 85004

D.I.E. Corp.

Daicel (U.S.A), Inc.
One Parker Plaza, 400 Kelby Street ,
Fort Lee, NJ, 7024
(201) 461-4466; Fax: (201) 461-2776
www.daicel.co.jp

Daicel Chem. Ind.
1, Teppo-cho, Sakai-sha,
Osaka, Japan, 590-8501
81 722 27 3111; Fax: 81 722 27
3000
www.daicel.co.jp

Daihachi Chem. Ind. Co. Ltd.
1-9 Nitto-cho, Handa-shi,
Aichi, Japan, 475-0033

Dajac Labs.
See: Monomer-Polymer & Dajac
Laboratories Inc.

Danbert Chemical Co.

Dead Sea Bromine
Israel Chemicals Ltd.
Beer Sheva, Israel

Dean AG
3600 North River Rd.,
Franklin Park, IL, 60131-2185

Decco
Vale Lane, Hartcliffe Way
Bedminster, Bristol
BS3 5RR, UK
+44 0117966-9364, Fax: +44 0117
953-5279

Degen Co.
200 Kellogg Street, PO Box 5240
Jersey City, NJ 07305
(201) 432-1192, Fax (201) 432-8483

Degussa AG
Weissfrauenstr. 9,
Frankfurt/M, Germany, 60287
49 69 21801, Fax 49 69 218-3218
www.degussa.de

Degussa-Hu+a5ls Corp.
65 Challenger Rd.,
Ridgefield Park, NJ 07660, USA
201-807-3224, 800-237-6745, Fax
201-807-3111
www.degussa.com

Degussa Ltd.
Earl Rd, Stanley Green,
Handforth, Wilmslow, Cheshire, SK9
3RL, UK
44-161-486-6211, Fax 44-161-485-
6445
www.degussa.com

Delavau J. W. S Co. Inc.
2140 Germantown Avenue
Philadelphia, Pa 19122
(215) 235-1100, Fax (215) 235-2202

Dermik Labs. Inc.
500 Arcola Road,
P.O. Box 1200,
Collegeville, PA 19426-0107, USA
312-687-7440

Devere Chemical Co.
PO Box 8444, 1923 Beloit Ave.,
Janesville, WI, 53545

Diachem SpA
via Tonale, 15-24061
Albano S.A., Italy
035 581228

Diamalt GmbH
Georg-Reismuller-Strasse 32,
Munich, Germany
49 89-81060; Fax: 49 89-8106513

Diverseylever Inc.
3630 E. Kemper Rd.,
Sharonville, Cincinnati, OH, 45241-
2046
(513) 554-4200; Fax: (513) 554-4330

Dover Chemical Corp.
PO Box 40, 3676 Davis Rd. NW,
Dover, OH, 44622
(330) 343-7711; Fax: (330) 364-1579
www.doverchem.com

Dow Agrosciences, Crop Protection
9330 Zionsville Rd., Bldg. 308,
Indianapolis, IN, 46268
(800) 352-6776; Fax: (800) 258-3033
www.dowagro.com

Dow Chemical North America
2040 Willard H. Dow Center,
Midland, MI 48674, USA
517-636-1000, 800-441-4DOW
Fax 517-636-9752
www.dow.com/

Dow Corning Corp.
PO Box 0994,
Midland, MI 48686-0994, USA
517-496-4000, 800-248-2481
Fax 517-496-4586
www.dowcorning.com

Dow Chemical Europe S.A.
Bachtobelstrasse 3,
CH-8810 Horgen, Switzerland
41-1-728-2095
Fax 41-1-728-3081
www.dow.com/

Dow Chemical Co. Ltd.
2 Heathrow Boulevard, 284 Bath
Road,,
W. Drayton, Middlesex, UK, UB7
ODQ
44 181 917 5000; Fax: 44 181 917
5400
www.dowcorning.com

DowElanco Ltd.
See: Dow Agrosciences, Crop
Protection

Dragon Chemical Corp.
PO Box 7311,
Roanoke, VA, 24019
(540) 362-3657
www.dragonchemical.com

Drew Industrial
Division of Ashland Inc.,
One Drew Plaza,
Boonton, NJ , 07005-1924
(973) 263-7636

Drexel Chemical Co.
PO Box 13327,
Memphis, TN, 38113
(901) 774-4370
www.drexchem.com

Dry Branch Kaolin Inc.
Highway 80 S,
Dry Branch, GA, 31020
(478) 945-3125

DSM Chemie Linz GmbH
Postfach 933
A-4021, Linz, Austria
+43 706916-3619, Fax +43
691663619

Manufacturer and Supplier Directory

DSM Fine Chemicals Inc.
Park 80 West, Plaza 2,
Saddle Brook, NJ, 07663-5817
(201) 845-4404; Fax: (201) 845-4406
www.dsm.com

DuPont Chemicals
1007 Market St.,
Wilmington, DE 19898, USA
800-441-7515
www.dupont.com

DuPont (UK) Ltd.
Wedgewood Way,
Stevenage, Herts, SG1 4QN, UK
44-1438-734026, Fax 44-1438-734379
www.dupont.com

DuPont Merck Pharmaceuticals
Barley Mill Plaza,
Wilmington, DE 19880-0025, USA
302-992-5000, 800-441-9861,
Fax 302-892-8530
www.dupont.com

Eagle-Picher Industries
PO Box 550,
C & Porter Sts.,
Joplin, MO 64801, USA
417-623-8000, Fax 417-782-1923

Earth Care
PO Box 282,
Palm Harbor, FL, 34682
(813) 937-1324

Eastman Chemical Products
PO Box 431,
Kingsport, TN 37662-5280, USA
423-229-2000, 800-EASTMAN,
Fax 423-229-1196,
www.eastman.com/

Eaton, J. T. & Co. Inc.
1393 E. Highland Road
Toledo, OH 44087
(330) 425-7801, Fax (330) 425-8353
www.jteaton.com

ECC International
5775 Peachtree-Dunwoody Rd.
NE,Suite 200G,
Atlanta, GA 30342, USA
404-303-4415, 800-843-3222
Fax 404-303-4384
www.ecc.com

ECC International Ltd
John Keay House,
St. Austell, Cornwall, PL25 4DJ, UK
44-1726-74482
Fax 44-1726-623019
www.ecc.com

Ecolab Inc.
370 N. Wabasha Street
St. Paul, MN 55102-2233
(651) 293-2233, Fax (651) 293-2092
www.ecolab.com

Eka Nobel Inc.
2622 Nashville Ferry Rd. East,
PO Box 2167,
Columbus, MS 39701, USA
770-956-2520, 800-821-9486,
www.ekanobel.com

Electrochem. Inc.
400 W. Cummings Park
Woburn, MA 01801
Elementis plc
Elementis House
56 Kingston Road
Staines TW18 4ES, UK
+44 (0) 1784 22 7000, Fax +44 (0)
1784 46 0731
www.elementis-eu.com

Elf Atochem North Am.
2000 Market St.,
Philadelphia, PA 19103-3222, USA
215-419-7000, 800-628-4453, Fax
215-419-7875
www.elf-atochem.com

Eli Lilly & Co.
Lilly Corporate Center,
Indianapolis, IN 46285, USA
317-276-2000, Fax 317-276-6876
www.lilly.com

Elkins-Sinn Pharmaceutical Co.
2 Esterbrook Lane
Cherry Hill, NJ 08002-4009
(610) 688-4400

Elliott, Thomas Ltd.

Ellis & Everard plc
Pine Street,
South Bank Road, Cargo Fleet,
Middlesbrough,TS3 8BD, UK
01642-227388, Fax 01642-242609
www.elliseverard.com

Elm Research Institute
867 Rt. 12, Unit 5,
Westmoreland, NH, 3467
(603) 358-6199
www.forelms.org

EM Industries, Inc.
5 Skyline Dr.,
Hawthorne, NY 10532, USA
914-592-4660
Fax 914-592-9469

Embetec Crop Protection Ltd.
See: Embetec BV

Embetec BV
Postbus 70, 4130 EB Vianen (ZH),
Netherlands, 4130
347-329891; Fax: 347-329893

EMD Chemicals Inc.
480 S. Democrat Road
Gibbstown, NJ 08027
(856) 423-6300, (800) 222-0342,
Fax (856) 423-4389
emdinfo@emdchemicals.com

EniChem America, Inc.
2000 West Loop South,
Suite 2010,
Houston, TX 77027, USA
713-940-0700, 800-441-3646
Fax 713-940-0761
www.eni.it

EniChem Elastomeri Srl
Strada 3, Palazzo B1, Milanofiori,
I-20090 Assago, Milan, Italy
39-2-5201, Fax 39-2-52026077
www.eni.it
Ensystex Inc.
2709 Breezewood Ave.,
Fayetteville, NC, 28302
(910) 484-6163
www.ensystex.com

Entek Corp.
6835 Deerpath Rd., Suite E,
Elkridge, MD, 21075

Environ Intercontinental Ltd.
410 Leonard Blvd., N., Unit #3,
Lehigh Acres, FL, 33971
(941) 303-0076

Environmental Laboratories, Inc.
2304 Gulf Life Tower,
Jacksonville, FL, 32207

Enzypharm BV
Industrieweg 17, 205
3762 EG Soest
Utrecht, Netherlands
+31 35 6030051, Fax +31 35
6029962
www.enzypharm.nl

Eprova AG
See: Merck-Eprova AG

Equistar Chemical Products
See: Lyondell Chemical Co.

Esprit Chem. Co.
7680 Matoaka Road
Sarasota, FL 34243
(941) 355-5100, Fax (941) 358-1339
www.espritchem.com

Estron Chemical, Inc.
409 N. Main Street
Calvert City, KY 42029
(270) 395-4195, Fax (270) 395-5070
www.estron.com

Ethicon Inc.
Rt 22, PO Box 151,
Somerville, NJ 08876, USA
908-218-0707

Manufacturer and Supplier Directory

Ethitek Pharmaceuticals Co.
7701 N. Austin Avenue
Skokie, IL 60077

Ethyl Corpn.
330 S. Fourth St.,
Richmond, VA 23217
(804) 788-5000

Evans Chemetics
See: Hampshire Chemicals

Evans Medical Ltd.
Evans House, Regents Park
Kingston Road, Leatherhead
Surrey KT22 7PQ, UK
+44 (01372) 364000

Expansia SA
BP-6, F-30390 Aramon
France
+33 66 57 01 01, Fax +33 66 57 01 48

Exxon Chemical Co.
PO Box 3272,
Houston, TX 77253-3272, USA
713-870-6000, 800-526-0749,
Fax 713-870-6661
www.exxon.com/exxonchemical

ExxonMobil Chem. Co.
4999 Scenic Highway
Baton Rouge, LA 70805-3359
www.exxonchemical.com

Fabrichem
2226 Black Rock Turnpike, #206
Fairfield, CT 06430
(203) 366-1820, Fax (203) 366-1850
www.fabricheminc.com

Fabriquimica
Calle 32, No. 3313
San Martin (B16501JA)
Buenos Aires, Argentina
(54-11) 4753-0894, Fax (54-11)
4755-7290
www.fabriquimica.com

Faesy & Besthoff Inc.
143 River Rd.,
Edgewater, NJ, 07020
(201) 945-6200, Fax (201) 945-6145

Fahlberg-List
See: Salutas Pharma GmbH

Fair Products
PO Box 386,
Cary, NC, 27512
(919) 467-8352

Fairmount Chemical Co.
117 Blanchard St.,
Newark, NJ 07105, USA
201-344-5790, 800-872-9999, Fax
201-690-5298

Faith, Keyes and Clark
See: Wiley, John

The Fanning Corp.
2450 W. Hubbard St.,
Chicago, IL 60612-1408, USA
312-563-1234, Fax 312-563-0087

Fargro Ltd.
Toddington Lane,
Littlehampton, W.Sussex, UK, BN17
7PP
44 1903 721591; Fax: 44 1903
730737
www.fargro.co.uk

Farleyway Chem. Ltd.
Ham Lane, Kingswinford
W. Midlands, DY6 7JU UK
+44 1384 400 222, Fax +44 1384
400 020

Farm Protection Ltd.

Farmers Crop Chemicals Ltd.
Thorn Farm, Evesham Rd, Inkberrow
Worcester, Worcestershire WR7 4LJ,
UK
44-01386-793401, Fax 44-01386-
793184

Farmitalia Carlo Erba Ltd.
Italia House
23 Grosvenor Road
St. Albans, Herts, AL1 3AW, UK
+44 (01727)-40041

Farnam Companies Inc.
PO Box 34820
Phoenix, AZ 85067-4820
(800) 234-2269
www.farnam.com

Fermenta
Terv u. 17/a,
Budapest, H1223, Hungary
36 424 0313; 36 424 0314; Fax: 36
424-0314

Fermenta Animal Health Co.
See: Boehring-Ingelheim Vetmedica
Inc.

Ferro Corp./Bedford
7050 Krick Rd.,
Bedford, OH 44146, USA
216-641-8580, 800-321-9946, Fax
216-439-7686
www.ferro.com

Ferro Corp./Grant Chemicals
111 W. Irene Rd.,
Zachary, LA 70791-9738, USA
www.ferro.com

Ferro Corp./Keil Chemicals
3000 Sheffield Ave.,
Hammond, IN 46320, USA
219-931-2630, 800-628-9079, Fax
219-931-0895
www.ferro.com

Ferro Corp./Transelco
Box 217, 1789 Transelco Dr.
Penn Yan, NY 14527, USA
315-536-3357, Fax 315-536-8091
www.ferro.com

Ferro-Pfanstiehl
Ferro Corp.
1000 Lakeside Avenue
Cleveland, OH 44114-7000
(216) 641-8580
www.ferro.com

Fina Chemicals
52 Rue de l'Industrie,
B-1040 Brussels, Belgium
32-2-288 9132
Fax 32-2-288-3322

Fina plc
Fina House, 1 Ashley Ave,
Epsom, Surrey, KT18 5AD, UK
44-1372-726226
Fax 44-1372-744520

Fine Agrochemicals Ltd.
Hill End House, Whittington
Worcester, UK, WR5 2R
www.fine-agrochemicals.com

Firmenich
Postfach 4160
D-50155 Kerpen, Germany
+49 2237 6 90 10, Fax +49 2237 69
01 69
www.firmenich.com

Fisher Scientific International Inc.
Liberty Lane, Hampton, NH 03842
(603) 926-5911, Fax (603) 929-2379
webmaster@nh.fishersci.com

Fisons plc, Pharmaceuticals
Weyside Park, Catteshall Lane,
Godalming, Surrey, GU7 1XE, UK
44-1483-410210, Fax 44-1483-
410220

Fisons plc, Horticultural Div.
177 Sanfordville Rd.,
Warwick, NY, 10990

Flexabar Corp.
1969 Rutgers University Blvd.,
Lakewood, NJ, 8701

Florida Treatt
4900 Lakeland Commerce Parkway
Lakeland, FL 33805
(863) 668-9500, Fax (863) 422-5930
www.treatt.com

Manufacturer and Supplier Directory

Florida Water Works, Inc.
505 Power Rd.,
Sanford, FL, 32771
(407) 324-5888

Fluka
See: Sigma-Aldrich Co.
www.sigma-aldrich.com

Fluorochem Ltd.
Wesley Street
Glossop SK13 7RY
Derbyshire, UK
+44 01457 868921, Fax +44 01457
869360
www.fluorochem.net

FMC Corp./Ag Chem Group
1735 Market St.,
Philadelpha, PA 19103, USA
215-299-6000, Fax 215-299-5999
www.fmc.com

**FMC Corp./Chemical Products
Group**
1735 Market St.,
Philadelphia, PA 19103, USA
215-299-6000, 800-346-5101,
Fax 215-299-5999
www.fmc.com

FMC Corp./Pharmaceuticals
1735 Market St.,
Philadelphia, PA 19103, USA
215-299-6000, 800-362-3773,
Fax 215-299-6821
www.fmc.com

Forest Pharm. Inc.
13600 Shoreline Drive
St. Louis, MO 63045
(800) 678-1605, (314) 493-7450
www.forestpharm.com

Fortune Biotech Ltd.
4937 Monaco Drive
Pleasanton, CA 94566
(925) 485-9849, Fax (925) 485-9831
www.fortunebiotech.com

Foseco (FS) Ltd.
Tamworth, Staffs, B78 3TL, UK
44-1827-289999, Fax 44-1827-
250806
www.foseco.com

Foseco Ltd./Metallurgy
Tamworth, Staffs, B78 3TL, UK
www.foseco.com

Franklin Mineral Products
PO Drawer 390,
Hartwell, GA 30643, USA
706-376-3174, Fax 706-376-3044

Frinton Chemicals
PO Box 2428
Vineland, NJ 08362
(877) 374-6866, Fax (856) 439-1977
www.frinton.com

Frowein, Kurt, GmbH & Co. KG
Lenneper Str. 130c
D-42289 Wuppertal
Germany
+49 0202 262 900, Fax +49 0202
262 9049

Frutarom Inc.
9500 Railroad Avenue
Bergen, NJ 07047
(201) 861-9500, (800) 526-7147,
Fax (201) 861-4323
www.frutarom.com

Frys Metals (Pty) Ltd.
PO Box 519
Germiston 1400 RSD
S. Africa
+27 11 8275413, Fax +27 11
8242232
www.frys.co.za

Fuji Chem. Ind.
7-B Marlen Drive
Robbinsville, NJ 08691
(609) 890-2490, Fax (609) 890-2495
www.fujichemusa.com

G.H.G. Co. Inc.
13345 57th Pl. S.,
Lake Worth, FL, 33467
(561) 371-9200

Galderma Labs Inc.
3000 Altamesa Blvd., Suite 300,
Fort Worth, TX 76133, USA
817-551-8664

Gattefosse Corp.
372 Kinderkamack Rd.,
Westwood, NJ 07675, USA
201-358-1700, Fax 201-358-4050

Gattefosse S.A.
36 Chemin de Genas,
BP 603,
69804 Saint Priest, C, France
33-78- 90-6311, Fax 33-78-90-4567

Gaylord Chemical Co.
PO Box 1209, 106 Galeria Blvd.,
Slidell, LA 70459-1209, USA
504-639-5633, 800-426-6620,
Fax 504-649-0068

GB Biosciences Corp.
See: Syngenta Crop Protection

General Electric Co.
One Plastics Ave.,
Pittsfield, MA 01201, USA
413-448-4808, 800-845-0600
www.ge.com

General Electric Co.
260 Hudson River Rd.,
Waterford, NY 12188, USA
518-237-3330, 800-255-8886, Fax
518-233-3931
www.ge.com

General Electric Co.
21800 Tungsten Rd,
Cleveland, OH 44117, USA
216-266-2451
Fax 216-266-3372
www.ge.com

GE Silicones

GE Specialities
1000 Morgantown Industrial Park
Morgantown, WV 26501
(304) 284-2353, Fax (304) 284-2307
www.ge.com/specialtychemicals

Gelest Inc.
612 William Leigh Drive
Tullytown, PA 19007-6308
(215) 547-1015, Fax (215) 547-2484
www.gelest.com

General Chemicals
13300 Foley Street
Detroit, MI 48227
(313) 491-0355, Fax (313) 491-8545
www.generalchem.com

Generic
422 N. 5th St.,
Milwaukee, WI, 53201

Genesee Polymers Corp.
G-5251 Fenton Rd.,
PO Box 7047,
Flint MI, 48507-0047, USA
810-238-4966, Fax 810-767-3016

Genstar Stone Products
Route 235
Hollywood, MD 20636
(410) 527-4000, Fax (410) 527-4535

Geoliquids Inc.
15 E. Palatine Rd., Suite 109,
Prospect Heights, IL 60070, USA
708-215-0938, 800-827-2411, Fax
708-215-9821

Georgia Gulf
Highway 405,
Plaquemine, LA, 70765
(504) 685-1200
www.ggc.com

Georgia Gulf Corp.
PO Box 105197,
400 Perimeter Center Terrace,
Atlanta, GA 30348, USA
504-389-2500

Georgia Marble Co.
1201 Roberts Blvd., Bldg. 100
Kennesaw, GA 30144-3619
(404) -421-6500, Fax (404) 421-6507

Georgia Pacific Corp.
133 Peachtree Street
Atl;anta, GA 30303
(404) 650-4000

Manufacturer and Supplier Directory

Georgia Pacific Resins
PO Box 105605,
Atlanta, GA, 30348
www.gp.com

Gharda USA Inc.
1116 Taylorsville Rd.,
Washington, Crossing, PA, 18977
(215) 321-9091

Gist-Brocades Intl.
Koestraat 3, PO Box 18
6160 MD Geleen
Netherlands

Gittens, Edward
Giulini Chemie GmbH,
Giulini-strasse 2,
67029 Ludwigshafen, Germany
49-621 570 901

Givaudan-Roure Corp.
100 Delawanna Ave.,
Clifton, NJ 07014, USA
201-365-8277, Fax 201-777-9304

Givaudan Iberica SA
100 Delawanna Ave.
Clifton, NJ, 07014
(201) 365-8000
www.givaudan.com

Glaxo Laboratories
Glaxo House, Berkeley Avenue,
Greenford, Middlesex, UB6 0NN, UK
44-171-493-4060
Fax 44-181-966-8330

Glaxo Wellcome Inc.
P.O. Box 13398,
Research Triangle Park, NC 27709,
USA
919-248-2100
www.glaxowellcome.com

Glaxo Welcome plc,
Glaxo Wellcome House,
Berkeley Ave,
Greenford, Middlesex, UB6 0NN, UK
0171-493-4060, Fax 0181-966-8330
www.glaxowellcome.co.uk

Glenwood Inc.
83 N. Summit St,
PO Box 518,
Tenafly, NJ 07670, USA
201-569-0050

Glucona
114 East Conde Street
Janesville, WI 53546
(888) GLUCONA, Fax (608) 752-7643
www.glucona.com

Gold Coast Chemical Products
Dora Industries Inc.
2790 S. Park Rd.
Pembroke Park, FL, 33009
(954) 921-9100

Goldschmidt AG,
Goldschmidtstrasse 100,
Postfach 101461,
45127 Essen, Germany
49-201-173-01, Fax 49-201-173-3000
www.goldschmidt.com

Goldschmidt Chemical
914 E. Randolph Rd.,
PO Box 1299,
Hopewell, VA 23860, USA
804-541-8658, 800-446-1809, Fax
804-541-2783
www.goldschmidt.com

Good Food
W. Main St., PO Box 160
Honey Brook, PA 19344
(610) 273-3776, Fax (610) 273-2087

Goodrich B.F. Co.
240 W. Emerling Avenue
Akron, OH 44301
330-374-2449
www.bfgoodrich.com

Goodyear Tire & Rubber
1144-East Market St.,
Akron, OH 44316, USA
216-796-3845, 800-321-2385
www.goodyear.com

Goodyear Tire & Rubber
1485 E. Archwood Ave.,
Akron, OH 44306-3299, USA
216-796-6400, 800-548-8107,
Fax 216-796-2617
www.goodyear.com

Gowan Co.
4885 W Riverside Dr.,
Yuma, AZ, 85366-5569
(800) 883-1844
www.gowanco.com

Grace W.R. & Co.
7500 Grace Drive
Columbia, MD 21044
(410) 531-4000, Fax (410) 531-4367
www.gracedavison.com

Grain Processing Corp.
1600 Oregon St.,
Muscatine, IA 52761, USA
319-264-4265
Fax 319-264-4289

Great Lakes Fine Chemicals
PO Box 2200,
One Great Lakes Blvd.,
W. Lafayette, IN 47906-0200, USA
317-497-6100, 800-621-9521,
Fax 317-497-6123
www.greatlakeschem.com

Great Lakes Fine Chemicals
LaVie Sakuragicho Bldg.,
5-26-3, Sakuragicho, Nishi-Ku,
Yokohama 220, Japan
81-45-212-9541, Fax 81-45-212-9539
www.greatlakeschem.com

Great Western
808 SW 15th Avenue
Portland, OR 97205
(503) 228-2600, Fax (503) 221-5752
www.gwchem.com

Greeff R.W. & Co.
777 West Putnam Avenue
Greenwich, CT 06830
(203) 532-2900, Fax (203) 532-2980
www.rwgreef.com

Griffin Corp.
PO Box 1847, Rock Ford Rd.,
Valdosta, GA 31603-1847, USA
912-242-8635, 800-237-1854, Fax
912-244-5978

Grindsted UK
Northern Way, Bury St. Edmunds,
Suffolk IP32 6NP, UK
+44 1284 769631, Fax +44 1284
760839

Grindsted Products Denmark
Edwin Rahrs Vej 38,
DK-8220 Brabrand, Denmark
45-86-25-3366, Fax 45-6-25-1077

Guaranteed Chemical Co.
230 Ouiski Bayou Dr.
Houma, LA, 70360
(985) 876-9276

Gustafson LLC
1400 Preston Rd., Suite 400
Planos, TX, 75093
(972) 985-8877
www.gustafson.com

H & S Chemicals Division
Lonza Inc.
17-17 Route 208,
Fair Lawn, NJ, 7410
(201) 794-2433
www.lonza.com

Haarmann & Reimer Co.
PO Box 175,
70 Diamond Road,
Springfield, NJ 07081, USA
201-467-5600,800-422-1559, Fax
201-467-3514

Haarmann & Reimer Co.
1127 Myrtle St., PO Box 932,
Elkhart, IN 46515, USA
219-262-7874, 800-348-7414, Fax
219-262-6747

Manufacturer and Supplier Directory

Haarmann & Reimer GmbH
Postfach 1253,
Rumohrtalstrasse 1,
37603 Holzminden, Germany
49-5531-900, Fax 49-5531-901649

Haarmann & Reimer Ltd.
Fieldhouse Lane,
Marlow, Bucks., SL7 1NA, UK
44-1628-472051, Fax 44-1628 890795

Hacco Inc.
PO Box 3211, PO Box 2312
Avondale Estates, GA, 30002

Hall C.P.
7300 South Central Avenue
Chicago, IL 60638-0428
(708) 594-6000, Fax (708) 458-0428

Hamari Chemicals Ltd.
1-4-29, Kunijima
Higashi Yodogawa-ku
Osaka 533, Japan
+81 06 323-9027

Hampshire
7491 Federal Hwy., Suite 136,
Boca Raton, FL, 33487
(407) 241-4264

Hampshire Chemicals
2 East Split Brook Road
Building 10
Nashua, NH 03060
(800) 447-4369, Fax (989) 832-1465

Harcros Chemicals Inc.
PO Box 2930,
5200 Speaker Rd.,
Kansas City, KS 66110-2930, USA
913-621-7749, Fax 913-621-7746
www.harcoschem.com

Harcros Durham Chemicals,
Birtley,
Chester-le-Street, Co. Durham, DH3 1QX, UK
44-1914-102361, Fax 44-1914-106005
www.harcoschem.com

Hardman Inc.
600 Cortlandt St.,
Belleville, NJ 07109, USA
201-751-3000, Fax 201-751-8407

Harris, Keith
7 Sefton Rd.,
Thornleigh, NSW, Australia, 2120
61 2 9484-1341; Fax: 61 2 9481-9306
www.keithharris.com.au

Hart Products Corp.
173 Sussex St.,
Jersey City, NJ 07302, USA
201-433-6632, Fax 201-435-7268

Hart Chemicals Ltd.
256 Victoria Rd. South,
Guelph, Ont., N1H 6K8, Canada
519-824-3280, Fax 519-824-0755

Harwick Chemical Corp.
60 S. Seiberling St., PO Box 9360,
Akron, OH 44305-0360, USA
216-798-9300, Fax 216-798-0214

Hatco Corp.
1020 King George Road
Fords, NJ 08863-0601
(908) 738-1000, Fax (908) 738-9385

Heico Chemicals Inc.
Route 611
Delaware Water Gap, PA, 18327
(800) 344-3426; (570) 420-3900;
Fax: (570) 421-9012
www.heicochemicals.com

Helena Chemical Co.
225 Schilling Blvd.,
Collierville, TN 38017
(901) 761-0050, Fax (901) 761-5754

Henkel Corp./Cospha,
300 Brookside Ave.,
Ambler, PA 19002, USA
215-628-1476, 800-955-1456,
Fax 215-628-1450
www.henkel.com

Henkel Corp./Emery
5051 Estecreek Rd.,
Cincinnati, OH 45232-1446, USA
513-530-7300, 800-543-7370
Fax 513-530-7581
www.henkel.com

Henkel Corp./Organic
300 Brookside Ave.,
Ambler, PA 19002, USA
215-628-1441, 800-634-2436,
Fax 215-628-1200
www.henkel.com

Henkel Corp./Fine Chemicals
5325 South 9th Ave.,
La Grange, IL 60525-3602, USA
708-579-6150, 800-328-6199,
Fax 708-579-6152
www.henkel.com

Henley & Co.

Herbert Laboratories
2525 Dupont Drive,
Irvine, CA 92713, USA
714-252-4500

Hercules B.V.
8 Veraartlaan,
NL-2288GM Rijswijk, The Netherlands
31-70-150-000, Fax 31-70-398-9893
www.herc.com

Hercules B.V./Aqualo
Postbus 5832,
NL-2280 HV Rijswijk, The Netherlands
31-70-315-0226, Fax 31-70-390-7560
www.herc.com

Hercules Europe S.A.
Avenue de Tervuren 300,
B-1150 Brussels, Belgium
32-2-761-5511
www.herc.com

Hercules Inc.
Hercules Plaza-6205SW,
Wilmington, DE 19894-0001, USA
302-594-5000, 800-247-4372,
Fax 302-594-5400
www.herc.com

Hess & Clark Inc.
7th. & Orange Streets,
Ashland, OH, 44805
(800) 992-3594

Heterene Chemical Co.
PO Box 247, 295 Vreeland Ave.,
Paterson, NJ 07543, USA
201-278-2000, Fax 201-278-7512

Hexcel Corp./Chemicals
215 N. Centennial St.,
Zeeland, MI 49464, USA
616-772-2193, Fax 616-772-7344

Hexcel Corp./Resins
20701 Nordhoff St.,
Chatsworth, CA 91311, USA
213-322-8050, 800-423-5451, Fax 818-709-0399

Hexcel Corp./Trevarn,
5794 W. Las Positas Blvd.,
Pleasanton, CA 94588, USA
510-847-9500, 800-444-3923, Fax 510-734-9688

Hightex
60-Igualada
Barcelona, Spain

Hill Brothers Chemical Co.
1675 N. Main St.
Orange, CA, 92867
(714) 998-8800
www.hillbrothers.com

Hilton Davis
2235 Langdon Farm Road
Cincinnati, OH 45237
(800) 477-1022, Fax (800) 477-4565

Hindustan Insecticides
c/o Noris Chemical Corp.
P.O. Box 1192
Lompoc, CA, 93438
(805) 736-2081

Manufacturer and Supplier Directory

Hisamitsu Pharm. Co. Ltd.
408 Daikan-machi
Tashiro, Tosu, Saga
841-8686 Japan
+81 0942-83-2101, Fax +81 0942-
83-6119
www.hisamitsu.co.jp

Hi-Yield Chemical Co.
PO Box 460
Bonham, TX, 75418
Hoechst AG
Entwicklung TH 1,
D-65926 Frankfurt/M, Germany
49-69-305-2298, Fax 49-69-318435
www.hcc.com/

Hoechst/Fine Chemicals
5200 77 Center Dr.,
Charlotte, NC 28217, USA
704-559-6000, 800-242-6222
x6183,
Fax 704-559-6153
www.hcc.com/

Hoechst/Int'l. Headquarters
PO Box 2500, Rt. 202-206
North Somerville, NJ 08876-1258,
USA
908-231-2000, 800-235-2637
www.hcc.com/

Hoechst Celanese
50 Meister Avenue
Somerville, NJ 08876

Hoechst-Roussel Pharmaceuticals
Route 202-206 North, PO Box 2500
Somerville, NJ 08876-1258, USA
201-231-2000, 800-451-4455
www.hcc.com/

Hoechst Japan Ltd.
New Hoechst Bldg.,
10-16, Akas,Minato-ku,
Tokyo 107, Japan
81-3-3479-5118, Fax 81-3-3479-
6715
www.hcc.com/

Hoechst Mitsubishi Kase,
Hoechst Bldg.,
10-33, Akasaka, Minato-ku,
Tokyo 107, Japan
81-3-3582-8452, Fax 81-3-3582-
2375
www.hcc.com/

Hoechst/Bulk Pharmacueticals
1601 West LBJ Freeway,
PO Box 819005,
Dallas, TX 75381-9005, USA
214-277-4783, Fax 214-277-3858
www.hcc.com/

Hoechst Roussel Veterinary
See: Akzo Nobel Chemicals

Hoechst Chemicals (UK)
Hoechst House, Salisbury Rd.,
Hounslow, Middlesex, TW4 6JH, UK
44-181-570-7712, Fax 44-181-577-
1854
www.hcc.com/

Hoffmann Mineral
Münchenerstrasse 75, PO Box 1460,
D-86633 Neuburg, Germany
49-84-31-53-0, Fax 49-84-31-53-330

Hoffmann-LaRoche Inc.
340 Kingsland St.,
Nutley, NJ 07110, USA
201-909-8332, 800-526-0189,
Fax 201-909-8414

Hoffmann-LaRoche S.A,.
Grenzacherstrasse 124,
CH-4002 Basle, Switzerland
41-61-688.1111, Fax 41-61-688
6590

Hommel GmbH
PO Box 1662
D-59336 Luedingshausen
Germany
+49 (0) 25 91 23 05 0; Fax +49 (0)
25 91 44 13
www.hommel-pharma.com

Honeywell
Morris Township, NJ

Honeywill & Stein
Times House, Throwleyway
Sutton, Surrey, SM1 4AF, UK
+44 2- 8770 7090, Fax +44 20 8770
7295

Honolulu Wood Treating Co.
91-291 Hanua St.
Kapolei, HI, 96707
(808) 682-5704

Hooker Chemical
See: Occidental Chemcal

Hopkins
PO Box 7532
Madison, WI 53707

Hortichem Ltd.
1b Mills Way Boscombe Down
Business Park
Amesbury Wilts SP4 7RX, UK, SP4
7RX
44 1980 676500; Fax: 44 1980
626555
hortichem@hortichem.co.uk

Howard Hall

Hüls America Inc.
220 Davidson Ave.,
Somerset, NJ 08873, USA
908-560-6345, 800-631-5275
www.huls.com

Huber, J. M. Corp.
333 Thornall St,
Edison, NJ 8818, USA
908-549-8600, Fax 908-549-2239
www.huber.com/

Huber, J. M. Corp.
PO Box 310,
701 Fontain Street,
Havre de Grace, MD 21078, USA
410-939-3500, Fax 410-939-7301
www.huber.com/

Huber, J. M. Corp./Carbon
PO Box 2831,
Borger, TX 79008-2831, USA
806-274-6331, 800-631-6331
www.huber.com/

Huber, J. M. Corp./Clay
One Huber Rd.,
Macon, GA 31298, USA
912-745-4751, Fax 912-745-1116
www.huber.com/

**Huber, J. M. Corp./Engineered
Materials**
1100 Penn Ave.,
PO Box 2831,
Borger, TX 79008-2831, USA
806-274-6331
www.huber.com/

**Huber, J. M. Corp./Engineered
Materials**
One Huber Rd.,
Macon, GA 31298, USA
912-745-4751, Fax 912-745-1116
www.huber.com/

**Huber, J. M. Corp./Engineered
Materials**
4940 Peachtree Industrial Blvd, Suite
340,
Norcross, GA 30071, USA
404-441-1301, Fax 404-368-9908
www.huber.com/

Humphrey

Huntington Laboratories
970 East Tipton St.,
Huntington, IN 46750, USA
219-356-8100, 800-537-5724, Fax
219-356-6485

Hydrite Chemical Co.
300 N Patrick Blvd.
Brookfield, WI, 53045
(414) 792-1450
www.hydrite.com

Hyland Div. Baxter
Hyland Therapeutics,
444-W Gelnoaks Blvd,
Glendale, CA 91202, USA

Manufacturer and Supplier Directory

Hynson Westcott & Dunning
See: Becton Dickinson

I.G. Farben
See: Bayer, Hoechst, BASF

IBC Manufacturing Co.
416 E. Brooks Rd.
Memphis, TN, 38109
(901) 344-5316

ICI Agrochemicals
See: ICI Plant Protection

ICI Americas Inc.
1000 Uniqema Blvd.,
New Castle, DE, 19720
(302) 574-1659
www.ici.com

ICI Chemicals & Polymers
475 Creamery Way,
Exton, PA 19341, USA
610-363-4737, 800-ICI-PTFE, Fax
610-363-4748
www.ici.com

ICI Chemicals & Polymers
PO Box 14, The Heath,
Runcorn, Cheshire, WA7 4QF, UK
44-1928-514444, Fax 44-1928-515555
www.ici.com

ICI Garden Products
Fernhurst,
Haslemere, Surrey, GU27 3JE, UK
44-1428-645454, Fax 44-1428-657222
www.ici.com

ICI Plant Protection
Jealott's Hill Station
Berkshire, UK
www.ici.com

ICI Specialty Chemicals
Concord Pike & New Murphy Rd.,
Wilmington, DE 19897, USA
302-886-3000, 800-822-8215, Fax
302-886-2972
www.ici.com

ICI Surfactants (Australia)
Newsom St.,
Ascot Vale, Vic., 3032, Australia
61-3-9272-5355
Fax 61-3-9272-5353
www.ici.com

ICI Surfactants (Belgium)
Everslaan 45,
B-3078 Everberg, Belgium
32-2-758-92 11, Fax 32-2-758-96 86
www.ici.com

ICI Surfactants America
Delaware Corporate Center 1,
1 Righter Pkwy.,
Wilmington, DE 19803, USA
302-887-3000, 800-822-8215,
Fax 302-887-3525
www.ici.com

ICN Pharmaceuticals Inc.
ICN Plaza,
3300 Hyland Ave,
Costa Mesa, CA 92626, USA
714-545-0100, 800-556-1937

ICN Biomedical Research Products
1263 S. Chilicothe Rd.,
Aurora, OH, 44202
(330) 562-1500
www.icnbiomed.com

IFF International Flavors &
Fragrances
600 Highway 36
Hazlet, NJ 07730
(732) 264-4500, Fax (732) 335-3551
www.imcglobal.com

IMC Fertilizer
IMC Corp.
Lake Forest, IL
www.imcglobal.com

Industrias Quimicas del Valles SA
Avenida Rafael de Casanova 81,
Mollet del Vallés,
E-08100 Barcelona, Spain
34-3-570-56-96, Fax 34-3-593-80-11

Industrie Chimche Caffaro
via Friuli, 55
Cesano Maderno, Italy, I

Inolex Chemical Co.
Jackson & Swanson Sts.,
Philadelphia, PA 19148-3497, USA
215-271-0800, 800-521-9891, Fax
215-289-9065

Inst. Gentili S.p.A.
Pisa, Italy

Integra Life Sciences
311 Enterprise Drive
Plainsboro, NJ 08536
(609) 275-0500, Fax (609) 275-3684

**Integrated Environments
International Inc.**
9119 Alondra Blvd.
Bellflower, CA, 90706

International Dioxcide
544-Ten Rod Road,
North Kingstown, RI 02852-4220,
USA
908-499-9660, 800-477-6071, Fax
908-388-3648

International Gallium GmbH

International Hormones

International Paint, Inc.
2270 Morris Ave.
Union, NJ, 7083
(908) 964-2288

Intervet Inc.
405 State St.
Millsboro, DE, 19966
(302) 934-8051
www.intervet.com

Ishihara Sangyo
660 White Plains Road, #340
Tarrytown NY 10591
(914) 333-7800, Fax (914) 333-7848
www.ishihara.com

Ishihara Sangyo Kaisha, Ltd.
3-15, Edobori 1-chome
Nishi-ku, Osaka 550 Japan
+81 444-1451, Fax +81 445-7798

ISP
1361 Alps Rd.,
Wayne, NJ 07470-3688, USA
201-628-4000, 800-522-4423,
Fax 201-628-4117
www.ispcorp.com

ISP (Österreich) GmbH
Belvederegasse 18/1,
A-1040 Vienna, Austria
43-1-504-76-21,Fax 43-1-505-89-44
www.ispcorp.com

ISP (Australasia) Pty.
73-75 Derby St., Silverwater,
Sydney N.S.W., 2141, Australia
61-2-648-5177, Fax 61-2-647-1608
www.ispcorp.com

ISP (Canada) Inc.
1075 The Queensway East,
Box 1740, Station B,
Mississauga, Ont., L4Y 4C1, Canada
905-277-0381, Fax 905-272-0552
www.ispcorp.com

ISP Asia Pacific Pte,
200 Cantonment Rd.,
Hex 06-07 Southpoint,
0208 Singapore
65-224-9406, Fax 65-226-0853
www.ispcorp.com

ISP Europe
40 Alan Turing Rd.,
Surrey Research Park,
Guildford, Surrey, GU2 5YF, UK
44-1483-301757, Fax 44-1483-302175
www.ispcorp.com

Manufacturer and Supplier Directory

ISP Van Dyk, Inc.
11 William St.,
Belleville, NJ 07109, USA
201-450-7722, Fax 201-751-2047
www.ispcorp.com

Israel Chemicals Ltd.
23 Aranha Street, Millennium Tower
PO Box 20245, Tel Aviv
Israel 61202
03-6844401, Fax 03-6844428
www.israelchemicals.com

Itochu International Inc.
335 Madison Avenue
New York, NY 10017

J.C. Chemical Co.
1725 S. Harvard Blvd.
Los Angeles, CA, 90006

Jan Dekker
Jan Dekker BV, PO Box 10
Wormeveer, Netherlands, 1520 AA
31 75 647-9999; 31 75 640-3030
www.jandekker.com

Janssen Chimica
See: Acros

Janssen Pharmaceuticals
1125 Trenton-Harbourton Road,
P.O. Box 200,
Titusville, NJ 08560-0200, USA
609-730-2000, 800-253-3682, Fax
609-730-3044

Janssen Pharmaceutical Belgium
Janssen Pharmaceuticalaan 3,
B-2440 Geel, Belgium
32-14-60-420, Fax 32-14-60-4220

Jarchem Industries Inc.
414 Wilson Avenue
Newark, NJ 07105
(973) 344-0600, Fax (973) 344-5743
www.jarchem.com

Jones, JCI, Chemicals, Inc.
100 Sunny Sol Blvd.
Caledonia, NY, 14423
(716) 538-2314
www.jcichem.com

Jin Hung
543-6, Kajwa 3-Dong
Seo-Ku, Inchon 404253
Republic of Korea

Johnson & Johnson-Merck Consumer Pharm.
Camp Hill Rd.,
Ft. Washington, PA 19034, USA
215-233-7700
Fax 215-233-8315
www.jnj-merck.com

Johnson Matthey Inc.
17370 N. Laurel Park Dr.,
Suite 400 East,
Livonia, MI 48152, USA
313-591-4031, Fax 313-591-4032

Johnson Matthey plc,
York Way,
Royston, Herts., SG8 5HJ, UK
44-1763-253200, Fax 44-1763-
253492

Johnson Matthey plc,
Elton House, North
Powell Street,
Birmingham B1 3DD, UK
44-121-693 3555, Fax 44-121-236-
3351

Johnson Matthey GmbH Alpha
Postfach 6540,
Zeppelinstrasse 7
D76185 Karlsruhe, Germany
49-6196-7038-21, Fax 49-6196-
7038-012

Johnson, S. C. & Son Co.
1525 Howe Street
Racine, WI 53403-5011
(800) 494-4855

Jonas
1037 Broadway
Denver, CO, 80203

Jungbunzlauer AG
St. Alban-Vorstadt 90
CH-4002 Basel, Switzerland
+41 61 295 5100, Fax +41 61 295
5108
www.jungbunzlauer.com

Jungle Lab.
P.O. Box 630,
Cibolo, TX, 78108
(210) 658-3503
www.junglelabs.com

Kao Corp.
14-10, Nihonbashi,
Kayabacho 1,
Chuo-ku,Tokyo 103, Japan
81-3-3660-7111
chemicals@kao.co.jp

Kao Corp. S.A.
Puig dels Tudons, 10,
08210 Barbera Del Valles,
Barcelona, Spain
34-3-729-0000, Fax 34-3-719 0534,

Kaopolite, Inc.
2444 Morris Ave.,
Union, NJ 07083, USA
908-789-0609, Fax 908-851-2974

Karlshamns AB
S 37482 Karlshamn, Sweden
46-454-82000, Fax 46-454-18453

Kay Chemical Co.
PO Box 18407
Greensboro, NC, 27419
(910) 668-7290

Kelco
8355 Aero Drive,
PO Box 23576,
San Diego, CA 92123-1718, USA
619-292-4900, 800-535-2656,
Fax 619-467-6520
www.monsanto.com

Kelco International
Tadworth Surrey,
Waterfield, KT20 5HQ, UK
44-1737-377000, Fax 44-1737-
377100

Kelco International
Les Mercuriales,
40 Rue Jean Jaures,
93176 Bagnolet Cedex, France
33-1-49-72-2800, Fax 33-1-43-62-
8038

Kemichrom
Union Derivan, SA (UNDESA)
Avda Generalitat 175-179
08840 Viladecans, Barcelona, Spain
34-93-637-3537; Fax: 34-93-659-
1902
www.undesa.com

Kemira Agro OY
PO Box 330, Porkkalankatu 3
Helsinki, Finland, FIN-00101
358 10 861 511; Fax: 358 10-862
1126
www.kemira-agro.com

KenoGard
Diputació, 279
08007 Barcelona, Spain
34-93 488 12 70; Fax: 34-93 487 38
45
www.kenogard.es

Kerr-McGee Chemical
PO Box 25861,
Oklahoma City, OK 73125
405-270-1313, 800654-3911,
Fax 405-270-3123
www.kerr-mcgee.com

Kerr-McGee Chemical Europe
Hohegrabenweg 87,
40667 Meerbusch-Buderich,
Düsseldorf, Germany
www.kerr-mcgee.com

Killgerm Chemicals Ltd.
Denholme Drive,
Ossett, West Yorkshire, WF5 9BW,
UK
44-1924-277631

Manufacturer and Supplier Directory

Kincaid Enterprises
PO Box 549, Plant Rd.,
Nitro, WV 25143, USA
304-755-3377, Fax 304-755-4547

KMG-Bernuth Inc.
KMG Chemicals
10611 Harwin Drive, #402
Houston, TX 77036-1534
(713) 988-9252, Fax (713) 988-9298

Knoll AG
Knoll Pharmaceutical,
30 N Jefferson Rd,
Whippany, NJ 07981, USA
201-887-8300, 800-526-0710

Koninklijke Nederlandsche Gist-En Spiritusfabriek
See: DSM

Kop-Coat Inc.
436 7th. Ave.
Pittsburgh, PA, 15219
(412) 826-3387
www.kop-coat.com

Koppers Industries Inc.
436 7th. Ave.
Pittsburgh, PA, 15219
(412) 227-2424
www.koppers.com/

Koster Keunen Inc
1021 Echo Lake Road
PO Box 69
Watertown, CT 06795-0069
(860) 945-3333, Fax (860) 945-0330
www.kosterkeunen.com

Kraeber GmbH & Co.
Waldhofstrasse 14
25474 Ellerbek, Germany
+49 4101 30530, Fax +49 4101
305390
www.kraeber.de

Kraft Foods Inc.
Northfield, IL
www.kraft.com

Krishi Rasayan
Block - A/11; 4th. Fl. FMC Fortuna,
234/3A
AJC Bose Rd. Calcutta,
W. Benzal, 700020
247-5719; 247-3760;247-5731; Fax:
247 1436
krishi@giascal.01.vsnl.net.in

Kunshan Chemical Material Co. Ltd
Zhangpu Town, Kunshan
Jiangsu Province, China 215321
+86 0512-57441852, Fax +86 0512
57446052
www.kunshanchem.com

KV Pharmaceutical Co.
2503 S. Hanley Road
St. Louis, MO 63144
(314) 645-6600, Fax (314) 644-2419
www.kvpharmaceutical.com

La Littorale
220-19 Quai Duport Neuf
Bezires, Cedex, 34502 France

La Quinoleine SA
4 Allee Annand Camus,
Rueil Malmaison Cedex,
France, F-92565

La-Co Industries Inc.
1201 Pratt Blvd.
Elk Grove Village, IL 60007
(847) 956-7600, Fax (847) 956-9885
www.laco.com

Laevosan GmbH
Estermannstr. 17
4021 Linz, Austria
+43 (0) 732 76510, Fax +43 (0) 732
782833

Lambson Ltd.
Aire & Calder Works,
Cinder Lane,
Castleford, West Yorkshire, WF10
1LU, UK
44-1977-510511, Fax 44-1977
603049

Lanaetex Products Inc.
151 3rd. Street
Elizabeth, NJ 07206-1841
(908) 351-9700

Lancashire Chemical
High Street West,
Glossop, Derbyshire, SK13 8ES, UK
44-1457-860006
Fax 44-1457-868394

Lancaster Synthesis Co.
See: Clariant Corp.
PO Box 1000,
Windham, NH 03087-9977
(603) 889-3306, Fax (603) 889-3326
www.clariant.com

Lanco Manufacturing Corp.
Aponte 6,
San Lorenzo, PR, 754

Langley Smith & Co.
36 Spital Square,
London, E1 6DY, UK
44-171-247-7473, Fax 44-171-375-
1470

LaRoche Industries Inc.
1100 Johnson Ferry Rd. NE,
Atlanta, GA 30342, USA
404-851-0300, Fax 404-851-0476
www.larocheind.com

Lawn & Garden Products
P O Box 5317,
Fresno, CA 93755, USA
209 225 4770, Fax 209 225 1319

Lederle Laboratories
One Cyanamid Plaza,
Wayne, NJ 07470, USA
914-735-2815
www.ahp.com

Lederle Laboratories
Professional Services Dept.,
Pearl River, NY 10965, USA
914-735-2815
www.ahp.com

Lever Industrial Ltd.
P O Box 20, Cressex Industrial,
High Wycombe, Bucks., HP12 3TL,
UK
44-1494-461234, Fax 44-1494-
462565

Lever Industrial Ltd.
45 River Rd..
Edgewater, NJ, 07020
(201) 943-7032

Lever Industriel
103 Rue DeParis,
9300 Bobigny, France

Lipha Pharm. Inc.

LiphaTech, Inc.
3600 W. Elm St.,
Milwaukee, WI 53209, USA
414-351-1476, 800-558-1003,
Fax 414-351-1847
www.liphatech.com

Lipo
800 Scholz Dr.
Vandalia, OH, 45377
(513) 264-1222

Livingston Group Inc.
4768 Hermitage Rd.
Virginia Beach, VA, 23455
(757) 460-3115

Lohmann

Lonza France SARL
55, rue Aristide Briand,
F-92309 Levallois-Per, France
33-1-40-89-9925, Fax 33-1-40-89-
9921
www.lonza.com

Lonza Inc.
17-17 Route 208,
Fair Lawn, NJ USA
201-794-2400, 800-777-1875,
Fax 201-794-2695
www.lonza.com

Manufacturer and Supplier Directory

Lonza Japan Ltd.
Kyowa Shinkawa Bldg., 8th Fl.,
20-8, Shinkawa 2-chome,
Chuo-ku,Tokyo 104, Japan
81-3-5566-0612, Fax 81-3-5566-
0619
www.lonza.com

Lonza UK Ltd.,
Imperial House,
Lypiatt Road,
Cheltenham, Gloucestershire, GL50
2QJ, UK
44-1242-513211, Fax 44-1242-
222294
www.lonza.com

Lonza SpA
via Vittor Pisani, 31,
I-20124 Milan, Italy
39-2-66-9991, Fax 39-2-66-98-7630
www.lonza.com

Lonza Ltd./Fine Chemicals
Münchensteinerstrasse 38,
CH-4002 Basle, Switzerland
49-41-61-316-8111, Fax 49-41-61-
316-8301
www.lonza.com

Lowi
Lucta SA
PO Box 533, Kobe Port
Japan
+81 78 2222313, Fax +81 78
2222312
www.lucta.com.cn

Ludger Ltd.
Oxford BioBusiness Center
Littlemore Park
Oxford OX4 4SS, UK
+44 (0) 870-085-7011,
Fax +44 (0) 870-163-4620
www.ludger.com

Lukens

Luxembourg Industries Inc.
PO Box 13, 27 Hamered St.,
Tel Aviv, 68125 Israel
+972 3 796-4300, Fax +972 3 510-
0474
www.luxpam.com
Luxembourg-Pamol Inc.
5100 Poplar Avenue
Clark Tower, Suite 2700, PMB 111
Memphis, TN 38137
(713) 661-8800, Fax (713) 661-3299
www.luxpam.com

Lyondell Chemical Co.
PO Box 3646
Houston, TX 77253-3646
www.equistar.com

M & S Research
11500 W. Hill Dr.
Rockville, MD, 20852
(301) 468-1668

M&T Harshaw

Maag
PO Box 6430
Vero Beach, FL, 32961
(407) 567-7506

Madis Dr. Labs

Maggioni Farmaceutici S.p.A.

Magnesia GmbH
Kurt Höbold Str. 6
21337 Lüneburg, Germany
+49 4131-8710-0, Fax +49 4131-
8710-55
www.magnesia.de

Magnesium Elektron
500 Point Breeze Rd.,
Flemington, NJ, 08822, USA
908-782-5800, 800-366-9596, Fax
908-782-7768

Makhteshim-Agan
245 5th. Ave. Suite 1901
New York, NY, 10016
(212) 661-9800
www.main.co.il

Makhteshim Chemical Works Ltd.
PO Box 60,
Beer-Sheva,
84100, Israel,
972-7-296611, Fax 972-7-280304
www.main.co.il

Mallinckrodt Inc.
Brakesplan Road South, Harefie,
Uxbridge, Middlesex, UB9 7LS, UK
www.mallchem.com

Mallinckrodt Canada
7500 Trans Canada Hwy.,
Pointe Claire, PQ, H9R 5H8, Canada
514-695-1220
www.mallchem.com

Mallinckrodt Laboratory Chemicals
Postfach 1268,
Industriestrasse 19-21,
D-64802 Dieburg, Germany
49-6071-20040, Fax 49-6071-
200444
www.mallchem.com

Mallinckrodt Laboratory Chemicals
2443 Warrenville Rd,
Lisle, IL 60532, USA
708-955-4555
www.mallchem.com

Mallinckrodt Laboratory Chemicals
4100 North Elston Ave,
Chicago, IL 60618, USA
312-478-1118
www.mallchem.com

Mallinckrodt Laboratory Chemicals
222 Red School Street,
Phillipsburg, NJ O8865, USA
908-859-6916, 800-354-2050,
Fax 908-859-6916
www.mallchem.com

Mallinckrodt Veterinary
PO 5840,
St Louis, MO 63134, USA
314-654-2000,1-888-744-1414,
Fax 314-654-8410
www.mallchem.com

Mallinckrodt, Inc.
Mallinckrodt & 2nd Street,
PO Box 5439,
St Louis, MO 63147, USA
314-895-2000, 800-325-8888,
Fax 314-539-1251
www.mallchem.com

Mandoval Ltd.
Douglas Drive, Catteshall Lane
Godalming, Surrey GU7 1JX, UK
+44 1483 425326, Fax +44 1483
413947
www.mandoval.co.uk

Manuel Vilaseca SA
Roselló 188
08008 Barcelona, Spain
+34 934 512 729

Marion Merrell Dow Inc.
9300 Ward Parkway,
P.O. Box 8480,
Kansas City, MO 64114-0480, USA
800-552-3656

Marjon & Associates
PO Box 791, Maricopa, CA, 93252
(661) 769-8517

Marks, A. H. & Co. Ltd,
Wyke Lane, Wyke,
Bradford, West Yorkshire, BD12 9EJ,
UK
44-1274-691234, Fax 44-1274-
691176
www.ahmarks.com

Mauget, J. J. Co.
5435 Peck Road
Arcadia, CA 91006-5847
(626) 444-1057

Maxicrop International Ltd.
SeedQuest
1353 El Centro Avnue
Oakland, CA 94602-1817
(510) 482-5560, Fax (510) 482-5696
www.seedquest.com

May & Baker Ltd.
Rainham Road South,
Dagenham, Essex, RM10 7XS, UK
44-118-592-3060

Manufacturer and Supplier Directory

McGaw Inc.
B. Braun Medical Inc.
2525 McGaw Avenue, PO Box
19791
Irvine, CA 92623-9791

McIntyre Group Ltd.
24601 Governors Hwy.,
University Park, IL 60466-4127, USA
708-534-6200, 800-645-6457,
Fax 708-534-6216
www.mcintyregroup.com

McIntyre Chemicals Ltd.
Blk 513, Bishan Town Centre St., 13
#01-500 Singapore 570513
65-354-3547, 65-354-0353, Fax 65-
353-9068
www.mcintyregroup.com

McKechnie Chemicals
PO Box 4, Tanhouse Lane,
Widnes, Cheshire, WA8 0PG, UK
44-151-424-2611, Fax 44-151-424-
4221

Mclaughlin Gormley King Co
8810 10th Ave. N.,
Minneapolis, MN 55427
612-544-0341, Fax 612-544-6437

McNeil Consumer Products
Camp Hill Rd,
Fort Washington, PA 19304
215-233-7000

McNeil Pharmaceuticals
Spring House, PA 19477-0776
215-628-5000

Mead Johnson & Co.
P.O. Box 4500,
Princeton, NJ 08543-4500, USA
609-897-2000, 800-321-1335

Mearl

Mechema Chemicals Ltd.
15 Ta-Tun First Road
Kung-Ying, Tao-Yuan Hsien
Taiwan
+886 3 4833788, Fax +886 3
4833799
www.mechema.com.tw

Medical Chemical Corp.
19430 Van Ness Ave.
Torrance, CA, 90501
(310) 787-6800
www.med-chem.com

Mendell, Edward

Merck & Co., Inc.
PO Box 4,
West Point, PA 19486-0004, USA
800-672-6372
www.merck.com

Merck, E., GmbH
Postfach 4119,
Frankfurter Strasse 250,
D-64293 Darmstadt, Germany
49-6151-72-0, Fax 49-6151-72-2000
www.merck.com

Merck-Eprova AG
Im Laternenacker 5
CH-8200 Schaffhausen
Switzerland
+41 (52) 630 72 72, Fax (+41 (52)
630 72 55
www.eprova.com

Merck & Co., Inc.
PO Box 2000,
Rahway, NJ 07065-0900, USA
908-594-4000, Fax 908-594-5431
www.merck.com

Merck Japan Ltd.
Arco Tower, 8-1,
Shimomeguro 1,
Meguro-ku, Tokyo 153, Japan
81-3-5434-4700, Fax 81-3-5434-
4705
www.merck.com

Merck KgaA,
Postfach 4119,
Frankfurter Strasse 250,
D-64293 Darmstadt, Germany
49-6151-72-0, Fax 49-6151-72-2000
www.merck.com

Merck Ltd.
Merck House,
Poole, Dorset, BH15 1TD, UK
44-1202-669700, Fax 44-1202-
665599
www.merck.com

Merck Pty. Ltd.
207 Colchester Rd.,
Kilsyth, Vic., 3137, Australia
61-3-728-5855, Fax 61-3-728-1351
www.merck.com

Merck Sharp & Dohme/Iso
4545 Oleatha Ave.,
St. Louis, MO 63116, USA
800-325-9034, Fax 314-353-3754
www.merck.com

**Merichem Chemicals & Refinery
Services LLC**
5455 Old Spanish Trail
Houston, TX, 77023-5013
(713) 428-5100
www.merichem.com

Mesa Specialty Gases & Equipment
3619 Pendleton Avenue, Suite C
Santa Ana, CA 92704
(714) 434-7102, Fax (714) 434-8006
www.mesagas.com

Metallgesellschaft GmbH

Metrex Research Corp.
1717 W. Collins Ave.
Orange, CA, 92867
(714) 516-7425
www.metrex.com

M. Michel & Co., Inc.
90 Broad St.,
New York, NY 10004, USA
212-344-3878, Fax 212-344-3880

Micro-Flo Co. LLC
PO Box 772099
Memphis, TN, 38117
(901) 432-5131

Microban Systems Inc.
1135 Braddock Ave.
Braddock, PA, 15104
(412) 461-8686

Midori Kagaku
1-25-1 Higashi-Ikebukuro
Toshimaku, Tokyo 170-0013, Japan
+81 03-3980-8808, Fax +81 03-
3980-8805
www.midori-kagaku.co.jp

Miles Inc.
Bayer Inc., Mobay Rd.,
Pittsburgh, PA 15205-9741, USA
412-777-2000, 800-662-2927,
Fax 412-777-2608

Bayer Inc.,
400 Morgan Lane,
West Haven, CT 06516-4175, USA
203-937-2000, 800-468-0894,
Fax 412-394-5578

Miles Inc./Polysar
2603 W. Market St.,
Akron, OH 44313, USA
216-836-0451, 800-321-0997,
Fax 216-836-0200

Miles Inc./Polyurethane
Mobay Rd.,
Pittsburgh, PA 15205-9741, USA
412-777-2000, 800-662-2927

Miles Laboratories Inc.
PO Box 390,
Shawnee, KS 66201, USA

Miles Ltd./Biotechnology
PO Box 37, Stoke Court,
Stoke Poges, Slough, Berks., SL2 4LY,
UK
44-1281-45151, Fax 44-1281-43893

Miljac Corp.
280 Elm Street
New Canaan, CT 06840
(203) 966-8777, Fax (203) 966-3577
www.miljac.com

Manufacturer and Supplier Directory

Miller, Frank & Sons, Inc.
13831 S. Emerald Avenue
Riverdale, IL 60827
(708) 201-7200, Fax (708) 841-8073
fmillers@worldnet.att.net

Milliken Chemical
PO Box 1927, M-400
Spartanburg, SC, 29304
(864) 503-2200, Fax: (864) 503-2430
www.millikenchemical.com

Milliken Chemical N.V.
18-24 Ham,
B-9000 Gent, Belgium
32-9-265-1082, Fax 32-9-265-1195

Mirachem Srl.
See: Reckitt Benckiser Inc.
(800) 847-3527

Mirfield Sales Services Ltd.
Moorend House, Moorend Lane,
Dewsbury, West Yorkshire, WF13
4QQ, UK
44-1484-842851, Fax 44-1484-
847066

Mission Pharmacal Co.
1325 East Durango,
San Antonio, TX 78210, USA
512-533-7118

Mitchell Cotts Chemical Ltd.
PO Box No 6, Steanard Lane,
Mirfield, West Yorkshire, WF14 8QB,
UK
44-1924-493861, Fax 44-1924-
490972

Mitsubishi Chemical
Mitsubishi Bldg.,
5-2, Marunou,
Chiyoda-ku,Tokyo 100, Japan
81-3-3283-6531, Fax 81-3-3283-
6658

Mitsubishi Chemical
Niederkasseler Lohweg 8,
D-40547 Düsseldorf, Germany
49-211-52392-0, Fax 49-211-591272

Mitsubishi Gas Chemical
Mitsubishi Bldg.,
5-2, Marunouchi, 2-chome
Chiyoda-ku,Tokyo 100, Japan
81-3-3283-5000, Fax 81-3283-5120

Mitsubishi Gas Chemical
520 Madison Ave., 17th Floor,
New York, NY 10022, USA
212-752-4620, Fax 212-758-4012
www.mgc-a.com

Mitsubishi International
520 Madison Ave.,
New York, NY 10022-4223, USA
212-605-2193, 800-442-6266,
Fax 212-605-1704

Mitsubishi Kasei Corp.
Mitsubishi Bldg.,
5-2, Marunou,
Chiyoda-ku,Tokyo 100, Japan
81-3-3283-6254

Mitsubishi Kasei Poly.
Mitsubishi Bldg.,
5-2, Marunou,
Chiyoda-ku,Tokyo 100, Japan
81-3-3283-4405, Fax 81-3-3283-4480

Mitsubisho Materials
1-5-1, Ohte-machi,
Chiyoda-ku,Tokyo 100, Japan
81-3-5252-5200, Fax 81-3-5252-
5270/1

Mitsubishi Oil Co.
Sanyu Bldg.,
2-4, Toranomon 1-,
Minato-ku, Tokyo 105, Japan
81-3-3595-7663, Fax 81-3-3508-
2521

Mitsubishi Petrochemical
Mitsubishi Bldg.,
5-2, Marunou,
Chiyoda-ku,Tokyo 100, Japan
81-3-3283-5700, Fax 81-3-3283-
5472

Mitsui Petrochemical
250 Park Ave., Suite 950,
New York, NY 10017, USA
212-682-2366, Fax 212-490-6694

Mitsui Petrochemical
Kasumigaseki Bldg.,
2-5, Kasum,
Chiyoda-ku,Tokyo 100, Japan
81-3-3580-3616, Fax 81-3-3593-
0028

Mitsui Toatsu Chemicals
Kasumigaseki Bldg.,
2-5, Kasum,
Chiyoda-ku,Tokyo 100, Japan
81-3-3592-4111, Fax 81-3-3592-
4267

Mitsui Toatsu Chemicals
2500 Westchester Ave., Suite 1,
Purchase, NY 10577, USA
914-253-0777, Fax 914-253-0790

MLG Enterprises
PO Box 52568
Turtle Creek Post Office
Mississauga, ON L5J 4S6
Canada
(905) 696-6947, Fax (905) 696-6955

Mobay
See: Bayer Corp.

Mobil Chemical Co.
PO Box 3029,
Edison, NJ 08818-3029, USA
908-321-6000
www.mobil.com

Mobil Chemical Co./Films
1150 Pittsford-Victor Rd.,
Pittsford, NY 14534, USA
800-654-3436, 800-828-6381
www.mobil.com

Mobil Chemical Co./Petrochemicals
15600 J.F. Kennedy Blvd., Suite 800,
Houston, TX 77032, USA
713-590-7700, Fax 713-590-7908
www.mobil.com

Mobil Chemical Co./Polystyrene
Rt. 27 & Vinyard Rd., Box 3029,
Edison, NJ 08818-3029, USA
908-321-3500, 800-922-0380,
Fax 908-321-3501
www.mobil.com

Mobil Mining & Minerals
P.O. Box 26683,
Richmond, VA 23261, USA
804-798-4291
www.mobil.com

Mobil Oil Corp., Spec. Products
3225 Gallows Rd.,7W004,
Fairfax, VA 22037-0001, USA
703-849-3609, 800-662-4525, Fax
703-849-6637
www.mobil.com

Mobil Plastics Europe
Zoning Industriel de Latour,
B-6761 Virton, Belgium
32-63-21 32 11, Fax 32-63-21 34 24
www.mobil.com

Molekula Ltd.
Technology Road
Poole, Dorset BH17 7DA, UK
+44 (0) 1202 330066, Fax +44 (0)
1202 330055
www.molekula.com

Molekula USA
9648 Olive Blvd., Suite 239
St. Louis, MO 63132-3002
(314) 503-2245, Fax ((314) 569-0893
www.molekula.com

Mona Industries Inc.
PO Box 425,76 E. 24th St.,
Paterson, NJ 07544, USA
201-345-8220, 800-553-6662,
Fax 201-345-3527
www.monaweb.com

**Monomer-Polymer & Dajac
Laboratories Inc.**
1675 Bustleton Pike,
Feasterville, PA 19053, USA
215-364-1155, Fax 215-364-1583

Manufacturer and Supplier Directory

Monsanto Australia Ltd.
600 St. Kilda Rd.,12th Floor,
Melbourne,Vic.a, 3004, Australia
61-3-522-7122, Fax 61-3-525-2253
www.monsanto.com

Monsanto Canada Inc.
2330 Argentia Rd.,Box 787,
Streetsville Postal Station,
Mississauga, Ont., L5M 2G4, Canada
905-826-9222, Fax 905-826-3119
www.monsanto.com

Monsanto Chemical Co.
800 N. Lindbergh Blvd.,
St. Louis, MO 63167, USA
314-694-1000, 800-325-4330,
Fax 314-694-7625
www.monsanto.com

Monsanto Commercial
Bosques de Durazno 61,
3R Piso,Bosques de las Lomas,
Mexico City, DF 11700, Mexico
52-5-251-2715, Fax 52-5-251-7923
www.monsanto.com

Monsanto Plastics
800 N. Lindbergh Blvd.,
St. Louis, MO 63167, USA
800-325-4330
www.monsanto.com

Monsanto plc, Agriculture
Thames Tower, Burleys Way,
Leicester, Leicestershire, LE1 3TP, UK
44-1162 620864, Fax 44-1162
530320
www.monsanto.com

Monsanto, Detergents   Morton
Rue Laid Burniat,
1348 Louvain la Neuve, France,
www.monsanto.com

Morflex Inc.
2110 High Point Rd.
Greensboro, NC, 27403
(919) 292-1781
www.morflex.com

Morton International
150 Andover St.,
Danvers, MA 01923-1480, USA
508-774-3100, Fax 508-750-9512
www.mortonintl.com

Morton International
100 North Riverside Plaza,
Chicago, IL 60606, USA
312-807-2000
www.mortonintl.com

Morton International
100 North Riverside Plaza,
Chicago, IL 60606-1598, USA
312-807-2000, Fax 312-807-3150
www.mortonintl.com

Morton International
Westward House,155-157 Staines
Rd.,
Hounslow, Middlesex, TW3 3JB, UK
44-181-570-7766, Fax 44-181-570-
6943
www.mortonintl.com

Morton International
7900-A Taschereau Blvd.,
Brossard, PQ, J4X 1C2, Canada
514-466-7764, Fax 514-466-7771
www.mortonintl.com

Morton International
2700 E. 170 St.,
Lansing, IL 60438, USA
708-474-7000, 800-323-3224,
Fax 708-868-7490
www.mortonintl.com

Morton International
10 S. Electric St.,
West Alexandria, OH 45381, USA
513-839-4612, 800-348-8846,
Fax 513-839-5615
www.mortonintl.com

Morton International
150 Andover St.,
Danvers, MA 01923-1480, USA
508-774-3100, Fax 508-750-9512
www.mortonintl.com

Morton International
100 North Riverside Plaza,
Chicago, IL 60606, USA
312-807-2000
www.mortonintl.com

Morton International
2000 West St.,
Cincinnati, OH 45215, USA
513-733-2100, Fax 513-733-2133
www.mortonintl.com

Morton International
PO Box 3089,
130 Mountain Creek Church Rd.,
Greenville, SC 29602, USA
803-292-5700, 800-845-6810, Fax
803-292-5713
www.mortonintl.com

Morton International
Greville House, Hibernia Road,
Hounslow, Middlesex, TW3 3RX, UK
44-181-570-7766, Fax 44-181-570-
6943
www.mortonintl.com

Morton International
Oppauer Str 43,
D-68305 Mannheim, Germany
49-621-76260
www.mortonintl.com

Mosselman
NV Organon
PO Box 20, Oss
Molenstraat 110 Oss
5340 BH, Netherlands

MTM Agrochemicals Ltd.
18 Liverpool Road, Great Sankey,
Warrington, Cheshire, WA5 1QR, UK
44-1925-33232, Fax 44-1925-52679

MTM Spec. Chem. Ltd.

Multi-Kem
(800) 462-4425, (800) 441-7405,
Fax (201) 941-5239

Multisorb Technologies Inc.
325 Harlem Rd.
Buffalo, NY, 14224
(716) 824-8900
www.multisorb.com

Murphy Chemical Co Ltd.
1530 Locust St.
Philadelphia, PA, 19102

Nalco Chemical B.V.
Postbus 5131,
NL-5004 EC Tilburg, The
Netherlands
31-13-63 55 55, Fax 31-13-67 45 45
www.nalco.com

Nalco Chemical Co.,
One Nalco Center,
Naperville, IL 60563-1198, USA
708-305-1041, 800-527-7753, Fax
708-305-2998
www.nalco.com

Nalco Diversified Technologies, Inc.
PO Box 200
Chagrin Falls, OH, 44022
(216) 247-5000

Napp Laboratories Ltd.
Cambridge Science Park, Milton,
Cambridge, Cambridgeshire, CB4
4BH, UK

National Chelating Co.
221 West Meats Ave.
Orange, CA, 92865-2621
(714) 637-5056

**National Cottonseed Products
Association**
www.cottonseed.com

National Lead Co.
NL Industries Inc.
5430 LBJ Freeway, Suite 1700
Dallas, TX 75240-2697
(972) 233-1700, Fax (972) 448-1445
www.nl-ind.com

Manufacturer and Supplier Directory

Naturex
The Broadway, Tolworth
Surrey KT6 7HR, UK
+44 800 277 818, Fax +44 800 277
8182
www.naturex.co.uk

Naugatuck

Neste UK
See: Perstorp UK Ltd.

Neudorff, W. GmbH KG
Postfach 1209, An der Muhle 3
Emmerthal, Germany, 31860
49 180-5638367; Fax: 49 5155-6010
www.neudorff.de

New Metals & Chems. Ltd.

Niacet, Inc.
400 47th. Street
Niagara Falls, NY 14304
(716) 285-1474, Fax (716) 285-1497
www.niacet.com

Nichia Kagaku Kogyo
491 Oka, Kaminaka-Cho
Anan-Shi, Tokushima 774-8601
Japan
+81 884-22-2311, Fax +81 884-21-
0148
www.nichi.co.jp

Nickerson Seeds Ltd.
JNRC, Rothwell,
Lincolnshire, LN7 6DT, UK
44-1472-89471, Fax 44-1472-89602

Nihon Kagaku Sangyo
20-5, Shitaya 2-chome,
Taito-ku,Tokyo 110, Japan
81-3-3876-3131, Fax 81-3-3876-
3278

Nihon Nohyaku Co., Ltd.
2-5, Nihonbashi 1-Chome,
Chuo-ku,Tokyo 103, Japan
81-3-3278-0461, Fax 81-3-3281-
2443

Niklor Chemical Co. Inc.
2060 E. 220th. St.
Long Beach, CA, 90810-1695
(310) 830-2253

Nipa Hardwicke Inc.
2411 Silverside Rd.,
104 Hagley Bldg.,
Wilmington, DE 19810, USA
302-478-1522, Fax 302-478-4097
www.nipa.com

Nipa Hardwicke Inc./Germany
Hans-Boeckler-Ring 9,
Norderstedt, D-22851, Germany
49-40-529-54-60, Fax 49-40-529-54-
630
www.nipa.com

Nipa Hardwicke Inc./UK
Llantwit Farde, Pontypridd,
Mid Glamorgan, CF38 2SN, Wales
UK
44-443-205311, Fax 44-443-207746
www.nipa.com

**Nippon Nyukazai
Sankyo Corp.**
Yoshizawa Bldg., 9-19,
Ginza 3-chome, Chuo-ku
Tokyo 104, Japan
+81 3543-8571, Fax +81 3546-3174

Nippon Rikagakuyakuhin Corp.
2-12, Nihonbashi-Honcho 4-chome,
Chuo-ku, Tokyo 103, Japan
+81 3241-3557, Fax +81 3242-3345

Nippon Soda Co. Ltd.
Shin Ohtemachi Bldg., 2-1,
Ohte-machi 2-chome, Chiyoda-ku,
Tokyo 101, Japan
+81 3245-6054, Fax +81 3242-2882

Nippon Zeon
Furukawa Building
2-6-1 Marunouchi
Chiyoda-ku, Tokyo 100-8323
Japan
+81-3-3216-1772, Fax +81-3-3216-
0501

Nissan Chem. Ind.
7-1, Kanda Nishiki-cho 3-chome
Chiyoda-ku, Tokyo 101-0054, Japan
+81 3296-8320, Fax +81 3296-8210
www.nissanchem.co.jp

Nissho Iwai American Corp.
700 S Flower St. Suite 1900
Los Angeles, CA, 90017
(202) 546-8424

Noah Technologies Corp.
1 Noah Park
San Antonio, TX 78249-3419
(210) 691-2000, Fax (210) 691-2600
www.noahtech.com

Norac
405 S. Motor Avenue
Azusa, CA 91702
(626) 334-2908, Fax (626) 334-3512
www.norac.com

Nor-Am
2711 Centreville Rd.
Wilmington, DE, 19808
(302) 892-3039

Norit Americas Inc.
1050 Crown Pointe Pkwy.,
Atlanta, GA 30338, USA
404-512-4610, 800-641-9245, Fax
404-512-4622

Norit N.V.
Postbus 105,
NL-3800 AC Amersfoort, The
Netherlands
31-64-8911, Fax 31-61-7429

Norit UK Ltd.
Clydesmill Place,
Cambuslang Industrial Estate,
Glasgow, G32 8RF, UK
44-141-641-8841, Fax 44-141-641-
0742

Norman, Fox & Co.
5511 S. Boyle Ave.,
PO Box 58727,
Vernon, CA 90058, USA
323-583-0016, 800-632-1777, Fax
323-583-9769
www.norfoxx.com

Norsk Hydro AS
Bygdoyalle 2,
N 0240 Oslo 2, Norway
47-2-243-2100
Fax 47-2-243-2725

Norsk Hydro Plast Oy,
Sabiansgatan 12 G49, PL 299,
358-00131 Helsinki, Finland
358-0-660446

Norton Chem. Process Prods.
Box 350
Akron, OH 44309-0350

Novartis Pharmaceutical Corp.
59 Route 10
East Hanover, NJ 07936-1011
USA
(908) 503-7500

Novo Nordisk A/S
Bi, Novo Allé,
DK-2880 Bagsvaerd, Denmark
45-44-44-8888, Fax 45-44-44-60 88
www.novo.dk

Novo Nordisk Bioindustries
33 Turner Rd.,
Danbury, CT 06813-1907, USA
800-251-6686, Fax 203-790-2748
www.novo.dk

Novo Nordisk Bioindustries
Makuhari Techno Garden CB-6,
1-3, Nakase 1-chome,
Chiba 261-01, Japan
81-43-296-6767, Fax 81-43-296-
6767
www.novo.dk

Novo Nordisk Biotech Inc.
1445 Drew Aveenue
Davis, CA 95616
www.novo.dk

Manufacturer and Supplier Directory

Novo Nordisk Pharmaceuticals Inc.
100 Overlook Center, #2
Princeton, NJ 08540-7814
(609) 987-5800
www.novo.dk

Nufarm Americas Inc.
14140 SW Frwy 250
Sugarland, TX, 77478
(281) 295-0100

Nycomed Christiaens S.A.
Gentsesteenweg 615
1080 Brussels, Belgium
+32 2 4640611, Fax +32 2 4640698

NYCO Minerals, Inc.
124 Mountain View Dr.,
PO Box 368,
Willsboro, NY 12996-0368, USA
518-963-4262
Fax 518-963-1110

NYCO Minerals, Inc.
Ordrupvej 24,
PO Box 88,
DK-2920 Charlottenlun, Denmark
45-39-64-3370,Fax 45-39-64-3710

Occidental Chemical
Occidental Tower, 5005 LBJ
Freeway,
Dallas, TX 75244, USA
214-404-3800, 800-752-5151,
Fax 214-404-3669
www.oxychem.com/

Occidental Chemical
Toranomon 34 Mori Bldg., 9th Floor,
25-5 Toranomon 1-Chome,
Minato-ku, Tokyo 105, Japan
81-3-3502-4651, Fax 81-3-3502-
4640
www.oxychem.com

Occidental Chemical
PO Box 344,
Niagara Falls, NY 14302-0344, USA
800-733-1165, Fax 716-278-7297
www.oxychem.com

Occidental Chemical
Suite 2, 16th Floor,
275 Alfred St.,
N. Sidney, NSW, 2060, Australia
61-2-9-957-6722, Fax 61-2-9-957-
1548
www.oxychem.com

Occidental Chemical
Tampa, FL
813-286-3800
www.oxychem.com

Occidental Chemical
PO Box 27702,
Houston, TX 77227-7702, USA
713-623-7615, 800-800-4373
www.oxychem.com

Occidental Chemical
Occidental Tower,
5005 LBJ Freeway,
Dallas, TX 75244, USA
214-404-3300, Fax 214-404-4815
www.oxychem.com

Occidental Chemical
PO Box 809050,
5005 LBJ Freeway,
Dallas, TX 75380-9050, USA
214-404-4932, 800-733-3339,
Fax 214-404-4981
www.oxychem.com

Occidental Chemical
Holidaystraat 5,
B-1831 Diegem, Belgium
32-725-4450, Fax 32-725-4676
www.oxychem.com

Oclassen Pharm. Inc.
See: Watson Pharmaceuticals

Octavius Hunt Ltd.
5 Dove Lane, Redfield,
Bristol, Avon, BS5 9NQ, UK
44-1179-555304, Fax 44-1272-
557875

Octel Chemicals Ltd.
Halebank,
Widnes, Cheshire, WA8 8NS, UK
44-151-424-3671, Fax 44-151-420-
1301

Olin Australia Ltd.
PO Box 141, Level 2, Suite 1,
601 Pacific Hwy.,
St. Leonards, N.S.W., 2065, Australia
61-2-906-4455, Fax 61-2-439-4198
www.olin.com

Olin Brasil Ltd.
Nacoes Unidas, 11.857, 12th Floor,
Sao Paulo, 04578-000, Brazil
55-11-505-0382, Fax 55-11 505-
1950
www.olin.com

Olin Chemicals
501 Merritt 7, PO Box 4500,
Norwalk, CT 06856-4500, USA
203-750-3429, 800-462-6546,
Fax 203-270-3752
www.olin.com

Olin Chemicals USA
PO Box 586,
Cheshire, CT 6410, USA
800-344-9168, Fax 203-271-4060
www.olin.com

Olin Chlor-Alkali Products
650 25th St., Suite 300,
Cleveland, OH 37311, USA
615-336-4850
www.olin.com

Olin Corp./Agriculture
PO Box 991,
Little Rock, AR 72203, USA
www.olin.com

Olin Far East Ltd.
80-6 Soosong-dong,
Suktan Bldg.,
Chongro-ku, Seoul, 110-140, Korea
82-2-737-2840, Fax 82-2-730-7387
www.olin.com

Olin Far East Ltd.
7F-2 No. 137,
Fu Shing S. Road Sec. 1,
Taipei, Taiwan, R.O.C.,
886-2-752-4413, Fax 886-2-741-
2113
www.olin.com

Olin Industrial H.K.
111 Peninsula Centre,
67 Mody Rd.,Tsim Sha Tsui-East,
Kowloon, Hong Kong
852-366-8303, Fax 852-367-1309
www.olin.com

Olin Japan Inc.
Shiozaki Bldg.,
7-1 Hirakawa-Cho 2-Chome, Chiyod,
Tokyo,102, Japan
81-3-3263-4615
www.olin.com

Olin Pte., Ltd.
501 Orchard Rd.,
08-03 Lane Crawford Rd.,
Singapore 0923
65-735-1268, Fax 65-735-1298
www.olin.com

Olin Pty. Ltd.
PO Box 114,
Bergvlei, 2012,
Rep. of S. Africa
27-11-444-2244, Fax 27-11-444-
2241
www.olin.com

Olin Quimica, S.A.
Campos Eliseos No. 385,
Piso 9, Col. Polanco, Delg. Miguel
Hidalgo,
11560, Mexico
52-5-281-2045, Fax 52-5-281-2037
www.olin.com

Olin Research Center
350 Knotter Dr.,
PO Box 586,
Cheshire, CT 06410-0586, USA
203-271-4000, 800-654-6763
Fax 203-271-4060,
www.olin.com

OMG Americas Inc.
811 Sharon Dr.
Westlake, OH, 44145
(216) 575-3679

Manufacturer and Supplier Directory

Ondeo Nalco Co.
Ondeo Nalco Ctr.
Naperville, IL, 60563
(630) 305-1000
www.ondeo-nalco.com

Ore & Chem.

Organon Inc.,
375 Mt Pleasant Ave,
West Orange, NJ 07052, USA
201-325 4500

Original Bradford Soapworks
200 Providence Street
West Warwick RI 02893

Ortho Pharmaceutical Corp.
Route 202
S. Raritan, NJ 08869
(908) 704-1500, Fax (908) 526-4997

Osaka Org. Chem. Ind.
Shin Toyama Bldg., 1-7-20 Azuchi-Machi,
Choku, Osaka, Japan, 541
81 6 6264-5071; Fax: 81 6 6264-1675

Osmose Inc.
980 Ellicott Street
Buffalo, NY 14209
(716) 882-5905, Fax (716) 882-5134

Otsuka Pharm. Co. Ltd.
2-9 Kanda Tsukasa-cho
Chiyoda-ku, Tokyo 101-8535
Japan
+81 3 3292-0021
www.otsuka.co.jp

Outokumpu Oy Riihitontuntie 7
PO Box 140
FI-02201 Espoo, Finland
www.outokumpu.com

Outokumpu Technology Inc.
10771 E. Easter Avenue, Suite 120
Centennial, CO 80112
(303) 792-3110
www.outokumpu.com

Oxiteno S.A.
Av. Brigadeiro Luis Antônio
1343 Bela Vista
Sâo Paulo SP 1350
Brasil
55-11 3177-6043, Fax (55-11 285-0327

Oxychem, Petrochemicals
5 Greenway Plaza, Suite 2400,
Houston, TX 77046, USA
713-623-2246, 800-448-2246,
Fax 713-968-6318
www.oxychem.com

Pace International LLC
1011 Western Ave. Suite 505
Seattle, WA, 98104
(206) 264-7599
www.paceint.com

Pacific Anchor Chemical
1224 Mindon Road,
Cumberland, RI 02864, USA
401-333-4100, Fax 401-333-4630
www.airproducts.com

Pacific Anchor Chemical
3305 E. 26th St.,
Los Angeles, CA 90023, USA
213-264-0311
www.airproducts.com

Pan Britannica Industries
Britannica House,
Waltham Cross, Herts., EN8 7DY, UK
44-1992 23691, Fax 44-1992-26452

Parke-Davis Inc.
201 Tabor Rd,
Morris Plains, NJ 07950, USA
201-540-2000, 800-223-0423,
Fax 201-540-2248
www.parke-davis.com

Parke-Davis Inc.
188 Howard Ave,
Holland, MI 49424, USA
616-392-2375, Fax 616-392-8914
www.parke-davis.com

Parke-Davis Ltd.
Mitchell House, Southampton Row,
Eastleigh, Hants., SO5 5RY, UK
www.parke-davis.com

Patco Additives
American Ingredients Co.
3947 Broadway
Kansas City, MO 64111
(816) 561-9050 x1325, Fax (816) 561-7778
www.patco-additives.com

Parry, E. I. D. (India) Ltd.
234 NSC Bose Road
Parry's Corner
Chennai 600 001, India
+91 044-25340251, Fax +91 044-25341609
www.eidparry.com

PBI/Gordon Corp.
1217 W. 12th. Street
Kansas City, MO 64101
(816) 421-4070, Fax (816) 474-0462

PBT International
13532 Hunting Hill Way
N. Potomac, MD, 20878
(301) 926-9873
www.pbtigroup.com

PCR Inc.
PO Box 1466,
Gainesville, FL 32602-1466, USA
904-376-8246, 352-376-8246, 800-331-6313
Fax 904-371-6246, 352-371-6246

Pennsylvania Engineering Co.
1119-21 N Howard St.
Philadelphia, PA, 19123
(215) 627-3636

Pennwalt Corp.
Elf Atochem N. America
900 First Ave.
King of Prussia, PA, 19406

Pennwalt Italia SpA
via del Porto,
I-28040 Marano-Ticino, Italy
321-9791, Fax 321-979246

Pennwalt Holland
See: Atofina

Penta Manufacturing
PO Box 1448,
Fairfield, NJ 07007-1448, USA
201-740-2300, Fax 201-740-1839

Pentagon Chemicals Ltd.
Northside,
Workington, Cumbria, CA14 1JJ, UK

Pentagon Urethanes Ltd.
Northside,
Workington, Cumbria, CA14 1JJ, UK
44-1900-604371, Fax 44-1900-66943

Penwest Pharmaceuticals Co.
Church House,
48 Church St.,
Reigate, Surrey, RH1 6YS, UK
44-1737-222 323

Penwest Pharmaceuticals Co.
2981 Rt. 22,
Patterson, NY 12563-9970, USA
914-878-3414, 800-431-2457, Fax 914-878-3484

Penwest Pharmaceuticals Co.
Postfach 1207,
25430 Uetersen, Germany
49-6135-951410, Fax 49-6135-951223

Permviro Systems Inc.
Trotter's Creek Ind. Park
3520 Trotter Dr.
Alpharetta, GA, 30004
(770) 667-9800

Manufacturer and Supplier Directory

Perstorp UK Ltd.
Cambridge House
37 Bramhall Lane S., Bramhall
Stockport, Ches., SK7 2DU, UK
+44 (0) 161 439 0900, Fax +44 (0)
161 439 0200

PET Chemicals
PO Box 18993
Memphis, TN, 38181
(901) 362-1950

Pfaltz & Bauer
172 E. Aurora Street
Waterbury CT 06708
(203) 574-0075, Fax (203) 574-3181
www.pfaltzandbauer.com

Pfanstiehl Labs Inc.
See: Ferro-Pfanstiehl

Pfizer Inc.
235 E 42nd. St.
New York, NY, 10017
(800) 832-5778; (212) 573-2323;
Fax: 212-573-1166
www.pfizer.com

Pfizer Food Science
235 E 42nd. St.
New York, NY, 10017
(800) 832-5778; (212) 573-2323;
Fax: 212-573-1166
www.pfizer.com

Pfizer Asia/Australia
PO Box 57,
West Ryde, NSW, 2114, Australia
61-2-858-9500
www.pfizer.com

Pfizer Canada
PO Box 800,
Point Claire/Dorval,
Montreal,PQ, H9R 4V2, Canada
514-695-0500
www.pfizer.com

Pfizer K.K.
3-22 Toranomon 2-chome,
Minato-ku,Tokyo 105, Japan
81-3-3503-0441, Fax (03) 3503-0447
www.pfizer.com

Pfizer S.A.
Principe de Vergara 109,
E-28002 Madrid, Spain
34-1-262 11 00,
www.pfizer.com

Pfizer/Dairy & Brewery
4215 N. Port Washington Rd.,
Milwaukee, WI 53212, USA
414-332-3545, 800-231-1590
www.pfizer.com

Pharmacia
See: Pfizer Inc.

Pharmacia & Upjohn
Pfizer Inc.
95 Corporate Drive
Bridgewater, NJ 08807-1265
(908) 306-4400, Fax (908) 306-4433

Pharmacia & Upjohn AB
Lindhagensgatan 133
SE-112 87 Stockholm, Sweden
+46 (08) 695-8000, Fax +46 (08)
618-8607

Pharmacia & Upjohn Inc.
301 Henrietta Street
Kalamazoo, MI 49001
(616) 323-4000, Fax (616) 323-4077

Phelps Dodge Corp.
2600 N. Central Avenue
Phoenix, AZ 85004
602-234-8100

Phibro-Tech Inc.
1 Parker Plaza
Fort Lee, NJ, 7024
(201) 944-6020
www.phibro-tech.com

Phillips Chemical Co.
PO Box 968,
Borger, TX 79008, USA
806-274-5236, 800-858-4327,
Fax 806-274-5230

Phillips Chemical Co.
101 ARB Plastics Technical Center,
Bartlesville, OK 74004, USA
918-661-9845, Fax 918-662-2929

Phillips Petroleum Chemical
Steenweg op Brussels 355,
B-3090 Overijse, Belgium
32-2-689-1211, Fax 32-2-689-1472

Phillips Petroleum Co.
Philips Quadrant, 35 Guildford,
Woking, Surrey, GU22 7QT, UK
44-1483-756666, Fax 44-1483-
752371

Phillips Petroleum Inc.
Shin-Tokyo Bldg.,
3-1, Marunou,
Chioyda-ku,Tokyo 100,Japan
81-3-3216-6951
Fax 81-3-3216-6960

Pilot Chemical Co.
11756 Burke St.,
Santa Fe Springs, CA 90670, USA
310-723-0036, 800-707-4568,
Fax 310-945-1877
www.pilotchemical.com

**Planters Products Cooperative
Marketing & Supply, Inc.**
Planters Products Bldg. 109 Esteban
St.
Legaspi Village, Makati City,
Phillipines
Fax: 816-4388

PMC Specialities International
65B Wigmore Street,
London, W1H 9LG, UK
44-171-935-4058, Fax 44-171-935
9895

PMC Specialties Group
501 Murray Rd.,
Cincinnati, OH 45217, USA
513-242-3300, 800-543-2466, Fax
513-482-7353

PMC Specialties Group
20525 Center Ridge Road,
Rocky River, OH 44116, USA
216-356-0700, 800-543-2466, Fax
216-356-2787

PMP Fermentation Prods. Inc.
900 NE Adams Street
Peoria, IL 61603
(309) 637-0400, Fax (309) 637-9302
www.pmpinc.com

PMS Specialities Group Inc.

Polymer Ag Inc.
744 Whitesbridge Rd.
Fresno, CA, 93706
(209) 495-0234

Polysar
Division of Bayer AG
PO Box 3001
1265 Vidal St., South
Sarnia, ON Canada N7T 7M2
(519) 337-8251, Fax (519) 339-7723

Polysciences Inc.
400 Valley Road,
Warrington, PA 18976-2590, USA
215-343-6484, 800-523-2575,
Fax 800-343-3291

Polysciences, Europe
Postfach 1130,
D-69208 Eppelheim, Germany
49-6221-765767, Fax 49-6221-
764620

Polyvel
100 Ninth Street
Hammonton, NJ 08037
(609) 567-0080, Fax (609) 567-9522
www.polyvel.com

Portman Agrochemicals Ltd.
Apex House, Grand Parade, Tally-Ho
Corner
London, UK, N12 0EH
44 20 8446-8383; Fax: 44 20 8445
6045

Manufacturer and Supplier Directory

Poythress Laboratories
16 N 22nd St,
PO Box 26946,
Richmond, VA 23261, USA

PPG Industries Inc.
Urb. Industrial Mario Julia,
A&B Street Caparra Heights,
San Juan, 920, Puerto Rico
www.ppg.com

PPG Industrial do Brasil Limitada
Rua Sampaio Viana,
277,13o Andar-Conj.132,
Sao Paulo-SP, Paraiso,
CEP 04004-000, Brazil
55-11-887-6522, Fax 551-887-2224
www.ppg.com

PPG Industries (UK)
Carrington Business Park,
Carrington,Urmston,
Manchester, M31 4DD, UK
44-161-777-9203, Fax 44-161-777-9064
www.ppg.com

PPG Industries Asia/Pacific
LTD,Takanawa Court, 5th Floor,
13-1, Takanawa 3 chome,
Minato-ku-Tokyo 108, Japan
03-3280-2861, Fax 03-3280-2920
www.ppg.com

PPG Industries Sales, Inc.
Immeuble Scor,
1 Avenue du President Wilson,
Paris LaDéfense Cedex, 92704,
France
33-1-4698-8100, Fax 33-1-4698-8263
www.ppg.com

PPG Industries Taiwan
Suite 601, Worldwide House, No. 131,
Ming East Rd., Sec. 3,
Taipei 105, Taiwan, R.O.C.
886-2-514-8052, Fax 886-2-514-7957
www.ppg.com

PPG Industries, Inc.
One PPG Place, 34 North,
Pittsburgh, PA 15272, USA
412-434-3131, 800-CHEM-PPG,
Fax 412-434-2891
www.ppg.com

PPG Industries, Inc.
961 Division St.,
Adrian, MI 49221, USA
517-263-7831, Fax 517-263-2552
www.ppg.com

PPG Industries Inc.
PO Box 98,
100 Station Ave.,
Stockertown, PA 18083-0098, USA
215-759-3690, Fax 215-759-3692
www.ppg.com

PPG Industries, Inc.
3938 Porett Dr.,
Gurnee, IL 60031, USA
847-244-3410, 800-323-0856,
Fax 847-244-9633
www.ppg.com

PPG Industries-Asia
Takanawa Court, 5th floor,
13-1 Takanawa 3-Chome,
Minato-Ku, Tokyo 108, Japan
81-3-3280-2911, Fax 81-3-3280-2920
www.ppg.com

Pratt Pharmaceuticals
See: Pfizer Inc.

Praxair Inc.
39 Old Ridgebury Rd. M-1
Danbury, CT, 6810
(203) 794-5946
www.praxair.com

Presperse Inc.
141 Ethel Rd W,
Piscataway, NJ 08854-5928, USA
732-819-8009, Fax 732-819-7155

Procter & Gamble Co.
Rm C2N09A, 11530 Reed Hartman
Hwy.
Cincinnati, OH 45241, USA
513-626-3701, 800-477-8899,
Fax 513-626-3145
www.pg.com/chemicals

Procter & Gamble Inc.
4711 Yonge St.,
PO Box 355, Station A,
Toronto, Ont., M5W 1C5, Canada
416-730-4064, Fax 416-730-4122
www.pg.com/chemicals

Procter & Gamble Inc.
1-17 Koyocho-naka 1-chome,
Higashinaka-ku,
Kobe 658, Japan
81-78-845-5328, Fax 81-78-845-6912
www.pg.com/chemicals

Procter & Gamble Ltd.
PO Box 9, Hayes Gate House,
27 Uxbridge Rd.,
Hayes, Middlesex, UB4 0JD, UK
44-181-242-2303, Fax 44-181-242-2294
www.pg.com/chemicals

Protameen Chemicals Inc.
375 Minnisink Road
Totowa, NJ 07511
(973) 256-4374, Fax (973) 256-6764
www.protameen.com

Protan AS
PO Box 420
Brakeroya
N3002 Drammen, Norway
+47 32 22 16 00, Fax +47 32 22 17 00

Protex Chemie Basel
Thannerstrasse 72,
CH-4054 Basel, Switzerland
41-61-302 0066, Fax 41-61-302 0076

Protex Chemicals Ltd.
Astley Lane Industrial Estate,
Astley Way, Swillington,
Leeds, L228 8XT, UK
44-113-2876002, Fax 44-113-2875003

Protex Extrosa
Postfach 1415,
79504 Lorrach, Germany
49-7621-84772, Fax 49-7621-12429

Protex Korea Co. Ltd.
PO Box 1193,
Seoul 1351010, Korea
82-2-548-6993, Fax 82-2-548-6564

Protex Nederland
Broekhovenseweg 130 R,
502 LJ Tilburg, The Netherlands
31-13-36-7477, Fax 31-13-36-6592

Protex S.A.
B.P. 177, 6 rue Barbès,
92305 Levallois-Paris, France
33-1-41-34-1400, Fax 33-1-41-34-1416

Protocol Analytical LLC
203 Norcross Avenue
Metuchen, NJ 08840
(732) 627-0500, Fax (732) 627-0979

Punda Mercantile Inc.
4115-610 Sherbrooke St. W.
Montreal, Quebec
Canada H3Z 1K9
(514) 931-7278, Fax (514) 931-7200
www.punda.com

Purac Biochem. BV
Arkelsedijk 46
4206 AC, Gorinchem
Netherlands
+31 0183-695695, Fax +31 0183-695600
www.purac.com

Manufacturer and Supplier Directory

Purdue Pharma L.P.
100 Connecticut Avenue
Norwalk, CT 06856
(203) 853-0123, Fax (203) 838-1576

Purecha Group
3 Giri Kunj
M. G. Cross Road
Exit 4, Kandivali West
Mumbai 400 067, India
2806 42 78
purecha@vsnl.com

QO Chemicals (Australia)
16 Princess St.,
Kew, Vic., 3101, Australia
61-3-853-6464, Fax 61-3-853-5929
QO Chemicals Europe
8, rue Bellini
75782 Paris Cedex 16, France
33-1-4503-1450, Fax 33-1-4704-3604
Quest Inte
QO Chemicals, Inc.
Industrieweg 12,
Haven 391,
2030 Antwerp, Belgium
32-3-541-21-65, Fax 32-3541-65-03

Qualco Inc.
225 Passaic St.
Passaic, NJ, 7055
(201) 473-1222
www.qualco.com

Qualis Inc.
4600 Park Ave.
Des Moines, IA, 50321
(515) 243-3000

Quantum/USI
Quantum Chemical Co.,
11500 Northlake Dr.,
PO Box 429550,
Cincinnati, OH 45249, USA
513-530-6500, 800-323-4905, Fax
513-530-6119

Quantum Chemical Co.
1 Pierce Place, Suite 250C,
Itasca, IL 60143, USA
708-285-1111, 800-323-7659, Fax
708-285-3316

Quantum Chemical Europe
Lange Bunder 7,
4854 MB Bavel, The Netherlands
31-1613-6600, Fax 31-1613-3500

Quest Chemical Corp.
12255 F.M. 529 Northwoods
Industrial Park
Houston, TX, 77041
(713) 896-8188
www.questchemicalcorp.com

Quest International
Postbus 2,
1400 CA Bussum, The Netherlands ·
31-2159-99111, Fax 31- 2159-46067

Quest International
5115 Sedge Blvd.,
Hoffman Estates, IL 60192, USA
800-621-4710, Fax 847-645-7061

Quest International
Bromborough Port,
Wirral, Merseyside, L62 4SU, UK
44-151-645-2060, Fax 44-151-645-6975

Quest International
12 Britton St.,
Smithfield, NSW, 2164, Australia
61-2 827-4000, Fax 61-2-604-7926

Quest International
Postfach 650170,
Poppenbütteler Chaussee 36,
D-22361 Hamburg 65, Germany
49-40-607970, Fax 49-40-6079710

Quest International
10 Painters Mill Rd.,
Owings Mills, MD 21117-3686, USA
410-363-2550, Fax 410-363-7514

Quest International
400 International Dr.,
Mt. Olive, NJ 07828, USA
201-691-7100, 800-598-5986, Fax
201-691-7479

Quest International
PO Box 630, Woods Corners,
Norwich, NY 13815, USA
607-334-9951, Fax 607-334-5022

Quest International
Kennington Road,
Ashford, Kent, TN24 0LT, UK
44-1233 644444, Fax 44-1233-644146

Quimica Estrella
Av. de los Constituyentes, Cap. Fed.
(1427)
Buenos Aires, Argentina, 2995
54-11-4524-8100; Fax: 54-11-4522-3022

Rallis India Ltd.
Ralli House
21 D. Sukhadwala Marg.
Mumbai 400-001
India
+91 22 5665 2700/750 /987,
Fax +91 22 2207 7755/6573, 2208
7295
www.rallis.co.in

Rasa

Raschig AG
Mundenheimer Strasse 100,
D-67061 Ludwigshafen, Germany
49-621-56180, Fax 49-621-5618661

Raschig Corp.
5000 Old Osborne Tpke.,
PO Box 7656,
Richmond, VA 23231, USA
804-222-9516, Fax 804-226-1569

Raschig France S.A.R
49, ave. de Versailles,
F-75016 Paris, France
33-1-45-24- 0636, Fax 33-1-45-20-8408

Raschig UK Ltd.
Dock Office, Trafford Rd.,
Salford Quays,
Salford, Lancs., M5 2XB, UK
44-161-877-3933, Fax 44-161-877-3944

Reade Advanced Materials
PO Drawer 15039
Providence, RI 02915-0039
(401) 433-7000, Fax (401) 433-7001
www.reade.com

Reckitt Benckiser Inc.
1655 Valley Rd.
Wayne, NJ, 7474
(973) 686-7391
www.reckitt.com

Reddick Fumigants, Inc.
Highway 64W
Williamston, NC, 27892
(919) 792-1613

Reheis Inc.
PO Box 609,
235 Snyder Ave.,
Berkeley Heights, NJ 07922, USA
908-464-1500, Fax 908-464-8094
www.reheis.com

Reheis Ireland
Kilbarrack Rd.
Dublin 5, Ireland,
353-1-8322621, Fax 353-1-8392205
www.reheis.com

Reilly Industries Inc.
1500 S. Tibbs Ave.
Indianapolis, IN 46241, USA
317-248-6411, 800-777-3536, Fax
317-248-6402
www.reillyind.com

Rentokil Ltd.
College Building, Windrush Park
Road,
Witney, Oxfordshire, UK, OX8 5DY
44 1993 778821; Fax: 44 1993
706654
www.rentokil-initial.co.uk

Rewo Chemische Werke GmbH

Rheox Inc.
See: Elementis plc

Manufacturer and Supplier Directory

Rhône-Poulenc ABM,
Poleacre Lane, Woodley,
Stockport, Cheshire, SK6 1PQ, UK
44-161-430-4391, Fax 44-161-430-8523
www.rhone-poulenc.com

Rhône-Poulenc Ag Co.
PO Box 12014, 2 TW Alexander Dr.,
Research Triangle Park, NC 27709,
USA
919-549-2000, 800-334-9745, F
ax 919-549-3924
www.rhone-poulenc.com

Rhône-Poulenc Agrochemicals
14-20 rue Pierre Baizet, B.P. 9163,
69263 Lyon Cedex 09, France
33-7-229-2525, Fax 33-7-229-2885
www.rhone-poulenc.com

Rhône-Poulenc Basic
One Corporate Dr., Box 881,
Shelton, CT 06484, USA
800-642-4200
Fax 203-925-3627,
www.rhone-poulenc.com

Rhône-Poulenc Chemicals
Woodley,
Stockport, Cheshire, SK6 1PQ, UK
44-161-430-4391, Fax 44-161-430-4364
www.rhone-poulenc.com

Rhône-Poulenc Chemicals
Staveley,
Chesterfield, Derbyshire, S43 2PB,
UK
44-1246-277251, Fax 44-1246-280090
www.rhone-poulenc.com

Rhône-Poulenc Chemicals
25 Quai Paul Doumer,
F-92408 Courbevoie Cedex 29,
France
33-47-68-1234, Fax 33-47-68-23-00
www.rhone-poulenc.com

Rhône-Poulenc Chemicals
Ketenislaan 1,
B-9130 Kallo, Belgium
32-37-551759, Fax 32-37-756995
www.rhone-poulenc.com

Rhône-Poulenc Chemicals
Oak House,Reeds Crescent,
Watford, Herts, WD1 1QH, UK
44-181 984 3342, Fax 44-181 984 1701
www.rhone-poulenc.com

Rhône-Poulenc/Coatings & Paint
CN 7500, Prospect Plains Rd.,
Cranbury, NJ 08512, USA
609-860-4600, 800-253-5052
www.rhone-poulenc.com

Rhône-Poulenc/Detergents
CN 7500, Prospect Plains Rd.,
Cranbury, NJ 08512, USA
609-860-4600, 800-253-5052
www.rhone-poulenc.com

**Rhône-Poulenc Environmental Prods.
Ltd.**
Bayer CropScience
Alfred-Nobel-Str. 50
D-40789 Mannheim am Rhein,
Germany
49 21 73-38 31 25

Rhône-Poulenc/Film Div.
2754 West Park Dr.,
Holcomb, NY 14469, USA
716-657-5800, Fax 716-657-5838
www.rhone-poulenc.com

Rhône-Poulenc Food Ingredients
CN 7500, Prospect Plains Rd.,
Cranbury, NJ 08512, USA
609-860-4600, 800-253-5052
www.rhone-poulenc.com

Rhône-Poulenc Gerona
via Milano 78,
Ospiate Di Bollate,
I-20021 Milano, Italy
39-2-38 33 41, Fax 39-2-3833 43 01
www.rhone-poulenc.com

Rhône-Poulenc, Inc.
CN7500, Prospect Plain Rd.,
Cranbury, NJ 08512-7500, USA
609-860-4000, 800-922-2189, Fax
609-860-0269
www.rhone-poulenc.com

**Rhône-Poulenc, Inc./Specialty
Chemicals**
PO Box 769,
Marietta, GA 30061, USA
404-422-1250, 800-677-7412
Fax 404-427-0874
www.rhone-poulenc.com

Rhône-Poulenc North America
One Corporate Dr.,
Shelton, CT 06484, USA
203-925-8164, Fax 203-925-3670
www.rhone-poulenc.com

Rhône-Poulenc N.V.
Kuhlmannkaai,
B-9020 Ghent, Belgium
32-91 44-8891
www.rhone-poulenc.com

Rhône-Poulenc/Perf. Resins
1525 Church St. Ext.,
Marietta, GA 30060, USA
404-422-1250, Fax 404-427-0874
www.rhone-poulenc.com

Rhône-Poulenc Rorer
Rainham Rd South,
Dagenham, Essex, RM10 7XS, UK
44-181 592 3060, Fax 44-181-593 2140
www.rhone-poulenc.com

Rhône-Poulenc Rorer
500 Arcola Road,
Collegeville, PA 19426-0107, USA
610-454-8000
www.rhone-poulenc.com

Rhône-Poulenc Silico
Cranbury, NJ 08512, USA
800-288-1175
www.rhone-poulenc.com

Rhône-Poulenc Silico
27 06/07 The Concourse,
300 Beach Road,
Singapore 0719
65-291-1921, Fax 65-296-6044
www.rhone-poulenc.com

Rhône-Poulenc Surfactants
Les Miroirs-Défense 3,
18 ave. d'Alsace, Cedex 29,
92097 Paris, LaDéfense, France
(33-1) 4768 1234, Fax (33-1) 4768 0900
www.rhone-poulenc.com

Rhône-Poulenc Surfactants
3265 Wolfdale Rd.,
Mississauga, Ont., L5C 1V8, Canada
905-270-5534, Fax 905-270-5816
www.rhoône-poulenc.com

Rhône-Poulenc Surfactants,CN 7500
Prospect Plains Rd.,
Cranberry, NJ 08512-7500, USA
609-860-4000, 800-922-2189,
Fax 609-860-0459
www.rhone-poulenc.com

Rhône-Poulenc/Textiles
1000 Hurricane Shoals Rd.,
Lawrenceville, GA 30243, USA
www.rhone-poulenc.com

Rhône-Poulenc/Tire & Rubber
CN 7500,
Cranbury, NJ 08512-7500, USA
609-860-3068, Fax 609-395-1632
www.rhone-poulenc.com

Rhône-Poulenc UK
Ongar, Essex, UK

Rhône-Poulenc/Water Solvents
CN 7500,
Cranberry, NJ 08512-7500, USA
609-860-4000, 800-922-2189,
Fax 609-860-0459
www.rhone-poulenc.com

Manufacturer and Supplier Directory

Rhône-Poulenc/Water Treatment
One Gatehall Dr.,
Parsippany, NJ 07054, USA
201-292-2900, 800-848-7659, Fax
201-292-5295
www.rhone-poulenc.com

Rhodia Europe
Rhodia SA
26 quai Alphonse Le Gallo
92512 Boulogne-Billancourt Cedex
France
+33 1 55 38 40 00
www.eu.rhodia.com

Rhodia USA
Rhodia Inc.
259 Prospect Plains Road
CN 7500
Cranbury, NJ 08512-7500
(609) 860-4000
www.eu.rhodia.com

Richman Chemical Inc.
768 N. Bethlehem Pike
Lower Gwynedd, PA 19002
(215) 628-2946, Fax (215) 628-4262
www.richmanchemical.com

Ridge Technologies
117 Lyons Rd.,
Basking Ridge, NJ 07920, USA
908-766-1915, Fax 908-204-0312

Riedel de Haen (Chinosolfabrik)
See: Sigma-Aldrich Co.

Rigby Taylor Ltd.
The Riverway Estate,
Portsmouth Road, Peasmarsh,
Guildford, Surrey, GU3 1LZ, UK
44-1483-35657, Fax 44-1483-34058
www.rigbytaylor.com

Rigby Taylor Ltd.
Garside St.
Bolton, Lancs. UK, BL1 4AE
44 1204 377777; Fax: 44 1204
377755
www.rigbytaylor.com

Rita Corp.
1725 Kilkenny Court
Woodstock, IL, 60098-7437
(800) 426-7759; (815) 337-2500
www.ritacorp.com

Rit-Chem Co. Inc.
109 Wheeler Ave.,
PO Box 435,
Pleasantville, NY 10570-0435, USA
914-769-9110, Fax 914-769-1408

Riverdale Chemical Co.
1333 Burr Ridge Parkway, #125A
Burr Ridge, IL 60521-0866
(708) 754-3330, Fax (708) 754-0314

Robert Koch Industries
4770 Harback Road
Bennett, CO 80102
(303) 644-3763, (303) 644-3045

Roberts Pharmaceutical Corp.
4 Industrial Way West
Eatontown, NJ 07724
(800) 828-2088

Robins, A. H. Co.
P.O. Box 26609, 1407 Cummings Dr.
Richmond, VA, 23261
(804) 257-2035

Roche Diagnostics Corp.
Roche Applied Science
PO Box 50414
9115 Hague Road
Indianapolis, IN 46250-0414
(800) 428-5433, Fax (800) 428-2883
www.roche-applied-science.com

Roche Diagnostics GmbH
Roche Applied Science
Sandhofer Str. 116
68305 Mannheim, Germany
0800 759-4152, Fax 0621/759-4136
www.roche-applied-science.com

Roche Laboratories
340 Kingsland St,
Nutley, NJ 07110, USA
201-235-3381, Fax 201-235-8023

Roche Vitamins
340 Kingsland St.,
Nutley, NJ 07110-1199, USA
201-235-5000, Fax 201-535-7606

Roche Puerto Rico
See: ICN Pharmaceuticals

Roebic Labs, Inc.
25 Connair Rd.
Orange, CT, 6477
(203) 795-1283
www.roebic.com

Roerig Div. Pfizer Pharm.

Rohm and Haas (Australia)
969 Burke Rd., PO Box 11,
Camberwell, Vic. 3124, Australia
www.rohmhaas.com

Rohm and Haas (UK) Ltd.
Lennig House,
2 Mason's Avenue, Croydon,
Greater London, CR9 3NB, UK
44-181-686-8844, Fax 44-181-686-
8329
www.rohmhaas.com

Rohm and Haas Canada
2 Manse Rd.,
West Hill, Ont., M1E 3T9, Canada
416-284-4711, 800-268-4201, Fax
416-284-2982
www.rohmhaas.com

Rohm and Haas Co.
100 Independence Mall West,
Philadelphia, PA 19106-2399, USA
215-592-3000, 800-323-4165, Fax
215-592-6909
www.rohmhaas.com

Rohm and Haas Co. Europe
Chesterfield House, Bloomsbury,
London, WC1A 2TP, UK
44-171-242-4455, Fax-44-171-404-
4126
www.rohmhaas.com

Rohm Tech.

Rona Laboratories
Harrier House
High St., W. Drayton
Middlesex UB7 7QG, UK

Rottapharm SpA
Via Valosa di Sopra, 9
20052 Monza, Milan
Italy
+39 039-7390-1, Fax +39 039-7390-371
www.rotta.com

Roussel Laboratories
Orbital,
Uxbridge, Middlesex, UB9 5HP, UK
44-1895- 834343, Fax 44-1895-
834578

Roussel Uclaf, Fine Chemicals
Tour Roussel Hoechst,
F-92080 Paris la Défense, France
33-40-81-4884, Fax 33-40-90-0032

Royal H. M.
PO Box 28, 689 Pennington Avenue
Trenton, NJ 08618
(609) 396-9176, Fax (609) 396-3185

Royal H. M.
PO Box 6798, 6880 8th. Street
Buena Park, CA 90620
(714) 670-1554, (714) 670-2707

Ruetgers-Nease Chemicals,
201 Struble Rd.,
State College, PA 16801, USA
814-238-2424, Fax 814-238-1567

Rutgers Organics GmbH
Postfach 310160,
Sanhofer Str. 95,
D-68261 Mannheim, Germany
49-621-7654-286, Fax 49-621-7654-
413

Ruger Chemical Co.
83 Cordier St.
Irvington, NJ, 07111-4035
(973) 926-0331; Fax: (973) 926-4921
www.rugerchemical.com

Manufacturer and Supplier Directory

Sachtleben Chemie GmbH
Huntingdon House, Princess St.,
Bolton, BL1 1EJ, UK
44-1204-363634, Fax 44-1204-36-
1144

Salomon L. A.

Salutas Pharma GmbH
Otto von Guericke Allee 1
39179 Barleben, Germany
+49 03 92 03/71 - 28 20, Fax +49 03
92 03/71 29 50
www.salutas.de

San Yuan Chem. Co. Ltd.
34 Kuang Fu Road
Chai Tai Ind.,
Taipao City, Chiayih, Taiwan
886 5 2379361, Fax 886 5 2379695

Sandoz AG
Lichtstrasse 35,
CH-4002 Basel, Switzerland
61-24-1111, Fax 61-24-8081

Sandoz Pharmaceuticals
Route 10,
East Hanover, NJ 07936, USA
201-503-7500

Sandoz Products Ltd.
Frimley Business Park, Frimley,
Camberley, Surrey, GU16 5SG, UK
44-1276-692255, Fax 44-1276-
692508

Sanitized Inc.
57 Litchfield Rd.
New Preston, CT, 6777
(860) 868-9491

Sankyo Corp.
Yoshizawa Bldg., 9-19,
Ginza 3-chome, Chuo-ku
Tokyo 104, Japan
+81 3543-8571, Fax +81 3546-3174

Sankyo Co. Ltd.
3-5-1 Nihonbashi Hon-cho
Chuo-ku, Tokyo 103-8426
Japan

Sanofi Bio-Industries
620 Progress Ave.,
PO Box 1609,
Waukesha, WI 53187, USA
414-547-5531

Sanofi Bio-Industries
8 Neshaminy Interplex,Suite 213,
Trevose, PA 19053, USA
215-638-7801, Fax 215-638-8168

Sanofi Winthrop Ltd.
One Onslow Street,
Guildford, Surrey, GU1 4YS, UK
44-1483-505515, Fax 44-1483-
35432

Sanofi Winthrop
301 Oxford Valley Road
Morrisville, PA 19067-7706
(215) 321-7560

Sartomer Company Inc/America
Oaklands Corp. Center,
502 Thomas Jones Way,
Exton, PA 19341, USA
610-363-4117, 800-SARTOMER, Fax
610-363-6849
www.sartomer.com

Sartomer Company Inc/Asia
331 Northbridge Rd,
#23-06 Odeon Towers,
188720, Singapore
65-334-2645, 65-334-2645, Fax 65-
334-2647
www.sartomer.com

Sartomer Company Inc/Europe
Le Diamant B,
92970 Paris La Défense-Cedex,
France
33-1-4135-68-21, Fax 33-1-4135-
62-70
www.sartomer.com

Savage Labs.
60 Baylis Road,
PO Box 2006,
Melville, NY 11747, USA

Schaeffer Salt & Chem.

Schenectady Chemical
319 Comstock Rd.,
Scarborough, Ont., M1L 2H3,
Canada

Schenectady Chemical
PO Box 1046,
Schenectady, NY 12301, USA
518-370-4200, Fax 518-346-3111

Schenectady-Midland
Four Ashes, Wolverhampton,
West Midlands, WV10 7BT, UK
44-1902 790555, Fax 44-1902
791640

Scher Chemicals, Inc.
Industrial West & Styertowne R,
PO Box 4317,
Clifton, NJ 07012, USA
201-471-1300, Fax 201-471-3783

Schering AG
Postfach 650311,
Müllerstrasse 170-178,
D-13303 Berlin 65, Germany
+49 30-4680, Fax +49 30-4685305
www.schering.com

Schering Agrochemical
Hauxton, Cambridge,
Cambridgeshire, CB2 5HU, UK
44-1223 870312, Fax 44-1223
872142
www.schering.com

Schering Agrochemical
Chesterford Park Industrial Research
Station,
Saffron Walden, Essex, CB10 1XL,
UK
44-1799-30123, Fax 44-1799-30991
www.schering.com

Schering Corp.
Galloping Hill Rd,
Kenilworth, NJ 07033, USA
908-298-4000, 800-526-4099
www.schering.com

Schering Health Care
The Brow, Burgess Hills,
West Sussex, RH15 9NE, UK
44-1444-232323, Fax 44-1444-
246613
www.schering.com

Schering Industrial
Gorsey Lane, Widnes,
Cheshire, WA8 0HE, UK
44-151-495-1989, Fax 44-151-495-
2003
www.schering.com

**Schering-Plough Veterinary
Operations Inc.**
1095 Morris Avenue
Union, NJ, 7083
(908) 298-4000
www.sch-plough.com

Schering Agrochemicals Ltd.
See: Aventis Crop Science

Schweizerhall Inc.
10 Corporate Place South,
Piscataway, NJ 08854, USA
908-981-8200, 800-243-6564, Fax
908-981-8282

SCI Natural Ingredients
4 Kings Road
Reading, BerksRG1 3AA, UK
+44 1734 580247, Fax +44 1734
589580

SCM Glidco Organics
PO Box 389
Jacksonville, FL 32201-0389
(904) 768-5800, Fax (904) 768-2200

**Scotts Company Ltd Professional
Products**
Paper Mill Lane, Bramford,
Ipswich, Suffolk, IP8 LBZ, UK
44-1473-830492, Fax 44-1473-
830386

Scotts-Sierra Crop Protection
14111 Scottslawn Road
Marysville, OH 43041

Seaboard Industries
185 Van Winkle Ave.
Hawthorne, NJ, 07506-2151
(973) 427-8500

Searle G.D. & Co.
5200 Old Orchard Road
Skokie, IL 60077
(847) 982-7000, Fax (847) 470-1480

Searle Ltd.
PO Box 53
Lane End Road, High Wycombe
Bucks., HP12 4HL, UK
44 (01494) 521124, Fax 44 (01494)
447872

Seimi Chem.
2-10, Chigasaki 3-chome
Chigasaki-shi, Kanagawa 253
Japan
+81 82-4131, Fax +81 86-2767

Selectokil Ltd (Velpar)
Nomix-Chipman Ltd,
Portland Building, Portland Street,
Staple Hill, Bristol BS16 4PS, UK
44 117 957 4574; Fax: 44 117 956
3461

Shell Canada Chemical
PO Box 100, Station M,
Calgary, Alberta, T2P 2H5, Canada
800-567-8728, Fax 800-567-8862
www.shell.com

Shell Chemical Co.
910 Louisiana Street,
Houston, TX 77252, USA
800-USA-SHELL, Fax 713-241-6367
www.shell.com

Shell Chemical Co.
One Shell Plaza, PO Box 2463,
Houston, TX 77252, USA
713-241-6161, 800-872-7435,
Fax 713-241-4043
www.shell.com

Shell Chemicals UK Ltd.
Heronbridge House, Chester Business
Park,
Wrexham Rd, Chester, Cheshire,
CH4 9QA, UK
44-1244-685678, Fax 44-1244-
685010
www.shell.com

Shen Hong
5F-206 Manking E. Rd.
Sect 2-Box 46
209 Taipei, Taiwan, ROC

Shephard Bros Inc.
503 S. Cypress St.
La Habra, CA, 90631
(562) 697-1366

Sherex Chemical Co.
PO Box 646,
5777 Frantz Rd.,
Dublin, OH 43017, USA
614-764-6500, 800-848-7370, Fax
614-764-6650
www.witco.com

Sherman Chem. Co. Ltd.
Shaftesbury, UK
+44 (0) 1747 823293,
Fax +44 (0) 1747 825383
www.sherchem.co.uk

Sherwin-Williams Co.
101 Prospect Ave.
Cleveland, OH, 44115
(216) 566-2000
www.sherwin-williams.com

Shinung
Shinung Bldg. 45 Wu Chuan Center
Taichung, Taiwan

Shionogi & Co. Ltd.

Showa Denko
13-9, Shiba Daimon 1-chome
Minato-ku, Tokyo 105-8518, Japan
www.sdk.co.jp

Siapa

Siber Hegner Ltd.
County House,
221-241 Beckenham Road,
Beckenham,
Kent, BR3 4UF, UK
44-181-659-2345, Fax 44-181-659-
1292

Sigma-Aldrich Co.
PO Box 2060
Milwaukee, WI, 53201
(800) 558-9160; (414) 273-3850;
Fax: (414) 273-4979
www.sigma-aldrich.com

Sigma-Aldrich Fine Chem.
3500 DeKalb St.
St. Louis, MO, 63118
(314) 771-5765
www.sigma-aldrich.com

Sintesul S. A. (Brazil)
Roselandia CP 294, CEP 93352-000,
Novo Hamburgo - RS, Brazil, 93352-
000
51 594-3266; Fax: 51 594-3266

Sipcam
300 Colonial Pkwy. Suite 230,
Roswell, GA, 30076
(770) 587-1032
www.sipcamagrousa.com

Smith Werner G.
1730 Train Avenue
Cleveland, OH 44113
(216) 861-3676, Fax (216) 861-3680

SmithKline Beecham plc,
Clarendon Road,
Worthing, West Sussex, BN14 8QH,
UK

Snia UK
36 Broadway, St. James
London SW1H 0BH, UK
+44 171 222 8696, Fax +44 171 222
8705

Soil Chemicals Corp.

Solaris ChemTech Ltd.
First India Place
Tower C, Mehrauli
Gurgaon Road 122002
Haryana, India
+91 124 6804242, Fax +91 124
6804263

Solutia, Inc.
320 Interstate N. Pkwy. Suite 500
Atlanta, GA, 30339
(770) 991-7600
www.solutia.com

Solutia Australia Limited
1/437 Canterbury Road,
Surrey Hills, Vic., 3127, Australia
61-3-98884539
www.solutia.com

Solutia Beijing,
C612 Bejing Lufthasan Center,
Beijing 100016, China
011-8610-463-8046
www.solutia.com

Solutia Canada
2330 Argentina Rd.,
Mississauga, Ont., L5M 2G4, Canada
905-826-9222
www.solutia.com

Solutia Europe SA
Parc Scientifique-Fleming,
Rue Laid Burniate, 3,
B-1348 Louvain-la-Neuve, Belgium
32-10-48-13-21, Fax 32-10-48-12-24
www.solutia.com

Solutia Japan Limited
Nihonbashi Daini Building, 41-12,
Nihonbashi Hakozaki-cho,
Chuo-ku, Tokyo 103, Japan
81-3-56441638
www.solutia.com

Solutia Mexico
Bosques de Duranzo 61,3er Piso,
Bosques de las Lomas,
D. F. 11700 MX,Mexico
www.solutia.com

Solutia Moscow
Volkov Lane 19,
Moscow, 123242, Russia
7-502-221-0000,
www.solutia.com

Manufacturer and Supplier Directory

Solutia Singapore
101 Thompson Rd,
#19-00 United Square,
Singapore 307591, Singapore
65-733-1611
www.solutia.com

Solutia USA
1460 Broadway,
New York, NY 10036,
212-382-9600
www.solutia.com

Solutia Venezuela
Avenida Francisco de Miranda,
Edificio Parque Cristal,
Torre Este Piso 8 Ofc. 8-12,
Caracas, Los Palos Grandes,1062,
Venezuela
58-2-285-0944
www.solutia.com

Solvay Enzymes GmbH
Postfach 690307,
Hans Böckler Allee 20,
30612 Hannover, Germany
49-511-8570, Fax 49-511-8572371
www.solvay.com

Solvay Enzymes, Inc.
PO Box 4859,
1230 Randolph St.,
Elkhart, IN 46514-0859, USA
219-523-3700, 800-342-2097,
Fax 219-523-3800
www.solvay.com

Solvay Interox Ltd.
PO Box 7, Warrington,
Cheshire, WA4 6HB, UK
44-1925-651277, Fax 44-1925-
655856
www.solvay.com

Solvay Polymers Inc.
3333 Richmond Ave.,
PO Box 27328,
Houston, TX 77227-7328, USA
713-525-4000, 800-231-6313,
Fax 713-522-7890
www.solvay.com

Solvay S.A.
33 rue du Prince Albert,
B-1050 Brussels, Belgium
32-509-6111, Fax 32-509-6617
www.solvay.com

Sorex Ltd.
St Michael's Industrial Estate,
Hale Road, Widnes,
Cheshire, WA8 8TJ, UK
44-151-420-7151, Fax 44-151-495-
1163

Southern Agricultural Insecticides Inc.
Palmetto, FL
(914) 722-3285
www.southernag.com

Southern Agricultural Insecticides Inc.
Hendersonville, NC
(828) 692-2233
www.southernag.com

Southern Agricultural Insecticides Inc.
Boone NC
(828) 264-8843
www.southernag.com

Southern Clay Products
1212 Church St., PO Box 44
Gonzales, TX 78629
(210) 672-2891, Fax (210) 672-3650

Spartan Flame Retardants
PO Box 395
Crystal Lake, IL 60039
(815) 459-8500, Fax (815) 459-8560
www.spartancompany.com

Spectrum Chem. Manufacturing
755 Jersey Avenue
New Brunswick NJ 08901
(310) 516-7512

Spectrum Organic Products, Inc.
1304 South Point Blvd. Suite 280.
Petaluma, CA, 94954
(707) 778-8900; Fax: (707) 765-8470
www.spectrumnaturals.com

Speer Products Inc.
4242 B.F. Goodrich Blvd.
Memphis, TN 38118
(901) 362-1950, Fax (901) 795-9525
www.speerproducts.com

Spice King
6009 Washington Blvd.
Culver City, CA 90232-7488
(213) 836-7770, Fax (213) 836-6454

Spiess-Urania Chemicals GmbH
Heidenkampsweg 77
Hamburg, Germany, 20097
49 40/23 65 20; Fax: 49 40/23 65 22 55
www.spiess-urania.com

Squibb E.R. & Sons
See: Bristol-Myers Squibb Co.

SSC Chemical Corp.
P.O. Box 919
Bellmore, NY, 11710

Staley A. E. Mfg.
2200 E. Eldorado St., PO Box 151
Decatur, IL 62525
(217) 423-4411, Fax (217) 421-2881

Stepan Canada
PO Box 307,
Orillia, Ont., L3V 6J6, Canada
705-326-7329, Fax 705-326-4523
www.stepan.com

Stepan Co.
22 West Frontage Rd.,
Northfield, IL 60093; USA
708-446-7500, 800-745-7837, Fax
708-501-2443
www.stepan.com

Stepan Europe
BP127,
38340 Voreppe, France
33-76-50-51-00, Fax 33-7656-7165
www.stepan.com

Stepan/PVO.
100 West Hunter Ave.,
Maywood, NJ 07607, USA
201-845-3030, Fax 201-845-6754
www.stepan.com

Steris Corp.
5035 Manchester Ave.
St. Louis, MO, 63110
(314) 535-1390
www.steris.com

Sterling Health U.S.A.
See: Sanofi-Winthrop

Sterling Research Laboratories,
Sterling-Winthrop House,Onslow
Street,
Guildford, Surrey, GU1 4YS, UK
44-1483-505515, Fax 44-1483-
35432

Sterling Winthrop Inc.
See: Sanofi-Winthrop

Stiefel Labs Inc.
255 Alhambra Circle
Coral Gables, FL 33134
(305) 443-3800, Fax (305) 443-3467

Süd-Chemie
Lenbachplatz 6, D-80333 Munchen
Germany
+49 89 5110-0, Fax +49 89 5110-
375
www.sued-chemie.de

Suchema
Haupstrasse 15
CH-8251 Kaltenbach
Switzerland
49-41-54-41-1265, Fax 49-41-54-41-
3234

Sumitomo Corp.
600 3rd. Ave.
New York, NY, 10016-1901
(212) 207-0700
www.sumitomo-chem.co.jp

Sundat (S) Pte. Ltd. (Singapore)
Singapore
65 6861 2460; Fax: 65 6862 0287

Manufacturer and Supplier Directory

Superfos Biosector A/S
PO Box 39,
2950 Vedbaek
Denmark
42-89-31-11, fax 42-89-15-95

Surety Laboratories Inc.
PO Box 1002, Cranford, NJ, 7016
(908) 755-6501
www.suretylabs.com

Sutton Labs.
International Specialty Products
1361 Alps Rd., Wayne, NJ , 07490
(800) 622-4423

Sybron Chemicals Canada
666 Appleby Line,
Burlington, Ont., L7L5Y3, Canada
905-637-3337, 909-637-6198,
www.sybronchemicals.com

Sybron Chemicals Inc.
PO Box 66, Birmingham Rd,
Birmingham, NJ 8011, USA
609-893-1100, 800-678-0020
www.sybronchemicals.com

Sybron Chemie Nederlands
Postbus 46,
NL-6710 BA Ede, The Netherlands
31-8380-70911, Fax 31-8380-30236
www.sybronchemicals.com

Syn Prods.

Synchemicals Ltd.
PO Box 18300
Greensboro, NC, 27419

Synchemicals Ltd.
Owen Street,
Coalville, Leicestershire, LE6 2DE,
UK
44-1530-510060, Fax 44-1530-510299

Syngenta Crop Protection
1800 Concord Pike
Wilmington, DE 19850

Syngenta Professional Products
PO Box 18300
Greensboro, NC, 27419
(336) 632-6000
www.syngentaprofessionalproducts.com

Syntetics Products Co

Synthetic Chemicals Ltd.
Four Ashes, Wolverhampton,
West Midlands, WV10 7BP, UK
44-1902-794000, Fax 44-1902-794300

Taiho
1-27 Kanda-Nishikicho
Chiyoda-ku, Tokyo, Japan
+81 3-3294-4527, Fax +81 3-3233-4057
www.taiho.co.jp

Taiwan Surfactant Co.
8 Fl., No 11, Sec. 1,
Chung Shan North Road,
Taipei, Taiwan, R.O.C.
886-2-2541-1122, Fax 886-2-2542-3773
www.taiwansurfactant.com.tw

Takasago Intl. Corp.
516 Xiangyang Road
Shanghai, China

Tam Ceramics
4511 Hyde Park Blvd.,
PO Box 0067
Niagara Falls, NY 14305
(716) 278-9400, fax (716) 2855-3026

Tanabe Seiyaku Co. Ltd.
2.10, Dosho-machi 3-chome
Cho-ku, Osaka 541, Japan
+81 205-5555, Fax +81 222-5262

Tanabe Research Laboratories U.S.A. Inc.
4540 Towne Centre Court
San Diego, CA 92121
(619) 558-9211

Texaco Chemical Co.
PO Box 27707,
Houston, TX 77227-7707, USA
713-961-3711, Fax 713-235-6437
www.texaco.com

Texaco Chemical Co./Oxides & Specialties
4800 Fournace Place, PO Box 430,
Bellaire, TX 77401, USA
www.texaco.com

Texaco France S.A.
5, rue Bellini, Tour Arago,
F-92806 Puteaux Cedex, France
33-1-47-17-2602, Fax 33-1-47-76-3050
www.texaco.com

Thomson Research Associates
95 King St. E. Suite 100
Toronto, Canada
(416) 955-1881
www.ultra-fresh.com

Thor Chemicals Ltd. (UK)
Earl Rd., Cheadle Hulme,
Cheshire, SK8 6QP, UK
44-161-486-2028, Fax 44-161-488-4155

Thor Chemicals, Inc.
Brook House, 37 North Ave.,
Norwalk, CT 06851, USA
203-846-8613, Fax 203-846-4810

Tiarco Chemical Div.
1300 Tiarco Drive,
Dalton, GA 30720, USA
706-277-1300, Fax 706-277-9039

TIC Gums
4609 Richlynn Drive, PO Box 369
Belcamp. MD 21017-0369
(800) 221-3953, (410) 273-6469
www.ticgums.com

Tokuyama Petrochemical
4980, Kasei-cho, Shin-Nanyo-shi
Yamaguchi 746, Japan
0834 62 4121, Fax 0834 62 3545

TopPro Specialties

Toray Ind. inc.
Toray Bldg., 2-1, Nihonbashi-Muromachi 2-chome
Chuo-ku, Tokyo 103-8666, Japan
+81 3 3245 5111, Fax +81 3 3245 5555
www.toray.co.jp

Tosoh Canada Ltd.
1200 Sheppard Ave. East, Suite 511,
Willowdale, Ont., M2K 2S5, Canada
416-756-2226, Fax 416-756-2750
www.tosoh.com

Tosoh Corporation
7-7 Akasaka 1-chome,
Minato-ku,Tokyo 107, Japan
81-3-3582-8120
www.tosoh.com

Tosoh Europe B.V.
World Trade Centre Amsterdam,
Strawinskylaan 1351,
1077 XX Amsterdam, The
Netherlands
31-20-644026, 31-20-623412
www.tosoh.com

Tosoh USA Inc.
1100 Circle 75 Pkwy., Suite 60,
Atlanta, GA 30339, USA
770-956-1100, Fax 770-956-7368
www.tosoh.com

Trace Chemicals LLC
2320 Lakecrest Dr.
Pekin, IL, 61554
(309) 347-2184
www.tracechemicals.com

Trical
PO Box 1327
8770 Highway 25
Hollister, CA 95023
(800) 655-7262

Manufacturer and Supplier Directory

Tri-K Industries
PO Box 128,
Northvale, NJ 07647-0128, USA
201-261-2800, 800-526-0372, Fax
201-261-1432

Tri-K Industries
151 Veterans Dr.
Northvale, NJ, 07647
(201) 750-1055; Fax: (201) 750-9785

Trinity Manufacturing Inc.
PO Box 1519
Hamlet, NC, 28345
(910) 582-5650

Tripart Farm Chemicals Ltd.

Troy Biosciences Inc.
PO Box 7012
Chandler, AZ, 85246
(480) 940-1745
www.troybiosciences.com

Troy Chemical Corp.
8 Vreeland Rd.
Florham Park, NJ, 7932
(973) 443-4200

Truchem Ltd.
Brook House, 30 Larwood Grove,
Sherwood, Nottingham, Notts., HG5
3JD,UK
44-1159-260762, Fax 44-1602-
671153

Trutec Industries Inc.
4700 Gateway Blvd,
Springfield, OH, 45502
(800) 933-8832

TSE Industries

U.S. Biochemical (UK)
25 Signet Court, Newmarket Rd,
Cambridge, Cambridgeshire, CB5
8LA,UK
44-1223-467064, Fax 44-1223-
60732

U.S. Biochemical Corp.
PO Box 22400,
Cleveland, OH 44122, USA
216-765-5000, 800-321-9322,
Fax 216-464-5075

U.S. Borax Inc.
26877 Tourney Rd.,
Valencia, CA 91355-1847, USA
805-287-5400, 800-US BORAX, Fax
800-626-4872

U.S. Fish and Wildlife Services
Arlington Square Bldg. MS 725
Washington, DC, 20240
(608) 783-6451
www.fws.gov

U.S. Gypsum Co.
125 S. Franklin St.
Chicago, IL, 60606
(312) 795-6831; Fax: (312) 606-4519
www.usg.com

U.S. Petrochemical Industries
921 Canal St. Suite 802
New Orleans, LA, 70112

UAP - Platte Chemical
PO Box 1286
Greeley, CO, 80632
(970) 356-4400
www.uap.com

UAP- West
7108 N. Fresno, #410,
Fresno, CA, 93720
(559) 437-5360; Fax: (559) 452-8505
www.uap.com

UCB Chemicals Corp.
2000 Lake Park Dr.,
Smyrna, GA 30080, USA
770-434-6188, 800-433-2873, Fax
770-434-8314
www.ucb.be

UCB Chemical Corp. Malaysia
PT 12701, Tuanku Jaafar Industrial
Park,
71450 Seremban, Negeri Sembilan
D.K.,
Malaysia
60-6-675-1112, Fax 60-6-675-1115
www.ucb.be

UCB N.V. Filmsektor
Ottergemsesteenweg 801, PO Box
369,
B-9000 Ghent, Belgium
091-40-32-11, Fax 091-40-88-00
www.ucb.be

UCB S.A./Chemical
33 rue d'Anderlecht,
B-1620 Drogenbos, Belgium
32-2-3714923, Fax 32-2-3714924
www.ucb.be

UCB S.A./Chemical
60, Allée de la Recherche
B-1070 Brussels, Belgium
32-2-641-1411, Fax 32-2-640-9860
www.ucb.be

Unichema Australia
164 Ingles St.,
Port Melbourne,Vic., 3207, Australia
61-3-647-9311, Fax 61-3-645-3001

Unichema Chemicals Ltd.
Bebington,
Wirral, Merseyside, L62 4UF, UK
44-151-645-2020, Fax 44-151-645-
9197

Unichema Chemie B.V.
Postbus 2,
2800 AA Gouda, The Netherlands
31-1820-42911, Fax 31-1820-42250

Unichema Chemie GmbH
Postfach 100963,
D-46249 Emmerich, Germany
49-2822-720, Fax 49-2822-72276

Unichema France S.A.
148 Boulevard Haussemann,
75008 Paris, France
33-1-44-95-0840, Fax 33-1- 42-
563188

Unichema International
Postbus 2,
2800 AA Gouda, The Netherlands
31-1820-42911, Fax 31-1820-42250

Unichema North America
4650 S. Racine Ave.,
Chicago, IL 60609, USA
312-376-9000, 800-833-2864,
Fax 312-376-0095

Unicorn Laboratories
12385 Automobile Blvd.
Clearwater, FL, 33762
(727) 578-4545

Union Camp Chemicals
Vigo Lane,
Chester-le-Street, Co. Durham, DH3
2RB, UK
44-91-410-2631, Fax 44-91-410-
9391
www.unioncamp.com

Union Camp Corp./Chemicals
1600 Valley Rd.,
Wayne, NJ 07470, USA
201-628-2290, 800-733-1374, Fax
201-628-2840
www.unioncamp.com

Union Camp Corp./Chemicals
PO Box 2668
Savannah, GA 31402, USA
912-238-6000
www.unioncamp.com

Union Carbide Canada
7400 Blvd des Galleries,
d'Anjou, PQ, H1M 3M2, Canada
514-493-2610, Fax 514-493-2619
www.unioncarbide.com

Union Carbide Corp.
39 Old Ridgebury Rd.,
Danbury,CT,06817-0Q01, USA
203-794-2000, 800-335-8550, Fax
203-794-3170
www.unioncarbide.com

Manufacturer and Supplier Directory

Union Carbide Corp.
PO Box 12014,
Research Triangle Park, NC 27709,
USA
www.unioncarbide.com

Union Carbide (Europe)
7, rue du Pre-Bouvier,
CH-1217 Meyrin (Geneve),
Switzerland
41-22-989 66 53, Fax 41-22-989
6545
www.unioncarbide.com

Union Carbide (UK)
93-95 High Street,
Rickmansworth, Herts., WD3 1RB,
UK
44-1923 720 366, Fax 44-1923-
896721
www.unioncarbide.com

Uniroyal Chemical Co.
199 Benson Road,
Middlebury, CT 06749, USA
203-573-2269, 800-322-3243, Fax
203-573-2165
www.uniroyal.com

Uniroyal Chemical Ltd.
Kennet House, 4 Langley Quay,
Slough, Berks., SL3 6EH, UK
44-1753-603000, Fax 44-1753-
603078
www.uniroyal.com

United Catalysts Inc.
PO Box 32370,
Louisville, KY 40232, USA
502-634-7200, Fax 502-637-3732

United Horticultural Supply
150 S. Main St.
Fremont, NE, 68025
(303) 356-4400
www.uhsonline.com

Unitex Chemical Corp.
PO Box 16344,
520 Broome Rd.,
Greensboro, NC 27406, USA
910-378-0965, Fax 910-272-4312

Unitex Ltd.
Halfpenny Lane, Knaresborough,
North Yorkshire, HG5 0PP, UK
44-1423-862677, Fax 44-1423-
868340

Universal Cooperatives Inc.
Eagan, MN

Universal Crop Protection Ltd.
1300 Corporate Ctr. Curve
Eagan, MN, 55121
(651) 239-1000

Unocal Hydrocarbon Sales
1701 Golf Rd., Suite 1-1101,
Rolling Meadows, IL 60008-4295,
USA
708-734-7622, 800-967-7601
Fax 708-734-7677
www.unocal.com

UOP Inc.
25 E. Algonquin Rd.
Desplaines, IL 60017-5017
(847) 391-2000, Fax (847) 391-2253
www.uop.com

Upjohn Co./Fine Chemicals
7000 Portage Rd.,
Kalamazoo, MI 49001, USA
616-323-5505, 800-253-8600,
Fax 616-329-3604
www.upjohn.com

Upjohn Ltd.
PO Box 8, Fleming Way,
Crawley, West Sussex, RH10 2NJ,
UK
44-1293-531133, Fax 44-1293-
548850
www.upjohn.com

US Chemical Corp.
18000 West Sarah Lane
Brookfield, WI, 53045
(414) 921-1450

Usines de Melle
See: Rhône Poulenc

Valent Biosciences Corp.
870 Technology Way
Libertyville, IL, 60048
(847) 968-4724
www.valentbiosciences.com

Valent USA Corp.
1170 W. Shaw Ave. Suite 103
Fresno, CA, 93711
(559) 244-3960; Fax: (559) 228-2019
www.valent.com

Value Gardens Supply LLC
PO Box 585
St. Joseph, MO, 64502
(540) 864-8100

Van Dyk

Van Waters & Rogers
6100 Carillon Point
Kirkland, WA 98033
(206) 889-3400, Fax (206) 889-4133

Vandemoortele Professional
Ottergemsesteenweg 806
B-9000 Gent Belgium
+32 91 401711, Fax +32 91 227264

Vanderbilt R.T. Co. Inc.
30 Winfield St., PO Box 5150
Norwalk, CT 06856
(203) 853-1400, Fax (203) 853-1452
www.rtvanderbilt.com

Veitsiluoto Oy/Fores,
PO Box 196,
SF-90101 Oulu 10, Finland
358-81-316 3111
Fax 358-81-378 5755

Velsicol Chemical Corp.
10400 W. Higgins Road, Suite 600
Rosemont, IL 60018
(800) 843-7759, Fax (847) 298-9014
www.velsicol.com

Verdugt BV
Postbus 60, Papesteeg 91,
NL-4000 AB Tiel, Netherlands
+31 3440-15224, Fax +31 3440-
11475

Verichem Inc.
3499 Grand Ave.
Pittsburgh, PA, 15225
(412) 331-7299

Vineland Labs.
2285 E. Landis Ave.
Vineland, NJ, 8360
(856) 691-2411

Vinings Industries Inc.
See: Kemira Chemicals

Virbac AH Inc.
PO Box 162059
Fort Worth, TX, 76161
(817) 831-5030

Vista Chemical Co.
900 Threadneedlem, PO Box 19029
Houston, TX 77224-9029
(713) 588-3000, Fax, (713) 588-3236

Vitax Ltd.
Owen St.
Coalville, UK, LE67 3DE
44 1530 510-060; Fax: 44 1530 510-
299
www.vitax.co.uk

Voluntary Purchasing Group Inc.
PO Box 460
Bonham, TX 75418
(903) 583-5501

Vulcan Plastics Ltd.
Hosey Hill, Westerham,
Kent, TN16 1TB, UK
44-1959-562304
Wacker Chemicals (US)
3301 Sutton Rd.,
Adrian, MI 49221, USA
517-264-8131, 800-485-3686, Fax
517-264-8795

Manufacturer and Supplier Directory

Wacker Chemicals Ltd.
The Clock Tower,
Mount Felix,Bridge Street,
Walton-on-Thames, Surrey, KT12
1AS, UK
44-1932-246111, Fax 44-1932-
240141

Wacker Silicones Corp.
3301 Sutton Rd.,
Adrian, MI 49221-9397, USA
517-264-8500, 800-248-0063, Fax
517-264-8246
www.wacker.silicones.com

Wacker-Chemie GmbH
Hanns-Seidel-Platz 4,
D-81737 München, Germany
49-89-6279-01, Fax 49-89-6279-
1770
www.wacker.de

Wako Chemicals GmbH
Nissanstrasse 2,
D-41468 Neuss, Germany
49-2131-3110, Fax 49-2131-31-1100
www.wako-chem.co.ip

Wako Chemicals USA
1600 Bellwood Rd.,
Richmond, VA 23237, USA
804-271-7677, Fax 804-271-7791
www.wako-chem.co.ip

Wako Pure Chemical Industries Ltd.
1,2-Doshomachi 3-Chome,
Chuo-ku, Osaka 541, Japan
81-6-203-3741, Fax 81-6-222-1203
www.wako-chem.co.ip

Walco-Linck Co.
PO Drawer 598
Valley Cottage, NY 10989
(845) 353-7600, (800) 338-2329

Wallace Laboratories
Cranbury, NJ 08512, USA
609-665-6000

Walling Water Management
PO Box 5208
Sioux Falls, SD, 57117
(605) 336-6710
www.wallingwater.com

Walter G. Legge Co Inc.
444 Central Avenue, PO Box 591
Peekskill, NY 10566
(800) 345-3443, Fax (914) 737-2636
www.leggesystems.com

Walton Pharmaceuticals Inc.
PO Box 76, East Horsley
Leatherhead KT24 5YW
Surrey, UK
+44 01483 280001, Fax +44 01483
280002

Ward Blenkinsop
See: Great Lakes Fine Chemicals

Warner-Jenkinson
2526 Baldwin St., PO Box 14538
St. Louis, MO 63178-0538
(314) 889-7600, Fax (314) 658-7318

Warner-Lambert
See: Pfizer Inc.

Waterbury Companies Inc.
32 Mattatuck Heights
PO Box 1812
Waterbury, CT 06722
(203) 597-1812, Fax (203) 574-1710

Watson Pharmaceuticals
311 Bonnie Circle
Corona, CA 92880
(909) 493-5300, Fax (909) 493-5836

Weiders Farmasoytiske A/S
Hausmannsgaten 6
0186 Oslo, Norway
+47 22 205415, Fax +47 22 364052

Wellcome Foundation Ltd.
See: GlaxoWellcome plc
www.wellcome.ac.uk

Wensleydale Foods

Wepak Corp.
PO Box 36803
Charlotte, NC, 28236
(704) 334-5781

West Agro Inc.
501 Santa Fe
Kansas City, MO, 64105
(816) 421-0366
www.westagro.com

Western Farm Service Inc.
(209) 547-2637, Fax (209) 464-4652
www.westernfarmservice.com

Westvaco Corp., Chemicals ·
PO Box 70848,
Charleston Hts., SC 29415-0848,
USA
803-740-2300, Fax 803-740-2329

Westvaco Corp., Custom Chemicals
PO Box 237, Hwy. 60 East,
Mulberry, FL 33860, USA
813-425-3043

Westwood-Squibb Pharm. Inc.
See: Bristol-Myers Squibb Co.

Whitecourt

Whitehall Laboratories
22-24 Torrington Place,
London, WC1E 7ET, UK
44-171-636-8080
Fax 44-171-580-6037

Whiting Peter Ltd.
1 Oil Mill Lane
London W6 9UA, UK
+44 020 8741 4025, Fax +44
0208741 1737
www.whiting-chemicals.com

Whitmire Micro-Gen Research Laboratories Inc
3568 Tree Court Ind. Blvd.
St Louis, MO 63122-0000
(800) 777-8570

Whittaker Clark & Daniels
1000 Coolidge Street
s. Plainfield, NJ 07080
(908) 561-6100, Fax (800) 833-8139
www.wcdinc.com

Wilbur Ellis Co.
East Mission
Cashmere Branch, WA 98815
(509) 663-8753

Wiley, John
111 River Street
Hoboken, NJ 07030-5774
(201) 748-6000, Fax (201) 748-6088

Willert Home Products Inc.
4044 Park Street
St. Louis, MO 63110
(314) 772-2822, (800) 325-9680

Witco (Europe) S.A.
7, rue du Pre-Bouvier,
CH-1217 Meyrin, Geneva,
Switzerland
41-22-989-2111, Fax 41-22-989-
2391
www.witco.com

Witco Asia Pacific Pte Ltd.
12 Science Park Drive,
#03-04 The Mendel,
Singapore Science Park, 118225,
Singapore
65-774-4800, Fax 65-770-5148
www.witco.com

Witco Canada Ltd.
565 Coronation Drive,
Westhill, Ont., M1E2K3, Canada
416-497-991, Fax 416-284-8141
www.witco.com

Witco Chemical Ltd.
Paragon Works, Baxenden,
Near Accrington,
Lancs., BB5 2SL, UK
44-125-439-8616
Fax 44-125-439-8586
www.witco.com

Witco Corp.
One American Lane,
Greenwich, CT 06831-2559, USA
203-552-2000, 800-779-4826
x6400, Fax 203-552-2010
www.witco.com

Manufacturer and Supplier Directory

Witco Corp./Allied-Kelite,
17050 Lathrop Ave.,
Harvey, IL 60426, USA
800-323-9784
www.witco.com

Witco Corp./Concarb.
10500 Richmond, Suite 116,
PO Box 42817,
Houston, TX 77242-2817
713-978-5745, 800-231-4591,
Fax 713-978-5728
www.witco.com

Witco Corp./Lubricants
10100 Santa Monica Blvd., Suite 1470,
Los Angeles, CA 90067-4183, USA
310-277-4511, 800-429-4826, Fax 310-201-0383
www.witco.com

Witco Corp./Organics
One American Lane,
Greenwich, CT 06831-2559, USA
203-552-2000, Fax 203-552-2010
www.witco.com

Witco do Brasil Ltda.
Rua Verbo Divino 16661 cj 64,
Sao Paulo, 04719-002, Brasil
55-11-5181-2799, Fax 55-11-5181-7972
www.witco.com

Witco Surfactants UK
Paragon Works,
Baxenden, Nr. Accrington,
Lancs., BB5 2SL, UK
44-1254-39-8616, Fax 44-1254-39-8586
www.witco.com

Witco UK
Union Lane, Droitwich,
Worcestershire, WR9 9BB, UK
44-190 579 4795, Fax 44-190 579 4002
www.witco.com

WWH (Witco Corp.)
One American Lane,
Greenwich, CT 06831-2559, USA
203-552-2000, 800-779-4826, Fax 203-552-2010
www.witco.com

WPC Brands Inc.
1 Repel Rd.
Jackson, WI, 53037-9583
(262) 677-4121
www.wpcbrands.com

Wrap Pack Inc.
1728 Presson Place
Yakima, WA, 98903-223
(509) 453-2830

Wuhan Youji Industries Co. Ltd.
Wyeth Laboratories
Huntercombe Lane South, Taplow,
Maidenhead, Berks., SL6 0PH, UK
44-1628-604377, Fax 44-1628-666368
www.ahp.com

Wyeth Laboratories Inc.
P.O. Box 8299,
Philadelphia, PA 19101, USA
215-688-4400
www.ahp.com

Yamanouchi Europe BV
PO Box 108
NL 2350, A C Leiderdorp
Netherlands
+31 7154 55745, Fax +31 7154 800

Yamanouchi Pharma
10, Pl. de la Coupole - BP 105
94223 Charenton le Pont Cedex
France
+33 (1) 46 76 64 00, Fax +43 (1) 46 76 64 99

Zeeland Chemicals, Inc.
215 N. Centennial St.,
Zeeland, MI 49464, USA
616-772-2193, 800-223-0453, Fax 616-772-7344
www.cambrex.com

Zen Brasal Pvt. Ltd.
2207 Ph 4
GIDC Vatva Ahmedabad
India

Zeneca Pharmaceuticals
See: ICI Plant Protection

Zeochem
PO Box 35940, 1600 W. Hill St.,
Louisvcille, KY 40232
(502) 634-7681, Fax (502) 634-8133
www.zeochem.com

ZEP Manufacturing Co.
4401 Northside Parkway
Atlanta, GA 30327
(877) I BUY ZEP, Fax (877) 428-9937

ZinderSpA